Chemistry

H.-D. Belitz · W. Grosch · P. Schieberle

Food Chemistry

4th revised and extended ed.

With 481 Figures, 923 Formulas and 634 Tables

 Springer

Professor Dr.Hans-Dieter Belitz †

Professor em. Dr. Werner Grosch
Ehem. apl. Professor für Lebensmittelchemie
an der Technischen Universität München
Ehem. stellvertr. Direktorder Deutschen Forschungsanstalt
für Lebensmittelchemie München
Lichtenbergstraße
485748 Garching

Professor Dr. Peter Schieberle
Ordinarius für Lebensmittelchemie
an der Technischen Universität München
Leiter des Instituts für Lebensmittelchemie
an der Technischen Universität München
Direktor der Deutschen Forschungsanstalt
für Lebensmittelchemie München
Lichtenbergstraße
485748 Garching

ISBN 978-3-540-69935-4   e-ISBN 978-3-540-69934-7

DOI 10.1007/978-3-540-69934-7

Library of Congress Control Number: 2008931197

Production: le-tex publishing services oHG, Leipzig
Typesetting: le-tex publishing services oHG, Leipzig
Cover design: KünkelLopka GmbH, Heidelberg, Germany

Printed on acid-free paper

9 8 7 6 5 4 3 2 1

springer.com

Preface

Preface to the First German Edition

The very rapid development of food chemistry and technology over the last two decades, which is due to a remarkable increase in the analytical and manufacturing possibilities, makes the complete lack of a comprehensive, teaching or reference text particularly noticeable. It is hoped that this textbook of food chemistry will help to fill this gap. In writing this volume we were able to draw on our experience from the lectures which we have given, covering various scientific subjects, over the past fifteen years at the Technical University of Munich.

Since a separate treatment of the important food constituents (proteins, lipids, carbohydrates, flavor compounds, etc.) and of the important food groups (milk, meat, eggs, cereals, fruits, vegetables, etc.) has proved successful in our lectures, the subject matter is also organized in the same way in this book.

Compounds which are found only in particular foods are discussed where they play a distinctive role while food additives and contaminants are treated in their own chapters. The physical and chemical properties of the important constituents of foods are discussed in detail where these form the basis for understanding either the reactions which occur, or can be expected to occur, during the production, processing, storage and handling of foods or the methods used in analyzing them. An attempt has also been made to clarify the relationship between the structure and properties at the level of individual food constituents and at the level of the whole food system.

The book focuses on the chemistry of foodstuffs and does not consider national or international food regulations. We have also omitted a broader discussion of aspects related to the nutritional value, the processing and the toxicology of foods. All of these are an essential part of the training of a food chemist but, because of the extent of the subject matter and the consequent specialization, must today be the subject of separate books. Nevertheless, for all important foods we have included brief discussions of manufacturing processes and their parameters since these are closely related to the chemical reactions occurring in foods.

Commodity and production data of importance to food chemists are mainly given in tabular form. Each chapter includes some references which are not intended to form an exhaustive list. No preference or judgement should be inferred from the choice of references; they are given simply to encourage further reading. Additional literature of a more general nature is given at the end of the book.

This book is primarily aimed both at students of food and general chemistry but also at those students of other disciplines who are required or choose to study food chemistry as a supplementary subject. We also hope that this comprehensive text will prove useful to both food chemists and chemists who have completed their formal education.

We thank sincerely Mrs. A. Mödl (food chemist), Mrs. R. Berger, Mrs. I. Hofmeier, Mrs. E. Hortig, Mrs. F. Lynen and Mrs. K. Wüst for their help during the preparation of the manuscript and its proofreading. We are very grateful to Springer Verlag for their consideration of our wishes and for the agreeable cooperation.

Garching,   *H.-D. Belitz*
July 1982   *W. Grosch*

Preface to the Fourth English Edition

The fourth edition of the "Food Chemistry" textbook is a translation of the sixth German edition of this textbook. It follows a general concept as detailed in the preface to the first edition given below. All chapters have been carefully checked and updated with respect to the latest developments, if required. Comprehensive changes have been made in Chapters 9 (Contaminants), 18 (Phenolic Compounds), 20 (Alcoholic Beverages) and 21 (Tea, Cocoa). The following topics were newly added:

- the detection of BSE and D-amino acids,
- the formation and occurrence of acrylamide and furan,
- compounds having a cooling effect,
- technologically important milk enzymes,
- the lipoproteins of egg yolk,
- the structure of the muscle and meat aging,
- food allergies,
- the baking process,
- the reactivity of oxygen species in foods,
- phytosterols,
- glycemic index,
- the composition of aromas was extended: odorants (pineapple, raw and cooked mutton, black tea, cocoa powder, whisky) and taste compounds (black tea, roasted cocoa, coffee drink).

The production data for the year 2006 were taken from the FAO via Internet. The volume of the book was not changed during the revision as some existing chapters were shortened.

We are very grateful to Dr. Margaret Burghagen for translating the manuscript. It was our pleasure to collaborate with her.

We would also like to thank Prof. Dr. Jürgen Weder and Dr. Rolf Kieffer for several valuable recommendations. We are also grateful to Sabine Bijewitz and Rita Jauker for assistance in completing the manuscript, and Christel Hoffmann for help with the literature and the index.

Garching,   *W. Grosch*
Mai 2008   *P. Schieberle*

Contents

0   **Water** 1
0.1   Foreword 1
0.2   Structure 1
0.2.1   Water Molecule 1
0.2.2   Liquid Water and Ice 2
0.3   Effect on Storage Life.. 3
0.3.1   Water Activity 3
0.3.2   Water Activity as an Indicator 5
0.3.3   Phase Transition of Foods Containing Water 5
0.3.4   WLF Equation 6
0.3.5   Conclusion 7
0.4   References 7

1   **Amino Acids, Peptides, Proteins** 8
1.1   Foreword 8
1.2   Amino Acids 8
1.2.1   General Remarks... 8
1.2.2   Classification, Discovery and Occurrence 9
1.2.2.1   Classification 9
1.2.2.2   Discovery and Occurrence 9
1.2.3   Physical Properties 12
1.2.3.1   Dissociation 12
1.2.3.2   Configuration and Optical Activity 13
1.2.3.3   Solubility 14
1.2.3.4   UV-Absorption 15
1.2.4   Chemical Reactions 16
1.2.4.1   Esterification of Carboxyl Groups................................... 16
1.2.4.2   Reactions of Amino Groups... 16
1.2.4.2.1 Acylation 16
1.2.4.2.2 Alkylation and Arylation 18
1.2.4.2.3 Carbamoyl and Thiocarbamoyl Derivatives 20
1.2.4.2.4 Reactions with Carbonyl Compounds................................... 21
1.2.4.3   Reactions Involving Other Functional Groups 22
1.2.4.3.1 Lysine.. 23
1.2.4.3.2 Arginine 23
1.2.4.3.3 Aspartic and Glutamic Acids 23
1.2.4.3.4 Serine and Threonine 24
1.2.4.3.5 Cysteine and Cystine 24
1.2.4.3.6 Methionine.. 24
1.2.4.3.7 Tyrosine 24

1.2.4.4	Reactions of Amino Acids at Higher Temperatures	25
1.2.4.4.1	Acrylamide	25
1.2.4.4.2	Mutagenic Heterocyclic Compounds	26
1.2.5	Synthetic Amino Acids Utilized for Increasing the Biological Value of Food (Food Fortification)	29
1.2.5.1	Glutamic Acid	32
1.2.5.2	Aspartic Acid...	32
1.2.5.3	Lysine	32
1.2.5.4	Methionine...	33
1.2.5.5	Phenylalanine	33
1.2.5.6	Threonine..	33
1.2.5.7	Tryptophan...	33
1.2.6	Sensory Properties	34
1.3	Peptides	34
1.3.1	General Remarks, Nomenclature.........................	34
1.3.2	Physical Properties	36
1.3.2.1	Dissociation	36
1.3.3	Sensory Properties	36
1.3.4	Individual Peptides	38
1.3.4.1	Glutathione	38
1.3.4.2	Carnosine, Anserine and Balenine	39
1.3.4.3	Nisin	39
1.3.4.4	Lysine Peptides	40
1.3.4.5	Other Peptides	40
1.4	Proteins	40
1.4.1	Amino Acid Sequence	41
1.4.1.1	Amino Acid Composition, Subunits	41
1.4.1.2	Terminal Groups	42
1.4.1.3	Partial Hydrolysis	43
1.4.1.4	Sequence Analysis	44
1.4.1.5	Derivation of Amino Acid Sequence from the Nucleotide Sequence of the Coding Gene	46
1.4.2	Conformation	48
1.4.2.1	Extended Peptide Chains	48
1.4.2.2	Secondary Structure (Regular Structural Elements)	49
1.4.2.2.1	β-Sheet..	50
1.4.2.2.2	Helical Structures	51
1.4.2.2.3	Reverse Turns	52
1.4.2.2.4	Super-Secondary Structures	52
1.4.2.3	Tertiary and Quaternary Structures	53
1.4.2.3.1	Fibrous Proteins	53
1.4.2.3.2	Globular Proteins	53
1.4.2.3.3	BSE	55
1.4.2.3.4	Quaternary Structures	56
1.4.2.4	Denaturation	56
1.4.3	Physical Properties	58
1.4.3.1	Dissociation	58
1.4.3.2	Optical Activity.......................................	60
1.4.3.3	Solubility, Hydration and Swelling Power	60
1.4.3.4	Foam Formation and Foam Stabilization	62
1.4.3.5	Gel Formation	62

1.4.3.6	Emulsifying Effect	63
1.4.4	Chemical Reactions	64
1.4.4.1	Lysine Residue	64
1.4.4.1.1	Reactions Which Retain the Positive Charge	64
1.4.4.1.2	Reactions Resulting in a Loss of Positive Charge	65
1.4.4.1.3	Reactions Resulting in a Negative Charge	65
1.4.4.1.4	Reversible Reactions	66
1.4.4.2	Arginine Residue	66
1.4.4.3	Glutamic and Aspartic Acid Residues	67
1.4.4.4	Cystine Residue (cf. also Section 1.2.4.3.5)	67
1.4.4.5	Cysteine Residue (cf. also Section 1.2.4.3.5)	68
1.4.4.6	Methionine Residue	69
1.4.4.7	Histidine Residue	69
1.4.4.8	Tryptophan Residue	70
1.4.4.9	Tyrosine Residue	70
1.4.4.10	Bifunctional Reagents	70
1.4.4.11	Reactions Involved in Food Processing	70
1.4.5	Enzyme-Catalyzed Reactions	74
1.4.5.1	Foreword	74
1.4.5.2	Proteolytic Enzymes	74
1.4.5.2.1	Serine Endopeptidases	74
1.4.5.2.2	Cysteine Endopeptidases	76
1.4.5.2.3	Metalo Peptidases	76
1.4.5.2.4	Aspartic Endopeptidases	76
1.4.6	Chemical and Enzymatic Reactions of Interest to Food Processing	79
1.4.6.1	Foreword	79
1.4.6.2	Chemical Modification	80
1.4.6.2.1	Acylation	80
1.4.6.2.2	Alkylation	82
1.4.6.2.3	Redox Reactions Involving Cysteine and Cystine	82
1.4.6.3	Enzymatic Modification	83
1.4.6.3.1	Dephosphorylation	83
1.4.6.3.2	Plastein Reaction	83
1.4.6.3.3	Cross-Linking	86
1.4.7	Texturized Proteins	87
1.4.7.1	Foreword	87
1.4.7.2	Starting Material	88
1.4.7.3	Texturization	88
1.4.7.3.1	Spin Process	88
1.4.7.3.2	Extrusion Process	89
1.5	References	89
2	**Enzymes**	93
2.1	Foreword	93
2.2	General Remarks, Isolation and Nomenclature	93
2.2.1	Catalysis	93
2.2.2	Specificity	94
2.2.2.1	Substrate Specificity	94
2.2.2.2	Reaction Specificity	95
2.2.3	Structure	95

2.2.4	Isolation and Purification	96
2.2.5	Multiple Forms of Enzymes	97
2.2.6	Nomenclature	97
2.2.7	Activity Units	98
2.3	Enzyme Cofactors	98
2.3.1	Cosubstrates	99
2.3.1.1	Nicotinamide Adenine Dinucleotide	99
2.3.1.2	Adenosine Triphosphate	102
2.3.2	Prosthetic Groups	102
2.3.2.1	Flavins	102
2.3.2.2	Hemin	103
2.3.2.3	Pyridoxal Phosphate	103
2.3.3	Metal Ions	104
2.3.3.1	Magnesium, Calcium and Zinc	104
2.3.3.2	Iron, Copper and Molybdenum	105
2.4	Theory of Enzyme Catalysis	106
2.4.1	Active Site	106
2.4.1.1	Active Site Localization	107
2.4.1.2	Substrate Binding	108
2.4.1.2.1	Stereospecificity	108
2.4.1.2.2	"Lock and Key" Hypothesis	109
2.4.1.2.3	Induced-fit Model	109
2.4.2	Reasons for Catalytic Activity	110
2.4.2.1	Steric Effects – Orientation Effects	111
2.4.2.2	Structural Complementarity to Transition State	112
2.4.2.3	Entropy Effect	112
2.4.2.4	General Acid–Base Catalysis	113
2.4.2.5	Covalent Catalysis	114
2.4.3	Closing Remarks	117
2.5	Kinetics of Enzyme-Catalyzed Reactions	117
2.5.1	Effect of Substrate Concentration	117
2.5.1.1	Single-Substrate Reactions	117
2.5.1.1.1	Michaelis–Menten Equation	117
2.5.1.1.2	Determination of K_m and V	120
2.5.1.2	Two-Substrate Reactions	121
2.5.1.2.1	Order of Substrate Binding	121
2.5.1.2.2	Rate Equations for a Two-Substrate Reaction	122
2.5.1.3	Allosteric Enzymes	123
2.5.2	Effect of Inhibitors	125
2.5.2.1	Irreversible Inhibition	126
2.5.2.2	Reversible Inhibition	126
2.5.2.2.1	Competitive Inhibition	126
2.5.2.2.2	Non-Competitive Inhibition	127
2.5.2.2.3	Uncompetitive Inhibition	128
2.5.3	Effect of pH on Enzyme Activity	128
2.5.4	Influence of Temperature	130
2.5.4.1	Time Dependence of Effects	131
2.5.4.2	Temperature Dependence of Effects	131
2.5.4.3	Temperature Optimum	133
2.5.4.4	Thermal Stability	134
2.5.5	Influence of Pressure	136

2.5.6	Influence of Water	137
2.6	Enzymatic Analysis	137
2.6.1	Substrate Determination	138
2.6.1.1	Principles	138
2.6.1.2	End-Point Method	138
2.6.1.3	Kinetic Method	140
2.6.2	Determination of Enzyme Activity	140
2.6.3	Enzyme Immunoassay	141
2.6.4	Polymerase Chain Reaction	142
2.6.4.1	Principle of PCR	143
2.6.4.2	Examples	144
2.6.4.2.1	Addition of Soybean	144
2.6.4.2.2	Genetically Modified Soybeans	144
2.6.4.2.3	Genetically Modified Tomatoes	144
2.6.4.2.4	Species Differentiation	144
2.7	Enzyme Utilization in the Food Industry	144
2.7.1	Technical Enzyme Preparations	145
2.7.1.1	Production	145
2.7.1.2	Immobilized Enzymes	145
2.7.1.2.1	Bound Enzymes	145
2.7.1.2.2	Enzyme Entrapment	145
2.7.1.2.3	Cross-Linked Enzymes	148
2.7.1.2.4	Properties	148
2.7.2	Individual Enzymes	149
2.7.2.1	Oxidoreductases	149
2.7.2.1.1	Glucose Oxidase	149
2.7.2.1.2	Catalase	149
2.7.2.1.3	Lipoxygenase	149
2.7.2.1.4	Aldehyde Dehydrogenase	149
2.7.2.1.5	Butanediol Dehydrogenase	149
2.7.2.2	Hydrolases	150
2.7.2.2.1	Peptidases	150
2.7.2.2.2	α- and β-Amylases	150
2.7.2.2.3	Glucan-1,4-α-D-Glucosidase (Glucoamylase)	151
2.7.2.2.4	Pullulanase (Isoamylase)	152
2.7.2.2.5	Endo-1,3(4)-β-D-Glucanase	152
2.7.2.2.6	α-D-Galactosidase	152
2.7.2.2.7	β-D-Galactosidase (Lactase)	152
2.7.2.2.8	β-D-Fructofuranosidase (Invertase)	152
2.7.2.2.9	α-L-Rhamnosidase	153
2.7.2.2.10	Cellulases and Hemicellulases	153
2.7.2.2.11	Lysozyme	153
2.7.2.2.12	Thioglucosidase	153
2.7.2.2.13	Pectolytic Enzymes	153
2.7.2.2.14	Lipases	154
2.7.2.2.15	Tannases	154
2.7.2.2.16	Glutaminase	154
2.7.2.3	Isomerases	154
2.7.2.4	Transferases	154
2.8	References	155

3	**Lipids**	**158**
3.1	Foreword	158
3.2	Fatty Acids	159
3.2.1	Nomenclature and Classification	159
3.2.1.1	Saturated Fatty Acids	159
3.2.1.2	Unsaturated Fatty Acids	162
3.2.1.3	Substituted Fatty Acids	164
3.2.2	Physical Properties	165
3.2.2.1	Carboxyl Group	165
3.2.2.2	Crystalline Structure, Melting Points	165
3.2.2.3	Urea Adducts	166
3.2.2.4	Solubility	167
3.2.2.5	UV-Absorption	167
3.2.3	Chemical Properties	167
3.2.3.1	Methylation of Carboxyl Groups	167
3.2.3.2	Reactions of Unsaturated Fatty Acids	168
3.2.3.2.1	Halogen Addition Reactions	168
3.2.3.2.2	Transformation of Isolene-Type Fatty Acids to Conjugated Fatty Acids	168
3.2.3.2.3	Formation of a π-Complex with Ag^+ Ions	168
3.2.3.2.4	Hydrogenation	169
3.2.4	Biosynthesis of Unsaturated Fatty Acids	169
3.3	Acylglycerols	169
3.3.1	Triacylglycerols (TG)	170
3.3.1.1	Nomenclature, Classification, Calorific Value	170
3.3.1.2	Melting Properties	171
3.3.1.3	Chemical Properties	172
3.3.1.4	Structural Determination	173
3.3.1.5	Biosynthesis	177
3.3.2	Mono- and Diacylglycerols (MG, DG)	177
3.3.2.1	Occurrence, Production	177
3.3.2.2	Physical Properties	178
3.4	Phospho- and Glycolipids	178
3.4.1	Classes	178
3.4.1.1	Phosphatidyl Derivatives	178
3.4.1.2	Glyceroglycolipids	180
3.4.1.3	Sphingolipids	181
3.4.2	Analysis	182
3.4.2.1	Extraction, Removal of Nonlipids	182
3.4.2.2	Separation and Identification of Classes of Components	182
3.4.2.3	Analysis of Lipid Components	183
3.5	Lipoproteins, Membranes	183
3.5.1	Lipoproteins	183
3.5.1.1	Definition	183
3.5.1.2	Classification	184
3.5.2	Involvement of Lipids in the Formation of Biological Membranes	185
3.6	Diol Lipids, Higher Alcohols, Waxes and Cutin	186
3.6.1	Diol Lipids	186
3.6.2	Higher Alcohols and Derivatives	186
3.6.2.1	Waxes	186

3.6.2.2	Alkoxy Lipids	187
3.6.3	Cutin	187
3.7	Changes in Acyl Lipids of Food	187
3.7.1	Enzymatic Hydrolysis	187
3.7.1.1	Triacylglycerol Hydrolases (Lipases)	188
3.7.1.2	Polar-Lipid Hydrolases	190
3.7.1.2.1	Phospholipases	190
3.7.1.2.2	Glycolipid Hydrolases	190
3.7.2	Peroxidation of Unsaturated Acyl Lipids	191
3.7.2.1	Autoxidation	191
3.7.2.1.1	Fundamental Steps of Autoxidation	192
3.7.2.1.2	Monohydroperoxides	193
3.7.2.1.3	Hydroperoxide-Epidioxides	195
3.7.2.1.4	Initiation of a Radical Chain Reaction	196
3.7.2.1.5	Photooxidation	196
3.7.2.1.6	Heavy Metal Ions	198
3.7.2.1.7	Heme(in) Catalysis	200
3.7.2.1.8	Activated Oxygen	201
3.7.2.1.9	Secondary Products	203
3.7.2.2	Lipoxygenase: Occurrence and Properties	207
3.7.2.3	Enzymatic Degradation of Hydroperoxides	209
3.7.2.4	Hydroperoxide–Protein Interactions	211
3.7.2.4.1	Products Formed from Hydroperoxides	211
3.7.2.4.2	Lipid–Protein Complexes	212
3.7.2.4.3	Protein Changes	213
3.7.2.4.4	Decomposition of Amino Acids	214
3.7.3	Inhibition of Lipid Peroxidation	214
3.7.3.1	Antioxidant Activity	215
3.7.3.2	Antioxidants in Food	215
3.7.3.2.1	Natural Antioxidants	215
3.7.3.2.2	Synthetic Antioxidants	218
3.7.3.2.3	Synergists	219
3.7.3.2.4	Prooxidative Effect	220
3.7.4	Fat or Oil Heating (Deep Frying)	220
3.7.4.1	Autoxidation of Saturated Acyl Lipids	221
3.7.4.2	Polymerization	223
3.7.5	Radiolysis	224
3.7.6	Microbial Degradation of Acyl Lipids to Methyl Ketones	225
3.8	Unsaponifiable Constituents	225
3.8.1	Hydrocarbons	227
3.8.2	Steroids	227
3.8.2.1	Structure, Nomenclature	227
3.8.2.2	Steroids of Animal Food	227
3.8.2.2.1	Cholesterol	227
3.8.2.2.2	Vitamin D	229
3.8.2.3	Plant Steroids (Phytosterols)	229
3.8.2.3.1	Desmethylsterols	229
3.8.2.3.2	Methyl- and Dimethylsterols	231
3.8.2.4	Analysis	232
3.8.3	Tocopherols and Tocotrienols	233
3.8.3.1	Structure, Importance	233

3.8.3.2   Analysis 234
3.8.4   Carotenoids 234
3.8.4.1   Chemical Structure, Occurrence 235
3.8.4.1.1 Carotenes 236
3.8.4.1.2 Xanthophylls 237
3.8.4.2   Physical Properties 240
3.8.4.3   Chemical Properties 241
3.8.4.4   Precursors of Aroma Compounds 241
3.8.4.5   Use of Carotenoids in Food Processing 244
3.8.4.5.1 Plant Extracts....................................... 244
3.8.4.5.2 Individual Compounds 244
3.8.4.6   Analysis 244
3.9   References 245

4   **Carbohydrates** 248
4.1   Foreword 248
4.2   Monosaccharides...................................... 248
4.2.1   Structure and Nomenclature 248
4.2.1.1   Nomenclature 248
4.2.1.2   Configuration.. 249
4.2.1.3   Conformation 254
4.2.2   Physical Properties 256
4.2.2.1   Hygroscopicity and Solubility 256
4.2.2.2   Optical Rotation, Mutarotation 257
4.2.3   Sensory Properties 258
4.2.4   Chemical Reactions and Derivatives.................... 261
4.2.4.1   Reduction to Sugar Alcohols 261
4.2.4.2   Oxidation to Aldonic, Dicarboxylic and Uronic Acids 262
4.2.4.3   Reactions in the Presence of Acids and Alkalis............... 263
4.2.4.3.1 Reactions in Strongly Acidic Media 263
4.2.4.3.2 Reactions in Strongly Alkaline Solution 266
4.2.4.3.3 Caramelization 270
4.2.4.4   Reactions with Amino Compounds (*Maillard* Reaction) 270
4.2.4.4.1 Initial Phase of the Maillard Reaction................ 271
4.2.4.4.2 Formation of Deoxyosones 272
4.2.4.4.3 Secondary Products of 3-Deoxyosones 274
4.2.4.4.4 Secondary Products of 1-Deoxyosones 276
4.2.4.4.5 Secondary Products of 4-Deoxyosones 280
4.2.4.4.6 Redox Reactions 282
4.2.4.4.7 *Strecker* Reaction 282
4.2.4.4.8 Formation of Colored Compounds 284
4.2.4.4.9 Protein Modifications 285
4.2.4.4.10 Inhibition of the Maillard Reaction................... 289
4.2.4.5   Reactions with Hydroxy Compounds (O-Glycosides) 289
4.2.4.6   Esters 290
4.2.4.7   Ethers 291
4.2.4.8   Cleavage of Glycols 292
4.3   Oligosaccharides..................................... 292
4.3.1   Structure and Nomenclature 292
4.3.2   Properties and Reactions 294
4.4   Polysaccharides...................................... 296

4.4.1	Classification, Structure	296
4.4.2	Conformation	296
4.4.2.1	Extended or Stretched, Ribbon-Type Conformation	296
4.4.2.2	Hollow Helix-Type Conformation	297
4.4.2.3	Crumpled-Type Conformation	298
4.4.2.4	Loosely-Jointed Conformation	298
4.4.2.5	Conformations of Heteroglycans	298
4.4.2.6	Interchain Interactions	298
4.4.3	Properties	300
4.4.3.1	General Remarks	300
4.4.3.2	Perfectly Linear Polysaccharides	300
4.4.3.3	Branched Polysaccharides	300
4.4.3.4	Linearly Branched Polysaccharides	301
4.4.3.5	Polysaccharides with Carboxyl Groups	301
4.4.3.6	Polysaccharides with Strongly Acidic Groups	302
4.4.3.7	Modified Polysaccharides	302
4.4.3.7.1	Derivatization with Neutral Substituents	302
4.4.3.7.2	Derivatization with Acidic Substituents	302
4.4.4	Individual Polysaccharides	302
4.4.4.1	Agar	302
4.4.4.1.1	Occurrence, Isolation	302
4.4.4.1.2	Structure, Properties	302
4.4.4.1.3	Utilization	303
4.4.4.2	Alginates	303
4.4.4.2.1	Occurrence, Isolation	303
4.4.4.2.2	Structure, Properties	303
4.4.4.2.3	Derivatives	304
4.4.4.2.4	Utilization	304
4.4.4.3	Carrageenans	304
4.4.4.3.1	Occurrence, Isolation	304
4.4.4.3.2	Structure, Properties	305
4.4.4.3.3	Utilization	306
4.4.4.4	Furcellaran	306
4.4.4.4.1	Occurrence, Isolation	306
4.4.4.4.2	Structure, Properties	307
4.4.4.4.3	Utilization	307
4.4.4.5	Gum Arabic	307
4.4.4.5.1	Occurrence, Isolation	307
4.4.4.5.2	Structure, Properties	307
4.4.4.5.3	Utilization	309
4.4.4.6	Gum Ghatti	309
4.4.4.6.1	Occurrence	309
4.4.4.6.2	Structure, Properties	309
4.4.4.6.3	Utilization	309
4.4.4.7	Gum Tragacanth	310
4.4.4.7.1	Occurrence	310
4.4.4.7.2	Structure, Properties	310
4.4.4.7.3	Utilization	310
4.4.4.8	Karaya Gum	310
4.4.4.8.1	Occurrence	310
4.4.4.8.2	Structure, Properties	310

4.4.4.8.3   Utilization 311
4.4.4.9   Guaran Gum 311
4.4.4.9.1   Occurrence, Isolation 311
4.4.4.9.2   Structure, Properties 311
4.4.4.9.3   Utilization 312
4.4.4.10   Locust Bean Gum 312
4.4.4.10.1   Occurrence, Isolation 312
4.4.4.10.2   Structure, Properties 312
4.4.4.10.3   Utilization 312
4.4.4.11   Tamarind Flour 312
4.4.4.11.1   Occurrence, Isolation 312
4.4.4.11.2   Structure, Properties 312
4.4.4.11.3   Utilization 313
4.4.4.12   Arabinogalactan from Larch 313
4.4.4.12.1   Occurrence, Isolation 313
4.4.4.12.2   Structure, Properties 313
4.4.4.12.3   Utilization 313
4.4.4.13   Pectin 314
4.4.4.13.1   Occurrence, Isolation 314
4.4.4.13.2   Structure, Properties 314
4.4.4.13.3   Utilization 315
4.4.4.14   Starch 315
4.4.4.14.1   Occurrence, Isolation 315
4.4.4.14.2   Structure and Properties of Starch Granules 316
4.4.4.14.3   Structure and Properties of Amylose 321
4.4.4.14.4   Structure and Properties of Amylopectin 323
4.4.4.14.5   Utilization 324
4.4.4.14.6   Resistant Starch 325
4.4.4.15   Modified Starches 325
4.4.4.15.1   Mechanically Damaged Starches 325
4.4.4.15.2   Extruded Starches 325
4.4.4.15.3   Dextrins 326
4.4.4.15.4   Pregelatinized Starch 326
4.4.4.15.5   Thin-Boiling Starch 326
4.4.4.15.6   Starch Ethers 326
4.4.4.15.7   Starch Esters 326
4.4.4.15.8   Cross-Linked Starches 327
4.4.4.15.9   Oxidized Starches 327
4.4.4.16   Cellulose 327
4.4.4.16.1   Occurrence, Isolation 327
4.4.4.16.2   Structure, Properties 328
4.4.4.16.3   Utilization 328
4.4.4.17   Cellulose Derivatives 328
4.4.4.17.1   Alkyl Cellulose, Hydroxyalkyl Cellulose 329
4.4.4.17.2   Carboxymethyl Cellulose 329
4.4.4.18   Hemicelluloses 330
4.4.4.19   Xanthan Gum 331
4.4.4.19.1   Occurrence, Isolation 331
4.4.4.19.2   Structure, Properties 331
4.4.4.19.3   Utilization 331
4.4.4.20   Scleroglucan 331

4.4.4.20.1   Occurrence, Isolation 331
4.4.4.20.2   Structure, Properties 331
4.4.4.20.3   Utilization 332
4.4.4.21   Dextran... 332
4.4.4.21.1   Occurrence... 332
4.4.4.21.2   Structure, Properties 332
4.4.4.21.3   Utilization 332
4.4.4.22   Inulin and Oligofructose.................................. 332
4.4.4.22.1   Occurrence... 332
4.4.4.22.2   Structure... 332
4.4.4.22.3   Utilization 332
4.4.4.23   Polyvinyl Pyrrolidone (PVP).............................. 333
4.4.4.23.1   Structure, Properties 333
4.4.4.23.2   Utilization 333
4.4.5   Enzymatic Degradation of Polysaccharides 333
4.4.5.1   Amylases 333
4.4.5.1.1 α-Amylase... 333
4.4.5.1.2 β-Amylase... 333
4.4.5.1.3   Glucan-1,4-α-D-glucosidase (Glucoamylase) 333
4.4.5.1.4 α-Dextrin Endo-1,6-α-glucosidase (Pullulanase) 334
4.4.5.2   Pectinolytic Enzymes 334
4.4.5.3   Cellulases 335
4.4.5.4   Endo-1,3(4)-β-glucanase 335
4.4.5.5   Hemicellulases 335
4.4.6   Analysis of Polysaccharides 335
4.4.6.1   Thickening Agents 335
4.4.6.2   Dietary Fibers 336
4.5   References 337

5   Aroma Compounds...................................... 340
5.1   Foreword 340
5.1.1   Concept Delineation 340
5.1.2   Impact Compounds of Natural Aromas 340
5.1.3   Threshold Value 341
5.1.4   Aroma Value 342
5.1.5   Off-Flavors, Food Taints 343
5.2   Aroma Analysis.. 345
5.2.1   Aroma Isolation.. 345
5.2.1.1   Distillation, Extraction 346
5.2.1.2   Gas Extraction 348
5.2.1.3   Headspace Analysis 348
5.2.2   Sensory Relevance 349
5.2.2.1   Aroma Extract Dilution Analysis (AEDA)................... 350
5.2.2.2   Headspace GC Olfactometry 350
5.2.3   Enrichment 351
5.2.4   Chemical Structure 353
5.2.5   Enantioselective Analysis................................. 353
5.2.6   Quantitative Analysis, Aroma Values 356
5.2.6.1   Isotopic Dilution Analysis (IDA) 356
5.2.6.2   Aroma Values (AV)....................................... 356
5.2.7   Aroma Model, Omission Experiments 357

5.3	Individual Aroma Compounds	359
5.3.1	Nonenzymatic Reactions	360
5.3.1.1	Carbonyl Compounds	361
5.3.1.2	Pyranones	361
5.3.1.3	Furanones	361
5.3.1.4	Thiols, Thioethers, Di- and Trisulfides	363
5.3.1.5	Thiazoles	367
5.3.1.6	Pyrroles, Pyridines	367
5.3.1.7	Pyrazines	371
5.3.1.8	Amines	373
5.3.1.9	Phenols	374
5.3.2	Enzymatic Reactions	374
5.3.2.1	Carbonyl Compounds, Alcohols	376
5.3.2.2	Hydrocarbons, Esters	379
5.3.2.3	Lactones	380
5.3.2.4	Terpenes	382
5.3.2.5	Volatile Sulfur Compounds	387
5.3.2.6	Pyrazines	388
5.3.2.7	Skatole, p-Cresol	388
5.4	Interactions with Other Food Constituents	389
5.4.1	Lipids	390
5.4.2	Proteins, Polysaccharides	391
5.5	Natural and Synthetic Flavorings	393
5.5.1	Raw Materials for Essences	393
5.5.1.1	Essential Oils	394
5.5.1.2	Extracts, Absolues	394
5.5.1.3	Distillates	394
5.5.1.4	Microbial Aromas	394
5.5.1.5	Synthetic Natural Aroma Compounds	395
5.5.1.6	Synthetic Aroma Compounds	395
5.5.2	Essences	396
5.5.3	Aromas from Precursors	397
5.5.4	Stability of Aromas	397
5.5.5	Encapsulation of Aromas	398
5.6	Relationships Between Structure and Odor	398
5.6.1	General Aspects	398
5.6.2	Carbonyl Compounds	398
5.6.3	Alkyl Pyrazines	399
5.7	References	400
6	**Vitamins**	403
6.1	Foreword	403
6.2	Fat-Soluble Vitamins	404
6.2.1	Retinol (Vitamin A)	404
6.2.1.1	Biological Role	404
6.2.1.2	Requirement, Occurrence	404
6.2.1.3	Stability, Degradation	406
6.2.2	Calciferol (Vitamin D)	406
6.2.2.1	Biological Role	406
6.2.2.2	Requirement, Occurrence	406
6.2.2.3	Stability, Degradation	407

6.2.3	α-Tocopherol (Vitamin E)	407
6.2.3.1	Biological Role	407
6.2.3.2	Requirement, Occurrence	407
6.2.3.3	Stability, Degradation	408
6.2.4	Phytomenadione (Vitamin K_1 Phylloquinone)	408
6.2.4.1	Biological Role	408
6.2.4.2	Requirement, Occurrence	408
6.2.4.3	Stability, Degradation	408
6.3	Water-Soluble Vitamins	411
6.3.1	Thiamine (Vitamin B_1)	411
6.3.1.1	Biological Role	411
6.3.1.2	Requirement, Occurrence	411
6.3.1.3	Stability, Degradation	412
6.3.2	Riboflavin (Vitamin B_2)	413
6.3.2.1	Biological Role	413
6.3.2.2	Requirement, Occurrence	413
6.3.2.3	Stability, Degradation	413
6.3.3	Pyridoxine (Pyridoxal, Vitamin B_6)	413
6.3.3.1	Biological Role	413
6.3.3.2	Requirement, Occurrence	414
6.3.3.3	Stability, Degradation	414
6.3.4	Nicotinamide (Niacin)	414
6.3.4.1	Biological Role	414
6.3.4.2	Requirement, Occurrence	414
6.3.4.3	Stability, Degradation	415
6.3.5	Pantothenic Acid	415
6.3.5.1	Biological Role	415
6.3.5.2	Requirement, Occurrence	415
6.3.5.3	Stability, Degradation	415
6.3.6	Biotin	415
6.3.6.1	Biological Role	415
6.3.6.2	Requirement, Occurrence	415
6.3.6.3	Stability, Degradation	415
6.3.7	Folic Acid	415
6.3.7.1	Biological Role	415
6.3.7.2	Requirement, Occurrence	416
6.3.7.3	Stability, Degradation	416
6.3.8	Cyanocobalamin (Vitamin B_{12})	416
6.3.8.1	Biological Role	416
6.3.8.2	Requirement, Occurrence	417
6.3.8.3	Stability, Degradation	417
6.3.9	L-Ascorbic Acid (Vitamin C)	417
6.3.9.1	Biological Role	417
6.3.9.2	Requirement, Occurrence	417
6.3.9.3	Stability, Degradation	418
6.4	References	420
7	**Minerals**	421
7.1	Foreword	421
7.2	Main Elements	421
7.2.1	Sodium	421

7.2.2	Potassium	423
7.2.3	Magnesium	424
7.2.4	Calcium	424
7.2.5	Chloride	424
7.2.6	Phosphorus	424
7.3	Trace Elements	424
7.3.1	General Remarks	424
7.3.2	Individual Trace Elements	424
7.3.2.1	Iron	424
7.3.2.2	Copper	425
7.3.2.3	Zinc	425
7.3.2.4	Manganese	425
7.3.2.5	Cobalt	426
7.3.2.6	Chromium	426
7.3.2.7	Selenium	426
7.3.2.8	Molybdenum	426
7.3.2.9	Nickel	426
7.3.2.10	Fluorine	426
7.3.2.11	Iodine	427
7.3.3	Ultra-trace Elements	427
7.3.3.1	Tin	427
7.3.3.2	Aluminum	427
7.3.3.3	Boron	427
7.3.3.4	Silicon	428
7.3.3.5	Arsenic	428
7.4	Minerals in Food Processing	428
7.5	References	428
8	**Food Additives**	429
8.1	Foreword	429
8.2	Vitamins	430
8.3	Amino Acids	430
8.4	Minerals	430
8.5	Aroma Substances	430
8.6	Flavor Enhancers	430
8.6.1	Monosodium Glutamate (MSG)	430
8.6.2	5′-Nucleotides	431
8.6.3	Maltol	431
8.6.4	Compounds with a Cooling Effect	431
8.7	Sugar Substitutes	432
8.8	Sweeteners	432
8.8.1	Sweet Taste: Structural Requirements	432
8.8.1.1	Structure–Activity Relationships in Sweet Compounds	432
8.8.1.2	Synergism	433
8.8.2	Saccharin	433
8.8.3	Cyclamate	434
8.8.4	Monellin	436
8.8.5	Thaumatins	437
8.8.6	Curculin and Miraculin	438
8.8.7	*Gymnema silvestre* Extract	438
8.8.8	Stevioside	438

8.8.9	Phyllodulcin	439
8.8.10	Glycyrrhizin	439
8.8.11	Dihydrochalcones	439
8.8.12	Ureas and Guanidines	439
8.8.12.1	Suosan	439
8.8.12.2	Guanidines	440
8.8.13	Oximes	440
8.8.14	Oxathiazinone Dioxides	440
8.8.15	Dipeptide Esters and Amides	441
8.8.15.1	Aspartame	441
8.8.15.2	Superaspartame	442
8.8.15.3	Alitame	442
8.8.16	Hernandulcin	442
8.9	Food Colors	443
8.10	Acids	443
8.10.1	Acetic Acid and Other Fatty Acids	443
8.10.2	Succinic Acid	443
8.10.3	Succinic Acid Anhydride	447
8.10.4	Adipic Acid	447
8.10.5	Fumaric Acid	447
8.10.6	Lactic Acid	448
8.10.7	Malic Acid	448
8.10.8	Tartaric Acid	448
8.10.9	Citric Acid	448
8.10.10	Phosphoric Acid	449
8.10.11	Hydrochloric and Sulfuric Acids	449
8.10.12	Gluconic Acid and Glucono-δ-lactone	449
8.11	Bases	449
8.12	Antimicrobial Agents	449
8.12.1	Benzoic Acid	449
8.12.2	PHB-Esters	450
8.12.3	Sorbic Acid	451
8.12.4	Propionic Acid	452
8.12.5	Acetic Acid	452
8.12.6	SO_2 and Sulfite	452
8.12.7	Diethyl (Dimethyl) Pyrocarbonate	453
8.12.8	Ethylene Oxide, Propylene Oxide	453
8.12.9	Nitrite, Nitrate	454
8.12.10	Antibiotics	454
8.12.11	Diphenyl	454
8.12.12	o-Phenylphenol	454
8.12.13	Thiabendazole, 2-(4-Thiazolyl)benzimidazole	454
8.13	Antioxidants	455
8.14	Chelating Agents (Sequestrants)	455
8.15	Surface-Active Agents	456
8.15.1	Emulsions	456
8.15.2	Emulsifier Action	457
8.15.2.1	Structure and Activity	457
8.15.2.2	Critical Micelle Concentration (CMC), Lyotropic Mesomorphism	458
8.15.2.3	HLB-Value	459

8.15.3	Synthetic Emulsifiers	460
8.15.3.1	Mono-, Diacylglycerides and Derivatives	460
8.15.3.2	Sugar Esters	462
8.15.3.3	Sorbitan Fatty Acid Esters	462
8.15.3.4	Polyoxyethylene Sorbitan Esters	462
8.15.3.5	Polyglycerol – Polyricinoleate (PGPR)	462
8.15.3.6	Stearyl-2-Lactylate	463
8.16	Substitutes for Fat	463
8.16.1	Fat Mimetics	463
8.16.1.1	Microparticulated Proteins	463
8.16.1.2	Carbohydrates	463
8.16.2	Synthetic Fat Substitutes	463
8.16.2.1	Carbohydrate Polyester	464
8.16.2.2	Retrofats	464
8.17	Thickening Agents, Gel Builders, Stabilizers	464
8.18	Humectants	464
8.19	Anticaking Agents	464
8.20	Bleaching Agents	464
8.21	Clarifying Agents	464
8.22	Propellants, Protective Gases	465
8.23	References	465
9	**Food Contamination**	467
9.1	General Remarks	467
9.2	Toxic Trace Elements	468
9.2.1	Arsenic	468
9.2.2	Mercury	468
9.2.3	Lead	468
9.2.4	Cadmium	469
9.2.5	Radionuclides	470
9.3	Toxic Compounds of Microbial Origin	470
9.3.1	Food Poisoning by Bacterial Toxins	470
9.3.2	Mycotoxins	472
9.4	Plant-Protective Agents (PPA)	475
9.4.1	General Remarks	475
9.4.2	Active Agents	476
9.4.2.1	Insecticides	476
9.4.2.2	Fungicides	476
9.4.2.3	Herbicides	483
9.4.3	Analysis	483
9.4.4	PPA Residues, Risk Assessment	485
9.4.4.1	Exceeding the Maximum Permissible Quantity	485
9.4.4.2	Risk Assessment	485
9.4.4.3	Natural Pesticides	486
9.5	Veterinary Medicines and Feed Additives	486
9.5.1	Foreword	486
9.5.2	Antibiotics	487
9.5.3	Anthelmintics	487
9.5.4	Coccidiostats	487
9.5.5	Analysis	487
9.6	Polychlorinated Biphenyls (PCBs)	489

9.7	Harmful Substances from Thermal Processes	490
9.7.1	Polycyclic Aromatic Hydrocarbons (PAHs)	490
9.7.2	Furan	490
9.7.3	Acrylamide	490
9.8	Nitrate, Nitrite, Nitrosamines	492
9.8.1	Nitrate, Nitrite	492
9.8.2	Nitrosamines, Nitrosamides	492
9.9	Cleansing Agents and Disinfectants	495
9.10	Polychlorinated Dibenzodioxins (PCDD) and Dibenzofurans (PCDF)	496
9.11	References	497
10	**Milk and Dairy Products**	498
10.1	Milk	498
10.1.1	Physical and Physico-Chemical	498
10.1.2	Composition	501
10.1.2.1	Proteins	501
10.1.2.1.1	Casein Fractions	502
10.1.2.1.2	Micelle Formation	508
10.1.2.1.3	Gel Formation	509
10.1.2.1.4	Whey Proteins	511
10.1.2.2	Carbohydrates	512
10.1.2.3	Lipids	513
10.1.2.4	Organic Acids	515
10.1.2.5	Minerals	515
10.1.2.6	Vitamins	515
10.1.2.7	Enzymes	516
10.1.2.7.1	Plasmin	516
10.1.2.7.2	Lactoperoxidase	517
10.1.3	Processing of Milk	517
10.1.3.1	Purification	518
10.1.3.2	Creaming	518
10.1.3.3	Heat Treatment	518
10.1.3.4	Homogenization	518
10.1.3.5	Reactions During Heating	519
10.1.4	Types of Milk	520
10.2	Dairy Products	521
10.2.1	Fermented Milk Products	521
10.2.1.1	Sour Milk	523
10.2.1.2	Yoghurt	523
10.2.1.3	Kefir and Kumiss	523
10.2.1.4	Taette Milk	524
10.2.2	Cream	524
10.2.3	Butter	524
10.2.3.1	Cream Separation and Treatment	525
10.2.3.2	Churning	525
10.2.3.3	Packaging	526
10.2.3.4	Products Derived from Butter	526
10.2.4	Condensed Milk	526
10.2.5	Dehydrated Milk Products	527
10.2.6	Coffee Whitener	528

10.2.7	Ice Cream	528
10.2.8	Cheese	529
10.2.8.1	Curd Formation	529
10.2.8.2	Unripened Cheese	530
10.2.8.3	Ripening	532
10.2.8.4	Processed Cheese	536
10.2.8.5	Imitation Cheese	536
10.2.9	Casein, Caseinates, Coprecipitate	536
10.2.10	Whey Products	537
10.2.10.1	Whey Powder	537
10.2.10.2	Demineralized Whey Powder	538
10.2.10.3	Partially Desugared Whey Protein Concentrates	538
10.2.10.4	Hydrolyzed Whey Syrups	539
10.2.11	Lactose	539
10.2.12	Cholesterol-Reduced Milk and Milk Products	539
10.3	Aroma of Milk and Dairy Products	539
10.3.1	Milk, Cream	539
10.3.2	Condensed Milk, Dried Milk Products	539
10.3.3	Sour Milk Products, Yoghurt	540
10.3.4	Butter	540
10.3.5	Cheese	541
10.3.6	Aroma Defects	543
10.4	References	544
11	**Eggs**	546
11.1	Foreword	546
11.2	Structure, Physical Properties and Composition	546
11.2.1	General Outline	546
11.2.2	Shell	547
11.2.3	Albumen (Egg White)	548
11.2.3.1	Proteins	548
11.2.3.1.1	Ovalbumin	548
11.2.3.1.2	Conalbumin (Ovotransferrin)	550
11.2.3.1.3	Ovomucoid	550
11.2.3.1.4	Lysozyme (Ovoglobulin G1)	550
11.2.3.1.5	Ovoglobulins G2 and G3	551
11.2.3.1.6	Ovomucin	551
11.2.3.1.7	Flavoprotein	551
11.2.3.1.8	Ovoinhibitor	551
11.2.3.1.9	Avidin	551
11.2.3.1.10	Cystatin (Ficin Inhibitor)	551
11.2.3.2	Other Constituents	551
11.2.3.2.1	Lipids	551
11.2.3.2.2	Carbohydrates	552
11.2.3.2.3	Minerals	552
11.2.3.2.4	Vitamins	552
11.2.4	Egg Yolk	553
11.2.4.1	Proteins of Granules	554
11.2.4.1.1	Lipovitellins	554
11.2.4.1.2	Phosvitin	554
11.2.4.2	Plasma Proteins	555

11.2.4.2.1   Lipovitellenin 555
11.2.4.2.2   Livetin 555
11.2.4.3   Lipids 555
11.2.4.4   Other Constituents 556
11.2.4.4.1   Carbohydrates 556
11.2.4.4.2   Minerals 556
11.2.4.4.3   Vitamins 556
11.2.4.4.4   Aroma Substances 557
11.2.4.4.5   Colorants 557
11.3   Storage of Eggs 557
11.4   Egg Products 557
11.4.1   General Outline 557
11.4.2   Technically-Important Properties 558
11.4.2.1   Thermal Coagulation 558
11.4.2.2   Foaming Ability 558
11.4.2.2.1   Egg White 558
11.4.2.2.2   Egg Yolk 559
11.4.2.3   Emulsifying Effect 559
11.4.3   Dried Products 559
11.4.4   Frozen Egg Products 560
11.4.5   Liquid Egg Products 561
11.5   References 561

12   **Meat** 563
12.1   Foreword 563
12.2   Structure of Muscle Tissue 564
12.2.1   Skeletal Muscle 564
12.2.2   Heart Muscle 567
12.2.3   Smooth Muscle 567
12.3   Muscle Tissue: Composition and Function 568
12.3.1   Overview 568
12.3.2   Proteins 568
12.3.2.1   Proteins of the Contractile Apparatus and Their Functions 568
12.3.2.1.1   Myosin 568
12.3.2.1.2   Titin 569
12.3.2.1.3   Actin 570
12.3.2.1.4   Tropomyosin and Troponin 571
12.3.2.1.5   Other Myofibrillar Proteins 571
12.3.2.1.6   Contraction and Relaxation 572
12.3.2.1.7   Actomyosin 573
12.3.2.2   Soluble Proteins 573
12.3.2.2.1   Enzymes 573
12.3.2.2.2   Myoglobin 573
12.3.2.2.3   Color of Meat 575
12.3.2.2.4   Curing, Reddening 576
12.3.2.3   Insoluble Proteins 577
12.3.2.3.1   Collagen 577
12.3.2.3.2   Elastin 584
12.3.3   Free Amino Acids 584
12.3.4   Peptides 584
12.3.5   Amines 584

12.3.6	Guanidine Compounds	585
12.3.7	Quaternary Ammonium Compounds	585
12.3.8	Purines and Pyrimidines	586
12.3.9	Organic Acids	586
12.3.10	Carbohydrates	586
12.3.11	Vitamins	586
12.3.12	Minerals	587
12.4	Post Mortem Changes in Muscle	587
12.4.1	Rigor Mortis	587
12.4.2	Defects (PSE and DFD Meat)	588
12.4.3	Aging of Meat	589
12.5	Water Holding Capacity of Meat	590
12.6	Kinds of Meat, Storage, Processing	592
12.6.1	Kinds of Meat, By-Products	592
12.6.1.1	Beef	592
12.6.1.2	Veal	592
12.6.1.3	Mutton and Lamb	593
12.6.1.4	Goat Meat	593
12.6.1.5	Pork	593
12.6.1.6	Horse Meat	593
12.6.1.7	Poultry	593
12.6.1.8	Game	593
12.6.1.9	Variety Meats	593
12.6.1.10	Blood	594
12.6.1.11	Glandular Products	595
12.6.2	Storage and Preservation Processes	595
12.6.2.1	Cooling	595
12.6.2.2	Freezing	595
12.6.2.3	Drying	597
12.6.2.4	Salt and Pickle Curing	597
12.6.2.5	Smoking	597
12.6.2.6	Heating	597
12.6.2.7	Tenderizing	598
12.7	Meat Products	598
12.7.1	Canned Meat	598
12.7.2	Ham, Sausages, Pastes	598
12.7.2.1	Ham, Bacon	598
12.7.2.1.1	Raw Smoked Hams	598
12.7.2.1.2	Cooked Ham	598
12.7.2.1.3	Bacon	599
12.7.2.2	Sausages	599
12.7.2.2.1	Raw Sausages	600
12.7.2.2.2	Cooked Sausages	600
12.7.2.2.3	Boiling Sausages	601
12.7.2.3	Meat Paste (Pâté)	601
12.7.2.3.1	Pastes	601
12.7.2.3.2	Pains	601
12.7.3	Meat Extracts and Related Products	601
12.7.3.1	Beef Extract	601
12.7.3.2	Whale Meat Extract	602
12.7.3.3	Poultry Meat Extract	602

12.7.3.4   Yeast Extract 602
12.7.3.5   Hydrolyzed Vegetable Proteins 602
12.8   Dry Soups and Dry Sauces 603
12.8.1   Main Components 603
12.8.2   Production 604
12.9   Meat Aroma 605
12.9.1   Taste compounds 605
12.9.2   Odorants 605
12.9.3   Process Flavors 607
12.9.4   Aroma Defects 608
12.10   Meat Analysis 608
12.10.1   Meat 608
12.10.1.1 Animal Origin 608
12.10.1.1.1 Electrophoresis 608
12.10.1.1.2 Sexual Origin of Beef 610
12.10.1.2 Differentiation of Fresh and Frozen Meat 610
12.10.1.3 Pigments 611
12.10.1.4 Treatment with Proteinase Preparations 611
12.10.1.5 Anabolic Steroids 612
12.10.1.6 Antibiotics 612
12.10.2   Processed Meats 612
12.10.2.1 Main Ingredients 613
12.10.2.2 Added Water 613
12.10.2.3 Lean Meat Free of Connective Tissue 613
12.10.2.3.1 Connective Tissue Protein 613
12.10.2.3.2 Added Protein 613
12.10.2.4 Nitrosamines 614
12.11   References 614

13   **Fish, Whales, Crustaceans, Mollusks** 617
13.1   Fish 617
13.1.1   Foreword 617
13.1.2   Food Fish 618
13.1.2.1   Sea Fish 618
13.1.2.1.1 Sharks 618
13.1.2.1.2 Herring 618
13.1.2.1.3 Cod Fish 621
13.1.2.1.4 Scorpaenidae 622
13.1.2.1.5 Perch-like Fish 622
13.1.2.1.6 Flat Fish 622
13.1.2.2   Freshwater Fish 623
13.1.2.2.1 Eels 623
13.1.2.2.2 Salmon 623
13.1.3   Skin and Muscle Tissue Structure 623
13.1.4   Composition 624
13.1.4.1   Overview 624
13.1.4.2   Proteins 624
13.1.4.2.1 Sarcoplasma Proteins 625
13.1.4.2.2 Contractile Proteins 625
13.1.4.2.3 Connective Tissue Protein 625
13.1.4.2.4 Serum Proteins 626

13.1.4.3   Other N-Compounds.................................... 626
13.1.4.3.1   Free Amino Acids, Peptides 626
13.1.4.3.2   Amines, Amine Oxides 626
13.1.4.3.3   Guanidine Compounds................................. 626
13.1.4.3.4   Quaternary Ammonium Compounds 626
13.1.4.3.5   Purines 627
13.1.4.3.6   Urea 627
13.1.4.4   Carbohydrates 627
13.1.4.5   Lipids 627
13.1.4.6   Vitamins... 628
13.1.4.7   Minerals 628
13.1.4.8   Aroma Substances.................................... 628
13.1.4.9   Other Constituents 629
13.1.5   Post mortem Changes 629
13.1.6   Storage and Processing of Fish and Fish Products 630
13.1.6.1   General Remarks..................................... 630
13.1.6.2   Cooling and Freezing 631
13.1.6.3   Drying 632
13.1.6.4   Salting 633
13.1.6.5   Smoking... 633
13.1.6.6   Marinated, Fried and Cooked Fish Products 634
13.1.6.7   Saithe 635
13.1.6.8   Anchosen 635
13.1.6.9   Pasteurized Fish Products............................. 635
13.1.6.10   Fish Products with an Extended Shelf Life 635
13.1.6.11   Surimi, Kamboko 635
13.1.6.12   Fish Eggs and Sperm 635
13.1.6.12.1   Caviar.. 635
13.1.6.12.2   Caviar Substitutes 636
13.1.6.12.3   Fish Sperm.. 636
13.1.6.13   Some Other Fish Products 636
13.2   Whales 636
13.3   Crustaceans 636
13.3.1   Shrimps 636
13.3.2   Crabs 637
13.3.3   Lobsters 637
13.3.4   Crayfish, Crawfish.................................... 637
13.4   Mollusks (*Mollusca*)................................. 638
13.4.1   Mollusks (*Bivalvia*) 638
13.4.2   Snails 638
13.4.3   Octopus, Sepia, Squid................................. 639
13.5   Turtles 639
13.6   Frogdrums 639
13.7   References 639

14   **Edible Fats and Oils** 640
14.1   Foreword 640
14.2   Data on Production and Consumption 640
14.3   Origin of Individual Fats and Oils..................... 640
14.3.1   Animal Fats 640
14.3.1.1   Land Animal Fats 640

14.3.1.1.1   Edible Beef Fat 642
14.3.1.1.2   Sheep Tallow 643
14.3.1.1.3   Hog Fat (Lard) 643
14.3.1.1.4   Goose Fat 644
14.3.1.2   Marine Oils 644
14.3.1.2.1   Whale Oil 644
14.3.1.2.2   Seal Oil 644
14.3.1.2.3   Herring Oil 644
14.3.2   Oils of Plant Origin 644
14.3.2.1   Fruit Pulp Oils 645
14.3.2.1.1   Olive Oil 645
14.3.2.1.2   Palm Oil 646
14.3.2.2   Seed Oils 647
14.3.2.2.1   Production 647
14.3.2.2.2   Oils Rich in Lauric and Myristic Acids 647
14.3.2.2.3   Oils Rich in Palmitic and Stearic Acids 648
14.3.2.2.4   Oils Rich in Palmitic Acid 649
14.3.2.2.5   Oils Low in Palmitic Acid and Rich in Oleic
              and Linoleic Acids 650
14.4   Processing of Fats and Oils 653
14.4.1   Refining 653
14.4.1.1   Removal of Lecithin 654
14.4.1.2   Degumming 654
14.4.1.3   Removal of Free Fatty Acids (Deacidification) 654
14.4.1.4   Bleaching 655
14.4.1.5   Deodorization 655
14.4.1.6   Product Quality Control 656
14.4.2   Hydrogenation 656
14.4.2.1   General Remarks 656
14.4.2.2   Catalysts 657
14.4.2.3   The Process 658
14.4.3   Interesterification 658
14.4.4   Fractionation 659
14.4.5   Margarine – Manufacturing and Properties 660
14.4.5.1   Composition 660
14.4.5.2   Manufacturing 661
14.4.5.3   Varieties of Margarine 661
14.4.6   Mayonnaise 661
14.4.7   Fat Powder 661
14.4.8   Deep-Frying Fats 661
14.5   Analysis 662
14.5.0   Scope 662
14.5.1   Determination of Fat in Food 662
14.5.2   Identification of Fat 663
14.5.2.1   Characteristic Values 663
14.5.2.2   Color Reactions 664
14.5.2.3   Composition of Fatty Acids and Triacylglycerides 664
14.5.2.4   Minor Constituents 665
14.5.2.5   Melting Points 666
14.5.2.6   Chemometry 666
14.5.3   Detection of Changes During Processing and Storage 667

14.5.3.1   Lipolysis 667
14.5.3.2   Oxidative Deterioration 667
14.5.3.2.1 Oxidation State 667
14.5.3.2.2 Shelf Life Prediction Test.............................. 668
14.5.3.3   Heat Stability 668
14.5.3.4   Refining 669
14.6   References 669

15   **Cereals and Cereal Products** 670
15.1   Foreword 670
15.1.1   Introduction 670
15.1.2   Origin 670
15.1.3   Production 671
15.1.4   Anatomy – Chemical Composition, a Review 671
15.1.5   Special Role of Wheat–Gluten Formation 673
15.1.6   Celiac Disease 674
15.2   Individual Constituents 674
15.2.1   Proteins 674
15.2.1.1   Differences in Amino Acid Composition 674
15.2.1.2   A Review of the Osborne Fractions of Cereals 675
15.2.1.3   Protein Components of Wheat Gluten 680
15.2.1.3.1 High-Molecular Group (HMW Subunits of Glutenin) 681
15.2.1.3.2 Intermediate Molecular Weight Group (ω5-Gliadins,
            ω1,2-Gliadins)................................... 684
15.2.1.3.3 Low-Molecular Group (α-Gliadins, γ-Gliadins,
            LMW Subunits of Glutenin) 685
15.2.1.4   Structure of Wheat Gluten 691
15.2.1.4.1 Disulfide Bonds.. 691
15.2.1.4.2 Contribution of Gluten Proteins to the Baking Quality ... 692
15.2.1.5   Puroindolins... 695
15.2.2   Enzymes.. 695
15.2.2.1   Amylases 695
15.2.2.2   Proteinases 696
15.2.2.3   Lipases 696
15.2.2.4   Phytase 696
15.2.2.5   Lipoxygenases 697
15.2.2.6   Peroxidase, Catalase 698
15.2.2.7   Glutathione Dehydrogenase 698
15.2.2.8   Polyphenoloxidases 698
15.2.2.9   Ascorbic Acid Oxidase.................................. 698
15.2.2.10   Arabinoxylan Hydrolases 698
15.2.3   Other Nitrogen Compounds 699
15.2.4   Carbohydrates 701
15.2.4.1   Starch 701
15.2.4.2   Polysaccharides Other than Starch 702
15.2.4.2.1 Pentosans 702
15.2.4.2.2 β-Glucan 703
15.2.4.2.3 Glucofructans 703
15.2.4.2.4 Cellulose 703
15.2.4.3   Sugars... 703
15.2.5   Lipids 703

15.3	Cereals – Milling	706
15.3.1	Wheat and Rye	706
15.3.1.1	Storage	707
15.3.1.2	Milling	707
15.3.1.3	Milling Products	708
15.3.2	Other Cereals	710
15.3.2.1	Corn	710
15.3.2.2	Hull Cereals	710
15.3.2.2.1	Rice	710
15.3.2.2.2	Oats	710
15.3.2.2.3	Barley	711
15.4	Baked Products	711
15.4.1	Raw Materials	711
15.4.1.1	Wheat Flour	711
15.4.1.1.1	Chemical Assays	711
15.4.1.1.2	Physical Assays	713
15.4.1.1.3	Baking Tests	714
15.4.1.2	Rye Flour	715
15.4.1.3	Storage	716
15.4.1.4	Influence of Additives/Minor Ingredients on Baking Properties of Wheat Flour	716
15.4.1.4.1	Ascorbic Acid	716
15.4.1.4.2	Bromate, Azodicarbonamide	719
15.4.1.4.3	Lipoxygenase	719
15.4.1.4.4	Cysteine	719
15.4.1.4.5	Proteinases (Peptidases)	720
15.4.1.4.6	Salt	721
15.4.1.4.7	Emulsifiers, Shortenings	721
15.4.1.4.8	α-Amylase	721
15.4.1.4.9	Milk and Soy Products	722
15.4.1.5	Influence of Additives on Baking Properties of Rye Flour	722
15.4.1.5.1	Pregelatinized Flour	722
15.4.1.5.2	Acids	722
15.4.1.6	Dough Leavening Agents	722
15.4.1.6.1	Yeast	723
15.4.1.6.2	Chemical Leavening Agents	723
15.4.2	Dough Preparation	723
15.4.2.1	Addition of Yeast to Wheat Dough	723
15.4.2.1.1	Direct Addition	723
15.4.2.1.2	Indirect Addition	723
15.4.2.2	Sour Dough Making	724
15.4.2.3	Kneading	725
15.4.2.4	Fermentation	726
15.4.2.5	Events Involved in Dough Making and Dough Strengthening	726
15.4.2.5.1	Dough Making	726
15.4.2.5.2	Dough Strengthening	730
15.4.3	Baking Process	731
15.4.3.1	Conditions	731
15.4.3.2	Chemical and Physical Changes – Formation of Crumb	731
15.4.3.3	Aroma	734
15.4.3.3.1	White Bread Crust	734

15.4.3.3.2   White Bread Crumb 736
15.4.3.3.3   Rye Bread Crust 737
15.4.4   Changes During Storage................................ 739
15.4.5   Bread Types 740
15.4.6   Fine Bakery Products 741
15.5   Pasta Products 741
15.5.1   Raw Materials 741
15.5.2   Additives 742
15.5.3   Production 742
15.6   References 742

16   **Legumes** 746
16.1   Foreword 746
16.2   Individual Constituents.............................. 746
16.2.1   Proteins 746
16.2.1.1   Glubulines 746
16.2.1.2   Allergens 751
16.2.2   Enzymes... 753
16.2.3   Proteinase and Amylase Inhibitors 754
16.2.3.1   Occurrence and Properties 754
16.2.3.2   Structure... 755
16.2.3.3   Physiological Function 756
16.2.3.4   Action on Human Enzymes 757
16.2.3.5   Inactivation 757
16.2.3.6   Amylase Inhibitors 757
16.2.3.7   Conclusions 758
16.2.4   Lectins 759
16.2.5   Carbohydrates 759
16.2.6   Cyanogenic Glycosides 760
16.2.7   Lipids 761
16.2.8   Vitamins and Minerals 762
16.2.9   Phytoestrogens 762
16.2.10   Saponins.. 763
16.2.11   Other Constituents 764
16.3   Processing 764
16.3.1   Soybeans and Peanuts.................................. 764
16.3.1.1   Aroma Defects 764
16.3.1.2   Individual Products 765
16.3.1.2.1   Soy Proteins.. 765
16.3.1.2.2   Soy Milk 766
16.3.1.2.3   Tofu 766
16.3.1.2.4   Soy Sauce (Shoyu) 766
16.3.1.2.5   Miso 767
16.3.1.2.6   Natto... 767
16.3.1.2.7   Sufu 767
16.3.2   Peas and Beans 768
16.4   References 768

17   **Vegetables and Vegetable Products** 770
17.1   Vegetables 770
17.1.1   Foreword 770

17.1.2	Composition	770
17.1.2.1	Nitrogen Compounds	770
17.1.2.1.1	Proteins	770
17.1.2.1.2	Free Amino Acids	770
17.1.2.1.3	Amines	786
17.1.2.2	Carbohydrates	786
17.1.2.2.1	Mono- and Oligosaccharides, Sugar Alcohols	786
17.1.2.2.2	Polysaccharides	786
17.1.2.3	Lipids	787
17.1.2.4	Organic Acids	787
17.1.2.5	Phenolic Compounds	788
17.1.2.6	Aroma Substances	788
17.1.2.6.1	Mushrooms	788
17.1.2.6.2	Potatoes	788
17.1.2.6.3	Celery Tubers	788
17.1.2.6.4	Radishes	789
17.1.2.6.5	Red Beets (28)	790
17.1.2.6.6	Garlic and Onions	790
17.1.2.6.7	Watercress	791
17.1.2.6.8	White Cabbage, Red Cabage and Brussels Sprouts	792
17.1.2.6.9	Spinach	792
17.1.2.6.10	Artichoke	792
17.1.2.6.11	Cauliflower, Broccoli	792
17.1.2.6.12	Green Peas	793
17.1.2.6.13	Cucumbers	793
17.1.2.6.14	Tomatoes	793
17.1.2.7	Vitamins	793
17.1.2.8	Minerals	793
17.1.2.9	Other Constituents	793
17.1.2.9.1	Chlorophyll	793
17.1.2.9.2	Betalains	796
17.1.2.9.3	Goitrogenic Substances	798
17.1.2.9.4	Steroid Alkaloids	798
17.1.3	Storage	799
17.2	Vegetable Products	799
17.2.1	Dehydrated Vegetables	799
17.2.2	Canned Vegetables	800
17.2.3	Frozen Vegetables	801
17.2.4	Pickled Vegetables	802
17.2.4.1	Pickled Cucumbers (Salt and Dill Pickles)	802
17.2.4.2	Other Vegetables	802
17.2.4.3	Sauerkraut	802
17.2.4.4	Eating Olives	803
17.2.4.5	Faulty Processing of Pickles	804
17.2.5	Vinegar-Pickled Vegetables	804
17.2.6	Stock Brining of Vegetables	804
17.2.7	Vegetable Juices	805
17.2.8	Vegetable Paste	805
17.2.9	Vegetable Powders	805
17.3	References	805

18   **Fruits and Fruit Products** 807
18.1   Fruits 807
18.1.1   Foreword 807
18.1.2   Composition 807
18.1.2.1   N-Containing Compounds 807
18.1.2.1.1   Proteins, Enzymes................................... 807
18.1.2.1.2   Free Amino Acids 809
18.1.2.1.3   Amines 812
18.1.2.2   Carbohydrates 817
18.1.2.2.1   Monosaccharides..................................... 817
18.1.2.2.2   Oligosaccharides 817
18.1.2.2.3   Sugar Alcohols 817
18.1.2.2.4   Polysaccharides..................................... 818
18.1.2.3   Lipids 818
18.1.2.3.1   Fruit Flesh Lipids (Other than Carotenoids and Triterpenoids) .. 818
18.1.2.3.2   Carotenoids 818
18.1.2.3.3   Triterpenoids 819
18.1.2.3.4   Fruit Waxes 820
18.1.2.4   Organic Acids 820
18.1.2.5   Phenolic Compounds 822
18.1.2.5.1   Hydroxycinammic Acids, Hydroxycoumarins
              and Hydroxybenzoic Acids 823
18.1.2.5.2   Flavan-3-ols (Catechins), Flavan-3,4-diols,
              and Proanthocyanidins (Condensed Tanning Agents) 827
18.1.2.5.3   Anthocyanidins 829
18.1.2.5.4   Flavanones 832
18.1.2.5.5   Flavones, Flavonols 834
18.1.2.5.6   Lignans... 835
18.1.2.5.7   Flavonoid Biosynthesis 835
18.1.2.5.8   Technological Importance of Phenolic Compounds 835
18.1.2.6   Aroma Compounds..................................... 837
18.1.2.6.1   Bananas 837
18.1.2.6.2   Grapes 837
18.1.2.6.3   Citrus Fruits....................................... 837
18.1.2.6.4   Apples, Pears....................................... 839
18.1.2.6.5   Raspberries 839
18.1.2.6.6   Apricots 839
18.1.2.6.7   Peaches... 840
18.1.2.6.8   Passion Fruit 840
18.1.2.6.9   Strawberries 841
18.1.2.6.10   Pineapples 841
18.1.2.6.11   Cherries, Plums 841
18.1.2.6.12   Litchi 842
18.1.2.7   Vitamins.. 842
18.1.2.8   Minerals 843
18.1.3   Chemical Changes During Ripening of Fruit 843
18.1.3.1   Changes in Respiration Rate 843
18.1.3.2   Changes in Metabolic Pathways 844
18.1.3.3   Changes in Individual Constituents.................. 845
18.1.3.3.1   Carbohydrates 845
18.1.3.3.2   Proteins, Enzymes................................... 846

18.1.3.3.3   Lipids 846
18.1.3.3.4   Acids 846
18.1.3.3.5   Pigments 846
18.1.3.3.6   Aroma Compounds 846
18.1.4   Ripening as Influenced by Chemical Agents 847
18.1.4.1   Ethylene 847
18.1.4.2   Anti-Senescence Agents 848
18.1.4.2.1 Polyamines 848
18.1.4.2.2 1-Methylcyclopropene (MCP) 848
18.1.5   Storage of Fruits 848
18.1.5.1   Cold Storage 848
18.1.5.2   Storage in a Controlled (Modified) Atmosphere 848
18.2   Fruit Products 849
18.2.1   Dried Fruits 849
18.2.2   Canned Fruits 850
18.2.3   Deep-Frozen Fruits 850
18.2.4   Rum Fruits, Fruits in Sugar Syrup, etc. 851
18.2.5   Fruit Pulps and Slurries 851
18.2.6   Marmalades, Jams and Jellies 851
18.2.6.1   Marmalades 851
18.2.6.2   Jams 852
18.2.6.3   Jellies 852
18.2.7   Plum Sauce (Damson Cheese) 852
18.2.8   Fruit Juices 852
18.2.8.1   Preparation of the Fruit 853
18.2.8.2   Juice Extraction 853
18.2.8.3   Juice Treatment 853
18.2.8.4   Preservation 854
18.2.8.5   Side Products 854
18.2.9   Fruit Nectars 854
18.2.10   Fruit Juice Concentrates 854
18.2.10.1   Evaporation 855
18.2.10.2   Freeze Concentration 855
18.2.10.3   Membrane Filtration 855
18.2.11   Fruit Syrups 855
18.2.12   Fruit Powders 856
18.3   Alcohol-Free Beverages 856
18.3.1   Fruit Juice Beverages 856
18.3.2   Lemonades, Cold and Hot Beverages 856
18.3.3   Caffeine-Containing Beverages 856
18.3.4   Other Pop Beverages 856
18.4   Analysis 856
18.4.1   Various Constituents 857
18.4.2   Species-Specific Constituents 858
18.4.3   Abundance Ratios of Isotopes 858
18.5   References 859

19   **Sugars, Sugar Alcohols and Honey** 862
19.1   Sugars, Sugar Alcohols and Sugar Products 862
19.1.1   Foreword 862
19.1.2   Processing Properties 862

19.1.3	Nutritional/Physiological Properties	866
19.1.3.1	Metabolism	866
19.1.3.2	Glycemic Index	867
19.1.3.3	Functional Food	867
19.1.4	Individual Sugars and Sugar Alcohols	867
19.1.4.1	Sucrose (Beet Sugar, Cane Sugar)	867
19.1.4.1.1	General Outline	867
19.1.4.1.2	Production of Beet Sugar	869
19.1.4.1.3	Production of Cane Sugar	872
19.1.4.1.4	Other Sources for Sucrose Production	873
19.1.4.1.5	Packaging and Storage	873
19.1.4.1.6	Types of Sugar	873
19.1.4.1.7	Composition of some Sugar Types	873
19.1.4.1.8	Molasses	874
19.1.4.2	Sugars Produced from Sucrose	874
19.1.4.3	Starch Degradation Products	875
19.1.4.3.1	General Outline	875
19.1.4.3.2	Starch Syrup (Glucose or Maltose Syrup)	875
19.1.4.3.3	Dried Starch Syrup (Dried Glucose Syrup)	876
19.1.4.3.4	Glucose (Dextrose)	876
19.1.4.3.5	Glucose-Fructose Syrup (High Fructose Corn Syrup, HFCS)	877
19.1.4.3.6	Starch Syrup Derivatives	877
19.1.4.3.7	Polydextrose	877
19.1.4.4	Milk Sugar (Lactose) and Derived Products	877
19.1.4.4.1	Milk Sugar	877
19.1.4.4.2	Products from Lactose	877
19.1.4.5	Fruit Sugar (Fructose)	878
19.1.4.6	L-Sorbose and Other L-Sugars	878
19.1.4.7	Sugar Alcohols (Polyalcohols)	878
19.1.4.7.1	Isomaltol (Palatinit)	878
19.1.4.7.2	Sorbitol	879
19.1.4.7.3	Xylitol	879
19.1.4.7.4	Mannitol	879
19.1.5	Candies	879
19.1.5.1	General Outline	879
19.1.5.2	Hard Caramel (Bonbons)	879
19.1.5.3	Soft Caramel (Toffees)	880
19.1.5.4	Fondant	880
19.1.5.5	Foamy Candies	880
19.1.5.6	Jellies, Gum and Gelatine Candies	881
19.1.5.7	Tablets	881
19.1.5.8	Dragées	881
19.1.5.9	Marzipan	881
19.1.5.10	Persipan	881
19.1.5.11	Other Raw Candy Fillers	881
19.1.5.12	Nougat Fillers	881
19.1.5.13	Croquant	882
19.1.5.14	Licorice and its Products	882
19.1.5.15	Chewing Gum	882
19.1.5.16	Effervescent Lemonade Powders	882
19.2	Honey and Artificial Honey	883

19.2.1	Honey	883
19.2.1.1	Foreword	883
19.2.1.2	Production and Types	884
19.2.1.3	Processing	884
19.2.1.4	Physical Properties	885
19.2.1.5	Composition	885
19.2.1.5.1	Water	885
19.2.1.5.2	Carbohydrates	885
19.2.1.5.3	Enzymes	886
19.2.1.5.4	Proteins	888
19.2.1.5.5	Amino Acids	888
19.2.1.5.6	Acids	888
19.2.1.5.7	Aroma Substances	889
19.2.1.5.8	Pigments	889
19.2.1.5.9	Toxic Constituents	889
19.2.1.6	Storage	889
19.2.1.7	Utilization	890
19.2.2	Artificial Honey	890
19.2.2.1	Foreword	890
19.2.2.2	Production	890
19.2.2.3	Composition	890
19.2.2.4	Utilization	890
19.3	References	891
20	**Alcoholic Beverages**	892
20.1	Beer	892
20.1.1	Foreword	892
20.1.2	Raw Materials	892
20.1.2.1	Barley	892
20.1.2.2	Other Starch- and Sugar-Containing Raw Materials	892
20.1.2.2.1	Wheat Malt	892
20.1.2.2.2	Adjuncts	892
20.1.2.2.3	Syrups, Extract Powders	894
20.1.2.2.4	Malt Extracts, Wort Concentrates	894
20.1.2.2.5	Brewing Sugars	894
20.1.2.3	Hops	894
20.1.2.3.1	General Outline	894
20.1.2.3.2	Composition	895
20.1.2.3.3	Processing	896
20.1.2.4	Brewing Water	897
20.1.2.5	Brewing Yeasts	897
20.1.3	Malt Preparation	898
20.1.3.1	Steeping	898
20.1.3.2	Germination	898
20.1.3.3	Kilning	898
20.1.3.4	Continuous Processes	898
20.1.3.5	Special Malts	899
20.1.4	Wort Preparation	899
20.1.4.1	Ground Malt	899
20.1.4.2	Mashing	899
20.1.4.3	Lautering	900

20.1.4.4   Wort Boiling and Hopping 900
20.1.4.5   Continuous Processes 900
20.1.5   Fermentation 900
20.1.5.1   Bottom Fermentation 900
20.1.5.2   Top Fermentation 901
20.1.5.3   Continuous Processes, Rapid Methods 901
20.1.6   Bottling 901
20.1.7   Composition 901
20.1.7.1   Ethanol 901
20.1.7.2   Extract 901
20.1.7.3   Acids 902
20.1.7.4   Nitrogen Compounds 902
20.1.7.5   Carbohydrates 902
20.1.7.6   Minerals 902
20.1.7.7   Vitamins 902
20.1.7.8   Aroma Substances 902
20.1.7.9   Foam Builders 902
20.1.8   Kinds of Beer 903
20.1.8.1   Top Fermented Beers 903
20.1.8.2   Bottom Fermented Beers 904
20.1.8.3   Diet Beers 904
20.1.8.4   Alcohol-Free Beers 904
20.1.8.5   Export Beers 904
20.1.9   Beer Flavor and Beer Defects 904
20.2   Wine 906
20.2.1   Foreword 906
20.2.2   Grape Cultivars 907
20.2.3   Grape Must 911
20.2.3.1   Growth and Harvest 911
20.2.3.2   Must Production and Treatment 913
20.2.3.3   Must Composition 914
20.2.3.3.1 Carbohydrates 915
20.2.3.3.2 Acids 915
20.2.3.3.3 Nitrogen Compounds 915
20.2.3.3.4 Lipids 915
20.2.3.3.5 Phenolic Compounds 915
20.2.3.3.6 Minerals 916
20.2.3.3.7 Aroma Substances 916
20.2.4   Fermentation 916
20.2.5   Cellar Operations After Fermentation; Storage 917
20.2.5.1   Racking, Storing and Aging 917
20.2.5.2   Sulfur Treatment 917
20.2.5.3   Clarification and Stabilization 918
20.2.5.4   Amelioration 918
20.2.6   Composition 919
20.2.6.1   Extract 919
20.2.6.2   Carbohydrates 919
20.2.6.3   Ethanol 919
20.2.6.4   Other Alcohols 920
20.2.6.5   Acids 920
20.2.6.6   Phenolic Compounds 920

20.2.6.7	Nitrogen Compounds	921
20.2.6.8	Minerals	921
20.2.6.9	Aroma Substances	921
20.2.7	Spoilage	924
20.2.8	Liqueur Wines	926
20.2.9	Sparkling Wine	926
20.2.9.1	Bottle Fermentation ("*Méthode Champenoise*")	927
20.2.9.2	Tank Fermentation Process ("*Produit en Cuve Close*")	927
20.2.9.3	Carbonation Process	928
20.2.9.4	Various Types of Sparkling Wines	928
20.2.10	Wine-Like Beverages	928
20.2.10.1	Fruit Wines	928
20.2.10.2	Malt Wine; Mead	929
20.2.10.3	Other Products	929
20.2.11	Wine-Containing Beverages	929
20.2.11.1	Vermouth	929
20.2.11.2	Aromatic Wines	929
20.3	Spirits	929
20.3.1	Foreword	929
20.3.2	Liquor	929
20.3.2.1	Production	929
20.3.2.2	Alcohol Production	930
20.3.2.3	Liquor from Wine, Fruit, Cereals and Sugar Cane	931
20.3.2.3.1	Wine Liquor (Brandy)	931
20.3.2.3.2	Fruit Liquor (Fruit Brandy)	931
20.3.2.3.3	Gentian Liquor ("*Enzian*")	932
20.3.2.3.4	Juniper Liquor (*Brandy*) and Gin	932
20.3.2.3.5	Rum	932
20.3.2.3.6	Arrack	933
20.3.2.3.7	Liquors from Cereals	933
20.3.2.4	Miscellaneous Alcoholic Beverages	935
20.3.3	Liqueurs (Cordials)	935
20.3.3.1	Fruit Sap Liqueurs	935
20.3.3.2	Fruit Aroma Liqueurs	935
20.3.3.3	Other Liqueurs	935
20.3.4	Punch Extracts	936
20.3.5	Mixed Drinks	936
20.4	References	936
21	**Coffee, Tea, Cocoa**	938
21.1	Coffee and Coffee Substitutes	938
21.1.1	Foreword	938
21.1.2	Green Coffee	939
21.1.2.1	Harvesting and Processing	939
21.1.2.2	Green Coffee Varieties	939
21.1.2.3	Composition of Green Coffee	940
21.1.3	Roasted Coffee	940
21.1.3.1	Roasting	940
21.1.3.2	Storing and Packaging	941
21.1.3.3	Composition of Roasted Coffee	942
21.1.3.3.1	Proteins	942

21.1.3.3.2   Carbohydrates 942
21.1.3.3.3   Lipids 942
21.1.3.3.4   Acids 943
21.1.3.3.5   Caffeine 943
21.1.3.3.6   Trigonelline, Nicotinic Acid 944
21.1.3.3.7   Aroma Substances 944
21.1.3.3.8   Minerals 946
21.1.3.3.9   Other Constituents 946
21.1.3.4   Coffee Beverages 946
21.1.4   Coffee Products 948
21.1.4.1   Instant Coffee 948
21.1.4.2   Decaffeinated Coffee 949
21.1.4.3   Treated Coffee 949
21.1.5   Coffee Substitutes and Adjuncts 949
21.1.5.1   Introduction 949
21.1.5.2   Processing of Raw Materials 949
21.1.5.3   Individual Products 950
21.1.5.3.1   Barley Coffee 950
21.1.5.3.2   Malt Coffee 950
21.1.5.3.3   Chicory Coffee 950
21.1.5.3.4   Fig Coffee 950
21.1.5.3.5   Acorn Coffee 950
21.1.5.3.6   Other Products 950
21.2   Tea and Tea-Like Products 951
21.2.1   Foreword 951
21.2.2   Black Tea 951
21.2.3   Green Tea 952
21.2.4   Grades of Tea 952
21.2.5   Composition 952
21.2.5.1   Phenolic Compounds 953
21.2.5.2   Enzymes 953
21.2.5.3   Amino Acids 954
21.2.5.4   Caffeine 954
21.2.5.5   Carbohydrates 954
21.2.5.6   Lipids 955
21.2.5.7   Pigments (Chlorophyll and Carotenoids) 955
21.2.5.8   Aroma Substances 955
21.2.5.9   Minerals 956
21.2.6   Reactions Involved in the Processing of Tea 956
21.2.7   Packaging, Storage, Brewing 958
21.2.8   Maté (Paraguayan Tea) 958
21.2.9   Products from Cola Nut 958
21.3   Cocoa and Chocolate 959
21.3.1   Introduction 959
21.3.2   Cacao 959
21.3.2.1   General Information 959
21.3.2.2   Harvesting and Processing 960
21.3.2.3   Composition 961
21.3.2.3.1   Proteins and Amino Acids 961
21.3.2.3.2   Theobromine and Caffeine 962
21.3.2.3.3   Lipids 962

21.3.2.3.4   Carbohydrates 962
21.3.2.3.5   Phenolic Compounds 962
21.3.2.3.6   Organic Acids 963
21.3.2.3.7   Volatile Compounds and Flavor Substances.................. 964
21.3.2.4   Reactions During Fermentation and Drying.................. 964
21.3.2.5   Production of Cocoa Liquor................................ 965
21.3.2.6   Production of Cocoa Liquor with Improved Dispersability 965
21.3.2.7   Production of Cocoa Powder by Cocoa Mass Pressing........ 966
21.3.3   Chocolate 966
21.3.3.1   Introduction 966
21.3.3.2   Chocolate Production 966
21.3.3.2.1   Mixing 966
21.3.3.2.2   Refining 966
21.3.3.2.3   Conching 966
21.3.3.2.4   Tempering and Molding 967
21.3.3.3   Kinds of Chocolate 967
21.3.4   Storage of Cocoa Products 969
21.4   References 969

22   **Spices, Salt and Vinegar** 971
22.1   Spices 971
22.1.1   Composition 971
22.1.1.1   Components of Essential Oils 971
22.1.1.2   Aroma Substances... 974
22.1.1.2.1   Pepper 975
22.1.1.2.2   Vanilla 976
22.1.1.2.3   Dill 976
22.1.1.2.4   Fenugreek 976
22.1.1.2.5   Saffron 977
22.1.1.2.6   Mustard, Horseradish 977
22.1.1.2.7   Ginger 977
22.1.1.2.8   Basil 977
22.1.1.2.9   Parsley 978
22.1.1.3   Substances with Pungent Taste 979
22.1.1.4   Pigments 981
22.1.1.5   Antioxidants 981
22.1.2   Products 981
22.1.2.1   Spice Powders 981
22.1.2.2   Spice Extracts or Concentrates (Oleoresins) 982
22.1.2.3   Blended Spices 982
22.1.2.4   Spice Preparations....................................... 982
22.1.2.4.1   Curry Powder 982
22.1.2.4.2   Mustard Paste 982
22.1.2.4.3   Sambal 982
22.2   Salt (Cooking Salt) 982
22.2.1   Composition 982
22.2.2   Occurrence... 983
22.2.3   Production 983
22.2.4   Special Salt 983
22.2.5   Salt Substitutes 983
22.3   Vinegar 983

22.3.1   Production 984
22.3.1.1   Microbiological Production 984
22.3.1.2   Chemical Synthesis..................................... 984
22.3.2   Composition 984
22.4   References 985

23   Drinking Water, Mineral and Table Water 986
23.1   Drinking Water 986
23.1.1   Treatment 986
23.1.2   Hardness 986
23.1.3   Analysis 987
23.2   Mineral Water 988
23.3   Table Water 988
23.4   References 988

Index 989

Introduction

Foods are materials which, in their naturally occurring, processed or cooked forms, are consumed by humans as nourishment and for enjoyment.

The terms "nourishment" and "enjoyment" introduce two important properties of foods: the nutritional value and the hedonic value. The former is relatively easy to quantify since all the important nutrients are known and their effects are defined. Furthermore, there are only a limited number of nutrients. Defining the hedonic value of a food is more difficult because such a definition must take into account all those properties of a food, such as visual appeal, smell, taste and texture, which interact with the senses. These properties can be influenced by a large number of compounds which in part have not even been identified. Besides their nutritional and hedonic values, foods are increasingly being judged according to properties which determine their handling. Thus, the term "convenience foods". An obvious additional requirement of a food is that it be free from toxic materials.

Food chemistry is involved not only in elucidating the composition of the raw materials and end-products, but also with the changes which occur in food during its production, processing, storage and cooking. The highly complex nature of food results in a multitude of desired and undesired reactions which are controlled by a variety of parameters. To gain a meaningful insight into these reactions, it is necessary to break up the food into model systems. Thus, starting from compositional analyses (detection, isolation and structural characterization of food constituents), the reactions of a single constituent or of a simple mixture can be followed. Subsequently, an investigation of a food in which an individual reaction dominates can be made. Inherently, such a study starts with a given compound and is thus not restricted to any one food or group of foods. Such general studies of reactions involving food constituents are supplemented by special investigations which focus on chemical processes in individual foods. Research of this kind is from the very beginning closely associated with economic and technological aspects and contributes, by understanding the basics of the chemical processes occurring in foods, both to resolving specific technical problems and to process optimization.

A comprehensive evaluation of foods requires that analytical techniques keep pace with the available technology. As a result a major objective in food chemistry is concerned with the application and continual development of analytical methods. This aspect is particularly important when following possible contamination of foods with substances which may involve a health risk. Thus, there are close links with environmental problems.

Food chemistry research is aimed at establishing objective standards by which the criteria mentioned above – nutritional value, hedonic value, absence of toxic compounds and convenience – can be evaluated. These are a prerequisite for the industrial production of high quality food in bulk amounts.

This brief outline thus indicates that food chemistry, unlike other branches of chemistry which are concerned either with particular classes of compounds or with particular methods, is a subject which, both in terms of the actual chemistry and the methods involved, has a very broad field to cover.

0 Water

0.1 Foreword

Water (moisture) is the predominant constituent in many foods (Table 0.1). As a medium water supports chemical reactions, and it is a direct reactant in hydrolytic processes. Therefore, removal of water from food or binding it by increasing the concentration of common salt or sugar retards many reactions and inhibits the growth of microorganisms, thus improving the shelf lives of a number of foods. Through physical interaction with proteins, polysaccharides, lipids and salts, water contributes significantly to the texture of food.

Table 0.1. Moisture content of some foods

Food	Moisture content (weight-%)	Food	Moisture content (weight-%)
Meat	65–75	Cereal flour	12–14
Milk	87	Coffee beans,	
Fruits,		roasted	5
vegetables	70–90	Milk powder	4
Bread	35	Edible oil	0
Honey	20		
Butter,			
margarine	16–18		

The function of water is better understood when its structure and its state in a food system are clarified. Special aspects of binding of water by individual food constituents (cf. 1.4.3.3, 3.5.2 and 4.4.3) and meat (cf. 12.5) are discussed in the indicated sections.

0.2 Structure

0.2.1 Water Molecule

The six valence electrons of oxygen in a water molecule are hybridized to four sp³ orbitals

that are elongated to the corners of a somewhat deformed, imaginary tetrahedron (Fig. 0.1). Two hybrid orbitals form O–H covalent bonds with a bond angle of 105° for H–O–H, whereas the other 2 orbitals hold the nonbonding electron pairs (n-electrons). The O–H covalent bonds, due to the highly electronegative oxygen, have a partial (40%) ionic character.

Each water molecule is tetrahedrally coordinated with four other water molecules through hydrogen bonds. The two unshared electron pairs (n-electrons or sp^3 orbitals) of oxygen act as H-bond acceptor sites and the H–O bonding orbitals act as hydrogen bond donor sites (Fig. 0.2). The dissociation energy of this hydrogen bond is about 25 kJ mole^{-1}.

The simultaneous presence of two acceptor sites and two donor sites in water permits association in a three-dimensional network stabilized by

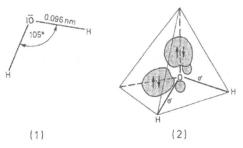

Fig. 0.1. Water. (**1**) Molecular geometry, (**2**) orbital model

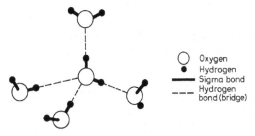

Fig. 0.2. Tetrahedral coordination of water molecules

H.-D. Belitz · W. Grosch · P. Schieberle, *Food Chemistry*
© Springer 2009

H-bridges. This structure which explains the special physical properties of water is unusual for other small molecules. For example, alcohols and compounds with iso-electric dipoles similar to those of water, such as HF or NH_3, form only linear or two-dimensional associations.

The above mentioned polarization of H–O bonds is transferred via hydrogen bonds and extends over several bonds. Therefore, the dipole moment of a complex consisting of increasing numbers of water molecules (multi-molecular dipole) is higher as more molecules become associated and is certainly much higher than the dipole moment of a single molecule. Thus, the dielectric constant of water is high and surpasses the value, which can be calculated on the basis of the dipole moment of a single molecule. Proton transport takes place along the H-bridges. It is actually the jump of a proton from one water molecule to a neighboring water molecule. Regardless of whether the proton is derived from dissociation of water or originates from an acid, it will sink into the unshared electron pair orbitals of water:

$$(0.1)$$

In this way a hydrated $H_3O^{\oplus}$ ion is formed with an exceptionally strong hydrogen bond (dissociation energy about $100\,\text{kJ}\,\text{mol}^{-1}$). A similar mechanism is valid in transport of $OH^{\ominus}$ ions, which also occurs along the hydrogen bridges:

$$(0.2)$$

Since the transition of a proton from one oxygen to the next occurs extremely rapidly ($v > 10^{12}\,\text{s}^{-1}$), proton mobility surpasses the mobilities of all other ions by a factor of 4–5, except for the stepwise movement of $OH^{\ominus}$ within the structure; its rate of exchange is only 40% less than that of a proton.

H-bridges in ice extend to a larger sphere than in water (see the following section). The mobility of protons in ice is higher than in water by a factor of 100.

0.2.2 Liquid Water and Ice

The arrangements of water molecules in "liquid water" and in ice are still under intensive investigation. The outlined hypotheses agree with existing data and are generally accepted.

Due to the pronounced tendency of water molecules to associate through H-bridges, liquid water and ice are highly structured. They differ in the distance between molecules, coordination number and time-range order (duration of stability). Stable ice-I is formed at $0\,°C$ and 1 atm pressure. It is one of nine known crystalline polymorphic structures, each of which is stable in a certain temperature and pressure range. The coordination number in ice-I is four, the $O–H\cdots O$ (nearest neighbor) distance is $0.276\,\text{nm}$ ($0\,°C$) and the H-atom between neighboring oxygens is $0.101\,\text{nm}$ from the oxygen to which it is bound covalently and $0.175\,\text{nm}$ from the oxygen to which it is bound by a hydrogen bridge. Five water molecules, forming a tetrahedron, are loosely packed and kept together mostly through H-bridges.

Table 0.2. Coordination number and distance between two water molecules

	Coordination number	$O–H\cdots O$ Distance
Ice ($0\,°C$)	4	$0.276\,\text{nm}$
Water ($1.5\,°C$)	4.4	$0.290\,\text{nm}$
Water ($83\,°C$)	4.9	$0.305\,\text{nm}$

When ice melts and the resultant water is heated (Table 0.2), both the coordination number and the distance between the nearest neighbors increase. These changes have opposite influences on the density. An increase in coordination number (i.e. the number of water molecules arranged in an orderly fashion around each water molecule) increases the density, whereas an increase in distance between nearest neighbors decreases the density. The effect of increasing coordination number is predominant during a temperature increase from 0 to $4\,°C$. As a consequence, water has an unusual property: its density in the liquid state at $0\,°C$ ($0.9998\,\text{g cm}^{-3}$) is higher than in the solid state (ice-I, $\rho = 0.9168\,\text{g cm}^{-3}$). Water is a structured liquid with a short time-range order. The water molecules, through H-bridges, form short-lived polygonal structures which are rapidly cleaved and then reestablished giving a dynamic equilibrium. Such fluctuations explain the lower viscosity of water, which otherwise could not be explained if H-bridges were rigid.

The hydrogen-bound water structure is changed by solubilization of salts or molecules with polar and/or hydrophobic groups. In salt solutions the n-electrons occupy the free orbitals of the cations, forming "aqua complexes". Other water molecules then coordinate through H-bridges, forming a hydration shell around the cation and disrupting the natural structure of water.

Hydration shells are formed by anions through ion-dipole interaction and by polar groups through dipole-dipole interaction or H-bridges, again contributing to the disruption of the structured state of water.

Aliphatic groups which can fix the water molecules by dispersion forces are no less disruptive. A minimum of free enthalpy will be attained when an ice-like water structure is arranged around a hydrophobic group (tetrahedral-four-coordination). Such ice-like hydration shells around aliphatic groups contribute, for example, to stabilization of a protein, helping the protein to acquire its most thermodynamically favorable conformation in water.

The highly structured, three-dimensional hydrogen bonding state of ice and water is reflected in many of their unusual properties.

Additional energy is required to break the structured state. This accounts for water having substantially higher melting and boiling points and heats of fusion and vaporization than methanol or dimethyl ether (cf. Table 0.3). Methanol has only one hydrogen donor site, while dimethyl ether has none but does have a hydrogen bond acceptor site; neither is sufficient to form a structured network as found in water.

Table 0.3. Some physical constants of water, methanol and dimethyl ether

	F_p (°C)	K_p (°C)
H_2O	0.0	100.0
CH_3OH	−98	64.7
CH_3OCH_3	−138	−23

0.3 Effect on Storage Life

Drying and/or storage at low temperatures are among the oldest methods for the preservation of food with high water contents. Modern food technology tries to optimize these methods. A product should be dried and/or frozen only long enough to ensure wholesome quality for a certain period of time.

Naturally, drying and/or freezing must be optimized for each product individually. It is therefore necessary to know the effect of water on storage life before suitable conditions can be selected.

0.3.1 Water Activity

In 1952, Scott came to the conclusion that the storage quality of food does not depend on the water content, but on water activity (a_w), which is defined as follows:

$$a_w = P/P_0 = ERH/100 \qquad (0.3)$$

P   = partial vapor pressure of food moisture at temperature T

P_0 = saturation vapor pressure of pure water at T

ERH = equilibrium relative humidity at T.

The relationship between water content and water activity is indicated by the sorption isotherm of a food (Fig. 0.3).

At a low water content (<50%), even minor changes in this parameter lead to major changes in water activity. For that reason, the sorption isotherm of a food with lower water content is shown with an expanded ordinate in Fig. 0.3b, as compared with Fig. 0.3a.

Figure 0.3b shows that the desorption isotherm, indicating the course of a drying process, lies slightly above the adsorption isotherm pertaining to the storage of moisture-sensitive food. As a rule, the position of the hysteresis loop changes when adsorption and desorption are repeated with the same sample. The effect of water activity on processes that can influence food quality is presented in Fig. 0.4. Decreased water activity retards the growth of microorganisms, slows enzyme catalyzed reactions (particularly involving hydrolases; cf. 2.2.2.1) and, lastly,

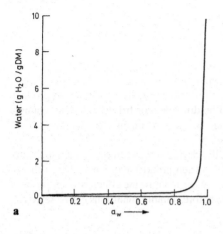

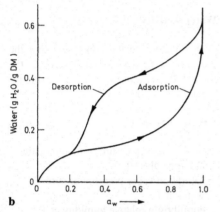

Fig. 0.3. Moisture sorption isotherm (according to *Labuza et al.*, 1970). **a** Food with high moisture content; **b** Food with low moisture content (DM: Dry matter)

Table 0.4. Water activity of some food

Food	a_w	Food	a_w
Leberwurst	0.96	Marmalades	0.82–0.94
Salami	0.82–0.85	Honey	0.75
Dried fruits	0.72–0.80		

retards non-enzymatic browning. In contrast, the rate of lipid autoxidation increases in dried food systems (cf. 3.7.2.1.4).

Foods with a_w values between 0.6 and 0.9 (examples in Table 0.4) are known as "intermediate moisture foods" (IMF). These foods are largely protected against microbial spoilage.

One of the options for decreasing water activity and thus improving the shelf life of food is to use additives with high water binding capacities (humectants). Table 0.5 shows that in addition to common salt, glycerol, sorbitol and sucrose

Table 0.5. Moisture content of some food or food ingredients at a water activity of 0.8

	Moisture content (%)		Moisture content (%)
Peas	16	Glycerol	108
Casein	19	Sorbitol	67
Starch		Saccharose	56
(potato)	20	Sodium chloride	332

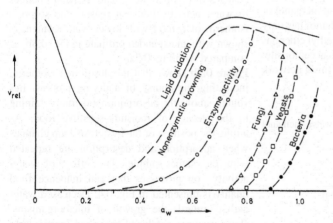

Fig. 0.4. Food shelf life (storage stability) as a function of water activity (according to *Labuza*, 1971)

have potential as humectants. However, they are also sweeteners and would be objectionable from a consumer standpoint in many foods in the concentrations required to regulate water activity.

0.3.2 Water Activity as an Indicator

Water activity is only of limited use as an indicator for the storage life of foods with a low water content, since water activity indicates a state that applies only to ideal, i.e. very dilute solutions that are at a thermodynamic equilibrium. However, foods with a low water content are non-ideal systems whose metastable (fresh) state should be preserved for as long as possible. During storage, such foods do not change thermodynamically, but according to kinetic principles. A new concept based on phase transition, which takes into account the change in physical properties of foods during contact between water and hydrophilic ingredients, is better suited to the prediction of storage life. This will be briefly discussed in the following sections (0.3.3–0.3.5).

0.3.3 Phase Transition of Foods Containing Water

The physical state of metastable foods depends on their composition, on temperature and on storage time. For example, depending on the temperature, the phases could be glassy, rubbery or highly viscous. The kinetics of phase transitions can be measured by means of differential scanning calorimetry (DSC), producing a thermogram that shows temperature T_g as the characteristic value for the transition from glassy to rubbery (plastic). Foods become plastic when their hydrophilic components are hydrated. Thus the water content affects the temperature T_g, for example in the case of gelatinized starch (Fig. 0.5).
Table 0.6 shows the T_g of some mono- and oligosaccharides and the difference between melting points T_m.
During the cooling of an aqueous solution below the freezing point, part of the water crystallizes, causing the dissolved substance to become enriched in the remaining fluid phase (unfrozen water). In the thermogram, temperature T'_g appears,

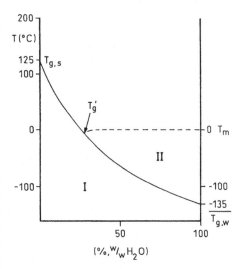

Fig. 0.5. State diagram, showing the approximate T_g temperatures as a function of mass fraction, for a gelatinized starch-water system (according to *Van den Berg*, 1986).
States: I = glassy; II = rubbery;
$T_{g,s}$ and $T_{g,w}$ = phase transition temperatures of dehydrated starch and water; T_m = melting point (ice)

Table 0.6. Phase transition temperature T_g and melting point T_m of mono- and oligosaccharides

Compound	T_g [°C]	T_m
Glycerol	−93	18
Xylose	9.5	153
Ribose	−10	87
Xylitol	−18.5	94
Glucose	31	158
Fructose	100	124
Galactose	110	170
Mannose	30	139.5
Sorbitol	−2	111
Sucrose	52	192
Maltose	43	129
Maltotriose	76	133.5

at which the glassy phase of the concentrated solution turns into a rubber-like state. The position of T'_g (−5 °C) on the T_g curve is shown by the example of gelatinized starch (Fig. 0.5); the quantity of unfrozen water W'_g at this temperature is 27% by weight. Table 0.7 lists the temperatures T'_g for aqueous solutions (20% by weight) of carbohydrates and proteins. In the case of oligosaccharides composed of three glucose molecules,

Table 0.7. T'_g and W'_g of aqueous solutions (20% by weight) of carbohydrates and proteins[a]

Substance	T'_g	W'_g
Glycerol	−65	0.85
Xylose	−48	0.45
Ribose	−47	0.49
Ribitol	−47	0.82
Glucose	−43	0.41
Fructose	−42	0.96
Galactose	−41.5	0.77
Sorbitol	−43.5	0.23
Sucrose	−32	0.56
Lactose	−28	0.69
Trehalose	−29.5	0.20
Raffinose	−26.5	0.70
Maltotriose	−23.5	0.45
Panose	−28	0.59
Isomaltotriose	−30.5	0.50
Potato starch (DE 10)	−8	
Potato starch (DE 2)	−5	
Hydroxyethylcellulose	−6.5	
Tapioca (DE 5)	−6	
Waxy corn (DE 0.5)	−4	
Gelatin	−13.5	0.46
Collagen, soluble	−15	0.71
Bovine serum albumin	−13	0.44
α-Casein	−12.5	0.61
Sodium caseinate	−10	0.64
Gluten	−5 to −10	0.07 to 0.41

[a] Phase transition temperature T'_g (°C) and water content W'_g (g per g of substance) of maximum freeze-concentrated glassy structure.

maltotriose has the lowest T'_g value in comparison with panose and isomaltotriose. The reason is probably that in aqueous solution, the effective chain length of linear oligosaccharides is greater than that of branched compounds of the same molecular weight.

In the case of homologous series of oligo- and polysaccharides, T_g and T'_g increase with the molecular weight up to a certain limit (Fig. 0.6). Table 0.8 lists the phase transition temperatures T'_g of some fruits and vegetables.

Table 0.8. Phase transition temperature T'_g of some fruits and vegetables

Fruit/vegetable	T'_g(°C)
Strawberries	−33 to −41
Peaches	−36.5
Bananas	−35
Apples	−42
Tomatoes	−41.5
Peas (blanched, frozen)	−25
Carrots	−25.5
Broccoli, stalks	−26.5
Broccoli, flower buds	−11.5
Spinach (blanched, frozen)	−17
Potatoes	−11

0.3.4 WLF Equation

The viscosity of a food is extremely high at temperature T_g or T'_g (about 10^{13} Pa.s). As the temperature rises, the viscosity decreases, which means that processes leading to a drop in quality will accelerate. In the temperature range of T_g to about ($T_g + 100$ °C), the change in viscosity does not follow the equation of *Arrhenius* (cf. 2.5.4.2), but a relationship formulated by *Williams, Landel* and *Ferry* (the WLF equation):

$$\log \frac{\eta}{\rho T} \bigg/ \frac{\eta_g}{\rho_t T_g} = -\frac{C_1(T - T_g)}{C_2 + (T - T_g)} \tag{0.4}$$

Viscosity (η) and density (ρ) at temperature T; viscosity (η_g) and density (ρ_g) at phase transition temperature T_g; C_1 and C_2: constants.

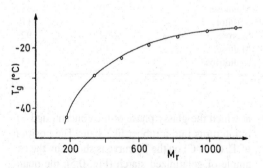

Fig. 0.6. Phase transition temperatures T'_g (aqueous solution, 20% by weight) of the homologous series glucose to maltoheptaose as a function of molecular weight M_r

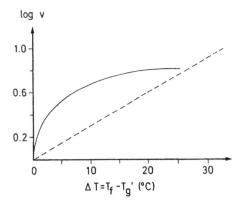

Fig. 0.7. Crystallization of water in ice cream (according to *Levine* and *Slade*, 1990).
v: crystallization velocity;
T_f: temperature in the freezer compartment;
T_g': phase transition temperature.
The *Arrhenius* kinetics (– – –) is shown for comparison

Table 0.9. Unwanted chemical, enzymatic and physical processes in the production and storage of foods, depending on phase transition temperature T_g or T_g' and delayed by the addition of starch partial hydrolysates (lower DE value)

Process
1. Agglomeration and lumping of foods in the amorphous state
2. Recrystallization
3. Enzymatic reaction
4. Collapse of structure in case of freeze-dried products
5. Non-enzymatic browning

According to the WLF equation, the rate of which in our example water crystallizes in ice cream at temperatures slightly above T_g' rises exponentially (Fig. 0.7). If the *Arrhenius* equation were to be valid, crystallization would accelerate linearly at a considerably slower rate after exceeding T_g' (Fig. 0.7).

0.3.5 Conclusion

In summary, we find that the rate of a food's chemical and enzymatic reactions as well as that of its physical processes becomes almost zero when the food is stored at the phase transition temperature of T_g or T_g'. Measures to improve storage life by increasing T_g or T_g' can include the extraction of water through drying and/or an immobilization of water by means of freezing, or by adding polysaccharides. Table 0.9 shows examples of how the drop in quality of certain foods can be considerably delayed when T_g or T_g' are increased by the addition of polysaccharides and approximated to the storage temperature.

0.4 References

Blanshard, J. M. V., Lillford, P. J., The glassy state in foods. Nottingham University press, 1983

Fennema, O. R., Water and ice. In: Principles of food science, Part I (Ed.: Fennema, O. R.), p. 13. New York: Marcel Dekker, Inc., 1976

Franks, F., Water, ice and solutions of simple molecules. In: Water relations of foods (Ed.: Duckworth, R. B.), p. 3. London: Academic Press, 1975

Hardman, T. M.(Ed.), Water and food quality. London: Elsevier Applied Science, 1989

Heiss, R., Haltbarkeit und Sorptionsverhalten wasserarmer Lebensmittel. Berlin: Springer-Verlag, 1968

Karel, M., Water activity and food preservation. In: Principles of food science, Part II (Editors: Karel, M., Fennema, O. R., Lund, D. B.), p. 237. New York: Marcel Dekker, Inc., 1975

Labuza, T. P., Kinetics of lipidoxidation in foods. Crit. Rev. Food Technol. *2*, 355 (1971)

Noel, T. R., Ring, S. G., Whittam, M. A., Glass transitions in low-moisture foods. Trends Food Sci. Technol. September 1990, pp. 62

Polesello, A., Gianciacomo, R., Application of near infrared spectrophotometry to the nondestructive analysis of food: a review of experimental results. Crit. Rev. Food Sci. Nutr., *18*, 203 (1982/83)

Slade, W., Levine, H., Beyond water activity – Recent advances based on an alternative approach to the assessment of food quality and safety. Crit. Rev. Food Sci. Nutr. *30*, 115 (1991)

1 Amino Acids, Peptides, Proteins

1.1 Foreword

Amino acids, peptides and proteins are important constituents of food. They supply the required building blocks for protein biosynthesis. In addition, they directly contribute to the flavor of food and are precursors for aroma compounds and colors formed during thermal or enzymatic reactions in production, processing and storage of food. Other food constituents, e. g., carbohydrates, also take part in such reactions. Proteins also contribute significantly to the physical properties of food through their ability to build or stabilize gels, foams, emulsions and fibrillar structures. The nutritional energy value of proteins (17 kJ/g or 4 kcal/g) is as high as that of carbohydrates.

The most important sources of protein are grain, oilseeds and legumes, followed by meat and milk. In addition to plants and animals, protein producers include algae (*Chlorella, Scenedesmus, Spirulina* spp.), yeasts and bacteria (single-cell proteins [SCP]). Among the C sources we use are glucose, molasses, starch, sulfite liquor, waste water, the higher n-alkanes, and methanol. Yeast of the genus *Candida* grow on paraffins, for example, and supply about 0.75 t of protein per t of carbohydrate. Bacteria of the species *Pseudomonas* in aqueous methanol produce about 0.30 t of protein per t of alcohol. Because of the high nucleic acid content of yeasts and bacteria (6–17% of dry weight), it is necessary to isolate protein from the cell mass. The future importance of single-cell proteins depends on price and on the technological properties.

In other raw materials, too, protein enrichment occurs for various reasons: protein concentration in the raw material may be too low for certain purposes, the sensory characteristics of the material (color, taste) may not be acceptable, or undesirable constituents may be present. Some products rich in protein also result from other processes, e. g., in oil and starch production. Enrichment results from the extraction of the constituents (protein concentrate) or from extraction and subsequent separation of protein from the solution, usually through thermal coagulation or isoelectric precipitation (protein isolate). Protein concentrates and protein isolates serve to enhance the nutritional value and to achieve the enhancement of the above mentioned physical properties of foods. They are added, sometimes after modification (cf. 1.4.6.1), to traditional foods, such as meat and cereal products, but they are also used in the production of novel food items such as meat, fish and milk substitutes. Raw materials in which protein enrichment takes place include:

- Legumes such as soybeans (cf. 16.3.1.2.1) and broad beans;
- Wheat and corn, which provide gluten as a byproduct of starch production;
- Potatoes; from the natural sap left over after starch production, proteins can be isolated by thermal coagulation;
- Eggs, which are processed into different whole egg, egg white and egg yolk products (cf. 11.4);
- Milk, which supplies casein (cf. 10.2.9 and whey protein (cf. 10.2.10);
- Fish, which supplies protein concentrates after fat extraction (cf. 13.1.6.13 and 1.4.6.3.2);
- Blood from slaughter animals, which is processed into blood meal, blood plasma concentrate (cf. 12.6.1.10) and globin isolate.
- Green plants grown for animal fodder, such as alfalfa, which are processed into leaf protein concentrates through the thermal coagulation of cell sap proteins.

1.2 Amino Acids

1.2.1 General Remarks

There are about 20 amino acids in a protein hydrolysate. With a few exceptions, their general

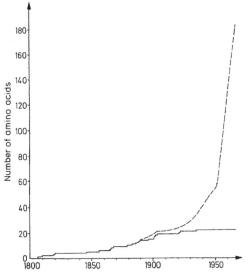

Fig. 1.1. Discovery of naturally occurring amino acids (according to *Meister*, 1965).--- Amino acids, total; — protein constituents

structure is:

R—CH—COOH
       |
      NH₂   (1.0)

In the simplest case, R=H (aminoacetic acid or glycine). In other amino acids, R is an aliphatic, aromatic or heterocyclic residue and may incorporate other functional groups. Table 1.1 shows the most important "building blocks" of proteins. There are about 200 amino acids found in nature (Fig. 1.1). Some of the more uncommon ones, which occur mostly in plants in free form, are covered in Chap. 17 on vegetables.

1.2.2 Classification, Discovery and Occurrence

1.2.2.1 Classification

There are a number of ways of classifiying amino acids. Since their side chains are the deciding factors for intra- and intermolecular interactions in proteins, and hence, for protein properties, amino acids can be classified as:

- Amino acids with nonpolar, uncharged side chains: e. g., glycine, alanine, valine, leucine, isoleucine, proline, phenylalanine, tryptophan and methionine.
- Amino acids with uncharged, polar side chains: e. g., serine, threonine, cysteine, tyrosine, asparagine and glutamine.
- Amino acids with charged side chains: e. g., aspartic acid, glutamic acid, histidine, lysine and arginine.

Based on their nutritional/physiological roles, amino acids can be differentiated as:

- Essential amino acids:
  Valine, leucine, isoleucine, phenylalanine, tryptophan, methionine, threonine, histidine (essential for infants), lysine and arginine ("semi-essential").
- Nonessential amino acids:
  Glycine, alanine, proline, serine, cysteine, tyrosine, asparagine, glutamine, aspartic acid and glutamic acid.

1.2.2.2 Discovery and Occurrence

Alanine was isolated from silk fibroin by *Weyl* in 1888. It is present in most proteins and is particularly enriched in silk fibroin (35%). Gelatin and zein contain about 9% alanine, while its content in other proteins is 2–7%. Alanine is considered nonessential for humans.

Arginine was first isolated from lupin seedlings by *Schulze* and *Steiger* in 1886. It is present in all proteins at an average level of 3–6%, but is particularly enriched in protamines. The arginine content of peanut protein is relatively high (11%). Biochemically, arginine is of great importance as an intermediary product in urea synthesis. Arginine is a semi-essential amino acid for humans. It appears to be required under certain metabolic conditions.

Asparagine from asparagus was the first amino acid isolated by *Vauguelin* and *Robiquet* in 1806. Its occurrence in proteins (edestin) was confirmed by *Damodaran* in 1932. In glycoproteins the carbohydrate component may be bound N-glycosidically to the protein moiety through the amide group of asparagine (cf. 11.2.3.1.1 and 11.2.3.1.3).

Table 1.1. Amino acids (protein building blocks) with their corresponding three ond one letter symbols

Structure	Name	Structure	Name	Structure	Name
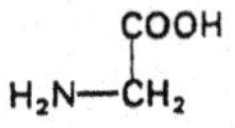 H_2N-CH_2, COOH	Glycine (Gly. G)	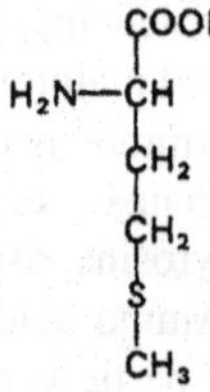 COOH, H_2N-CH, CH_2, CH_2, S, CH_3	L-Methionine (Met. M)	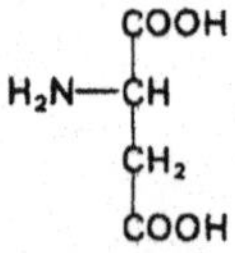 COOH, H_2N-CH, CH_2, COOH	L-Aspartic acid (Asp. D)
COOH, H_2N-CH, CH_3	L-Alanine (Ala. A)				
		COOH, H_2N-CH, CH_2OH	L-Serine (Ser. S)	COOH, H_2N-CH, CH_2, CH_2, COOH	L-Glutamic acid (Glu. E)
COOH, H_2N-CH, CH, H_3C CH_3	L-Valine (Val. V)				
		COOH, H_2N-CH, $HC-OH$, CH_3	L-Threonine (Thr. T)		
COOH, H_2N-CH, CH_2, CH, H_3C CH_3	L-Leucine (Leu. L)			COOH, H_2N-CH, CH_2, CH_2, CH_2, CH_2NH_2	L-Lysine (Lys. K)
		COOH, H_2N-CH, CH_2SH	L-Cysteine (Cys. C)		
COOH, H_2N-CH, H_3C-CH, CH_2, CH_3	L-Isoleucine (Ile. I)	COOH, HN ring, OH	L-4-Hydroxy-proline	COOH, H_2N-CH, CH_2, CH_2, $HO-CH$, CH_2NH_2	L-5-Hydroxy-lysine
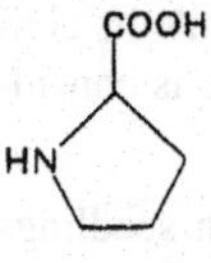 COOH, HN ring	L-Proline (Pro. P)	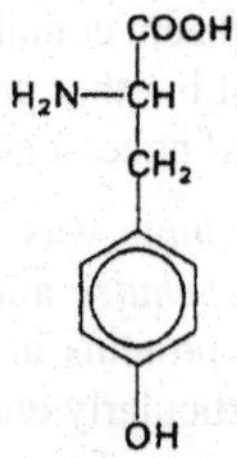 COOH, H_2N-CH, CH_2, (phenol ring), OH	L-Tyrosine (Tyr. Y)	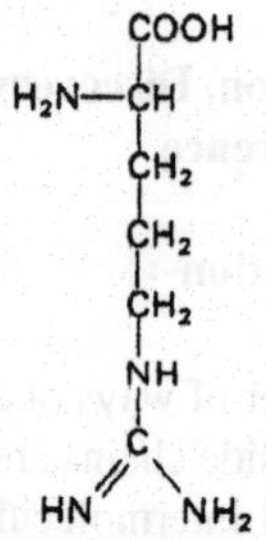	
				COOH, H_2N-CH, CH_2, (imidazole ring), HN N	L-Histidine (His. H)
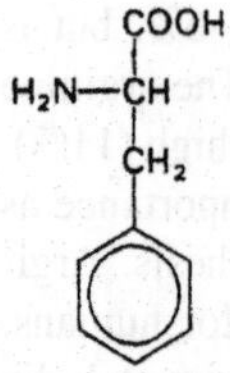 COOH, H_2N-CH, CH_2, (phenyl ring)	L-Phenylalanine (Phe. F)	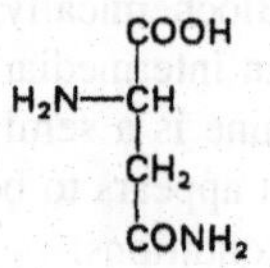 COOH, H_2N-CH, CH_2, $CONH_2$	L-Asparagine[a] (Asn. N)		
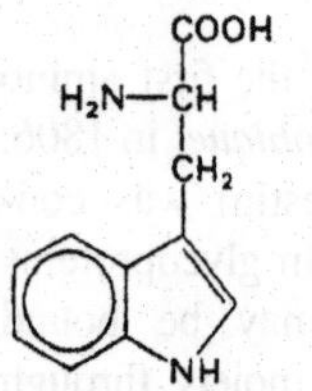 COOH, H_2N-CH, CH_2, (indole ring), NH	L-Tryptophan (Trp. W)	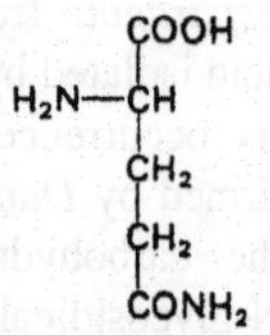 COOH, H_2N-CH, CH_2, CH_2, $CONH_2$	L-Glutamine[a] (Gln. Q)	COOH, H_2N-CH, CH_2, CH_2, CH_2, NH, C, HN NH_2	L-Arginine (Arg. R)

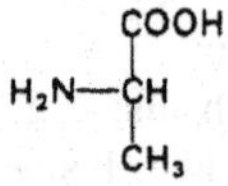

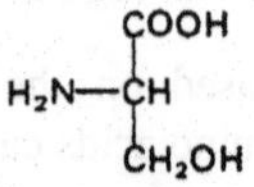

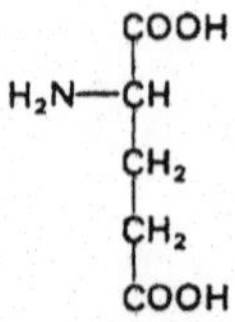

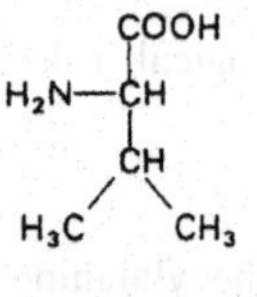

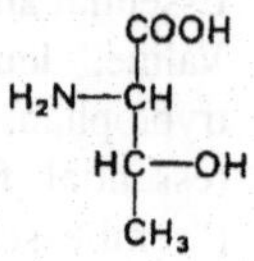

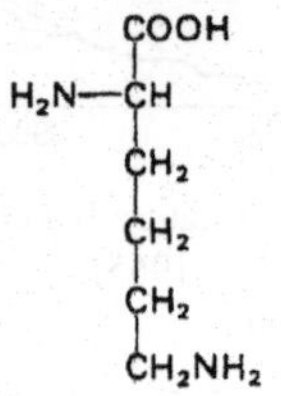

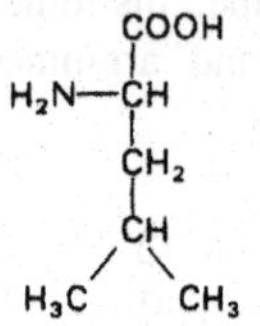

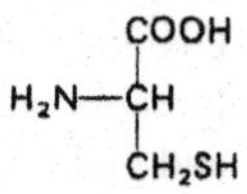

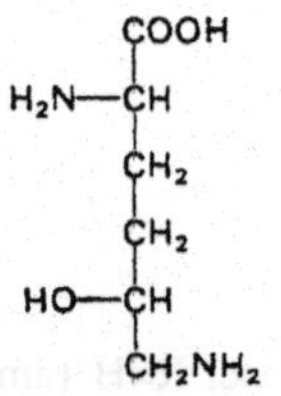

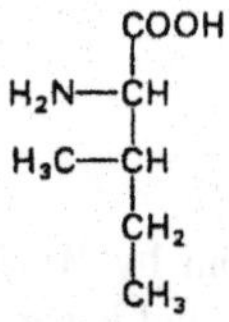

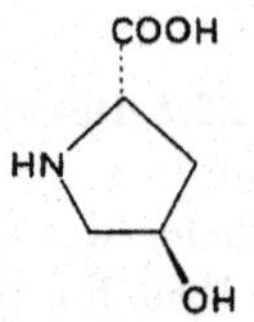

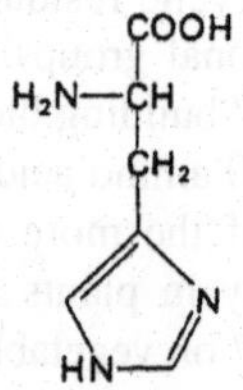

[a] When no distinction exists between the acid and its amide then the symbols (Asx, B) and (Glx, Z) are valid.

Aspartic Acid was isolated from legumes by *Ritthausen* in 1868. It occurs in all animal proteins, primarily in albumins at a concentration of 6–10%. Alfalfa and corn proteins are rich in aspartic acid (14.9% and 12.3%, respectively) while its content in wheat is low (3.8%). Aspartic acid is nonessential.

Cystine was isolated from bladder calculi by *Wolaston* in 1810 and from horns by *Moerner* in 1899. Its content is high in keratins (9%). Cystine is very important since the peptide chains of many proteins are connected by two cysteine residues, i.e. by disulfide bonds. A certain conformation may be fixed within a single peptide chain by disulfide bonds. Most proteins contain 1–2% cystine. Although it is itself nonessential, cystine can partly replace methionine which is an essential amino acid.

Glutamine was first isolated from sugar beet juice by *Schulze* and *Bosshard* in 1883. Its occurrence in protein (edestin) was confirmed by *Damodaran* in 1932. Glutamine is readily converted into pyrrolidone carboxylic acid, which is stable between pH 2.2 and 4.0, but is readily cleaved to glutamic acid at other pH's:

$$CH_2\!-\!CH_2\!-\!CONH_2$$
$$CH\!-\!NH_2 \longrightarrow$$
$$COOH$$

$$(1.1)$$

Glutamic Acid was first isolated from wheat gluten by *Ritthausen* in 1866. It is abundant in most proteins, but is particularly high in milk proteins (21.7%), wheat (31.4%), corn (18.4%) and soya (18.5%). Molasses also contains relatively high amounts of glutamic acid. Monosodium glutamate is used in numerous food products as a flavor enhancer.

Glycine is found in high amounts in structural protein. Collagen contains 25–30% glycine. It was first isolated from gelatin by *Braconnot* in 1820. Glycine is a nonessential amino acid although it does act as a precursor of many compounds formed by various biosynthetic mechanisms.

Histidine was first isolated in 1896 independently by *Kossel* and by *Hedin* from protamines occurring in fish. Most proteins contain 2–3% histidine. Blood proteins contain about 6%. Histidine is essential in infant nutrition.

5-Hydroxylysine was isolated by *van Slyke et al.* (1921) and *Schryver et al.* (1925). It occurs in collagen. The carbohydrate component of glycoproteins may be bound O-glycosidically to the hydroxyl group of the amino acid (cf. 12.3.2.3.1).

4-Hydroxyproline was first obtained from gelatin by *Fischer* in 1902. Since it is abundant in collagen (12.4%), the determination of hydroxyproline is used to detect the presence of connective tissue in comminuted meat products. Hydroxyproline is a nonessential amino acid.

Isoleucine was first isolated from fibrin by *Ehrlich* in 1904. It is an essential amino acid. Meat and ceral proteins contain 4–5% isoleucine; egg and milk proteins, 6–7%.

Leucine was isolated from wool and from muscle tissue by *Braconnot* in 1820. It is an essential amino acid and its content in most proteins is 7–10%. Cereal proteins contain variable amounts (corn 12.7%, wheat 6.9%). During alcoholic fermentation, fusel oil is formed from leucine and isoleucine.

Lysine was isolated from casein by *Drechsel* in 1889. It makes up 7–9% of meat, egg and milk proteins. The content of this essential amino acid is 2–4% lower in cereal proteins in which prolamin is predominant. Crab and fish proteins are the richest sources (10–11%). Along with threonine and methionine, lysine is a limiting factor in the biological value of many proteins, mostly those of plant origin. The processing of foods results in losses of lysine since its ε-amino group is very reactive (cf. *Maillard* reaction).

Methionine was first isolated from casein by *Mueller* in 1922. Animal proteins contain 2–4% and plant proteins contain 1–2% methionine. Methionine is an essential amino acid and in many biochemical processes its main role is as a methyl-donor. It is very sensitive to oxygen and heat treatment. Thus, losses occur in many food processing operations such as drying, kiln-drying, puffing, roasting or treatment with oxidizing agents. In the bleaching of flour with NCl_3 (nitrogen trichloride), methionine is converted to the toxic methionine sulfoximine:

$$H_3C-\underset{\underset{NH}{\|}}{\overset{\overset{O}{\|}}{S}}-CH_2-CH_2-\underset{\underset{NH_2}{|}}{CH}-COOH \tag{1.2}$$

Phenylalanine was isolated from lupins by *Schulze* in 1881. It occurs in almost all proteins (averaging 4–5%) and is essential for humans. It is converted *in vivo* into tyrosine, so phenylalanine can replace tyrosine nutritionally.

Proline was discovered in casein and egg albumen by *Fischer* in 1901. It is present in numerous proteins at 4–7% and is abundant in wheat proteins (10.3%), gelatin (12.8%) and casein (12.3%). Proline is nonessential.

Serine was first isolated from sericin by *Cramer* in 1865. Most proteins contain about 4–8% serine. In phosphoproteins (casein, phosvitin) serine, like threonine, is a carrier of phosphoric acid in the form of O-phosphoserine. The carbohydrate component of glycoproteins may be bound O-glycosidically through the hydroxyl group of serine and/or threonine [cf. 10.1.2.1.1 (κ-casein) and 13.1.4.2.4].

Threonine was discovered by *Rose* in 1935. It is an essential amino acid, present at 4.5–5% in meat, milk and eggs and 2.7–4.7% in cereals. Threonine is often the limiting amino acid in proteins of lower biological quality. The "bouillon" flavor of protein hydrolysates originates partly from a lactone derived from threonine (cf. 5.3.1.3).

Tryptophan was first isolated from casein hydrolysates, prepared by hydrolysis using pancreatic enzymes, by *Hopkins* in 1902. It occurs in animal proteins in relatively low amounts (1–2%) and in even lower amounts in cereal proteins (about 1%). Tryptophan is exceptionally abundant in lysozyme (7.8%). It is completely destroyed during acidic hydrolysis of protein. Biologically, tryptophan is an important essential amino acid, primarily as a precursor in the biosynthesis of nicotinic acid.

Tyrosine was first obtained from casein by *Liebig* in 1846. Like phenylalanine, it is found in almost all proteins at levels of 2–6%. Silk fibroin can have as much as 10% tyrosine. It is converted

through dihydroxyphenylalanine by enzymatic oxidation into brown-black colored melanins.

Valine was first isolated by Schutzenberger in 1879. It is an essential amino acid and is present in meat and cereal proteins (5–7%) and in egg and milk proteins (7–8%). Elastin contains notably high concentrations of valine (15.6%).

1.2.3 Physical Properties

1.2.3.1 Dissociation

In aqueous solution amino acids are present, depending on pH, as cations, zwitterions or anions:

$$\underset{\underset{NH_3^\oplus}{|}}{R-CH}-COOH \xrightleftharpoons[+H^\oplus]{-H^\oplus} \underset{\underset{NH_3^\oplus}{|}}{R-CH}-COO^\ominus$$

$$\xrightleftharpoons[+H^\oplus]{-H^\oplus} \underset{\underset{NH_2}{|}}{R-CH}-COO^\ominus \tag{1.3}$$

With the cation denoted as ^+A, the dipolar zwitterion as $^+A^-$ and the anion as A^-, the dissociation constant can be expressed as:

$$\frac{[^\oplus A^\ominus][H^\oplus]}{[^\oplus A]} = K_1 \qquad \frac{[A^\ominus][H^\oplus]}{[^\oplus A^\ominus]} = K_2 \tag{1.4}$$

At a pH where only dipolar ions exist, i.e. the isoelectric point, pI, $[^+A] = [A^-]$:

$$[^\oplus A] = \frac{[^\oplus A^\ominus][H^\oplus]}{K_1} = [A^\ominus] = \frac{[^\oplus A^\ominus]K_2}{[H^\oplus]}$$

$$[H^\oplus] = (K_1 \cdot K_2)^{0,5}$$

$$pI \approx 0{,}5\,(pK_1 + pK_2) \tag{1.5}$$

The dissociation constants of amino acids can be determined, for example, by titration of the acid. Figure 1.2 shows titration curves for glycine, histidine and aspartic acid. Table 1.2 lists the dissociation constants for some amino acids. In amino acids the acidity of the carboxyl group is higher and the basicity of the amino group lower than in the corresponding carboxylic acids and amines (cf. pK values for propionic acid, 2-propylamine and alanine). As illustrated by the comparison of pK values of 2-aminopropionic acid (alanine) and 3-aminopropionic acid (β-alanine), the pK is influenced by the distance between the two functional groups.

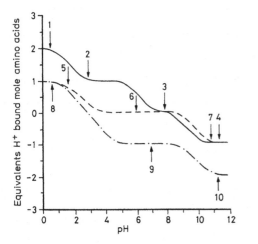

Fig. 1.2. Calculated titration curves for glycine (---), histidine (—) and aspartic acid (-·-·-). Numerals on curves are related to charge of amino acids in respective pH range: 1 $^{++}$His, 2 $^{++}$His$^-$, 3 $^+$His$^-$, 4 His$^-$, 5 $^+$Gly, 6 $^+$Gly$^-$, 7 Gly$^-$, 8 $^+$Asp. 9 $^+$Asp^{--}, 10 Asp^{--}

Table 1.2. Amino acids: dissociation constants and isoelectric points at 25 °C

Amino acid	pK$_1$	pK$_2$	pK$_3$	pK$_4$	pI
Alanine	2.34	9.69			6.0
Arginine	2.18	9.09	12.60		10.8
Asparagine	2.02	8.80			5.4
Aspartic acid	1.88	3.65	9.60		2.8
Cysteine	1.71	8.35	10.66		5.0
Cystine	1.04	2.10	8.02	8.71	5.1
Glutamine	2.17	9.13			5.7
Glutamic acid	2.19	4.25	9.67		3.2
Glycine	2.34	9.60			6.0
Histidine	1.80	5.99	9.07		7.5
4-Hydroxyproline	1.82	9.65			5.7
Isoleucine	2.36	9.68			6.0
Leucine	2.36	9.60			6.0
Lysine	2.20	8.90	10.28		9.6
Methionine	2.28	9.21			5.7
Phenylalanine	1.83	9.13			5.5
Proline	1.99	10.60			6.3
Serine	2.21	9.15			5.7
Threonine	2.15	9.12			5.6
Tryptophan	2.38	9.39			5.9
Tyrosine	2.20	9.11	10.07		5.7
Valine	2.32	9.62			6.0
Propionic acid	4.87				
2-Propylamine	10.63				
β-Alanine		3.55	10.24		6.9
γ-Aminobutyric acid		4.03	10.56		7.3

The reasons for this are probably as follows: in the case of the cation → zwitterion transition, the inductive effect of the ammonium group; in the case of the zwitterion → anion transition, the stabilization of the zwitterion through hydration caused by dipole repulsion (lower than in relation to the anion).

$$(\oplus\!-\!\ominus,\ \text{zwitterion};\ \longleftrightarrow \text{water dipole}) \qquad (1.6)$$

1.2.3.2 Configuration and Optical Activity

Amino acids, except for glycine, have at least one chiral center and, hence, are optically active. All amino acids found in proteins have the same configuration on the α-C-atom: they are considered L-amino acids or (S)-amino acids[*] in the *Cahn-Ingold-Prelog* system (with L-cysteine an exception; it is in the (R)-series). D-amino acids (or (R)-amino acids) also occur in nature, for example, in a number of peptides of microbial origin:

$$(1.7)$$

L-Amino acid (S)-Amino acid   D-Amino acid (R)-Amino acid

Isoleucine, threonine and 4-hydroxyproline have two asymmetric C-atoms, thus each has four isomers:

$$(1.8)$$

L-Isoleucine (2S:3S)- Isoleucine (Common in proteins)   D-Isoleucine (2R:3R)- Isoleucine   L-allo-Isoleucine (2S:3R)- Isoleucine   D-allo-Isoleucine (2R:3S)- Isoleucine

---

[*] As with carbohydrates, D,L-nomenclature is preferred with amino acids.

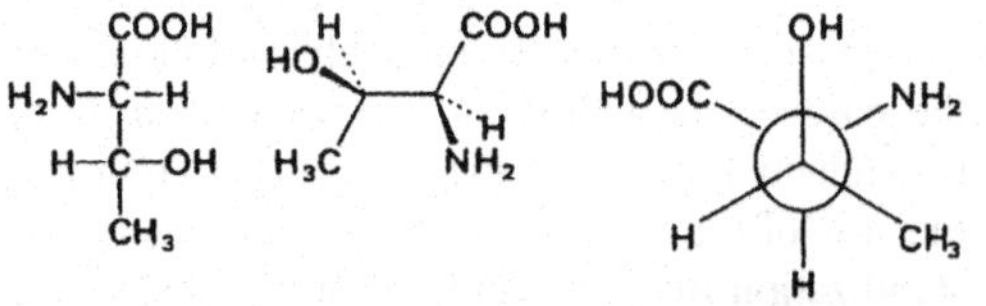

(Fischer-   (dotted line-   (Newman-
projection)   wedge)   projection)

L-Threonine, (2S : 3R)-Threonine (Common in proteins)

(1.9)

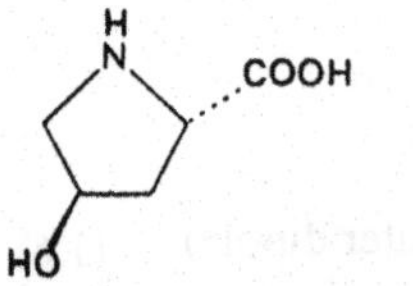

L-4-Hydroxyproline, (2S: 4R)-Hydroxyproline
(Common in proteins)   (1.10)

The specific rotation of amino acids in aqueous solution is strongly influenced by pH. It passes through a minimum in the neutral pH range and rises after addition of acids or bases (Table 1.3). There are various possible methods of separating the racemates which generally occur in amino acid synthesis (cf. 1.2.5). Selective crystallization of an over-saturated solution of racemate after seeding with an enantiomer is used, as is the fractioned crystallization of diastereomeric salts or other derivatives,

Table 1.3. Amino acids: specific rotation ($[\alpha]_D^t$)

Amino acid	Solvent system	Temperature (°C)	$[\alpha]_D$
L-Alanine	0.97 M HCl	15	+14.7°
	water	22	+ 2.7°
	3 M NaOH	20	+ 3.0°
L-Cystine	1.02 M HCl	24	−214.4°
L-Glutamic acid	6.0 M HCl	22.4	+31.2°
	water	18	+11.5°
	1M NaOH	18	+10.96°
L-Histidine	6.0 M HCl	22.7	+13.0°
	water	25.0	−39.01°
	0.5 M NaOH	20	−10.9°
L-Leucine	6.0 M HCl	25.9	+15.1°
	water	24.7	−10.8°
	3.0 M NaOH	20	+7.6°

such as (S)-phenylethylammonium salts of N-acetylamino acids. With enzymatic methods, asymmetric synthesis is used, e. g., of acylamino acid anilides from acylamino acids and aniline through papain:

$$D,L-R-CO-NH-CHR^1-COOH \xrightarrow[\text{Papain}]{\text{Aniline}}$$

$$L-R-CO-NH-CHR^1-CO-NH-C_6H_5$$
$$+ D-R-CO-NH-CHR^1-COOH \qquad (1.11)$$

or asymmetric hydrolysis, e. g., of amino acid esters through esterases, amino acid amides through amidases or N-acylamino acids through aminoacylases:

$$D,L-H_2N-CHR-COOR^1 \xrightarrow{\text{Esterase}}$$
$$L-H_2NCHR-COOH + D-H_2N-CHR-COOR^1$$

$$D,L-H_2N-CHR-CONHR^1 \xrightarrow{\text{Amidase}}$$
$$L-H_2N-CHR-COOH + D-H_2N-CHR-CONHR^1$$

$$D,L-R-CO-NH-CHR^1-COOH \xrightarrow{\text{Acylase}}$$
$$L-H_2N-CHR^1-COOH$$
$$+ D-R-CO-NH-CHR^1-COOH$$

(1.12)

The detection of D-amino acids is carried out by enantioselective HPLC or GC of chiral amino acid derivatives. In a frequently applied method, the derivatives are produced in a precolumn by reaction with o-phthalaldehyde and a chiral thiol (cf. 1.2.4.2.4). Alternatively, the amino acids can be transformed into trifluoroacetylamino acid-2-(R,S)-butylesters. Their GC separation is shown in Fig. 1.3.

1.2.3.3 Solubility

The solubilities of amino acids in water are highly variable. Besides the extremely soluble proline, hydroxyproline, glycine and alanine are also quite soluble. Other amino acids (cf. Table 1.4) are significantly less soluble, with cystine and tyrosine having particularly low solubilities. Addition of acids or bases improves the solubility through salt formation. The presence of other amino acids, in general, also brings about

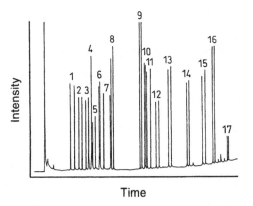

Fig. 1.3. Gas chromatogram of N-pentafluoropropanoylDL-amino acid isopropylesters on Chirasil-Val (N-propionyl-L-valine-tert-butylamide-polysiloxane) (1: D-, L-Ala, 2: D-, L-Val, 3. D-, L-Thr, 4: Gly, 5: D-, L-Ile, 6: D-, L-Pro, 7: D-, L-Leu, 8: D-, L-Ser, 9: D-, L-Cys, 10: D-, L-Asp, 11: D-, L-Met, 12: D-, L-Phe, 13: D-, L-Glu, 14: D-, L-Tyr, 15: D-, L-Orn, 16: D-, L-Lys, 17: D-, L-Trp; according to *Frank* et al., 1977)

an increase in solubility. Thus, the extent of solubility of amino acids in a protein hydrolysate is different than that observed for the individual components.

The solubility in organic solvents is not very good because of the polar characteristics of the amino acids. All amino acids are insoluble in ether. Only cysteine and proline are relatively soluble in ethanol (1.5 g/100 g at 19 °C). Methionine, arginine, leucine (0.0217 g/100 g; 25 °C), glutamic acid (0.00035 g/100 g; 25 °C), phenylalanine, hydroxy-proline, histidine and tryptophan are sparingly soluble in ethanol. The solubility of isoleucine in hot ethanol is relatively high (0.09 g/100 g at 20 °C; 0.13 g/100 g at 78–80 °C).

1.2.3.4 UV-Absorption

Aromatic amino acids such as phenylalanine, tyrosine and tryptophan absorb in the UV-range of the spectrum with absorption maxima at 200–230 nm and 250–290 nm (Fig. 1.4). Dissociation of the phenolic HO-group of tyrosine shifts the absorption curve by about 20 nm towards longer wavelengths (Fig. 1.5).

Table 1.4. Solubility of amino acids in water (g/100 g H₂O)

Amino acid	Temperature (°C)				
	0	25	50	75	100
L-Alanine	12.73	16.51	21.79	28.51	37.30
L-Asparatic acid	0.209	0.500	1.199	2.875	6.893
L-Cystine	0.005	0.011	0.024	0.052	0.114
L-Glutamic acid	0.341	0.843	2.186	5.532	14.00
Glycine	14.18	24.99	39.10	54.39	67.17
L-Histidine	–	4.29	–	–	–
L-Hydroxy-proline	28.86	36.11	45.18	51.67	–
L-Isoleucine	3.791	4.117	4.818	6.076	8.255
L-Leucine	2.270	2.19	2.66	3.823	5.638
D,L-Methionine	1.818	3.381	6.070	10.52	17.60
L-Phenyl-alanine	1.983	2.965	4.431	6.624	9.900
L-Proline	127.4	162.3	206.7	239.0	–
D,L-Serine	2.204	5.023	10.34	19.21	32.24
L-Tryptophan	0.823	1.136	1.706	2.795	4.987
L-Tyrosine	0.020	0.045	0.105	0.244	0.565
L-Valine	8.34	8.85	9.62	10.24	–

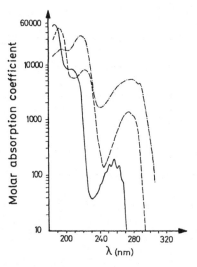

Fig. 1.4. Ultraviolet absorption spectra of some amino acids. (according to *Luebke, Schroeder* and *Kloss*, 1975). -----Trp. ---Tyr. —Phe

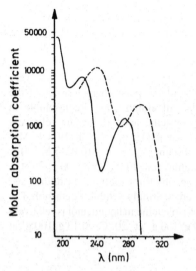

Fig. 1.5. Ultraviolet absorption spectrum of tyrosine as affected by pH. (according to *Luebke, Schroeder* and *Kloss*, 1975) — 0.1 mol/l HCl, --- 0.1 mol/l NaOH

Absorption readings at 280 nm are used for the determination of proteins and peptides. Histidine, cysteine and methionine absorb between 200 and 210 nm.

1.2.4 Chemical Reactions

Amino acids show the usual reactions of both carboxylic acids and amines. Reaction specificity is due to the presence of both carboxyl and amino groups and, occasionally, of other functional groups. Reactions occurring at 100–220 °C, such as in cooking, frying and baking, are particularly relevant to food chemistry.

1.2.4.1 Esterification of Carboxyl Groups

Amino acids are readily esterified by acid-catalyzed reactions. An ethyl ester hydrochloride is obtained in ethanol in the presence of HCl:

$$R-\underset{\underset{NH_3^{\oplus}Cl^{\ominus}}{|}}{CH}-COOH + R'-OH \xrightarrow{H^{\oplus}} R-\underset{\underset{NH_3^{\oplus}Cl^{\ominus}}{|}}{CH}-COOR' + H_2O$$

$$(1.13)$$

The free ester is released from its salt by the action of alkali. A mixture of free esters can then be separated by distillation without decomposition. Fractional distillation of esters is the basis of a method introduced by *Emil Fischer* for the separation of amino acids:

$$R-\underset{\underset{NH_3^{\oplus}X^{\ominus}}{|}}{CH}-COOR' \xrightarrow{B} R-\underset{\underset{NH_2}{|}}{CH}-COOR' + BH^{\oplus}X^{\ominus}$$

$$(1.14)$$

Free amino acid esters have a tendency to form cyclic dipeptides or open-chain polypeptides:

$$(1.15)$$

$$-NH-\underset{\underset{R}{|}}{CH}-\overset{O}{\underset{OR'}{C}} + H_2N-\underset{\underset{R}{|}}{CH}-\overset{O}{C}$$

$$\longrightarrow -NH-\underset{\underset{R}{|}}{CH}-CO-NH-\underset{\underset{R}{|}}{CH}-CO- + R'OH$$

$$(1.16)$$

tert-butyl esters, which are readily split by acids, or benzyl esters, which are readily cleaved by HBr/glacial acetic acid or catalytic hydrogenation, are used as protective groups in peptide synthesis.

1.2.4.2 Reactions of Amino Groups

1.2.4.2.1 Acylation

Activated acid derivatives, e. g., acid halogenides or anhydrides, are used as acylating agents:

$$R'-COX + H_2N-\underset{\underset{R}{|}}{CH}-COO^{\ominus} + OH^{\ominus}$$

$$\longrightarrow R'-CO-NH-\underset{\underset{R}{|}}{CH}-COO^{\ominus} + X^{\ominus} + H_2O$$

$$(1.17)$$

N-acetyl amino acids are being considered as ingredients in chemically-restricted diets and for fortifying plant proteins to increase their biological value. Addition of free amino acids to food which must be heat treated is not problem free. For example, methionine in the presence of a reducing sugar can form methional by a *Strecker* degradation mechanism, imparting an off-flavor to food. Other essential amino acids, e. g., lysine or threonine, can lose their biological value through similar reactions. Feeding tests with rats have shown that N-acetyl-L-methionine and N-acetyl-L-threonine have nutritional values equal to those of the free amino acids (this is true also for humans with acetylated methionine). The growth rate of rats is also increased significantly by the α- or ε-acetyl or α,ε-diacetyl derivatives of lysine.

Some readily cleavable acyl residues are of importance as temporary protective groups in peptide synthesis.

The trifluoroacetyl residue is readily removed by mild base-catalyzed hydrolysis:

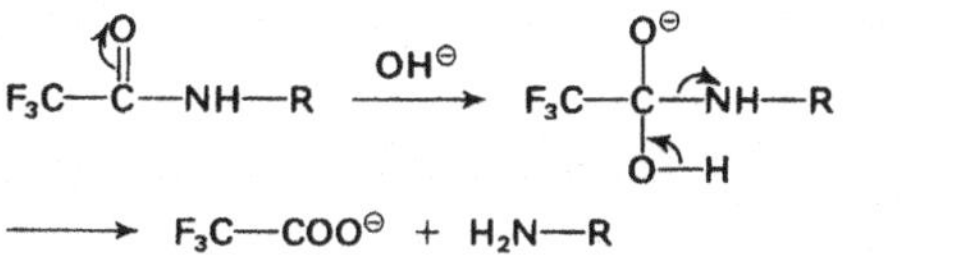

$$(1.18)$$

The phthalyl residue can be readily cleaved by hydrazinolysis:

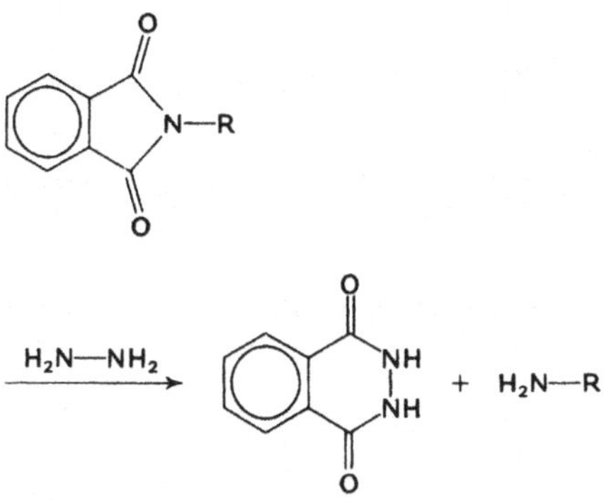

$$(1.19)$$

The benzyloxycarbonyl group can be readily removed by catalytic hydrogenation or by hydrolysis with HBr/glacial acetic acid:

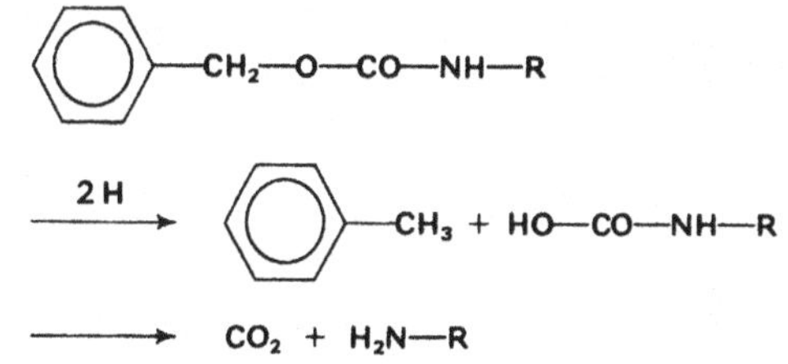

$$(1.20)$$

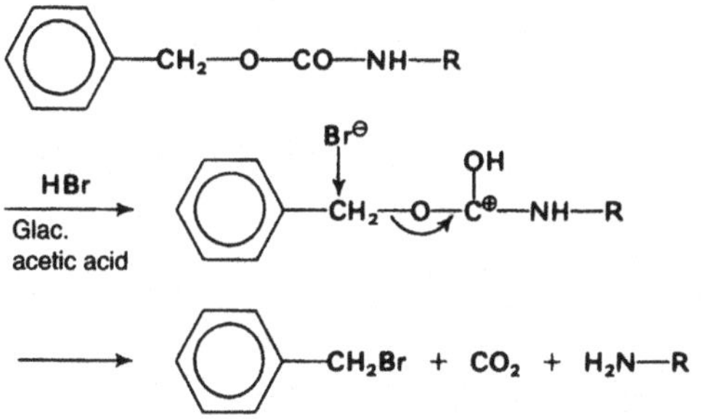

$$(1.21)$$

The *tert*-alkoxycarbonyl residues, e. g., the *tert*-butyloxycarbonyl groups, are cleaved under acid-catalyzed conditions:

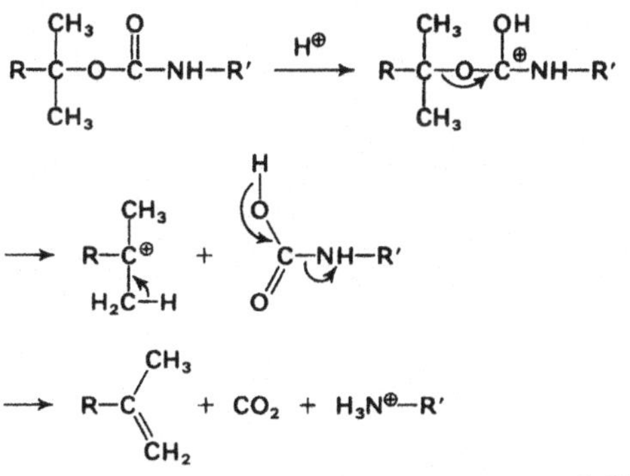

$$(1.22)$$

N-acyl derivatives of amino acids are transformed into oxazolinones (azlactones) by elimination of water:

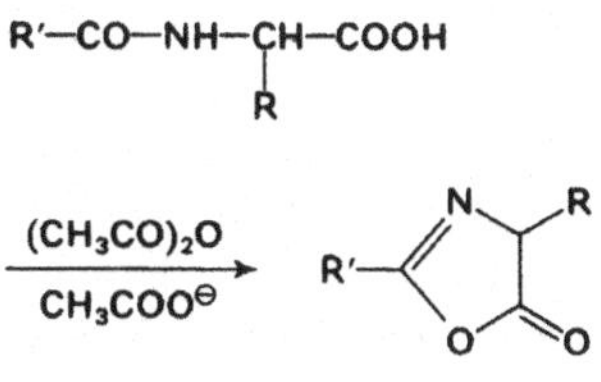

$$R'-CO-NH-CH-COOH$$
$$|$$
$$R$$

$$\xrightarrow[CH_3COO^\ominus]{(CH_3CO)_2O}$$

$$(1.23)$$

These are highly reactive intermediary products which form a mesomerically stabilized anion. The anion can then react, for example, with aldehydes. This reaction is utilized in amino acid synthesis with glycine azlactone as a starting compound:

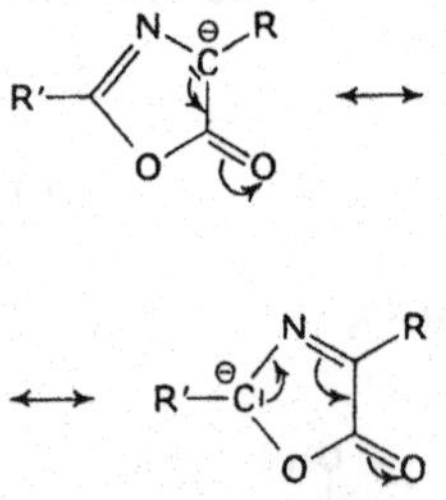

$$(1.24)$$

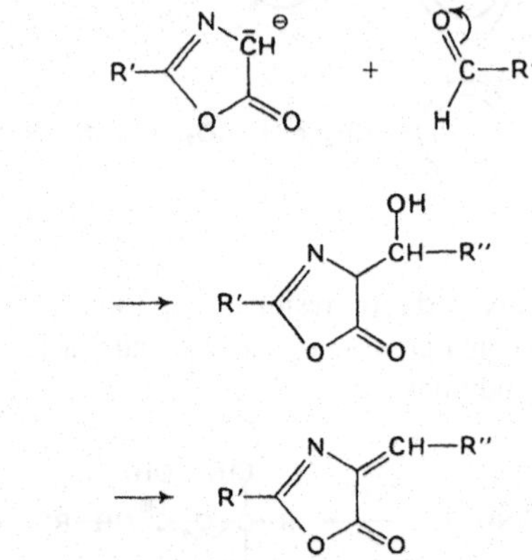

$$(1.25)$$

Acylation of amino acids with 5-dimethylaminonaphthalene-1-sulfonyl chloride (dansyl chloride, DANS-Cl) is of great analytical importance:

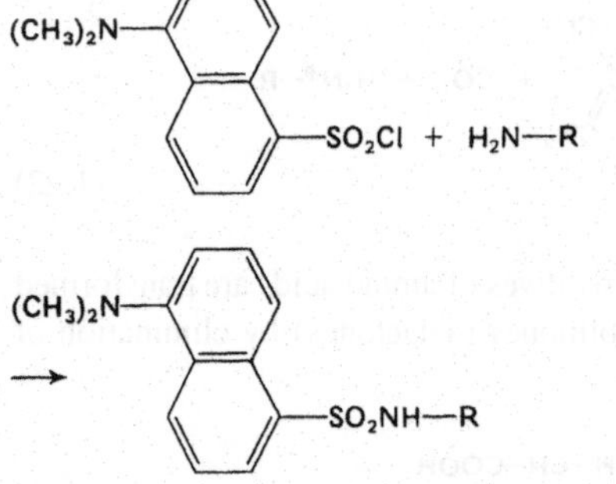

$$(1.26)$$

The aryl sulfonyl derivatives are very stable against acidic hydrolysis. Therefore, they are suitable for the determination of free N-terminal amino groups or free ε-amino groups of peptides or proteins. Dansyl derivatives which fluoresce in UV-light have a detection limit in the nanomole range, which is lower than that of 2,4-dinitrophenyl derivatives by a factor of 100. Dimethylaminoazobenzenesulfonylchloride (DABS-Cl) and 9-fluoroenylmethylchloroformate (FMOC) detect amino acids (cf. Formula 1.27 and 1.28) including proline and hydroxyproline. The fluorescent derivatives can be quantitatively determined after HPLC separation.

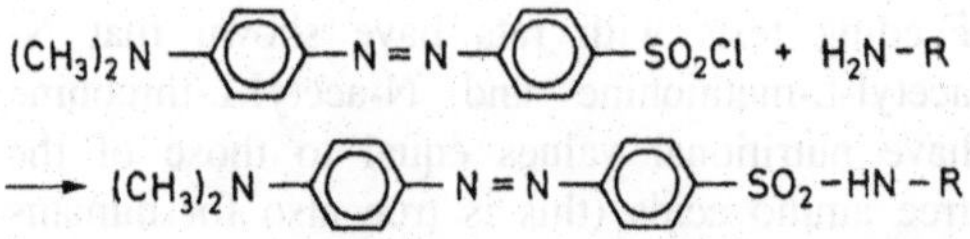

$$(1.27)$$

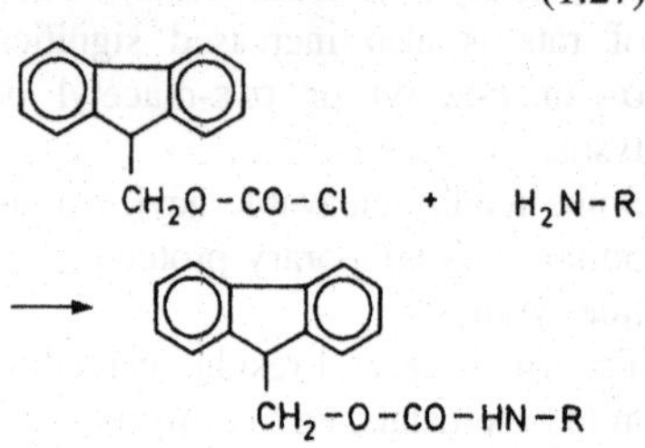

$$(1.28)$$

1.2.4.2.2 Alkylation and Arylation

N-methyl amino acids are obtained by reaction of the N-tosyl derivative of the amino acid with methyl iodide, followed by removal of the tosyl substituent with HBr:

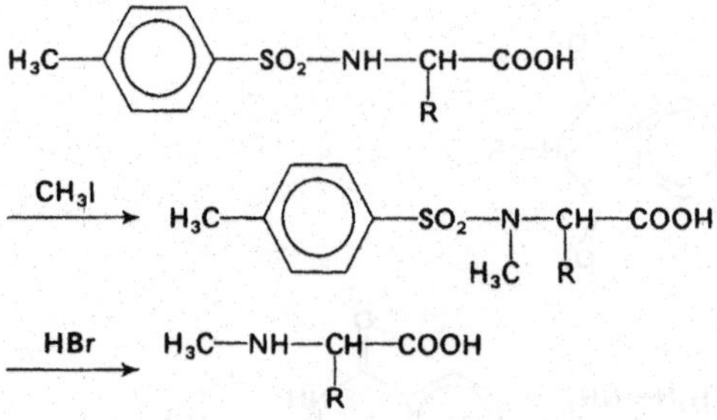

$$(1.29)$$

The N-methyl compound can also be formed by methylating with HCHO/HCOOH the benzylidene derivative of the amino acid, formed initially by reaction of the amino acid with benzaldehyde. The benzyl group is then eliminated by hydrogenolysis:

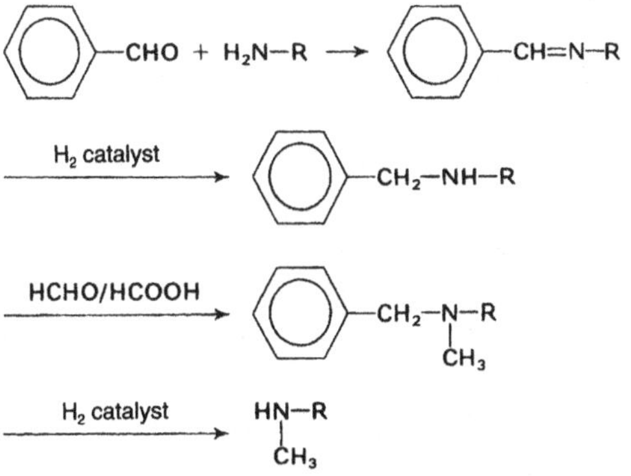

(1.30)

Dimethyl amino acids are obtained by reaction with formaldehyde, followed by reduction with sodium borohydride:

$$2\,HCHO + H_2N{-}R \xrightarrow[\text{pH 9, 0 °C}]{NaBH_4} (CH_3)_2N{-}R \qquad (1.31)$$

The corresponding reactions with proteins are being considered as a means of protecting the ε-amino groups and, thus, of avoiding their destruction in food through the *Maillard* reaction (cf. 1.4.6.2.2).

Direct reaction of amino acids with methylating agents, e. g. methyl iodide or dimethyl sulfate, proceeds through monomethyl and dimethyl compounds to trimethyl derivatives (or generally to N-trialkyl derivatives) denoted as betaines:

$$H_2N{-}\underset{R}{\underset{|}{C}}H{-}COOH \xrightarrow{CH_3I} (CH_3)_3N^{\oplus}{-}\underset{R}{\underset{|}{C}}H{-}COO^{\ominus} \qquad (1.32)$$

As shown in Table 1.5, betaines are widespread in both the animal and plant kingdoms.

Derivatization of amino acids by reaction with 1-fluoro-2,4-dinitrobenzene (FDNB) yields N-2,4-dinitrophenyl amino acids (DNP-amino acids), which are yellow compounds and crystallize readily. The reaction is important for labeling N-terminal amino acid residues and free ε-amino groups present in peptides and proteins; the DNP-amino acids are stable under conditions of acidic hydrolysis (cf. Reaction 1.33).

Table 1.5. Occurrence of trimethyl amino acids $(CH_3)_3N^+$-CHR-COO^- (betaines)

Amino acid	Betaine	Occurrence
β-Alanine	Homobetaine	Meat extract
γ-Amino-butyric acid	Actinine	Mollusk (shell-fish)
Glycine	Betaine	Sugar beet, other samples of animal and plant origin
Histidine	Hercynine	Mushrooms
β-Hydroxy-γ-amino-butyric acid	Carnitine	Mammals muscle tissue, yeast, wheat germ, fish, liver, whey, mollusk (shell-fish)
4-Hydroxy-proline	Betonicine	Jack beans
Proline	Stachydrine	Stachys, orange leaves, lemon peel, alfalfa, Aspergillus oryzae

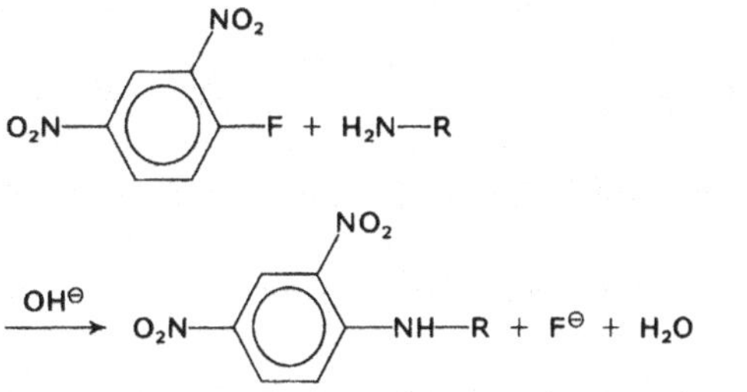

(1.33)

Another arylation reagent is 7-fluoro-4-nitrobenzo-2-oxa-1,3-diazol (NBD-F), which is also used as a chlorine compound (NBD-Cl) and which leads to derivatives that are suited for an amino acid analysis through HPLC separation:

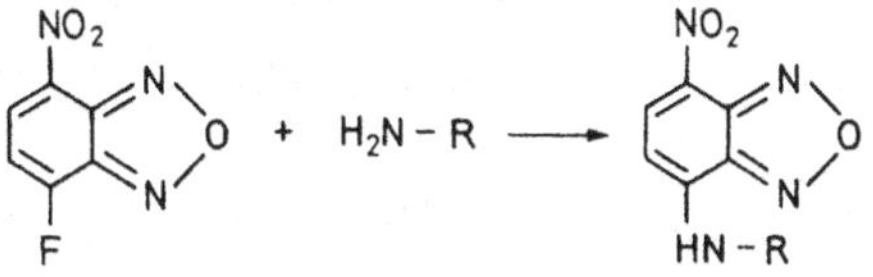

(1.34)

Reaction of amino acids with triphenylmethyl chloride (tritylchloride) yields N-trityl derivatives, which are alkali stable. However, the derivative is cleaved in the presence of acid, giving a stable triphenylmethyl cation and free amino acid:

$$(C_6H_5)_3C-Cl + H_2N-R \xrightarrow{\ OH^\ominus\ } (C_6H_5)_3C-NH-R$$

$$\xrightarrow{\ H^\oplus\ } (C_6H_5)_3C^\oplus + H_2N-R$$

(1.35)

The reaction with trinitrobenzene sulfonic acid is also of analytical importance. It yields a yellow-colored derivative that can be used for the spectrophotometric determination of protein:

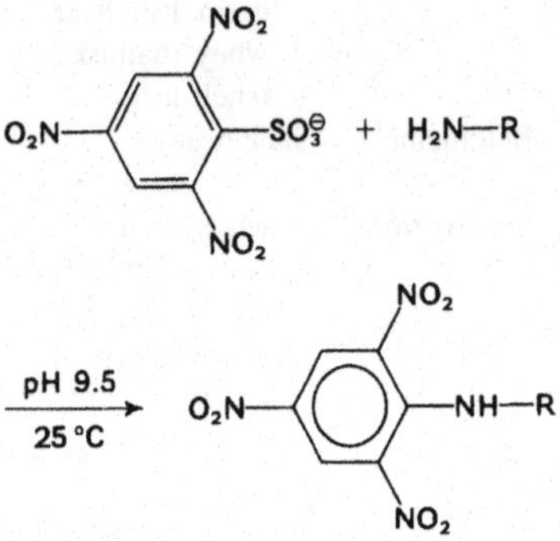

$$\xrightarrow[25\,°C]{pH\ 9.5}$$

(1.36)

The reaction is a nucleophilic aromatic substitution proceeding through an intermediary addition product (*Meisenheimer* complex). It occurs under mild conditions only when the benzene ring structure is stabilized by electron-withdrawing substituents on the ring (cf. Reaction 1.37).

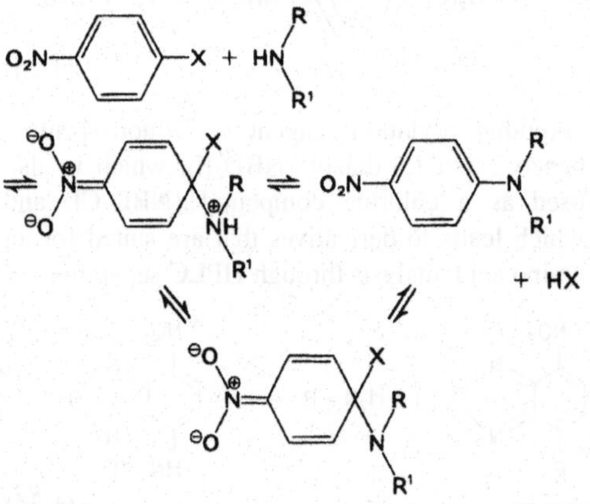

(1.37)

The formation of the *Meisenheimer* complex has been verified by isolating the addition product from the reaction of 2,4,6-trinitroanisole with potassium ethoxide (cf. Reaction 1.38).

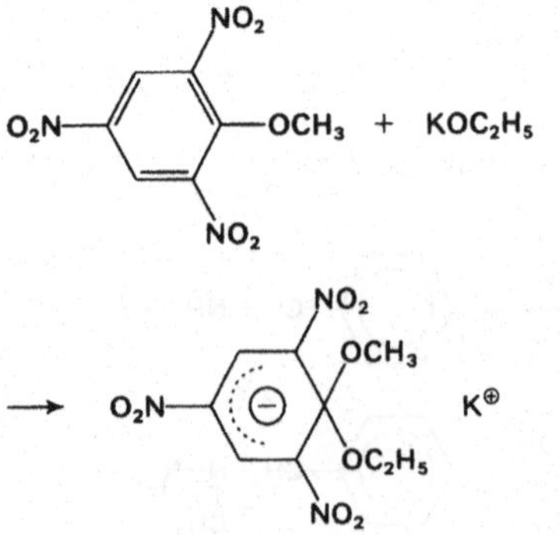

(1.38)

An analogous reaction occurs with 1,2-naphthoquinone-4-sulfonic acid (*Folin* reagent) but, instead of a yellow color (cf. Formula 1.36), a red color develops:

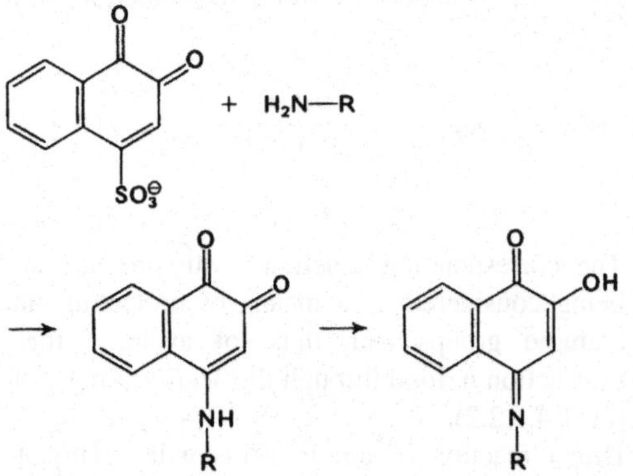

(1.39)

1.2.4.2.3 Carbamoyl and Thiocarbamoyl Derivatives

Amino acids react with isocyanates to yield carbamoyl derivatives which are cyclized into 2,4-dioxoimidazolidines (hydantoins) by boiling in an acidic medium:

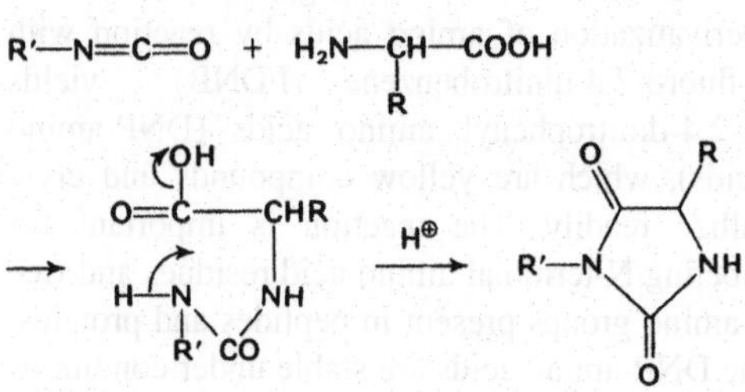

$$R'-N{=}C{=}O + H_2N-CH-COOH$$
$$\hphantom{R'-N{=}C{=}O + H_2N-}R$$

(1.40)

A corresponding reaction with phenylisothiocyanate can degrade a peptide in a stepwise fashion (*Edman* degradation). The reaction is of great importance for revealing the amino acid sequence in a peptide chain. The phenylthiocarbamoyl derivative (PTC-peptide) formed in the first step (coupling) is cleaved non-hydrolytically in the second step (cleavage) with anhydrous trifluoroacetic acid into anilinothiazolinone as derivative of the N-terminal amino acid and the remaining peptide which is shortened by the latter. Because of its instability, the thiazolinone is not suited for an identification of the N-terminal amino acid and is therefore – after separation from the remaining peptide, in the third step (conversion) – converted in aqueous HCl via the phenylthiocarbamoylamino acid into phenyl-thiohydantoin, while the remaining peptide is fed into a new cycle.

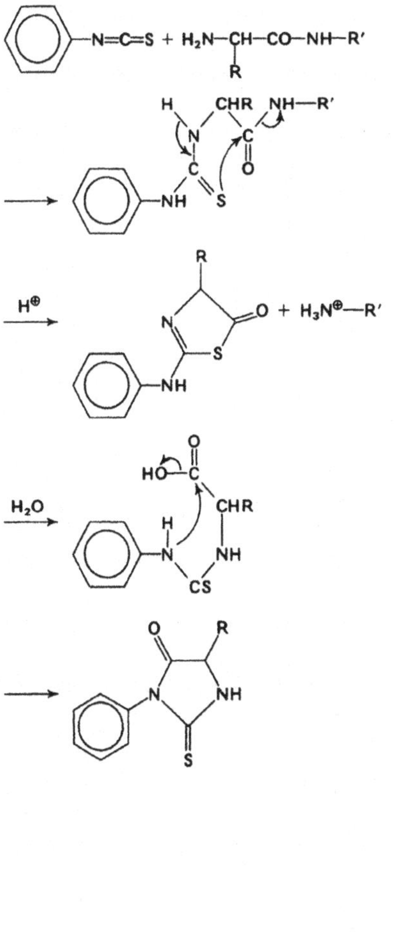

(1.41)

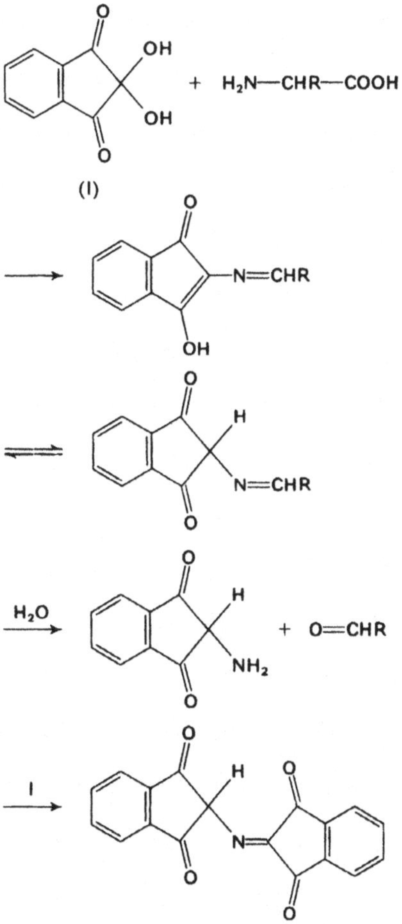

(1.42)

1.2.4.2.4 Reactions with Carbonyl Compounds

Amino acids react with carbonyl compounds, forming azomethines. If the carbonyl compound has an electron-withdrawing group, e.g., a second carbonyl group, transamination and decarboxylation occur. The reaction is known as the *Strecker* degradation and plays a role in food since food can be an abundant source of dicarbonyl compounds generated by the *Maillard* reaction (cf. 4.2.4.4.7). The aldehydes formed from amino acids (*Strecker* aldehydes) are aroma compounds (cf. 5.3.1.1). The ninhydrin reaction is a special case of the *Strecker* degradation. It is an important reaction for the quantitative determination of amino acids using spectrophotometry (cf. Reaction 1.42). The detection limit lies at 1–0.5 nmol. The resultant blue-violet color has an absorption maximum at 570 nm. Proline yields a yellow-colored compound with $\lambda_{max} = 440$ nm (Reaction 1.43):

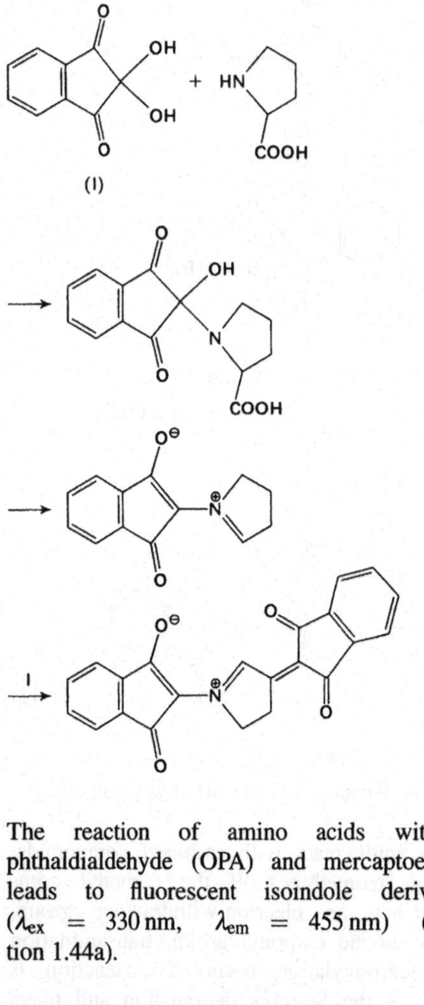

(1)

(1.43)

The reaction of amino acids with o-phthaldialdehyde (OPA) and mercaptoethanol leads to fluorescent isoindole derivatives ($\lambda_{ex} = 330$ nm, $\lambda_{em} = 455$ nm) (Reaction 1.44a).

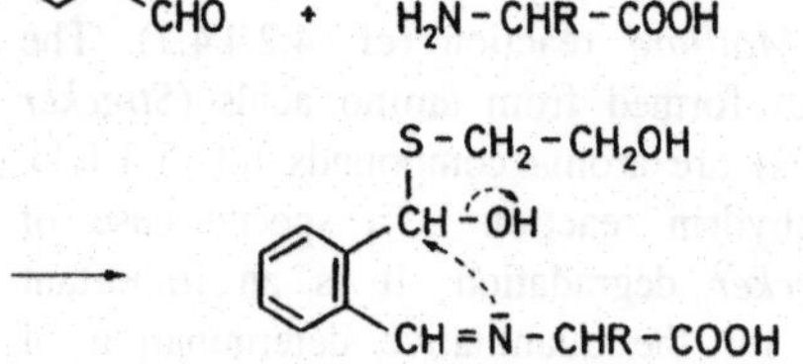

(1.44a)

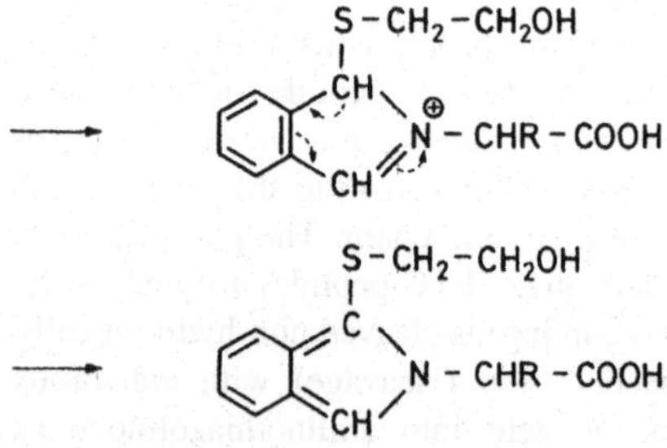

(1.44b)

The derivatives are used for amino acid analysis via HPLC separation. Instead of mercaptoethanol, a chiral thiol, e.g., N-isobutyryl-L-cysteine, is used for the detection of D-amino acids. The detection limit lies at 1 pmol. The very fast racemizing aspartic acid is an especially suitable marker. One disadvantage of the method is that proline and hydroxyproline are not detected. This method is applied, e.g., in the analysis of fruit juices, in which high concentrations of D-amino acids indicate bacterial contamination or the use of highly concentrated juices. Conversely, too low concentrations of D-amino acids in fermented foods (cheese, soy and fish sauces, wine vinegar) indicate unfermented imitations.

Fluorescamine reacts with primary amines and amino acids – at room temperature under alkaline conditions – to form fluorescent pyrrolidones ($\lambda_{ex} = 390$ nm, $\lambda_{em} = 474$ nm). The detection limit lies at 50–100 pmol:

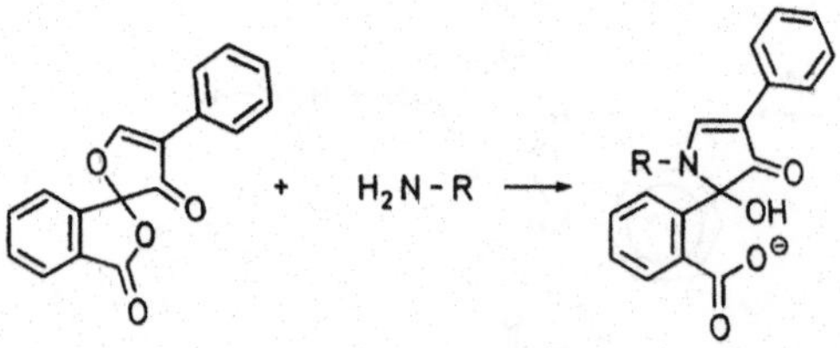

(1.45)

The excess reagent is very quickly hydrolyzed into water-soluble and non-fluorescent compounds.

1.2.4.3 Reactions Involving Other Functional Groups

The most interesting of these reactions are those in which α-amino and α-carboxyl groups are blocked, that is, reactions occurring with peptides and proteins. These reactions will be covered in detail in sections dealing with modification of proteins (cf. 1.4.4 and 1.4.6.2). A number of reactions of importance to free amino acids will be covered in the following sections.

1.2.4.3.1 Lysine

A selective reaction may be performed with either of the amino groups in lysine. Selective acylation of the ε-amino group is possible using the lysine-Cu^{2+} complex as a reactant:

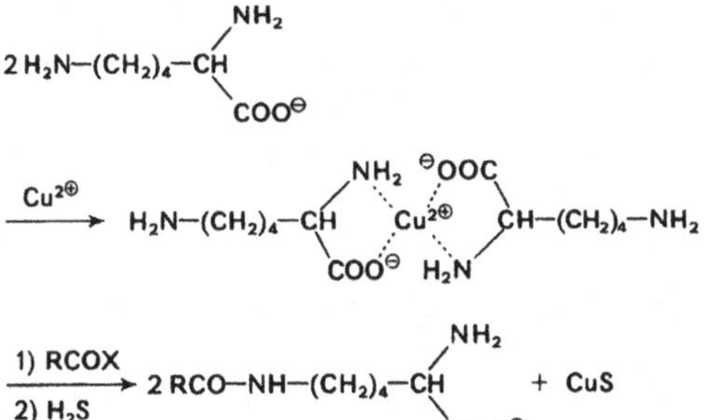

$$2\,H_2N-(CH_2)_4-\overset{\displaystyle NH_2}{\underset{\displaystyle COO^{\ominus}}{CH}}$$

$$\xrightarrow{Cu^{2\oplus}} H_2N-(CH_2)_4-\underset{COO^{\ominus}}{\overset{NH_2}{CH}}\,\,\underset{H_2N}{\overset{\ominus OOC}{Cu^{2\oplus}}}\,\,CH-(CH_2)_4-NH_2$$

$$\xrightarrow[\text{2) H}_2\text{S}]{\text{1) RCOX}} 2\,RCO-NH-(CH_2)_4-\underset{COO^{\ominus}}{\overset{NH_2}{CH}} + CuS$$

(1.46)

Selective reaction with the α-amino group is possible using a benzylidene derivative:

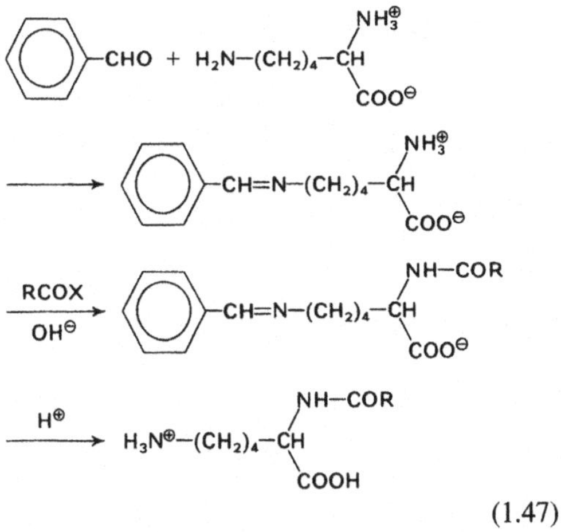

ε-N-benzylidene-L-lysine and ε-N-salicylidene-L-lysine are as effective as free lysine in growth

feeding tests with rats. Browning reactions of these derivatives are strongly retarded, hence they are of interest for lysine fortification of food.

1.2.4.3.2 Arginine

In the presence of α-naphthol and hypobromite, the guanidyl group of arginine gives a red compound with the following structure:

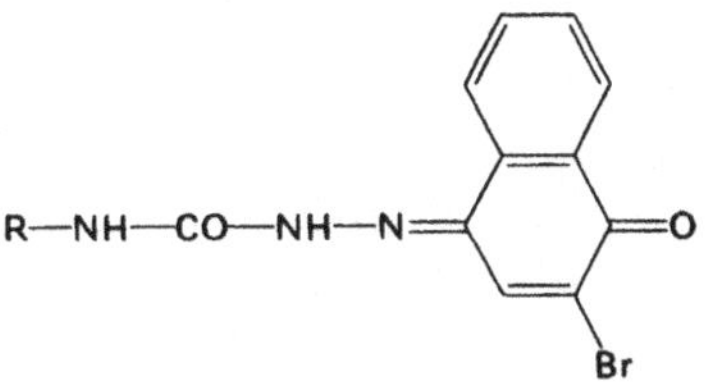

$$R-NH-CO-NH-N=$$

(1.48)

1.2.4.3.3 Aspartic and Glutamic Acids

The higher esterification rate of β- and γ-carboxyl groups can be used for selective reactions. On the other hand the β- and γ-carboxyl groups are more rapidly hydrolyzed in acid-catalyzed hydrolysis since protonation is facilitated by having the ammonium group further away from the carboxyl group. Alkali-catalyzed hydrolysis of methyl or ethyl esters of aspartic or glutamic acids bound to peptides can result in the formation of isopeptides.

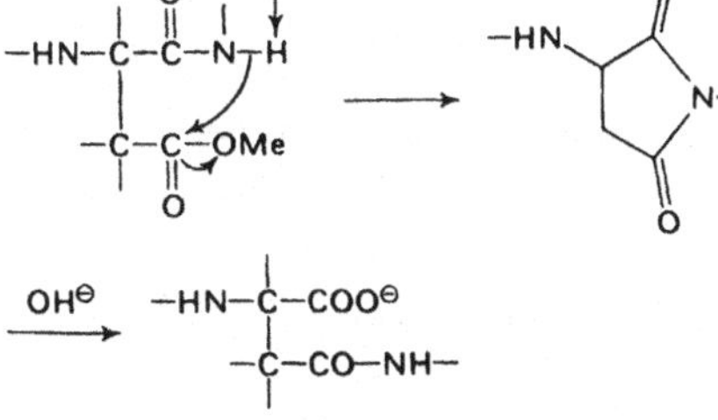

(1.49)

Decarboxylation of glutamic acid yields γ-aminobutyric acid. This compound, which also occurs in wine (cf. 20.2.6.9), tastes sour and produces a dry feeling in the mouth at concentrations above its recognition threshold (0.02 mmol/l).

1.2.4.3.4 Serine and Threonine

Acidic or alkaline hydrolysis of protein can yield α-keto acids through β-elimination of a water molecule:

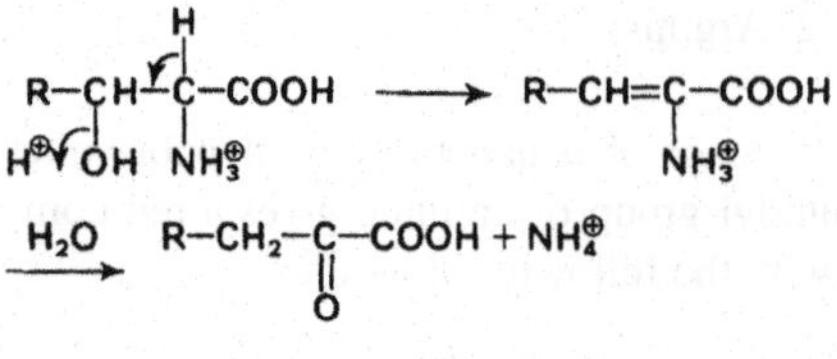

$$R-\underset{\underset{\text{OH}}{\overset{\oplus}{H}}}{CH}-\underset{NH_3^{\oplus}}{\overset{\overset{H}{|}}{C}}-COOH \longrightarrow R-CH=\underset{NH_3^{\oplus}}{C}-COOH$$

$$\xrightarrow{H_2O} R-CH_2-\underset{\overset{\|}{O}}{C}-COOH + NH_4^{\oplus}$$

$$(1.50)$$

In this way, α-ketobutyric acid formed from threonine can yield another amino acid, α-aminobutyric acid, via a transamination reaction. Reaction 1.51 is responsible for losses of hydroxy amino acids during protein hydrolysis.

Reliable estimates of the occurrence of these amino acids are obtained by hydrolyzing protein for varying lengths of time and extrapolating the results to zero time.

1.2.4.3.5 Cysteine and Cystine

Cysteine is readily converted to the corresponding disulfide, cystine, even under mild oxidative conditions, such as treatment with I_2 or potassium hexacyanoferrate (III). Reduction of cystine to cysteine is possible using sodium borohydride or thiol reagents (mercaptoethanol, dithiothreitol):

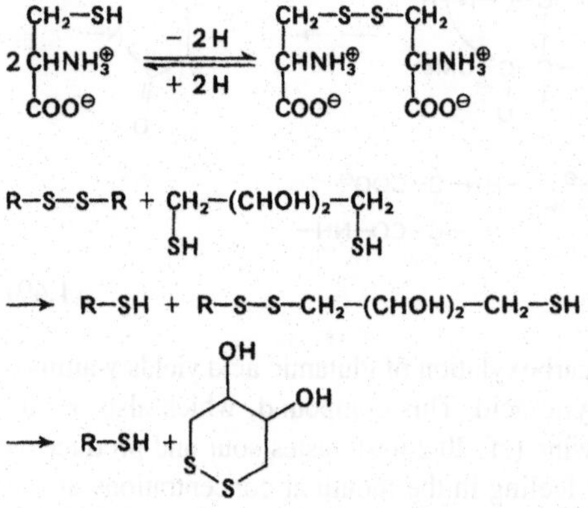

$$2 \underset{\underset{COO^{\ominus}}{|}}{\overset{\overset{CH_2-SH}{|}}{CHNH_3^{\oplus}}} \underset{+2H}{\overset{-2H}{\rightleftharpoons}} \underset{\underset{COO^{\ominus}}{|}}{\overset{\overset{CH_2-S-S-CH_2}{|}}{CHNH_3^{\oplus}}} \underset{\underset{COO^{\ominus}}{|}}{CHNH_3^{\oplus}}$$

$$R-S-S-R + \underset{\underset{SH}{|}}{CH_2}-(CHOH)_2-\underset{\underset{SH}{|}}{CH_2}$$

$$\longrightarrow R-SH + R-S-S-CH_2-(CHOH)_2-CH_2-SH$$

$$\longrightarrow R-SH + \text{(cyclic structure with OH, OH, S, S)}$$

$$(1.51)$$

The equilibrium constants for the reduction of cystine at pH 7 and 25 °C with mercaptoethanol or dithiothreitol are 1 and 10^4, respectively.

Stronger oxidation of cysteine, e. g., with performic acid, yields the corresponding sulfonic acid, cysteic acid:

$$\left.\begin{array}{c} R-SH \\ \\ R-S-S-R \end{array}\right\} \xrightarrow{HCOOOH} R-SO_3H$$

$$(1.52)$$

Reaction of cysteine with alkylating agents yields thioethers. Iodoacetic acid, iodoacetamide, dimethylaminoazobenzene iodoacetamide, ethylenimine and 4-vinylpyridine are the most commonly used alkylating agents:

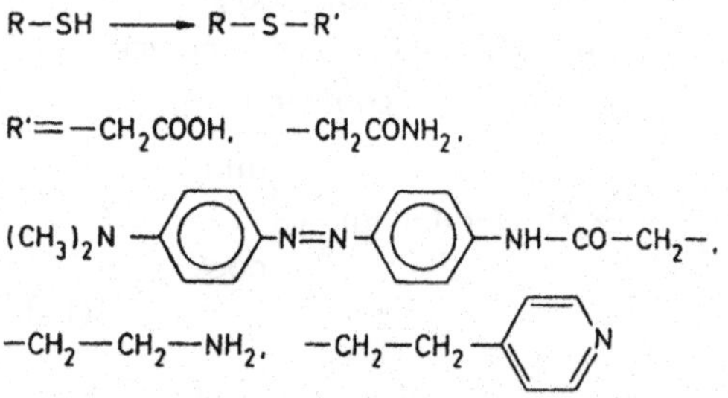

$$R-SH \longrightarrow R-S-R'$$

$$R'=-CH_2COOH, \quad -CH_2CONH_2,$$

$$(CH_3)_2N-\bigcirc\!\!\!\!-N=N-\bigcirc\!\!\!\!-NH-CO-CH_2-,$$

$$-CH_2-CH_2-NH_2, \quad -CH_2-CH_2-\bigcirc\!\!\!\!-N$$

$$(1.53)$$

1.2.4.3.6 Methionine

Methionine is readily oxidized to the sulfoxide and then to the sulfone. This reaction can result in losses of this essential amino acid during food processing:

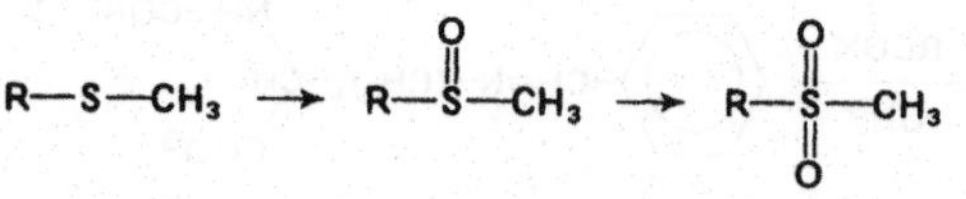

$$R-S-CH_3 \longrightarrow R-\underset{\overset{\|}{O}}{S}-CH_3 \longrightarrow R-\underset{\overset{\|}{O}}{\overset{\overset{\|}{O}}{S}}-CH_3$$

$$(1.54)$$

1.2.4.3.7 Tyrosine

Tyrosine reacts, like histidine, with diazotized sulfanilic acid (*Pauly* reagent). The coupled-reaction product is a red azo compound:

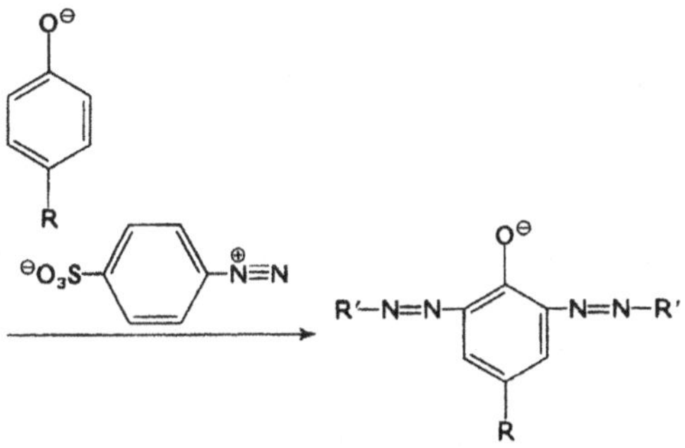

$$(1.55)$$

1.2.4.4 Reactions of Amino Acids at Higher Temperatures

Reactions at elevated temperatures are important during the preparation of food. Frying, roasting, boiling and baking develop the typical aromas of many foods in which amino acids participate as precursors. Studies with food and model systems have shown that the characteristic odorants are formed via the *Maillard* reaction and that they are subsequent products, in particular of cysteine, methionine, ornithine and proline (cf. 12.9.3).

1.2.4.4.1 Acrylamide

The toxic compound acrylamide is one of the volatile compounds formed during the heating of food (cf. 9.7.3). Model experiments have shown that it is produced in reactions of asparagine with reductive carbohydrates or from the resulting cleavage products (e. g., 2-butanedione, 2-oxopropanal).

The formation is promoted by temperatures >100 °C and/or longer reaction times. Indeed, model experiments have shown that the highest yields based on asparagine are ca. 0.1–1 mol%. Cysteine and methionine also form acrylamide in the presence of glucose, but the yields are considerably lower than those from asparagine. The thermal reaction of acrolein with ammonia also produces acrylamide, but again only in small amounts.

Although from a purely stoichiometric standpoint, it would be possible that the degradation of asparagine by the cleavage of CO_2 und NH_3 directly produces acrylamide, the course of formation is quite complex. Indeed, various proposals exist for the mechanism of this formation. It was shown that considerable amounts of 3-aminopropionamide are produced in the reaction of asparagine with α-dicarbonyl compounds with the formation of the *Schiff* base and subsequent decarboxylation and hydrolysis in the sense of a *Strecker* reaction (Fig. 1.6). It could be shown in model studies and in additional experiments with foods (cocoa, cheese) that the splitting-off of ammonia from 3-aminopropionamide occurs relatively easily at higher temperatures and even in the absence of carbohydrates results in very high yields of acrylamide (>60 mol%). Therefore, 3-aminopropionamide, which is to be taken as the biogenic amine of asparagine, represents a transient intermediate in the formation of acrylamide in foods. In the meantime, this compound has also been identified in different foods.

Another mechanism (Fig. 1.7, right) starts out from the direct decomposition of the *Schiff* base obtained from a reductive carbohydrate and asparagine via instable analytically undetectable intermediates. It is assumed that the ylide formed by the decarboxylation of the *Schiff* base directly decomposes on cleavage of the

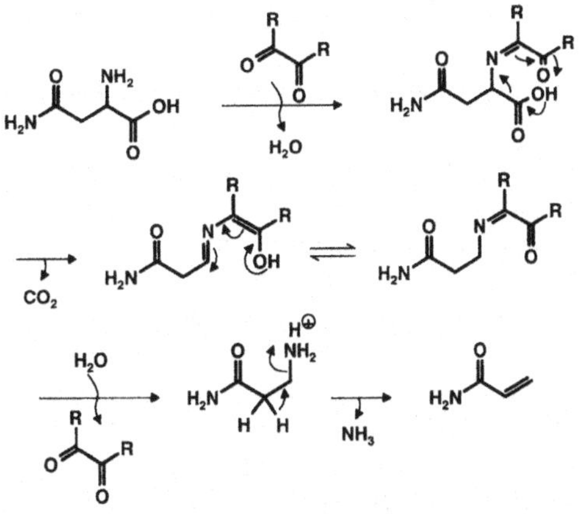

Fig. 1.6. Formation of 3-aminopropionamide (3-APA) from the Strecker reaction of asparagine and subsequent deamination to acrylamide (according to *Granvogl* et al., 2006)

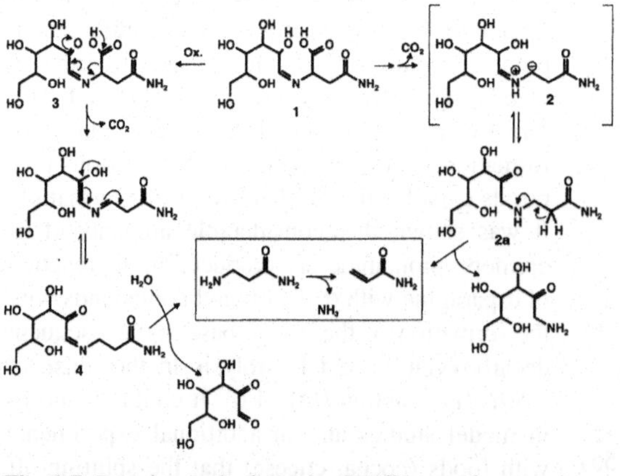

Fig. 1.7. Reaction paths from the *Schiff* base of asparagine and glucose which result in acrylamide (according to *Stadler* et al., 2004 and *Granvogl* et al., 2006)

C-N bond to give acrylamide and a 1-amino-2-hexulose. Another proposed mechanism (Fig. 1.7, left) is the oxidation of the *Schiff* base and subsequent decarboxylation. Here, an intermediate is formed which can decompose to 3-aminopropionamide after enolization and hydrolysis. 3-Aminopropionamide can then be

meat extract, deep-fried meat, grilled fish and heated model mixtures on the basis of creatine, an amino acid (glycine, alanine, threonine) and glucose. For the most part they were imidazoquinolines and imidazoquinoxalines. The highest concentrations (μ/kg)

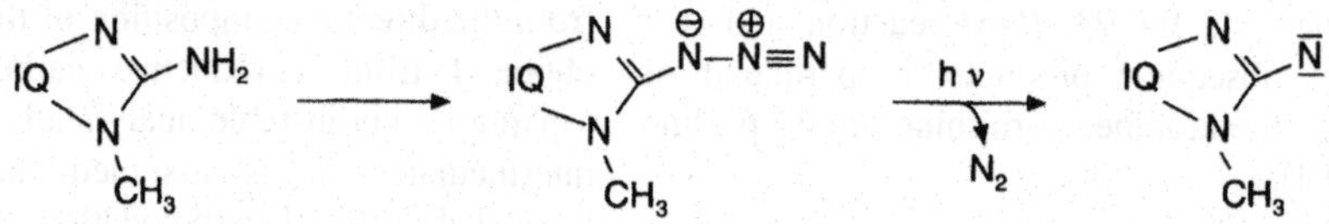

$$\tag{1.56}$$

converted to acrylamide after the splitting-off of ammonia.

1.2.4.4.2 Mutagenic Heterocyclic Compounds

In the late 1970s it was shown that charred surface portions of barbecued fish and meat as well as the smoke condensates captured in barbecuing have a highly mutagenic effect in microbial tests (*Salmonella typhimurium* tester strain TA 98). In model tests it could be demonstrated that pyrolyzates of amino acids and proteins are responsible for that effect. Table 1.6 lists the mutagenic compounds isolated from amino acid pyrolyzates. They are pyridoindoles, pyridoimidazoles and tetra-azafluoroanthenes.

At the same time, it was found that mutagenic compounds of amino acids and proteins can also be formed at lower temperatures. The compounds listed in Table 1.7 were obtained from

were found in meat extract: IQ (0−15), MeIQ (0−6), MeIQx (0−80). A model experiment directed at processes in meat shows that heterocyclic amines are detectable at temperatures around 175 °C after only 5 minutes. It is assumed that they are formed from creatinine, subsequent products of the *Maillard* reaction (pyridines, pyrazines, cf. 4.2.4.4.3) and amino acids as shown in Fig. 1.8.

The toxicity is based on the heteroaromatic amino function. The amines are genotoxic after oxidative metabolic conversion to a strong electrophile, e. g., a nitrene. Nitrenes of this type are synthesized for model experiments as shown in Formula 1.56. According to these experiments, MeIQ, IQ and MeIQx have an especially high genotoxic potential. The compounds listed in Table 1.6 can be deaminated by nitrite in weakly acid solution and thus inactivated.

The β-carbolines norharmane (I, R=H) and harmane (I, R=CH₃) are well known as components

Table 1.6. Mutagenic compounds from pyrolysates of amino acids and proteins

Mutagenic compound	Short form	Pyrolized compound	Structure
3-Amino-1,4-dimethyl-5H-pyrido[4,3-b]indole	Trp-P-1	Tryptophan	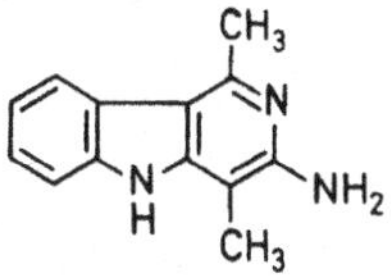
3-Amino-1-methyl-5H-pyrido[4,3-b]indole	Trp-P-2	Tryptophan	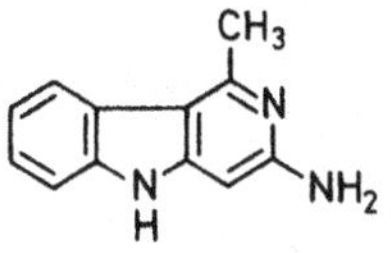
2-Amino-6-methyldipyrido [1,2-a:3',2'-d]imidazole	Glu-P-1	Glutamic acid	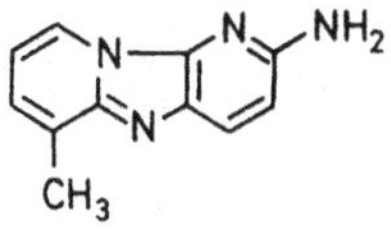
2-Aminodipyrido[1,2-a:3',2'-d] imidazole	Glu-P-2	Glutamic acid	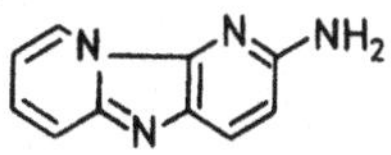
3,4-Cyclopentenopyrido[3,2-a] carbazole	Lys-P-1	Lysine	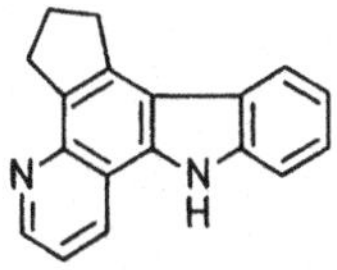
4-Amino-6-methyl-1H-2,5,10, 10b-tetraazafluoranthene	Orn-P-1	Ornithine	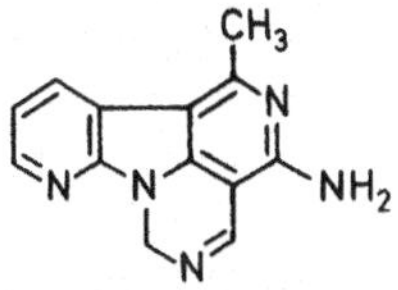
2-Amino-5-phenylpyridine	Phe-P-1	Phenylalanine	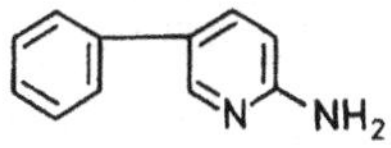
2-Amino-9H-pyrido[2,3-b] indole	AαC	Soya globulin	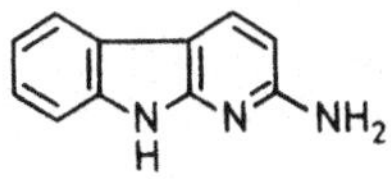
2-Amino-3-methyl-9H-pyrido[2,3-b] indole	MeAαC	Soya globulin	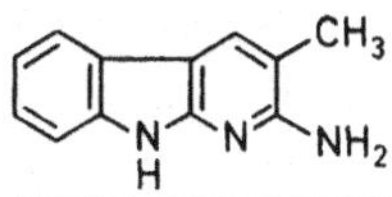

of tobacco smoke. They are formed by a reaction of tryptophan and formaldehyde or acetaldehyde:

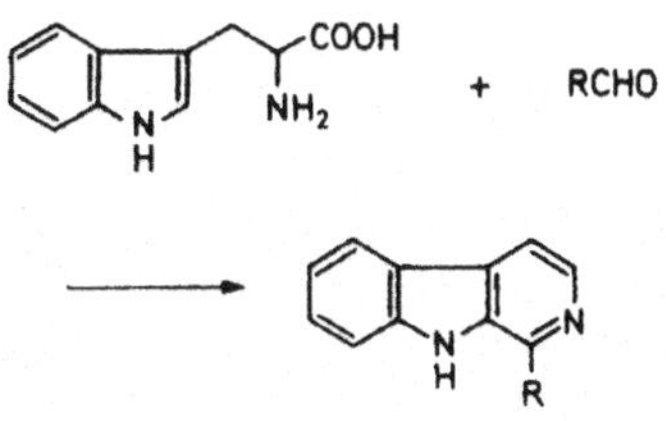

$$(1.57)$$

Table 1.7. Mutagenic compounds from various heated foods and from model systems

Mutagenic compound	Short form	Food Model system[a]	Structure
2-Amino-3-methylimidazo-[4,5-*f*]quinoline	IQ	1,2,3	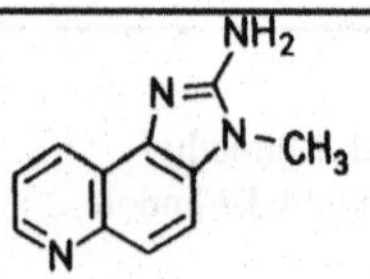
2-Amino-3,4-dimethylimidazo-[4,5-*f*]quinoline	MelQ	3	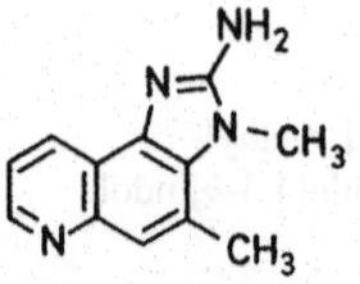
2-Amino-3-methylimidazo-[4,5-*f*]quinoxaline	IQx	2	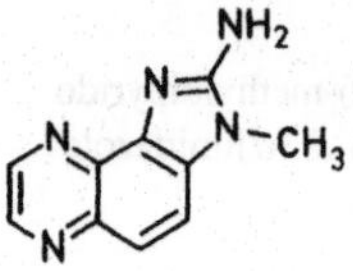
2-Amino-3,8-dimethylimidazo-[4,5-*f*]quinoxaline	MelQ2x	2,3	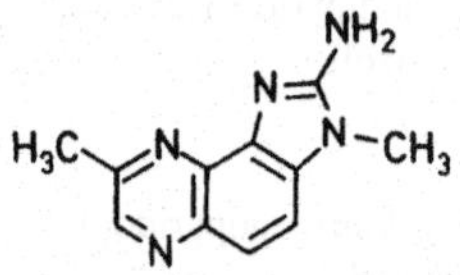
2-Amino-3,4,8-trimethyl-imidazo-[4,5-*f*]quinoxaline	4,8-Di MelQx	2,3,5,6	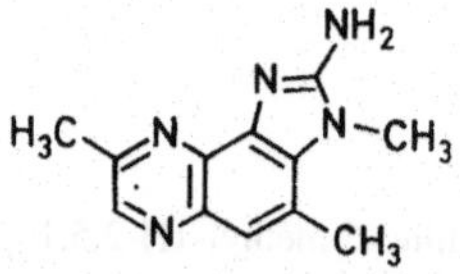
2-Amino-3,7,8-trimethyl-imidazo-[4,5-*f*]quinoxaline	7,8-Di MelQx	4	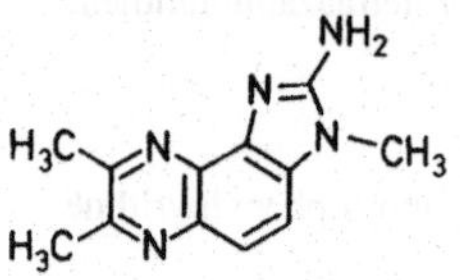
2-Amino-1-methyl-6-phenyl-imidazo[4,5-*b*]pyridine	PhIP	2	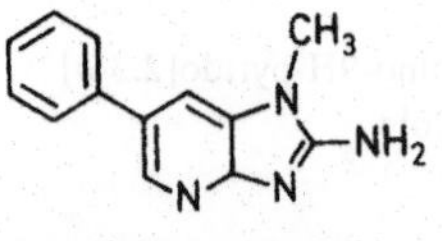

[a] 1: Meat extract; 2: Grilled meat; 3: Grilled fish; 4: Model mixture of creatinine, glycine, glucose; 5: as 4, but alanine; 6: as 4, but threonine

Tetrahydro-β-carboline-3-carboxylic acid (II) and (1S, 3S)-(III) and (1R, 3S)-methyltetrahydro-β-carboline-3-carboxylic acid (IV) were detected in beer (II: 2–11 mg/L, III + IV: 0.3–4 mg/L) and wine (II: 0.8–1.7 mg/L, III + IV: 1.3–9.1 mg/L). The ratio of diastereomers III and IV (Formula 1.58) was always near 2:1:

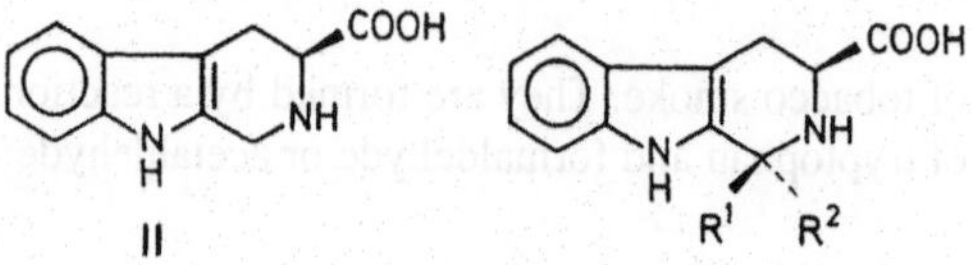

III $R^1 = CH_3$, $R^2 = H$
IV $R^1 = H$, $R^2 = CH_3$

(1.58)

The compounds are pharmacologically active.

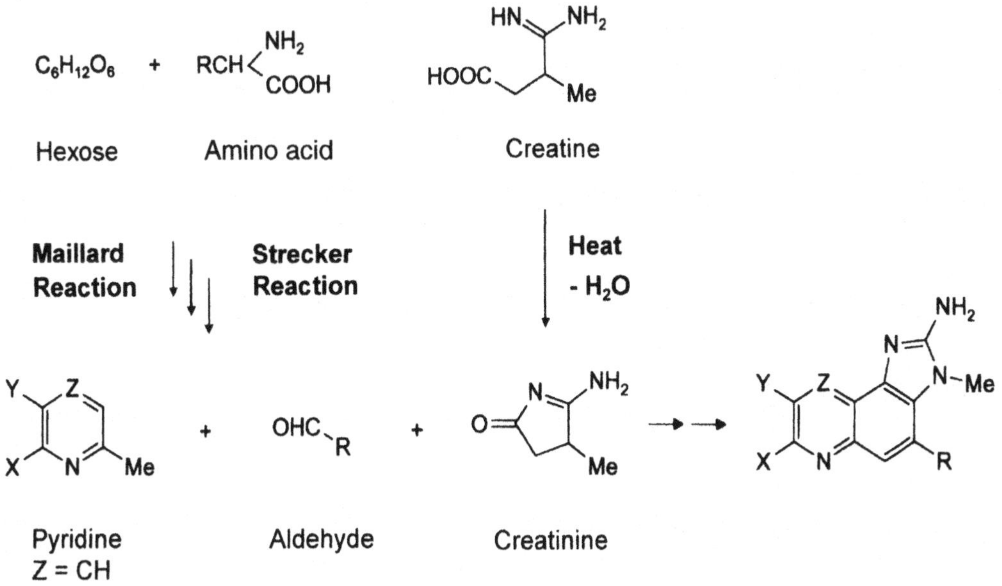

$C_6H_{12}O_6$ + RCH⟨$^{NH_2}_{COOH}$

Hexose   Amino acid   Creatine

Maillard Reaction   **Strecker Reaction**   **Heat - H₂O**

Pyridine
Z = CH

Pyrazine
Z = N

Pyridine
Z = CH   Aldehyde   Creatinine

	X	Y	Z	R
IQ	H	H	CH	H
MeIQ	H	H	CH	Me
MeIQx	H	Me	N	H
7,8-DiMeIQx	Me	Me	N	H
4,8-DiMeIQx	H	Me	N	Me

Fig. 1.8. Formation of heterocyclic amines by heating a model system of creatine, glucose and an amino acid mixture corresponding to the concentrations in beef (according to *Arvidsson* et al., 1997). For abbreviations, see Table 1.7

1.2.5 Synthetic Amino Acids Utilized for Increasing the Biological Value of Food (Food Fortification)

The daily requirements of humans for essential amino acids and their occurrence in some important food proteins are presented in Table 1.8. The biological value of a protein (g protein formed in the body/100 g food protein) is determined by the absolute content of essential amino acids, by the relative proportions of essential amino acids, by their ratios to nonessential amino acids and by factors such as digestibility and availability. The most important (more or less expensive) *in vivo* and *in vitro* methods for determining the biological valence are based on the following principles:

• Replacement of endogenous protein after protein depletion.
  The test determines the amount of endogenous protein that can be replaced by 100 g of food protein. The test person is given a nonprotein diet and thus reduced to the absolute N minimum. Subsequently, the protein to be examined is administered, and the N balance is measured. The biological valence (BV) follows from

$$BV = \frac{\text{Urea-N(non-protein diet)} + \text{N balance}}{\text{N intake}}$$

$$\times 100, \qquad (1.59)$$

"Net protein utilization" (NPU) is based on the same principle and is determined in animal experiments. A group of rats

Table 1.8. Adult requirement for essential amino acids and their occurrence in various food

Amino acid	1	2	3	4	5	6	7	8	9
Isoleucine	10–11	3.5	4.0	4.6	3.9	3.6	3.4	5.0	3.5
Leucine	11–14	4.2	5.3	7.1	4.3	5.1	6.5	8.2	5.4
Lysine	9–12	3.5	3.7	4.9	3.6	4.4	2.0	3.6	5.4
Methionine + Cystine	11–14	4.2	3.2	2.6	1.9	2.1	3.8	3.4	1.9
Methionine		2.0	1.9	1.9	1.2	0.9	1.4	2.2	0.8
Phenylalanine + Tyrosine	13–14	4.5	6.1	7.2	5.8	5.5	6.7	8.9	6.0
Phenylalanine		2.4	3.5	3.5	3.1	3.3	4.6	4.7	2.5
Threonine	6–7	2.2	2.9	3.3	2.9	2.7	2.5	3.7	3.8
Tryptophan	3	1.0	1.0	1.0	1.0	1.0	1.0	1.0	1.0
Valine	11–14	4.2	4.3	5.6	3.6	3.3	3.8	6.4	4.1
Tryptophan[a]			1.7	1.4	1.4	1.5	1.1	1.0	1.3

1: Daily requirement in mg/kg body weight.
2–8: Relative value related to Trp = 1 (pattern).
2: Daily requirements, 3: eggs, 4: bovine milk, 5: potato, 6: soya, 7: wheat flour, 8: rice, and 9: *Torula*-yeast.
[a] Tryptophan (%) in raw protein.

is fed a non-protein diet (Gr 1), while the second group is fed the protein to be examined (Gr 2). After some time, the animals are killed, and their protein content is analyzed. The biological valence follows from

$$NPU = \frac{\text{Protein content Gr 2} - \text{protein content Gr 1}}{\text{Protein intake}} \times 100$$

- Utilization of protein for growth.
  The growth value (protein efficiency ratio = PER) of laboratory animals is calculated according to the following formula:

$$PER = \frac{\text{Weight gain (g)}}{\text{Available protein (g)}}$$

- Maintenance of the N balance.
- Plasma concentration of amino acids.
- Calculation from the amino acid composition.
- Determination by enzymatic cleavage *in vitro*.

Table 1.9 lists data about the biological valence of some food proteins, determined according to different methods.

The highest biological value observed is for a blend of 35% egg and 65% potato proteins. The biological value of a protein is generally limited by:

- Lysine: deficient in proteins of cereals and other plants

- Methionine: deficient in proteins of bovine milk and meat

Table 1.9. Biological valence of some food proteins determined according to different methods[a]

Protein from	Biological valence			Limiting amino acid
	BV	NPU	PER	
Chicken egg	94	93	3.9	
Cow's milk	84	81	3.1	Met
Fish	76	80	3.5	Thr
Beef	74	67	2.3	Met
Potatoes	73	60	2.6	Met
Soybeans	73	61	2.3	Met
Rice	64	57	2.2	Lys, Tyr
Beans	58	38	1.5	Met
Wheat flour (white)	52	57	0.6	Lys, Thr

[a] The methods are explained in the text.

Table 1.10. Increasing the biological valence (PER[a]) of some food proteins through the addition of amino acids

Protein from	Addition(%)					
	with out	0.2 Lys	0.4 Lys	0.4 Lys 0.2 Thr	0.4 Lys 0.07 Thr	0.4 Lys 0.07 Thr 0.2 Thr
Casein (Reference)	2.50					
Wheat flour	0.65	1.56	1.63	2.67		
Corn	0.85		1.08		2.50	2.59

[a] The method is explained in the text.

- Threonine: deficient in wheat and rye
- Tryptophan: deficient in casein, corn and rice.

Since food is not available in sufficient quantity or quality in many parts of the world, increasing its biological value by addition of essential amino acids is gaining in importance. Illuminating examples are rice fortification with L-lysine and L-threonine, supplementation of bread with L-lysine and fortification of soya and peanut protein with methionine. Table 1.10 lists data about the increase in biological valence of some food proteins through the addition of amino acids. Synthetic amino acids are used also for chemically defined diets which can be completely absorbed and utilized for nutritional purposes in space travel, in pre-and post-operative states, and during therapy for maldigestion and malabsorption syndromes.

The fortification of animal feed with amino acids (0.05–0.2%) is of great significance. These demands have resulted in increased production of amino acids. Table 1.11 gives data for world production in 1982. The production of L-glutamic acid, used to a great extent as a flavor enhancer, is exceptional. Production of methionine and lysine is also significant.

Four main processes are distinguished in the production of amino acids: chemical synthesis, isolation from protein hydrolysates, enzymatic and microbiological methods of production, which is currently the most important. The following sections will further elucidate the important

Table 1.11. World production of amino acids, 1982

Amino acid	t/year	Process[a]				Mostly used as
		1	2	3	4	
L-Ala	130		+		+	Flavoring compound
D,L-Ala	700	+				Flavoring compound
L-Arg	500			+	+	Infusion Therapeutics
L-Asp	250		+		+	Therapeutics Flavoring compound
L-Asn	50				+	Therpeutics
L-CySH	700				+	Baking additive Antioxidant
L-Glu	270,000			+		Flavoring compound flavor enhancer
L-Gln	500			+		Therapeutics
Gly	6.000	+				Sweetener
L-His	200			+	+	Therapeutics
L-Ile	150			+	+	Infusion
L-Leu	150			+	+	Infusion
L-Lys	32,000		+	+		Feed ingredient
L-Met	150		+			Therapeutics
D,L-Met	110,000	+				Feed ingredient
L-Phe	150		+	+		Infusion
L-Pro	100			+	+	Infusion
L-Ser	50			+	+	Cosmetics
L-Thr	160			+	+	Food additive
L-Trp	200		+	+		Infusion
L-Tyr	100			+		Infusion
L-Val	150		+		+	Infusion

[a] 1: Chemical synthesis, 2: protein hydrolysis, 3: microbiological procedure, 4: isolation from raw materials.

industrial processes for a number of amino acids.

1.2.5.1 Glutamic Acid

Acrylnitrile is catalytically formylated with CO/H_2 and the resultant aldehyde is transformed through a *Strecker* reaction into glutamic acid dinitrile which yields D,L-glutamic acid after alkaline hydrolysis. Separation of the racemate is achieved by preferential crystallization of the L-form from an oversaturated solution after seeding with L-glutamic acid:

$$H_2C{=}CH{-}CN \xrightarrow[\text{[Co(CO)}_4]_2]{CO/H_2} OHC{-}CH_2{-}CH_2{-}CN$$

$$\xrightarrow{HCN/NH_3} \underset{\underset{NH_2}{|}}{NC{-}CH{-}CH_2{-}CH_2{-}CN}$$

$$\xrightarrow{OH^{\ominus}} \text{D,L-Glu}$$

$$(1.60)$$

A fermentation procedure with various selected strains of microorganisms (*Brevibacterium flavum, Brev. roseum, Brev. saccharolyticum*) provides L-glutamic acid in yields of 50 g/l of fermentation liquid:

$$CH_3COONH_4 \text{ (20 g/l)} \xrightarrow[\text{pH 7.5}]{MO} \text{L-Glu (50 g/l)}$$

$$(1.61)$$

1.2.5.2 Aspartic Acid

Aspartic acid is obtained in 90% yield from fumaric acid by using the aspartase enzyme:

$$\text{Fumaric acid} \xrightarrow[NH_3]{Aspartase} \text{L-Asp}$$

$$(1.62)$$

1.2.5.3 Lysine

A synthetic procedure starts with caprolactam, which possesses all the required structural features, except for the α-amino group which is in-

troduced in several steps:

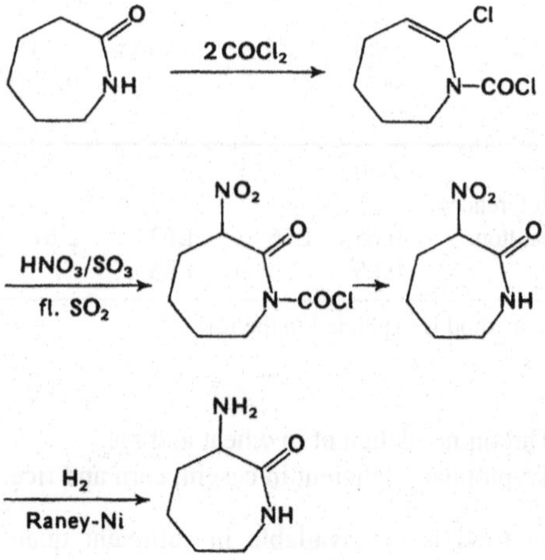

$$(1.63)$$

Separation of isomers is done at the α-amino caprolactam (Acl) step through the sparingly soluble salt of the L-component with L-pyrrolidone carboxylic acid (Pyg):

$$\text{D,L-Acl + L-Pyg} \longrightarrow \text{D-Acl + L,L-salt}$$

Base

$$\longrightarrow \text{L-Acl} \xrightarrow{HCl} \text{L-Lys · HCl} \qquad (1.64)$$

More elegant is selective hydrolysis of the L-enantiomer by an L-α-amino-ε-caprolactamase which occurs in several yeasts, for example in *Cryptococcus laurentii*. The racemization of the remaining D-isomers is possible with a racemase of *Achromobacter obae*. The process can be performed as a one-step reaction: the racemic aminocaprolactam is incubated with intact cells of *C. laurentii* and *A. obae*, producing almost 100% L-lysine.

In another procedure, acrylnitrile and ethanal react to yield cyanobutyraldehyde which is then transformed by a *Bucherer* reaction into cyanopropylhydantoin. Catalytic hydrogenation of the nitrile group, followed by alkaline hydrolysis yields D,L-lysine.

The isomers can be separated through the sparingly soluble L-lysine sulfanilic acid salt:

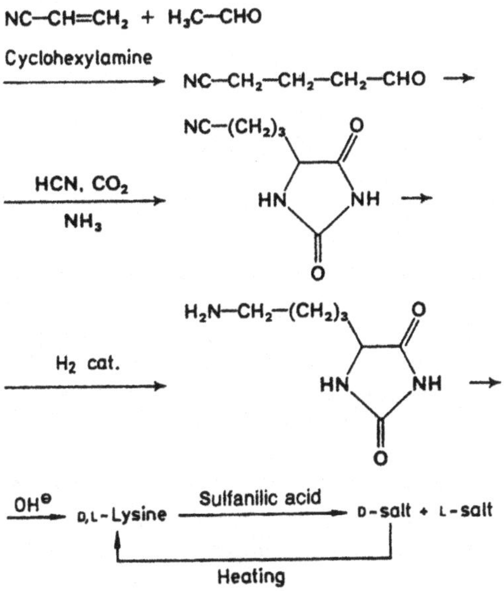

$$NC-CH=CH_2 + H_2C-CHO$$

$$\xrightarrow{\text{Cyclohexylamine}} NC-CH_2-CH_2-CH_2-CHO \rightarrow$$

$$\xrightarrow[NH_3]{HCN, CO_2}$$

$$\xrightarrow{H_2 \text{ cat.}}$$

$$\xrightarrow{OH^{\ominus}} \text{D,L-Lysine} \xrightarrow{\text{Sulfanilic acid}} \text{D-salt + L-salt}$$

Heating

(1.65)

Fermentation with a pure culture of *Brevibacterium lactofermentum* or *Micrococcus glutamicus* produces L-lysine directly:

$$CH_3COONH_4 \xrightarrow[pH\ 7-8.5]{MO} \text{L-Lys} \quad (40-90 \text{ g/l})$$
$$(<15 \text{ g/l})$$

(1.66)

1.2.5.4 Methionine

Interaction of methanethiol with acrolein produces an aldehyde which is then converted to the corresponding hydantoin through a *Bucherer* reaction. The product is hydrolyzed by alkaline catalysis. Separation of the resultant racemate is usually not carried out since the D-form of methionine is utilized by humans via transamination:

$$CH_3SH + H_2C=CH-CHO \rightarrow H_3C-S-CH_2CH_2-CHO$$

$$\xrightarrow[2)\ OH^{\ominus}]{1)\ HCN,\ CO_2,\ NH_3} \text{D,L-Met}$$

(1.67)

1.2.5.5 Phenylalanine

Benzaldehyde is condensed with hydantoin, then hydrogenation using a chiral catalyst gives a product which is about 90% L-phenylalanine:

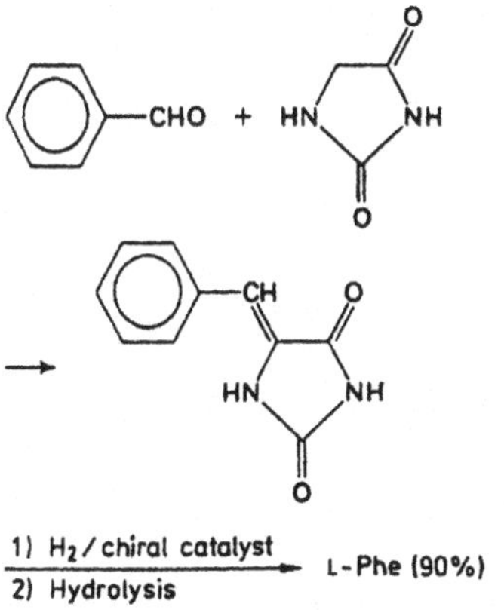

$$\xrightarrow[2)\ \text{Hydrolysis}]{1)\ H_2/\text{chiral catalyst}} \text{L-Phe (90\%)}$$

(1.68)

1.2.5.6 Threonine

Interaction of a copper complex of glycine with ethanal yields the *threo* and *erythro* isomers in the ratio of 2:1. They are separated on the basis of their differences in solubility:

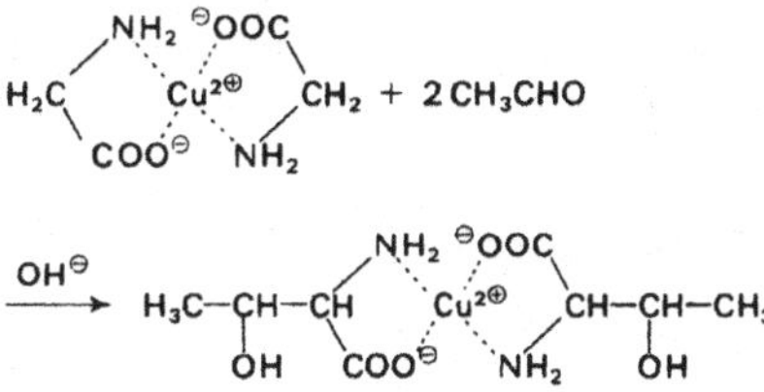

D,L-threonine is separated into its isomers through its N-acetylated form with the help of an acylase enzyme.
Threonine is also accessible via microbiological methods.

1.2.5.7 Tryptophan

Tryptophan is obtained industrially by a variation of the *Fischer* indole synthesis. Addition of hydrogen cyanide to acrolein gives 3-cyanopropanal which is converted to hydantoin through a *Bucherer* reaction. The nitrile group is then reduced to an aldehyde group. Reaction with phenylhydrazine produces an indole derivative. Lastly, hydantoin is saponified with alkali:

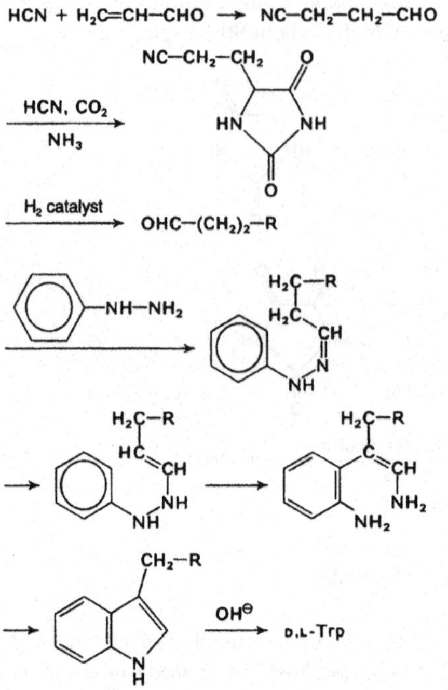

$$HCN + H_2C{=}CH{-}CHO \longrightarrow NC{-}CH_2{-}CH_2{-}CHO$$

(1.70)

L-Tryptophan is also produced through enzymatic synthesis from indole and serine with the help of tryptophan synthase:

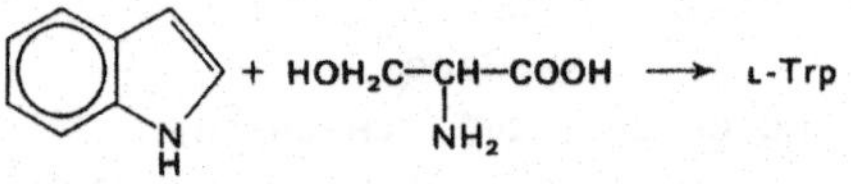

(1.71)

1.2.6 Sensory Properties

Free amino acids can contribute to the flavor of protein-rich foods in which hydrolytic processes occur (e. g. meat, fish or cheese).

Table 1.12 provides data on taste quality and taste intensity of amino acids. Taste quality is influenced by the molecular configuration: sweet amino acids are primarily found among members of the D-series, whereas bitter amino acids are generally within the L-series. Consequently amino acids with a cyclic side chain

(1-aminocycloalkane-1-carboxylic acids) are sweet and bitter.

The taste intensity of a compound is reflected in its recognition threshold value. The recognition threshold value is the lowest concentration needed to recognize the compound reliably, as assessed by a taste panel. Table 1.12 shows that the taste intensity of amino acids is dependent on the hydrophobicity of the side chain.

L-Tryptophan and L-tyrosine are the most bitter amino acids with a threshold value of $c_{t\,bitter} = 4{-}6\,mmol/l$. D-Tryptophan, with $c_{t\,sweet} = 0.2{-}0.4\,mmol/l$, is the sweetest amino acid. A comparison of these threshold values with those of caffeine ($c_{t\,bi} = 1{-}1.2\,mmole/l$) and sucrose ($c_{t\,sw} = 10{-}12\,mmol/l$) shows that caffeine is about 5 times as bitter as L-tryptophan and that D-tryptophan is about 37 times as sweet as sucrose.

L-Glutamic acid has an exceptional position. In higher concentrations it has a meat broth flavor, while in lower concentrations it enhances the characteristic flavor of a given food (flavor enhancer, cf. 8.6.1). L-Methionine has a sulfur-like flavor.

The bitter taste of the L-amino acids can interfere with the utilization of these acids, e. g., in chemically defined diets.

1.3 Peptides

1.3.1 General Remarks, Nomenclature

Peptides are formed by binding amino acids together through an amide linkage.

On the other hand peptide hydrolysis results in free amino acids:

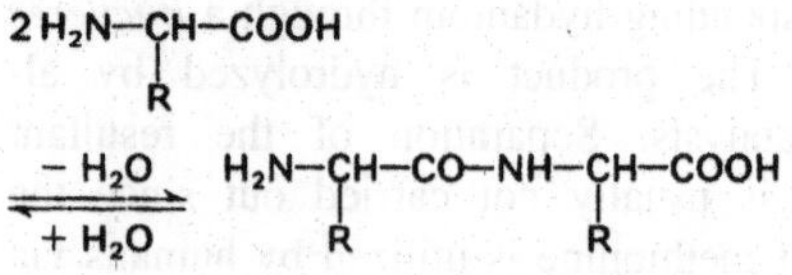

(1.72)

Functional groups not involved in the peptide synthesis reaction should be blocked. The protecting or blocking groups must be removed after synthesis under conditions which retain the stability of the newly formed peptide bonds:

Table 1.12. Taste of amino acids in aqueous solution at pH 6–7 sw – sweet, bi – bitter, neu – neutral

Amino acid	Taste			
	L-Compound		D-Compound	
	Quality	Intensity[a]	Quality	Intensity[a]
Alanine	sw	12–18	sw	12–18
Arginine	bi		neu	
Asparagine	neu		sw	3–6
Aspartic acid	neu		neu	
Cystine	neu		neu	
Glutamine	neu		sw	
Glutamic acid	meat broth like (3.0)		neu	
Glycine[b]	sw	25–35		
Histidine	bi	45–50	sw	2–4
Isoleucine	bi	10–12	sw	8–12
Leucine	bi	11–13	sw	2–5
Lysine	sw		sw	
	bi	80–90		
Methionine	sulphurous		sulphurous	
			sw	4–7
Phenylalanine	bi	5–7	sw	1–3
Proline	sw	25–40	neu	
	bi	25–27		
Serine	sw	25–35	sw	30–40
Threonine	sw	35–45	sw	40–50
Tryptophan	bi	4–6	sw	0.2–0.4
Tyrosine	bi	4–6	sw	1–3
1-Aminocycloalkane-1-carboxylic acid[b]				
Cyclobutane derivative	sw	20–30		
Cyclopentane derivative	sw	3–6		
	bi	95–100		
Cyclohexane derivative	sw	1–3		
	bi	45–50		
Cyclooctane derivative	sw	2–4		
	bi	2–5		
Caffeine	bi	1–1.2		
Saccharose	sw	10–12		

[a] Recognition threshold value (mmol/l).
[b] Compounds not optically active.

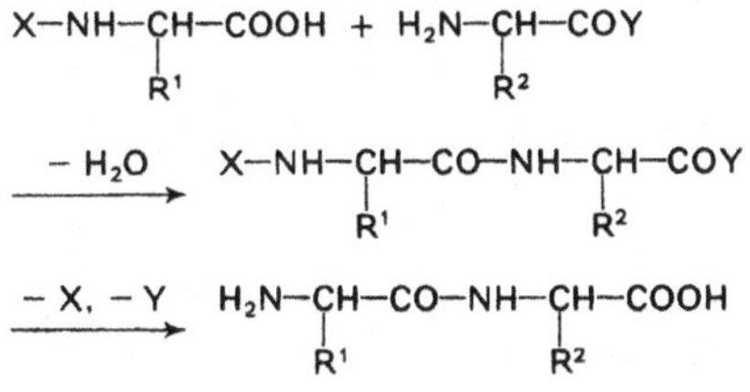

$$X-NH-\underset{R^1}{CH}-COOH + H_2N-\underset{R^2}{CH}-COY$$

$$\xrightarrow{-H_2O} X-NH-\underset{R^1}{CH}-CO-NH-\underset{R^2}{CH}-COY$$

$$\xrightarrow{-X, -Y} H_2N-\underset{R^1}{CH}-CO-NH-\underset{R^2}{CH}-COOH \quad (1.73)$$

Peptides are denoted by the number of amino acid residues as di-, tri-, tetrapeptides, etc., and the term "oligopeptides" is used for those with 10 or less amino acid residues. Higher molecular weight peptides are called polypeptides. The transition of "polypeptide" to "protein" is rather undefined, but the limit is commonly assumed to be at a molecular weight of about 10 kdal, i.e., about 100 amino acid residues are needed in the chain for it to be called a protein. Peptides are interpreted as acylated amino acids:

H₂N—CH—CO—NH—CH—CO—NH—CH₂—COOH
$\quad\quad$|$\quad\quad\quad\quad\quad$|
$\quad\quad$CH₃$\quad\quad\quad\quad$CH₂OH

$\quad$Alanyl$\quad$−$\quad$seryl$\quad$−$\quad$glycine

$$(1.74)$$

The first three letters of the amino acids are used as symbols to simplify designation of peptides (cf. Table 1.1). Thus, the peptide shown above can also be given as:

Ala—Ser—Gly   or   ASG $\qquad$ (1.75)

One-letter symbols (cf. Table 1.1) are used for amino acid sequences of long peptide chains.

D-Amino acids are denoted by the prefix D-. In compounds in which a functional group of the side chain is involved, the bond is indicated by a perpendicular line. The tripeptide glutathione (γ-glutamyl-cysteinyl-glycine) is given as an illustration along with its corresponding disulfide, oxidized glutathione:

Glu $\qquad\qquad$ Glu
$\;$└—Cys—Gly $\qquad$ └—Cys—Gly
$\qquad\qquad\qquad\qquad\qquad$|
$\qquad\qquad\qquad\quad$┌—Cys—Gly
$\qquad\qquad\qquad\;$Glu $\qquad\qquad$ (1.76)

By convention, the amino acid residue with the free amino group is always placed on the left. The amino acids of the chain ends are denoted as N-terminal and C-terminal amino acid residues. The peptide linkage direction in cyclic peptides is indicated by an arrow, i.e., —CO → NH-.

1.3.2 Physical Properties

1.3.2.1 Dissociation

The pK values and isoelectric points for some peptides are listed in Table 1.13. The acidity of the free carboxyl groups and the basicity of the free amino groups are lower in peptides than in the corresponding free amino acids. The amino acid sequence also has an influence (e.g., Gly-Asp/Asp-Gly).

Table 1.13. Dissociation constants and isoelectric points of various peptides (25 °C)

Peptide	pK₁	pK₂	pK₃	pK₄	pK₅	pI
Gly-Gly	3.12	8.17				5.65
Gly-Gly-Gly	3.26	7.91				5.59
Ala-Ala	3.30	8.14				5.72
Gly-Asp	2.81	4.45	8.60			3.63
Asp-Gly	2.10	4.53	9.07			3.31
Asp-Asp	2.70	3.40	4.70	8.26		3.04
Lys-Ala	3.22	7.62	10.70			9.16
Ala-Lys-Ala	3.15	7.65	10.30			8.98
Lys-Lys	3.01	7.53	10.05	11.01		10.53
Lys-Lys-Lys	3.08	7.34	9.80	10.54	11.32	10.93
Lys-Glu	2.93	4.47	7.75	10.50		6.10
His-His	2.25	5.60	6.80	7.80		7.30

1.3.3 Sensory Properties

While the taste quality of amino acids does depend on configuration, peptides, except for the sweet dipeptide esters of aspartic acid (see below), are neutral or bitter in taste with no relationship to configuration (Table 1.14). As

Table 1.14. Taste threshold values of various peptides: effect of configuration and amino acid sequence (tested in aqueous solution at pH 6–7); bi – bitter

Peptide[a]	Taste	
	Quality	Intensity[b]
Gly-Leu	bi	19–23
Gly-D-Leu	bi	20–23
Gly-Phe	bi	15–17
Gly-D-Phe	bi	15–17
Leu-Leu	bi	4–5
Leu-D-Leu	bi	5–6
D-Leu-D-Leu	bi	5–6
Ala-Leu	bi	18–22
Leu-Ala	bi	18–21
Gly-Leu	bi	19–23
Leu-Gly	bi	18–21
Ala-Val	bi	60–80
Val-Ala	bi	65–75
Phe-Gly	bi	16–18
Gly-Phe	bi	15–17
Phe-Gly-Phe-Gly	bi	1.0–1.5
Phe-Gly-Gly-Phe	bi	1.0–1.5

[a] L-Configuration if not otherwise designated.
[b] Recognition threshold value in mmol/l.

Table 1.15. Bitter taste of dipeptide A–B: dependence of recognition threshold value (mmol/l) on side chain hydrophobicity (0: sweet or neutral taste)

A/B	Asp	Glu	Asn	Gln	Ser	Thr	Gly	Ala	Lys	Pro	Val	Leu	Ile	Phe	Tyr	Trp
	0	0	0	0	0	0	0	0	85	26	21	12	11	6	5	5
Gly	0[a]	–	–	–	–	–	0	0	–	45	75	21	20	16	17	13
Ala	0	–	–	–	–	–	0	0	–	–	70	20	–	–	–	–
Pro	26	–	–	–	–	–	–	–	–	–	–	6	–	–	–	–
Val	21	–	–	–	–	–	65	70	–	–	20	10	–	–	–	–
Leu	12	–	–	–	–	–	20	20	–	–	–	4.5	–	–	3.5	0.4
Ile	11	43	43	33	33	33	33	21	21	23	4	9	5.5	5.5	–	0.9
Phe	6	–	–	–	–	–	17	–	–	2	–	1.4	–	0.8	0.8	–
Tyr	5	–	–	–	–	–	–	–	–	–	–	4	–	–	–	–
Trp	5	–	28	–	–	–	–	–	–	–	–	–	–	–	–	–

[a] Threshold of the amino acid (cf. Table 1.12).

with amino acids, the taste intensity is influenced by the hydrophobicity of the side chains (Table 1.15). The taste intensity does not appear to be dependent on amino acid sequence (Table 1.14). Bitter tasting peptides can occur in food after proteolytic reactions. For example, the bitter taste of cheese is a consequence of faulty ripening. Therefore, the wide use of proteolytic enzymes to achieve well-defined modifications of food proteins, without producing a bitter taste, causes some problems. Removal of the bitter taste of a partially hydrolyzed protein is outlined in the section dealing with proteins modified with enzymes (cf. 1.4.6.3.2).

The sweet taste of aspartic acid dipeptide esters (I) was discovered by chance in 1969 for α-L-aspartyl-L-phenylalanine methyl ester ("Aspartame", "NutraSweet"). The corresponding peptide ester of L-aminomalonic acid (II) is also sweet.

A comparison of structures I, II and III reveals a relationship between sweet dipeptides and

$$
\begin{array}{ccc}
\text{COO}^{\ominus} & \text{COO}^{\ominus} & \text{COO}^{\ominus} \\
\text{CH}_2 & \text{H–C–NH}_3^{\oplus} & \text{H–C–NH}_3^{\oplus} \\
\text{H–C–NH}_3^{\oplus} & \text{CO} & \text{R} \\
\text{CO} & \text{NH} & \\
\text{NH} & \text{R}^1\text{–C–R}^3 & \text{III} \\
\text{R}^1\text{–C–R}^3 & \text{R}^2 & \\
\text{R}^2 & & \\
\text{I} & \text{II} &
\end{array}
$$

(1.77)

chain substituent, R, is found only in peptide types I and II.

Since the discovery of the sweetness of compounds of type I, there has been a systematic study of the structural prerequisites for a sweet taste.

The presence of L-aspartic acid was shown to be essential, as was the peptide linkage through the α-carboxyl group.

R^1 may be an H or CH_3 group[2], while the R^2 and R^3 groups are variable within a certain range. Several examples are presented in Table 1.16. The sweet taste intensity passes through a maximum with increasing length and volume of the R^2 residue (e. g., COO-fenchyl ester is $22-23 \times 10^3$ times sweeter than sucrose). The size of the R^3 substituent is limited to a narrow range. Obviously, the R^2 substituent has the greatest influence on taste intensity.

The following examples show that R^2 should be relatively large and R^3 relatively small: L-Asp-L-Phe-OMe (aspartame, $R^2—CH_2C_6H_5$, $R_3 =$ COOMe) is almost as sweet ($f_{sac,g}(1) = 180$) as L-Asp-D-Ala-OPr ($f_{sac,g}(0.6) = 170$), while L-Asp-D-Phe-OMe has a bitter taste.

In the case of acylation of the free amino group of aspartic acid, the taste characteristics depend on the introduced group. Thus, D-Ala-L-Asp-L-Phe-OMe is sweet ($f_{sac,g}(0.6) = 170$), while L-Ala-L-Asp-L-Phe-OMe is not. It should be noted that superaspartame is extremely sweet (cf. 8.8.15.2).

sweet D-amino acids. The required configuration of the carboxyl and amino groups and the side

---

[2] Data are not yet available for compounds with $R^1 >$ CH_3.

Table 1.16. Taste of dipeptide esters of aspartic acid[a] and of amino malonic acid[b]

R^2	R^3	Taste[c]
Asparagin acid derivate		
$COOCH_3$	H	8
$n-C_3H_7$	$COOCH_3$	4
$n-C_4H_9$	$COOCH_3$	45
$n-C_4H_9$	$COOC_2H_5$	5
$n-C_6H_{13}$	CH_3	10
$n-C_7H_{15}$	CH_3	neutral
$COOCH(CH_3)_2$	$n-C_3H_7$	17
$COOCH(CH_3)_2$	$n-C_4H_9$	neutral
$COOCH_3$	$CH_2C_6H_5$	bitter
$CH(CH_3)C_2H_5$	$COOCH_3$	bitter
$CH_2CH(CH_3)_2$	$COOCH_3$	bitter
$CH_2C_6H_5$	$COOCH_3$	140
COO-2-methyl-cyclohexyl	$COOCH_3$	5–7000
COO-fenchyl	$COOCH_3$	22–33,000
D,L-Aminomalon acid derivate		
$COOiC_3H_7$	CH_3	58
CH_3	$COOiC_3H_7$	neutral

[a] Formula 1.77 I, $R^1 = H$.
[b] Formula 1.77 II, $R^1 = H$.
[c] For sweet compounds the factor $f_{sac, g}$ is given, related to the threshold value of a 10% saccharose solution (cf. 8.8.1.1).

The intensity of the salty taste of Orn-β-Ala depends on the pH (Table 1.18). Some peptides exhibit a salty taste, e. g. ornithyl-β-alanine hydrochloride (Table 1.17) and may be used as substitutes for sodium chloride.

Table 1.17. Peptides with a salty taste

Peptide[a]	Taste	
	Threshold (mmol/l)	Quality[b]
Orn-βAla.HCl	1.25	3
Orn-γAbu.HCl	1.40	3
Orn-Tau.HCl	3.68	4
Lys-Tau.HCl	5.18	4
NaCl	3.12	3

[a] Abbreviations: Orn, ornithine; β-Ala, β-alanine, γ-Abu, γ-aminobutyric acid; Tau, taurine.
[b] The quality of the salty taste was evaluated by rating it from 0 to 5 on a scale in comparison with a 6.4 mmol/L NaCl solution (rated 3); 4 is slightly better, 5 clearly better than the control solution.

Table 1.18. Effect of HCl on the salty taste of Orn-β-Ala[a]

Equivalents HCl	pH	Taste	
		salty[b]	sour[c]
0	8.9	0	
0.79	7.0	0	
0.97	6.0	1	
1.00	5.5	2	
1.10	4.7	3	+/−
1.20	4.3	3.5	+
1.30	4.2	3	++

[a] Peptide solution: 30 mmol/L.
[b] The values 1, 3 and 5 correspond in intensity to 0.5%, 0.25% and 0.1% NaCl solutions respectively.
[c] Very weak (+) and slightly sour (++).

1.3.4 Individual Peptides

Peptides are widespread in nature. They are often involved in specific biological activities (peptide hormones, peptide toxins, peptide antibiotics). A number of peptides of interest to food chemists are outlined in the following sections.

1.3.4.1 Glutathione

Glutathione (γ-L-glutamyl-L-cysteinyl-glyci-ne) is widespread in animals, plants and microorganisms. Beef (200), broccoli (140), spinach (120), parsley (120), chicken (95), cauliflower (74), potatoes (71), paprika (49), tomatoes (49) and oranges (40) are especially rich in glutathione (mg/kg). A noteworthy feature is the binding of glutamic acid through its γ-carboxyl group. The peptide is the coenzyme of glyoxalase.

$$(1.78)$$

It is involved in active transport of amino acids and, due to its ready oxidation, is also involved in many redox-type reactions. It influences the rheological properties of wheat flour dough through thiol-disulfide interchange with wheat gluten. High concentrations of reduced glutathione in flour bring about reduction of protein disulfide bonds and a corresponding decrease in molecular weight of some of the protein constituents of dough gluten (cf. 15.4.1.4.1).

1.3.4.2 Carnosine, Anserine and Balenine

These peptides are noteworthy since they contain a β-amino acid, β-alanine, bound to L-histidine or 1-methyl- or 3-methyl-L-histidine, and are present in meat extract and in muscle of vertebrates (cf. Formula 1.79).

Data on the amounts of these peptides present in meat are given in Table 1.19. Carnosine is predominant in beef muscle tissue, while anserine is predominant in chicken meat. Balenine is a characteristic constituent of whale muscle, although it appears that sperm whales do not have

probably due to the presence of meat from other whale species. These peptides are used analytically to identify the meat extract. Their physiological roles are not clear. Their buffering capacity in the pH range of 6–8 may be of some importance. They may also be involved in revitalizing exhausted muscle, i. e. in the muscle regaining its excitability and ability to contract. Carnosine may act as a neurotransmitter for nerves involved in odor perception.

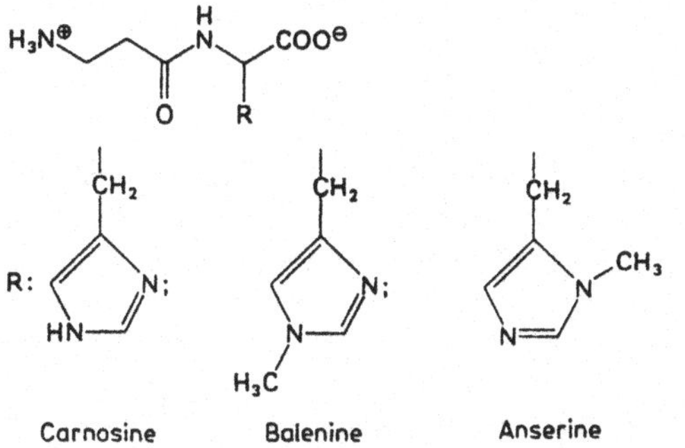

1.3.4.3 Nisin

This peptide is formed by several strains of *Streptococcus lactis* (Langfield-N-group). It contains a number of unusual amino acids, namely dehydroalanine, dehydro-β-methyl-alanine, lanthionine, β-methyl-lanthionine, and therefore also five thioether bridges (cf. Formula 1.80).

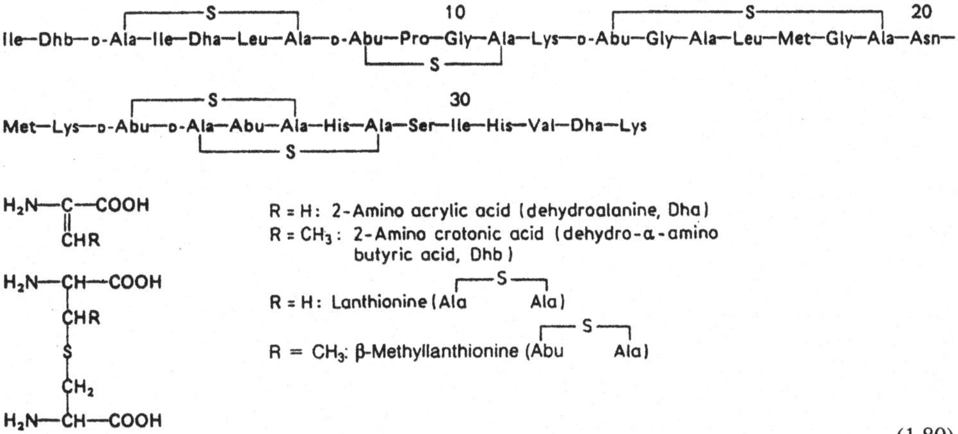

(1.80)

this dipeptide. The amounts found in commercial sperm whale meat extract are

Carnosine   Balenine   Anserine

(1.79)

The peptide subtilin is related to nisin. Nisin is active against Gram-positive microorganisms (lactic acid bacteria, *Streptococci*, *Bacilli*, *Clostridia* and other anaerobic spore-forming microorganisms). Nisin begins to act against the cytoplasmic membrane as soon as the spore has germinated. Hence, its action is more pronounced against spores than against vegetative cells. Nisin is permitted as a preservative in several countries. It is used to suppress anaerobes in cheese and cheese products, especially in hard cheese and processed cheese to inhibit butyric acid fermentation. The use of nisin in the canning of vegetables allows mild sterilization conditions.

Table 1.19. Occurrence of carnosine, anserine and balenine (%) in meat[a]

Meat	Carnosine	Anserine	Balenine	Σ[b]
Beef muscle				0.2–0.4
tissue	0.15–0.35	0.01–0.05		
Beef meat				4.4–6.2
extract	3.1–5.7	0.4–1.0		
Chicken meat[c]	0.01–0.1	0.05–0.25		
Chicken meat				
extract	0.7–1.2	2.5–3.5		
Whale meat				ca. 0.3
Whale meat				
extract a[d]	3.1–5.9	0.2–0.6	13.5–23.0	16–30
Whale meat				
extract b[e]	2.5–4.5	1.2–3.0	0–5.2	3.5–12

[a] The results are expressed as % of the moist tissue weight, or of commercially available extracts containing 20% moisture.
[b] β-Alanine peptide sum.
[c] Lean and deboned chiken meat.
[d] Commercial extract mixture of various whales.
[e] Commercial extract mixture, with sperm whale prevailing.

1.3.4.4 Lysine Peptides

A number of peptides, such as:

Gly—Lys,   Ala—Lys,   Glu—Lys,   Lys,
                                                    Gly⌐

    Lys,   Lys,   Gly—Lys
Glu⌐ | Gly⌐
            Glu

(1.81)

have been shown to be as good as lysine in rat growth feeding tests. These peptides substantially retard the browning reaction with glucose (Fig. 1.9), hence they are suitable for lysine fortification of sugar-containing foods which must be heat treated.

1.3.4.5 Other Peptides

Other peptides occur commonly and in variable levels in protein rich food as degradation products of proteolytic processes.

1.4 Proteins

Like peptides, proteins are formed from amino acids through amide linkages. Covalently bound

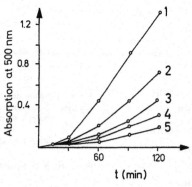

Fig. 1.9. Browning of some lysine derivatives (0.1 M lysine or lysine derivative, 0.1 M glucose in 0.1 M phosphate buffer pH 6.5 at 100 °C in sealed tubes, (according to *Finot* et al., 1978.) 1 Lys, 2 Ala-Lys, 3 Gly–Lys, 4 Glu–Lys, 5 Lys)
Gly⌐   Glu⌐

hetero constituents can also be incorporated into proteins. For example, phosphoproteins such as milk casein (cf. 10.1.2.1.1) or phosvitin of egg yolk (cf. 11.2.4.1.2) contain phosphoric acid esters of serine and threonine residues.
The structure of a protein is dependent on the amino acid sequence (the primary structure)

which determines the molecular conformation (secondary and tertiary structures). Proteins sometimes occur as molecular aggregates which are arranged in an orderly geometric fashion (quaternary structure). The sequences and conformations of a large number of proteins have been elucidated and recorded in several data bases.

R : H, CH$_3$;   R^1 : H, Sugar residue;   Ac : Acetyl   (1.82)

Glycoproteins, such as ϰ-casein (cf. 10.1.2.1.1), various components of egg white (cf. 11.2.3.1) and egg yolk (cf. 11.2.4.1.2), collagen from connective tissue (cf. 12.3.2.3.1) and serum proteins of some species of fish (cf. 13.1.4.2.4), contain one or more monosaccharide or oligosaccharide units bound O-glycosidically to serine, threonine or δ-hydroxylysine or N-glycosidically to asparagine (Formula 1.82). In glycoproteins, the primary structure of the protein is defined genetically. The carbohydrate components, however, are enzymatically coupled to the protein in a co- or post-transcriptional step. Therefore, the carbohydrate composition of glycoproteins is inhomogeneous (microheterogeneity).

1.4.1 Amino Acid Sequence

1.4.1.1 Amino Acid Composition, Subunits

Sequence analysis can only be conducted on a pure protein. First, the amino acid composition is determined after acidic hydrolysis. The procedure (separation on a single cation-exchange resin column and color development with ninhydrin reagent or fluorescamine) has been standardized and automated (amino acid analyzers). Figure 1.10 shows a typical amino acid chromatogram.

As an alternative to these established methods, the derivatization of amino acids with the subsequent separation and detection of derivatives is possible (pre-column derivatization). Various derivatization reagents can be selected, such as:

- 9-Fluorenylmethylchloroformate (FMOC, cf. 1.2.4.2.1)
- Phenylisothiocyanate (PITC, cf. 1.2.4.2.3)
- Dimethylaminoazobenzenesulfonylchloride (DABS-Cl, cf. 1.2.4.2.1)
- Dimethylaminonaphthalenesulfonylchloride (DANS-Cl, cf. 1.2.4.2.1)
- 7-Fluoro-4-nitrobenzo-2-oxa-1,3-diazole (NBDF, cf. 1.2.4.2.1)
- 7-Chloro-4-nitrobenzo-2-oxa-1,3-diazole (NBDCl, cf. 1.2.4.2.1)
- o-Phthaldialdehyde (OPA, cf. 1.2.4.2.4)

It is also necessary to know the molecular weight of the protein. This is determined by gel column chromatography, ultracentrifugation or SDS-PAG electrophoresis. Furthermore, it is necessary to determine whether the protein is a single molecule or consists of a number of identical or different polypeptide chains (subunits) associated through disulfide bonds or noncovalent forces. Dissociation into subunits can be accomplished by a change in pH, by chemical modification of the protein, such as by succinylation, or with denaturing agents (urea, guanidine hydrochloride, sodium dodecyl sulfate). Disulfide bonds, which are also found in proteins which consist of only one peptide chain, can be cleaved by oxidation of cystine to cysteic acid or by reduction to cysteine with subsequent alkylation of the thiol group

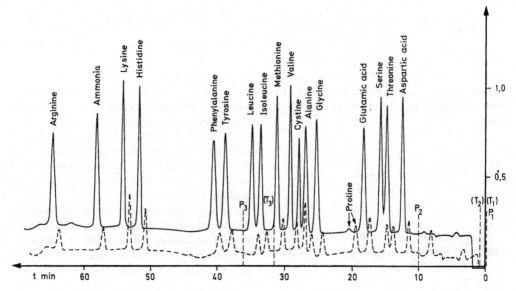

Fig. 1.10. Amino acid chromatogram. Separation of a mixture of amino acids (10 nmol/amino acid) by an amino acid analyzer. Applied is a single ion exchange column: Durrum DC-4A, 295×4 mm buffers $P_1/P_2/P_3$: 0.2 N Na-citrate pH 3.20/0.2 N Na-citrate pH 4.25/1.2 N Na-citrate and NaCl of pH 6.45. Temperatures $T_1/T_2/T_3$: 48/56/80 °C. Flow rate: 25 ml/h; absorbance reading after color development with ninhydrin at 570/440 nm: —/– – – –

(cf. 1.2.4.3.5) to prevent reoxidation. Separation of subunits is achieved by chromatographic or electrophoretic methods.

1.4.1.2 Terminal Groups

N-terminal amino acids can be determined by treating a protein with 1-fluoro-2,4-dinitrobenzene (*Sanger's* reagent; cf. 1.2.4.2.2) or 5-dimethylaminonaphthalene-1-sulfonyl chloride (dansyl chloride; cf. 1.2.4.2.1). Another possibility is the reaction with cyanate, followed by elimination of the N-terminal amino acid in the form of hydantoin, and separation and recovery of the amino acid by cleavage of the hydantoin (cf. 1.2.4.2.3). The N-terminal amino acid (and the amino acid sequence close to the N-terminal) is accessible by hydrolysis with aminopeptidase, in which case it should be remembered that the hydrolysis rate is dependent on amino acid side chains and that proline residues are not

cleaved. A special procedure is required when the N-terminal residue is acylated (N-formyl- or N-acetyl amino acids, or pyroglutamic acid). Determination of C-terminal amino acids is possible via the hydrazinolysis procedure recommended by *Akabori:*

$$H_2N-\underset{\underset{R_1}{|}}{C}H-CO-(HN-\underset{\underset{R_{2-n}}{|}}{C}H-CO-)HN-\underset{\underset{R_m}{|}}{C}H-COOH$$

$$\xrightarrow[100\,°C]{H_2N-NH_2} H_2N-\underset{\underset{R_{1-n}}{|}}{C}H-CO-NH-NH_2$$

$$+ \; H_2N-\underset{\underset{R_m}{|}}{C}H-COOH$$

$$(1.83)$$

The C-terminal amino acid is then separated from the amino acid hydrazides, e. g., by a cation exchange resin, and identified. It is possible to mark the C-terminal amino acid through selective titration via oxazolinone:

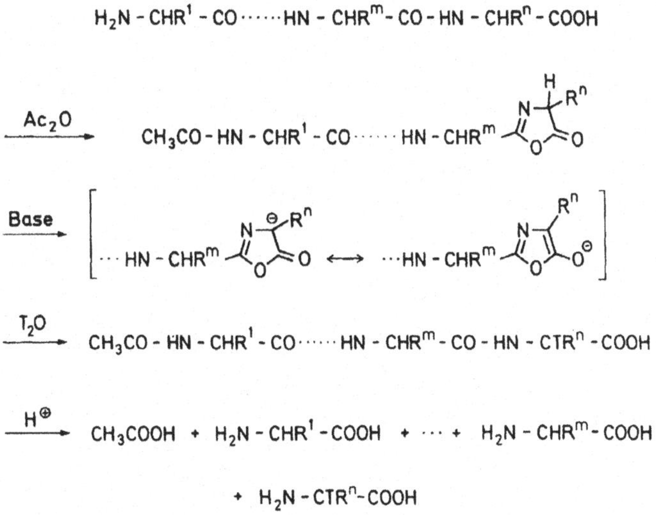

$$H_2N - CHR^1 - CO \cdots\cdots HN - CHR^m - CO - HN - CHR^n - COOH$$

$$\xrightarrow{Ac_2O} \quad CH_3CO - HN - CHR^1 - CO \cdots\cdots HN - CHR^m \underset{O}{\overset{H}{\bigwedge}}\overset{R^n}{\underset{O}{}}$$

$$\xrightarrow{Base} \quad \left[\cdots HN - CHR^m \underset{O}{\overset{N^{\ominus}}{\bigwedge}}\overset{R^n}{\underset{O}{}}O \longleftrightarrow \cdots HN - CHR^m \underset{O}{\overset{N}{\bigwedge}}\overset{R^n}{\underset{O^{\ominus}}{}} \right]$$

$$\xrightarrow{T_2O} \quad CH_3CO - HN - CHR^1 - CO \cdots\cdots HN - CHR^m - CO - HN - CTR^n - COOH$$

$$\xrightarrow{H^{\oplus}} \quad CH_3COOH + H_2N - CHR^1 - COOH + \cdots + H_2N - CHR^m - COOH$$

$$+ H_2N - CTR^n - COOH$$

$$(1.84)$$

The C-terminal amino acids can be removed enzymatically by carboxypeptidase A which preferentially cleaves amino acids with aromatic and large aliphatic side chains, carboxypeptidase B which preferentially cleaves lysine, arginine and amino acids with neutral side chains or carboxypeptidase C which cleaves with less specificity but cleaves proline.

1.4.1.3 Partial Hydrolysis

Longer peptide chains are usually fragmented. The fragments are then separated and analyzed individually for amino acid sequences. Selective enzymatic cleavage of peptide bonds is accomplished primarily with trypsin, which cleaves exclusively Lys-X- and Arg-X-bonds, and chymotrypsin, which cleaves peptide bonds with less specificity (Tyr-X, Phe-X, Trp-X and Leu-X). The enzymatic attack can be influenced by modification of the protein. For example, acylation of the ε-amino group of lysine limits tryptic hydrolysis to Arg-X (cf. 1.4.4.1.3 and 1.4.4.1.4), whereas substitution of the SH-group of a cysteine residue with an aminoethyl group introduces a new cleavage position for trypsin into the molecule "pseudolysine residue"):

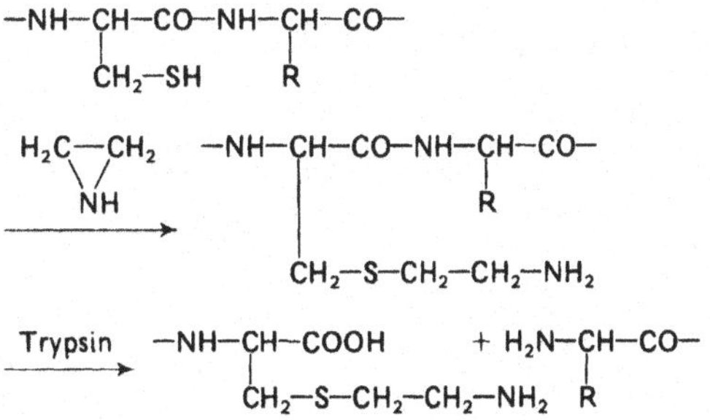

$$-NH-CH-CO-NH-CH-CO-$$
$$\underset{CH_2-SH}{|} \quad \underset{R}{|}$$

$$\underset{\underset{NH}{\backslash\;/}}{H_2C-CH_2} \quad -NH-CH-CO-NH-CH-CO-$$
$$\xrightarrow{} \quad \underset{R}{|}$$
$$\underset{CH_2-S-CH_2-CH_2-NH_2}{|}$$

$$\xrightarrow{Trypsin} \quad -NH-CH-COOH \quad + \; H_2N-CH-CO-$$
$$\underset{CH_2-S-CH_2-CH_2-NH_2}{|} \quad \underset{R}{|}$$

$$(1.85)$$

Also suited for the specific enzymatic hydrolysis of peptide chains is the endoproteinase Glu-C from *Staphylococcus aureus* V8. It cleaves Glu-X bonds (ammonium carbonate buffer pH 7.8 or ammonium acetate buffer pH 4.0) as well as Glu-X plus Asp-X bonds (phosphate buffer pH 7.8). The most important chemical method for selective cleavage uses cyanogen bromide (BrCN) to attack Met-X-linkages (Reaction 1.86).

Hydrolysis of proteins with strong acids reveals a difference in the rates of hydrolysis of peptide bonds depending on the adjacent amino acid side chain. Bonds involving amino groups of serine and threonine are particularly susceptible to hydrolysis. This effect is due to

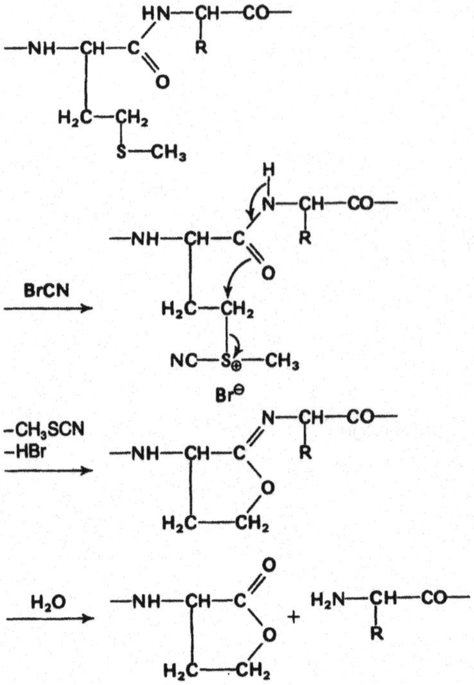

$$HN—CH—CO—$$
$$—NH—CH—C\;\;\;\;R$$
$$\;\;\;\;\;\;\;\;\;\;\|\;\;\;\;\\O$$
$$H_2C—CH_2$$
$$\;\;\;\;\;\;\;S—CH_3$$

BrCN →

$$\;\;\;\;\;\;\;\;\;\;N—CH—CO—$$
$$—NH—CH—C\;\;\;\;R$$
$$\;\;\;\;\;\;\;\;\;\;\\O$$
$$H_2C—CH_2$$
$$\;\;\;\;\;NC—S^{\oplus}—CH_3$$
$$\;\;\;\;\;\;\;\;\;\;Br^{\ominus}$$

−CH₃SCN
−HBr →

$$\;\;\;\;\;\;\;\;\;\;N—CH—CO—$$
$$—NH—CH—C\;\;\;\;R$$
$$\;\;\;\;\;\;\;\;\;\;\\O$$
$$H_2C—CH_2$$

H₂O →

$$\;\;\;\;\;\;\;\;\;O$$
$$\;\;\;\;\;\;\;\;\;\\\\$$
$$—NH—CH—C\;\;\;\;+\;\;H_2N—CH—CO—$$
$$\;\;\;\;\;\;\;\;\;\\O\;\;\;\;\;\;\;\;\;\;\;\;R$$
$$H_2C—CH_2$$

(1.86)

$N \rightarrow O$-acyl migration via the oxazoline and subsequent hydrolysis of the ester bond:

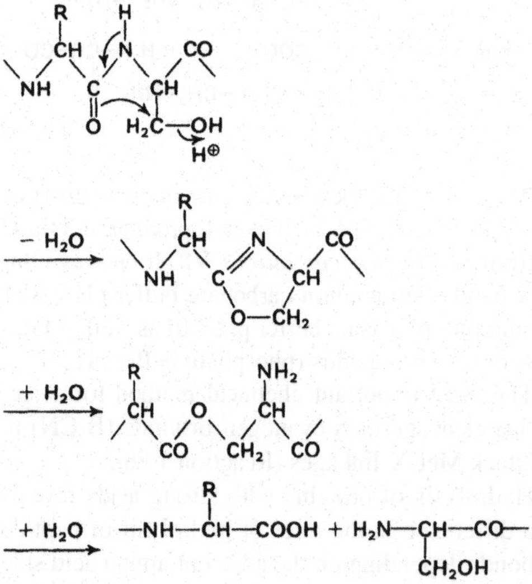

$$\begin{array}{c} R \;\;\; H \\ CH \;\; N \;\;\; CO \\ NH \;\; C \;\; CH \\ O \;\; H_2C—OH \\ H^{\oplus} \end{array}$$

− H₂O →

$$\begin{array}{c} R \\ CH \;\;\; N \;\;\; CO \\ NH \;\; C \;\; CH \\ O—CH_2 \end{array}$$

+ H₂O →

$$\begin{array}{c} R \;\;\;\;\;\; NH_2 \\ CH \;\; O \;\; CH \\ CO \;\; CH_2 \;\; CO \end{array}$$

+ H₂O →

$$—NH—CH—COOH + H_2N—CH—CO—$$
$$\;\;\;\;\;\;\;R\;\;\;\;\;\;\;\;\;\;\;\;\;\;\;\;\;CH_2OH$$

(1.87)

Hydrolysis of proteins with dilute acids preferentially cleaves aspartyl-X-bonds.

Separation of peptide fragments is achieved by gel and ion-exchange column chromatography using a volatile buffer as eluent (pyridine, morpholine acetate) which can be removed by freeze-drying of the fractions collected. The separation of peptides and proteins by reversed-phase HPLC has gained great importance, using volatile buffers mixed with organic, water-soluble solvents as the mobile phase.

The fragmentation of the protein is performed by different enzymic and/or chemical techniques, at least by two enzymes of different specifity. The arrangement of the obtained peptides in the same order as they occur in the intact protein is accomplished with the aid of overlapping sequences. The principle of this method is illustrated for subtilisin BPN′ as an example in Fig. 1.11.

1.4.1.4 Sequence Analysis

The classical method is the *Edman* degradation reaction. It involves stepwise degradation of peptides with phenylisothiocyanate (cf. 1.2.4.2.3) or suitable derivatives, e. g. dimethylaminoazobenzene isothiocyanate (DABITC). The resultant phenylthiohydantoin is either identified directly or the amino acid is recovered. The stepwise reactions are performed in solution or on peptide bound to a carrier, i. e. to a solid phase. Both approaches have been automated ("sequencer"). Carriers used include resins containing amino groups (e. g. amino polystyrene) or glass beads treated with amino alkylsiloxane:

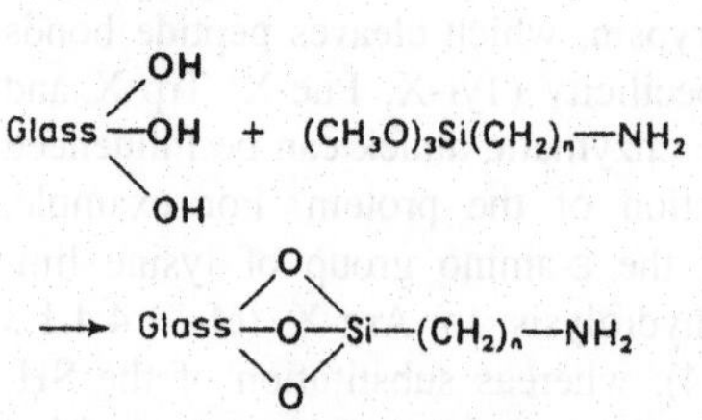

$$\text{Glass}\begin{array}{c}—OH \\ —OH \\ —OH\end{array} + (CH_3O)_3Si(CH_2)_n{-}NH_2$$

$$\longrightarrow \text{Glass}\begin{array}{c}—O \\ —O \\ —O\end{array}Si{-}(CH_2)_n{-}NH_2$$

(1.88)

The peptides are then attached to the carrier by carboxyl groups (activation with carbodiimide or carbonyl diimidazole, as in peptide synthesis) or by amino groups. For example, a peptide

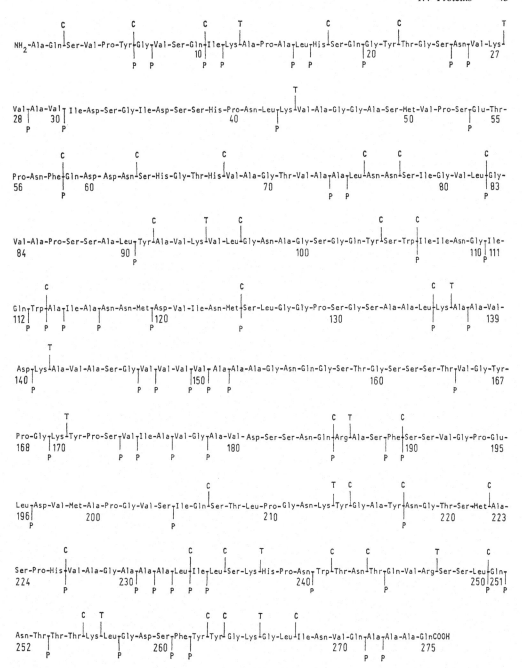

Fig. 1.11. Subtilisin BPN; peptide bonds hydrolyzed by trypsin (T), chymotrypsin (C), and pepsin (P)

segment from the hydrolysis of protein by trypsin has lysine as its C-terminal amino acid. It is attached to the carrier with p-phenylene-diisothiocyanate through the α- and ε-amino groups. Mild acidic treatment of the carrier under conditions of the *Edman* degradation splits the first peptide bond. The *Edman* procedure is then performed on the shortened peptide through second, third and subsequent repetitive reactions:

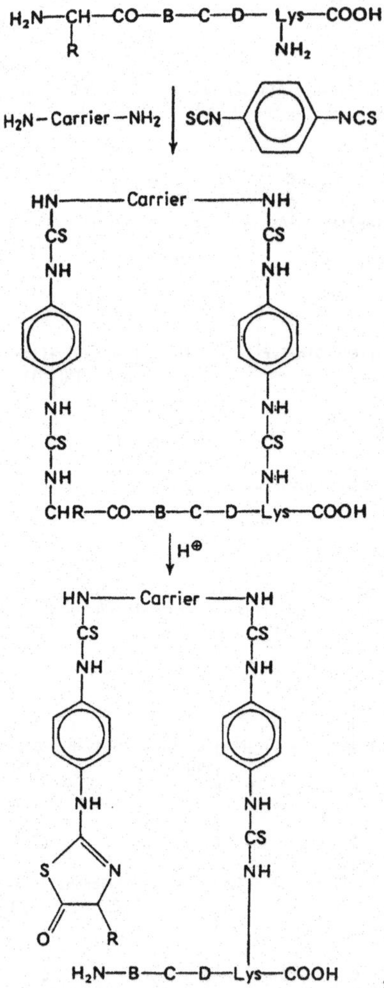

$$(1.89)$$

Microvariants allow working in the picomole range. In the reaction chamber, the protein is fixed on a glass-fiber disc, and the coupling and cleaving reagents are added and removed in a carrier gas stream (vapour-phase sequentiation).

Apart from the *Edman* degradation, other methods can give valuable additional information on sequence analysis. These methods include the hydrolysis with amino- and carboxypeptidases as mentioned in the case of end group analysis and the fragmentation of suitable volatile peptide derivatives in a mass spectrometer.

1.4.1.5 Derivation of Amino Acid Sequence from the Nucleotide Sequence of the Coding Gene

The number of proteins for which the coding gene in the genome has been characterized is increasing steadily. However, a considerable part of the amino acid sequences known today has already been derived from the nucleotide sequences in question.

The background of this process will be briefly described here. The nucleotides consist of four different bases as well as 2-deoxyribose and phosphoric acid. They are the building blocks of the high-molecular deoxyribonucleic acid (DNA). The nucleotides are linked via 2-deoxyribose and phosphoric acid as $3' \rightarrow 5'$-diesters. In DNA, two polynucleotide strands are linked together in each case via hydrogen bridge bonds to give a double helix. The bases thymine and adenine as well as cytosine and guanine are complementary (cf. Formula 1.90). DNA is the carrier of the genetic information which controls protein biosynthesis via transcription to messenger ribonucleic acid (RNA). In translation into proteins, the sequence of bases codes the primary sequence of amino acids. Here, three of the four bases adenine, guanine, cytosine and thymine (abbreviated AGCT) in each case determines one amino acid, e. g., UGG codes for tryptophan (cf. Fig. 1.12).

UUU Phe	UCU Ser	UAU Tyr	UGU Cys
UUC Phe	UCC Ser	UAC Tyr	UGC Cys
UUA Phe	UCA Ser	UAA Stop	UGA Stop
UUG Phe	UCG Ser	UAG Stop	UGG Trp
CUU Leu	CCU Pro	CAU His	CGU Arg
CUC Leu	CCC Pro	CAC His	CGC Arg
CUA Leu	CCA Pro	CAA Gln	CGA Arg
CUG Leu	CCG Pro	CAG Gln	CGG Arg
AUU Ile	ACU Thr	AAU Asn	AGU Ser
AUC Ile	ACC Thr	AAC Asn	AGC Ser
AUA Ile	ACA Thr	AAA Lys	AGA Arg
AUG Met	ACG Thr	AAG Lys	AGG Arg
GUU Val	GCU Ala	GAU Asp	GGU Gly
GUC Val	GCC Ala	GAC Asp	GGC Gly
GUA Val	GCA Ala	GAA Glu	GGA Gly
GUG Val	GCG Ala	GAG Glu	GGG Gly

Fig. 1.12. The genetic code

Therefore, the primary sequence of a protein can be derived from the nucleotide (base) sequence.

For the sequencing of DNA, the method of choice is the dideoxy process (chain termination process) introduced by *Fred Sanger* in 1975. The principle is based on the specific termination of the enzymatic synthesis of a DNA strand by means of DNA polymerase by using a $2',3'$-dideoxynucleotide, i.e., to prevent polymerization with the formation of the $3' \rightarrow 5'$-phosphodiester at the position of the base in question. For instance, if the $2',3'$-dideoxynucleotide of guanine is used, the biosynthesis is stopped at guanine in each case. To detect all the guanine residues, only about 0.5 mol% of the dideoxynucleotide in question (based on the 2-deoxynucleotide) is used. In this way, DNA fragments of varying length are obtained which all have the same $5'$-terminal and thus mark the position of the base.

The starting material is a hybrid of the single-stranded DNA to be sequenced and a primer consisting of about 20 nucleotides. This is lengthened with the help of DNA polymerase and a mixture of the 4 nucleotides and one $2',3'$-dideoxynucleotide in each case. The primer serves as a defined starting position and also as an initiator for the start of the synthesis of the complementary DNA strand. The DNA fragments of different length obtained in four experiments are then separated electrophoretically according to molecular size. For detection, either the primer can be labelled with four different fluorescent dyes (TAG) or the four dideoxynucleotides are labelled with different fluorescent dyes. In the former case, 4 series of experiments are carried out with differently labelled primers and one of the 4 dideoxynucleotides in each case. The charges are combined and electrophoretically separated together. The primary sequence is determined from the signals measured at different wave lengths (Fig. 1.13). When 4 differently labelled dideoxynucleotides are used, the primer is not labelled. Alternatively, the dideoxynucleotides can also be radioactively labelled (e. g., with ^{32}P). In this case, four separate DNA syntheses are also required.

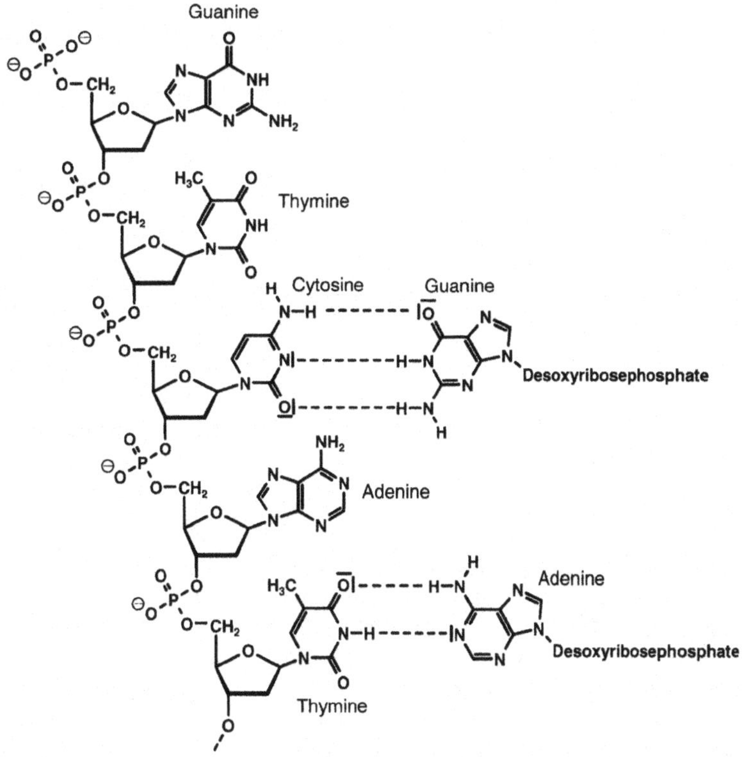

(1.90)

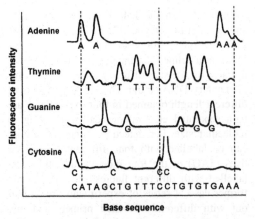

Base sequence

Fig. 1.13. Fluorescence detection of electrophoreti-cally separated DNA fragments obtained by using the dideoxy method (according to *Smith* et al., 1986)

Fig. 1.15. Structure of an elongated peptide chain. ● Carbon, ○ oxygen, ○ nitrogen, ○ hydrogen and ® side chain

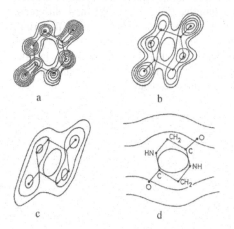

Fig. 1.14. Electron density distribution patterns for 2,5-dioxopiperazine with varying resolution extent. **a** 0.11 nm, **b** 0.15 nm, **c** 0.20 nm, **d** 0.60 nm (after *Pe-rutz*, 1962)

1.4.2 Conformation

Information about conformation is available through X-ray crystallographic analysis of protein crystals and by measuring the distance (≤ 30 nm) between selected protons of the peptide chain (NH_i–NH_{i+1}, NH_{i+1}–$C_\alpha H_i$, NH_{i+1}–$C_\beta H_i$, $C_\alpha H_i$–$C_\alpha H_{i+1}$, $C_\alpha H_i$–$C_\beta H$) by means of H-NMR spectroscopy in solution. This assumes that, in many cases, the conformation of the protein in crystalline form is similar to that of the protein in solution. As an example the calculated electron

density distributions of 2,5-dioxopiperazine based on various degrees of resolution are presented in Fig. 1.14. Individual atoms are well revealed at 0.11 nm. Such a resolution has not been achieved with proteins. Reliable localization of the C_α-atom of the peptide chain requires a resolution of less than 0.3 nm.

1.4.2.1 Extended Peptide Chains

X-ray structural analysis and other physical mea-surements of a fully extended peptide chain re-veal the lengths and angles of bonds (see the "ball and stick" representation in Fig. 1.15). The pep-tide bond has partial (40%) double bond char-acter with π electrons shared between the C'—O and C'—N bonds. The resonance energy is about 83.6 kJ/mole:

$$\begin{array}{ccc} & & \\ \end{array} \tag{1.91}$$

Normally the bond has a trans-configuration, i. e. the oxygen of the carbonyl group and the hydro-gen of the NH group are in the trans-position; a cis-configuration which has 8 kJ mol^{-1} more energy occurs only in exceptional cases (e. g. in small cyclic peptides or in proteins before proline residues).

Thus in ribonuclease A, two X-Pro bonds have trans-conformation (Pro-42 and Pro-117), and two have cis-conformation (Pro-93 and Pro-114). The equilibrium between the two isomers is catalyzed by specific enzymes (peptidyl-prolyl-cis/trans-isomerases). This accelerates the folding of a peptide chain (cf. 1.4.2.3.2), which in terms of the biosynthesis occurs initially in all-trans-conformation.

Six atoms of the peptide bonds, C_i^{α}, C_i', O_i, N_{i+1}, C_{i+1}^{α} and H_{i+1}, lie in one plane (cf. Fig. 1.16). For a trans-peptide bond, ω_i is $180°$. The position of two neighboring planes is determined by the numerical value of the angles ψ_i (rotational bond between a carbonyl carbon and an α-carbon) and ϕ_i (rotational bond between an amide-N and an α-carbon). For an extended peptide chain, $\psi_i = 180°$ and $\phi_i = 180°$. The position of side chains can also be described by a series of angles χ_i^{1-n}.

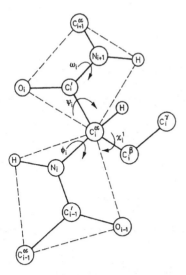

Fig. 1.16. Definitions for torsion angles in a peptide chain

$\omega_i = 0°$ for $C_i^{\alpha}-C_i'/N_{i+1}-C_{i+1}^{\alpha} \rightarrow$ cis,

$\psi_i = 0°$ for $C_i^{\alpha}-N_i/C_i'-O_i \rightarrow$ trans,

$\phi_i = 0°$ for $C_i'-C_i'/N_i-H \rightarrow$ trans,

$\chi_i = 0°$ for $C_i^{\alpha}-N_iC_i^{\beta}-C_i \rightarrow$ cis.

The angles are positive when the rotation is clockwise and viewed from the N-terminal side of a bond or (for X) from the atom closer to the main chain respectively. (according to *Schulz* and *Schirmer*, 1979)

1.4.2.2 Secondary Structure (Regular Structural Elements)

The primary structure gives the sequence of amino acids in a protein chain while the secondary structure reveals the arrangement of the chain in space. The peptide chains are not in an extended or unfolded form ($\psi_i, \phi_i = 180°$). It can be shown with models that ψ_i and ϕ_i, at a permissible minimum distance between non-bonding atoms (Table 1.20), can assume only particular angles. Figure 1.17 presents the

Table 1.20. Minimal distances for nonbonded atoms (Å)

	C	N	O	H
C	3.20[a]	2.90	2.80	2.40
	(3.00)[b]	(2.80)	(2.70)	(2.20)
N		2.70	2.70	2.40
		(2.60)	(2.60)	(2.20)
O			2.70	2.40
			(2.60)	(2.20)
H				2.00
				(1.90)

[a] Normal values.
[b] Extreme values.

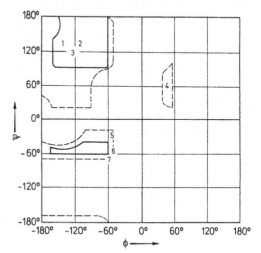

Fig. 1.17. ϕ, ψ-Diagram (*Ramachandran* plot). Allowed conformations for amino acids with a C^{β}-atom obtained by using normal ($-$) and lower limit (- - -) contact distances for non-bonded atoms, from Table 1.20. β-Sheet structures: antiparallel (1); parallel (2), twisted (3). Helices: α-, left-handed (4), 3_{10} (5), α, right-handed (6), π (7)

Table 1.21. Regular structural elements (secondary structures) in polypeptides

Structure	Φ (°)	ψ (°)	n^a	d^b (Å)	r^c (Å)	Comments
β-Pleated sheet, parallel	−119	+113	2.0	3.2	1.1	Occurs occasionally in neighbouring chain sectors of globular proteins
β-Pleated sheet, antiparallel	−139	+135	2.0	3.4	0.9	Common in proteins and synthetic polypeptides
3_{10}-Helix	−49	−26	2.3	2.0	1.9	Observed at the ends of α-helixes
α-Helix, left-handed coiling	+57	+47	3.6	1.5	2.3	Common in globular proteins, as α "coiled coil" in fibrous proteins
α-Helix, right-handed coiling	−57	−47	3.6	1.5	2.3	Poly-D-amino acids poly-(β-benzyl)-L-aspartate
π-Helix	−57	−70	4.4	1.15	2.8	Hypothetical
Polyglycine II	−80	+150	3.0	3.1		Similar to antiparallel β-pleated-sheet formation
Polyglycine II, left-handed coiling	+80	−150	3.0	3.1		Synthetic polyglycine is a mixture of right- and left-handed helices; in some silk fibroins, the left-handed helix occurs
Poly-L-proline I	−83	+158	3.3	1.9		Synthetic poly-L-proline, only cis-peptide bonds
Poly-L-proline II	−78	+149	3.0	3.1		As left-handed polyglycine II, as triple helix in collagen

[a] Amino acid residues per turn.
[b] The rise along the axis direction, per residue.
[c] The radius of the helix.

permissible ranges for amino acids other than glycine (R ≠ H). The range is broader for glycine (R = H). Figure 1.18 demonstrates that most of 13 different proteins with a total of about 2500 amino acid residues have been shown empirically to have values of ψ, ϕ-pairs within the permissible range. When a multitude of equal ψ, ϕ-pairs occurs consecutively in a peptide chain, the chain acquires regular repeating structural elements. The types of structural elements are compiled in Table 1.21.

1.4.2.2.1 β-Sheet

Three regular structural elements (pleated-sheet structures) have values in the range of $\phi = -120\,°C$ and $\psi = +120\,°C$. The peptide chain is always lightly folded on the C_α atom (cf. Fig. 1.19), thus the R side chains extend perpendicularly to the extension axis of the chain, i. e. the side chains change their projections alternately from +z to −z. Such a pleated structure is stabilized when more chains are present.

Subsequently, adjacent chains interact along the x-axis by hydrogen bonding, thus providing the cross-linking required for stability. When adjacent chains run in the same direction, the peptide chains are parallel. This provides a stabilized,

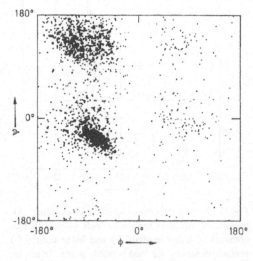

Fig. 1.18. ϕ, ψ-Diagram for observed values of 13 different proteins containing a total of 2500 amino acids. (according to *Schulz* and *Schirmer*, 1979)

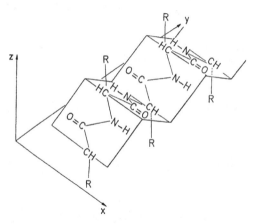

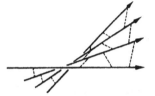

Fig. 1.21. Diagrammatic presentation of a twisted sheet structure of parallel peptide chains (according to *Schulz* and *Schirmer*, 1979)

Fig. 1.19. A pleated sheet structure of a peptide chain

planar, parallel sheet structure. When the chains run in opposite directions, a planar, antiparallel sheet structure is stabilized (Fig. 1.20). The lower free energy, twisted sheet structures, in which the main axes of the neighboring chains are arranged at an angle of 25 °C (Fig. 1.21), are more common than planar sheet structures.

The β structures can also be regarded as special helix with a continuation of 2 residues per turn. With proline, the formation of a β structure is not possible.

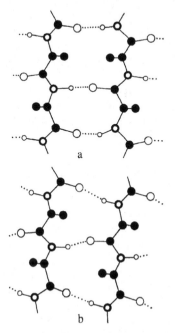

Fig. 1.20. Diagrammatic presentation of antiparallel (**a**) and parallel (**b**) peptide chain arrangements

1.4.2.2.2 Helical Structures

There are three regular structural elements in the range of $\phi = -60°$ and $\psi = -60°$ (cf. Fig. 1.17) in which the peptide chains are coiled like a threaded screw. These structures are stabilized by intrachain hydrogen bridges which extend almost parallel to the helix axis, crosslinking the CO and NH groups, i.e., the CO group of amino acid residue i with the NH group of residue $i+3$ (3_{10}-helix), $1+4$ (α-helix) or $i+5$ (π-helix).

The most common structure is the α-helix and for polypeptides from L-amino acids, exclusively the right-handed α-helix (Fig. 1.22). The left-handed α-helix is energetically unfavourable for L-amino acids, since the side chains here are in close contact with the backbone. No α-helix is possible with proline. The 3_{10}-helix was observed only at the ends of α-helices but not as an independent regular structure. The π-helix is hypothetical. Two helical conformations are known of polyproline (I and II). Polyproline I contains only cispeptide bonds and is right-handed, while polyproline II contains trans-peptide bonds and is left-handed. The stability of the two conformations depends on the solvent and other factors. In water, polyproline II predominates. Polyglycine can also occur in two conformations. Polyglycine I is a β-structure, while polyglcine II corresponds largely to the polyproline II-helix. A helix is characterized by the angles ϕ and ψ, or by the parameters derived from these angles: n, the number of amino acid residues per turn; d, the rise along the main axis per amino acid residue; and r, the radius of the helix. Thus, the equation for the pitch, p, is $p = n \cdot d$. The parameters n and d are presented within a ϕ, ψ plot in Fig. 1.23.

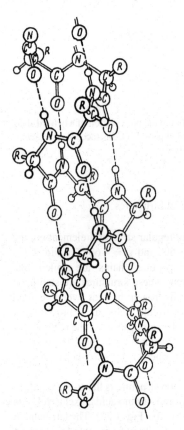

Fig. 1.22. Right-handed α-helix

Table 1.22. β-Turns in the peptide chain of egg white lysozyme

Residue Number	Sequence			
20–23	Y	R	G	Y
36–39	S	N	F	N
39–42	N	T	Q	A
47–50	T	D	G	S
54–57	G	I	L	E
60–63	S	R	W	W
66–69	D	G	R	T
69–72	T	P	G	S
74–77	N	L	C	N
85–88	S	S	D	I
100–103	S	D	G	D
103–106	D	G	M	N

1.4.2.2.3 Reverse Turns

An important conformational feature of globular proteins are the reverseturns β-turns and β-bends. They occur at "hairpin" corners, where the peptide chain changes direction abruptly. Such corners involve four amino acid residues often including proline and glycine. Several types of turns are known; of greatest importance are type I (42% of 421 examined turns), type II (15%) and type III (18%); see Fig. 1.24.

In type I, all amino acid residues are allowed, with the exception of proline in position 3. In type II, glycine is required in position 3. In type III, which corresponds to a 3_{10}-helix, all amino acids are allowed. The sequences of the β-bends of lysozyme are listed in Table 1.22 as an example.

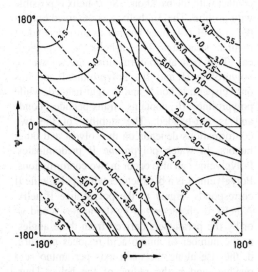

Fig. 1.23. φ,ψ-Diagram with marked helix parameters n (---) and d (—). (according to *Schulz* and *Schirmer*, 1979)

1.4.2.2.4 Super-Secondary Structures

Analysis of known protein structures has demonstrated that regular elements can exist in combined forms. Examples are the coiled-coil α-helix (Fig. 1.25, a), chain segments with antiparallel β-structures (β-meander structure; Fig. 1.25, b) and combinations of α-helix and β-structure (e. g., βαβαβ; Fig. 1.25 c).

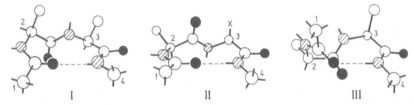

Fig. 1.24. Turns of the peptide chains (β-turns), types I–III. ○ = carbon, ⊘ = nitrogen, ● = oxygen. The α-C atoms of the amino acid residues are marked 1–4. X = no side chain allowed

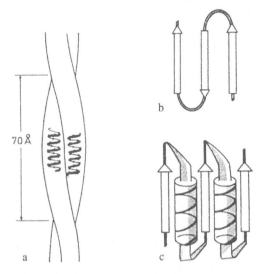

Fig. 1.25. Superhelix secondary structure (according to *Schulz* and *Schirmer*, 1979). **a** coiled-coil α-helix, **b** β-meander, **c** βαβαβ-structure

1.4.2.3 Tertiary and Quaternary Structures

Proteins can be divided into two large groups on the basis of conformation: (a) fibrillar (fibrous) or scleroproteins, and (b) folded or globular proteins.

1.4.2.3.1 Fibrous Proteins

The entire peptide chain is packed or arranged within a single regular structure for a variety of fibrous proteins. Examples are wool keratin (α-helix), silk fibroin (β-sheet structure) and collagen (a triple helix). Stabilization of these structures is achieved by intermolecular bonding (electrostatic interaction and disulfide linkages, but primarily hydrogen bonds and hydrophobic interactions).

1.4.2.3.2 Globular Proteins

Regular structural elements are mixed with randomly extended chain segments (randomly coiled structures) in globular proteins. The proportion of regular structural elements is highly variable: 20–30% in casein, 45% in lysozyme and 75% in myoglobin (Table 1.23). Five structural subgroups are known in this group of proteins: (1) α-helices

Table 1.23. Proportion of "regular structural elements" present in various globular proteins

Protein	α-Helix	β-Structure	n_G	n	%
Myoglobin	3–16[a]			14	
	20–34			15	
	35–41			7	
	50–56			7	
	58–77			20	
	85–93			9	
	99–116			18	
	123–145			23	
			151	173	75
Lysozyme	5–15			11	
	24–34			11	
		41–54		14	
	80–85			6	
	88–96			9	
	97–101			5	
	109–125			7	
			129	63	49
α_{s1}-Casein				199	ca. 30
β-Casein				209	ca. 20

[a] Position number of the amino acid residue in the sequence.
n_G: Total number of amino acid residues.
n: Amino acid residues within the regular structure.
%: Percentage of the amino acid residues present in regular structure.

occur only; (2) β-structures occur only; (3) α-helical and β-structural portions occur in separate segments on the peptide chain; (4) α-helix and β-structures alternate along the peptide chain; and (5) α-helix and β-structures do not exist.

The process of peptide chain folding is not yet fully understood. It begins spontaneously, probably arising from one center or from several centers of high stability in larger proteins. The tendency to form regular structural elements shows a very different development in the various amino acid residues. Table 1.24 lists data which were derived from the analysis of globular proteins of known conformation. The data indicate, for example, that Met, Glu, Leu and Ala are strongly helix-forming. Gly and Pro on the other hand show a strong helix-breaking tendency. Val, Ile and Leu promote the formation of pleated-sheet structures, while Asp, Glu and Pro prevent them.

Table 1.24. Normalized frequencies[a] of amino acid residues in the regular structural elements of globular proteins

Amino acid	α-Helix (P_α)	Pleated sheet (P_β)	β-Turn (P_t)
Ala	1.29	0.90	0.78
Cys	1.11	0.74	0.80
Leu	1.30	1.02	0.59
Met	1.47	0.97	0.39
Glu	1.44	0.75	1.00
Gln	1.27	0.80	0.97
His	1.22	1.08	0.69
Lys	1.23	0.77	0.96
Val	0.91	1.49	0.47
Ile	0.97	1.45	0.51
Phe	1.07	1.32	0.58
Tyr	0.72	1.25	1.05
Trp	0.99	1.14	0.75
Thr	0.82	1.21	1.03
Gly	0.56	0.92	1.64
Ser	0.82	0.95	1.33
Asp	1.04	0.72	1.41
Asn	0.90	0.76	1.28
Pro	0.52	0.64	1.91
Arg	0.96	0.99	0.88

[a] Shown is the fraction of an amino acid in a regular structural element, related to the fraction of all amino acids of the same structural element. P = 1 means random distribution; P > 1 means enrichment, P < 1 means depletion. The data are based on an analysis of 66 protein structures.

Pro and Gly are important building blocks of turns. Arginine does not prefer any of the three structures. By means of such data it is possible to forecast the expected conformations for a given amino acid sequence.

Folding of the peptide chain packs it densely by formation of a large number of intermolecular noncovalent bonds. Data on the nature of the bonds involved are provided in Table 1.25.

The H-bonds formed between main chains, main and side chains and side-side chains are of particular importance for folding. The portion of polar groups involved in H-bond buildup in proteins of $Mr > 8.9$ kdal appears to be fairly constant at about 50%.

The hydrophobic interaction of the nonpolar regions of the peptide chains also plays an important role in protein folding. These interactions are responsible for the fact that nonpolar groups are folded to a great extent towards the interior of the protein globule. The surface areas accessible to water molecules have been calculated for both unfolded and native folded forms for a number of monomeric proteins with known conformations. The proportion of the accessible surface in the stretched state, which tends to be buried in the interior of the globule as a result of folding, is a simple linear function of the molecular weight (M). The gain in free energy for the folded surface is

Table 1.25. Bond-types in proteins

Type	Examples	Bond strength (kJ/mole)
Covalent bonds	–S–S–	ca. −230
Electrostatic bonds	–COO–H_3N^+–	−21
	>C=O O=C<	+ 1.3
Hydrogen bonds	–O–H ⋯ O<	− 16.7
	>N–H ⋯ O=C<	− 12.5
Hydrophobic bonds	$-CH\begin{smallmatrix}CH_3\;H_3C\\ \\CH_3\;H_3C\end{smallmatrix}CH-$	0.01[b]
	–Ala ⋯ Ala–	−3
	–Val ⋯ Val–	−8
	–Leu ⋯ Leu–	−9
	–Phe ⋯ Phe–	−13
	–Trp ⋯ Trp–	−19

[a] For $\varepsilon = 4$.
[b] Per Å^2-surface area.

$10 \, \mathrm{kJnm}^{-2}$. Therefore, the total hydrophobic contribution to free energy due to folding is:

$$\Delta G_{HP} = 88 \, M + 79 \cdot 10^{-5} \, M^2 \; [\mathrm{J \cdot mol^{-1}}] \qquad (1.92)$$

This relation is valid for a range of $6108 \leq M \leq 34{,}409$, but appears to be also valid for larger molecules since they often consist of several loose associations of independent globular portions called structural domains (Fig. 1.26).

Proteins with disulfide bonds fold at a significantly slower rate than those without disulfide bonds. Folding is not limited by the reaction rate of disulfide formation. Therefore the folding process of disulfide-containing proteins seems to proceed in a different way. The reverse process, protein unfolding, is very much slowed down by the presence of disulfide bridges which generally impart great stability to globular proteins. This stability is particularly effective against denaturation. An example is the *Bowman-Birk* inhibitor from soybean (Fig. 1.27) which inhibits the activity of trypsin and chymotrypsin. Its tertiary structure is stabilized by seven disulfide bridges. The reactive sites

of inhibition are $\mathrm{Lys^{16}\text{-}Ser^{17}}$ and $\mathrm{Leu^{43}\text{-}Ser^{44}}$, i.e. both sites are located in relatively small rings, each of which consists of nine amino acid residues held in ring form by a disulfide bridge. The thermal stability of this inhibitor is high.

As examples of the folding of globular proteins, Fig. 1.28 shows schematically the course of the peptide chains in the β-chain of hemoglobin, in triosephosphate isomerase and carboxypeptidase. Other protein conformations are shown in the following figures:

– Fig. 8.7 (cf. 8.8.4): Thaumatin and monellin (two-dimensional)
– Fig. 8.8 (cf. 8.8.5): Thaumatin and monellin (three-dimensional)
– Fig. 11.3 (cf. 11.2.3.1.4): Lysozyme

1.4.2.3.3 BSE

The origination of transmissible spongiform encephalopathies (TESs) is explained by a change in the protein conformation. (The name refers to the spongy deformations which occur in the brain in this disease. The resulting defects interrupt the transmission of signals). One of the TESs is bovine spongiform encephalopathy (BSE). According to the current hypothesis, TESs are caused by pathogenic prion proteins (PrPp), which can be present in the animal meal used as feed. PrPp are formed from normal prion proteins (PrPn) found in all mammalian cells.

Fig. 1.26. Globular protein with two-domain structure (according to *Schulz* and *Schirmer*, 1979)

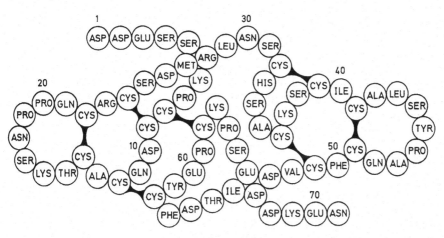

Fig. 1.27. *Bowman-Birk* inhibitor from soybean (according to *Ikenaka* et al., 1974)

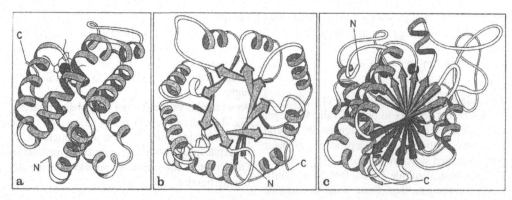

Fig. 1.28. Tertiary structures (schematic: spiral: α-helix, arrow: pleated sheet) of the β-chain of hemoglobin (**a**), of triosephosphate isomerase (**b**) and carboxypeptidase (**c**). (according to *Walton*, 1981)

However, a PrPp forces the pathogenic conformation on a PrPn. The stability towards serine proteinase K from the fungus *Tritirachium album* is used to differentiate between PrPp and PrPn. Serine proteinase K, which attacks the carboxyl side of hydrophobic amino acids, largely hydrolyzes PrPn while a characteristic peptide (M_r 27-30 kDa) is released from PrPp. This marker can be identified using the sandwich ELISA (cf. 2.6.3).

1.4.2.3.4 Quaternary Structures

In addition to the free energy gain by folding of a single peptide chain, association of more than one peptide chain (subunit) can provide further gains in free energy. For example, hemoglobin (4 associated peptide chains) $\Delta G^0 = -46\,\text{kJ mole}^{-1}$ and the trypsin-trypsin inhibitor complex (association of 2 peptide chains) $\Delta G^0 = -75.2\,\text{kJ mole}^{-1}$. In principle such associations correspond to the folding of a larger peptide chain with several structural domains without covalently binding the subunits. Table 1.26 lists some proteins which partially exhibit quaternary structures.

1.4.2.4 Denaturation

The term denaturation denotes a reversible or irreversible change of native conformation (tertiary structure) without cleavage of covalent bonds (except for disulfide bridges). Denaturation is possible with any treatment that cleaves hydrogen

Table 1.26. Examples of globular proteins

Name	Origin	Molecular weight (Kdal)	Number of subunits
Lysozyme	Chicken egg	14.6	1
Papain	*Papaya latex*	20.7	1
α-Chymotrypsin	Pancreas (beef)	23	1
Trypsin	Pancreas (beef)	23.8	1
Pectinesterase	Tomato	27.5	
Chymosin	Stomach (calf)	31	
β-Lactoglobulin	Milk	35	2
Pepsin A	Stomach (swine)	35	1
Peroxidase	Horseradish	40	1
Hemoglobin	Blood	64.5	4
Avidin	Chicken egg	68.3	4
Alcohol-dehydrogenase	Liver (horse)	80	2
	Yeast	150	4
Hexokinase	Yeast	104	2
Lactate dehydrogenase	Heart (swine)	135	4
Glucose oxidase	*P. notatum*	152	
Pyruvate kinase	Yeast	161	8
	A. niger	186	
β-Amylase	Sweet potato	215	4
Catalase	Liver (beef)	232	4
	M. lysodeikticus	232	
Adenosine triphosphatase	Heart (beef)	284	6
Urease	Jack beans	483	6
Glutamine synthetase	*E. coli*	592	12
Arginine decarboxylase	*E. coli*	820	10

bridges, ionic or hydrophobic bonds. This can be accomplished by: changing the temperature, adjusting the pH, increasing the interface area, or adding organic solvents, salts, urea, guanidine hyrochloride or detergents such as sodium dodecyl sulfate. Denaturation is generally reversible when the peptide chain is stabilized in its unfolded state by the denaturing agent and the native conformation can be reestablished after removal of the agent. Irreversible denaturation occurs when the unfolded peptide chain is stabilized by interaction with other chains (as occurs for instance with egg proteins during boiling). During unfolding reactive groups, such as thiol groups, that were buried or blocked, may be exposed. Their participation in the formation of disulfide bonds may also cause an irreversible denaturation.

An aggregation of the peptide chains caused by the folding of globular proteins is connected with reduced solubility or swellability. Thus the part of wheat gluten that is soluble in acetic acid diminishes as heat stress increases (Fig. 1.29). As a result of the reduced rising capacity of gluten caused by the pre-treatment, the volume of bread made of recombined flours is smaller (Fig. 1.30). In the case of fibrous proteins, denaturation, through destruction of the highly ordered structure, generally leads to increased solubility or

rising capacity. One example is the thermally caused collagen-to-gelatin conversion, which occurs when meat is cooked (cf. 12.3.2.3.1).

The thermal denaturation of the whey proteins β-lactoglobulin and α-lactalbumin has been well-studied. The data in Table 1.27 based on reaction kinetics and the *Arrhenius* diagram (Fig. 1.31) indicate that the activation energy of the overall reaction in the range of 80–90 °C changes. The higher E_a values at lower temperatures must be attributed to folding, which is the partial reaction that determines the reaction rate at temperatures <90 °C. At higher temperatures (>95 °C), the aggregation to which the lower activation energy corresponds predominates.

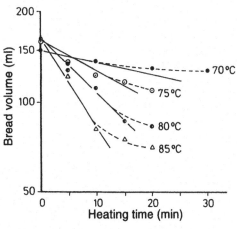

Fig. 1.30. Volume of white bread of recombined flours using thermally treated liquid gluten (wheat) (according to *Pence* et al., 1953)

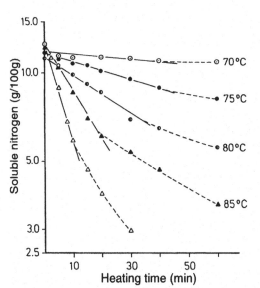

Fig. 1.29. Solubility of moist gluten (wheat) in diluted acetic acid after various forms of thermal stress (according to *Pence* et al., 1953)

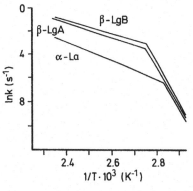

Fig. 1.31. *Arrhenius* diagram for the denaturation of the whey proteins β-lactoglobulin A, β-lactoglobulin B and α-lactalbumin B (according to *Kessler*, 1988)

Table 1.27. Denaturation of β-lactoglobulins A and B (β-LG-A, β-LG-B) and of α-lactalbumin (a-LA)

Protein	n	ϑ (°C)	E_a (kJ mol^{-1})	$\ln(k_o)$ (s^{-1})	$\Delta S^{\neq}$ (kJ mol^{-1}) K^{-1}
β-LG-A	1.5	70–90	265.21	84.16	0.445
		95–150	54.07	14.41	−0.136
β-LG-B	1.5	70–90	279.96	89.43	0.487
		95–150	47.75	12.66	−0.150
α-LA	1.0	70–80	268.56	84.92	0.452
		85–150	69.01	16.95	−0.115

n: reaction order, δ: temperature, E_a: activation energy, k_o: reaction rate constant, $\Delta S^{\neq}$: activation entropy.

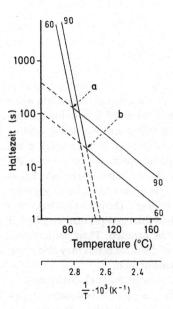

Fig. 1.32. Lines of equal denaturation degrees of β-lactoglobulin B. [The *steeper lines* correspond to the folding (60%, 90%), the *flatter lines* to aggregation (60%, 90%); at point a, 60% are folded and 90% can be aggregated, corresponding to an overall reaction of 60%; at point b, 90% are folded and 60% can be aggregated, corresponding to an overall reaction of 60%; according to *Kessler*, 1988]

The values in Table 1.27 determined for activation entropy also support the above mentioned attribution. In the temperature range of 70–90 °C, $\Delta S^{\#}$ is always positive, which indicates a state of greater disorder than should be expected with the predominance of the folding reaction. On the other hand, the negative $\Delta S^{\#}$ values at 95–105 °C indicate a state of greater order than should be expected considering that aggregation predominates in this temperature range. Detailed studies of the kind described above allow optimal control of thermal processes. In the case of milk processing, the data have made it possible, for example, to avoid the separation of whey proteins in heating equipment and to optimize the properties of yogurt gels (cf. 10.1.3.3 and 10.2.1.2).

Figure 1.32 shows the denaturation of β-LG in a diagram that combines the heating period with the temperature (cf. 2.5.4.3) in the form of straight lines of equal denaturation degrees. This allows us to read directly the time/temperature combinations required for a certain desired effect. At 85 °C/136 s for example, only 60% of the β-LG-B are folded, so that only 60% can aggregate, although 90% would be potentially able to aggregate: at this temperature, the folding determines the overall reaction, as shown above. Conversely, 90% of the protein is potentially folded at 95 °C/21 s while only 60% can be aggregated. At this temperature, aggregation determines the overall reaction.

Denaturation of biologically active proteins is usually associated with loss of activity. The fact that denatured proteins are more readily digested by proteolytic enzymes is also of interest.

1.4.3 Physical Properties

1.4.3.1 Dissociation

Proteins, like amino acids, are amphoteric. Depending on pH, they can exist as polyvalent cations, anions or zwitter ions. Proteins differ in their α-carboxyl and α-amino groups – since these groups are linked together by peptide bonds, the uptake or release of protons is limited to free terminal groups. Therefore, most of the dissociable functional groups are derived from side chains. Table 1.28 lists pK values of some protein groups. In contrast to free amino acids, these values fluctuate greatly for proteins since the dissociation is influenced by neighboring groups in the macromolecule. For example, in lysozyme the γ-carboxyl group of Glu[35] has a pK

Table 1.28. pK values of protein side chains

Group	pK (25 °C)	Group	pK (25 °C)
α-Carboxyl-	3–4	Imidazolium-	4–8
β, γ-Carboxyl-	3–5	Hydroxy-	
α-Ammonium-	7–8	(aromatic)	9–12
ε-Ammonium-	9–11	Thiol	8–11
Guanidinium-	12–13		

of 6–6.5, while the pK of the β-carboxyl group of Asp^{66} is 1.5–2, of Asp^{52} is 3–4.6 and of Asp^{101} is 4.2–4.7.

The total charge of a protein, which is the absolute sum of all positive and negative charges, is differentiated from the so-called net charge which, depending on the pH, may be positive, zero or negative. By definition the net charge is zero and the total charge is maximal at the isoelectric point. Lowering or raising the pH tends to increase the net charge towards its maximum, while the total charge always becomes less than at the isolectric point.

Since proteins interact not only with protons but also with other ions, there is a further differentiation between an isoionic and an isoelectric point. The isoionic point is defined as the pH of a protein solution at infinite dilution, with no other ions present except for H^+ and HO^-. Such a protein solution can be acquired by extensive dialysis (or, better, electrodialysis) against water. The isoionic point is constant for a given substance while the isoelectric point is variable depending on the ions present and their concentration. In the presence of salts, i. e. when binding of anions is stronger than that of cations, the isoelectric point is lower than the isoionic point. The reverse is true when cationic binding is dominant. Figure 1.33 shows the shift in pH of an isoionic serum albumin solution after addition of various salts. The shift in pH is consistently positive, i. e. the protein binds more anions than cations.

The titration curve of β-lactoglobulin at various ionic strengths (Fig. 1.34) shows that the isoelectric point of this protein, at pH 5.18, is independent of the salts present. The titration curves are, however, steeper with increasing ionic strength, which indicates greater suppression of the electrostatic interaction between protein molecules.

At its isoelectric point a protein is the least soluble and the most likely to precipitate ("isoelectric precipitation") and is at its maximal crystallization capacity. The viscosity of solubilized proteins and the swelling power of insoluble proteins are at a minimum at the isoelectric point.

When the amino acid composition of a protein is known, the isoelectric point can be estimated according to the following formula:

$$pI = -10 \log Q_{pI} + 7.0 \qquad (1.93)$$

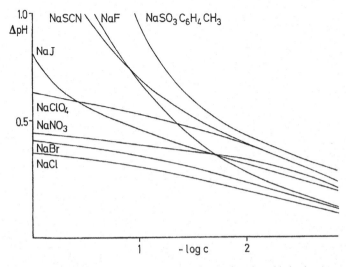

Fig. 1.33. pH-shift of isoionic serum albumin solutions by added salts. (according to *Edsall* and *Wymann*, 1958)

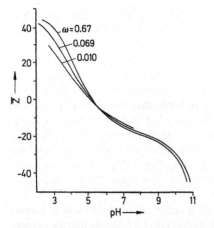

Fig. 1.34. Titration curves for β-lactoglobulin at various ionic strengths ω. (according to *Edsall* and *Wyman*, 1958)

where QpI is the sum of deviations of the isoelectric points of all participating amino acids from the neutral point:

$$Q_{pI} = \frac{4.2 \cdot n \, Asp + 3.8m \, Glu}{3.8q \, Arg + 2.6r \, Lys + 0.5s \, His} \qquad (1.94)$$

The formula fails when acid or alkaline groups occur in masked form.

1.4.3.2 Optical Activity

The optical activity of proteins is due not only to asymmetry of amino acids but also to the chirality resulting from the arrangement of the peptide chain. Information on the conformation of proteins can be obtained from a recording of the optical rotatory dispersion (ORD) or the circular dichroism (CD), especially in the range of peptide bond absorption wavelengths (190–200 nm). The *Cotton* effect occurs in this range and reveals quantitative information on secondary structure. An α-helix or a β-structure gives a negative *Cotton* effect, with absorption maxima at 199 and 205 nm respectively, while a randomly coiled conformation shifts the maximum to shorter wavelengths, i.e. results in a positive *Cotton* effect (Fig. 1.35).

1.4.3.3 Solubility, Hydration and Swelling Power

Protein solubility is variable and is influenced by the number of polar and apolar groups and their arrangement along the molecule. Generally, proteins are soluble only in strongly

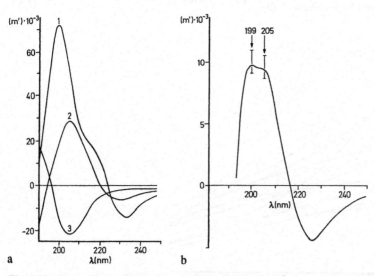

Fig. 1.35. *Cotton* effect. **a** Polylysine α-helix (1, pH 11–11.5) β-sheet structure (2, pH 11–11.3 and heated above 50 °C) and random coiled (3, pH 5–7). **b** Ribonuclease with 20% α-helix, 40% β-sheet structure and 40% random coiled region. (according to *Luebke, Schroeder,* and *Kloss,* 1975)

polar solvents such as water, glycerol, formamide, dimethylformamide or formic acid.
In a less polar solvent such as ethanol, proteins are rarely noticeably soluble (e. g.
prolamines). The solubility in water is dependent on pH and on salt concentration.
Figure 1.36 shows these relationships for
β-lactoglobulin.

At low ionic strengths, the solubility rises with
increase in ionic strength and the solubility minimum (isoelectric point) is shifted from pH 5.4 to
pH 5.2. This shift is due to preferential binding of
anions to the protein.

If a protein has enough exposed hydrophobic
groups at the isoelectric point, it aggregates
due to the lack of electrostatic repulsion
via intermolecular hydrophobic bonds, and
(isoelectric) precipitation will occur. If on
the other hand, intermolecular hydrophobic interactions are only poorly developed,
a protein will remain in solution even at the
isoelectric point, due to hydration and steric
repulsion.

As a rule, neutral salts have a two-fold effect on
protein solubility. At low concentrations they increase the solubility ("salting in" effect) by suppressing the electrostatic protein-protein interaction (binding forces).

The log of the solubility (S) is proportional
to the ionic strength (μ) at low concentrations
(cf. Fig. 1.36.):

$$\log S = k \cdot \mu. \tag{1.95}$$

Protein solubility is decreased ("salting out" effect) at higher salt concentrations due to the ion
hydration tendency of the salts. The following relationship applies (S_0: solubility at $\mu = 0$; K: salting out constant):

$$\log S = \log S_0 - K \cdot \mu. \tag{1.96}$$

Cations and anions in the presence of the same
counter ion can be arranged in the following orders (*Hofmeister* series) based on their salting out
effects:

$$K^+ > Rb^+ > Na^+ > Cs^+ > Li^+ > NH_4^+;$$
$$SO_4^{2-} > citrate^{2-} > tratrate^{2-} > acetate^-$$
$$> Cl^- > NO_3^- > Br^- > J^- > CNS^-. \tag{1.97}$$

Multivalent anions are more effective than monovalent anions, while divalent cations are less effective than monovalent cations.

Since proteins are polar substances, they are
hydrated in water. The degree of hydration
(g water of hydration/g protein) is variable.
It is 0.22 for ovalbumin (in ammonium sulfate), 0.06 for edestin (in ammonium sulfate),
0.8 for β-lactoglobulin and 0.3 for hemoglobin.
Approximately 300 water molecules are sufficient to cover the surface of lysozyme (about
$6000 \, Å^2$), that is one water molecule per
$20 \, Å^2$.

The swelling of insoluble proteins corresponds
to the hydration of soluble proteins in that
insertion of water between the peptide chains
results in an increase in volume and other
changes in the physical properties of the protein. For example, the diameter of myofibrils
(cf. 12.2.1) increases to 2.5 times the original
value during rinsing with 1.0 mol/L NaCl,
which corresponds to a six-fold volume increase (cf. 12.5). The amount of water taken
up by swelling can amount to a multiple of
the protein dry weight. For example, muscle
tissue contains 3.5–3.6 g water per g protein dry
matter.

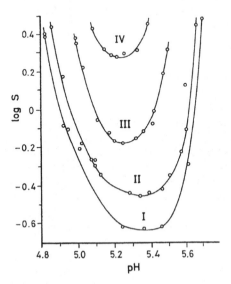

Fig. 1.36. β-Lactoglobulin solubility as affected by pH
and ionic strength I. 0.001, II. 0.005, III. 0.01, IV. 0.02

The water retention capacity of protein can be estimated with the following formula:

$$a = f_c + 0.4 f_p + 0.2 f_n \qquad (1.98)$$

(a: g water/g protein; f_c, f_p, f_n: fraction of charged, polar, neutral amino acid residues).

1.4.3.4 Foam Formation and Foam Stabilization

In several foods, proteins function as foam-forming and foam-stabilizing components, for example in baked goods, sweets, desserts and beer. This varies from one protein to another. Serum albumin foams very well, while egg albumin does not. Protein mixtures such as egg white can be particularly well suited (cf. 11.4.2.2). In that case, the globulins facilitate foam formation. Ovomucin stabilizes the foam, egg albumin and conalbumin allow its fixation through thermal coagulation.

Foams are dispersions of gases in liquids. Proteins stabilize by forming flexible, cohesive films around the gas bubbles. During impact, the protein is adsorbed at the interface via hydrophobic areas; this is followed by partial unfolding (surface denaturation). The reduction of surface tension caused by protein adsorption facilitates the formation of new interfaces and further gas bubbles. The partially unfolded proteins associate while forming stabilizing films.

The more quickly a protein molecule diffuses into interfaces and the more easily it is denatured there, the more it is able to foam. These values in turn depend on the molecular mass, the surface hydrophobicity, and the stability of the conformation.

Foams collapse because large gas bubbles grow at the expense of smaller bubbles (disproportionation). The protein films counteract this disproportionation. That is why the stability of a foam depends on the strength of the protein film and its permeability for gases. Film strength depends on the adsorbed amount of protein and the ability of the adsorbed molecules to associate. Surface denaturation generally releases additional amino acid side chains which can enter into intermolecular interactions. The stronger the cross-linkage, the more stable the film.

Since the smallest possible net charge promotes association, the pH of the system should lie in the range of the isoelectric points of the proteins that participate in film formation.

In summary, the ideal foam-forming and foam-stabilizing protein is characterized by a low molecular weight, high surface hydrophobicity, good solubility, a small net charge in terms of the pH of the food, and easy denaturability.

Foams are destroyed by lipids and organic solvents such as higher alcohols, which due to their hydrophobicity displace proteins from the gas bubble surface without being able to form stable films themselves. Even a low concentration of egg yolk, for example, prevents the bursting of egg white. This is attributed to a disturbance of protein association by the lecithins.

The foam-forming and foam-stabilizing characteristics of proteins can be improved by chemical and physical modification. Thus a partial enzymatic hydrolysis leads to smaller, more quickly diffusing molecules, better solubility, and the release of hydrophobic groups. Disadvantages are the generally lower film stability and the loss of thermal coagulability. The characteristics can also be improved by introducing charged or neutral groups (cf. 1.4.6.2) and by partial thermal denaturation (e. g. of whey proteins). Recently, the addition of strongly alkaline proteins (e. g. clupeines) is being tested, which apparently increases the association of protein in the films and allows the foaming of fatty systems.

1.4.3.5 Gel Formation

Gels are disperse systems of at least two components in which the disperse phase in the dispersant forms a cohesive network. They are characterized by the lack of fluidity and elastic deformability. Gels are placed between solutions, in which repulsive forces between molecules and the disperse phase predominate, and precipitates, where strong intermolecular interactions predominate. We differentiate between two types of gel, the *polymeric networks* and the *aggregated dispersions*, although intermediate forms are found as well.

Examples of *polymeric networks* are the gels formed by gelatin (cf. 12.3.2.3.1) and polysaccharides such as agarose (cf. 4.4.4.1.2) and carrageenan (4.4.4.3.2). Formation of a three-dimensional network takes place through the aggregation of unordered fibrous molecules via partly ordered structures, e. g. while double helices are formed (cf. 4.4.4.3.2, Fig. 4.14, Fig. 12.21). Characteristic for gels of this type is the low polymer concentration ($\sim$1%) as well as transparency and fine texture. Gel formation is caused by setting a certain pH, by adding certain ions, or by heating/cooling. Since aggregation takes place mostly via intermolecular hydrogen bonds which easily break when heated, polymeric networks are thermo-reversible, i. e. the gels are formed when a solution cools, and they melt again when it is heated.

Examples of *aggregated dispersions* are the gels formed by globular proteins after heating and denaturation. The thermal unfolding of the protein leads to the release of amino acid side chains which may enter into intermolecular interactions. The subsequent association occurs while small spherical aggregates form which combine into linear strands whose interaction establishes the gel network. Before gel can be formed in the unordered type of aggregation, a relatively high protein concentration (5–10%) is necessary. The aggregation rate should also be slower than the unfolding rate, since otherwise coarse and fairly unstructured gels are formed, such as in the area of the iso-electric point. The degree of denaturation necessary to start aggregation seems to depend on the protein. Since partial denaturation releases primarily hydrophobic groups, intermolecular hydrophobic bonds generally predominate, which results in the thermoplastic (thermo-irreversible) character of this gel type, in contrast to the thermoreversible gel type stabilized by hydrogen bonds. Thermoplastic gels do not liquefy when heated, but they can soften or shrink. In addition to hydrophobic bonds, disulfide bonds formed from released thiol groups can also contribute to cross-linkage, as can intermolecular ionic bonds between proteins with different isoelectric points in heterogeneous systems (e. g. egg white).

Gel formation can be improved by adding salt. The moderate increase in ionic strength increases interaction between charged macro-molecules or molecule aggregates through charge shielding without precipitation occurring. An example is the heat coagulation of soybean curd (tofu,

cf. 16.3.1.2.3) which is promoted by calcium ions.

1.4.3.6 Emulsifying Effect

Emulsions are disperse systems of one or more immiscible liquids. They are stabilized by emulsifiers – compounds which form interface films and thus prevent the disperse phases from flowing together (cf. 8.15). Due to their amphipathic nature, proteins can stabilize o/w emulsions such as milk (cf. 10.1.2.3). This property is made use of on a large scale in the production of food preparations.

The adsorption of a protein at the interface of an oil droplet is thermodynamically favored because the hydrophobic amino acid residues can then escape the hydrogen bridge network of the surrounding water molecules. In addition, contact of the protein with the oil droplet results in the displacement of water molecules from the hydrophobic regions of the oil-water boundary layer. Therefore, the suitability of a protein as an emulsifier depends on the rate at which it diffuses into the interface and on the deformability of its conformation under the influence of interfacial tension (surface denaturation). The diffusion rate depends on the temperature and the molecular weight, which in turn can be influenced by the pH and the ionic strength. The adsorbability depends on the exposure of hydrophilic and hydrophobic groups and thus on the amino acid profile, as well as on the pH, the ion strength and the temperature. The conformative stability depends in the amino acid composition, the molecular weight and the intramolecular disulfide bonds. Therefore, a protein with ideal qualities as an emulsifier for an oil-in-water emulsion would have a relatively low molecular weight, a balanced amino acid composition in terms of charged, polar and nonpolar residues, good water solubility, well-developed surface hydrophobicity, and a relatively stable conformation. The β-casein molecule meets these requirements because of less pronounced secondary structures and no crosslinks due to the lack of SH groups (cf. 10.1.2.1.1). The apolar "tail" of this flexible molecule is adsorbed by the oil phase of the boundary layer and the polar "head", which projects into the aqueous medium, prevents coalescence.

The solubility and emulsifying capacity of some proteins can be improved by limited enzymatic hydrolysis.

1.4.4 Chemical Reactions

The chemical modification of protein is of importance for a number of reasons. It provides derivatives suitable for sequence analysis, identifies the reactive groups in catalytically active sites of an enzyme, enables the binding of protein to a carrier (protein immobilization) and provides changes in protein properties which are important in food processing. In contrast to free amino acids and except for the relatively small number of functional groups on the terminal amino acids, only the functional groups on protein side chains are available for chemical reactions.

1.4.4.1 Lysine Residue

Reactions involving the lysine residue can be divided into several groups, (a) reactions leading to a positively charged derivative, (b) reactions eliminating the positive charge, (c) derivatizations introducing a negative charge, and (d) reversible reactions. The latter are of particular importance.

1.4.4.1.1 Reactions Which Retain the Positive Charge

Alkylation of the free amino group of lysine with aldehydes and ketones is possible, with a simultaneous reduction step:

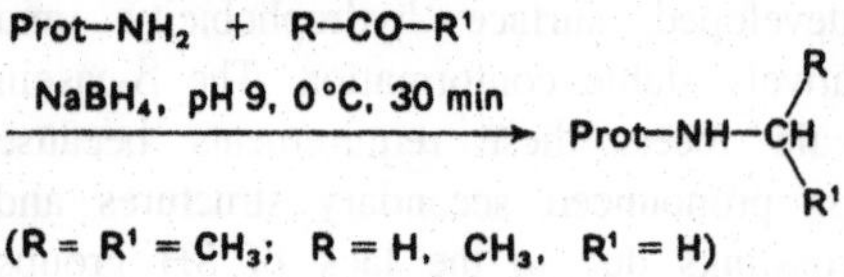

$$\text{Prot-NH}_2 \ + \ \text{R-CO-R}^1$$

$$\xrightarrow{\text{NaBH}_4, \ \text{pH 9, 0°C, 30 min}} \text{Prot-NH-CH}\overset{\displaystyle R}{\underset{\displaystyle R^1}{}}$$

$$(R = R^1 = CH_3; \ R = H, \ CH_3, \ R^1 = H)$$

$$(1.99)$$

A dimethyl derivative [Prot−N(CH$_3$)$_2$] can be obtained with formaldehyde (R=R$_1$=H) (cf. 1.2.4.2.2).

Guanidination can be accomplished by using O-methylisourea as a reactant. α-Amino groups react at a much slower rate than ε-amino groups:

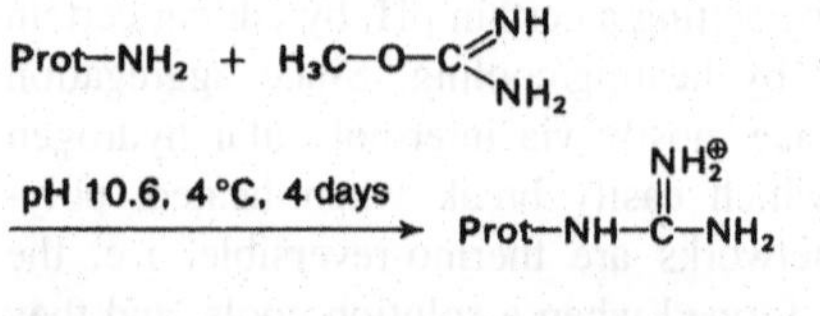

$$\text{Prot-NH}_2 \ + \ \text{H}_3\text{C-O-C}\overset{\displaystyle NH}{\underset{\displaystyle NH_2}{}}$$

$$\xrightarrow{\text{pH 10.6, 4°C, 4 days}} \text{Prot-NH-}\overset{\displaystyle NH_2^{\oplus}}{\underset{}{C}}\text{-NH}_2 \qquad (1.100)$$

This reaction is used analytically to assess the amount of biologically available ε-amino groups and for measuring protein digestibility.

Derivatization with imido esters is also possible. The reactant is readily accessible from the corresponding nitriles:

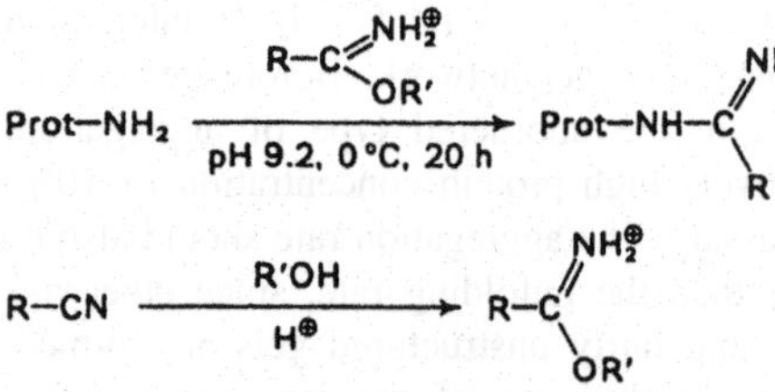

$$\text{Prot-NH}_2 \ \xrightarrow[\text{pH 9.2, 0°C, 20 h}]{R-C\overset{NH_2^{\oplus}}{\underset{OR'}{}}} \ \text{Prot-NH-C}\overset{NH_2^{\oplus}}{\underset{R}{}}$$

$$\text{R-CN} \ \xrightarrow[\text{H}^{\oplus}]{\text{R'OH}} \ \text{R-C}\overset{NH_2^{\oplus}}{\underset{OR'}{}}$$

$$(1.101)$$

Proteins can be cross-linked with the use of a bifunctional imido ester (cf. 1.4.4.10).

Treatment of the amino acid residue with amino acid carboxyanhydrides yields a polycondensation reaction product:

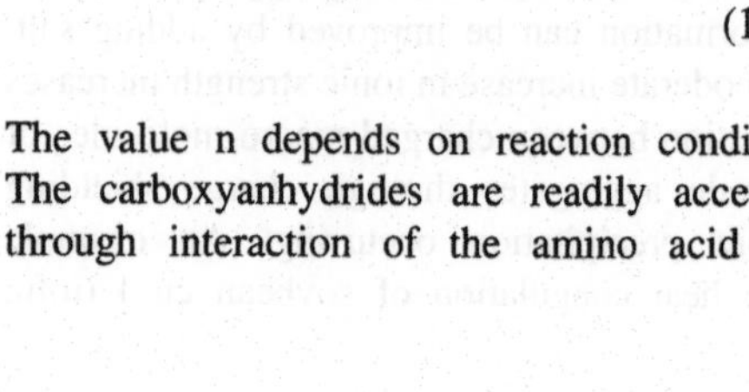

$$\text{Protein-NH}_2 \ + \ \longrightarrow$$

$$\text{Protein-NH-[CO-CHR-NH]}_n\text{-CO-CHR-NH}_2$$

$$(1.102)$$

The value n depends on reaction conditions. The carboxyanhydrides are readily accessible through interaction of the amino acid with phosgene:

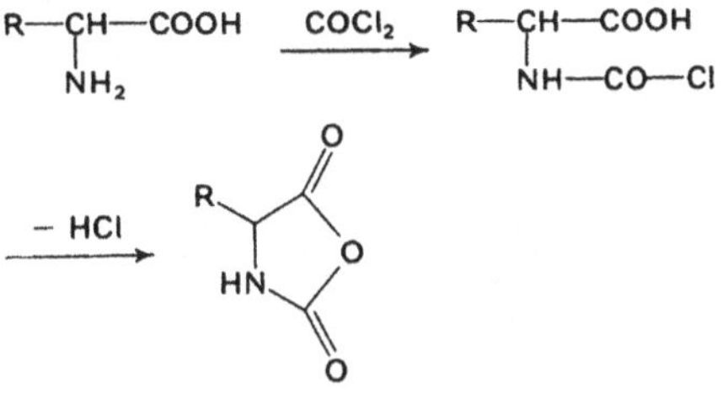

R—CH—COOH   COCl₂   R—CH—COOH
   | |
   NH₂   NH—CO—Cl

$$\xrightarrow{-\ HCl}$$

(structure: oxazolidine-2,5-dione with R)

(1.103)

1.4.4.1.2 Reactions Resulting in a Loss of Positive Charge

Acetic anhydride reacts with lysine, cysteine, histidine, serine, threonine and tyrosine residues. Subsequent treatment of the protein with hydroxylamine (1 M, 2 h, pH 9, 0 °C) leaves only the acetylated amino groups intact:

$$\text{Prot—NH}_2 \xrightarrow[\text{pH 7--9.5, 0 °C}]{(CH_3CO)_2O} \text{Prot—NH—CO—CH}_3$$

(1.104)

Carbamoylation with cyanate attacks α- and ε-amino groups as well as cysteine and tyrosine residues. However, their derivatization is reversible under alkaline conditions:

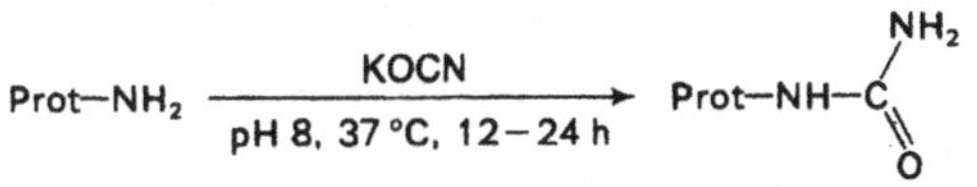

$$\text{Prot—NH}_2 \xrightarrow[\text{pH 8, 37 °C, 12--24 h}]{KOCN} \text{Prot—NH—C}\overset{NH_2}{\underset{O}{<}}$$

(1.105)

Arylation with 1-fluoro-2,4-dinitrobenzene (*Sanger's* reagent; FDNB) and trinitrobenzene sulfonic acid was outlined in Section 1.2.4.2.2. FDNB also reacts with cysteine, histidine and tyrosine.
4-Fluoro-3-nitrobenzene sulfonic acid, a reactant which has good solubility in water, is also of interest for derivatization of proteins:

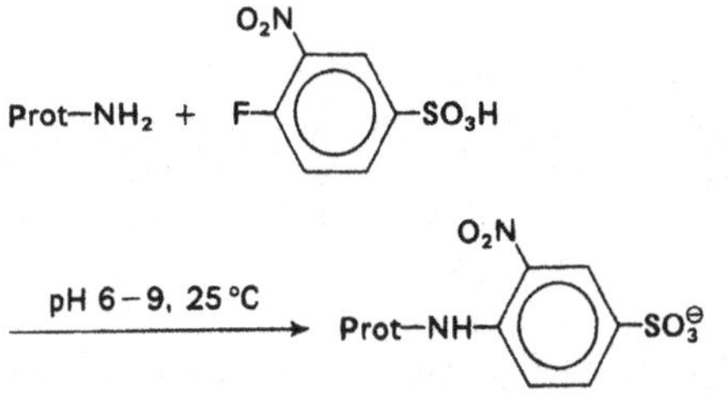

Prot—NH₂ + F—(O₂N-phenyl)—SO₃H

$$\xrightarrow[\text{pH 6--9, 25 °C}]{} \text{Prot—NH—(O}_2\text{N-phenyl)—SO}_3^{\ominus}$$

(1.106)

Deamination can be accomplished with nitrous acid:

$$\text{Prot—NH}_2 \xrightarrow[\text{pH 4.35, 0 °C}]{HNO_2} \text{Prot—OH + N}_2$$

(1.107)

This reaction involves α- and ε-amino groups as well as tryptophan, tyrosine, cysteine and methionine residues.

1.4.4.1.3 Reactions Resulting in a Negative Charge

Acylation with dicarboxylic acid anhydrides, e. g. succinic acid anhydride, introduces a carboxyl group into the protein:

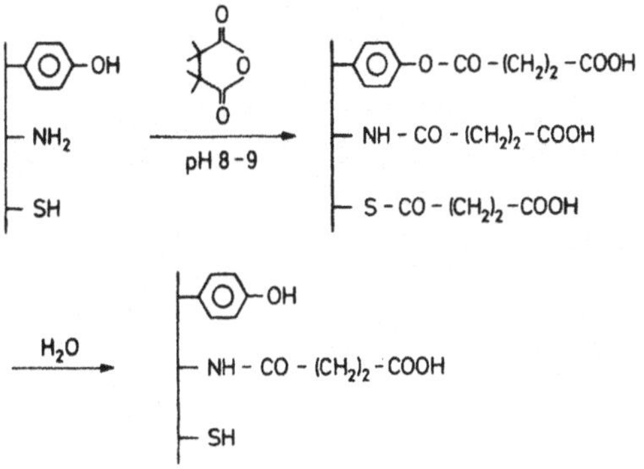

(reaction scheme, pH 8-9):

—OH
—NH₂   →   —O-CO-(CH₂)₂-COOH
—SH   —NH-CO-(CH₂)₂-COOH
               —S-CO-(CH₂)₂-COOH

$$\xrightarrow{H_2O}$$

—OH
—NH-CO-(CH₂)₂-COOH
—SH

(1.108)

Introduction of a fluorescent acid group is possible by interaction of the protein with pyridoxal phosphate followed by reduction of the intermediary *Schiff* base:

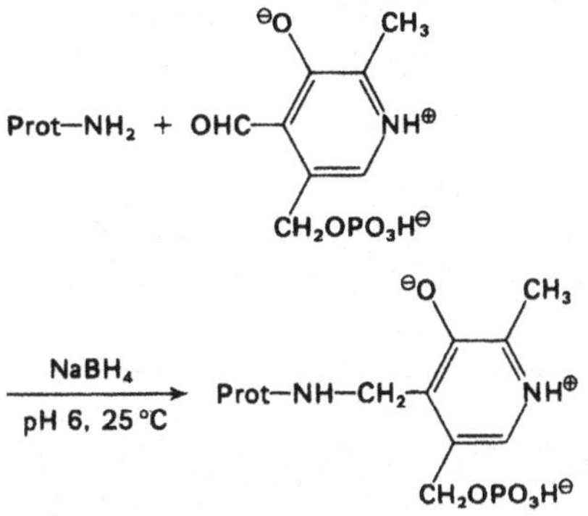

Prot—NH₂ + OHC—(pyridoxal phosphate structure)

$$\xrightarrow[\text{pH 6, 25 °C}]{NaBH_4} \text{Prot—NH—CH}_2\text{—(reduced pyridoxal phosphate structure)}$$

(1.109)

1.4.4.1.4 Reversible Reactions

N-Maleyl derivatives of proteins are obtained at alkaline pH by reaction with maleic acid anhydride. The acylated product is cleaved at pH < 5, regenerating the protein:

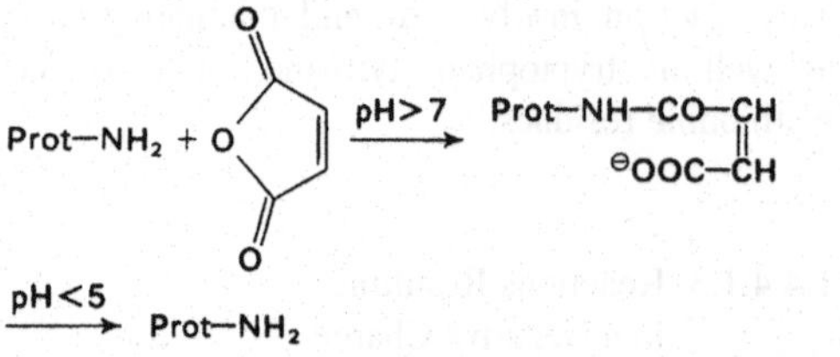

$$
\text{Prot}-\text{NH}_2 + \text{O} \xrightarrow{\text{pH} > 7} \text{Prot}-\text{NH}-\text{CO}-\text{CH} \atop {}^{\ominus}\text{OOC}-\text{CH}
$$

$$
\xrightarrow{\text{pH} < 5} \text{Prot}-\text{NH}_2
$$

(1.110)

The half-life (τ) of ε-N-maleyl lysine is 11 h at pH 3.5 and 37 °C. More rapid cleavage is observed with the 2-methyl-maleyl derivative (τ < 3 min at pH 3.5 and 20 °C) and the 2,2,3,3-tetrafluoro-succinyl derivative (τ very low at pH 9.5 and 0 °C). Cysteine binds maleic anhydride through an addition reaction. The S-succinyl derivative is quite stable. This side reaction is, however, avoided when protein derivatization is done with exo-cis-3,6-end-oxo-1,2,3,6- tetrahydrophthalic acid anhydride:

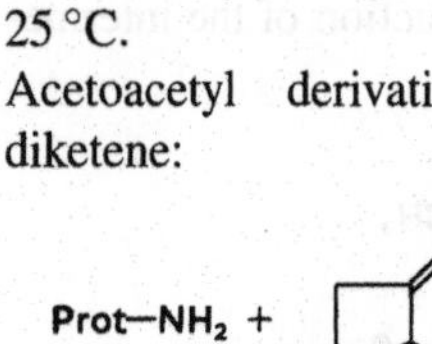

(1.111)

For ε-N-acylated lysine, $\tau = 4-5$ h at pH 3 and 25 °C.

Acetoacetyl derivatives are obtained with diketene:

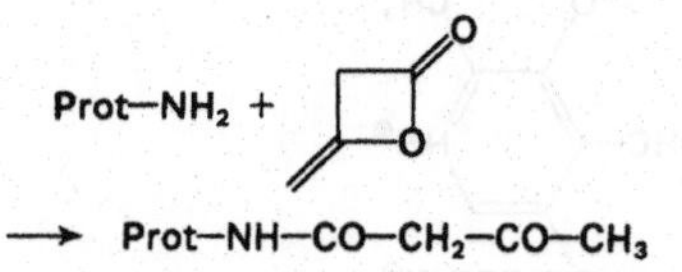

$$
\text{Prot}-\text{NH}_2 +
$$

$$
\longrightarrow \text{Prot}-\text{NH}-\text{CO}-\text{CH}_2-\text{CO}-\text{CH}_3
$$

(1.112)

This is type of reaction also occurs with cysteine and tyrosine residues. The acyl group is readily split from tyrosine at pH 9.5. Complete release of protein from its derivatized form is possible by treatment with phenylhydrazine or hydroxylamine at pH 7.

1.4.4.2 Arginine Residue

The arginine residue of proteins reacts with α- or β-dicarbonyl compounds to form cyclic derivatives:

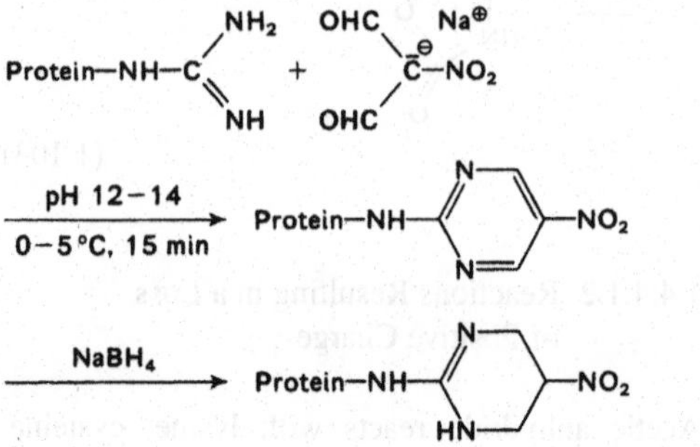

$$
\text{Protein}-\text{NH}-\text{C} {\overset{\text{NH}_2}{\underset{\text{NH}}{}}} + {\overset{\text{OHC}}{\underset{\text{OHC}}{}}} \text{C}-\text{NO}_2 \quad \text{Na}^{\oplus}
$$

$$
\xrightarrow[0-5\,°C,\ 15\ min]{\text{pH } 12-14} \text{Protein}-\text{NH} \quad \text{—NO}_2
$$

$$
\xrightarrow{\text{NaBH}_4} \text{Protein}-\text{NH} \quad \text{—NO}_2
$$

(1.113)

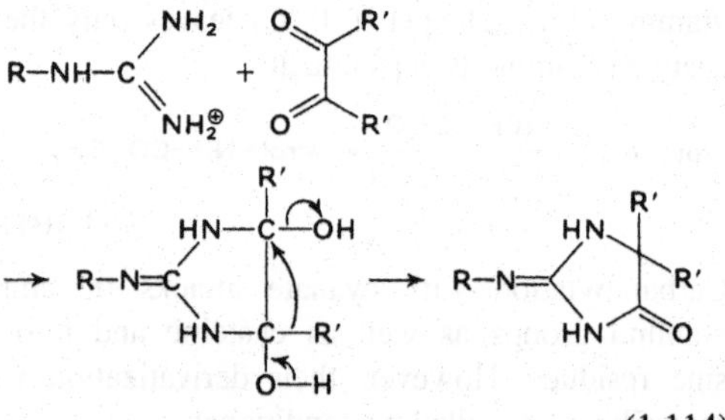

$$
\text{R}-\text{NH}-\text{C} {\overset{\text{NH}_2}{\underset{\text{NH}_2^{\oplus}}{}}} + {\overset{\text{O}}{\underset{\text{O}}{}}} {\overset{\text{R}'}{\underset{\text{R}'}{}}}
$$

$$
\longrightarrow \text{R}-\text{N}=\text{C} \longrightarrow \text{R}-\text{N}=
$$

(1.114)

$$
\text{R}-\text{NH}-\text{C} {\overset{\text{NH}_2}{\underset{\text{NH}_2^{\oplus}}{}}} +
$$

$$
\xrightarrow[25-40\,°C]{\text{pH } 8-9} \text{R}-\text{N}=
$$

$$
\xrightarrow[\text{pH } 7]{\text{H}_2\text{N}-\text{OH}} \text{R}-\text{NH}-\text{C} {\overset{\text{NH}_2}{\underset{\text{NH}_2^{\oplus}}{}}}
$$

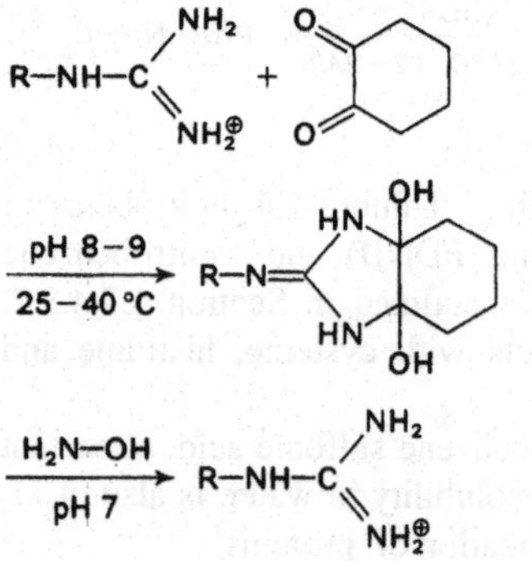

(1.115)

The nitropyrimidine derivative absorbs at 335 nm. The arginyl bond of this derivative is not cleaved by trypsin but it is cleaved in its tetrahydro form, obtained by reduction with NaBH$_4$ (cf. Reaction 1.113). In the reaction with benzil, an iminoimidazolidone derivative is obtained after a benzilic acid rearrangement (cf. Reaction 1.114).

Reaction of the arginine residue with 1,2-cyclohexanedione is highly selective and proceeds under mild conditions. Regeneration of the arginine residue is again possible with hydroxylamine (cf. Reaction 1.115).

1.4.4.3 Glutamic and Aspartic Acid Residues

These amino acid residues are usually esterified with methanolic HCl. There can be side reactions, such as methanolysis of amide derivatives or N,O-acyl migration in serine or threonine residues:

$$\text{Protein—COOH} \xrightarrow[\text{0 °C}]{\text{CH}_3\text{OH/HCl}} \text{Protein—COOCH}_3$$

$$(1.116)$$

Diazoacetamide reacts with a carboxyl group and also with the cysteine residue:

$$\text{Protein—COOH} + \text{N}_2\text{CH—CONH}_2$$
$$\longrightarrow \quad \text{R—COOCH}_2\text{CONH}_2 \qquad (1.117)$$

Amino acid esters or other similar nucleophilic compounds can be attached to a carboxyl group of a protein with the help of a carbodiimide:

$$\text{Protein—COOH} + \text{H}_2\text{N—CH}_2\text{—COOCH}_3$$

$$\xrightarrow{\text{R—N=C=N—R}} \text{Protein—CO—NH—CH}_2\text{—COOCH}_3$$

$$(1.118)$$

Amidation is also possible by activating the carboxyl group with an isooxazolium salt (*Woodward* reagent) to an enolester and its conversion with an amine.

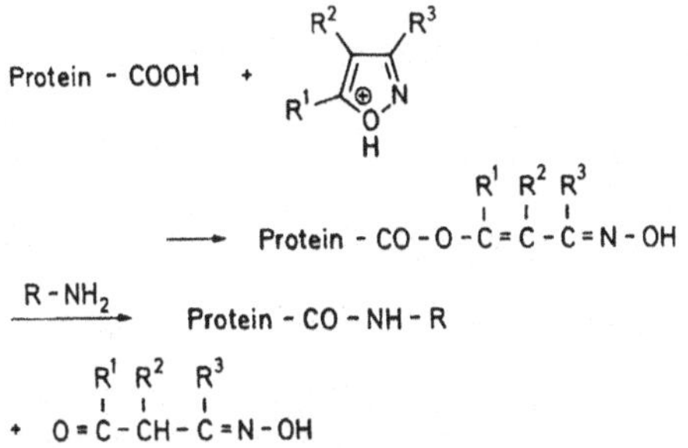

$$\text{Protein - COOH} + $$

$$\longrightarrow \text{Protein - CO - O - } \overset{R^1}{\underset{}{C}} = \overset{R^2}{\underset{}{C}} - \overset{R^3}{\underset{}{C}} = \text{N - OH}$$

$$\xrightarrow{\text{R - NH}_2} \text{Protein - CO - NH - R}$$

$$+ \quad \overset{R^1}{\underset{}{O}} = \overset{R^2}{\underset{}{C}} - \overset{}{\underset{}{CH}} - \overset{R^3}{\underset{}{C}} = \text{N - OH}$$

$$(1.119)$$

1.4.4.4 Cystine Residue (cf. also Section 1.2.4.3.5)

Cleavage of cystine is possible by a nucleophilic attack:

$$\text{Protein—S—S—Protein} + \text{Y}^\ominus$$
$$\longrightarrow \quad \text{Protein—S—Y} + \text{Protein—S}^\ominus \qquad (1.120)$$

The nucleophilic reactivity of the reagents decreases in the series: hydride > arsenite and phosphite > alkanethiol > aminoalkanethiol > thiophenol and cyanide > sulfite > OH⁻ > p-nitrophenol > thiosulfate > thiocyanate. Cleavage with sodium borohydride and with thiols was covered in Section 1.2.4.3.5. Complete cleavage with sulfite requires that oxidative agents (e. g. Cu^{2+}) be present and that the pH be higher than 7:

$$\text{RSSR} + \text{SO}_3^{2\ominus} \longrightarrow \text{RSSO}_3^\ominus + \text{RS}^\ominus$$

$$2\,\text{RS}^\ominus \xrightarrow{Cu^{2\oplus}} \text{RSSR} \qquad (1.121)$$

The resultant S-sulfo derivative is quite stable in neutral and acidic media and is fairly soluble in water. The S-sulfo group can be eliminated with an excess of thiol reagent.

Cleavage of cystine residues with cyanides (nitriles) is of interest since the thiocyanate formed in the reaction is cyclized to a 2-iminothiazolidine derivative with cleavage of the N-acyl bond:

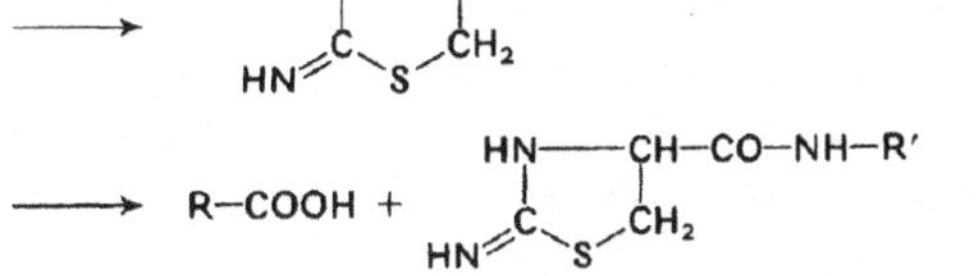

$$(1.122)$$

This reaction can be utilized for the selective cleavage of peptide chains. Initially, all the disulfide bridges are reduced with dithiothreitol, and then are converted to mixed disulfides through

reaction with 5,5′-dithio-bis-(2-nitro-benzoic acid). These mixed disulfides are then cleaved by cyanide at pH 7.
Electrophilic cleavage occurs with Ag^+ and Hg^+ or Hg^{2+} as follows:

$$\begin{aligned} &2\,Ag^{\oplus} + 2\,RSSR \longrightarrow 2\,RSAg + 2\,RS^{\oplus} \\ &2\,RS^{\oplus} + 2\,OH^{\ominus} \longrightarrow 2\,RSOH \longrightarrow RSO_2H + RSH \\ &RSH + Ag^{\oplus} \longrightarrow RSAg + H^{\oplus} \\ \hline &3\,Ag^{\oplus} + 2\,RSSR + 2\,OH^{\ominus} \longrightarrow 3\,RSAg + RSO_2H + H^{\oplus} \end{aligned} \quad (1.123)$$

Electrophilic cleavage with H^+ is possible only in strong acids (e. g. 10 mol/L HCl). The sulfenium cation which is formed can catalyze a disulfide exchange reaction:

$$\begin{aligned} &RSSR + H^{\oplus} \longrightarrow RSH + RS^{\oplus} \\ &RS^{\oplus} + R'SSR' \longrightarrow RSSR' + R'S^{\oplus} \end{aligned} \quad (1.124)$$

In neutral and alkaline solutions a disulfide exchange reaction is catalyzed by the thiolate anion:

$$\begin{aligned} &RSSR + OH^{\ominus} \rightleftharpoons R{-}SOH + RS^{\ominus} \\ &R'SSR' + RS^{\ominus} \rightleftharpoons R'SSR + R'S^{\ominus} \end{aligned} \quad (1.125)$$

1.4.4.5 Cysteine Residue (cf. also Section 1.2.4.3.5)

A number of alkylating agents yield derivatives which are stable under the conditions for acidic hydrolysis of proteins. The reaction with ethylene imine giving an S-aminoethyl derivative and, hence, an additional linkage position in the protein for hydrolysis by trypsin, was mentioned in Section 1.4.1.3. Iodoacetic acid, depending on the pH, can react with cysteine, methionine, lysine and histidine residues:

$$\text{Protein{-}SH} \xrightarrow{ICH_2COOH} \text{Protein{-}S{-}}CH_2{-}COOH \quad (1.126)$$

The introduction of methyl groups is possible with methyl iodide or methyl isourea, and the introduction of methylthio groups with methylthiosulfonylmethane:

$$\text{Protein-SH} \xrightarrow{CH_3I} \text{Protein{—}S{—}}CH_3 \quad (1.127)$$

$$\text{Protein{—}SH} + CH_3{—}O{—}C(NH_2){=}NH_2^{\oplus} \longrightarrow \text{Protein{—}S{—}}CH_3 + O{=}C(NH_2)_2 \quad (1.128)$$

$$\text{Protein{—}SH} + CH_3{—}S{—}SO_2{—}CH_3 \longrightarrow \text{Protein{—}S{—}S{—}}CH_3 + CH_3SO_3H \quad (1.129)$$

Maleic acid anhydride and methyl-p-nitrobenzene sulfonate are also alkylating agents:

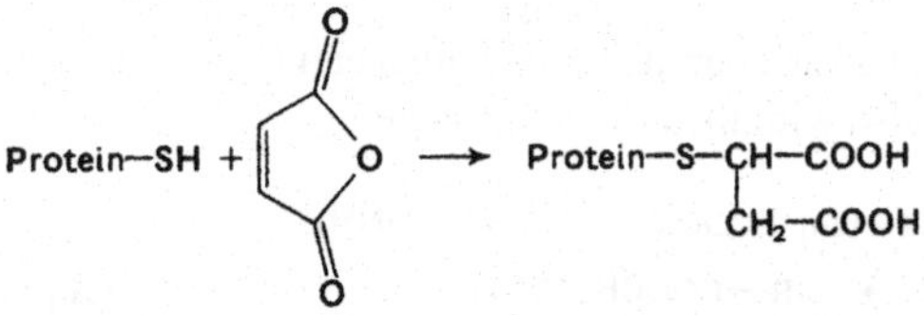

(1.130)

(1.131)

A number of reagents make it possible to measure the thiol group content spectrophotometrically. The molar absorption coefficient, ε, for the derivative of azobenzene-2-sulfenylbromide, ε_{353}, is 16,700 $M^{-1}cm^{-1}$ at pH 1:

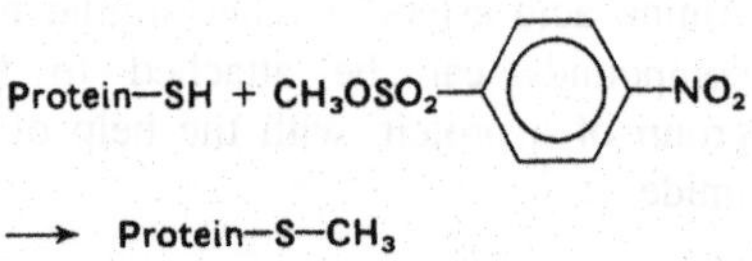

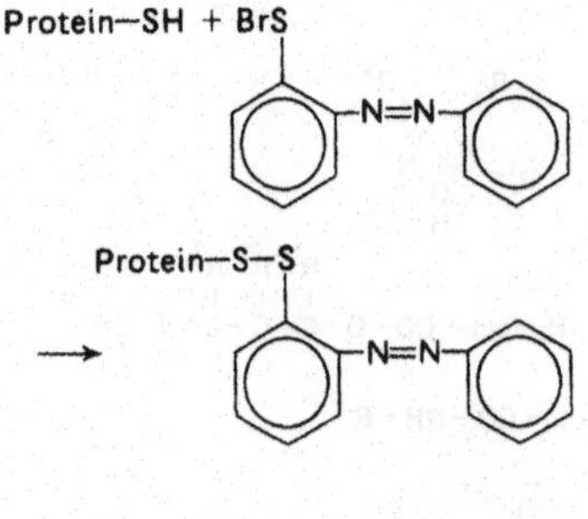

(1.132)

5,5′-Dithiobis-(2-nitrobenzoic acid) has a somewhat lower ε_{412} of 13,600 at pH 8 for its product, a thionitrobenzoate anion:

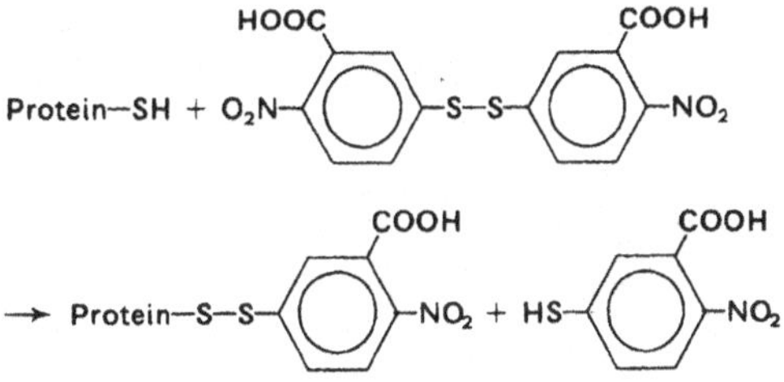

(1.133)

The derivative of p-hydroxymercuribenzoate has an ε_{250} of 7500 at pH 7, while the derivative of N-ethylmaleic imide has an ε_{300} of 620 at pH 7:

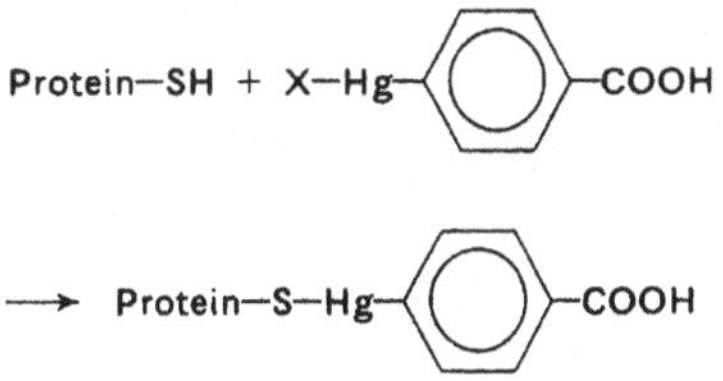

(1.134)

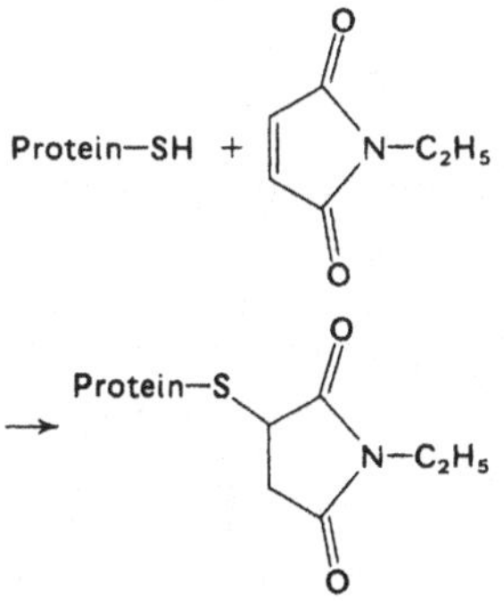

(1.135)

Especially suitable for the specific isolation of cysteine-containing peptides of great sensitivity is N-dimethylaminoazobenzenemaleic acid imide (DABMA).

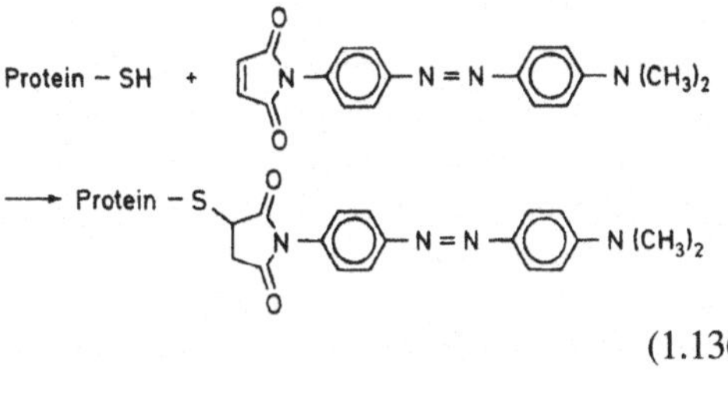

(1.136)

1.4.4.6 Methionine Residue

Methionine residues are oxidized to sulfoxides with hydrogen peroxide. The sulfoxide can be reduced, regenerating methionine, using an excess of thiol reagent (cf. 1.2.4.3.6). α-Halogen carboxylic acids, β-propiolactone and alkyl halogenides convert methionine into sulfonium derivatives, from which methionine can be regenerated in an alkaline medium with an excess of thiol reagent:

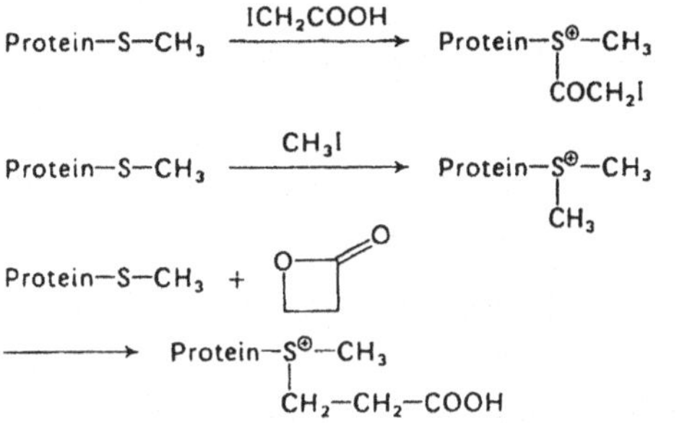

(1.137)

Reaction with cyanogen bromide (BrCN), which splits the peptide bond on the carboxyl side of the methionine molecule, was outlined in Section 1.4.1.3.

1.4.4.7 Histidine Residue

Selective modification of histidine residues present on active sites of serine proteinases is possible. Substrate analogues such as halogenated methyl ketones inactivate such enzymes (for example, 1-chloro-3-tosylamido-7-aminoheptan-2-one inactivates trypsin and 1-chloro-3-tosylamido-4-phenylbutan-2-one inactivates chymotrypsin) by N-alkylation of the histidine residue:

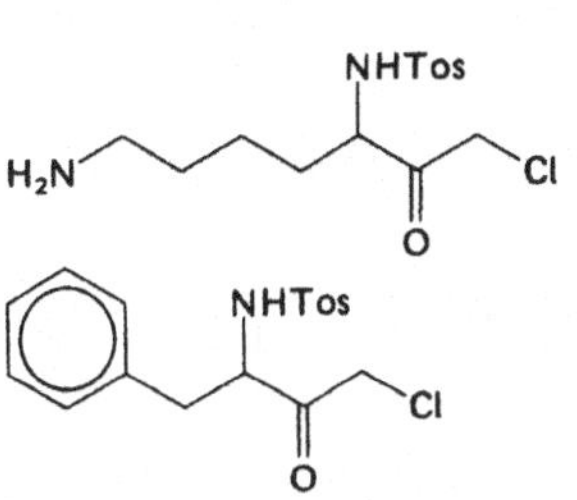

(1.138)

1.4.4.8 Tryptophan Residue

N-Bromosuccinimide oxidizes the tryptophan side chain and also tyrosine, histidine and cysteine:

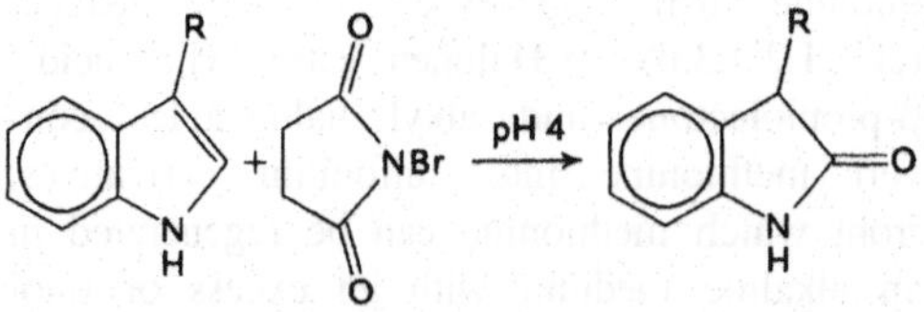

$$(1.139)$$

The reaction is used for the selective cleavage of peptide chains and the spectrophotometric determination of tryptophan.

1.4.4.9 Tyrosine Residue

Selective acylation of tyrosine can occur with acetylimidazole as a reagent:

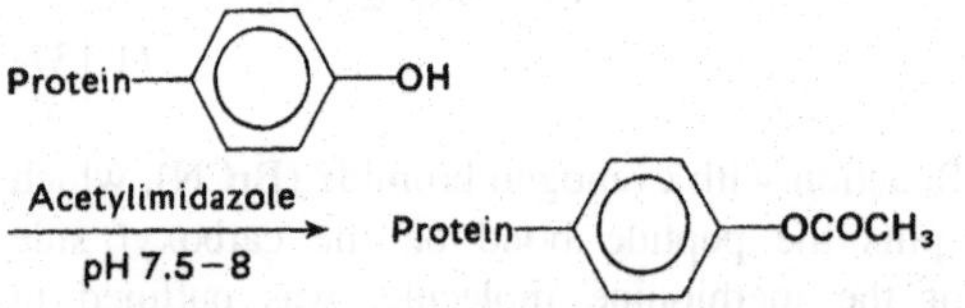

$$(1.140)$$

Diazotized arsanilic acid reacts with tyrosine and with histidine, lysine, tryptophan and arginine:

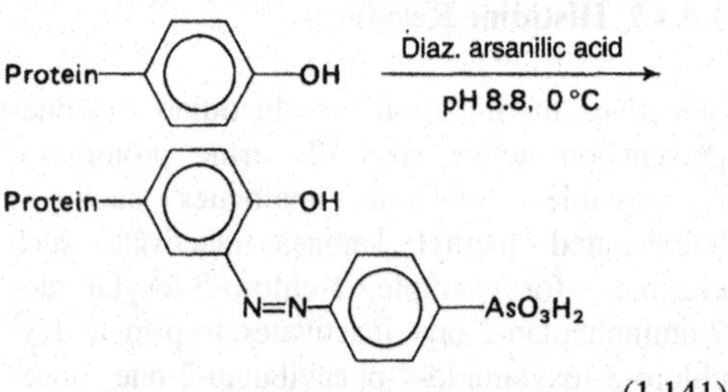

$$(1.141)$$

Tetranitromethane introduces a nitro group into the ortho position:

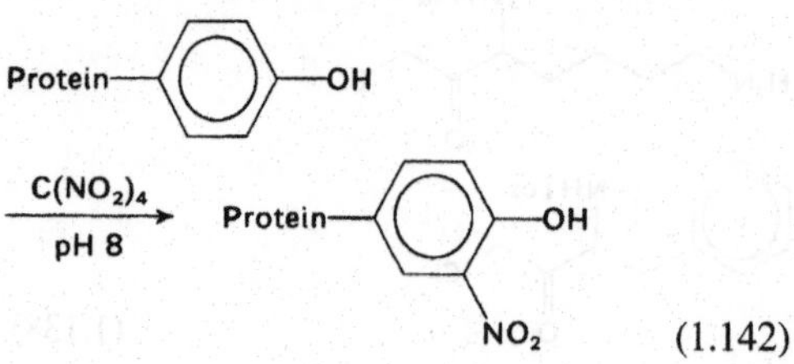

$$(1.142)$$

1.4.4.10 Bifunctional Reagents

Bifunctional reagents enable intra- and inter-molecular cross-linking of proteins. Examples are bifunctional imidoester, fluoronitrobenzene, isocyanate derivatives and maleic acid imides:

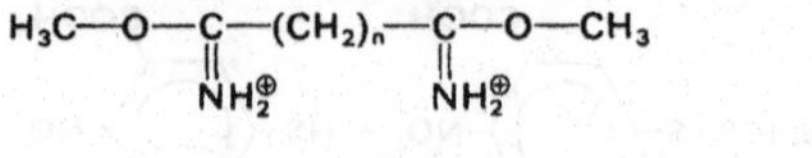

$$(1.143)$$

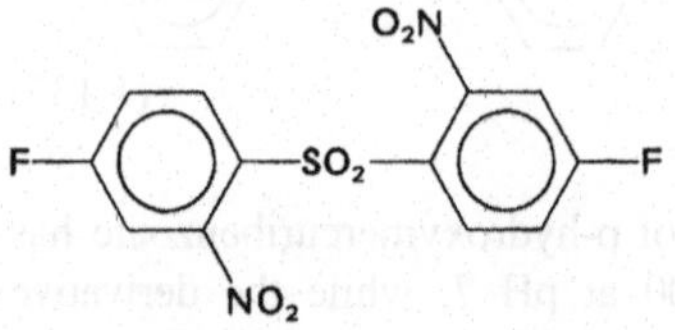

$$(1.144)$$

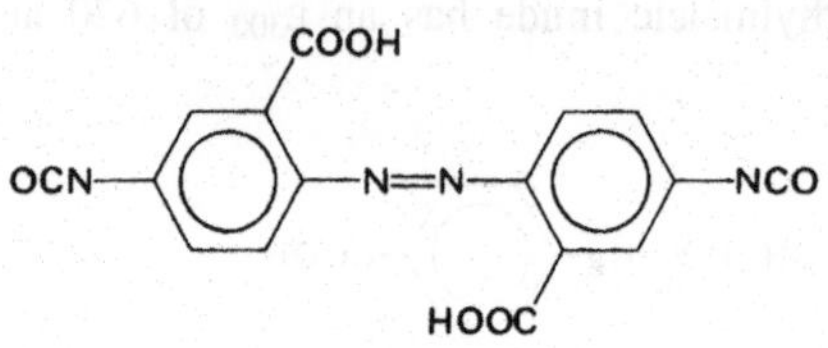

$$(1.145)$$

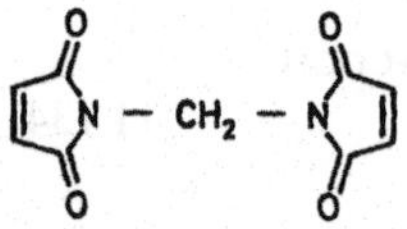

$$(1.146)$$

1.4.4.11 Reactions Involved in Food Processing

The nature and extent of the chemical changes induced in proteins by food processing depend on a number of parameters, for example, composition of the food and processing conditions, such as temperature, pH or the presence of oxygen. As a consequence of these reactions, the biological value of proteins may be decreased:

- Destruction of essential amino acids
- Conversion of essential amino acids into derivatives which are not metabolizable
- Decrease in the digestibility of protein as a result of intra- or interchain cross-linking.

Formation of toxic degradation products is also possible. The nutritional/physiological and toxicological assessment of changes induced by processing of food is a subject of some controversy and opposing opinions.

The *Maillard* reaction of the ε-amino group of lysine prevails in the presence of reducing sugars, for example, lactose or glucose, which yield protein-bound ε-N-deoxylactulosyl-1-lysine or ε-N-deoxyfructosyl-1-lysine, respectively. Lysine is not biologically available in these forms. Acidic hydrolysis of such primary reaction products yields lysine as well as the degradation products furosine and pyridosine in a constant ratio (cf. 4.2.4.4):

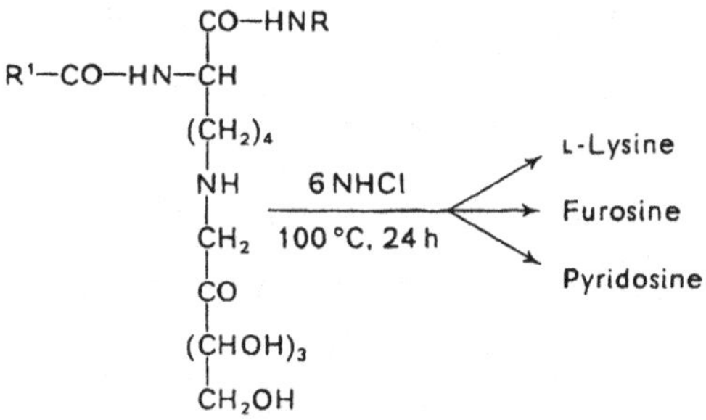

$$
\begin{array}{l}
\text{CO—HNR} \\
\text{R'—CO—HN—CH} \\
\quad\quad\quad\quad (\text{CH}_2)_4 \\
\quad\quad\quad\quad \text{NH} \quad\quad \text{6 NHCl} \\
\quad\quad\quad\quad \text{CH}_2 \quad 100\,°\text{C, 24 h} \\
\quad\quad\quad\quad \text{CO} \\
\quad\quad\quad\quad (\text{CHOH})_3 \\
\quad\quad\quad\quad \text{CH}_2\text{OH}
\end{array}
\longrightarrow
\begin{array}{l}
\text{L-Lysine} \\
\text{Furosine} \\
\text{Pyridosine}
\end{array}
$$

(1.147)

A nonreducing sugar (e. g. sucrose) can also cause a loss of lysine when conditions for sugar hydrolysis are favorable.

Losses of available lysine, cystine, serine, threonine, arginine and some other amino acids occur at higher pH values. Hydrolysates of alkali-treated proteins often contain some unusual compounds, such as ornithine, β-aminoalanine, lysinoalanine, ornithinoalanine, lanthionine, methyllanthionine and D-alloiso-leucine, as well as other D-amino acids.

The formation of these compounds is based on the following reactions: 1,2-elimination in the case of hydroxy amino acids and thio amino acids results in 2-amino-acrylic acid (dehydroalanine) or 2-aminocrotonic acid (dehydro-aminobutyric acid):

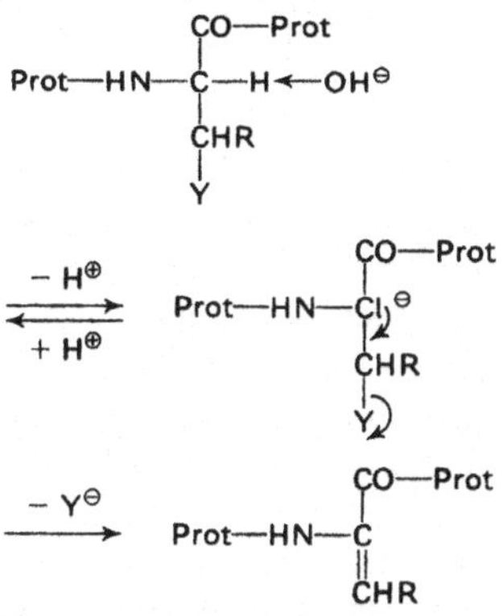

$$
\begin{array}{l}
\quad\quad\quad \text{CO—Prot} \\
\text{Prot—HN—C—H} \leftarrow \text{OH}^\ominus \\
\quad\quad\quad \text{CHR} \\
\quad\quad\quad\quad Y
\end{array}
$$

$$
\underset{+\,H^\oplus}{\overset{-\,H^\oplus}{\rightleftharpoons}}
\begin{array}{l}
\quad\quad\quad \text{CO—Prot} \\
\text{Prot—HN—C}^\ominus \\
\quad\quad\quad \text{CHR} \\
\quad\quad\quad\quad Y
\end{array}
$$

$$
\overset{-\,Y^\ominus}{\longrightarrow}
\begin{array}{l}
\quad\quad\quad \text{CO—Prot} \\
\text{Prot—HN—C} \\
\quad\quad\quad \text{CHR}
\end{array}
$$

R = H, CH₃; Y = OH, OPO₃H₂, SH, SR', SSR'   (1.148)

In the case of cystine, the eliminated thiolcysteine can form a second dehydroalanine residue:

$$
\begin{array}{l}
-\overset{|}{\underset{|}{\text{C}}}-\text{CH}_2-\text{S}-\text{S}^\ominus \\
\quad\, \text{H}
\end{array}
\longrightarrow
\begin{array}{l}
-\overset{|}{\underset{|}{\text{C}}}-\text{CH}_2-\text{S}^\ominus + \text{S}^0 \\
\text{HO}^\ominus \rightarrow \text{H}
\end{array}
$$

$$
\longrightarrow \quad \text{C}=\text{CH}_2 + \text{SH}^\ominus + \text{OH}^\ominus
$$

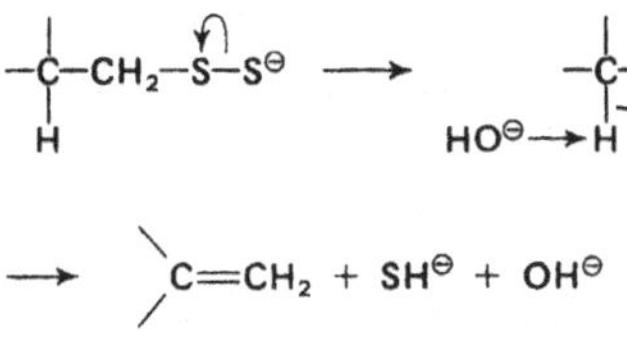

(1.149)

Alternatively, cleavage of the cystine disulfide bond can occur by nucleophilic attack on sulfur, yielding a dehydroalanine residue via thiol and sulfinate intermediates:

$$
\text{R—S—S—R} + 2\,\text{OH}^\ominus \longrightarrow
$$
$$
\text{R—S—O}^\ominus + \text{R—S}^\ominus + \text{H}_2\text{O} \tag{1.150}
$$

$$
\text{R—S—O}^\ominus + \text{R—S—S—R}
$$
$$
\longrightarrow \quad \underset{\underset{O}{\|}}{\text{R—S}}\text{—S—R} + \text{R—S}^\ominus \tag{1.151}
$$

$$
\underset{\underset{O}{\|}}{\text{R—S}}\text{—S—R} + 2\,\text{OH}^\ominus
$$
$$
\longrightarrow \quad \text{R—SO}_2^\ominus + \text{R—S}^\ominus + \text{H}_2\text{O} \tag{1.152}
$$

Intra- and interchain cross-linking of proteins can occur in dehydroalanine reactions involving additions of amines and thiols. Ammonia may also react via an addition reaction:

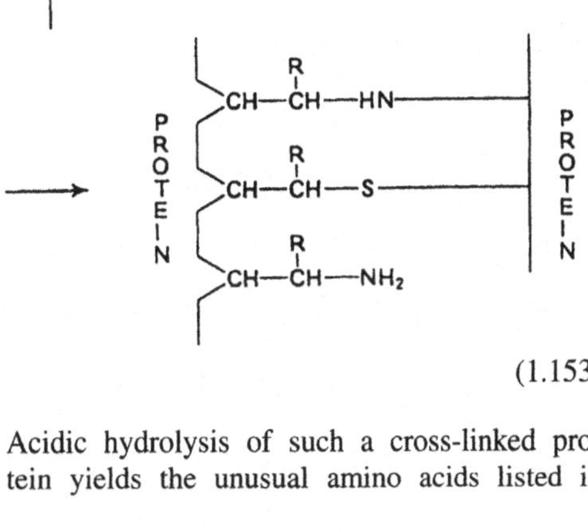

$$
\begin{array}{cc}
\text{PROTEIN}
\left\{
\begin{array}{l}
\text{C}=\text{C}\overset{H}{\underset{R}{}} \\
\text{C}=\text{C}\overset{H}{\underset{R}{}} \\
\text{C}=\text{C}\overset{H}{\underset{R}{}}
\end{array}
\right.
&
\begin{array}{l}
\text{H}_2\text{N}- \\
\text{HS}- \\
+\ \text{NH}_3
\end{array}
\left.\begin{array}{}\end{array}\right\}\text{PROTEIN}
\end{array}
$$

$$
\longrightarrow
\begin{array}{l}
\text{PROTEIN}
\left\{
\begin{array}{l}
\text{CH—CH—HN—} \\
\text{CH—CH—S—} \\
\text{CH—CH—NH}_2
\end{array}
\right.
\end{array}
\left.\begin{array}{}\end{array}\right\}\text{PROTEIN}
$$

(1.153)

Acidic hydrolysis of such a cross-linked protein yields the unusual amino acids listed in

Table 1.29. Ornithine is formed during cleavage of arginine (Reaction 1.54).

$$
\begin{array}{c}
\underset{\substack{\\}}{\text{COOH}}\\
\underset{\substack{\\}}{\text{CHNH}_2}\\
\underset{\substack{\\}}{(\text{CH}_2)_3\ \ \ \text{NH}}\\
\text{NH}-\text{C}\diagup^{\text{NH}_2}
\end{array}
\xrightarrow{\text{OH}^{\ominus}}
\begin{array}{c}
\text{COOH}\\
\text{CHNH}_2\\
(\text{CH}_2)_3\\
\text{NH}_2
\end{array}
+\ \ \text{O}{=}\text{C}\diagup^{\text{NH}_2}_{\diagdown\text{NH}_2}
$$

(1.154)

Formation of D-amino acids occurs through abstraction of a proton via a C2-carbanion. The reaction with L-isoleucine is particularly interesting. L-Isoleucine is isomerized to D-alloisoleucine which, unlike other D-amino acids, is a diastereoisomer and so has a retention time different from L-isoleucine, making its determination possible directly from an amino acid chromatogram:

$$
\begin{array}{c}
\text{CO}-\\
-\text{HN}-\overset{|}{\text{C}}-\text{H}\\
\text{H}_3\text{C}-\overset{|}{\text{C}}-\text{H}\\
\overset{|}{\text{C}_2\text{H}_5}
\end{array}
\underset{+\,\text{H}^{\oplus}}{\overset{-\,\text{H}^{\oplus}}{\rightleftharpoons}}
\begin{array}{c}
\text{CO}-\\
-\text{HN}-\overset{|}{\text{C}}{}^{\ominus}\\
\text{H}_3\text{C}-\overset{|}{\text{C}}-\text{H}\\
\overset{|}{\text{C}_2\text{H}_5}
\end{array}
$$

L-Isoleucine

$$
\underset{-\,\text{H}^{\oplus}}{\overset{+\,\text{H}^{\oplus}}{\rightleftharpoons}}
\begin{array}{c}
\text{CO}-\\
\text{H}-\overset{|}{\text{C}}-\text{NH}-\\
\text{H}_3\text{C}-\overset{|}{\text{C}}-\text{H}\\
\overset{|}{\text{C}_2\text{H}_5}
\end{array}
$$

D-allo-Isoleucine   (1.155)

Heating proteins in a dry state at neutral pH results in the formation of isopeptide bonds between the ε-amino groups of lysine residues and the β- or γ-carboxamide groups of asparagine and glutamine residues:

$$
\begin{array}{ccc}
\overset{P}{\underset{N}{\overset{R}{\underset{I}{\overset{O}{\underset{E}{T}}}}}}
\left\{\begin{array}{l}-\text{NH}_2\\[6pt]-\text{NH}_2\end{array}\right.
&
\begin{array}{l}\text{H}_2\text{NOC}-\text{CH}_2-\\[6pt]\text{H}_2\text{NOC}-(\text{CH}_2)_2-\end{array}
&
\left.\begin{array}{r}\\ \\ \end{array}\right\}\overset{P}{\underset{N}{\overset{R}{\underset{I}{\overset{O}{\underset{E}{T}}}}}}
\end{array}
$$

$$
\xrightarrow{-\,\text{NH}_3}
\begin{array}{ccc}
\overset{P}{\underset{N}{\overset{R}{\underset{I}{\overset{O}{\underset{E}{T}}}}}}
&
\begin{array}{l}-\text{NH}-\text{OC}-\text{H}_2\text{C}-\\[6pt]-\text{NH}-\text{OC}-(\text{H}_2\text{C})_2-\end{array}
&
\overset{P}{\underset{N}{\overset{R}{\underset{I}{\overset{O}{\underset{E}{T}}}}}}
\end{array}
$$

(1.156)

Table 1.29. Formation of unusual amino acids by alkali treatment of protein

Name	Formula	
3-N^6-Lysinoalanine (R = H) 3-N^6-Lysino-3-methyl-alanine(R = CH$_3$)	COOH | CHNH$_2$ | CHR—NH—(CH$_2$)$_4$	COOH | CHNH$_2$ |
3-N^5-Ornithinoalanine (R = H) 3-N^5-Ornithino-3-methylalanine (R = CH$_3$)	COOH | CHNH$_2$ | CHR—NH—(CH$_2$)$_3$	COOH | CHNH$_2$ |
Lanthionine (R = H) 3-Methyllanthionine (R = CH$_3$)	COOH | CHNH$_2$ | CHR——S——CH$_2$	COOH | CHNH$_2$ |
3-Aminoalanine (R =H) 2,3-Diamino butyric acid (R = CH$_3$)	COOH | CHNH$_2$ | CHRNH$_2$	

These isopeptide bonds are cleaved during acidic hydrolysis of protein and, therefore, do not contribute to the occurrence of unusual amino acids. A more intensive heat treatment of proteins in the presence of water leads to a more extensive degradation.

Oxidative changes in proteins primarily involve methionine, which relatively readily forms methionine sulfoxide:

$$
\begin{array}{c}
\text{CO}-\\
-\text{HN}-\overset{|}{\text{C}}-\text{H}\\
\overset{|}{(\text{CH}_2)_2}\\
\overset{|}{\text{S}}\\
\overset{|}{\text{CH}_3}
\end{array}
\longrightarrow
\begin{array}{c}
\text{CO}-\\
-\text{HN}-\overset{|}{\text{C}}-\text{H}\\
\overset{|}{(\text{CH}_2)_2}\\
\overset{|}{\text{S}{=}\text{O}}\\
\overset{|}{\text{CH}_3}
\end{array}
$$

(1.157)

The formation of methionine sulfoxide was observed in connection with lipid peroxidation, phenol oxidation and light exposure in the presence of oxygen and sensitizers such as riboflavin.

After in vivo reduction to methionine, protein-bound methionine sulfoxide is apparently biologically available.

Figure 1.37 shows the effect of alkaline treatment of a protein isolate of sunflower seeds. Serine, threonine, arginine and isoleucine concentrations

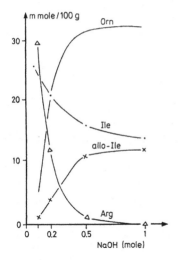

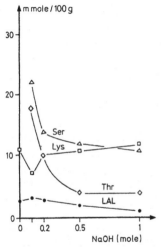

Fig. 1.37. Amino acid contents of a sunflower seed protein isolate heated in sodium hydroxide solutions at 80 °C for 16 h. (according to *Mauron*, 1975)

Table 1.30. Formation of D-amino acids by alkali treatment of proteins[a] (1% solution in 0.1 N NaOH, pH ∼ 12.5, temperature 65 °C)

Protein	Heating time (h)	D-Asp (%)	D-Ala	D-Val	D-Leu	D-Pro	D-Glu	D-Phe
Casein	0	2.2	2.3	2.1	2.3	3.2	1.8	2.8
	1	21.8	4.2	2.7	5.0	3.0	10.0	16.0
	3	30.2	13.3	6.1	7.0	5.3	17.4	22.2
	8	32.8	19.4	7.3	13.6	3.9	25.9	30.5
Wheat	0	3.3	2.0	2.1	1.8	3.2	2.1	2.3
gluten	3	29.0	13.5	3.9	5.6	3.2	25.9	23.3
Promine D								
(soya	0	2.3	2.3	2.6	3.3	3.2	1.8	2.3
protein)	3	30.1	15.8	6.6	8.0	5.8	18.8	24.9
Lactal-	0	3.1	2.2	2.9	2.7	3.1	2.9	2.3
bumin	3	22.7	9.2	4.8	5.8	3.6	12.2	16.5

[a] Results in % correspond to D- + L-amino acids = 100%.

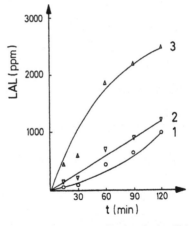

Fig. 1.38. Formation of lysinoalanine (LAL) by heating casein (5% solution at 100 °C) (according to *Sternberg* and *Kim*, 1977) 1 pH 5.0, 2 pH 7.0, 3 pH 8.0

are markedly decreased with increasing concentrations of NaOH. New amino acids (ornithine and alloisoleucine) are formed. Initially, lysine concentration decreases, but increases at higher concentrations of alkali. Lysinoalanine behaves in the opposite manner. The extent of formation of D-amino acids as a result of alkaline treatment of proteins is shown in Table 1.30.

Data presented in Figs. 1.38 and 1.39 clearly show that the formation of lysinoalanine is influenced not only by pH but also by the protein source. An extensive reaction occurs in casein even at pH 5.0 due to the presence of phosphory-

lated serine residues, while noticeable reactions occur in gluten from wheat or in zein from corn only in the pH range of 8–11. Figure 1.40 illustrates the dependence of the reaction on protein concentration.

Table 1.31 lists the contents of lysinoalanine in food products processed industrially or prepared under the "usual household conditions".

The contents are obviously affected by the food type and by the processing conditions.

In the radiation of food, o-hydroxyphenylalanine called o-tyrosine is formed through the re-

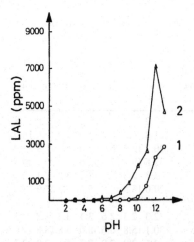

Fig. 1.39. Lysinoalanine (LAL) formation from wheat gluten (2) and corn gluten (1). Protein contents of the glutens: 70%; heated as 6.6% suspension at 100 °C for 4 h. (according to *Sternberg* and *Kim*, 1977)

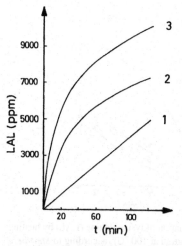

Fig. 1.40. Lysinoalanine (LAL) formation as influenced by casein concentration. (1): 5%, (2): 15%, and (3) 20% all at pH 12.8. (according to *Sternberg* and *Kim*, 1977)

action of phenylalanine with OH-radicals. In hydrolysates, the compound can be detected with the help of HPLC (fluorescence detection or electrochemical detection). It is under discussion as an indicator for food radiation. The amount formed depends on the irradiated dose and on the temperature. In samples of chicken and pork, fish and shrimps, <0.1 mg/kg (non-radiated controls), 0.5–0.8 mg/kg (5 kGy, −18 °C) and 0.8–1.2 mg/kg (5 kGy, 20 °C) were found.

1.4.5 Enzyme-Catalyzed Reactions

1.4.5.1 Foreword

A great number and variety of enzyme-catalyzed reactions are known with protein as a substrate. These include hydrolytic reactions (cleavage of peptide bonds or other linkages, e. g., the ester linkage in a phosphoprotein), transfer reactions (phosphorylation, incorporation of acyl residues, sugar residues and methyl groups) and redox reactions (thiol oxidation, disulfide reduction, amino group oxidation or incorporation of hydroxyl groups). Table 1.32 is a compilation of some examples. Some of these reactions are covered in Section 1.4.6.3 or in the sections related to individual foodstuffs. Only enzymes that are involved in hydrolysis of peptide bonds (proteolytic enzymes, peptidases) will be covered in the following sections.

1.4.5.2 Proteolytic Enzymes

Processes involving proteolysis play a role in the production of many foods. Proteolysis can occur as a result of proteinases in the food itself, e. g., autolytic reactions in meat, or due to microbial proteinases, e. g., the addition of pure cultures of selected microorganisms during the production of cheese.

This large group of enzymes is divided up as shown in Table 1.33. The two subgroups formed are: peptidases (exopeptidases) that cleave amino acids or dipeptides stepwise from the terminal ends of proteins, and proteinases (endopeptidases) that hydrolyze the linkages within the peptide chain, not attacking the terminal peptide bonds. Further division is possible, for example, by taking into account the presence of a given amino acid residue in the active site of the enzyme. The most important types of proteolytic enzymes are presented in the following sections.

1.4.5.2.1 Serine Endopeptidases

Enzymes of this group, in which activity is confined to the pH range of 7–11, are denoted as alkaline proteinases. Typical representatives

Table 1.31. Lysinoalanine content of various foods

Food	Origin/Treatment		Lysinoalanine (mg/kg protein)
Frankfurter	CP[a]	Raw	0
		Cooked	50
		Roasted in oven	170
Chicken drums	CP	Raw	0
		Roasted in oven	110
		Roasted in micro wave oven	200
Egg white, fluid	CP		0
Egg white		Boiled	
		(3 min)	140
		(10 min)	270
		(30 min)	370
		Baked	
		(10 min/150 °C)	350
		(30 min/150 °C)	1100
Dried egg white	CP		160–1820[b]
Condensed milk, sweetened	CP		360–540
Condensed milk, unsweetened	CP		590–860
Milk product for infants	CP		150–640
Infant food	CP		<55 150
Soya protein isolate	CP		0–370
Hydrolyzed vegetable protein	CP		40–500
Cocoa powder	CP		130–190
Na-caseinate	CP		45–6900
Ca-caseinate	CP		250–4320

[a] Commercial product.
[b] Variation range for different brand name products.

from animal sources are trypsin, chymotrypsin, elastase, plasmin and thrombin. Serine proteinases are produced by a great number of bacteria and fungi, e. g. *Bacillus cereus, B. firmus, B. licheniformis, B. megaterium, B. subtilis, Serratia marcescens, Streptomyces fradiae, S. griseus, Trititrachium album, Aspergillus flavus, A. oryzae* and *A. sojae*.

These enzymes have in common the presence of a serine and a histidine residue in their active sites (for mechanism, see 2.4.2.5).

Inactivation of these enzymes is possible with reagents such as diisopropylfluorophosphate (DIFP) or phenylmethanesulfonylfluoride (PMSF). These reagents irreversibly acylate the serine residue in the active site of the enzymes:

$$E - CH_2OH + FY \rightarrow E - CH_2OY + HF$$
$$(Y : -PO(iC_3H_7O)_2 - SO_2 - CH_2C_6H_6)$$

(1.158)

Irreversible inhibition can also occur in the presence of halogenated methyl ketones which alkylate the active histidine residue (cf. 2.4.1.1), or as a result of the action of proteinase inhibitors, which are also proteins, by interaction with the enzyme to form inactive complexes. These natural inhibitors are found in the organs of animals and plants (pancreas, colostrum, egg white, potato tuber and seeds of many legumes; cf. 16.2.3). The specificity of serine

Table 1.32. Enzymatic reactions affecting proteins

Hydrolysis
 – Endopeptidases
 – Exopeptidases
Proteolytic induced aggregation
 – Collagen biosynthesis
 – Blood coagulation
 – Plastein reaction
Cross-linking
 – Disulfide bonds
   Protein disulfide isomerase
   Protein disulfide reductase (NAD(P)H)
   Protein disulfide reductase (glutathione)
   Sulfhydryloxidase
   Lipoxygenase
   Peroxidase
 – ε(γ-Glutamyl)lysine
 – Transglutaminase
 – Aldol-, aldimine condensation and subsequent
   reactions (connective tissue)
   Lysyloxidase
Phosphorylation, dephosphorylation
 – Protein kinase
 – Phosphoprotein phosphatase
Hydroxylation
 – Proline hydroxylase
 – Lysine hydroxylase
Glycosylation
 – Glycoprotein-β-galactosyltransferase
Methylation and demethylation
 – Protein(arginine)-methyl-transferase
 – Protein(lysine)-methyl-transferase
 – Protein-O-methyl-transferase
[2pt] Acetylation, deacetylation
 – ε-N-Acetyl-lysine

endopeptidases varies greatly (cf. Table 1.34). Trypsin exclusively cleaves linkages of amino acid residues with a basic side chain (lysyl or arginyl bonds) and chymotrypsin preferentially cleaves bonds of amino acid residues which have aromatic side chains (phenylalanyl, tyrosyl or tryptophanyl bonds). Enzymes of microbial origin often are less specific.

1.4.5.2.2 Cysteine Endopeptidases

Typical representatives of this group of enzymes are: papain (from the sap of a tropical, melonlike fruit tree, *Carica papaya*), bromelain (from the sap and stem of pineapples, *Ananas comosus*), ficin (from *Ficus latex* and other *Ficus spp.*) and a Streptococcus proteinase. The range of activity of these enzymes is very wide and, depending on the substrate, is pH 4.5–10, with a maximum at pH 6–7.5.

The mechanism of enzyme activity appears to be similar to that of serine endopeptidases. A cysteine residue is present in the active site. A thioester is formed as a covalent intermediary product. The enzymes are highly sensitive to oxidizing agents. Therefore, as a rule they are used in the presence of a reducing agent (e. g., cysteine) and a chelating agent (e. g., EDTA).

Inactivation of the enzymes is possible with oxidative agents, metal ions or alkylating reagents (cf. 1.2.4.3.5 and 1.4.4.5). In general these enzymes are not very specific (cf. Table 1.34).

1.4.5.2.3 Metalo Peptidases

This group includes exopeptidases, carboxypeptidases A and B, aminopeptidases, dipeptidases, prolidase and prolinase, and endopeptidases from bacteria and fungi, such as *Bacillus cereus*, *B. megaterium*, *B. subtilits*, *B. thermoproteolyticus* (thermolysin), *Streptomyces griseus* (pronase; it also contains carboxy- and aminopeptidases) and *Aspergillus oryzae*.

Most of these enzymes contain one mole of Zn^{2+} per mole of protein, but prolidase and prolinase contain one mole of Mn^{2+}. The metal ion acts as a *Lewis* acid in carboxypeptidase A, establishing contact with the carbonyl group of the peptide bond which is to be cleaved. Figure 1.41 shows the arrangement of other participating residues in the active site, as revealed by X-ray structural analysis of the enzyme-substrate complex.

The enzymes are active in the pH 6–9 range; their specificity is generally low (cf. Table 1.34).

Inhibition of these enzymes is achieved with chelating agents (e. g. EDTA) or sodium dodecyl sulfate.

1.4.5.2.4 Aspartic Endopeptidases

Typical representatives of this group are enzymes of animal origin, such as pepsin and rennin

Table 1.33. Classification of proteolytic enzymes (peptidases)

EC-No.[a]	Enzyme group	Comments	Examples
	Exopeptidases	Cleave proteins/peptides stepwise from N- or C-terminals	
3.4.11.	Aminopeptidases	Cleave amino acids from N-terminal	Various aminopeptidases
3.4.13.	Dipeptidases	Cleave dipeptides	Various dipeptidases (carnosinase, anserinase)
3.4.14.	Dipeptidyl- and tripeptidylpeptidases	Cleave di- and tripeptides from N-terminal	Cathepsin C
3.4.15.	Peptidyl-dipeptidases	Cleave dipeptides from C-terminal	Carboxycathepsin,
3.4.16.	Serine carboxypeptidases	Cleave amino acids from C-terminal, serine in the active site	Carboxypeptidase C, cathepsin A
3.4.17.	Metalocarboxypeptidases	Cleave amino acids from C-terminal, Zn^{2+} or Co^{2+} in the active site	Carboxypeptidases A and B
3.4.18.	Cysteine carboxypeptidases	Cleave amino acids from C-terminal, cysteine in the active site	Lysosomal carboxypeptidase B
	Endopeptidases	Cleave protein/peptide bonds other than terminal ones	
3.4.21.	Serine endopeptidase	Serine in the active site	Chymotrypsins A, B and C, peptidase B alkaline proteinases α-and β-trypsin,
3.4.22.	Cysteine endopeptidase	Cysteine in the active site	Papain, ficin, bromelain, cathepsin B
3.4.23.	Aspartic endopeptidase	Aspartic acid (2 residues) in the active site	Pepsin, cathepsin D, rennin (chymosin)
3.4.24.	Metaloendopeptidase	Metal ions in the active site	Collagenase, microbial neutral proteinases

[a] cf. 2.2.7

Table 1.34. Specificity of proteolytic enzymes [based on cleavage of oxidized B chain of bovine insulin; strong cleavage: ↓, weak cleavage (↓)]

No.[a]	Phe	Val	Asn	Gln	His	Leu	Cys	Gly	Ser	His	Leu	Val	Glu	Ala	Leu	Tyr	Leu	Val	Cys	Gly	Glu	Arg	Gly	Phe	Phe	Tyr	Thr	Pro	Lys	Ala
Serine peptidases
1																						→								→
2				→		→										(↓)	→								→	→				
3			→		→	→			→	(↓)					→	→	→↓	→							→	→	→			→
4								(↓)				→																		
5																						→								→
6				(↓)	→	→			→	(↓)			→	→	→	→	→		→		→	→			→	→	→			
7																					(↓)									→
8														→										(↓)						
Cysteine peptidases
9							(↓)	(↓)	(↓)	(↓)	→		(↓)	→	→	→	→	→	→	→	→	→	→	→	→→	→				
10									→	→	→		→	→	→	→	→								→→ (↓)	→ (↓)				
11							(↓)		→				→	→	→	→	→					→			→	→				
Metallopeptidases
12	→		→						→→	→↓					→	→	→→	→					→		→	→→				
13														→			→						→↓	→						
14														→		→	→		→	→		→	→↓	→	→					
Aspartic peptidases
15	(↓)		(↓)		→→	→			(↓)		→	→↓	(↓)	(↓)	→	→	→	→	→	→	→	(↓)	→↓	→						
16	(↓)					(↓)								(↓)	→↓	→	→						→							
17			→						→	→	→↓		→	→	→	→		→					→							
18		→			(↓)	→↓			→→	→	→		→	→	→	→	→↓	→				(↓)	→	→						
19		→				(↓)				→	→	→			→	→	(↓)	→				→		→	→	(↓)				→

---

[a] **Enzymes.**
1) Trypsin (bovine)
2) Chymotrypsin A (bovine)
3) Chymotrypsin B (porcine)
4) Aspergillopeptidase C (*Aspergillus oryzae*)
5) Endopeptidase from *Streptomyces griseus* (trypsin-like)
6) Subtilisin BPN'
7) Endopeptidase from *Aspergillus oryzae*
8) Endopeptidase from *Aspergillus flavus*
9) Papain (*Papaya carica*)
10) Ficin III (*Ficus glabrata*)
11) Chymopapain (*Charica papaya*)
12) Endopeptidase II from (*Aspergillus oryzae*)
13) Endopeptidase from *Bacillus subtilis*
14) Thermolysin (*Bacillus thermoproteolyticus Rokko*)
15) Pepsin A (porcine)
16) Renin (calf)
17) Endopeptidase from *Candida albicans*
18) Endopeptidase from *Mucor miehei*
19) Endopeptidase from *Rhizopus chinensis*

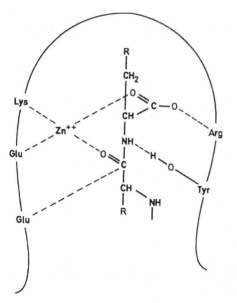

Fig. 1.41. Carboxypeptidase A active site. (according to *Lowe* and *Ingraham*, 1974)

(called Lab-enzyme in Europe), active in the pH range of 2–4, and cathepsin D, which has a pH optimum between 3 and 5 depending on the substrate and on the source of the enzyme. At pH 6–7 rennin cleaves a bond of κ-casein with great specificity, thus causing curdling of milk (cf. 10.1.2.1.1).

Aspartic proteinases of microbial origin can be classified as pepsin-like or rennin-like enzymes. The latter are able to coagulate milk. The pepsin-like enzymes are produced, for example, by *Aspergillus awamori*, *A. niger*, *A. oryzae*, *Penicillium spp.* and *Trametes sanguinea*. The rennin-like enzymes are produced, for example, by *Aspergillus usamii* and *Mucor spp.*, such as *M. pusillus*.

There are two carboxyl groups, one in undissociated form, in the active site of aspartic proteinases. The mechanism postulated for cleavage of peptide bonds is illustrated in Reaction 1.159. The nucleophilic attack of a water molecule on the carbonyl carbon atom of the peptide bond is catalyzed by the side chains of Asp-32 (basic catalyst) and Asp-215 (acid catalyst). The numbering of the amino acid residues in the active site applies to the aspartic proteinase from *Rhizopus chinensis*.

$$(1.159)$$

Inhibition of these enzymes is achieved with various diazoacetylamino acid esters, which apparently react with carboxyl groups on the active site, and with pepstatin. The latter is isolated from various *Streptomyces* as a peptide mixture with the general formula (R: isovaleric or n-caproic acid; AHMHA: 4-amino-3-hydroxy-6-methyl heptanoic acid):

$$\text{R} - \text{Val} - \text{Val} - \text{AHMHA} - \text{Ala} - \text{AHMAH} \tag{1.160}$$

The specifity of aspartic endopeptidases is given in Table 1.34.

1.4.6 Chemical and Enzymatic Reactions of Interest to Food Processing

1.4.6.1 Foreword

Standardization of food properties to meet nutritional/physiological and toxicological demands and requirements of food processing operations is a perennial endeavor. Food production is similar to a standard industrial fabrication process: on the one hand is the food commodity with all its required properties, on the other hand are the components of the product, each of which supplies a distinct part of the required properties. Such considerations have prompted investigations into the relationships in food between macroscopic physical and chemical properties and the structure and reactions at the molecular level. Reliable understanding of such relationships is a fundamental prerequisite for the design and operation of a process, either to optimize the process or to modify the food components to meet the desired properties of the product.

Modification of proteins is still a long way from being a common method in food processing, but it is increasingly being recognized as essential, for two main reasons:

Firstly, proteins fulfill multipurpose functions in food. Some of these functions can be served better by modified than by native proteins.

Secondly, persistent nutritional problems the world over necessiate the utilization of new raw materials.

Modifying reactions can ensure that such new raw materials (e. g., proteins of plant or microbial origin) meet stringent standards of food safety, palatability and acceptable biological value. A review will be given here of several protein modifications that are being used or are being considered for use. They involve chemical or enzymatic methods or a combination of both. Examples have been selected to emphasize existing trends. Table 1.35 presents some protein properties which are of interest to food processing. These properties are related to the amino acid composition and sequence and the conformation of proteins. Modification of the properties of proteins is possible by changing the amino acid composition or the size of the molecule, or by removing or inserting hetero constituents. Such changes can be accomplished by chemical and/or enzymatic reactions. From a food processing point of view, the aims of modification of proteins are:

- Blocking the reactions involved in deterioration of food (e. g., the *Maillard* reaction)
- Improving some physical properties of proteins (e. g., texture, foam stability, whippability, solubility)
- Improving the nutritional value (increasing the extent of digestibility, inactivation of toxic or other undesirable constituents, introducing essential ingredients such as some amino acids).

1.4.6.2 Chemical Modification

Table 1.36 presents a selection of chemical reactions of proteins that are pertinent to and of current importance in food processing.

1.4.6.2.1 Acylation

Treatment with succinic anhydride (cf. 1.4.4.1.3) generally improves the solubility of protein.

Table 1.35. Properties of protein in food

Properties with	
nutritional/physiological relevance	processing relevance
Amino acid composition	Solubility, dispersibility
Availability of amino acids	Ability to coagulate
	Water binding/holding capacity
	Gel formation
	Dough formation, extensibility, elasticity
	Viscosity, adhesion, cohesion
	Whippability
	Foam stabilization
	Emulsifying ability
	Emulsion stabilization

Table 1.36. Chemical reactions of proteins significant in food

Reactive group	Reaction	Product
$-NH_2$	Acylation	$-NH-CO-R$
$-NH_2$	Reductive alkylation with HCHO	$-N(CH_3)_2$
$-CONH_2$	Hydrolysis	$-COOH$
$-COOH$	Esterification	$-COOR$
$-OH$	Esterification	$-O-CO-R$
$-SH$	Oxidation	$-S-S-$
$-S-S-$	Reduction	$-SH$
$-CO-NH-$	Hydrolysis	$-COOH + H_2N-$

For example, succinylated wheat gluten is quite soluble at pH 5 (cf. Fig. 1.40). This effect is related to disaggregation of high molecular weight gluten fractions (cf. Fig. 1.41). In the case of succinylated casein it is obvious that the modification shifts the isoelectric point of the protein (and thereby the solubility minimum) to a lower pH (cf. Fig. 1.42). Succinylation of leaf proteins improves the solubility as well as the flavor and emulsifying properties.

Succinylated yeast protein has not only an increased solubility in the pH range of 4–6, but is also more heat stable above pH 5. It has better emulsifying properties, surpassing many other proteins (Table 1.37), and has increased whippability.

Introduction of aminoacyl groups into protein can be achieved by reactions involving amino acid carboxy anhydrides (Fig. 1.44), amino acids and carbodiimides (Fig. 1.46) or by BOC-amino acid hydroxysuccinimides with subsequent removal of the aminoprotecting group (BOC) (cf. 1.161):

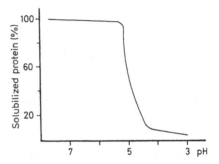

Fig. 1.42. Solubility of succinylated wheat protein as a function of pH (0.5% solution in water). (according to *Grant*, 1973)

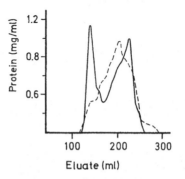

Fig. 1.43. Gel column chromatography of an acetic acid (0.2 mol/L) wheat protein extract. Column: Sephadex G-100 (— before and - - - after succinylation). (according to *Grant*, 1973)

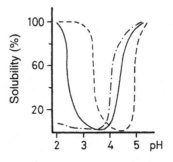

Fig. 1.44. Solubilities of native (—) and succinylated casein (— 50% and — · — · — 76%) as a function of pH. (according to *Schwenke* et al., 1977)

$$\text{BOC} - \text{NH} - \underset{\underset{R}{|}}{\text{CH}} - \text{CO} - \text{O} - \text{N} \qquad + \qquad \text{H}_2\text{N}- \qquad \text{HO}-$$

$$\xrightarrow{\text{pH 9}}$$

$$\text{BOC} - \text{NH} - \underset{\underset{R}{|}}{\text{CH}} - \text{CO} - \text{HN} -$$

$$\text{BOC} - \text{NH} - \underset{\underset{R}{|}}{\text{CH}} - \text{CO} - \text{O} -$$

$$\xrightarrow[\text{NH}_2\text{OH}]{\text{pH 8}}$$

$$\text{BOC} - \text{NH} - \underset{\underset{R}{|}}{\text{CH}} - \text{CO} - \text{HN} -$$

$$\text{HO}-$$

$$\xrightarrow[\text{TFA}]{\text{Anhydrous}}$$

$$\text{H}_2\text{N} - \underset{\underset{R}{|}}{\text{CH}} - \text{CO} - \text{HN} -$$

$$\text{HO}-$$

$$\text{H}_2\text{N} - \underset{\underset{R}{|}}{\text{CH}} - \text{COOH} : \text{Ala, Trp, Gly, Met}$$

(1.161)

Feeding tests with casein with attached methionine, as produced by the above method, have demonstrated a satisfactory availability of methionine (Table 1.38). Such covalent attachment of essential amino acids to a protein may avoid the problems associated with food supplementation with free amino acids: losses in

processing, development of undesired aroma due to methional, etc.

Table 1.39, using β-casein as an example, shows to what extent the association of a protein is affected by its acylation with fatty acids of various chain lengths.

Table 1.37. Emulsifying property of various proteins[a]

Protein	Emulsifying Activity Index $(m^2 \times g^{-1})$	
	pH 6.5	pH 8.0
Yeast protein (88%) succinylated	322	341
Yeast protein (62%) succinylated	262	332
Sodium dodecyl sulfate (0.1%)	251	212
Bovine serum albumin	–	197
Sodium caseinate	149	166
β-Lactoglobulin	–	153
Whey protein powder A	119	142
Yeast protein (24%) succinylated	110	204
Whey protein powder B	102	101
Soya protein isolate A	41	92
Hemoglobin	–	75
Soya protein isolate B	26	66
Yeast protein (unmodified)	8	59
Lysozyme	–	50
Egg albumin	–	49

[a] Protein concentration: 0.5% in phosphate buffer of pH 6.5.

Table 1.38. Feeding trial (rats) with modified casein: free amino acid concentration in plasma and PER value

Diet	μmole/100 ml plasma				
	Lys	Thr	Ser	Gly	Met
Casein	101	19	34	32	5
Met-casein[a]	96	17	33	27	39

	PER[b]
Casein (10%)	2.46
Casein (10%) + Met (0.2%)	3.15
Casein (5%) + Met-casein[a] (5%)	2.92

[a] Covalent binding of methionine to ε-NH_2 groups of casein.
[b] Protein Efficiency Ratio (cf. 1.2.5).

Table 1.39. Association of acylated β-casein A

Protein	SD[a] (%)	Mono-mer (%)	Poly-mer (%)	$S^0_{20,w}$ $(S \cdot 10^{13})$	$S^{1\%}_{20,w}$ $(S \cdot 10^{13})$
β-Casein A (I)	–	11	89	12.6	6.3
Acetyl-I	96	41	59	4.8	4.7
Propionyl-I	97	24	76	10.5	5.4
n-Butyryl-I	80	8	92	8.9	8.3
n-Hexanoyl-I	85	0	100	7.6	11.6
n-Octanoyl-I	89	0	100	6.6	7.0
n-Decanoyl-I	83	0	100	5.0	6.5

[a] Substitution degree.

Gliadin (1.4% Ala)

Edestin (3.8% Ala)

$$+ \quad H_3C-CH$$

pH 6.8, 48 h → Polyalanyl-gliadin (23.6% Ala)
pH 6.8, 0.005 M SDS 72 h ↘ Polyalanyl-edestin (6.2% Ala)

Fig. 1.45. Hydrolysis of a reductively methylated casein by bovine α-chymotrypsin. Modification extents: a 0%, b 33%, and c 52%. (according to *Galembeck* et al., 1977)

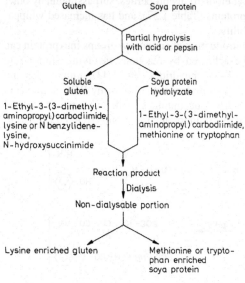

Fig. 1.46. Properties of modified wheat gluten. (according to *Lasztity*, 1975)

1.4.6.2.2 Alkylation

Modification of protein by reductive methylation of amino groups with formaldehyde/NaBH₄ retards *Maillard* reactions. The resultant methyl derivative, depending on the degree of substitution, is less accessible to proteolysis (Fig. 1.47). Hence, its value from a nutritional/physiological point of view is under investigation.

1.4.6.2.3 Redox Reactions Involving Cysteine and Cystine

Disulfide bonds have a strong influence on the properties of proteins. Wheat gluten can be

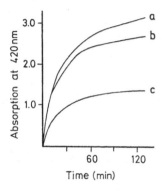

Fig. 1.47. Viscosity curves during reduction of different wheat glutens. For sample designation see Fig. 1.44. (according to *Lasztity*, 1975)

modified by reduction of its disulfide bonds to sulfhydryl groups and subsequent reoxidation of these groups under various conditions (Fig. 1.48). Reoxidation of a diluted suspen-

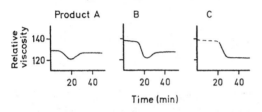

A: readily soluble, soft, adhesive, non-elastic
B: cohesive, elastic
C: sparingly soluble, strong, cohesive and non-elastic

Fig. 1.48. Reaction of proteins with D,L-alanine carboxy anhydride. (according to *Sela* et al., 1962 and *St. Angelo* et al., 1966)

Fig. 1.49. Covalent binding of lysine to gluten (according to *Li-Chan* et al., 1979) and of methionine or tryptophan to soya protein (according to *Voutsinas* and *Nakai*, 1979), by applying a carbodiimide procedure

sion in the presence of urea results in a weak, soluble, adhesive product (gluten A), whereas reoxidation of a concentrated suspension in the presence of a higher concentration of urea yields an insoluble, stiff, cohesive product (gluten C). Additional viscosity data have shown that the disulfide bridges in gluten A are mostly intramolecular while those in gluten C are predominantly intermolecular (Fig. 1.49).

1.4.6.3 Enzymatic Modification

Of the great number of enzymatic reactions with protein as a substrate (cf. 1.4.5), only a small number have so far been found to be suitable for use in food processing.

1.4.6.3.1 Dephosphorylation

Figure 1.50 uses β-casein as an example to show that the solubility of a phosphoprotein in the presence of calcium ions is greatly improved by partial enzymatic dephosphorylation.

1.4.6.3.2 Plastein Reaction

The plastein reaction enables peptide fragments of a hydrolysate to join enzymatically through peptide bonds, forming a larger polypeptide of

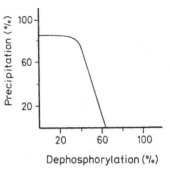

Fig. 1.50. Solubility of β-casein, partially dephosphorylated by phosphoprotein phosphatase: Precipitation: pH 7.1: 2.5 mg/ml protein: 10 mmol/L $CaCl_2$: 35 °C; 1 h. (according to *Yoshikawa* et al., 1974)

about 3 kdal:

$$R-CO-NH-R^1 + E-OH$$
$$\rightleftharpoons R-CO-O-E + H_2N-R^1$$

$$\xrightarrow{H_2O} R-COOH + E-OH$$

$$\xrightarrow{H_2N-R^2} R-CO-NH-R^2 + E-OH$$

$$\tag{1.162}$$

The reaction rate is affected by, among other things, the nature of the amino acid residues. Hydrophobic amino acid residues are preferably linked together (Fig.1.51). Incorporation of amino acid esters into protein is affected by the alkyl chain length of the ester. Short-chain alkyl esters have a low rate of incorporation, while the long-chain alkyl esters have a higher rate of incorporation. This is especially important for the incorporation of amino acids with a short side chain, such as alanine (cf. Table 1.40).

The plastein reaction can help to improve the biological value of a protein. Figure 1.52 shows the plastein enrichment of zein with tryptophan, threonine and lysine. The amino acid composition of such a zein-plastein product is given in Table 1.41.

Enrichment of a protein with selected amino acids can be achieved with the corresponding amino

Table 1.40. Plastein reaction catalyzed by papain: rate of incorporation of amino acid esters[a]

Aminoacyl residue	OEt	OnBu	OnHex	OnOct
L-Ala	0.016	0.054	0.133	0.135
D-Ala	0.0	–	0.0	–
α-Methylala	0.0	–	0.0	–
L-Val	0.005	–	0.077	–
L-Norval	0.122	–	0.155	–
L-Leu	0.119	–	0.140	–
L-Norleu	0.125	–	0.149	–
L-Ile	0.005	–	0.048	–

[a] μ mole $\times$ mg papain$^{-1} \times$ min^{-1}.

acid esters or, equally well, by using suitable partial hydrolysates of another protein.

Figure 1.53 presents the example of soya protein enrichment with sulfur-containing amino acids through "adulteration" with the partial hydrolysate of wool keratin. The PER (protein efficiency ratio) values of such plastein products are significantly improved, as is seen in Table 1.42.

Figure 1.54 shows that the production of plastein with an amino acid profile very close to that recommended by FAO/WHO can be achieved from very diverse proteins.

The plastein reaction also makes it possible to improve the solubility of a protein, for example, by increasing the content of glutamic acid (Fig. 1.55). A soya protein with 25% glutamic acid yields a plastein with 42% glutamic acid.

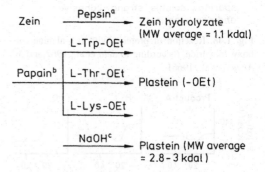

Fig. 1.52. Zein enrichment with Trp, Thr, and Lys by a plastein reaction. (according to *Aso* et al., 1974)
[a] 1% substrate, E/S = 1/50, pH 1.6 at 37 °C for 72 h
[b] 50% substrate, hydrolyzate/AS-OEt = 10/1, E/S = 3/100 at 37 °C for 48 h
[c] 0.1 mol/L in 50% ethanol at 25 °C for 5 h

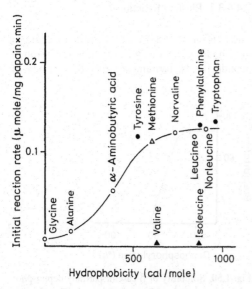

Fig. 1.51. Plastein reaction with papain: incorporation rates of amino acid esters as function of side chain hydrophobicity. (according to *Arai* et al., 1978)

Table 1.41. Amino acid composition of various plasteins (weight-%)

	1	2	3	4	5	6
Arg	1.56	1.33	1.07	1.06	1.35	1.74
His	1.07	0.95	0.81	0.75	0.81	1.06
Ile	4.39	6.39	6.58	5.49	6.23	5.67
Leu	20.18	23.70	23.05	23.75	25.28	23.49
Lys	0.20	0.20	0.24	2.14	3.24	0.19
Phe	6.63	7.26	6.82	7.34	7.22	6.98
Thr	2.40	2.18	9.23	2.36	2.46	2.13
Trp	0.38	9.71	0.25	0.40	0.42	0.33
Val	3.62	5.23	5.77	5.53	6.18	6.20
Met	1.58	1.87	1.67	1.89	2.06	2.04
Cys	1.00	0.58	0.88	0.81	0.78	0.92
Ala	7.56	7.51	8.05	7.97	7.93	8.77
Asp	4.61	3.38	3.42	3.71	3.60	3.91
Glu	21.70	12.48	14.03	14.77	12.95	13.02
Gly	1.48	1.15	1.23	1.29	1.27	1.52
Pro	10.93	8.42	9.10	9.73	9.14	9.37
Ser	4.42	3.40	3.89	3.93	3.74	4.28
Tyr	4.73	5.35	4.97	5.00	6.08	5.54

1) Zein hydrolyzate; 2) Trp-plastein; 3) Thr-plastein; 4) Lys-plastein; 5) Ac-Lys-plastein; 6) Control without addition of amino acid ethyl esters.

Table 1.42. PER-values for various proteins and plasteins

Protein	PER value (rats)
Casein	2.40
Soya protein (I)	1.20
Plastein SW[a] + I (1:2)	2.86
Plastein-Met[b] + I (1:3)	3.38

[a] From hydrolyzate I and wool keratin hydrolyzate.
[b] From hydrolyzate I and Met-OEt. PER (cf. 1.2.5).

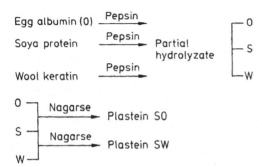

Fig. 1.53. Protein enrichment with sulfur amino acids applying plastein reaction. (according to *Yamashita et al.*, 1971)

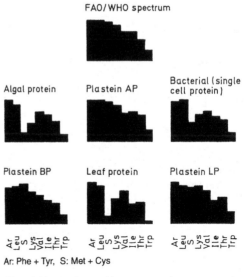

Ar: Phe + Tyr, S: Met + Cys

Fig. 1.54. Amino acid patterns of some proteins and their corresponding plasteins. (according to *Arai et al.*, 1978)

Soya globulin $\xrightarrow{\text{Pepsin}^a}$ Partial hydrolyzate

$\xrightarrow[\text{Papain}^b]{\text{Glu-}\alpha,\gamma\text{-(OEt)}_2}$ Plastein (-OEt)

$\xrightarrow{\text{NaOH}^c}$ Plastein

Fig. 1.55. Soy globulin enrichment with glutamic acid by a plastein reaction. (according to *Yamashita et al.*, 1975)
[a] pH 1.6
[b] Partial hydrolyzate/Glu-α-γ-(OEt)$_2$ = 2:1, substrate concentration: 52.5%, E/S = 1/50, pH 5.5 at 37 °C for 24 h; sample contains 20% acetone
[c] 0.2 mol/L at 25 °C for 2 h

Soya protein has a pronounced solubility minimum in the pH range of 3–6. The minimum is much less pronounced in the case of the unmodified plastein, whereas the glutamic acidenriched soya plastein has a satisfactory solubility over the whole pH range (Fig. 1.56) and is also resistant to thermal coagulation (Fig. 1.57).

Proteins with an increased content of glutamic acid show an interesting sensory effect: partial hydrolysis of modified plastein does not result in a bitter taste, rather it generates a pronounced "meat broth" flavor (Table 1.43).

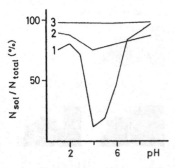

Fig. 1.56. Effect of pH on solubility of soy protein and modified products (1 g/100 ml water). 1 Soy protein, 24.1% Glu; 2 Plastein 24.8% Glu; 3 Glu-plastein with 41.9% Glu. (according to *Yamashita* et al., 1975)

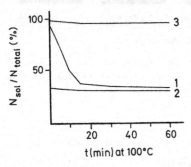

Fig. 1.57. Solubility of soy protein and modified products (800 mg/10 ml water) as a function of heating time at 100 °C. 1 Soy protein 24.1% Glu; 2 Plastein 24.8% Glu; 3 Glu-plastein, 41.9% Glu. (according to *Yamashita* et al., 1975)

Elimination of the bitter taste from a protein hydrolysate is also possible without incorporation of hydrophilic amino acids. Bitter-tasting peptides, such as Leu-Phe, which are released by partial hydrolysis of protein, react preferentially in the subsequent plastein reaction and are incorporated into higher molecular weight peptides with a neutral taste.

The versatility of the plastein reaction is also demonstrated by examples wherein undesired amino acids are removed from a protein. A phenylalanine-free diet, which can be prepared by mixing amino acids, is recommended for certain metabolic defects. However, the use of a phenylalanine-free higher molecular weight peptide is more advantageous with respect to sensory and osmotic properties. Such peptides can be prepared from protein by the plastein reaction. First, the protein is partially hydrolyzed with pepsin. Treatment with pronase under

Table 1.43. Taste of glutamic acid enriched plasteins

Enzyme	pH	Sub-strate[a]	Hydro-lysis[b]	Taste[c] bitter	meat broth type
Pepsin	1.5	G	67	1	1.3
		P	73	4.5	1.0
α-Chymo-	8.0	G	48	1	1.0
trypsin		P	72	4.5	1.0
Molsin	3.0	G	66	1.0	5.0
		P	74	1.3	1.3
Pronase	8.0	G	66	1.0	4.3
		P	82	1.3	1.2

[a] G: Glu-plastein, P: plastein; 1 g/100 ml.
[b] N_{sol} (10% TCA)/N_{total} (%).
[c] 1: no taste, 5: very strong taste.

suitable conditions then preferentially releases amino acids with long hydrophobic side chains. The remaining peptides are separated by gel chromatography and then subjected to the plastein reaction in the presence of added tyrosine and tryptophan (Fig. 1.58). This yields a plastein that is practically phenylalanine-free and has a predetermined ratio of other amino acids, including tyrosine (Table 1.44).

The plastein reaction can also be carried out as a one-step process (Fig. 1.59), thus putting these reactions to economic, industrial-scale use.

1.4.6.3.3 Cross-Linking

Cross-linking between protein molecules is achieved with transglutaminase (cf. 2.7.2.4) and with peroxidase (cf. 2.3.2.2). The cross-linking occurs between tyrosine residues when a protein is incubated with peroxidase/H_2O_2 (cf. Reaction 1.163).

(1.163)

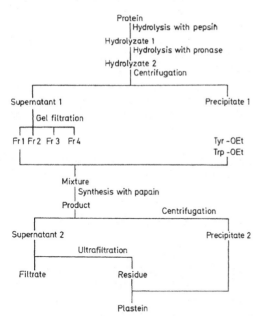

Fig. 1.58. Production of plasteins with high tyrosine and low phenylalanine contents. (according to *Yamashita* et al., 1976)

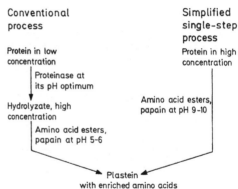

Fig. 1.59. An outline for two- and single-step plastein reactions. (according to *Yamashita* et al., 1979)

Incubation of protein with peroxidase/H_2O_2/catechol also results in cross-linking. The reactions in this case are the oxidative deamination of lysine residues, followed by aldol and aldimine condensations, i. e. reactions analogous to those catalyzed by lysyl oxidase in connective tissue:

$$ (1.164) $$

Table 1.44. Amino acid composition (weight-%) of plasteins with high tyrosine and low phenylalanine contents from fish protein concentrate (FPC) and soya protein isolate (SPI)

Amino acid	FPC	FPC-Plastein	SPI	SPI-Plastein
Arg	7.05	4.22	7.45	4.21
His	2.31	1.76	2.66	1.41
Ile	5.44	2.81	5.20	3.83
Leu	8.79	3.69	6.73	2.43
Lys	10.68	10.11	5.81	3.83
Thr	4.94	4.20	3.58	4.39
Trp	1.01	2.98	1.34	2.80
Val	5.88	3.81	4.97	3.24
Met	2.80	1.90	1.25	0.94
Cys	0.91	1.41	1.78	1.82
Phe	4.30	0.05	4.29	0.23
Tyr	3.94	7.82	3.34	7.96
Ala	6.27	4.82	4.08	2.56
Asp	11.13	13.67	11.51	18.00
Glu	17.14	27.17	16.94	33.56
Gly	4.42	3.94	4.88	3.89
Pro	3.80	4.25	6.27	2.11
Ser	4.59	3.58	5.45	4.67

Table 1.45 presents some of the proteins modified by peroxidase/H_2O_2 treatment and includes their ditryrosine contents.

1.4.7 Texturized Proteins

1.4.7.1 Foreword

The protein produced for nutrition in the world is currently about 20% from animal sources and 80% from plant sources. The plant proteins are primarily from cereals (57%) and oilseed meal (16%). Some nonconventional sources of protein (single cell proteins, leaves) have also acquired some importance.

Table 1.45. Content of dityrosine in some proteins after their oxidation with horseradish peroxidase/H_2O_2 (pH 9.5, 37 °C, 24 h. Substrate/enzyme = 20:1)

Protein	Tyrosine content prior to oxidation (g/100 g protein)	Tyrosine decrease (%)	Dityrosine content (g/100 g protein)
Casein	6.3	21.8	1.37
Soyamine[a]	3.8	11.5	0.44
Bovine serum albumin	4.56	30.7	1.40
Gliadin	3.2	5.4	0.17

[a] Protein preparation from soybean.

Proteins are responsible for the distinct physical structure of a number of foods, e. g. the fibrous structure of muscle tissue (meat, fish), the porous structure of bread and the gel structure of some dairy and soya products.

Many plant proteins have a globular structure and, although available in large amounts, are used to only a limited extent in food processing. In an attempt to broaden the use of such proteins, a number of processes were developed in the mid-1950's which confer a fiber-like structure to globular proteins. Suitable processes give products with cooking strength and a meat-like structure. They are marketed as meat extenders and meat analogues and can be used whenever a lumpy structure is desired.

1.4.7.2 Starting Material

The following protein sources are suitable for the production of texturized products: soya; casein; wheat gluten; oilseed meals such as from cottonseed, groundnut, sesame, sunflower, safflower or rapeseed; zein (corn protein); yeast; whey; blood plasma; or packing plant offal such as lungs or stomach tissue.

The required protein content of the starting material varies and depends on the process used for texturization. The starting material is often a mixture such as soya with lactalbumin, or protein plus acidic polysaccharide (alginate, carrageenan or pectin).

The suitability of proteins for texturization varies, but the molecular weight should be in the range of 10–50 kdal. Proteins of less than 10 kdal are weak fiber builders, while those higher than 50 kdal are disadvantageous due to their high viscosity and tendency to gel in the alkaline pH range. The proportion of amino acid residues with polar side chains should be high in order to enhance intermolecular binding of chains. Bulky side chains obstruct such interactions, so that the amounts of amino acids with these structures should be low.

1.4.7.3 Texturization

The globular protein is unfolded during texturization by breaking the intramolecular binding forces. The resultant extended protein chains are stabilized through interaction with neighboring chains. In practice, texturization is achieved in one of two ways:

- The starting protein is solubilized and the resultant viscous solution is extruded through a spinning nozzle into a coagulating bath (spin process).
- The starting protein is moistened slightly and then, at high temperature and pressure, is extruded with shear force through the orifices of a die (extrusion process).

1.4.7.3.1 Spin Process

The starting material (protein content >90%, e. g. a soya protein isolate) is suspended in water and solubilized by the addition of alkali. The 20% solution is then aged at pH 11 with constant stirring. The viscosity rises during this time as the protein unfolds. The solution is then pressed through the orifices of a die (5000–15,000 orifices, each with a diameter of 0.01–0.08 mm) into a coagulating bath at pH 2–3. This bath contains an acid (citric, acetic, phosphoric, lactic or hydrochloric) and, usually, 10% NaCl. Spinning solutions of protein and acidic polysaccharide mixtures also contain earth alkali salts. The protein fibers are extended further (to about 2- to 4-times the original length) in a "winding up" step and are bundled into thicker fibers with diameters of 10–20 mm. The molecular interactions are enhanced during stretching of the fiber, thus increasing the mechanical strength of the fiber bundles.

The adherent solvent is then removed by pressing the fibers between rollers, then placing them in a neutralizing bath ($NaHCO_3$ + NaCl) of pH 5.5–6 and, occasionally, also in a hardening bath (conc. NaCl).

The fiber bundles may be combined into larger aggregates with diameters of 7–10 cm.

Additional treatment involves passage of the bundles through a bath containing a binder and other additives (a protein which coagulates when heated, such as egg protein; modified starch or other polysaccharides; aroma compounds; lipids). This treatment produces bundles with improved thermal stability and aroma. A typical bath for fibers which are to be processed into a meat analogue might consist of 51% water, 15% ovalbumin, 10% wheat gluten, 8% soya flour, 7% onion powder, 2% protein hydrolysate, 1% NaCl, 0.15% monosodium glutamate and 0.5% pigments.

Finally, the soaked fiber bundles are heated and chopped.

1.4.7.3.2 Extrusion Process

The moisture content of the starting material (protein content about 50%, e. g., soya flour) is adjusted to 30–40% and additives (NaCl, buffers, aroma compounds, pigments) are incorporated. Aroma compounds are added in fat as a carrier, when necessary, after the extrusion step to compensate for aroma losses. The protein mixture is fed into the extruder (a thermostatically controlled cylinder or conical body which contains a polished, rotating screw with a gradually decreasing pitch) which is heated to 120–180 °C and develops a pressure of 30–40 bar. Under these conditions the mixture is transformed into a plastic, viscous state in which solids are dispersed in the molten protein. Hydration of the protein takes place after partial unfolding of the globular molecules and stretching and rearrangement of the protein strands along the direction of mass transfer.

The process is affected by the rotation rate and shape of the screw and by the heat transfer and viscosity of the extruded material and its residence time in the extruder. As the molten material exits from the extruder, the water vaporizes, leaving behind vacuoles in the ramified protein strands.

The extrusion process is more economical than the spin process. However, it yields fiber-like particles rather than well-defined fibers. A great number and variety of extruders are now in operation. As with other food processes, there is a trend toward developing and utilizing high-temperature/short-time extrusion cooking.

1.5 References

Aeschbach, R., Amado, R., Neukom, H.: Formation of dityrosine cross-links in proteins by oxidation of tyrosine residues. Biochim. Biophys. Acta *439*, 292 (1976)

Arai, S., Yamashita, M., Fujimaki, M.: Nutritional improvement of food proteins by means of the plastein reaction and its novel modification. Adv. Exp. Med. Biol. *105*, 663 (1978)

Aso, K., Yamashita, M., Arai, S., Suzuki, J., Fujimaki, M.: Specificity for incorporation of α-amino acid esters during the plastein reaction by papain. J. Agric. Food Chem. *25*, 1138 (1977)

Belitz, H.-D., Wieser, H.: Zur Konfigurationsabhängigkeit des süßen oder bitteren Geschmacks von Aminosäuren und Peptiden. Z. Lebensm. Unters. Forsch. *160*, 251 (1976)

Biochemistry, 5th edition, Berg, J.M., Tymoczko, J.L., Stryer, L., W.H. Freeman and Company, New York, 2002

Bodanszky, M.: Peptide Chemistry. Springer-Verlag: Berlin, 1988

Boggs, R.W.: Bioavailability of acetylated derivatives of methionine, threonine and lysine. Adv. Exp. Med. Biol. *105*, 571 (1978)

Bosin, T.R., Krogh, S., Mais, D.: Identification and quantitation of 1,2,3,4-Tetrahydro-β-carboline-3-carboxylic acid and 1-methyl-1,2,3,4-tetrahydro-β-carboline-3-carboxylic acid in beer and wine. J. Agric. Food Chem. *34*, 843 (1986)

Bott, R.R., Davies, D.R.: Pepstatin binding to *Rhizopus chinensis* aspartyl proteinase. In: Peptides: Structure and function (Eds.: Hruby, V.J., Rich, D.H.), p. 531. Pierce Chemical Co.: Rockford, Ill. 1983

Brückner, H., Päzold, R.: Sind D-Aminosäuren gute molekulare Marker in Lebensmitteln? Pro und Kontra. Lebensmittelchemie *60*, 141 (2006)

Chen, C., Pearson, A.M., Gray, J.I.: Meat Mutagens. Adv. Food Nutr. Res. *34*, 387, (1990)

Cherry, J.P. (Ed.): Protein functionality in foods. ACS Symposium Series 147, American Chemical Society: Washington, D.C. 1981

Creighton, T.E.: Proteins: structures and molecular properties. W.H. Freeman and Co.: New York. 1983

Croft, L.R.: Introduction to protein sequence analysis, 2nd edn., John Wiley and Sons, Inc.: Chichester. 1980

Dagleish, D.G.: Adsorptions of protein and the stability of emulsions. Trends Food Sci. Technol. *8*, 1 (1997)

Dickinson, E.: Towards more natural emulsifiers. Trends Food Sci. Technol. *4*, 330 (1993)

Einsele, A.: Biomass from higher n-alkanes. In: Biotechnology (Eds.: Rehm, H.-J., Reed, G.), Vol. 3, p. 43, Verlag Chemie: Weinheim. 1983

Faust, U., Präve. P.: Biomass from methane and methanol. In: Biotechnology (Eds.: Rehm, H.-J., Reed, G.), Vol. 3, p. 83, Verlag Chemie: Weinheim. 1983

Felton, J.S., Pais, P., Salmon, C.P., Knize, M.G.: Chemical analysis and significance of heterocyclic aromatic amines. Z. Lebensm. Unters. Forsch. A *207*, 434 (1998)

Finot, P.-A., Mottu, F., Bujard, E., Mauron, J.: N-substituted lysines as sources of lysine in nutrition. Adv. Exp. Med. Biol. *105*, 549 (1978)

Friedman, M., Granvogl, M., Schieberle, P.: Thermally generated 3-aminopropionamide as a transient intermediate in the formation of acrylamide. J. Agric. Food Chem. *54*, 5933-6938 (2006)

Galembeck, F., Ryan, D.S., Whitaker, J.R., Feeney, R.E.: Reactions of proteins with formaldehyde in the presence and absence of sodium borohydride. J. Agric. Food Chem. *25*, 238 (1977)

Glazer, A.N.: The chemical modification of proteins by group-specific and site-specific reagents. In: The proteins (Eds.: Neurath, H., Hill, R.L., Boeder, C.-L.), 3rd edn., Vol. II, p. 1, Academic Press: New York-London. 1976

Grant, D.R.: The modification of wheat flour proteins with succinic anhydride. Cereal Chem. *50* 417 (1973)

Gross, E., Morell, J.L.: Structure of nisin. J. Am. Chem. Soc. *93*, 4634 (1971)

Herderich, M., Gutsche, B.: Tryptophan-derived bioactive compounds in food. Food Rev. Int. *13*, 103 (1997)

Hudson, B.J.F. (Ed.): Developments in food proteins-1. Applied Science Publ.: London. 1982

Ikenaka, T., Odani, S., Koide, T: Chemical structure and inhibitory activities of soybean proteinase inhibitors. Bayer-Symposium V "Proteinase inhibitors", p. 325, Springer-Verlag: Berlin-Heidelberg. 1974

Jägerstad, M., Skog, K., Arvidson, P., Solyakov, A.: Chemistry, formation and occurrence of genotoxic heterocyclic amines identified in model systems and cooked foods. Z. Lebensm. Unters. Forsch A *207*, 419 (1998)

Kasai, H., Yamaizumi, Z., Shiomi, T., Yokoyama, S., Miyazawa, T., Wakabayashi, K., Nagao, M., Sugimura, T., Nishimura, S.: Structure of a potent mutagen isolated from fried beef. Chem. Lett. *1981*, 485

Kasai, H., Yamaizumi, Z., Wakabayashi, K., Nagao, M., Sugimura, T., Yokoyama, S., Miyazawa, T., Spingarn, N.E., Weisberger, J.H., Nishimura, S.: Potent novel mutagens produced by broiling fish under normal conditions. Proc. Jpn. Acad. Ser. B56, 278 (1980)

Kessler, H.G.: Lebensmittel- und Bioverfahrenstechnik; Molkereitechnologie. 3rd edn., Verlag A. Kessler: Freising, 1988

Kinsella, J.E.: Functional properties of proteins in foods: A survey. Crit. Rev. Food Sci. Nutr. *7*, 219 (1976)

Kinsella, J.E.: Texturized proteins: Fabrication, flavoring, and nutrition. Crit. Rev. Food Sci. Nutr. *10*, 147 (1978)

Kleemann. A., Leuchtenberger, W., Hoppe, B., Tanner. H.: Amino acids. In: Ullmann's encyclopedia of industrial chemistry, 5ᵗʰ Edition, Volume A2, p. 57, Verlag VCA. Weinheim, 1986

Klostermeier, H., et al.: Proteins. In: Ullmann's encyclopedia of industrial chemistry, 5ᵗʰ Edition, Volume A22, p. 289, Verlag VCH, Weinheim, 1993

Kostka, V. (Ed.): Aspartic proteinases and their inhibitors. Walter de Gruyter: Berlin. 1985

Lásztity, R.: Rheologische Eigenschaften von Weizenkleber und ihre Beziehungen zu molekularen Parametern. Nahrung *19*, 749 (1975)

Lottspeich, F., Henschen, A., Hupe, K.-P. (Eds.): High performance liquid chromatography in protein and peptide chemistry. Walter de Gruyter: Berlin. 1981

Lübke, K., Schröder, E., Kloss, G.: Chemie und Biochemie der Aminosäuren, Peptide und Proteine. Georg Thieme Verlag: Stuttgart. 1975

Masters, P.M., Friedman, M.: Racemization of amino acids in alkali-treated food proteins. J. Agric. Food Chem. *27*, 507 (1979)

Mauron, J.: Ernährungsphysiologische Beurteilung bearbeiteter Eiweißstoffe. Dtsch. Lebensm. Rundsch. *71*, 27 (1975)

Mazur, R.H., Goldkamp, A.H., James, P.A., Schlatter, J.M.: Structure-taste relationships of aspartic acid amides. J. Med. Chem. *13*, 1217 (1970)

Mazur, R.H., Reuter, J.A., Swiatek, K.A., Schlatter, J.M.: Synthetic sweetener. 3. Aspartyl dipeptide esters from L- and D-alkylglycines. J. Med. Chem. *16*, 1284 (1973)

Meister, A.: Biochemistry of the amino acid. 2nd edn., Vol. I, Academic Press: New York-London. 1965

Morrissay, P.A., Mulvihill, D.M., O'Neill, E.M.: Functional properties of muscle proteins. In: Development in Food Proteins – 5; (Ed.: Hudson, B.J.F.), p. 195, Elsevier Applied Science: London, 1987

Mottram, D.S., Wedzicha, B.L., Dodson, A.T.: Acrylamide is formed in the Maillard reaction. Nature 419, 448 (2002)

Nagao, M., Yahagi, T., Kawachi, T., Seino, Y., Honda, M., Matsukura, N., Sugimura, T., Wakabayashi, K., Tsuji, K., Kosuge, T.: Mutagens in foods, and especially pyrolysis products of protein. Dev. Toxicol. Environ. Sci. 2nd (Prog. Genet. Toxicol.), p. 259 (1977)

Nakai, S., Modler, H.W.: Food proteins. Properties and characterization. Verlag Chemie, Weinheim, 1996

Oura, E.: Biomass from carbohydrates. In: Biotechnology (Eds.: Rehm, H.-J., Reed, G.), Vol. 3, p. 3, Verlag Chemie: Weinheim, 1983

Pence, J.W., Mohammad, A., Mecham, D.K.: Heat denaturation of gluten. Cereal Chem. 30, 115 (1953)

Pennington, S.R., Wilkins, M.R., Hochstrasser, D.F., Dunn, M.J.: Proteome analysis: from protein characterization to biological function. Trends in Cell Biology 7, 168 (1997)

Perutz, M.F.: Proteins and nucleic acids. Elsevier Publ. Co.: Amsterdam. 1962

Phillips, E.G., Whitehead, D.M., Kinsella, J.: Structure-function properties of food proteins. Academic Press, London, 1994

Poindexter, E.H., Jr., Carpenter, R.D.: Isolation of harmane and norharmane from cigarette smoke. Chem. Ind. 1962, 176

Puigserver, A.J., Sen, L.C., Clifford, A.J., Feeney, R.E., Whitaker, J.R.: A method for improving the nutritional value of food proteins: Covalent attachment of amino acids. Adv. Exp. Med. Biol. 105, 587 (1978)

Repley, J.A., Careri, G.: Protein hydration and function. Adv. Protein Chem. 41, 38 (1991)

Richmond, A.: Phototrophic microalgae. In: Biotechnology (Eds.: Rehm, H.-J., Reed G.). Vol. 3. p. 109, Verlag Chemie: Weinheim. 1983

Schmitz, M.: Möglichkeiten und Grenzen der Homoarginin-Markierungsmethode zur Messung der Proteinverdaulichkeit beim Schwein. Dissertation, Universität Kiel. 1988

Schulz, G.E., Schirmer, R.H.: Principle of protein structure. Springer-Verlag: Berlin-Heidelberg. 1979

Schwenke, K.D.: Beeinflussung funktioneller Eigenschaften von Proteinen durch chemische Modifizierung. Nahrung 22, 101 (1978)

Seki, T., Kawasaki, Y., Tamura, M., Tada. M., Okai, H.: Further study on the salty peptide ornithyl-β-alanine. Some effects of pH and additive ions on saltiness. J. Agric. Food Chem. 38, 25 (1990)

Severin, Th., Ledl, F.: Thermische Zersetzung von Cystein in Tributyrin. Chem. Mikrobiol. Technol. Lebensm. 1, 135 (1972)

Shinoda, I., Tada, M., Okai, H.: A new salty peptide, ornithyl-β-alanine hydrochloride. Pept. Chem. 21st. Proceeding of the Symposium on Peptide Chemistry 1983, p. 43 (1984)

Soda, K., Tanaka, H., Esaki, N.: Amino acids. In: Biotechnology (Eds.: Rehm, H.-J. Reed. G.). Vol. 3, p. 479, Verlag Chemie: Weinheim. 1983

Stadler, R.H., Robert, F., Riediker, S., Varga, N., Davidek, T., Devaud, S., Goldmann, T., Hau, J., Blank I.: In-depth mechanistic study on the formation of acrylamide and other vinylogous compounds in the Maillard reaction. J. Agric. Food Chem. 52, 5550 (2004)

Sternberg, M., Kim, C.Y.: Lysinoalanine formation in protein food ingredients. Adv. Exp. Med. Biol. 86B, 73 (1977)

Sugimura, T., Kawachi, T., Nagao, M., Yahagi, T., Seino, Y., Okamoto, T., Shudo, K., Kosuge, T., Tsuji, K. et al.: Mutagenic principle(s) in tryptophan and phenylalanine pyrolysis products. Proc. Jpn. Acad. 33, 58 (1977)

Sulser, H.: Die Extraktstoffe des Fleisches. Wissenschaftliche Verlagsgesellschaft mbH: Stuttgart. 1978

Traub, W., Piez, K.A.: The chemistry and structure of collagen. Adv. Protein Chem. 25, 267 (1971)

Treleano, R., Belitz, H.-D., Jugel, H., Wieser, H.: Beziehungen zwischen Struktur und Geschmack bei Aminosäuren mit cyclischen Seitenketten. Z. Lebensm. Unters. Forsch. 167, 320 (1978)

Tschesche, H. (Ed.): Modern Methods in Protein Chemistry, Vol. 2, Walter de Gruyter: Berlin, 1985

Voutsinas, L.P., Nakai, S.: Covalent binding of methionine and tryptophan to soy protein. J. Food Sci. 44, 1205 (1979)

Walton, A.G.: Polypeptides and protein structure, Elsevier North Holland, Inc., New York-Oxford. 1981

Watanabe, M., Arai, S.: The plastein reaction and its applications. In: Developments in Food Proteins – 6; (Ed.: Hudson, B.J.F.), p. 179, Elsevier Applied Science: London. 1988

Whitaker, J.R., Fujimaki, M. (Eds.): Chemical deterioration of proteins, ACS Symposium Series 123, American Chemical Society: Washington D.C. 1980

Wieser, H., Belitz, H.-D.: Zusammenhänge zwischen Struktur und Bittergeschmack bei Aminosäuren und Peptiden. I. Aminosäuren und verwandte Verbindungen. Z. Lebensm. Unters. Forsch. 159, 65 (1975)

Wieser, H., Belitz, H.-D.: Zusammenhänge zwischen Struktur und Bittergeschmack bei Aminosäuren und Peptiden. II. Peptide und Peptidderivate. Z. Lebensm. Unters. Forsch. *160*, 383 (1976)

Wieser, H., Jugel, H., Belitz, H.-D.: Zusammenhänge zwischen Struktur und Süßgeschmack bei Aminosäuren. Z. Lebensm. Unters. Forsch. *164*, 277 (1977)

Wild, D., Kerdar, R.S.: The inherent genotoxic potency of food mutagens and other heterocyclic and carboxylic aromatic amines and corresponding azides. Z. Lebensm. Unters. Forsch. A *207*, 428 (1998)

Wittmann-Liebold, B., Salnikow, J., Erdmann, V.A. (Eds.): Advanced Methods in Protein Microsequence Analysis. Springer-Verlag: Berlin. 1986

Yamashita, M., Arai, S., Fujimaki, M.: A lowphenylalanine, high-tyrosine plastein as an acceptable dietetic food. J. Food Sci. *41*, 1029 (1976)

Yamashita, M., Arai, S., Amano, Y., Fujimaki, M.: A novel one-step process for enzymatic incorporation of amino acids into proteins: Application to soy protein and flour for enhancing their methionine levels. Agric. Biol. Chem. *43*, 1065 (1979)

Yamashita, M., Arai, S., Tsai, S.-J., Fujimaki, M.: Plastein reaction as a method for enhancing the sulfur-containing amino acid level of soybean protein. J. Agric. Food Chem. *19*, 1151 (1971)

Yamashita, M., Arai, S., Kokubo, S., Aso, K., Fujimaki, M.: Synthesis and characterization of a glutamic acid enriched plastein with greater solubility. J. Agric. Food Chem. *23*, 27 (1975)

Yoshikawa, M., Tamaki, M., Sugimoto, E., Chiba, H.: Effect of dephosphorylation on the self-association and the precipitation of β-Casein. Agric. Biol. Chem. *38*, 2051 (1974)

2 Enzymes

2.1 Foreword

Enzymes are proteins with powerful catalytic activity. They are synthesized by biological cells and in all organisms, they are involved in chemical reactions related to metabolism. Therefore, enzyme-catalyzed reactions also proceed in many foods and thus enhance or deteriorate food quality. Relevant to this phenomenon are the ripening of fruits and vegetables, the aging of meat and dairy products, and the processing steps involved in the making of dough from wheat or rye flours and the production of alcoholic beverages by fermentation technology.

Enzyme inactivation or changes in the distribution patterns of enzymes in subcellular particles of a tissue can occur during storage or thermal treatment of food. Since such changes are readily detected by analytical means, enzymes often serve as suitable indicators for revealing such treatment of food. Examples are the detection of pasteurization of milk, beer or honey, and differentiation between fresh and deep frozen meat or fish.

Enzyme properties are of interest to the food chemist since enzymes are available in increasing numbers for enzymatic food analysis or for utilization in industrial food processing. Examples of both aspects of their use are provided in this chapter in section 2.6.4 on food analysis and in section 2.7, which covers food processing.

Details of enzymes which play a role in food science are restricted in this chapter to only those enzyme properties which are able to provide an insight into the build-up or functionality of enzymes or can contribute to the understanding of enzyme utilization in food analysis or food processing and storage.

2.2 General Remarks, Isolation and Nomenclature

2.2.1 Catalysis

Let us consider the catalysis of an exergonic reaction:

$$A \underset{k_{-1}}{\overset{k_1}{\rightleftharpoons}} P \tag{2.1}$$

with a most frequently occurring case in which the reaction does not proceed spontaneously. Reactant A is metastable, since the activation energy, E_A, required to reach the activated transition state in which chemical bonds are formed or cleaved in order to yield product P, is exceptionally high (Fig. 2.1).

The reaction is accelerated by the addition of a suitable catalyst. It transforms reactant A into intermediary products (EA and EP in Fig. 2.1), the transition states of which are at a lower energy level than the transition state of a noncatalyzed reaction (A$\neq$ in Fig. 2.1). The molecules of the

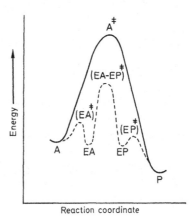

Fig. 2.1. Energy profile of an exergonic reaction A → P; — without and - - - with catalyst E

Table 2.1. Examples of catalyst activity

Reaction	Catalyst	Activation energy $(kJ \cdot mol^{-1})$	$k_{rel}(25 \,°C)$
1. $H_2O_2 \rightarrow H_2O + 1/2\ O_2$	Absent	75	1.0
	$I^{\oplus}$	56.5	$\sim 2.1 \cdot 10^3$
	Catalase	26.8	$\sim 3.5 \cdot 10^8$
2. Casein + n $H_2O \rightarrow$	$H^{\ominus}$	86	1.0
(n+1) Peptides	Trypsin	50	$\sim 2.1 \cdot 10^6$
3. Ethylbutyrate	$H^{\ominus}$	55	1.0
$+H_2O \rightarrow$ butyric acid+ethanol	Lipase	17.6	$\sim 4.2 \cdot 10^6$
4. Saccharose + $H_2O \rightarrow$	$H^{\ominus}$	107	1.0
Glucose+Fructose	Invertase	46	$\sim 5.6 \cdot 10^{10}$
5. Linoleic acid	Absent	150–270	1.0
$+O_2 \rightarrow$ Linoleic acid	Cu^{2+}	30–50	$\sim 10^2$
hydroperoxide	Lipoxygenase	16.7	$\sim 10^7$

species A contain enough energy to combine with the catalyst and, thus, to attain the "activated state" and to form or break the covalent bond that is necessary to give the intermediary product which is then released as product P along with free, unchanged catalyst. The reaction rate constants, k_{+1} and k_{-1}, are therefore increased in the presence of a catalyst. However, the equilibirum constant of the reaction, i.e. the ratio $k_{1+}/k_{-1} = K$, is not altered.

Activation energy levels for several reactions and the corresponding decreases of these energy levels in the presence of chemical or enzymatic catalysts are provided in Table 2.1. Changes in their reaction rates are also given. In contrast to reactions 1 and 5 (Table 2.1) which proceed at measurable rates even in the absence of catalysts, hydrolysis reactions 2, 3 and 4 occur only in the presence of protons as catalysts. However, all reaction rates observed in the case of inorganic catalysts are increased by a factor of at least several orders of magnitude in the presence of suitable enzymes. Because of the powerful activity of enzymes, their presence at levels of 10^{-8} to 10^{-6} mol/l is sufficient for *in vitro* experiments. However, the enzyme concentrations found in living cells are often substantially higher.

2.2.2 Specificity

In addition to an enzyme's ability to substantially increase reaction rates, there is a unique enzyme property related to its high specificity for both the compound to be converted (substrate specificity) and for the type of reaction to be catalysed (reaction specificity).

The activities of allosteric enzymes (cf. 2.5.1.3) are affected by specific regulators or effectors. Thus, the activities of such enzymes show an additional regulatory specificity.

2.2.2.1 Substrate Specificity

The substrate specificity of enzymes shows the following differences. The occurrence of a distinct functional group in the substrate is the only prerequisite for a few enzymes, such as some hydrolases. This is exemplified by nonspecific lipases (cf. Table 3.21) or peptidases (cf. 1.4.5.2.1) which generally act on an ester or peptide covalent bond.

More restricted specificity is found in other enzymes, the activities of which require that the substrate molecule contains a distinct structural feature in addition to the reactive functional group. Examples are the proteinases trypsin and chymotrypsin which cleave only ester or peptide bonds with the carbonyl group derived from lysyl or arginyl (trypsin) or tyrosyl, phenylalanyl or tryptophanyl residues (chymotrypsin). Many enzymes activate only one single substrate or preferentially catalyze the conversion of one substrate while other substrates are converted into products with a lower reaction rate (cf. ex-

Table 2.2. Substrate specificity of a legume α-glucosidase

Substrate	Relative activity (%)	Substrate	Relative activity (%)
Maltose	100	Cellobiose	0
Isomaltose	4.0	Saccharose	0
Maltotrisose	41.5	Phenyl-α-	
Panose	3.5	glucoside	3.1
Amylose	30.9	Phenyl-α-	
Amylopectin	4.4	maltoside	29.7

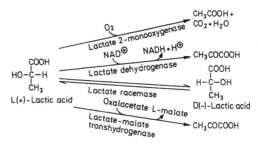

Fig. 2.2. Examples of reaction specificity of some enzymes

amples in Table 2.2 and 3.24). In the latter cases a reliable assessment of specificity is possible only when the enzyme is available in purified form, i. e. all other accompanying enzymes, as impurities, are completely removed.

An enzyme's substrate specificity for stereoisomers is remarkable. When a chiral center is present in the substrate in addition to the group to be activated, only one enantiomer will be converted to the product. Another example is the specificity for diastereoisomers, e. g. for cis-trans geometric isomers.

Enzymes with high substrate specificity are of special interest for enzymatic food analysis. They can be used for the selective analysis of individual food constituents, thus avoiding the time consuming separation techniques required for chemical analyses, which can result in losses.

2.2.2.2 Reaction Specificity

The substrate is specifically activated by the enzyme so that, among the several thermodynamically permissible reactions, only one occurs. This is illustrated by the following example: L(+)-lactic acid is recognized as a substrate by four enzymes, as shown in Fig. 2.2, although only lactate-2-monooxygenase decarboxylates the acid oxidatively to acetic acid. Lactate dehydrogenase and lactate-malate transhydrogenase form a common reaction product, pyruvate, but by different reaction pathways (Fig. 2.2). This may suggest that reaction specificity should be ascribed to the different cosubstrates, such as NAD^+ or oxalacetate. But this is not the case since a change in cosubstrates stops the reaction. Obviously, the enzyme's reaction specificity as

well as the substrate specificity are predetermined by the structure and chemical properties of the protein moiety of the enzyme.

Of the four enzymes considered, only the lactate racemase reacts with either of the enantiomers of lactic acid, yielding a racemic mixture.

Therefore, enzyme reaction specificity rather than substrate specificity is considered as a basis for enzyme classification and nomenclature (cf. 2.2.6).

2.2.3 Structure

Enzymes are globular proteins with greatly differing particle sizes (cf. Table 1.26). As outlined in section 1.4.2, the protein structure is determined by its amino acid sequences and by its conformation, both secondary and tertiary, derived from this sequence. Larger enzyme molecules often consist of two or more peptide chains (subunits or protomers, cf. Table 1.26) arranged into a specified quaternary structure (cf. 1.4.2.3). Section 2.4.1 will show that the three dimensional shape of the enzyme molecule is actually responsible for its specificity and its effective role as a catalyst. On the other hand, the protein nature of the enzyme restricts its activity to a relatively narrow pH range (for pH optima, cf. 2.5.3) and heat treatment leads readily to loss of activity by denaturation (cf. 1.4.2.4 and 2.5.4.4).

Some enzymes are complexes consisting of a protein moiety bound firmly to a nonprotein component which is involved in catalysis, e. g. a "prosthetic" group (cf. 2.3.2). The activities of other enzymes require the presence of a cosubstrate which is reversibly bound to the protein moiety (cf. 2.3.1).

2.2.4 Isolation and Purification

Most of the enzyme properties are clearly and reliably revealed only with purified enzymes. As noted under enzyme isolation, prerequisites for the isolation of a pure enzyme are selected protein chemical separation methods carried out at 0–4 °C since enzymes are often not stable at higher temperatures.

Tissue Disintegration and Extraction. Disintegration and homogenization of biological tissue requires special precautions: procedures should be designed to rupture the majority of the cells in order to release their contents so that they become accessible for extraction. The tissue is usually homogenized in the presence of an extraction buffer which often contains an ingredient to protect the enzymes from oxidation and from traces of heavy metal ions. Particular difficulty is encountered during the isolation of enzymes which are bound tenaciously to membranes which are not readily solubilized. Extraction in the presence of tensides may help to isolate such enzymes. As a rule, large amounts of tissue have to be homogenized because the enzyme content in proportion to the total protein isolated is low and is usually further diminished by the additional purification

of the crude enzyme isolate (cf. example in Table 2.3).

Enzyme Purification. Removal of protein impurities, usually by a stepwise process, is essentially the main approach in enzyme purification. As a first step, fractional precipitation, e. g. by ammonium sulfate saturation, is often used or the extracted proteins are fractionated by molecular weight e. g., column gel chromatography. The fractions containing the desired enzyme activity are collected and purified further, e. g., by ion-exchange chromatography. Additional options are also available, such as various forms of preparative electrophoresis, e. g. disc gel electrophoresis or isoelectric focusing. The purification procedure can be substantially shortened by using affinity column chromatography. In this case, the column is packed with a stationary phase to which is attached the substrate or a specific inhibitor of the enzyme. The enzyme is then selectively and reversibly bound and, thus, in contrast to the other inert proteins, its elution is delayed.

Control of Purity. Previously, the complete removal of protein impurities was confirmed by crystallization of the enzyme. This "proof" of pu-

Table 2.3. Isolation of a glucosidase from beans (*Phaseolus vidissimus*)

No. Isolation step	Protein (mg)	α-Glucosidase			
		Activity (µcat)	Specific activity (µcat/mg)	Enrichment (-fold)	Yield (%)
1. Extraction with 0.01 mol/L acetate buffer of pH 5.3					
2. Saturation to 90% with ammonium sulfate followed by solubilization in buffer of step 1	44,200	3840	0.087	1	100
3. Precipitation with polyethylene glycol (20%). Precipitate is then solubilized in 0.025 mol/L Tris-HCl buffer of pH 7.4	7610	3590	0.47	5.4	93
4. Chromatography on DEAE-cellulose column, an anion exchanger	1980	1650	0.83	9.5	43
5. Chromatography on SP-Sephadex C-50, a cation exchanger	130	845	6.5	75	22
6. Preparative isoelectric focusing	30	565	18.8	216	15

rity can be tedious and is open to criticism. To-day, electrophoretic methods of high separation efficiency or HPLC are primarily used.

The behavior of the enzyme during chromatographic separation is an additional proof of purity. A purified enzyme is characterized by a symmetrical elution peak in which the positions of the protein absorbance and enzyme activity coincide and the specific activity (expressed as units per amount of protein) remains unchanged during repeated elutions.

During a purification procedure, the enzyme activities are recorded as shown in Table 2.3. They provide data which show the extent of purification achieved after each separation step and show the enzyme yield. Such a compilation of data readily reveals the undesired separation steps associated with loss of activity and suggests modifications or adoption of other steps.

2.2.5 Multiple Forms of Enzymes

Chromatographic or electrophoretic separations of an enzyme can occasionally result in separation of the enzyme into "*isoenzymes*", i. e. forms of the enzyme which catalyze the same reaction although they differ in their protein structure. The occurrence of multiple enzyme forms can be the result of the following:

a) Different compartments of the cell produce genetically independent enzymes with the same substrate and reaction specificity, but which differ in their primary structure. An example is glutamate-oxalacetate transaminase occurring in mitochondria and also in muscle tissue sarcoplasm. This is the indicator enzyme used to differentiate fresh from frozen meat (cf. 12.10.1.2).

b) Protomers associate to form polymers of differing size. An example is the glutamate dehydrogenase occurring in tissue as an equilibrium mixture of molecular weights $M_r = 2.5 \cdot 10^5 - 10^6$.

c) Different protomers combine in various amounts to form the enzyme. For example, lactate dehydrogenase is structured from a larger number of subunits with the reaction specificity given in Fig. 2.2. It consists of five forms (A_4, A_3B, A_2B_2, AB_3 and B_4), all derived from two protomers, A and B.

2.2.6 Nomenclature

The Nomenclature Commitee of the "International Union of Biochemistry and Molecular Biology" (IUBMB) adopted rules last amended in 1992 for the systematic classification and designation of enzymes based on reaction specificity. All enzymes are classified into six major classes according to the nature of the chemical reaction catalyzed:

1. Oxidoreductases.
2. Transferases.
3. Hydrolases.
4. Lyases (cleave C−C, C−O, C−N, and other groups by elimination, leaving double bonds, or conversely adding groups to double bonds).
5. Isomerases (involved in the catalysis of isomerizations within one molecule).
6. Ligases (involved in the biosynthesis of a compound with the simultaneous hydrolysis of a pyrophosphate bond in ATP or a similar triphosphate).

Each class is then subdivided into subclasses which more specifically denote the type of reaction, e. g. by naming the electron donor of an oxidation-reduction reaction or by naming the functional group carried over by a transferase or cleaved by a hydrolase enzyme.

Each subclass is further divided into sub-subclasses. For example, sub-subclasses of oxidoreductases are denoted by naming the acceptor which accepts the electron from its respective donor.

Each enzyme is classified by adopting this system. An example will be analyzed. The enzyme ascorbic acid oxidase catalyzes the following reaction:

$$\text{L} - \text{Ascorbic acid} + \tfrac{1}{2} O_2 \rightleftharpoons$$
$$\text{L} - \text{Dehydroascorbic acid} + H_2O \qquad (2.2)$$

Hence, its systematic name is L-ascorbate: oxygen oxidoreductase, and its systematic number is E.C. 1.1.10.3.3 (cf. Formula 2.3). The systematic names are often quite long. Therefore, the short, trivial names along with the systematic numbers are often convenient for enzyme designation. Since enzymes of different biological origin often differ in their properties, the source and, when known, the subcellular fraction used for isolation are specified in addition to the name of the enzyme preparation; for example, "ascorbate oxidase (E.C. 1.1.10.3.3) from cucumber". When known, the subcellular fraction of origin (cytoplasmic, mitochondrial or peroxisomal) is also specified.

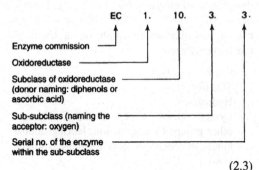

$$\text{EC} \quad 1. \quad 10. \quad 3. \quad 3.$$

Enzyme commission

Oxidoreductase

Subclass of oxidoreductase
(donor naming: diphenols or
ascorbic acid)

Sub-subclass (naming the
acceptor: oxygen)

Serial no. of the enzyme
within the sub-subclass

$$(2.3)$$

A number of enzymes of interest to food chemistry are described in Table 2.4. The number of the section in which an enzyme is dealt with is given in the last column.

2.2.7 Activity Units

The catalytic activity of enzymes is exhibited only under specific conditions, such as pH, ionic strength, buffer type, presence of cofactors and suitable temperature. Therefore, the rate of substrate conversion or product formation can be measured in a test system designed to follow the enzyme activity. The International System of Units (SI) designation is $mol\,s^{-1}$ and its recommended designation is the "katal" (kat[*]). Decimal units are formed in the usual way, e. g.:

$$\mu kat = 10^{-6}\,kat = \mu mol \cdot s^{-1} \qquad (2.4)$$

Concentration of enzymatic activity is given as $\mu kat\,l^{-1}$. The following activity units are derived from this:

a) The *specific catalytic activity*, i. e. the activity of the enzyme preparation in relation to the protein concentration.

---

[*] The old definition in the literature may also be used: 1 enzyme unit (U) $\triangleq$ 1 $\mu mol\,min^{-1}(1U \triangleq 16.67 \cdot 10^{-9}kat)$.

b) The *molar catalytic activity*. This can be determined when the pure enzyme with a known molecular weight is available. It is expressed as "katal per mol of enzyme" $(kat\,mol^{-1})$. When the enzyme has only one active site or center per molecule, the molar catalytic activity equals the "turnover number", which is defined as the number of substrate molecules converted per unit time by each active site of the enzyme molecule.

2.3 Enzyme Cofactors

Rigorous analysis has demonstrated that numerous enzymes are not pure proteins. In addition to protein, they contain metal ions and/or low molecular weight nonprotein organic molecules. These nonprotein hetero constituents are denoted as cofactors which are indispensable for enzyme activity.

According to the systematics (Fig. 2.3), an apoenzyme is the inactive protein without a cofactor. Metal ions and coenzymes participating in enzymatic activity belong to the cofactors which are subdivided into prosthetic groups and cosubstrates. The prosthetic group is bound firmly to the enzyme. It can not be removed by, e. g. dialysis, and during enzyme catalysis it remains attached to the enzyme molecule. Often, two substrates are converted by such enzymes, one substrate followed by the other, returning the prosthetic group to its original state. On the other

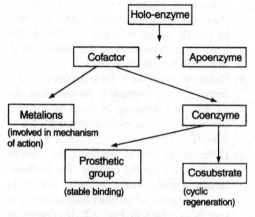

Fig. 2.3. Systematics of cofactor-containing enzymes (according to *Schellenberger*, 1989)

hand, during metabolism, the cosubstrate reacts with at least two enzymes. It transfers the hydrogen or the functional group to another enzyme and, hence, is denoted as a "transport metabolite" or as an "intermediary substrate". It is distinguished from a true substrate by being regenerated in a subsequent reaction. Therefore the concentration of the intermediary substrates can be very low. In food analysis higher amounts of cosubstrates are often used without regeneration.

Only those cofactors with enzymatic activities of importance in enzymatic analysis of food and/or in food processing will be presented. Some cofactors are related to water-soluble vitamins (cf. 6.3). The metal ions are dealt with separately in section 2.3.3.

2.3.1 Cosubstrates

2.3.1.1 Nicotinamide Adenine Dinucleotide

Transhydrogenases (e. g. lactate dehydrogenase, alcohol dehydrogenase) dehydrogenate or hydrogenate their substrates with the help of a pyridine cosubstrate (Fig. 2.4); its nicotinamide residue accepts or donates a hydride ion (H^-) at position 4:

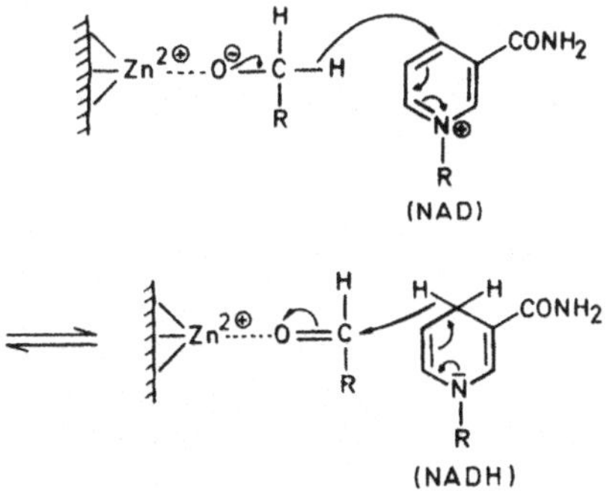

$$(2.5)$$

The reaction proceeds stereospecifically (cf. 2.4.1.2.1); ribose phosphate and the $-CONH_2$ group force that pyridine ring of the cosubstrate to become planar on the enzyme surface. The role of Zn^{2+} ions in this catalysis is outlined in section 2.3.3.1. The transhydrogenases differ according to the site on the pyridine ring involved in or accessible to H-transfer. For example, alcohol and lactate dehydrogenases transfer

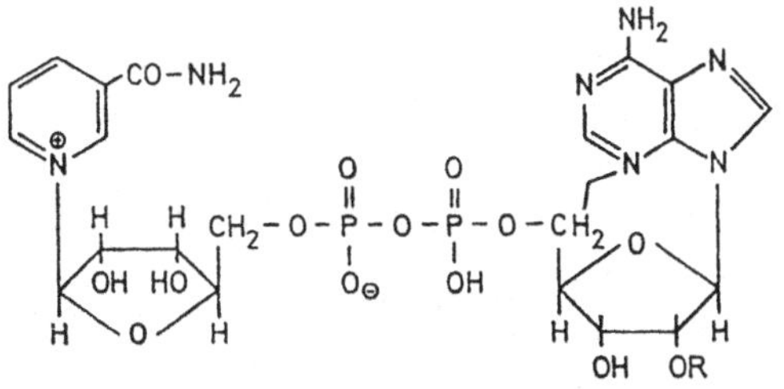

Fig. 2.4. Nicotinamide adenine dinucleotide (NAD) and nicotinamide adenine dinucleotide phosphate (NADP); R = H: NAD; R = PO_3H_2: NADP

the pro-R-hydrogen from the A* side, whereas glutamate or glucose dehydrogenases transfer the pro-S-hydrogen from the B* side*.

The oxidized and reduced forms of the pyridine cosubstrate are readily distinguished by absorbance readings at 340 nm (Fig. 2.5). Therefore, whenever possible, enzymatic reactions which are difficult to measure directly are coupled with an NAD(P)-dependent indicator reaction (cf. 2.6.1.1) for food analysis.

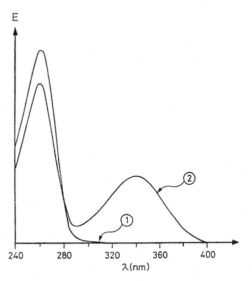

Fig. 2.5. Electron excitation spectra of NAD (1) and NADH (2)

---

* Until the absolute configuration of the chiral center is determined, the two sides of the pyridine ring are denoted as A and B.

Table 2.4. Systematic classification of some enzymes of importance to food chemistry

Class/subclass		Enzyme	EC-Number	In text found under
1.	*Oxidoreductases*			
1.1	CH−OH as donor			
1.1.1	With NAD$^\oplus$ or NADP$^\oplus$ as acceptor	Alcohol dehydrogenase	1.1.1.1	2.6.1
		Butanediol dehydrogenase	1.1.1.4	2.7.2.1.5
		L-Iditol 2-dehydrogenase	1.1.1.14	2.6.1
		L-Lactate dehydrogenase	1.1.1.27	2.6.1
		3-Hydroxyacyl-CoA dehydrogenase	1.1.1.35	12.10.1.2
		Malate dehydrogenase	1.1.1.37	2.6.1
		Galactose 1-dehydrogenase	1.1.1.48	2.6.1
		Glucose-6-phosphate 1-dehydrogenase	1.1.1.49	2.6.1
1.1.3	With oxygen as acceptor	Glucose oxidase	1.1.3.4 2.7.2.1.1	2.6.1 and
		Xanthine oxidase	1.1.3.22	2.3.3.2
1.2	Aldehyde group as donor			
1.2.1	With NAD$^\oplus$ or NADP$^\oplus$ as acceptor	Aldehyde dehydrogenase	1.2.1.3	2.7.2.1.4
1.8	S-Compound as donor			
1.8.5	With quinone or related compound as acceptor	Glutathione dehydrogenase (ascorbate)	1.8.5.1	15.2.2.7
1.10	Diphenol or endiol as donor			
1.10.3	With oxygen as acceptor	Ascorbate oxidase	1.10.3.3	2.2.6
1.11	Hydroperoxide as acceptor	Catalase	1.11.1.6	2.7.2.1.2
		Peroxidase	1.11.1.7	2.3.2.2 and 2.5.4.4
1.13	Acting on single donors			
1.13.11	Incorporation of molecular oxygen	Lipoxygenase	1.13.11.12	2.5.4.4 and 3.7.2.2
1.14	Acting on paired donors			
1.14.18	Incorporation of one oxygen atom	Monophenol monooxygenase (Polyphenol oxidase)	1.14.18.1	2.3.3.2
2.	*Transferases*			
2.3	Transfer of acyl groups			
2.3.2	Aminoacyl transferases	Transglutaminase	2.3.2.13	2.7.2.4
2.7	Transfer of phosphate			
2.7.1	HO-group as acceptor	Hexokinase	2.7.1.1	2.6.1
		Glycerol kinase	2.7.1.30	2.6.1
		Pyruvate kinase	2.7.1.40	2.6.1
2.7.3	N-group as acceptor	Creatine kinase	2.7.3.2	2.6.1
3.	*Hydrolases*			
3.1	Cleavage of ester bonds			
3.1.1	Carboxylester hydrolases	Carboxylesterase	3.1.1.1	3.7.1.1
		Triacylglycerol lipase	3.1.1.3	2.5.4.4 and 3.7.1.1
		Phospholipase A$_2$	3.1.1.4	3.7.1.2
		Acetylcholinesterase	3.1.1.7	2.4.2.5
		Pectinesterase	3.1.1.11	4.4.5.2
		Phospholipase A$_1$	3.1.1.32	3.7.1.2
3.1.3	Phosphoric monoester hydrolases	Alkaline phosphatase	3.1.3.1	2.5.4.4

Table 2.4. Continued

Class/subclass		Enzyme	EC-Number	In text found under
3.1.4	Phosphoric diester hydrolases	Phospholipase C	3.1.4.3	3.7.1.2
		Phospholipase D	3.1.4.4	3.7.1.2
3.2	Hydrolyzing O-glycosyl compounds			
3.2.1	Glycosidases	α-Amylase	3.2.1.1	4.4.5.1.1
		β-Amylase	3.2.1.2	4.4.5.1.2
		Glucan-1,4-α-D-glucosidase (Glucoamylase)	3.2.1.3	4.4.5.1.3
		Cellulase	3.2.1.4	4.4.5.3
		Polygalacturonase	3.2.1.15	2.5.4.4 and 4.4.5.2
		Lysozyme	3.2.1.17	2.7.2.2.11 and 11.2.3.1.4
		α-D-Glucosidase (Maltase)	3.2.1.20	2.6.1
		β-D-Glucosidase	3.2.1.21	2.6.1
		α-D-Galactosidase	3.2.1.22	
		β-D-Galactosidase (Lactase)	3.2.1.23	2.7.2.2.7
		β-Fructofuranosidase (Invertase, saccharase)	3.2.1.26	2.7.2.2.8
		1,3-β-D-Xylanase	3.2.1.32	2.7.2.2.10
		α-L-Rhamnosidase	3.2.1.40	2.7.2.2.9
		Pullulanase	3.2.1.41	4.4.5.1.4
		Exopolygalacturonase	3.2.1.67	4.4.5.2
3.2.3	Hydrolysing S-glycosyl compounds	Thioglucosidase (Myrosinase)	3.2.3.1	2.7.2.2.12
3.4	Peptidases[a]			
3.4.21	Serine endopeptidases[a]	Microbial serine endopeptidases e. g. Subtilisin	3.4.21.62	1.4.5.2.1
3.4.22	Cysteine endopeptidases[a]	Papain	3.4.22.2	1.4.5.2.2
		Ficin	3.4.22.3	1.4.5.2.2
		Bromelain	3.4.22.33	1.4.5.2.2
3.4.23	Aspartic acid endopeptidases[a]	Chymosin (Rennin)	3.4.23.4	1.4.5.2.4
3.4.24	Metalloendopeptidases[a]	Thermolysin	3.4.24.27	1.4.5.2.3
3.5	Acting on C−N bonds, other than peptide bonds			
3.5.2	In cyclic amides	Creatininase	3.5.2.10	2.6.1
4.	Lyases			
4.2	C−O-Lyases			
4.2.2	Acting on polysaccharides	Pectate lyase	4.2.2.2	4.4.5.2
		Exopolygalacturonate lyase	4.2.2.9	4.4.5.2
		Pectin lyase	4.2.2.10	4.4.5.2
5.	*Isomerases*			
5.3	Intramolecular oxidoreductases			
5.3.1	Interconverting aldoses and ketoses	Xylose isomerase	5.3.1.5	2.7.2.3
		Glucose-6-phosphate isomerase	5.3.1.9	2.6.1

[a] cf. Table 1.33.

2.3.1.2 Adenosine Triphosphate

The nucleotide adenosine triphosphate (ATP) is an energy-rich compound. Various groups are cleaved and transferred to defined substrates during metabolism in the presence of ATP. One possibility, the transfer of orthophosphates by kinases, is utilized in the enzymatic analysis of food (cf. Table 2.16).

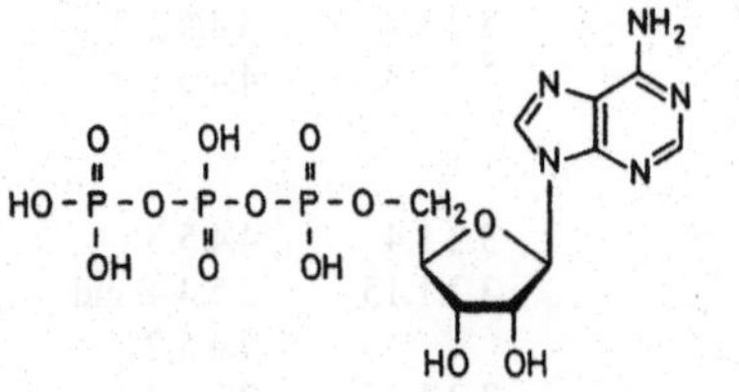

(2.6)

$$ATP + H_2O \longrightarrow ADP + H_3PO_4$$

$$(\Delta G^0 \text{ at pH 7} = -50 \text{ kJ mol}^{-1})$$

(2.7)

2.3.2 Prosthetic Groups

2.3.2.1 Flavins

Riboflavin (7,8-dimethyl-10-ribityl-isoalloxazine), known as vitamin B_2 (cf. 6.3.2), is the building block of flavin mononucleotide (FMN) and flavin adenine dinucleotide (FAD). Both act as prosthetic groups for electron transfer reactions in a number of enzymes.

Due to the much wider redox potential of the flavin enzymes, riboflavin is involved in the transfer of either one or two electrons. This is different from nicotinamides which participate in double electron transfer only. Values between $+0.19$ V (stronger oxidizing effect than $NAD^{\oplus}$) and -0.49 V (stronger reducing effect than NADH) have been reported.

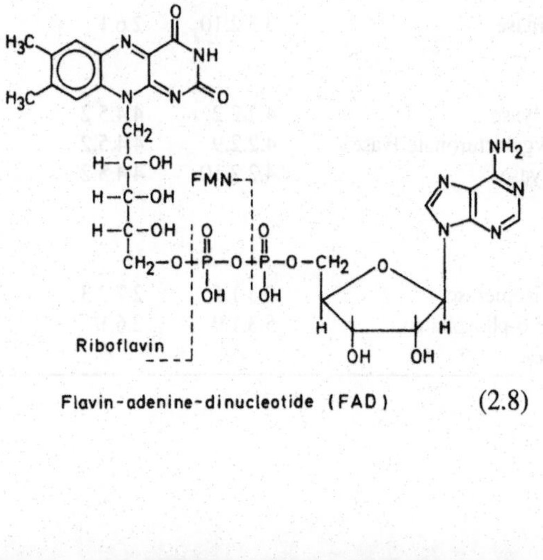

Flavin-adenine-dinucleotide (FAD)   (2.8)

An example for a flavin enzyme is glucose oxidase, an enzyme often used in food processing to trap residual oxygen (cf. 2.7.2.1.1). The enzyme isolated and purified from *Aspergillus niger* is a dimer ($M_r = 168,000$) with two noncovalently bound FAD molecules. In contrast to xanthine oxidase (cf. 2.3.3.2), for example, this enzyme has no heavy metal ion. During oxidation of a substrate, such as the oxidation of β-D-glucose to δ-D- gluconolactone, the flavoquinone is reduced by two single electron transfers:

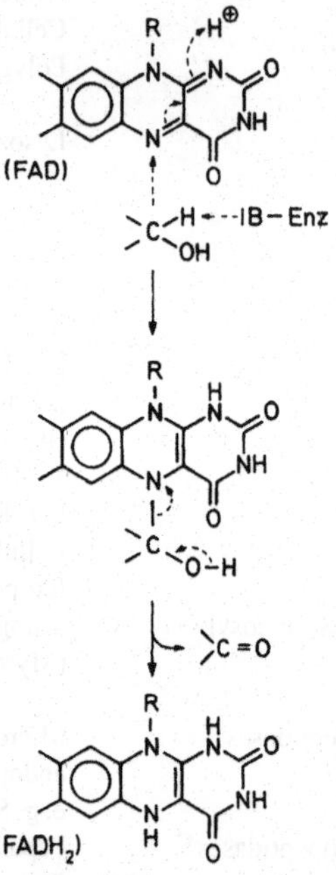

(2.9)

Like glucose oxidase, many flavin enzymes transfer the electrons to molecular oxygen, forming H_2O_2 and flavoquinone. The following intermediary products appear in this reaction:

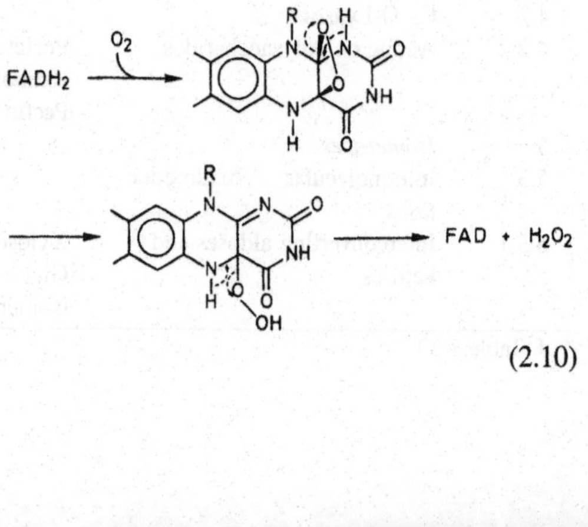

(2.10)

2.3.2.2 Hemin

Peroxidases from food of plant origin and several catalases contain ferri-protoporphyrin IX (hemin, cf. Formula 2.11) as their prosthetic group and as the chromophore responsible for the brown color of the enzymes:

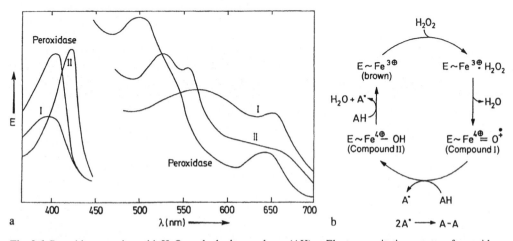

$$(2.11)$$

In catalytic reactions there is a change in the electron excitation spectra of the peroxidases (Fig. 2.6a) which is caused by a valence change of the iron ion (Fig. 2.6b). Intermediary compounds I (green) and II (pale red) are formed during this change by reaction with H_2O_2 and reducing agent AH. The reaction cycle is completed by another single electron transfer.

Some verdoperoxidases, which are green in color (as suggested by their name) and found in various foods of animal origin, e. g. milk, contain an unidentified Fe-protoporphyrin as their prosthetic group.

2.3.2.3 Pyridoxal Phosphate

Pyridoxal phosphate (Formula 2.12) and pyridoxamine (Formula 2.13), derived from it, are designated as vitamin B_6 (cf. 6.3.3) and are essential ingredients of food:

$$(2.12)$$

$$(2.13)$$

Coupled to the enzyme as a prosthetic group through a lysyl residue, pyridoxal phosphate is involved in conversion reactions of amino acids. In the first step of catalysis, the amino group of the amino acid substrate displaces the 6-amino group of lysine from the aldimine linkage (cf. Reaction 2.14). The positively charged pyridine ring then exerts an electron shift towards the α-C-atom of the amino acid substrate; the shift being supported by the release of one substituent of the α-C-atom. In Fig. 2.7 is shown how the ionization of the proton attached to the α-C-atom leads to *transamination*

Fig. 2.6. Peroxidase reaction with H_2O_2 and a hydrogen donor (AH). **a** Electron excitation spectra of peroxidase and intermediates I and II; **b** mechanism of catalysis

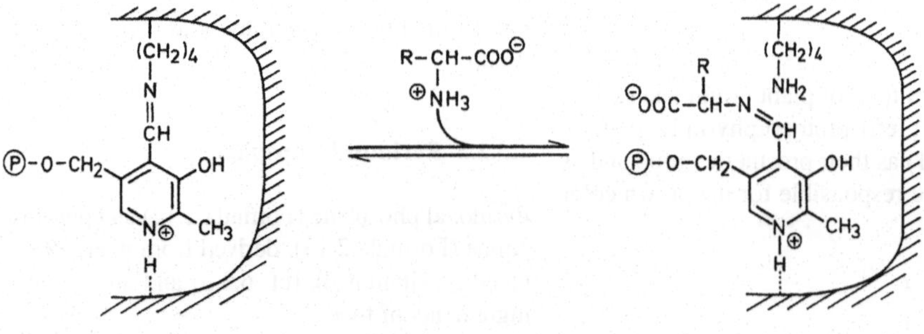

$$(2.14)$$

of the amino acid with formation of an α-keto acid. The reaction may also proceed through a *decarboxylation* (Fig. 2.7) and yield an amine. Which of these two pathway options will prevail is decided by the structure of the protein moiety of the enzyme.

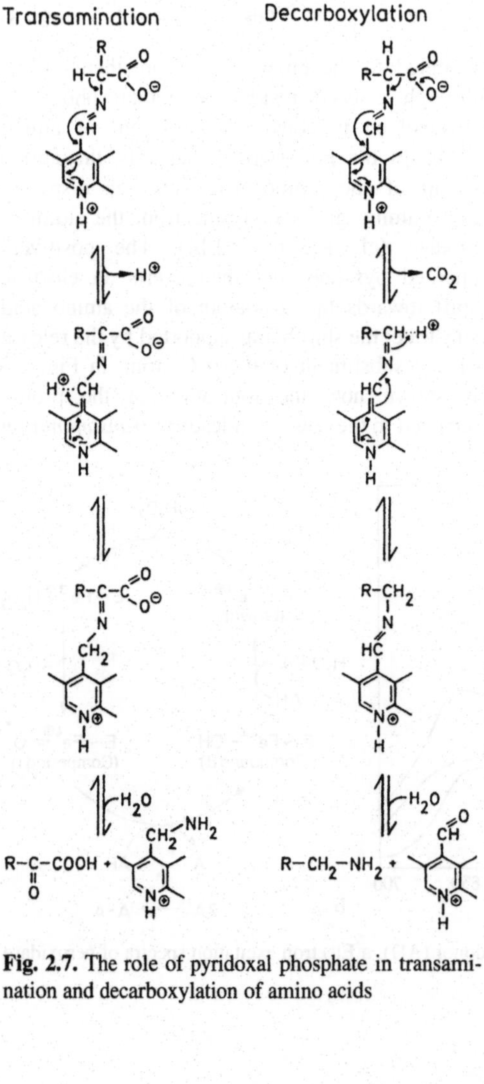

Transamination   **Decarboxylation**

Fig. 2.7. The role of pyridoxal phosphate in transamination and decarboxylation of amino acids

2.3.3 Metal Ions

Metal ions are indispensable cofactors and stabilizers of the conformation of many enzymes. They are especially effective as cofactors with enzymes converting small molecules. They influence the substrate binding and participate in catalytic reactions in the form of a *Lewis* acid or play the role of an electron carrier. Only the most important ions will be discussed.

2.3.3.1 Magnesium, Calcium and Zinc

$Mg^{2\oplus}$ ions activate some enzymes which hydrolyze phosphoric acid ester bonds (e. g. phosphatases; cf. Table 2.4) or transfer phosphate residues from ATP to a suitable acceptor (e. g. kinases; cf. Table 2.4). In both cases, $Mg^{2\oplus}$ ions act as an electrophilic *Lewis* acid, polarize the P–O-linkage of the phosphate residue of the substrate or cosubstrate and, thus, facilitate a nucleophilic attack (water with hydrolases; ROH in the case of kinases). An example is the hexokinase enzyme (cf. Table 2.16) which, in glycolysis, is involved in catalyzing the phosphorylation of glucose to glucose-6-phosphate with ATP as cosubstrate. The effect of a $Mg^{2\oplus}$ ion within the enzyme-substrate complex is obvious from the following formulation:

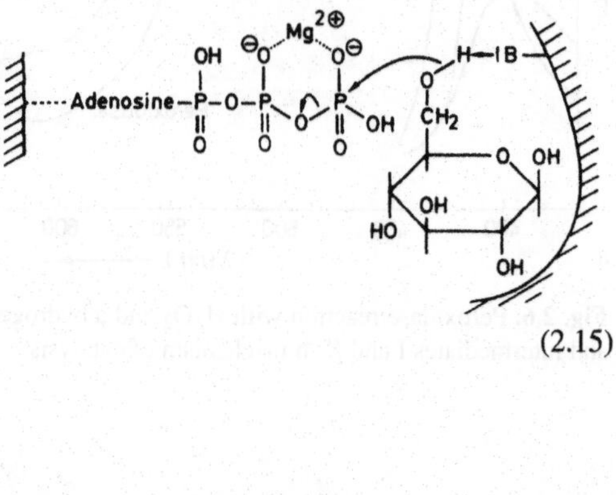

$$(2.15)$$

$Ca^{2\oplus}$ ions are weaker *Lewis* acids than $Mg^{2\oplus}$ ions. Therefore, the replacement of $Mg^{2\oplus}$ by $Ca^{2\oplus}$ may result in an inhibition of the kinase enzymes. Enhancement of the activity of other enzymes by $Ca^{2\oplus}$ is based on the ability of the ion to interact with the negatively charged sites of amino acid residues and, thus, to bring about stabilization of the enzyme conformation (e. g. α-amylase; cf. 4.4.4.5.1). The activation of the enzyme may be also caused by the involvement of the $Ca^{2\oplus}$ ion in substrate binding (e. g. lipase; cf. 3.7.1.1).

The $Zn^{2\oplus}$ ion, among the series of transition metals, is a cofactor which is not involved in redox reactions under physiological conditions. As a *Lewis* acid similar in strength to $Mg^{2\oplus}$, $Zn^{2\oplus}$ participates in similar reactions. Hence, substituting the $Zn^{2\oplus}$ ion for the $Mg^{2\oplus}$ ion in some enzymes is possible without loss of enzyme activity. Both metal ions can function as stabilizers of enzyme conformation and their direct participation in catalysis is readily revealed in the case of alcohol dehydrogenase. This enzyme isolated from horse liver consists of two identical polypeptide chains, each with one active site. Two of the four $Zn^{2\oplus}$ ions in the enzyme readily dissociate. Although this dissociation has no effect on the quaternary structure, the enzyme activity is lost. As described under section 2.3.1.1, both of these $Zn^{2\oplus}$ ions are involved in the formation of the active site. In catalysis they polarize the substrate's $C-O$ linkage and, thus, facilitate the transfer of hydride ions from or to the cosubstrate. Unlike the dissociable ions, removal of the two residual $Zn^{2\oplus}$ ions is possible only under drastic conditions, namely disruption of the enzyme's quaternary structure which is maintained by these two ions.

2.3.3.2 Iron, Copper and Molybdenum

The redox system of $Fe^{3\oplus}/Fe^{2\oplus}$ covers a wide range of potentials (Table 2.5) depending on the attached ligands. Therefore, the system is exceptionally suitable for bridging large potential differences in a stepwise electron transport system. Such an example is encountered in the transfer of electrons by the cytochromes as members of the respiratory chain (cf. textbook of biochemistry) or in the biosynthesis of unsaturated fatty acids (cf. 3.2.4), and by some individual enzymes.

Iron-containing enzymes are attributed either to the heme (examples in 3.3.2.2) or to the non-heme Fe-containing proteins. The latter case is exemplified by lipoxygenase, for which the mechanism of activity is illustrated in section 3.7.2.2, or by xanthine oxidase.

Xanthine oxidase from milk ($M_r = 275,000$) reacts with many electron donors and acceptors. However, this enzyme is most active with substrates such as xanthine or hypoxanthine as electron donors and molecular oxygen as the electron acceptor. The enzyme is assumed to have two active sites per molecule, with each having 1 FAD moiety, 4 Fe-atoms and 1 Mo-atom. During the oxidation of xanthine to uric acid:

$$(2.16)$$

oxygen is reduced by two one-electron steps to H_2O_2 by an electron transfer system in which the following valence changes occur:

$$(2.17)$$

Under certain conditions the enzyme releases a portion of the oxygen when only one electron transfer has been completed. This yields $O_2^{\ominus}$, the superoxide radical anion, with one unpaired electron. This ion can initiate lipid peroxidation by a chain reaction (cf. 3.7.2.1.8).

Polyphenol oxidases and ascorbic acid oxidase, which occur in food, are known to have a $Cu^{2\oplus}/Cu^{1\oplus}$ redox system as a prosthetic group. Polyphenol oxidases play an important role in the quality of food of plant origin because they cause the "*enzymatic browning*" for example in potatoes, apples and mushrooms. Tyrosinases, catecholases, phenolases or cresolases are enzymes that react with oxygen and a large range of mono and diphenols.

Table 2.5. Redox potentials of Fe^{3+}/Fe^{2+} complex compounds at pH 7 (25 °C) as affected by the ligand

Redox-System	E_0' (Volt)
$[Fe^{III}(o\text{-phen}^a)_3]^{3\oplus}/[Fe^{II}(o\text{-phen})_3]^{2\oplus}$	+1.10
$[Fe^{III}(OH_2)_6]^{3\oplus}/[Fe^{II}(OH_2)_6]^{2\oplus}$	+0.77
$[Fe^{III}(CN)_6]^{3\ominus}/[Fe^{II}(CN)_6]^{4\ominus}$	+0.36
Cytochrome a($Fe^{3\oplus}$)/Cytochrome a ($Fe^{2\oplus}$)	+0.29
Cytochrome c ($Fe^{3\oplus}$)/Cytochrome c ($Fe^{2\oplus}$)	+0.26
Hemoglobin ($Fe^{3\oplus}$)/Hemoglobin ($Fe^{2\oplus}$)	+0.17
Cytochrome b ($Fe^{3\oplus}$)/Cytochrome b ($Fe^{2\oplus}$)	+0.04
Myoglobin ($Fe^{3\oplus}$)/Myoglobin ($Fe^{2\oplus}$)	0.00
$(Fe^{III}EDTA)^{1\ominus}/(Fe^{II}EDTA)^{2\ominus}$	−0.12
$(Fe^{III}(oxin^b)_3)/(Fe^{II}(oxin)_3)^{1\ominus}$	−0.20
Ferredoxin ($Fe^{3\oplus}$)/Ferredoxin ($Fe^{2\oplus}$)	−0.40

[a] o-phen: o-Phenanthroline.
[b] oxin: 8-Hydroxyquinoline.

Polyphenol oxidase catalyzes two reactions: first the hydroxylation of a monophenol to o-diphenol (EC 1.14.18.1, monophenol monooxygenase) followed by an oxidation to o-quinone (EC 1.10.3.1, o-diphenol: oxygen oxidoreductase). Both activities are also known as cresolase and catecholase activity. At its active site, polyphenol oxidase contains two $Cu^{1\oplus}$ ions with two histidine residues each in the ligand field. In an "ordered mechanism" (cf. 2.5.1.2.1) the enzyme first binds oxygen and later monophenol with participation of the intermediates shown in Fig. 2.8. The Cu ions change their valency ($Cu^{1\oplus} \rightarrow Cu^{2\oplus}$). The newly formed complex ([] in Fig. 2.8) has a strongly polarized

O−O=bonding, resulting in a hydroxylation to o-diphenol. The cycle closes with the oxidation of o-diphenol to o-quinone.

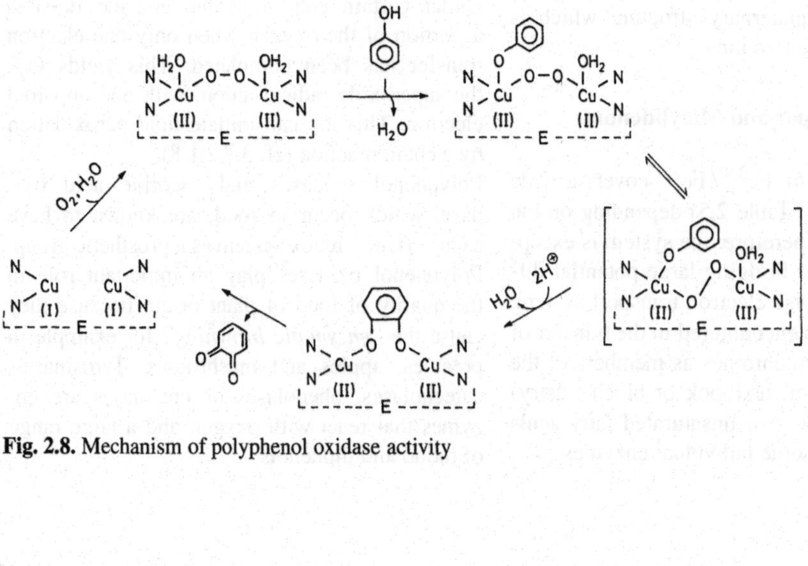

2.4 Theory of Enzyme Catalysis

It has been illustrated with several examples (Table 2.1) that enzymes are substantially better catalysts than are protons or other ionic species used in nonenzymatic reactions. Enzymes invariably surpass all chemical catalysts in relation to substrate and reaction specificities.

Theories have been developed to explain the exceptional efficiency of enzyme activity. They are based on findings which provide only indirect insight into enzyme catalysis. Examples are the identification of an enzyme's functional groups involved in catalysis, elucidation of their arrangement within the tertiary structure of the enzyme, and the detection of conformational changes induced by substrate binding. Complementary studies involve low molecular weight model substrates, the reactions of which shed light on the active sites or groups of the enzyme and their coordinated interaction with other factors affecting enzymatic catalysis.

2.4.1 Active Site

An enzyme molecule is, when compared to its substrate, often larger in size by a factor

Fig. 2.8. Mechanism of polyphenol oxidase activity

of several orders of magnitude. For example, glucose oxidase ($M_r = 1.5 \cdot 10^5$) and glucose ($M_r = 180$). This strongly suggests that in catalysis only a small locus of an active site has direct contact with the substrate. Specific parts of the protein structure participate in the catalytic process from the substrate binding to the product release from the so-called *active site*. These parts are amino acid residues which bind substrate and, if required, cofactors and assist in conversion of substrate to product.

Investigations of the structure and function of the active site are conducted to identify the amino acid residues participating in catalysis, their steric arrangement and mobility, the surrounding micro-environment and the catalysis mechanism.

2.4.1.1 Active Site Localization

Several methods are generally used for the identification of amino acid residues present at the active site since data are often equivocal. Once obtained, the data must still be interpreted with a great deal of caution and insight.

The influence of pH on the activity assay (cf. 2.5.3) provides the first direct answer as to whether dissociable amino acid side chains, in charged or uncharged form, assist in catalysis. The data readily obtained from this assay must again be interpreted cautiously since neighboring charged groups, hydrogen bonds or the hydrophobic environment of the active site can affect the extent of dissociation of the amino acid residues and, thus, can shift their pK values (cf. 1.4.3.1).

Selective labeling of side chains which form the active site is also possible by chemical modification. When an enzyme is incubated with reagents such as iodoacetic acid (cf. 1.2.4.3.5) or dinitrofluorobenzene (cf. 1.2.4.2.2), resulting in a decrease of activity, and subsequent analysis of the modified enzyme shows that only one of the several available functional groups is bound to reagent (e. g. one of several −SH groups), then this group is most probably part of the active site. Selective labeling data when an inhibiting substrate analogue is used are more convincing. Because of its similarity to the chemical structure of the substrate, the analogue will be bound covalently to the enzyme but not converted into product. We will consider the following examples:

N-tosyl-L-phenylalanine ethyl ester (Formula 2.18) is a suitable substrate for the proteinase chymotrypsin which hydrolyzes ester bonds. When the ethoxy group is replaced by a chloromethyl group, an inhibitor whose structure is similar to the substrate is formed (N_α-tosyl-L-phenylalanine chloromethylketone, TPCK).

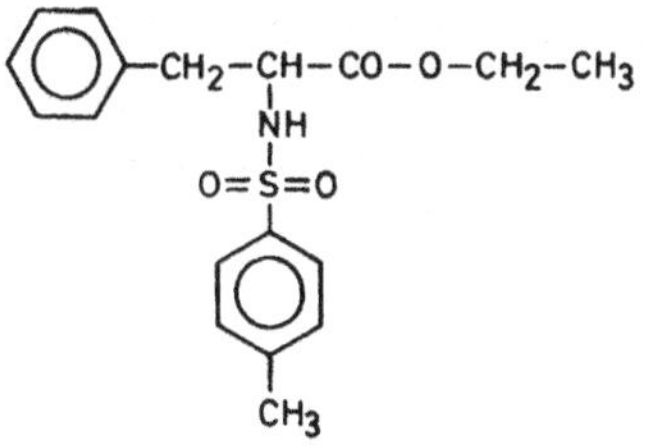

$$(2.18)$$

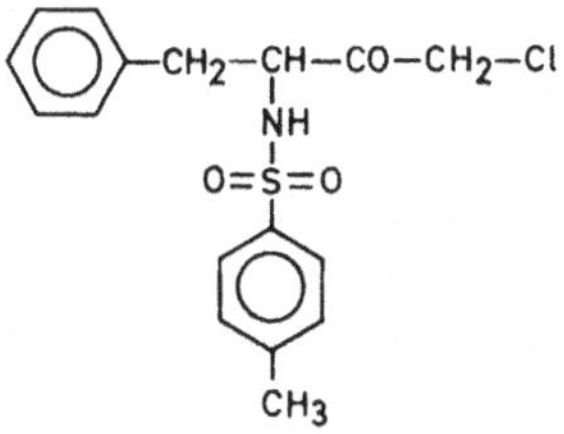

$$(2.19)$$

Thus, the substrate analogue binds specifically and irreversibly to the active site of chymotrypsin. Analysis of the enzyme inhibitor complex reveals that, of the two histidine residues present in chymotrypsin, only His[57] is alkylated at one of its ring nitrogens. Hence, the modified His residue is part of the active site (cf. mechanism of chymotrypsin catalysis, Fig. 2.17). TPCK binds highly specifically, thus the proteinase trypsin is not inhibited. The corresponding inhibiting substrate analogue, which binds exclusively to trypsin, is N-tosyl-L-lysine chloromethylketone (TLCK):

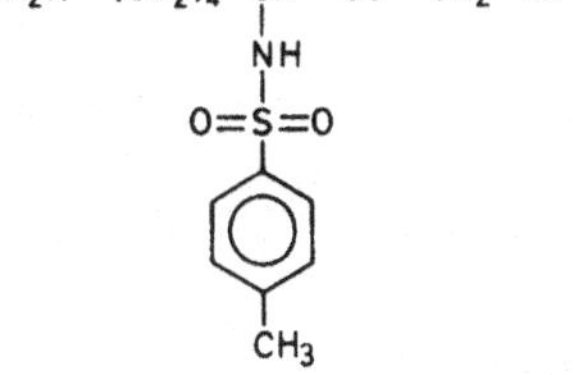

$$(2.20)$$

Reaction of diisopropylfluorophosphate (DIFP)

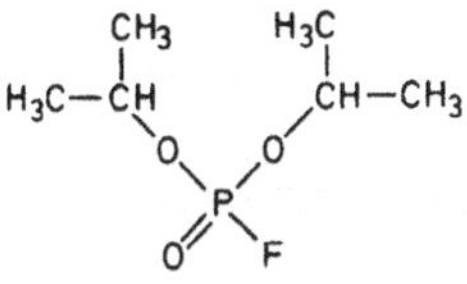

$$(2.21)$$

with a number of proteinases and esterases alkylates the unusually reactive serine residue at the active site. Thus, of the 28 serine residues present in chymotrypsin, only Ser^{195} is alkylated, while the other 27 residues are unaffected by the reagent. It appears that the reactivity of Ser^{195} is enhanced by its interaction with the neighboring His^{57} (cf. mechanism of catalysis in Fig. 2.17). The participation of a carboxyl group at the active site in β-glucosidase catalysis has been confirmed with the help of conduritol B-epoxide, an inhibiting substrate analogue:

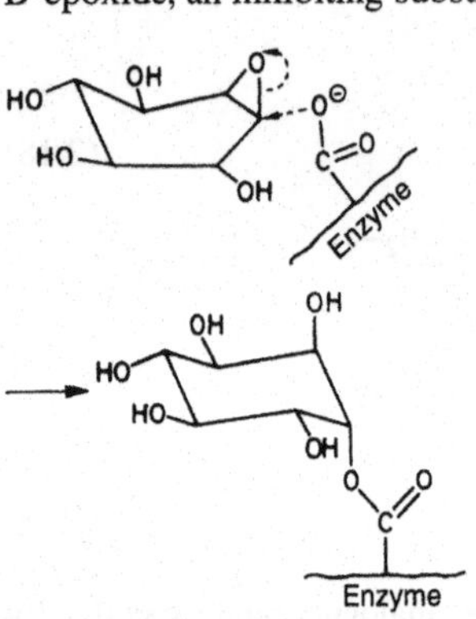

$$(2.22)$$

A lysine residue is involved in enzyme catalysis in a number of lyase enzymes and in enzymes in which pyridoxal phosphate is the cosubstrate. An intermediary *Schiff* base product is formed between an ε-amino group of the enzyme and the substrate or pyridoxal phosphate (cf. 2.3.2.3). The reaction site is then identified by reduction of the *Schiff* base with $NaBH_4$.

An example of a "lysine" lyase is the aldolase enzyme isolated from rabbit muscle. The intermediary product formed with dihydroxyacetone phosphate (cf. mechanism in Fig. 2.19) is detected as follows:

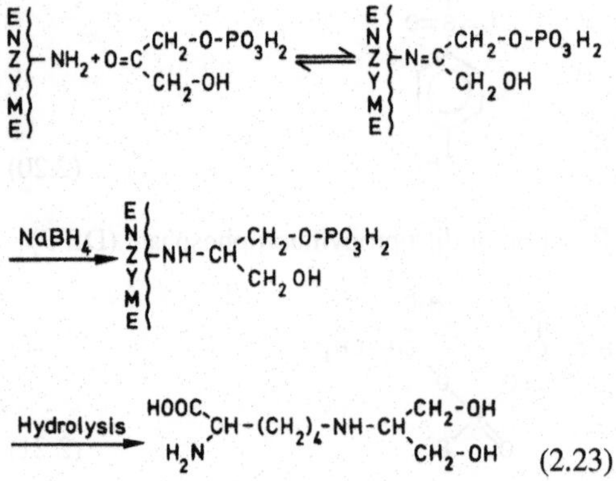

$$(2.23)$$

2.4.1.2 Substrate Binding

2.4.1.2.1 Stereospecificity

Enzymes react stereospecifically. Before being bound to the binding locus, the substrates are distinguished by their cis, trans-isomerism and also by their optical antipodes. The latter property was illustrated by the reactions of L(+)-lactic acid (Fig. 2.2). There are distinct recognition areas on the binding locus. Alcohol dehydrogenase will be used to demonstrate this. This enzyme removes two hydrogen atoms, one from the methylene group and the other from the hydroxyl group, to produce acetaldehyde. However, the enzyme recognizes the difference between the two methylene hydrogens since it always stereospecifically removes the same hydrogen atom. For example, yeast alcohol dehydrogenase always removes the pro-R-hydrogen from the C-1 position of a stereospecifically deuterated substrate and transfers it to the C-4 position of the nicotinamide ring of NAD:

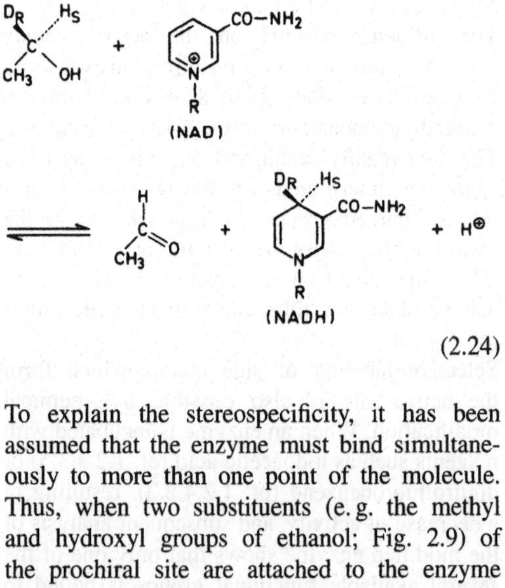

$$(2.24)$$

To explain the stereospecificity, it has been assumed that the enzyme must bind simultaneously to more than one point of the molecule. Thus, when two substituents (e. g. the methyl and hydroxyl groups of ethanol; Fig. 2.9) of the prochiral site are attached to the enzyme surface at positions A and B, the position of the third substituent is fixed. Therefore, the same substituent will always be bound to reactive position C, e. g. one of the two methylene hydrogens in ethanol. In other words, the two equal substituents in a symmetrical molecule are differentiated by asymmetric binding to the enzyme.

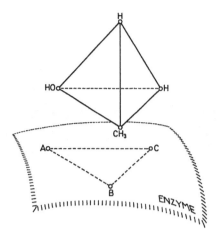

Fig. 2.9. A model for binding of a prochiral substrate (ethanol) by an enzyme

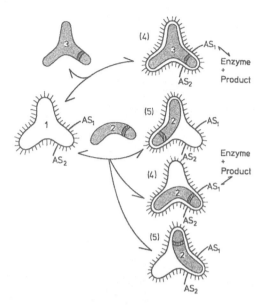

Fig. 2.10. Binding of a good (3) and of a bad substrate (2) by the active site (1) of the enzyme (according to *Jencks*, 1969). (4) A productive enzyme-substrate complex; (5) a nonproductive enzyme substrate complex. As_1 and As_2: reactive amino acid residues of the enzyme involved in conversion of substrate to product

2.4.1.2.2 "Lock and Key" Hypothesis

To explain substrate specificity, E. *Fischer* proposed a hypothesis a century ago in which he depicted the substrate as being analogous to a key and the enzyme as its lock. According to this model, the active site has a geometry which is complementary only to its substrate (Fig. 2.10). In contrast, there are many possibilities for a "bad" substrate to be bound to the enzyme, but only one provides the properly positoned enzyme-substrate complex, as illustrated in Fig. 2.10, which is converted to the product.

The proteinases chymotrypsin and trypsin are two enzymes for which secondary and tertiary structures have been elucidated by x-ray analysis and which have structures supporting the lock and key hypothesis to a certain extent. The binding site in chymotrypsin and trypsin is a three-dimensional hydrophobic pocket (Fig. 2.11). Bulky amino acid residues such as aromatic amino acids fit neatly into the pocket (chymotrypsin, Fig. 2.11a), as do substrates with lysyl or arginyl residues (trypsin, Fig. 2.11b). Instead of Ser^{189}, the trypsin peptide chain has Asp^{189} which is present in the deep cleft in the form of a carboxylate anion and which attracts the positively charged lysyl or arginyl residues of the substrate. Thus, the substrate is stabilized and realigned by its peptide bond to face the enzyme's Ser^{195} which participates in hydrolysis (transforming locus).

The peptide substrate is hydrolyzed by the enzyme elastase by the same mechanism as for chymotrypsin. However, here the pocket is closed to such an extent by the side chains of Val^{216} and Thr^{226} that only the methyl group of alanine can enter the cleft (Fig. 2.11c). Therefore, elastase has specificity for alanyl peptide bonds or alanyl ester bonds.

2.4.1.2.3 Induced-fit Model

The conformation of a number of enzymes is changed by the binding of the substrate. An example is carboxypeptidase A, in which the Try^{248} located in the active site moves approximately 12 Å towards the substrate, glycyl-L-phenylalanine, to establish contact. This and other observations support the dynamic induced-fit model proposed by *Koshland* (1964). Here, only the substrate has the power to induce a change in the tertiary structure to the active form of the enzyme. Thus, as the substrate molecule approaches the enzyme surface, the amino acid

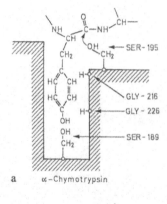

a   α-Chymotrypsin

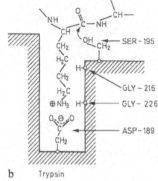

b   Trypsin

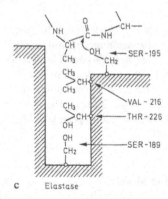

c   Elastase

Fig. 2.11. A hypothesis for substrate binding by α-chymotrypsin, trypsin and elastase enzymes (according to *Shotton*, 1971)

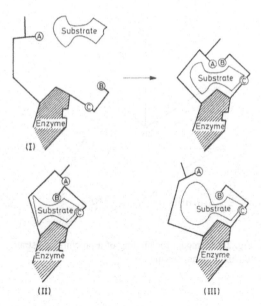

Fig. 2.12. A schematic presentation of "induced-fit model" for an active site of an enzyme (according to *Koshland*, 1964).
— Polypeptide chain of the enzyme with catalytically active residues of amino acids, A and B; the residue C binds the substrate

residues A and B change their positions to conform closely to the shape of the substrate (I, in Fig. 2.12). Groups A and B are then in the necessary position for reaction with the substrate. Diagrams II and III (Fig. 2.12) illustrate the case when the added compound is not suitable as substrate. Although group C positioned the substrate correctly at its binding site, the shape of the compound prevents groups A and B from being aligned properly in their active positions and, thus, from generating the product.

In accordance with the mechanisms outlined above, one theory suitable for enzymes following the lock and key mechanism and the other theory for enzymes operating with the dynamic induced-fit model, the substrate specificity of any enzyme-catalyzed reaction can be explained satisfactorily.

In addition, the relationship between enzyme conformation and its catalytic activity thus outlined also accounts for the extreme sensitivity of the enzyme as catalyst. Even slight interferences imposed on their tertiary structure which affect the positioning of the functional groups result in loss of catalytic activity.

2.4.2 Reasons for Catalytic Activity

Even though the rates of enzymatically catalyzed reactions vary, they are very high compared to the effectiveness of chemical catalysts (examples in Table 2.2). The factors responsible for the high increase in reaction rate are outlined below. They are of different importance for the individual enzymes.

2.4.2.1 Steric Effects – Orientation Effects

The specificity of substrate binding contributes substantially to the rate of an enzyme-catalyzed reaction.

Binding to the active site of the enzyme concentrates the reaction partners in comparison with a dilute substrate solution. In addition, the reaction is now the favored one since binding places the substrate's susceptible reactive group in the proximity of the catalytically active group of the enzyme.

Therefore the contribution of substrate binding to the reaction rate is partially due to a change in the molecularity of the reaction. The intermolecular reaction of the two substrates is replaced by an intramolecular reaction of an enzyme-substrate complex. The consequences can be clarified by using model compounds which have all the reactive groups within their molecules and, thus, are subjected to an intramolecular reaction. Their reactivity can then be compared with that of the corresponding bimolecular system and the results expressed as a ratio of the reaction rates of the intramolecular (k_1) to the intermolecular (k_2) reactions. Based on their dimensions, they are denoted as "effective molarity". As an example, let us consider the cleavage of p-bromophenylacetate in the presence of acetate ions, yielding acetic acid anhydride:

$$(2.25)$$

Intramolecular hydrolysis is substantially faster than the intermolecular reaction (Table 2.6). The effective molarity sharply increases when the reactive carboxylate anion is in close proximity to the ester carbonyl group and, by its presence, retards the mobility of the carbonyl group. Thus, the effective molarity increases (Table 2.6) as the C–C bond mobility decreases. Two bonds can rotate in a glutaric acid ester, whereas only one can

rotate in a succinic acid ester. The free rotation is effectively blocked in a bicyclic system. Hence, the reaction rate is sharply increased. Here, the rigid steric arrangement of the acetate ion and of the ester group provides a configuration that imitates that of a transition state.

In contrast to the examples given in Table 2.6, examples should be mentioned in which substrates are not bound covalently by their enzymes. The following model will demonstrate that other interactions can also promote close positioning of the two reactants. Hydrolysis of p-nitrophenyldecanoic acid ester is catalyzed by an alkylamine:

$$(2.26)$$

The reaction rate in the presence of decylamine is faster than that in the presence of ethylamine by a factor of 700. This implies that the reactive amino group has been oriented very close to the susceptible carbonyl group of the ester by the establishment of a maximal number of hydrophobic contacts. Correspondingly, there is a decline in the reaction rate as the alkyl amine group is lengthened further.

Table 2.6. Relative reaction rate for the formation of acid anhydrides

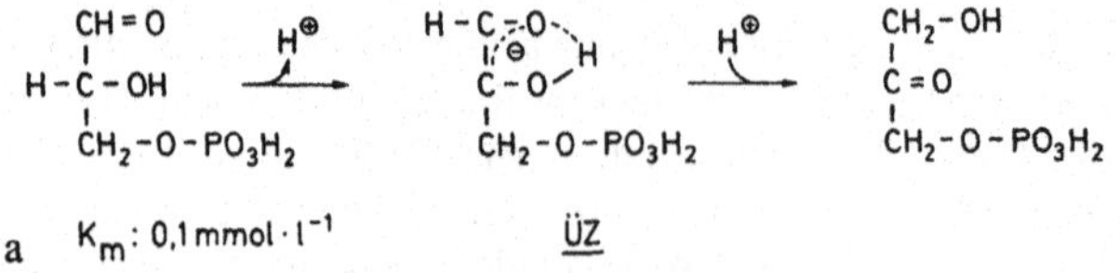

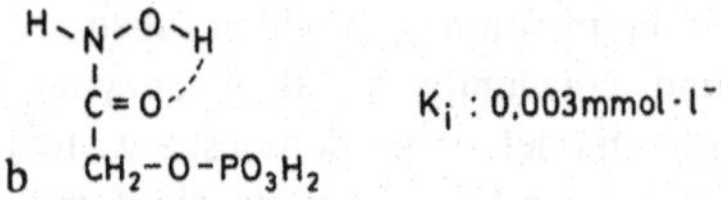

$$
\begin{array}{ccc}
\begin{array}{l}
CH=O \\
| \\
H-C-OH \\
| \\
CH_2-O-PO_3H_2
\end{array}
&
\xrightarrow{H^{\oplus}}
&
\begin{array}{l}
H-C \cdot . O \cdots \\
\quad \! \! \! \! \mathop{.}\limits_{.}^{.} \, {}^{\oplus}\! \! H \\
C-O \diagup \\
| \\
CH_2-O-PO_3H_2
\end{array}
\xrightarrow{H^{\oplus}}
\begin{array}{l}
CH_2-OH \\
| \\
C=O \\
| \\
CH_2-O-PO_3H_2
\end{array}
\end{array}
$$

a K_m: 0,1 mmol · l^{-1} ÜZ

$$
\begin{array}{l}
H \diagdown N \diagup O \diagdown H \\
\quad | \\
\quad C=O \diagup \\
\quad | \\
CH_2-O-PO_3H_2
\end{array}
\qquad K_i : 0,003 \, mmol \cdot l^{-1}
$$

b

Fig. 2.13. Example of a transition state analog inhibitor **a** reaction of triosephosphate isomerase, TT: postulated transition state; **b** inhibitor

2.4.2.2 Structural Complementarity to Transition State

It is assumed that the active conformation of the enzyme matches the transition state of the reaction. This is supported by affinity studies which show that a compound with a structure analogous to the transition state of the reaction ("*transition state analogs*") is bound better than the substrate. Hydroxamic acid, for example, is such a transition state analog which inhibits the reaction of triosephosphate isomerase (Fig. 2.13). Comparisons between the *Michaelis* constant and the inhibitor constant show that the inhibitor has a 30 times higher affinity to the active site than the substrate.

The active site is complementary to the transition state of the reaction to be catalyzed. This assumption is supported by a reversion of the concept. It has been possible to produce catalytically active monoclonal antibodies directed against transition state analogs. The antibodies accelerate the reaction approximating the transition state of the analog. However, their catalytic activity is weaker compared to enzymes because only the environment of the antibody which is complementary to the transistion state causes the acceleration of the reaction.

Transition state analog inhibitors were used to show that in the binding the enzyme displaces the hydrate shell of the substrates. The reaction rate can be significantly increased by removing the hydrate shell between the participants.

Other important factors in catalytic reactions are the distortion of bonds and shifting of charges. The substrate's bonds will be strongly polarized by the enzyme, and thus highly reactive, through the precise positioning of an acid or base group or a metall ion (*Lewis* acid, cf. 2.3.3.1) (example see Formula 2.15). These hypotheses are supported by investigations using suitable transition state analog inhibitors.

2.4.2.3 Entropy Effect

An interpretation in thermodynamic terms takes into account that a loss of entropy occurs during catalysis due to the loss of freedom of rotation and translation of the reactants. This entropy effect is probably quite large in the case of the formation of an enzyme-substrate complex since the reactants are fairly rigidly positioned before the transition state is reached. Consequently, the conversion of the enzyme-substrate complex to the transition state is accompanied by little or no change of entropy. As an example, a reaction running at 27 °C with a decrease in entropy of $140 \, J K^{-1} \, mol^{-1}$ is considered. Calculations indicate that this decrease leads to a reduction in free activation energy by about 43 kJ. This value falls in the range of the amount by which the activation energy of a reaction is lowered by an enzyme (cf. Table 2.1) and which can have the effect of increasing the reaction rate by a factor of 10^8

The catalysis by chymotrypsin, for example, shows how powerful the entropy effects can be. In section 2.4.2.5 we will see that this catalysis is a two-step event proceeding through an acylated enzyme intermediate. Here we will consider only the second step, deacylation, thereby distinguishing the following intermediates:

a) N-acetyl-L-tyrosyl-chymotrypsin
b) Acetyl-chymotrypsin.

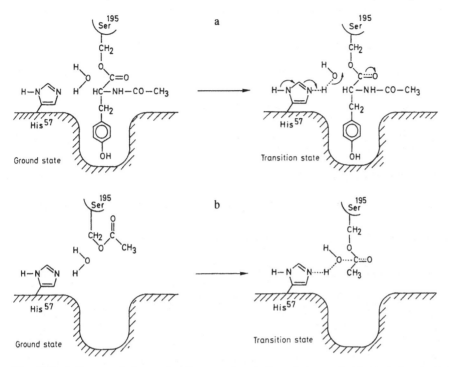

Fig. 2.14. Influence of the steric effect on deacylation of two acyl-chymotrypsins (according to *Bender* et al., 1964). **a** N-acetyl-L-tyrosyl-chymotrypsin, **b** acetyl-chymotrypsin

In case a) deacylation is faster by a factor of 3540 since the carbonyl group is immobilized by insertion of the bulky N-acetyl-L-tyrosyl group into a hydrophobic pocket on the enzyme (Fig. 2.14a) at the correct distance from the attacking nucleophilic $OH^{\ominus}$ ion derived from water (cf. 2.4.2.5). In case b) the immobilization of the small acetyl group is not possible (Fig. 2.14b) so that the difference between the ground and transition states is very large. The closer the ground state is to the transition state, the more positive will be the entropy of the transition state, $\Delta S^{\#}$; a fact that as mentioned before can lead to a considerable increase in reaction rate. The thermodynamic data in Table 2.7 show that the difference in reaction rates depends, above all, on an entropy effect; the enthalpies of the transition states scarcely differ.

Table 2.7. Thermodynamic data for transition states of twoacyl-chymotrypsins

Acyl-enzyme	$\Delta G^{\#}$ $(kJ \cdot mol^{-1})$	$\Delta H^{\#}$ $(kJ \cdot mol^{-1})$	$\Delta S^{\#}$ $(J \cdot K^{-1} \cdot mol^{-1})$
N-Acetyl-L-tyrosyl	59.6	43.0	−55.9
Acetyl	85.1	40.5	−149.7

2.4.2.4 General Acid–Base Catalysis

When the reaction rate is affected by the concentration of hydronium ($H_3O^{\oplus}$) or $OH^{\ominus}$ ions from water, the reaction is considered to be specifically acid or base catalyzed. In the so-called general acid or base catalysis the reaction rate is affected by prototropic groups located on the side chains of the amino acid residues. These groups involve proton donors (denoted as general acids) and proton acceptors (general bases). Most of the amino acids located on the active site of the enzyme influence the reaction rate by general acid-base catalysis.

As already mentioned, the amino acid residues in enzymes have prototropic groups which have the potential to act as a general acid or as a general base. Of these, the imidazole ring of histidine is of special interest since it can perform both functions simultaneously:

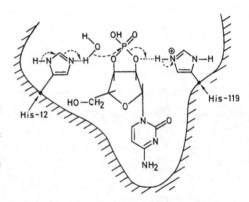

$$
(2.27)
$$

The imidazole ring ($pK_2 = 6.1$) can cover the range of the pH optima of many enzymes.

Thus, two histidine residues are involved in the catalytic activity of ribonuclease, a phosphodiesterase. The enzyme hydrolyzes pyrimidine-$2',3'$-cyclic phosphoric acids. As shown in Fig. 2.15, cytidine-$2',3'$-cyclic phosphoric acid is positioned between two imidazole groups at the binding locus of the acitive site.

His^{12} serves as a general base, removing the proton from a water molecule. This is followed by nucleophilic attack of the intermediary $OH^{\ominus}$ ions on the electrophilic phosphate group. This attack is supported by the concerted action of the general acid His^{119}.

Another concerted general acid-base catalysis is illustrated by triose phosphate isomerase, an enzyme involved in glycolysis. Here, the concerted action involves the carboxylate anion of a glutamic acid residue as a general base with a general acid which has not yet been identified:

$$
(2.28)
$$

The endiol formed from dihydroxyacetone-3-phosphate in the presence of enzyme isomerizes into glyceraldehyde-3-phosphate.

These two examples show clearly the significant differences to chemical reactions in solutions. The enzyme driven acid-base catalysis takes place selectively at a certain locus of the active site. The local concentration of the amino acid residue acting as acid or base is fairly high due to the perfect position relative to the substrate. On the other hand, in chemical reactions in solutions all reactive groups of the substrate are nonspecifically attacked by the acid or base.

Fig. 2.15. Hydroylysis of cytidine-$2',3'$-phosphate by ribonuclease (reaction mechanism according to *Findlay*, 1962)

2.4.2.5 Covalent Catalysis

Studies aimed at identifying the active site of an enzyme (cf. 2.4.1.1) have shown that, during catalysis, a number of enzymes bind the substrate by covalent linkages. Such covalent linked enzyme-substrate complexes form the corresponding products much faster than compared to the reaction rate in a non-catalyzed reaction.

Examples of enzyme functional groups which are involved in covalent bonding and are responsible for the transient intermediates of an enzyme-substrate complex are compiled in Table 2.8. Nucleophilic catalysis is dominant (examples 1–6, Table 2.8), since amino acid residues are present in the active site of these enzymes, which

Table 2.8. Examples of covalently linked enzymesubstrate intermediates

Enzyme	Reactive functional group	Intermediate
1. Chymotrypsin	HO-(Serine)	Acylated enzyme
2. Papain	HS-(Cysteine)	Acylated enzyme
3. β-Amylase	HS-(Cysteine)	Maltosylenzyme
4. Aldolase	ε-H_2N-(Lysine)	*Schiff* base
5. Alkaline phosphatase	HO-(Serine)	Phosphoenzyme
6. Glucose-6-phosphatase	Imidazole-(Histidine)	Phosphoenzyme
7. Histidine decarboxylase	O=C < (Pyruvate)	*Schiff* base

only react with substrate by donating an electron pair (nucleophilic catalysis). Electrophilic reactions occur mostly by involvement of carbonyl groups (example 7, Table 2.8) or with the help of metal ions.

A number of peptidase and esterase enzymes react covalently in substitution reactions by a two-step nucleophilic mechanism. In the first step, the enzyme is acylated; in the second step, it is deacylated. Chymotrypsin will be discussed as an example of this reaction mechanism. Its activity is dependent on His^{57} and Ser^{195}, which are positioned in close proximity within the active site of the enzyme because of folding of the peptide chain (Fig. 2.16).

Because Asp^{102} is located in hydrophobic surroundings, it can polarize the functional groups in close proximity to it. Thus, His^{57} acts as a strong general base and abstracts a proton from the OH-group of the neighboring Ser^{195} residue (step 'a', Fig. 2.17). The oxygen remaining on Ser^{195} thus becomes a strong nucleophile and attacks the carbon of the carbonyl group of the peptide bond of the substrate. At this stage an amine (the first product) is released (step 'b', Fig. 2.17) and the transient covalently-bound acyl enzyme is formed. A deacylation step follows. The previous position of the amine is occupied by a water molecule. Again, His^{57}, through support from Asp^{102}, serves as a general base, abstracting the proton from water (step 'c', Fig. 2.17). This is followed by nucleophilic attack of the resultant $OH^{\ominus}$ ion on the carbon of the carbonyl group of the acyl enzyme (step 'd', Fig. 2.17), resulting

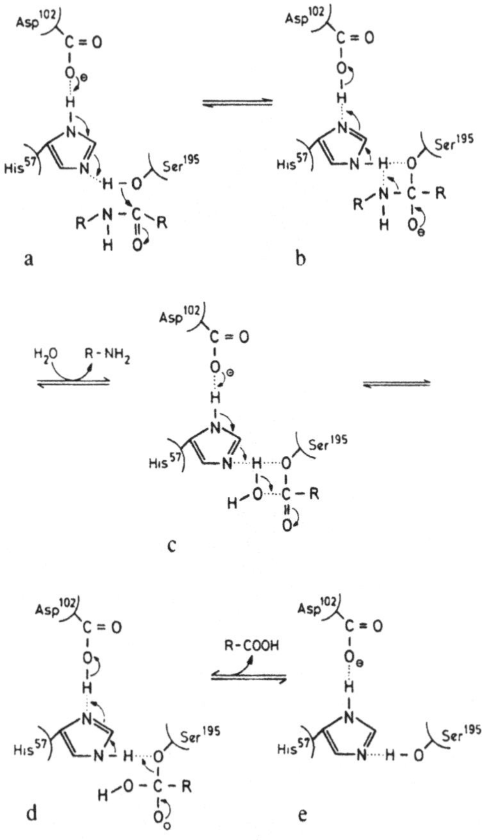

Fig. 2.17. Postulated reaction mechanism for chymotrypsin activity (according to *Blow* et al., 1969)

in free enzyme and the second product of the enzymic conversion.

An exceptionally reactive serine residue has been identified in a great number of hydrolase enzymes, e. g., trypsin, subtilisin, elastase, acetylcholine esterase and some lipases. These enzymes appear to hydrolyze their substrates by a mechanism analogous to that of chymotrypsin. Hydrolases such as papain, ficin and bromelain, which are distributed in plants, have a cysteine residue instead of an "active" serine residue in their active sites. Thus, the transient intermediates are thioesters.

Enzymes involved in the cleavage of carbohydrates can also function by the above mechanism. Figure 2.18 shows that amylose hydrolysis by β-amylase occurs with the help of four functional groups in the active site. The enzyme-substrate complex is subjected to a nucleophilic attack

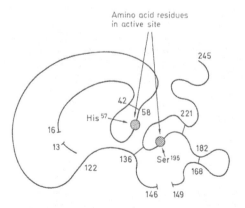

Fig. 2.16. Polypeptide chain conformation in the chymotrypsin molecule (according to *Lehninger*, 1977)

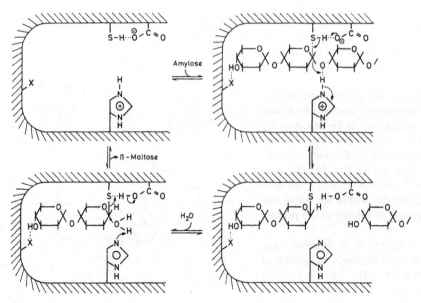

Fig. 2.18. Postulated mechanism for hydrolysis of amylose by β-amylase

by an SH-group on the carbon involved in the α-glycosidic bond. This transition step is facilitated by the carboxylate anion in the role of a general base and by the imidazole ring as an acid which donates a proton to glycosidic oxygen. In the second transition state the imidazole ring, as a general base in the presence of a water molecule, helps to release maltose from the maltosylenzyme intermediate.

Lysine is another amino acid residue actively involved in covalent enzyme catalysis (cf. 2.4.1.1). Many lyases react covalently with a substrate containing a carbonyl group. They catalyze, for example, aldol or retroaldol condensations important for the conversion and cleavage of monosaccharides or for decarboxylation reactions of β-keto acids. As an example, the details of the reaction involved will be considered for aldolase (Fig. 2.19). The enzyme-substrate complex is first stabilized by electrostatic interaction between the phosphate residues of the substrate and the charged groups present on the enzyme. A covalent intermediate, a *Schiff* base, is then formed by nucleophilic attack of the ε-amino group of the "active" lysine on a carbonyl group of the substrate. The *Schiff* base cation facilitates the retroaldol cleavage of the substrate, whereas a negatively charged group on the enzyme (e.g. a thiolate or carboxylate anion) acts as a general base, i.e. binds the free proton. Thus,

Fructose-1,6-diphosphate + Enzyme

$\downarrow\uparrow$ — H_2O

HC-O-(P)
    | H
   C=N —   "active" lysine
    |⊕
HO-C-H
    |
H-C-O-H ← |B —
    |
H-C-OH
    |
CH₂-O-(P)

$\downarrow\uparrow$ — Glyceraldehyde -3 -(P)

H₂C-O-(P)
    | H
   C-N —   "active" lysine
   H⊕
H⊕   HO-C-H

$\downarrow\uparrow$ — H_2O

Dihydroxyacetone -(P) + enzyme

Fig. 2.19. Aldolase of rabbit muscle tissue. A model for its activity; P: PO_3H_2

the first product, glyceraldehyde-3-phosphate, is released. An enamine rearrangement into a ketimine structure is followed by release of dihydroxyacetone phosphate.

This is the mechanism of catalysis by aldolases which occur in plant and animal tissues (lysine aldolases or class I aldolases). A second group of these enzymes often produced by microorganisms contains a metal ion (metallo-aldolases). This group is involved in accelerating retroaldol condensations through electrophilic reactions with carbonyl groups:

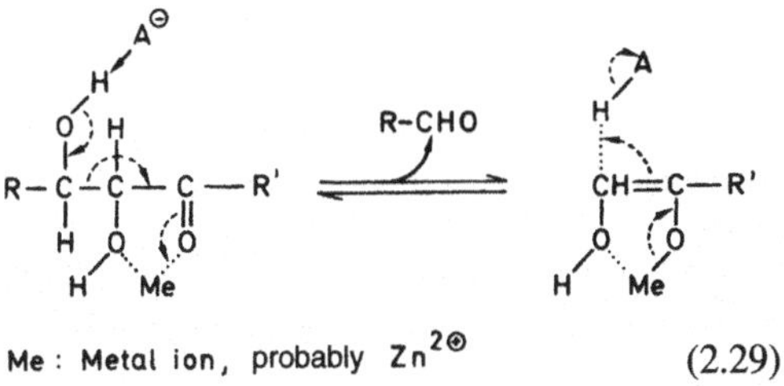

Me : Metal ion, probably $Zn^{2\oplus}$ (2.29)

Other examples of electrophilic metal catalysis are given under section 2.3.3.1. Electrophilic reactions are also carried out by enzymes which have an α-keto acid (pyruvic acid or α-keto butyric acid) at the transforming locus of the active site. One example of such an enzyme is histidine decarboxylase in which the N-terminal amino acid residue is bound to pyruvate. Histidine decarboxylation is initiated by the formation of a *Schiff* base by the reaction mechanism in Fig. 2.20.

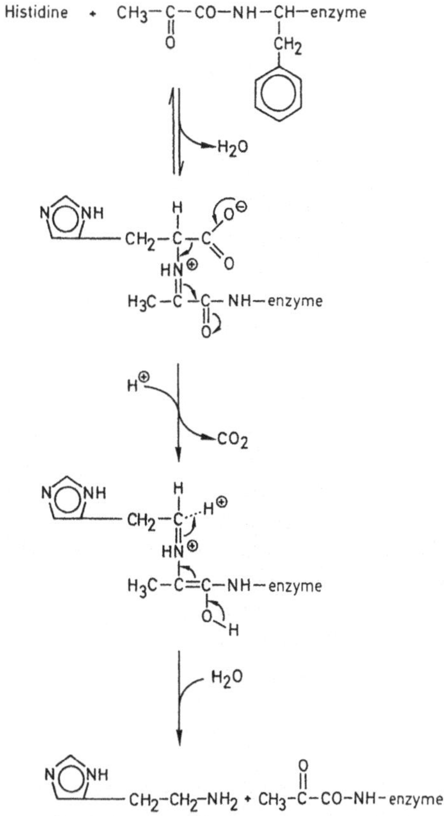

Fig. 2.20. A proposed mechanism for the reaction of histidine decarboxylase

analyze the parameters which influence or determine the rate of an enzyme-catalyzed reaction. The reaction rate is dependent on the concentrations of the components involved in the reaction. Here we mean primarily the substrate and the enzyme. Also, the reaction can be influenced by the presence of activators and inhibitors. Finally, the pH, the ionic strength of the reaction medium, the dielectric constant of the solvent (usually water) and the temperature exert an effect.

2.4.3 Closing Remarks

The hypotheses discussed here allow some understanding of the fundamentals involved in the action of enzymes. However, the knowledge is far from the point where the individual or combined effects which regulate the rates of enzyme-catalyzed reactions can be calculated.

2.5 Kinetics of Enzyme-Catalyzed Reactions

Enzymes in food can be detected only indirectly by measuring their catalytic activity and, in this way, differentiated from other enzymes. This is the rationale for acquiring knowledge needed to

2.5.1 Effect of Substrate Concentration

2.5.1.1 Single-Substrate Reactions

2.5.1.1.1 Michaelis–Menten Equation

Let us consider a single-substrate reaction. Enzyme E reacts with substrate A to form an intermediary enzyme-substrate complex, EA. The complex then forms the product P and releases the free enzyme:

$$E + A \xrightleftharpoons[k_{-1}]{k_1} EA \xrightarrow{k_2} E + P \tag{2.30}$$

In order to determine the catalytic activity of the enzyme, the decrease in substrate concentration or the increase in product concentration as a function of time can be measured. The activity curve obtained (Fig. 2.21) has the following regions:

a) The maximum activity which occurs for a few msec until an equilibrium is reached between the rate of enzyme-substrate formation and rate of breakdown of this complex.

Measurements in this pre-steady state region which provide an insight into the reaction steps and mechanism of catalysis are difficult and time consuming. Hence, further analysis fo the pre-steady state will be ignored.

b) The usual procedure is to measure the enzyme activity when a steady state has been reached. In the steady state the intermediary complex concentration remains constant while the concentration of the substrate and end product are changing. For this state, the following is valid:

$$\frac{dEA}{dt} = -\frac{dEA}{dt} \tag{2.31}$$

c) The reaction rate continuously decreases in this region in spite of an excess of substrate. The decrease in the reaction rate can be considered to be a result of:

Enzyme denaturation which can readily occur, continuously decreasing the enzyme concentration in the reaction system, or the product formed increasingly inhibits enzyme activity or, after the concentration of the product increases, the reverse reaction takes place, converting the product back into the initial reactant.

Since such unpredictable effects should be avoided during analysis of enzyme activities, as a rule the initial reaction rate, v_0, is measured as soon as possible after the start of the reaction.

The basics of the kinetic properties of enzymes in the steady state were given by Briggs and

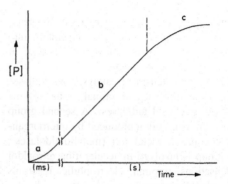

Fig. 2.21. Progress of an enzyme-catalyzed reaction

Haldane (1925) and are supported by earlier mathematical models proposed by Michaelis and Menten (1913).

The following definitions and assumptions should be introduced in relation to the reaction in Equation 2.30:

$[E_0]$ = total enzyme concentration available at the start of the catalysis.

$[E]$ = concentration of free enzyme not bound to the enzyme-substrate complex, EA, i. e. $[E] = [E_0] - [EA]$.

$[A_0]$ = total substrate concentration available at the start of the reaction. Under these conditions, $[A_0] \gg [E_0]$. Since in catalysis only a small portion of A_0 reacts, the substrate concentration at any time, $[A]$, is approximately equal to $[A_0]$.

When the initial reaction rate, v_0, is considered, the concentration of the product, $[P]$, is 0. Thus, the reaction in Equation 2.30 takes the form:

$$\frac{dP}{dt} = v_0 = k_2(EA) \tag{2.32}$$

The concentration of enzyme-substrate complex, $[EA]$, is unknown and can not be determined experimentally for Equation 2.32. Hence, it is calculated as follows: The rate of formation of EA, according to Equation 2.30, is:

$$\frac{dEA}{dt} = k_1(E)(A_0) \tag{2.33}$$

and the rate of EA breakdown is:

$$-\frac{dEA}{dt} = k_{-1}(EA) * k_2(EA) \tag{2.34}$$

Under steady-state conditions the rates of breakdown and formation of EA are equal (cf. Equation 2.31):

$$k_1(E)(A_0) = (k_{-1} * k_2)(EA) \tag{2.35}$$

Also, the concentration of free enzyme, [E], can not be readily determined experimentally. Hence, free enzyme concentration from the above relationship ([E] = [E_0] − [EA]) is substituted in Equation 2.35:

$$k_1[(E_0) − (EA)](A_0) = (k_{-1} * k_2)(EA) \tag{2.36}$$

Solving Equation 2.36 for the concentration of the enzyme-substrate complex, [EA], yields:

$$(EA) = \frac{(E_0)(A_0)}{\frac{k_{-1}+k_2}{k_1} + (A_0)} \tag{2.37}$$

The quotient of the rate constants in Equation 2.37 can be simplified by defining a new constant, K_m, called the *Michaelis* constant:

$$(EA) = \frac{(E_0)(A_0)}{K_m + (A_0)} \tag{2.38}$$

Substituting the value of [EA] from Equation 2.38 in Equation 2.32 gives the *Michaelis–Menten* equation for v_0 (initial reaction rate):

$$v_0 = \frac{k_2(E_0)(A_0)}{K_m + (A_0)} \tag{2.39}$$

Equation 2.39 contains a quantity, [E_0], which can be determined only when the enzyme is present in purified form. In order to be able to make kinetic measurements using impure enzymes, *Michaelis* and *Menten* introduced an approximation for Equation 2.39 as follows. In the presence of a large excess of substrate, $[A_0] \gg K_m$ in the denominator of Equation 2.39. Therefore, K_m can be neglected compared to $[A_0]$:

$$v_0 = \frac{k_2(E_0)(A_0)}{(A_0)} = V \tag{2.40}$$

Thus, a zero order reaction rate is obtained. It is characterized by a rate of substrate breakdown or product formation which is independent of substrate concentration, i. e. the reaction rate, V, is dependent only on enzyme concentration. This rate, V, is denoted as the maximum velocity.

From Equation 2.40 it is obvious that the catalytic activity of the enzyme must be measured in the presence of a large excess of substrate.

To eliminate the [E_0] term, V is introduced into Equation 2.39 to yield:

$$v_0 = \frac{V(A_0)}{K_m * (A_0)} \tag{2.41}$$

If $[A_0] = K_m$, the following is derived from Equation 2.41:

$$v_0 = \frac{V}{2} \tag{2.42}$$

Thus, the *Michaelis* constant, K_m, is equal to the substrate concentration at which the reaction rate is half of its maximal value. K_m is independent of enzyme concentration. The lower the value of K_m, the higher the affinity of the enzyme for the substrate, i. e. the substrate will be bound more tightly by the enzyme and most probably will be more efficiently converted to product. Usually, the values of K_m, are within the range of 10^{-2} to 10^{-5} mol · 1^{-1}. From the definition of K_m:

$$K_m = \frac{k_{-1}+k_2}{k_1} \tag{2.43}$$

it follows that K_m approaches the enzymesubstrate dissociation constant, K_s, only if

$k_{+2} \ll k_{-1}$.

$$k_2 \ll k_{-1} \frown K_m \approx \frac{k_{-1}}{k_1} = K_s \tag{2.44}$$

Some values for the constants k_{+1}, k_{-1}, and k_0 are compiled in Table 2.9. In cases in which the catalysis proceeds over more steps than shown in Equation 2.30 the constant k_{+2} is replaced by k_0. The rate constant, k_{+1}, for the formation of the enzyme-substrate complex has values in the order of 10^6 to 10^8: in a few cases it approaches the maximum velocity ($\sim 10^9 1 \cdot mol^{-1} s^{-1}$), especially when small molecules of substrate readily diffuse through the solution to the active site of the enzyme. The values for k_{-1} are substantially lower in most cases, whereas k_0 values are in the range of 10^1 to 10^6 s^{-1}.

Table 2.9. Rate constants for some enzyme catalyzed reactions

Enzyme	Substrate	k_1 ($l \cdot mol^{-1} s^{-1}$)	K_{-1} (s^{-1})	k_0 (s^{-1})
Fumarase	Fumarate	$> 10^9$	$4.5 \cdot 10^4$	10^3
Acetylcho- line esterase	Acetyl- choline	10^9		10^3
Alcohol dehydro- genase (liver)	NAD NADH Ethanol	$5.3 \cdot 10^5$ $1.1 \cdot 10^7$ $> 1.2 \cdot 10^4$	74 3.1 > 74	10^3
Catalase	H_2O_2	$5 \cdot 10^6$		10^7
Peroxidase	H_2O_2	$9 \cdot 10^6$	< 1.4	10^6
Hexokinase	Glucose	$3.7 \cdot 10^6$	$1.5 \cdot 10^3$	10^3
Urease	Urea	$> 5 \cdot 10^6$		10^4

Another special case to be considered is if $[A_0] \ll K_m$, which occurs at about $[A_0] < 0.05 \, K_m$. Here $[A_0]$ in the denominator of Equation 2.39 can be neglected:

$$v_0 = \frac{k_2 (E_0) (A_0)}{K_m} \qquad (2.45)$$

and, considering that $k_2[E_0] = V$, it follows that:

$$v_0 = \frac{V}{K_m}(A_0) \qquad (2.46)$$

In this case the *Michaelis–Menten* equation reflects a first-order reaction in which the rate of substrate breakdown depends on substrate concentration. In using a kinetic method for the determination of substrate concentration (cf. 2.6.1.3), the experimental conditions must be selected such that Equation 2.46 is valid.

2.5.1.1.2 Determination of K_m and V

In order to determine values of K_m and V, the catalytic activity of the enzyme preparation is measured as a function of substrate concentration. Very good results are obtained when $[A_0]$ is in the range of $0.1 \, K_m$ to $10 \, K_m$.

A graphical evaluation of the result is obtained by inserting the data into Equation 2.41. As can be seen from a plot of the data in Fig. 2.22, the equation corresponds to a rectangular hyperbola. This graphical approach yields correct values for K_m

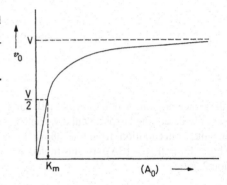

Fig. 2.22. Determination of *Michaelis* constant, K_m, according to Equation (2.41)

only when the maximum velocity, V, can be accurately determined.

For a more reliable extrapolation of V, Equation 2.41 is transformed into a straight-line equation. Most frequently, the *Lineweaver–Burk* plot is used which is the reciprocal form of Equation 2.41:

$$\frac{1}{v_0} = \frac{K_m}{V} \cdot \frac{1}{(A_0)} + \frac{1}{V} \qquad (2.47)$$

Figure 2.23 graphically depicts a plot of $1/v_0$ versus $1/[A_0]$. The values V and K_m are obtained from the intercepts of the ordinate $(1/V)$ and of the abscissa $(-1/K_m)$, respectively. If the data do not fit a straight line, then the system deviates from the required steady-state kinetics; e. g., there is inhibition by excess substrate or the system is influenced by allosteric effects (cf. 2.5.1.3; allosteric enzymes do not obey *Michaelis–Menten* kinetics).

A great disadvantage of the *Lineweaver–Burk* plot is the possibility of departure from a straight line since data taken in the region of saturating substrate concentrations or at low substrate concentrations can be slightly inflated. Thus, values taken from the straight line may be somewhat overestimated.

A procedure which yields a more uniform *distribution* of the data on the straight line is that proposed by *Hofstee* (the *Eadie–Hofstee* plot). In this procedure the *Michaelis–Menten* equation, 2.41, is algebraically rearranged into:

$$v_0(A_0) + v_0 K_m = V \cdot (A_0) \qquad (a)$$

$$v_0 + \frac{v_0}{(A_0)} \cdot K_m = V \qquad (b)$$

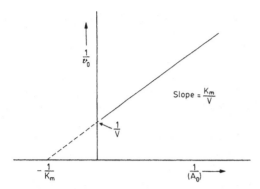

Fig. 2.23. Determination of K_m and V (according to *Lineweaver* and *Burk*)

$$v_0 = -K_m \frac{v_0}{(A_0)} \cdot V \qquad \text{(c)} \qquad (2.48)$$

When Equation 2.48c is plotted using the substrate-reaction velocity data, a straight line with a negative slope is obtained (Fig. 2.24) where y is v_0 and x is $v_0/[A_0]$. The y and x intercepts correspond to V and V/K_m, respectively.

Single-substrate reactions, for which the kinetics outlined above (with some exceptions, cf. 2.5.1.3) are particularly pertinent, are those catalyzed by lyase enzymes and certain isomerases. Hydrolysis by hydrolase enzymes can also be considered a single-substrate reaction when the water content remains unchanged, i.e., when it is present in high concentration (55.6 mol/μ). Thus, water, as a reactant, can be disregarded.

Characterization of an enzyme-substrate system by determining values for K_m and V is import-

ant in enzymatic food analysis (cf. 2.6.4) and for assessment of enzymatic reactions occurring in food (e.g. enzymatic browning of sliced potatoes, cf. 2.5.1.2.1) and for utilization of enzymes in food processing, e.g., aldehyde dehydrogenase (cf. 2.7.2.1.4).

2.5.1.2 Two-Substrate Reactions

For many enzymes, for examples, oxidoreductase and ligase-catalyzed reactions, two or more substrates or cosubstrates are involved.

2.5.1.2.1 Order of Substrate Binding

In the reaction of an enzyme with two substrates, the binding of the substrates can occur sequentially in a specific order. Thus, the binding mechanism can be divided into catalysis which proceeds through a ternary adsorption complex (enzyme + two substrates) or through a binary complex (enzyme + one substrate), i.e. when the enzyme binds only one of the two available substrates at a time.

A ternary enzyme-substrate complex can be formed in two ways. The substrates are bound to the enzyme in a random fashion ("random mechanism") or they are bound in a well-defined order ("ordered mechanism").

Let us consider the reaction

$$A + B \underset{}{\overset{E}{\rightleftharpoons}} P + Q \qquad (2.49)$$

If the enzyme reacts by a "random mechanism", substrates A and B form the ternary enzyme-substrate complex, EAB, in a random fashion and the P and Q products dissociate randomly from the ternary enzyme-product complex, EPQ:

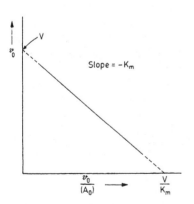

$$(2.50)$$

Creatine kinase from muscle (cf. 12.3.6) is an example of an enzyme which reacts by a random

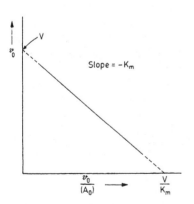

Fig. 2.24. Determination of K_m and V (according to *Hofstee*)

mechanism:

Creatine + ATP

$\rightleftharpoons$ Creatine-phosphate + ADP   (2.51)

In an "ordered mechanism" the binding during the catalyzed reaction according to equation 2.49 is as follows:

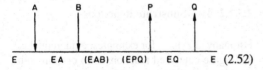

E   E A   (EAB) (EPQ)   E Q   E   (2.52)

Alcohol dehydrogenase reacts by an "*ordered mechanism*", although the order of the binding of substrates NAD^+ and ethanol is decided by the ethanol concentration. NAD^+ is absorbed first at low concentrations (< 4 mmol/l):

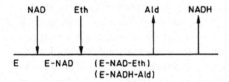

Eth : Ethanol
Ald : Acetaldehyde   (2.53)

When the ethanol concentration is increased to 7–8 mmol/l, ethanol is absorbed first, followed by the cosubstrate. The order of removal of products (acetaldehyde and NADH) is, however, not altered.

Polyphenol oxidase from potato tubers also reacts by an "*ordered mechanism*". Oxygen is absorbed first, followed by phenolic substrates. The main substrates are chlorogenic acid and tyrosine. Enzyme affinity for tyrosine is greater and the reaction velocity is higher than for chlorogenic acid. The ratio of chlorogenic acid to tyrosine affects enzymatic browning to such an extent that it is considered to be the major problem in potato processing. The deep-brown colored melanoidins are formed quickly from tyrosine but not from chlorogenic acid. In assessing the processing quality of potato cultivars, the differences in phenol oxidase activity and the content of ascorbic acid in the tubers should also be considered in relation to "enzymatic browning". Ascorbic acid retards formation of melanoidins by its ability to reduce o-quinone, the initial product of enzymatic oxidation (cf. 18.1.2.5.8).

In enzymatic reactions where functional group transfers are involved, as a rule only binary enzyme-substrate complexes are formed by the so-called "*ping pong mechanism*".

A substrate is adsorbed by enzyme, E, and reacts during alteration of the enzyme (a change in the oxidation state of the prosthetic group, a conformational change, or only a change in covalent binding of a functional group). The modified enzyme, which is denoted F, binds the second substrate and the second reaction occurs, which regenerates the initial enzyme, E, and releases the second product:

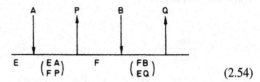

(2.54)

The glycolytic enzyme hexokinase reacts by a "ping pong mechanism":

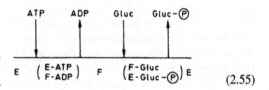

(2.55)

2.5.1.2.2 Rate Equations for a Two-Substrate Reaction

Here the reaction rate is distinguished by its dependence on two reactants, either two molecules of the same compound or two different compounds. The rate equations can be derived by the same procedures as used for single-substrate catalysis. Only the final forms of the equations will be considered.

When the catalysis proceeds through a ternary enzyme-substrate complex, EAB, the general equation is:

$$v_0 = \frac{V}{1 + \dfrac{K_a}{(A_0)} + \dfrac{K_b}{(B_0)} + \dfrac{K_{ja} \cdot k_b}{(A_0)(B_0)}}   \quad (2.56)$$

When compared to the rate equation for a single-substrate reaction (Equation 2.41), the difference becomes obvious when the equation for a single-substrate reaction is expressed in the following form:

$$v_0 = \frac{V}{1 + \dfrac{K_a}{(A_0)}} \qquad (2.57)$$

The constants K_a and K_b in Equation 2.56 are defined analogously to K_m, i.e. they yield the concentrations of A or B for $v_0 = V/2$ assuming that, at any given moment, the enzyme is saturated by the other substrate (B or A). Each of the constants, like K_m (cf. Equation 2.43), is composed of several rate constants. K_{ia} is the inhibitor constant for A.

When the binding of one substrate is not influenced by the other, each substrate occupies its own binding locus on the enzyme and the substrates form a ternary enzyme-substrate complex in a defined order ("*ordered mechanism*"), the following is valid:

$$K_{ia} \cdot K_b = K_a \cdot K_b \qquad (2.58)$$

or from Equation 2.56:

$$v_0 = \frac{V}{1 + \dfrac{K_a}{(A_0)} + \dfrac{K_b}{(B_0)} + \dfrac{K_a \cdot K_b}{(A_0)(B_0)}} \qquad (2.59)$$

However, when only a binary enzyme-substrate complex is formed, i.e. one substrate or one product is bound to the enzyme at a time by a "ping pong mechanism", the denominator term $K_{ia} \cdot K_b$ must be omitted since no ternary complex exists. Thus, Equation 2.56 is simplified to:

$$v_0 = \frac{V}{1 + \dfrac{K_a}{(A_0)} + \dfrac{K_b}{(B_0)}} \qquad (2.60)$$

For the determination of rate constants, the initial rate of catalysis is measured as a function of the concentration of substrate B (or A) for several concentrations of A (or B). Evaluation can be done using the *Lineweaver–Burk* plot. Reshaping Equation 2.56 for a "*random mechanism*" leads to:

$$\frac{1}{v_0} = \left[\frac{K_b}{V} + \frac{K_{ia} \cdot K_b}{(A_0)V} \right] \frac{1}{(B_0)} + \left[1 + \frac{K_a}{(A_0)} \right] \frac{1}{V} \qquad (2.61)$$

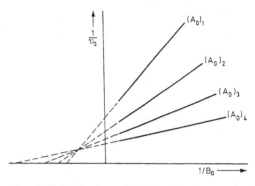

Fig. 2.25. Evaluation of a two-substrate reaction, proceeding through a ternary enzyme-substrate complex (according to *Lineweaver* and *Burk*). $[A_0]_4 > [A_0]_3 > [A_0]_2 > [A_0]_1$

First, $1/v_0$ is plotted against $1/[B_0]$. The corresponding slopes and ordinate intercepts are taken from the straight lines obtained at various values for $[A_0]$ (Fig. 2.25):

$$\text{Slope} = \frac{K_b}{V} + \frac{K_{ia}K_b}{V} \cdot \frac{1}{(A_0)}$$

$$\text{Ordinate intercept} = \frac{1}{V} + \frac{K_a}{V} \cdot \frac{1}{(A_0)} \qquad (2.62)$$

and are then plotted against $1/[A_0]$. In this way two straight lines are obtained (Fig. 2.26a and b), with slopes and ordinate intercepts which provide data for calculating constants K_a, K_b, K_{ia}, and the maximum velocity, V. If the catalysis proceeds through a "ping pong mechanism", then plotting $1/v_0$ versus $1/[B_0]$ yields a family of parallel lines (Fig. 2.27) which are then subjected to the calculations described above.

A comparison of Figs. 2.25 and 2.27 leads to the conclusion that the dependence of the initial catalysis rate on substrate concentration allows the differentiation between a ternary and a binary enzyme-substrate complex. However, it is not possible to differentiate an "ordered" from a "random" reaction mechanism by this means.

2.5.1.3 Allosteric Enzymes

We are already acquainted with some enzymes consisting of several protomers (cf. Table 1.26). When the protomer activities are independent

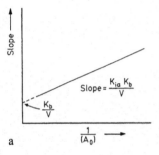

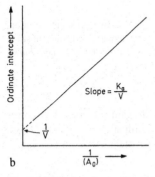

Fig. 2.26. Plotting slopes (**a**) and ordinate intercepts (**b**) from Fig. 2.25 versus $1/[A_0]$

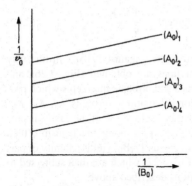

Fig. 2.27. Evaluation of a two-substrate reaction, proceeding through a binary enzyme-substrate complex (according to *Lineweaver* and *Burk*). $[A_0]_4 > [A_0]_3 > [A_0]_2 > [A_0]_1$

of each other in catalysis, the *Michaelis–Menten* kinetics, as outlined under sections 2.5.1.1 and 2.5.1.2, are valid. However, when the subunits cooperate, the enzymes deviate from these kinetics. This is particularly true in the case of positive cooperation when the enzyme is activated by the substrate. In this kind of plot, v_0 versus $[A_0]$ yields not a hyperbolic curve but a saturation curve with a sigmoidal shape (Fig. 2.28).

Thus, enzymes which do not obey the *Michaelis–Menten* model of kinetics are allosterically regulated. These enzymes have a site which reversibly binds the allosteric regulator (substrate, cosubstrate or low molecular weight compound) in addition to an active site with a binding and transforming locus. Allosteric enzymes are, as a rule, engaged at control sites of metabolism. An example is tetrameric phosphofructokinase, the key enzyme in glycolysis. In glycolysis and alcoholic fermentation it catalyzes the phosphorylation of fructose-6-phosphate to fructose-1,6-diphosphate. The enzyme is activated by its substrate in the presence of ATP. The prior binding of a substrate molecule which enhances the binding of each succeeding substrate molecule is called positive cooperation. The two enzyme-catalyzed reactions, one which obeys *Michaelis–Menten* kinetics and the other which is regulated by allosteric effects, can be reliably distinguished experimentally by comparing the ratio of the substrate concentration needed to obtain the observed value of 0.9 V to that needed to obtain 0.1 V. This ratio, denoted as R_s, is a measure of the cooperativity of the interaction.

$$R_S = \frac{(A_0)_{0.9\,V}}{(A_0)_{0.1\,V}} \tag{2.63}$$

For all enzymes which obey *Michaelis–Menten* kinetics, $R_s = 81$ regardless of the value of K_m or V. The value of R_s is either lower or higher than 81 for allosteric enzymes. $R_s < 81$ is

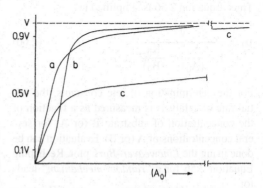

Fig. 2.28. The effect of substrate concentration on the catalytic reaction rate. **a** Enzyme obeying *Michaelis–Menten* kinetics; **b** allosterically regulated enzyme with positive cooperativity; **c** allosterically regulated enzyme with negative cooperativity

indicative of positive cooperation. Each substrate molecule, often called an effector, accelerates the binding of succeeding substrate molecules, thereby increasing the catalytic activity of the enzyme (case b in Fig. 2.28). When $R_s > 81$, the system shows negative cooperation. The effector (or allosteric inhibitor) decreases the binding of the next substrate molecule (case c in Fig. 2.28). Various models have been developed in order to explain the allosteric effect. Only the symmetry model proposed by *Monod, Wyman* and *Changeux* (1965) will be described in its simplified form: specifically, when the substrate acts as a positive allosteric regulator or effector. Based on this model, the protomers of an allosteric enzyme exist in two conformations, one with a high affinity (R-form) and the other with a low affinity (T-form) for the substrate. These two forms are interconvertible. There is an interaction between protomers. Thus, binding of the allosteric regulator by one protomer induces a conformational change of all the subunits and greatly increases the activity of the enzyme.

Let us assume that the R- and T-forms of an enzyme consisting of four protomers are in an equilibrium which lies completely on the side of the T-form:

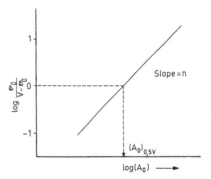

The equation says that the catalytic rate increases by the nth power of the substrate concentration when $[A_0]$ is small in comparison to K. The *Hill* coefficient, n, is a measure of the sigmoidal character of the curve and, therefore, of the extent of the enzyme's cooperativity. For $n = 1$ (Equation 2.65) the reaction rate is transformed into the *Michaelis–Menten* equation, i.e. in which no cooperativity factor exists.

In order to assess the experimental data, Equation 2.65 is rearranged into an equation of a straight line:

$$\log \frac{v_0}{V - v_0} = n \log(A_0) - \log K' \qquad (2.66)$$

The slope of the straight line obtained by plotting the substrate concentration as $\log[A_0]$ versus $\log[v_0/(V - v_0)]$ is the *Hill* coefficient, n (Fig. 2.29). The constant K incorporates all the individual K_m values involved in all the steps of substrate binding and transformation. The value of K_m is obtained by using the substrate concentration, denoted as $[A_0]_{0.5\,v}$, at which $v_0 = 0.5\,V$. Under these conditions, the following is derived from Equation 2.66):

$$\log \frac{0.5\,V}{0.5\,V} = 0 = n \cdot \log(A_0)_{0.5\,v} - \log K' \qquad (a)$$

$$K' = (A_0)_{0.5\,v}^n \qquad (b)$$
$$(2.67)$$

Fig. 2.29. Linear presentation of *Hill's* equation

Addition of substrate, which here is synonymous to the allosteric effector, shifts the equilibrium from the low affinity T-form to the substantially more catalytically active R-form. Since one substrate molecule activates four catalytically active sites, the steep rise in enzyme activity after only a slight increase in substrate concentration is not unexpected. In this model it is important that the RT conformation is not permitted. All subunits must be in the same conformational state at one time to conserve the symmetry of the protomers. The equation given by *Hill* in 1913, derived from the sigmoidal absorption of oxygen by hemoglobin, is also suitable for a quantitative description of allosteric enzymes with sigmoidal behavior:

$$v_0 = \frac{V(A_0)^n}{K' + (A_0)^n} \qquad (2.65)$$

2.5.2 Effect of Inhibitors

The catalytic activity of an enzyme, in addition to substrate concentration, is affected by the type and concentration of inhibitors, i. e. compounds which decrease the rate of catalysis, and activators, which have the opposite effect. Metal ions and compounds which are active as prosthetic groups or which provide stabilization of the enzyme's conformation or of the enzyme-substrate complex (cf. 2.3.2 and 2.3.3) are activators. The effect of inhibitors will be discussed in more detail in this section.

Inhibitors are found among food constituents. Proteins which specifically inhibit the activity of certain peptidases (cf. 16.2.3), amylases or β-fructofuranosidase are examples. Furthermore, food contains substances which nonselectively inhibit a wide spectrum of enzymes. Phenolic constituents of food (cf. 18.1.2.5) and mustard oil (cf. 17.1.2.6.5) belong to this group. In addition, food might be contaminated with pesticides, heavy metal ions and other chemicals from a polluted environment (cf. Chapter 9) which can become inhibitors under some circumstances. These possibilities should be taken into account when enzymatic food analysis is performed.

Food is usually heat treated (cf. 2.5.4) to suppress undesired enzymatic reactions. As a rule, no inhibitors are used in food processing. An exception is the addition of, for example, SO_2 to inhibit the activity of phenolase (cf. 8.12.6).

Much data concerning the mechanism of action of enzyme inhibitors have been compiled in recent biochemical research. These data cover the elucidation of the effect of inhibitors on funtional groups of an enzyme, their effect on the active site and the clarification of the general mechanism involved in an enzymecatalyzed reaction (cf. 2.4.1.1).

Based on kinetic considerations, inhibitors are divided into two groups: inhibitors bound *irreversibly* to enzyme and those bound *reversibly*.

2.5.2.1 Irreversible Inhibition

In an irreversible inhibition the inhibitor binds mostly covalently to the enzyme; the EI complex formed does not dissociate:

$$E + I \xrightarrow{k_1} EI \qquad (2.68)$$

The rate of inhibition depends on the reaction rate constant k_1 in Equation 2.68, the enzyme concentration, [E], and the inhibitor concentration, [I]. Thus, irreversible inhibition is a function of reaction time. The reaction cannot be reversed by diluting the reaction medium. These criteria serve to distinguish irreversible from reversible inhibition.

Examples of irreversible inhibition are the reactions of SH-groups of an enzyme with iodoacetic acid:

$$Enz\text{-}SH + ICH_2COOH$$
$$\longrightarrow Enz\text{-}S\text{-}CH_2\text{-}COOH + HI \qquad (2.69)$$

and other reactions with the inhibitors described in section 2.4.1.1.

2.5.2.2 Reversible Inhibition

Reversible inhibition is characterized by an equilibrium between enzyme and inhibitor:

$$E + I \rightleftharpoons EI \quad (a)$$

$$\frac{(E) \cdot (I)}{(EI)} = K_i \quad (b) \qquad (2.70)$$

The equilibrium constant or dissociation constant of the enzyme-inhibitor complex, K_i, also known as the inhibitor constant, is a measure of the extent of inhibition. The lower the value of K_i, the higher the affinity of the inhibitor for the enzyme. Kinetically, three kinds of reversible inhibition can be distinguished: competitive, non-competitive and uncompetitive inhibition (examples in Table 2.10). Other possible cases, such as allosteric inhibition and partial competitive or partial non-competitive inhibition, are omitted in this treatise.

2.5.2.2.1 Competitive Inhibition

Here the inhibitor binds to the active site of the free enzyme, thus preventing the substrate from binding. Hence, there is competition between substrate and inhibitor:

$$E + I \rightleftharpoons EI \ (a) \quad E + A \rightleftharpoons EA \ (b) \qquad (2.71)$$

According to the steady-state theory for a single-substrate reaction, we have:

$$v_0 = \frac{V(A_0)}{K_m \left(1 + \frac{(I)}{K_i}\right) + (A_0)} \qquad (2.72)$$

Table 2.10. Examples of reversible enzyme inhibition

Enzyme	EC-Number	Sustrate	Inhibitor	Inhibition type[a]	K_i (mmol/1)
Glucose dehydrogenase	1.1.1.47	Glucose/NAD	Glucose-6-phosphate	C	$4.4 \cdot 10^{-5}$
Glucose-6-phosphate dehydrogenase	1.1.1.49	Glucose-6-phosphate/NADP	Phosphate	C	$1 \cdot 10^{-1}$
Succinate dehydrogenase	1.3.99.1	Succinate	Fumarate	C	$1.9 \cdot 10^{-3}$
Creatine kinase	2.7.3.2	Creatine/ATP	ADP	NC	$2 \cdot 10^{-3}$
Glucokinase	2.7.1.2	Glucose/ATP	D-Mannose	C	$1.4 \cdot 10^{-2}$
			2-Deoxyglucose	C	$1.6 \cdot 10^{-2}$
			D-Galactose	C	$6.7 \cdot 10^{-1}$
Fructose-biphosphatase	3.1.3.11	D-Fructose-1, 6-biphosphate	AMP	NC	$1.1 \cdot 10^{-4}$
α-Glucosidase	3.2.1.20	p-Nitrophenyl-α-D-glucopyranoside	Saccharose	C	$3.7 \cdot 10^{-2}$
			Turanose	C	$1.1 \cdot 10^{-2}$
Cytochrome c oxidase	1.9.3.1	Ferrocytochrome c	Azide	UC	

[a] C: competitive, NC: noncompetitive, and UC: uncompetitive.

In the presence of inhibitors, the *Michaelis* constant is apparently increased by the factor:

$$1 + \frac{(I)}{K_i} \qquad (2.73)$$

Such an effect can be useful in the case of enzymatic substrate determinations (cf. 2.6.1.3). When inhibitor activity is absent, i.e. $[I] = 0$, Equation 2.72 is transformed into the *Michaelis–Menten* equation (Equation 2.41). The *Lineweaver–Burk* plot (Fig. 2.30a) shows that the intercept $1/V$ with the ordinate is the same in the presence and in the absence of the inhibitor, i.e. the value of V is not affected although the slopes of the lines differ. This shows that the inhibitor can be fully dislodged by the substrate from the active site of the enzyme when the substrate is present in high concentration. In other words, inhibition can be overcome at high substrate concentrations (see application in Fig. 2.49). The inhibitor constant, K_i, can be calculated from the corresponding intercepts with the abscissa in Fig. 2.30a by calculating the value of K_m from the abscissa intercept when $[I] = 0$.

2.5.2.2.2 Non-Competitive Inhibition

The non-competitive inhibitor is not bound to the active site of the enzyme but to some other site. Therefore, the inhibitor can react equally with free enzyme or with enzyme-substrate complex. Thus, three processes occur in parallel:

$$E + A \rightleftharpoons EA \text{ (a)} \qquad E + I \rightleftharpoons EI \text{ (b)}$$
$$EA + I \rightleftharpoons EAI \text{ (c)} \qquad (2.74)$$

Postulating that EAI and EI are catalytically inactive and the dissociation constants K_i and K_{EAi} are numerically equal, the following equation is obtained by rearrangement of the equation for a single-substrate reaction into its reciprocal form:

$$\frac{1}{v_0} = \frac{K_m}{V}\left(1 + \frac{(I)}{K_i}\right)\frac{1}{(A_0)} + \frac{1}{V}\left(1 + \frac{(I)}{K_i}\right) \qquad (2.75)$$

The double-reciprocal plot (Fig. 2.30b) shows that, in the presence of a noncompetitive inhibitor; K_m is unchanged whereas the values of V are decreased such that V becomes $V/(1 + [I]/K_i)$, i.e. non-competitive inhibition can not be overcome by high concentrations of substrate.

This also indicates that, in the presence of inhibitor, the amount of enzyme available for catalysis is decreased.

2.5.2.2.3 Uncompetitive Inhibition

In this case the inhibitor reacts only with enzyme-substrate complex:

$$E + A \rightleftharpoons EA \begin{array}{c} \nearrow E + P \\ \searrow \\ EAI \end{array} \qquad (2.76)$$

Rearranging Equation 2.76 into an equation for a straight line, the reaction rate becomes:

$$\frac{1}{v_0} = \frac{K_m}{V} \frac{1}{(A_0)} + \frac{1}{V}\left(1 + \frac{(1)}{K_i}\right) \qquad (2.77)$$

The double reciprocal plot (Fig. 2.30c) shows that in the presence of an uncompetitive inhibitor, both the maximum velocity, V, and K_m are changed but not the ratio of K_m/V. Hence the slopes of the lines are equal and in the presence of increasing amounts of inhibitor, the lines plotted are parallel. Uncompetitive inhibition is rarely found in single-substrate reactions. It occurs more often in two-substrate reactions.

In conclusion, it can be stated that the three types of reversible inhibition are kinetically distinguishable by plots of reaction rate versus substrate concentration using the procedure developed by *Lineweaver* and *Burk* (Fig. 2.30).

2.5.3 Effect of pH on Enzyme Activity

Each enzyme is catalytically active only in a narrow pH range and, as a rule, each has a pH optimum which is often between pH 5.5 and 7.5 (Table 2.11).

The optimum pH is affected by the type and ionic strength of the buffer used in the assay. The rea-

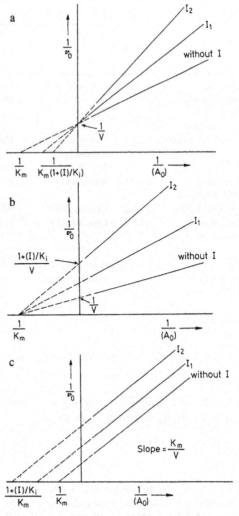

Fig. 2.30. Evaluation of inhibited enzyme-catalyzed reaction according to *Lineweaver* and *Burk*, $[I_1] < (I_2)$. **a** Competitive inhibition, **b** noncompetitive inhibition, **c** uncompetitive inhibition

sons for the sensitivity of the enzyme to changes in pH are two-fold:

a) sensitivity is associated with a change in protein structure leading to irreversible denaturation,

b) the catalytic activity depends on the quantity of electrostatic charges on the enzyme's active site generated by the prototropic groups of the enzyme (cf. 2.4.2.4).

In addition the ionization of dissociable substrates as affected by pH can be of importance to

Table 2.11. pH Optima of various enzymes

Enzyme	Source	Substrate	pH Optimum
Pepsin	Stomach	Protein	2
Chymotrypsin	Pancreas	Protein	7.8
Papain	Tropical plants	Protein	7–8
Lipase	Microorganisms	Olive oil	5–8
α-Glucosidase (maltase)	Microorganisms	Maltose	6.6
β-Amylase	Malt	Starch	5.2
β-Fructofuranosidase (invertase)	Tomato	Saccharose	4.5
Pectin lyase	Microorganisms	Pectic acid	9.0–9.2
Xanthine oxidase	Milk	Xanthine	8.3
Lipoxygenase, type I[a]	Soybean	Linoleic acid	9.0
Lipoxygenase, type II[a]	Soybean	Linoleic acid	6.5

[a] See 3.7.2.2.

the reaction rate. However, such effects should be determined separately. Here, only the influences mentioned under b) will be considered with some simplifications.

An enzyme, E, its substrate, A, and the enzyme-substrate complex formed, EA, depending on pH, form the following equilibria:

$$E^{n+1} \underset{-H^\oplus}{\overset{-H^\oplus}{\rightleftharpoons}} E^n \underset{-H^\oplus}{\overset{-H^\oplus}{\rightleftharpoons}} E^{n-1}$$

$$E^{n+1}A \underset{-H^\oplus}{\overset{-H^\oplus}{\rightleftharpoons}} E^nA \underset{-H^\oplus}{\overset{-H^\oplus}{\rightleftharpoons}} E^{n-1}A \qquad (2.78)$$

Which of the charged states of E and EA are involved in catalysis can be determined by following the effect of pH on V and K_m.

a) Plotting K_m versus pH reveals the type of prototropic groups involved in substrate binding and/or maintaining the conformation of the enzyme. The results of such a plot, as a rule, resemble one of the four diagrams shown in Fig. 2.31.
Figure 2.31a: K_m is independent of pH in the range of $4 - 9$. This means that the forms E^{n+1}, E^n, and E^{n-1}, i.e. enzyme forms which are neutral, positively or negatively charged on the active site, can bind substrate. Figures 2.31b and c: K_m is dependent on one prototropic group, the pK value of which is below (Fig. 2.31b) or above (Fig. 2.31c) neutrality. In the former case, E^n and E^{n-1} are the active forms, while in the latter, E^{n+1} and E^n are the active enzyme forms in substrate binding.

Figure 2.31d: K_m is dependent on two prototropic groups; the active form in substrate binding is E^n.

b) The involvement of prototropic groups in the conversion of an enzyme-substrate complex into product occurs when the enzyme is saturated with substrate, i.e. when equation 2.40 which defines V is valid ($[A_0] \gg K_m$). Thus, a plot of V versus pH provides essentially the same four possibilities presented in Fig. 2.31, the difference being that, here, the prototropic groups of EA, which are involved in the conversion to product, are revealed.

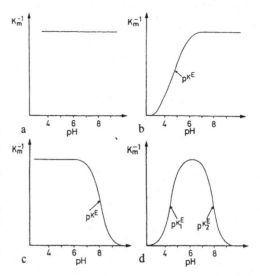

Fig. 2.31. The possible effects of pH on the *Michaelis* constant, K_m

In order to better understand the form of the enzyme involved in catalysis, a hypothetical enzyme-substrate system will be assayed and interpreted. We will start from the assumption that data are available for v_0 (initial velocity) as a function of substrate concentration at several pH's, e. g., for the *Lineweaver* and *Burk*. The values for K_m and V are obtained from the family of straight lines (Fig. 2.32) and plotted against pH. The diagram of $K_m^{-1} = f(pH)$ depicted in Fig. 2.33a corresponds to Fig. 2.31c which implies that neutral (E^n) and positively charged (E^{n+1}) enzyme forms are active in binding the substrate.

Figure 2.33b: V is dependent on one prototropic group, the pK value of which is below neutrality. Therefore, of the two enzyme-substrate complexes, $E^{n+1}A$ and E^nA, present in the equilibrium state, only the latter complex is involved in the conversion of A to the product.

In the example given above, the overall effect of pH on enzyme catalysis can be illustrated as follows:

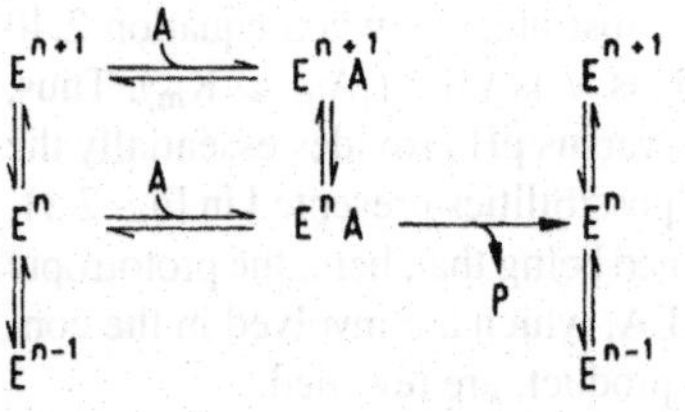

$$
\begin{array}{ccc}
E^{n+1} \xrightleftharpoons{\quad A \quad} & E^{n+1}A & E^{n+1} \\
\Updownarrow \qquad & \Updownarrow & \Updownarrow \\
E^{n} \xrightleftharpoons{\quad A \quad} & E^{n}A \xrightarrow{\ P\ } & E^{n} \\
\Updownarrow \qquad & & \Updownarrow \\
E^{n-1} & & E^{n-1}
\end{array}
\tag{2.79}
$$

This schematic presentation is also in agreement with the diagram of $V/K_m = f(pH)$ (Fig. 2.33c)

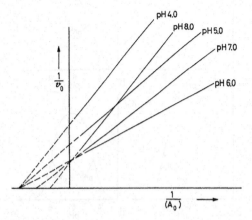

Fig. 2.32. Determination of V and K_m at different pH values

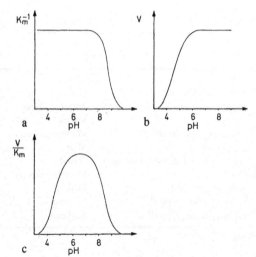

Fig. 2.33. Evaluation of K_m and V versus pH for a hypothetical case

which reveals that, overall, two prototropic groups are involved in the enzymecatalyzed reaction.

An accurate determination of the pK values of prototropic groups involved in enzyme-catalyzed reactions is possible using other assays (cf. *J.R. Whitaker*, 1972). However, identification of these groups solely on the basis of pK values is not possible since the pK value is often strongly influenced by surrounding groups. Pertinent to this claim is our recollection that the pH of acetic acid in water is 4.75, whereas in 80% acetone it is about 7. Therefore, the enzyme activity data as related to pH have to be considered only as preliminary data which must be supported and verified by supplementary investigations.

2.5.4 Influence of Temperature

Thermal processes are important factors in the processing and storage of food because they allow the control of chemical, enzymatic and microbial changes. Undesired changes can be delayed or stopped by refrigerated storage. Heat treatment may either accelerate desirable chemical or enzymatic reactions or inhibit undesirable changes by inactivation of enzymes or microorganisms. Table 2.12 informs about quality dete-

Table 2.12. Thermal inactivation of enzymes to prevent deterioration of food quality

Food product	Enzyme	Quality loss
Potato products, apple products	Monophenol oxidase	Enzymatic browning
Semi-ripe peas	Lipoxygenase, peroxidase	Flavor defects; bleaching
Fish products	Proteinase, thiaminase	Texture (liquefaction), loss of vitamine B_1
Tomato purée	Polygalacturonase	Texture (liquefaction)
Apricot products	β-Glucosidase	Color defects
Oat flakes	Lipase, lipoxygenase	Flavor defects (bitter taste)
Broccoli Cauliflower	Cystathionine β-Lyase (cystine-lyase)	Off-flavor

rioration caused by enzymes which can be eliminated e. g., by thermal inactivation.

Temperature and time are two parameters responsible for the effects of a thermal treatment. They should be selected carefully to make sure that all necessary changes, e. g., killing of pathogens, are guaranteed, but still all undesired changes such as degradation of vitamins are kept as low as possible.

2.5.4.1 Time Dependence of Effects

The reaction rates for different types of enzymatic reactions have been discussed in section 2.5.1. The inactivation of enzymes and the killing of microorganisms can be depicted as a reaction of 1st order:

$$c_t = c_0 e^{-kt} \tag{2.80}$$

with c_0 and c_t = concentrations (activities, germ counts) at times 0 and t, and k = rate constant for the reaction. For c_t and t follows from equation 2.80:

$$\log c_t = -\frac{k}{2,3} \cdot t + \log c_0 \tag{2.81}$$

$$t = \frac{2,3}{k} \log \frac{c_0}{c_t} \tag{2.82}$$

$c_0/c_t = 10$ gives:

$$t = \frac{2,3}{k} = D \tag{2.83}$$

The co-called "D-value" represents the time needed to reduce the initial concentration (activity, germ count) by one power of ten. It refers to a certain temperature which has to be stated in each case. For example: *Bacillus cereus* $D_{121°C} = 2.3$ s, *Clostridium botulinum* $D_{121°C} = 12.25$ s. For a heat treatment process, the D-value allows the easy determination of the holding time required to reduce the germ count to a certain level. If the germ count of *B. cereus* or *Cl. botulinum* in a certain food should be reduced by seven powers of ten, the required holding times are $2.3 \times 7 = 16.1$ s and $12.25 \times 7 = 85.8$ s.

2.5.4.2 Temperature Dependence of Effects

A relationship exists for the dependence of reaction rate on temperature. It is expressed by an equation of *Arrhenius:*

$$k = A \cdot e^{-E_a/RT} \tag{2.84}$$

with k = rate constant for the reaction rate, E_a = activation energy, R = general gas constant and A = Arrhenius factor. For the relationship between k and T, the *Arrhenius* equation is only an approximation. According to the theory of the transition state (cf. 2.2.1), A is transferred via the active state $A^{\#}$ into P. A and $A^{\#}$ are in equilibrium.

$$A \; \underset{k_{-1}}{\overset{k_1}{\rightleftharpoons}} \; A^{\#} \longrightarrow P \tag{2.85}$$

For the reaction rate follows:

$$k = M \cdot \frac{A}{A^{\neq}} = M \cdot \frac{k_1}{k_{-1}} = M \cdot K^{\neq} \qquad (2.86)$$

with

$$M = \frac{K_B \cdot T}{h} = \frac{R \cdot T}{N_A \cdot h} \qquad (2.87)$$

($K^{\#}$ equilibrium constant, k_B *Boltzmann* constant, h: *Planck* constant, N_A: *Avogadro* number).
For the equilibrium constant follows:

$$K^{\neq} = e^{-\Delta G^{\neq}/RT} \qquad (2.88)$$

Resulting for the equilibrium constant in:

$$k = \frac{k_B \cdot T}{h} e^{-\Delta G^{\neq}/RT} \qquad (2.89)$$

and for the free activation enthalpy:

$$\Delta G^{\neq} = -RT \ln \frac{k \cdot h}{k_B \cdot T} \qquad (2.90)$$

If k is known for any temperature, $\Delta G^{\neq}$ can be calculated according to equation 2.90. Furthermore, the following is valid:

$$\Delta G^{\neq} = \Delta H^{\neq} - T \Delta S^{\neq} \qquad (2.91)$$

A combination with equation 2.90 results in:

$$-RT \ln \frac{k \cdot h}{k_B \cdot T} = \Delta H^{\neq} - T \Delta S^{\neq} \qquad (2.92)$$

and

$$\log \frac{k}{T} = -\log \frac{h}{k_B} - \frac{\Delta H^{\neq}}{2.3RT} + \frac{T \Delta S^{\neq}}{2.3R} \qquad (2.93)$$

It is possible to determine $\Delta H^{\neq}$ graphically based on the above equation if k is known for several temperatures and $\log k/T$ is plotted against $1/T$. If $\Delta G^{\neq}$ and $\Delta H^{\neq}$ are known, $\Delta S^{\neq}$ can be calculated from equation 2.91.
The activation entropy is contained in the *Arrhenius* factor A as can be seen by comparing the empirical *Arrhenius* equation 2.84 with equation 2.89 which is based on the transition state hypothesis:

$$k = A \cdot e^{-E_a/RT} \qquad (2.94a)$$

$$k = \frac{k_B}{h} \cdot e^{-\Delta S^{\neq}/R} \cdot T \cdot e^{-\Delta H^{\neq}/RT} \qquad (2.94b)$$

Activation energy E_a and activation enthalpy $\Delta H^{\neq}$ are linked with each other as follows:

$$\frac{d \ln k}{dT} = \frac{E_a}{RT^2} \qquad (2.95)$$

$$\frac{d \ln k}{dT} = \frac{1}{T} + \frac{\Delta H^{\neq}}{RT^2} = \frac{RT + \Delta H^{\neq}}{RT^2} \qquad (2.96)$$

$$E_a = \Delta H^{\neq} + RT \qquad (2.97)$$

Using plots of $\log k$ against $1/T$, the activation energy of the *Arrhenius* equation can be determined. For enzyme catalyzed reactions, E_a is 10–60, for chemical reactions this value is 50–150 and for the inactivation of enzymes, the unfolding of proteins, and the killing of microorganisms, 250–350 kJ/mol are required.
For enzymes which are able to convert more than one substrate or compound into product, the activation energy may be dependent on the substrate. One example is alcohol dehydrogenase, an important enzyme for aroma formation in semiripened peas (Table 2.13). In this case the activation energy for the reverse reaction is only slightly influenced by substrate.
Under consideration of the temperature dependence of the rate constant k in equation 2.80, the implementation of the expression from *Arrhenius* equation 2.84 leads to:

$$c_1 = c_0 \cdot e^{-k_0 \cdot t \cdot e^{-E_a/RT}} \qquad (2.98)$$

For a constant effect follows:

$$\frac{c_t}{c_0} = \text{const.} = e^{-k_0 \cdot t \cdot e^{-E_a/RT}} \qquad (2.99)$$

Table 2.13. Alcohol dehydrogenase from pea seeds: activation energy of alcohol dehydrogenation and aldehyde reduction

Alcohol	E_a (kJ · mole^{-1})	Aldehyde	E_a (kJ · mole^{-1})
Ethanol	20		
n-Propanol	37	n-Propanal	20
2-Propenol	18		
n-Butanol	40	n-Butanal	21
n-Hexanol	37	n-Hexanal	18
2-trans-hexenol	15	2-trans-Hexenal	19
		2-trans-Heptenal	18

and

$$\ln t = +\frac{E_a}{RT} + \text{const.} \qquad (2.100)$$

When plotting $\ln t$ against $1/T$, a family of parallel lines results for each of different activation energies E_a with each line from a family corresponding to a constant effect c_t/c_0 (cf. equation 2.99) (Fig. 2.34).

For very narrow temperature ranges, sometimes a diagram representing $\log t$ against temperature δ (in °C) is favourable. It corresponds to:

$$\log \frac{t}{t_B} = -\frac{E_a}{2.3R \cdot T_B \cdot T}(\vartheta - \vartheta_B) = \frac{1}{z}(\vartheta - \vartheta_B) \qquad (2.101)$$

with t_B as reference time and T_B or δ_B as reference temperature in K respectively °C. For $\log t/t_B$ the following is valid:

$$z = \frac{2.3R \cdot T_B \cdot T}{E_a} \qquad (2.102)$$

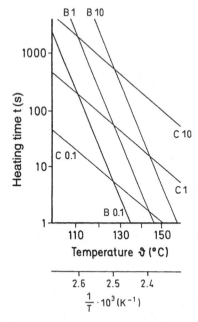

Fig. 2.34. Lines of equal microbiological and chemical effects for heat-treated milk (lines B10, B1, and BO.1 correspond to a reduction in thermophilic spores by 90, 9, and 1 power of ten compared to the initial load; lines C10, C1, and CO.1 correspond to a thiamine degradation of 30%, 3%, and 0.3%; according to *Kessler*, 1988)

This z-value, used in practice, states the temperature increase in °C required to achieve a certain effect in only one tenth of the time usually needed at the reference temperature. However, due to the temperature dependence of the z-value (equation 2.101), linearity can be expected for a very narrow temperature range only. A plot according to equation 2.100 is therefore more favourable.

In the literature, the effect of thermal processes is often described by the Q_{10} value. It refers to the ratio between the rates of a reaction at temperatures $\delta + 10(°C)$ and $\delta(°C)$:

$$Q_{10} = \frac{k_{\vartheta+10}}{k_{\vartheta}} = \frac{t_{\vartheta}}{t_{\vartheta+10}} \qquad (2.103)$$

The combination of equations 2.101 and 2.103 shows the relationship between the Q_{10} value and z-value:

$$\frac{\log Q_{10}}{10} = \frac{E_a}{2.3RT^2} = \frac{1}{z} \qquad (2.104)$$

2.5.4.3 Temperature Optimum

Contrary to common chemical reactions, enzyme-catalyzed reactions as well as growth of microorganisms show a so-called temperature optimum, which is a temperature-dependent maximum resulting from the overlapping of two counter effects with significantly different activation energies (cf. 2.5.4.2):

- increase in reaction or growth rate
- increase in inactivation or killing rate

For starch hydrolysis by microbial α-amylase, the following activation energies, which lie between the limits stated in section 2.5.4.2, were derived from e. g. the *Arrhenius* diagram (Fig. 2.35):

- E_a (hydrolysis) $= 20 \, \text{kJ} \cdot \text{mol}^{-1}$
- E_a (inactivation) $= 295 \, \text{kJ} \cdot \text{mol}^{-1}$

As a consequence of the difference in activation energies, the rate of enzyme inactivation is substantially faster with increasing temperature than the rate of enzyme catalysis. Based on activation energies for the above example, the following relative rates are obtained (Table 2.14). Increasing δ from 0 to 60 °C increases the hydrolysis rate by a factor of 5, while the rate of inactivation is accelerated by more than 10 powers of ten.

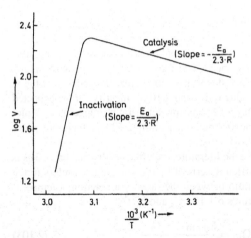

Fig. 2.35. Fungal α-amylase. Amylose hydrolysis versus temperature. *Arrhenius* diagram for assessing the activation energy of enzyme catalysis and enzyme inactivation; V = total reaction rate

Table 2.14. α-Amylase activity as affected by temperature: relative rates of hydrolysis and enzyme inactivation

Temperature (°C)	Relative rate[a]	
	hydrolysis	inactivation
0	1.0	1.0
10	1.35	$1.0 \cdot 10^2$
20	1.8	$0.7 \cdot 10^4$
40	3.0	$1.8 \cdot 10^7$
60	4.8	$1.5 \cdot 10^{10}$

[a] Activation energies of 20 kJ · mole^{-1} for hydrolysis and 295 kJ · mole^{-1} for enzyme inactivation were used for calculation according to *Whitaker* (1972).

The growth of microorganisms follows a similar temperature dependence and can also be depicted according to the *Arrhenius* equation (Fig. 2.36) by replacing the value k by the growth rate and assuming E_a is the reference value µ of the temperature for growth.

For maintaining food quality, detailed knowledge of the relationship between microbial growth rate and temperature is important for optimum production processes (heating, cooling, freezing).

The highly differing activation energies for killing microorganisms and for normal chemical reactions have triggered a trend in food technology towards the use of high-temperature short-time (HTST) processes in production. These are based on the findings that at higher

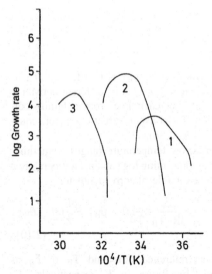

Fig. 2.36. Growth rate and temperature for 1) psychrophilic (*Vibrio AF-1*), 2) mesophilic (*E. coli K-12*) and 3) thermophilic (*Bacillus cereus*) microorganisms (according to *Herbert*, 1989)

temperatures the desired killing rate of microorganisms is higher than the occurrence of undesired chemical reactions.

2.5.4.4 Thermal Stability

The thermal stability of enzymes is quite variable. Some enzymes lose their catalytic activity at lower temperatures, while others are capable of withstanding – at least for a short period of time – a stronger thermal treatment. In a few cases enzyme stability is lower at low temperatures than in the medium temperature range.

Lipase and alkaline phosphatase in milk are thermolabile (Fig. 2.37), whereas acid phosphatase is relatively stable. Therefore, alkaline phosphatase is used to distinguish raw from pasteurized milk because its activity is easier to determine than that of lipase. Of all the enzymes in the potato tuber (Fig. 2.38), peroxidase is the last one to be thermally inactivated. Such inactivation patterns are often found among enzymes in vegetables. In such cases, peroxidase is a suitable indicator for controlling the total inactivation of all the enzymes e. g., in assessing the adequacy of a blanching process. However, newer developments aim to limit the enzyme inactivation to

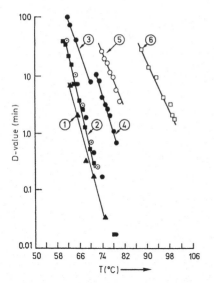

Fig. 2.37. Thermal inactivation of enzymes of milk. 1 Lipase (inactivation extent, 90%), 2 alkaline phosphatase (90%), 3 catalase (80%), 4 xanthine oxidase (90%), 5 peroxidase (90%), and 6 acid phosphatase (99%)

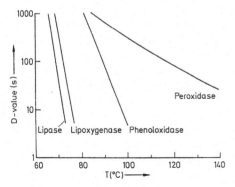

Fig. 2.38. Thermal inactivation (90%) of enzymes present in potato tuber

such enzymes responsible for quality deterioration during storage. For example semiripened pea seeds in which lipoxygenase is responsible for spoilage. However, lipoxygenase is more sensitive than peroxidase, thus a sufficient but gentle blanching requires the inactivation of lipoxygenase only. Inactivation of peroxidase is not necessary.

All the changes which occur in proteins outlined in section 1.4.2.4 also occur during the heating of enzymes. It the case of enzymes the consequences are even more readily observed since a slight con-

formational change at the active site can result in total loss of activity.

The inactivation or killing rates for enzymes and microorganisms depend on several factors. Most significant is the pH. Lipoxygenase isolated from pea seeds (Fig. 2.39) denatures most slowly at its isoelectric point (pH 5.9) as do many other enzymes.

Table 2.21 contains a list of technically useful proteinases and their thermal stability. However, these data were determined using isolated enzymes. They may not be transferrable to the same enzymes in food because in its natural environment an enzyme usually is much more stable. In additional studies, mostly related to heat transfer in food, some successful procedures to calculate the degree of enzyme inactivation based on thermal stabilty data of isolated enzymes have been developed. An example for the agreement between calculated and experimental results is presented in Fig. 2.40.

Peroxidase activity can partially reappear during storage of vegetables previously subjected to a blanching process to inactivate enzymes. The reason for this recurrence, which is also observed for alkaline phosphatase of milk, is not known yet.

Enzymes behave differently below the freezing point. Changes in activity depend on the type of enzyme and on a number of other factors which are partly contrary. The activity is positively influenced by increasing the concentration of enzyme

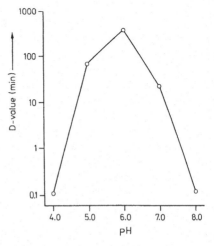

Fig. 2.39. Pea seed lipoxygenase. Inactivation extent at 65 °C as affected by pH

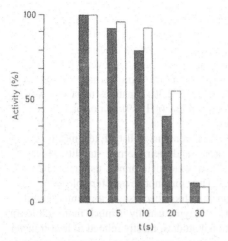

Fig. 2.40. Blanching of semiripened peas at 95 °C; lipoxygenase inactivation (according to *Svensson*, 1977). ■ Experimentally found, □ calculated

and substrate due to formation of ice crystals. A positive or negative change might be caused by changes in pH. Viscosity increase of the medium results in negative changes because the diffusion of the substrate is restricted. In completely frozen food (T < phase transition temperature T'_g, cf. 0.3.3 and Table 0.8), a state reached only during deep-freezing, the catalytic activity stops temporarily. Relatively few enzymes are irreversibly destroyed by freezing.

2.5.5 Influence of Pressure

The application of high pressures can inhibit the growth of microorganisms and the activity of enzymes. This allows the protection of sensitive nutrients and aroma substances in foods. Some products preserved in this gentle way are now marketable. Microorganisms are relatively sensitive to high pressure because their growth is inhibited at pressures of 300–600 MPa and lower pH values increase this effect. However, bacterial spores withstand pressures of >1200 MPa.

In contrast to thermal treatment, high pressure does not attack the primary structure of proteins at room temperature. Only H-bridges, ionic bonds and hydrophobic interactions are disrupted. Quaternary structures are dissociated into subunits by comparatively low pressures (<150 MPa). Higher pressures (>1200 MPa) change the tertiary structure and very high pressures disrupt the H-bridges which stabilize the secondary structure. The hydration of proteins is also changed by high pressure because water molecules are pressed into cavities which can exist in the hydrophobic interior of proteins. In general, proteins are irreversibly denatured at room temperature by the application of pressures above 300 MPa while lower pressures cause only reversible changes in the protein structure.

In the case of enzymes, even slight changes in the steric arrangement and mobility of the amino acid residues which participate in catalysis can lead to loss of activity. Nevertheless, a relatively high pressure is often required to inhibit enzymes. But the pressure required can be reduced by increasing the temperature, as shown in Fig. 2.41 for α-amylase. While a pressure of 550 MPa is required at 25 °C to inactivate the enzyme with a rate constant (first order reaction) of $k = 0.01$ min^{-1}, a pressure of only 340 MPa is required at 50 °C.

It is remarkable that enzymes can also be activated by changes in the conformation of the polypeptide chain, which are initiated especially by low pressures around 100 MPa. In the application of the pressure technique for the production of stable food, intact tissue, and not isolated enzymes, is exposed to high pressures. Thus, the enzyme activity can increase instead of decreasing when cells or membranes are disintegrated with the release of enzyme and/or substrate.

Some examples are presented here to show the pressures required to inhibit the enzyme activity which can negatively effect the quality of foods.

– Pectin methylesterase (EC 3.1.1.11) causes the flocculation of pectic acid (cf. 2.7.2.2.13) in orange juices and reduces the consistency of tomato products. In orange juice, irreversible enzyme inactivation reaches 90% at a pressure of 600 MPa. Even though the enzyme in tomatoes is more stable, increasing the temperature to 59–60 °C causes inactivation at 400 MPa and at 100 MPa after the removal of Ca^{2+} ions.
– Peroxidases (EC 1.11.1.3) induce undesirable aroma changes in plant foods. In green beans, enzyme inactivation reached 88% in 10 min after pressure treatment at 900 MPa. At pressures above 400 MPa (32 °C), the activity of this enzyme in oranges fell continuously

to 50%. However, very high pressures increased the activity at 32–60 °C. It is possible that high pressure denatures peroxidase to a heme(in) catalyst (cf. 3.7.2.1.7).

– Lipoxygenase from soybeans (cf. 3.7.2.2). This enzyme was inactivated in 5 min at pH 8.3 by pressures up to 750 MPa and temperatures in the range 0–75 °C. The pressure stability was reduced by gassing with CO_2 and reducing the pH to 5.4.

– Polyphenol oxidases (cf. 0.3.3) in mushrooms and potatoes require pressures of 800–900 MPa for inactivation. The addition of glutathione (5 mmol/l) increases the pressure sensitivity of the mushroom enzyme. In this case, the inactivation is obviously supported by the reduction of disulfide bonds.

2.5.6 Influence of Water

Up to a certain extent, enzymes need to be hydrated in order to develop activity. Hydration of e. g. lysozyme was determined by IR and NMR spectroscopy. As can be seen in Table 2.15, first the charged polar groups of the side chains hydrate, followed by the uncharged ones. Enzymatic activity starts at a water content of 0.2 g/g protein, which means even before a monomolecular layer of the polar groups with water has taken place. Increase in hydration resulting in

Table 2.15. Hydration of Lysozyme

g Water / g Protein	Hydration sequence	Molecular changes
0.0	Charged groups	Relocation of protons
	Uncharged, polar groups (formation of clusters)	New orientation of disulfide bonds
0.1	Saturation of COOH groups Saturation of polar groups in side chains	Change in conformation
0.2	Peptide-NH	Start of enzymatic activity
0.3	Peptide-CO Monomolecular hydration of polar groups Apolar side chains	
0.4	Complete enzyme hydration	

a monomolecular layer of the whole available enzyme surface at 0.4 g/g protein raises the activity to a limiting value reached at a water content of 0.9 g/g protein. Here the diffusion of the substrate to the enzyme's active site seems to be completely guaranteed.

For preservation of food it is mandatory to inhibit enzymatic activity completely if the storage temperature is below the phase transition temperature T_g or T'_g (cf. 0.3.3). With help of a model system containing glucose oxidase, glucose and water as well as sucrose and maltodextrin (10 DE) for adjustment of T'_g values in the range of -9.5 to -32 °C, it was found that glucose was enzymatically oxidized only in such samples that were stored for two months above the T'_g value and not in those kept at storage temperatures below T'_g.

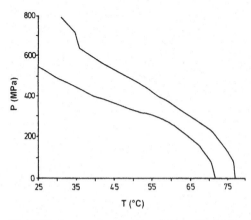

Fig. 2.41. Pressure–temperature diagram for the inactivation kinetics of α-amylase from *Bacillus subtilis* at pH 8.6 (according to *Ludikhuyze* et al., 1997). Range of the rate constants: k = 0.01 min^{-1} (lower line) to k = 0.07 min^{-1} (upper line)

2.6 Enzymatic Analysis

Enzymatic food analysis involves the determination of food constituents, which can be both substrates or inhibitors of enzymes, and the determination of enzyme activity in food.

2.6.1 Substrate Determination

2.6.1.1 Principles

Qualitative and quantitative analysis of food constituents using enzymes can be rapid, highly sensitive, selective and accurate (examples in Table 2.16). Prior purification and separation steps, as a rule, are not necessary in the enzymatic analysis of food.

In an enzymatic assay, spectrophotometric or electrochemical determination of the reactant or the product is the preferred approach. When this is not applicable, the determination is performed by a coupled enzyme assay. The coupled reaction includes an auxiliary reaction in which the food constituent is the reactant to be converted to product, and an indicator reaction which involves an indicator enzyme and its reactant or product, the formation or breakdown of which can be readily followed analytically. In most cases, the indicator reaction follows the auxiliary reaction:

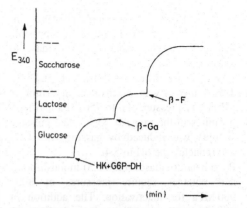

Fig. 2.42. Enzymatic determination of glucose, saccharose and lactose in one run. After adding cosubstrates, ATP and NADP, the enzymes are added in the order: hexokinase (HK), glucose-6-phosphate dehydrogenase (G6P-DH), β-galactosidase (β-Ga) and β-fructosidase (β-F)

$$\text{Glucose} + \text{ATP} \xrightarrow{\text{HK}} \text{Glucose-6P} \qquad\qquad (a)$$

$$\text{Glucose-6P} + \text{NADP}^\oplus \xrightarrow{\text{G6P-DH}} \text{6-Phosphogluconate} + \text{NADPH} + \text{H}^\oplus \text{ (b)}$$

$$\text{Lactose} \xrightarrow{\beta\text{-Ga}} \text{Glucose} + \text{Galactose} \qquad (c)$$

$$\text{Saccharose} \xrightarrow{\beta\text{-F}} \text{Glucose} + \text{Fructose} \qquad (d) \qquad\qquad (2.105)$$

$$\text{A} + \text{B} \underset{\text{Auxiliary reaction}}{\rightleftharpoons} \text{P} + \text{Q}$$

$$\text{P} + \text{C} \underset{\text{Indicator reaction}}{\rightleftharpoons} \text{R} + \text{S} \qquad (2.106)$$

Reactant A is the food constituent which is being analyzed. C or R or S is measured. The equilibrium state of the coupled indicator reaction is concentration dependent. The reaction has to be adjusted in some way in order to remove, for example, P from the auxiliary reaction before an equilibrium is achieved. By using several sequential auxiliary reactions with one indicator reaction, it is possible to simultaneously determine several constituents in one assay. An example is the analysis of glucose, lactose and saccharose (cf. Reaction 2.105).

First, glucose is phosphorylated with ATP in an auxiliary reaction (a). The product, glucose-6-phosphate, is the substrate of the NADP-dependent indicator reaction (b). Add-

ition of β-galactosidase starts the lactose analysis (c) in which the released glucose, after phosphorylation, is again measured through the indicator reaction [(b) of Reaction 2.105 and also Fig. 2.42]. Finally, after addition of β-fructosidase, saccharose is cleaved (d) and the released glucose is again measured through reactions (a) and (b) as illustrated in Fig. 2.42.

2.6.1.2 End-Point Method

This procedure is reliable when the reaction proceeds virtually to completion. If the substrate is only partly consumed, the equilibrium is displaced in favor of the products by increasing the concentration of reactant or by removing one of the products of the reaction. If it is not possible to achieve this, a standard curve must

Table 2.16. Examples of enzymatic analysis of food constituents[a]

Constituent	Auxiliary reaction	Indicator reaction
Glucose	$\beta\text{-D-Glucose}^b + O_2 \xrightarrow{\text{Glucose oxidase}} \delta\text{-D-Gluconolactone} + H_2O_2(a_H)$	o-Dianisidine $+ H_2O_2 \xrightarrow{\text{Peroxidase}}$ Oxid. o-dianisidine (a_1)
	Glucose $+ ATP \xrightarrow{\text{Hexokinase}}$ Glucose-6P (b_H)	Glucose-6P $+ NADP^{\oplus} \xrightarrow{\text{Glucose-6P dehydrogenase}}$ Gluconate-6P $+ NADPH + H^{\oplus} (b_1)$
Fructose	Fructose $+ ATP \xrightarrow{\text{Hexokinase}}$ Fructose-6P	As glucose-6P (b_1)
	Fructose-6P $\xrightarrow{\text{Glucosephosphate isomerase}}$ Glucose-6P	
Sorbitol	D-Sorbitol $+ NAD \xrightarrow{\text{Sorbitol dehydrogenase}}$ Fructose $+ NADH + H^{\oplus}$	As glucose $(b_H + b_1)$
Maltose	Maltose $+ H_2O \xrightarrow{\alpha\text{-Glucosidase}}$ 2 Glucose	As glucose $(b_H + b_1)$
Starch	Starch $+ (n-1) H_2O \xrightarrow{\text{Amyloglucosidase}}$ n-Glucose	As glucose $(b_H + b_1)$
Galactose	β-D-Galactose $+ NAD^{\oplus} \xrightarrow{\text{Galactose dehydrogenase}}$ D-Galactono-γ-lactone $+ NADH + H^{\oplus}$	
Ethanol	Ethanol $+ NAD^{\oplus} \xrightarrow{\text{Alcohol dehydrogenase}}$ Acetaldehyde $+ NADH + H^{\oplus}$	
Glycerol	Glycerol $+ ATP \xrightarrow{\text{Glycerol kinase}}$ sn-Glycerol-3P $+ ATP$	ADP + Phosphoenolpyruvate $\xrightarrow{\text{Pyruvate kinase}} ATP + $ Pyruvate (c)
		Pyruvate $+ NADH + H^{\oplus} \xrightarrow{\text{Lactate dehydrogenase}}$ Lactate $+ NAD^{\oplus}$ (d)
Lactate	L-Lactate assay is achieved by a reversed reaction of d), and D-lactate assay with a dehydrogenase specific for D-enantiomer.	
Creatinine and Creatine	Creatinine $+ H_2O \xrightarrow{\text{Creatininase}}$ Creatine	
	Creatine $+ ATP \xrightarrow{\text{Creatine kinase}}$ Creatine-P $+ ADP$; ADP is determined through c) and d)	
Individual amino acids	R-CH(NH$_2$)COOH $\xrightarrow{\text{Amino acid decarboxylase}^d}$ R-CH$_2$-NH$_2$ + CO$_2$	
L-Malate	L-Malate $+ NAD^{\oplus} \xrightarrow{\text{Malate dehydrogenase}}$ Oxalacetate $+ NADH + H^{\oplus}$	

[a] For saccharose and lactose see Fig. 2.42.
[b] The content of α-anomeric form is accessible through mutarotation.
[c] After hydrolysis this method is suitable for the assay of acylglycerols.
[d] Specific decarboxylases are availabe as exemplified by those for L-tryosine, L-lysine, L-glutamic acid, L-aspartic acid, or L-arginine.

Table 2.17. Enzyme concentrations used in the end-point method of enzymatic food analysis

Substrate	Enzyme	K_m (mol/l)	Enzyme concentration (μcat/l)
Glucose	Hexokinase	$1.0 \cdot 10^{-4}$ (30 °C)	1.67
Glycerol	Glycerol kinase	$5.0 \cdot 10^{-5}$ (25 °C)	0.83
Uric acid	Urate oxidase	$1.7 \cdot 10^{-5}$ (20 °C)	0.28
Fumaric acid	Fumarase	$1.7 \cdot 10^{-6}$ (21 °C)	0.03

be prepared. In contrast to kinetic methods (see below), the concentration of substrate which is to be analyzed in food must not be lower than the *Michaelis* constant of the enzyme catalyzing the auxiliary reaction. The reaction time is readily calculated when the reaction rate follows first-order kinetics for the greater part of the enzymatic reaction.

In a two-substrate reaction the enzyme is saturated with the second substrate. Since Equation 2.41 is valid under these conditions, the catalytic activity of the enzyme needed for the assay can be determined for both one- and two-substrate reactions. The examples shown in Table 2.17 suggest that enzymes with low K_m values are desirable in order to handle the substrate concentrations for the end-point method with greater flexibility.

Data for K_m and V are needed in order to calculate the reaction time required. A prerequisite is a reaction in which the equilibrium state is displaced toward formation of product with a conversion efficiency of 99%.

2.6.1.3 Kinetic Method

Substrate concentration is obtained using a method based on kinetics by measuring the reaction rate. To reduce the time required per assay, the requirement for the quantitative conversion of substrate is abandoned. Since kinetic methods are less susceptible to interference than the endpoint method, they are advantageous for automated methods of enzymatic analysis.

The determination of substrate using kinetic methods is possible only as long as Equation 2.46 is valid. Hence, the following is required to perform the assay:

a) For a two-substrate reaction, the concentration of the second reactant must be so high that the rate of reaction depends only on the concentration of the substrate which is being analyzed.

b) Enzymes with high *Michaelis* constants are required; this enables relatively high substrate concentrations to be determined.

c) If enzymes with high *Michaelis* constants are not available, the apparent K_m is increased by using competitive inhibitors.

In order to explain requirement c), let us consider the example of the determination of glycerol as given in Table 2.16. This reaction allows the determination of only low concentrations of glycerol since the K_m values for participating enzymes are low: 6×10^{-5} mol/l to 3×10^{-4} mol/l. In the reaction sequence the enzyme pyruvate kinase is competitively inhibited by ATP with respect to ADP. The expression $K_m(1 + (I)/K_I)$ (cf. 2.5.2.2.1) may in these circumstances assume a value of 6×10^{-3} mol/l, for example. This corresponds to an apparent increase by a factor of 20 for the K_m of ADP (3×10^{-4} mol/l). The ratio $(S)/K_m(1 + [I]/K_I)$ therefore becomes 1×10^{-3} to 3×10^{-2}. Under these conditions, the auxilary reaction (Table 2.16) with pyruvate kinase follows pseudo-first-order kinetics with respect to ADP over a wide range of concentrations and, as a result of the inhibition by ATP, it is also the rate-determining step of the overall reaction. It is then possible to kinetically determine higher concentrations of glycerol.

2.6.2 Determination of Enzyme Activity

In the foreword of this chapter it was emphasized that enzymes are suitable indicators for identifying heat-treated food. However, the determination of enzyme activity reaches far beyond this possibility: it is being used to an increasing extent for the evaluation of the quality of raw food and for optimizing the parameters of particular food processes. In addition, the activities of enzyme preparations have to be controlled prior to use in processing or in enzymatic food analysis.

The measure of the catalytic activity of an enzyme is the rate of the reaction catalyzed by the enzyme. The conditions of an enzyme activity assay are optimized with relation to: type and ionic strength of the buffer, pH, and concentrations of substrate, cosubstrate and activators used. The closely controlled assay conditions, including the temperature, are critical because, in contrast to substrate analysis, the reliability of the results in this case often can not be verified by using a weighed standard sample.

Temperature is a particularly important parameter which strongly influences the enzyme assay. Temperature fluctuations significantly affect the reaction rate (cf. 2.5.4); e. g., a $1\,°C$ increase in temperature results in about a 10% increase in activity. Whenever possible, the incubation temperature should by maintained at $25\,°C$.

The substrate concentration in the assay is adjusted ideally so that Equation 2.40 is valid, i. e. $[A_0] \gg K_m$. Difficulties often arise while trying to achieve this condition: the substrate's solubility is limited; spectrophotometric readings become unreliable because of high light absorbance by the substrate; or the high concentration of the substrate inhibits enzyme activity. For such cases procedures exist to assess the optimum substrate concentration which will support a reliable activity assay.

2.6.3 Enzyme Immunoassay

Food compounds can be determined specifically and sensitively by immunological methods. These are based on the specific reaction of an antibody containing antiserum with the antigen, the substance to be determined. The antiserum is produced by immunization of rabbits for example. Because only compounds with a high molecular weight ($M_r > 5000$) display immunological activity, for low molecular compounds (haptens) covalent coupling to a protein is necessary. The antiserum produced with the "conjugate" contains antibodies with activities against the protein as well as the hapten.

Prior to the application, the antiserum is tested for its specificity against all proteins present in the food to be analyzed. As far as possible all un-specificities are removed. For example, it is possible to treat an antiserum intended to be used for the determination of peanut protein with proteins from other nuts in such a way that it specifically reacts with peanut protein only. However, there are also cases in which the specificity could not be increased because of the close immunochemical relationship between the proteins. This happens, for example, with proteins from almonds, peach and apricot kernels.

The general principle of the competitive immunoassay is shown in Fig. 2.43. Excess amounts of marked and unmarked antigens compete for the antibodies present. The concentration of the unmarked antigen to be determined is the only variable if the concentration of the marked antigen and the antibody concentration are kept on a constant level during the examination. Following the principle of mass action, the unknown antigen concentration can be calculated indirectly based on the proportion of free marked antigen. Older methods still require the formation of a precipitate for the detection of an antibody–antigen reaction (cf. 12.10.2.3.2). Immunoassays are much faster and more sensitive.

Radioisotopes (^3H, ^{14}C) and enzymes are used to mark antigens. Furthermore, fluorescent and luminescent dyes as well as stable radicals are important. Horseradish peroxidase, alkaline phosphatase from calf stomach, and β-D-galactosidase from E. coli are often used as indicator enzymes because they are available in high purity, are very stable and their activity can be determined sensitively and precisely. Enzymes are bound to antigens or haptens by covalent bonds, e. g., by reaction with glutaraldehyde or carbodiimide.

Enzyme immunoassays are increasingly used in food analysis (examples see Table 2.18). Laboratories employing these methods need no specific

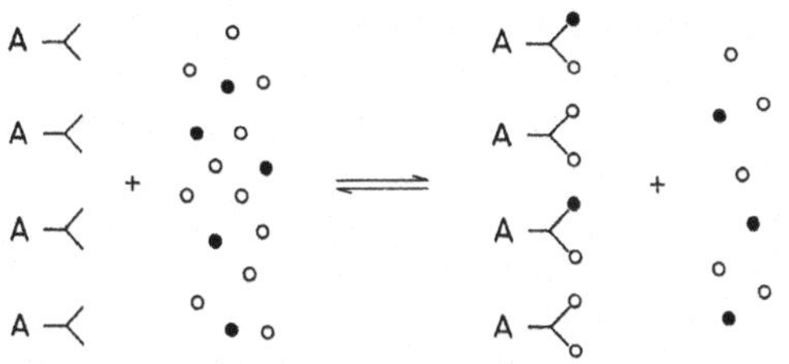

Fig. 2.43. Principle of an immunoassay. Marked antigens (●) and unmarked antigens (○) compete for the binding sites of the antibodies A

Table 2.18. Examples for application of enzyme immunoassay in food analysis

Detection and quantification
Type of meat
Soya protein in meat products
Myosin in muscle meat
Cereal proteins as well as papain in beer
Gliadins (absence of gluten in foods)
Veterinary drugs and fattening aids, e. g. penicillin in milk, natural or synthetic estrogens in meat
Toxins (aflatoxins, enterotoxins, ochratoxins) in food
Pesticides (atrazine, aldicarb, carbofuran)
Glycoalkaloids in potatoes

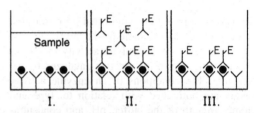

Fig. 2.44. Principle of non-competitive ELISA (sandwich ELISA)

Y   Immobilized antibody,

●   antigen,

$\curlyvee^E$ enzyme-marked antibody

equipment contrary to use of radio immunological methods (RIA). Furthermore, for radio immunoassays free antigens have always to be separated from the ones bound to antibodies (heterogeneous immunoassay) while an enzyme immunoassay is suitable for homogeneous tests if the activity of the indicator enzyme is inhibited by the formation of an antigen–antibody-complex.

In food analysis, the ELISA test (*enzyme linked immunosorbent assay*) is the most important immunochemical method. In fact, two experimental procedures are applied: the competitive ELISA, as shown in Fig. 2.43, and the sandwich ELISA. While the competitive ELISA is directed at the detection of low-molecular substances, the sandwich ELISA is suitable only for analytes (antigens) larger than a certain minimum size. The antigen must have at least two antibody binding sites (epitopes) which are spatially so far apart that it can bind two different antibodies. The principle of the sandwich ELISA is shown in Fig. 2.44. A plastic carrier holds the antibodies, e. g. against a toxin, by adsorption. When the sample is added, the toxin (antigen) reacts with the excess amount of antibodies (I in Fig. 2.44). The second antibody marked with an enzyme (e. g. alkaline phosphatase, peroxidase, glucose oxidase or luciferase) and with specificity for the antigen forms a sandwich complex (II). Unbound enzyme-marked antibodies are washed out. The remaining enzyme activity is determined (III). It is directly proportional to the antigen concentration in the sample which can be calculated based on measured standards and a calibration curve.

2.6.4 Polymerase Chain Reaction

With the polymerase chain reaction (PCR), a few molecules of any DNA sequence can be multiplied by a factor of 10^6 to 10^8 in a very short time. The sequence is multiplied in a highly specific way until it becomes visible electrophoretically. Based on PCR, analytical techniques have been developed for species identification in the case of animal and plant foods and microorganisms. It is

Table 2.19. Examples of approved genetically modified crops (as of 2003)[a]

Crop	Property
Cauliflower	Herbicide tolerance
Broccoli	Herbicide tolerance
Chicory	Herbicide tolerance
Cucumber	Fungal resistance
Potato	Insect and virus resistance
Pumpkin	Virus resistance
Corn	Herbicide tolerance, insect resistance
Melon	Virus resistance, delayed ripening
Papaya	Virus resistance
Paprika	Virus resistance
Rape seed	Higher concentrations of 12:0 und 14:0, herbicide resistance
Rice	Virus resistance
Red Bean	Insect resistance
Soybean	Altered fatty acid spectrum (cf. 14.3.2.2.5), herbicide tolerance
Tomato	Delayed ripening, increased pectin content
Wheat	Herbicide tolerance
Sugar beet	Herbicide tolerance

[a] The crop is approved in at least one country.

of special interest that PCR allows the detection of genetically modified food (genetically modified organism, GMO). Thus, it is possible to control the labeling of GMOs, which is required by law. In fact, the number of GMOs among food crops is increasing steadily (cf. Table 2.19); cf. survey by Anklam et al. (2002).

2.6.4.1 Principle of PCR

The first steps of a PCR reaction are shown schematically in Fig. 2.45. First, the extract

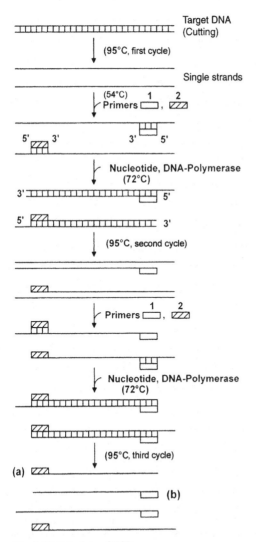

Fig. 2.45. Principle of PCR

which contains, among other substances, the DNA fragment (analyte) to be identified is briefly heated to 95 °C. This causes denaturation and separation into single strands. After cooling to 54 °C, two oligodeoxynucleotides (primer 1 and 2 having base sequences complementary to the target DNA) which flank the DNA sequence to be multiplied are added in excess. These primers, which are 15–30 nucleotides long and made with a synthesizer, hybridize with the complementary segments on the single strands. The temperature is increased to 72 °C and a mixture of the four deoxynucleoside 5′-triphosphates (dATP, dCTP, dGTP, dTTP, for structures of the bases cf. Formula 2.107) and a thermostable DNA polymerse, e.g., *Taq polymerase* from *Thermus aquaticus*, are added. The polymerase synthesizes new complementary strands starting from the primers in the 5′ → 3′ direction using the deoxynucleotides. In the subsequent heating step, these strands are separated, in addition to the denatured target DNA which is no longer shown in Fig. 2.45. In the second cycle, the primers hybridize with the single strands which end with the nucleotide sequence of the other primer in each case. The PCR yields two DNA segments (a and b in Fig. 2.45) which are bounded by the nucleotide sequences of the primer. The DNA segment is amplified by repeating the steps denaturation – addition of primer – PCR 20 to 30 times, and is then electophoretically analyzed.

Base:

Cytosine (C)   Thymine (T)

Adenine (A)   Guanine (G)

$$(2.107)$$

In comparison with protein analysis, DNA analysis is more sensitive by several orders of magnitude due to amplification which is millionfold after 20 cycles and billionfold after 30 cycles. Heated food can be analyzed because DNA is considerably more stable than proteins. It is also possible to detect GMOs which do not contain altered or added proteins identifiable by chemical methods. Acidic foods can cause problems when they are strongly heated, e. g., tomato products. In this case, the DNA is hydrolyzed to such an extent that the characteristic sequences are lost. The exceptional sensitivity of this method can also give incorrect positive results. For this reason, it is important that the PCR is quantitatively evaluated, especially when controlling limiting values. A known amount of a synthetic DNA is added to the sample and amplified competitively with the analyte. For calibration, mixtures of the target and competing DNA are subjected to PCR analysis.

2.6.4.2 Examples

2.6.4.2.1 Addition of Soybean

The addition of soybean protein to meat and other foods can be detected with the help of the primers GMO3 (5′-GCCCTCTACTCCA-CCCCCATCC-3′) and GMO4 (5′-GCCCATC-TGCAAGCCTTTTTGTG-3′). They label a small but still sufficiently specific sequence of 118 base pairs (bp) of the gene for a lectin occurring in soybean. A small amplicon is of advantage since the DNA gets partially fragmented when meat preparations are heated.

2.6.4.2.2 Genetically Modified Soybeans

Genetically modified soybeans are resistant to the herbicide glyphosphate (cf. 9.4.3), which inhibits the key enzyme, 5-enolpyruvylshiki-mi-3-phosphate synthase (EPSPS), in the metabolism of aromatic amino acids in plants. However, glyphosphate is inactive against the EPSPS of bacteria. Hence transgenic soybeans contain a genetic segment which codes for an EPSPS from *Agrobacterium sp.* and a peptide for the transport of this enzyme. To detect this segment and, consequently, genetically modified soybeans, primers are used which induce the amplification of a segment of 172 bp in the PCR.

2.6.4.2.3 Genetically Modified Tomatoes

During ripening and storage, tomatoes soften due to the activity of an endogenous enzyme poly-galacturonase (PG). The expression of the gene for PG is specifically inhibited in a particular tomato, resulting in extended storage life and better aroma. PCR methods have been developed to detect these transgenic tomatoes. However, this detection can fail if the DNA is too strongly hydrolyzed on heating the tomato products.

2.6.4.2.4 Species Differentiation

If specific primers fail, a PCR with universal primers can be applied in certain cases, followed by an RFLP analysis (restriction fragment length polymorphism). The DNA of a meat sample is first determined with a primer pair which exhibits a high degree of correspondence in its binding sites to the DNA of many animal species. In the case of various animal species, the PCR yields equally long products which should be relatively large (ca. 300–500 bp). The amplicon is cleaved in the subsequent RFLP analysis with different restriction endonucleases. After electrophoretic separation, the pattern of the resulting DNA fragments can be assigned to individual animal species. This method is suitable for samples of one type of meat. Preparations containing meat of several animal species or DNA which is more strongly fragmented on heating can be reliably analyzed only with animal species-specific primers.

2.7 Enzyme Utilization in the Food Industry

Enzyme-catalyzed reactions in food processing have been used unintentionally since ancient times. The enzymes are either an integral part of the food or are obtained from microorganisms. Addition of enriched or purified enzyme preparations of animal, plant or, especially, microbial origin is a recent practice. Most of these enzymes come from microorganisms, which have been genetically modified in view of their economic production. Such intentionally used additives provide a number of advantages in food processing: exceptionally pronounced substrate specificity (cf. 2.2.2), high reaction rate under mild reaction conditions (temperature, pH), and a fast and continuous, readily controlled reaction process with generally modest operational costs and investment. Examples for the application of microbial enzymes in food processing are given in Table 2.20.

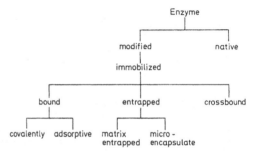

Fig. 2.46. Forms of immobilized enzymes

2.7.1 Technical Enzyme Preparations

2.7.1.1 Production

The methods used for industrial-scale enzyme isolation are outlined in principle under section 2.2.4. In contrast to the production of highly purified enzymes for analytical use, the production of enzymes for technical purposes is directed to removing the interfering activities which would be detrimental to processing and to staying within economically acceptable costs. Selective enzyme precipitation by changing the ionic strength and/or pH, adsorption on inorganic gels such as calcium phosphate gel or hydroxyl apatite, chromatography on porous gel columns and ultrafiltration through membranes are among the fractionation methods commonly used. Ionexchange chromatography, affinity chromatography (cf. 2.2.4) and preparative electrophoresis are relatively expensive and are seldom used. A few temperature-stable enzymes are heat treated to remove the other contaminating and undesired enzyme activities.

Commercial enzyme preparations are available with defined catalytic activity. The activity is usually adusted by the addition of suitable inert fillers such as salts or carbohydrates. The amount of active enzyme is relatively low, e. g., proteinase preparations contain 5–10% proteinase, whereas amylase preparations used for treamtent of flour contain only 0.1% pure fungal α-amylase.

2.7.1.2 Immobilized Enzymes

Enzymes in solution are usually used only once. The repeated use of enzymes fixed to a carrier is more economical. The use of enzymes in a continous process, for example, immobilized enzymes used in the form of a stationary phase which fills a reaction column where the reaction can be controlled simply by adjustment of the flow rate, is the most advanced technique. Immobilized enzymes are produced by various methods (Fig. 2.46).

2.7.1.2.1 Bound Enzymes

An enzyme can be bound to a carrier by covalent chemical linkages, or in many cases, by physical forces such as adsorption, by charge attraction, H-bond formation and/or hydrophobic interactions. The covalent attachment to a carrier, in this case an activated matrix, is usually achieved by methods employed in peptide and protein chemistry. First, the matrix is activated. In the next step, the enzyme is coupled under mild conditions to the reactive site on the matrix, usually by reaction with a free amino group. This is illustrated by using cellulose as a matrix (Fig. 2.47). Another possibility is a process of copolymerization with suitable monomers. Generally, covalent attachment of the enzyme prevents leaching or "bleeding".

2.7.1.2.2 Enzyme Entrapment

An enzyme can be entrapped or enclosed in the cavities of a polymer network by polymerization of a monomer such as acrylamide or N,N'-methylene-bis-acrylamide in the presence of enzyme, and still remain accessible to substrate through the network of pores. Furthermore,

Table 2.20. Examples for the use of microbial enzymes in food processing

EC Number	Enzyme[a]	Biological Origin	Application[b]
Oxidoreductases			
1.1.1.39	Malate dehydrogenase (decarboxylating)	*Leuconostoc oenos*	10
1.1.3.4	Glucose oxidase	*Aspergillus niger*	7, 10, 16
1.11.1.6	Catalase	*Micrococcus lysodeicticus*	
		Aspergillus niger	1, 2, 7, 10, 16
Transferases			
2.7.2.4	Transglutaminase	*Streptoverticillium*	5, 8
Hydrolases			
3.1.1.1	Carboxylesterase	*Mucor miehei*	2, 3
3.1.1.3	Triacylglycerol lipase	*Aspergillus niger, A. oryzae, Candida lipolytica, Mucor javanicus, M. miehei, Rhizopus arrhizus, R. niveus*	2, 3
3.1.1.11	Pectinesterase	*Aspergillus niger*	9, 10, 17
3.1.1.20	Tannase	*Aspergillus niger, A. oryzae*	10
3.2.1.1	α-Amylase	*Bacillus licheniformis, B. subtilis, Aspergillus oryzae*	3, 8, 9, 10, 12 14, 15
		Aspergillus niger, Rhizopus delemar, R. oryzae	8, 9, 10, 12, 14, 15
3.2.1.2	β-Amylase	*Bacillus cereus,*	8, 10
3.2.1.3	Glucan-1,4-α-D-gluco-sidase (glucoamylase)	*B. magatherium, B. subtilis Aspergillus oryzae*	3, 9, 10, 12, 14 15, 18
		Aspergillus niger, Rhizopus arrhizus, R. delemar, R. niveus, R. oryzae, Trichoderma reesei	9, 10, 12, 14, 15, 18
3.2.1.4	Cellulase	*Aspergillus niger, A. oryzae, Rhizopus delemar, R. oryzae, Sporotrichum dimorphosporum, Thielavia terrestris, Trichoderma reesei*	9, 10, 18
3.2.1.6	Endo-1,3(4)-β-D-glucanase	*Bacillus circulans, B. subtilis, Aspergillus niger, A. oryzae, Penicillum emersonii, Rhizopus delemar, R. oryzae*	10
3.2.1.7	Inulinase	*Kluyveromyces fragilis*	12
3.2.1.11	Dextranase	*Klebsiella aerogenes, Penicillium funicolosum, P. lilacinum*	12
3.2.1.15	Polygalacturonase	*Aspergillus niger, Penicillium simplicissimum, Trichoderma reesei*	3, 9, 10, 17
		Aspergillus oryzae, Rhizopus oryzae	3, 9, 10
		Aspergillus niger	9, 10, 17

Table 2.20. (continued)

EC Number	Enzyme[a]	Biological Origin	Application[b]
3.2.1.20	α-D-Glucosidase	*Aspergillus niger, A. oryzae, Rhizopus oryzae*	8
3.2.1.21	β-D-Glucosidase	*Aspergillus niger, Trichoderma reesei*	9
3.2.1.22	α-D-Galactosidase	*Aspergillus niger, Mortierella vinacea sp., Saccharomyces carlsbergensis*	12
3.2.1.23	β-D-Galactosidase	*Aspergillus niger, A. oryzae, Kluyveromyces fragilis, K. lactis*	1, 2, 4, 18
3.2.1.26	β-D-Fructofuranosidase	*Aspergillus niger, Saccharomyces carlsbergensis, S. cerevisiae*	14
3.2.1.32	Xylan endo-1,3-β-D-xylosidase	*Streptomyces sp., Aspergillus niger, Sporotrichum dimorphosporum*	8, 10, 13
3.2.1.41	α-Dextrin endo-1,6-α-glucosidase (pullunanase)	*Bacillus acidopullulyticus* *Klebsiella aerogenes*	8, 10, 12, 14, 15 8, 10, 12
3.2.1.55	α-L-Arabinofuranosidase	*Aspergillus niger*	9, 10, 17
3.2.1.58	Glucan-1,3-β-D-glucosidase	*Trichoderma harzianum*	10
3.2.1.68	Isoamylase	*Bacillus cereus*	8, 10
3.2.1.78	Mannan endo-1,4-β-D-mannanase	*Bacillus subtilis, Aspergillus oryzae, Rhizopus delemar, R. oryzae, Sporotrichum dimor-phosporum, Trichoderma reesei* *Aspergillus niger*	13 13, 17
3.5.1.2	Glutaminase	*Bacillus subtilis*	5
3.4.21.14	Serine endopeptidase[c]	*Bacillus licheniformis*	5, 6, 10, 11
3.4.23.6	Aspartic acid endopeptidase	*Aspergillus melleus, Endothia parasitica, Mucor miehei, M. pusillus* *Aspergillus oryzae*	2 2, 5, 6, 8, 9, 10, 11, 15, 18
3.4.24.4	Metalloendopeptidase	*Bacillus cereus, B. subtilis*	10, 15

Lyases 4.2.2.10	Pectin lyase	*Aspergillus niger*	9, 10, 17

Isomerases 5.3.1.5	Xylose isomerase[d]	*Streptomyces murinus S. olivaceus, S. olivochromogenes, S. rubiginosus*	8, 9, 10, 12

[a] Principal activity.

[b] 1) Milk, 2) Cheese, 3) Fats and oils, 4) Ice cream, 5) Meat, 6) Fish, 7) Egg, 8) Cereal and starch, 9) Fruit and vegetables, 10) Beverages (soft drinks, beer, wine), 11) Soups and broths, 12) Sugar and honey, 13) Cacao, chocolate, coffee, tea, 14) Confectionery, 15) Bakery, 16) Salads, 17) Spices and flavors, 18) Diet food.

[c] Similar to Subtilisin.

[d] Some enzymes also convert D-glucose to D-fructose, cf. 2.7.2.3.

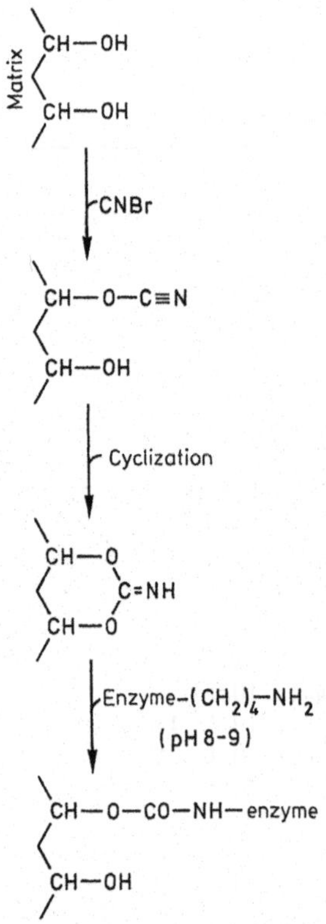

Matrix

$$\begin{matrix} \diagdown \\ CH{-}OH \\ \diagup\diagdown \\ CH{-}OH \\ \diagup \end{matrix}$$

↓ CNBr

$$\begin{matrix} \diagdown \\ CH{-}O{-}C{\equiv}N \\ \diagup\diagdown \\ CH{-}OH \\ \diagup \end{matrix}$$

↓ Cyclization

$$\begin{matrix} \diagdown \\ CH{-}O \\ \diagup\diagdown\quad C{=}NH \\ CH{-}O \\ \diagup \end{matrix}$$

↓ Enzyme-$(CH_2)_4^-$NH$_2$
(pH 8-9)

$$\begin{matrix} \diagdown \\ CH{-}O{-}CO{-}NH{-}enzyme \\ \diagup\diagdown \\ CH{-}OH \\ \diagup \end{matrix}$$

Fig. 2.47. Enzyme immobilization by covalent binding to a cellulose matrix

suitable processes can bring about enzyme encapsulation in a semipermeable membrane (microencapsulation) or confinement in hollow fiber bundles.

2.7.1.2.3 Cross-Linked Enzymes

Derivatization of enzymes using a bifunctional reagent, e. g. glutaraldehyde, can result in cross-linking of the enzyme and, thus, formation of large, still catalytically active insoluble complexes. Such enzyme preparations are relatively unstable for handling and, therefore, are used mostly for analytical work.

2.7.1.2.4 Properties

The properties of an immobilized enzyme are often affected by the matrix and the methods used for immobilization.

Kinetics. As a rule, higher substrate concentrations are required for saturation of an entrapped enzyme than for a free, native enzyme. This is due to a decrease in the concentration gradient which takes place in the pores of the polymer network. Also, there is an increase in the "apparent" *Michaelis* constant for an enzyme bound covalently to a matrix carrying an electrostatic charge. This is also true when the substrate and the functional groups of the matrix carry the same charge. On the other hand, opposite charges bring about an increase of substrate affinity for the matrix. Consequently, this decreases the "apparent" K_m.

pH Optimum. Negatively charged groups on a carrier matrix shift the pH optimum of the covalently bound enzyme to the alkaline region, whereas positive charges shift the pH optimum towards lower pH values. The change in pH optimum of an immobilized enzyme can amount to one to two pH units in comparison to that of a free, native enzyme.

Thermal Inactivation. Unlike native enzymes, the immobilized forms are often more heat stable (cf. example for β-D-glucosidase, Fig. 2.48). Heat stability and pH optima changes induced by immobilization are of great interest in the industrial utilization of enzymes.

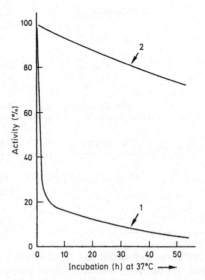

Fig. 2.48. Thermal stabilities of free and immobilized enzymes (according to *Zaborsky*, 1973). 1 β-D-glucosidase, free, 2 β-D-glucosidase, immobilized

2.7.2 Individual Enzymes

2.7.2.1 Oxidoreductases

Broader applications for the processing industry, besides the familiar use of glucose oxidase, are found primarily for catalase and lipoxygenase, among the many enzymes of this group. A number of oxidoreductases have been suggested or are in the experimental stage of utilization, particularly for aroma improvement (examples under 2.7.2.1.4 and 2.7.2.1.5).

2.7.2.1.1 Glucose Oxidase

The enzyme produced by fungi such as *Aspergillus niger* and *Penicillium notatum* catalyzes glucose oxidation by consuming oxygen from the air. Hence, it is used for the removal of either glucose or oxygen (Table 2.20). The H_2O_2 formed in the reaction is occasionally used as an oxidizing agent (cf. 10.1.2.7.2), but it is usually degraded by catalase.

Removal of glucose during the production of egg powder using glucose oxidase (cf. 11.4.3) prevents the *Maillard* reaction responsible for discoloration of the product and deterioration of its whippability. Similar use of glucose oxidase for some meat and protein products would enhance the golden-yellow color rather than the brown color of potato chips or French fries which is obtained in the presence of excess glucose.

Removal of oxygen from a sealed package system results in suppression of fat oxidation and oxidative degradation of natural pigments. For example, the color change of crabs and shrimp from pink to yellow is hindered by dipping them into a glucose oxidase/catalase solution. The shelf life of citrus fruit juices, beer and wine can be prolonged with such enzyme combinations since the oxidative reactions which lead to aroma deterioration are retarded.

2.7.2.1.2 Catalase

The enzyme isolated from microorganisms is important as an auxiliary enzyme for the decomposition of H_2O_2:

$$2\,H_2O_2 \rightleftharpoons 2\,H_2O + O_2 \qquad (2.108)$$

Hydrogen peroxide is a by-product in the treatment of food with glucose oxidase. It is added to food in some specific canning procedures. An example is the pasteurization of milk with H_2O_2, which is important when the thermal process is shut down by technical problems. Milk thus stabilized is also suitable for cheesemaking since the sensitive casein system is spared from heat damage. The excess H_2O_2 is then eliminated by catalase.

2.7.2.1.3 Lipoxygenase

The properties of this enzyme are described under section 3.7.2.2 and its utilization in the bleaching of flour and the improvement of the rheological properties of dough is covered under section 15.4.1.4.3.

2.7.2.1.4 Aldehyde Dehydrogenase

During soya processing, volatile degradation compounds (hexanal, etc.) with a "bean-like" aroma defect are formed because of the enzymatic oxidation of unsaturated fatty acids. These defects can be eliminated by the enzymatic oxidation of the resultant aldehydes to carboxylic acids. Since the flavor threshold values of these acids are high, the acids generated do not interfere with the aroma improvement process.

$$\text{n-Hexanal} + \text{NAD}^{\oplus}$$
$$\rightleftharpoons \quad \text{Caproic acid} + \text{NADH} + \text{H}^{\oplus} \qquad (2.109)$$

Of the various aldehyde dehydrogenases, the enzyme from beef liver mitochondria has a particularly high affinity for n-hexanal (Table 2.21). Hence its utilization in the production of soya milk is recommended.

2.7.2.1.5 Butanediol Dehydrogenase

Diacetyl formed during the fermentation of beer can be a cause of a flavor defect. The enzyme from *Aerobacter aerogenes*, for example, is able to correct this defect by reducing the diketone to the flavorless 2,3-butanediol:

$$\text{CH}_3-\text{CO}-\text{CO}-\text{CH}_3 + \text{NADH} + \text{H}^{\oplus}$$
$$\rightleftharpoons \quad \underset{\text{OH OH}}{\text{CH}_3-\text{CH}-\text{CH}-\text{CH}_3} + \text{NAD}^{\oplus} \qquad (2.110)$$

Table 2.21. *Michaelis* constants for aldehyde dehydrogenase (ALD) from various sources

Substrate	K_m (µmol/l)			
	ALD (bovine liver)			ALD
	Mitochondria	Cytosol	Microsomes	Yeast
Ethanal	0.05	440	1500	30
n-Propanal	–	110	1400	–
n-Butanal	0.1	< 1	–	–
n-Hexanal	0.075	< 1	< 1	6
n-Octanal	0.06	< 1	< 1	–
n-Decanal	0.05	–	–	–

Such a process is improved by the utilization of yeast cells which, in addition to the enzyme and NADH, contain a system able to regenerate the cosubstrate. In order to prevent contamination of beer with undesirable cell constituents, the yeast cells are encapsulated with gelatin.

2.7.2.2 Hydrolases

Most of the enzymes used in the food industry belong to the class of hydrolase enzymes (cf. Table 2.20).

2.7.2.2.1 Peptidases

The mixture of proteolytic enzymes used in the food industry contains primarily endopeptidases (specificity and classification under section 1.4.5.2). These enzymes are isolated from animal organs, higher plants or microorganisms, i.e. from their fermentation media (Table 2.22). Examples of their utilization are as follows. Proteinases are added to wheat flour in the production of some bakery products to modify rheological properties of dough and, thus, the firmness of the endproduct. During such dough treatment, the firm or hard wheat gluten is partially hydrolyzed to a soft-type gluten (cf. 15.4.1.4.5).
In the dairy industry the formation of casein curd is achieved with chymosin or rennin (cf. Table 2.20) by a reaction mechanism described under section 10.1.2.1.1. Casein is also precipitated through the action of other proteinases by a mechanism which involves secondary proteolytic activity resulting in diminished curd yields and lower curd strength. Rennin is essentially free of other undesirable proteinases and is, therefore, especially suitable for cheesemaking. However, there is a shortage of rennin since it has to be isolated from the stomach of a suckling calf. However, it is now possible to produce this enzyme using genetically engineered microorganism. Proteinases from *Mucor miehei, M. pusillus* and *Endothia parasitica* are a suitable replacement for rennin.
Plant proteinases (cf. Table 2.22) and also those of microorganisms are utilized for ripening and tenderizing meat. The practical problem to be solved is how to achieve uniform distribution of the enzymes in muscle tissue. An optional method appears to be injection of the proteinase into the blood stream immediately before slaughter, or rehydration of the freeze-dried meat in enzyme solutions.
Cold turbidity in beer is associated with protein sedimentation. This can be eliminated by hydrolysis of protein using plant proteinases (cf. Table 2.22). Utilization of papain was suggested by *Wallerstein* in 1911. Production of complete or partial protein hydrolysates by enzymatic methods is another example of an industrial use of proteinases. This is used in the liquefaction of fish proteins to make products with good flavors.
One of the concerns in the enzymatic hydrolysis of proteins is to avoid the release of bitter-tasting peptides and/or amino acids (cf. 1.2.6 and 1.3.3). Their occurrence in the majority of proteins treated (an exception is collagen) must be expected when the molecular weight of the peptide fragments falls below 6000. Bitter-tasting peptides, e. g., those which are formed in the ripening of cheese, can be converted to a hydrolyzate which is no longer bitter by adding a mixture of endo- and exopeptidases from Latobacilli.

2.7.2.2.2 α- and β-Amylases

Amylases are either produced by bacteria or yeasts (Table 2.20) or they belong to the components of malt preparations. The high temperature-resistant bacterial amylases, particularly those of *Bac. licheniformis* (Fig. 2.49) are of interest for the hydrolysis of corn starch (gelatinization at 105–110 °C). The hydrolysis rate of these enzymes can be enhanced further

Table 2.22. Peptidases (proteinases) utilized in food processing

Name	Source	pH optimum	Optimal stability pH range
	A. *Peptidases of animal origin*		
Pancreatic proteinase[a]	Pancreas	9.0[b]	3–5
Pepsin	Gastric lining of swine or bovine	2	
Chymosin	Stomach lining of calves or genetically engineered microorganisms B. *Peptidases of plant origin*	6–7	5.5–6.0
Papain	Tropical melon tree (*Carica papaya*)	7–8	4.5–6.5
Bromelain	Pineapple (fruit and stalk)	7–8	
Ficin	Figs (*Ficus carica*) C. *Bacterial peptidases*	7–8	
Alkaline proteinases e. g. subtilisin	*Bacillus subtilis*	7–11	7.5–9.5
Neutral proteinases e. g. thermo-lysin	*Bacillus thermoproteolyticus*	6–9	6–8
Pronase	*Streptomyces griseus* D. *Fungal peptidases*		
Acid proteinase	*Aspergillus oryzae*	3.0–4.0[d]	5
Neutral proteinase	*Aspergillus oryzae*	5.5–7.5[d]	7.0
Alkaline proteinase	*Aspergillus oryzae*	6.0–9.5[d]	7–8
Proteinase	*Mucor pussillus*	3.5–4.5[d]	3–6
Proteinase	*Rhizopus chinensis*	5.0	3.8–6.5

[a] A mixture of trypsin, chymotrypsin, and various peptidases with amylase and lipase as accompanying enzymes.
[b] With casein as a substrate.
[c] A mixture of various endo- and exopeptidases including amino- and carboxypeptidases.
[d] With hemoglobin as as substrate.

by adding $Ca^{2\oplus}$ ions. α-Amylases added to the wort in the beer production process accelerate starch degradation. These enzymes are also used in the baking industry (cf. 15.4.1.4.8).

2.7.2.2.3 Glucan-1,4-α-D-Glucosidase (Glucoamylase)

Glucoamylase cleaves β-D-glucose units from the non-reducing end of an 1,4-α-D-glucan. The α-1,6-branching bond present in amylo-pectin is cleaved at a rate about 30 times slower than the α-

1,4-linkages occurring in straight chains. The enzyme preparation is produced from bacterial and fungal cultures. The removal of transglucosidase enzymes which catalyze, for example, the transfer of glucose to maltose, thus lowerung the yield of glucose in the starch saccharification process, is important in the production of glucoamy-lase.

The starch saccharification process is illustrated in Fig. 2.50. In a purely enzymatic process (left side of the figure), the swelling and gelatinization and liquefaction of starch can occur in a single step using heat-stable bacterial α-amylase (cf. 2.7.2.2.2). The action of amylases

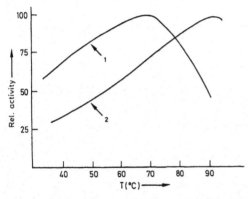

Fig. 2.49. The activity of α-amylase as influenced by temperature. 1 α-amylase from *Bacillus subtilis*, 2 from *Bacillus licheniformis*

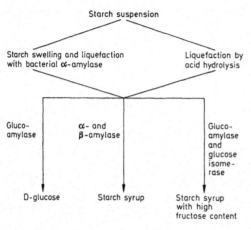

Fig. 2.50. Enzymatic starch degradation

yields starch syrup which is a mixture of glucose, maltose and dextrins (cf. 19.1.4.3.2).

2.7.2.2.4 Pullulanase (Isoamylase)

Pullulanase (cf. 4.4.5.1.4) is utilized in the brewing process and in starch hydrolysis. In combination with β-amylase, it is possible to produce a starch sirup with a high maltose content.

2.7.2.2.5 Endo-1,3(4)-β-D-Glucanase

In the brewing process, β-glucans from barley increase wort viscosity and impede filtration. Enzymatic endo-hydrolysis reduces viscosity.

2.7.2.2.6 α-D-Galactosidase

This and the following enzymes (up to and including section 2.7.2.2.9) attack the non-reducing ends of di-, oligo- and polysaccharides with release of the terminal monosaccharide. The substrate specificity is revealed by the name of the enzyme, e. g., α-D-galactosidase:

$$(2.111)$$

In the production of sucrose from sugar beets (cf. 19.1.4.1.2), the enzymatic preparation from *Mortiella vinacea* hydrolyzes raffinose and, thus, improves the yield of granular sugar in the crystallization step. Raffinose in amounts >8% effectively prevents crystallization of sucrose.

Gas production (flatulence) in the stomach or intestines produced by legumes originates from the sugar stachyose (cf. 16.2.5). When this tetrasaccharide is cleaved by α-D-galactosidase, flatulency from this source is eliminated.

2.7.2.2.7 β-D-Galactosidase (Lactase)

Enzyme preparations from fungi (*Aspergillus niger*) or from yeast are used in the dairy industry to hydrolyze lactose. Immobilized enzymes are applied to produce milk suitable for people suffering from lactose malabsorption. Milk treated in this way can also be used to make products like skim milk concentrate or ice cream, thus avoiding interference by lactose due to its low solubility.

2.7.2.2.8 β-D-Fructofuranosidase (Invertase)

Enzyme preparations isolated from special yeast strains are used for saccharose (sucrose) inversion in the confectionery or candy industry. Invert sugar is more soluble and, because of the presence of free fructose, is sweeter than saccharose.

2.7.2.2.9 α-L-Rhamnosidase

Some citrus fruit juices and purées (especially those of grapes) contain naringin, a dihydrochalcone with a very bitter taste. Treatment of naringin with combined preparations of α-L-rhamnosidase and β-D-glucosidase yields the nonbitter aglycone compound naringenin (cf. 18.1.2.5.4).

2.7.2.2.10 Cellulases and Hemicellulases

The baking quality of rye flour and the shelf life of rye bread can be improved by partial hydrolysis of the rye pentosans. Technical pentosanase preparations are mixtures of β-glycosidases (1,3- and 1,4-β-D-xylanases, etc.).

Solubilization of plant constituents by soaking in an enzyme preparation (maceration) is a mild and sparing process. Such preparations usually contain exo- and endo-cellulases, α- and β-mannosidases and pectolytic enzymes (cf. 2.7.2.2.13). Examples of the utilization are: production of fruit and vegetable purées (mashed products), disintegration of tea leaves, or production of dehydrated mashed potatoes. Some of these enzymes are used to prevent mechanical damage to cell walls during mashing and, thus, to prevent excessive leaching of gelatinized starch from the cells, which would make the purée too sticky.

Glycosidases (cellulases and amylases from *Aspergillus niger*) in combination with proteinases are recommended for removal of shells from shrimp. The shells are loosened and then washed off in a stream of water.

2.7.2.2.11 Lysozyme

The cell walls of gram-positive bacteria are formed from peptidoglycan (synonymous with murein). Peptidoglycan consists of repeating units of the disaccharide N-acetylglucosamine (NAG) and N-acetylmuramic acid (NAM) connected by β-1,4-glycosidic linkages, a tetrapeptide and a pentaglycine peptide bridge. The NAG and NAM residues in peptidoglycan alternate and form the linear polysaccharide chain.

Lysozyme (cf. 11.2.3.1.4) solubilizes peptidoglycan by cleaving the 1,4-β-linkage between NAG and NAM. Combination preparations containing both lysozyme and nisin (cf. 1.3.4.3) are recommended for the preservation of meat preparations, salad dressings and cheese preparations. They are more effective than the components.

2.7.2.2.12 Thioglucosidase

Proteins from seeds of the mustard family (*Brassicaceae*), such as turnip, rapeseed or brown or black mustard, contain glucosinolates which can be enzymatically decomposed into pungent mustard oils (esters of isothiocyanic acid, $R-N=C=S$). The oils are usually isolated by steam distillation. The reactions of thioglycosidase and a few glucosinolates occurring in *Brassicaceae* are covered in section 17.1.2.6.5.

2.7.2.2.13 Pectolytic Enzymes

Pectolytic enzymes are described in section 4.4.5.2. Pectic acid which is liberated by pectin methylesterases flocculates in the presence of Ca^{2+} ions. This reaction is responsible for the undesired "cloud" flocculation in citrus juices. After thermal inactivation of the enzyme at about 90 °C, this reaction is not observable. However, such treatment brings about deterioration of the aroma of the juice. Investigations of the   pectin esterase of orange peel have shown that the enzyme activity is affected by competitive inhibitors: oligogalacturonic acid and pectic acid (cf. Fig. 2.51). Thus, the increase in turbidity of citrus juice can be prevented by the addition of such compounds.

Pectinolytic enzymes are used for the clarification of fruit and vegetable juices. The mechanism of clarification is as follows: the core of the turbidity causing particles consists of carbohydrates and proteins (35%). The prototropic groups of these proteins have a positive charge at the pH of fruit

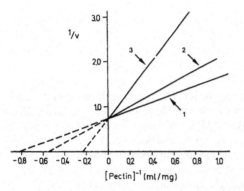

Fig. 2.51. Pectin esterase (orange) activity as affected by inhibitors (according to *Termote*, 1977). 1 Without inhibitor, 2 hepta- and octagalacturonic acids, 3 pectic acid

juice (3.5). Negatively charged pectin molecules form the outer shell of the particle. Partial pectinolysis exposes the positive core. Aggregation of the polycations and the polyanions then follows, resulting in flocculation. Clarification of juice by gelatin (at pH 3.5 gelatin is positively charged) and the inhibition of clarification by alginates which are polyanions at pH 3.5 support this suggested model.

In addition, pectinolytic enzymes play an important role in food processing, increasing the yield of fruit and vegetable juices and the yield of oil from olive fruits.

2.7.2.2.14 Lipases

The mechanism of lipase activity is described under section 3.7.1.1. Lipase from microbial sources (e. g. *Candida lipolytica*) is utilized for enhancement of aromas in cheesemaking.

Limited hydrolysis of milk fat is also of interest in the production of chocolate milk. It enhances the "milk character" of the flavor.

The utilization of lipase for this commodity is also possible.

Staling of bakery products is retarded by lipase, presumably through the release of mono- and diacylglycerols (cf. 15.4.4). The defatting of bones, which has to be carried out under mild conditions in the production of gelatin, is facilitated by using lipase-catalyzed hydrolysis.

2.7.2.2.15 Tannases

Tannases hydrolyze polyphenolic compounds (tannins):

$$Digallate \xrightarrow{\text{H}_2\text{O}} Gallate \qquad (2.112)$$

For example, preparations from *Aspergillus niger* prevent the development of turbidity in cold tea extracts.

2.7.2.2.16 Glutaminase

This enzyme catalyzes the hydrolysis of glutamine (Formula 2.113). For meat preparations, the addition of an enzyme preparation from *Bacillus subtilis* is under discussion. It increases the concentration of glutamic acid, which substantially contributes to the taste of meat.

$$
\begin{array}{c}
\text{COOH} \\
| \\
\text{H}_2\text{N - CH - CH}_2\text{ - CH}_2\text{ - CONH}_2
\end{array}
\xrightarrow[\text{NH}_3]{\text{H}_2\text{O}}
$$

$$
\begin{array}{c}
\text{COOH} \\
| \\
\text{H}_2\text{N - CH - CH}_2\text{ - CH}_2\text{ - COOH}
\end{array}
\qquad (2.113)
$$

2.7.2.3 Isomerases

Of this group of enzymes, glucose isomerse, which is used in the production of starch syrup with a high content of fructose (cf. 19.1.4.3.5), is very important. The enzyme used industrially is of microbial origin. Since its activity for xylose isomerization is higher than for glucose, the enzyme is classified under the name "xylose isomerase" (cf. Table 2.4).

2.7.2.4 Transferases

Protein glutamine-γ-glutamyl transferase (transglutaminase, TGase) catalyzes the acyl transfer between the γ-carboxyamide group of peptide-bound glutamine (acyl donor) and primary amines (acyl acceptor, I in Formula 2.114), e. g., peptide-bound lysine (II in Formula 2.114). Free

I. $[\overset{|}{\underset{|}{Gln}}] - CO - NH_2 \;+\; R - NH_2 \;\longrightarrow\; [\overset{|}{\underset{|}{Gln}}] - CO - NH - R \;+\; NH_3$

II. $[\overset{|}{\underset{|}{Gln}}] - CO - NH_2 \;+\; H_2N - [\overset{|}{Lys}] \;\longrightarrow\; [\overset{|}{\underset{|}{Gln}}] - CO - NH - [\overset{|}{Lys}] \;+\; NH_3$

III. $[\overset{|}{\underset{|}{Gln}}] - CO - NH_2 \;+\; HOH \;\longrightarrow\; [\overset{|}{\underset{|}{Gln}}] - CO - OH \;+\; NH_3$

$$(2.114)$$

acid amides and amino acids also react. Proteins or peptides are cross linked in this way. If amines are absent, TGase can catalyze the deamination of glutamine residues in proteins with H_2O as the acyl acceptor (III in Formula 2.114).

TGases play an important role in the metabolism of animals and plants. For the production of protein gels (cf. 1.4.6.3.3), the TGase from the actinomycete *Streptoverticillum mobaraense* is of special interest. In contrast to the TGases from mammals, the activity of this enzyme, which is released in large amounts by the microorganisms into the nutrient medium, does not depend on $Ca^{2\oplus}$. This enzyme consists of 331 amino acids (M_r: 37,842) of known sequence. A cysteine residue is probably at the active center. The pH optimum of TGase activity is between 5 and 8. This enzyme can also be used at low temperatures and is rapidly denatured at 70 °C. Proteins are cross linked by the formation of ε-(γ-glutamyl)lysine isopeptide bonds. However, the biological availability of lysine is not appreciably reduced. The viscoelastic properties of the resulting protein gels depend not only on the type of proteins and the catalytic conditions (TGase concentration, pH, temperature, time), but also on the pretreatment of the protein, e. g., heat denaturation.

Possible applications of TGase in the production of food are shown in Table 2.23.

2.8 References

Table 2.23. Possible applications of transglutaminase

Raw material	Application
Meat	Restructured meat from small pieces Partial replacement of cutter aids in the production of boiling sausage ("Brühwurst").
Fish	Production of fish gel (surimi, cf. 13.1.6.11) Reducing water loss in the thawing of frozen fish.
Milk	Texture control of low-fat yoghurt to produce the palate feeling of a whole-fat product Increasing the solubility of casein in the presence of $Ca^{2\oplus}$ ions or at a lower pH, e. g., for beverages. Cross linking of casein with whey proteins to increase the protein yield in cheese making.
Wheat	"Hardening" of soft wheat flour for the production of pasta.

Acker, L., Wiese, R.: Über das Verhalten der Lipase in wasserarmen Systemen. Z. Lebensm. Unters. Forsch. *150*, 205 (1972)

Anklam, E., Gadani, F., Heinze, P., Pijnenberg, H., Van Den Eede, G.: Analytical methods for detection and determination of genetically modified organisms in agrictultural crops and plant-derived food products. Eur. Food Res. Technol. *214*, 2 (2002)

Baudner, S., Dreher, R.M.: Immunchemische Methoden in der Lebensmittelanalytik – Prinzip und Anwendung. Lebensmittelchemie *45*, 53 (1991)

Bender, M.L., Bergeron, R.J., Komiyama, M.: The bioorganic chemistry of enzymatic catalyis. John Wiley & Sons: New York. 1984

Bergmeyer, H.U., Bergmeyer, J., Graßl, M.: Methods of enzymatic analysis. 3rd edn., Vol. 1 ff., Verlag Chemie: Weinheim. 1983ff

Bergmeyer, H.U., Gawehn, K.: Grundlagen der enzymatischen Analyse. Verlag Chemie: Weinheim. 1977

Betz, A.: Enzyme. Verlag Chemie: Weinheim. 1974

Birch, G.G., Blakebrough, N., Parker, K.J.: Enzyme and food processing. Applied Science Publ.: London. 1981

Blow, D.M., Birktoft, J.J., Harley, B.S.: Role of a buried acid group in the mechanism of action of chymotrypsin: Nature *221*, 337 (1969)

Dreher, R.M., Märtlbauer, E.: Moderne Methoden in der Lebensmittelanalytik – Enzymimmunoassays und DNS-Hybridsierungstests. Lebensmittelchemie *49*, 1 (1995)

Eriksson, C.E.: Enzymic and non-enzymic lipid degradation in foods. In: Industrial aspects of biochemistry (Ed.: Spencer, B.), Federation of European Biochemical Societies, Vol. 30, p. 865, North Holland/American Elsevier: Amsterdam. 1974

Fennema, O.: Actvity of enzymes in partially frozen aqueous systems. In: Water relations of foods (Ed.: Duckworth, R.B.), p. 397, Academic Press: London–New York–San Francisco. 1975

Gachet, E., Martin, G.G., Vigneau, F., Meyer, G.: Detection of genetically modified organisms (GMOs) by PCR: a brief review of methodologies available. Trends Food Sci. Technol. *9*, 380 (1999)

Gray, C.J.: Enzyme-catalysed reactions. Van Nostrand Reinhold Comp.: London. 1971

Guibault, G.G.: Analytical uses of immobilized enzymes. Marcel Dekker, Inc.: New York. 1984

Hendrickx, M., Ludikhuyze, L., Van den Broek, I., Weemaes, C.: Effects of high pressure on enzymes related to food quality. Trends Food Sci. Technol. *9*, 197 (1998)

International Union of Biochemistry and Molecular Biology: Enzyme nomenclature 1992. Academic Press: New York–San Francisco–London. 1992

Kessler, H.G.: Lebensmittel- und Bioverfahrenstechnik, Molkereitechnologie, 3. Auflage, Verlag A. Kessler, Freising, 1988

Kilara, A., Shahani, K. A.: The use of immobilized enzymes in the food industry: a review. Crit. Rev. Food Sci. Nutr. *12*, 161 (1979)

Koshland, D.E.: Conformation changes at the active site during enzyme action. Fed. Proc. *23*, 719 (1964)

Koshland, D.E., Neet, K.E.: The catalytic and regulatory properties of proteins. Ann. Rev. Biochem. *37*, 359 (1968)

Lehninger, A.L.: Biochemie. Verlag Chemie: Weinheim. 1977

Levine, H., Slade, L.: A polymer physico-chemical approach to the study of commercial starch hydrolysis products (SHPs). Carbohydrate Polymers *6*, 213 (1986)

Matheis, G.: Polyphenoloxidase and enzymatic browning of potatoes (*Solanum*). Chem. Microbiol. Technol. Lebensm. *12*, 86 (1989)

Meyer, R., Candrian, U.: PCR-based DNA analysis for the identification and characterization of food components. Lebensm. Wiss. Technol. *29*, 1 (1996)

Morris, B.A., Clifford, M.N. (Eds.): Immunoassays in food analysis. Elsevier Applied Science Publ.: London. 1985

Morris, B.A., Clifford, M.N., Jackman, R. (Eds.): Immunoassays for veterinary and food analysis. Elsevier Applied Science, London, 1988

Motoki, M., Seguro, K.: Transglutaminase and its use for food processing. Trends Food Sci. Technol. *9*, 204 (1998)

Page, M.I., Williams, A. (Eds.): Enzyme mechanisms. The Royal Society of Chemistry, London, 1987

Palmer, T.: Understanding enzymes. 2nd edn., Ellis Horwood Publ.: Chichester. 1985

Perham, R.N. et al.: Enzymes. In: Ullmann's encyclopedea of industrial chemistry, 5th Edition, Volume A23, p. 341, Verlag VCH, Weinheim, 1987

Phipps, D.A.: Metals and metabolism. Clarendon Press: Oxford. 1978

Plückthun, A.: Wege zu neuen Enzymen: Protein Engineering und katalytische Antikörper. Chem. unserer Zeit *24*, 182 (1990)

Potthast, K., Hamm, R., Acker, L.: Enzymic reactions in low moisture foods. In: Water relations of foods (Ed.: Duckworth, R.B.), p. 365, Academic Press: London–New York–San Francisco. 1975

Reed, G.: Enzymes in food processing. 2nd edn., Academic Press: New York–London. 1975

Richardson, T.: Enzymes. In: Principles of food science, Part I (Ed.: Fennema, O.R.), Marcel Dekker, Inc.: New York–Basel. 1977

Schellenberger, A. (Ed.): Enzymkatalyse, Springer-Verlag, Berlin, 1989

Schwimmer, S.: Source book of food enzymology. AVI Publ. Co.: Westport, Conn. 1981

Scrimgeour, K.G.: Chemistry and control of enzyme reactions. Academic Press: London–New York. 1977

Segel, I.H.: Biochemical calculations. John Wiley and Sons, Inc.: New York. 1968

Shotton, D.: The molecular architecture of the serine proteinases. In: Proceedings of the international research conference on proteinase inhibitors (Eds.: Fritz, H., Tschesche, H.), p. 47, Walter de Gruyter: Berlin–New York. 1971

Straub, J.A., Hertel, C., Hammes, W.P.: Limits of a PCR-based detection method for genetically modified soya beans in wheat bread production. Z. Lebensm. Unters. Forsch. A *208*, 77 (1999)

Suckling, C.J. (Eds.): Enzyme chemistry. Chapman and Hall: London. 1984

Svensson, S.: Inactivation of enzymes during thermal processing. In: Physical, chemical and biological changes in food caused by thermal processing (Eds.: Hoyem, T., Kvalle, O.), p. 202, Applied Science Publ.: London. 1977

Termote, F., Rombouts, F.M., Pilnik, W.: Stabilization of cloud in pectinesterase active orange juice by pectic acid hydrolysates. J. Food Biochem. *1*, 15 (1977)

Teuber, M.: Production of chymosin (EC 3.4.23) by microorganisms and its use for cheesemaking. Int. Dairy Federation, Ann. Sessions Copenhagen 1989, B-Doc 162

Uhlig, H.: Enzyme arbeiten für uns – Technische Enzyme und ihre Anwendung. Carl Hanser Verlag, München, 1991

Whitaker, J.R.: Principles of enzymology for the food sciences. 2nd edn Marcel Dekker, Inc.: New York. 1993

Whitaker J.R., Sonnet, P.E. (Eds.): Biocatalysis in Agricultural Biotechnology. ACS Symp. Ser. 389, American Chemical Society, Washington DC, 1989

Zaborsky, O.R.: Immobilized enzymes. CRC-Press: Cleveland, Ohio. 1973

3 Lipids

3.1 Foreword

Lipids are formed from structural units with a pronounced hydrophobicity. This solubility characteristic, rather than a common structural feature, is unique for this class of compounds. Lipids are soluble in organic solvents but not in water. Water insolubility is the analytical property used as the basis for their facile separation from proteins and carbohydrates. Some lipids are surface-active since they are amphiphilic molecules (contain both hydrophilic and hydrophobic moieties). Hence, they are polar and thus distinctly different from neutral lipids. The two approaches generally accepted for lipid classification are presented in Table 3.1.

The majority of lipids are derivatives of fatty acids. In these so-called acyl lipids the fatty acids are present as esters and in some minor lipid groups in amide form (Table 3.1). The acyl residue influences strongly the hydrophobicity and the reactivity of the acyl lipids.

Some lipids act as building blocks in the formation of biological membranes which surround cells and subcellular particles. Such lipids occur in all foods, but their content is often less than 2% (cf. 3.4.1). Nevertheless, even as minor food constituents they deserve particular attention, since their high reactivity may strongly influence the organoleptic quality of the food.

Primarily triacylglycerols (also called triglycerides) are deposited in some animal tissues and organs of some plants. Lipid content in such storage tissues can rise to 15–20% or higher and so serve as a commercial source for isolation of triacylglycerols. When this lipid is refined, it is available to the consumer as an edible oil or fat. The nutritive/physiological importance of lipids is based on their role as fuel molecules ($37\,\mathrm{kJ/g}$ or $9\,\mathrm{kcal/g}$ triacylglycerols) and as a source of essential fatty acids and vitamins. Apart from these roles, some other lipid properties are indispensable in food handling or processing.

Table 3.1. Lipid classification

A. Classification according to "acyl residue" characteristics

I. Simple lipids (not saponifiable)

Free fatty acids, isoprenoid lipids (steroids, carotenoids, monoterpenes), tocopherols

II. Acyl lipids (saponifiable)	Constituents
Mono-, di-, triacyl-glycerols	Fatty acid, glycerol
Phospholipids (phosphatides)	Fatty acid, glycerol or sphingosine, phosphoric acid, organic base
Glycolipids	Fatty acid, glycerol or sphingosine, mono-, di- or oligosaccharide
Diol lipids	Fatty acid, ethane, pro-pane, or butane diol
Waxes	Fatty acid, fatty alcohol
Sterol esters	Fatty acid, sterol

B. Classification according to the characteristics "neutral–polar"

Neutral lipids	Polar (amphiphilic) lipids
Fatty acids ($>C_{12}$)	Glycerophospholipid
Mono-, di-, triacyl-glycerols	Glyceroglycolipid
Sterols, sterol esters	Sphingophospholipid
Carotenoids	Sphingoglycolipid
Waxes	
Tocopherols[a]	

[a] Tocopherols and quinone lipids are often considered as "redox lipids".

These include their melting behavior and the pleasant creamy or oily taste that is recognized by a receptor, which has recently been identified. Therefore, there are all together six taste qualities (cf. 8.6.1). – Fats also serve as solvents for certain taste substances and numerous odor substances. On the whole, fats enrich the nutritional quality and are of importance in food to achieve the desired texture, specific mouthfeel and aroma, and a satisfactory aroma retention. In addition, foods can be prepared by deep frying, i. e. by dipping the food into fat or oil heated to a relatively high temperature.

The lipid class of compounds also includes some important food aroma substances or precursors which are degraded to aroma compounds. Some lipid compounds are indispensable as food emulsifiers, while others are important as fat- or oil-soluble pigments or food colorants.

3.2 Fatty Acids

3.2.1 Nomenclature and Classification

Acyl lipid hydrolysis releases aliphatic carboxylic acids which differ in chemical structure. They can be divided into groups according to chain length, number, position and configuration of their double bonds, and the occurrence of additional functional groups along the chains. The fatty acid distribution pattern in food is another criterion for differentiation.

Table 3.2 compiles the major fatty acids which occur in food. Palmitic, oleic and linoleic acids frequently occur in higher amounts, while the other acids listed, though widely distributed, as a rule occur only in small amounts (major vs minor fatty acids). Percentage data of acid distribution make it obvious that unsaturted fatty acids are the predominant form in nature.

Fatty acids are usually denoted in the literature by a "shorthand description", e. g. 18:2 (9, 12) for linoleic acid. Such an abbreviation shows the number of carbon atoms in the acid chain and the number, positions and configurations of the double bonds. All bonds are considered to be cis; whenever trans-bonds are present, an additional "tr" is shown. As will be outlined later in a detailed survey of lipid structure, the carbon skeleton of lipids should be shown as a *zigzag* line (Table 3.2).

3.2.1.1 Saturated Fatty Acids

Unbranched, straight-chain molecules with an even number of carbon atoms are dominant among the saturated fatty acids (Table 3.6). The short-chain, low molecular weight fatty acids ($< 14:0$) are triglyceride constituents only in fat and oil of milk, coconut and palmseed. In the free form or esterified with low molecular weight alcohols, they occur in nature only in small amounts, particularly in plant foods and in foods processed with the aid of microorganisms, in which they are aroma substances.

Odor and taste threshold values of fatty acids are compiled in Table 3.3 for cream, butter and cocoa fat. The data for cream and coconut fat indicate lower odor than taste threshold values of C_4- and C_6-fatty acids, while it is the reverse for C_8- up to C_{14}-fatty acids.

The aroma threshold increases remarkably with higher pH-values (Table 3.4) since only the undissociated fatty acid molecule is aroma active.

Table 3.2. Structures of the major fatty acids

Abbreviated designation	Structure[a]	Common name	Proportion (%)[b]
14:0	∿∿∿∿COOH	Myristic acid	2
16:0	∿∿∿∿COOH	Palmitic acid	11
18:0	∿∿∿∿COOH	Stearic acid	4
18:1(9)	∿∿∿=∿∿COOH	Oleic acid	34
18:2(9,12)	∿∿=∿=∿∿COOH	Linoleic acid	34
18:3(9, 12, 15)	∿=∿=∿=∿∿∿COOH	Linolenic acid	5

[a] Numbering of carbon atoms starts with carboxyl group-C as number 1.
[b] A percentage estimate based on world production of edible oils.

Table 3.3. Aroma threshold values (odor and/or taste) of free fatty acids in different food items

Fatty acid	Aroma threshold (mg/kg) in				
	Cream		Sweet cream butter[a]	Coconut fat	
	Odor	Taste		Odor	Taste[b]
4:0	50	60	40		35 160
6:0	85	105	15		25 50
8:0	200	120	455	> 1000	25
10:0	> 400	90	250	> 1000	15
12:0	> 400	130	200	> 1000	35
14:0	> 400	> 400	5000	> 1000	75
16:0	n.d.	n.d.	10,000	n.d.	n.d.
18:0	n.d.	n.d.	15,000	n.d.	n.d.

[a] Odor/taste not separated.
[b] Quality of taste: 4:0 rancid, 6:0 rancid, like goat, 8:0 musty, rancid, soapy, 10:0, 12:0 and 14:0 soapy
n.d.: not determined.

Table 3.4. Threshold values[a] of fatty acids depending on the pH-value of an aqueous solution

Fatty acids	Threshold (mg/kg) at pH		
	3.2	4.5	6.0
4:0	0.4	1.9	6.1
6:0	6.7	8.6	27.1
8:0	2.2	8.7	11.3
10:0	1.4	2.2	14.8

[a] Odor and taste.

Additive effects can be observed in mixtures: examples No. 1 and 2 in Table 3.5 demonstrate that the addition of a mixture of C_4-C_{12} fatty

acids to cream will produce a rancid soapy taste if the capryl, capric and lauryl acid contents rise from 30 to 40% of their threshold value concentration. A further increase of these fatty acids to about 50% of the threshold concentration, as in mixture No. 3, results in a musty rancid odor.

Some high molecular weight fatty acids (>18:0) are found in legumes (peanut butter). They can be used, like lower molecular weight homologues, for identification of the source of triglycerides (cf. 14.5.2.3). Fatty acids with odd numbers of carbon atoms, such as valeric (5:0) or enanthic (7:0) acids (Table 3.6) are present in food only in traces. Some of these short-chain homologues are important as food aroma constituents. Pentadecanoic and heptadecanoic acids are odd-numbered fatty acids present in milk and a number of plant oils. The common name "margaric acid" for 17:0 is an erroneous designation. *Chevreul* (1786–1889), who first discovered that fats are glycerol esters of fatty acids, coined the word "margarine" to denote a product from oleomargarine (a fraction of edible beef tallow), believing that the product contained a new fatty acid, 17:0. Only later was it clarified that such margarine or "17:0 acid" was a mixture of palmitic and stearic acids.

Branched-chain acids, such as iso- (with an isopropyl terminal group) or anteiso- (a secondary butyl terminal group) are rarely found in food. Pristanic and phytanic acids have been detected in milk fat (Table 3.6). They are isoprenoid acids obtained from the degradation of the phytol side chain of chlorophyll.

Table 3.5. Odor and taste of fatty acid mixtures in cream

No.	Fatty acid mixtures of					Odor	Taste
	4:0	6:0	8:0	10:0	12:0		
	Concentration in % of aroma threshold[a]						
1	28	17	29	31	30	n.O.	n.T.
2	28	17	40	42	37	n.O.	rancid, soapy
3	28	17	52	53	45	musty, rancid	rancid, soapy
4	48	29	29	31	30	musty, rancid	n.T.
5	48	29	40	42	37	musty, rancid	rancid, soapy

[a] The concentration of each fatty acid is based on the threshold values indicated in Table 3.3 for odor for 4:0 and 6:0 and for taste for 8:0–12:0.
n.O. = no difference in odor from that of cream.
n.T. = no difference in taste from that of cream.

Table 3.6. Saturated fatty acids

Abbreviated designation	Structure	Systematic name	Common name	Melting point (°C)
A. Even numbered straight chain fatty acids				
4:0	$CH_3(CH_2)_2COOH$	Butanoic acid	Butyric acid	−7.9
6:0	$CH_3(CH_2)_4COOH$	Hexanoic acid	Caproic acid	−3.9
8:0	$CH_3(CH_2)_6COOH$	Octanoic acid	Caprylic acid	16.3
10:0	$CH_3(CH_2)_8COOH$	Decanoic acid	Capric acid	31.3
12:0	$CH_3(CH_2)_{10}COOH$	Dodecanoic acid	Lauric acid	44.0
14:0	$CH_3(CH_2)_{12}COOH$	Tetradecanoic acid	Myristic acid	54.4
16:0	$CH_3(CH_2)_{14}COOH$	Hexadecanoic acid	Palmitic acid	62.9
18:0	$CH_3(CH_2)_{16}COOH$	Octadecanoic acid	Stearic acid	69.6
20:0	$CH_3(CH_2)_{18}COOH$	Eicosanoic acid	Arachidic acid	75.4
22:0	$CH_3(CH_2)_{20}COOH$	Docosanoic acid	Behenic acid	80.0
24:0	$CH_3(CH_2)_{22}COOH$	Tetracosanoic acid	Lignoceric acid	84.2
26:0	$CH_3(CH_2)_{24}COOH$	Hexacosanoic acid	Cerotic acid	87.7
B. Odd numbered straight chain fatty acids				
5:0	$CH_3(CH_2)_3COOH$	Pentanoic acid	Valeric acid	−34.5
7:0	$CH_3(CH_2)_5COOH$	Heptanoic acid	Enanthic acid	−7.5
9:0	$CH_3(CH_2)_7COOH$	Nonanoic acid	Pelargonic acid	12.4
15:0	$CH_3(CH_2)_{13}COOH$	Pentadecanoic acid		52.1
17:0	$CH_3(CH_2)_{15}COOH$	Heptadecanoic acid	Margaric acid	61.3
C. Branched chain fatty acids				
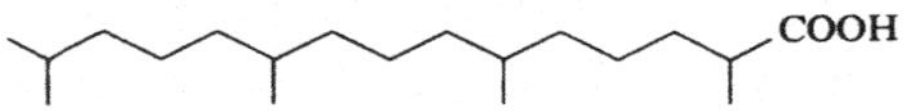		2,6,10,14-Tetra-methyl-penta-decanoic acid	Pristanic acid	
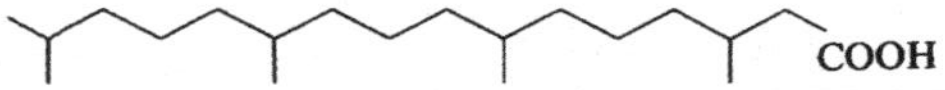		3,7,11,15-Tetra-methyl-hexa-decanoic acid	Phytanic acid	

3.2.1.2 Unsaturated Fatty Acids

The unsaturated fatty acids, which dominate lipids, contain one, two or three allyl groups in their acyl residues (Table 3.7). Acids with isolated double bonds (a methylene group inserted between the two cis-double bonds) are usually denoted as isolene-type or nonconjugated fatty acids.

The structural relationship that exists among the unsaturated nonconjugated fatty acids derived from a common biosynthetic pathway is distinctly revealed when the double bond position is determined by counting from the methyl end of the chain (it should be emphasized that position designation using this method of counting requires the suffix "ω" or "n"). Acids with the same methyl ends are then combined into groups. Thus, three family groups exist: ω3 (linolenic type), ω6 (linoleic type) and ω9 (oleic acid type; Table 3.7). Using this classification, the common structural features abundantly found in C_{18} fatty acids (Table 3.2) are also found in less frequently occurring fatty acids. Thus, erucic acid (20:1) occurring only in the mustard family of seeds (*Brassicaceae*, cf. 14.3.2.2.5), belongs to the ω9 group, arachidonic acid (20:4), occurring in meat, liver, lard and lipids of chicken eggs, belongs to the ω6 group, while the $C_{20}-C_{22}$ fatty acids with 5 and 6 double bonds, occurring in fish lipids, belong to the ω3 group (cf. 13.1.4.5 and 14.3.1.2).

Linoleic acid can not be synthesized by the human body. This acid and other members of the ω6 family are considered as essential fatty acids required as building blocks for biologically active membranes. α-Linolenic acid, which belongs to the ω3 family and which is synthesized only by plants, also plays a nutritional role as an essential fatty acid.

A formal relationship exists in some olefinic unsaturated fatty acids with regard to the position of the double bond when counted from the carboxyl end of the chain. Oleic, palmitoleic and myristoleic acids belong to such a Δ9 family (cf. Table 3.7); the latter two fatty acids are minor constituents in foods of animal or plant origin.

Unsaturated fatty acids with an unusual structure are those with one trans-double bond and/or conjugated double bonds (Table 3.7). They are formed in low concentrations on biohydrogenation in the stomach of ruminants and are consequently found in meat and milk (cf. 10.1.2.3). Such trans-unsaturated acids are formed as artifacts in the industrial processing of oil or fat (heat treatment, oil hardening). Since trans-fatty acids are undesirable, their content in German margarines has been lowered from 8.5% (1994) to 1.5% (1996) by improving the production process. Conjugated linoleic acids (CLA) are of special interest because they are attributed to have an anticarcinogenic effect. In fact, C_{18} fatty acids with two double bonds which differ in position and geometry belong to the group CLA. The occurrence of CLA in foods is shown in Table 3.8. Up to nine isomers have been identified in lipids and, apart from exceptions, 18:2 (9c, 11tr) predominates (Table 3.8). Conjugated fatty acids with diene, triene or tetraene systems also occur frequently in several seed oils, but do not play a role in human nutrition. Table 3.7 presents, as an example, two naturally occurring acids with conjugated triene systems which differ in the configuration of one double bond at position 9 (cis, trans).

Unsaturated fatty acids emulsified in water taste bitter with a relatively low threshold value for α-linolenic acid (Table 3.9). Thus an off-taste can be present due to fatty acids liberated, as indicated in Table 3.9, by the enzymatic hydrolysis of unsaturated triacyl glycerides which are tasteless in an aqueous emulsion.

Table 3.8. Conjugated linoleic acids in food

Food	Total CLA[a] (g/kg fat)	18:2(c9,tr11) (% of CLA[a])
Milk	2–30	90
Butter	9.4–11.9	91
Cheese	0.6–7.1	17–90
Processed cheese	3.2–8.9	17–90
Ice cream	3.8–4.9	73–76
Sour cream	7.5	78
Yoghurt	5.1–9.0	82
Beef, roasted	3.1–9.9	60
Plant oils, marine oils	0.2–0.5	45

[a] CLA, conjugated linoleic acid.

Table 3.7. Unsaturated fatty acids

Abbreviated designation	Structure	Common name	Melting point (°C)
A. Fatty acids with nonconjugated cis double bonds			
	ω9-Family		
18:1 (9)	$CH_3-(CH_2)_7-CH=CH-CH_2-(CH_2)_6-COOH$	Oleic acid	13.4
22:1 (13)	$-(CH_2)_{10}-COOH$	Erucic acid	34.7
24:1 (15)	$-(CH_2)_{12}-COOH$	Nervonic acid	42.5
	ω6-Family		
18:2 (9, 12)	$CH_3-(CH_2)_4-(CH=CH-CH_2)_2-(CH_2)_6-COOH$	Linoleic acid	−5.0
18:3 (6,9,12)	$-(CH=CH-CH_2)_3-(CH_2)_3-COOH$	γ-Linolenic acid	
20:4 (5,8,11,14)	$-(CH=CH-CH_2)_4-(CH_2)_2-COOH$	Arachidonic acid	−49.5
	ω3-Family		
18:3 (9, 12, 15)	$CH_3-CH_2-(CH=CH-CH_2)_3-(CH_2)_6-COOH$	α-Linolenic acid	−11.0
20:5 (5,8,11,14,17)	$-(CH=CH-CH_2)_5-(CH_2)_2-COOH$	EPA[a]	
22:6 (4,7,10,13,16,19)	$-(CH=CH-CH_2)_6-CH_2-COOH$	DHA[a]	
	Δ9-Family		
18:1 (9)	$CH_3-(CH_2)_7-CH=CH-CH_2-(CH_2)_6-COOH$	Oleic acid	13.4
16:1 (9)	$CH_3-(CH_2)_5-$	Palmitoleic acid	0.5
14:1 (9)	$CH_3-(CH_2)_3-$	Myristoleic acid	
B. Fatty acids with nonconjugated trans-double bonds			
18:1 (tr9)	$CH_3-(CH_2)_7-CH\overset{tr}{=}CH-(CH_2)_7-COOH$	Elaidic acid	46
18:2 (tr9, tr12)	$CH_3-(CH_2)_4-CH\overset{tr}{=}CH-CH_2-CH\overset{tr}{=}CH-(CH_2)_7-COOH$	Linolelaidic acid	28
C. Fatty acids with conjugated double bonds			
18:2 (9, tr11)	$CH_3-(CH_2)_5-CH\overset{tr}{=}CH-CH\overset{c}{=}CH-(CH_2)_7- COOH$		
18:3 (9, tr11, tr13)	$CH_3-(CH_2)_3-CH\overset{tr}{=}CH-CH\overset{tr}{=}CH-CH\overset{c}{=}CH-(CH_2)_7-COOH$	α-Eleostearic acid	48
18:3 (tr9, tr11, tr13)	$CH_3-(CH_2)_3-CH\overset{tr}{=}CH-CH\overset{tr}{=}CH-CH\overset{tr}{=}CH-(CH_2)_7-COOH$	β-Eleostearic acid	71.5
18:4 (9, 11, 13, 15)[b]	$CH_3-CH_2-(CH=CH)_4-(CH_2)_7-COOH$	Parinaric acid	85

[a] EPA: Eicosapentanoic acid, DHA: Docosahexanoic acid.
[b] Geometry of the double bond was not determined.

Table 3.9. Taste of unsaturated fatty acids emulsified in water

Compound	Threshold (mmol/1)	Quality
Oleic acid	9–12	bitter, burning, pungent
Elaidic acid	22	slightly burning
Linoleic acid	4–6	bitter, burning, pungent
Linolelaidic acid	11–15	bitter, burning, scratchy
γ-Linolenic acid	3–6	bitter, burning, pungent
α-Linolenic acid	0.6–1.2	bitter, burning, pungent, like fresh walnut
Arachidonic acid	6–8	bitter, repugnant off-taste

3.2.1.3 Substituted Fatty Acids

Hydroxy Fatty Acids. Ricinoleic acid is the best known of the straight-chain hydroxy fatty acids. Its structure is 12-OH, 18:1 (9). It is an optically active acid with a D(+)-configuration:

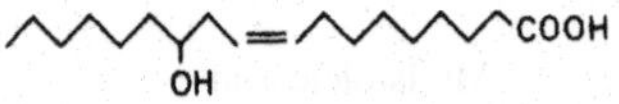

$$(3.1)$$

Ricinoleic acid is the main acid of castor bean oil, comprising up to 90% of the total acids. Hence, it can serve as an indicator for the presence of this oil in edible oil blends.

D-2-Hydroxy saturated 16:0 to 25:0 fatty acids with both even and odd numbers of carbons in their chains occur in lipids in green leaves of a great number of vegetables. γ- or ∂-Lactones are obtained from 4- and 5-hydroxycarboxylic acids (C_8 to C_{16}) by the elimination of water. ∂-Lactones have been found in milk fat and fruits. They are very active aroma components (cf. 5.3.2.3).

Oxo Fatty Acids. Natural oxo (or keto) acids are less common than hydroxy acids. About 1% of milk fat consists of saturated (C_{10}–C_{24}) oxo fatty

acids, with an even number of carbon atoms, in which the carbonyl group is located on C-5 to C-13: One of 47 identified compounds of this substance class has the following structure:

$$(3.2)$$

Furan Fatty Acids. These occur in fish liver oil in a range of 1–6% and up to 25% in some freshwater fish. Furan fatty acids are also part of the minor constituents of some plant oils and butter (Table 3.10). They are also present in fruits (lemon, strawberry), vegetables (cabbage, potato) and mushrooms (champignons). Two of these acids have the following formulas

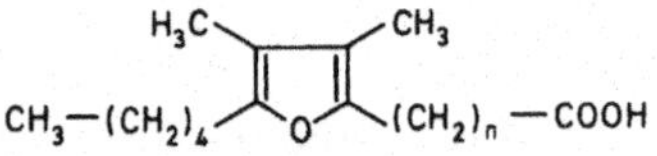

I: n = 8;   II: n = 10

$$(3.3)$$

Photooxidation (cf. 3.7.2.1.4) of these acids can deteriorate especially the quality of soybean oil. Substituted fatty acids are also derived by autoxidation or enzymatic peroxidation of unsaturated fatty acids, which will be dealt with in more detail in 3.7.2.3 and 3.7.2.4.1.

Table 3.10. Examples for the occurrence of furan fatty acids I and II

Oil	Concentration (mg/kg)	
	I[a]	II[a]
Soya oil	120–170	130–230
Wheat germ oil	100–130	105–150
Rapeseed oil	6–16	7–20
Corn oil	8–11	9–13
Butter	13–139	24–208
Leaves of the tea shrub[b]	50	713
Green tea[b]	4	80–100
Black tea[b]	10	159
Spinach[b]	86	733

[a] I: 10,13-epoxy-11,12-dimethyloctadeca-10,12-dienoic acid.
II: 12,15-epoxy-13,14-dimethyleicosa-12,14-dienoic acid (Formula 3.3).
[b] Values based on dry weight.

3.2.2 Physical Properties

3.2.2.1 Carboxyl Group

Carboxylic acids have a great tendency to form dimers which are stabilized by hydrogen bonds:

$$R-C\underset{\substack{\\ O-H\cdots O}}{\overset{\substack{O\cdots H-O\\ }}{\diagup}}C-R \tag{3.4}$$

The binding energy of the acid dimer dissolved in hexane is 38 kJ/mole. Also, the fatty acid molecules are arranged as dimers in the crystalline lattice (cf. Fig. 3.2).

The acidic character of the carboxyl group is based on proton dissociation and on the formation of the resonance-stabilized carboxylate anion:

$$R-C\overset{O}{\underset{OH}{\diagup}} \rightleftharpoons H^{\oplus} + \left[R-C\overset{O}{\underset{O^{\ominus}}{\diagup}} \longleftrightarrow R-C\overset{O^{\ominus}}{\underset{O}{\diagup}} \right] \tag{3.5}$$

The pK$_s$ values for the C$_2$–C$_9$ short-chain acid homologues range from 04.75–4.95. The pK$_s$ of 7.9 for linoleic acid deviates considerably from this range. This unexpected and anomalous behavior, which has not yet been clarified, is clearly illustrated in the titration curves for propionic, caprylic and linoleic acids recorded under identical conditions (Fig. 3.1).

3.2.2.2 Crystalline Structure, Melting Points

Melting properties of fats depend on the arrangement of the acyl residues in the crystal lattice in addition to other factors attributed solely to the structure of triglycerides.

Calculations of the energy content of the carbon chain conformation have revealed that at room temperature 75% of the C–C bonds of a saturated fatty acid are present in a fully staggered *zigzag* or "trans" conformation and only 25% in the energetically slightly less favorable skew conformation.

The unsaturated fatty acids, because of their double bonds, are not free to rotate and hence have rigid kinks along the carbon chain skeleton. However, a molecule is less bent by a trans than by

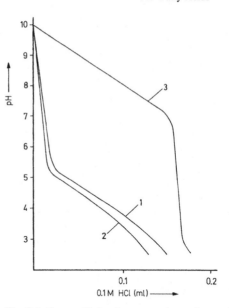

Fig. 3.1. Fatty acid titration curves (according to *Bild* et al., 1977). Aqueous solutions (0.1 mol/l) of Na-salts of propionic (1), caprylic (2) and linoleic acids (3) were titrated with 0.1 mol/l HCl

a cis double bond. Thus, this cis-configuration in oleic acid causes a bending of about 40°:

$$\tag{3.6}$$

The corresponding elaidic acid, with a trans-configuration, has a slightly shortened C-chain, but is still similar to the linear form of stearic acid:

$$\tag{3.7}$$

The extent of molecular crumpling is also increased by an increase in the number of cis double bonds. Thus, the four cis double bonds in arachidonic acid increase the deviation from a straight line to 165°:

$$\tag{3.8}$$

When fatty acids crystallize, the saturated acids are oriented as depicted by the simplified pattern in Fig. 3.2. The dimer molecular arrangement is thereby retained. The principal reflections of the X-ray beam are from the planes (c) of high electron density in which the carboxyl groups are situated. The length of the fatty acid molecule can be determined from the "main reflection" site intervals (distance d in Fig. 3.2). For stearic acid (18:0), this distance is 2.45 nm.

The crystalline lattice is stabilized by hydrophobic interaction along the acyl residues. Correspondingly, the energy and therefore the temperature required to melt the crystal increase with an increased number of carbons in the chain.

Odd-numbered as well as unsaturated fatty acids can not be uniformly packed into a crystalline lattice as can the saturated and even-numbered acids. The odd-numbered acids are slightly interfered by their terminal methyl groups.

The consequence of less symmetry within the crystal is that the melting points of even-numbered acids (C_n) exceed the melting points of the next higher odd-numbered (C_{n+1}) fatty acids (cf. Table 3.6).

The molecular arrangement in the crystalline lattice of unsaturated fatty acids is not strongly influenced by trans double bonds, but is strongly influenced by cis double bonds. This difference, due to steric interference as mentioned above, is reflected in a decrease in melting points in the fatty acid series 18:0, 18:1 (tr9) and 18:1 (9). However, this ranking should be considered as reliable only when the double bond positions within the molecules are fairly comparable. Thus, when a cis double bond is at the end of the carbon chain, the

deviation from the form of a straight extended acid is not as large as in oleic acid. Hence, the melting point of such an acid is higher. The melting point of cis-2-octadecenoic acid is in agreement with this rule; it even surpasses the 9-trans isomer of the same acid (Table 3.11).

The melting point decreases with an increasing number of isolated cis-double bonds (Table 3.11). This behavior can be explained by the changes in the geometry of the molecules, as can be seen when comparing the geometric structures of oleic and arachidonic acid.

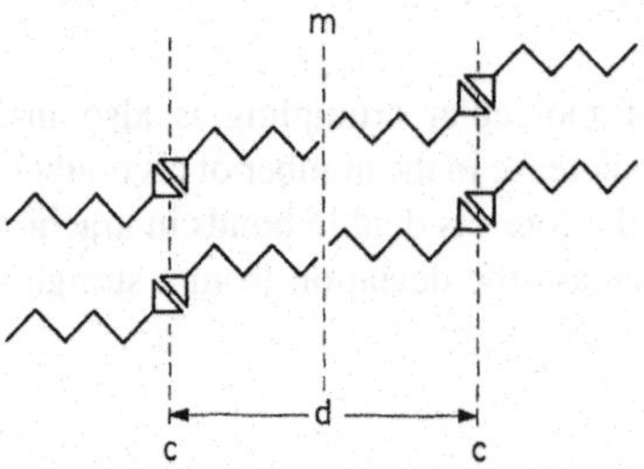

3.2.2.3 Urea Adducts

When urea crystallizes, channels with a diameter of 0.8–1.2 nm are formed within its crystals and can accomodate long-chain hydrocarbons. The stability of such urea adducts of fatty acids parallels the geometry of the acid molecule. Any deviation from a straight-chain arrangement brings about weakening of the adduct. A tendency to form inclusion compounds decreases in the series 18:0 > 18:1 (9) > 18:2 (9, 12).

A substitution on the acyl chain prevents adduct formation. Thus, it is possible to separate branched or oxidized fatty acids or their methyl esters from the corresponding straightchain compounds on the basis of the formation of urea adducts. This principle is used as a method for preparative-scale enrichment and separation of branched or oxidized acids from a mixture of fatty acids.

Fig. 3.2. Arrangement of caproic acid molecules in crystal (according to *Mead* et al., 1965). Results of a X-ray diffraction analysis reveal a strong diffraction in the plane of carboxyl groups (c) and a weak diffraction at the methyl terminals (m): d: identity period

Table 3.11. The effect of number, configuration and double bond position on melting points of fatty acids

Fatty acid		Melting point (°C)
18:0	Stearic acid	69
18:1 (tr9)	Elaidic acid	46
18:1 (2)	cis-2-Octadecenoic acid	51
18:1 (9)	Oleic acid	13.4
18:2 (9, 12)	Linoleic acid	−5
18:2 (tr9, tr12)	Linolelaidic acid	28
18:3 (9, 12, 15)	α-Linolenic acid	−11
20:0	Arachidic acid	75.4
20:4 (5,8,11,14)	Arachidonic acid	−49.5

3.2.2.4 Solubility

Long-chain fatty acids are practically insoluble in water; instead, they form a floating film on the water surface. The polar carboxyl groups in this film are oriented toward the water, while the hydrophobic tails protrude into the gaseous phase. The solubility of the acids increases with decreasing carbon number; butyric acid is completely soluble in water.

Ethyl ether is the best solvent for stearic acid and other saturated long-chain fatty acids since it is sufficiently polar to attract the carboxyl groups. A truly nonpolar solvent, such as petroleum benzine, is not suitable for free fatty acids.

The solubility of fatty acids increases with an increase in the number of cis double bonds. This is illustrated in Fig. 3.3 with acetone as a solvent. The observed differences in solubility can be utilized for separation of saturated from unsaturated fatty acids. The mixture of acids is dissolved at room temperature and cooled stepwise to $-80\,°C$. However, the separation efficiency of such a fractional crystallization is limited since, for example, stearic acid is substantially more soluble in acetone containing oleic acid than in pure acetone. This mutual effect on solubility has not been considered in Fig. 3.3.

3.2.2.5 UV-Absorption

All unsaturated fatty acids which contain an isolated cis double bond absorb UV light at a wavelength close to 190 nm. Thus, the acids can not be distinguished spectrophotometrically.

Conjugated fatty acids absorb light at various wavelenths depending on the length of conjugation and configuration of the double bond system. Figure 3.4 illustrates such behavior for several fatty acids. See 3.2.3.2.2 for the conversion of an isolene-type fatty acid into a conjugated fatty acid.

3.2.3 Chemical Properties

3.2.3.1 Methylation of Carboxyl Groups

The carboxyl group of a fatty acid must be depolarized by methylation in order to facilitate gas chromatographic separation or separation by fractional distillation. Reaction with diazomethane is preferred for analytical purposes. Diazomethane is formed by alkaline hydrolysis of N-nitroso-N-methyl-p-toluene sulfonamide.

The gaseous CH_2N_2 released by hydrolysis is swept by a stream of nitrogen into a receiver containing the fatty acid solution in ether-methanol (9:1 v/v). The reaction:

$$R-COOH+CH_2N_2 \longrightarrow R-COOCH_3+N_2$$

$$(3.9)$$

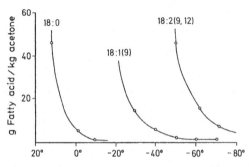

Fig. 3.3. Fatty acid solubility in acetone (according to *Mead* et al., 1965)

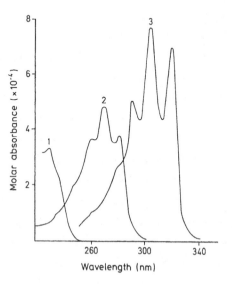

Fig. 3.4. Electron excitation spectra of conjugated fatty acids (according to *Pardun*, 1976). 1 9,11-isolinoleic acid, 2 α-eleostearic acid, 3 parinaric acid

proceeds under mild conditions without formation of by-products. Further possibilities for methylation include: esterification in the presence of excess methanol and a *Lewis* acid (BF$_3$) as a catalyst; or the reaction of a fatty acid silver salt with methyl iodide:

$$R - COOAg + CH_3I \longrightarrow R - COOCH_3 + AgI,$$
$$(3.10)$$

3.2.3.2 Reactions of Unsaturated Fatty Acids

A number of reactions which are known for olefinic hydrocarbons play an important role in the analysis and processing of lipids containing unsaturated fatty acids.

3.2.3.2.1 Halogen Addition Reactions

The number of double bonds present in an oil or fat can be determined through their iodine number (cf. 14.5.2.1). The fat or oil is treated with a halogen reagent which reacts only with the double bonds. Substitution reactions generating hydrogen halides must be avoided. IBr in an inert solvent, such as glacial acetic acid, is a suitable reagent:

$$(3.11)$$

The number of double bonds is calculated by titrating the unreacted IBr reagent with thiosulfate.

3.2.3.2.2 Transformation of Isolene-Type Fatty Acids to Conjugated Fatty Acids

Allyl systems are labile and are readily converted to a conjugated double bond system in the presence of a base (KOH or K-tertbutylate):

$$-CH = CH - CH = CH - CH_2 -$$
$$(3.12)$$

During this reaction, an equilibrium is established between the isolene and the conjugated forms of the fatty acid, the equilibrium state being dependent on the reaction conditions. This isomerization is used analytically since it provides a way to simultaneously determine linoleic, linolenic and arachidonic acids in a fatty acid mixture. The corresponding conjugated diene, triene and tetraene systems of these fatty acids have a maximum absorbance at distinct wavelengths (cf. Fig. 3.4). The assay conditions can be selected to isomerize only the naturally occurring cis double bonds and to ignore the trans fatty acids formed, for instance, during oil hardening (cf. 14.4.2).

3.2.3.2.3 Formation of a π-Complex with Ag$^+$ Ions

Unsaturated fatty acids or their triacylglycerols, as well as unsaturated aldehydes obtained through autoxidation of lipids (cf. 3.7.2.1.5), can be separated by "argentation chromatography". The separation is based on the number, position and configuration of the double bonds present. The separation mechanism involves interaction of the π-electrons of the double bond with Ag$^+$ ions, forming a reversible π-complex of variable stability:

$$(3.13)$$

The complex stability increases with increasing number of double bonds. This means a fatty acid with two cis double bonds will not migrate as far as a fatty acid with one double bond on a thin-layer plate impregnated with a silver salt. The R$_f$ values increase for the series 18:2 (9, 12) < 18:1 (9) < 18:0. Furthermore, fatty acids with isolated double bonds form a stronger Ag$^+$ complex than those with conjugated bonds. Also, the complex is stronger with a cis- than with a trans-configuration. The complex is also more stable, the further the double bond is from the end of the chain. Finally, a separation of nonconjugated from conjugated fatty acids and of isomers that differ only in their double bond configuration is possible by argentation chromatography.

3.2.3.2.4 Hydrogenation

In the presence of a suitable catalyst, e. g. Ni, hydrogen can be added to the double bond of an acyl lipid. This heterogeneous catalytic hydrogenation occurs stereo selectively as a cis-addition. Catalyst-induced isomerization from an isolene-type fatty acid to a conjugated fatty acid occurs with fatty acids with several double bonds:

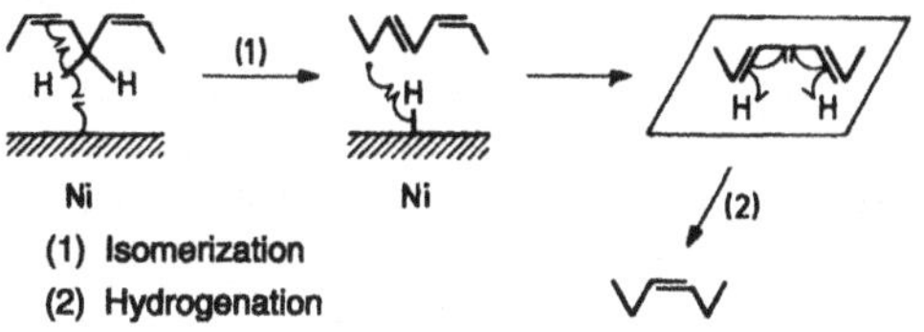

(1) Isomerization
(2) Hydrogenation

$$(3.14)$$

Since diene fatty acids form a more stable complex with a catalyst than do monoene fatty acids, the former are preferentially hydrogenated. Since nature is not an abundant source of the solid fats which are required in food processing, the partial and selective hydrogenation, just referred to, plays an important role in the industrial processing of fats and oils (cf. 14.4.2).

3.2.4 Biosynthesis of Unsaturated Fatty Acids

The biosynthetic precursors of unsaturated fatty acids are saturated fatty acids in an activated form (cf. a biochemistry textbook). These are aerobiocally and stereospecifically dehydrogenated by dehydrogenase action in plant as well as animal tissues. A flavoprotein and ferredoxin are involved in plants in the electron transport system which uses oxygen as a terminal electron acceptor (cf. Reaction 3.15).

To obtain polyunsaturated fatty acids, the double bonds are introduced by a stepwise process. A fundamental difference exists between mammals and plants. In the former, oleic acid synthesis is possible, and, also, additional double bonds can be inserted towards the carboxyl end of the fatty acid molecule. For example, γ-linolenic acid can be formed from the essential fatty acid linoleic acid and, also, arachidonic acid (Fig. 3.5) can be formed by chain elongation of γ-linolenic acid. In a diet deficient in linoleic acid, oleic acid is dehydrogenated to isolinoleic acid and its derivatives (Fig. 3.5), but none of these acquire the physiological function of an essential acid such as linoleic acid.

Plants can introduce double bonds into fatty acids in both directions: towards the terminal CH_3-group or towards the carboxyl end. Oleic acid (oleoyl-CoA ester or β-oleoyl-phosphatidylcholine) is thus dehydrogenated to linoleic and then to linolenic acid. In addition synthesis of the latter can be achieved by another pathway involving stepwise dehydrogenation of lauric acid with chain elongation reactions involving C_2 units (Fig. 3.5).

3.3 Acylglycerols

Acylglycerols (or acylglycerides) comprise the mono-, di- or triesters of glycerol with fatty acids (Table 3.1). They are designated as neutral lipids. Edible oils or fats consist nearly completely of triacylglycerols.

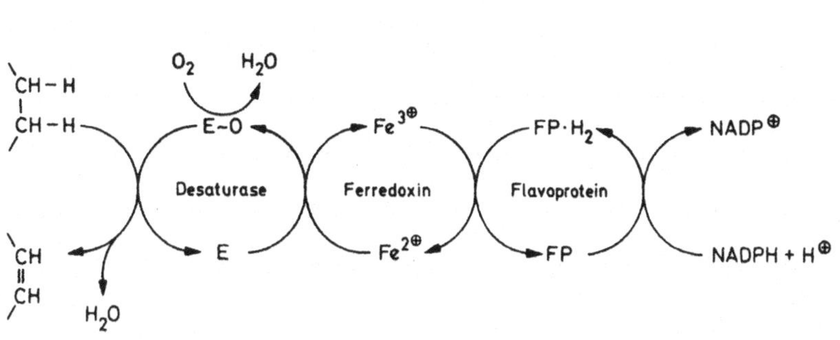

$$(3.15)$$

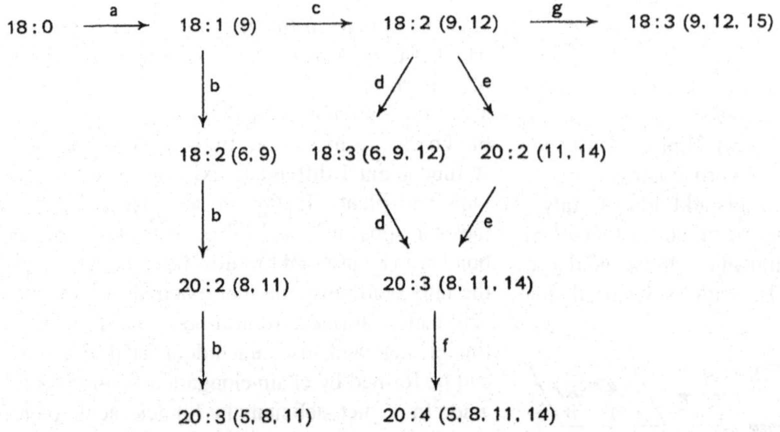

$$18:0 \xrightarrow{a} 18:1\,(9) \xrightarrow{c} 18:2\,(9,\,12) \xrightarrow{g} 18:3\,(9,\,12,\,15)$$

$$18:2\,(6,\,9) \quad 18:3\,(6,\,9,\,12) \quad 20:2\,(11,\,14)$$

$$20:2\,(8,\,11) \qquad 20:3\,(8,\,11,\,14)$$

$$20:3\,(5,\,8,\,11) \qquad 20:4\,(5,\,8,\,11,\,14)$$

Fig. 3.5. Biosynthesis of unsaturated fatty acids. Synthesis pathways: a, c, g in higher plants; a, c, g and a, c, d, f in algae; a, b and d, f (main pathway for arachidonic acid) or e, f in mammals

3.3.1 Triacylglycerols (TG)

3.3.1.1 Nomenclature, Classification, Calorific Value

Glycerol, as a trihydroxylic alcohol, can form triesters with one, two or three different fatty acids. In the first case a triester is formed with three of the same acyl residues (e. g. tripalmitin; P_3). The mixed esters involve two or three different acyl residues, e. g., dipalmito-olein (P_2O) and palmito-oleo-linolein (POL). The rule of this shorthand designation is that the acid with the shorter chain or, in the case of an equal number of carbons in the chain, the chain with fewer double bonds, is mentioned first. The Z number gives the possible different triacylglycerols which can occur in a fat (oil), where n is the number of different fatty acids identified in that fat (oil):

$$Z = \frac{n^3 + n^2}{2} \tag{3.16}$$

For n = 3, the possible number of triglycerols (Z) is 18. However, such a case where a fat (oil) contains only three fatty acids is rarely found in nature. One exception is Borneo tallow (cf. 14.3.2.2.3), which contains essentially only 16:0, 18:0 and 18:1 (9) fatty acids.

Naturally, the Z value also takes into account the number of possible positional isomers within a molecule, for example, by the combination of POS, PSO and SOP. When only positional isomers are considered and the rest disregarded, Z is reduced to Z′:

$$Z' = \frac{n^3 + 3n^2 + 2n}{6} \tag{3.17}$$

Thus, when n = 3, Z′ = 10.
A chiral center exists in a triacylglycerol when the acyl residues in positions 1 and 3 are different:

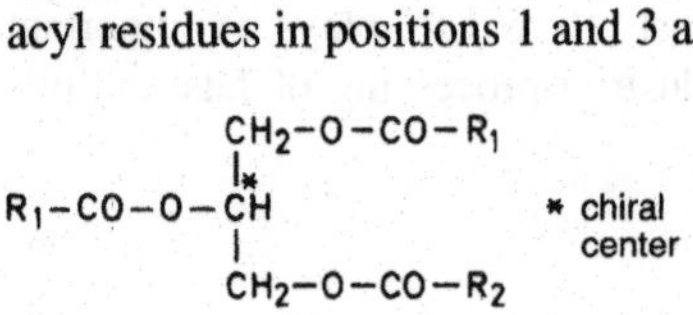

$$
\begin{array}{l}
\text{CH}_2\text{-O-CO-R}_1 \\
\quad | * \\
\text{R}_1\text{-CO-O-CH} \qquad\qquad * \text{ chiral} \\
\quad | \qquad\qquad\qquad\qquad\;\; \text{center} \\
\text{CH}_2\text{-O-CO-R}_2
\end{array}
\tag{3.18}
$$

In addition enantiomers may be produced by 1-monoglycerides, all 1,2-diglycerides and 1,3-diglycerides containing unlike substituents.

In the stereospecific numbering of acyl residues (prefix sn), the L-glycerol molecule is shown in the *Fischer* projection with the secondary HO-group pointing to the left. The top carbon is then denoted C-1. Actually, in a *Fischer* projection, the horizontal bonds denote bonds in front and the vertical bonds those behind the plane of the page:

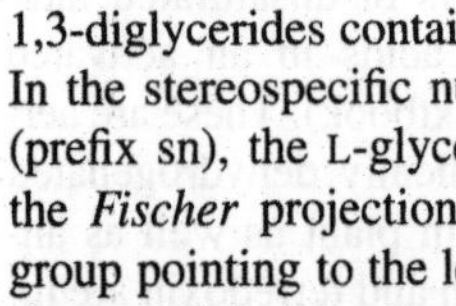

$$
\begin{array}{ll}
\text{CH}_2\text{-OH} & \text{sn-1} \\
\quad | & \\
\text{HO}\!-\!\!\text{C}\!-\!\!\text{H} & \text{sn-2} \\
\quad | & \\
\text{CH}_2\text{-OH} & \text{sn-3}
\end{array}
\tag{3.19}
$$

For example, the nomenclature for a triacylglycerol which contains P, S and O:

sn-POS = sn-1-Palmito-2-oleo-3-stearin .

This assertion is only possible when a stereospecific analysis (cf. 3.3.1.4) provides information on the fatty acids at positions 1, 2 and 3. rac-POS = sn-POS and sn-SOP in the molar ratio 1:1, i.e. the fatty acid in position 2 is fixed while the other two acids are equally distributed at positions 1 and 3.

POS = mixture of sn-POS, sn-OPS, sn-SOP,

sn-PSO, sn-OSP and sn-SPO

The physiological calorific value of TGs depends on the fatty acid composition. In the case of TGs with fatty acids of medium chain length (6–10 C atoms), the calorific value decreases from 9 to 7 kcal/g and in the case of asymmetric TGs, e. g., a combination of 2:0, 3:0 or 4:0 with 18:0, it decreases to 5 kcal/g. These special TGs, which are available only synthetically, are classified as fat substitutes (cf. 8.16).

3.3.1.2 Melting Properties

TG melting properties are affected by fatty acid composition and their distribution within the glyceride molecule (Table 3.12).
Mono-, di- and triglycerides are polymorphic, i.e. they crystallize in different modifications, denoted as α, β' and β. These forms differ in their melting points (Table 3.12) and crystallographic properties.
During the cooling of melted acylglycerols, one of the three polymorphic forms is yielded. This depends also on the temperature gradient cho-

Table 3.12. Triacylglycerols and their polymorphic forms

Compound	Melting point (°C) of polymorphic form		
	α	β'	β
Tristearin	55	63.2	73.5
Tripalmitin	44.7	56.6	66.4
Trimyristin	32.8	45.0	58.5
Trilaurin	15.2	34	46.5
Triolein	−32	−12	4.5–5.7
1,2-Dipalmitoolein	18.5	29.8	34.8
1,3-Dipalmitoolein	20.8	33	37.3
1-Palmito-3-stearo-2-olein	18.2	33	39
1-Palmito-2-stearo-3-olein	26.3	40.2	
2-Palmito-1-stearo-3-olein	25.3	40.2	
1,2-Diacetopalmitin	20.5	21.6	42.3

sen. The α-form has the lowest melting point. This modification is transformed first into the β'-form upon heating and then into the β-form. The β-form is the most stable and, hence, also has the highest melting point (Table 3.12). These changes are typically monotropic, i.e. they proceed in the order of lower to higher stability.
Crystallization of triglycerides from a solvent system generally yields β-form crystals.
X-ray analysis as well as measurements by Raman spectroscopy revealed that saturated triglycerols in their crystalline state exist in a "chair form" (Fig. 3.6a): The "tuning fork" configuration for the β'-modification was not verified. The different properties of the three forms are based on the crystallization in different systems.

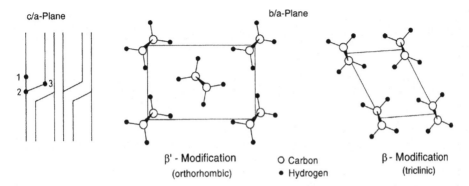

Fig. 3.6. Arrangement of the β'- and β-form of saturated triacylglycerols in the crystalline lattice (Cartesian coordinates a, b, c)

Table 3.13. Crystallization patterns of edible fats or oils

β-Type	β'-Type	β-Type	β'-Type
Coconut oil	Cottonseed oil	Peanut	
Corn germ oil	Butter	Sunflower oil	Whale oil
Olive oil	Palm oil	Lard	
Palm seed oil	Rapeseed oil		

α-form: hexagonal system; the melting point is relatively low, since areas of the methyl ends are freely arranged as in liquid crystals.

β'-form: (Fig. 3.6b): orthorhombic system; the carbon chains are perpendicular to each other.

β-form: (Fig. 3.6c): triclinic system; parallel arrangement of the carbon chains.

Unsaturated fatty acids interfere with the orderly packing of molecules in the crystalline lattice, thereby causing a decrease in the melting point of the crystals.

TG such as 1,3-diaceto-palmitin, i. e. a triglyceride with one long and two short-chain fatty acids, exists in the exceptionally stable α-form. Since films of such TGs can expand by 200 to 300 times their normal length, they are of interest for application as protective coating for fat-containing foods. In edible fats and oils, more than the three mentioned polymorphic forms can be present, e. g., 4–6 forms are being discussed for cocoa butter. In order to classify fats and oils, that form is used that is predominant after solidification (Table 3.13).

3.3.1.3 Chemical Properties

Hydrolysis, methanolysis and interesterification are the most important chemical reactions for TGs.

Hydrolysis. The fat or oil is cleaved or saponified by treatment with alkali (e. g. alcoholic KOH):

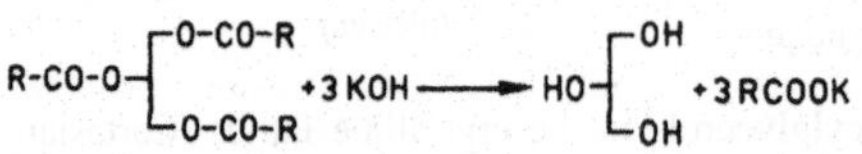

$$(3.20)$$

After acidification and extraction, the free fatty acids are recovered as alkali salts (commonly called soaps). This procedure is of interest for analysis of fat or oil samples. Commercially, the free fatty acids are produced by cleaving triglycerides with steam under elevated pressure and temperature and by increasing the reaction rate in the presence of an alkaline catalyst (ZnO, MgO or CaO) or an acidic catalyst (aromatic sulfonic acid).

Methanolysis. The fatty acids in TG are usually analyzed by gas liquid chromatography, not as free acids, but as methyl esters. The required transesterification is most often achieved by Na-methylate (sodium methoxide) in methanol and in the presence of 2,2-dimethoxypropane to bind the released glycerol. Thus, the reaction proceeds rapidly and quantitatively even at room temperature.

Interesterification. This reaction is of industrial importance (cf. 14.4.3) since it can change the physical properties of fats or oils or their mixtures without altering the chemical structure of the fatty acids. Both intra- and inter-molecular acyl residue exchanges occur in the reaction until an equilibrium is reached which depends on the structure and composition of the TG molecules. The usual catalyst for interesterification is Na-methylate.

The principle of the reaction will be elucidated by using a mixture of tristearin (SSS) and triolein (OOO) or stearodiolein (OSO). Two types of interesterification are recognized:

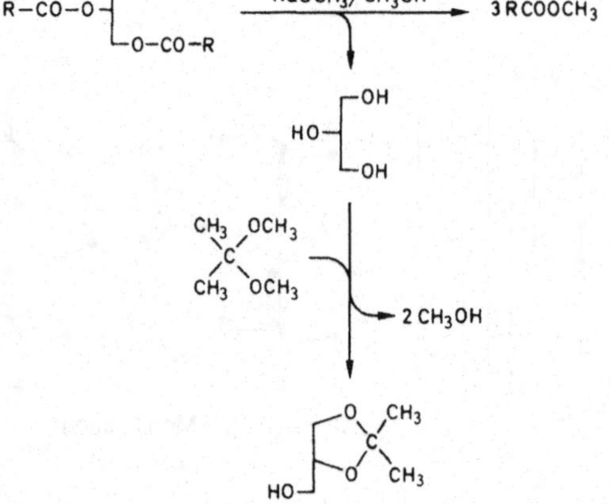

$$(3.21)$$

a) A single-phase interesterification where the acyl residues are randomly distributed:

$$
\begin{array}{c}
\underset{(50\%)}{S\,S\,S} \cdot \underset{(50\%)}{0\,0\,0} \\
\downarrow (NaOCH_3) \\
\underset{(12.5\%)}{S\,S\,S} \quad \underset{(12.5\%)}{S\,0\,S} \quad \underset{(25\%)}{0\,S\,S} \quad \underset{(25\%)}{S\,0\,0} \quad \underset{(12.5\%)}{0\,S\,0} \quad \underset{(12.5\%)}{0\,0\,0}
\end{array}
\tag{3.22}
$$

b) A directed interesterification in which the reaction temperature is lowered until the higher melting and least soluble TG molecules in the mixture crystallize. These molecules cease to participate in further reactions, thus the equilibrium is continuously changed. Hence, a fat (oil) can be divided into high and low melting point fractions, e. g.:

$$
\begin{array}{c}
0\,S\,0 \\
\downarrow (NaOCH_3) \\
\underset{(33.3\%)}{S\,S\,S} \quad \underset{(66.7\%)}{0\,0\,0}
\end{array}
\tag{3.23}
$$

3.3.1.4 Structural Determination

Apart from identifying a fat or oil from an unknown source (cf. 14.5.2), TG structural analysis is important for the clarification of the relationship existing between the chemical structure and the melting or crystallization properties, i.e. the consistency.

An introductory example: cocoa butter and beef tallow, the latter used during the past century for adulteration of cocoa butter, have very similar fatty acid compositions, especially when the two main saturated fatty acids, 16:0 and 18:0, are considered together (Table 3.14). In spite of their compositions, the two fats differ significantly in their melting properties. Cocoa butter is hard and brittle and melts in a narrow temperature range (28–36 °C). Edible beef tallow, on the other hand, melts at a higher temperature (approx. 45 °C) and over a wider range and has a substantially better plasticity. The melting property of cocoa butter is controlled by the presence of a different pattern of triglycerols: SSS, SUS and SSU (cf. Table 3.14). The chemical composition of Borneo

Table 3.14. Average fatty acid and triacylglycerol composition (weight-%) of cocoa butter, tallow and Borneo tallow (a cocoa butter substitute)

	Cocoa butter	Edible beef tallow	Borneo tallow[a]
16:0	25	36	20
18:0	37	25	42
20:0	1		1
18:1 (9)	34	37	36
18:2 (9,12)	3	2	1
SSS[b]	2	29	4
SUS	81	33	80
SSU	1	16	1
SUU	15	18	14
USU		2	
UUU	1	2	1

[a] cf. 14.3.2.2.3
[b] S: Saturated, and U: unsaturated fatty acids.

tallow (Tenkawang fat) is so close to that of cocoa butter that the TG distribution patterns shown in Table 3.14 are practically indistinguishable. Also, the melting properties of the two fats are similar, consequently, Borneo tallow is currently used as an important substitute for cocoa butter. Analysis of the TGs present in fat (oil) could be a tedious task, when numerous TG compounds have to be separated. The composition of milk fat is particularly complex. It contains more than 150 types of TG molecules.

The separation by HPLC using reverse phases is the first step in TG analysis. It is afforded by the chain length and the degree of unsaturation of the TGs. As shown in Fig. 3.7 the oils from different plant sources yield characteristic patterns in which distinct TGs predominate.

TGs differing only in the positions of the acyl residues are not separated. However, in some cases it is possible to separate positional isomeric triglycerols after bromination of the double bonds because triglycerols with a brominated acyl group in β-position are more polar compared to those in α-position.

The separation capacity of the HPLC does not suffice for mixtures of plant oils with complex triglycerol composition. Therefore it is advisable to perform a preseparation of the triglycerols according to their number of double bonds by "argentation chromatography" (cf. 3.2.3.2.3).

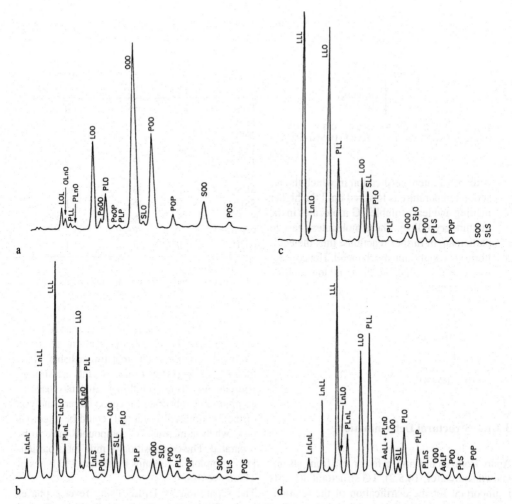

Fig. 3.7. Composition of triacylglycerols present in edible fats or oils as determined by HPLC. **a** Olive oil, **b** soybean oil, **c** sunflower oil, **d** wheat germ oil. Fatty acids: P palmitic, S stearic, O oleic, L linoleic, Ln linolenic, Ao eicosanoic

Various hypotheses have been advanced, supported by results of TG biosynthesis, to predict the TG composition of a fat or oil when all the fatty acids occurring in the sample are known. The values calculated with the aid of the *1,3-random-2-random* hypothesis agree well with values found experimentally for plant oil or fat. The hypothesis starts with two separated fatty acid pools. The acids in both pools are randomly distributed and

used as such for TG biosynthesis. The primary HO-groups (positions 1 and 3 of glycerol) from the first pool are esterified, while the secondary HO-group is esterified in the second pool. The proportion of each TG is then determined (as mole-%):

$$\beta\text{-XYZ(mol-\%)} = 2 \cdot \begin{bmatrix} \text{mol-\% X in} \\ 1,3\text{-Position} \end{bmatrix} \cdot \begin{bmatrix} \text{mol-\% Y in} \\ 2\text{-Position} \end{bmatrix} \cdot \begin{bmatrix} \text{mol-\% Z in} \\ 1,3\text{-Position} \end{bmatrix} \cdot 10^{-4}$$

(3.24)

The data required in order to apply the formula are obtained as follows: after partial hydrolysis of fat (oil) with pancreatic lipase (cf. 3.7.1.1), the

Table 3.15. Triacylglycerol composition (mole-%) of a sunflower oil.
A comparison of experimental values with calculated values based on a 1,3-random-2-random hypothesis

Triacylglycerol[a]	Found	Calculated	Triacylglycerol[a]	Found	Calculated
β-StOSt	0.3	0.5	β-OStL	0.5	0.2
β-StStO	0.2	trace	β-OOL	8.1	6.5
β-StOO	2.3	1.6	β-OLO	3.1	4.2
β-OStO	0.1	trace	β-StLL	13.2	14.0
β-StStL	0.3	0.2	β-LStL	1.3	0.3
β-StLSt	2.2	1.7	β-OLL	20.4	21.9
OOO	1.3	1.2	β-LOL	8.4	8.7
β-StOL	4.4	4.2	LLL	28.1	28.9
β-StLO	4.0	5.3	Others	0.9	0.9

St: Stearic, O: oleic, and L: linoleic acid.
[a] Prefix β: The middle fatty acid is esterified at the β- or sn-2-position, the other two acids are at the sn-1 or sn-3 positions.

fatty acids bound at positions 1 and 3 are determined. The fatty acids in position 2 are calculated from the difference between the total acids and those acids in positions 1 and 3.

Table 3.15 illustrates the extent of agreement for the TG composition of sunflower oil obtained experimentally and by calculation using the 1,3-random-2-random hypothesis. However, both approaches disregard the differences between positions 1 and 3. In addition, the hypothesis is directed to plants, of which the fats and oils consist of only major fatty acids.

Stereospecific Analysis. Biochemically, the esterified primary OH-groups of glycerol can be differentiated from each other; thus, the determination of fatty acids in positions 1, 2 and 3 is possible. The reactions carried out are presented in Fig. 3.8. First, the TG (I) is hydrolyzed under controlled conditions to a diacylglycerol using pancreatic lipase (cf. 3.7.1.1). Phosphorylation with a diacylglycerol kinase follows. The enzyme reacts stereospecifically since it phosphorylates only the 1,2- or (S)- but not the 2,3-diglycerol. Subsequently, compound I is hydrolyzed to a monoacylglycerol (III). The distribution of the acyl residues in positions 1, 2 and 3 is calculated from the results of the fatty acid analysis of compounds I, II and III.

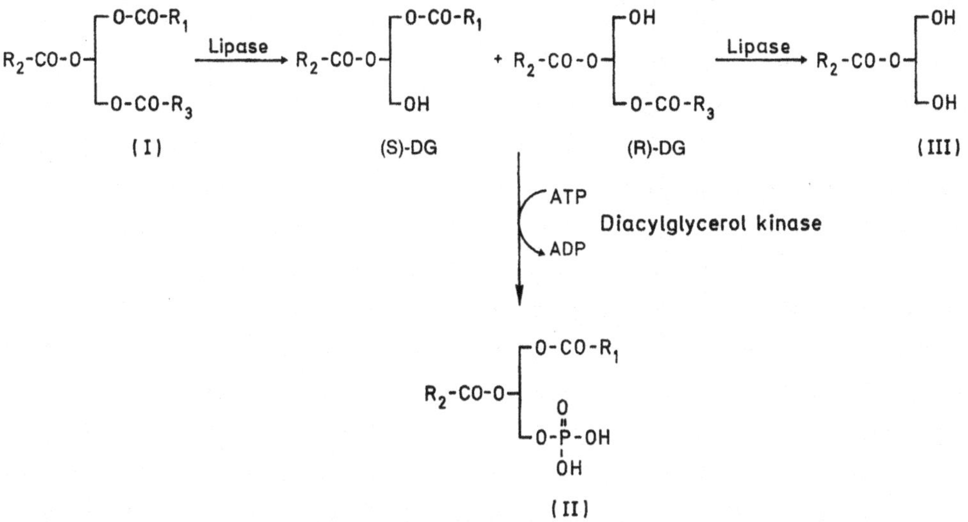

Fig. 3.8. Enzymatic stereospecific analysis of triacylglycerols

Alternatively, the stereo-specific analysis can be carried out chemically. The TGs are partially hydrolyzed in the presence of ethyl magnesium bromide. The resulting diacylglycerols are isolated and their OH groups converted to urethane with (S)-1-(1-naphthyl)ethylisocyanate. The sn-1,3- and the diastereomeric sn-1,2- and 2,3-di-acylglycerol urethane derivatives are separated in a subsequent HPLC step. The fatty acid analysis of the urethanes show the distribution of the acyl residues in positions 1, 2 and 3.

Individual TGs or their mixtures can be analyzed with these procedures. Based on these results (some are presented in Table 3.16), general rules for fatty acid distribution in plant oils or fats can be deduced:

• The primary HO-groups in positions 1 and 3 of glycerol are preferentially esterified with saturated acids.
• Oleic and linolenic acids are equally distributed in all positions, with some exceptions, such as cocoa butter (cf. Table 3.16).
• The remaining free position, 2, is then filled with linoleic acid.

Results compiled in Table 3.16 show that for oil or fat of plant origin, the difference in acyl residues between positions 1 and 3 is not as great as for TGs of animal origin (e. g., chicken egg). Therefore, the 1,3-random-2-random hypothesis can provide results that agree well with experimental findings.

The fatty acid pattern in animal fats is strongly influenced by the fatty acid composition of animal feed. A steady state is established only after 4–6 months of feeding with the same feed composition. The example of chicken egg (Table 3.16) indicates that positions 1 and 3 in triglycerides of animal origin show much greater variability than in fats or oils of plant origin. Therefore, any prediction of TG types in animal fat should be calculated from three separate fatty acid pools (*1-random-2-random-3-random* hypothesis).

The specific distribution of saturated fatty acids in the triglycerols of fats and oils of plant origin serves as an evidence of *ester oils*.

Ester oils are produced by esterification of glycerol with purified fatty acids obtained from olive oil residues. In this case the saturated acyl groups are equally distributed between all three positions of the glycerol molecule, whereas

Table 3.16. Results of stereospecific analysis of some fats and oils[a]

Fat/Oil	Position	16:0	18:0	18:1 (9)	18:2 (9,12)	18:3 (9,12,15)
Peanut	1	13.6	4.6	59.2	18.5	–
	2	1.6	0.3	58.5	38.6	–
	3	11.0	5.1	57.3	18.0	–
Soya	1	13.8	5.9	22.9	48.4	9.1
	2	0.9	0.3	21.5	69.7	7.1
	3	13.1	5.6	28.0	45.2	8.4
Sun-flower	1	10.6	3.3	16.6	69.5	–
	2	1.3	1.1	21.5	76.0	–
	3	9.7	9.2	27.6	53.5	–
Olive	1	15.2	2.9	68.6	11.0	–
	2	2.5	0.6	81.0	14.6	–
	3	19.6	5.2	62.6	9.4	–
Palm	1	60.1	3.4	26.8	9.3	–
	2	13.3	0.2	67.9	17.5	–
	3	71.9	7.6	14.4	3.2	–
Cocoa	1	34.0	50.4	12.3	1.3	–
	2	1.7	2.1	87.4	8.6	–
	3	36.5	52.8	8.6	0.4	–
Chicken egg	1	68.2	6.0	12.4	2.3	–
	2	4.8	0.3	60.8	31.3	–
	3	8.9	7.7	69.4	5.4	–

[a] Values in mol%. In order to simplify the Table other fatty acids present in fat/oil are not listed.

in olive oil saturated acyl groups are attached to position 1 and 3. As proof, the amount of 2-MG containing palmitic acid is determined after hydrolysis of the triglycerols with a lipase (pancreas). Values above 2% are indicative of an adulteration of the olive oil with an ester oil.

The positional specific distribution of palmitic acid is unfavorable for the use of fats and oils of plant origin in infant food, as this acid is liberated by lipolysis in the gastric tract. Palmitic acid then forms insoluble salts with Ca^{2+}-ions from the food, possibly resulting in severe bilious attacks. The fatty acids of human milk consist of up to 25% of palmitic acid; 70% are bound to the 2-position of the triglycerols. During lipolysis 2-monopalmitin is formed that is easily resorbed.

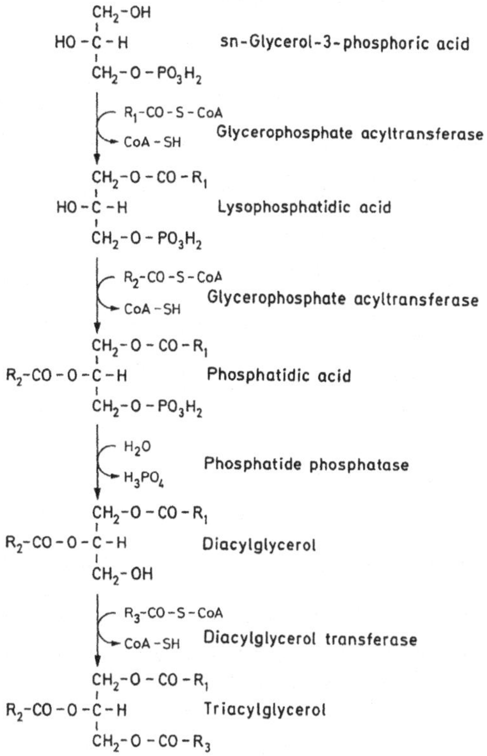

3.3.1.5 Biosynthesis

A TG molecule is synthesized in the fat cells of mammals and plants from L-glycerol-3-phosphate and fatty acid-CoA esters (Fig. 3.9). The L-glycerol-3-phosphate supply is provided by the reduction of dihydroxy acetone phosphate by NAD^+-dependent glycerol phosphate dehydrogenase. The dihydroxy acetone phosphate originates from glycolysis.

The lipid bodies (oleosomes, spherosomes) synthesized are surrounded by a membrane and are deposited in storage tissues.

The TG fatty acid composition within a plant species depends on the environment, especially the temperature. A general rule is that plants in cold climates produce a higher proportion of unsaturated fatty acids. Obviously, the mobility of TGs is thus retained. In the sunflower (cf. Fig. 3.10), this rule is highly pronounced; whereas in safflower, only a weak response to temperature variations is observed (Fig. 3.10).

3.3.2 Mono- and Diacylglycerols (MG, DG)

3.3.2.1 Occurrence, Production

The occurrence of MG and DG in edible oils or fats or in raw food is very low. However, their levels may be increased by the action of hydrolases during food storage or processing. MG and DG are produced commercially by fat glycerolysis (200 °C, basic catalyst)

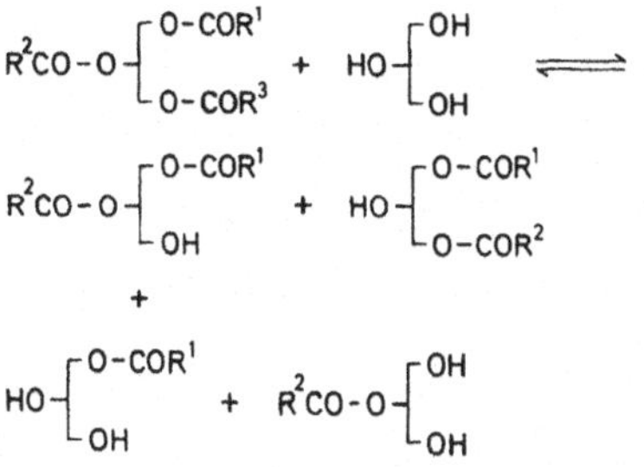

Fig. 3.9. Biosynthesis of triacylglycerols

$$(3.25)$$

From the equilibrium (cf. Formula 3.25) that contains 40–60% MG, 45–35% DG and 15–5% TG, the MG are separated by distillation under high vacuum. The amount of 1-MG (90–95%) is predominant over the amount of 2-MG.

3.3.2.2 Physical Properties

MG and DG crystallize in different forms (polymorphism; cf. 3.3.1.2). The melting point of an ester of a given acid increases for the series 1,2-DG < TG < 2-MG < 1,3-DG < 1-MG:

	Melting Point (°C) β-form
Tripalmitin	65.5
1,3-Dipalmitin	72.5
1,2-Dipalmitin	64.0
1-Palmitin	77.0
2-Palmitin	68.5

MG and DG are surface-active agents. Their properties can be further modified by esterification with acetic, lactic, fumaric, tartaric or citric acids. These esters play a significant role as emulsifiers in food processing (cf. 8.15.3.1).

3.4 Phospho- and Glycolipids

3.4.1 Classes

Phospho- and glycolipids, together with proteins, are the building blocks of biological membranes. Hence, they invariably occur in all foods of animal and plant origin. Examples are compiled in Table 3.17. As surface-active compounds, phospho- and glycolipids contain hydrophobic moieties (acyl residue, N-acyl sphingosine) and hydrophilic portions (phosphoric acid, carbohydrate). Therefore, they are capable of forming orderly structures (micelles or planar layers) in aqueous media; the bilayer structures are found in all biological membranes. Examples for the composition of membrane lipids are listed in Table 3.18.

3.4.1.1 Phosphatidyl Derivatives

The following phosphoglycerides are derived from phosphatidic acid. Phosphatidyl choline or lecithin (phosphate group esterified with the

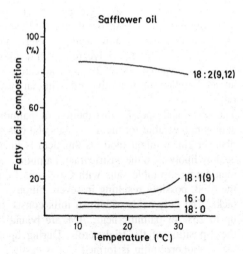

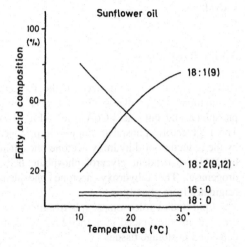

Fig. 3.10. The effect of climate (temperature) on the fatty acid composition of triacylglycerols

Table 3.17. Composition of lipids of various foods[a]

	Milk	Soya	Wheat	Apple
Total lipids	3.6	23.0	1.5	0.088
Triacylglycerols	94	88	41	5
Mono-, and diacylglycerols	1.5		1	
Sterols	< 1		1	15
Sterol esters			1	2
Phospholipids	1.5	10	20	47
Glycolipids		1.5	29	17
Sulfolipids			1	
Others		0.54	7	15

[a] Total lipids as %, while lipid fractions are expressed as percent of the total lipids.

Table 3.18. Occurrence of phosphatidyl derivates

Food	Lipid (g/kg)	P-containing lipids[a] (g/kg)	Phosphatidyl derivatives[b] (mg/kg)			
		PC	PS	PE	PI	
Milk	37.8	0.35	120	10	100	2
Egg	113	35.1	27,000	¾	5810	¾
Meat (beef)	19	8.3	4290	690	1970	¾
Meat (chicken)	62	6.6	3320	850	1590	¾
Tuna fish	155	19.4	6410	1940	5030	¾
Potato	1.1	0.56	280	10	160	90
Rice	6.2	0.89	320	30	350	¾
Soybean	183	17.8	7980	¾	4660	2500

[a] Phosphatidyl derivatives and other P-containing lipids, e. g., plasmalogens, sphingomyelins.
[b] The abbreviations correspond to Formulas 3.26 – 3.29.

OH-group of choline, PC):

$$
\begin{array}{l}
\text{(1) } CH_2-O-CO-R_1\\
R_2-CO-O-CH \text{ (2)}\quad O \qquad\qquad CH_3\\
\text{(3) } CH_2-O-\overset{\;}{\underset{O^{\ominus}}{P}}-O-CH_2-CH_2-\overset{\oplus}{N}-CH_3\\
\qquad\qquad\qquad\qquad\qquad\qquad\quad CH_3
\end{array}
\tag{3.26}
$$

Phosphatidyl serine (phosphate group esterified with the HO-group of the amino acid serine, PS):

$$
\begin{array}{l}
CH_2-O-CO-R_1\\
R_2-CO-O-CH \qquad O\\
\qquad CH_2-O-\overset{O}{\underset{O^{\ominus}\; X^{\oplus}}{P}}-O-CH_2-\underset{COO^{\ominus}}{CH}-\overset{\oplus}{N}H_3
\end{array}
\tag{3.27}
$$

Phosphatidyl ethanolamine (phosphate group esterified with ethanolamine, PE):

$$
\begin{array}{l}
CH_2-O-CO-R_1\\
R_2-CO-O-CH \qquad O\\
\qquad CH_2-O-\overset{O}{\underset{O^{\ominus}}{P}}-O-CH_2-CH_2-\overset{\oplus}{N}H_3
\end{array}
\tag{3.28}
$$

Phosphatidyl inositol (phosphate group esterified with inositol, PI):

$$
\begin{array}{l}
CH_2-O-CO-R_1\\
R_2-CO-O-CH \qquad O \qquad\quad OH\\
\qquad CH_2-O-P-O-\text{(inositol ring: OH, OH, OH, HO, OH)}
\end{array}
\tag{3.29}
$$

A mixture of phosphatidyl serine and phosphatidyl ethanolamine was once referred to as cephalin.

Examples of foods which contain phosphatidyl derivatives are shown in Table 3.18. The differences to the data in Table 3.17 are caused by the biological range of variations.

Only one acyl residue is cleaved by hydrolysis (cf. 3.7.1.2.1) with phospholipase A. This yields the corresponding lyso-compounds from lecithin or phosphatidyl ethanolamine. Some of these lyso-derivatives occur in nature, e. g., in cereals. Phosphatidyl glycerol is invariably found in green plants, particularly in chloroplasts:

$$
\begin{array}{ll}
CH_2-O-CO-R_1 & CH_2-OH\\
R_2-CO-O-CH \qquad O & CH-OH\\
\qquad CH_2-O-\overset{O}{\underset{O^{\ominus}\; X^{\oplus}}{P}}-O-CH_2
\end{array}
$$

L-α-Phosphatidyl-D-glycerol   (3.30)

Cardiolipin, first identified in beef heart, is also a minor constituent of green plant lipids. Its chemical structure is diphosphatidyl glycerol:

$$
\begin{array}{lll}
CH_2-O-CO-R_1 & CH_2-O-\overset{O}{\overset{||}{P}}-O-CH_2 &\\
R_2-CO-O-CH \qquad O & CH-OH \quad O^{\ominus} X^{\oplus} \quad CH-O-CO-R_3 &\\
\qquad CH_2-O-\underset{O^{\ominus}\,X^{\oplus}}{P}-O-CH_2 & R_4-CO-O-CH_2 &
\end{array}
$$

Diphosphatidyl glycerol (cardiolipin)   (3.31)

The plasmalogens occupy a special place in the class of phospho-glycerides. They are phosphatides in which position 1 of glycerol is linked to a straight-chain aldehyde with 16 or 18 carbons. The linkage is an enolether type with a double bond in the cis-configuration. Plasmalogens of 1-O-(1-alkenyl)-2-O-acylglycerophospholipid type occur in small amounts in animal muscle tissue and also in milk fat. The enol-ether linkage, unlike the ether bonds of a 1-O-alkylglycerol (cf. 3.6.2), is readily hydrolyzed even by weak acids.

$$
\begin{array}{l}
CH_2-O-CH=CH-R_1 \\
R_2-CO-O-CH \qquad O \\
\qquad CH_2-O-\overset{\overset{\displaystyle O}{\|}}{P}-O-CH_2-CH_2-\overset{\oplus}{N}H_3 \\
\qquad\qquad\quad \underset{\ominus}{O}
\end{array}
$$

Plasmalogen   (3.32)

Phospholipids are sensitive to autoxidation since they contain an abundance of linoleic acid. The other acid commonly present is palmitic acid.

Phospholipids are soluble in chloroform-methanol and poorly soluble in water-free acetone. The pK_s value of the phosphate group is between 1 and 2. Phosphatidyl choline and phosphatidyl ethanolamine are zwitter-ions at pH 7.

Phospholipids can be hydrolyzed stepwise by alcoholic KOH. Under mild conditions, only the fatty acids are cleaved, whereas, with strong alkalies, the base moiety is released. The bonds between phosphoric acid and glycerol or phosphoric acid and inositol are stable to alkalies, but are readily hydrolyzed by acids. Phosphatidyl derivatives, together with triacylglycerols and sterols, occur in the lipid fraction of lipoproteins (cf. 3.5.1).

Lecithin. Lecithin plays a significant role as a surface-active agent in the production of emulsions. "Raw lecithin", especially that of soya and that isolated from egg yolk, is available for use on a commercial scale. "Raw lecithins" are complex mixtures of lipids with phosphatidyl cholines, ethanolamines and inositols as main components (Table 3.19).

The major phospholipids of raw soya lecithin are given in Table 3.19. The manufacturer often separates lecithin into ethanol-soluble and ethanol-insoluble fractions.

Pure lecithin is a W/O emulsifier with a HLB-value (cf. 8.15.2.3) of about 3. The HLB-value rises to 8–11 by hydrolysis of lecithin to lysophosphatidyl choline; an o/w emulsifier is formed. Since the commercial lecithins are complex mixtures of lipids their HBL-values lie within a wide range. The ethanol-insoluble

Table 3.19. Composition of glycerol phospholipids in soya "raw lecithin" and in the resulting fractions[a]

	Unfrac-tionated	Ethanol soluble fraction	Ethanol insoluble fraction
Phosphatidyl ethanolamine	13–17	16.3	13.3
Phosphatidyl choline	20–27	49	6.6
Phosphatidyl inositol	9	1	15.2

[a] Values in weight%.

fraction (Table 3.19) is suitable for stabilization of W/O emulsions and the ethanol-soluble fraction for O/W emulsions. To increase the HLB-value, "hydroxylated lecithins" are produced by hydroxylation of the unsaturated acyl groups with hydrogen peroxide in the presence of lactic-, citric- and tartaric acid.

3.4.1.2 Glyceroglycolipids

These lipids consist of 1,2-diacylglycerols and a mono-, di- or, less frequently, tri- or tetrasaccharide bound in position 3 of glycerol. Galactose is predominant as the sugar component among plant glycerolipids. The glyceroglycolipids occurring in wheat influence the baking properties (cf. 15.2.5).

(3.33)

Monogalactosyl diacylglycerol (MGDG)
(1,2-diacyl-3-β-D-galactopyranosyl-L-glycerol)

(3.34)

Digalactosyl diacylglycerol (DGDG)

(1,2-diacyl-3-(α-D-galactopyranosyl-1,6-β-D-galactopyranosyl)-L-glycerol)
6-O-acyl-MGDG and 6-0-acyl-DGDG are minor components of plant lipids.
Sulfolipids are glyceroglycolipids which are highly soluble in water since they contain a sugar moiety esterified with sulfuric acid. The sugar moiety is 6-sulfochinovose. Sulfolipids occur in chloroplasts but are also detected in potato tubers:

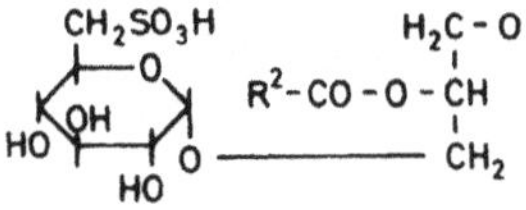

(3.35)

Sulfolipid(1,2-diacyl-(6-sulfo-α-D-chinovosyl-1,3)-L-glycerol)

3.4.1.3 Sphingolipids

Sphingolipids contain sphingosine, an amino alcohol with a long unsaturated hydrocarbon chain (D-*erythro*-1,3-dihydroxy-2-amino-trans-4-octadecene) instead of glycerol:

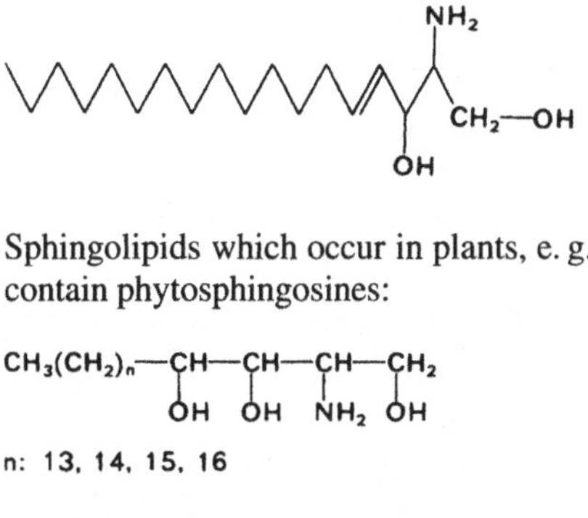

(3.36)

Sphingolipids which occur in plants, e. g., wheat, contain phytosphingosines:

$$CH_3(CH_2)_n—\underset{OH}{CH}—\underset{OH}{CH}—\underset{NH_2}{CH}—\underset{OH}{CH_2}$$

n: 13, 14, 15, 16 (3.37)

The amino group in sphingolipids is linked to a fatty acid to form a carboxy amide, denoted as ceramide. The primary hydroxyl group is either esterified with phosphoric acid (sphingophospholipid: ceramide-phosphate-base) or bound glycosidically to a mono- di-, or oligosaccharide (sphingoglycolipid: ceramide-phosphate-sugar$_n$). In the third group of sphingolipids the ceramide moiety is linked by a phosphate residue to the carbohydrate building blocks. These compounds are also referred to as phytoglycolipids.

Sphingophospholipids. Sphingomyelin is one example of a sphingophospholipid. It is the most abundant sphingolipid and is found in myelin, the fatty substance of the sheath around nerve fibers. The structure of sphingomyelin is:

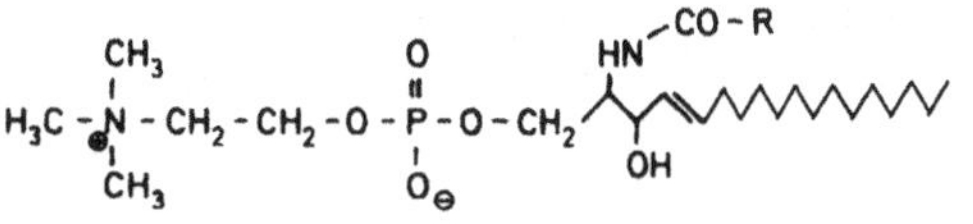

(3.38)

Sphingoglycolipids are found in tissue of animal origin, milk and in plants (especially cereals). Based on structural properties of the carbohydrate building blocks, one differentiates neutral and acid glycosphingolipids. The sulfatides and gangliosides also belong to this group.
Lactosylceramide in milk and the ceramide glycosides of wheat are examples of neutral glycosphingolipids that contain, next to glucose and mannose, also saturated (14:0–28:0) and monounsaturated (16:1–26:1) 2-hydroxy- or 2,3-dihydroxy fatty acids.
Formula 3.40 a depicts a sphingoglycolipid of wheat.
Gangliosides contain sialic acid (N-acetylneuraminic acid; cf. Formula 3.40 b). In the ganglioside fraction of milk monosialosyllactosyl-ceramide (cf. Formula 3.41) was identified.

Phytosphingolipids. These lipids also have a complex structure. Total hydrolysis yields phytosphingosine, inositol, phosphoric acid and various monosaccharides (galactose, arabinose, mannose, glucosamine, glucuronic acid).

$$\text{D-Glc}p\text{-}\beta\text{-}(1\rightarrow4)\text{-D-Man}p\text{-}\beta\text{-}(1\rightarrow4)\text{-D-Man}p\text{-}\beta\text{-}(1\rightarrow4)\text{-D-Glc}p\text{-}\beta\text{-}(1\rightarrow1)\text{-}O\text{-}CH_2\text{-}\underset{}{CH}(NH\text{-}CO\text{-}CH(OH)\text{-}(CH_2)_{21}\text{-}CH_3)\text{-}\underset{OH}{CH}\text{-}\underset{OH}{CH}\text{-}(CH_2)_{13}\text{-}CH_3$$

(3.39)

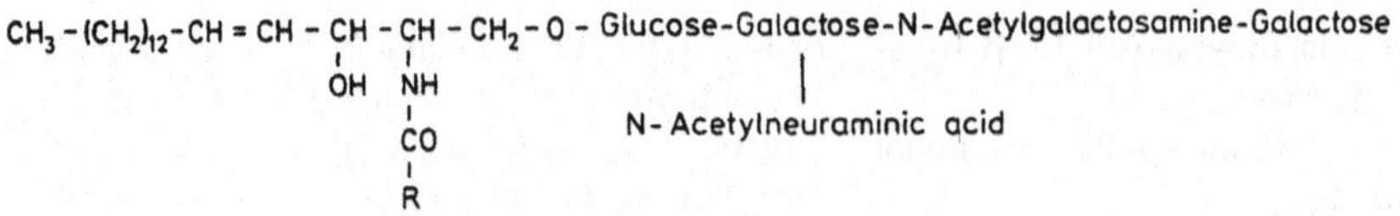

$$CH_3-(CH_2)_{12}-CH=CH-CH-CH-CH_2-O-Glucose-Galactose-N-Acetylgalactosamine-Galactose$$

with substituents OH, NH below, then CO, then R; and N-Acetylneuraminic acid branching.

(3.40a)

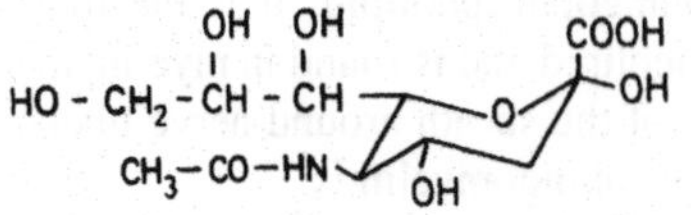

(3.40b)

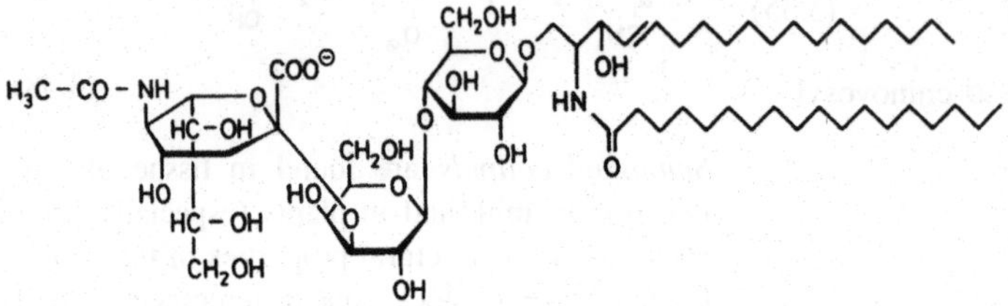

(3.41)

$$CH_3-(CH_2)_{13}-CH-CH-CH-CH_2-O-P-O-Inositol-Glucuronic\ acid-Glucosamine$$

with OH, OH, NH; CO; R$_1$; and $O^\ominus$ $X^\oplus$; bracket to Galactose, Arabinose, Mannose.

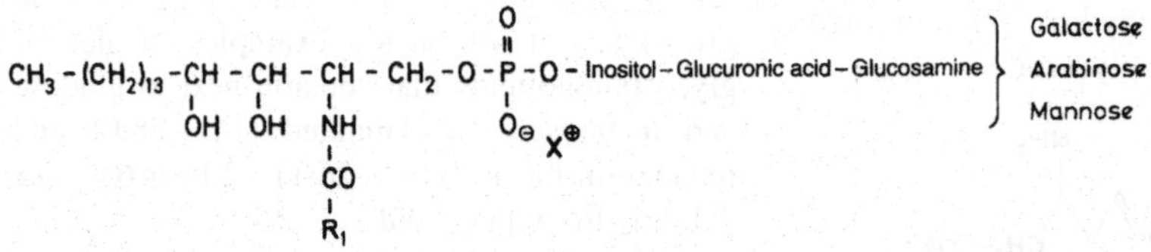

(3.42)

Phytosphingolipids are isolated from soya and peanuts (cf. Formula 3.42).

3.4.2 Analysis

3.4.2.1 Extraction, Removal of Nonlipids

A solvent mixture of chloroform/methanol $(2 + 1\,v/v)$ is suitable for a quantitative extraction of lipids. Addition of a small amount of BHA (cf. 3.7.3.2.2) is recommended for the stabilization of lipids against autoxidation.

Nonlipid impurities were earlier removed by shaking the extracts with a special salt solution under rather demanding conditions. An improved procedure, by which emulsion formation is avoided, is based on column chromatography with dextran gels.

3.4.2.2 Separation and Identification of Classes of Components

Isolated and purified lipids can be separated into classes by thin layer chromatography using developing solvents of different polarity. Figure 3.12 shows an example of the separation of neutral and polar lipids. Identification of polar lipids is based on using spraying reagents which react on the plates with the polar moiety of the lipid molecules. For example, phosphoric acid is identified by the molybdenum-blue reaction, monosaccharides by orcinol-FeCl$_3$, choline by bismuth iodide (*Dragendorff* reagent), ethanolamine and serine by the ninhydrin reaction, and sphingosine by a chlorine-benzidine reagent.

When sufficient material is available, it is advisable to perform a preliminary separation of lipids by column chromatography on magnesium sili-

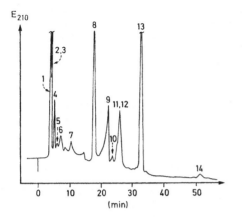

E_{210}

Fig. 3.11. HPLC-analysis of soy "raw lethicin" (according to *Sotirhos* et al. 1986) 1 Triacylglycerols, 2 free fatty acids, 3 phosphatidyl glycerol, 4 cerebrosides, 5 phytosphingosine, 6 diphosphatidyl glycerol, 7 digalactosyldiacyl glycerol, 8 phosphatidyl ethanolamine, 9 phosphatidyl inositol, 10 lysophosphatidyl ethanolamine, 11 phosphatidic acid, 12 phosphatidyl serine, 13 phosphatidyl choline, 14 lysophosphatidyl choline

cate (florisil), silicic acid, hydrophobic dextran gel or a cellulose-based ion-exchanger, such as DEAE-cellulose.

Also the HPLC-analysis of phospho- and glycolipids is of growing importance. Figure 3.11 for example demonstrates the separation of soya "raw lecithin".

3.4.2.3 Analysis of Lipid Components

Fatty acid composition is determined after methanolysis of the lipid. For positional analysis of acyl residues (positions 1 or 2 in glycerol), phosphatidyl derivatives are selectively hydrolyzed with phospholipases (cf. 3.7.1.2.1) and the fatty acids liberated are analyzed by gas chromatography.

The sphingosine base can also be determined by gas chromatography after trimethylsilyl derivatization. The length of the carbon skeleton, of interest for phytosphingosine, can be determined by analyzing the aldehydes released after the chain has been cleaved by periodate:

$$CH_3-(CH_2)_n-CH-CH-CH-CH_2-OH$$
$$\quad\quad\quad\quad\quad\quad OH\ OH\ NH_2$$

$$\downarrow\ IO_4^{\ominus}$$

$$CH_3-(CH_2)_n-CHO\ +\ NH_3\ +\ 3\,CH_2O \tag{3.43}$$

The monosaccharides in glycolipids can also be determined by gas chromatography. The lipids are hydrolyzed with trifluoroacetic acid and then derivatized to an acetylated glyconic acid nitrile. By using this sugar derivative, the chromatogram is simplified because of the absence of sugar anomers (cf. 4.2.4.6).

3.5 Lipoproteins, Membranes

3.5.1 Lipoproteins

3.5.1.1 Definition

Lipoproteins are aggregates, consisting of proteins, polar lipids and triacylglycerols, which are water soluble and can be separated into protein and lipid moieties by an extraction procedure using suitable solvents. This indicates that only noncovalent bonds are involved in the formation of lipoproteins. The aggregates are primarily stabilized by hydrophobic interactions between the apolar side chains of hydrophobic regions of the protein and the acyl residues of the lipid. In addition, there is a contribution to stability by ionic forces between charged amino acid residues and charges carried by the phosphatides. Hydrogen bonds, important for stabilization of the secondary structure of protein, play a small role in binding lipids since phosphatidyl derivatives have only a few sites available for such linkages. Hydrogen bonds could exist to a greater extent between proteins and glycolipids; however, such lipids have not yet been found as lipoprotein components, but rather as building blocks of biological membranes. An exception may be their occurrence in wheat flour, where they are responsible for gluten stability of dough. Here, the lipoprotein complex consists of prolamine and glutelin attached to glycolipids by hydrogen bonds and hydrophobic forces.

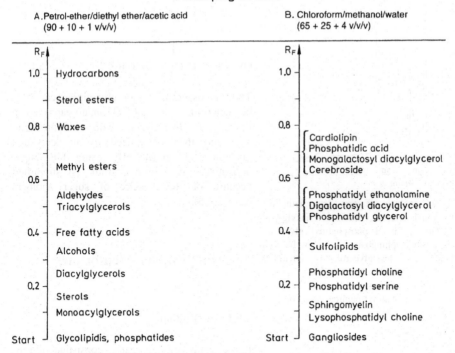

Fig. 3.12. Separation of lipid classes by thin layer chromatography using silicagel as an adsorbent. R_f values in two solvent systems

Thus, lipoproteins are held together only by non-covalent bonds.

3.5.1.2 Classification

Lipoproteins exist as globular particles in an aqueus medium. They are solubilized from biological sources by buffers with high ionic strength, by a change of pH or by detergents in the isolating medium. The latter, a more drastic approach, is usually used in the recovery of lipoproteins from membranes.

Lipoproteins are characterized by ultracentrifugation. Since lipids have a lower density (0.88–0.9 g/ml) than proteins (1.3–1.35 g/ml), the separation is possible because of differences in the ratios of lipid to protein within a lipoprotein complex. The lipoproteins of blood plasma have been thoroughly studied. They are separated by a stepwise centrifugation in solutions of NaCl into three fractions with different densities (Fig. 3.13). The "very low density lipoproteins" (VLDL; density <1.006 g/ml), the "low density lipoproteins" (LDL; 1.063 g/ml) and the "high density lipoproteins" (HDL; 1.21 g/ml) float, and the sediment contains the plasma proteins. The VLDL fraction can be separated further by electrophoresis into chylomicrons (the lightest lipoprotein, density <1.000 g/ml) and pre-β-lipoprotein.

Lipoproteins in the LDL fraction from an electrophoretic run have a mobility close to that of blood plasma β-globulin. Therefore, the LDL fraction is denoted as β-lipoprotein. An analogous designation of α-lipoprotein is assigned to the HDL fraction.

Chylomicrons, the diameters of which range from 1000–10,000 Å, are small droplets of triacylglycerol stabilized in the aqueous medium by a membrane-like structure composed of protein, phosphatides and cholesterol. The role of chylomicrons in blood is to transport triacylglycerols to various organs, but preferentially from the intestines to adipose tissue and the liver. The milk fat globules (cf. 10.1.2.3) have a structure

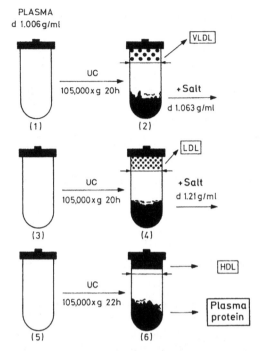

PLASMA
d 1.006 g/ml

UC
105,000 × g 20h

+ Salt
d 1.063 g/ml

VLDL

(1)   (2)

UC
105,000 × g 20h

+ Salt
d 1.21 g/ml

LDL

(3)   (4)

UC
105,000 × g 22h

HDL

Plasma
protein

(5)   (6)

Fig. 3.13. Plasma protein fractionation by a preparative ultracentrifugation (UC) method (according to *Seidel*, 1971)

similar to that of chylomicrons. The composition of some lipoproteins is presented in Table 3.20.

Certain diseases related to fat metabolism (hyperlipidemias) can be clinically diagnosed by the content and composition of the plasma lipoprotein fractions.

Electron microscopy studies have revealed that the fat globules in milk have small particles attached to their membranes; these are detached by detergents and have been identified as LDL (cf. Table 3.20).

3.5.2 Involvement of Lipids in the Formation of Biological Membranes

Membranes that compartmentalize the cells and many subcellular particles are formed from two main building blocks: proteins and lipids (phospholipids and cholesterol). Differences in membrane structure and function are reflected by the compositional differences of membrane proteins and lipids (see examples in Table 3.18).

Studies of membrane structure are difficult since the methods for isolation and purification profoundly change the organization and functionality of the membrane.

Model membranes are readily formed. The major forces in such events are the hydrophobic interactions between the acyl tails of phospholipids, providing a bilayer arrangement. In addition, the amphipathic character of the lipid molecules makes membrane formation a spontaneous process. The acyl residues are sequestered and oriented in the nonpolar interior of the bilayer, whereas the polar hydrophilic head groups are oriented toward the outer aqueous phase.

Another arrangement in water that satisfies both the hydrophobic acyl tails and the hydrophilic polar groups is a globular micelle. Here, thehydrocarbon tails are sequestered inside, while the polar groups are on the surface of the sphere. There is no bilayer in this arrangement.

Table 3.20. Composition of typical lipoproteins

Source	Lipoprotein	Particle weight (kdal)	Protein (%)	Glycerophospholipids (%)	Cholesterol free (%)	Cholesterol esterified (%)	Triacylglycerols (%)
Human blood serum	Chylomicron	10^9–10^{10}	1–2	4	2.5–3	3–4	85–90
	Pre-β-lipoprotein	5–$100 \cdot 10^6$	8.3	19.2	7.4	11.1	54.2
	LDL (β-lipoprotein)	$2.3 \cdot 10^6$	22.7	27.9	8.5	28.8	10.5
	HDL (α-lipoprotein)	1–$4 \cdot 10^5$	58.1	24.7	2.9	9.2	5.9
Egg yolk (chicken)	β-Lipovitellin	$4 \cdot 10^5$	78	12	0.9	0.1	9
	LDL	2–$10 \cdot 10^6$	18	22	3.5	0.2	58
Bovine milk	LDL	$3.9 \cdot 10^6$	12.9	52	0	0	35.1

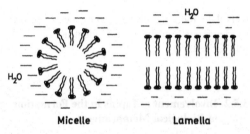

Fig. 3.14. Arrangement of polar acyl lipids in aqueous medium. ∞ Polar lipid tails; ≈ hydrophobic lipid tails

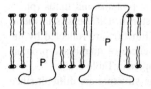

Fig. 3.15. Fluid mosaic model of a biological membrane. The protein (P) is not fixed but is mobile in the phospholipid phase

The favored structure for most phospho- and glycolipids in water is a bimolecular arrangement, rather than a micelle. Two model systems can exist for such bimolecular arrangements. The first is a lipid vesicle, known as a liposome, the core of which is an aqueous compartment surrounded by a lipid bilayer, and the second is a planar, bilayer membrane. The latter, together with the micellar model, is presented in Fig. 3.14. Globular proteins, often including enzymes, are found in animal cell membranes and are well embedded or inserted into the bimolecular layer. Some of these so-called integral membrane proteins protrude through both sides of the membrane (fluid mosaic model, Fig. 3.15). Although integral proteins interact extensively with the hydrophobic acyl tails of membrane lipids they are mobile within the lipid membrane.

3.6 Diol Lipids, Higher Alcohols, Waxes and Cutin

3.6.1 Diol Lipids

The diol lipids which occur in both plant and animal tissues are minor lipid constituents. The diol content is about 1% of the content of glycerol. Exceptions are sea stars, sea urchins and mollusks, the lipids of which in summer contain 25–40% diol lipids. This proportion decreases sharply in winter and spring. Neutral and polar lipids derived from ethylene glycol, propane-(1,2 and 1,3)-diol and butane-(1,3; 1,4- and 2,3)-diol have been identified in the diol lipid fraction. Several of those isolated from corn oil have the following structures:

$$
\begin{array}{ll}
\text{H}_2\text{C}-\text{O}-\text{CO}-\text{R} & \qquad \text{H}_3\text{C} \\
\text{H}_2\text{C}-\text{O}-\text{CO}-\text{R} & \text{HC}-\text{O}-\text{CO}-\text{R} \\
& \text{HC}-\text{O}-\text{CO}-\text{R} \\
& \qquad \text{H}_3\text{C} \qquad\qquad (3.44)
\end{array}
$$

$$
(3.45)
$$

In a glycodiol lipid one hydroxyl group of ethylenediol is esterified with a fatty acid.
Diol lipids with structures analogous to phosphatidyl choline or plasmalogen have also been identified.

3.6.2 Higher Alcohols and Derivatives

3.6.2.1 Waxes

Higher alcohols occur either free or bound in plant and animal tissues. Free higher alcohols are abundant in fish oil and include:

Cetyl alcohol	$C_{16}H_{33}OH$
Stearyl alcohol	$C_{18}H_{37}OH$
Oleyl alcohol	$C_{18}H_{35}OH$

Waxes are important derivatives of higher alcohols. They are higher alcohols esterified with long-chain fatty acids. Plant waxes are usually found on leaves or seeds. Thus, cabbage leaf wax consists of the primary alcohols C_{12} and C_{18}–C_{28} esterified with palmitic acid and other acids. The dominant components are stearyl and ceryl alcohol ($C_{26}H_{53}OH$). In addition to primary alcohols, esters of secondary alcohols, e. g., esters of nonacosane-15-ol, are present:

$$
\text{H}_3\text{C}-(\text{CH}_2)_{13}-\underset{\underset{\text{OH}}{|}}{\text{CH}}-(\text{CH}_2)_{13}-\text{CH}_3
$$

$$
(3.46)
$$

The role of waxes is to protect the surface of plant leaves, stems and seeds from dehydration and infections by microorganisms. Waxes are removed together with oils by solvent extraction of nondehulled seeds. Waxes are oil-soluble at elevated temperatures but crystallize at room temperature, causing undesired oil turbidity. Ceryl cerotate (ceryl alcohol esterified with cerotic acid, $C_{25}H_{51}COOH$)

$$H_3C-(CH_2)_{24}-CO-O-CH_2-(CH_2)_{24}-CH_3 \quad (3.47)$$

is removed from seed hulls during extraction of sunflower oil. Waxes are removed by an oil refining winterization step during the production of clear edible oil.

Waxes are also components of the mass used to cover fruit to protect it from drying out.

Waxes are present in fish oils, especially in sperm whale blubber and whale head oil, which contain a "reservoir" of spermaceti wax.

3.6.2.2 Alkoxy Lipids

The higher alcohols, 16:0, 18:0 and 18:1 (9), form mono- and diethers with glycerol. Such alkoxy-lipids are widely distributed in small amounts in mammals and sea animals. Examples of confirmed structures are shown in Formula 3.48. The elucidation of ether lipid structure is usually accomplished by cleavage by concentrated HI at elevated temperatures.

Alkyl glycerol ether   1,2-Dialkyl-glycerol ether

O-Alkyl-diacylglycerol   O,O-Dialkyl-monoacylglycerol

(3.48)

Common names of some deacylated alkoxy lipids (1-O-alkylglycerol) are the following:

$$CH_3-(CH_2)_{14}-CH_2-O-CH_2-CH-CH_2$$
$$\qquad\qquad\qquad\qquad OH\ \ OH$$

Chimyl alcohol   (3.49)

$$CH_3-(CH_2)_{16}-CH_2-O-CH_2-CH-CH_2$$
$$\qquad\qquad\qquad\qquad OH\ \ OH$$

Batyl alcohol   (3.50)

$$CH_3-(CH_2)_7-CH=CH-(CH_2)_7-CH_2-O-CH_2-CH-CH_2$$
$$\qquad\qquad\qquad\qquad\qquad\qquad\qquad OH\ \ OH$$

Selachyl alcohol   (3.51)

3.6.3 Cutin

Plant epidermal cells are protected by a suberized or waxy cuticle. An additional layer of epicuticular waxes is deposited above the cuticle in many plants. The waxy cuticle consists of cutin. This is a complex, high molecular weight polyester which is readily solubilized in alkali. The structural units of the polymer are hydroxy fatty acids. The latter are similar in structure to the compounds given in 3.7.2.4.1. A segment of the postulated structure of cutin is presented in Fig. 3.16.

3.7 Changes in Acyl Lipids of Food

3.7.1 Enzymatic Hydrolysis

Hydrolases, which cleave acyl lipids, are present in food and microorganisms. The release of short-chain fatty acids ($<C_{14}$), e.g., in the hydrolysis of milk fat, has a direct effect on food aroma. Lipolysis is undesirable in fresh milk

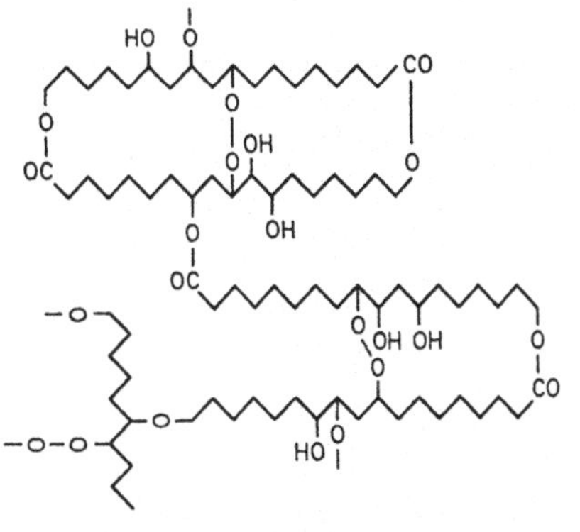

Fig. 3.16. A structural segment of cutin (according to *Hitchcock* and *Nichols*, 1971)

since the free C_4–C_{12} fatty acids (cf. Tables 3.3 to 3.5 for odor threshold values) are responsible for the rancid aroma defect. On the other hand, lipolysis occurring during the ripening of cheese is a desired and favorable process because the short-chain fatty acids are involved in the build-up of specific cheese aromas. Likewise, slight hydrolysis of milk fat is advantageous in the production of chocolate.

Linoleic and linoleiic acid released by hydrolysis and present in emulsified form affect the flavor of food even at low concentrations. They cause a bitter-burning sensation (cf. Table 3.9). In addition, they decompose by autoxidation (cf. 3.7.2.1) or enzymatic oxidation (cf. 3.7.2.2) into compounds with an intensive odor. In fruits and vegetables enzymatic oxidation in conjunction with lipolysis occur, as a rule, at a high reaction rate, especially when tissue is sliced or homogenized (an example for rapid lipolysis is shown in Table 3.21). Also, enzymatic hydrolysis of a small amount of the acyl lipids present can not be avoided during disintegration of oil seeds. Since the release of higher fatty acids promotes foaming, they are removed during oil refining (cf. 14.4.1).

Enzymes with lipolytic activity belong to the carboxyl-ester hydrolase group of enzymes (cf. 2.2.6).

3.7.1.1 Triacylglycerol Hydrolases (Lipases)

Lipases (cf 2.2.6) hydrolyze only emulsified acyl lipids; they are active on a water/lipid interface.

Lipases differ from esterase enzymes since the latter cleave only water-soluble esters, such as triacetylglycerol.

Lipase activity is detected, for example, in milk, oilseeds (soybean, peanut), cereals (oats, wheat), fruits and vegetables and in the digestive tract of mammals. Many microorganisms release lipase-type enzymes into their culture media.

As to their specificity, lipases are distinguished according to the criteria presented in Table 3.22. The lipase secreted by the swine pancreas has been the most studied. Its molecular weight is $M_r = 48,000$. The enzyme cleaves the following types of acyl glycerols with a decreasing rate of hydrolysis: triacyl- >diacyl- ≫monoacylglycerols. Table 3.22 shows that pancreatic lipase reacts with acyl residues at positions 1 and 3. The third acyl residue of a triacylglycerol is cleaved (cf. Reaction 3.52) only after acyl migration, which requires a longer incubation time.

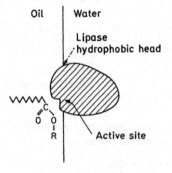

$$(3.52)$$

The smaller the size of the oil droplet, the larger the oil/water interface and, therefore, the higher the lipase activity. This relationship should not

Table 3.21. Lipid hydrolysis occurring during potato tuber homogenization

	μmoles/g[a]	
	Acyl lipids	Free fatty acids
Potato	2.34	0.70
Homogenate[b]	2.04	1.40
Homogenate[b] kept for 10 min at 0 °C	1.72	1.75
Homogenate kept for 10 min at 25 °C	0.54	2.90

[a] Potato tissue fresh weight.
[b] Sliced potatoes were homogenized for 30 sec at 0 °C.

Oil | Water

Fig. 3.17. A hypothetical model of pancreatic lipase fixation of an oil/water interphase (according to *Brockerhoff*, 1974)

Table 3.22. Examples of the specificity of lipases

Specificity	Lipase from
Substrate specific	
Monoacylglycerides	Rats (adipose tissue)
Mono- and diacyl-glycerides	*Penicillium camembertii*
Triacylglycerides	*Penicillium* sp.
Regiospecific	
1,3-Regioselective	Pancreas, milk, *Aspergillus niger*
sn-2-Regioselective	*Candida antarctia*
Non-regiospecific	
	Oats, castor, *Aspergillus flavus*
Acyl residue-specific	
Short chain fatty acids	*Penicillium roqueforti*
cis-9-Unsaturated fatty acids	*Geotrichum candidum*
Long chain fatty acids	*Botrytis cinerea*
Stereospecific[a]	
sn-1	*Pseudomonas neruginosa*
sn-3	Rabbit (digestive tract)

[a] Lipases differentiate between the sn-1 and sn-3 position in TGs.

be ignored when substrate emulsions are prepared for the assay of enzyme activities.

A model for pancreatic lipase has been suggested to account for the enzyme's activity on the oil/water interface (Fig. 3.17). The lipase's "hydrophobic head" is bound to the oil droplet by hydrophobic interactions, while the enzyme's active site aligns with and binds to the substrate molecule. The active site resembles that of serine proteinase. The splitting of the ester bond occurs with the involvement of Ser, His and Asp residues on the enzyme by a mechanism analogous to that of chymotrypsin (cf. 2.4.2.5). The dissimilarity between pancreatic lipase and serine proteinase is in the active site: lipase has a leucine residue within this site in order to establish hydrophobic contact with the lipid substrate and to align it with the activity center.

Lipase-catalyzed reactions are accelerated by Ca^{2+} ions since the liberated fatty acids are precipitated as insoluble Ca-salts.

The properties of milk lipase closely resemble those of pancreatic lipase.

Lipases of microbiological origin are often very heat stable. As can be seen from the exam-

ple of a lipase of *Pseudomonas fluorescence* (Table 3.23), such lipases are not inactivated by pasteurization, ultra high temperature treatment, as well as drying procedures, e. g., the production of dry milk. These lipases can be the cause of decrease in quality of such products during storage.

A lipase of microbial origin has been detected which hydrolyzes fatty acids only when they have a cis-double bond in position 9 (Table 3.22). It is used to elucidate triacylglyceride structure. The use of lipases in food processing was outlined under 2.7.2.2.14.

Lipase activities in foods can be measured very sensitively with fluorochromic substrates, e. g., 4-methyl umbelliferyl fatty acid esters. Of course it is not possible to predict the storage stability of a food item with regard to lipolysis based only on such measurements. The substrate specificity of the lipases, which can vary widely as shown in Table 3.22, is of essential importance for the aroma quality. Therefore, individual fatty acids can increase in different amounts even at the same lipase activity measured against a standard substrate. Since the odor and taste threshold values of the fatty acids differ greatly (cf. Tables 3.3–3.5), the effects of the lipases on the aroma are very variable. It is not directly possible to predict the point of time when rancid aroma notes will be present from the determination of the lipase activity. More precise information about the changes to be expected is obtained through storage experiments during which the fatty acids are quantitatively determined by gas chromatographic analysis. Table 3.24 shows the change in the concentrations of free fatty acids in sweet cream butter together with the resulting rancid aroma notes.

Table 3.23. Heat inactivation of a lipase of *Pseudomonas fluorescence* dissolved in skim milk

Temperature °C	D-value[a] (min)
100	23.5
120	7.3
140	2.0
160	0.7

[a] Time for 90% decrease in enzyme activity (cf. 2.5.4.1).

Table 3.24. Free fatty acids in butter (sweet cream) samples of different quality

Fatty acid	Butter A (mg/kg)	B	C	D	E
4:0	0	5	38	78	119
6:0	0	4	28	25	46
8:0	8	22	51	51	86
10:0	38	58	104	136	229
12:0	78	59	142	137	231
14:0	193	152	283	170	477
Aroma[a]	2.3	2.8	3.0	4.6	5.4

[a] Classification: 2 not rancid, 3 slightly rancid, 4 rancid, 5 very rancid.

3.7.1.2 Polar-Lipid Hydrolases

These enzymes are denoted as phospholipases, lysophospholipases or glycolipid hydrolases, depending on the substrate.

3.7.1.2.1 Phospholipases

Phospholipase A_1. The enzyme is present together with phospholipase A_2 in many mammals and bacteria. It cleaves specifically the sn-1 ester bonds of diacylphosphatides (Formula 3.53).

Phospholipase A_2. Enzymes with sn-2 specificity isolated form snake and bee venoms. They are very stable, are activated by Ca^{2+}-ions and are amongst the smallest enzyme molecules (molecular weight about 14,000).

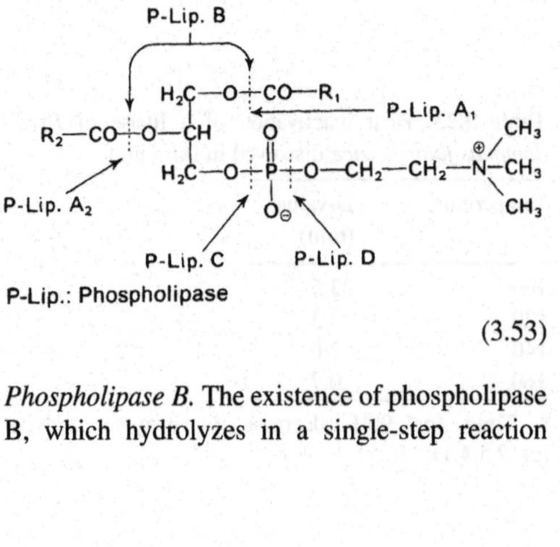

P-Lip.: Phospholipase

$$(3.53)$$

Phospholipase B. The existence of phospholipase B, which hydrolyzes in a single-step reaction

both acyl groups in diacylphosphatides, is controversial. Other than the phospholipases A_1, A_2, C and D, the B-type could not be isolated in its pure form. A phospholipase B has been enriched from germinating barley. However, the B-specificity appears to be only a secondary activity because the enzyme hydrolyzes the acyl residue of lysolecithin considerably faster than the acyl residues of lecithin.

Phospholipase C. It hydrolyzes lecithin to a 1,2-diacylglyceride and phosphoryl choline. The enzyme is found in snake venom and in bacteria.

Phospholipase D. This enzyme cleaves the choline group in the presence of water or an alcohol, such as methanol, ethanol or glycerol, yielding free or esterified phosphatidic acid. For example:

Phosphatidylcholine $+ ROH$

$\longrightarrow$ Phosphatidyl-OR $+$ Choline

$R : H, CH_3, CH_2, CH_2(OH) - CH(OH) - CH_2$

$$(3.54)$$

Phospholipase D cannot cleave phosphatidyl inositol. The enzyme is present in cereals, such as rye and wheat, and in legumes. It was isolated and purified from peanuts.

Lysophospholipases. The enzymes, hydrolyzing only lysophosphatides, are abundant in animal tissue and bacteria. There are lysophospholipases that split preferentially 1-acylphosphatides while others prefer 2-acylphosphatides, and a third group doesn't differentiate at all between the two lysophosphatide types.

3.7.1.2.2 Glycolipid Hydrolases

Enzymes that cleave the acyl residues of mono- and digalactosyl-diacylglycerides are localized in green plants. A substrate specificity study for such a hydrolase from potato (Table 3.25) shows that plants also contain enzymes that are able to hydrolyze polar lipids in general. The potato enzyme preferentially cleaves the acyl residue from monoacylglycerols and lysolecithin, whereas triacylglycerols, such as triolein, are not affected.

Table 3.25. Purified potato acyl hydrolase: substrate specificity

Substrate	Relative activity (%)	Substrate	Relative activity (%)
Monolein	100	Lecithin	13
Diolein	21	Monogalactosyl-	
Triolein	0.2	diacylglycerol	31
Methyloleate	28	Digalactosyl-	
Lysolecithin	72	diacylglycerol	17

3.7.2 Peroxidation of Unsaturated Acyl Lipids

Acyl lipid constituents, such as oleic, linoleic and linolenic acids, have one or more allyl groups within the fatty acid molecule (cf. Table 3.7) and thus are readily oxidized to hydroperoxides. The latter, after subsequent degradation reaction, yield a great number of other compounds. Therefore, under the usual conditions of food storage, unsaturated acyl lipids cannot be considered as stable food constituents.

Autoxidation should be distinguished from *lipoxygenase catalysis* in the process denoted as *lipid peroxidation*. Both oxidations provide hydroperoxides, but the latter occurs only in the presence of the enzyme.

Lipid peroxidation provides numerous volatile and nonvolatile compounds. Since some of the volatiles are exceptionally odorous compounds, lipid peroxidation is detected even in food with unsaturated acyl lipids present as minor constituents, or in food in which only a small portion of lipid was subjected to oxidation.

Induced changes in food aroma are frequently assessed by consumers as objectionable, for example, as rancid, fishy, metallic or cardboardlike, or as an undefined old or stale flavor. On the other hand, the fact that some volatile compounds, at a level below their off-flavor threshold values, contribute to the pleasant aroma of many fruits and vegetables and to rounding-off the aroma of many fator oil-containing foods should not be neglected.

3.7.2.1 Autoxidation

Autoxidation is quite complex and involves a great number of interrelated reactions of intermediates. Hence, autoxidation of food is usually imitated by the study of a model system in which, for example, changes of one unsaturated fatty acid or one of its intermediary oxidation products are recorded in the presence of oxygen under controlled experimental conditions.

Model system studies have revealed that the rate of autoxidation is affected by fatty acid composition, degree of unsaturation, the presence and activity of pro- and antioxidants, partial pressure of oxygen, the nature of the surface being exposed to oxygen and the storage conditions (temperature, light exposure, moisture content, etc.) of fat/oil-containing food. The position of the unsaturated fatty acid in the triacylglyceride molecule also influences the rate of autoxidation. TGs with an unsaturated fatty acid in the 1- or 3-position oxidize faster than TGs with an unsaturated acyl residue in the more protected 2-position.

The oxygen uptake of an unsaturated fatty acid as a function of time is shown in Fig. 3.18. Studying this figure helps in the understanding of the elementary steps involved in autoxidation. The extreme case 1 demonstrates what has invariably been found in food: the initial oxidation products are detectable only after a certain elapsed storage time. When this *induction period*, which is typical for a given autoxidation process, has expired, a steep rise occurs in the reaction rate. The proox-

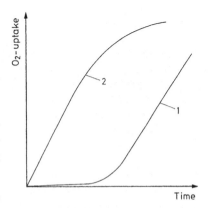

Fig. 3.18. Autoxidation of unsaturated acyl lipids. Prooxidant concentration: 1 low, 2 high

Table 3.26. Induction period and relative rate of oxidation for fatty acids at 25 °C

Fatty acid	Number of allyl groups	Induction period (h)	Oxidation rate (relative)
18:0	0		1
18:1 (9)	1	82	100
18:2 (9, 12)	2	19	1,200
18:3 (9, 12, 15)	3	1.34	2500

idant concentration is high in some foods. In these cases, illustrated in Fig. 3.18-2, the induction period may be nonexistent.

3.7.2.1.1 Fundamental Steps of Autoxidation

The length of the induction period and the rate of oxidation depend, among other things, on the fatty acid composition of the lipid (Table 3.26); the more allyl groups present, the shorter the induction period and the higher the oxidation rate.
Both phenomena, the induction period and the rise in reaction rate in the series, oleic, linoleic and linolenic acid can be explained as follows: Oxidation proceeds by a sequential free radical chain-reaction mechanism. Relatively stable radicals that can abstract H-atoms from the activated methylene groups in an olefinic compound are formed. On the basis of this assumption and, in addition, on the fact that the oxidation rate is exponential, *Farmer* et al. (1942) and *Bolland* (1949) proposed an autoxidation mechanism for olefinic compounds and, thus, also for unsaturated fatty acids. This mechanism has several fundamental steps. As shown in Fig. 3.19, the oxidation process is essentially a radical-induced chain reaction divided into initiation (start), propagation, branching and termination steps. Autoxidation is initiated by free radicals of frequently unknown origin.
Measured and calculated reaction rate constants for the different steps of the radical chain reaction show that due to the stability of the peroxy free radicals (ROO$^\bullet$), the whole process is limited by the conversion of these free radicals into monohydroperoxide molecules (ROOH). This reaction is achieved by abstraction of an H-atom

from a fatty acid molecule [reaction step 2 (RS-2 in Fig. 3.19)]. The H-abstraction is the slowest and, hence, the rate limiting step in radical (R$^\bullet$) formation. Peroxidation of unsaturated fatty acids is accelerated autocatalytically by radicals generated from the degradation of hydroperoxides by a monomolecular reaction mechanism (RS-4 in Fig. 3.19). This reaction is promoted by heavy metal ions or heme(in)-containing molecules (cf. 3.7.2.1.7). Also, degradation of hydro-peroxides is considered to be a starting point for the formation of volatile reaction products (cf. 3.7.2.1.9).
After a while, the hydroperoxide concentration reaches a level at which it begins to generate free radicals by a bimolecular degradation mechanism (RS-5 in Fig. 3.19). Reaction RS-5 is exothermic, unlike the endothermic monomolecular decomposition of hydroperoxides (RS-4 in Fig. 3.19) which needs approx. 150 kJ/mol. However, in most foods, RS-5 is of no relevance since fat (oil) oxidation makes a food unpalatable well before reaching the necessary hydroperoxide level for the RS-5 reaction step to occur. RS-4 and RS-5 (Fig. 3.19) are the branching reactions of the free radical chain.

Start: Formation of peroxy (RO$_2^\bullet$), alkoxy (RO$^\bullet$) or alkyl (R$^\bullet$) radicals

Chain propagation:

(1) R$^\bullet$ + O$_2$ ⟶ RO$_2^\bullet$ k_1: 10^9 l mol^{-1} s^{-1}

(2) RO$_2^\bullet$ + RH ⟶ ROOH + R$^\bullet$ k_2: 10–60 l mol^{-1} s^{-1}

(3) RO$^\bullet$ + RH ⟶ ROH + R$^\bullet$

Chain branching:

(4) ROOH ⟶ RO$^\bullet$ + $^\bullet$OH

(5) 2 ROOH ⟶ RO$_2^\bullet$ + RO$^\bullet$ + H$_2$O

Chain termination:

(6) 2 R$^\bullet$ ⟶ ⎫

(7) R$^\bullet$ + RO$_2^\bullet$ ⟶ ⎬ Stable products

(8) 2 RO$_2^\bullet$ ⟶ ⎭

Fig. 3.19. Basic steps in the autoxidation of olefins

At room temperature, a radical may inititiate the formation of 100 hydroperoxide molecules before chain termination occurs. In the presence of air (oxygen partial pressure >130 mbar), all alkyl radicals are transformed into peroxy radicals through the rapid radical chain reaction 1 (RS-1, Fig. 3.19). Therefore, chain termination occurs through collision of two peroxy radicals (RS-8, Fig. 3.19).

Termination reactions RS-6 and RS-7 in Fig. 3.19 play a role when, for example, the oxygen level is low, e. g. in the inner portion of a fatty food.

The hypothesis presented in Fig. 3.19 is valid only for the initiation phase of autoxidation. The process becomes less and less clear with increasing reaction time since, in addition to hydroperoxides, secondary products appear that partially autoxidize into tertiary products. The stage at which the process starts to become difficult to survey depends on the stability of the primary products. It is instructive here to compare the difference in the structures of monohydroperoxides derived from linoleic and linolenic acids.

3.7.2.1.2 Monohydroperoxides

The peroxy radical formed in RS-1 (Fig. 3.19) is slow reacting and therefore it selectively abstracts the most weakly bound H-atom from a fat molecule. It differs in this property from, for example, the substantially more reactive hydroxy (HO•) and alkoxy(RO•) radicals (cf. 3.7.2.1.8). RS-2 in Fig. 3.19 has a high reaction rate only when the energy for H-abstraction is clearly lower than the energy released in binding H to O during formation of hydroperoxide groups (about 376 kJ mol^{-1}).

Table 3.27 lists the energy inputs needed for H-abstraction from the carbon chain segments or groups occurring in fatty acids. The peroxy radical abstracts hydrogen more readily from a methylene group of a 1,4-pentadiene system than from a single allyl group. In the former case, the 1,4-diene radical that is generated is more effectively stabilized by resonance, i. e. electron delocalization over 5 C-atoms. Such considerations explain the difference in rates of autoxidation for unsaturated fatty acids and show why, at room temperature, the unsaturated fatty

Table 3.27. Energy requirement for a H-atom abstraction

	D_{R-H}, (kJ/mole)
$\overset{\displaystyle H}{\overset{\displaystyle \mid}{CH_2-}}$	422
$\overset{\displaystyle H}{\overset{\displaystyle \mid}{CH_3-CH-}}$	410
$\overset{\displaystyle H}{\overset{\displaystyle \mid}{-CH-CH=CH-}}$	322
$\overset{\displaystyle H}{\overset{\displaystyle \mid}{-CH=CH-CH-CH-}}$	272

acids are attacked very selectively by peroxy radicals while the saturated acids are stable.

The general reaction steps shown in Fig. 3.19 are valid for all unsaturated fatty acids. In the case of oleic acid, H-atom abstraction occurs on the methylene group adjacent to the double bond, i. e. positions 8 and 11 (Fig. 3.20). This would give rise to four hydroperoxides. In reality, they have all been isolated and identified as autoxidation products of oleic acid. The configuration of the newly formed double bond of the hydroperoxides is affected by temperature. This configuration has 33% of cis and 67% of the more stable trans-configuration at room temperature.

Oxidation of the methylene group in position 11 of linoleic acid is activated especially by the two neighboring double bonds. Hence, this is the initial site for abstraction of an H-atom (Fig. 3.21). The pentadienyl radical generated is stabilized by formation of two hydroperoxides at positions 9 and 13, each retaining a conjugated diene system. These hydroperoxides have an UV maximum absorption at 235 nm and can be separated by high performance liquid chromatography as methyl esters, either directly or after reduction to hydroxydienes (Fig. 3.22).

The monoallylic groups in linoleic acid (positions 8 and 14 in the molecule), in addition to the bis-allylic group (position 11), also react to a small extent, giving rise to four hydroperoxides (8-, 10-, 12- and 14-OOH), each isomer having two isolated double bonds. The proportion of these minor monohydroperoxides is about 4% of the total (Table 3.28).

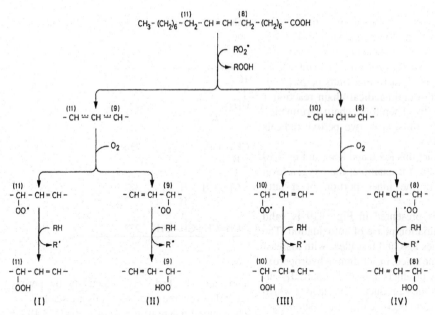

Fig. 3.20. Autoxidation of oleic acid. Primary reaction products: I 11-Hydroperoxyoctadec-9-enoic acid; II 9-hydroperoxyoctadec-10-enoic acid, III 10-hydroperoxyoctadec-8-enoic acid, IV 8-hydroperoxyoctadec-9-enoic acid

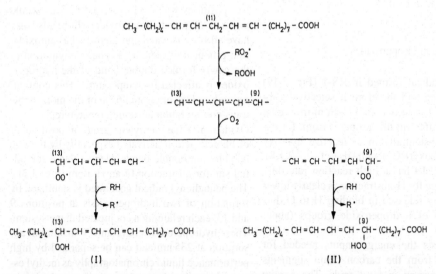

Fig. 3.21. Autoxidation of linoleic acid. Primary reaction products: I 13-Hydroperoxyoctadeca-9,11-dienoic acid, II 9-hydroperoxyoctadeca-10,12-dienoic acid

Autoxidation of linolenic acid yields four monohydroperoxides (Table 3.28). Formation of the monohydroperoxides is easily achieved by H-abstraction from the bis-allylic groups in positions 11 and 14. The resultant two pentadiene radicals then stabilize analogously to linoleic

acid oxidation (Fig. 3.21); each radical corresponds to two monohydroperoxides. However, the four isomers are not formed in equimolar amounts; the 9- and 16-isomers predominate (Table 3.28). The configuration of the conjugated double bonds again depends on the reaction

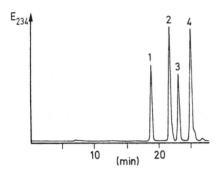

Fig. 3.22. Autoxidation of linoleic acid methyl ester. Analysis of primary products (after reduction of the hydroperoxy group) by HPLC (according to *Chan* and *Levett*, 1977). 1 13-Hydroxy-cis-9, trans-11-octadecadienoic acid methyl ester, 2 13-hydroxytrans-9,trans-11-octadecadienoic acid methyl ester, 3 9-hydroxy-trans-10, cis-12-octadecadienoic acid methyl ester, 4 9-hydroxy-trans-10, trans-12-octadecadienoic acid methyl ester

Table 3.28. Monohydroperoxides formed by autoxidation ($^3O_2$) and photooxidation ($^1O_2$) of unsaturated fatty acids

Fatty acid	Monohydroperoxide		Proportion (%)	
	Position of			
	HOO- group	double bond	$^3O_2$	$^1O_2$
Oleic acid	8	9	27	
	9	10	23	48
	10	8	23	52
	11	9	27	
Linoleic acid	8	9,12	1.5	
	9	10,12	46.5	32
	10	8,12	0.5	17
	12	9,13	0.5	17
	13	9,11	49.5	34
	14	9,12	1.5	
Linolenic acid	9	10,12,15	31	23
	10	8,12,15		13
	12	9,13,15	11	12
	13	9,11,15	12	14
	15	9,12,16		13
	16	9,12,14	46	25

conditions. Cis-hydroperoxides are the main products at temperatures <40 °C. Competition exists between conversion of the peroxy radical to monohydroperoxide and reactions involving β-fragmentation and cyclization. Allyl

peroxy radicals can undergo β-fragmentation which results, after a new oxygen molecule is attached, in a peroxy radical positional isomer, e.g. rearrangement of an oleic acid peroxy radical:

$$-\overset{(11)}{CH} - CH = \overset{(9)}{CH} - \quad \longleftrightarrow \quad -\overset{(11)}{CH} = CH - \overset{(9)}{CH} -$$
$$\underset{O-O\cdot}{|} \qquad\qquad\qquad\qquad \underset{\cdot O-O}{|}$$

$$-\overset{(10)}{CH} - CH = \overset{(8)}{CH} - \quad \longleftrightarrow \quad -\overset{(10)}{CH} = CH - \overset{(8)}{CH} -$$
$$\underset{O-O\cdot}{|} \qquad\qquad\qquad\qquad \underset{\cdot O-O}{|} \qquad\qquad (3.55)$$

However, hydroperoxides can also be isomerized by such a reaction pathway. When they interact with free radicals (H-abstraction from –OOH group) or with heavy metal ions (cf. Reaction 3.64), they are again transformed into peroxy radicals. Thus, the 13-hydroperoxide of linoleic acid isomerizes into the 9-isomer and vice versa:

$$-\overset{(13)}{CH}-CH{=}CH-CH{=}\overset{(9)}{CH}-$$
$$\underset{OOH}{|}$$

$$\text{RO}^\cdot$$
$$\text{ROH} \qquad O^\cdot$$

$$-\overset{}{CH}-CH{=}CH-CH{=}CH- \cdots O$$
$$\underset{OO^\cdot}{|}$$

$$\text{RH}$$
$$\text{R}^\cdot$$

$$-\overset{(13)}{CH}{=}CH-CH{=}CH-\overset{(9)}{CH}-$$
$$\underset{HOO}{|} \qquad\qquad\qquad (3.56)$$

3.7.2.1.3 Hydroperoxide-Epidioxides

Peroxy radicals which contain isolated β, γ double bonds are prone to cyclization reactions in competition with reactions leading to mono-hydroperoxides. A hydroperoxide-epidioxide results through attachment of a second oxygen molecule and abstraction of a hydrogen atom:

$$(3.57)$$

Peroxy radicals with isolated β,γ double bonds are formed as intermediary products after autoxidation and photooxidation (reaction with singlet O_2) of unsaturated fatty acids having two or more double bonds.

For this reason the 10- and 12-peroxy radicals obtained from linoleic acid readily form hydroperoxy-epidioxides. While such radicals are only minor products in autoxidation, in photooxidation they are generated as intermediary products in yields similar to the 9- and 13-peroxy radicals, which do not cyclize. Ring formation by 10- and 12-peroxy radicals decreases formation of the corresponding monohydroperoxides (Table 3.28; reaction with $^1O_2$).

Among the peroxy radicals of linolenic acid which are formed by autoxidation, the isolated β,γ double bond system exists only for the 12- and 13-isomers, and not for the 9- and 16-isomers. Also, the tendency of the 12- and 13-peroxy radicals of linolenic acid to form hydroperoxy-epidioxides results in the formation of less monohydroperoxide of the corresponding isomers as opposed to the 9- and 16-isomers (Table 3.28).

Peroxy radicals interact rapidly with antioxidants which may be present to give monohydroperoxides (cf. 3.7.3.1). Thus, it is not only the chain reaction which is inhibited by antioxidants, but also β-fragmentation and peroxy radical cyclization.

Fragmentation occurs when a hydroperoxideepidioxide is heated, resulting in formation of aldehydes and aldehydic acids. For example,

$$R^1 \diagup\diagdown\diagup\diagdown\diagup R^2$$
$$O - O^\bullet$$

$$\longrightarrow \quad R^1 \diagup\diagdown\diagup\diagdown\diagup R^2$$
$$O{\vdash}O \quad O{\vdash}OH$$

$$\longrightarrow \quad R^1 \diagup\diagdown\diagup CHO \; + \; OCH - R^2 \; + \cdots$$

$$R^1: CH_3-(CH_2)_3 \; ; \quad R^2: (CH_2)_7-COOH \qquad (3.58)$$

hydroperoxide-epidioxide fragments derived from the 12-peroxy radical of linoleic acid are formed as shown in Reaction 3.58.

Peroxy radicals formed from fatty acids with three or more double bonds can form bicy-

cloendoperoxides with an epidioxide radical as intermediate. This is illustrated in Reaction 3.74.

3.7.2.1.4 Initiation of a Radical Chain Reaction

Since autoxidation of unsaturated acyl lipids frequently results in deterioration of food quality, an effort is made to at least decrease the rate of this deterioration process. However, pertinent measures are only possible when more knowledge is acquired about the reactions involved during the induction period of autoxidation and how they trigger the start of autoxidation.

In recent decades model system studies have revealed that two fundamentally different groups of reactions are involved in initiating autoxidation. The first group is confined to the initiating reactions which overcome the energy barrier required for the reaction of molecular oxygen with an unsaturated fatty acid. The most important is photosensitized oxidation (photooxidation) which provides the "first" hydroperoxides. These hydroperoxides are then converted further into radicals in the second group of reactions. Heavy metal ions and heme(in) proteins are involved in this second reaction group. Some enzymes which generate the superoxide radical anion can be placed in between these two delineated reaction groups since at least H_2O_2 is necessary as reactant for the formation of radicals.

The following topics will be discussed here:

- Photooxidation
- Effect of heavy metal ions
- Heme(in) catalysis
- Activated oxygen from enzymatic reactions.

3.7.2.1.5 Photooxidation

In order to understand photooxidation and to differentiate it from autoxidation, the electronic configuration of the molecular orbital energy levels for oxygen should be known. As presented in Fig. 3.23, the allowed energy levels correspond to $^3\Sigma^-$ g, $^1\Delta$ g and $^1\Sigma^+$ g.

The notation for the molecular orbital of O_2 is $(\sigma \, 2s)^2 \, (\sigma^* \, 2s)^2 \, (\sigma \, 2p)^2 \, (\pi \, 2p)^4 \, (\pi^* \, 2p)^2$.

In the ground state, oxygen is a triplet ($^3O_2$). As seen from the above notation, the term $(\pi^* \, 2p)^2$

accounts for two unpaired electrons in the oxygen molecule. These are the two antibonding π orbitals available: π^*2p_y and π^*2p_z. The two electrons occupy these orbitals alone. The net angular momentum of the unpaired electrons has three components, hence the term "triplet". When the electrons are paired, the angular momentum can not be split into components and this represents a singlet state. In the triplet state, oxygen reacts preferentially with radicals, i.e. molecules having one unpaired electron. In contrast, direct reactions of tripletstate oxygen with molecules which have all electrons paired, as in the case of fatty acids, are prevented by spin barriers. For this reason the activation energy of the reaction

$$RH + {}^3O_2 \longrightarrow ROOH \qquad (3.59)$$

is so high (146–273 kJ/mole) that it does not occur without some assistance.

Oxygen goes from the ground state to the short-lived 1-singlet-state ($^1O_2$) by the uptake of 92 kJ/mole of energy (Fig. 3.23). The previously unpaired single electrons are now paired on the π^*2p_y antibonding orbital. The reactivity of this molecule resembles ethylenic or general olefinic π electron pair reactions, but it is more electrophilic. Hence, in the reaction with oleic acid, the 1-singlet-state oxygen attacks the 9−10 double bond, generating two monohydroperoxides, the 9- and 10-isomers (cf. Table 3.28). The second singlet-state of oxygen ($^1\Sigma^+$ g) has

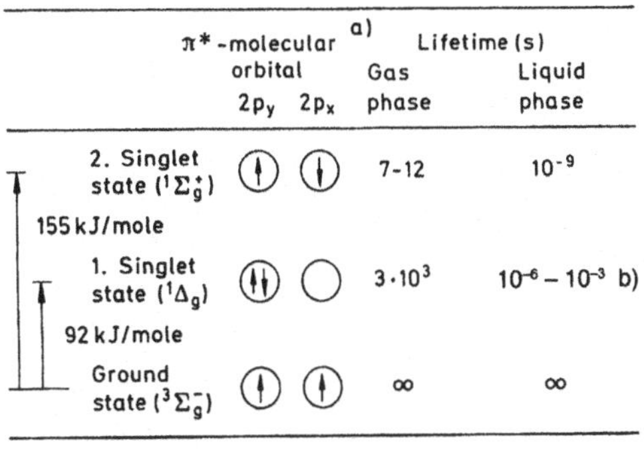

a much shorter life than the 1-singlet-state and plays no role in the oxidation of fats or oils.

For a long time it has been recognized that the stability of stored fat (oil) drops in the presence of light. Light triggers lipid autoxidation. Low amounts of some compounds participate as sensitizers.

According to *Schenk* and *Koch* (1960), there are two types of sensitizers. Type I sensitizers are those which, once activated by light (sen*), react directly with substrate, generating substrate radicals. These then trigger the autoxidation process. Type II sensitizers are those which activate the ground state of oxygen to the $^1O_2$ singlet state.

Type I and II photooxidation compete with each other. Which reaction will prevail depends on the structure of the sensitizer but also on the concentration and the structure of the substrate available for oxidation.

Table 3.28 shows that the composition of hydroperoxide isomers derived from an unsaturated acid by autoxidation ($^3O_2$) differs from that obtained in the reaction with $^1O_2$. The isomers can be separated by analysis of hydroperoxides using high performance liquid chromatography and, thus, one can distinguish Type I from Type II photooxidation. Such studies have revealed that sensitizers, such as chlorophylls a and b, pheophytins a and b and riboflavin, present in food, promote the Type II oxidation of oleic and linoleic acids.

As already stated, the Type II sensitizer, once activated, does not react with the substrate but with ground state triplet oxygen, transforming it with an input of energy into 1-singlet-state oxygen:

$$Sen \xrightarrow{h\nu} Sen^*$$
$$Sen^* + O_2 \longrightarrow Sen + {}^1O_2 \qquad (3.60)$$

The singlet $^1O_2$ formed now reacts directly with the unsaturated fatty acid by a mechanism of "cyclo-addition":

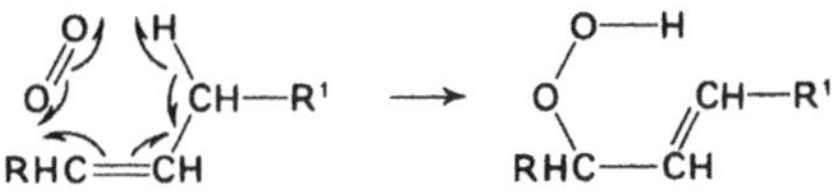

$$\qquad (3.61)$$

The fact that the number of hydroperoxides formed are double the number of isolated double bonds present in the fatty acid molecule is in

π^*-molecular orbital		Lifetime (s)	
2p_y	2p_x [a]	Gas phase	Liquid phase
2. Singlet state ($^1\Sigma_g^+$)	↕ ↕	7–12	10^{-9}
155 kJ/mole			
1. Singlet state ($^1\Delta_g$)	↕↓ ◯	$3 \cdot 10^3$	$10^{-6} - 10^{-3}$ [b]
92 kJ/mole			
Ground state ($^3\Sigma_g^-$)	↕ ↕	∞	∞

Fig. 3.23. Configuration of electrons in an oxygen molecule
[a] Electrons in $2p_x$ and $2p_y$ orbitals
[b] Dependent on solvent, e.g. 2 μs in water, 20 μs in D_2O and 7 μs in methanol

agreement with the above reaction mechanism. The reaction is illustrated in Fig. 3.24 for the oxidation of linoleic acid. In addition to the two hydroperoxides with a conjugated diene system already mentioned (Fig. 3.21), two hydroperoxides are obtained with isolated double bonds.

The mechanism reveals that double bonds of unsaturated fatty acids containing more than one double bond behave as isolated C=C units rather than 1,4-diene systems. Consequently, the difference in the reaction rates of linoleic acid and oleic acid with singlet oxygen is relatively small. In model reactions linoleic acid was photooxidized only 2 to 3 times faster. This is in contrast to autoxidation where linoleic acid is oxidized at least 12 times faster (cf. 3.7.2.1.1). However, furan fatty acids (cf. 3.2.1.3) react much faster with $^1O_2$ than linoleic or linolenic acid. An endo peroxide is the main reaction product. In addition, diacetyl, 3-methyl-2,4-nonandione (MND) and 2,3-octandione are formed. MND contributes to the aroma of tea (cf. 21.2.5.8) and, with diacetyl, to the aroma changes in stored soybean oil (cf. 14.3.2.2.5). The formation of MND is explained by the secondary reactions shown in Fig. 3.25. First, the allyl group of C-9 to C-11 is oxidized to an 11-hydroperoxide. The following β-cleavage of hydroperoxide results in the formation of a carbonyl group at C-11 and a hydroxyl radical which combines with C-13 after homolysis of the furan ring. The furan fatty acid thus is split into MND and a fragment of unknown struc-

ture. The odor threshold of MND in water is fairly low (Table 3.32), even in comparison with that of octandione.

Formation of 1-singlet oxygen ($^1O_2$) is inhibited by carotenoids (car):

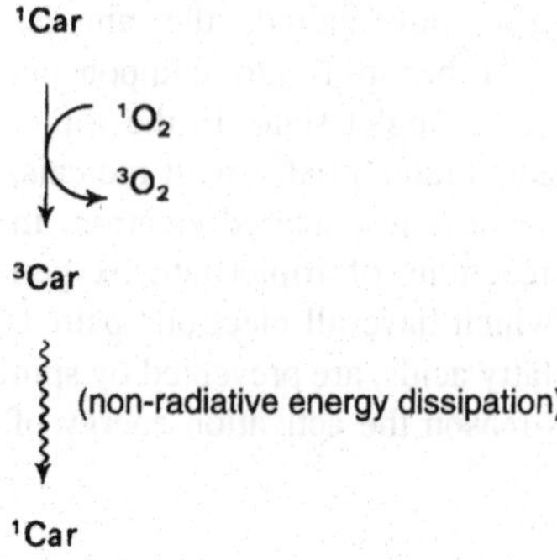

^{1}**Car**

$^1O_2$

$^3O_2$

^{3}**Car**

(non-radiative energy dissipation)

^{1}**Car**   (3.62)

The quenching effect of carotenoids (transition of $^1O_2$ to $^3O_2$) is very fast ($k = 3 \times 10^{10}$ 1 mole^{-1}s^{-1}). They also prevent energy transfer from excited-state chlorophyll to $^3O_2$. Therefore, carotenoids are particularly suitable for protecting fat (oil)-containing food from Type II photooxidation.

3.7.2.1.6 Heavy Metal Ions

These ions are involved in the second group of initiation reactions, namely, in the decomposition of initially-formed hydroperoxides into radicals

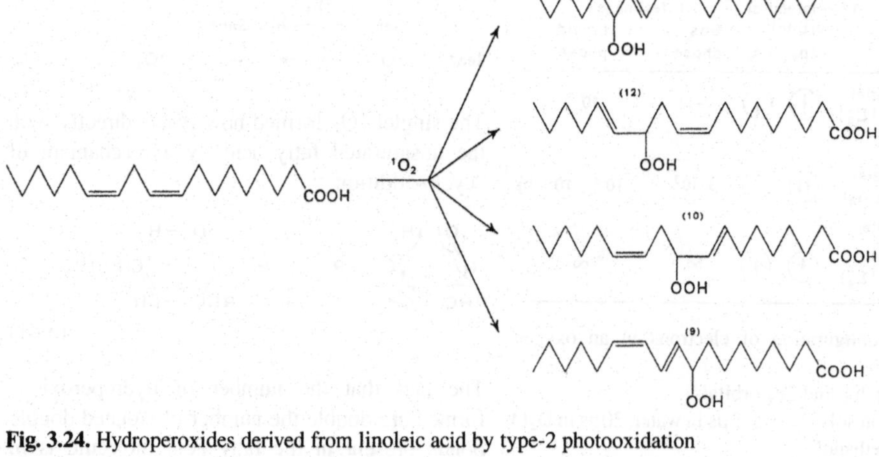

Fig. 3.24. Hydroperoxides derived from linoleic acid by type-2 photooxidation

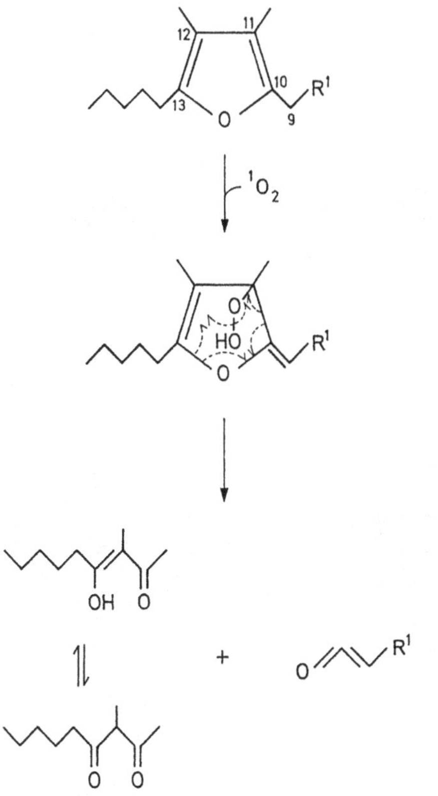

Fig. 3.25. Side reaction of a branched furan fatty acid with singlet oxygen (R^1: $(CH_2)_7COOH$)

which then propel the radical chain reaction of the autooxidation process. Fats, oils and foods always contain traces of heavy metals, the complete removal of which in a refining step would be uneconomical. The metal ions, primarily Fe, Cu and Co, may originate from:

- Raw food. Traces of heavy metal ions are present in many enzymes and other metal-bound proteins. For example, during the crushing and solvent extraction of oilseeds, metal bonds dissociate and the free ions bind to fatty acids.
- From processing and handling equipment. Traces of heavy metals are solubilized during the processing of fat (oil). Such traces are inactive physiologically but active as prooxidants.
- From packaging material. Traces of heavy metals from metal foils or cans or from wrapping paper can contaminate food and diffuse into the fat or oil phase.

The concentration of heavy metal ions that results in fat (oil) shelf-life instability is dependent on the nature of the metal ion and the fatty acid composition of the fat (oil). Edible oils of the linoleic acid type, such as sunflower and corn germ oil, should contain less than 0.03 ppm Fe and 0.01 ppm Cu to maintain their stability. The concentration limit is 0.2 ppm for Cu and 2 ppm for Fe in fat with a high content of oleic and/or stearic acids, e. g. butter.

Heavy metal ions trigger the autoxidation of unsaturated acyl lipids only when they contain hydroperoxides. That is, the presence of a hydroperoxide group is a prerequisite for metal ion activity, which leads to decomposition of the hydroperoxide group into a free radical:

$$Me^{n\oplus} + ROOH \rightarrow Me^{(n+1)\oplus} + RO^{\cdot} + OH^{\ominus}$$
$$(3.63)$$

$$Re^{(n+1)\oplus} + ROOH \rightarrow RO_2^{\cdot} + H^{\oplus} + Me^{n\oplus} \quad (3.64)$$

Reaction rate constants for the decomposition of linoleic acid hydroperoxide are given in Table 3.29. As seen with iron, the lower oxidation state (Fe^{2+}) provides a ten-fold faster decomposition rate than the higher state (Fe^{3+}). Correspondingly, Reaction 3.63 proceeds much faster than Reaction 3.64 in which the reduced state of the metal ion is regenerated. The start of autoxidation then is triggered by radicals from generated hydroperoxides.

The decomposition rates for hydroperoxides emulsified in water depend on pH (Table 3.29). The optimal activity for Fe and Cu ions is in the pH range of 5.5–6.0. The presence of ascorbic acid, even in traces, accelerates the decomposition. Apparently, it sustains the reduced state of the metal ions.

The direct oxidation of an unsaturated fatty acid to an acyl radical by a heavy metal ion

$$RH + Me^{(n+1)\oplus} \rightarrow R^{\cdot} + H^{\oplus} + Me^{n\oplus} \quad (3.65)$$

proceeds, but at an exceptionally slow rate. It seems to be without significance for the initiation of autoxidation.

The autoxidation of acyl lipids is also influenced by the moisture content of food. The reaction rate is high for both dehydrated and water-containing food, but is minimal at a water activity (a_w) of 0.3 (Fig. 0.4). The following hypotheses are discussed to explain these differences: The high reaction rate in dehydrated food is due to metal

Table 3.29. Linoleic acid hydroperoxides[a]: decomposition by heavy metal or heme compounds at 23 °C. Relative reaction rates k_{rel} are given at two pH's[a]

Heavy metal ion[b]	k_{rel}		Heme compound[b]	k_{rel}	
	pH 7	pH 5.5		pH 7	pH 5.5
Fe^{3+}	1	10^2	Hematin	4.10^3	4.10^4
Fe^{2+}	14	10^3	Methemoglobin	5.10^3	$7.6.10^3$
Cu^{2+}	0.2	1.5	Cytochrome C	$2.6.10^3$	$3.9.10^3$
Co^{3+}	6.10^2	1	Oxyhemoglobin	$1.2.10^3$	
Mn^{2+}	0	0	Myoglobin	$1.1.10^3$	
			Catalase	1	
			Peroxidase	1	

[a] Linoleic acid hydroperoxide is emulsified in a buffer.
[b] Reaction rate constant is related to reaction rate in presence of Fe^{3+} at pH 7 ($k_{rel} = 1$).

ions with depleted hydration shells. In addition, ESR spectroscopic studies show that food drying promotes the formation of free radicals which might initiate lipid peroxidation. As the water content starts to increase, the rate of autoxidation decreases. It is assumed that this decrease in rate is due to hydration of ions and also of radicals. Above an a_w of 0.3, free water is present in food in addition to bound water. Free water appears to enhance the mobility of prooxidants, thus accounting for the renewed increase in autoxidation rate that is invariably observed at high moisture levels in food.

3.7.2.1.7 Heme(in) Catalysis

Heme (Fe^{2+}) and hemin (Fe^{3+}) proteins are widely distributed in food. Lipid peroxidation in animal tissue is accelerated by hemoglobin, myoglobin and cytochrome C. These reactions are often responsible for rancidity or aroma defects occurring during storage of fish, poultry and cooked meat. In plant food the most important heme(in) proteins are peroxidase and catalase. Cytochrome P_{450} is a particularly powerful catalyst for lipid peroxidation, although it is not yet clear to what extent the compound affects food shelf life "in situ".

During heme catalysis, a Fe^{2+} protoporphyrin complex ($P-Fe^{2+}$), like in myoglobin, will be oxidized by air to $P-Fe^{3+}$ as indicated in Formula 3.66. The formed superoxide radical anion O_2^-, whose properties are discussed below, will further react yielding H_2O_2. Hydrogen peroxide will then oxidize $P-Fe^{3+}$ to the oxene species P–Fe=O. The reaction with H_2O_2 is accelerated by acid/base catalysis, facilitating the loss of the water molecule; the hemin protein and one carboxylic group of the protoporphyrin system acts as proton acceptor and proton donor respectively.

Oxene is the active form of the hemin catalyst. It oxidizes two fatty acid hydroperoxide molecules to peroxy radicals that will then initiate lipid peroxidation.

In comparison with iron ions, some heme(in) compounds degrade the hydroperoxides more rapidly by several orders of magnitude (cf. Table 3.29). Therefore they are more effective as initiators of lipid peroxidation. Their activity is also negligibly influenced by a decrease in the pH-value.

However, the activity of a heme(in) protein towards hydroperoxides is influenced by its steric accessibility to fatty acid hydroperoxides. Hydroperoxide binding to the Fe-porphyrin moiety of native catalase and peroxidase molecules is obviously not without interferences. The prosthetic group is free to promote hydroperoxide decomposition only after heat denaturation of the enzymes. Indeed, a model experiment with peroxidase showed that the peroxidation of linoleic acid increased by a factor of 10 when the enzyme was heated for 1 minute to 140 °C. As expected, the enzymatic activity of peroxidase decreased and was only 14%. Similar results were obtained in reaction systems containing catalase.

$$P\text{-}Fe^{2\oplus} + O_2 \longrightarrow P\text{-}Fe^{3\oplus} + O_2^{\ominus}$$

$$2O_2^{\ominus} + 2H^{\oplus} \longrightarrow H_2O_2 + O_2$$

$$P\text{-}Fe^{3\oplus} + H_2O_2 \longrightarrow P\overset{\oplus}{\underset{\underset{Fe\cdots O^{\ominus}\text{-}O\text{-}H}{}}{\overset{H}{\diagup}}} \longrightarrow \begin{array}{c} \overset{3\oplus}{P\text{-}Fe}=O \\ \updownarrow \\ \overset{4\oplus}{P\text{-}Fe}\text{-}O^{\ominus} \end{array} + H_2O$$

$$\overset{4\oplus}{P\text{-}Fe}\text{-}O^{\ominus} + ROOH \longrightarrow \overset{4\oplus}{P\text{-}Fe}\cdots\overset{\ominus}{OH} + ROO^{\bullet}$$

$$\overset{4\oplus}{P\text{-}Fe}\cdots\overset{\ominus}{O}\text{-}H + ROOH \longrightarrow \overset{3\oplus}{P\text{-}Fe} + ROO^{\bullet} + H_2O \qquad (3.66)$$

Suppression of peroxidase and catalase activity is of importance for the shelf life of heat-processed food. As long as the protein moiety has not been denatured, it is the lipoxygenase enzyme which is the most active for lipid peroxidation (cf. 3.7.2.2). After lipoxygenase activity is destroyed by heat denaturation, its role is replaced by the heme(in) proteins. As already suggested, an assay of heme(in) protein enzyme activity does not necessarily reflect its prooxidant activity.

3.7.2.1.8 Activated Oxygen

In enzymatic reactions oxygen can form three intermediates, which differ greatly in their activities and which are all ultimately reduced to water:

$$O_2 \xrightarrow{e} O_2^{\ominus} \xrightarrow[2H^{\oplus}]{e} H_2O_2$$

$$\xrightarrow[H^{\oplus}\ H_2O]{e} OH^{\bullet} \xrightarrow[H^{\oplus}]{e} H_2O$$

e: Electron

$$(3.67)$$

Oxygen takes up one electron to form the superoxide radical anion ($O_2^{\ominus}$). This anion radical is a reducing agent with chemical properties dependent on pH, according to the equilibrium:

$$O_2^{\ominus} + H^{\oplus} \rightleftharpoons HO_2^{\bullet}(pK_s : 4.8) \qquad (3.68)$$

Based on its pK_s value under physiological conditions, this activated oxygen species occurs as an anion with its radical character suppressed. It acts as a nucleophilic reagent (e. g. it promotes phospholipid hydrolysis within the membranes) under such conditions, but is not directly able to abstract an H-atom and to initiate lipid peroxidation. The free radical activity of the superoxide anion appears only in acidic media, wherein the perhydroxy radical form ($HO_2^{\bullet}$) prevails. Some reactions of ($HO_2^{\bullet}$) are presented in Table 3.30. $O_2^{\ominus}$ is comparatively inactive (Table 3.30). As shown in Reaction 3.69, it dismutates at a rate that is dependent on the pH, e.g., pH 7: $k = 5.105\,l\,mol^{-1}\,s^{-1}$, pH 11: $k = 102\,l\,mol^{-1}\,s^{-1}$.

$$2O_2^{\ominus} + 2H^{\oplus} \longrightarrow H_2O_2 + O_2 \qquad (3.69)$$

An enzyme with superoxide dismutase activity which significantly accelerates ($k = 2 \times 10^9\,l\,mol^{-1}\,s^{-1}$) Reaction 3.69 occurs in numerous animal and plant tissues. The superoxide radical anion, O_2^-, is generated especially by flavin enzymes, such as xanthine oxidase (cf. 2.3.3.2). The involvement of this enzyme in the development of milk oxidation flavor has been questioned for a long time. The superoxide radical anion reacts at an exceptionally high rate ($k = 1.9 \times 10^{10}\,l\,mol^{-1}\,s^{-1}$) with nitrogen oxide (NO), the monomer being present as the free radical, to give peroxy nitrite ($ONOO^{\ominus}$). NO is formed in animal and plant foods from arginine by nitrogen oxide synthase (cf. 9.8.1). It is relatively stable with a half life of $400\,s$ (H_2O). Peroxy nitrite is a versatile oxidant; it oxidizes unsaturated fatty acids, ascorbic acid, tocopherols, uric acid and amino acids, among

Table 3.30. Rate constants of reactions of reactive oxygen species with food constituents

Constituent	$^1O_2$	$HO^\bullet$	$O_2^\ominus$	$HOO^\bullet$
Lipid		$k\ (l \times mol^{-1} \times s^{-1})$		
Oleic acid	5.3×10^4			
Linoleic acid	7.3×10^4			No reaction
Linolenic acid	1.0×10^5			1.2×10^3
Arachidonic acid				1.7×10^3
Cholestererol	2.5×10^8			3.1×10^3
Amino acids				
Histidine	4.6×10^7	4.8×10^9	<1.0	<95
Tryptophane	1.3×10^7	1.3×10^{10}	<24	
Cysteine	5.0×10^7	1.9×10^{10}	<15	<600
Cystine		2.1×10^9	$<4.0 \times 10^{-1}$	
Methionine	1.3×10^7	7.4×10^9	$<3.3 \times 10^{-1}$	<49
Sugar				
Glucose	1.4×10^4	1.5×10^9		
Fructose		1.6×10^9		
Sucrose	2.5×10^4	2.3×10^9		
Maltose		2.3×10^9		
Vitamins				
β-Carotene	5.0×10^9			
Riboflavin	6.0×10^7	1.2×10^{10}		
Ascorbic acid	1.1×10^7	8.2×10^9		1.6×10^4
Vitamin D	2.3×10^7			
α-Tocopherol	13.2×10^7		No reaction	2.0×10^5

other substances. It easily decomposes with the formation of radicals, which can start lipid peroxidation.

Hydrogen peroxide, H_2O_2, is the second intermediate of oxygen reduction. In the absence of heavy metal ions, energy-rich radiation including UV light and elevated temperatures, H_2O_2 is a rather indolent and sluggish reaction agent. On the other hand, the hydroxy radical ($HO^\bullet$) derived from it is exceptionally active. During the abstraction of an H-atom,

$$R\!-\!H + HO^\bullet \longrightarrow R^\bullet + H_2O \qquad (3.70)$$

the energy input in the HO-bond formed is $497\ kJ/mol$, thus exceeding the dissociation energy for abstraction of hydrogen from each C–H bond by at least $75\ kJ/mol$ (cf. Table 3.27). Therefore, the $HO^\bullet$ radical reacts non-selectively with all organic constituents of food at an almost diffusion-controlled rate. Consequently, it can directly initiate lipid peroxidation. However, in

a complex system such as food, the following question is always pertinent: "Has the $HO^\bullet$ radical actually reached the unsaturated acyl lipid, or was it trapped prior to lipid oxidation by some other food ingredient?".

The reaction of the superoxide radical anion with hydrogen peroxide should be emphasized in relation to initiation of autoxidation. This is the so-called *Fenton* reaction in particular of an Fe-complex:

$$
\begin{array}{ccc}
O_2^\ominus & \text{ADP-Fe}^{3\oplus} & HO^\bullet + OH^\ominus \\
O_2 & \text{ADP-Fe}^{2\oplus} & H_2O_2
\end{array}
$$

$$(3.71)$$

The Fe-complex (e. g. with ADP) occurs in food of plant and animal origin. The Fe^{2+} obtained by reduction with O_2^- can then reduce the H_2O_2 present and generate free $HO^\bullet$ radicals.

3.7.2.1.9 Secondary Products

The primary products of autoxidation, the mono-hydroperoxides, are odorless and tasteless (such as linoleic acid hydroperoxides; cf. 3.7.2.4.1). Food quality is not affected until volatile compounds are formed. The latter are usually powerfully odorous compounds and, even in the very small amounts in which they occur, affect the odor and flavor of food.

From the numerous volatile secondary products of lipid peroxidation the following compounds will be discussed in detail

- odor-active carbonyl compounds
- malonic dialdehyde
- alkanes, alkenes

Odor-Active Monocarbonyl Compounds. Model expriments showed that the volatile fractions formed during the autoxidation of oleic, linoleic and linolenic acid contain mainly aldehydes and ketones (Table 3.31). Linoleic acid, a component of all lipids sensitive to autoxidation, is a precursor of hexanal that is predominant in

the volatile fraction. Therefore this substance, since it can easily be determined by headspace analysis, is used as an indicator for the characterization of off-flavors resulting from lipid peroxidation.

A comparison of the sensory properties (Table 3.32) shows that some carbonyl compounds, belonging to side components of the volatile fractions, may intensively contribute to an off-flavor due to their low threshold values. Food items containing linoleic acid, especially (E)-2-nonenal, trans-4,5-epoxy-(E)-2-decenal and 1-octen-3-one, are very aroma active.

The rapid deterioration of food containing linolenic acid should not be ascribed solely to the preferential oxidation of this acid but also to the low odor threshold values of the carbonyl compounds formed, such as (Z)-3-hexenal, (E,Z)-2,6-nonadienal and (Z)-1,5-octadien-3-one (Table 3.32). Aldehydes with exceptionally strong aromas can be released in food by the autoxidation of some fatty acids, even if they are present in low amounts. An example is octadeca-(Z,Z)-11, 15-dienoic acid (the precursor for

Table 3.31. Volatile compounds formed by autoxidation of unsaturated fatty acids $(\mu g/g)^a$

Oleic acid		Linoleic acid		Linolenic acid	
Heptanal	50	Pentane[b]	+[c]	Propanal[b]	
Octanal	320	Pentanal	55	1-Penten-3-one	30
Nonanal	370	Hexanal	5,100	(E)-2-Butenal	10
Decanal	80	Heptanal	50	(E)-2-Pentenal	35
(E)-2-Decenal	70	(E)-2-Heptenal	450	(Z)-2-Pentenal	45
(E)-2-Undecenal	85	Octanal	45	(E)-2-Hexenal	10
		1-Octen-3-one	2	(E)-3-Hexenal	15
		1-Octen-3-hydroperoxide	+[c]	(Z)-3-Hexenal	90
		(Z)-2-Octenal	990	(E)-2-Heptenal	5
		(E)-2-Octenal	420	(E,Z)-2,4-Heptadienal	320
		(Z)-3-Nonenal	30	(E,E)-2,4-Heptadienal	70
		(E)-3-Nonenal	30	(Z,Z)-2,5-Octadienal	20
		(Z)-2-Nonenal	+[c]	3,5-Octadien-2-one	30
		(E)-2-Nonenal	30	(Z)-1,5-Octadien-3-one	+[c]
		(Z)-2-Decenal	20	(Z)-1,5-Octadien-3-hydroperoxide	+[c]
		(E,E)-2,4-Nonadienal	30	(E,Z)-2,6-Nonadienal	10
		(E,Z)-2,4-Decadienal	250	2,4,7-Decatrienal	85
		(E,E)-2,4-Decadienal	150		
		trans-4,5-Epoxy-(E)-2-decenal	+[c]		

[a] Each fatty acid in amount of 1 g was autoxidized at 20 °C by an uptake of 0.5 mole oxygen/mole fatty acid.
[b] Major compound of autoxidation.
[c] Detected, but not quantified.

Table 3.32. Sensory properties of aroma components resulting from lipid peroxidation

| Compound | Flavor quality | Odor threshold (μg/kg) | | in water |
| | | in oil | | |
		nasal	retronasal	nasal
Aldehydes				
2:0	fruity, pungent	0.22	7.1	–
3:0	fruity, pungent	9.4	68	–
5:0	pungent, like bitter almonds	240	150	18
6:0	tallowy, green leafy	320	75	12
7:0	oily, fatty	3200	50	5
8:0	oily, fatty, soapy	55	515	0.7
9:0	tallowy, soapy-fruity	13,500	260	1.0
10:0	orange peel like	300	75	5
5:1 (E-2)	pungent, apple	2300	600	–
6:1 (E-2)	apple	420	250	316
6:1 (Z-3)	green leafy	1.7	1.2	0.03
7:1 (E-2)	fatty, bitter almond	14,000	400	51
7:1 (Z-4)	cream, putty	2	1	0.8
8:1 (Z-2)	walnut	–	50	–
8:1 (E-2)	fatty, nutty	7000	125	4
9:1 (Z-2)	fatty, green leafy	4.5	0.6	0.02
9:1 (E-2)	tallowy, cucumber	900	65	0.25
9:1 (Z-3)	cucumber	250	35	–
10:1 (E-2)	tallowy, orange	33,800	150	–
7:2 (E,Z-2,4)	frying odor, tallowy	4000	50	–
7:2 (E,E-2,4)	fatty, oily	10,000	30	–
9:2 (E,E-2,4)	fatty, oily	2500	460	–
9:2 (E,Z-2,6)	like cucumber	4	1.5	–
9:2 (Z,Z-3,6)	fatty, green	–	–	0.05[a]
9:3 (E,E,Z-2,4,6)	Oat flakes	–	–	0.026
10:2 (E,Z-2,4)	frying odor	10	–	–
10:2 (E,E-2,4)	frying odor	180	40	0.2
10:3 (E,Z,Z-2,4,7)	cut beans	–	24	–
trans-4,5-Epoxy-(E)-2-decenal	metallic	1.3	3	–
Ketones				
1-Penten-3-one	hot, fishy	0.73	3	–
1-Octen-3-one	like mushrooms, fishy	10	0.3	0.05
1-Nonen-3-one	like mushrooms, earthy	–	–	8×10^{-6}
(Z)-1,5-Octadien-3-one	like geraniums, metallic	0.45	0.03	1.2×10^{-3}
(E,E)-3,5-Octadien-2-one	fatty, fruity	300	–	–
(E,Z)-3,5-Octadien-2-one	fatty, fruity	200	–	–
3-Methyl-2,4-nonanedione	like straw, fruity, like butter	23	1.5	0.01
Miscellaneous compounds				
1-Octen-3-hydroperoxide	metallic	240	–	–
2-Pentylfuran	like butter, like green beans	2000	–	–

[a] retronasal.

(Z)-4-heptenal), which occurs in beef and mutton and often in butter (odor threshold in Table 3.32). Also, the processing of oil and fat can provide an altered fatty acid profile. These can then provide new precursors for a new set of carbonyls. For example, (E)-6-nonenal, the precursor of which is octadeca-(Z,E)-9,15-dienoic acid, is a product of the partial hydrogenation of linolenic acid. This aldehyde can be formed during storage of partially hardened soya and linseed oils. The aldehyde, together with other compounds, is responsible for an off-flavor denoted as "hardened flavor". Several reaction mechanisms have been suggested to explain the formation of volatile carbonyl compounds. The most probable mechanism is the β-scission of monohydroperoxides with formation of an intermediary short-lived alkoxy radical (Fig. 3.26). Such β-scission is catalyzed by heavy metal ions or heme(in) compounds (cf. 3.7.2.1.7).

There are two possibilities for β-scission of each hydroperoxide fatty acid (Fig. 3.26). Option "B", i.e. the cleavage of the C–C bond located further away from the double bond position, is the energetically preferred one since it leads to resonance-stabilized "oxoene" or "oxo-diene" compounds. Applying this β-scission mechanism ("B") to both major monohydroperoxide isomers of linoleic acid gives the products shown in Formula 3.72 and 3.73.

From the volatile autoxidation products which contain the methyl end of the linoleic acid molecule, the formation of 2,4-decadienal and pentane can be explained by reaction 3.72.

The formation of hexanal among the main volatile compounds derived from linoleic acid (cf. Table 3.31) is still an open question. The preferential formation of hexanal in aqueous

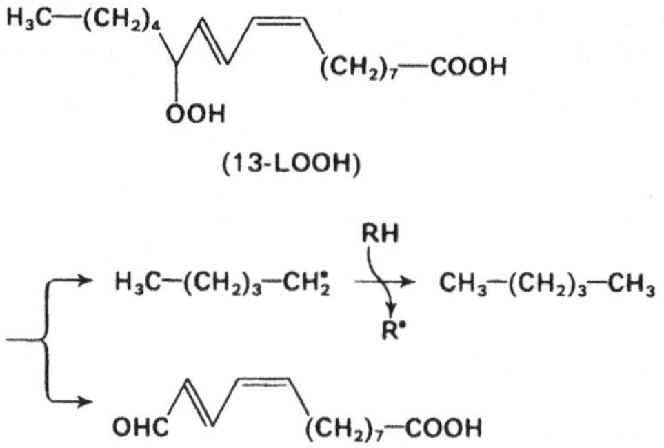

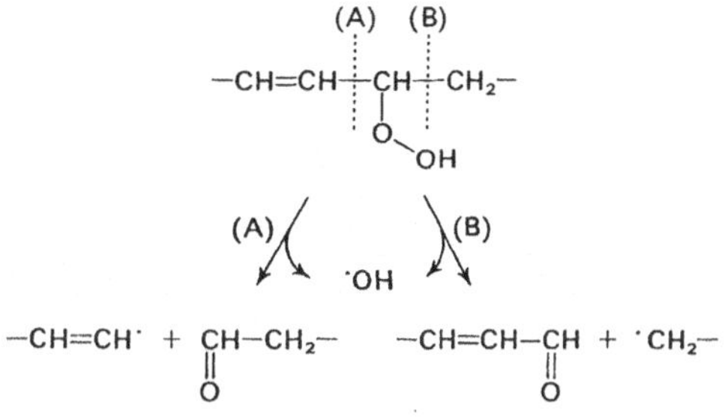

Fig. 3.26. β-Scission of monohydroperoxides (according to *Badings*, 1970)

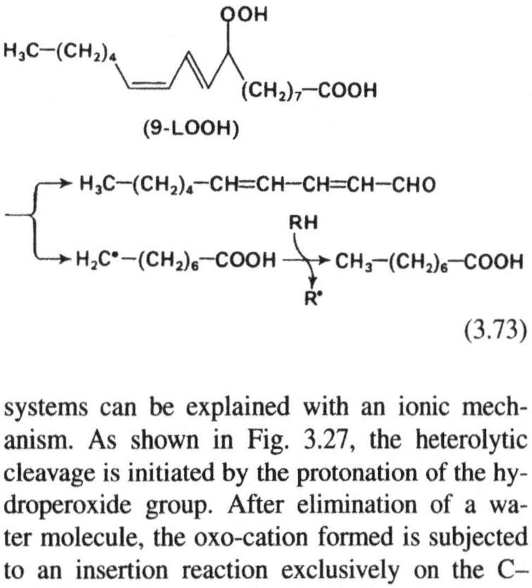

(3.73)

systems can be explained with an ionic mechanism. As shown in Fig. 3.27, the heterolytic cleavage is initiated by the protonation of the hydroperoxide group. After elimination of a water molecule, the oxo-cation formed is subjected to an insertion reaction exclusively on the C–C linkage adjacent to the double bond. The carbonium ion then splits into an oxo-acid and hexanal. The fact that linoleic acid 9-hydroperoxide gives rise to 2-nonenal is in agreement with this outline.

However, in the water-free fat or oil phase of food, the homolytic cleavage of hydroperoxides presented above is the predominant reaction mechanism. Since option "A" of the cleavage reaction is excluded (Fig. 3.26), some other reactions should be assumed to occur to account for formation of hexanal and other aldehydes from linoleic acid. The further oxidation reactions of monohydroperoxides and carbonyl compounds are among the possibilities.

The above assumption is supported by the finding that 2-alkenals and 2,4-alkadienals are oxidized substantially faster than the unsaturated

(3.72)

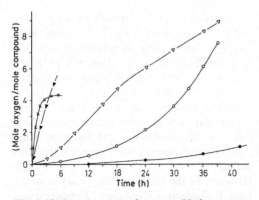

R₁—CHO

Hexanal

R_1: $CH_3(CH_2)_4$ R_2: $(CH_2)_7$—COOH

Fig. 3.27. Proton-catalyzed cleavage of linoleic acid 13-hydroperoxide (according to *Ohloff*, 1973)

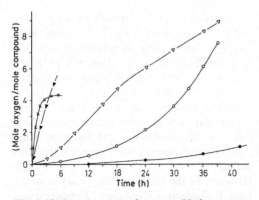

Fig. 3.28. Reaction rate of an autoxidation process (according to *Lillard* and *Day*, 1964).
–∇–∇– Linolenic acid methyl ester, –o–o– linoleic acid methyl ester, ×–×– 2-nonenal, ▼–▼–▼ 2,4-heptadienal, –•–•– nonanal

fatty acids (Fig. 3.28). In addition, the autoxidation of 2,4-decadienal yields hexanal and other volatiles which coincide with those obtained from linoleic acid. Since saturated aldehydes oxidize slowly, as demonstrated by nonanal (Fig. 3.28), they will enrich the oxidation products and become predominant.

Also the delayed appearance of hexanal during the storage of linoleic acid containing fats and oils compared to pentane and 2,4-decadienal, supports the hypothesis that hexanal is not directly formed by a β-scission of the 13-hydroperoxide. It is mainly produced in a tertiary reaction, e. g., during the autoxidation of 2,4-decadienal.

Other studies to elucidate the multitude of aldehydes which arise suggest that the decomposition of minor hydroperoxides formed by autoxidation of linoleic acid (cf. Table 3.28) contribute to the profile of aldehydes. This suggestion is supported by pentanal, which originates from the 14-hydroperoxide.

The occurrence of 2,4-heptadienal (from the 12-hydroperoxide isomer) and of 2,4,7-decatrienal (from the 9-hydroperoxide isomer) as oxidation products is, thereby, readily explained by accepting the fragmentation mechanism outlined above (option "B" in Fig. 3.26) for the autoxidation of α-linolenic acid. The formation of other volatile carbonyls can then follow by autoxidation of these two aldehydes or from the further oxidation of labile monohydroperoxides.

Malonic Aldehyde. This dialdehyde is preferentially formed by autoxidation of fatty acids with three or more double bonds. The compound is odorless. In food it may be bound to proteins by a double condensation, crosslinking the proteins (cf. 3.7.2.4.3). Malonic aldehyde is formed from α-linolenic acid by a modified reaction pathway, as outlined under the formation of hydroperoxide-epidioxide (cf. 3.7.2.1.3). However, a bicyclic compound is formed here as an intermediary product that readily fragments to malonic aldehyde:

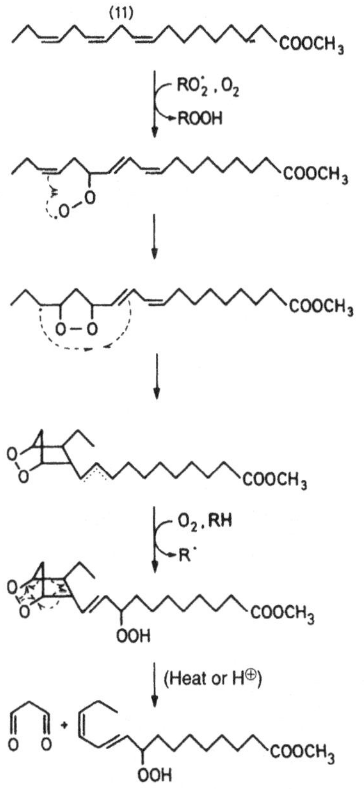

$$(3.74)$$

Alkanes, Alkenes. The main constituents of the volatile hydrocarbon fraction are ethane and pentane. Since these hydrocarbons are readily quantitated by gas chromatography using head-space analysis, they can serve as suitable indicators for *in vivo* detection of lipid peroxidation. Pentane is probably formed from

the 13-hydroperoxide of linoleic acid by the β-scission mechanism (cf. reaction 3.72). The corresponding pathway for 16-hydroperoxide of linolenic acid should then yield ethane.

3.7.2.2 Lipoxygenase: Occurrence and Properties

A lipoxygenase (linoleic acid oxygen oxidoreductase, EC 1.13.11.12) enzyme occurs in many plants and also in erythrocytes and leucocytes. It catalyzes the oxidation of some unsaturated fatty acids to their corresponding monohydroperoxides. These hydroperoxides have the same structure as those obtained by autoxidation. Unlike autoxidation, reactions catalyzed by lipoxygenase are characterized by all the features of enzyme catalysis: substrate specificity, peroxidation selectivity, occurrence of a pH optimum, susceptibility to heat treatment and a high reaction rate in the range of 0–20 °C. Also, the activation energy for linoleic acid peroxidation is rather low: 17 kJ/mol (as compared to the activation energy of a noncatalyzed reaction, see 3.7.2.1.5). Lipoxygenase oxidizes only fatty acids which contain a 1-cis,4-cis-pentadiene system. Therefore, the preferred substrates are linoleic and linolenic acids for the plant enzyme, and arachidonic acid for the animal enzyme; oleic acid is not oxidized.

Lipoxygenase is a metal-bound protein with an Fe-atom in its active center. The enzyme

Table 3.33. Regio- and stereospecificity of lipoxygenases (LOX)

Origin (isoenzyme)		Hydroperoxide from 18:2 (9,12)[a]					PH
		13S	13R	9S	9R	8R	
Soybean, seed	(LOX I)	94	2	2	2		10.5
	(LOX II)	77	3	18	2		7
Pea, seed	(LOX I)	23	16	32	29		6.8
	(LOX II)	87	2	6	5		6.8
Corn, germ		3.5	3.5	89	4		6.5
Tomato, fruit		13	2	84	<1		5.5
Potato, tuber		1.6	2.4	96	0		6.8
Barley, seed				92	3		7.0
Wheat, germ		10	5	83	2		6.8
Gaeumannomyces graminis						100	7.4
Marchantia polymorpha		89	2				9.0

[a] Composition of the hydroperoxide fraction in %.

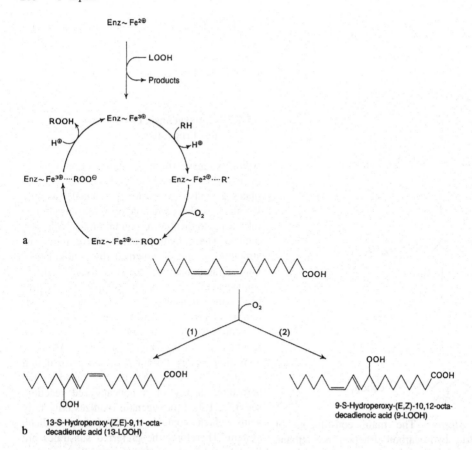

Fig. 3.29. Lipoxygenase catalysis

a Proposed mechanism of reaction (according to *Veldink*, 1977); RH: linoleic acid; LOOH: linoleic acid hydroperoxide

b Regio- and stereospecificity for linoleic acid oxidation. (1) Lipoxygenase from soybean (LOX 1; cf. Table 3.33); (2) lipoxygenase from tomato (cf. Table 3.33)

is activated by its product and during activation, Fe^{2+} is oxidized to Fe^{3+}. The catalyzed oxidation pathway is assumed to have the following reaction steps (cf. Fig. 3.29a): abstraction of a methylene H-atom from the substrate's 1,4-pentadiene system and oxidation of the H-atom to a proton. The pentadienyl radical bound to the enzyme is then rearranged into a conjugated diene system, followed by the uptake of oxygen. The peroxy radical formed is then reduced by the enzyme and, after attachment of a proton, the hydroperoxide formed is released.

In the rate-limiting step of catalysis, the isoenzyme LOX 1 from soybeans abstracts the pro-(S)-hydrogen from the n-8 methylene group[a] of linoleic acid. Molecular oxygen is then introduced into the fatty acid present as a pentadienyl radical from the opposite side at n-6 with the formation of the 13S-hydroperoxide (Fig. 3.29b). Another group of LOX, to which the enzyme from tomatoes belongs, abstracts the pro-(R)-hydrogen. This results in the formation of a 9S-hydroperoxide (Fig. 3.29b) if the oxygen coming from the opposite side docks onto C-9.

Lipoxygenases from plants mostly exhibit 9- or 13-regiospecificity. A LOX with C-8 specificity has been found in a mushroom (Table 3.33).

---

[a] "n": the C atoms are counted from the methyl end of the fatty acid.

Non-specific LOX occur in legumes, e. g., LOX 1 in peas (Table 3.33) and LOX III in soybeans (pH optimum: 6.5). These enzymes oxidize linoleic acid to mixtures of 9- and 13-hydroperoxides, which approach racemic proportions. In addition, oxo fatty acids and volatile compounds are formed, i. e., the product spectrum resembles that formed by the autoxidation of linoleic acid. Moreover, they also react with esterified substrate fatty acids. In contrast to specific LOX, they do not require prior release of fatty acids by a lipase enzyme for activity in food.

The non-specific lipoxygenases can cooxidize carotenoids and chlorophyll and thus can degrade these pigments to colorless products. This property is utilized in flour "bleaching" (cf. 15.4.1.4.3). The involvement of LOX in cooxidation reactions can be explained by the possibility that the peroxy radicals are not as rapidly and fully converted to their hydroperoxides as in the case of specifically reacting enzymes. Thus, a fraction of the free peroxy radicals are released by the enzyme. It can abstract an H-atom either from the unsaturated fatty acid present (pathway 2a in Fig. 3.30) or from a polyene (pathway 2b in Fig. 3.30).

The non-specific lipoxygenases present in legumes produces a wide spectrum of volatile aldehydes from lipid substrates. These aldehydes, identical to those of a noncatalyzed autoxidation, can be further reduced to their alcohols, depending on the status of NADH-NAD$^{\oplus}$.

3.7.2.3 Enzymatic Degradation of Hydroperoxides

Animals and plants degrade fatty acid hydroperoxides differently. In animal tissue, the enzyme glutathione peroxidase (cf. 7.3.2.8) catalyzes a reduction of the fatty acid hydroperoxides to the corresponding hydroxy acids, while in plants and mushrooms, hydroperoxide lyase (HPL), hydroperoxide isomerase, allene oxide synthase (AOS) and allene oxide cyclase (AOC) are active. The HPL reaction is highly interesting with regard to food chemistry since the hydroperoxides, which are formed by lipoxygenase catalysis of linoleic and linolenic acid, are precursors of odorants. Those are important for fruits, vegetables and mushrooms, like the green-grassy or cucumberlike smelling aldehydes hexanal, (Z)-3-hexenal ("leafy aldehyde"), (Z,Z)-3,6-nonadienal and the mushroomlike (R)-1-octen-3-ol (Table 3.34). The suggested mechanism is a β-cleavage of the hydroperoxide (Fig. 3.31).

The difference in volatile products in plants (aldehydes) and mushrooms (allyl alcohols) is due to the different substrate and reaction specificity of HPL. In the first case, in hydroperoxides with conjugated diene systems (Fig. 3.31a), the bond between the C-atom bearing the HOO-group and the C-atom of the diene system is cleaved. In the second case (Fig. 3.31b), cleavage of hydroperoxides with isolated double bonds occurs in the opposite direction between the C-atom

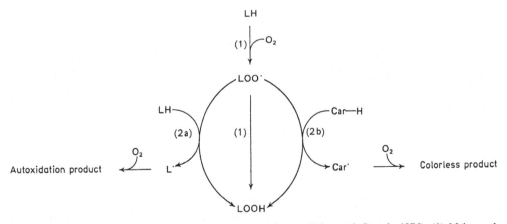

Fig. 3.30. Reactions of non-specific lipoxygenase (according to *Weber* and *Grosch*, 1976). (1) Main catalysis pathway; (2a) and (2b) cooxidation pathways. LH: linoleic acid; Car-H: carotenoid; LOOH: linoleic acid hydroperoxide

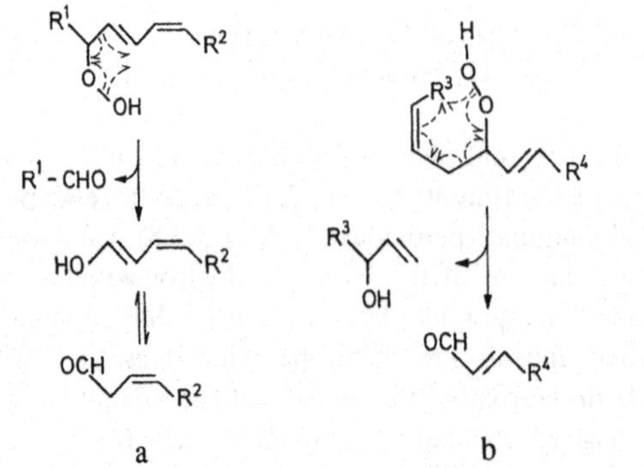

	R^1	R^2
13 - LOOH	$CH_3(CH_2)_4$	$(CH_2)_7 COOH$
13 - LnOOH	$CH_3CH_2CH = CHCH_2$	$(CH_2)_7 COOH$
9 - LOOH	$HOOC(CH_2)_7$	$(CH_2)_4 CH_3$
9 - LnOOH	$HOOC(CH_2)_7$	$CH_2CH = CH CH_2CH_3$
	R^3	R^4
10 - LOOH	$CH_3(CH_2)_4$	$(CH_2)_6 COOH$
10 - LnOOH	$CH_3CH_2CH = CHCH_2$	$(CH_2)_6 COOH$

Fig. 3.31. Mechanism of the cleavage of hydroperoxides by lyases (according to *Wurzenberger and Grosch*, 1986)
a in plants, **b** in mushrooms

with the OOH-group and the C-atom with the adjacent methylene group. The (Z)-3-aldehydes in plants formed by the splitting reaction can transform themselves into the respective (E)-2-aldehydes. Isomerases that catalyze this reaction were identified in cucumbers, apples and tea chloroplasts.

The widespread presence of the C6- and C9-aldehydes in fruits and vegetables as well as the C8-alcohols in mushrooms (Table 3.34) permits the conclusion to be drawn that enzymatic-oxidative cleavage of linoleic and linolenic acid with the enzymes lipoxygenase, hydroperoxide-lyase and, if necessary, an aldehyde-isomerase generally contributes to the formation of aroma in these food items. This process is intensified when oxygen can permeate the cells freely by destruction of tissue (during the chopping of fruits and vegetables).

Allene oxide synthase (AOS) is a cytochrome P_{450} enzyme which was first found in flaxseed. It catalyzes the degradation of hydroperoxides to very instable allene oxides ($t_{1/2}$ at 0 °C: 33 s). On

hydrolysis, the allene oxide originating from the 13-hydroperoxide of linoleic acid gives rise to α- or γ-ketol fatty acids depending on whether the OH-ion attacks at C-13 or C-9 (Formula 3.75). The allene oxide can also react with other nucleophiles, e. g., ROH, RSH, as well as the anion of linoleic acid. Allene oxide cyclase (AOC) competes with these non-enzymatic reactions, cyclizing the allene oxide to 15,16-dihydro-12oxophyto-dienoic acid (Formula 3.75). AOS and AOC convert the 13-hydroperoxide of linolenic acid to 12-oxophytadienoic acid.

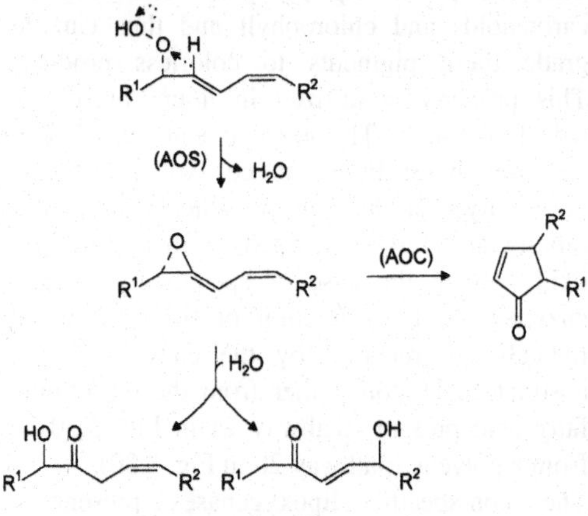

R^1 : $CH_3- (CH_2)_4$; R^2 : $(CH_2)_7- COOH$

(3.75)

Linoleic acid 9-hydroperoxide formed by lipoxygenase in potato is changed enzymatically by elimination of water into a fatty acid with a dienyl-ether structure:

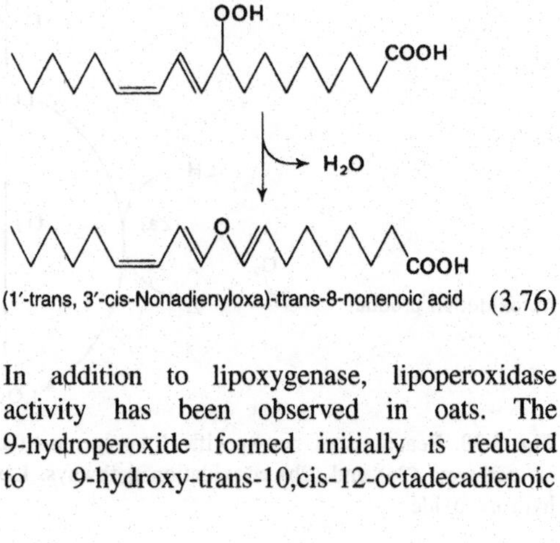

(1′-trans, 3′-cis-Nonadienyloxa)-trans-8-nonenoic acid   (3.76)

In addition to lipoxygenase, lipoperoxidase activity has been observed in oats. The 9-hydroperoxide formed initially is reduced to   9-hydroxy-trans-10,cis-12-octadecadienoic

Table 3.34. Occurrence and properties of various hydroperoxide-lyases

Occurrence	Substrate	Products of the catalyses
Apple, tomato, cucumber, tea leaf (chloroplasts), soy beans, grape	13(S)-hydroperoxy-9-cis,11-trans-octadecadienoic acid (13-LOOH)	hexanal + 12-oxo-9-cis-dodecenoic acid
Apple, tomato, cucumber, tea leaf (chloroplasts), soy beans, grape	13(S)-hydroperoxy-9-cis,11-trans, 15-cis-octadecatrienoic acid (13-LnOOH)	(Z)-3-hexenal + 12-oxo-9-cis-dodecenoic acid
Cucumber, pear	9(S)-hydroperoxy-10-trans, 12-cis-octadecadienoic acid (9-LOOH)	(Z)-3-nonenal + 9-oxo-nonanoicacid
Cucumber, pear	9(S)-hydroperoxy-8-trans, 12-cis, 15-cis-octadecatrienoic acid (9-LnOOH)	(Z,Z)-3,6-nonadienal + 9-oxononanoic acid
Champignon	10(S)-hydroperoxy-10-trans, 12-cis-octadecadienoic acid (10-LOOH)	1-octen-3(R)-ol + 10-oxo-8-trans-dcenoic acid (Z)-1,5-octadien-3(R)-ol+
Champignon	10(S)-hydroperoxy-8-trans,12-cis-15-cis-octadecatrienoic acid (10-LnOOH)	10-oxo-8-trans-decenoicacid

acid. Since hydroxy but not hydroperoxy acids taste bitter, this reaction should contribute to the bitter taste generated during the storage of oats (cf. 15.2.2.3).

3.7.2.4 Hydroperoxide–Protein Interactions

3.7.2.4.1 Products Formed from Hydroperoxides

Hydroperoxides formed enzymatically in food are usually degraded further. This degradation can also be of a nonenzymatic nature. In nonspecific reactions involving heavy metal ions, heme(in) compounds or proteins, hydroperoxides are transformed into oxo, expoxy, mono-, di- and trihydroxy carboxylic acids (Table 3.35). Unlike hydroperoxides, i.e. the primary products of autoxidation, some of these derivatives are characterized as having a bitter taste (Table 3.35). Such compounds are detected in legumes and cereals. They may play a role in other foods rich in unsaturated fatty acids and proteins, such as fish and fish products.

In order to clarify the formation of the compounds presented in Table 3.35, the reaction sequences given in Fig. 3.32 have been assumed to occur. The start of the reaction is from the alkoxydiene radical generated from the 9- or 13-hydroperoxide by the catalytic action of heavy metal ions or heme(in) compounds (cf. 3.7.2.1.7). The alkoxydiene radical may disproportionate into a hydroxydiene and an oxodiene fatty acid. Frequently this reaction is only of secondary importance since the alkoxydiene radical rearranges immediately to an epoxyallylic radical which is susceptible to a variety of radical combination reactions. Under aerobic conditions the epoxyallylic radical combines preferentially with molecular oxygen. The epoxyhydroperoxides formed are, in turn, subject to homolysis via an oxyradical. A disproportionation reaction leads to epoxyoxo and epoxyhydroxy compounds. Under anaerobic conditions the epoxyallylic radical combines with other radicals, e. g. hydroxy radicals (Fig. 3.32) or thiyl radicals (Fig. 3.33).

Of the epoxides produced, the allylic epoxides are known to be particularly susceptible to hydrolysis in the presence of protons. As shown in Fig. 3.32 trihydroxy fatty acids may result from the hydrolysis of an allylic epoxyhydroxy compound.

Table 3.35. Products obtained by non-enzymic degradation of linoleic acid hydroperoxides

Product[a]	Hydroperoxide interaction with				
	Fe^{3+} cysteine	Hemo-globin	Soya homogenate	Pea homogenate	Wheat flour
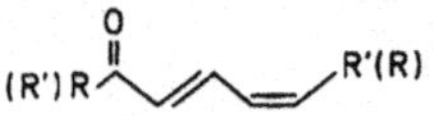 (R')R–(C=O)–CH=CH–R'(R)	+	+	+	+	
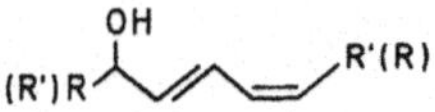 (R')R–CH(OH)–CH=CH–R'(R)	+	+	+	+	+
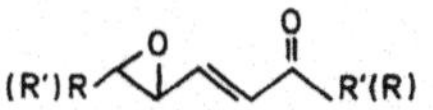 (R')R–(epoxy)–CH=CH–(C=O)–R'(R)	+		+		
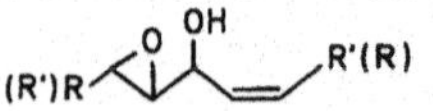 (R')R–(epoxy,OH)–CH=CH–R'(R)	+	+	+		
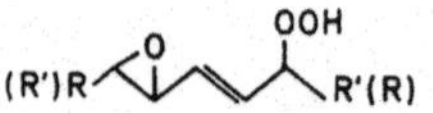 (R')R–(epoxy)–CH=CH–CH(OOH)–R'(R)	+		+		
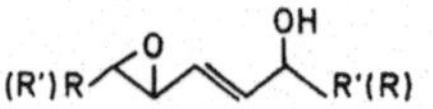 (R')R–(epoxy)–CH=CH–CH(OH)–R'(R)		+		+	+
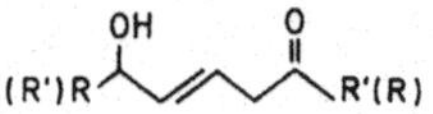 (R')R–CH(OH)–CH=CH–(C=O)–R'(R)	+				
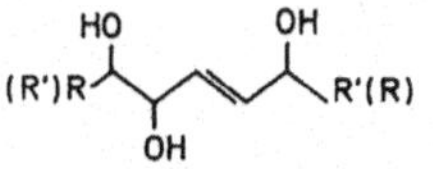 (R')R–CH(HO)–CH=CH–CH(OH)–R'(R), OH	+		+	+	+

[a] As a rule a mixture of two isomers are formed with R: CH$_3$(CH$_2$)$_4$ and R': (CH$_2$)$_7$COOH.

3.7.2.4.2 Lipid–Protein Complexes

Studies related to the interaction of hydroperoxides with proteins have shown that, in the absence of oxygen, linoleic acid 13-hydroperoxide reacts with N-acetylcysteine, yielding an adduct of which one isomer is shown:

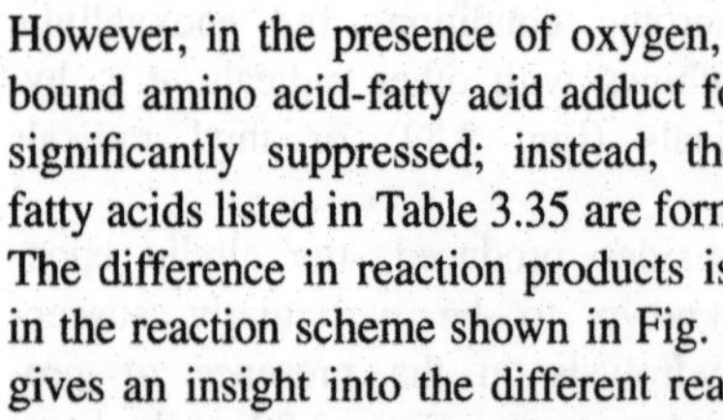

$$(3.77)$$

However, in the presence of oxygen, covalently bound amino acid-fatty acid adduct formation is significantly suppressed; instead, the oxidized fatty acids listed in Table 3.35 are formed. The difference in reaction products is explained in the reaction scheme shown in Fig. 3.33 which gives an insight into the different reaction pathways. The thiyl radical, derived from cysteine by abstraction of an H-atom, is added to the epoxyallyllic radical only in the absence of oxygen (pathway 2 in Fig. 3.33). In the presence

of oxygen, oxidation of cysteine to cysteine oxide and of fatty acids to their more oxidized forms (Fig. 3.32) occur with a higher reaction rate than in the previous reaction.

Table 3.36. Taste of oxidized fatty acids

Compound	Threshold value for bitter taste (mmol/1)
13-Hydroperoxy-cis-9,trans-11-octadecadienoic acid	not bitter[a]
9-Hydroperoxy-trans-10,cis-12-octadecadienoic acid	not bitter[a]
13-Hydroxy-cis-9, trans-11-octadecadienoic acid	7.6–8.5[a]
9-Hydroxy-trans-10,cis-12-octadecadienoic acid	6.5–8.0[a]
9,12,13-Trihydroxy-trans-10-octadecenoic acid	0.6–0.9[b]
9, 10, 13-Trihydroxy-trans-11-octadecenoic acid	

[a] A burning taste sensation.
[b] A blend of the two trihydroxy fatty acids was assessed.

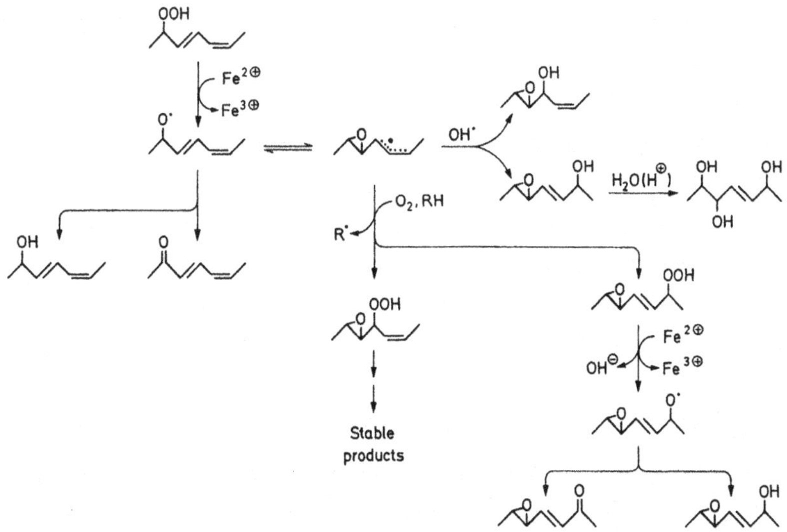

Fig. 3.32. Degradation of linoleic acid hydroperoxides to hydroxy-, epoxy- and oxo-fatty acids. The postulated reaction sequence explains the formation of identified products. Only segments of the structures are presented (according to *Gardner*, 1985)

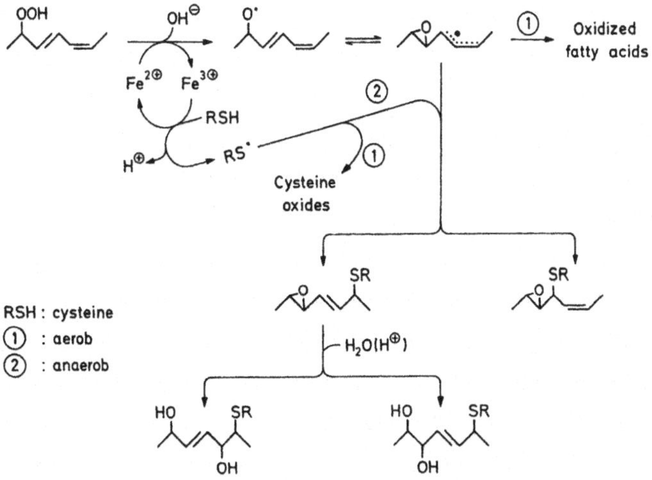

Fig. 3.33. Interaction of linoleic acid hydroperoxides with cysteine. A hypothesis to explain the reaction products obtained. Only segments of the structures are presented (according to *Gardner*, 1985)

As a consequence, a large portion of the oxidized lipid from protein-containing food stored in air does not have lipid–protein covalent bonds and, hence, is readily extracted with a lipid solvent such as chloroform/methanol (2:1).

3.7.2.4.3 Protein Changes

Some properties of proteins are changed when they react with hydroperoxides or their degradation products. This is reflected by changes in food texture, decreases in protein solubility (formation of cross-linked proteins), color (browning) and changes in nutritive value (loss of essential amino acids).

The radicals generated from hydroperoxides (cf. Fig. 3.33) can abstract H-atoms from protein (PH), preferentially from the amino acids Trp, Lys, Tyr, Arg, His and cysteine, in which the phenolic HO-, S- or N-containing groups react:

$$RO^\bullet + PH \longrightarrow P^\bullet + ROH \qquad (3.78)$$

$$2\,P^\bullet \longrightarrow P\text{–}P \qquad (3.79)$$

In Reaction 3.79. protein radicals combine with each other, resulting in the formation of a protein network. Malonaldehyde is generated (cf. 3.7.2.1.9) under certain conditions during lipid peroxidation. As a bifunctional reagent, malonaldehyde can crosslink proteins through a *Schiff* base reaction with the ε-NH_2 groups of two lysine residues:

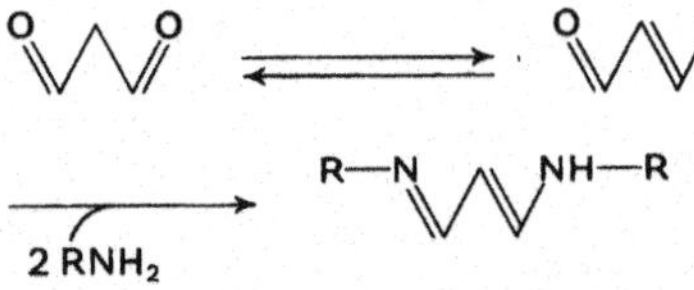

$$(3.80)$$

The *Schiff* base adduct is a conjugated fluorochrome that has distinct spectral properties (λ_{max} excitation ~ 350 nm; λ_{max} emission ~ 450 nm). Hence, it can be used for detecting lipid peroxidation and the reactions derived from it with the protein present.

Reactions resulting in the formation of a protein network like that oulined above also have practical implications, e. g., they are responsible for the decrease in solubility of fish protein during frozen storage.

Also, the monocarbonyl compounds derived from autoxidation of unsaturated fatty acids readily condense with protein-free NH_2 groups, forming *Schiff* bases that can provide brown polymers by repeated aldol condensations (Fig. 3.34). The brown polymers are often N-free since the amino compound can be readily eliminated by hydrolysis. When hydrolysis occurs in the early stages of aldol condensations (after the first or second condensation; cf. Fig. 3.34) and the released aldehyde, which has a powerful odor, does not

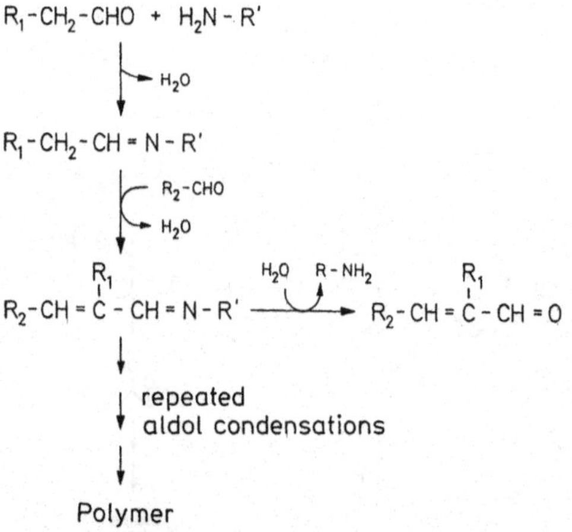

Fig. 3.34. Reaction of volatile aldehydes with protein amino groups

reenter the reaction, the condensation process results not only in discoloration (browning) but also in a change in aroma.

3.7.2.4.4 Decomposition of Amino Acids

Studies of model systems have revealed that protein cleavage and degradation of side chains, rather than formation of protein networks, are the preferred reactions when the water content of protein/lipid mixtures decreases. Several examples of the extent of losses of amino acids in a protein in the presence of an oxidized lipid are presented in Table 3.37. The strong dependence of this loss on the nature of the protein and reaction conditions is obvious. Degradation products obtained in model systems of pure amino acids and oxidized lipids are described in Table 3.38.

3.7.3 Inhibition of Lipid Peroxidation

Autoxidation of unsaturated acyl-lipids can be retarded by:

- Exclusion of oxygen. Possibilities are packaging under a vacuum or addition of glucose oxidase (cf. 2.7.2.1.1).

Table 3.37. Amino acid losses occurring in protein re-action with peroxidized lipids

Reaction system		Reaction conditions		Amino acids lost
protein	lipid	time	T (°C)	(% loss)
Cyto-chrome C	Linolenic acid	5 h	37	His(59),Ser(55), Pro(53),Val(49), Arg(42),Met(38), Cys(35)[a]
Trypsin	Linoleic acid	40 min	37	Met(83),His(12)[a]
Lysozyme	Linoleic acid	8 days	37	Trp(56),His(42), Lys(17),Met(14), Arg(9)
Casein	Linoleic acid ethyl ester	4 days	60	Lys(50), Met(47), Ile(30),Phe(30), Arg(29),Asp(29), Gly(29),His(28), Thr(27),Ala(27), Tyr(27)[a,b]
Oval-bumin	Linoleic acid ethyl ester	24 h	55	Met(17),Ser(10), Lys(9),Ala(8), Leu(8)[a,b]

[a] Trp analysis was not performed.
[b] Cystine analysis was not performed.

Table 3.38. Amino acid products formed in reaction with peroxidized lipid

Reaction system		Compounds formed from amino acids
amino acid	lipid	
His	Methyl linoleate	Imidazolelactic acid, Imidazoleacetic acid
Cys	Ethyl arachidonate	Cystine, H_2S, cysteic acid, alanine, cystine-disulfoxide
Met	Methyl linoleate	Methionine-sulfoxide
Lys	Methyl linoleate	Diaminopentane, aspartic acid, glycine, alanine, α-aminoadipic acid, pipecolinic acid, 1,10-diamino-1,10-dicarboxydecane

- Storage at low temperature in the dark. The autoxidation rate is thereby decreased substantially. However, in fruits and vegetables which contain the lipoxygenase enzyme, these precautions are not applicable. Food deterioration is prevented only after in activation of the enzyme by a blanching process (cf. 2.6.4).

- Addition of antioxidants to food.

3.7.3.1 Antioxidant Activity

The peroxy and oxy free radicals formed during the propagation and branching steps of the autoxidation radical chain (cf. Fig. 3.19) are scavenged by antioxidants (AH; cf. Fig. 3.35). Antioxidants containing a phenolic group play the major role in food. In reactions 1 and 2 in Fig. 3.35, they form radicals which are stabilized by an aromatic resonance system. In contrast to the acyl peroxy and oxy free radicals, they are not able to abstract a H-atom from an unsaturated fatty acid and therefore cannot initiate lipid peroxidation. The end-products formed in reactions 3 and 4 in Fig. 3.35 are relatively stable and in consequence the autoxidation radical chains are shortened.

The reaction scheme (Fig. 3.35) shows that one antioxidant molecule combines with two radicals. Therefore, the maximum achievable stoichiometric factor is $n = 2$. In practice, the value of n is between 1 and 2 for the antioxidants used. Antioxidants, in addition to their main role as radical scavengers, can also partially reduce hydroperexides to hydroxy compounds.

3.7.3.2 Antioxidants in Food

3.7.3.2.1 Natural Antioxidants

The unsaturated lipids in living tissue are relatively stable. Plants and animals have the necessary complement of antioxidants and of enzymes, for instance, glutathione peroxidase and superoxide dismutase, to effectively prevent lipid oxidation.

During the isolation of oil from plants (cf. 3.8.3), tocopherols are also isolated. A sufficient level is retained in oil even after refining, thus, toco-

$$RO_2^{\cdot} + AH \longrightarrow ROOH + A^{\cdot} \quad (1)$$

$$RO^{\cdot} + AH \longrightarrow ROH + A^{\cdot} \quad (2)$$

$$RO_2^{\cdot} + A^{\cdot} \longrightarrow ROOA \quad (3)$$

$$RO^{\cdot} + A^{\cdot} \longrightarrow ROA \quad (4)$$

Fig. 3.35. Activity of an antioxidant as a radical scavenger. AH: Antioxidant

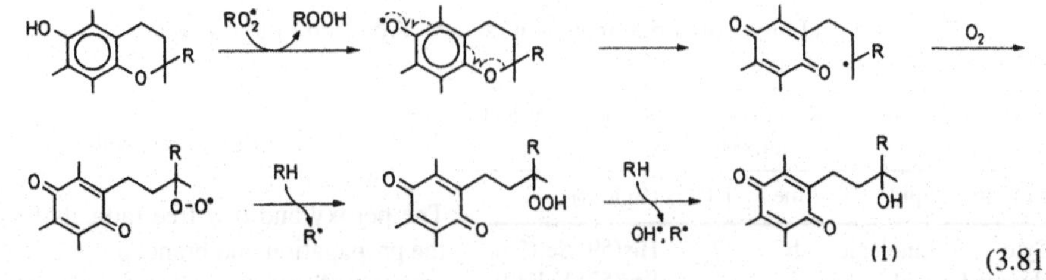

$$(3.81)$$

pherols secure the stability of the oil end-product. Soya oil, due to its relatively high level of furan fatty acids and linolenic acid (cf. 14.3.2.2.5), is an exception. The tocopherol content of animal fat is influenced by animal feed.

The antioxidant activity of tocopherols increases from α → δ. It is the reverse of the vitamin E activity (cf. 6.2.3) and of the rate of reaction with peroxy radicals. Table 3.39 demonstrates that α-tocopherol reacts with peroxy radicals faster than the other tocopherols and the synthetic antioxidants DBHA and BHT.

The higher efficiency of γ-tocopherol in comparison to α-tocopherol is based on the higher stability of γ-tocopherol and on different reaction products formed during the antioxidative reaction.

After opening of the chroman ring system, α-tocopherol is converted into an alkyl radical which in turn oxidizes to a hydroxy-alkylquinone (I in Formula 3.81). α-Tocopherol is a faster scavenger for peroxy radicals formed during autoxidation than γ-tocopherol (Table 3.39), but α-tocopherol then generates an alkyl radical which, in contrast to the slow reacting chromanoxyl radical, can start autoxidation of unsaturated fatty acids. Therefore, the peroxidation rate of an unsaturated fatty acid increases with higher α-tocopherol concentrations after

going through a minimum. This prooxidative effect is smaller in the case of γ-tocopherol because in contrast to α-tocopherol, no opening of the chroman ring takes place but formation of diphenylether and biphenyl dimers occurs. The supposed explanation for these reaction products is: The peroxy radical of a fatty acid abstracts a hydrogen atom from γ-tocopherol (Formula 3.82). A chromanoxyl radical (I) is formed, that can transform into a chromanyl radical (II). Recombination of (I) and (II) results in the diphenylether dimer (III) and recombination of two radicals (II) into the biphenyl dimer (IV). Unlike p-quinone from the reaction of α-tocopherol, the dimer structures (III) and (IV) possess one or two phenolic OH-groups that are also antioxidatively active.

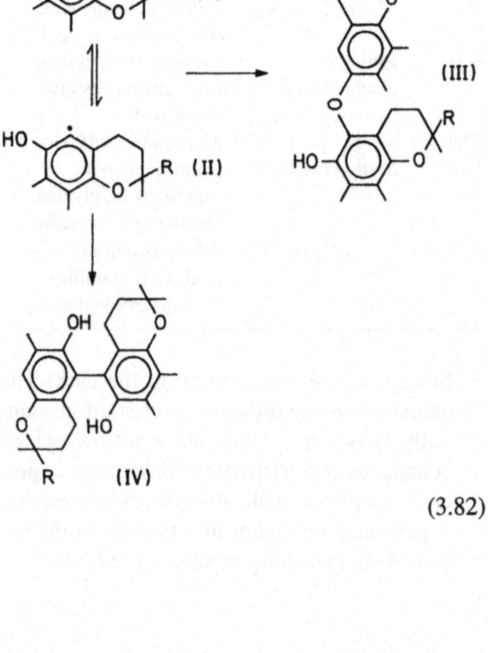

$$(3.82)$$

Table 3.39. Rate constants of tocopherols and BHT for reaction 2 in Fig. 3.35 at 30°C

Antioxidant	$k(1 \cdot mol^{-1} \cdot s^{-1}) \cdot 10^{-5}$
α-Tocopherol	23.5
β-Tocopherol	16.6
γ-Tocopherol	15.9
δ-Tocopherol	6.5
2,6-Di-*tert*-butyl-4-hydroxyanisole (DBHA)	1.1
2,6-Di-tert-butyl-p-cresol (BHT)	0.1

In the reaction with peroxyl radicals, the higher rate of tocopherols compared with DBHA (cf. Table 3.39) is based on the fact that the chromanoxyl radical formed on H-abstraction is more stable than the phenoxyl radical. Both types of radical are stabilized by the following resonance:

$$(3.83)$$

This resonance effect is the highest when the orbital of the 2p electron pair of the ether oxygen and the half occupied molecule orbital of the radical oxygen are aligned parallel to each other, i. e., vertical to the plane of the aromatic ring. Any deviation lowers the stability, slowing down H-abstraction. Due to incorporation into a six-membered ring present in the *half-chair* conformation, the ether oxygen is so strongly fixed in the chromanoxyl radical that the deviation is only 17°. The methoxy group in the DBHA phenoxyl radical is freely rotatable so that the orbital of the 2p electron pair is oriented to the plane of the aromatic ring; thus the deviation is ~90°. BHT reacts even slower than DBHA (Table 3.39) because there is no ether oxygen.

Ascorbic acid (cf. 6.3.9) is active as an antioxidant in aqueous media, but only at higher concentrations ($\sim 10^{-3}$ mol/l). A prooxidant activity is observed at lower levels (10^{-5} mol/l), especially in the presence of heavy metal ions. The effect of tocopherols is enhanced by the addition of fat soluble ascorbyl palmitate or ascorbic acid in combination with an emulsifier (e. g. lecithin) since the formed tocopherol radical from reaction 2 in Fig. 3.35 is rapidly reduced to α-tocopherol by vitamin C.

Carotinoids also can act as scavengers for alkyl radicals. Radicals stabilized by resonance are formed (Formula 3.84), unable to initiate lipid peroxidation. β-Carotenes are most active at a concentration of $5 \cdot 10^{-5}$ mol/l, while at higher concentrations the prooxidative effect is predominant. Also the partial pressure of oxygen is critical, it should be below 150 mm Hg.

Phenolic compounds (cf. 18.1.2.5) which are widely distributed in plant tissues, act as natural antioxidants. The protective effect of several herbs, spices (e. g. sage or rosemary) and tea extracts against fat (oil) oxidation is based on the presence of such natural antioxidants (cf. 21.2.5.1 and 22.1.1.4). The antioxidative effect of phenols depends on the pH. It is low in an acidic medium (pH 4) and high in an alkaline medium (pH 8) when phenolation occurs. In the protection of linoleic acid micelles, the antioxidative activity of quercetin is approximately as high as that of α-tocopherol (Table 3.40). The activity of the two synthetic dihydroxyflavones at 70% and 63% is also high. Therefore, it is not only the number of OH-groups in the molecule, but the presence of OH-groups in the ortho position that is important. But this characteristic feature is not enough to explain the high activity of quercetin compared with that of catechin, which is four times less active (Table 3.40) although the OH patterns correspond. Obviously the carbonyl group, which is absent in catechin, increases the stability of the phenoxyl radical by electron attraction which

Table 3.40. Relative antioxidative activity (RAA) of flavonoids, cumarins and hydroxycinnamic acids[a,b]

Compound	RAA × 100
α-Tocopherol	100
Quercetin (cf. Formula 18.32)	90
Cyanidin	90
Catechin (cf. Formula 18.20)	22
6,7-Dihydroxyflavone	70
7,8-Dihydroxyflavone	63
7,8-Dihydroxycumarin	3.3
Ferulic acid	< 0.1
Caffeic acid	< 0.1

[a] Test system: linoleic acid micelles stabilized with Na dodecyl sulfate (pH 7.4, T: 50 °C).
[b] RAA with reference to the activity of α-tocopherol.

$$(3.84)$$

includes the 2,3-double bond. The phenoxyl radical arises from the trapping reaction of peroxyl radicals with quercetin. Another factor is the lipophilicity. Catechin is more hydrophobic than caffeic acid, which is present as an anion at pH 7.4 and, consequently, cannot penetrate the linoleic acid micelle. For this reason, although caffeic acid has OH-groups in ortho position, it is not an antioxidant under the conditions given in Table 3.40.

Polyphenols in wood, such as lignin, undergo thermal cracking, resulting in volatile phenols, during the generation of smoke by burning wood or, even more so, sawdust. These phenols deposit on the food surface during smoking and then penetrate into the food, thus acting as antioxidants.

It was also demonstrated that vanillin, in food items where its aroma is desired, plays an important role as an antioxidant. Finally, some of the *Maillard* reaction products, such as reductones (cf. 4.2.4.4), should be considered as naturally active antioxidants.

3.7.3.2.2 Synthetic Antioxidants

In order to be used as an antioxidant, a synthetic compound has to meet the following requirements: it should not be toxic; it has to be highly active at low concentrations (0.01–0.02%); it has to concentrate on the surface of the fat or oil phase. Therefore, strongly lipophilic antioxidants are particularly suitable (with low HLB values, e. g. BHA, BHT or tocopherols, dodecylgallate) for o/w emulsions. On the other hand, the more polar antioxidants, such as TBHQ and propyl gallate, are very active in fats and oils since they are enriched at the surface of fat and come in contact with air. Antioxidants should be stable under the usual food processing conditions. This stability is denoted as the "carry through" effect. Some of the synthetic antioxidants used worldwide are:

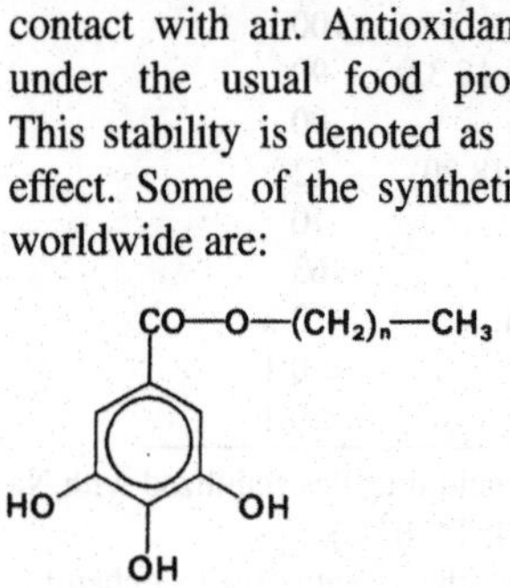

$$(3.85)$$

Propyl (n = 2); octyl (n = 7) and dodecyl (n = 11) gallate

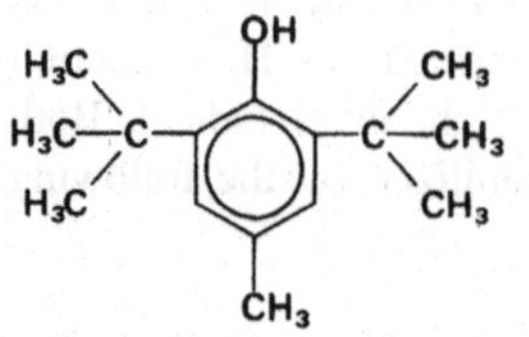

$$(3.86)$$

2,6-Di-tert-butyl-p-hydroxytoluene (BHT)

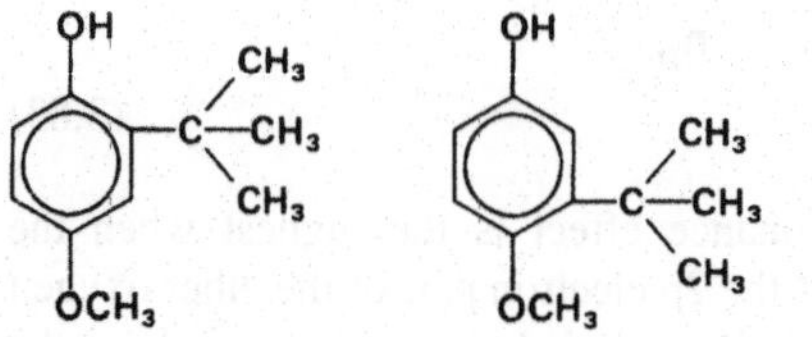

$$(3.87)$$

tert-Butyl-4-hydroxyanisole (BHA)
Commercial BHA is a mixture of two isomers, 2- and 3-tert-butyl-4-hydroxyanisole

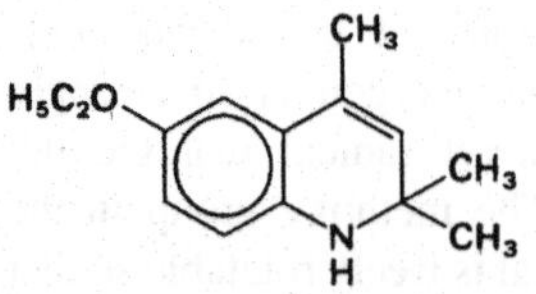

$$(3.88)$$

6-Ethoxy-1,2-dihydro-2,2,4-trimethylquinoline (ethoxyquin)

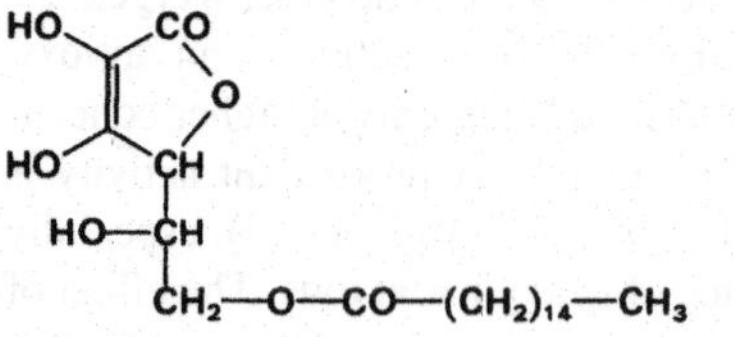

$$(3.89)$$

Ascorbyl palmitate
ESR spectroscopy has demonstrated that a large portion of ethoxyquin is present in oil as a free radical

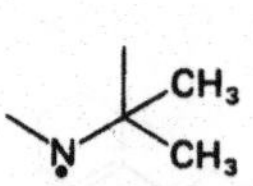

$$(3.90)$$

and stabilization by dimerization of the radical occurs. The radical, and not the dimer, is the active antioxidant.

tert-Butylhydroquinone (TBHQ) is a particularly powerful antioxidant used, for example, for stabilization of soya oil. The "carry through" properties are of importance in the use of BHA, TBHQ and BHT in food processing. All three antioxidants are steam distillable at higher temperatures. Utilization of antioxidants is often regulated by governments through controls on the use of food additives. In North America incorporation of antioxidants is permitted at a maximum level of 0.01% for any one antioxidant, and a maximum of 0.02% for any combination. The regulations related to permitted levels often vary from country to country.

The efficiency of an antioxidant can be evaluated by a comparative assay, making use of an "antioxidative factor" (AF):

$$AF = I_A/I_0 \qquad (3.91)$$

where I_A = oxidation induction period for a fat or oil (cf. 3.7.2.1.1) in the presence of an antioxidant and I_0 = oxidation induction period of a fat or oil without an antioxidant.

Hence, the efficiency of an antioxidant increases with an increase in the AF value. As illustrated by the data in Table 3.41, BHA in comparison with BHT shows a higher efficiency in a lard sample. This result is understandable since in BHT both tertiary butyl substituents sterically hinder the reaction with radicals to a certain extent (reaction 1 in Fig. 3.35). The effect on antioxidants depends not only on the origin of fat or oil but, also, on the processing steps used in the isolation and refining procedures. Hence, data in Table 3.41 serve only as an illustration.

BHA and BHT together at a given total concentration are more effective in extending shelf-life of a fat or oil than either antioxidant alone at the same level of use (Table 3.41).

To explain this, it is suggested that BHA, by participating in reaction 1 (Fig. 3.35), provides

Table 3.41. Antioxidative factor (AF) values of some antioxidants (0.02%) in refined lard

Antioxidant	AF	Antioxidant	AF
d-α-Tocopherol	5	Octyl gallate	6
dl-γ-Tocopherol	12	Ascorbyl palmitate	4
BHA	9.5	BHA and	
BHT	6	BHT[a]	12

[a] Each compound is added in amount of 0.01%.

a phenoxy radical (I):

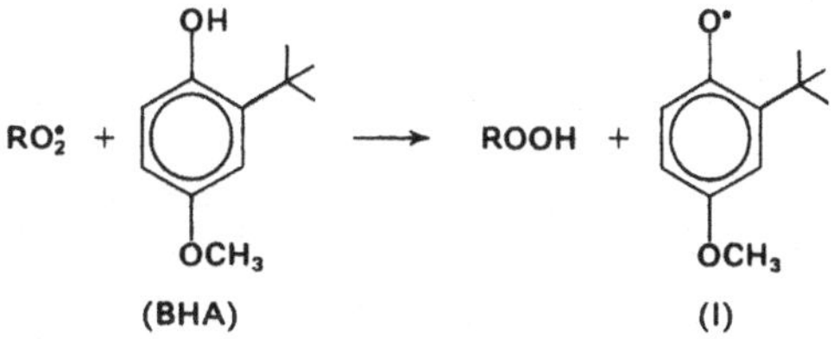

(3.92)

which is then regenerated into the original molecule by rapid interaction with BHT:

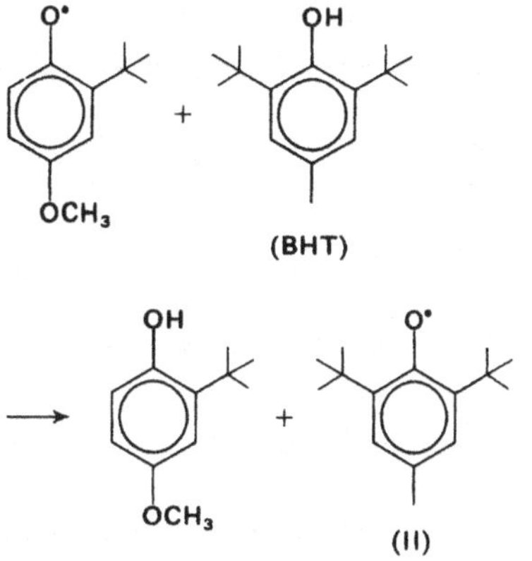

(3.93)

On the other hand, the phenoxy radical (II) derived from BHT can react further with an additional peroxy radical:

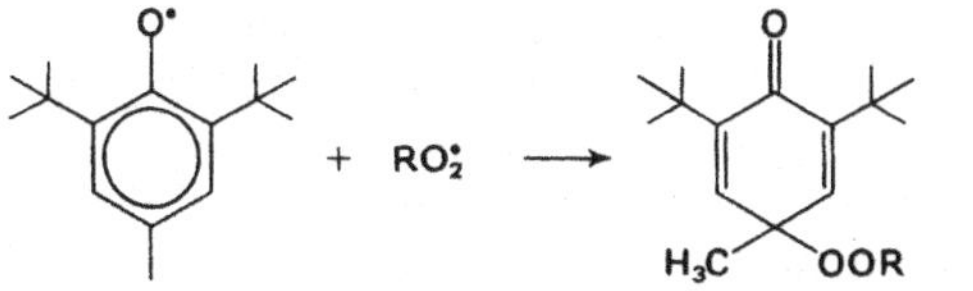

(3.94)

Propyl gallate (PG) increases the efficiency of BHA, but not that of BHT. Ascorbyl palmitate, which is by itself a rather weak antioxidant, substantially sustains the antioxidative activity of γ,d,l-tocopherol.

3.7.3.2.3 Synergists

Substances which enhance the activity of antioxidants are called synergists. The main examples are lecithin, amino acids, citric, phosphoric, citraconic and fumaric acids, i. e. compounds which complex heavy metal ions (chelating agents, sequesterants or scavengers of trace metals). Thus, initiation of heavy metal-catalyzed lipid autoxidation can be prevented (cf. 3.7.2.1.6). Results compiled in Table 3.42 demonstrate the synergistic activities of citric and phosphoric acids in combination with lauryl gallate. Whereas citric acid enhances the antioxidant effectiveness in the presence of all three metal ions, phosphoric acid is able to do so with copper and nickel, but not with iron. Also, use of citric acid is more advantageous since phosphoric acid promotes polymerization of fat or oil during deep frying.

The synergistic effect of phospholipids is different. Addition of dipalmitoylphosphatidylethanolamine (0.1–0.2 weight %) to lard enhances the antioxidative activity of α-tocopherol, BHA, BHT and propyl gallate, while phosphatidylcholine shows no activity.

The reaction of ascorbic acid with tocopherol radicals as described in 3.7.3.2 is a synergistic effect.

3.7.3.2.4 Prooxidative Effect

The activity of antioxidants reverses under certain conditions: they become prooxidants. One way in which α-tocopherol can become peroxidatively active is shown in Formula 3.81. Another way is through the formation of the chromanoxyl radical in concentrations high enough to overcome the inertness mentioned in 3.7.3.1 and abstract H-atoms from unsaturated acyl lipids to a definite extent, starting lipid peroxidation. This activity reversion, which is also undesirable from a nutritional and physiological point of view, is prevented by co-antioxidants, e.g., vitamin C (cf. 3.7.3.2.1), which can reduce the chromanoxyl radical to α-tocopherol.

In the presence of heavy metal ions, e.g., $Fe^{3\oplus}$, ascorbic acid becomes a peroxidant. It reduces $Fe^{3\oplus}$ to $Fe^{2\oplus}$, which can produce superoxide radical anions or hydroxyl radicals with oxygen or H_2O_2 (*Fenton* reaction, cf. 3.7.2.1.8) Prooxidative effects have also been observed with carotenoids and flavonoids at higher concentrations.

3.7.4 Fat or Oil Heating (Deep Frying)

Deep frying is one of the methods of food preparation used both in the home and in industry. Meat, fish, doughnuts, potato chips or french fries are dipped into fat (oil) heated to about 180 °C. After several minutes of frying, the food is sufficiently tender to be served.

The frying fat or oil changes substantially in its chemical and physical properties after prolonged use. Data for a partially hydrogenated soybean oil

Table 3.42. Synergistic action of citric (C) and phosphoric acids (P) in combination with lauryl gallate (LG) on oxidation of fats and oils

Added to fat/oil	AF value after addition of				
	0.01% C	0.01% P	0.01% LG	0.01%LG + 0.01% C	0.01% LG + 0.01% P
0.2 ppm Cu	0.3	0.2	0.9	4.7	4.1
2 ppm Fe	0.6	0.5	0.1	5.7	0.2
2 ppm Ni	0.5	0.6	3.0	7.0	4.4

Table 3.43. Characteristics of partially hydrogenated soybean oil before and after simulated deep fat frying[a]

Characteristics	Fresh oil	Heated oil
Iodine number	108.9	101.3
Saponification number	191.4	195.9
Free fatty acids[b]	0.03	0.59
Hydroxyl number	2.25	9.34
DG	1.18	2.73
Composition of fatty acids (weight %)		
14:0	0.06	0.06
16:0	9.90	9.82
18:0	4.53	4.45
18:1 (9)	45.3	42.9
18:2 (9, 12)	37.0	29.6
18:3 (9, 12, 15)	2.39	1.67
20:0	0.35	0.35
22:0	0.38	0.38
Other	0.50	0.67

[a] The oil was heated for 80 h (8 h/day) at 195 °C. Batches of moist cotton balls containing 75% by weight of water were fried at 30-min intervals (17 frying operations/day) in order to simulate the deep frying process.
[b] Weight % calculated as oleic acid.

compiled in Table 3.43 indicate that heating of oil causes reactions involving double bonds. This will result in a decrease in iodine number. As can be deduced from changes in the composition of fatty acids (Table 3.43), in the case of soybean oil, linoleic and linolenic acid are the most affected. Peroxides formed at elevated temperatures fragment immediately with formation of hydroxy compounds thus increasing the hydroxyl number (Table 3.43). Therefore, determination of peroxide values to evaluate the quality of fat or oil in deep frying is not appropriate.

Unsaturated TG polymerize during heating thus increasing the viscosity of the fat. Di- and trimeric TG are formed. The increase of these components can be monitored by means of gel permeation chromatography (GPC) (Fig. 3.36).

Before or after methanolysis of the oil sample, GPC is a valuable first tool to analyze the great number of reaction products formed during deep frying. Monomeric methyl esters are further fractionated via the urea adducts, while the cyclic fatty acids enrich themselves in the supernatant. Dimeric methyl esters can be preseparated by RP-HPLC and further analyzed by GC/MS after silylation of the OH-groups. A great number of volatile and nonvolatile products are obtained during deep frying of oil or fat. The types of reactions involved in and responsible for changes in oil and fat structures are compiled in Table 3.44. Some of the reactions presented will be outlined in more detail.

3.7.4.1 Autoxidation of Saturated Acyl Lipids

The selectivity of autoxidation decreases above 60 °C since the hydroperoxides formed are subjected to homolysis giving hydroxy and alkoxy radicals (Reaction RS-4 in Fig. 3.19) which, due to their high reactivity, can abstract H-atoms even from saturated fatty acids.

Numerous compounds result from these reactions. For example, Table 3.45 lists a series of aldehydes and methyl ketones derived preferentially from tristearin. Both classes of compounds are also formed by thermal degradation of free fatty acids. These acids are formed by triglyceride hydrolysis or by the oxidation of aldehydes.

Table 3.44. A review of reactions occurring in heat treated fats and oils

Fat/oil heating	Reaction	Products
1. Deep frying without food	Autoxidation Isomerization Polymerization	Volatile acids aldehydes esters alcohols Epoxides Branched chain fatty acids Dimeric fatty acids Mono- and bicyclic compounds Aromatic compounds Compounds with trans double bonds Hydrogen, CO_2
2. Deep frying with food added	As under 1. and in addition hydrolysis	As under 1. and in addition free fatty acids, mono- and diacylglycerols and glycerol

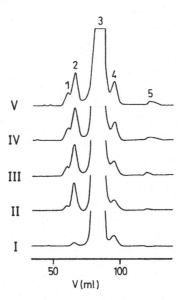

Fig. 3.36. Gel permeation chromatography of heated soybean oil (according to *Rojo* and *Perkins*, 1987). Oil samples (composition and heating conditions see Table 3.41) were analyzed immediately (I), as well as after 8 h (II), 24 h (III), 48 h (IV) and 80 h (V), *1* Trimeric TG, *2* Dimeric TG, *3* TG, *4* DG, *5* free fatty acids

Table 3.45. Volatile compounds formed from heat-treated tristearin[a]

Class of compound	Portion	C-number	Major compounds
Alcohols	2.7	4–14	n-Octanol
			n-Nonanol
			n-Decanol
γ-Lactones	4.1	4–14	γ-Butyrolactone
			γ-Pentalactone
			γ-Heptalactone
Alkanes	8.8	4–17	n-Heptadecane
			n-Nonane
			n-Decane
Acids	9.7	2–12	Caproic acid
			Valeric acid
			Butyric acid
Aldehydes	36.1	3–17	n-Hexanal
			n-Heptanal
			n-Octanal
Methyl ketones	38.4	3–17	2-Nonanone
			2-Heptanone
			2-Decanone

[a] Tristearin is heated in air at 192 °C.

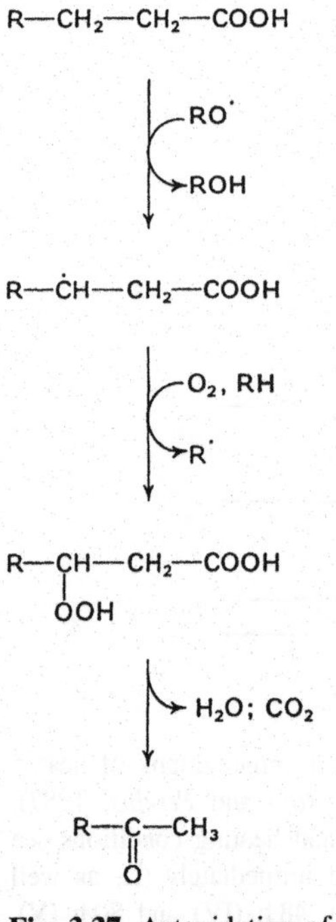

$$R-CH_2-CH_2-COOH$$

$$\downarrow \begin{array}{l} RO^{\cdot} \\ ROH \end{array}$$

$$R-\overset{\cdot}{C}H-CH_2-COOH$$

$$\downarrow \begin{array}{l} O_2, RH \\ R^{\cdot} \end{array}$$

$$R-\underset{\underset{OOH}{|}}{C}H-CH_2-COOH$$

$$\downarrow H_2O; CO_2$$

$$R-\underset{\underset{O}{\|}}{C}-CH_3$$

Fig. 3.37. Autoxidation of saturated fatty acids. Postulated reaction steps involved in formation of methyl ketones

Methyl ketones are obtained by thermally induced β-oxidation followed by a decarboxylation reaction (Fig. 3.37). Aldehydes are obtained from the fragmentation of hydroperoxides by a β-scission mechanism (Fig. 3.38) occurring nonselectively at elevated temperatures (compare the difference with 3.7.2.1.9).

Unsaturated aldehydes with a double bond conjugated to the carbonyl group are easily degraded during the deep frying process (Formula 3.95). Addition of water results in the formation of a 3-hydroxyaldehyde that is split by retro aldol condensation catalyzed by heat. Examples of this mechanism are the degradation of (E,Z)-2,6-nonadienal to (Z)-4-heptenal and acetaldehyde, as well as the cleavage of 2,4-decadienal into 2-octenal and acetaldehyde.

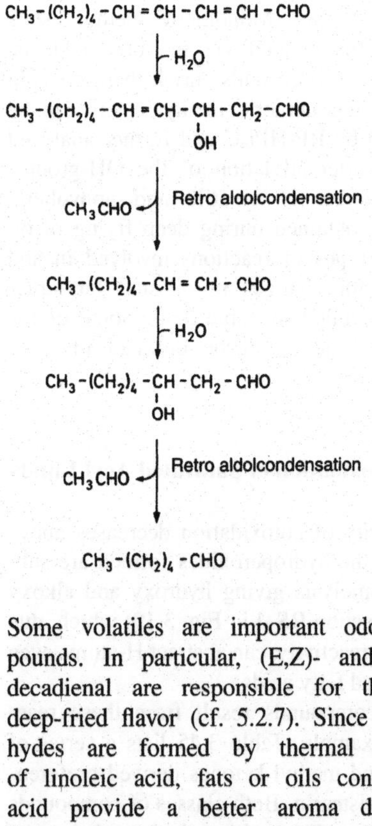

$$CH_3-(CH_2)_4-CH=CH-CH=CH-CHO$$

$$\downarrow H_2O$$

$$CH_3-(CH_2)_4-CH=CH-\underset{\underset{OH}{|}}{C}H-CH_2-CHO$$

$$CH_3CHO \dashv \quad \text{Retro aldolcondensation}$$

$$CH_3-(CH_2)_4-CH=CH-CHO$$

$$\downarrow H_2O$$

$$CH_3-(CH_2)_4-\underset{\underset{OH}{|}}{C}H-CH_2-CHO$$

$$CH_3CHO \dashv \quad \text{Retro aldolcondensation}$$

$$CH_3-(CH_2)_4-CHO \qquad (3.95)$$

Some volatiles are important odorous compounds. In particular, (E,Z)- and (E,E)-2,4-decadienal are responsible for the pleasant deep-fried flavor (cf. 5.2.7). Since these aldehydes are formed by thermal degradation of linoleic acid, fats or oils containing this acid provide a better aroma during deep frying than hydrogenated fats. However, if a fat is heated for a prolonged period of time, the volatile compounds produce an off-flavor.

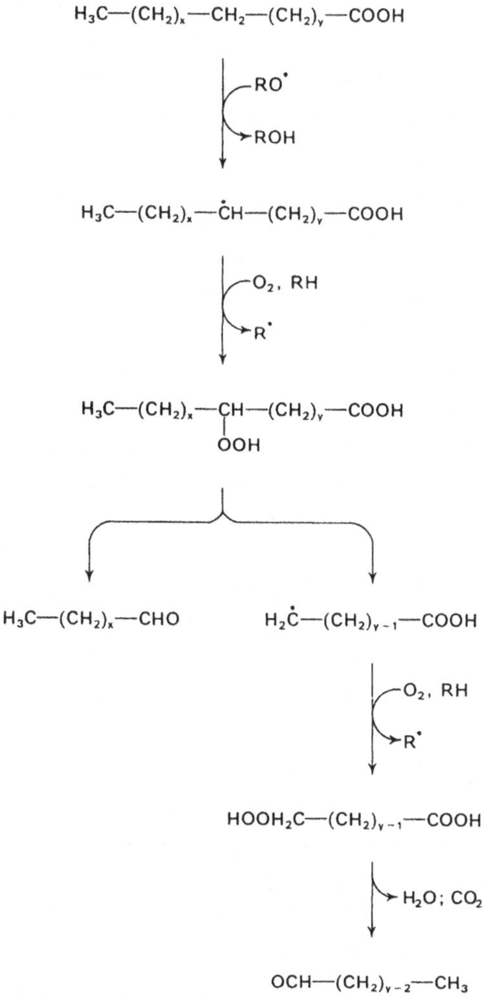

H_3C—$(CH_2)_x$—CH_2—$(CH_2)_y$—COOH

$\downarrow$ —RO$^\bullet$

$\rightarrow$ROH

H_3C—$(CH_2)_x$—$\overset{\bullet}{C}H$—$(CH_2)_y$—COOH

$\downarrow$ —O_2, RH

$\rightarrow$R$^\bullet$

H_3C—$(CH_2)_x$—$\underset{OOH}{CH}$—$(CH_2)_y$—COOH

H_3C—$(CH_2)_x$—CHO $H_2\overset{\bullet}{C}$—$(CH_2)_{y-1}$—COOH

$\downarrow$ —O_2, RH

$\rightarrow$R$^\bullet$

HOOH$_2$C—$(CH_2)_{y-1}$—COOH

$\downarrow$ $\rightarrow$$H_2O$; CO_2

OCH—$(CH_2)_{y-2}$—CH_3

Fig. 3.38. Autoxidation of saturated fatty acids. Hypothetical reactions involved in formation of volatile aldehydes

3.7.4.2 Polymerization

Under deep frying conditions, the isolenic fatty acids are isomerized into conjugated fatty acids which in turn interact by a 1,4-cycloaddition, yielding so-called *Diels–Alder* adducts (cf. Reaction 3.96).

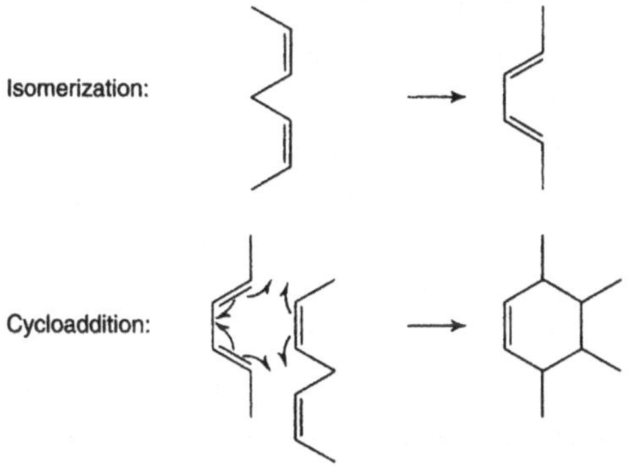

Isomerization:

Cycloaddition:

(3.96)

The side chains of the resultant tetra-substituted cyclohexene derivatives are shortened by oxidation to oxo, hydroxy or carboxyl groups. In addition, the cyclohexene ring is readily dehydrogenated to an aromatic ring, hence compounds related to benzoic acid can be formed. The fatty acid or triacylglycerol radicals formed by H-abstraction in the absence of oxygen can dimerize and then form a ring structure:

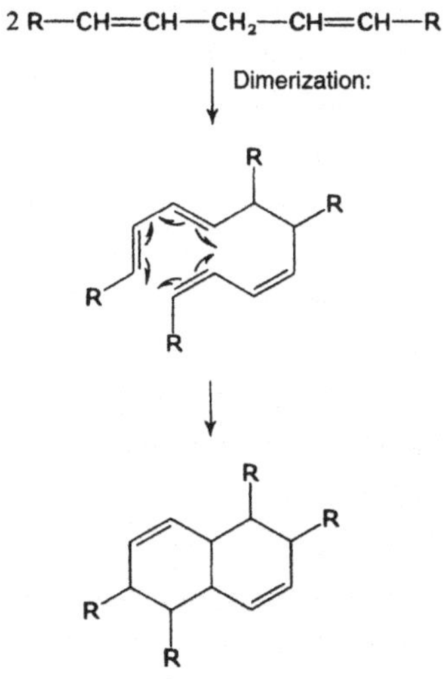

2 R—CH=CH—CH_2—CH=CH—R

$\downarrow$ Dimerization:

(3.97)

On the other hand, polymers with ether and peroxide linkages are formed in the presence of oxygen. They also may contain hydroxy, oxo or epoxy groups. The following structures, among others, have been identified:

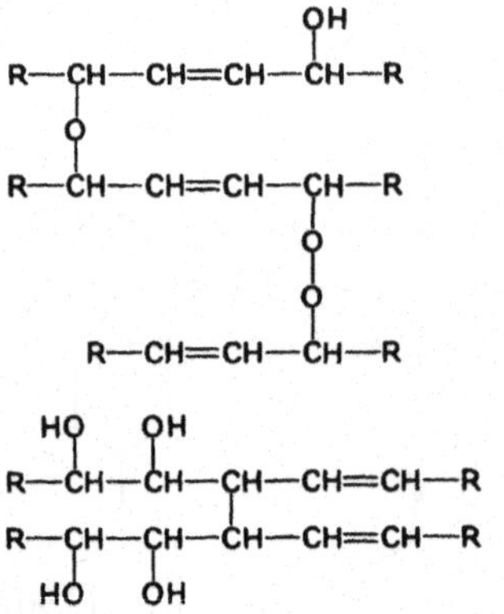

$$R—CH—CH=CH—CH—R$$

(Formula structures)

$$(3.98)$$

Such compounds are undesirable in deep-fried oil or fat since they permanently diminish the flavoring characteristics of the oil or fat and, because of their HO-groups, behave like surface-active agents, i. e. they foam.

Disregarding the odor or taste deficiencies developed in a fat or oil heated for a prolonged period of time, the oil is considered spoiled when its petroleum ether-insoluble oxidized fatty acids reach a level $\geq 1\%$ (or $\geq 0.7\%$ at the decreased smokepoint temperature of $\leq 170\,^{\circ}C$). The fats or oils differ in their heat stability (Table 3.46). The stability is increased by hydrogenation of the double bonds.

3.7.5 Radiolysis

Alkyl and acyloxy radicals are formed during radiolysis of acyl lipids. These will further react to form volatile compounds. The formation of alkanes and alkenes, that lack one or two C-atoms, from the original acyl residue are of interest for the detection of irradiation (Fig. 3.39).

The proposed indicators for the irradiation of meat are the hydrocarbons 14:1, 15:0, 16:1, 16:2, 17:0 and 17:1 which are formed during radiolysis of palmitic, oleic and stearic acid. It was demonstrated that their concentrations in fat increased depending on the radiation dose, e. g., in chicken meat (Fig. 3.40).

Alkyl cyclobutanones are another group of compounds which can be used as indicators of irradi-

ation. They are produced from triacylglycerides (cf. Formula 3.99), but are not formed on heating, e. g., meat, or by microorganisms. On the irradiation of chicken meat with a dose of 1 kGy, $0.72\,\mu g$ of 2-dodecyl cyclobutanone per g of lipid were detected. The indicator is stable because this value fell by only 15% in 18 days.

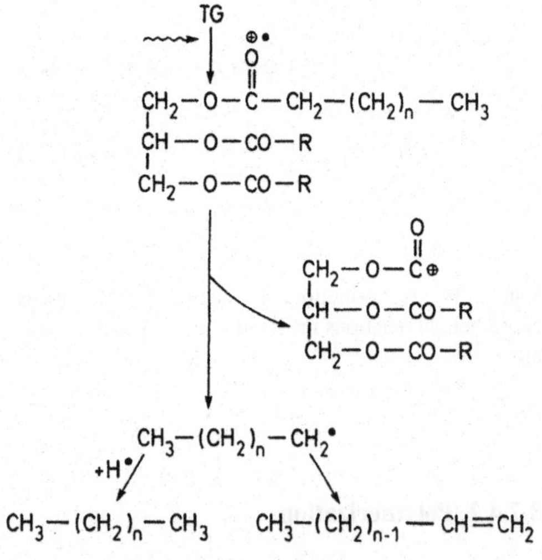

Table 3.46. Relative stability of various fats and oils on deep-frying (RSDF)

Oil/fat	RSDF	Oil/fat	RSDF
Sunflower	1.0	Coconut	2.4
Rapeseed	1.0	Edible beef tallow	2.4
Soya	1.0	Soya oil,	
Peanut	1.2	hydrogenated	2.3
Palm	1.5	Peanut oil,	
Lard	2.0	hydrogenated	4.4
Butter fat	2.3		

Fig. 3.39. Formation of alkanes and 1-alkenes during radiolysis of saturated triacyl glycerols

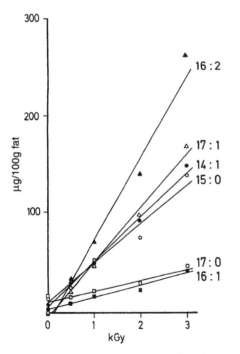

$$(3.99)$$

The dose used to decontaminate meat and spices is in the range 3–30 kGy. Therefore, the chemical reactions mentioned here are to be expected. Considerably lower doses are used for the treatment of fruit and vegetables as shown in the following examples: disinfection of fruit and nuts

Fig. 3.40. Increase in the concentration of hydrocarbons relative to the radiation dose during irradiation of chicken meat (according to *Nawar* et al., 1990)

(0.15–0.5 kGy), inhibition of the germination of potatoes, onions and garlic (0.2–0.15 kGy), delayed ripening of bananas, mangoes and papayas (0.12–0.75 kGy). Consequently, the chemical detection of the irradiation of these foods is quite difficult.

3.7.6 Microbial Degradation of Acyl Lipids to Methyl Ketones

Fatty acids of short and medium chain lengths present in milk fat, coconut and palm oils are degraded to methyl ketones by some fungi. A number of *Penicillium* and *Aspergillus* species, as well as several *Ascomycetes, Phycomycetes* and *fungi imperfecti* are able to do this.

The microorganisms first hydrolyze the triglycerides enzymatically (cf. 3.7.1) and then they degrade the free acids by a β-oxidation pathway (Fig. 3.41). The fatty acids $<C_{14}$ are transformed to methyl ketones, the C-skeletons of which have one C-atom less than those of the fatty acids. Apparently, the thiohydrolase activity of these fungi is higher than the β-ketothiolase activity. Hence, ester hydrolysis occurs instead of thioclastic cleavage of the thioester of a β-keto acid (see a textbook of biochemistry). The β-keto acid released is rapidly decarboxylated enzymatically; a portion of the methyl ketones is reduced to the corresponding secondary alcohols.

The odor threshold values for methyl ketones are substantially higher than those for aldehydes (cf. Tables 3.32 and 3.47). Nevertheless, they act as aroma constituents, particularly in flavors of mold-ripened cheese (cf. 10.2.8.3). However, methyl ketones in coconut or palm oil or in milk fat provide an undesirable, unpleasant odor denoted as "perfume rancidity".

3.8 Unsaponifiable Constituents[*]

Disregarding a few exceptions, fats and oils contain an average of 0.2–1.5% unsaponifiable compounds (Table 3.48). They are isolated from

---

[*] The free higher alcohols described under 3.6.2 and the deacylated alkoxy-lipids belong to this class.

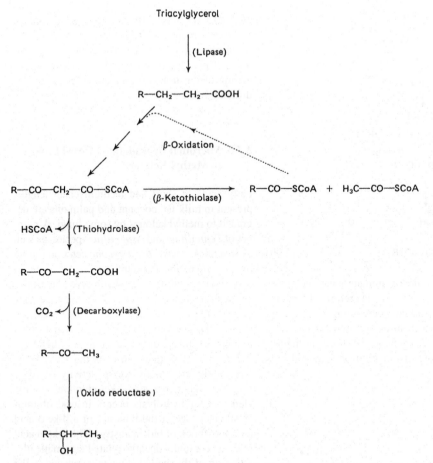

Fig. 3.41. Fungal degradation of triacylglycerols to methyl ketones (according to *Kinsella* and *Hwang*, 1976)

Table 3.47. Sensory properties of methyl ketones

Compound	Odor description	Odor threshold (ppb; in water)
2-Pentanone	Fruity, like bananas	2300
2-Hexanone		930
2-Heptanone	Fragrant, herbaceous	650
2-Octanone	Flowery, refreshing	190
2-Nonanone	Flowery, fatty	190

Table 3.48. Content of unsaponifiables in various fats and oils

Fat/oil	Unsaponifiable (weight-%)	Fat/oil	Unsaponifiable (weight-%)
Soya	0.6–1.2	Shea	3.6–10.0
Sunflower	0.3–1.2	Lard	0.1–0.2
Cocoa	0.2–0.3	Shark	15–17
Peanut	0.2–4.4	(refined)	
Olive	0.4–1.1	Herring	0.7–1.0
Palm	0.3–0.9	(refined)	
Rapeseed	0.7–1.1		

a soap solution (alkali salts of fatty acids) by extraction with an organic solvent.

The unsaponifiable matter contains hydrocarbons, steroids, tocopherols and carotenoids. In addition, contaminants or fat or oil additives, such as mineral oil, plasticizers or pesticide residues, can be found.

Each class of compounds in the unsaponifiable matter is represented by a number of components, the structures and properties of which have been thoroughly elucidated in the past decade or two, thus reflecting the advance in the analytical chemistry of fats and oils.

Studies aimed at elucidating the constituents, and their structures, of unsaponifiable matter are motivated by a desire to find compounds which can serve as a reliable indicator for the identity of a fat or an oil.

3.8.1 Hydrocarbons

All edible oils contain hydrocarbons with an even or an odd C-number (C_{11} to C_{35}). Olive, rice and fish oils are particularly rich in this class of compounds. The main hydrocarbon constituent of olive oil (1–7 g/kg) and rice oil (3.3 g/kg) is a linear triterpene (C_{30}), squalene:

(3.100)

This compound is used as an analytical indicator for olive oil (cf. Table 14.24).

Squalene is present in a substantially higher concentration in fish liver oil. For example, shark liver oil has up to 30% squalene, and 7% pristane (2,6,10,14-tetramethylpentadecane) and some phytane (3,7,11,15-tetramethylhexadecane).

3.8.2 Steroids

The unsaponifiable part of edible fats contains a series of cyclic triterpenes which have structures related to that of steroids. Quantitatively, the 3β-hydroxysteroids are the sterols which are to the fore. Especially diverse is the sterol spectrum of plant fats which contain not only desmethyl, but also 4-methyl and 4,4-dimethyl sterols.

3.8.2.1 Structure, Nomenclature

The steroid skeleton is made up of four condensed rings; A, B, C and D. The first three are in the chair conformation, whereas ring D is usually pla-

nar. While rings B and C, and C and D are fused in a trans-conformation, rings A and B can be fused in a trans- or in a cis-conformation.

(3.101)

Conformational isomers introduced by fusing rings A and B in cholest-5-ene-3-β-ol (cholesterol; cf. Formula 3.101) are not possible since the C-5 position has a double bond.

By convention, the steric arrangement of substituents and H-atoms is related to the angular methyl group attached at C-10. When the plane containing the four rings is assumed to be the plane of this page, the substituent at C-10, by definition, is above the plane; all substituents below the plane are denoted by dashed or dotted lines. They are said to be α-oriented and have a trans-conformation. Substituents above the plane are termed β-oriented and are shown by solid line bonds and, in relation to the angular C-10 methyl group, are of cis-conformation.

In cholesterol (cf. Formula 3.101) the HO-group, the angular methyl group at C-13, the side chain on C-17 and the H-atom on C-8 are β-oriented (cis), whereas the H-atoms at C-9, C-14 and C-17 are α-oriented (trans). These steroids that are not methylated at position C-4 are also denoted as desmethyl steroids.

3.8.2.2 Steroids of Animal Food

3.8.2.2.1 Cholesterol

Cholesterol (cf. Formula 3.101) is obtained biosynthetically from squalene (see a textbook of biochemistry). It is the main steroid of mammals and occurs in lipids in free form or esterified with saturated and unsaturated fatty acids. The content of cholesterol in some foods is illustrated by the data in Table 3.49.

Autoxidation of cholesterol, which is accelerated manyfold by 18:2 and 18:3 fatty acid

Table 3.49. Cholesterol content of some food

Food	Amount (mg/100 g)
Calf brain	2000
Egg yolk[a]	1010
Pork kidney	410
Pork liver	340
Butter	215–330
Pork meat, lean	70
Beef, lean	60
Fish (Halibut;	
Hypoglossus vulgaris)	50

[a] Egg white is devoid of cholesterol.

Table 3.50. Progesterone in foods

Food	Progesterone (µg/kg)
Beef, male[a]	0.01–5
Beef, female[a]	0.5–40
Pork (muscle)	1.1–1.8
Chicken	0.24
Turkey	8.18
Chicken egg	12.5–43.6
Skim milk (0.1% fat)	1.3–4.6
Whole milk (3.5% fat)	9.5–12.5
Cream (32% fat)	42–73
Butter (82% fat)	133–300
Cheese (Gouda, 29% fat)	44
Potatoes	5.1
Wheat	0.6–2.9
Corn germ oil	0.3
Safflower oil	0.7

[a] edible parts.

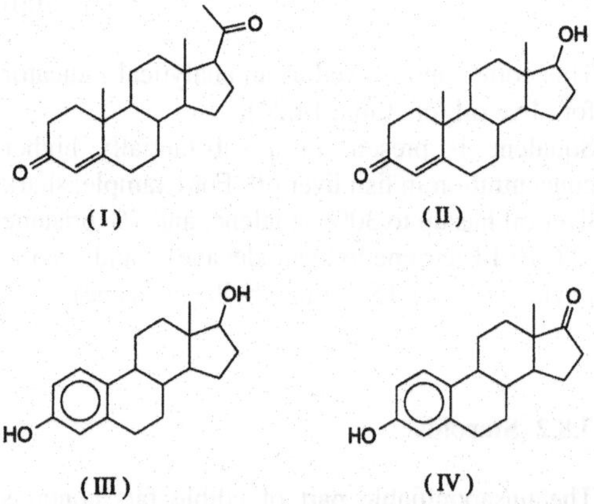

peroxy radicals, proceeds through the intermediary 3β-hydroxycholest-5-en-7α- and 7β-hydroperoxides, of which the 7β-epimer is more stable because of its quasi-equatorial conformation and, hence, is formed predominantly. Unlike autoxidation, the photosensitized oxidation (reaction with a singlet oxygen) of cholesterol yields 3β-hydroxycholest-6-en-5α-hydroperoxide. Among the many derivatives obtained by the further degradation of the hydroperoxides, cholest-5-en-3β,7α-diol, cholest-5en-3β,7β-diol, 3β-hydroxycholest-5-en-7-one, 5,6β-epoxy-5β-cholestan-3β-ol and 5α-cholestan-3β,5,6β-triol have been identified as major products. These so-called "oxycholesterols" have been detected as side components in some food items (dried egg yolk, whole milk powder, butter oil and heated meat). It is difficult to quantify these oxidation products because significant losses can occur in the clean-up of the analyte, e.g., in the case of polar cholestantriol. In addition, artifacts are easily formed. For this reason, quantitative values found in the literature are frequently only approximations.

In the animal organism, cholesterol is the starting point for the synthesis of other steroids, such as sex hormones and bile acids. In fact, GC-MS analyses and radio immunoassays show that among the sex hormones, progesterone (I in Formula 3.102) appears most often in animal food. It is enriched in the fat phase, leading to relatively high concentrations in butter (Table 3.50). Traces of this steroid also occur in plant foods. Testosterone (II in Formula 3.102), 3,17-estradiol (III) and 17-estrone (IV) are other sex hormones which have been identified as

natural trace components of meat, milk and their products.

(3.102)

Products of cholesterol metabolism include C_{19}-sterols which produce the specific smell of boar in boar meat. Five aroma components (Table 3.51) were identified; 5α-androst-16-en-3α-ol (Formula 3.103) has also been detected in truffels (cf. 17.1.2.6.1).

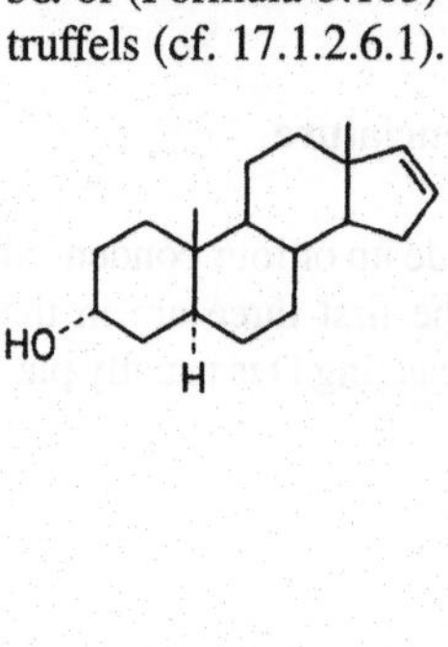

(3.103)

Table 3.51. Odor-active C_{19}-steroids

Compound	Odor threshold (mg/kg ; oil)
5α-Androst-16-en-3-one	0.6
5α-Androst-16-en-3α-ol	0.9
5α-Androst-16-en-3β-ol	1.2
4,16-Androstadien-3-one	7.8
5,16-Androstadien-3β-ol	8.9

3.8.2.2.2 Vitamin D

Cholecalciferol (vitamin D_3) is formed by photolysis of 7-dehydrocholesterol, a precursor in
cholesterol biosynthesis. As shown in Fig. 3.42,
UV radiation opens the B-ring. The precalciferol formed is then isomerized to vitamin D_3 by
a rearrangement of the double bond which is influenced by temperature. Side-products, such as
lumi- and tachisterol, have no vitamin D activity. Cholecalciferol is converted into the active
hormone, 1,25-dihydroxy-cholecalciferol, by hydroxylation reactions in liver and kidney.

7-Dehydrocholesterol, the largest part of which is
supplied by food intake and which accumulates
in human skin, is transformed by UV light into
vitamin D_3. The occurrence and the physiological significance of the D vitamins are covered in
Sect. 6.2.2.

Ergosterol (ergosta-5,7,22-trien-3β-ol), which
occurs in yeast, moulds and algae, is provitamin D_2. It can serve as an indicator for fungal
contamination. A tolerance limit of 15 mg/kg
solid has been proposed for tomato products.

3.8.2.3 Plant Steroids (Phytosterols)

The sterols and stanols (hydrogenation products
of sterols) occurring in plants are known as phytosterols. The best known representatives are the
desmethylsterols shown in 3.8.2.3.1.

The phytosterols are of interest from a nutritional
and physiological point of view because they
lower the concentration of cholesterol and LDL in
the blood plasma (cf. 3.5.1.2). The absorption of
cholesterol is inhibited, a significant effect being
reached with an intake of 1 g/day of phytosterol.
Since the normal dietary intake amounts to only
200–400 mg/day of phytosterol, margarines are
enriched with phytosterols. However, as the free
sterols are only poorly soluble in the fat phase,
sterol esters are used in the production of margarine. Sterol esters are hydrolysed in the digestive tract. The starting material for the extraction
of phytosterols is plant oils and tall oil (Swedish
"tall" = pine), which accumulates as a by-product
in the production of paper and pulp. Tall oil is
rich in phytostanols, mainly β-sitostanol.

3.8.2.3.1 Desmethylsterols

Cholesterol, long considered to be an indicator of
the presence of animal fat, also occurs in small
amounts in plants (Table 3.52). Campe-, stigma-
and sitosterol, which are predominant in the sterol
fraction of some plant oils, are structurally related to cholesterol; only the side chain on C-17 is
changed. The following structural segments (only

Table 3.52. Average sterol composition of plant oils[a]

Component	Sun-flower	Peanut	Soya	Cotton-seed	Corn	Olive	Palm
Cholesterol	0.5	6.2	0.5	0.5	0.5	0.5	0.5
Brassicasterol	0.5	0.5	0.5	0.5	0.5	0.5	0.5
Campesterol	242	278	563	276	2655	19	88
Stigmasterol	236	145	564	17	499	0.5	42
β-Sitosterol	1961	1145	1317	3348	9187	732	252
Δ^5-Avenasterol	163	253	46	85	682	78	0.5
Δ^7-Stigmasterol	298	0.5	92	0.5	96	0.5	51
Δ^7-Avenasterol	99	34	63	18	102	30	0.5
24-Methylene-cycloartenol	204	0.5	53	0.5	425	580	0.5

[a] Values in mg/kg.

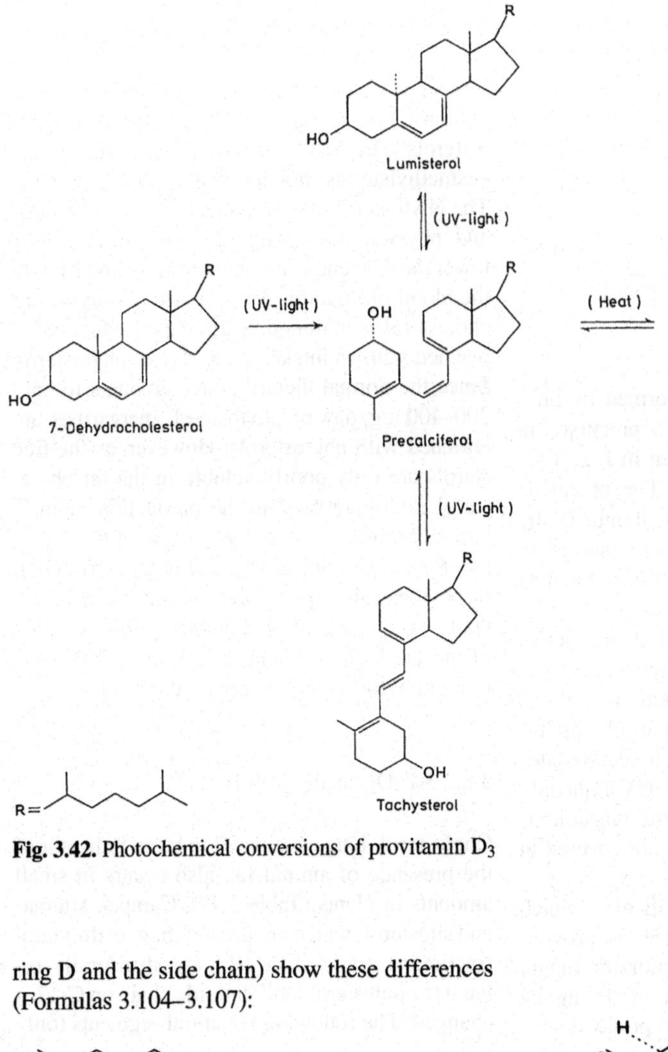

Fig. 3.42. Photochemical conversions of provitamin D₃

ring D and the side chain) show these differences
(Formulas 3.104–3.107):

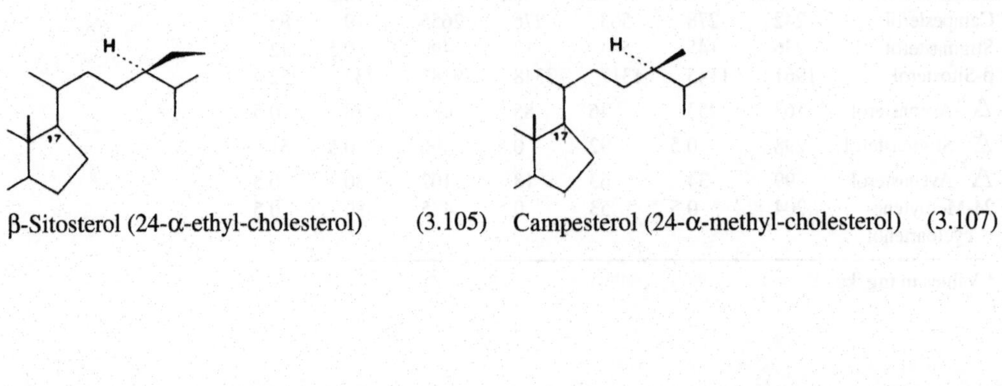

Cholesterol (3.104)

Stigmasterol (double bond at C-22) (3.106)

β-Sitosterol (24-α-ethyl-cholesterol) (3.105)

Campesterol (24-α-methyl-cholesterol) (3.107)

Δ^5-Avenasterol is a sitosterol derivative:

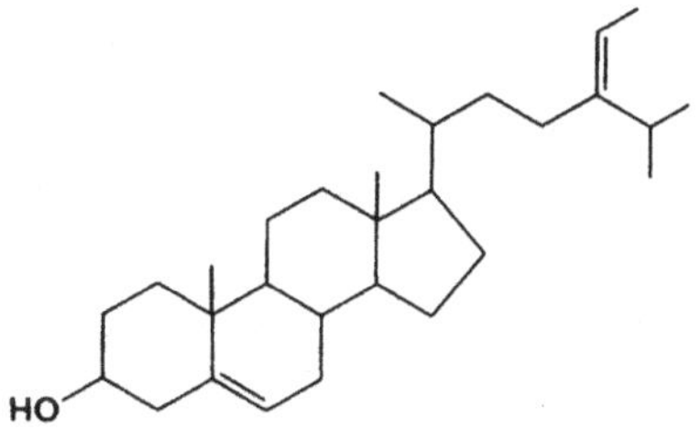

Δ^5-Avenasterol   (3.108)

Steroids which, like the avenasterols, contain an ethylidene group have antioxidative activity at the temperatures used in deep frying because under these conditions, a peroxyl radical can abstract an H-atom from this group.

In addition to Δ^5-sterols, Δ^7-sterols occur in plant lipids; for example:

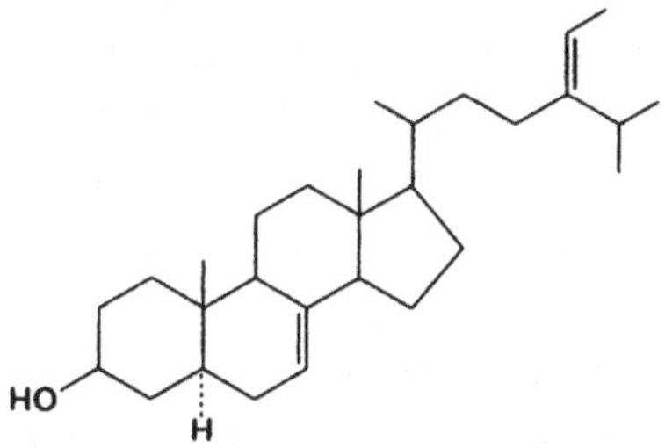

Δ^7-Avenasterol   (3.109)

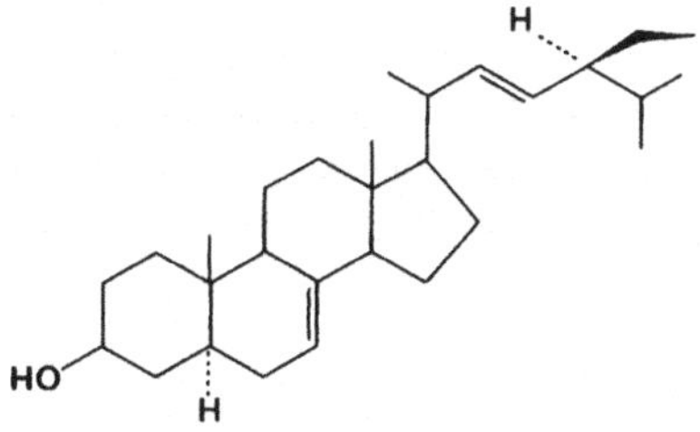

Δ^7-Stigmasterol   (3.110)

Plant lipids contain 0.15–0.9% sterols, with sitosterol as the main component (Table 3.52). In order to identify blends of fats (oils), the data on the predominant steroids are usually expressed as a quotient. For example, the ratio of stigmasterol/campesterol is determined in order to detect adulteration of cocoa butter. As seen from Table 3.53, this ratio is significantly lower in a number of cocoa butter substitutes than in pure cocoa butter. The phytosterol fraction (e. g. sito- and campesterol) has to be determined in order to detect the presence of plant fats in animal fats.

Table 3.53. Ratio of stigmasterol to campesterol in various fats

Fat	Sterols (g/kg)	% Stigmasterol / % Campesterol
Cocoa butter	1.8	2.8–3.5
Tenkawang[a]	2.15	0.42–0.55
Sal oil[a]	3	0.98
Illexao 30 – 90[b]	1.15	–[d]
Palm oil	0.67	0.43
Palm seed oil	0.81	1.28
Choclin[c]		0.38
Kaobien[c]		0.56
Coconut oil	0.75	1.47
Peanut oil, hydrogenated		0.72
Coberine[c]		0.31–0.60

[a] Cocoa butter substitutes (cf. 14.3.2.2.3).
[b] Trade name for sheasterol.
[c] Trade name for cocoa butter substitutes from the middle fraction of palm oil and shea butter.
[d] Contains no stigmasterol.

3.8.2.3.2 Methyl- and Dimethylsterols

Sterols with α-oriented C-4 methyl groups occur in oils of plant origin. The main compounds are:

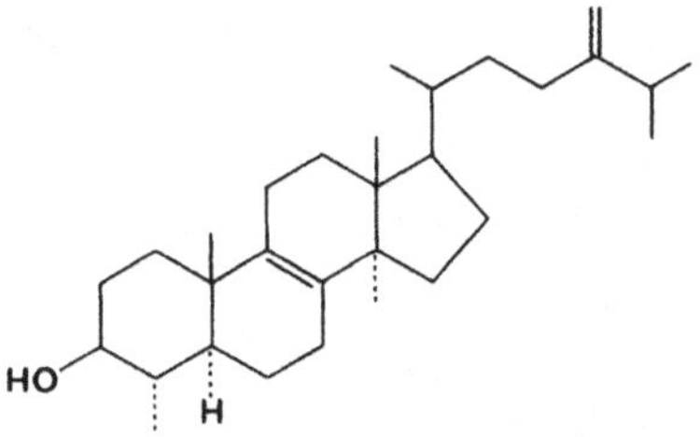

(3.111)

4α,14α-Dimethyl-24-methylene-5α-cholest-8-en-3β-ol (Obtusifoliol)

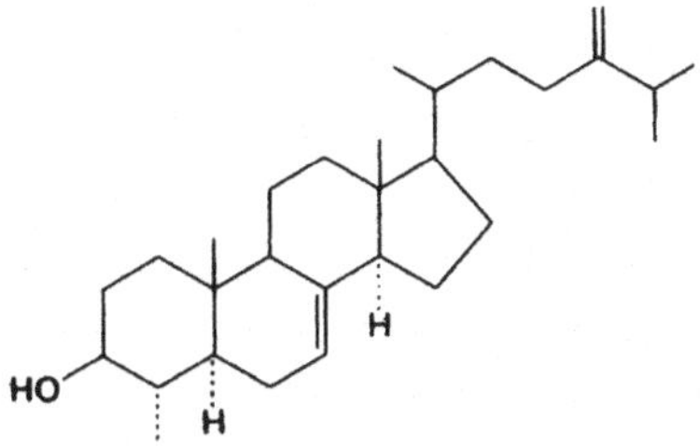

(3.112)

4α-Methyl-24-methylene-5α-cholest-7-en-3β-ol (Gramisterol)

Gas chromatographic-mass spectrometric studies have also revealed the presence of 4,4-dimethylsterols in the steroid fraction of many plant oils:

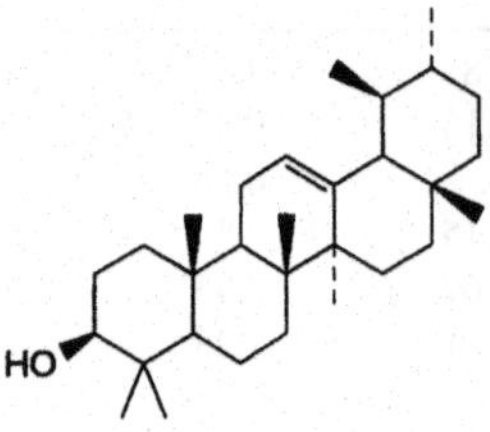

α-Amyrine   (3.113a)

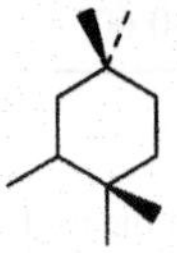

β-Amyrine   (3.113b)

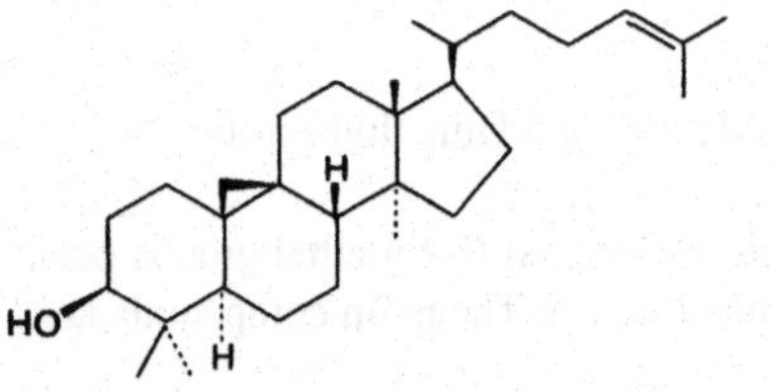

Cycloartenol   (3.114)

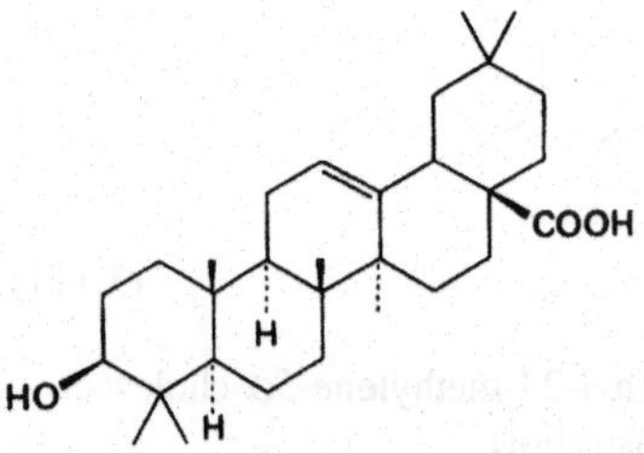

Oleanolic acid   (3.115)

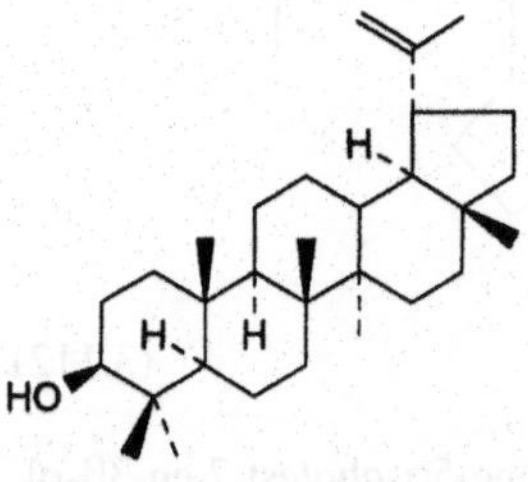

Lup-20(29)-en-3β-ol (Lupeol)   (3.116)

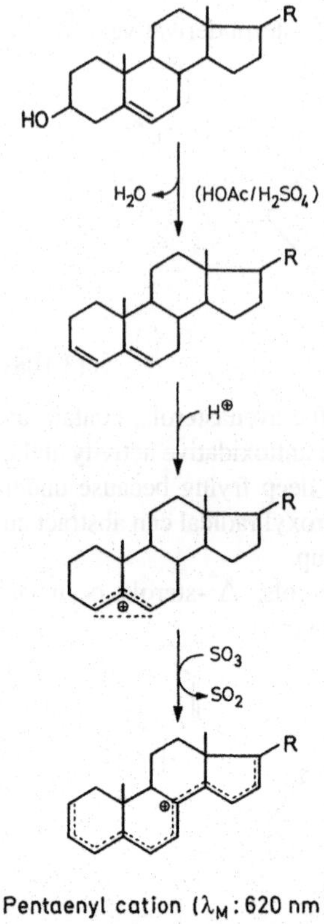

Pentaenyl cation (λ_M: 620 nm)

Fig. 3.43. Sterol detection according to *Lieberman–Burchard.* Reactions involved in color development

Oleanolic acid has long been known as a constituent of olive oil. Methyl- and dimethylsterols are important in identifying fats and oils (cf. Fig. 3.44).

3.8.2.4 Analysis

Qualitative determination of sterols is conducted using the *Liebermann–Burchard* reaction, in which a mixture of glacial acetic and concentrated sulfuric acids reacts directly with the fat or oil or the unsaponifiable fraction. Several modifications of this basic assay have been developed which, depending on the steroid and the oxidizing agent used, result in the production of a green or red color. The reaction is more sensitive when the SO_3 oxidizing agent is replaced by the Fe^{3+} ion. The conversion of sterols into a chromophore is based on the

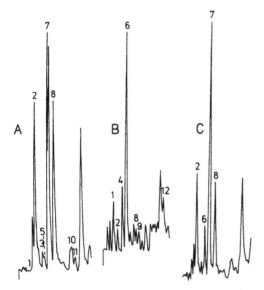

Fig. 3.44. Gas chromatographic separation of the triterpene alcohol fraction from coberine (**A**), cocoa butter (**B**) and cocoa butter +5% coberine (**C**) (according to *Gegiou and Staphylakis*, 1985). *1*, lanosterol; *2*, β-amyrine; *3*, butyrospermol; *4*, 24-methylene lanostenol; *5*, parkeol; *6*, cycloartenol; *7*, α-amyrine; *8*, lup-20(29)-en-3β-ol; *9*, 24-methylene cycloartenol; *10*, ψ-taraxasterol; *11*, taraxasterol; *12*, cyclobranol

reaction sequence given in Fig. 3.43. As shown, the assay is applicable only to sterols containing a double bond, such as in the B ring of cholesterol.

Sterols are separated as 3,5-dinitrobenzoic acid derivatives by thin layer chromatography and, after reaction with 1,3-diaminopropane, are determined quantitatively with high sensitivity in the form of a *Meisenheimer* adduct. Sterols and triterpene alcohols are silylated and then analysed by gas chromatography. One application of this method is illustrated by the detection of 5% coberine in cocoa butter (Fig. 3.44). The compounds α-amyrine and lup-20(29)-en-3β-ol (Formula 3.113a and 3.116) serve as indicators. They are present in much higher concentrations in some cocoa butter substitutes than in cocoa butter. Coberine is a cocoa butter substitute made by blending palm oil and shea butter (the shea is an African tree with seeds that yield a thick white fat, shea butter).

The content of egg (more accurately, the yolk) in pasta products or cookies can be calculated af-

ter the cholesterol content has been determined, usually by gas chromatography or HPLC. Vitamin D determination requires specific procedures in which precautions are taken with regard to the compound's sensitivity to light. A chemical method uses thin layer chromatographic separation of unsaponifiables, elution of vitamin D from the plate and photometric reading of the color developed by antimony (III) chloride. An alternative method recommends the use of HPLC.

3.8.3 Tocopherols and Tocotrienols

3.8.3.1 Structure, Importance

The methyl derivatives of tocol [2-methyl-2(4′, 8′, 12′-trimethyltridecyl)-chroman-6-ol] are denoted tocopherols. In addition the corresponding methyl derivatives of tocotrienol occur in food.

All four tocopherols and tocotrienols, with the chemical structures given in Fig. 3.45, are found primarily in cereals (especially wheat germ oil), nuts and rapeseed oils. These redox-type lipids are of nutritional/physiological and analytical interest. As antioxidants (cf. 3.7.3.2.1), they prolong the shelf lives of many foods containing fat or oil. The significance of tocopherols such as vitamin E is outlined in 6.2.3.

About 60–70% of the tocopherols in oilseeds are retained during the oil extraction and refining process (cf. 14.4.1 and Table 3.54). Some oils with very similar fatty acid compositions can be distinguished by their distinct tocopherol spectrum. To illustrate this, two examples are provided. The amount of β-tocopherol in wheat germ oil is quite high (Table 3.54), hence it serves as an indicator of that oil. The blending of soya oil with sunflower oil is detectable by an increase in the content of linolenic acid (cf. 14.5.2.3). However, it is possible to make a final conclusive decision about the presence and quantity of soya oil in sunflower oil only after an analysis of the composition of the tocopherols.

The tocopherol pattern is also different in almond and apricot kernel oil (Table 3.54) whose fatty acid compositions are very similar. Therefore adulteration of marzipan with persipan can be detected by the analysis of the tocopherols.

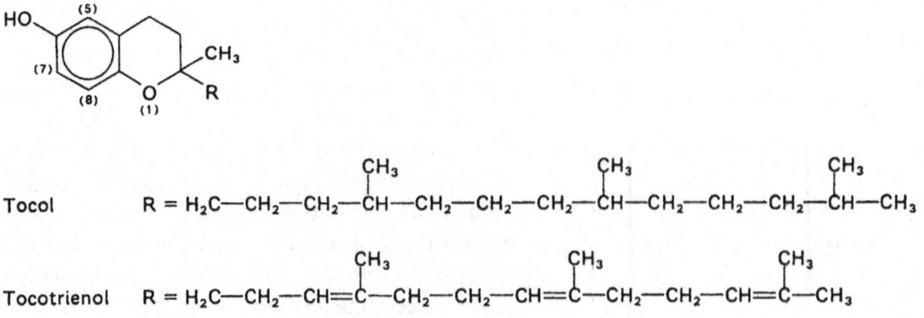

HO (5) CH₃ / R (7) (8) O (1)

CH₃ CH₃ CH₃

Tocol   R = H₂C—CH₂—CH₂—CH—CH₂—CH₂—CH₂—CH—CH₂—CH₂—CH₂—CH—CH₃

CH₃ CH₃ CH₃

Tocotrienol   R = H₂C—CH₂—CH=C—CH₂—CH₂—CH=C—CH₂—CH₂—CH=C—CH₃

Substitution	Tocopherols (T)	Tocotrienols (T-3)
5,7,8-Trimethyl	α-T	α-T-3
5,8-Dimethyl	β-T	β-T-3
7,8-Dimethyl	γ-T	γ-T-3
8-Methyl	δ-T	δ-T-3

Fig. 3.45. Tocopherols and tocotrienols present in food

3.8.3.2 Analysis

Isolation of tocopherols is accompanied by losses due to oxidation. Therefore, the edible oil is dissolved in acetone at 20–25 °C in the presence of ascorbyl palmitate as an antioxidant. The major portion of triacylglycerols is separated by crystallization at −80 °C. Tocopherols remaining in solution are then analyzed by thin layer or gas chromatography (after silylation of the phenolic HO-group) or by HPLC (cf. Fig. 3.46). UV spectrophotometry is also possible. However, the fluorometric method based on an older colorimetric procedure developed by *Emmerie* and *Engel* is even more sensitive. It involves reduction of the Fe (III) ion to Fe (II) by tocopherols and the reaction of the reduced iron with 2,2′-bipyridyl to form an intensive red colored complex.

3.8.4 Carotenoids

Carotenoids are polyene hydrocarbons biosynthesized from eight isoprene units (tetraterpenes) and, correspondingly, have a 40-C skeleton.

Table 3.54. Tocopherols and Tocotrienols in plant oil[a]

Oil	α-T	α-T-3	β-T	β-T-3	γ-T	γ-T-3	∂-T	∂-T-3
Sunflower	56.4	< 0.02	2.45	0.2	0.4	0.02	0.09	
Peanut	14.1	< 0.02	0.4	0.4	13.1	0.03	0.92	
Soya	17.9	< 0.02	2.8	0.4	60.4	0.08	37.1	
Cottonseed	40.3	< 0.02	0.2	0.9	38.3	0.09	0.5	
Corn	27.2	5.4	0.2	1.1	56.6	6.2	2.5	
Olive	9.0	< 0.02	0.2	0.4	0.5	0.03	0.04	
Palm (raw)	20.6	39.2	< 0.1	2.5	< 0.1	42.6	2.6	10.1
Wheat germ	133.0	< 2.6	71.0	18.1	26.0		27.1	
Almond	20.7		0.3		0.9			
Apricot kernel	0.5				22.4		0.3	
Peach kernel	6.4		1.3		1.0			
Cocoa butter	0.3		< 0.1		5.3		< 0.1	
Palm oil, middle fraction	< 0.1		< 0.1		0.43		< 0.1	
Shea fat stearin	< 0.1		< 0.1		0.43		< 0.1	

[a] Average composition; indicated in mg/100 g.

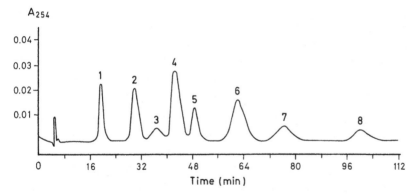

Fig. 3.46. Tocopherol and tocotrienol analysis by HPLC (according to *Cavins* and *Inglett*, 1974). 1 α-Tocopherol, 2 α-tocotrienol, 3 β-tocopherol, 4 γ-tocopherol, 5 β-tocotrienol, 6 γ-tocotrienol, 7 δ-tocopherol, and 8 δ-tocotrienol

They provide the intensive yellow, orange or red color of a great number of foods of plant origin (Table 3.55; cf. also 17.1.2.3 and 18.1.2.3.2). They are synthesized only by plants (see a textbook of biochemistry). However, they reach animal tissues via the feed (pasture, fodder) and can be modified and deposited there.

A well known example is the chicken egg yolk, which is colored by carotenoids. The carotenoids in green plants are masked by chlorophyll. When the latter is degraded, the presence of carotenoids is readily revealed (e. g. the green pepper becomes red after ripening).

3.8.4.1 Chemical Structure, Occurrence

Other carotenoids are derived by hydrogenation, dehydrogenation and/or cyclization of the basic structure of the C_{40}-carotenoids (cf. Formula 3.117). The cyclization reaction can occur at one or both end groups. The differences in C_9-end groups are denoted by Greek letters (cf. Formula 3.118).

$$(3.117)$$

A semisystematic nomenclature used at times has two Greek letters as a prefix for the generic

$$(3.118)$$

Table 3.55. Carotenoids in various food

Food	Concentration (ppm)[a]	Food	Concentration (ppm)[a]
Carrots	54	Peaches	27
Spinach	26–76	Apples	0.9–5.4
Tomatoes	51	Peas	3–7
Apricots	35	Lemons	2–3

[a] On dry weight basis.

name "carotene", denoting the structure of both C_9- end groups (cf. Formulas III, IV, VI or X: cf. Formulas 3.120, 3.121, 3.122 and 3.128, respectively). Designations such as α-, β- or γ-carotene are common names.

Carotenoids are divided into two main classes: carotenes and xanthophylls. In contrast to carotenes, which are pure polyene hydrocarbons, xanthophylls contain oxygen in the form of hydroxy, epoxy or oxo groups. Some carotenoids of importance to food are presented in the following sections.

3.8.4.1.1 Carotenes

Acyclic or aliphatic carotenes

Carotenes I, II and III (cf. Formulas 3.119–3.122) are intermediary or precursor compounds which, in biosynthesis after repeated dehydrogenizations, provide lycopene (IV; see a textbook of biochemistry). Lycopene is the red color of the tomato (and also of wild rose hips). In yellow tomato cultivars, lycopene precursors are present together with β-carotene (Table 3.56).

Phytoene (I)   (3.119)

Phytofluene (II)   (3.120)

ξ-Carotene (7,8,7′,8′-tetrahydro-ψ,ψ-carotene) (III)   (3.121)

Lycopene (ψ,ψ-carotene) (IV)   (3.122)

Table 3.56. Carotenes (ppm) in some tomato cultivars

Cultivar	Phytoene (I)	Phytofluene (II)	β-Carotene (VII)	ξ-Carotene (III)	γ-Carotene (V)	Lycopene (IV)
Campbell	24.4	2.1	1.4	0	1.1	43.8
Ace Yellow	10.0	0.2	trace	0	0	0
High Beta	32.5	1.7	35.6	0	0	0
Jubilee	68.6	9.1	0	12.1	4.3	5.1

Monocyclic Carotenes

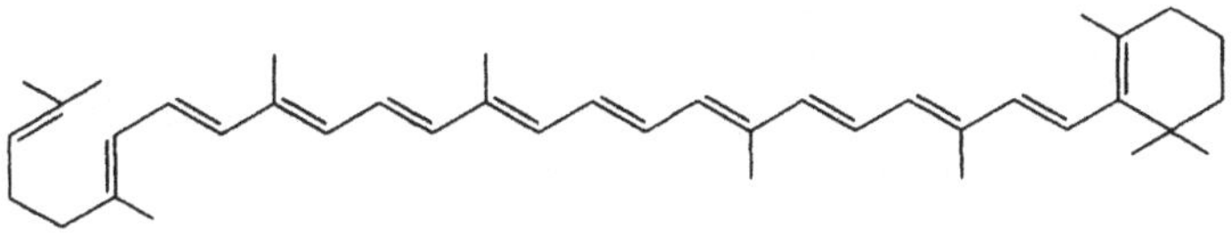

γ-Carotene (ψ, β-carotene) (V)   (3.123)

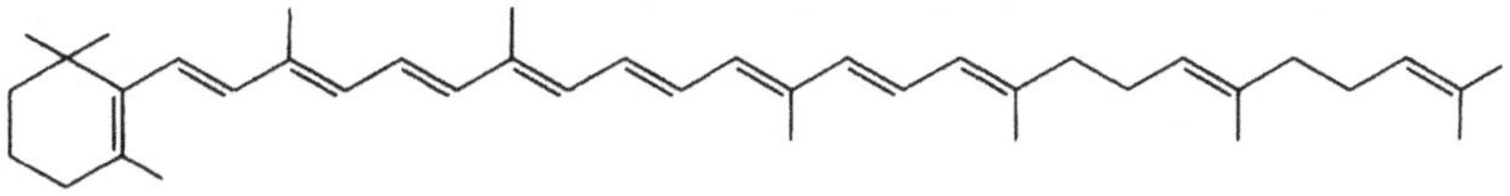

β-Zeacarotene (Va)   (3.124)

Bicyclic Carotenes

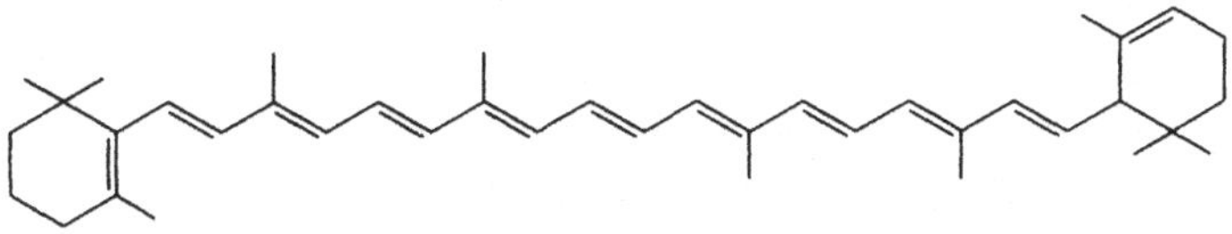

α-Carotene (β, ε-carotene) (VI)   (3.125)

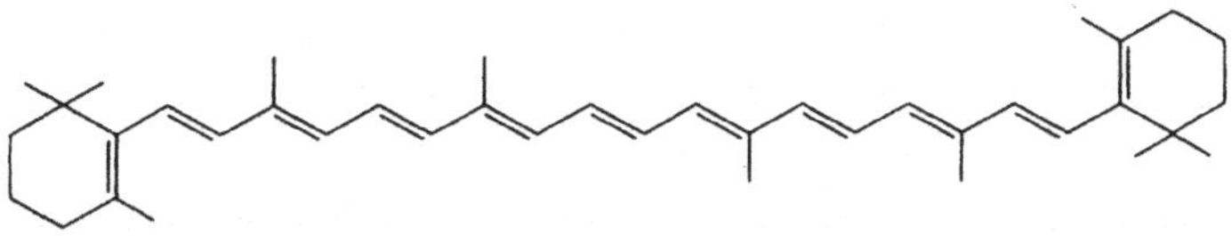

β-Carotene (β, β-carotene) (VII)   (3.126)

The importance of β-carotene as provitamin A is
covered under 6.2.1.

3.8.4.1.2 Xanthophylls

Hydroxy Compounds

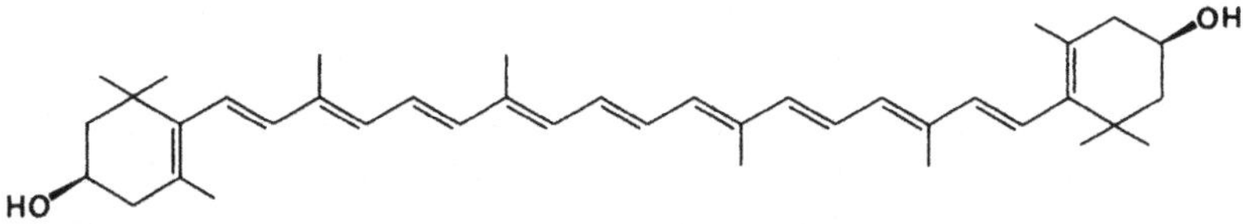

Zeaxanthin (β, β-carotene-3, 3'-diol) (VIII)   (3.127)

This xanthophyll is present in corn (*Zea mays*).

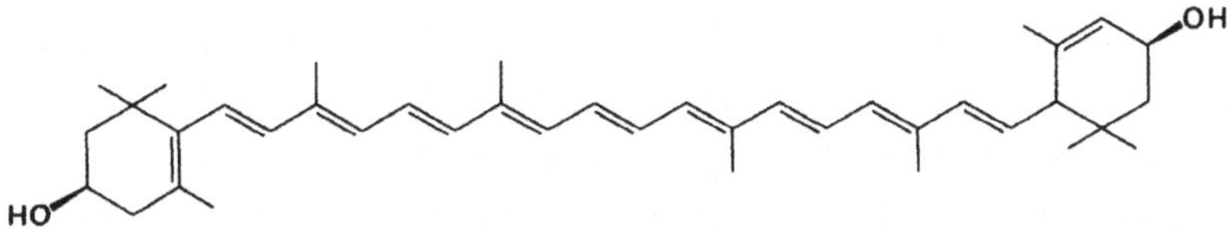

Lutein (β, ε-carotene-3, 3'-diol) (IX)   (3.128)

This xanthophyll occurs in green leaves and in egg yolk.

Keto Compounds

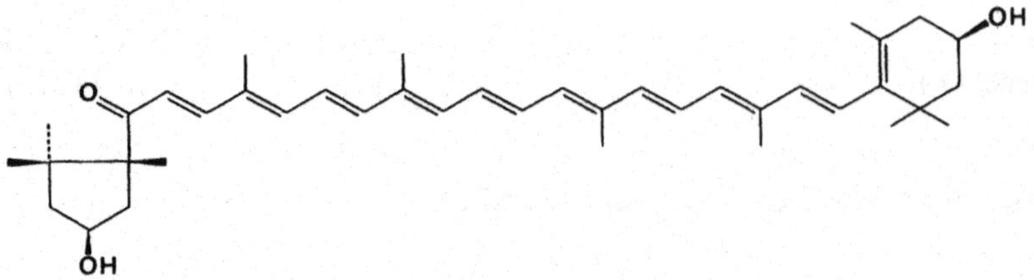

Capsanthin (3, 3′-dihydroxy-β, ϰ-carotene-6′-one) (X)   (3.129)

This xanthophyll is the major carotene of paprika peppers.

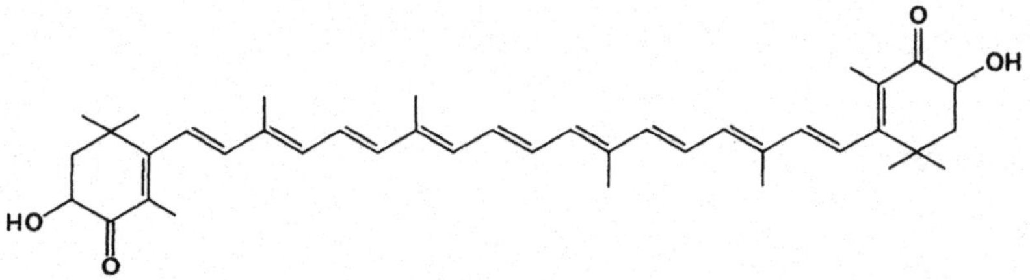

Astaxanthin (XI)   (3.130)

Astaxanthin is present in crab and lobster shells and, in combination with proteins, provides three blue hues (α-, β- and γ-crustacyanin) and one yellow pigment. During the cooking of crabs and lobsters, the red astaxanthin is released from a green carotenoid-protein complex. Astaxanthin usually occurs in lobster shell as an ester, e. g., di-palmitic ester.

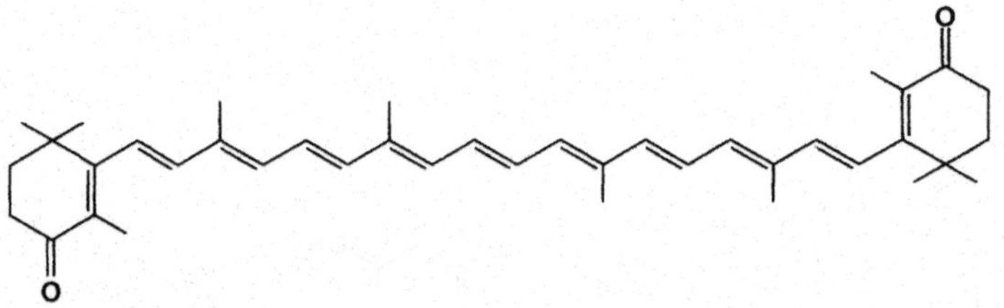

Canthaxanthin (XII)   (3.131)

This xanthophyll is used as a food colorant (cf. 3.8.4.5).

Epoxy Compounds

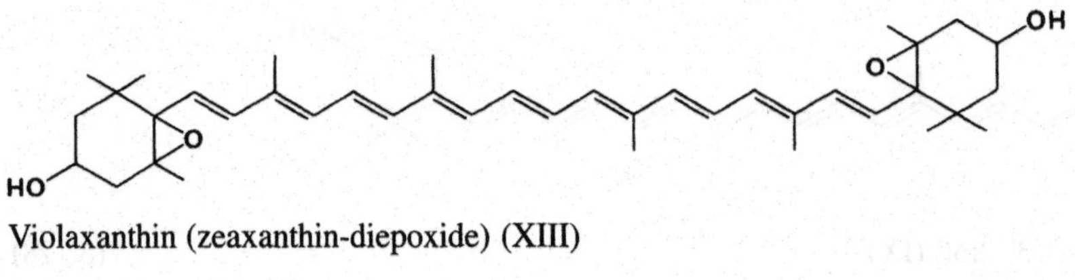

Violaxanthin (zeaxanthin-diepoxide) (XIII)   (3.132)

Violaxanthin is present in orange juice (cf. Table 3.57) and it also occurs in green leaves.

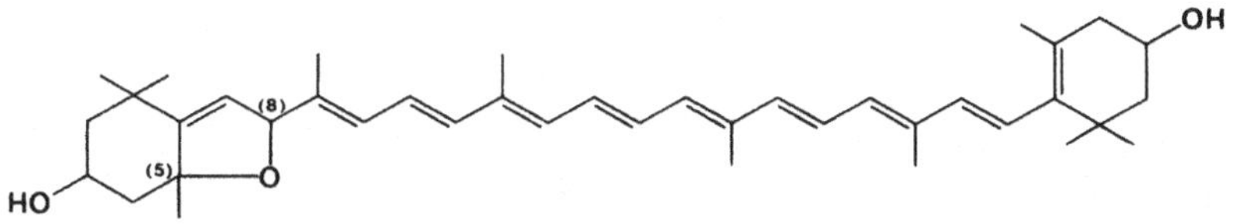

Mutatoxanthin (5,8-epoxy-5,8-dihydro-β, β-carotene-3, 3′-diol) (XVI)   (3.133)

This epoxy carotenoid is present in oranges (cf. Table 3.57).

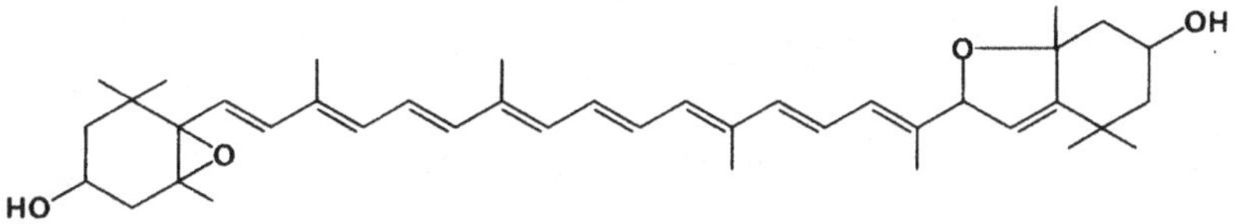

Luteoxanthin (XIV)   (3.134)

Luteoxanthin is the major carotenoid of oranges (cf. Table 3.57).

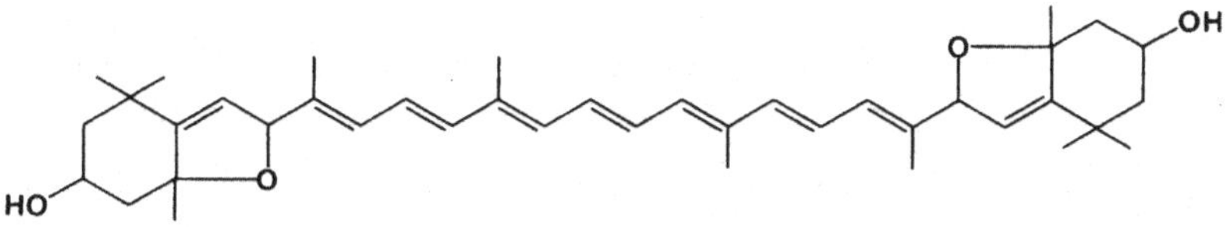

Auroxanthin (XV)   (3.135)

This carotenoid is a constituent of oranges (cf. Table 3.57).

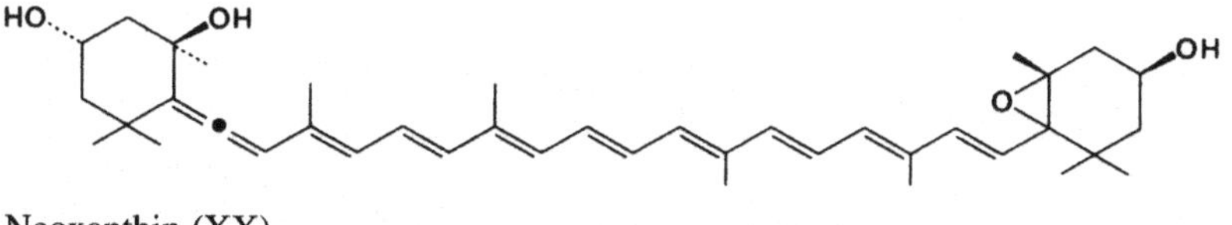

Neoxanthin (XX)   (3.136)

Dicarboxylic Acids and Esters

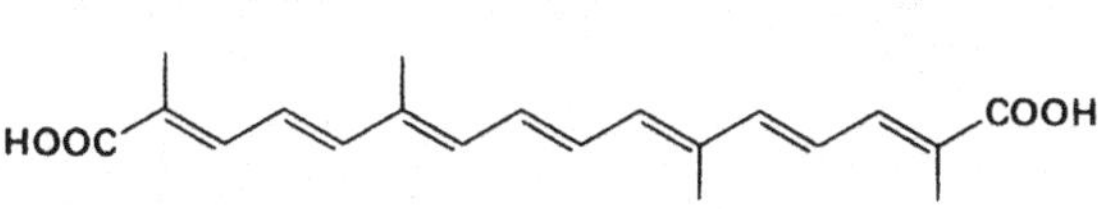

Crocetin (XVII)   (3.137)

This carboxylic acid carotenoid is the yellow pigment of saffron. It occurs in plants as a diester, i. e. glycoside with the disaccharide gentiobiose. The diester, called crocin, is therefore water-soluble.

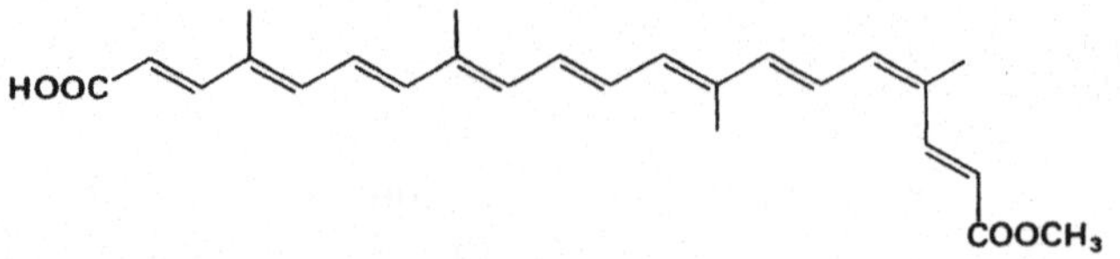

Bixin (XVIII)   (3.138)

Bixin is the main pigment of annato extract. Annato originates from the West Indies and the pigment is isolated from the seed pulp of the tropical bush *Bixa orellana*. Bixin is the monomethyl ester of norbixin, a dicarboxylic acid homologous to crocetin.

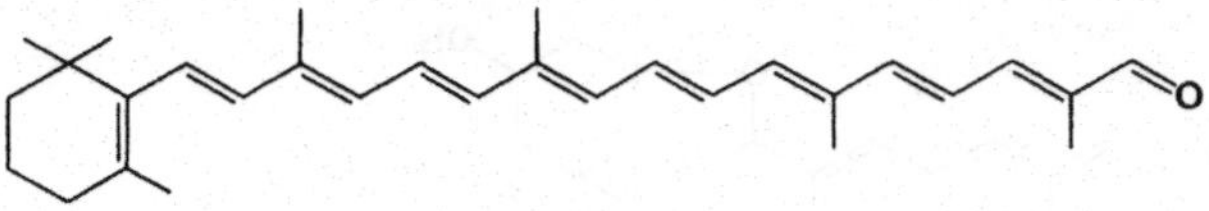

3.8.4.2 Physical Properties

Carotenoids are very soluble in apolar solvents, including edible fats and oils, but they are not soluble in water. Hence, they are denoted "lipochromes". Carotenoids are readily extracted

β-apo-8′-carotenal[*] (XIX)   (3.139)

Carotenoids are, as a rule, present in plants as a complex mixture. For example, the orange has more than 50 well characterized compounds, of which only those that exceed 5% of the total carotenoids are presented in Table 3.57.

Hydroxy-carotinoids are often present as esters of fatty acids; e. g., orange juice contains 3-hydroxy-β-carotene (cryptoxanthin) esterified with lauric, myristic and palmitic acid. The quantitative analysis of this ester fraction is used as proof of an adulteration of orange juice with mandarin juice.

from plant sources with petroleum ether, ether or benzene. Ethanol and acetone are also suitable solvents.

The color of carotenoids is the result of the presence of a conjugated double bond system in the molecules. The electron excitation spectra of such systems are of interest for elucidation of their structure and for qualitative and quantitative analyses.

Carotenoids show three distinct maxima in the visible spectrum, with wavelength positions dependent on the number of conjugated double

Table 3.57. Major carotenoid components in orange juice

Carotenoid	As percent of total carotenoids
Phytoene (I)	13
ξ-Carotene (III)	5.4
Cryptoxanthin (3-Hydroxy-β-carotene)	5.3
Antheraxanthin (5,6-Epoxyzeaxanthin)	5.8
Mutatoxanthin (XVI)	6.2
Violaxanthin (XIII)	7.4
Luteoxanthin (XIV)	17.0
Auroxanthin (XV)	12.0

Table 3.58. Absorption wavelength maxima for some carotenoids

Compound	Conjugated double bonds	Wavelength, nm (petroleum ether)		
A. Effect of the number of conjugated double bonds				
Phytoene (I)	3	275	285	296
Phytofluene (II)	5	331	348	367
ξ-Carotene (III)	7	378	400	425
Neurosporene	9	416	440	470
Lycopene (IV)	11	446	472	505
B. Effect of the ring structure				
γ-Carotene (V)	11	431	462	495
β-Carotene (VII)	11	425[a]	451	483

[c] The prefix "apo" indicates a compound derived from a carotenoid by removing part of its structure.

[a] Maximum absorption wavelength is not unequivocal (cf. Fig. 3.47).

bonds (Table 3.58). The fine structure of the spectrum is better distinguished in the case of acyclic lycopene (IV) than bicyclic β-carotene, since the latter is no longer a fully planar molecule. The methyl groups positioned on the rings interfere with those on the polyenic chain. Such steric effects prevent the total overlapping of π orbitals; consequently, a hypsochromic shift (a shift to a shorter wavelength) is observed for the major absorption bands (Fig. 3.47a).

Oxo groups in conjugation with the polyene system shift the major absorption bands to longer wavelengths (a bathochromic effect) with a simultaneous quenching of the fine structure of the spectrum (Fig. 3.47b). The hydroxyl groups in the molecule have no influence on the spectra.

A change of solvent system alters the position of absorption maxima. For example, replacing hexane with ethanol leads to a bathochromic shift.

Most of the carotenoids in nature and, thus, in food are of the trans-double bond configuration. When a mono-cis- or di-cis-compound occurs, the prefix "neo" is used. When one bond of all trans-double bonds is rearranged into this cis-configuration, there is a small shift in absorption maxima with a new minor "cis band" shoulder on the side of the shorter wavelength.

3.8.4.3 Chemical Properties

Carotenoids are highly sensitive to oxygen and light. When these factors are excluded, carotenoids in food are stable even at high temperatures. Their degradation is, however, accelerated by intermediary radicals occurring in food due to lipid peroxidationo (cf. 3.7.2). The cooxidation phenomena in the presence of lipoxygenase (cf. 3.7.2.2) are particularly visible. Changes in extent of coloration often observed with dehydrated paprika and tomato products are related to oxidative degradation of carotenoids. Such discoloration is desirable in flours (flour bleaching; cf. 15.4.1.4.3).

The color change in paprika from red to brown, as an example, is due partly to a slow *Maillard* reaction, but primarily to oxidation of capsanthin (Fig. 3.48) and to some as yet unclear polymerization reactions.

3.8.4.4 Precursors of Aroma Compounds

Aroma compounds are formed during the oxidative degradation of carotenoids. Such compounds, their precursors and the foods in which they occur are listed in Table 3.59. The mentioned ionones and β-damascenone

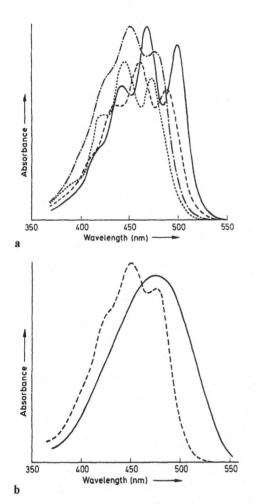

Fig. 3.47. Electron excitation spectra of carotenoids (according to *Isler*, 1971). **a** — Lycopene (IV), – – – – γ-carotene (V), ······ α-carotene (VI), – · – · – β-carotene (VII); **b** Canthaxanthin (XII) before — and after – – – – oxo groups reduction with NaBH₄

Table 3.59. Aroma compounds formed in oxidative degradation of carotenoids

Precursor[a]	Aroma compound	Odor threshold ($\mu/1$, water)	Occurrence
Lycopene (I)	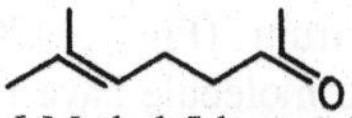 6-Methyl-5-hepten-2-one	50	Tomato
	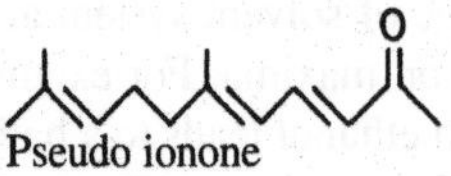 Pseudo ionone	800	Tomato
Dehydrolycopene	 6-Methyl-3,5-heptadien-2-one	380	Tomato
α-Carotene (VI)	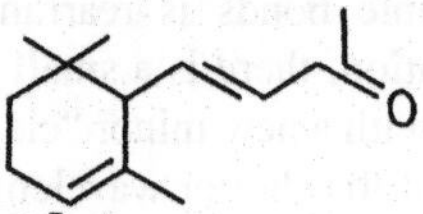 α-Ionone	R(+): 0.5–5 S(−): 20–40	Raspberry, black tea carrots, vanilla
β-Carotene (VII)	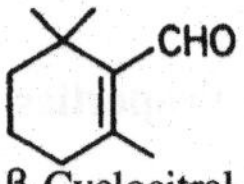 β-Cyclocitral	5	Tomato
	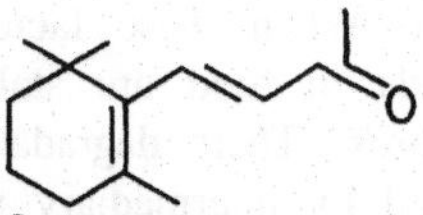 β-Ionone	0.007	Tomato, raspberry, blackberry, passion fruit, black tea
Neoxanthin (XX)	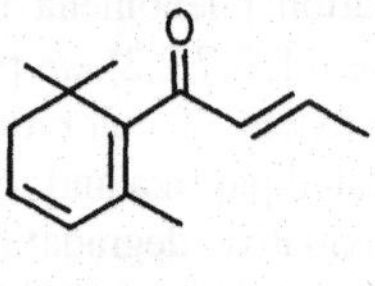 β-Damascenone	0.002	Tomato, coffee, black tea, wine, beer, honey, apple
	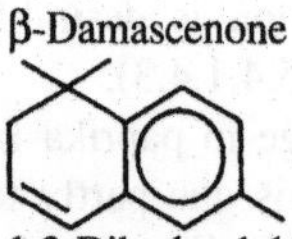 1,2-Dihydro-1,1,6-trimethylnaphthalene	2	Wine, peach, strawberry

[a] Roman numerals refer to the chemical structures presented in 3.8.4.1.

belong to the class of C_{13}-norisoprenoides. Other than β-ionone, α-ionone is a chiralic aroma compound whose R-enantiomer is present almost in optical purity in the food items listed in Table 3.59. α- and β-Damascone (Formula 3.140), present in black tea are probably derived from α- and β-carotene. Chiro-specific analysis (cf. 5.2.4) indicated that α-damascone occurs as racemate.

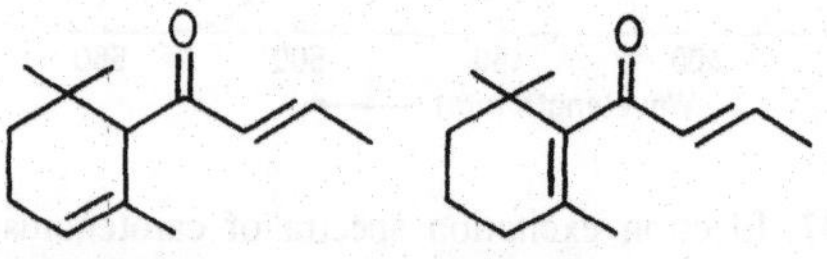

α-Damascone   β-Damascone

(3.140)

The odor thresholds of the R- and S-form (about 1 µg/kg, water) differ rarely.

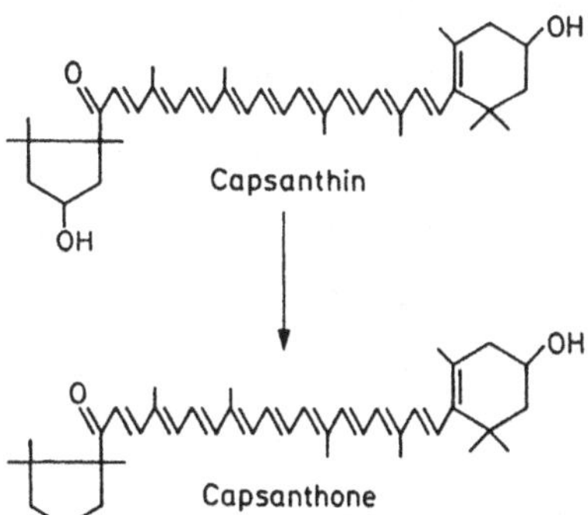

Capsanthin

Capsanthone

β-Citraurin

3-Keto-β-apo-8'-carotinal

Short-chain compounds

Fig. 3.48. Oxidative degradation of capsanthin during storage of paprika (according to *Philip* and *Francis*, 1971)

Of all C_{13}-norisoprenoids, β-damascenone and β-ionone, smelling like honey and violets respectively, have the lowest odor threshold values (Table 3.59). Precursor of β-damascenone is neoxanthine, out of which the *Grasshopper ketone* (I in Formula 3.142) is formed by oxi-

dative cleavage. The oxygen function migrates from the C-9 to the C-7 position by reduction of I to form an allentriol, elimination and attachment of HO-ions. In acid medium, 3-hydroxy-β-damascone and β-damascenone result from the intermediate (II).

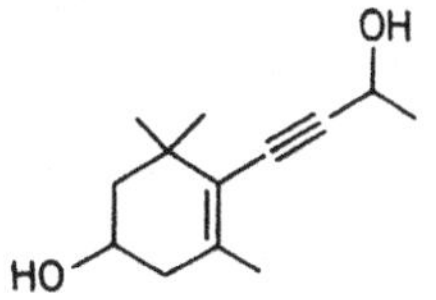

(3.141)

Besides the *Grasshopper ketone*, another enindiol (Formula 3.141) was identified in grape juices. When heated (pH 3), this enindiol yields 3-hydroxy- β-damascone as main and β-damascenone as minor product.

Hydroxylated C_{13}-norisoprenoids (i. e. *Grasshopper ketone*, 3-hydroxy-β-damascone) often occur in plants as glycosides, and can be liberated from these by enzymatic or acid hydrolysis and then transformed into aroma compounds. Therefore the aroma profile changes when fruits are heated during the production of juice or marmalade. An example is the formation of vitispirane (II in Formula 3.143) by hydrolysis of glycosidic bound 3-hydroxy-7,8-dihydro-β-ionol (I) in wine. The odor threshold of vitispirane is relatively high (800 μg/kg, wine) but is clearly exceeded in some port varieties.

1,2-Dihydro-1,1,6-trimethyl naphthalene (Table 3.59) can be formed by a degradation of neoxanthin and other carotinoids during the storage of wine. It smells like kerosene (threshold 20 μg/kg, wine). It is thought that this odorant contributes considerably to the typical aroma of white wine that was stored for a long period in the bottle. The compound may cause an off-flavor in pasteurized passion fruit juice.

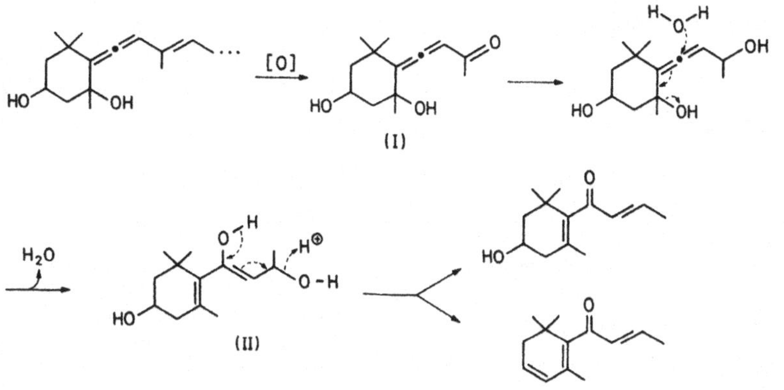

(3.142)

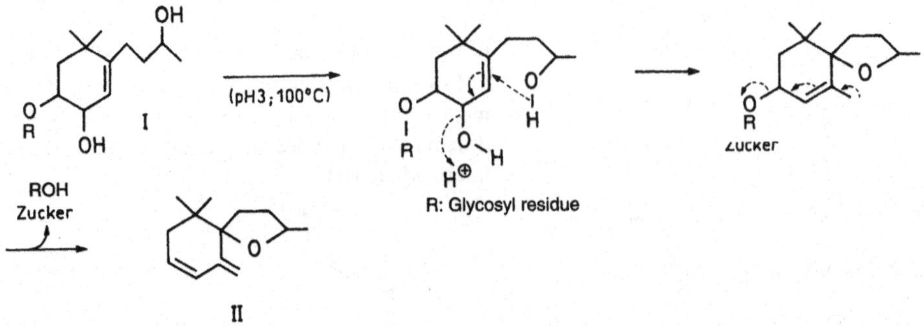

$$(3.143)$$

3.8.4.5 Use of Carotenoids in Food Processing

Carotenoids are utilized as food pigments to color margarine, ice creams, various cheese products, beverages, sauces, meat, and confectionery and bakery products. Plant extracts and/or individual compounds are used.

3.8.4.5.1 Plant Extracts

Annato is a yellow oil or aqueous alkaline extract of fruit pulp of *Raku* or *Orleans* shrubs or brushwood (*Bixa orellana*). The major pigments of annato are bixin (XVIII) and norbixin, both of which give dicarboxylic acids upon hydrolysis. Oleoresin from paprika is a red, oil extract containing about 50 different pigments. The aqueous extract of saffron (more accurately, from the pistils of the flower *Crocus sativus*) contains crocin (XVII) as its main constituent. It is used for coloring beverages and bakery products.

Raw, unrefined palm oil contains 0.05–0.2% carotenoids with α- and β-carotenes, in a ratio of 2:3, as the main constituents. It is of particular use as a colorant for margarine.

3.8.4.5.2 Individual Compounds

β-Carotene (VII), canthaxanthin (XII), β-apo-8′-carotenal (XIX) and the carboxylic acid ethyl ester derived from the latter are synthesized for use as colorants for edible fats and oils. These carotenoids, in combination with surface-active agents, are available as micro-emulsions (cf. 8.15.1) for coloring foods with a high moisture content.

3.8.4.6 Analysis

The total lipids are first extracted from food with ispropanol/petroleum ether (3:1 v/v) or with acetone. Alkaline hydrolysis follows, removing the extracted acyl-lipids and the carotenoids from the unsaponifiable fraction. This is the usual procedure when alkali-stable carotenoids are analyzed. Although carotenoids are generally alkali stable, there are exceptions. When alkali-labile carotenoids are present, the acyl lipids are removed instead by a saponification method using column chromatography as the separation technique.

A preliminary separation of the lipids into classes of carotenoids is carried out when a complex mixture of carotenoids is present. For example, column chromatography is used with Al_2O_3 as an adsorbent (Table 3.60). Additional separation into classes or individual compounds is achieved

Table 3.60. Separation of carotenoids into classes by column chromatography using neutral aluminum oxide (6% moisture) as an adsorbent P: Petroleum ether, D: diethyl ether

Elution with	Carotenoids in effluent
100% P	Carotenes
5% D in P	Carotene-epoxides
20–59% D in P	Monohydroxy-carotenoids
100% D	Dihydroxy-carotenoids
5% Ethanol in D	Dihydroxy-epoxy-carotenoids

by HPLC and thin layer chromatography. Thin layers made of MgO or ZnCO$_3$ are suitable. These adsorbent layers permit separation of carotenoids into classes according to the number, position and configuration of double bonds.

Identification of carotenoids is based on chromatographic data and on electron excitation spectra (cf. 3.8.4.2), supplemented when necessary with tests specific to each group. For example, a hypsochromic effect after addition of NaBH$_4$ suggests the presence of oxo or aldehyde groups, whereas the same effect after addition of HCl suggests the presence of a 5,6-epoxy group. The latter "blue hue shift" is based on a rearrangement reaction:

(5,6-Epoxide)   (5,8-Epoxide)

$$(3.144)$$

Such rearrangements can also occur during chromatographic separations of carotenoids on silicic acid. Hence, this adsorbent is a potential source of artifacts.

Epoxy group rearrangement in the carotenoid molecule can also occur during storage of food with a low pH, such as orange juice.

Elucidation of the structure of carotenoids requires, in addition to VIS/UV spectrophotometry, supplemental data from mass spectrometry and IR spectroscopy. Carotenoids are determined photometrically with high sensitivity based on their high molar absorbancy coefficients. This is often used for simultaneous qualitative and quantitative analysis. New separation methods based on high performance liquid chromatography have also proved advantageous for the qualitative and quantitative analysis of carotenoids present as a highly complex mixture in food.

3.9 References

Allen, J.C., Hamilton, R.J.: Rancidity in food. 3rd edition. Blackie Academic & Professional, London, 1996

Amorati, R., Pedulli, G.F., Cabrini, L., Zambonin, L., Laudi, L.: Solvent and pH effects on the antioxidant activity of caffeic and other phenolic acids. J. Agric. Food. Chem. 54, 2932 (2006)

Andersson, R.E., Hedlund, C.B., Jonsson, U.: Thermal inactivation of a heat-resistant lipase produced by the psychrotrophic bacterium Pseudomonas fluoreszens. J. Dairy Sci. 62, 361 (1979)

Badings, H.T.: Cold storage defects in butter and their relation to the autoxidation of unsaturated fatty acids. Ned. Melk Zuiveltijdschr. 24, 147 (1970)

Barnes, P.J.: Lipid composition of wheat germ and wheat germ oil. Fette Seifen Anstrichm. 84, 256 (1982)

Bergelson, L.D.: Diol lipids. New types of naturally occurring lipid substances. Fette Seifen Anstrichm. 75, 89 (1973)

Brannan, R.G., Conolly, B.J., Decker, E.A.: Peroxynitrite: a potential initiator of lipid oxidation in food. Trends Food Sci. Technol. 12, 164 (2001)

Burton, G.W., Ingold, K.U.: Vitamin E: Application of the principles of physical organic chemistry to the exploration of its structure and function. Acc. Chem. Res. 19, 194 (1986)

Chan, H.W.-S. (Ed.): Autoxidation of unsaturated lipids. Academic Press: London. 1987

Choe, E., Min, D.B.: Chemistry and reactions of reactive oygen species in foods. Crit. Rev. Food Sci. Nutr. 46, 1 (2006)

Christie, W.W.: Lipid analysis. 2. Aufl. Pergamon Press: Oxford. 1982

Christie, W.W.: High-performance liquid chromatography and lipids. Pergamon Press: Oxford. 1987

Christie, W.W., Nikolova-Damyanova, B., Laakso, P., Herslof, B.: Stereospecific analysis of triacylsnglycerols via resolution of diastereomeric diacylglycerol derivatives by high-performance liquid chromatography on silica. J. Am. Oil Chem. Soc. 68, 695 (1991)

Christopoulou, C.N., Perkins, E.G.: Isolation and characterization of dimers formed in used soybean oil. J. Am. Oil Chem. Soc. 66, 1360 (1989)

Dennis, E.A.: Phospholipases. In: The enzymes (Ed.: Boyer, P.D.) 3rd edn., Vol. XVI, p. 307, Academic Press: New York. 1983

Dionisi, F., Golay, P.A., Aeschlimann, J.M., Fay, L.B.: Determination of cholesterol oxidation products in milk powders: methods comparison and validation. J. Agric. Food Chem. 46, 2227 (1998)

Fedeli, E.: Lipids of olives. Prog. Chem. Fats Other Lipids 15, 57 (1977)

Foote, C.S.: Photosensitized oxidation and singlet oxygen: Consequences in biological systems. In: Free radicals in biology (Ed.: Pryor, W.A.), Vol. II, p. 85, Academic Press: New York. 1976

Foti, M., Piattelli, M., Baratta, M.T., Ruberto, G.: Flavonoids, coumarins, and cinnamic acids as antioxidants in a micellar system. Structure-activity relationship. J. Agric. Food Chem. *44*, 497 (1996)

Frankel, E.N.: Recent advances in lipid oxidation. J. Sci. Food Agric. *54*, 495 (1991)

Fritsche, J., Steinhart, H.: Trans fatty acid content in German margarines. Fett/Lipid *99*, 214 (1997)

Fritsche, S., Steinhart, H.: Occurrence of hormonally active compounds in food: a review. Eur. Food Res. Technol. *209*, 153 (1999)

Fritz, W., Kerler, J., Weenen, H.: Lipid derived flavours. In: Current topics in flavours and fragrances (Ed.: K.A.D. Swift,) Kluwer Academic Publishers, Dortrecht, 1999

Galliard, T., Mercer, E.I. (Eds.): Recent advances in the chemistry and biochemistry of plant lipids. Academic Press: London. 1975

Gardner, H.W.: Recent investigations in to the lipoxygenase pathway of plants. Biochim. Biophys. Acta *7084*, 221 (1991)

Gardner, H.W.: Lipoxygenase as versatile biocatalyst. J. Am. Oil Chem. Soc. *73*, 1347 (1996)

Garti, N., Sato, K.: Crystallization and polymorphism of fats and fatty acids. Marcel Dekker: New York. 1988

Gertz, C., Herrmann, K.: Zur Analytik der Tocopherole und Tocotrienole in Lebensmitteln. Z. Lebensm. Unters. Forsch. *174*, 390 (1982)

Gortstein, T., Grosch, W.: Model study of different antioxidant properties of α- and γ-tocopherol in fats. Fat Sci. Technol. *92*, 139 (1990)

Grosch, W.: Reaction of hydroperoxides – Products of low molecular weight. In: Autoxidation of unsaturated lipids (Ed.: H.W.-S. Chan), p. 95, Academic Press, London, 1987

Gunstone, F.D.: Fatty acid and lipid chemistry. Blackie Academic & Professional, London, 1996

Guth, H., Grosch, W.: Detection of furanoid fatty acids in soya-bean oil. – Cause for the light-induced off-flavour. Fat Sci. Technol. *93*, 249 (1991)

Guth, H., Grosch, W.: Deterioration of soya-bean oil: Quantification of primary flavour compounds using a stable isotope dilution assay. Lebensm. Wiss. u. Technol. *23*, 513 (1990)

Hamilton, R.J., Bhati, A.: Recent advances in chemistry and technology of fats and oils. Elsevier Applied Science: London. 1987

Haslbeck, F., Senser, F., Grosch, W.: Nachweis niedriger Lipase-Aktivitäten in Lebensmitteln. Z. Lebensm. Unters. Forsch. *181*, 271 (1985)

Hicks, B.; Moreau, R.A.: Phytosterols and phytostanols: Functional Food Cholesterol Busters. Food Technol. *55*, 62 (2001)

Homberg, E., Bielefeld, B.: Sterine und Methylsterine in Kakaobutter und Kakaobutter-Ersatzfetten. Dtsch. Lebensm. Rundsch. *78*, 73 (1982)

Hudson, B.J.F. (Eds.): Food antioxidations. Elsevier Applied Science: London. 1990

Institute of Food Science & Technology: Information statement on Phytosterl Esters (Plant Sterol and Stanol Esters), www.ifst.org./hottop29.htm (2002)

Isler, O. (Ed.): Carolenoids. Birkhäuser Verlag: Basel. 1971

Jeong, T.M., Itoh, T., Tamura, T., Matsumoto, T.: Analysis of methylsterol fractions from twenty vegetable oils. Lipids *10*, 634 (1975)

Johnson, A.R., Davenport, J.B.: Biochemistry and methodology of lipids. John Wiley and Sons: New York. 1971

Kadakal, C., Artiik, N.: A new quality parameter in tomato and tomato products: Ergosterol. Crit. Rev. Food Sci. & Nutr. *44*, 349 (2004)

Kinsella, J.E., Hwang, D.H.: Enzyme *penicillium roqueforti* involved in the biosynthesis of cheese flavour. Crit. Rev. Sci. Nutr. *8*, 191 (1976)

Kinsella, J.E.: Food lipids and fatty acids: Importance in food quality, nutrition, and health. Food Technol. *42* (10), 124 (1988)

Kinsella, J.E.: Seafoods and fish oils in human health and disease. Marcel Dekker: New York. 1987

Kochler, S.P.: Stable and healthful frying oil for the 21st century. Inform *11*, 642 (2000)

Korycka-Dahl, M.B., Richardson, T.: Active oxygen species and oxidation of food constituents. Crit. Rev. Food Sci. Nutr. *10*, 209 (1978)

Laakso, P.: Analysis of triacylglycerols – Approaching the molecular composition of natural mixtures. Food Rev. Int. *12*, 199 (1996)

Meijboom, P.W., Jongenotter, G.A.: Flavor perceptibility of straight chain, unsaturated aldehydes as a function of double-bond position and geometry. J. Am. Oil Chem. Soc. 680 (1981)

Min, D.B., Smouse, T.H. (Eds.): Flavor chemistry of lipid foods. American Oil Chemists' Society: Champaign. 1989

Nawar, W.W.: Volatiles from food irradiation. Food Rev. Internat. *2*, 45 (1986)

O'Shea, M., Lawless, F., Stanton, C., Devery, R.: Conjugated linoleic acid in bovine milk fat: a food-based approach to cancer chemoprevention. Trends Food Sci. Technol. *9*, 192 (1998)

Pardun, H.: Die Pflanzenlecithine. Verlag für chemische Industrie H. Ziolkowsky KG: Augsburg. 1988

Perkins, E.G. (Ed.): Analysis of lipids and lipoproteins. American Oil Chemists' Society: Champaign, Ill. 1975

Philip, T., Francis, F.J.: Oxidation of capsanthin. J. Food Sci. *36*, 96 (1971)

Podlaha, O., Töregard, B., Püschl, B.: TG-type composition of 28 cocoa butters and correlation between some of the TG-type components. Lebensm. Wiss. Technol. *17*, 77 (1984)

Porter, N.A., Lehman, L.S., Weber, B.A., Smith, K.J.: Unified mechanism for polyunsaturated fatty acid autoxidation. Competition of peroxy radical hydrogen atom abstraction, β-scission, and cyclization. J. Am. Chem. Soc. *103*, 6447 (1981)

Porter, N.A., Caldwell, S.E., Mills, K.A.: Mechanisms of free radical oxidation of unsaturated lipids. Lipids *30*, 277 (1995)

Pryde, E.H. (Ed.): Fatty acids. American Oil Chemists' Society: Champaign, Ill. 1979

Rietjens, I.M.C.M. et al.: The pro-oxidant chemistry of the natural antioxidants vitamin C, vitamin E, carotenoids and flavonoids. Environmental Toxicology and Pharmacology *11*, 321 (2002)

Rojo, J.A., Perkins, E.G.: Cyclic fatty acid monomer formation in frying fats. I. Determination and structural study. J. Am. Oil Chem. Soc. *64*, 414 (1987)

Schieberle, P., Haslbeck, F., Laskawy, G., Grosch, W.: Comparison of sensitizers in the photooxidation of unsaturated fatty acids and their methyl esters. Z. Lebensm. Unters. Forsch. 179, 93 (1984

Shahidi, F., Wanasundara, P.K.J.P.D.: Phenolic antioxidants. Crit. Rev. Food Sci. Nutr. *32*, 67 (1992)

Sherwin, E.R.: Oxidation and antioxidants in fat and oil processing. J. Am. Oil Chem. Soc. *55*, 809 (1978)

Simic, M.G., Karel, M. (Eds.): Autoxidation in food and biological systems. Plenum Press: New York. 1980

Slower, H.L.: Tocopherols in foods and fats. Lipids *6*, 291 (1971)

Smith, L.L.: Review of progress in sterol oxidations: 1987–1995. Lipids *31*, 453 (1996)

Sotirhos, N., Ho, C.-T., Chang, S.S.: High performance liquid chromatographic analysis of soybean phospholipids. Fette Seifen Anstrichm. *88*, 6 (1986)

Szuhaj, B.F., List, G.R.: Lecithins. American Oil Chemists' Society: Champaign, 1985

Thiele, O.W.: Lipide, Isoprenoide mit Steroiden. Georg Thieme Verlag: Stuttgart. 1979

Van Niekerk, P.J., Burger, A.E.C.: The estimation of the composition of edible oil mixtures. J. Am. Oil Chem. Soc. *62*, 531 (1985)

Veldink, G.A., Vliegenthart, J.F.G., Boldingh, J.: Plant lipoxygenases. Prog. Chem. Fats Other Lipids *15*, 131 (1977)

Wada, S., Koizumi, C.; Influence of the position of unsaturated fatty acid esterified glycerol on the oxidation rate of triglyceride. J. Am. Oil Chem. Soc. *60*, 1105 (1983)

Wagner, R.K., Grosch, W.: Key odorants of French fries. J. Am. Oil. Chem. Soc. *75*, 1385 (1998)

Wanasundara, P.K.J.P.D., Shahidi, F., Shukla, V.K.S.: Endogenous antioxidants from oilseeds and edible oils. Food Rev. Int. *13*, 225 (1997)

Werkhoff, P., Bretschneider, W., Güntert, M., Hopp, R., Surburg, H.: Chirospecific analysis in flavor and essential oil chemistry. Part B. Direct enantiomer resolution of trans-α-ionone and trans-α-damascone by inclusion gas chromatography. Z. Lebensm. Unters. Forsch. *192*, 111 (1991)

Winterhalter, P., Schreier, P.: Natural precursors of thermally induced C_{13} norisoprenoids in quince. In: Thermal generation of aromas (Eds.: T.H. Parliment, R.J. McGorrin, C.-T. Ho) p. 320, ACS Symposium Series 409, American Chemical Society: Washington D.C. 1989

Wolfram, G.: ω-3- und ω-6-Fettsäuren – Biochemische Besonderheiten und biologische Wirkungen. Fat Sci. Technol. *91*, 459 (1989)

Woo, A.H., Lindsay, R.C: Statistical correlation of quantitative flavour intensity assessments and individual free acid measurements for routine detection and prediction of hydrolytic rancidity off-flavours in butter. J. Food Sci. *48*, 1761 (1983)

4 Carbohydrates

4.1 Foreword

Carbohydrates are the most widely distributed and abundant organic compounds on earth. They have a central role in the metabolism of animals and plants. Carbohydrate biosynthesis in plants starting from carbon dioxide and water with the help of light energy, i.e., photosynthesis, is the basis for the existence of all other organisms which depend on the intake of organic substances with food.

Carbohydrates represent one of the basic nutrients and are quantitatively the most important source of energy. Their nutritional energy value amounts to 17 kJ/g or kcal/g. Even the nondigestible carbohydrates, acting as bulk material, are of importance in a balanced daily nutrition. Other important functions in food are fulfilled by carbohydrates. They act for instance as sweetening, gel- or paste-forming and thickening agents, stabilizers and are also precursors for aroma and coloring substances, especially in thermal processing.

The term carbohydrates goes back to times when it was thought that all compounds of this class were hydrates of carbon, on the basis of their empirical formula, e.g. glucose, $C_6H_{12}O_6$ (6C+6H$_2$O). Later, many compounds were identified which deviated from this general formula, but retained common reactions and, hence, were also classed as carbohydrates. These are exemplified by deoxysugars, amino sugars and sugar carboxylic acids. Carbohydrates are commonly divided into monosaccharides, oligosaccharides and polysaccharides. Monosaccharides are polyhydroxy-aldehydes or -ketones, generally with an unbranched C-chain. Well known representatives are glucose, fructose and galactose. Oligosaccharides are carbohydrates which are obtained from <10 carbohydrate units, which formally polymerize from monosaccharides with the elimination of water to give full acetals, e.g. by the reaction:

$$n\ C_6H_{12}O_6 \xrightarrow{-(n-1)\ H_2O} C_{6n}H_{10n+2}O_{5n+1} \quad (4.1)$$

Well known representatives are the disaccharides saccharose (sucrose), maltose and lactose, and the trisaccharide raffinose, and the tetrasaccharide stachyose.

In polysaccharides, consisting of n monosaccharides, the number n is as a rule >10. Hence, the properties of these high molecular weight polymers differ greatly from other carbohydrates. Thus, polysaccharides are often considerably less soluble in water than mono- and oligosaccharides. They do not have a sweet taste and are essentially inert. Well known representatives are starch, cellulose and pectin.

4.2 Monosaccharides

4.2.1 Structure and Nomenclature

4.2.1.1 Nomenclature

Monosaccharides are polyhydroxy-aldehydes (aldoses), formally considered to be derived from glyceraldehyde, or polyhydroxyketones (ketoses), derived from dihydroxyacetone by inserting CHOH units into the carbon chains. The resultant compounds in the series of aldoses are denoted by the total number of carbons as trioses, for the starting glyceraldehyde, and tetroses, pentoses, hexoses, etc. The ketose series begins with the simplest ketose, dihydroxyacetone, a triulose, followed by tetruloses, pentuloses, hexuloses, etc. The position of the keto group is designated by a numerical prefix, e.g. 2-pentulose, 3-hexulose.

When a monosaccharide carries a second carbonyl group, it is denoted as a -dialdose (2 aldehyde groups), -osulose (aldehyde and keto groups) or -diulose (2 keto groups). Substitution of an HO-group by an H-atom gives rise to a deoxy sugar, and by an H$_2$N-group to aminodeoxy compounds (cf. Formula 4.2). Analogous to 4- or 5-hydroxypentanal, aldoses (starting from tetroses) and ketoses (starting from 2-pentuloses) undergo intramolecular cyclization with hemiacetal formation to form lactols (Formula 4.3). With the exception of erythrose, monosaccharides are crystallized in these cyclic forms and, even in solution, there is an equilibrium between the open chain carbonyl form and cyclic hemiacetals, with the latter predominating. The tendency to cyclize is, compared to hydroxyaldehydes, even more pronounced in monosaccharides, as shown by $\Delta G°$-values and equilibrium concentrations in 75% aqueous ethanol (cf. Formula 4.3).

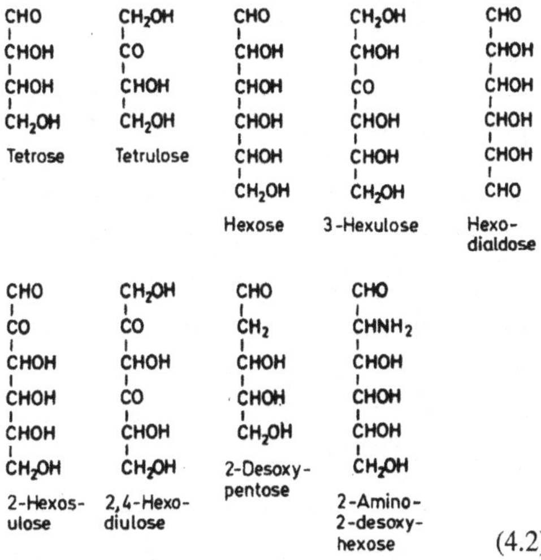

4.2.1.2 Configuration

Glyceraldehyde, a triose, has one chiral center, so it exists as an enantiomer pair, i.e. in D- and L-forms. According to the definition, the secondary hydroxyl group is on the right in D-glyceraldehyde and on the left in L-glyceraldehyde. Although this assignment was at first arbitrarily made, it later proved to be correct. It is possible by cyanhydrin synthesis to obtain from each enantiomer a pair of diastereomeric tetroses (*Kiliani–Fischer synthesis*):

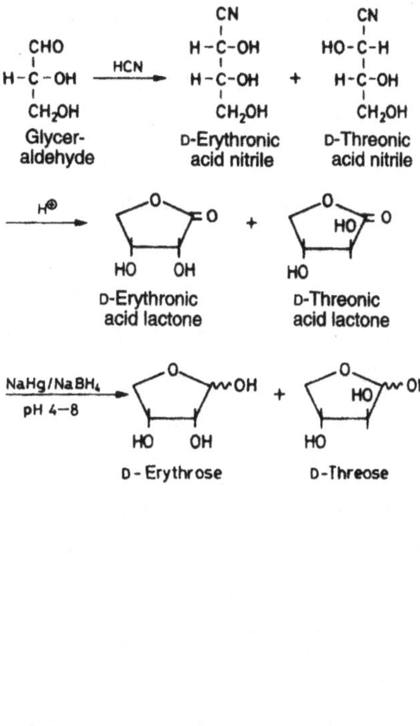

(4.2)

(4.4)

Lactols can be considered as tetrahydrofuran or tetrahydropyran derivatives, hence, they are also denoted as furanoses or pyranoses.

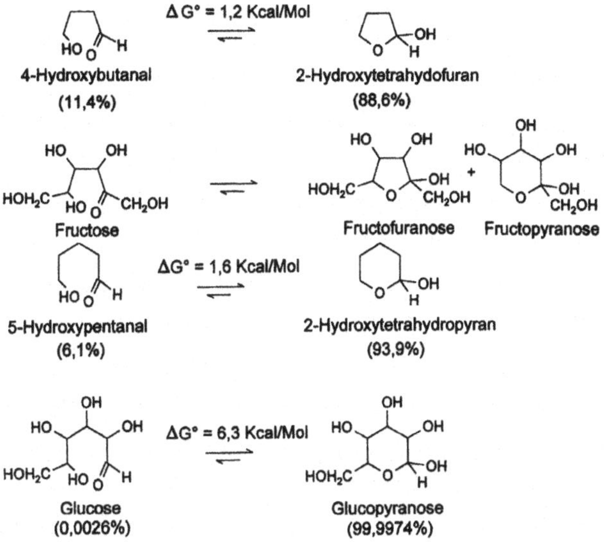

(4.3)

Correspondingly, L-erythrose and L-threose are obtained from L-glyceraldehyde:

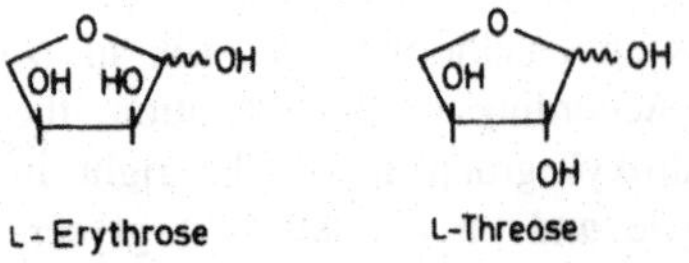

L-Erythrose   L-Threose   (4.5)

The nitriles can also be directly reduced to diastereomeric aldoses with $PdO/BaSO_4$, by passing the lactone intermediate stage. Another reaction for the formation of monosaccharides is the nitroalkane synthesis. The epimeric nitro compounds, obtained by the reaction of an aldose with nitromethane as anions, are separated and converted to the corresponding aldoses by an acinitroalkane cleavage (*Nef*-reaction):

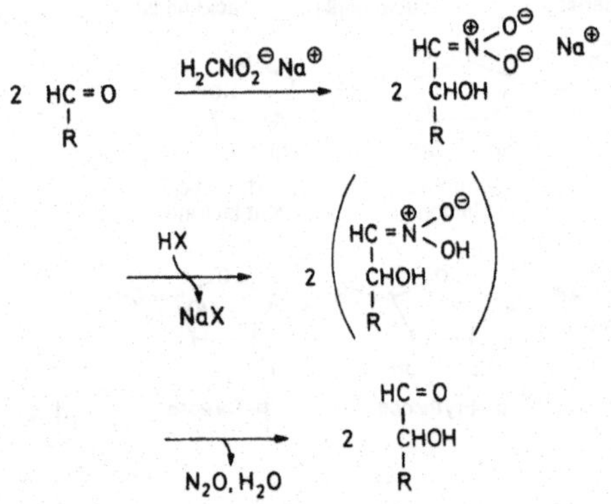

(4.6)

After repeated cyanhydrin reactions, four tetroses will provide a total of eight pentoses (each tetrose provides a pair of new diastereomers with one more chiral center), which can then yield sixteen stereoisomeric hexoses. The compounds derived from D-glyceraldehyde are designated as D-aldoses and those from L-glyceraldehyde as L-aldoses.

An important degradation reaction of aldoses proceeds via the disulfone of the dithioacetal:

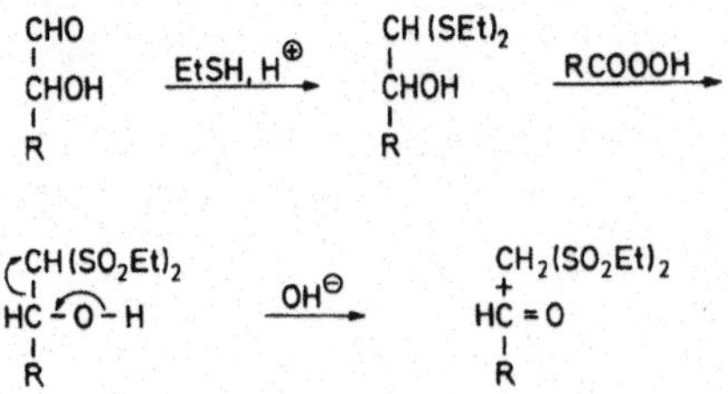

(4.7)

Figure 4.1 shows the formulas and names for D-aldoses using simplified *Fischer* projections. The occurrence of aldoses of importance in food is compiled in Table 4.1. Epimers are monosaccharides which differ in configuration at only one chiral C-atom. D-Glucose and D-mannose are 2-epimers. D-glucose and D-galactose are 4-epimers.

The enantiomers D- and L-tetrulose, by formally inserting additional CHOH-groups between the keto and existing CHOH-groups, form a series of D- and L-2-ketoses. Figure 4.2 gives D-2-ketoses in their simplified *Fischer* projections.

Data are provided in Table 4.2 on the occurrence of ketoses of interest in food.

For the simplified presentation of structures, abbreviations are used which usually consist of the first letters of the name of the monosaccharide. Figure 4.1 gives the configuration prefix derived from the trivial names, representing a specified configuration applied in monosaccharide classification. Thus, systematic names for D-glucose and D-fructose are D-gluco-hexose

Table 4.1. Occurrence of aldoses

Name, structure	Occurrence
Pentoses	
D-Apiose (3-C-Hydroxy-methyl-D-glycero-tetrose)	Parsley, celery seed
L-Arabinose	Plant gums, hemicelluloses, pectins, glycosides
2-Deoxy-D-ribose	Deoxyribonucleic acid
D-Lyxose	Yeast-nucleic acid
2-O-Methyl-D-xylose	Hemicelluloses
D-Ribose	Ribonucleic acid
D-Xylose	Xylanes, hemicelluloses plant gums, glycosides
Hexoses	
L-Fucose (6-Deoxy-L-galactose)	Human milk, seaweed (algae), plant gums and mucilage
D-Galactose	Widespread in oligo- and polysaccharides
D-Glucose	Widespread in plants and animals
D-Mannose	Widespread as polysaccharide building blocks
L-Rhamnose (6-Deoxy-L-mannose)	Plant gums and mucilage, glycosides

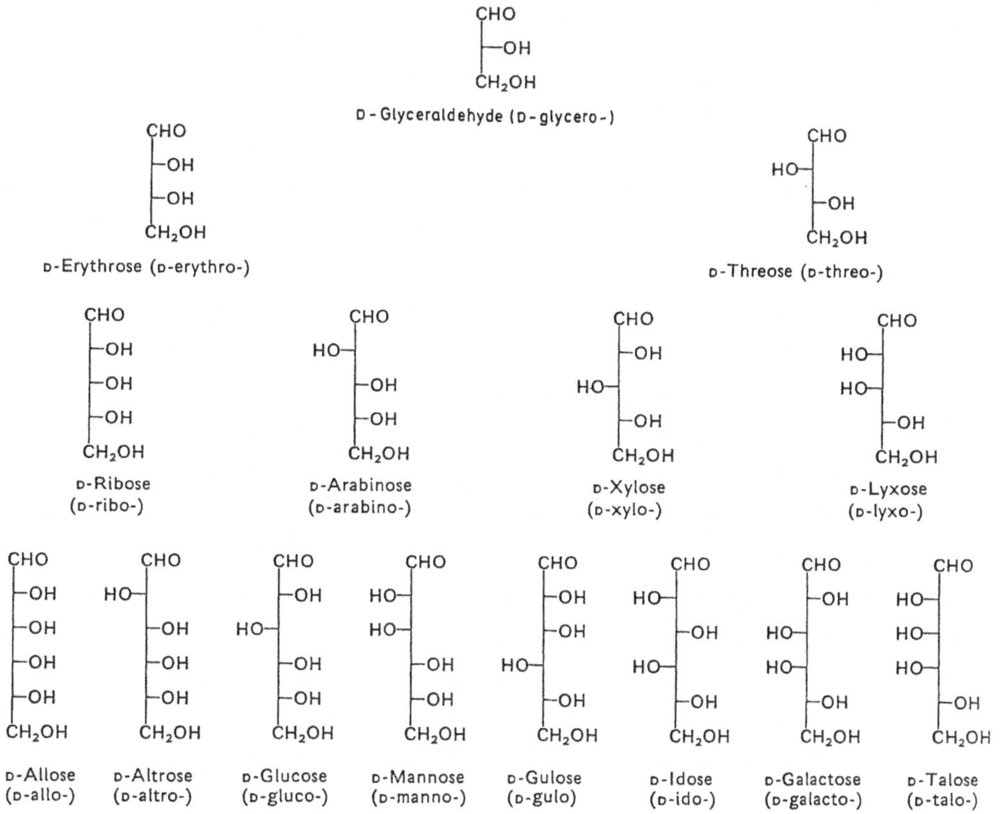

```
                              CHO
                             ─OH
                              CH₂OH
                    D-Glyceraldehyde (D-glycero-)

       CHO                                              CHO
      ─OH                                            HO─
      ─OH                                               ─OH
       CH₂OH                                            CH₂OH
  D-Erythrose (D-erythro-)                     D-Threose (D-threo-)
```

```
   CHO           CHO              CHO              CHO
  ─OH         HO─              ─OH           HO─
  ─OH            ─OH         HO─              HO─
  ─OH            ─OH            ─OH              ─OH
   CH₂OH         CH₂OH          CH₂OH            CH₂OH
 D-Ribose      D-Arabinose     D-Xylose        D-Lyxose
 (D-ribo-)     (D-arabino-)    (D-xylo-)       (D-lyxo-)
```

```
 CHO      CHO     CHO      CHO     CHO      CHO      CHO      CHO
─OH    HO─      ─OH    HO─     ─OH    HO─      ─OH    HO─
─OH       ─OH  HO─    HO─        ─OH       ─OH  HO─     HO─
─OH       ─OH     ─OH      ─OH HO─     HO─   HO─      HO─
─OH       ─OH     ─OH      ─OH    ─OH       ─OH     ─OH      ─OH
 CH₂OH    CH₂OH    CH₂OH    CH₂OH    CH₂OH    CH₂OH    CH₂OH    CH₂OH
D-Allose  D-Altrose D-Glucose D-Mannose D-Gulose D-Idose D-Galactose D-Talose
(D-allo-) (D-altro-) (D-gluco-) (D-manno-) (D-gulo) (D-ido-) (D-galacto-) (D-talo-)
```

Fig. 4.1. D-Aldoses in *Fischer* projection

and D-arabino-2-hexulose. Such nomenclature makes it possible to systematically denote all monosaccharides that contain more than four

Table 4.2. Occurrence of ketoses

Name, structure	Occurrence
Hexulose	
D-Fructose	Present in plants and honey
D-Psicose	Found in residue of fermented molasses
Heptulose	
D-manno-2-Heptulose	Avocado fruit
Octulose	
D-glycero-D-manno-2-Octulose	Avocado fruit
Nonulose	
D-erythro-L-gluco-2-Nonulose	Avocado fruit

chiral centers. According to this procedure the portion of the molecule adjacent to the carbonyl group is given the maximal possible prefix, while the portion furthest from the carbonyl group is denoted first. In the case of ketoses, the two portions of the molecule separated by the keto group are given. In a combined prefix designation, as with aldoses, the portion which has the C-atom furthest from the keto group is mentioned first. However, when a monosaccharide does not have more than four chiral centers, a designation in the ketose series may omit the two units separated by the keto group. The examples in Formula 4.8 illustrate the rule.

Lactol formation provides a new chiral center. Thus, there are two additional diastereomers for each pyranose and furanose. These isomers are called anomers and are denoted as α and β-forms. Formula 4.9 illustrates the two anomeric D-glucose molecules in *Tollens* ring formulas and *Haworth* projections.

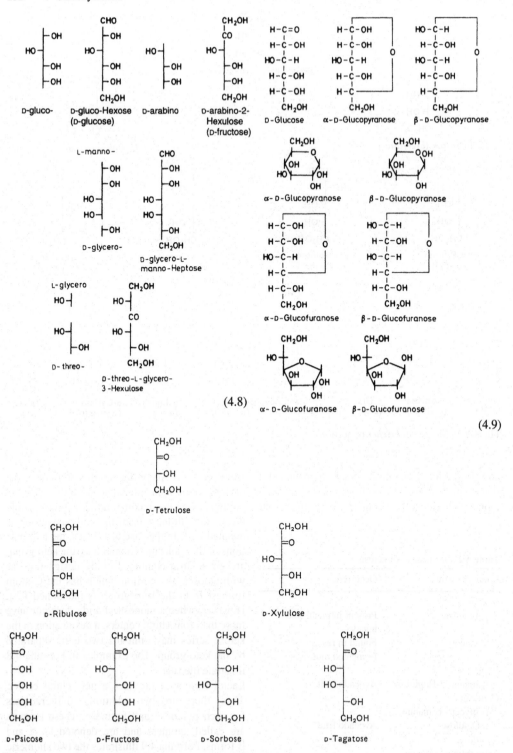

(4.8)

(4.9)

Fig. 4.2. D-Ketoses in *Fischer* projection

The cis-arrangement of the two adjacent HO-groups in positions C-1 and C-2 of α-D-glucopyranose, unlike its β-anomer, increases the conductivity of boric acid. A borate complex is formed which is a stronger acid than boric acid itself (cf. Formula 4.10).

In *Tollens* ring formula, in the D-series, the α-anomer has the hydroxyl group at C-1 on the right and the β-anomer has this OH-group on the left.

In *Haworth* projections the HO-group of α/β-anomers of the D-series usually occurs below/above the pyranose or furanose ring planes, while in the L-series the reverse is true (cf. Formula 4.11).

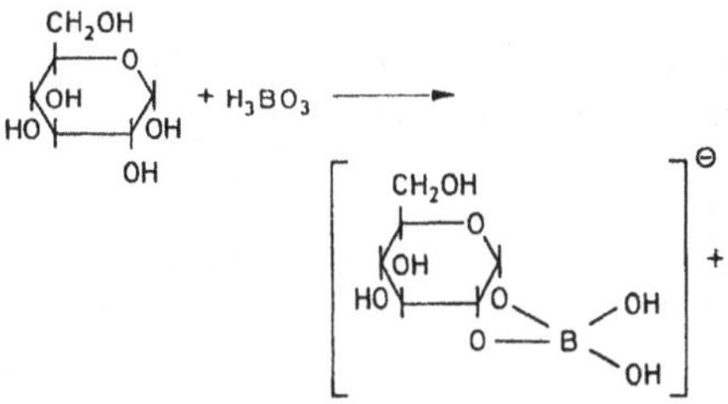

(4.10)

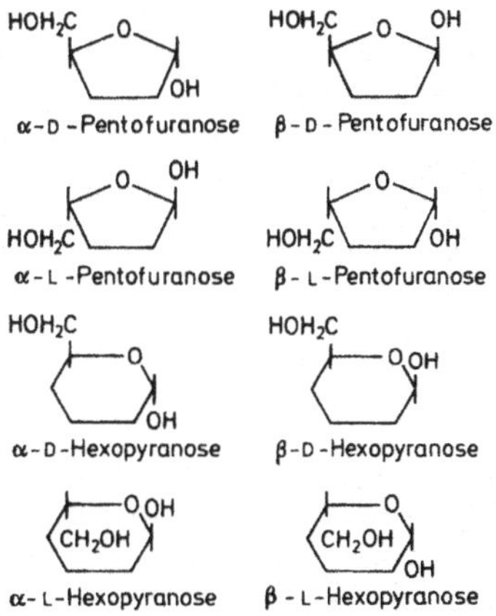

(4.11)

equilibrium state varies greatly among sugars (cf. Table 4.6).

The transition into the different hemiacetal forms, called mutarotation, proceeds via the open-chain carbonyl compound. The acid- and base-catalyzed ring opening is the rate limiting step of the reaction:

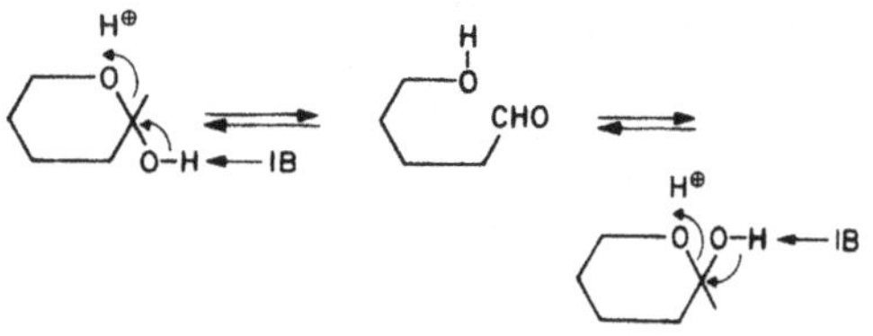

(4.12)

2,3,4,6-Tetramethyl-D-glucose reaches equilibrium in benzene rapidly through the concerted action of cresol and pyridine as acid–base catalysts (Table 4.3). Bifunctional reagents, like 2-pyridone and benzoic acid, are especially efficient acid–base catalysts in both polar and nonpolar solvents:

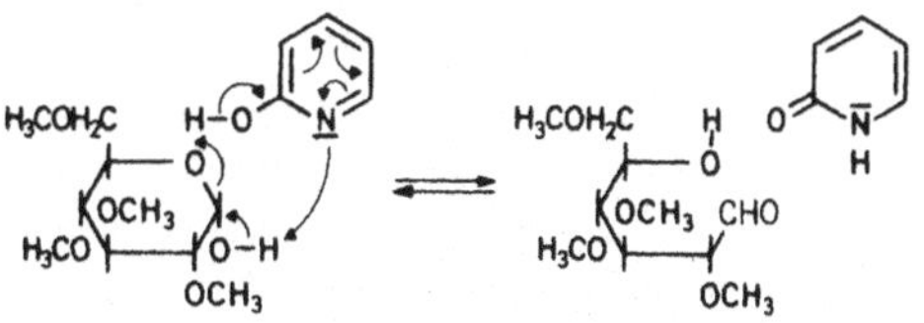

(4.13)

Water can also be a bifunctional catalyst:

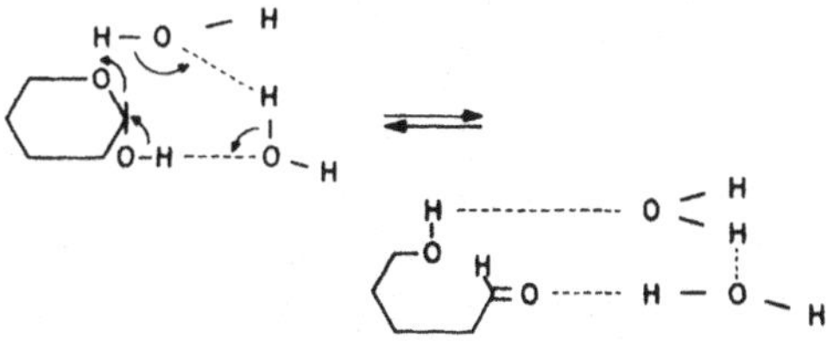

(4.14)

Each monosaccharide can exist in solution together with its open chain molecule in a total of five forms. Due to the strong tendency towards cyclization, the amount of the open chain form is greatly reduced. The contribution of the different cyclic forms to the equilibrium state in a solution depends on the conformation. An aqueous D-glucose solution is nearly exclusively the two pyranoses, with 36% α- and 64% β-anomer, while the furanose ring form is less than 1%. The

Table 4.3. Mutarotation rate of 2,3,4,6-tetramethyl-D-glucose (0.09 mol/l) in benzene

Catalyst	$(k\,min^{-1})$	k_{rel}
–	7.8×10^{-5}	1.0
Pyridine (0.1 mol/l)	3.7×10^{-4}	4.7
p-Cresol (0.1 mol/l)	4.2×10^{-4}	5.4
Pyridine + p-cresol (0.1 mol/l)	7.9×10^{-3}	101
2-Pyridone (0.1 mol/l)	1.8×10^{-1}	2307
Benzoic acid (0.1 mol/l)	2.2	28,205

The reaction rate for the conversion of the α-and β-forms has a wide minimum in an aqueous medium in a pH range of 2–7, as illustrated in section 10.1.2.2 with lactose, and the rate increases rapidly beyond this pH range.

4.2.1.3 Conformation

A series of physicochemical properties of monosaccharides can be explained only by the conformation formulas (*Reeves* formulas).

The preferred conformation for a pyranose is the so-called chair conformation and not the twisted-boat conformation, since the former has the highest thermodynamic stability. The two chair C-conformations are $^4C_1$ (the superscript corresponds to the number of the C-atom in the upper position of the chair and the subscript to that in the lower position; often designated as C1 or "O-outside") and $^1C_4$ (often designated as 1C, the mirror image of C1, and C-1 in upper and C-4 in lower positions, or simply the "O-inside" conformer). The $^4C_1$-conformation is preferred in the series of D-pyranoses, with most of the bulky groups, e. g., HO and, especially, CH_2OH, occupying the roomy equatorial positions. The interaction of the bulky groups is low in such a conformation, hence the conformational stability is high. This differs from the C_4-conformation, in which most of the bulky groups are crowded into axial positions, thus imparting a thermodynamic instability to the molecule (Table 4.4).

β-D-Glucopyranose in the $^4C_1$-conformation is an exception. All substituents are arranged equatorially, while in the $^1C_4$ all are axial (Formula 4.15). α-D-Glucopyranose in the $^4C_1$-conformation has one axial group at C-1 and is also lower in energy by far (cf. Table 4.5).

β-D-Glucose $(^4C_1)$ α-D-Glucose $(^4C_1)$

β-D-Glucose $(^1C_4)$ α-D-Glucose $(^1C_4)$ (4.15)

Table 4.4. Free energies of unfavorable interactions between substituents on the tetrahydropyran ring

Interaction	Energy kJ/mole [a]
$H_{ax} - O_{ax}$	1.88
$H_{ax} - C_{ax}$	3.76
$O_{ax} - O_{ax}$	6.27
$O_{ax} - C_{ax}$	10.45
$O_{eq} - O_{eq}/O_{ax} - O_{eq}$	1.46
$O_{eq} - C_{eq}/O_{ax} - C_{eq}$	1.88
Anomeric effect[b]	
for O_{eq}^{c2}	2.30
for O_{ax}^{c2}	4.18

[a] Aqueous solution, room temperature.
[b] To be considered only for an equatorial position of the anomeric HO-group.

The arrangement of substituents differs e. g., in α-D-idopyranose. Here, all the substituents are in axial positions in the $^4C_1$-conformation (axial HO-groups at 1, 2, 3, 4), except for the CH_2OH-group, which is equatorial. However, the $^1C_4$-conformation is thermodynamically more favorable despite the fact that the CH_2OH-group is axial (cf. Table 4.5):

α-D-Idose $(^4C_1)$ α-D-Idose $(^1C_4)$ (4.16)

A second exception (or rather an extreme case) is α-D-altropyranose. Both conformations (O-outside and O-inside) have practically the same stability in this sugar (cf. Table 4.5).

The free energy of the conformers in the pyranose series can be calculated from partial interaction energies (derived from empirical data). Only the 1,3-diaxial interactions (with exception of the interactions between H-atoms), 1,2-gauche or staggered (60°) interactions of two HO-groups and that between HO-groups and the CH_2OH-group will be considered. The partial interaction energies are compiled in Table 4.4, the relative free enthalpies $\Delta G°$ calculated from these data for various conformers are presented in Table 4.5. In addition to the interaction energies an effect is considered

Table 4.5. Relative free enthalpies for hexopyranoses

Hexopyranose	Conformation	$\Delta G°$ (kJ/mole)
α-D-Glucose	$^4C_1$	10.03
	$^1C_4$	27.38
β-D-Glucose	$^4C_1$	8.57
	$^1C_4$	33.44
α-D-Mannose	$^4C_1$	10.45
β-D-Mannose	$^4C_1$	12.33
α-D-Galactose	$^4C_1$	11.91
β-D-Galactose	$^4C_1$	10.45
α-D-Idose	$^4C_1$	18.18
	$^1C_4$	16.09
β-D-Idose	$^4C_1$	16.93
	$^1C_4$	22.36
α-D-Altrose	$^4C_1$	15.26
	$^1C_4$	16.09

The other substituents also influence the anomeric effect, particularly the HO-group in C-2 position. Here, due to an antiparallel dipole formation, the axial position enhances stabilization better than the equatorial position. Correspondingly, in an equilibrium state in water, D-mannose is 67% in its α-form, while α-D-glucose or α-D-galactose are only 36% and 32%, respectively (Table 4.6). The anomeric effect (dipole–dipole interaction) increases as the dielectric constant of the solvent system decreases e. g., when water is replaced by methanol.

Alkylation of the lactol HO-group also enhances the anomeric effect (Table 4.7). A reduction of the dielectric constant of the solvent (e. g., transition from water to methanol), resulting in an increase in the dipole–dipole interaction, also enhances the anomeric effect. Conformational isomers of furanose also occur since its ring is not planar. There are two basic conformations, the envelope (E) and the twist (T), which are the most stable; in solu-

which destabilizes the anomeric HO-group in the equatorial position, while it stabilizes this group in the axial position. This is called the *anomeric effect* and is attributed to the repulsion between the parallel dipoles. If the 1-OH group (β-anomer) is in the equatorial position (cf. Formula 4.17), repulsion results from the polarized bonds carbon atom 5 – ring oxygen and carbon atom 1 – oxygen of the anomeric OH-group. The repulsion forces the anomeric HO-group to take up the more stable axial or α-position:

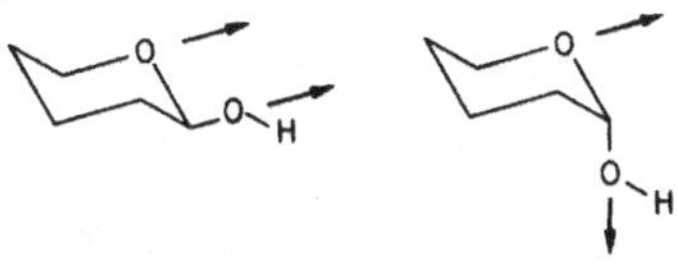

(4.17)

Table 4.7. Methylglucoside isomers in methanol (1% HCl) at equilibrium state[a]

Compound	α-Pyranoside	β-Pyranoside	α-Furanoside	β-Furanoside
Methyl-D-glucoside	66	32.5	0.6	0.9
Methyl-D-mannoside	94	5.3	0.7	0
Methyl-D-galactoside	58	20	6	16
Methyl-D-xyloside	65	30	2	3

[a] Values in %.

Table 4.6. Equilibrium composition[a] of aldoses and ketoses in aqueous solution

Compound	T (°C)	α-Pyranose	β-Pyranose	α-Furanose	β-Furanose
D-Glucose	20	36	64	–	–
D-Mannose	20	67	33	–	–
D-Galactose	20	32	64	1	3
D-Idose	60	31	37	16	16
D-Ribose	40	20	56	6	18
D-Xylose	20	35	65	–	–
D-Fructose	20	–	76	4	20

[a] Values in %.

tion a mixture exists of conformers similar in energy (cf. Formula 4.18).

$$^1E \qquad\qquad ^4T_3 \qquad\qquad (4.18)$$

An anomeric effect preferentially forces the anomeric HO-group into the axial position. This is especially the case when the HO-group attached to C-2 is also axial. When pyranose ring formation is prevented or blocked, as in 5-O-methyl-D-glucose, the twisted $^3T_2$-conformer becomes the dominant form:

5-O-Methyl-β-D-glucose ($^3T_2$)   (4.19)

A pyranose is generally more stable than a furanose, hence, the former and not the latter conformation is predominant in most monosaccharides (Table 4.6).

The composition of isomers in aqueous solution, after equilibrium is reached, is compiled for a number of monosaccharides in Table 4.6. Evidence for such compositions is obtained by polarimetry, by oxidation with bromine, which occurs at a much higher reaction rate with β- than α-pyranose and, above all, by nuclear magnetic resonance spectroscopy (^{1}H-NMR).

In proton magnetic resonance spectroscopy of sugars, the protons bound to oxygen, which complicate the spectrum, are replaced by derivatization (O-acyl derivatives) or are exchanged for deuterium when the sugar is in D_2O solution. The chemical shift of the retained protons bound covalently to carbon varies. Due to less shielding by the two oxygens in α position, the proton on the anomeric carbon atom appears at a lower magnetic field than other protons, the chemical shift increasing in the order pyranoses < furanoses in the range of $\delta = 4.5-5.5$ (free monosaccharides). As a result of the coupling

with the H-atom at C-2, the anomeric proton appears as a doublet. Furthermore, an axial proton (β-form of D-series) appears at higher field than an equatorial proton (α-form of D-series). The sugar conformation is elucidated from the coupling constant of neighboring protons: equatorial–equatorial, equatorial–axial (small coupling constants) or axial-axial (larger coupling constants).

The proton resonance spectrum of D-glucose ($^1C_4$-conformation) in D_2O is shown in Fig. 4.3. The figure first shows the signals of the protons at C-2 to C-6 in the range of 3.2–3.9 ppm. The large coupling constant of the doublet at δ 4.62 (7.96 Hz) shows a diaxial position of the H-atoms at C-1/C-2 and, thus, the equatorial position of the hydroxy group at C-1. This indicates the β-D-glucopyranose conformation. The equatorial proton in α-D-glucopyranose (5.2 ppm) appears at lower field (higher ppm). The smaller coupling constant of the doublet at δ 5.2 (J = 3.53 Hz) confirms the axial/equatorial arrangement of the H-atoms at C-1/C-2 of α-D-glucopyranose.

The content of both anomers in aqueous solution can be calculated from the signal areas. α- and β-Glucofuranoses are not present in aqueous solution (Table 4.6).

Elucidation of sugar conformation can also be achieved by ^{13}C-nuclear magnetic resonance spectroscopy. Like ^{1}H-NMR, the chemical shifts differ for different C-atoms and are affected by the spatial arrangement of ring substituents.

4.2.2 Physical Properties

4.2.2.1 Hygroscopicity and Solubility

The moisture uptake of sugars in crystallized form is variable and depends, for example, on the sugar structure, isomers present and sugar purity. Solubility decreases as the sugars cake together, as often happens in sugar powders or granulates. On the other hand, the retention of food moisture by concentrated sugar solutions, e. g., glucose syrup, is utilized in the baking industry.

The solubility of mono- and oligosaccharides in water is good. However, anomers may differ substantially in their solubility, as exemplified by α- and β- lactose (cf. 10.1.2.2). Monosaccharides

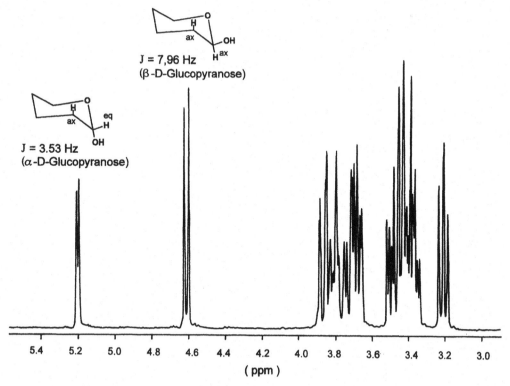

J = 7,96 Hz
(β-D-Glucopyranose)

J = 3.53 Hz
(α-D-Glucopyranose)

Fig. 4.3. Proton resonance spectrum of D-glucose (in D_2O)

are soluble to a small extent in ethanol and are insoluble in organic solvents such as ether, chloroform or benzene.

4.2.2.2 Optical Rotation, Mutarotation

Specific rotation constants, designated as $[\alpha]$ for sodium D-line light at 20–25°C, are listed in Table 4.8 for some important mono- and oligosaccharides. The specific rotation constant $[\alpha]_\lambda^t$ at a selected wavelength and temperature is calculated from the angle of rotation, α, by the equation:

$$[\alpha]_\lambda^t = \frac{\alpha}{l \cdot c} \tag{4.20}$$

where l is the polarimeter tube length in decimeters and c the number of grams of the optically active sugar in 100 ml of solution. The molecular rotation, [M], is suitable for comparison of the rotational values of compounds with differing molecular weights:

$$[M]_\lambda^t = M [\alpha]_\lambda^t \tag{4.21}$$

where M represents the compound's molecular weight. Since the rotational value differs for anomers and also for pyranose and furanose conformations, the angle of rotation for a freshly prepared solution of an isomer changes until an equilibrium is established. This phenomenon is known as mutarotation. When an equilibrium exists only between two isomers, as with glucose (α- and β-pyranose forms), the reaction rate follows first order kinetics:

$$-\frac{dc_\alpha}{dt} = k \cdot c_\alpha - k' \cdot c_\beta \tag{4.22}$$

A simple mutarotation exists in this example, unlike complex mutarotations of other sugars, e.g., idose which, in addition to pyranose, is also largely in the furanose form. Hence, the order of its mutarotation kinetics is more complex.

Table 4.8. Specific rotation of various mono- and oligosaccharides

Compound	$[\alpha]_D{}^a$	Compound	$[\alpha]_D{}^a$
Monosaccharides		*Oligosaccharides*	
L-Arabinose	+105	(continued)	
α-	+55.4	Kestose	+28
β-	+190.6	Lactose	+53.6
D-Fructose	−92	β-	+34.2
β-	−133.5	Maltose	+130
D-Galactose	+80.2	α-	+173
α-	+150.7	β-	+112
β-	+52.8	Maltotriose	+160
D-Glucose	+52.7	Maltotetraose	+166
α-	+112	Maltopentaose	+178
β-	+18.7	Maltulose	+64
D-manno-2-		Manninotriose	+167
Heptulose	+29.4	Melezitose	+88.2
D-Mannose	+14.5	Melibiose	+ 143
α-	+29.3	β-	+ 123
β-	−17	Palatinose	+ 97.2
D-Rhamnose	−7.0	Panose	+ 154
D-Ribose	−23.7	Raffinose	+ 101
D-Xylose	+18.8	Saccharose	+ 66.5
α-	+93.6	α-*Schardinger-*	
		Dextrin	+151
Oligosaccharides		β-*Schardinger-*	
(including disaccharides)		Dextrin	+ 162
Cellobiose	+34.6	γ-*Schardinger-*	
β-	+14.2	Dextrin	+180
Gentianose	+33.4	Stachyose	+146
Gentiobiose	+10		
α-	+31		
β-	−3		

^{a} Temperature: 20–25 °C.

4.2.3 Sensory Properties

Mono- and oligosaccharides and their corresponding sugar alcohols, with a few exceptions, are sweet. β-D-Mannose has a sweet-bitter taste, and some oligosaccharides are bitter, e. g. gentiobiose.

The most important sweeteners are saccharose (sucrose), starch syrup (a mixture of glucose, maltose and malto-oligosaccharides) and glucose. Invert sugar, fructose-containing glucose syrups (high fructose corn syrup), fructose, lactose and sugar alcohols, such as sorbitol, mannitol and xylitol, are also of importance. The sugars differ in quality of sweetness and taste intensity. Saccharose is distinguished from other sugars by its pleasant taste even at high concentrations. The taste intensity of oligosaccharides drops regularly as the chain length increases.

The taste intensity can be measured by determining the recognition threshold of the sugar (the lowest concentration at which sweetness is still perceived) or by comparison with a reference substance (isosweet concentrations). The threshold value is related to the affinity of sweet-taste chemoreceptors for the sweet substance and is of importance in elucidation of reltionships between the chemical structure of a compound and its taste. For practical purpose, the use of a reference substance is of greater importance: taste intensity is dependent on concentration and varies greatly among sweet compounds.

Saccharose is the reference substance usually chosen. Tables 4.9, 4.10 and 4.11 list some sugar sweetness threshold values and relative sweet-

Table 4.9. Taste threshold values of sugars in water

Sugar	Recognition threshold		Detection threshold	
	mol/1	%	mol/1	%
Fructose	0.052	0.94	0.02	0.24
Glucose	0.090	1.63	0.065	1.17
Lactose	0.116	4.19	0.072	2.60
Maltose	0.080	2.89	0.038	1.36
Saccharose	0.024	0.81	0.011	0.36

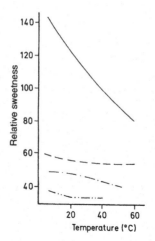

Fig. 4.4. Temperature dependence of the relative sweetness of some sugars (based on saccharose $\triangleq 100$ at each temperature; — D-fructose, – – – D- glucose, –·–·– D-galactose, –··–··– maltose) (according to *Shallenberger* and *Birch*, 1975)

ness values. Only mean values are given with deviations omitted. The recognition threshold values for saccharose cited in the literature vary from 0.01 to 0.037 mol/1.

Taste quality and intensity are dependent not only on a compound's structure but on other taste reception parameters: temperature, pH and the presence of additional sweet or non-sweet compounds.

The temperature dependence of the taste intensity is especially pronounced in the case of D-fructose (Fig. 4.4). It is based on the varying intensity of sweetness of the different isomers:

β-D-fructopyranose is the sweetest isomer, and its concentration decreases as the temperature increases (Fig. 4.5).

Table 4.10. Relative sweetness of sugars and sugar alcohols to sucrose[a]

Sugar/ sugar alcohol	Relative sweetness	Sugar/ sugar alcohol	Relative sweetness
Saccharose	100	D-Mannitol	69
Galactitol	41	D-Mannose	59
D-Fructose	114	Raffinose	22
D-Galactose	63	D-Rhamnose	33
D-Glucose	69	D-Sorbitol	51
Invert sugar	95	Xylitol	102
Lactose	39	D-Xylose	67
Maltose	46		

[a] 10% aqueous solution.

Table 4.11. Concentration (%) of iso-sweet aqueous solutions of sugars and sugar alcohols

D-Fructose	D-Glucose	Lactose	Saccharose	D-Sorbitol	Xylitol
0.8	1.8	3.5	1.0		
1.7	3.6	6.5	2.0		
4.2	8.3	15.7	5.0	9.3	8.5
8.6	13.9	25.9	10.0	17.2	9.8
13.0	20.0	34.6	15.0		
16.1	27.8		20.0	28.2	18.5
	39.0		30.0		25.4
	48.2		40.0	48.8	

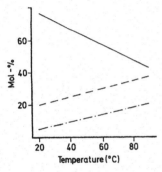

Fig. 4.5. Temperature dependence of the mutarotation equilibrium of D-fructose, (— β-D-fructo-pyranose, – – – β-D-fructofuranose, —·—·— α-D-fructofuranose) (according to *Shallenberger* and *Birch*, 1975)

Furthermore, an interrelationship exists between the sugar content of a solution and the sensory assessment of the volatile aroma compounds present. Even the color of the solution might affect taste evaluation. Figures 4.6–4.9 clarify these phenomena, with fruit juice and canned fruits as selected food samples.

The overall conclusion is that the composition and concentration of a sweetening agent has to be adjusted for each food formulation to provide optimum sensory perception.

A prerequisite for a compound to be sweet is the presence in its structure of a proton

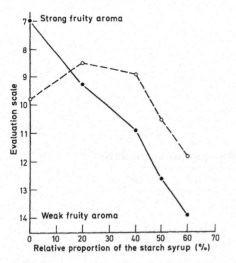

Fig. 4.6. Sensory evaluation of the "fruity flavor" of canned peaches at different ratios of saccharose/starch syrup (•—•60°, ○—○50° Brix) (according to *Pangborn*, 1959)

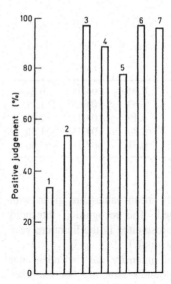

Fig. 4.7. Sensory evaluation of canned cherries prepared with different sweeteners 1, 2, 3: 60, 50, 40% saccharose, 4: 0.15% cyclamate, 5: 0.05% saccharin, 6: 10% saccharose + 0.10% cyclamate, 7: 10% saccharose + 0.02% saccharin (according to *Salunkhe*, 1963)

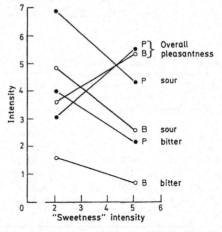

Fig. 4.8. Sensory evaluation of the categories "overall pleasantness", "sour" and "bitter" versus sweetness intensity. B bilberry (○—○) and P (•—•) cranberry juice (according to *Sydow*, 1974)

donor/acceptor system (AH/B-system), which may be supplemented by a hyrophobic site X. This AH/B/X-system interacts with a complementary system of the taste receptor site located on the taste buds. Based on studies of the taste quality of sugar derivatives and deoxy sugars, the following AH/B/X-systems have been proposed

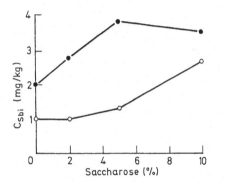

Fig. 4.9. Bitter taste threshold values of limonin (o—o) and naringin ($\times 10^{-1}$ •—•) in aqueous saccharose solution (according to *Guadagni*, 1973)

for β-D-glucopyranose and β-D-fructopyranose respectively:

(4.23)

β-D-glucopyranose and β-L-glucopyranose are sweet. Molecular models show that the AH/B/X-system of both sugar components fits equally well with the complementary receptor system $AH_R/B_R/X_R$ (Formula 4.24):

β-D-Glucose   β-L-Glucose

(4.24)

With asparagine enantiomers, the D-form is sweet, while the L-form is tasteless. Here, unlike D- and L-glucose, only the D-form interacts with the complementary $AH_R/B_R/X_R$-system:

D-Asparagine   L-Asparagine   (4.25)

As described in 8.8.1.1, the AH/B/X-system has been extended to explain the large differences in structure and sweetening strength which can exist in compounds of different substance classes.

4.2.4 Chemical Reactions and Derivatives

4.2.4.1 Reduction to Sugar Alcohols

Monosaccharides can be reduced to the corresponding alcohols by $NaBH_4$, electrolytically or via catalytic hydrogenation. Two new alcohols are obtained from ketoses, e. g., fructose, since a new chiral center is formed:

(4.26)

The alcohol name is derived from the sugar name in each case by replacing the suffix -ose or -ulose with the suffix -itol. The sugar alcohols of importance in food processing are xylitol, the only one of the four pentitols (mesoribitol, D,L-arabitol, mesoxylitol) used, and only D-glucitol (sorbitol) and D-mannitol of the ten stereoisomeric forms of hexitols [meso-allitol, meso-galactitol (dulcitol), D,L-glucitol (sorbitol), D,L-iditol, D,L-mannitol and D,L-altritol]. They are used as sugar substitutes in dietetic food formulations to decrease water activity in "intermediate moisture foods", as softeners, as crystallization inhibitors and for improving the rehydration characteristics of dehydrated food. Sorbitol is found in nature in many fruits, e. g., pears, apples and plums. Palatinitol (a mixture of glucopyranosyl glucitol and glucopyranosyl mannitol) is a sugar alcohol. It is produced using biotechnological methods by the rearrangement of sucrose ($1 \rightarrow 2$ to $1 \rightarrow 6$), followed by reduction. Maltitol, the reduction product of the disaccharide maltose, is being considered for wider use in food formulations.

4.2.4.2 Oxidation to Aldonic, Dicarboxylic and Uronic Acids

Under mild conditions, e. g., with bromine water in buffered neutral or alkaline media, aldoses are oxidized to aldonic acids. Oxidation involves the lactol group exclusively. β-Pyranose is oxidized more rapidly than the α-form. Since the β-form is more acidic (cf. 4.2.1.3), it can be considered that the pyranose anion is the reactive form. The oxidation product is the δ-lactone which is in equilibrium with the γ-lactone and the free form of aldonic acid. The latter form prevails at pH > 3.

(4.27)

The transition of lactones from δ- to γ-form and vice versa probably proceeds through an intermediary bicyclic form.
The acid name is obtained by adding the suffix–onic acid (e. g. aldose $\rightarrow$ aldonic acid).
Glucono-δ-lactone is utilized in food when a slow acid release is required, as in baking powders, raw fermented sausages or dairy products.

Treatment of aldose with more vigorous oxidizing agents, such as nitric acid, brings about oxidation of the C-1 aldehyde group and the CH_2OH-group, resulting in formation of a dicarboxylic acid (nomenclature: stem name of the parent sugar + the suffix -aric acid, e. g. aldose $\rightarrow$ aldaric acid). Thus, galactaric acid (common or trivial name: mucic acid) is obtained from galactose:

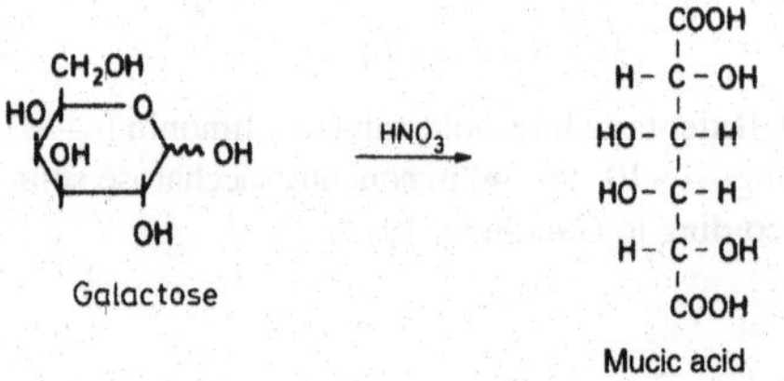

(4.28)

The dicarboxylic acid can, depending on its configuration, form mono- or dilactones.
Oxidation of the CH_2OH-group by retaining the carbonyl function at C-1, with the aim of obtaining uronic acids (aldehydocarboxylic acids), is possible only by protecting the carbonyl group during oxidation. A suitable way is to temporarily block the vicinal HO-groups by ketal formation which, after the oxidation at C-6 is completed, are deblocked under mild acidic conditions:

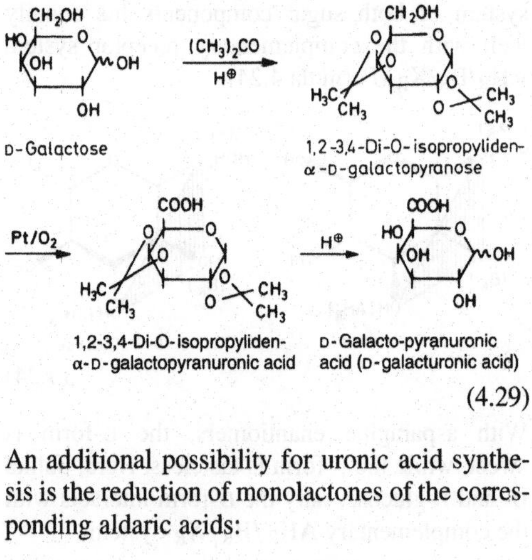

(4.29)

An additional possibility for uronic acid synthesis is the reduction of monolactones of the corresponding aldaric acids:

(4.30)

Another industrially applied glucuronic acid synthesis involves first oxidation then hydrolysis of starch:

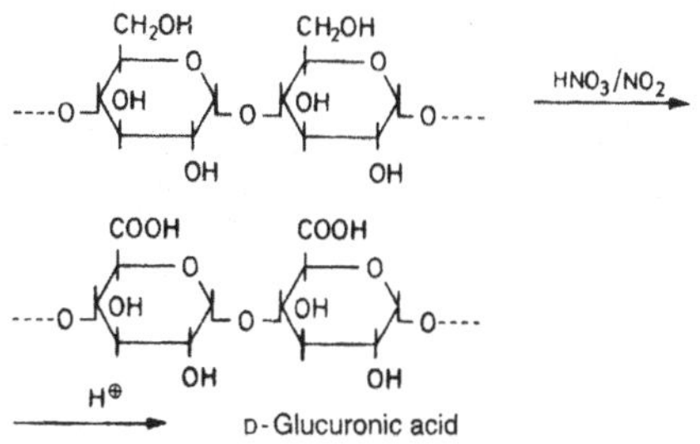

$$(4.31)$$

Depending on their configuration, the uronic acids can form lactone rings in pyranose or furanose forms.

A number of uronic acids occur fairly abundantly in nature. Some are constituents of polysaccharides of importance in food processing, such as gel-forming and thickening agents, e. g. pectin (D-galacturonic acid) and sea weed-derived alginic acid (D-mannuronic acid, L-guluronic acid).

4.2.4.3 Reactions in the Presence of Acids and Alkalis

In the absence of amine components, monosaccharides are relatively stable in the pH range 3–7. Beyond these pH limits, more or less extensive conversions occur, depending on the conditions. Enolizations and subsequent elimination of water with retention of the C-chain predominate in an acidic medium. In a basic medium, enolizations with subsequent fragmentation (*retro*-aldol reactions) and secondary reactions of the fragments (aldol additions) predominate.

4.2.4.3.1 Reactions in Strongly Acidic Media

The reverse of glycoside hydrolysis (cf. 4.2.4.5) i.e. re-formation of glycosides, occurs in dilute mineral acids. All the possible disaccharides and higher oligosaccharides, but preferentially isoma-

ltose and gentiobiose, are obtained from glucose:

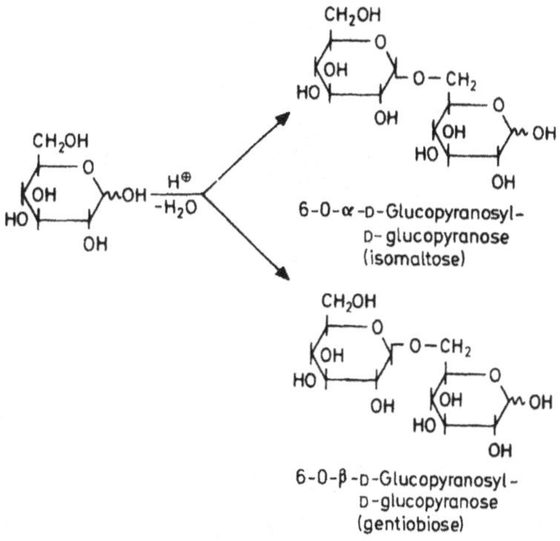

$$(4.32)$$

Such reversion-type reactions are observed, e. g., in the acidic hydrolysis of starch.

In addition to the formation of intermolecular glycosides, intramolecular glycosidic bonds can be readily established when the sugar conformation is suitable. β-Idopyranose, which occurs in the $^1C_4$-conformation, is readily changed to 1,6-anhydroidopyranose, while the same reaction with β-D-glucopyranose ($^4C_1$-conformation) occurs only under more drastic conditions, e. g., during pyrolysis of glucose, starch or cellulose. Heating glucose syrup above 100 °C can form 1,6-anhydrogluco-pyranose, but only in traces:

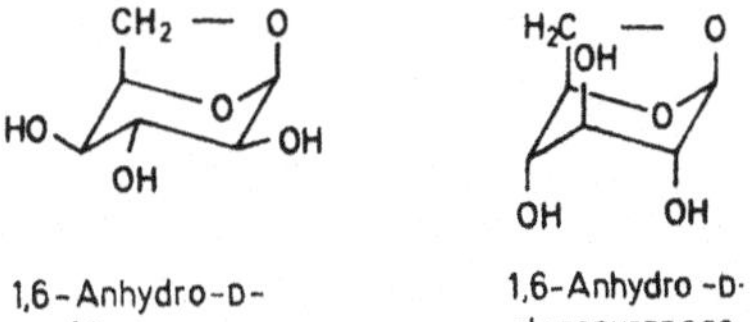

1,6-Anhydro-D-idopyranose

1,6-Anhydro-D-glucopyranose

$$(4.33)$$

In the formation of reversion products, it is assumed that in the presence of strong acids an oxonium cation is formed which, as an alkylating agent, reacts with the nucleophilic hydroxy groups with the cleavage of H^+ (Formula 4.34).

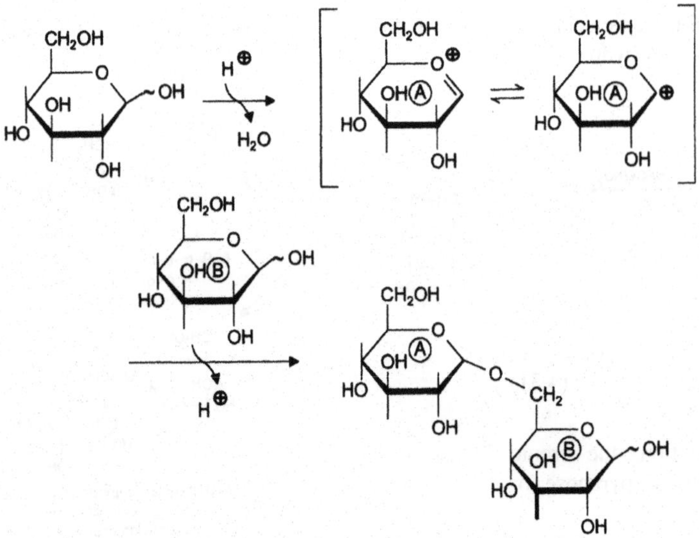

(4.34)

Dehydrating Reactions. The heating of monosaccharides under acidic conditions, e. g., during pasteurization of fruit juices and baking of rye bread, gives rise to a large number of furan and pyran compounds (examples in Formula 4.35). The formation of these compounds can be explained by enolizations and dehydrating reactions of the carbohydrates. It is noticeable that in some compounds, the aldehyde group of an aldose is formally retained at C-1 (furfural, 5-hydroxymethyl furfural, 5-methylfurfural) and in other components, the aldehyde group is reduced to a methyl group. As explained later, this indicates the course of formation in each case.

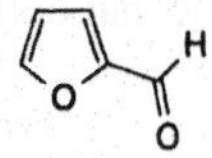

The reaction pathway in acid starts slowly with enolization to important intermediates called enediols. Glucose gives rise to 1,2-enediol, and fructose to 2,3-enediol as well (Formula 4.36). Starting with the enediols, the further course of the reaction is shown below.

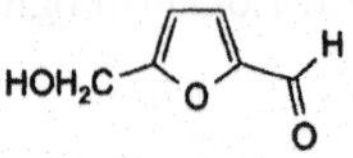

The steps in the formation of 5-hydroxymethyl furfural (HMF) from 1,2-enediol is shown in Formula 4.37. HMF is also used as an indicator for the heating of carbohydrate containing food, e. g., honey. The (*retro-Michael* addition) water elimination at C-3 and subsequently at C-4 leads to a 1,2-diulose (3,4-dideoxyosone), which after cyclization to a hemiacetal, a dihydrofuran, releases another molecule of water, producing HMF. In the same way, e. g., furfural can be made from pentoses and 5-methylfurfural from the 6-methylpentose rhamnose. 2-Hydroxyacetylfuran, which is preferentially formed from fructose, can be obtained starting from the corresponding 2,3-enediol by water elimination at C-4, followed by C-5 (Formula 4.38).

Furfural

5-Hydroxymethylfurfural

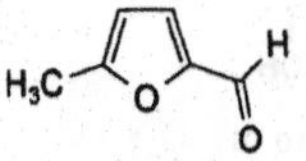

5-Methylfurfural

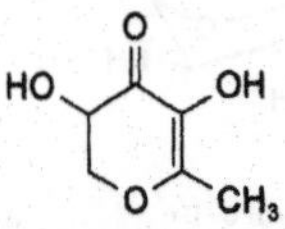

3,5-Dihydroxy-2-methyl-5,6-dihydropyran-4-one

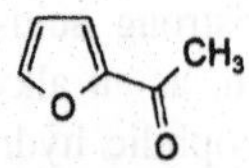

2-Acetylfuran

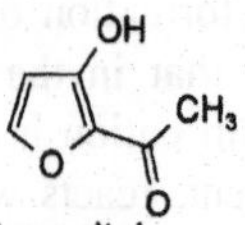

Isomaltol

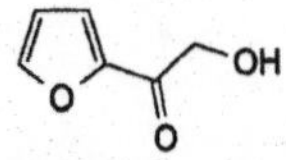

2-Hydroxyacetylfuran

(4.35)

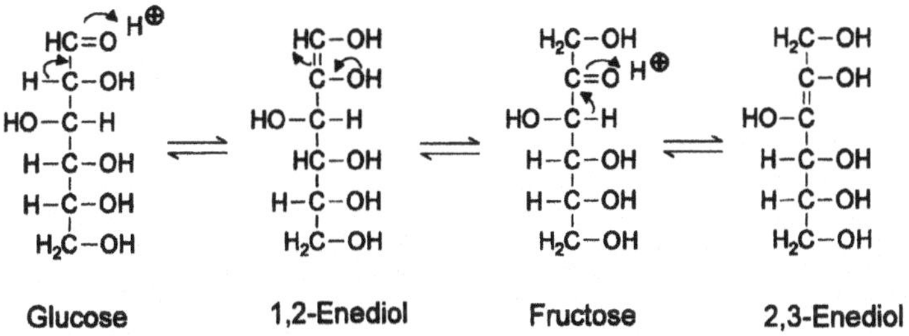

Glucose	1,2-Enediol	Fructose	2,3-Enediol	(4.36)

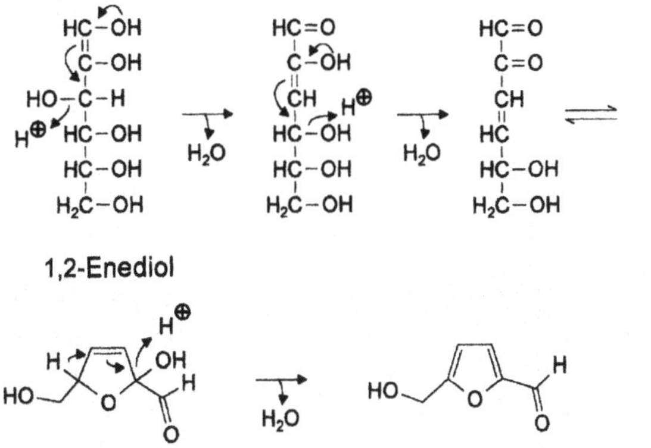

1,2-Enediol

(4.37)

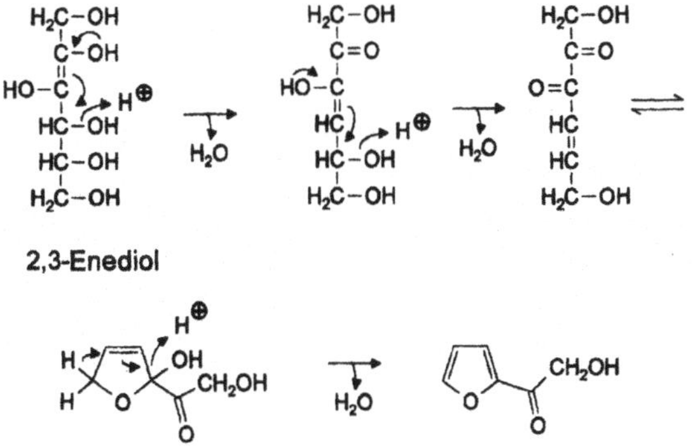

2,3-Enediol

(4.38)

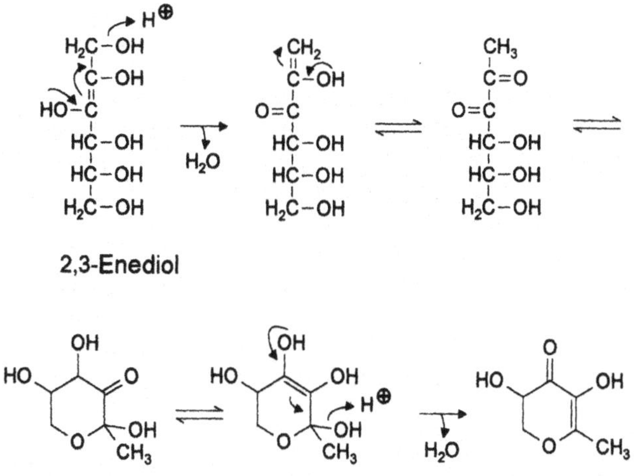

2,3-Enediol

(4.39)

With the 2,3-enediol, not only water elimination at C-4, but also the elimination of the hydroxyl group at C-1 is possible (Formula 4.39). This reaction pathway gives, among other compounds, 3,5-dihydroxy-2-methyl-5,6-dihydropyran-4-one, which is also used as an indicator for the heating of food. The formation of two different enediols is the reason for the wider product spectrum from ketoses, like fructose, than from aldoses. In the presence of amino compounds, all the reactions mentioned here proceed very easily also in the pH range 3–7. Since free amino acids are present in many foods, the reactions shown here also occur in connection with the pathways discussed in 4.2.4.4.

4.2.4.3.2 Reactions in Strongly Alkaline Solution

Alkaline reaction conditions occur in food, e. g., in the isolation of sucrose from sugar beet and in

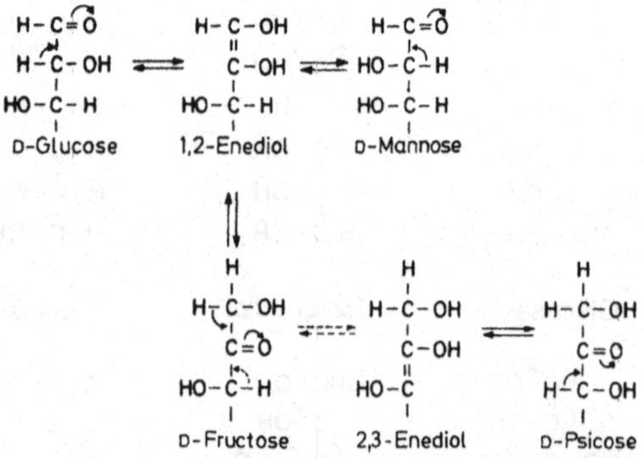

(4.40)

In this isomerization reaction, known as the *Lobry de Bruyn–van Ekenstein* rearrangement, one type of sugar can be transformed to another sugar in widely differing yields. The reaction plays a role in transforming an aldose to a ketose. For example, in the presence of sodium aluminate as a catalyst, lactose (4-O-β-D-galactopyranosido-D-glucopyranose, I) is rearranged into lactulose (4-O-β-D-galacto-pyranosido-D-fructose):

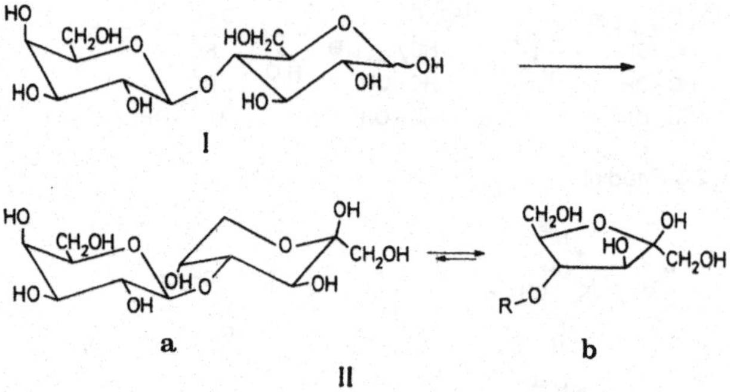

(4.41)

the production of alkali-baked goods. Apart from enolization reactions, which can also occur under acidic conditions but proceed much faster under alkaline conditions, the degradation of the carbohydrate skeleton is an important characteristic of base-catalyzed degradation reactions. Glucose, mannose and fructose are in equilibrium through the common 1,2-enediol. Also present is a small amount of psicose, which is derived from fructose by 2,3-enolization:

In this disaccharide, fructose is present mainly as the pyranose (IIa) and, to a small extent, as the furanose (IIb).

Lactulose utilization in infant nutrition is under consideration since it acts as a bifidus factor and prevents obstipation.

Oxidation of the enediol occurs in the presence of oxygen or other oxidizing agents, e. g., Cu^{2+}, resulting in carboxylic acids. In such a reaction with glucose, the main products are D-arabinonic and

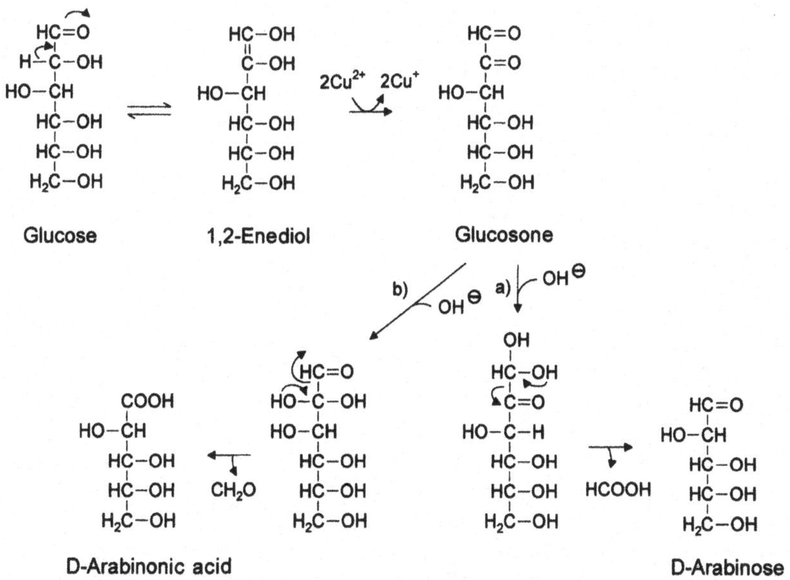

$$
\begin{array}{ccc}
\text{HC=O} & \text{HC-OH} & \text{HC=O} \\
\text{H-C-OH} & \text{C-OH} & \text{C=O} \\
\text{HO-CH} & \text{HO-CH} & \text{HO-CH} \\
\text{HC-OH} & \text{HC-OH} & \text{HC-OH} \\
\text{HC-OH} & \text{HC-OH} & \text{HC-OH} \\
\text{H}_2\text{C-OH} & \text{H}_2\text{C-OH} & \text{H}_2\text{C-OH} \\
\text{Glucose} & \text{1,2-Enediol} & \text{Glucosone}
\end{array}
\qquad 2Cu^{2+} \;\; 2Cu^{+}
$$

$$
\begin{array}{ccccc}
 & \text{HC=O} & \text{HC-OH} & & \text{HC=O} \\
\text{COOH} & \text{HO-C-OH} & \text{C=O} & & \text{HO-CH} \\
\text{HO-CH} & \text{HO-CH} & \text{HO-C-H} & & \text{HC-OH} \\
\text{HC-OH} & \text{HC-OH} & \text{HC-OH} & & \text{HC-OH} \\
\text{HC-OH} & \text{HC-OH} & \text{HC-OH} & \text{HCOOH} & \text{H}_2\text{C-OH} \\
\text{H}_2\text{C-OH} & \text{H}_2\text{C-OH} & \text{H}_2\text{C-OH} & & \\
\text{D-Arabinonic acid} & & & & \text{D-Arabinose}
\end{array}
\qquad (4.42)
$$

formic acids in addition to formaldehyde and D-arabinose (Formula 4.42). Depending on reaction conditions, particularly the type of alkali present, further hydroxyacids are also formed due to enolization occurring along the molecule.

The nonstoichiometric sugar oxidation process in the presence of alkali is used for both qualitative and quantitative determination of reducing sugars (*Fehling's* reaction with alkaline cupric tartrate; *Nylander's* reaction with alkaline trivalent bismuth tartrate; or using *Benedict's* solution, in which cupric ion complexes with citrate ion).

Hydroxyaldehydes and hydroxyketones are formed by chain cleavage due to retroaldol reaction under nonoxidative conditions using dilute alkali at elevated temperatures or concentrated alkali even in the cold.

For example, fructose can yield glyceraldehyde and dihydroxyacetone (Formula 4.43), and the latter easily undergoes water elimination to give 2-oxopropanal (methylglyoxal). Starting from 1-deoxy-2,3-hexodiulose, several degradation pathways leading to short-chain compounds are possible (Formula 4.44). Among other compounds, 2-oxopropanal, monohydroxyacetone, acetic acid, glyceraldehyde or glyceric acid can be formed by *retro*-aldol reactions (a), α-dicarbonyl cleavages (b) and β-dicarbonyl cleavages (c).

Since enolization is not restricted to any part of the molecule and since water elimination and redox reactions are not restricted in amount, even the spectrum of primary cleavage products is great. These primary products are highly

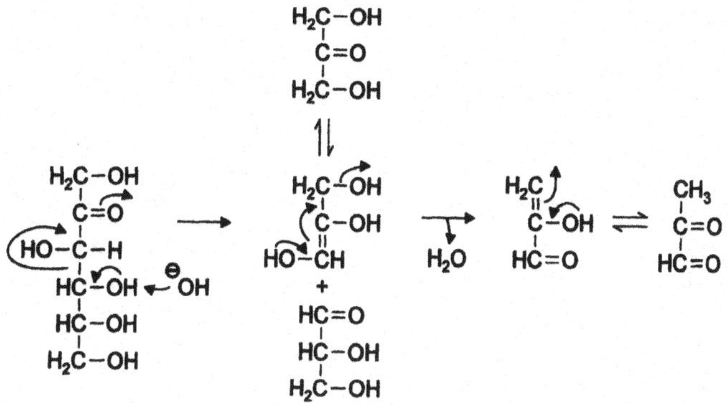

$$
\begin{array}{c}
\text{H}_2\text{C-OH} \\
\text{C=O} \\
\text{H}_2\text{C-OH}
\end{array}
$$

$$
\begin{array}{ccccc}
\text{H}_2\text{C-OH} & & \text{H}_2\text{C-OH} & & \\
\text{C=O} & \longrightarrow & \text{C-OH} & & \text{H}_2\text{C} \qquad \text{CH}_3 \\
\text{HO-C-H} & & \text{HO-CH} & \quad \text{H}_2\text{O} \quad & \text{C-OH} \rightleftharpoons \text{C=O} \\
\text{HC-OH} \;\; \text{OH} & & + & & \text{HC=O} \qquad \text{HC=O} \\
\text{HC-OH} & & \text{HC=O} & & \\
\text{H}_2\text{C-OH} & & \text{HC-OH} & & \\
 & & \text{H}_2\text{C-OH} & &
\end{array}
\qquad (4.43)
$$

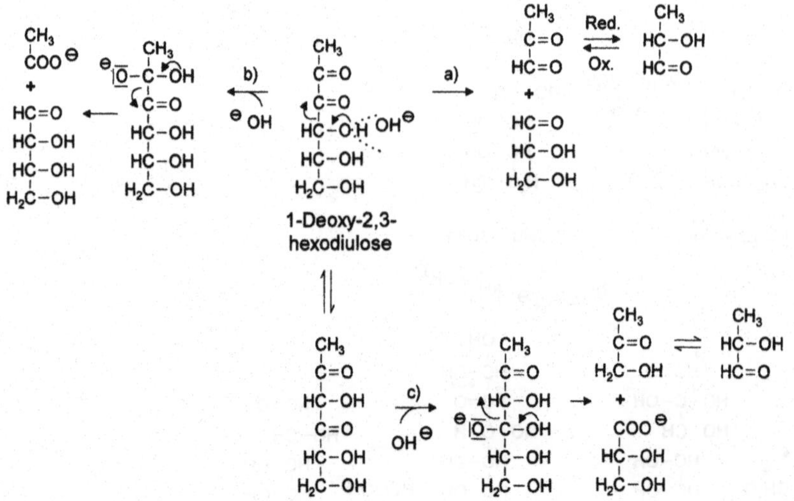

1-Deoxy-2,3-hexodiulose

$$(4.44)$$

reactive and provide a great number of secondary products by aldol condensations (*retro*-reactions) and intramolecular *Cannizzaro* reactions.

The compounds formed in fructose syrup of pH 8–10 heated for 3 h are listed in Table 4.12. The cyclopentenolones are typical caramel-like aroma substances. The formation of 2-hydroxy-3-methyl-2-cyclopenten-1-one as an example is shown in Formula 4.46. The compound 1,3,5-trideoxy-2,5-hexulose is formed as an intermediate via the aldol condensation

Table 4.12. Volatile reaction products obtained from fructose by an alkali degradation (pH 8–10)

Acetic acid
Hydroxyacetone
1-Hydroxy-2-butanone
3-Hydroxy-2-butanone
4-Hydroxy-2-butanone
Furfuryl alcohol
5-Methyl-2-furfuryl alcohol
2,5-Dimethyl-4-hydroxy-3-(2H)-furanone
2-Hydroxy-3-methyl-2-cyclopenten-1-one
3,4-Dimethyl-2-hydroxy-2-cyclopenten-1-one
3,5-Dimethyl-2-hydroxy-2-cyclopenten-1-one
3-Ethyl-2-hydroxy-2-cyclopenten-1-one
γ-Butyrolactone

of 2 molecules of monohydroxyacetone (or the isomer 2-hydroxypropanone). The linking of the C-atoms 6 and 2 then leads to a carbocycle which yields the target compound on water elimination. Analogously, the substitution of 1-hydroxy-2-butanone or 3-hydroxy-2-butanone for one molecule of hydroxyacetone can give rise to 2-hydroxy-3-ethyl- or 2-hydroxy-3,4-dimethyl-2-cyclopenten-1-one.

Among other compounds, C-3 fragments are also precursors for the formation of 4-hydroxy-2,5-dimethyl-3(2H)-furanone (II, furaneol), which can be formed, e. g., from 2-hydroxypropanal and its oxidation product, 2-oxopropanal (I in Formula 4.47). The substitution of 2-oxobutanal for 2-oxopropanal yields the homologous ethylfuraneol. In a similar way, 3-hydroxy-4,5-dimethyl-2(5H)-furanone (sotolon) can be formed from 2,3-butanedione (diacetyl) and glycolaldehyde (Formula 4.48).

Saccharinic acids are specific reaction products of monosaccharides in strong alkalies, particularly of alkaline-earth metals. They are obtained in each case as diastereomeric pairs by benzilic acid rearrangements from deoxy-hexodiuloses according to Formula 4.48a. In fact, 1-deoxy-2,3-hexodiulose yields saccharinic acid, 3-deoxy-1,2-hexodiulose yields metasaccharinic acid and 4-deoxy-2,3-hexodiulose yields isosaccharinic acid (Formula 4.48b).

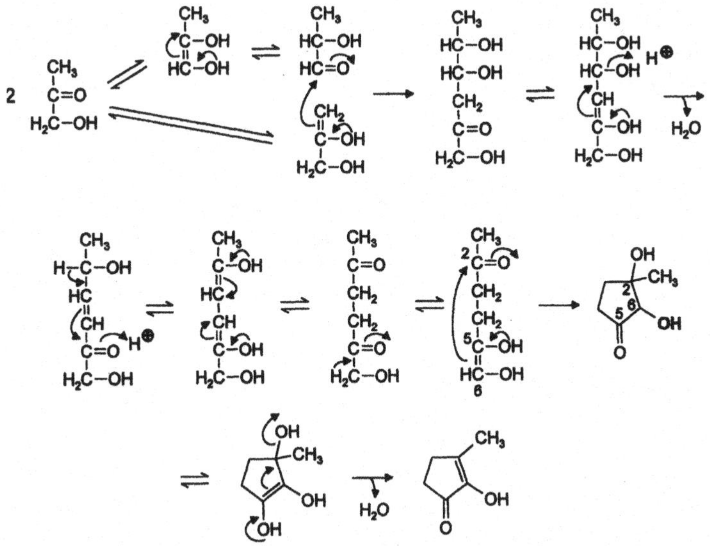

(4.46)

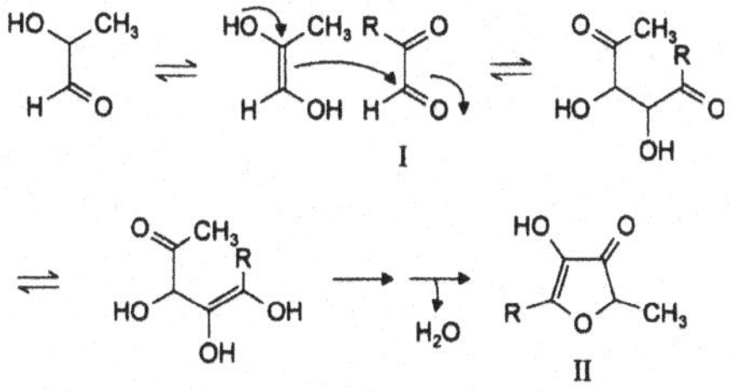

R = Methyl (I : 2-Oxopropanal; II : Furaneol)
R = Ethyl (I : 2-Oxobutanal; II : Ethylfuraneol)

(4.47)

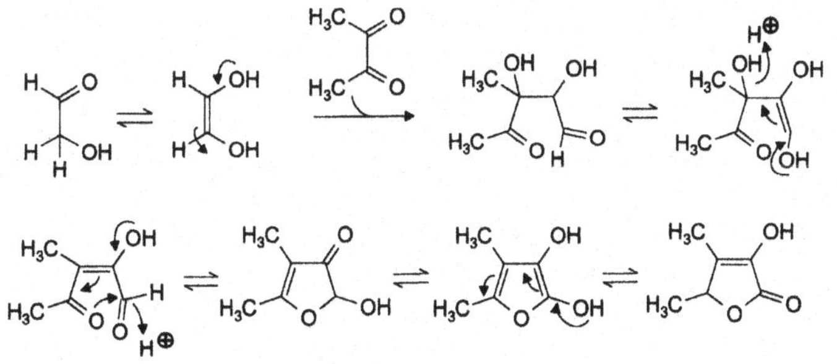

(4.48)

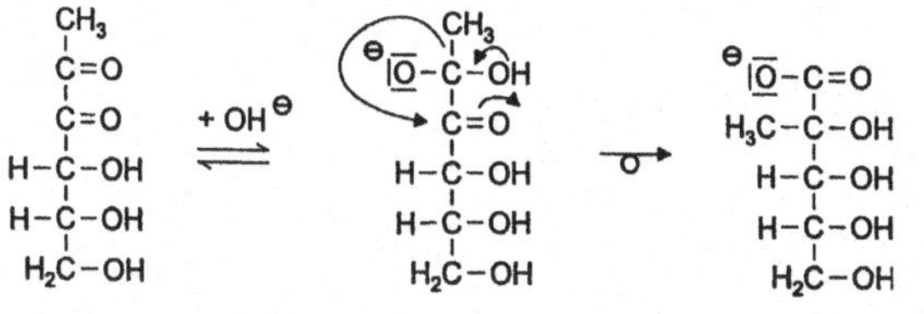

(4.48 a)

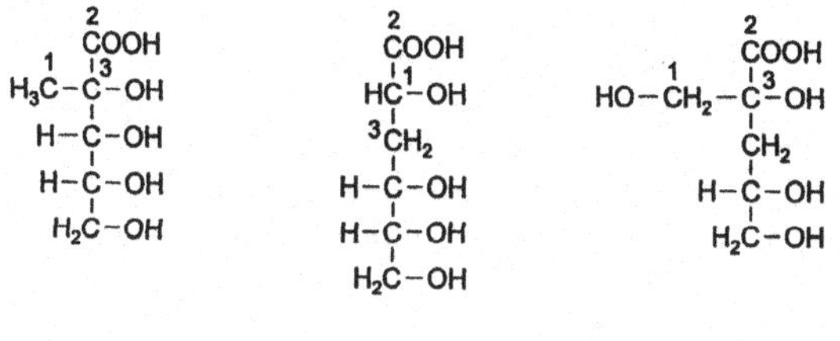

(4.48 b)

Ammonia catalyzes the same reactions as alkali and alkaline earths. Reactive intermediary products can polymerize further into brown pigments ("sucre couleur") or form a number of imidazole, pyrazine and pyridine derivatives.

4.2.4.3.3 Caramelization

Brown-colored products with a typical caramel aroma are obtained by melting sugar or by heating sugar syrup in the presence of acidic and/or alkaline catalysts. The reactions involved were covered in the previous two sections. The process can be directed more towards aroma formation or more towards brown pigment accumulation. Heating of saccharose syrup in a buffered solution enhances molecular fragmentation and, thereby, formation of aroma substances. Primarily dihydrofuran-ones, cyclopentenolones, cyclohexenolones and pyrones are formed (cf. 4.2.4.3.2). On the other hand, heating glucose syrup with sulfuric acid in the presence of ammonia provides intensively colored polymers ("*sucre couleur*"). The stability and solubility of these polymers are enhanced by bisulfite anion addition to double bonds:

$$
\begin{array}{c}
\overset{\diagdown}{\underset{\diagup}{\overset{\displaystyle C}{\underset{\displaystyle C}{\parallel}}}}
\end{array}
\quad \xrightarrow{\;SO_3H^{\ominus}\;} \quad
\begin{array}{c}
\overset{\diagdown}{CH} \\
\underset{\diagup}{C}\text{-}SO_3^{\ominus}
\end{array}
\qquad (4.50)
$$

4.2.4.4 Reactions with Amino Compounds (*Maillard* Reaction)

In this section, the formation of N-glycosides as well as the numerous consecutive reactions classed under the term *Maillard* reaction or nonenzymatic browning will be discussed.

N-Glycosides are widely distributed in nature (nucleic acids, NAD, coenzyme A). They are formed in food whenever reducing sugars occur together with proteins, peptides, amino acids or amines. They are obtained more readily at a higher temperature, low water activity and on longer storage.

On the sugar side, the reactants are mainly glucose, fructose, maltose, lactose and, to a smaller extent, reducing pentoses, e. g., ribose. On the side of the amino component, amino acids with a primary amino group are more important than those with a secondary because their concentration in foods is usually higher. Exceptions are, e. g., malt and corn products which have a high proline content. In the case of proteins, the ε-amino groups of lysine react predominantly. However, secondary products from reactions with the guanidino group of arginine are also known. In fact, imidazolin-ones and pyrimidines, which are formed from arginine and reactive α- and β-dicarbonyl compounds obtained from sugar degradation, have been detected.

The consecutive reactions of N-glycosides partially correspond to those already outlined for acid/base catalyzed conversions of monosaccharides. However, starting with N-containing intermediates, which with the nitrogen function possess a catalyst within the molecule, these reactions proceed at a high rate under substantially milder conditions, which are present in many foods.

$$
\begin{array}{cccccc}
\text{HC=O} & & \overset{H^{\oplus}}{\underset{|}{\overset{OH}{H-C-N-R}}} & \overset{H^{\oplus}}{HC=N-R} & \overset{H}{\underset{|}{HC-N-R}} & \overset{H}{\underset{|}{H_2C-N-R}} \\
\text{H-C-OH} & & \text{H-C-OH} & \text{H-C-OH} & \overset{|}{C-OH} & \text{C=O} \\
\text{HO-C-H} & R\text{-}NH_2 & \text{HO-C-H} & \text{HO-C-H} & \text{HO-C-H} & \text{HO-C-H} \\
\text{H-C-OH} & \xrightarrow{\quad} & \text{H-C-OH} & \text{H-C-OH} & \text{H-C-OH} & \text{H-C-OH} \\
\text{H-C-OH} & & \text{H-C-OH} \quad H_2O & \text{H-C-OH} & \text{H-C-OH} & \text{H-C-OH} \\
\text{H}_2\text{C-OH} & & \text{H}_2\text{C-OH} & \text{H}_2\text{C-OH} & \text{H}_2\text{C-OH} & \text{H}_2\text{C-OH} \\
\text{Glucose} & & & \text{Imine} & \text{1,2-Eneaminol} & \text{1-Amino-1-deoxy-} \\
& & & & & \text{ketose (\textit{Amadori}} \\
& & & & & \text{compound)} \qquad (4.50)
\end{array}
$$

These reactions result in:

- Brown pigments, known as melanoidins, which contain variable amounts of nitrogen and have varying molecular weights and solubilities in water. Little is known about the structure of these compounds. Studies have been conducted on fragments obtained after *Curie* point pyrolysis or after oxidation with ozone or sodium periodate. Browning is desired in baking and roasting, but not in foods which have a typical weak or other color of their own (condensed milk, white dried soups, tomato soup).
- Volatile compounds which are often potent aroma substances. The *Maillard* reaction is important for the desired aroma formation accompanying cooking, baking, roasting or frying. It is equally significant for the generation of off-flavors in food during storage, especially in the dehydrated state, or on heat treatment for the purpose of pasteurization, sterilization and roasting.
- Flavoring matter, especially bitter substances, which are partially desired (coffee) but can also cause an off-taste, e. g., in grilled meat or fish (roasting bitter substances).
- Compounds with highly reductive properties (reductones) which can contribute to the stabilization of food against oxidative deterioration.
- Losses of essential amino acids (lysine, arginine, cysteine, methionine).
- Compounds with potential mutagenic properties.
- Compounds that can cause cross-linkage of proteins. Reactions of this type apparently also play a role *in vivo* (diabetes).

4.2.4.4.1 Initial Phase of the Maillard Reaction

D-Glucose will be used as an example to illustrate the course of reactions occurring in the early phase of the *Maillard* reaction. The open-chain structures will be used for simplification although the hemiacetal forms are predominantly present in solution.

Nucleophilic compounds like amino acids or amines easily add to the carbonyl function of reducing carbohydrates with the formation of imines (*Schiff* bases). As a result of the hydroxyl group present in the α-position (Formula 4.50), the imines can rearrange via the 1,2-eneaminols corresponding to the 1,2-enediol (cf. Formula 4.36). This rearrangement leads to an aminoketose called an *Amadori* compound (1-amino-1-deoxyketose) after its discoverer. If fructose reacts in a corresponding way with an amine or an amino acid (Formula 4.51), an aminoaldose, called a *Heyns* compound (2-amino-2-deoxyaldose), is formed. Since the addition of the amine to fructose or the addition of the H-atom to the intermediate aminoenol can proceed from two sides, an enantiomeric pair is obtained in each case.

Amadori compounds with different amino acid residues have been detected in many heated and stored foods, e. g., in dried fruit and vegetables, milk products, cocoa beans or soy sauce. *Amadori* compounds are also found in the blood serum, especially of patients suffering from Diabetes mellitus. As secondary amino acids, *Amadori* and *Heyns* compounds can be analytically detected by amino acid analysis (cf. protein section).

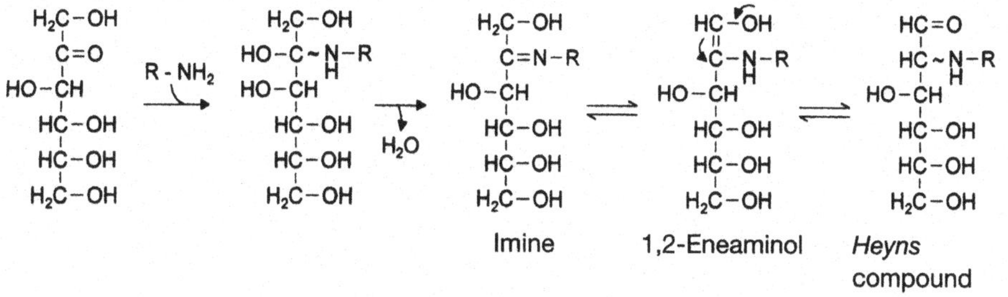

$$(4.51)$$

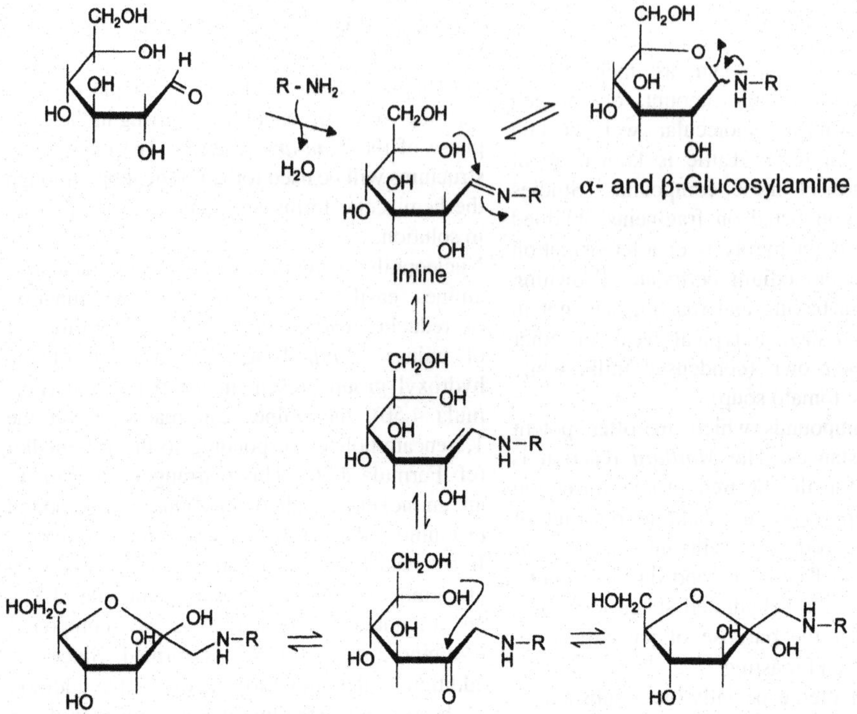

$$\alpha\text{- and }\beta\text{-Glucosylamine}$$

$$\text{Imine}$$

(4.52)

The reasons for the partial stability of such *Amadori* compounds in comparison with imines can be explained by the cyclic molecular structures.

The imine (Formula 4.52) formed by the reaction of glucose with the amine is easily converted to the cyclic hemiaminal, α- and β-glucosylamine. However, N-glycosides of this type are relatively instable because they very easily mutarotate, i. e., they are easily hydrolyzed via the open-chain imine or are converted to the respective α- and β-anomer. However, the *Amadori* rearrangement leads to furanose, which as a hemiacetal, has a stability to mutarotation comparable with that of carbohydrates.

The *Amadori* compounds can react further with a second sugar molecule, resulting in glycosylamine formation and subsequent conversion to di-D-ketosylamino acids ("diketose amino acids") by an *Amadori* rearrangement:

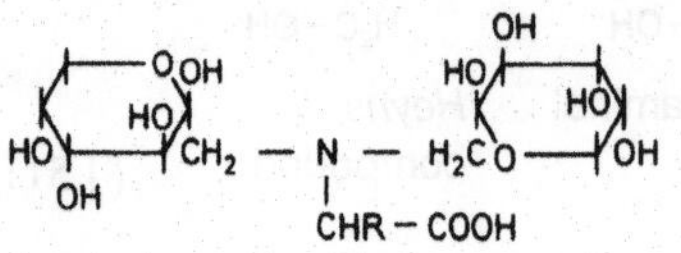

Di-D-fructosylamino acid   (4.54)

4.2.4.4.2 Formation of Deoxyosones

Amadori products are only intermediates formed in the course of the *Maillard* reaction. In spite of their limited stability, these products can be used under certain conditions as an analytical indicator of the extent of the heat treatment of food. Unlike the acidic (pH <3) and alkaline (pH >8) sugar degradation reactions, the *Amadori* compounds are degraded to 1-, 3-, and 4-deoxydicarbonyl compounds (deoxyosones) in the pH range 4–7. As reactive α-dicarbonyl compounds, they yield many secondary products. Formulas 4.54–4.57 summarize the degradation reactions starting with the *Amadori* compound.

Analogous to fructose (cf. Formula 4.37), amino-1-deoxyketose can be converted to 2,3-eneaminol as well as 1,2-eneaminol (Formula 4.54) by enolization. Analogous to the corresponding 1,2-enediol, water elimination and hydrolysis of the imine cation gives 3-deoxy-1,2-diulose, also called 3-deoxyosone (Formula 4.55).

Like the corresponding 2,3-enediol, the 2,3-eneaminol has two different β-elimination options. Formula 4.56 shows the elimination of the amino acid by a *retro-Michael* reaction with formation of 1-deoxy-2,3-diulose, also called 1-deoxyosone. In addition, 4-deoxy-2,3-diulose, also called 4-deoxyosone, can be formed by water elimination at C-4 of 2,3-eneaminol (Formula 4.57). In contrast to the previously mentioned pathways, the amino acid residue remains bound to the carbohydrate in this reaction path. As shown in the reaction scheme, all three deoxyosones occur in different cyclic hemiacetal forms.

As in the case of the deoxyosones, the concentrations of *Amadori* and *Heyns* compounds vary, depending on the reaction conditions (pH value, temperature, time, type and concentration of the educts). As a result, there is a change in the product spectrum and, consequently, in the color, taste, odor, and other properties of the food in each case.

Like all α-dicarbonyl compounds, deoxyosones can be stabilized as quinoxalines by a trapping reaction with o-phenylenediamine (Formula 4.58) and subsequently quantitatively determined using liquid chromatographic techniques. In this way, deoxyosones were detected for the first time as intermediates in carbohydrate degradation.

The stable secondary products of the *Maillard* reaction, that are isolated from many different reaction mixtures and have known structures, can be generally assigned to a definite deoxyosone by a series of plausible reaction steps (enolization, elimination of water, retroaldol cleavage, substitution of an amino function for a hydroxy function etc.).

Of the large number of secondary products known today, a few typical examples will be dealt with here for each deoxyosone.

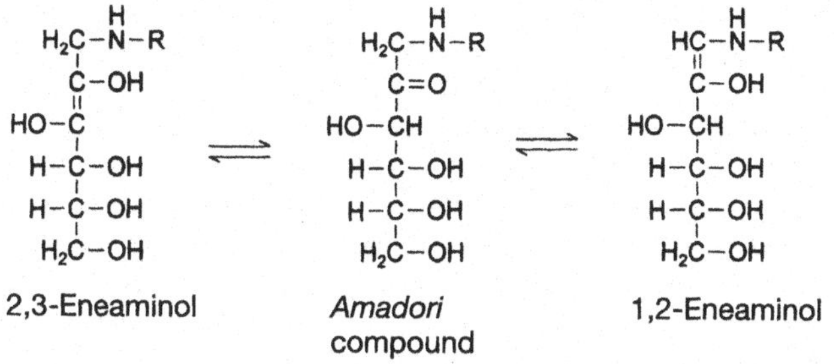

$$(4.54)$$

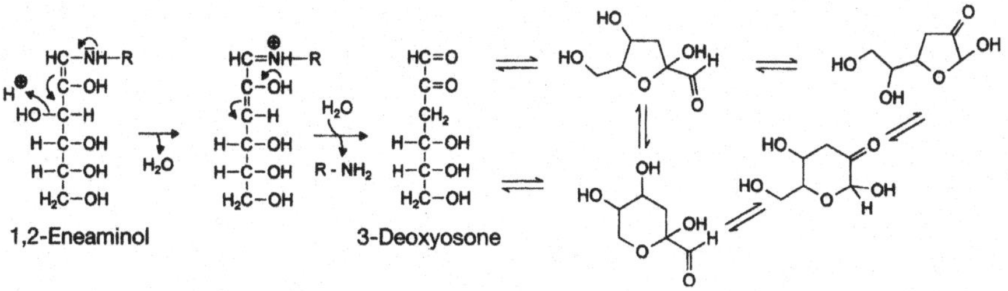

$$(4.55)$$

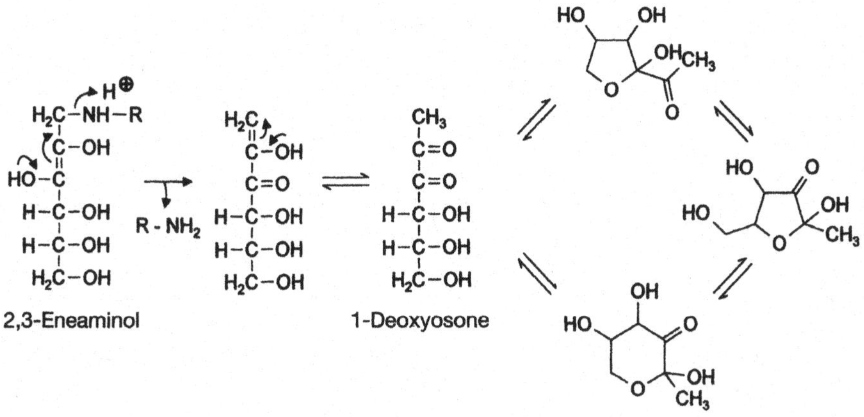

$$(4.56)$$

H₂C–N–R (H)
C–OH
HO–C (H⊕)
H–C–OH
H–C–OH
H₂C–OH

2,3-Enaminol

$\xrightarrow{H_2O}$

H₂C–N–R (H)
C=O
HO–C
H–C
H–C–OH
H₂C–OH

$\rightleftharpoons$

H₂C–N–R (H)
C=O
C=O
CH₂
H–C–OH
H₂C–OH

4-Deoxyosone

$$\text{(4.57)}$$

HC=O
C=O
H₂C
HC–OH
HC–OH
H₂C–OH

+

NH₂
NH₂

$\xrightarrow{2\,H_2O}$

[quinoxaline structure]

H—OH
H—OH
H₂C–OH

$$\text{(4.58)}$$

4.2.4.4.3 Secondary Products of 3-Deoxyosones

Formula 4.59 shows examples of products obtained on the decomposition of 3-deoxyosones. The best known compounds are 5-hydroxymethylfurfural from hexoses (HMF, II in Formula 4.59) and furfural from pentoses (I in Formula 4.59). Taking the furanoid structures of 3-deoxyosone as a basis (Formula 4.55), 3,4-dideoxyosone is obtained after ring opening, enolization, and water elimination (Formula 4.60). Water elimination from the hemiacetal form of 3,4-dideoxyosone directly yields HMF. Taking into account the water elimination required to form 3-deoxyosone (cf. Formula 4.55), 5-hydroxymethylfurfural is formed from hexose by the stoichiometric elimination of 3 mols of water.

In the presence of higher concentrations of ammonia, primary amines or amino acids, 3-deoxyosone preferentially gives rise to 2-formyl-5-hydroxymethylpyrrole (III in Formula 4.59) or the corresponding N-alkylated derivatives rather than to HMF. The most important reaction intermediate is 3,4-dideoxyosone (cf. Formula 4.60), which can react with amino compounds with the elimination of water to give the corresponding pyrrole (Formula 4.61) or pyridine derivatives (Formula 4.62). The reaction

with ammonia plays a role, especially in the production of sugar couleur.

If pyrrole formation occurs with an amino acid, this product can react further (Formula 4.63) to yield a bicyclic lactone (V in Formula 4.59). Other secondary products of 3-deoxyosone are compounds with a pyranone structure. In fact, β-pyranone (VI in Formula 4.59) is under discussion as the most important intermediate. It can be formed from the pyranose hemiacetal form of 3-deoxyosone (Formula 4.64). This compound has been identified only in the full acetal form (e.g., with carbohydrates on drying) because only this structure makes a relatively stable end product possible. The compounds mentioned have acidic hydrogen atoms in position 4, easily allowing condensation reactions with aldehydes and polymerization or the formation of brown dyes.

Another compound obtained from 3-deoxyosone via a relatively complex reaction is maltoxazine (VII in Formula 4.59), which has been identified in malt and beer. This compound could be formed from 3,4-dideoxyosone, which first undergoes a *Strecker* reaction with the secondary amino acid proline with decarboxylation to give the 1-pyrroline derivative (Formula 4.65). Enolization, formation of a five-membered carbocyclic compound and nucleophilic addition of the hydroxymethyl group to the pyrroline cation yields the tricyclic maltoxazine. In general, the formation of such carbocyclic compounds is favored in the presence of secondary amino acids like proline.

3-Deoxyosones predominantly form pyrazines and imidazoles with ammonia. The following compounds were isolated from sugar coloring (cf. Formula 4.66).

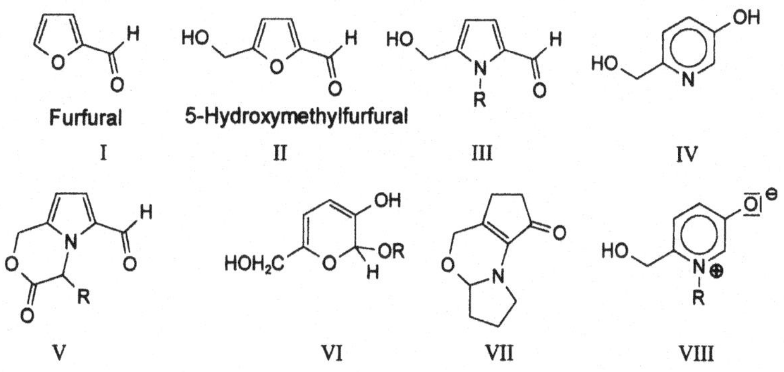

Furfural   5-Hydroxymethylfurfural

I   II   III   IV

V   VI   VII   VIII   (4.59)

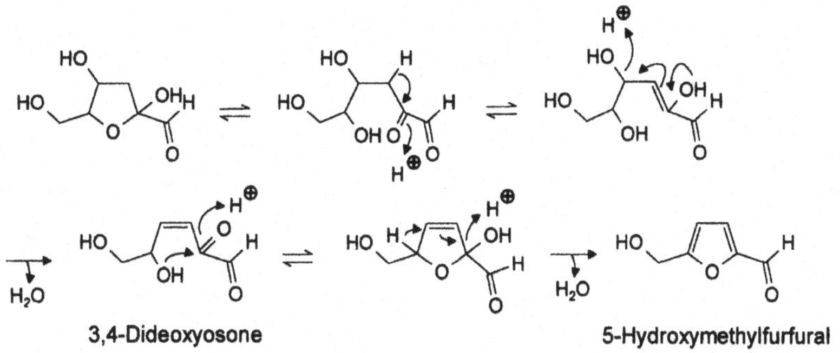

3,4-Dideoxyosone   5-Hydroxymethylfurfural   (4.60)

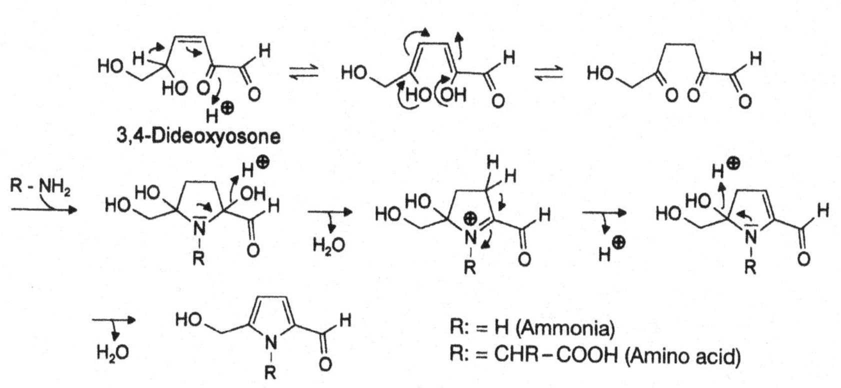

3,4-Dideoxyosone

R–NH₂

R: = H (Ammonia)
R: = CHR–COOH (Amino acid)   (4.61)

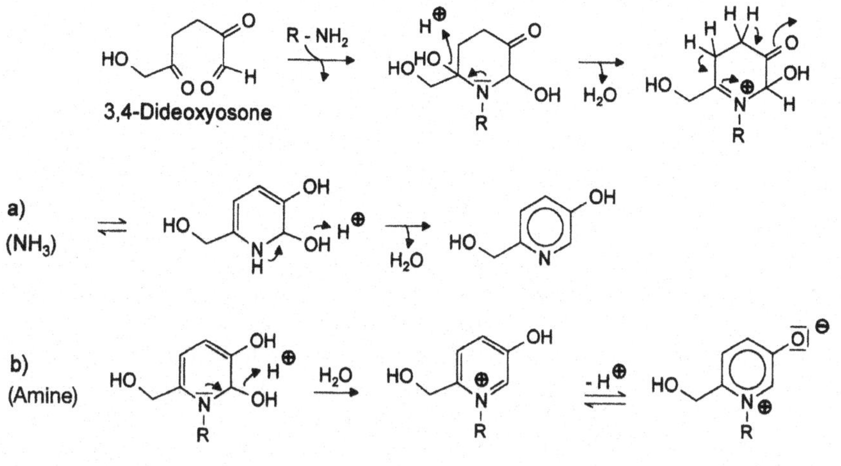

3,4-Dideoxyosone   R–NH₂

a)
(NH₃)

b)
(Amine)   (4.62)

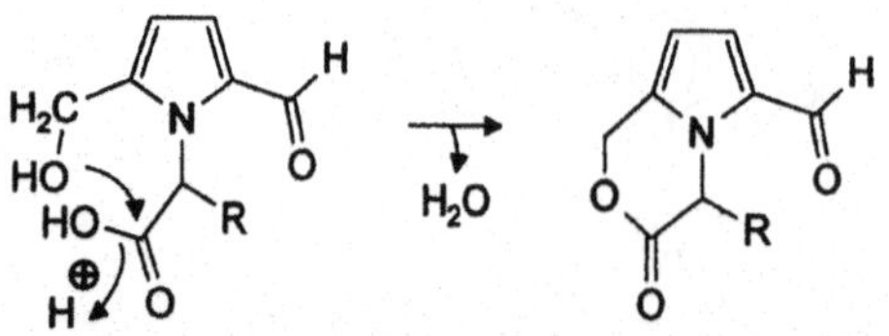

(4.63)

by reduction at C-1 of the carbohydrate (cf. Formula 4.56), all these compounds contain a methyl or acetyl group at position 2 of the furan or pyran derivatives. The product structures show that apart from the water elimination at C-1 leading to 1-deoxyosone, other dehydrations occurred at C-2, C-5 and/or C-6.

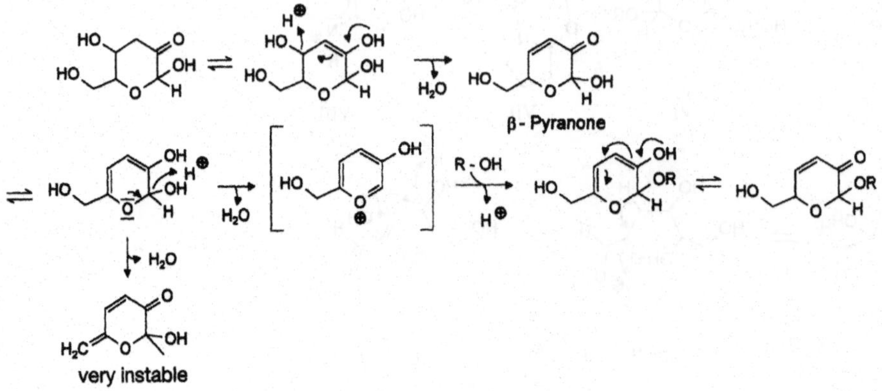

(4.64)

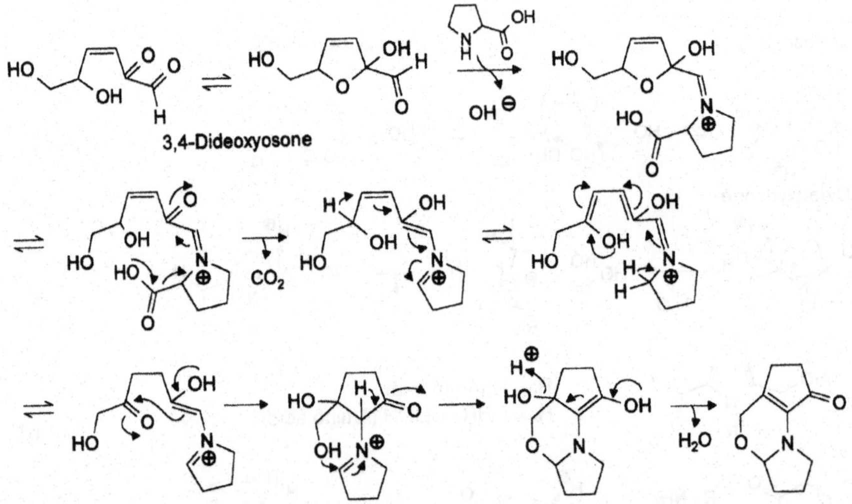

(4.65)

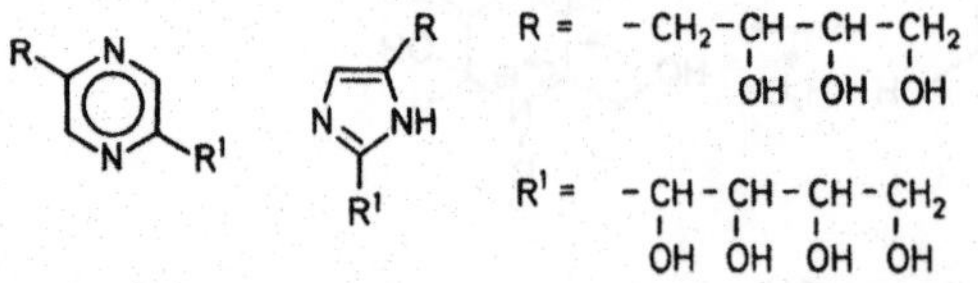

$$R = -CH_2-CH-CH-CH_2$$
$$\quad\quad\quad\;\; OH\;\; OH\;\; OH$$

$$R^1 = -CH-CH-CH-CH_2$$
$$\quad\quad\quad OH\;\; OH\;\; OH\;\; OH$$

(4.66)

4.2.4.4.4 Secondary Products of 1-Deoxyosones

Unlike the 3-deoxyosones which have been studied for a long time, the 1-deoxyosones were detected only a few years ago. Formula 4.67 shows known compounds derived from 1-deoxyosone. Since 1-deoxyosones formally occur

In the reaction yielding 3-hydroxy-5-hydroxy-methyl-2-methyl-(5H)-furan-4-one (IV in Formula 4.67), this compound can be directly formed by water elimination from the furanoid hemiacetal of 1-deoxyosone (Formula 4.68). It was found that isomerization to the 4-hydroxy-3-oxo compound does not occur under the conditions relevant to food. On the other hand, it is interesting that significant degradation to norfuraneol occurs (Formula 4.68). Norfuraneol is also formed as the main reaction product on the degradtion of the 1-deoxyosones of pentoses.

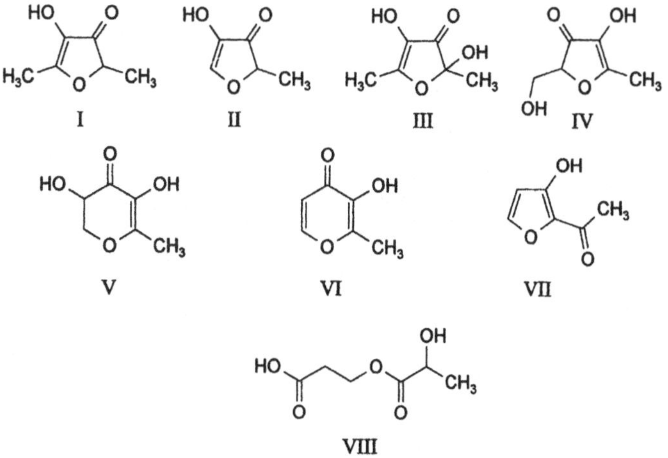

(4.67)

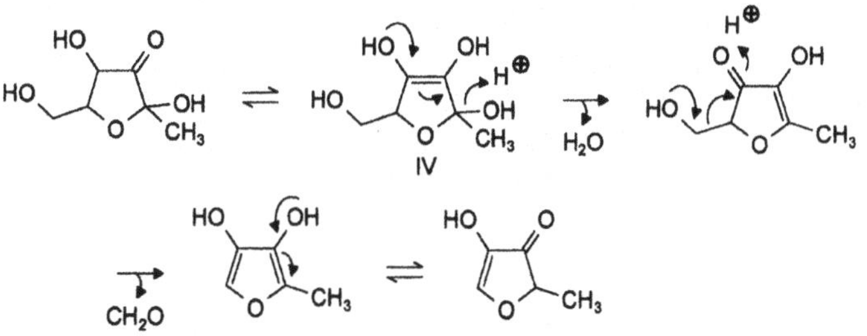

(4.68)

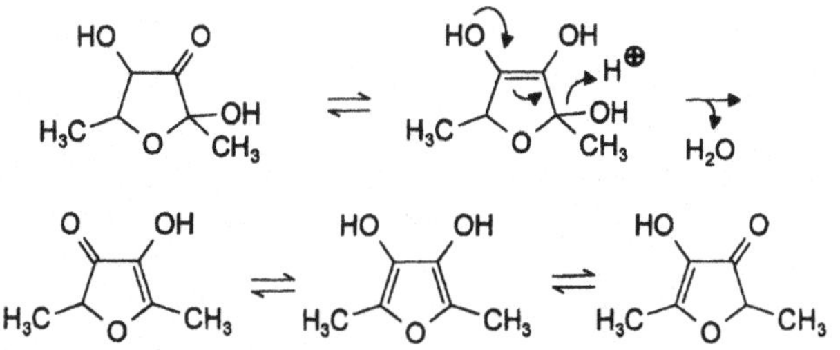

(4.69)

The compound 4-hydroxy-2,5-dimethyl-3(2H)-furanone (furaneol, I in Formula 4.67) is the corresponding degradation product from the 6-deoxy-L-mannose (rhamnose) (Formula 4.69). Furaneol can also be formed from hexose phosphates under reducing conditions (cf. 4.2.4.4.6) and from C-3 fragments (cf. Formula 4.47). With a relatively low odor threshold value, furaneol has an intensive caramel-like odor. It is interesting that furaneol is also biosynthesized in plants, e. g., in strawberries (cf. 18.1.2.6.9) and pineapples (cf. 18.1.2.6.10).

The formation of another degradation product of 1-deoxyosone, acetylformoin (III in Formula 4.67) is shown in Formula 4.70. In comparison with the formation of furanone in the synthesis of acetylformoin (cf. Formula 4.68),

the water elimination at C-6 of the carbohydrate skeleton occurs *before* cyclization to the furan derivative. Although further water elimination is no longer easy, it is suggested to explain the formation of methylene reductic acid (Formula 4.70).

As a result of the presence of an enediol structure element in the α-position to the oxo function in the open-chain structures of acetylformoin, this compound belongs to the group of substances called reductones. Substances of this type, e. g., also vitamin C (ascorbic acid), are weakly acidic (Formula 4.71), reductive (Formula 4.72) and exhibit antioxidative properties. The latter are attributed to the possible formation of resonance-stabilized radicals (Formula 4.73) and also to the disproportionation of two radicals with

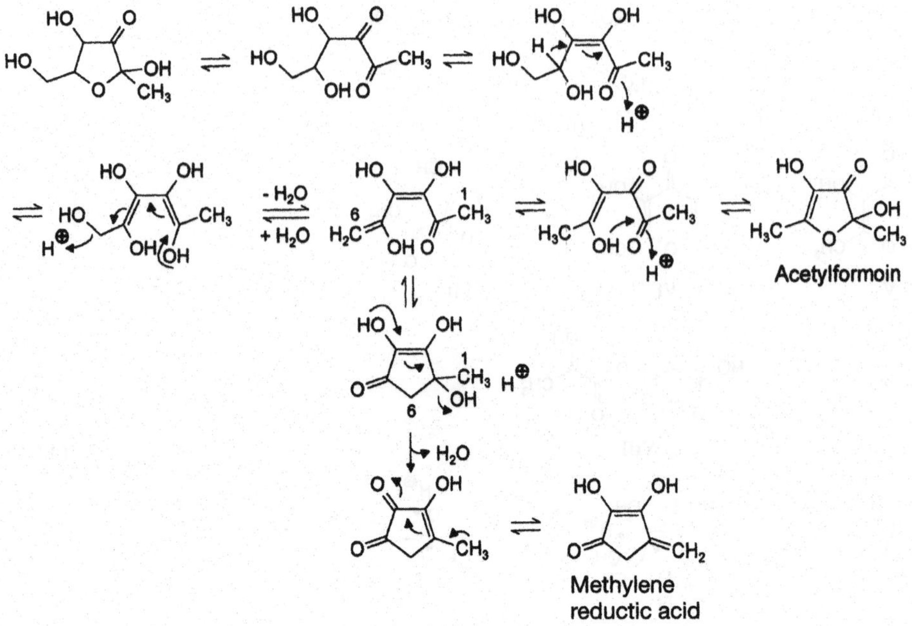

Acetylformoin

Methylene
reductic acid

(4.70)

re-formation of the reductone structure (Formula 4.74). Reductones reduce $Ag^{\oplus}$, $Au^{3\oplus}$, $Pt^{4\oplus}$ to the metals, $Cu^{2\oplus}$ to $Cu^{\oplus}$, $Fe^{3\oplus}$ to $Fe^{2\oplus}$ and Br_2 or I_2 to $Br^{\ominus}$ or $I^{\ominus}$ respectively. Reductones are present as mono-anions at pH values <6. The di-anion occurring under alkaline conditions is easily oxidized in the presence of O_2.

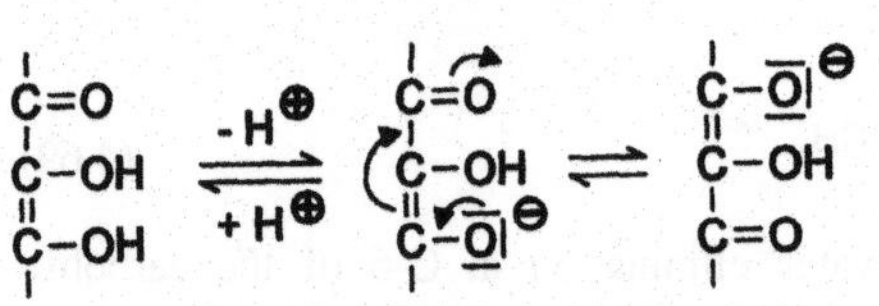

(4.71)

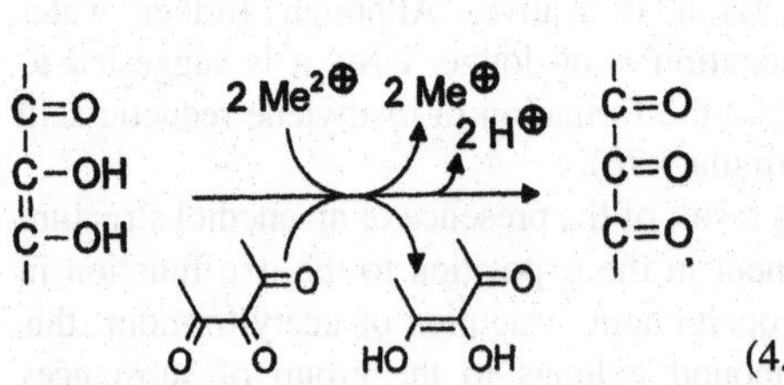

(4.72)

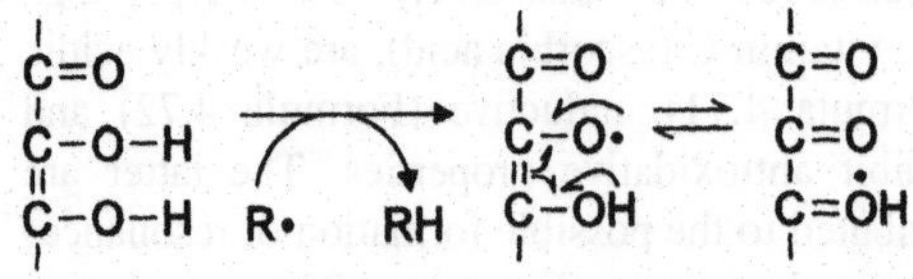

(4.73)

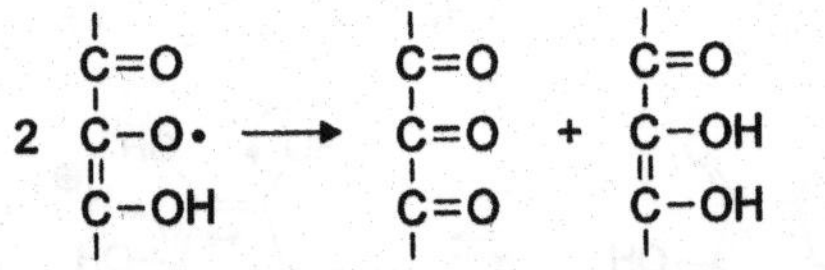

$$2 \begin{array}{c} | \\ C=O \\ | \\ C-O\bullet \\ | \\ C-OH \\ | \end{array} \longrightarrow \begin{array}{c} | \\ C=O \\ | \\ C=O \\ | \\ C=O \\ | \end{array} + \begin{array}{c} | \\ C=O \\ | \\ C-OH \\ | \\ C-OH \\ | \end{array}$$

(4.74)

The compound 3,5-dihydroxy-2-methyl-5,6-dihydropyran-4-one (V in Formula 4.67) is also formed from the pyranoid hemiacetals of 1-deoxy-2,3-hexodiulose (Formula 4.75). In comparison, maltol is preferentially formed from disaccharides like maltose or lactose (Formula 4.76) and not from dihydroxypyranone by water elimination. The formation of maltol from monosaccharides is negligible. A comparison of the decomposition of 1-deoxyosones from the corresponding cyclic pyranone structure clearly shows (cf. Formula 4.75 and 4.76) that the glycosidically bound carbohydrate in the disaccharide directs the course of water elimination in another direction (Formula 4.76). It is the stabilization of the intermediates to quasi-aromatic maltol which makes possible the cleavage of the glycosidic bond with the formation of maltol. Parallel to the formation of maltol, isomaltol derivatives which still contain the second carbohydrate molecule are also formed from disaccharides (Formula 4.77). Indeed, the formation of free isomaltol is possible by the hydrolysis of the glycosidic bond. β-Galactosyl-isomaltol was detected as the main product in heated milk. However, glucosyl-isomaltol is formed from maltose in much lower amounts. In this case, the formation of maltol dominates. The galactosyl residue clearly favors the formation of furanoid 1-deoxyosone from lactose, whereas the pyranoid 1-deoxyosone is preferentially formed from maltose (cf. Formula 4.76 and 4.77).

An open-chain compound, the lactic acid ester of β-hydroxypropionic acid (VIII in Formula 4.67) can also be formulated from 1-deoxyosone via 1,5-dideoxyosone. Hydration of this β-dicarbonyl compound and subsequent cleavage of the bond between C-2 and C-3 directly yield the lactic acid ester (Formula 4.78). Among the compounds mentioned, acetylformoin is one of the comparatively

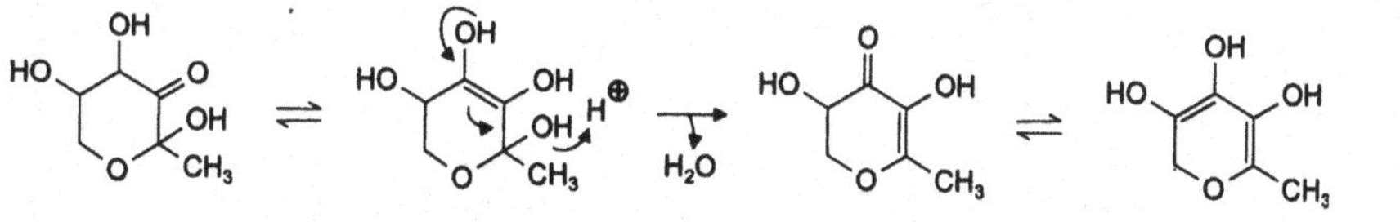

(4.75)

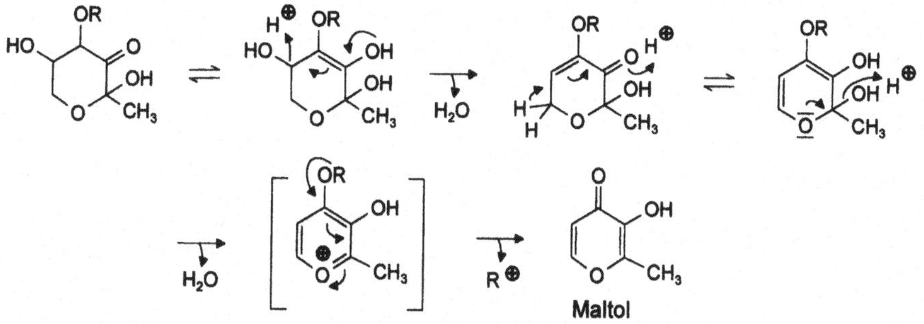

Maltol

(4.76)

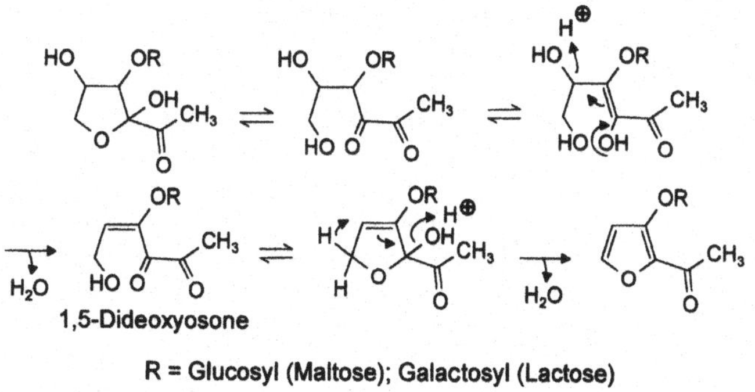

1,5-Dideoxyosone

R = Glucosyl (Maltose); Galactosyl (Lactose)

(4.77)

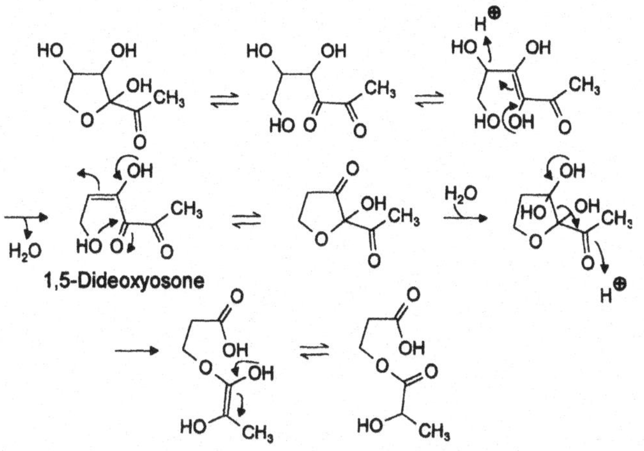

1,5-Dideoxyosone

(4.78)

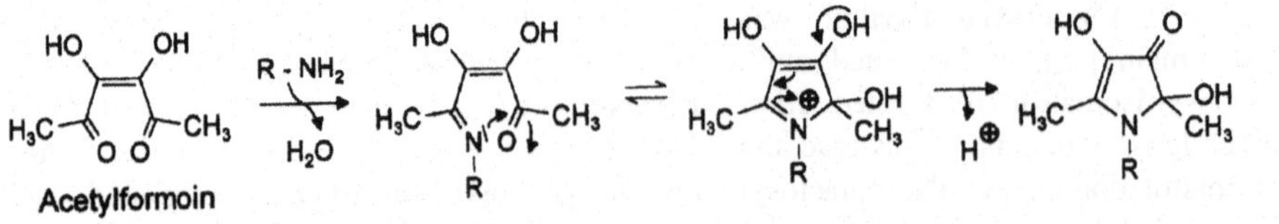

Acetylformoin

(4.79)

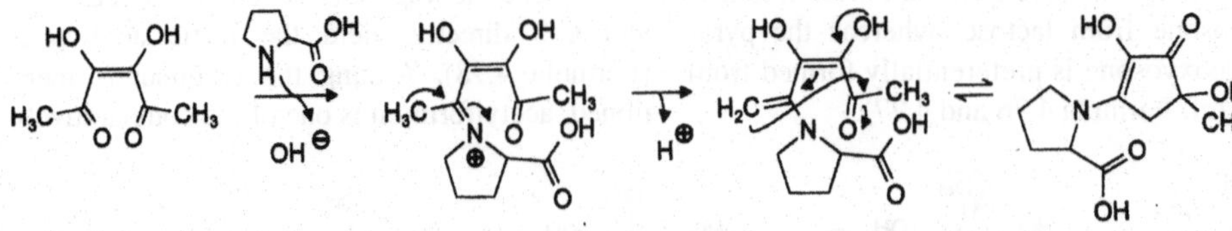

(4.80)

instable compounds which undergo reactions with other amino components. In the presence of mainly primary amines (amino acids), aminoreductones are formed (pyrrolinones, Formula 4.79) and in the presence of secondary amino acids, relatively stable carbocyclic compounds (Formula 4.80).

The following compounds were detected in reaction mixtures containing proline and hydroxyproline. Their formation must proceed via the 1-deoxyosones as well:

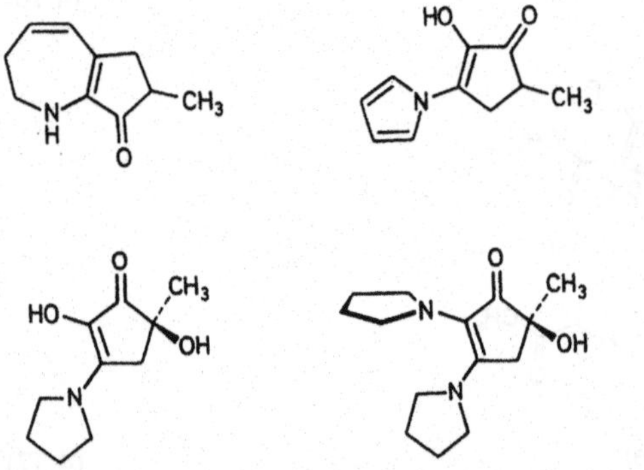

(4.81)

The pyrrolidino- and dipyrrolidinohexose reductones were characterized as bitter substances obtained from heated proline/saccharose mixtures (190 °C, 30 min, molar ratio 3:1; c_{sbi}: 0.8 and 0.03 mmol/l).

4.2.4.4.5 Secondary Products of 4-Deoxyosones

As shown in Formula 4.38, 2-hydroxyacetylfuran is one the reaction products of 4-deoxyosone. However, this compound is preferentially formed in carbohydrate degradation in the absence of amine components. If it is formed from the *Amadori* product in accordance with Formula 4.57, the amino acid remains in the reaction product, producing furosine (Formula 4.82). In the presence of higher concentrations of primary amines, the formation of 2-hydroxyacetylfuran (and also furosine) is significantly suppressed in favor of the corresponding pyrrole and pyridiniumbetaine (Formula 4.84). The reason for this is that the triketo structures formed from 4,5-dideoxyosone by enolization react with primary amines, amino acids or ammonia to give pyrrole and pyridine derivatives (Formula 4.83).

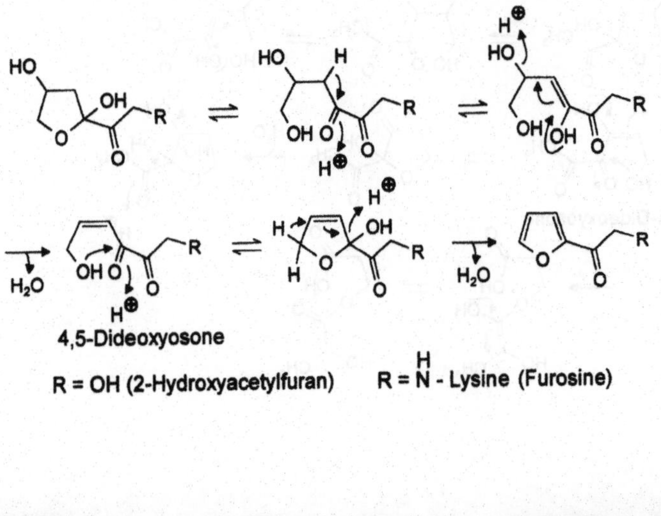

4,5-Dideoxyosone

R = OH (2-Hydroxyacetylfuran)   **R = N-H - Lysine (Furosine)**   (4.82)

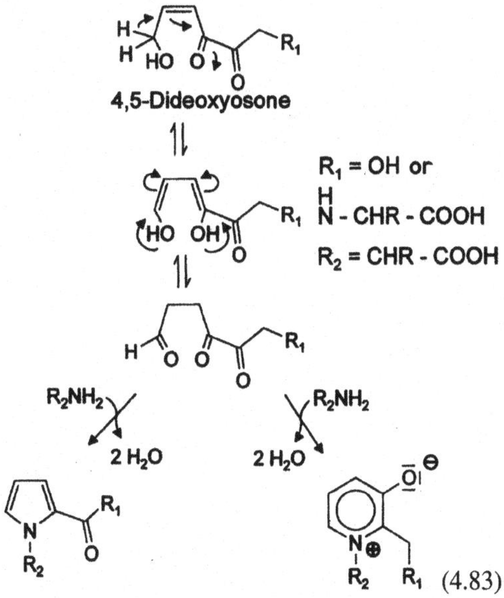
4,5-Dideoxyosone
R_1 = OH or
H
N - CHR - COOH
R_2 = CHR - COOH
R_2NH_2
2 H_2O

(4.83)

With ammonia, aminoacetylfuran is very easily converted to 2-(2-furoyl)-5-(2-furyl)-1 Himidazole, known as FFI, which was previously isolated from acid hydrolysates from protein/glucose reaction mixtures:

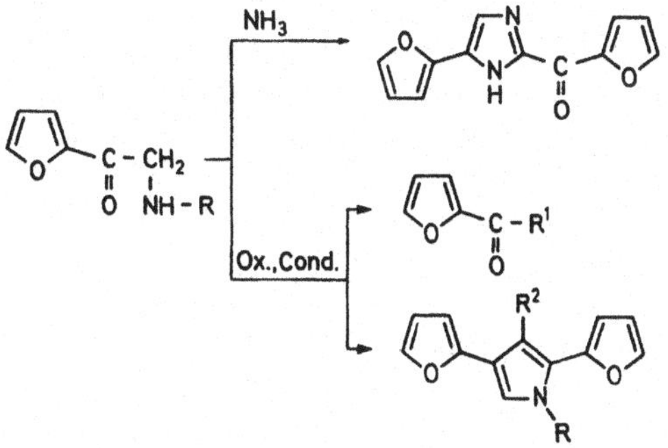
NH_3
Ox.,Cond.

(4.84)

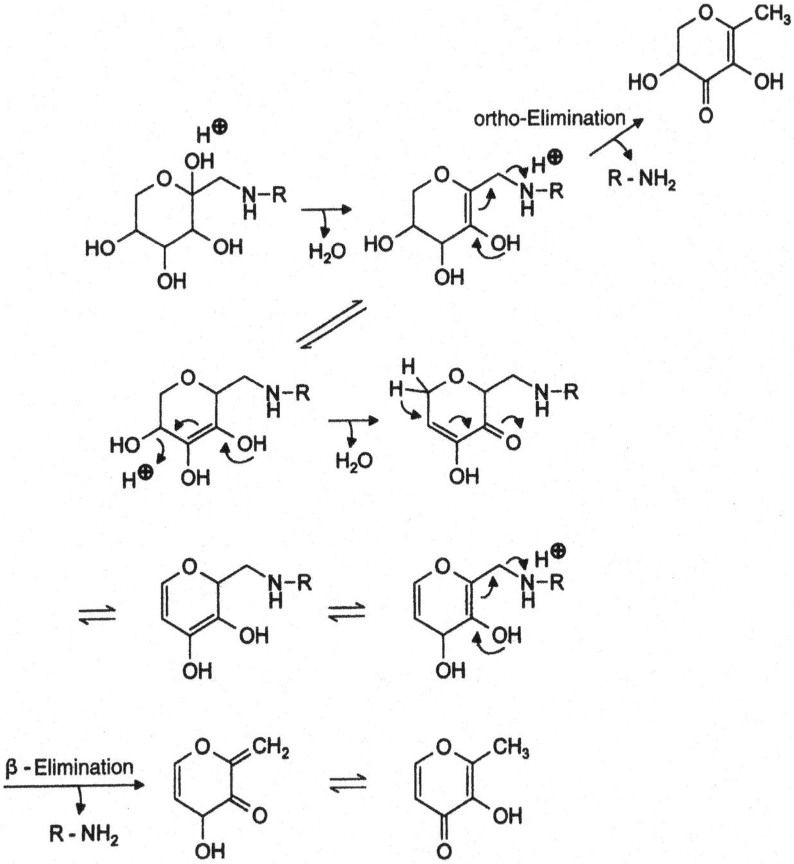
ortho-Elimination
R - NH_2
H_2O
β - Elimination

Maltol

(4.85)

Various oxidation and condensation products (R^1 = OH, CONHR; R^2 = OH, NHR) were isolated from a heated, neutral solution. It can be assumed from the structures shown that protein cross-linkages are possible if, e. g., R_1 and R_2 represent the ε-amino group of lysine (Formula 4.83).

In conclusion, it should be mentioned here that more recent studies also assume the direct elimination of water from the cyclic hemiacetals. This is shown in Formula 4.85 for the direct formation of maltol from the *Amadori* product.

4.2.4.4.6 Redox Reactions

In the course of the *Maillard* reaction, deoxyosones and reductones, e. g., acetylformoin (cf. III, Formula 4.67), are formed. They can react to give enol and triketo compounds via an addition with disproportionation (Formula 4.86). Redox reactions of this type can explain the formation of products which are not possible according to the reactions described till now. In fact, it has recently been found that, for example, glucose 6-phosphate and fructose-1,6-diphosphate, which occur in baker's yeast and muscle, form 4-hydroxy-2,5-dimethyl-3(2H)-furanone to a large extent. Since the formation from hexoses (or hexose phosphates) is not explainable, reduction of the intermediate acetylformoin (Formula 4.87) must have occurred. As shown, this reduction can proceed through acetylformoin itself or other reductones, e. g., ascorbic acid. Such re-

dox reactions also play an important role in the formation of the aroma substances 2-acetyl-1-pyrroline and 2-acetyltetrahydropyridine, which have been shown to exhibit slight oxidizability of α-eneaminols (cf. 5.3.1.6). Furthermore, such processes also play a part in the formation of carboxymethyllysine (cf. 4.2.4.4.9).

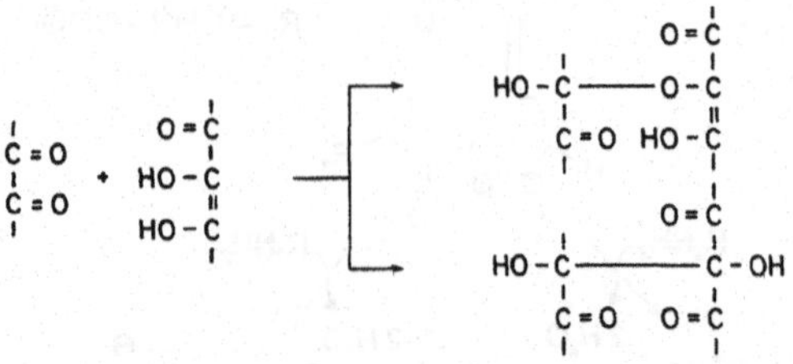

(4.86)

4.2.4.4.7 *Strecker* Reaction

The reactions between α-dicarbonyl compounds, like the deoxyosones obtained in the *Maillard* reaction, and amino acids are classed under the term *Strecker* reaction. This reaction leads to the formation of aldehydes (*Strecker* aldehydes), CO_2 and α-aminoketones on oxidative decarboxylation of the α-amino acids (Formula 4.88). It occurs in foods at higher concentrations of free amino acids and under more drastic reaction conditions, e. g., at higher temperatures or under pressure.

The aldehydes, which have one C-atom less than the amino acids, possess a considerable aroma potential, depending on the amino acid degraded.

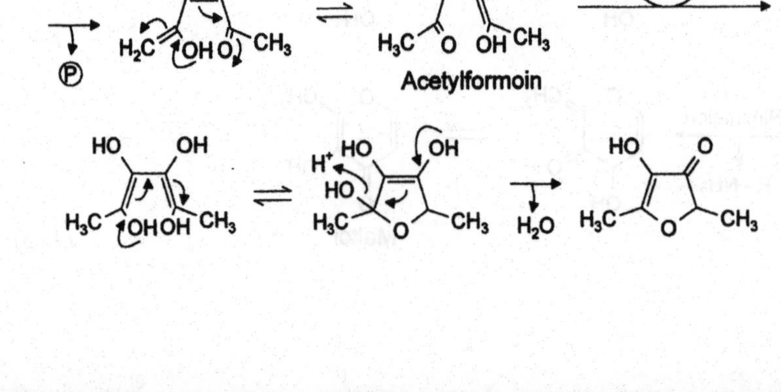

Acetylformoin

(4.87)

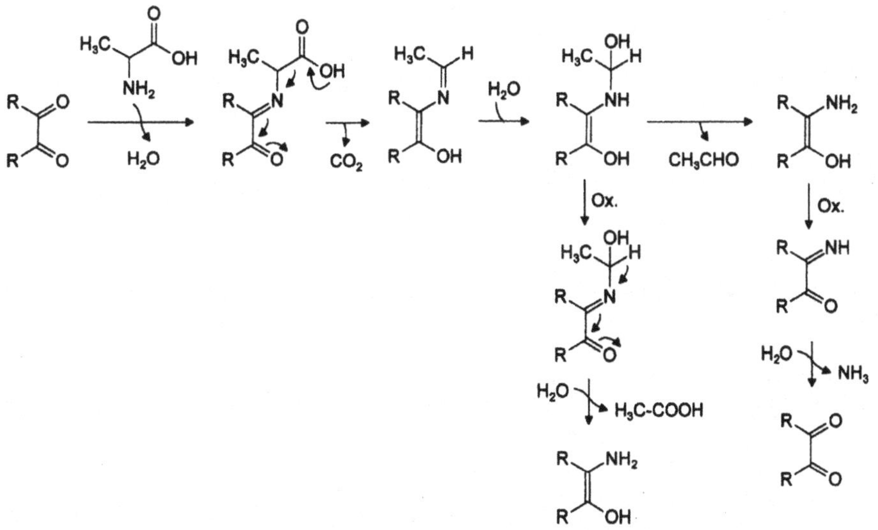

(4.88)

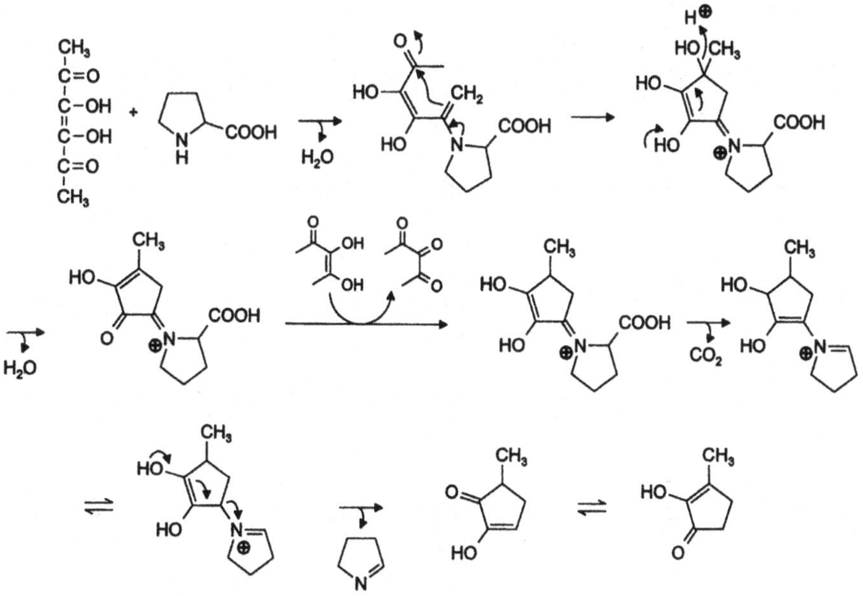

(4.89)

Strecker aldehydes which are important for their aroma are methional, phenylacetaldehyde, 3- and 2-methylbutanal and methylpropanal (cf. 5.3.1.1). Other compounds which are formed via the *Strecker* degradation and influence the aroma of food are H_2S, NH_3, 1-pyrroline (cf. 5.3.1.6) and cysteamine (cf. 5.3.1.4). Recently, the corresponding *Strecker* acids have also been found, especially in the presence of oxygen. They can be formed via the oxidation of the intermediate eneaminol (Formula 4.88). All the α-dicarbonyl compounds obtained on car-

bohydrate degradation as well as the reductones can undergo *Strecker* reactions. The product spectrum is significantly increased due to the redox reactions of the resulting intermediates. The complex course of a *Strecker* reaction is represented in Formula 4.89 with the formation of 2-hydroxy-3-methyl-cyclopent-3-enone. Apart from the pathway shown in Formula 4.46, this reaction is also of importance. When amino acids with functional groups in the side chain are involved, even more complex reactions are possible (cf. 5.3.1.4–5.3.1.8).

4.2.4.4.8 Formation of Colored Compounds

As a result of the mostly brown colors (bread crust, meat) formed by the heating of reducing carbohydrates with amine components, the *Maillard* reaction is also called nonenzymatic browning. Clinical biochemical studies have recently shown that these browning products partly exhibit antimutagenic and anticarcinogenic properties. As a result of the complex course of the reaction, however, it has only rarely been possible to identify colored compounds till now. One of the first colored compounds identified in model reactions of xylose/amines and furfural/norfuraneol is compound I in Formula 4.90. It is under discussion that this compound is formed via condensation reactions of the CH-acidic compound norfuraneol with the aldehyde group of furfural. Similar condensation reactions of 3-deoxyosone with furfural and of acetylformoin with furfural in model reactions led to the formation of the yellow products II and III (Formula 4.91).

However, both compounds could be stabilized only as the full acetal, e. g., in alcoholic solutions. In general, it is assumed today that condensation reactions between nucleophilic/electrophilic intermediates of the *Maillard* reaction result in the formation of the colored components, which are also called *melanoidins*.

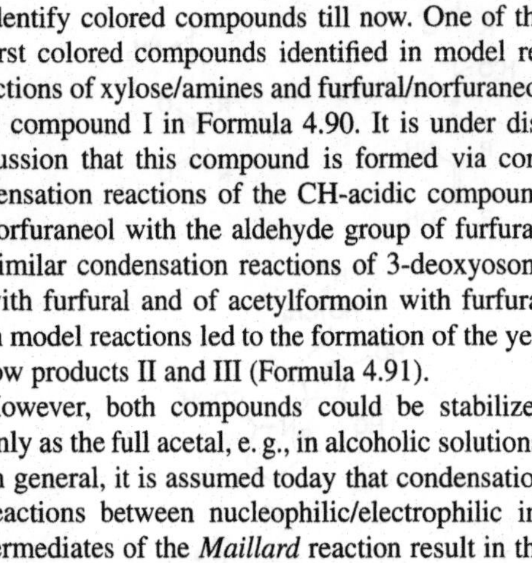

(4.90)

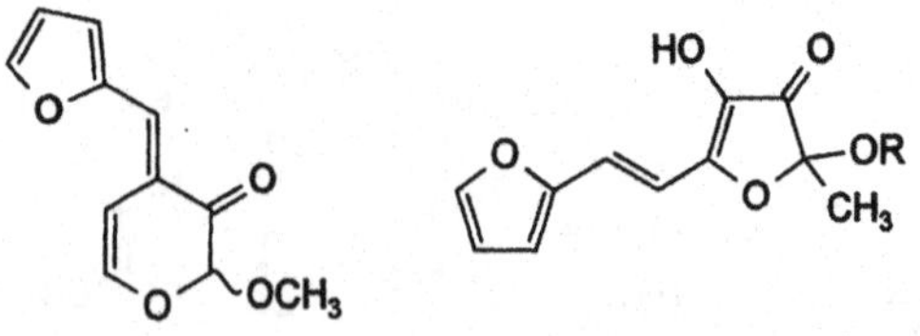

(4.91)

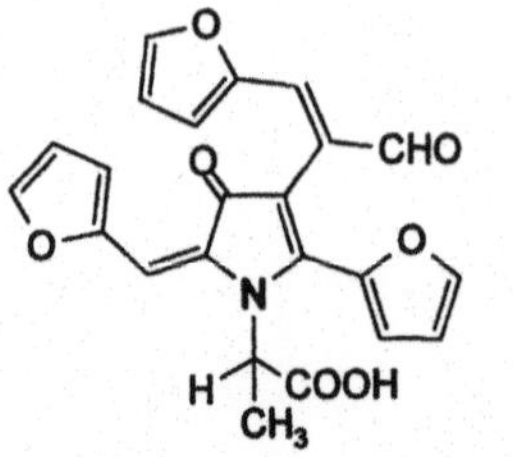

(4.92)

A red colored pyrroline dye (IV, Formula 4.92) could be identified in a model reaction of furfural and alanine. This dye is formed from 4 molecules of furfural and 1 molecule of alanine. Labeling experiments with ^{13}C showed that one open-chain molecule of furfural is inserted into the pyrrolinone structure. The proline/furfural reaction system indicated further that ring opening proceeds via a cyanine dye with the structure illustrated in V, Formula 4.93. Other colored compounds could be obtained by the condensation of 3,5-dihydroxy-2-methyl-5,6-dihydropyran-4-one with furfural (VI, Formula 4.94) and of 3-hydroxy-4-methyl-3-cyclopenten-1,2-dione (methylene reductic acid) with furfural (VII, Formula 4.94). Both dyes were also obtained by heating the *Amadori* product of proline and glucose in the presence of furfural. The orange colored compound VIII and the red compound IX were identified in a reaction system containing xylose/alanine/furfural (Formula 4.94).

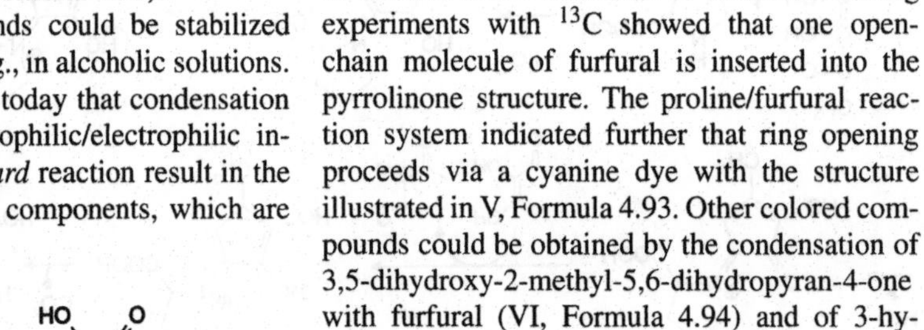

(4.93)

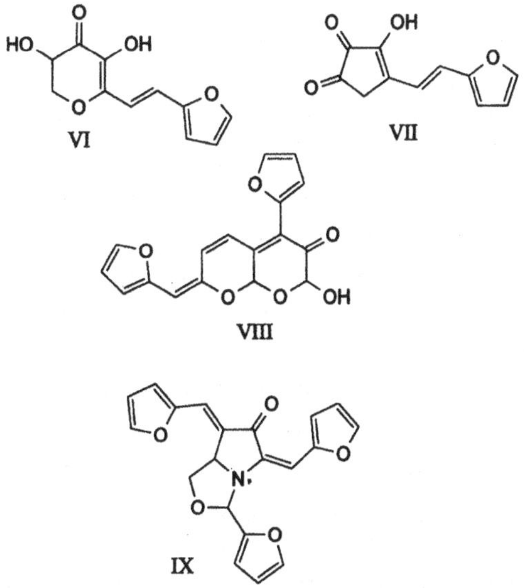

VI

VII

VIII

IX

(4.94)

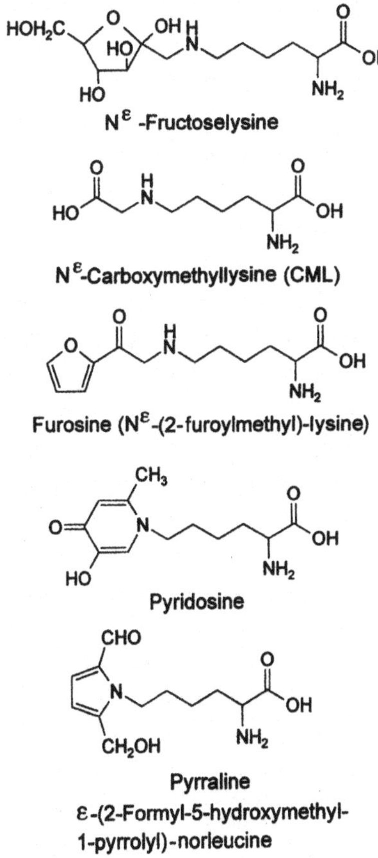

HOH₂C.

N^ε -Fructoselysine

N^ε-Carboxymethyllysine (CML)

Furosine (N^ε-(2-furoylmethyl)-lysine)

CH₃

Pyridosine

CHO

CH₂OH

Pyrraline
ε-(2-Formyl-5-hydroxymethyl-
1-pyrrolyl)-norleucine

(4.95)

4.2.4.4.9 Protein Modifications

The side chains of proteins can undergo post-translational modification in the course of thermal processes. The reaction can also result in the formation of protein cross-links. A known reaction which mainly proceeds in the absence of carbohydrates, for example, is the formation of dehydroalanine from serine, cysteine or serine phosphate by the elimination of H_2O, H_2S or phosphate. The dehydroalanine can then lead to protein cross-links with the nucleophilic side chains of lysine or cysteine (cf. 1.4.4.11). In the presence of carbohydrates or their degradation products, especially the side chains of lysine and arginine are subject to modification, which is accompanied by a reduction in the nutritional value of the proteins. The structures of important lysine modifications are summarized in Formula 4.95. The best known compounds are the *Amadori* product N^ε-fructoselysine and furosine, which can be formed from the former compound via the intermediate 4-deoxyosone (Formula 4.96). To detect of the extent of heat treatment, e. g., in the case of heat treated milk products, furosine is released by acid hydrolysis of the proteins and quantitatively determined by amino acid analysis. In this process, all the intermediates which lead to furosine are degraded and an unknown portion of already existing furosine is destroyed. Therefore, the hydrolysis must occur under standardized conditions or preferably by using enzymes. Examples showing the concentrations of furosine in food are presented in Table 4.13.

Table 4.13. Concentration of furosine in heated milk products

Product	Furosine (mg/kg protein)
Raw milk	35–55
Milk (pasteurized)	48–75
Milk (ultrahigh heated)	500–1800
Sterile milk	5000–12,000
Milk powder	1800–12,000
Baby food (powder)	9300–18,900
Noodles	400–8500
Bakery products	200–6000

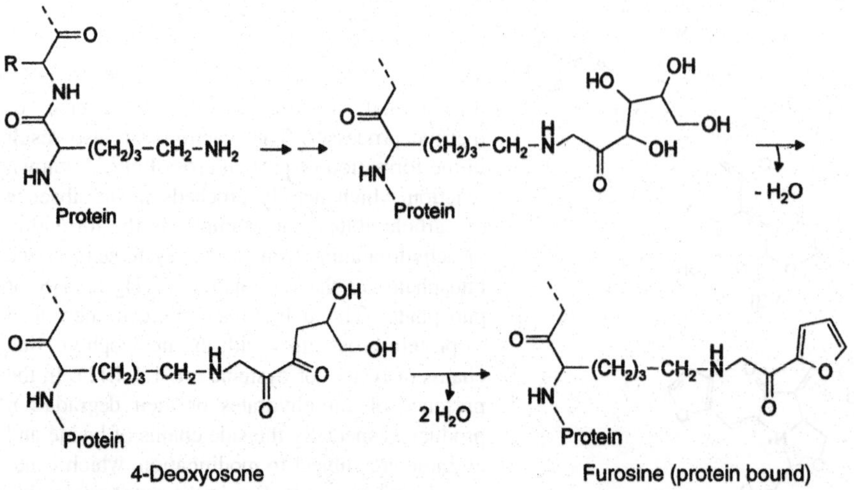

4-Deoxyosone   Furosine (protein bound)   (4.96)

Pyridosine (Formula 4.95) is also formed by the degradation of N^ε-fructoselysine, but in lower amounts than furosine (ratio ca. 3:1). It is assumed that 1-deoxyosone is the precursor.

Carboxymethyllysine is also formed from N^ε-fructoselysine and is also used as an indicator of the degree of thermal treatment of protein containing foods. This compound can be produced in different ways starting with oxidized N^ε-fructoselysine or the reaction of glyoxal with the lysine side chain. The reaction path shown in Formula 4.97 takes into account general mechanisms of carbohydrate degradation with cleavage of β-dicarbonyl compounds.

Pyrraline (Formula 4.95) is also formed as a modification of the amino acid lysine in proteins. The reaction partner is, according to Formula 4.61, 3,4- dideoxyosone obtained from 4-deoxyosone. Pyrraline is found in high concentrations especially in foods that have been subjected to strong thermal treatment, e. g., biscuits and pastries (Table 4.14). The concentrations in milk are clearly less than those of furosine (Table 4.13). The cross-linkage of proteins is also possible via the pyrrole residues of 2 molecules of pyrraline. The corresponding dimers have already been detected in model reactions. (cf. Formula 4.100).

The amino acid arginine can also be modified at the guanidino group, e. g., by reaction with α-dicarbonyl compounds from carbohydrate degradation. The compounds characterized were, among others, those formed from the reaction with methylglyoxal (I, Formula 4.98), 3-deoxyosone (II, Formula 4.98), a pentan-

dione (III, Formula 4.98) and glyoxal (IV, Formula 4.98). The synthesis of GLARG is displayed in Formula 4.99. It is interesting that glyoxal reacts with the N-atoms 1 and 2 of the guanidino group (Formula 4.99), whereas methylglyoxal bridges N-2 and N-3.

Among the identified compounds, only ornithino-imidazolinone (I, Formula 4.98) has been quanti-

Table 4.14. Concentrations of pyrraline in foods

Foods	Pyrraline (mg/kg protein)
UHT milk	<2–5
Sterile milk	60–80
Condensed milk	30–135
Pretzels	220–230
White bread crust	540–3680
White bread crumb	25–110
Nibbling biscuits	970–1320

Table 4.15. Concentrations of ornithino-imidazolinone (OIZ) in foods

Foods	OIZ (mg/kg protein)	Arginine loss (%)
Alkali-baked products	9000–13,000	20–30
Pretzel crust	25,000–28,000	60–70
Coffee beans (roasted)	7000–9000	20–25
Nibbling biscuits	6000–20,000	15–40

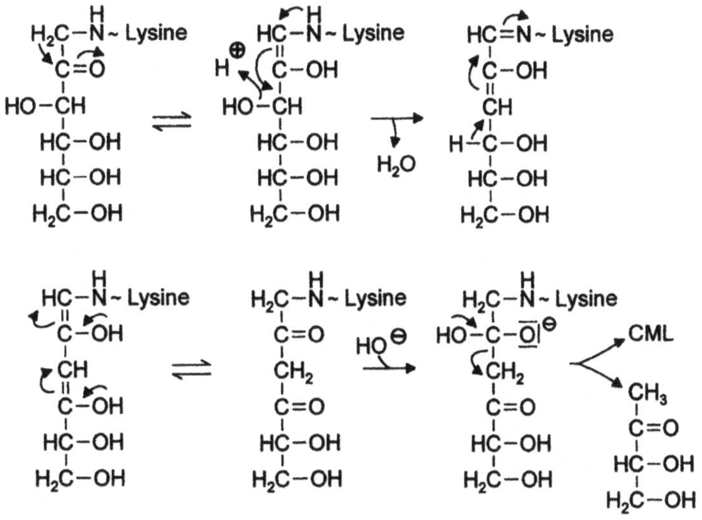

(4.97)

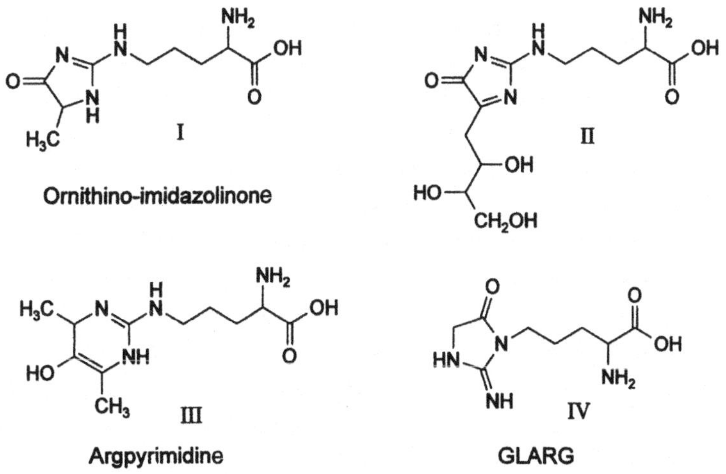

I

Ornithino-imidazolinone

II

III

Argpyrimidine

IV

GLARG

(4.98)

tatively determined in different foods. The data (Table 4.15) show that especially in alkali-baked products, about 60–70% of the arginine present in the flour reacts to imidazolinone.

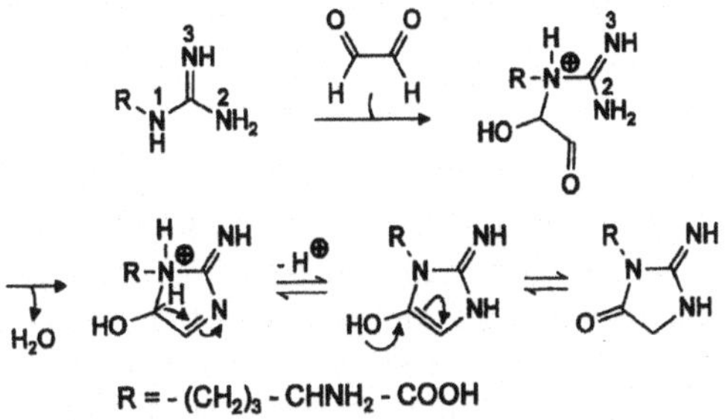

$R = - (CH_2)_3 - CHNH_2 - COOH$

(4.99)

Apart from the modification of amino acid side chains in individual protein strands, cross-linkage of two protein chains can also occur. Some of the structures are shown in Formula 4.100. Pentosidine was first found in physiological protein. It strongly fluoresces and is formed by bridging an arginine residue with a lysine residue via a pentose. The concentrations of pentosidine in food are comparatively low (Table 4.16). The formation of pentosidine is assumed to be as shown in Formula 4.101. Formation of the *Amadori* product with the ε-amino group of lysine is followed by water elimination at C-2 and C-3 of pentose with the formation of the 4,5-diulose, which condenses with the

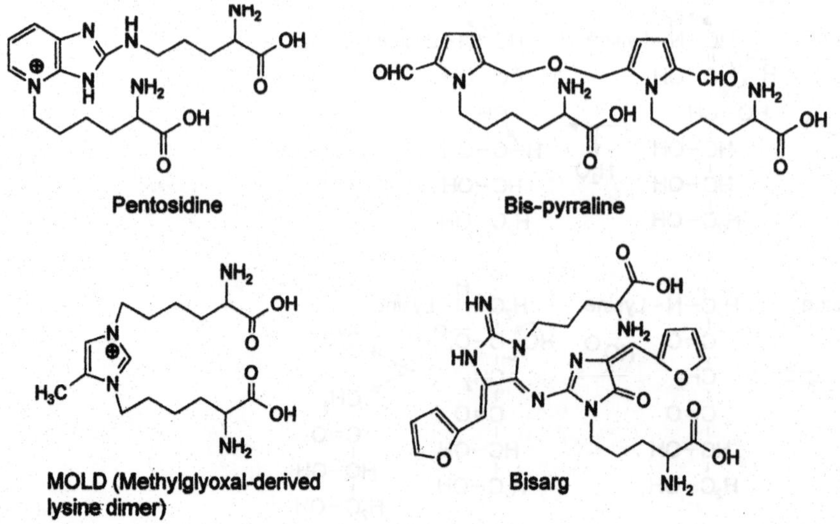

Pentosidine

Bis-pyrraline

MOLD (Methylglyoxal-derived lysine dimer)

Bisarg

(4.100)

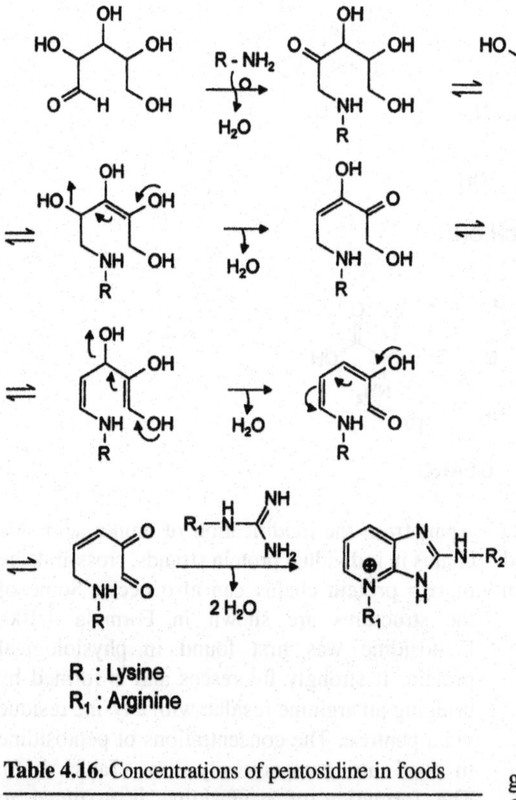

R : Lysine
R₁ : Arginine

(4.101)

Table 4.16. Concentrations of pentosidine in foods

Foods	Concentration (mg/kg protein)
Sterile milk	0.1–2.6
Condensed milk	0.3–0.6
Bread crust	0.4–2.6
Salt pretzel	9.3–22.8
Roasted coffee	10.8–39.9

guanidino group of arginine. The recently identified compound Bisarg (Formula 4.100) is a condensation product of 2 molecules each of arginine, glyoxal and furfural. Apart from the protein cross-linking properties, the intensive brown-orange color of Bisarg should be mentioned.

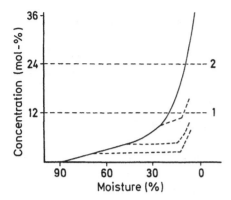

Fig. 4.10. Increase of *Amadori* compounds in two stage air drying of carrots as influenced by carrot moisture content. — 10, 20, 30 min at 110 °C; – – – 60 °C; sensory assessment: 1) detection threshold 2) quality limit (according to *Eichner* and *Wolf*, in *Waller* and *Feather*, 1983)

4.2.4.4.10 Inhibition of the Maillard Reaction

Measures to inhibit the *Maillard* reaction in cases where it is undesirable involve lowering of the pH value, maintenance of lowest possible temperatures and avoidance of critical water contents (cf. 0.3.2) during processing and storage, use of nonreducing sugars, and addition of sulfite. Figure 4.10 demonstrates by the example of carrot dehydration the advantages of running a two-stage process to curtail the *Maillard* reaction.

4.2.4.5 Reactions with Hydroxy Compounds (O-Glycosides)

The lactol group of monosaccharides heated in alcohol in the presence of an acid catalyst is substituted by an alkoxy or aryloxy group, denoted as an aglycone (*Fischer* synthesis), to produce alkyl- and arylglycosides. It is assumed that the initial reaction involves the open form. With the majority of sugars, the furanosides are formed in the first stage of reaction. They then equilibrate with the pyranosides. The transition from furanoside to pyranoside occurs most probably through an open carboxonium ion, whereas pyranoside isomerization is through a cyclic one (cf. Reaction 4.102). Furanosides are obtainable by stopping the reaction at a suitable time. The equilibrium state in alcohol is, as in water, de-

pendent on conformational factors. The alcohol as solvent and its R-moiety both increase the anomeric effect and thus α-pyranoside becomes a more favorable form than was α-pyranose in aqueous free sugar solutions (Table 4.7). In the system D-glucose/methanol in the presence of 1% HCl, 66% of the methylglucoside is present as α-pyranoside, 32.5% as β-pyranoside, and only 0.6% and 0.9% are in α- and β-furanoside forms. Under the same conditions, D-mannose and D-galactose are 94% and 58% respectively in α-pyranoside forms.

A highly stereospecific access to glycosides is possible by C-1 bromination of acetylated sugars.

$$(4.102)$$

$$(4.103)$$

In the reaction of peracetylated sugar with HBr, due to the strong anomeric effect, α-halogenide is formed almost exclusively (cf. Formula 4.103).

This then reacts, probably through its glycosyl cation form. Due to the steric influence of the acetylated group on C-2, the 1,2-trans-glycoside is preferentially obtained, e. g., in the case of D-glucose, β-glucoside results.

Acetylglycosyl halogenides are also used for a highly stereoselective synthesis of α-glycosides. The compound is first dehalogenated into a glycal. Then, addition of nitrosylchloride follows, giving rise to 2-deoxy-2-nitrosoglycosylchloride. The latter, in the presence of alcohol, eliminates HCl and provides a 2-deoxy-2-oximino-α-glycoside. Reaction with ethanal yields the 2-oxo compound, which is then reduced to α-glycoside:

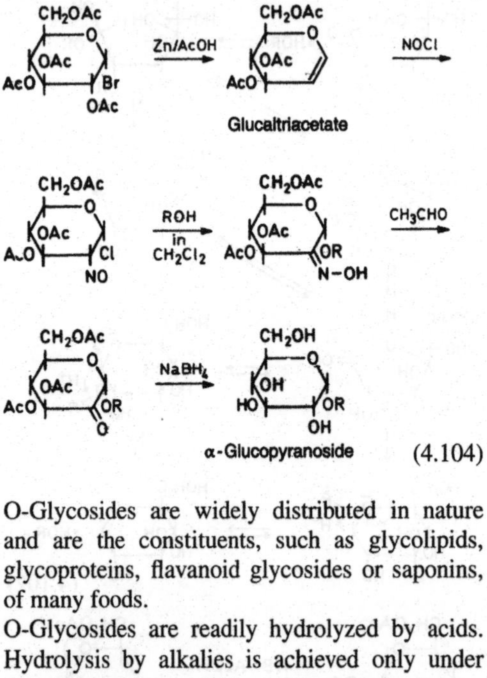

Glucaltriacetate

α-Glucopyranoside   (4.104)

O-Glycosides are widely distributed in nature and are the constituents, such as glycolipids, glycoproteins, flavanoid glycosides or saponins, of many foods.

O-Glycosides are readily hydrolyzed by acids. Hydrolysis by alkalies is achieved only under drastic conditions which simultaneously decompose monosaccharides.

The acid hydrolysis is initiated by glycoside protonation. Alcohol elimination is followed by addition of water:

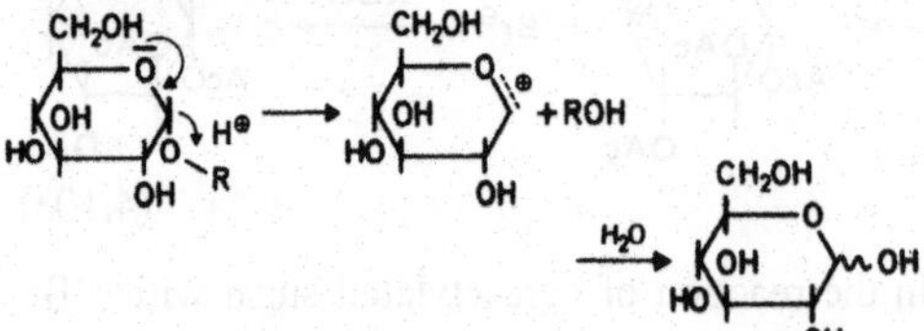

(4.105)

Table 4.17. Relative rate of hydrolysis of glycosides (a: 2 mol/l HCl, 60 °C; b: 0.5 mol/l HCl, 75 °C)

Compound	Hydrolysis condition	K_{rel}
Methyl-α-D-glucopyranoside	a	1.0
Metyhl-β-D-glucopyranoside	a	1.8
Phenyl-α-D-glucopyranoside	a	53.7
Phenyl-β-D-glucopyranoside	a	13.2
Methyl-α-D-glucopyranoside	b	1.0
Methyl-β-D-glucopyranoside	b	1.9
Methyl-α-D-mannopyranoside	b	2.4
Methyl-β-D-mannopyranoside	b	5.7
Methyl-α-D-galactopyranoside	b	5.2
Methyl-β-D-galactopyranoside	b	9.2

The hydrolysis rate is dependent on the aglycone and the monosaccharide itself. The most favored form of alkylglycoside, α-pyranoside, usually is the isomer most resistant to hydrolysis. This is also true for arylglycosides, however, due to steric effects, the β-pyranoside isomer is synthesized preferentially and so the β-isomer better resists hydrolysis.

The influence of the sugar moiety on the rate of hydrolysis is related to the conformational stability. Glucosides with high conformational stability are hydrolyzed more slowly (cf. data compiled in Table 4.17).

4.2.4.6 Esters

Esterification of monosaccharides is achieved by reaction of the sugar with an acyl halide or an acid anhydride. Acetylation, for instance with acetic anhydride, is carried out in pyridine solution:

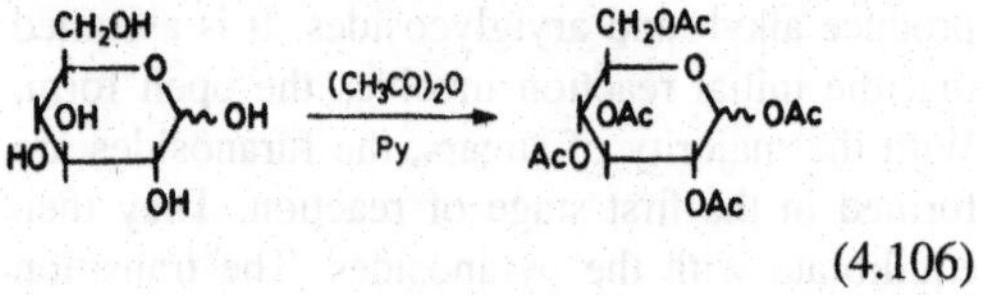

(4.106)

Acyl groups have a protective role in some synthetic reactions. Gluconic acid nitrile acetates (aldonitrile acetates) are analytically suitable sugar derivatives for gas chromatographic separation and identification of sugars. An advantage of these compounds is that they simplify a chromatogram since there are no anomeric peaks:

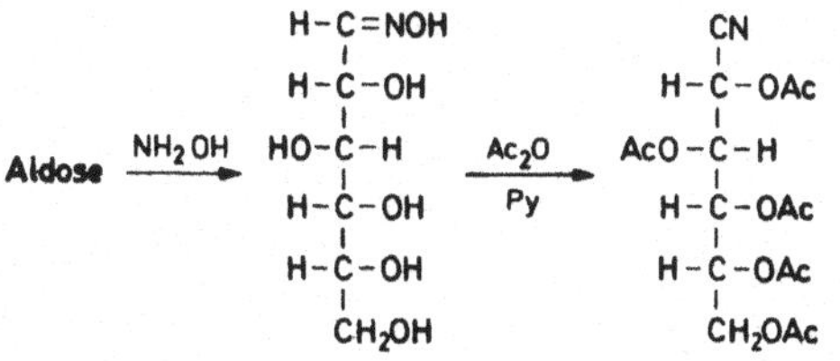

$$ \text{Aldose} \xrightarrow{NH_2OH} \begin{array}{c} H-C=NOH \\ H-C-OH \\ HO-C-H \\ H-C-OH \\ H-C-OH \\ CH_2OH \end{array} \xrightarrow[Py]{Ac_2O} \begin{array}{c} CN \\ H-C-OAc \\ AcO-C-H \\ H-C-OAc \\ H-C-OAc \\ CH_2OAc \end{array} $$

$$ (4.107) $$

Selective esterification of a given HO-group is also possible. For example, glucose can be selectively acetylated in position 3 by reacting 1,2,5,6-di-O-isopropylidene-α-D-glucofuranose with acetic acid anhydride, followed by hydrolysis of the diketal:

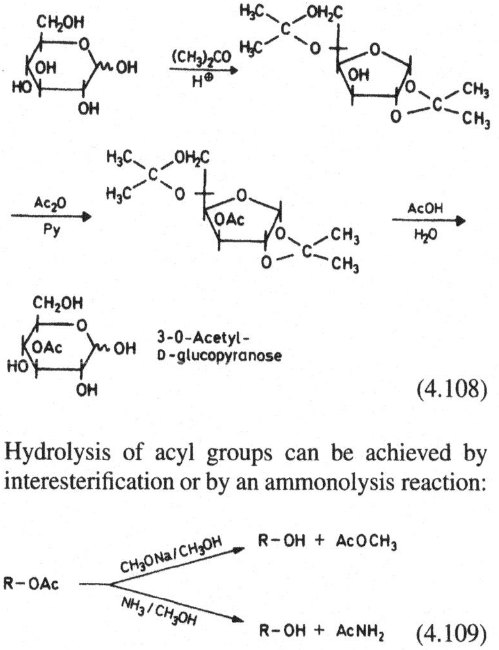

3-0-Acetyl-
D-glucopyranose

$$ (4.108) $$

Hydrolysis of acyl groups can be achieved by interesterification or by an ammonolysis reaction:

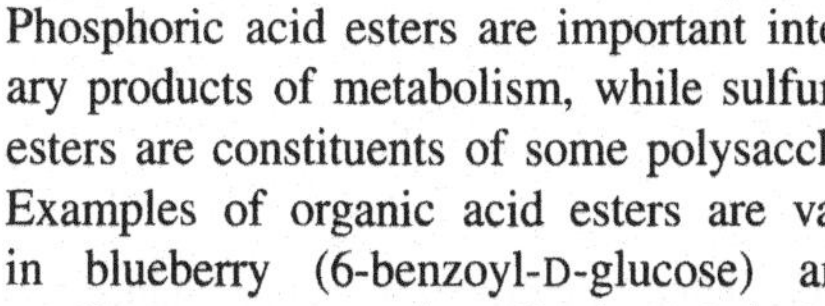

$$ (4.109) $$

Sugar esters are also found widely in nature. Phosphoric acid esters are important intermediary products of metabolism, while sulfuric acid esters are constituents of some polysaccharides. Examples of organic acid esters are vacciniin in blueberry (6-benzoyl-D-glucose) and the tannintype compound, corilagin (1,3,6-trigalloyl-D-glucose):

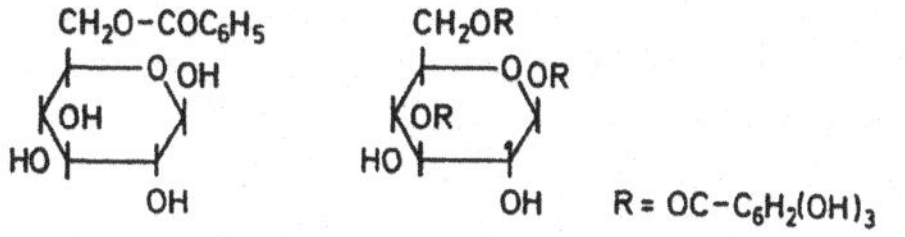

$$ R = OC-C_6H_2(OH)_3 $$

$$ (4.110) $$

Sugar esters or sugar alcohol esters with long chain fatty acids (lauric, palmitic, stearic and oleic) are produced industrially and are very important as surface-active agents. These include sorbitan fatty acid esters (cf. 8.15.3.3) and those of saccharose (cf. 8.15.3.2), which have diversified uses in food processing.

4.2.4.7 Ethers

Methylation of sugar HO-groups is possible using dimethylsulfate or methyliodide as the methylating agent. Methyl ethers are of importance in analysis of sugar structure since they provide data about ring size and linkage positions.

Permethylated saccharose, for example, after acid hydrolysis provides 2,3,4,6-tetra-O-methyl-D-glucose and 1,3,4,6-tetra-O-methyl-D-fructose. This suggests the presence of a 1,2'-linkage between the two sugars and the pyranose and furanose structures for glucose and fructose, respectively:

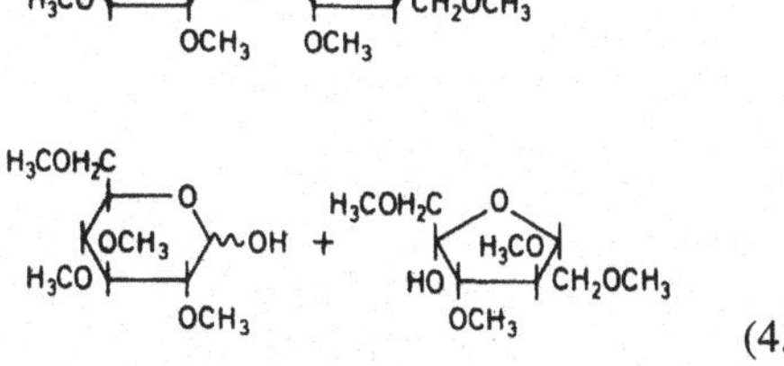

$$ (4.111) $$

Trimethylsilyl ethers (TMS-ethers) are unstable against hydrolysis and alcoholysis, but have remarkable thermal stability and so are suitable for gas chromatographic sugar analysis. Treatment of a sugar with hexamethyldisilazane and trimethylchlorosilane, in pyridine as solvent, provides a sugar derivative with all HO-groups silylated:

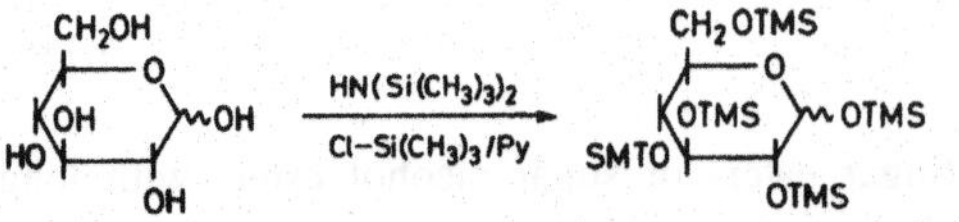

$$\text{(4.112)}$$

4.2.4.8 Cleavage of Glycols

Oxidative cleavage of vicinal dihydroxy groups or hydroxy-amino groups of a sugar with lead tetraacetate or periodate is of importance for structural elucidation. Fructose, in a 5-membered furanose form, consumes 3 moles of periodate (splitting of each α-glycol group requires 1 mole of oxidant) while, in a pyranose ring form, it consumes 4 moles of periodate.

Saccharose consumes 3 moles (cf. Reaction 4.113) and maltose 4 moles of periodate. The final conclusion as to sugar linkage positions and ring structure is drawn from the periodate consumption, the amount of formic acid produced (in the case of saccharose, 1 mole; maltose, 2 moles) and the other carbonyl fragments which are oxidized additionally by bromine to stable carboxylic acids and then released by hydrolysis. The glycol splitting reaction should be considered an optional or complementary method to the permethylation reaction applied in structural elucidation of carbohydrates.

4.3 Oligosaccharides

4.3.1 Structure and Nomenclature

Monosaccharides form glycosides (cf. 4.2.4.5). When this occurs between the lactol group of one monosaccharide and any HO-group of a second monosaccharide, a disaccharide results.

Compounds with up to about 10 monosaccharide residues are designated as oligosaccharides. When a glycosidic linkage is established only between the lactol groups of two monosaccharides, then a *nonreducing disaccharide* is formed, and when one lactol group and one alcoholic HO-group are involved, a *reducing disaccharide* re-

sults. The former is denoted as a glycosylglycoside, the latter as a glycosylglycose, with additional data for linkage direction and positions. Examples are saccharose and maltose:

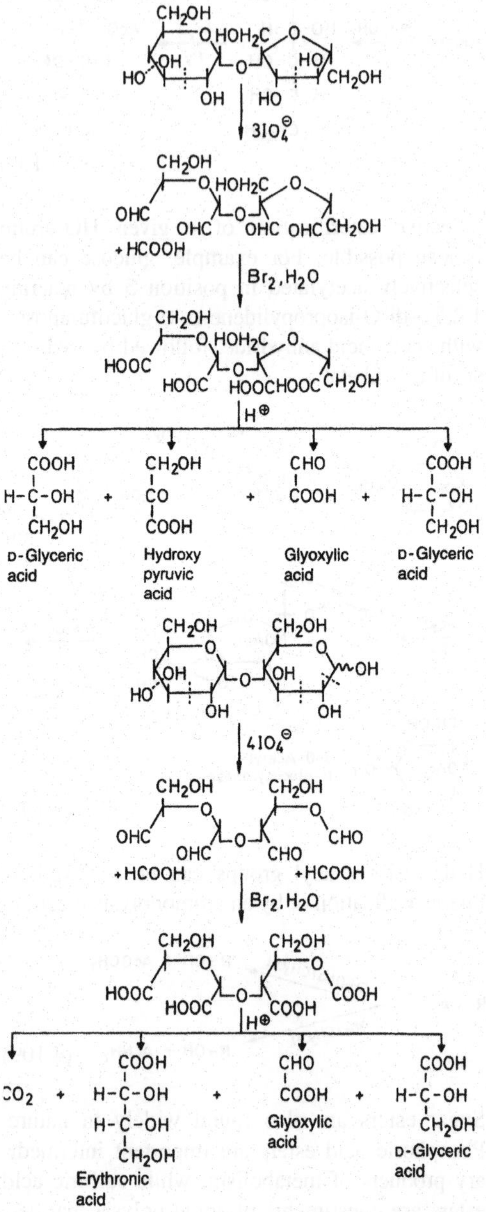

$$\text{(4.113)}$$

An abbreviated method of nomenclature is to use a three letter designation or symbol for a monosaccharide and suffix *f* or *p* for furanose or pyranose. For example, saccharose and maltose can be written as O-β-D-Fru$f(2 \rightarrow 1)$α-D-Glc$p$ and O-α-D-Glc$p(1 \rightarrow 4)$D-Glc$p$, respectively.

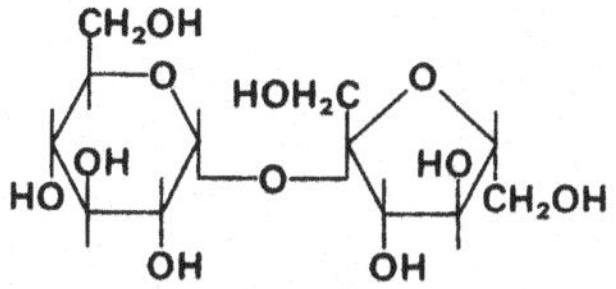

β-D-Fructofuranosyl-α-D-
glucopyranoside
(saccharose)

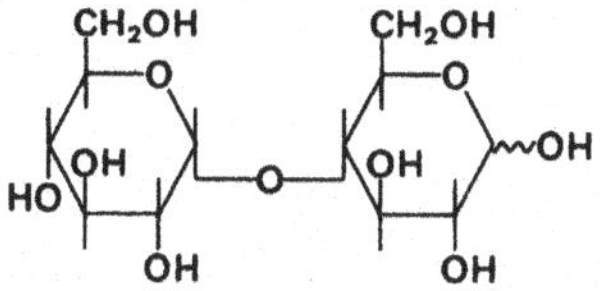

O-α-D-Glucopyranosyl-(1 → 4)-
D-glucopyranose
(maltose)   (4.114)

Branching also occurs in oligosaccharides. It results when one monosaccharide is bound to two glycosyl residues. The name of the second glycosyl residue is inserted into square brackets. A trisaccharide which represents a building block of the branched chain polysaccharides amylopectin and glycogen is given as an example:

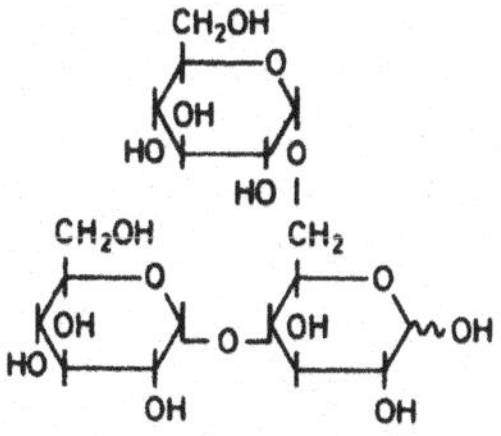

O-α-D-Glucopyranosyl-(1→4)-0-[α-D-
glucopyranosyl-(1→6)]-D-glucopyranose   (4.115)

The abbreviated formula for this trisaccharide is as follows:

α-D-Glc$p$ (1 → 4) Glc$p$
                (6
                ↑
            1)-α-D-Glc$p$ (4.116)

The conformations of oligo- and polysaccharides, like peptides, can be described by providing the angles Φ and ψ:

                                                 (4.117)

A calculation of conformational energy for all conformers with allowed Φ, ψ pairs provides a Φ, ψ diagram with lines corresponding to iso-conformational energies. The low-energy conformations calculated in this way agree with data obtained experimentally (X-ray diffraction, NMR, ORD) for oligo- and polysaccharides.

H-bonds fulfill a significant role in conformer stabilization. Cellobiose and lactose conformations are well stabilized by an H-bond formed between the HO-group of C-3 in the glucose residue and the ring oxygen of the glycosyl residue. Conformations in aqueous solutions appear to be similar to those in the crystalline state:

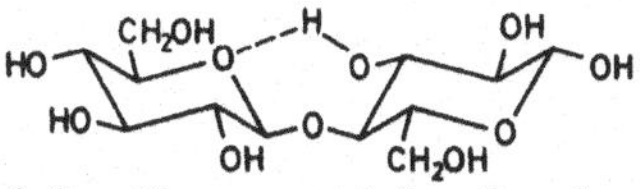

O-β-D-Glucopyranosyl-(1→4)-D-glucopyranose (cellobiose)

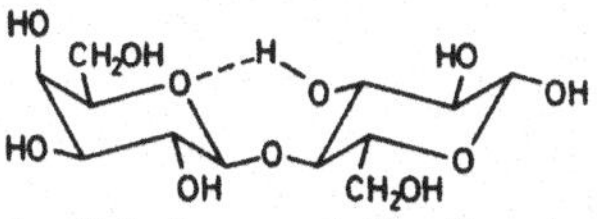

O-β-D-Galactopyranosyl-(1→4)-D-glucopyranose (lactose)

                                                 (4.118)

In crystalline maltose and in nonaqueous solutions of this sugar, a hydrogen bond is established between the HO-groups on C-2 of the glucosyl and on C-3 of the glucose residues (4.119). However, in aqueous solution, a conformer partially present is stabilized by H-bonds established between the CH$_2$OH-group of the glucosyl residue and the HO-group of C-3 on the glucose residue (4.120). Both conformers correspond to the two energy minima in the Φ, ψ diagram.

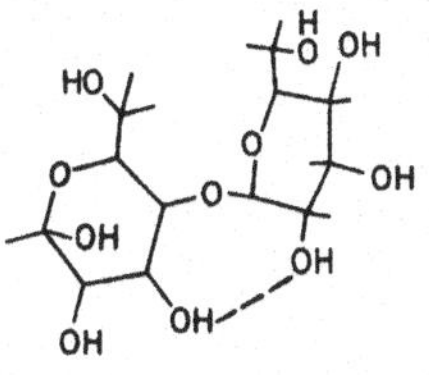

                                                 (4.119)

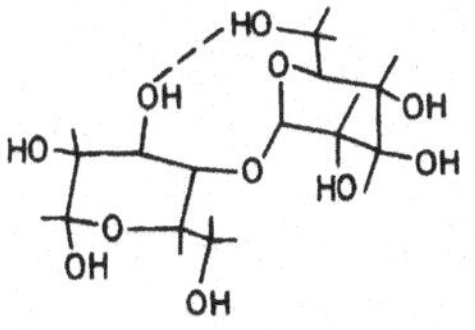

                                                 (4.120)

Two H-bonds are possible in saccharose, the first between the HO-groups on the C-1 of the fructose and the C-2 of the glucose residues, and the second between the HO-group on the C-6 of the fructose residue and the ring oxygen of the glucose residue:

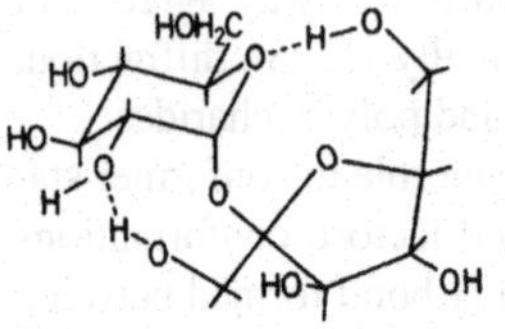

β -D-Fructofuranosyl -α-D-glucopyranoside (saccharose)

(4.121)

4.3.2 Properties and Reactions

The oligosaccharides of importance to food, together with data on their occurrence, are compiled in Table 4.18. The physical and sensory properties were covered with monosaccharides, as were reaction properties, though the difference between reducing and nonreducing oligosaccharides should be mentioned. The latter do not have a free lactol group and so lack reducing properties, mutarotation and the ability to react with alcohols and amines.

As glycosides, oligosaccharides are readily hydrolyzed by acids, while they are relatively stable against alkalies. Saccharose hydrolysis is denoted as an inversion and the resultant equimolar mixture of glucose and fructose is called invert sugar. The term is based on a change of specific rotation during hydrolysis. In saccharose the rotation is positive, while it is negative in the hydrolysate, since D-glucose rotation to the right (hence its name dextrose) is surpassed by the value of the left-rotating fructose (levulose):

Saccharose $\xrightarrow{H^{\oplus}}$ D-Glucose + D-Fructose

$[\alpha]_D = +66.5°$ $[\alpha]_D = +52.7°$ $[\alpha]_D = -92.4°$

$[\alpha]_D = -19.8°$

(4.122)

Conclusions can be drawn from mutarotation, which follows hydrolysis of reducing disaccharides, about the configuration on the anomeric C-atom. Since the α-anomer has a higher specific

rotation in the D-series than the β-anomer, cleavage of β-glycosides increases the specific rotation while cleavage of α-glycosides decreases it:

Cellobiose (β) $\longrightarrow$ 2 D-Glucose

$[\alpha]_D = +34.6°$ $[\alpha]_D = +52.7°$

Maltose (α) $\longrightarrow$ 2 D-Glucose

$[\alpha]_D = +130°$ $[\alpha]_D = +52.7°$

Lactose (β) $\longrightarrow$ D-Galactose + D-Glucose

$[\alpha]_D = +52.3°$ $[\alpha]_D = +80.2°$ $[\alpha]_D = +52.7°$

$[\alpha]_D = +66.3°$

(4.123)

Enzymatic cleavage of the glycosidic linkage is specified by the configuration on anomeric C-1 and also by the whole glycosyl moiety, while the aglycone residue may vary within limits.

The methods used to elucidate the linkage positions in an oligosaccharide (methylation, oxidative cleavage of glycols) were outlined under monosaccharides.

The cyclodextrins   listed in Table 4.18 are prepared by the action of cyclomaltodextrin glucanotransferase (E. C. 2.4.1.19), obtained from *Bacillus macerans*, on maltodextrins. Maltodextrins are, in turn, made by the degradation of starch with α-amylase. This glucanotransferase splits the α-1,4-bond, transferring glucosyl groups to the nonreducing end of maltodextrins and forming cyclic glucosides with 6-12 glucopyranose units. The main product, β-cyclodextrin, consists of seven glucose units and is a non-hygroscopic, slightly sweet compound:

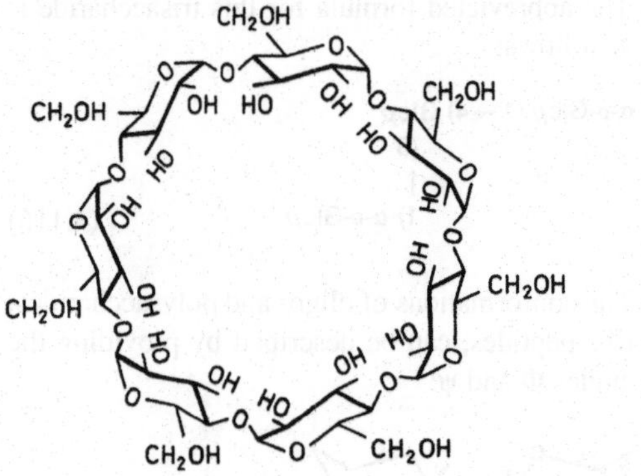

(4.124)

Table 4.18. Structure and occurrence of oligosaccharides

Name	Structure	Occurrence
Disaccharides		
Cellobiose	O-β-D-Glc*p*-(1 → 4)-D-Glc*p*	Building block of cellulose
Gentiobiose	O-β-D-Glc*p*-(1 → 6)-D-Glc*p*	Glycosides (amygdalin)
Isomaltose	O-α-D-Glc*p*-(1 → 6)-D-Glc*p*	Found in mother liquor during glucose production from starch
Lactose	O-β-D-Gal*p*-(1 → 4)-D-Glc*p*	Milk
Lactulose	O-β-D-Gal*p*-(1 → 4)-D-Fru*p*	Conversion product of lactose
Maltose	O-α-D-Glc*p*-(1 → 4)-D-Glc*p*	Building block of starch, sugar beet, honey
Maltulose	O-α-D-Glc*p*-(1 → 4)-D-Fru*f*	Conversion product of maltose, honey, beer
Melibiose	O-α-D-Gal*p*-(1 → 6)-D-Glc*p*	Cacao beans
Neohesperidose	O-α-L-Rha*p*-(1 → 2)-D-Glc*p*	Glycosides (naringin, neohesperidin)
Neotrehalose	O-α-D-Glc*p*-(1 → 1)-β-D-Glc*p*	Koji extract
Nigerose	O-α-D-Glc*p*-(1 → 3)-D-Glc*p*	Honey, beer
Palatinose	O-α-D-Glc*p*-(1 → 6)-D-Fru*f*	Microbial product of saccharose
Rutinose	O-α-L-Rha*p*-(1 → 6)-D-Glc*p*	Glycosides (hesperidin)
Saccharose	O-β-D-Fru*f*-(2 → 1)-α-D-Glc*p*	Sugar beet, sugar cane, spread widely in plants
Sophorose	O-β-D-Glc*p*-(1 → 2)-D-Glc*p*	Legumes
Trehalose	O-α-D-Glc*p*-(1 → 1)-α-D-Glc*p*	Ergot (*Claviceps purpurea*), young mushrooms
Trisaccharides		
Fucosidolactose	O-α-D-Fuc*p*-(1 → 2)-O-β-α-Gal*p*-(1 → 4)-D-Gal*p*	Human milk
Gentianose	O-β-D-Glc*p*-(1 → 6)-O-α-D-Glc*p*-(1 → 2)-β-D-Fru*f*	Gentian rhizome
Isokestose (1-Kestose)	O-α-D-Glc*p*-(1 → 2)-O-β-D-Fru*f*-(1 → 2)-β-D-Fru*f*	Product of saccharase action on saccharose as a substrate
Kestose (6-Kestose)	O-α-D-Glc*p*-(1 → 2)-O-β-D-Fru*f*-(6 → 2)-β-D-Fru*f*	Saccharose subjected to yeast saccharase activity, honey
Maltotriose	O-α-D-Glc*p*-(1 → 4)-O-α-D-Glc*p*-(1 → 4)-D-Glc*p*	Degradation product of starch, starch syrup
Manninotriose	O-α-D-Gal*p*-(1 → 6)-O-α-D-Gal*p*-(1 → 6)-D-Glc*p*	Manna
Melezitose	O-α-D-Glc*p*-(1 → 3)-O-β-D-Fru*f*-(2 → 1)-α-D-Glc*p*	Manna, nectar
Neokestose	O-β-D-Fru*f*-(2 → 6)-O-α-D-Glc*p*-(1 → 2)-β-D-Fru*f*	Product of saccharase action on saccharose as a substrate
Panose	O-α-D-Glc*p*-(1 → 6)-O-α-D-Glc*p*-(1 → 4)-D-Glc*p*	Degradation product of amylopectin, honey
Raffinose	O-α-D-Gal*p*-(1 → 6)-O-α-D-Glc*p*-(1 → 2)-β-D-Fru*f*	Sugar beet, sugar cane, widely distributed in plants
Umbelliferose	O-α-D-Gal*p*-(1 → 2)-O-α-D-Glc*p*-(1 → 2)-β-D-Fru*f*	Umbelliferae roots
Tetrasaccharides		
Maltotetraose	O-α-D-Glc*p*-(1 → 4)-O-α-D-Glc*p*-(1 → 4)-O-α-D-Glc*p*-(1 → 4)-D-Glc*p*	Starch syrup
Stachyose	O-α-D-Gal*p*-(1 → 6)-O-α-D-Gal*p*-(1 → 6)-O-α-D-Glc*p*-(1 → 2)-β-D-Fru*f*	Widespread in plants (artichoke, soybean)
Higher oligosaccharides		
Maltopentaose	[O-α-D-Glc*p*-(1 → 4)]₄-D-Glc*p*	Starch syrup
α-*Schardinger*-Dextrin, Cyclohexaglucan (α, 1 → 4)		Growth of
β-*Schardinger*-Dextrin, Cycloheptaglucan (α, 1 → 4)		*Bacillus macerans*
γ-*Schardinger*-Dextrin, Cyclooctaglucan (α, 1 → 4)		on starch syrup

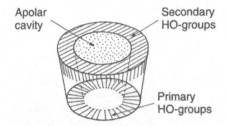

Apolar cavity

Secondary HO-groups

Primary HO-groups

Fig. 4.11. Schematic representation of the hollow cylinder formed by β-cyclodextrin

The β-cyclodextrin molecule is a cylinder (Fig. 4.11) which has a primary hydroxyl (C6) rim on one side and a secondary hydroxyl (C2, C3) rim on the other. The surfaces made of pyranose rings are hydrophobic. Indeed, the water of hydration is very easily displaced from this hydrophobic cavity by sterically suitable apolar compounds, which are masked in this way. In food processing, β-cyclodextrin is therefore a suitable agent for stabilizing lipophilic vitamins and aroma substances and for neutralizing the taste of bitter substances

4.4 Polysaccharides

4.4.1 Classification, Structure

Polysaccharides, like oligosaccharides, consist of monosaccharides bound to each other by glycosidic linkages. Their acidic hydrolysis yields monosaccharides. Partial chemical and enzymatic hydrolysis, in addition to total hydrolysis, are of importance for structural elucidation. Enzymatic hydrolysis provides oligosaccharides, the analysis of which elucidates monosaccharide sequences and the positions and types of linkages. Polysaccharides (glycans) can consist of one type of sugar structural unit (homoglycans) or of several types of sugar units (heteroglycans). The monosaccharides may be joined in a linear pattern (as in cellulose and amylose) or in a branched fashion (amylopectin, glycogen, guaran). The frequency of branching sites and the length of side chains can vary greatly (glycogen, guaran). The monosaccharide residue sequence may be periodic, one period containing one or several alternating structural units (cellulose, amylose or hyaluronic acid), the sequence may

contain shorter or longer segments with periodically arranged residues separated by nonperiodic segments (alginate, carrageenans, pectin), or the sequence may be nonperiodic all along the chain (as in the case of carbohydrate components in glycoproteins).

4.4.2 Conformation

The monosaccharide structural unit conformation and the positions and types of linkages in the chain determine the chain conformation of a polysaccharide. In addition to irregular conformations, regular conformations are known which reflect the presence of at least a partial periodic sequence in the chain. Some typical conformations will be explained in the following discussion, with examples of glucans and some other polysaccharides.

4.4.2.1 Extended or Stretched, Ribbon-Type Conformation

This conformation is typical for 1,4-linked β-D-glucopyranosyl residues (Fig. 4.12 a), as occur, for instance, in cellulose fibers:

$$(4.125)$$

This formula shows that the stretched chain conformation is due to a *zigzag* geometry of monomer linkages involving oxygen bridging. The chain may be somewhat shortened or compressed to enable formation of H-bonds between neighboring residues and thus contribute to conformational stabilization. In the ribbon-type, stretched conformation, with the number of monomers in turn denoted as n and the pitch (advancement) in the axial direction per monomer unit as h, the range of n is from 2 to ±4, while h is the length of a monomer unit. Thus, the chain given in Fig. 4.12 a has $n = -2.55$ and $h = 5.13$ Å.

A strongly pleated, ribbon-type conformation might also occur, as shown by a segment of

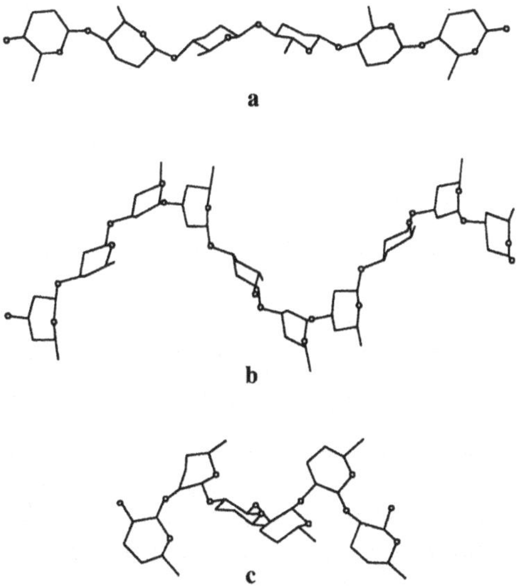

a

b

c

Fig. 4.12. Conformations of some β-D-glucans. Linkages: **a** 1 → 4, **b** 1 → 3, **c** 1 → 2 (according to *Rees*, 1977)

a pectin chain (1,4-linked α-D-galactopyranosyl-uronate units):

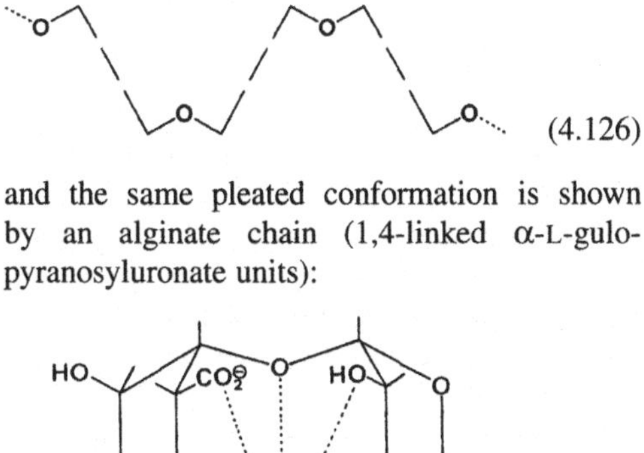

(4.126)

and the same pleated conformation is shown by an alginate chain (1,4-linked α-L-gulo-pyranosyluronate units):

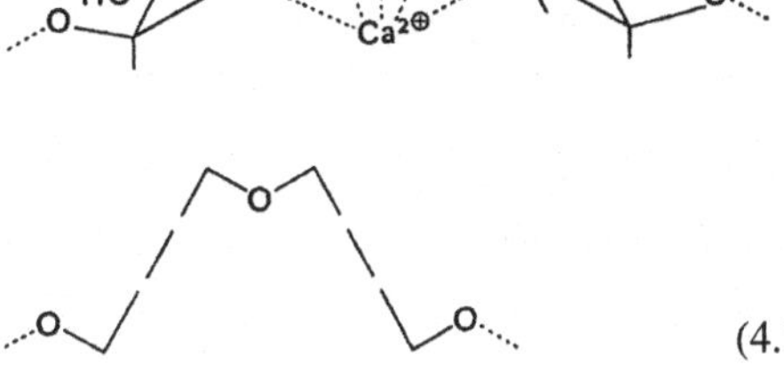

(4.127)

Ca^{2+} ions can be involved to stabilize the conformation. In this case, two alginate chains are assembled in a conformation which resembles an egg box (*egg box type of conformation*):

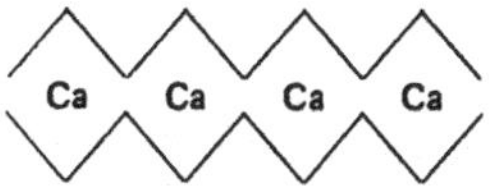

(4.128)

It should be emphasized that in all examples the linear, ribbon-type conformation has a zigzag geometry as a common feature.

4.4.2.2 Hollow Helix-Type Conformation

This conformation is typical for 1,3-linked β-D-glucopyranose units (Fig. 4.12, b), as occur in the polysaccharide lichenin, found in moss-like plants (lichens):

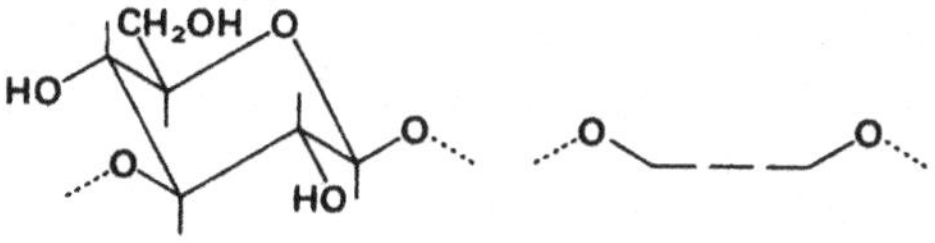

(4.129)

The formula shows that the helical conformation of the chain is imposed by a U-form geometry of the monomer linkages. Amylose (1,4-linked α-D-glucopyranosyl residues) also has such a geometry, and hence a helical conformation:

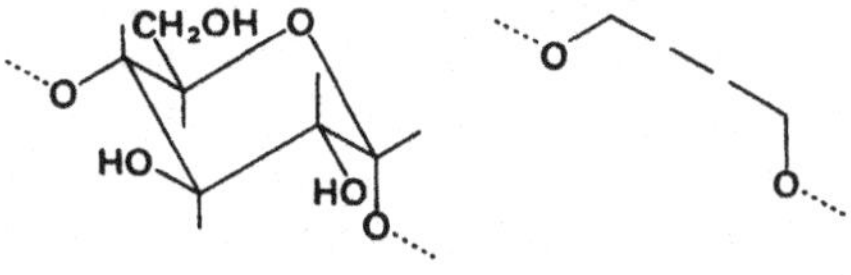

(4.130)

The number of monomers per turn (n) and the pitch in the axial direction per residue (h) is highly variable in a hollow helical conformation. The value of n is between 2 and ±10, whereas h can be near its limit value of 0. The conformation of a $\beta(1 \rightarrow 3)$-glucan, with $n = 5.64$ and $h = 3.16$ Å, is shown in Fig. 4.12, b. The helial conformation can be stabilized in various ways. When the helix diameter is large, inclusion (clathrate) compounds can be formed (Fig. 4.13, a; cf. 4.4.4.14.3). More extended or stretched chains, with smaller helix diameter,

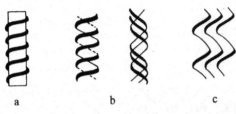

a   b   c

Fig. 4.13. Stabilizatioin of helical conformations. **a** Clathrate compounds, **b** coiled double or triple helices, **c** "nesting" (according to *Rees*, 1977)

can form double or triple stranded helices (Fig. 4.13, b; cf. 4.4.4.3.2 and 4.4.4.14.3), while strongly-stretched chains, in order to stabilize the conformation, have a zigzag, pleated association and are not stranded (Fig. 4.13, c).

4.4.2.3 Crumpled-Type Conformation

This conformation occurs with, for example, 1,2-linked β-D-glucopyranosyl residues (Fig. 4.12, c). This is due to the wrinkled geometry of the monomer O-bridge linkages:

$$(4.131)$$

Here, the n value varies from 4 up to −2 and h is 2–3 Å. The conformation reproduced in Fig. 4.12, c has $n = 2.62$ and $h = 2.79$ Å. The likelihood of such a disorderly form associating into more orderly conformations is low. Polysaccharides of this conformational type play only a negligible role in nature.

4.4.2.4 Loosely-Jointed Conformation

This is typical for glycans with 1,6-linked β-D-glucopyranosyl units, because they exhibit a particularly great variability in conformation.

The great flexibility of this glycan-type conformation is based on the nature of the connecting bridge between the monomers. The bridge has three free rotational bonds and, furthermore, the sugar residues are further apart:

$$(4.132)$$

4.4.2.5 Conformations of Heteroglycans

The examples considered so far have demonstrated that a prediction is possible for a homoglycan conformation based on the geometry of the bonds of the monomer units which maintain the oxygen bridges. It is more difficult to predict the conformation of a heteroglycan with a periodic sequence of several monomers, which implies different types of conformations. Such a case is shown by ι-carrageenan, in which the β-D-galactopyranosyl-4-sulfate units have a U-form geometry, while the 3,6-anhydro-α-D-galactopyranosyl-2-sulfate residues have a zigzag geometry:

$$(4.133)$$

Calculations have shown that conformational possibilities vary from a shortened, compressed ribbon band type to a stretched helix type. X-ray diffraction analyses have proved that a stretched helix exists, but as a double stranded helix in order to stabilize the conformation (cf. 4.4.4.3.2 and Fig. 4.19).

4.4.2.6 Interchain Interactions

It was outlined in the introductory section (cf. 4.4.1) that the periodically arranged monosaccharide sequence in a polysaccharide can be interrupted by nonperiodic segments. Such sequence interferences result in conformational disorders. This will be explained in more detail with ι-carrageenan, mentioned above, since it will shed light on the gel-setting mechanism of macromolecules in general.

Initially, a periodic sequence of altering units of β-D-galactopyranose- 4-sulfate (I, conformation $^4C_1$) and α-D-galactopyranose-2,6-disulfate (II, conformation $^4C_1$) is built up in carrageenan biosynthesis:

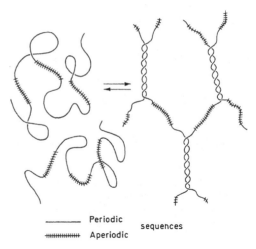

I   II

III   (4.134)

When the biosynthesis of the chain is complete, an enzyme-catalyzed reaction eliminates sulfate from most of α-D-galactopyranose-2,6-disulfate (II), transforming the unit to 3,6-anhydro-α-D-galactopyranose-2-sulfate (III, conformation $^1C_4$). This transformation is associated with a change in linkage geometry. Some II-residues remain in the sequence, acting as interference sites. While the undisturbed, ordered segment of one chain can associate with the same segment of another chain, forming a double helix, the nonperiodic or disordered segments can not participate in such associations (Fig. 4.14).

In this way, a gel is formed with a three-dimensional network in which the solvent is immobilized. The gel properties, e.g., its strength, are influenced by the number and distribution of α-D-galactopyranosyl-2,6-disulfate residues, i.e. by a structural property regulated during polysaccharide biosynthesis.

The example of the τ-carrageenan gel-building mechanism, involving a chain–chain interaction of sequence segments of orderly conformation, interrupted by randomly-coiled segments corresponding to a disorderly chain sequence, can be applied generally to gels of other macromolecules. Besides a sufficient chain length, the

——— Periodic
++++++ Aperiodic sequences

Fig. 4.14. Schematic representation of a gel setting process (according to *Rees*, 1977)

structural prerequisite for gel-setting ability is interruption of a periodic sequence and its orderly conformation. The interruption is achieved by insertion into the chain of a sugar residue of a different linkage geometry (carrageenans, alginates, pectin), by a suitable distribution of free and esterified carboxyl groups (glycuronans) or by insertion of side chains. The interchain associations during gelling (network formation), which involve segments of orderly conformation, can then occur in the form of a double helix (Fig. 4.15,a); a multiple bundle of double helices (Fig. 4.15,b); an association between stretched ribbon-type conformations, such as an egg box model (Fig. 4.15,c); some other similar associations (Fig. 4.15,d); or, lastly, forms consisting of double helix and ribbon-type combinations (Fig. 4.15,e).

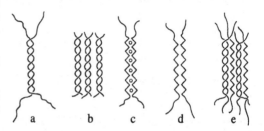

a   b   c   d   e

Fig. 4.15. Interchain aggregation between regular conformations. **a** Double helix, **b** double helix bundle, **c** egg-box, **d** ribbon–ribbon, and **e** double helix, ribbon interaction

4.4.3 Properties

4.4.3.1 General Remarks

Polysaccharides are widely and abundantly distributed in nature, fulfilling roles as:

- Structure-forming skeletal substances (cellulose, hemicellulose and pectin in plants; chitin, mucopolysaccharides in animals).
- Assimilative reserve substances (starch, dextrins, inulin in plants; glycogen in animals).
- Water-binding substances (agar, pectin and alginate in plants; mucopolysaccharides in animals).

As a consequence, polysaccharides occur in many food products and even then they often retain their natural role as skeletal substances (fruits and vegetables) or assimilative nutritive substances (cereals, potatoes, legumes). Isolated polysaccharides are utilized to a great extent in food processing, either in native or modified form, as: thickening or gel-setting agents (starch, alginate, pectin, guaran gum); stabilizers for emulsions and dispersions; film-forming, coating substances to protect sensitive food from undesired change; and inert fillers to increase the proportion of indigestible ballast substances in a diet (cf. 15.2.4.2). Table 4.19 gives an overview of uses in food technology.

The outlined functions of polysaccharides are based on their highly variable properties. They vary from insoluble forms (cellulose) to those with good swelling power and solubility in hot and cold water (starch, guaran gum). The solutions may exhibit low viscosities even at very high concentrations (gum arabic), or may have exceptionally high viscosities even at low concentrations (guaran gum). Some polysaccharides, even at a low concentration, set into a thermoreversible gel (alginates, pectin). While most of the gels melt at elevated temperatures, some cellulose derivatives set into a gel.

These properties and their utilization in food products are described in more detail in section 4.4.4, where individual polysaccharides are covered. Here, only a brief account will be given to relate their properties to their structures in a general way.

4.4.3.2 Perfectly Linear Polysaccharides

Compounds with a *single* neutral monosaccharide structural unit and with *one* type of linkage (as occurs in cellulose or amylose) are denoted as perfectly linear polysaccharides. They are usually insoluble in water and can be solubilized only under drastic conditions, e. g. at high temperature, or by cleaving H-bonds with alkalies or other suitable reagents. They readily precipitate from solution (example: starch retrogradation). The reason for these properties is the existence of an optimum structural prerequisite for the formation of an orderly conformation within the chain and also for chain–chain interaction. Often, the conformation is so orderly that a partial crystallinity state develops. Large differences in properties are found within these groups of polysaccharides when there is a change in structural unit, linkage type or molecular weight. This is shown by properties of cellulose, amylose or β-1,3-glucan macromolecules.

4.4.3.3 Branched Polysaccharides

Branched polysaccharides (amylopectin, glycogen) are more soluble in water than their perfectly linear counterparts since the chain–chain interaction is less pronounced and there is a greater extent of solvation of the molecules. Solutions of branched polysaccharides, once dried, are readily rehydrated. Compared to their linear counterparts of equal molecular weights and equal concentrations, solutions of branched polysaccharides have a lower viscosity. It is assumed that the viscosity reflects the "effective volume" of the macromolecule. The "effective volume" is the volume of a sphere with diameter determined by the longest linear extension of the molecule. These volumes are generally larger for linear than for branched molecules (Fig. 4.16). Exceptions are found with highly pleated linear chains. The tendency of branched polysaccharides to precipitate is low. They form a sticky paste at higher concentrations, probably due to side chain–side chain interactions (interpenetration, entanglement). Thus, branched polysaccharides are suitable as binders or adhesives.

Table 4.19. Examples of uses of polysaccharides in foods

Area of application/food	Suitable polysaccharides
Stabilization of emulsions/suspensions in condensed milk and chocolate milk	Carrageenan, algin, pectin, carboxymethylcellulose
Stabilization of emulsions in coffee whiteners, low-fat margarines	Carrageenan
Stabilization of ice cream against ice crystal formation, melting, phase separation; improvement of consistency (smoothness)	Algin, carrageenan, agar, gum arabic, gum tragacanth, xanthan gum, guaran gum, locust bean flour, modified starches, carboxymethylcellulose, methylcellulose
Water binding, improvement of consistency, yield increase of soft cheese, cream cheese, cheese preparations	Carrageenan, agar, gum tragacanth, karaya gum, guaran gum, locust bean flour, algin, carboxymethylcellulose
Thickening and gelation of milk in puddings made with and without heating, creams; improvement of consistency	Pectin, algin, carrageenan, guaran gum, locust bean flour, carboxymethylcellulose, modified starches
Water binding, stabilization of emulsions in meat products (corned beef, sausage)	Agar, karaya gum, guaran gum, locust bean flour
Jellies for meat, fish, and vegetable products	Algin, carrageenan, agar
Stabilization and thickening, prevention of synaeresis, freeze-thaw stability of soups, sauces, salad dressing, mayonnaise, ketchup; obtaining "body" in low-fat and low-starch products	Gum tragacanth, algin, karaya gum, xanthan gum, guaran gum, locust bean flour, carboxymethyl-cellulose, propylene glycol alginate, modified starches
Stabilization of protein foam in beer, whipped cream, meringues, chocolate marshmallows	Algin, carrageenan, agar, gum arabic, karaya gum, xanthan gum
Prevention of starch retrogradation in bread and cakes, water binding in dough	Agar, guaran gum, locust bean flour, carrageenan, xanthan gum
Thickening and gelation of fruit pulp (confiture, jams, jellies, fruit pulp for ice cream and yoghurt)	Pectin, algin
Gelation of jelly candies, jelly beans, glaze, icing, water-dessert jellies	Pectin, algin, carrageenan, agar, gum arabic, modified starches
Sediment stabilization in fruit juices, obtaining "body" in beverage powders	Algin, pectin, propylene glycol alginate, gum arabic, xanthan gum, guaran gum, methylcellulose
Stabilization of powdery aroma emulsions, encapsulation of aroma substances	Gum arabic, gum ghatti, xanthan gum

4.4.3.4 Linearly Branched Polysaccharides

Linearly branched polysaccharides, i.e. polymers with a long "backbone" chain and with many short side chains, such as guaran or alkyl cellulose, have properties which are a combination of those of perfectly linear and of branched molecules. The long "backbone" chain is responsible for high solution viscosity. The presence of numerous short side chains greatly weakens interactions between the molecules, as shown by the good solubility and rehydration rates of the molecules and by the stability even of highly concentrated solutions.

4.4.3.5 Polysaccharides with Carboxyl Groups

Polysaccharides with carboxyl groups (pectin, alginate, carboxymethyl cellulose) are very soluble as alkali salts in the neutral or alkaline pH range. The molecules are negatively charged due to carboxylate anions and, due to their repulsive charge forces, the molecules are relatively stretched and resist intermolecular associations. The solution viscosity is high and is pH-dependent. Gel setting or precipitation occurs at pH ≤ 3 since electrostatic repulsion ceases to exist. In addition, undissociated carboxyl groups dimerize through

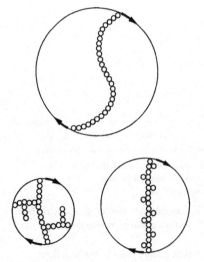

Fig. 4.16. Schematic representation of the "effective volumes" of linear, branched and linearly branched types of polysaccharides

H-bridges. However, a divalent cation is needed to achieve gel setting in a neutral solution.

4.4.3.6 Polysaccharides with Strongly Acidic Groups

Polysaccharides with strongly acidic residues, present as esters along the polymer chains (sulfuric, phosphoric acids, as in furcellaran, carrageenan or modified starch), are also very soluble in water and form highly viscous solutions. Unlike polysaccharides with carboxyl groups, in strongly acidic media these solutions are distinctly stable.

4.4.3.7 Modified Polysaccharides

Modification of polysaccharides, even to a low substitution degree, brings about substantial changes in their properties.

4.4.3.7.1 Derivatization with Neutral Substituents

The solubility in water, viscosity and stability of solutions are all increased by binding neutral sub-

stituents to linear polysaccharide chains. Thus the properties shown by methyl, ethyl and hydroxypropyl cellulose correspond to those of guaran and locust bean gum. The effect is explained by interference of the alkyl substituents in chain interactions, which then facilitates hydration of the molecule. An increased degree of substitution increases the hydrophobicity of the molecules and, thereby, increases their solubility in organic solvents.

4.4.3.7.2 Derivatization with Acidic Substituents

Binding acid groups to a polysaccharide (carboxymethyl, sulfate or phosphate groups) also results in increased solubility and viscosity for reasons already outlined. Some derivatized polysaccharides, when moistened, have a pasty consistence.

4.4.4 Individual Polysaccharides

4.4.4.1 Agar

4.4.4.1.1 Occurrence, Isolation

Agar is a gelatinous product isolated from seaweed (red algae class, *Rhodophyceae*), e. g., *Gelidium spp.*, *Pterocladia spp.* and *Gracilaria spp.*, by a hot water extraction process. Purification is possible by congealing the gel.

4.4.4.1.2 Structure, Properties

Agar is a heterogenous complex mixture of related polysaccharides having the same backbone chain structure. The main components of the chain are β-D-galactopyranose and 3,6-anhydro-α-L-galactopyranose, which alternate through 1 → 4 and 1 → 3 linkages:

$$(4.135)$$

The chains are esterified to a low extent with sulfuric acid. The sulfate content differentiates between the agarose fraction (the main gelling component of agar), in which close to every tenth galactose unit of the chain is esterified, and the agaropectin fraction, which has a higher sulfate esterification degree and, in addition, has pyruvic acid bound in ketal form [4,6-(l-carboxyethylidene)-D-galactose]. The ratio of the two polymers can vary greatly. Uronic acid, when present, does not exceed 1%. Agar is insoluble in cold water, slightly soluble in ethanolamine and soluble in formamide. Agar precipitated by ethanol from a warm aqueous dispersion is, in its moist state, soluble in water at 25 °C, while in the dried state it is soluble only in hot water. Gel setting occurs upon cooling. Agar is a most potent gelling agent as gelation is perceptible even at 0.04%. Gel setting and stability are affected by agar concentration and its average molecular weight. A 1.5% solution sets to a gel at 32–39 °C, but does not melt below 60–97 °C. The great difference between gelling and melting temperatures, due to hysteresis, is a distinct and unique feature of agar.

4.4.4.1.3 Utilization

Agar is widely used, for instance in preparing nutritive media in microbiology. Its application in the food industry is based on its main properties: it is essentially indigestible, forms heat resistant gels, and has emulsifying and stabilizing activity. Agar is added to sherbets (frozen desserts of fruit juice, sugar, water or milk) and ice creams (at about 0.1%), often in combination with gum tragacanth or locust (carob) bean gum or gelatin. An amount of 0.1–1% stabilizes yoghurt, some cheeses and candy and bakery products (pastry fillings). Furthermore, agar retards bread staling and provides the desired gel texture in poultry and meat canning. Lastly, agar has a role in vegetarian diets (meat substitute products) and in desserts and pretreated instant cereal products.

4.4.4.2 Alginates

4.4.4.2.1 Occurrence, Isolation

Alginates occur in all brown algae (*Phaeophyceae*) as a skeletal component of their cell walls. The major source of industrial production is the giant kelp, *Macrocystis pyrifera*. Some species of *Laminaria, Ascophyllum* and *Sargassum* are also used. Algae are extracted with alkalies. The polysaccharide is usually precipitated from the extract by acids or calcium salts.

4.4.4.2.2 Structure, Properties

Alginate building blocks are β-D-mannuronic and α-L-guluronic acids, joined by $1 \rightarrow 4$ linkages:

$$(4.136)$$

The ratio of the two sugars (mannuronic/guluronic acids) is generally 1.5, with some deviation depending on the source. Alginates extracted from *Laminaria hyperborea* have ratios of 0.4–1.0. Partial hydrolysis of alginate yields chain fragments which consist predominantly of mannuronic or guluronic acid, and also fragments where the two uronic acid residues alternate in a 1:1 ratio. Alginates are linear copolymers consisting of the following structural units:

$$[\rightarrow 4)\text{-}\beta\text{-}D\text{-}ManpA(1 \rightarrow 4)\text{-}\beta\text{-}D\text{-}ManpA(1 \rightarrow]_n$$
$$[\rightarrow 4)\text{-}\alpha\text{-}L\text{-}GulpA(1 \rightarrow 4)\text{-}\alpha\text{-}L\text{-}GulpA(1 \rightarrow]_m$$
$$[\rightarrow 4)\text{-}\beta\text{-}D\text{-}ManpA(1 \rightarrow 4)\text{-}\alpha\text{-}L\text{-}GulpA(1 \rightarrow]_p$$

$$(4.137)$$

The molecular weights of alginates are 32–200 kdal. This corresponds to a degree of polymerization of 180–930. The carboxyl group pK-values are 3.4–4.4. Alginates are water soluble in the form of alkali, magnesium, ammonia or amine salts. The viscosity of alginate solutions is influenced by molecular weight and the counter ion of the salt. In the absence of di- and trivalent cations or in the presence of a chelating agent, the viscosity is low ("long flow" property). However, with a rise in multivalent cation levels (e. g., calcium) there is a parallel rise in viscosity ("short flow"). Thus, the viscosity can be adjusted as desired. Freezing and thawing of a Na-alginate solution containing Ca^{2+} ions can result in a further rise in viscosity. The curves in Fig. 4.17 show the effect on viscosity of the concentrations of three alginate preparations: low, moderate and high viscosity types. These data reveal that a 1% solution, depending on the type of alginate, can have a viscosity range of 20–2000 cps. The viscosity is unaffected in a pH range of 4.5–10. It rises at a pH below 4.5, reaching a maximum at pH 3–3.5.

Gels, fibers or films are formed by adding Ca^{2+} or acids to Na-alginate solutions. A slow reaction is

needed for uniform gel formation. It is achieved by a mixture of Na-alginate, calcium phosphate and glucono-δ-lactone, or by a mixture of Na-alginate and calcium sulfate.

Depending on the concentration of calcium ions, the gels are either thermoreversible (low concentration) or not (high concentration). Figure 4.18 shows a schematic section of a calcium alginate gel.

4.4.4.2.3 Derivatives

Propylene glycol alginate is a derivative of economic importance. This ester is obtained by the reaction of propylene oxide with partially neutralized alginic acid. It is soluble down to pH 2 and, in the presence of Ca^{2+} ions, forms soft, elastic, less brittle and syneresisfree gels.

4.4.4.2.4 Utilization

Alginate is a powerful thickening, stabilizing and gel-forming agent. At a level of 0.25–0.5% it improves and stabilizes the consistency of fillings for baked products (cakes, pies), salad dressings and milk chocolates, and prevents formation of larger ice crystals in ice creams during storage. Furthermore, alginates are used in a variety of gel products (cold instant puddings, fruit gels, dessert gels, onion rings, imitation caviar) and are applied to stabilize fresh fruit juice and beer foam.

4.4.4.3 Carrageenans

4.4.4.3.1 Occurrence, Isolation

Red sea weeds (*Rhodophyceae*) produce two types of galactans: agar and agar-like polysaccharides, composed of D-galactose and 3,6-anhydro-L-galactose residues, and carrageenans and related polysaccharides, composed of D-galactose and 3,6-anhydro-D-galactose which are partially sulfated as 2-, 4- and 6-sulfates and 2,6-disulfates. Galactose residues are alternatively linked by $1 \rightarrow 3$ and $1 \rightarrow 4$ linkages. Carrageenans are isolated from *Chondrus* (*Chondrus crispus*, the Irish moss), *Eucheuma*, *Gigartina*, *Gloiopeltis* and *Iridaea* species by hot

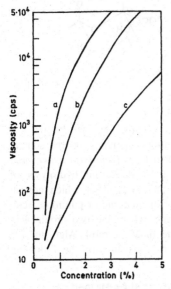

Fig. 4.17. Viscosity of aqueous alginate solutions. Alginate with (*a*) high, (*b*) medium, and (*c*) low viscosity

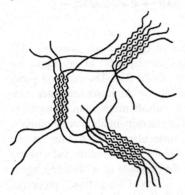

Fig. 4.18. Schematic representation of a calcium alginate gel (cross-linkage by egg box formation, cf. Formula 4.120; according to *Franz*, 1991)

water extraction under mild alkaline conditions, followed by drying or isolate precipitation.

4.4.4.3.2 Structure, Properties

Carrageenans are a complex mixture of various polysaccharides. They can be separated by fractional precipitation with potassium ions. Table 4.20 compiles data on these fractions and their monosaccharide constituents. Two major fractions are $\varkappa$ (gelling and K^+-insoluble fraction) and λ (nongelling, K^+-soluble).

Table 4.20. Building blocks of carrageenans

Carrageenan	Monosaccharide building block
$\varkappa$-Carrageenan	D-Galactose-4-sulfate, 3,6-anhydro-D-galactose-2-sulfate
$\varkappa$-Carrageenan	D-Galactose-4-sulfate, 3,6-anhydro-D-galactose
λ-Carrageenan	D-Galactose-2-sulfate, D-galactose-2,6-disulfate
μ-Carrageenan	D-Galactose-4-sulfate, D-galactose-6-sulfate, 3,6-anhydro-D-galactose
ν-Carrageenan	D-Galactose-4-sulfate, D-galactose-2,6-disulfate, 3,6-anhydro-D-galactose
Furcellaran	D-Galactose-D-galactose-2-sulfate, D-galactose-4-sulfate, D-galactose-6-sulfate, 3,6-anhydro-D-galactose

$\varkappa$-Carrageenan is composed of D-galactose, 3,6-anhydro-D-galactose and ester-bound sulfate in a molar ratio of 6:5:7. The galactose residues are essentially fully sulfated in position 4, whereas the anhydrogalactose residues can be sulfated in position 2 or substituted by α-D-galactose-6-sulfate or -2,6-disulfate. A typical sequence of $\varkappa$- (or ι-)carrgeenan is:

(4.138)

The sequence favors the formation of a doublestranded helix (Fig. 4.19). λ-Carrageenan contains as the basic building block β-D-Gal$p$-$(1 \rightarrow 4)$-α-D-Gal$p$ (cf. Formula 4.139), which is joined through a 1,3-glycosidic linkage to the polymer. Position 6 of the second galactose residue is esterified with sulfuric acid as is ca. 70% of position 2 of both residues. The high sulfate content favors the formation of a zigzag ribbon-shaped conformation.

(4.139)

The molecular weights of $\varkappa$- and λ-carrageenans are 200–800 kdal. The water solubility increases

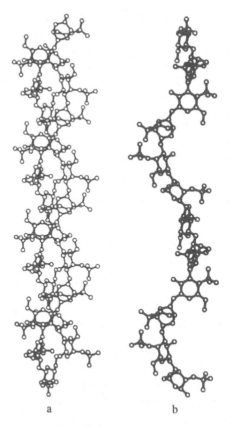

a   b

Fig. 4.19. ι-Carrageenan conformation. **a** Double helix, **b** single coil is presented to clarify the conformation (according to *Rees*, 1977)

as the carrageenan sulfate content increases and as the content of anhydrosugar residue decreases. The viscosity of the solution depends on the carrageenan type, molecular weight, temperature, ions present and carrageenan concentration.

As observed in all linear macromolecules with charges along the chain, the viscosity increases exponentially with the concentration (Fig. 4.20). Aqueous ϰ-carrageenan solutions, in the presence of ammonium, potassium, rubidium or caesium ions, form thermally reversibly gels. This does not occur with lithium and sodium ions.

This strongly suggests that gel-setting ability is highly dependent on the radius of the hydrated counter ion. The latter is about 0.23 nm for the former group of cations, while hydrated lithium (0.34 nm) and sodium ions (0.28 nm) exceed the limit. The action of cations is visualized as a zipper arrangement between aligned segments of linear polymer sulfates, with low ionic radius cations locked between alternating sulfate residues. Gel-setting ability is probably also due to a mechanism based on formation of partial double helix structures between various chains. The extent of intermolecular double helix formation, and thus the gel strength, is greater, the more uniform the chain sequences are. Each substitution of a 3,6-anhydrogalactose residue by another residue, e. g., galactose-6-sulfate, results in a kink within the helix and, thereby, a decrease in gelling strength. The helical conformation is also affected by the position of sulfate groups. The effect is more pronounced with sulfate in

the 6-position, than in 2- or 4-positions. Hence, the gel strength of ϰ-carrageenan is dependent primarily on the content of esterified sulfate groups in the 6-position.

The addition of carubin, which is itself non-gelling, to ϰ-carrageenan produces more rigid, more elastic gels that have a lower tendency towards synaeresis. Carubin apparently prevents the aggregation of ϰ-carrageenan helices.

The 6-sulfate group can be removed by heating carrageenans with alkali, yielding 3,6-anhydrogalactose residues. This elimination results in a significantly increased gel strength.

Carrageenans and other acidic polysaccharides coagulate proteins when the pH of the solution is lower than the proteins' isoelectric points. This can be utilized for separating protein mixtures.

4.4.4.3.3 Utilization

Carrageenan utilization in food processing is based on the ability of the polymer to gel, to increase solution viscosity and to stabilize emulsions and various dispersions. A level as low as 0.03% in chocolate milk prevents fat droplet separation and stabilizes the suspension of cocoa particles. Carrageenans prevent synerisis in fresh cheese and improve dough properties and enable a higher amount of milk powder incorporation in baking. The gelling property in the presence of K$^+$ salt is utilized in desserts and canned meat. Protein fiber texture is also improved. Protein sedimentation in condensed milk is prevented by carrageenans which, like κ-casein, prevent milk protein coagulation by calcium ions. Carrageenans are also used to stabilize ice cream and clarify beverages.

4.4.4.4 Furcellaran

4.4.4.4.1 Occurrence, Isolation

Furcellaran (Danish agar) is produced from red sea weed (algae *Furcellaria fastigiata*). Production began in 1943 when Europe was cut off from its agar suppliers. After alkali pretreatment of algae, the polysaccharide is isolated using hot water. The extract is then concentrated under vacuum and seeded with 1–1.5% KCl solution. The

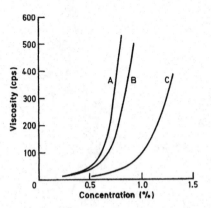

Fig. 4.20. Viscosity curves of carrageenan aqueous solutions. A: *Eucheuma spinosum*, C: *Chondrus crispus*, B: A and C in a ratio of 2:1, 40 °C, 20 rpm (according to *Whistler*, 1973)

separated gel threads are concentrated further by freezing, the excess water is removed by centrifugation or pressing and, lastly, the polysaccharide is dried. The product is a K-salt and contains, in addition, 8–15% occluded KCl.

4.4.4.4.2 Structure, Properties

Furcellaran is composed of D-galactose (46–53%), 3,6-anhydro-D-galactose (30–33%) and sulfated portions of both sugars (16–20%).
The structure of furcellaran is similar to χ-carrageenan. The essential difference is that χ-carrageenan has one sulfate ester per two sugar residues, while furcellaran has one sulfate ester residue per three to four sugar residues. Sugar sulfates identified are: D-galactose-2-sulfate, -4-sulfate and -6-sulfate, and 3,6-anhydro-D-galactose-2-sulfate. Branching of the polysaccharide chain can not be excluded. Furcellaran forms thermally reversible aqueous gels by a mechanism involving double helix formation, similar to χ-carrageenan.
The gelling ability is affected by the polysaccharide polymerization degree, amount of 3,6-anhydro-D-galactose, and by the radius of the cations present. K^+, NH_4^+, Rb^+ and Cs^+ from very stable, strong gels. Ca^{2+} has a lower effect, while Na^+ prevents gel setting. Addition of sugar affects the gel texture, which goes from a brittle to a more elastic texture.

4.4.4.4.3 Utilization

Furcellaran, with milk, provides good gels and therefore it is used as an additive in puddings. It is also suitable for cake fillings and icings. In the presence of sucrose, it gels rapidly and retains good stability, even against food grade acids. Furcellaran has the advantage over pectin in marmalades since it allows stable gel setting at a sugar concentration even below 50–60%. The required amount of polysaccharide is 0.2–0.5%, depending on the marmalade's sugar content and the desired gel strength. To keep the hydrolysis extent low, a cold aqueous 2–3% solution of furcellaran is mixed into a hot, cooked slurry of fruits and sugar.

Furcellaran is also utilized in processed meat products, such as spreadable meat pastes and pastry fillings. It facilitates protein precipitation during brewing of beer and thus improves the final clarification of the beer.

4.4.4.5 Gum Arabic

4.4.4.5.1 Occurrence, Isolation

Gum arabic is a tree exudate of various *Acacia* species, primarily *Acacia Senegal*, and is obtained as a result of tree bark injury. It is collected as air-dried droplets with diameters from 2–7 cm. The annual yield per tree averages 0.9–2.0 kg. The major producer is Sudan, with 50–60,000 t/annum, followed by several other African countries. Gum arabic has been known since ancient Egypt as "kami", an adhesive for pigmented paints.

4.4.4.5.2 Structure, Properties

Gum arabic is a mixture of closely related polysaccharides, with an average molecular weight range of 260–1160 kdal. The main structural units, with molar proportions for the gum exudate *A. senegal* given in brackets, are L-arabinose (3.5), L-rhamnose (1.1), D-galactose (2.9) and D-glucuronic acid (1.6). The proportion varies significantly depending on the Acacia species. Gum arabic has a major core chain built of β-D-galactopyranosyl residues linked by $1 \rightarrow 3$ bonds, in part carrying side chains attached at position 6 (cf. Formula 4.140).
Gum arabic occurs neutral or as a weakly acidic salt. Counter ions are Ca^{2+}, Mg^{2+} and K^+. Solubilization in 0.1 mol/l HCl and subsequent precipitation with ethanol yields the free acid. Gum arabic exhibits marked emulsifying and film-forming properties, which are caused not only by its structure, but also by the slight admixture (ca. 2%) of a protein. The serine and threonine residues of this protein are thought to be covalently bound to the carbohydrate.
The interfacial activity of gum arabic is low compared to that of proteins. The proportion of gum arabic to oil used in formulations has to be ap-

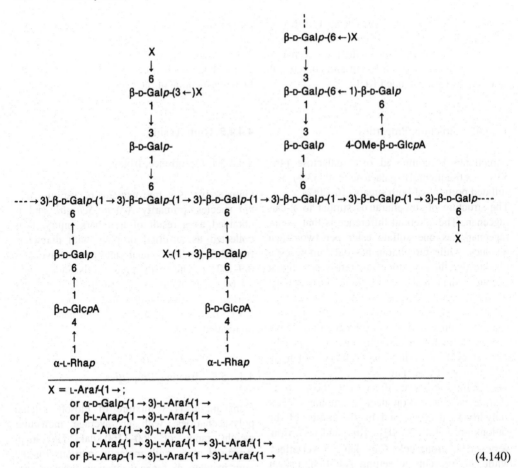

$$X = \text{L-Ara}f\text{-}(1 \rightarrow;$$
$$\text{or } \alpha\text{-D-Gal}p\text{-}(1 \rightarrow 3)\text{-L-Ara}f\text{-}(1 \rightarrow$$
$$\text{or } \beta\text{-L-Ara}p\text{-}(1 \rightarrow 3)\text{-L-Ara}f\text{-}(1 \rightarrow$$
$$\text{or }\quad \text{L-Ara}f\text{-}(1 \rightarrow 3)\text{-L-Ara}f\text{-}(1 \rightarrow$$
$$\text{or }\quad \text{L-Ara}f\text{-}(1 \rightarrow 3)\text{-L-Ara}f\text{-}(1 \rightarrow 3)\text{-L-Ara}f\text{-}(1 \rightarrow$$
$$\text{or } \beta\text{-L-Ara}p\text{-}(1 \rightarrow 3)\text{-L-Ara}f\text{-}(1 \rightarrow 3)\text{-L-Ara}f\text{-}(1 \rightarrow \qquad (4.140)$$

proximately 1:1. In contrast, a protein oil ratio of about 1:10 is used in an emulsion stabilized by milk proteins.

Gum arabic is very soluble in water and solutions of up to 50% gum can be prepared. The solu-

tion viscosity starts to rise steeply only at high concentrations (Fig. 4.21). This property is unlike that of many other polysaccharides, which provide highly viscous solutions even at low concentrations (Table 4.21).

Table 4.21. Viscosity (mPas) of polysaccharides in aqueous solution as affected by concentration (25 °C)

Concentration (%)	Gum arabic	Tragacanth	Carrageenan	Sodium alginate	Methyl cellulose	Locust bean gum	Guaran gum
1		54	57	214	39	59	3025
2		906	397	3760	512	1114	25,060
3		10,605	4411	29,400	3850	8260	111,150
4		44,275	25,356		12,750	39,660	302,500
5	7	111,000	51,425		67,575	121,000	510,000
6		183,500					
10	17						
20	41						
30	200						
40	936						
50	4163						

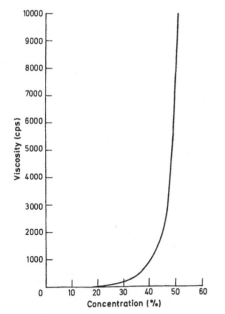

Fig. 4.21. Viscosity curve of an aqueous gum arabic solution (according to *Whistler*, 1973) (25.5 °C, Brook-field viscometer)

4.4.4.5.3 Utilization

Gum arabic is used as an emulsifier and stabilizer, e.g., in baked products. It retards sugar crystallization and fat separation in confectionery products and large ice crystal formation in ice creams, and can be used as a foam stabilizer in beverages. Gum arabic is also applied as a flavor fixative in the production of encapsulated, powdered aroma concentrates. For example, essential oils are emulsified with gum arabic solution and then spray-dried. In this process, the polysaccharide forms a film surrounding the oil droplet, which then protects the oil against oxidation and other changes.

4.4.4.6 Gum Ghatti

4.4.4.6.1 Occurrence

Gum ghatti is an exudate from the tree *Anogeissus latifolia* found in India and Ceylon.

4.4.4.6.2 Structure, Properties

The building blocks are L-arabinose, D-galactose, D-mannose, D-xylose, and D-glucuronic acid. L-Rhamnose has also been detected. The sugars are partially acetylated (5.5% acetyl groups based on dry weight). Three characteristic structural elements have been detected (cf. Formula 4.141). This acidic polysaccharide occurs as a Ca/Mg salt. Gum ghatti is soluble in water to the extent of ca. 90% and dispersible. Although it produces solutions that are more viscous than gum arabic, it is less soluble.

4.4.4.6.3 Utilization

Like gum arabic, gum ghatti can be used for the stabilization of suspensions and emulsions.

4 Carbohydrates

$$
\begin{array}{l}
\text{--- → 4)-D-GlcpA-(1 → 6)-D-Galp-(1 → ---} \qquad\qquad\qquad \text{R = -L-Ara} \\
\qquad\qquad\qquad\qquad\qquad 3 \qquad\qquad\qquad\qquad\qquad\qquad\qquad\quad \text{-L-Rha} \\
\qquad\qquad\qquad\qquad\qquad \uparrow \qquad\qquad\qquad\qquad\qquad\qquad\qquad\quad \text{-[-L-Ara]}_n\text{-L-Ara} \\
\qquad\qquad\qquad\qquad\qquad \text{R}
\end{array}
$$

$$
\begin{array}{l}
\qquad\quad \text{R} \qquad\qquad\qquad\qquad\qquad\qquad \text{R} \\
\qquad\quad \downarrow \qquad\qquad\qquad\qquad\qquad\qquad \downarrow \\
\qquad\quad 6 \qquad\qquad\qquad\qquad\qquad\qquad 6 \\
\text{---D-GlcpA-(1 → 2)-D-Manp-(1 → 2)-D-Manp-(1 → ---} \\
\qquad\qquad\qquad\qquad\qquad\qquad\qquad\qquad 3 \\
\qquad\qquad\qquad\qquad\qquad\qquad\qquad\qquad \uparrow \\
\qquad\qquad\qquad\qquad\qquad\qquad\qquad\qquad \text{R} \\
\qquad\qquad\qquad\qquad\qquad\qquad\qquad\qquad\quad\; 4 \\
\text{--- → 6)-D-Galp-(1 → 6)-D-Galp-(1 → 6)-D-Galp-(1 → 3)-L-Arap-(1 → ---} \\
\qquad\qquad\qquad\qquad 3 \qquad\qquad\qquad\qquad\qquad 3 \\
\qquad\qquad\qquad\qquad \uparrow \qquad\qquad\qquad\qquad\qquad \uparrow \\
\qquad\qquad\qquad\qquad \text{R} \qquad\qquad\qquad\qquad\qquad \text{R}
\end{array}
$$

(4.141)

4.4.4.7 Gum Tragacanth

4.4.4.7.1 Occurrence

Gum tragacanth is a plant exudate collected from *Astragalus* species shrubs grown in the Middle East (Iran, Syria, Turkey).

4.4.4.7.2 Structure, Properties

Gum tragacanth consists of a water-soluble fraction, the so-called tragacanthic acid, and the insoluble swelling component, bassorin. Tragacanthic acid contains 43% of D-galacturonic acid, 40% of D-xylose, 10% of L-fucose, and 4% of D-galactose. Like pectin, it is composed of a main polygalacturonic acid chain which bears side chains made of the remaining sugar

residues (Formula 4.142). Bassorin consists of 75% of L-arabinose, 12% of D-galactose, 3% of D-galacturonic acid methyl ester, and L-rhamnose.

Its molecular weight is about 840 kdal. The molecules are highyl elongated (450 × 1.9 nm) in aqueous solution and are responsible for the high viscosity of the solution (Table 4.21). As shown in Fig. 4.22, the viscosity is highly dependent on shear rate.

4.4.4.7.3 Utilization

Gum tragacanth is used as a thickening agent and a stabilizer in salad dressings (0.4–1.2%) and in fillings and icings in baked goods. As an additive in ice creams (0.5%), it provides a soft texture.

4.4.4.8 Karaya Gum

4.4.4.8.1 Occurrence

Karaya gum, also called Indian tragacanth, is an exudate from an Indian tree of the *Sterculia ureus* and other *Sterculia* species.

4.4.4.8.2 Structure, Properties

The building blocks are D-galactose, L-rhamnose, D-galacturonic acid, and L-glucuronic acid. The sugars are partially acetylated (13% acetyl groups based on dry weight). The molecule consists of three main chains which are polymers of different disaccharide units (cf. Formula 4.143). The main chains carry side chains and are also covalently linked via the side chains.

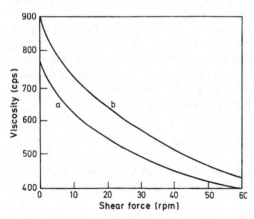

Fig. 4.22. The effect of shear rate on viscosity of aqueous tragacanth solutions. **a** Flake form tragacanth, 1%; **b** tragacanth, ribbon form, 0.5% (according to *Whistler*, 1973)

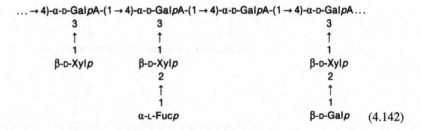

$$(4.142)$$

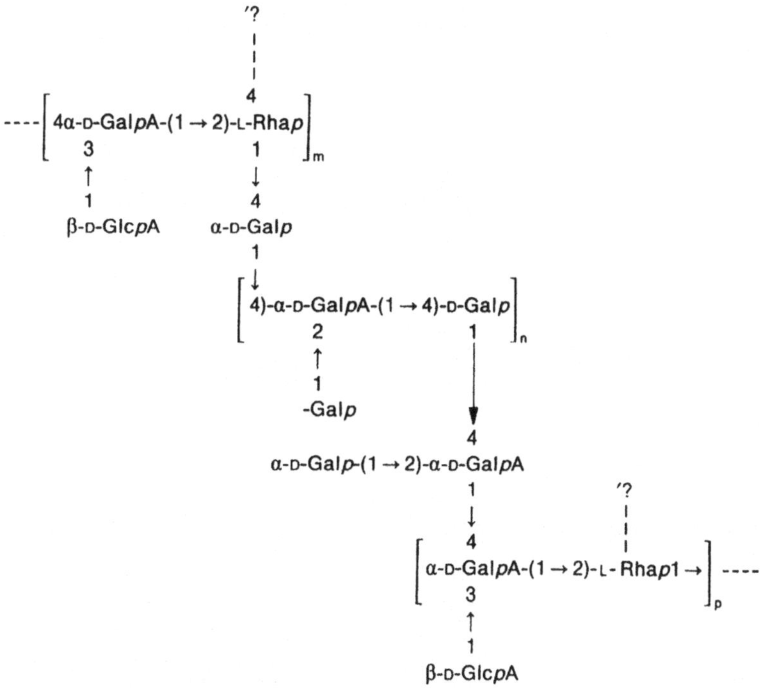

$$
\begin{array}{c}
'? \\
| \\
| \\
4 \\
----\left[4\alpha\text{-D-Gal}p\text{A-}(1\to 2)\text{-L-Rha}p\right]_m \\
\end{array}
$$

$$
\begin{array}{cc}
3 & 1 \\
\uparrow & \downarrow \\
1 & 4 \\
\beta\text{-D-Glc}p\text{A} & \alpha\text{-D-Gal}p \\
& 1
\end{array}
$$

$$
\left[\cdot\downarrow \atop 4)\text{-}\alpha\text{-D-Gal}p\text{A-}(1\to 4)\text{-D-Gal}p\right]_n
$$

$$
\begin{array}{c}
2 \quad\quad\quad 1 \\
\uparrow \\
1 \\
\text{-Gal}p
\end{array}
$$

$$
\begin{array}{c}
4 \\
\alpha\text{-D-Gal}p\text{-}(1\to 2)\text{-}\alpha\text{-D-Gal}p\text{A} \\
1 \quad\quad\quad\quad '? \\
\downarrow \quad\quad\quad\quad | \\
4 \quad\quad\quad\quad | \\
\left[\alpha\text{-D-Gal}p\text{A-}(1\to 2)\text{-L-Rha}p1\to\right]_p ---- \\
3 \\
\uparrow \\
1 \\
\beta\text{-D-Glc}p\text{A}
\end{array}
$$

(4.143)

As a result of the strong cross-linkage, this polysaccharide is insoluble in water and resistant to enzymes and microorganisms. However, it swells greatly even in cold water. Suspensions have a pasty consistency at concentrations of more than 3%.

4.4.4.8.3 Utilization

Karaya gum is used as a water binder (soft cheese), a binding agent (meat products like corned beef, sausages), a stabilizer of protein foams (beer, whipped cream, meringues), and as a thickener (soups, sauces, salad dressings, mayonnaise, ketchup). It increases the freeze-thaw stability of products, prevents synaeresis of gels, and provides "body".

4.4.4.9 Guaran Gum

4.4.4.9.1 Occurrence, Isolation

Guar flour is obtained from the seed endosperm of the leguminous plant *Cyamopsis tetragonoloba*. The seed is decoated and the germ removed. In addition to the polysaccharide guaran, guar flour

contains 10–15% moisture, 5–6% protein, 2.5% crude fiber and 0.5–0.8 ash. The plant is cultivated for forage in India, Pakistan and the United States (Texas).

4.4.4.9.2 Structure, Properties

Guaran gum consists of a chain of β-D-mannopyranosyl units joined by $1 \to 4$ linkages. Every second residue has a side chain, a D-galactopyranosyl residue that is bound to the main chain by an $\alpha(1 \to 6)$ linkage (cf. Formula 4.126).

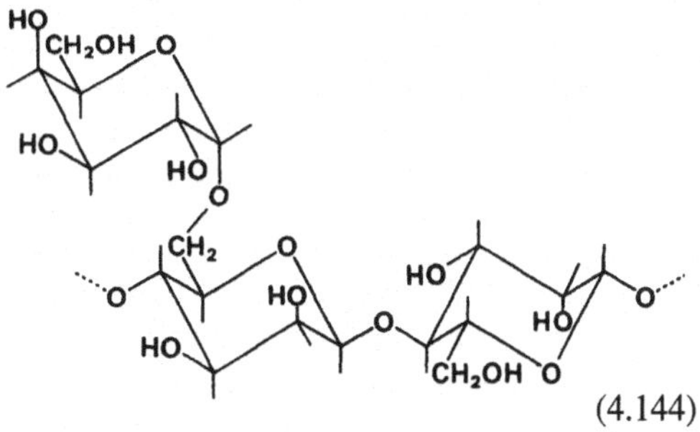

(4.144)

Guaran gum forms highly viscous solutions (Table 4.21), the viscosity of which is shear rate dependent (Fig. 4.23).

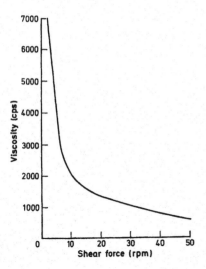

Fig. 4.23. Viscosity of 1% aqueous guar solution at 25 °C versus shear rate (rpm.). Viscometer: Haake rotovisco (according to *Whistler*, 1973)

4.4.4.9.3 Utilization

Guaran gum is used as a thickening agent and a stabilizer in salad dressings and ice creams (application level 0.3%). In addition to the food industry, it is widely used in paper, cosmetic and pharmaceutical industries.

4.4.4.10 Locust Bean Gum

4.4.4.10.1 Occurrence, Isolation

The locust bean (carob bean; St. John's bread) is from an evergreen cultivated in the Mediterranean area since ancient times. Its long, edible, fleshy seed pod is also used as fodder. The dried seeds were called "carat" by Arabs and served as a unit of weight (approx. 200 mg). Even today, the carat is used as a unit of weight for precious stones, diamonds and pearls, and as a measure of gold purity (1 carat = 1/24 part of pure gold). The locust bean seeds consist of 30–33% hull material, 23–25% germ and 42–46% endosperm. The seeds are milled and the endosperm is separated and utilized like the guar flour described above. The commercial flour contains 88% galactomannoglycan, 5% other polysaccharides, 6% protein and 1% ash.

4.4.4.10.2 Structure, Properties

The main locust bean polysaccharide is similar to that of guaran gum: a linear chain of $1 \rightarrow 4$ linked β-D-mannopyranosyl units, with α-D-galactopyranosyl residues $1 \rightarrow 6$ joined as side chains. The ratio mannose/galactose is 3 to 6; this indicates that, instead of every second mannose residue, as in guaran gum, only every 4th to 5th is substituted at the C-6 position with a galactose molecule.

The molecular weight of the galactomannan is close to 310 kdal. Physical properties correspond to those of guar gum, except the viscosity of the solution is not as high (cf. Table 4.21).

4.4.4.10.3 Utilization

Locust bean flour is used as a thickener, binder and stabilizer in meat canning, salad dressings, sausages, soft cheeses and ice creams. It also improves the water binding capacity of dough, especially when flour of low gluten content is used.

4.4.4.11 Tamarind Flour

4.4.4.11.1 Occurrence, Isolation

Tamarind is one of the most important and widely grown trees of India (*Tamarindus indica*; date of India). Its brown pods contain seeds which are rich in a polysaccharide that is readily extracted with hot water and, after drying, recovered in a powdered form.

4.4.4.11.2 Structure, Properties

The polysaccharide consists of D-galactose (1), D-xylose (2) and D-glucose (3), with respective molar ratios given in brackets. L-Arabinose is also present. The suggested structure is presented in Formula 4.145.

The polysaccharide forms a stable gel over a wide pH range. Less sugar is needed to achieve a desired gel strength than in corresponding pectin gels (Fig. 4.24). The gels exhibit only a low syneresis phenomenon.

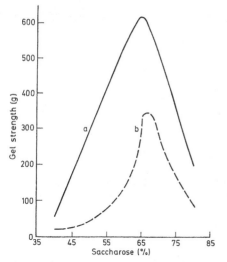

Fig. 4.24. Gel strength of (*a*) tamarind flour and (*b*) pectin from lemons versus saccharose concentration (according to *Whistler*, 1973)

4.4.4.11.3 Utilization

The tamarind seed polysaccharide is a suitable substitute for pectin in the production of marmalades and jellies. It can be used as a thickening agent and stabilizer in ice cream and mayonnaise production.

4.4.4.12 Arabinogalactan from Larch

4.4.4.12.1 Occurrence, Isolation

Coniferous larch-related woods (*Larix* species; similar to pine, but shed their needle-like leaves)

contain a water-soluble arabinogalactan of 5–35% of the dry weight of the wood. It can be isolated from chipped wood by a counter-current extraction process, using water or dilute acids. The extract is then usually drum dried.

4.4.4.12.2 Structure, Properties

The polysaccharide consists of straight chain β-D-galactopyranosyl units joined by $1 \rightarrow 3$ linkages and, in part, has side chains of galactose and arabinose residues bound to positions 4 and 6. The suggested structure is given in Formula 4.146.

In general, the polysaccharide is highly branched. The molecular weight is 50–70 kdal. The molecule is nearly spherical in shape, so its aqueous solution behaves like a *Newton*ian fluid. The viscosity is exceptionally low. At a temperature of 20 °C, the viscosity of a 10% solution is 1.74 cps, a 30% solution 7.8 cps at pH 4 or 8.15 cps at pH 11, and a 40% solution 23.5 cps. These data show that the viscosity is practically unaffected by pH. The solution acquires a thick paste consistency only at concentrations exceeding 60%.

4.4.4.12.3 Utilization

Arabinogalactan, due to its good solubility and low viscosity, is used as an emulsifier and stabilizer, and as a carrier substance in essential oils, aroma formulations, and sweeteners.

$$
\left[\begin{array}{c}
\rightarrow 4)\text{-}\beta\text{-}D\text{-}Glc\,p\,(1 \rightarrow 4)\text{-}\beta\text{-}D\text{-}Glc\,p\,(1 \rightarrow 4)\text{-}\beta\text{-}D\text{-}Glc\,p\,(1 \rightarrow 4)\text{-}\beta\text{-}D\text{-}Glc\,p\,(1 \rightarrow \\
6 \qquad\qquad\qquad 6 \qquad\qquad\qquad 6 \\
\uparrow \qquad\qquad\qquad \uparrow \qquad\qquad\qquad \uparrow \\
1 \qquad\qquad\qquad 1 \qquad\qquad\qquad 1 \\
\alpha\text{-}D\text{-}Xyl\,p \qquad\quad \alpha\text{-}D\text{-}Xyl\,p \qquad\quad L\text{-}Ara\,f \\
2 \\
\uparrow \\
1 \\
\beta\text{-}D\text{-}Gal\,p
\end{array} \right]_n \qquad (4.145)
$$

$$
\begin{array}{ccc}
\beta\text{-}L\text{-}Ara\,p\,(1 \rightarrow 3)_L\text{-}Ara\,f & L\text{-}Ara\,f\,(1 \rightarrow 3)_D\text{-}Gal\,p & D\text{-}Gal\,p\,(1 \rightarrow 6)_D\text{-}Gal\,p \\
1 & 1 & 1 \\
\downarrow & \downarrow & \downarrow \\
6 & 6 & 6 \\
\end{array}
$$

$$
\rightarrow 3)\text{-}\beta\text{-}D\text{-}Gal\,p\,(1 \rightarrow 3)\text{-}\beta\text{-}D\text{-}Gal\,p\,(1 \rightarrow 3)\text{-}\beta\text{-}D\text{-}Gal\,p\,(1 \rightarrow 3)\text{-}\beta\text{-}D\text{-}Gal\,p\,(1 \rightarrow 3)\text{-}\beta\text{-}D\text{-}Gal\,p\,(1 \rightarrow 3)\text{-}\beta\text{-}D\text{-}Gal\,p\,(1 \rightarrow
$$

$$
\begin{array}{c}
4 \\
\uparrow \\
1 \\
\beta\text{-}D\text{-}Gal\,p\,(1 \rightarrow 6)_D\text{-}Gal\,p
\end{array} \qquad (4.146)
$$

4.4.4.13 Pectin

4.4.4.13.1 Occurrence, Isolation

Pectin is widely distributed in plants. It is produced commercially from peels of citrus fruits and from apple pomace (crushed and pressed residue). It is 20–40% of the dry matter content in citrus fruit peel and 10–20% in apple pomace. Extraction is achieved at pH 1.5–3 at 60–100 °C. The process is carefully controlled to avoid hydrolysis of glycosidic and ester linkages. The extract is concentrated to a liquid pectin product or is dried by spray- or drum-drying into a powered product. Purified preparations are obtained by precipitation of pectin with ions which form insoluble pectin salts (e.g. Al^{3+}), followed by washing with acidified alcohol to remove the added ions, or by alcoholic precipitation using isopropanol and ethanol.

4.4.4.13.2 Structure, Properties

Pectin is a polysaccharide mixture with a complicated structure containing at least 65% of galacturonic acid (GalA). Three structural elements are involved in the make-up of a pectin molecule: a homogalacturonan (cf. Formula 4.147) consisting of $(1 \rightarrow 4)$ linked α-D-GalA, a galacturonan with differently arranged side chains (building blocks: apiose, fucose, arabinose, xylose), and a rhamnogalacturonan with a backbone consisting of the disaccharide units $[\rightarrow 4)$-α-D-GalA-$(1 \rightarrow 2)$-α-L-Rha-$(1 \rightarrow]$ and with its rhamnose residues linked by arabinan and galactan chains. In pectins, the GalA residues

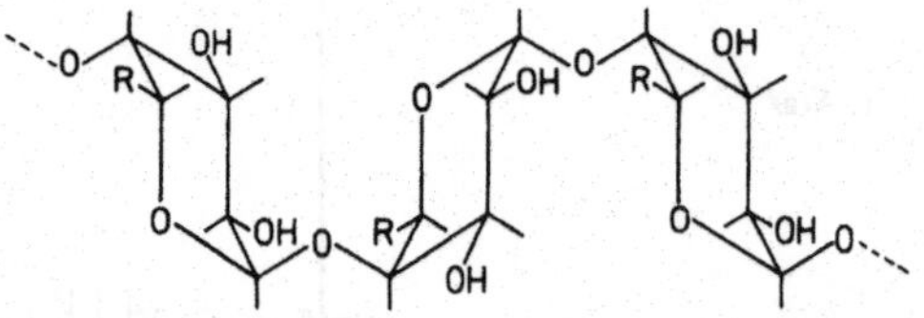

R: $COO^{\ominus}$, $COOCH_3$ $\qquad$ (4.147)

are esterified to a variable extent with methanol, while the HO-groups in 2- and 3-positions may be acetylated to a small extent. Pectin stability is highest at pH 3–4. The glycosidic linkage hydrolyzes in a stronger acidic medium. In an alkaline medium, both linkages, ester and glycosidic, are split to the same extent, the latter by an elimination reaction (cf. Formula 4.148).

The elimination reaction occurs more readily with galacturonic acid units having an esterified carboxyl group, since the H-atom on C-5 is more acidic than with residues having free carboxyl groups.

At a pH of about 3, and in the presence of Ca^{2+} ions also at higher pH's, pectin forms a thermally reversible gel. The gel-forming ability, under comparable conditions, is directly proportional to the molecular weight and inversely proportional to the esterification degree. For gel formation, low-ester pectins require very low pH values and/or calcium ions, but they gelatinize in the presence of a relatively low sugar content. High-ester pectins require an increasing amount of sugar with rising esterification degree. The gelsetting time for high ester pectins is longer than that for pectin products of low esterification degree (Table 4.22).

Apart from the degree of esterification, gel formation is also influenced by the distribution of the ester groups in the pectin molecule.

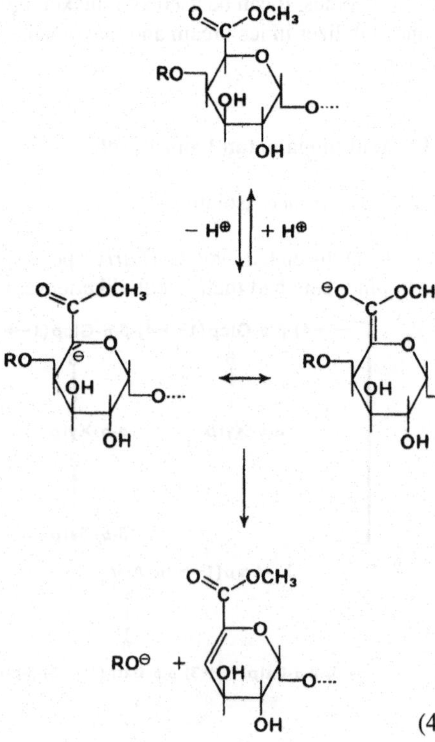

(4.148)

Table 4.22. Gelling time of pectins with differing degrees of esterification

Pectin type	Esterification degree (%)	Gelling time[a] (s)
Fast gelling	72–75	20–70
Normal	68–71	100–135
Slow gelling	62–66	180–250

[a]Difference between the time when all the prerequisites for gelling are fulfilled and the time of actual gel setting.

4.4.4.13.3 Utilization

Since pectin can set into a gel, it is widely used in marmalade and jelly production. Standard conditions to form a stable gel are, for instance: pectin content <1%, sucrose 58–75% and pH 2.8–3.5. In low-sugar products, low-ester pectin is used in the presence of Ca^{2+} ions. Pectin is also used to stabilize soured milk beverages, yoghurts and ice creams.

4.4.4.14 Starch

4.4.4.14.1 Occurrence, Isolation

Starch is widely distributed in various plant organs as a storage carbohydrate. As an ingredient of many foods, it is also the most important carbohydrate source in human nutrition. In addition, starch and its derivatives are important industrially, for example, in the paper and textile industries.

Starch is isolated mainly from the sources listed in Table 4.23. Starch obtained from corn, potatoes, cassava, and wheat in the native and modified form accounted for 99% of the world production in 1980. Some other starches are also available commercially. Recently, starches obtained from legumes (peas, lentils) have become more interesting because they have properties which appear to make them a suitable substitute for chemically modified starches in a series of products.

Starches of various origin have individual, characteristic properties which go back to the shape, size, size distribution, composition, and crystallinity of the granules. The existing connections are not yet completely understood on a molecular basis.

Table 4.23. Raw materials for starch

Raw material	Starch production 1980[a]
Raw materials of industrial importance	
Corn	77
Waxy corn	
Potato	10
Cassava	8
Wheat	4
Rice	
Waxy rice	
Other raw materials	
Sago palm	Kouzu
Sweet potato	Water chestnut
Arrowroot	Edible canna
Negro corn	Mungo bean
Lotus root	
Taro	Lentil

[a] % of the world production.

In some cases, e. g., potato tubers, starch granules occur free, deposited in cell vacuoles; hence, their isolation is relatively simple. The plant material is disintegrated, the starch granules are washed out with water, and then sedimented from the "starch milk" suspension and dried. In other cases, such as cereals, the starch is embedded in the endosperm protein matrix, hence granule isolation is a more demanding process. Thus, a counter-current process with water at 50 °C for 36–48 h is required to soften corn (steeping step of processing). The steeping water contains 0.2% SO_2 in order to loosen the protein matrix and, thereby, to accelerate the granule release and increase the starch yield. The corn grain is then disintegrated. The germ, due to its high oil content, has a low density and is readily separated by flotation. It is the source for corn oil isolation (cf. 14.3.2.2.4). The protein and starch are then separated in hydrocyclones. The separation is based on density difference (protein < starch).

The protein by-product is marketed as animal feed or used for production of a protein hydrolysate (seasoning). The recovered starch is washed and dried.

In the case of wheat flour, a dough is made first, from which a raw starch suspension is washed out. After separation of fiber particles from this suspension, the starch is fractionated by centrifugation. In addition to the relatively pure primary starch, a finer grained secondary starch is obtained which contains pentosans. The starch is then dried and further classified. The residue, gluten (cf. 15.1.5), serves, e. g., as a raw material in the production of food seasoning (cf. 12.7.3.5) and in the isolation of glutamic acid. If dried gently, it retains its baking quality and is used as a flour improver. In the case of rye, the isolation of starch is impeded by the relatively high content of swelling agents. Starch isolated from the tubers of various plants in tropical countries is available on the market under a variety of names (e. g., canna, maranta, and tacca starch). The real sago is the product obtained from the pith of the sago palm.

Starch is a mixture of two glucans, amylose and amylopectin (cf. 4.4.4.14.3 and 4.4.4.14.4).

Most starches contain 20–30% amylose (Table 4.24). New corn cultivars (amylomaize) have been developed which contain 50–80% amylose. The amylose can be isolated from starch, e. g., by crystallization of a starch dispersion, usually in the presence of salts ($MgSO_4$) or by precipitation with a polar organic compound (alcohols, such as n-butanol, or lower fatty acids, such as caprylic or capric), which forms a complex with amylose and thus enhance its precipitation.

Normal starch granules contain 70–80% amylopectin, while some corn cultivars and millet, denoted as waxy maize or waxy millet, contain almost 100% amylopectin.

4.4.4.14.2 Structure and Properties of Starch Granules

Starch granules are formed in the amyloplasts. These granules are simple or compound and consist of concentric or eccentric layers of varying density. They are of varying size (2–150 μm),

size distribution, and shape (Table 4.24). In addition to amylose and amylopectin, they usually contain small amounts of proteins and lipids. They are examined by using various physical methods, including light microscopy, small-angle light scattering, electron microscopy, X-ray diffraction, small-angle neutron scattering, and small-angle X-ray scattering. On the basis of X-ray diffraction experiments, starch granules are said to have a semicrystalline character, which indicates a high degree of orientation of the glucan molecules. About 70% of the mass of a starch granule is regarded as amorphous and ca. 30% as crystalline (Table 4.24). The amorphous regions contain the main amount of amylose, but also a considerable part of the amylopectin. The crystalline regions consist primarily of amylopectin. Although this finding was surprising at first because of the branched structure of amylopectin (cf. 4.4.4.14.4), it was deduced from the fact that amylose can be dissolved out of the granule without disturbing the crystalline character and that even amylose-free starches, like waxy corn starch, are semicrystalline. The degree of crystallinity depends on the water content. It is 24% for air-dried potato starch (19.8% of water), 29–35% for the wetted product (45–55% of water), and only 17% for starch dried via P_2O_5 and subsequently rehydrated.

On the basis of results obtained from different physical methods, the model shown in Fig. 4.25 is under discussion for the crystalline regions of the starch granule. It contains double helices of amylopectin (cf. 4.4.4.14.4), mixed amylose/amylopectin double helices, V helices of amylose with enclosed lipid (cf. 4.4.4.14.3), free amylose, and free lipid.

With the aid of X-ray diffraction diagrams, native starches can be divided into types A, B, and C. An additional form, called the V-type, occurs in swollen granules (Fig. 4.26). While types A and B are real crystalline modifications, the C-type is a mixed form. The A-type is largely present in cereal starches, and the B-type in potatoes, amylomaize, and in retrograded starches (resistant starch, cf. 4.4.4.16.3). The C-type is not only observed in mixtures of corn and potato starches, but it is also found in various legume starches.

When suspended in cold water, air-dried starch granules swell with an increase in diameter of 30–40%. If this suspension is heated, irreversible

Table 4.24. Shape, composition, and properties of different starch granules

Source	Shape[a]	Diameter (μm)	Crystallinity (%)	Gelatinization temperature (°C)	Swelling at 95°C[b]	Amylose Percentage (%)[c]	Amylose Polymerisation degree	Amylopectin Iodine binding constant[d]	Amylopectin Polymerisation degree[e]
Cereal									
Wheat	l,p	2–38	36	53–65	21	22–28	2100	0.21	19–20
Rye	l	12–40		57–70		28		0.74	26
Barley	l	2–5		56–62		22–29	1850		26
Corn	p	5–25		62–70	24	28	940	0.91	25–26
Amylomaize			20–25	67–87		52–80	1300	0.11	23
Waxy corn	p		39	63–72	64	0–1			20–22
Oats		5–15		56–62		27			20
Rice	p	3–8	38	61–78	19	14–32	1300	0.59	
Waxy rice				55–65	56	1			
Millet	p,s	4–12		}69–75[g]	}22[g]				
Sorghum	p,s	4–24		68–74	49	21–34			
Waxy sorghum									
Legumes									
Horsebean	s,o	17–31		64–67		32–34	1800	1.03	23
Smooth pea	n(si)	5–10		}57–70[h]		33–35	1300	1.66	26
Wrinkled pea	n(c)	30–40				63–75	1100	0.91	27
Roots and tubers									
Potato	e	15–100	25	58–66		23	3200	0.58	24
Cassava	semi-s,s	5–35	38[f]	52–64		17		1.06	

a l = lenticular, p = polyhedral, s = spherical, o = oval, n = kidney-shaped, el = elliptical, si = simple, c = compound.
b Weight of swollen starch, based on its dry weight; loss of soluble polysaccharides is considered.
c Based on the cum of amylose and amylopectin.
d mg iodine/100 mg starch.
e Cleavage degree of polymerization, determined by degradation of branches with pullulanase or isoamylase.
f Tapioca.
g Millet.
h Pea.

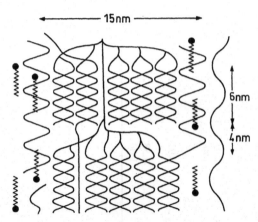

Fig. 4.25. Model of a crystalline region in a starch granule (according to *Galliard*, 1987). Amylopectin double helix XXXXX; mixed double helix of amylose and amylopectin XXX; V-helix of amylose and enclosed lipid VV; free lipid wwww; free amylose

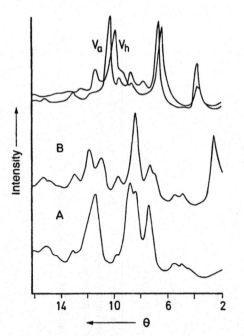

Fig. 4.26. X-ray diffraction diagrams of starches: A-type (cereals), B-type (legumes) and V-type (swollen starch, V_a: water free, V_h: hydrated) (according to *Galliard*, 1987)

changes occur starting at a certain temperature, which is characteristic of each type of starch (50–70 °C, cf. Table 4.24), called the gelatinization temperature. The starch granules absorb 20–40 g

of water/g of starch, the viscosity of the suspension rising steeply. At the same time, a part of the amylose diffuses out of the granule and goes into solution. Finally, the granule bursts. In the first step of gelatinization, the starch crystallites melt and form a polymer network. This network breaks up at higher temperatures (ca. 100 °C), resulting in a solution of amylose and amylopectin. In gelatinization, water first diffuses into the granule, crystalline regions then melt with the help of hydration, and, finally, swelling gives rise to a solution through further diffusion of water. In this process, hydrogen bridges between glucose chains in the crystallites are primarily disrupted, and perhaps some of those in the amorphous regions as well. It is probable that the swelling of the amorphous regions facilitates the dissolving out of amylose from the crystallites, which are thereby destabilized. As with heating in water, the same effect occurs when starch is suspended in other solvents, e. g., liquid ammonia or dimethyl sulfoxide, or mechanically damaged, e. g., by dry grinding.

The course of gelatinization depends not only on the botanical origin of the starch and the temperature used, but also on the water content of the suspension (Fig. 4.27). Thus, dried starch with 1–3% of water undergoes only slight changes up to a temperature of 180 °C, whereas starch with 60% of water completely gelatinizes at temperatures as low as 70 °C.

If an aqueous starch suspension is maintained for some time at temperatures below the gelatinization temperature, a process known as tempering, the gelatinization temperature is increased, apparently due to the reorganization of the structure of the granule. Treatment of starch at low water contents and higher temperatures results in the stabilization of the crystallites and, consequently, a decrease in the swelling capacity. Figure 4.28 shows the resulting change in the X-ray diffraction spectrum from type B to type A, using potato starch as an example.

The changes in the physical properties caused by treating processes of this type can, however, vary considerably, depending on the botanical origin of the starches. This is shown in Table 4.25 for potato and wheat starch. On wet heating, the swelling capacity of both starches decreases, although to different extents. On the other hand, there is a decrease in solubility only of

Table 4.25. Physicochemical properties of starches before (1) and after (2) heat treatment in the wet state ($T = 100\,°C$, $t = 16\,h$, $H_2O = 27\%$)

Property	Wheat starch		Potato starch	
	1	2	1	2
Start of gelatinization (°C)	56.5	61	60	60.5
End of gelatinization (°C)	62	74	68	79
Swelling capacity at 80 °C (ratio)	7.15	5.94	62.30	19.05
Solubility at 80 °C (%)	2.59	5.93	31.00	10.10
Water binding capacity (%)	89.1	182.6	102.00	108.7
Enzymatic vulnerability (% dissolved)	0.44	48.55	0.57	40.35

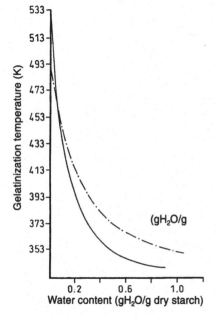

Fig. 4.27. Gelatinization temperature of differently hydrated starches (— potato starch, — · — wheat starch, determined by differential thermal analysis, differential calorimetry, and double refraction) (according to *Galliard*, 1987)

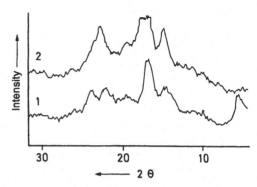

Fig. 4.28. X-ray diffraction diagrams of potato starch before (1) and after (2) thermal treatment (102 °C/16 h) at a water content of 40%. The pattern of native starch (18.3% of water) corresponds to the B-type and that of treated starch (24.2% of water) to the A-type (according to *Galliard*, 1987)

potato starch, while that of wheat starch clearly increases. The explanation put forward to explain these findings is that the amorphous amylose of potato starch is converted to an ordered, less soluble state, while the amylose of cereal starch, present partially in the form of helices with enclosed lipids, changes into a more easily leachable state.

A gelatinization curve for potato starch is presented in Fig. 4.29. The number of gelatinized

starch granules was determined by microscopy. Another way to monitor gelatinization as a function of temperature is to measure the viscosity of a starch suspension. The viscosity curves in Fig. 4.30 show that, as mentioned above, the viscosity initially increases due to starch granule swelling. The subsequent disintegration of the swollen granule is accompanied by a drop in viscosity. The shape of the curve varies greatly for different starches.

Potato starch shows a very high maximum (∼4000 *Brabender* units), followed by a steep drop. Waxy corn starch exhibits similar behavior, except that the maximum is not as high. In normal corn starch, the maximum is still lower, but the following drop is slight, i. e., the granules are more stable. Under these conditions, amylomaize starch does not swell, even though ca. 35% of

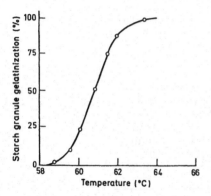

Fig. 4.29. Potato starch gelatinization curve (according to *Banks* and *Muir*, 1980)

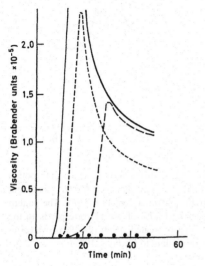

Fig. 4.30. Gelatinization properties of various starches (according to *Banks* and *Muir*, 1980). Brabender visco-amylograph. 40 g starch/460 ml water, temperature programming: start at 50 °C, heated to 95 °C at a rate of 1.5° C/min. Held at 95 °C for 30 min — potato, - - - waxy corn, – – – normal corn, and ••• amylomaize starch

the amylose goes into solution. The viscosity of a starch paste generally increases on rapid cooling with mixing, while a starch gel is formed on rapid cooling without mixing.

Amylose gels tend to retrograde. This term denotes the largely irreversible transition from the solubilized or highly swollen state to an insoluble, shrunken, microcrystalline state (Fig. 4.31). This state can also be directly achieved by slowly cooling a starch paste. The tendency towards retrogradation is enhanced at low temperatures, es-

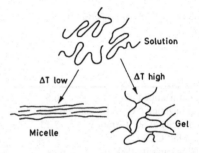

Fig. 4.31. Behavior of amylose molecules during cooling of a concentrated aqueous solution

pecially near 0 °C, neutral pH values, high concentration, and by the absence of surface active agents. It also depends on the molecular weight and on the type of starch, e. g., it increases in the series potato < corn < wheat. The transitions described from very water-deficient starting states via very highly swollen states or solutions to more or less shrunken states are linked to changes in the interactions between the glucans and to conformational changes. At present, these changes cannot be fully described because they greatly depend on the conditions in each case, e. g., even on the presence of low molecular compounds.

It is known that the gelatinization temperature is increased by polyhydroxy compounds (glycerol, sugar) and decreased by salts (NaCl, $CaCl_2$), as presented in Fig. 4.32 (top) as a function of water activity, which is lowered by the dissolved substances (a_w, cf. 0.3.1). Apart from the activity of the solvent water, if its volume fraction (v_1), which changes in reverse order to the volume fraction of the solute, is considered and if the gelatinization temperature is plotted against ln a_w/V_1, instead of a_w, the effect of the different dissolved substances is unified (Fig. 4.32, bottom). The reason is that polyhydroxy compounds cause a large change in v_e and a small change in a_w, while a small change in v_e is linked to a large change in a_w in the case of the salts.

Lipids also influence the properties of starch. Like free amino acids, monoglycerides or fatty acid esters of hydroxy acids, lipids form inclusion compounds with amylose (cf. 4.4.4.14.3). Like di- and triglycerides, they also reduce the swelling capacity and solubility by inhibiting water diffusion. Therefore, both degreasing as well as lipid addition are of importance as physical modification methods of starches.

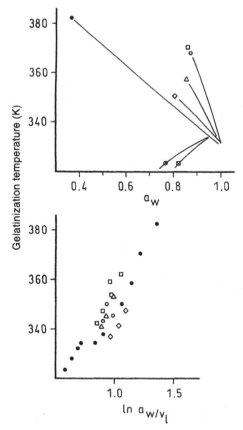

Fig. 4.32. Gelatinization temperature of potato starch as a function of water activity a_w (*top*) and of the natural logarithm of the quotient of activity a_w to volume fraction v_l of water (*bottom*); ● glycerol, ○ maltose, □ saccharose, △ glucose, ◇ ribose, ⊗ NaCl, ⊠ CaCl$_2$ (according to *Galliard*, 1987)

4.4.4.14.3 Structure and Properties of Amylose

Amylose is a chain polymer of α-D-gluco-pyranosyl residues linked $1 \rightarrow 4$:

(4.149)

Enzymatic hydrolysis of the chain is achieved by α-amylase, β-amylase and glucoamylase. Often, β-amylase does not degrade the molecule completely into maltose, since a very low branching is found along the chain with $\alpha(1 \rightarrow 6)$ linkages. The molecular size of amylose is variable. The polymerization degree in wheat starch lies between 500 and 6000, while in potatoes it can rise up to 4500. This corresponds to a molecular weight of 150–750 kdal. X-ray diffraction experiments conducted on oriented amylose fibers make possible the assignment of the types of starch mentioned above to definite molecular structural elements. Oriented fibers of the A-type were obtained by cutting and stretching thin films of acetylamylose at 150 °C, deacetylation in alcoholic alkali, and conditioning at 80% relative air humidity and 85 °C. Type B fibers were obtained in a corresponding manner by conditioning the deacetylated material at room temperature for three days at 80% and for another three days at 100% relative air humidity, followed by aftertreatment in water at 90 °C for 1 h. The diffraction patterns obtained with these oriented fibers corresponded to those of types A and B given by native starch powders, allowing the development of structural models.

The structural elements of type B are left-hand double helices (Fig. 4.34a), which are packed in a parallel arrangement (Fig. 4.33). One turn of the double helix is 2.1 nm long, which corresponds to 6 glucose residues, i.e., three residues from each glucan chain. Hydrogen bridges between the amylose molecules stabilize the double helix. The central channel surrounded by six double helices is filled with water (36 H$_2$O/unit cell). The A-type is very similar to the B-type, except that the central channel is occupied by another double helix, making the packing more close. In this type, only eight molecules of water per unit cell are inserted between the double helices. The transition from type B to type A achieved by wet heating has been described already (4.4.4.14.2, Fig. 4.28). It is difficult to bring the postulated antiparallel arrangement of the double helices into line with the requirements of biosynthesis, where a parallel arrangement can be expected. It is possible that the present experimental data do not exclude such an arrangement.

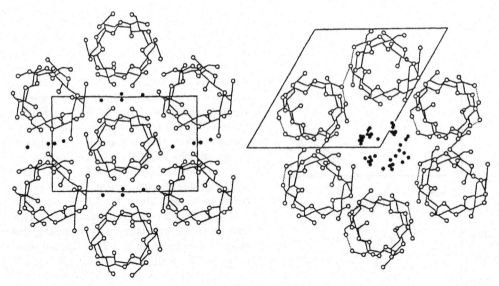

Fig. 4.33. Unit cells and arrangement of double helices (*cross section*) in A-amylose (*left*) and B-amylose (*right*) (according to *Galliard*, 1987)

The double helix mentioned above and shown in Fig. 4.34 can, depending on conditions, change into other helical conformations.

In the presence of KOH, for instance, a more extended helix results with 6 glucose residues per helical turn (Fig. 4.34, b) while, in the presence of KBr, the helix is even more stretched to 4 residues per turn (Fig. 4.34, c). Inclusion (clathrate) compounds are formed in the presence of small molecules and stabilize the V starch conformation (Fig. 4.34, d); it also has 6 glucose residues per helical turn. Stabilization may be achieved by H-bridges between O-2 and O-3 of neighboring residues within the same chain and between O-2 and O-6 of the residues i and $i+6$ neighbored on the helix surface. Many molecules, such as iodine, fatty acids, fatty acid esters of hydroxycarboxylic acids (e. g., stearyllactate), monoglycerides, phenols, arylhalogenides, n-butanol, t-butanol, and cyclohexane are capable of forming clathrate compounds with amylose molecules. The helix diameter, to a certain extent, conforms to the size of the enclosed guest molecule; it varies from 13.7 Å to 16.2 Å. While the iodine complex and that of n-butanol have 6 glucose residues per turn in a V conformation, in a complex with t-butanol the helix turn is enlarged to 7 glucose residues/turn (Fig. 4.34, e).

It is shown by an α-naphthol clathrate that up to 8 residues are allowed (Fig. 4.34, f). Since the helix is internally hydrophobic, the enclosed "guest" has also to be lipophilic in nature. The enclosed molecule contributes significantly to the stability of a given conformation. For example, it is observed that the V conformation, after "guest" compound removal, slowly changes in a humid atmosphere to a more extended B conformation. Such a conformational transition also occurs during staling of bread or other bakery products. Freshly baked bread shows a V spectrum of gelatinized starch, but aged bread typically has the retrograded starch B spectrum. Figure 4.35 illustrates both conformations in the form of cylinder projections. While in V amylose, as already outlined, O-2 of residues i and O-6 of residues $i+6$ come into close contact through H-bridges, in the B pattern the inserted water molecules increase the double-strand distance along the axis of progression (h) from 0.8 nm for the V helix to 1 nm for the B helix.

Cereal starches are stabilized by the enclosed lipid molecules, so their swelling power is low. The swelling is improved in the presence of alcohols (ethanol, amyl alcohol, tert-amyl alcohol). Obviously, these alcohols are dislodging and removing the "guest" lipids from the helices.

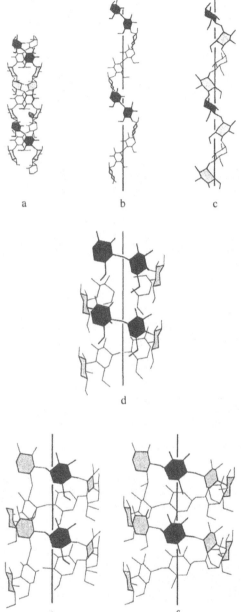

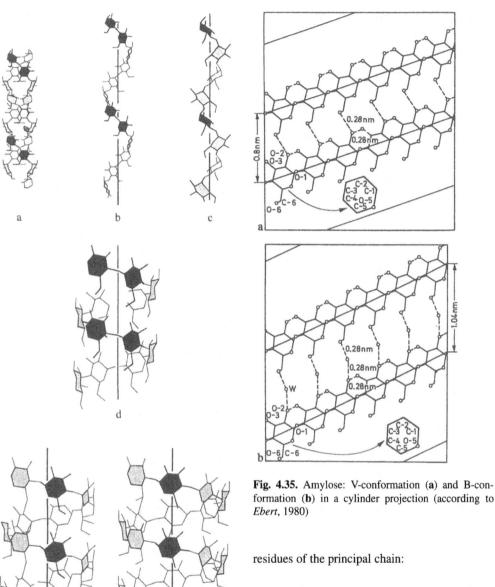

a

b

c

d

e

f

Fig. 4.34. Amylose conformation (for explanation see text) (according to *Rees*, 1977)

Fig. 4.35. Amylose: V-conformation (**a**) and B-conformation (**b**) in a cylinder projection (according to *Ebert*, 1980)

residues of the principal chain:

$$\text{(4.150)}$$

4.4.4.14.4 Structure and Properties of Amylopectin

Amylopectin is a branched glucan with side chains attached in the 6-position of the glucose

An average of 20–60 glucose residues are present in short chain branches and each of these branch chains is joined by linkage of C-1 to C-6 of the next chain. The proposed structural models (Fig. 4.36) suggest that amylopectin also has double helices organized in parallel. As mentioned above, the main portion of a starch granule's crystalline structure is apparently derived from amylopectin. The structural modell II in Fig. 4.36 clearly shows from left to right the sequence of more compact (crystalline) and less compact (amorphous) sections. In this model, a distinction is made between shorter A-chains that are free of side chains and longer B chains that bear side chains. In the B chains, sections with compact successive side chains (cluster) alternate with branch-free sections.

The degree of polymerization of amylopectin (wheat) lies in the range of $3 \times 10^5 - 3 \times 10^6$ glucose units, which corresponds to a molecular mass of $5 \times 10^7 - 5 \times 10^8$. One phosphoric acid residue is found for an average of 400 glucose residues.

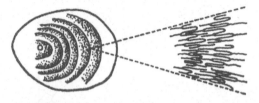

Fig. 4.37. Arrangement of amylopectin molecules in a starch granule

The organization of amylopectin molecules in starch granules is shown in Fig. 4.37: it is radial, the reducing end being directed outwards.

Enzymatic degradation of amylopectin is similar to that of amylose. The enzyme β-amylase degrades the molecule up to the branching points. The remaining resistant core is designated as "limit-dextrin".

Amylopectin, when heated in water, forms a transparent, highly viscous solution, which is ropy, sticky and coherent. Unlike with amylose, there is no tendency toward retrogradation. There are no staling or aging phenomena and no gelling, except at very high concentrations. However, there is a rapid viscosity drop in acidic media and on autoclaving or applying stronger mechanical shear force.

4.4.4.14.5 Utilization

Starch is an important thickening and binding agent and is used extensively in the production of puddings, soups, sauces, salad dressings, diet food preparations for infants, pastry filling, mayonnaise, etc. Corn starch is the main food starch and an important raw material for the isolation of starch syrup and glucose (cf. 19.1.4.3).

A layer of amylose can be used as a protecting cover for fruits (dates or figs) and dehydrated and candied fruits, preventing their sticking together. Amylose treatment of French fries decreases their susceptibility to oxidation. The good gelling property of a dispersable amylose makes it a suitable ingredient in instant puddings or sauces. Amylose films can be used for food packaging, as edible wrapping or tubing, as exemplified by a variety of instant coffee or tea products. Amylopectin utilization is also diversified. It is used to a large extent as a thickener or stabilizer

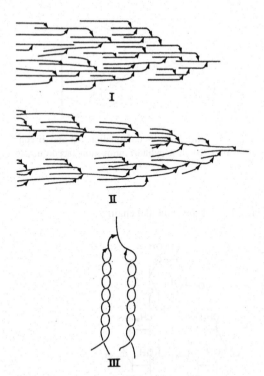

Fig. 4.36. Structural models (I, II) for amylopectin with parallel double helices. III is an enlarged segment of I or II (according to *Banks* and *Muir*, 1980)

Table 4.26. Utilization of amylopectin and its derivatives

Starch	Utilization
Unmodified waxy starch (also in blend with normal starch and flours)	Salad dressing, sterilized canned and frozen food, soups, broth, puffed cereals, and snack food
Pregelatinized waxy starch or isolated amylopectin	Baked products, paste (pâté) fillings, sterilized bread, salad dressing, pudding mixtures
Thin boiling waxy starch	Protective food coatings
Cross-linked waxy starch	Paste fillings, white and brown sauces, broth, sterilized or frozen canned fruit, puddings, salad dressing, soups, spreadable cream products for sandwiches, infant food
Waxy starch, hydroxypropyl ether	Sterilized and frozen canned food
Waxy starch, carboxymethyl ether	Emulsion stabilizer
Waxy starch acetic acid ester	Sterilized and frozen canned food, infant food
Waxy starch succinic- and adipic acid esters	Sterilized and frozen canned food, aroma encapsulation
Waxy starch sulfuric acid ester	Thickenig agent, emulsion stabilizer, ulcer treatment (pepsin inhibitor)

and as an adhesive or binding agent. Table 4.26 lists the range of its applications.

4.4.4.14.6 Resistant Starch

Starch and its degradation products which are not absorbed in the small intestine are called resistant starch (RS). RS can, however, be metabolized by the bacteria of the colon. Acetic acid, propionic acid and butyric acid are formed, stimulating the growth of the cells of the intestinal epithelium. Especially butyric acid has been found to positively affect health. A distinction is made between 4 forms of RS: Type I, starch enclosed in cells (e. g., coarse-ground grain or legumes), Type II, native starch granules (e. g., in bananas, potatoes), Type III, starch fractions produced on retrogradation (e. g., in boiled potatoes, bread crumb), and

Type IV, starch modified by the *Maillard* reaction or caramelization (formation of glycosidic bonds which are not hydrolyzed by α-amylase).

Only amylose, and not amylopectin, is involved in Type III RS. The formation of RS depends on the temperature and on the water and lipid content. Indeed, 20% of RS is obtained on autoclaving corn starch. The yield can be raised to about 40% by heating under pressure and cooling (ca. 20 cycles). The optimal amylose/water ratio is 1:3.5 (g/g). Lipids complexed by amylose inhibit RS formation (cf. 15.2.4.1).

Type III RS consists of 60–70% of double helical $\alpha(1–4)$polyglucan aggregates and only 25–30% of crystalline structures. It is assumed that the high content of the double helical conformation, which is similar to that of amylose type B, limits the activity of α-amylases.

Various methods have been proposed for the determination of RS, e. g., RS equals total starch minus digestible starch. The results are only comparable if the incubation conditions and the α-amylases used correspond.

4.4.4.15 Modified Starches

Starch properties and those of amylose and amylopectin can be improved or "tailored" by physical and chemical methods to fit or adjust the properties to a particular application or food product.

4.4.4.15.1 Mechanically Damaged Starches

When starch granules are damaged by grinding or by application of pressure at various water contents, the amorphous portion is increased, resulting in improved dispersibility and swellability in cold water, a decrease in the gelatinization temperature by 5–10 °C, and an increase in enzymatic vulnerability. In bread dough made from flour containing damaged starch, for instance, the uptake of water is faster and higher and amylose degradation greater.

4.4.4.15.2 Extruded Starches

The X-ray diffraction diagram changes on extrusion of starch. The V-type appears first, followed by its conversion to an E-type at higher temperatures ($>185\,°C$), and reformation of the V-type on cooling. The E-type apparently differs from the V-type only in the spacing of the V helices of amylose.

Extruded starches are easily dispersible, better soluble, and have a lower viscosity. The partial degradation of appropriately heated amylose shows that chemical changes also occur at temperatures of $185–200\,°C$. Apart from maltose, isomaltose, gentiobiose, sophorose, and 1,6-anhydroglucopyranose appeared.

4.4.4.15.3 Dextrins

Heating of starch ($<15\%$ of water) to $100–200\,°C$ with small amounts of acidic or basic catalysts causes more or less extensive degradation. White and yellow powders are obtained which deliver clear or turbid, highly sticky solutions of varying viscosity. These products are used as adhesives in sweets and as fat substitutes.

4.4.4.15.4 Pregelatinized Starch

Heating of starch suspensions, followed by drying, provides products that are swellable in cold water and form pastes or gels on heating. These products are used in instant foods, e. g., pudding, and as baking aids (cf. Table 4.26).

4.4.4.15.5 Thin-Boiling Starch

Partial acidic hydrolysis yields a starch product which is not very soluble in cold water but is readily soluble in boiling water. The solution has a lower viscosity than the untreated starch, and remains fluid after cooling. Retrogradation is slow. These starches are utilized as thickeners and as protective films (cf. Table 4.26).

4.4.4.15.6 Starch Ethers

When a 30–40% starch suspension is reacted with ethylene oxide or propylene oxide in the presence of hydroxides of alkali and/or alkali earth metals (pH 11–13), hydroxyethyl- or hydroxypropyl-derivates are obtained ($R' = H, CH_3$):

$$R{-}OH + \underset{O}{\diagdown\!\diagup}^{R'} \xrightarrow{OH^{\ominus}} R{-}O{-}CH_2{-}\underset{OH}{CHR'}$$

(4.151)

The derivatives are also obtained in reaction with the corresponding epichlorohydrins. The substitution degree can be controlled over a wide range by adjusting process parameters. Low substitution products contain up to 0.1 mole alkyl group/mole glucose, while those with high substitution degree have 0.8–1 mole/mole glucose. Introduction of hydroxyalkyl groups, often in combination with a small extent of cross-linking (see below) greatly improves starch swelling power and solubility, lowers the gelatinization temperature and substantially increases the freeze–thaw stability and the paste clarity of highly-viscous solutions. Therefore, these products are utilized as thickeners for refrigerated foods (apple and cherry pie fillings, etc), and heat-sterilized canned food (cf. Table 4.26).

Reaction of starch with monochloroacetic acid in an alkaline solution yields carboxymethyl starch:

$$R{-}OH + ClCH_2COO^{\ominus} \xrightarrow{OH^{\ominus}} R{-}O{-}CH_2{-}COO^{\ominus}$$

(4.152)

These products swell instantly, even in cold water and in ethanol. Dispersions of 1–3% carboxymethyl starch have an ointment-like (pomade) consistency, whereas 3–4% dispersions provide a gel-like consistency. These products are of interest as thickeners and gelforming agents.

4.4.4.15.7 Starch Esters

Starch monophosphate ester is produced by dry heating of starch with alkaline orthophosphate or alkaline tripolyphosphate at $120–175\,°C$:

$$R{-}OH \xrightarrow[POCl_3/\text{Alkali phosphate}]{OH^{\ominus}} R{-}OPO_3H^{\ominus}$$

(4.153)

Starch organic acid esters, such as those of acetic acid, longer chain fatty acids ($C_6–C_{26}$), succinic, adipic or citric acids, are obtained in reactions with the reactive derivatives (e. g., vinyl acetate) or by heating the starch with free acids or their salts. The thickening and paste clarity properties of the esterified starch are better than in the corresponding native starch.

In addition, esterified starch has an improved freeze–thaw stability. These starches are utilized as thickeners and stabilizers in bakery products, soup powders, sauces, puddings, refrigerated food, heat-sterilized canned food and in margarines. The starch esters are also suitable as protective coatings, e. g., for dehydrated fruits or for aroma trapping or encapsulation (cf. Table 4.26).

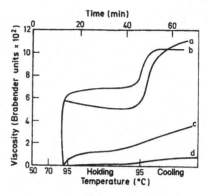

Fig. 4.38. Corn starch viscosity curves as affected by crosslinking degree. Instruments: Brabender amylograph; *a* control, *b* crosslinked with 0.05%, *c* 0.10%, *d* 0.15% epichlorohydrin (according to *Pigman*, 1970)

4.4.4.15.8 Cross-Linked Starches

Cross-linked starches are obtained by the reaction of starch (R—OH) with bi- or polyfunctional reagents, such as sodium trimetaphosphate, phosphorus oxychloride, epichlorohydrin or mixed anhydrides of acetic and dicarboxylic acids (e. g., adipic acid):

$$2\,R{-}OH + R'CO{-}O{-}CO{-}(CH_2)_n{-}CO{-}O{-}COR'$$
$$\rightarrow \quad R{-}O{-}CO{-}(CH_2)_n{-}CO{-}O{-}R \qquad (4.154)$$

$$\xrightarrow{OH^\ominus} R{-}O{-}CH_2{-}CH{-}CH_2Cl$$
$$\qquad\qquad\qquad\underset{OH}{|}$$

$$\xrightarrow{OH^\ominus} R{-}O{-}CH_2$$

$$\xrightarrow[OH^\ominus]{ROH} R{-}O{-}CH_2{-}CH{-}CH_2{-}O{-}R$$
$$\qquad\qquad\qquad\qquad\underset{OH}{|} \qquad\qquad (4.155)$$

$$2\,R{-}OH \xrightarrow{(NaPO_3)_3} R{-}O{-}\overset{O}{\underset{O^\ominus}{\overset{\|}{P}}}{-}O{-}R$$
$$\qquad\qquad\qquad\qquad\qquad (4.156)$$

The starch granule gelatinization temperature increases in proportion to the extent of cross-linking, while the swelling power decreases (Fig. 4.38). Starch stability remains high at extreme pH values (as in the presence of food acids) and under conditions of shear force. Cross-linked starch derivatives are generally used when high starch stability is demanded.

4.4.4.15.9 Oxidized Starches

Starch hydrolysis and oxidation occur when aqueous starch suspensions are treated with sodium hypochlorite at a temperature below the starch gelatinization temperature range. The products obtained have an average of one carboxyl group per 25–50 glucose residues:

$$(4.157)$$

Oxidized starch is used as a lower-viscosity filler for salad dressings and mayonnaise. Unlike thin-boiling starch, oxidized starch does not retrograde nor does it set to an opaque gel.

4.4.4.16 Cellulose

4.4.4.16.1 Occurrence, Isolation

Cellulose is the main constituent of plant cell walls, where it usually occurs together with hemicelluloses, pectin and lignin. Since cellulase enzymes are absent in the human digestive tract, cellulose, together with some other inert polysaccharides, constitute the indigestible carbohydrate of plant food (vegetables, fruits or cereals), referred to as dietary fiber. Cellulases are also absent in the digestive tract of animals, but herbivorous animals can utilize cellulose because of the rumen microflora (which hydrolyze the cellulose). The importance of dietary fiber in human nutrition appears mostly to be the maintenance of intestinal motility (peristalsis).

4.4.4.16.2 Structure, Properties

Cellulose consists of β-glucopyranosyl residues joined by 1 → 4 linkages (cf. Formula 4.158). Cellulose crystallizes as monoclinic, rod-like crystals. The chains are oriented parallel to the fiber direction and form the long b-axis of the unit cell (Fig. 4.39). The chains are probably somewhat pleated to allow intrachain hydrogen bridge formation between O-4 and O-6, and between O-3 and O-5 (cf. Formula 4.159). Intermolecular hydrogen bridges (stabilizing the parallel chains) are present in the direction of the a-axis while hydrophobic interactions exist in the c-axis direction. The crystalline sections comprise an average of 60% of native cellulose. These sections are interrupted by amorphous gel regions, which can become crystalline when moisture is removed. The acid- or alkali-labile bonds also apparently occur in these regions. Microcrystalline cellulose is formed when these bonds are hydrolyzed. This partially depolymerized cellulose product with a molecular weight of 30–50 kdal, is still water insoluble, but does not have a fibrose structure.

Cellulose has a variable degree of polymerization (denoted as DP; number of glucose residues per chain) depending on its origin. The DP can range from 1000 to 14,000 (with corresponding molecular weights of 162 to 2268 kdal). Because of its high molecular weight and crystalline structure, cellulose is insoluble in water. Also, its swelling power or ability to absorb water, which depends partly on the cellulose source, is poor or negligible.

4.4.4.16.3 Utilization

Microcrystalline cellulose is used in low-calorie food products and in salad dressings, desserts and ice creams. Its hydration capacity and dispersibility are substantially enhanced by adding it in combination with small amounts of carboxymethyl cellulose.

4.4.4.17 Cellulose Derivatives

Cellulose can be alkylated into a number of derivatives with good swelling properties and improved solubility. Such derivatives have a wide field of application.

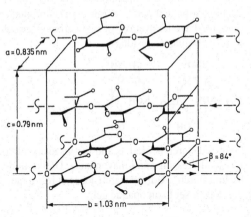

Fig. 4.39. Unit cell of cellulose (according to *Meyer* and *Misch*)

(4.158)

(4.159)

4.4.4.17.1 Alkyl Cellulose, Hydroxyalkyl Cellulose

The reaction of cellulose with methylchloride or propylene oxide in the presence of a strong alkali introduces methyl or hydroxypropyl groups into cellulose (cf. Reaction 4.160). The degree of substitution (DS) is dependent on reaction conditions.

Mixed substituted products are also produced, e. g., methylhydroxypropyl cellulose or methylethyl cellulose. The substituents interfere with the normal crystalline packing of the cellulose chains, thus facilitating chain solvation. Depending on the nature of the substituent (methyl, ethyl, hydroxymethyl, hydroxyethyl or hydroxypropyl) and the substitution degree, products are obtained with variable swelling powers and water solubilities. A characteristic property for methyl cellulose and double-derivatized methylhydroxypropyl cellulose is their initial viscosity drop with rising temperature, setting to a gel at a specific temperature. Gel setting is reversible. Gelling temperature is dependent on substitution type and degree. Figure 4.40 shows the dependence of gelling temperature on the type of substitution and the concentration of the derivatives in water. Hydroxyalkyl substituents stabilize the hydration layer around the macromolecule and, thereby, increase the gelling temperature. Changing the proportion of methyl to hydroxypropyl substituents can vary the jelling temperature within a wide range.

The above properties of cellulose derivatives permit their diversified application (Table 4.27). In baked products obtained from gluten-poor or gluten-free flours, such as those of rice, corn or rye, the presence of methyl and methylhydroxypropyl celluloses decreases the crumbliness and friability of the product, enables a larger volume of water to be worked into the dough and, thus, improves the extent of starch swelling during oven baking. Since differently substituted celluloses offer a large choice of gelling temperatures, each application can be met by

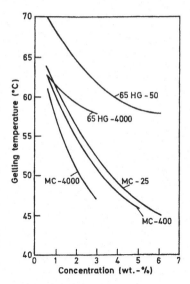

Fig. 4.40. Gelling behavior of alkyl celluloses (according to *Balser*, 1975). MC: methyl cellulose, HG: hydroxypropylmethyl cellulose with a hydroxypropyl content of about 6.5%. The numerical suffix is the viscosity (cps) of a 2% solution

using the most suitable derivative. Their addition to batter or a coating mix for meats (panure) decreases oil uptake in frying. Their addition to dehydrated fruits and vegetables improves rehydration characteristics and texture upon reconstitution. Sensitive foods can be preserved by applying alkyl cellulose as a protective coating or film. Cellulose derivatives can also be used as thickening agents in low calorie diet foods. Hydroxypropyl cellulose is a powerful emulsion stabilizer, while methylethyl cellulose has the property of a whipping cream: it can be whipped into a stable foam consistency.

4.4.4.17.2 Carboxymethyl Cellulose

Carboxymethyl cellulose is obtained by treating alkaline cellulose with chloroacetic acid.

$$(4.160)$$

Table 4.27. Utilization of cellulose derivatives (in amounts of 0.01 to 0.8%)

Food product	Cellulose derivative[a]			Effect								
	1	2	3	A[b]	B	C	D	E	F	G	H	I
Baked products	+	+			+		+		+			
Potato products	+	+			+		+					+
Meat and fish	+	+		+		+						+
Mayonnaise, dressings	+	+		+	+			+				
Fruit jellies	+			+	+	+						
Fruit juices	+			+								
Brewery	+	+								+	+	
Wine	+	+								+	+	
Ice cream, cookies	+			+	+			+				
Diet food	+	+	+	+								

[a] 1: Carboxymethyl cellulose, Na-salt; 2: methyl cellulose; 3: hydroxypropyl methyl cellulose.
[b] A: Thickening effect; B: water binding/holding; C: cold gel setting; D: gel setting at higher temperatures; E: emulsifier; F: suspending effect; G: surface activity; H: adsorption; and I: film-forming property.

The properties of the product depend on the degree of substitution (DS; 0.3–0.9) and of polymerization (DP; 500–2000). Low substitution types (DS ≤0.3) are insoluble in water but soluble in alkali, whereas higher DS types (>0.4) are water soluble. Solubility and viscosity are dependent on pH.

Carboxymethyl cellulose is an inert binding and thickening agent used to adjust or improve the texture of many food products, such as jellies, paste fillings, spreadable process cheeses, salad dressings and cake fillings and icings (Table 4.27). It retards ice crystal formation in ice cream, stabilizing the smooth and soft texture. It retards undesired saccharose crystallization in candy manufacturing and inhibits starch retrogradation or the undesired staling in baked goods. Lastly, Carboxymethyl cellulose improves the stability and rehydration characteristics of many dehydrated food products.

4.4.4.18 Hemicelluloses

The term hemicelluloses refers to substances which occupy the spaces between the cellulose fibrils within the cell walls of plants. Various studies, e.g., on apples, potatoes, and beans, show that xyloglucans dominate in the class *Dicotyledoneae*. A section of the structure of a xyloglucan from runner beans is presented in Formula 4.161.

In the class *Monocotyledoneae*, the composition of the hemicelluloses in the endosperm tissue varies greatly, e.g., wheat and rye contain mainly arabinoxylans (pentosans, cf. 15.2.4.2.1), while β-glucans (cf. 15.2.4.2.2) predominate in barley and oats.

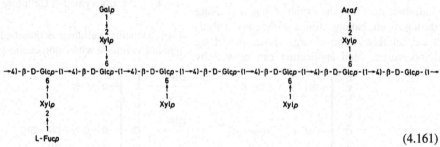

(4.161)

4.4.4.19 Xanthan Gum

4.4.4.19.1 Occurrence, Isolation

Xanthan gum, the extracellular polysaccharide from *Xanthomonas campestris* and some related microorganisms, is produced on a nutritive medium containing glucose, NH_4Cl, a mixture of amino acids, and minerals. The polysaccharide is recovered from the medium by isopropanol precipitation in the presence of KCl.

4.4.4.19.2 Structure, Properties

Xanthan gum can be regarded as a cellulose derivative. The main chain consists of 1,4 linked β-glucopyranose residues. On an average, every second glucose residue bears in the 3-position a trisaccharide of the structure β-D-Man*p*-(1 → 4)-β-D-Glc*p*A(1 → 2)-α-D-Man*p* as the side chain. The mannose bound to the main chain is acetylated in position 6 and ca. 50% of the terminal mannose residues occur ketalized with pyruvate as 4,6-O-(1-carboxyethylidene)-D-mannopyranose (cf. Formula 4.162; Glc*p*A: glucuronic acid).

$$[4)\text{-}\beta\text{-}D\text{-}Glcp\text{-}(1 \to 4)\text{-}\beta\text{-}D\text{-}Glcp\text{-}(1 \to]_n$$

$$3$$
$$\uparrow$$
$$1$$

β-D-Man*p*-(1 → 4)-β-D-Glc*p*A-(1 → 2)-α-D-Man*p*-6-OAc

```
          /  \
         4    6
          \  /
           C
          / \
       H₃C   COOH
```
$$\qquad (4.162)$$

The molecular weight of xanthan gum is $>10^6$ dal. In spite of this weight, it is quite soluble in water. The highly viscous solution exhibits a pseudoplastic behavior (Fig. 4.41). The viscosity is to a great extent, independent of temperature. Solutions, emulsions and gels, in the presence of xanthan gums, acquire a high freeze–thaw stability.

4.4.4.19.3 Utilization

The practical importance of xanthan gum is based on its emulsion-stabilizing and particle-

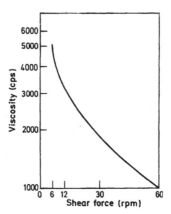

Fig. 4.41. Viscosity of aqueous xanthan gum solution as affected by shear rate (according to *Whistler*, 1973). Viscometer: Brookfield model LVF

suspending abilities (turbidity problems, essential oil emulsions in beverages). Due to its high thermal stability, it is useful as a thickening agent in food canning. Xanthan gum addition to starch gels substantially improves their freeze-thaw stability.

Xanthan gum properties might also be utilized in instant puddings: a mixture of locust bean flour, Na-pyrophosphate and milk powder with xanthan gum as an additive provides instant jelly after reconstitution in water. The pseudoplastic thixotropic properties, due to intermolecular association of single-stranded xanthan gum molecules, are of interest in the production of salad dressings, i.e. a high viscosity in the absence of a shear force and a drop in viscosity to a fluid state under a shear force.

4.4.4.20 Scleroglucan

4.4.4.20.1 Occurrence, Isolation

Sclerotium species, e.g., *S. glucanicum*, produce scleroglucan on a nutritive medium of glucose, nitrate as N-source and minerals. The polysaccharide is recovered from the nutritive medium by alcoholic precipitation.

4.4.4.20.2 Structure, Properties

The "backbone" of scleroglucan is a β-1,3-glucan chain that, on an average, has an attached glucose as a side chain on every third sugar residue (cf. Formula 4.163).

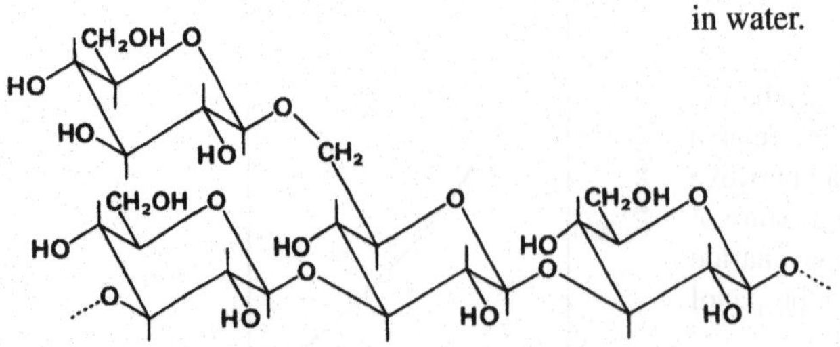

(4.163)

The polysaccharide has a molecular weight of about 130 kdal and is very soluble in water. Solutions have high viscosities and exhibit pseudoplastic thixotropic properties.

4.4.4.20.3 Utilization

Scleroglucan is used as a food thickener and, on the basis of its good film-forming property, is applied as a protective coating to dried foods.

4.4.4.21 Dextran

4.4.4.21.1 Occurrence

Leuconostoc mesenteroides, Streptobacterium dextranicum, Streptococcus mutans and some other bacteria produce extracellular dextran from saccharose with the help of α-1,6-glucan: D-fructose-2-glucosyl transferase (dextran sucrase, EC 2.4.1.5).

4.4.4.21.2 Structure, Properties

Dextran is an α-1,6-glucan (Formula 4.164; molecular weight $M_r = 4$–5×10^7 dal) with several glucose side chains, which are bound to the main chain of the macromolecule primarily through 1,3-linkages but, in part, also by 1,4- and 1,2-linkages.

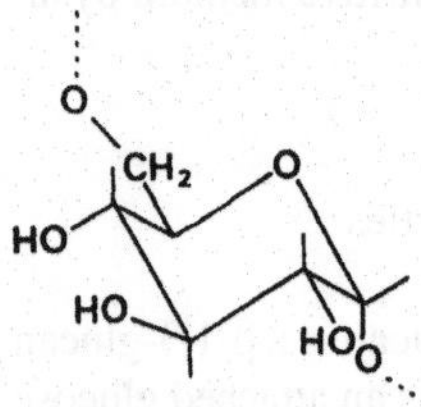

(4.164)

On an average, 95% of the glucose residues are present in the main chain. Dextran is very soluble in water.

4.4.4.21.3 Utilization

Dextran is used mostly in medicine as a blood substitute. In the food industry it is used as a thickening and stabilizing agent, as exemplified by its use in baking products, confections, beverages and in the production of ice creams.

4.4.4.22 Inulin and Oligofructose

4.4.4.22.1 Occurrence

Inulin occurs as a reserve carbohydrate in many plant families, e. g., scorzonera, topinambur, chicory, rye, onion and dahlia bulb.

4.4.4.22.2 Structure

Inulin contains about 30 furanoid D-fructose units in a β-1,2-linkage. This linear polysaccharide has α-glucose residues in 2,1-bonding at its ends. Individual α-glucose residues in 1,3-bonding have also been detected in the interior of the polysaccharide. Inulin (M, 5000–6000) is soluble in warm water and resistant to alkali.

4.4.4.22.3 Utilization

Inulin is nondigestible in the small intestine, but is degraded by the bacteria in the large intestine. It can be used in many foods as a sugar and fat substitute (cf. 8.16.1.2), e. g., biscuits, yoghurt, desserts and sweets. Inulin yields D-fructose on acid or enzymatic hydrolysis. Oligofructans have a slightly sweet taste due to the lower degree of polymerization.

4.4.4.23 Polyvinyl Pyrrolidone (PVP)

4.4.4.23.1 Structure, Properties

This compound is used as if it were a poly-saccharide-type additive. Therefore, it is described here. The molecular weight of PVP can range from 10–360 kdal.

$$(4.165)$$

It is quite soluble in water and organic solvents. The viscosity of the solution is related to the molecular weight.

4.4.4.23.2 Utilization

PVP forms insoluble complexes with phenolic compounds and, therefore, is applied as a clarifying agent in the beverage industry (beer, wine, fruit juice). Furthermore, it serves as a binding and thickening agent, and as a stabilizer, e. g., of vitamin preparations. Its tendency to form films is utilized in protective food films (particle solubility enhancement and aroma fixation in instant tea and coffee production).

4.4.5 Enzymatic Degradation of Polysaccharides

Enzymes that cleave polysaccharides are of interest for plant foods. Examples are processes that occur in the ripening of fruit (cf. 18.1.3.3.2), in the processing of flour to cakes and pastries (cf. 15.2.2.1), and in the degradation of cereals in preparation for alcoholic fermentation (cf. 20.1.4). In addition, enzymes of this type are used in food technology (cf. 2.7.2.2) and in carbohydrate analysis (cf. Table 2.16 and 4.4.6). The following hydrolases are of special importance.

4.4.5.1 Amylases

Amylases hydrolyze the polysaccharides of starch.

4.4.5.1.1 α-Amylase

α-Amylase hydrolyzes starch, glycogen, and other 1,4-α-glucans. The attack occurs inside the molecule, i. e., this enzyme is comparable to endopeptidases. Oligosaccharides of 6–7 glucose units are released from amylose. The enzyme apparently attacks the molecule at the amylose helix (cf. 4.4.4.14.3) and hydrolyzes "neighboring" glycoside bonds that are one turn removed. Amylopectin is cleaved at random; the branch points (cf. 4.4.4.14.4) are overjumped. α-Amylase is activated by Ca^{2+} ions (cf. 2.3.3.1 and 2.7.2.2.2).

The viscosity of a starch solution rapidly decreases on hydrolysis by α-amylase (starch liquefaction) and the iodine color disappears. The dextrins formed at first are further degraded on longer incubation, reducing sugars appear and, finally, α- maltose is formed. The activity of the enzyme decreases rapidly with decreasing degree of polymerization of the substrate.

Catalysis is accelerated by the gelatinization of starch (cf. 4.4.4.14.2). For example, the swollen substrate is degraded 300 times faster by a bacterial amylase and 10^5 times faster by a fungal amylase than is native starch.

4.4.5.1.2 β-Amylase

This enzyme catalyzes the hydrolysis of 1,4-α-D-glucosidic bonds in polysaccharides (mechanism, 2.4.2.5), effecting successive removals of maltose units from the nonreducing end. Hydrolysis is linked to a Walden inversion at C-1, giving rise to β-maltose. This inversion, which can be detected polarimetrically, represents a definite characteristic of an exoglycanase.

In contrast to amylose, amylopectin is not completely hydrolyzed. All reaction stops even before branch points are reached.

4.4.5.1.3 Glucan-1,4-α-D-glucosidase (Glucoamylase)

This glucoamylase starts at the nonreducing end of 1,4-α-D-glucans and successively liberates β-D-glucose units. In amylopectin, α-1,6-branches are cleaved ca. 30 times slower than α-1,4-bonds.

4.4.5.1.4 α-Dextrin Endo-1,6-α-glucosidase (Pullulanase)

This enzyme hydrolyzes 1,6-α-D-glucosidic bonds in polysaccharides, e. g., in amylopectin, glycogen, and pullulan. Linear amylose fragments are formed from amylopectin.

4.4.5.2 Pectinolytic Enzymes

Pectins (cf. 4.4.4.13) in plant foods are attacked by a series of enzymes. A distinction is made between:

- Pectin esterases which occur widely in plants and microorganisms and demethylate pectin to pectic acid (Formula 4.166).
- Enzymes which attack the glycosidic bond in polygalacturonides (Table 4.28). These include hydrolases and lyases which catalyze a transelimination reaction (see Formula 4.167). The double bond formed in the product of the last mentioned reaction results in an increase in the absorption at 235 nm.

The second group can be further subdivided according to the substrate (pectin or pectic acid) and to the site of attack (endo-/exo-enzyme), as shown in Table 4.28. The endo-enzymes strongly depolymerize and rapidly reduce the viscosity of a pectin solution.

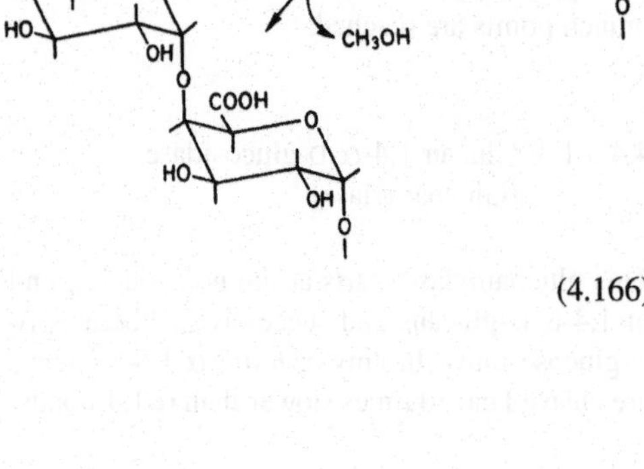

(4.166)

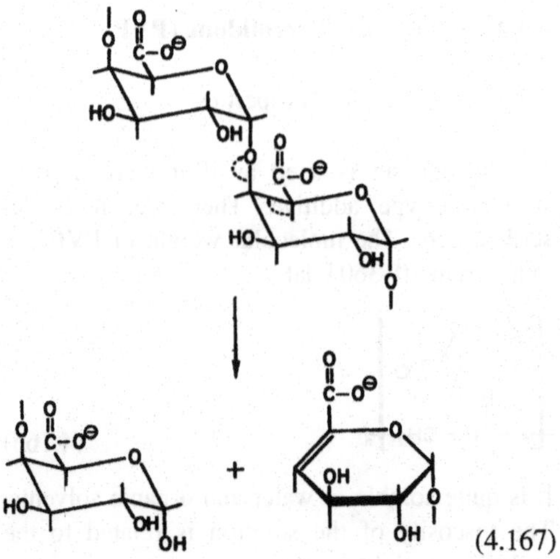

(4.167)

Table 4.28. Enzymes that cleave pectin and pectic acid

Enzyme	EC No.	Substrate
Polygalacturonase	3.2.1.15	
Endo-polymethyl galacturonase		Pectin
Endo-polygalacturonase		Pectic acid
Exo-polygalacturonase	3.2.1.67	
Exo-polymethyl galacturonase		Pectin
Exo-polygalacturonase		Pectic acid
Pectin lyase	4.2.2.10	
Endo-polymethyl galacturonlyase		Pectin
Pectate lyase	4.2.2.2	
Endo-polygalacturonate lyase		Pectic acid
Exo-polygalacturonate lyase	4.2.2.9	Pectic acid

Polygalacturonases occur in plants and microorganisms. They are activated by NaCl and some by Ca^{2+} ions as well.

Pectin and pectate lyases are only produced by microorganisms. They are activated by Ca^{2+} ions and differ in the pH optimum (pH 8.5–9.5) from the polygalacturonases (pH 5–6.5).

4.4.5.3 Cellulases

Hydrolysis of completely insoluble, microcrystalline cellulose is a complicated process. For this purpose, certain microorganisms produce particles called cellusomes (particle weight ca. 10^6). During isolation, these particles readily disintegrate into enzymes, which synergistically perform cellulose degradation, and components, which, among other things, support substrate binding. At least three enzymes are involved in the degradation of cellulose to cellobiose and glucose:

$$\text{Cellulose} \xrightarrow[\text{C}_1]{\text{C}_x} \text{Cellobiose} \xrightarrow{\text{Cellobiase}} \text{Glucose}$$

$$(4.168)$$

As shown in Table 4.29, the C_1 and C_x factors, which were found to be endo- and exo-1,4-β-glucanases respectively, hydrolyze cellulose to cellobiose. Since the C_1 factor is increasingly inhibited by its product, a cellobiase is needed so that cellulose breakdown is not rapidly brought to a standstill. However, cellobiase is also subject to product inhibition. Therefore, complete cellulose degradation is possible only if cellobiase is present in large excess or the glucose formed is quickly eliminated.

4.4.5.4 Endo-1,3(4)-β-glucanase

This hydrolase is also called laminarinase and hydrolyzes 1,3(4)-β-glucans. This enzyme occurs together with cellulases, e.g., in barley malt, and is involved in the degradation of β-glucans (cf. 15.2.4.2.2) in the production of beer.

4.4.5.5 Hemicellulases

The degradation of hemicelluloses also proceeds via endo- and exohydrolases. The substrate specificity depends on the monosaccharide building blocks and on the type of binding, e.g., endo-1,4-β-D-xylanase, endo-1,5-α-L-arabinase. These enzymes occur in plants and microorganisms, frequently together with cellulases.

4.4.6 Analysis of Polysaccharides

The identification and quantitative determination of polysaccharides plays a role in the examination of thickening agents, balast material etc.

4.4.6.1 Thickening Agents

First, thickening agents must be concentrated. The process used for this purpose is to be modified depending on the composition of the food. In general, thickening agents are extracted from the defatted sample with hot water. Extracted starch is digested by enzymatic hydrolysis (α-amylase, glucoamylase), and proteins are separated by precipitation (e. g., with sulfosalicylic acid). The polysaccharides remaining in the solution are separated with ethanol. An electropherogram of the polysaccharides dissolved in a borate buffer provides an initial survey of the thickening agents present. It is sometimes difficult to identify and, consequently, differentiate between the added polysaccharides and those that are endogenously present in many foods. In simple cases, it is sufficient if the electropherogram is supported by structural analysis. Here, the polysaccharides are permethylated (cf. 4.2.4.7), then subjected to acid hydroysis, reduced with sodium borohydride (cf. 4.2.4.1) and converted to partially methylated alditol acetates by acetylation of the OH-groups (cf. 4.2.4.6).

These derivatives of the monosaccharide structural units are then qualitatively and quantitatively analyzed by gas chromatography on capillary columns. In more difficult cases, a preliminary separation of acidic and neutral polysaccharides on an ion exchanger is recommended. Methanolysis or hydrolysis of polysaccharides containing uronic acids and anhydro sugars are critical due to losses of these labile building blocks.

Reductive cleavage of the permethylated polysaccharide is recommended as a gentle alternative to hydrolysis. In this process, partially methylated anhydroalditolacetates are formed as shown in Fig. 4.42, using a galactomannan as an example. Conclusions about the structure of the polysaccharide can be drawn from the result of the qualitative and quantitative analysis, which is carried out by gas chromatography/mass

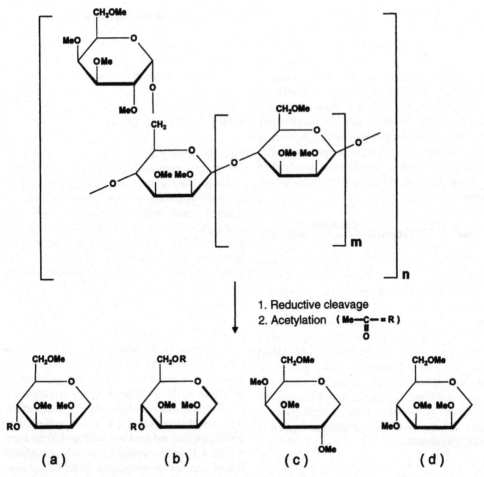

Fig. 4.42. Reductive depolymerization of a permethylated galactomannan (according to *Kiwitt-Haschemie* et al., 1996) 1. Reductive cleavage with triethylsilane and trimethylsilylmethanesulfonate/boron trifluoride 2. Acetylation with acetic anhydride and N-methylimidazole

spectrometry. In the example presented here, the derivative 4-O-acetyl-1,5-anhydro-2,3,6-tri-O-methyl-D-mannitol (*a* in Fig. 4.42) results from the 1,4-linked D-mannose, the structural unit of the main chain. The derivative 4,6-di-O-acetyl-1,5-anhydro-2,3-di-O-methyl-D-mannitol (*b*) indicates the structural unit which forms the branch and the derivative 1,5-anhydro-2,3,4,6-tetra-O-methyl-D-galactitol (*c*) indicates the terminal D-galactopyranose of the side chain. The derivative 1,5-anhydro-2,3,4,6-tetra-O-methyl-D-mannitol (*d*) produced in small amounts shows the end of the main chain formed by D-mannopyranose. The appearance of glucose in the structural analysis indicates glucans or

modified glucans, e. g., modified starches or celluloses. The identification of thickening agents of this type is achieved by the specific detection of the hetero-components, e. g., acetate or phosphate.

4.4.6.2 Dietary Fibers

Gravimetric methods are the methods of choice for the determination of dietary fibers (cf. 15.2.4.2). In the defatted sample, the digestible components (1,4-α-glucans, proteins) are enzymatically hydrolyzed (heat-stable α- amylase, glucoamylase, proteinase). After centrifugation,

Table 4.29. Cellulases

EC No.	Name	Synonym	Reaction
3.2.1.4	Cellulase	C_x factor CMCase[a], endo-1,4-β-glucanase	Endohydrolysis of 1,4-β-D-glucosidic bonds
3.2.1.91	Cellulose 1,4-β-cellobiosidase	C_1 factor avicellase	Exohydrolysis of 1,4-β-D-glucosidic bonds with formation of cellobiose from cellulose or 1,4-β-glucooligo-saccharides. The attack proceeds from the nonreducing end.
3.2.1.21	β-Glucosidase	Cellobiase amygdalase	Hydrolysis of terminal β-D-glucose residues in β-glucans

[a] CMC: carboxymethylcellulose; the enzyme activity can be measured via the decrease in viscosity of a CMC solution.

the insoluble fibers remain in the residue. The water soluble fibers in the supernatant are isolated by precipitation with ethanol, ultrafiltration or dialysis. The protein and mineral matter still remaining with the soluble and insoluble dietary fibers is deducted with the help of correction factors.

4.5 References

Angyal, S.J.: Zusammensetzung und Konformation von Zuckern in Lösung. Angew. Chem. *81*, 172 (1969)

Balser, K.: Celluloseäther. In: Ullmanns Encyklopädie der technischen Chemie, 4. edn., Vol. *9*, S. 192. Verlag Chemie: Weinheim. 1975

Banks, W., Muir, D.D.: Structure and chemistry of the starch granule. In: The biochemistry of plants (Eds.: Stumpf, P.K., Conn, E.E.), Vol. *3*, p. 321, Academic Press: New York. 1980

Brouns, F., Kettlitz, B., Arrigoni, E.: Resistanz starch and "the butyrate revolution". Trends Food Sci. Technol. *13*, 251 (2002)

Birch, G.G. (Ed.): Developments in food carbohydrate-1 ff, Applied Science Publ.: London. 1977 ff.

Birch, G.G., Parker, K.J. (Eds.): Nutritive sweeteners. Applied Science Publ.: London. 1982

Birch, G.G., Parker, K.J. (Eds.): Dietary fibre. Applied Science Publ.: London. 1983

Birch, G.G. (Ed.): Analysis of food carbohydrate. Elsevier Applied Science Publ.: London. 1985

Blanshard, J.M.V., Mitchell, J.R. (Eds.): Polysaccharides in food. Butterworths: London. 1979

Büser, W., Erbersdobler, H.F.: Carboxymethyllysine, a new compound of heat damage in milk products. Milchwissenschaft *41*, 780 (1986)

Davidson, R.L. (Ed.): Handbook of water-soluble gums and resins. McGraw-Hill Book Co.: New York. 1980

Ebert, G.: Biopolymere. Dr. Dietrich Steinkopff Verlag: Darmstadt. 1980

Erlingen, R.C., Delcour, J.A.: Formation, analysis, structure and properties of type III enzyme resistant starch. J. Cereal Sci. *22*, 129 (1995)

Friedman, M.: Food browning and its prevention: An Overview. J. Agric. Food. Chem. *44*, 631 (1996)

Galliard, T. (Ed.): Starch: Properties and Potential. John Wiley and Sons: Chichester. 1987

Gidley, M.J., Cooke, D., Darke, A.H., Hoffmann, R.A., Russell, A.L., Greenwell, P.: Molecular order and structure in enzyme-resistant retrograded starch. Carbohydrate Polymers *28*, 23 (1995)

Glomb, M.A., Monnier, V.M.: Mechanism of protein modification by glyoxal and glycolaldehyde, reactive intermediates of the Maillard reaction. J. Biol. Chem. *270*, 10017 (1995)

Guadagni, D.G., Maier, V.P., Turnbaugh, J.G.: Effect of some citrus juice constituents on taste thresholds for limonine and naringin bitterness. J. Sci. Food Agric. *24*, 1277 (1973)

Henle, T., Zehetner, G., Klostermeyer, H.: Fast and sensitive determination of furosine. Z. Lebensm. Unters. Forsch. *200*, 235 (1995)

Henle, T., Schwarzenbolz, U., Klostermeyer, H.: Detection and quantification of pentosidine in foods. Z. Lebensm. Unters. Forsch. *204*, 95 (1997)

Hill, R.D., Munck, L. (Eds.): New approaches to research on cereal carbohydrates. Elsevier Science Publ.: Amsterdam. 1985

Hofmann, T.: Characterization of the chemical structure of novel colored Maillard reaction products from furan-2-carboxaldehyde and amino acids. J. Agric. Food Chem. *46*, 932 (1998)

Hofmann, T.: 4-Alkylidene-2-imino-5-[4-alkylidene-5-oxo-1,3-imidazol-2-inyl]azamethylidene-1,3-imidazolidine. A novel colored substructure in melanoidins formed by Maillard reactions of bound arginine with glyoxal and furan-2-carboxaldehyde. J. Agric. Food Chem. *46*, 3896 (1998)

Hofmann, P., Münch, P., Schieberle, P.: Quantitative model studies on the formation of aroma-active aldehydes and acids by Strecker type reactions. J. Agric. Food Chem. *48*, 434 (2000)

Hough, L., Phadnis, S.P.: Enhancement in the sweetness of sucrose. Nature *263*, 800 (1976)

Jenner, M.R.: Sucralose. How to make sugar sweeter. ACS Symposium Series *450*, p. 68 (1991)

Kiwitt-Haschemie, K., Renger, A., Steinhart, H.: A comparison between reductive-cleavage and standard methylation analysis for determining structural features of galactomannans. Carbohydrate Polymers *30*, 31 (1996)

Ledl, F., Severin, T.: Untersuchungen zur *Maillard*-Reaktion. XIII. Bräunungsreaktion von Pentosen mit Aminen. Z. Lebensm. Unters. Forsch. *167*, 410 (1978)

Ledl, F., Severin, T.: Untersuchungen zur *Maillard*-Reaktion. XVI. Bildung farbiger Produkte aus Hexosen. Z. Lebensm. Unters. Forsch. *175*, 262 (1982)

Ledl, F., Krönig, U., Severin, T., Lotter, H.: Untersuchungen zur *Maillard*-Reaktion. XVIII. Isolierung N-haltiger farbiger Verbindungen. Z. Lebensm. Unters. Forsch. *177*, 267 (1983)

Ledl, F.: Low molecular products, intermediates and reaction mechanisms. In: Amino-carbonyl reactions in food and biological systems (Eds.: Fujimaki, M., Namiki, M., Kato, H.), p. 569, Elsevier: Amsterdam. 1986

Ledl, F., Fritul G., Hiebl, H., Pachmayr, O., Severin, T.: Degradation of *Maillard* products. In: Aminocarbonyl reactions in food and biological systems (Eds.: Fujimaki, M., Namiki, M., Kato, H.). p. 173, Elsevier: Amsterdam. 1986

Ledl, F.: Chemical Pathways of the *Maillard* Reaction. In: The *Maillard* Reaction in Food Processing, Human Nutrition and Physiology (Eds.: Finot, P.A. et al.) p. 19, Birkhäuser Verlag: Basel. 1990

Ledl, F., Schleicher, E.: Die *Maillard*-Reaktion in Lebensmitteln und im menschlichen Körper – neue Ergebnisse zu Chemie, Biochemie und Medizin. Angewandte Chemie *102*, 597 (1990)

Ledl, F., Glomb, M., Lederer, M.: Nachweis reaktiver *Maillard*-Produkte. Lebensmittelchemie *45*, 119 (1991)

Lehmann, J.: Chemie der Kohlenhydrate. Georg Thieme Verlag: Stuttgart. 1976

Loewus, F.A., Tanner, W. (Eds.): Plant carbohydrates I, II. Springer-Verlag: Berlin. 1981/82

Nedvidek, W., Noll, P., Ledl, F.: Der Einfluß des *Strecker*abbaus auf die *Maillard*-Reaktion. Lebensmittelchemie *45*, 119 (1991)

Pagington, J.S.: β-Cyclodextrin and its uses in the flavour industry. In: Developments in food flavours (Eds.: Birch, G.G., Kindley, M.G.), p. 131, Elsevier Applied Science: London. 1986

Pigman, W., Horton, D. (Eds.): The Carbohydrates. 2nd edn., Academic Press: New York. 1970–1980

Pilnik, W., Voragen, F., Neukom, H., Nittner, E.: Polysaccharide. In: Ullmanns Encyklopädie der technischen Chemie, 4. edn., Vol. *19*, S. 233, Verlag Chemie: Weinheim. 1980

Preuß, A., Thier, H.-P.: Isolierung natürlicher Dickungsmittel aus Lebensmitteln zur capillargaschromatographischen Bestimmung. Z. Lebensm. Unters. Forsch. *176*, 5 (1983)

Radley, J.A. (Ed.): Starch production technology. Applied Science Publ.: London. 1976

Rees, D.A.: Polysaccharide shapes. Chapman and Hall: London. 1977

Reilly, P.J.: Xylanases: structure and function. Basic Life Sci. *18*, 111 (1981)

Rodriguez, R., Jimenez, A., Fernandez-Bolanos, J., Guillen, R., Heredia, A.: Dietary fibre from vegetable products as source of functional ingredients. Trends Food Sci. Technol. *17*, 3 (2006)

Salunkhe, D.K., McLaughlin, R.L., Day, S.L., Merkley, M.B.: Preparation and quality evaluation of processed fruits and fruit products with sucrose and synthetic sweeteners. Food Technol. *17*, 203 (1963)

Scherz, H., Mergenthaler, E.: Analytik der als Lebensmittelzusatzstoffe verwendeten Polysaccharide. Z. Lebensm. Unters. Forsch. *170*, 280 (1980)

Scherz, H.: Verwendung der Polysaccharide in der Lebensmittelverarbeitung. In: Polysaccharide. Eigenschaften und Nutzung (Hrsg.: Burchard, W.), S. 142, Springer-Verlag: Berlin. 1985

Scherz, H., Bonn, G.: Analytical Chemistry of Carbohydrates, Georg Thieme Verlag, Stuttgart, 1998

Schweizer, T.F.: Fortschritte in der Bestimmung von Nahrungsfasern. Mitt. Geb. Lebensmittelunters. Hyg. *75*, 469 (1984)

Shallenberger, R.S., Birch, G.G.: Sugar chemistry. AVI Publ. Co.: Westport, Conn. 1975

Shallenberger, R.S.: Advanced sugar chemistry, principles of sugar stereochemistry. Ellis Horwood Publ.: Chichester. 1982

Sutherland, I.W.: Extracellular polysaccharides. In: Biotechnology (Eds.: Rehm, H.-J., Reed, G.), Vol. 3, p. 531, Verlag Chemie: Weinheim. 1983

Szente, L., Szeijtli, J.: Cyclodestrins as food ingredients. Trends Food Sci Technol. *15*, 137 (2004)

Von Sydow, E., Moskowitz, H., Jacobs, H., Meiselman, H.: Odor – Taste interaction in fruit juices. Lebensm. Wiss. Technol. *7*, 18 (1974)

Voragen, A.G.J.: Technological aspects of functional food-related carbohydrates. Trends Food Sci. Technol. *9*, 328 (1998)

Waller, G.R., Feather, M.S. (Eds.): The *Maillard* reaction in foods and nutrition. ACS Symposium Series 215, American Chemical Society: Washington, D.C. 1983

Wells-Knecht, K.J., Brinkmann, E., Baynes, J.W.: Characterization of an imidazolium salt formed from glyoxal and N-hippuryllysine: a model for Maillard reaction crosslinks in proteins. J. Org. Chem. *60*, 6246 (1995)

Whistler, R.L. (Ed.): Industrial gums. 2nd edn., Academic Press: New York. 1973

Willats, W.G.T., Knox, J.P., Mikkelsen, J.D.: Pectin: new insights into an old polymer are starting to gel. Trends Food Sci. Technol. *17*, 97 (2006)

Yaylayan, V.: In search of alternative mechanisms for the *Maillard* reaction, Trends Food Sci Technol *1*, 20 (1990)

5 Aroma Compounds

5.1 Foreword

5.1.1 Concept Delineation

When food is consumed, the interaction of taste, odor and textural feeling provides an overall sensation which is best defined by the English word "flavor". German and some other languages do not have an adequate expression for such a broad and comprehensive term. Flavor results from compounds that are divided into two broad classes: Those *responsible for taste* and those *responsible for odors*, the latter often designated as aroma substances. However, there are compounds which provide both sensations.

Compounds *responsible for taste* are generally nonvolatile at room temperature. Therefore, they interact only with taste receptors located in the taste buds of the tongue. The four important basic taste perceptions are provided by: sour, sweet, bitter and salty compounds. They are covered in separate sections (cf., for example, 8.10, 22.3, 1.2.6, 1.3.3, 4.2.3 and 8.8). Glutamate stimulates the fifth basic taste (cf. 8.6.1).

Aroma substances are volatile compounds which are perceived by the odor receptor sites of the smell organ, i. e. the olfactory tissue of the nasal cavity. They reach the receptors when drawn in through the nose (orthonasal detection) and via the throat after being released by chewing (retronasal detection). The concept of aroma substances, like the concept of taste substances, should be used loosely, since a compound might contribute to the typical odor or taste of one food, while in another food it might cause a faulty odor or taste, or both, resulting in an off-flavor.

5.1.2 Impact Compounds of Natural Aromas

The amount of volatile substances present in food is extremely low (ca. 10–15 mg/kg). In general, however, they comprise a large number of components. Especially foods made by thermal processes, alone (e. g., coffee) or in combination with a fermentation process (e. g., bread, beer, cocoa, or tea), contain more than 800 volatile compounds. A great variety of compounds is often present in fruits and vegetables as well.

All the known volatile compounds are classified according to the food and the class of compounds and published in a tabular compilation (*Nijssen, L. M.* et al., 1999). A total of 7100 compounds in more than 450 foods are listed in the 1999 edition, which is also available as a database on the internet.

Of all the volatile compounds, only a limited number are important for aroma. Compounds that are considered as aroma substances are prima-

Table 5.1. Examples of key odorants

Compound	Aroma	Occurrence
(R)-Limonene	Citrus-like	Orange juice
(R)-1-p-Menthene-8-thiol	Grapefruit-like	Grapefruit juice
Benzaldehyde	Bitter almond-like	Almonds, cherries, plums
Neral/geranial	Lemon-like	Lemons
1-(p-Hydroxy-phenyl)-3-butanone (raspberry ketone)	Raspberry-like	Raspberries
(R)-(−)-1-Octen-3-ol	Mushroom-like	Champignons, Camembert cheese
(E,Z)-2,6-Nonadienal	Cucumber-like	Cucumbers
Geosmin	Earthy	Beetroot
trans-5-Methyl-2-hepten-4-one (filbertone)	Nut-like	Hazelnuts
2-Furfurylthiol	Roasted	Coffee
4-Hydroxy-2,5-dimethyl-3(2H)-furanone	Caramel-like	Biscuits, dark beer, coffee
2-Acetyl-1-pyrroline	Roasted	White-bread crust

rily those which are present in food in concentrations higher than the odor and/or taste thresholds (cf. "Aroma Value", 5.1.4). Compounds with concentrations lower than the odor and/or taste thresholds also contribute to aroma when mixtures of them exceed these thresholds (for examples of additive effects, see 3.2.1.1, 20.1.7.8, 21.1.3.4).

Among the aroma substances, special attention is paid to those compounds that provide the characteristic aroma of the food and are, consequently, called key odorants (character impact aroma compounds). Examples are given in Table 5.1.

In the case of important foods, the differentiation between odorants and the remaining volatile compounds has greatly progressed. Important findings are presented in the section on "Aroma" in the corresponding chapters.

5.1.3 Threshold Value

The lowest concentration of a compound that is just enough for the recognition of its odor is called the odor threshold (recognition threshold). The detection threshold is lower, i. e., the concentration at which the compound is detectable but the aroma quality still cannot be unambiguously established. The threshold values are frequently determined by smelling (orthonasal value) and by tasting the sample (retronasal value). With a few exceptions, only the orthonasal values are given in this chapter. Indeed, the example of the carbonyl compounds shows how large the difference between the ortho- and retronasal thresholds can be (cf. 3.7.2.1.9).

Threshold concentration data allow comparison of the intensity or potency of odorous substances. The examples in Table 5.2 illustrate that great differences exist between individual aroma compounds, with an odor potency range of several orders of magnitude.

In an example provided by nootkatone, an essential aroma compound of grapefruit peel oil (cf. 18.1.2.6.3), it is obvious that the two enantiomers (optical isomers) differ significantly in their aroma intensity (cf. 5.2.5 and 5.3.2.4) and, occasionally, in aroma quality or character.

The threshold concentrations (values) for aroma compounds are dependent on their vapor pressure, which is affected by both temperature and

Table 5.2. Odor threshold values in water of some aroma compounds (20 °C)

Compound	Threshold value (mg/l)
Ethanol	100
Maltol	9
Furfural	3.0
Hexanol	2.5
Benzaldehyde	0.35
Vanillin	0.02
Raspberry ketone	0.01
Limonene	0.01
Linalool	0.006
Hexanal	0.0045
2-Phenylethanal	0.004
Methylpropanal	0.001
Ethylbutyrate	0.001
(+)-Nootkatone	0.001
(-)-Nootkatone	1.0
Filbertone	0.00005
Methylthiol	0.00002
2-Isobutyl-3-methoxypyrazine	0.000002
1-p-Menthene-8-thiol	0.00000002

medium. Interactions with other odor-producing substances can result in a strong increase in the odor thresholds. The magnitude of this effect is demonstrated in a model experiment in which the odor thresholds of compounds in water were determined in the presence and absence of 4-hydroxy-2,5-dimethyl-3(2H)-furanone (HD3F). The results in Table 5.3 show that HD3F does not influence the threshold value of 4-vinylguaiacol. However, the threshold values of the other odor-

Table 5.3. Influence of 4-hydroxy-2,5-dimethyl-3(2H)-furanone (HD3F) on the odor threshold of aroma substances in water

Compound	Threshold value (µg/1)		Ratio
	I[a]	II[b]	II to I
4-Vinylguaiacol	100	90	≈1
2,3-Butanedione	15	105	7
2,3-Pentanedione	30	150	5
2-Furfurylthiol	0.012	0.25	20
β-Damascenone	2×10^{-3}	0.18	90

[a] I, odor threshold of the compound in water.
[b] II, odor threshold of the compound in an aqueous HD3F solution having a concentration (6.75 mg/1, aroma value A = 115) as high as in a coffee drink.

Table 5.4. Comparison of threshold values[a] in water and beer

Compound	Threshold (mg/kg) in	
	Water	Beer
n-Butanol	0.5	200
3-Methylbutanol	0.25	70
Dimethylsulfide	0.00033	0.05
(E)-2-Nonenal	0.00008	0.00011

[a] Odor and taste.

ants increase in the presence of HD3F. This effect is the greatest in the case of β-damascenone, the threshold value being increased by a factor of 90. Other examples in this book which show that the odor threshold of a compound increases when it is influenced by other odor-producing substances are a comparison of the threshold values in water and beer (cf. Table 5.4) as well as in water and in aqueous ethanol (cf. 20.2.6.9).

5.1.4 Aroma Value

As already indicated, compounds with high "aroma values" may contribute to the aroma of foods. The "aroma value" A_x of a compound is calculated according to the definition:

$$A_x = \frac{c_x}{a_x} \tag{5.1}$$

(c_x: concentration of compound X in the food, a_x: odor threshold (cf. 5.1.3) of compound X in the food). Methods for the identification of the corresponding compounds are described under Section 5.2.2.

The evaluation of volatile compounds on the basis of the aroma value provides only a rough pattern at first. The dependence of the odor intensity on the concentration must also be taken into account. In accordance with the universally valid law of *Stevens* for physiological stimuli, it is formulated as follows:

$$E = k \cdot (S - S_o)^n \tag{5.2}$$

E: perception intensity, k: constant, S: concentration of stimulant, S_o: threshold concentration of stimulant.

The examples presented in Fig. 5.1 show that the exponent n and, therefore, the dependency of the odor intensity on the concentration can vary substantially. Within a class of compounds, the range of variations is not very large, e. g., $n = 0.50 - 0.63$ for the alkanals $C_4 - C_9$.

In addition, additive effects that are difficult to assess must also be considered. Examinations of mixtures have provided preliminary information. They show that although the intensities of compounds with a similar aroma note add up, the intensity of the mixture is usually lower than the sum of the individual intensities (cf. 3.2.1.1). For substances which clearly differ in their aroma note, however, the odor profile of a mixture is composed of the odor profiles of the components added together, only when the odor intensities are approximately equal. If the concentration ratio is such that the odor intensity of one component predominates, this component then largely or completely determines the odor profile.

Examples are (E)-2-hexenal and (E)-2-decenal which have clearly different odor profiles (cf. Fig. 5.2 a and 5.2 f). If the ratio of the odor intensities is approximately one, the odor notes of both aldehydes can be recognized in the odor profile of the mixture (Fig. 5.2 d). But if the dominating odor intensity is that of the decenal (Fig. 5.2 b), or of the hexenal (Fig. 5.2 e), that particular note determines the odor profile of the mixture.

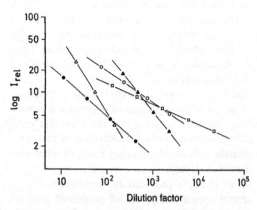

Fig. 5.1. Relative odor intensity I_{rel} (reference: n-butanol) as a function of the stimulant concentration (according to *Dravnieks*, 1977).
Air saturated with aroma substance was diluted. ● – ● – ● α-pinene, ○ – ○ – ○ 3-methylbutyric acid methyl ester, △ – △ – △ hexanoic acid, ◆ – ◆ – ◆ 2,4-hexadienal, □ – □ – □ hexylamine

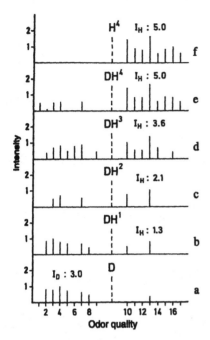

Fig. 5.2. Odor profiles of (E)-2-decenal (D), (E)-2-hexenal (H) and mixtures of both aldehydes (according to *Laing* and *Willcox*, 1983). The following concentrations (mg/kg) dissolved in di-2-ethylhexyl-phthalate were investigated: 50 (D); 2 (H^1); 3.7 (H^2); 11 (H^3) and 33 (H^4).
I_D and I_H: Odor intensity of each concentration of (E)-2-decenal and (E)-2-hexenal. Odor quality: **1**, warm; **2**, like clean washing; **3**, cardboard; **4**, oily, fatty; **5**, stale; **6**, paint; **7**, candle; **8**, rancid; **9**, stinkbug; **10**, fruity; **11**, apple; **12**, almond; **13**, herbal, green; **14**, sharp, pungent; **15**, sweet; **16**, banana; **17**, floral. The broken line separates the aroma qualities of (E)-2-decenal (*left side*) and (E)-2-hexenal

The mixture in Fig. 5.2, c gives a new odor profile because definite features of the decenal (stale, paint-like, rancid) and the hexenal (like apples, almonds, sweet) can no longer be recognized in it. The examples show clearly that the aroma profiles of foods containing the same aroma substances can be completely dissimilar owing to quantitative differences. For example, changes in the recipe or in the production process which cause alterations in the concentrations of the aroma substances can interfere with the balance in such a way that an aroma profile with unusual characteristics is obtained.

5.1.5 Off-Flavors, Food Taints

An off-flavor can arise through foreign aroma substances, that are normally not present in a food, loss of key odorants, or changes in the concentration ratio of individual aroma substances. Figure 5.3 describes the causes for aroma defects in food. In the case of an odorous contaminant, which enters the food via the air or water and then gets enriched, it can be quite difficult to determine its origin if the limiting concentration for odor perception is exceeded only on enrichment. Examples of some off-flavors that can arise during food processing and storage are listed in Table 5.5. Examples of microbial metabolites wich may be involved in pigsty-like and earthy-muddy off-flavors are skatole (I; faecal-like, $10\,\mu\mathrm{g/kg}^*$), 2-methylisoborneol (II; earthy-muddy, $0.03\,\mu\mathrm{g/kg}^*$) and geosmin (III; earthy, $(-): 0.01\,\mu\mathrm{g/kg}^*$; $(+): 0.08\,\mu\mathrm{g/kg}^*$):

$$(5.3)$$

2,4,6-Trichloroanisole (IV) with an extremely low odor threshold (mouldy-like: $3 \cdot 10^{-5}\,\mu\mathrm{g/kg}$, water) is an example of an off-flavor substance (cf. 20.2.7) which is produced by fungal degradation and methylation of pentachlorophenol fungicides.
To a certain extent, unwanted aroma substances are concealed by typical ones. Therefore, the threshold above which an off-flavor becomes noticeable can increase considerably in food compared to water as carrier, e.g., up to $0.2\,\mu\mathrm{g/kg}$ 2,4,6-trichloroanisole in dried fruits.

---

*Odor threshold in water.

Table 5.5. "Off-flavors" in food products

Food product	Off-flavor	Cause
Milk	Sunlight flavor	Photooxidation of methionine to methional (with riboflavin as a sensitizer)
Milk powder	Bean-like	The level of O_3 in air too high: ozonolyis of 8,15- and 9,15-isolinoleic acid to 6-trans-nonenal
Milk powder	Gluey	Degradation of tryptophan to o-amino-acetophenone
Milk fat	Metallic	Autoxidation of pentaene- and hexaene fatty acids to octa-1,cis-5-dien-3-one
Milk products	Malty	Faulty fermentation by *Streptococcus lactis, var. maltigenes*; formation of phenylacetaldehyde and 2-phenylethanol from phenylalanine
Peas, deep froze	Hay-like	Saturated and unsaturated aldehydes, octa-3,5-dien-2-one, 3-alkyl-2-methoxypyrazines, hexanal
Orange juice	Grapefruit note	Metal-catalyzed oxidation or photooxidation of valencene to nootkatone 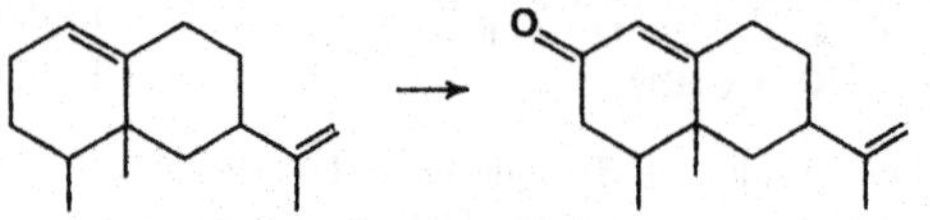

Orange juice	Terpene note	d-Limonene oxidation to carvone 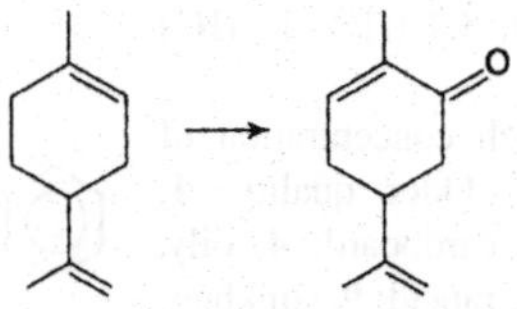

Passion fruit juice	Aroma flattening during pasteurization	Oxidation of (6-trans-2'-trans)-6-(but-2'-enyliden)-1,5,5-trimethylcyclohex-1-ene to 1,1,6-trimethyl-1,2-dihydronaphthalene: 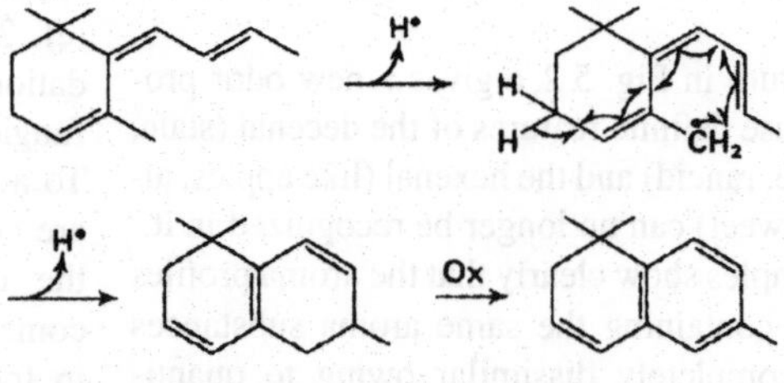

Beer	Sunlight flavor	Photolysis of humulone: reaction of one degradation product with hydrogen sulfide yielding 3-methyl-2-buten-1-thiol
Beer	Phenolic note	Faulty fermentation: hydrocinnamic acid decarboxylation by *Hafnia protea*

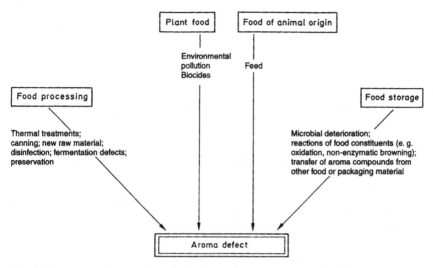

Fig. 5.3. The cause of aroma defects in food

5.2 Aroma Analysis

The aroma substances consist of highly diversified classes of compounds, some of them
being highly reactive and are present in food
in extremely low concentrations. The difficulties usually encountered in qualitative and
quantitative analysis of aroma compounds are
based on these features. Other difficulties are
associated with identification of aroma compounds, elucidation of their chemical structure
and characterization of sensory properties.

The results of an aroma analysis can serve as
an objective guide in food processing for assessing the suitability of individual processing steps,
and for assessing the quality of raw material,
intermediate- and endproducts. In addition, investigation of food aroma broadens the possibility of
food flavoring with substances that are prepared
synthetically, but are chemically identical to those
found in nature, i. e. the so-called "nature identical flavors" (cf. 5.5).

The elucidation of the aroma of any food is carried out stepwise; the following instrumental and
sensory analyses are conducted:

- Isolation of the volatile compounds
- Differentiation of the aroma substances from
  the remaining components of the volatile frac-
  tion by dilution analyses
- Concentration and identification
- Quantification and calculation of aroma values

- Simulation of the aroma on the basis of the
  analytical results
- Omission experiments

5.2.1 Aroma Isolation

The amount of starting material must be selected
to detect even those aroma substances which are
present in very low concentrations (ppb range),
but contribute considerably to the aroma because
of still lower odor thresholds. The volatile compounds should be isolated from food using gentle
methods because otherwise artifacts can easily be
produced by the reactions listed in Table 5.6.

Additional difficulties are encountered in foods
which retain fully-active enzymes, which can
alter the aroma. For example, during the homogenization of fruits and vegetables, hydrolases split
the aroma ester constituents, while lipoxygenase,
together with hydroperoxide lyase, enrich the
aroma with newly-formed volatile compounds.
To avoid such interferences, tissue disintegration
is done in the presence of enzyme inhibitors,
e. g., $CaCl_2$ or, when possible, by rapid sample
preparation. It is useful in some cases to inhibit
enzyme-catalyzed reactions by the addition of
methanol or ethanol. However, this can result
in a change in aroma due to the formation of
esters and acetals from acids and aldehydes
respectively.

Table 5.6. Possible changes in aromas during the isolation of volatile compounds

Reaction

Enzymatic

1. Hydrolysis of esters (cf. 3.7.1)
2. Oxidative cleavage of unsaturated fatty acids (cf. 3.7.2.3)
3. Hydrogenation of aldehydes (cf. 5.3.2.1)

Non-enzymatic

4. Hydrolysis of glycosides (cf. 5.3.2.4 and 3.8.4.4)
5. Lactones from hydroxy acids
6. Cyclization of di-, tri-, and polyols (cf. 5.3.2.4)
7. Dehydration and rearrangement of tert-allyl alcohols
8. Reactions of thiols, amines, and aldehydes (cf. 5.3.1.4) in the aroma concentrate
9. Reduction of disulfides by reductones from the *Maillard* reaction
10. Fragmentation of hydroperoxides

At the low pH values prevalent in fruit, non-enzymatic reactions, especially reactions 4–7 shown in Table 5.6, can interfere with the isolation of aroma substances by the formation of artifacts. In the concentration of isolates from heated foods, particularly meat, it cannot be excluded that reactive substances, e. g., thiols, amines and aldehydes, get concentrated to such an extent that they condense to form heterocyclic aroma substances, among other compounds (Reaction 8, Table 5.6).

In the isolation of aroma substances, foods which owe their aroma to the *Maillard* reaction should not be exposed to temperatures of more than 50 °C. At higher temperatures, odorants are additionally formed, i. e., thiols in the reduction of disulfides by reductones. Fats and oils contain volatile and non-volatile hydroperoxides which fragment even at temperatures around 40 °C.

An additional aspect of aroma isolation not to be neglected is the ability of the aroma substances to bind to the solid food matrix. Such binding ability differs for many aroma constituents (cf. 5.4).

The aroma substances present in the vapor space above the food can be very gently detected by *headspace analysis* (cf. 5.2.1.3). However, the amounts of substance isolated in this process are so small that important aroma substances, which are present in food in very low concentrations, give no detector signal after gas chromatographic separation of the sample. These substances can be determined only by sniffing the carrier gas stream. The difference in the detector sensitivity is clearly shown in Fig. 5.4, taking the aroma substances of the crust of white bread as an example. The gas chromatogram does not show, e. g., 2-acetyl-1-pyrroline and 2-ethyl-3,5-dimethylpyrazine, which are of great importance for aroma due to high FD factors in the FD chromatogram (definition in 5.2.2.1). These aroma substances can be identified only after concentration from a relatively large amount of the food.

5.2.1.1 Distillation, Extraction

The volatile aroma compounds, together with some water, are removed by vacuum distillation from an aqueous food suspension. The highly volatile compounds are condensed in an efficiently cooled trap. The organic compounds contained in the distillate are separated from the water by extraction or by adsorption to a hydrophobic matrix and reversed phase chromatography and then prefractionated.

The apparatus shown in Fig. 5.5 is recommended for the gentle isolation of aroma substances from aqueous foods by means of distillation. In fact, a condensate can be very quickly obtained because of the short distances. As in all distillation processes, the yield of aroma substances decreases if the food or an extract is fatty (Table 5.7).

After application of high vacuum ($\approx$5 mPa) the distillation procedure is started by dropping the

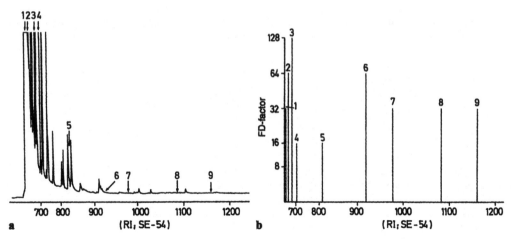

Fig. 5.4. Headspace analysis of aroma substances of white-bread crust. **a** Capillary gas chromatogram (the *arrows* mark the positions of the odorants), **b** FD chromatogram. Odorants: *1* methylpropanal, *2* diacetyl, *3* 3-methylbutanal, *4* 2,3-pentanedione, *5* butyric acid, *6* 2-acetyl-1-pyrroline, *7* 1-octen-3-one, *8* 2-ethyl-3,5-dimethylpyrazine, *9* (E)-2-nonenal (according to *Schieberle* and *Grosch*, 1992)

Table 5.7. Yields of aroma substances on distillation under vacuum[a]

Aroma substance (amount)[a]	Yield[b] (%)	
	Model I	Model II
3-Methylbutyric acid (1.9 µg)	91	31
Phenylacetaldehyde (4.2 µg)	84	21
3-Hydroxy-4,5-dimethyl-2(5H)-furanone (2.2 µg)	100	3.3
2-Phenylethanol (3.7 µg)	100	10.7
(E,E)-2,4-Decadienal (1.4 µg)	100	3.4
(E)-β-Damascenone (0.9 µg)	100	2.8
Vanillin (3.7 µg)	100	0.4

[a] Amount in the model solution: I in diethylether (50 ml), II in a mixture of diethylether (50 ml) and triglycerides (50 ml)
[b] Distillation in the apparatus shown in Fig. 5.5 at 35 °C

liquid food or the extract from the funnel (*1* in Fig. 5.5) into the distillation flask which is heated to 35–40 °C in a water bath (*2*). The volatiles including the solvent vapor are transferred into the distillation head (*3*). The distillate is condensed by liquid nitrogen in the receiver (*4*). The *Dewar* flask (*5*) protects the vacuum pump (reduced pressure 10^{-3} Pa).

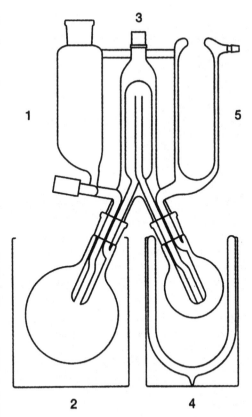

Fig. 5.5. Apparatus for the distillation of aroma substances from foods (for explanation, see text. According to *Engel* et al., 1999)

Solid foods are first extracted, the addition of water may be required to increase the yield of aroma substances.

An extraction combined with distillation can be achieved using an apparatus designed by *Likens–Nickerson* (Fig. 5.6).

In this process, low-boiling solvents are usually used to make subsequent concentration of the aroma substances easier. Therefore, this process is carried out at normal pressure or slightly reduced pressure. The resulting thermal treatment of the food can lead to reactions (examples in Table 5.6) that change the aroma composition. The example in Table 5.8 shows the extent to which some aroma substances are released from glycosides during simultaneous distillation/extraction.

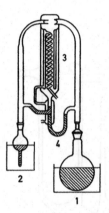

Fig. 5.6. Apparatus according to *Likens* and *Nickerson* used for simultaneous extraction and distillation of volatile compounds.
1 Flask with heating bath containing the aqueous sample, *2* flask with heating bath containing the solvent (e. g. pentane), *3* cooler, *4* condensate separator: extract is the upper and water the lower phase

Table 5.8. Isolation of odorants from cherry juice – Comparison of distillation in vacuo (I) with simultaneous distillation and extraction (II)

Odorant	I	(μg/l)	II
Benzaldehyde	202		5260
Linalool	1.1		188

Table 5.9. Relative retention time (t_{rel}) of some compounds separated by gas chromatography using Porapak Q as stationary phase (Porapak: styrene divinylbenzene polymer; T: 55 °C)

Compound	t_{rel}	Compound	t_{rel}
Water	1.0	Methylthiol	2.6
Methanol	2.3	Ethylthiol	20.2
Ethanol	8.1	Dimethylsulfide	19.8
Acetaldehyde	2.5	Formic acid	
Propanal	15.8	ethyl ester	6.0

5.2.1.2 Gas Extraction

Volatile compounds can be isolated from a solid or liquid food sample by purging the sample with an inert gas (e. g., N_2, CO_2, He) and adsorbing the volatiles on a porous, granulated polymer (Tenax GC, Porapak Q, Chromosorb 105), followed by recovery of the compounds. Water is retarded to only a negligible extent by these polymers (Table 5.9). The desorption of volatiles is usually achieved stepwise in a temperature gradient. At low temperatures, the traces of water are removed by elution, while at elevated temperatures, the volatiles are released and flushed out by a carrier gas into a cold trap, usually connected to a gas chromatograph.

5.2.1.3 Headspace Analysis

The headspace analysis procedure is simple: the food is sealed in a container, then brought to the desired temperature and left for a while to establish an equilibrium between volatiles bound to the food matrix and those present in the vapor phase. A given volume of the headspace is withdrawn with a gas syringe and then injected into a gas chromatograph equipped with a suitable separation column (static headspace analysis). Since the water content and an excessively large volume of the sample substantially reduce the separation efficiency of gas chromatography, only the major volatile compounds are indicated by the detector. The static headspace analysis makes an important contribution when the positions of the aroma substances in the chromatogram are determined by olfactometry (cf. 5.2.2.2).

More material is obtained by dynamic headspace analysis or by solid phase microextraction (SPME). In the dynamic procedure the headspace volatiles are flushed out, adsorbed and thus concentrated in a polymer, as outlined in 5.2.1.2. However, it is difficult to obtain a representative sample by this flushing procedure, a sample that would match the original headspace composition. A model system assay (Fig. 5.7) might clarify the problems. Samples (e) and (f) were obtained by adsorption on different polymers. They are different from each other and differ from sample (b), which was obtained

directly for headspace analysis. The results might agree to a greater extent by varying the gas flushing parameters (gas flow, time), but substantial differences would still remain. A comparison of samples (a) and (g) in Fig. 5.7 shows that the results obtained by the distillation-extraction procedure give a relatively good representation of the composition of the starting solution, with the exception of ethanol. However, the formation of artifacts is critical (cf. 5.2.1.1).

SPME is based on the partitioning of compounds between a sample and a coated fiber immersed in it. The odorants are first adsorbed onto the fiber (e. g. nonpolar polydimethylsilo-xane or polar polyacrylate) immersed in a liquid food, a food extract or in the headspace above a food sample for a certain period of time. After adsorption is completed, the compounds are thermally desorbed into a GC injector block for further analysis.

Particularly in food applications headspace SPME is preferred to avoid possible contamination of the headspace system by non-volatile food components. Also SPME analysis is quite sensitive to experimental conditions. In addition to the stationary phase, sample, volume concentration of odorants, sample matrix and uniformity as well as temperature and time of the adsorption and desorption processes influence the yield. In quantitative SPME analysis these influences are eliminated by the use of labelled internal standards (cf. 5.2.6.1).

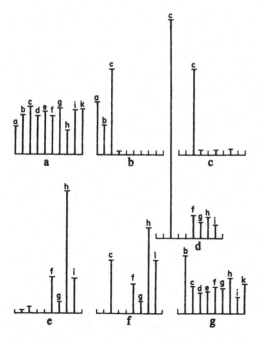

Fig. 5.7. A comparison of some methods used for aroma compound isolation (according to *Jennings* and *Filsoof*, 1977).

a *a* Ethanol, *b* 2-pentanone, *c* heptane, *d* pentanol, *e* hexanol, *f* hexyl formate, *g* 2-octanone, *h* d-limonene, *i* heptyl acetate and *k* γ-heptalactone. **b** Headspace analysis of aroma mixture **a**. **c** From aroma mixture 10 μl is dissolved in 100 ml water and the headspace is analyzed. **d** As in **c** but the water is saturated with 80% NaCl. **e** As in **c** but purged with nitrogen and trapped by Porapak Q. **f** As in **c** but purged with nitrogen and trapped by Tenax GC. **g** As in **e** but distilled and extracted (cf. Fig. 5.6)

5.2.2 Sensory Relevance

In many earlier studies on the composition of aromas, each volatile compound was regarded as an aroma substance. Although lists with hundreds of compounds were obtained for many foods, it was still unclear which of the volatiles were really significant as odorants and to what extent important odorants occurring in very low concentrations were detected.

The studies meanwhile concentrate on those compounds which significantly contribute to aroma. The positions of these compounds in the gas chromatogram are detected with the help of dilution analyses. Here, both of the following methods based on the aroma value concept (cf. 5.1.4) find application.

5.2.2.1 Aroma Extract Dilution Analysis (AEDA)

In AEDA, the concentrate of the odorants obtained by distillation is separated by gas chromatography on a capillary column. To determine the retention times of the aroma substances, the carrier gas stream is subjected to sniffing detection after leaving the capillary column (GC/olfactometry). The sensory assessment of a single GC run, which is often reported in the literature, is not very meaningful because the perception of aroma substances in the carrier gas stream depends on limiting quantities which have nothing to do with the aroma value, e. g., the amount of food analysed, the degree of concentration of the volatile fraction, and the amount of sample separated by gas chromatography.

These limitations are eliminated by the stepwise dilution of the volatile fraction with solvent, followed by the gas chromatographic/olfactometric analysis of each dilution. The process is continued until no more aroma substance can be detected by GC olfactometry. In this way, a dilution factor 2^n (n = number of $1 + 1$ dilutions) is determined for each aroma substance that appears in the gas chromatogram. It is designated as the flavor dilution factor (FD factor) and indicates the number of parts of solvent required to dilute the aroma extract until the aroma value is reduced to one.

Another more elaborate variant of the dilution analysis requires, in addition, that the duration of each odor impression is recorded by a computer and CHARM values are calculated (CHARM: acronym for combined hedonic response measurement), which are proportional to aroma values. The result of an AEDA can be represented as a diagram. The FD factor is plotted against the retention time in the form of the retention index (RI) and the diagram is called a FD chromatogram.

The FD chromatograms of the volatile compounds of white bread and French fries are presented in Fig. 5.4 and 5.8, respectively.

The identification experiments concentrate on those aroma substances which appear in the FD chromatogram with higher FD factors. To detect all the important aroma substances, the range of FD factors taken into account must not be too narrowly set at the lower end because

differences in yield shift the concentration ratios. Labile compounds can undergo substantial losses and when distillation processes are used, the yield decreases with increasing molecular weight of the aroma substances.

In the case of French fries (Fig. 5.8), 19 aroma substances appearing in the FD-factor range 2^1–2^7 were identified (cf. legend of Fig. 5.8). Based on the high FD factors, the first approximation indicates that methional, 2-ethyl-3,5-dimethylpyrazine, 2,3-diethyl-5-methylpyrazine and (E,E)-2,4-decadienal substantially contribute to the aroma of French fries.

5.2.2.2 Headspace GC Olfactometry

In the recovery of samples for AEDA, highly volatile odorants are lost or are covered by the solvent peak in gas chromatography, e. g., methanethiol and acetaldehyde. For this reason, in addition to AEDA, a sample is drawn from the gas space above the food, injected into the gas chromatograph, transported by the carrier gas stream into a cold trap and concentrated there, as shown in Fig. 5.9. After quick evaporation, the sample is flushed into a capillary column by the carrier gas and chromatographed. At the end of the capillary, the experimentor sniffs the carrier gas stream and determines the positions of the chromatogram at which the odorants appear. The gas chromatogram is simultaneously monitored by a detector.

To carry out a dilution analysis, the volume of the headspace sample is reduced stepwise until no odorant is detectable by gas chromatography/olfactometry. In our example with French fries (Fig. 5.10), e. g., the odors of methanethiol, methylpropanal and dimethyltrisulfide were detectable in the sixth dilution, but only methanethiol was detectable in the seventh. The eighth dilution was odorless. Further experiments showed that methanethiol does in fact belong to the key odorants of French fries.

In GC/olfactometry, odor thresholds are considerably lower than in solution because the aroma substances are subjected to sensory assessment in a completely vaporized state. The examples given in Table 5.10 show how great the differences can be when compared to solutions of the aroma substances in water.

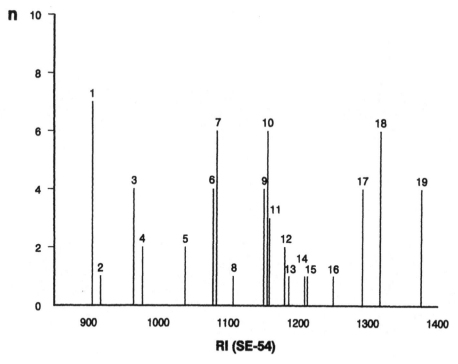

Fig. 5.8. FD chromatogram of the volatile fraction of French fries. Ordinate: n, number of 1 + 1 dilutions. Abscissa: retention index (RI) on the capillary SE-54. The following odorants were identified: *1* methional, *2* 2-acetyl-1-pyrroline, *3* dimethyltrisulfide, *4* 1-octen-3-one, *5* phenylacetaldehyde, *6* 2-ethyl-3,6-dimethyl-pyrazine, *7* 2-ethyl-3,5-dimethylpyrazine, *8* nonanal, *9* (Z)-2-nonenal, *10* 2,3-diethyl-5-methylpyrazine, *11* (E)-2-nonenal, *12* 2-ethenyl-3-ethyl-5-methylpyrazine, *13* 2-isobutyl-3-methoxypyrazine, *14* dimethyltetrasulfide, *15* (E,E)-2,4-nonadienal, *16* (Z)-2-decenal, *17* (E,Z)-2,4-decadienal, *18* (E,E)-2,4-decadienal, *19* trans-4,5-epoxy-(E)-2-decenal (according to *Wagner* and *Grosch*, 1997)

Table 5.10. Odor thresholds of aroma substances in air and water

Compound	Odor thresholds in		
	Air (a) (ng/I)	Water (b) (μg/l)	b/a
β-Damascenone	0.003	0.002	6.7×10^2
Methional	0.12	0.2	1.6×10^3
2-Methylisoborneol	0.009	0.03	3.3×10^3
2-Acetyl-1-pyrroline	0.02	0.1	5×10^3
4-Vinylguaiacol	0.6	5	8.3×10^3
Linalool	0.6	6	1.0×10^4
Vanillin	0.9	20	2.2×10^4
4-Hydroxy-2,5-dimethyl-3(2H)-furanone (furaneol)	1.0	30	3×10^4

5.2.3 Enrichment

When an aroma concentrate contains phenols, organic acids or bases, preliminary separation of these compounds from neutral volatiles by extraction with alkali or acids is advantageous.

The acidic, basic and neutral fractions are individually analyzed. The neutral fraction by itself consists of so many compounds that in most cases not even a gas chromatographic column with the highest resolving power is able to separate them into individual peaks. Thus, separation of the neutral fraction is advisable and is usually achieved by liquid chromatography, or preparative gas or high performance liquid chromatography. A preliminary separation of aroma extracts is achieved

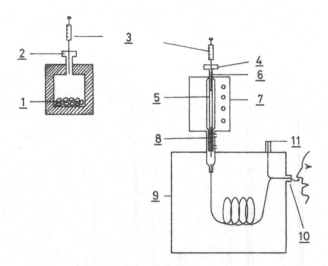

Fig. 5.9. Apparatus for the gas chromatography–olfactometry of static headspace samples. *1* Sample in thermostated glass vessel, *2* septum, *3* gastight syringe, *4* injector, *5* hydrophobed glass tube, *6* carrier gas, e. g., helium, *7* purge and trap system, *8* cold trap, *9* gas chromatograph with capillary column, *10* sniffing port, *11* flame ionization detector (according to *Guth* and *Grosch*, 1993)

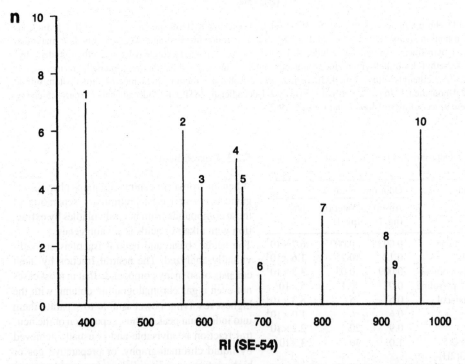

Fig. 5.10. FD chromatogram of static headspace samples of French fries. Ordinate: n, number of 1 + 1 dilutions. Abscissa: retention index (RI) on the capillary SE-54. The following odorants were identified: *1* methanethiol, *2* methylpropanal, *3* 2,3-butanedione, *4* 3-methylbutanal, *5* 2-methylbutanal, *6* 2,3-pentanedione, *7* hexanal, *8* methional, *9* 2-acetyl-1-pyrroline, *10* dimethyltrisulfide (according to *Wagner* and *Grosch*, 1997)

by chromatography on silica gel, as shown in Table 5.11 for coffee aroma. To localize the aroma substances each of the four fractions is analyzed by gas chromatography and olfactometry. Some volatile compounds are aroma active in such low concentrations that even enrichment by column chromatography does not allow identification, e. g., 3-methyl-2-butenethiol (Fraction A in Table 5.11) and the two methoxypyrazines (Fraction B) in coffee. In most cases, further concentration is achieved with the help of multidimensional gas chromatography (MGC). The fraction which contains the unknown aroma

Table 5.11. Column chromatographic preliminary separation of an aroma extract of roasted coffee

Fraction[a]	Aroma substance
A	2-Methyl-3-furanthiol, 2-furfurylthiol, bis(2-methyl-3-furyl)disulfide, 3-methyl-2-butenethiol
B	2,3-Butanedione, 3-methylbutanal, 2,3-pentanedione, trimethylthiazole, 3-mercapto-3-methylbutylformiate, 3-isopropyl-2-methoxypyrazine, phenylacetaldehyde, 3-isobutyl-2-methoxypyrazine, 5-methyl-5(H)-cyclopentapyrazine, p-anisaldehyde, (E)-β-damascenone
C	Methional, 2-ethenyl-3,5-dimethylpyrazine, linalool, 2,3-diethyl-5-methylpyrazine, guaiacol, 2-ethenyl-3-ethyl-5-methylpyrazine, 4-ethylguaiacol, 4-vinylguaiacol
D	2-/3-Methylbutyric acid, trimethylpyrazine, 3-mercapto-3-methyl-1-butanol, 5-ethyl-2,4-dimethylthiazole, 2-ethyl-3,5-dimethylpyrazine, 3,4-dimethyl-2-cyclopentenol-1-one, 4-hydroxy-2,5-dimethyl-3(2H)-furanone, 5-ethyl-4-hydroxy-2-methyl-3(2H)-furanone, 3-hydroxy-4,5-dimethyl-2(5H)-furanone, 5-ethyl-3-hydroxy-4-methyl-2(5H)-furanone, vanillin

[a] Chromatography at 10–12 °C on a silica gel column (24 × 1 cm, deactivated with 7% water); elution with mixtures of pentane-diethylether (50 ml, 95 + 5, v/v, fraction A; 30 ml, 75 × 25, v/v, Fraction B; 30 ml, 1 + 1, v/v, Fraction C) and with diethylether (100 ml, Fraction D).

substance is first subjected to preliminary separation on a polar capillary. The eluate containing the substance is cut out, rechromatographed on a non-polar capillary and finally analyzed by mass spectrometry. The MGC is also used in quantitative analysis for the preliminary purification of analyte and internal standard (cf. 5.2.6.1).

5.2.4 Chemical Structure

In the structure elucidation of aroma substances, mass spectrometry has become an indispensable tool because the substance amounts eluted by gas chromatography are generally sufficient for an evaluable spectrum. If the corresponding reference substance is available, identification of the aroma substance is based on agreement of the mass spectrum, retention indices on at least two capillary columns of different polarity, and of odor thresholds, which are compared by gas chromatography/olfactometry. If the reference substance is not available, the following procedure is suitable for the identification of the odorant:

The analyte is concentrated until a ^{1}H-NMR spectrum and, if necessary, a ^{13}C-spectrum can be measured. An example is the identification of the characteristic odorant of roasted hazelnuts. The mass spectrum of this substance (Fig. 5.11a) indicates an unsaturated carbonyl compound with a molar mass of 126. In conjunction with the structural elements shown by the ^{1}H-NMR spectrum (Fig. 5.11b), it was proposed that the odorant is 5-methyl-(E)-2-hepten-4-one (Filbertone).

It goes without saying that the synthesis of the proposed aroma substance was part of the identification. It was also guaranteed that its chromatographic and sensory properties correspond with those of the unknown odorant.

5.2.5 Enantioselective Analysis

In the case of chiral aroma substances, elucidation of the absolute configuration and determination of the enantiomeric ratio, which is usually given as the enantiomeric excess (ee), are of especial interest because the enantiomers of a compound can differ considerably in their odor quality and threshold. The compound 3a,4,5,7a-

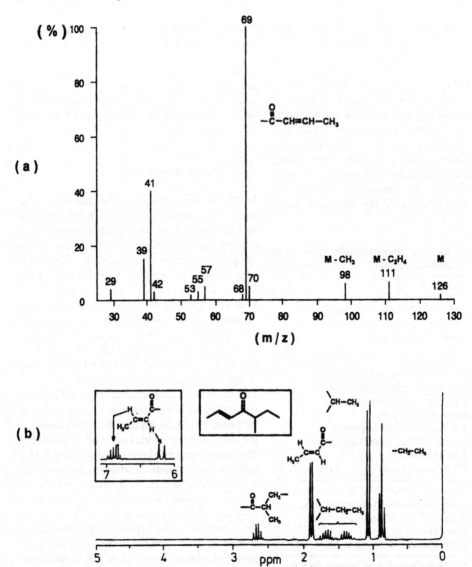

Fig. 5.11. Instrumental analysis of 5-methyl-(E)-2-hepten-4-one (according to *Emberger*, 1985) **(a)** mass spectrum, **(b)** ^{1}H-NMR spectrum (for discussion, see text)

tetrahydro-3,6-dimethyl-2(3H)-benzofuranone (wine lactone) represents an impressive example which shows how much the odor activity of enantiomers can vary. The four enantiomeric pairs of this compound have been separated by gas chromatography on a chiral phase (Fig. 5.12). The 3S,3aS,7aR-enantiomer (No. 6 in Table 5.12) has the lowest odor threshold of the eight diastereomers. The identification of this substance in wine (cf. 20.2.6.9) led to the name wine

lactone. Two diastereomers (No. 3 and 8) are odorless.

The determination of the ee value can be used to detect aromatization with a synthetic chiral aroma substance because in many cases one enantiomer is preferentially formed in the biosynthesis of chiral aroma substances (examples in Table 5.13). In contrast to biosynthesis, chemical synthesis gives the racemate which is usually not separated for economic reasons. The addition of an aroma sub-

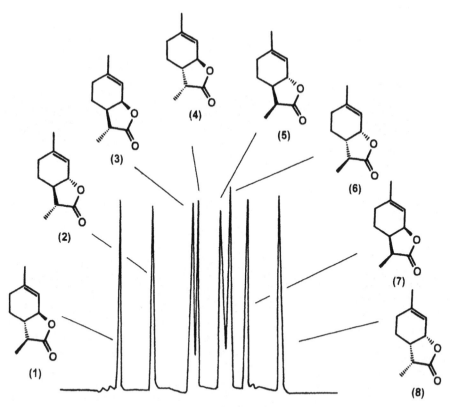

Fig. 5.12. Gas chromatogram of the diastereomers of 3a,4,5,7a-tetrahydro-3,6-dimethyl-2(3H)-benzofuranone (wine lactone) on a chiral phase (according to *Guth*, 1997)

Table 5.12. Odor threshold values of diastereomeric 3a,4,5,7a-tetrahydro-3,6-dimethyl-2(3H)-benzofuranone

No.[a]	Stereoisomer-conformation	Odor threshold (ng/l air)
1	(3S,3aS,7aS)	0.007–0.014
2	(3R,3aR,7aR)	14–28
3	(3R,3aR,7aS)	>1000
4	(3R,3aS,7aS)	8–16
5	(3S,3aR,7aR)	0.05–0.2
6	(3S,3aS,7aR)	0.00001–0.00004
7	(3S,3aR,7aS)	80–160
8	(3R,3aS,7aR)	>1000

[a] Numbering as in Fig. 5.12.

ing, e. g., that of filbertone decreases during the roasting of hazelnuts (cf. Table 5.13).

Table 5.13. Enantiomeric excess (ee) of chiral aroma substances in some foods

Aroma substance	Food	ee (%)
R(+)-γ-Decalactone	Peach, apricot, mango, strawberry pineapple, maracuya	>80
R(+)-δ-Decalactone	Milk fat	60
R(+)-trans-α-Ionone	Raspberry	92.4
	Carrot	90.0
	Vanilla bean	94.2
R(-)-1-Octen-3-ol	Mushroom, chanterelle	>90
S(+)-E-5-Methyl-2-hepten-4-one (filbertone)	Hazelnut, raw	60–68
	Hazelnut, roasted	40–45
R-3-Hydroxy-4,5-dimethyl-2(5H)-furanone (sotolon)	Sherry	ca. 30

stance of this type can be determined by enantioselective analysis if safe data on the enantiomeric excess of the compound in the particular food are available. It should also be taken into account that the ee value can change during food process-

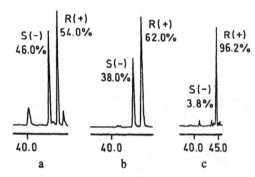

Fig. 5.13. Enantioselective gas chromatographic analysis of trans-α-ionone in aroma extracts of different raspberry fruit juice concentrates (according to *Werkhoff* et al., 1990): **a** and **b** samples with nature identical aroma, **c** natural aroma

The method frequently applied to determine ee values is the enantioselective gas chromatographic analysis of the aroma substance on a chiral phase, e. g., peralkylated cyclodextrins. This method was used, e. g., to test raspberry fruit juice concentrates for unauthorized aromatization with trans-α-ionone. The gas chromatograms of trans-α-ionone from two different samples are shown in Fig. 5.13. The low excesses of the R-enantiomer of ee = 8% (concentrate A) and ee = 24% (B) can probably be put down to the addition of synthetic trans-α-ionone racemate to the fruit juice concentrate because in the natural aroma (C), the ee value is 92.4%.

5.2.6 Quantitative Analysis, Aroma Values

5.2.6.1 Isotopic Dilution Analysis (IDA)[*]

The quantitative analysis of aroma substances using conventional methods often gives incorrect values. The high vapor pressure, the poor extractability especially of polar aroma substances from hydrous foods and the instability of important aroma substances, e. g., thiols, can cause unforeseeable losses in the purification of the samples and in gas chromatography.

The results of quantitative analyses are exact (standard deviation <10%) and reproducible if the chemical structure of the internal standard is very similar to the structure of the analyte. An

_____
[*] Most of the quantitative data on aroma substances in this book come from IDAs.

isotopomer of the analyte is the most similar. In this case, the physical and chemical properties of both substances correspond, except for a small isotope effect which can lead to partial separation in capillary gas chromatography.

The examples given in Fig. 5.14 show that for economic reasons, mostly internal standards labelled with deuterium are synthesized for IDA. The considerably more expensive carbon isotope 13 is introduced into the odorant (examples are the internal standards No. 11 and 12 in Fig. 5.14) only if a deuterium/protium exchange can occur in the course of analysis. This exchange would falsify the result. Another advantage of this isotope is the completely negligible isotope effect compared to deuterium.

It is easy to conduct an IDA because losses of analyte in the distillative recovery (cf. 5.2.1.1) and in purification do not influence the result since the standard suffers the same losses. These advantages of IDA are used in food chemistry for other analytes as well, e. g., pantothenic acid (cf. 6.3.5.2) or for the mycotoxin patulin (cf. 9.2.3).

The quantification of the odorants 2-furfurylthiol (FFT), 2-methyl-3-furanthiol (MFT) and 3-mercapto-2-pentanone (3M2P) in boiled meat will be considered as an example. Especially MFT and 3M2P are very instable, so after the addition of the deuterated standards d-FFT, d-MFT and d-3M2P (No. 13 in Fig. 5.14) to the extract, it is advisable to concentrate via a trapping reaction for thiols which is performed with p-hydroxymercuribenzoic acid. The analytes and their standards are displaced from the derivatives by cysteine in excess, separated by gas chromatography, and analyzed by mass spectrometry. In this process, mass chromatograms for the ions are monitored in which the analyte and its isotopomer differ (Fig. 5.15). After calibration, the mass chromatograms are evaluated via a comparison of the areas of analyte and standard. 2-Mercapto-3-pentanone (2M3P) is also identified in this analysis. However, this compound is of no importance for the aroma of boiled meat because of its lower concentration and higher odor threshold compared to those of 3M2P.

5.2.6.2 Aroma Values (AV)

To approach the situation in food aroma values (definition cf. 5.1.4) are calculated. It is assumed

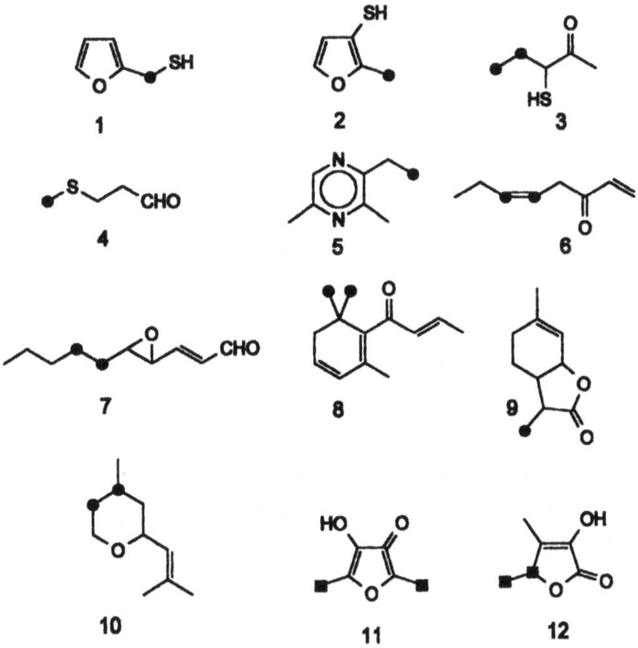

Fig. 5.14. Odorants labelled with deuterium (●) or carbon-13 (■) as internal standard substances for isotopic dilution analyses of the corresponding unlabelled odorants.
1 2-[α-^{2}H$_2$]furfurylthiol, *2* 2-[^{2}H$_3$]methyl-3-furanthiol, *3* 3-mercapto-2-[4,5-^{2}H$_2$]pentanone, *4* [4-^{2}H$_3$]methio-nal, *5* 2-[^{2}H$_3$]ethyl-3,5-dimethylpyrazine, *6* (Z)-1,5-[5,6-^{2}H$_2$]octadien-3-one, *7* trans-4,5-epoxy-(E)-2-[6,7^{2}H$_4$] decenal, *8* 1-(2,6,6-[6,6-^{2}H$_6$]trimethyl-1,3-cyclohexadienyl)-2-buten-1-one (β-damascenone), *9* 3a,4,5,7atetra-hydro-3,6-[3-^{2}H$_3$]dimethyl-2(3H)-benzofuranone (wine lactone), *10* tetrahydro-4-methyl-2-(2-methylpropenyl)-2H-[3,4-^{2}H$_3$]pyran (sotolon), *11* 4-hydroxy-2,5-[^{13}C$_2$]dimethyl-3(2H)-furanone, *12* 3-hydroxy-4,5-[4-^{13}C]di-methyl-2(5H)-[5-^{13}C]furanone (rose oxide)

that the odorants showing higher AVs contribute strongly to the aroma of the food. For this purpose, the odor thresholds of the compounds dissolved in water, in oil or applied to starch are used, depending on which of these materials dominates in the food.

An example are the AVs of the odorants of French fries based on their odor thresholds in an oil (Table 5.14). Methanethiol, methional, methylpropanal and 2-methylbutanal exhibit the highest aroma values. Consequently, they should belong to the most important odorants of French fries.

5.2.7 Aroma Model, Omission Experiments

Finally, the identified odorants must actually produce the aroma in question. To test this, the determined concentrations of the odorants are dis-

solved in a suitable medium, which is not difficult in the case of liquid foods. The solvent for the recombination mixture called the aroma model can be adapted to the food. An ethanol/water mixture, for example, is suitable for wine. In the case of solid foods, however, compromises have to be accepted.

The aroma profile of the model is then compared to that of the food. In the example of French fries discussed in detail here, a very good approximation of the original aroma was achieved.

The selection of odorants by dilution analyses (cf. 5.2.2) does not take into account additive (cf. 20.1.7.8) or antagonistic effects (example in Fig. 5.2) because the aroma substances, after separation by gas chromatography, are sniffed individually. Therefore, in view of the last mentioned effect, the question arises whether all the compounds occurring in the aroma model really contribute to the aroma in question. To answer

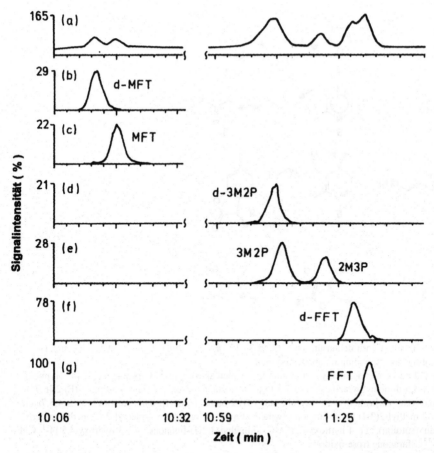

Fig. 5.15. Isotopic dilution analysis of 2-furfurylthiol (FFT), 2-methyl-3-furanthiol (MFT) and 3-mercapto-2-pentanone (3M2P).
(a) Gas chromatogram, (b–g) mass chromatograms of the analytes and the deuterated (d) internal standards; traces of the ions shown in brackets were monitored: d-MFT (m/z 118), MFT (m/z 115), d-3M2P (m/z 121), 3M2P and 2M3P (m/z 119), d-FFT (m/z 83), FFT (m/z 81) (according to *Kerscher* and *Grosch*, 1998)

this question, one or several aroma substances are omitted in the model and a triangle test is used to examine which of three samples (two complete and one reduced aroma model) offered to the testers in random order differs in aroma from the others. If a significant number of testers determine a difference in the reduced model, it can be assumed that the odorants lacking in the reduced model contribute to the aroma and, consequently, belong to the key odorants of the food.

Some omission experiments, e.g., conducted with the aroma model for French fries, are shown in Table 5.15.

If methanethiol and the two decadienal isomers are missing in Experiments 1 and 2, the aroma

has no similarity to that of French fries. All five testers were in agreement. The *Strecker* aldehydes with the malt odor (Exp. 3), 4,5-epoxydecenal (Exp. 4) and both pyrazines (Exp. 5) are also important for the aroma because their absence was noticed by four of the five testers. 1-Octen-3-one, (Z)-2- and (E)-2-nonenal are of no importance for the aroma (Exp. 6). Surprisingly, this also applies to methional (Exp. 7) although it has the second highest aroma value (cf. Table 5.14) and smells of boiled potatoes. It is obvious that methional is masked by other odorants occurring in the aroma model. In French fries, the odor note "like boiled potatoes" is probably produced by methanethiol in combination with pyrazines.

Table 5.14. Volatile compounds with high aroma values in French fries[a]

Compound	Concentration[b] (µg/kg)	Odor threshold[c] (µg/kg)	Aroma value[d]
Methanethiol	1240	0.06	2×10^4
Methional	783	0.2	3.9×10^3
Methylpropanal	5912	3.4	1.7×10^3
2-Methylbutanal	10599	10	1.1×10^3
trans-4,5-Epoxy-(E)-2-decenal	771	1.3	592
3-Methylbutanal	2716	5.4	503
(E,Z)-2,4-Decadienal	1533	4	383
4-Hydroxy-2,5-dimethyl-3 (2H)-furanone	2778	25	111
2,3-Diethyl-5-methylpyrazine	41	0.5	83
(E,E)-2,4-Decadienal	6340	180	35
2,3-Butanedione	306	10	31
2-Ethyl-3,5-dimethylpyrazine	42	2.2	19
2-Ethenyl-3-ethyl-5-methylpyrazine	5.4	0.5	11
3-Isobutyl-2-methoxypyrazine	8.6	0.8	11
2-Ethyl-3,6-dimethylpyrazine	592	57	10

[a] Potato sticks deep-fried in palm oil.
[b] Results of IDA.
[c] Odor threshold of the compound dissolved in sunflower oil.
[d] Quotient of concentration and odor threshold.

Table 5.15. Aroma model for French fries as affected by the absence of one or more odorants[a]

Exp. No.	Odorant omitted in the model	Number[b]
1	Methanethiol	5
2	(E,Z)-2,4-Decadienal and (E,E)-2,4-decadienal	5
3	Methylpropanal, 2- and 3-methylbutanal	4
4	trans-4,5-Epoxy-(E)-2-decenal	4
5	2-Ethyl-3,5-dimethylpyrazine and 3-ethyl-2,5-dimethylpyrazine	4
6	1-Octen-3-one, (Z)-2- and (E)-2-nonenal	1
7	Methional	0

[a] Models lacking in one or more components were each compared to the model containing the complete set of 19 odorants.
[b] Number of the assessors detecting an odor difference in triangle tests, maximum 5.

The instrumental and sensory methods presented in the French fries example have also been successfully applied in the elucidation of other aromas. The results are presented in the book for some individual foods.

5.3 Individual Aroma Compounds

The results of dilution analyses and of aroma simulation experiments show that only 5% of the more than 7000 volatile compounds identified in foods contribute to aromas. The main reason for the low number of odorants in the volatile fraction is the marked specificity of the sense of smell (for examples, cf. 5.6).

Important odorants grouped according to their formation by nonenzymatic or enzymatic reactions and listed according to classes of compounds are presented in the following sections. Some aroma compounds formed by both enzymatic and nonenzymatic reactions are covered in sections 5.3.1 and 5.3.2. It should be noted that the reaction pathways for each aroma compound are differentially established. Frequently, they are dealt with by using hypothetical reaction pathways which lead from the precursor to the odorant. The reaction steps and the intermediates of the pathway are postulated

Table 5.16. Some *Strecker* degradation aldehydes[a]

Amino acid precursor	Strecker-aldehyde			Odor threshold value
	Name	Structure	Aroma description	(µg/l; water)
Gly	Formaldehyde	CH_2O	Mouse-urine, ester-like	50×10^3
Ala	Ethanal	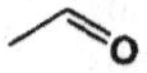	Sharp, penetrating, fruity	10
Val	Methylpropanal	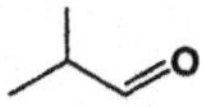	Malty	1
Leu	3-Methylbutanal	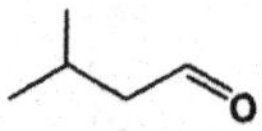	Malty	0.2
Ile	2-Methylbutanal	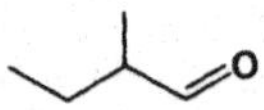	Malty	4
Phe	2-Phenylethanal	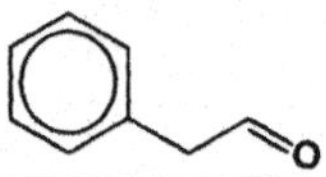	Flowery, honey-like	4

[a] Methional will be described in 5.3.1.4.

by using the general knowledge of organic chemistry or biochemistry. For an increasing number of odorants, the proposed formation pathway can be based on the results of model experiments. Postulated intermediates have also been confirmed by identification in a numbers of cases. However, studies on the formation of odorants are especially difficult since they involve, in most cases, elucidation of the side pathways occurring in chemical or biochemical reactions, which quantitatively are often not much more than negligible.

5.3.1 Nonenzymatic Reactions

The question of which odorants are formed in which amounts when food is heated depends on the usual parameters of a chemical reaction. These are the chemical structure and concentration of the precursors, temperature, time and environment, e. g., pH value, entry of oxygen and the water content. Whether the amounts formed are really sufficient for the volatiles to assert themselves in the aroma depend on their odor thresholds and on interactions with other odorants.

Aroma changes at room temperature caused by nonenzymatic reactions are observed only after prolonged storage of food. Lipid peroxidation (cf. 3.7.2.1), the *Maillard* reaction and the related *Strecker* degradation of amino acids (cf. 4.2.4.4.7) all play a part. These processes are greatly accelerated during heat treatment of food. The diversity of aroma is enriched at the higher temperatures used during roasting or frying. The food surface dries out and pyrolysis of carbohydrates, proteins, lipids and other constituents, e. g., phenolic acids, takes place generating odorants, among other compounds.

The large number of volatile compounds formed by the degradation of only one or two constituents is characteristic of nonenzymatic reactions. For example, 41 sulfur-containing compounds, including 20 thiazoles, 11 thiophenes, 2 dithiolanes and 1 dimethyltrithiolane, are obtained by heating cysteine and xylose in tributyrin at 200 °C. Nevertheless, it should not be overlooked that even under these drastic conditions, most of the volatile compounds are only formed in concentrations which are far less than the often relatively high odor thresholds (cf. 5.6). For this reason, only a small fraction of the many volatile compounds formed in heated foods is aroma active.

5.3.1.1 Carbonyl Compounds

The most important reactions which provide volatile carbonyl compounds were presented in sections 3.7.2.1.9 (lipid peroxidation), 4.2.4.3.3 (caramelization) and 4.2.4.4.7 (amino acid decomposition by the *Strecker* degradation mechanism).

Some *Strecker* aldehydes found in many foods are listed in Table 5.16 together with the corresponding aroma quality data. Data for carbonyls derived from fatty acid degradation are found in Table 3.32. Carbonyls are also obtained by degradation of carotenoids (cf. 3.8.4.4).

5.3.1.2 Pyranones

Maltol (3-hydroxy-2-methyl-4H-pyran-4-one) is obtained from carbohydrates as outlined in 4.2.4.4.4 and has a caramel-like odor. It has been found in a series of foods (Table 5.17), but in concentrations that were mostly in the range of the relatively high odor threshold of 9 mg/kg (water).

Maltol enhances the sweet taste of food, especially sweetness produced by sugars (cf. 8.6.3), and is able to mask the bitter flavor of hops and cola.

Ethyl maltol [3-hydroxy-2-ethyl-4H-pyran-4-one] enhances the same aroma but is 4- to 6-times more powerful than maltol. It has not been detected as a natural constituent in food. Nevertheless, it is used for food aromatization.

5.3.1.3 Furanones

Among the great number of products obtained from carbohydrate degradation, 3(2H)- and 2(5H)-furanones belong to the most striking aroma compounds (Table 5.18).

Table 5.17. Occurrence of maltol in food

Food product	mg/kg	Food product	mg/kg
Coffee, roasted	20–45	Chocolate	3.3
Butter, heated	5–15	Beer	0–3.4
Biscuit	19.7		

Compounds I–III, V and VI in Table 5.18, as well as maltol and the cyclopentenolones (cf. 4.2.4.3.2), have a planar enol-oxo-configuration

$$(5.4)$$

and a caramel-like odor, the odor threshold of aqueous solutions being influenced by the pH.

In Table 5.19, the examples furanone I and II show that the threshold value decreases with decreasing pH. As with the fatty acids (cf. 3.2.1.1), the vapor pressure and, consequently, the concentration in the gas phase increase with decreasing dissociation. The fact that furanone I does not appreciably contribute to food aromas is due to its high odor threshold. However, this compound is of interest as a precursor of 2-furfurylthiol (cf. 5.3.1.4). If the hydroxy group in furanone II is methylated to form IV, the caramel-like aroma note disappears.

A list of foods in which furanone II has been identified as an important aroma substance is given in Table 5.20.

As the furanones are secondary products of the *Maillard* reaction, their formation is covered in 4.2.4.3.2, 4.2.4.4.4 and 4.2.4.4.6. Whether the furanone II detected in fruit, which is partly present as the β-glycoside (e. g., in tomatoes, cf. Formula 5.5), is formed exclusively

$$(5.5)$$

by nonenzymatic reactions favored by the low pH is still not clear. Furanone V (sotolon) is a significant contributor to the aroma of, e. g., sherry, French white wine, coffee (drink) and above all of seasonings made on the basis of a protein hydrolysate (cf. 12.7.3.5). It is a chiral compound having enantiomers that differ in their odor threshold (Table 5.18) but not in their odor quality. It is formed in the *Maillard* reaction (cf. 4.2.4.4), but can also be produced from 4-hydroxyisoleucine (e. g., in fenugreek seeds, cf. 22.1.1.2.4). Furanone VI (abhexon) has

Table 5.18. Furanones in food

Structure	Substituent/trivial name or trade name (odor threshold in µg/kg, water)	Aroma quality	Occurrence
	A. 3(2H)-Furanones		
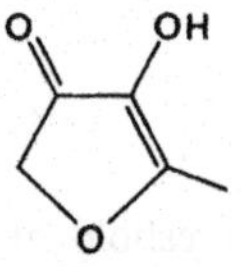 (I)	4-Hydroxy-5-methyl *Norfuraneol* (nasal: 23,000)	Roasted chicory-like, caramel	Meat broth
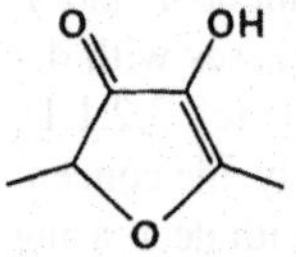 (II)	4-Hydroxy-2,5-dimethyl *Furaneol* (nasal: 60; retronasal: 25)	Heat-treated strawberry, pineapple-like, caramel	cf. Table 5.20
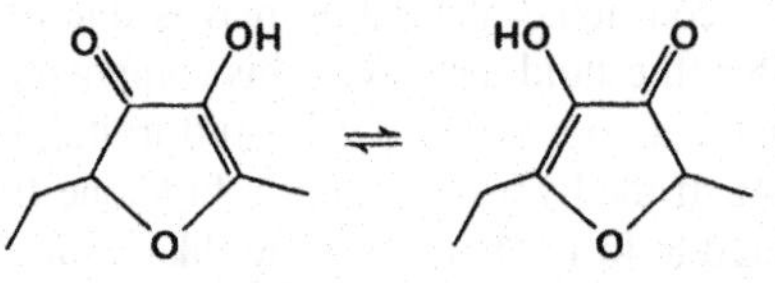 (III)	2-(5)-Ethyl-4-hydroxy-5-(2)-methyl[a] *Ethylfuraneol* (nasal: 7.5)	Sweet, pastry, caramel	Soya sauce Emmental cheese
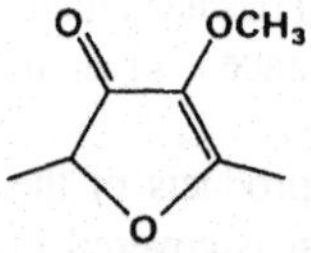 (IV)	4-Methoxy-2,5-dimethyl *Mesifuran* (nasal: 3400)	Sherry-like	Strawberry, raspberry[b]
	B. 2(5H)-Furanones		
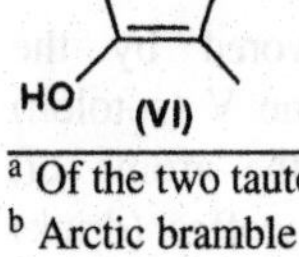 (V)	3-Hydroxyl-4,5-dimethyl *Sotolon* (nasal, R-form 90, recemate, retronasal: 3)	Caramel, protein hydrolysate S-form 7	Coffee, sherry, seasonings, fenugreek seeds
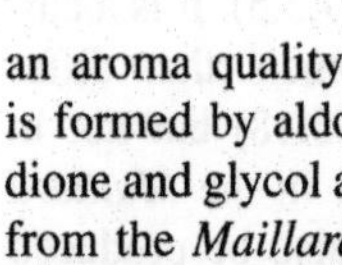 (VI)	5-Ethyl-3-hydroxy- *Abhexon* hydrolysate (nasal: 30, retronsal: 3)	Caramel, 4-methyl protein	Coffee, seasonings

[a] Of the two tautomeric forms, only the 5-ethyl-4-hydroxy-2-methyl isomer is aroma active.
[b] Arctic bramble (*Rubus arcticus*).

an aroma quality similar to that of sotolon and is formed by aldol condensation of 2,3-pentanedione and glycol aldehyde, which can be obtained from the *Maillard* reaction, or by aldol condensation of 2 molecules of α-oxobutyric acid, a degradation product of threonine (Fig. 5.16).

Quantitative analysis of furanones is not very easy because due to their good solubility in water, they are extracted from aqueous foods with poor yields and easily decompose, e. g., sotolon (cf. Formula 5.6). Correct values are obtained by IDA.

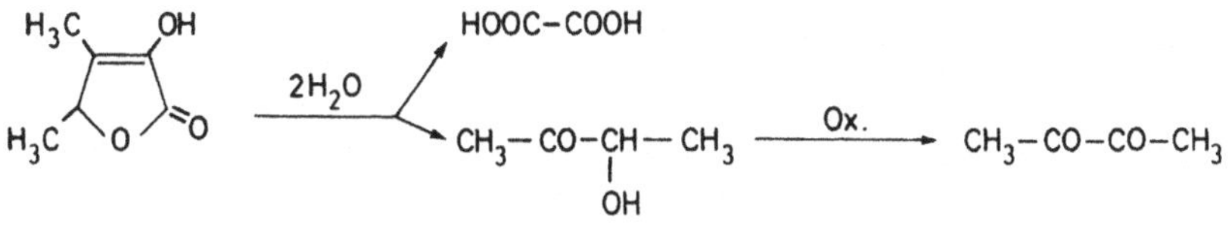

$$(5.6)$$

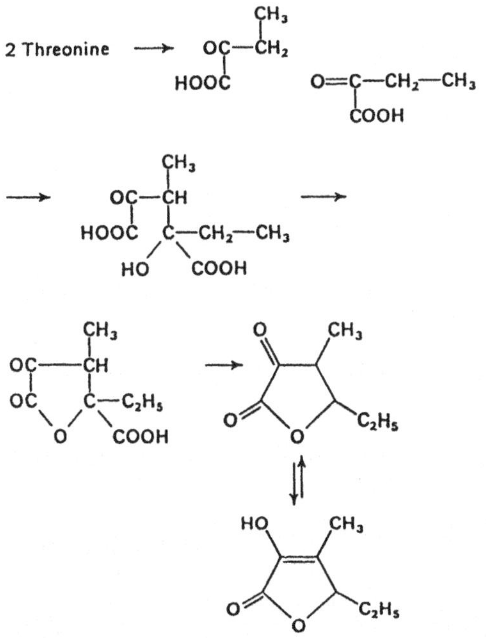

Fig. 5.16. Formation of 5-ethyl-3-hydroxy-4-methyl-2(5H)-furanone from threonine by heating

Table 5.19. Odor thresholds of 4-hydroxy-5-methyl-(I) and 4-hydroxy-2,5-dimethyl-3(2H)-furanone (II) as a function of the pH value of the aqueous solution

pH	Threshold (μg/l)	
	I	II
7.0	23,000	60
4.5	2100	31
3.0	2500	21

5.3.1.4 Thiols, Thioethers, Di- and Trisulfides

An abundance of sulfurous compounds is obtained from cysteine, cystine, monosaccharides, thiamine and methionine by heating food. Some

Table 5.20. Occurrence of 4-hydroxy-2,5-dimethyl-3(2H)-furanone

Food	mg/kg
Beer, light	0.35
Beer, dark	1.3
White bread, crust	1.96
Coffee drink[a]	1.5–7
Emmental cheese	1.2
Beef, boiled	9
Strawberry	1–30
Pineapple	1.6–35

[a] Coffee, medium roasted, 54 g/l water.

are very powerful aroma compounds (Table 5.21) and are involved in the generation of some delightful but also some irritating, unpleasant odor notes.

Thiols are important constituents of food aroma because of their intensive odor and their occurrence as intermediary products which can react with other volatiles by addition to carbonyl groups or to double bonds.

Hydrogen sulfide and 2-mercaptoacetaldehyde are obtained during the course of the *Strecker* degradation of cysteine (Fig. 5.17). In a similar way, methionine gives rise to methional, which releases methanethiol by β-elimination (Fig. 5.18). Dimethylsulfide is obtained by methylation during heating of methionine in the presence of pectin:

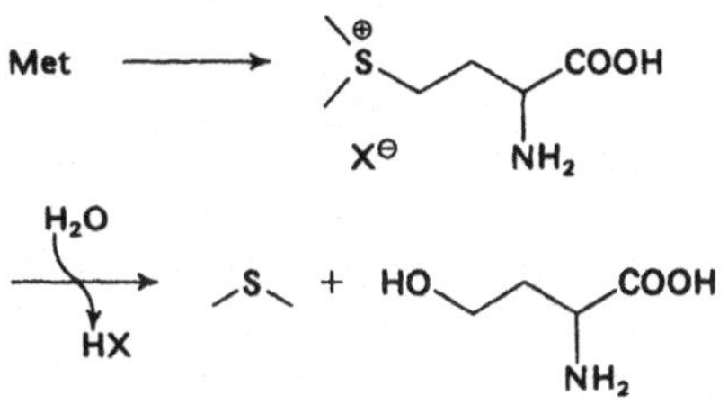

$$(5.7)$$

Methanethiol oxidizes easily to dimethyldisulfide, which can disproportionate to dimethylsulfide and dimethyltrisulfide (Formula 5.8).

Table 5.21. Sensory properties of volatile sulfur compounds

Compound	Odor	
	Quality	Threshold (µg/l)[a]
Hydrogen sulfide	Sulfurous, putrid	10
Methanethiol	Sulfurous, putrid	0.02
Dimethylsulfide	Asparagus, cooked	1.0
Dimethyldisulfide	Cabbage-like	7.6
Dimethyltrisulfide	Cabbage-like	0.01
Methional	Potatoes, boiled	0.2
Methionol	Sulfurous	5.0
3-Methyl-2-butenethiol	Animal	0.0003
3-Mercapto-2-butanone	Sulfurous	3.0
3-Mercapto-2-pentanone	Sulfurous	0.7
2-Mercapto-3-pentanone	Sulfurous	2.8
2-Furfurylthiol	Roasted, like coffee	0.012
2-Methyl-3-furanthiol	Meat, boiled	0.007
Bis(2-methyl-3-furyl)disulfide	Meat-like	0.00002
3-Mercapto-2-methylpentan-1-ol	Meat-like, like onions	0.0016

[a] In water.

Due to its very low odor threshold (Table 5.21), the trisulfide is very aroma active and is frequently found in dilution analyses as a companion substance of methanethiol. For the moment, it is unknown whether it is derived from food or whether it is an artifact obtained in the isolation and concentration of volatile compounds.

Except for the exceptionally reactive 2-mercaptoethanal, the sulfur compounds mentioned above have been identified in practically all protein-containing foods when they are heated or stored for a prolonged period of time.

The addition of H_2S to α-diketones, which are produced in the *Maillard* reaction (cf. 4.2.4.3.2 and 4.2.4.4), the elimination of water and a reaction called reductive sulfhydrylation result in mercaptoalkanes (Formula 5.9). Here, two position isomers 2-mercapto-3-pentanone (2M3P) and 3-mercapto-2-pentanone (3M2P) are produced from 2,3-pentanedione, 3M2P being an important contributor to the aroma of meat (cf. 12.9.2). Model experiments with various monosaccharides (cf. 12.9.3) show that ribose yields more 2M3P and 3M2P than glucose, the optimal pH being 5.0. The optimum probably results from the fact that while the liberation of H_2S from cysteine is favored at low pH, the fragmentation of the monosaccharides to α-diketones is favored at higher pH values.

2-Furfurylthiol (FFT) is the key odorant of roasted coffee (cf. 21.1.3.3.7). It also plays a role in meat aromas and in the aroma of rye bread crust (cf. 12.9.2 and 15.4.3.3.3). It appears on toasting when white bread is baked with a higher amount of yeast. The precursor of FFT is furfural which, according to the hypothesis, adds hydrogen sulfide to give a thiohemiacetal (Formula 5.10). Water elimination and reductive sulfhydrylation then yield FFT. On the other hand, FFT can also be formed from furfuryl alcohol after the elimination of water and addition of hydrogen sulfide. Furfuryl alcohol is one of the volatile main products of the *Maillard* reaction. Roasted coffee contains FFT and other volatile thiols not only in the free state, but also bound via disulfide bridges to cysteine, SH-peptides and proteins. The thiols can be released by reduction, e. g., with dithioerythritol.

An isomer of FFT, 2-methyl-3-furanthiol (MFT), has a similarly low odor threshold (Table 5.21), but differs in the odor quality. MFT smells like boiled meat, being one of its key odorants (cf. 12.9.2). The SH-group of MFT is considerably more instable than that of FFT because in an H-abstraction, a thiyl radical can be generated which is stabilized by resonance with the aromatic ring (Formula 5.11). The thiyl radicals dimerize to bis(2-methyl-3-furyl)

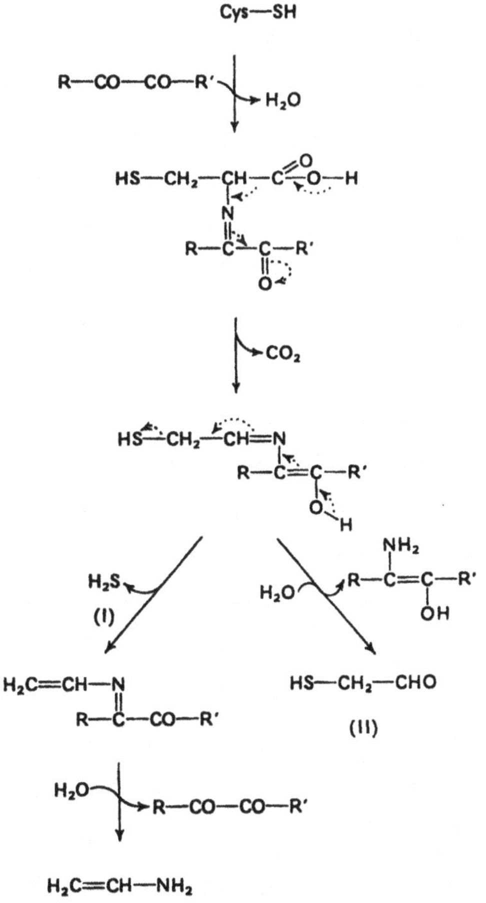

Fig. 5.17. Cysteine decomposition by a *Strecker* degradation mechanism: formation of H_2S (I) or 2-mercaptoethanal (II)

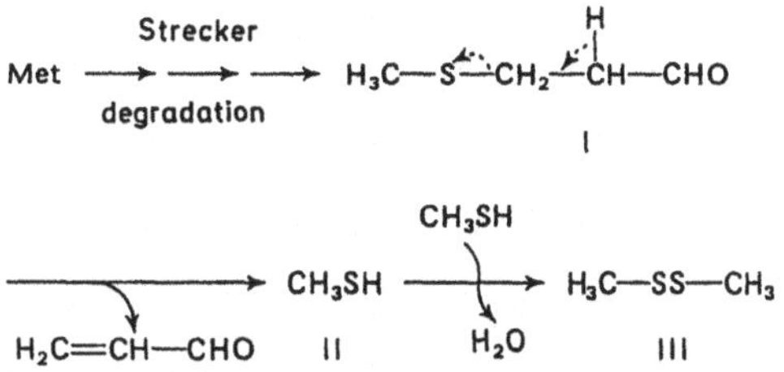

Fig. 5.18. Methionine degradation to methional, methanethiol and dimethylsulfide

disulfide, which is cleaved again at a higher temperature (Formula 5.11), e. g., during cooking. If constituents which have H-atoms abstractable by thiyl radicals, e. g., reductones, are present in food, MFT is regenerated. This is desirable because although the disulfide of MFT has a very low odor threshold (Table 5.21), its meat-like odor has a medical by-note and, unlike MFT, its *Stevens* curve is much flatter (cf. 5.1.4), i. e., the odor is not very intensive even in a higher concentration range.

Norfuraneol (I in Table 5.18) is under discussion as the precursor of MFT. As proposed in Formula 5.12, the addition of hydrogen sulfide leads to 4-mercapto-5-methyl-3(2H)-furanone, which yields MFT after reduction, e. g., by reductones from the *Maillard* reaction, and water elimination. MFT can also be formed in meat by the hydrolysis of thiamine (Fig. 5.19). The postulated intermediate is the very reactive 5-hydroxy-3-mercaptopentan-2-one.

(5.8)

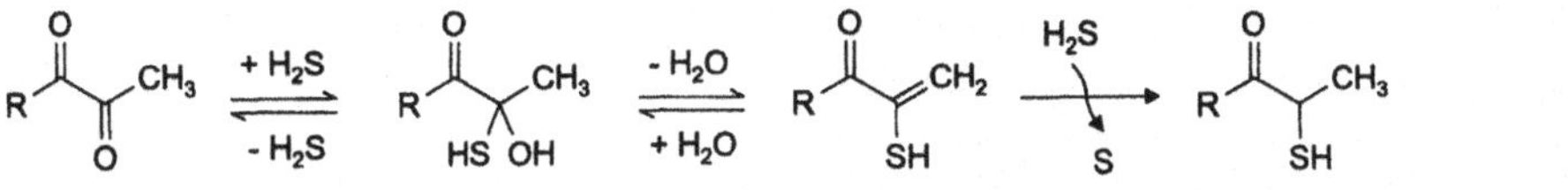

(5.9)

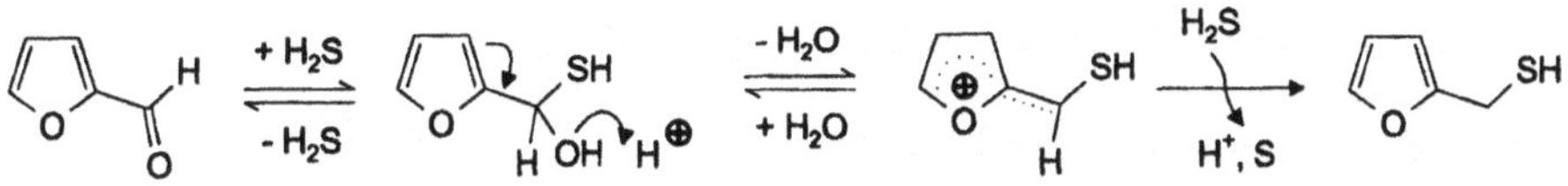

(5.10)

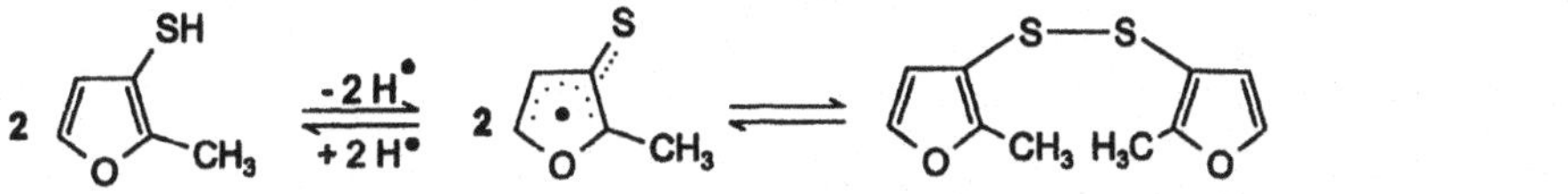

(5.11)

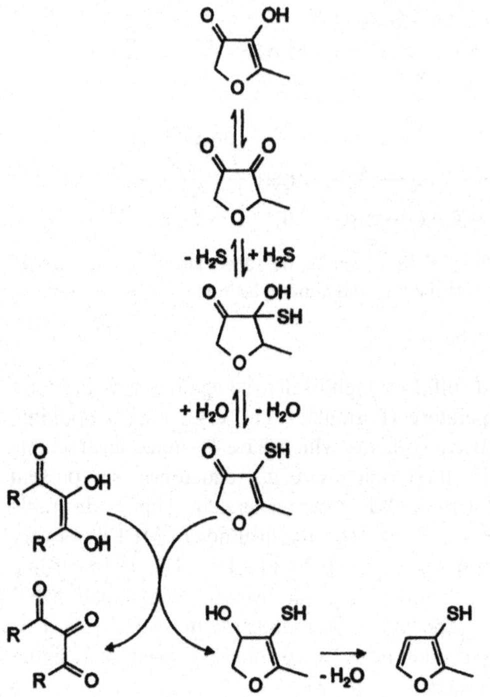

(5.12)

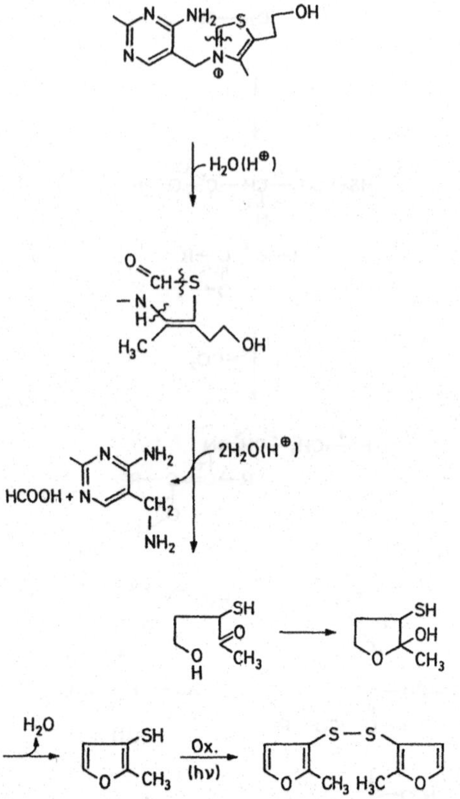

Fig. 5.19. Formation of 2-methyl-3-furanthiol and bis(2-methyl-3-furyl)disulfide from thiamine

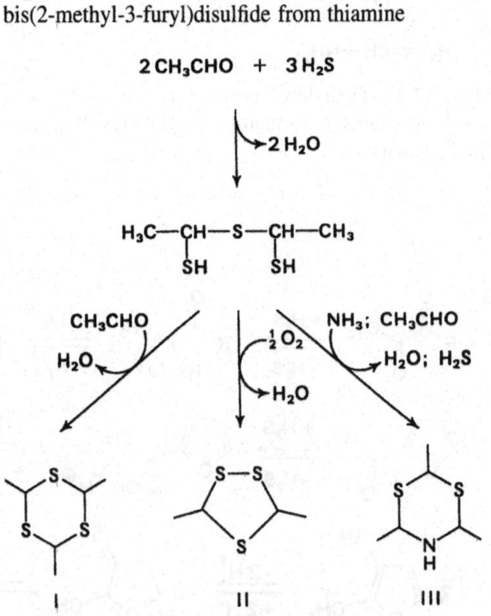

Some reaction systems, which have been described in the patent literature for the production of meat aromas, regard thiamine as precursor. 3-Methyl-2-butene-1-thiol is one of the roast odorants of coffee (cf. 21.1.3.3.7) and can cause on off-flavor in beer (cf. Table 5.5). In general, only very small amounts are formed which are still aroma active on account of the very low odor threshold (Table 5.21). The formation of the thiol is explained by the fact that the 3-methyl-2-butene radical is formed from terpenes by photolysis (beer) or under the drastic conditions of the roasting process (coffee). This radical then meets a SH•-radical formed from cysteine under these conditions. In the case of beer, humulons (cf. 20.1.2.3.2) are under discussion as the source of the alkyl radical. In coffee 3-methyl-2-butene-1-ol (prenyl alcohol) is also a possible precursor, which yields the thiol after water elimination and hydrogen sulfide addition.

It is unclear whether sulfides I–III in Fig. 5.20 and trithioacetone, analogous to trithioacetaldehyde (I), are really formed during the cooking of meat or whether these compounds are artifacts that are produced on concentration of the volatile fraction in the course of analysis (cf. 5.2.1).

Fig. 5.20. Formation of 2,4,6-trimethyl-s-trithiane (I), 3,5-dimethyl-1,2,4-trithiolane (II) and 2,4,6-trimethyl-5,6-dihydro-1,3,5-dithiazine (III)

5.3.1.5 Thiazoles

Thiazole and its derivatives are detected in foods such as coffee, boiled meat, boiled potatoes, heated milk and beer. Aroma extract dilution analyses show that among the compounds I–III in Table 5.22, 2-acetyl-2-thiazoline (II) contributes most intensively to the aroma of quick fried beef. Model experiments showed that cysteamine, formed by the decarboxylation of cysteine, and 2-oxopropanal are the precursors. It was also found that higher yields of II are obtained at pH 7.0 compared to pH 5.0. The intermediates in the reaction path to thiazoline II (Fig. 5.21) were identified as the odorless 2-(1-hydroxyethyl)-4,5-dihydrothiazole (a) and 2-acetylthiazolidine (b), which are in tautomeric equilibrium, presumably with 2-(1-hydroxyethylene)thiazolidine (c) as the intermediate compound (Fig. 5.21). The intermediates a and b are oxidized to thiazoline II by atmospheric oxygen in the presence of catalytic amounts of heavy metals. It is assumed that the

metal ion, e. g., Cu^{2+}, oxidizes the eneaminol c to a resonance-stabilized radical d in a one-electron reaction (Fig. 5.22). This radical then traps an oxygen molecule with the formation of a peroxy radical (e). H-Abstraction from the eneaminol c results in the conversion of e to 2-acetyl-2-thiazolinehydroperoxide (f), which decomposes to thiazoline I and H_2O_2. H_2O_2 can oxidize the metal ion and regenerate it for a new cycle.

In the conversion of the precursor b, only the limitation of the reaction time to 10 minutes in the temperature range 50–100 °C results in the highest yield of thiazolidine II (Fig. 5.23). This is in accord with the aroma formation during the frying of beef. The concentration of II in meat, decreases again if heating continues.

Thiazole IV (Table 5.22) can occur in milk when it is heated, and is responsible for a "stale" off-flavor. Thiazole V (Table 5.22) is a constituent of tomato aroma. The aroma of tomato products is usually enhanced by the addition of 20–50 ppb of thiazole V (for the biosynthesis of the compound, see Section 5.3.2.5).

5.3.1.6 Pyrroles, Pyridines

The volatile compounds formed by heating food include numerous pyrrole and pyridine derivatives. Of special interest are the N-heterocyclic compounds with the following structural feature:

$$\diagdown N\!\!\diagup{}^{C}\!-\!\underset{\overset{\|}{O}}{C}\!-\!R \tag{5.13}$$

This characteristic feature appears to be required for a roasted odor. In fact, all the pyrrolines and pyridines listed in Table 5.23 as well as 2-acetyl-thiazole, 2-acetylthiazoline (cf. Table 5.22) and acetylpyrazine (cf. Table 5.23) contain this structural element and have a roasted or cracker-like odor. However, the thresholds of these compounds vary greatly. The lowest values were found for 2-acetyl-and 2-propionyl-1-pyrroline.

The length of the alkanoyl group also influences the aroma activity because in the transition from 2-propionyl- to 2-butanoyl-1-pyrroline, the roasted note suddenly disappears and the odor threshold increases by several powers of ten.

2-Acetyl-1-pyrroline (Apy) is responsible for the typical aroma of the crust of white bread and it

Table 5.22. Thiazoles and thiazolines in food

Name	Structure	Aroma quality	Odor threshold (μg/kg, H_2O)
2-Acetyl-thiazole	I	Cereal, popcorn	10
2-Acetyl-2-thiazoline	II	Popcorn	1
2-Propionyl-2-thiazoline	III	Popcorn	1
Benzo-thiazole	IV	Quino-line, rubber	
2-Isobutyl-thiazole	V	Green, tomato, wine	3

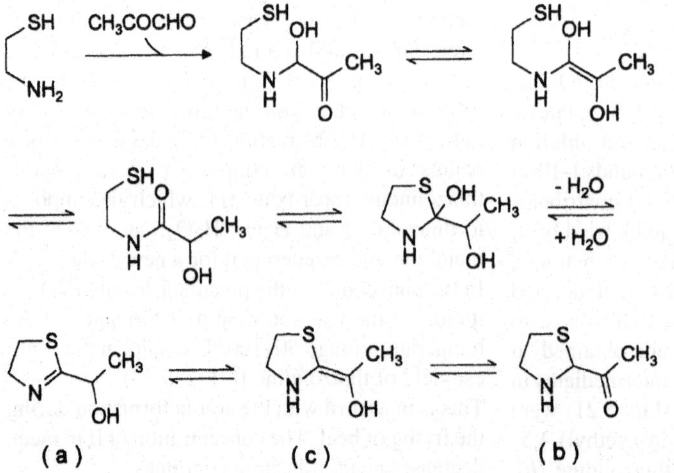

Fig. 5.21. Formation of precursors of 2-acetyl-2-thiazoline (according to *Hofmann* and *Schieberle*, 1995)

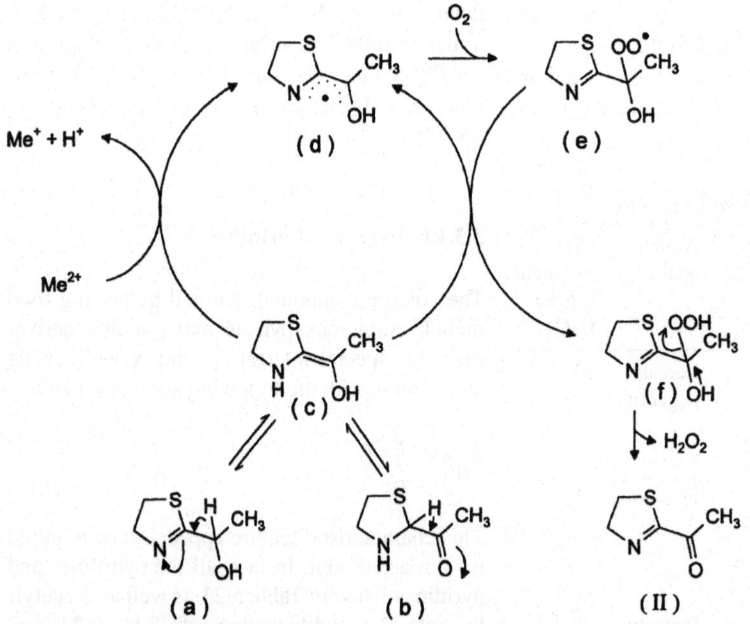

Fig. 5.22. Metal catalyzed oxidation of 2-(1-hydroxyethyl)-4,5-dihydrothiazole and 2-acetylthiazolidine (according to *Hofmann* and *Schieberle*, 1995)

produces the pleasant popcorn aroma of certain types of rice consumed mainly in Asia. In gas chromatography, Apy appears predominantly in the imine form shown in Table 5.23, whereas 2-acetyltetrahydropyridine (ATPy) appears as the eneamine and imine tautomers. Model experiments show that 1-pyrroline is the precursor of Apy and ATPy. 1-Pyrroline is formed by the

Strecker degradation of both proline (cf. Formula 5.14) and ornithine (cf. Formula 5.15). In the baking of white bread, ornithine comes from yeast where it is found in a concentration about four times that of free proline.

In addition, triose phosphates occurring in yeast have been identified as precursors. They yield on heating, e.g., 2-oxopropanal from di-

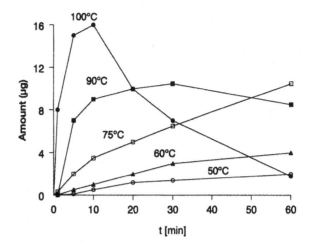

Fig. 5.23. Dependence on time and temperature of the formation of 2-acetyl-2-thiazoline from 2-(1-hydroxyethyl)-3,5-dihydrothiazole (according to *Hofmann* and *Schieberle*, 1996)

Table 5.23. Pyrrole and pyridine derivatives with a roasted aroma

Name	Structure	Odor threshold (μg/kg, water)	Occurrence
2-Acetyl-1-pyrroline (APy)		0.1	White-bread crust, rice, cooked meat, popcorn
2-Propionyl-1-pyrroline		0.1	Popcorn, heated meat
2-Acetyltetra-hydropyridine (ATPy)		1.6	White-bread crust, popcorn
2-Acetylpyridine		19	White-bread crust

hydroxyacetone phosphate (cf. Formula 5.16), which is involved in the *Strecker* degradation (cf. Formula 5.14). Another source of 2-oxopropanal is the retroaldol condensation of 3-deoxy-1,2-dicarbonyl compounds in the course of the *Maillard* reaction (cf. 4.2.4.4.2). The reaction route which can explain the formation of Apy is based on an investigation of the model 1-pyrroline/2-oxopropanal and on labelling experiments. They show that in the reaction of proline with [^{13}C]$_6$-glucose under roasting conditions, two ^{13}C atoms are inserted into the Apy molecule. As a start in the reaction sequence to Apy, it is assumed that 2-oxopropanal (cf. 4.2.4.3.2), which is formed in the degradation of glucose, is present as a hydrate and participates

in a nucleophilic attack on 1-pyrroline (Fig. 5.24). The resulting 2-(1,2-dioxopropyl)pyrrolidine is sensitive to oxygen and, consequently, rapidly oxidizes to 2-(1,2-dioxopropyl)pyrroline. After hydration, decarboxylation takes place in accord with the labelling experiment. This is followed by rearrangement and oxidation to Apy.

Hydroxy-2-propanone, which is formed by the *Strecker* degradation of amino acids, e. g., proline (cf. Formula 5.14), is in the enolized form the reaction partner of 1-pyrroline in the formation of ATPy (Fig. 5.25). The aldol addition of the two educts gives 2-(1-hydroxy-2-oxopropyl)-pyrrolidine (HOP) which undergoes ring opening to yield 5,6-dioxoheptylamine. The subsequent *Schiff* reaction to a 6-ring results in ATPy.

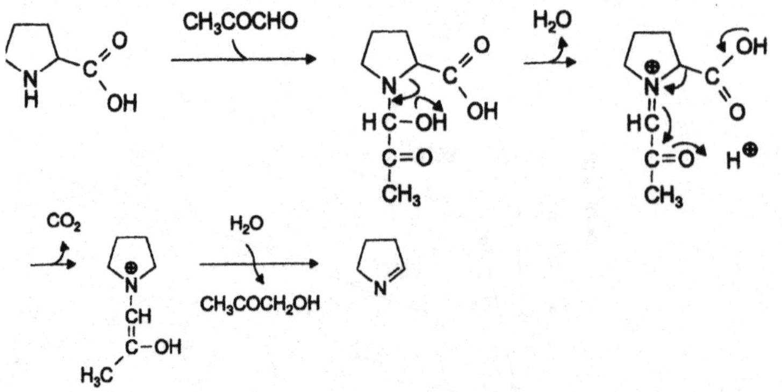

$$(5.14)$$

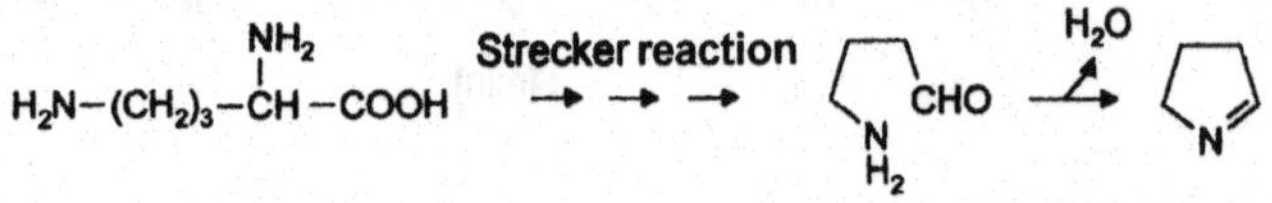

$$(5.15)$$

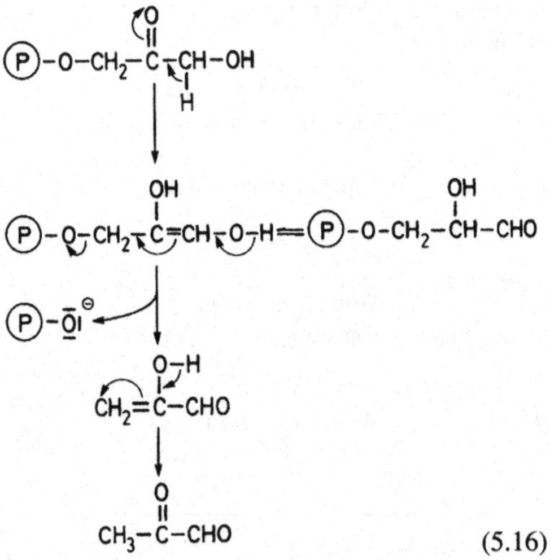

$$(5.16)$$

The reaction pathway shown in Fig. 5.25 can be based on the identification of HOP as an intermediate in the formation of ATPy and on a model experiment in which 2-methyl-1-pyrroline was used instead of 1-pyrroline. 2-Acetyl-3-methyl-3,4,5,6-tetrahydropyridine (cf. Formula 5.17) was produced, i. e., a displacement of the methyl group from position 2 in the 5-ring of the starting compound to position 3 in the 6-ring of the product. This shift can only be explained by the ring enlargement mechanism (Fig. 5.25).

A comparison of the reaction paths in Fig. 5.24 and Fig. 5.25 allows the conclusion that the concentration ratio of 2-oxopropanal to hydroxy-2-propanone in food decides whether Apy or ATPy is preferentially formed from proline. If free amino acids are present in the food and the *Strecker* degradation dominates, then the formation of ATPy predominates. This could explain the preference for ATPy (430 µg/kg) compared to Apy (24 µg/kg) in the production of popcorn.

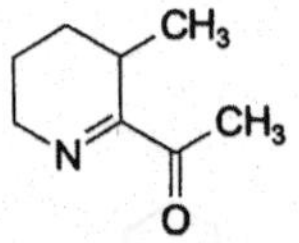

$$(5.17)$$

Although the odor threshold increases by about a factor of 10, the popcorn-like aroma note remains on oxidation of ATPy to 2-acetyl-pyridine. Substantially greater effects on the aroma are obtained by the oxidation of APy to 2-acetylpyrrole, which has an odor threshold that is more than 5 powers of ten higher and no longer smells roasted.

2-Pentylpyridine contributes to the smell of roasting lamb fat (greasy, suety odor; threshold: 0.12 µg/kg water); it produces an aroma defect in soybean products (cf. 16.3.1.1). The precursors identified were ammonia from the pyrolysis of asparagine and glutamine and 2,4-decadienal:

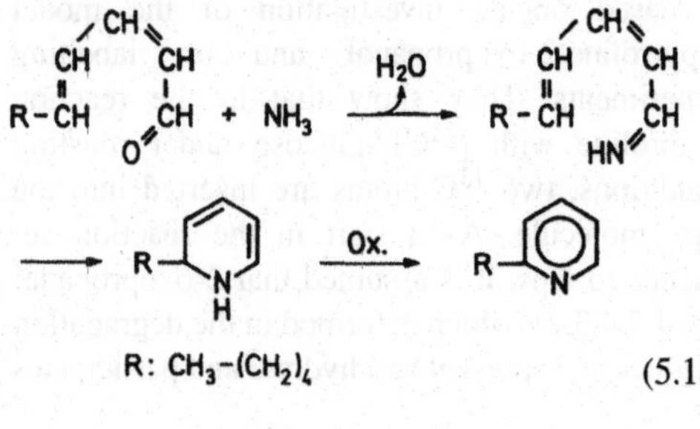

R: $CH_3-(CH_2)_4$

$$(5.18)$$

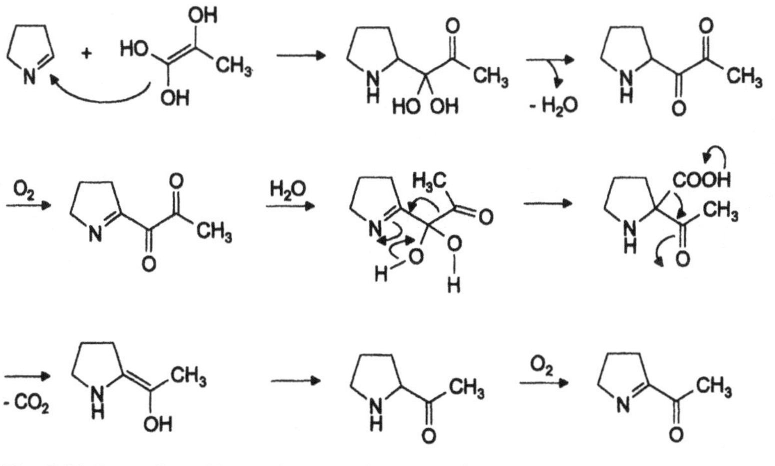

Fig. 5.24. Formation of 2-acetyl-1-pyrroline (according to *Hofmann* and *Schieberle*, 1998)

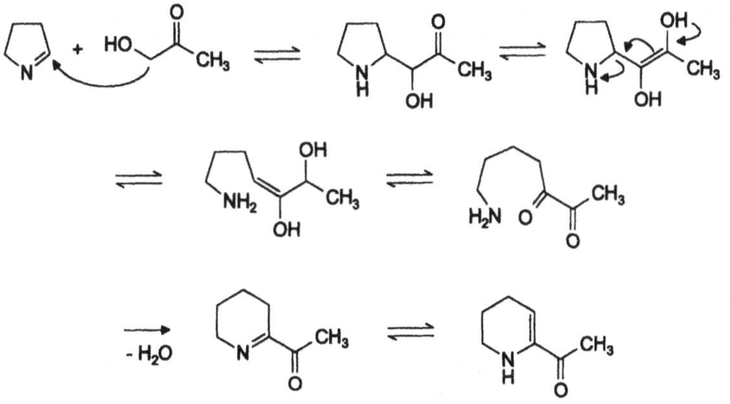

Fig. 5.25. Formation of 2-acetyltetrahydropyridine (according to *Hofmann* and *Schieberle*, 1998)

5.3.1.7 Pyrazines

A large number of volatile pyrazines are formed on heating food. Seventy compounds are known alone in the group of alkyl pyrazines consisting only of the elements C, H and N. In dilution analyses, e. g., of coffee, bread crust, fried meat and cocoa liquor, only the first six compounds in Table 5.24 were detected; pyrazine II and V reached the highest FD factors.

According to gas chromatographic–olfactometric studies, pyrazines II, III, V and VI (Table 5.24) have the lowest odor thresholds (0.07 pmol/l air) that have ever been measured for alkyl pyrazines (cf. 5.6.3). Of these four pyrazines, II and V are produced in food in higher concentrations than III and VI (cf. example coffee, 21.1.3.3.7). As a re-

sult of this favorable ratio of concentration to odor threshold, the aroma activities of II and V exceed those of the other alkyl pyrazines.

Although the odor thresholds of pyrazines I and IV are much higher than those of pyrazines II, III, V and VI (Table 5.24), they are still detected in dilution analyses because they are formed in much higher concentrations on heating food and, consequently, can partially compensate for their "aroma weakness". 2-Oxopropanal and alanine are the precursors of pyrazines II, IV and V as well as 2-ethyl-5,6-dimethylpyrazine, which is odorless in the concentrations present in food. In accord with the formation of pyrazine in food, pyrazine IV is the main compound in model experiments (Table 5.25), followed by II and V. To explain the formation of II and IV,

Table 5.24. Pyrazines in food

Structure	Substituent	Aroma quality	Odor threshold value (µg/l; water)
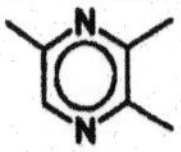 (I)	Trimethyl-	Earthy	90
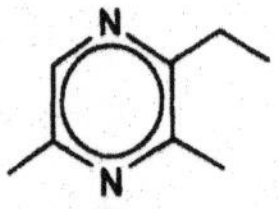 (II)	2-Ethyl-3,5-dimethyl-	Earthy, roasted	0.04
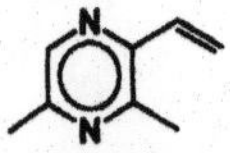 (III)	2-Ethenyl-3,5-dimethyl-	Earthy, roasted	0.1
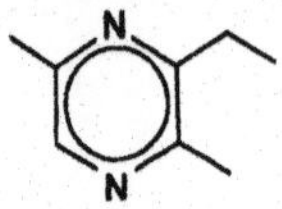 (IV)	2-Ethyl-3,6-dimethyl-	Earthy, roasted	9
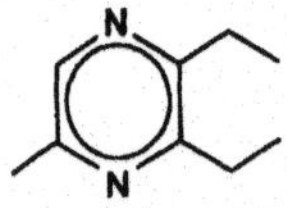 (V)	2,3-Diethyl-5-methyl-	Earthy, roasted	0.09
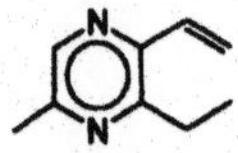 (VI)	2-Ethenyl-3-ethyl-5-methyl-	Earthy, roasted	0.1
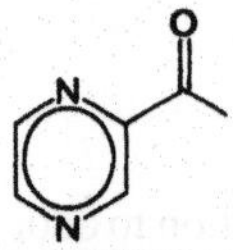 (VII)	Acetyl-	Roasted corn	62
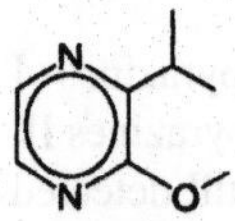 (VIII)	2-Isopropyl-3-methoxy-	Potatoes	0.002
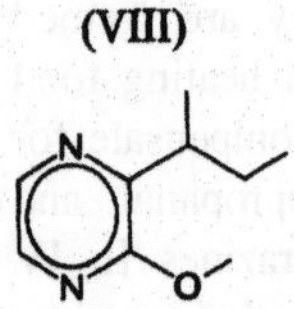 (IX)	2-sec-Butyl-3-methoxy-	Earthy	0.001
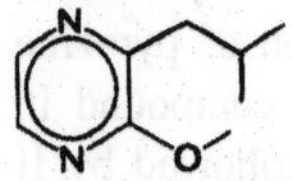 (X)	2-Isobutyl-3-methoxy-	Hot paprika (red pepper)	0.002

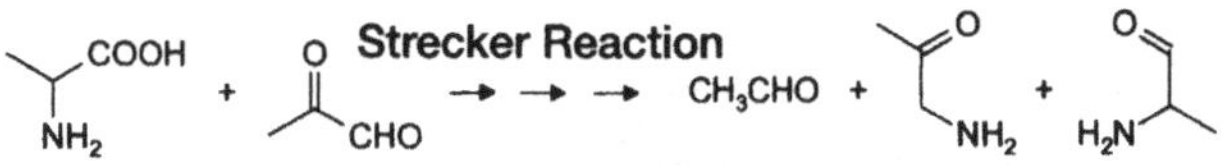

(5.19)

Table 5.25. Formation of aroma active alkyl pyrazines on heating alanine and 2-oxopropanal[a]

Pyrazine[b]	Amount (μg)
2-Ethyl-3,5-dimethyl-(II)	27
2-Ethyl-3,6-dimethyl-(IV)	256
2-Ethyl-5,6-dimethyl-	2.6
2,3-Diethyl-5-methyl-(V)	18

[a] The mixture of educts (2 mmol each; pH 5.6) was heated for 7 min to 180 °C.
[b] Roman numerals refer to Table 5.24.

it is postulated that the *Strecker* reaction of alanine and 2-oxopropanal represents the start, resulting in acetaldehyde, aminoacetone and 2-aminopropanal (cf. Formula 5.19). The precursor of pyrazine IV, 3,6-dimethyl-dihydropyrazine, is formed by the condensation both of two molecules of aminoacetone as well as two molecules of 2-aminopropanal (cf. Formula 5.20). The nucleophilic attack by dihydropyrazine on the carbonyl group of acetaldehyde and water elimination yield pyrazine IV. This mechanism also explains the formation of pyrazine II if 3,5-dimethyldihydro-pyrazine, which is produced by the condensation

of aminoacetone and 2-aminopropanal (cf. Formula 5.21), is assumed to be the intermediate. The preferential formation of pyrazine IV in comparison with II can be explained by the fact that the *Strecker* reaction produces less 2-aminopropanal than aminoacetone because the aldehyde group in 2-oxopropanal is more reactive than the keto group. However, both aminocarbonyl compounds are required to the same extent for the synthesis of pyrazine II (cf. Formula 5.21). The powerfully odorous pyrazines VIII–X (Table 5.24) appear as metabolic by-products in some plant foods and microorganisms (cf. 5.3.2.6). Since they are very stable, they withstand, e. g., the roasting process in coffee (cf. 21.1.3.3.7).

5.3.1.8 Amines

Not only aldehydes (cf. 5.3.1.1), but also amines are formed in the *Strecker* reaction (cf. 4.2.4.4.7). The odor thresholds of these amines (examples in Table 5.26) are pH dependent. The enzymatic decarboxylation of amino acids produces the same amines as the *Strecker* reaction; the precursors are shown in Table 5.26. Both reactions take place e. g. in the production of cocoa, but the *Strecker*

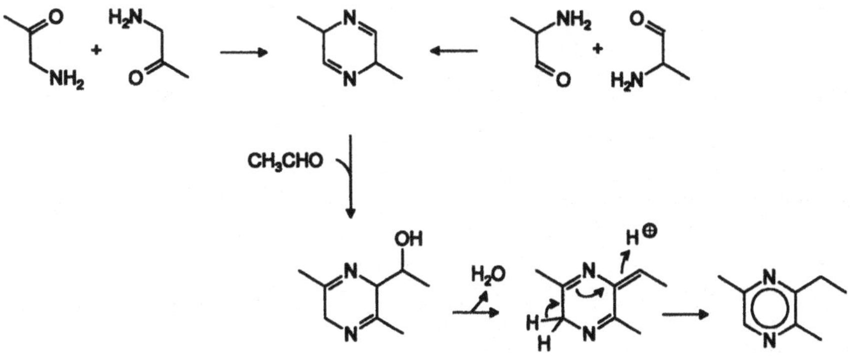

(IV)   (5.20)

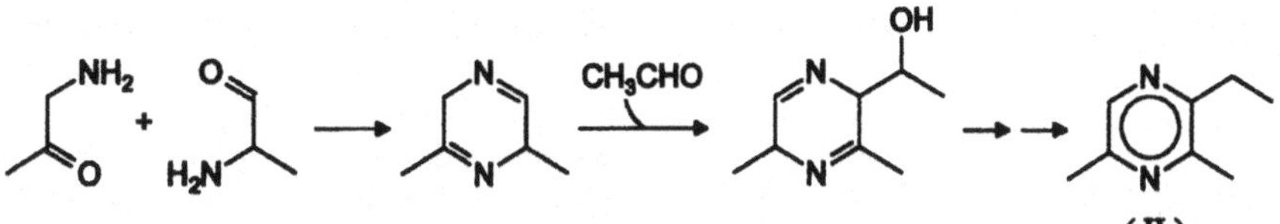

(II)   (5.21)

Table 5.26. Precursors and sensory properties of amines

Amine	Amino acid precursor	Odor Quality	Threshold (mg/l) Water[a]	Oil
2-Methylpropyl	Val	Fishy, amine-like, malty	8.0	48.3
2-Methylbutyl	Ile	Fishy, amine-like, malty	4.9	69.7
3-Methylbutyl	Leu	Fishy, amine-like, malty	3.2	13.7
2-Phenylethyl	Phe	Fishy, amine-like, honey-like	55.6	89.7
3-(Methylthio)propyl	Met	Fishy, amine-like, boiled potato	0.4	0.3

[a] pH 7.5.

reaction predominates. An especially odor intensive amine, trimethylamine, is formed in the degradation of choline (cf. 11.2.4.4.4).

5.3.1.9 Phenols

Phenolic acids and lignin are degraded thermally or decomposed by microorganisms into phenols, which are then detected in food. Some of these compounds are listed in Table 5.27.

Smoke generated by burning wood (lignin pyrolysis) is used for cold or hot smoking of meat and fish products. This is a phenol enrichment process since phenol vapors penetrate the meat or fish muscle tissue. Also, some alcoholic beverages, such as Scotch whiskey, and also butter have low amounts of some phenols, the presence of which is needed to roundoff their typical aromas. Ferulic acid was identified as an important precursor in model experiments. 4-Vinylguaiacol is formed as the main product in pyrolysis, the secondary products being 4-ethylguaiacol, vanillin and guaiacol. To explain such a reaction which, for example, accompanies the process of roasting coffee or the kiln drying of malt, it has to be assumed that thermally formed free radicals regulate the decomposition pattern of phenolic acids (cf., for example, heat decomposition of ferulic acid, Fig. 5.26). In the pasteurization of orange juice, p-vinyl-guaiacol can also be formed from ferulic acid, producing a stale taste at concentrations of 1 mg/kg.

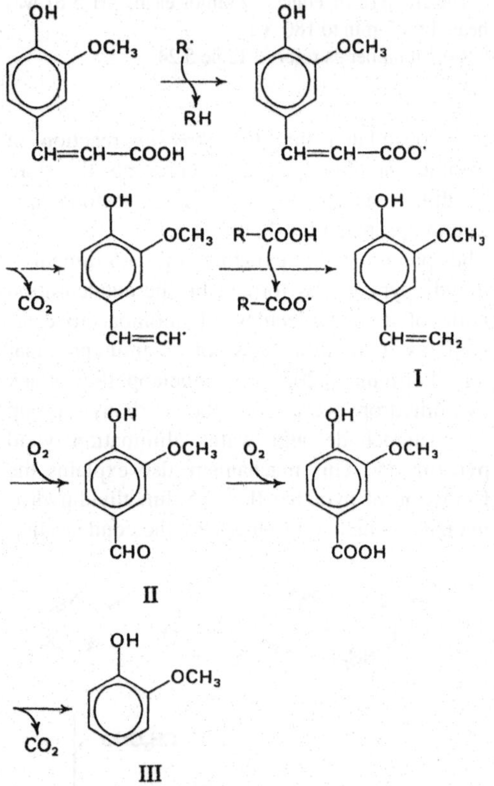

5.3.2 Enzymatic Reactions

Aroma compounds are formed by numerous reactions which occur as part of the normal metabolism of animals, plants and microorganisms. The enzymatic reactions triggered by tissue disruption, as experienced during disintegration or slicing of fruits and vegetables, are of particular importance. Enzymes can also be involved indirectly in aroma formation by providing the preliminary stage of the process, e. g. by releasing

Fig. 5.26. Thermal degradation of ferulic acid. 4-Vinylguaiacol (I), vanillin (II), and guaiacol (III) (according to *Tressl* et al., 1976)

Table 5.27. Phenols in food

Name	Structure	Aroma quality	Odor threshold (μg/kg, water)	Occcurrence
p-Cresol	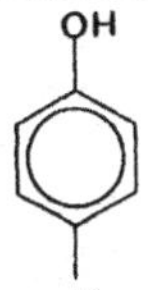 (I)	Smoky	55	Coffee, sherry, milk, roasted peanuts, asparagus
4-Ethylphenol	 (II)	Woody		Milk, soya souce, roasted peanuts, tomatoes, coffee
Guaiacol	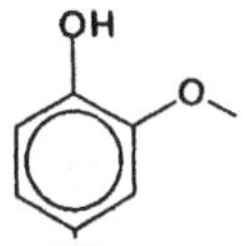 (III)	Smoky, burning, sweet	1	Coffee, milk, crisp-bread, meat (fried)
4-Vinylphenol	 (IV)	Harsh, smoky	10	Beer, milk, roasted peanuts
2-Methoxy-4-vinylphenol	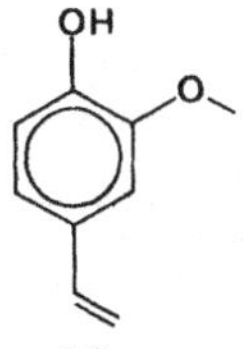 (V)	Clove-like	5	Coffee, beer, apple (cooked), asparagus
Eugenol	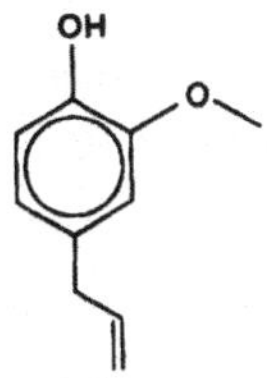 (VI)	Spicy	1	Tomato paste, brandy, plums, cherries
Vanillin	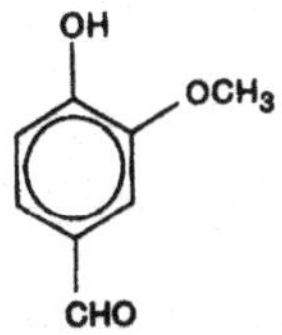 (VII)	Vanilla	20	Vanilla, rum, coffee, asparagus (cooked), butter

Table 5.28. Pyrolysis products of some phenolic acids (T: 200 °C; air)

Phenolic acid	Product	Distribution (%)
Ferulic acid	4-Vinylguaiacol	79.9
	Vanillin	6.4
	4-Ethylguaiacol	5.5
	Guaiacol	3.1
	3-Methoxy-4-hydroxy-acetophenone (Acetovanillone)	2.6
	Isoeugenol	2.5
Sinapic acid	2,6-Dimethoxy-4-vinylphenol	78.5
	Syringaldehyde	13.4
	2,6-Dimethoxyphenol	4.5
	2,6-Dimethoxy-4-ethylphenol	1.8
	3,5-Dimethoxy-4-hydroxy-acetophenone (Acetosyringone)	1.1

amino acids from available proteins, sugars from polysaccharides, or ortho-quinones from phenolic compounds. These are then converted into aroma compounds by further nonenzymatic reactions. In this way, the enzymes enhance the aroma of bread, meat, beer, tea and cacao.

5.3.2.1 Carbonyl Compounds, Alcohols

Fatty acids and amino acids are precursors of a great number of volatile aldehydes, while carbohydrate degradation is the source of ethanal only. Due to its aroma activity at higher concentrations ethanal is of great importance for the fresh note, e. g., in orange and grapefruit juice.

Linoleic and linolenic acids in fruits and vegetables are subjected to oxidative degradation by lipoxygenase alone or in combination with a hydroperoxide lyase, as outlined in sections 3.7.2.2 and 3.7.2.3. The oxidative cleavage yields oxo acids, aldehydes and allyl alcohols. Among the aldehydes formed, hexanal, (E)-2-hexenal, (Z)-3-hexenal and/or (E)-2-nonenal, (Z)-3-nonenal, (E,Z)-2,6-nonadienal and (Z,Z)-3,6-nonadienal are important for aroma.

Frequently, these aldehydes appear soon after the disintegration of the tissue in the presence of oxygen. A part of the aldehydes is enzymatically reduced to the corresponding alcohols (see below). In comparison, lipoxygenases and hydroperoxide lyases from mushrooms exhibit a different reaction specificity. Linoleic acid, which predominates in the lipids of champignon mushrooms, is oxidatively cleaved to R(–)-1-octen-3-ol and 10-oxo-(E)-8-decenoic acid (cf. 3.7.2.3). The allyl alcohol is oxidized to a small extent by atmospheric oxygen to the corresponding ketone. Owing to an odor threshold that is about hundred times lower (cf. Table 3.32), this ketone together with the alcohol accounts for the mushroom odor of fresh champignons and of Camembert.

Aldehydes formed by the *Strecker* degradation (cf. 5.3.1.1; Table 5.16) can also be obtained as metabolic by-products of the enzymatic transamination or oxidative deamination of amino acids. First, the amino acids are converted enzymatically to α-keto acids and then to aldehydes by decarboxylation in a side reaction:

$$R\text{—}\underset{\underset{\displaystyle CO_2}{}}{\underset{\displaystyle NH_2}{CH}}\text{—}COOH \longrightarrow R\text{—}\underset{\displaystyle O}{C}\text{—}COOH$$

$$\longrightarrow R\text{—}CHO \qquad (5.22)$$

Unlike other amino acids, threonine can eliminate a water molecule and, by subsequent decarboxylation, yield propanal:

$$H_3C\text{—}\underset{\displaystyle OH}{CH}\text{—}\underset{\displaystyle NH_2}{CH}\text{—}COOH$$

$$\downarrow H_2O$$

$$H_3C\text{—}CH\text{=}\underset{\displaystyle NH_2}{C}\text{—}COOH$$

$$\rightleftharpoons H_3C\text{—}CH_2\text{—}\underset{\displaystyle NH}{C}\text{—}COOH$$

$$\downarrow H_2O,\ NH_3$$

$$H_3C\text{—}CH_2\text{—}CO\text{—}COOH$$

$$\downarrow CO_2$$

$$H_3C\text{—}CH_2\text{—}CHO \qquad (5.23)$$

Many aldehydes derived from amino acids occur in plants and fermented food.

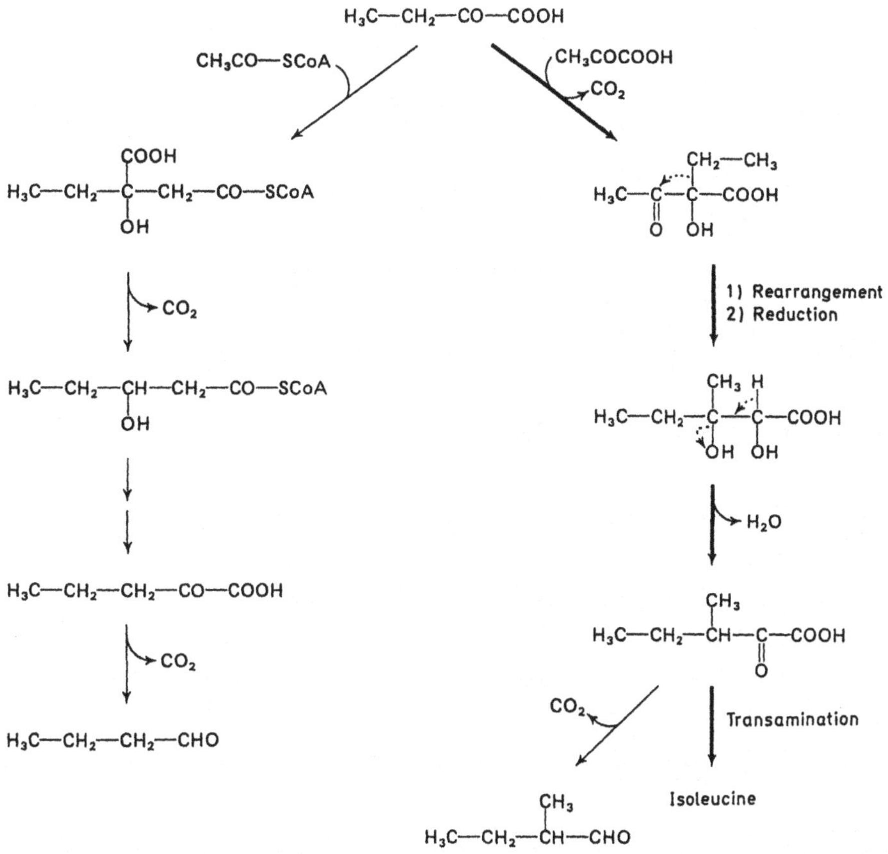

$$H_3C-CH_2-CO-COOH$$

CH$_3$CO—SCoA

CH$_3$COCOOH
CO$_2$

$$H_3C-CH_2-\overset{\displaystyle COOH}{\underset{\displaystyle OH}{C}}-CH_2-CO-SCoA$$

$$H_3C-\overset{\displaystyle CH_2-CH_3}{\underset{\displaystyle OH}{C}}\overset{}{\underset{\displaystyle O}{C}}-COOH$$

CO$_2$

1) Rearrangement
2) Reduction

$$H_3C-CH_2-\overset{}{\underset{\displaystyle OH}{CH}}-CH_2-CO-SCoA$$

$$H_3C-CH_2-\overset{\displaystyle CH_3}{\underset{\displaystyle OH}{C}}\overset{\displaystyle H}{\underset{\displaystyle OH}{C}}-COOH$$

H$_2$O

$$H_3C-CH_2-CH_2-CO-COOH$$

$$H_3C-CH_2-\overset{\displaystyle CH_3}{CH}-\overset{}{\underset{\displaystyle O}{C}}-COOH$$

CO$_2$

$$H_3C-CH_2-CH_2-CHO$$

CO$_2$

Transamination

Isoleucine

$$H_3C-CH_2-\overset{\displaystyle CH_3}{CH}-CHO$$

Fig. 5.27. Formation of aldehydes during isoleucine biosynthesis (according to *Piendl*, 1969). → main pathway → side pathway of the metabolism

A study involving the yeast *Saccharomyces cerevisiae* clarified the origin of methylpro-panal and 2- and 3-methylbutanal. They are formed to a negligible extent by decomposition but mostly as by-products during the biosynthesis of valine, leucine and isoleucine.

Figure 5.27 shows that α-ketobutyric acid, derived from threonine, can be converted into isoleucine. Butanal and 2-methylbutanal are formed by side-reaction pathways.

2-Acetolactic acid, obtained from the condensation of two pyruvate molecules, is the intermediary product in the biosynthetic pathways of valine and leucine (Fig. 5.28). However, 2-acetolactic acid can be decarboxylated in a side reaction into acetoin, the precursor of diacetyl. At α-keto-3-methylbutyric acid, the metabolic pathway branches to form methylpropanal and branches again at α-keto-4-methyl valeric acid to form 3-methylbutanal (Fig. 5.28).

The enzyme that decarboxylates the α-keto-carboxylic acids to aldehydes has been detected in oranges. Substrate specificity for this decarboxylase is shown in Table 5.29.

Table 5.29. Substrate specificity of a 2-oxocarboxylic acid decarboxylase from orange juice

Substrate	V_{rel} (%)
Pyruvate	100
2-Oxobutyric acid	34
2-Oxovaleric acid	18
2-Oxo-3-methylbutyric acid	18
2-Oxo-3-methylvaleric acid	18
2-Oxo-4-methylvaleric acid	15

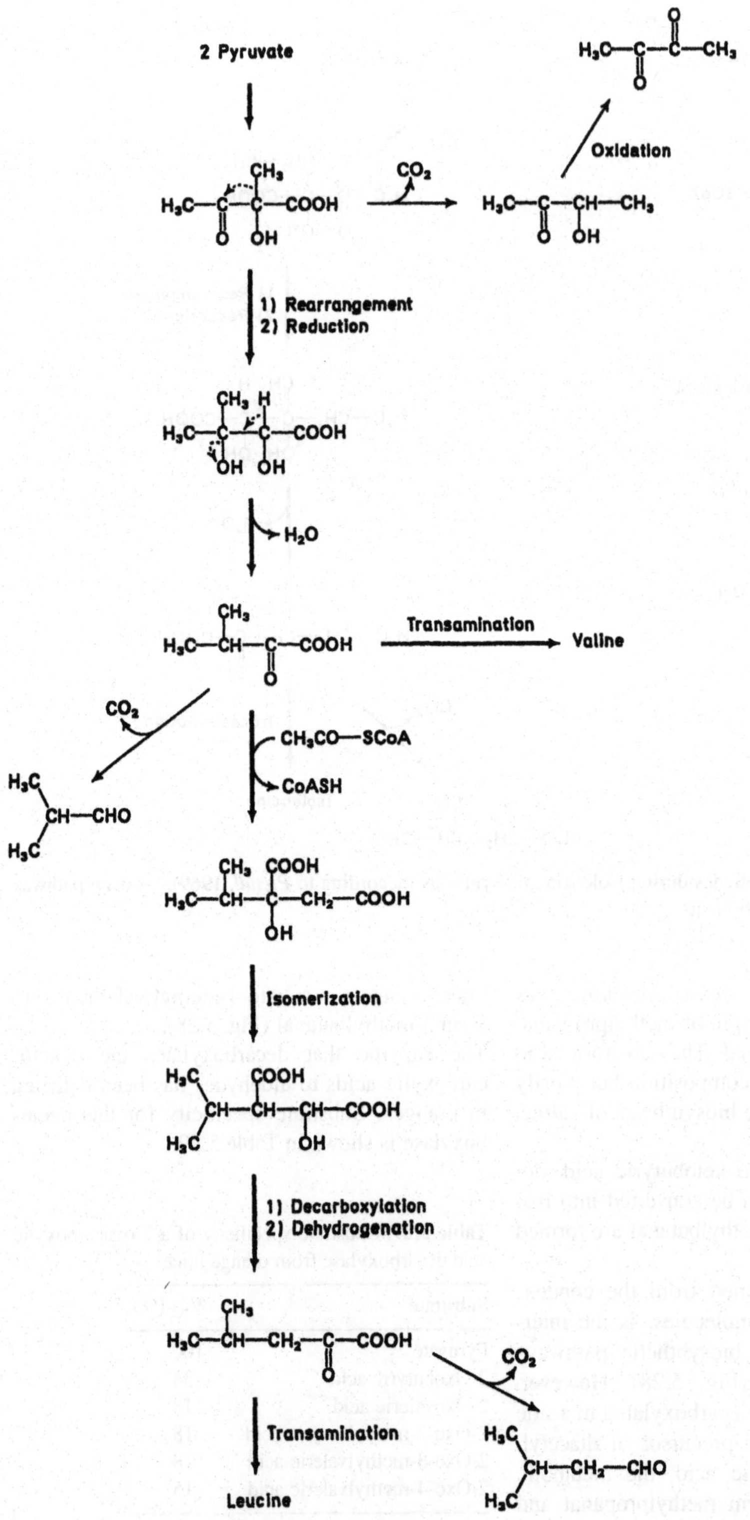

Fig. 5.28. Formation of carbonyl compounds during valine and leucine biosynthesis (according to *Piendl*, 1969). → main pathway → side pathway of the metabolism

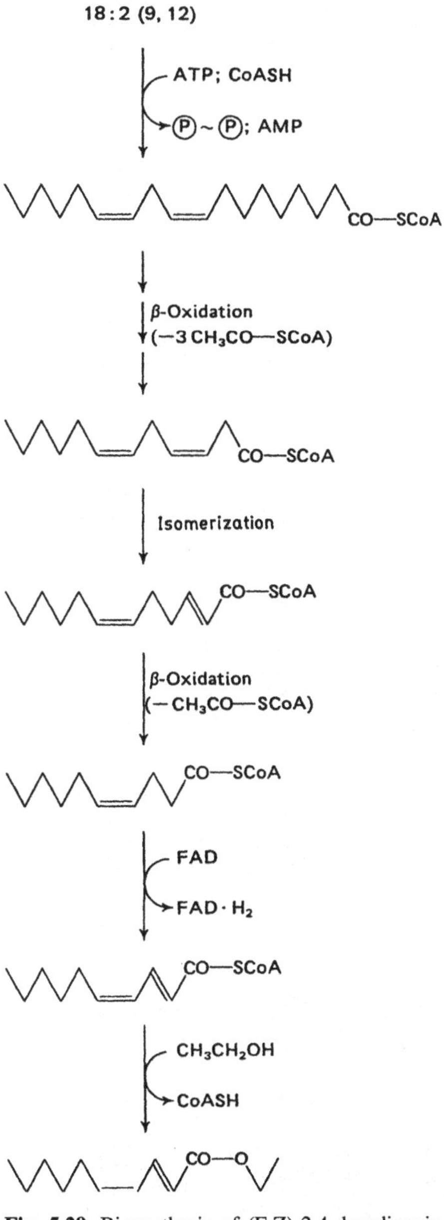

18:2 (9, 12)

Fig. 5.29. Biosynthesis of (E,Z)-2,4-decadienoic acid ethyl ester in pears (according to *Jennings* and *Tressl*, 1974)

Alcohol dehydrogenases (cf. 2.3.1.1) can reduce the aldehydes derived from fatty acid and amino acid metabolism into the corresponding alcohols:

$$R—CH_2—OH + NAD^{\oplus}$$
$$\rightleftharpoons R—CHO + NADH + H^{\oplus} \qquad (5.24)$$

Alcohol formation in plants and microorganisms is strongly favoured by the reaction equilibrium and, primarily, by the predominance of NADH over NAD^+. Nevertheless, the enzyme specificity is highly variable. In most cases aldehydes $>C_5$ are only slowly reduced; thus, with aldehydes rapidly formed by, for example, oxidative cleavage of unsaturated fatty acids, a mixture of alcohols and aldehydes results, in which the aldehydes predominate.

5.3.2.2 Hydrocarbons, Esters

Fruits and vegetables (e.g., pineapple, apple, pear, peach, passion fruit, kiwi, celery, parsley) contain unsaturated C_{11} hydrocarbons which play a role as aroma substances. Of special interest are (E,Z)-1,3,5-undecatriene and (E,Z,Z)-1,3,5,8-undecatetraene, which with very low threshold concentrations have a balsamic, spicy, pine-like odor. It is assumed that the hydrocarbons are formed from unsaturated fatty acids by β-oxidation, lipoxygenase catalysis, oxidation of the radical to the carbonium ion and decarboxylation. The hypothetical reaction pathway from linoleic acid to (E,Z)-1,3,5-undecatrieneis shown in Formula 5.25.

$$R—CO—SCoA + R'—OH$$
$$\rightarrow R—CO—O—R' + CoASH \qquad (5.25)$$

Esters are significant aroma constituents of many fruits. They are synthetized only by intact cells:

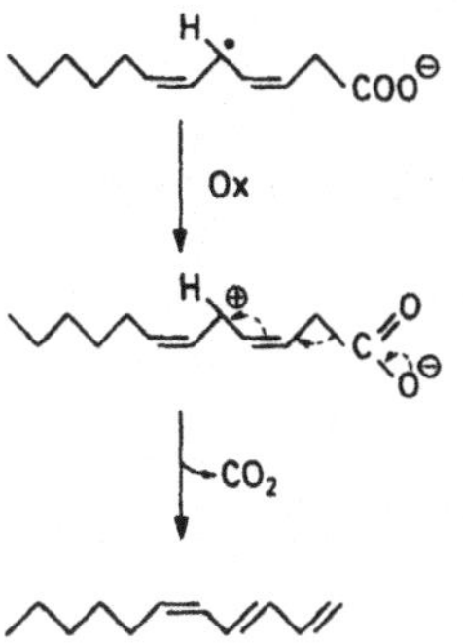

(5.26)

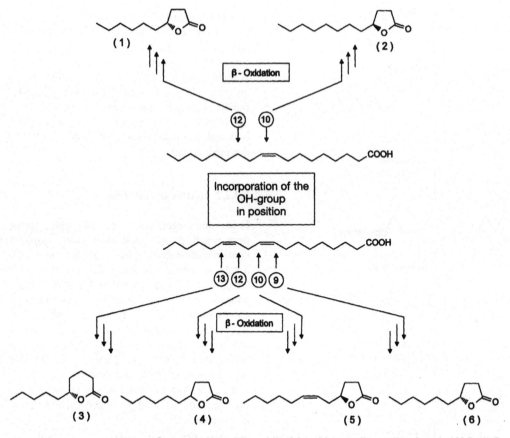

Fig. 5.30. Biosynthesis of γ- and δ-lactones from oleic and linoleic acid (according to *Tressl et al.*, 1996) (1) R-γ-decalactone, (2) S-δ-dodecalactone, (3) R-δ-decalactone, (4) γ-decalactone, (5) R-(Z)-6-γ-dodecenelactone, (6) R-γ-nonalactone

Acyl-CoA originates from the β-oxidation of fatty acids and also occasionally from amino acid metabolism. Figure 5.30 shows an example of how ethyl (E,Z)-2,4-decadienoate, an important aroma constituent of pears, is synthesized from linoleic acid.

Table 5.30 gives information on the odor thresholds of some esters. Methyl branched esters, from the metabolism of leucine and isoleucine, were found to have very low values. The odor thresholds of the acetates are higher than those of the corresponding ethylesters.

When fruits are homogenized, such as in the processing of juice, the esters are rapidly hydrolyzed by the hydrolase enzymes present, and the fruit aroma flattens.

5.3.2.3 Lactones

Numerous lactones are found in food. Some of the representatives which belong to the typical aroma substances of butter, coconut oil, and various fruits are presented in Table 5.31.

Since the aroma of lactones is partly very pleasant, these substances are also of interest for commercial aromatization of food. In the homologous series of γ- and δ-lactones, the odor threshold decreases with increasing molecular weight (Table 5.32).

The biosynthesis of lactones was studied using the yeast *Sporobolomyces odorus* and it was shown that the results are valid for animal and plant foods. Labelling with deuterium indicates

Table 5.30. Odor thresholds of esters

Compound	Odor threshold (µg/kg, water)
Methylpropionic acid methyl ester	7
2-Methylbutyric acid methyl ester	0.25
Methylpropionic acid ethyl ester	0.1
(S)-2-Methylbutyric acid ethyl ester	0.06
Butyric acid ethyl ester	0.1
Isobutyric acid ethyl ester	0.02
3-Methylbutyric acid ethyl ester	0.03
Caproic acid ethyl ester	5
Cyclohexanoic acid ethyl ester	0.001
(R)-3-Hydroxyhexanoic ethyl ester	270
Caprylic acid ethyl ester	0.1
(E,Z)-2,4-Decadienoic acid ethyl ester	100
trans-Cinnamic acid ethyl ester	0.06
Benzoic acid ethyl ester	60
Salicylic acid methyl ester	40
Butyl acetate	58
2-Methylbutyl acetate	5
3-Methylbutyl acetate	3
Pentyl acetate	38
Hexyl acetate	101
(Z)-3-Hexenyl acetate	7.8
Octyl acetate	12
2-Phenylethyl acetate	20

Table 5.32. Odor thresholds of lactones

Compound	Odor threshold (µg/kg, water)
γ-Lactones	
γ-Hexalactone	1600
γ-Heptalactone	400
γ-Octalactone	7
γ-Nonalactone	30–65
γ-Decalactone	11
γ-Dodecalactone	7
δ-Lactones	
δ-Octalactone	400
δ-Decalactone	100
6-Pentyl-α-pyrone	150

that the precursors oleic and linoleic acid are regio- and stereospecifically oxidized to hydroxy acids (Fig. 5.30), which are shortened by β-oxidation and cyclized to lactones. The individual steps in the biosynthesis are represented in Fig. 5.31 using (R)-δ-decalactone, a key odorant of butter (cf. 10.3.4).

Linoleic acid is metabolized by cows with the formation of (Z)-6-dodecen-γ-lactone as a secondary product (Fig. 5.30). Its sweetish odor enhances the aroma of butter. On the other hand, it is undesirable in meat.

Table 5.31. Lactones in food

Name	Structure	Aroma quality	Occurrence
4-Nonanolide (γ-nonalactone)	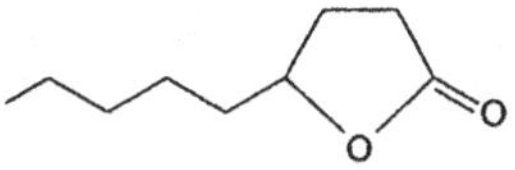	Reminiscent of coconut oil, fatty	Fat-containing food, crispbread, peaches
4-Decanolide (γ-decalactone)	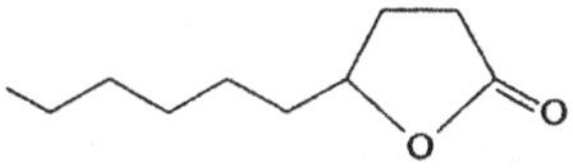	Fruity, peaches	Fat-containing food, cf. Table 5.13
5-Decanolide (δ-decalactone)	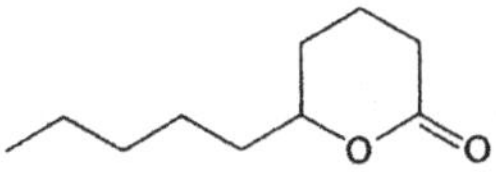	Oily, peaches	Fat-containing food, cf. Table 5.13
(Z)-6-Dodecen-γ-lactone	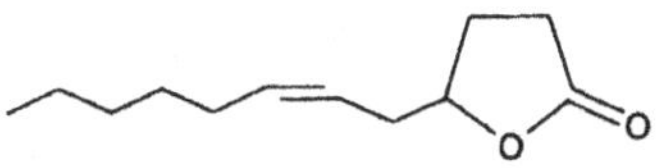	Sweet	Milk fat, peaches
3-Methyl-4-octanolide (whisky- or quercus lactone)	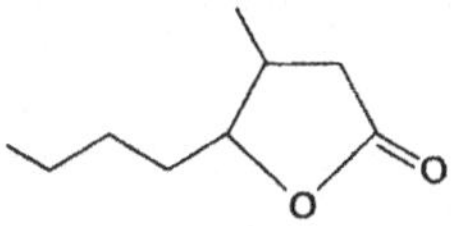	Coconut-like	Alcoholic beverages

The whisky or oak lactone is formed when alcoholic beverages are stored in oak barrels. 3-Methyl-4-(3,4-dihydroxy-5-methoxybenzo)octanoic acid is extracted from the wood. After elimination of the benzoic acid residue, this compound cyclizes to give the lactone. The odor thresholds of the two cis-oak lactones (3R, 4R and 3S, 4S) are about ten times lower than those of the trans diastereomers (3S, 4R and 3R, 4S).

5.3.2.4 Terpenes

The mono- and sesquiterpenes in fruits (cf. 18.1.2.6) and vegetables (cf. 17.1.2.6), herbs and spices (cf. 22.1.1.1) and wine (cf. 20.2.6.9) are presented in Table 5.33. These compounds stimulate a wide spectrum of aromas, mostly perceived as very pleasant (examples in Table 5.34). The odor thresholds of terpenes vary greatly (Table 5.34). Certain terpenes occur in flavoring plants in such large amounts that in spite of relatively high odor thresholds, they can act as character impact compounds, e. g., S(+)-α-phellandrene in dill.

Monoterpenes with hydroxy groups, such as linalool, geraniol and nerol, are present in

fruit juice at least in part as glycosides. Linalool-β-rutinoside (I) and linalool-6-0-α-L-arabinofuranosyl-β-D-glucopyranoside (II) have been found in wine grapes and in wine (cf. 20.2.6.9):

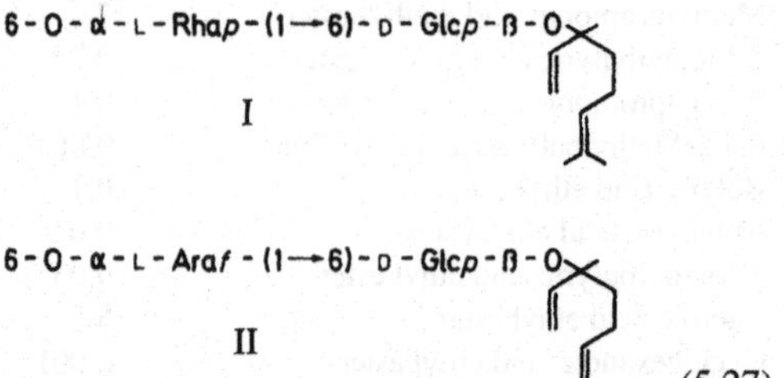

$$6 - O - \alpha - L - Rha p - (1 \rightarrow 6) - D - Glc p - \beta - O \qquad I$$

$$6 - O - \alpha - L - Ara f - (1 \rightarrow 6) - D - Glc p - \beta - O \qquad II \qquad (5.27)$$

Terpene glycosides hydrolyze, e. g., in the production of jams (cf. 18.1.2.6.11), either enzymatically (β-glucosidase) or due to the low pH of juices. The latter process is strongly accelerated by a heat treatment. Under these conditions, terpenes with two or three hydroxyl groups which are released undergo further reactions, forming hotrienol (IV) and neroloxide (V) from 3,7-dimethylocta-1,3-dien-3,7-diol (cf. Formula 5.28) in grape juice, or cis- and trans-furanlinalool oxides (VIa and VIb) from 3,7-dimethylocta-1-en-3,6,7-triol in grape juice and peach sap (cf. Formula 5.29).

Table 5.33. Terpenes in food

Monoterpenes

Acyclic (including cyclic derivatives)

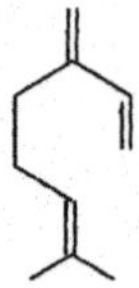

Myrcene (I)

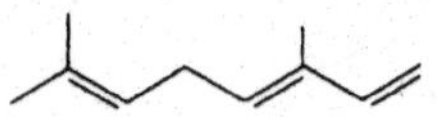

trans-Ocimene (II)

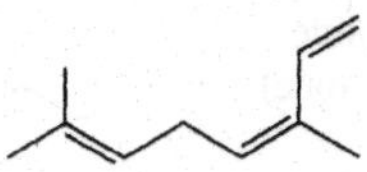

cis-Ocimene (III)

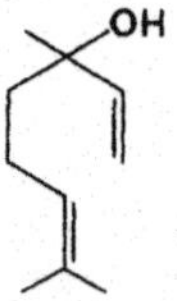

Linalool (IV)

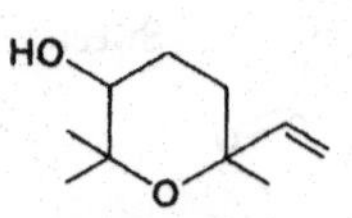

2,6,6-Trimethyl-2-vinyl-5-hydroxytetrahydropyran[a] (IVa)

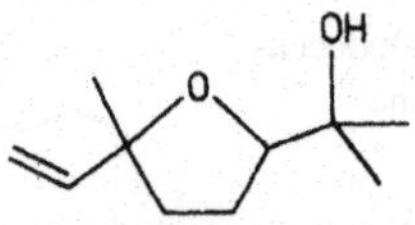

2-Methyl-2-vinyl-5-hydroxyisopropyltetrahydrofuran[a] (IVb)

Table 5.33. (*Continued*)

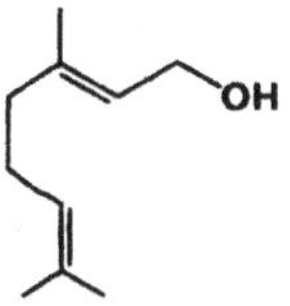

Geraniol[b] (V)

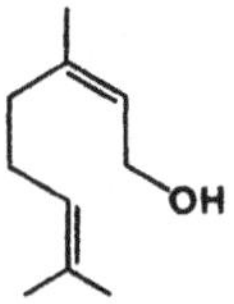

Nerol[b] (VI)

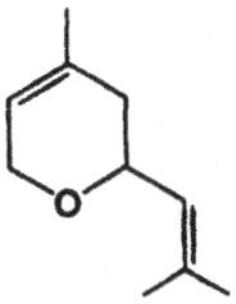

Neroloxide (VIa)

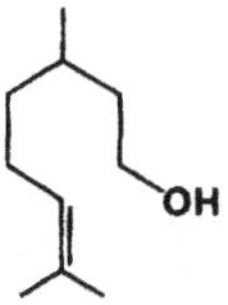

Citronellol[b] (VII)

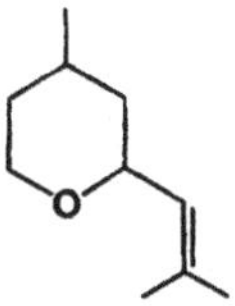

Rose oxide (VIIa)

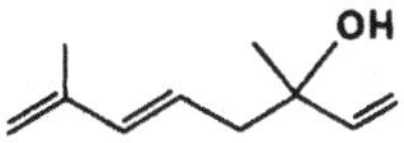

Hotrienol[c] (VIII)

Monocyclic

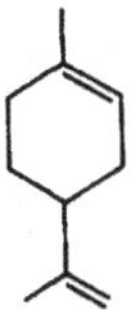

Limonene (IX)

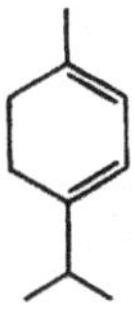

α-Terpinene (X)

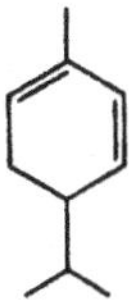

α-Phellandrene (XI)

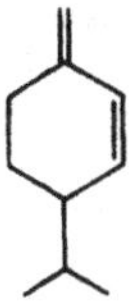

β-Phellandrene (XII)

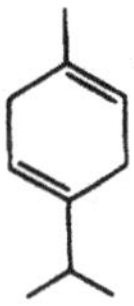

γ-Terpinene (XIII)

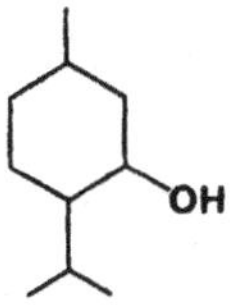

Menthol (XIV)

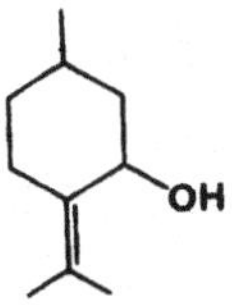

Pulegol (XV)

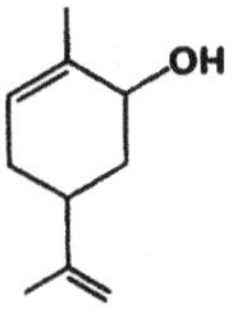

Carveol (XVI)

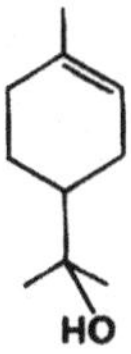

α-Terpineol (XVII)

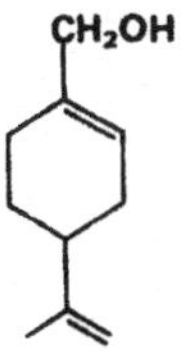

CH₂OH

Perilla alcohol (XVIII)

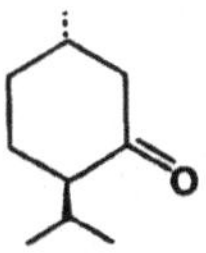

Menthone (XIX)

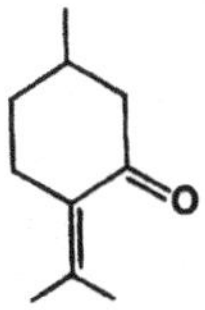

Pulegone (XX)

Table 5.33. (*Continued*)

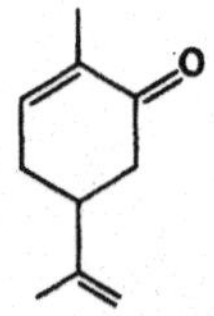

Carvone (**XXI**)

1,3-p-Menthadien-7-al
(**XXII**)

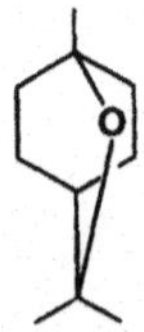

1,8-Cineole (**XXIII**)

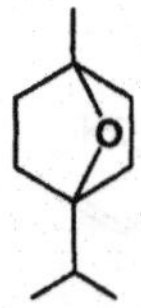

1,4-Cineole (**XXIV**)

Bicyclic

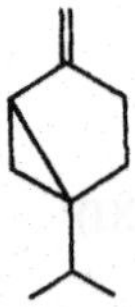

Sabinene (**XXV**)

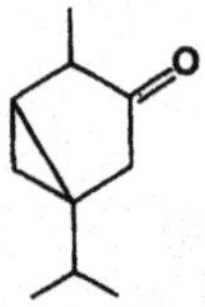

Thujone (**XXVI**)

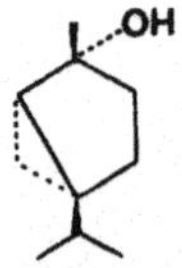

(+)-cis-Sabinene hydrate
(**XXVII**)

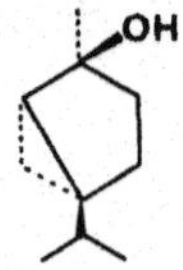

(+)-trans-Sabinene hydrate
(**XXVIII**)

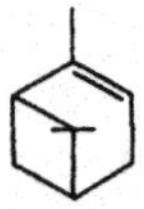

α-Pinene (**XXIX**)

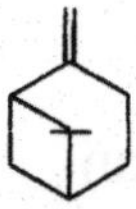

β-Pinene (**XXX**)

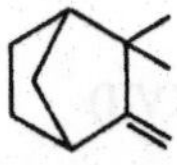

Camphene (**XXXI**)

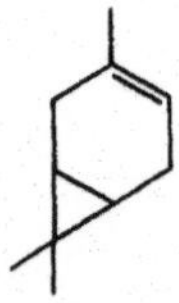

Δ³-Carene (**XXXII**)

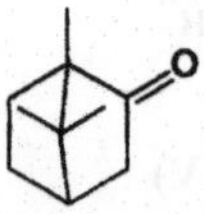

Camphor (**XXXIII**)

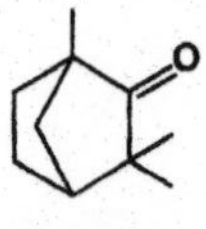

Fenchone (**XXXIV**)

Table 5.33. (Continued)

Sesquiterpenes
Acyclic

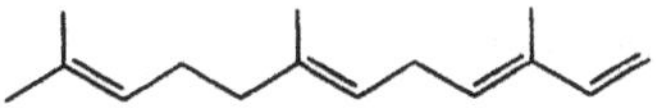

trans-α-Farnesene (XXXV)

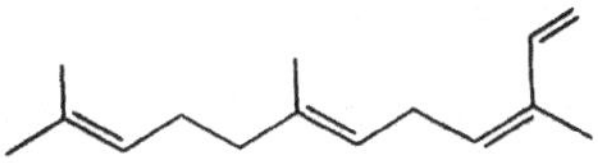

cis-α-Farnesene (XXXVI)

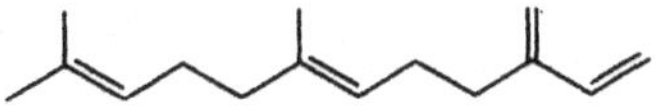

β-Farnesene (XXXVII)

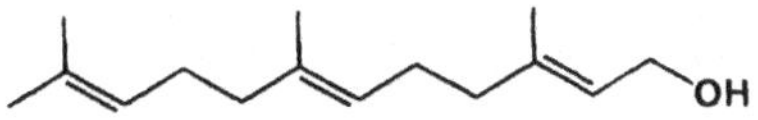

Farnesol (XXXVIII)

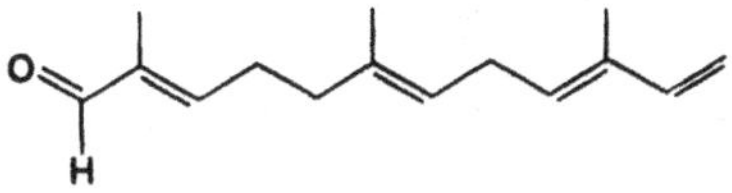

(all-trans)-α-Sinensal (XXXIX)

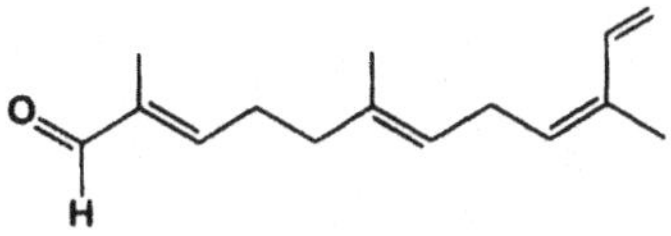

(trans,trans,cis)-α-Sinensal (XL)

Monocyclic

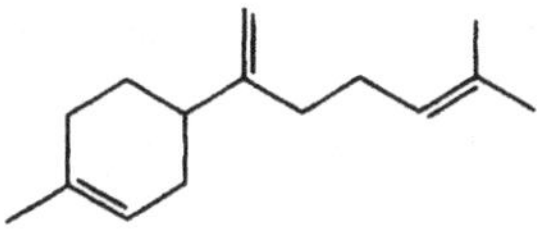

β-Bisabolene (XLI)

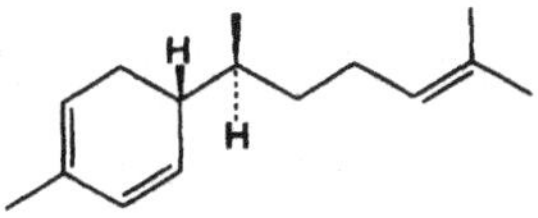

(−)-Zingiberene (XLII)

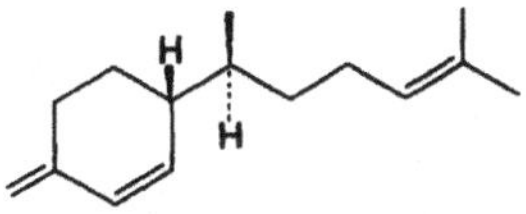

(−)-Sesquiphellandrene (XLIII)

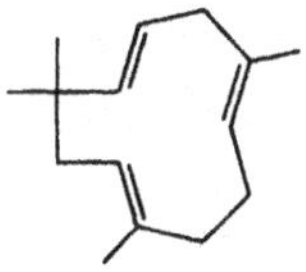

Humulene (XLIV)

Bicyclic

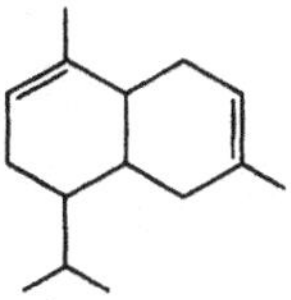

β-Cadinene (XLV)

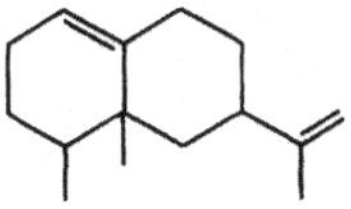

Valencene (XLVI)

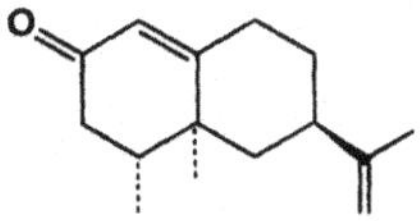

(+)-Nootkatone (XLVII)

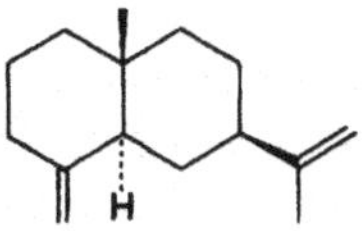

β-Selinene (XLVIII)

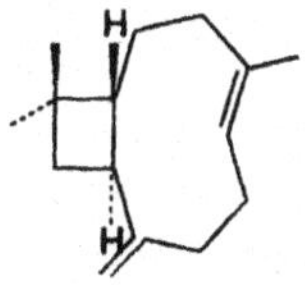

β-Caryophyllene (XLIX)

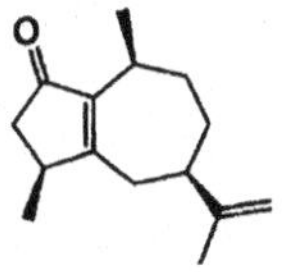

L (-) - Rotundone (L)

[a] Compounds IVa and IVb are also denoted as pyranlinalool and furanlinalool oxide, respectively.
[b] Corresponding aldehydes geranial (Va), neral (VIb) and citronellal (VIIa) also occur in food. Citral is a mixture of neral and geranial.
[c] (−)-3,7-Dimethyl-1,5,7-octatrien-3-ol (hotrienol) is found in grape, wine and tea aromas.

Table 5.34. Sensory properties of some terpenes

Compound[a]	Aroma quality	Odor threshold (µg/kg, water)
Myrcene (I)	Herbaceous, metallic	14
Linalool (IV)	Flowery	6
cis-Furanlinalool oxide (IVb)	Sweet-woody	6000
Geraniol (V)	Rose-like	7.5
Geranial (Va)	Citrus-like	32
Nerol (VI)		300
Citronellol (VII)	Rose-like	10
cis-Rose oxide (VIIa)	Geranium-like	0.1
R(+)-Limonene (IX)	Citrus-like	200
R(−)-α-Phellandrene (XI)	Terpene-like, medicinal	500
S(−)-α-Phellandrene (XI)	Dill-like, herbaceous	200
α-Terpineol (XVII)	Lilac-like, peach-like	330
(R)-Carvone (XXI)		50
1,8-Cineol (XXIII)	Spicy, camphor-like	12
(all-E)-α-Sinensal (XXXIX)	Orange-like	0.05
(−)-β-Caryophyllene (XLIX)	Spicy, dry	64
(−)-Rotundone (L)	Peppery	0.008

[a] The numbering of the compounds refers to Table 5.33.

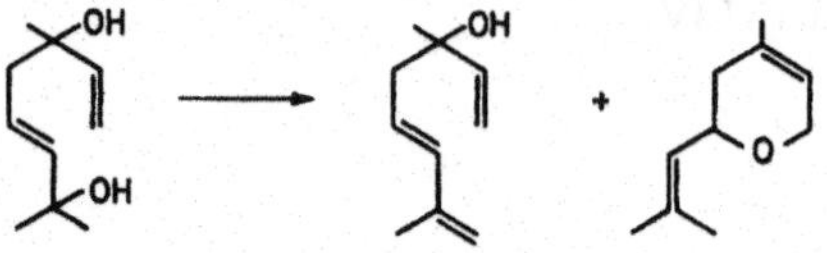

IV   V   (5.28)

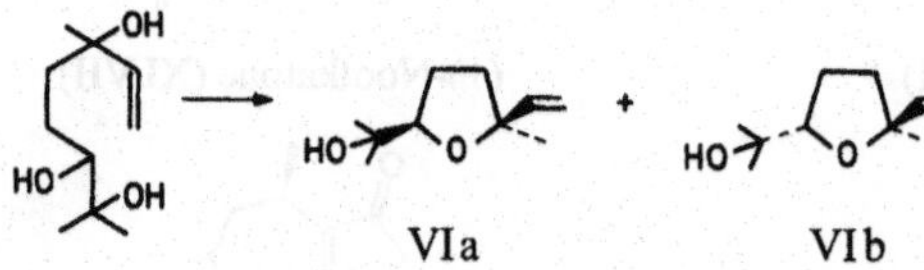

VIa   VIb

(5.29)

Most terpenes contain one or more chiral centers. Of several terpenes, the optically inactive form and the l- and d-form occur in different plants. The enantiomers and diastereoisomers differ regularly in their odor characteristics. For example, menthol (XIV in Table 5.33) in the l-form

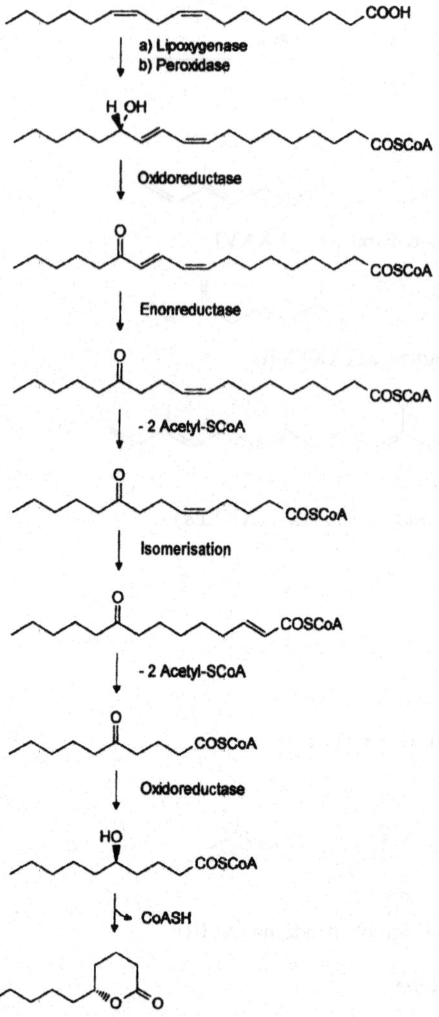

Fig. 5.31. Formation of R-δ-decalactone from linoleic acid (according to *Tressl et al.*, 1996)

(1R, 3R, 4S) which occurs in peppermint oil, has a clean sweet, cooling and refreshing peppermint aroma, while in the d-form (1S, 3S, 4R) it has remarkable, disagreeable notes such as phenolic, medicated, camphor and musty. Carvone (XXI in Table 5.33) in the R(−)-form has a peppermint odor. In the S(+)-form it has an aroma similar to caraway. Other examples that show the influence of stereochemistry on the odor threshold of terpenes are 3a,4,5,7a-tetrahydro-3,6-dimethyl-2(3H)-benzofuranone (cf. 5.2.5) and 1-p-menthene-8-thiol (cf. 5.3.2.5).

Some terpenes are readily oxidized during food storage. Examples of aroma defects resulting from oxidation are provided in Table 5.5 and Section 22.1.1.1.

5.3.2.5 Volatile Sulfur Compounds

The aroma of many vegetables is due to volatile sulfur compounds obtained by a variety of enzymatic reactions. Examples are the vegetables of the plant families *Brassicacea* and *Liliaceae*; their aroma is formed by decomposition of glucosinolates or S-alkyl-cysteine-sulfoxides (cf. 17.1.2.6.7).

2-Isobutylthiazole (compound V, Table 5.22) contributes to tomato aroma (cf. 17.1.2.6.13). It is probably obtained as a product of the secondary metabolism of leucine and cysteine (cf. postulated Reaction 5.30).

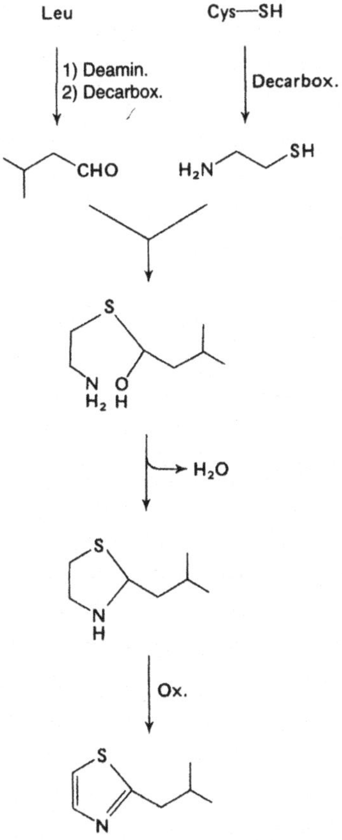

(5.30)

Isobutyric acid is the precursor of asparagus acid (1,2-dithiolane-4-carboxylic acid) found in

asparagus. It is dehydrogenated to give methylacrylic acid which then adds on an unknown S-containing nucleophile (see Formula 5.31). During cooking, asparagus acid is oxidatively decarboxylated to a 1,2-dithiocyclopentene (see Formula 5.32), which contributes to the aroma of asparagus.

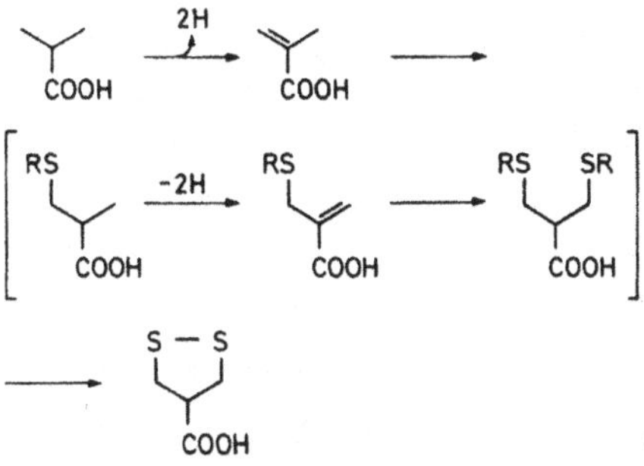

(5.31)

Volatile sulfur compounds formed in wine and beer production originate from methionine and are by-products of the microorganism's metabolism. The compounds formed are methional (I), methionol (II) and acetic acid-3-(methylthio)-propyl ester (III, cf. Reaction 5.33).

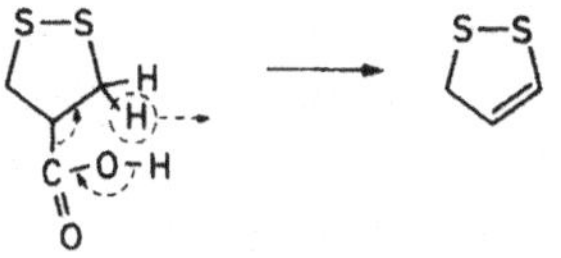

(5.32)

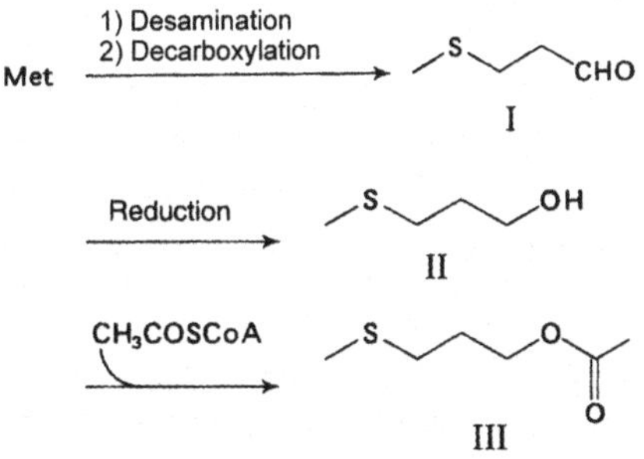

(5.33)

Tertiary thiols (Table 5.35) are some of the most intensive aroma substances. They have a fruity odor at the very low concentrations in which they occur in foods. With increasing concentration, they smell of cat urine and are called *catty odorants*. Tertiary thiols have been detected in some fruits, olive oil, wine (*Scheurebe*) and roasted coffee (Table 5.35). They make important contributions to the aroma and are possibly formed by the addition of hydrogen sulfide to metabolites of isoprene metabolism. In beer,

Table 5.35. Tertiary thiols in food

Name	Structure	Odor threshold (µg/kg, water)	Occurrence
4-Mercapto-4-methyl-2-pentanone	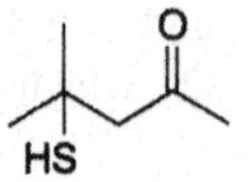	0.0001	Basil, wine (*Scheurebe*), Grapefruit
4-Methoxy-2-methyl-2-butanethiol	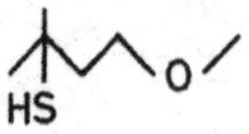	0.00002	Olive oil (cf. 14.3.2.1.1), black currants
3-Mercapto-3-methylbutylformate	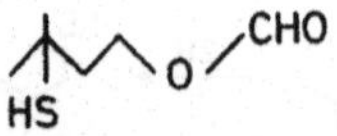	0.003	Roasted coffee
1-p-Menthen-8-thiol	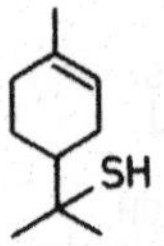	0.00002	Grapefruit

3-mercapto-3-methylbutylformate is undesirable because it causes off-flavor at concentrations as low as 5 ng/l. 1-p-Menthene-8-thiol, which contributes to grapefruit aroma, is a chiral compound. The (R)-enantiomer exhibits an extremely low odor threshold shown in Table 5.35. The (S)-enantiomer has a weak and unspecific odor.

5.3.2.6 Pyrazines

Paprika pepper (*Capsicum annum*) and chillies (*Capsicum frutescens*) contain high concentrations of 2-isobutyl-3-methoxypyrazine (X in Table 5.24 for structure). Its biosynthesis from leucine is assumed to be through the pathway shown in Formula 5.34.
The compound 2-sec-butyl-3-methoxy-pyrazine is one of the typical aroma substances of carrots.

Pyrazines are also produced by microorganisms. For example, 2-isopropyl-3-methoxypyrazine has been identified as a metabolic byproduct of *Pseudomonas perolans* and *Pseudomonas taetrolens*. This pyrazine is responsible for a musty/earthy off-flavor in eggs, dairy products and fish.

5.3.2.7 Skatole, p-Cresol

The amino acids tryptophan and tyrosine are degraded by microorganisms to skatole and p-cresol respectively (cf. Formula 5.40).
The odor thresholds of skatole have been determined in sunflower oil (15.6 µg/kg) and on starch (0.23 µg/kg). This compound plays a role in the aroma of Emmental cheese (cf. 10.3.5) and causes an aroma defect in white pepper (cf. 22.1.1.2.1). It can probably also be formed nonenzymatically from tryptophan by the

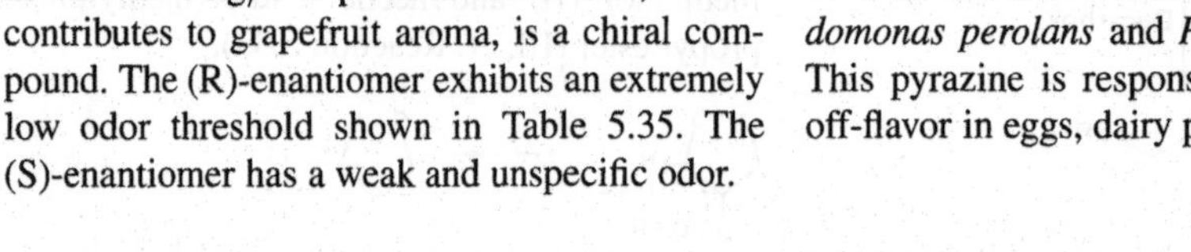

(5.34)

Strecker degradation, oxidation to indolylacetic acid and decarboxylation. The oxidative cleavage of skatole yields o-aminoacetophenone (cf. Formula 5.36), which has an animal odor and is the key aroma substance of tortillas and taco shells made of corn treated with lime (*Masa corn*). In the case of milk dry products, o-aminoacetophenone causes an aroma defect (cf. 10.3.2). Its odor threshold of $0.2\,\mu g/kg$ (water) is very low. On the other hand, p-amino-acetophenone has an extremely high odor threshold of $100\,mg/kg$ (water).

R−CH₂−CH−COOH
     |
     NH₂

↓ Oxidative deamination

↓ Decarboxylation

R−CH₂−CHO

↓ Oxidation

R−CH₂−COOH

↳ CO₂   R : [indole structure] (Trp)

R−CH₃   [Tyr structure] HO—◯ (Tyr)

(5.35)

p-Cresol (odor threshold on starch $130\,\mu g/kg$) has been detected as an accompanying substance of skatole in samples of white pepper having an aroma defect. It is also formed in citrus oil and juice by the degradation of citral (cf. 5.5.4).

[skatole structure with CH₃]

↓ Oxidation

[o-acetyl formamide structure]

↓ Oxidation
  Decarboxylation

[o-aminoacetophenone structure]

(5.36)

5.4 Interactions with Other Food Constituents

Aroma interactions with lipids, proteins and carbohydrates affect the retention of volatiles within the food and, thereby, the levels in the gaseous phase. Consequently, the interactions affect the intensity and quality of food aroma. Since such interactions cannot be clearly followed in a real food system, their study has been transferred to model systems which can, in essence, reliably imitate the real systems. Consider the example of emulsions with fat contents of 1%, 5% and 20%, which have been aromatized with an aroma cocktail for mayonnaise consisting of diacetyl, (Z)-3-hexenol, (E,Z)-2,6-nonadienol, allyl isothiocyanate and allyl thiocyanate. The sample with 20% of fat has the typical and balanced odor of mayonnaise (Fig. 5.32 a). If the fat content decreases, the aroma changes drastically. The emulsion with 5% of fat has an untypical creamy and pungent odor since there is a decrease in the intensities of the buttery and fatty notes in the aroma profile (Fig. 5.32 b). In the case of 1% of fat, pungent, mustard-like aroma notes dominate (Fig. 5.32 c).

Headspace analyses show that the drastic change in the aroma of the emulsions is based on the fact that the concentrations of the fatsoluble aroma substances (Z)-3-hexenol, allyl isothiocyanate and allyl thiocyanate in the gas phase increase with decreasing fat content (Fig. 5.33). Only the water-soluble diacetyl remains unaffected (Fig. 5.33).

The concentration of the very aroma active (E,Z)-2,6-nonadienol (cf. 10.3.6) in the head space is below the detection limit. However, this odorant can be detected by headspace GC-olfactometry (cf. 5.2.2.2). The results in Table 5.36 show that this alcohol as well as (Z)-3-hexenol no longer contribute to the aroma in the 20% fat emulsion. In the emulsion with 1% of fat, (E,Z)-2,6-nonadienol, allyl isothiocyanate and allyl thiocyanate predominate and produce the green, mustard-like aroma (Table 5.36).

A knowledge of the binding of aroma to solid food matrices, from the standpoint of food aromatization, aroma behavior and food processing and storage, is of great importance.

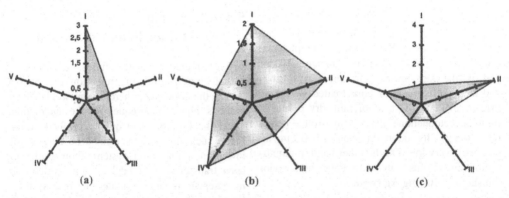

Fig. 5.32. Influence of the fat content on the aroma profile of emulsions; a) 20% fat, b) 5% fat, c) 1% fat. The intensities of the aroma qualities buttery (I), pungent, sharp (II), fatty (III), sweet (IV) and green (V) were evaluated as 1 (weak) to 4 (strong) (according to *Widder* and *Fischer*, 1996)

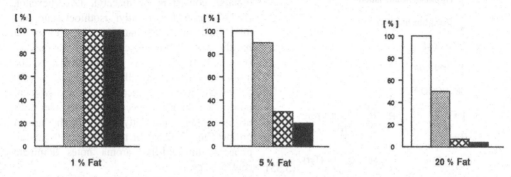

Fig. 5.33. Influence of the fat content of an emulsion on the concentration of aroma substances in the gas phase (according to *Widder* and *Fischer*, 1996).
☐ diacetyl, ■ (Z)-3-hexenol, ⊠ allyl isothiocyanate, ■ allyl thiocyanate

Table 5.36. Mayonnaise model: gas chromatography/olfactometry of headspace samples

Aroma substance[a]	Odor quality	Odor intensity[b]		
		1% fat	5% fat	20% fat
Diacetyl	Buttery	3	4	>4
(Z)-3-Hexenol	Green	2	1	0
(E,Z)-2-6-Nonadienol	Green, fatty	4	<1	0
Allyl isothiocyanate	Pungent, mustard-like	4	3	<1
Allyl thiocyanate	Pungent, mustard-like	4	3	<1

[a] Components of an aroma cocktail to which an oil emulsion was added.
[b] Intensity on sniffing the carrier gas stream 1 (weak)–4 (strong).

5.4.1 Lipids

In an o/w emulsion (cf. 8.15.1), the distribution coefficient, K, for aroma compounds is related to aroma activity:

$$K = \frac{C_o}{C_w} \qquad (5.37)$$

where C_o is the concentration of the aroma compound in the oil phase, and C_w the concentration of the aroma compound in the aqueous phase.

In a homologous series, e. g., n-alkane alcohols (cf. Fig. 5.34), the value of K increases with increasing chain length. The solubility in the fat or oil phase rises proportionally as the hydrophobic-

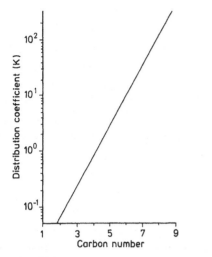

Fig. 5.34. Distribution of n-alkanols in the system oil/water (according to *McNulty* and *Karel*, 1973)

ity imposed by chain length increases. The vapor pressure behavior is exactly the reverse; it drops as the hydrophobicity of the aroma compounds increases. The vapor pressure also drops as the volume of the oil phase increases, and the odor threshold value increases at the same time. This is well clarified in Fig. 5.35. The solubility of 2-heptanone is higher in whole milk than

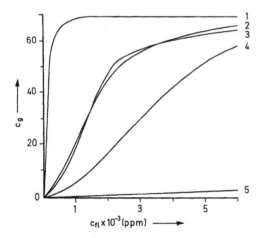

Fig. 5.35. Influence of the medium on 2-heptanone concentration in the gas phase (according to *Nawar*, 1966). 2-Heptanone alone (*1*), in water (*2*), in skim milk (*3*), in whole milk (*4*), in oil (*5*). c_{fl}: concentration in liquid; c_g: concentration in gas phase (detection signal height from headspace analysis)

in skim milk which, in this case, behaves as an aqueous phase. When this phase is replaced by oil (Fig. 5.35), 2-heptanone concentration in the gas phase is the lowest.

Experiments with n-alcohols demonstrate that, with increasing chain length of volatile compounds, the migration rate of the molecules from oil to water phase increases. An increase in oil viscosity retards such migration.

5.4.2 Proteins, Polysaccharides

The sorption characteristics of various proteins for several volatile compounds are presented in Fig. 5.36. Ethanol is bound to the greatest extent, probably with the aid of hydrogen bonds. The binding of the nonpolar aroma compounds probably occurs on the hydrophobic protein surface regions. A proposal for the evaluation of data on the sorption of aroma volatiles on a biopolymer (protein, polysaccharide) is based on the law of mass action. When a biopolymer, B, has a group which attracts and binds the aroma molecule, A, then the following equation is valid:

$$K = \frac{(BA)}{c_f(B)} \tag{5.38}$$

where K = a single binding constant; and c_f = concentration of free aroma compound molecules.

$$[BA] = K \cdot c_f \cdot (B) \tag{5.39}$$

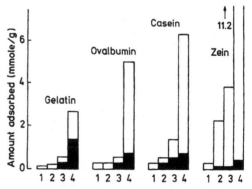

Fig. 5.36. Sorption of volatile compounds on proteins at 23 °C (according to *Maier*, 1974). Hexane (*1*), ethyl acetate (*2*), acetone (*3*), ethanol (*4*). □ plus ■: maximal sorption, ■: after desorption

To calculate the average number of aroma molecules bound to a biopolymer, the specific binding capacity, r, has to be introduced:

$$r = \frac{(BA)}{(B) + (BA)} \qquad (5.40)$$

The concentration of the complex BA from Equation 5.39 is substituted in Equation 5.40:

$$r = \frac{K \cdot c_f(B)}{(B) + K(B)c_f} = \frac{Kc_f}{1 + Kc_f} \qquad (5.41)$$

When a biopolymer binds not only one molecular species (as A in the above case) but has a number (n) of binding groups (or sites) equal in binding ability and independent of each other, then r has to be multiplied by n, and Equation 5.41 acquires the form:

$$r = \frac{n \cdot Kc_f}{1 + Kc_f} \qquad (5.42)$$

$$\frac{r}{c_f} = K \cdot n - K \cdot r = K' - K \cdot r \qquad (5.43)$$

where K' = overall binding constant.
The evaluation of data then follows Equation 5.43 presented in graphic form, i.e. a diagram of $r/C_f = f(r)$. Three extreme or limiting cases should be observed:

a) A straight line (Fig. 5.37, a) indicates that only one binding region on a polymer, with n binding sites (all equivalent and independent from each other) is involved. The values n and K' are obtained from the intersection of the straight line with the abscissa and the ordinate, respectively.

b) A straight line parallel to the abscissa (Fig. 5.37, b) is obtained when the single binding constant, K, is low and the value of n is very high. In this special case, Equation 5.48 has the form:

$$r = K' \cdot c_f \qquad (5.44)$$

c) A curve (Fig. 5.37, c) which in approximation is the merging of two straight lines, as shown separately (Fig. 5.37, d). This indicates two binding constants, K_1', and K_2', and their respective binding groups, n_1 and n_2, which are equivalent and independent of each other.

By plotting r versus c_f, values of K' are obtained from the slope of the curve. An example for a model system with two binding regions (case c) is given by aroma binding to starch. It should be remembered that starch binds the volatiles only after gelatinization by trapping the volatiles in its helical structure, and that starch is made up of two constituents, amylose and amylopectin. The binding parameters are listed for some aroma compounds in Table 5.37. Numerous observations indicate that K_1' and binding region n_1 are related to the inner space of the helix, while K_2' and the n_2 region are related to the outer surface of the helix. K_1' is larger than K_2', which shows that, within the helix, the aroma compounds are bound more efficiently to glucose residues of the helix. The fraction $1/n$ is a measure of the size of the binding region. It decreases, as expected, with

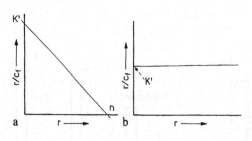

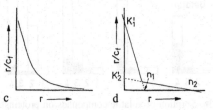

Fig. 5.37. Binding of aroma compounds by biopolymers. Graphical determination of binding parameters according to *Solms*, 1975

Table 5.37. Binding of aroma compounds by potato starch

Compounds	Binding constant			
	K_1'	n_1	K_2'	n_2
1-Hexanol	$5.45 \cdot 10^1$	0.10	–	–
1-Octanol	$2.19 \cdot 10^2$	0.05	$2.15 \cdot 10^1$	0.11
1-Decanol	$1.25 \cdot 10^2$	0.04	$1.29 \cdot 10^1$	0.11
Capric acid	$3.30 \cdot 10^2$	0.07	$4.35 \cdot 10^1$	0.19
Menthone	$1.84 \cdot 10^2$	0.012	8.97	0.045
Menthol	$1.43 \cdot 10^2$	0.007	–	–
β-Pinene	$1.30 \cdot 10^1$	0.027	1.81	0.089

Table 5.38. Binding of aroma compounds by proteins (0.4% solutions at pH 4.5)

Aroma compound	Total binding constant $K' \cdot 10^3$ (lmol^{-1})			
	Bovine serum albumin		Soya protein	
	20 °C	60 °C	20 °C	60 °C
Butanal	9.765	11.362	10.916	9.432
Benzaldehyde	6.458	6.134	5.807	6.840
2-Butanone	4.619	5.529	4.975	5.800
1-Butanol	2.435	2.786	2.100	2.950
Phenol	3.279	3.364	3.159	3.074
Vanillin	2.070	2.490	2.040	2.335
2,5-Dimethyl pyrazine	0.494			
Butyric acid	0.			

Table 5.39. Use of aromas in the production of foods

Product group	Percentage (%)[a]
Non-alcoholic beverages	38
Confectionery	14
Savoury products[b], snacks	14
Bread and cakes	7
Milk products	6
Desserts	5
Ice cream	4
Alcoholic beverages	4
Others	8

[a] Approximate values.
[b] Salty product line like vegetables, spices, meat.

increasing molecular weight of alkyl alcohols, but it is still larger within the helix than on the outer surface. Altogether, it should be concluded that, within a helix, the trapped compound cannot fulfill an active role as an aroma constituent.

An unlimited number of binding sites exist in proteins dissolved or dispersed in water (case b). K' values for several aroma compounds are given in Table 5.38. The value of the constant decreases in the order of aldehydes, ketones, alcohols, while compounds such as dimethylpyrazine and butyric acid are practically unable to bind. In the case of aldehydes, it must be assumed that they can react with free amino- and SH-groups. The high values of K' can reflect other than secondary forces.

Bovine serum albumin and soya proteins are practically identical with regard to the binding of aroma compounds (Table 5.38). Since both proteins have a similar hydrophobicity, it is apparent that hydrophobic rather than hydrophilic interactions are responsible for aroma binding in proteins.

5.5 Natural and Synthetic Flavorings

Aromatized food has been produced and consumed for centuries, as exemplified by confectionery and baked products, and tea or alcoholic beverages. In recent decades the number of aromatized foods has increased greatly. In Germany, these foods account for about 15–20% of the total food consumption. A significant reason for this development is the increase in industrially produced food, which partly requires aromatization because certain raw materials are available only to a limited extent and, therefore, expensive or because aroma losses occur during production and storage. In addition, introduction of new raw materials, e. g., protein isolates, to diversify or expand traditional food sources, or the production of food substitutes is promising only if appropriate aromatization processes are available. This also applies to the production of nutraceuticals (cf. 19.1.3).

Aroma concentrates, essences, extracts and individual compounds are used for aromatization. They are usually blended in a given proportion by a flavorist; thus, an aroma mixture is "composed". The empirically developed "aroma formulation" is based primarily on the flavorist's experience and personal sensory assessment and is supported by the results of a physico-chemical aroma analysis. Legislative measures that regulate food aromatization differ in various countries.

At present, non-alcoholic beverages occupy the first place among aromatized foods (Table 5.39). Of the different types of aroma, citrus, mint and red fruit aromas predominate (Table 5.40).

5.5.1 Raw Materials for Essences

In Germany, up to about 60% of the aromas used for food aromatization are of plant origin and,

Table 5.40. Types of aroma used

Aroma type	Percentage (%)[a]
Citrus	20
Mint	15
Red fruits	11
Vanilla	10.5
Meat	10.5
Spices	8.5
Chocolate	8.5
Cheese	5.5
Nut	2.5
Others	8

[a] Approximate values.

thus, designated as "natural aroma substances". The rest of the aroma compounds are synthetic, but 99% of this portion is chemically identical to their natural counterparts. Only 1% are synthetic aroma compounds not found in nature.

5.5.1.1 Essential Oils

Essential (volatile) oils are obtained preferentially by steam distillation of plants (whole or parts) such as clove buds, nutmeg (mace), lemon, caraway, fennel, and cardamon fruits (cf. 22.1.1.1). After steam distillation, the essential oil is separated from the water layer, clarified and stored. The pressure and temperature used in the process are selected to incur the least possible loss of aroma substances by thermal decomposition, oxidation or hydrolysis.

Many essential oils, such as those of citrus fruits, contain terpene hydrocarbons which contribute little to aroma but are readily autooxidized and polymerized ("resin formation"). These undesirable oil constituents (for instance, limonene from orange oil) can be removed by fractional distillation. Fractional distillation is also used to enrich or isolate a single aroma compound. Usually, this compound is the dominant constituent of the essential oil. Examples of single aroma compounds isolated as the main constituent of an essential oil are: 1,8-cineole from eucalyptus, 1(−)-menthol from peppermint, anethole from anise seed, eugenol from clove, or citral (mixture of geranial and neral, the pleasant odorous

compounds of lemon or lime oils) from litseacuba.

5.5.1.2 Extracts, Absolues

When the content of essential oil is low in the raw material or the aroma constituents are destroyed by steam distillation or the aroma is lost by its solubility in water, then the oil in the raw material is recovered by an extraction process. Examples are certain herbs or spices (cf. 22.1.1.1) and some fruit powders. Hexane, triacetin, acetone, ethanol, water and/or edible oil or fat are used as solvents. Good yields are also obtained by using liquid CO_2. The volatile solvent is then fully removed by distillation. The oil extract (resin, absolue) often contains volatile aroma compounds in excess of 10% in addition to lipids, waxes, plant pigments and other substances extractable by the chosen solvent. Extraction may be followed by chromatographic or counter-current separation to isolate some desired aroma fractions. If the solvent used is not removed by distillation, the product is called an extract. The odor intensity of the extract, compared to the pure essential oil, is weaker for aromatization purposes by a factor of 10^2 to 10^3.

5.5.1.3 Distillates

The aroma constituents in fruit juice are more volatile during the distillation concentration process than is the bulk of the water. Hence, the aroma volatiles are condensed and collected (cf. 18.2.10). Such distillates yield highly concentrated aroma fractions through further purification steps.

5.5.1.4 Microbial Aromas

Cheese aroma concentrates offered on the market have an aroma intensity at least 20-fold higher than that of normal cheese. They are produced by the combined action of lipases and *Penicillium roqueforti* using whey and fats/oils of plant origin as substrates. In addition to C_4–C_{10} fatty acids, the aroma is determined by the presence of 2-heptanone and 2-nonanone.

5.5.1.5 Synthetic Natural Aroma Compounds

In spite of the fact that a great number of food aroma compounds have been identified, economic factors have resulted in only a limited number of them being synthesized on a commercial scale. Synthesis starts with a natural compound available in large amounts at the right cost, or with a basic chemical. Several examples will be considered below.

A most important aroma compound worldwide, vanillin, is obtained primarily by alkaline hydrolysis of lignin (sulfite waste of the wood pulp industry), which yields coniferyl alcohol. It is converted to vanillin by oxidative cleavage:

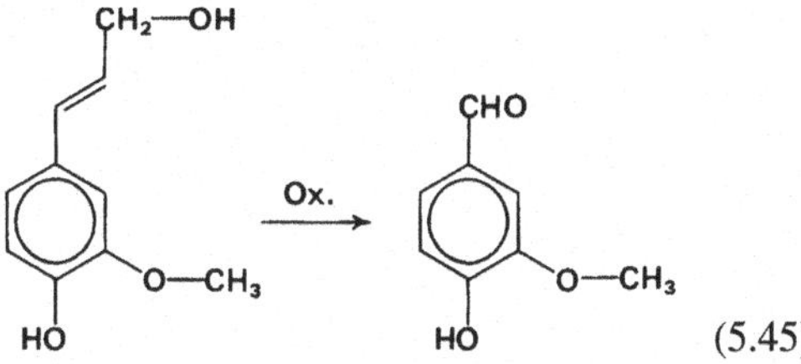

$$(5.45)$$

A distinction can be made between natural and synthetic vanillin by using quantitative ^{13}C analysis (cf. 18.4.3). The values in Table 5.41 show that the ^{13}C distribution in the molecule is more meaningful than the ^{13}C content of the entire molecule. The most important source of citral, used in large amounts in food processing, is the steamdistilled oil of lemongrass (*Cymbopogon flexuosus*). Citral actually consists of two geometrical isomers: geranial (I) and neral (II). They are isolated from the oil in the form of bisulfite adducts:

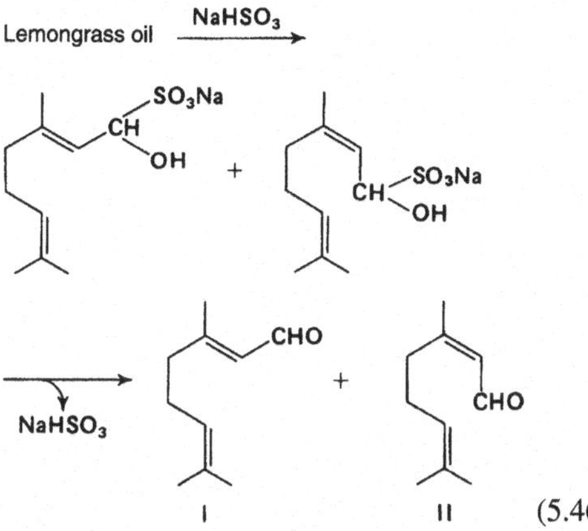

$$(5.46)$$

The aroma compound menthol is primarily synthesized from petrochemically obtained m-cresol.

Table 5.41. Site-specific ^{13}C isotopic analysis of vanillin from different sources

Source	R (%)[a] in			R(%)[a] total
	CHO	Benzene ring	OCH$_3$	
Vanilla	1.074	1.113	1.061	1.101
Lignin	1.062	1.102	1.066	1.093
Guaiacol	1.067	1.102	1.026	1.089

[a] R ($^{13}C/^{12}C$) was determined by site-specific natural isotope fractionation NMR (SNF-NMR). Standard deviation: 0.003–0.007.

Thymol is obtained by alkylation and is then further hydrogenated into racemic menthol:

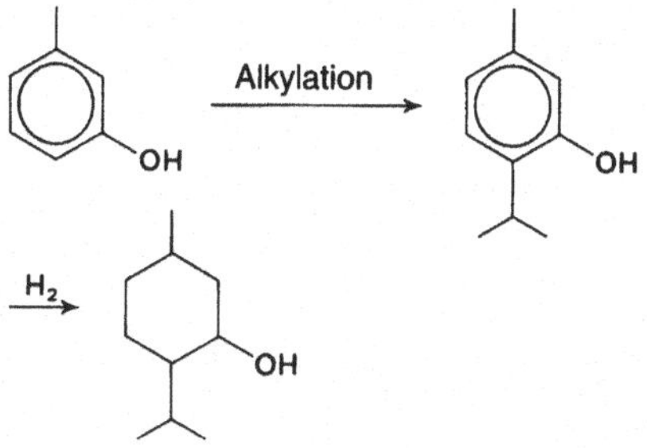

$$(5.47)$$

A more expensive processing step then follows, in which the racemic form is separated and 1(–)-menthol is recovered. The d-optical isomer substantially decreases the quality of the aroma (cf. 5.3.2.4).

The purity requirement imposed on synthetic aroma substances is very high. The purification steps usually used are not only needed to meet the stringent legal requirements (i. e. beyond any doubt safe and harmless to health), but also to remove undesirable contaminating aroma compounds. For example, menthol has a phenolic off-flavor note even in the presence of only 0.01% thymol as an impurity. This is not surprising since the odor threshold value of thymol is lower than that of 1(–)-menthol by a factor of 450.

5.5.1.6 Synthetic Aroma Compounds

Some synthetic flavorings which do not occur in food materials are compiled in Table 5.42. Except for ethyl vanillin, they are of little importance in the aromatization of foods.

Table 5.42. Synthetic Flavoring Materials (not naturally occurring in food)

Name	Structure	Aroma description
Ethyl vanillin	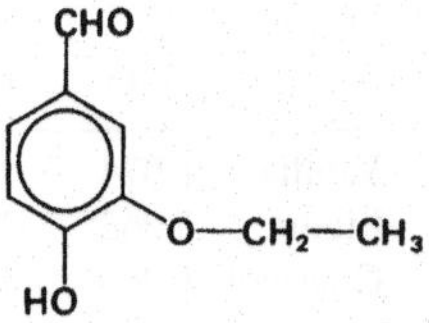	Sweet like vanilla (2 to 4-times stronger than vanillin)
Ethyl maltol	cf. 5.3.1.2	Caramel-like
Musk ambrette	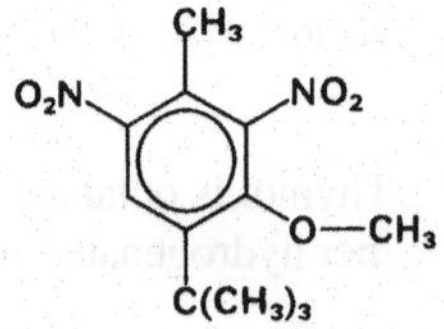	Musk-like
Allyl phenoxyacetate		Fruity, pineapple-like
α-Amyl cinnamic-aldehyde	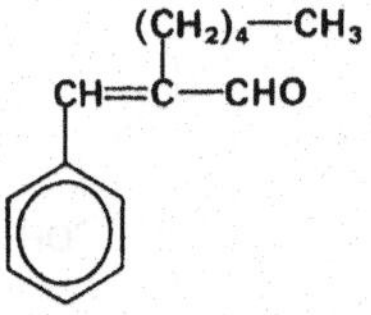	Floral, jasmin and lilies
Hydroxycitronellal	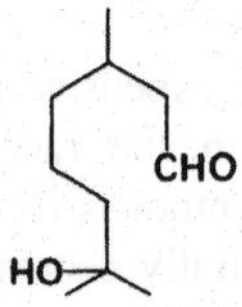	Sweet, flowery, liliaceous
6-Methyl coumarin	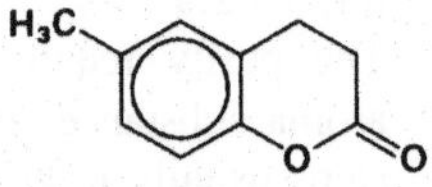	Dry, herbaceous
Propenylguaethol (vanatrope)	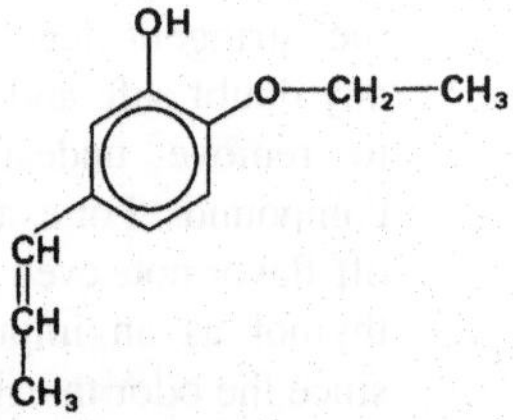	Phenolic, anise-like
Piperonyl isobutyrate	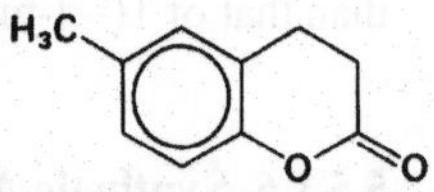	Sweet, fruity, like berry fruits

5.5.2 Essences

The flavorist composes essences from raw materials. In addition to striving for an optimal aroma, the composition of the essence has to meet food processing demands, e. g., compensation for possible losses during heating. The "aroma formulation" is an empirical one, developed as a result of long experience dealing with many problems, disappointments and failures, and is rigorously guarded after the "know-how" is acquired.

5.5.3 Aromas from Precursors

The aroma of food that has to be heated, in which the impact aroma compounds are generated by the *Maillard* reaction, can be improved by increasing the levels of precursors involved in the reaction. This is a trend in food aromatization. Some of the precursors are added directly, while some are generated within the food by the preliminary release of the reaction components required for the *Maillard* reaction (cf. 4.2.4.4). This is achieved by adding protein and polysaccharide hydrolases to food.

5.5.4 Stability of Aromas

Aroma substances can undergo changes during the storage of food. Aldehydes and thiols are especially sensitive because they are easily oxidized to acids and disulfides respectively. Moreover, unsaturated aldehydes are degraded by reactions which will be discussed using (E)-2-hexenal and citral as examples. These two aldehydes are important aromatization agents for leaf green and citrus notes. (Z)-3-Hexenal, an important contributor to the aroma of freshly pressed juices, e. g., orange and grapefruit (cf. 18.1.2.6.3), is considerably more instable than (E)-2-hexenal (Table 5.43) and, consequently, hardly finds application in aromatization.

Table 5.43. Half-life periods in the degradation of C_6 and C_7 aldehydes in different solvents at 38 °C[a]

Aldehyde	Water/ Ethanol (8+2, v/v)	Buffer[b]/ Ethanol (8+2, v/v)	Triacetin
n-Hexanal	100	91	86
(E)-2-Hexenal	256	183	71
(Z)-3-Hexenal	42	36	26
n-Heptanal	79	76	73
(E)-2-Heptenal	175	137	57
(Z)-4-Heptenal	200	174	64

[a] The half-life period is given in hours.
[b] Na-citrate buffer of pH 3.5 (0.2 mol/l).

In an apolar solvent, e. g., a triacylglycerol, (E)-2-hexenal decreases much more rapidly than in a polar medium in which its stability exceeds that of hexanal (Table 5.43). It oxidizes mainly to (E)-2-hexenoic acid, with butyric acid, valeric acid and 2-penten-1-ol being formed as well. The reaction pathway to the C_{6-} and C_{5-} acids is shown in Formula 5.48.

$$CH_3-(CH_2)_2-CH=CH-CHO \ (RH)$$

$$CH_3-(CH_2)_2-CH=CH-C\overset{\displaystyle O}{\underset{\displaystyle O-OH}{}}$$

$$2 CH_3-(CH_2)_2-CH=CH-COOH$$

$$CH_3-(CH_2)_2-CH=CH-OH$$

$$CH_3-(CH_2)_2-CH_2-CHO$$

$$CH_3-(CH_2)_2-CH_2-COOH$$

In[•]:
Start radical

(5.48)

At the acidic pH values found in fruit, autoxidation decreases, (E)-2-hexenal preferentially adds water with the formation of 3-hydroxy-hexanal. In addition, the double bond is iso-merized with the formation of low concentrations of (Z)-3-hexenal. As a result of its low threshold value, (Z)-3-hexenal first influences the aroma to a much greater extent than 3-hydroxyhexanal which has a very high threshold (cf. 18.1.2.6.3). Citral is also instable in an acidic medium, e. g., lemon juice. At citral equilibrium, which consists of the stereoisomers geranial and neral in the ratio of 65:35, neral reacts as shown in Formula: 5.49. It cyclizes to give the labile p-menth-1-en-3,8-diol which easily eliminates water, forming various p-menthadien-8-ols. This is followed by aromatization with the formation of p-cymene, p-cymen-8-ol, and α,p-dimethylstyrene. p-Methylacetophenone is formed from the last mentioned compound by oxidative cleavage of the Δ^8-double bond. Together with p-cresol, p-methylacetophenone contributes appreciably to the off-flavor formed on storage of lemon juice. Citral is also the precursor of p-cresol.

In citrus oils, limonene and γ-terpinene are also attacked in the presence of light and oxygen. Carvone and a series of limonene hydroperoxides are formed as the main aroma substances.

5.5.5 Encapsulation of Aromas

Aromas can be protected against the chemical changes described in 5.5.4 by encapsulation. Materials suitable for inclusion are polysaccharides, e. g., gum arabic, maltodextrins, modified starches, and cyclodextrins. The encapsulation proceeds via spray drying, extrusion or formation of inclusion complexes. For spray drying, the aroma substances are emulsified in a solution or suspension of the polysaccharide, which contains solutizer in addition to the emulsifying agent.

In preparation for extrusion, a melt of wall material, aroma substances, and emulsifiers is produced. The extrusion is conducted in a cooled bath, e. g., isopropanol.

β-Cyclodextrins, among other compounds, can be used for the formation of inclusion complexes (cf. 4.3.2). Together with the aroma substances, they are dissolved in a water/ethanol mixture by heating. The complexes precipitate out of the cooled solution and are removed by filtration and dried. Criteria for the evaluation of encapsulated aromas are: stability of the aroma, concentration of aroma substance, average diameter of the capsules and, amount of aroma substance adhering to the surface of the capsule.

5.6 Relationships Between Structure and Odor

5.6.1 General Aspects

The effect of stimulants on the peripheral receptors of an organism results in responses that are characterized by their *quality* and their *intensity*. The intensity is quantifiable, e. g., by determining odor threshold values (cf. 5.1.3). The quality can be described only by comparison. Odor stimulants can be grouped into those of the same or similar qualities, e. g., compounds with a caramel-like odor (cf. 5.3.1.2 and 5.3.1.3) or roasted smelling N-heterocyclic compounds (cf. 5.3.1.6). The dependence of the odor threshold on the structure is of great interest since specificity of the odor detection is the reason why aroma substances are only a fraction of the volatile compounds occurring in foods (cf. 5.3). The specificity of the sense of smell will be elucidated by using two classes of compounds as an example. Studies have shown how the odor thresholds in these classes change when the structures are systematically varied. Only odor thresholds in air are used because the influence of a solvent or a solid carrier does not have to be considered.

5.6.2 Carbonyl Compounds

In the series of saturated aldehydes C_5–C_{10}, the odor threshold reaches a minimum with octanal (Table 5.44). An E-configured double bond in the 2-position raises the odor threshold in the case of the alkenals 5:1 to 8:1 compared with the corresponding alkanals. (E)-2-Nonenal is the ex-

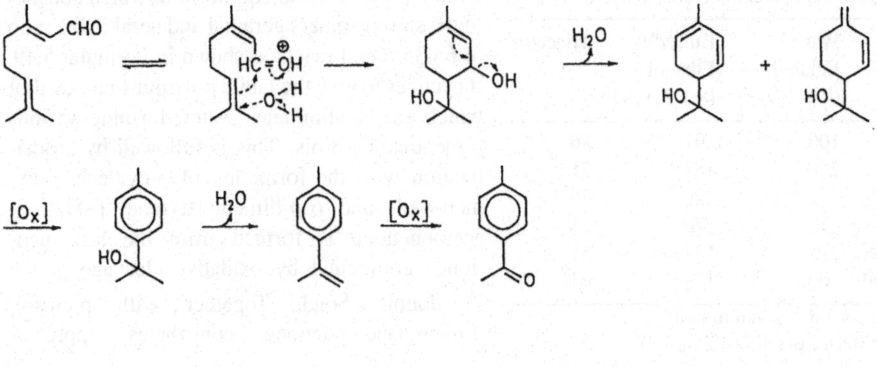

(5.49)

Table 5.44. Odor thresholds (T) in air of alkanals and (E)-2-alkenals

Alkanal	T (pmol/ stimulant)	(E)-2-Alkenal	T (pmol/ stimulant)
5:0	125	5:1	1600
6:0	80	6:1	900
7:0	66	7:1	1250
8:0	4	8:1	100
9:0	7	9:1	0.4
10:0	30	10:1	25

ception with an odor threshold 17.5 times lower than that of nonanal. Decanal and (E)-2-decenal have similar intensive odors. In chiral compounds (cf. 5.2.5 and 5.3.2.4) as well as cis/trans isomers, e. g., C_6 and C_9 aldehydes with a double bond, the molecule geometry influences the odor intensity and quality (Table 5.45).

Except for the pair (E/Z)-6-nonenal, the odor threshold of the E-isomer exceeds that of the corresponding Z-isomer. In particular, the values for (E)- and (Z)-3-hexenal differ greatly. Some of the aldehydes listed in Table 5.44 and 5.45 are formed by the peroxidation of unsaturated fatty acids (cf. 3.7.2.1.9). However, they play a role in aromas only when they are produced in foods in a concentration higher than their odor threshold concentration. The aroma active aldehydes usually include hexanal, which appears as the main product in the volatile fraction of peroxidized linoleic acid and, therefore, can surmount the relatively high odor threshold (Table 5.44). (E)-2-Nonenal also belongs to this

Table 5.45. Dependence of the odor thresholds (T) of C_6–C_9 aldehydes on the position and geometry of the double bond

C_6-Aldehyde	T (pmol/ stimulant)	C_9-Aldehyde	T (pmol/ stimulant)
(E)-2-6:1	900	(E)-2-9:1	0.4
(Z)-2-6:1	600	(Z)-2-9:1	0.014
(E)-3-6:1	>400	(E)-3-9:1	0.5
(Z)-3-6:1	1.4	(Z)-3-9:1	0.2
(E)-4-6:1	77	(E)-4-9:1	9
		(Z)-4-9:1	1.6
		(E)-5-9:1	70
		(E)-6-9:1	0.05
		(Z)-6-9:1	1.3

group. Although it is formed in considerably lower concentrations than hexanal, it can prevail in aromas due to its very low odor threshold. This also applies to (Z)-3-hexenal which is enzymatically formed from linolenic acid (cf. 3.7.2.3) and has a very low odor threshold. Consequently, it plays a much larger role in aromas, e. g., of fruit and vegetables, olive oil and fish, than its quantitatively more dominant companion substance (E)-2-hexenal.

5.6.3 Alkyl Pyrazines

The following example illustrates how pronounced the specificity of the sense of smell can be in cyclic compounds. The relationship between structure and odor activity was tested with 80 alkyl pyrazines. A part of the results is shown in Fig. 5.38.

In the series of mono-, di-, tri- and tetramethylpyrazines P1–P6, trimethylpyrazine (P5) shows the highest aroma activity. In the transition from dimethylpyrazines to trimethylpyrazine, the odor quality changes from nutty to earthy/roasted. If the methyl group in the ring position 2 of P5 is substituted by an ethyl group, P7 is formed, which has an odor threshold approximately 6000 times lower and an unchanged odor quality. If the ethyl group moves to the 3- (P8) or 5-position (P9), the odor threshold increases substantially. It increases even more if the ethyl group is substituted by a propyl group (P10–P12). An ethenyl group in position 2 instead of the ethyl group gives P13, but the odor threshold remains as low as with P7. If the ethenyl group moves round the ring (P14, P15), the threshold value again increases substantially. The insertion of a second ethyl group in position 3 of P7 and P13 changes neither the threshold value nor the odor quality in P16 and P17 respectively. However, if the methyl group in position 2 of P14 or in position 3 of P15 is replaced by an ethyl group, the resulting pyrazines P18 and P19 have very high threshold values. A comparison between P17 and P18 shows that whether the ethenyl group is in position 2 or 3 of ethenyl-ethyl-5-methylpyrazines is very important for the contact of the alkyl pyrazines with the odor receptor. If the methyl and ethyl group in P19 exchange positions, P20 is formed and the

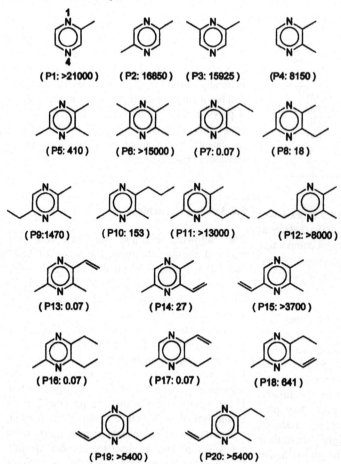

Fig. 5.38. Odor thresholds of alkyl pyrazines (according to *Wagner* et al., 1999). The odor threshold in pmol/l air is given in brackets

odor threshold remains very high. Seventy alkyl pyrazines have been identified in foods. However, in dilution analyses, the compounds which appear with a high odor intensity are only P7 and P16 in addition to P5, P13 and P17 (cf. 5.3.1.7). This is explainable by the specificity of odor detection of alkyl pyrazines discussed here.

5.7 References

Acree, T. E., Barnard, J., Cunningham, D.G.: A procedure for the sensory analysis of gas chromatographic effluents. Food Chem. *14*, 273 (1984)

Acree, T. E., Teranishi, R. (Eds.): Flavor science – Sensible principles and techniques. Am. Chem. Soc., Washington DC, 1993

Beets, M.G.J.: Structure–activity relationships in human chemoreception. Applied Science Publ.: London. 1978

Berger, R.G.: Microbial flavors. In: Flavor chemistry. Thirty years of progress. (Eds.: R. Teranishi, E.L. Wick, I. Hornstein) p. 229, Kluwer Academic/Plenum Publ., 1999

Beyeler, M., Solms, J.: Interaction of flavour model compounds with soy protein and bovine serum albumin. Lebens. Wiss. Technol. *7*, 217 (1974)

Blank, I.; Sen, A., Grosch, W.: Potent odorants of the roasted powder and brew of Arabica coffee. Z. Lebensm. Unters. Forsch. *195*, 239 (1992)

Blank, I., Milo, C., Lin, J., Fay, L.B.: Quantification of aroma-impact components by isotope dilution assay – Recent developments. In: Flavor chemistry. Thirty years of progress. (Eds.: R. Teranishi, E.L. Wick, I. Hornstein) p. 63, Kluwer Academic/Plenum Publ., 1999

Buttery, R.G., Haddon, W.F., Seifert, R.M., Turnbaugh, J.G.: Thiamin odor and bis(2-methyl-3-furyl)disulfide. J. Agric. Food Chem. *32*, 674 (1984)

Buttery, R.G.: Flavor chemistry and odor thresholds. In: Flavor chemistry. Thirty years of progress (Ed.: R. Teranishi, E.L. Wick, I. Hornstein) p. 353, Kluver Academic/Plenum Publ., New York, 1999

Buttery, R.G., Ling, L.C.: 2-Ethyl-3,5-dimethylpyrazine and 2-ethyl-3,6-dimethylpyrazine: odor thresholds in water solution. Lebensm. Wiss. Technol. *30*, 109 (1997)

Buttery, R.G., Ling. L.C.: Importance of 2-aminoacetophenone to the flavor of masa corn flour products. J. Agric. Food Chem. *42*, 1 (1994)

Cerny, C., Grosch, W.: Precursors of ethyldimethylpyrazine isomers and 2,3-diethyl-5-methylpyrazine formed in roasted beef. Z. Lebensm. Unters. Forsch. *198*, 210 (1994)

Eisenreich, W., Rohdich, F., Bacher, A.: The deoxyxylulose phosphate pathways to terpenoids. Trends in Plant Science 2001, 78

Engel, W., Bahr, W., Schieberle, P.: Solvent assisted flavour evaporation – a new and versatile technique for the careful and direct isolation of aroma compounds from complex food matrices. Eur. Food Res. Technol. *209*, 237 (1999)

Engel, K.-H., Flath, R. A., Buttery, R. G., Mon, T.R., Ramming, D.W., Teranishi, R.: Investigation of volatile constituents in nectarines. 1. Analytical and sensory characterization of aroma components in some nectarine cultivars. J. Agric. Food Chem. *36*, 549 (1988)

Etiévant, P.X.: Artifacts and contaminants in the analysis of food flavor. Crit. Rev. Food Sci. Nutr. *36*, 733 (1996)

Frijters, J.E.R.: A critical analysis of the odour unit number and its use. Chem. Senses Flavour *3*, 227 (1978)

Grosch, W.: Analyse von Aromastoffen. Chem. unserer Zeit *24*, 82 (1990)

Grosch, W.: Review – Determination of potent odorants in foods by aroma extract dilution analysis (AEDA) and calculation of odour activity values. Flavour Fragrance J. *9*, 147 (1994)

Grosch, W.: Evaluation of the key odorants of foods by dilution experiments, aroma models and omission. Chem. Senses, *26*, 533 (2001)

Guichard, E., Fournier, N.: Enantiomeric ratios of sotolon in different media and sensory differentiation of the pure enantiomers. Poster at the EURO FOOD, CHEM VI, Hamburg, 1991

Guth, H.: Determination of the configuration of wine lactone. Helv. Chim. Acta *79*, 1559 (1997)

Hofmann, T., Schieberle, P.: 2-Oxopropanal, hydroxy-2-propanone, and 1-pyrroline – important intermediates in the generation of the roast-smelling food

flavor compounds 2-acetyl-1-pyrroline and 2-acetyltetrahydropyridine. J. Agric. Food Chem. *46*, 2270 (1998)

Jennings, W.G., Filsoof, M.: Comparison of sample preparation techniques for gas chromatographic analyses. J. Agric. Food Chem. *25*, 440 (1977)

Kim, Y.-S., Hartman, T.G., Ho, C.-T.: Formation of 2-pentylpyridine from the thermal interaction of amino acids and 2,4-decadienal. J. Agric. Food Chem. *44*, 3906 (1996)

Laing, D.G., Wilcox, M.E.: Perception of components in binary odour mixtures. Chem. Senses *7*, 249 (1983)

Land, D.G., Nursten, H. E. (Eds.): Progress in flavour research. Applied Science Publ.: London. 1979

Larsen, M., Poll, L.: Odour thresholds of some important aroma compounds in raspberries. Z. Lebensm. Unters. Forsch. *191*, 129 (1990)

Leffingwell, J.C., Leffingwell, D.: GRAS flavor chemicals-detection thresholds. Perfumer & Flavorist *16* (1), 1 (1991)

Maarse, H., Belz, R. (Eds.): Isolation, separation and identification of volatile compounds in aroma research. Akademie-Verlag: Berlin. 1981

Maier, H.G.: Zur Bindung flüchtiger Aromastoffe an Proteine. Dtsch. Lebensm. Rundsch. *70*, 349 (1974)

McNulty, P.B., Karel, M.: Factors affecting flavour release and uptake in O/W-emulsions. J. Food Technol, *8*, 319 (1973)

Mosandl, A.: Capillary gas chromatography in quality assessment of flavours and fragrances. J. Chromatogr. *624*, 267 (1992)

Mosandl, A.: Analytical authentication of genuine flavor compounds – Review and preview. In: Flavor chemistry. Thirty years of progress. (Eds.: R. Teranishi, E.L. Wick, I. Hornstein) p. 31, Kluwer Academic/Plenum Publ., 1999

Naim, M., Striem, B.J., Kanner, J., Peleg, H.: Potential of ferulic acid as a precursor to off-flavors in stored orange juice. J. Food Sci. *53*, 500 (1988)

Nijssen, L.M., Visscher, C. A., Maarse, H., Willemsens, L.C., Boelens, M.H.: Volatile compounds in food. Qualitative and quantitative data. 7th Edition and supplements 1 and 2. TNO Nutrition and Food Research Institute, Zeist, The Netherlands, 1999

Ohloff, G.: Riechstoffe und Geruchssinn. Springer-Verlag, Berlin. 1990

Piendl, A.: Brauereitechnologie und Molekularbiologie. Brauwissenschaft *22*, 175 (1969)

Reineccius, G.A.: Off-flavors in foods. Crit. Rev. Food Sci. Nutr. *29*, 381 (1991)

Risch, S.J., Reineccius, G.A. (Eds.): Flavor encapsulation. ACS Symp. Ser. *370* (1988)

Rizzi, P.: The biogenesis of food-related pyrazines. Food Rev. Internat. *4*, 375 (1988)

Saxby, M.J. (Ed.): Food taints and off-flavours, 2nd edition. Blackie Academic & Professional, London, 1996

Schieberle, P.: Primary odorants in popcorn. J. Agric. Food Chem. *39*, 1141 (1991)

Schieberle, P.: The role of free amino acids present in yeast as precursors of the odorants 2-acetyl-1-pyrroline and 2-acetyltetrahydropyridine in wheat bread crust. Z. Lebensm. Unters. Forsch. *191*, 206 (1990)

Schieberle, P.: New developments on methods for the analysis of volatile flavor compounds and their precursors. In: Characterization of food-emerging methods. Ed. A.G. Goankar, Elsevier, Amsterdam, p. 403–443, 1995

Schreier, P.: Chromatographic studies of biogenesis of plant volatiles. Dr. Alfred Hüthig Verlag: Heidelberg. 1984

Silberzahn, W.: Aromen, in: Taschenbuch für Lebensmittelchemiker und -technologen (Frede, W., Ed.), Band 1, Springer-Verlag, Berlin. 1991

Solms, J.: Aromastoffe als Liganden. In: Geruch und Geschmackstoffe (Ed.: Drawert, F.), S. 201, Verlag Hans Carl: Nürnberg. 1975

Teranishi, R., Buttery, R.G., Shahidi, F. (Eds.): Flavor chemistry – Trends and Developments. ACS Symp. Ser. *388* (1989)

Tressl, R., Haffner, T., Lange, H., Nordsiek, A.: Formation of γ- and δ-lactones by different biochemical pathways. In: Flavour Science – recent developments (Eds.: A.J. Taylor, D.S. Mottram) p. 141, The Royal Society of Chemistry, Cambridge, 1996

Tressl, R., Helak, B., Martin, N., Kersten, E.: Formation of amino acid specific *Maillard* products and their contribution to thermally generated aromas. ACS Symp. Ser. *409*, 156 (1989)

Tressl, R., Kersten, E., Nittka, C., Rewicki, D.: Formation of sulfur-containing flavor compounds from [^{13}C]-labeled sugars, cysteine and methionine. In: Sulfur compounds in foods (Eds.: C.J. Mussinan, M.E. Keelan) p. 224, ACS Symp. Ser. 564, American Chemical Society, Washington, 1994

Tressl, R., Rewicki, D.: Heated generated flavors and precursors. In: Flavor chemistry. Thirty years of progress. (Eds.: R. Teranishi, E.L. Wick, I. Hornstein, I.) p. 305, Kluwer Academic/Plenum Publ., 1999

von Ranson, C., Belitz, H.-D.: Untersuchungen zur Struktur-Aktivitätsbeziehung bei Geruchsstoffen. 2. Mitteilung: Wahrnehmungs- und Erkennungsschwellenwerte sowie Geruchsqualitäten gesättigter und ungesättigter aliphatischer Aldehyde. Z. Lebensm. Unters. Forsch. *195*, 515 (1992)

von Ranson, C., Schnabel, K.-O., Belitz, H.-D.: Untersuchungen zur Struktur-Aktivitätsbeziehung bei Geruchsstoffen. 4. Mitteilung: Struktur und Geruchsqualität bei aliphatischen, alicyclischen und aromatischen Aldehyden. Z. Lebensm. Unters. Forsch. *195*, 527 (1992)

Vermeulen, C., Gijs, L., Collin, S.: Sensorial contribution and formation pathways of thiols in food. Food Rev. Internat. *21*, 69 (2005)

Wagner, R., Grosch, W.: Evaluation of potent odorants of French fries. Lebensm. Wiss. Technol. *30*, 164 (1997)

Wagner, R., Czerny, M., Bieloradsky, J., Grosch, W.: Structure–odour–activity relationships of alkylpyrazines. Z. Lebensm. Unters. Forsch. *A208*, 308 (1999)

Wagner, R., Grosch, W.: Key odorants of French fries. J. Am. Oil Chem. Soc. *75*, 1385 (1998)

Werkhoff, P., Bretschneider, W., Herrmann, H.-J., Schreiber, K.: Fortschritte in der Aromastoff-Analytik (1–9). Labor Praxis (1989) S. 306, 426, 514, 616, 766, 874, 1002, 1121 and (1990) S. 51

Winterhalter, P., Knapp, H., Straubinger, M.: Watersoluble aroma precursors: Analysis, structure, and reactivity. In: Flavor chemistry. Thirty years of progress. (Eds.: R. Teranishi, E.L. Wick, I. Hornstein) p. 255, Kluwer Academic/Plenum Publ., 1999

Williams, P.J., Strauss, C.R., Wilson, B., Massy-Westropp, R. A.: Novel monoterpene disaccharide glycosides of *Vitis vinifera* grapes and wines. Phytochemistry *21*, 2013 (1982)

Wüst, M., Mosandl, A.: Important chiral monoterpenoid esters in flavours and essential oils – enantioselective analysis and biogenesis. Eur. Food Res. Technol. *209*, 3 (1999)

Ziegler, E., Ziegler, H.: Flavourings. Wiley-VCH, Weinheim, 1998

6 Vitamins

6.1 Foreword

Vitamins are minor but essential constituents of food. They are required for the normal growth, maintenance and functioning of the human body. Hence, their preservation during storage and processing of food is of far-reaching importance. Data are provided in Tables 6.1 and 6.2 to illustrate vitamin losses in some preservation methods for fruits and vegetables. Vitamin losses can occur through chemical reactions which lead to inactive products, or by extraction or leaching, as in the case of water-soluble vitamins during blanching and cooking.

The vitamin requirement of the body is usually adequately supplied by a balanced diet. A defi-

Table 6.1. Vitamin losses (%) through processing/canning of vegetables

Processed/ canned product	Samples of vegetable analyzed	Vitamin losses as % of freshly cooked and drained product				
		A	B_1	B_2	Niacin	C
Frozen products (cooked and drained)	10[a]	12[c]	20	24	24	26
		0–50[d]	0–61	0–45	0–56	0–78
Sterilized products (drained)	7[b]	10	67	42	49	51
		0–32	56–83	14–50	31–65	28–67

[a] Asparagus, lima beans, green beans, broccoli, cauliflower, green peas, potatoes, spinach, brussels sprouts, and baby corn-cobs.
[b] As under a) with the exception of broccoli, cauliflower and brussels sprouts; the values for potato include the cooking water.
[c] Average values.
[d] Variation range.

Table 6.2. Vitamin loss (%) through processing/canning of fruits

Processed/canned product	Fruit samples analyzed	Vitamin losses as % of fresh product				
		A	B_1	B_2	Niacin	C
Frozen products (not thawed)	8[a]	37[c]	29	17	16	18
		0–78[d]	0–66	0–67	0–33	0–50
Sterilized products (including the cooking water)	8[b]	39	47	57	42	56
		0–68	22–67	33–83	25–60	11–86

[a] Apples, apricots, bilberries, sour cherries, orange juice concentrate (calculated for diluted juice samples), peaches, raspberries and strawberries.
[b] As under a) except orange juice and not its concentrate was analyzed.
[c] Average values.
[d] Variation range.

ciency can result in hypovitaminosis and, if more severe, in avitaminosis. Both can occur not only as a consequence of insufficient supply of vitamins by food intake, but can be caused by disturbances in resorption, by stress and by disease.

An assessment of the extent of vitamin supply can be made by determining the vitamin content in blood plasma, or by measuring a biological activity which is dependent on the presence of a vitamin, as are many enzyme activities.

Vitamins are usually divided into two general classes: the fat-soluble vitamins, such as A, D, E and K_1, and the water-soluble vitamins, B_1, B_2, B_6, nicotinamide, pantothenic acid, biotin, folic acid, B_{12} and C.

Data on the desirable human daily intake of some vitamins are presented by age group in Table 6.3.

6.2 Fat-Soluble Vitamins

6.2.1 Retinol (Vitamin A)

6.2.1.1 Biological Role

Retinol (I, in Formula 6.1) is of importance in protein metabolism of cells which develop from the ectoderm (such as skin or mucouscoated linings of the respiratory or digestive systems). Lack of retinol in some way negatively affects epithelial tissue (thickening of skin, hyperkeratosis) and also causes night blindness.

$$(6.1)$$

Furthermore, retinol, in the form of 11-cis-retinal (II), is the chromophore component of the visual cycle chromoproteins in three types of cone cells, blue, green and red (λ_{max} 435, 540 and 565 nm, respectively) and of rods of the retina.

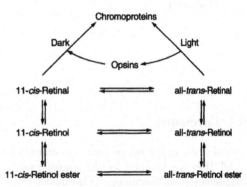

Fig. 6.1. Schematic representation of the visual cyle

The chromoproteins (rhodopsins) are formed in the dark from the corresponding proteins (opsins) and 11-cis-retinal, while in the light the chromoproteins dissociate into the more stable all-trans-retinal and protein. This conformational change triggers a nerve impulse in the adjacent nerve cell. The all-trans-retinal is then converted to all-trans-retinol and through an intermediary, 11-cis-retinol, is transformed back into 11-cis-retinal (see Fig. 6.1 for the visual cycle reactions).

6.2.1.2 Requirement, Occurrence

The daily requirement of vitamin A (Table 6.3) is provided to an extent of 75% by retinol intake (as fatty acid esters, primarily retinyl palmitate), while the remaining 25% is through β-carotene and other provitaminactive carotenoids. Due to the limited extent of carotenoid cleavage, at least 6 g of β-carotene are required to yield 1 g retinol. Vitamin A absorption and its storage in the liver occur essentially in the form of fatty acid esters. Its content in liver is 250 µg/g fresh tissue, i.e. a total of about 240–540 mg is stored. The liver supplies the blood with free retinol, which then binds to proteins in blood. The plasma concentration of retinol averages 1.78 µmol/l in women and 2.04 µmol/l in men.

A hypervitaminosis is known, but the symptoms disappear if the intake of retinal is decreased.

Vitamin A occurs only in animal tissue; above all in fish liver oil, in livers of mammals, in milk fat and in egg yolk. Plants are devoid of vitamin A but do contain carotenoids which yield vitamin A

Table 6.3. Recommended daily intake of vitamins

Age group (years)	A (mg Retinol)[a]	D (µg)[b]	E (mg)[c]	K (µg)[d]	C (mg)	B$_1$ (mg)	B$_2$ (mg)	Niacin[e] (mg)	B$_6$ (µg)	Folic acid[f] (mg)	Pantothenic acid (mg)	Biotin (µg)	B$_{12}$ (µg)
<1	0.5–0.6	10	3–4	4–10	50–55	0.2–0.4	0.3–0.4	2–5	0.1–0.3	60–80	2–3	5–10	0.4–0.8
1–4	0.6	5	6	15	60	0.6	0.7	7	0.4	200	4	10–15	1.0
4–10	0.7–0.8	5	8–10	20–30	70–80	0.8–1.0	0.9–1.1	10–12	0.5–0.7	300	4–5	15–20	1.5–1.8
10–15	0.9–1.1	5	10–14	40–50	90–100	1.0–1.3	1.2–1.6	13–18	1.0–1.4	400	5–6	20–35	2.0–3.0
15–25	0.9–1.1	5	15	60–70	100	1.0–1.3	1.2–1.5	13–17	1.2–1.6	400	6	30–60	3.0
25–51	0.8–1.0	5	14	60–70	100	1.0–1.2	1.2–1.4	13–16	1.2–1.5	400	6	30–60	3.0
52–65	0.8–1.0	5	13	80	100	1.0–1.1	1.2–1.3	13–15	1.2–1.5	400	6	30–60	3.0
>65	0.8–1.0	10	12	80	100	1.0	1.2	13	1.2–1.4	400	6	30–60	3.0
Pregnant women	1.1	5	13	60	100	1.2	1.5	15	1.9	600	6	30–60	3.5
Lactating women	1.5	5	17	60	150	1.4	1.6	17	1.9	600	6	30–60	4.0

[a] 1 mg retinol = 1 mg retinol equivalent = 6 mg all-trans-β-carotene = 12 mg other provitamin A carotinoids = 1.15 mg all-trans-retinyl acctate = 1.83 mg all-trans-retinyl palmitate (IU = 0.34 µg retinol).

[b] Ergocalciferol (D$_2$) or cholecalciferol (D$_3$) (1 IU = 0.025 µg).

[c] Tocopherol equivalent (cf. 6.2.3.1).

[d] Phylloquinone (cf. 6.2.4).

[e] 1 mg niacin equivalent = 60 mg tryptophan.

[f] 1 µg folate equivalent = 1 µg food folate = 0.5 µg folic acid (PGA, cf. 6.3.7.1).

by cleavage of the centrally located double bond (provitamins A). Carotenoids are present in almost all vegetables but primarily in green, yellow and leafy vegetables (carrots, spinach, cress, kale, bell peppers, paprika peppers, tomatoes) and in fruit, outstanding sources being rose hips, pumpkin, apricots, oranges and palm oil, which is often used for yellow coloring. Animal carotenoids are always of plant origin, derived from feed.

Table 6.7 gives the vitamin A content of some common foods. These values can vary greatly with cultivar, stage of ripeness, etc. An accurate estimate of the vitamin A content of a food must include a detailed analysis of its carotenoids.

6.2.1.3 Stability, Degradation

Food processing and storage can lead to 5–40% destruction of vitamin A and carotenoids. In the absence of oxygen and at higher temperatures, as experienced in cooking or food sterilization, the main reactions are isomerization and fragmentation. In the presence of oxygen, oxidative degradation leads to a series of products, some of which are volatile (cf. 3.8.4.4). This oxidation often parallels lipid oxidation (cooxidation process). The rate of oxidation is influenced by oxygen partial pressure, water activity, temperature, etc. Dehydrated foods are particularly sensitive to oxidative degradation.

6.2.2 Calciferol (Vitamin D)

6.2.2.1 Biological Role

Cholecalciferol (vitamin D_3, I) is formed from cholesterol in the skin through photolysis of 7-dehydrocholesterol (provitamin D_3) by ultraviolet light ("sunshine vitamin"; cf. 3.8.2.2.2). Similarly, vitamin D_2 (ergocalciferol, II; cf. Formula 6.2) is formed from ergosterol.

Vitamin D_2 and D_3 are hydroxylated first in the liver to the prohormone 25-hydroxycholecalciferol (calcidiol) and subsequently in the kidney to the vitamin D hormone $1\alpha,25$-dihydroxycholecalciferol (calcitriol). Calcitriol acts as an inductor of proteins in various organs. It promotes calcium resorption in the intestine and

an optimal calcium concentration in the kidney and in the bones, it induces the synthesis of proteins involved in the structure of the bone matrix and in calcification.

Vitamin D deficiency can result in an increased excretion of calcium and phosphate and, consequently, impairs bone formation through inadequate calcification of cartilage and bones (childhood rickets). Vitamin D deficiency in adults leads to osteomalacia, a softening and weakening of bones. Hypercalcemia is a result of excessive intake of vitamin D (>50 µg/day), causing calcium carbonate and calcium phosphate deposition disorders involving various organs.

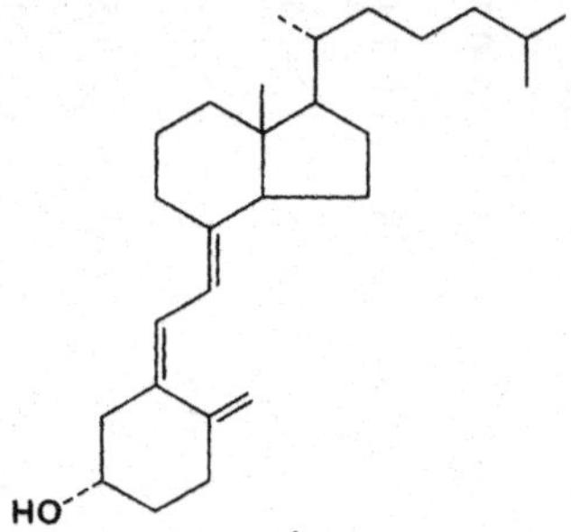

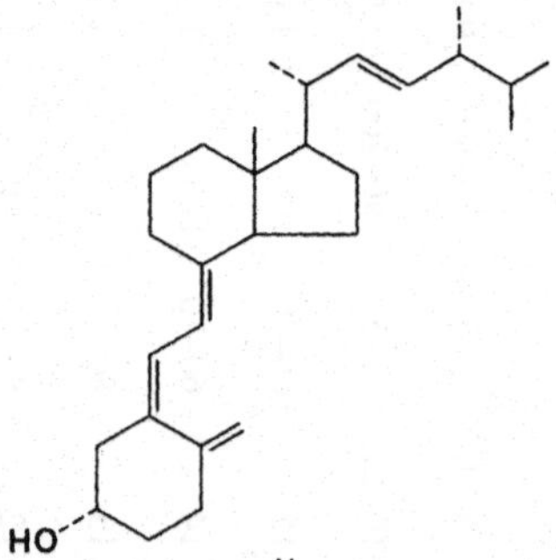

$$(6.2)$$

6.2.2.2 Requirement, Occurrence

The daily requirement is shown in Table 6.3. Indicators of deficiency are the concentration of the metabolite 25-hydroxycholecalciferol in plasma and the activity of alkaline serum phosphatase, which increases during vitamin deficiency.

Most natural foods have a low content of vitamin D_3. Fish liver oil is an exceptional source of vitamin D_2. The D-provitamins, ergosterol and 7-dehydrocholesterol, are widely distributed in the animal and plant kingdoms. Yeast, some mushrooms, cabbage, spinach and wheat germ oil are particularly abundant in provitamin D_2. Vitamin D_3 and its provitamin are present in egg yolk, butter, cow's milk, beef and pork liver, mollusks, animal fat and pork skin. However, the most important vitamin D source is fish oil, primarily liver oil. The vitamin D requirement of humans is best supplied by 7-dehydrocholesterol. Table 6.7 gives data on vitamin D occurrence in some foods. However, these values can vary widely, as shown by variations in dairy cattle milk (summer or winter), caused by feed or frequency of pasture grazing and exposure to the ultraviolet rays of sunlight.

6.2.2.3 Stability, Degradation

Vitamin D is sensitive to oxygen and light. Its stability in food is not a problem, because adults usually obtain a sufficient supply of this vitamin.

6.2.3 α-Tocopherol (Vitamin E)

6.2.3.1 Biological Role

The various tocopherols differ in the number and position of the methyl groups on the ring. α-Tocopherol (Formula 6.3; the configuration at the three asymmetric centers, 2, 4′ and 8′, is R) has the highest biological activity (Table 6.4). Its activity is based mainly on its antioxidative properties, which retard or prevent lipid oxidation (cf. 3.7.3.1). Thus, it not only contributes to the stabilization of membrane structures, but also stabilizes other active agents (e. g., vitamin A, ubiquinone, hormones, and enzymes) against oxidation. Vitamin E is involved in the conversion of arachidonic acid to prostaglandins and slows down the aggregation of blood platelets. Vitamin E deficiency is associated with chronic disordes (sterility in domestic and experimental animals, anemia in monkeys, and muscular dystrophy in chickens). Its mechanism of action is not fully elucidated.

Table 6.4. Biological activity of some tocopherols

Tocopherol (T)	Vitamin E activity	
	In IU/mg[a]	Conversion factor[b]
2R,4′R,8′R-α-T	1.49	1.00
2S,4′R,8′R-α-T	0.46	0.31
2R,4′R,8′S-α-T	1.34	0.90
2S,4′R,8′S-α-T	0.55	0.37
2R,4′S,8′S-α-T	1.09	0.73
2S,4′S,8′R-α-T	0.31	0.21
2R,4′S,8′R-α-T	0.85	0.57
2S,4′S,8′S-α-T	1.10	0.60
2R,4′R,8′R-α- Tocopheryl acetate	1.36	0.91
all-rac-α-T	1.10	0.74
all-rac-α- Tocopheryl acetate	1.00	0.67
all-rac-β-T	0.30	0.20
all-rac-γ-T	0.15	0.10
all-rac-δ-T	0.01	

[a] International units (IU) per mg substance.
[b] Conversion factor from mg substance to mg α-tocopherol equivalents.

Table 6.5. Requirement of tocopherol equivalents (TE) on supply of unsaturated fatty acids

Fatty acid	TE (mg/g fatty acid)
Monoene acids	0.06
Diene acids	0.4
Triene acids	0.8
Tetraene acids	1.0
Pentaene acids	1.2
Hexaene acids	1.45

6.2.3.2 Requirement, Occurrence

The daily requirement is given in Table 6.3. It increases when the diet contains a high content of unsaturated fatty acids (cf. Table 6.5). A normal supply results in a tocopherol concentration of 12–46 µmol/l in blood plasma.
A level less than 0.4 mg/100 ml is considered a deficiency.
Section 3.8.3.1 and Table 6.7 provide data on the tocopherol content in some foods. The main sources are vegetable oils, particularly germ oils of cereals.

(6.3)

6.2.3.3 Stability, Degradation

Losses occur in vegetable oil processing into margarine and shortening. Losses are also encountered in intensive lipid autoxidation, particularly in dehydrated or deep fried foods (Table 6.6).

Table 6.6. Tocopherol stability during deep frying

	Toco-pherol total (mg/ 100 g)	Loss (%)
Oil before deep frying	82	
after deep frying	73	11
Oil extracted from potato chips		
immediately after production	75	
after 2 weeks storage		
at room temperature	39	48
after 1 month storage		
at room temperature	22	71
after 2 months storage		
at room temperature	17	77
after 1 month kept at $-12\,°C$	28	63
after 2 months kept at $-12\,°C$	24	68
Oil extracted from French fries		
immediately after production	78	
after 1 month kept at $-12\,°C$	25	68
after 2 months kept at $-12\,°C$	20	74

6.2.4 Phytomenadione (Vitamin K$_1$ Phylloquinone)

6.2.4.1 Biological Role

The K-group vitamins are naphthoquinone derivatives which differ in their side chains. The structure of vitamin K$_1$ is shown in Formula 6.4. The configuration at carbon atoms

$7'$ and $11'$ is R and corresponds to that of natural phytol. Racemic vitamin K$_1$ synthesized from optically inactive isophytol has the same biological activity as the natural product. Vitamin K is involved in the post-translational synthesis of γ-carboxyglutamic acid (Gla) in vitamin K-dependent proteins. It is reduced to the hydroquinone form (Formula 6.4) which acts as a cofactor in the carboxylation of glutamic acid. In this process, it is converted to the epoxide from which vitamin K is regenerated. Blood clotting factors (prothrombin, proconvertin, Christmas and Stuart factor) as well as proteins which perform other functions belong to the group of vitamin K-dependent proteins which bind Ca^{2+} ions at Gla. Deficiency of this vitamin causes reduced prothrombin activity, hypothrombinemia and hemorrhage.

6.2.4.2 Requirement, Occurrence

The activity is given in vitamin equivalents (VE): 1 VE = 1 µg phylloquinone. The daily requirement of vitamin K$_1$ is shown in Table 6.3. It is covered by food (cf. Table 6.7). The bacteria present in the large intestine form relatively high amounts of K$_2$. However, it is uncertain whether they appreciably contribute to covering the requirement.

Vitamin K$_1$ occurs primarily in green leafy vegetables (spinach, cabbage, cauliflower), but liver (veal or pork) is also an excellent source (Table 6.7).

6.2.4.3 Stability, Degradation

Little is known about the reactions of vitamin K$_1$ in foods. The vitamin K compounds are destroyed by light and alkali. They are relatively stable to atmospheric oxygen and exposure to heat.

Table 6.7. Vitamin content of some food products[a]

Food product	Carotene[b] mg	A mg	D μg	E mg	K mg	B1 mg	B2 mg	NAM[c] mg	PAN[d] mg	B6 mg	BIO[e] μg	FOL[f] μg	B12 μg	C mg
Milk and milk products														
Bovine milk	0.018	0.028	0.088	0.07	0.0003	0.04	0.18	0.09	0.35	0.04	3.5	8.0	0.4	1.7
Human milk	0.003	0.054	0.07	0.28	0.0005	0.02	0.04	0.17	0.21	0.01	0.6	8.0	0.05	6.5
Butter	0.38	0.59	1.2	2.2	0.007	0.005	0.02	0.03	0.05	0.005				0.2
Cheese														
Emmental	0.12	0.27	1.1	0.53	0.003	0.05	0.34	0.18	0.40	0.11	3.0	9.0	3.0	0.5
Camembert (60% fat)	0.29	0.50		0.77		0.04	0.37	0.95	0.7	0.2	2.8	38	2.4	
Camembert (30% fat)	0.1	0.2	0.17	0.30		0.05	0.67	1.2	0.9	0.3	5.0	66	3.1	
Eggs														
Chicken egg yolk	0.29	0.88	5.6	5.7		0.29	0.40	0.07	3.7	0.3	53	208	2.0	0.3
Chicken egg white						0.02	0.32	0.09	0.14	0.012	7	9.2	0.1	0.3
Meat and meat products														
Beef, muscle			0.02	0.41	0.48	0.013	0.08	0.26	7.5	0.31	0.24	3.0	3.0	5.0
Pork, muscle		0.006		0.24	0.018	0.90	0.23	5.0	0.70	0.4			0.8	
Calf liver		28.0	0.33	0.60	0.09	0.28	2.61	15.0	7.9	0.17	80	240	60	35
Pork liver		36		0.5	0.06	0.31	3.2	15.7	6.8	0.6	30	220	40	23
Chicken liver		33	1.3	0.45	0.08	0.32	2.49	11.6	7.2	0.8		380	20	28
Pork kidneys						0.34	1.8	8.4	3.1	0.6			50	16
Blood sausage	0.02	0.06				0.07	0.13	1.2						
Fish and fish products														
Herring		0.04	27	1.5		0.04	0.22	3.8	0.9	0.5	4.5	5	8.5	0.5
Eel		0.98	20	8		0.18	0.32	2.6	0.3			13	1	1.8
Cod-liver oil				3.26										
Cereals and cereal products														
Wheat, whole kernel	0.02			1.4		0.48	0.09	5.1	1.2	0.27	6	58		
Wheat flour, type 550				0.34		0.11	0.03	0.5	0.4	0.10	1.1	16		
Wheat flour, type 1050						0.43	0.07	1.4	0.63	0.24	1.1	30		
Wheat germ	0.06			27.6	0.13	2.01	0.72	4.5	1.0	0.5	17	520		
Rye whole kernel				2.0		0.35	0.17	1.8	1.5	0.23	5.0	35		
Rye flour, type 997						0.19	0.11	0.8				33		
Corn whole kernel	1.3			2.0	0.04	0.36	0.20	1.5	0.7	0.4	6	31		

Table 6.7. Continued

Food product	Carotene[b] mg	A mg	D µg	E mg	K mg	B$_1$ mg	B$_2$ mg	NAM[c] mg	PAN[d] mg	B$_6$ mg	BIO[e] µg	FOL[f] µg	B$_{12}$ µg	C mg
Corn (breakfast cereal, corn flakes)	0.17			0.18		0.60	0.15	1.4	0.2	0.07	20	6		
Oat flakes				1.5	0.063	0.59		1.0	1.1	0.16	12	67		
Rice, unpolished				0.74		0.41	0.09	5.2	1.7	0.28	3.0	16		
Rice, polished				0.18		0.06	0.03	1.3	0.6	0.15		11		
Vegetables														
Watercress	4.9					0.09	0.17	0.7		0.07				96
Mushrooms (cultivated)	0.01		1.94	0.12	0.02	0.10	0.44	5.2	2.1	0.05	16	25		4.9
Chicory	3.4					0.06	0.03	0.24			4.8	50		8.7
Endive	1.7					0.05	0.12	0.4				109		9.4
Lamb's lettuce	3.9			0.6	0.82	0.07	0.08	0.4		0.25		145		35
Kale	5.2			1.7	0.002	0.1	0.25	2.1	0.4	0.3	0.5	187		105
Potatoes	0.005			0.05		0.11	0.05	1.2	0.1	0.31	0.4	15		17
Kohlrabi	0.2					0.05	0.05	1.8	0.1	0.1	2.7	70		63
Head lettuce	1.1			0.6	0.2	0.06	0.08	0.3	1.4	0.06	1.9	41		13
Lentils, dried	0.1					0.48	0.26	2.5	0.3	0.6		168		7.0
Carrots	12			0.47	0.015	0.07	0.05	0.6	0.3	0.27	5	17		7.1
Brussels sprouts	0.4			0.6	0.24	0.13	0.14	0.7		0.3	0.4	101		114
Spinach	4.8			2.5	0.4	0.09	0.20	0.6		0.22	6.9	145		52
Edible mushroom (*Boletus edulis*)			3.1	0.63	0.006	0.03	0.37	4.9	2.7	0.1				2.5
Tomatoes	0.59			0.81		0.06	0.04	0.5	0.3	0.1	4	33		19
White cabbage	0.07			1.7	0.07	0.05	0.04	0.3	0.3	0.19	3.1	31		48
Fruits														
Apple	0.05			0.49	0.004	0.035	0.032	0.3	0.1	0.1	0.0045	5		12
Orange	0.1			0.32		0.08	0.04	0.3	0.2	0.1	2.3	22		50
Apricot	1.8			0.5	0.02	0.04	0.05	0.8	0.3	0.1		4		9.4
Strawberry	0.02			0.12		0.03	0.05	0.5	0.3	0.06	4	43		64
Grapefruit	0.01			0.30		0.05	0.02	0.24	0.25	0.03	0.4	10		44
Rose hips	4.8			4.2		0.09	0.06	0.48		0.05				1250
Red currants	0.03			0.71		0.04	0.03	0.23	0.06	0.05	2.6	11		36
Black currants	0.08			1.9		0.05	0.04	0.28	0.4	0.08	2.4	8.8		177
Sour cherries	0.24			0.13		0.05	0.06	0.4				29		12
Plums	0.41			0.86		0.07	0.04	0.4	0.2	0.05	0.1	2		5.4
Sea buckthorn	1.5			3.2		0.03	0.21	0.3	0.2	0.11	3.3	10		450
Yeast														
Baker's yeast, pressed						1.43	2.31	17.4	3.5	0.68	33	293		
Brewer's yeast, dried						12.0	3.8	44.8	7.2	4.4	20			

[a] Values are given in mg or µg per 100 g of edible portion. [b] Total carotenoids with vitamin A activity. [c] Nicotinamide. [d] Pantothenic acid. [e] Biotin. [f] Folic acid.

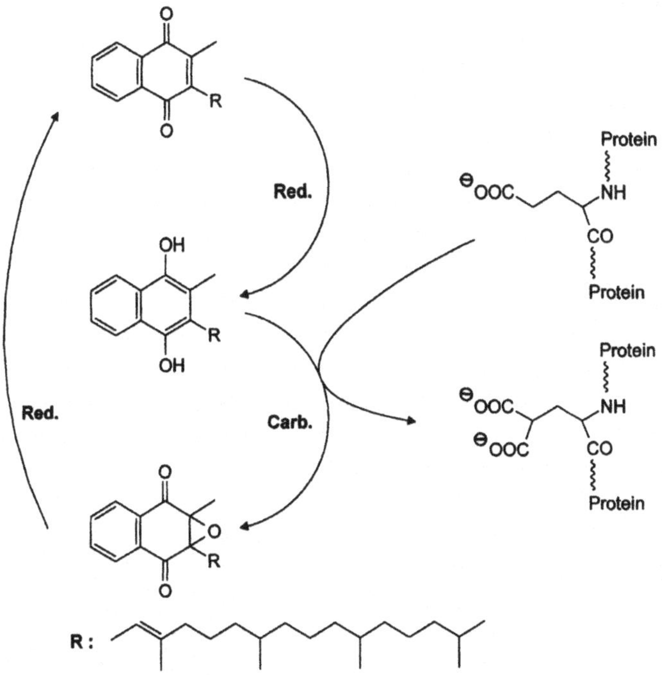

Red.: Reductase; Carb.: Carboxylase (6.4)

In the hydrogenation of oils, the double bond in residue R (cf. Formula 6.4) is attacked. Although hydrogenated vitamin K (2′,3′-dihydrophylloquinone) is absorbed, it is apparently no longer as active as the natural form.

6.3 Water-Soluble Vitamins

6.3.1 Thiamine (Vitamin B$_1$)

6.3.1.1 Biological Role

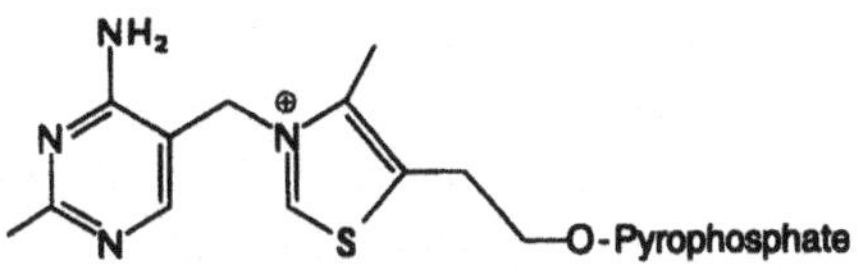

Thiamine, in the form of its pyrophosphate, such as pyruvate dehydrogenase, transketolase, phosphoketolase and α-ketoglutarate dehydrogenase, in reactions involving the transfer of an activated aldehyde unit (D: donor; A: acceptor):

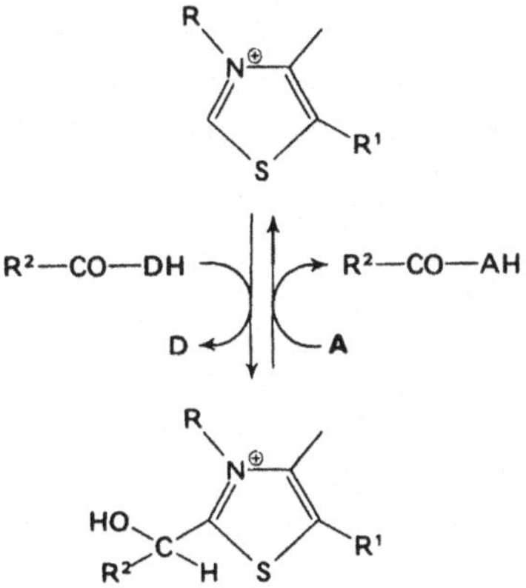

Vitamin B$_1$ deficiency is shown by a decrease in activity of the enzymes mentioned above. The disease known as beri-beri, which has neurological and cardiac symptoms, results from a severe dietary deficiency of thiamine.

6.3.1.2 Requirement, Occurrence

The daily requirement is shown in Table 6.3. Since thiamine is a key substance in carbohydrate metabolism, the requirement increases in a carbohydrate-enriched diet. The assay of transketolase activity in red blood cells or the extent of transketolase reactivation on addition of thiamine pyrophosphate can be used as indicators for sufficient vitamin intake in the diet.

Vitamin B_1 is found in many plants. It is present in the pericarp and germ of cereals, in yeast, vegetables (potatoes) and shelled fruit. It is abundant in pork, beef, fish, eggs and in animal organs such as liver, kidney, brain and heart. Human milk and cow's milk contain vitamin B_1. Whole grain bread and potatoes are important dietary sources. Since vitamin B_1 is localized in the outer part of cereal grain hulls, flour milling with a low extraction grade or rice polishing remove most of the vitamin in the bran (cf. 15.3.1.3 and 15.3.2.2.1). Table 6.7 lists data on the occurrence of thiamine.

6.3.1.3 Stability, Degradation

Thiamine stability in aqueous solution is relatively low. It is influenced by pH (Fig. 6.2), temperature (Table 6.8), ionic strength and metal ions. The enzyme-bound form is less stable than free thiamine (Fig. 6.2). Strong nucleophilic reagents, such as HSO_3^- or OH^-, cause rapid decomposition by forming 5-(2-hydroxyethyl)-4-methylthiazole and 2-methyl-4-amino-5(methyl-sulfonic acid)-pyrimidine, or 2-methyl-4-amino-5-hydroxymethylpyrimidine (see Reactions 6.7). Thermal degradation of thiamine, which also initially yields the thiazole and pyrimidine

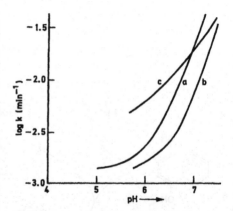

Fig. 6.2. Inactivation rate of thiamine as affected by pH **a** Thiamine in phosphate buffer, **b** thiamine in wheat or oat flour, **c** thiamine pyrophosphate in flour

Table 6.8. Thiamine losses in food during storage (12 months)

Food	Thiamine loss, %	
	1.5 °C	38 °C
Apricots	28	65
Orange juice	0	22
Peas	0	32
Green beans	24	92
Tomato juice	0	40

derivatives mentioned above, is involved in the formation of meat-like aroma in cooked food (cf. 5.3.1.4).

Thiamine is inactivated by nitrites, probably through reaction with the amino group attached to the pyrimidine ring.

$$\text{(6.7)}$$

Strong oxidants, such as H_2O_2 or potassium ferricyanide, yield the fluorescent thiochrome. This reaction is often used in chemical determination of the thiamine content in food (see Reaction 6.8).

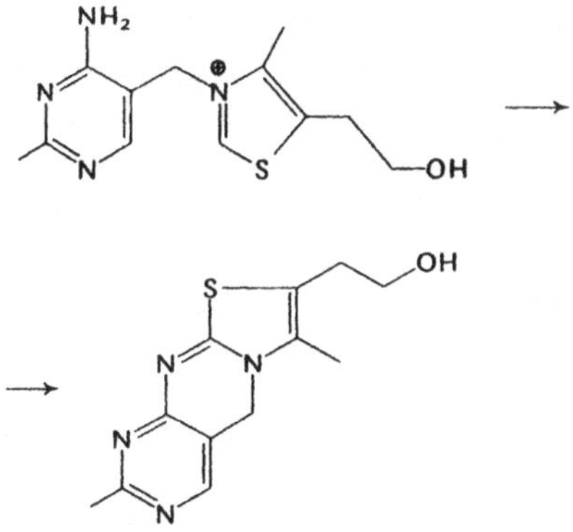

Riboflavin deficiency will lead to accumulation of amino acids. A specific deficiency symptom is the decrease of glutathione reductase activity in red blood cells.

6.3.2.2 Requirement, Occurrence

The daily requirement is given in Table 6.3. Deficiency symptoms are rarely observed with a normal diet and, since the riboflavin pool in the body is very stable, even in a deficient diet it is not depleted by more than 30–50%. The riboflavin content of urine is an indicator of riboflavin supply levels. Values above 80 µg riboflavin/g creatinine are normal; 27–79 µg/g are low; and less than 27 µg/g strongly suggests a vitamin-deficient diet. Glutathione reductase activity assay can provide similar information.

The most important sources of riboflavin are milk and milk products, eggs, various vegetables, yeast, meat products, particularly variety meats such as heart, liver and kidney, and fish liver and roe. Table 6.7 provides data about the occurrence of riboflavin in some common foods.

(6.8)

The following losses of thiamine can be expected: 15–25% in canned fruit or vegetables stored for more than a year; 0–60% in meat cooked under household conditions, depending on temperature and preparation method; 20% in salt brine pickling of meat and in baking of white bread; 15% in blanching of cabbage without sulfite and 40% with sulfite. Losses caused by sulfite are pH dependent. Practically no thiamine degradation occurs in a stronger acidic medium (e. g. lemon juice).

6.3.2 Riboflavin (Vitamin B$_2$)

6.3.2.1 Biological Role

Riboflavin (Formula 6.9) is the prosthetic group of flavine enzymes, which are of great importance in general metabolism and particularly in metabolism of protein.

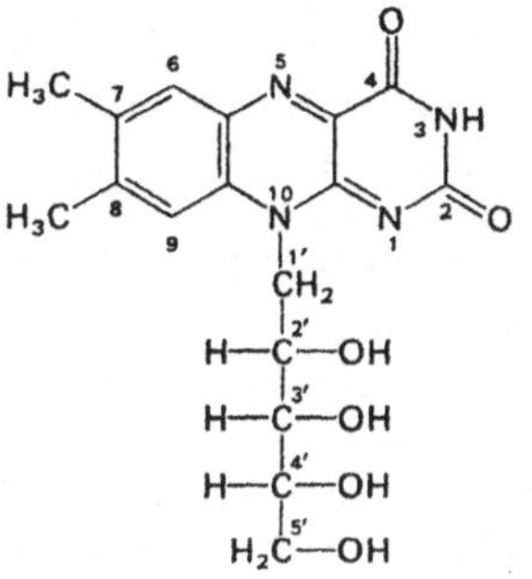

6.3.2.3 Stability, Degradation

Riboflavin is relatively stable in normal food handling processes. Losses range from 10–15%. Exposure to light, especially in the visible spectrum from 420–560 nm, photolytically cleaves ribitol from the vitamin, converting it to lumiflavin:

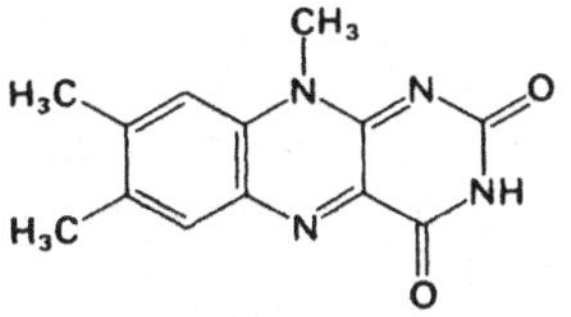

(6.10)

6.3.3 Pyridoxine (Pyridoxal, Vitamin B$_6$)

6.3.3.1 Biological Role

Vitamin B$_6$ activity is exhibited by pyridoxine (Formula 6.11) or pyridoxol (R = CH_2OH), pyridoxal (R = CHO) and pyridoxamine (R = CH_2NH_2). The metabolically active form,

(6.9)

pyridoxal phosphate, functions as a coenzyme (cf. 2.3.2.3) of amino acid decarboxylases, amino acid racemases, amino acid dehydrases, amino transferases, serine palmitoyl transferase, lysyl oxidase, δ-aminolevulinic acid synthase, and of enzymes of tryptophan metabolism. Furthermore, it stabilizes the conformation of phosphorylases.

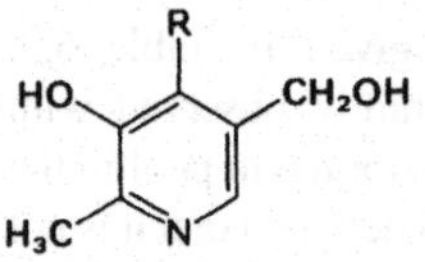

$$(6.11)$$

The intake of the vitamin occurs usually in the forms of pyridoxal or pyridoxamine.

Pyridoxine deficiency in the diet causes disorders in protein metabolism, e. g., in hemoglobin synthesis. Hydroxykynurenine and xanthurenic acid accumulate, since the conversion of tryptophan to nicotinic acid, a step regulated by the kynureninase enzyme, is interrupted.

6.3.3.2 Requirement, Occurrence

The daily requirement is given in Table 6.3. An indicator of sufficient supply is the activity of glutamate oxalacetate transaminase, an enzyme present in red blood cells. This activity is decreased in vitamin deficiency. The occurrence of pyridoxine in food is outlined in Table 6.8.

6.3.3.3 Stability, Degradation

The most stable form of the vitamin is pyridoxal, and this form is used for vitamin fortification of food. Vitamin B_6 loss is 45% in cooking of meat and 20–30% in cooking of vegetables. During milk sterilization, a reaction with cysteine transforms the vitamin into an inactive thiazolidine derivative (Formula 6.12). This reaction may account for vitamin losses also in other heat-treated foods.

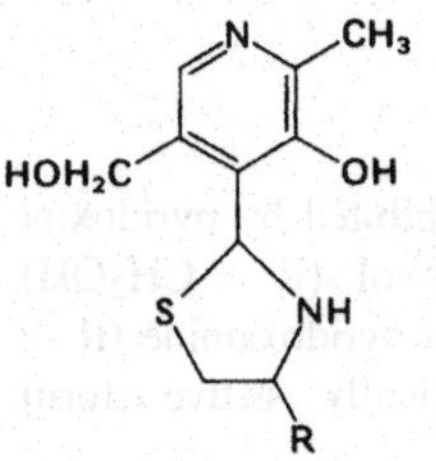

$$(6.12)$$

6.3.4 Nicotinamide (Niacin)

6.3.4.1 Biological Role

Nicotinic acid amide (I), in the form of nicotinamide adenine dinucleotide (NAD^+, cf. 2.3.1.1), or its phosphorylated form ($NADP^+$), is a coenzyme of dehydrogenases. Its excretion in urine is essentially in the form of N^1-methylnicotinamide (trigonelline amide, II), N^1-methyl-6-pyridone-3-carboxamide (III) and N^1-methyl-4-pyridone-3-carboxamide (IV):

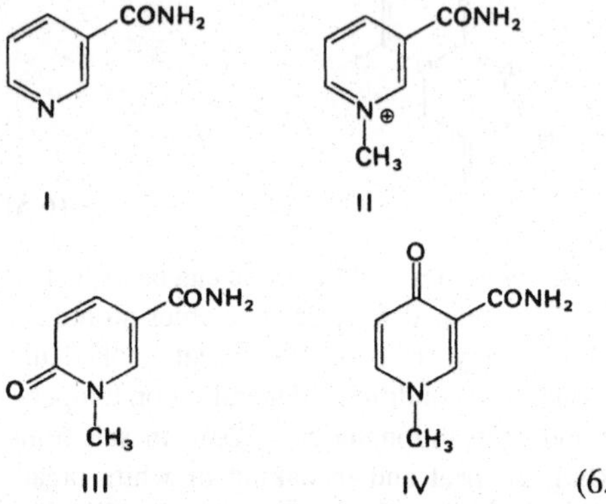

$$(6.13)$$

Vitamin deficiency is observed initially by a drop in concentration of NAD^+ and $NADP^+$ in liver and muscle, while levels remain normal in blood, heart and kidney. The classical deficiency disease is pellagra, which affects the skin, digestion and the nervous system (dermatitis, diarrhea and dementia). However, the initial deficiency symptoms are largely non-specific.

6.3.4.2 Requirement, Occurrence

The daily requirement (cf. Table 6.3) is covered to an extent of 60–70% by tryptophan intake. Hence, milk and eggs, though they contain little niacin, are good foods for prevention of pellagra because they contain tryptophan. It substitutes for niacin in the body, with 60 mg L-tryptophan equalling 1 mg nicotinamide. Indicators of sufficient supply of niacin in the diet are the levels of metabolites II (cf. Formula 6.13) in urine or III and IV in blood plasma.

The vitamin occurs in food as nicotinic acid, either as its amide or as a coenzyme. Animal organs, such as liver, and lean meat, cereals, yeast and mushrooms are abundant sources of niacin. Table 6.7 provides data on its occurrence in food.

6.3.4.3 Stability, Degradation

Nicotinic acid is quite stable. Moderate losses of up to 15% are observed (cf. Tables 6.1 and 6.2) in blanching of vegetables. The loss is 25–30% in the first days of ripening of meat.

6.3.5 Pantothenic Acid

6.3.5.1 Biological Role

Pantothenic acid (Formula 6.14) is the building unit of coenzyme A (CoA), the main carrier of acetyl and other acyl groups in cell metabolism. Acyl groups are linked to CoA by a thioester bond. Pantothenic acid occurs in free form in blood plasma, while in organs it is present as CoA. The highest concentrations are in liver, adrenal glands, heart and kidney.

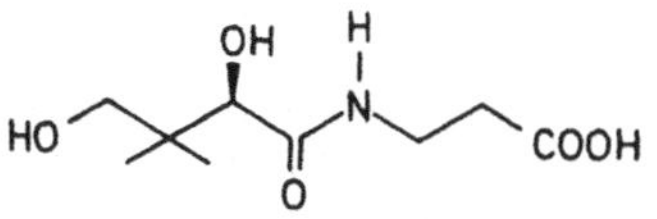

$$(6.14)$$

Only the R enantiomer occurs in nature and is biologically active. A normal diet provides an adequate supply of the vitamin.

6.3.5.2 Requirement, Occurrence

The daily requirement is 6–8 mg. The concentration in blood is 10–40 µg/100 ml and 2–7 mg/day are excreted in urine.

Pantothenic acid in food is determined with microbiological or ELISA (cf. 2.6.3) techniques. A gas chromatographic method using a ^{13}C-isotopomer of pantothenic acid as the internal standard is very accurate and much more sensitive. Table 6.7 lists data on pantothenic acid occurrence in food.

6.3.5.3 Stability, Degradation

Pantothenic acid is quite stable. Losses of 10% are experienced in processing of milk. Losses of 10–30%, mostly due to leaching, occur during cooking of vegetables.

6.3.6 Biotin

6.3.6.1 Biological Role

Biotin is the prosthetic group of carboxylating enzymes, such as acetyl-CoA-carboxylase, pyruvate carboxylase and propionyl-CoA-carboxylase, and therefore plays an important role in fatty acid biosynthesis and in gluconeogenesis. The carboxyl group of biotin forms an amide bond with the ε-amino group of a lysine residue of the particular enzyme protein. Only the (3aS, 4S, 6aR) compound, D-(+)-biotin, is biologically active:

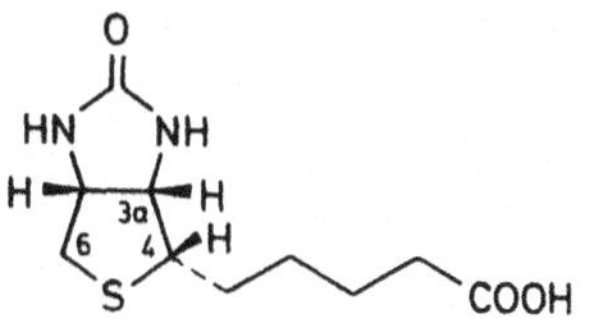

$$(6.15)$$

Biotin deficiency rarely occurs. Consumption of large amounts of raw egg white might inactivate biotin by its specific binding to avidin (cf. 11.2.3.1.9).

6.3.6.2 Requirement, Occurrence

The daily requirement is shown in Table 6.3. An indicator of sufficient biotin supply is the excretion level in the urine, which is normally 30–50 µg/day. A deficiency is indicated by a drop to 5 µg/day.

Biotin is not free in food, but is bound to proteins. Table 6.7 provides data on its occurrence in food.

6.3.6.3 Stability, Degradation

Biotin is quite stable. Losses during processing and storage of food are 10–15%.

6.3.7 Folic Acid

6.3.7.1 Biological Role

The tetrahydrofolate derivative (Formula (6.16), II) of folic acid (I, pteroylmonoglutamic acid, PGA) is the cofactor of enzymes which trans-

$$(6.16)$$

fer single carbon units in various oxidative states, e. g., formyl or hydroxymethyl residues. In transfer reactions the single carbon unit is attached to the N^5- or N^{10}-atom of tetrahydrofolic acid.

Folic acid deficiency caused by insufficient supply in the diet or by malfunction of absorptive processes is detected by a decrease in folic acid concentration in red blood cells and plasma, and by a change in blood cell patterns. There are clear indications that a congenital defect (neural tube defect) and a number of diseases are based on a deficiency of folate.

6.3.7.2 Requirement, Occurrence

The requirement shown in Table 6.3 is not often reached. In some countries, cereal products are supplemented with folic acid in order to avoid deficiency, e. g., with 1.4 mg/kg in the USA. Correspondingly, positive effects on consumer health have been observed.

In cooperation with vitamin B_{12}, folic acid methylates homocysteine to methionine. Therefore, homocysteine is a suitable marker for the supply of folate. In the case of a deficiency, the serum concentration of this marker is clearly raised compared with the normal value of 8–10 μmol/ml, resulting in negative effects on health because higher concentrations of homocysteine are toxic.

In food folic acid is mainly bound to oligo-γ-L-glutamates made up of 2–6 glutamic acid residues. Unlike free folic acid, the absorption of this conjugated form is limited and occurs only after the glutamic acid residues are cleaved by folic acid conjugase, a γ-glutamyl-hydrolase, to give the monoglutamate compound. Since certain constituents can reduce the absorption of folates, the average bioavailability is estimated at 50%. The folic acid content of foods varies. Data on folic acid occurrence in food are compiled in Table 6.7.

6.3.7.3 Stability, Degradation

Folic acid is quite stable. There is no destruction during blanching of vegetables, while cooking of meat gives only small losses. Losses in milk are apparently due to an oxidative process and parallel those of ascorbic acid. Ascorbate added to food preserves folic acid.

6.3.8 Cyanocobalamin (Vitamin B_{12})

6.3.8.1 Biological Role

Cyanocobalamin (Formula 6.17) was isolated in 1948 from *Lactobacillus lactis*. Due to its stability and availability, it is the form in which the vitamin is used most often. In fact, cyanocobalamin is formed as an artifact in the processing of biological materials. Cobalamins occur naturally as adenosylcobalamin and methylcobalamin, which instead of the cyano group contain a 5′-deoxyadenosyl residue and a methyl group respectively.

Adenosylcobalamin (coenzyme B_{12}) participates in rearrangement reactions in which a hydrogen atom and an alkyl residue, an acyl group or an electronegative group formally exchange places on two neighboring carbon atoms. Reactions of this type play a role in the metabolism of a series of bacteria. In mammals and bacteria a rearrangement reaction that depends on vitamin B_{12} is the conversion of methylmalonyl CoA to succinyl CoA (cf. 10.2.8.3). Vitamin B_{12} deficiency results in the excretion of methylmalonic acid in the urine.

Another reaction that depends on adenosylcobalamin is the reduction of ribonucleoside triphosphates to the corresponding 2′-deoxy compounds, the building blocks of deoxyribonucleic acids.

Methylcobalamin is formed, e. g., in the methylation of homocysteine to methionine with N^5-

methyltetrahydrofolic acid as the intermediate stage. The enzyme involved is a cobalamin-dependent methyl transferase.

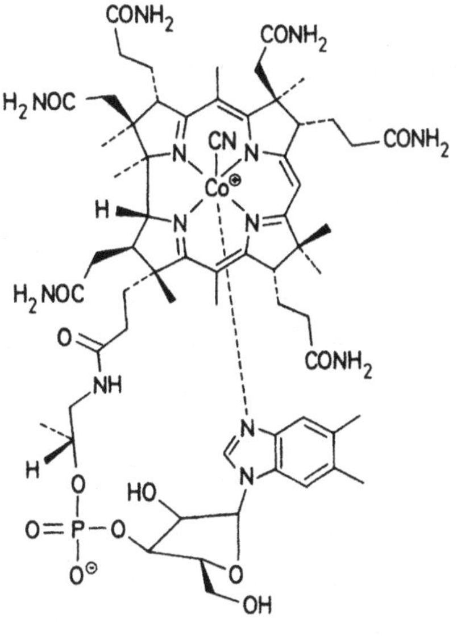

(6.17)

The absorption of cyanocobalamin is achieved with the aid of a glycoprotein, the "intrinsic factor" formed by the stomach mucosa. Deficiency of vitamin B_{12} is usually caused by impaired absorption due to inadequate formation of "intrinsic factor" and results in pernicious anemia.

6.3.8.2 Requirement, Occurrence

The daily requirement of vitamin B_{12} is shown in Table 6.3. The plasma concentration is normally 450 pg/ml.

The ability of vitamin B_{12} to promote growth alone or together with antibiotics, for example in young chickens, suckling pigs and young hogs, is of particular importance. This effect appears to be due to the influence of the vitamin on protein metabolism, and it is used in animal feeding. The increase in feed utilization is exceptional with underdeveloped young animals. Vitamin B_{12} is of importance also in poultry operations (enhanced egg laying and chick hatching). The use of vitamin B_{12} in animal feed vitamin fortification is obviously well justified.

Liver, kidney, spleen, thymus glands and muscle tissue are abundant sources of vitamin B_{12} (Table 6.7). Consumption of internal organs (variety meats) of animals is one method of alleviating vitamin B_{12} deficiency symptoms in humans.

6.3.8.3 Stability, Degradation

The stability of vitamin B_{12} is very dependent on a number of conditions. It is fairly stable at pH 4–6, even at high temperatures. In alkaline media or in the presence of reducing agents, such as ascorbic acid or SO_2, the vitamin is destroyed to a greater extent.

6.3.9 L-Ascorbic Acid (Vitamin C)

6.3.9.1 Biological Role

Ascorbic acid (L-3-keto-threo-hexuronic acid-γ-lactone, Formula 6.18, I) is involved in hydroxylation reactions, e.g., biosynthesis of catecholamines, hydroxyproline and corticosteroids (11-β-hydroxylation of deoxycorticosterone and 17-β-hydroxylation of corticosterone). Vitamin C is fully absorbed and distributed throughout the body, with the highest concentration in adrenal and pituitary glands.

About 3% of the body's vitamin C pool, which is 20–50 mg/kg body weight, is excreted in the urine as ascorbic acid, dehydroascorbic acid (a combined total of 25%) and their metabolites, 2,3-diketo-L-gulonic acid (20%) and oxalic acid (55%). An increase in excreted oxalic acid occurs only with a very high intake of ascorbic acid. Scurvy is caused by a dietary deficiency of ascorbic acid.

6.3.9.2 Requirement, Occurrence

The daily requirement is shown in Table 6.3. An indicator of insufficient vitamin supply in the diet is a low level in blood plasma (0.65 mg/100 ml). Vitamin C is present in all animal and plant cells, mostly in free form, and it is probably bound to protein as well. Vitamin C is particularly abundant in rose hips, black and red currants, strawberries, parsley, oranges, lemons (in peels more than in pulp), grapefruit, a variety of cabbages and potatoes. Vitamin C loss during storage of vegetables from winter through late spring can be as high as 70%.

Table 6.7 provides data on vitamin C occurrence in a variety of foods.

Ascorbic acid is chemically synthesized. However, the synthesis by means of genetically modified microorganisms (GMO vitamin C) is more cost effective. Therefore, the largest proportion is synthesized by these means.

6.3.9.3 Stability, Degradation

Ascorbic acid (I) has an acidic hydroxyl group ($pK_1 = 4.04$, $pK_2 = 11.4$ at 25 °C). Its UV absorption depends on the pH value (Table 6.9). Ascorbic acid is readily and reversibly oxidized to dehydroascorbic acid (II), which is present in aqueous media as a hydrated hemiketal (IV). The biological activity of II is possibly weaker than that of I because the plasma and tissue concentrations of II are considerably lower after the administration of equal amounts of I and II. The activity is completely lost when the dehydroascorbic acid lactone ring is irreversibly opened, converting II to 2,3-diketogulonic acid (III), cf. Formula 6.18:

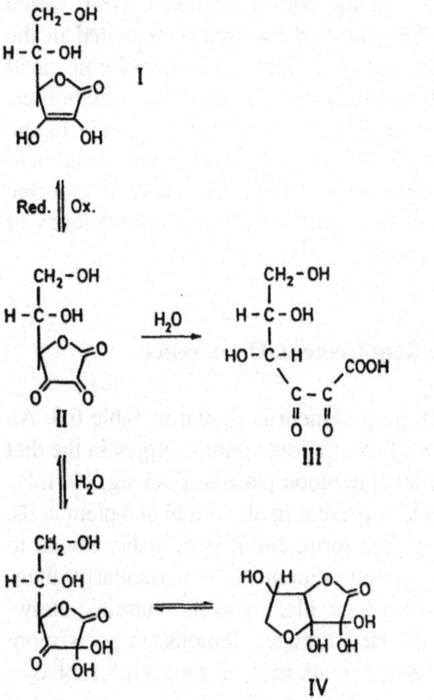

(6.18)

Table 6.9. Effect of pH on ultraviolet absorption maxima of ascorbic acid

pH	λ max (nm)
2	244
6–10	266
>10	294

The oxidation of ascorbic acid to dehydroascorbic acid and its further degradation products depends on a number of parameters. Oxygen partial pressure, pH, temperature and the presence of heavy metal ions are of great importance. Metal-catalyzed destruction proceeds at a higher rate than noncatalyzed spontaneous autoxidation. Traces of heavy metal ions, particularly Cu^{2+} and Fe^{3+}, result in high losses.

The principle of metal catalysis is schematically presented in Reaction 6.19 (Me = metal ion).

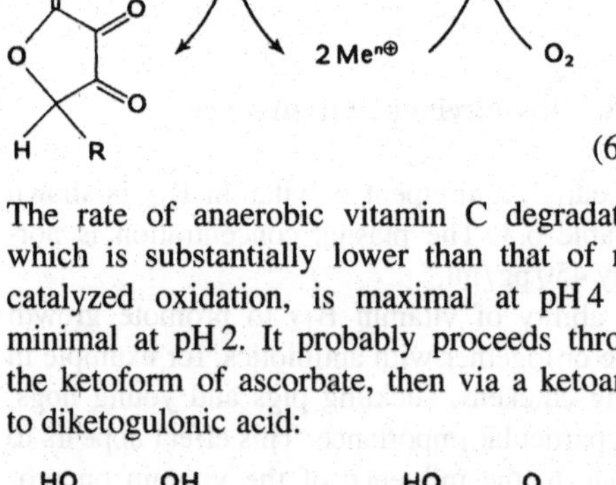

(6.19)

The rate of anaerobic vitamin C degradation, which is substantially lower than that of non-catalyzed oxidation, is maximal at pH 4 and minimal at pH 2. It probably proceeds through the ketoform of ascorbate, then via a ketoanion to diketogulonic acid:

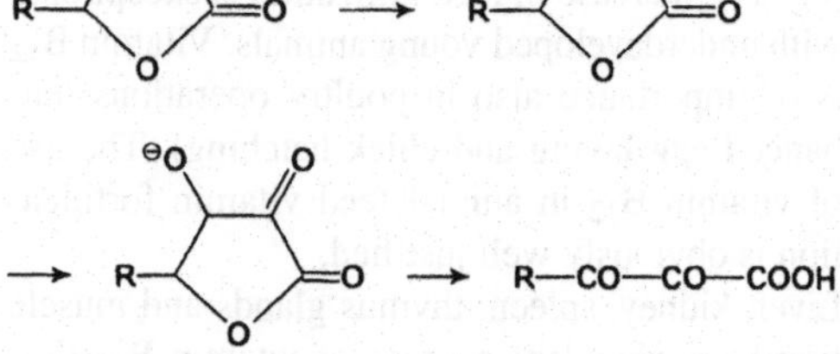

(6.20)

Diketogulonic acid degradation products, xylosone and 4-deoxypentosone (Formula 6.21), are then converted into ethylglyoxal, various reductones (cf. 4.2.4.3.1), furfural and furancarboxylic acid.

juices and dried fruits. The intermediates that have been identified are scorbamic acid (I in Formula 6.22), which is produced by *Strecker* degradation with an amino acid, and a red pigment (II). A wealth of data is available on ascorbic acid losses during preservation, storage and processing of food. Tables 6.1 and 6.2 and Figs. 6.3 and 6.4 present several examples. Ascorbic acid degradation is often used as a general indicator of changes occurring in food.

$$(6.21)$$

In the presence of amino acids, ascorbic acid, dehydroascorbic acid and their degradation products might be changed further by entering into *Maillard*-type browning reactions (cf. 4.2.4.4).

An example is the reaction of dehydroascorbic acid with amino compounds to give pigments, which can cause unwanted browning in citrus

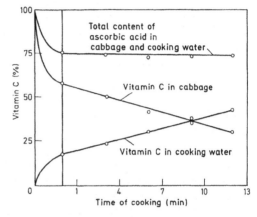

Fig. 6.3. Ascorbic acid losses as a result of cooking of cabbage (according to *Plank*, 1966)

$$(6.22)$$

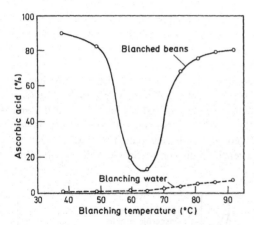

Fig. 6.4. Ascorbic acid losses in green beans versus blanching temperature (according to *Plank*, 1966)

6.4 References

Bässler, K.H.: On the problematic nature of vitamin E requirements: net vitamin E.Z. Ernährungswiss. *30*, 174 (1991)

Bässler, K.H., Lang, K.: Vitamine. Dr. Dietrich Steinkopff Verlag: Darmstadt. 1975

Booth, S.L., Mayer, J.: A hydrogenated form of vitamin K in food supply. Inform *11*, 258 (2000)

Counsell, J.N., Hornig, D.H. (Eds.): Vitamin C (Ascorbic Acid). Applied Science Publ.: London 1981

Deutsche Gesellschaft für Ernährung: Referenzwerte für die Nährstoffzufuhr. 1. Aufl. Umschau Braus Verlagsgesellschaft, Frankfurt a.M., 2000

Farrer, K.T.H.: The thermal destruction of vitamin B_1 in foods. Adv. Food Res. *6*, 257 (1955)

Gregory, J.F., Quinlivan, E.P., Davis, S.R.: Integrating the issues of folate bioavailability, intake and metabolism in the era of fortification. Trends Food Sci. Technol. *16*, 229 (2005)

Isler, O., Brubacher, G.: Vitamine I. Fettlösliche Vitamine. Georg Thieme Verlag: Stuttgart. 1982

Isler, O., Brubacher, G., Ghisla, S., Kräutler, B.: Vitamine II. Wasserlösliche Vitamine. Georg Thieme Verlag: Stuttgart, 1988

Körner, W.F., Völlm, J.: Vitamine. In: Klinische Pharmakologie und Pharmakotherapie. 3. Aufl. (Eds.: Kuemmerle, H.P., Garrett, E.R., Spitzy, K.H.), p. 361, Urban u. Schwarzenberg: München. 1976

Labuza, T.P., Riboh, D.: Theory and application of *Arrhenius* kinetics to the prediction of nutrient losses in foods. Food Technol. *36* (10), 66 (1982)

Liao, M.-L., Seib, P.A.: Chemistry of L-ascorbic acid related to foods. Food Chem. *30*, 289 (1988)

Lund, D.B.: Effect of commercial processing on nutrients. Food Technol. *33* (2), 28 (1979)

Machlin, L.J. (Ed.): Handbook of Vitamins. Marcel Dekker: New York. 1984

Ogiri, Y., Sun, F., Hayami, S., Fujimura, A., Yamamoto, K.; Yaita, M., Kojo, S.: The low vitamin C activity of orally administered L-dehydroascorbic acid. J. Agric. Food Chem. *50*, 227 (2002)

Plank, R.: Handbuch der Kältetechnik. Bd. XIV, p. 475, Springer-Verlag: Berlin. 1966

Rehner, G., Daniel, H.: Biochemie der Ernährung. Spektrum Akademischer Verlag, Heidelberg, 1999

Reiff, F., Paust, J., Meier, W., Fürst, A., Ernst, H.G., Kuhn, W., Härtner, H., Florent, J., Suter, C., Pollak, P.: Vitamine. Ullmanns Encyklopädie der technischen Chemie, 4th edn. Vol. 23, p. 621, Verlag Chemie: Weinheim. 1983

Rychlik, M.: Quantification of free and bound pantothenic acid in foods and blood plasma by stable isotope dilution assay. J. Agric. Food Chem. *48*, 1175 (2000)

Seib, P.A., Tolbert, B.M. (Eds.): Ascorbic acid: Chemistry, metabolism, and uses. Advances in Chemistry Series 200, American Chemical Society: Washington, D.C. 1982

7 Minerals

7.1 Foreword

Minerals are the constituents which remain as ash after the combustion of plant and animal tissues. Minerals are divided into:

- main elements,
- trace elements and
- ultra-trace elements

The main elements (Na, K, Ca, Mg, Cl, P) are essential for human beings in amounts >50 mg/day. Sulfur also belongs to this group. However, it will not be discussed here because sulfur requirements are met by the intake of sulfur-containing amino acids. Trace elements (Fe, I, F, Zn, Se, Cu, Mn, Cr, Mo, Co, Ni) are essential in concentrations of <50 mg/day; their biochemical actions have been elucidated. Ultra-trace elements (Al, As, Ba, Bi, B, Br, Cd, Cs, Ge, Hg, Li, Pb, Rb, Sb, Si, Sm, Sn, Sr, Tl, Ti, W) are elements whose essentiality has been tested in animal experiments over several generations and deficiency symptoms have been found under these extreme conditions. For one of these elements, if it is possible to detect a biochemical function in a vital tissue or organ, the element is assigned to the trace elements.

Main and trace elements have very varied functions, e.g., as electrolytes, as enzyme constituents (cf. 2.3.3) and as building materials, e.g., in bones and teeth. Table 7.1 summarizes the content of main elements in the human body. Table 7.2 shows the content of sodium, potassium, calcium, iron, and phosphorus in some foods. In the same food raw material, the mineral content can fluctuate greatly depending on genetic and climatic factors, agricultural procedures, composition of the soil, and ripeness of the harvested crops, among other factors. This applies to both main and trace elements. Changes in the mineral content usually occur also in the processing of raw materials, e.g., in thermal processes and material separations. Table 7.3

Table 7.1. Main elements in the human body

Element	Content g/kg
Calcium	10–20
Phosphorus	6–12
Potassium	2–2.5
Sodium	1–1.5
Chlorine	1–1.2
Magnesium	0.4–0.5

shows data on mineral losses in food processing. Mineral supply depends not only on the intake in food but primarily on the bioavailability, which is essentially related to the composition of the food. Thus, the redox potential and pH value determine the valency state, solubility, and, consequently, absorption. A series of food constituents, e.g., proteins, peptides, amino acids, polysaccharides, sugars, lignin, phytin, and organic acids, bind minerals and enhance or inhibit their absorption. The importance of minerals as food ingredients depends not only on their nutritional and physiological roles. They contribute to food flavor and activate or inhibit enzyme-catalyzed and other reactions, and they affect the texture of food.

7.2 Main Elements

7.2.1 Sodium

The sodium content of the body is 1.4 g/kg. Sodium is present mostly as an extracellular constituent and maintains the osmotic pressure of the extracellular fluid. In addition, it activates some enzymes, such as amylase. Sodium absorption is rapid; it starts 3–6 min after intake and is completed within 3 h. Daily intake of sodium averages 2.5 g (females) to 3.3 g (males); the adult's average requirement ranges from

Table 7.2. Mineral content (Na, K, Ca, Fe, and P) of some foods

Food product	Na	K	Ca	Fe	P
Milk and dairy products					
Bovine milk,					
raw, high quality	48	157	120	0.046	92
Human milk	16	53	31	0.06	15
Butter	5	16	13	0.02–0.2	21
Cheese					
Emmental (45% fat)	275	95	1020	0.35	636
Camembert (60% fat)	709	95	90	0.13	310
Camembert (30% fat)	669	120	600	0.17	385
Eggs					
Chicken egg yolk	51	138	140	7.2	590
Chicken egg white	170	154	11	0.2	21
Meat and meat products					
Beef, whole carcass, lean	66	342	5.7	2.6	190
Pork, whole carcass, lean	69	397	5	1.0	192
Calf liver	87	316	8.7	7.9	306
Pork liver	77	363	7.6	18	407
Chicken liver	68	218	18	7.4	240
Pork kidney	173	242	7	7.3	260
Blood sausage	680	38	6.5	6.4	22
Fish and fish products					
Herring	117	360	34	1.1	250
Eel	65	259	17	0.9	334
Cereals and cereal products					
Wheat, whole kernel	7.8	381	33	3.3	341
Wheat flour, type 550	2.0	146	15	1.0	108
Wheat flour, type 1050	3.0	203	24	2.2	208
Wheat germ	5	993	49	8.5	1100
Rye, whole kernel	3.8	530	37	2.8	337
Rye flour, type 997	1	285	25	1.9	189
Corn, whole kernel	6	294	8	1.5	213
Breakfast cereals					
(corn flakes)	915	120	13	2.0	59
Oat flakes	6.8	374	48	5.4	415
Rice, unpolished	10	238	16	3.2	282
Rice, polished	3.9	103	6	0.8	114
Vegetables					
Watercress	12	276	180	3.1	64
Mushrooms (cultivated)	8	390	11	1.26	123
Chicory	4.4	192	26	0.74	26
Endive	43	346	54	1.4	54
Peas, green	2	274	24	1.7	113
Lamb's lettuce	4	421	35	2.0	49
Kale	35	451	212	1.9	87
Potatoes	3.2	418	6.4	0.43	50
Kohlrabi	20	322	68	0.48	50
Head lettuce	7.5	179	22	0.34	23
Lentils, dried	6.6	837	65	8.0	412
Carrots	60	321	37	0.39	35
Brussels sprout	7	451	31	1.1	84
Spinach	65	554	117	3.8	46
Edible mushroom (*Boletus edulis*)	6	341	4.2	1.0	85

Table 7.2. continued

Food product	Na	K	Ca	Fe	P
Vegetables					
Tomato	3.3	242	9.4	0.3	22
White cabbage	13	255	46	0.4	36
Fruits					
Apple	1.2	122	5.8	0.25	12
Orange	1.4	165	42	0.19	23
Apricots	2	278	16	0.65	21
Strawberry	1.4	161	21	0.64	29
Grapefruit	1.1	148	24	0.17	17
Rose hips	24	291	257	0.52	258
Currants-red	1.4	257	29	0.91	27
Currants-black	1.5	310	46	1.29	40
Cherries-sour	2	114	8	0.6	19
Plums	1.7	177	8.3	0.26	18
Sea buckthorn	3.5	133	42	0.44	9
Yeast					
Baker's yeast, pressed	34	649	28	3.5	473
Brewer's yeast, dried	77	1410	50	17.6	1900

[a] Data are in mg/100 g edible portion (average values).

Table 7.3. Mineral losses in food processing

Raw material	Product	Loss (%)						
		Cr	Mn	Fe	Co	Cu	Zn	Se
Spinach	Canned		87		71		40	
Beans	Canned						60	
Tomatoes	Canned						83	
Carrots	Canned				70			
Beetroot	Canned				67			
Green beans	Canned				89			
Wheat	Flour		89	76	68	68	78	16
Rice	Polished rice	75	26			45	75	

1.3–1.6 g/day (equal to 3.3–4.0 g/day NaCl). The intake of too little or too much sodium can result in serious disorders. From a nutritional standpoint, only the excessive intake of sodium is of importance because it can lead to hypertension. A low intake of sodium can be achieved by a nonsalty diet or by using diet salt (common salt substitutes, cf. 22.2.5). Table 7.2 provides values for the sodium content of some foods.

7.2.2 Potassium

The concentration of potassium in the body is 2 g/kg. At a concentration of 140 mmol/l, it is the most common cation in the intracellular fluid. Potassium is localized mostly within the cells. It regulates the osmotic pressure within the cell, is involved in cell membrane transport and also in the activation of a number of glycolytic and respiratory enzymes. The potassium intake in a normal diet is 2–5.9 g/day. The minimum daily requirement is estimated to be 782 mg. Potassium deficiency is associated with a number of symptoms and may be a result of undernourishment or predominant consumption of potassium-deficient foods, e. g., white bread, fat or oil. The potassium content in food is summarized in Table 7.2. Potatoes and molasses are particularly abundant sources.

7.2.3 Magnesium

The concentration of magnesium in the body is 250 mg/kg. The daily requirement is 300–400 mg. In a normal diet, the daily intake is 300–500 mg. As a constituent and activator of many enzymes, particularly those associated with the conversion of energy-rich phosphate compounds, and as a stabilizer of plasma membranes, intracellular membranes, and nucleic acids, magnesium is a life-supporting element. Because of its indispensable role in body metabolism, magnesium deficiency causes serious disorders.

7.2.4 Calcium

The total amount of calcium in the body is about 1500 g. Because of the large amounts of calcium all over the body, it is one of the most important minerals. It is abundant in the skeleton and in some body tissues. Calcium is an essential nutrient because it is involved in the structure of the muscular system and controls essential processes like muscle contraction (locomotor system, heartbeat) blood clotting, activity of brain cells and cell growth. Calcium deficiency causes serious disorders. The desirable calcium intake (g/day) is stipulated as: birth to 6 months (0.4), 6 to 12 months (0.6), 1 to 5 years (0.8), 6 to 10 years (0.8–1.2), 11 to 24 years and pregnant women (1.2 to 1.5), 25 to 65 years (1.0) and above 65 years (1.5). The main source of calcium is milk and milk products, followed at a considerable distance by fruit and vegetables, cereal products, meat, fish and eggs. Table 7.2 provides data on the calcium content of some foods. An adequate supply of vitamin D is required for the absorption of calcium.

7.2.5 Chloride

The chloride content of human tissue is 1.1 g/kg body weight and the plasma concentration is 98–106 mmol/l. Chloride serves as a counter ion for sodium in extracellular fluid and for hydrogen ions in gastric juice. Chloride absorption is as rapid as its excretion in the urine. The minimum intake of chloride largely corresponds on a molar basis to the sodium requirement.

7.2.6 Phosphorus

The total phosphorus content in the body is about 700 g. The daily requirement is about 0.8–1.2 g. The Ca/P ratio in food should be about 1. Phosphorus, in the form of phosphate, free or bound as an ester or present as an anhydride, plays an important role in metabolism and, as such, is an essential nutrient. The organic forms of phosphorus in food are cleaved by intestinal phosphatases and, thereby, absorption occurs mostly in the form of inorganic phosphate. Polyphosphates, used as food additives, are absorbed only after prior hydrolysis into orthophosphate. The extent of hydrolysis is influenced by the degree of condensation of the polyphosphates. Table 7.2 includes a compilation of the phosphorus content of some foods.

7.3 Trace Elements

7.3.1 General Remarks

There are 11 trace elements present in hormones, vitamins, enzymes and other proteins which have distinct biological roles. A deficiency in the trace elements results in metabolic disorders that are primarily associated with the absence or decreased activity of metabolic enzymes.

7.3.2 Individual Trace Elements

7.3.2.1 Iron

The iron content of the body is 4–5 g. Most of it is present in the hemoglobin (blood) and myoglobin (muscle tissue) pigments. The metal is also present in a number of enzymes (peroxidase, catalase, hydroxylases and flavine enzymes), hence it is an essential ingredient of the daily diet. The iron requirement depends on the age and sex of the individual, it is about 1.5–2.2 mg/day. Iron supplied in the diet must be in the range of 15 mg/day in order to meet this daily requirement. The large variation in intake can be explained by different extents of absorption of the various forms of iron present in food (organic iron compounds *vs* simple

Table 7.4. Trace elements in the human body and their daily intake[a]

Element	Content (mg/kg body weight)	Adequate Intake[b] (mg/day)
Fe	60	15
F	37	2.9–3.8
Zn	33	10–15
Cu	1.0	1.0–1.5
Se	0.2	0.03–0.07
Mn	0.2	2–5
I	0.2	0.2–0.26
Ni	0.1	0.025–0.03
Mo	0.1	0.05–0.1
Cr	0.1	0.003–0.1
Co	0.02	0.002–0.1

[a] Average values.
[b] Estimated for adults.

salts). The most utilizable source is iron in meat, for which the extent of absorption is 20–30%. The absorption is much less from liver (6.3%) and fish (5.9%), or from cereals, vegetables and milk, from which iron absorption is the lowest (1.0–1.5%). Eggs decrease and ascorbic acid increases the extent of absorption. Bran interferes with iron absorption due to the high content of phytate. Apparently, the absorption of iron present in food is, in a healthy organism, regulated by the requirement of the organism. Nevertheless, in order to provide a sufficient supply of iron to persons who require higher amounts (children, women before menopause and pregnant or nursing women), cereals (flour, bread, rice, pasta products) fortified with iron to the extent of 55–130 mg/kg are recommended. Extensive feeding tests with chickens and rats have shown that $FeSO_4$ is the most suitable form of iron, but ferrous gluconate and ferrous glycerol phosphate are also efficiently absorbed. Two food processing problems arising from mineral fortification are the increased probability that oxidation will occur and, in the case of wheat flour, decreased baking quality. Generally, iron is an undesirable element in food processing; for example, iron catalyzes the oxidation of fat or oil, increases turbidity of wine and, as a constituent of drinking water, it supports the growth of iron-requiring bacteria. The iron content of various foods is shown in Table 7.2.

7.3.2.2 Copper

The amount of copper in the body is 80–100 mg. Copper is a component of a number of oxidoreductase enzymes (cytochrome oxidase, superoxide dismutase, tyrosinase, uricase, amine oxidase). In blood plasma, it is bound to ceruloplasmin, which catalyzes the oxidation of Fe^{2+} to Fe^{3+}. This reaction is of great significance since it is only the Fe^{3+} form in blood which is transported by the transferrin protein to the iron pool in the liver. The daily copper requirement is 1–1.5 mg and it is supplied in a normal diet. Copper is even less desirable than iron during food processing and storage since it catalyzes many unwanted reactions. Cu^{2+}-Ions are taste bearing. The threshold value 2.4–3.8 mg/l was determined with aqueous solutions of $CuSO_4$ or $CuCl_2$.

7.3.2.3 Zinc

The total zinc content in adult human tissue is 2–4 g. The daily requirement of 5–10 mg is provided by a normal diet (6–22 mg zinc/day). Zinc is a component of a number of enzymes (e. g., alcohol dehydrogenase, lactate dehydrogenase, malate dehydrogenase, glutamate dehydrogenase, carboxypeptidases A and B, and carbonic anhydrase). Other enzymes, e. g., dipeptidases, alkaline phosphatase, lecithinase and enolase, are activated by zinc and by some other divalent metal ions. Zinc deficiency in animals causes serious disorders, while high zinc intake by humans is toxic. Zinc poisoning has been reported as a result of consumption of soured food kept in zinc-plated metal containers (e. g., potato salad from institutional catering services).

7.3.2.4 Manganese

The body contains a total of 10–40 mg of manganese. The daily requirement, 2–5 mg, is met by the normal daily food intake (2–48 mg manganese/day). Manganese is the metal activator for pyruvate carboxylase and, like some other divalent metal ions, it activates various enzymes, such as arginase, amino peptidase, alkaline phosphatase, lecithinase or enolase. Manganese, even in higher amounts, is relatively nontoxic.

7.3.2.5 Cobalt

The total cobalt content of the body is 1–2 mg. Since it was discovered that vitamin B_{12} contains cobalt as its central atom, the nutritional importance of cobalt has been emphasized and it has been assigned the status of an essential element. Its requirement is met by normal nutrition.

7.3.2.6 Chromium

The chromium content of the body varies considerably depending on the region; the range is 6–12 mg. The daily intake also varies greatly from 5 to 200 µg. The supply is considered suboptimal. Chromium is important in the utilization of glucose. For instance, it activates the enzyme phosphoglucomutase and increases the activity of insulin; therefore, chromium deficiency causes a decrease in glucose tolerance. And the risk of cardiovascular disease increases. Chromium, as the chromate ion, proved to be nontoxic when used at 25 ppm in a long-term feeding experiment with rats.

7.3.2.7 Selenium

The selenium content in humans is 10–15 mg, while the daily intake is 0.05–0.1 mg. Depending on the region, it can vary greatly because of the varying content of selenium in the soil. Selenium is an antioxidant and can enhance tocopherol activity. The enzyme glutathione peroxidase contains selenium. It catalyzes the following reaction, protecting membranes from oxidative destruction:

$$ROOH + 2GSH \rightarrow ROH + H_2O + GSSG \quad (7.1)$$

Selenium toxicity, for example, its strong carcinogenic activity, is well known from numerous animal feeding studies and from diseases of cattle grazing in pastures on selenium-rich soil.

For adults, an adequate intake is estimated at 30–70 µg Se/day.

7.3.2.8 Molybdenum

The body contains 8–10 mg of molybdenum. Daily intake in food is approx. 0.3 mg. It is a component of aldehyde oxidase and xanthine oxidase. The bacterial nitrate reductase involved in meat curing and pickling processes contains molybdenum. High levels of the metal are toxic, as has been shown by cattle grazing on molybdenum-enriched soil. The grass on such soil contains 20–100 µg molybdenum/g dry matter.

7.3.2.9 Nickel

Nickel is an activator of a number of enzymes, e. g., alkaline phosphatase and oxalacetate decarboxylase, which can also be activated by other divalent metal ions. Nickel also enhances insulin activity. The essential role of nickel has been established by inducing deficiency symptoms in feeding experiments with chickens and rats. These symptoms include changes in the liver mitochondria. The daily intake in food amounts to 150–700 µg. The nickel requirement is estimated to be 35–500 µg/day.

7.3.2.10 Fluorine

The body contains 2.6 g fluorine. It plays an essential role, as indicated by feeding experiments with rats and mice – deficient diets containing less than 2.5 ppm and 0.1–0.3 ppm respectively, resulted in disorders in growth and reproduction. The positive effect of fluorine on teeth caries is well established. The addition to drinking water of 0.5–1.5 ppm fluorine in the form of NaF or $(NH_4)_2SiF_6$ inhibits tooth decay. Its beneficial effect appears to be in retarding solubilization of tooth enamel and inhibiting the enzymes involved in development of caries. Toxic effects of fluorine appear at a level of 2 ppm. Therefore, the beneficial effects of fluoridating drinking water are disputed by some and it is a controversial topic of mineral nutrition.

7.3.2.11 Iodine

The content of iodine in the body is about 10 mg, of which the largest portion (70–80%) is covalently bound in the thyroid gland. Iodine absorption from food occurs exclusively and rapidly as iodide and is utilized in the thyroid gland in the biosynthesis of the hormone thyroxine (tetraiodothyronine) and its less iodized form, triiodothyronine. In this process, the iodide ion is first oxidized, then iodization of the tyrosine residues of thyroglobulin occurs. Diiodotyrosine condenses with itself or with monoiodotyrosine to form thyroglobulin-bound thyroxine or triiodothyronine. Both active hormones are released from thyroglobulin by the action of a proteinase. Also released are several peptides which, however, lack activity. The iodine requirement of humans is 100–200 µg/day; pregnant and nursing women require 230 and 260 µg/day respectively. Iodine deficiency results in enlargement of the thyroid gland (iodine-deficiency induced goiter). There is little iodine in most food. Good sources are milk, eggs and, above all, seafood. Drinking water contributes little to the body's iodine supply. In areas where goiter is found, the water has 0.1–2.0 µgI/l, while in goiter-free districts, 2–15 µgI/l are present in drinking water. To avoid diseases caused by low iodine supply, some countries with iodine-deficient districts employ prophylactic measures to combat the deficiency symptoms. This involves iodization of common salt with potassium iodate, with 100 µg iodine added to 1–10 g NaCl. Higher amounts of iodine are toxic and, as shown with rats, disturb the animal's normal reproduction and lactation. In humans, diseases of the thyroid can develop.

7.3.3 Ultra-trace Elements

7.3.3.1 Tin

Tin occurs in all humans organs. Although a growth-promoting effect was detected in rats, it is disputed. The natural level of tin in food is very low, but it can be increased in the case of foods canned in tinplate cans. Very acidic foods can often dissolve substantial amounts of tin. Thus, the concentration of tin in pineapple and grapefruit juice transported in poorly tin plated cans was 2 g/l. The tin content of foods in tinplate cans is generally below 50 mg/kg and should not exceed 250 mg/kg. In the form of inorganic compounds, tin is resorbed only to a low extent and, therefore, it is only slightly toxic. In comparison, organic tin compounds can be very toxic.

7.3.3.2 Aluminum

The body contains 50–150 mg of aluminum. Higher levels are found in aging organisms. The daily average intake of aluminum is 2–10 mg. It is resorbed in only negligible amounts by the gastrointestinal tract. The largest portion is eliminated in feces. Excretion of aluminum in urine is less than 0.1 mg/day. It is not secreted in milk. Animal feeding tests with high levels of aluminum in the diet through several generations showed that aluminum is nontoxic. This seems to be true also for humans. Hence, the reluctance to use aluminum cookware in food processing is unfounded. Some recent studies, however, have revealed that a pathologically caused accumulation of aluminum in humans can cause significant damage to the cells of the central nervous system.

7.3.3.3 Boron

Boron is found in humans and animals. The concentrations in the organs and tissues vary. In human beings, the highest concentrations are found in the heart (28 mg/kg), followed by the ribs (10 mg/kg), spleen (2.6 mg/kg) and liver (2.3 mg/kg). Muscle tissue contains only 0.1 mg/kg. Boron seems to be an essential nutrient, which promotes bone formation by interaction with calcium, magnesium and vitamin D. In addition, there are indications that boron is involved in the hydroxylation of steroids, e. g., in the synthesis of 17β-estradiol and testosterone. The daily requirement is estimated to be 1–2 mg. Apples (40), soy flour (28), grapes (27), tomatoes (27), celery (25) und broccoli (22) are rich in boron (mg/kg solids). Important sources also include wine (8) and water.

7.3.3.4 Silicon

Silicon, as soluble silicic acid, is rapidly absorbed. The silicon content of the body is approx. 1 g. The main source is cereal products. Silicon promotes growht and thus has a biological role. The toxicity of silicic acid is apparent only at concentrations ≥ 100 mg/kg. The intake in food amounts to 21–46 mg/day.

7.3.3.5 Arsenic

Arsenic was shown to be an essential trace element for the growth of chickens, rats, and goats. Its metabolic role is not yet understood. It appears to be involved in the metabolism of methionine. Choline can be replaced by arsenocholine in some of its functions. The possible human requirement is estimated to be 12–25 µg/day. The intake in food amounts to 20–30 µg/day. The main source is fish.

7.4 Minerals in Food Processing

The contribution of minerals to the nutritive/physiological value and the physical state of food has been covered in the Foreword of this Chapter and under the individual elements.

However, there are metal ions, derived from food itself or acquired during food processing and storage, which interfere with the quality and visual appearance of food. They can cause discoloration of fruit and vegetable products (cf. 18.1.2.5.8) and many metal-catalyzed reactions are responsible for losses of some essential nutrients, for example, ascorbic acid oxidation (cf. 6.3.9.3). Also, they are responsible for taste defects or off-flavors, for example, as a consequence of fat oxidation (cf. 3.7.2.1.6). Therefore, the removal of many interfering metal ions by chelating agents (cf. 8.14) or by other means is of importance in food processing.

7.5 References

Bohl, C.H., Volpe, S.L.: Magnesium and exercise. Crit. Rev. Food Sci. Nutr. *42*, 533 (2002)

Deutsche Gesellschaft für Ernährung (DGE): Referenzwerte für die Nährstoffzufuhr. 1. Auflage. Umschau Braus Verlagsgesellschaft, Frankfurt a.M., 2000

Devirian, T.A., Volpe, S.L.: The physiological effects of dietary boron. Crit. Rev. Food Sci. Nutr. *43*, 219 (2003)

Lang, K.: Biochemie der Ernährung, 4. Aufl., Dr. Dietrich Steinkopff Verlag: Darmstadt. 1979

Pfannhauser, W.: Essentielle Spurenelemente in der Nahrung. Springer Verlag: Berlin, 1988

Smith, K.T.: Trace Minerals in Foods. Marcel Dekker: New York, 1988

Wolfram, G., Kirchgeßner, M. (Eds.): Spurenelemente und Ernährung. Wissenschaftl. Verlagsgesellschaft: Stuttgart, 1990

8 Food Additives

8.1 Foreword

A food additive is a substance (or a mixture of substances) which is added to food and is involved in its production, processing, packaging and/or storage without being a major ingredient. Additives or their degradation products generally remain in food, but in some cases they may be removed during processing. The following examples illustrate and support the use of additives to enhance the:

- *Nutritive Value of Food*
  Additives such as vitamins, minerals, amino acids and amino acid derivatives are utilized to increase the nutritive value of food. A particular diet may also require the use of thickening agents, emulsifiers, sweeteners, etc.
- *Sensory Value of Food*
  Color, odor, taste and consistency or texture, which are important for the sensory value of food, may decrease during processing and storage. Such decreases can be corrected or readjusted by additives such as pigments, aroma compounds or flavor enhancers. Development of "off-flavor", for instance, derived from fat or oil oxidation, can be suppressed by antioxidants. Food texture can be stabilized by adding minerals or polysaccharides, and by many other means.
- *Shelf Life of Food*
  The current forms of food production and distribution have increased the demand for longer shelf life. Furthermore, the world food supply situation requires preservation by avoiding deterioration as much as possible. The extension of shelf life involves protection against microbial spoilage, for example, by using antimicrobial additives and by using active agents which suppress and retard undesired chemical and physical changes in food. The latter is achieved by stabilization of pH using buffering additives or stabilization of texture with thickening or gelling agents, which are polysaccharides.

- *Practical Value*
  The common trend towards foods which are easy and quick to prepare (convenience foods) can also necessitate the increasing use of additives.

It is implicitly understood that food additives and their degradation products should be non-toxic at their recomended levels of use. This applies equally to acute and to chronic toxicity, particularly the potential carcinogenic, teratogenic (causing a malformed fetus) and mutagenic effects. It is generally recognized that additives are applied only when required for the nutritive or sensory value of food, or for its processing or handling. The use of additives is regulated by Food and Drug or Health and Welfare administrations in most countries. The regulations differ in part from country to country but there are endeavors under way to harmonize them on the basis of both current toxicological knowledge and the requirements of modern food technology. The most important groups of additives are outlined in the following sections. No reference

Table 8.1. Utilization of food additives in United States (1965 as % of total additives used)[a]

Additives, class	% of total	Additives, class	% of total
Aroma compounds	42.5	Chelating agents	2.6
		Colors	2.1
Natural aroma substances	21	Chemical preservatives	1.8
Nutritional fortifiers	6.9	Stabilizers	1.8
		Antioxidants	1.7
Surface active agents (tensides)	5	Maturing and bleaching agents	1.4
Buffering substances,		Sweeteners	0.5
acids, bases	3.5	Other additives	9.4

[a] In 1965 a total of 1696 substances (= 100%) were utilized.

is made to legislated regulations or definitions provided therein. A compilation of the relative importance of various groups of additives is presented in Table 8.1.

8.2 Vitamins

Many food products are enriched or fortified with vitamins to adjust for processing losses or to increase the nutritive value. Such enrichment is important, particularly for fruit juices, canned vegetables, flour and bread, milk, margarine and infant food formulations. Table 8.2 provides an overview of vitamin enrichment of food.

Several vitamins have some desirable additional effects. Ascorbic acid is a dough improver, but can play a role similar to tocopherol as an antioxidant. Carotenoids and riboflavin are used as coloring pigments, while niacin improves the color stability of fresh and cured and pickled meat.

8.3 Amino Acids

The increase in the nutritive value of food by addition of essential amino acids and their derivatives is dealt with in sections 1.2.5 and 1.4.6.3.

Table 8.2. Examples of vitamin fortification of food

Vitamin	Food product
B_1	Cocoa powder and its products, beverages and concentrates, confectionary and other baked products
B_2	Baked products, beverages
B_6	Baked and pasta products
B_{12}	Beverages, etc.
Pantothenic acid	Baked products
Folic acid	Cereals (cf. 6.3.7.2)
C	Fruit drinks, desserts, dairy products, flour
A	Skim milk powder, breakfast cereals (flakes), beverage concentrates, margarine, baked products, etc.
D	Milk, milk powder, etc.
E	Various food products, e. g. margarine

8.4 Minerals

Food is usually an abundant source of minerals. Fortification is considered for iron, which is often not fully available, and for calcium, magnesium, copper and zinc. Iodization of salt is of importance in iodine deficient areas (cf. 22.2.4).

8.5 Aroma Substances

The use of aroma substances of natural or synthetic origin is of great importance (cf. Table 8.1). The aroma compounds are dealt with in detail in Chapter 5 and in individual sections covering some food commodities.

8.6 Flavor Enhancers

These are compounds that enhance the aroma of a food commodity, though they themselves have no distinct odor or taste in the concentrations used. An enhancer's effect is apparent to the senses as "feeling", "volume", "body" or "freshness" (particularly in thermally processed food) of the aroma, and also by the speed of aroma perception ("time factor potentiator").

8.6.1 Monosodium Glutamate (MSG)

Glutamic acid was isolated by *Ritthausen* (cf. 1.2.2.2). In 1908 *Ikeda* found that MSG is the beneficial active component of the algae *Laminaria japonica*, used for a long time in Japan as a flavor improver of soup and similarly prepared food. The consumption of MSG in 1978 was 200,000 tonnes worldwide.

The taste of MSG cannot be explained by a combination of sweet, salty, sour and bitter tastes. It is, as the fifth quality, of an elementary nature. This assumption, which was made as early as 1908 by a Japanese researcher to explain the special taste called *umami*, was confirmed by the identification of a taste receptor for MSG. The sixth quality of taste is "fatty" (cf. 3.1). Indeed, MSG is one of the most important taste-bearing substances in meat (cf. 12.9) and cheese ripened

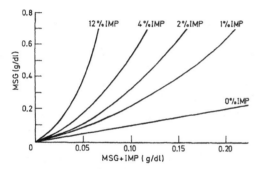

Fig. 8.1. Synergistic activities of Na-glutamate (MSG) and disodium-inosine monophosphate (IMP). The *curves* give the concentrations of MSG and MSG + IMP in water that are rated as being sensory equivalent by a taste panel

Synergistic flavor-enhancing effects are experienced with simultaneous use of MSG and IMP or GMP (Fig. 8.1). A mixture of MSG (59 mmol) and GMP (2.75 mmol) can replace 1230 mmol of MSG.

8.6.3 Maltol

Maltol (3-hydroxy-2-methyl-4-pyrone, cf. 5.3.1.2) has a caramel-like odor (melting point 162–164 °C). It enhances the perception of sweetness in carbohydrate-rich food (e. g. fruit juices, marmalades, fruit jelly). Addition of 5–75 ppm maltol allows a decrease of sugar content by about 15%, while retaining the sweetness intensity.

8.6.4 Compounds with a Cooling Effect

A cooling feeling in the mouth is produced by both fats (cf. 14.3.2.2.2), which melt on consumption, as well as low-molecular compounds which are capable of stimulating receptors for cold perception. Menthol is well known (cf. 5.3.2.4). Its threshold for the cooling effect is 9 µmol/kg of water. In comparison with the cooling effect, however, the retronasal threshold for the characteristic menthol odor is lower by a factor of 9.5, which is a disadvantage for the wider application of menthol.

In dark roasted malt, α-keto enamines with a cooling effect have been identified as products of the *Maillard* reaction. The odorless 5-methyl-4-(1-pyrrolidinyl)-3(2H)-furanone (Formula 8.1) was especially active. Its threshold for the cooling effect was 13.5 µmol/l, which is comparable with that of menthol. Studies on the relationship

for longer periods of time (cf. 10.3.5). Reports by Japanese researchers that glutamyl peptides, e. g., Glu-Glu, also taste like MSG have not been confirmed.

The taste of MSG is intensified by certain nucleotides (Fig. 8.1). Glutamate promotes sensory perception, particularly of meat-like aroma notes, and is frequently used as an additive in frozen, dehydrated or canned fish and meat products. MSG is added in the concentration range of 0.2–0.8%. The intake of larger amounts of MSG by some hypersensitive persons can trigger a "Chinese restaurant syndrome", which is characterized by temporary disorders such as drowsiness, headache, stomach ache and stiffening of joints. These disappear after a short time.

8.6.2 5'-Nucleotides

5'-Inosine monophosphate (IMP, disodium salt) and 5'-guanosine monophosphate (GMP, disodium salt) have properties similar to MSG but heightened by a factor of 10–20. Their flavor enhancing ability at 75–500 ppm is good in all food (e. g. soups, sauces, canned meat or tomato juice). However, some other specific effects, besides the "MSG effect", have been described for nucleotides. For example, they imprint a sensation of higher viscosity in liquid food. The sensation is often expressed as "freshness" or "naturalness", the expressions "body" and "mouthfeel" being more appropriate for soups.

$$\text{(8.1)}$$

between structure and effect showed that the substitution of a methylene group for the oxygen in the five-membered ring resulted in an increase in the threshold to 218 µmol/l, i. e., 16 fold. However, shifting the ring oxygen from position 4 to 5 resulted in an odourless, extremely cooling-intensive α-keto enamine (Formula 8.2) having a threshold of 0.24 µmol/l, which is 38 times lower than that of menthol.

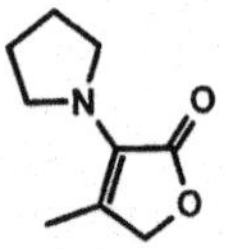

(8.2)

8.7 Sugar Substitutes

Sugar substitutes are those compounds that are used like sugars (sucrose, glucose) for sweetening, but are metabolized without the influence of insulin. Important sugar substitutes are the sugar alcohols, sorbitol, xylitol and mannitol and, to a certain extent, fructose (cf. 19.1.4.5–19.1.4.7).

8.8 Sweeteners

Sweeteners are natural or synthetic compounds which imprint a sweet sensation and possess no or negligible nutritional value ("nonnutritive sweeteners") in relation to the extent of sweetness.

There is considerable interest in new sweeteners. The rise in obesity in industrialized countries has established a trend for calorie-reduced nutrition. Also, there is an increased discussion about the safety of saccharin and cyclamate, the two sweeteners which were predominant for a long time. The search for new sweeteners is complicated by the fact that the relationship between chemical structure and sweetness perception is not yet satisfactorily resolved. In addition, the safety of suitable compounds has to be certain. Some other criteria must also be met, for example, the compound must be adequately soluble and stable over a wide pH and temperature range, have a clean sweet taste without side or post-flavor effects, and provide a sweetening effect as cost-effectively as does sucrose. At present, some new sweeteners are on the market (e. g., acesulfame and aspartame). The application of a number of other compounds will be discussed here.

The following sections describe several sweeteners, irrespective of whether they are approved, banned or are just being considered for future commercial use.

8.8.1 Sweet Taste: Structural Requirements

8.8.1.1 Structure–Activity Relationships in Sweet Compounds

A sweet taste can be derived from compounds with very different chemical structures. *Shallenberger* and *Acree* consider that for sweetness, a compound must contain a proton donor/acceptor system (AH_s/B_s-system), which has to meet some steric requirements and which can interact with a complementary receptor system (AH_r/B_r-system) by involvement of two hydrogen bridges (Fig. 8.2). The expanded model of *Kier* has an additional hydrophobic interaction with a group, X, present at a distinct position of the molecule (Fig. 8.3). The examples in Figs. 8.2 and 8.3 show that these models are applicable to many sweet compounds from highly different classes. An enlarged model substitutes a nucleophilic/electrophilic system (n_s/e_s system) for the AH_s/B_s system and an extended hydrophobic contact for the localized contact with group X.

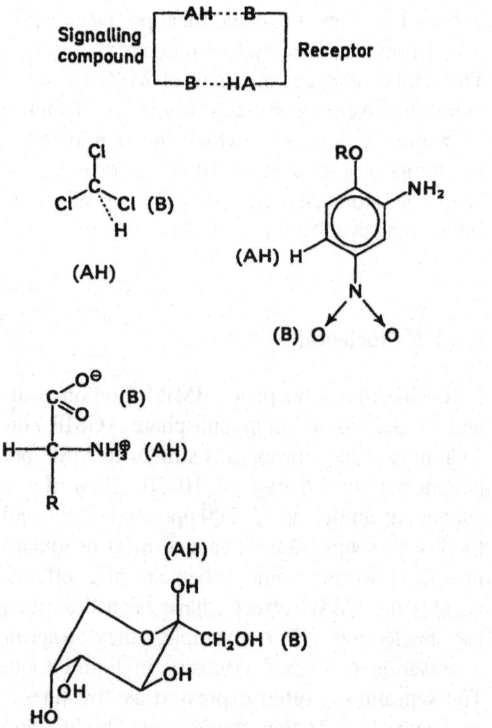

Fig. 8.2. AH/B-systems of various sweet compounds

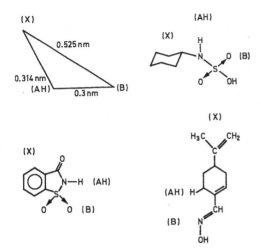

Fig. 8.3. AH/B/X-systems of various sweet compounds

Thus, a receptor for sweet compounds is to be depicted schematically as a hydrophobic pocket, containing a complementary n_r/e_r system.

It has been shown with numerous compounds that as the hydrophobicity and the space-filling properties of hydrophobic groups increase, the sweetening strength increases, passes through a maximum, and finally reaches a limit beyond which the sweet taste is either quenched or changes into a bitter taste.

According to *Nofre* and *Tinti*, even the AH/B/X system does not adequately explain the effect of hyperpotent sweeteners, e. g., guanidine (cf. 8.8.12.2). They propose a sweetness receptor which should make the large differences in the structure and sweetening strength understandable.

It is postulated that at least eight amino acid residues form the recognition sites B, AH, XH, G1, G2, G3, G4 and D in the sweetness receptor (Fig. 8.4 a). With the exception of D, two functional groups of an amino acid residue can interact with the sweet substance in each case through H-bridges, ionic relationships and *van der Waals* contacts. The last mentioned interactions involve G1–G4 (Fig. 8.4 a). The OH group of a serine or threonine residue located in the neighborhood of the phenyl ring of a phenylalanine residue is assumed for D. According to this theory, substances with weak sweetening strength, e. g., glucose (Fig. 8.4 b), make contact with only two or three amino acid residues. On

the other hand, sucrose makes contact with seven, but not with D (Fig. 8.4 c). A functional group, e. g., a CN group, which accepts a H-bridge involving D and the appropriate steric orientation towards the groups G1, G2 and G4 of the receptor are characteristic of hyperpotent sweeteners, e. g., lugduname (Fig. 8.4 d), which is 230,000 times sweeter than sucrose.

The sweetening strength of a compound can be measured numerically and expressed as:

- Threshold detection value, c_{tsv} (the lowest concentration of an aqueous solution that can still be perceived as being sweet).
- Relative sweetening strength of a substance X, related to a standard substance S, which is the quotient of the concentrations c (w/w per cent or mol/l) of isosweet solutions of S and X:

$$f(c_s) = \frac{c_s}{c_x} \quad \text{for} \rightarrow c_s \text{ isosweet } c_x \quad (8.3)$$

Saccharose in a 2.5 or 10% solution usually serves as the standard substance $(f_{sac, g})$. Since the sweetening strength is concentration dependent (cf. Fig. 8.5), the concentration of the reference solution must always be given $(f(c_s))$. When the sweetening strength of a substance is expressed as $f_{sac, g}(10) = 100$, this means, e. g., that the substance is 100 times sweeter than a 10% saccharose solution or a 0.1% solution of this substance is isosweet with a 10% saccharose solution.

8.8.1.2 Synergism

In mixtures of sweet tasting substances, synergistic intensification of taste occurs, i. e., the sweetness intensity is higher than the calculated value. An example is the intensification of sweetness in acesulfame–aspartame mixtures (Fig. 8.6).

8.8.2 Saccharin

Saccharin is an important sweetener $(f_{sac, g}(10) = 550)$ and is mostly used in the form of the water-soluble Na salt, which is not so sweet $(f_{sac, g}(10) = 450)$. At higher concentrations, this compound has a slightly metallic to bitter after-taste. The present stipulated ADI value

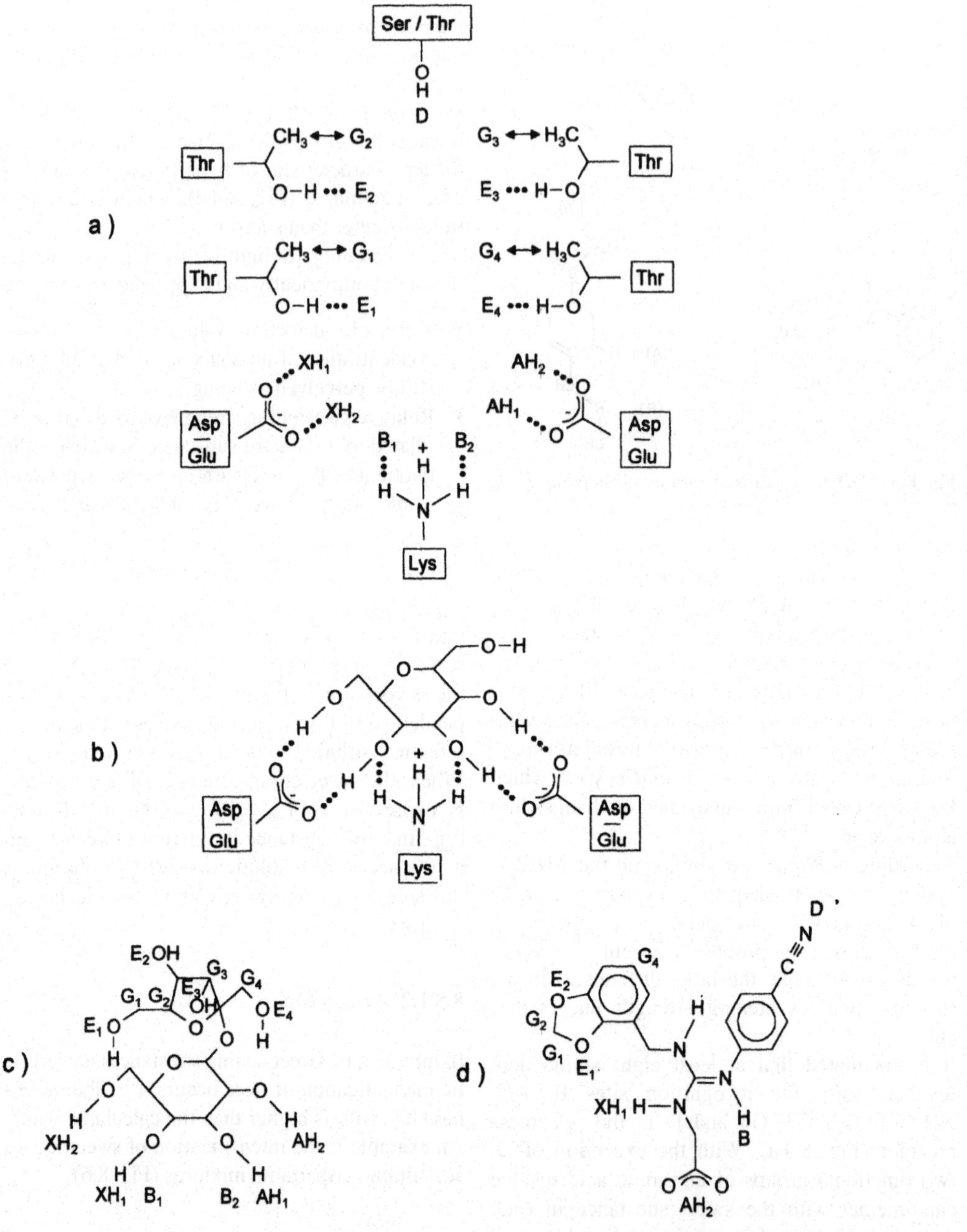

Fig. 8.4. Model of a sweetness receptor according to *Nofre* and *Tinti* (1996). **a)** Possible interactions of a sweet substance with the receptor. Interactions of the receptor with **b)** glucose, **c)** sucrose and **d)** lugduname

is 0–2.5 mg/kg of body weight. The synthesis of saccharin usually starts with toluene (*Remsen/Fahlberg* process, Formula 8.4) or sometimes with the methyl ester of anthranilic acid (*Maumee* process, Formula 8.5).

8.8.3 Cyclamate

Cyclamate is a widespread sweetener and is marketed as the Na- or Ca-salt of cyclohexane sulfamic acid. The sweetening strength is substan-

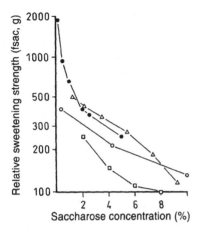

Fig. 8.5. Relative sweetening strength of some sweeteners as a function of the saccharose concentration (● neohesperidin dihydrochalcone, △ saccharin, ○ aspartame, ☐ acesulfame K) (according to *Bär* et al. 1990)

tially lower than that of saccharin and

$$\text{(8.4)}$$

is $f_{sac, g}(10) = 35$. It has no bitter after-taste. Overall, the sweet taste of cyclamate is not as pleasant as that of saccharin. The present stipulated ADI value of the acid is $0-11$ mg/kg of body weight. The synthesis of the compound is based on sulfonation of cyclohexylamine:

$$\text{(8.5)}$$

$$\text{(8.6)}$$

Table 8.3 shows several homologous compounds, illustrating the dependence of sweetness intensity on cycloalkyl ring size. The larger the ring size, the higher the sweetness, i. e. the lower the sweetness threshold value.

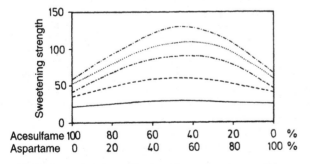

Fig. 8.6. Synergistic intensification of sweetness in acesulfame–aspartame mixtures (according to *v. Rymon Lipinski*, 1994) Ordinate: sweetening strength by comparison with a sucrose solution (g/l). Sweetener mixture: (—) 100 mg/l, (- - - - -) 200 mg/l, (—· ·—) 300 mg/l, (······) 400 mg/l, (—·—) 500 mg/l

Table 8.3. Taste threshold values of some alicyclic sulfamic acids (Na-salts)

R	c_{tsw} (mmol/l)	R	c_{tsw} (mmol/l)
Cyclobutyl	100	Cycloheptyl	0.5–0.7
Cyclopentyl	2–4	Cyclooctyl	0.5–0.8
Cyclohexyl	1–3		

8.8.4 Monellin

The pulp of *Dioscoreophyllum cumminsii* fruit contains monellin, a sweet protein with a molecular weight of 11.5 kdal. It consists of two peptide chains, A and B, which are not covalently bound. Their amino acid sequences are shown in Table 8.4.

The conformation is known (Fig. 8.7 and 8.8). As a result of cross reactions with an antiserum against thaumatin (cf. 8.8.5), sequence Y(13)ASD in a β-turn is regarded as the site of contact with the sweetness receptor. It corresponds to sequence Y(57)FD of thaumatin. The separated individual chains are not sweet. When the chains are recombined, a sweet taste is restored slowly, but the sweetness intensity of the original native protein is not reached. This strongly suggests that peptide chain separation results in irreversible conformational changes. However, combination of synthesized A and B chains gave a product with the same sweetening strength as natural monellin. The thermal stability of the protein was increased

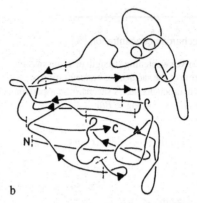

Fig. 8.7. A two dimensional representation of the conformation of the peptide chains of monellin (**a**) and thaumatin (**b**). (β-structure: →| ; α-helix: ∂; β-turn: ⊃; N, C, or N_A, N_B, C_A, C_B: N and C termini of the chains) (according to *Kim* et al., 1991)

by covalently bonding the two peptide chains via the amino acid residues A2 and B50 (cf. Fig. 8.9). For this purpose, a synthetic gene

Table 8.4. Amino acid sequences of the A and B chains of monellin. The sequence YASD shown in *bold type*, which is localized in a β-turn, is regarded as a part of the structure responsible for the cross reaction of monellin with antibodies against thaumatin as well as for making contact with the sweetness receptor (cf. Table 8.5 and Fig. 8.7 and 8.8)

					5					10					15					20
A-chain:	F[a]	R	E	I	K	G	Y	E	Y	Q	L	Y	V	**Y**	**A**	**S**	**D**	K	L	F	R
		A	D	I	S	E	D	Y	K	T	R	G	R	K	L	L	R	F	N	G	P
		V	P	P	P																

					5					10					15					20
B-chain	T[b]	G	E	W	E	I	I	D	I	G	P	F	T	Q	N	L	G	K	F	A	V
		D	E	E	N	K	I	G	Q	Y	G	R	L	T	F	N	K	V	I	R	P
		C	M	K	K	T	I	Y	E	N	E										

[a] Ca. 10% of the A chains also contain phenylalanine at the N-terminal (Phe-A-chain).
[b] Ca. 19% of the B chains also contain threonine at the N-terminal (Thr-B-chain) and N-terminal glycine is absent in ca. 24% (de-gly[1]-B-chain).

Fig. 8.8. A stereoscopic representation of the conformation of the peptide chains of monellin (*top*) and thaumatin (*bottom*) (the location of tryptic peptides that cross react with heterologous antibodies is indicated by *thicker lines*) (according to *Kim* et al., 1991)

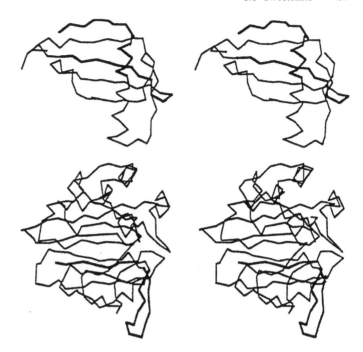

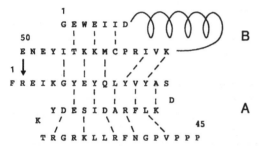

Fig. 8.9. Monellin: schematic representation of the A and B chains, showing the intra- and intermolecular hydrogen bonds (— —). Using genetic engineering techniques, the two chains were linked via a peptide bond (→) between the amino acids residues E (B50) and R (A2) (according to *Kim* et al., 1991)

was cloned and expressed in *E. coli* and yeast. The protein (I) thus obtained was as sweet as natural monellin (II). While the sweet taste of II was completely quenched at pH 2 by heating to 50 °C, I exhibited its full sweetness at room temperature even after being heated to 100 °C.

The threshold value is $f_{sac, g} = 3000$. Based on its low stability, slow triggering and slow fading away of taste perception, monellin probably will not succeed as a commercial sweetener.

8.8.5 Thaumatins

The fruit of *Thaumatococcus daniellii* contains two sweet proteins: thaumatin I and II, with $f_{sac, g} \sim 2000$. There are also low amounts of three other sweet proteins (thaumatin a, b and c). The complete amino acid sequence and the conformation (Fig. 8.7 and 8.8) of thaumatin I, a peptide chain with 207 amino acid residues, has been established (Table 8.5). As a result of cross reactions with an anti-serum against monellin (cf. 8.8.4), sequence Y(57)FD in a β-turn is regarded as the site of contact with the sweetness receptor. It corresponds to sequence Y(A13)ASD of monellin.

The sweet taste gets weaker with the increasing acetylation of the 11 ε-amino groups of the protein, being lost already with four acetyl groups. In contrast, up to 7 ε-amino groups could be modified by reductive methylation with HCHO/NaBH$_4$ without reduction in the taste intensity. The isoelectric point of the protein is obviously of importance for its activity. Thaumatin which is regarded as toxicologically safe is used, e. g., in chewing gum and milk products. Synergistic effects have been observed when thaumatin is used in combination with saccharin and acesulfame.

Table 8.5. Amino acid sequence of thaumatin I. (Disulfide bonds: 9–204, 55–66, 71–77, 121–193, 126–177, 134–145, 149–158, 159–164; the sequence YFD shown in *bold type*, which is localized in a β-turn, is regarded as a part of the structure responsible for the cross reaction of thaumatin with antibodies against monellin as well as for making contact with the sweetness receptor (cf. Table 8.4 and Fig. 8.7 and 8.8)

	5	10	15	20	25	30	35	40
	A T F E I	V N R C S	Y T V W A	A A S K G	D A A L D	A G G R Q	L N S G E	S W T I N
41	V E P G T	N G G K I	W A R T D	**C Y F D D**	S G S G I	C K T G D	C G G L L	R C K R F
81	G R P P T	T L A E F	S L N Q Y	G K D Y I	D I S N I	K G F N V	P M N F S	P T T R G
121	C R G V R	C A A D I	V G Q C P	A K L K A	P G G G C	N D A C T	V F Q T S	E Y C C T
161	T G K C G	P T E Y S	R F F K R	L C P D A	F S Y V L	D K P T T	V T C P G	S S N Y R
201	V T F C P	T A						

8.8.6 Curculin and Miraculin

Curculin is a sweet protein ($f_{sac, g}(6.8) = 550$) of known sequence (Table 8.6). It occurs in the fruit of *Curculigo latifolia*. The sweet taste induced by this protein disappears after a few minutes, only to reappear with the same intensity on rinsing with water. It is assumed that $Ca^{2\oplus}$ and/or $Mg^{2\oplus}$ ions in the saliva suppress the sweet taste. Rinsing with citric acid (0.02 mol/l) considerably enhances the impression of sweetness ($f_{sac, g}(12) = 970$). Thus, like miraculin, curculin acts as a taste modifier.

Miraculin is a glycoprotein present in the fruit of *Synsepalum dulcificum* (a tropical fruit known as miracle berry). Although it is tasteless, it has the property of giving sour solutions a sweet taste and therefore it is called a taste modifier. Thus, lemon juice seems sweet when the mouth is first rinsed with a solution of miraculin. The molecular weight of this taste modifier is 42–44 kdal.

8.8.7 *Gymnema silvestre* Extract

The extract from *Gymnema silvestre* is related to the taste modifier miraculin. It has the property of eliminating the ability to perceive sweetness for a few hours, without interfering with the perception of other taste qualities. The active substance has not yet been identified.

8.8.8 Stevioside

Leaves of *Stevia rebaudiana* contain approx. 6% stevioside ($f_{sac, g}(4) \sim 300$). Its structure is shown in Formula 8.7. This compound is of interest as a sweetener, however its toxic properties are unclear.

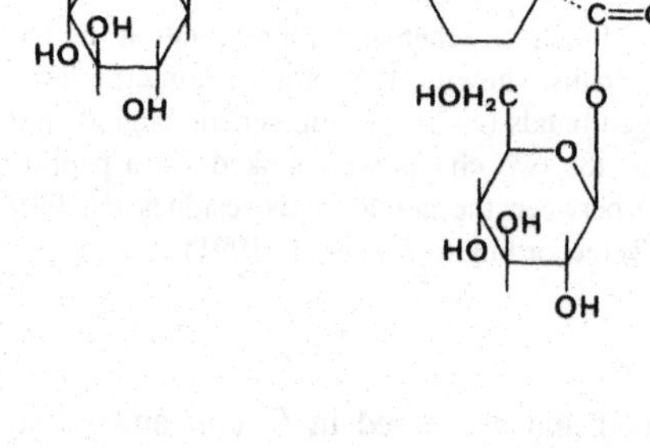

$$(8.7)$$

Table 8.6. Amino acid sequence of curculin

	5	10	15	20	25	30	35	40
	D N V L L	S G Q T L	H A D H S	L Q A G A	Y T L T I	Q N N C N	L V K Y Q	N G R Q I
41	W A S N T	D R R G S	G C R L T	L L S D G	N L V I Y	D H N N N	D V N G S	A C C G D
81	A G K Y A	L V L Q K	D G R F V	I Y G P V	L W S L G	P N G C R	R V N G	

8.8.9 Phyllodulcin

The leaves of *Hydrangea macrophylla* contain a 3,4-dihydroisocoumarin derivative, phyllodulcin (Formula 8.8). Its sweetness matches that of dihydrochalcones and of licorice root.

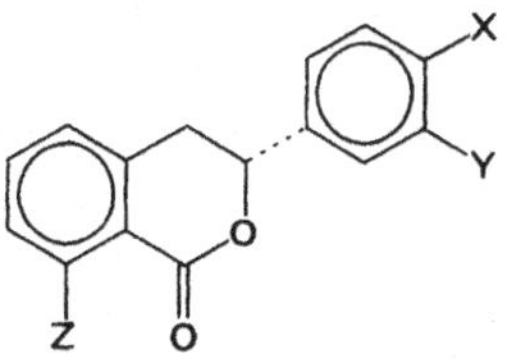

Z = OH, X = OMe, Y = OH   (8.8)

The taste perception builds relatively slowly and also fades away slowly. The sweetening strength is $f_{sac}(5) = 250$. A study of a number of related isocoumarin derivatives shows that taste quality and strength are very much dependent on the substitution pattern of the molecule (cf. Table 8.7).

8.8.10 Glycyrrhizin

The active substance from licorice root (*Glycyrrhiza glabra*) is a β, β'-glucuronido-glucuronide of glycyrrhetic acid:

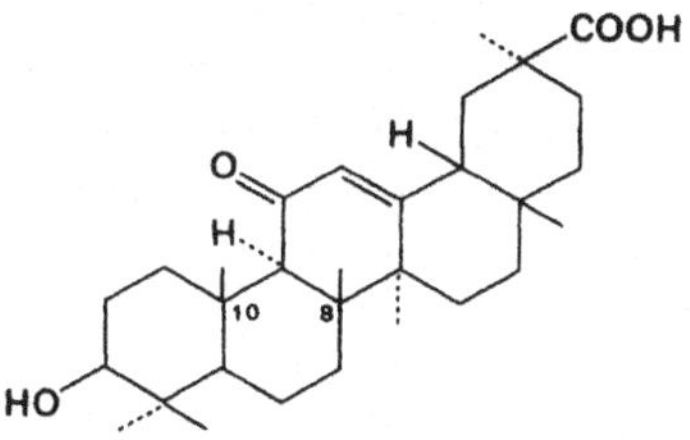

                                    (8.9)

The sweetening strength is $f_{sac, g}(4) = 50$. The compound is utilized for production of licorice (also spelled as liquorice). Its cortisone-like side effect limits its wide application.

8.8.11 Dihydrochalcones

Some dihydrochalcones are derived from flavanones (cf. 18.1.2.5.4) and have a relatively clean sweet taste that is slowly perceived but persists for some time. The sweetening strength of

Table 8.7. Sensory properties of some 2,3-dihydro-isocoumarins

Compound[a]			
X	Y	Z	Taste
OMe	OH	OH	very sweet
OMe	OMe	OH	bitter
OMe	OMe	OMe	no taste
OMe	OAc	OAc	slightly sweet
OH	OH	OH	no taste
OH	H	OH	no taste
OH	OH	H	no taste
OMe	OH	H	very sweet
OH	OMe	H	no taste

[a] Formula 8.8.

β-neohesperidin dihydrochalcone is $f_{sac, g} = 1100$ (threshold value) or $f_{sac, g}(10) = 667$ (R = β-neohesperidosyl in Formula 8.10).

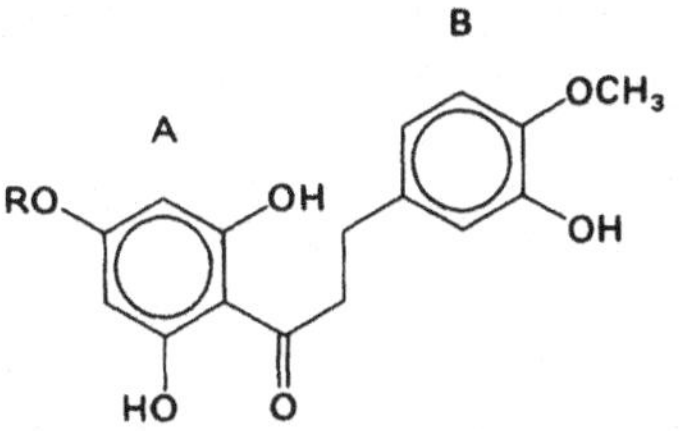

                                    (8.10)

In different countries, this compound is used in chewing gum, mouthwashes, beverages, and various types of candy. The quality and strength of the sweet taste of dihydrochalcone are related particularly to the substitution pattern in ring B. The prerequsite for a sweet taste is the presence in ring B of at least one hydroxy group, but not three adjacent hydroxy and alkoxy substituents.

8.8.12 Ureas and Guanidines

8.8.12.1 Suosan

Suosan, N-[(p-nitrophenyl)carbamoyl]-β-alanine (Formula 8.11), $f_{sac, g}(2) = 700$, is clearly sweeter than saccharin. The e/n system could be the NH/COO$^-$ system of β-alanine, which corresponds to the e/n system of aspartame (cf. 1.3.3 and 8.8.15). The p-cyanophenyl compound ($f_{sac, g}(2) = 450$), the N-glycine homolog and the thiocarbamoyl compound are also sweet.

$$O_2N-\langle\bigcirc\rangle-NH-CO-NH-CH_2-CH_2-COOH$$

$$(8.11)$$

8.8.12.2 Guanidines

Derivatives of guanidinoacetic acid (Formula 8.12) are among the sweetest compounds known until now (Table 8.8).

$$RN \overset{\displaystyle \overset{NHR^1}{|}}{\underset{}{\diagup}}{}^C\diagdown_{NHR^2}$$

$$(8.12)$$

Replacement of the carboxyl group by a tetrazole residue results in loss of sweetening strength (Table 8.8).

The guanidines can be synthesized, e. g., via the isothiocyanates:

$$R-N=C=S \xrightarrow{R^1NH_2} R-NH-\overset{\displaystyle \underset{\|}{S}}{C}-NHR^1$$

$$\xrightarrow{MeI} R-N=C-NHR^1$$
$$\underset{SMe}{|}$$

$$\xrightarrow{R^2NH_2} R-N=C-NHR^1$$
$$\underset{NHR^2}{|}$$

$$(8.13)$$

Table 8.8. Taste of some guanidines (Formula 8.12)

R	R^1	R^2	$f_{sac,\,g}(2)$
p-Cyanophenyl	H	Carboxy-methyl	2700
	Benzyl		30,000
	Phenylsulfonyl		45,000
	1-Naphthyl		60,000
	Cyclohexyl		12,000
	Cyclooctyl		170,000
	Cyclononyl		200,000
3,5-Dichloro-	Benzyl		80,000
phenyl	Cyclooctyl		60,000
p-Cyanophenyl	Cyclohexyl	Tetrazolyl-methyl	400[a]
	Cyclooctyl		5000[b]

[a] $f_{sac,\,g}(4)$.
[b] $f_{sac,\,g}(5)$.

8.8.13 Oximes

It has long been known that the anti-aldoxime of perillaldehyde (discovered in the essential oil of *Perilla nankinensis*) has an intensive sweet taste ($f_{sac,\,g} \sim 2000$). For its structure see Formula 8.14 (I).

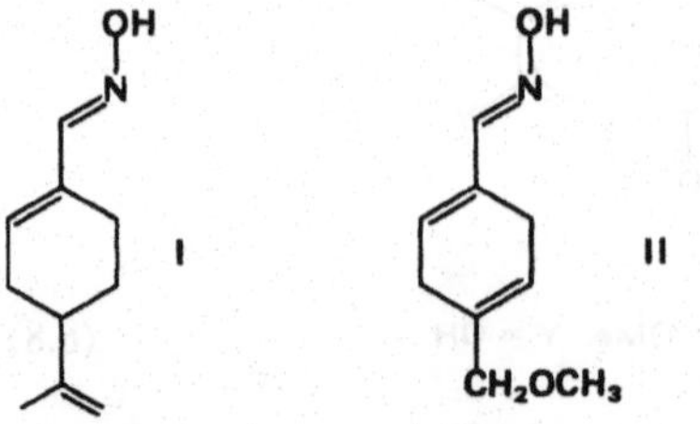

$$(8.14)$$

In the meantime, a related compound, (II), with improved solubility, has been reported, but its sweetness is just moderately high ($f_{sac,\,g} \sim 450$).

8.8.14 Oxathiazinone Dioxides

These compounds (Formula 8.15) belong to a class of sweeteners with an AH/B-system corresponding to that of saccharin. Based on their properties and present toxicological data, they are suitable for use. The sweetening strength depends on substituents R^1 and R^2 and is $f_{ac,\,g}(10) = 200$ for acesulfame K (Table 8.9).
The ADI value stipulated for the potassium salt of acesulfame is 0–9 mg/kg of body weight.

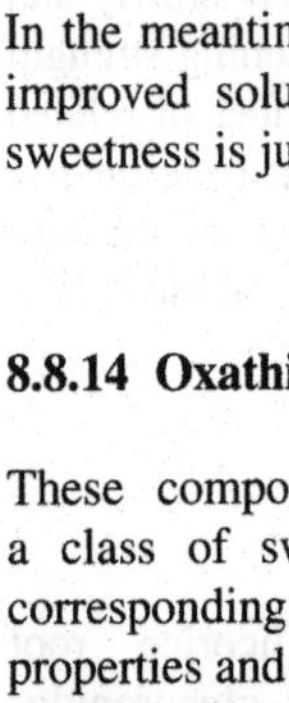

$$(8.15)$$

Oxathiazinone dioxides are obtained from fluorosulfonyl isocyanate and alkynes, or from com-

Table 8.9. Sweetness of some oxathiazinone dioxides (Na-salts) (Formula 8.15)

R^1	R^2	$f_{sac,\,g}$	R^1	R^2	$f_{sac,\,g}$
H	H	10	Et	H	20
H	Me	130[a]	Et	Me	250
Me	H	20	Pr	Me	30
Me	Me	130	i-Pr	Me	50
H	Et	150			

[a] Acesulfame.

pounds with active methylene groups, as exemplified by 1,3-diketones, 3-oxocarboxylic acids and 3-oxocarboxylic acid esters:

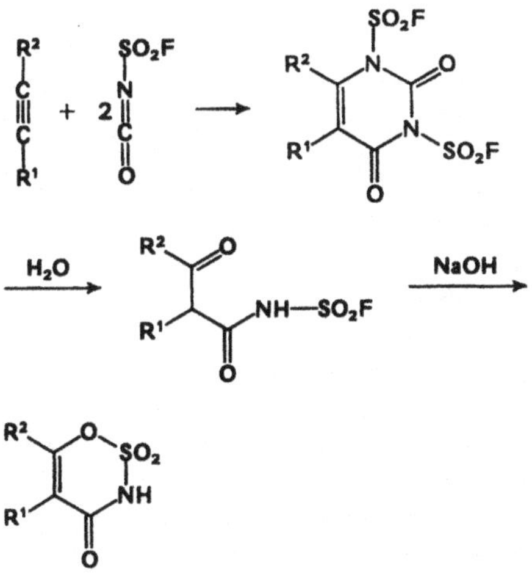

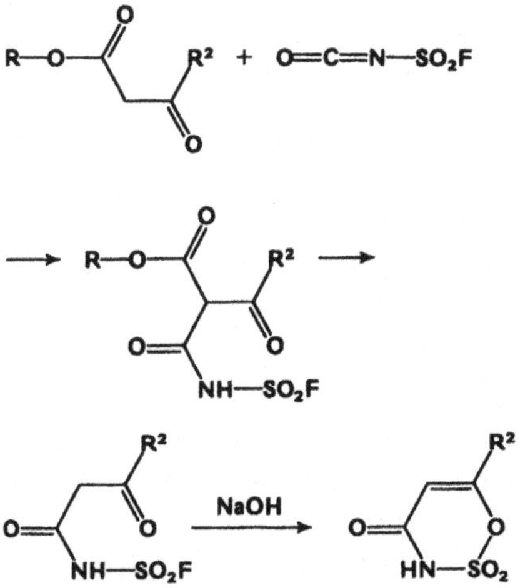

The sweetness of acesulfame is perceived quickly and this substance is practically stable in foods under the common processing and storage conditions. It is used in a large number of different products.

8.8.15 Dipeptide Esters and Amides

8.8.15.1 Aspartame

A dipeptide, L-aspartyl-L-phenylalanine methyl ester (L-Asp-L-Phe-OMe), has recently been

Table 8.10. Comparison of sweetening strengths of aspartame and saccharose (concentrations of isosweet aqueous solutions in %)

Saccharose	Aspartame	$f_{sac,g}$
0.34[a]	0.001[a]	340
4.3	0.02	215
10.0	0.075	133
15.0	0.15	100

[a] Threshold value.

approved for use as a sweetener in North America (aspartame, "NutraSweet"). It is as sweet as a number of other dipeptide esters of L-aspartic acid and D,L-aminomalonic acid. The relationship between chemical structure and taste among these compounds was outlined in more detail in 1.3.3. The sweetening strength relative to saccharose is concentration dependent (Table 8.10). Aspartame is used worldwide, though its stability is not always satisfactory. Unlike sweetening of drinks (coffee or tea) which are drunk immediately, problems arise in the use of aspartame in food which has to be heated or in sweetened drinks which have to be stored for longer periods of time. Possible degradation reactions involve α/β-rearrangement, hydrolysis

Table 8.11. Taste of some dipeptide amides of the type L-Asp-D-Ala-NHR

R	$f_{sac,g}(10)$
Cyclopentyl	50
Cyclohexyl	90
(2,2,5,5-tetramethyl)-cyclopentyl	800
(2,2,6,6-tetramethyl)-cyclohexyl	300
(Diethyl)-methyl	100
(Diisopropyl)-methyl	250
(Di-tert-butyl)-methyl	450
(Di-cyclopropyl)-methyl	1200
(Cyclopropyl)-(tert-butyl)-methyl	1200
(Cyclopropyl)-(methyl)-methyl	100
(2,2,4,4-Tetramethyl)-cyclobutyl	300
(2,2,4,4-Tetramethyl)-cyclobutan-3-onyl	240
(3-Hydroxy-2-2,2,4,4-tetramethyl)-cyclobutyl	125
3-(2,2,4,4-Tetramethyl)-thietanyl	2000[a]
3-(1-cis-Oxo-2,2,4,4-tetramethyl)-thietanyl	300
3-(1-trans-Oxo-2,2,4,4-tetramethyl)-thietanyl	350
3-(1,1-Dioxo-2,2,4,4-tetramethyl)-thietanyl	805

[a] Alitame.

into amino acid constituents and cyclization to the 2,5-dioxopiperazine derivative:

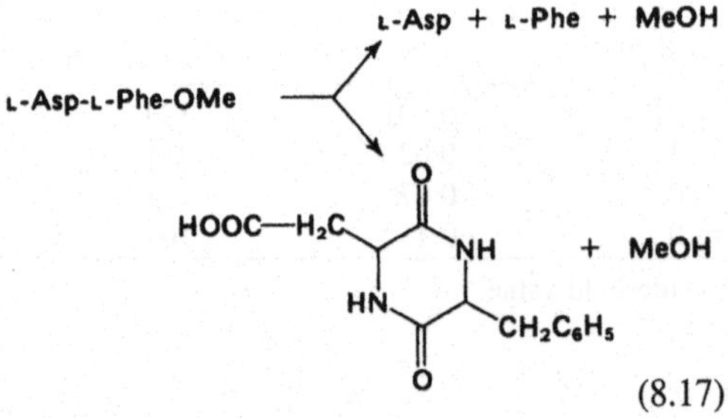

$$L\text{-Asp} + L\text{-Phe} + MeOH$$

L-Asp-L-Phe-OMe

HOOC—H$_2$C ... NH   + MeOH
HN ... CH$_2$C$_6$H$_5$

(8.17)

The ADI values stipulated for aspartame and diketopiperazine are 0–40 mg/kg of body weight and 0–7.5 mg/kg of body weight.
Aspartame synthesis on a large scale is achieved by the following reactions:

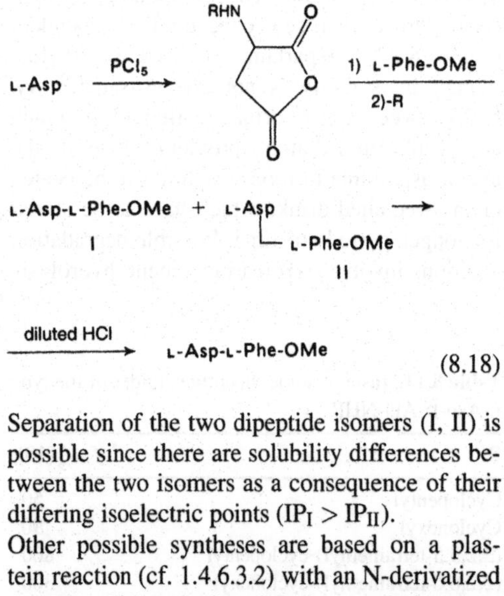

RHN

L-Asp $\xrightarrow{PCl_5}$ $\xrightarrow[\text{2)-R}]{\text{1) }L\text{-Phe-OMe}}$

L-Asp-L-Phe-OMe + L-Asp
I   L-Phe-OMe
                     II

$\xrightarrow{\text{diluted HCl}}$ L-Asp-L-Phe-OMe

(8.18)

Separation of the two dipeptide isomers (I, II) is possible since there are solubility differences between the two isomers as a consequence of their differing isoelectric points (IP$_I$ > IP$_{II}$).
Other possible syntheses are based on a plastein reaction (cf. 1.4.6.3.2) with an N-derivatized aspartic acid and phenylalanine methylester or on bacterial synthesis of an Asp–Phe polymer, achieved by genetic engineering techniques, enzymatic cleavage of the polymer to Asp–Phe, followed by acid or enzyme catalyzed esterification of the dipeptide with methanol.

8.8.15.2 Superaspartame

Substitution of a (p-cyanophenyl)carbamoyl residue for the free amino group of aspartame produces a compound called superaspartame (For-

mula 8.19), $f_{sac,g}(2) = 14{,}000$, which is sweeter than aspartame by about two powers of ten. This molecule contains structural elements of aspartame and of cyanosuosan.

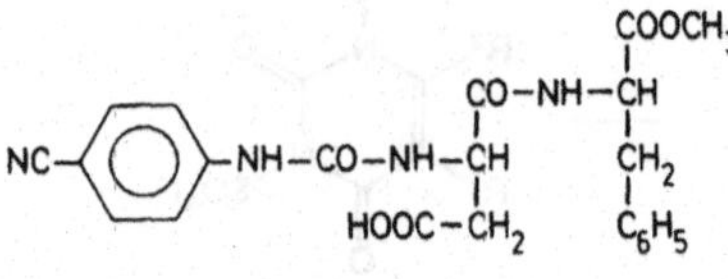

NC—⟨ ⟩—NH—CO—NH—CH   CH$_2$
HOOC—CH$_2$ C$_6$H$_5$
CO—NH—CH
COOCH$_3$

(8.19)

8.8.15.3 Alitame

Amides of dipeptides consisting of L-aspartic acid and D-alanine are sweet (Table 8.11). The compound alitame is the N-3-(2,2,4,4-tetra-methyl)-thietanylamide of L-Asp-D-Ala (Formula 8.20) and with $f_{sac,g}(10) = 2000$, it is a potential sweetener.

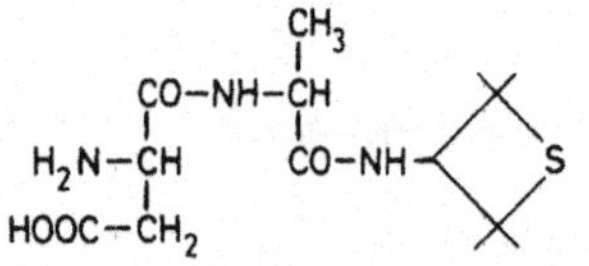

H$_2$N—CH   CO—NH—
HOOC—CH$_2$ CO—NH—CH
CH$_3$
S

(8.20)

Since the second amino acid has a D configuration, its side chain must be small, corresponding to the structure activity relationships discussed for dipeptide esters of the aspartame type (cf. 1.3.3). On the other hand, the carbonyl group should carry the largest possible hydrophobic residue.
The stability of dipeptide amides of the alitame type is substantially higher than that of dipeptide esters of the aspartame type. Therefore, alitame can also be used in bread and confectionery.
Like aspartame, alitame also undergoes α/β-rearrangement. Both isomers hydrolyze slowly to give L-aspartic acid and D-alanine amide, which is excreted either directly or as the glucuronide. A small part is oxidized to sulfoxides and sulfone. Cyclization to diketopiperazine which is typical of dipeptide methylesters does not occur.

8.8.16 Hernandulcin

(+)-Hernandulcin is a sweet sesquiterpene from *Lippia dulcis Trev.* (*Verbenaceae*), with the structure 6-(1,5-dimethyl-1-hydroxy-hex-4-enyl)

-3-methyl-cyclohex-2-enone:

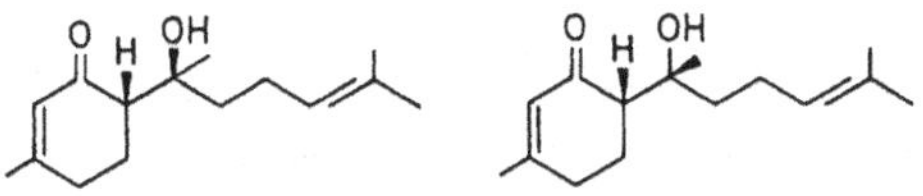

I   II   (8.21)

In comparison with sucrose, the sweetening strength of this compound is $f_{sac, mol}(0.25) = 1250$. Hernandulcin is somewhat less pleasant in taste than sucrose and exhibits some bitterness.

The racemic compound was synthesized via a directed aldol-condensation reaction by adding 6-methyl-5-hepten-2-one to a mixture of 3-methyl-2-cyclohexen-1-one and lithium di-isopropylamide in tetrahydrofuran, followed by chromatographic separation of (±)-hernandulcin (I, 95%) from the diastereomeric counterpart (±)-epihernandulcin (II, 5%). Whereas I is sweet, II exhibits no sweet taste.

The carbonyl and hydroxyl groups, which are located about 0.26 nm apart in the preferred conformation, are considered as an AH/B-system. The sweet taste is lost when these groups are modified (reduction of the carbonyl group to an alcohol, or acetylation of the hydroxy group).

8.9 Food Colors

A number of natural colors are available and used to adjust or correct food discoloration or color change during processing or storage. Carotenoids (cf. 3.8.4.5) are used the most, followed by red beet pigment and brown colored caramels. The number of approved synthetic dyes is low. Table 8.12 lists the pigments of importance in food coloring. Yellow and red colors are used the most. Food products which are often colored are confections, beverages, dessert powders, cereals, ice creams and dairy products.

8.10 Acids

The acid taste is caused only by the $H^\oplus$ ion. The intensity depends on the potential and not on the actual $H^\oplus$-ion concentration, which indicates the pH. Consequently, the solution of a weak acid,

which is not completely dissociated, tastes as sour as the solution of a strong acid of the same concentration. Therefore, the first step in the detection of an acid is comparable with an acid–base titration, the receptor for the sour taste functioning as the base.

Apart from the taste effect and antimicrobial activities, acids have a number of other functions in foods. The most important acids used in food processing and storage are outlined in this section.

8.10.1 Acetic Acid and Other Fatty Acids

Acetic, propionic and sorbic acids are dealt with under antimicrobial agents (8.10). Other short chain fatty acids, such as butyric and higher homologues, are used in aroma formulations.

8.10.2 Succinic Acid

The acid ($pK_1 = 4.19$; $pK_2 = 5.63$) is applied as a plasticizer in dough making. Succinic acid monoesters with glycerol are used as emulsifiers in the baking industry. The acid is synthesized by catalytic hydrogenation of fumaric or maleic acids.

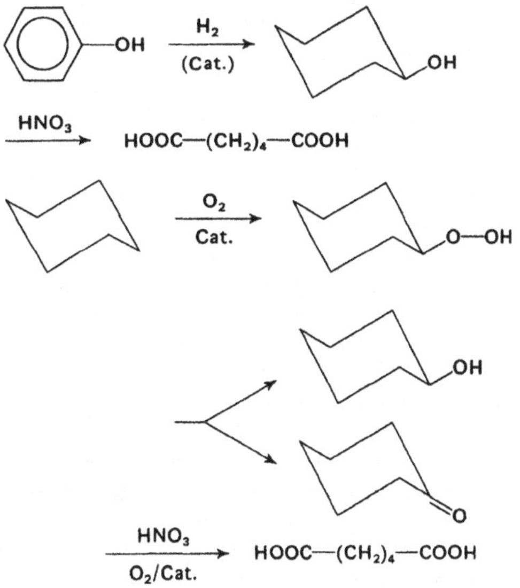

(8.22)

Table 8.12. Examples of food colorants (natural and synthetic)

Number	Name	FD & C (USA)	EU No.	Color	λ max (nm) (solvent)[b]	Formula[a]	Examples for utilization in food processing
1	Tartrazine	Yellow No 5	E 102	lemon-yellow (W)	426 (W)	I	Pudding powders, confectionary and candies, ice creams, pop (effervescent) drinks
2	Riboflavin		E 101	yellow (W)	445 (W)		Mayonnaise, soups, puddings, desserts, confectionary and candy products
3	Curcumin		E 100	yellow-red (E)	426 (E)	II	Mustard
4	Zeaxanthin			yellow (oil)	455–460 (CH)		Fat, hot and cold drinks, puddings, water
5	Sunset Yellow FCF	Yellow No 6	E 110	orange (W)	485 (W)	III	Beverages, fruit preserves, confectionary and candy products, honey-like products, sea salmon, crabs
6	β-Carotene		E 160a	orange (oil)	453–456 (CH)		Fat, beverages, soups, pudding, water, confectionary and candy products, yoghurt
7	Bixin		E 160b	orange (oil)	471/503 (CHCl$_3$)		Fat, mayonnaise
8	Lycopene		E 160d	orange (oil)	478 (H)		Mayonnaise, ketchup, sauces
9	Canthaxanthin	Food Orange 8	E 161g	orange (oil)	485 (CHCl$_3$)		Sea salmon, beverages, tomato products
10	Astaxanthin			orange (oil)	488 (CHCl$_3$)		Beverages, tomato products, confectionary and candy products
11	β-Apo-8'- carotenal		E 160e	orange (oil)	460–462 (CH)		Sauces, beverages, confectionary and candy products
12	Carmoisine		E 122	red with bluish tint (W)	516 (W)	IV	Beverages, confectionary and candy products, ice cream, pudding, fruit preserves
13	Amaranth	Red No 2	E 123	red with bluish tint (W)	520 (W)	V	Beverages, fruit preserves, confectionary and candy products, jams
14	Ponceau 4R		E 124	scarlet-red (W)	505 (W)	VI	Beverages, candy products (bonbons), sea salmon, cheese coatings

Table 8.12. (*Continued*)

Number	Name	FD & C (USA)	EU No.	Color	λ max (nm) (solvent)[b]	Formula[a]	Examples for utilization in food processing
15	Carmine		E 120	bright-red	518 (W ammonia solution)	VII	Alcoholic beverages
16	Anthocyanidin (from red grape pomace)		E 163a–f	red-violet[c] (W) (M + 0.01% HCl)	520–546		Jams, pop (effervescent) drinks
17	Erythrosine	Red No 3	E 127	cherry-red (W) red with bluish tint	527 (W) 532 (W)	VIII	Fruits, jams, confectionary and candy products
18	Red 2G					IX	Confectionary and candy products
19	Indigo Carmine (Indigotine)	Blue No 2	E 132	purple blue (W)	610 (W)	X	Also in combination with yellow colorant for confectionary and candy products and liqueurs
20	Patent Blue V		E 131	blue with a greenish tint (W)	638 (W)	XI	Mostly in combination with yellow colorants for confectionary and candy products, beverages
21	Brilliant blue FCF	Blue No 1		blue with a greenish tint (W)	630 (W)	XII	Mostly in combination with yellow colorants for confectionary and candy products, beverages
22	Chlorophyll		E 140	green	412 (CHCl$_3$)		Edible oils
23	Chlorophyllin copper complex		E 141	green (W)	405 (W)		Confectionary and candy products, liqueurs, jellies, cream food products
24	Green S (Brilliant Green BS)		E 142	green (W)	632 (W)	XIII	
25	Black BN		E 151	violet with bluish tint	570 (W)	XIV	Fish roe coloring, confectionary and candy products

[a] Formulas in Table 8.13;

[b] Solvent W: water, CH: cyclohexane, M: methanol, H: hexane, and E: ethanol;

[c] Color is pH dependent.

Table 8.13. Structures of the synthetic food colorants listed in Table 8.12

I

II

III

IV

V

VI

VII

VIII

IX

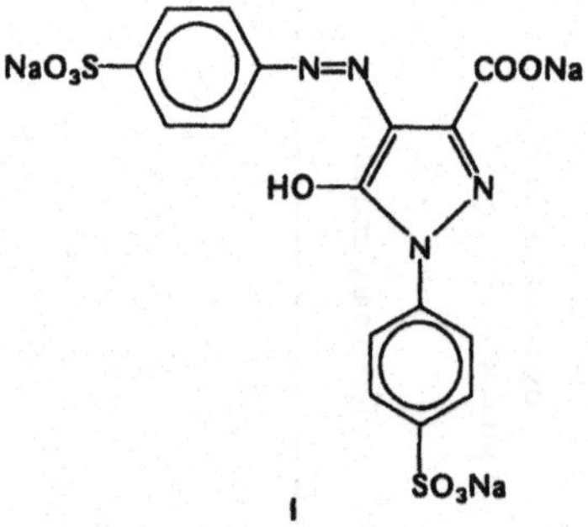

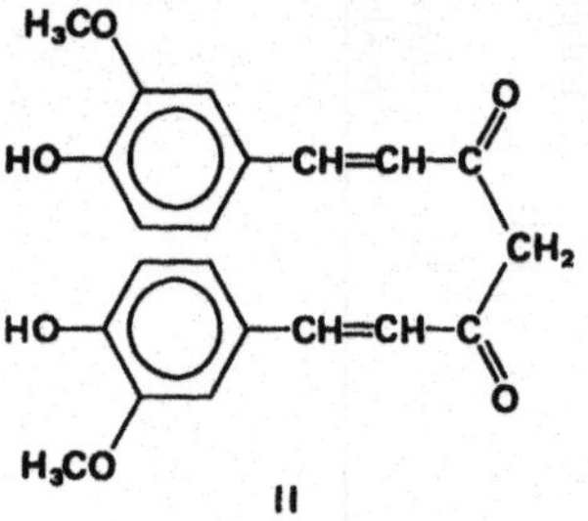

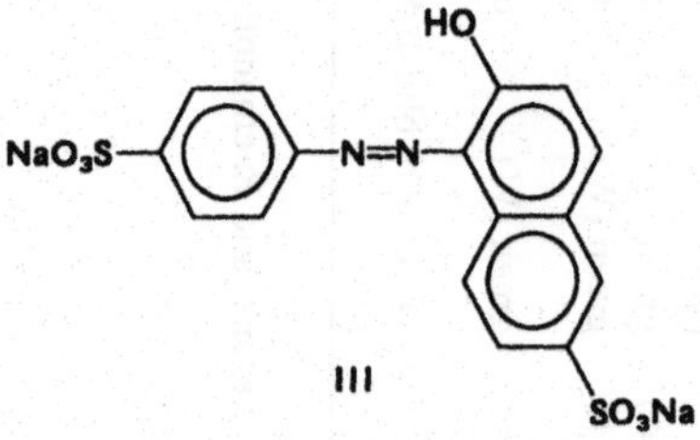

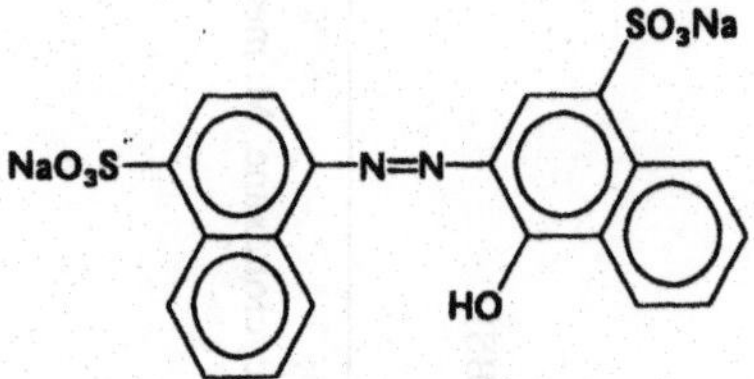

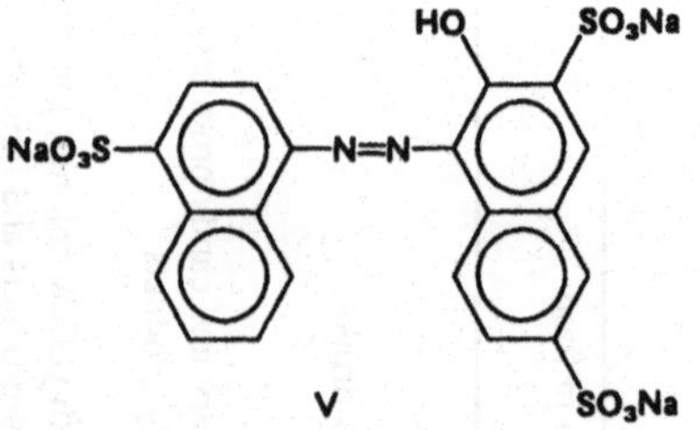

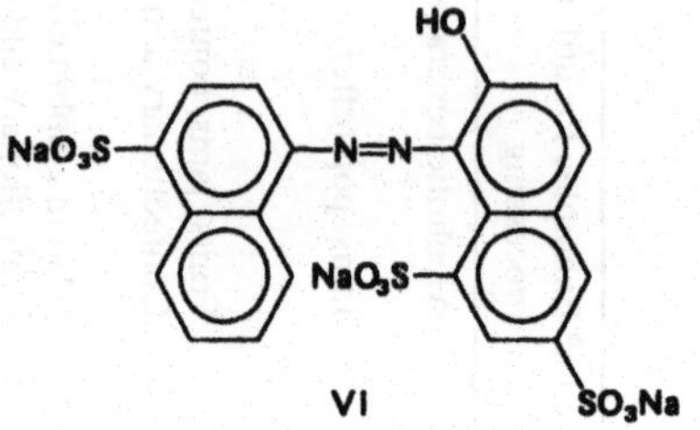

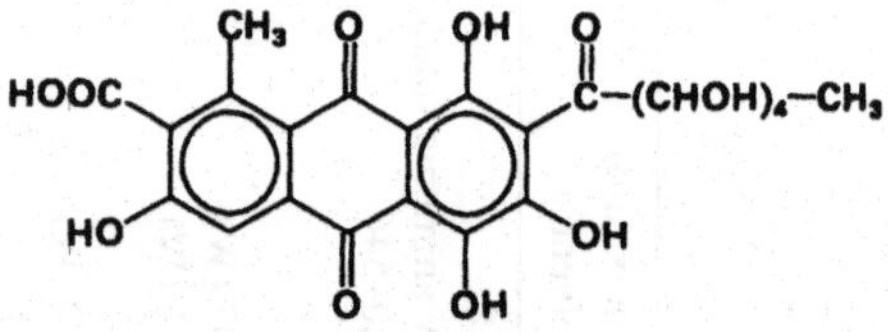

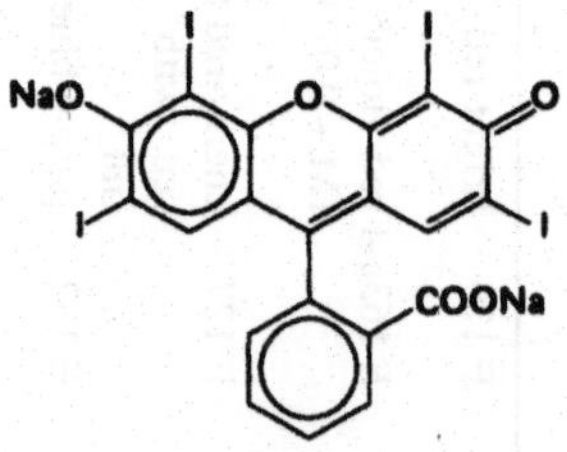

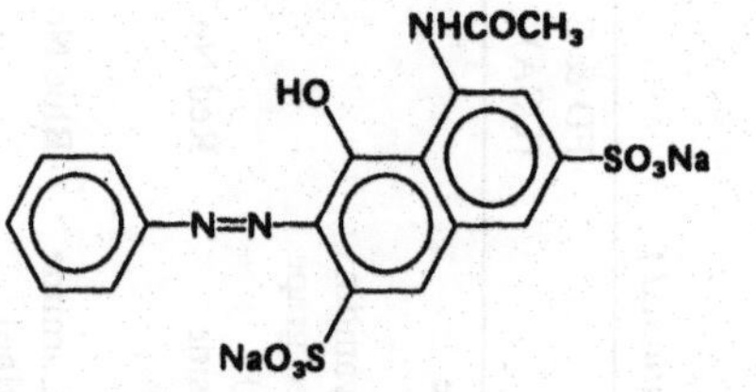

Table 8.13. (Continued)

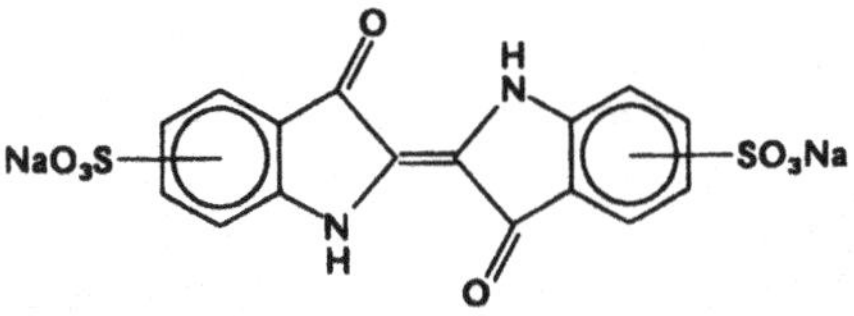

NaO₃S—[...structure X...]—SO₃Na

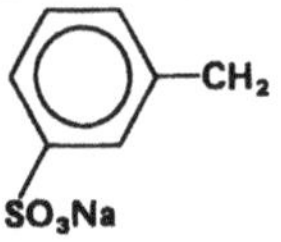

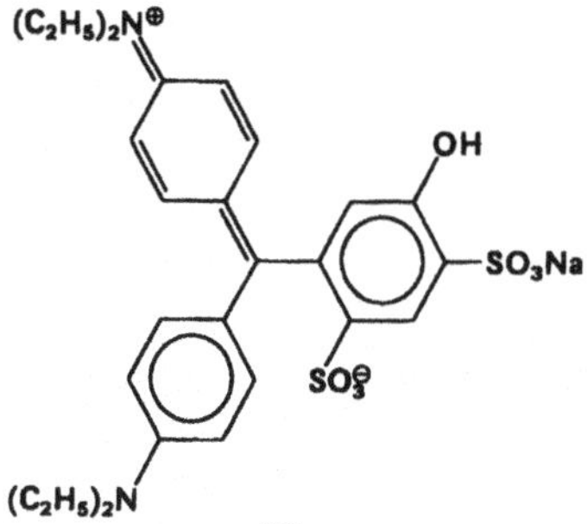

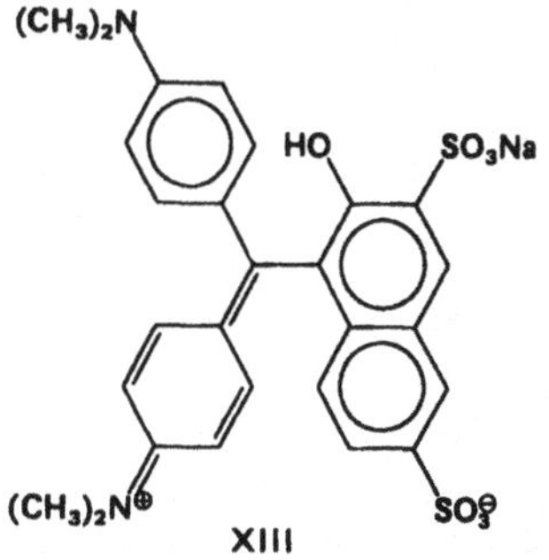

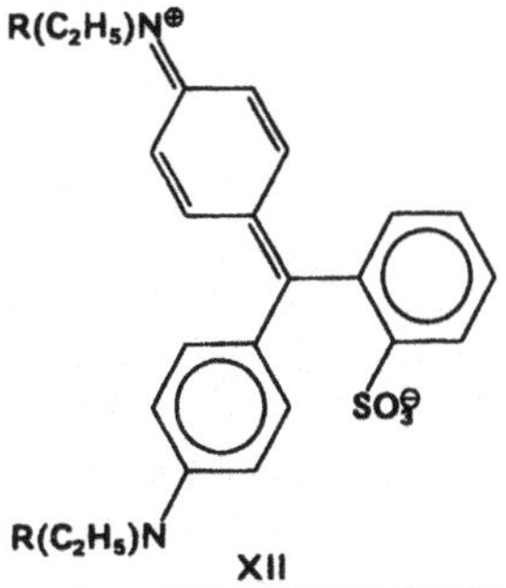

X

(C₂H₅)₂N⊕ [...structure XI...] OH, —SO₃Na, SO₃⊖, (C₂H₅)₂N

XI

R: [...]—CH₂, SO₃Na

(CH₃)₂N [...structure XIII...] HO, SO₃Na, (CH₃)₂N⊕, SO₃⊖

XIII

R(C₂H₅)N⊕ [...structure XII...] SO₃⊖, R(C₂H₅)N

XII

NaO₃S—[...]—N=N—[...]—N=N—[...] NHCOCH₃, HO, SO₃Na, NaO₃S, SO₃Na

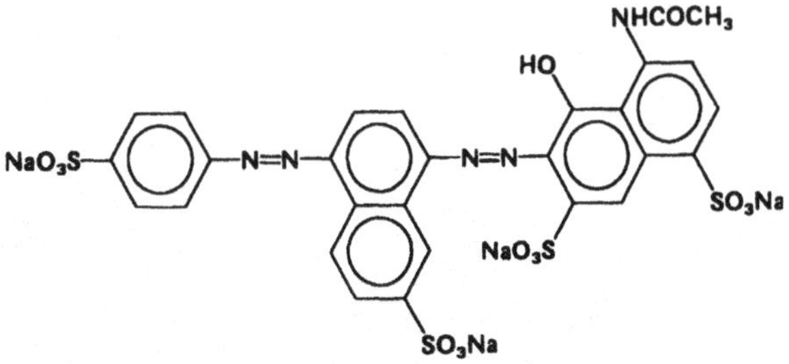

XIV

8.10.3 Succinic Acid Anhydride

This is the only acid anhydride used as a food additive. The hydrolysis proceeds slowly, hence the compound is suitable as an agent in baking powders and for binding of water in some dehydrated food products.

8.10.4 Adipic Acid

Adipic acid ($pK_1 = 4.43$; $pK_2 = 5.62$) is used in powdered fruit juice drinks, for improving the gelling properties of marmalades and fruit jellies, and for improving cheese texture. It is syn-

thesized from phenol or cyclohexane (cf. Reactions 8.22).

8.10.5 Fumaric Acid

Fumaric acid ($pK_1 = 3.00$; $pK_2 = 4.52$) increases the shelf life of some dehydrated food products (e. g. pudding and jelly powders). It is also used to lower the pH, usually together with food preservatives (e. g. benzoic acid), and as an additive promoting gel setting. Fumaric acid is synthesized via maleic acid anhydride which is obtained by catalytic oxidation of butene or benzene (cf. Reaction 8.23) or is produced from molasses by fermentation using *Rhizopus spp.*

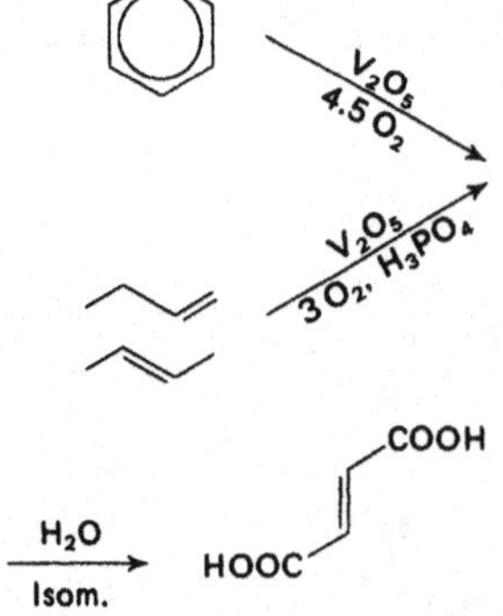

$$\xrightarrow[\text{Isom.}]{H_2O} \quad (8.23)$$

8.10.6 Lactic Acid

D,L- or L-lactic acid (pK = 3.86) is utilized as an 80% solution. A specific property of the acid is its formation of intermolecular esters, providing oligomers or a dimer lactide:

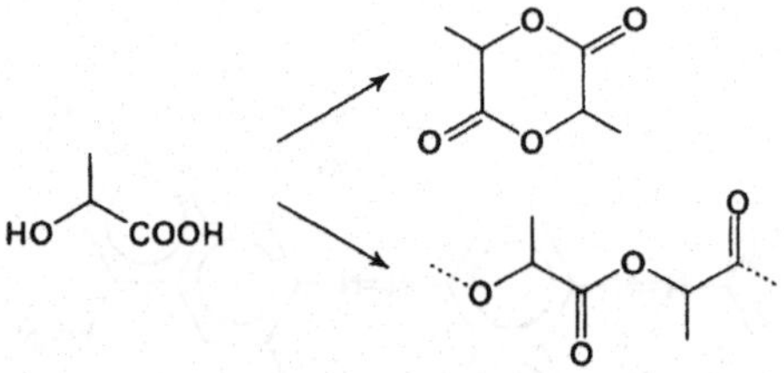

$$(8.24)$$

Such intermolecular esters are present in all lactic acid solutions with an acid concentration higher than 18%. More dilute solutions favor complete hydrolysis to lactic acid. The lactide can be utilized as an acid generator. Lactic acid is used for improving egg white whippability (pH adjustment to 4.8–5.1), flavor improvement of beverages and vinegarpickled vegetables, prevention of discoloration of fruits and vegetables and, in the form of calcium lactate, as an additive in milk powders.

Lactic acid production is based on synthesis starting from ethanal, leading to racemic D,L-lactic acid (Formula 8.25) or on homofermentation (*Lactobacillus delbruckii, L. bulgaricus, L. leichmannii*) of carbohydrate-containing raw material, which generally provides L-but also D,L-lactic acids under the conditions of fermentation.

$$CH_3CHO \xrightarrow{HCN} CH_3CHOHCN \longrightarrow Lactate$$
$$(8.25)$$

8.10.7 Malic Acid

Malic acid (pK$_1$ = 3.40; pK$_2$ = 5.05) is widely utilized in the manufacturing of marmalades, jellies and beverages and canning of fruits and vegetables (e. g. tomato). The monoesters with fatty alcohols are effective antispattering agents in cooking and frying fats and oils.

Malic acid synthesis, which provides the racemic D,L-form, is achieved by addition of water to maleic/fumaric acid. L-Malic acid can be synthesized enzymatically from fumaric acid with fumarase (*Lactobacillis brevis, Paracolobactrum spp.*) and from other C sources (paraffins) by fermentation with *Candida spp.*

8.10.8 Tartaric Acid

Tartaric acid (pK$_1$ = 2.98; pK$_2$ = 4.34) has a "rough", "hard" sour taste. It is used for the acidification of wine, in fruit juice drinks, sour candies, ice cream, and because of its formation of metal complexes, as a synergist for antioxidants. The production of (2R,3R)-tartaric acid is achieved from wine yeast, pomace, and cask tartar, which contain a mixture of potassium hydrogentartrate and calcium tartrate. This mixture is first converted entirely to calcium tartrate, from which tartaric acid is liberated by using sulfuric acid. Racemic tartaric acid is obtained by cis-epoxidation of maleic acid, followed by hydrolysis:

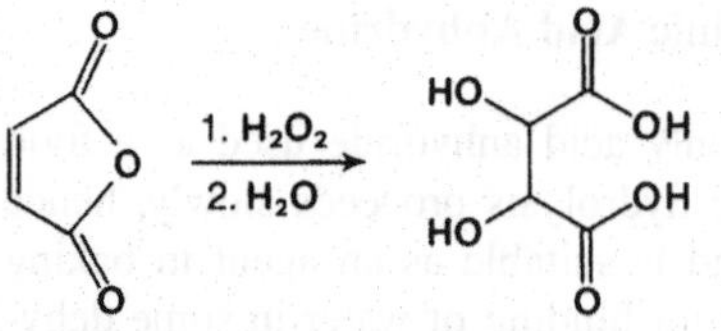

$$(8.26)$$

8.10.9 Citric Acid

Citric acid (pK$_1$ = 3.09; pK$_2$ = 4.74; pK$_3$ = 5.41) is utilized in candy production, fruit juice, ice cream, marmalade and jelly manufacturing, in vegetable canning and in dairy products such as processed cheese and buttermilk (aroma improver). It is also used to suppress browning in fruits and vegetables and as a synergetic compound for antioxidants. Its production is based on microbial fermentation of molasses by *Aspergillus niger*. The yield of citric acid is 50–70% of the fermentable sugar content.

8.10.10 Phosphoric Acid

Phosphoric acid ($pK_1 = 2.15$; $pK_2 = 7.1$; $pK_3 \sim 12.4$) and its salts account for 25% of all the acids used in food industries. The bulk of the acids (salts) used in the industry is citric acid (about 60%), while the use of other acids accounts for only 15%. The main field of use of phosphoric acid is the soft drink industry (cola drinks). It is also used in fruit jellies, processed cheese and baking powder and as an active buffering agent or pH-adjusting ingredient in fermentation processes. Acid salts, e.g., $Ca(H_2PO_4)_2 \cdot H_2O$ (fast activity), NaH_{14}-Al_3 $(PO_4)_8 \cdot 4 H_2O$ (slow activity) and Na_2H_2-P_2O_7 (slow activity) are used in baking powders as components of the reaction to slowly or rapidly release the CO_2 from $NaHCO_3$.

8.10.11 Hydrochloric and Sulfuric Acids

Both acids are used in starch and sucrose hydrolyses. Hydrochloric acid is also used in protein hydrolysis in industrial production of seasonings.

8.10.12 Gluconic Acid and Glucono-δ-lactone

Gluconic acid is obtained by the oxidation of glucose, which proceeds either by metal catalysis or enzymatically (*Aspergillus niger, Guconobacter suboxidans*).
Gluconic acid is used, e.g., in the production of invert sugar, beverages, and candies. The δ-lactone is produced by evaporating a gluconic acid solution at 35–60 °C. Glucono-δ-lactone slowly hydrolyzes, releasing protons. Hence, it is applied as an additive when slow acidification is needed, as with baking powder, raw sausage ripening and several sour milk products.

8.11 Bases

NaOH and a number of alkaline salts, such as $NaHCO_3$, Na_{2CO_3}, $MgCO_3$, MgO, $Ca(OH)_2$, Na_{2HPO_4} and Na-citrate, are used in food processing for various purposes, for example:
Ripe olives are treated with 0.25–2% NaOH to eliminate the bitter flavor and to develop the desired dark fruit color.
In alkali-baked goods (bread and cakes that keep) molded dough pieces are dipped into 1.25% NaOH at 85–88 °C or, in the case of larger fresh alkali-baked goods, into 3.5% NaOH at room temperature in the baking process in order to form the typical smooth, light to dark brown surface.
In chocolate manufacturing, $NaHCO_3$ enhances the *Maillard* reaction, providing dark bitter chocolates.
In production of molten processed cheese, the pH rise needed to improve the swelling of casein gels is achieved by addition of alkali salts.

8.12 Antimicrobial Agents

Elimination of microflora by physical methods is not always possible, therefore, antimicrobial agents are needed. The spectrum of compounds used for this purpose has hardly changed for a long time. It is not easy to find new compounds with wider biological activity, negligible toxicity for mammals and acceptable cost.
In the use of weak acids as preservatives, their pK value and the pH value of the food are very important for the application because only the undissociated molecule can penetrate into the inside of the microbial cell. Accordingly, weak acids are suitable preferably for acidic foods.

8.12.1 Benzoic Acid

Benzoic acid activity is directed both to cell walls and to inhibition of citrate cycle enzymes (α-ketoglutaric acid dehydrogenase, succinic acid dehydrogenase) and of enzymes involved in oxidative phosphorylation.
The acid is used in its alkali salt form as an additive, since solubility of the free acid is unsatisfactorily low. However, the undissociated acid

(pK$_a$ = 4.19) is predominantly active. As in the case of sorbic acid and propionic acid, a certain activity is also attributed to the anion.

Benzoic acid usually occurs in nature as a glycoside (in cranberry, bilberry, plum and cinnamon trees and cloves). Its activity is primarily against yeasts and molds, less so against bacteria. Figures 8.10 and 8.12 show the pH-dependent activity of the acid against *Escherichia coli, Staphylococcus aureus* and *Aspergillus niger*.

The LD$_{50}$ (rats; orally) is 1.7–3.7 g/kg body weight; the LD$_{100}$ (guinea pig, cat, dog, rabbit; orally) is 1.4–2 g/kg. A daily intake of <0.5 g Na-benzoate is tolerable for humans. No dangerous accumulation of the acid occurs in the body even at a dosage of as much as 4 g/day. It is eliminated by excretion in the urine as hippuric acid while, at higher levels of intake, the glucuronic acid derivative is also excreted.

Benzoic acid (0.05–0.1%) is often used in combination with other preservatives and, on the basis of its higher activity at acidic pH's, it is used for preservation of sour food (pH 4–4.5 or lower), beverages with carbon dioxide, fruit salads, marmalades, jellies, fish preserves, margarine, paste (pâté) fillings and pickled sour vegetables. A change in aroma, occurring mostly in fruit products, may result as a consequence of benzoic acid esterification.

8.12.2 PHB-Esters

The alkyl esters of p-hydroxybenzoic acid (PHB; parabens) are quite stable. Their solubility in water decreases with increasing alkyl chain length (methyl → butyl). The esters are mostly soluble in 5% NaOH.

The esters are primarily antifungal agents and are also active against yeasts but less so against bacteria, especially those which are gram-negative. The activity rises with increasing alkyl chain length (Fig. 8.11). Nevertheless, lower members of the homologous series are preferred because of better solubility.

The LD$_{50}$ (mice; orally) is >8 g/kg body weight. In a feeding experiment over 96 weeks using 2% PHB-ester, no weight decrease was observed, while a slight decrease was found at the 8% level. In humans, the compounds are excreted in urine as p-hydroxybenzoic acid or its glycine or glucuronic acid conjugates.

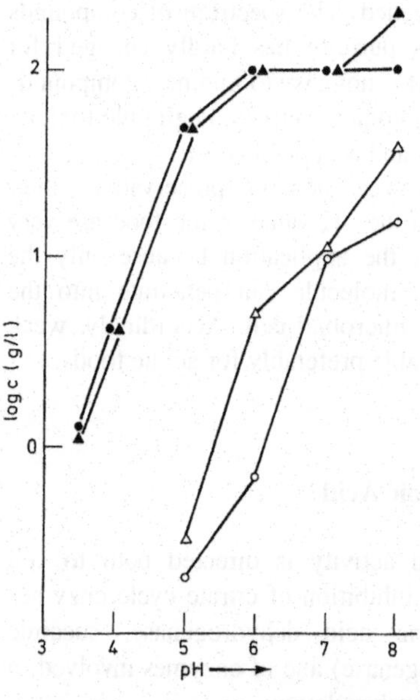

Fig. 8.10. The effect of benzoic acid on *Escherichia coli* (○ bacteriostatic. ● bactericidal activity) and *Staphylococcus aureus* (△ bacteriostatic and ▲ bactericidal activity)

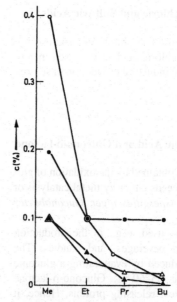

Fig. 8.11. Inhibition of *Salmonella typhosa* (●), *Aspergillus niger* (△), *Staphylococcus aureus* (○), and *Saccharomyces cerevisiae* (▲) by PHB-esters

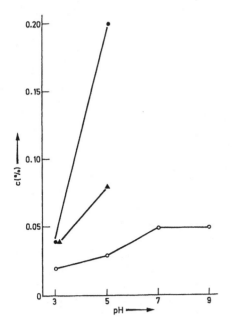

Fig. 8.12. Growth inhibition of *Aspergillus niger* by benzoic acid (•), p-hydroxybenzoic acid propyl ester (○) and sorbic acid (▲)

8.12.3 Sorbic Acid

The antimycotic effect of straight chain carboxylic acids has long been known. In particular the unsaturated acids, for example crotonic acid and its homologues, are very active. Sorbic acid (2-trans, 4-trans-hexadienoic acid; pK = 4.76) has the advantage that it is odorless and tasteless at the levels of use (0.3% or less). The acid is obtained by several syntheses:

- From parasorbic acid [(S)-2-hexen-5-olide); cf. Reaction 8.27]. The acid is present in berries of the mountain ash tree (*Sorbus aucuparia*).

$$\text{(parasorbic acid)} \xrightarrow[\text{EtOH}]{\text{HCl in}} \text{(hexadienoate)} COOEt$$

$$\xrightarrow{H^{\oplus}} \text{(sorbic acid)} COOH \tag{8.27}$$

- From ethanal:

$$3\,CH_3CHO \xrightarrow[\text{acetate}]{\text{Piperidine}} \text{(dienal)} CHO + 2\,H_2O$$

$$\xrightarrow[\text{Ag/O}_2/30\,°C]{\text{pH}>12.5} \text{(dienoic acid)} COOH \tag{8.28}$$

- From crotonaldehyde obtained from ethanal (cf. Reaction 8.29).

Unlike benzoic acid, the esters can be used over a wide pH range since their activity is almost independent of pH (cf. Fig. 8.12). As additives, they are applied at 0.3–0.06% as aqueous alkali solutions or as ethanol or propylene glycol solutions in fillings for baked goods, fruit juices, marmalades, syrups, preserves, olives and pickled sour vegetables.

$$2\,CH_3CHO \longrightarrow \overset{OH}{\underset{}{\text{...}}}CHO \xrightarrow{-H_2O} \text{...}CHO$$

$$CH_2CO$$

a: BF₃, T<0°C   b: Diethylene glycol, Zn-Isobutyrate, T>25°C   (8.29)

The third synthesis is the most important.

The microbial activity of sorbic acid is primarily against fungi and yeasts, less so against bacteria. The activity is pH dependent (Fig. 8.12). Its utilization is possible up to pH 6.5, the proportion of undissociated acid being still 1.8%.

The LD_{50} (rats) of sorbic acid is ca. 10 g/kg body weight. Feeding experiments with rats for more than 90 days, with 1–8% sorbic acid in the diet, had no effect, while only 60% of the animals survive an 8% level of benzoic acid.

Sorbic acid is degraded biochemically like a fatty acid, i. e. by a β-oxidation mechanism. A small portion of the acid is degraded by ω-oxidation, yielding trans, trans-muconic acid (cf. Reaction 8.30).

$$\text{(8.30)}$$

Some microorganisms, such as *Penicillium roqueforti*, have the ability to decarboxylate sorbic acid and thus convert it into 1,3-pentadiene, which has no antimicrobial activity and in addition may contribute to an off-flavor in cheeses:

$$\text{(8.31)}$$

Sorbic acid or its salts are effective antifungal agents in baked products, cheeses, beverages (fruit juices, wines), marmalades, jellies, dried fruits and in margarine.

8.12.4 Propionic Acid

Propionic acid is found in nature where propionic acid fermetnation occurs, e. g., in Emmental cheese, in which it is present up to 1%.

Its antimicrobial activity is mostly against molds, less so against bacteria. Propionic acid has practically no effect against yeast. Its activity is pH dependent. It is recommended and used up to pH 5 and only occasionally up to pH 6.

Propionic acid is practically nontoxic. It is used as an additive in baked products for inhibition of molds, and to prevent ropiness caused by the action of *Bacillus mesentericus*. It is added to flour at 0.1–0.2% as its Ca-salt and is used in cheese manufacturing by dipping the cheese into an 8% solution of the acid.

8.12.5 Acetic Acid

The preserving activity of vinegar (cf. 22.3) has been known from ancient times. The acid has a two-fold importance: as a preservative and as a seasoning agent. It is more active against yeasts and bacteria than against molds. It is used as the free acid, Na- and Ca-salts, or as Na-diacetate ($CH_3COOH \cdot CH_3COONa \cdot 1/2H_2O$), in ketchup, mayonnaise, acid-pickled vegetables, bread and other baked products.

8.12.6 SO$_2$ and Sulfite

The activity of these preserving agents covers yeasts, molds and bacteria. The activity increases with decreasing pH and is mostly derived from undissociated sulfurous acid, which predominates at a pH < 3.

Toxicity is negligible at the levels usually applied. Possible mutagenic activity is under investigation. Excretion in the urine occurs as sulfate.

Sulfite reacts with a series of food constituents, e. g., proteins with cleavage of disulfide bonds (cf. 1.4.4.4), with various cofactors like $NAD^{\oplus}$, folic acid, pyridoxal, and thiamine (cf. 6.3.1.3) and with ubiquinone:

$$\text{(8.32)}$$

The pyrimidines in nucleic acids can also react, e. g., cytosine and uracil (cf. Formula 8.33). Anthocyanins are bleached (cf 18.1.2.5.3).

$$\text{(8.33)}$$

SO$_2$ is used in the production of dehydrated fruits and vegetables, fruit juices, syrups, concentrates or purée. The form of application is SO_2, Na_2SO_3, K_2SO_3, $NaHSO_3$, $Na_2S_2O_5$ and $K_2S_2O_5$ at levels of 200 ppm or less.

SO_2 is added in the course of wine making prior to must fermentation to eliminate interfering microorganisms. During wine fermentation with selected pure yeast cultures, SO_2 is used at a level of 50–100 ppm, while 50–75 ppm are used for wine storage.

SO_2 is not only antimicrobially active, but inhibits discoloration by blocking compounds with a reactive carbonyl group (*Maillard* reaction; nonenzymatic browning) or by inhibiting oxidation of phenols by phenol oxidase enzymes (enzymatic browning).

8.12.7 Diethyl (Dimethyl) Pyrocarbonate

Diethyl pyrocarbonate (DEPC or diethyl dicarbonate) is a colorless liquid of fruit-like or ester odor. Its antimicrobial activity covers yeasts (10–100 ppm), bacteria (*Lactobacilli*: 100–170 ppm) and molds (300–800 ppm). The levels of the compound required for a clear inhibition are given in brackets. Diethyl pyrocarbonate readily hydrolyzes to yield carbon dioxide and ethanol:

$$\longrightarrow 2\,C_2H_5OH + 2\,CO_2 \qquad (8.34)$$

or it reacts with food ingredients. In alcoholic beverages it yields a small amount of diethyl carbonate:

$$\xrightarrow{C_2H_5OH} \qquad + CO_2 + C_2H_5OH \qquad (8.35)$$

In the presence of ammonium salts, DEPC can form ethyl urethane in a pH-dependent reaction:

$$+ \ NH_3$$

$$\longrightarrow \qquad + CO_2 + C_2H_5OH \qquad (8.36)$$

Since diethyl carbonate may be a teratogenic agent and ethyl urethane is a carcinogen, the use of diethyl pyrocarbonate is discussed under toxicological aspects. The compound should be replaced by dimethyl pyrocarbonate, since methyl urethane, unlike ethyl urethane, is not carcinogenic.

Both compounds are used in cold pasteurization of fruit juices, wine and beer at a concentration of 120–300 ppm.

8.12.8 Ethylene Oxide, Propylene Oxide

These compounds are active against all microorganisms, particularly vegetative cells and spores, and also against viruses. Propylene oxide is somewhat less reactive than ethylene oxide.

Since they are efficient alkylating agents, the pure compounds are very toxic. After application, all the residual amounts must be completely removed. The glycols resulting from their hydrolysis are not as toxic (ethylene glycol: LD_{50} for rats is 8.3 g/kg body weight). Toxic reaction products can be formed, as exemplified by chlorohydrin obtained in the presence of chloride:

$$\xrightarrow{Cl^{\ominus}} \qquad (8.37)$$

In addition, some essential food constituents react with formation of biologically inactive derivatives. Examples are riboflavin, pyridoxine, niacin, folic acid, histidine or methionine. However, these reactions are not of importance under the conditions of the normal application of ethylene oxide or propylene oxide.

Both compounds are used as gaseous sterilants (ethylene oxide, boiling point 10.7 °C; propylene oxide, 35 °C) against insects and for gaseous sterilization of some dehydrated foods for which other methods, e.g. heat sterilization, are not suitable. Examples are the gaseous sterilization of walnuts, starches, dehydrated foods (fruits and vegetables) and, above all, spices, in which a high spore count (and plate count in general) is often a sanitary problem. The sterilization is carried out in pressure chambers in a mixture with an inert gas (e.g. 80–90% CO_2). The need to remove the residual unreacted gas (vacuum, "gaseous rinsing") has already been empha-sized. An alternative method of sterilization for the above-mentioned food products is high energy irradiation (UV-light, X-ray, or gamma irradiation).

8.12.9 Nitrite, Nitrate

These additives are used primarily to preserve the red color of meat (cf. 12.3.2.2.2). However, they also have antimicrobial activity, particularly in a mixture with common salt. Of importance is their inhibitive action, in nonsterilized meat products, against infections by *Clostridium bo-tulinum* and, consequently, against accumulation of its toxin. The activity is dependent on the pH and is proportional to the level of free HNO_2. In-deed, 5–20 mg of nitrite per kg are considered sufficient to redden meat, 50 mg/kg for the pro-duction of the characteristic taste, and 100 mg/kg for the desired antimicrobial effects. Acute tox-icity has been found only at high levels of use (formation of methemoglobin). A problem is the possibility of the formation of nitrosamines, com-pounds with powerful carcinogenic activity. Nu-merous animal feeding tests have demonstrated tumor occurrence when the diet contained amines (sensitive to nitroso substitution) and nitrite. Con-sequently, the trend is to exclude or further re-duce the levels of nitrate and nitrite in food. No suitable replacement has been found for nitrite in meat processing.

8.12.10 Antibiotics

The use of antibiotics in food preservation raises a problem since it might trigger development of more resistant microorganisms and thus create medical/therapeutic difficulties.

Of some importance is nisin, a polypeptide antibiotic, produced by some *Lactococcus lactis* strains. It is active against Gram-positive microorganisms and all spores, but is not used in human medicine. This heat-resistant peptide is applied as an additive for sterilization of dairy products, such as cheeses or condensed or evaporated milk (cf. 1.3.4.3). Natamycin (pimaricin, Formula 8.38), which is produced by

Streptomyces natalensis and *S. chattanogensis*, is active at 5–100 ppm against yeasts and molds and is used as an additive in surface treatment of cheeses. It also finds application for suppressing the growth of molds on ripening raw sausages.

$$R = COOH \tag{8.38}$$

The possibility of incorporating the wide spec-trum antibiotics chlortetracycline and oxytetracy-cline into fresh meat, fish and poultry, in order to delay spoilage, is still under investigation.

8.12.11 Diphenyl

Diphenyl, due to its ability to inhibit growth of molds, is used to prevent their growth on peels of citrus fruits (lemon, orange, lime, grapefruit). It is applied by impregnating the wrapping paper and/or cardboard packaging material (1–5 g diphenyl/m^2).

8.12.12 o-Phenylphenol

This compound, at a level of 10–50 ppm and a pH range of 5–8, inhibits the growth of molds. The inhibition effect, which increases with increasing pH, is utilized in the preservation of citrus fruits. It is applied by dipping the fruits into a 0.5–2% solution at pH 11.7.

8.12.13 Thiabendazole, 2-(4-Thiazolyl)benzimidazole

This compound (Formula 8.39) is particularly powerful against molds which cause the socalled blue mold rots, e. g., *Penicillium italicum* (blue-green-spored "contact mold") and *Penicillium digitatum* (green-spored mold). It is used for pre-serving the peels of citrus fruits and bananas. The application mode is by dipping or spraying the fruit with a wax emulsion containing 0.1–0.45% thiabendazole.

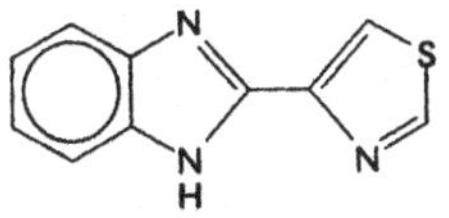

$$(8.39)$$

8.13 Antioxidants

Since lipids are widely distributed in food and since their oxidation yields degradation products of great aroma impact, their degradation is an important cause of food deterioration by generation of undesirable aroma. Lipid oxidation can be retarded by oxygen removal or by using antioxidants as additives. The latter are mostly phenolic compounds, which provide the best results often as a mixture and in combination with a chelating agent. The most important antioxidants, natural or synthetic, are tocopherols, ascorbic acid esters, gallic acid esters, tert-butylhydroxyanisole (BHA) and di-tert-butylhydroxytoluene (BHT). They are covered in 3.7.3.2.2.

8.14 Chelating Agents (Sequestrants)

Chelating agents have acquired greater importance in food processing. Their ability to bind metal ions has contributed significantly to stabilization of food color, aroma and texture. Many natural constituents of food can act as chelating agents, e. g., carboxylic acids (oxalic, succinic), hydroxy acids (lactic, malic, tartaric, citric), polyphosphoric acids (ATP, pyrophosphates), amino acids, peptides, proteins and porphyrins.

Table 8.14 lists the chelating agents utilized by the food industry, while Table 8.15 gives the stability constants for some of their metal complexes.

Traces of heavy metal ions can act as catalysts for fat or oil oxidation. Their binding by chelating agents increases antioxidant efficiency and inhibits oxidation of ascorbic acid and fat-soluble vitamins. The stability of the aroma and color of canned vegetables is substantially improved.

Table 8.14. Chelating agents used as additives in food processing. (Compounds given in brackets are utilized only as salts or derivatives)

(Acetic acid)	Na-, K-, Ca-salts
Citric acid	Na-, K-, Ca-salts, monoisopropyl ester, monoglyceride ester, triethyl ester, monostearyl ester,
EDTA	Na-, Ca-salts
(Gluconic acid)	Na-, Ca-salts
Oxystearin	
Orthophosphoric acid	Na-, K-, Ca-salts
(Pyrophosphoric acid)	Na-salt
(Triphosphoric acid)	Na-salt
(Hexametaphosphoric acid, 10–15 residues)	Na-, Ca-salts
(Phytic acid)	Ca-salt
Sorbitol	
Tartaric acid	Na-, K-salts
(Thiosulfuric acid)	Na-salt

Table 8.15. Stability constants (pK-values) of some metal complexes

Chelating agent	Ca^{2+}	Co^{2+}	Cu^{2+}	Fe^{2+}	Fe^{3+}	Mg^{2+}	Zn^{2+}
Acetate	0.5	2.2				0.5	1.0
Glycine	1.4	5.2	8.2	4.3	10.0	3.5	5.2
Citrate	3.5	4.4	6.1	3.2	11.9	2.8	4.5
Tartrate	1.8		3.2		7.5	1.4	2.7
Gluconate	1.2		18.3			0.7	1.7
Pyrophosphate	5.0		6.7		22.2	5.7	8.7
ATP	3.6	4.6	6.1			4.0	4.3
EDTA	10.7	16.2	18.8	14.3	25.7	8.7	16.5

In the production of herb and spice extracts, the combination of an antioxidant and a chelating agent provides an improved extract quality. Chelating agents are also used in dairy products, wherein their deaggregating activity for the casein complexes is often utilized; in blood recovery processes to prevent clotting; and in the sugar industry to facilitate sucrose crystallization, a process which is otherwise retarded by sucrose-metal complexes.

8.15 Surface-Active Agents

Naturally occurring and synthetic surface-active agents (tensides), some of which are listed in Table 8.16, are used in food processing when a decrease in surface tension is required e. g., in production and stabilization of all kinds of dispersions (Table 8.17).

Dispersions include emulsions, foams, aerosols and suspensions (Table 8.18). In all cases an *outer, continuous* phase is distinct from an *inner*, discontinuous, *dispersed* phase. Emulsions are of particular importance and they will be outlined in more detail.

8.15.1 Emulsions

Emulsions are dispersed systems, usually of two immiscible liquids. When the outer phase consists of water and the inner of oil, it is considered as an

Table 8.16. Surfactants (surface active agents) in food

I. Naturally occurring:
A. Ions: proteins (cf. 1.4.3.6), hydrocolloids
   (gum arabic, cf. 4.4.4.5.2), phospholipids
   (lecithin, cf. 3.4.1.1), bile acids
B. Neutral substances: glycolipids (cf. 3.4.1.2),
   saponins
II. Synthetic:
A. Ions: stearyl-2-lactylate, Datem, Citrem
   (cf. Table 8.24)
B. Neutral substances: mono-, diacylglycerols
   and their acetic- and lactic
   acid esters, saccharose fatty acid esters,
   sorbitan fatty acid esters, polyoxyethylene
   sorbitan fatty acid esters
   Polyglycerol-polyricinoleate (PGPR)

Table 8.17. Examples of surfactant utilization in the food industry

Utilization in production of	Effect
Margarine	Stabilization of a w/o emulsion
Mayonnaise	Stabilization of an o/w emulsion
Ice cream	Stabilization of an o/w emulsion, achievement of a "dry" consistency
Sausages	Prevention of fat separation
Bread and other baked products	Improvement of crumb structure, baked product volume, inhibition of starch retrogradation (bread staling)
Chocolate	Improvement of rheological properties, inhibition of "fat blooming"
Instant powders	Solubilization
Spice extracts	Solubilization

Table 8.18. Dispersion systems

Type	Inner phase	Outer phase
Emulsion	liquid	liquid
Foam	gaseous	liquid
Aerosol	liquid or solid	gaseous
Suspension	solid	liquid

"oil in water" (o/w) type of emulsion. When this is reversed, i. e., water is dispersed in oil, a w/o emulsion exists. Examples of food emulsions are: milk (o/w), butter (w/o) and mayonnaise (o/w).

The visual appearance of an emulsion depends on the droplet diameter. If the diameter is in the range of 0.15–100 µm, the emulsion appears milky-turbid. In comparison, micro-emulsions (diameter: 0.0015–0.15 µm) are transparent and considerably more stable because the sedimentation rate depends on the droplet diameter (Table 8.19).

Table 8.19. Sedimentation rate (v) as a function of droplet diameter (d)

d (µm)	v (cm/24 h)
0.02	3.75×10^{-4}
0.2	3.76×10^{-2}
2	3.76
20	3.76×10^{2}
200	3.76×10^{4}

Each emulsifier can disperse a limited amount of an inner phase, i. e. it has a fixed *capacity*. When the limit is reached, further addition of outer phase breaks down the emulsion. The capacity and other related parameters differ among emulsifiers and can be measured accurately under standardized conditions.

8.15.2 Emulsifier Action

8.15.2.1 Structure and Activity

Emulsions are made and stabilized with the aid of a suitable tenside, usually called an emulsifier. Its activity is based on its molecular structure. There is a lipophilic or hydrophobic part with good solubility in a nonaqueous phase, such as an oil or fat, and a polar or hydrophilic part, soluble in water. The hydrophobic part of the molecule is generally a long-chain alkyl residue, while the hydrophilic part consists of a dissociable group or of a number of hydroxyl or polyglycolether groups.

In an immiscible system such as oil/water, the emulsifier is located on the interface, where it decreases interfacial tension. Thus, even in a very low concentration, it facilitates a fine distribution of one phase within the other. The emulsifier also prevents droplets, once formed, from aggregating and coalescing, i. e. merging into a single, large drop (Fig. 8.13).

Ionic tensides stabilize o/w emulsions in the following way (Fig. 8.14a): at the interface, their alkyl residues are solubilized in oil droplets, while the charged end groups project into the aqueous phase. The involvement of counter ions forms an electrostatic double layer, which prevents oil droplet aggregation.

Nonionic, neutral tensides are oriented on the oil droplet's surface with the polar end of the tenside projecting into the aqueous phase. The coalescence of the droplets of an o/w emulsion is prevented by an anchored "hydrate shell" built around the polar groups.

The coalescence of water droplets in a w/o emulsion first requires that water molecules break through the double-layered hydrophobic region of emulsifier molecules (Fig. 8.14 b). This escape is only possible when sufficient energy is supplied to rupture the emulsifier's hydrophobic interaction.

The stability of an emulsion is increased when additives are added which curtail droplet mobility. This is the basis of the stabilization effect of hydrocolloids (cf. 4.4.3) on o/w emulsions since they increase the viscosity of the outer, aqueous phase.

A rise in temperature negatively affects emulsion stability, and can be applied whenever an emulsion has to be destroyed. Elevated temperatures are used along with shaking, agitation or pressure

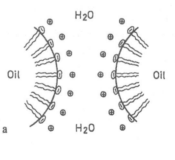
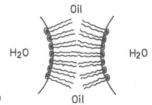

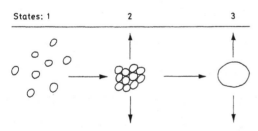

Fig. 8.13. Changes in an emulsion. *1* The droplets are dispersed in a continuous phase. *2* The droplets form aggregates. An increase in particle diameter results in acceleration of their flotation or sedimentation. *3* Coalescence: the aggregated droplets merge into larger droplets. Finally, two continuous phases are formed; the emulsion is destroyed

Fig. 8.14. Stabilization of an emulsion. **a** Activity of an ionic emulsifier in an o/w emulsion. **b** Activity of a nonpolar emulsifier in w/o emulsion. ○ Polar groups, ∼ apolar tails of the emulsifier

(mechanical destruction of interfacial films as, for example, in butter manufacturing, cf. 10.2.3.3). Other ways of decreasing the stability of an emulsion are addition of ions which collapse the electrostatic double layer, or hydrolysis to destroy the emulsifier.

8.15.2.2 Critical Micelle Concentration (CMC), Lyotropic Mesomorphism

The surface tension of an aqueous solution of an o/w emulsifier decreases down to the critical micelle concentration (CMC) as a function of the emulsifier concentration. Above this limiting value, the emulsifier aggregates reversibly to give spherical micelles, the surface tension changing only slightly. The CMC is a characteristic value of the emulsifier, which decreases as the hydrophobic part of the molecule increases. It is also influenced by the temperature, pH value, and electrolyte concentration.

The temperature at which the solubility of an emulsifier reaches the CMC is called the critical micelle temperature (Tc, Krafft point). Crystals, micelles, and the dissolved emulsifier are in equilibrium at the Tc (Fig. 8.15). An emulsifier cannot form micelles below the Tc which, e. g., depends on the structure of the fatty acid residues in lecithin (Table 8.20).

Emulsifiers are lyotropic mesomorphous, i. e., they form one of the following liquid crystalline mesophases depending on the water content

Table 8.20. Effect of fatty acid residues on the critical micelle temperature T_c of lecithins

Fatty acid	T_c (°C)
12:0	0
14:0	23
16:0	41
18:0	58
18:1	−20

and the temperature (shown schematically in Fig. 8.16):

Hexagonal I: Cylindrical aggregates of emulsifier molecules; the polar groups are oriented towards the outer water phase.

Lamellar: Emulsifier bilayers which are separated by thin water zones.

Hexagonal II (Inverse Hexagonal): Cylindrical aggregates of emulsifier molecules; the polar groups are oriented towards the inner water phase.

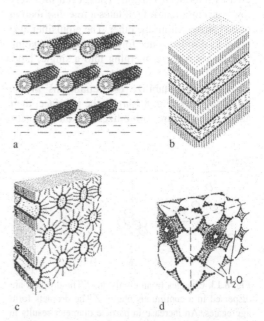

a   b

c   d

Fig. 8.16. Lyotropic mesophases of emulsifiers (according to *Schuster*, 1985). **(a)** Hexagonal I, **(b)** Lamellar, **(c)** Hexagonal II, **(d)** Cubic ● ●〜〜 Emulsifier ‑‑‑‑‑ Water

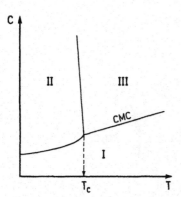

Fig. 8.15. Solubility of an emulsifier in water. *Ordinate*: concentration, *abscissa*: temperature. I: Solution, II: Crystals, III: Micelles, T_c: Critical micelle temperature

Cubic: Cubic space- and face-centered water aggregates in a matrix of emulsifier molecules; the polar groups are oriented towards the water.

Phase diagrams show the mesophase present as a function of water content and temperature. In the phase diagram of the o/w emulsifier lysolecithin (Fig. 8.17, a), micelles, a lamellar, and a hexagonal phase appear. The w/o emulsifier 1-monoelaidin (Fig. 8.17, b) crystallizes at

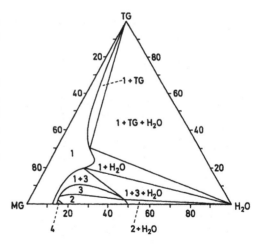

Fig. 8.18. Ternary phase diagram for the system monoglycerides (from sunflower oil)/water/soybean oil at 40 °C (according to *Larsson* and *Dejmek*, 1990) *1* Microemulsion, *2* Cubic, *3* Hexagonal II, *4* Lamellar

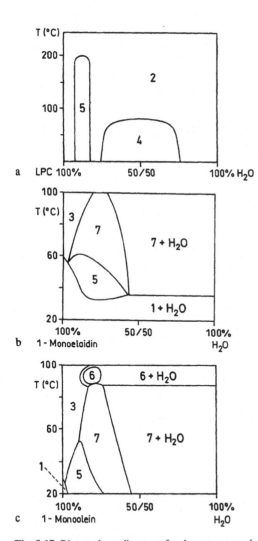

Fig. 8.17. Binary phase diagrams for the system emulsifier/water (according to *Krog*, 1990) **(a)** Lysolecithin, **(b)** 1-Monoelaidin, **(c)** 1-Monoolein
1 Crystals, *2* Micelles, *3* Microemulsion, *4* Hexagonal I, *5* Lamellar, *6* Hexagonal II, *7* Cubic

temperatures below 30 °C. The β-modification formed first is converted to the more stable β-form, which unlike the α-form, has no emulsifying properties. The melted 1-monoelaidin forms a microemulsion with little water and lamellar and cubic mesophases with much water. 1-Monoolein (Fig. 8.17, c) melts at lower temperatures and an inverse hexagonal mesophase appears.

The phases of simply constituted food emulsions, which are present at a certain temperature depending on the composition, show a ternary phase diagram, e. g., in Fig. 8.18.

8.15.2.3 HLB-Value

A tenside with a relatively strong lipophilic group and weak hydrophilic group is mainly soluble in oil and preferentially stabilizes a w/o emulsion, and vice versa. This fact led to the development of a standard with which the relative strength or "activity" of the hydrophilic and lipophilic groups of emulsifiers can be evaluated. It is called the *HLB value* (hydrophilic–lipophilic balance). It can be determined, e. g., from dielectric constants or from the chromatographic behavior of the surface-active substance. The HLB value of the fatty acid esters of polyhydroxy alcohols can also be calculated as follows (SV =

saponification number of the emulsifier, AV = acid value of the separated acid):

$$HLB = 20\left(1 - \frac{SV}{AV}\right) \qquad (8.40)$$

On the basis of experimental group numbers (Table 8.21), the HLB value can be calculated using the formula:

$$HLB = \sum(\text{hydrophilic group number}) -$$
$$\sum(\text{hydrophobic group number}) + 7 \quad (8.41)$$

Some examples listed in Table 8.22 show very good correspondence between calculated and experimentally found HLB values.

Table 8.21. Group number N_H and N_L for HLB calculation

Hydrophilic group	N_H	Lipophilic group	N_L
$-OSO_3^-$, Na^+	38.7	$-\overset{\mid}{C}H-$	0.475
$-SO_{3-}$, Na^+	37.4	$-CH_2-$	0.475
$-COO^-$, Na^+	21.1	$-CH_3$	0.475
$-COO^-$, K^+	19.1	$=CH-$	0.475
Sorbitan ring	6.8	$-\overset{\mid}{C}H-CH_2-O-$	0.15
Ester	2.4	CH_3	
$-COOH$	2.1		
$-OH$ (free)	1.9	Benzene ring	1.662
$-O-$	1.3		
$-(CH_2-CH_2-O)-$	0.33		

Table 8.22. Hydrophilic lipophilic balance (HLB) values of some surfactants

Compound	HLB-value	
	Found	Calculated
Oleic acid	1.0	
Sorbitol tristearate	2.1	2.1
Stearyl monoglyceride	3.4	3.8
Sorbitol monostearate	4.7	4.7
Sorbitol monolaurate	8.6	
Gelatin	9.8	
Polyoxyethylene sorbitol tristearate	10.5	10
Methylcellulose	10.5	
Polyoxyethylene sorbitol monostearate	14.9	
Polyoxyethylene sorbitol monooleate	15.0	15
Sodium oleate	18.0	
Potassium oleate	20.0	

Table 8.23. HLB-values related to their industrial application

HLB-range	Application
3–6	w/o-Emulsifiers
7–9	Humectants
8–18	o/w-Emulsifiers
15–18	Turbidity stabilization

The HLB values indicated the first industrial applications (Table 8.23). For a detailed characterization, however, comprehensive knowledge of possible interactions of the emulsifier with the many components of a food emulsion is still lacking. Hence, emulsifiers are mainly used in accordance with empirical considerations. It has been observed with neutral emulsifiers that the degree of hydration of the polar groups decreases with a rise in temperature and the influence of the lipophilic groups increases. Phase inversion occurs o/w → w/o. The temperature at which inversion occurs is called the phase conversion temperature.

8.15.3 Synthetic Emulsifiers

Today, 150,000–200,000 t of emulsifiers are produced worldwide. Of this amount, mono-and diacylglycerides and their derivatives account for the largest part, i.e. about 75%. Synthetic emulsifiers include a series of nonionic compounds. Unlike the ionic compounds, the nonionic emulsifiers are not in danger of decreasing in interfacial activity through salt formation with food constituents. The utilization of emulsifiers is legislated and often differently regulated in some countries. The synthetic emulsifiers described below are used worldwide.

8.15.3.1 Mono-, Diacylglycerides and Derivatives

Mono- and diacylglycerides, which are mostly used as mixtures, are produced as described in 3.3.2. Other emulsifiers with special activities are obtained by derivatization (cf. Table 8.24). As a result of the diverse reaction possibilities

Table 8.24. Emulsifiers from mono- and diacylglyceride mixtures

Name			Production by conversion of mixtures of mono- and
Mono- and diglycerides esterified with		EU-number	diacylglycerides with
Acetic acid (acetylated mono- and diglycerides)	*Acetem*	E472a	Acetic anhydride
Lactic acid	*Lactem*	E472b	Lactic acid
Citric acid	*Citrem*	E472c	Citric acid
Monoacetyl- and diacetyltartaric acid	*Datem*	E472e anhydride	Tartaric acid and acetic

of the starting compounds, complex products are obtained in this process. An example is represented by the diacetyltartaric acid ester of monoglycerides (DATEM). At concentrations of ca. 0.3% (based on the amount of flour), this ester increases the volume of wheat biscuits. For the production of this emulsifier, acetic anhydride and tartaric acid are heated, diacetyltartaric acid anhydride (I in Formula 8.43) being formed on removal of acetic acid by distillation. Compound I is converted to DATEM

with monoacylglycerides (II). In the series 6:0 to 22:0 as well as 18:1 (9) and 18:2 (9,12), the baking activity of DATEM is the highest with stearic acid as the acyl residue. DATEMs on the basis of diacylglycerides containing the acyl residues 10:0 or 18:0 exhibit only slight activity. Of the 10 components of a DATEM preparation which quantitatively appeared, the main product (III in Formula 8.43) gave the largest increase in volume of white bread, closely followed by compounds IV and V.

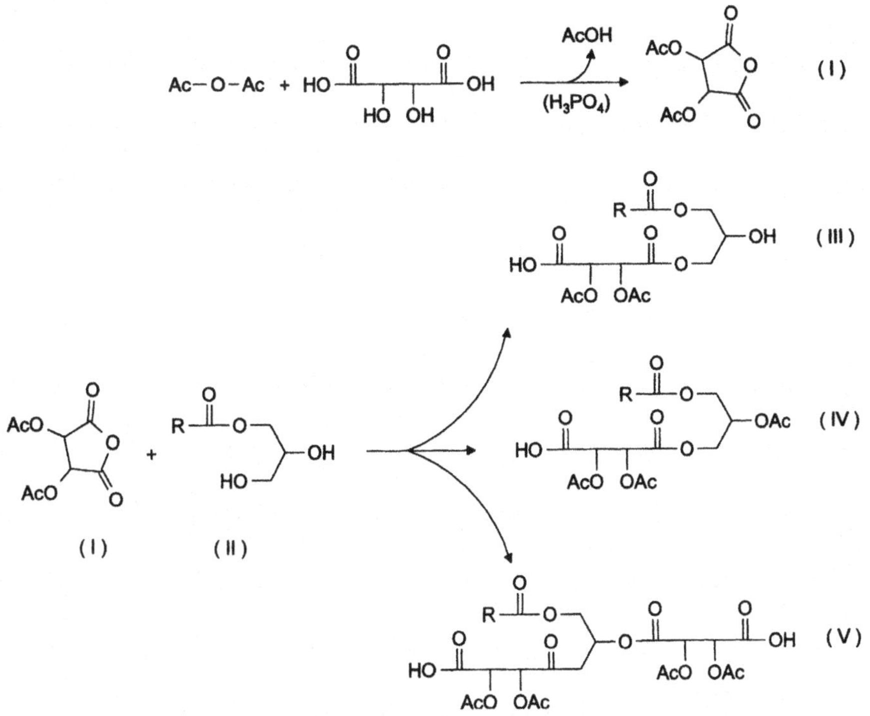

(8.43)

Unlike *acetem* and *lactem, citrem* is an acid (cf. Formula 8.43).

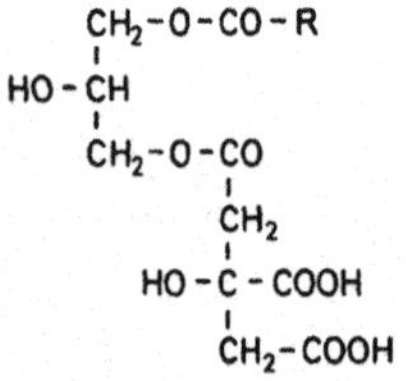

$$
\begin{array}{l}
CH_2\text{-}O\text{-}CO\text{-}R \\
| \\
HO\text{-}CH \\
| \\
CH_2\text{-}O\text{-}CO \\
\qquad\quad | \\
\qquad\quad CH_2 \\
\qquad\quad | \\
\quad HO\text{-}C\text{-}COOH \\
\qquad\quad | \\
\qquad\quad CH_2\text{-}COOH
\end{array}
\tag{8.43}
$$

8.15.3.2 Sugar Esters

They are obtained, among other methods, by transesterification of fatty acid methyl esters (14:0, 16:0, 18:0 and/or 18:1, double bond position 9) with sucrose and lactose. The resultant mono- and diesters are odorless and tasteless. Depending on their structure, they cover an HLBrange of 7–13, and are used in stabilization of o/w emulsions, or in stabilization of some instant dehydrated and powdered foods.

8.15.3.3 Sorbitan Fatty Acid Esters

Esters of sorbitan (cf. 19.1.4.6) with fatty acids (*Span's*) serve the stabilization of w/o emulsions. Sorbitan tristearate is used in the production of chocolate to delay the fat bloom formation.

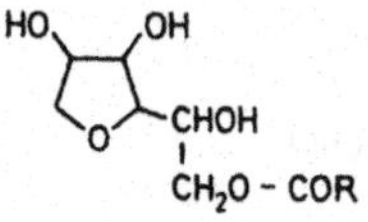

$$ \tag{8.44} $$

8.15.3.4 Polyoxyethylene Sorbitan Esters

Polyoxyethylene groups are introduced into the molecules to increase the hydrophilic property of sorbitan esters:

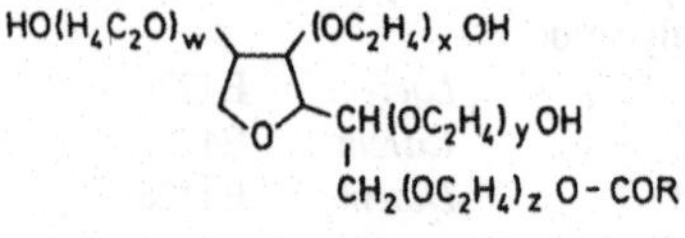

$$ \tag{8.45} $$

Polyoxyethylene sorbitan monoesters (examples in Table 8.22) are used to stabilize o/w emulsions.

8.15.3.5 Polyglycerol – Polyricinoleate (PGPR)

In the production of the emulsifier PGPR (cf. Formula 8.46), oligomeric glycerol is made by attachment of 2,3-epoxy-1-propanol (glycid) to glycerol and at the same time ricinoleic acids are esterified with each other under controlled heating conditions. In a third step, the oligomeric glycerol is esterified with polyesterricinoleic acid.

The emulsifier has a very complicated composition: apart from different types of esters, oligomeric glycerol and free ricinoleic acid are present.

Together with lecithin, PGPR is used in the production of chocolate. It completely eliminates the flow point of a molten chocolate mass, but hardly lowers the viscosity.

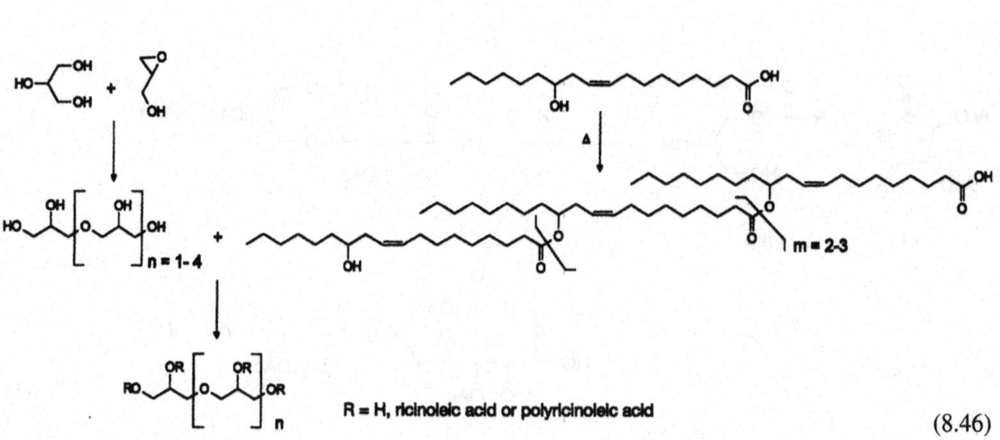

R = H, ricinoleic acid or polyricinoleic acid

$$ \tag{8.46} $$

8.15.3.6 Stearyl-2-Lactylate

In the presence of sodium or calcium hydroxide, esterification of stearic acid with lactic acid gives a mixture of stearyl lactylates (Na or Ca salt), the main component being stearyl-2-lactylate:

$$CH_3-(CH_2)_{16}-CO-O-\overset{\overset{\displaystyle CH_3}{|}}{CH}-CO-O-\overset{\overset{\displaystyle CH_3}{|}}{CH}-COO^{\ominus}NA^{\oplus}$$

$$(8.47)$$

The free acid acts as a w/o emulsifier and the salts as o/w emulsifiers. The HLB-value of the sodium salt is 8–9, and that of the calcium salt, 6–7. The sodium salt is used to stabilize an o/w emulsion which is subjected to repeated cycles of freezing and thawing.

8.16 Substitutes for Fat

In the industrial, highly developed countries, the intake of energy with food is higher than the physiological requirements. To avoid the consequences manifested by, e. g., overweight and adipositas, attempts are made to substitute the fat, the main source of energy. However, fat has many functions in food which cannot be completely taken on by a substitute. For this reason, various substances are offered which make partial solutions possible. They are divided into two groups depending on their origin:

- natural (fat mimetics)
- synthetic (fat substitutes, fat replacers)

8.16.1 Fat Mimetics

8.16.1.1 Microparticulated Proteins

The mouth feeling of substances depends on their chemical composition and on the particle size. Protein particles with a diameter of more than 8 µm are experienced as sandy, those in the range of 3–8 µm as powdery, 0.1–3 µm as creamy, and less than 0.1 µm as watery. Therefore, by means of microparticulation of protein concentrates to particles of 0.1–3 µm, it is possible to achieve the melt-in-the-mouth feeling produced by fat globules. In this process, concentrates of ovalbumin, casein and whey protein are exposed to varying pressures and temperatures, the proteins being ground by high shear forces. Rapid cooling to $4-1\,°C$ yields a thick cream.

These substitutes are suitable for milk products (ice cream, desserts etc.) which are not strongly heated. In fact, 3 g of fat can be replaced by 3 g of swollen substitute (1 g of protein + 2 g of water) or 27 kcal by 4 kcal.

8.16.1.2 Carbohydrates

Polymeric carbohydrates are used as fat substitutes. They are nondigestible in the small intestine and classed as fiber. However, a number of these substances are degraded by the bacteria in the large intestine with the formation of short-chain acids (2:0, 3:0, 4:0). These acids are absorbed, the gain in energy at 2 kcal/g being half as much as with digestible carbohydrates. The energy (kcal/g) provided by fiber substances which can be used to replace fat are: wheat bran (1.5), barley brans (0.9), oat bran (0.1), apple fibers (1.6), soybean bran (0.7) and pea fibers (0.2). In the production of foods, attention must be paid to the taste of the preparations. The carbohydrate-based fat substitutes include the resistant starches (cf. 4.4.4.14.6), which can be formed during starch retrogradation, but also occur in some fruits, e. g., bananas. Fructose polymers (cf. 4.4.4.22.1), pectin (cf. 4.4.4.13), modified starch and cellulose, e. g., carboxymethyl cellulose (4.4.4.17.2), also play a role.

From corn starch, e. g., non-sweet oligosaccharides (maltodextrins, DE5) which dissolve completely in hot water are obtained. When this solution is cooled, a gel is formed which has the texture of edible oil. It can partially replace fat, e. g., in margarine, allowing a 35% reduction of the energy content.

8.16.2 Synthetic Fat Substitutes

Energetically inefficient fat substitutes can basically be made as follows:

- replacement of glycerol with other alcohols,
- replacement of the usual fatty acids with branched, polybasic or especially long-chain carboxylic acids,

- introduction of inverse ester bonds (retrofats),
- use of ether instead of ester bonds.

8.16.2.1 Carbohydrate Polyester

Mono-, oligo- and polysaccharides yield fatlike products when esterified with fatty acids. In general, the starting material is sucrose as the acetate, which is melted with fatty acid alkylesters in the presence of alkali metals. The degree of esterification of sucrose should be high because otherwise the ester bonds are hydrolyzed in the gastrointestinal tract. In the best known product, *Olestra*®, 6–8 OH groups are esterified with fatty acids 8:0–12:0. This product is tasteless and thermally stable so that it can be as strongly heated as an edible fat during baking and frying.

8.16.2.2 Retrofats

These are esters of polybasic acids (e. g., malonic acid, citric acid, propane 1,2,3-tricar-boxylic acid, butane 1,2,3,4-tetracarboxylic acid) with long-chain alcohols.

8.17 Thickening Agents, Gel Builders, Stabilizers

A number of polysaccharides and their modified forms, even at low concentrations, are able to increase a system's viscosity, to form gels and to stabilize emulsions, suspensions or foams. These compounds are also active as crystallization inhibitors (e. g. in confections, ice creams) and are suitable for aroma encapsulation, as is often needed for dehydrated food. These properties make polysaccharides important additives in food processing and storage. The compounds of importance, together with their properties and utilization, were described in detail in the chapter on carbohydrates. Among proteins, gelatin is an important gel-forming agent used widely in food products (cf. 12.3.2.3.1).

8.18 Humectants

Some polyols (1,2-propanediol, glycerol, mannitol, sorbitol) have distinct hygroscopic properties and act as humectants, i. e. additives for retaining food moisture and softness and inhibiting crystallization. They are often required in a confectionery product. When glycerol or sorbitol is added to mashed vegetables or fruits or in the production of other powdered foods before the final drying stage, the dehydrated products have improved rehydration characteristics.

8.19 Anticaking Agents

Some food products, such as common salt, seasoning salt (e. g. a mixture of onion or garlic powder with common salt), dehydrated vegetable and fruit powders, soup and sauce powders and baking powder, tend to cake into a hard lump. Lumping can be avoided by using any of a number of compounds that either absorb water or provide protective hydrophobic films. Anticaking compounds include sodium, potassium and calcium hexacyanoferrate (II), calcium and magnesium silicate, tricalcium phosphate and magnesium carbonate.

8.20 Bleaching Agents

Bleaching is used primarily in flour production. The removal of yellow carotenoids by oxidation can be achieved by a number of compounds that, in addition to bleaching, improve the baking quality of flour. Examples of some approved common bleaching agents are Cl_2, ClO_2, $NOCl$, NO_2 and N_2O_4. Lipoxygenase enzyme also has an efficient bleaching activity.

8.21 Clarifying Agents

In some beverages, such as fruit juices, beer or wine, turbidity and sediment formation can occur with the involvement of phenolic compounds, pectins and proteins. These defects can be corrected by: (a) partial enzymatic degradation of pectins and proteins; (b) removal of phenolic compounds with the aid of gelatin, polyamide or polyvinyl pyrrolidone powders; and (c) by protein removal with bentonite or tannin. Bentonite consists of hydrous aluminium silicate,

$Al_2SiO_9(OH)_x$, and changing amounts of iron, calcium and magnesium salts.

8.22 Propellants, Protective Gases

Food sensitive to oxidation and/or microbial spoilage can be stored in an atmosphere of protective gas or a gas mixture (N_2, CO_2, CO, etc.; modified or controlled atmosphere storage). This is often a suitable method for lengthening the shelf life of food.

Liquid food can be filled into pressurized containers and, when needed, using a propellant, discharged in the form of a cream or paste (e. g. cream cheese, ketchup), a foam (whipping cream) or a mist (herb or spice extracts in oil; liquid barbecue smoke). Propellants used are N_2, N_2O, and CO_2.

Due to its low solubility in water, fat and oil, N_2 is used preferentially as a propellant when foam formation is not desired (ketchup). On the other hand, gases such as N_2O and CO_2 are preferred for foam formation (whipped cream) due to their good solubility in water.

8.23 References

Anonymus: Fat substitute update. Food Technol. *44* (3), 92 (1990)

Ariyoshi, Y., Kohmura, M., Hasegawa, Y., Ota, M., Nio, N.: Sweet peptides and proteins: Synthetic studies. ACS Symposium Series *450*, p. 41 (1991)

Bär, A., Borrego, F., Castillo, J., del Rio, J. A.: Neohesperidin dihydrochalcone: Properties and applications. Lebensm. Wiss. Technol. *23*, 371 (1990)

Beets, M.G.J.: Structure–activity relationships in human chemoreception. Applied Science Publ.: London. 1978

Belitz, H.-D., Chen, W., Jugel H., Treleano, R., Wieser, H., Gasteiger, J., Marsili, M.: Sweet and bitter compounds: Structure and taste relationship. In: Food taste chemistry (Ed.: Boudreau, J.C.), ACS Symposium Series 115, p. 93, American Chemical Society: Washington, D.C. 1979

Belitz, H.-D., Chen, W, Jugel, H., Stempfl, W., Treleano, R., Wieser, H.: Structural requirements for sweet and bitter taste. In: Flavour '81 (Ed.: Schreier, P.), p. 741, Walter de Gruyter: Berlin. 1981

Belitz, H.-D., Chen, W., Jugel, H., Stempfl, W, Treleano, R., Wieser, H.: QSAR of bitter tasting compounds. Chem. Ind. *1983*, 23

Birch, G.G., Brennan, J.G., Parker, K.J. (Eds.): Sensory properties of foods. Applied Science Publ.: London. 1977

Branen, A.L., Davidson, P.M., Salminen, S. (Eds.): Food Additives. Marcel Dekker: New York. 1990

Buchta, K.: Lactic acid. In: Biotechnology (Eds.: Rehm, H.-J., Reed, G.), Vol. 3, p. 409, Verlag Chemie: Weinheim. 1983

Compadre, C.M., Pezzuto, J.M., Kinghorn, A.D.: Hernandulcin: an intensely sweet compound discovered by review of ancient literature. Science *227*, 417 (1985)

Deutsche Forschungsgemeinschaft: Bewertung von Lebensmittelzusatz- und -inhaltsstoffen. VCH Verlagsgesellschaft: Weinheim. 1985

Furia, T.E. (Ed.): Handbook of food additives. 2nd edn., CRC Press: Cleveland, Ohio. 1972

Gardner, D.R., Sanders, R.A.: Isolation and characterization of polymers in heated olestra and olestra/triglyceride blend. J. Am. Oil Chem. Soc. *67*, 788 (1990)

Glandorf, K.K., Kuhnert, P., Lück, E.: Handbuch Lebensmittelzusatzstoffe. Behr's Verlag, Hamburg, 1999

Glowaky, R.C., Hendrick, M.E., Smiles, R.E., Torres, A.: Development and uses of Alitame: A novel dipeptide amide sweetener. ACS Symposium Series *450*, p. 57 (1991)

Gould, G.W. (Ed.): Mechanisms of action of food preservation procedures. Elsevier Applied Science: London (1989)

Griffin, W.C., Lynch, M.J.: Surface active agents. In: Handbook of food additives, 2nd edn. (Ed.: Furia, T.E.), p. 397, CRC Press: Cleveland, Ohio. 1972

Haberstroh, H.-J., Hustede, H.: Weinsäure. Ullmanns Encyklopädie der technischen Chemie, 4. edn., vol. 24, p. 431. Verlag Chemie: Weinheim. 1983

Herbert, R.A.: Microbial growth at low temperatures. In: Gould, G.W. (Ed.) Mechanism of action of food preservation procedures, p. 71. Elsevier Applied Science: London (1989)

Heusch, R.: Emulsionen. In: Ullmanns Encyklopädie der technischen Chemie, 4. edn., vol. 10, p. 449, Verlag Chemie: Weinheim. 1975

Iyengar, R. B., Smits, P., Van der Ouderas, F., Van der Wel, H., Van Brouwershaven, J., Ravestein, P., Richters, G., von Wassenaar, P.D.: The complete amino-acid sequence of the sweet protein Thaumatin I. Eur. J. Biochem. *96*, 193 (1979)

Jenner, M.R.: Sucralose: How to make sugar sweeter. ACS Symposium Series *450*, p. 68 (1991)

Kaneko, R., Kitabatake, N.: Structure-sweetness relationship in thaumatin: importance of lysine residues. Chem. Senses *26*, 167 (2001)

Ottinger, H., Soldo, T., Hofmann, T.: Systematic studies on structure and physiological activity of cyclic α-keto enamines, a novel class of "cooling" compounds. J. Agric. Food Chem. *49*, 5383 (2001)

Kim, S.-H., Kang, C.-H., Cho, J.-M.: Sweet proteins: Biochemical studies and genetic engineering. ACS Symposium Series *450*, p. 28 (1991)

Köhler, P.: Synthetische Emulgatoren für Backwaren. Untersuchungen zur Wirksamkeit von DATEM und seinen Komponenten. Habilitations-schrift, TU München, 1999

Köhler, P., Grosch, W.: Study of the effect of DATEM. 1. Influence of fatty acid chain length on rheology and baking. J. Agric. Food Chem. *47*, 1863 (1999)

Kommission der Europäischen Gemeinschaften: Berichte des Wissenschaftlichen Lebensmittelausschusses (Sechzehnte Folge) – Süßstoffe: Brüssel. 1985

Lancet, D., Ben-Arie, N.: Sweet taste transduction. A molecular-biological analysis. ACS Symposium Series *450*, p. 226 (1991)

Lange, H., Kurzendorfer, C.-P.: Zum Mechanismus der Stabilisierung von Emulsionen. Fette Seifen Anstrichm. *76*, 120 (1974)

Larsson, K., Friberg, S.E.: Food emulsions. 2. edn., Marcel Dekker, Inc.: New York, 1990

Lucca, P.A., Tepper, B.J.: Fat replacers and the functionality of fat in foods. Trends Food Sci. Technol. *5*, 12 (1994)

Lück, E.: Chemische Lebensmittelkonservierung. Springer Verlag: Berlin. 1977

Maga, J.A.: Flavor potentiators. Crit. Rev. Food Sci. Nutr. *18*, 231 (1982/83)

Nofre, C., Tinti, J.-M.: Sweetness reception in man: the multipoint attachment theory. Food Chem. *56*, 263 (1996)

Owens, W.H., Kellogg, M.S., Klade, C.A., Madigan, D.L., Mazur, R.H., Muller, G.W.: Tetrazoles as carboxylic acid surrogates: High-potency sweeteners. ACS Symposium Series *450*, p. 100 (1991)

Powrie, W.D., Tung, M.A.: Food dispersions. In: Principles of food science, part I (Ed.: Fennema, O.R.), p. 539, Marcel Dekker, Inc.: New York. 1976

Röhr, M., Kubicek, C.P., Kominek, J.: Citric acid. In: Biotechnology (Eds.: Rehm, H.-J., Reed, G.), Vol. 3, p. 419, Verlag Chemie: Weinheim. 1983

Röhr, M., Kubicek, C.P., Kominek, J.: Gluconic acid. In: Biotechnology (Eds.: Rehm, H.-J., Reed, G.), Vol. 3, p. 455, Verlag Chemie: Weinheim. 1983

Rohse, H., Belitz, H.-D.: Shape of sweet receptors studied by computer modeling. ACS Symposium Series *450*, p. 176 (1991)

Rosival, L., Engst, R., Szokolay, A.: Fremd- und Zusatzstoffe in Lebensmitteln. VEB Fachbuchverlag: Leipzig. 1978

Rymon Lipinski, G.-W. v.: Süßstoffe und Zuckeraustauschstoffe. Lebensmittelchemie *48*, 34 (1994)

Schuster, G. (Ed.): Emulgatoren für Lebensmittel. Springer-Verlag: Berlin. 1985

Schwall, H.: Milchsäure. In: Ullmanns Encyklopädie der technischen Chemie, 4. ed., vol. 17, p. 1, Verlag Chemie: Weinheim. 1979

Shallenberger, R. S., Acree, T.E.: Chemical structure of compounds and their sweet and bitter taste. In: Handbook of sensory physiology, Vol. IV, Part 2 (Ed.: Beidler, L.M.), p. 221, Springer-Verlag: Berlin. 1971

Shallenberger, R. S.: The AH, B glycophore and general taste chemistry. Food Chem. *56*, 209 (1996)

Souci, S.W., Mergenthaler, E.: Fremdstoffe in Lebensmitteln. J.F. Bergmann Verlag: München. 1958

Tinti, J.-M., Nofre, C.: Design of sweeteners: A rational approach. ACS Symposium Series *450*, p. 88 (1991)

Walters, D.E., Orthoefer, F.T., DuBois, G.E. (Eds.): Sweeteners – Discovery, Molecular Design and Chemoreception. ACS Symposium Series *450*, American Chemical Society: Washington, D.C. 1991

Van der Wel, H., Bel, W. J.: Effect of acetylation and methylation on the sweetness intensity of Thaumatin I. Chem. Senses Flavor *2*, 211 (1976)

Yamashita, H., Theerasilp, S., Aiuchi T., Nakaya, K., Nakamura, Y., Kurihara, Y.: Purification and complete amino acid sequence of a new type of sweet protein with taste-modifying activity; curulin. J. Biol. Chem. *265*, 15770 (1990)

Zunft, H.-J.F., Ragotzky, K.: Strategien zur Fettsubstitution in Lebensmitteln. Fett/Lipid *99*, 204 (1997)

9 Food Contamination

9.1 General Remarks

Special attention must be paid to the possibility of contamination of food with toxic compounds. They may be present incidentally and may be derived in various ways. Examples of such contaminants are:

- Pollutants derived from burning of fossil fuels, radionuclides from fallout, or emissions from industrial processing (toxic trace elements, radionuclides, polycyclic aromatic hydrocarbons, dioxins).
- Components of packaging material and of other frequently used products (monomers, polymer stabilizers, plasticizers, polychlorinated biphenyls, cleansing/washing agents and disinfectants).
- Toxic metabolites of microorganisms (enterotoxins, mycotoxins).
- Residues of plant-protective agents (PPA).
- Residues from livestock and poultry husbandry (veterinary medicinals and feed additives).

Toxic food contaminants might also be formed within the food itself or within the human digestive tract by reactions of some food ingredients and additives (e. g. nitrosamines). Measures required to prevent contamination include:

- Extensive analytical control of food.
- Determination of the sources of contamination.
- Legislation (legal standards to permit, ban, curtail or control the use of potent food contaminants, and the processes associated with them) to establish permissible levels of contaminants.

Toxicological assessment of a contaminant may, for various reasons, be a difficult task. Firstly, sufficient data are not available for all compounds. Also, the possibility of synergistic effects of various substances, often including their degradation products, should not be excluded. Further,

the hazard might be influenced by age, sex, state of health and by habitual consumption. Based on these considerations, any nutritional statement about the "tolerable concentration" must take sufficient safety factors into account.

Toxicity assay involves the determination of:

- Acute toxicity, designated as LD_{50} (the dose that will kill 50% of the animals in a test series).
- Subacute toxicity, determined by animal feeding tests lasting four weeks.
- Chronic toxicity, assessed by animal feeding tests lasting 6 months to 2 years.

In chronic toxicity tests attention is especially given to the occurrence of carcinogenic, mutagenic and teratogenic symptoms. The tests are conducted with at least two animal species, one of which is not a rodent.

The upper dosage level for a substance, fed to test animals over their life span and observed for several generations, which does not produce any effect, is designated as the "No Observed Adverse Effect Level" (NOAEL, mg/kg body weight of the animal tested per day or mg/kg feed per day). This level can be used as a basis for estimating the hazard for humans in all cases in which a correlation between dose and effect has been observed. The NOAEL is multiplied by a safety factor (SF: 10^{-1} to 5×10^{-4}, mostly 10^{-2}), with which especially sensitive persons, extreme deviations from the average consumption and other unknown factors are taken into account, giving the toxicologically acceptable dose. It is expressed as the acute RfD (Reference Dose in mg/kg body weight (BW)/day) or as the Acceptable Daily Intake (ADI).

RfD: the acute RfD is the estimated amount of a chemical compound present in food which, related to the body weight and based on all the facts known at the time of the estimation, can be taken in over a short period of time (usually during a meal or a day) without posing a discernible risk to the consumer's health.

ADI: the ADI value denotes the amount of substance which the consumer can take in every day and lifelong with food without discernible injury to his health.

Taking the consumption habits into account, the RfD can be used to calculate the tolerable concentrations (TC) of substances for individual foods:

$$TC = \frac{NOAEL \times FV}{SF} \times \frac{BW}{CA \times ASF}$$

In this formula, TC is the toxicological tolerable concentration for a particular food (expressed in mg/kg food); NOAEL, no observed adverse effect level (mg/kg feed); FV, daily intake of feed by test animals (kg feed/kg body weight); SF, safety factor (10–2000, but usually 100); BW, body weight of an adult (50–80 kg); CA, amount in kg consumed per day of the food for which the TC is being calculated; and ASF, additional safety factor (up to 10) for particularly sensitive persons, such as children or the sick. The maximum concentrations of contaminants (MRL, maximum residue limit in mg/kg food) allowed by legislation are often still well below toxicological tolerance concentrations because other parameters such as "good agricultural practice" are taken into account.

The ADI value is compared with the NEDI or IEDI value (national or international estimated daily intake) to check the risk which comes from contaminants, e.g., pesticides. If the last mentioned value is higher than the ADI, tests are conducted to find out if there is in fact a risk, which would then lead to further measures, if necessary. In many countries, food monitoring is carried out for the early detection of possible danger due to undesirable substances like plant-protective agents (PPA), heavy metals and other contaminants. Foods belonging to the most important groups of goods are repeatedly tested for the presence of certain contaminants. The results are published on the internet; cf. www.bvl.bund.de.

9.2 Toxic Trace Elements

9.2.1 Arsenic

From the viewpoint of its frequency in the environment, toxic activity and the probability of

man's exposure to the substance, arsenic was first on the list of dangerous substances compiled in the USA in 1999. Arsenic was followed by lead, mercury, vinyl chloride, benzene, PCBs, cadmium and benzo[α]pyrene (source: Agency for Toxic Substances and Disease Registry, ATSDR). The amount of arsenic which is probably not dangerous when taken orally is estimated at 0.3 µg/kg body weight/day.

9.2.2 Mercury

Mercury poisoning caused by food intake is derived from organomercury compounds, e.g., dimethyl mercury ($CH_3-Hg-CH_3$), methyl mercury salts (CH_3-Hg; X = chloride or phosphate), and phenyl mercury salts (C_6H_5-Hg-X; X = chloride or acetate). These highly toxic compounds are lipid soluble, readily absorbed and accumulate in erythrocytes and the central nervous system. Some are used as fungicides and for treating seeds (seed dressing). Methyl mercury compounds are also synthesized by microflora from inorganic mercury salt sediments found on lake and river bottoms. Hence, the content of these compounds might rise in fish and other organisms living in water.

The natural mercury level in the environment appears to have stabilized in the last 50 years. Poisonings recorded in Japan appear to have been caused by consumption of fish caught in waters heavily contaminated by mercury-containing industrial waste water, and in Iraq by milling and consuming seed cereals dressed with mercury, which were intended for sowing. The tolerable dose for an adult of 70 kg is 0.35 mg Hg per week, of which a maximum of 0.2 mg may be derived from the highly toxic methyl mercury. The average mercury intake with food, most of which consumed fish, is shown in Table 9.1

9.2.3 Lead

The contamination of the environment with lead is increased by industrialization and by emissions from cars running on leaded gasoline. Tetraethyllead [(C_2H_5)$_4$Pb], an antiknocking additive used to increase the octane value of gasoline, is converted by combustion into PbO, $PbCl_2$ and other inorganic lead compounds. The major part of these compounds is found in an approx. 30 m wide band along roads or highways; the lead level sharply decreases beyond this distance. At a distance of 100 m from a road with heavy traffic, the lead level in the atmosphere decreases by a factor of 10 and that in soil and plants by a factor of 20 from the level found at or close to the road. A decrease in the level of lead in gasoline and increased use of unleaded gasoline has resulted in a drop in the extent of contamination. Environmental lead contamination has not, however, significantly increased the level of lead in food. The lead in soil is rather immobilized; thus the increase in the lead level of plants is not proportional to the extent of soil contamination. Vegetables with larger surface areas (spinach, cabbage) may contain higher levels of lead when cultivated near the lead emission source. When contaminated plants are fed to animals, the body does not absorb much lead since most is excreted in feces.

Further sources of contamination are leadcontaining tin cookware and soldered metal cans and lead-containing enamels. This is particularly so in contact with sour food. These sources of contamination are of lesser importance.

1.75 mg of lead are considered as the tolerable weekly dose for adults of 70 kg. The lead content of food is shown in Table 9.1.

Hair and bone analyses have revealed that lead contamination of humans in preindustrialized times was apparently higher than today. This might be due to the use in those days of lead pipes for drinking water, lead-containing tinware, and excessive use of lead salts for heavily glazed pottery used as kitchenware.

9.2.4 Cadmium

Cadmium ions, unlike Pb^{2+} and Hg^{2+}, are readily absorbed by plants and distributed uniformly in all their tissues, thus decontamination by dehulling or by removal of outer leaves, as with lead, is not possible. Certain wild mushrooms (horse mushrooms, giant mushrooms etc.), peanuts and linseed can contain larger amounts of cadmium. The finer the linseed is ground, the

Table 9.1. Intake of lead, mercury and cadmium through food consumption[a]

Country	Year[b]	µg/Person × Week
Lead		
Germany	1988–92	85–544
Finland	1975–78	460
	about 1990	85
Great Britain	1994	170
The Netherlands	1988–89	168–175
Sweden	1987	119
USA	1990–91	29
Mercury		
Germany	1986	117
	1988	<70
	1988	
	1988	8, 61[c]
Great Britain	1994	28
The Netherlands	1984–86	5
Sweden	1987	13
USA	1986–1991	19,5
Cadmium		
Germany	1986	192
	1988–91	49–99
Great Britain	1994	96
Japan	1992	189–245
The Netherlands	1988–89	84–112
Sweden	1987	84
USA	1986–91	90

[a] Source: J.F. Diehl (cf. Literature, Chap. 9).
[b] Year of the investigation.
[c] Daily consumption of fish.

higher the intake of cadmium on consumption. The contamination sources are industrial waste water and the sludge from plant clarifiers, which is often used as fertilizer. The cadmium content of food is compiled in Table 9.1.

A prolonged intake of cadmium results in its accumulation in the human organism, primarily in liver and kidney. A level of 0.2–0.3 mg Cd/g kidney cortex causes damage of the tubuli. The tolerable weekly dose for an adult (70 kg) is considered to be 0.49 mg of cadmium. On the whole, the concentrations of the toxic trace elements lead, mercury and cadmium in food show a clearly decreasing tendency, especially in recent studies. This is partly due to improvements in trace analyses, but also due to a real decrease in food.

9.2.5 Radionuclides

It is estimated that the average radiation exposure in FR Germany in 1975 was 172 mrad, of which 21 mrad were ascribed to internal radiation by natural radionuclides incorporated in the body (about 90% from ^{40}K, the rest from ^{14}C) and less than 1 mrad by nuclides acquired as a result of atmospheric fallout from nuclear explosion tests (50% from ^{137}Cs, a radionuclide with a half life of 30 years, but quickly excreted by the body; approx. 50% from ^{90}Sr, a most dangerous radioisotope, capable of inducing leukemia and bone cancer; and traces of ^{14}C and tritium). ^{137}Cs and ^{90}Sr are the escort elements of potassium and calcium, respectively. Food contamination with radionuclides in FR Germany had its peak in 1964/65, when the intake in food per day per person was 240 pCi of ^{137}Cs and 30 pCi of ^{90}Sr. Up to the Chernobyl reactor accident in April 1986, the intake was less than 10% of previous values as a result of the moratorium on atmospheric testing of atomic weapons. Radionuclide residues in food were not a health hazard.

For 1986, the accident in Chernobyl caused an additional intake of radionuclides with food that is estimated at (children up to the age of 1 year/adults) 1779/4598 Bq/year of ^{131}J, 986/1758 Bq/year of ^{134}Cs, and 1849/3399 Bq/year of ^{137}Cs. The resulting additional effective equivalent dose for people in the FR Germany is estimated at 0.06–0.22 mSv. In comparison, natural radiation exposure is about 2 mSv per year, of which 0.38 mSv/year is caused by radionuclides in food. As a precaution, maximum activity values of 500 Bq/l and 250 Bq/kg have been stipulated for milk and vegetables respectively. In comparison, the activity of natural radionuclides (mainly ^{40}K) in food is: milk 40–50 Bq/kg, milk powder 400–500 Bq/kg, fruit juice concentrate 600–800 Bq/kg and soluble coffee (powder) >1000 Bq/kg.

The level of tritium infiltrating the biosphere is expected to rise further due to increasing nuclear plant operation worldwide.

9.3 Toxic Compounds of Microbial Origin

9.3.1 Food Poisoning by Bacterial Toxins

Most (60–90%) cases of food poisoning are bacterial in nature. They are distinguished by food intake causing:

- Intoxication (poisoning, e. g., by *Clostridium botulinum, Staphylococcus aureus*).
- Diseases caused by massive pollution with facultative pathogenic spores, e. g., *Clostridium perfringens, Bacillus cereus*.
- Infections by *Salmonella* spp. or *Shigella* spp., *Escherichia coli*.
- Diseases of unclear etiology, such as those from *Proteus* spp., *Pseudomonas* spp.

The harmful activity of these bacteria in the digestive tract is ascribed to enterotoxins, which are classified into two groups: exotoxins (toxins excreted by microorganisms into the surrounding medium) and endotoxins (retained by the microorganism cells but released when the cell disintegrates).

Exotoxins are released primarily by gram-positive bacteria during their growth. They consist mostly of proteins which are antigenic and very poisonous. They become active after a latent period. This group includes the toxins released by *Clostridium botulinum* (botulin toxin, a globular protein neurotoxin), *Cl. perfringens* and *Staphylococcus aureus*. Table 9.2 gives some important data for these microorganisms, including harmful effects. Intoxications with *St. aureus* are the most frequent cause of food poisoning. Symptoms are vomiting, diarrhea and stomach ache and are caused primarily by food of animal origin (meat and meat products, poultry, cheese, potato salad, pastry).

Endotoxins are produced primarily by gramnegative bacteria. They act as antigens, are firmly bound to the bacterial cell wall and are complex in nature. They have protein, poly-saccharide and lipid components. Endotoxins are relatively heat stable and are in general active without a latent period. The toxins causing typhoid and paratyphoid fevers, salmonellosis and bacterial dysentery are in this group. Salmonellosis is very serious. It is an infection by toxins of about 300 different but closely related organisms. The infec-

Table 9.2. Food poisoning by bacterial toxins

Microorganism:	Staphylococcus aureus	Clostridium botulinum	Bacillus cereus perfringens	Clostridium	E. coli 0157:H7
Growth conditions					
temperature range	10–45 °C	4–35 °C	10–45 °C	12–52 °C	8–45 °C
pH range	4.5	5		5–8.5	4.0–7.5
Toxin					
type	Protein	Protein	Lipid (?)	Protein	Protein
effective amount	0.5–1 μg	0.1–1 μg	10^8 spores/g	10^6 spores/g	
stability	Relatively thermostable	Thermolabile, inactivation at 80 °C/30 min or 100 °C/5 min			
Incubation time	2–6 h	1–3 days	1–12 h	8–24 h	1 day
Duration of disease	1–3 days	Death after 1–8 days, with survivors ill 6–8 months	0.5–1 day	0.5–1 day	1–3 days first symptoms 2 days – ? disease
Symptoms	Vomiting, diarrhea, abdominal pain	Paralysis of the nerve centres of the medulla oblongata	Abdominal pain, nausea, diarrhea, vomiting	Diarrhoe, abdominal convulsions, nausea, loss of appetite	Severe diarrhoe, destruction of erythrocytes
Foods usually accounting for poisoning	Cold meat and cheese slices, mildly acidic salads (meat, poultry, sausage, cheese, potatoes), mayonnaise, cream fillings in baked products	Homemade canned meat, hind bony ham, sliced sausages, trout fillets, canned green beans	Institutional/community catering: Heated and warmed dishes, cereal containing dishes (corn, rice)	Institutional/communal catering: Heated and warmed meat dishes, warm desserts, puddings, cream fillings in baked products	Insufficiently heated meat, raw milk, unpasteurized apple juice, unwashed fruit and vegetables

tion is characterized by enteric fever, gastroen-teritis and salmonella septicemia. Sources of in-fections are egg products, frozen poultry, ground or minced beef, confectionery products and co-coa.

Although the bacterium *E. coli* first served only as an indicator of fecal contamination, it has meanwhile received special attention. This bac-terium also includes enterotoxic strains, e. g., the especially dangerous strain 0157:H7 discovered in 1983 (Table 9.2).

9.3.2 Mycotoxins

There are more than 200 mycotoxins produced under certain conditions by about 120 fungi or molds. Table 9.3 presents data on mycotoxins of particular interest to food preservation and stor-age. The chemical structures of these toxins are presented in Fig. 9.1.

Infections of rye and, to a lesser extent, of other cereal grains with *Claviceps purpurea* (ergot, or rooster's spur) are responsible for the disease called ergotism (symptoms: gangrene and convulsions). The disease was important in the past when bread from infected rye grain was eaten. It has practically ceased to exist due to seed treatment with fungistatic agents and grain cleaning prior to milling.

Most mycotoxin data are on the genera *As-pergillus spp.* and the aflatoxins they produce

during growth. These are the most common and highly toxic fungal toxins, e. g., aflatoxin B_1, the most powerful carcinogen known. In animal feeding tests with rats, its carcinogenic effect is revealed at a daily dose of only 10 µg/kg body weight. In a comparative study, the carcinogenic property of the highly toxic dimethylnitrosamine was revealed only at a daily dose of 750 µg/kg body weight. It is primarily plant material (particularly nuts and fruit) that is contaminated with aflatoxins. Aflatoxin passes from moldy feed to animal products, primarily milk. The dairy cow's metabolism converts the B-group aflatoxins to those of the M-group ("M" stands for metabolite), which are also carcinogenic. Nephrotoxic ochratoxin A passes from fodder cereals mainly to the blood and kidney tissue of pigs, but it is also found in the muscles, liver and adipose tissue.

In the course of food monitoring between 1995 and 2002, more than 40 foods were tested for the presence of aflatoxins, deoxynivalenol, fumosins, patulin, ochratoxin A und zearalenone. Individual mycotoxins were detected in 21% of the samples; pistachios were especially conspicuous.

An assessment of the health hazards caused by mycotoxins is not meaningful when applied to the aflatoxins because these substances damage DNA, are carcinogenic and have no threshold be-low which no harmful effects are observed. An assessment was possible in the case of deoxyni-valenol und ochratoxin A, with the reservation

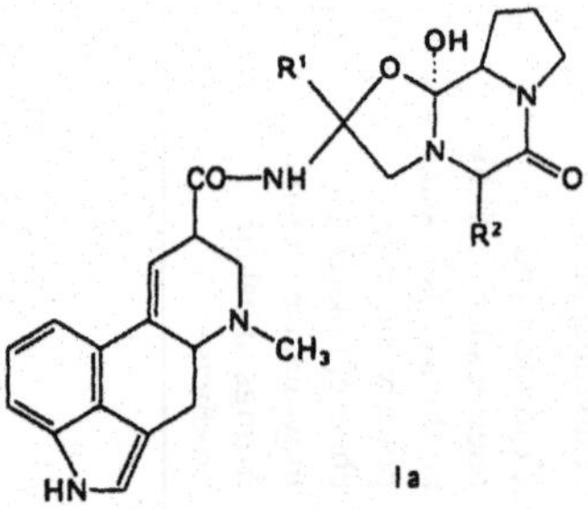

I: Alkaloids of Ergot
Ia: Ergocristine (R^1 = CH(CH$_3$)$_2$,
R^2= H$_2$C–C$_6$H$_5$), Ergostine (R^1= C$_2$H$_5$,
R^2= H$_2$C–C$_6$H$_5$), Ergotamine (R^1= CH$_3$,
R^2= H$_2$C–C$_6$H$_5$),Ergocryptine (R^1= CH(CH$_3$)$_2$,
R^2= H$_2$C–CH(CH$_3$)$_2$), Ergocornine (R^1= CH(CH$_3$)$_2$
R^2= CH(CH$_3$)$_2$), Ergosine (R^1= CH$_3$,
R^2= CH$_2$–CH(CH$_3$)$_2$) Ib: Ergometrine

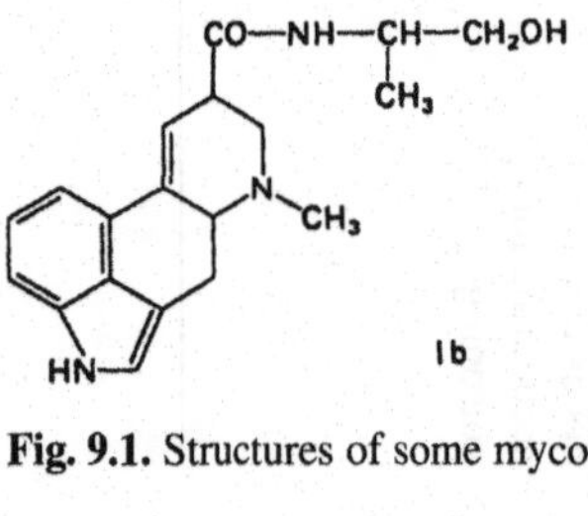

Fig. 9.1. Structures of some mycotoxins (cf. Table 9.3)

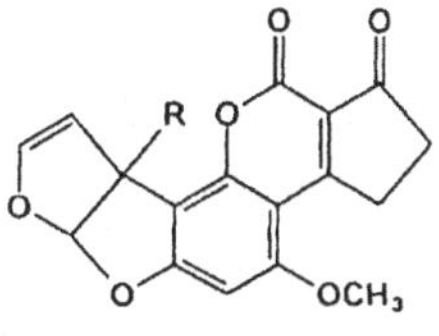

II a

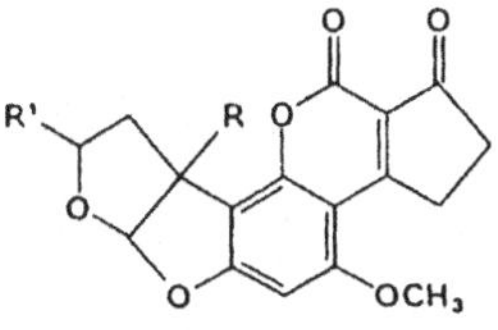

II b

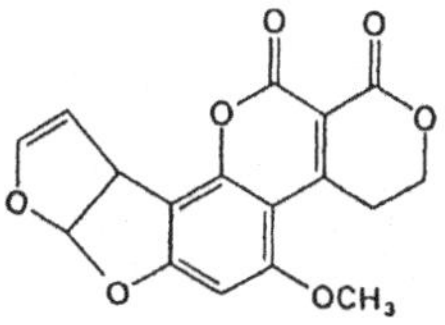

II c

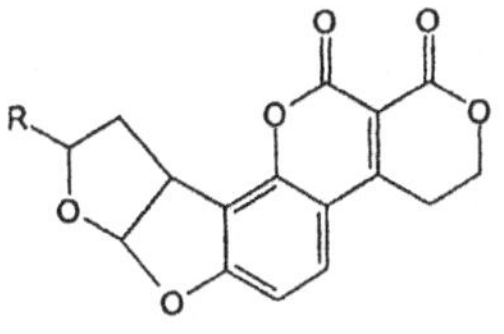

II: Aflatoxins
II a: Aflatoxin B$_1$ (R=H),
Aflatoxin M$_1$ (R=OH), II b: Aflatoxin B$_2$
(R, R^1=H), Aflatoxin M$_2$ (R=OH, R^1=H),
Aflatoxin B$_{2a}$ (R=H, R^1=OH), II c: Aflatoxin G$_1$,
II d: Aflatoxin G$_2$ (R=H), Aflatoxin G$_{2a}$ (R=OH)

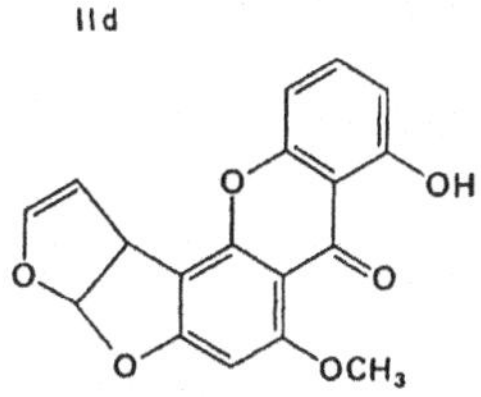

II d

III: Sterigmatocystine
IV: Patuline

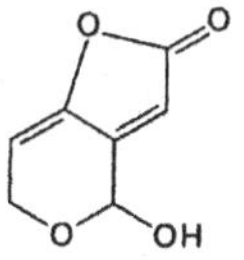

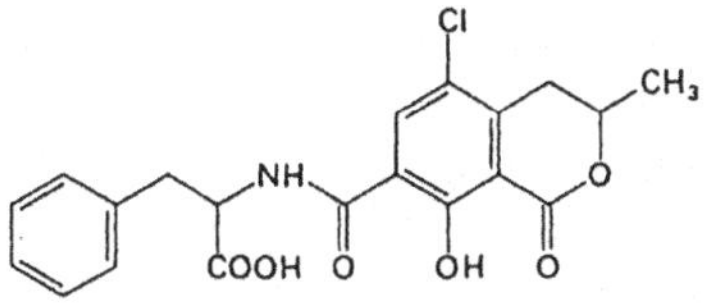

III

IV

V

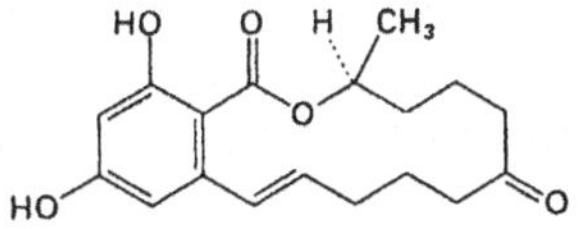

VI

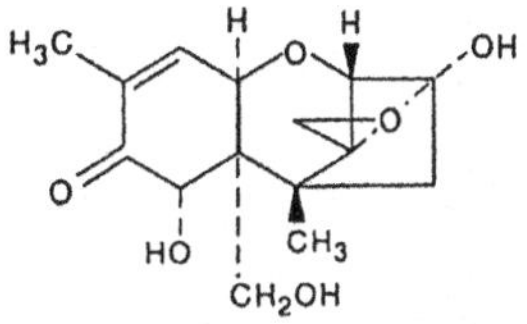

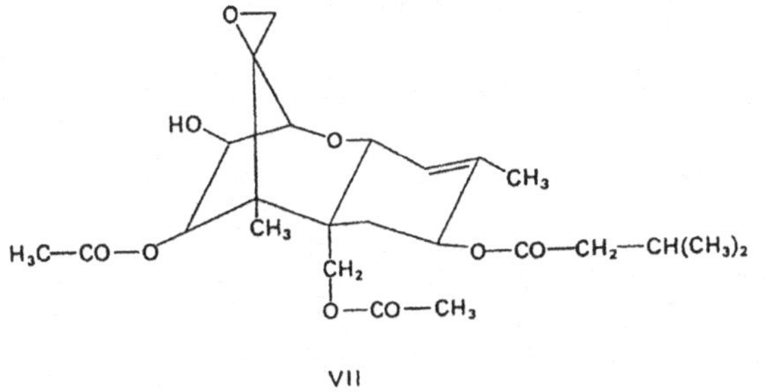

VII

VIII

V: Ochratoxin A
VI: trans-Zearalenone
VII: Fusariotoxin T$_2$
VIII: Vomitoxin

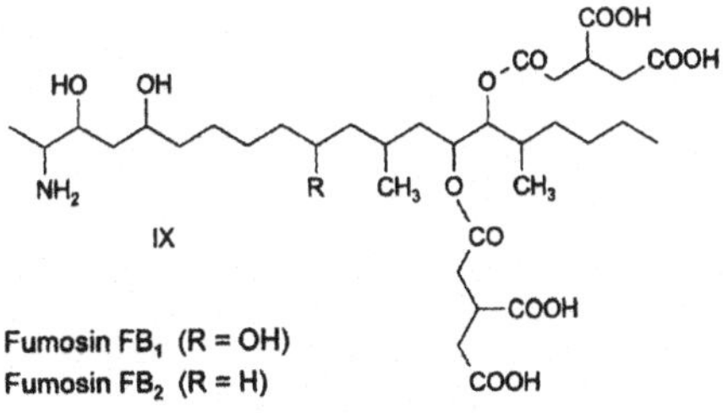

IX

IX: Fumosin FB$_1$ (R = OH)
Fumosin FB$_2$ (R = H)

Fig. 9.1. (Continued)

Table 9.3. Mycotoxins

Fungus/mold	Toxin[a]	Toxicity[b]	Effect	Occurrence
Claviceps purpurea	Ergot alkaloids (I)		Ergotism (gangrenous convulsions)	Mainly rye, to a lesser extent wheat
Aspergillus flavus A. parasiticus	Aflatoxins (II)	7.2 mg/kg (rat, orally)	Liver cirrhosis, liver cancer	Groundnuts and other nuts (almond, Brasil nut) corn and other cereals, animal feed, milk
Aspergillus versicolor A. nidulans	Sterigmatocystin (III)	120 mg/kg (rat, orally)	Liver cancer	Corn, wheat, animal feed
Penicillium expansum P. urticae	Patulin (IV)	35 mg/kg (mouse, orally)	Cellular poison	Putrifying fruits, fruit juices
Byssochlamis nivea, B. fulva				
Aspergillus ochraceus A. melleus	Ochratoxin A (V)	20 mg/kg (rat, orally)	Fatty liver and kidney damage	Barley, corn
Fusarium graminearum	Zearalenone (VI) (Fusariotoxin F$_2$)	0.1 mg/kg over 5 days (swine, orally[c])	Estrogen, infertility	Corn and other cereals, animal feed
Fusarium oxysporum F. tricinctum	Fusariotoxin T$_2$ (VII)	3.8 mg/kg (rat, orally)	Toxic aleukia, hemorrhagic syndrome	Cereals, animal feed
Fusarium roseum F. graminearum	Deoxynivalenol (VIII) (Vomitoxin)	70 mg/kg (mouse)	Vomiting	Cereals, animal feed
Fusarium moniliforme	Fumosin FB$_1$ and FB$_2$ (IX)		Liver cancer	Corn

[a] Roman numerals refer to the structural formulas in Fig. 9.1.
[b] Acute toxicity (LD$_{50}$).
[c] Estrogenic activity.

Table 9.4. Reference values and their utilization for two mycotoxins (food monitoring 1995–2002)

Substance	Reference[a]	Reference value (µg/kg kg/d)	Utilization (%)
Deoxyni-valenol	PMTDI	1	34.1–82.5
Ochra-toxin A	PTDI	0.005	7.4–16.1

[a] P(M)TDI: provisional (maximum) tolerable daily intake

that (provisional) defined reference values were assumed. The reservation referred to the fact that the data base for a sound evaluation of the effects on human health is still too limited.

The food intake must be known in order to calculate the utilization of the reference values. For this purpose, a large national study on consumption was carried out in the FRG between 1985 and 1988. With regard to the preferred foods, the amount consumed and the average body weight of the test persons were evaluated. For a differentiated presentation of the results, the test persons were divided into a total of 10 different age, sex and consumption groups, e.g., children, men, women. Among the men and women, a distinction was also made between the meat, fish and fruit eaters (cf. 9.4.4.2).

Table 9.4 shows that in the case of deoxynivalenol, the utilization of the reference value is relatively high at 34.1–82.5%. The upper value was calculated for 4–6 year old children and the lower value for women (fish eaters). Cereal products are mainly contaminated with deoxynivalenol. Ochratoxin A is also most frequently taken in by children. In addition to cereal products, especially fruit juices play a role as a source of this substance.

In comparison with the usual HPLC/UV method, it has been shown in the analysis of mycotoxins using patulin (IV in Fig. 9.1) that the detection limit decreases by a factor of 100 if an isotopic dilution assay (cf. 5.2.6) is carried out with $[^{13}C_2]$ patulin as the internal standard. After silylation and gas chromatographic/mass spectrometric measurement of analyte and isotopomer, 5.7–26.0 µg/l of patulin were found in apple juices.

9.4 Plant-Protective Agents (PPA)

9.4.1 General Remarks

The term PPA includes all the compounds used in agricultural food production to protect cultivated plants from plant- and insect-caused diseases, parasites or weeds, or from detrimental microorganisms. The most important groups of PPA are: (1) *herbicides* to protect the plant from weeds; (2) *fungicides* to suppress the growth of undesired fungi or molds; and (3) *insecticides* to protect the plants from damage caused by insects. In addition to these main groups, there are *acaricides* to control mites, *nematocides* to control worms or nematodes, *molluscicides* to protect the plant from snails and slugs, *rodenticides* to control rodents (mice or rats), and plant *growth regulators* (cf. 18.1.4). In Germany in 2003, the herbicides had the largest market share at 43%, followed by the fungicides (28%), insecticides (18%) and the growth regulators (9%). The remaining agents accounted for 2%.

The use of PPA is rewarding since it reduces losses in crop yield and stocks. It has also contributed to the control or eradication of insect-spread diseases such as malaria. Without pest control, the harvest losses of rice, which is especially susceptible, would be 24%. The use of PPA reduces rice losses to 14% and wheat, soybean and corn losses to 7–10%. Apart from losses during cultivation, about 15% of the world harvest is lost during storage due to pests in barns and silos. The substances applied must be effective against pests but safe for the user, consumer and the environment. Accordingly, these substances are evaluated with regard to their toxicological and ecological properties in the registration procedure. Their influence on, e. g., beneficial organisms, aquatic organisms, birds and mammals is tested and their degradability in the soil or in the plant is determined. Since the registration of pesticides has to be renewed at certain times and the costs of the re-evaluation have to be carried mainly by the manufacturer, only those substances which are successful on the market are put through this procedure.

PPA are applied in various forms and by various means: dusting as powder, fumigation, spraying as a liquid, or pad or furrow irrigation. The strict observance of directions for use, waiting the recommended time between final application and harvest, and restricting the application to the necessary dose, is required to maintain the residual pesticide levels in food at a minimum. Legal regulations with bans on use, stipulated maximum permissible quantities (MRL) etc. emphasize these requirements.

Contamination of food of plant origin can occur directly by treating the crop before storage and distribution (fruit and vegetable treatment with fungicides, cereal treatment with insecticides). It can occur indirectly by uptake from the soil of residual PPA by the subsequent crop, from the atmosphere or drifting from neighboring fields, or from a storage space pretreated with PPA.

Contamination of food of animal origin occurs by ingestion of feed containing stall- and barn-cleansing agents (fungicides, insecticides), by coming in contact with wooden studs and boards preserved with fungistatic agents, and veterinary medicines and, occasionally, by use of disinfected corn as fodder.

The structures of the PPA mentioned in the following sections are shown in Table 9.5 and Fig. 9.2. Comprehensive details on PPA are given on the internet at www.hclrss.demon.co.uk/index.html and http://extoxnet.orst.edu/pips/ghindex.html and by Tomlin (cf. Literature), who has listed more than 800 compounds.

9.4.2 Active Agents

9.4.2.1 Insecticides

Organophosphate compounds (e.g., IX, XXX, XXXVII, XXXIX in Table 9.5), carbamates (e.g., VII) and pyrethroids (e.g., XIII) have been used as insecticides for many years. The pyrethroids are synthetic modifications of pyrethrin I (cf. Formula 9.1), the main active agent of pyrethrum. Pyrethrum is isolated from the capitulum of different varieties of chrysanthemums and used as a natural insecticide.

(9.1)

Chlorinated hydrocarbons like dichlorodiphenyltrichloroethane (DDT, XII) und lindane (XXVIII) belong to those pesticides which are no longer approved in the EU and the USA. Exceptions are made worldwide only for the use against mosquitoes to control malaria and only if no alternatives are available. The reason for the rejection of chlorinated hydrocarbons is their persistence. They are stable and accumulate in human beings and animals (in the fat phase) and in the environment. The half life DT_{50} (time for 50% dissipation of the initial concentration) of DDT is 4–30 years. In comparison, the DT_{50} values of the organophosphates, carbamates and pyrethroids are in the range of days to a few months, e.g., chlorpyriphos (IXa, b) 10–120 days, carbofuran (VII) 30–60 days and deltamethrin (XIII) <23 days. In Germany, the ban on DDT has resulted in a decrease in the concentration of DDT and its degradation product DDE (cf. 9.4.3) in human milk (mg/kg milk fat) from 1.83 (1979–81) to 0.132 (2002).

Pest populations which are resistant to the active agents develop on longer application. Therefore, new active agents have to be continually synthesized to calculate this resistance. Examples are indoxacarb (XXV) and tebufenozide (XLIII).

Insecticides are mainly nerve poisons. In particular, the older active agents, e.g., parathion (XXXVII, introduced in 1946), chlorpyriphos (IXa, 1965) and methidation (XXXIV, 1965) are also very toxic to mammals (compare acute toxicity LD_{50} in Table 9.5). The toxicity falls when the ethyl groups in IXa are replaced by methyl groups (compare LD_{50} of IXa and b in Table 9.5). Nerve poisons inhibit acetylcholine esterase, bind to receptors which are controlled by the neurotransmitter acetylcholine or interfere with the neurotransmission in the nervous system by modifying the ion canals (examples in Table 9.5). Other insecticides damage the respiratory chain. As mitochondrial uncouplers, they prevent the formation of a proton gradient (Table 9.5). Another mechanism of action is directed at the development of the pests, which can be prevented, e.g., by inhibiting the biosynthesis of chitin.

9.4.2.2 Fungicides

Real fungi (Ascomycetes, Basidomycetes, Deuteromycetes) and lower fungi (Oomycetes)

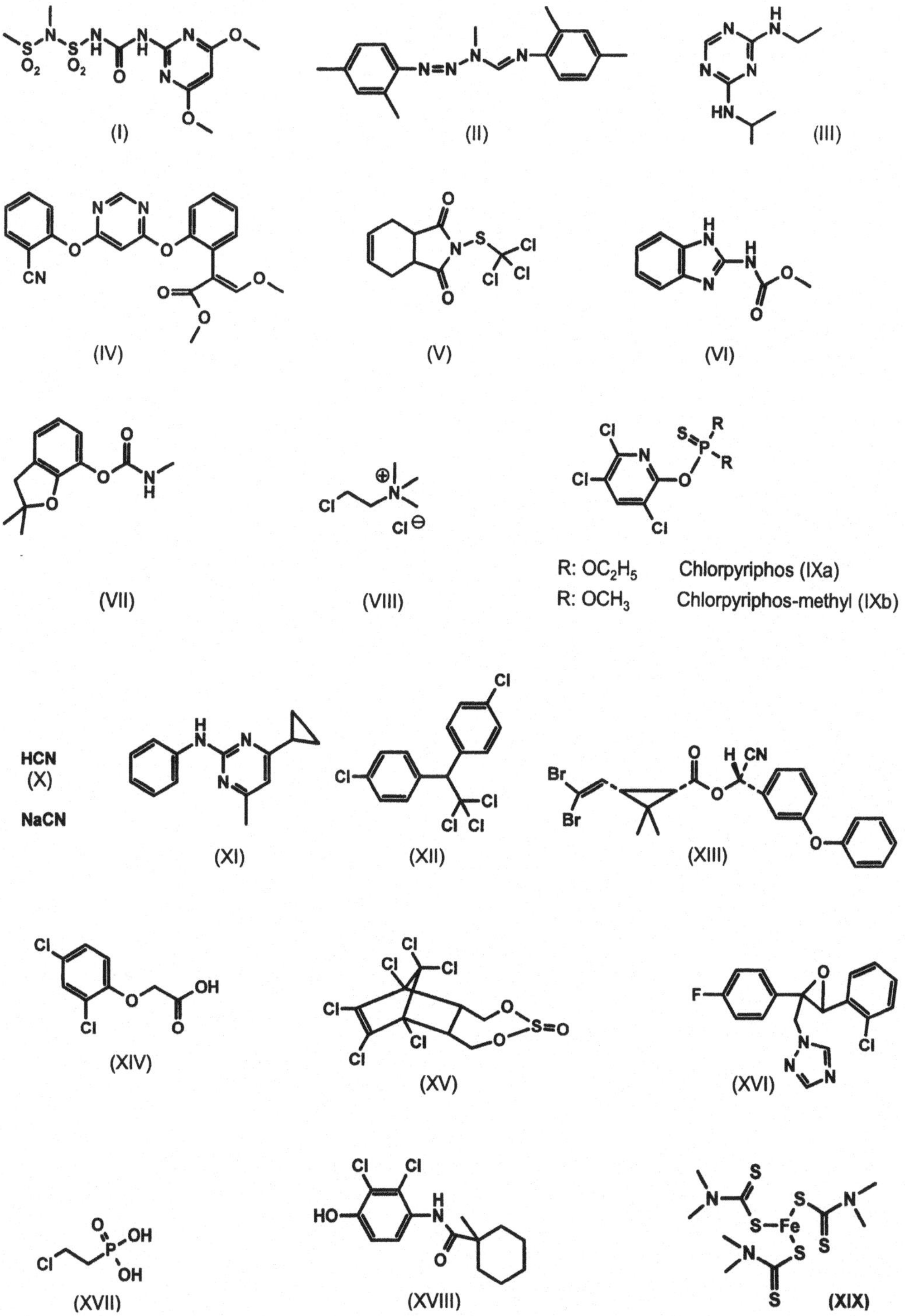

Fig. 9.2. Structures of some selected PPA. The Roman numerals refer to Table 9.5 – sheet 1

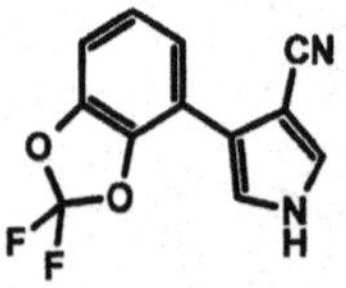

(XX)

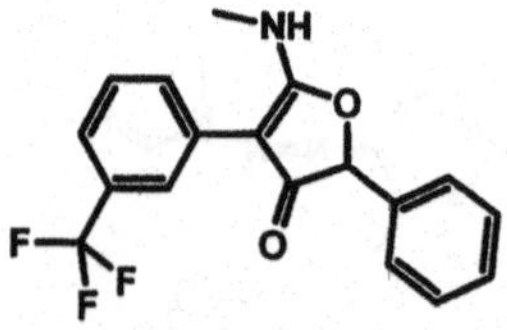

(XXI)

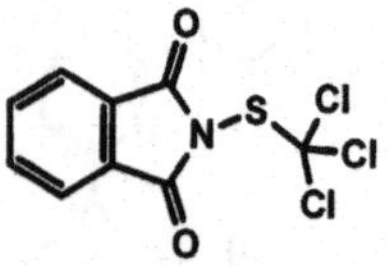

(XXII)

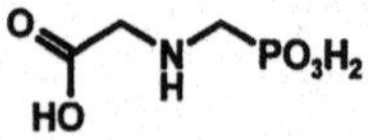

(XXIII)

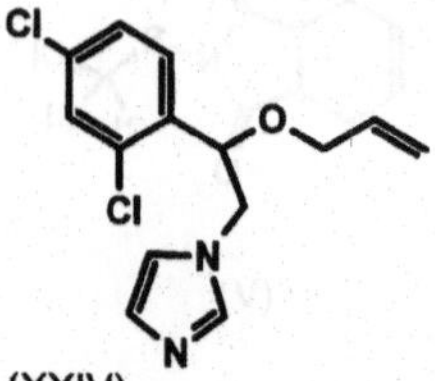

(XXIV)

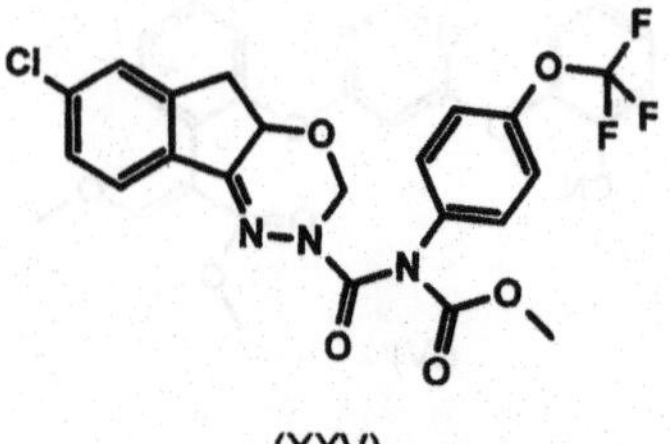

(XXV)

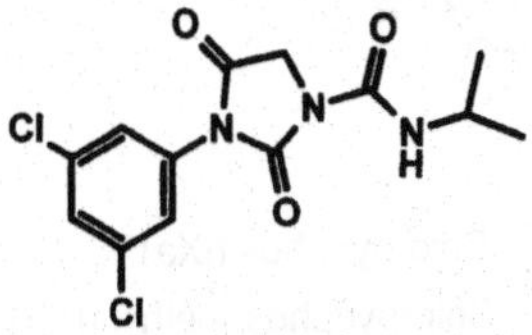

(XXVI)

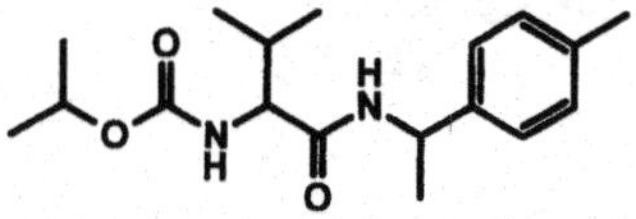

(XXVII)

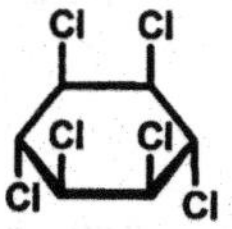

(XXVIII)

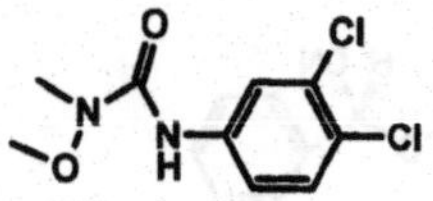

(XXIX)

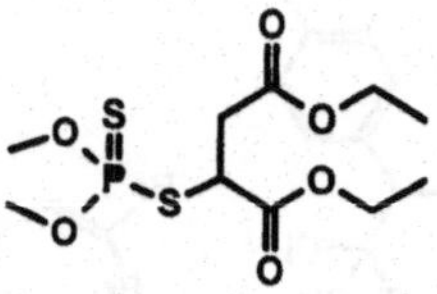

(XXX)

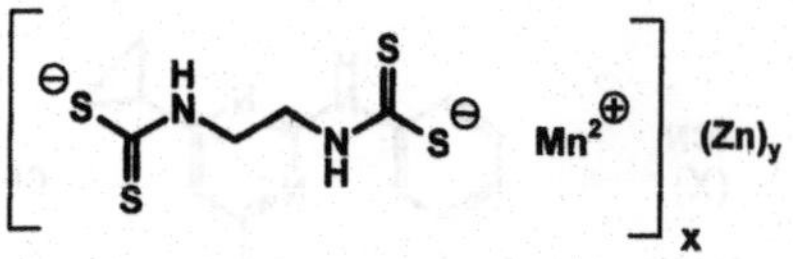

(XXXIa)

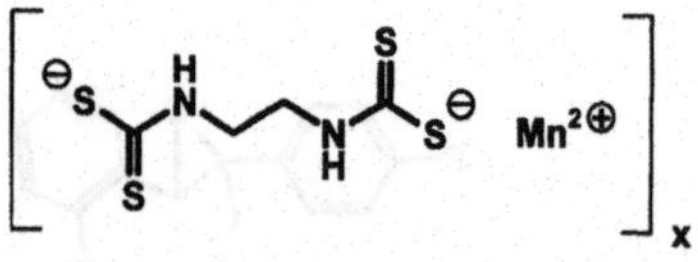

(XXXIb)

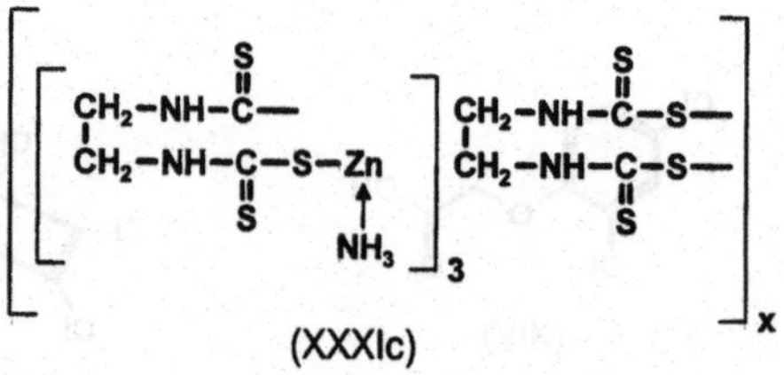

(XXXIc)

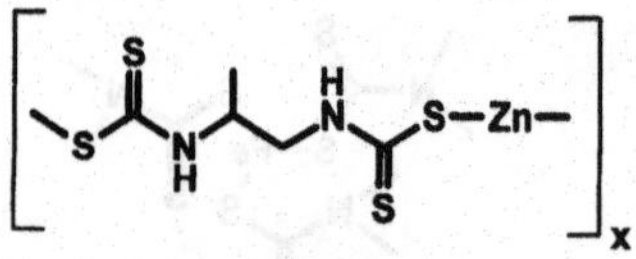

(XXXId)

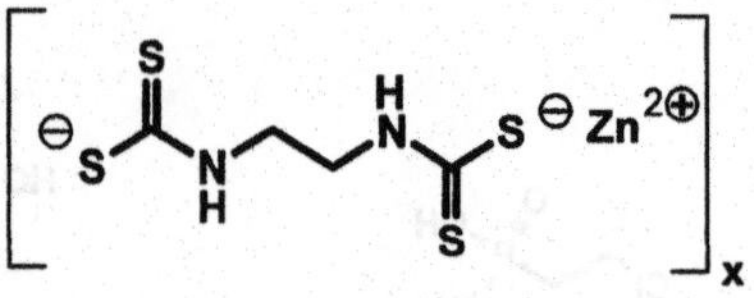

(XXXIe)

Fig. 9.2. (Continued)

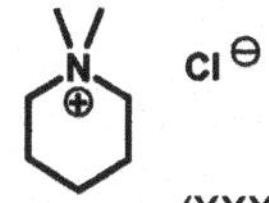

(XXXII)

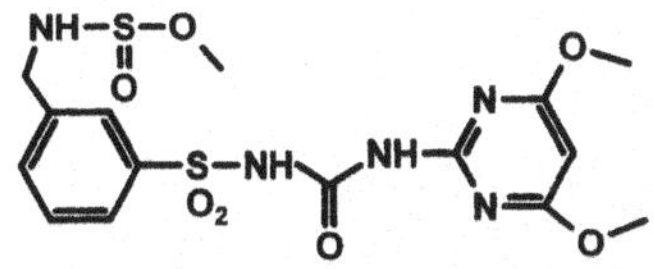

(XXXIII)

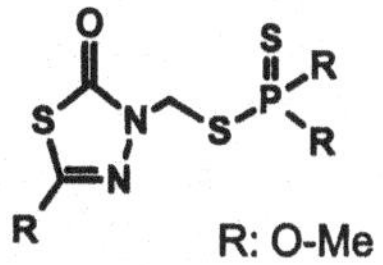

(XXXIV)

CH_3-Br

(XXXV)

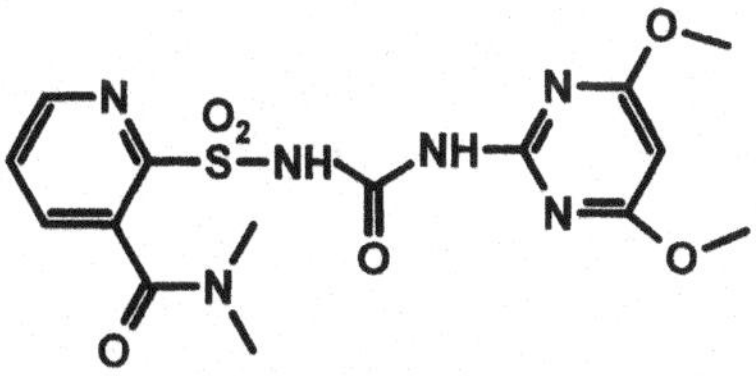

(XXXVI)

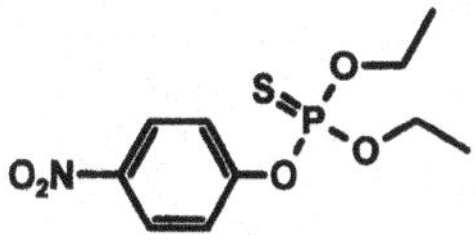

(XXXVII)

PH_3

(XXXVIII)

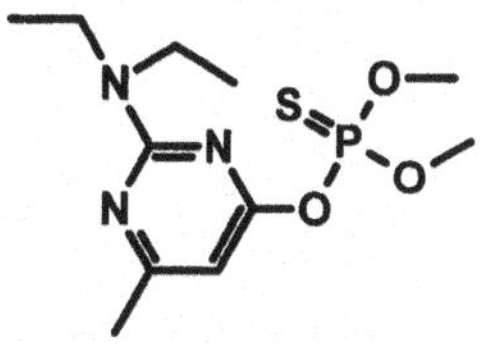

(XXXIX)

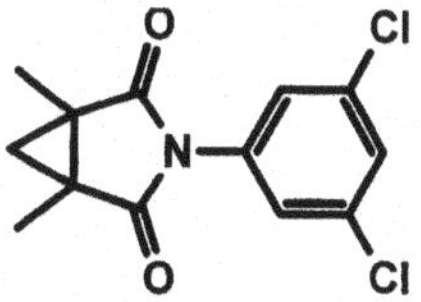

(XL)

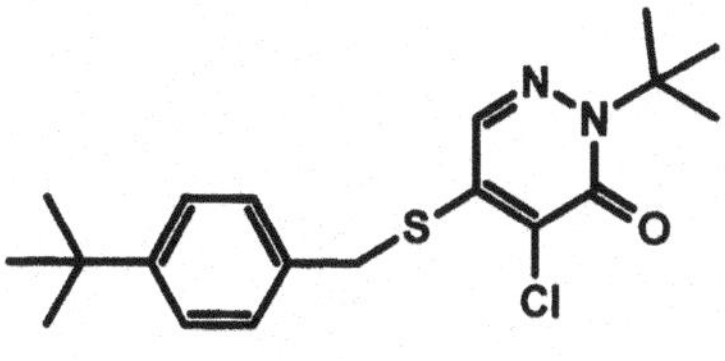

(XLI)

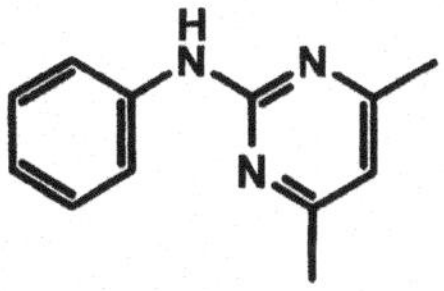

(XLII)

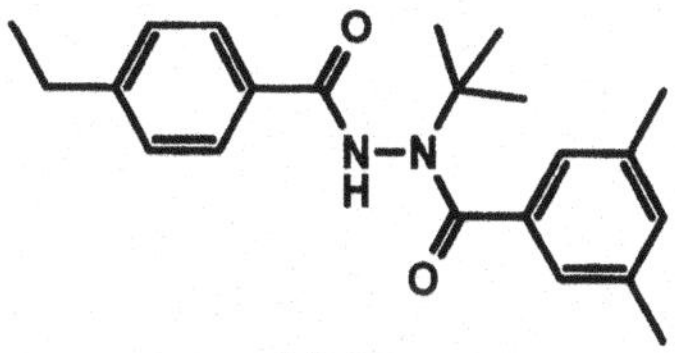

(XLIII)

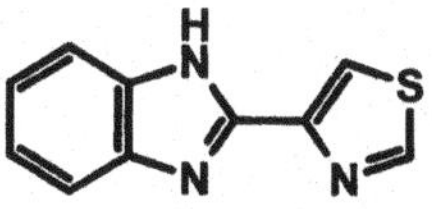

(XLIV)

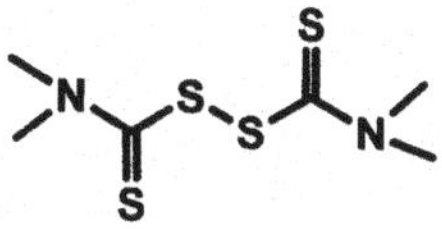

(XLV)

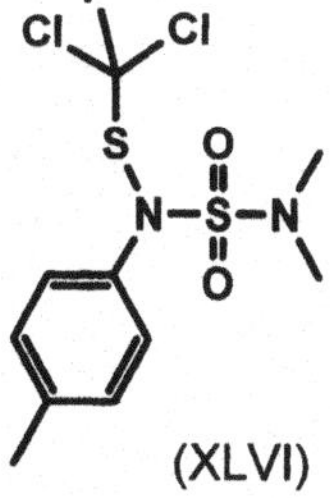

(XLVI)

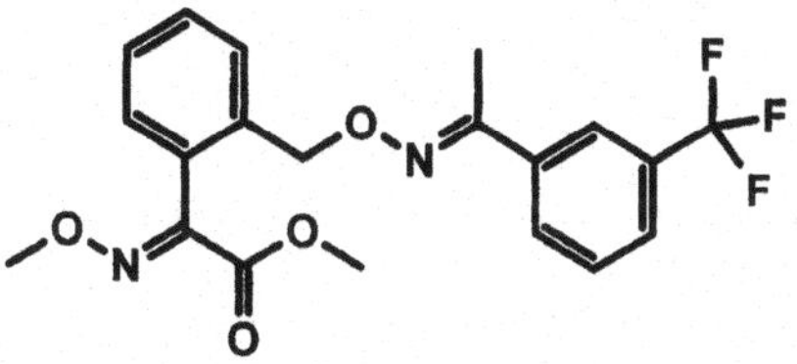

(XLVII)

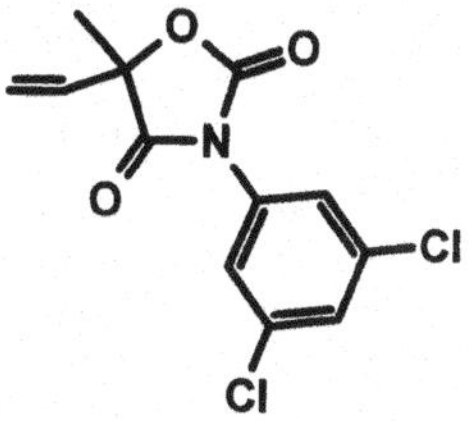

(XLVIII)

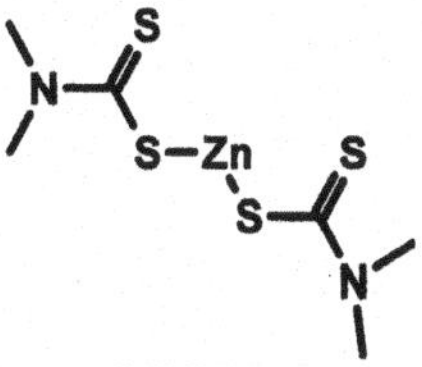

(XLIX)

Fig. 9.2. (Continued)

Table 9.5. Selected PPA

No.	Name	Application[a]	Biochemical activity	LD$_{50}$ (mg/kg)[b]
I	Amidosulfuron	H	Inhibits synthesis of branched AA by inhibition of acetolactate synthase	$\geq$5000 (R, M)
II	Amitraz	I, A	Nerve poison	650 (R)
III	Atrazine	H	Inhibits electron transport in photosystem II	1869–3090 (R)
IV	Azoxystrobin	F	Inhibits the respiratory chain	>5000 (R, M)
V	Captan	F	Inhibits respiration	9000 (R)
VI	Carbenedazine	F	Inhibits β-tubulin biosynthesis	6400 (R), >2500 (H)
VII	Carbofuran	I, N	Choline esterase inhibitor	8 (R), 15 (H)
VIII	Chloromequat	Pgr	Inhibits gibberellin biosynthesis	807–966 (R)
IXa	Chlorpyriphos	I	Choline esterase inhibitor	135–163 (R), 1000–2000 (K)
IXb	Chlorpyriphos methyl	I, A	Choline esterase inhibitor	>3000 (R), 1100–2250 (M)
X	Cyanide	I, R	Prevents O_2 transport by hemoglobin into the cell	6–15 (R)
XI	Cyprodinil	F	Probably inhibitor of methionine synthesis	>2000 (R)
XII	DDT	See text	Nerve poison	113–118 (R), 500–570 (H)
XIII	Deltamethrin	I	Nerve poison: inhibits the function of Na$^+$ ion channels	>2000 (R), >300 (H)
XIV	2,4-Dichlorophenoxyacetic acid (2,4-D)	H	Inhibits growth	639–764 (R)
XV	Endosulfan	I, A	Antagonist for the GABA receptor[c]	70–240 (R)
XVI	Epoxiconazole	F	Inhibits sterol biosynthesis	>5000 (R)
XVII	Ethephon	Pgr	Decomposes to ethylene	3030 (R)
XVIII	Fenhexamide	F	Inhibits sterol biosynthesis	>5000 (R)
XIX	Ferbam	F		>4000 (R)
XX	Fludioxonil	F	Inhibits MAP kinase[d]	>5000 (R)
XXI	Flurtamone	H	Blocks carotene biosynthesis	500 (R)
XXII	Folpet	F	Inhibits respiration	>9000 (R)
XXIII	Glyphosates	H	Inhibits biosynthesis of aromatic compounds	5600 (R), 3530 (goat)
XXIV	Imazalil	F	Inhibits ergosterol biosynthesis	227–343 (R), >640 (H)
XXV	Indoxacarb	I	Blocks Na channels in nerve cells	1732 (male R), 268 (female R)
XXVI	Iprodion	F	Inhibits germination of spores and growth of mycelium	>2000 (R, M)
XXVII	Iprovalicarb	F	Inhibits growth of Oomycetes fungi	>5000 (R)

Table 9.5. (Continued)

No.	Name	Application[a]	Biochemical activity	LD_{50} (mg/kg)[b]
XXVIII	Lindane (γ-HCH)	See text	Antagonist for the GABA receptor	88–270 (R)
XXIX	Linuron	H	Inhibits electron transport in photosystem II	1500–4000 (R)
XXX	Malathion	I, A	Inhibitor of choline esterase	1375–5500 (R)
XXXI	Maneb group	F	Inactivates SH groups, inhibits respiration	
	a. Mancozeb			>5000 (R)
	b. Maneb			>5000 (R)
	c. Metiram			>5000 (R)
	d. Propineb			>5000 (R)
	e. Zineb			>5200 (R)
XXXII	Mepiquat	Pgr	Inhibits biosynthesis of gibberellic acid	464
XXXIII	Mesosulfuron methyl	H	Inhibits biosynthesis of branched AA like I	>5000
XXXIV	Methidathion	I, A	Inhibits choline esterase	25–54 (R), 25–70 (M)
XXXV	Methylbromide	I, A, R		See legend[e]
XXXVI	Nicosulfuron	H	Inhibits biosynthesis of branched AA like I	>5000 (R)
XXXVII	Parathion	I, A	Inhibits choline esterase	$\approx$2 (R), 12 (M)
XXXVIII	Phosphide/PH3	I, R	Inhibits respiration	8,7 (R)[f]
XXXIX	Pirimiphos methyl	I, A	Inhibits choline esterase	1414 (R), 1180 (M)
XL	Procymidone	F	Inhibits triglyceride biosynthesis in molds	6800 (R)
XLI	Pyridaben	I, A	Inhibits electron transport in the mitochondria	1350 (R)
XLII	Pyrimethanil	F	Inhibits methionine biosynthesis (assumption)	4150–5971 (R)
XLIII	Tebufenozid	I	Antagonist of the insect hormone ecdyson	>5000 (R)
XLIV	Thiabendazol	F	Inhibits mitosis by binding to tubulin	3100 (R)
XLV	Thiram	F		2600 (R), 210 (K)
XLVI	Tolylfluanid	F	Inhibits respiration	>5000 (R)
XLVII	Trifloxystrobin	F	Inhibits the respiratory chain	>5000 (R)
XLVIII	Vinclozolin	F	Prevents germination of spores	>15000 (R, M)
XLIX	Ziram	F	Inactiviates the SH-group	>2000 (R), 100–300 (K)

[a] A, acaricide; H, herbicide; I, insecticide; R, rodenticide; Pgr, plant growth regulator.
[b] Mean lethal dose (LD_{50}); experimental animal: H, dog; M, Mouse, R, rat; K, rabbit.
[c] GABA: γ-aminobutyric acid.
[d] MAP kinase: nitrogen-activated protein kinase (participates in mitosis).
[e] Toxic with threshold value for inhalation of 0.019 mg/l air.
[f] LD_{50} for aluminium phosphide.

can destroy entire harvests. These fungi and their spores are controlled by fungicides so that mildew, rust, leaf blight, stem rot, botrytis and other plant diseases do not occur. Depending on the mode of action, a distinction is made between the contact fungicides, which act on the surface of the plant preventing germination and/or penetration of the fungus into the plant, and systemic fungicides, which penetrate into the plant and eliminate hidden seats of disease.

The active agents can be divided into inorganic, organometallic and organic compounds. Inorganic fungicides include Bordeaux mixture, copper chloride oxide, lime sulfur and colloidal sulfur. Examples of organometallic compounds are dithiocarbamates of zinc and manganese (Maneb group; XXXIa–e in Table 9.5), which are relatively often encountered as residues in foods (cf. 9.4.4). Most of the fungicides are, however, metal-free organic compounds (examples in Table 9.5).

In the case of the fungicides, too, the development of resistance necessitates the continual development of new active agents. A special innovation was the introduction of synthetic modifications of the fungal constituent strobilurin A (Formula 9.2), which has antibiotic and fungicidal activity. Examples are azoxystrobin (IV) and epoxyconazole (XVI).

Another new class of substances are the valinamides, which has given rise to the active agent iprovalicarb (XXVII). Information on the toxic activity of these fungicides is given in Table 9.5.

$$(9.2)$$

The development of active agents with increased fungicidal activity but constant relatively low toxicity for mammals has led to a considerable reduction of the dose required. The examples in Table 9.6 show that this trend has also been observed for insecticides and herbicides.

PPA residues in foods have most often been found in fruit and vegetables (Table 9.7). Among the identified active agents, the fungicides play the biggest part (Table 9.8). Therefore, special attention has been paid to them in Table 9.5.

Table 9.6. Amounts used of some plant protection agents

Active agent	Introduced in the year	Dose (g/ha)
A. Insecticides		
Chlorpyriphos methyl (IXb)	1966	250–1000
Deltamethrin (XIII)	1984	5–20
Indoxacarb (XXV)	1996	12.5–125
B. Fungicides		
Mancozeb (XXXIa)	1961	1500–3000
Azoxystrobin (IV)	1992	100–375
Epoxyconazole (XVI)	1993	125
C. Herbicides		
2,4-D (XIV)	1942	300–2300
Atrazine (III)	1957	≤1500
Nicosulferon (XXXVI)	1990	35–70

Table 9.7. Foods with pesticide residues which exceeded the permissible upper limit (investigated in 2003 in Germany)[a,b]

Food	N	N_O	N_R	N_H	N_H (%)
A. Cereal					
Barley	23	11	11	1	4.3
Rice	159	126	31	2	1.3
Wheat	301	186	113	2	0.7
B. Food of animal origin					
Poultry meat	583	322	259	2	0.3
Cheese and curd	273	100	172	1	0.4
Mutton	24	11	11	2	8.3
Bird's eggs	324	125	197	2	0.6
C. Fruit and vegetables					
Pineapple	64	23	28	13	20.3
Apple	456	161	277	18	3.9
Apricot	159	54	88	17	10.7
Aubergine	185	122	51	12	6.5
Pear	426	139	243	44	10.3
Cauliflower	123	62	58	3	2.4
Bean with hull	109	62	40	7	6.4
Broccoli	14	12	1	1	7.1
Blackberry	8	0	6	2	25.0
Chinese cabbage	32	18	8	6	18.8
Peas without pods	122	45	66	11	9.0
Strawberry	894	173	663	58	6.5
Fig (dried)	6	3	2	1	16.7
Fennel	10	7	1	2	20.0
Grapefruit	51	22	28	1	2.0
Kale	11	6	2	3	27.3
Cucumber	381	214	140	27	7.1 (11.4)
Hazelnut	39	36	2	1	2.6

Table 9.7. (Continued)

Food	N	N_O	N_R	N_H	N_H (%)
Blueberry	54	41	12	1	1.9
Raspberry	59	27	27	5	8.5
Currant	107	25	67	15	14.0 (17.2)
Cherry	173	101	69	3	1.7
Kiwi	102	47	54	1	1.0
Turnip	64	46	14	4	6.3
Mandarin	233	26	187	20	8.6
Mango	45	25	17	3	6.7
Melon	32	13	18	1	3.1
Orange	209	34	164	11	5.3
Papaya	24	13	4	7	29.2
Bell pepper	922	367	403	152	16.5 (21.5)
Parsley	30	13	12	5	16.7
Peach	271	110	125	36	13.3
Plum	158	93	62	3	1.9
Leek	13	12	0	1	7.7
Rocket	80	8	52	20	25.0 (30.1)
Salad	451	153	255	43	9.5
Asparagus	135	104	29	2	1.5
Spinach	87	72	11	4	4.6
Tomato	691	333	311	47	6.8
Grape	933	157	645	131	14.0 (12.9)
Lemon	300	124	149	27	9.0
Zucchini (courgette)	89	52	34	3	3.4

[a] N: number of samples; N_O: number of samples without detectable residues; N_R: number of samples with residues including the maximum permissible quantity; N_H: number of samples with residues above the maximum quantity; N_H in percent (with reference to: N).
[b] In brackets are the values for 2004.

9.4.2.3 Herbicides

A distinction is made between non-specific and specific herbicides. The former inhibit the growth of both cultivated plants and weeds. For this reason, they can only be used before sowing. The introduction of resistance genes in soybean, corn and rape seed, among others, allows their weeds to be controlled by non-specific herbicides even during growth.

Selective herbicides inhibit the growth of weeds while protecting the cultivated plants. This selective action is achieved, e. g., because, unlike the weed, the cultivated plant quickly degrades the herbicide. One of the first selective herbicides was 2,4-dichlorophenoxyacetic acid (XIV), which eliminates only dicotyledon weeds but not monocotyledon cereal plants.

Newer selectively acting herbicides are the compounds amidosulfuron (I), mesosulfuron methyl (XXXIII) and nicosulfuron (XXXVI), which belong to the class of sulfonyl ureas. Since they are very active, the amount used is very small (compare atrazine (III) and nicosulfuron (XXXVI) in Table 9.6). As in the case of the other PPA, the biochemical mechanism of action of most of the herbicides is known (examples in Table 9.5). They frequently target a reaction site in the chloroplasts.

9.4.3 Analysis

The purpose of the analyses is to detect PPA which are not registered and to expose cases where the stipulated maximum permissible amounts have been exceeded. In addition, it is necessary to continually measure the contamination of food with PPA (monitoring, cf. 9.4.4).

The analysis of PPA residues is difficult because the number of active agents which can be taken into consideration is very large. For example, 255 compounds were registered in Germany in 2003 and maximum permissible amounts were stipulated by law for 600 compounds. However, we have to reckon with the use of about 1000 active agents worldwide. The analysis is made more difficult by the large differences in the chemical structures and the requirement for an exact quantification, the maximum permissible quantities being in the trace range, i. e., between 0.001 and 10 mg/kg. The following example gives an insight into the most important steps of a method (multimethod) with which a series of pesticides are identified.

A sample of a fruit or vegetable (ca. 10 g) is homogenized with the solvent (acetonitrile) which contains the internal standard (triphenyl phosphate). Anhydrous $MgSO_4$ and NaCl are added to bind water. After centrifugation, an SPE (solid phase extraction) sorbent is stirred in an aliquot of the supernatant to bind organic acids, pigments and sugar. After a second centrifugation, the PPA and the internal standard are identified

Table 9.8. Number of food samples with reference to the individual active agent (investigated in 2003 in Germany)[a]

Active agent	N_1	N_2	N_2 (%)	Concentration (mg/kg)
A. Cereal				
Hydrogen cyanide (V)	40	40	100	0.1
Bromide	80	62	77.5	0.25
Chlormequat (VIII)	46	20	43.5	0.01
Ethephon (XVII)	37	12	32.4	0.2
Primiphos methyl (XXXIX)	195	19	9.7	0.02
Flurtamone (XXI)	20	1	5.0	0.005
Phosphide/PH3 (XXXVIII)	28	1	3.6	0.01
DDT (XII)[b]	99	3	3.0	0.2
B. Fruit and vegetables				
Bromide	562	254	45.2	0.01
Chlormequat (VIII)	1416	338	23.9	0.01
Maneb group (XXXI)[c]	1890	367	19.4	0.01
Amitraz (II)	315	49	15.6	0.01
Chlorpyriphos (IXa)	5412	617	11.4	0.01
Carbenedazine (VI)	2608	273	10.5	0.005
Cyprodinil (XI)	3894	396	10.2	0.05
Procymidone (XL)	5264	524	10.0	0.01
Thiabendazole (XLIV)	2621	250	9.5	0.005
Fenhexamide (XVIII)	1460	129	8.8	0.03
Endosulfan (XV)	5363	417	7.8	0.005
Imazalil (XXIV)	4387	314	7.2	0.05
Mepiquat (XXVII)	1127	71	6.3	0.01
Fludioxonil (XX)	3715	216	5.8	0.005
Tolylfluanid (XLVI)	5052	287	5.7	0.001
Captan (V)/Folpet (XXII)	5041	261	5.2	0.01
Pyridaben (XLI)	1692	72	4.3	0.005
Pyrimethanil (XLII)	3399	146	4.3	0.005
Vinclozolin (XLVIII)	5196	199	3.8	0.01
Methidation (XXXIV)	5389	184	3.4	0.02
Trifloxystrobin (XLVII)	1079	36	3.3	0.005

[a] Of the 373 (cereal) or 399 (fruit and vegetables) active agents analyzed, those with $N_2 \geq 3\%$ are listed.
N_1: number of samples; N_2: number of samples with an active agent concentration equal to or higher than the concentration given in the column on the right. N_2 (%): N_2 with reference to N_1.
[b] Including degradation products.
[c] Calculated as CS_2.

and quantified by gas chromatography–mass spectrometry (MS). Alternatively, LC (liquid chromatography)–MS processes are also used to detect the analytes. The LC–MS process is the method of choice in the case of thermolabile active agents. Some pesticides can be accurately identified only by MS–MS measurements.

A series of PPA cannot be identified by a multimethod because their polarity and structural differences are too large. For the analysis of such active agents, special processes have been elaborated, with isotope dilution analyses (principle, cf. 5.2.6.1) also playing a part.

If metabolites of PPA are toxicologically relevant, they are also identified in the analysis, e. g., the degradation product 2,4-dimethylaniline together with the active agent amitraz (II), and dichlorodiphenyldichloroethane (DDD) and dichlorodiphenyldichloroethylene (DDE) with DDT (XII).

On the plant, PPA are exposed to light and can undergo photolysis. An example is parathion (XXXVII). Nitrosoparathion is formed, among other compounds, and reacts with components of the cuticle. The resulting product is insoluble and is not detected by the conventional PPA analysis. It can be identified directly on the cuticle with the help of the ELISA technique (principle, cf. 2.6.3).

9.4.4 PPA Residues, Risk Assessment

9.4.4.1 Exceeding the Maximum Permissible Quantity

In Germany, 12,874 samples were tested for the presence of PPA in 2003. 2515 samples came from the monitoring program (cf. 9.4.4.2) and 10,359 from official food monitoring. No PPA were found in 42.9% of these samples. In 50.1% of the samples, residues were present which were equal to or less than the maximum permissible quantity. Altogether 890 samples (6.9%) had residue content higher than the maximum permissible quantity.

Table 9.7 shows the foodstuffs that were most affected. Thus, the permissible upper limit was exceeded in more than 10% of the samples of pineapples, apricots, pears, blackberries, Chinese cabbage, dried figs, fennel, kale, currants, papaya, bell pepper, parsley, peaches, rocket and grapes. The figures for 2004 are also given for a few foods which were more contaminated.

Table 9.8 shows how often the individual active agents occurred. Hydrogen cyanide, followed by bromide and chloromequat were detected in every sample of cereal. The two last mentioned active agents were found most frequently in fruit and vegetables. The maneb group, amitraz, chlorpyriphos, carbenedazine, cyprodinil and procymidones also accounted for proportions of 10% and more of the samples of these foods. The frequency of bromide is due to the fact that it is a naturally and ubiquitously occurring substance. Higher concentrations can indicate the use of bromine-containing fumigants, e. g., methyl bromide (XXXV), in soil treatment or in storage.

The EU coordinated a pesticide monitoring program with the participation of Norway, Iceland and Liechtenstein in 2002. They investigated the occurrence of a total of 41 pesticides in pears, bananas, oranges/mandarins, peaches/nectarines, beans, potatoes, carrots and spinach.

Residue concentrations higher than the permissible upper limit were found most frequently in spinach (13% of the tested samples), followed by beans (7%), oranges/mandarins (4%) and peaches/nectarines (3%). Of the active agents, residues of the maneb group most often exceeded the permissible upper limit (1.19% of all the tested samples).

In the USA, a monitoring program for insecticides in 2002 showed that DDT (0.0001–0.025 mg/kg) was the most frequently found active agent. It was found in 21% of the samples. Furthermore, chlorpyriphos methyl (17%, 0.0002–0.59 mg/kg), malathion (15%, 0.0007–0.071 mg/kg), endosulfan (14%, 0.0001–0.166 mg/kg) and dieldrin (11%, 0.0001–0.010 mg/kg) occurred in more than 10% of the tested samples.

9.4.4.2 Risk Assessment

The results of the food monitoring program in 1995–2002 were used for the risk assessment. In this time period, 30,682 samples from 130 different foods were analyzed for the presence of 160 PPA. No active substances or only traces were found in 45.7% of the samples. Concentrations exceeding the maximum permissible amounts were determined in 2.6% of the samples.

The PPA selected for the risk assessment were those which were quantified in at least 5% of the samples of three or more foodstuffs. Table 9.9 shows the selection. The calculation of the food intake was based on the results of the national consumption study, which was carried out in the FRG between 1985 and 1988 (cf. 9.3.2).

The overview of the results in Table 9.9 shows that the reference values of the selected PPA are mostly utilized to less than 1%. The dithiocarbamates are the only exception with a utilization of 7.7 to 18.3%. The lower value was found for the group of men (women 9.6%) and the upper value for children (4–6 years of age). The reason for this difference is probably the consumption of fruit because men eat less fruit than children. Fruit can contain residues of dithiocarbamates used for controlling fungal diseases.

Table 9.9. Utilization of the reference value[a] (food monitoring 1995–2002)

Active agent	Reference ADI[b] (μg/kg kg/d)	Utilization (%)
Bromide	1000	0.7–1.3
Carbenedazine (VI)	30	0.2–0.5
Captan (V)/Folpet (XXII)	100	0.03–0.10
Chlorpyrifos (IXa)	10	0.1–0.3
Dithiocarbamates[c]	3[d]	7.7–18.3
Iprodion (XXVI)	60	0.2–0.4
Pirimiphos methyl (XXXIX)	30	0.1–0.3
Procymidon (XL)	100	0.03–0.09
Thiabendazole (XLIV)	100	0.1–0.4
Vinclozolin (XLVIII)	10	0.8–3.5

[a] The utilization (%) is defined as the ratio of the estimated daily intake (μg/d) of an active agent to the appropriate reference value.
[b] ADI: definition cf. 9.1.
[c] Dithiocarbamates: mancozeb (XXXIa), maneb (XXXIb), metiram (XXXIc), zineb (XXXIe), thiram (XLV), ferbam (XIX), ziram (XLIX) and propineb (XXXId).
[d] The ADI value of the dithiocarbamates lies between 3 and 30 μg/kg kg/d. The lowest value was taken as a basis for the calculation.

The German Federal Office for Consumer Protection and Food Safety came to the following conclusion based on the monitoring results: "The utilization of the toxicological reference values (permissible or tolerable daily intake) is without exception very small, being only about 1% for the plant-protective agents and the persistent organochloro compounds tested. A substantial exposure was observed only with the dithiocarbamates. However, it should be taken into account that the results can be partly superimposed by the residues of naturally occurring sulfur-containing substances. The results obtained up to now do not represent the possible total contamination of the consumers with residues of plant protection agents, however a distinct section because the analysis spectrum did not include all the conceivable active substances of plant protection agents and their metabolites, but about 160 substances relevant to foodstuffs."

The latest results of the food monitoring program are to be found in the annually published report on food safety at www.bvl.bund.de

9.4.4.3 Natural Pesticides

The natural pesticides should not be disregarded in the risk assessment. Ames and Gold (2000) have shown that in comparison with the synthetic pesticides, the pesticides occurring naturally in food are of much greater importance. Substances of this type are produced by the plant for protection against microorganisms and insects, e.g., allylisothiocyanate, benzaldehyde, caffeic acid, capsaicin, catechin, estragol, d-limonene and 4-methylcatechin.

These substances include as high a percentage of compounds which are carcinogenic as among the synthetic pesticides. It has been estimated for the USA (the conditions in Europe should be similar) that the population consumes with food on average 1500 mg per person and day of natural pesticides, but only 0.09 mg of synthetic pesticides. In summary, it can be concluded that a risk of impairment of consumer health is not discernible in the case of the proper and controlled application of registered PPA.

9.5 Veterinary Medicines and Feed Additives

9.5.1 Foreword

The current practice in animal husbandry is the wide use of veterinary medicines, which serve not

Table 9.10. Animal medicines (selected structural formulas are presented in Fig. 9.3)

Number	Class of compounds	Example
Antibiotics		
I	Sulfonamides	Sulfapyridine (Ia), Sulfathiazole (Ib)
II	β-Lactams	Amipicillin (IIa), Amoxycillin (IIb)
III	Tetracyclines	Tetraycline (IIIa), Oxytetracycline (IIIb)
IV	Aminoglycosides	Streptomycin (IVa), Dihydrostreptomycin (IVb)
V	Macrolides	Tiamulin (Vt)
VI	Crinolines & Fluorochinolones	Ciprofloxacin (VIa), Marbofloxacin (VIb)
Anthelmintics		
VII	Benzimidazoles	Fenbendazole (VIIa), Mebendazole (VIIb)
VIII	Tetrahydroimidathiazoles	Levamisol (VIIIa), Morantel (VIIIb)
IX	Avermectins	Ivermectin (IX)
Coccidiostats		
X	Different classes	Dicoquinat (Xa), Clopidol (Xb), Lasalocid (Xc)

only for therapy, but to a large extent for prophylaxis and economic aims (e.g. to shorten animal growth or feeding time; to abate the risk of losses). Veterinary preparation residues in food are ingested by humans in low amounts but continuously and, hence, could be a health hazard. This possibility was, for a long time, not carefully examined. Therefore, as in the field of pesticides, supporting and maintaining appropriate measures (imposing legally binding regulations, analytical control or supervision, elucidation of toxicological problems) has the ultimate aim of protecting human health.

A brief outline of some important groups of veterinary medicines follows. Table 9.10 and Fig. 9.3 provide a review of their use and chemical structures.

9.5.2 Antibiotics

Antibiotics are used prophylactically to stem infections, e.g., in intensive mass animal farming and in the therapy of infectious diseases. Since they inhibit the growth of the microflora, which is present in the digestive tract of livestock, feed utilization is improved. The animals gain weight faster. This application of antibiotics as growth promoters is regarded critically in the EU and has led to bans. A constant intake of antibiotics, even at low doses, is a risk to human health since some microorganisms may become resistant and allergic reactions may develop.

9.5.3 Anthelmintics

In meadows and sheds, animals come into contact with their excrements and subsequently with worms in all developmental stages. Anthelmintics are used against the resulting diseases caused by worms.

9.5.4 Coccidiostats

The compounds of this class are added to animal feed to combat coccidiosis diseases (such as enteritis or cachexie) caused by protozoans living as parasites in intestines. Poultry and rabbits are the animals most often affected. Residues may be found in eggs.

9.5.5 Analysis

The aim of these analyses is:

- to detect medicines which are banned or not approved, e.g., chloramphenicol (XIt), nitrofurans (derivatives of 2-nitrofuran, e.g., nitrofurantoin, XIIt), fattening aids with estrogenic activity such as 17-estradiol (XIIIt), diethylstilbestrol (XIVt), zeranol (XVt).
- to check if the residue of an approved therapeutic agent is still within the permissible upper limit (MRL, cf. 9.1).

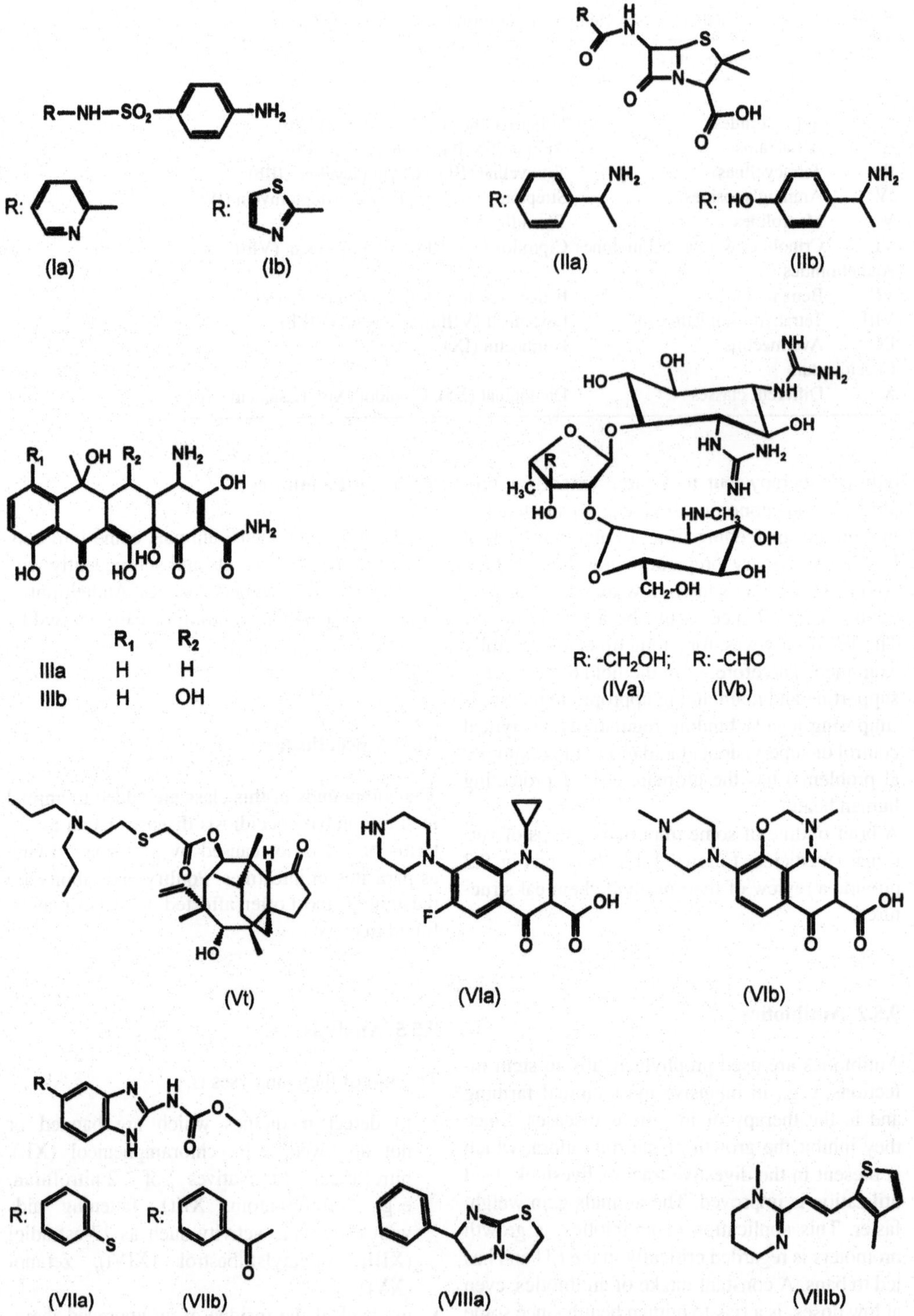

Fig. 9.3. Structures of some selected veterinary medicines. The Roman numerals refer to Table 9.10 and to the text (cf. 9.5.5)

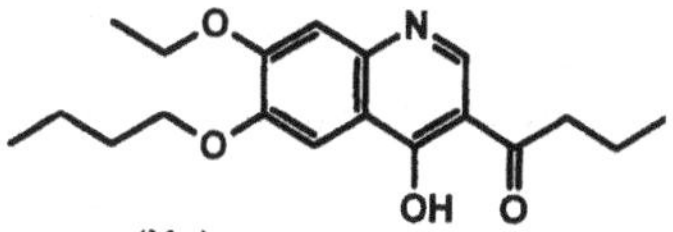

(Xa)

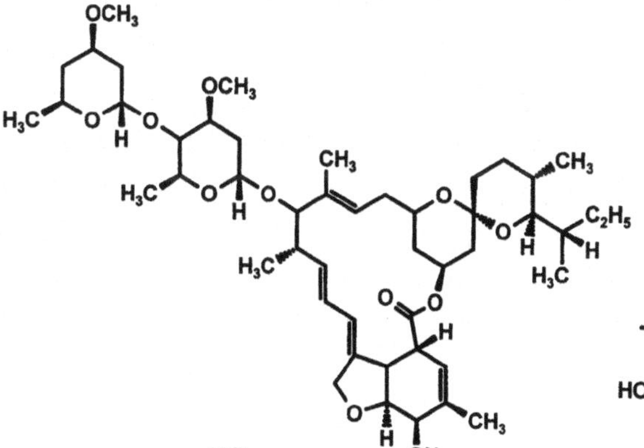

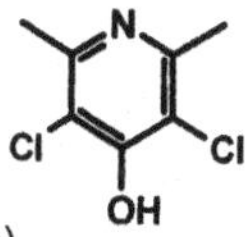

(Xb)

(IX)

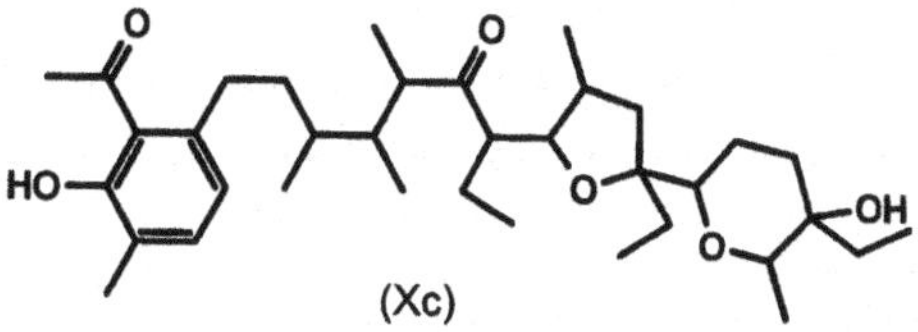

(Xc)

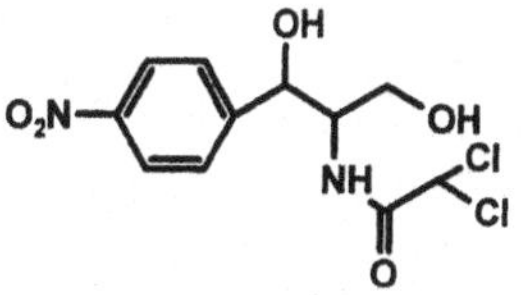

(XIt)

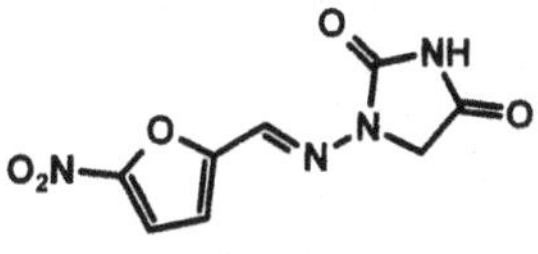

(XIIt)

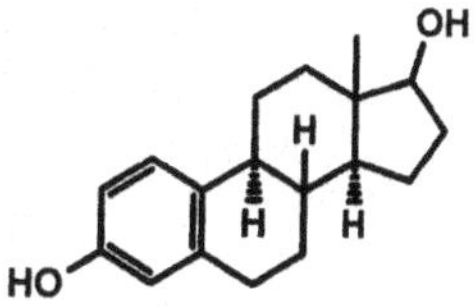

(XIIIt)

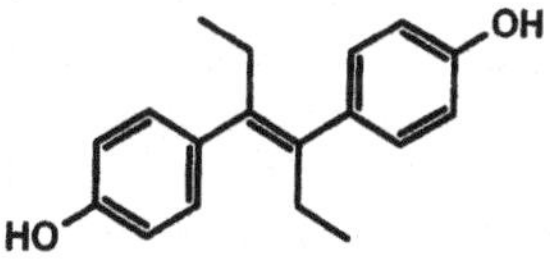

(XIVt)

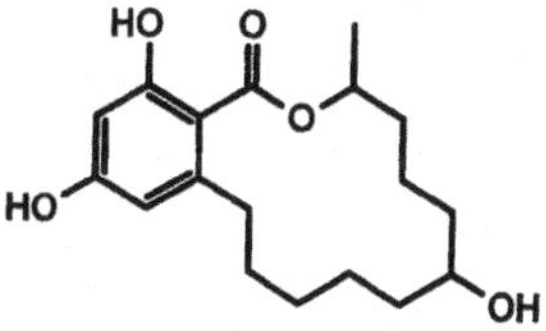

(XVt)

Fig. 9.3. (Continued)

The analysis starts with screening. Antibiotics can be detected with the help of the bacteria whose growth they inhibit. Their differentiation can be further improved by means of an electrophoretic preliminary separation.

In principle, the same isolation and separation methods and mass spectrometric techniques are used for the unambiguous identification of veterinary medicines as for pesticides (cf. 9.4.3). Enzyme immunoassays are also used (cf. 2.6.3).

9.6 Polychlorinated Biphenyls (PCBs)

The PCBs are complex mixtures of substances which have been on the market since 1950. They are widely used, e. g., as transformer oil, hydraulic fluid, heat exchange medium, dielectric fluid in condensers, plasticizer and additive for printing ink. Formula 9.3 shows 2,2',5,5'-tetrachlorobiphenyl (I) and 2,2',4,5,5'-pentachlorobiphenyl (II) as examples.

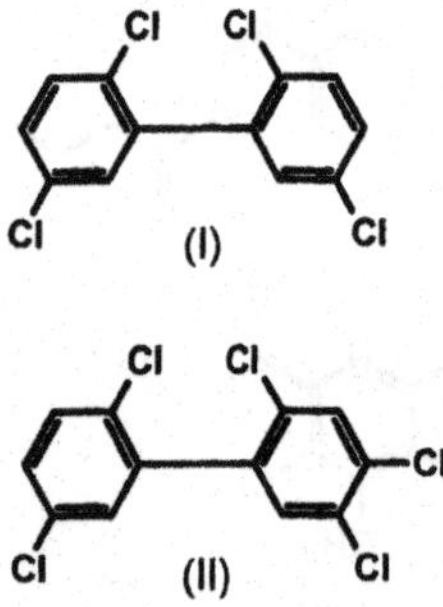

(I)

(II)   (9.3)

As a result of their widespread use, the PCBs also came into contact with food. Because of their persistence and solubility in fat, they accumulated like in the case of DDT (cf. 9.4.2.1). Therefore, they have been increasingly identified in fatty foods since their discovery. This and the fact that PCBs can produce highly toxic dioxins (cf. 9.10) in the combustion process led to the banning of the production and application of PCBs in 1989. In Germany, the contamination with PCB, e. g., in milk fat (mg/kg) has subsequently fallen on average: 0.012 (1986), 0.007 (1992), 0.003 (2001).

9.7 Harmful Substances from Thermal Processes

9.7.1 Polycyclic Aromatic Hydrocarbons (PAHs)

Burning of organic materials, such as wood (wood smoke and its semi-dry distillation product, the wood smoke vapor phase), coal or fuel oil, results in pyrolytic reactions which yield a great number of polycyclic aromatic hydrocarbons (abouot 250 have been identified) with more than three linearly or angularly fused benzene rings, that are carcinogenic to varying extents. The quantity and diversity of compounds generated is affected by the conditions of the burning process. Benzo[a]pyrene (Bap) (Formula 9.4) usually serves as an indicator compound.

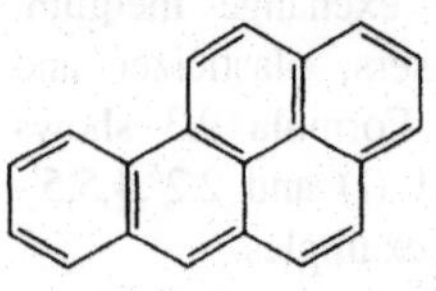

(9.4)

Contamination of food with polycyclic compounds can be caused by fall-out from the atmosphere (as often occurs with fruit and leafy vegetables in industrial districts), by direct drying of cereals with combustion gases, by smoking or roasting of food (barbecuing or charcoal broiling; smoking of sausage, ham or fish; roasting of coffee). PAHs accumulate in high-fat tissues. The content in meat and processed meat products should not exceed 1 µg/kg end-product measured as Bap. A reduction of Bap contamination to this limiting value has been achieved by the use of modern smoking techniques. A maximum of 5 µg/kg Bap is tolerated in smoked fish. Values less than 1.6 µg/kg were found in 95% of the samples tested in food monitoring in 2005.

9.7.2 Furan

Furan is possibly a carcinogenic substance. It occurs in heated food, especially in roasted coffee. Isotopic dilution analyses with $[^2H_4]$-furan as the internal standard yielded 2.5–4.3 mg/kg furan in differently produced coffee powders. Baby food, e.g., carrot mash and potato/spinach mash contained 74 and 75 µg/kg respectively. Furan is formed from amino acids which yield acetaldehyde and glycolaldehyde on thermal degradation (Fig. 9.4). Aldol condensation, cyclization and elimination of water are the reaction steps. Other precursors of furan are carbohydrates, polyunsaturated fatty acids and carotinoids (Fig. 9.4). Furan can also be formed from the thermolysis of ascorbic acid.

9.7.3 Acrylamide

Polyacrylamide, produced from monomeric acrylamide (2-propenamide), has been used for decades in various industrial processes, e. g., as a flocculant in the treatment of drinking water. Especially for reasons of occupational health and safety, numerous toxicological studies on acrylamide have already been conducted. These studies have shown above all that on high exposure, acrylamide (i) binds to hemoglobin in the blood, (ii) is metabolized to reactive epoxide glycidamide and (iii) is carcinogenic on chronic exposure in animal tests. For this reason, acrylamide was put about 20 years ago

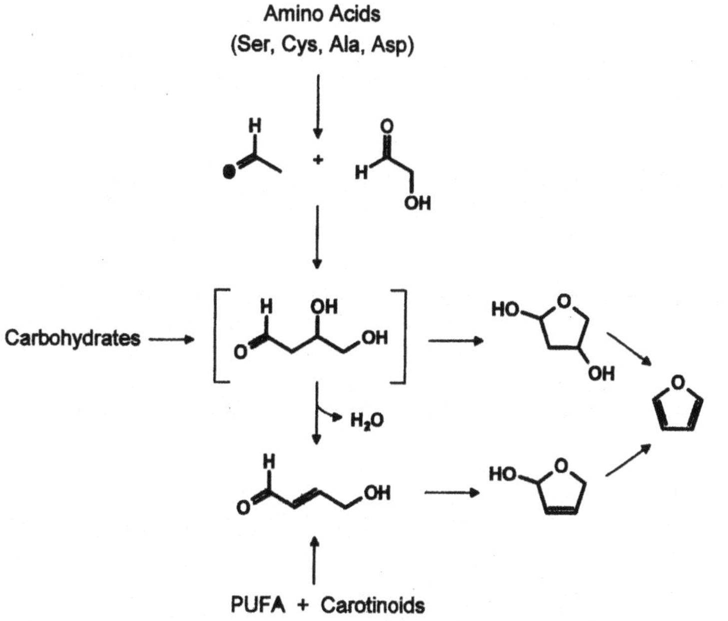

Amino Acids
(Ser, Cys, Ala, Asp)

Carbohydrates ⟶

H₂O

PUFA + Carotinoids

Fig. 9.4. Formation of furan on heating food (according to *Yaylayan*, 2006)

into Category III A2 of carcinogenic working substances. According to EU guidelines for drinking water, the concentration of acrylamide in water should not exceed 0.1 μg/l.

The occurrence of relatively high concentrations of acrylamide in tobacco smoke has been known for a long time, but in 2002 this compound was described for the first time also as a constituent in various thermally treated foods. In particular, processed potato products such as chips, but also fine breads and cakes contain relatively high concentrations (Table 9.11). Today, mainly stable isotope assays in combination with GC–MS or LC–MS with the use of deuterium- or carbon-

13-labelled acrylamide are used for the quantitative determination. The large range of variations in the concentrations measured in food indicates that the raw material and the process conditions exert a significant influence on the formation of acrylamide. Thus, it could be shown, e. g., that the formation of acrylamide in potato products clearly fluctuates depending on the variety of potato (Fig. 9.5) and the concentrations of acryl-

Table 9.11. Maximum concentrations and variation ranges of acrylamide in selected foods

Food	Concentration (μg/kg)	
Gingerbread	7800	(80–7800)
Potato chips	3700	(100–3700)
Crispbread	2800	(25–2800)
Roasted nuts	2000	(10–2000)
Ground coffee	500	
Roasted meat	50	
Bread	40	

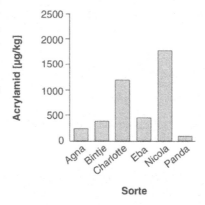

Fig. 9.5. Formation of acrylamide during the frying of potato strips from different varieties of potatoes (according to *Amrein* et al., 2003)

amide can also be distinctly reduced by lowering the heating temperature, e. g., in deep frying. As presented in 1.2.4.4.1, acrylamide is preferentially formed by the reaction of the amino acid asparagine with reductive carbohydrates (or their degradation products). In fact, studies with isotopically labelled asparagine have shown that the carbon skeleton of the amino acid is retained in acrylamide. Nevertheless, the formation of acrylamide, e. g., in the case of potatoes, correlates after heating considerably better with the concentration of fructose and glucose in fresh potatoes than with the concentration of free asparagine although potatoes have a very high concentration of free asparagine at 2–4 g/kg dry weight. In the case of gingerbread, in addition to the concentration of free asparagine, the NH_4HCO_3 used as baking powder was identified as a promoter in the formation of acrylamide.

Apart from the enzymatic hydrolysis of asparagine with asparaginase, the ways in which the content of acrylamide in food can be reduced include the use of various additives, lowering the pH value and reducing the heating temperature. According to more recent calculations, the daily intake of acrylamide from foodstuffs in Germany is assumed to be about 0.57 µg/kg body weight.

9.8 Nitrate, Nitrite, Nitrosamines

9.8.1 Nitrate, Nitrite

The plants which belong to Group A in Table 9.12 can store very much more nitrate than those in Groups B and C, their nitrate content depending

Table 9.12. Nitrate concentrations in vegetables

A. *High concentration (1000–4000 mg/kg fresh weight)*
Chinese cabbage, endivie, corn salad, lettuce, fennel, kohlrabi, beetroot, radish, rocket, spinach
B. *Moderate concentration (500–1000 mg/kg fresh weight)*
Aubergine, white cabbage, cauliflower, kale, red cabbage and savoy cabbage, leek, carrots, celery, zucchini
C. *Low concentration (<500 mg/kg fresh weight)*
Peas, cereal, green beans, cucumber, potatoes, garlic, fruit, bell peppers, Brussels sprouts, tomatoes, onions

among other things on the N supplied on fertilization. Apart from the properties of the soil, even light plays a role because some plants store more nitrate when there is a lack of light. The foods of animal origin listed in Table 9.13 and drinking water (cf. 23.1.3) are a further source of nitrate. It was calculated on the basis of a national consumption study (cf. 9.3.2) that the intake of nitrate is highest in 4–6 year old children (Table 9.14), followed by women and men who prefer fruit and vegetables in their diet rather than meat and fish. The ADI value for nitrate is utilized to 23–40% by the population.

Nitrite mainly comes from cured meat and meat products (Table 9.12). The daily supply amounts to about 0.25 mg $NO_2^{\ominus}$.

It is remarkable that the amount of nitrate formed every day in the human organism, about 1 mg/kg body weight, is just as much as the intake in the diet. The precursor is arginine which is cleaved to give NO and citrullin (cf. 3.7.2.1.8). NO is oxidized to N_2O_3, which reacts with water to give nitrite. Hemoglobin oxidizes nitrite to nitrate, giving rise to methemoglobin which cannot transport oxygen. Therefore, nitrite is toxic, especially for infants (cyanosis) because their methemoglobin reductase still has low activity. This enzyme reduces methemoglobin to hemoglobin.

The toxicity of nitrate starts from the bacterial reduction to nitrite. In the human organism, about 25% of the nitrate absorbed from the food is eliminated with the saliva and up to 7% is reduced to nitrite in the mouth cavity within 24 hours by the action of bacterial nitrate reductases and transported to the stomach. About 90% of the total amount of nitrite which reaches the digestive tract comes from nitrate reduction.

The bacterial formation of nitrite led to the assumption that toxic nitrosamines can arise endogenously by the nitrosation of amines (cf. 9.8.2). This danger has been overestimated. Endogenous nitrosation was described as "practically insignificant" in the nutrition report as early as 1996.

9.8.2 Nitrosamines, Nitrosamides

Nitrosamines and nitrosamides are powerful carcinogens. They are obtained from secondary

Table 9.13. Nitrate and nitrite in milk, cheese and meat products (mg/kg fresh substance)

Food		Nitrate			Nitrite	
	n^a	Mean	Variation	n^a	Mean	Variation
Milk	16	1.4	1.0–4.1			
Cheese				39	0.3	0.2–1.3
Meat	110	7.6	1.0–49.5			
Uncooked smoked pork ribs	73	68.6	5.0–425.5	47	27.9	0.2–94.1
Uncooked smoked black forest ham	23	351.0	21.6–1384.3	20	12.3	1.2–80.2
Uncooked smoked ham				23	10.7	0.9–44.2
Uncooked sausages, firm	20	208.4	7.0–1042.0			
Cooked smoked shoulder ham				44	15.7	0.8–91.0
Salami				76	5.1	0.3–48.7
Fresh soft sausage				35	6.9	0.2–45.6
Fried sausage				108	3.5	0.2–41.5
Finely minced pork sausage				32	7.8	0.2–18.6
Calf-liver sausage, finely grained				19	5.4	1.9–12.3
Salted herring filet	154	27.4	1.0–405.0			
Herring titbit	103	74.7	19.0–276.0			

[a] Number of samples.

Table 9.14. Intake of nitrate and utilization of the reference value (food monitoring 1995–2002)

Group of persons	Intake		Utilization
	(mg/d)	(mg/kg kg/d)	%-ADI[a]
Children, 4–6 years	30.6	1.465	40.1
Children, 7–10 years	37.7	1.220	33.4
Men	64.6	0.830	22.7
Men, fish eaters	69.0	0.862	23.6
Men, meat eaters	73.3	0.921	25.2
Men, fruit and vegetable eaters	90.8	1.155	31.6
Women	61.0	0.950	26.0
Women, fish eaters	67.8	1.034	28.3
Women, meat eaters	70.0	1.064	29.2
Women, fruit and vegetable eaters	86.2	1.328	36.4

[a] The ADI value for nitrate is 3.65 mg/kg kg/d.

amines, N-substituted amides and nitrous acid:

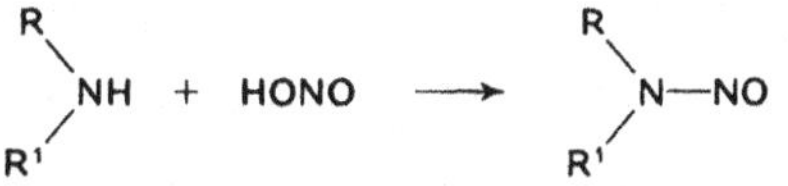

$$R_{R'}NH + HONO \longrightarrow R_{R'}N{-}NO \quad (9.5)$$

$$R{-}CO{-}NHR' + HONO \longrightarrow R{-}CO{-}\underset{NO}{N}{-}R' \quad (9.6)$$

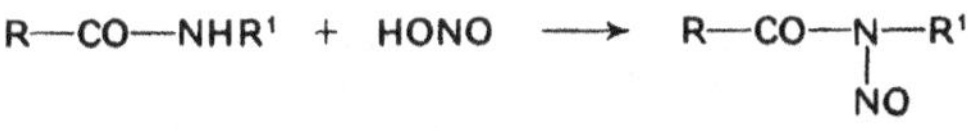

The nitrosonium ion, NO^+, or a nitrosyl halogenide, XNO, is the reactive intermediate:

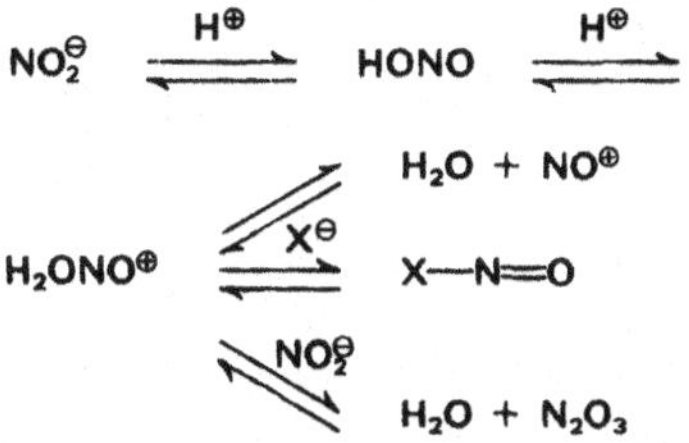

$$NO_2^\ominus \underset{}{\overset{H^\oplus}{\rightleftharpoons}} HONO \overset{H^\oplus}{\rightleftharpoons}$$

$$H_2ONO^\oplus \underset{NO_2^\ominus}{\overset{X^\ominus}{\rightleftharpoons}} \begin{cases} H_2O + NO^\oplus \\ X{-}N{\equiv}O \\ H_2O + N_2O_3 \end{cases} \quad (9.7)$$

Nitrosamine formation is also possible from primary amines:

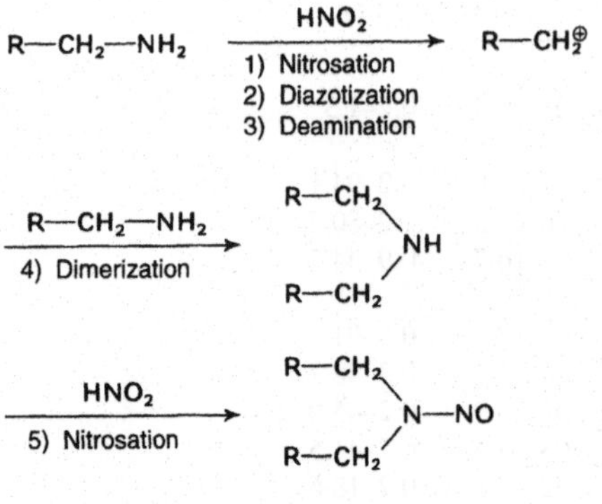

$$R-CH_2-NH_2 \xrightarrow[\substack{1)\ Nitrosation \\ 2)\ Diazotization \\ 3)\ Deamination}]{HNO_2} R-CH_2^{\oplus}$$

$$R-CH_2-NH_2 \xrightarrow[4)\ Dimerization]{} \begin{array}{c} R-CH_2 \\ \diagdown \\ NH \\ \diagup \\ R-CH_2 \end{array}$$

$$\xrightarrow[5)\ Nitrosation]{HNO_2} \begin{array}{c} R-CH_2 \\ \diagdown \\ N-NO \\ \diagup \\ R-CH_2 \end{array} \qquad (9.8)$$

from diamines:

$$H_2N-(CH_2)_n-\underset{R}{CH}-NH_2 \xrightarrow[N_2]{HNO_2}$$

$$H_2N-(CH_2)_n-\underset{R}{CH}^{\oplus} \xrightarrow{Cyclization}$$

$$\begin{array}{c} (CH_2)_n \\ \diagdown \\ CHR \\ \diagup \\ NH \end{array} \xrightarrow{HNO_2} \begin{array}{c} (CH_2)_n \\ \diagdown \\ CHR \\ \diagup \\ \underset{NO}{N} \end{array}$$

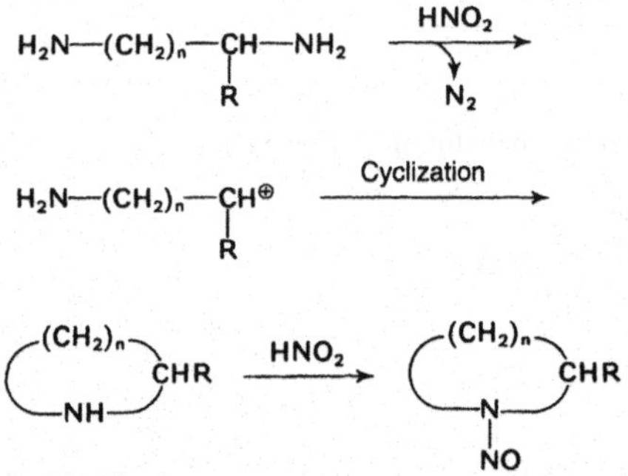

R = H: diamine
R = COOH: diamino carboxylic acid   (9.9)

and from tertiary amines:

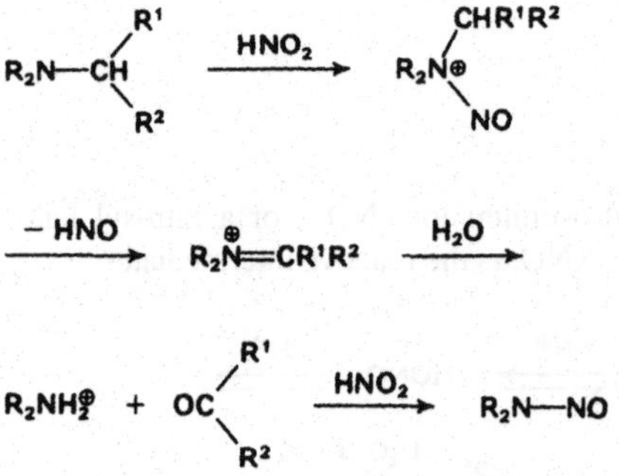

$$R_2N-\underset{R^2}{\overset{R^1}{CH}} \xrightarrow{HNO_2} R_2N^{\oplus} \overset{CHR^1R^2}{\underset{NO}{}}$$

$$\xrightarrow{-\ HNO} R_2N^{\oplus}{=}CR^1R^2 \xrightarrow{H_2O}$$

$$R_2NH_2^{\oplus} + OC\overset{R^1}{\underset{R^2}{}} \xrightarrow{HNO_2} R_2N-NO$$
$$(9.10)$$

$$2\ HNO \longrightarrow H_2N_2O_2 \longrightarrow H_2O + N_2O$$
$$(9.11)$$

Table 9.15. Nitrosamines in food

Food product	Compound[a]	Content µg/kg
Frankfurter (hot dog)	NDMA	0–84
Fish (raw)	NDMA	0–4
Fish, smoked and pickled with nitrites or nitrates	NDMA	4–26
Fish, fried	NDMA	1–9
Cheese (Danish, Blue, Gouda, Tilsiter, goatmilk cheese)	NDMA	1–4
Salami	NDMA	10–80
Bacon (hog's hind leg) smoked meat	NDMA	1–60
Pepper-coated ham, raw and roasted	NPIP	4–67
	NPYR	1–78

[a] NDMA: N-Nitrosodimethylamine, NPIP: N-nitrosopiperidine, NPYR: N-nitrosopyrrolidine.

Nitrosamines are detected in variable amounts in many foods (Table 9.15). The most common compound is dimethylnitrosamine, which is also a most powerful carcinogen. Some activity has been ascribed to nitrosopiperidine and nitrosopyrrolidine. In meat products cured and treated with pickling salt, 30% of the samples contained nitrosodimethylamine (NDMA; 0.5–15 µg/kg) and 13% nitrosopyrrolidine (NPYR; >0.5 µg/kg). About 25% of the cheese samples analyzed were contaminated (0.5–4.9 µg/kg).

Nitrosopyrrolidine is formed from the amino acid proline by nitrosation followed by decarboxylation at elevated temperatures, such as in roasting or frying:

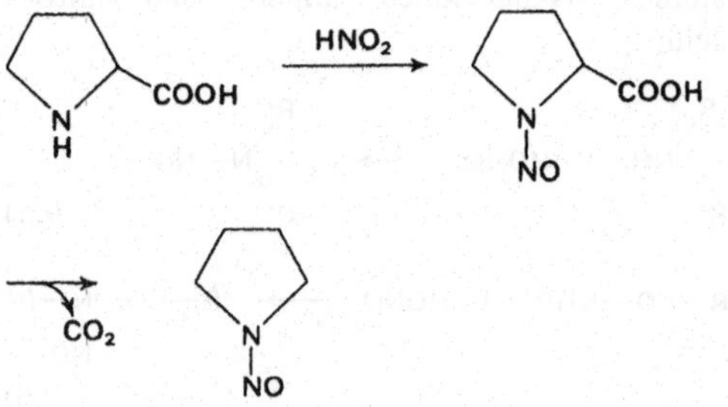

$$(9.12)$$

Table 9.16. Amines in food (mg/kg)

Compound	Spinach	Cabbage (kale) red	green	Carrots	Red beet	Celery	Lettuce	Rhu-barb	Herring salted	smoked	in oil	Tilsiter	Camem-bert	Lim-burger
Ammonia	18.280	11.060	15.260	3.970	8.800	19.600	10.260	6.340	2.928	270	–	164.400	–	–
Methylamine	12	22.7	16.6	3.8	30	64	37.5	–	3.4	–	7	–	12	3
Ethylamine	8.4	1.3	–	1	–	–	3.3	0.1	0.4	0.4	–	–	4	1
Dimethylamine	–	2.8	5.5	–	–	51	7.2	–	7.8	6.3	45	–	–	–
Methylethylamine	–	0.9	–	7	–	–	7.5	–	–	–	1	–	–	–
n-Propylamine	–	–	–	–	–	–	–	–	–	–	–	8.7	2	2
Diethylamine	15	–	–	–	–	–	–	–	1.9	5.2	–	–	–	–
n-Butylamine	–	–	7	–	–	–	–	–	–	–	–	3.7	–	–
i-Butylamine	–	–	–	–	–	–	–	–	–	0.3	–	–	0.2	0.2
Pyrroline	–	–	–	–	–	–	–	–						
n-Pentylamine	0.3	0.6	0.4	–	–	0.8	3	–	–	–	17	1.2	–	–
i-Pentylamine	3.8	–	0.5	–	–	–	–	3.9	–	–	–	–	0.2	tr[a]
Pyrrolidine	2.5	–	–	–	–	0.4	–	–	–	–	–	19.9	1	0.1
Di-n-propylamine									–	–	–	8.4	–	–
Piperidine									0.7	0.2	–	–	–	tr
Aniline	–	1.0	0.7	30.9	0.6	0.7	0.6	5						
N-Methylaniline	3.4	0.3	–	0.8	–	0.5	–	–	–	–	–	37.9	–	–
N-Methyl-benzylamine	–	–	–	16.5	–	–	10	–	–	–	2	–	–	–
Toluidine	–	–	1.1	7.2	–	1.1	–	–						
Benzylamine	6.1	3.3	3.8	2.8	0.1	3.4	11.5	2.9						
Phenylethylamine	1.1	8.6	3	2	0.5	–		3.2						
N-Methylphenyl-ethylamine	2.4	3.7	2	2	0.4	0.5	0.4	2.6	–	–	0.1	2.6	–	–

[a] Traces.

The nitrosopyrrolidine (1.5 µg/kg) in meat products increases almost ten fold (to 15.4 µg/kg) during roasting and frying. An estimate of the average daily intake of nitrosamines ranges from 0.1 µg nitrosodimethylamine and 0.1 µg nitrosopyrrolidine to a total of 1 µg.

The concentrations in food of ammonia and amines which can possibly undergo nitrosation are presented in Table 9.16.

Inhibition of a nitrosation reaction is possible, e. g., with ascorbic acid, which is oxidized by nitrite to its dehydro form, while nitrite is reduced to NO. Similarly, tocopherols and some other food constituents inhibit substitution reactions. Representative suitable measures to decrease exo- and endogenic nitrosamine hazards are:

- Decreasing nitrite and nitrate incorporation into processed meat. However, to completely relinquish the use of nitrite is a great health hazard due to the danger from bacterial intoxication (especially by botulism).
- Addition of inhibitors (ascorbic acid, tocopherols).
- Decreasing the nitrate content of vegetables.

9.9 Cleansing Agents and Disinfectants

Residues introduced by large-scale animal husbandry and use of milking machines are gaining importance. The residues in meat and processed meat products originate from the surfaces of processing equipment, while residues in milk arise from measures involved in disinfection of the udder. Iodine-containing disinfectants, including udder dipping or soaking agents, may be an additional source of contamination of milk with iodine.

Also in the case of fruit and vegetables, measures have to be taken to kill pathogenic microbes. Although chlorine is normally used, it does not kill all the bacteria in the permitted concentrations. In addition, chlorine can convert pesticide residues to substances of unknown biological activity. *Ozone* is recommended as an alternative. In disinfection, it is 1.5 times as active as chlorine, kills microorganisms which are not attacked by chlorine and destroys pesticide residues. It decomposes to molecular oxygen with a half life (aqueous ozone solution) of 20 minutes at room temperature. To disinfect process water, 0.5–5 mg/kg of ozone are used for <5 minutes. Apart from disinfection, ozone also delays the ripening of fruit (cf. 18.1.4.1).

9.10 Polychlorinated Dibenzodioxins (PCDD) and Dibenzofurans (PCDF)

Polychlorinated dibenzo-p-dioxins (PCDD) and polychlorinated dibenzofurans (PCDF), informally called "dioxins", occur as companion compounds or impurities in a large number of bromine- and chlorine-containing chemicals.

(9.13)

Furthermore, they are formed in many thermal processes (600 °C > T ≥ 200 °C) in the presence of chlorine or other halogens in inorganic or organic form. Consequently, they are widely distributed in the environment. The number of isomers (congeners) is large, For rodents, 2,3,7,8-tetrachlorodibenzodioxin (2,3,7,8-TCDD, "Seveso dioxin") has proved to be expecially toxic (LD_{50} = 0.6 µg/kg, guinea-pig) and carcinogenic. With the exception of PnCDD, the toxicity of other congeners is lower and is generally expressed as toxicity equivalent factors (TEFs), based on 2,3,7,8-TCDD (TEF = 1) (Table 9.17). With the help of these values, 2,3,7,8-TCDD equivalents (TCDD equivalents, TEQ) can be calculated, which are a measure of the total exposure to corresponding compounds (cf. Table 9.17 and 9.18).
The daily intake of dioxin in industrial countries is estimated at 1–3 pg TEQ/kg body weight. The half life and the absorption rate are assumed to be 7.5 years and 50% respectively. In 1997, the WHO set a value of 1–4 pg TEQ per kg body weight and day as a tolerable, lifelong intake (tolerable daily intake, TDI). Table 9.18 shows the estimated dioxin intake with food. Due to the concentration of dioxins in the fat phase, mother's milk in the industrial countries has on average

Table 9.17. Risk assessment of dibenzo-p-dioxins and dibenzofurans

Congener	TEF[a]
Dibenzo-p-dioxins	
2,3,7,8-TCDD	1
1,2,3,7,8-PnCDD	1
1,2,3,4,7,8-HxCDD	0.1
1,2,3,6,7,8-HxCDD	0.1
1,2,3,7,8,9-HxCDD	0.1
1,2,3,4,6,7,8-HpCDD	0.01
OCDD	0.0001
Dibenzofurans	
2,3,7,8-TCDF	0.1
1,2,3,7,8-PnCDF	0.05
2,3,4,7,8-PnCDF	0.5
1,2,3,4,7,8-HxCDF	0.1
1,2,3,6,7,8-HxCDF	0.1
1,2,3,7,8,9-HxCDF	0.1
2,3,4,6,7,8-HxCDF	0.1
1,2,3,4,6,7,8-HpCDF	0.01
1,2,3,4,7,8,9-HpCDF	0.01
OCDF	0.0001

[a] Toxicity equivalent factors (2,3,7,8-TCDD = 1).

Table 9.18. Average daily intake of 2,3,7,8-tetrachlorodibenzo-p-dioxin (2,3,7,8-TCDD) and related compounds with the food (pg/day)[a]

	2,3,7,8-TCDD	ΣTEQ[b]
Meat products (including poultry)	7	23.5
Milk	6.2	28.5
Eggs	0.8	4.2
Fish	8.6	33.3
Vegetable oil	<0.2[c]	<0.6
Vegetables	<2.4[c]	<2.4[c]
Fruits	<1.4[c]	<2.6[c]
Sum:	24.6	93.5[d]

[a] Based on an "average food basket".
[b] Sum of the compounds taken in, expressed as toxicity equivalents TEQ (cf. text).
[c] These numbers are included in the sum with 50%.
[d] At present, the TDI value (cf. text) is 1–4 pg/kg body weight and day. In outdoor air that is not directly contaminated, it can be assumed that the intake through breathing is 0.03 pg TEQ/kg body weight and day.

10–35 pg TEQ per g of fat and in the developing countries, 10 pg TEQ per g of fat.

9.11 References

Ames, B.N., Gold, L.S.: Paracelsus to parascience: the environmental cancer distraction. Mutation Research *447*, 3 (2000)

Amrein, T., Bachman, S., Noti, A., Biedermann, M., Barbarosa M.M., Biedermann-Brem, S., Grob, K., Keiser, A., Realini P., Scher, E., Amado, R.: Potential of acrylamide formation, sugars, and free asparagine in potatoes: a comparison of cultivars and farming systems. J. Agric. Food Chem. *51*, 5556 (2004)

Anastassiades, M., Lehotay, S.J., Stajnbaher, D., Schenk, F.J.: Fast and easy multiresidue method employing acetonitrile extraction/partitioning and "dispersive solid-phase extraction" for the determination of pesticide residues in produce. JAOAC internat *86*, 412 (2003)

Atreya, N.: Does the mere presence of a pesticide residue in food indicate a risk? J. Envir. Monit. *2*, 53N (2000)

Ballschmiter, K.: Chemie und Vorkommen der halogenierten Dioxine und Furane. Nachr. Chem. Tech. Lab. *39*, 988 (1991)

Beckmann, M., Haack, K.-J.: Insektizide für die Landwirtschaft. Chemie in unserer Zeit *37*, 88 (2003)

Buchanan, R.L., Doyle, M.P.: Food borne disease significance of Escherichia coli 0157:H7 and other enterohemorrhagic E. coli. Food Technol. *51* (no. 10), 69 (1997)

Bundesamt für Verbraucherschutz und Lebensmittelsicherheit: Ergebnisse des bundesweiten Lebensmittel-Monitorings der Jahre 1995–2002. www.bvl.bund.de/dl/monitoring/monitoring-1995-2002.pdf

Bundesforschungsanstalt für Ernährung: Radioaktivität in Lebensmitteln – Tschernobyl und die Folgen. Bericht BFE-R-86-04, Karlsruhe, Dezember 1986

Cikryt, P.: Die Gefährdung der Menschen durch Dioxine und verwandte Verbindungen. Nachr. Chem. Tech. Lab. *39*, 648 (1991)

Diehl, J.F.: Chemie in Lebensmitteln – Rückstände, Verunreinigungen, Inhalts- und Zusatzstoffe. Wiley-VCH-Verlag, Weinheim, 2000

Dixon, S.N.: Veterinary drug residues. In: Food chemical safety. Vol. 1: Contaminants (Ed. D.H. Watson). Woodhead Publishing, Cambridge, England, 2001, p. 109

Eisenbrand, G.: N-Nitrosoverbindungen in Nahrung und Umwelt. Wissenschaftliche Verlagsgesellschaft mbH, Stuttgart, 1981

Hathway, D.E.: Molecular aspects of toxicology. The Royal Society of Chemistry, Burlington House, London, 1984

Henningsen, M.: Moderne Fungizide. Chemie in unserer Zeit *37*, 98 (2003)

Macholz, R., Lewerenz, H.: Lebensmitteltoxikologie. Springer-Verlag, Berlin, 1989

National Research Council (U.S.A.): Regulating Pesticides in Food. National Academy Press, Washington, D.C., 1987

Petz, M.: Tandem-MS; Biosensoren und weitere analytische Trends sowie jüngste Erkenntnisse bei Tierarzneimittelrückständen. Lebensmittelchemie *55*, 1 (2001)

Petz, M.: Toxikologische Aspekte der Ernährung. In: Ernährungsbericht 2004. Deutsche Gesellschaft für Ernährung (ed.), p. 119

Rychlik, M., Schieberle, P.: Quantification of the mycotoxin Patulin by a stable isotope dilution assay. J. Agric. Food Chem. *47*, 3749 (1999)

Schrenk, D., Fürst, P.: WHO setzt Werte für die tolerierbare tägliche Aufnahme an Dioxinen neu fest. Nachr. Chem. Techn. Lab. *47*, 313 (1999)

Schwack, W., Anastassiades, M., Scherbaum, E.: Rückstandsanalytik von Pflanzenschutzmitteln. Chemie in unser Zeit *37*, 324 (2003)

Seitz, T., Hoffmann, M.G., Krähmer, H.: Herbizide für die Landwirtschaft. Chemie in unserer Zeit *37*, 112 (2003)

U.S. Food and Drug Administration: Pesticide Program. Residue Monitoring 2002. www.cfsan.fda.gov/dms/pres02rep.html

Von Hoof, N., de Wasch, K., Hoppe, H., Poelmans, S., de Brabander, H.F.: In: Rapid and on-line instrumentation for food quality assurance (Ed.: I.E. Tothill). Woodhead Publishing, Cambridge, 2002, p. 91

Wettach, J.W., Rung, B., Schwack, W.: Detection of photochemically induced cuticle-bound residues of parathion by immunoassay. Food Agric. Immunol. *14*, 5 (2002)

Xu, L.: Use of ozone to improve the safety of fresh fruits and vegetables. Food Technol. *53* (10), 58 (1999)

10 Milk and Dairy Products

10.1 Milk

Milk is the secreted fluid of the mammary glands of female mammals. It contains nearly all the nutrients necessary to sustain life. Since the earliest times, mankind has used the milk of goats, sheep and cows as food. Today the term "milk" is synonymous with cow's milk. The milk of other animals is spelled out, e. g., sheep milk or goat milk, when supplied commercially.

In Germany, the yield of milk per cow in kg/year has increased steadily as a result of selective breeding and improvements in feed. The yield was 1260 kg per cow in 1812, 2163 kg in 1926, 3800 kg in the FRG in 1970, 4181 kg in 1977 and 6537 kg in 2003. In the EU in 2003, Swedish cows were the best performers at 8073 kg, followed by Danish and Dutch animals at 7889 kg and 7494 kg respectively.

In some countries it is permitted to increase the yield of milk by injection of the growth hormone bovine somatropin (BST). The recombinant BST (rBST) used is identical in activity to natural BST. This is done by taking, from the DNA of cows, the specific gene sequence that carries the instructions for preparing BST and inserting it into *E. coli*, which can then produce large amounts of rBST. Natural BST consists of 190 or 191 amino acids. rBST may differ slightly in that a few extra amino acids may be attached at the N-terminal end of the BST molecule. Due to differences in the molecular mass it is possible to distinguish between rBST and natural BST. Milk production in various countries, its processing into dairy products and its consumption are summarized in Tables 10.1–10.3.

10.1.1 Physical and Physico-Chemical Properties

Milk is a white or yellow-white, opaque liquid. The color is influenced by scattering and absorption of light by milk fat globules and protein micelles. Therefore, skim milk also retains its white color. A yellowish, i. e. yellow-green, color is derived from carotene (ingested primarily during pasture grazing) present in the fat phase and from riboflavin present in the aqueous phase. Milk tastes mildly sweet, while its odor and flavor are normally quite faint.

Milk fat occurs in the form of droplets or globules, surrounded by a membrane and emulsified in milk serum (also called whey). The fat globules (called cream) separate after prolonged storage or after centrifugation. The fat globules float on the skim milk. Homogenization of milk so finely divides and emulsifies the fat globules that cream separation does not occur even after prolonged standing.

Proteins of various sizes are dispersed in milk serum. They are called micelles and consist mostly of calcium salts of casein molecules. Furthermore, milk contains lipoprotein particles, also called milk microsomes, which consist of the residues of cell membranes, microvilli, etc., as well as somatic cells, which are mainly leucocytes (10^8/l of milk). Some of the properties of the main structural elements of milk are listed in Table 10.4.

Various proteins, carbohydrates, minerals and other ingredients are solubilized in milk serum. The specific density of milk decreases with increasing fat content, and increases with increasing amounts of protein, milk sugar and salts. The specific density of cow's milk ranges from 1.029 to 1.039 (15 °C). Defatted (skim) milk has a higher specific density than whole milk. From the relationships given by *Fleischmann*:

$$m = 1.2f + \frac{266.5(s-1)}{s} \qquad (10.1)$$

and by *Richmond*:

$$m = 0.25s + 1.21f + 0.66 \qquad (10.2)$$

H.-D. Belitz · W. Grosch · P. Schieberle, *Food Chemistry*
© Springer 2009

Table 10.1. Production of milk, 2006 (1000 t)

Continent	Cow milk	Buffalo milk	Sheep milk	Goat milk
World	549,693	80,094	8723	13,801
Africa	24,674	2300	1719	3129
America, Central-	14,179	–	–	–
America, North-	90,564	–	–	–
America, South- and Caribbean	66,030	–	36	164
Asia	134,170	77,571	4006	7821
Europe	209,441	222	2963	2479
Oceania	24,814	–	–	–

Country	Cow milk	Country	Buffalo milk	Country	Sheep milk
USA	82,463	India	52,100	China	1091
India	39,775	Pakistan	21,136	Turkey	790
China	32,249	China	2850	Greece	752
Russian Fed.	31,074	Egypt	2300	Syria	604
Germany	28,453	Nepal	927	Italy	554
Brazil	25,333	Iran	232	Romania	545
France	24,195	Italy	215	Iran	534
UK	14,577	Myanmar	171	Sudan	487
New Zealand	14,498	Turkey	38	Spain	403
Ukraine	12,988	Viet Nam	31	France	263
Poland	11,982	$\Sigma\,(\%)^a$	100	Algeria	210
Italy	11,013			Mali	128
Netherlands	10,532			Bulgaria	108
Australia	10,250			Portugal	100
Mexico	10,029			$\Sigma\,(\%)^a$	75
Turkey	10,026				
Pakistan	9404				
Japan	8134				
Argentina	8100				
Canada	8100				
Colombia	6770				
$\Sigma\,(\%)^a$	75				

Country	Goat milk
India	3790
Sudan	1519
Bangladesh	1416
Pakistan	676
France	583
Greece	511
Spain	423
Iran	365
China	262
Ukraine	258
Russian Fed.	256
Turkey	254
$\Sigma\,(\%)^a$	75

^{a} World production $= 100\%$.

Table 10.2. Production of dairy products in 2004 (1000 t)

Continent	Cheese	Butter[a]	Condensed milk	Whole milk powder	Skim milk powder[b]	Whey powder
World	17,824	7968	3892	2702	3455	2038
Africa	915	226	64	21	11	2
America, North-, Central-	4944	646	1112	140	796	542
America, South-	668	191	377	768	64	–
Asia	1090	3678	559	83	239	4
Europe	9558	2622	1760	946	1699	1386
Oceania	649	605	21	744	647	105

Country	Cheese	Country	Butter[a]	Country	Condensed milk
USA	4357	India	2500	USA	797
Germany	1852	Pakistan	557	Germany	505
France	1840	USA	525	The Netherlands	291
Italy	1320	New Zealand	473	Peru	274
The Netherlands	670	Germany	440	Russian Fed.	193
Egypt	661	France	420	Thailand	179
Poland	520	Russian Fed.	262	Malaysia	164
Russian Fed.	483	Poland	180	Mexico	158
UK	370	UK	160	UK	139
Australia	364	Iran	150	China	114
Argentina	360	Ireland	142	Ukraine	80
Canada	360	Australia	130	Canada	78
Denmark	335	Italy	125	Σ (%)[c]	76
Σ (%)[c]	76	Σ (%)[c]	76		

Country	Whole milk powder	Country	Skim milk powder[b]	Country	Whey powder
New Zealand	557	USA	674	France	610
Brazil	420	New Zealand	425	USA	493
France	220	France	271	Germany	262
Australia	187	Germany	250	The Netherlands	219
Argentina	165	Russian Fed.	243	Australia	82
The Netherlands	112	Australia	222	UK	56
Mexico	105	Japan	180	Canada	49
UK	90	Poland	140	Denmark	39
Russian Fed.	85	Ukraine	117	Finnland	32
Denmark	80	Canada	102	Ireland	30
Σ (%)[c]	75	Σ (%)[c]	76	Σ(%)[c]	92

[a] Including fat from buffalo milk (ghee)
[b] Including butter-milk powder
[c] World production $= 100\%$

the dry matter content of milk, m, in percent, can be calculated from the percent fat content (f), knowing the specific density (s).

The freezing point of milk is -0.53 to $-0.55\,°C$. This rather constant value is a suitable test for detection of watering of milk.

The pH of fresh milk is 6.5–6.75, while the acid degree according to *Soxhlet–Henkel* (°SH) is 6.5–7.5.

The refractive index (n_D^{20}) is 1.3410–1.3480, and the specific conductivity at $25\,°C$ is 4–$5.5 \times 10^{-3}\ \mathrm{ohm^{-1} cm^{-1}}$.

Table 10.3. Consumption of milk and dairy products in FR Germany (in kg/capita and year)

	1996	2003	2005
Consumer milk	66.7	66	67
Fresh milk products (without yoghurt)	9.9	12.2	12
Yoghurt	13.1	15.3	16.8
Cream and cream products	7.6	7.4	7.4
Butter	7.3	6.6	6.5

Table 10.5. Composition of human milk and milk of various mammals (%)

Milk	Protein	Casein	Whey protein	Sugar	Fat	Ash
Human	0.9[a]	0.4	0.5	7.1	4.5	0.2
Cow (bovine)	3.2	2.6	0.6	4.6	3.9	0.7
Donkey	2.0	1.0	1.0	7.4	1.4	0.5
Horse	2.5	1.3	1.2	6.2	1.9	0.5
Camel	3.6	2.7	0.9	5.0	4.0	0.8
Zebu	3.2	2.6	0.6	4.7	4.7	0.7
Yak	5.8			4.6	6.5	0.9
Buffalo	3.8	3.2	0.6	4.8	7.4	0.8
Goat	3.2	2.6	0.6	4.3	4.5	0.8
Sheep	4.6	3.9	0.7	4.8	7.2	0.9
Reindeer	10.1	8.6	1.5	2.8	18.0	1.5
Cat	7.0	3.8	3.2	4.8	4.8	0.6
Dog	7.4	4.8	2.6			
Rabbit	10.4					

[a] After the 15-th day of the breast feeding period the protein content is increased to 1.6%.

The measurement of redox potentials of milk and its products can also be of value. The redox potential is +0.30 V for raw and +0.10 V for pasteurized milk, +0.05 V for processed cheese, −0.15 V for yoghurt and −0.30 V for Emmental cheese.

10.1.2 Composition

The composition of dairy cattle milk varies to a fairly significant extent. Table 10.5 provides some data. In all cases water is the main ingredient of milk at 63–87%. In the following sections, only cow's milk will be dealt with in detail since it is the main source of our dairy foods.

10.1.2.1 Proteins

In 1877 *O. Hammarsten* distinguished three proteins in milk: casein, lactalbumin and lactoglobulin. He also outlined a procedure for their separation: skim milk is diluted then acidified with acetic acid. Casein flocculates, while the whey proteins stay in solution. This established a specific property of casein: it is insoluble in weakly acidic media. It was later revealed that the milk protein system is much more complex. In 1936 *Pedersen* used ultra-centrifugation to demonstrate the nonhomogeneity of casein, while in 1939 *Mellander* used electrophoresis to prove that casein consists of three fractions, i. e. α-, β- and γ-casein. The most important proteins of milk are listed in Table 10.7. The casein fraction forms the main portion. Major constituents of whey proteins, β-lactoglobulin A and B and α-lactalbumin, can be differentiated genetically. Other protein constituents, e. g., enzymes, are present in much lower quantities; they are not listed in Table 10.7.

Table 10.4. Main structural elements of milk

Name	Type of dispersion	Percentage	Number (1^{-1})	Diameter (mm)	Surface $(m^2/1$ milk$)$	Specific density[a] (g/ml)
Fat globules	Emulsion	3.8	10^{13}	100–10,000	70	0.92
Casein micelles	Suspension	2.8	10^{17}	10–300	4000	1.11
Globular proteins (whey proteins)	Colloidal solution	0.6	10^{20}	3–6	5000	1.34
Lipoprotein particles	Colloidal suspension	0.01	10^{17}	10	10	1.10

[a] 20 °C.

Table 10.6. Amino acid composition (g AA/100 g protein) of the total protein, casein, and whey protein of bovine milk

Amino acid	Total protein	Casein	Whey protein
Alanine	3.7	3.1	5.5
Arginine	3.6	4.1	3.3
Aspartic acid	8.2	7.0	11.0
Cystine	0.8	0.3	3.0
Glutamic acid	22.8	23.4	15.5
Glycine	2.2	2.1	3.5
Histidine	2.8	3.0	2.4
Isoleucine	6.2	5.7	7.0
Leucine	10.4	10.5	11.8
Lysine	8.3	8.2	9.6
Methionine	2.9	3.0	2.4
Phenylalanine	5.3	5.1	4.2
Proline	10.2	12.0	4.4
Serine	5.8	5.5	5.5
Threonine	4.8	4.4	8.5
Tryptophan	1.5	1.5	2.1
Tyrosine	5.4	6.1	4.2
Valine	6.8	7.0	7.5

The amino acid composition of the total protein, casein, and whey protein of bovine milk is presented in Table 10.6.

10.1.2.1.1 Casein Fractions

The main constituents of this milk protein fraction have been fairly well investigated.
Their amino acid sequences are summarized in Table 10.8. Data showing the genetic variations are provided in Table 10.9. Caseins are not denaturable because of the lacking tertiary structure.

α_s-Caseins. The B variant of α_{s1}-casein consists of a peptide chain with 199 amino acid residues and has a molecular weight of 23 kdal. The sequence contains 8 phosphoserine residues, 7 of which are localized in positions 43–80, and these positions have an additional 12 carboxyl groups. Thus these positions are extremely polar acidic segments along the peptide chain. Proline is uni-

Table 10.7. Bovine milk proteins

Fraction	Genetic variants	Portion[a]	Isoionic point	Molecular weight[b] (kdal)	Phosphorus content (%)
Caseins		80	–	–	0.9
α_{s1}-Casein	A, B, C, D, E	34	4.92–5.35	23.6[f]	1.1
α_{s2}-Casein	A, B, C, D	8		25.2[g]	1.4
κ-Casein	A, B, C, E	9	5.77–6.07	19[h]	0.2
β-Casein	A^1, A^2, A^3, B, C, D, E	25	5.20–5.85	24	0.6
γ-Casein		4	5.8–6.0	12–21	0.1
γ_1-Casein	A^1, A^2, A^3, B			20.5	
γ_2-Casein	$A^1/A^2, A^3$, B			11.8	
γ_3-Casein	$A^1/A^2/A^3$, B			11.6	
Whey proteins		20	–	–	
β-Lactoglobulin	A, B, C, D, E, F, G	9	5.35–5.41	18.3	
α-Lactalbumin	A, B, C	4	4.2–4.5[e]	14.2	
Serum albumin	A	1	5.13	66.3	
Immunoglobulin		2			
IgG1			5.5–6.8	162	
IgG2			7.5–8.3	152	
IgA			–	400[c]	
IgM			–	950[d]	
FSC(s)[i]				80	
Proteose-Peptone		4	3.3–3.7	4–41	

[a] As % of skim milk total protein, [b] monomers, [c] dimer, [d] pentamer, [e] isoelectric point, [f] Variant B, [g] Variant A, [h] Variant A^2, [i] Free secretory component

Table 10.8. Amino acid sequences of bovine milk proteins

α_{s1}-*Casein B-8P*

R	P	K	H	P	I	K	H	Q	G	L	P	Q	E	V	L	N	E	N	L
L	R	F	F	V	A	P	F	P	Q	V	F	G	K	E	K	V	N	Q	L
S	K	D	I	G	S^a	E	S^a	T	E	D	Q	A	M	E	D	I	K	E	M
E	A	E	S^a	I	S^a	S^a	S^a	E	E	I	V	P	N	S^a	V	Q	E	K	H
I	Q	K	E	D	V	P	S	E	R	Y	L	G	Y	L	E	Q	L	L	R
L	K	K	Y	K	V	P	Q	L	E	I	V	P	N	S^a	A	E	E	R	L
H	S	M	K	E	G	I	H	A	Q	Q	K	E	P	M	I	G	V	N	Q
E	L	A	Y	F	Y	P	E	L	F	R	Q	F	Y	Q	L	D	A	Y	P
S	G	A	W	Y	Y	V	P	L	G	T	Q	Y	T	D	A	P	S	F	S
D	I	P	N	P	I	G	S	E	N	S	E	K	T	T	M	P	L	W	

α_{s2}-*Casein A-11P*

K	N	T	M	E	H	V	S^a	S^a	S^a	E	E	S	I	I	S^a	Q	E	T	Y
K	Q	E	K	N	M	A	I	N	P	S	K	E	N	L	C	S	T	F	C
K	E	V	V	R	N	A	N	E	E	E	Y	S	I	G	S^a	S^a	S^a	E	E
S^a	A	E	V	A	T	E	E	V	K	I	T	V	D	D	K	H	Y	Q	K
A	L	N	E	I	N	E	F	Y	Q	K	F	P	Q	Y	L	Q	Y	L	Y
Q	G	P	I	V	L	N	P	W	D	Q	V	K	R	N	A	V	P	I	T
P	T	L	N	R	E	Q	L	S^a	T	S^a	E	E	N	S	K	K	T	V	D
M	E	S^a	T	E	V	F	T	K	K	T	K	L	T	E	E	E	K	N	R
L	N	F	L	K	K	I	S	Q	R	Y	Q	K	F	A	L	P	Q	Y	L
K	T	V	Y	Q	H	Q	K	A	M	K	P	W	I	Q	P	K	T	K	V
I	P	Y	V	R	Y	L													

β-*Casein A²-5P*

R	E	L	E	E	L	N	V	P	G	E	I	V	E	S^a	L	S^a	S^a	S^a	E
E	S	I	T	R	I	N	K	K	I	E	K	F	Q	S^a	E	E	Q	Q	Q
T	E	D	E	L	Q	D	K	I	H	P	F	A	Q	T	Q	S	L	V	Y
P	F	P	G	P	I	P	N	S	L	P	Q	N	I	P	P	L	T	Q	T
P	V	V	V	P	P	F	L	Q	P	E	V	M	G	V	S	K	V	K	E
A	M	A	P	K	H	K	E	M	P	F	P	K	Y	P	V	Q	P	F	T
E	S	Q	S	L	T	L	T	D	V	E	N	L	H	L	P	P	L	L	L
Q	S	W	M	H	Q	P	H	Q	P	L	P	P	T	V	M	F	P	P	Q
S	V	L	S	L	S	Q	S	K	V	L	P	V	P	E	K	A	V	P	Y
P	Q	R	D	M	P	I	Q	A	F	L	L	Y	Q	Q	P	V	L	G	P
V	R	G	P	F	P	I	I	V											

κ-*Casein B-1P*

Z^d	E	Q	N	Q	E	Q	P	I	R	C	E	K	D	E	R	F	F	S	D
K	I	A	K	Y	I	P	I	Q	Y	V	L	S	R	Y	P	S	Y	G	L
N	Y	Y	Q	Q	K	P	V	A	L	I	N	N	Q	F	L	P	Y	P	Y
Y	A	K	P	A	A	V	R	S	P	A	Q	I	L	Q	W	Q	V	L	S
D	T	V	P	A	K	S	C	Q	A	Q	P	T	T	M	A	R	H	P	H
P	H	L	S	F	M	A	I	P	P	K	K	N	Q	D	K	T	E	I	P
T	I	N	T	I	A	S	G	E	P	T^b	S	T^b	P	T^b	I	E	A	V	E
S	T	V	A	T	L	E	A	S^a	P	E	V	I	E	S	P	P	E	I	N
T	V	Q	V	T	S	T	A	V											

formly distributed along the chain and apparently to a great extent hinders the formation of a regular structure. A portion of the chain, up to 30%, is assumed to have regular conformations. Amino acid residues 100–199 are distinctly apolar and are responsible for strong association tendencies,

Table 10.8. continued

α-*Lactalbumin B*[c]																
E	Q	L	T	K	C	E	V	F	R	E	L	K	D	L	K	G	Y	G	G
V	S	L	P	E	W	V	C	T	T	F	H	T	S	G	Y	D	T	E	A
I	V	E	N	N	Q	S	T	D	Y	G	L	F	Q	I	N	N	K	I	W
C	K	N	D	Q	D	P	H	S	S	N	I	C	N	I	S	C	D	K	F
L	N	N	D	L	T	N	N	I	M	C	V	K	K	I	L	D	K	V	G
I	N	Y	W	L	A	H	K	A	L	C	S	E	K	L	D	Q	W	L	C
E	K	L																	

β-*Lactoglobulin B*[e]																
L	I	V	T	Q	T	M	K	G	L	D	I	Q	K	V	A	G	T	W	Y
S	L	A	M	A	A	S	D	I	S	L	L	D	A	Q	S	A	P	L	R
V	Y	V	E	E	L	K	P	T	P	E	G	D	L	E	I	L	L	Q	K
W	E	N	G	E	C	A	Q	K	K	I	I	A	E	K	T	K	I	P	A
V	F	K	I	D	A	L	N	E	N	K	V	L	V	L	D	T	D	Y	K
K	Y	L	L	F	C	M	E	N	S	A	E	P	E	Q	S	L	A	C	Q
C	L	V	R	T	P	E	V	D	D	E	A	L	E	K	F	D	K	A	L
K	A	L	P	M	H	I	R	L	S	F	N	P	T	Q	L	E	E	Q	C
H	I																		

[a] The serine residue is phosphorylated.

[b] These threonine residues can be glycosylated.

[c] Disulfide bonds: 6–120, 28–111, 61–77, 73–91.

[d] Pyrrolidone carboxylic acid.

[e] Disulfide bonds: 66–160 and apparently either 106–119 or 106–121. Accordingly, the free thiol group is either Cys-119 or Cys-121.

which are limited by the repulsing forces of phosphate groups. In the presence of Ca^{2+} ions, in the levels found in milk, $α_{s1}$-casein forms an insoluble Ca-salt. In the A variant of the molecule, amino acid residues 14–26 are missing; in the C variant the glutamic acid in position 192 (Glu-192) is replaced by Gly-192; and in the D variant Pth-53 (phosphothreonine) replaces Ala-53.

$α_{s2}$-*Casein* (M_r 25,000) consists of 207 amino acid residues, has a pronounced dipolar structure with a concentration of anionic groups in the region of the N-terminus and cationic groups in the region of the C-terminus. It contains 11 phosphoserine and 2 cysteine residues and is even more easily precipitable with $Ca^{2⊕}$ than $α_{s1}$-casein. Other proteins, previously known as $α_{s3}$-, $α_{s4}$-, $α_{s5}$-, and $α_{s6}$-caseins, appear to be members of the $α_{s2}$ family and to differ in the degree of phosphorylation. Dimers linked via disulfide bridges also appear to be present.

β-*Caseins.* The A^2 variant is a peptide chain consisting of 209 residues and has a molecular weight of 24.0 kdal. Five phosphoserine residues

are localized in positions 1–40; these positions contain practically all of the ionizing sites of the molecule. Positions 136–209 contain mainly residues with apolar side chains. On the whole, β-casein is the most hydrophobic casein. The molecule has a structure with a "polar head" and an "apolar tail", thus resembling a "soaplike" molecule. Indeed, CD measurements have shown that β-casein contains about 9% of α-helix structure and about 25% of β-structure. An increase in temperature results in an increase in the β-structure at the cost of the aperiodic part. The self-association of β-casein is an endothermic process. Like $α_{s1}$-casein, β-casein contains no cysteine. The protein precipitates in the presence of Ca^{2+} ions at the levels found in milk. However, at temperatures at or below 1 °C the calcium salt is quite soluble.

κ-*Caseins.* The B variant consists of a peptide chain with 169 residues and has a molecular weight of 18 kdal. The monomer, which contains 1 phosphoserine and 2 cysteine residues, is accessible only under reducing conditions.

Table 10.9. Amino acid sequences[a] of genetic variants of bovine milk proteins

Protein	Variant	Frequency[b]	14–26	53	59	192
α_{s1}-Casein	A	s	14–26 are lacking			
(199 AS)	B	w		Ala	Glu	Glu
	C	i				Gly
	D	s		ThrP		
	E	s			Lys	Gly

Protein	Variant	Frequency[b]	33	47	50–58	130
α_{s2}-Casein	A		Glu	Ala		Thr
(207 AS)	B					
	C		Gly	Thr		Ile
	D				lacking	

Protein	Variant	Frequency[b]	18	35	36	37	67	106	122
β-Casein	A^1						His		
(209 AS)	A^2	w, i	SerP	SerP	Glu	Glu	Pro	His	Ser
	A^3						Gln		
	B	s					His		Arg
	C	s		Ser			Lys	His	
	D	s	Lys						
	E	s					Lys		

Protein	Variant	Frequency[b]	97	136	148	155
κ-Casein	A	x		Thr	Asp	Ser
(169 AS)	B	w, i	Arg	Ile	Ala	
	C			His		
	E					Gly

Protein	Variant	Frequency[b]	10
α-Lactalbumin	A	i	Gln
(123 AS)	B	w	Arg

Protein	Variant	Frequency[b]	45	50	59	64	78	118	130	158
β-Lactoglobulin	A	x				Asp		Val		
(162 AS)	B	w, i	Glu	Pro	Gln	Gly	Ile	Ala	Asp	Gln
	C				His					
	D		Gln							
	E									Gly
	F			Ser					Tyr	Gly
	G						Met			Gly

[a] cf. Table 10.8.

[b] w: predominant in the western world (*Bos taurus*), i: predominant in India (*Bos indicus, Bos grunniens*), s: rare, x: not predominant, but not rare.

Normally, κ-casein occurs as a trimer or as a higher oligomer in which the formation of disulfide bonds is probably involved. The protein contains varying amounts of carbohydrates (average values: 1% galactose, 1.2% galactosamine, 2.4% N-acetyl neuramic acid) that are bound to the peptide chain through Thr-131, 133, 135 or (in variant A) 136. κ-Casein is separated electrophoretically into various components that have the same composition of amino acids, but differ in their carbohydrate moiety, e. g., per protein molecule they contain 0–3 moles N-acetyl neuramic acid, 0–4 moles galatose and 0–3 moles galactosamine. Three different glycosyl residues could be isolated, one of which has the structure shown in Formula 10.3.

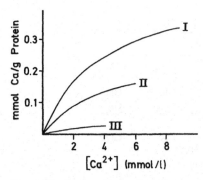

Fig. 10.1. Calcium binding by I: α_{s1}-casein (0.38), II: β-casein (0.21) and III: κ-casein (0.05). The bound phosphate residues in mmol/g of casein are given in brackets (according to *Walstra* and *Jenness*, 1984)

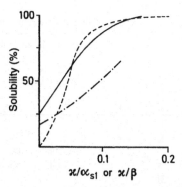

Fig. 10.2. Influence of κ-casein on the solubility of κ_{s1}-casein. (-2.5 mg/ml) and β-casein (-1.5 mg/ml; -6 mg/ml) at pH 7.0, 30 °C, 100 mmol/l $CaCl_2$ (according to *Walstra* and *Jenness*, 1984)

In the other two oligosaccharide units, one of the two N-acetylneuraminic acid residues is lacking in each case.

κ-Casein is the only main constituent of casein which remains soluble in the presence of Ca^{2+} ions in the concentrations found in milk (Fig. 10.1). Aggregation of α_{s1}- and β-caseins with κ-casein prevents their coagulation in the presence of Ca^{2+} ions (Fig. 10.2). This property of κ-casein is of utmost importance for formation and maintenance of stable casein complexes and casein micelles, as occur in milk. Chymosin (rennet, rennin cf. 1.4.5.2) selectively cleaves the peptide chain of κ-casein at $-Phe^{105} - Met^{106}-$ into two fragments: para-κ-casein and a glycopeptide (Pyg = pyroglutamic acid, i. e. pyrrolidone carboxylic acid):

The released glycopeptide is soluble, while para-κ-casein precipitates in the presence of Ca^{2+} ions. In this way κ-casein loses its protective effect; the casein complexes and casein micelles coagulate (curdle formation) from the milk. The specificity of rennin is high, as is shown in Table 10.10. If Met^{106} in κ-casein is replaced with Phe^{106} by genetic engineering techniques, the rate of catalysis is increased by 80%. The sugar moiety of κ-casein is not essential for rennin action, nor for the stabilizing property of its protein portion. However, the sugar moiety delays protein cleavage by rennin. Also, it appears that the stability of α_s- and κ-casein mixtures in the presence of Ca^{2+} ions is influenced by the carbohydrate content of κ-casein.

(10.3)

1	105	106	169
Pyg $\cdots\cdots$	Phe $-$	Met $\cdots\cdots$	Val

κ-Casein

$\xrightarrow{\text{Lab}}$

1	105	106	169
Pyg $\cdots\cdots$	Phe $+$	Met $\cdots\cdots$	Val

para-κ-Casein   Glycopeptide   (10.4)

Table 10.10. Chymosin specificity: relative rate of hydrolysis of peptides from the κ-casein amino acid sequence

Substrate	V_{rel}^a
105 106 Phe-Met	0.00
104 108 Ser-Phe-Met-Ala-Ile	0.04
109 Ser-Phe-Met-Ala-Ile-Pro	0.11
103 Leu-Ser-Phe-Met-Ala-Ile	21.6
102 His-Leu-Ser-Phe-Met-Ala-Ile	31
110 Leu-Ser-Phe-Met-Ala-Ile-Pro-Pro	100
101 Pro-His-Leu-Ser-Phe-Met-Ala-Ile	100
98 112 His-Pro-His-Pro-His-Leu-Ser-Phe-Met-Ala-Ile-Pro-Pro-Lys-Lys	2500

^{a} Relative rate: $kcat/K_m$.

In the C variant of κ-casein, Arg^{97} is replaced with His^{97} (Table 10.9), which has a weaker positive charge. As a result, chymosin is not as strongly bound as in the case of the B variant; the rate of catalysis decreases. Therefore, C variant milk is less suitable for the production of sweet-milk cheese than B variant milk.

γ-Caseins. These proteins are degradation products of the β-caseins, formed by milk proteases, e. g., $γ_1$-casein is obtained by cleavage of the residues 1–28. The peptide released is identical to the proteose-peptone PP8F which has been found in milk. Correspondingly, $γ_2$- and $γ_3$-caseins are formed by hydrolysis of the amino acid residues 1–105 and 1–107 respectively. According to more recent nomenclature recommendations, β-casein fragments should be described by the position numbers. Thus, $γ_1$-casein from any β-casein variant X is called, e. g., β-casein X (f29–209) and the corresponding proteose peptone PF8F β-casein X (f1–28).

λ-Caseins. The λ-casein fraction consists mainly of fragments of the $α_{s1}$-caseins. In vitro the α-caseins are formed by incubation of the $α_{s1}$-caseins with bovine plasmin.
The molar ratio of the main components $α_{s1}/β + γ/κ/α_{s2}$ is on an average 8/8/3/2. All casein forms contain phosphoric acid, which always occurs in a tripeptide sequence pattern (Pse = phos-

phoserine):

$$Pse-X-Glu \quad or \quad Pse-X-Pse \quad (10.5)$$

in which X is any amino acid, including phosphoserine and glutamic acid. Examples are:

$α_{s1}$-Casein :	Pse-Glu-Pse
	Pse-Ile-Pse-Pse-Pse-Glu
	Pse-Val-Glu
	Pse-Ala-Glu
β-Casein :	Pse-Leu-Pse-Pse-Pse-Glu
	Pse-Glu-Glu
κ-Casein :	Pse-Pro-Glu (10.6)

Most probably this regular pattern originates from the action of a specific protein kinase. The various distribution of polar and apolar groups of the individual proteins outlined above are summarized in Table 10.11. The hydrophobicity values listed are average hydrophobicity values $\overline{H}$ of the amino acid side chains present in the sequence of the given segments, and are calculated as follows:
A measure of the hydrophobicity of a compound is the free energy, F_t, needed to transfer the compound from water into an organic solvent, and is given as the ratio of the compound's solubility in water (N_w, as mole fraction) and in the organic solvent (N_{org}, as mole fraction), involving the ac-

Table 10.11. Distribution of amino acid residues with ionizing side chains (net charge) and with nonpolar side chains (hydrophobicity) in α_{s1}-casein and β-casein

Residue	α_{s1}-Casein		Residue	β-Casein	
	1	2		1	2
1–40	+3	1340	1–43	−16	783
41–80	−22.5	641	44–92	−3.5	1429
81–120	0	1310	93–135	+2	1173
121–160	−1	1264	136–177	+3	1467
161–199	−2.5	1164	178–209	+2	1738

1 Net charge.
2 Hydrophobicity $\overline{H}$ (Cal/mole; cf. text).

tivity coefficients (γ_w, γ_{org}):

$$\Delta F_t = RT \ln \frac{N_w \cdot \gamma_w}{N_{org} \cdot \gamma_{org}} \tag{10.7}$$

The corresponding free energy of transfer of the side chain of an amino acid $H\Phi_i$ is obtained from the following relationship:

$$H\Phi_i = \Delta F_t(\text{amino acid } i) - \Delta F_i(\text{glycine})$$

The average hydrophobicity of a sequence segment of a polypeptide chain with n amino acid residues is then:

$$\overline{H} = \frac{\Sigma H\Phi_i}{n} \tag{10.8}$$

The higher the $H\Phi_i$, i.e. $\overline{H}$, the higher is the hydrophobicity of individual side chains, i.e. the sequence segment. Data provided in Table 10.11 are related to the ethanol/water system.

10.1.2.1.2 Micelle Formation

Only up to 10% of the total casein fraction is present as monomers. They are usually designated as serum caseins and the concentration ratio $c_\beta > c_\kappa > c_{\alpha sl}$ is quite valid. However, the main portion is aggregated to casein complexes and casein micelles. This aggregation is regulated by a set of parameters, as presented in Fig. 10.3. Dialysis of casein complexes against a chelating agent might shift the equilibrium completely to

Monomers (soluble caseins)

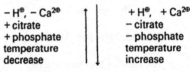

Casein complex

Micella
(calcium caseinate + calcium phosphate)

Fig. 10.3. Casein complex and casein micelle formation

monomers, while against high Ca^{2+} ion concentrations the shift would be to large micelles.

From Fig. 10.4 it follows that the diameter of the micelles in skim milk varies from 50–300 nm, with a particle distribution peak at 150 nm. Using an average diameter of 140 nm, the micelle volume is 1.4×10^6 nm^3 and the particle weight is 10^7–10^9 dal. This corresponds to 25,000 monomers per micelle. Casein micelles are substantially smaller than fat globules, which have diameters between 0.1–10 μm. Scanning electron micrographs of micelles are shown in Fig. 10.5 and compositional data are provided in Table 10.12. The ratio of monomers in micelles varies to a great extent (Table 10.13), depending on dairy cattle breed, season and fodder, and is influenced also by micellular size (Table 10.14). The micelles are not tightly packed and so are of variable density. They are strongly solvated

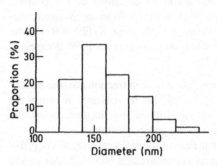

Fig. 10.4. Particle size distribution of casein micelles in skim milk (fixation with glutaraldehyde)

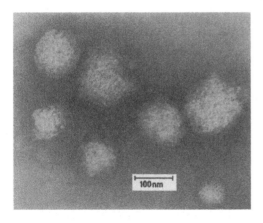

Fig. 10.5. Electron micrograph of the casein micelles in skim milk (according to *Webb*, 1974). The micelles are fixed with glutaraldehyde and then stained with phosphomolybdic acid

Table 10.12. Composition of casein micelles (%)

Casein	93.2	Phosphate	
Ca	2.9	(organic)	2.3
Mg	0.1	Phosphate	
Na	0.1	(inorganic)	2.9
K	0.3	Citrate	0.4

Table 10.13. Typical distribution of components in casein micelles

Component	Ratio numbers			
α_{s1}	3	6	9	12
β	1	1	4	4
γ		1	1	1
κ	1	3	3	3

Table 10.14. Composition and size of casein micelles isolated by centrifugation

Centrifugation time (min)[a]	Composition of the sediment(%)			
	α_{s1}	β	κ	Others
0[b]	50	32	15	3
0–7.5	47	34	16	3
7.5–15	46	32	18	4
15–30	45	31	20	4
39–60	42	29	26	3
Serum casein	39	23	33	5

[a] Centrifugation speed $10^5 \times g$.
[b] Isoelectric casein.

which consist of ca. 30 different casein monomers and aggregate to large micelles via calcium phosphate bridges. Two types of subunits apparently exist: one type contains κ-casein and the other does not. The κ-casein molecules are arranged on the surface of the corresponding submicelles. At various positions, their hydrophilic C-termini protrude like hairs from the surface, preventing aggregation. Indeed, aggregation of the submicelles proceeds until the entire surface of the forming micelle is covered with κ-casein, i.e., covered with "hair", and, therefore, exhibits steric repulsion. The effective density of the hair layer is at least 5 nm. A small part of the κ-casein is also found inside the micelle.

10.1.2.1.3 Gel Formation

The micelle system, can be destabilized by the action of rennin or souring. Rennin attacks κ-casein, eliminating not only the C-terminus in the form of the soluble glycopeotide 106–169, but also the cause of repulsion. The remaining paracasein micelles first form small aggregates with an irregular and often long form, which then assemble with gel formation to give a three dimensional network with a pore diameter of a few µm. The fat globules present are included in this network with pore enlargement. It is assumed that dynamic equilibria exist between casein monomers and submicelles, dissolved and bound calcium phosphate, and submicelles and micelles.

(1.9 g water/g protein) and hence are porous. The monomers are kept together with:

* Hydrophobic interactions that are minimal at a temperature less than 5 °C.
* Electrostatic interactions, mostly as calcium or calcium phosphate bridges between phosphoserine and glutamic acid residues (Fig. 10.6).
* Hydrogen bonds.

On a molecular level different micelle models have been proposed which to a certain extent explain the experimental findings. The most probable model is shown in Fig. 10.7. This model comprises subunits (submicelles, $M_r \sim 760,000$)

Fig. 10.6. Peptide chain bridging with calcium ions

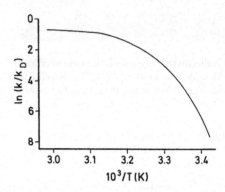

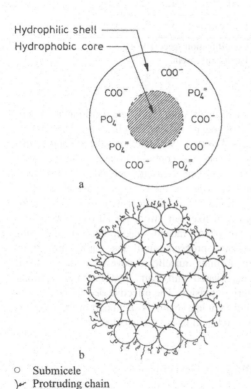

Hydrophilic shell
Hydrophobic core

○ Submicele
}⌒ Protruding chain
— Calcium phosphate

Fig. 10.7. Schematic model of a casein micelle; (**a**) a subunit consisting of α_{s1}-, β-, γ-, κ-caseins, (**b**) Micelle made of subunits bound by calcium phosphate bridges (according to *Webb*, 1974)

The rate of gel formation increases with increasing temperature (Fig. 10.8). It is slow at $T < 25\,°C$ and proceeds almost under diffusion control at $T \sim 60\,°C$. It follows that hydrophobic interactions, especially due to the very hydrophobic para-κ-casein remaining on the surface after the action of rennin, are the driving force for

Fig. 10.8. Temperature dependency of the aggregation rate of para-casein micelles (rate constant k in fractions of the diffusion-controlled rate k_D; according to *Dalgleish*, 1983)

gel formation. In addition, other temperature-dependent reactions play a role, like the binding of calcium ions and of β-casein to the micelles, and the change in solubility of colloidal calcium phosphate.

Acid coagulation of casein is also definitely caused by hydrophobic interactions, as shown by the dependency of the coagulation rate on the temperature and pH value (Fig. 10.9). On acidification, the micelle structure changes due to the migration of calcium phosphate and monomeric casein. Since the size of the micelle remains practically constant, this migration of components must be associated with swelling. During coagulation, dissolved casein reassociates with the micelles, forming a gel network.

The gel structure can be controlled via changes in the hydrophobicity of the micelle surface. A decrease in hydrophobicity is possible, e. g., by heating milk ($90\,°C/10\,min$). Covalent bonding of denatured β-lactoglobulin to κ-casein (cf. 10.1.3.5) occurs, burying hydrophobic groups.

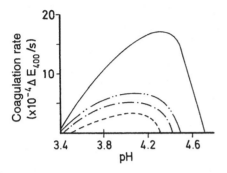

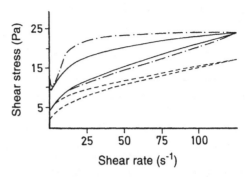

Fig. 10.9. Rate of coagulation of casein micelles as a function of temperature and pH value (——— 25 °C, — ·· — 15 °C, — · — 10 °C, – – – 5 °C, according to *Bringe, Kinsella*, 1986)

Fig. 10.11. Flow curves of stiff skim-milk yoghurt subjected to defined prestirring as a function of the rate of denaturation of β-lactoglobulin B (temperature/time/denaturation rate 90 °C/2.2 s/10%: – – –, 90 °C/21 s/60%: ———, 90 °C/360 s/99%: – . –; according to *Kessler*, 1988)

Due to weaker interactions, stable, rigid gels with a chain-like structure are formed on acidification. These gels exhibit no syneresis and are desirable, e. g., in yoghurt (10.2.1.2). Figure 10.10 shows that the firmness of stiff yoghurt is highest when the denaturation of β-lactoglobulin is 90–99%. If this rate of denaturation is achieved at lower temperatures (e. g., 85 °C), gels are formed that are more rigid and coarser than those formed by heating to higher temperatures (e. g., 130 °C), which results in a soft, smooth gelatinous mass. The gel stability of whole-milk yoghurt is lower than that of skim-milk yoghurt because the protein network is interrupted by included fat globules.

Flow curves of skim-milk yoghurt as a function of the rate of denaturation of β-lactoglobulin are presented in Fig. 10.11. The yield point is a measure of the elastic properties of the gel and the area enclosed by the hysteresis loop is a measure of the total energy required to destroy the gel. Both parameters increase with increasing rate of denaturation, which is a sign of increasing gel stability. In contrast to yoghurt production, syneresis of the gel is desirable in the production of cottage cheese, so that the typical texture is attained. For this reason, the milk is only slightly heat treated and the surface hydrophobicity is increased by the addition of chymosin before acidification.

10.1.2.1.4 Whey Proteins

β-*Lactoglobulin* occurs in genetic variants A, B and C of the Jersey dairy cattle breed, and variant D of the Montbeliarde dairy cow. Two other A_{Dr} and B_{Dr} variants of Australian drought master cows are identical to variants A and B apart from the carbohydrate content.

Table 10.9 shows the corresponding changes in the amino acid composition of β-lactoglobulin.

The monomeric β-lactoglobulin has a molecular weight of 18 kdal and consists of 162 amino acids, whose sequence is shown in Table 10.8. It exhibits a reversible, pH-dependent oligomerization, as represented by the equation:

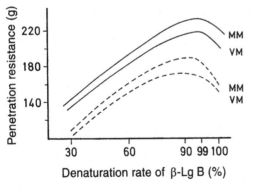

Fig. 10.10. Firmness of yoghurt as a function of the rate of denaturation of β-lactoglobulin B (the final value of the penetration resistance of a conical test piece in stiff yoghurt is given; heating temperature 85 °C: ———, 130 °C: – – –, WM: whole milk with 3.5% fat, SM: skim milk; according to *Kessler*, 1988)

$$A \rightleftarrows A_2 \rightleftarrows (A_2)_4 \rightleftarrows A_2 \rightleftarrows A$$
$$pH < 3.5 \quad 3.7 < pH < 5.1 \quad pH > 7.5 \qquad (10.9)$$

Hence, the monomer is stable only at a pH less than 3.5 or above 7.5. The octamer occurs with variant A, but not with variants B and C. Irreversible denaturation occurs at a pH above 8.6 as well as by heating or at higher levels of Ca^{2+} ions. β-lactoglobulin has 5 cysteine residues, one (Cys^{121}, Table 10.8) of which being free. In the native protein, however, this cysteine is buried within the structure. This SH group is exposed on partial denaturation and can participate either in protein dimerization via disulfide bridge formation or in reactions with other milk proteins, especially with κ-casein and α-lactalbumin, which proceed during the heating of milk.

α-Lactalbumin (M_r 14,200). This protein exists in two genetic forms, A and B (Gln → Arg). It has 8 cysteine residues. Its amino acid sequence (Table 10.8), which is similar to that of lysozyme, has been elucidated. Disulfide bonds and a $Ca^{2⊕}$ ion participate in the stabilization of the tertiary structure. α-Lactalbumin has a biological function since it is the B subunit of the enzyme lactose synthetase. The enzyme subunit A is a nonspecific UDP-galactosyl transferase; the subunit B makes sure that the transfer of the galactose residue can occur at the low glucose concentration present in mammals. The affinity of the transferase alone for glucose is too low ($K_m = 2$ mol/l). It is increased 1000 fold by cooperation with α-lactalbumin.

10.1.2.2 Carbohydrates

The main sugar in milk is lactose, an O-β-D-galactopyranosyl-(1 → 4)-D-glucopyranose, which is 4–6% of milk.
The most stable form is α-lactose monohydrate, $C_{12}H_{22}O_{11} \cdot H_2O$. Lactose crystallizes in this form from a supersaturated aqueous solution at T < 93.5 °C. The crystals may have a prism- or pyramid-like form, depending on conditions. Vacuum drying at T > 100 °C yields a hygroscopic α-anhydride. Crystallization from aqueous solutions above 93.5 °C provides water-free β-lactose (β-anhydride, cf. Formula 10.10). Rapid drying of a lactose solution, as in milk powder production, gives a hygroscopic and amorphous equilibrium mixture of α- and β-lactose.

β-Lactose   (10.10)

Some physical data of lactose are summarized in Table 10.15. The ratio of anomers is temperature dependent. As temperature increases, the β-form decreases. The mutarotation rate is temperature ($Q_{10} = 2.8$) and pH dependent (Fig. 10.12). The rise in mutarotation rate at pH < 2 and pH > 7 originates from the rate-determining step of ring opening, which is catalyzed by both H^+ and OH^- ions:

(10.11)

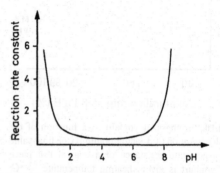

Fig. 10.12. Mutarotation rate of lactose as affected by pH

Table 10.15. Some physical characteristics of lactose

	α-Lactose	β-Lactose	Equilibrium mixture
Melting point (°C)	223[a]	252.2[a]	
Spec. rotation $[\alpha]_D^{20}$	89.4	35.0	
Equilibrium in aqueous solution[b]			
0 °C	1.00	1.80	
20 °C	1.00	1.68	
50 °C	1.00	1.63	
Solubility in water[c]			
0 °C	5.0	45.1	11.9
25 °C	8.6		21.6
39 °C	12.6		31.5
100 °C	70	94.7	157.6

[a] Anhydrous. [b] Relative concentration.
[c] g Lactose/100 g water.

The great solubility difference between the two anomers is noteworthy. The sweetness of lactose is significantly lower than that of fructose, glucose or sucrose (Table 10.16). For people who suffer under lactose intolerance, dietetic milk products are produced by treatment with β-1,4-galactosidase (cf. 2.7.2.2.7). Glucose and some other amino sugars and oligosaccharides are present in small amounts in milk.

Lactulose is found in heated milk products. It is a little sweeter and clearly more soluble than lactose. For example, condensed milk contains up to 1% of lactulose, corresponding to an isomerization of ca. 10% of the lactose present. The formation proceeds via the *Lobry de Bruyn–van Ekenstein* rearrangement (cf. 4.2.4.3.2) or via *Schiff* base. Traces of epilactose (4-O-β-D-glacto-

Table 10.16. Relative sweetness of saccharose, glucose, fructose and lactose[a]

Saccharose	Glucose	Fructose	Lactose
0.5	0.9	0.4	1.9
5.0	8.3	4.2	15.7
10.0	12.7	8.7	20.7
20.0	21.8	16.7	33.3

[a] Results are expressed as concentration % for isosweet aqueous sugar solutions.

pyranosyl-D-mannose) are also formed on heating milk.

10.1.2.3 Lipids

The composition of milk fat is presented in Table 10.17. Milk fat contains 95–96% triglycerols. Its fatty acid composition is given in Table 10.18. The relatively high content of low molecular weight fatty acids, primarily of butyric acid, is characteristic of milk. Although linoleic acid dominates in the lipids occurring in feed, the content of this fatty acid is very low in milk fat (Table 10.17). The reason was found to be that microorganisms living in the rumen hydrogenate the linoleic acid to oleic acid and stearic acid with the formation of conjugated linoleic acid (CLA, cf. 3.2.1.2) and vaccenic acid as intermediates, as shown in Fig. 10.13. It is possible to increase the concentration of linoleic

Table 10.17. Milk lipids

Lipid fraction	Percent of the total lipid
Triacylglycerols	95–96
Diacylglycerols	1.3–1.6
Monoacylglycerols	0.02–0.04
Keto acid glycerides	0.9–1.3
Hydroxy acid glycerides	0.6–0.8
Free fatty acids	0.1–0.4
Phospholipids	0.8–1.0
Sphingolipids	0.06
Sterols	0.2–0.4

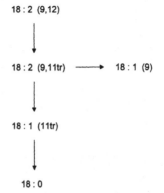

Fig. 10.13. Biohydrogenation of linoleic acid in the rumen of ruminants

Table 10.18. Fatty acid composition of milk fat[a]

Fatty acid	Weight-%
Saturated, straight chain	
Butyric acid	2.79
Caproic acid	2.34
Caprylic acid	1.06
Capric acid	3.04
Lauric acid	2.87
Myristic acid	8.94
Pentadecanoic acid	0.79
Palmitic acid	23.8
Heptadecanoic acid	0.70
Stearic acid	13.2
Nonadecanoic acid	0.27
Arachidic acid	0.28
Behenic acid	0.11
Saturated, branched chain	
12-Methyltetradecanoic acid	0.23
13-Methyltetradecanoic acid	0.14
14-Methylpentadecanoic acid	0.20
14-Methylhexadecanoic acid	0.23
15-Methylhexadecanoic acid	0.36
3,7,11,15-Tetramethylhexadecanoic acid	0.12–0.18
Unsaturated	
9-Decenoic acid	0.27
9-cis-Tetradecenoic acid	0.72
9-cis-Hexadecenoic acid	1.46
9-cis-Heptadecenoic acid	0.19
8-cis-Octadecenoic acid	0.45
Oleic acid	25.5
11-cis-Octadecenoic acid	0.67
9-trans-Octadecenoic acid	0.31
10-trans-Octadecenoic acid	0.32
11-trans-Oxtadecenoic acid	1.08
12-trans-Octadecenoic acid	0.12
13-trans-Octadecenoic acid	0.32
14-trans-Octadecenoic acid	0.27
15-trans-Octadecenoic acid	0.21
16-trans-Octadecenoic acid	0.23
Linoleic acid	2.11
Linolenic acid	0.38

[a] Only acids with a content higher than 0.1% are listed.

acid in milk fat, e. g., by adding plant fats of the appropriate composition in encapsulated form to the feed. The disadvantage of such a nutritionally/physiologically interesting approach is the changed physico-chemical properties of the dairy product, e. g., an increased susceptibility to oxidation and the formation of unsaturated lactones (γ-dodec-cis-6-enolactone from linoleic acid) which influences the flavor of milk and meat. In addition to the main straight-chain fatty acids, small amounts of odd-C-number, branched-chain and oxo-fatty acids (cf. 3.2.1.3) are present.

Phospholipids are 0.8–1.0% of milk fat and sterols, mostly cholesterol, are 0.2–0.4%. Butterfat melting properties, as affected by season and fodder, are listed in Table 10.19. Milk fat is very finely distributed in plasma. The diameter of fat globules is $0.1–10\,\mu m$, but for the main part in the range of $1–5\,\mu m$. During homogenization, milk at 50–75 °C is forced through small passages under pressure of up to 35 MPa, the diameter of the globules lowers to $<1\,\mu m$, depending on the pressure. The fat droplets are surrounded by a membrane that consists of phospholipids and a double layer of proteins and accounts for about 2% of the total mass of the droplet. The layer thickness is on average 8–9 nm, but is not uniform. Membrane compositional data are given in Table 10.20.

About 40 proteins with M_r 15–240 kdal (milk fat globule membrane proteins, MFGM proteins) are involved in the make-up of the membrane. Although their nutritional value with regard to the caseins and whey proteins is low, they receive

Table 10.19. Melting characteristics of butterfat

Temperature (°C)	Solid content (%)	Temperature (°C)	Solid content (%)
5	43–47	30	6–8
10	40–43	35	1–2
20	21–22	40	0

Table 10.20. Membrane composition of milkfat globules

Constituent	Proportion (%)
Protein	41
Phospho- and glycolipids	30
Cholesterol	2
Neutral glycerides	14
Water	13

attention because they can be detrimental to the health of sensitized persons. Casein proteins enter and participate in membrane formation when the fat globule surface area is expanded 4- to 6-fold during homogenization of milk. Six of the eight dominant proteins in the MFGM are glycoproteins, e. g., xanthine oxidoreductase. Other enzymes which are present in the membrane are acetylcholine esterase as well as alkaline and acidic phosphatase (cf. Table 10.24). A very active lipoprotein lipase, a glycoprotein (8.3% carbohydrate, molecular weight 48.3 kdal), occurs in the casein micelles. However, if the milking and storage procedures are appropriate, the raw milk can be kept for several days without the development of a rancid off-flavor. It is likely that the membranes of the fat globules prevent lipolysis. Disruption of the organized structure of the membrane, for instance by homogenization, allows the lipase to bind to the fat globules and to hydrolyze the triacylglycerols at a high rate (1 µmole fatty acid per min per ml milk, pH 7, 37 °C). The milk becomes unpalatable within a few minutes. Therefore the lipoprotein lipase has to be inactivated by pasteurization prior to milk homogenization.

The small amounts of gangliosides that occur in milk (5.6 µmol/l, calculated as ganglioside-bound sialic acid) are of interest for the analytical differentiation of skim-milk and butter-milk powder. As structural elements of the membrane of the fat globules, the cream gets enriched with gangliosides during skimming and only about 8% remain in the skimmed milk. During butter-making, the membrane of the fat globules is mechanically destroyed and the highly polar gangliosides pass almost completely into the buttermilk. Therefore, unlike skim-milk powder, butter-milk powder is rich in gangliosides (ca. 480 µmol/kg, calculated as sialic acid).

10.1.2.4 Organic Acids

Citric acid (1.8 g/l) is the predominant organic acid in milk. During storage it disappears rapidly as a result of the action of bacteria. Other acids (lactic, acetic) are degradation products of lactose. The occurrence of orotic acid (73 mg/l), an intermediary product in biosynthesis of pyrimidine nucleotides, is specific for milk:

Table 10.21. Indicators for the proportion of milk in foods

Compound	Whole milk powder	Skim milk powder
Orotic acid		
photometric	50.6	66.4
polarographic	46.6	58.1
Total creatinine	66.3	84.4
Uric acid	12.4	15.3

Expressed as mg/100 g solids.

Dihydro orotic acid   Orotic acid

$$\text{Uridine-} \atop \text{Cytidine-} \atop \text{Thymidine-}\Bigg\}\text{5-phosphate}$$

(10.12)

Orotic acid as well as total creatinine and uric acid are suitable indicators for the determination of the proportion of milk in foods. The average values for whole-milk and skim-milk powder given in Table 10.21 can serve as reference values.

10.1.2.5 Minerals

Minerals, including trace elements, in milk are compiled in Table 10.22.

10.1.2.6 Vitamins

Milk contains all the vitamins in variable amounts (Table 10.23). During processing, the fat-soluble vitamins are retained by the cream, while the water-soluble vitamins remain in skim milk or whey.

Table 10.22. Mineral composition of milk

Constituent	mg/l	Constituent	µg/l
Potassium	1500	Zinc	4000
Calcium	1200	Aluminum	500
Sodium	500	Iron	400
Magnesium	120	Copper	120
Phosphate	3000	Molybdenum	60
Chloride	1000	Manganese	30
Sulfate	100	Nickel	25
		Silicon	1500
		Bromine	1000
		Boron	200
		Fluorine	150
		Iodine	60

Table 10.23. Vitamin content of milk

Vitamin	mg/l	Vitamin	mg/l
A (Retinol)	0.4	Nicotinamide	1
D (Calciferol)	0.001	Pantothenic acid	3.5
E (Tocopherol)	1.0	C (Ascorbic acid)	20
B_1 (Thiamine)	0.4	Biotin	0.03
B_2 (Riboflavin)	1.7	Folic acid	0.05
B_6 (Pyridoxine)	0.6		
B_{12} (Cyano-cobalamine)	0.005		

10.1.2.7 Enzymes

Milk contains a great number of enzymes which are not only of analytical importance for the detecton of heat-treated milk, but can also influence the processing properties. The rate of heat inactivation of the enzymes indicates the type and extent of heating (cf. 2.5.4).

Hydrolases identified include: amylases, lipases, esterases, proteinases and phosphatases. Proteinase inhibitors have also been found. Important oxidoreductases in milk are aldehyde dehydrogenase (xanthine oxidase), lactoperoxidase and catalase. A general idea of the occurrence and localization of enzymes in boving milk is given in Table 10.24

10.1.2.7.1 Plasmin

The serine endopeptidase plasmin is of special interest in milk technology.

Table 10.24. Enzymes in bovine milk

EC Number	Name	Localization[a]
1.1.1.27	L-Lactate dehydrogenase	P
1.1.1.37	Malate dehydrogenase	
1.1.3.22	Xanthine oxidase	F
1.4.3.6	Amine oxidase (copper-containing)	
1.6.99.3	NADH dehydrogenase	F
1.8	Sulfhydryl oxidase[b]	S
1.8.1.4	Dihydrolipoamide dehydrogenase	F
1.11.1.6	Catalase	L
1.11.1.7	Lactoperoxidase	S
1.15.1.1	Superoxide dismutase	
2.3.2.2	γ-Glutamyltransferase	F
2.4.1.22	Lactose synthase	S
2.4.99.1	β-Galactoside-α-2,6-sialytransfersase	
2.6.1.1	Aspartate aminotransferase	P
2.6.1.2	Alanine aminotransferase	
2.7.1.26	Riboflavin kinase	
2.7.1.30	Glycerol kinase	
2.7.7.2	FMN adenyl transferase	
2.8.1.1	Thiosulfate sulfurtransferase	
3.1.1.1	Carboxylesterase	S
3.1.1.2	Arylesterase	
3.1.1.7	Acetylcholine esterase	F
3.1.1.8	Choline esterase	S
3.1.1.34	Lipoproteinlipase	C
3.1.3.1	Alkaline phosphatase	F
3.1.3.2	Acid phosphatase	F
3.1.3.5	5'-Nucleotidase	F
3.1.3.9	Glucose-6-phosphatase	F
3.1.3.16	Phosphoprotein phosphatase	P
3.1.4.1	Phosphodiesterase I	F
3.1.27.5	Pancreatic ribonuclease	S
3.2.1.1	α-Amylase	S
3.2.1.2	β-Amylase	
3.2.1.17	Lysozyme	S
3.2.1.24	α-Mannosidase	
3.2.1.31	β-Glucuronidase	
3.2.1.52	β-N-Acetylglucosaminidase	
3.4	Acid peptidases	
3.4.21.7	Plasmin	C
3.6.1.3	Adenosinetriphosphatase	F
3.6.1.9	Nucleotide pyrophosphatase	
4.1.2.13	Fructosebiphosphate aldolase	
5.3.1.9	Glucose-6-phosphate isomerase	

[a] C: casein micelle, F: fat-globule membrane, L: leucocytes, P: plasma, S: serum.
[b] Not thiol oxidase EC 1.8.3.2.

Plasmin, its precursor plasminogen and the plasminogen activator (PA) are present in milk associated to the casein micelles and the membranes of the fat globules. A shift of the pH value to the acidic range (pH 4.7) promotes the release of plasmin from casein. The concentration of plasmin is 0.3–2.5 µg/ml milk. It increases with the age of the cow and during the lactation period. Plasmin is in the pH range 7.5–8.0 most active at 37 °C. It preferentially hydrolyzes β-casein and at a lower rate also α-casein. κ-Casein is resistant just like the whey proteins α-lactalbumin and β-lactoglobulin. Like trypsin, it attacks the carboxyl side of L-lysine as well as L-arginine on hydrolysis. The activity of plasmin is controlled by the PA inhibitor and the plasmin inhibitor.

The plasmin activity is reduced by only 10–17% under the conditions of pasteurization, e. g., 72 °C for 15 s. Storage of pasteurized milk indirectly promotes plasmin activity because the inhibitors of PA are inactivated. Complete thermal inactivation of plasmin is achieved at 120 °C in 15 min and at 142 °C in 18 s.

Plasmin influences the ripening process, e.g., in Camembert. It is accelerated and aroma formation is improved. In the recovery of caseinates, on the other hand, the separation of plasmin is absolutely necessary.

10.1.2.7.2 Lactoperoxidase

In the preservation of raw milk, the antimicrobial lactoperoxidase (LP) system present in milk is of interest. This LP system offers an alternative, especially in countries where it is not possible to cool the milk to protect it from spoilage.

The system consists of LP (EC 1.11.1.7) and the substrates thiocyanate and H_2O_2. The enzyme, a glycoprotein (carbohydrate content 10%), consists of 612 amino acid residues (M_r 78,000, IP 9.6) and Fe-protoporphyrin IX, which as the prosthetic group carries out the catalysis, as described in 2.3.2.2. Thiocyanate takes part in this reaction process as the electron donor AH. Lactoperoxidase is one of the heat stable enzymes of milk, especially when the structure is fixed by $Ca^{2\oplus}$ ions. It is still active after 30 min at 63 °C (neutral pH) and after 15 s at 72 °C, but it is inactive after only 2.5 s at 80 °C. Acidification (pH 5.3) accelerates the inactivation by liberating the $Ca^{2\oplus}$ ions. After

xanthine oxidase, LP is the most common enzyme in milk: ca. 30 mg/l.

The thiocyanate concentration in milk depends on the feed, e. g., on the occurrence of glucosinolates (cf. 17.1.2.9.3). H_2O_2 is not a component of milk, but is supplied by bacteria, e. g., lactic acid bacteria.

Hypothiocyanite is the main product formed from the hydrogen donor thiocyanate in LP catalysis.

$$SCN^{\ominus} + H_2O_2 \xrightarrow{LP} OSCN^{\ominus} + H_2O$$

This compound is a bactericidal agent because it can oxidize the SH groups of structure-forming proteins and enzymes in bacteria. The LP system is used to prolong the shelf life of raw milk and pasteurized milk. H_2O_2 is produced here by the addition of glucose/glucose oxidase (cf. 2.7.2.1.1) and the concentration of thiocyanate is increased by addition.

In the production of fermented milk products, the LP system can inhibit the development of starter cultures.

10.1.3 Processing of Milk

Only a small amount of milk is sold to the consumer without processing (certified raw milk). The main part is subjected to a processing procedure shown schematically in Fig. 10.14.

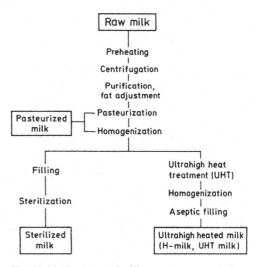

Fig. 10.14. Treatment of milk

10.1.3.1 Purification

The milk is usually delivered in the cooled tank (at least $-8\,°C$ of a milk truck. For purification, it is fed into a clarifier (self-cleaning disk separator) via a deaerating vessel. These separators can process either cold or warm milk (40 °C) at speeds of 4500–8400 rpm with throughput capacities of up to 50,000 l/h.

10.1.3.2 Creaming

After heating to about 40 °C (increase in creaming efficiency by lowering the viscosity), the milk is separated into cream and skimmed milk in a cream separator. Cream separators have a nominal capacity of up to 25,000 l/h at speeds of 4700–6500 rpm. The fat content of the milk can be standardized by careful back-mixing.

10.1.3.3 Heat Treatment

The fluid milk is heated to improve its durability and to kill disease-causing microorganisms. Heat treatments used are (cf. Fig. 10.15):

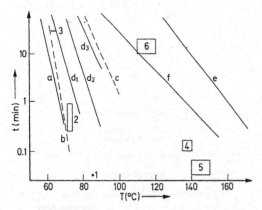

Fig. 10.15. Heating of milk. *1–3* Pasteurization: *1* high temperature treatment, *2* short time and *3* long time heat treatment; *4* and *5* UHT treatment: *4* indirect and *5* direct; *6* sterilization. *a*: Killing pathogenic microorganisms (*Tubercle bacilli* as labelling organism), *b/c*: inactivation of alkaline/acid phosphatase. d_1, d_2, d_3 denaturation (5, 40, 100%) of whey proteins. *e*: casein heat coagulation, *f*: start of milk browning

- Thermization.
  The process involves heating under conditions that are milder than those of pasteurization, e. g., 57–68 °C. The number of bacteria is reduced, e. g., for the production of cheese. The taste of the milk and the coagulation time during treatment with rennet are not impaired.
- Pasteurization.
  The milk is treated: at high temperature (85 °C for 2–3 s) in a short-time, flash process (72–75 °C/15–30 s) in plate heaters; or by the low temperature or holder process, in which it is heated at 63–66 °C for at least 30–32 min, with stirring, and is then cooled.
- Ultrahigh Temperature (UHT) Treatment.
  The process involves indirect heating by coils or plates at 136–138 °C for 5–8 s, or direct heating by live steam injection at 140–145 °C for 2–4 s, followed by aseptic packaging.
  To prevent dilution or concentration of the milk, the amount of injected steam must be controlled in such a way that it corresponds to the amount of water withdrawn during expansion under vacuum.
- Bactotherm Process.
  This is a combination of centrifugal sterilization in bactofuges (65 to 70 °C) and UHT heating of the separated sediment (2–3% of the milk), followed by recombination. Since the total amount of milk is not heated in this process, the taste is improved. The storability is ca. 8–10 days.
- Sterilization.
  Milk in retail packages is heated in autoclaves at 107–115 °C/20–40 min, 120–130 °C/8–12 min.

10.1.3.4 Homogenization

Homogenization is conducted to stabilize the emulsion milk by reducing the size of the fat globules. This is achieved by high-pressure homogenization (up to 35 MPa, 50–75 °C). In principle, the high-pressure homogenizer is a high-pressure pump which presses the product through a homogenizing valve. The fat globules are reduced in size to a diameter of <1 µm by turbulence, cavitation and shear forces, resulting in a ca. 10 fold increase in the surface area. The membranes of the reduced fat globules are formed by the uptake of caseins and whey proteins.

10.1.3.5 Reactions During Heating

Heat treatment affects several milk constituents. Casein, strictly speaking, is not a heat-coagulable protein; it coagulates only at very high temperatures (cf. Fig. 10.15). Heating at 120 °C for 5 h dephosphorylates sodium or calcium caseinate solutions (100% and 85%, respectively) and releases 15% of the nitrogen in the form of low molecular weight fragments.

However, temperature and pH strongly affect casein association and cause changes in micellular structure (cf. 10.1.2.1.2 and 10.1.2.1.3). An example of such a change is the pH-dependent heat coagulation of skim milk. The coagulation temperature drops with decreasing pH (Fig. 10.16 and 10.9). Salt concentration also has an influence, e. g., the heat stability of milk decreases with a rise in the content of free calcium.

All pasteurization processes supposedly kill the pathogenic microorganisms in milk. The inactivation of the alkaline phosphatase is used in determining the effectiveness of pasteurization. At higher temperatures or with longer heating times, the whey proteins start to denature – this coincides with the complete inactivation of acid phosphatase. Denatured whey proteins, within the pH-range of their isoelectric points, cease to the soluble and coagulate together with casein due to souring or chymosin action of the milk. Such coprecipitation of the milk proteins is of importance in some milk processing (as in cottage cheese production). The thermal stability of whey proteins is illustrated in Fig. 10.17.

Heat treatment of milk activates thiol groups; e. g., a thiol-disulfide exchange reaction occurs between κ-casein and β-lactoglobulin. This reaction reduces the vulnerability of κ-casein to chymosin, resulting in a more or less strong retardation of the rennet coagulation of heated milk.

Further changes induced by heating of milk are:

• Calcium phosphate precipitation on casein micelles.
• *Maillard* reactions between lactose and amino groups (e. g. lysine) which, in a classical sterilization process, causes browning of milk and formation of hydroxymethyl furfural (HMF).
• δ-Lactone and methyl ketone formation from glycerides esterified with hydroxy- or keto-fatty acids.
• Degradation of vitamin B_1, B_6, B_{12}, folic acid and vitamin C. Losses of 10–30% in the production of UHT milk are possible. Sterilization destroys ca. 50% of the vitamins B_1, B_6 and folic acid and up to 100% of vitamin C and B_{12}.

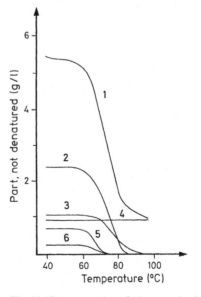

Fig. 10.17. Denaturation of whey proteins by heating at various temperatures for 30 min. *1* Total whey protein, *2* β-lactoglobulin, *3* α-lactalbumin, *4* proteose peptone, *5* immunoglobulin, *6* serum albumin

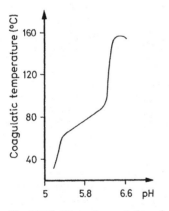

Fig. 10.16. Thermal coagulation of skim milk

- Changes in membranes of milk fat globules, which affect the cream separation property of the globules.

Detailed studies have shown that the rate of several reactions which occur during heating of milk, e. g., thiamine and lysine degradations, formation of HMF and nonenzymic browning, can be calculated over a great temperature-time range (including extended storage) by application of a second-order rate law. Assuming an average activation energy of $E_a = 102 \, kJ/mole$, a "chemical effect" $C^* = 1$ has been defined which gives a straight line in a log t vs. T^{-1} diagram, from which the thiamine loss is seen to be approx. 0.8 mg/l (Fig. 10.18). Other lines in Fig. 10.18 represent a power of ten of heat treatment and chemical reactions ($C^* = 10^{-1}, 10^{-2}, \ldots$, or $10^1, 10^2, \ldots$). The pigments formed in a browning reaction become visible only in the range of $C^* = 10$.

Quality deterioration in the form of nutritional degradation, changes in color or development of off-flavor have also been predicted for other foods by application of a suitable mathematical model.

In most cases the loss of quality fits a zero- or first-order rate law. Knowledge of the rate constant allows one to predict the extent of reaction for any time.

The influence of temperature on the reaction rate follows the *Arrhenius* equation (cf. 2.5.4). Thus by studying a reaction and measuring the rate constants at two or three high temperatures, one could then extrapolate with a straight line to a lower temperature and predict the rate of the reaction at the desired lower temperature. However, these data allow only a prediction of the shelf life when the physical and chemical properties of the components of a food do not alter with temperature. For example, as temperature rises a solid fat goes into a liquid state. The reactants may be mobile in the liquid fat and not in the solid phase. Thus, shelf life will be underestimated for the lower temperature.

10.1.4 Types of Milk

Milk is consumed in the following forms:

- *Raw fluid milk* (high quality milk), which has to comply with strict hygienic demands.
- *Whole milk* is heat-treated and contains at least 3% fat. It can be a standardized whole milk adjusted to a predetermined fat content, in which case the fat content has to be at least 3.5%.
- *Low-fat milk* is heat-treated and the cream is separated. The fat content is 1.5–2%.
- *Skim milk* is heat-treated and the fat content is less than 0.3%.
- *Reconstituted milk* is most common in regions where milk production is not feasible (e. g. many Japanese cities). For production, melted butter fat is emulsified in a suspension of skim milk powder at 45 °C. The "cream" with a fat content of 20–30% is subjected to two-stage homogenization (20 and 5 MPa, 55–60 °C) and then diluted with the skim milk suspension.
- *Filled milk* is less expensive because the butter fat is replaced with a plant fat.
- *Toned milk* is a blend of a fat-rich fresh milk and reconstitued skim milk in which the nonfat solids are "toned up". Addition of water "tones down" the fat and nonfat solids.

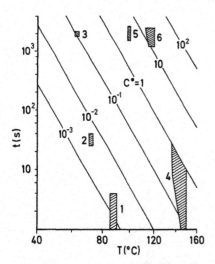

Fig. 10.18. Chemical reactions in heat-treated milk. ("Chemical effect" $C^* = 1$: losses of approx. 3% thiamine and approx. 0.7% lysine and formation of approx. 0.8 mg/l HMF); commonly used heat treatments: *1* high heat, *2* short time heating, *3* prolonged heating, *4* UHT treatment, *5* boiling, *6* sterilization (according to *Kessler*, 1983).
HMF: Hydroxymethylfurfural

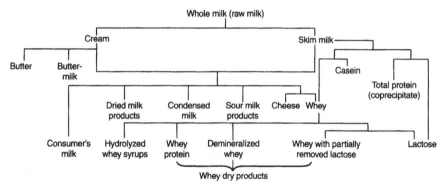

Fig. 10.19. Schematic presentation of milk processing

10.2 Dairy Products

Milk processing is illustrated schematically in Fig. 10.19.

10.2.1 Fermented Milk Products

All sour milk products have undergone fermentation, which can involve not only lacticacid bacteria, but also other microorganisms, e. g., yeasts. To the lactic acid bacteria count the genera *Lactobacillus, Lactococcus, Leuconostoc*, and *Pediococcus*. The most important species are presented in Table 10.25.

Depending on the microorganisms involved, fermentation proceeds via the glycolysis pathway with the almost exclusive formation of lactic acid (homofermentation), via the pentose phosphate pathway with formation of lactic acid, acetic acid (ethanol), and possibly CO_2 (heterofermentation) or via both pathways. These metabolic pathways are shown in Fig. 10.20. Organisms that are obligatorily homofermentative have fructose-bisphosphate aldolase, but not glucose 6-phosphate dehydrogenase and 6-phosphogluconate dehydrogenase. However, organisms that are obligatorily heterofermentative have both dehydrogenases, but no aldolase. Facultatively homofermentative organisms have all three enzymes and can use both metabolic pathways.

Apart from the type of fermentation, the configuration of the lactic acid formed also depends on

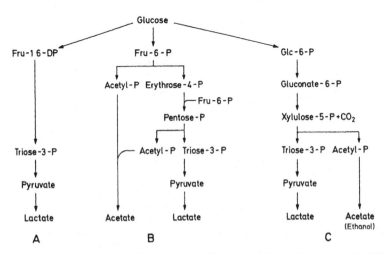

Fig. 10.20. Glucose metabolism in lactic acid bacteria. A: homofermentation, B: Bifidus pathway, and C: heterofermentation (6-phosphogluconate pathway)

Table 10.25. Lactic acid bacteria

Microorganisms	L-Lactic acid[a] (%)	Remarks
Lactobacillus *bulgaricus*	0.6–4	thermophilic,
L. lactis	0	homofermentative,
L. leichmanii		D-, L- or
L. delbrueckii		D,L-Lactic acid
L. helveticus	70	
L. jugurti		
L. acidophilus	60	
L. casei subsp. casei		mesophilic,
L. casei subsp. alactosus		homofermentative,
L. casei subsp. pseudo plantarum		D-, L- or D,L-Lactic acid
L. casei subsp. rhamnosus		
L. casei subsp. fusiformis		
L. casei subsp. tolerans		
L. plantarum		
L. curvatus		
L. fermentum		heterofermentative,
L. cellobiosus		D,L-Lactic acid
L. brevii		
L. hilgardii		
L. vermiformis		
L. reuteri		
Streptococcus *thermophilus*	99	thermophilic, homofermentative
S. faecium		homofermentative
Lactococcus lactis *subsp. lactis*	92–99	mesophilic, homofermentativ
Lactococcus lactis *subsp. cremoris*	99	
Leuconostoc cremoris		heterofermentative,
L. mesenteroides		D-Lactic acid
L. dextranicum		
L. lactis		
Pediococcus acidilactici		thermophilic, homofermentative, D,L-Lactic acid

[a] Orientation values; the proportion of L-lactic acid depends on the bacterial strain and on the culture conditions.

the microorganisms involved. As shown in Table 10.25, both enantiomers are formed in varying amounts. Table 10.26 lists the content of total lactic acid and L-lactic acid in various dairy products.

In human metabolism, L-lactic acid is formed exclusively and only one L-lactate dehydrogenase is available. Therefore, the intake of larger amounts of D-lactic acid can result in enrichment in the blood and hyperacidity of the urine. For this reason, the WHO recommends a limitation of the intake of D-lactic acid to 100 mg per day and kg of body weight. Apart from the main products mentioned, various aroma substances are formed in the course of fermentation (cf. 10.3.3). In addition, proteolytic and lipolytic processes occur to a certain extent. During proteolysis, peptides can be formed which have opiate activity and hypotensive, immune-stimulating or antimicrobial effects (cf. Literature).

According to the consistency, a distinction is made between stiff, gel-like products, stirred, creamy products, and drinkable, flowable products. The thermal pretreatment of milk influences the rheological properties of the products as described in section 10.1.2.1.3. The keeping time of sour milk products can be increased if they are produced and filled under aseptic conditions or produced under normal conditions but subsequently heat treated.

Table 10.26. Total lactic acid and L-lactic acid in some dairy products

Product	Total lactic acid (%)	L-Lactic acid (%)[b]
Sour milk	0.97	88
Buttermilk	0.86	87
Sour cream	0.86	96
Joghurt	1.08	54
Curd	0.59	94
Cottage cheese	0.34	92
Emmental	0.27	76
Sbrinz	1.53	58
Tilset cheese	1.27	52

[a] Average values. [b] Based on total lactic acid.

10.2.1.1 Sour Milk

Sour milk is the product obtained by the fermentation of milk, which occurs either by spontaneous souring caused by various lactic-acid-producing bacteria or on addition of mesophilic microorganisms (*Lactococcus lactis, L. cremoris, L. diacetylactis, Leuconostoc cremoris*) to heated milk at 20 °C. As fermentation proceeds, lactose is transformed into lactic acid, which coagulates casein at pH 4–5. The thick, sour-tasting curdled milk is manufactured from whole milk (at least 3.5% milk fat), low-fat milk (1.5–1.8% fat) or from skim milk (at most 0.3% fat), often by blending with skim milk powder to increase the total solids content and to improve the resultant protein gel structure. Sour milk contains 0.5–0.9% of lactic acid. In some countries sheep, water buffalo, reindeer or mare's milk are also processed. Sour cream is produced by a process very similar to that used in sour milk manufacture except that coffee grade cream is used as the raw material.

10.2.1.2 Yoghurt

The production of yoghurt is presented schematically in Fig. 10.21. Yoghurt cultures consist of thermophilic lactic acid bacteria that live together symbiotically (*Streptococcus thermophilus* and *Lactobacillus bulgaricus*). Incubation is conducted on addition of 1.5–3% of the operating culture at 42–45 °C for about 3 h. The final product has a pH value of about 4–4.2 and contains 0.7–1.1% of lactic acid. Functional foods include yoghurts which have been incubated with probiotics. Probiotics are defined, cultured strains of lactic acid bacteria, which have been isolated from human intestinal flora, e. g., certain lactobacilli and bifidobacteria. On consumption, they are supposed to reach the large intestine and contribute to the formation of an optimal intestinal flora.

The variety of products is increased by the addition of fruits and fruit pastes to yoghurt.

The addition of fruit or fruit pastes and sugar yields special products (fruit yoghurts).

An essential part of the specific yoghurt aroma comes from carbonyl compounds, predominantly acetaldehyde and diacetyl. In addition to 1-octen-

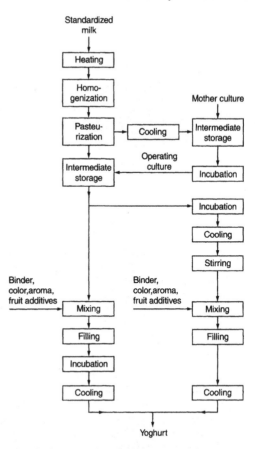

Fig. 10.21. Production of different types of yoghurt

3-one, 1-nonen-3-one has also been detected as an important odorant, which has an exceptionally low odor threshold (cf. 3.7.2.1.9). An autoxidation product of linoleic acid, (E)-2-nonenal (Formula 10.13), is thought to be the precursor.

$$(10.13)$$

10.2.1.3 Kefir and Kumiss

Kefir and kumiss are sparkling, carbonated alcoholic beverages. The microflora of kefir include *Torula* yeast (responsible for alcoholic fermentation) and *Lactobacillus brevis, L. casei, Leuconostoc mesenteroides, Streptococcus durans, Saccharomyces delbrueckii, S. cerevisiae* and *Acetobacter aceti*. The kefir bacillus causes a buildup of "kefir grains", which resemble cauliflower heads when wet and brownish seeds when dry, and are particles of clotted milk plus the kefir organisms. Their addition to fluid milk produces kefir. Kumiss is made of mare's or goat's milk fermented by the obligatory pure kumiss culture.

Both dairy beverages are indigenous to the Caucasus and steppes of Turkestan. Kefir contains lactic acid (0.5–1.0%), noticeable amounts of alcohol (0.5–2.0%) and carbon dioxide, and some products of casein degradation resulting from proteolytic action of yeast. Normal kumiss contains 1.0–3.0% of alcohol. The production is similar to that of yoghurt.

10.2.1.4 Taette Milk

Taette (Lapp's milk) is a specially fermented, sour cow's milk product consumed in Sweden, Norway and Finland. Its thread-like, viscous structure is due to the formation of slimy substances at the low fermentation temperatures used. Mesophilic microorganisms (*Lactococcus* and *Leuconostoc spp.*) are involved in this process.

10.2.2 Cream

Milk is practically completely defatted (remaining fat content 0.03–0.06%) in hermetic, self-cleaning or hermetic/self-cleaning creaming separators. The cream products are subsequently standardized by back-mixing. Whipping cream contains at least 30% milk fat, coffee cream at least 10% and butter cream 25–82%. Cream is utilized in many ways, either by direct consumption or for production of butter and ice creams.

Whippability and stability of the whipped foam products are necessary whipping cream properties. For the best quality cream, a volume increase of at least 80% is expected and a standard cone with 100 g load must penetrate 3 cm deep in >10 s. No serum separation should occur at 18 °C after 1 h.

Fat droplets accumulate during whipping on the surface of large air bubbles which form the froth. An increased build-up of smaller bubbles tears apart the membrane of the droplet and enlarges the fat interphase area, thus resulting in gel setting of the lamella separating the individual air bubbles. Sour cream is the product of progressive lactic acid fermentation of cream.

10.2.3 Butter

Butter is a water-in-oil emulsion (w/o emulsion) made from cream by phase inversion occurring in the butter-making process. According to its manufacturing process, three types exist:

- Butter from sour cream (cultured-cream butter).
- Butter from nonsoured, sweet cream (sweet cream butter).
- Butter from sweet cream, which is soured in a subsequent step (soured butter).

Butter contains 81–85% fat, 14–16% water, 0.5–4.0% fat-free solids and 1.2% NaCl in the case of salted butter. The composition generally must meet legal standards. Butter is an emulsion with a continuous phase of liquid milk fat in which are trapped crystallized fat grains, water droplets

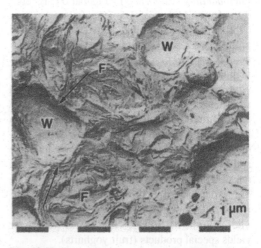

Fig. 10.22. Freeze-fracture micrograph of butter (F: fat globule, W: water droplet; according to *Juriaanse* and *Heertje*, 1988)

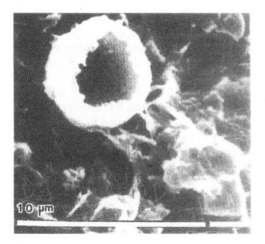

Fig. 10.23. Crystalline shell of a fat grain, as found in butter, which was obtained by eliminating the included oil; (according to *Juriaanse* and *Heertje*, 1988)

and air bubbles. A freeze-fracture micrograph of butter showing the continuous fat phase with included fat globules and water droplets is shown in Fig. 10.22. Butter consistency is determined by the ratio of free fluid fat to that of solidified fat. Due to seasonal variations in the unsaturated fatty acid content of milk fat, the solid/fluid fat ratio fluctuates at 24 °C between 1.0 in summer and 1.5 in winter. Equalization of these ratios is achieved by a preliminary cream-tempering step in a cream-ripening process, then churning and kneading the cream, which influences the

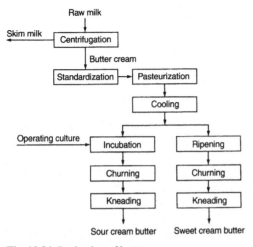

Fig. 10.24. Production of butter

extent of fluid fat inclusion into the solidified "fat grains". Figure 10.23 shows the crystalline shell of a cut fat grain, from which the liquid fat was removed during preparation.

A general idea of the most important processing steps involved in butter making is given in Fig. 10.24.

10.2.3.1 Cream Separation and Treatment

Cream is separated from whole milk by high-efficiency separators (cf. 10.1.3.2 and 10.2.2). The cream, depending on the subsequent churning process, should contain 25–82% milk fat. The cream is then pasteurized at 90–110 °C.

Cream ripening and souring are the most important steps in the production of sour cream butter. The process is performed in a cream ripener or vat, with suitable mixing and temperature control. Soon after the cream has filled the ripener, a "starter culture" is added, followed by incubation for 12–24 h at 8–19 °C. The pH falls to 4.6–5.0. The "starter culture" consists of various strains of lactic acid bacteria (primarily *Lactococcus lactis subsp. lactis, Lactococus lactis subsp. cremoris, Lactococus lactis subsp. diacetylactis* and *Lactococcus lactis subsp. cremoris bv. citrovorum*). The subsequent ripening at 8–19 °C proceeds for 12 to 24 h.

The formation of fat crystals can be influenced by suitable temperature control during the cream ripening process. Consequently, the consistency of the butter can be influenced and corrected. The souring step is omitted in the production of sweet cream butter. The pasteurized cream is cooled for about 3 h at 4–6 °C to induce the crystallization of fat in the fat globules. It is then stored for about 5 h at a temperature which is 1–2 °C higher than the melting range (17–19 °C) of the low-melting milk fat fraction. As a result, a mixture of crystalline higher-melting TG and liquid low-melting TG is formed, which is easy to spread. The cream then ripens for at least 10 h at 10–14 °C.

10.2.3.2 Churning

Churning is essentially strong mechanical cream shearing which tears the membranes of the fat globules and facilitates coalescence of the globules. The cream "breaks" and tiny granules of butter appear. Prolonged churning results in a continuous fat phase. Foam build-up is also desirable since the tiny air bubbles, with their large surface area, attract some membrane materials. Some membrane phospholipids are transferred into the aqueous phase. Buttermilk, a milky, turbid liquid, separates out initially (it is later drained off), followed by the butter granules of approx. 2 mm diameter. These granules still contain 30% of the aqueous phase. This is reduced to 15–19% by churning. The finely distributed water droplets (diameter 10 μm or less) are retained by the fat phase.

Churning is mainly carried out in stainless steel vessels of different forms which rotate nonsymmetrically. Continuously operated churns are also used with cream having a fat content of 32–38% (sour cream butter) or 40–50% (sweet cream butter). The machines are divided into churning, separation, and kneading compartments. In the churning compartment, a rotating impact wave causes butter granule formation. The separation compartment is divided into two parts. The butter is first churned further, resulting in the formation of butter granules of a larger diameter. Subsequently, the buttermilk is separated and the butter is washed, if necessary. The cooled kneading compartment consists of transport screws and kneading elements for further processing the butter. Both kneading compartments are operated under vacuum conditions to reduce the air content of the butter to less than 1%. The final salt and water content of the butter is adjusted by apportioning.

In the continuous *Alfa*-process the phase conversion is achieved in a screw-type cooler, using previously pasteurized (90%) 82% cream by repeated chilling to 8–13 °C, without the aqueous phase being separated.

The *Booser* process and the *NIZO* process allow a subsequent souring of butter from sweet cream. Both processes are of economic interest, because they yield a more aromatic sour butter and sweet buttermilk, which is a more useful by-product than sour butter-milk.

During the *Booser* process 3–4% of starter cultures are incorporated into the butter granules (water content: 13.5–14.5%) obtained from sweet cream.

Lactic acid and a flavor concentrate are obtained by separate fermentations during the first step of the *NIZO* process. In a second step they are mixed and incorporated into the butter granules from sweet cream.

Lactobacillus helveticus cultivated on whey produces the lactic acid, which is then separated by ultrafiltration and concentrated in vacuum up to about 18%. The flavor concentrate is obtained by growing starter cultures and *Lactococcus lactis subsp. diacetylactis* on skim milk of about 16% dry matter.

10.2.3.3 Packaging

After the butter is formed, it is cut by machine into rectangular blocks and is wrapped in waxed or grease-proof paper or metallic (aluminium) foil laminates (coated within with polyethylene).

10.2.3.4 Products Derived from Butter

- Melted butter consists of at least 99.3% milk fat. The aqueous phase is removed by decantation of the melted butter or by evaporation.
- Fractionated butterfat. The butter is separated by fractional crystallization into high- and low-melting fractions, and is utilized for various purposes (e. g. consistency improvement of whipping creams and butters).
- Spreadable blends with vegetable oils ("butterine").

10.2.4 Condensed Milk

Condensed milk is made from milk by the partial removal of water and addition of saccharose, if necessary (sweetened condensed milk). It is used, diluted or undiluted, like milk. Nonsweetened condensed milk is mainly available with a fat content of 7.5% or 10% and in some countries up to 15%. The solids content is 25–33%.

The production process (Fig. 10.25) starts with milk of the desired fat content. The milk is first heated, e.g., to 120 °C for 3 minutes to separate albumin, kill germs, and reduce the danger of delayed thickening. Subsequently, it is evap-

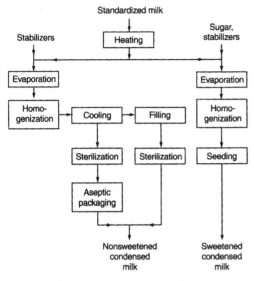

Fig. 10.25. Production of condensed milk

45–50% of the weight of the end-product. Homogenization and sterilization steps are omitted. To avoid graininess caused by lactose crystallization – the solubility limit of lactose is exceeded after sucrose addition – the condensed milk is cooled rapidly, then seeded with finely pulverized α-lactose hydrate. Seeding ensures that the lactose crystal size is 10 μm or less.

The critical quality characteristics of condensed milk are the degree of heat damage (lysine degradation), prevention of separation during the storage life, absence of coarse crystallized lactose, as well as color and taste. These criteria are influenced not only by the process management (heat treatment during evaporation and sterilization and suitable selection of the homogenization temperature and pressure), but also by the source of the milk (feed) and the producer's ability to maintain hygienic conditions.

orated in a continuously operated vacuum evaporator at 40 to 60 °C. In comparison with previously used circulation, riser, and flat-plate evaporators, film evaporators are mainly employed today. Several units (up to seven stages) are usually connected in series, each unit being heated by the vapor from the previous stage. The temperature and pressure decrease from stage to stage. Optimal energy utilization is achieved by mechanical or thermal vapor compression. Fat separation is prevented by homogenization at 40–60 °C (12.5–25 MPa). The resulting evaporated milk, with a solid content of 24–31% or more, is homogenized, poured into lacquer (enamel)-coated cans made of white metal sheets, and is sterilized in an autoclave at 115–120 °C for 20 min. Continuous flow sterilization followed by aseptic packaging is also used. To prevent coagulation during processing and storage, Na-hydrogen carbonate, disodium phosphate and trisodium citrate are incorporated into the condensed milk. These additives have a dual effect: pH correction and adjustment of free Ca^{2+} ion concentration, both aimed at preventing casein aggregation (cf. Fig. 10.3). The additives are in the range of 0.2–0.8 g/l. They are controlled by law.

In the production of *sweetened condensed milk*, after a preheating step (short-time heating at 110–130 °C), sucrose is added to a concentration of

10.2.5 Dehydrated Milk Products

Skim milk powder and whole milk powder are used either for the reconstitution of milk in countries that for climatic reasons have no dairy farming or as intermediate products for further processing into infant milk products, milk chocolate etc. The quality of these instant products depends on the durability, redissolution capacity (cold and warm), taste, microbiological characteristics, and preservation of essential constituents (proteins, vitamins) during production.

The main drying process used is spray drying. However, drum drying (with and without vacuum) and fluid-bed drying (foaming with inert gas N_2 or CO_2) are used for special purposes. Freeze drying offers no particular advantages over the less expensive spray drying process and is only of interest for special products.

Using film evaporating systems, the milk is first preconcentrated to 30–55% solids.

In drum drying, the liquid (30–40% solids) is applied in a thin layer to a heated drying cylinder (100–130 °C) and, after a defined residence time (rotation, 2–3 s), removed with a scraping knife. The liquid film can be applied in various ways. In drum drying, relatively large particles are obtained. The thermal exposure (temperature, time) is considerably higher than in spray drying, which is consequently preferred. The solubility is poor due to the denaturation of whey proteins. The product is clearly brown owing to the *Maillard* reaction.

In spray drying, the milk concentrate (30–55% solids) is finely dispersed in the spray tower by centrifugal atomization or by nozzle atomization and dried with hot air (150–220 °C) cocurrently or countercurrently. The water content drops to 6–7% in 0.5–1 s. A further decrease to 3–4% is achieved by after drying in a vibration fluid bed with hot air (130–140 °C).

Particles with a diameter in the range of 5 to 100 μm consist of a continuous mass of amorphous lactose and other low-molecular components, which includes fat globules, casein micelles, whey proteins and usually vacuoles. When the powder absorbs water, lactose crystallizes at $a_w > 0.4$, causing agglomeration. During drying, the temperature of the particles normally does not rise above 70 °C. Therefore, the whey proteins do not denature and remain soluble. Many enzymes are still active. Storage problems are caused by the *Maillard* reaction and by fat oxidation in the case of fat-containing powders. Foam dried products can have excellent properties (aroma, solubility).

Other dehydrated dairy products, in addition to whole milk or skim milk powders, are manufactured by similar processes. Products include dehydrated malted milk powder, spray- or roller-dried creams with at least 42% fat content of their solids and a maximum 4% moisture, and butter or cream powders with 70–80% milk fat. Dehydrated buttermilk and lactic acid-soured milk are utilized as children's food.

Adaptation of infant milk product formulation to approximate mothers' milk can be achieved, for example, by addition of whey proteins, sucrose, whey or lactose, vegetable oil, vitamins and trace elements and by reduction of minerals, i. e., by a shift of the Na/K ratio.

The compositions of some dehydrated dairy products are illustrated in Table 10.27.

10.2.6 Coffee Whitener

Coffee whiteners are products that are available in liquid, but more often in dried instant form. They are used like coffee cream or condensed milk. A formulation typical of these products is shown in Table 10.28. In contrast to milk products, plant fats are used in the production of coffee whiteners. Caseinates are usually the protein component. The most important process steps in the production are: preemulsification of the constituents at temperatures of up to 90 °C, high-pressure homogenization (cf. 10.1.3.4), spray drying, and instantization (cf. 10.2.5).

10.2.7 Ice Cream

Ice cream is a frozen mass which can contain whole milk, skim milk products, cream or butter, sugar, vegetable oil, egg products, fruit and fruit ingredients, coffee, cocoa, aroma substances and approved food colors. A typical formulation is 10% milk fat, 11% fat-free milk solids, 14% saccharose, 2% glucose syrup-solids, 0.3% emulsifiers, 0.3% thickener, and 62% water. The thickeners, mostly polysaccharides (cf. Table 4.15), increase the viscosity and the emulsifiers destabilize the fat globules, favoring their aggregation during the freezing process.

Table 10.27. Composition of dried milk products (%)

	1	2	3	4
Water	2.7	3	4.6	3.3
Protein	26.5	38.2	13	91.4
Fat	27.4	0.9	1.1	0.9
Lactose	37.7	49.6	73	0.2
Minerals	5.7	8.2	8.2	4.1

1: whole milk powder; 2: skim milk powder; 3: whey powder; 4: caseinate

Table 10.28. Typical formulation of coffee whiteners

Constituent	Amount (%)
Glucose syrup	52.6
Fat	30.0
Sodium caseinate	12.0
Water	3.15
Emulsifiers	1.6
K_2HPO_4	0.6
Carrageenan	0.05
Color and aroma substances	

For the production of ice cream, the mixture of components is subjected to high-temperature short-time pasteurization (80–85 °C, 20–30 s), high-pressure homogenization (150–200 bar) and cooling to ca. 5 °C. Air is then mixed into the mixture (60–100 vol%) while it is frozen at temperatures of up to −10 °C and then hardened. The freezers used are mainly continuously working systems furnished with coolants which evaporate at −30 °C to −40 °C. The process is controlled in such a way that the core temperature of the ice cream production is ca. −18 °C.

The structural elements of ice cream are ice crystals (∼50 μm), air bubbles (60–150 μm), fat globules (<2 μm), and aggregated fat globules (5–10 μm). The fat is mostly attached to the air bubbles. The air bubbles have a three fold function: they reduce the nutritional value, soften the product, and prevent a strong cold sensation during consumption.

10.2.8 Cheese

Cheese is obtained from curdled milk by removal of whey and by curd ripening in the presence of special microflora (Table 10.29). The great abundance of cheese varieties, about 2000 worldwide, can be classified from many viewpoints, e. g., according to:

- Milk utilized (cow, goat or sheep milk).
- Curd formation (using acids, rennet extract or a combination of both).
- Texture or consistency, or water content (%) in fat-free cheese. Following the latter criterion, the more important cheese groups are (water content in %):

  Extra hard: <51%
  Hard: 49–56%
  Semihard: 54–63%
  Semisolid: 61–69%
  Soft: >67%

- Fat content (% dry matter). By this criterion, the more important groups are:

  Double cream cheese (60–85% fat);
  Cream cheese (≥50);
  Whole fat cheese (≥45);
  Fat cheese (≥40);
  Semi fat cheese (≥20);
  Skim cheese (max. 10).

Within each group, individual cheeses are characterized by aroma. A small selection of the more important cheese varieties is listed in Table 10.30. Cheese manufacturing essentially consists of curd formation and ripening (Fig. 10.26).

10.2.8.1 Curd Formation

The milk fat content is adjusted to a desired level and, when necessary, the protein content is also adjusted. Additives include calcium salts to improve protein coagulation and cheese texture, nitrates to inhibit anaerobic spore-forming microflora, and color pigments. The prepared raw or pasteurized milk is mixed at 18–50 °C in a vat with a starter culture (cf. Table 10.29) (lactic acid or propionic acid bacteria; molds, such as *Penicillium camemberti, P. candidum, P. roqueforti*; red- or yellow-smearing cultures, such as *Bacterium linens* with *cocci* and yeast). The milk coagulates into a soft, semi-solid mass, the curd, after lactic acid fermentation (sour milk cheese, pH 4.9–4.6), or by addition of rennet (sweet milk cheese, pH 6.6–6.3), or some other combination, the most common being combined acid and rennet treatment. This protein gel is cut into cubes while being heated and is then gently stirred. The whey is drained off while the retained fat-containing curd is subjected to a firming process (syneresis). The firming gets more intense as the mechanical input and the applied temperature increase. The process and the starter culture (pH) determine the curd properties. When the desired curd consistency has been achieved, curd and whey separation is accomplished either by draining off the whey or by pressing off the curd while simultaneously molding it.

New methods of cheese making aim at including the whey proteins in the curd, instead of removing them with the whey. Apart from giving higher yields (12–18%), these processes help to economize on waste water costs or elaborate whey treatments (cf. 10.2.10).

The use of ultrafiltration steps as compared with conventional cheese making is shown in Fig. 10.26. Alternatively, conventionally produced whey can be concentrated by ultrafiltration and then added to the curd or milk can be soured with starter culture and/or rennet

Table 10.29. Characteristic microflora of some types of cheese

Type of cheese	Starter cultures	Other species
Parmigiano-Reggiano	*Streptococcus thermophilus*	
	Lactobacillus helveticus	
	L. bulgaricus	
Emmental	*Lactococcus lactis*	*Propionibacterium*
	subsp. lactis	*freudenreichii*
	Lactococcus lactis	*P. freudenreichii*
	subsp. cremoris	*subsp. shermanii*
	S. thermophilus	
	Lactobacillus helveticus	
	L. bulgaricus	
Cheddar	*Lactococcus lactis*	*None*
	subsp. cremoris	
	(Lactococcus lactis	
	subsp. lactis)	
Roquefort	*Lactococcus lactis*	*Penicillium roqueforti*
	subsp. lactis	
	Lactococcus lactis	
	subsp. cremoris	
	Lactococcus lactis	
	subsp. diacetylactis	
	Leuconostoc cremoris	
Limburger	*Lactococcus lactis*	*Brevibacterium linens*
	subsp. lactis	*Micrococcus spp.*
	Lactococcus lactis	*Yeasts*
	subsp. cremoris	
Edamer, Gouda	*Lactococcus lactis*	*Brevibacterium linens*
	subsp. lactis	*Micrococcus spp.*
	Lactococcus lactis	*Yeasts*
	subsp. cremoris	
	Lactococcus lactis	
	subsp. diacetylactis	
	Leuconostoc cremoris	
Camembert, Brie	*Lactococcus lactis*	*Penicillium camemberti*
	Lactococcus lactis	
	subsp. cremoris	*P. caseicolum*
	Lactococcus lactis	*Brevibacterium linens*
	subsp. diacetylactis	*Micrococcus spp. Yeasts*

addition and then concentrated by ultrafiltration. To reduce the cost of enzymes in the casein precipitation step with chymosin (rennet or usually microbial rennet substitutes), processes using carrier-bound enzymes are being tested. Here, the enzyme reaction proceeds in the cold and precipitation occurs subsequently on heating the milk. In this way, clogging of the enzyme bed is avoided.

The individual process steps in cheese making are being increasingly mechanized and automated.

The equipment used includes discontinuously operated cheesemakers (vats or tanks with stirring and cutting devices) and coagulators for continuous curd formation with subsequent fully automatic whey separation and molding.

10.2.8.2 Unripened Cheese

Unripened cheeses have a soft (quark), gelatinous (layer cheese), or grainy (cottage cheese) consis-

Table 10.30. Cheese Varieties

Unripened Cheeses (F: <10–70, T: 39–44, R: unripened)

Quark. Neuchâtel, Petit Suisse, Demi Sel, Cottage Cheese
Schichtkäse (layers of different fat content)
Rahm-, Doppelrahmfrischkäse, Demi Suisse, Gervais, Carré-frais, Cream Cheese
Mozzarella (plastic curd by heating to >60 °C within the whey), Scamorze

Ripened Cheese

Hard Cheeses (F: 30–50, T: 58–63, R: 2–8 M)

Chester, Cheddar, Cheshire, Cantal
Emmental, Alpkäse, Bergkäse, Gruyère, Herrgωrdskäse, Samsoe
Gruyère (Greyerzer), l'Emmental française, Beaufort, Gruyère de Comte
Parmigiano-Reggiano (granular structure, very hard, grating type), Grana, Bagozzo, Sbrinz
Provolone (plastic curd by heating to >60 °C in the whey: Pasta filata),
    Cacciocavallo

Slicing Cheeses (F: 30–60, T: 44–57, R: 3–5 W)

Edam, Geheimratskäse, Brotkäse, Molbo, Thybo Gouda, Fynbo, Naribo
Pecorino (from ewe's milk), Aunis Brinsenkäse
Port Salut, St. Paulin, Esrom, Jerome, Deutscher Trappistenkäse
Tilsiter, Appenzeller, Danbo, Steppenkäse, Svecia-Ost

Semi-solid slicing Cheeses (F: 30–40, T: 44–55, R: 3–5 W)

Butterkäse, Italico, Bel Paese, Klosterkäse
Roquefort (from ewe's milk), Bleu d'Auvergne, Bresse Bleu, Bleu du Haut-Jura, Gorgonzola, Stracchino,
    Stilton, Blue Dorset, Blue Cheese, Danablue
Steinbuscher
Weiψlacker, Bierkäse

Soft Cheeses (F: 20–60, T 35–52, R: 2 W)

Chevre (from goat's milk), Chevret, Chevretin, Nicolin, Cacciotta, Rebbiola,
    Pinsgauer Käse
Brie, Le Coulommiers
Camembert, veritable Camembert, Petit Camembert
Limburger, Backsteinkäse, Allgäuer Stangenkäse
Münsterkäse, Mainauer, Mondseer, Le Munster, Gérômè
Pont l'Eveque, Angelot, Maroilles
Romadour, Kümmelkäse, Weinkäse, Limburger

Low-fat Cheeses (F: <10, T: 35, R: 1–2 W)

Harzer Käse, Mainzer Käse (ripened with *Bact. linens*, different cocci and yeasts)
Handkäse, Korbkäse, Stangenkäse, Spitzkäse (ripened with *Bact. linens*, different
    cocci and yeasts, or with *Penicillium camemberti*), Gamelost

Cooking cheese (from Cottage Cheese by heating with emulsifying agents, F: <10 60)

[a] Related types are grouped together. For the classes average values are given for
- fat content in the dry matter: F (%)
- dry matter: T(%)
- ripening time: R in months (M) or weeks (W).

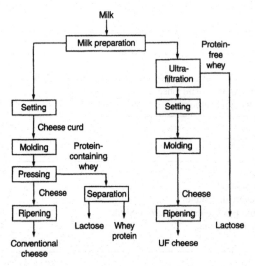

Fig. 10.26. Cheese making (conventional or with ultra-filtration)

tency. In the production of quark, the whey is usually separated after souring. Cottage cheese is generally produced in continuously operated coagulators with special temperature regulation. After whey separation via a filter band, the curd grain can be washed in a screw vat, cooled, and dried via another drying band.

10.2.8.3 Ripening

The molded cheese mass is placed in a salt bath for some time, dried, and then left to ripen in air-conditioned rooms. Ripening or curing is dependent on cheese mass composition, particularly the water content, the microflora and the external conditions, such as temperature and humidity in the curing rooms.

The ripening of soft cheeses proceeds inwards, so in the early stages there is a ripened rind and an unripe inner core. This nonuniform ripening is due to the high whey content which causes increased formation of lactic acid and a pH drop at the start of ripening. In the rind, special molds that grow more favorably at higher pH values contribute to a pH increase by decarboxylating amino acids.

Ripening in hard cheeses occurs uniformly throughout the whole cheese mass. Rind formation is the result of surface drying, so it can be avoided by packaging the cheese mass in suitable plastic foils before curing commences. The duration of curing varies and lasts several days for soft cheeses and up to several months or even a couple of years for hard cheeses. The yield per 100 kg fluid milk is 8 kg for hard cheeses and up to 12 kg for soft cheeses.

All cheese ingredients are degraded biochemically to varying extents during curing.

Lactose is degraded to lactic acid by homofermentation. In cheddar cheese, for example, the pH drops from 6.55 to 5.15 from the addition of the starter culture to the end of mold pressing. In the presence of propionic acid bacteria (as in the case of Emmental cheese), lactic acid is converted further to propionic and acetic acids and CO_2, according to the reaction:

$$3\ CH_3CHOHCOOH \rightarrow 2\ CH_3CH_2COOH$$
$$+CH_3COOH + CO_2 + H_2O \qquad (10.14)$$

The ratio of propionic to acetic acid is influenced by the redox potential of the cheese, and in the presence of nitrates, for example, the ratio is lower. Propionic acid fermentation is shown in Fig. 10.27. The crucial step is the reversible rearrangement of succinyl-CoA into methylmalonyl-CoA:

$$\begin{array}{ccc}
& CO-SCoA & CO-SCoA \\
& | & *CH_2 \\
H_3C^*-CH & \longrightarrow & | \\
& | & CH_2 \\
& CO_2^\ominus & CO_2^\ominus
\end{array} \qquad (10.15)$$

The catalysis is mediated by adenosyl-B_{12}, which is a coenzyme for transformations of the general type:

$$\begin{array}{ccc}
X-C-H & & H-C-H \\
| & \rightarrow & | \\
Y-C-H & & Y-C-X
\end{array} \qquad (10.16)$$

Based on a study of a coenzyme B_{12}-analogue, it is obvious that a nonclassical carbanion mechanism is involved:

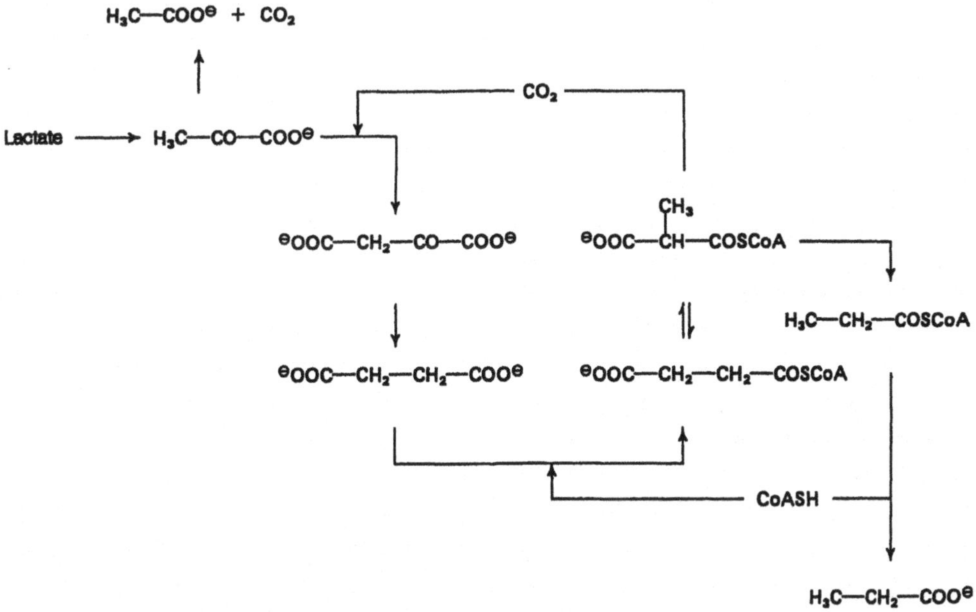

$$H_3C—COO^\ominus + CO_2$$

Lactate $\longrightarrow$ $H_3C—CO—COO^\ominus$

CO$_2$

$^\ominus OOC—CH_2—CO—COO^\ominus$ $^\ominus OOC—\overset{\overset{\displaystyle CH_3}{|}}{CH}—COSCoA$

$H_3C—CH_2—COSCoA$

$^\ominus OOC—CH_2—CH_2—COO^\ominus$ $^\ominus OOC—CH_2—CH_2—COSCoA$

CoASH

$H_3C—CH_2—COO^\ominus$

Fig. 10.27. Scheme for propionic acid fermentation

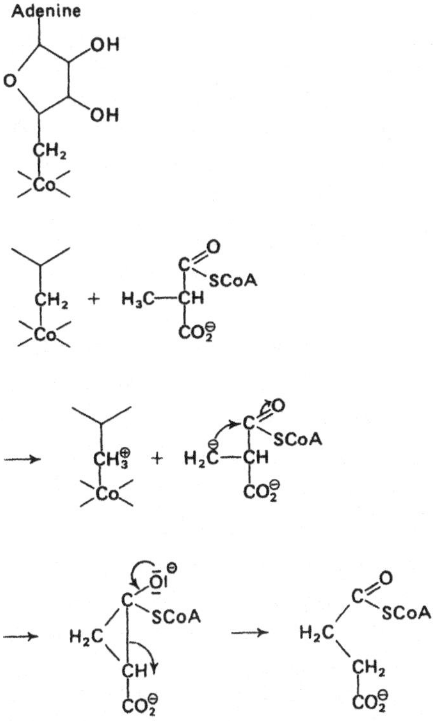

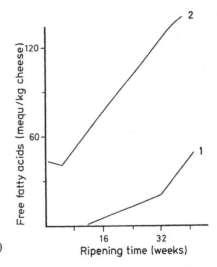

ing. In most types of cheese, as little lipolysis as possible is a prerequisite for good aroma. Exceptions are varieties like Roquefort, Gorgonzola, and Stilton, that are characterized by a marked degradation of fat.

Lipolysis is strongly enhanced by homogenization of the milk (Fig. 10.28). The release of fat-

$$(10.17)$$

The mode and extent of milk *fat* degradation depend on the microflora involved in cheese ripen-

Fig. 10.28. Lipolysis during ripening of blue cheese: *1* untreated milk, *2* homogenized milk

ty acids, especially those that affect cheese aroma, depends on the specificity of the lipases (Table 10.31). In addition to free fatty acids, 2-alkanones and 2-alkanols are formed as by-products of the β-oxidation of the fatty acids (cf. 3.7.5).

Molds, particularly *Penicillium roqueforti*, utilize β-ketoacyl-CoA deacylase (thiohydrolase) and β-ketoacid decarboxylase to provide the compounds typical for the aroma of semi-soft cheeses e. g., the blue-veined cheese (Roquefort, Stilton, Gorgonzola, cf. Table 10.32).

Protein degradation to amino acids occurs through peptides as intermediary products. Depending on the cheese variety, 20–40% of casein is transformed into soluble protein derivatives, of which 5–15% are amino acids. A pH range of 3–6 is optimum for the activity of peptidases from *Penicillium roqueforti*. Proteolysis is strongly influenced by the water and salt content of the cheese. The amino acid content is 2.8–9% of the cheese solids. Of the amino acids released, glutamic acid is of special importance to cheese taste (cf. 10.3.5). Ripening defects can produce bitter-tasting peptides.

The amino acids are transformed further. In early stages of cheese ripening, at a lower pH, they are decarboxylated to amines. In later stages, at a higher pH, oxidation reactions prevail:

$$R-CH-NH_2 \quad\diagdown\quad \begin{array}{l} R-CH_2-NH_2 \;+\; CO_2 \\[2ex] R-CO-COOH \end{array}$$

$$\downarrow$$

$$R-CHO \;+\; CO_2$$

$$(10.18)$$

Proteolysis contributes not only to aroma formation, but it affects cheese texture. In overripening of soft cheese, proteolysis can proceed almost to liquefaction of the entire cheese mass.

The progress of proteolysis can be followed by electrophoretic and chromatographic methods, e. g., via the peptide pattern obtained with the help of RP-HPLC (Fig. 10.29) and via changes in concentration of individual peptides which correspond to certain casein sequences (Table 10.33) and can serve as an indicator of the degree of cheese ripening.

The decarboxylation of amino acids (name in brackets) leads to the biogenic amines phenyl-ethylamine (phenylalanine), tyramine (tyrosine), tryptamine (tryptophan), histamine (histidine), putrescine (ornithine) and cadaverine (lysine). The content of these compounds in some types of cheese is presented in Table 10.34. These values can fluctuate greatly depending on the degree of ripening. On average, 350–500 µmol per person per day are consumed. Apart from cheese,

Table 10.31. Substrate specificity of a lipase from *Penicillium roqueforti*

Substrate	Hydrolysis (V_{rel})
Tributyrin	100
Tripropionin	25
Tricaprylin	75
Tricaprin	50
Triolein	15

Table 10.32. 2-Alkanones in blue cheese

2-Alkanone n^a	mg/100 g cheese (dry matter)
3	0.5–0.8
5	1.4–4.1
7	3.8–8.0
9	4.4–17.6
11	1.2–5.9

[a] Number of C-atoms.

Table 10.33. Amino acid sequences of some small peptides from Cheddar cheese

Peptide[a]	Sequence	Corresponding casein sequence	
30	A P F P E	α_{s1}B	$26-30^b$
37	D K I (H) P F	βA^2	47–52
39	L P Q E (V L)	α_{s1}B	11–16
46	L Q D K I (H) P (F)	βA^2	45–52
58	Y P F P G P I P N	βA^2	60–68
60	A P F P E (V F)	α_{s1}B	$26-32^b$

[a] Numbering cf. Fig. 10.29.
[b] In the literature, Q represents position 30 of α_{s1}-casein B, and E the corresponding position of the precursor protein. () Added on the basis of the amino acid composition.

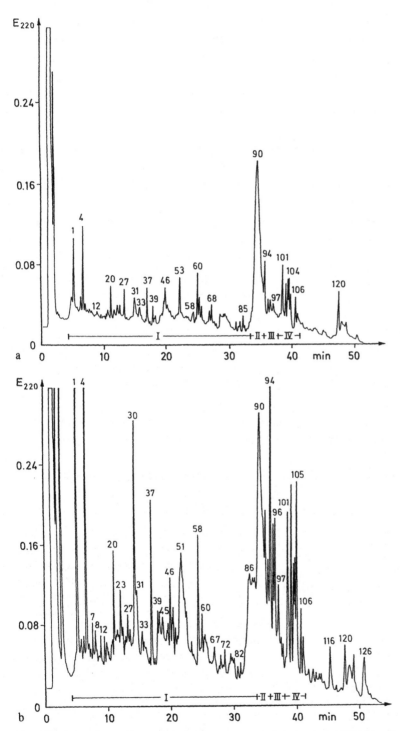

Fig. 10.29. RP-HPLC of the pH 4.6-soluble fraction of a citrate extract of Cheddar cheese after 3 (**a**) and 24 (**b**) weeks; ripening at 10 °C (according to *Kaiser* et al., 1992)

Table 10.34. Biogenic amines in cheese (mg/100 g)

Cheese	Phenylethyl-amine	Tyramine	Tryptamine	Histamine	Putrescine[a]	Cadaverine[b]
Cheddar	0–30	6–112	0–0.2	2.4–140	0–100	0–88
Emmentaler	0–23.4	3.3–40	0–1.3	0.4–250	0–15	0–8
Gruyere		6.4–9.9		0–20	0.5	2.5
Parmesan		0.4–2.9		0–58		
Provolone					1–20	2–20
Edamer	0–1.3		0–0.4	1.4–6.5		0.5–9.4
Gouda		0–110		3.5–18	2–20	2.5
Tilsiter	0–14.8	0–78	0–7.1	0–95.3	0–31.3	0–31.8
Gorgonzola					0–75	0–430
Roquefort		2.7–110	0–160	1–16.8	1.5–3.3	7.1–9.3
Camembert		2–200	2	0–48	0.7–3.3	1.2–3.7

[a] Butane-1,4-diamine
[b] Pentane-1,5-diamine

fruits (cf. 18.1.4.2.1) and meat (cf. 12.3.5) are important sources.

10.2.8.4 Processed Cheese

Processed (or melted) cheese is made from natural, very hard grating or hard cheeses by shredding and then heating the shreds to 75–95 °C in the presence of 2–3% melting salts (lactate, citrate, phosphate) and, when required, utilizing other ingredients, such as milk powder, cream, aromas, seasonings and vegetable and/or meat products. The cheese can be spreadable or made firm and cut as desired. The shelf life of processed cheese is long due to thermal killing of microflora.

The heating process is carried out batchwise by steam injection in a double-walled pressure vessel equipped with a mixer, usually under a slight vacuum. Continuous processes are conducted in double-walled cylinders with agitator shafts.

10.2.8.5 Imitation Cheese

Imitation cheese (analogue cheese) is mainly found in North America. They are made of protein (mostly milk protein), fat (mostly hardened vegetable fat), water, and stabilizers by using processed cheese technology. A typical formulation is shown in Table 10.35.

Table 10.35. Typical formulation of imitation cheese (Mozarella type)

Component	Amount (%)
Water	51.1
Ca/Na caseinate	26.0
Vegetable oil (partially hydrogenated)	8.0
Glucono-δ-lactone	2.8
Salt	2.0
Color and aroma substances	

10.2.9 Casein, Caseinates, Coprecipitate

The production of casein, caseinates, and coprecipitate is shown schematically in Fig. 10.30.

Coagulation and separation of casein from milk is possible by souring the milk by lactic acid fermentation, or by adding acids such as HCl, H_2SO_4, lactic acid or H_3PO_4. Another way to achieve coagulation is to add proteinase enzymes, such as chymosin and pepsin. The acid coagulation is achieved at 35–50 °C and pH 4.2–4.6 (isoelectric point of casein is pH 4.6–4.7). Casein precipitates out as coarse grains and is usually separated in sedimentation centrifuges, washed, and dried (whirlwind drier). The enzymatic process involves heating to 65 °C after precipitation in whey.

Increasing the level of Ca^{2+} ions (addition to milk of 0.24% $CaCl_2$) causes casein and whey proteins to coagulate when the temperature is

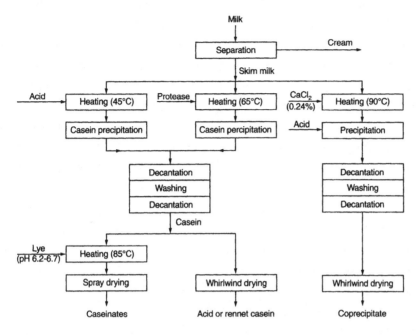

Fig. 10.30. Production of casein, caseinates, and coprecipitate

at 90 °C. Joint coagulation of proteins can also be achieved by first heat-denaturing the whey proteins, then acidifying the milk. Washing followed by drying of the curd gives a coprecipitate which contains up to 96% of the total proteins of the milk. When casein dispersions, 20–25%, are treated with alkali [NaOH or Ca(OH)$_2$, alkali or alkaline-earth carbonates or citrates] at 80–90 °C and pH 6.2–6.7, and then the solubilized product is spray-dried, a soluble or readily dispersable casein product is obtained (caseinate, disintegrated milk protein).

Caseins and whey proteins are also concentrated by ultrafiltration and reverse osmosis. Since the molar masses of the whey proteins and casein micelles are in the range of 10^3–10^4 and 10^7–10^9 respectively, membranes with a pore diameter of 5–50 nm are suitable for the separation of these proteins.

Casein and caseinate are utilized as food and also have nonfood uses. In food manufacturing they are used for protein enrichment and/or to achieve stabilization of some physical properties of processed meats, baked products, candies, cereal products, ice creams, whipping creams, coffee whiteners, and some dietetic food products and drugs.

The nonfood uses involve wide application of casein/caseinate as a sizing (coating) for better quality papers (for books and journals, with a surface suitable for fine printing), in glue manufacturing, as a type of waterproof glue (alkali caseinate with calcium components as binder); in the textile industry (dye fixing, water-repellent impregnations); and for casein paints and production of some plastics (knobs, piano keyboards, etc.).

10.2.10 Whey Products

Whey accumulates in considerable amounts in the production of cheese and casein.

The composition of whey and whey products is presented in Table 10.36. Whey and whey products are used in animal feed, dietetic foods (infant food), bread, confectionery, candies, and beverages.

10.2.10.1 Whey Powder

In dairy farming, two process variants are applied for the drying of whey:

Table 10.36. Protein, lactose and mineral contents of whey products[a]

Product	DM[b] (%)	Protein (%)	Lactose (%)	Minerals (%)
Skim milk	9.0	36	53	7
Whey (from coagulating with rennet)	6.0–6.4	13	75	8
Whey (from coagulating with acid)	5.8–6.2	12	67	14
Demineralized whey powder		12–13	85	1–2
Whey protein powder[c]				
I		47	44	9
II		74	20	6

[a] Average values are expressed as % of dry matter.
[b] Dry matter.
[c] After one (I) and two (II) ultrafiltrations.

- Preliminary concentration of the whey to 50–55% dry matter in falling-film evaporating systems (thermal or mechanical vapor compression), followed by spray drying (one step or two step with subsequent vibraton fluid bed).
- Preliminary concentration of the whey to 21–25% dry matter by reverse osmosis (hyperfiltration), followed by concentration to 50–55% dry matter via falling-film evaporators and spray drying.

The composition of whey powder is presented in Tables 10.27 and 10.36.

10.2.10.2 Demineralized Whey Powder

In the applications of whey powder, the minerals can interfere with the taste. The production of demineralized whey powder proceeds via ion exchange or, preferentially, electrodialysis (1.5–4.5V/cell; current density 5–20 mA/cm^2 membrane area, Fig. 10.31). The course of demineralization is shown in Fig. 10.32.

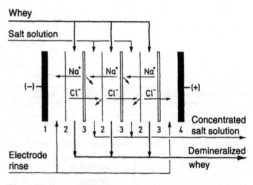

Fig. 10.31. Principle of electrodialysis of whey. *1* cathode, *2* cation membrane, *3* anion membrane, *4* anode

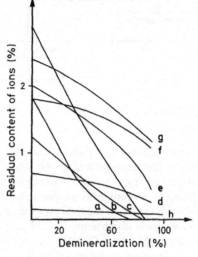

Fig. 10.32. Whey demineralization. Ions of *a* chloride, *b* sodium, *c* potassium, *d* calcium, *e* phosphate, *f* lactate, *g* citrate, and *h* magnesium

10.2.10.3 Partially Desugared Whey Protein Concentrates

In the ultrafiltration of whey, protein concentrates depleted of lactose to various extents are obtained, depending on the number of stages and amount of wash water. Another, less gentle method involves the heating of whey (95 °C, 3–4 min) by direct steam injection, followed by precipitation of the denatured proteins at pH 4.5, separation in a sedimentation centrifuge (2000–4000 min^{-1}), and drying.

10.2.10.4 Hydrolyzed Whey Syrups

The production of sweet whey syrups is becoming increasingly important due to the use of carrier-bound lactase (β-galactosidase, EC. 3.2.1.23). In these syrups, lactose is hydrolyzed to glucose and galactose. Concentration to 60–75% solids is achieved by evaporation.

10.2.11 Lactose

For lactose production the whey is evaporated to 55–65% solid content, and the concentrate is then seeded and cooled slowly to induce sugar crystallization. The raw lactose (food quality) is recrystallized to yield a raffinade (pharmaceutical-grade lactose). Lactose is used in manufacturing of drugs (tablet filler), dietetic food products, baked products, dehydrated foods, cocoa products, beverages and ice creams.

10.2.12 Cholesterol-Reduced Milk and Milk Products

In the production of milk products with a reduced cholesterol content, more than 90% of the cholesterol is removed from water-free milk fat by extraction with supercritical carbon dioxide or by steam distillation. The fat is then recombined with skim milk to give low-cholesterol milk, which is used to make the usual milk products. The extent of cholesterol reduction in a series of products is listed in Table 10.37.

Recombined milk does not have the same properties as the original milk because, e. g., the membrane composition of the fat globule changes in the process. Cheese made from milk of this type can exhibit texture defects. Since skim milk with a fat content of 0.2% still contains about 18 mg/l of cholesterol, skim milk must also be freed of cholesterol for the production of cholesterol-free products.

10.3 Aroma of Milk and Dairy Products

10.3.1 Milk, Cream

Raw or gently pasteurized milk has a mild but characteristic taste.

In the AEDA of UHT milk (Table 10.38), δ-decalactone, which contributes to the aroma of butter (Table 10.40) as well as unripened and ripened cheese (cf. 10.3.5), is the predominant aroma substance. Apart from other lactones, 2-acetyl-1-pyrroline, methional, 2-acetyl-2-thioazoline and 4,5-epoxy-2-decenal are among the identified aroma substances.

A higher thermal exposure of milk, e. g., by sterilization, allows the accumulation of *Maillard* products, such as methylpropanal, 2- and 3-methyl butanal and 4-hydroxy-2,5-dimethyl-3(2H)-furanone.

10.3.2 Condensed Milk, Dried Milk Products

During the concentration and drying of milk, reactions that are similar to those described for heat-treated milk (cf. 10.1.3.5 and 10.3.1) occur, but to a greater extent. Therefore, like the aroma of UHT milk (cf. 10.3.1 and Table 10.38), the aroma of condensed milk is also caused by *Maillard* reaction products. The stale flavor that appears when condensed milk is stored for longer periods is due especially to the presence of the degradation product of tryptophan, o-aminoacetophenone, which is aroma active in concentrations $\geq 1\,\mu g/kg$. A rubbery aroma defect results from higher concentrations of benzothiazole.

Table 10.37. Effects of a 90% reduction of cholesterol in butter oil on the cholesterol content of recombined milk and its products

Food	Fat (%)	Cholesterol (mg/kg) I[a]	II[a]
Whole milk	3.3	135	26
Butter	81	2400	300
Yoghurt	3.5	124	26
Ice cream	10.8	450	41
Cottage cheese	4.6	150	12
Mozzarella	21.6	786	68
Brie	20.8	1000	75
Camembert	24.6	714	57
Roquefort	30.6	929	107
Cheddar	33.1	1071	114

[a] Product before (I) and after (II) cholesterol reduction.

Table 10.38. Potent aroma substances of UHT milk[a]. Result of an AEDA

Compound	Odor quality	FD Factor
2-Acetyl-1-pyrroline	Rusty	8
(Z)-4-Nonenal	Fatty	1
Methional	Boiled potatoes	8
2,3-Diethyl-5-methylpyrazine	Earthy	1
Unknown	Fatty, cardboard	1
Butyric acid	Sweaty	4
Unknown	Mint	2
2-Acetyl-2-thiazoline	Rusty	8
Caproic acid	Sweaty	2
d-Octalactone	Coconut	16
trans-4,5-Epoxy-(E)-2-decenal	Metallic	16
Caprylic acid	Sweaty	1
δ-Nonalactone	Coconut	1
Unknown	Musty	2
δ-Decalactone	Coconut	128
Unknown	Musty	8
Capric acid	Sweaty	2
Unknown	Coconut	1
γ-Dodecalactone	Coconut	16
γ-(Z)-6-Dodecenol-actone	Coconut	16
Unknown	Woody	8
Vanillin	Vanilla	16

[a] In bottles

The content of free butyric and caprylic acid as well as (Z)-3-hexenal rises when cream is whipped (Table 10.39). Pasteurization results in the formation of 2-acetyl-2-thiazoline in whipped cream and the content of (E,Z)-2,6-nonadienal is greatly increased. A model corresponding to Table 10.39 (without No. 12, 14, 17 and 20) approaches the aroma of whipped pasteurized cream and reproduces especially the "creamy" note.

Maillard reaction products are also characteristic of the aroma of milk powder. The development of aroma defects during the storage of whole milk powder is due to products of lipid peroxidation, e. g., (Z)- and (E)-2-nonenal.

10.3.3 Sour Milk Products, Yoghurt

Metabolic products of lactic acid bacteria, such as diacetyl, ethanal, dimethylsulfide, acetic acid and lactic acid contribute to this aroma. Carbon dioxide also appears to be important. In good

Table 10.39. Aroma substances in raw cream (I), whipped raw cream (II) and in whipped pasteurized raw cream (III)

No.	Aroma substance	Concentration (µg/kg)		
		I	II	III
1	Butyric acid	4400	8000	2000
2	Caprylic acid	4200	7500	1800
3	δ-Dedecalactone	1100	1400	1200
4	δ-Decalactone	300	300	250
5	γ-Dodecalactone	63	99	63
6	δ-Octalactone	28	37	26
7	3-Methylbutyric acid	18	18	17
8	(Z)-6-Dodecen-g-lactone	7.5	10	9.2
9	3-Methylindol	3.4	3.1	3.4
10	(Z)-3-Hexenal	1.6	3.3	7.7
11	(E)-2-Nonenal	1.3	1.7	0.8
12	trans-4,5-Epoxy-(E)-2-decenal	1	0.97	0.29
13	2-Phenylethanol	0.57	0.58	0.51
14	(E)-2-Ddodecenal	0.37	0.37	0.4
15	1-Octen-3-one	0.33	0.19	0.11
16	(E,Z)-2,6-Nonadienal	0.11	0.2	1.4
17	2-Aminoacetophenone	0.13	0.15	0.13
18	1-Hexen-3-one	0.1	0.1	0.21
19	Methional	0.07	0.06	0.07
20	2-Acetyl-1-pyrroline	0.03	0.05	0.07
21	2-Acetyl-2-thiazoline	n.d.	n.d.	0.06
22	Methanthiol	n.d.	n.a.	27
23	Dimethylsulfide	10	n.a.	13

[a] Cream (fat content: 30%); n.d.: not detected, n.a.: not analyzed

sour milk products, the concentration ratio of diacetyl/ethanal should be ca. 4. At values of ≤3, a green taste appears, which is to be regarded as an aroma defect. Diacetyl is formed from citrate (Fig. 10.33). The conversion of acetolactate to diacetyl is disputed. It should occur spontaneously or be catalyzed by an α-acetolactate oxidase.

Ethanal greatly contributes to the aroma of yoghurt. Concentrations of 13–16 µg/kg are characteristic of good products.

10.3.4 Butter

Only the three compounds listed in Table 10.40 make an appreciable contribution to the aroma of butter. A comparison of the aroma profiles of five samples of butter (Table 10.41) with the results of a quantitative analysis (Table 10.40) show that

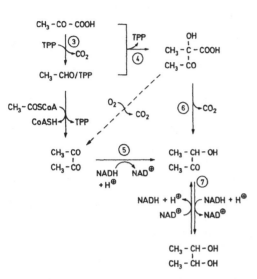

Fig. 10.33. Formation of diacetyl and butanediol from citrate by Streptococci. *1* citratase, *2* oxaloacetate decarboxylase, *3* pyruvate decarboxylase, *4* α-acetolactate synthase, *5* diacetyl reductase, *6* α-acetolactate decarboxylase, *7* 2,3-butanediol dehydrogenase

Table 10.40. Concentrations of the key aroma substances in five samples of butter[a]

Aroma substance	Concentration (mg/kg) in sample number				
	1	2	3	4	5
Diacetyl	0.62	0.34	0.11	0.32	<0.01
(R)-δ-Decalactone	5.0	4.91	3.06	2.15	3.8
Butyric acid	4.48	3.63	2.66	94.5	2.48

[a] The aroma profiles of the samples are presented in Table 10.41.

the concentrations of these three odorants, which are present in Samples 1 and 2, produce an intensive butter aroma. In Samples 3 and 4, especially the diacetyl content is too low and in Sample 4, the excessively high butyric acid concentration stimulates a rancid aroma defect. Lactic acid is primarily involved in the taste of sour cream butter.

If butter contains lipases, fatty acids are released on storage. Above certain limiting concentrations

Table 10.41. Aroma profile of samples of butter

No.	Sample	Odor quality	Intensity[a]
1	Sour cream	Buttery, creamy, sweet	3
2	Sour cream	Buttery, creamy	2–3
3	Sour cream	Slightly buttery, mild, sour	1–2
4	Sour cream	Rancid, like butyric acid	3
5	Sweet cream	Mild, somewhat sour	1

[a] Evaluation: 1, weak; 2, medium; 3, strong.

(cf. 3.2.1.1), these fatty acids cause a rancid off-flavor.

Rancid, soapy aroma defects, which occur in butter samples with very low concentrations of free fatty acids, can be due to contamination with anionic detergents (sodium dodecyl sulfate, sodium dodecyl benzosulfonate). Detergents of this type are used to disinfect the udder and the milking machine.

10.3.5 Cheese

The aroma profile of unripened cheese, e. g., Mozzarella, consists of butter-like, sweetish, salty and sour notes produced by 1, δ-decalactone, NaCl and lactic acid. The characteristic odor and taste of the type of cheese are formed during ripening, whereby the composition of the microflora and the storage conditions (temperature, air humidity, time) have the greatest influence. For a soft cheese (Camembert) and a hard cheese (Emmentaler), the compounds mainly responsible for the odor and taste in the ripened product will be discussed here.

The butter-like note of unripened cheese can still be detected in Camembert and Emmentaler, but the intensity is lower, because other aroma substances formed during ripening become evident. Thus, Camembert also has mushroom-like, sulfurous and flowery notes and Emmentaler, nutty, sweet and fruity notes. In comparison with unripened cheese, the taste profile is extended to include a glutamate note and in the case of Emmentaler, an additional and characteristic sour/pungent impression.

Among the odorants in Camembert (Table 10.42), 1-octen-3-ol is responsible for the mushroom-like note, which should be intensified by 1-octen-3-one. Although its concentration is only 2.1 µg/kg, it is aroma-active in Camembert because its odor threshold is 100 times lower than that of the alcohol. Methanethiol, methional, dimethylsulfide and methylene-bis(methylsulfide) produce the sulfurous and phenylethylacetate produces the flowery note. The relatively high concentration of glutamate (Table 10.42) is noticeable in the taste. Methylketones contribute to the typical aroma of blue cheese (Roquefort). It is unknown which additional compounds are important.

Table 10.43 shows the odorants and taste compounds of Emmental. The high concentration of more than 6 g/kg of propionic acid produces the

Table 10.42. Odorants and taste compounds of Camembert[a]

Compound/Ion	Concentration[b] (mg/kg)
Odorant	
Diacetyl	0.074–0.110
3-Methylbutanal	0.094–0.142
Methional	0.027–0.125
1-Octen-3-ol	0.075–0.130
1-Octen-3-one	0.0021
Phenethylacetate	0.250–0.320
2-Undecanone	0.180–0.700
δ-Decalactone	0.910–1.08
Methanthiol	0.260–0.275
Dimethylsulfide	0.250–0.410
Acetaldehyde	0.015–0.025
Hexanal	0.124–0.144
Dimethyltrisulfide	0.008–0.010
Methylen-bis(methylsulfide)	0.250–0.360
2-Acetyl-1-pyrroline	0.001–0.003
2-Phenylethanol	0.130–0.137
Taste compound	
Acetic acid	59–92
Butyric acid	122–130
3-Methylbutyric acid	3.4–4.5
Caprylic acid	62–70
Glutamic acid[c]	2690–4381
Lactic acid	88–174
Succinic acid[c]	535–892
Ammonia[c]	448–632
Sodium[c]	12,190–13,570
Potassium[c]	665–743
Calcium[c]	761–802
Magnesium[c]	61–97
Chloride[c]	12,053–14,180
Phosphate[c]	1330–1425

[a] Fat: 45% solids, Water: 52%.
[b] Reference: fresh weight.
[c] Concentration in aqueous extract from 1 kg cheese.

Table 10.43. Oderants and taste compounds of Emmentaler[a]

Compound/Ion	Concentration[b]
Odorant (µg/kg)	
Diacetyl	431
2-Methylbutanal	181
3-Methylbutanal	145
Butyric acid ethyl ester	27
3-Methylbutyric acid ethyl ester	0.40
2-Heptanone	522
Dimethyltrisulfide	0.11
Methional	67
Caproic acid ethyl ester	51
1-Octen-3-one	0.06
4-Hydroxy-2,5-dimethyl-3(2H)-furanone (HD3F)	1186
5-Ethyl-4-hydroxy-2-methyl-3(2H)-furanone (EHM3F)	253
2-*sec*-Butyl-3-methoxypyrazine	0.07
Skatole	47
δ-Decalactone	3751
Taste compound (mg/kg)	
Acetic acid	3830
Propionic acid	6750
Butyric acid	70
3-Methylbutyric acid	20
Glutamic acid[c]	5380
Lactic acid[c]	9150
Succinic acid[c]	1320
Ammonia[c]	560
Sodium[c]	5150
Potassium[c]	1280
Calcium[c]	6650
Magnesium[c]	680
Chloride[c]	3730
Phosphate[c]	10570

[a] Ripening time 3 months.
[b] Reference: dry matter (water content: 36%, fat: 50% solids).
[c] Concentration in aqueous extract from 1 kg cheese.

characteristic sour/pungent taste, which is intensified by lactic acid. The two furanones HD3F and EHM3F probably contribute to the nutty note. Model experiments show that lactic acid bacteria (*Lactobacillus helveticus, Lactobacillus delbrueckii*) are involved in the formation of HD3F.

At the Emmentaler pH of 5.6, magnesium (threshold value: 3.5 mmol/kg) and calcium propionate (7.1 mmol/kg) taste sweet. Consequently, it is assumed that these propionates contribute to the sweet note. On the other hand, glutamic acid is an important taste substance, which has the additional function of neutralizing the bitter taste of amino acids and peptides. Only if the concentrations of these constituents climb too high on longer ripening of Emmentaler, the effect of glutamic acid is no longer sufficient and the bitter taste appears. An off-flavor can also be formed if there is a greater increase in the fatty acids 4:0–12:0.

The caseins are increasingly degraded during longer ripening. Water-soluble peptides and amino acids are formed which bind a part of the ions. Thus, when chewing a cheese ripened for a long time, the water-soluble portion of the ions increases, possibly causing an intensification of the salty taste.

It is probable that not only peptides, but also other amides are responsible for the bitter taste of cheese. For example, the presence of bitter N-isobutyl acetamide has been detected in Camembert cheese.

10.3.6 Aroma Defects

As already indicated, aroma defects can arise in milk and milk products either by absorption of aroma substances from the surroundings or by formation of aroma substances via thermal and enzymatic reactions.

Exogenous aroma substances from the feed or cowshed air enter the milk primarily via the respiratory or digestive tract of the cow. Direct absorption apparently plays only a minor role. Metabolic disorders of the cow can cause aroma defects, e. g., the acetone content of milk is increased in ketosis.

The oxidation of lipids is involved in the endogenous formation of aroma defects. While very low concentrations of certain carbonyl compounds, e. g., (Z)-4-heptenal (1 μg/kg), 1-octen-3-one, and hexanal, appear to contribute to the full creamy taste, increased concentrations of these and other compounds produce cardboard-like, metallic, and green aroma notes. In butter, for instance, the phospholipids of the fat globule membrane are especially susceptible to oxidation. The subsequent products get distributed in the entire fat fraction and cause taste defects which range from metallic to fatty and from fishy to tallowy.

Light can cause the degradation of methionine to 3-methylthiopropanal via riboflavin as sensitizer. Together with other sulfides and methanethiol, this sulfur compound produces the aroma defect of milk and milk products called "light taste".

A series of aroma defects are caused by enzymatic reactions. These include:

- An unclean taste due to an increased concentration of dimethylsulfide produced by psychotropic microorganisms.
- A fruity taste due to the formation of ethyl esters produced by psychotropic microorganisms, e. g., *Pseudomonas fragii*.
- A malty taste due to increased formation of 3-methylbutanal, 2-methylbutanal, and methylpropanal by *Strept. lactis var. maltigenes*.
- A metallic taste in buttermilk due to (E,Z)-2,6-nonadienol in concentrations >1.3 μg/l. The precursor is the triglycerol-bound α-linolenic acid which is oxidized to 9-hydro-peroxy-10,12,15-octadecatrienoic acid by oxygenases from the starter culture. Proton catalysis liberates (E,Z)-2,6-nonadienal which is reduced to the corresponding alcohol by lactic acid bacteria.
- A phenolic taste due to spores of *Bacillus circulans*.
- A rancid taste due to the release of lower fatty acids (C_4–C_{12}) by milk lipases or bacterial lipases.
- A bitter taste can occur due to proteolytic activity, e. g., on storage of UHT milk. The milk proteinase plasmin is inactivated on intensive heating (142 °C, >16 s). However, some bacterial proteinases can still be active even after much longer exposure to heat (142 °C, 6 min).

10.4 References

Aimutis, W.R., Eigel, W.N.: Identification of λ-casein as plasmin-derived fragments of bovine α-casein. J. Dairy Sci. *65*, 175 (1982)

Badings, H.T., De Jong, C., Dooper, R.P.M., De Nijs, R.C.M.: Rapid analysis of volatile compounds in food products by purge- and cold-trapping/capillary gas chromatography. In: Progress in flavor research 1984 (Ed.: Adda, J.), p. 523, Elsevier Science Publ., Amsterdam. 1985

Bottazzi, V.: Other fermented dairy products. In: Biotechnology (Eds.: Rehm, H.-J., Reed, G.), Vol. 5, p. 315, Verlag Chemie: Weinheim. 1983

Bringe, N.A., Kinsella, J.E.: The effects of pH, calcium chloride and temperature on the rate of acid coagulation of casein micelles. Am. Dairy Sci., Annual Meeting, 23–26 June, Davis, CA. (1986)

Burghalter, G., Steffen, C., Puhan, Z.: Cheese, processed cheese, and whey. In: Ullmann's Enzyclopedia of Industrial Chemistry. 5ᵗʰ Edition, Volume A6 (Eds.: F.T. Campbell, R. Pfefferkorn, J.F. Rounsaville) Verlag Chemie, Heidelberg, 1986, pp. 163

Creamer, L.K., Richardson, T., Parry, D.A.D.: Secondary structure of bovine α_{s1}- and β-casein in solution. Arch. Biochem. Biophys. *211*, 689 (1981)

Czerny, M., Schieberle, P.: Influence of the polyethylene packaging on the adsorption of odour-active compounds from UHT-milk. Eur. Food Res. Technol. 2007, in press

Dalgleish, D.G.: Coagulation of renneted bovine casein micelles: dependence on temperature, calcium ion concentration and ionic strength. J. Dairy Res. *50*, 331 (1983)

Davies, F.L., Law, B.A.: Advances in the microbiology and biochemistry of cheese and fermented milk. Elsevier Appl. Science Publ., London. 1984

Dirks, U., Reimerdes, E.H.: Analytik von Gangliosiden unter besonderer Berücksichtigung der Milch. Z. Lebensm. Unters. Forsch. *186*, 99 (1988)

Eigel, W.N., Butler, J.E., Ernstrom, C.A., Farrell Jr., H.M., Harwalkar, V.R., Jenness, R., Whitney, R.McL.: Nomenclature of Proteins of Cow's Milk: Fifth Revision, J. Dairy Sci. *67*, 1599 (1984)

Gobbetti, M., Stepaniak, L., De Angelis, M., Corsetti, A., Di Cagno, R.: Latent bioactive peptides in milk proteins: Proteolytic activation and significance in dairy processing. Crit. Rev. Food Sci. Nutr. *42*, 223 (2002)

Groψklaus, R.: Pro- und Präbiotika. In: Workshop "Moderne Ernährung-Lifestyle". Werkstattbericht 5 der Stockmeyer Stiftung für Lebensmittelforschung, Eigenverlag Kuratorium der Stockmeyer Stiftung, Sassenberg-Füchtorf, 1999, p. 21

Henle, T., Walter, H., Krause, I., Klostermeyer, H.: Efficient determination of individual *Maillard* compounds in heat-treated milk products by amino acid analysis. Int. Dairy Journal *1*, 125 (1991)

Juriaanse, A.C., Herrtje, J.: Microstructure of shortenings, margarine and butter – a review. Food Microstructure *7*, 181 (1988)

Kaiser, K.-P., Belitz, H.-D., Fritsch, R.J.: Monitoring Cheddar cheese ripening by chemical indices of proteolysis. 2. Peptide mapping of casein fragments by reverse-phase high-performance liquid chromatography. Z. Lebensm. Unters. Forsch. *195*, 8 (1992)

Kessler, H.G.: Lebensmittel- und Bioverfahrenstechnik. Molkereitechnologie. 3. Auflage. Verlag A. Kessler: Freising. 1988

Kinsella, J.E., Hwang, D.H.: Enzymes of Penicillium roqueforti involved in the biosynthesis of cheese flavor. Crit. Rev. Food Sci. Nutr. *8*, 191 (1977)

Kosikowski, F.V.: "Cholesterol-free" milks and milk products: Limitations in production and labeling. Food Technol. *44* (11), 130 (1990)

Kubickova, J., Grosch, W.: Quantification of potent odorants in Camembert cheese and calculation of their odour activity values. Int. Dairy J. *8*, 17 (1998)

Moran, N., Schieberle, P.: On the role of aroma compounds for the creaminess of dairy products. Deutsche Forschungsanstalt für Lebensmittelchemie, Report 2006, p. 186

Nielsen, S.S.: Plasmin system and mircrobial proteases in milk: characteristics, roles and relationship. J. Agric. Food Chem. *50*, 6628 (2002)

Nursten, H.: The flavour of milk and dairy products: I. Milk of different kinds, milk powder, butter and cream. Int. J. Dairy Technol. *50*, 48 (1997)

Oh, S., Richardson, T.: Genetic engeneering of bovine χ-casein to enhance proteolysis by chymosin. In: Interactions of Food Proteins (Eds.: N. Parris, R. Barford) ACS Symp. Ser. *454*, 195 (1991)

Olivecrona, T., Bengtsson, G.: Lipase in milk. In: Lipases (Eds.: Borgström, B., Brockman, H.L.), p. 205, Elsevier Science Publ., Amsterdam. 1984

Ott, A.; Fay, L.B., Chaintreau, A.: Determination and origin of the aroma impact compounds of yoghurt flavor. In: Flavour perception. Aroma evaluation (Eds.: H.P. Kruse, M. Rothe) Universität Potsdam, 1997, p. 203

Payens, T.A.J.: Casein micelles: the colloid-chemical approach. J. Dairy Res. *46*, 291 (1979)

Preininger, M., Warmke, R., Grosch, W.: Identification of the character impact flavour compounds of Swiss cheese by sensory studies of models. Z. Lebensm. Unters. Forsch. *202*, 30 (1996)

Riccio, P.: The proteins of the milk fat globule membrane. Trends Food Sci. Technol. *15*, 458 (2004)

Schieberle, P., Gassenmeier, K., Guth, H., Sen, A., Grosch, W.: Character impact odour compounds of different kinds of butter. Lebensm. Wiss. Technol. *26*, 347 (1993).

Schieberle, P., Heiler, C.: Influence of processing and storage on the formation of the metallic smelling (E,Z)-2,6-nonadienol in buttermilk. In: Flavour perception – Aroma evaluation (Eds.: H.P. Kruse, M. Rothe), Universität Potsdam, 1997, pp. 213

Seifu, E., Buys, E.M., Donkin, E.F.: Significance of the lactoperoxidase system in the dairy industry and its potential applications: a review. Trends Food Sci. Technol. *16*, 137 (2005)

Stein, J.L., Imhof, K.: Milk and dairy products. In: Ullmann's Encyclopedia of Industrial Chemistry, 5th Edition, Volume A16 (Eds.: B. Elvers, S. Hawkins, G. Schulz) Verlag Chemie, Heidelberg, 1990, pp. 589

Swaisgood, H.E.: The Caseins. Crit. Rev. Food Technol. *3*, 375 (1972/73)

Thomas, M.E.C., Scher, J., Desobry-Banon, S., Desobry, S.: Milch powders ageing: Effect on physical and functional properties. Crit. Rev. Food Sci. Nutr. *44*, 297 (2005)

Timm, F.: Speiseeis. Verlag Paul Parey: Berlin. 1985

Vedamuthu, E.R., Washam, C.: Cheese. In: Biotechnology (Eds.: Rehm, H.-J., Reed, G.), Vol. 5, p. 231. Verlag Chemie: Weinheim. 1983

Walstra, P., Jenness, R.: Dairy chemistry and physics. John Wiley & Sons. New York. 1984

11 Eggs

11.1 Foreword

Eggs have been a human food since ancient times. They are one of nature's nearly perfect protein foods and have other high quality nutrients. Eggs are readily digested and can provide a significant portion of the nutrients required daily for growth and maintenance of body tissues. They are utilized in many ways both in the food industry and the home. Chicken eggs are the most important. Those of other birds (geese, ducks, plovers, seagulls, quail) are of lesser significance. Thus, the term "eggs", without a prefix, generally relates to chicken eggs and is so considered in this chapter. Table 11.1 gives some data on the production of eggs.

11.2 Structure, Physical Properties and Composition

11.2.1 General Outline

The egg (Fig. 11.1) is surrounded by a 0.2–0.4 mm thick calcareous and porous shell. Shells of chicken eggs are white-yellow to brown, duck's are greenish to white, and those of most wild birds are characteristically spotted. The inside of the shell is lined with two closely adhering membranes (inner and outer). The two membranes separate at the large end of the egg to form an air space, the so-called air cell. The air cell is approx. 5 mm in diameter in fresh eggs and increases in size during storage, hence it can be used to determine the age of eggs. The egg white (albumen) is an aqueous, faintly straw-tinted, gel-like liquid, consisting of three fractions that differ in viscosity. The inner portion of the egg, the yolk, is surrounded by albumen. A thin but very firm layer of albumen (chalaziferous layer) closely surrounds the yolk and it branches on opposite sides of the yolk

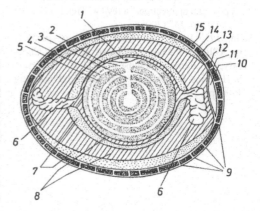

Fig. 11.1. Cross-section of a chicken egg – a schematic representation. Egg yolk: *1* germinal disc (blastoderm), *2* yolk membrane, *3* latebra, *4* a layer of light colored yolk, *5* a layer of dark colored yolk, *6* chalaza, *7* egg white (albumen) thin gel, *8* albumen thick gel, *9* pores, *10* air cell, *11* shell membrane, *12* inner egg membrane, *13* shell surface cemented to the mammillary layer, *14* cuticle, and *15* the spongy calcareous layer

into two chalazae that extend into the thick albumen.

The chalazae resemble two twisted rope-like cords, twisted clockwise at the large end of the egg and counterclockwise at the small end. They serve as anchors to keep the yolk in the center. In an opened egg the chalazae remain with the yolk. The germinal disc (blastoderm) is located at the top of a clubshaped latebra on one side of the yolk. The yolk consists of alternate layers of dark- and light-colored material arranged concentrically.

The average weight of a chicken egg is 58 g. Its main components are water ($\sim$74%), protein ($\sim$12%), and lipids ($\sim$11%). The proportions of the three main egg parts, yolk, white and shell, and the major ingredients are listed in Table 11.2. Table 11.3 gives the amino acid composition of whole egg, white and yolk.

H.-D. Belitz · W. Grosch · P. Schieberle, *Food Chemistry*
© Springer 2009

Table 11.1. Production of eggs, 2006 (1000 t)[a]

Continent	Chicken egg	Other egg
World	61,111	5421
Africa	2224	7
America, Central-	2302	–
America, North-	5760	–
America, South- and Caribbean	5715	76
Asia	37,162	5256
Europe	10,021	79
Oceania	230	3

Country	Chicken egg	Country	Other egg
China	25,326	China	4529
USA	5360	Thailand	310
India	2604	Indonesia	202
Japan	2497	Brazil	75
Russian Fed.	2100	Philippines	72
Mexico	2014	Uzbekistan	48
Brazil	1675	Russian Federation	31
Indonesia	932	Korea, Republic of	28
France	850	Bangladesh	26
Spain	850	United Kingdom	16
Ukraine	819	$\Sigma(\%)^b$	98
Turkey	753		
$\Sigma(\%)^b$	75		

[a] Including egg for hatching.
[b] World production = 100%.

Table 11.2. Average composition of chicken eggs

Fraction	Percent of the total weight	Dry matter (%)	Protein (%)	Fat (%)	Carbo-hydrates (%)	Minerals (%)
Shell	10.3	98.4	3.3[a]			95.1
Egg white	56.9	12.1	10.6	0.03	0.9	0.6
Egg yolk	32.8	51.3	16.6	32.6	1.0	1.1

[a] A protein mucopolysaccharide complex.

11.2.2 Shell

The shell consists of calcite crystals embedded in an organic matrix or framework of interwoven protein fibers and spherical masses (protein-mucopolysaccharide complex) in a proportion of 50:1. There are also small amounts of magnesium carbonate and phosphates.

The shell structure is divided into four parts: the cuticle or bloom, the spongy layer, the mammillary layer and the pores. The outermost shell coating is an extremely thin (10 μm), transparent, mucilaginous protein layer called the cuticle, or bloom. The spongy, calcareous layer, i.e. a matrix comprising two-thirds of the shell thickness, is below the thin cuticle. The mammillary layer consists of a small layer of compressed, knob-like particles, with one side firmly cemented to the spongy layer and the other side adhering closely to the outer surface of the

Table 11.3. Amino acid composition of whole egg, egg white and yolk (g/100 g edible portion)

Amino acid	Whole egg	Egg white	Egg yolk
Ala	0.71	0.65	0.82
Arg	0.84	0.63	1.13
Asx	1.20	0.85	1.37
Cys	0.30	0.26	0.27
Glx	1.58	1.52	1.95
Gly	0.45	0.40	0.57
His	0.31	0.23	0.37
Ile	0.85	0.70	1.00
Leu	1.13	0.95	1.37
Lys	0.68	0.65	1.07
Met	0.40	0.42	0.42
Phe	0.74	0.69	0.72
Pro	0.54	0.41	0.72
Ser	0.92	0.75	1.31
Thr	0.51	0.48	0.83
Trp	0.21	0.16	0.24
Tyr	0.55	0.45	0.76
Val	0.95	0.84	1.12

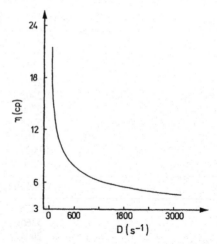

Fig. 11.2. Egg white viscosity, η, as affected by shear rate D, at 10 °C. (according to *Stadelman*, 1977)

11.2.3.1 Proteins

Table 11.4 lists the most important albumen proteins in order of their abundance in egg white.

The carbohydrate moieties of the glycoprotein constituents are presented in Table 11.5. Several albumen proteins have biological activity (Table 11.4), i.e., as enzymes (e. g., lysozyme), enzyme inhibitors (e. g., ovomucoid, ovoinhibitor) and complex-forming agents for some coenzymes (e. g., flavoprotein, avidin). The biological activities may be related to protection of the egg from microbial spoilage. Egg white protein separation is relatively easy: the albumen is treated with an equal volume of saturated ammonium sulfate; the globulin fraction precipitates together with lysozyme, ovomucin and other globulins; while the major portion of the egg white remains in solution. This albumen fraction consists of ovalbumin, conalbumin and ovomucoid. Further separation of these fractions is achieved by ion-exchange chromatography.

shell membrane. The shell membrane is made of two layers (48 and 22 μm, respectively), each an interwoven network of protein polysaccharide fibers. The outer layer adheres closely to the mammillary layer. Tiny pore canals which extend through the shell are seen as minute pores or round openings (7000–17,000 per egg). The cuticle protein partially seals the pores, but they remain permeable to gases while restricting penetration by microorganisms.

11.2.3 Albumen (Egg White)

Albumen is a 10% aqueous solution of various proteins. Other components are present in very low amounts. The thick, gel-like albumen differs from thin albumen (cf. Fig. 11.1) only in its approx. four-fold content of ovomucin. Albumen is a pseudoplastic fluid. Its viscosity depends on shearing force (Fig. 11.2). The surface tension (12.5% solution, pH 7.8, 24 °C) is 49.9 dynes cm^{-1}. The pH of albumen of freshly laid egg is 7.6–7.9 and rises to 9.7 during storage due to diffusion of solubilized CO_2 through the shell. The rise is time and temperature dependent. For example, a pH of 9.4 was recorded after 21 days of storage at 3–35 °C.

11.2.3.1.1 Ovalbumin

This is the main albumen protein, crystallized by *Hofmeister* in 1890. It is a glycophospho-protein with 3.2% carbohydrates (Table 11.5) and 0–2 moles of serine-bound phosphoric acid per mole of protein (ovalbumin components A_3, A_2 and A_1, approx. 3, 12 and 85%, respectively).

Table 11.4. Proteins of egg white

Protein	Percent of the total protein[a]	Denaturation temperature (°C)	Molecular weight (kdal)	Isoelectric point (pH)	Comments
Ovalbumin	54	84.5	44.5	4.5	
Conalbumin (Ovotransferrin)	12	61.5	76	6.1	binds metal ions
Ovomucoid	11	70.0	28	4.1	proteinase inhibitor
Ovomucin	3.5		$5.5\text{--}8.3 \times 10^6$	4.5–5.0	inhibits viral hemagglutination
Lysozyme (Ovoglobulin G$_1$)	3.4	75.0	14.3	10.7	N-acetylmuramidase
Ovoglobulin G$_2$	4	92.5	30–45	5.5 ⎱	good foam builders
Ovoglobulin G$_3$	4			5.8 ⎰	
Flavoprotein	0.8		32	4.0	binds riboflavin
Ovoglycoprotein	1.0		24	3.9	
Ovomacroglobulin	0.5		760–900	4.5	inhibits serine and cysteine proteinases
Ovoinhibitor	1.5		49	5.1	proteinase inhibitor
Avidin	0.05		68.3[b]	9.5	binds biotin
Cystatin (ficin inhibitor)	0.05		12.7	5.1	inhibits cysteine peptidases

[a] Average values are presented.
[b] Four times 15.6 kdal + approx. 10% carbohydrate.

Table 11.5. Carbohydrate composition of some chicken egg white glycoproteins

Protein	Carbo-hydrate (%)	Components (moles/mole protein)				
		Gal	Man	GlcN	GalN	Sialic acid
Ovalbumin	3.2		5	3		
Ovomucoid	23	2	7	23		1
α-Ovomucin[a]	13	21	46	63	6	7
Ovoglyco-protein	31	6	12	19		2
Ovoinhibitor (A)	9.2		10[b]	14		0.2
Avidin[c]	10		4(5)	3		

[a] In addition to carbohydrate, it contains 15 moles of esterified sulfuric acid per mole protein.
[b] Sum of Gal + Man.
[c] Data per subunit (16 kdal).

Ovalbumin consists of a peptide chain with 385 amino acid residues. It has a molecular weight $M_r = 42,699$ and contains four thiol and one disulfide group. The phosphoric acid groups are at Ser-68 and Ser-344. During the storage of eggs, the more heat-stable S-ovalbumin (coagulation temperature 92.5 °C) is formed from the native protein (coagulation temperature 84.5 °C) probably by a thiol-disulfide exchange. The content of S-ovalbumin increases from 5% in fresh eggs to 81% in eggs cold stored for 6 months. The carbohydrate moiety is bound to Asn-292 in the sequence:

–Glu–Lys–Thr–Asn–Leu–Thr–Ser– with a probable structure as follows:

$$(\beta GlcNAc)_{0-1} \rightarrow \alpha Man \rightarrow (\alpha Man)_3 (1 \rightarrow$$

$$\uparrow$$
$$(\alpha Man)_{0-1}$$
$$|$$

$$\rightarrow 3)\beta GlcNAc(1-4)\beta GlcNAc \rightarrow Asn$$
$$\underset{\uparrow}{4} \qquad\qquad\qquad |$$
$$1$$
$$\uparrow$$
$$\beta Man$$
$$\uparrow$$
$$(\beta GlcNAc)_{0-2}$$

$$(11.1)$$

Ovalbumin is relatively readily denatured, for example, by shaking or whipping its aqueous solution. This is an interphase denaturation which occurs through unfolding and aggregation of protein molecules.

11.2.3.1.2 Conalbumin (Ovotransferrin)

Conalbumin and serum transferrin are identical in the chicken. This protein, unlike ovalbumin, is not denatured at the interphase but coagulates at lower temperatures. Conalbumin consists of one peptide chain and contains one oligosaccharide unit made of four mannose and eight N-acetylglucosamine residues.
Binding of metal ions (2 moles Mn^{3+}, Fe^{3+}, Cu^{2+} or Zn^{2+} per mole of protein) at pH 6 or above is a characteristic property of conalbumin. Table 11.6 lists the absorption maxima of several complexes. The occasional red discoloration of egg products during processing originates from a conalbumin-iron complex. The complex is fully dissociated at a pH less than 4. Tyrosine and

Table 11.6. Metal complexes of conalbumin

Metal ion	λ max (nm)	ε ($l\,mol^{-1}\,cm^{-1}$)	Complex color
Fe^{3+}	470	3280	pinkish
Cu^{2+}	440	2500	yellow
	670	350	
Mn^{3+}	429	4000	yellow

histidine residues are involved in metal binding. Alkylation of 10 to 14 histidine residues with bromoacetate or nitration of tyrosine residues with tetranitromethane removes its iron-binding ability. Conalbumin has the ability to retard growth of microorganisms.

11.2.3.1.3 Ovomucoid

Ion-exchange chromatography or electrophoresis reveals 2 or 3 forms of this protein, which apparently differ in their sialic acid contents. The carbohydrate moiety (Table 11.5) consists of three oligosaccharide units bound to protein through asparagine residues. The protein has 9 disulfide bonds and, therefore, stability against heat coagulation. Hence, it can be isolated from the supernatants of heatcoagulated albumen solutions, and then precipitated by ethanol or acetone. Ovomucoid inhibits bovine but not human trypsin activities. The proportion of regular structural elements is high (26% of α-helix, 46% of β-structure, and 10% of β-turn).

11.2.3.1.4 Lysozyme (Ovoglobulin G1)

Lysozyme is widely distributed and is found not only in egg white but in many animal tissues and secretions, in latex exudates of some plants and in some fungi. This protein, with three known components, is an N-acetylmuramidase enzyme that hydrolyzes the cell walls of Gram-positive bacteria (murein; AG = N-acetyl-glucosamine; AMA = N-acetylmuramic acid; $\rightarrow$ = lysozyme attack):

$$\downarrow \qquad\qquad \downarrow$$
$$-\beta AG(1-4)\beta AMS(1-4)\beta AG(1\rightarrow$$
$$|$$
$$Peptide$$

$$\qquad\qquad\qquad \downarrow$$
$$\rightarrow 4)\beta AMS(1-4)\beta AG-$$
$$|$$
$$Peptide$$
$$| \qquad\qquad (11.2)$$

Lysozyme consists of a peptide chain with 129 amino acid residues and four disulfide bonds, Its primary (Table 11.7) and tertiary structures have been elucidated (Fig. 11.3).

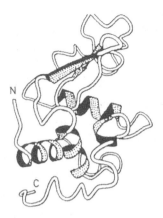

Fig. 11.3. Tertiary structure of lysozyme from chicken egg white (according to *McKenzie* and *White*, 1991)

11.2.3.1.5 Ovoglobulins G2 and G3

These proteins are good foam builders.

11.2.3.1.6 Ovomucin

This protein, of which three components are known, can apparently form fibrillar structures and so contribute to a rise in viscosity of albumen, particularly of the thick, gel-like egg white (see egg structure, Fig. 11.1), where it occurs in a four-fold higher concentration than in fractions of thin albumen.

Ovomucin has been separated into a low-carbohydrate (carbohydrate content ca. 15%) α-fraction and a high-carbohydrate (carbohydrate content ca. 50%) β-fraction. It appears to be associated with polysaccharides. The compositions of its carbohydrate moieties are given in Table 11.5. Ovomucin is heat stable. It forms a water-insoluble complex with lysozyme. The dissociation of the complex is pH dependent. Presumably it is of importance in connection with the thinning of egg white during storage of eggs.

11.2.3.1.7 Flavoprotein

This protein binds firmly with riboflavin and probably functions to facilitate transfer of this coenzyme from blood serum to egg.

11.2.3.1.8 Ovoinhibitor

This protein is, like ovomucoid, a proteinase inhibitor. It inhibits the activities of trypsin, chymotrypsin and some proteinases of microbial origin. Its carbohydrate composition is given in Table 11.5.

11.2.3.1.9 Avidin

Avidin is a basic glycoprotein (Table 11.5). Its amino acid sequence has been determined. Noteworthy is the finding that 15 positions (12% of the total sequence, Table 11.7) are identical with those of lysozyme. Avidin is a tetramer consisting of four identical subunits, each of which binds one mole of biotin. The dissociation constant of the avidin-biotin complex at pH 5.0 is $k_{-1}/k_1 = 1.3 \times 10^{-15}$ mol/l, i.e., it is extremely low. The free energy and free enthalpy of complex formation are $\Delta G = -85$ kJ/mole and $\Delta H = -90$ kJ/mole, respectively. Avidin, in its form in egg white, is practically free of biotin, and presumably fulfills an antibacterial role. Of interest is the occurrence of a related biotin-binding protein (streptavidin) in *Streptomyces* spp., which has antibiotic properties.

11.2.3.1.10 Cystatin (Ficin Inhibitor)

Chicken egg cystatin C consists of one peptide chain with a ca. 120 amino acid residues (M_r 12,700). The two isomers known differ in their isoelectric point (pI 5.6 and pI 6.5) and their immunological properties. This inhibitor inhibits cysteine endopeptidases such as ficin and papain. In fact, cathepsins B, H, and L and dipeptidyl peptidase I are also inhibited.

11.2.3.2 Other Constituents

11.2.3.2.1 Lipids

The lipid content of albumen is negligible (0.03%).

Table 11.7. Amino acid sequences of avidin (1) and lysozyme (2)[a]

1)		Ala	Arg	Lys	*Cys*	Ser	*Leu*	Thr	Gly	Lys	Trp
2)	Lys Val	Phe	Gly	Arg	*Cys*	Glu	*Leu*	Ala	Ala	Ala	Met
1)		Thr	Asn	Asp	Leu	Gly	Ser	*Asn*[b]	Met	Thr	Ile
2)		Lys	Arg	His	Gly	Leu	Asp	*Asn*[b]	Tyr	Arg	Gly
1)		Gly	Ala	Val	Asn	Ser	Arg	Gly	Glu	Phe	Thr
2)		Tyr	Ser	Leu	Gly	Asn	Trp	Val	Cys	Ala	Ala
1)		Gly	Thr	Tyr	Ile	Thr	Ala	Val	*Thr*	Ala	Thr
2)		Lys	Phe	Glu	Ser	Asn	Phe	Asn	*Thr*	Glu	Ala
1)		Ser	*Asn*	Glu	Ile	Lys	Glu	Ser	Pro	Leu	His
2)		Thr	*Asn*	Arg	Asn	Thr	Asp	Gly	Ser	Thr	Asp
1)		Gly	Thr	Glu	Asn	Thr	*Ile*	Asn	Lys	*Arg*	Thr
2)		Tyr	Gly	Ile	Leu	Glu	*Ile*	Asn	Ser	*Arg*	Trp
1)		Gln	Pro	Thr	Phe	*Gly*	Phe	*Thr*	Val	Asn	Trp
2)		Trp	Cys	Asn	Asp	*Gly*	Arg	*Thr*	Pro	Gly	Ser
1)		Lys	Phe	Ser	Glu	Ser	Thr	Thr	Val	Phe	Thr
2)		Arg	Asn	Leu	Cys	Asp	Ile	Pro	Cys	Ser	Ala
1)		Gly	Gln	Cys	Phe	Ile	Asp	Arg	Asn	Gly	Lys
2)		Leu	Leu	Ser	Ser	Asp	Ile	Thr	Ala	Ser	Val
1)		Glu	Val	Leu	*Lys*	Thr	Met	Trp	Leu	Leu	Arg
2)		Asn	Cys	Ala	*Lys*	Lys	Ile	Val	Ser	Asp	Gly
1)		Ser	Ser	Val	*Asn*	Asp	Ile	Gly	Asp	Asp	*Trp*
2)		Asp	Glu	Met	*Asn*	Ala	Trp		Val	Ala	*Trp*
1)		Lys	Ala	Thr	*Arg*	Val	*Gly*	Ile	Asn	Ile	Phe
2)		Arg	Asn	Arg	*Cys*	Lys	*Gly*	Thr	Asp	Val	Gln
1)		Thr	Arg	Leu	*Arg*	Thr	Gln	Lys	Glu		
2)		Ala	Trp	Ile	*Arg*	Gly	Cys	Arg	Leu		

[a] Italics: Identical amino acid in 1 and 2.
[b] Binding site for carbohydrates.

11.2.3.2.2 Carbohydrates

Carbohydrates (approx. 1%) are partly bound to protein (approx. 0.5%) and partly free (0.4–0.5%). Free carbohydrates include glucose (98%) and mannose, galactose, arabi-nose, xylose, ribose and deoxyribose, totaling 0.2–2.0 mg/100 g egg white. There are no free oligosaccharides or polysaccharides. Bound carbohydrates were covered previously with proteins (cf. 11.2.3.1 Table 11.5). Mannose, galactose and glucosamine are predominant, and sialic acid and galactosamine are also present.

11.2.3.2.3 Minerals

The mineral content of egg white is 0.6%. Its composition is listed in Table 11.8.

Table 11.8. Mineral composition of eggs

	Egg white (%)	Egg yolk (%)
Sulfur	0.195	0.016
Phosphorus	0.015–0.03	0.543–0.980
Sodium	0.161–0.169	0.026–0.086
Potassium	0.145–0.167	0.112–0.360
Magnesium	0.009	0.016
Calcium	0.008–0.02	0.121–0.262
Iron	0.0001–0.0002	0.0053–0.011

11.2.3.2.4 Vitamins

Data on vitamins found in egg white are summarized in Table 11.12.

11.2.4 Egg Yolk

Yolk is a fat-in-water emulsion with about 50% dry weight. It consists of 65% lipids, 31% proteins and 4% carbohydrates, vitamins and minerals. The main components of egg yolk are LDL (68%; cf. 3.5.1.2), HDL (16%), livetins (10%) and phosvitins (4%). Water transfer from egg white drops the solid content of the yolk by 2–4% during storage for 1–2 weeks. Egg yolk is a pseudoplastic non-*Newtonian* fluid with a viscosity which depends on the shear forces applied. Its surface tension is $0.044\ \mathrm{Nm}^{-1}$ (25 °C), while its pH is 6.0 and, unlike egg white, increases only slightly (to 6.4–6.9) even after prolonged storage. Yolk contains particles of differing size that can be classified into two groups:

- *Yolk droplets* of highly variable size, with a diameter range of 20–40 μm. They resemble fat droplets, consist mostly of lipids, and some have protein membranes. They are a mixture of lipoproteins with a low density (LDL, cf. 3.5.1.2).
- *Granules* that have a diameter of 1.0–1.3 μm, i.e., they are substantially smaller than yolk droplets, and are more uniform in size but less uniform in shape. They have a substructure and consist of proteins but also contain lipids and minerals.

The first steps in the analysis of the proteins present in egg yolk are orientated towards the method used to classify lipoproteins (cf. 3.5.1.2). First, the granules are separated by the centrifugation of the diluted yolk (Fig. 11.4). They consist of 70% HDL, 12% LDL, which is very similar to the plasma LDL, and 16% phosvitin. After raising the density by adding salt, as shown in Fig. 11.4, the plasma is separated by ultracentrifugation into floating LDL (85% of the plasma), sedimenting γ-livetin and the α- und β-livetins remaining in solution, which are then precipitated with alginate.

Electrophoretic analyses of the lipid-free samples provide an insight into the proteins and apoproteins (lipoproteins after removal of the lipids, e.g., by extraction with acetone) present in egg yolk and its fractions. A relevant experiment, the results of which are presented in Table 11.9, shows 20 protein zones in the molecular weight

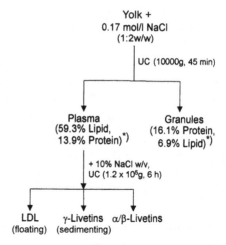

Fig. 11.4. A schematic representation of the fractionation of egg yolk.
UC: ultracentrifuge
*) Numbers: proportions of the yolk dry weight

Table 11.9. Proteins and apoproteins[a] identified in egg yolk, plasma (P) and granules (G) after electrophoretic separation (SDS-PAGE)

MW (kdal)	P/G	RV	S	Protein/Apoprotein
221	P	2.9	2	Apovitellenin VIa
203	P	8.7	1	γ-Livetin + Apovitellenin VI
122	P	7.7	2	Apovitellenin Va
110	G	21.4	1	Apovitellin 3 + 4
93	P	0.6	2	Apovitellenin Vb
85	P	1.6	2	Apovitellenin V
78	G	4.5	1	Apovitellin 5 + 6
73	P	1.5	2	α-Livetin
68	P	3.6	1	Apovitellenin IV
62	P	0.4	1	Apovitellenin IIIa
59	G	1.3		Phosvitin
55	P	10.7	1	α-Livetin/Apovitellenin III
47	G	4.8	2	Apovitellin 7
36	P	2.9	0	β-Livetin
33	P	4.8	0	β-Livetin
31	G	7.6	1	Apovitellin 8
21	P	0.3	2	Apovitellenin IIa
20	P	1.2	2	Apovitellenin II
17	P	9.6	1	Apovitellenin I
5	P	3.3	1	Apolipoprotein CII

[a] The samples were defatted before SDS-PAGE
MW: molecular weight; RV: relative volume of the protein zone; S: stability on heating (egg yolk diluted 1:5 (w/w) with 1% (v/v) NaCl solution was heated at 74 °C for 15 min): 0, thermostable; 1, partial damage; 2, thermolabile

range of 5 to 221 kdal. Fifteen zones which are mainly due to the apovitellins come from the plasma (Table 11.9). The granules are separated into phosvitin and four apovitellins, which appear at the molecular weights 31, 47, 78 and 110 kdal. However, it should be taken into account that the molecular weights of proteins determined using physical methods (e. g., electrophoresis, chromatography) represent only approximate values. The exact values can be calculated only after the amino acid sequence has been determined. The apovitellins 3 and 4 are quantitatively conspicuous. However, the quantities stated in Table 11.9 are only rough estimations because when the electropherogram is dyed, the proteins do not react with the same color yield, e. g., phosvitin is greatly undervalued in Table 11.9. Table 11.9 also indicates the differences in the thermostability of the proteins. On heating, most of the apovitellenins and α-livetin become insoluble and are, consequently, no longer visible in the electropherogram. Some lipoproteins and proteins will now be characterized more closely.

11.2.4.1 Proteins of Granules

11.2.4.1.1 Lipovitellins

The lipovitellin fraction represents high density lipoproteins (HDL). Its lipid moiety is 22% of dry matter and consists of 35% triglycerides, approx. 60% phospholipids and close to 5% cholesterol and cholesterol esters (cf. 3.5.1). The lipovitellins can be separated by electrophoretic and chromatographic methods into their α- and β-components, which differ in their protein-bound phosphorus content (0.39 and 0.19% P, respectively). α-Lipovitellin consists of two polypeptide chains (M_r 111,000 and 85,000), but β-lipovitellin has only one chain (M_r 110,000). The vitellins are covalently bound to oligosaccharides made up of mannose, galactose, glucosamine and sialic acid. The stronger acidic character of α-lipovitellin is based not only on the higher phosphoric acid content, but also on the higher content of sialic acid. The two lipovitellins form a quaternary structure (M_r 420,000), which decomposes into subunits above pH 9. The amino acid composition is shown in Table 11.10. In the yolk, lipovitellins are present as a complex with

Table 11.10. Amino acid composition of phosvitin and α- and β-lipovitellins (mole %)

Amino acid	Phosvitin[a]	α-Lipovitellin	β-Lipovitellin
Gly	2.7	5.0	4.6
Ala	3.6	8.0	7.5
Val	1.3	6.2	6.6
Leu	1.3	9.2	9.0
Ile	0.9	5.6	6.2
Pro	1.3	5.5	5.5
Phe	0.9	3.2	3.3
Tyr	0.5	3.3	3.0
Trp	0.5	0.8	0.8
Ser	54.5	9.0	9.0
Thr	2.2	5.2	5.6
Cys	0.0	2.1	1.9
Met	0.5	2.6	2.6
Asx	6.2	9.6	9.3
Glx	5.8	11.4	11.6
His	4.9	2.2	2.0
Lys	7.6	5.7	5.9
Arg	5.3	5.4	5.6

phosvitin, with about two phosvitin molecules (M_r 32,000) for each lipovitellin molecule (M_r 420,000). The lipovitellins are heat stable. However, they lose this property if the lipids are separated.

11.2.4.1.2 Phosvitin

Phosvitin is a glycophosphoprotein with an exceptionally high amount of phosphoric acid bound to serine residues. For this reason, it behaves like a polyelectrolyte (polyanion) in aqueous solution. On electrophoresis, two components are obtained, α- and β-phosvitin, which are protein aggregates with molecular weights of 160,000 and 190,000. In the presence of sodium dodecyl sulfate, α-phosvitin dissociates into three different subunits (M_r = 37,500, 42,500 and 45,000) and β-phosvitin into only one subunit (M_r = 45,000). Its amino acid composition is given in Table 11.10. The partial specific volume of 0.545 ml/g is very low, probably due to the large repulsive charges of the molecule. The frictional ratio suggests the presence of a long, mostly stretched molecular form.

A partial review of its amino acid sequence shows that sequences of 6–8 phosphoserine residues, interrupted by basic and other amino acid residues, are typical of this protein:

$$\cdots Asp—(Pse)_6—Arg—Asp\cdots$$
$$\cdots His—His—Arg—(Pse)_6—Arg—His—Lys\cdots$$
$$(11.3)$$

The carbohydrate moiety is a branched oligosaccharide, consisting of mannose (3 residues), galactose (also 3 residues), N-acetylglucosamine (5) and N-acetylneuraminic acid (2). The oligosaccharide is bound by an N-glycosidic linkage to asparagine. The amino acid sequence in the vicinity of the linkage position is shown in Formula 11.4.

$$\cdots Ser—Asn^{169}—Ser—Gly—(Pse)_8—Arg—Ser—$$
$$|$$

    Carbohydrate

$$—Val—Ser—His—His\cdots \qquad (11.4)$$

Phosvitin is relatively heat stable. In fact, no changes can be electrophoretically detected after 10 minutes at 110 °C. Phosphate is eliminated at 140 °C. Coagulating egg yolk is frequently enclosed in coagulates of other proteins.

Phosvitin very strongly binds multivalent cations, the type of metal and the pH being of importance. The iron in eggs, present as $Fe^{3\oplus}$, is bound to an extent of 95% to phosvitin and so strongly that its availability for nutrition is greatly limited. The $Fe^{3\oplus}$ complex is monomeric and phosvitin is saturated with iron at a molar ratio of Fe/P = 0.5; this strongly suggests formation of a chelate complex involving two phosphate groups from the same peptide chain per iron. Since phosvitin traps iron and other heavy metal ions, it can synergistically support antioxidants.

11.2.4.2 Plasma Proteins

11.2.4.2.1 Lipovitellenin

Lipovitellenin is obtained as a floating, low density lipoprotein (LDL) by ultracentrifugation of diluted yolk. Several components with varying densities can be separated by fractional centrifugation. The lipid moiety represents 84–90% of the dry matter and consists of 74% triglycerides and 26% phospholipids. The latter contain predominantly phosphatidyl choline (approx. 75%), phosphatidyl ethanolamine (approx. 18%) as well as sphingomyelin and lysophospholipids (approx. 8%). Eleven bands appear on electrophoresis of the apoproteins (Table 11.9).

11.2.4.2.2 Livetin

The water-soluble globular protein fraction can be separated electrophoretically into α-, β- and γ-livetins (Table 11.9). These have been proven to correspond to chicken blood serum proteins, i.e. serum albumin, α_2-glycoprotein and γ-globulin.

11.2.4.3 Lipids

Egg yolk contains 32.6% of lipid whose composition is given in Table 11.11. These lipids occur as the lipoproteins described above and, as such, are closely associated with the proteins occurring in yolk.

The fatty acid composition of the lipids depends on that of the feed (Table 11.12). However, the extent to which individual fatty acids are incorporated varies greatly. The addition of fats rich in linoleic acid to the feed, e.g., soy oil, leads to a great increase in this fatty acid. In comparison, only traces of the main fatty acid 10:0 of coconut oil is recovered (Table 11.12). Highly unsaturated ω-3-fatty acids (20:5, 22:6) from fish oils do appear in egg lipids, but not in proportion to their

Table 11.11. Egg yolk lipids

Lipid fraction	a	b
Triacylglycerols	66	
Phospholipids	28	
Phosphatidyl choline		73
Phosphatidyl ethanolamine		15.5
Lysophosphatidyl choline		5.8
Sphingomyelin		2.5
Lysophosphatidyl ethanolamine		2.1
Plasmalogen		0.9
Phosphatidyl inositol		0.6
Cholesterol, cholesterol esters and other compounds	6	

[a] As percent of total lipids.
[b] As percent of phospholipid fraction.

Table 11.12. Fatty acid composition of the lipids in egg yolk – Influence of feed[a]

Fatty acid	Fish oil (3%)[b]		Soy oil (12%)		Coconut oil (10%)	
	I	II	I	II	I	II
10:0	–	–	–	–	7.9	–
12:0	–	–	–	–	40.0	1.0
14:0	–	–	0.1	0.2	18.5	7.5
16:0	23.0	26.8	11.4	24.0	12.5	25.5
16:1	7.9	4.9	–	1.6	–	4.6
18:0	7.0	18.2	5.0	8.6	3.6	8.1
18:1	15.8	31.7	24.5	38.1	10.9	39.3
18:2	4.9	11.3	50.3	33.1	4.9	9.0
18:3	4.4	0.4	7.5	1.4	0.3	0.2
20:4	4.9	1.3	–	1.2	–	–
20:5	9.5	0.4	–	–	–	–
20:1	–	–	0.6	0.2	–	–
22:4	6.2	0.2	–	–	–	–
22:5	7.5	0.5	–	–	–	–
22:6	8.0	4.1	–	0.5	–	–

[a] Fatty acid distribution (weight %) in the lipids of the feed (I) and in the yolk (II).
[b] Fat content in the feed.

content in the feed. This also applies to 22:5 (Table 11.12). Furthermore, it has been observed that the fatty acid pattern of the feed is reflected more clearly in the triglyceride fraction of egg lipids than in the polar lipids.

About 4% of the egg lipids consists of sterols. The main component is cholesterol (96%), ca. 15% of which is esterified with fatty acids. The cholesterol content is 2.5%, based on the egg yolk solids. Disregarding mammalian brain, this level exceeds by far that in all other foods (cf. 3.8.2.2.1) and, therefore, serves as an indicator of the addition of eggs. Cholestanol, 7-cholestenol, campesterol, β-sitosterol, 24-methylene cholesterol and lanosterol are other components of the sterol fraction. The quality of egg products is endangered by autoxidation of cholesterol (cf. 3.8.2.2.1).

11.2.4.4 Other Constituents

11.2.4.4.1 Carbohydrates

Egg yolk carbohydrates are about 1% of the dry matter, with 0.2% bound to proteins. The free carbohydrates present in addition to glucose are the same monosaccharides identified in egg white (cf. 11.2.3.2.2).

11.2.4.4.2 Minerals

The minerals in egg yolk are listed in Table 11.8.

11.2.4.4.3 Vitamins

The vitamins in egg yolk are presented in Table 11.13.

Table 11.13. Vitamin content of whole egg, egg white and yolk (mg/100 g edible portion)

Vitamin	Whole egg	Egg white	Egg yolk
Retinol (A)	0.22	0	1.12
Thiamine	0.11	0.022	0.29
Riboflavin	0.30	0.27	0.44
Niacin	0.1	0.1	0.065
Pyridoxine (B$_6$)	0.08	0.012	0.3
Pantothenic acid	1.59	0.14	3.72
Biotin	0.025	0.007	0.053
Folic acid	0.051	0.009	0.15
Tocopherols	2.3	0	6.5
α-Tocopherol	1.9		5.4
Vitamin D	0.003		0.0056
Vitamin K	0.009		

11.2.4.4.4 Aroma Substances

The typical aroma substances of egg white and egg yolk are still unknown. The "fishy" aroma defect that can occur in eggs is caused by trimethylamine TMA, which has an odor threshold that depends on the pH (25 μg/kg, pH 7.9) because only the undissociated form is odor active. TMA is formed by the microbial degradation of choline, e. g., on feeding fish meal or soy meal. Normally, TMA does not interfere because it is enzymatically oxidized to odorless TMA oxide. However, in feed, e. g., soy meal, substances exist which could inhibit this reaction.

11.2.4.4.5 Colorants

The color of the yolk, which is produced by carotenoids in the feed, is considered to be a quality characteristic. Normally, xanthophylls (cf. 3.8.4.1.2) are absorbed from the feed, preferably lutein, followed by luteinmono- and diester, 3′-oxolutein and zeaxanthin. The color of the yolk can be intensified by the appropriate feed composition. The substances available dissolved in an oil are, e. g., β-*apo*-8′-carotene ethyl ester, citranaxanthin (5′,6′-dihydro-5′-*apo*-β-carotene-6′-one, Formula 11.5) and canthaxanthin.

$$(11.5)$$

11.3 Storage of Eggs

A series of changes occurs in eggs during storage. The diffusion of CO_2 through the pores of the shell, which starts soon after the egg is laid, causes a sharp rise in pH, especially in egg white. The gradual evaporation of water through the shell causes a decrease in density (initially approx. $1.086\,g/cm^3$; the daily reduction coefficient is about $0.0017\,g/cm^3$) and the air cell enlarges. The viscosity of the egg white drops. The yolk is compact and upright in a fresh egg, but it flattens during storage. After the egg is cracked and the contents are released onto a level surface, this flattening is expressed as

yolk index, the ratio of yolk height to diameter. Furthermore, the vitellin membrane of the yolk becomes rigid and tears readily once the egg is opened. Of importance for egg processing is the fact that several properties change, such as egg white whipping behavior and foam stability. In addition, a "stale" flavor develops.

These changes are used for determination of the age of an egg, e. g., in the floating test (change in egg density), flash candling (egg yolk form and position), egg white viscosity test, measurement of air cell size, refractive index, and sensory assay of the "stale" flavor (performed mostly with softboiled eggs). The lower the storage temperature and the lower the losses of CO_2 and water, the lower the quality loss during storage of eggs. Therefore, cold storage is an important part of egg preservation. A temperature of 0 to $-1.5\,°C$ (common chilled storage or subcooling at $-1.5\,°C$) and a relative humidity of 85–90% are generally used. A coating (oiling) of the shell surface with light paraffin-base mineral oil quite efficiently retards CO_2 and vapor escape, but a tangible benefit is derived only if oil is applied soon (1 h) after laying, since at this time the CO_2 loss is the highest. Controlled atmosphere storage of eggs (air or nitrogen with up to 45% CO_2) has been shown to be a beneficial form of egg preservation. Cold storage preserves eggs for 6–9 months, with a particularly increased shelf life with subcooled storage at $-1.5\,°C$. Egg weight loss is 3.0–6.5% during storage.

Microbial spoilage is indicated by an increase in lactic acid and succinic acid to values above $1\,g/kg$ and $25\,mg/kg$ of dry matter respectively. 3-Hydroxybutyric acid serves as an indicator of fertilized eggs ($>10\,mg/kg$ of egg mass).

11.4 Egg Products

11.4.1 General Outline

Egg products, in liquid, frozen or dried forms, are made from whole eggs, white or yolk. They are utilized further as semi-end products in the manufacturing of baked goods, noodles, confectionery, pastry products, mayonnaise and other salad dressings, soup powders, margarine, meat products, ice creams and egg liqueurs. Figure 11.5 gives an overview of the main

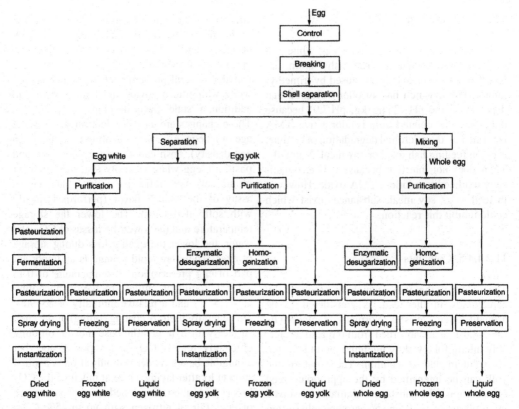

Fig. 11.5. Schematic presentation of the production of egg products

processing steps involved in manufacturing of egg products.

11.4.2 Technically-Important Properties

The many uses of egg products are basically a result of three properties of eggs: coagulation when heated; foaming ability (whippability); and emulsifying properties. The coloring ability and aroma of egg should also be mentioned.

11.4.2.1 Thermal Coagulation

Egg white begins to coagulate at 62 °C and egg yolk at 65 °C. The coagulation temperature is influenced by pH. At a pH at or above 11.9 egg white gels or sets even at room temperature, though after a while the gel liquiefies. All egg proteins coagulate, except ovomucoid and phosvitin. Conalbumin is particularly sensitive, but can be stabilized by complexing it with metal ions. Due to their ability to coagulate, egg products are important food-binding agents.

11.4.2.2 Foaming Ability

11.4.2.2.1 Egg White

Whipping of egg white builds a foam which entraps air and hence is used as a leavening agent in many food products (baked goods, angel cakes, biscuits, soufflés, etc.).

Due to a large surface area increase in the liquid/air interphase, proteins denature and aggregate during whipping. In particular, ovomucin forms a film of insoluble material between the liquid lamella and air bubble, thereby stabilizing the foam. Egg globulin also contributes to this effect by increasing the fluid viscosity and by decreasing the surface tension, both effects of importance in the initial stage of the whipping process. In angel cake, egg white without ovomucin and globulins leads to long whipping times and cakes with reduced volumes. An excessive ovomucin content decreases the elasticity of the ovomucin film and thus decreases the thermal stability (expansion of air bubbles) of the foam. The stability of the foam is increased by polymers of conalbumin and ovalbumin, which are stable to sodium dodecyl sulfate and 2-mercaptoethanol.

The whipping properties of dried egg white can be improved by the addition of whey proteins, casein and bovine serum albumin. The foaming ability is also increased by weak proteolysis and partial hydrolysis of the oligosaccharides in the glycoproteins by treatment with amylases.

The whippability of egg white can be assayed by measurement of foam volume and foam stability (amount of liquid released from the foam in a given time).

Small amounts of yolk (0.1%) considerably reduce the foam formation.

11.4.2.2.2 Egg Yolk

Egg yolk can be whipped into stable foam only at higher temperatures (optimum 72 °C), the volume increasing about six fold in the process. Above the critical temperature, the volume falls and the proteins coagulate. The protein coagulation is prevented by reducing the pH value, e. g., by the addition of acetic acid. This effect is used in the production of highly stable sauces.

11.4.2.3 Emulsifying Effect

The emulsifying effect of whole egg or egg yolk alone is utilized, for example, in the production of mayonnaise and of creamy salad dressings (cf. 14.4.6). Phospholipids, LDLs and proteins are responsible for the emulsifying action of eggs.

11.4.3 Dried Products

The eggs are stored at 15 °C for up to 2 days because the content of the egg can be easily separated from the shell at this temperature. In some countries, eggs are disinfected with an aqueous chlorine solution (200 mg/l) before they are broken open.

The liquid content of eggs is mixed or churned either immediately or only after egg white and yolk separation. This homogenization is followed by a purification step using centrifuges (separators), and then by a pasteurization step (Fig. 11.5).

Since the egg white coagulates at 55 °C and the yolk and the whole egg coagulate between 62 °C and 65 °C, pasteurization requires lower temperatures than those used for milk (cf. 10.1.3.3), e. g., 52 °C/7 min for the egg white, 62 °C/6 min for the yolk and 64.5 °C/6 min for the whole egg.

The sugars are removed prior to egg drying to prevent reaction between amino compoents (proteins, phosphatidyl ethanolamines) and reducing sugars (glucose), thereby avoiding undesired brown discoloration and faulty aroma.

Glucose is removed from egg white after pasteurization (cf. 11.4.5), usually by microbiological sugar fermentation. The pH of the pasteurized egg liquid is shifted from 9.0–9.3 to pH 7.0–7.5 using citric or lactic acid, and then is incubated at 30–33 °C with suitable microorganisms (*Streptococcus* spp., *Aerobacter* spp.). The sugar in whole egg homogenate or yolk is removed in part by yeasts (e. g. *Saccharomyces cerevisiae*) or mainly by glucoseoxidase/catalase enzymes (cf. 2.7.2.1.1 and 2.7.2.1.2), which oxidize glucose to gluconic acid. Addition of hydrogen peroxide releases oxygen and accelerates the process.

Spray drying is the most important egg drying process. The yolk, which has a relatively high solids content, is dried directly. The egg white and the whole egg are concentrated from 11 to 18% and from 24 to 32% solids respectivey using membrane filtration, which saves energy in the drying process. The whole egg can also be concentrated by film boiling in vacuum. After heating to 45–50 °C, egg white is usually dried by high-pressure dispersion in a stream of air at 165 °C. In this process, the egg white heats up to 50–60 °C. The product is then stored in heat-maintaining rooms (post-pasteurization) for at least 7 days at 55 °C to kill pathogenic microbes.

The whole egg and egg yolk are brought to 64.5 and 63 °C respectively to reduce germs, fol-

Table 11.14. Composition of dried egg products (values in %)

Constituent	Whole egg	Egg white	Egg yolk
Moisture[a]	5.0	8.0	5.0
Fat[b]	40.0	0.12	57.0
Protein[b]	45.0	80.0	30.0
Ash	3.7	5.7	3.4

[a] Maximum values.
[b] Minimum values.

lowed by spray drying at high pressure or with a centrifugal atomizer. The temperatures of the hot air blown in are 185 °C (whole egg) or 165 °C (yolk). The temperatures of the air drop to 50–60 °C in the drying process, which results in a water content of 4–5%. The products are quickly cooled in a cold air stream to 25–30 °C to prevent lipid peroxidation. There is no post-pasteurization. Other egg drying processes, e. g., freeze drying, are hardly ever applied commercially.

Dried instant powder can be made in the usual way: rewetting and additionally drying the agglomerated particles. Egg white agglomerization is facilitated by addition of sugar (sucrose or lactose).

The shelf life of dried egg white is essentially unlimited. Whole egg powder devoid of sugar has a shelf life of approx. 1 year at room temperature, while sugarless yolk lasts 8 months at 20–24 °C and more than a year in cold storage. The shelf-life of powders containing egg yolk is limited by aroma defects which develop gradually from oxidation of yolk fat. The compositions of dried egg products are given in Table 11.14.

11.4.4 Frozen Egg Products

The eggs are pretreated as described above (cf. 11.4.3 and Fig. 11.5). The homogenate is pasteurized at 63 °C for 1 min (cf. 11.4.5) to lower the germ count and is then frozen quickly at −40 °C. The shelf-life of the frozen eggs is up to 12 months at a storage temperature of −15 to −18 °C.

Frozen egg white thickens negligibly after thawing, while the viscosity of egg yolk rises

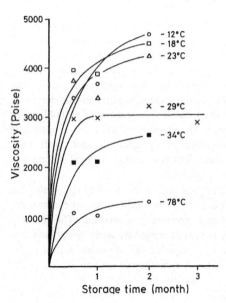

Fig. 11.6. Egg yolk viscosity after frozen storage. (According to *Palmer* et al., 1970)

irreversibly when freezing and storage temperatures are below −6 °C (Fig. 11.6). The egg yolk has a gel-like consistency after thawing, which hampers further utilization by dosage metering or mixing. Thawed whole egg gels can cause similar problems, but to a lesser extent than yolk. Pretreatment of yolk with proteolytic enzymes, such as papain, and with phospholipase A prevents gel formation. Mechanical treatments after thawing of yolk can result in a drop in viscosity. Gel formation can also be prevented by adding 2–10% common salt or 8–10% sucrose to egg yolk, cf. Fig. 11.7. Good results are also obtained with a solution containing glucose and fructose in the ratio of 45:55. The egg yolk is diluted with 70.3% and the whole egg with 45.2% of this solution. Although salted and sugar-sweetened yolk is of limited acceptability to some manufacturers, this process is of great importance.

The consistency of the frozen egg products is influenced by the temperature gradients during freezing and thawing, and also by storage duration and temperature. Rapid freezing and thawing are best.

The molecular events leading to gel formation by freezing are poorly understood. Apparently, the formation of ice crystals causes a partial dehydration of protein, coupled with a rearrangement of

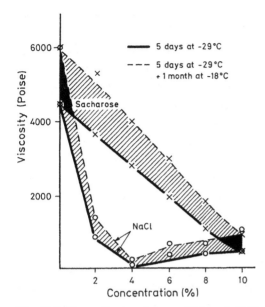

Fig. 11.7. Egg yolk viscosity on addition of NaCl or saccharose and after frozen storage. (according to *Palmer* et al., 1970)

Table 11.15. Composition of frozen and liquid egg products (values in %)

Constituent	Whole egg	Egg white	Egg yolk
Moisture	75.3	88.0	57.0
Fat	11	<0.03[a]	27.2
Protein	12	10.5	13.5
Reducing sugars	0.7	0.8	0.7

[a] Proportion of egg yolk (weight-%).

lipoprotein. This probably induces formation of entangled protein strands.

The whippability of egg white can be enhanced by various additives, such as glycerol, starch syrup and triethyl citrate. Typical compositional data for frozen egg products are provided in Table 11.15.

11.4.5 Liquid Egg Products

Eggs are pretreated as described earlier (cf. 11.4.3 and Fig. 11.5). Despite sanitary conditions at plants, eggs cannot be entirely protected from microorganisms. Pasteurization is difficult due to the heat sensitivity of egg protein and the need to kill the pathogens under specific conditions. It is especially important to eliminate *Salmonella* spp., which have varying resistances to heat. The most resistant are *S. senftenberg, S. oranienburg* and *S. paratyphi B*. Inactivation of α-amylase occurs as the temperature lethal to *S. senftenberg* is approached; hence, this enzyme can be used as an indicator to monitor the adequacy of the heat treatment. The heating conditions differ for different liquid egg products (cf. 11.4.3).

Most of the egg white proteins are relatively stable at pH 7, so normal pasteurization conditions do not negatively affect processing properties such as whippability. An exception is conalbumin, but addition of metal ions (e. g. Al-lactate) can stabilize even this protein. Addition of Na-hexametaphosphate can also improve the stability of conalbumin.

Pasteurized liquid egg products are generally also preserved by chemical means, e. g., addition of sorbic or benzoic acid.

The compositions of liquid egg products are presented in Table 11.15.

11.5 References

Chang, P.K., Powrie, W.D., Fennema, O.: Effect of heat treatment on viscosity of yolk. J. Food Sci. *35*, 864 (1970)

Green, N.M.: Avidin. Adv. Protein Chem. *29*, 85 (1975)

Guilmineau, F., Krause, I., Kulozik, U.: Efficient analysis of egg yolk proteins and their thermal sensitivity using sodium dodecyl sulfate polyacrylamide gel electrophoresis under reducing and nonreducing conditions. J. Agric. Food Chem. *53*, 9329 (2005)

Janssen, H.J.L.: Neuere Entwicklungen bei der Fabrikation von Eiprodukten. Alimenta *10*, 121 (1971)

Kovacs-Nolan, J., Phillips, M., Mine, Y.: Advances in the value of eggs and egg components for human health. Review. J. Agric. Food Chem. *53*, 8421 (2005)

Li-Chan, E.C.Y., Powrie, W.D., Nakai, S.: The chemistry of eggs and egg products. In Egg Science and Technology (eds.: W.J. Stadelman, O.J. Cotterill), 4th edition, Food Products Press, New York, 1995, pp. 105

Maga, J.A.: Egg and egg product flavour. J. Agric. Food Chem. *30*, 9 (1982)

McKenzie, H.A., White jr., F.H.: Lysozyme and α-lactalbumin: Structure, function, and interrelationships. Adv. Protein Chem. *41*, 174 (1991)

Palmer, H.H., Ijichi, K., Roff, H.: Partial thermal reversal of gelation in thawed egg yolk products. J. Food Sci. *35*, 403 (1970)

Spiro, R.G.: Glycoproteins. Adv. Protein Chem. *27*, 349 (1973)

Stadelmann, W.J., Cotterill, O.J. (Eds.): Egg science and technology. 2nd edn., AVI Publ. Cp.: Westport, Conn, 1977

Taborsky, G.: Phosphoproteins. Adv. Protein Chem. *28*, 1 (1974)

Ternes, W., Acker, L., Scholtyssek, S.: Ei und Eiprodukte, Verlag Paul Parey, Berlin, 1994

Tunmann, P., Silberzahn, H.: Über die Kohlenhydrate im Hühnerei. I. Freie Kohlenhydrate. Z. Lebensm. Forsch. *115*, 121 (1961)

12 Meat

12.1 Foreword

Much evidence from many civilizations has verified that the meat of wild and domesticated animals has played a significant role in human nutrition since ancient times. In addition to the skeletal muscle of warm-blooded animals, which in a strict sense is "meat", other parts are also used: fat tissue, some internal organs and blood. Definitions of the term "meat" can vary greatly, corresponding to the intended purpose. From the aspect of food legislation for instance, the term meat includes all the parts of warm-blooded animals, in fresh or processed form, which are suitable for human consumption. In the colloquial language the term meat means skeletal muscle tissue containing more-or-less adhering fat. Some data concerning meat production and consumption are compiled in Tables 12.1–12.3.

Table 12.1. World meat production in 2006 (1000 t)[a]

Continent	(Beef/Veal) Cattle meat	Buffalo	(Mutton/Lamb) Sheep meat	Goat
World	61,033	3183	8633	4945
Africa	4565	270	1167	932
America, Central-	1992	—	49	43
America, North-	13,301	—	101	21
America, South- and Caribbean	15,672	—	293	138
Asia	13,664	2911	4653	3706
Europe	11,033	1	1294	128
Oceania	2797	—	1126	20

Continent	Pork	Horse	Chicken meat	Duck meat
World	105,604	772	73,057	3846
Africa	838	11	3472	57
America, Central-	1225	84	3103	20
America, North-	11,448	44	16,941	93
America, South- and Caribbean	6011	191	16,567	37
Asia	62,013	333	24,226	3226
Europe	24,767	169	10,908	423
Oceania	526	22	944	10

Continent	Meat, grand total
World	272,884
Africa	12,528
America, Central-	6542
America, North-	45,574
America, South- and Caribbean	39,628
Asia	118,103
Europe	51,204
Oceania	5846

Table 12.1. Continuation 1

Country	Beef/Veal	Country	Buffalo	Country	(Mutton/Lamb) Sheep meat
USA	11,910	India	1488	China	2540
Brazil	7774	Pakistan	571	Australia	626
China	7173	China	351	New Zealand	500
Argentina	2980	Egypt	270	Iran, Islamic Rep of	389
Australia	2077	Nepal	142	United Kingdom	331
Russian Fed.	1755	Philippines	70	Turkey	272
Mexico	1602	Thailand	63	India	239
France	1473	Indonesia	40	Spain	227
Canada	1391	Myanmar	24	Syrian Arab Republic	200
India	1334	Iran	14	Algeria	185
Germany	1167	$\Sigma(\%)^b$	95	Pakistan	172
Italy	1109			Sudan	148
South Africa	804			Russian Federation	136
Colombia	792			South Africa	117
UK	762			Morocco	112
New Zealand	700			Kazakhstan	106
Spain	671			Nigeria	103
Ukraine	592			France	99
$\Sigma(\%)^b$	75			$\Sigma(\%)^b$	75

Country	Goat	Country	Pork (swine)	Country	Horse
China	2161	China	52,927	China	200
India	475	USA	9550	Mexico	79
Pakistan	392	Germany	4500	Russian Fed.	56
Sudan	186	Spain	3230	Kazakhstan	56
Nigeria	147	Brazil	3140	Argentina	56
Bangladesh	137	Vietnam	2446	Italy	41
Iran	105	Poland	2092	Mongolia	41
Greece	57	France	2011	USA	26
Indonesia	53	Canada	1898	Australia	21
Mali	51	Denmark	1749	Brazil	21
$\Sigma(\%)^b$	75	$\Sigma(\%)^b$	75	$\Sigma(\%)^b$	75

[a] Slaughtering in each region is independent of the origin of the animals.
[b] World production = 100%

12.2 Structure of Muscle Tissue

12.2.1 Skeletal Muscle

Skeletal muscle (Fig. 12.1) consists of long, thin, parallel cells arranged into fiber bundles. Each of these muscle fibers exists as a separate entity surrounded by connective tissue, the endomysium. Numbers of these primary muscle fibers are held together in a bundle which is surrounded by a larger sheet of thin connective tissue, the perimysium. Many such primary bundles are then held together and wrapped by an outer, large, thick layer of connective tissue called the epimysium. Figure 12.2 shows a cross-section of rabbit *Psoas major* muscle in which the endomysium and perimysium are readily recognized.

Table 12.1. Continuation 2

Country	Chicken meat	Country	Duck meat	Country	Meat, grand total
USA	15,945	China	2673	China	81,733
China	10,701	France	233	USA	41,081
Brazil	8507	Malaysia	105	Brazil	19,783
Mexico	2411	Viet Nam	86	Germany	6868
India	2000	USA	86	India	6103
Russian Fed.	1534	Thailand	85	Mexico	5331
Japan	1337	Korea, Rep.	67	Spain	5293
Indonesia	1333	India	65	France	5206
UK	1331	Myanmar	60	Russian Fed.	5153
Argentina	1156	UK	42	Argentina	4537
Iran	1153			Canada	4493
Thailand	1100	$\Sigma(\%)^b$	91	Australia	3941
Spain	1048			Italy	3915
Canada	997			Poland	3490
South Africa	971			UK	3389
Poland	960			Viet Nam	3151
Turkey	937				
Malaysia	914			$\Sigma (\%)^b$	75
France	819				
$\Sigma (\%)^b$	75				

[a] Data refer to slaughtered animals irrespective of the possibility of being imported as live animals.
[b] World production $= 100\%$

Table 12.2. Annual meat consumption in FR Germany (kg/person)

Year	Beef/Veal	Pork	Poultry	Total
1960	18.7	29.3	4.2	59.0
1964/65	19.2	33.9	6.0	66.5
1970	22.9	36.1	7.4	72.0
1972/73	20.5	42.0	9.0	79.0
1974/75	21.0	44.6	8.8	82.5
1976/77	21.6	45.5	9.1	84.9
1993	19.7	56.1	12.4	95.2
1995	16.6	54.9	13.4	92.0
1997	14.5	53.8	14.7	89.9
2003	12.8	55.1	18.2	90.7

Table 12.3. Meat consumption in selected countries (kg/person/year)

Country/ region	Year	Beef/ Veal	Pork	Poultry	Total
EEC (European Economic Community)	1960	19.9	19.2	5.2	52.2
	1970	25.2	23.7	8.9	65.7
EU-15	1995	20.1	41.0	20.0	93.2
	1997	19.0	40.8	20.7	92.3
	2003	20.0	43.4	23.4	96.6
France	1960	29.2	19.8	8.6	74.9
	1970	35.9	24.8	11.3	89.3
	1995	28.1	35.9	22.6	107.9
	1997	26.7	35.4	23.9	106.5
	2003	27.1	36.4	24.5	109.6
Italy	1960	12.9	7.2	3.6	30.0
	1970	18.9	9.1	9.2	43.5
	1995	25.9	33.1	18.4	88.7
	1997	24.2	34.4	18.6	88.2
	2003	25.1	39.4	18.0	91.5
UK	1995	17.5	23.1	25.8	133.2
	1997	16.6	23.3	26.6	128.1
	2003	20.0	22.1	28.5	83.0
USA	2004	30.0	22.3	15.3	100.4

The membrane surrounding each individual muscle fiber is called the sarcolemma (thickness ca. 75 nm). It consists of three layers: the endomysium, a middle amorphous layer and an inner plasma membrane. Muscle fibers are polynuclear cells. The nuclei are surrounded by the sarcoplasm and by other cellular elements (mitochondria, sarcoplasmic reticulum, lyso-

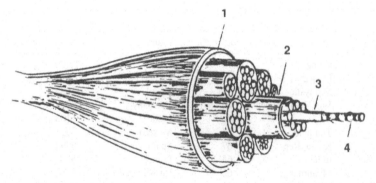

Fig. 12.1. The structural element of skeletal muscle (according to *Gault*, 1992). *1* epimysium, *2* perimysium, *3* endomysium, *4* muscle fiber

Fig. 12.2. A cross-section of *M. psoas* rabbit muscle. (From *Schultz, Anglemier*, 1964)

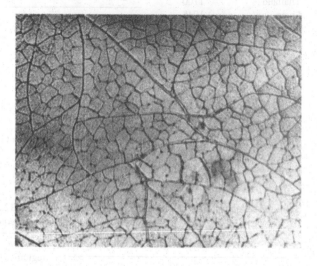

somes). Under aerobic conditions, the bulk of the cellular energy is produced in the form of ATP in the mitochondria. The lysosomes are the source of the endopeptidases which participate in the aging of meat (cf. 12.4.3). The muscle fibers or muscle cells have a diameter of 0.01 to 0.1 mm and attain a length of 150 mm and more.

The main components of muscle cells are the myofibrils, each of which has a diameter of 1–2 µm. Up to 1000 myofibrils arranged parallel to each other stretch through the muscle cell in the direction of the fiber.

White muscle (birds, poultry), which has a high ratio of myofibrils to sarcoplasm, contracts rapidly but tires quickly. It can be distinguished from red muscle, which is poor in myofibrils but rich in sarcoplasm. Red muscles are used for slow, long-lasting contractions and do not tire quickly. Figure 12.3 shows a cross-section

of a muscle fiber with numerous myofibrils. A greatly magnified, oblique view of a fiber of this type is presented in Fig. 12.4 and Fig. 12.5 shows separated myofibrils.

The organization of the muscle contractile apparatus is revealed in a longitudinal section of the muscle fiber. The characteristic crossbondings ("striations"; Fig. 12.6) of skeletal muscle are due to the regularly overlapping anisotropic A bands, which double refract polarized light, and the isotropic I bands. The dark bands, the Z line, are in the middle of the light I bands and perpendicular to the axis of the fiber. The dark A bands are crossed in the middle with light H bands, while the dark M line is situated in the middle of the H bands (Fig. 12.7). A single contractile unit of a myofibril, called the sarcomere, stretches from one Z line to the next and consists of thick and thin filaments. Figure 12.8a shows

Fig. 12.3. A cross-section of a muscle fiber. (from *Schultz, Anglemier*, 1964)

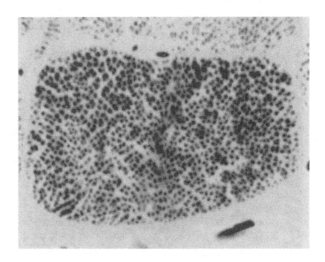

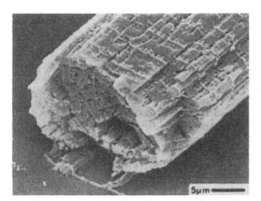

Fig. 12.4. An oblique view of a fractured muscle fiber; scanning electron microscopy at −180 °C (*Sargent*, 1988)

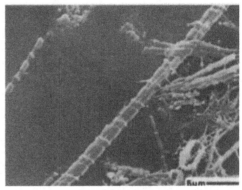

Fig. 12.5. Separated myofibrils; scanning electron microscopy at −180 °C (*Sargent*, 1988)

a schematic representation of the structure of a sarcomere derived from Fig. 12.6 and 12.7. The thick filaments are formed from the protein myosin. They stretch through the entire A band and are fixed in a hexagonal arrangement by the bulge at the center (M line) (Fig. 12.8 a, IV). Thin filaments consist mainly of actin. They original from the Z line, pass across the I band and between the thick filaments, and penetrate the A bands (Figs. 12.7 and 12.8). During muscle contraction, the mechanism of which is explained in section 12.3.2.1.5, the thick filaments penetrate into the H zones and the Z lines move closer to each other. Thus, the width of the I band gradually decreases and finally disappears. Figure 12.8b schematically presents these changes which take place during muscle contraction.

12.2.2 Heart Muscle

The structure of heart muscle is similar to striated skeletal muscle but has significantly more mitochondria and sarcoplasm.

12.2.3 Smooth Muscle

The smooth muscle cells are distinguished by their centrally located cell nuclei and optically uniform myofibrils which do not have crossstri-

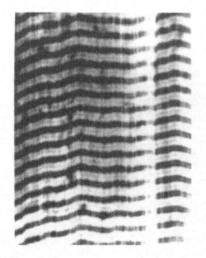

Fig. 12.6. A longitudinal section of two adjacent muscle fibers. (from *Schultz* and *Anglemier*, 1964)

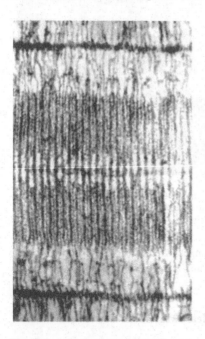

Fig. 12.7. A longitudinal section of a sarcomere. (from *Schultz, Anglemier*, 1964)

ations. Smooth muscles occur in mucous linings, the spleen, lymphatic glands, epidermis and intestinal tract. Smooth muscle fibers are useful in the examination of meat products; preferentially for the detection of pharynx (esophagus), stomach or calf pluck (heart, liver and lungs).

12.3 Muscle Tissue: Composition and Function

12.3.1 Overview

Muscles freed from adhering fat contain on the average 76% moisture, 21.5% N-substances, 1.5% lipids and 1% minerals. In addition, variable amounts of carbohydrates (0.05–0.2%) are present. Table 12.4 provides data on the average composition of some cuts of beef, pork and chicken.

12.3.2 Proteins

Muscle proteins can be divided into three large groups (cf. Table 12.5):

- Proteins of the contractile apparatus, extractable with concentrated salt solutions (actomyosin, together with tropomyosin and troponin).
- Proteins soluble in water or dilute salt solutions (myoglobin and enzymes).
- Insoluble proteins (connective tissue and membrane proteins).

12.3.2.1 Proteins of the Contractile Apparatus and Their Functions

About 20 different myofibrillar proteins are known. Myosin, actin and titin quantitatively predominate, acounting for 65–70% of the total protein. The remaining proteins are the tropomyosins and troponins, which are important for contraction, and various cytoskeletal proteins, which are involved in the stabilization of the sarcomere.

12.3.2.1.1 Myosin

Myosin molecules form the thick filaments and make up about 50% of the total proteins present in the contractile apparatus. Myosin can be isolated from muscle tissue with a high ionic strength buffer, for example, $0.3 \, mol/l \, KCl/0.15 \, mol/l$ phosphate buffer, pH 6.5. The molecular weight of myosin is approx. 500 kdal. Myosin consists

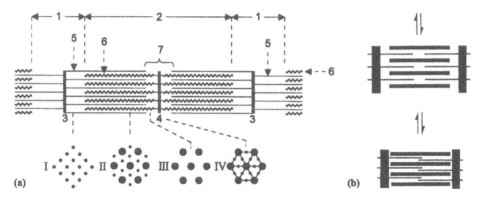

Fig. 12.8. Schematic representation of a sarcomere in the relaxed (**a**) and contracted (**b**) state (according to *Gault*, 1992). *1* I band, *2* A band, *3* Z line, *4* M line, *5* thin filament, *6* thick filament, *7* H zone. Cross section: I thin filaments near the Z line, II overlapping thick and thin filaments, III thick filaments, IV M line

Table 12.4. Average composition of meat (%)

Meat	Cut	Moisture	Protein	Fat	Ash
Pork	Boston butt (*M. subscapularis*)	74.9	19.5	4.7	1.1
	Loin (*M. psoas maior*)	75.3	21.1	2.4	1.2
	Cutlets, chops[a]	54.5	15.2	29.4	0.8
	Ham	75	20.2	3.6	1.1
	Side cuts	60.3	17.8	21.1	0.85
Beef	Shank	76.4	21.8	0.7	1.2
	Sirloin steak[a]	74.6	22.0	2.2	1.2
Chicken[b]	Hind leg (thigh + drum stick)	73.3	20.0	5.5	1.2
	Breast	74.4	23.3	1.2	1.1

[a] With adhering adipose tissue.
[b] Without skin.

of two large (M_r ca. 200,000) and four small (M_r ca. 20,000) subunits and is a very long molecule (measurements 140 × 2 nm). The two large subunits form a long, double-stranded α-helical rod with a double head of globular protein, both heads being joined at the same end of the coil (head dimensions, 5 × 20 nm) (Fig. 12.9a). The myosin ATPase activity is localized in the heads and is required for the interaction of the heads with actin, the protein constituent of the thin filaments. Myosin is cleaved by trypsin into two fragments: light (LMM, M_r 150,000) and heavy meromyosin (HMM, M_r 340,000). The HMM fraction contains the globularheaded region and has the ATPase activity and the ability to react with actin. Further proteolysis of HMM yields two subfragments S1 and S2, which correspond to the actual head and neck. The four smaller subunits mentioned above are found in the head region.

Up to 400 myosin molecules are arranged in the thick filaments ($l \sim 1500$ nm, $d \sim 12$ nm). By bringing the tails together, a major cord is formed and on its surface the heads are spirally located. The distance between two adjacent heads on such a spiral is 14.3 nm, and that between the two repeating heads in the same row or line is 42.9 nm. Their association is reversible under certain conditions. Myosin is stabilized by titin during muscle contraction (cf. Fig. 12.9b).

12.3.2.1.2 Titin

Apart from actin and myosin, titin is the third filament in the sarcomere (Fig. 12.9b). It connects the myosin filaments with the Z line and forms an "elastic" region with actin. Therefore, titin is the "backbone" of the sarcomere. As a result of its size ($M_r = 3 \times 10^6$), it moves very slowly

Fig. 12.9. Schematic repre-
sentations of (**a**) a myosin
molecule (according to
Lehninger, 1975), (**b**) arrangement of the Z line *(1)*,
actin *(2)*, titin *(3)* and
myosin *(4)* in the sarcomere; (**I**) stretched muscle,
(**II**) contracted muscle

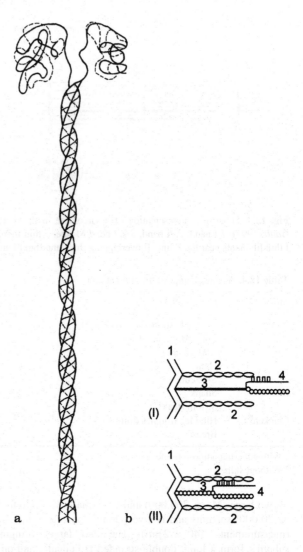

in an electric field and, consequently, was not
noticed for a long time in electrophoretic separations of muscle proteins. Titin is the largest
known protein until now. Its sequence consists
of 26,926 amino acid residues. In fact, 90% of
the molecule consists of globular domains, a large
part of which bind to other proteins, especially to
myosin.

12.3.2.1.3 Actin

Actin is the main constituent of the thin filament.
It makes up ca. 22% of the total protein of the
contractile apparatus. It is substantially less
soluble than myosin, probably because it is

fixed to substances in the Z line. Actin can be
isolated, for example, by extraction of pulverized,
acetone-dried muscle tissue with an aqueous ATP
solution.

The monomer G-actin (globular actin) consists
of 375 amino acids, has a molecular weight
of 42,000 and is able to bind ATP and a doubly
charged cation. G-actin exists only at low ionic
strengths. The addition of singly and doubly
charged cations starts the polymerization to
F-actin (fibrillar actin) with the cleavage of
ATP to ADP, which remains in the bound
state.

F-actin in the thin filaments ($1 \sim 1000$ nm, $d \sim$
8 nm) is in the form of a double-stranded helix
in which the G-actin beads are stabilized by two

Table 12.5. Muscle proteins

Proteins	Percentage[a]	
Myofibrillar proteins	60.5	
Myosin (H-, L-meromyosin, various associated components)		29
Actin		13
Titin		3.7
Tropomyosins		3.2
Troponins C, I, T		3.2
α, β-, γ-Actinins		2.6
Myomesin, N-line proteins etc.		3.7
Desmin etc.		2.1
Sarcoplasma proteins	29	
Glyceraldehydephosphate dehydrogenase		6.5
Aldolase		3.3
Creatine kinase		2.7
Other glycolytic enzymes		12.0
Myoglobin		1.1
Hemoglobin, other extracellular proteins		3.3
Connective tissue proteins		
Proteins from organelles	10.5	
Collagen		5.2
Elastin		0.3
Mitochondrial proteins (including cytochrome c and insoluble enzymes)		5.0

[a] Average percentage of the total protein of a typical mammalian muscle after rigor mortis and before other post mortem changes.

tropomyosin fibrils (cf. 12.3.2.1.3), as shown in Fig. 12.10. Altogether, it is a four-stranded filament. Six F-actin strands surround a thick filament; consequently, each F-actin strand adheres to the heads of three thick filaments (Fig. 12.8a, II).

12.3.2.1.4 Tropomyosin and Troponin

Tropomyosin (ca. 5% of the contractile proteins) is a highly elongated molecule (2×45 nm) with a molecular weight of about 68,000, and is assumed to be a double-stranded α-helix. Although each chain contains the same number of amino acids, their sequences differ in 39 positions. Tropomyosin contains no di-sulfide bridges and no proline. Indeed, 100% of tropomyosin is an α-helix. The monomer readily forms polymeric fibrils which are bound to F-actin on the thin filament.

Troponin (ca. 5% of the contractile proteins) sits on the actin filaments (cf. Fig. 12.10) and controls the contact between the filaments of myosin and actin during muscle contraction by means of a $Ca^{2\oplus}$ concentration-dependent change in conformation (cf. 12.3.2.1.5). It is a complex of three components, T, I, and C. Troponin T consists of a peptide chain with 259 amino acid residues and binds to tropomyosin. Troponin I (179 amino acid residues) binds to actin and inhibits various enzyme activities (ATPase). Troponin C (158 amino acid residues) binds $Ca^{2\oplus}$ ions reversibly through a change in conformation.

12.3.2.1.5 Other Myofibrillar Proteins

Apart from the main components of sarcomeres, myosin and actin, a series of cytoskeletal proteins exist that are responsible for the stabilization of the structure of the sarcomeres. The most important component is *connectin* (ca. 10% of the contractile proteins), an insoluble protein (M_r ca. 2,000,000) capable of forming fine filaments (g-filaments, $d = 2$ nm) which start at the Z line and proceed between the thick filaments of neighboring sarcomeres. These g-filaments greatly contribute to the elasticity and firmness of meat because they can be stretched from 1.1 μm to a length of 3.0 μm. Another protein of the cell skeleton is *myomesin* (subunit: $M_r = 165,000$). As the main component of the M line, myomesin is involved in fixing the thick filaments of the A band and in connecting neighboring myofibrils. Since myomesin strongly binds to myosin, it is possibly involved in the packing and cohesion of the myosin molecules in the thick filaments as well.

Among other proteins, α-*actinin* ($M_r = 200,000$), *desmin* ($M_r = 55,000$), *vimentin* ($M_r = 58,000$) and *synemin* ($M_r = 23,000$) are localized in the Z lines. Desmin appears to connect neighboring myofibrils.

A *N line protein* ($M_r = 60,000$) has been isolated from the N lines which run parallel to the Z lines on both sides and through the I bands.

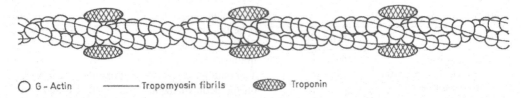

○ G – Actin   ——— Tropomyosin fibrils   ▨▨▨ Troponin

Fig. 12.10. A schematic representation of a thin filament. (according to *Karlsson*, 1977)

12.3.2.1.6 Contraction and Relaxation

Muscle stimulation by a nerve impulse triggers depolarization of the outer membrane of the muscle cell and thus release of Ca^{2+} ions from the sarcoplasmic reticulum. The Ca^{2+} concentration in the sarcoplasm of the resting muscle increases quickly from 10^{-7} to 10^{-5} mole/l. The binding of this Ca^{2+} to the troponin complex causes a conformational change in this protein. As a consequence, displacement of the tropomyosin fibrils occurs along the F-actin filament. Thus, the sterically hindered sites on the actin units are exposed for interaction with the myosin heads. The energy required for the shifting of the unbound myosin heads is obtained from the hydrolysis of ATP. The hydrolysis products of ATP, ADP and inorganic phosphate (P_i), remain on the myosin heads, which then bind to the actin monomers (Fig. 12.11a). Consequently, the myosin heads, now bound to actin, are forced to undergo a conformational change, which forces the thin filament to move relative to the thick filament (Fig. 12.11b).

The thin filaments and the heads of the thick filaments reverses half way between the Z lines. Therefore, the two thin filaments which interact with one thick filament are drawn toward each other, resulting in a shortening of the distance between the Z lines.

When the myosin heads release ADP and P_1 and become detached from the thin filaments (Fig. 12.11c), the heads are ready to take up a fresh charge of ATP (Fig. 12.11d). If the Ca^{2+} concentration in the sarcoplasm remains high, the ATP will again hydrolyze and the interaction of the myosin heads with the thin filament is repeated (Fig. 12.11a). However, if the Ca^{2+} concentration drops in the meantime, no ATP hydrolysis occurs, tropomyosin again blocks the access of myosin heads to the actin binding sites

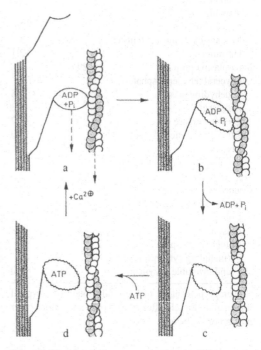

Fig. 12.11. Molecular processes involved in muscle contraction (see text; according to *Karlsson*, 1977)

and the muscle returns to its resting state. The decrease in Ca^{2+} concentration when muscle excitation has ceased, as well as the increase in Ca^{2+} during stimulation, i.e. the flow of calcium ions, is controlled by the sarcoplasmic reticulum. The Ca^{2+} concentration is low in the sarcoplasm of the resting muscle, while it is high within the sarcoplasmic reticulum. When the ATP level is low, detachment of the myosin and actin filaments does not occur. The muscle remains in a stiff, contracted state called rigor mortis (cf. 12.4). Hence, relaxation of muscle depends on the presence of regenerated ATP.

12.3.2.1.7 Actomyosin

Solutions of F-actin and myosin at high ionic strength ($\mu = 0.6$) *in vitro* form a complex called actomyosin. The formation of the complex is reflected by an increase in viscosity and occurs in a definite molar ratio: 1 molecule of myosin per 2 molecules of G-actin, the basic unit of the double-helical F-actin strand. It appears that a spike-like structure is formed, which consists of myosin molecules embedded in a "backbone" made of the F-actin double helix. Addition of ATP to actomyosin causes a sudden drop in viscosity due to dissociation of the complex. When this addition of ATP is followed by addition of Ca^{2+}, the myosin ATPase is activated, ATP is hydrolyzed and the actomyosin complex again restored after the ATP concentration decreases.

Upon spinning of an actomyosin solution into water, fibers are obtained which, analogous to muscle fibers, contract in the presence of ATP. Glycerol extraction of muscle fibers removes all the soluble components and abolishes the semipermeability of the membrane. Such a model muscle system shows all the reactions of *in vivo* muscle contraction after the readdition of ATP and Ca^{2+}. This and similar model studies demonstrate that the muscle contraction mechanism is understood in principle, although some molecular details are still not clarified.

12.3.2.2 Soluble Proteins

Soluble proteins make up 25–30% of the total protein in muscle tissue. They consist of ca. 50 components, mostly enzymes and myoglobin (cf. Table 12.5). The high viscosity of the sarcoplasm is derived from a high concentration of solubilized proteins, which can amount to 20–30%. The glycolytic enzymes are bound to the myofibrillar proteins *in vivo*.

12.3.2.2.1 Enzymes

Sarcoplasm contains most of the enzymes needed to support the glycolytic pathway and the pentosephosphate cycle. Glyceraldehyde-3-phosphate dehydrogenase can make up more than 20% of the total soluble protein. A series of enzymes involved in ATP metabolism, e. g., creatine phosphokinase and ADP-deaminase (cf. 12.3.6 and 12.3.8) are also present.

12.3.2.2.2 Myoglobin

Muscle tissue dry matter contains an average of 1% of the purple-red pigment myoglobin. However, the amounts in white and red meat vary considerably.

Myoglobin consists of a peptide chain (globin) of molecular weight of 16.8 kdal. It has known primary and tertiary structures (Fig. 12.12). The pigment component is present in a hydrophobic pocket of globin and is bound to a histidyl (His[93]) residue of the protein. The pigment, heme, is the same as that in hemoglobin (blood pigment), i. e. Fe^{2+}-protoporphyrin (Fig. 12.13).

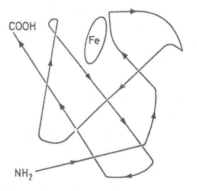

Fig. 12.12. Molecular model of myoglobin (**a**) and a schematic representation of peptide chain course (**b**). (from *Schormueller*, 1965)

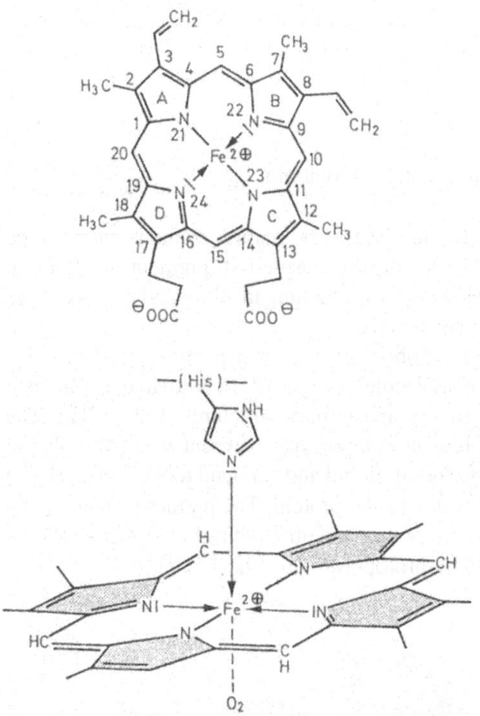

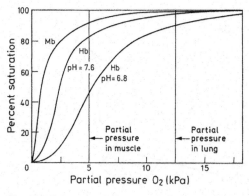

Fig. 12.13. Octahedral environment of Fe^{2+}-protoporphyrin with the imidazole ring of a globin histidine residue and oxygen (according to *Karlsson*, 1977)

Myoglobin supplies oxygen because of its ability to bind oxygen reversibly. Comparison of the oxygen binding curves for hemoglobin and myoglobin (Fig. 12.14) shows that at low p_{O_2}, such as exists in muscle, hemoglobin releases

oxygen to myoglobin. The sigmoidal shape of the O_2-binding curve for hemoglobin is due to its quaternary structure. It consists of four polypeptide chains, with one pigment molecule bound to each. The binding of O_2 to the four pigment molecules occurs cooperatively because of allosteric effects. Therefore, the degree of saturation, S, is expressed by the following equation (p_{O_2} = oxygen partial pressure; k = dissociation constant for the O_2-protein complex):

$$S = \frac{k \cdot p_{O_2}^n}{1 + k \cdot p_{O_2}^n} \tag{12.1}$$

For hemoglobin, $n \sim 2.8$ (sigmoidal saturation curve), and for myoglobin, $n = 1$ (hyperbolic saturation curve). The efficiency of O_2 transfer from hemoglobin to myoglobin is further enhanced by a decrease in pH since oxygen binding is pH-dependent (the *Bohr* effect).

While in the living animal approx. 10% of the total iron is bound to myoglobin, 95% of all the iron in well-bled beef muscle is bound to myoglobin. Unlike myoglobin, hemoglobin contributes little to the color of meat. The contribution of other pigments, such as the cytochromes, is negligible. However, attention must be paid to the fact that the visual appearance of a cut of meat is influenced not only by the light absorption of pigments, i.e. primarily myoglobin, but also by light scattering by the surface of muscle fiber. A bright red color is obtained when the coefficient of absorption is high and that of light scattering is low. Myoglobin (Mb) is purple ($\lambda_{max} = 555$ nm); oxymyoglobin (MbO$_2$), a covalent complex of ferrous Mb and O_2, is bright red ($\lambda_{max} = 542$ and 580 nm); and the oxidation product of Mb in the ferric state, metmyoglobin (MMb$^+$), is brown ($\lambda_{max} = 505$ and 635 nm). Some other ligands, such as electron pair donors (e.g. CO, NO, N_3^-, CN^-), like O_2, bind covalently, giving rise to low-spin complexes with similar absorption spectra and hence to a color similar to MbO$_2$. Figure 12.15 shows several absorption spectra of myoglobins.

Heme devoid of globin (free heme, Fe^{2+}-protoporphyrin) does not form the O_2-adduct, but oxidizes rapidly to hemin (Fe^{3+}-protoporphyrin). A prerequisite for reversible O_2 binding is the presence of an effective donor ligand on the iron's axial site, which is bound by formation of a quadratic-pyramidal complex. The imidazole

Fig. 12.14. Oxygen binding curves of myoglobin and hemoglobin

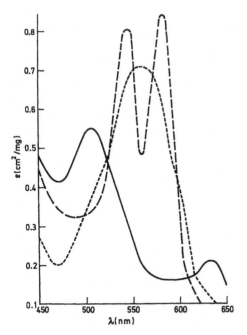

Fig. 12.15. Absorption spectra of myoglobin - - -, oxymyoglobin – – – and metmyoglobin ——. (according to *Fennema*, 1976)

side chain of His[93] of myoglobin has this function. Upon interaction with this fifth ligand, iron is raised above the heme plane by about 0.05 nm:

Binding of the sixth ligand moves the iron to its original position in the heme plane. Since the Fe–N bond distance (His[93]) remains constant, dislocation of the fifth ligand occurs (His[93], proximal His), i. e. a conformational change of the globin takes place.

The basicity of the fifth ligand affects the binding of the sixth ligand. The imidazole ring of His[93] is a good π-donor and, hence, stabilizes the O_2-adduct. A weaker base would enhance oxidation of the iron rather than adduct formation, while a stronger base would increase the stability of the adduct and diminish the probability of iron oxidation. From a biochemical viewpoint, the latter effect is rated as (O_2 supplier) negative; while from a food science point of view, it is desirable and positive (stable, bright red meat color).

As mentioned above, His[93] is located in a hydrophobic pocket of the myoglobin molecule. The electron density and, therefore, the oxidation state of the iron are regulated by protonation and deprotonation of the imidazole ring. With an increase in pH, there is an increase in basicity and, hence, an increase in binding of O_2 (the *Bohr* effect; cf. Fig. 12.14). A second histidine residue of myoglobin, His[64] (distal His), contributes to heme-O_2-complex stabilization by formation of a hydrogen bridge or ionic bond between N and O (cf. Formula 12.2).

12.3.2.2.3 Color of Meat

The color of fresh meat is determined by the ratios of myoglobin (Mb), oxymyoglobin (MbO_2) and metmyoglobin (MMb^+):

$$(12.3)$$

Stable MbO_2 is formed at a high partial pressure of oxygen. Fresh cuts of meat, to a depth of about 1 cm, acquire a bright cherry-red color which is considered a mark of quality. A slow and continuous oxidation to MMb^+ occurs at a low partial pressure of O_2. The change of $Fe^{2+} \rightarrow Fe^{3+}$ is reflected in the change in color from red to brown. MMb^+ does not form an O_2-adduct, since Fe^{3+} appears to be a less efficient π donor than Fe^{2+}. With better donor ligands than O_2 (CN^-,

$$(12.2)$$

NO, N_3^-), low-spin complexes are formed, the spectra of which are similar to those of MbO_2.

The change of $Fe^{2+} \rightarrow Fe^{3+}$ is designated as autoxidation:

"Dioxygen-Fe" "Fe$^{3\oplus}$-superoxide"
(ionic)

$$(12.4)$$

The oxygen molecule dissociates from the heme, taking along an electron from the iron, after protonation of its outer, more negative oxygen atom to form a hydroperoxy radical, the conjugate acid of the superoxide anion (cf. 3.7.2.1.4). The proton may originate from the distal histidine residue or other globin residues or from the surrounding medium. Autoxidation is accelerated by a drop in pH. The reason is the increased dissociation of the protein-pigment complex:

$$Globin + Heme \underset{k_{-1}}{\overset{k_1}{\rightleftharpoons}} Myoglobin \qquad (12.5)$$

Soon after slaughter, the meat has a pH at or near 7, at which the equilibrium constant of the above reaction is $K = k_1/k_{-1} = 10^{12} - 10^{15}\,mole^{-1}$. Since, during post-rigor, glycolysis decreases the pH of the meat to 5–6, myoglobin becomes increasingly susceptible to autoxidation.

The stability of MbO_2 is also highly dependent on temperature. Its half-life, τ, at pH 5 is 2.8 h at 25 °C and 5 days at 0 °C. Fresh meat has a system which can reduce MMb^+ back to Mb. Although various enzymatic and non-enzymatic reactions take part in this system, their contribution to the preservation of the color of fresh meat is unclear. One proposal suggests the participation of a NADH-cytochrome b_5-reductase. This enzyme which was detected in meat reduces $MMb^{\oplus}$ and $MHb^{\oplus}$. In addition, a series of non-specific reductases, also known as diaphorases, are supposed to play a role in this system. The

slow formation of MMb^+ can be reversed at the low partial pressure of O_2 which is found inside the cut of meat or in packaged, sealed meat. Therefore, for color stabilty, packaging of meat in O_2-permeable materials is not suitable since, after a time, its reduction capacity is fully exhausted. A non-O_2-permeable material is suitable for packaging meat. All of the pigment is present as Mb and is transformed to the bright red MbO_2 only when the package is opened.

Stabilization of the color of meat is also possible under controlled atmosphere packaging. A gaseous mixture of CO and air appears to be advantageous.

Copper ions promote autoxidation of heme to a great extent, while other metal ions, such as Fe^{3+}, Zn^{2+} or Al^{3+}, are less active.

12.3.2.2.4 Curing, Reddening

Color stabilization by the addition of nitrate or nitrite (meat curing) plays an important role in meat processing. Nitrite initially oxidizes myoglobin to metmyoglobin:

$$Mb + NO_2^{\ominus} \rightarrow MMb^{\oplus} + NO \qquad (12.6)$$

The resulting NO forms bright-red, highly stable complexes with Mb and MMb^+, MbNO and MMb^+NO:

$$(12.7)$$

Reducing agents, such as ascorbate, thiols or NADH, accelerate the *reddening* by reducing nitrite to NO and MMb^+ to Mb. Nitrosylmyoglobin, MbNO, is formed, which is highly stable when O_2 is absent. However, in the presence of O_2, the NO released by dissociation of MbNO is oxidized to NO_2.

The color of cured meat is heat stable. Denatured nitrosylmyoglobin is present in heated meat, or, due to dissociation of the proteinpigment complex, heme occurs with NO ligands present in

Fig. 12.16. Myoglobin reactions (Mb: myoglobin, MMb$^+$: metmyoglobin, MbO$_2$: oxymyoglobin, MbNO: nitrosylmyoglobin, MMb$^+$NO: nitrosylmetmyoglobin)

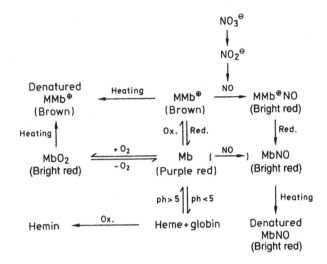

both axial binding sites:

$$O^{\oplus}N-Fe^{2\oplus}-NO^{\ominus}$$

(12.8)

MbNO antioxidatively protects the meat against lipid peroxidation. As shown in Formula 12.9, it traps fatty acid peroxyl radicals ROO$^{\bullet}$ with the formation of myoglobin and nitrite. MbNO is reformed in the presence of the above mentioned reducing agents.

$$ROO^{\bullet} + MbNO + H_2O \rightarrow ROOH + Mb + HNO_2$$
$$ROOH + 2MbNO + H_2O \rightarrow ROH + 2Mb + 2HNO_2$$
(12.9)

A color change to brown is observed when non-cured meat is heated. A Fe^{3+} complex is present which has its fifth and sixth coordination sites occupied by histidine residues of denatured meat proteins.

The myoglobin reactions relevant to meat color are presented schematically in Fig. 12.16.

12.3.2.3 Insoluble Proteins

The main fraction of proteins insoluble in water or salt solutions are the proteins of connective tissue. Membranes and the insoluble portion of the contractile apparatus are included in this group (cf. 12.3.2.1.4).

Connective tissue contains various types of cells. These cells synthesize many intercellular amorphous substances (carbohydrates, lipids, proteins) in which the collagen fibers are embedded.

Lipoproteins are present mostly in membranes. The lipids make up 3–4% of muscle tissue and are located in membranes. They consist of phospholipids, triacylglycerols and cholesterol. The phospholipid portion varies greatly: it makes up 50% of the plasma membrane and 90% of the mitochondrial membrane.

12.3.2.3.1 Collagen

Collagen constitutes 20–25% of the total protein in mammals. Table 12.6 shows data on its amino acid composition. The high contents of glycine and proline and the occurrence of 4-hydroxyproline and 5-hydroxylysine are characteristic. Since the occurrence of hydroxyproline is confined to connective tissue, its determination may provide quantitative data on the extent of connective tissue incorporation into a meat product.

Collagen also contains carbohydrates (glucose and galactose). These are attached to hydroxylysine residues of the peptide chain by O-glycosidic bonds. The presence of 2-O-α-D-glucosyl-O-β-D-galactosyl-hydroxylysine and of O-β-D-galactosyl-hydroxylysine has been confirmed.

Various types of collagen are known. They are characteristic of different organs and also of dif-

Table 12.6. Amino acid composition of muscle proteins (values are in g/16 g N)

Amino acid	Beef muscle total	Poultry muscle total[a]	Myosin	Actin (calf skin)	Collagen	Elastin
Aspartic acid	9.7–9.9	9.7–11.0	10.9	10.4	5.4	1.0
Threonine	4.8	3.5–4.5	4.7	6.7	2.1	1.1
Serine	4.1–4.5	–	4.1	5.6	2.9	0.9
Glutamic acid	15.8–16.2	16–18	21.9	14.2	9.7	2.4
Proline	3.0–4.1	–	2.4	4.9	13.0	11.6
Hydroxyproline					10.5	1.5
Glycine	4.6–6.1	4.6–6.7	2.8	4.8	22.5	25.5
Alanine	6.1–6.3	–	6.7	6.1	8.2	21.1
Cystine	1.3–1.5	–	1.0	1.3	0	0.3
Valine	4.8–5.5	4.7–4.9	4.7	4.7	2.9	16.5
Methionine	4.1–4.5	–	3.1	4.3	0.7	Trace
Isoleucine	5.2	4.6–5.2	5.3	7.2 ⎫	4.8[b]	3.7
Leucine	8.1–8.7	7.3–7.8	9.9	7.9 ⎭		8.6
Tyrosine	3.8–4.0	–	3.1	5.6	1.2	1.3
Phenylalanine	3.8–4.5	3.7–3.9	4.5	4.6	2.2	5.9
Lysine	9.2–9.4	8.3–8.8	11.9	7.3	3.9	0.5
Hydroxylysine					1.1	–
Histidine	3.7–3.9	2.2–2.3	2.2	2.8	0.7	0.1
Arginine	5.3–5.5	5.7–6.1	6.8	6.3	7.6	1.2
Tryptophan	–	–	0.8	2.0	0	–

[a] Chicken, duck, turkey: average values.
[b] Sum of isoleucine and leucine.

ferent connective tissue layers of muscular tissue (cf. 12.2.1). An overview is presented in Table 12.7. The amino acid sequence of an α^1-chain of collagen type I of mammalian skin is shown in Table 12.8. It is typical that every third residue in this sequence is glycine. Deviations from this regularity have been observed only at the ends of a chain. A frequently recurring sequence is:

Gly – Pro – Hyp – .

As a result of the specificity of the hydroxylating enzyme in vertebrates, hydroxyproline is always located, as shown in the sequence (Table 12.8) before glycine.

Collagen consists of three peptide chains which can be different or identical, depending on the type (cf. Table 12.7). The three peptide chains, each of which has a helical structure, form together a triple-stranded helix which has a structure corresponding to that of polyglycine II. A triple helix of this type is shown in Fig. 12.17. The basic structural unit of collagen fibers is called tropocollagen. It has a molecular weight of approx. 30 kdal. With a length of approx. 280 nm and a diameter of 1.4–1.5 nm, collagen is one

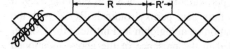

Fig. 12.17. Schematic representation of the conformation of tropocollagen (period R = 8.7 nm, pseudoperiod R' = 2.9 nm)

of the longest proteins. Tropocollagen fibers associate in a specific way to form collagen fibers, as presented schematically in Fig. 12.18. The association of adjacent rows is not in register, but is displaced by about one-fourth of the tropocollagen length (a "quarter staggered" array). This is responsible for cross-striations in the collagen fibers.

Fig. 12.18. Build up of a collagen fiber (**b**) from tropocollagen (**a**) molecules

Table 12.7. Types of collagen

Type	Peptide chains[a]	Molecular composition	Occurrence
I	α^1, α^2	$[\alpha^1(I)]_2\ \alpha^2(I)$	Skin, tendons, bones, muscle (epimysium)
II	α^1	$[\alpha^1(II)]_3$	Cartilage
III	α^1	$[\alpha^1(III)]_3$	Fetal skin, cardiovascular system, synovial membranes, inner organs, muscle (perimysium)
IV	α^1, α^2	$[\alpha^1(IV)]_3(?)^b$	Basal membranes, capsule of lens, glomeruli
		(?)	Placental membrane, lung, muscle (endomysium)
V	$\alpha A, \alpha B,$ αC (?)	$[\alpha B]_2\ \alpha A$ or $(\alpha B)_3 +$ $(\alpha A)_3$	Placental membrane, cardiovascular system, lung, muscle (endomysium),
		$(\alpha C)_3$ (?)	secondary component of many tissues

[a] Since the α chains of various types of collagen differ, they are called $\alpha^1(I)$, $\alpha^1(II)$, αA etc.
[b] (?) Not completely elucidated.

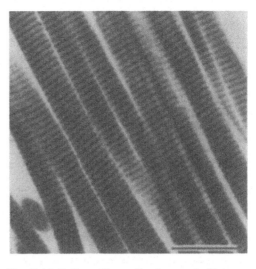

Fig. 12.19. Collagen fibers of bovine muscle (*Extensor carpi radialis*); transmission electron microscopy; sample fixed with 2% glutaraldehyde/paraformaldehyde, bar: 0.5 μm. (according to *Elkhalifa* et al., 1988)

Figure 12.19 shows an electron micrograph of collagen fibers of bovine muscle.

During maturation or aging, collagen fibers strengthen and are stabilized, primarily by covalent cross-linkages. Thus, cross-links confer mechanical strength to collagen fibers.

Cross-link formation involves the following reactions:

- Enzymatic oxidation of lysine and hydroxylysine to the corresponding ω-aldehydes.
- Conversion of these aldehydes to aldols and aldimines.
- Stabilization of these primary products by additional reduction or oxidation reactions.

It appears that only certain residues undergo reaction, mainly in the terminal, non-helical regions of the peptide chains.

Lysine and hydroxylysine residues within the peptide chain are oxidized by an enzyme that requires Cu^{2+} and pyridoxal phosphate for its activity and which is related to amine oxidase.

This reaction yields an α-aminoadipic acid semialdehyde residue bound to the existing peptide chain (R = H or OH):

$$\begin{array}{l} \text{NH} \\ | \\ \text{CH---CH}_2\text{---CH}_2\text{---CHR---CH}_2\text{NH}_2 \\ | \\ \text{CO} \end{array}$$

$$\longrightarrow \begin{array}{l} \text{NH} \\ | \\ \text{CH---CH}_2\text{---CH}_2\text{---CHR---C} \overset{\text{O}}{\underset{\text{H}}{}} \\ | \\ \text{CO} \end{array} \qquad (12.10)$$

The two aldehyde-containing chains may interact through an aldol condensation followed by elimination of water, forming a cross-link:

$$\begin{array}{cc} \text{Protein} & \text{Protein} \\ | & | \\ \text{H}_2\text{C---CHO} + & \text{H}_2\text{C---CHO} \longrightarrow \\ | & | \end{array}$$

$$\begin{array}{cc} \text{Protein} \quad \text{Protein} & \text{Protein} \quad \text{Protein} \\ | \qquad\quad | & | \qquad\quad | \\ \text{H}_2\text{C---CHOH---CH---CHO} \longrightarrow & \text{H}_2\text{C---CH}=\text{C---CHO} \\ & \qquad\qquad\qquad \| \end{array}$$

$$(12.11)$$

Table 12.8. Sequence of mammalian skin collagen, α^1-chain[a]

G P M G P S G P R G P R G L P* G P G Z* M S Y G Y D E K S A G V P
G E P* G A S G P M G P P* G P Q* G P R G P P* A P P* A P* L P* G E K P
G R P* G F S G Q R G A K G P P* Q* G P T G P R R A Z* P* G G G E K H R
G F Q G A A G P L G P R G L P* P* G P Q K N T G P A R P* A R A G M G A P
G A A G A D G A P G P T G P A G P V R G E R P* A P* D A G G E G P N Q*
G L P* G E R G S P* G A A G S E G P K* P* L N S R P* E P* G G D A P P S K
G L T G S P* G S P* G P D G K T R A G G G Z* Z* N N A A A
G P V G A A G A R G S P* A P* T P P K P A S P* G R A A R A A
G E P* G E Q* G A L P* R Q N A K* K D P* K D A P* G A A B A R A
G S Q K G P A G P A G A P D V D Q P* K A G G A K A P* A R E
G P K A G Q* P* G R D M A L L P* P I D D A P G A D P* E A
G D F G P A G E R G P A P* K A P* K N A G P A G P D P P* A
G A T G M P* G F P* G K S V T K V R P* A G D G S P* A V V S
G K E K G A G G A K A P D D Q T A Y P* A A S E A
G E L G Q P* G R P* A G K R E A A P A G A N Z A E P*
G E S G R D G P M S P* L P* P* A K Q* Z S A A
G S P* G V G G S P* P* B K* A G P A G P A A
G A R G L Q A G A K* L N E Z P* P L I G A P V R
G F S G A G A Q P* D R* G P I I R G P G H R
G D A G Q A G S G A G P P* K D D P* P* G P S G P R T
P Q P Q G Z K A H D G G R Y Y G A Y G P P L S F L

Z*: Pyrrolidone carboxylic acid, P*: 4-hydroxyproline, K*: 5-hydroxylysine, P‡: 3-hydroxyproline.
[a] The sequence is derived from very similar sequences of skin collagen of various mammals.

A polypeptide chain with an aldehyde residue (I) can interact with a lysine residue of the adjacent chain to form an aldimine, which can be further reduced to peptide-bound lysinonorleucine (III):

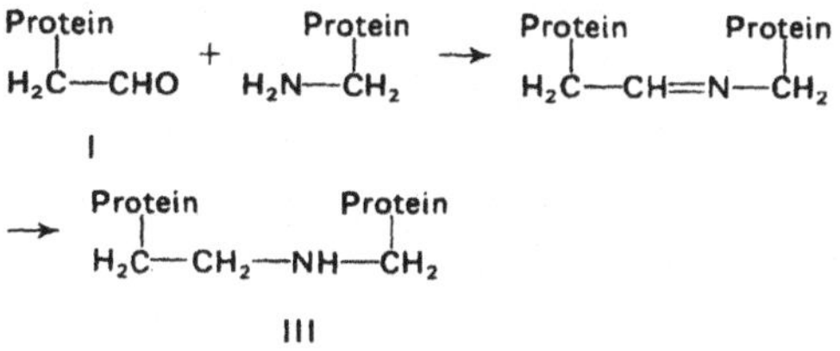

$$
\begin{array}{cc}
\text{Protein} & \text{Protein} \\
| & | \\
H_2C\!-\!CHO & H_2N\!-\!CH_2
\end{array}
+
\rightarrow
\begin{array}{cc}
\text{Protein} & \text{Protein} \\
| & | \\
H_2C\!-\!CH\!=\!N\!-\!CH_2
\end{array}
$$

I

$$
\rightarrow
\begin{array}{cc}
\text{Protein} & \text{Protein} \\
| & | \\
H_2C\!-\!CH_2\!-\!NH\!-\!CH_2
\end{array}
$$

III

$$(12.12)$$

If hydroxylysine is involved, an aldimine formed initially can be converted to a more stable β-aminoketone by *Amadori* rearrangement (cf. 4.2.4.4.1):

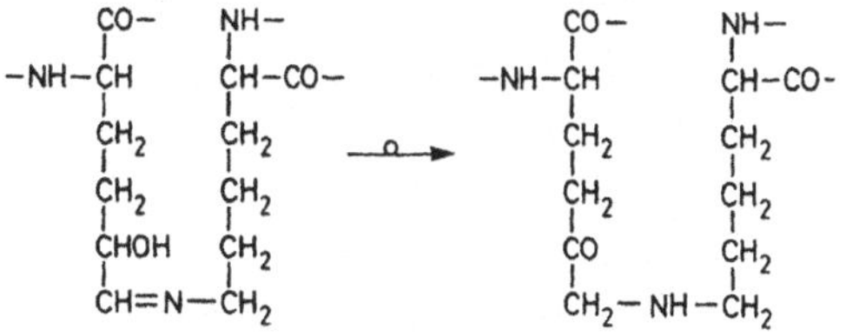

$$(12.13)$$

Likewise, aldehyde II can interact with one lysine residue through the intermediary dehydromerodesmosine (IV) to merodesmosine (V) and, thus, provide cross-links between the three adjacent polypeptide chains:

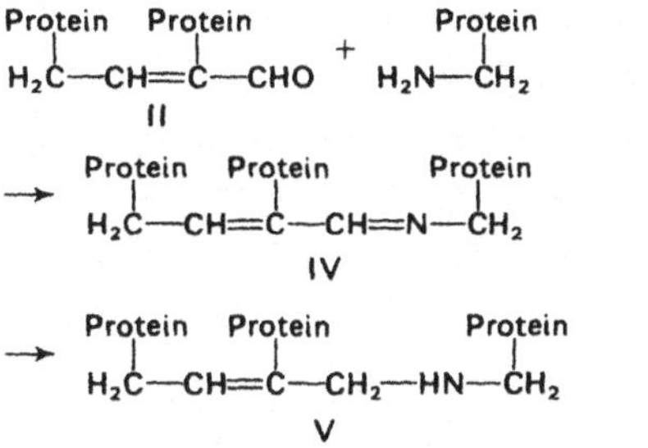

$$
\begin{array}{ccc}
\text{Protein} & \text{Protein} & \text{Protein} \\
| & | & | \\
H_2C\!-\!CH\!=\!C\!-\!CHO & & H_2N\!-\!CH_2
\end{array}
$$

II

$$
\rightarrow
\begin{array}{ccc}
\text{Protein} & \text{Protein} & \text{Protein} \\
| & | & | \\
H_2C\!-\!CH\!=\!C\!-\!CH\!=\!N\!-\!CH_2
\end{array}
$$

IV

$$
\rightarrow
\begin{array}{ccc}
\text{Protein} & \text{Protein} & \text{Protein} \\
| & | & | \\
H_2C\!-\!CH\!=\!C\!-\!CH_2\!-\!HN\!-\!CH_2
\end{array}
$$

V

$$(12.14)$$

During the reaction of three aldehyde molecules of type I with a lysine residue (actually a total of four lysine side chains are involved), a pyridine derivative is formed which, depending on the extent of reduction, yields desmosine (VI), dihydro- (VII) and tetrahydrodesmosine (VIII):

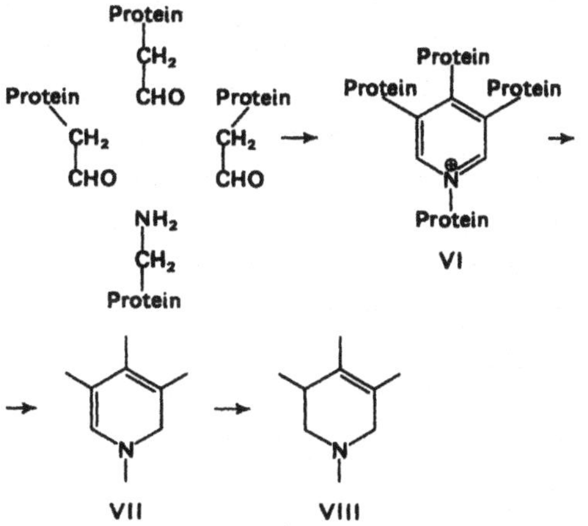

VI

VII   VIII

$$(12.15)$$

Depending on the kind of condensation, in addition to desmosine VI, designated as an A-type condensation product, rings with other substitution patterns are observed, i. e. B- and C-type condensation products:

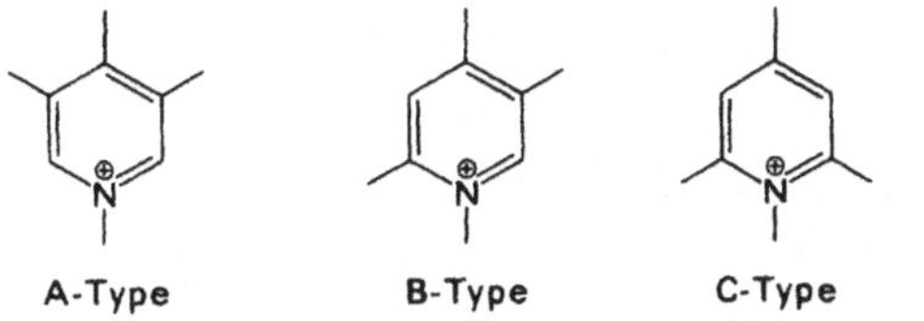

A-Type   B-Type   C-Type

$$(12.16)$$

Pyridinolines have also been detected. They are probably formed from β-aminoketones and the ω-aldehyde of a hydroxylysine residue:

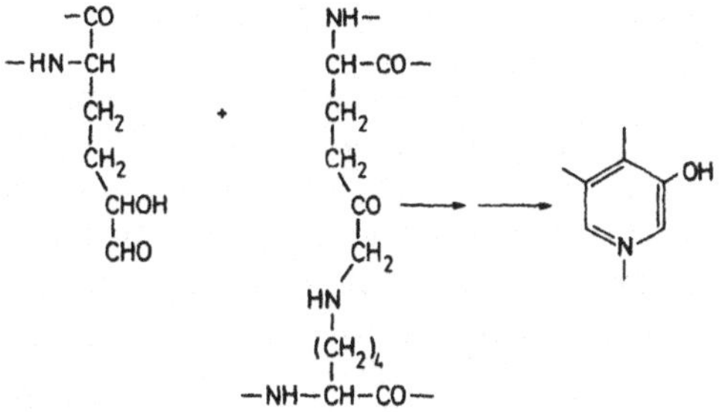

$$(12.17)$$

Studies of bovine muscle collagens have shown that the pyridinoline content increases with increasing age of the animal and, like the collagen content, negatively correlates with the tenderness. In intensively fattened cows, the pyridinoline content was higher than in extensively fattened animals.

Histidine can also be involved in cross-linking reactions, as shown by the detection of histidino-hydroxylysino-norleucine:

$$\text{(12.18)}$$

The amino acid pentosidine was also obtained from collagen, which indicates the bonding of lysine and arginine with the participation of a pentose:

$$\text{(12.19)}$$

The outlined reactions can also occur with hydroxylysine residues present on collagen fibers. Of all the compounds mentioned, hydroxylysino-norleucine and dihydroxylysino-norleucine have been isolated from collagen in significant amounts.

In the case of type I, collagen biosynthesis (Fig. 12.20a–h) involves first the synthesis of pro-α^1- and pro-α^2-precursor chains. The N-terminus of these precursors contains up to 25% of extended α^1- and α^2-chains (a). Immediately after the chains are released from polysomes, hydroxylation of the proline and lysine residues occurs (cf. reactions under 12.20).

Realignment of the chains follows: two strands of pro-α^1 and one chain of pro-α^2 are joined to form a triple-stranded helix (b–d). The extended peptides at the N-terminus appear to play a distinct role in these reactions. Disulfide bridging occurs between the strands at this stage in order to stabilize the structure. The procollagen thus formed will cross the membrane of the cell in which it was synthesized (e). The N-terminal peptides are removed by limited proteolysis (f) and the pro-

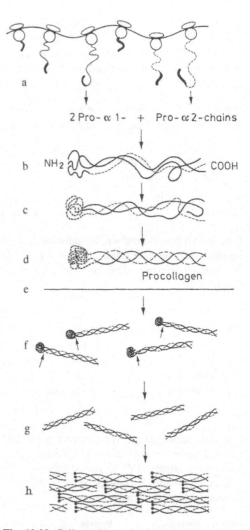

Fig. 12.20. Collagen biosynthesis (according to *Bornstein*, 1974). **a** Polysome, **b** hydroxylation, **c** chain straightening, **d** disulfide bond formation, **e** cell membrane, **f** membrane crossing, **g** a limited hydrolysis to tropocollagen, **h** collagen fiber formation, cross-linking

collagen is converted to tropocollagen (g). Finally, the tropocollagen is realigned to form collagen fibers (h). At this stage, collagen maturation, which coincides with strengthening of collagen fibers by covalent cross-linking along the peptide strands, begins. The maturation is initiated by oxidation of lysine and is followed by the reactions described above.

Collagen swells but does not solubilize. Enzymatically, it can be hydrolyzed to various extents with

$$\text{(proline structure)} + O_2 \xrightarrow[\text{(Fe}^{2\oplus}\text{, ascorbate)}]{\text{Proline hydroxylase}}$$

$$\text{(hydroxyproline structure)} + 2\text{-oxoglutarate} \longrightarrow$$

$$\longrightarrow \begin{array}{c} COO^{\ominus} \\ CH_2 \\ CH_2 \\ COO^{\ominus} \end{array} + CO_2 + \text{(hydroxyproline structure)}$$

$$\begin{array}{c} H_2C-NH_2 \\ H_2C \\ R \end{array} \xrightarrow[\text{(Fe}^{2\oplus}\text{, ascorbate)}]{\text{Lysine hydroxylase}} \begin{array}{c} H_2C-NH_2 \\ CHOH \\ R \end{array}$$

$$(12.20)$$

a series of collagenases from different sources and with different specificities. A vertebrate animal collagenase, which is a metal proteinase, splits a special bond in native collagen while the collagenase from *Clostridium histolyticum*, also a metal proteinase, cleaves collagen preferentially at glycine residues, forming tripeptides:

$$-Pro-X-\overset{\downarrow}{G}ly-Pro-X-\overset{\downarrow}{G}ly-Pro- \tag{12.21}$$

Collagenase enzymes which are serine proteinases are also known.

Denatured collagen, as formed post-mortem by the action of lactic acid, can also be cleaved by lysosomal enzymes, e. g., lysosomal collagenase and cysteine proteinase cathepsin B_1. Thermally denatured collagen is attacked by pepsin and trypsin.

One characteristic of the intact collagen fiber is that it shrinks when heated (cooking or roast-

ing). The shrinkage temperature (T_s) is different for different species. For fish collagen, the T_s is 45 °C and for mammals, 60–65 °C. When native or intact collagen is heated to $T > T_s$, the triple-stranded helix is destroyed to a great extent, depending on the cross-links. The disrupted structure now exists as random coils which are soluble in water and are called gelatin. Depending on the concentration of the gelatin solution and of the temperature gradient, a transition into organized structures occurs during cooling. Figure 12.21 schematically summarizes these transitions. At low concentrations, intramolecular back-pleating occurs preferentially with single-strands. At higher concentrations and slow rates of cooling, a structure is rebuilt which resembles the original native structure. At even higher concentrations and rapid cooling, structures are obtained in which the helical segments alternate with randomly coiled portions of the strand. All these structures can immobilize a large amount of water and form gelatin gels.

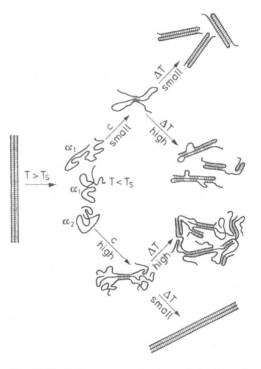

Fig. 12.21. Collagen conversion into gelatin. (according to *Traub* and *Piez*, 1971). T_s: shrinkage temperature, T: temperature, c: concentration; (see text)

The transition of collagen to gelatin outlined above occurs during the cooking and roasting of meat. The extent of gelatinization is affected by the collagen cross-linking as determined by the age of the animal and the amount of heat applied (temperature, time, pressure).

Gelatin plays a role as a gelling agent. It is produced on a large scale from animal bones or skin by treatment with alkali or acid, followed by a water extraction step. Depending on the process, products are obtained which differ in molecular weight and, consequently, in their gelling properties. Some brands are used as food gelatins, others play an important role in industry (film emulsions, glue manufacturing).

12.3.2.3.2 Elastin

Elastin is found in lower amounts in connective tissue along with collagen. It is a nonswelling, highly stable protein (M_r 70,000) which forms elastic fibers. The protein has rubber-like properties. It can stretch and then return to its original length or shape. Large amounts of elastin are present in ligaments and the walls of blood vessels. The ligament located in the neck of grazing animals is an exceptionally rich source of this protein. Table 12.6 shows that the amino acid composition is different from that of collagen. The amount of basic and acidic amino acids is low, e. g., hydroxylysine is absent, while the content of amino acids with nonpolar side chains (Ala, Val) is greatly increased. This difference explains that, unlike collagen, elastin lacks the capability of swelling on heating in water. The elastic properties are based on very strong cross links, which involve the desmosines described in the section on collagen (cf. Formula 12.15).

Elastin is hydrolyzed by the serine proteinase elastase, which is excreted by the pancreas. This enzyme preferentially cleaves peptide bonds at sites where the carbonyl residue has a nonaromatic, nonpolar side chain.

12.3.3 Free Amino Acids

Fresh beef muscle contains 0.1–0.3% free amino acids (fresh weight basis). All amino acids are detectable in low amounts (<0.005), with alanine (0.01–0.05%) and glutamic acid (0.01–0.05%) being most predominant.

The free amino acid fraction also contains 0.02–0.1% taurine (I). As such, taurine should be regarded as a major constituent of this fraction. It is obtained biosynthetically from cysteine through cysteic acid and/or from a side pathway involving cysteamine and hypotaurine (II):

$$\begin{array}{ccccc}
CH_2-SH & & CH_2-SO_2H & & CH_2-SO_3H \\
CH-NH_2 & \longrightarrow & CH-NH_2 & \longrightarrow & CH-NH_2 \\
COOH & & COOH & & COOH \\
\downarrow CO_2 & & \downarrow CO_2 & & \downarrow CO_2 \\
CH_2-SH & & CH_2-SO_2H & & CH_2-SO_3H \\
CH_2-NH_2 & \longrightarrow & CH_2-NH_2 & \longrightarrow & CH_2-NH_2 \\
& II & & & I
\end{array}$$

(12.22)

The biochemical role of taurine includes derivatization of bile acids (taurocholic and taurodeoxycholic acids). A neurotransmitting function has also been ascribed to this compound.

12.3.4 Peptides

The characteristic β-alanyl histidine peptides, carnosine, anserine and balenine, of muscle are described in section 1.3.4.2. Their contribution to taste is discussed in 12.9.1.

12.3.5 Amines

Methylamine in fresh beef muscle is present at 2 mg/kg, while the other volatile aliphatic amines (dimethyl-, trimethyl-, ethyl-, diethyl- and isopropylamine) are detected only in trace amounts.

Of the biogenic amines produced on the decarboxylation of amino acids (cf. 10.2.8.3), histamine, tyramine, putrescine and cadaverine have been identified in beef and pork. Since these substances are microbial metabolic products, they were the proposed indicators of microbial quality and were listed in the biogenic amine index (BAI = concentration of the four amines in mg/kg). A BAI value of <5 indicates clean meat, 5–20 acceptable (early stages of microbial infestation), 20–50 inferior quality and >50 spoiled meat. The BAI values of fermented meat products are naturally higher; a limit of 500 mg/kg was proposed for salami.

Other biogenic amines are spermidine [N-(3-aminopropyl)-1,4-butandiamine] and spermine [N, N'-bis-(3-aminopropyl)-1,4-butandiamine], which are biogenetically formed from putrescine and belong to the constituents of meat. The main compound is spermine, with a concentration in the range of 25–65 mg/kg.

12.3.6 Guanidine Compounds

Creatine and creatinine (I and II, respectively; cf. Formula 12.23) are characteristic constituents of muscle tissue and their assay is used to detect the presence of meat extract in a food product. Creatine is present in fresh beef at 0.3–0.6% and creatinine at 0.02–0.04%.

In living muscle, 50–80% of creatine is in the phosphorylated form, creatine phosphate (III, cf. Formula 12.23), which is in equilibrium with ATP. The reaction rate is highly influenced by the enzyme creatine phosphokinase. Creatine phosphate serves as an energy reservoir (free energy of hydrolysis, $\Delta G^0 = -42.7$ kJ/mole; of ATP: $\Delta G^0 = -29.7$ kJ/mole). Creatine phosphate has a higher phosphoryl group transfer potential than ATP. Hence, when muscle is stimulated for a prolonged period in the absence of glycolysis or respiration, the supply of creatine phosphate will become depleted within a couple of hours by maintaining the ATP concentration. This is especially the case in post-mortem muscle, when the ATP supply has declined significantly through oxidative respiration.

12.3.7 Quaternary Ammonium Compounds

Choline and carnitine are present in muscle tissue at 0.02–0.06% and 0.05–0.2%, respectively (on a fresh weight basis). Choline is synthesized from serine with colamine as an intermediary product (cf. Reactions 12.24) and carnitine is obtained from lysine through ε-N-trimethyllysine and butyrobetaine (cf. Reactions 12.25).

The carnitine fatty acid esters, which are in equilibrium with long chain acyl-CoA molecules in living muscle tissue, are of biochemical importance. The carnitine fatty acid ester, but not the acyl-CoA ester, can traverse the inner mitochondrial membrane. After the fatty acid is oxidized within the mitochondria, carnitine is instrumental in transporting the generated acetic acid out of the mitochondria.

$$
\begin{array}{l}
CH_2OH \\
| \\
CHNH_2 \\
| \\
COOH
\end{array}
\xrightarrow[CO_2]{}
\begin{array}{l}
CH_2OH \\
| \\
CH_2NH_2
\end{array}
\longrightarrow
\begin{array}{l}
CH_2OH \\
| \\
CH_2N^\oplus(CH_3)_3
\end{array}
$$

(12.24)

$$
\begin{array}{l}
CH_2{-}NH_2 \\
| \\
CH_2 \\
| \\
CH_2 \\
| \\
CH_2 \\
| \\
CHNH_2 \\
| \\
COOH
\end{array}
\longrightarrow
\begin{array}{l}
CH_2N^\oplus(CH_3)_3 \\
| \\
CH_2 \\
| \\
CH_2 \\
| \\
CH_2 \\
| \\
CHNH_2 \\
| \\
COOH
\end{array}
\longrightarrow
\begin{array}{l}
CH_2N^\oplus(CH_3)_3 \\
| \\
CH_2 \\
| \\
CH_2 \\
| \\
COOH
\end{array}
\longrightarrow
$$

$$
\xrightarrow[\text{Fe}^{2\oplus},\ \text{Ascorbate}]{\text{O}_2,\ 2\text{-Ketoglutarate}}
\begin{array}{l}
CH_2N^\oplus(CH_3)_3 \\
| \\
CHOH \\
| \\
CH_2 \\
| \\
COOH
\end{array}
$$

(12.25)

Formula/structures (12.23):

I: creatine; II: creatinine; III: creatine phosphate

$$
\underset{\text{I}}{\underset{\underset{CH_3}{|}}{\underset{N{-}CH_2{-}COOH}{HN{=}C{\overset{NH_2}{<}}}}}
+ ATP \rightleftharpoons
\underset{\text{III}}{\underset{\underset{CH_3}{|}}{\underset{N{-}CH_2{-}COOH}{HN{=}C{\overset{NH{-}PO_3H_2}{<}}}}}
+ ADP
$$

II:

$$
HN{=}C
\begin{array}{c}
\overset{H}{\underset{}{N}}{-}C{=}O \\
| \quad\quad | \\
N{-}CH_2 \\
| \\
CH_3
\end{array}
$$

(12.23)

12.3.8 Purines and Pyrimidines

The total content of purines in fresh beef muscle tissue is 0.1–0.25% (on a fresh weight basis). ATP, present predominantly in living tissue, breaks down to inosine-5'-monophosphate (IMP) in the post-mortem stages. The breakdown rate is influenced by the condition of the animal and by temperature. IMP is then slowly decomposed through successive steps to hypoxanthine, with inosine as an intermediary product:

$$ATP \quad \rightleftarrows ADP + P_{in} \text{ (Myosin-ATPase)}$$
$$2ADP \quad \rightleftarrows ATP + AMP \text{ (Myokinase)}$$
$$AMP \quad \rightarrow IMP + NH_3 \text{ (Adenylate-deaminase)}$$
$$IMP \quad \rightarrow Inosine + P_{in} \text{ (5' - Nucleotidase)}$$
$$Inosine \rightarrow Hypoxanthine + Ribose \text{ (Nucleosidase)}$$
$$(12.26)$$

Post-mortem data on the *Psoas major* rabbit muscle are given in Table 12.10. They relate to nucleotide breakdown and to other important muscle tissue constituents.

The changes in water holding capacity of meat resulting from ATP transition to IMP are dealt with in Section 12.5. Unlike purines, pyrimidine nucleotide content in muscle is very low (Table 12.9).

Table 12.9. Purines and pyrimidines in fresh-beef muscle

Compound	Content (%)
Inosine-5'-phosphate	0.02–0.2[a]
Inosine	Trace
Hypoxanthine	0.01–0.03
Adenosine-5'-phosphate	0.001–0.01
Adenosine-5'-diphosphate	<0.3[b]
Adenosine-5'-triphosphate	
Nicotinamide-adenine-dinucleotide	0.1
Guanosine-5'-phosphate	0.002
Cytidine-5'-phosphate	0.001
Uridine-5'-phosphate	0.002

[a] Until approx. 1 h post-mortem no IMP is found in muscle.

[b] There is a fairly rapid decrease in post-mortem concentration influenced by cooling and other muscle handling conditions.

Table 12.10. Post-mortem changes in the concentration of some constituents of rabbit muscle (*M. psoas*)

Compound	μmol/g Fresh tissue	
	living muscle	post-rigor muscle
Total acid-soluble phosphorus	68	68
Inorganic phosphorus	<12	>48
Adenosine triphosphate (ATP)	9	<1
Adenosine diphosphate (ADP)	1	<1
Inosine monophosphate	<1	9
Creatine phosphate	20	<1
Creatine	23	42
NAD/NADP	2	1
Glycogen	50	<10
Glucose-1-phosphate	<1	<1
Glucose-6-phosphate	5	6
Fructose-1,6-bisphosphate	<1	<1
Lactic acid	10	100

12.3.9 Organic Acids

The predominant acid in muscle tissue is the lactic acid formed by glycolysis (0.2–0.8% on a fresh meat weight basis), followed by glycolic (0.1%) and succinic acids (0.05%). Other acids of the *Krebs* cycle are present in negligible amounts.

12.3.10 Carbohydrates

The glycogen content of muscle varies greatly (0.02–1.0% on a fresh tissue weight basis) and is influenced by the age and condition of the animal prior to slaugher. The rate of the post-mortem decrease in glycogen als varies. Sugars are only 0.1–0.15% of the weight of fresh muscle, of which 0.1% is shared by glucose-6-phosphate and other phosphorylated sugars. The free sugars present are glucose (0.009–0.09%), fructose and ribose.

12.3.11 Vitamins

Table 12.11 provides data on water-soluble vitamins in beef muscle.

Table 12.11. Vitamins in beef muscle

Compound	mg/kg Fresh tissue
Thiamine	0.6–1.6
Riboflavin	1–34
Nicotinamide	40–120
Pyridoxine, pyridoxal, pyridoxamine	1–4
Pantothenic acid	4–10
Folic acid	0.03
Biotin	0.05
Cyanocobalamine (B_{12})	0.01–0.02
α-Tocopherol	4.8
Retinol	0.2
Vitamin K	0.13

12.3.12 Minerals

Table 12.12 provides data on minerals in meat. Table 12.13 provides data on the occurrence of soluble and insoluble iron in meat of different animals. The other trace elements, which are 1 mg/kg fresh meat tissue, are not listed individually.

Table 12.12. Minerals in beef muscle

Element	% in fresh tissue	Element	%in fresh tissue
K	0.25–0.4	Zn	0.001–0.008
Na	0.07–0.2	P (as	
Mg	0.015–0.035	P_2O_5)	0.30–0.55
Ca	0.005–0.025	Cl	0.04–0.1
Fe	0.001–0.005		

12.4 Post Mortem Changes in Muscle

Immediately after death, the muscle is soft, limp, and dry and can be reversibly extended by using a low load (5–15 kPa). Cadaveric rigidity (rigor mortis) occurs after a few hours. The muscle can then be extended only by using a heavy load (>200 kPa) and becomes moist or wet. Rigor can occur in various stages of contraction or stretching. It subsides after some time and the muscle can be easily extended, but irreversibly. Meat with a more or less tender consistency is formed from the muscle. This process is caused by complicated proteolytic reactions, which are discussed in 12.4.3.

12.4.1 Rigor Mortis

Cessation of blood circulation ends the O_2 supply to muscle. Anaerobic conditions start to develop. The energy-rich phosphates, such as creatine phosphate, ATP and ADP, are degraded. The glycolysis process, which is pH and temperature dependent and which is influenced by the presence of glycogen, is the sole remaining energy source. The lactic acid formed remains in the muscle, thereby decreasing the muscle pH from 6.5 to less than 5.8.

Table 12.10 gives an example of post mortem changes in rabbit *Psoas major* as related to concentrations of some of the more important muscle tissue constituents. The data shown in Fig. 12.22 illustrate the post mortem decreases in pH, creatine phosphate and ATP in beef *Longissimus dorsi* and *Psoas major* muscles and emphasizes that the changes are dependent on the type of muscle.

Table 12.13. Occurrence of iron in meat of different animal species

Animal species	Concentration ($\mu g/g$)[a]		Distribution of soluble iron (%)[a]			
	Insoluble Fe	Soluble Fe	Ferritin	Hemoglobin	Myoglobin	Free Fe
Beef (rump steak)	5.9	20.0	1.6	6.0	89.0	3.4
Pork (loin)	3.0	3.6	8.4	22.2	64.0	5.4
Lamb (loin)	5.9	12.3	7.3	13.0	74.0	5.7
Chicken (leg)	4.7	3.4	26.4	55.7	12.1	5.8

[a] Average value of three meat samples.

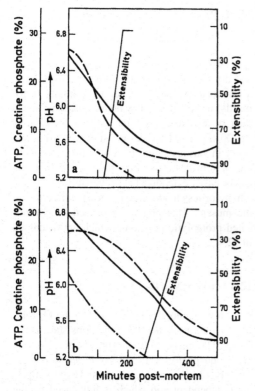

Fig. 12.22. Post-mortem changes in beef muscle. **a** *M. longissimus dorsi*; **b** *M. psoas*; ———: pH value; ---: ATP as % of the total acid-soluble phosphate; ·····: creatine phosphate as % of the total acid soluble phosphate. (according to *Hamm*, 1972)

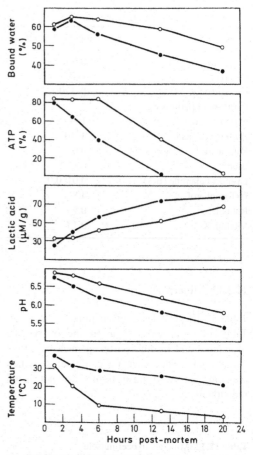

Fig. 12.23. The effect of temperature on post mortem changes in beef muscle.

M. semimembranosus ●—●: normal cooling, animal carcass kept for the first hour post-mortem at 2–4 °C then posterior hind quarters cut and kept at 14 °C for 10 h followed by 2 °C; o—o: cooling in ice, hind quarters 11 h in crushed ice, followed by 2 °C. Temperature measurement of the meat at 4 cm depth; bound water as percent of total water; lactic acid results are on fresh weight basis and ATP expressed as percent of total nucleotides. (according to *Disney* et al., 1967)

Although muscle tissue is soft and flexible and dry on the surface immediately following death, its flexibility or extensibility is lost very rapidly. ATP breaks down (Fig. 12.22). The muscle tissue becomes stiff and rigid (death's stiffening, rigor mortis; cf. 12.3.2.1.5 and 12.3.2.1.6) and, as the rigor proceeds, the muscle tissue surface becomes wetter. The depletion of the energy reserves results in the distribution of the calcium ions, which are stored in the mitochondria and in the sarcoplasmic reticulum, throughout the entire intracellular matrix.

The onset of rigor mortis occurs in beef muscle within 10–24 h; in pork, 4–18 h; and in chicken, 2–4 h.

The rate of decrease in pH and the final pH value of meat are of significance for water holding capacity and, therefore, for meat quality. Figure 12.23 shows that a more rapid and intensive cooling of the post mortem muscle results in meat with a noticeably higher water holding capacity than that of muscle cooled slowly.

12.4.2 Defects (PSE and DFD Meat)

Rapid drops in ATP and pH (Fig. 12.24) cause pork muscle to become pale and soft and to undergo extensive drip loss because of lowered wa-

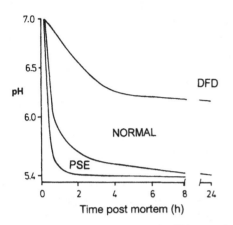

Fig. 12.24. Post mortem decrease in pH in normal meat, PSE meat and DFD meat in the case of pork (according to *Moss*, 1992)

Table 12.14. Some differences between normal and faulty meat[a]

	Quality	pH (1 h)	pH (24 h)	ATP	Gly-cogen	Lac-tate
Normal meat		6.5	5.8	2.2	6.2	4.7
PSE-Meat	Pale, exudative, loose soft texture	5.6	5.6	0.3	1.9	9.0
DFD-Meat	Dark, sticky, firm texture	6.5	6.3	1.1	1.5	4.0

[a] Pork meat: *M. longissimus dorsi*. Values are averages expressed as mg/g muscle 1 h post mortem; pH 1 h (initial) and pH 24 h (final) values post mortem.

ter holding capacity (PSE meat: pale, soft, exudative). PSE meat has a low tensile strength and loses a substantial amount of weight when hung and, when thawed, drip losses occur. Such defects are typical of hogs with an inherited sensitivity towards stress, such as fear prior to slaughter, anxiety during transport, exposure to temperature changes, etc. Immediately prior to or during slaughtering, an abnormally rapid ATP breakdown occurs and, consequently, the rate of glycolysis is accelerated. The pH value falls rapidly and the body temperature, which would normally drop (from 38 °C to 36 °C in 45 minutes post mortem), rises to 40–41 °C as a result of the intensified metabolism.

The falling pH value and the high temperature cause protein denaturation. Soluble proteins precipitate and scatter light. Consequently, the meat appears paler in spite of the unchanged myoglobin content. At the same time, cell membranes disintegrate and water loss increases. In fact, PSE pork incurs up to 15% drip loss in 3 days and normal meat only ca. 4%.

The occurrence of dark and firm pork meat (DFD meat: dark, firm and dry) is likewise characteristic of a stress-impaired hog. Since glycogen is largely used up due to stress, only a little lactic acid is formed after slaughtering and the pH hardly falls (Fig. 12.24). The microfibrils which are more swollen at higher pH values bind more water (dry texture). As a result of this effect and the higher stability of oxymyoglobin at higher pH values (cf. 12.3.2.2.3), the color

appears darker than normal meat. The relatively high pH value makes DFD meat susceptible to microbial infection and, therefore, not suitable for raw meat products.

Data relating to normal and faulty cuts of meat are summarized in Table 12.14. Both defects mentioned may occur in different muscles of the same animal. The PSE effect is not significant in beef muscle tissue since energy is available from fat oxidation so glycogen breakdown can occur slowly. These meat defects may be avoided in hog muscles by careful handling of stress-sensitive animals and by rapid cooling of carcasses.

12.4.3 Aging of Meat

Rigor mortis in beef muscle tissue is usually resolved 2–3 days post mortem. By this time, the meat again becomes soft and tender (aging). Further aging of the meat to improve tenderness and to form aroma requires various amounts of time, depending on the temperature. At temperatures around 3 °C (−1 °C to +7 °C), aging of poultry requires at least 36 h, pork 60 h, veal 7 days and beef 14 days. Apart from the animal species, the age of the animal (degree of cross linkage of collagen) and the released enzymes influence the duration of aging. A slight rise in pH is observed with aging, the water holding capacity is increased somewhat and, also, fluid loss from heat-treated meat is slightly decreased.

Table 12.15. Endopeptidases involved in the aging of meat

Origin	Enzyme	M_r	pH range	Hydrolysis
Sarcoplasma	μ-Calpain[a]	110,000	6.5–7.5	Z line proteins
	m-Calpain[a]	110,000	6.5–7.5	Z line proteins
Lysosomes	Cathepsin B[a]	25,000	3.5–6.0	
	Cathepsin L[a]	28,000	3.0–6.0	Myosin, actin,
	Cathepsin D[a]	42,000	3.0–6.0	troponin, collagen

[a] Cysteine endopeptidase.
[b] Aspartic acid endopeptidase.

Maturation or aging is accompanied by morphological changes which primarily affect the cytoskeleton. Microexaminations show that the Z lines, which as cross structures (cf. 12.2.1) separate the individual sarcomeres in the muscle fibril, are broken up during aging. In addition, the fibrillar proteins titin and desmin are degraded. In comparison, the contractile proteins myosin and actin are stable. They are attacked only at temperatures above 25 °C. The connective tissue present outside the muscle cells also remains intact.

The degradation of the myofibrillar proteins is catalyzed by endopeptidases. The participation of the enzymes listed in Table 12.15 is under discussion. Special attention is directed to μ-calpain, which, like m-calpain, is activated by the Ca ions liberated during the rigor phase. Both calpains are cysteine endopeptidases, which consist of a large (80 kdal) and a small (28 kdal) subunit. The large subunit contains the active center. These two calpains can be distinguished by the Ca concentration required for their activation. μ-Calpain requires about 30 μmol/l and m-calpain 250–270 μmol/l. The activity of the calpains is regulated, among other factors, by their endogenous inhibitor calpastatin. It has been proposed that the calpains synergistically cooperate in the aging of meat with the cathepsins shown in Table 12.15. Most of the cathepsins are also cysteine endopeptidases, which are similar to papain. Their endogenously occurring inhibitors are the cystatins. On the whole, the processes involved in the aging of meat are so unclear that it has not been possible to define markers which can predict the development of tenderness in meat.

12.5 Water Holding Capacity of Meat

Muscle tissue contains 20–25% protein and approx. 74–76% water, i.e. 350–360 g water per 100 g protein. Of this total water not more than 5% is bound directly to hydrophilic groups on the proteins. The rest of the water in the muscle tissue, i.e. 95%, is held by capillary forces between the thick and thin filaments. When a larger amount of water is bound to the network, the muscle is more swollen and the meat is softer and juicier. Hence, water holding capacity, protein swelling and meat consistency are intimately interrelated. The extent of water holding by the protein gel network depends on the abundance of cross-links among the peptide chains. These links may be hydrophobic bonds, hydrogen and ionic bonds and may involve divalent metal ions. A decrease in the number of these cross-linkages results in swelling, whereas an increase in the number of cross-linkages results in shrinkage (syneresis) of the protein gel. The transversal swelling of the myofibrils caused by NaCl has been visualized by phase-contrast microscopy. On washing with 40.6–1.0 mol/L of NaCl, first the centers of the myosin polymer A bands (thick filaments) (cf. 12.3.1 and 12.3.2.1.1) are extracted, and, with increasing concentration, the entire bands are extracted. There is a 2.5 fold increase in the diameter of the myofibrils, corresponding to a 6 fold increase in volume. The cause of these changes is attributed to the depolymerization of the thick filaments to give soluble myosin molecules and the weakening of the bonding of myosin heads to actin. Furthermore, weakening of transversal structural ele-

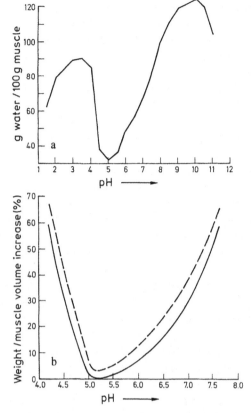

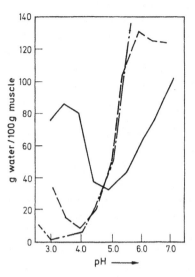

Fig. 12.26. Swelling of meat as affected by salts. Beef muscle homogenate; the ionic strength of the salt added to homogenate is $\mu = 0.20$; — control, – – – NaCl, —·—·— NaSCN (according to *Hamm*, 1972)

The water holding capacity of muscle tissue soon after slaughter is high because the muscle

Fig. 12.25. Swelling of meat as affected by pH. **a** Beef muscle homogenate, 5 days post mortem, **b** beef muscle cut in cubes 3 mm edge length, – – – weight increase, — volume increase (according to *Hamm*, 1972)

ments (M line, Z line, cf. 12.3.2.1.4) probably occurs, which facilitates the extension of myofibrils. The water holding capacity of meat is of great practical importance for meat processing and is affected by pH and the ion environment of the proteins (cf. 1.4.3.1 and 1.4.3.3).

The total charge on the proteins and, hence, their electrostatic interactions are the highest at their isoelectric points. Therefore, meat swelling is minimal in the pH range of 5.0–5.5 (Fig. 12.25). Addition of salt to meat shifts the isoelectric point and, hence, the corresponding swelling minimum to lower pH values, due to the prefered binding of the anion. This means that, in the presence of salts, water holding is increased at all pH's higher than the isoelectric point of the unsalted meat (Fig. 12.26).

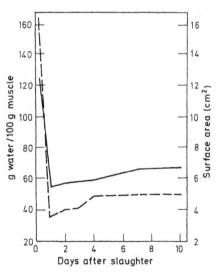

Fig. 12.27. Water holding capacity and rigidity of beef muscle. — Water holding capacity, – – – rigidity (stiffness) expressed as the surface area acquired by homogenate after being pressed between filter papers, under standardized conditions (according to *Hamm*, 1972)

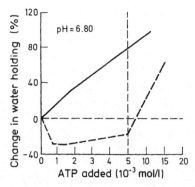

Fig. 12.28. Swelling of meat as affected by ATP addition. Beef muscle homogenate; pH 6.8; — 2 h post-mortem, – – – 4 days post-mortem (according to *Hamm*, 1972)

is still warm and due to the presence of high concentrations of ATP. After the onset of rigor mortis ATP breaks down, the rigidity of the tissue increases and the water holding capacity starts to decrease (Fig. 12.27). Addition of ATP to muscle tissue homogenates prior to the onset of rigor mortis brings about a rise in tissue swelling (Fig. 12.28). Addition of low levels of ATP (to about 1×10^{-3} molar) during post-rigor brings about tissue contraction or shrinkage, while higher levels of ATP cause tissue swelling (Fig. 12.28). This influence on swelling, however, is of short duration since, as ATP breaks down, contraction and shrinkage take place. Nevertheless, these studies amply illustrate the softening effect of ATP and, as already mentioned, the ability of ATP to dissociate actinmyosin complexes (cf. 12.3.2.1.5 and 12.3.2.1.6). Thus, because of high ATP levels and high pH, the slaughtered muscle which is still warm has a high water holding capacity, whereas post-rigor meat, with low ATP and low pH, has a low water holding capacity.

12.6 Kinds of Meat, Storage, Processing

Modern slaughterhouses are higly automated. After the delivered animals are stunned either electrically, or by using a bolt apparatus, or with CO_2, they are bled. The blood (3–4% of the live weight) can be processed into plasma (60–70%) and blood concentrate (30–40%, hemoglobin). The animal bodies are then passed to skinning machines via scalding vats and unhairing machines. Subsequently, the animals are disemboweled, the red organs and the stomach/intestine package are separated for further processing. The animals sides are passed through a shock tunnel (air temperature −4 to −10 °C, 1–2 h). They are stored in the cold until they are cut up on conveyor belts. During processing, accumulating fat is fed to the grease boiler. All discarded materials and bones are processed into meat and bone meal in carcass processing plants. The waste water is treated using specific processes.

12.6.1 Kinds of Meat, By-Products

12.6.1.1 Beef

The most important categories are:

- Young bull meat from full-grown animals (18–22 months, live weight >300 kg): fine fibered, well marbled.
- Cow meat from animals (>2 years) which have already calved: medium red to brown red, moderately fine to coarse fibered, yellow fat, marbled.
- Heifer meat from young, full-grown female animals (15 to 24 months) which have not calved: red, fine fibered, white fat.

The meat of bulls (>5 years) and oxen (2–3 years) is of little economic importance. The average amount of waste from slaughterhouse oxen is 40–55%; that from cows, 42–66%. Beef carcasses are hung for 4–8 days before being cut up for soup meat, and 10–14 days for roasts or steaks.

12.6.1.2 Veal

Meat from young cattle (ca. 4 months) with a body weight up to 150 kg when slaughtered. Color: pale red. The meat aroma is weaker than that of beef. The meat is hung for 8 days before use.

12.6.1.3 Mutton and Lamb

Depending on the age of the animals, the meat has a light, brick, or dark red color and is generally interpersed with fat tissue. The most important types are:

- Lamb from animals not older than 6 months (milk lamb) or 12 months (fattened lamb).
- Mutton from male, castrated and female animals not older than 2 years. The meat of older animals is called *sheep meat.*
- Sheep meat

The odor and taste of mutton and sheep meat are specific.

12.6.1.4 Goat Meat

Goat meat is generally from young animals (2–4 months).

12.6.1.5 Pork

The meat is from very young animals (sucking pig) or from 5–7 month old animals. It exhibits a fairly soft consistency and is fine fibered with a pale pink, pink or whitish grey color. The meat should be hung for 3–4 days before use. The meat becomes greyish-white when cooked, making it different from all other meats. Pork is interpersed and entwined with fat.

12.6.1.6 Horse Meat

The meat of a young horse is bright red, whereas that of older horses is dark or reddish-brown or, when exposed to air, darkens to a reddish-black color. The consistency of the meat is firm and compact and the muscle tissue is not marbled with fat. During cooking, the white fat (melting point 30 °C) appears as yellow droplets on the surface of the broth. The characteristic sweet flavor and taste of the meat are derived from the high glycogen content. In addition to the determination of glycogen, an immunoassay (cf. 2.6.3) and fatty acid analysis can be used to detect horse meat. Horse fat is characterized by a higher content of linolenic acid than beef or pork lard.

12.6.1.7 Poultry

The color of poultry meat differs according to age, breed and body part (breast meat is light, thighs and drumsticks are dark). Species of poultry which have dark meat (geese, ducks, pigeons) can be distinguished from those with light meat (chickens, turkeys, peacocks). The age, breed and feeding of the bird influence meat quality. Poultry fat tends to become rancid because of its high content of unsaturated fatty acids.

12.6.1.8 Game

Wild game can be divided into fur-bearing animals: deer (antelope, caribou, elk, white-tailed deer), wild boars (wild pigs) and other wild game (hare, rabbit, badger, beaver, bear); and birds or fowl (heathcock, partridge, pheasant, snipe, etc.). The meat of wild game consists of fragile fibers with a firm consistency. The meat remains red to red-brown in color. It has low amounts of connective and adipose tissues. The taste and flavor of each type of wild meat is characteristic. Aging of the meat requires a longer time than meat from domestic animals because of the thick and compact muscle tissue structure. The meat then becomes dark-brown to black-red.

12.6.1.9 Variety Meats

Meats of various animal organs are called variety meats. They include tongue, heart, liver, kidney, spleen, brains, retina, intestines, tripe (the first and second stomachs of ruminants), bladder, pork crackling (skin), cow udders, etc. Many of these variety meats, such as liver, kidney or heart, are highly-valued foods because they contain vitamins and trace elements as well as high quality protein. Liver contributes the specific aroma of liver sausage and pastes (goose liver). Liver is also consumed as such. Heart, kidney, lungs, pork or beef stomach, calf giblets and cow's udders are incorporated into sausages: spleen is also made into sausage. Tongues are cooked, pickled and smoked, used for the production of better-quality sausages, and canned or sold as fresh meat. Calf brain and sweetbreads (thymus glands) are especially valued as food for patients. The compo-

Table 12.16. Average composition of some internal organs and blood (g/100 g edible portion)

Organ	Moisture	Protein	Fat	Carbohydrate	Caloric value (kJ)
Heart					
beef	75.5	16.8	6.0	0.56	517
pork	76.8	16.9	2.6	0.4	390
Kidney					
beef	76.1	16.6	5.1	–	471
pork	76.3	16.5	3.8	0.80	435
Liver					
beef	69.9	19.7	3.1	5.90	550
pork	71.8	20.1	4.9	1.14	542
Spleen					
beef	76.7	18.5	2.9		422
pork	77.4	17.2	3.6	–	426
Tongue, beef	66.8	16.0	15.9	0.4	867
Lung, pork	79.1	13.5	6.7	–	477
Brain, veal	80.4	9.8	7.6	0.8	461
Thymus, veal	77.7	17.2	3.4	–	418
Blood					
beef	80.5	17.8	0.13	0.065	309
pork	79.2	18.5	0.11	0.06	319

sitions of some variety meats are shown in Table 12.16.

Intestines, with their high content of elastin, make excellent natural sausage casings. These and beef stomach are specialty dishes.

Pork skin is an ingredient of jellied meat and blood sausage. It is also consumed directly and is a good source of vitamin D. Cartilage and bones contain tendons and ligaments which are collagen- and elastin-type proteins. Cartilage and bones are similar in composition, with the exception of their mineral content; the former contains 1% minerals and the later averages 22% minerals, ranging from 20–70%. The fat content of bones can be as high as 30% and commonly varies between 10–25%. Spinal cord and ribs, when boiled in water, release gelatin-type substances and fat and, therefore, both are used in soup preparations (bouillon, clear broth or bouillon cubes or concentrated stock).

12.6.1.10 Blood

The blood which drains from a slaughtered animal is, on the average, about 3–4% of the live weight (oxen, cows, calves) but is particularly high for horses (9.98%) and low for hogs (3.3%). Blood has been used since ancient times for making blood and red sausages and other food products.

Blood consists of protein-rich plasma in which the cells or corpuscles are suspended. They are the red and white blood cells (erythrocytes and leucocytes, respectively) and the platelets (thrombocytes). The red blood cells do not have nuclei and are flexible round or elliptical discs with indented centers. The diameters of red blood cells vary (in µm: 4 in goat; 6 in pig; 10 in whale; and up to 50 or more in birds, amphibians, reptiles and fish). Red blood cells contain hemoglobin, the red blood pigment. White blood cells contain nuclei but no pigments, are surrounded by membranes, are 4–14 µm in diameter and are fewer in number than red blood cells. In addition to salts (potassium phosphate, sodium chloride and lesser amounts of Ca-, Mg- and Fe-salts), various proteins, such as albumins, globulins and fibrinogen, are present in blood.

The N-containing low molecular weight substances ("residual nitrogen") of blood comprise primarily urea and lesser amounts of amino acids, uric acid, creatine and creatinine. During coagulation or clotting of blood, the soluble fibrinogen in the plasma is converted to insoluble fibrin fibers which separate as a clot. Coagulation is a complex reaction catalyzed by the enzyme thrombin, the precursor of which is prothrombin. Thrombin reacts with fibrinogen to form insoluble fibrin. The mesh of long fibrin fibers traps and holds blood cells (platelets, erythrocytes and leucocytes). Hence, the clot is colored red. The remaining fluid, which contains albumins and globulins, is the serum. Blood plasma contains 0.3–0.4% fibrinogen and 6.5–8.5% albumin plus globulin in the ratio of 2.9:2.0.

The composition of blood is given in Table 12.16. Blood clotting requires the presence of Ca^{2+} ions. Hence, Ca^{2+}-binding agents, such as citrate, phosphate, oxalate and fluoride, prevent blood coagulation. In the processing of blood into food, coagulation is occasionally retarded by stirring the blood with metal rods onto which the fibrin deposits. Currently, blood clotting is inhibited by using Ca^{2+}-complexing salts. After centrifugation, blood stabilized in this way yields about 70% of plasma containing 7–8%

protein. The proteins can be processed further by spray-drying into powdered plasma. Recovery of liquid plasma is permitted only from the blood of cattle (excluding calves) and hogs. Addition of dried and liquid plasma to processed meats is legal. Citrate and/or phosphate are used as calcium-binding agents.

12.6.1.11 Glandular Products

Animal glands, such as the adrenal, pancreas, pineal, mammary, ovary, pituitary and thyroid glands, provide useful by-products for pharmaceutical use. Some of these products are adrenalin, cortisone, epinephrine, insulin, progesterone, trypsin and thyroid gland extract.

12.6.2 Storage and Preservation Processes

Meat must be appropriately treated to allow storage.

12.6.2.1 Cooling

Refrigeration (cooling or freezing the meat) is an important process for prolonged preservation of fresh meat. Carcasses in the form of sides or quarters are cooled. Cooling is performed slowly (e. g., with a blast of air at $0.5\,\mathrm{m/s}$ at $4\,°\mathrm{C}$) or quickly (e. g., stepwise for $3\,\mathrm{h}$ with a $3.5\,\mathrm{m/s}$ blast of air at $-10\,°\mathrm{C}$, for $19\,\mathrm{h}$ with a blast of air at $1.2\,\mathrm{m/s}$ at $2\,°\mathrm{C}$, and over $18\,\mathrm{days}$ with air at $4\,°\mathrm{C}$). The shelf-life of meat at $0\,°\mathrm{C}$ is 3 to 6 weeks. Weight loss due to moisture evaporation is low at high relative humidities, and decreases as the water holding capacity increases.

If meat is cooled to cold storage temperatures ($<10\,°\mathrm{C}$) before rigor occurs, it shrinks and becomes tough. This is due to the fact that at low temperatures, binding of $Ca^{2\oplus}$ by the sarcoplasmic reticulum and mitochondria is reduced, the $Ca^{2\oplus}$ concentration in the intracellular space is increased, inducing contraction (cf. 12.3.2.1.5). To prevent this phenomenon, meat is kept at $15–16\,°\mathrm{C}$ for $16–24\,\mathrm{h}$ and cooled after rigor has occurred. Electrical stimulation is also possible. This process causes rigor by accelerating

glycolysis and a decline in pH. The same effect is achieved by stunning the animals with CO_2.

As long as the meat is stored in the cold in large cuts, lipid oxidation is very slow. Only chopping or mincing or warming of the muscle tissue causes a high rate of peroxidation. Muscle disintegration results in a low but significant release of highly unsaturated membrane phospholipids and Fe^{2+} ions from myoglobin. This non-heme iron is an effective catalyst of lipid peroxidation. Its concentration increases during cooking, as shown for beef in Table 12.18. Even after short cold storage of heated meat and subsequent warming, a rancid off-flavor may develop (warmed over flavor, WOF) (cf. 12.6.2.6 and 12.9.4).

Curing prevents WOF. Myoglobin is stabilized by nitrite, therefore, no additional non-heme iron is formed during cooking (Table 12.18). In addition, the MbNO formed has an antioxidative effect (cf. 12.3.2.2.4). Lipid peroxidation does not occur and new aroma substances are formed that are characteristic of cured meat.

12.6.2.2 Freezing

The shelf life of meat is substantially lengthened by freezing. Freezing can be performed in a single step (direct freezing) or in a two step process (initial cooling followed by freezing) using an air

Table 12.17. Loss of quality of frozen chicken from producer to consumer[a]

Frozen food chain	Average storage temperature (°C)	Shelf-life (day)	Quality loss (% per day)	Average storage time (day)	Quality loss (%)
Producer	−23	540	0.186	40	7.5
Transport	−20	420	0.239	2	0.5
Wholesaler	−22	520	0.196	190	37.1
Transport	−16	370	0.370	1	0.4
Retailer	−20	420	0.239	30	7.2
	−14[b]	210	0.476	3	1.4
Transport	−7	60	1.67	0.17	0.3
Consumer	−12	150	0.666	14	9.3
Σ				280	63.7

[a] For definition cf. Fig. 12.29.
[b] A temperature estimate for food storage on the surface of open freezers.

Table 12.18. Oxidative fat deterioration in cooked beef

	Beef			
	Raw	Cooked	Cured[a]	Cured and cooked[a]
Non-heme iron ($\mu g/g$)	6.62	10.8	6.65	6.80

Storage at 4 °C	Malonic aldehyde	(mg/kg)[b]
0. Day	0.58	0.56
5. Day	1.55	0.48
12. Day	2.78	0.47
21. Day	2.83	0.54

[a] Cured with 156 mg/kg nitrite.
[b] Determined with the thiobarbituric acid test.

blast freezer with an air temperature of $-40\,°C$ and an air stream velocity of 3–10 m/s. The shelf life for storage at $-18\,°C$ to $-20\,°C$ and 90% relative humidity is 9 to 15 months. The shelf life of frozen chicken, as affected by storage temperature, is presented in Fig. 12.29, while Table 12.17 shows the deterioration of frozen chicken as it is shipped from producer to consumer. The shelf life is largely determined by oxidative changes affecting the lipids, which take place more readily in poultry (ducks, geese, chickens) and pork than in beef or mutton.

The water holding capacity of frozen meat increases as the freezing temperature decreases.

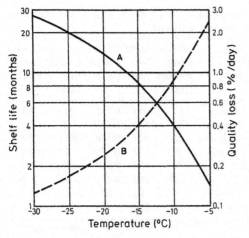

Fig. 12.29. Shelf life of frozen chicken as affected by storage temperature. — Shelf life A, – – – quality loss B = 100/A (according to *Gutschmidt*, 1974)

Water holding capacity also remains high when freezing is performed rapidly. Under these conditions the formation of large ice crystals is suppressed and damage to membranes and the irreversible change in myofibrillar proteins caused by temporary high salt concentrations are avoided.

Freezing meat immediately after slaughtering, without precooling (single-stage process), causes substantial shortening and high fluid losses on thawing, if the meat freezes completely before rigor occurs. However, this is possible only with smaller cuts. The reason for this phenomenon is an extremely fast ATP breakdown with a corresponding decrease in the water holding capacity. The sudden high rate of ATP breakdown is initiated by release of Ca^{2+} ions from the sarcoplasmic reticulum, which triggers the high activity of myosin-ATPase ("thaw rigor"). This "thaw rigor", which is associated with toughness, can be avoided if the warm meat is frozen and then minced in the frozen state after addition of NaCl. Thaw rigor may also be avoided by disintegrating warm meat in the presence of NaCl and then freezing it.

Freezing matured meat results in lower fluid losses than freezing meat in a prerigor or rigor state. However, this process is not widely used for economic reasons. Rigor can be induced before freezing by using electrical stimulation. Long storage of frozen meat results in a decrease in water holding capacity. Solubility changes and shifts in the isoelectric point of proteins of the sarcoplasm and contractile apparatus are observed.

Slow thawing of frozen meat is generally considered more favorable than rapid thawing, although some opposing data exist. Obviously, freezing, storage and thawing should be considered as related process steps, which should be coordinated.

12.6.2.3 Drying

Drying is an ancient method of meat preservation. Drying is frequently used in combination with salting, curing, and smoking. Some processes are: drying in a stream of hot air (40–60 °C), drying in vacuum under variable conditions, e. g., in hot fat, and freeze-drying, the most gentle process. The moisture content of the end product is usually 3–10%. Important quality criteria of such dried meat products are the rehydration capacity, which can be determined by water uptake under standard conditions, and the fraction of firmly-bound water. The drying process should not affect the water holding and aroma characteristics of the meat. The shelf life of dried meat products is limited by the development of off-flavors due to fat oxidation and by discoloration due to the *Maillard* reaction. Dried beef and chicken are important ingredients of many soup powders. In addition to pieces of meat, minced meat, with or without binders, and processed meats, e. g., meat balls or dumplings, are also dried for this purpose.

12.6.2.4 Salt and Pickle Curing

Salt in high concentrations inhibits the growth of microorganisms and curtails activity of meat enzymes. Hence, salt is considered as a meat preservative. Salting meat at a level up to 5% NaCl causes swelling (cf. Fig. 12.26). Higher salt concentrations (10–20%) induce shrinkage in meat and its products, causing a decrease in moisture to a level below that of untreated meat. The meat retains its natural color, usually dark red, since the myoglobin concentration increases due to the moisture loss. The color of such meat changes upon cooking to grayish-brown.

Salting by the addition of sodium nitrite and/or nitrate (curing or pickling) produces products with highly stable color (cf. 12.3.2.2.4). Since nitrite reacts faster and less is required for color stabilization, it is widely used in place of nitrate. Salt curing is done either by rubbing salt on the meat surface (dry curing or pickling), by submerging the meat in 15–20% brine (wet pickle curing), or by injection of brine in special automats.

Additives, such as sugar or spices, which favorably affect the red color and formation of meat aroma, are often added to pickling salts. The aroma of cured meat is specific and differs from that of noncured meat. Aroma formation is enhanced by the microflora (*Micrococcus* spp. and *Achromobacter* spp.) of curing brine, which are simultaneously involved in reduction of nitrate (NO_3^-) and nitrite (NO_2^-) ions and thereby contribute to the stabilization of the pinkish or red color of cured meat.

12.6.2.5 Smoking

Smoking of meat is usually associated with salting. Depending on the smoking procedure, the moisture drops 10–40%. Compounds present in smoke with bactericidal and antioxidative properties are deposited on and penetrate into the meat. Important smoke ingredients include phenols, acids, and carbonyl compounds. The concentration of polycyclic hydrocarbons in smoke depends on the type of smoke generation and can be largely suppressed by suitable process management, e. g., by external smoke generation with cleaning of the smoke via cold traps, showers, or filters. A distinction is made between hot smoking (50–85 °C) over a period ranging from less than one hour to several hours (e. g., used for cooked and boiling sausages), warm (25–50 °C) and cold smoking (12–25 °C) over a period ranging from two days to several weeks (e. g., used for raw sausage and ham). Special smoking processes include wet smoking processes, electrostatic processes, and the use of smoke condensates.

12.6.2.6 Heating

Heat treatment is an important finishing process and also serves for the production of canned meat. Typical changes involved in heat treatment are: development of grayish-brown color, protein coagulation, release of juices due to decrease in water holding capacity (Fig. 12.30), increase in

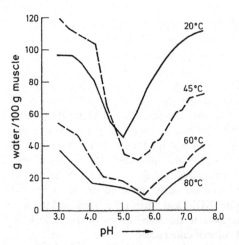

Fig. 12.30. Water holding capacity of beef muscle versus heat treatment and pH. (according to *Hamm*, 1972)

Table 12.19. Effect of can size and product on required heating time of canned meat (time in min to reach 121 °C at the center of the can)

Canned meat	400 g	850 g	2500 g
Beef	47	57	80
Pork	58	98	120
Liver sausage	90	130	
Blood sausage	106	113	130

pH, development of a typical cooked or roasted meat aroma and, finally, softening induced by the shrinking and partial conversion of collagen to gelatin (cf. 12.3.2.3.1).

Refrigerated storage of heated meat and reheating may lead to WOF (cf. 12.6.2.1 and 12.9.4).

12.6.2.7 Tenderizing

Plant enzyme preparations (ficin, papain, bromelain) are used to tenderize meat. These substances are either sprayed onto the meat cuts or are distributed via the blood vessels of the animal either shortly before or after slaughtering.

12.7 Meat Products

Canned meat, ham, and sausages, and meat extracts are produced from meat.

12.7.1 Canned Meat

Examples of canned meat are beef and pork in their own juice, corned beef, luncheon meat, cooked sausages, jellied meat, and cured and pickled hams. In order to achieve sterile canned

meat, the required heating time and temperature depend on the size and content of the can since heat penetration is highly variable (Table 12.19). Therefore, mathematical models have been developed which allow the temperature to be controlled in such a way that even the coldest point in the contents is heated to a temperature high enough and for long enough to kill pathogenic bacteria and microbes responsible for spoilage. Correspondingly controlled processes are also used in the production of cooked and boiling sausages.

12.7.2 Ham, Sausages, Pastes

12.7.2.1 Ham, Bacon

12.7.2.1.1 Raw Smoked Hams

After the center ham has been cut (longitudinal or circular), *ham on the bone* is dry, then wet cured (4–7 weeks), matured (reddened) for 2–3 weeks by dry storing, followed by washing, drying and exposure to cold smoke for 4–7 weeks. In *rolled ham*, the bone is taken out, and it is subsequently processed like ham on the bone, except that the curing time is shorter. *Lightly-salted lean hams* are made from cutlet or chop meats by a mild curing process, filled into casings and warm smoked.

12.7.2.1.2 Cooked Ham

Bone-free ham is cured for 2–3 weeks, stored dry to mature, washed and warm smoked. It is subsequently cooked by gently simmering. In the cook-in process, the cured meat is first packed in foil that is resistant to boiling, then cooked and smoked. *Praha ham* is a special cooked ham which is often baked in a bread dough.

12.7.2.1.3 Bacon

Back fat from the pig is salted, washed, dried, and cold smoked.

12.7.2.2 Sausages

Sausage manufacturing consists of grinding, mincing or chopping the muscle tissue and other organs and blending them with fat, salts, seasonings (herbs and spices) and, when necessary, with binders or extenders. The sausage mix or dough is then stuffed into cylindrical synthetic or cellulose casings or tubings of traditional sausage shape or, often, natural casings, such as hog or sheep intestines or the hog's bun (for liver sausage) are used. They are sold as raw, precooked or cooked, and/or smoked sausages. The composition of ham and sausage products is shown in Table 12.20. The different types of sausages have in common that a continuous, hydrophilic salt/protein/water matrix stabilizes a disperse phase (coarse meat/fat particles, fat

Table 12.20. Protein and fat content of ham and sausage products

Product	Moisture %	Protein %	Fat %	Caloric value (kJ) (kJ/100 g)
Salami				
(German style)	40	21	33	1578
Cervelat sausage	41	20	34	1598
Knackwurst	60	12	26	1166
Bratwurst (pork)	57	12	29	1277
Hunter's sausage	64	16	16	864
Gelbwurst	58	11	27	1186
Munich Weisswurst				
(white sausage				
Munich style)	62	11	25	1112
Bockwurst	59	12	25	1129
Liver sausage	52	12	29	1351
Rotwurst	56	12	29	1277
Ham, raw	43	18	33	1527
Ham, cooked	70	23	4	539
Bacon, marbled	20	9	65	2558

globules, insoluble proteins, connective tissue, and seasoning particles). The stability of systems of this type is influenced by the pH value, ionic strength, melting range of the lipids, and by the protein content. In finely ground systems with emulsion character, the grinding temperature is also important for stability. A temperature of 14 °C is regarded as optimal, unstable products resulting at T > 20 °C.

In the emulsions mentioned above, a monomolecular protein film is formed around the fat globules present (Fig. 12.31). The importance of the different protein components as film formers decreases in the following order: myosin > actomyosin > sarcoplasma proteins > actin. The hydrophobic heads of the myosin molecules evidently dip into the fat globules, while the tails interact with actomyosin in the continuous phase. The monomolecular myosin layer formed in this way should have a thickness of ∼130 nm. On the outside, there is probably a multimolecular actomyosin layer which binds water and contributes to the stabilization of the emulsion because of its viscous, elastic, and cohesive properties. Higher temperatures, which lead to destabilization (see above), probably cause increased protein/protein interactions in the actomyosin layer which, in turn, result in a decrease in the water binding, elasticity losses, and disturbances in the myosin film.

While the formation of myosin films on fat globules is responsible for the stabilization of raw sausages with emulsion character, protein/protein interactions and gel formation are important for the stabilization of fat and water in the system in the case of cooked and boiling sausages.

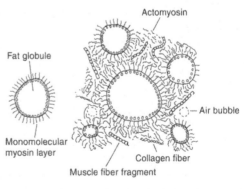

Fig. 12.31. Schematic representation of a sausage emulsion (according to *Morrissey* et al., 1987)

12.7.2.2.1 Raw Sausages

Typical products are Cervelat sausage, salami and the German Mettwurst. They are made of raw, ground skeletal muscle (pH < 5.8), fat and spices. They are cured by the addition of 2.8–3.2% of nitrite curing salt (cf. 22.2.4) or common salt and sodium nitrate (max. 300 mg/kg), max. 2% of sugar (sucrose, glucose, maltose, lactose) and other curing aids (D-gluconic acid-5-lactone, ascorbic acid etc.). Starter cultures are normally added to optimize ripening. The production of raw sausage is schematically presented in Fig. 12.32.

The grinding is preferably carried out in a cutter in which the following steps are conducted: cutting, distribution, mixing, very fine grinding, and emulsification. A cutter consists essentially of a dish rotating around a highspeed cutter head. Products of varying texture can be produced by varying the form and speed of the cutter, the size of the cutting chamber (possible insertion of stowing rings), plate speed, and cutting duration. Operation under a vacuum has advantages. Apart from batch cutters with a dish content of up to 750 l, continuously operating dish cutters are also available today.

In the case of firm types of raw sausages, frozen material is used for grinding (−20 °C) and the temperature is kept below 4 °C during the grinding process by cooling. After the mass has been stuffed, the sausage is ripened in air conditioned rooms. At first, the temperature is 20–26 °C (air humidity 90–95%) to increase lactobacilii; it is later reduced to 10–15 °C (air humidity 75%).

The specific aroma is formed in the course of ripening by microorganisms present (micrococci and lactobacilli, often added in the form of starter cultures). Reddening (cf. 12.3.2.2.4) also plays a big role. The drop in pH due to lactic acid formation (5.2–4.8) results in shrinkage of the protein gel. The sausages become stable, firm and suitable for slicing after vaporization of the water released (20–40% weight loss). Ripening takes 2–3 weeks (fast processes) or 7–8 weeks (slow processes). The raw sausage is subsequently smoked, e. g., cold smoking at 16–28 °C. The white layer on various types of salami is due to mold mycelia or, in cheaper products, a layer of lime milk dip.

12.7.2.2.2 Cooked Sausages

Cooked sausages are made from cooked starting materials. Typical products are liver sausage, blood sausage, and jellied sausage. The production of liver sausage is shown schematically in Fig. 12.33. Modern plants generally use cooking cutters in which the following steps are conducted in one machine: preliminary grinding, cooking, mixing, and cutting. In comparison with boiling sausages, cooked sausages can be cut only when cold.

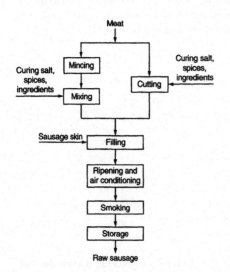

Fig. 12.32. Production of raw sausage

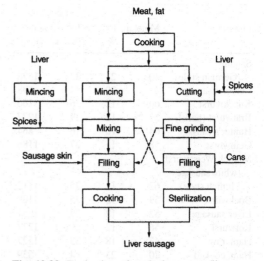

Fig. 12.33. Production of cooked sausage (liver sausage)

12.7.2.2.3 Boiling Sausages

Boiling sausages are made from raw meat (beef, pork, veal) by boiling, baking or frying. During processing in the cutter, the water holding capacity is increased by the addition of common salt and cutter aids (condensed phosphates, lactate, acetate, tartrate, and citrate). The swelling resulting from added salts is caused by an increase in the pH of the meat slurry and by the complexing of divalent cations which, in free form, suppress swelling.

The temperature during grinding/chopping has to be kept low (addition of ice or ice-cold water) since higher temperatures decrease the water holding capacity. Water retention increases as the fat component of the meat slurry is increased as long as the fat:protein ratio does not exceed 2.8 to 1. As a consequence the salt concentration is increased. After chopping and stuffing, the sausages are hot smoked and scalded at 72–78 °C. At this temperature, coagulation of protein gel, which holds the water, forms the broken texture so typical of these sausages.

Typical products are bockwurst, wieners, white and hunter's sausage and mortadella. Figure 12.34 schematically shows the production of boiling sausages.

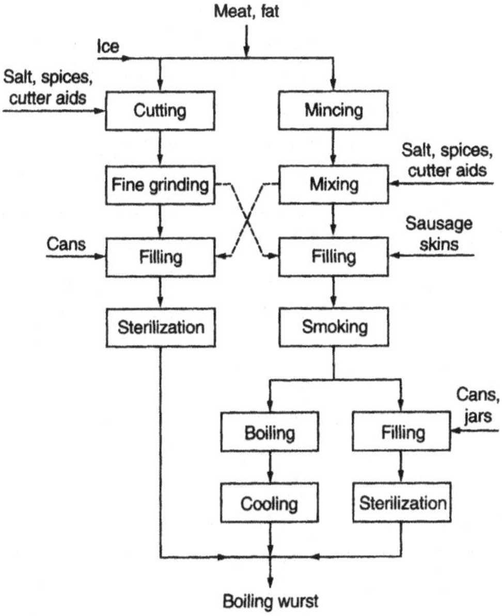

12.7.2.3 Meat Paste (Pâté)

12.7.2.3.1 Pastes

Meat pastes are delicately cooked meat products made primarily from meat and fat of calves and hogs and, often, from poultry (e. g. goose liver paste) or wild animal meat (hare, deer or boar). Unlike sausages, pastes contain quality meat and are free of slaughter scrapings or other inferior by-products. A portion of meat or the whole meat used is present as finely comminuted spreadable paste.

12.7.2.3.2 Pains

Pains usually consist of larger pieces of meat (especially game and poultry), which are processed into a cooked sausage-like mass with fat, truffle, and various spices.

12.7.3 Meat Extracts and Related Products

12.7.3.1 Beef Extract

Meat extract is a concentrate of water-soluble beef ingredients devoid of fat and proteins. Its preparation dates back to *Liebig's* work in Munich in 1847. Comminuted beef is counter-currently extracted with water at 90 °C. After removal of fat by separators and subsequent filtration, the extract containing 1.5–5% solids is concentrated to 45–65% solids in a multiple stage vacuum evaporator which operates in a decreasing temperature gradient (a range of 92 to 46 °C). The final evaporation to 80–83% solids is then carried out under atmospheric pressure at a temperature of 65 °C or higher or under vacuum on a belt dryer.

In the same way, the cooking water recovered during the production of corned beef can be processed into meat extract. Only this latter source of meat extract is of economic significance. The yield is 4% of fresh meat weight.

Fig. 12.34. Production of boiling sausage ("Brüh-wurst")

Table 12.21. Chemical composition of beef extract

	%
Organic matter	56–64
Amino acids, peptides	15–20
Other N-compounds	10–15
Total creatinine	5.4–8.2
Ammonia	0.2–0.4
Urea	0.1–0.3
N-free compounds	10–15
Total lipids	>1.5
Pigments	10–20
Minerals	18–24
Sodium chloride	2.5–5
Moisture	15–23
pH-value of a 10% aqueous solution	approx. 5.5

The composition of the extract is given in Table 12.21. For addition to soup powders and sauce powders, the thick pasty meat extract is blended with a carrier substance and vacuumor spray-dried.

12.7.3.2 Whale Meat Extract

This product is obtained from meat of various whales (blue, finback, sei, humpback and sperm) in a process similar to that used for beef extract.

12.7.3.3 Poultry Meat Extract

Chicken extract is obtained by evaporation of chicken broth or by extraction of chicken halves with water at 80 °C, followed by a concentration step under vacuum to an end-product of 70–80% solids.

12.7.3.4 Yeast Extract

Yeast cells (*Saccharomyces* and *Torula* spp.) are forced to undergo shrinking of protoplasm by addition of salt, which causes loss of cell water and solutes (plasmolysis), or the cells are steamed or subjected to autolysis. Cells treated in this way are extracted with water and the extract is concentrated to yield a brown paste. Yeast ex-

tract is rich in the B-vitamins. The concentrations of thiamine and thiamine diphosphate are above their taste threshold values and may contribute to the product's unpleasant flavor. On the other hand, the spicy flavor of the paste is essentially due to 5′-nucleotides freed during hydrolysis and to amino acids, particularly glutamic acid.

12.7.3.5 Hydrolyzed Vegetable Proteins

The production of this protein hydrolysate is schematically presented in Fig. 12.35. According to the given formulation, the different plant protein-containing raw materials, such as wheat and rice gluten and roughly ground soybeans, palm kernels or peanuts, are automatically delivered from raw material silos, weighed, and fed to a hydrolysis boiler (double-walled,

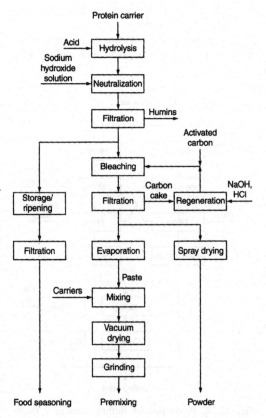

Fig. 12.35. Production of hydrolyzed vegetable protein

pressure-stable stirred tank). Hydrolysis proceeds at temperatures above 100 °C and the appropriate pressure with hydrochloric acid or sulfuric acid (salt-free seasoning).

The hydrolysate is subsequently neutralized to pH 5.8 with sodium or calcium carbonate or with sodium hydroxide solution. In this process, the pH range of 2.5–4 must be passed through as quickly as possible to repress the formation of pyrrolidone carboxylic acid from glutamic acid.

The hydroysate is filtered and the filtrate (seasoning) stored. The filtration residue is washed with water and refiltered, if necessary. The diluted filtrate is evaporated and added to the seasoning obtained in the first step.

The seasoning is subsequently stored; it is filtered several times before filling. Apart from liquid food seasoning, seasoning in paste and powder form and mixtures for use in dry soups and sauces are produced. These products are partly bleached with activated carbon and the taste is neutralized.

The compound 3-hydroxy-4,5-dimethyl-2(5H)-furanone (HD2F, cf. 5.3.1.3) is responsible for the intensive, typical seasoning aroma. The products have a meat- or bouillon-like odor and taste. It was found in 1978 that genotoxic compounds are formed in hydrochloric acid hydrolysates of protein-containing raw materials. Thus, 3-chloropropane-1,2-diol, 2-chloropropane-1,3-diol, 1,3-dichloropropane-2-ol, 1,2-dichloropropane-3-ol, and 3-chloropropane-1-ol have been identified as secondary products of lipids in amounts of 0.1 to >100 ppm in commercial protein hydrolysates and products derived from them. In feeding experiments on rats, these dichloro compounds were found to be carcinogenic. The testing of the monochloro compounds is still in progress. The chlorinated glycerols, which are partly also present as fatty acid esters, have half life periods of several hundred days in the hydrolysates. The N-(2,3-dihydroxypropyl) derivatives of the amino acids serine and threonine as well as 3-aminopropane-1,2-diol have been detected as aminolysis products.

Chlorinated steriods, e. g., 3-chloro-5-cholestene (Formula 12.27a), 3-chloro-24-methyl-5,22-cholestadiene (Formula 12.27b) and 3-chloro-24-ethyl-5,22-cholestadiene (Formula 12.27c), have been identified in the insoluble residue of the corresponding products.

a R =

b R = , R¹ = Me

c R as in b, R¹ = Et

(12.27)

Moreover, there have been indications of the presence of chlorinated *Maillard* compounds in hydrochloric acid hydrolysates, e. g., 5-(chloromethyl)furfural.

To avoid or minimize the unwanted compounds mentioned above, the production process has been or is being modified, e. g., in the form of an additional alkali treatment of the hydrochloric acid hydrolysate. Thus, concentrations of <1 ppm of 3-chloro-1,2-propanediol were found in the majority of samples tested in 1990, which is clearly less than it was in previous years.

12.8 Dry Soups and Dry Sauces

Meat extract, hydrolysates of vegetable proteins, and yeast autolysate are used to a large extent in the production of dry soups amd dry sauces. For this reason, these substances will be described here. The industrial production of these products for use in home and canteen kitchens has become increasingly important in the past 20 years. In particular, a special pretreatment of the raw materials made possible the development of products which, after quick rehydration, give ready-to-consume complete meals (dry stews), snacks between meals (dry soups, instant soups), or sauces.

12.8.1 Main Components

Not only meat extracts, protein hydrolysates, and yeast autolysates, but also glutamate, ribonucleotides (inosinate/guanylate), and reaction aromas are used as the taste-bearing substance (cf. 12.9.3). These substances are dried with and without a carrier (belt vacuum drying, spray drying). Flour (wheat, rice, corn), legume flour (peas, lentils, beans), and starches (potato, rice and corn) serve as binding agent. Apart from native flour or starch, swelling flour or instant starch that is pregelatinized by drum drying or boil extrusion is used. In fact, especially good swelling and dispersing properties are achieved by agglomeration. Legumes are precooked in pressure vessels for up to several hours before drying. The rehydration time can be reduced to 4–5 minutes by freeze drying. Standard products are normally air dried on belt dryers. Pasta is subjected to a precooking process by means of steam and/or water or used in a fat dried form, like in the Far East.

Rice is added in a pre-cooked, freeze-dried form or as reformed rice (dried rice flour extrudate). After the appropriate pretreatment (e. g., blanching), vegetables and mushrooms are dried (drum, spray, and freeze drying). Products with instant character are obtained by centrifugal fluidized bed drying. In this process, which is used on a large scale for carrots and rice, the products in a perforated and basket-shaped rotating cylinder are dried with hot air of ca. 130 °C with simultaneous puffing. The fats used are mainly beef fine tallow, hardened plant fats, chicken fat, and milk fat. These fats are often applied in powder form (cf. 14.4.7). The meat additives are primarily beef and chicken which are air dried or freeze dried. To perfect the taste, salt and spices are used as ground natural spices or in the form of spice extracts.

To improve the technological properties, dry soups and sauces contain a series of other ingredients, e. g., milk products, egg products, sugar, and maltodextrin, acids, soybean protein, sugar coloring, and antioxidants.

12.8.2 Production

The production of dry soups and sauces essentially involves mixing the preproduced raw materials. The process steps are shown in Fig. 12.36. Weighing of individual components from the raw material silos and their pneumatic dosing

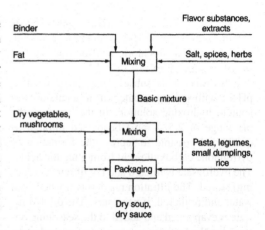

Fig. 12.36. Production of dry soups and sauces

into the mixer are conducted automatically. In soup mixtures that contain breakable components, such as pasta and dry vegetables, a basic mixture of the powdery components (binder, fat powder, extract powder etc.) is first produced in high-speed mixers. The breakable components are gently mixed in a second slow mixing step. The mixtures are agglomerated for special uses (instant soups and sauces); they generally have no coarse components. This is usually conducted in batchwise or continuously operated fluid bed spray granulators. In continuous agglomeration plants (Fig. 12.37), extract substances and fat are dosed in separated systems. Alternatively, finished soup/sauce mixtures are agglomerated by back wetting with steam or water and dried

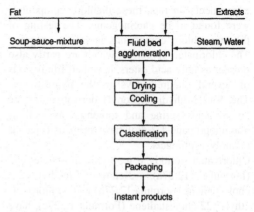

Fig. 12.37. Production of instant products by agglomeration

via a separate fluid bed. The packaging materials used protect the dry mixture from light, air, and moisture.

12.9 Meat Aroma

Raw meat has only a weak aroma. Numerous intensive aroma variations arise from heating, the character of the aroma being dependent on the type of meat and the method of preparation (stewing, cooking, pressure cooking, roasting or broiling-barbecuing). The preparation effects are based on reaction temperatures and reactant concentrations. Thus, a carefully dried, cold aqueous meat extract provides a roasted meat aroma when heated, while an extract heated directly, without drying, provides a bouillon aroma.

12.9.1 Taste compounds

Meat aroma consists of: (a) nonvolatile taste substances, (b) taste enhancers and (c) aroma constituents. The latter compounds or their precursors originate essentially from the water-soluble fraction. The constituents listed in Table 12.22 have been identified as the taste substances of beef broth and roasted meat juice. Solutions of these substances in the given concentrations (Table 12.22) give the typical taste profiles, which are composed of sweet, sour, salty, and glutamate-like (umami) notes. The meat note is produced by odorants.

12.9.2 Odorants

Dilution analyses were used to elucidate the potent odorants (Table 12.23) of boiled beef and pork and of the meat and skin of fried chicken. Omission experiments (cf. 5.2.7) show that octanal, nonanal, (E,E)-2,4-decadienal, methanethiol, methional, 2-furfurylthiol, 2-methyl-3-furanthiol, 3-mercapto-2-pentanone and HD3F are the key aroma substances of boiled beef. These compounds are also present in boiled pork and chicken, but species-specific differences

Table 12.22. Taste compounds in beef broth and pot roast gravy

Compound/Ion	Concentration (mmol /l)	
	Broth[a]	Roast gravy[b]
Aspartic acid	0.05	0.18
Alanine	–[c]	9.41
Glutamic acid	0.3	1.71
Cysteine	–[c]	0.48
5'-AMP	0.14	0.64
5'-IMP	0.4	7.82
Hypoxanthine	–[c]	3.62
Carnosine	6.2	23.4
Anserine	0.7	–[c]
Lactic acid	25.6	155
Succininc acid	–[c]	2.16
Carnitine	2.0	–[c]
Pyroglutamic acid	2.6	–[c]
Creatinine	–[c]	43.3
Creatine	–[c]	20.3
Sodium	2.3	35.6
Potassium	31.3	170
Magnesium	3.0	12.1
Calcium	1.0	–[c]
Chloride	3.1	18.9
Phosphate	10.1	49.4

[a] Ground meat (500 g) suspended in 1 l of water and boiled for 2 h, followed by fat separation and filtration.
[b] Meat (2 kg) fried for 20 min and braised for 4 h after the addition of 1 l of water. The meat juice or gravy is poured off.
[c] Does not contribute to taste in the sample.

in concentration exist. The meaty/caramel-like note typical of beef is produced by 2-furfurylthiol,2-methyl-3-furanthiol and HD3F, which occur in relatively high concentrations in this meat. In comparison, the lower concentration of HD3F in pork is due to the considerably lower contents of the precursors glucose 6-phosphate and fructose 6-phosphate.

The aroma of boiled pork is not as intensive as that of beef and the fatty note is more pronounced. The concentrations of the fatty smelling carbonyl compounds, e. g., hexanal, octanal and nonanal, are lower in pork, but in proportion to the concentrations of 2-furfurylthiol, 2-methyl-3-furanthiol and HD3F, they are higher than in beef. This difference appears to favor the intensity of the fatty note in the odor profile of pork. In chicken, the fatty notes become even more noticeable due to

Table 12.23. Concentrations of odorants in boiled beef and pork and in fried chicken

| | Concentration (µg/kg) | | | |
	Beef[a]	Pork[a]	Chicken[b] Meat	Skin
Acetaldehyde	1817	3953	3815	3287
Methylpropanal	117	90	83	538
2-Methylbutanal	n.a.	n.a.	8	455
3-Methylbutanal	26	27	17	668
Hexanal	345	173	283	893
Octanal	382	154	190	535
1-Octen-3-one	9.4	4.8	7.2	10.8
Nonanal	1262	643	534	832
(Z)-2-Nonenal	6.2	1.4	5.5	10.5
(E)-2-Nonenal	32	15	23	147
(E,E)-2,4-Decadienal	27	7.4	11	711
12-Methyltridecanal	961	n.a.	n.a.	n.a.
Hydrogen sulfide	n.a.	n.a.	290	n.a.
Methanethiol	311	278	202	164
Dimethylsulfide	105	n.a.	n.a.	n.a.
Methional	36	11	53	97
2-Furfurylthiol	29	9.5	0.1	1.9
2-Methyl-3-furanthiol	24	9.1	0.4	4.1
3-Mercapto-2-pentanone	69	66	29	27
4-Hydroxy-2,5-dimethyl-3 (2H)-furanone(HD3F)	9075	2170	50	395
2-Acetyl-2-thiazoline	1.4	1.6	2.6	5.8
2-Acetyl-1-pyrroline	1.1	n.a.	0.2	2.9
2-Propionyl-1-pyrroline	n.a.	n.a.	n.a.	0.8
2-Ethyl-3,5-dimethylpyrazine	n.a.	n.a.	n.a.	4.3
2,3-Diethyl-5-methylpyrazine	n.a.	n.a.	n.a.	2.5
p-Cresol	5.9	n.a.	3.4	1.1
Guaiacol	4.3	n.a.	4.3	<1
Butyric acid	7024	17,200	8119	4817

[a] Beef (8.8% fat) and pork (1.7% fat) were boiled for 45 min at a pressure of 80 kPa at 116 °C.
[b] Chicken was fried for 1 h in coconut fat at 180 °C, breast meat (1% fat) and skin (46% fat) were analyzed separately.
n.a: not analyzed

the low formation of the two sulfur-containing odorants and HD3F.

The aroma of fried chicken is primarily caused by the *Strecker* aldehydes methyl propanal, 2-and 3-methyl butanal as well as the roast aroma substances 2-acetyl-2- thiazoline, 2-acetyl-1-pyrroline and the two alkyl pyrazines. The thiazoline and the pyrroline are also formed in lower concentrations during the boiling of meat. 2-Acetyl- 2-thiazoline is the most important roast aroma substance in meat fried for only a few min-utes. It decreases on longer heating (cf. 5.3.1.5), while the stable alkylpyrazines increase further.

If beef is heated for a longer period of time, 12-methyltridecanal (MT) appears as an important odorant. Especially in a pot roast, this substance is one of the indispensable aroma substances, which develops its full effect on retronasal detection and increases mouth feeling. The precursors of MT are plasmalogens which occur in the membrane lipids of muscle and slowly hydrolyze on heating (cf. Formula 12.28).

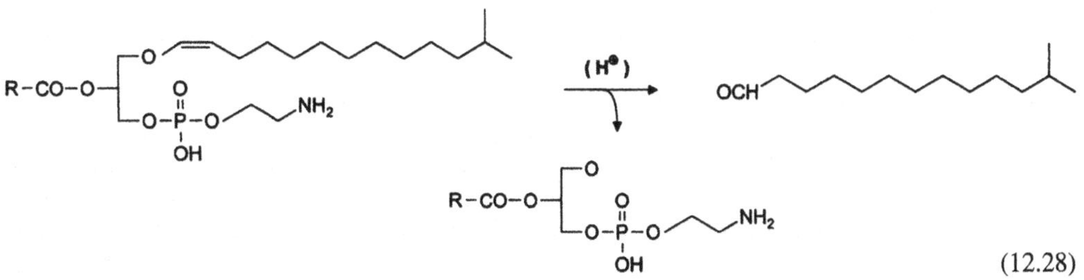

R–CO–O⟶ ... O–P–O ... NH$_2$ ... OH   (H$^\bullet$) ⟶ OCH ...

R–CO–O ... O–P–O ... NH$_2$ ... OH

$$(12.28)$$

Larger amounts of MT are released only on hydrolysis of the lipids in the meat of ruminants, but not from the lipids of prok and poultry meat, as shown in Table 12.24. In fact, there are indications that microorganisms present in the stomach of ruminants produce MT which is then incorproated into plasmalogens.

The MT concentration in the phospholipids of bovine muscle increases with increasing age. The studies conducted until now indicate a linear relationship, which could be of interest for the determination of the age of beef.

Important aroma substances of raw and cooked mutton are listed in Table 12.25. A special feature is the two branched fatty acids, which are already present in the raw meat and produce the "mutton" odor. (E)-2-Nonenal and the other odor substances from lipid peroxidation are also present in not inconsiderable concentrations in the raw meat. Only HD3F is formed during cooking.

Table 12.24. Release of 12-methyltridecanal (MT) on hydrolysis of lipids from different animal species

Animal species	Lipid (g/kg)	MT (µg/g lipid)
Beef[a]	14–22	55–149
Beef[b]	n.a.	44–63
Veal[a]	12	19
Red deer[a]	25	5
Springbok[a]	14	16
Pork[a]	15–19	1.3–2.7
Pork[b]	n.a.	1.6
Chicken[b]	n.a.	0.3
Turkey[b]	n.a.	1.6

The samples were refluxed: [a] with HCl (1 h), [b] at pH 5.7 (4 h).
n.a.: not analyzed

12.9.3 Process Flavors

Aromas obtained by heating aroma precursors are used in the aromatization of foods. An important aim of process flavors is the production of odor qualities similar to those of meat. This is achieved especially on heating cysteine with ribose, as shown in Table 12.26. Glucose is less effective and rhamnose promotes the formation of HD3F.

For economic reasons, attempts are made to replace individual precursors with inexpensive materials, e. g., a relatively inexpensive protein hy-

Table 12.25. Comparison of the aroma substances of raw (I) and cooked, lean mutton (II)

Compound	Amount (µg/kg)	
	I	II
4-Ethyloctanoic acid	255	217
4-Methyloctanoic acid	278	502
(E)-2-Nonenal	27	21
(E,E)-2,4-Decadienal	2.9	4.6
(E,E)-2,4-Nonadienal	1.4	3.8
(Z)-1,5-Octadien-3-one	0.8	2.1
4-Hydroxy-2,5-dimethyl -3(2H)-furanone (HD3F)	<50	9162

Table 12.26. Formation of relevant aroma substances on heating cysteine with ribose, glucose or rhamnose[a]

Compound	Amount (µg/kg)		
	Ribose	Glucose	Rhamnose
2-Furfurylthiol	12.1	2.8	0.8
2-Methyl-3-furanthiol	19.8	1.9	0.8
3-Mercapto-2-pentanone	59.9	13.9	7.3
4-Hydroxy-2,5-dimethyl- 3(2H)-furanone (HD3F)	18.5	79.4	19,800

[a] Mixtures of cysteine (3.3 mmol) and the monosaccharide (10 mmol) dissolved in phosphate buffer (100 ml; 0.5 mol/l; pH 5.0) were heated to 145 °C in 20 min.

drolysate is used as the amino acid source and other important precursors of meat aromas like thiamine and monosaccharide phosphates are applied in the form of yeast autolysates.

Fats or oils are added to produce the carbonyl compounds which contribute to the animal species specific note of meat aroma.

12.9.4 Aroma Defects

If cooked meat is stored for a short time, e. g., 48 h at ca. 4 °C, an aroma defect develops, which becomes unpleasantly noticeable especially after heating and is characterized by the terms metallic, green, musty and pungent. This aroma defect, also called warmed over flavor (WOF), is caused by lipid peroxidation (cf. 12.6.2.1). The indicator of this aroma defect is hexanal, which increases as shown in Table 12.27.

Other changes which contribute to the aroma defect are the increase in metallic/musty smelling epoxydecenal, which, like hexanal, is formed in the peroxidiation of linoleic acid (cf. 3.7.2.1.9), and the decrease in HD3F. The latter is probably due to the reaction of its enolic OH group with peroxy radicals.

The WOF appears in chicken much faster because its linoleic acid content is about 10 times higher than that in beef. Apart from the changes in concentration of the odorants listed in Table 12.27, the degradation of 2,4-decadienal, which is typical of an advanced lipid peroxidation

(cf. 3.7.2.1.9), has an additional negative effect on the aroma.

The WOF is inhibited by additives which bind Fe ions, e. g., polyphosphates, phytin, and EDTA. In comparison, antioxidants are almost ineffective. Therefore, it is assumed that a site specific mechanism is involved in the formation of WOF. The Fe ions liberated in the cooking process are bound by the phospholipids via the negatively charged phosphate residues and, consequently, adjoin the unsaturated acyl residues of these lipids. Radicals from the *Fenton* reaction of Fe ions with hydroperoxides (cf. 3.7.2.1.8) attack only the unsaturated acyl residues, starting their peroxidation. This hypothesis can also explain the observation that multivalent ions (Ca^{3+}, Al^{3+}) inhibit WOF as they probably displace the Fe ions from the phospholipids.

12.10 Meat Analysis

12.10.1 Meat

The determination of the kind of animal, the origin of meat, differentiation of fresh meat from that kept frozen and then thawed, and the control of veterinary medicines is of interest. The latter include antibiotics (penicillin, streptomycin, tetracyclines, etc.) used to treat dairy cattle infected with mastitis, and other chemicals, including diethyl stilbestrol, used for cattle to increase the efficiency of conversion of feed into meat.

12.10.1.1 Animal Origin

The animal origin of the meat can be determined by immunochemical and/or electrophoretic methods of analysis as well as by PCR. The PCR method is described in 2.6.4.2.2. Electrophoretic protein analysis will be discussed here. The sexual origin of a meat sample can also be of interest, as discussed here for beef.

12.10.1.1.1 Electrophoresis

To determine the animal or plant origin of the food, electrophoretic procedures have often

Table 12.27. Changes in the concentrations of important aroma substances on cold storage and reheating of roasted beef

Compound	Concentration (µg/kg)	
	I[a]	II[b]
Hexanal	269	2329
trans-4,5-Epoxy-(E)-2-decenal	1.5	10.7
4-Hydroxy-2,5-dimethyl-3(2H) -furanone (HD3F)	1108	665

[a] I: hamburgers were fried for 7 min.
[b] II: as in I, then storage at 4 °C for 48 h and heated at 70 °C for 45 min until a core temperature of 60–65 °C is reached.

proved to be valuable when the electrophero-grams of the protein extracts reveal protein zones or bands specific for the protein source. Thus, in meat analysis, such a method allows for the differentiation between more than 40 animal species, e. g., beef, pork, horse, buffalo, sheep, game, and poultry (cf. Fig. 12.38).

To carry out an analyis, the sarcoplasm proteins are extracted with water. The electrophoretic separation is predominantly conducted on poly-acrylamide gels, previously also on starch and agarose gels. The application of a pH gradient (isoelectric focusing) provides excellent protein patterns. The first assignment is achieved directly after the electrophoretic separation on the basis of two red myoglobin zones (Fig. 12.38a). The ratio of the intensities of these zones, which

represent met- and oxymyoglobin, changes with the storage time of the meat or extract and is not important for evaluation. The myoglobin and hemoglobin zones can be intensified by treatment with o-dianisidine/H_2O_2 (Fig. 12.38b) and subsequent staining with coomassie brilliant blue makes all the proteins visible (12.38 c).

Some animal species can be recognized via the myoglobin bands (e. g., beef, buffalo, pork, horse, red and grey kangaroo) and others are assigned to groups. The identification is achieved with electropherograms stained with coomassie blue (Fig. 12.39). A differentiation within the families Cervidae (deer) and Bovidae (horned animals) is difficult, with the exception of the subfamily Bovinae (cattle), e. g., between roe deer, fallow buck, elk, reindeer, kudu, springbok,

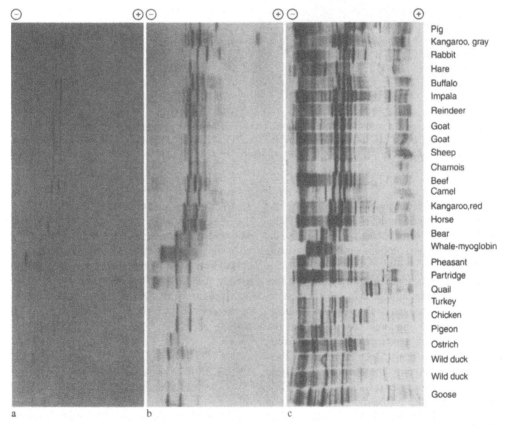

Pig
Kangaroo, gray
Rabbit
Hare
Buffalo
Impala
Reindeer
Goat
Goat
Sheep
Chamois
Beef
Camel
Kangaroo, red
Horse
Bear
Whale-myoglobin
Pheasant
Partridge
Quail
Turkey
Chicken
Pigeon
Ostrich
Wild duck
Wild duck
Goose

Fig. 12.38. Separation of sarcoplasm proteins of various warm blooded animals (mammals and fowl) by iso-electric focusing on polyacrylamide gels (PAGIF, PAGplate, pH range 3.5–9.5). (according to *Kaiser*, 1988). **a** Myoglobin (and hemoglobin) zones without staining; **b** Myoglobin and hemoglobin zones after treatment with o-dianisidine/H_2O_2; **c** Protein zones after staining with coomassie brilliant blue

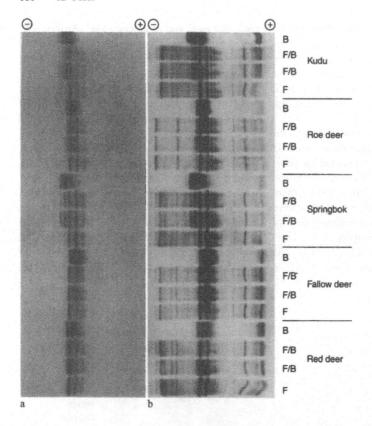

Fig. 12.39. Animal species with the same myoglobin patterns. Separation by isoelectric focusing of water soluble muscle proteins [F], blood [B], and mixtures of both [F/B] (cf. Fig. 12.38, according to *Kaiser*, 1988). **a** Myoglobin and hemoglobin zones after treatment with o-dianisidine/H_2O_2; **b** Protein zones after staining with coomassie brilliant blue

impala, sheep, goat, and chamois. Here, the hemoglobins can be used if the meat contains sufficient blood components, as is usually the case with game, or if blood is separately available (Fig. 12.39a).

The analyses mentioned above are largely limited to raw meat because protein denaturation occurs in heat treated meat. Denaturation increases with temperature and time and makes the immunochemical and electrophoretic identification more and more difficult. Since DNA is more thermostable than proteins, the PCR is a promising alternative in these cases (cf. 2.6.4.2.2).

From the intensities of the indicator zones in an electropherogram, it is possible to estimate the proportion of one kind of meat in a meat mix. This is illustrated in Fig. 12.40 using a mixture of ground beef and pork.

12.10.1.1.2 Sexual Origin of Beef

The sexual origin of beef can be determined by an analysis of the steroid hormones. Since the concentrations of individual compounds vary too greatly, the ratio of progesterone/pregnenolone obtained from GC/MS is used. This value is on average 0.5 for oxen and bulls and 7.9 for heifers.

12.10.1.2 Differentiation of Fresh and Frozen Meat

The isoenzyme patterns of cell organelles, for instance mitochondria and microsomes, differ often from those of cytoplasm. When the organelle membranes are damaged by a physical or chem-

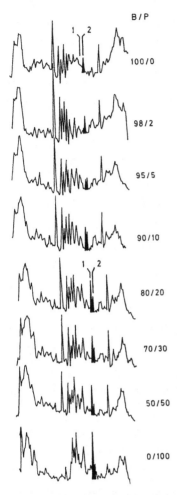

Fig. 12.40. Blended beef and pork meat: Densitograms of sarcoplasm proteins separated by PAGIF and PAG-plate; pH range 3.5–9.5. B/P (beef/pork) blend ratios in weight %. (according to *Kaiser*, 1980b)

B / P

100 / 0

98 / 2

95 / 5

90 / 10

80 / 20

70 / 30

50 / 50

0 / 100

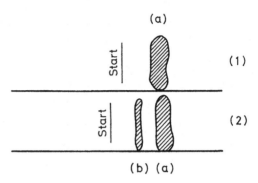

Fig. 12.41. Differentiation of fresh liver (1) from frozen and thawed liver (2) by electrophoretic separation of glutamate-oxalacetate transaminases (a) GOT sarcoplasm, (b) GOT mitochondria (according to *Hamm* and *Mašić*, 1975)

ical process, isoenzyme blending will occur in the cytoplasm.

Such membrane damage has been observed by freezing and thawing of tissue, for example, of muscle tissue, in which the isoenzymes of glutamate oxalacetate transaminase (GOT) bound to mitochondrial membranes are partially released and found in the sarcoplasm. The pressed sap collected from fresh unfroze meat has only sarcoplasm enzymes, while the frozen and thawed meat has, in addition, the isoenzymes derived from mitochondria. The GOT isoenzymes can be separated by electrophoresis (Fig. 12.41). This procedure is also applicable to fish.

The enzyme β-hydroxyacyl-CoA-dehydrogenase (HADH, EC 1.1.1.35) is also suitable for the detection of frozen meat or fish. In the oxidation of fatty acids, HADH catalyzes the reaction shown in Formula 12.29. This enzyme is bound to the inner membrane of mitochondria and is liberated in the freeze/thaw process. Its activity can then be measured in the issuing sap with acetoacetyl CoA or with the artificial substrate N-acetylacetoacetylcysteamine.

$$3\text{-Oxoacyl-CoA} + NADH + H^{\oplus} \rightarrow$$
$$(S)\text{-3-Hydroxyacyl-CoA} + NAD^{\oplus} \qquad (12.29)$$

12.10.1.3 Pigments

Pigment analysis is carried out for the evaluation of meat freshness. The individual pigments, such as myoglobin (purple-red), oxymyoglobin (red) and metmyoglobin (brown), are determined.

12.10.1.4 Treatment with Proteinase Preparations

Proteinases injected intramuscularly or through blood vessels degrade the structural proteins and, hence, proteolytic enzymes can be used to soften or tenderize meat. The enzymes are of plant or microbial origin and are used in the meat and poultry industries, while some are also used in the household as meat tenderizers. Analytical determination of proteinases is relatively difficult.

A possible assay may be based on disc gel electrophoresis of meat extracts, prepared in the presence of urea and SDS. The band intensities of the lower molecular weight collagen fragments increase in proteinase-treated meat.

12.10.1.5 Anabolic Steroids

Anabolic compounds present in animal feed as an additive increase muscle tissue growth. Owing to a potential health hazard, some of these compounds are banned in many countries. Their detection can be achieved by the mouse uterus test or by a radioimmunoassay. Special receptor proteins which have the property of binding strongly to estrogens are isolated from rabbit or cattle uterus. The hormone-receptor complex is in equilibrium with its components:

$$\text{Receptor} + \text{estrogen} \rightleftharpoons \text{Receptor-estrogen complex} \tag{12.30}$$

The nonlabelled estrogens bound to receptor in the test sample will be competitively displaced by the addition of 17-β-estradiol labelled with tritium for radiochemical assay.

To reach equilibrium, a suitable amount of receptor protein and a constant amount of labelled estradiol are incubated together with the test sample. The amount of the radioactive ^{3}H-estradiol receptor complex will decrease in the presence of competitive estrogens from the meat extract. The binding affinity of the estrogen receptor depends on the type of estrogen present (Fig. 12.42). Hence, detection limits differ and range from 0.3 to 50 ppb (mg per metric ton).

Anabolic compounds can be further separated by gas–liquid chromatography after derivatization of the polar functional groups, and identified by mass spectrometry. This method allows the determination of weak or nonestrogenic components too, but in the past it suffered from high losses in sample preparation and could not compete with radioimmunoassay in sensitivity. In the meantime disadvantages of the method have been eliminated.

12.10.1.6 Antibiotics

Antibiotics are used as part of therapy to treat animal diseases and, sometimes in low concentrations, as constitutents of animal feed to increase feed utilization and to accelerate animal growth.

Detection of antibiotics is usually achieved by the inhibition of the growth of bacteria ("inhibitor test"). *Bacillus subtilis*, strain BGA, is one of the recommended test organisms.

Chemical methods must be used in order to identify and quantify the antibiotics and other veterinary medical residues. The principal method is chromatographic separation coupled with mass spectrometry. The tetracyclines, which are common antibiotics, can be determined relatively easily by fluorometric measurement of adequately prepared and purified meat extracts.

12.10.2 Processed Meats

Besides the estimation of the animal species and the control of additives, the analysis of processed meats is associated with verifying composition. Here the emphasis is on the content of extraneous added water, carbohydrate-containing thickeners and binders, nonmeat protein additives and fat. In addition, the determination of nitrites, nitrates, nitrosamines and, for enhancing the

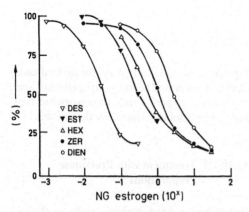

Fig. 12.42. Relative binding affinity of estrogen compounds to estrogen receptor. 50% binding achieved by: 0.034 ng diethylstilbestrol (DES), 0.33 ng 17-β-estradiol (EST), 0.6 ng hexestrol (HEX), 1.2 ng zeranol (ZER), 2.9 ng dienestrol (DIEN). (according to *Ingerowski* and *Stan*, 1978)

pinkish-red color of processed meat, L-ascorbic acid is of importance in pickle-cured meat products. Other analytical problems involve the detection of condensed phosphates, citric acid and glucono-δ-lactone, as well as the detection of polycyclic aromatic compounds in smoked meats, of mycotoxins in products with desirable or undesirable growth of molds and of chlorine compounds in seasonings.

12.10.2.1 Main Ingredients

The first insight into the composition of processed meat, i. e. whether it contains an excess of fat or carbohydrate, which would lower the protein content and thus lower the value of the processed meat, is obtained by proximate analysis of the product's main ingredients: moisture, raw protein, fat and ash content. If their sum is less than $100 \pm 0.5\%$ of the sample weight, then the presence of carbohydrate binders should be verified. A positive finding should be further investigated since incorporation of liver into processed meat may provide glycogen. Hence, thorough carbohydrate analysis is required.

12.10.2.2 Added Water

Moisture content is related to protein content and is relatively constant. *Feder's* method of analysis of water added to chopped or ground meat or to emulsion-type sausages is based on these findings. The method uses the empirical equation:

Water added $(\%)$ = Moisture $(\%)$
    − Protein $(\%) \times$ F   (12.31)

F for beef and pork = 4.0 ;

F for poultry leg = 3.9 and breast = 3.6

This indirect method for assessing the amount of added water has been repeatedly criticized. In spite of this, no better method has yet been developed. Moreover, the calculated water content is never used alone to evaluate a meat product. Other significant data, such as muscle protein content and the proportion of fat to protein, are also included.

12.10.2.3 Lean Meat Free of Connective Tissue

A measure of meat quality is expressed as the amount of lean meat free of connective tissue, which corresponds to meat proteins devoid of connective tissue protein (MPDCP). To obtain this value, the meat sample is analyzed for connective tissue proteins (CP), added extraneous protein (EP) and nonprotein-nitrogen (NPN), e. g., glutamate, purine and pyrimidine derivatives, urea. These values are then deducted from the value for total protein (TP):

$$MPDPC = TP - (CP + EP + NPN) \qquad (12.32)$$

Another method still being tested is based on drastic treatment (heating to 130 °C at pH 9) of a meat sample. Under these conditions, extraneous proteins, collagen and blood plasma proteins solubilize, while the residual protein is calculated as MPDCP using a constant factor.

12.10.2.3.1 Connective Tissue Protein

The amino acid 4-hydroxyproline is a marker compound for connective tissue. It occurs only in connective tissue protein. The amount of 4-hydroxyproline is determined in the acidhydrolysate of the sample or the isolated protein using an amino acid analyzer, or color-imetrically using a specific color reaction. The latter, accepted widely in practice, is a direct photometric procedure based on the oxidation of hydroxyproline in alkaline solution by H_2O_2 or N-chloro-p-toluenesulfonamide (chloramine-T). The oxidation yields a pyrrole derivative which is then condensed with p-dimethylaminobenzaldehyde to form a red pigment. The connective tissue content of meat is calculated by multiplying the hydroxyproline value by a factor of 8, which corresponds to an average of 12.4% hydroxyproline content of connective tissue.

12.10.2.3.2 Added Protein

In order to extend or improve the water holding capacity of processed meat, the product may contain milk, egg or soy proteins. These proteins can be detected immunochemically, e. g., with the ELISA technique (cf. 2.6.3), with high sensitivity. The PCR technique (cf. 2.6.4.2) is even more sensitive and suitable for heated meat preparations. The addition of other proteins is usually limited by law so that a quantitative evaluation is required, which is fairly difficult.

12.10.2.4 Nitrosamines

Not only does the question of the content of nitrite or nitrate in pickle-cured meat arise, but also whether nitrosamines are formed and to what extent they occur in meat (cf. 9.8).

In general, nitrosamines arise only in very low concentrations. Since some of these compounds are a great health hazard, they should be detectable below 0.1 ppm in food for human consumption. The same procedures are available for identifying volatile nitrosamines which have been described earlier for the analysis of aroma constituents (cf. 5.2). However, precautions should be taken during the isolation step. Isolation of nitrosamines should not proceed at low pH since an acid medium in the presence of residual meat nitrites promotes further *de novo* synthesis of nitrosamines. Since the isolated fraction of neutral volatile compounds, which also includes nitrosamines, is highly complex in composition, reliable nitrosamines identification by gas chromatographic retention data is not possible. Additional mass spectrometric data are needed to verify the chemical structure.

12.11 References

Bailey, A. J. (Ed.): Recent advances in the chemistry of meat. The Royal Society of Chemistry, Burlington House: London. 1984

Bailey, A. J.: The Chemistry of Collagen Cross-links and their role in meat texture. Proc. 42nd, Annual Reciprocal Meat Conf., 127 (1989) publ. 1990

Bekhit, A.E.D., Faustman, C.: Metmyoglobin reducing activity. Review. Meat Science 71, 407 (2005)

Belloque, J., Garcia, M.C., Torre, M., Marina, M.L.: Analysis of soybean proteins in meat products: A review. Crit. Rev. Food Sci. Nutri. 42, 507 (2002)

Bornstein, P., Sage, H.: Structurally distinct collagen types. Annu. Rev. Biochem. 49, 957 (1980)

Brander, J., Eyring, G., Richter, B.: Würzen. In: Ullmanns Encyklopädie der technischen Chemie, 4. edn., vol. 24, p. 507, Verlag Chemie: Weinheim. 1983

Cerny, C., Grosch, W.: Quantification of character-impact odour compounds of roasted beef. Z. Lebensm. Unters. Forsch. 196, 417 (1993)

Drumm, T.D., Spanier, A.M.: Changes in the content of lipid autoxidation and sulfur-containing compounds in cooked beef during storage. J. Agric. Food Chem. 39, 336 (1991)

Elkhalifa, E.A., Marriott, N.G., Grayson, R.L., Graham, P.P., Perkins, S.K.: Ultrastructural and textural properties of restructured beef treated with a bacterial culture and splenic pulp. Food Microstructure 7, 137 (1988)

Gault, N. F. S.: Structural aspects of raw meat. In: The chemistry of muscle-based foods (Eds.: D.E. Johnston, M.K. Knight, D.A. Ledward) Royal Society of Chemistry, Cambridge, 1992, pp. 79

Geesink, G.H., Kuchay, S. Chrishti, A.H., Koohmaraie, M.: μ-Calpain is essential for postmortem proteolysis of muscle proteins. J. Anim. Sci. 84, 2834 (2006)

Giddings, G.G.: The basis of color in muscle foods. Crit. Rev. Food Sci. Nutr. 9, 81 (1977)

Graf, E., Panter, S.S.: Inhibition of WOF development by polyvalent cations. J. Food Sci. 56, 1055 (1991)

Grundhöfer, F.: Fleisch und Erzeugnisse aus Fleisch. In: Taschenbuch für Lebensmittelchemiker und -technologen, Band 1 (Ed.: W. Frede) Springer-Verlag, Berlin, 1991, pp. 249

Guth, H., Grosch, W.: Dependence of the 12-methyltridecanal concentration in beef on the age of the animal. Z. Lebensm. Unters. Forsch. 201, 25 (1995)

Gutschmidt, J.: The storage life of frozen chicken with regard to the temperature in the cold chain. Lebensm. Wiss. Technol. 7, 139 (1974)

Hamm, R.: Kolloidchemie des Fleisches. Verlag Paul Parey: Berlin. 1972

Hamm, R., Masic, D.: Routinemethode zur Unterscheidung zwischen frischer Leber und aufgetauter Gefrierleber. Fleischwirtschaft 55, 242 (1975)

Hawkes, R.: Identification of Concanavalin A-binding proteins after sodium dodecyl sulfate-gelelectrophoresis and protein blotting. Anal. Biochem. 123, 143 (1982)

Herrera-Mendez, C.H., Becila, S., Boudjellal, A. Quali, A.: Meat aging: Reconsideration of the current concept. Trends Food Sci. Technol. 17, 394 (2006)

Herrmann, Ch., Thoma, H., Kotter, L.: Zur direkten Bestimmung von Muskeleiweiß in Fleischerzeugnissen. Fleischwirtschaft 56, 87 (1976)

Hornung, H.H.: Schlachtvieh. In: Lebensmitteltechnologie (Ed.: R. Heiss) Springer-Verlag, Berlin, 1988, pp. 46

Honikel, K.O.: Vom Fleisch zum Produkt. Fleischwirtschaft *84*, 228 (2004)

Igene, J. O., Yamauchi, K., Pearson, A. M., Gray, J. I., Aust, S.D.: Mechanisms by which nitrite inhibits the development of warmed-over flavour (WOF) in cured meat. Food Chem. *18*, 1 (1985)

Ingerowski, G.H., Stan, H.-J.: Nachweis von Östrogen-Rückständen in Fleisch mit Hilfe des cytoplasmatischen Östrogenrezeptors aus Rinderuterus. Dtsch. Lebensm. Rundsch. *74*, 1 (1978)

Johnston, D.E., Knight, M.K., Ledward, D.A.: The chemistry of muscle-based foods. Royal Society of Chemistry, Cambridge, 1992

Kaiser, K.-P., Matheis G., Kmita-Dürrmann, Ch., Belitz, H.-D.: Identifizierung der Tierart bei Fleisch, Fisch und abgeleiteten Produkten durch Proteindifferenzierung mit elektrophoretischen Methoden. I. Rohes Fleisch und roher Fisch. Z. Lebensm. Unters. Forsch. *170*, 334 (1980 a)

Kaiser, K.-P., Matheis G., Kmita-Dürrmann, Ch., Belitz, H.-D.: Proteindifferenzierung mit elektrophoretischen Methoden bei Fleisch, Fisch und abgeleiteten Produkten. II. Qualitative and quantitative Analyse roher binärer Fleischmischungen durch isoelektrische Fokussierung in Polyacrylamidgel. Z. Lebensm. Unters. Forsch. *171*, 415 (1980b)

Karlson, P.: Kurzes Lehrbuch der Biochemie. 10. edn., Georg Thieme Verlag: Stuttgart. 1977

Kerler, J., Grosch, W.: Odorants contributing to warmed-over flavor (WOF) of refrigerated cooked beef. J. Food. Sci. *61*, 1271 (1996)

Kerscher, R., Grosch, W.: Comparison of the aromas of cooked beef, pork and chicken. In: Frontiers of Flavour Science (eds.: P. Schieberle, K.-H. Engel) Deutsche Forschungsanstalt für Lebensmittelchemie, Garching, 2000, pp. 17

Kerscher, R., Nürnberg, K., Voigt, J., Schieberle, P., Grosch, W.: Occurrence of 12-methyltridecanal in microorganisms and physiological samples isolated from beef. J. Agric. Food Chem. *48*, 2387 (2000)

Ladikos, D., Lougovois, V.: Lipid oxidation in muscle foods: a review. Food Chem. *35*, 295 (1990)

Lawrie, R. A.: The conversion of muscle to meat. In: Recent advances in food science (Eds.: Hawthorn, J., Leitch, J.M.), Vol. I, p. 68, Butterworth: London. 1962

Lawrie, R.A.: Meat science, 4th edn., Pergamon Press: Oxford. 1985

Ledward, D. A.: A note on the nature of the haematin pigments present in freeze dried and cooked beef. Meat Sci. *21*, 231 (1987)

Lehninger, A. L.: Biochemie. 2nd edn., Verlag Chemie: Weinheim. 1977

Manley, C.H., Ahmedi, S.: The development of process flavors. Trends Food Sci. Technol. *6*, 46 (1995)

Mottram, D.S.: Some observations on the role of lipids in meat flavour. In: Sensory quality in foods and beverages (Eds.: Williams. A.A., Atkin, R.K.), p. 394, Ellis Horwood Ltd.: Chichester. 1983

Müller, W.-D.: Fleischverarbeitung. In: Taschenbuch für Lebensmittelchemiker und -technologen, Band 2 (Ed.: D. Osteroth) Springer-Verlag, Berlin, 1991, pp. 387

Müller, W.-D.: Erhitzen und Räuchern von Kochwurst und Kochpökelware. Fleischwirtschaft *69*, 308 (1989)

Pearson, A.M., Young, R.B.: Muscle and meat biochemistry. Academic Press: San Diego, CA. 1989

Piasecki, A., Ruge, A., Marquardt, H.: Malignant transformation of mouse M2-fibroblasts by glycerol chlorohydrines contained in protein hydrolysates and commercial food. Arzneim.-Forsch./Drug. Res. *40*, 1054 (1990)

Potthast, K., Hamm, R.: Biochemie des DFD-Fleisches. Fleischwirtschaft *56*, 978 (1976)

Price, J.F., Schweigert, B. S. (Eds.): The science of meat and meat products. 2nd edn., W. H. Freeman: San Francisco, 1971

Ruiz-Capillas, C., Jimenez-Colmenero, F.: Biogenic amines in meat and meat products. Crit. Rev. Food Sci. Nutr. *44*, 489 (2004)

Sargent, J. A.: The application of cold stage scanning electron microscopy to food research. Food Microstructure *7*, 123 (1988)

Scanlan, R.A.: N-Nitrosamines in foods. Crit. Rev. Food Technol. *5*, 357 (1974)

Schlichtherle-Cerny, H., Grosch, W.: Evaluation of taste compounds of stewed beef juice. Z. Lebensm. Unters. Forsch. A *207*, 369 (1998)

Schormüller, J. (Ed.): Handbuch der Lebensmittelchemie. Vol. I, Springer-Verlag: Berlin. 1965

Schultz, H.W., Anglemier, A.F. (Eds.): Proteins and their reactions. AVI Publ. Co.: Westport, Conn. 1964

Schwägele, F.: Struktur und Funktion des Muskels. Fleischwirtschaft *84*, 168 (2004).

Shahidi, F.: Flavor of meat and meat products. Blackie Academic & Professional, London, 1994

Silhankova, L., Smid, F., Cerna, M., Davidek, J., Velisek, J.: Mutagenicity of glycerol chlorohydrines and of their esters with higher fatty acids present in protein hydrolysates. Mutation Research *103*, 77 (1982)

Steinhart, H.: Chemische Kriterien zur Beurteilung der Fleischqualität. Lebensmittelchemie *46*, 61 (1992)

Thanh, V.H., Shibasaki, K.: Major proteins of soybean seeds. A straightforward fractionation and their characterization. J. Agric. Food Chem. *24*, 1117 (1976)

Tóth, L.: Chemie der Räucherung. Verlag Chemie: Weinheim, 1982

Traub, W., Piez, K. A.: The chemistry and structure of collagen. Adv. Protein Chem. *25*, 243 (1971)

Van Rillaer, W., Beernaert, H.: Determination of residual 1,3-dichloro-2-propanol in protein hydrolysates by capillary gas chromatography. Z. Lebensm. Unters. Forsch. *188*, 343 (1989)

Velisek, J., Davidek, J., Davidek, T., Hamburg, A.: 3-Chloro-1,2-propanediol derived amino alcohol in protein hydrolysates. J. Food Sci. *56*, 136 (1991)

Velisek, J., Davidek, J., Kubelka, V, Janisek, G., Svobodava, Z., Simicova, Z.: New chlorine-containing organic compounds in protein hydrolysates. J. Agric. Food Chem. *28*, 1142 (1980)

Velisek, J., Davidek, J., Kubelka, V.: Formation of $\Delta^{3.5}$-diene and 3-chloro Δ^{5}-ene analogues of sterols in protein hydrolysates. J. Agric. Food Chem. *34*, 660 (1986)

Warendorf, T., Belitz, H.-D.: Zum Geschmack von Fleischbrühe. 2. Sensorische Analyse der Inhaltsstoffe und Imitation einer Brühe. Z. Lebensm. Unters. Forsch. *195*, 209 (1992)

Weenen, H., Kerler, J., van der Ven, J. G. M.: The Maillard reaction in flavour formation. In: Flavours and fragrances (Ed. K. A. Swift) The Royal Society of Chemistry, Cambridge, 1997, pp. 153

Wittkowski, R.: Phenole im Räucherrauch. VCH Verlagsgesellschaft: Weinheim. 1985

Wittmann, R.: Chlorpropandiole in Lebensmitteln. Lebensmittelchemie *45*, 89 (1991)

Wykle, B., Gillett, T.A., Addis, P.B.: Myoglobin heterogeneity in pigs with PSE and normal muscle by an improved isoelectric focusing technique. J. Animal. Sci. *47*, 1260 (1978)

13 Fish, Whales, Crustaceans, Mollusks

13.1 Fish

13.1.1 Foreword

Fish and fish products fulfill an important role in human nutrition as a source of biologically valuable proteins, fats and fat-soluble vitamins. Fish can be categorized in many different ways, e. g., according to:

- The environment in which the fish lives: sea fish (herring, cod, saithe) and freshwater fish (pike, carp, trout), or those which can live in both environments, e. g., eels and salmon. Sea fish can be subdivided into groundfish and pelagial fish.
- The body form: round (cod, saithe) or flat (common sole, turbot or plaice).

Commercial fishing takes place in the open sea, coastal and freshwater areas. Conservation programs and hatcheries to rebuild stocks play important roles in the management of fresh and saltwater fish resources.

Table 13.1. Fish, crustaceans, mullusks (fished in 2001)

Continent	1000 t	Country	1000 t
World	129,943	China	44,063
		Peru	7996
Africa	7292	India	5965
America, North,		Japan	5521
Central	8885	USA	5405
America, South	15,817	Indonesia	5068
Asia	78,763	Chile	4363
Europe	17,875	Russian Fed.	3718
Oceania	1163	Thailand	3606
		Norway	3199
		Philippines	2380
		Korea	2282
		Vietnam	2010
		Iceland	1985
		Σ (%)[a]	75

[a] World production = 100%

Table 13.2. Catch of fish by German fishery in 2007

Name	Metric ton
Herring	24,515
Mackerel	11,043
Cod	8064
Haddock	136
Pollack	2247
Red fish	263
Crab	11,385
Mussel	5797
Others	18,317
Sum	81,767

Table 13.3. Average per capita consumption of fish and shellfish (2001–2003)[a]

Country	kg/year
Iceland	91.4
Japan	66.9
Portugal	57.1
Norway	49.5
Spain	44.5
France	33.5
Finland	32.1
Sweden	29.5
Luxembourg	28.6
China	25.7
Italy	24.3
Denmark	22.9
United States	22.6
Belgium	22.3
Greece	21.9
Netherlands	21.7
United Kingdom	20.4
Ireland	20.4
Switzerland	15.7
Germany	14.3; 15.5[b]
Ukraine	13.6
Austria	11.5
Czech Republic	10.2
Poland	8.9
India	4.8

[a] Source: FAO
[b] 2006.

The fishing industry catch has risen sharply in tonnage during this century. In 1900 the catch was approx. 4 million t, while it had increased to 129 million t by 2001. Table 13.1 shows the catch in tonnage of the leading countries engaged in fishing. This includes shellfish products, i. e. lobsters and crustaceans such as crabs, crayfish and shrimps, and mollusks such as clams, oysters, scallops, squid, etc., which are not true fish but are harvested from the sea by the fishing industry. Table 13.2 lists the catch of fish by the German fishery. Table 13.3 gives information on the per capita consumption of fish and shellfish. Because of the high nutritional value of ω-polyunsaturated fatty acids (cf. 3.2.1.3) fish consumption is recommended as an important part of the human diet.

13.1.2 Food Fish

Table 13.4 shows the important food fish. In general, a predatory fish is better tasting than a non-predaceous fish, a fatty fish better than a non-fatty fish. The fishbone-rich species, such as carp, perch, pike and fench, are often less in demand than fish with fewer bones.

Some important food fish will be described in more detail.

13.1.2.1 Sea Fish

13.1.2.1.1 Sharks

Dogfish (*Squalus acanthias*) about 1 m long are marinated or smoked before marketing. In North America fish of the family *Squalidae* are generally referred to as dogfish sharks. Other names are spiny, spring or piked dogfish and rock salmon. Trade names used in th U.K. are flake, huss or rig. In Germany the name used is Dornhai (Dornfisch), and the smoked dorsal muscle is sold as "Seeaal", while the hot-smoked, skimmed belly walls are called "Schillerlocken". Mackerel sharks of the family *Lamnidae* are also in this group. The main species of this family are: (a) porbeagle, blue dog or Beaumaris shark (*Lamna nasus*); and (b) salmon, (c) mako and (d) white sharks. The blue shark is found in the

Atlantic Ocean and the North Sea as an escort of herring schools. It possesses a meat similar to veal and is known in the trade as sea or wild sturgeon, or calf-fish. Due to a high content of urea (cf. 13.1.4.3.6), the meat of these fish is often tainted with a mild odor of ammonia. Endeavors to popularize shark and related fish meat are well justified since the meat is highly nutritious; however consumer acceptance would be hampered by the word "shark", therefore other terms are commonly preferred in the trade. Shark fins are a favorite dish in China and are imported to Europe as a specialty food.

13.1.2.1.2 Herring

The herring (*Clupea harengus*) is a fatty fish (fat content >10%, cf. Table 13.5). It is one of the most processed and most important food fish and a source of raw materials for meal and marine oil. Herrings are categorized according to the season of the catch (spring or winter herring), spawning time or stage of development (e. g. matje, the young fatty herring with roe only slightly developed, is cured and packed in half barrels), or according to the ways in which the fish were caught: drag or drift nets, trawling nets, gill and trammel nets or by seining (purse seining), the most important form of snaring.

The herring averages 12–35 cm in length and migrates in large swarms or schools throughout the nothern temperate and cold seas.

Herring is marketed cold or hot smoked (kippers, buckling), frozen, salt-cured, dried and spiced, jellied, marinated and canned in a large variety of sauces, creams, vegetable oils, etc. Sprat (*Sprattus sprattus phalericus*), called brisling in Scandinavia and Sprotte in Germany, is processed into an "Appetitsild" (skinned fillets or spice-cured sprats packed in vinegar, salt, sugar and seasonings). Canned brisling is packed in edible oil, tomato and mustard sauces, etc. and is sold as brisling sardines. Brisling is often lightly smoked and marketed as such. Sprats are also processed into a delicatessen product called "Anchosen", which consists mostly of small sprats, sometimes mixed with cured matje, and preserved in salt and sugar, with or without spices and sodium nitrate. Anchovies (*Engraulis encrasicolus*; German term "Sardellen"; found in Atlantic and Pacific

Table 13.4. Major commercial fish species – quality and utilization

Name	Family	Genus, sp.	Comments on quality and processing
Sea fish			
Pleurotremata (sharks)			
Dogfish	Squalidae	*Squalus acanthias (Acanthias vulgaris)*	
Rajiformes (skates)			
Skates, e. g., thronback, common skate	Rajidae	*Raja clavata, R. batis*	Used are the wing shaped body widenings, the pectoral part and breast fins as a delicacy; it is fried, smoked or jellied
Acipenseriformes (sturgeons)			
Sturgeon	Acipenseridae	*Acipenser sturio*	Exceptionally delicate when smoked, caviar is made from its roe
Clupeiformes (herrings)			
Herring	Clupeidae	*Clupea harengus*	Valuable fish with fine white meat, fried and grilled; industrially processed, for example into Bismarck herrings, rollmops and brat-herring
Sprat	Clupeidae	*Sprattus sprattus*	Mostly cold or warm smoked; anchovies
Sardine	Clupeidae	*Sardina pilchardus*	Mostly steam cooked and canned in oil; along sea coast grilled and fried
Anchovy (Anchovis)	Engraulidae	*Engraulis encrasicolus*	Pleasant, aromatic fragrant, cured in brine, made into rings and paste
Lophiiformes (anglers)			
Angler, allmouth	Lophiidae	*Lophius piscatorius*	White, good and firm meat, poached
Gadiformes (cods)			
Ling	Gadidae	*Molva molva*	Tasty firm white meat
Cod	Gadidae	*Gadus morhua*	Meat is prone to fracturing, used fresh, filleted, salted and frozen, dried (stock- and klipfish), cooked, poached; oil is produced from liver
Haddock	Gadidae	*Melanogrammus aeglefinus*	Very fine in taste, processed as fresh, pickled, or marinated, smoked, fried, roasted, cooked or poached, or used for fish salad
Coalfish pollack, black coor or Boston bluefish	Gadidae	*Pollachius virens, P. pollachius*	Meat is lightly tinted grayish-brown, it is filleted, smoked, sliced as cutlets or chops and processed in oil (used forsalmon substitute)
Whithing	Merlangius	*Merlangius merlangius*	Good meat, easily digested, but very sensitive, fried or deep fried roasted or smoked, used for fish stuffings
Hake	Merlucciidae	*Merluccius merluccius*	Fresh or frozen, all processing methods are used
Scorpaeniformes			
Red fish, ocean perch	Scorpaenidae	*Sebastes marinus*	Tasty meat, fattier than cod, it is filleted or smoked
Gurnard, sea robin (gray gurnard, red gurnard)	Triglidae	*Trigla gurnardus, T. lucerna*	White firm meat (red sp. is of higher quality), used fresh or smoked
Lumpfish, sea hen	Cyclopteridae	*Cyclopterus lumpus*	Smoked, its roe is processed into caviar substitute

Table 13.4. continued

Name	Family	Genus, sp.	Comments on quality and processing
Perciformes (percoid fishes)			
Red mullet	Mullidae	*Mullus barbatus*	White fine and a piquant delicius meat, mostly grilled
Catfish, wolffish	Anarhicha-didae	*Anarhichas lupus, A. minor*	Fine white fragrant meat, poached grilled, doughtype crust coated
Mackerel	Scombridae	*Scomber scombrus*	Highly valued fish, tasty reddish meat, fried, grilled, smoked or canned; its meat is not readily digestible
Tuna	Scombridae	*Thunnus thynnus*	Reddish meat of exceptional taste, it is fried, roasted, smoked, or canned in oil or processed into paste sausages or rolls
Pleuronectiformes (flat fishes)			
Turbot (butt or britt)	Scopthalmidae	*Psetta maxima (Rhombus maximum)*	Apart from common sole, the highest valued flat fish, meat is snow-white firm and piquant, it is cooked, grilled or poached
Halibut	Pleuronectidae	*Hippoglossus hippoglossus*	Tasty meat, it is poached, fried or smoked
Plaice	Pleuronectidae	*Pleuronectes platessa*	Tasty meat, fried or filleted and poached
Flounder	Pleuronectidae	*Platichthys flesus*	Good white meat, it is poached, fried or smoked
Common sole	Soleidae	*Solea solea*	It is the finest flat fish, it is poached, fried, grilled or roasted
Freshwater fish			
Petromyzones (lampreys)			
Lamprey	Petromyzoni-dae	*Lampetra fluviatilis*	It is industrially processed
Anguilliformes (eel sp.)			
Eel	Angullidae	*Anguilla anguilla*	Tasty meat, good quality when not exceeding 1 kg in weight, the fresh fish is fried, roasted, or it is smoked, marinated of jellied
Salmoniformes (salmons)			
Salmon	Salmonidae	*Salmo salar*	High quality fish (5–10 kg), it is poached, grilled, cured or smoked, also pickled
River trout	Salmonidae	*Salmo trutta*	High quality fish, no fishbone, bluish tinted when cooked, roasted a la meuniere
Rainbow trout (lake- or steelhead trout)	Salmonidae	*Salmo gairdnerii*	
Brook trout	Salmonidae	*Salvelinus fontinalis*	A worthy fish, meat is pale pinkish, processed as trout but mostly fried
Whitefish	Salmonidae	*Coregonus sp.*	Processed as trout
Coregonus, whitefish	Salmonidae	*Coregonus sp.*	White tender and tasty meat, though somewhat dry, it is fried or deep fried
Smelt	Osmeridae	*Osmerus eperlanus*	A fishbone rich meat which is mostly deep fried
Pike (jackfish)	Esocidae	*Esox lucius*	Young pikes (best quality 2–3 kg) are soft tender and tasty, the meat is well rated though it is bone rich, steam cooked, cooked or fried

Table 13.4. continued

Cypriniformes (carps)			
Roach	Cyprinidae	*Rutilus rutilus*	It has a tasty meat though fishbone rich
Bream	Cyprinidae	*Abramis brama*	Tasty meat but fishbone rich
Tench	Cyprinidae	*Tinca tinca*	Tender fatty meat, tasty, bluish when cooked, mostly steam cooked
Carp	Cyprinidae	*Cyprinus carpio*	Soft meat, readily disgestible, a valuable fish food, bluish when cooked
Crucian	Cyprinidae	*Carassim carassim*	A good food fish, but not as good as carp, the meat is bone rich
Perciformes (perchoid fishes)			
Perch	Percidae	*Perca fluviatilis*	Firm, white and very tasty meat, best quality is below 1 kg (25–40 cm), it is fried, filleted and/or steam cooked
Zander	Percidae	*Stizostedion lucioperca*	White, soft, juicy and tasty meat, 40–50 cm, fried or steam cooked, it is the best quality freshwater fish
Ruffle	Percidae	*Gymno-cephaluscernua*	Exceptionally tasty meat

Oceans) should also be included in the herring group. Anchovies are usually salted (cured in brine in barrels until the flesh has reddened). They are also canned in glass jars, marketed as a paste or cream, smoked or dried. Sardines (*Sardina pilchardus*), from the Mediterranean or Atlantic (France, Spain, Portugal) or from Africa's west coast, are often marketed steamed, fried or grilled, or canned in oil or tomato sauce. The fully grown sardine is known in the trade as pilchard (Californian, Chilean, Japanese). It is also salt-cured and pressed in barrels or canned in edible oil or in sauce. "Russian sardines" or "Kron-sardine" are actually marinated small herrings or sprats caught in the Baltic Sea. Also in the herring group is the allis shad (*Alosa alosa* or *Clupea alosa*), which is sold fresh, smoked or canned.

13.1.2.1.3 Cod Fish

These nonfatty fish (fat content <1%, cf. Table 13.5) are usually marketed fresh, whole and gutted, and many have the head and/or skin removed or be filleted. The Atlantic cod (*Gadus morhua*) is considered the most important food fish of Northern Europe. Classified according to

size, designations are: small codling, codling, sprag and cod in the U.K. and Iceland; and scrod in the United States.

Saithe is also known as coalfish or pollack (*Pollachius virens*) or by names such as black cod or Boston bluefish. After salt-curing and slicing, it is lightly smoked and packed in edible oil. Saithe is marketed in Germany as a salmon substitute called "Seelachs". Rolled in balls and canned, it is called "side boller" in Norway.

Whiting (*Merlangius merlangius*), known as mer-lan in France, is a North Atlantic, North Sea fish, marketed in many forms.

Haddock (*Melanogrammus aeglefinus*) is a North Atlantic and Arctic Sea fish. Small haddock are called gibbers or pingers, and large ones are jum-bos. The annual haddock catch is lower than those of the above-mentioned fish, i.e. anchovy, her-ring, cod, sardine and pollack. Hake (*Merluccius merluccius*) is an Atlantic and North Sea fish. Its various subspecies are the Cape, Chilean, North-east Pacific, Mediterranean and North American east coast white hake. The annual catch is some-what higher than that of haddock. Even higher than both is the catch of menhaden (*Brevoortia tyrannus*), which accounts for almost 38% of the fish tonnage in the United States.

Table 13.5. Average chemical composition of fish

Fish	Moisture[a]	Protein[a]	Fat[a]	Minerals[a]	Edible portion[b]
Freshwater fish					
Eel	61	15	26	1.0	70
Perch	80	18	0.8	1.3	38
Zander	78	19	0.7	1.2	50
Carp	75	19	4.8	1.3	55
Tench	77	18	0.8	1.8	40
Pike	80	18	0.9	1.1	55
Salmon	66	20	14	1.0	64
River trout	78	19	2.7	1.2	50
Smelt	80	17	1.7	0.9	48
Sea fish					
Cod	82	17	0.64	1.2	75
Haddock	81	18	0.61	1.1	57
Ling	79	19	0.6	1.0	68
Hake	79	17	2.5	1.1	58
Red fish (ocean perch)	78	19	3	1.4	52
Catfish	80	16	2.0	1.1	52
Plaice	79	17	1.9	1.4	56
Flounder	81	17	0.7	1.3	45
Common sole	80	18	1.4	1.1	71
Halibut (butt)	75	19	1.7	1.3	80
Turbot (britt)	80	17	1.7	0.7	46
Herring (Atlantic)	63	17	18	1.3	67
Herring (Baltic Sea)	71	18	9	1.3	65
Sardine	74	19	5	1.6	59
Mackerel	68	19	12	1.3	62
Tuna	62	22	16	1.1	61

[a] As % of edible portion.
[b] As % of the whole fish weight.

13.1.2.1.4 Scorpaenidae

Red fish of the North Atlantic and arctic regions (*Sebastes marinus* and other species), which are known as red fish or ocean perch (U.K.) or rose-fish or Norway haddock (U.S.A.), have gained in importance in recent decades. Red fish meat is rich in vitamins, firm and moderately fat (fat content 1–10%, cf. Table 13.5). It is marketed fresh or frozen, whole or as fillets; as cold or hot smoked steaks; and roasted or cooked.

13.1.2.1.5 Perch-like Fish

The bluefin tuna (*Thunnus thynnus*) is one of the several *Thunnus* spp. It has a red, beef-like muscle tissue and is caught primarily in the North and Mediterranean Seas and the Atlantic Ocean. It is a fatty fish (cf. Table 13.5), which is marketed salted and dried, smoked, or canned in edible oil, brine or tomato sauce. Tuna meat is also a common delicatessen item (tuna paste, sausages, rolls, etc.). The Atlantic mackerel (*Scomber scombrus*) is of great importance, as are the chub or Pacific mackerel (*S. japonicus*) and the blue Australian mackerel (*S. australasicus*). Mackerel are sold whole, gutted or ungutted; or filleted, frozen, smoked, salted, pickle-salted (Boston mackerel), etc.

13.1.2.1.6 Flat Fish

This group includes: plaice or hen fish (*Pleuronectes platessa*); flounder (*Platichthys flesus*, also known as fluke); Atlantic halibut or butt (*Hippoglossus hippoglossus*); common dab (*Pleuronectes limanda*); brill (*Rhombus laevis*); Atlantic and North Sea common sole (*Solea solea*, "Dover" sole); and turbot (*Psetta maxima*, also called butt or britt). These fish and haddock (cf. 13.1.2.1.3) are the sea fish most popular with consumers.

13.1.2.2 Freshwater Fish

Some important freshwater fish are; eels; carp; tench; roach; silver bream; pike, jackfish or pickerel; perch, pike-perch or blue pike; salmon; rainbow, river or brown trout; and pollan (freshwater herring or white fish). Unlike sea fish, the catch of freshwater fish is of little economic importance (cf. Table 13.2), although it does offer an important source of biologically valuable proteins.

13.1.2.2.1 Eels

Freshwater and sea eels (*Anguilla anguilla*, *A. rostrata*, *Conger conger*, etc.) are sold fresh, marinated, jellied, frozen or smoked as unripe summer (yellow or brown eel) or ripe winter eels (bright or silver eel). Due to their high fat content (cf. Table 13.5), eels are not readily digestible.

13.1.2.2.2 Salmon

Salmon (*Salmo salar*) and sea trout (*Salmo trutta*) are migratory. Salted or frozen fish are supplied to the European market by Norway and by imports from Alaska and the Pacific Coast of Canada. Also included in this group are: river trout (*Salmo trutta f. fario*) and lake trout (*Salmo gairdnerii*), which is commonly called steelhead trout in North America when it journeys between the sea and inland lakes.

13.1.3 Skin and Muscle Tissue Structure

As in other backboned animals, fish skin consists of two layers: the outer epidermis and the inner derma (cutis or corium). The outer epidermis is not horny but is rich in water, has numerous gland cells and is responsible for the slimy surface. Mucopolysaccharides are major components of this mucous, with galactosamine and glucosamine as the main sugars. The derma is permeated with connective tissue fibers and has various pigment cells, among them guanophores, which contain silverywhite glistening guanine crystals. Scales protrude from the derma. Their number, size and kind differ from species to species. This is of importance in fish processing since it determines whether a fish can be processed with or without skin. The nature and state of fish skin affects shelf life and flavor. The spreading of skin microflora after death is the main cause of the rapid decay of fish. The skin contains numerous spores resistant to low temperatures; they can grow even at $< -10\,°C$ (psychrophiles or psychrotolerant microorganisms). The decay is also enhanced by bacteria present in fish intestines.

The fish body is fully covered by muscle tissue. It is divided dorsoventrally by spinous processes and fin rays and in the horizontal direction by septa. Corresponding to the number of vertebra, the rump muscle tissue is divided into muscle sections (myomeres) (Fig. 13.1), which are separated from each other by connective tissue envelopes. The transversal envelopes are called myocommata, the horizontal ones myosepta. While myosepta are arranged in a straight line, myocommata are pleated in a zig-zag fashion. Since cooking gelatinizes the connective tissue, the muscle tissue is readily disintegrated into flake-like segments.

The muscle fibers (muscle cells), which are enclosed by the sarcolemma, contain 1000–2000 myofibrils (Fig. 13.1), the cell nucleus, sarcoplasm, mitochondria and the sarcoplasmic reticulum. The myofibrils are divided into sarcomeres which, as in mammalian muscle (cf. 12.2.1), consists of thick and thin filaments. Depending on the myoglobin content (cf. 13.1.4.2.1), fish flesh is dark or light colored. The dark muscles are similar to the heart muscle. They lie directly under the skin (Fig. 13.1) and allow persistent swimming. In comparison, the light muscles allow sudden exertion. The proportion of dark muscles is correspondingly low in sea bed fish, e. g., flounder. On the other hand, it is relatively high in fish which constantly swim, e. g., herring and mackerel. Unlike the light muscles, the dark muscles

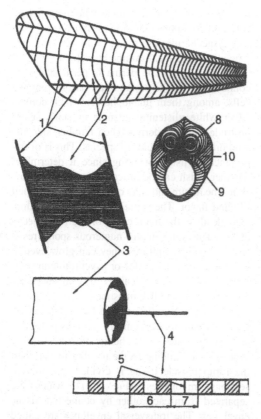

Fig. 13.1. Schematic representation of the muscle structure of fish (according to *Tülsner*, 1994) *1* myosepta, *2* myomeres, *3* muscle fibers, *4* myofibrils, *5* Z line, *6* A band, *7* I band, *8* light muscle (back part), *9* light muscle (stomach part), *10* dark side muscle

are richer in lipids, nucleic acids and B vitamins. The light muscles get their energy from glycolysis and exhibit a higher ATPase activity.

13.1.4 Composition

13.1.4.1 Overview

The edible portion of a fish body is less than in warm-blooded animals. The total waste might approach 50% and 10–15% after head removal. Fish meat and that of land animals are readily digestible, but fish is digested substantially faster and has therefore a much lower nutritive saturation value. The cooking loss is approx. 15% with

fish, which is significantly less than that of beef. The biological value of fish proteins is similar to that of land animals. While the crude protein content of fish is about 17–20%, the fat and water contents vary widely. Some are distinctly non-fatty, with fat contents of only 0.1–0.4% (haddock or cod), while some are very fatty (eels, herring or tuna), with fat contents of 16–26%. Many fish species have fat contents between these extreme values. Table 13.5 provides data on the basic composition of fish.

13.1.4.2 Proteins

The protein-N content of fish muscle tissue is between 2–3%. The amino acid composition, when compared to that of beef or milk casein (Table 13.6), reveals the high nutritional value of fish proteins. The sarcoplasma protein accounts for 20–30% of the muscle tissue total protein. The contractile apparatus accounts for 65–75% protein; the connective tissue of teleosts is 3%; and of elasmobranchs, such as sharks and rays (skate or rocker), is up to 10%. The individual protein groups and their functions in muscle tissue of mammals (cf. 12.3.2) also apply to fish.

Table 13.6. Amino acid composition of fish muscle, beef muscle and casein (amino acid-N as % of total-N)

	Casein	Beef muscle	Cod muscle
Aspartic acid	4.7	4.0	6.8
Threonine	3.6	3.7	3.4
Serine	5.3	4.6	3.6
Glutamic acid	13.3	9.3	8.8
Proline	7.5	4.3	3.4
Glycine	3.2	6.0	5.8
Alanine	3.0	4.9	5.9
Cystine	0.2	0.8	2.5
Valine	5.4	3.7	2.5
Methionine	1.8	2.2	2.0
Isoleucine	4.1	4.2	2.7
Leucine	6.1	5.1	5.1
Tyrosine	3.0	2.1	1.7
Phenylalanine	2.7	2.7	2.1
Tryptophan	1.0	1.2	1.1
Lysine	9.8	9.8	11.7
Histidine	5.3	4.9	3.5
Arginine	8.2	14.5	13.2

13.1.4.2.1 Sarcoplasma Proteins

Fish sarcoplasma proteins consist largely of enzymes. The enzymes correspond to those of mammalian muscle tissue. When these proteins are separated electrophoretically, specific patterns are obtained for each fish species. This is a helpful chemical means of fish taxonomy. The pigments are concentrated in the dark muscle. For instance, 3.9 g/kg of myoglobin, 5.8 g/kg of hemoglobin and 0.13 g/kg of cytochrome C are present in the dark muscle of a type of mackerel (*Scomber japonicus*). The light muscles contain only 0.1 g/kg of hemoglobin and myoglobin. Hemoglobin is absent in some mollusks and in Antarctic fish with colorless blood. The amino acid composition of fish and mammalian myoglobins is clearly different. Fish myoglobin contains, e. g., a cysteine residue which is absent in mammals. In strongly pigmented fish (e. g. tuna), pigment degradation reactions can induce meat discoloration (e. g. observable "greening" in canned tuna meat).

13.1.4.2.2 Contractile Proteins

The proportion of myofibrillar proteins in fish total protein is higher than in mammalian muscle tissue, however the proportions among individual components (Table 13.7) are similar (cf. 12.3.2.1). The heat stability of fish proteins is lower than that of mammals, the protein denaturation induced by urea occurs more readily, and protein hydrolysis by trypsin is faster (Fig. 13.2). These properties provide additional evidence of the good digestibility of fish proteins. Mollusks contain paramyosin. The percentage of this protein in smooth muscles, e. g., of oysters, is 38%.

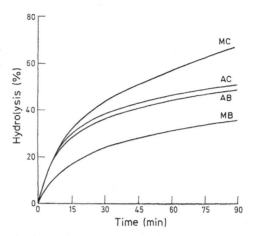

Fig. 13.2. Tryptic hydrolysis of myofibrils (M) and actin (A) from cod fish (C) and beef (B) under the same conditions. (according to *Connell*, 1964)

The amino acid composition shows relatively high amounts of Arg (12%) and Lys (9%) and little Pro. The paramyosin molecule consists of two peptide chains (M_r) 95,000–125,000), each of which is 120 nm long, has a helical structure and is twisted to a rod. In fact, two disulfide bonds contribute to the stability of the molecule. It forms the core in the thick filaments and is surrounded by myosin. In the production of gels, it influences the rheological properties and is the reason why gels made from mollusk meat are more elastic and more cohesive than gels made from fish protein.

13.1.4.2.3 Connective Tissue Protein

The content of connective tissue protein (1.3%) in fish muscle is lower than in mammalian mus-

Table 13.7. Myofibrillar proteins of fish

Protein	Content (%)	Molar mass
Myosin	50–58	Two long (200,000 and 240,000) and two short (15,000 and 28,000) peptide chains
Actin G	15–20	41,785
Tropomyosin	4–6	70,000
Troponin	4–6	72,000
Paramyosin	2–19[a]	200,000–258,000

[a] High percentages in the smooth muscle of mussels and squid.

cle. Collagen is the main component with a content of up to 90% and the remainder is elastin. The shrinkage temperature, T_s, is about 45 °C in fish collagen, i. e. much lower than in mammalian collagen (60–65 °C). These two factors make fish meat more tender than mammalian meat.

13.1.4.2.4 Serum Proteins

The congealing temperature of the blood serum of polar fish of Arctic or Antarctic regions (e. g. *Trematomus borchgrevinski, Dissostichus mawsoni, Boreogadus saida*) is about −2 °C and thus is significantly lower than that of other fish (−0.6 to −0.8 °C). Antifreeze gly-coproteins account for such low values. The amino acid sequence of this class of proteins is characterized by high periodicity:

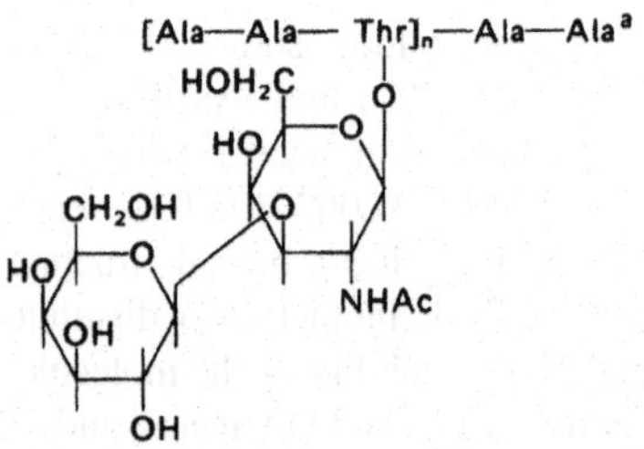

[Ala—Ala— Thr]$_n$—Ala—Ala[a]

[a] C-terminal has one or two alanine residues   (13.1)

The molecular weight range is 10.5–27 kdal, while the conformation is generally stretched, with several *a*-helical regions. These glycoproteins are hydrated to a great extent in solution. The antifreezing effects are attributed to the disaccharide residues as well as to the methyl side chains of the peptide moiety.

13.1.4.3 Other N-Compounds

The nonprotein-N content is 9–18% of the total nitrogen content in teleosts and 33–38% in elasmobranchs.

13.1.4.3.1 Free Amino Acids, Peptides

Histidine is the predominant free amino acid in fish with dark-colored flesh (tuna, mackerel). Its content in the flesh is 0.6–1.3% fresh weight and can even exceed 2%. During bacterial decay of the flesh, a large amount of histamine is formed from histidine. Fish with light colored flesh contain only 0.005–0.05% free histidine. Free 1-methylhistidine is also present in fish muscle tissue. Anserine and carnosine contents are 25 mg/kg fresh tissue. Taurine content is high (500 mg/kg).

13.1.4.3.2 Amines, Amine Oxides

Sea fish contain 40–120 mg/kg of trimethylamine oxide, which is involved in the regulation of the osmotic pressure. After death, this compound is reduced by bacteria to "fishy" smelling trimethylamine (cf. 13.1.4.8). On the other hand, fresh-water fish contain only very low amounts of trimethylamine (0–5 mg/kg). On storage of fish, a part of the trimethylamine is enzymatically broken down to dimethylamine and formaldehyde. The latter then undergoes cross-linking reactions with proteins, which reduce the solubility (cf. 13.1.6.2) and make the fish tougher. In addition to trimethylamine, the amine fraction contains dimethyl- and monomethylamines and ammonia, and some other biogenic amines derived from amino acid decarboxylation. The concentration of volatile nitrogen bases increases after death, the increase being influenced by storage duration and conditions. The level of volatile amines can be used as an objective measure of fish freshness (Fig. 13.3).

13.1.4.3.3 Guanidine Compounds

Guanidine compounds, such as creatine, are 600–700 mg/kg fresh fish muscle tissue. In crustaceans, the role of creatine is taken over by arginine.

13.1.4.3.4 Quaternary Ammonium Compounds

Glycine betaine and γ-butyrobetaine are present in low amounts in fish flesh.

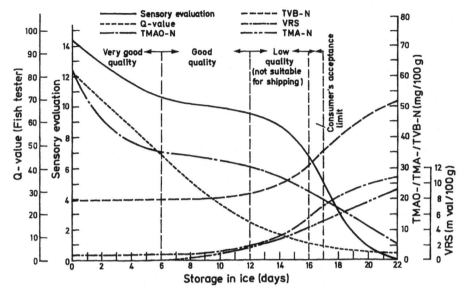

Fig. 13.3. Cod fish quality change during storage (according to *Ludorff*, 1973). Sensory evaluation: in total 15 points are given, 5 for visual appearance and 10 for odor, taste and texture; Q-value: electric resistance of the fish tissue as recorded by a "fish tester"; Q40: quality class S, Q = 30–40: A, Q = 20–30: B, Q20: C and worse; TMAO-N: trimethylamine oxide-N; TVB-N; total volatile base-N; VRS: volatile reducing substances, TMA-N: trimethylamine-N

13.1.4.3.5 Purines

The purine content in fish muscle tissue is about 300 mg/kg.

13.1.4.3.6 Urea

The fairly high content of urea in muscle tissue (1.3–2.1 g/kg) is characteristic of elasmobranchs (rays, sharks). The compound is decomposed to ammonia by bacterial urease during fish storage.

13.1.4.4 Carbohydrates

Glycogen is the principal carbohydrate. Its content (up to 0.3%) is generally lower than in mammalian muscle tissue.

13.1.4.5 Lipids

The fat (oil) content of fish is highly variable. It is influenced not only by the kind of fish but by

the maturity, season, food availability and feeding habit. Fat deposition occurs in muscle tissue (e. g. carp, herring), in liver (cod, haddock, saithe) or in intestines (blue pike, pike, perch).

Fish is an important source of ω-3-polyenic acids with 5 and 6 double bonds (cf. Table 13.8), which are considered valuable from a physiological and nutritional viewpoint. In contrast to the high content of unsaturated fatty acids, the level of anti-

Table 13.8. The content of ω-3-fatty acids in fish (g/100 g of fillet)

Type of fish	EPA (20:5)[a]	DHA (22:6)[a]
Mackerel	0.65	1.10
Salmon (Atlantic)	0.18	0.61
Salmon (red)	1.30	1.70
Trout	0.22	0.62
Tuna	0.63	1.70
Cod	0.08	0.15
Flounder	0.11	0.11
Perch	0.17	0.47
Haddock	0.05	0.10
Sole	0.09	0.09

[a] Structure (cf. 3.2.1.2).

Table 13.9. Minerals in fish muscle

Element	Content (mg/kg)	Element	Content (mg/kg)
Ca	48–420	Fe	5–248
Mg	240–310	Cu	0.4–1.7
P	1730–2170	I	0.1–1.0

oxidatively active tocopherols is relatively low. Therefore, the lipids of fish represents a major problem in preservation because of its easy per-oxidation (cf. 13.1.4.8).

13.1.4.6 Vitamins

Fish fat and liver (liver oil) are significant sources of fat-soluble vitamins, A and D. Also present are vitamins E (tocopherol) and K. The water-soluble vitamins, thiamine, riboflavin and niacin, occur in relatively high amounts, while others are present only in low amounts.

13.1.4.7 Minerals

The average content of major minerals in fish muscle tissue is compiled in Table 13.9.

13.1.4.8 Aroma Substances

Aroma substances are formed by the enzymatic oxidative degradation of the highly unsaturated fatty acids with the participation of lipoxygenases of varying specificity. These aroma substances contribute to the mild green-metallic-mushroom aroma of fresh fish. Dilution analyses have shown that these substances are acetaldehyde, propanal, 1-octen-3-one, (Z)-1,5-octadien-3-one, (E,Z)-2,6-nonadienal, (Z,Z)-3,6-nonadienal and (E,E)-2,4-decadienal.

Using salmon as an example, Table 13.10 (compare LO with LI) shows how the concentrations of important odorants change in the cooking process. The highly volatile substances acetaldehyde and propanal decrease and hexanal, (Z)-4-hep-

Table 13.10. Influence of the storage temperature of the raw material on the formation of odorants during cooking of salmon and cod[a]

Aroma substance	Salmon			Cod	
	LO[b]	LI[c]	LII[c]	KI[c]	KII[c]
Acetaldehyde	3700	2300	2500	1300	2400
Propanal	3500	1700	4700	n.a.	n.a.
2,3-Butandione	57	52	n.a.	200	596
2,3-Pentandione	141	234	318	86	26
Hexanal	35	58	148	n.a.	28
(Z)-3-Hexenal	2.6	3.9	50	1.3	4.3
(Z)-4-Heptenal	3.0	6.0	47	1.6	2.8
Methional	3.0	8.0	4.4	11	10
1-Octen-3-one	0.5	0.4	0.4	0.7	0.2
(Z)-1,5-Octadien-3-one	0.4	0.3	0.5	0.1	0.16
(Z,Z)-3,6-Nonadienal	n.a.	5.7	49	1.3	4.2
(E,Z)-2,6-Nonadienal	9.3	9.7	26	3.5	2.8
(E)-2-Nonenal	2.0	2.7	6.4	n.a.	n.a.
(E,E)-2,4-Nonadienal	2.2	2.6	3.7	3.2	2.0
(E,E)-2,4-Decadienal	4.8	6.0	18	3.5	2.2
Methanethiol	n.a.	n.a.	n.a.	100	130
2-Methylbutanal	n.a.	n.a.	n.a.	20	270
3-Methylbutanal	n.a.	n.a.	n.a.	51	620

[a] Concentrations given in µg/kg raw or cooked fish; n.a.: not analyzed.
[b] LO: raw salmon (*Salmo salar*), stored for 14 weeks at −60 °C.
[c] Salmon and cod (*Gadus morhua*) were each stored for 14 weeks at −60 °C (LI, KI) and −13 °C (LII, KII), then covered with aluminium foil and heated in a boiling water bath for 15 min.

tenal and methional increase. In the aroma profile of boiled cod, "mild fishy" and "boiled potato" notes can be detected. If methional and (Z)-1,5-octadien-3-one are dissolved in water in the concentrations that are formed during the cooking of cod (sample KI in Table 13.10), the mixture has a fishy and boiled-potato odor (Table 13.11). In addition, a geranium-like odor is detectable, which is produced by the pure octadienone. Consequently, these two carbonyl compounds are primarily responsible for the cooking aroma of this low-fat fish.

It is well known that on cold storage, aroma defects can appear faster in a high-fat fish than in a low-fat fish. This is clearly shown in an experiment in which salmon and cod were stored for 14 weeks at different temperatures and then boiled. While the aroma of the fish stored at $-60\,°C$ was perfect, the relatively low temperature of $-13\,°C$ had a negative effect on the aroma. The salmon had an intensive fatty/train oil odor and, in comparison, the low-fat cod had only a more intensively malty odor. The aroma defect of the fatty fish, which becomes very unpleasantly noticeable, is based on the peroxidation of polyunsaturated ω-3 fatty acids, which results in a 13 fold increase in (Z)-3-hexenal (compare LII with LI in Table 13.10), an 8 fold increase in (Z)-4-heptenal and a 9 fold increase in (Z,Z)-3,6-nonadienal. In low-fat cod, these aldehydes increased at most by a factor of 3.

The change in the aroma of cod was produced by an increase in malty smelling 2- and 3-methylbutanal, which had 12 to 13 times higher concentrations in KII than in KI (Table 13.10).

The compound 2,6-dibromophenol, which has a very low aroma threshold of 0.5 ng/kg, also contributes to the aroma of fresh sea fish. In higher concentrations, it produces an iodoform-like odor defect which has been observed in

shrimps. The meaty aroma note of cooked tuna is caused by the formation of 2-methyl-3-furanthiol (cf. 12.9).

Trimethylamine also has a fishy odor. However, its odor threshold at the pH of fish meat is very much higher than that of the potent lipid peroxidation products, e.g., (Z)-1,5-octadien-3-one (cf. 3.7.2.1.8 and 11.2.4.4.4). Therefore, it plays a role as an off-flavor substance only on stronger bacterial infection of fish at temperatures $>0\,°C$.

13.1.4.9 Other Constituents

More than 500 tropical fish species (barracuda, sting ray, fugu, globefish), including some valuable food fish, are known to be passively poisonous. Poisoning can occur as a result of their consumption. The toxicity can vary with the season, and can extend to the whole body or be localized in individual organs (gonads, i.e. ovaries and testicles, liver, intestines, blood). Cooking can destroy some of these toxic substances. They consist of peptides, proteins and other compounds. Some of their structures have been elucidated. There are also actively poisonous fish, with prickles or tiny needle-like spines used as the poisoning apparatus. These are triggered as a weapon in defence or attack. This group of fish includes the species *Dasytidae, Scorpaenidae* and *Trachinidae*. The latter, known as lesser weever (Trachinus viperd), is a fish of the Atlantic Ocean and the Mediterranean Sea.

13.1.5 Post mortem Changes

After death, fish muscle tissue is subjected to practically the same spontaneous reactions as mammalian muscle tissue (cf. 12.4.3). Due to the low glycogen content of fish muscle, its pH drop is small. Generally, pH values of 6.2 are obtained. The duration and extent of rigor mortis depend on the type and physiological condition of the fish. In some species, several days can pass before the rigor subsides as a result of the activity of endogenous fish proteinases. The enzymes hydrolyze the Z region of the myofibrils, releasing α-actinin and converting the high molecular α-connectin, which is arranged

Table 13.11. Aroma profile of a mixture of methional (1) and (Z)-1,5-octadien-3-one (2) dissolved in water[a]

Odor quality	Intensity[b]
Fishy	2
Potatoes, boiled	1
Geranium-like	2.5

[a] Concentration: 1 (10 µg/kg), 2 (0.16 µ/kg).
[b] Scale: 0 (not detectable) – 3 (strong).

lengthwise along the myofibrils, to the β-form. Collagen is attacked by collagenase; myosin and actin are not degraded. Deep freezing inhibits the proteolysis, which starts again on thawing. This can cause water loss and undesirable changes in texture in the production of filets. Therefore, deep frozen filets should be made from killed off fish or fish which have passed through rigor mortis. In the case of fresh fish kept in ice, attempts are made to keep the fishing time short so that the fish are stabilized until consumption by rapid and continuous rigor mortis.

Because of the particular structure of fish muscle, the tendency to generate an alkaline pH reaction in muscle, and a high probability of microbial infection during fishing and fish dressing, conditions are highly favorable for rapid spoilage of fish. Therefore, bacteriological supervision and control, from the market to processing plants and during distribution, are of utmost importance.

There are various physical and chemical criteria for assessing fish meat freshness.

The pH of fresh fish is 6.0–6.5. The suitability limit for consumption is pH 6.8, while spoiled fish meat has a pH of 7.0 or above due to the formation of ammonia and amines (cf. 13.1.4.3.2).

The specific resistance of fish muscle changes with storage duration. Soon after catching it is 440–460 ohms, after 4-days approx. 280 ohms, and after 12 days it drops to 260 ohms. The suitability limit for consumption is reached after 16 days, when the resistance is 220 ohms.

The refractive index (n) of fish eye fluid is affected by storage duration. In cod of very good quality, n ranges from 1.3347 to 1.3366. Fish with a n of 1.3394 or higher is not suitable for marketing. The decrease in TMAO concentration and a concomitant increase in volatile N-containing substances, such as trimethyl-amine and several volatile reducing compounds, belong to the chemical criteria for fish quality assessment. Figure 13.3 provides data on the usefulness of some quality criteria, with cod stored in ice taken as an example. In addition to chemical and physical data, sensory evaluation data are included. Another method is based on the observation that the post mortem degradation of ATP, which gives ino-sine in some fish and hypoxanthine in others, proceeds analogously to the loss of freshness. The K value serves to objectify this development. It has been defined as the ratio of the concentrations of inosine plus hypoxanthine to the total concentration of the ATP metabolites. The nucleotides and their degradation products are determined by using HPLC. In practice, not only is the relatively elaborate analysis a disadvantage, but also the dependency of the K value on a series of variables, e. g., the fish species.

A promising development is the gas sensor which can quickly register off-flavor substances or at least indicate an increase in volatile compounds that accompany the decrease in quality on storage.

13.1.6 Storage and Processing of Fish and Fish Products

13.1.6.1 General Remarks

Today, the fishing grounds are not only further and further away and the fishing trips longer, but the fishing ships must also be economically utilized. Therefore, as a result of the easy deterioration of fish, it has become increasingly necessary to process the fish on accompanying factory ships. An overview of the steps involved in fish processing is given in Fig. 13.4. The manual operation steps of the past, such as bleeding, gutting, washing, cutting off the heads, skinning, and filleting, have now been replaced to a significant extent by machines.

The fish waste that accumulates on processing, which accounts for up to 50% of the whole fish, is economically utilized by processing into fish meal on board and on land (cf. 13.1.6.13).

The ready decomposition or spoilage of fish flesh is the result of the special structure of the muscle tissue and the diverse ways in which microbial contamination occurs while handling fish, from catching, through processing and during distribution. From the earliest times, fish handling methods, like those for land animals, have been designed to increase the shelf life or storage ability. Fish are usually initially cooled or frozen, or are dried, salted and smoked, followed by pickling in vinegar or in gelatin with vinegar added. They may also be deep fried in oil, or pickled with or without vinegar and soaked in a sauce in an airtight, sealed container. The expected shelf life of such products determines if they are consid-

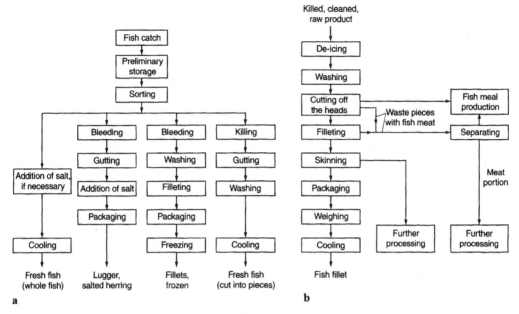

Fig. 13.4. Fish processing on board (**a**) and on land (**b**)

ered fully preserved, canned or semi-preserved products. Semi-preserves may contain additives against microbial spoilage. The compositions of some fish products are given in Table 13.12.

13.1.6.2 Cooling and Freezing

Preservation of freshness by refrigeration is the most modern and effective way to retain the wholesomeness and nutritional value of food. Refrigeration also enables fishing fleets to range the oceans for months in search of fish. Refrigeration permits stockpiling of fish, thus making fish processing plants more economical and better able to respond to market demand and supply.

Fish deteriorates rapidly at temperatures only slightly above 0 °C. Therefore, immediately after catching fish are packed in ice on board the ship. The ice used may be sprinkled with a bactericidal substance. Freezing, which may also be used on ships, is suitable for whole fish (gutted or ungutted, with or without head or skin removal), as is the case with flat fish, tuna, mackerel or herring, or for fish fillets (cod, haddock, saithe, red fish). Only quick freezing is used (−30 to −40 °C; cf. Fig. 13.5), so the critical temperature range of −0.5 to −5 °C is rapidly passed over. Apart

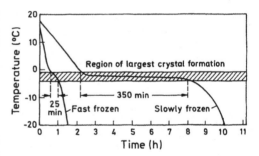

Fig. 13.5. Temperature course during fish fillet freezing process

from air and contact freezing processes, cryogen frosters are being increasingly used, especially for sensitive and high quality product (crustaceans).

In air freezing, freezing takes place in a cold current of air in differently arranged, usually continuously operated systems (tunnel, spiral band etc.). In the contact freezing processes used, the fish are pressed and frosted between two contact plates that are cooled by a flow of coolant. The blocks obtained by this process can be portioned into slabs or sticks using band saws. They can be sold to the consumer as such or breaded and prefried (170 °C, 20 s). Waste pieces (8–12%) are

Table 13.12. Average chemical composition of processed fish

Product	Moisture[a]	Protein[a]	Fat[a]	NaCl[a]	Edible portion[b]
Salted fish					
Matje herring	54	18	18	10	68
Salt cured herring	48	21	16	15	68
Dried fish					
Stockfish	15	79	2.5	3	64
Klipfish	34	45	0.7	13	99
Smoked fish					
Buckling	58	23	16	3	62
Smoked sprats	62	17	20	2	60
Eel	53	19	26	1	73
Mackerel	61	21	16	1	70
Schillerlocken					
(smoked haddock filet)	53	21	24		100
Semi-preserved fish					
Bismarck herring	60	20	17	3	95
Bratherring (fried and					
pickled herring)	62	17	15	4	92
Herring, jellied	56	29	13		55
Anchovies	69	13	5	1	100
Herring tidbit	62	15	10	3	100

[a] As % of edible portion.
[b] As % of the whole fish weight.

used for fishburgers and similar products. In comparison with conventional freezing systems, the food comes into direct contact with the refrigerant (liquid nitrogen or liquid carbon dioxide) in shock freezing. The spatial arrangement of the freezing systems essentially corresponds to those in air freezing.

During freezing of fish, problems associated with drip or sap losses, discoloration and rancidity due to lipid peroxidation and, consequently, fish weight loss, poor visual appearance and flat taste may arise and must be avoided by suitable processing. Cold storage should proceed at high air humidity (90%) and with stationary, noncirculating air. Data on the storage properties of some frozen fish are provided in Table 13.13.

Thawing is carried out in a stream of air saturated with water vapor at 20–25 °C or by spraying with water. Fish must be processed immediately after thawing because it loses juices rapidly and decays. Fish muscle enzymes have noticeable activity even at −10 °C. Excessively long storage or insufficient cooling, especially of fatty fish, results in a rancid off-flavor and an unattractive,

Table 13.13. Shelf life of frozen fish, crustaceans and mollusks as influenced by storage temperature

Product	Shelf life (months) at		
	−18 °C	−25 °C	−30 °C
Fatty fish	4	8	12
Nonfatty fish	8	18	24
Lobsters, cray-			
and crawfish	6	12	15
Crabs	6	12	12
Oysters	4	10	12

yellow colored fish surface. Antioxidants and associated synergistic compounds, such as ascorbic and citric acids, are used to inhibit fat deterioration. Changes in muscle texture are primarily due to changes in protein solubility (Fig. 13.6).

13.1.6.3 Drying

Fish can be preserved by drying naturally in the sun or in drying installations.

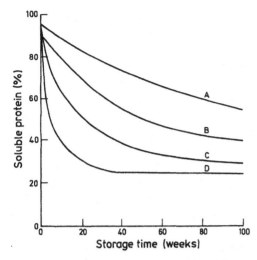

Fig. 13.6. Changes in fish muscle protein solubility as a result of cold storage (−14 °C). A: plaice, B: halibut, C: dogfish, D: cod. (according to *Connell*, 1964)

Stock fish, primarily nonfatty fish (cod, saithe, haddock, ling or tuck, which is often called cusk in North America) with head removed, split and gutted, is spread outdoors to dry in sea air (water content 12–18%). It is an unsalted fish product that is primarily produced in Norway and Iceland. Alternatively, machine-cut, headless and tailless fish which has been belly-clipped ("clipped" fish) is salted, either directly or in brine, and then put through a drying process (salt content 18–20%, water content < 40%). This is most often done with cod or other non-fatty fish species in Canada and Norway.

13.1.6.4 Salting

Salted fish (whole or parts) are obtained by salting fresh, deep frozen or frozen fish. Salt is the most important and oldest preservative for fish. Rubbing or sprinkling of fish with salt or immersion in brine, often followed by smoking, is called fish curing. Pickling in vinegar might be an additional preservation step. Salted products include: herring, anchovies, saithe, cod, salmon, tuna, and roe or caviar. It should be taken into account that if salting is used as the sole method of preservation without further processing (like matje, marinaded fish, smoked products etc.), complete bacterial

protection is not provided because halophilic microorganisms can cause spoilage.

In dry salting, fish and salt are alternately stacked in open piles and the resulting brine can drain off. In wet salting, the fish are put into more or less concentrated salt solutions. In heavily salted fish there are at least 20 g of salt in 100 g tissue fluid; in a medium-salted fish the salt content is 12–20 g.

Salting of herring is of special importance. There are the mildly salted matje (8–10% NaCl), the medium-salted "scotch cure", and the heavily salted herring, i. e. dry salting up to 25% NaCl. Herrings are also dressed, salted and packaged, both at sea and on land. The shelf life of salted herring is several months. Matje herring (immature sea herring, often wrongly called "sardines") must be consumed soon after they are removed from refrigeration. Salting might provide a finished end-product, but it is often used as a form of fast, temporary preservation, yielding semi-finished products which are later to be processed further. After salting, herring pass through a maturation process which generates a typical flavor. The proteolytic enzyme of the fish are involved in such "gibbed" herring maturation. During gibbing (i. e. a process of removing gills, long gut and stomach), the milt (male fish) or roe (female fish) and some of the pyloric cacea are left in the fish. These organs release enzymes which contribute to the maturation of the fish. If all organs are removed, no maturation occurs. Salting causes cell shrinkage and denaturation of muscle proteins, which manifests itself in a decrease in solubility (Fig. 13.7). This is used to convert a finely ground mass of low-fat fish meat to firm products. Similar in importance to herring are salted cod (Atlantic, Pacific and Greenland cod), which are salted dry or in brine as split or boneless fillets, and salted saithe, pollack (Dover hake) and some other saltwater fish of the *Gadus* species. Salted achovies (Mediterranean or Scandinavian) are also of importance.

13.1.6.5 Smoking

Smoked fish are obtained from fresh, deep frozen, frozen or salted fish which have been dressed in various ways. The whole fish body or fish por-

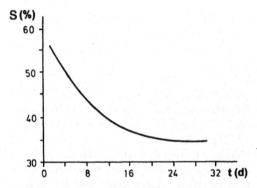

Fig. 13.7. Decrease in solubility of proteins during the salting of cod (according to *Tülsner*, 1994) *Ordinate*: S, solubility in % of total protein. *Abscissa*: duration of action of NaCl

tions are exposed to freshly generated sawdust smoke.

Cold smoking is performed for 1–3 days below 25 °C, generally at 18–25 °C, and is most often used with cooked and mature fatty fish (large size herring, salmon, cod, tuna). A product called "Buckling" is a large, fatty herring, sometimes nobbed (with head), that has been smoked. Delicatessen Buckling are made from gutted herring. Kippered herring (Newcastle Kipper) are obtained from fresh, fatty herring with the back split from head to tail. The dressed fish are then lightly brined and cold smoked. In the United States the term "kipper" corresponds to products hot smoked on trays (e. g. kippered black cod). Salted or frozen salmon are smoked in North America. The shelf life is 2 weeks.

Hot smoking is performed for 1–4 h at temperatures between 70 and 150 °C. It is used with whole, gutted or descaled fish such as herring (Buckling), sprat ("Kieler Sprotten" in Germany), plaice, flounder, halibut (with or without skin), eels, mackerel, tuna, haddock, whiting (merlan), saithe, cod, red fish (ocean perch), dog-fish, sturgeon and shad. In the process, the fish are cooked by air cooking. In some places, a minimum temperature of 85 °C is stipulated to kill microorganisms. Unlike cold-smoked fish, hot-smoked fish have only a limited shelf life, 3–10 days, which can be extended only by cold storage. Hot-smoked caviar (cod or saithe) is available. Smoking is conducted in smoking installations which consist of a smoke generator and a smoking chamber. In this chamber, the fish are dried and cooked and smoked.

13.1.6.6 Marinated, Fried and Cooked Fish Products

Marinade is vinegar or wine, or a mixture of both, usually spiced and salted, in which fish are soaked or steeped before use or before being pickled and stored for a longer time. The fish used might be fresh, deep frozen or frozen, or salted whole fish or fish portions. Marinating tenderizes muscle tissue without heat treatment. Fish preservation by pickling in this manner is based on the combined action of salt and vinegar. Vinegar-packed herring, called simply marinades, are a popular German fish food packaged in retail glass jars. Pickled fish can be packaged together with some plant extracts, sauces, gravy, creams, mayonnaise or related products, or they can be immersed in an edible oil (although oilpacked fish are not called pickled fish). Some of these products might contain chemical preservatives.

Marinated fish are packaged in cans, jars, etc., and may be handled without packaging. Fish marinades have only a limited shelf life (they are semipreserves), and even chemical preservatives can not prevent their eventual decay. Marinated fish considered as delicatessen items are "Kronsardines", Bismarck herring, rollmops and pickled herring.

Fried fish products are prepared from variously dressed fresh, deep frozen or frozen whole fish or fish portions, with or without further dressing in eggs and bread crumbs (batter formulations, such as "shake and bake"). They are then made tender by frying, baking, roasting or barbecuing. These products may be packaged or canned in the presence of vinegar, sauces, gravy or an edible oil, often with a chemical preservative added. Examples of these products are fried marinated fish sticks, "Brathering", "Bratrollmops", balls, etc.

Cooked fish products are processed in a similar manner. Tenderization is achieved by cooking or steaming. Processing also involves the use of vinegar or wine, addition of salt and the use of a preservative. The cooked products can be solidified, with or without plant ingredients, into a jelly (herring in jelly) or packaged with other extracts, sauces or gravy. Cooked fish products include herring marinades, rollmops, bacon rollmops in jelly, sea eel (dogfish) in jelly, or broths made from disintegrated saltwater fish meat. The occasional liquefaction of cooked fish jelly ("jelly disease") indicates microbial proteolysis.

13.1.6.7 Saithe

Saithe, often called coalfish, coley, pollack or Boston bluefish (trade name "Dover hake"), are processed into fillets, saltcured, dyed or tinted, and smoked. They are then cut into slices or cutlets and covered with edible oil. The product has a good shelf life.

13.1.6.8 Anchosen

Anchosen are made from fresh, frozen or deep-frozen small sprats and herring, preserved with salt in the presence of added sugar or sugars derived from starch saccharification, spiced and biologically ripened with sodium nitrate. Flavors are also added. Proteinases are also used to accelerate ripening.

Anchosen can be packed in sauce (gravy), creams or in edible oils, garnished with plant ingredients and a chemical preservative may be added. Examples of anchosen are appetit-sild, cut spiced herring and spiced herring. Appetit-sild is a product consisting of skinned fillets of spice-cured sprats, cured and packed in vinegar, salt, sugar and spices.

13.1.6.9 Pasteurized Fish Products

Pasteurized fish products made of fresh, deep-frozen or frozen fish or fish portions have shelf lives, even without cold storage, of at least 6 months. These products are prepared by prolonged heat treatment of fish at temperatures below 100 °C. They are then tightly sealed in a container. Such products are salted or soaked in vinegar prior to pasteurization.

13.1.6.10 Fish Products
with an Extended Shelf Life

Canned fish products of extended shelf life are products made from fresh fish, frozen or deep-frozen whole fish or fish portions. The shelf life without special cold storage of at least one year is achieved by adequate heat treatment in gas-tight containers.

These products with an extended shelf life can normally be kept indefinitely (in practice about 5 years). Special can materials have to be chosen when the fish is canned with corrosive ingredients such as tomato or mustard sauce, vinegar or lemon juice. The can is usually made of a lacquer-coated tinplate or inert aluminum.

Products with extended shelf lives are in their own juice or in added oil, or in some sauce or cream (e. g. "sardine" pilchards, *Sardina pilchardus*, packed in olive or soya oil, tomato mustard, or lemon juice). Also available are fish paste, meat balls or "Frikadellen" (Germany), i. e. flesh of white fish made into rissoles using flour, eggs and spices, which are then roasted, deep fried and used ready-to-serve, as hors d'oeuvres, and fish salad. The latter products are canned or packed in glass jars, and may be packed under controlled atmosphere.

13.1.6.11 Surimi, Kamboko

Surimi is a concentrate of insoluble muscle proteins (ca. 20%). It forms a solid cohesive gel with water (ca. 80%), which solidifies when warm. For production, lean fish meat is ground at 5–10 °C and extracted with water until basically only myosin, actin, actomyosin and small amounts of collagen remain. The addition of paramyosin (cf. 13.1.4.2.2) intensifies the structure of the gel. In the further processing of Surimi to Kamboko, starch (ca. 5%), egg white, flavor enhancers, colorants and aroma substances are added, whereby an attempt is made to imitate crab or mussel meat. The resulting mixture is solidified by denaturation of the proteins first at 40–50 °C and then at 80–90 °C. Fibrous structures are produced by extrusion.

13.1.6.12 Fish Eggs and Sperm

13.1.6.12.1 Caviar

Specially prepared sturgeon eggs (roe) are called caviar. The roe ("hard roe") are detached from the fish ovary gland. The roe are washed in cold water, salted and left to ripen until they become transparent. They are then drained from the brine slime and are marketed for the wholesale market in small metal or glass containers or in barrels. Occasionally, the caviar is pasteurized. Two basic types are marketed: grainy caviar, where eggs are readily detached from roe, and pressed caviar, where the ovarian membrane and the excess fluid are removed by gentle pressing. Caviar is made from various sturgeon species (beluga, stoer or sevruga). The roe of these sturgeon species caught in winter, when mildly salted (below 6% NaCl), give a high quality caviar called "Malossol". The beluga (the largest of the three sturgeons mentioned) provides the most valuable caviar.

Pressed caviar is obtained from all species. Salmon caviar (such as Amur and Keta caviars from Siberian salmon roe) is processed using less than 8.5% salt. American whitefish caviar is a mixture of roe from salmon, whitefish, carp and some other fish. Scandinavian caviar is from cod and lumpfish.

Sturgeon caviar is gray or brown to black in color. Salmon caviar is yellow-red or red. Most caviar is imported from the Russian Fed. and Iran (Caspian Sea caviars). It readily decays and so must be kept refrigerated. A medium-size beluga sturgeon can provide 15–20 kg caviar.

13.1.6.12.2 Caviar Substitutes

Caviar substitutes are made of roe of various sea and freshwater fish. Germany produces the dyed caviar of lumpfish (lumpsuckers), and also cod and herring caviars. The roe are soured, salted, spiced, dyed black, treated with traganth gum and, occasionally, a preservative is added.

13.1.6.12.3 Fish Sperm

Fish sperm are a product of the gonads of male fish and are often called milt or soft roe. Salted sperm from sea and freshwater fish, particularly herring, are most commonly marketed.

13.1.6.13 Some Other Fish Products

These include the nutritional products and seasonings derived from fish protein hydrolysates; insulin from shark pancreas; proteins recovered from saltwater fish fillet cutting; fish meal used

as feed for young animals, poultry and pond fish; and, lastly, fish fat (oil), as mentioned in 14.3.1.2. Of increasing importance is the production of fish protein concentrates and, when necessary, their modified products (cf. 1.4.6.3.2 and Table 1.44).

13.2 Whales

Although a whale is in a true sense a mammal and not a fish, it will be covered here. The blue (*Balaenoptera muculus*) and the finback whale (*B. physalus*) are the two most important whales, each growing up to 30 m in length and up to 150 tons in weight. Also caught are the humpback (*Megaptera nodosa*), the sperm (*Physeter macrocephalus*) and the sei whale (*Balaenoptera borealis*). Whale meat is similar to big game meat or beef. It has long and coarse muscle fibers arranged in bundles and colored gray-reddish. The color of the meat is affected by the age of the whale, and may be bright red or dark red, while frozen whale meat becomes dark black-brown in color. Freezing imparts a rough, firm texture to the meat. The fresh meat has a pleasant flavor but, due to the fast rate of fat oxidation, the shelf life is very short. For this reason bulk whale meat is not readily accepted by food wholesalers and the retail market. Whale meat extracts are also produced (cf. 12.7.3.2).

13.3 Crustaceans

Crustaceans have no backbone; their body is divided into sections, each bearing a pair of joint-legs. An armor-like shell covers and protects the body. Included are shrimp, crayfish (also called crabfish), crabs (e. g. freshwater, edible green shore crab) and lobster. Compositional data are provided in Table 13.14.

13.3.1 Shrimps

The most important shrimps are the common or brown shrimp from the North Sea (*Crangon crangon*), the Baltic Sea shrimp (*Palaemon adspersus fabricii*), the deep sea shrimp (*Pan-dalus bo-*

Table 13.14. Average chemical composition of crustaceans and mollusks

Crustaceans/ mollusks	Moisture[a]	Protein[a]	Fat[a]	Minerals[a]	Edible portion[b]
Shrimps	78	19	2	1.4	41
Lobsters	80	16	2	2.1	36
Crayfish	83	15	0.5	1.3	23
Oysters	83	9	1.2	2.0	10
Scallop	80	16	0.1	1.4	44
Mussel	83	10	1.3	1.7	18

[a] As % of edible portion.
[b] As % of the whole fish weight.

realis) and the larger species in tropical waters, such as blue Brasilian (*Penaeus* spp.) or the royal red shrimps (*Hymenopenaeus robustus*). Larger species are called prawn.

Shrimps are marketed soon after catch as: live, fresh with shell, with or without head, cooked in brine, or cooked without shell. They have very short shelf lives. Shrimps are also sold canned, deep frozen or as an extract or a salad ingredient. Canned shrimps are heated (pasteurized) at just 80–90 °C so as not to affect their flavor; hence, they are semi-preserves with a limited shelf life.

13.3.2 Crabs

Crabs live in shallow or deep water along the sea coast or in freshwater. Blue crab (*Call-inectes sapidus*) is the most common crab of the Atlantic coast of North America. Other important species are the common shore crab (also called green shore crab); the edible crab of Europe (*Cancer pagurus*), which lives in sandy, shallow water; the king crab of Alaska (*Pralithodes camchaticus*), also called Japanese crab; and the dungeness crab (*Cancer magister*) from the shallow waters from California to Alaska. These crabs differ in shape and size of their big claws, but all have no tail. The color and shape of the body varies, as does the ability to swim or to run sideways.

When crabs shed their shell and the new shell has not yet hardened, they are at their tastiest and are marketed as "extra choice soft" crabs. The forms sold are: live, fresh, frozen and canned. Crab paste, canned soup and crab cakes similar to deepfried fish cakes are delicatessen sea foods. In the trade, the term crab meat means white muscle meat, colored red only in leg muscles and chelae, and is distinguished from brown crab meat obtained from crab liver and gonads. The latter are usually processed into crab paste. All crab products are of limited shelf life.

13.3.3 Lobsters

The European lobster (*Homarus gammarus*) caught in the Atlantic is the largest in Europe. It reaches a length of 35–90 cm and a maximum weight of 10 kg. The major area of catch is Helgoland, the north and west sea coast of Europe, the Mediterranean and the Black Sea. The tastiest lobster meat is that from the breast shell. The American or northern lobster (*Homarus americanus*) is closely related to the European lobster. Lobsters are marketed live (remain alive up to 36 h after catch), whole boiled, or canned as cooked meat in its own juice or as soup (cream of lobster, lobster chowder). Lobster paste is also available. Cooking of lobster changes its color to red. The color change involves the release of astaxanthin from ovoverdin, a brown-green chromoprotein (cf. 3.8.4.1.2).

The Norway lobster (*Nephrops norvegicus*; also called Langoustine) also belongs to the lobster family. It is marketed fresh, frozen, semi-preserved, as in salad, or canned, as soup, paste or mildly-brined meat in its own juice.

13.3.4 Crayfish, Crawfish

Crayfish are freshwater crustaceans considered as a delicacy in Europe. The major crayfish of Europe belongs to Astacus spp. (*Astacus astacus* or *fluviatilis*). Its meat is the most tasteful in May–August when it sheds its shell and the new shell is still soft. The eastern part of North America has the freshwater crayfish of *Cambarus* spp. The Australian crayfish belongs to *Enastacus serratus*.

The cray(craw)fish die when they are dropped into boiling water. Their tail curls up – this is a sign that they were cooked fresh or alive. For the color change during cooking see above (cf. 13.3.3).

The seawater species of crayfish are called crawfish. They include *Palinurus, Panulirus* and *Jasus* spp. The most important crawfish are the European spiny lobster (*Palinurus vulgaris*), the Pacific North American counterpart (*Panulinus interruptus*) and others from Africa, Australia and Japan. The European spiny lobster is 30–40 cm long, up to 6 kg in weight, has rudimentary front legs shaped into sharp claws and has a knobby shell covering the body. It is often caught in the Mediterranean Sea, the west and south coasts of England, and along the coast of Ireland. The rock lobster (*Jasus lalandei*) and the Mediterranean crawfish (*Palinurus elephas*) are also available on European markets. The meat of these crawfish is rather coarse and fiberlike and is colored yellow to yellow-red.

The cray- and crawfishes are marketed fresh live, raw or cooked, and canned in differet forms: meat, butter (precooked meat mixed with butter), soup and soup powders, soup extracts (these are crawfish butter, spiced, salted and blended with flour) and crayfish bisque (French purée or thick soup of crawfish and lobsters).

13.4 Mollusks (*Mollusca*)

13.4.1 Mollusks (*Bivalvia*)

The bivalve mollusks include clams, oysters, mussels and scallops. The common oyster (also called flat native or European oyster) and the blue or common mussel are the most often processed molluscan shellfish.

Oysters (*Ostreidae*, e. g. the European oyster, *Ostrea edulis*) live in colonies along the sea coast or river banks, or are cultivated in ponds ("oys-

ter farms") which are often connected with the sea. Oysters are usually sold unshelled. Only the adductor muscle is consumed; the pleated gills and the digestive system are discarded. In addition to the common oyster, the Portuguese oyster (*Gryphea angulata*) and the American blue point oyster, (*Crassostrea virginica*), used most commonly for canning, are of importance. The best meat is obtained from oysters harvested when they are 3–5 years old, with the top quality harvested between September and April (an old saying is: oysters should be eaten in months which have "r" in their names).

The blue or common mussel (*Mytilus edulis*) lives in shallow, sandy freshwater, while the sea mussel lives in ocean water or is cultivated in ponds or lakes. The shell, 7–15 cm long, is bluish black and the body meat is yellowish. The meat is rich in protein (16.8%) and also in vitamin A and the vitamin B-complex. The meat is eaten cooked, fried or marinated. The major mussel growing areas in Germany are the Kiel Bay and the East Friesian Islands.

In addition to common mussel, numerous other mussels are eaten, mostly canned in vegetable oil, e. g., Pacific Bay or Cape Cod scallops (*Pectinidae*) and cockles (*Cardiae*).

Due to rapid spoilage, mussels are marketed live or canned. They are eaten soon after being caught or after the can is opened, and are avoided in warm seasons. Moreover, they should originate from uncontaminated clear waters.

13.4.2 Snails

Snails are univalve mollusks, i. e. they have only a single, coiled shell. They are eaten preferentially in Italy, France and Germany, and are nearly exclusively the large Helix garden snail (*Helix pomatid*). Snails are sometimes collected wild in South or Central Germany and in France, but most are supplied by snail gardens and feeding lots where lettuce and cabbage leaves are the food source, or in damp shady cellars, where wheat bran and leafy vegetable leaves (e. g. cabbage) are used as a feed. The meat is considered a delicacy. Since the shelf life of the meat is very limited, snails are marketed live (with the shell plugged) or canned. Marine snails of various kinds are fried, steamed, baked or cooked in soups, and are also considered a delicacy.

13.4.3 Octopus, Sepia, Squid

Octopus, sepia and calmar (*Cephalopoda*) are softbodied mollusks with eight or ten arms, and without an outside shell.

The sepia or cuttlefish (*Sepia officinalis*), the squid or calmar (*Loligo loligo*) and the octopus or devilfish (*Octopus vulgaris*) are caught in the Mediterranean region, mostly in Italy, and other parts of the world (Atlantic and Pacific Oceans, e. g. the North American poulp, Japanese *Polypus* spp., etc.). They are consumed deep fried in oil, baked, cooked in wine, pickled in vinegar after being boiled, cooked in soups, in salads, stewed or canned.

13.5 Turtles

Turtles, tortoises or terrapin (for American fresh-water turtles) are reptiles with a shell used as a "house". The logger head and green sea turtles are caught commercially for their meat. In Germany turtle is mostly eaten in soup or stew. The meat of the so-called soup turtle (*Chelonia mydas*) is faintly red to bright red, and is marketed canned. An imitation or mock turtle soup is prepared from edible parts of heads of calves and has no relation to turtles except for the name.

13.6 Frogdrums

The thigh portion (frogdrum) of a frog's hinged leg is sold as a delicacy. Frogs providing frog-drums are the common bullfrog (*Rana catesbo-niand*), the leopard frog (*Rana pipiens*) and oth-ers (*Rana arvalis, Rana tigrena, Rana esculenta*).

The meat is soft in texture, white in color and tasty; however, it has a very limited shelf life as it readily deteriorates. Frogdrums are eaten cooked, roasted or stewed.

13.7 References

Borgstrom, G. (Ed.): Fish as food. Vol. I–III. Academic Press: New York. 1961–1965

Connell, J.J.: Fish muscle proteins and some effects on them of processing. In: Proteins and the reactions (Eds.: Schultz, H.W., Anglemier, A.F.), p. 255, AVI Publ. Co.: Westport, Conn. 1964

Connell, J.J. (Ed): Advances in fish science and tech-nology. Fishing New Books Ltd: Farnham, Surrey, England. 1980

Feeney, R.E., Yin Yeh: Antifreeze proteins from fish bloods. Adv. Protein Chem. *32*, 191 (1978)

Fernandez, M., Mano, S., Garcia de Fernando, G.D., Ordonez, J.A., Hoz, L.: Use of β-hydroxyacyl-CoA-dehydrogenase (HADH) activity to differentiate frozen from unfrozen fish and shellfish. Eur. Food Res. Technol. *209*, 205 (1999)

Habermehl, G.: Gift-Tiere und ihre Waffen. 2. edn., Springer-Verlag: Berlin. 1977

Lindsay, R.C.: Fish flavors. Food Reviews International *6*, 437 (1990)

Ludorff, W., Meyer, V.: Fische und Fischerzeugnisse. 2. edn., Verlag Paul Parey: Berlin. 1973

Milo, C., Grosch, W.: Changes in the odorants of boiled Salmon and cod as affected by the storage of the raw material. J. Agric. Food Chem. *44*, 2366 (1996)

Multilingual dictionary of fish and fish products (pre-pared by the OECD), 2nd edn., Fishing News Books Ltd., Farnham, Surrey, England. 1978

Olafsdottir, G. et al.: Methods to evaluate fish freshness in research and industry. Trends Food Sci. Technol. *8*, 258 (1997)

Sikorksi, Z.E., Sun Pan, B., Shahidi, F.: Seafood pro-teins. Chapman & Hall, New York, 1994

Tülsner, M.: Fischverarbeitung. Band 1. Rohstoff-eigenschaften und Grundlagen der Verarbeitungs-prozesse. Behr's Verlag, Hamburg, 1994

Venugopal, V., Shahidi, F.: Structure and composition of fish muscle. Food Rev. Int. *12*, 175 (1996)

Whitfield, F.B.: Flavor of prawns and lobsters. Food Reviews International *6*, 505 (1990)

14 Edible Fats*and Oils

14.1 Foreword

Most fats and oils consist of triacylglycerides (recently also denoted as triacylglycerols; cf. 3.3.1) which differ in their fatty acid compositions to a certain extent. Other constituents which make up less than 3% of fats and oils are the unsaponifiable fraction (cf. 3.8) and a number of acyl lipids; e. g., traces of free fatty acids, mono- and diacylglycerols.

The term "fat" generally designates a solid at room temperature and "oil" a liquid. The designations are rather imprecise, since the degree of firmness is dependent on climate and, moreover, many fats are neither solid nor liquid, but are semi-solid. Nevertheless, in this chapter, unless specifically emphasized, these terms based on consistency will be retained.

14.2 Data on Production and Consumption

Data on the production of oilseeds and other crops are summarized in Table 14.0. The world production of vegetable fats has multiplied since the time before the Second World War (Table 14.1). There has been a significant rise in production since 1964 of soybean, palm and sunflower oils, as well as rapeseed oil. Soybean oil, butter and edible beef fat and lard are most commonly produced in FR Germany (Table 14.1). The per capita consumption of plant oils in Germany has increased in the past years (Table 14.2).

---

* Butter is dealt with in Chapter 10.2.3.

14.3 Origin of Individual Fats and Oils

14.3.1 Animal Fats

14.3.1.1 Land Animal Fats

The depot fats and organ fats of domestic animals, such as cattle and hogs, and milk fat, which was covered in Chapter 10, are important animal raw materials for fat production. The role of sheep fat, however, is not significant. The major fatty acids of these three sources are oleic, stearic and palmitic (Table 14.3).

It should be noted that the fatty acid composition of individual fat samples may vary greatly. The fat composition of land animals is affected by the kind and breed of animal and by the feed. The composition of plant fats depends on the cultivar and growth environment, i. e. climate and geographical location of the oilseed or fruit plant (cf. Fig. 3.3.1.5). Therefore only average values are given in the following tables dealing with fatty acid composition.

In contrast to oil from plant tissue, the recovery of animal fat is not restricted by rigid cell walls or sclerenchyma supporting tissue. Only heating is needed to release fat from adipose tissue (dry or wet rendering with hot water or steam). The fat expands when heated, tearing the adipose tissue cell membrane and flowing freely. Further fat separation is simple and does not pose a technical problem (Fig. 14.1).

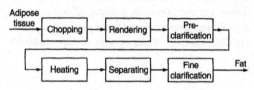

Fig. 14.1. Steps involved in wet rendering

Table 14.0. Production of major oilseeds, 2006 (1,000)[a]

Continent	(Castorbean) Castor oil seed	Sunflower seed	Rapeseed	Sesame seed
World	1140	31,332	48,974	3338
Africa	35	868	95	976
America, Central	1	–	–	54
America, North	–	1118	9823	–
America, South and Caribbean	107	4218	228	153
Asia	998	5325	21,421	2207
Europe	1	19,709	16,963	1
Oceania	–	95	444	–

Continent	Linseed	Safflower seed	Cottonseed	Copra
World	2570	583	44,173	5370
Africa	161	11	2887	194
America, Central	–	72	219	208
America, North	1321	89	6666	–
America, South and Caribbean	70	90	2513	263
Asia	764	356	30,665	4714
Europe	245	1	597	–
Oceania	10	36	844	198

Continent	Palm kernel	Palm oil	Olives	Olive oil
World	10,641	37,291	16,926	2710
Africa	1685	2359	2636	340
America, Central	164	503	13	1
America, North	–	–	45	1
America, South and Caribbean	599	1791	193	24
Asia	8256	32,755	2536	285
Europe	–	–	11,493	2060
Oceania	101	386	23	–

Country	(Castorbean) Castor oil seed	Country	Sunflower seed	Country	Rapeseed
India	730	Russian Fed.	6753	China	12,649
China	240	Ukraine	5324	Canada	9105
Brazil	92	Argentina	3798	India	8130
Ethiopia	15	China	1820	Germany	5337
Thailand	11	Romania	1526	France	4144
Paraguay	11	France	1440	UK	1870
Vietnam	5	Bulgaria	1197	Poland	1652
South Africa	5	Hungary	1165	Czech. Rep.	880
Philippines	4	India	1120	USA	718
Angola	4	Turkey	1118	Russian Fed.	522
Σ (%)[b]	98	Σ (%)[b]	80	Σ (%)[b]	92

Table 14.0. continued

Country	Sesame seed	Country	Linseed	Country	Safflower seed
China	666	Canada	1041	India	230
India	628	China	480	USA	87
Myanmar	580	USA	280	Kazakhstan	75
Sudan	200	India	210	Mexico	72
Uganda	166	Ethiopia	128	Australia	36
Ethiopia	160	Argentina	54	China	30
Nigeria	100	Bangladesh	50	Argentina	18
Bangladesh	50	UK	49	Kyrgyzstan	14
Paraguay	50	France	43	Ethiopia	6
Tanzania	48	Russian Fed.	36	Tanzania	5
Σ (%)[b]	79	Σ (%)[b]	92	Σ (%)[b]	98

Country	Cottonseed	Country	Copra	Country	Palm kernel
China	13,460	Philippines	2200	Malaysia	4125
India	7128	Indonesia	1310	Indonesia	3860
USA	6666	India	750	Nigeria	1250
Pakistan	4065	Vietnam	243	Thailand	191
Uzbekistan	2376	Mexico	204	Colombia	178
Brazil	1785	Sri Lanka	65	Brazil	122
Turkey	1350	Thailand	65	Ecuador	95
Australia	844	Malaysia	51	Cameroon	75
Syrian Arab Rep.	664	Mozambique	46	Côte d'Ivoire	74
Greece	520	Côte d'Ivoire	45	China	62
Σ (%)[b]	88	Σ (%)[b]	93	Σ (%)[b]	94

Country	Palm oil	Country	Olives	Country	Olive oil
Indonesia	15,900	Spain	5032	Spain	955
Malaysia	15,880	Italy	3424	Italy	627
Nigeria	1287	Greece	2661	Greece	439
Colombia	711	Turkey	1600	Tunisia	205
Thailand	685	Tunisia	1000	Syrian Arab Rep.	130
Côte d'Ivoire	330	Morocco	750	Turkey	90
Ecuador	291	Syrian Arab Rep.	501	Morocco	85
China	230	Algeria	365	Algeria	32
Congo	175	Egypt	310	Jordan	27
Honduras	175	Portugal	276	Portugal	26
Σ (%)[b]	96	Σ (%)[b]	94	Σ (%)[b]	97

[a] Soybean and peanuts are presented in Table 16.1.
[b] World production = 100%.

14.3.1.1.1 Edible Beef Fat

Edible beef fat is obtained from bovine adipose tissue covering the abdominal cavity and surrounding the kidney and heart and from other compact, undamaged fat tissues. The fat is light-yellow due to carotenoids derived from animal feed. It is of a friable, brittle consistency and melts between 45 and 50 °C.

The fatty acid composition of beef fat (Table 14.3) is not influenced greatly by feed intake, but that of hog fat (lard) is. The composition of edible beef fat triacylglycerols is given in 3.3.1.4. The following commercial products are prepared from beef fat: *Prime Beef Fat ("premier jus")* is obtained by melting fresh and selected fat trimmings in water heated to 50–55 °C. The acid value resulting from lipolytic action (cf. 14.5.3.1)

Table 14.1. World production of plant fats (in 10^6 t)

Fat/oil	1935/39	1965	1981	2004
Soya oil	1.23	4.86	12.5	31.9
Sunflower oil	0.56	2.38	4.5	9.5
Cottonseed oil	1.56	2.57	3.25	3.9
Peanut oil	1.51	3.17	2.95	4.8
Rapeseed (Canola) oil	1.21	1.47	3.75	13.1
Palm kernel and palm oil	1.33	1.6	6.16	31.6
Coconut oil	1.93	2.23	2.93	3.3
Olive oil	0.87	1.95[a]	1.33[b]	2.8

[a] Production data for 1964
[b] An estimate for 1982

Table 14.2. Consumption of plant oils and margarine in Germany (kg per capita per year)

Year	Plant oils	Margarine
1993	9.7	7.7
1995	11.4	7.1
1997	13.2	7.3

Table 14.3. Average fatty acid composition of some animal fats (weight-%)

Fatty acid	Beef tallow	Sheep tallow	Lard	Goose fat
12:0	0	0.5	0	0
14:0	3	2	2	0.5
14:1 (9)	0.5	0.5	0.5	0
16:0	26	21	24	21
16:1 (9)	3.5	3	4	2.5
18:0	19.5	28	14	6.5
18:1 (9)	40	37	43	58
18:2 (9, 12)	4.5	4	9	9.5
18:3 (9, 12, 15)	0	0	1	2[a]
20:0	0	0.5	0.5	0
20:1				
20:2	0	0.5	2	
Other	3	3	0	0

[a] It includes fatty acid 20:1.

is not allowed to exceed 1.3 (corresponding to approx. 0.65% free fatty acid).

This beef fat, when heated to 30–34 °C, yields two fractions: oleomargarine (liquid) and oleostearine (solid). Oleomargarine is a soft fat with a consistency similar to that of melted butter. It is used by the margarine and baking industries. Oleostearine (pressed tallow) has a high melting point of 50–56 °C and is used in the production of shortenings (cf. Table 14.18). *Edible Beef Fat (secunda beef fat)* is obtained by melting fat in water at 60–65 °C, followed by a purification step. It has a typical beef fat odor and taste and a free fatty acid content not exceeding 1.5%. Lower quality tallow has only industrial or technical importance, for example, as raw material for the soap and detergent industries.

14.3.1.1.2 Sheep Tallow

The unpleasant odor adhering to sheep tallow is difficult to remove, hence it is not used as an edible fat. Sheep tallow is harder and more brittle or friable than beef tallow. The fatty acid composition of sheep tallow is presented in Table 14.3.

14.3.1.1.3 Hog Fat (Lard)

Hog (swine) fat, called lard, is obtained from fat tissue covering the belly (belly trimmings) and other parts of the body. The back fat is mainly utilized for manufacturing bacon. After tallow and butter, lard is currently the animal fat which is consumed the most (Table 14.1). Its grainy and oily consistency is influenced by the breed and feeding of hogs.

Some commercial products are:

Lard obtained exclusively from belly trimmings (abdominal wall fat). This is the highest quality *neutral lard*. It has a mild flavor, is white in color and its acid value is not more than 0.8.

Lard from other organs and from the back is rendered using steam. The maximum acid value is 1.0.

Lard obtained from all the dispersed fat tissues, including the residues left after the recovery of neutral lard, is rendered in an autoclave with steam (120–130 °C). This type of lard has a maximum acid value of 1.5.

In contrast to the composition of triacylglycerols found in beef fat (Table 3.13), lard contains fewer triacylglycerols of the type SSS and more of the types SUU, USU and UUU (S = saturated; U = unsaturated fatty acid). As a consequence, lard melts at lower temperatures and over a range of temperatures rather than sharply at a single temperature, and its shelf life is not particularly long. In comparison with beef, pig depot fat contains the saturated fatty acids mainly in the sn-2 position. This difference can be used in the detection of lard, e. g., to control exports to Islamic countries.

14.3.1.1.4 Goose Fat

As the only kind of poultry fat produced, goose fat is a delicacy. Its production is insignificant in quantity. The fatty acid composition of goose fat is given in Table 14.3.

14.3.1.2 Marine Oils

Sea mammals, whales and seals, and fish of the herring family serve as sources of marine oils. These oils typically contain highly unsaturated fatty acids with 4–6 allyl groups, such as (double bond positions are given in brackets): 18:4 (6, 9, 12, 15); 20:5 (5, 8, 11, 14, 17); 22:5 (7, 10, 13, 16, 19); and 22:6 (4, 7, 10, 13, 16, 19) (Table 14.4). Since these acids are readily susceptible to autoxidation, marine oils are not utilized directly as edible oils, but only after hydrogenation of double bonds and refining.

Of analytical interest is the occurrence in marine oils of about 1% branched methy-

lated fatty acids, for example, 12-methyl- and 13-methyltetradecanoic acids or 14-methylhexadecanoic acid. These acids are also readily detectable in hardened marine oils.

14.3.1.2.1 Whale Oil

There are two suborders of whales: Baleen whales which have horny plates rather than teeth, and whales which have teeth. The blue and the finback whales, both live on plankton, and belong to the Baleen suborder. The oils from these whales do not differ substantially in their fatty acid compositions.

A blue whale, weighing approx. 130 t, yields 25–28 t of oil, which is usually recovered by a wet rendering process. The ruthless exploitation of the sea has nearly wiped out the whale population, hence their raw oil has become a rare product.

14.3.1.2.2 Seal Oil

The composition of seal oil is similar to that of whale oil (Table 14.4).

14.3.1.2.3 Herring Oil

The following members of the herring fish family are considered to be satisfactory sources of oil: herring, sardines (Californian or Japanese pilchards, etc.), sprat or brisling, anchovies (German Sardellen or Swedish sardell) and the Atlantic menhaden. The fatty acid compositions of the various fish oils differ from each other (Table 14.4).

14.3.2 Oils of Plant Origin

All the edible oils (with the sole exception of oleomargarine-type products) are of plant origin. With regard to the processes used to recover plant oils, it is practical to divide them into fruit and oilseed oils. While only two fruits are of economic importance in oil production, the number of oilseed sources is enormous.

Table 14.4. Average fatty acid composition of some marine oils (weight-%)

Fatty acid	Blue whale	Seal	Herrin (*Clupea harengus*)	Pilchard[a]	Menhaden (*Brevoortia tyrannus*)
14:0	5	4	7.5	7.5	8
16:0	8	7	18	16	29
16:1	9	16	8	9	8
18:0	2	1	2	3.5	4
18:1	29	28	17	11	13
18:2	2	1	1.5	1	1
18:3	0.5		0.5	1	1
18:4	0.4		3	2	2
20:1	22	12	9.5	3	1
20:4	0.5		0.5	1.5	1
20:5	2.5	5	9	17	10
22:1	14	7	11	4	2
22:5	1.5	3	1.5	2.5	1.5
22:6	3	6	7.5	13	13

[a] Trade name of grown sardines (*Sardinóps caerulea*).

The oils are sold and consumed as pure oil from a single oilseed plant or fruit plant, for example, olive, sunflower or corn oils, or are marketed and used as blended oils, which are generally designated as edible, cooking, frying, table or salad oil.

14.3.2.1 Fruit Pulp Oils

The oils obtained from the fruits of the olive tree and several oil palm species are of great economic importance. The fatty acid compositions of the oils of these fruits are summarized in Table 14.5. Due to the high enzymatic activity in fruit pulp, particularly of lipases, the shelf life of fruit oil is severely limited.

14.3.2.1.1 Olive Oil

Olive oil is obtained from the pulp of the stone fruit of the olive tree (*Olea europaea sativa*).

Table 14.5. Characteristics of olives (fruits/oil) and oil palm

	Olive (*Olea europaea sativa*)	Oil palm (*Elaeis guineensis*)
Fruits		
Length (cm)	2–3	3–5
Width (cm)	2–3	2–4
Fruit pulp (weight-%)	78–84	35–85
Fruit seed (weight-%)	14–16	15–65
Fruit pulp (mesocarp)		
Oil (weight-%)	38–58	30–55
Moisture (weight-%)	to 60	35–45
Fruit pulp oil		
Solidification point (°C)	−5 to −9	27–38
Average fatty acid composition (weight-%)		
	Olive oil	Palm oil
14:0	0	1
16:0	11.5	43.8
16:1	1.5	0.5
18:0	2.5	5
18:1 (9)	75.5	39
18:2 (9,12)	7.5	10
18:3 (9,12,15)	1.0	0.2
20:0	0.5	0.5

More than 90% of the world's olive harvest comes from the Mediterranean region, primarily in Italy and Spain (cf. Table 14.0). Olive tree plantations are found to a smaller extent in Japan, Australia, California and South America.

Oil Production. The disintegrated fruit is kneaded to release the oil droplets from the pulp, occasionally by adding common salt.

The oil is then pressed out or separated by gravity decantation. The initial cold pressing provides virgin oil (provence oil). This is then followed by warm pressing at about 40 °C.

In addition to the conditions used for oil recovery, the quality of olive oil is affected by the ripeness of the fruit (overripe fruit is not preferred) and length of storage. In virgin oils there is a relationship between sensory properties and the content of free fatty acids:

- Virgin olive oil extra (*extra vierge*): Pleasant aromatic taste; up to 0.8% free fatty acids, calculated as oleic acid
- Virgin olive oil (*fine vierge*): Slightly less aromatic in taste; up to 2% free fatty acids
- Lampante oil: Much less taste; more than 2% free fatty acids

For the isolation of refined olive oil, the oil cake is extracted with a solvent. The resulting extraction oil ("sansa" oil) is refined so that it contains at most 0.3% of free fatty acids.

From time to time, the expensive *extra vierge* oils are adulterated with refined "lampante" oils or extraction oils. In particular, the concentrations of waxes, sterol esters, and of the triterpene alcohols erythrodiol (I) and uvaol (II, cf. Formula 14.1) are used for an analytical differentiation (Table 14.6 and 14.5.2.4).

On storage or on thermal treatment, e. g., with steam, 1,2-diacylglycerides (DG) isomerize to 1,3-DG. After separation of the DG fraction on a silica gel column and silylation, 1,2- and 1,3-DG can be determined by gas chromatography. The ratio 1,2-/1,3-DG is calculated from their areas; a value of less than 45% 1,2-DG is regarded as critical and indicates a loss of quality resulting from a longer storage period. Fresh oils contain more than 70% 1,2-DG, the content falling by about 1% per month. If the cold index is also raised, then heat treatment has occurred. The ratio of pyropheophytin A (cf. 17.1.2.9.1) to pheophytin A is known as the cold index.

Table 14.6. Olive oils: concentrations of the minor constituents

Type	Alcohols[a]	Waxes[a,b]	Sitosterol[a] free	esterified	Erythrodiol[a]	Erythrodiol + Uvaol (%)[c]
Extra vierge oil	67	40	914	219	13	1
"*Lampante*" oil, raw	84	292	945	877	10	0.6
"*Lampante*" oil, refined	44	180	692	544	8	0.8
Extraction oil, raw	725	3294	1234	2702	283	13.5
Extraction oil, refined	75	3277	659	2624	116	5.6

[a] Values in mg/kg.
[b] Sum of the wax esters C_{40}–C_{46}.
[c] Percentage of the sum of sterols and triterpene dialcohols.

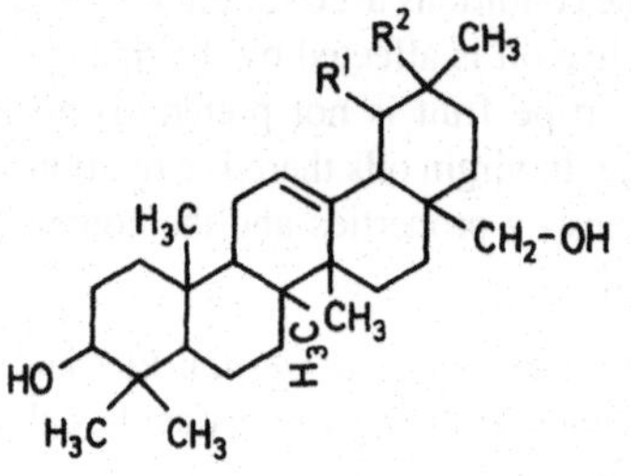

$$I : R^1 : H, \quad R^2 : CH_3$$
$$II : R^1 : CH_3, \quad R^2 : H \tag{14.1}$$

The aroma of natural oils is of special interest. The most important aroma substances of two *extra vierge* olive oils with different aromas are shown in Table 14.7. In oil I, green, fruity, and fatty notes are predominant while in oil II, the compounds with a green odor are concealed by an aroma substance with a "blackcurrant" odor, possibly due to a special variety of the olives. The compound involved is the extremely intensive odorant 4-methoxy-2-methyl-2-butanethiol (cf. Table 5.34), which has the highest aroma value of all the volatile compounds in oil II (Table 14.7).

The fatty acid composition of tea seed oil is very similar to that of olive oil. However, these two oils can be differentiated by using the *Fitelson* Test (cf. Table 14.21).

14.3.2.1.2 Palm Oil

This oil is obtained from oil palm, the utilization of which is constantly increasing (cf. Table 14.1). Palm plantations are found primarily in western Malaysia and Indonesia (cf. Table 14.0). The fruits provide two different oils, the first from the pulp and the second from the seeds.

Oil Production. The fruit cluster, which contains about 3000–6000 fruits, is first steam-treated to inactivate the high lipase activity and to separate the pulp from the seed. The oil is recovered by pressing the disintegrated pulp. The crude

Table 14.7. Important aroma substances of two *extra vierge* olive oils[a]

Aroma substance	Aroma quality	Oil I C	Oil I A_x	Oil II C	Oil II A_x
Isobutyric acid ethyl ester	Fruity	4.9	7	14	19
2-Methylbutyric acid ethyl ester	Fruity	3.9	5	14	19
Cyclohexanoic acid ethyl ester	Fruity	1.6	4.2	4.3	11
(Z)-3-Hexenal	Green	33	12	53	19
(Z)-2-Nonenal	Green, fatty	9	15	10	17
Acetic acid	Like vinegar	10,490	10	6680	6
4-Methoxy-2-methyl-2-butanethiol	Like blackcurrants	n.d.		1.8	40

[a] Oil I is from Italy, oil II from Spain; concentration C is in µg/kg and the aroma value A_x is calculated on the basis of the odor thresholds (retronasal) in an oil; n.d.: not detectable (C < 0.05 µg/kg).

oil is then clarified by centrifugation. Washing with hot water, followed by drying, provides a crude oil product that has a high carotene content (cf. 3.8.4.5) and, hence, the color of the oil is yellow to red. During refining (cf. 14.4.1), the palm oil color is destroyed by bleaching and the free fatty acids are removed. Palm fruit characteristics and oil composition are given in Table 14.5.

Adulteration of palm oil with palm stearin increases the ratio of the triacylglycerides PPP to MOP, which is usually between 3.5 and 4.5.

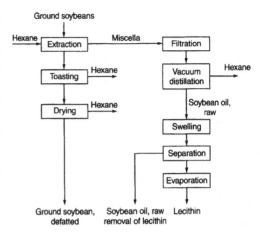

Fig. 14.2. Production of oil and lecithin from soybeans

14.3.2.2 Seed Oils

Some oilseeds have acquired great significance in the large-scale industrial production of edible oils. After a general review of their production, some individual oils, grouped together according to their characteristic fatty acid compositions, will be discussed.

14.3.2.2.1 Production

Conditioning. The ground or flaked seeds are heated with live steam of about 90 °C to facilitate oil recovery. This treatment ruptures all the cells, partly denatures the proteins and inactivates most of the enzymes. The temperature is regulated to avoid formation of undesirable colors and aromas.

After conditioning and moisture adjustment to about 3%, the oil is obtained by pressing and/or solvent extraction. The choice between these two processes depends on the oil content of the seed. Solvent extraction is the only economic choice for seeds with an oil content below 25%.

Pressing. The oil is removed by pressure from an expeller or screw press. The residual oil in the resultant meal flakes is 4–7%. It is, however, more economical to apply lower pressures and to leave 15–20% of the oil in the flakes, and then to remove this oil by a solvent extraction process ("prepress solvent extraction" process).

Extraction. The ground seeds are rolled into thin flakes by passing them between smooth steel rollers. This flaking step provides the enlarged surface area needed for efficient sol-

vent extraction. The extraction is performed using petroleum ether, i. e. technical hexane, as a solvent (boiling point 60–70 °C). In addition to n-hexane, it contains 2- and 3-methylpentane and 2,3-dimethylbutane and is free of aromatic compounds.

Solvent removal from the raw oil-solvent mixture, called miscella, is achieved by distillation. The maximum amount of solvent remaining in the oil is 0.1%. The oil-free flakes are then steamed to remove the solvent ("*desolventizing*") and, after dry heating ("*toasting*"), cooled and sold as protein-rich feed for cattle. The production of oil from soybeans is schematically shown in Fig. 14.2.

The crude oil obtained either by pressing or solvent extraction contains suspended plant debris, protein and mucous substances. These impurities are removed by filtration.

14.3.2.2.2 Oils Rich in Lauric and Myristic Acids

The most important representatives of this group of oils are coconut, palm seed and babassu oils. The acceptable shelf life stability of these oils is reflected in their fatty acid compositions (Table 14.8). Since linoleic acid is present in negligible amounts, autoxidative changes in these oils do not occur. However, when these oils are used in preparations containing water, microbiological deterioration may occur; this involves release of free C_8–C_{12} fatty acids and their partial degra-

Table 14.8. Characteristics of palm kernel oils

	Oil palm (Elaeis guineensis)	Coconut palm (Cocos nucifera)	Babassu palm (Orbignya speciosa)
Kernel oil content (weight-%)	40–52	63–70	67–69
Fat/oil melting range (°C)	23–30	20–28	22–26
Average fatty acid composition (weight-%)			
8:0	6	8	4.5
10:0	4	6	7
12:0	47	47	45
14:0	16	18	16
16:0	8	9	7
18:0	2.5	2.5	4
18:1 (9)	14	7	14
18:2 (9,12)	2.5	2.5	2.5

dation to methyl ketones ("perfume scent rancidity", cf. 3.7.6).

Coconut and palm seed oils are important ingredients of vegetable margarines which are solid at room temperature. However, they melt in the mouth with a significant heat uptake, producing a cooling effect.

Coconut oil is obtained from the stone fruit of the coconut palm, which grows throughout the tropics. The moisture content of the oil-containing endosperm, when dried, decreases from 50% to about 5–7%. Such crushed and dried coconut endosperm is called "copra" and is sold under this name as a raw material for oil production around the world.

Palm kernel oil is obtained from the kernels of the fruit of the oil palm. The kernels are separated from the fruit pulp, then removed from the stone shells and dried prior to recovery of the oil. Babassu oil is obtained from seeds of the babassu palm, which is native to Brazil. This oil is rarely found on the world market and is mainly consumed in Brazil.

14.3.2.2.3 Oils Rich in Palmitic and Stearic Acids

Cocoa butter and fats (solid at room temperature) belong to this group, with the latter referred to as cocoa butter substitutes ("cocoa butter interchangeable fats"). They are relatively hard and can crystallize in several polymorphic forms (cf. 3.3.1.2). Their melting points are between 30 and 40 °C. The relatively narrow melting range for cocoa butter, as well as for some other types of butter, is to be expected (Table 14.9). When cocoa butter melts in the mouth, a pleasant, cooling sensation is experienced (cf. 14.3.2.2.2). This is characteristic of only a few types of triacylglycerols present in fats which contain predominantly palmitic, oleic and stearic acids. This fatty acid composition is also reflected by the resistance of these fats to autoxidation and microbiological deterioration (Table 14.9). These fats are utilized preferentially in the manufacturing of chocolates, candy and confections.

Table 14.9. Fatty acid composition of cocoa butter and cocoa butter substitutes

Trade name	Cocoa butter	Illipè butter (Mowrah butter)	Borneo tallow (Tengkawang fat, Illipè butter)	Shea butter (Kerité fat)
Source	Cacao tree (Theobroma cacao)	Madhuca logifolia	Shorea stenoptera	Butyrospermum parkii
Fat, melting range (°C)[a]	28–36	24.5–28.5	28–37	23–42
Average fatty acid composition (weight-%)				
16:0	25	28	20	7
18:0	37	14	42	38
18:1 (9)	34	49	36	50
18:2 (9, 12)	3	9	<1	5

[a] The melting ranges reflect a pronounced fat polymorphism (cf. 3.3.1.2); the highest temperature given represents the melting point of the stable fat modification.

Cocoa butter is the fat from cocoa beans. The seed germ contains up to 50–58% of the fat, which is recovered as a by-product during cocoa manufacturing (cf. 21.3.2.7). It is light yellow and has the pleasant, mild odor of cocoa. Cocoa butter contains 1,3-dipalmito-2-olein, 1-palmito-3-stearo-2-olein, and 1,3-distearo-2-olein in an almost constant ratio of 22:46:31 (% peak area). Since cocoa butter substitutes clearly differ in the content of these TGs, the amount of cocoa butter can be determined by HPLC of the TGs. Bromination of the double bonds (cf. 3.3.1.4) improves the separation of the three TGs. Other indicators of cocoa butter are given in 3.8.2.3.1 and 3.8.2.4. The denotation of the "cocoa butter interchangeable fats" may be confusing since fats from diverse sources are sometimes marketed under a collective name such as Illipè butter. Confusion can be avoided by providing the Latin name of the plant, i. e. the source of the fat.

Shea Butter (Kerité Fat) is obtained from seeds of a tree which grows in western Africa and the cultivation of which appears to be uneconomical. The high content of unsaponifiable matter (up to 11%) in this kind of butter is of interest.

Borneo Tallow (Illipè Butter) is obtained from the seeds of a plant native to Java, Borneo, the Philippines and India. It serves as a valuable edible fat in the Tropics. *Mowrah butter* (often marketed as Illipè butter) is derived from a different plant

(Madhuca longifolia) and is also indigenous to the Asian tropics.

14.3.2.2.4 Oils Rich in Palmitic Acid

Oils in this group contain more than 10% palmitic acid along with oleic and linoleic acids (cf. Table 14.10).

Cottonseed Oil is obtained from seeds of many cotton plant cultivars. The plant is widely cultivated (cf. Table 14.0). The raw oil is dark, usually dark red, and has a unique odor. It contains a poisonous phenolic, gossypol,

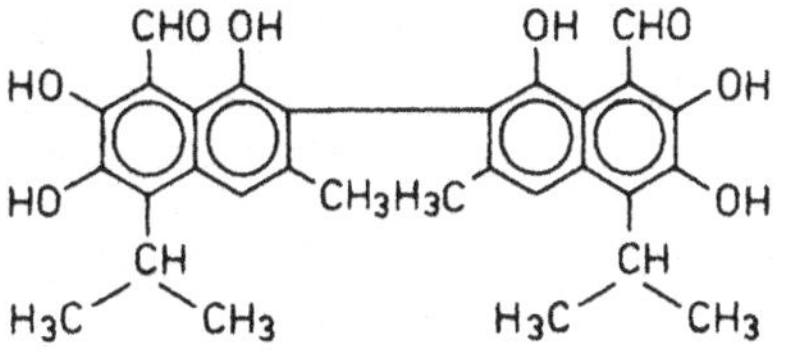

$$\text{(14.2)}$$

which is removed during refining. Another substance present in this oil is malvalia acid,

$$CH_3-(CH_2)_7-C=C-(CH_2)_6-COOH$$
$$\quad\quad\quad\quad\quad \backslash / $$
$$\quad\quad\quad\quad\quad CH_2 \quad\quad\quad\quad\quad \text{(14.3)}$$

which survives refining, but not hydrogenation of the oil. This substance is responsible for detection

Table 14.10. Oils rich in palmitic acid

	Cottonseed (Gossypium)	Corn germ (Zea mays)	Wheat germ[a] (Triticum aestivum)	Pumpkin seed (Cucurbitapepo)
Seed oil content (weight-%)	22–24	3.5–5[b]		35
Solidification point (°C)	0 to +4	−10 to −18		−15 to −16
Average fatty acid composition (weight-%)				
14:0	1.5	0	0	0
16:0	22	10.5	17	16
18:0	5	2.5	1	5
20:0	1	0.5	0	0
16:1 (9)	1.5	0.5	0	0.5
18:1 (9)	16	32.5	20	24
18:2 (9,12)	55	52	52	54
18:3 (9,12,15)	0	1	10	0.5

[a] Oil content in germ amounts to 8–11 weight-%.
[b] Of the seed oil content 80% is located in germ and the rest in seed endosperm.

or identification of the oil by the *Halphen* reaction (cf. Table 14.21).

At temperatures below +8 °C, cottonseed oil becomes turbid due to crystallization of high melting point triacylglycerols. Such undesirable low temperature characteristics are avoided using a "winterization" process (cf. 14.4.4).

Cereal Germ Oils. All cereals contain significant amounts of oil in the germ. It is available after the germ is separated during grain processing. Corn (maize) oil is the most important. Germ separation is achieved during dry or wet processing of the kernels into corn meal and starch (cf. 4.4.4.14.1). The oil is recovered from the germ collected as a by-product by pressing and solvent extraction. After crude oil refining, the corn waxes which originate from the skin-like coating of the epidermis (the cuticle), are removed by a winterization process (cf. 14.4.4).

Corn oil is suitable for manufacture of margarine and mayonnaise (creamy salad dressing), but is used preferentially as salad and cooking oil.

The oil present in wheat and rice is also concentrated in the germ. This oil can be recovered by pressing and/or solvent extraction of the germ. Wheat germ oil has a high content of tocopherol and, therefore, additional nutritive value. Rice germ oil is consumed to a minor extent in Asia. Pumpkin oil is obtained by pressing dehulled pumpkin seeds. In southern Europe it is utilized as an edible oil. It is brown in color and has a nut-like taste.

14.3.2.2.5 Oils Low in Palmitic Acid and Rich in Oleic and Linoleic Acids

A large number of oils from diverse plant families belong to this group (cf. Table 14.11). These oils are important raw materials for manufacturing margarine.

The sunflower is the most cultivated oilseed plant in Europe. Data on the production of the sunflower in regions and countries are given in Table 14.0. Prepressing of dehulled sunflower seeds yields a light yellow oil with a mild flavor. The oil is suitable for consumption once it is clarified mechanically. Refined oils are used in large amounts as salad oil or as frying oil and as a raw material for margarine production. The refining of the oil includes a wax-removal step.

Two legume oils, soybean and peanut (or ground nut), are of great economic significance (cf. Table 14.1). Soybean oil (fatty acid composition in Table 14.11) is currently at the top of the world production of edible oils of plant origin. It is cultivated mostly in the United States, Brazil and China. The refined oil is light yellow and has a mild flavor. It contains in low concentrations (Table 3.9) branched furan fatty acids which are rapidly oxidized on exposure to light. In fact, two of these fatty acids, which differ only in the length of the carboxyl ends (see Formula 3.3), produce the intensive aroma substances 3-methyl-2,4-nonandione (MND) and diacetyl in a side reaction with singlet oxygen. These aroma substances are significantly involved in the "bean-like, buttery, hay-like" aroma defect called *reversion flavor*. In the case of the soybean oils listed in Table 14.12, the two furan fatty acids were almost completely oxidized after 48 hours. However, the amounts of MND formed were very different. This is put down to differences in the stability of the intermediate hydroperoxide (cf. Fig. 3.25).

Other experiments have shown that the hydroperoxides formed from furan fatty acids on exposure to light fragment to the dione, even if the soybean oil is subsequently stored in the dark. In the complete absence of light, soybean oil is relatively stable. The shelf life of the oil is also improved significantly by partial hydrogenation to give a melting point range of 22–28 °C or 36–43 °C. Such oils are utilized as raw materials for the manufacture of margarine and shortening (semi-solid vegetable fats used in baked products, such as pastry, to make them crisp or flaky).

Cultivation using traditional and genetic engineering techniques has made it possible to develop soybean genotypes which have a fatty acid composition which meets the different demands made on edible oils. Table 14.13 shows the extent to which the fatty acid composition of soybean oil has been changed. The genotypes *low linolenic* and *high oleic* are considerably more stable to oxidation than the normal line. In addition, the composition of high oleic corresponds to that of a salad oil, partial hydrogenation is no longer required. Palmitic acid is decreased in the types *low palmitic* and *low saturate*, as it is involved in the increase in cholesterol in LDL (cf. 3.5.1.2)

Table 14.11. Oils low in palmitic acid and rich in oleic and linoleic acids

	Sunflower (Helianthus annuus)	Soya (Glycine max.)	Peanut[a] (Arachis hypogaea)	Rapeseed[b] (Brassica napus)	Sesame (Sesamum indicum)	Safflower (Carthamus tinctorius)	Linseed (Linum usitatissimum)	Poppy (Papaver somniferum)	Walnut (Juglans regia)
Seed oil content (weight-%)	25–30	18–23	42–52	ca. 40	45–55	32–43	32–43	40–51	58–71
Solidification point (°C)	−18 to −20	−8 to −18	−2 to +3	0 to −2	−3 to −6	−13 to −20	−18 to −27	−15 to −20	−15 to −20
Average fatty acid composition (weight-%)									
16:0	6.5	10	10	4	8.5	6	6.5	9.5	8
18:0	5	5	3	1.5	4.5	2.5	3.5	2	2
20:0	0.5	0.5	1.5	0.5	0.5	0.5	0	0	1
22:0	0	0	3	0	0	0	0	0	0
18:1 (9)	23	21	41	63	42	12	18	10.5	16
18:2 (9,12)	63	53	35.5	20	44.5	78	14	76	59
18:3 (9,12,15)	<0.5	8	0	9	0	0.5	58	1	12
20:1 & 20:2	1	3.5	1	1	0	0.5	0	0	0
22:1 (13)	0	0	–	0.5	0	0	0	0	0

[a] African peanut oil. [b] Canola-type oil (practically free of erucic acid).

Table 14.12. Oxidation of furan fatty acids I and II and formation of 3-methyl-2,4-nonandione in three refined soybean oils[a]

Compound[a]	Time[b]	Soybean oil		
		A	B	C
		(mg/kg)		
Furan fatty acid I	0 h	143	148	131
Furan fatty acid I	48 h	5	5	3
Furan fatty acid II	0 h	152	172	148
Furan fatty acid II	48 h	5	5	5
		(μ/kg)		
MND	0 h	<1	<1	<1
MND	48 h	89	3.4	<1
MND	30 d	721	204	43

[a] I: 10,13-Epoxy-11,12-dimethyloctadeca-10,12-dienoic acid;
II: 12,15-Epoxy-13,14-dimethyleicosa-12,14-dienoic acid. MND: 3-Methyl-2,4-nonandione.
[b] The soybean oils were stored at room temperature at a window facing north.

The fatty acid composition of peanut oil is greatly influenced by the region in which the peanuts are grown. In contrast to the peanut oils produced in Africa (Senegal or Nigeria), the peanut oils from South America are enriched in linoleic acid (41% vs 25%, w/w; see fatty acid composition, Table 14.11) at the expense of oleic acid (37% vs 55%, w/w). The contents of arachidic (20:0), eicosenoic (20:1), behenic (22:0), erucic (22:1) and lignoceric (24:0) acids are characteristic of peanut oil. Their glycerols readily crystallize below 8 °C.

Peanut Butter is a spreadable paste made from roasted and ground peanuts by the addition of peanut oil and, occasionally, hydrogenated peanut oil.

Rapeseed Oil. This oil is produced from seeds of two *Brasica* species: *Brassica napus var. napus* and *Brassica rapa var. Silvestris.* The latter plants yield slightly less oil, are shorter (approx. 80 cm), but mature more quickly. They are more tolerant to frost and have improved resistance to pest and diseases. Old rape and turnip rape cultivars contained high levels of erucic acid (45–50 by weight), which is hazardous in human nutrition. "Zero" erucic acid cultivars (22:1 <5% by weight), called *Canola*, have been developed and, recently, "double zero" cultivars, with low levels of erucic in the oil and goitrogenic compounds in the seed meal, have been developed. The major rapeseed-cultivating regions and countries are listed in Table 14.0.

The above-mentioned plants, as *Brassicacea*, contain mustard oil glucosides (glucosinolates, cf. 17.1.2.6.5) which, immediately after seed crushing, are hydrolyzed to esters of isothiocyanic acid. The hydrolysis is dependent on seed moisture and is catalyzed by a thioglucosidase enzyme called myrosinase (EC 3.2.3.1). In the presence of the enzyme, some of the isocyanates are isomerized into thiocyanates (esters of normal thiocyanic acid or rhodanides) and, in part, are decomposed into nitrile compounds which do not contain sulfur. All these compounds are volatile and, when dissolved in oil, are hazardous to health and detrimental to oil flavor. Moreover, they interfere with hydrogenation of the oil by acting as Ni-catalyst poisons (cf. 14.4.2.2). Therefore, in the production of rapeseed oil, a dry conditioning step is used (without live steam) to

Table 14.13. Changes in the composition of soybean oil through cultivation or modification[a] using genetic engineering techniques

	Genotype					
Fatty acid	Normal	Low linolenic	High oleic[b]	Low palmitic	Low saturate[b]	High stearic
16:0	11.2	10.1	6.4	5.9	3.0	9.2
18:0	3.4	5.3	3.3	3.7	1.0	20.5
18:1(9)	21.5	41.1	85.6	40.4	31.0	21.7
18:2 (9,12)	55.8	41.2	1.6	43.4	57.0	43.2
18:3 (9,12,15)	8.0	2.2	2.2	6.6	9.0	2.8

[a] Expressed in weight %.
[b] Developed by genetic engineering techniques.

inactivate the myrosinase enzyme and only then is the seed ground and subjected to prepress and solvent extraction processes.

Despite these precautions, small amounts of volatile sulfur compounds are formed. However, they are removed during the refining process. Irrespective of technical achievements in rapeseed production and processing, the selection and breeding of rapeseed "double zero" cultivars is being continued.

Rapeseed (*Canola*) oil is used as an edible oil. It is susceptible to autoxidation because of its relatively high content of linolenic acid. It is saturated by hyrogenation to a melting point of 32–34 °C and, with its stability and melting properties, resembles coconut oil.

Turnip rape oil has practically the same composition as the *B. napus* oil. It may contain at most 5% of erucic acid, because this can damage the heart muscle in high concentrations.

Sesame Oil is obtained from an ancient oilseed crop (*Sesamum indicum*, L.), which is widely cultivated in India, China, Burma and east Africa (cf. Table 14.0). In its refined form the oil is nearly crystal clear and has a good shelf life. In addition to a considerable amount of tocopherols, it contains another phenolic antioxidant, sesamol, which is derived from hydrolysis of sesamolin (Fig. 14.3).

Sesame oil can be readily identified with great reliability (cf. Table 14.21). Therefore, in some countries, blending this oil into margarine is required by law in order to identify the product as margarine.

Safflower Oil is obtained from a thistle-like plant (*Carthamus tinctorius*) grown in the arid regions

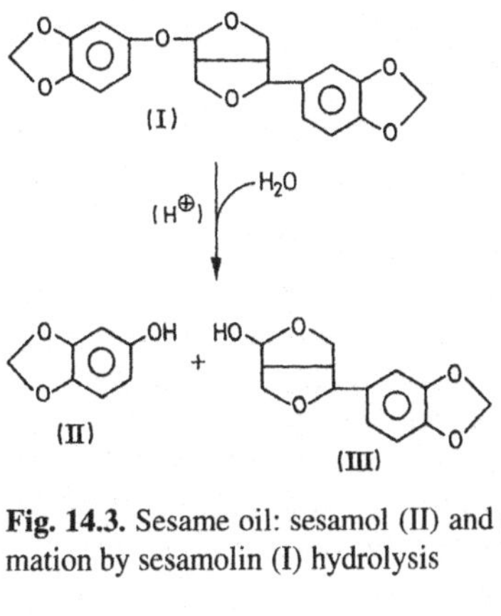

Fig. 14.3. Sesame oil: sesamol (II) and samin (III) formation by sesamolin (I) hydrolysis

of North America and India (cf. Table 14.0). New cultivars have been bred with oil compositions which deviate greatly from those listed in Table 14.11. These new cultivars contain 80% by weight oleic acid (18:1) and 15% by weight linoleic acid (18:2; 9,12).

Linseed Oil. Flax, used for fiber and seed production and the subsequent processing of the seed into linseed oil, is grown mainly in Canada, China and India (cf. Table 14.0). Due to its high content of linolenic acid (cf. Table 14.11), the oil readily autoxidizes, one of the processes by which some bitter substances are created. Since autoxidation involving polymerization reactions proceeds rapidly, the oil solidifies ("*fast drying oil*"). Therefore, it is used as a base for oil paints, varnishes and linoleum manufacturing, etc. A comparatively negligible amount, particularly of the coldpressed oil, is utilized as an edible oil.

Poppy Oil is very rich in linoleic acid (Table 14.11). The cold-pressed oil from flawless seeds is colorless to light yellow and can be used directly as an edible oil.

Walnut Oil has a pleasant odor and a nut-like taste. It contains relatively high concentrations of linolenic acid (Table 14.11) and, consequently, has a very limited shelf life.

14.4 Processing of Fats and Oils

14.4.1 Refining

Apart from some oils obtained by cold pressing (examples in 14.3.2.1), most of the oils obtained using expeller, screw or hydraulic presses, solvent extraction or by melting at elevated temperatures are not suitable for immediate consumption. Depending on the raw material and the oil recovery process, the oil contains polar lipids, especially phospholipids, free fatty acids, some odor- and taste-imparting substances, waxes, pigments (chlorophyll, carotenoids and their degradation products), sulfur-containing compounds (e. g. thioglucosides in rapeseed oils), phenolic compounds, trace metal ions, contaminants (pesticides or polycyclic hydrocarbons) and autoxidation products.

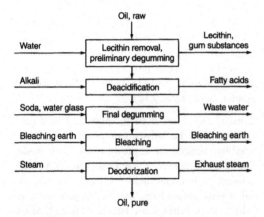

Fig. 14.4. Refining of oils

In a refining process comprising the following steps:

- Lecithin removal
- Degumming
- Free fatty acid removal
- Bleaching
- Deodorization,

all the undesired compounds and contaminants are removed. An overview of the refining process is given in Fig. 14.4. In practice the refining steps used depend on the quality of the crude oil and its special constituents (e. g. carotene in palm oil or gossypol in cottonseed oil). The following precautionary measures are taken during refining in order to avoid undesirable autoxidation and polymerization reactions:

- Absence of oxygen (also required during transport or storage)
- Avoidance of heavy metal contamination
- Maintaining the processing temperatures as low and duration as short as possible.

14.4.1.1 Removal of Lecithin

This processing step is of special importance for rapeseed and soybean oils. Water (2–3%) is added to crude oil, thereby enriching the phospholipids in the oil/water interface. The emulsion thus formed is heated up to 80 °C and then separated or clarified by centrifugation. The "crude lecithin" (cf. 3.4.1.1) is isolated from the aqueous phase and is recovered as crude vegetable lecithin after evaporating the water in a vacuum.

14.4.1.2 Degumming

Finely dispersed protein and carbohydrates are coagulated in oil by addition of phosphoric acid (0.1% of oil weight). A filtering aid is then added and the oil is clarified by filtration. This also removes the residual phospholipids from the previous processing step.

14.4.1.3 Removal of Free Fatty Acids (Deacidification)

Several methods exist for deacidification of fat or oil. The choice depends on the amount of free fatty acids present in crude fat or oil. The removal of fatty acids with 15% sodium hydroxide (alkali refining) is the most frequently used method. Technically, this is not very simple since fat hydrolysis has to be avoided and, moreover, the sodium soap (the "soap-stock"), which tends to form stable emulsions, has to be washed out by hot water. After vacuum drying, the fat or oil may contain only about 0.05% free fatty acids and 60 to 70 ppm of sodium soaps. When the fat or oil is treated with diluted phosphoric acid, the content of sodium soaps decreases to 20 ppm and part of the trace heavy-metal ions is removed.

Fats (oils) with a high content of free fatty acids require relatively high amounts of alkali for extraction, resulting in an unavoidably high loss of neutral fats (oils) due to alkaline hydrolysis. Therefore, extraction with alkali is frequently replaced by deacidification by distillation in these cases (14.4.1.5).

In special cases, a selective fluid/fluid extraction is of interest. Ethanol extracts free fatty acids (above a level of 3%) from triacylglycerols in crude oils – this is a suitable way to treat oils with exceptionally high amounts of free acids. At a given temperature, furfural can extract only the polyunsaturated triacylglycerols. On the other hand, propane under pressure preferentially solubilizes the saturated triacylglycerols and leaves behind the unsaturated ones, together with unsaponifiable matter. Pressurized propane is utilized in marine oil fractionation, e. g., in the production of vitamin A concentrates.

14.4.1.4 Bleaching

In order to remove the plant pigments (chlorophyll, carotenoids) and autoxidation products, the fat or oil is stirred for 30 min in the presence of Al-silicates (bleaching or *Fuller's* earth) in a vacuum at 90 °C. The silicate has to be activated prior to use – a suspension in water is treated with hydrochloric acid, followed by thorough washing with water, then drying. The amount of silicate used is 0.5–2% of the fat (oil) weight. It is often used together with 0.1–0.4% activated charcoal. The bleached oil is removed from the adsorbent by filtration. The oil retained by the adsorbent can be recovered by hexane extraction and recycled into the refining process. The residual alkali soaps, gums, part of the unsaponifiable matter and the heavy metal ions are also removed during the bleaching process.

After bleaching, some oils or fats which contain polyunsaturated fatty acids show an increase in absorbance at 270 nm. This is due to decomposition of hydroperoxides, formed by autoxidation, into oxo-dienes (I) and fatty acids with three double bonds (II):

$$R_1-CH=CH-CH=CH-\underset{\underset{O}{\parallel}}{C}-R_2$$

$$(I)$$

$$\uparrow\;\vdash\!\!-H_2O$$

$$R_1-CH=CH-CH=CH-\underset{\underset{OOH}{\mid}}{CH}-R_2$$

$$(Hydroperoxide)$$

$$\vdash\!\!-\;^{\bullet}OH$$

$$R_1-CH=CH-CH=CH-\underset{\underset{O\bullet}{\mid}}{CH}-R_2$$

$$\vdash\!\!-RH$$

$$\vdash\!\!-R^{\bullet}$$

$$R_1-CH=CH-CH=CH-\underset{\underset{OH}{\mid}}{CH}-R_2$$

$$\vdash\!\!-H_2O$$

$$R_1-CH=CH-CH=CH-CH=CH-R_3$$

$$(II) \qquad\qquad (14.4)$$

Table 14.14. Removal of chloro pigments in the refining of rapeseed oil (values in mg/kg)

		Amounts after		
	Raw oil	Deacidification	Bleaching	Deodorization
Chlorophyll A	2.62	0.89	0.028	0.007
Chlorophyll B	2.92	0.08	0.059	0.023
Pheophytin A	35.6	31.5	0.235	0.108
Pheophytin B	4.99	6.85	0.071	0.036
Sum	46.1	39.3	0.393	0.174

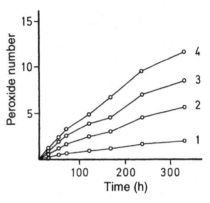

Fig. 14.5. Oxidation of soybean oil on exposure to room light (according to *Usuki* et al., 1984). (1), (2), (3), and (4) contain 39, 233, 425, and 623 µg/kg respectively of the mixture chlorophyll A/chlorophyll B/pheophytin A/pheophytin B (1:3:10:3)

As shown in Table 14.14 with rapeseed oil, most of the chlorophylls and their degradation products are removed during bleaching. However, the rest, which is in the range of 70–1200 µg/kg in refined plant oils, could still accelerate photooxidation (Fig. 14.5). As a result of their high stability on exposure to light, the pheophytins are stronger photooxidants than the chlorophylls. As shown in Table 14.14, the pheophytins are the predominating pigments.

14.4.1.5 Deodorization

Deodorization is essentially vacuum steam distillation (190–230 °C, 0.5–10 mbar). The volatile compounds, together with undesirable odorants present in the fat or oil, are separated in this refining step. Deodorization takes from 20 min to 6 h, depending on the type of fat or oil and the content of volatile compounds.

The processing loss in this refining step is 0.2%. This is negligible since the fat or oil droplets carried by the steam are caught by baffles or are intercepted by an external trap system.

Deodorization can been combined with deacidification by distillation when the oil has a low content of accompanying substances or if they have been largely removed by degumming and bleaching, e. g., after reduction of the phospholipids to less than 5 mg/kg. Since fatty acids are less steam-volatile than the odorous substances, higher temperatures (up to 270 °C) are used than in deodorization. Carotenoids are decomposed, so that, e. g., palm oil is thermally bleached in this process. The combination of deodorization with distillative deacidification is called physical refining; it is the process of choice in the case of higher acid concentrations (>0.7–1%). The relatively large amounts of waste water which have to be disposed off after alkaline extraction are avoided here. In addition, the fatty acids accumulated as by-products are of higher quality then the "refining fatty acids" obtained by alkali extraction (cf. 14.4.1.3).

In steaming or physical refining, the double bonds of linoleic and linolenic acid isomerize to a small extent. For this reason, an HPLC determination of isomeric linoleic acids is used to distinguish between refined and natural plant oils (cf. 14.5.3.4).

14.4.1.6 Product Quality Control

In addition to sensory evaluation, free fatty acid analysis (the content is usually below 0.05%) and analysis of possible contaminants are carried out. The data given in Table 14.15 illustrate the amounts of pesticides and polycyclic aromatic compounds removed by deodorization. However, this refining step also removes the highly desirable aroma substances which are characteristic of some cold-pressed oils such as olive oil.

The composition of phytosterols and tocopherols does not change appreciably during refining. Therefore, an analysis of these compounds is suitable for the identification of the type of fat. On the other hand, cholesterol can increase during steaming, e. g., due to the cleavage of

Table 14.15. Removal of endrin and polycyclic hydrocarbons during edible oil refining ($\mu g/kg$)

Compound	Content in raw oil	Content in oil after		
		deacidification	bleaching	steaming
Endrin	620[a]	590	510	<30
Anthracene	10.1[b]	5.8	4.0	0.4
Phenanthrene	100[b]	68	42	15
1,2-Benzanthracene	14[b]	7.8	5.0	3.1
3,4-Benzpyrene	2.5[b]	1.6	1.0	0.9

[a] Soybean oil.
[b] Rapeseed oil.

glycosides. In palm oil, for instance, the percentage of cholesterol in the sterol fraction increased from 2.8% to 8.8%.

14.4.2 Hydrogenation

14.4.2.1 General Remarks

Liquid oils are supplied mostly from natural sources. However, a great demand exists for fats which are solid or semi-solid at room temperature. To satisfy this demand, *W. Normann* developed a process in 1902 to convert liquid oil into solid fat, based on the hydrogenation of unsaturated triacylglycerols using Ni as catalyst; a process designated as "fat hardening". The process rapidly gained great economic importance; even marine oils became suitable for human consumption after the hardening process. Today more than 4 million tons of fat per year are produced worldwide by hydrogenation of oils; most is consumed as food.

The unsaturated triacylglycerols can be fully hydrogenated, providing high-melting cooking, frying and baking fats or partially hydrogenated, providing products such as:

• Oils rich in fatty acids with one double bond. They are stable and resistant to autoxidation and have a shelf life similar to olive oil. They are used as salad oil or as shortening.
• Products in which the linolenic acid is hydrogenated, but most of the essential fatty acid, linoleic acid, is left intact. An example is soy-

bean oil, hydrogenated selectively to increase its stability against oxidation.

- Fats that melt close to 30 °C and have a plastic or spreadable consistency at room temperature.

Fully or partially hydrogenated oils are important raw materials for margarine manufacturing and serve as deep-frying fats (cf. 14.4.8).

Table 14.16. Properties of hydrogenation catalysts

Catalyst	Selectivity		trans-Fatty acids
	S_{32}	S_{21}	(weight-%)[a]
Nickel-contact	2–3	40	40
Ni_3S_2-contact	1–2	75	90
Copper-contact	10–12	50	10

[a] trans-Fatty acids as monoenoic acids total content is calculated as elaidic acid.

14.4.2.2 Catalysts

The principle of the heterogeneous catalytic hydrogenation of unsaturated acylglycerols was outlined under 3.2.3.2.4. The most widely used catalyst is carrier-bound nickel. Raney nickel, copper, and noble metals serve special purposes. The choice of catalyst is made according to:

- Reaction specificity.
- Extent of trans-isomer formation
- Duration of activity and cost.

To determine the specificity of a catalyst, the reaction rates for each individual hydrogenation step must be determined. Simplified, there are three reaction rate constants (k) involved (AG = acylglycerol):

$$\text{Triene-AR} \xrightarrow{k_3} \text{Diene-AR} \xrightarrow{k_3} \text{Monoene-AR}$$
$$\downarrow k_1$$
$$\text{AR-acylresidue} \qquad \text{Saturated-AR}$$
$$(14.5)$$

The catalytic reactions considered here require that $k_3 > k_2 > k_1$. The following equations determine the specificity "S":

$$S_{32} = \frac{k_3}{k_2}; \quad S_{21} = \frac{k_2}{k_1}; \quad S_{31} = \frac{k_3}{k_1} \qquad (14.6)$$

That means, the greater the value of "S", the faster the hydrogenation at this step. Therefore, specificity (or selectivity) is proportional to the value of "S". For the three catalysts mentioned, Table 14.16 shows that the hydrogenation of diene → monoene by Ni_3S_2 and the hydrogenation of triene → monoene by copper become accelerated with marked specificity. Copper is particularly suitable for decreasing the linolenic acid

content in soybean and rapeseed oils. However, copper catalysts are not sufficiently economical, since they can not be used more than twice. Their complete removal, which is necessary because this is a prooxidatively active metal, is relatively tedious.

Although noble metals are up to 100 times more effective than nickel catalysts, they are not popular because of their high costs. It is of great advantage that the nickel catalyst can be used repeatedly for up to 50 times under the following conditions: the plant oil must be deacidified, freed of gum ingredients and contain no sulfur compounds (cf. rapeseed oil, 14.3.2.2.5). The favorable ratio of duration of activity to cost places the nickel catalysts ahead with advantages not readily surpassed by any other catalyst. For the production of nickel-carrier catalysts, kieselguhr or zeolite is impregnated with nickel hydroxide, which is precipitated out of a solution of nickel nitrate with sodium hydroxide or carbonate. After drying, nickel hydroxide is reduced to nickel with hydrogen at 350–500 °C.

For the production of carrier-free nickel catalysts, nickel formate is suspended in a fat and then decomposed:

$$Ni(HCOO)_2 \xrightarrow[(200-250\,°C)]{} Ni + 2CO_2 + H_2 \qquad (14.7)$$

The catalysts obtained with a Ni content of 22–25% are pyrophoric. For this reason it is embedded in fat and handled and marketed in this form. To evaluate the catalysts, calculation programs were developed for the determination of the actual selectivity of a catalyst based on the fatty acid composition of the starting material and of the hydrogenated product.

During hydrogenation, linolenic acid yields, among others, isolinoleic and isooleic acids (cf. Reaction 14.8).

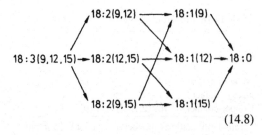

$$(14.8)$$

The diversity of the reaction products present in partially hydrogenated fat is increased further by the positional- and stereoisomers of the double bonds. Hydrogenation of soybean oil in the presence of a copper catalyst gives, for example, a number of trans-monoene fatty acids (Table 14.17). The extent of isomerization is affected, among other factors, by the type of catalyst used in hydrogenation. Efforts are being made to reduce the formation of trans-fatty acids during hydrogenation because these acids have a detrimental effect on the composition of the blood lipids. Also, the presence of a hydrogenated fat in a mixture is easily detected via the identification of trans-fatty acids, e. g., by IR spectroscopy or gas chromatography (cf. 14.5.2.3).

A further drawback of the partial hydrogenation of an oil is the pattern of linoleic acid isomersformed. The two isomers formed during hydrogenation, linolelaidic acid 18:2 (9 trans, 12 cis) and 18:2 (9 cis, 12 trans) are, unlike linoleic acid, not essential fatty acids (cf. 3.2.1.2).

Table 14.17. Fatty acid composition of a soya oil before and after hydrogenation with a copper catalyst

Fatty acid	Hydrogenation	
	before (weight-%)	after (weight-%)
16:0	10.0	10.0
18:0	4.2	4.2
18:1(9)	26.0	30.4
18:1[a]	0	5.5
18:2(9,12)	52.5	42.5
18:2(conjugated)[b]	0	0.7
18:2[c]	0	5.2
18:3(9,12,15)	7.3	0.7

[a] This fraction contains eight trans fatty acids: 18:1 (7 tr)–18:1 (14 tr); major components are 18:1 (10 tr) and 18:1(11 tr).
[b] It consists of various conjugated fatty acids.
[c] Isolinoleic and isolinolelaidic acids.

14.4.2.3 The Process

The hydrogen required can be obtained by electrolysis of dilute aqueous KOH, through water-to-gas conversion:

$$H_2O + C \rightarrow H_2 + CO$$
$$CO + H_2O \rightarrow H_2 + CO_2 \qquad (14.9)$$

or by the decomposition of natural gas with steam:

$$CH_3(CH_2)_x CH_3 + H_2O \rightarrow H_2 + CO$$
$$CO + H_2O \rightarrow H_2 + CO_2 \qquad (14.10)$$

in the latter two processes, the poisonous by products, H_2S and CO, have to be completely removed.

Oil hydrogenation is performed in an autoclave equipped with a stirrer under hydrogen gas pressure of 1–5 bar and a temperature of 150–220 °C. More recent developments aim at a continuous process, e. g., in fixed-bed reactors. A newer hydrogenation process uses a recycling hydrogenation unit equipped with a spraying nozzle, external heat exchanger and recycling pump.

The process conditions have a significant effect on the composition and therefore on the consistency of the end-product. Selective hydrogenation of double bonds is favored by a high concentration of catalyst (which, depending on the Ni activity, is 200–800 g Ni/t fat), a high temperature and low pressure of hydrogen gas. After hydrogenation, the fat is filtered, then deacidified, bleached and deodorized during further refining (cf. 14.4.1.3–14.4.1.5).

Some constituents of the unsaponifiable matter are also affected by the hydrogenation process. Carotenoids, including vitamin A, are hydrogenated extensively. Some of the chlorine-containing pesticide contaminants are hydrogenated. Sterols, under the usual operating conditions, are not affected. The ratios and levels of tocopherols are essentially unchanged.

14.4.3 Interesterification

Natural fats and oils are subjected to extensive interesterification during processing. This involves a rearrangement or randomization of acyl residues in triacylglycerols and thus provides fats or oils with new properties. By choosing the raw material and processing parameters, the interesterification can be controlled to obtain a fat with melting characteristics and consistency which match the intended use ("tailored fats").

The basics of the interesterification process are outlined under 3.3.1.3. Sodium methylate is used almost exclusively as the catalyst. The dried, bleached and deacidified fat (or oil) is stirred at 70–100 °C under a vacuum in the presence of alcoholate (0.1–0.3% of fat weight). When the reaction is completed, the catalyst is destroyed by adding water, then the degraded catalyst together with the resultant soaps are removed from the fat (oil) by repeated washing with water. The interesterified product is then bleached (cf. 14.4.1.4) and deodorized (cf. 14.4.1.5).

Table 14.18 illustrates the changes in triacylglycerols brought about by the process and its influence on fat melting points.

The baking properties of lard (improvement of volume and softness of the baked goods) are improved by interesterification. The uniform distribution of palmitic acid in the triacylglycerols accounts for such an improvement.

Furthermore, interesterification is of importance in the manufacturing of different varieties of margarine with a given composition, for example:

- Vegetable margarine with 30% w/w of 18:2 (9, 12) fatty acid may be produced by interesterification of partially hardened sunflower oil blended with its natural liquid oil.

- Interesterification of palm oil with palm seed or coconut oil (2:1) and the use of 6 parts of this product with 4 parts of sunflower oil provides a margarine which contains 20–25% w/w of linoleic acid and does not contain hydrogenated fat.

14.4.4 Fractionation

The undesirable fat ingredients are removed or the desirable triacylglycerols (TG) are enriched by fractional crystallization. The rising demand of food processors for special fats with standardized properties has led to large-scale isolation of special fractions, particularly from palm oil and the fats and oils listed under 14.3.2.2.2. The following procedures are used for the fractional crystallization of fats: The melted fat is slowly cooled until the high melting TG selectively crystallize, i.e. without forming mixed crystals of low and high melting TG. A sharp separation into two or more fractions is assumed to be satisfactory when the melting points of the fractions differ by at least 10 °C. The separated crystals are either removed by filtration or are washed out with a tenside solution. In the latter case, the fat crystals adsorb a water soluble surface-active agent, such as sodium dodecyl sulfate, and thus acquire hydrophilic properties. The crystals are then transferred to the aqueous phase. The isolated aqueous suspension is then heated and the TG recovered as liquid fat.

An even sharper fractional crystallization procedure may be achieved by solubilization of fat in hexane or some other suitable solvent. However, solvent distillation and recovery are rather time consuming, so the use of this procedure is justified only in very special cases. In the processing step of "winterization" of rapeseed (Canola), cottonseed or sunflower oil, small amounts of higher melting TG or waxes are removed which would otherwise cause turbidity during refrigeration. The basis of winterization is the fractional crystallization by slowly cooling the oil, as outlined above. Other procedures for the production of cold-stable oils are based on the use of crystallization inhibitors. These are mono- and diacylglycerols, esters of succinic acid, etc.

Table 14.18. Changes in the pattern of triacylglycerols in a partially hydrogenated palm oil by interesterification

Melting point	Prior to interesterification	Single phase interesterification	Directed interesterification
Melting point (°C)	41	47	52
	Triacylglycerols[a] in mole-%		
S_3	7	13	32
S_2U	49	38	13
SU_2	38	37	31
U_3	6	12	24

[a] S: Saturated, U: unsaturated fatty acids.

The application of the fractional extraction of fat or oil, instead of crystallization, has been outlined under 14.4.1.3.

14.4.5 Margarine – Manufacturing and Properties

The inventor of margarine, *Mège Mouries*, described in his patent issued in 1869 a process for the production of spreadable fat from beef fat which would substitute for and imitate the scarce and costly dairy butter. Based on the assumption that margaric acid (17:0) is the predominant fatty acid of beef fat, the name "margarine" was suggested for the new product. The assumption was, however, proven to be incorrect (cf. Table 14.3). Nevertheless, the name remained.

Margarine, which is produced worldwide in amounts exceeding 7 million t/a, is a water in oil emulsion. Its stability is achieved by an increase in viscosity of the continuous fat phase due to partial crystallization and through emulsifiers. The fat crystals form a three dimensional network. They should be present in the β'-modification; a conversion $\beta' \rightarrow \beta$ is undesirable because the β-modification causes a "sandy" texture defect. Hydrogenated fats, which are frequently used as raw materials, crystallize in the β'-modification when the lengths of the acyl residues differ. The erucic acid-rich and partially hydrogenated rapeseed fat used in the past crystallizes in the β'-form. The cultivation of rapeseed with a low content of erucic acid at first produced a fat that, after partial hydrogenation, consisted to almost 90% of 18:0 and 18:1 and, as a result of this homogeneity, crystallized in the β-form. By means of cultivation, 16:0 was increased from 5 to 12% at the cost of 18:1, which is sufficient for the stabilization of the β'-form.

14.4.5.1 Composition

The properties of margarine, such as nutritional value, spreadability, plasticity, shelf life and melting properties, resemble those of butter and are influenced essentially by the varieties and properties of the main fat ingredients. Since choice of ingredients is large, numerous varieties of margarine are produced (cf. Table 14.19).

The fat in margarine, which by regulation is 80% by weight (diet margarine is 39–41% fat), contains about 18% of emulsified water. The W/O emulsion is stabilized by a mixture of mono- and diacylglycerols (approx. 0.5%) and

Table 14.19. Examples of margarine types

Type	Comments
A. Household margarine	
Standard product	At least 50% of the fat is vegetable oil, the rest being animal fat.
Vegetable margarine	At least 98% of the fat is vegetable oil; contains at least 15% linoleic acid.
Linoleic acid enriched margarine	At least 30% linoleic acid, otherwise as vegetable margarine.
B. Semi-fat margarine	The fat content is halved. This type is not suitable for baking and frying.
C. Molten or fused margarine	Practically free of water and protein. It is aromatized with diacetyl and butyric acid; soft consistency; with large TG crystals it has a grainy structure; applied in cooking, frying and baking.
D. Special types for	
industrial processing	
Baking margarine	Strongly aromatized with heat stable compounds that contribute to baked products' aroma; mainly moderately melting TG's.
Margarine for pastry production	This margarine is strongly aromatized; its high melting TG's are embedded in oil phase; suitable for dough extension into thin sheets ("strudel dough") used in flaky pastry production.
Creamy margarine	It is not or only slightly aromatized; has a soft consistency; contains high content of coconut oil and approx. 10 vol-% of air.

crude lecithin (approx. 0.25%). Diet margarines have higher levels of emulsifiers. Skim milk or skim milk powder suspended in water (milk proteins, 1%; 2% in semi-fat margarine) is added in the production of high quality retail brands of margarine. The casein assists the action of the emulsifiers and, together with lactose, provides the desired browning when heated.

The aqueous phase of the margarine acquires a pH of 4.2–4.5 by addition of citric and lactic acids. This not only affects the flavor, but protects against microbial spoilage. In addition, traces of heavy metal ions are complexed. Margarine also contains the aroma substances typical of butter, which can be produced by microbiological souring (cf. 10.2.3.2). Readily available synthetic compounds, such as diacetyl, butyric acid, lactones of C_8–C_{14} hydroxy-fatty acids (cf. 5.3.1.4) and (Z)-4-heptenal, may also be used for aromatization. Common salt (0.1–0.2%) is used to round-off the flavor. Margarine is colored with β-carotene or with gently refined, unbleached palm oil. Attention is also given to maintaining the presence of 1 mg of α-tocopherol per g of linoleic acid. High quality products are vitaminized by the addition of about 25 IU/g vitamin A and 1 IU/g vitamin D_2. The authenticity of margarine is verified in some countries by an indicator substance added to it. This is required by legislation. Gently refined sesame oil (for its detection, see Table 14.22) is one of these substances.

14.4.5.2 Manufacturing

Margarine is manufactured continuously by a process consisting essentially of three steps:

- Emulsification of water within the continuous oil phase.
- Chilling and mechanical handling of the emulsion.
- Crystallization, preserving the type of w/o emulsion by efficient removal of the released heat of crystallization.

The triacylglycerols should preferentially crystallize in their β′-forms (cf. 3.3.1.2). The higher melting β-forms are not desired since they cause a "sandy" texture. The transitioin β′ → β-form is inhibited by addition of 1% saturated diacylglycerols.

14.4.5.3 Varieties of Margarine

The characteristic features of some varieties of margarine are summarized in Table 14.19.

14.4.6 Mayonnaise

Mayonnaise is an "oil in water" or o/w emulsion (cf. 8.15.1) consisting of 50–85% edible oil, 5–10% egg yolk, vinegar, salt and seasonings (cf. 11.4.2.3). The emulsion is stabilized by egg yolk phospholipids. Products with a lower oil content (<50%) may contain thickening agents such as starch, pectin, traganth, agar-agar, alginate, carboxymethylcellulose, milk proteins or gelatin. Sorbic acid, benzoic acid or the ethyl ester of p-hydroxybenzoic acid are added as preservatives. The stable emulsion is produced in a combinator with a homogenizer and then packed.

14.4.7 Fat Powder

In contrast to fats and oils, fat powders have better stability against autoxidation and, in some food products such as dehydrated soup powders or broths, are easier to handle. They are manufactured from natural or hardened plant fats, sometimes with the addition of emulsifiers and protein carriers. Butter and cream powders are also produced.

Two basic flow diagrams of the production of fat powders are shown in Fig. 14.6.

In a cold-spray process, the melted fat is sprayed under high pressure into a cooled (−35°C) air-blast crystallization chamber, where the fat particles solidify. After being recrystallized, the particles are coated to avoid clumping.

In a spray-drying process, the fat is homogenized with emulsifiers, water and skim milk, spray dried and subsequently crystallized.

14.4.8 Deep-Frying Fats

Traditionally, the fats used for deep frying are those whose stability against autoxidation have

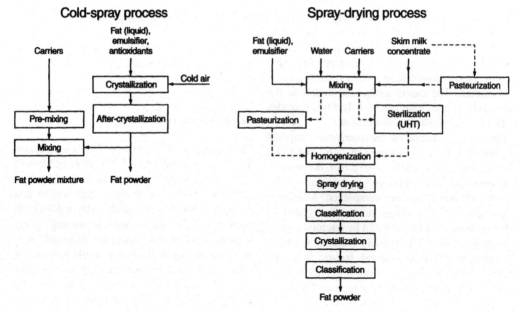

Fig. 14.6. Production of fat powder

been increased by hydrogenation. However, trans-fatty acids are produced in this process (cf. 14.4.2) and these acids are nutritionally and physiologically undesirable. In contrast, fat mixtures containing 70% oleic and 10% linoleic acid offer an improved nutritional value and a pleasant aroma (cf. 3.7.4.1). Small amounts of sesame oil (cf. 14.3.2.2.5) and oil isolated from rice bran are added to provide protection against autoxidation. This oil contains sterol esters of ferulic acid which are called γ-oryzanal. The ferulic acid (cf. 18.1.2.5) is the antioxidatively active component of these esters.

14.5 Analysis

14.5.0 Scope

The problems and scope of fat or oil analysis include identification of the type, determination of the composition of the blend, detection of additives, antioxidants, color pigments, and extraneous contaminants (solvent residues, pesticides, trace metals, mineral oils, plasticizers). In addition, the scope of analysis encompasses determination of other quality parameters, such

as the extent of lipolysis, autoxidation or thermal treatments. Also of interest is the extent of refining which the fats and oils have been subjected to as well as detection of hardened fat and products which were interesterified.

14.5.1 Determination of Fat in Food

The methods used for determination of fat or oil in food are often based on extraction with either ethyl ether or petroleum ether and gravimetric determination of the extraction residue. These methods may provide unreliable or incorrect results, particularly with food of animal origin. As shown in Table 14.20, where a corned beef sample was analyzed, the amount and composition of fatty acids in the fat residue were influenced greatly by the analytical methods used. In addition to the accessible free lipids, the emulsifiers present and the changes induced by autoxidation affect the amount of extractable lipids and the lipid-to-nonlipid ratio in the residue. The use of a standard method still does not eliminate the disadvantages shown by analytical methods of fat analysis. Therefore, in questionable cases, quantitative determination of fatty acids and/or glycerol is recommended.

Table 14.20. Determination of the fat content of canned corned beef

Analytical method	Fat content (%)[a]	Fatty acid composition (g/100 g)			
		Saturated acids	18:1 (9)	18:2 (9,12)	18:3 (9,12,15)
1. Dried sample is extracted with ethyl ether	7.9	3.98	2.06	0.05	0.08
2. Sample is homogenized in 95% ethanol and then extracted with ether	15.8	4.0	2.60	0.77	0.32
3. Sample is hydrolyzed with 4 mol/1 HCl (at 60 °C for 30 min), then extracted with ether	12.3	5.66	3.94	0.95	0.71
4. Sample is hydrolyzed with conc. HCl (at 100 °C for 1 h), methanol added and then extracted with carbon tetrachloride	13.9	2.45	1.68	0.34	0.21
5. Sample is homogenized in chloroform methanol mixture (2:1 v/v), washed with water and then the chloroform phase recovered	11.2	4.89	3.31	0.85	0.39

[a] The fat is determined gravimetrically after the solvent is evaporated.

A rapid and accurate determination of fats or oils in food is achieved by IR- (cf. 15.3.1) and ^{1}H-NMR spectrometry. The method is based on the fact that hydrogen nuclei in fluids respond to substantially higher magnetic resonance effects than do immobilized hydrogen atoms of solid substances. Thus, the ^{1}H-NMR signal of a fluid, such as an oil, differs from that of a nonoil matrix, such as carbohydrate, protein or firmly-bound water. The intensity of the signal is directly proportional to the oil content. This method is also of great value in oilseed selection or breeding research, since it permits determination of the oil content in a single kernel without damaging it by grinding or drying, i.e. retaining its ability to germinate.

The proportion of solid to fluid triacylglycerols in fat can also be determined using ^{1}H-NMR spectrometry.

14.5.2 Identification of Fat

14.5.2.1 Characteristic Values

For both, the identification and the determination of the quality of a fat or oil, the older lipid chemistry defines a series of characteristic values in which the reagent uptake is used to quantitatively estimate the selected functional groups or calculate the constituents of a fat or oil. The introduction of new analytical methods, such as gas chromatography of fatty acids and the HPLC of triacylglycerols (cf. 3.3.1.4), has made many of these measures obsolete. The values which are still used to differentiate fats or oils are:

Saponification Value (SV). This is the weight of KOH (in mg) needed to hydrolyze 1 g of fat or oil under standardized conditions. The higher the SV, the lower the average molecular weight of the fatty acids in the triacylglycerols (for examples, see Table 14.21).

Acid Value (AV). This value is important for a first quick characterization of the quality of a fat. It is the number of milligrams of KOH needed to neutralize the organic acids present in 1 g of fat.

Iodine Value (IV). This number is the number of grams of halogen, calculated as iodine, which bind to 100 g fat (cf. 3.2.3.2.1). The halogen uptake by fat or oil is affected by the contents of oleic (IV: 89.9), linoleic (IV: 181) and linolenic (IV: 273) acids. Examples of iodine numbers are provided in Table 14.21.

Table 14.21. Iodine (IV) and saponification numbers (SV) of various edible fats and oils

Oil/fat	IV	SV	Oil/fat	IV	SV
Coconut	256	9	Rapeseed		
Palm kernel	250	17	(turnip)	225	30
Cocoa	194	37	Sunflower	190	132
Palm	199	55	Soya	192	134
Olive	190	84	Butter	225	30
Peanut	192	156			

Hydroxyl Value (OHV). This number reflects the content of hydroxy fatty acids, fatty alcohols, mono- and diacylglycerols and free glycerol.

14.5.2.2 Color Reactions

Some oils give specific color reactions caused by particular ingredients. Examples are summarized in Table 14.22. Since many specific nonfat components are removed from oils by refining, these tests are negative when applied to refined oils.

14.5.2.3 Composition of Fatty Acids and Triacylglycerides

The acyl residues of an acylglycerol are released as methyl esters (cf. 3.3.1.3) and are analyzed as such by gas chromatography. However, free fatty acid analysis is also possible by using specially selected stationary solid phases. Capillary-column gas chromatography should be used to

Table 14.22. Color reactions for fat and oil identification

Reaction according to[a]	Identification of
Baudouin (furfural and hydrochloric acid)	Sesame oil
Halphen (sulfur and carbondisulfide)	Cottonseed oil
Fitelson[b] (sulfuric acid and acetic acid anhydride)	Teaseed oil

[a] Reagents are listed in brackets.
[b] It is a modification of *Liebermann–Burchard* reaction for sterols (cf. 3.8.2.4).

differentiate between cis and trans fatty acids, which is required for the detection of partially hydrogenated fats. The fatty acids indicative of the identity or type of fat or oil are summarized in Table 14.23. An enrichment step must precede gas chromatographic separation when fatty acids of analytical significance are present as minor constituents.

Prior to the enrichment step, specific techniques such as "argentation" chromatography (cf. 3.2.3.2.3) or fractionation by urea-adduct formation (cf. 3.2.2.3) are carried out in addition to the usual preparative chromatographic procedure. The methyl branched fatty acids in marine oils are an example. These acids are enriched by the urea-adduct inclusion method since, unlike straight-chain acids, they are unable to form inclusion compounds. These branched-chain fatty acids do not change during hydrogenation, hence they can be used as marine oil indicators, i.e. to reveal the presence of marine oil in hydrogenated vegetable oils such as margarine. Another example is the use of gas chromatography to determine furan fatty acids in soybean oils (cf. Table 3.9), which is possible only after enrichment in an urea filtrate.

In the interpretation of the results of fatty acid analyses, it should be taken into account that the fatty acid composition is subject to considerable variations. It depends on the breed and feed in the case of animal fats, and on the plant variety, geographic location of the area of cultivation, and the climate in the case of plant fats. Therefore, guide values have been set for individual oils and fats (cf. Table 14.24), which can differ from country to country.

The ratio of the content of a fatty acid in position 2 of the triacylglycerides to its total content (E factor, E = *enrichment*) is independent of the origin of the plant oil. After hydrolysis of the fat with pancreatic lipase, separation of the 2-monoglycerides, and their methanolysis, the concentration in position 2 is determined by gas chromatography and the E-factor calculated (examples for linoleic acid are shown in Table 14.25). Adulteration of olive oils with ester oils is shown by an increased E-factor for palmitic acid (cf. 3.3.1.4).

In many cases, the triacylglyceride pattern is more expressive than the fatty acid composition. This pattern can be quickly and easily determined

Table 14.23. Fatty acid indicators suitable for determination of fat and oil origin

Fatty acid	Content (%)[a]	Indicator of
4:0	3.7	Milk fat
12:0	45	Coconut-, palm kernel-, and babassu fat
18:1 (9)	65–85[b]	Teaseed-, olive- and hazelnut oil
18:3 (9, 12, 15)	9	Soya-, rapeseed (also erucic acid free) oil
18:2 (9, 12)	50–70	Sunflower-, corn germ-, cottonseed-, wheat germ-, and soya oil
22:0	3	Peanut oil
20:4 (5, 8, 11, 14)	0.1–0.6	Animal fat
18:1 (9, 12-OH)	80	Castor bean oil
Trans-fatty acids		Partially or fully hydrogenated oil/fat[c]
Methyl-branched fatty acids	0.2–1.6	Animal fat[d]

[a] When value range is omitted fatty acid percentage composition is given as an average value.
[b] A high percentage of this acid is a characteristic indicator.
[c] Here precautions are needed: animal fat, e. g. from beef, might contain up to 10% trans fatty acids.
[d] It is relatively high in marine oils (approx. 1%).

Table 14.24. Fatty acid composition of sunflower oil

Fatty acid	Per cent by weight	
	Average	Variation range[a]
16:0	6.2	3.0–10.0
16:1	0.08	<0.1
18:0	4.75	1.0–10.0
18:1 (9)	19.8	14–65
18:2 (9, 12)	67.0	20–75
18:3 (9, 12, 15)	0.08	<0.7
20:0	0.34	<1.5
20:1	0.15	<0.5
22:0	0.89	<1.0

[a] German guide values.

Table 14.25. E-factor of various oils for linoleic acid

Oils/fats	E-factor
Sunflower	1.2
Corn	1.3
Soybean	1.3
Rapeseed	1.7
Peanut	1.7

with the help of HPLC and GC (cf. 3.3.1.4). An example is the detection of foreign fat in milk fat. From extensive data on the triacylglyceride composition (GC differentiation according to the C-number), formulas have been developed which permit the detection of all important plant and animal fats up to a limiting value of 2–5 percent by weight. The older method, which is based

on a decrease in the butyric acid concentration due to the foreign fat, does not safely detect an addition of 20 percent by weight because of the biological variation (3.5–4.5 w/w percent 4:0).

14.5.2.4 Minor Constituents

Some fats which can not be unequivocally distinguished by their fatty acid or triacylglyceride composition may be identified by analysis of the unsaponifiable minor constituents. Examples are given in Table 14.26.

Table 14.26. Fat or oil identification by analysis of unsaponifiable constituents

Analysis	Identification
Squalene	Olive or rice oil and fish liver oil
Campesterol/stigmasterol[a] (cf. 3.8.2.3.1)	Cocoa butter substitutes
Carotene (cf. 3.8.4.5)	Raw palm oil
γ-/β-Tocopherol[b] (cf. Table 3.51)	Corn oil
γ-Tocopherol (cf. Table 3.51)	Wheat germ oil
α-/γ-Tocopherol[b] (cf. Table 3.51)	Sunflower oil
γ-/δ-Tocopherol[b] (cf. Table 3.51)	Soybean oil
Cholesterol[c] (cf. 3.8.2.2.1)	Animal fat

[a] Concentration ratios are characteristic.
[b] Concentration of individual compounds and their concentration ratios are characteristic.
[c] Cholesterol concentration must exceed by 5% the total sterol fraction.

The detection of adulteration of oils and fats has been improved further by coupled HPLC and GC of the minor constituents. The saponification of the sample is not required, free and esterified compounds being detected separately.

An example is the differentiation between the olive oil qualities "*extra vierge*" and "*lampante*". After esterification of the free OH-groups with pivalic acid, the free fat alcohols, wax esters, free acids, triterpene alcohols and esters are eluted in a relatively narrow fraction in HPLC and separated from the triacylglycerides. The eluate is transferred to a gas chromatograph

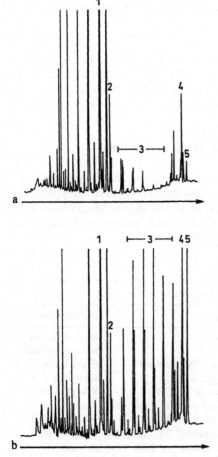

Fig. 14.7. On-line HPLC-GC of sterol and wax fractions of olive oils. **a** "*Extra vierge*" oil, **b** "*lampante*" oil. Peak 1: sitosterol, peak 2: 24-methylene cycloartenol, peak group 3: wax esters, peak 4: sitosterol ester, peak 5: 24-methylene cycloartenolester (according to *Grob* et al., 1991)

and analyzed on an apolar capillary column. As shown in Fig. 14.7, a clear distinction is made between "*lampante*" oils and "*extra vierge*" oils because the former have high contents of wax and sterol esters (sitosterol, 24-methylene cycloartenol) (cf. 14.3.2.1.1).

14.5.2.5 Melting Points

In addition to specific density, index of refraction, color and viscosity, the melting properties can be used to identify fats and oils.

The composition and the crystalline forms (cf. 3.3.1.2) of triacylglycerols present in fat determine the melting points and the temperature range over which melting occurs. The onset, flow point and end point of melting are of interest. They are determined by standardized procedures. The melting properties of fat are more accurately determined by differential thermal analysis. The temperature difference is measured between the fat sample and a blank, i.e. a thermally inert substance, as a function of the heating temperature (Fig. 14.8). In this way the temperatures at which polymorphic transitions of fat occur are detectable. In addition, the content of solid triacylglycerols can be assessed from the heat absorbed during melting at various temperatures. Thus, the solid triacylglycerol (TG) portion of coconut oil at $-3\,°C$ can be calculated using data from the recorded curve (Fig. 14.8) and the following formula:

$$\% \, \text{Coconut} \, (\text{solid TG}) = \frac{\text{Area (BCDE)}}{\text{Area (AEDA)}} \cdot 100$$

$$(14.11)$$

The solid: liquid ratio of acylglycerols is of importance in fat hydrogenation and interesterification processes (cf. 14.4.3). This ratio can also be assessed using the Solid Fat Index by measuring the expansion of the fat, i.e. the volume increase of a fat during its transition from solid to liquid and by ^{1}H-NMR spectroscopy (cf. 14.5.1).

14.5.2.6 Chemometry

To solve difficult problems in food chemistry, e.g., the detection of the authenticity of olive

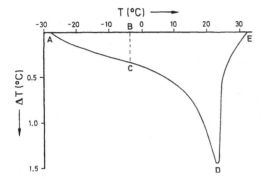

Fig. 14.8. Differential thermal analysis of a coconut fat

oils, the test procedures include chemometry. The measurements are planned and evaluated with the help of mathematical or statistical methods in order to gain the maximum amount of chemical information. An example is the addition of hazelnut oil to olive oil, *Raman* spectra being evaluated for the detection.

14.5.3 Detection of Changes During Processing and Storage

Processes used in recovery and refining and subsequent storage conditions are the main factors affecting the quality of edible fat or oil. A number of analytical methods are available for assessing the quality and deterioration of fat or oil.

14.5.3.1 Lipolysis

The extent of lipolysis (cf. 3.7.1) is determined by the free fatty acid content (FFA or Acid Value). Oils with FFA content exceeding 1% are commonly designated as crude oils, while lard with this level of free acids is considered spoiled. An exception is olive oil, which is still considered suitable for direct consumption even with a 3% FFA content. The FFA content is lowered to less than 0.1% by refining of oil or fat.

There is no relationship between the sensory perception of quality deterioration and the levels of FFA in fats which contain low-molecular acyl residues (e. g., milk, coconut, and palm kernel fats) because among the free fatty acids, the sensory-relevant compounds (C number <14)

usually take second place. A better correlation is provided by the analysis of the low-molecular free fatty acids (cf. 3.7.1.1). They are first separated from the fat, e. g., with an ion exchanger, released from the exchanger by esterification with ethyl iodide, and then determined by gas chromatography.

14.5.3.2 Oxidative Deterioration

Fats and oils deteriorate rather rapidly by autoxidation of their unsaturted acyl residues (cf. 3.7.2.1). A number of analytical methods have been developed to determine the extent of such deterioration and to predict the expected shelf life of a fat or oil.

14.5.3.2.1 Oxidation State

Peroxide Value. The method for determination of peroxide concentration is based on the reduction of the hydroperoxide group with HI or Fe^{2+}. The result of the iodometric titration is expressed as the peroxide value. The Fe^{2+} method is more suitable for determining a low hydroperoxide concentration since the amount of the resultant Fe^{3+}, in the form of the ferrithiocyanate (rhodanide) complex, is determined photometrically with high sensitivity (Fe-test in Table 14.27). The peroxide concentration reveals the extent of oxidative deterioration of the fat, nevertheless, no relationship exists between the peroxide value and aroma defects, e. g. rancidity (already existing or anticipated). This is because hydroperoxide degradation into odorants is influenced by so many factors (cf. 3.7.2.1.9) which make its retention by fat or oil or its further conversion into volatiles unpredictable.

Carbonyl Compounds. The analysis of the compounds responsible for the rancid aroma defect of a fat or oil is of great value. Volatile carbonyls (cf. 3.7.2.1.9) are among such compounds.

In a simple test, such as benzidine, anisidine or heptanal values, the volatile aldehydes are not separated from fat or oil, rather the reaction with the group-specific reagents is carried out in the fat or oil. In addition to the odorous aldehydes, the flavorless oxo-acylglycerols and oxo-acids can be

Table 14.27. Analytical aspects related to the determination of the extent of oxidation of unsaturated fatty acids: relative sensitivities of spectrophoto-metric procedures[a]

Method		Autoxidized fatty acid methyl esters	
		18:2 (9,12)[b]	18:3 (9,12,15)[c]
UV-Absorption	234 nm	1.0	1.0
	270 nm	0.1	0.3
Fe^{2+}/Thiocyanate (rhodanide)		9.4	6.3
Thiobarbituric	452 nm	0.1	0.5
acid test	530 nm	0.1	1.0
Kreis-test		0.1	0.1
Anisidine value		0.3	0.75
Heptanal value		0.1	0.1

[a] Related to UV absorption at 234 nm.
[b] Peroxide value: 475.
[c] Peroxide value: 450.

determined. Since the proportion of aroma-active and sensory neutral carbonyls is not known, any correlation found between the carbonyl value and aroma defects is clearly coincidental.

The *thiobarbituric acid test* (TBA) is a preferred method for detecting lipid peroxidation in biological systems. However, the reaction is nonspecific since a number of primary and secondary products of lipid peroxidation form malonaldehyde which in turn reacts in the TBA test. In food containing oleic and linoleic acids, the TBA-test is not as sensitive as the Fe^{2+}-test outlined above.

The *gas chromatographic determination* of individual carbonyl compounds appears to be a method suitable for comparison with findings of sensory panel tests. Analytical methods for the odorants causing aroma defects is still in the early stages of development because only a few fats or fat-containing foods have been examined in such detail that the aroma substances involved are clearly identified.

The well studied warmed-over flavor of cooked meat (cf. 12.6.2.1) is an example. It can be controlled relatively easily because the easy-to-determine hexanal has been identified as the most important off-flavor substance. On the other hand, the easily induced rancid aroma defect of rapeseed oil is primarily caused by the volatile hy-

droperoxides (1-octen-3-hydroperoxide, (Z)-1,5-octadiene-3-hydroper-oxide) and (Z)-2-nonenal which can be quantitatively detected only by using isotopic dilution analyses (cf. 5.2.6). This limitation also applies to 3-methyl-2,4-nonandione, which appears as the most important off-flavor substance in soybean oil on exposure to light.

To simplify the analytical procedure, individual aldehydes (e. g., hexanal, 2,4-decadienal), which are formed in larger amounts during lipid peroxidation, have been proposed as indicators. In most of the cases, however, it was not tested whether the indicator increases proportionally to the off-flavor substances which cause the aroma defect.

14.5.3.2.2 Shelf Life Prediction Test

To estimate susceptibility to oxidation, the fat or oil is subjected to an accelerated oxidation test under standardized conditions so that the signs of deterioration are revealed within several hours or days. Examples of such tests are the *Schaal test* (fat maintained at 60 °C) and the *Swift stability test* (fat kept at 97.8 °C and aerated continuously). The extent of oxidation is then measured by sensory and chemical tests such as peroxide value (cf. 3.7.2), ultraviolet absorption (suitable for fats and oils containing linoleic or linolenic acids) or oxygen uptake. There are also methods based on the fact that in the process of triglycerol oxidation, when the initiation period is terminated, large amounts of low molecular weight acids are released. They are then determined electro-chemically. During oxidation of a given fat or oil sample, a good correlation exists between the length of the induction period and the shelf-life.

14.5.3.3 Heat Stability

The behavior of a frying oil, when heated, is assessed from the content of oxidized fatty acids which are insoluble in petroleum ether and from the smoke point (cf. 3.7.4) of the fat or oil. The smoke point of a fat or oil is the temperature at which its triacylglycerols start to decompose in the presence of air. Smoke is the sign of decomposition. The smoke point of a fat or oil is normally in the range of 200–230 °C during pro-

Table 14.28. Indicators of refined oils and fats

Refining step	Indicator	Remarks
Bleaching (cf.14.4.1.4)	a. Fatty acids with conjugated triene systems	Determination of "b" is more reliable than the UV measurement of "a"
	b. Disterylether (>0.5 mg/kg)	
Deodorization (cf. 14.4.1.5)	a. Dimeric and oligomeric triacylglycerides	Unlike "a", the indicators "b" appear even on relatively gentle deodorization
	b. Position and substitution isomers of linoleic acids	

longed frying, and it decreases in the presence of decomposition products. When it falls below 170 °C, the fat is considered to be spoiled. At this point, the amount of fatty acids which are insoluble in petroleum ether exceeds 0.7%. However, this petroleum ether method is not reproducible. Fat separation by column chromatography is more reliable. The heated fat or oil is separated into a polar and a nonpolar fraction using silicic acid as an adsorbent. The value of 0.7% oxidized petroleum ether-insoluble fat corresponds, in this chromatographic separation, to 73% nonpolar and 27% polar fractions.

14.5.3.4 Refining

The addition of a refined oil to natural plant oil is detected by the determination of substances which can be formed during bleaching and deodorization (cf. Table 14.28).

14.6 References

Baltes, J.: Gewinnung und Verarbeitung von Nahrungsfetten. Verlag Paul Parey: Berlin. 1975

Bockisch, M.: Nahrungsfette und -öle. Verlag Eugen Ulmer, Stuttgart, 1993

DGF-Einheitsmethoden. Ed. Deutsche Gesellschaft für Fettwissenschaft e.V., Münster/Westf., Wissenschaftliche Verlagsgesellschaft mbH: Stuttgart. 1950–1979

Gertz, C.: Native und nicht raffinierte Speisefette und -öle. Fat Sci. Technol. *93*, 545 (1991)

Gertz, C., Fiebig, H.-J.: Statement on the applicability of methods for the determination of pyropheophytin A and isomeric diacylglycerols in virgin olive oils. German Society for Fat Science (DGF), Division Analysis and Quality Assurance, Münster. 2005

Grob, K., Artho, A., Mariani, C.: Gekoppelte LC-GC für die Analyse von Olivenölen. Fat Sci. Technol. *93*, 494 (1991)

Grosch, W., Tsoukalas, B.: Analysis of fat deterioration – Comparison of some photometric tests. J. Am. Oil Chem. Soc. *54*, 490 (1977)

Guhr, G., Waibel, J.: Untersuchungen an Fritierfetten; Zusammenhänge zwischen dem Gehalt an petrolätherunlöslichen Fettsäuren und dem Gehalt an polaren Substanzen bzw. dem Gehalt an polymeren Triglyceriden. Fette Seifen Anstrichm. *80*, 106 (1978)

Hamilton, R. J., Rossell, J.B. (Eds.): Analysis of oils and fats. Elsevier Applied Science Publ.: London. 1986

Hoffmann, G.: The chemistry and technology of edible oils and fats and their high fat products. Academic Press: London. 1989

Kiritsakis, A., Markakis, P.: Olive oil – a review. Adv. Food Res. *31*, 453 (1987)

Li-Chan, E.: Developments in the detection of adulteration of olive oil. Trends Food Sci. Technol. 5, 3 (1994)

Official methods and recommended practices of the American Oil Chemists' Society, Firestone, D. (Ed.), 4th edn. American Oil Chemists' Society, Champaign, 1989

Reid, L.M., O'Donnell, C.P., Downey, G.: Recent technological advances for the determination of food authenticity. Trends Food Sci. Technol. *17*, 344 (2006)

Schulte, E., Weber, N.: Disterylether in gebleichten Fetten und Ölen – Vergleich mit den konjugierten Trienen. Fat Sci. Technol. *93*, 517 (1991)

Sheppard, A. J., Hubbard, W. D., Prosser, A. R.: Evaluation of eight extraction methods and their effects upon total fat and gas liquid chromatographic fatty acid composition analyses of food products. J. Am. Oil Chem. Soc. *51*, 417 (1974)

Stansby, M. E.: Fish oils. AVI Publ. Co.: Westport, Conn. 1967

Swern, D. (Ed.): Bailey's industrial oil and fat producis. 4th edn., Vol. 1, John Wiley and Sons: New York. 1979

Usuki, R., Suzuki, T., Endo, Y., Kaneda, T.: Residual amounts of chlorophylls and pheophytins in refined edible oils. J. Am. Oil Chem. Soc. *61*, 785 (1984)

15 Cereals and Cereal Products

15.1 Foreword

15.1.1 Introduction

Cereal products are amongst the most important staple foods of mankind. Nutrients provided by bread consumption in industrial countries meet close to 50% of the daily requirement of carbohydrates, one third of the proteins and 50–60% of vitamin B. Moreover, cereal products are also a source of minerals and trace elements.

The major cereals are wheat, rye, rice, barley, millet and oats. Wheat and rye have a special role since only they are suitable for bread-making.

15.1.2 Origin

The genealogy of the cereals begins with wild grasses (*Poaceae*), as shown in Fig. 15.1. Barley (*Hordeum vulgare*), probably one of the first cereals grown systematically, was known as early as 5000 B.C. in Egypt and Babylon. Also, the bearded wheat cultivars from the groups Einkorn (*Triticum monococcum*) and Emmer (*T. dicoccum*), with diploid (genome formula: AA, $2n = 14$) and tetraploid (AABB, $2n = 28$) sets of chromosomes, (the chromosome number of the wheat genome is $n = 7$), were found among cultivated plants that were widely spread in temperate zones of Euroasia during the neolithic period. These cultivars are becoming extinct. Only the durum form of Emmer (*T. turgidum durum*, hard wheat, AABB), at 10% of the total wheat grown, has a significant role. The hexaploid (AABBDD, $2n = 42$) wheats derived from spelt are grown worldwide as bread cereals. The A genome of the spelts is closely related to that of Einkorn (*T. monococcum*). The origin of the B genome is unknown. It probably comes from species of the genus *Aegilops* and the D genome from *Aegilops squarrosa*.

Two varieties are derived from the spelts, bare wheat (soft wheat, *T. aestivum*) and bearded spelt (*T. spelta*). The low yield and the additional dehusking procedure led to the fact that soft wheat (called "wheat" below) gained more acceptance than spelt. As late as the middle of the 19th century, 15–20 times more spelt than wheat was grown in Southern Germany. Since the 1980s, however, demand for spelt has increased, especially in the natural food market. To compensate for the disadvantages with respect to the yield and the baking properties, wheat cultivars are crossed with the spelt. Such varieties differ from pure spelt in their gliadin pattern, which can be determined by HPLC (cf. 15.2.1.3.1).

Rice (*Oryza sativa*) and corn (*Zea mays*) have been cultivated for 5000 years, first in tropical Southeast Asia and then in Central and South

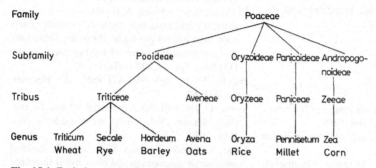

Fig. 15.1. Evolutionary development (phylogeny) of cereals

America. Cereals designated as millet have had a role from antiquity in subtropical and tropical regions of Asia and Africa. True millet from the subfamilies *Eragrostoideae* and *Panicoideae*, to which many regionally important cultivars belong (for instance, *Eragrostis tef, Eleusine coracan, Echinochloa frumentacea, Pennisetum glaucum, Setaria italica*), is distinguished from sorghum (*Sorghum bicolor*), which belongs to the subfamily *Andropogonoideae* and is cultivated worldwide.

Rye (*Secale cereale*) and oats (*Avena sativa*) are so-called secondary culture plants. Initially hardy and unwanted escorts of cultivated plants, they prospered and established themselves in northern regions with unfavorable climates. Their high tolerance for unfavorable climates surpassed that of both wheat and barley. Rye and oats have been cultivated for millenia.

Breeders have for many years attempted to combine the baking quality of wheat with the hardiness of rye. *Triticale*, the man-made hybrid of wheat and rye, does not yet fulfill this aim, hence its economic significance is low.

15.1.3 Production

Cereals are of great importance as raw materials for production of food and feed. Accordingly, they are grown on close to 60% of cultivated land in the world. Wheat production takes up the greatest part of land cultivated with cereals (Table 15.1) and wheat is produced in the largest

Table 15.1. Land cultivated with a cereal crop as % of the world total area under cereal cultivation (1979: 7.6×10^8 ha)

Cereal	1966	1976	1984	1988	1990	1996
Wheat	30.6	31.5	34.5	31.2	32.7	32.4
Rice	18.8	19.2	21.9	20.9	20.6	21.2
Corn	15.5	15.9	19.3	18.2	18.3	19.7
Millet	15.4[a]	15.6[a]	12.3[a]	5.7	5.3	5.1
Sorghum				6.6	6.3	6.6
Barley	12.2	11.9	11.7	10.8	10.1	9.4
Oats	4.5	3.8	3.8	3.2	3.1	2.4
Rye	2.4	2.1	2.6	2.3	2.3	1.6

[a] Sum of millet and sorghum.

quantity (Table 15.2). Wheat surplus producers are the USA, Canada, Argentina, Australia, and France. In the Federal Republic of Germany (FRG), winter wheat (92%) and spring wheat (8%) are both cultivated. The rise of cereal production in the world is shown in Table 15.3. The yields per hectare vary greatly from one country to another (Table 15.4). Due to an intensive effort in breeding and crop production programs, the yields per hectare in the FRG are very high and are surpassed by only a few countries, e. g., Holland. The FRG utilized 25.7×10^6 tons of cereals in 1976/77, of which 38% was bread and 62% feed cereals.

15.1.4 Anatomy – Chemical Composition, a Review

Cereals, in contrast to forage grasses, form a relatively large fruit, termed a caryopsis, in which the fruit shell is strongly bound to the seed shell. The kernel size, which is expressed as grams per 1000 kernels (Table 15.5), is not only dependent on the kind of cereal but on the cultivar and crop production techniques, hence it varies widely.

In oats, barley, and rice the front and back husks are fused together with the fruit. In contrast, threshing separates wheat and rye kernels from the husks as bare seed.

The major constituents of seven kinds of cereal are fairly uniform (Table 15.6). Noteworthy variations are the higher lipid content in oats and a lower fiber content in millet and rice. The available carbohydrates consist mainly of starch. Oats are especially rich in nonstarch polysaccharides (cf. 15.2.4.2). These cereals also differ in their vitamin B content (Table 15.6).

Fruit and seed coats enclose the nutrient tissue (endosperm) and germ in the kernel (Fig. 15.2). Botanically the endosperm consists of the starchy endosperm (70–80% of the kernel; Table 15.7) and the aleurone layer, which, with exception of barley, is a single cell layer.

The aleurone layer is rich in protein and also contains fat, enzymes and vitamins (Table 15.8 and 15.9). The proteins, of which half are water-soluble, appear as granules in the aleurone cells. They have no influence on the baking properties

Table 15.2. Cereal production in 2006 (1000 t)

Continent	Wheat	Rice, paddy	Barley	(Corn) Maize	Rye
World	605,946	634,606	138,643	695,228	13,261
Africa	25,096	21,131	6133	46,260	39
America, Central	3345	1179	875	24,788	–
America, North	84,575	8787	13,925	276,866	484
America, South and Caribbean	22,636	24,564	3024	91,778	38
Asia	272,185	576,518	22,441	203,025	1107
Europe	191,378	3459	89,048	76,742	11,574
Oceania	10,075	148	4072	558	20

Continent	Oats	Millet	Sorghum	Cereals, grand total
World	23,101	31,781	56,485	2,221,119
Africa	226	17,788	26,113	145,892
America, Central	153	1	5831	36,176
America, North	4963	300	7050	397,456
America, South and Caribbean	1247	16	10,973	154,677
Asia	1630	12,891	10,691	1,102,274
Europe	14,377	751	658	403,644
Oceania	658	35	1001	17,176

[a] World production = 100%.

Country	Wheat	Country	Rice	Country	Barley
China	104,470	China	184,070	Russian Fed.	18,154
India	69,350	India	136,510	Germany	11,967
USA	57,298	Indonesia	54,400	Ukraine	11,316
Russian Fed.	45,006	Bangladesh	43,729	France	10,412
France	35,367	Vietnam	35,827	Canada	10,005
Canada	27,277	Thailand	29,269	Turkey	9551
Germany	22,428	Myanmar	25,200	Spain	8318
Pakistan	21,277	Philippines	15,327	UK	5239
Turkey	20,010	Brazil	11,505	USA	3920
UK	14,735	Japan	10,695	Australia	3722
Iran	14,500			China	3430
Argentina	14,000	$\Sigma(\%)^a$	86	Denmark	3270
Ukraine	14,000			Poland	3161
				Iran	3000
$\Sigma(\%)^a$	76			$\Sigma(\%)^a$	76

Country	(Corn) Maize	Country	Rye	Country	Oats
USA	267,598	Russian Fed.	2965	Russian Fed.	4880
China	145,625	Germany	2644	Canada	3602
Brazil	42,632	Poland	2622	USA	1361
Mexico	21,765	Belarus	1072	China	1160
India	14,710	Ukraine	920	Poland	1035
Argentina	14,446	China	783	Finland	1029
France	12,902	Canada	302	Spain	918
Indonesia	11,611	Turkey	246	Germany	830
Italy	9671	USA	183	UK	728
Canada	9268	Spain	159	Ukraine	700
				Sweden	635
$\Sigma(\%)^a$	79	$\Sigma(\%)^a$	90	Australia	633
				$\Sigma(\%)^a$	76

Table 15.2. Continued

Country	Millet	Country	Sorghum	Country	Cereals, grand total
India	10,100	Nigeria	9866	China	445,355
Nigeria	7705	India	7240	USA	346,562
Niger	3200	USA	7050	India	239,130
China	1821	Mexico	5487	Russian Fed.	76,866
Burkina Faso	1199	Sudan	5203	Indonesia	66,011
Mali	1060	China	2490	France	61,813
Sudan	792	Argentina	2328	Brazil	59,017
Uganda	687	Ethiopia	2313	Canada	50,895
Russian Fed.	600	Brazil	1556	Bangladesh	45,010
Chad	590	Burkina Faso	1554	Germany	43,475
				Vietnam	39,648
$\Sigma(\%)^a$	87	$\Sigma(\%)^a$	80	Turkey	34,598
				Ukraine	33,698
				Argentina	33,556
				Thailand	33,146
				Pakistan	32,839
				Mexico	31,959
				$\Sigma(\%)^a$	75

Table 15.3. World production of cereals 1948–2006 (10^6 t)

Year	Amount	Year	Amount
1948	683	1988	1742
1956	789	1989	1881
1964	1019	1990	1955
1968	1180	1996	2050
1976	1456	2004	2239
1984	1802	2006	2221

of wheat. Millers regard the aleurone layer as part of the bran.

The starchy endosperm is the source of flour. Its thin-walled cells are packed with starch granules which lie imbedded in a matrix which is largely protein. A portion of these proteins, the gluten proteins, is responsible for the baking properties of wheat. The concentrations of the proteins and some other constituents (vitamins and minerals) decrease from outer to inner cells of the endosperm. The germ is separated from the endosperm by the scutellum. The germ is rich in enzymes and lipids (Table 15.8). Table 15.9 shows that wheat milling, when starchy endosperm cells are separated from germ and bran, results in a substantial loss of B-vitamins and minerals.

15.1.5 Special Role of Wheat–Gluten Formation

After addition of water a viscoelastic cohesive dough can be kneaded only from wheat flour. The resulting gluten, which can be isolated as a residue after washing out the dough with water, removing starch and other ingredients, is responsible for plasticity and dough stability.

Gluten consists of 90% protein (cf. 15.2.1.3), 8% lipids and 2% carbohydrates. The latter are primarily the water-insoluble pentosans (cf. 15.2.4.2.1), which are able to bind and hold a significant amount of water, while the lipids (cf. 15.2.5) form a lipoprotein complex with certain gluten proteins. In addition, enzymes such as proteinases and lipoxygenase are detectable in freshly isolated gluten.

The gluten proteins, in association with lipids, are responsible for the cohesive and viscoelastic flow properties of dough. Such rheological properties give the dough gas-holding capacity during leavening and provide a porous, spongy product with an elastic crumb after baking.

Rye and other cereals can not form gluten. The baking quality of rye is due to pentosans and to some proteins which swell after acidification (cf. 15.4.2.2) and contribute to gas-holding properties.

Table 15.4. Yield per hectare in year 2000/2001/2002/2003/2004/2006 (dt/ha)

	Wheat	Rye	Corn	Rice
Germany	72.8/78.8/68.1/65.0/81.7/72.0	49.3/61.3/50.3/42.9/61.3/49.1	92.1/88.4/93.8/72.4/90.9/80.3	-.-/-.-/-.-/-.-/-.-/-.-
Argentina	24.9/22.4/20.3/25.3/25.4/25.5	14.4/13.3/14.0/ 9.5/ 9.5/11.5	54.3/54.5/61.7/64.7/ 64.6/59.0	47.8/57.0/57.5/54.0/62.8/70.6
Australia	18.2/21.1/ 9.1/20.7/17.0/ 8.8	6.4/ 6.4/ 6.4/ 6.4/ 5.7/ 5.7	49.4/46.6/55.1/59.4/ 49.6/50.0	82.6/92.8/79.5/95.2/82.3/63.0
China	37.4/38.1/37.8/39.3/42.0/44.6	15.3/13.4/14.5/22.7/21.0/35.1	46.0/47.0/49.3/48.1/ 51.5/53.7	62.6/61.5/61.9/60.6/62.6/62.6
France	71.2/66.2/74.5/62.5/75.8/67.4	46.1/40.8/48.6/40.3/50.5/45.7	90.8/85.6/89.8/71.2/ 89.8/85.6	58.4/53.6/56.9/56.1/57.1/55.3
India	27.8/27.1/27.6/26.2/27.1/26.2	-.-/-.-/-.-/-.-/-.-/-.-	18.2/20.0/16.4/19.8/ 20.0/19.4	28.5/31.2/28.9/30.8/30.5/31.2
Russian Fed.	16.1/20.6/20.7/17.0/19.8/19.5	15.8/18.9/19.9/18.6/15.4/17.1	21.2/18.1/28.6/32.2/ 40.4/36.3	34.9/34.9/37.7/31.5/37.7/43.9
Turkey	22.4/20.3/21.0/20.9/22.3/21.5	17.7/15.7/17.0/17.1/18.5/16.5	41.4/40.0/42.0/50.0/ 42.9/58.6	60.3/61.0/60.0/57.2/50.0/87.0
USA	28.3/27.1/23.7/29.7/29.0/28.3	17.8/17.2/15.5/17.0/16.9/16.5	85.9/86.7/81.6/89.2/100.7/93.6	70.4/72.8/73.7/74.5/77.8/76.9
World	27.2/27.5/26.8/27.1/29.1/28.0	20.4/23.6/23.0/21.8/25.2/22.1	42.8/44.2/43.5/44.4/ 49.1/48.2	38.9/39.4/39.1/39.1/40.0/41.1

Table 15.5. Average thousand kernel weight of cereals (g)

Wheat	37	Oats	32
Rye	30	Barley	37
Corn	285	Millet	23
Rice	27		

15.1.6 Celiac Disease

Wheat, rye and barley can cause celiac disease in genetically predisposed persons; the role of oats in this disease is uncertain. Celiac disease affects both infants and adolescents, and in adults it is also called sprue. It is associated with a loss of villous structure of the intestinal mucosa; epithelial cells exhibit degenerative changes and nutrient absorption functions are severly impaired. Incidence of the disease varies, e. g., 0.1% of the children are affected in central Europe and 0.3% in Ireland. The prolamin fractions of wheat, barley or rye are the cause of the disease, which is therefore eliminated by a change of diet to rice, millet or corn.

15.2 Individual Constituents

The role of constituents is of particular interest in the processing of wheat and rye into bakery products.

15.2.1 Proteins

15.2.1.1 Differences in Amino Acid Composition

The proteins of different cereal flours vary in their amino acid composition (Table 15.10). Lysine content is low in all cereals. Methionine is also low, particularly in wheat, rye, barley, oats and corn. Both amino acids are significantly lower in flour than in muscle, egg or milk proteins. By breeding, attempts are being made to improve the content of all essential amino acids. This approach has been successful in the case of high-lysine barley and several corn cultivars.

Table 15.6. Chemical composition of cereals (average values)

	Wheat	Rye	Corn	Barley	Oats	Rice	Millet
	weight %						
Moisture	13.2	13.7	12.5	11.7	13.0	13.1	12.1
Protein (N × 6.25)	11.7	9.5	9.2	10.6	12.6	7.4	10.6
Lipids	2.2	1.7	3.8	2.1	7.1	2.4[a]	4.05
Available carbohydrates	59.6	60.7	64.2	63.3	55.7	74.1	68.8
Fiber	13.3	13.2	9.7	9.8	9.7	2.2	3.8
Minerals	1.5	1.9	1.30	2.25	2.85	1.2	1.6
	mg/kg						
Thiamine	5.5	4.4	4.6	5.7	7.0	3.4	4.6
Niacin	63.6	15.0	26.6	64.5	17.8	54.1	48.4
Riboflavin	1.3	1.8	1.3	2.2	1.8	0.55	1.5
Pantothenic acid	13.6	7.7	5.9	7.3	14.5	7.0	12.5

[a] Polished rice: 0.8%.

Table 15.7. Fractions of various cereals separated by milling (average weight-%)

Cereal variety	Husk	Bran	Germ	Endosperm
Wheat	0	15.0	2.0	83.0
Corn	0	7.2	11.0	81.8
Oats	20	8.0	2.0	70.0
Rice	20	8.0	2.0	70.0
Millet	0	7.9	9.8	82.3

15.2.1.2 A Review of the Osborne Fractions of Cereals

In 1907 *Osborne* separated wheat proteins, on the basis of their solubility, into four fractions. Sequential extraction of a flour sample yielded: water-soluble albumins, salt-soluble (e. g., 0.4 mol/l NaCl) globulins, and 70% aqueous ethanol-soluble prolamins. The glutelins remained in the flour residue. They can be separated into two sub-fractions. For this purpose, all the proteins remaining in the residue are first dissolved in 50% aqueous 1-propanol at 60 °C with reduction of the disulfide bonds, e. g., with dithioerythritol. The high-molecular (HMW) subunits (cf. 15.2.1.3.1) precipitate out on increasing the propanol concentration to 60%, while the low-molecular (LMW) subunits (cf. 15.2.1.3.3) remain in solution.

Further separation of the *Osborne* fractions and subfractions into the components is possible analytically with electrophoretic methods (cf. Fig. 15.3, 15.4) and analytically and preparatively with RP-HPLC (cf. Figs. 15.5–15.8).

Table 15.8. Chemical composition of anatomical parts of a wheat kernel (average weight-% on dry weight basis)

	Ash	Crude protein (N × 6.25)	Lipids	Crude fiber[a]	Cellulose	Pentosans	Starch
Longitudinal cells	1.3	3.9	1.0	27.7	32.1	50.1	–
Cross- and tube cells	10.6	10.7	0.5	20.7	22.9	38.9	–
Fruit and seed coatings	3.4	6.9	0.8	23.9	27.0	46.6	–
Aleurone cells[b]	10.9	31.7	9.1	6.6	5.3	28.3	–
Germ[b]	5.8	34.0	27.6	2.4	–	–	–
Starchy endosperm	0.6	12.6	1.6	0.3	0.3	3.3	80.4

[a] Crude fiber includes parts of cellulose and pentosans.
[b] Data for carbohydrates are incomplete.

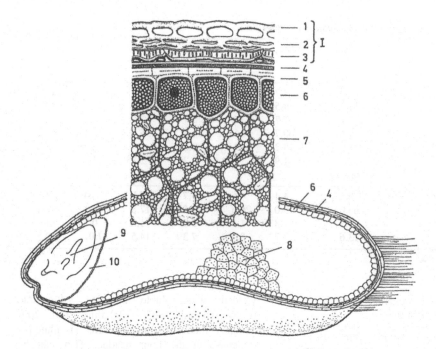

Fig. 15.2. Longitudinal section of a wheat grain. *1* Pericarp, *1* epidermis (epicarp), *2* hypodermis, *3* tube cells, *4* seed coat (testa), *5* nucellar tissue, *6* aleurone layer, *7* outer starchy endosperm cells, *8* inner starchy endosperm cells, *9* germ and *10* scutellum

Table 15.9. Mineral and vitamin distribution as % in kernel fractions of wheat

Fraction	Minerals	Thiamine	Riboflavin	Niacin	Pyridoxal phosphate	Pantothenic acid
Fruit coat	7	1	5	4	12	9
Germ	12	64	26	2	21	7
Aleurone layer	61	32	37	82	61	41
Starchy endosperm	20	3	32	12	6	43

In the literature, *Osborne* fractions derived from different cereals are often designated by special names (cf. review Table 15.11). The various designations may result in confusion and incorrect conclusions with regard to protein homogeneity. Therefore, it is better to preferentially use the general designations of the *Osborne* fractions and specify the protein source, e. g., wheat glutelin instead of glutenin.

Albumins and globulins are derived mostly from cytoplasmic residues and other subcellular fractions which are part of the kernel. Thus, enzymes are present in the first two *Osborne* fractions. Prolamins and glutelins, on the other hand, are storage proteins.

Cereals contain variable levels of *Osborne* fractions (Table 15.12). Wheat has the highest content of prolamin, corn has the second highest. The albumin fraction is the highest in rye and the lowest in corn. The content of albumin in oats is comparable to that in rye. Oats and rice have a higher content of glutelin than wheat, while rye, millet and corn have a much lower glutelin content. The amino acid composition of only the prolamins (Table 15.13) can be correlated to the botanical genealogy of cereals as shown in Fig. 15.1. In general, the amino acid composition is similar for wheat, of rye and barley. The prolamin composition of oats is intermediate between *Triticeae* and the other cereals. The amount of glutamic acid in

Table 15.10. Amino acid composition of the total proteins (mole-%)[a] of flours from various cereals

Amino acid	Wheat	Rye	Barley	Oats	Rice	Millet	Corn
Asx	4.2	6.9	4.9	8.1	8.8	7.7	5.9
Thr	3.2	4.0	3.8	3.9	4.1	4.5	3.7
Ser	6.6	6.4	6.0	6.6	6.8	6.6	6.4
Glx	31.1	23.6	24.8	19.5	15.4	17.1	17.7
Pro	12.6	12.2	14.3	6.2	5.2	7.5	10.8
Gly	6.1	7.0	6.0	8.2	7.8	5.7	4.9
Ala	4.3	6.0	5.1	6.7	8.1	11.2	11.2
Cys	1.8	1.6	1.5	2.6	1.6	1.2	1.6
Val	4.9	5.5	6.1	6.2	6.7	6.7	5.0
Met	1.4	1.3	1.6	1.7	2.6	2.9	1.8
Ile	3.8	3.6	3.7	4.0	4.2	3.9	3.6
Leu	6.8	6.6	6.8	7.6	8.1	9.6	14.1
Tyr	2.3	2.2	2.7	2.8	3.8	2.7	3.1
Phe	3.8	3.9	4.3	4.4	4.1	4.0	4.0
His	1.8	1.9	1.8	2.0	2.2	2.1	2.2
Lys	1.8	3.1	2.6	3.3	3.3	2.5	1.4
Arg	2.8	3.7	3.3	5.4	6.4	3.1	2.4
Trp	0.7	0.5	0.7	0.8	0.8	1.0	0.2
Amide group	31.0	24.4	26.1	19.2	15.7	22.8	19.8

[a] Mol amino acid per 100 mol amino acids.

Table 15.11. Designations of *Osborne*-fractions

Fraction	Wheat	Rye	Oats	Barley	Corn	Rice	Millet
Albumins	Leukosin						
Globulins	Edestin		Avenalin				
Prolamins	Gliadin	Secalin	Avenin	Hordein	Zein	Oryzin	Cafirin
Glutelins	Glutenin	Secalinin		Hordenin	Zeanin	Oryzenin	

Table 15.12. Protein distribution (%)[a] in *Osborne*-fractions[b]

Fraction	Wheat	Rye	Barley	Oats	Rice	Millet	Corn
Albumins	14.7	44.4	12.1	20.2	10.8	18.2	4.0
Globulins	7.0	10.2	8.4	11.9	9.7	6.1	2.8
Prolamins	32.6	20.9	25.0	14.0	2.2	33.9	47.9
Glutelins[c]	45.7	24.5	54.5	53.9	77.3	41.8	45.3

[a] Calculated from amino acid analyses.
[b] Ash content of the flours (% based on dry weight), wheat (0.55), rye (0.97), barley (0.96), oats (1.87), rice (1.0), millet (1.10), and corn (0.33).
[c] Protein residue after extraction of prolamins.

oat prolamins is similar to that of the *Triticeae*, whereas amounts of proline and leucine in oat prolamins are lower and higher, respectively, than those found in the *Triticeae*; this is also the case in comparison with rice, millet and corn. The amino acid compositions of rice, millet and corn are not related to the *Pooideae*.

The *Triticeae*, in which the prolamin amino acid compositions are closely related, can also cause *Celiac* disease (cf. 15.1.6). In comparison to other

Table 15.13. Amino acid composition (mole %) of the *Osborne* fractions of various cereals

Amino acid	Albumins						
	Wheat	Rye	Barley	Oats	Rice	Millet	Corn
Asx	9.7	8.8	10.2	10.2	9.9	11.0	16.7
Thr	3.8	4.0	4.7	4.4	4.6	5.0	4.4
Ser	6.2	6.2	6.4	8.9	6.5	6.3	6.2
Glx	20.9	22.1	13.8	12.4	14.2	12.1	12.4
Pro	9.3	12.0	7.4	6.1	4.6	5.1	8.6
Gly	6.9	6.6	9.7	12.6	9.8	10.0	9.7
Ala	6.9	6.5	8.2	7.6	9.4	10.5	10.0
Cys	3.2	2.3	3.8	6.8	1.9	1.4	1.8
Val	6.0	5.2	6.3	4.7	6.3	6.4	4.8
Met	1.6	1.3	2.0	1.2	1.7	2.0	1.1
Ile	3.3	3.4	3.3	2.7	3.7	3.3	3.0
Leu	6.4	6.3	5.9	5.5	6.9	6.5	5.1
Tyr	2.8	2.4	3.4	3.2	3.1	3.0	3.8
Phe	3.1	3.9	2.6	2.7	3.2	3.1	2.0
His	1.8	1.7	1.7	1.6	2.3	2.3	2.1
Lys	3.0	2.9	4.3	4.5	5.0	5.6	3.9
Arg	4.0	3.6	4.5	3.7	6.1	5.7	3.9
Trp	1.1	0.8	1.8	1.2	0.8	0.7	0.5
Amide groups	21.3	23.4	14.0	14.4	11.9	13.4	20.4

Amino acid	Globulins						
	Wheat	Rye	Barley	Oats	Rice	Millet	Corn
Asx	7.7	6.8	8.6	7.9	6.5	7.8	9.1
Thr	4.6	4.6	4.8	4.3	2.9	4.5	5.2
Ser	6.6	6.9	6.5	6.9	7.0	8.1	7.5
Glx	15.2	17.0	12.9	16.0	14.6	12.1	10.7
Pro	6.9	7.8	6.8	5.3	5.6	5.2	5.6
Gly	8.3	8.7	9.5	9.4	10.2	9.3	10.3
Ala	7.5	7.6	8.3	7.4	8.0	9.7	10.7
Cys	3.6	2.1	3.0	2.4	4.1	3.5	3.2
Val	6.8	6.3	6.8	6.5	6.1	6.3	6.2
Met	2.0	1.5	1.4	1.3	4.4	0.9	1.5
Ile	3.8	3.9	3.1	4.1	2.6	3.6	4.1
Leu	7.3	6.9	7.5	7.0	6.2	6.8	6.5
Tyr	2.9	2.3	2.7	2.7	3.7	3.0	2.6
Phe	3.1	3.6	3.3	4.0	2.8	3.3	3.2
His	2.4	2.5	2.2	2.4	2.2	2.9	2.3
Lys	4.0	4.3	4.7	4.4	2.4	4.0	4.6
Arg	6.4	6.5	7.0	7.3	9.8	8.2	6.0
Trp	0.9	0.7	0.9	0.7	0.9	0.8	0.7
Amide groups	13.9	14.6	9.6	14.5	10.4	11.3	11.2

Table 15.13. (continued)

Amino acid	Prolamins						
	Wheat	Rye	Barley	Oats	Rice	Millet	Corn
Asx	2.7	2.4	1.7	2.3	7.3	6.8	4.9
Thr	2.3	2.6	2.1	2.3	2.9	3.8	3.1
Ser	5.9	6.6	4.6	3.8	7.5	6.4	6.9
Glx	37.1	35.4	35.3	34.1	19.6	21.8	19.4
Pro	16.6	18.4	23.0	10.2	5.1	7.8	10.2
Gly	2.9	4.5	2.2	2.7	5.8	1.5	2.6
Ala	2.8	3.0	2.3	5.5	9.1	13.5	13.6
Cys	2.2	2.2	1.9	3.3	0.8	1.1	1.0
Val	4.2	4.4	3.9	7.7	6.9	6.4	4.0
Met	1.1	1.0	0.9	2.1	0.5	1.7	1.1
Ile	4.1	3.0	3.6	3.3	4.6	5.2	3.9
Leu	6.9	5.8	6.1	10.6	11.8	13.4	18.5
Tyr	2.0	1.7	2.3	1.7	6.1	2.1	3.6
Phe	4.6	4.5	5.8	5.3	4.8	4.9	4.9
His	1.7	1.2	1.2	1.1	1.5	1.3	1.1
Lys	0.8	1.0	0.5	1.0	0.5	0.0	0.0
Arg	1.7	1.9	2.0	2.7	4.7	0.8	1.2
Trp	0.4	0.4	0.6	0.3	0.5	1.5	0.0
Amide groups	37.5	34.7	34.9	31.6	23.3	28.6	23.0

Amino acid	Glutelins[a]						
	Wheat	Rye	Barley	Oats	Rice	Millet	Corn
Asx	3.7	7.1	4.9	9.3	9.5	7.6	5.5
Thr	3.6	4.7	4.2	4.2	4.2	5.1	4.2
Ser	7.3	6.9	6.7	6.6	6.7	5.9	6.1
Glx	30.1	19.7	24.2	19.0	15.5	16.8	16.0
Pro	11.9	9.4	14.2	5.5	5.1	8.4	11.1
Gly	7.9	9.2	6.4	7.9	7.4	6.9	6.9
Ala	4.4	7.3	5.6	6.5	7.9	10.1	9.4
Cys	1.4	0.8	0.5	1.2	1.2	1.7	1.8
Val	4.8	5.9	7.2	6.2	7.0	6.6	6.1
Met	1.3	1.6	1.3	1.3	2.4	1.6	2.8
Ile	3.5	3.7	4.0	4.6	4.5	4.1	3.4
Leu	6.9	7.4	7.5	7.8	8.4	9.1	10.9
Tyr	2.4	2.3	1.7	2.8	3.6	2.9	2.9
Phe	3.6	3.8	4.0	4.8	4.3	3.7	3.3
His	1.8	2.0	2.0	2.4	2.1	2.3	3.3
Lys	2.1	4.0	2.8	3.2	3.3	3.1	2.4
Arg	2.7	3.8	2.5	6.0	6.1	3.5	3.2
Trp	0.6	0.4	0.3	0.7	0.8	0.6	0.3
Amide groups	31.0	21.3	23.6	20.2	16.6	17.0	16.4

[a] Protein residue after extraction of prolamins.

cereals, *Triticeae* prolamins contain significantly higher levels of glutamic acid and proline. This suggests that the difference in prolamin composition, induced by these amino acids, may be responsible for *Celiac* disease.

15.2.1.3 Protein Components of Wheat Gluten

Wheat protein fractionation by the *Osborne* method provides prolamins and glutelins in a ratio of 1:1. Both fractions, in hydrated form, have different effects on the rheological characteristics of dough: prolamins are responsible, preferentially, for viscosity, and glutelins for dough strength and elasticity.

The genes for the gluten proteins occur at nine different complex loci in the wheat genome. The high molecular weight glutenin subunits are coded by the loci Glu-A1, Glu-B1 and Glu-D1, which are carried on the long arms of the chromosomes 1A, 1B and 1D. The low molecular weight glutenin subunits, the ω-gliadins and the γ-gliadins are coded by the loci Gli-A1, Gli-B1 and Gli-D1, which occur on the short arms of the chromosomes 1A, 1B and 1D. The α-gliadins are coded by the loci Gli-A2, Gli-B2 and Gli-D2 on the short arms of the group G chromosomes. It is presumed that the variation seen in different varieties is due to the presence of allelic genes at each of the nine storage protein loci. The relative importance of different alleles for gluten quality seems to be Glu-1 > Gli-1 > Gli-2.

A fractionation of gluten proteins is possible by two-dimensional electrophoresis. Figure 15.3 provides a schematic overview of the position of the most important protein groups in a two-dimensional electropherogram. The pattern of glutenins of two wheat cultivars are shown in Fig. 15.4.

Gluten proteins can be separated into their components on an analytical and micropreparative scale by using RP-HPLC. In general, this separation starts with the *Osborne* fractions or subfractions.

In this way, the *prolamines* of wheat can be separated into ω-, α-, γ-gliadins (Fig. 15.5), different varieties of wheat giving different patterns, e. g., the cultivars Clement and Maris Huntsman

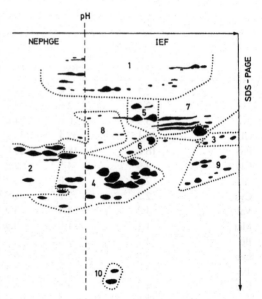

Fig. 15.3. Wheat endosperm proteins (cultivar "*Chinese Spring*"). Simplified schematic representation of a two-dimensional electrophoretic separation.

1st. Dimension: isoelectric focussing (IEF) and non-equilibrium pH gradient electrophoresis (NEPHGE). The electropherograms obtained by both methods are put together at the broken line in such a way that a continuous pH gradient is formed.

2nd. Dimension: polyacrylamide gel electrophoresis in the presence of sodium dodecyl sulfate and mercaptoethanol (SDS-PAGE).

The following protein fractions can be recognized: high-molecular glutenin subunits (1); basic (2) and acidic (3) low-molecular glutenin subunits; γ- (4) and ω-gliadins (5); subunits of the triplet band (6, 10); high-molecular albumins (7); globulins (8) and nonreserve proteins (9). (according to *Payne* et al., 1985)

known to produce sticky dough have a characteristically high ω-gliadin content.

The prolamin patterns of other cereals (Fig. 15.6) differ greatly from that of wheat. In *rye*, the hydrophilic ω-secalins are followed by the hydrophobic γ-secalins. And unlike wheat (α-gliadins), the area of moderate hydrophobicity is not occupied. In *barley*, a hydrophilic fraction is missing: the C-hordeins eluted in the middle area are followed by the hydrophobic B-hordeins. The chromatogram of *oats* is characterized by two hydrophobic fractions that are close to each other. The *low-molecular subunits of wheat glutelins* also give a chromatogram rich in components

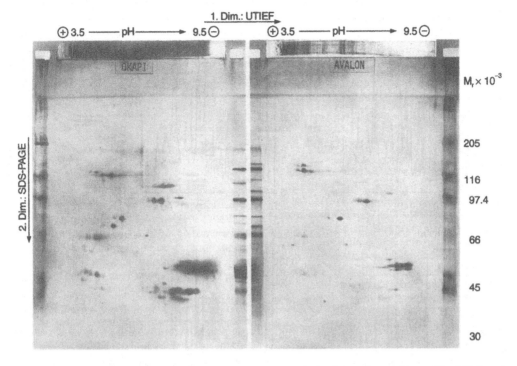

Fig. 15.4. Two-dimensional electrophoretic separation of glutenins[a] of the wheat cultivars *Okapi* (B4) and *Avalon* (A6)[b]. (according to *Krause* et al., 1988)

1st. Dimension: isoelectric focussing in ultrathin (0.25 mm) layer (UTIEF), pH 3.5–9.5; 8 mol/l urea.

2nd. Dimension: polyacrylamide gel electrophoresis in the presence of sodium dodecyl sulfate and mercaptoethanol (SDS-PAGE).

[a] Residue after extraction of defatted flour with water, salt solution, and aqueous ethanol.

[b] The baking quality class is given in brackets after the variety (bread volume yield for A6: average to high, for B4: low to average)

(Fig. 15.7). This chromatogram also contains the ω5-, ω1,2-, and γ-gliadins, which are not separated during pre-fractionation because of varying solubilities (cf. 15.2.1.2).

The *high-molecular subunits of wheat glutelins* show a protein pattern typical of the cultivar (Fig. 15.8).

Based on the data available on the structure of gluten proteins, three main groups can be formed which consist of several subunits. A *high-molecular group* with the HMW subunits of glutenins, a *group of intermediate molecular weights* with the ω5- and ω1,2-gliadins, and a *low-molecular group* with the α- and γ-gliadins as well as the LMW subunits of the glutenins. The properties of the protein groups mentioned are summarized in Table 15.14 and the amino acid composition is given in Table 15.15.

15.2.1.3.1 High-Molecular Group (HMW Subunits of Glutenin)

As shown in Table 15.15, the HMW subunits of glutenin are the only gluten proteins in which Gly (ca. 19%), and not Pro (ca. 12%), takes second place in the order of amino acids after Glx (ca. 36%). Furthermore, the proteins are characterized by the highest content of Tyr (ca. 6%) and Thr (ca. 3.5%) and the lowest content of Phe (ca. 0.3%) and Ile (ca. 0.8%). In the N-terminal amino acid sequence presented in Table 15.16, the sequence EGEAS-RQLQC is valid for all known HMW subunits and varies only in position 6 (E, K, G). From the total sequences known until now, it can be deduced that the HMW subunits consist of three segments (A–C in Table 15.17). The N- and C-terminal segments

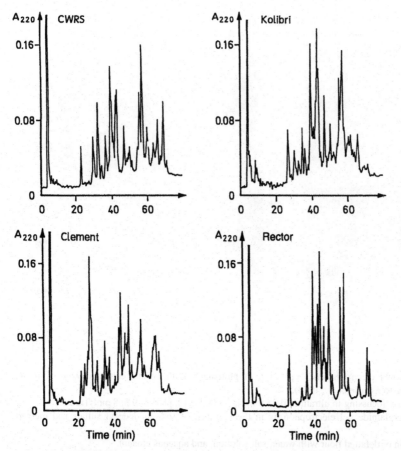

Fig. 15.5. RP-HPLC of the gliadin fractions of various wheat cultivars[a] on Synchro Pac C_{18} (50 °C, aqueous 2-propanol/trifluoroacetic acid/acetonitrile; 22–34 min: ω-gliadins, 33–51 min: α-gliadins, 52–72 min: γ-gliadins; according to *Wieser* et al., 1987)
[a] CWRS (Canadian Western Red Spring) is a mark of origin.

contain no recurring sequences and are characterized by the occurrence of Cys and amino acids with charged side chains. The middle segment consists of recurring sequences with the peptide unit QQPGQG as the backbone and insertions with the sequences YYPTSP, QQG, and QPG. It largely determines the unusual amino acid composition (high Gly and Tyr content). The individual HMW subunits differ mainly in the substitution of individual amino acid residues and in the number and arrangement of recurring peptide units.

The relative molecular masses (M_r) calculated from the known total sequences are 67,000–88,000, while the molecular masses derived from SDS-PAGE are 35–40% higher (Table 15.14).

Based on typical differences in the N-terminals and middle sequence segments, the HMW subunits can be assigned to two subgroups (x-type, M_r = 83,000–88,000; y-type, M_r = 67,000–74,000) (Tables 15.14 and 15.17). These proteins result from the fact that two genes are localized on each of the chromosomes of group 1 (1A, 1B, 1C; cf. 15.1.2). These genes code for the HMW subunits of type x and type y, e. g., ID contains the genes 1Dx and 1Dy. However, not all the genes are expressed in wheat cultivars. The allele pairs 1Dx2 and 1Dy12 as well as 1Dx5 and 1Dy10 are common. The pairs 1Bx6 and 1By8 as well as 1Bx7 and 1By9 are also frequently found. The occurrence of the corresponding HMW subunits 2, 5, 6–10,

Table 15.14. Classification and properties of gluten proteins

| Group | HMW | | MMW | | LMW | | |
| | HMW Subunits | | ω-Gliadins | | | | |
	x-Type	y-Type	ω5	ω1,2	α-Gliadins	γ-Gliadins	LMW Subunits
Mr × 10⁻³ (SDS-PAGE)[a]	104–124	90–102	66–79	55–65	32	38–42	36–44
Mr × 10⁻³ (sequence)[b]	83–88	67–74	44–55[c]	34–44[c]	28–35	31–35	32–39
Number of amino acid residues	770–827	627–684	n.a.	n.a.	262–298	272–308	281–333
Content of gluten proteins	4–9%	3–4%	3–6%	4–7%	28–33%	23–31%	19–25%
Number of cysteine residues	4	7	0	0	6	8	8
μmol Cys/g flour	0.3	0.3	0	0	6.0	6.7	5.0

[a] Result of electrophoresis.
[b] Calculated from the amino acid sequence.
[c] Determined with MALDI-TOF mass spectrometry.
n.a.: not analyzed.

Table 15.15. Amino acid composition[a] of protein groups of wheat gluten (cultivar Rektor)

	HMW subunits of glutenin	ω5-Gliadins	ω1,2-Gliadins	LMW subunits of glutenin	α-Gliadins	γ-Gliadins
Asx	0.7–0.9	0.3–0.5	0.5–1.3	0.7–1.5	2.7–3.3	1.9–4.0
Thr	3.2–3.8	0.4–0.6	0.8–2.3	1.8–2.9	1.5–2.3	1.6–2.4
Ser	6.4–8.4	2.6–3.3	5.8–6.3	7.7–9.5	5.3–6.6	4.9–6.8
Glx	35.9–37.0	55.4–56.0	42.5–44.9	38.0–41.9	35.8–40.4	34.2–39.1
Pro	11.2–12.8	19.7–19.8	24.8–27.4	14.0–16.2	15.0–16.6	15.8–18.4
Gly	18.2–19.8	0.6–0.8	0.9–2.1	2.3–3.2	1.9–3.2	2.0–3.0
Ala	2.9–3.5	0.2–0.3	0.3–1.3	1.7–2.3	2.6–4.1	2.8–3.5
Cys	0.6–1.3	0.0	0.0	1.9–2.6	1.9–2.2	2.2–2.8
Val	1.6–2.7	0.3	0.6–1.4	3.8–5.3	4.2–4.9	4.4–5.4
Met	0.1–0.3	0.0	0.0–0.3	1.2–1.6	0.4–0.9	1.2–1.6
Ile	0.7–1.1	4.3–4.7	1.9–3.5	3.6–4.4	3.6–4.6	4.0–4.6
Leu	3.1–4.3	2.7–3.3	3.9–5.3	5.3–7.5	6.5–8.7	6.4–7.3
Tyr	5.1–6.4	0.6–0.7	0.8–1.5	0.9–1.9	2.3–3.2	0.6–1.4
Phe	0.2–0.5	9.0–9.5	7.6–8.1	3.8–5.5	2.9–3.9	4.7–5.6
His	0.8–1.9	1.3–1.4	0.6–1.1	1.3–1.8	1.4–2.8	1.1–1.5
Lys	0.7–1.1	0.4–0.5	0.3–0.6	0.2–0.6	0.2–0.6	0.4–0.9
Arg	1.6–2.1	0.5–0.6	0.5–1.4	1.5–2.1	1.7–2.9	1.2–2.9

[a] mol % (without Trp).

and 12 in a series of wheat cultivars is shown in Figs. 15.8 and 15.9. The amino acid sequences of the subunits 1Dx5 and 1By9 are shown in Table 15.17.

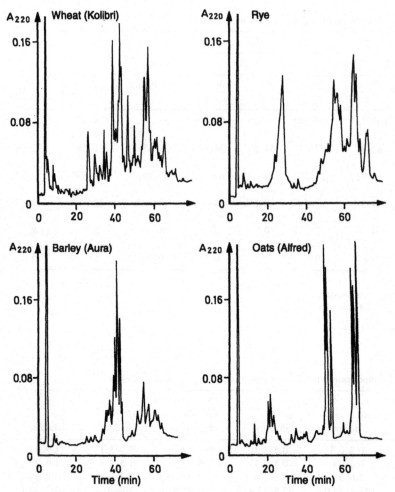

Fig. 15.6. RP-HPLC of the prolamin fractions of different varieties of cereal (conditions as in Fig. 15.5. *Wheat* 26–30 min: ω-gliadins, 32–50 min: α-gliadins, 54–71 min: α-gliadins. *Rye* 21–37 min: γ-secalins, 45–77 min: γ-secalins. *Barley* 32–44 min: C-hordeins, 46–66 min: B-hordeins. *Oats* 49–55 min/62–69 min: avenins; according to *Wieser, Belitz* et al., 1989)

15.2.1.3.2 Intermediate Molecular Weight Group (ω5-Gliadins, ω1,2-Gliadins)

This group of ω-gliadins is characterized by high values of Glx, Pro, and Phe (Table 15.15). The proportion of most of the other amino acids is less than in the other groups and the sulfur-containing amino acids Cys and Met are either absent or present only in traces. Total sequences have not yet been published, but some information is available on partial sequences. The ω-gliadins can be divided into two subgroups, the ω5-type and the ω1,2-type. This nomenclature is based on the varying mobility on acidic PAGE.

The ω5-*gliadins* are characterized by an extremely high content of Glx (ca. 56%) and a relatively high content of Phe (ca. 9%). Although the content of Pro (ca. 20%) is lower than in the ω1,2-type, it is clearly higher than in the other groups. These three amino acids account for about 85% of the total protein. ω5-Gliadins are free of sulfur-containing amino acids and the content of the remaining amino acids is comparatively low. It is noticeable that only this protein type has more Ile (ca. 4.5%) than Leu (ca. 3%).

The N-terminal sequences consist of the sequence SRLLSPRGKELHT and are typical of this type

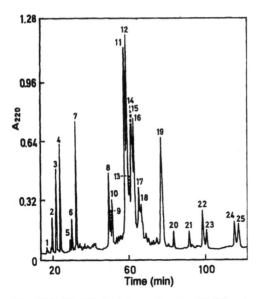

Fig. 15.7. RP-HPLC of low-molecular (LMW) sub-units of wheat glutelin of the cultivar Rektor on Nu-cleosil C_{18} (60 °C, aqueous 2-propanol/trifluoroacetic acid/acetonitrile. After the prolamins, the protein fraction was extracted from the residue with 70% aqueous ethanol/0.5% dithioerythritol at pH 7.6 and 4 °C; peaks 2–4, 5–7: ω5-, ω1,2-gliadins. Peaks 8–17: LMW subunits, peaks 20–25: γ-gliadins or related proteins, according to *Wieser* et al., 1990)

Table 15.16. N-terminal sequences of the protein groups of wheat gluten

	Position	
	5	10
HMW	E G E A S R Q L Q C	
subunits of	E	
glutenin	G	
	K	
ω5-Gliadins	S R L L S P R G K E	
	Q	
ω1,2-Gliadins	– – – – – – – – K E L Q S	
	A R Q L N P S N K E	
LMW-s[a]:	S H I P G L E R P S	
	C	
LMW-m[a]:	M E T S H I P G L E	
	C	
α-Gliadins	V R V P V P Q L Q P	
	F	
		P
γ-Gliadins	N M Q V D P S G Q V	
	I	
		S
	A	

[a] The LMW type is named after the first amino acid in the sequence, serine (s) or methionine (m).

of protein. Recurring sequences with the peptide unit PQQQF evidently start from position 14. This represents a clear difference to the ω1,2-type, which explains the typical differences in the amino acid composition of the two subgroups.

The ω5-gliadins have a higher mobility than the ω1,2-gliadins in acidic PAGE and a lower mobility in SDS-PAGE, M_r being between 44,000 and 55,000 (Table 15.14).

In comparison with the ω5-gliadins, ω1,2-gliadins have lower values for Glx (ca. 43%), Phe (ca. 7.5%) and Ile (ca. 3%) (Table 15.14). Most of the other values are higher, especially the content of Pro (ca. 26%), which is the highest within the gluten groups of proteins. In the case of the N-terminal sequences, two basic variants apparently exist a: ARQLNPSNKELQS; b: KELQS, which with varying length are homologous and lead into a recurring sequence. The variant a was found more frequently in ω2-gliadins and the variant b more frequently in ω1-gliadins. The N-terminal sequence of the

ω5-gliadins corresponds to that of variant a in 6 positions. The immediately following recurring sequence consists of the peptide unit PQQPY, while the dominating recurring peptide unit in this protein type seems to be the sequence PQQPFPQQ.

In acidic PAGE, ω1,2-gliadins have the lowest mobility. The molecular masses are in the range $M_r = 34,000 - 44,000$.

15.2.1.3.3 Low-Molecular Group (α-Gliadins, γ-Gliadins, LMW Subunits of Glutenin)

The quantitatively predominant low-molecular protein group in gluten (Table 15.14) has the best balanced amino acid composition. Most of the values lie between those of the high and intermediate molecular groups. Only the content of Cys, Val, Met, and Leu is higher (Table 15.15).

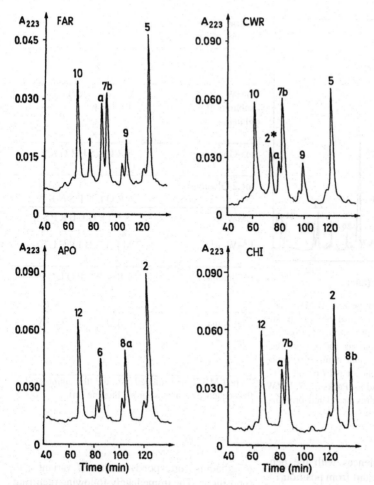

Fig. 15.8. RP-HPLC of the high-molecular (HMW) subunits of the glutelins of different varieties of wheat on Nucleosil C_8 (60 °C, urea/trifluoroacetic acid/acetonitrile/dithioerythritol; numbering of the peaks: 1–7 (x-type), 8–12 (y-type)); (FAR: Farmer, CWR: Canadian Western Red Spring, APO: Apollo, CHI: Chinese spring; according to *Seilmeier* et al., 1991)

A large number of partial and total sequences are found in the literature. With one exception (A-gliadin), the total sequences have all been derived from the corresponding nucleic acids. Based on present data, the low-molecular gluten proteins can be assigned to three subgroups (α-gliadins, γ-gliadins, LMW subunits of glutenin). As shown in Table 15.18 with three examples, the total sequences consist of up to seven differently structured segments: the N- and C-terminal sequence, segment 1–V. The individual proteins differ in the N- and C-terminal sequences, in the recurring sequences (segment I) and in the Gln-rich sequences (segment IV).

On the other hand, they exhibit long homologous sequence segments that are low in Pro. These segments are characterized by the frequent occurrence of amino acids with charged side chains. In addition, with a few exceptions, they contain all the sulfurous amino acids. The M_r of this group of proteins lies in the range of 28,000–39,000 (Table 15.14).

As shown in Table 15.15, the amino acid composition of the α-gliadins differs on the whole only slightly from that of the γ-gliadins and LMW subunits of glutenin. In the case of individual amino acids, however, significant differences are exhibited. The content of Tyr (ca. 3%) is considerably

Table 15.17. Sequence comparison of the HMW subunits of glutenin

a) x-Typ (HMW-subunit 1Dx5)*

N-terminal sequence (A[a])

Pos	Sequence
1	EGEASEQLQ**C**ERELQELQERELKA**C**QQVMDQQLRDISPE**C**
41	HPVVVSPVAGQYEQQIVVPPKGGSFYPGETTPPQQLQQRI
81	FWGIPALLKR

Recurring sequences (B)

Pos			Pos		
91	YYPSVTCP	Q-QVS	448		QQPGQGQQG
103	YYPGQASP	Q-RPGQG	457		QQPGQGQQG
117		Q-QPGQGQQG	466		QQPGQGQPG
126	YYP--TSP	Q-QPGQW	475	YYPTSP	QQSGQG
138		Q-QPEQGQPR	487		QQPGQW
147	YYP--TSP	Q-QSGQL	493		QQPGQGQPG
159		Q-QPAQG	502	YYPTSP	LQPGQGQPG
165		Q-QPGQGQQG	517	YDPTSP	QQPGQG
174		Q-QPGQGQPG	529		QQPGQL
183	YYP--TSS	QLQPGQL	535		QQPAQGQQG
196		Q-QPAQGQQG	544		QQLAQGQQG
205		Q-QPGQAQQG	553		QQPAQVQQG
214		Q-QPGQG	562		QRPAQGQQG
220		Q-QPGQGQQG	571		QQPGQGQQG
229		Q-QPGQG	580		QQLGQGQQG
235		Q-QPGQGQQG	589		QQPGQGQQG
244		Q-QLGQGQQG	598		QQPAQGQQG
253	YYP--TSL	Q-QSGQGQPG	607		QQPGQGQQG
268	YYP--TSL	Q-QLGQGQSG	616		QQPGQGQQG
283	YYP--TSP	Q-QPGQG	625		QQPGQG
295		Q-QPGQL	631		QESGQGQPW
301		Q-QPAQG	640	YYPTSP	QQPGQW
307		Q-QPGQGQQG	652		QQPGQGQPG
316		Q-QPGQGQQG	658		LQLGQGQQG
325		Q-QPGQG	667	YYLTSP	QQPGQGQQG
331		Q-QPGQGQPG	682	YYPTSL	QQPGQW
340	YYP--TS	PQ-QSGQGQPG	694		QQSGQGQHW
355	YYP--TSS	Q-QPTQS	700		QLSGQG
367		Q-QPGQGQQG	709	YYPTSP	QRPGQW
376		Q-QVGQGQQA	721		LQPGQGQQG
385		Q-QPGQG	727		QQPGQG
391		Q-QPGQGQPG	736	YYPTSP	QQLGQW
400	YYP--TSP	Q-QSGQGQPG	748		LQPGQGQQG
415	YYL--TSP	Q-QSGQG	754		QQTGQG
427		Q-QPGQL	763		QQSGQGQQG
433		Q-QSAQGQKG	775	YYPTSL	
442		Q-QPGQG	784	YY	

C-terminal sequence (C)

Pos	Sequence
786	SSYHVSVEHQAASLKVAKAQQLAAQLPAM**C**RLEGGDAL
824	SASQ

Table 15.17. (continued)

*b) γ-Type (HMW subunit 1 By 9)**

N-terminal sequence (A[a])

Pos	Sequence
1	EGEASRQLQ**C**ERELQESSLEA**C**RQVVDQQLAGRLPWSTGL
41	QMR**CC**QLRDVSAK**C**RPVAVSQVVRQYEQTVVPPKGGSFY
81	PGETTPLQQLQQVIFWGTSSQTVQG

Recurring sequence (B)

Left block (positions 106–402):

Pos	Segment
106	YYPSVSSPQQGP
118	YYPGQASP
132	
138	
147	YYP--TSL
162	YYP--SSL
174	
183	YYP--TSL
195	
204	YYP--TSP
216	
222	
228	
234	
240	
249	YYP--TSP
261	
267	
276	YYP--TSQ
291	QYP--ASQ
306	QYP--ASQ
321	QYP--ASQ
336	HYL--ASQ
351	HYP--ASL
366	HYT--ASL
381	HYP--ASL
402	

Right block (positions 408–641):

Pos	Seg 1	Seg 2	Seg 3
408	QQPGQG		
414	QQPGQG	QQTRQG	
420	QQPGKW	QQLEQG	
426	QELGQGQQG	QQTRQG	
432	HQSGQGQQG	QQLEQG	
438	QQPGQG	QQPGQGGQQG	
447	QQIGQGQQG	YYPTSP	QQSGQG
459	QQPGQG		QQPGQS
465	QQIGQGQQG		QQPGQGGQQG
474	QHPGQR	YYSSSL	QQPGQGGLQG
489	QQPGQG	HYPASL	QQPGQG
501	QQIGQG		HPGQR
506	QQLGQG		QQPGQG
512	QQPEQG		QQPEQG
518	RQIGQG		QQPGQGQQG
527	QQSGQGQQG	YYPTSP	QQPGQG
539	QQLGQG		KQLGQGQQG
548	QQPGQW		QQPGQG
560	QQSGQGQQG	YPTSPQ	QQPGQGGQQG
569	QQPGQGQQG	H**C**PTSP	QQTGQA
581	QQPGQGQQG		QQPGQG
587	QQPGQGQQG		QQIGQV
593	QQPGQGQQR		QQPGQGGQQG
602	QQPGQGQQG	YYPISL	QQSGQG
614	QQPGQGQQG		QQSGQG
620	QQVGQG		QQSGQG
626	QQIGQLGQR		HQLGQG
632	QQPGQG		QQSGQEQQG
641	YD		

C-terminal sequence (C)

Pos	Sequence
643	NPYHVNTEQQTASPKVAKVQQPATQLPIM**C**RMEGGDAL
681	SASQ

[a] The capital letters mark sequence segments (cf. Fig. 15.12).

The numbers give the positions that the amino acids at the beginning of the line occupy in the total sequence. The segments of recurring sequences are arranged according to the best possible homology. (–:space to maximize homology) Cysteine residues (**C**) are marked in bold type.

* The numbering of the HMW subunits (5 and 9) corresponds to Figs. 15.8 and 15.9.

Table 15.18. Sequence comparison[a] of an α-gliadin (clone 1235), γ-gliadin (clone genesA), and a LMW subunit (clone LMWG-1D1)

N-terminal sequences

α-	1	VRVPVPQLQPQNPSQQQPQEQVPLMQQQQQFPG
γ-	1	NMQVDPFGQVQWP-QQQPVLL
LMW	1	R**C**IPGLERPW

Recurring sequences 1[b]

	α-Gliadin		γ-Gliadin		LMW
34	QQEQ-FP-PQQPYP	21	PQQPFSQ	11	QQQPLPP
46	HQQP-FP-SQQPYP	28	QPQQTFPQ	18	QQT-FP
58	QPQP-FP-PQLPYP	36	PQQTFPH	23	QQPLFS
70	QTQP-FP-PQQPYP	43	QPQQQFPQ	29	QQQQQQL-FP
82	QPQPQYPQPQQPIS	51	PQPQQQFLQ	38	QQPSFS
		61	PQQPFPQ	44	QQQPPFW
		68	QPQQPYPQ	51	QQQPPFS
		76	QPQQPFPQ	58	QQQPILP
		84	TQQPQQLFPQ	65	QQPPFS
		94	SQQPQQPYPQ	71	QQQQLVLP
		104	QPQQQPYPQ	79	QQPPFS
		112	TQQPQQQFPQ	84	QQQQPVLP
		122	SQQPQ-PFPQ	93	PQQSP-FP
		131	PQQPQQSFPQ	100	QQQQQH
		141	QQPS	106	QQLV
				110	QQQ-1-P

Poly-Gln sequences (II)

α-	96	QQQAQQQQQQQQ

Sequences low in Pro III

α-	108	TLQQILQQQLIP**C**RDVVLQQHNIAHASSQVL----QQSSY
γ-	145	FIQPSLQQQLNP**C**KNLLLQQ**C**RPVSLVSSLW-SMIWPQSA**C**
LMW	115	VVQPSILQQLNP**C**KVFLQQQ**C**SPVAMPQRLARSQMLQQSS**C**
α-		QQLQQL**CC**QQLFQIPEQSR**C**QAIHNVVHAIIL
γ-		QVMRQQ**CC**QQLAQIPQQLQ**C**AAIHSVVHSISM
LMW		HVMQQQ**CC**QQLPQIPQQSRYEAIRAIIYSIIL

Sequences high in Gln IV

α-	176	HHHQQQQQQQPSSQVSYQQPQEQYPSGQVSFQSSQQN
γ-	216	QEQQQQQQQQQQQQQQQQQGMRILLPLYQQQQVGQGTL
LMW	188	QEQQQVQGSIQSQQQQPQQLGQCVSQPQQSQQQLGQQPQQQQ

Sequences low in Pro V

α-	212	PQAQGSVQPQQLPQFQEIRNLALQTLPAM**C**NVYIPPY**C**STTIAPFG
γ-	253	VQGQGIIQPQQPAQLEAIRSLVLQTLPTM**C**NVYVPPE**C**SIIKAPFA
LMW	231	LAQGTFLQPHQIAQLEVMTSIALRILPTM**C**SVNVPLYRTTTSVPFG

C-terminal sequences

α-	258	IFGTN
γ-	299	SIVTGIGGQ
LMW	277	VGTGVGAY

[a] The segments of the recurring and low-proline sequences are arranged according to the best possible homology (-: space to maximize homology). Cysteine residues (**C**) are in bold type.

[b] Roman numerals: division into sequence segments (cf. Fig. 15.10)

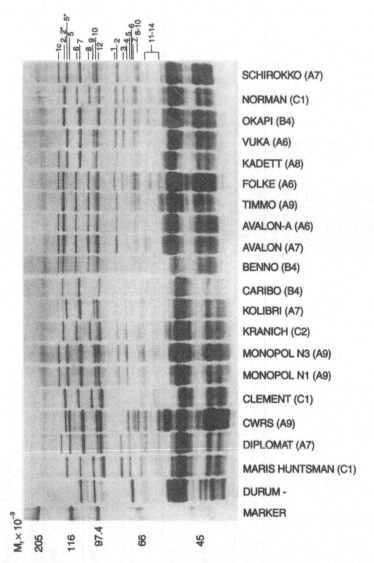

Fig. 15.9. Electrophoretic separation of the glutenins[a,b] of various wheat cultivars[c] on polyacrylamide gel in the presence of sodium dodecyl sulfate and mercaptoethanol (SDS-PAGE) (according to *Krause* et al., 1988)

---

[a] Residue after the extraction of defatted flour with water, salt solution, and aqueous ethanol.

[b] The numbering of the HMW subunits of glutenin differs from that in the original publication, Payne et al., 1981b: Band 1–7: (x-type), 8–12 (y-type).

[c] After each variety, the baking quality class is given in brackets (bread volume yield very high: A9, high to very high: A8, high: A7, average to high: A6, average: B5, low to average: B4, low: B3, very low to low: C2, very low: Cl).

higher and the content of Met (ca. 0.7%) and Phe (ca. 3.4%) is lower.

The N-terminal sequences determined directly by *Edman* degradation correspond to those derived from the nucleic acids. Apart from a few variations, the individual amino acid residues VRVPVPQLQPQN have been found for these N-terminal sequences (Table 15.16). The recurring sequences consist of the peptide unit QPQPFPPQQPYP, which usually occurs five times and varies in individual amino acids (Table 15.18). The balanced Tyr/Phe ratio of the α-gliadins is based on this domain. Deviating from the γ-gliadins and the LMW subunits of glutenin, α-gliadin contains a poly-Gln sequence between the recurring and low-Pro sequence segments.

In comparison with the α-gliadins, the *γ-gliadins* exhibit higher values for Phe (ca. 5%) and Met (ca. 1.4%) and lower values for Tyr (ca. 1%) (Table 15.15).

The most common N-terminal sequence that is directly determined or derived from the nucleic acids is NMQVDPSGQV. Individual positions are modified, e.g., position 2 with I (Table 15.16). The recurring sequences consist of the peptide units PQQPFPQ, in which Q, TQQ, LQQ or PQQ can be inserted. There are up to 15 repetitions of such peptide units which can vary in individual residues (Table 15.18). The absence of Tyr in the recurring sequence segments shifts the Tyr/Phe ratio to ca. 1.5 (Table 15.15).

The *LMW subunits of glutenin* differ from the α- and γ-gliadins by having higher values for Ser (ca. 9%) and lower values for Ala (ca. 2%) and Asx (ca. 1%). The values for the other amino acids coincide (Table 15.15).

The N-terminal sequences of the LMW subunits were found to be SHIPGL or SCISGL (s-type) and METSCI or METSHI (m-type). The known total sequences show that the LMW subunits of glutenin have typical N-terminal, C-terminal, Gln-rich and recurring sequence segments (Table 15.18). The remaining sequence segments correspond largely to those of α- and γ-gliadins. The recurring peptide units usually consist of the sequence Q_nPPFS with $n = 2-10$. These units are repeated up to 20 times and the hydrophobic tripeptide PPF is partly varied (e.g., by PVL, PLP). In comparison with the α- and γ-gliadins, the high Ser content in the total composition is due to the recurring sequences.

15.2.1.4 Structure of Wheat Gluten

15.2.1.4.1 Disulfide Bonds

α-Gliadin and γ-gliadin are mainly monomeric proteins which, consequently, contain only intramolecular disulfide bonds. In comparison, the glutenins are protein aggregates of HMW and LMW subunits with molar masses from ca. 200,000 to a few million, which are stabilized by intermolecular disulfide bonds, hydrophobic interactions and other forces.

It has been possible to elucidate the position and type of the disulfide bonds so that the structure of the gluten proteins is discernible. First, α-gliadin, γ-gliadin and the LMW subunits will be discussed based on two complementary schemes. Figure 15.10 shows the cysteine residues involved in intra- and intermolecular disulfide bridges. Figure 15.11 shows the extent of the relationship in the disulfide structures.

In Fig. 15.10 and Fig. 15.12, the cysteine residues (C) at the N-terminal of the sequence are marked with the first letters of the alphabet and those at the C-terminal with the last letters of the alphabet. Homologous cysteine residues have the same letters. In α- and γ-gliadins, disulfide bonds are found in segments III–V; in LMW, also in segment I (Fig. 15.10). In γ-gliadin, four intramolecular disulfide bridges are concentrated in a relatively small sequence section so that a compact structural element is formed from the three small rings A, B, C and one large ring D (Fig. 15.11a). The disulfide structure of α-gliadin is related to that of γ-gliadin. Since the disulfide bond C^d/C^e is lacking (Fig. 15.10), the small rings A and B open to give a larger ring AB (Fig. 15.11b). LMW subunits do contain the small rings A and B, but since the disulfide bond C^w/C^z is lacking (Fig. 15.10), ring D expands to give ring CD (Fig. 15.11c).

For steric reasons, the cysteine residues C^{b*} and C^x in the LMW subunits cannot form intramolecular disulfide bonds, but are available for intermolecular disulfide bridges, preferably with other LMW and HMW subunits (Fig. 15.10).

HMW subunits of the x-type contain four and those of the y-type contain seven cysteine residues (Table 15.17). Except for the residue C^y in the y-type, all the residues are in the segments A and C (Fig. 15.12). In the x-type, the residues C^a and C^b form an intramolecular disulfide bridge (Fig. 15.12) and C^d and C^z are available for intermolecular bonding. The y-type contains five cysteine residues in segment A and one in each of the segments B and C (Fig. 15.12). Until now, intermolecular disulfide bonds to other

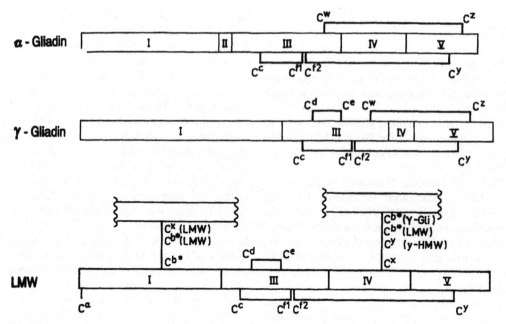

Fig. 15.10. Schematic representation of the disulfide structures of α-gliadins, γ-gliadins and LMW subunits (according to *Köhler* et al., 1993). Segments I–V (cf. Table 15.18)

HMW subunits of the y-type as well as to LMW subunits could be detected (Fig. 15.12).

It is noticeable that small amounts of α-, γ- and ω-gliadins are not extractable from flour with aqueous alcohol, but remain with the glutenins. It is assumed that these proteins contain an odd number of cysteine residues due to point mutations, one residue being available for intermolecular disulfide bonding. In fact, it has been observed that LMW subunits are linked to γ-gliadins which have 9 instead of 8 cysteine residues, via a bridge from C^x to C^{b*} (Fig. 15.10).

15.2.1.4.2 Contribution of Gluten Proteins to the Baking Quality

Investigations of the structure and amount of gluten proteins in wheat cultivars with varying dough and baking properties allow an estimation of the contributions of individual gluten proteins to quality. An important feature is the suitability for forming, as far a possible, high molecular protein aggregates. The x-type of the HMW subunits appears to be especially predestined for this because, e. g., it can form linear polymers via

the cysteine residues C^d and C^z (Fig. 15.12). This is expressed in the close relationship (correlation coefficient r > 0.8) between its amount and the strength of the dough (Fig. 15.13).

The considerably lower coefficient in the case of y-type (r < 0.3; Fig. 15.13) indicates that cross linkage via C^{c1} and C^{c2} or of C^y with LMW subunits (Fig. 15.12) does not have an especially positive effect on the consistency of the dough.

Apart from the HMW subunits of the x-type, the LMW subunits also make a positive contribution to the strength of dough and gluten (r = 0.58−0.85). It is to be assumed that the tendency of C^{b*} and C^x to polymerize is responsible for this effect (Fig. 15.10). However, about twice the amount of LMW subunits in flour is required for the same effect. The reason for this could be that the bonds of C^{b*} and C^x are not firmly directed but variable. Thus, C^x also binds to γ-gliadin with an odd number of cysteine residues (cf. 15.2.1.4.1) which would lead to chain breakage in the polymerization of gluten proteins, which possibly occurs during dough making.

Monomeric gliadins (cysteine-free ω-gliadins, α- and γ-gliadins with an even number of cysteine residues) are regarded as "solvent" or

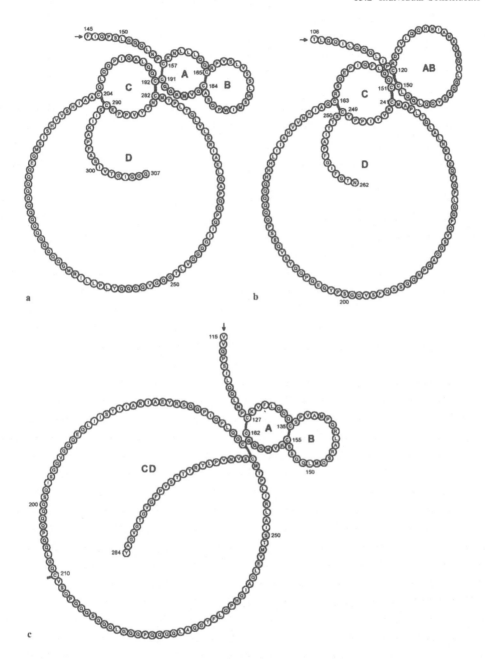

Fig. 15.11. Schematic two-dimensional structures of the C-terminal segments of γ-gliadin (**a**), α-gliadin (**b**) and LMW subunits (**c**) (according to *Müller* and *Wieser*, 1997)

"lubricant" for aggregated glutenins and made responsible for the viscosity of dough and gluten. Accordingly, it is not the absolute amount of gliadins that is correlated with the rheological properties of dough and gluten, but their quan- titative ratio to the glutenins (Fig. 15.14). The homologous arrangement of the intramolecular disulfide bonds of α- and γ-gliadins is reflected in their similar contributions to the rheological properties of dough and gluten.

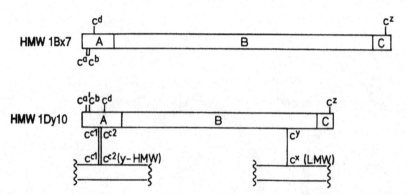

Fig. 15.12. Schematic representation of the disulfide structures of HMW subunits of the x-type and y-type (according to *Köhler* et al., 1993). Segments A–C (cf. Table 15.17). Nomenclature of the cysteine residues as in Fig. 15.10.

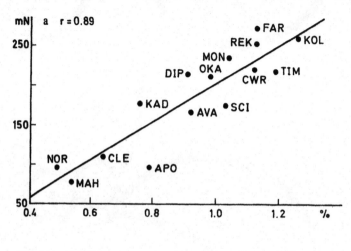

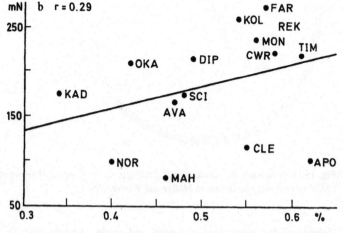

Fig. 15.13. Correlation between the maximal resistance to stretching of dough from various wheat cultivars in a micro-scale extension test and the concentration of HMW subunits (% based on flour) of x- (**a**) and y-type (**b**). (According to *Wieser* et al., 1992)

In summary, in the dough preparation and gluten formation phase, the competing processes of chain formation and termination are substantial factors for the properties of dough and significantly depend on the type of disulfide bonds. For high dough and gluten strength, a sufficient

F(N)

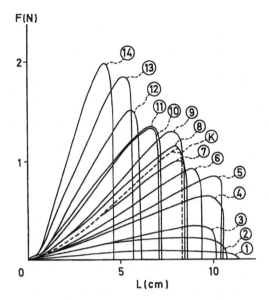

Fig. 15.14. Tensile tests of glutens with varying content of gliadin (gluten K from retail wheat flour was extracted with 70% aqueous ethanol. The extracted gliadin and the remaining glutenin were freeze dried, remixed in different proportions, and then hydrated. Gliadin content of the glutens: K) 33.9%, 1) 55.9%, 14) 22.6%; the gliadin contents of the other samples are in between, according to *Kim* et al., 1988)

amount of polymerizable gluten proteins (HMW subunits of the x-type, LMW subunits) is required with the lowest possible amounts of terminators (low molecular thiol compounds, gliadins with an odd number of cysteine residues, possibly also HMW subunits of the y-type).

15.2.1.5 Puroindolins

The wheat endosperm contains two basic, cysteine-rich proteins, puroindolin a and b (PIN-a and -b). The name is derived from the presence of tryprophan-rich segments in the amino aicd sequences: Trp-Arg-Trp-Trp-Lys-Trp-Trp-Lys in PIN-a and Trp-Pro-Thr-Trp-Trp-Lys in PIN-b. PIN-a consists of a peptide chain with 115 residues (M_r 12,479) and five disulfide bridges. The peptide chains of PIN-a and PIN-b are homologous to an extent of 60%. It has been shown that the puroindolins are identical to the basic friabilins which have been discovered on the surface of starch granules. PIN-a and -b are positively charged and bind negatively charged phospholipids with a high affinity. PIN-a also forms stable complexes with glycolipids while PIN-b is less suitable. It is assumed that the indolyl residues of the tryptophan-rich segments are involved in the stabilization of the complexes, hydrogen bridges being formed between the indole-NH and the OH-groups of the glycolipids. Thus, the higher stability of PIN-a compared with PIN-b complexes is based on the longer tryprophan-rich segment.

Foams of PIN-a and, to a smaller extent, of PIN-b are stabilized especially in the presence of polar wheat lipids. In this respect, PIN-a is clearly superior to egg white proteins as shown by the following comparison. After a drip off time of 5 min, a foam density of 0.028 was obtained with 0.3 mg of PIN-a/ml, while 1.25 mg/ml of egg white proteins were required for this purpose.

For the baking process, it is expected that the puroindolins protect the foam-like texture of the dough from destabilization by lipids.

15.2.2 Enzymes

Of the enzymes present in cereal kernels, those which play a role in processing or are involved in the reactions which are decisive for the quality of a cereal product will be described.

15.2.2.1 Amylases

α- and β-amylases (for their reactions, see 4.4.5.1) are present in all cereals. Wheat and rye amylases are of particular interest; their optimum activities are desirable in dough making in the presence of yeast (cf. 15.4.1.4.8). In mature kernels, α-amylase activity is minimal, while it increases abruptly during sprouting or germination. Unlike the situation with wheat, dormancy in rye is not very pronounced. Unfavorable harvest conditions (high moisture and temperature) favor premature germination ("*sprouting*"), not visible externally. During this time, α-amylase activity rises, resulting in extensive starch degradation during the baking process. Bread faults appear, as mentioned under 15.4.1.2.

Table 15.19. Amylases in wheat

Properties	α-Amylase I	α-Amylase II	β-Amylase
pH optimum	3.6–5.75	5.5–5.7	5.4–6.2
Molar mass	37,000[a]	21,000[a]	64,200[b]
Isoelectric point	4.65–5.11	6.05–6.20	4.1–4.9

[a] gel chromatography, [b] ultracentrifugation.

Two α-amylases, α-AI and α-AII, have been isolated from wheat by affinity chromatography and chromatofocussing. These two enzymes produce a series of multiple forms on SDS-PAGE electrophoresis. The ratio of the concentrations of the two α-amylases depends on the stage of development. After flowering, α-AI appears first in the outer layers of the kernel, then decreases with increasing ripeness. Low activities of α-AII are detectable even before dormancy, but they greatly increase during germination. The two α-amylases differ in their pH optimum, molar mass, and isoelectric point (Table 15.19). α-AII is more temperature resistant.

The pH optimum of α-amylase in germinating rye lies in a range similar to that of α-AII of wheat. Therefore, α-amylase is partially inhibited by the decrease in pH in sour dough (cf. 15.4.2.2).

The properties of wheat β-amylase are shown in Table 15.19.

15.2.2.2 Proteinases

Acid endopeptidases with pH optima of 4–5 occur in wheat, rye and barley. Their substrate specificity has been determined. The possibility that the wheat proteinases are involved in cleavage of gluten bonds, thereby affecting softening or mellowing of gluten during baking, is still disputed.

15.2.2.3 Lipases

These enzymes occur in various concentrations in all cereals. Carboxylester hydrolase, readily isolated from wheat germ, is not considered a lipase but an esterase (cf. 3.7.1.1). The activity in dormant seeds is low, but increases greatly on germination and can be detected with great sensitivity with a fluorochrome substrate, e. g., fluorescein dibutyrate. Therefore, this forms the basis of a method for the quick detection of "sprouting" in wheat and rye.

In addition to the esterase, a wheat lipase occurs enriched in the bran. A rise in free fatty acids observable during flour storage also involves lipases from metabolism of microorganisms present in flour.

In comparison to other cereals, oats contain a significant level of lipase. Its high activity is released once the oat kernel is disintegrated, crushed or squeezed. Linoleic acid is released from the acyl lipids that are present. It is then converted into hydroxy fatty acids by lipoxygenase and hydroperoxidase enzymes, giving rise to off-flavors (Fig. 15.11). All these enzymes are inactivated by heat treatment and thus quality deterioration can be avoided (cf. 15.3.2.2.2).

It should be taken into account in lipid extraction that the phospholipase D activities are relatively high in ripe cereals and this enzyme transfers the phosphatidyl residue of phospholipids to alcohols, which are used to extract lipids (cf. 3.7.1.2.1). The enzyme is inactivated during extraction with boiling water-saturated butanol. A phospholipase that hydrolyzes both acyl residues in the lecithin molecule ("phospholipase B") has been found in germinating cereal. It influences the foam stability in beer (cf. 20.1.7.9).

In the production and storage of egg dough products, phopholipases B and D can lower the phospholipid content.

15.2.2.4 Phytase

Cereals contain about 1% of phytate [myoinositol (1,2,3,4,5,6) hexakisphosphate], which binds about 70% of the phosphorus in the grain. Since it occurs mainly in the aleurone layer, the content of phytate in flour depends on the extent of grinding (Table 15.20). A part of it is hydrolyzed in stages to myo-inositol during dough making.

The phytases originate in cereals (Table 15.21), but are also synthesized by microorganisms, e. g., yeast. Therefore, if the baking process

Table 15.20. Phytate content in wheat flour

Degree of grinding	Phytate $(mg/kg)^a$
70%	53
85%	451
92%	759

[a] Based on solids.

Table 15.21. Phytase activity and phytate content in cereals

Type of cereal	Phytase activity[a]	Phytate content[b]
Wheat	180	12.4
Triticale	650	12.9
Rye	2800	11.8
Barley	350	11.9
Oats	48	11.3
Corn	9	9.2

[a] Activity: units/g cereal.
[b] Content: mg/g cereal.

takes 1 hour, 85–90% of the phytate is degraded in white bread made of flour with 1.2 g of phytate/kg. In rye whole grain bread (10 g of phytate/kg), 25–35% is degraded and 40–50% if the baking process is extended to 4 hours.

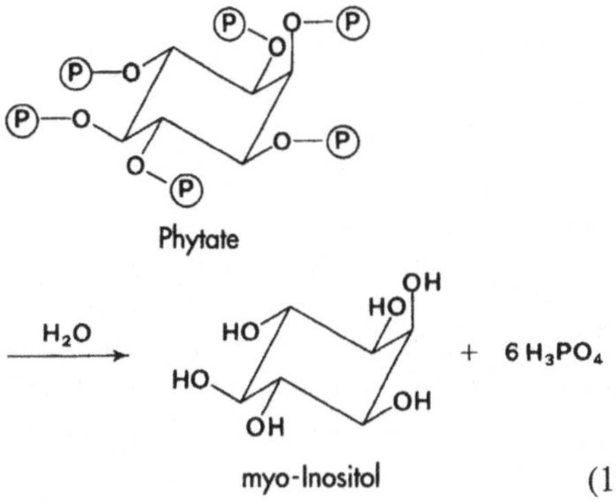

Phytate

$$\xrightarrow{H_2O} \text{myo-Inositol} + 6\,H_3PO_4 \qquad (15.1)$$

Partial hydolysis of phytate to myo-inositoltetrakis- and -triphosphate is desirable from a nutritional physiological point of view. In comparison with phytate, these less phosphory-lated myo-inositols do not form such stable complexes with cations. Consequently, the absorption of zinc, iron, calcium and magnesium ions is not

impeded. On the other hand, they still possess the positive nutritive properties of phytate.

15.2.2.5 Lipoxygenases

Cereals contain lipoxygenases (cf. 3.7.2.2) which, with the exception of the enzyme from rye, oxidize linoleic acid preferentially to 9-hydroperoxy acids. The rye lipoxygenase forms mainly the 13-hydroperoxide isomer. Though the enzyme from wheat belongs to the specifically reacting LOX (cf. Table 3.33) and thus cooxidizes carotenoids at a slow rate, it can still bring about a loss of yellow color in pasta products. This is the reason for the inactivation of wheat lipoxygenase during the preparation of pasta products (cf. 15.5).

The involvement of endogenous lipoxygenase in the baking of wheat flour is not clear. However, by addition of lipoxygenase-active soy flour, a significant improvement of the flour quality is achieved (cf. 15.4.1.4.3).

As shown in Fig. 15.11, oats contain a lipoxygenase with lipoperoxidase activity. This activity reduces the hydroperoxides initially formed, in the presence of phenolic compounds as H-donors, to the corresponding hydroxy acids:

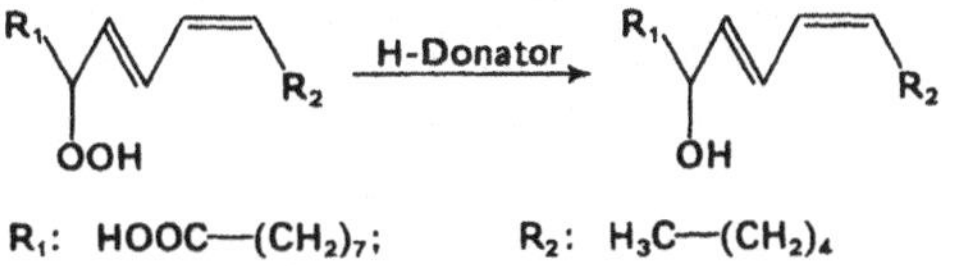

R_1: $HOOC\!-\!(CH_2)_7$; R_2: $H_3C\!-\!(CH_2)_4$

$$(15.2)$$

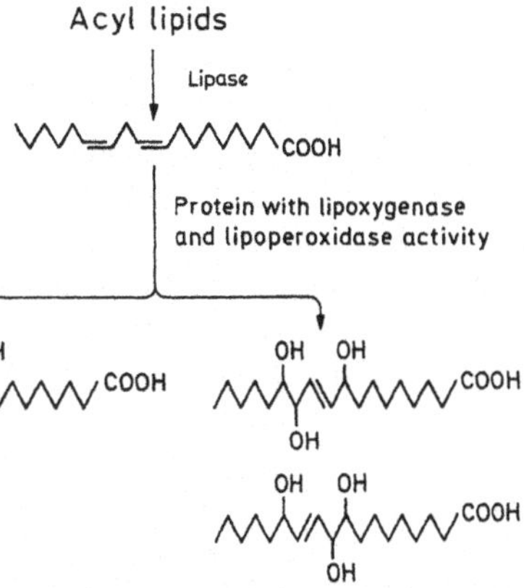

Fig. 15.15. Formation of bitter tasting compounds in oats (taste threshold values cf. 3.7.2.4.1)

15.2.2.6 Peroxidase, Catalase

Both enzymes are widely distributed among cereals. The pH-activity curves of the enzymes from wheat show that at the normal pH values of a dough, about 6.3, catalase still has 40–50% and peroxidase less than 10% of its activity at the pH optimum (peroxidase pH 4.5; catalase pH 7.5). Therefore, it is unlikely that the oxidative cross-linkage of pentosans (Fig. 15.19), which is catalyzed by peroxidase, plays an essential role in dough.

As heme catalysts they accelerate the nonenzymatic oxidation of ascorbic acid to the dehydro form. The involvement of both enzymes in the action of ascorbic acid as an improver will be discussed (cf. 15.4.1.4.1).

15.2.2.7 Glutathione Dehydrogenase

This enzyme catalyzes the oxidation of glutathione (GSH) in the presence of dehydroascorbic acid as an H-acceptor:

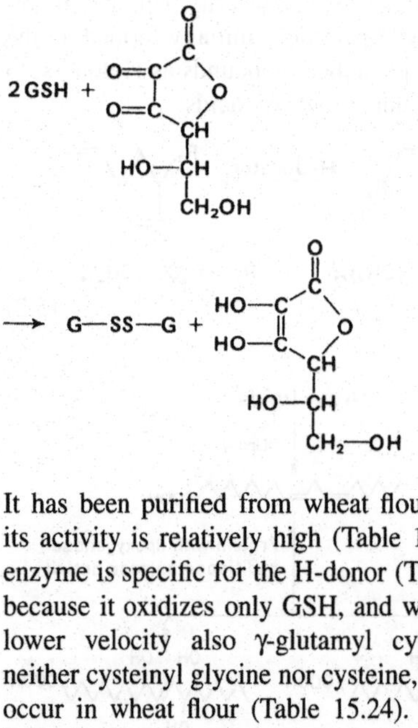

$$(15.3)$$

It has been purified from wheat flour in which its activity is relatively high (Table 15.22). The enzyme is specific for the H-donor (Table 15.23) because it oxidizes only GSH, and with a much lower velocity also γ-glutamyl cysteine, but neither cysteinyl glycine nor cysteine, which also occur in wheat flour (Table 15.24). The specificity for the H-acceptor is not so pronounced. As shown in Table 15.23, all four diastereomeric forms of dehydroascorbic acid are converted, but with different velocities. The substrate specificity corresponds to the varying activity of

Table 15.22. Activity of glutathione dehydrogenase (GSH-DH) in wheat flour

Wheat cultivar	GSH-DH[a]
Kranich	17.3
Kolibri	13.2
Benno	15.4
Mephisto	16.1
Diplomat	13.2
Jubilar	16.1
Caribo	12.5

[a] Activity at pH 6.5 (25 °C): μmol of L-*threo*-ascorbic acid per minute and g of flour.

the diastereomeric dehydroascorbic acids in flour improvement (cf. 15.4.1.4.1).

15.2.2.8 Polyphenoloxidases

In cereals, polyphenoloxidases preferably occur in the outer layers of the kernels. Wheat enzymes that exhibited cresol activity only (cf. 2.3.3.2) and were known as tyrosinases have been separated from polyphenoloxidases by chromatography and preparative gel electrophoresis.

Polyphenoloxidases can cause browning in whole-meal flours.

15.2.2.9 Ascorbic Acid Oxidase

An ascorbic acid oxidase (AO) occurs in wheat flour (Table 15.25), which oxidizes L-*threo*- and D-*erythro*-ascorbic acid at comparable rates. In addition, a substance has been found in flour extracts which oxidizes L-*threo*-ascorbic acid at pH 10 at a maximal rate. In comparison with AO, this activity does not decrease on incubation with proteases nor is it inhibited by the addition of the AO inhibitors KCN and NaF. It obviously catalyzes a nonenzymatic oxidation of ascorbic acid.

15.2.2.10 Arabinoxylan Hydrolases

In aqueous extracts of wheat flour, arabinoxylan hydrolases have been detected with the synthetic

Table 15.23. Substrate specificity of glutathione dehydrogenase (GSH-DH) from wheat

Substrate	Relative activity (%)	Kinetic constants	
		V_M (nkat/ml)	K_M(mmol/l)
H-Donor			
Glutathione (GSH)	100	362[a]	1.8
Cysteine	0	0	n.a.
Cysteinyl glycine (Cys-Gly)	0	1.3[b]	n.a.
γ-Glutamyl cysteine (Glu-Cys)	n.a.	37[a]	5.5
H-Acceptor[c]			
Dehydroascorbic acid (DHAsc)[d]			
L-*threo*	100	275[e]	0.14
L-*erythro*	67	n.a.	n.a.
D-*erythro*	60	305	1.2
D-*threo*	16	n.a.	n.a.

[a] Concentration of L-*threo*-DHA: 0.5 mol/l.
[b] Reaction system: L-*threo*-DHA 0.5 mmol/l, Cys-Gly 34 mmol/l.
[c] Reaction system: H-acceptor 0.29 mmol/l, GSH 0.5 mmol/l.
[d] Structures of the corresponding ascorbic acid diastereomers, cf. 15.4.1.4.1.
[e] Concentration of GSH: 3 mmol/l.
n.a.: not analyzed.

Table 15.24. Occurrence of low-molecular thiols in wheat flour[a]

Thiol	Concentration (nmol/g flour)
Glutathione (GSH)	100
Glu-Cys	17
Cys-Gly	5
Cysteine	13

[a] Origin: DNS (ash 0.78%).

Table 15.25. Activity of ascorbic acid oxidase (AO) in wheat flour

Wheat cultivar	AO[a]
Domino	60
Otane	40
Norseman	39
Amethyst	30
Sapphire	21
Brock	18

[a] Activity at pH 6.2 (25 °C): nmol L-*threo*-ascorbic acid per minute and g of flour.

substrates p-nitrophenyl-β-D-xylopyranoside and p-nitrophenyl-α-L-arabinofuranoside. A water soluble arabinoxylan gave arabinose and xylose on incubation as the main products and xylobiose and xylotetraose as the side products. The results show that low activities of arabinofuranosidase, xylosidase and endo-xylanase are found in wheat flour.

15.2.3 Other Nitrogen Compounds

Wheat contains glutathione and cysteine in the free state as thiol compounds (GSH, CSH), in the oxidized forms (GSSG, CSSC) and in the protein-bound forms (GSSP and CSSP) (Table 15.26). Reduction of GSSP and CSSP releases GSH and CSH respectively, e. g., with dithioerythritol.

It has been shown that glutathione is predominantly localized in the germ and in the aleurone layer. Therefore its concentration in flour increases as the extraction grade increases (Table 15.27).

During dough making, GSH reacts very quickly undergoing disulfide interchange with the flour proteins PSSP:

$$GSH + PSSP \rightleftharpoons PSSG + PSH \qquad (15.4)$$

If high-molecular gluten proteins are cleaved, the viscosity of the dough drops. Rheological meas-

Table 15.26. Reduced (GSH), oxidized (GSSG) and protein-bound glutathione (PSSG), total glutathione (GSS) as well as cysteine (CSH) and CSS[a]

Wheat cultivar	Ash (wt. %)	Concentration (nmol/g flour)					
		GSH	GSSG	PSSG	GSS	CSH	CSS
DNS	0.78	100	n.a.	n.a.	279	13	159
Maris Huntsman	0.68	81	n.a.	n.a.	232	9	145
Kanzler	0.62	35	n.a.	n.a.	180	8	118
Fresco[b]	n.a.	31	24	131	210	n.a.	n.a.
Norman[b]	n.a.	74	15	73	177	n.a.	n.a.
Mercia[b]	n.a.	74	27	102	230	n.a.	n.a.
Haven[b]	n.a.	18	20	89	147	n.a.	n.a.

[a] CSS consists of free cysteine, cystine and cysteine, which is linked with wheat proteins only via disulfide bridges but not via peptide bonds.
[b] Degree of grinding 64–68%.
n.a.: not analyzed.

Table 15.27. Glutathione concentration as a function of the degree of grinding

Wheat cultivar	Ash (w/w%)	Glutathione[a]	
		GSH	Total[b]
CWRS[c]	0.54	16	172
	0.71	35	348
	1.44	60	575
DNS[c]	0.59	41	175
	0.78	110	345
	1.57	215	657
Maris Huntsman	0.55	20	185
	0.68	94	273
	1.73	210	435

[a] Calculated as GSH in nmol/g based on the dry weight.
[b] Sum of GSH, GSSG, and GSSP.
[c] Marks of origin: Canadian Western Red Spring (CWRS), Dark Northern Spring (DNS).

urements of flour/water dough show the effect of GSH (Fig. 15.16). At 124 nmol/g, the concentration of GSH in the sample of flour is relatively high. The strength of the dough decreases with the addition of 100 nmol/g GSH. GSSG is also active but not as much as GSH (Fig. 15.16) because it first has to be reduced to GSH, e. g., by proteins with free SH groups (PSH, cf. Formula 15.5), before it can depolymerize aggregated gluten proteins as shown in Formula 15.4.

$$GSSG + PSH \rightleftharpoons GSSP + GSH \qquad (15.5)$$

Cysteine is also rheologically active and cystine after disulfide exchange, e. g., with PSH. To identify the disulfide bonds in gluten proteins, which are cleaved by endogenous GSH by disulfide exchange, flour was kneaded with added [35S]-GSH. The use of high specific radioactivity allowed the additive to be kept very low in comparison to the GSH present in flour. The disulfide bonds labelled by [35S]-GSH are cleaved by the endogenous GSH according to the reaction given

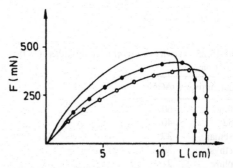

Fig. 15.16. Influence of reduced (GSH) and oxidized glutathione (GSSG) on the rheological properties of wheat dough (according to *Hahn* and *Grosch*, 1998). Tensile tests with dough made of 10 g of DNS flour (0.76% ash, 15.5% protein, 124 nmol/g GSH), water, 2% NaCl and the following additives (nnol/1 flour): GSH (100; o–o), GSSG (50; •–•), control without additive (—). F: force, L: distance

in Formula 15.4. It was found that GSH reacts very specifically in dough making because the intermolecular disulfide bridges involving C^{b*} and C^x in LMW (Fig. 15.10) are reduced to an extent of 47% in each case. Intramolecular disulfide bonds are hardly attacked. Since only one cleavage of intermolecular disulfide bonds can weaken the gluten and the dough, the strong rheological effect of GSH is understandable. The specificity of GSH is very remarkable since per g of flour, 50–100 nmol of GSH are faced with about 9000 nmol/g PSSP, which contain intermolecular disulfide bonds only to an extent of 10%.

15.2.4 Carbohydrates

15.2.4.1 Starch

The major carbohydrate storage form of cereals, starch (cf. Table 15.6) occurs only in the en- dosperm cells. The size and form of the starch granules are specific for different cereals. The polysaccharide molecules in starch granules are radially organized. Due to the presence of alternating amorphous (mainly amylose) and semicrystalline layers (amylopectin), differences in indices of refraction can be observed under a microscope.

Starch granules swell when heated in water suspension. At the end of swelling, they lose their form; i.e. they gelatinize. The temperature range in which these changes occur and also the extent of swelling at a given temperature are characteristic (cf. 4.4.4.14.1) and may be used for starch source identification. Starch absorbs ca. 45% of water in the baking process.

Cereal starches consist of about 25% amylose and 75% amylopectin (cf. Table 4.20). The chemical structures of these polysaccharides are presented in 4.4.4.14.3 and 4.4.4.14.4. Starch granules in some cultivars, for instance waxy corns, contain only amylopectin, while some cultivars

Table 15.28. Lipids in various cereal starches

	Wheat	Corn[a] (maize)	Amylomaize[a]	Waxy maize[a]
		(% or mg/100 g)[b]		
Nonpolar lipids	6%	60%	73%	88%
Sterol esters	2	3	9	7
Triacylglycerols	15	5	16	12
Diacylglycerols	7	3	16	6
Monoacylglycerols	8	12	13	5
Free fatty acids	27	380	650	105
Glycolipids	5%	1%	5%	6%
Sterol glycosides	3	7	13 ⎫	3
Monogalactosyldiacylglycerols	4		⎬	1
Monogalactosylmonoacylglycerols	10		18 ⎭	
Digalactosyldiacylglycerols	11			2
Digalactosylmonoglycerols	24		17	3
Phospholipids	89%	39%	22%	6%
Lyso-phosphatidyl ethanolamines	104	17	16	1
Lyso-phosphatidyl glycerols	23	6	7	trace
Lyso-phosphatidyl cholines	783	226	183	8
Lyso-phosphatidyl serines	26	8	6	trace
Lyso-phosphatidyl inositols				
Total lipids	1,047	667	964	153

[a] Amylose content in starch amounts to 23% (corn), 70% (amylomaize) and 5% (waxy maize cultivars).
[b] Results for lipid classes are expressed as % of total lipids present in starch, and for individual lipid compounds as mg/100 g starch dry matter.

are rich in amylose (cf. Table 4.20). Waxy corn starch swells considerably on heating, while granules with amylose swell only slightly (cf. Table 4.20 and Fig. 4.31).

Lipids (Table 15.28) and proteins (about 0.5%) are among the heterogeneous constituents of starch granules. Lipids are enclosed within the amylose helices. In wheat starch, they consist predominantly of lysolecithins (Table 15.28). They are extractable from partially gelatinized starch by using hot water-saturated butanol. During extraction, the lipid in the amylose helix is replaced by butanol.

The lipids complexed within the starch granules retard swelling and increase their gelatinization temperatures; thus they influence the baking behavior of cereals and the properties of the baked products.

15.2.4.2 Polysaccharides Other than Starch

Cereals also contain polysaccharides other than starch. In endosperm cells their content is much less than that of starch (cf. Table 15.29). They include pentosans, cellulose, β-glucans and glucofructans. These polysaccharides are primarily constituents of cell walls, and are more abundant in the outer portions than the inner portions of the kernel. Therefore, their content in flour increases as the degree of fineness increases (cf. rye as an example in Table 15.36).

From a nutritional and physiological viewpoint, soluble and insoluble polysaccharides other than starch and lignin (cf. 18.1.2.5.1) are also called dietary fiber. The most important fiber sources are cereals and legumes, while their content in fruits and vegetables is relatively low.

15.2.4.2.1 Pentosans

The pentosan content of cereals varies. Rye flour is exceptionally rich (6–8%) in comparison to wheat flour (1.5–2.5%). A portion of pentosans, 25–33% in wheat and 15–25% in rye, is water-soluble.

Unlike the water-soluble proteins of cereals, the soluble pentosans are able to absorb 15–20 times more water and thus can form highly viscous solutions. This soluble fraction consists mainly (ca. 85%) of a linear arabinoxylan and a soluble highly branched arabinogalactan peptide. A chain of D-xylopyranose units is typical of the structure of arabinoxylan (Ws-AX), which is extractable with water. The OH groups in the 2- and 3-position of this chain are glycosidically linked to L-arabinofuranose (e. g. 3-position in Fig. 15.17). The arabinose residues can be cleaved by mild acid hydrolysis or treatment with an α-L-arabinofuranosidase, giving water-insoluble xylan. Although a part of the arabinoxylan is insoluble in water (Wi-AX) as a result of cross-linking of the chains, it can become soluble by means of alkaline or enzymatic hydrolysis. The backbone of the arabinogalactan peptide is made of $\beta(1 \rightarrow 3)$ and $\beta(1 \rightarrow 6)$ linked galactopyranose units. It is α-glycosidically bound and contains, in addition, arabinofuranose residues. The bonding to the peptide is achieved via 4-transhydroxyproline.

The Ws-AX cause up to 25% of the water binding in dough. They increase the viscosity and, consequently, the stability of the gas bubbles. In contrast, the action of Wi-AX is considered to be unfavourable. They form physical barriers against the gluten and destabilize the gas bubbles.

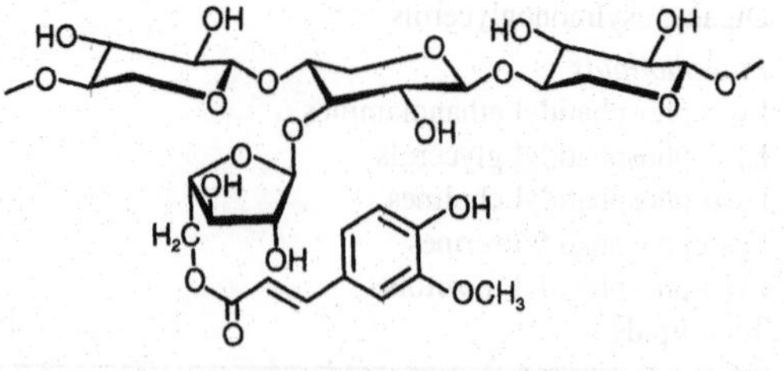

Table 15.29. Distribution of carbohydrates in wheat (%)

	Endosperm	Germ	Bran
Pentosans and hemicelluloses	2.4	15.3	43.1
Cellulose	0.3	16.8	35.2
Starch	95.8	31.5	14.1
Sugars	1.5	36.4	7.6

Fig. 15.17. A section of the structure of a water soluble arabinoxylan from wheat. A xylose in the $(1 \rightarrow 4)$-β-xylan section is linked in position 3 with a 5-O-*trans*-feruloyl-α-L-arabinofuranose

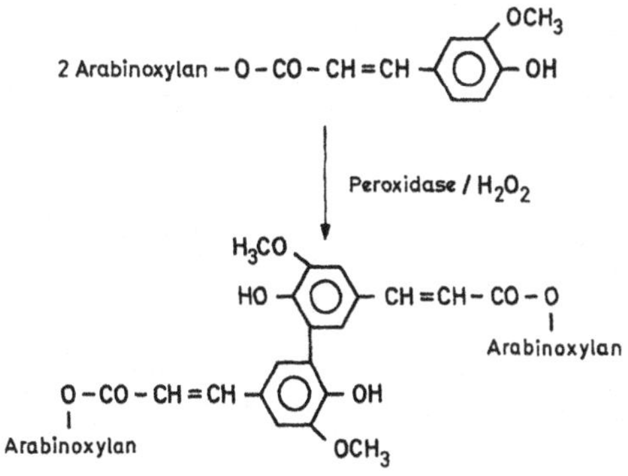

2 Arabinoxylan – O – CO – CH = CH –⟨◯⟩– OCH₃, OH

Peroxidase / H₂O₂

H₃CO, HO –⟨◯⟩– CH = CH – CO – O | Arabinoxylan

O – CO – CH = CH –⟨◯⟩– OH | Arabinoxylan   OCH₃

Fig. 15.18. Oxidative crosslinking of cereal pentosans

Accordingly, the baking result is positively influenced by endoxylanases which preferentially hydrolyze Wi-AX. Since inhibitors are present in wheat which inhibit the activity of added endoxylanases, attempts are being made with the help of molecular engineering techniques to produce microbial enzymes which do not react with these inhibitors.

The insoluble portion of pentosans from rye swells extensively in water. This portion is responsible for the rheological properties of dough and the baking behaviour of rye, and increases the crumb juiciness and chewability of baked products. An optimum starch-pentosan ratio is 16:1 (by weight) for rye flour.

Pentosan solutions gel when treated with hydrogen peroxide/peroxidase. This is due to the presence of low levels of ferulic acid (ca. 0.2%). An enzymic phenol oxidation occurs (cf. Fig. 15.18), which causes polymerization. This results in build-up of a network which, along with the low content of branched arabinofuranose, is responsible for the lack of solubility of most pentosans.

15.2.4.2.2 β-Glucan

The β-glucan content of cereals varies: barley 3–7%, oats 3.5–4.9%, wheat and rye kernels only 0.5–2%. These are linear polysaccharides with D-glucopyranose units joined by β-1,3 and β-1,4 linkages. Polysaccharides of the β-glucan type are also called lichenins. At 38 °C, 38–69% of the β-glucans of barley dissolve in 2 hours and 65–90% of the β-glucans of oats. β-Glucans are slimy mucous substances which provide a high viscosity to water solutions. In beer production

from barley β-glucans can interfere in wort filtration.

15.2.4.2.3 Glucofructans

Wheat flour contains 1% water-soluble, nonreducing oligosaccharides of molecular weight up to 2 kdal. They consist of D-glucose and D-fructose. Glucofructan, which predominates in durum wheats, probably has the following structure:

$$\text{ß-D-Fru}_f\,(2{\rightarrow}6)\text{ß-D-Fru}_f\,(2{\rightarrow}6)\text{ß-D-Fru}_f\,(2{\rightarrow}1)\text{α-D-Glc}_p$$
$$\begin{array}{c} 1 \\ \uparrow \\ 2 \\ \text{ß-D-Fru}_f \end{array} \qquad (15.6)$$

15.2.4.2.4 Cellulose

Cellulose is a minor constituent of the carbohydrate fraction obtained from starchy endosperm cells (cf. Table 15.29).

15.2.4.3 Sugars

Mono-, di- and trisaccharides, as well as other low molecular weight degradation products of starch, occur in wheat and other cereals in relatively low concentrations (Table 15.30). When starch degradation occurs during dough making, their levels increase (cf. 15.4.2.5). Mono-, di- and trisaccharides are of importance for dough leavening in the presence of yeast (cf. 15.4.1.6.1).

15.2.5 Lipids

Cereal kernels contain relatively low levels of lipids; nevertheless, differences occur among cer-

Table 15.30. Mono- and oligosaccharides in wheat flour

Compound	(%)
Raffinose	0.05–0.17
Glucodifructose	0.20–0.30
Maltose	0.05–0.10
Saccharose	0.10–0.40
Glucose	0.01–0.09
Fructose	0.02–0.08
Oligosaccharides[a]	1.2–1.3

[a] Fraction soluble in 80% ethanol.

eals (cf. Table 15.6). The endosperm cells of oats contain a higher level of lipids (6–8%) than wheat (1.6%). For this reason, the overall lipid content of oats is higher than in wheat and in other cereals.

The lipids are preferentially stored in the germ which, in the case of corn and wheat, serves as a source for oil production (cf. 14.3.2.2.4). Lipids are stored to a smaller extent in the aleurone layer. Cereal lipids do not differ significantly in their fatty acid composition (Table 15.31). Linoleic acid always predominates. Close attention has been given to wheat lipids since they greatly influence baking quality and they have therefore been studied thoroughly.

A wheat kernel weighs 30–42 mg and contains 0.92–1.24 µg of lipid. The germ and the aleurone cells are rich in triglycerides, which are present as spherosomes, while phospholipids and glycolipids predominate in the endosperm.

Wheat flour contains 1.5–2.5% lipids, depending on milling extraction rate. Part of this lipid is nonstarch lipid. This portion is extracted with a polar solvent, water-saturated butanol, at room temperature. Nonstarch lipid comprises about 75% of the total lipid of flour (Fig. 15.19). The residual lipids (25%) are bound to starch (cf. 15.2.4.1).

Nonstarch- and starch-bound lipids in wheat differ in their composition (cf. Table 15.28 and Table 15.32). In nonstarch-bound lipids the major constituents are the triacylglycerides and digalactosyl diacylglycerides, while in starch-bound lipids, the major constituents are lysophosphatides in which the acyl residue is located primarily in position 1. A decrease in amylose content is accompanied by a decrease in the lipid content (Table 15.28). The ratios of nonstarch-bound lipid classes are dependent

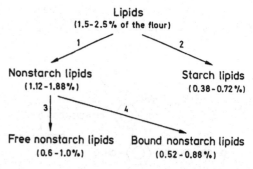

Fig. 15.19. Differentiation of wheat flour lipids by their solubility. *1* Flour extraction with water-saturated butanol (WSB) at room temperature, *2* with WSB at 90–100 °C, *3* with petroleum-ether, and subsequently, *4* with WSB

on the flour extraction grade. An increase in extraction grade increases the triacylglyceride content, since more of the germ is transferred into the flour.

The rheological dough properties are affected by nonstarch-bound lipids which are separated into free and bound lipids when extracted with solvents of different polarity. The free lipid fraction contains 90% of the total nonpolar lipids and 20% of the total polar lipids listed in Table 15.32. By kneading the flour into dough, the glycolipids become completely bound to gluten, while other lipids are only 70–80% bound. The extent of binding of triacylglycerides depends on dough handling. Intensive oxygen aeration and, particularly, addition of lipoxygenase (cf. 15.4.1.4.3) increase the fraction of free lipids.

The increased binding of lipids in the transition of flour to dough, which is expressed in their decreasing extractability, is explained by the following hypothesis.

Table 15.31. Average fatty acid composition of acyl lipids of cereals (weight-%)

	14:0	16:0	16:1	18:0	18:1	18:2	18:3
Wheat		20	1.5	1.5	14	55	4
Rye		18	<3	1	25	46	4
Corn		17.7		1.2	29.9	50.0	1.2
Oats	0.6	18.9		1.6	36.4	40.5	1.9
Barley	2	22	<1	<2	11	57	5
Millet		14.3	1.0	2.1	31.0	49.0	2.7
Rice	1	<28	6	2	35	39	3

Table 15.32. Nonstarch lipids in wheat flour

	mg/100g[a]
Nonpolar lipids (59%)	
Sterol lipids	43
Triacylglycerols (TG)	909
Diacylglycerols (DG)	67
Monoacylglycerols (MG)	53
Free fatty acids (FFA)	64
Glycolipids (26%)	
Sterol glycosides	18
Monogalactosyldiacylglycerols (MGDG)	115
Monogalactosylmonoacylglycerols (MGMG)	17
Digalactosyldiacylglycerols (DGDG)	322
Digalactosylmonoacylglycerols (DGMG)	52
[3pt] *Phospholipids* (15%)	
N-Acyl-phosphatidyl ethanolamines	95
N-Acyl-lyso-phosphatidyl ethanolamines	33
Phosphatidyl ethanolamines	
Phosphatidyl glycerols	19
Phosphatidyl cholines	96
Phosphatidyl serines	
Phosphatidyl inositols	9
Lyso-phosphatidyl glycerols	5
Lyso-phosphatidyl cholines	29

[a] Based on dry matter.

The neutral lipids are present in flour in the form of spherosomes and their membranes are formed by a part of the phospholipids. The spherosomes can be extracted with nonpolar solvents. The other phospholipids and all the glycolipids form inverse hexagonal phases (cf. 8.15.2.2), which are only partly extractable. During dough making, the water added results in the conversion of the inverse hexagonal to a laminar phase, which in turn stabilizes a microemulsion of the neutral lipids. The microemulsion vesicles are enclosed by the network of gluten proteins and, consequently, difficult to extract. If the dough is suspended in water, the lipids appear in the aqueous phase that separates on ultracentrifugation only when the framework of gluten proteins has been destroyed by reduction, e. g., with dithiothreitol.

Other hypotheses which explain the decreasing extractability of free lipids by selective binding, e. g., of glycolipids to starch and gluten, have not been confirmed.

The gas-holding capacity of doughs and, after passing through a minimum, the baking volume (Fig. 15.20) are positively influenced by polar lipids. Two effects are assumed in explanation. The polar lipids get concentrated in the boundary layer gas/liquid and stabilize the gas bubbles against coalescence. In addition, the lipid vesicles seal the pores which are formed in the protein films on kneading. On the other hand, the nonpolar lipids generally negatively influence the backing result with most varieties of wheat (Fig. 15.20).

Carotenoids and tocopherols belong to the minor components of the cereal lipid fraction. Wheat flour has a carotenoid content averaging 5.7 mg/kg. In durum wheats, which have a more intense yellow color, the carotenoids are 7.3 mg/kg of flour.

The major carotenoid, lutein (cf. 3.8.4.1.2), is present in free or esterified form (either mono or diester) with the fatty acids listed in Table 15.31. The following carotenoid pigments are also present: β-carotene, β-*apo*-carotenal, cryptoxanthin, zeaxanthin and antheraxanthin (for structures see 3.8.4.1). Carotenoid content of corn, depending on the cultivar, is 0.6–57.9 mg/kg, with lutein and zeaxanthin being the major constituents.

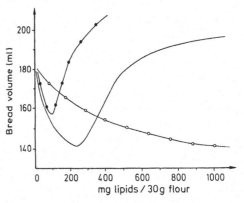

Fig. 15.20. The effect of free nonstarch lipids on the baking quality of defatted wheat flour (according to *Morrison*, 1976). — Lipids (total), −o−o− nonpolar lipids, −•−•− polar lipids

Table 15.33. Tocopherol content of parts of the wheat kernel

Part of kernel	Tocopherols in mg/kg			
	α-T	β-T	α-T-3	β-T-3
Germ	256	114	n.d.	n.d.
Aleurone layer	0.5	n.d.	10	69
Endosperm	0.07	0.10	0.45	13.5

n.d., not detectable.

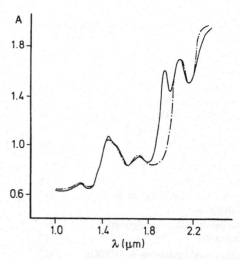

Fig. 15.21. Absorption of ground wheat in the near IR region. Sample dried (–·–·–) and with 9 w/w % water (——)

The composition of the tocopherols of wheat (Table 15.33) shows that the proportions of germ and aleurone lipids in nonstarch lipids can be determined by using β-T and β-T-3 as markers. Values of ca. 25% have been found, but they can fluctuate greatly depending on the milling process and extraction grade.

15.3 Cereals – Milling

15.3.1 Wheat and Rye

Quality control of the raw materials and milling products usually includes the determination of water, protein, and minerals. The absorption bands of food in the near infrared region (0.8–2.6 μm) are suitable for a quick basic analysis (water, protein, fat, carbohydrates etc.).

The overtones of CH, OH, and NH valence vibrations appear in the near IR region. Therefore, foods give a large number of absorption bands that can be assigned to definite components and have intensities that correlate with the amounts of the constituents. As an example, Fig. 15.21 shows the absorption of wheat in the near IR region. The sample containing water absorbs at 1.94 μm in addition. Therefore, after subtraction of the absorption of dried wheat and after calibration, the water content can be determined. Other consituents which can be determined in food by near-infrared (NIR) spectrophotometry are listed in Table 15.34.

In the development of methods for these materials, the measurement of IR reflection was at first given the most attention because it is technically

Table 15.34. Examples of quantitative analysis of foods by near-infrared (NIR) spectrophotometry

Component	Food
Water	Meat, cereals, control of fruit and vegetable drying processes, chocolate, coffee
Protein	Meat, cereals, milk and milk products
Fat	Meat, cereals, milk and milk products, oil seeds
Minerals	Cereals, meat
Starch	Cereals
Pentosans	Wheat
β-Glucans	Barley
Lysine	Wheat, barley

easier to perform. Since reproducible results can be obtained only if the surface and granulation of the samples are constant, sources of error arise here. In the meantime, however, technical improvements allow food, e. g., cereal kernels, to be irradiated in the range of 0.8–1.1 μm. Thus, the water and protein content in unground samples can also be determined by measuring the transmission. In food technology, measurements in the near IR region are widely used for the quick quality control of raw materials (Table 15.34).

15.3.1.1 Storage

Cereals can be stored without loss of quality for 2 to 3 years, provided that the kernel moisture content, which is 20–24% after threshing, is reduced to at least 14%. The low moisture content prevents microbial spoilage, especially by mycotoxin-forming organisms, and it also lowers kernel respiration, i. e., metabolism.

The water is slowly removed from grains by ripple-type dryers in a stream of hot air or burned gas at 60–80 °C (to the extent of 4% per passage) to avoid damage to kernels by uncontrolled shrinkage. Grains with high moisture content can be stored for short periods of time in the cold without quality deterioration. Stored grains are fumigated for pest control. Aluminum and magnesium phosphides are introduced. At 20 °C and 75% relative humidity, they decompose into gaseous PH_3. HCN or ethylene oxide fumigants are also used.

Wheat and rye are suitable for the production of bakery products, especially bread, and are called bread cereals. Other cereals serve only as additives for bakery products and are mainly used in other ways, e. g., for porridge and pancakes.

15.3.1.2 Milling

The aim of milling is to obtain preferentially a flour in which the constituents of the endosperm cells predominate. The outer part of the kernel, including the germ and aleurone layer (cf. Fig. 15.2) is removed. Such a requirement is not easy to accomplish since the kernel's groove and the unequal sizes of aleurone cells in cereals do not facilitate simple dehulling. Therefore, the grain has to be carefully broken, the particles sorted and separated by size and, only then, further disintegrated.

In a preliminary step to milling, the grain is cleaned of impurities such as weed seeds, straw, soil particles, spoiled decayed grains, dust, etc. This cleaning step is based on the cereal's kernel size and specific gravity. Washing with water is rarely done, since it promotes the growth of microorganisms.

The next step is grain wetting or steeping in water for 3–24 h, since an increased moisture content to 15–17% facilitates the separation of starchy endosperm cells from germ and hull. An alternative procedure is wheat conditioning at elevated temperatures up to 65 °C; it is faster than steeping and also favorably affects the baking quality. The kernels are disintegrated stepwise. Each passage through rollers involves particle size reduction by pressure and shear forces, followed by flour separation according to particle size using sieves in the form of flat sifters (Fig. 15.22). Rollers are matched to the product needed. Their size, surface flutes, rotation velocity, gap between pairs of rollers rotating in opposite directions at dissimilar speeds – all can be selected or adjusted. Wheat and rye are milled differently because of structural differences in the kernels. The wheat kernel is rather brittle; the rye kernel is gluey or sticky. Therefore, rye is less suitable for coarse grist milling than wheat. The wheat milling process can be adjusted so that the first passages provide the grist and the following ones provide the flour.

The germ of the rye kernel, because of its loose attachment, falls off readily during the cleaning

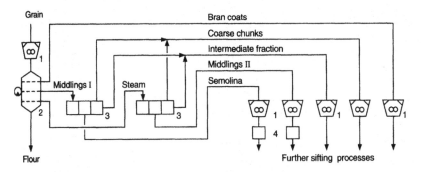

Fig. 15.22. Milling of cereal (*1*: roller mill, *2*: sifters, *3* and *4*: purifiers)

step, while the wheat germ is removed only on sifters. The hull and a substantian part of the aleurone layer is removed in the form of bran.

A portion (ca. 5–8%) of the starch granules is mechanically damaged during milling. The extent depends both on the type and intensity of milling and on the hardness of the kernel. The harder the structure of the kernel, the greater the damage. Since the rate of water absorption during dough making and the enzymatic degradation of starch increase with increasing damage, they are important for the baking process and desirable to a limited extent. To measure starch damage, the amylose extractable with a sodium sulfate solution is determined. Alternatively, the amount of starch degradable without gelatinization, e. g., at 30 °C by a- and/or β-amylase is determined. The starch damage expected during the milling process can also be estimated by determining the hardness of the kernel, e. g., by NIR reflectance spectrophotometry (cf. Table 15.34).

15.3.1.3 Milling Products

A miller distinguishes the end-products of milling on the basis of particle size or diameter, e. g., >500 μm for grist; 200–500 μm for semolina from durum or farina from bread wheats; 120–200 μm for "*dunst*"; and 14–120 μm for flour. The larger flour particles can be felt between the fingers (*graspable flour*), as opposed to smooth or polished flours in which the average particle size is 40–50 μm.

Differently milled flours vary considerably in baking quality. Flours obtained also differ greatly from cultivar to cultivar. This is especially the case with wheat cultivars (cf. 15.4.1.1).

In addition, quality depends on whether the milled flour comes from the inner or outer parts of the endosperm. Therefore, milled flour is controlled in the plant for its baking properties and blended or mixed to yield a commercial product based on present market standards (see also below). The characteristics of a few milling products and their applications are listed in Table 15.35.

The chemical composition of the flour depends on the milling extraction rate, e. g., flour weight obtained from 100 parts by weight of grain. Examples are given in Table 15.36. Increasing the rate of flour extraction decreases the pro-

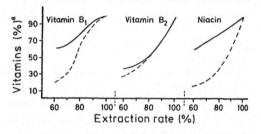

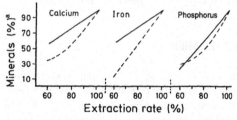

Fig. 15.23. Content of B vitamins and minerals in flour as affected by milling extraction rate (according to Lebensmittellexikon, 1979). —— rye, --- wheat
[a] Calculated as percent of the total content present in grain.

Table 15.35. Wheat and rye milling products

All purpose flour	Commercially available (retail market) flour for household preparations of baked products.
Special flour	It is used for special baked products, e. g., strong gluten wheat flour for toast bread, wheat flour with weak gluten for baked goods of loose tender or crispy structure as pastry etc.
Compounded (ready to use) flour	Special flour that contains other ingredients such as milk or egg powder, sugar etc., required by formulation of a selected baked product.
Groats (grist)	Coarsely ground dehulled cereal (devoid of germ and seed hull).
Whole grain groats	Ground from whole kernel (including germ).

Table 15.36. Average composition of wheat and rye flours[a]

A. Wheat flour

	Type				
	405	550	812	1050	1700[b]
	Flour extraction rate[c]				
	40–56%	64–71%	76–79%	82–85%	100%
Starch	82.3	81.8	78.1	77.8	69.2
Protein (N × 5.8)	11.7	12.3	13.0	12.9	12.7
Lipids	1.0	1.2	1.5	2.0	2.3
Dietary fiber[d]	4.7	5.0	5.6	6.0	13.4
Minerals (ash)	0.41	0.55	0.81	1.05	1.7

B. Rye flour

	Type				
	815	997	1150	1370	1740
	Flour extraction rate[c]				
	69–72%	75–78%	79–83%	84–87%	90–95%
Starch	74.8	73.5	71.3	71.1	68.6
Protein (N × 5.8)	7.2	8.03	9.6	9.6	11.7
Lipids	n.a.	1.3	1.5	1.7	1.8
Dietary fiber[d]	7.6	10.1	9.3	10.5	16.2
Minerals (ash)	0.82	1.0	1.15	1.37	1.74

[a] Weight-% per dry matter of wheat and rye flours. Flour average moisture content is 13 weight-%.
[b] Whole wheat flour.
[c] Approximate data.
[d] Indigestible carbohydrates (water soluble and insoluble), lignin. n.a.: not analyzed.

portion of starch and increases the amount of kernel-coating constituents such as minerals, vitamins and crude fiber (cf. Tables 15.8 and 15.9). Comparing products of the same extraction rate, rye flour contains higher proportions of both minerals and vitamins than wheat flour (Fig. 15.23). It should be pointed out that in the case of some B-vitamins, such as niacin, this difference is well-balanced by the higher concentrations in wheat in comparison to rye kernels (cf. Table 15.6). Consequently the concentrations of such vitamins are similar in rye and wheat flour.

Bread flours are standardized on the basis of their ash content in Europe and, particularly, Germany. The type of flour = ash content (weight %) × 1000 corresponds to the extraction grade. Examples are provided in Table 15.36 for wheat and rye flours and their chemical composition is detailed. Protein and starch contents are also related to flour particle size (cf. Table 15.37). Because of the variable particle sizes and densities of protein and starch, a flour sample can be separated by air classification into a fraction enriched in protein and starch. These are the so-called special purpose flours.

Table 15.37. Protein content of wheat flours as affected by flour particle size

Particle size (μm)	As portion of flour (weight %)	Protein content (weight %)
0–13	4	19
13–17	8	14
17–22	18	7
22–28	18	5
28–35	9	7
>35	43	11.5

The commercial product semolina ("*griess*") is made from endosperm cells of hard durum wheats. Semolina keeps its integrity during cooking and is used mostly for pasta production. Since semolina is a milled flour of low extraction rate, it contains few minerals and vitamins.

Table 15.38. Vitamin content of raw, white and parboiled rice

	B-vitamins (mg/kg)		
	Thiamine	Riboflavin	Niacin
Raw rice	3.4	0.55	54.1
White rice	0.5	0.19	16.4
Parboiled rice	2.5	0.38	32.2

15.3.2 Other Cereals

15.3.2.1 Corn

Corn endosperm, with the germ removed, is ground to grist for corn porridge (*Polenta*) and into corn flour for flat cakes (*tortillas*). Corn flakes are made from cooked and sweetened corn slurry, by drying, flaking and toasting. Similar products are made from millet, rice and oats.

15.3.2.2 Hull Cereals

Dehulling of rice, oats and barley requires special processes (cf. 15.1.4).

15.3.2.2.1 Rice

Rice milling involves the following processing steps: rough rice (paddy rice) → hull removal → brown rice → polishing to remove the bran coats (fruit and seed coats), the silvery cuticle, the germ and the aleurone layer → rubbing-off or rice polishing to obtain the end-product, white rice. Undamaged rice (45–55%), broken kernels or flour (20–35%) and a husk/hull fraction (20–24%) are obtained.

Polished white rice is made from this cleaned rice by additional treatment of the kernels with talc (a magnesium silicate) and 50% glucose solution. This imparts a glossy, transparent coating to the kernels.

White rice, in comparison to rough or brown rice, is low in vitamin content (cf. Table 15.38) and in minerals. A nutritionally improved product may be obtained by a parboiling process, orginally developed to facilitate seed coat removal. About 25% of the world's rice harvest is treated by the following process: raw rice → steeping in hot water, steaming in autoclaves, followed by drying and polishing → parboiled rice.

This treatment causes the following changes: the starch gelatinizes, but partly retrogrades again during drying. Enzymes are inactivated by the heat, causing inhibition of the enzymatic hydrolysis of lipids during storage of rice. The oil droplets (cf. 3.3.1.5) are broken and lipids partly migrate from the endosperm to the outer layers of the rice kernels. Since antioxidants are simultaneously destroyed, parboiled rice is more susceptible to lipid peroxidation. In contrast, minerals and vitamins diffuse from the outer layers to the inner endosperm and remain there after the separation of the aleurone layer (Table 15.38). The changes in starch mentioned above result in reduced cooking time.

Unlike in Europe and USA, some rice varieties popular in Asia develop a popcorn-like aroma on cooking. This is due to the formation of 2-acetyl-1-pyrroline, which is present in concentrations of 550–750 µg/kg in aromatic varieties of rice (cooked) and <8 µg/kg in lowaroma varieties.

15.3.2.2.2 Oats

Oat flakes are produced by the following processing steps: the kernels (12–16% water content) are steamed and then the moisture content is decreased to 7–10% in 2–3 h by heating at 90–100 °C. The hull (fruit and seed coats) is removed, i.e., the kernel is polished. This is followed by repeated steaming, squeezing between drum rollers, and drying of the moist flakes till the water content is 10–11%. The yield is 55–65%. This hydrothermic process also inactivates the oat enzymes involved in offflavor development. (E,E,Z)-2,4,6-Nonadienal produces the cereal-like, sweet aroma of oat flakes. It has an extremely low odor threshold and is formed from linolenic acid.

15.3.2.2.3 Barley

Removal of hull (fruit and seed coatings) yields groats which, after grinding, provide marketable products of large or fine particle size.

15.4 Baked Products

Baked products (for a review, see Table 15.39) are made from milled wheat, rye and, to a lesser extent, other cereals by the addition of water, salt, a leavening agent and other ingredients (shortening, milk, sugar, eggs, etc.). The following operations are involved:

- Selection and preparation of the raw materials
- Dough making and handling
- Baking
- Measures for quality preservation

15.4.1 Raw Materials

Among the ingredients involved in a formulation, only flour and those additives which affect dough rheological and/or baking properties will be cov-

Table 15.39. Classification of baked products

Bread including small baked products (rolls, buns)	Made entirely or mostly from cereal flours; moisture content on average 15%. Addition of sugar, milk and/or shortenings amounts to less than 10%. Small baked products differ from bread only by their size, form and weight.
Fine baked goods, including long term or extended shelf life products such as biscuits, crackers, cookies etc.	Made of cereal flours with at least 10% shortening and/or sugar, as well as other added ingredients. In baked goods for long shelf life the moisture content is greatly reduced.

ered. Flour improvers and dough leavening agents will be emphasized.

Characterization of the raw materials and additives is, in practice, made by assessing the dough rheological properties and by baking tests. Basic research endeavors to understand the nature of flour constituents and the reactions which affect their behavior in dough handling and baking.

15.4.1.1 Wheat Flour

A flour of optimal baking properties is required and chosen to match the quality of the desired product (cf. Table 15.35). The baking quality of wheat is strongly influenced by the cultivar (cf. Table 15.41) and also by conditions of growth and cultivation (climate, location), and subsequently by flour storage conditions and duration. Prior quality control is of importance to assess the overall baking quality of wheat flour. Flour particle size and color are assessed by sensory analysis. Graspable flours (cf. 15.3.1.3) are made from hard gluten-rich cultivars. Water uptake is slow when compared to smooth flour, and they make dry doughs.

The color difference is important, and is assessed with a wetted flour sample on a black background (*Pekar*-test).

15.4.1.1.1 Chemical Assays

Flour *acidity* (ml of 0.1 mol/l NaOH/10 g, titrated in the presence of phenolphthalein) depends upon the extraction rate of the flour and ranges between 2.0 ml/g (flour type 450) and 5.5 ml/g (flour type 1800). Too low acidity often reflects poorly aged flour. Acidity above 7.0 suggests microbial spoilage.

The *gluten content*, which is the residue left after the dough is washed (10 g flour kneaded into a dough with 6 ml of 2% NaCl, then washed with tap water), provides an indication of flour quality. A very low gluten content (<20%) frequently results in dough deterioration when machine-handled and also in baking faults. A higher content of gluten will not guarantee good baking quality (see "Maris Huntsman"

Table 15.40. Concentration of SH- and SS-groups in flour of different wheat cultivars

Cultivar	SH	SS	SS/SH
	μmole	per g	flour
Kolibri	1.15	12.5	10.9
Caribo topfit	0.88	12.2	13.9
Strong Canadian wheats	0.95	13.4	14.1
Inland wheat I[a]	0.75	10.2	13.6
Inland II[a]	1.05	12.6	12.0
Canadian Western Red Spring Wheat (CWRS)	1.26	12.9	10.2

[a] Marketed flour blendings.

cultivar, Table 15.41). Gluten swelling power is assessed by a *sedimentation value* as recommended by *Zeleny*. In this test, flour is suspended in an aliquot of a mixture of lactic acid (3.8 g), isopropanol (200 ml) and water (800 ml). The higher the volume of the sedimented gluten and starch, the better should be the baking quality of the flour.

For a given wheat cultivar, grown under similar climatic and soil conditions, the baking volume correlates with the *protein content* of the flour (Fig. 15.24). A similar linear relationship is not readily attainable for flours from different cultivars, as evidenced by the very different slopes of the regression lines.

The parameters involved here are those described in Section 15.2.1.4 as responsible for the properties of gluten. These include type, amount, and degree of polymerization of the HMW and LMW subunits of glutenin as well as the ratio gliadin/glutenin.

On the whole, the structure of wheat gluten has been so far evaluated to be able to describe variety-specific differences in the technological properties.

Wheat cultivars differ in the content of their *thiol* and *disulfide groups* (Table 15.40). This implies that the stability of a dough may be strongly influenced by a SH/SS exchange between a low molecular weight SH-peptide and gluten proteins. This also implies that a positive correlation

Table 15.41. Baking quality data of some wheat flours

	Wheat cultivar[a]		
	Monopol	Nimbus	Maris Huntsman
Protein (% dry matter)[b]	13.2	11.6	11.8
Wet gluten (%)	35.1	24.7	34.3
Farinogram[c]			
Water absorption (%)	59.2	54.8	59.8
Dough development time (min)	5.0	1.0	2.0
Dough stability (min)	5.0	1.5	0.5
Mixing Tolerance Index[d] (FU)	30	80	130
Extensogram[e]			
Area (dough strength, cm^2)	143	75	17
Resistance of the dough to extension (EU)	700	680	110
Extensibility (mm)	170	92	100
Baking test			
Dough surface	somewhat wet to normal	normal	wet, gluey
Dough elasticity	normal	somewhat short	weak
Baking volume (ml)	738	630	510

[a] Wheat cultivars with breadmaking quality corresponding to very good ("Monopol"), average ("Nimbus") and poor ("Maris Huntsman").
[b] Factor $N \times 5.7$.
[c] Explanation in Fig. 15.26; dough consistency: 500 FU.
[d] Measured after 10 min in Farinogram units (FU).
[e] Explanation in Fig. 15.29.

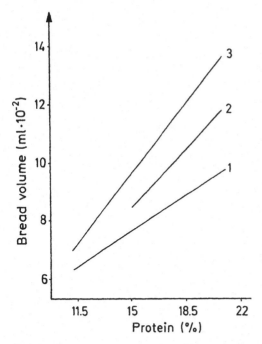

Fig. 15.24. Examples of relationship between protein content of flour and bread volume (according to *Pomeranz*, 1977). United States winter wheat cultivars: *1* Chiefkan, *2* Blackhull, *3* Nebred. The regression lines are based on numerous sample analyses

between the contents of SH- and SS-groups in flour, or their ratios, would be reflected in baking quality. However, low correlation coefficients of about 0.6 have been found. This corresponds to the observation (15.4.1.4.1) that the relationships are much more complex and cannot be grasped by means of an easily determinable characteristic quantity.

Of all enzymes in flour, quality control is aimed at the determination of amylase activity. The *Falling Number* test (*Hagberg* and *Perten*) serves this aim. A piston-type mixer falls through an aqueous flour paste. The falling time of the piston is measured for a given distance under standard conditions. The results are related, among other things, to starch granule stability in the presence of amylase enzymes. The *dextrin value* should be determined to assess amylase activity specifically. In a method developed by *Lemmerzahl*, the extent of standard dextrin hydrolysis in the presence of flour extract is measured. The fermentation power of a flour (cf. 15.4.1.6.1) involves determination

of the *maltose value* (diastatic activity). This is a quantitative determination of reducing sugars prior to and after incubation of a flour suspension at 27 °C for 1 h. Flours with a maltose content of <1.0% are regarded as weak fermentation promoters; values above 2.5% are flours from sprouted kernels. They provide poor baking quality.

15.4.1.1.2 Physical Assays

The instruments widely used in practice for the determination of the rheological properties of dough can be divided into recording dough kneaders and tensile testers. Dough development is followed with a Brabender farinograph (Fig. 15.25), with which measurement is made of the volume of water absorbed by the flour in order to make a dough of predetermined consistency (normal consistency). A plot of dough consistency versus time is recorded, as shown in Fig. 15.26.

In addition to the water absorption, the shape of the farinogram is used to characterize a flour. Var-

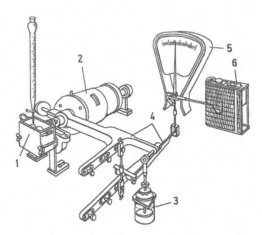

Fig. 15.25. Farinograph (according to *Rohrlich* and *Thomas*, 1967). The apparatus consists of a thermostated mixer or kneader *(1)*, its blades are driven by an electromotor *(2)*. The reaction torque acts through a lever system *(4)* of analytical balance precision on the indicator scale *(5)* simultaneously recorded on a strip chart recorder *(6)*. The movement of the lever system is damped by an oil dash pot *(3)*. The farinogram is a diagram of force versus time

ious indices have been defined (cf. Fig. 15.26); usually they refer to doughs with a maximum consistency of 500 FU.

Flours with strong gluten absorb more water and show longer dough development and stability times than do flours with weak gluten (Table 15.41). Corresponding results are obtained with the Swanson and Working mixographs.

A standardized piece of dough is stretched with the hook of a Brabender extensograph until the piece breaks (Fig. 15.27). As shown in Fig. 15.29, a graph of force (resistance to extension) versus stretching distance (extensibility) provides information about the stability of a dough, its gasholding capacity and fermentation tolerance. Of the examples given in Table 15.41, the "Monopol" cultivar obviously has strong gluten. The "Nimbus" cultivar has short gluten, as reflected by its low extensibility. The "Maris Huntsman" cultivar has a very weak gluten, as shown by the low resistance of its dough to extension and also by its low extensibility, and very small extension area.

Similar results are obtained with the Chopin extensograph or alveograph used widely in France. A piece of dough mounted on a perforated plate is blown into a ball. The pressure in the ball of dough is plotted against the time (cf. Fig. 15.28). In contrast to the Brabender extensograph, the dough is extended in two dimensions. As in the extensogram, the resistance of the dough to extension and its extensibility are obtained from the maximal height and width of the alveogram.

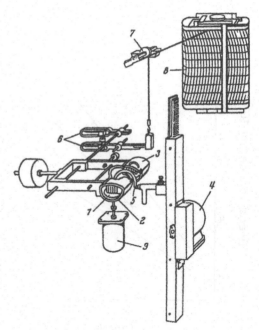

Fig. 15.27. Extensograph (according to *Rohrlich* and *Thomas*, 1967). The cylindrical piece of dough *(1)* is fixed by dough clamps *(3)* and placed on the balance fork *(2)*. The motor *(4)* of the stretching unit *(5)* is then started. The arm moves downward into the dough and extends it at constant speed. Simultaneously, the forces opposing the stretching action are transmitted through the lever system *(6)* to the balance system *(7)*. This is coupled to a recording arm of the strip chart recorder *(8)*. The fork of the balance system is coupled to an oil damper *(9)* to reduce the recoil

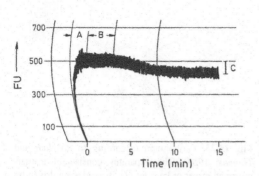

Fig. 15.26. Farinogram. The following data are pertinent for quality assessment of flour: *A* dough development time, *B* dough stability (dough consistency does not change), *C* decrease in dough consistency after a given time, here 12 min. FU: farinogram units

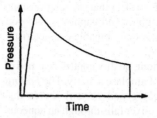

Fig. 15.28. Alveogram (cf. text)

15.4.1.1.3 Baking Tests

Direct information about the baking quality of a flour is obtained from baking tests under standardized conditions. Baking volume (cf. Ta-

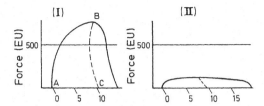

Fig. 15.29. Extensograms of a normal *(I)* and weak dough *(II)*. For quality assessment the following parameters are determined: resistance to extension, height of the curve at its peak (B–C) given in extensogram units (EU); extensibility, abscissa length between A–C in mm; extension area (A–B–C–A, cm^2) is related to energy input required to reach the maximum resistance; extensogram number (overall dough quality) is the ratio of extension resistance to extensibility

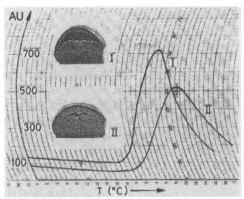

Fig. 15.30. Amylograms of two rye flours (according to *H. Stephan*, 1976)

	Gelatinization maximum (peak)	Gelatinization temperature	α-Amylase
Flour I	720 AU	67 °C	high
Flour II	520 AU	73.5 °C	low

AU: amylogram units.

ble 15.41), form, crumb structure and elasticity, and the taste of the baked product are evaluated. A baking test is performed with 1000 g flour for each product.

When the effects of expensive and not readily available flour constituents and/or additives are tested or a new cultivar is assessed, of which only several hundred kernels are available, a *"micro baking test"* is used, with 10 g flour for each baked product (cf. Fig. 15.35). If even less material is available, 2 g are sufficient. The sample is then kneaded in a mixograph and baked in a capsule.

15.4.1.2 Rye Flour

The Falling Number test (cf. 15.4.1.1.1) and an amylographic assay are the most important tests to assess the baking properties of rye flour. These tests depend to a great extent on gelatinization properties of starches and the presence of α-amylase. The higher the α-amylase activity, the lower the Falling Number.

An amylograph is a rotational torsion viscometer. It measures the viscosity change of an aqueous suspension of flour as a function of temperature. The recorded curve, called an amylogram (Fig. 15.30), shows that with increasing temperature there is an initial small fall followed by a steep rise in viscosity to a maximum value. The steep rise is due to intensive starch gelatinization.

The viscosity value and temperature at maximum viscosity (i. e., the temperature reflecting the end of gelatinization) are then read.

In rye flour with balanced baking properties, an optimal relationship should exist between α-amylase activity and starch quality. The extent of enzymatic starch degradation influences the stabilty of the gas-cell membranes which are formed by gas released in the dough and which consolidate during baking into an elastic crumb structure. These membranes contain pentosans, proteins and intact starch granules in addition to gelatinized and partially hydrolyzed starch. High α-amylase activity in rye or a large difference between the temperatures needed for enzyme inactivation (close to 75 °C) and those required for termination of starch gelatinization will produce poor bread since too much starch will be degraded during breadmaking. The gas-cell membranes are liquefied to a great extent; so the gas can escape. This gas will then be trapped in a hollow space below the bread crust (I in Fig. 15.30). Low α-amylase activity, especially in conjunction with low starch gelatinization, leads to a firm and brittle crumb structure.

15.4.1.3 Storage

Rye flour acquires optimal baking properties after 1–2 weeks of storage after milling. Wheat flour requires 3–4 weeks. This storage period is the flour "maturation time". In wheat the time is needed for oxidative processes to occur and thus provide a stronger (shorter) gluten. In this time, the concentrations of endogenous glutathione (GSH, GSSG), which reduces the stability of gluten in dough making (cf. 15.2.3), and PSSG decrease, the rate depending on the wheat cultivar.

Flour with a moisture content of <12% may be stored at 20 °C and a relative humidity of <70% for more than 6 months without significant change in baking quality.

Flour fumigation with $Cl_2, ClO_2, NOCl, N_2O_4$ or NO, or treatment with dibenzoyl or acetone peroxide results in carotenoid destruction. The flour becomes bleached. Other reactions, not yet elucidated, are involved with $Cl_2, NOCl, ClO_2$ and acetone peroxide treatment since they provide simultaneous improvement in baking quality of flours which have weak gluten.

15.4.1.4 Influence of Additives/Minor Ingredients on Baking Properties of Wheat Flour

The baking properties of wheat flours differ widely (cf. Table 15.41). In small traditional plants, a baker can use his experience to compensate for changes in the quality of raw materials: flexibility in formulations, dough handling and baking – all these parameters can be adjusted in order to obtain the desired end-product.

In a large-scale automated bakery, economic production demands uniform raw materials with uniform properties. Additives are used when necessary to adjust the flour characteristics to match the baking process (for instance, shortened dough handling time with low energy input). Additives are also used to ensure that the end-product meets existing standards. Incorporation of ascorbic acid, alkali bromates or enzyme-active soy flour improves the quality of weak gluten flour – e. g., in bread or bun baking. In these cases the dough becomes drier and there are increases in dough resistance to extension, mixing tolerance and fermentation stability. In addition, baking volume will increase and the crumb structure will improve. Ascorbic acid and lipoxygenase require oxygen for their actions; hence their beneficial role is very dependent on the intensity of dough mixing, which traps oxygen from the air.

In contrast, opposite effects may be observed by adding cysteine or proteinases, the result being gluten softening. Biscuits are made from such mellowed, softened doughs, which are made with little energy input. Additives which affect the rheological quality of the dough and/or the quality of baked products include emulsifiers, shortenings, salt, milk, soy flour, α-amylase and proteinase preparations and starch syrups.

15.4.1.4.1 Ascorbic Acid

The improver effect of ascorbic acid (Asc) was recognized by *Jorgensen* as early as 1935. He found that small amounts (2–10 g Asc per 100 kg flour) caused an improvement in flour. The dough becomes stronger (Fig. 15.31) and drier and the bread volume increases in most cases. The oxidation product of Asc, dehydroascorbic acid (DHAsc) is also effective (Table 15.42), but its use would be uneconomical. In the example in Fig. 15.31, the addition of 40 mg/kg of Asc has a greater strengthening effect on dough than 20 mg/kg. A further increase in Asc to 80 or even 160 mg/kg no longer increases the effect. But in comparison with

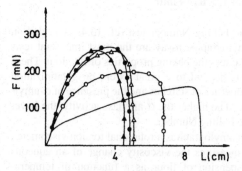

Fig. 15.31. Rheological properties of wheat dough as a function of different concentrations of added L-*threo*-ascorbic acid (Asc) (according to *Kieffer*, unpublished). Tensile tests with dough made of 10 g of flour of the variety Flair. Addition of Asc (mg/kg): 20 (o–o), 40 (•–•), 80 and 120 (▲–▲), 160 (Δ–Δ). Control without additive: —

oxidizing agents like bromate, no overdosage is observed.

In dough making, the Asc added to the flour oxidizes very rapidly to DHAsc (Fig. 15.32). Diastereomers of Asc are converted at the same rate. In contrast, the four diastereomers of Asc (stereochemistry in Fig. 15.33) as well as the corresponding DHAsc differ in their effect as flour improvers. As shown in Table 15.42, L-*threo*-Asc (vitamin C) has the highest dough strengthening effect. The two *erythro*-Asc have a weaker effect and D-*threo*-Asc is almost ineffective. Since these

Table 15.42. Effect of additives on the rheological properties of wheat dough

Additive (0.15 μmol/g flour)	Resistance to extension (%)	Extensibility[a] (%)
Control (without additive)	100	100
Cysteine	63	106
Glutathione (reduced form)	56	105
L-*threo*-Ascorbic acid	147	58
D-*erythro*-Ascorbic acid	122	86
L-*erythro*-Ascorbic acid	118	93
D-*threo*-Ascorbic acid	94	88
L-*threo*-Dehydroascorbic acid	145	56

[a] Relative values.

differences correspond with the substrate specificity of the GSH dehydrogenase found in flour (cf. 15.2.2.7), it is assumed that this enzyme is involved in flour improvement on the addition of Asc as shown in Fig. 15.34.

The atmospheric oxygen kneaded into the dough first oxidizes Asc to DHAsc (Reaction a in Fig. 15.34). The reaction is accelerated by ascorbic acid oxidase (cf. 15.2.2.9) and other factors. Subsequently (Reaction b), the GSH present in flour is oxidized to the disulfide. This reaction proceeds very rapidly because it

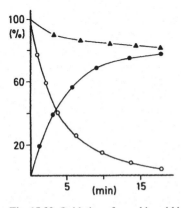

Fig. 15.32. Oxidation of ascorbic acid in dough making from wheat flour (according to *Nicolas* et al., 1980) o—o ascorbic acid, •—• dehydroascorbic acid, ▲—▲ sum of ascorbic and dehydroascorbic acids

$$\text{L-}threo\text{-Asc} + {}^{1}\!/_{2}\,O_2 \xrightarrow{\text{(AO)}} \text{L-}threo\text{-DHAsc} + H_2O \quad \text{(a)}$$

$$\text{L-}threo\text{-DHAsc} + 2\,\text{GSH} \xrightarrow{\text{(GSH-DH)}} \text{L-}threo\text{-Asc} + \text{GSSG} \quad \text{(b)}$$

$$\text{GSSG} + \text{PSH} \rightarrow \text{PSSG} + \text{GSH} \quad \text{(c)}$$

$$\text{PSSP} + \text{GSH} \rightarrow \text{PSSG} + \text{PSH} \quad \text{(d)}$$

$$\text{CSSC} + \text{GSH} \rightarrow \text{CSH} + \text{GSSC} \quad \text{(e)}$$

$$\text{PSSP} + \text{CSH} \rightarrow \text{PSSC} + \text{PSH} \quad \text{(f)}$$

$$\text{CSSC} + \text{PSH} \rightarrow \text{PSSC} + \text{CSH} \quad \text{(g)}$$

Fig. 15.34. Reactions involved in flour improvement by ascorbic acid (according to *Grosch* and *Wieser*, 1999) Asc, ascorbic acid; DHAsc, dehydroascorbic acid; AO, ascorbic acid oxidase; GSH-DH, glutathione dehydrogenase; GSH, reduced glutathione; GSSG, oxidized glutathione; CSH, cysteine; CSSC, cystine; PSSP, gluten proteins

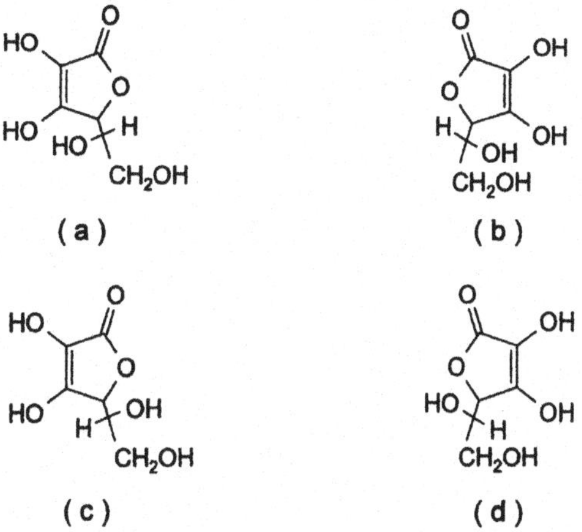

Fig. 15.33. Chemical structures of stereoisomeric ascorbic acids (Asc).
(a) L-*threo*-Asc (vitamin C), (b) D-*threo*-Asc,
(c) D-*erythro*-Asc (isoascorbic acid), (d) L-*erythro*-Asc

is catalyzed by enzyme GSH-DH (cf. 15.2.2.7) which requires DHAsc as a cofactor. Ascorbic acid is reformed which explains that relatively small amounts of Asc are sufficient for flour improvement. The GSSG formed in Reaction (b) can undergo an SH/SS exchange with gluten proteins (Reaction c), which has been shown to proceed especially rapidly on addition of Asc. Thus, GSH is incorporated into gluten proteins as a terminator of polymerization reactions via the intermediate GSSG. On the other hand, GSH formed in Reaction (c) is immediately oxidized. The reaction sequence (a→b→c) stops when all the GSH is present as GSSG or incorporated into gluten proteins. Consequently, GSH is largely withdrawn from the dough before it can depolymerize the gluten proteins by SH/SS interchange. If it is not withdrawn from the dough, Reactions (d)–(f) proceed without any interference. In comparison with Reaction (c), Reaction (d) can result in the softening of the

dough because GSH very specifically cleaves intermolecular disulfide bonds of the gluten proteins (cf. 15.2.3).

The results summarized in Tables 15.43 and 15.44 show that free cysteine increases rapidly in dough making. This is explained by Reaction (e) in Fig. 15.34, which shows that GSH reacts with the cystine present in flour. Cysteine increases and can, in turn, depolymerize gluten proteins (Reaction f). If, however, L-*threo*-Asc is added to the dough, GSH is so rapidly oxidized (Reaction b) that the comparatively slow Reaction (e) is strongly inhibited and cysteine increases only slightly (Tables 15.43 and 15.44). Corresponding to the substrate specificity of GSH-DH (cf. 15.2.2.7), D-*erythro*-Asc is almost ineffective; cysteine increases and GSH decreases as in the experiment without an additive (Table 15.43). The reaction scheme shown in Fig. 15.34 also explains the baker's experience that the consistency of dough can be reduced by the addition of cysteine almost independently of the presence of L-*threo*-Asc. Reaction (f) is not prevented due to the substrate specificity of GSH-DH which is only directed at GSH. Reaction (g) is promoted by the addition of L-*threo*-Asc because the reduction of cystine (Reaction e) is inhibited by the trapping of GSH. Consequently, cystine can undergo SS/SH interchange with gluten proteins according to Reaction (g) so that PSSC increases, as demonstrated in model experiments.

The reaction scheme in Fig. 15.34 also explains the observation that Asc cannot be overdosed. The effect of Asc stops at the moment when all the GSH is bound as GSSP and GSSG. An increase in Asc has no further effect.

Table 15.43. Influence of L-*threo*- and D-*erythro*-ascorbic acid (Asc) on the concentration of cysteine (CSH) and glutathione in wheat dough

Sample[a]	Additive (30 mg/kg)	CSH (nmol/g)	GSH (nmol/g)
Flour	Without	13	100
Dough	Without	42	44
Dough	L-*threo*-Asc	28	20
Dough	D-*erythro*-Asc	41	39

[a] Production of the dough: DNS flour (0.78% ash, 10 g) and water (6.5 ml) are kneaded at 30 °C for 2 min; liquid nitrogen is poured over the dough which is then freeze dried; CSH and GSH are determined by isotopic dilution analysis.

Table 15.44. Changes in concentration of glutathione (GSH) and cysteine (CSH) in dough on addition of L-*threo*-ascorbic acid or bromate

Additive[a]	GSH (nmol/g)			CSH (nmol/g)		
	Kneading time (min) at 30 °C					
	3	9	9 + 20[b]	3	9	9 + 20[b]
Without	57	22	17	68	28	31
L-*threo*-Asc (30 mg/kg)	11	6	2	26	17	19
KBrO₃ (50 mg/kg)	40	20	11	52	26	27

[a] The DNS flour used contained 124 nmol/g GSH and 22 nmol/g cysteine.
[b] The dough was allowed to rest for 20 min.

15.4.1.4.2 Bromate, Azodicarbonamide

Addition of alkali bromates to flour also prevents excessive softening of gluten during dough making. The reaction involves oxidation of endogenous glutathione to its disulfide. Bromate reacts slower than ascorbic acid (Table 15.44). After a kneading time of 3 minutes, GSH decreases from 124 nmol/g to a concentration of 40 nmol/g and cysteine increases from 22 nmol/g to 52 nmol/g. These values are relatively close to the corresponding values in dough without additives. On the other hand, only 11 nmol/g of GSH remain and cysteine increases only slightly after the addition of Asc. The reactions of bromate in dough have not yet been elucidated. Model experiments indicate that it can link gluten proteins by the formation of intermolecular disulfide bonds. Then the oxidation of GSH would not be the decisive step in flour improvement. In comparison with Asc, bromate can be overdosed, which also shows that another mechanism must be involved here. During baking, bromates are completely reduced to bromides with no bromination of flour constituents. Azodicarbonamide is of interest as a flour improver

$$H_2N-CO-N=N-CO-NH_2 \qquad (15.7)$$

since it improves not only the dough properties of weak gluten flour, but also lowers the energy input in dough mixing (cf. Fig. 15.40). Details of the reactions involved are unknown.

15.4.1.4.3 Lipoxygenase

The addition of a small amount of enzymeactive soy flour to a wheat dough increases the mixing tolerance, improves the rheological properties and may increase the bread volume. The effect on dough rheology is shown only with high-power mixing in the presence of air. The carotenoid pigments of wheat flour are bleached by the addition of enzyme-active soy flour. This is desirable in the production of white bread. The amount of enzyme-active soy flour is restricted to approximately 1% since higher levels may generate off-flavors.

It was demonstrated that nonspecific lipoxygenase (cf. 3.7.2.2) is responsible for the

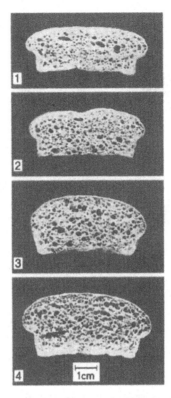

Fig. 15.35. Wheat flour quality improvement by the nonspecific lipoxygenase enzyme of soybean[a] (according to *Kieffer* and *Grosch*, 1979). Additions: **1** control (no addition, bread volume 31 ml), **2** extract of defatted soya meal in which lipoxygenase was thermally inactivated (31 ml), **3** extract of a defatted soya meal with 290 units of lipoxygenase[b] (35 ml), **4** purified type-II enzyme with 285 activity units (37 ml).
[a] Results in small-scale baking, 10 g flour cv. Clement.
[b] One enzyme unit = 1 μmole · min^{-1} oxygen uptake with linoleic acid as substrate

improver action (Fig. 15.35) and the bleaching effect caused by the enzyme-active soy flour. This enzyme, in contrast to endogenous wheat flour lipoxygenase, releases peroxy radicals which cooxidize carotenoids and other flour constituents.

15.4.1.4.4 Cysteine

Cysteine, in its hydrochloride form, softens gluten due to a SH/SS interchange with the glutenin fraction. The resistance to extension

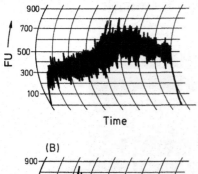

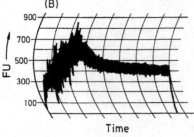

Fig. 15.36. Farinograms. Effect of L-cysteine hydrochloride on a flour with strong gluten (according to *Finney* et al., 1971). **A** control (no addition), **B** cysteine added (120 ppm)

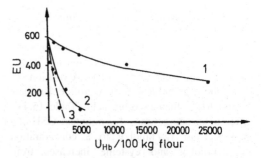

Fig. 15.37. The effect of a proteinase preparation on resistance to extension (in extensogram units) of a wheat flour dough (according to *Sproessler*, 1980). Proteinase preparation: *1* fungal, *2* papain, and *3* bacterial. U_{Hb} proteinase activity units determined with hemoglobin as a substrate

of the dough decreases and the extensibility increases (cf. Table 15.42). Decreases in dough development time and dough stability, as shown in farinograms (Fig. 15.36), clearly reveal the addition of cysteine. Flours with strong gluten and with optimum levels of cysteine also show a favorable increase in baking volume since, prior to baking, the gas trapped within the dough can develop a more spongy dough. The action of sodium sulfite is similar to that of cysteine.

15.4.1.4.5 Proteinases (Peptidases)

Proteinase preparations of microbial or plant origin are used for dough softening (cf. 2.7.2.2.1). Their action involves protein hydrolysis, i.e., gluten-protein endo-hydrolysis. Their effect on dough rheology, therefore, depends on the nature of the enzymes and the activity of the preparations towards gluten proteins. This is shown in Fig. 15.37. Despite equal hydrolase activities with hemoglobin as a test substrate, a fungal proteinase degrades gluten to a lesser extent and consequently causes a smaller decrease in dough resistance to extension in comparison to a bacterial enzyme preparation. Also, the latter is more effective than papain.

Fungal proteinases, because of their low enzyme activity and, therefore, high dosage tolerance, are suitable for optimization of flours containing strong gluten, used for bread and buns. However, bacterial enzymes are preferred in production of biscuits and wafers since they degrade gluten to a greater extent, providing accurate flat dough pieces with high form stability. Bacterial enzymes are also preferred for the desirable end product qualities of porosity and breaking strength.

Data are shown in Table 15.45 for white bread prepared with and without papain. There is a rise in the content of both free amino acids in the crumb and volatile carbonyl compounds in the crust when proteinase is used. As long as proteinases are active in a baking process, they release amino acids from flour proteins, which are then changed via *Strecker* degradation

Table 15.45. Effects of papain addition in white bread making (values in μmole/g dry matter)

Constituent		Without papain	With papain
Free amino acids	Dough	183	186
	Crumb	182	272
	Crust	10	15
Volatile carbonyl compounds	Crust	158	217

(cf. 4.2.4.4.7) into volatile carbonyl compounds in the crust. Bread aroma is enhanced, as is the crust color, by a build-up of melanoidin compounds from nonenzymatic browning reactions.

15.4.1.4.6 Salt

The taste of bread is rounded-off by the addition to dough of about 1.5% NaCl. As with other salts with small cations (e. g., sodium fumarate or phytate), the addition of NaCl increases dough stability. It is assumed that this is due to the ions masking the repulsion between one charged gluten protein molecule and another of like charge. This allows a sufficiently close approach of one molecule to another, thus hydrophobic and hydrophilic interactions can occur.

15.4.1.4.7 Emulsifiers, Shortenings

Flour baking quality is positively correlated to the content of polar lipids, particularly glycolipids (cf. 15.2.5). Further improvements in dough properties, baking results and end-product freshness or shelf life (cf. 15.4.4) are gained by adding emulsifiers to the dough, e. g., crude lecithin (cf. 3.4.1.1), mono- and diacylglycerides or their derivatives in which the OH-group(s) is esterified with acetic, tartaric, lactic, monoacetyl or diacetyl tartaric acid (cf. 3.3.2 and 8.15.3.1). The hypotheses presented in 15.2.5 are under discussion to explain this effect in the baking process.

Addition of triacylglycerides (shortenings) generally reduces the end-product volume, but there

Table 15.46. The effect of shortening on baking volume

Wheat flour	Baking volume (ml)[a]	
	Without shortening	With 3% shortening
I	64.5	81.0
II	73.3	71.8
III[b]	51.6	46.3

[a] Baking test performed on a small scale (10 g flour).
[b] Flour of poor baking quality.

are exceptions depending on the wheat variety. As illustrated by flour I in Table 15.46, addition of 3 % shortening provides a substantial increase in baking volume. Emulsifiers are also added to the dough to delay the aging of the crumb (cf. 15.4.4).

15.4.1.4.8 α-Amylase

Flours contain very small amounts of sugars which are metabolizable by yeast (cf. Table 15.30). Addition of sucrose or starch syrup at 1–2% to dough is advisable to maintain favorable growth of yeast and therefore to provide CO_2 needed for dough leavening. Uniform leavening over an extended time improves the quality of many baked end-products; the crumb structure acquires finer and more uniform porosity, while the crust has greater elasticity.

Flours derived from wheat without sprouted grains have some β- but very little α-amylase activity (cf. 15.2.2.1). Thus, only a small amount of starch is degraded to fermentable maltose by handling dough. An insight into the extent of starch degradation is provided by the maltose value (cf. 15.4.1.1.1). Addition of α-amylase in the form of malt flour or as a microbial preparation increases the flour capacity to hydrolyze the starch.

The activity of α-amylase as well as the levels of maltose and glucose increase in the germination of cereals; hence, addition of flour from malted grains enhances the growth of yeast in dough. However, the addition of malt to flours with weak gluten may not be expedient because of the proteolytic activity of the malt. α-Amylase preparations free of proteolytic activity are available from microorganisms (cf. 2.7.2.2).

Examples in Table 15.47 illustrate the effects of α-amylase from various sources on baking quality. While malt and fungal amylases show similar effects, the heat-stable α-amylase from *Bacillus subtilis*, with its prolonged activity even in the oven, may be easily used to excess. Products formed by the activities of α- and β-amylases are also available as reactants for nonenzymatic browning reactions. This favorably affects the aroma and color of the crust. α-Amylases are added to flour not only to standardize the baking properties, but also to delay the aging of the crumb (cf. 15.4.4).

Table 15.47. The effect of α-amylase preparations on baking results

α-Amylase preparation		White bread		
Origin	Activity[a] (units)	Volume (ml)	Crumb	
			pores	structure
Without addition		2400	average	average
Wheat malt	140	2790	good	good
	560	3000	good	good
	1120	2860	average	good
Aspergillus	140	2750	very good	very good
oryzae	560	2900	good	good
	1120	2950	average	average
Bacillus	7	2600	good	good
subtilis	35	2600	good	average
	140	2640	poor	very poor

[a] α-Amylase units in 700 g flour.

15.4.1.4.9 Milk and Soy Products

Dairy products such as skim milk, buttermilk, whey and casein are added to flour in combination with the ingredients or additives mentioned so far. These dairy products are used in either powdered or liquid form as well as either whole or in the form of defatted powder. In such cases, the proteins added to the dough increase its water binding capacity and provide a juicy crumb.

15.4.1.5 Influence of Additives on Baking Properties of Rye Flour

Rye flour often requires an improved water binding capacity. For this purpose, 2–4% of pregelatinized flour is added. In addition, artificial acidification of the rye dough is practiced; hence both aspects will be covered.

15.4.1.5.1 Pregelatinized Flour

Pregelatinized flour is made from ground cereals such as wheat, rye, rice, millet, etc. by cooking and steaming in autoclaves followed by drying and repeated milling. Such pregelatinized flours are sometimes blended with guar flour or alginates.

15.4.1.5.2 Acids

Rye flour is used in bread baking with sour dough fermentation (cf. 15.4.2.2).

Artificial acidification can be achieved by the addition of lactic, acetic, tartaric or citric acid to rye dough or by adding acidic forms of sodium and calcium salts of ortho- and/or pyrophosphoric acids.

Other preparations for acidification, the so-called dry or instant acids, consist of pregelatinized flour blended with a sour dough concentrate or of cereal mash prefermented by lactic bacteria. The acid values (for definitions see 15.4.1.1.1) vary from 100–1000.

15.4.1.6 Dough Leavening Agents

Dough consisting only of flour and water gives a dense flat cake. Baked products with a porous crumb, such as bread, are obtained only after the dough is leavened. This is achieved for wheat dough by addition of yeast while, for fine baked products, baking powders are used. Rye dough leavening is achieved by a sour dough formulation which includes lactic and acetic acid bacteria.

15.4.1.6.1 Yeast

A given amount (Table 15.48) of surface-fermenting yeast, *Saccharomyces cerevisiae*, is used. While normal yeasts preferentially degrade sucrose rather than maltose, special rapidly fermenting yeasts are used which metabolize both disaccharides at the same rate, shortening the fermentation time.

Yeasts differ in their growth temperature optima (24–26 °C) and their fermentation temperature optima (28–32 °C). The optimum pH for growth is 4.0–5.0. In addition to CO_2 and ethanol, which raise the dough, the yeast forms a variety of aroma compounds (cf. 5.3.2.1). Whether other compounds released by the growth of yeast would affect the dough rheology is unclear; there appears to be no effect of yeast proteinase and GSH.

15.4.1.6.2 Chemical Leavening Agents

The interaction of water, acid, heat and chemical leavening agents (baking powders) releases CO_2. The release of gas may occur in the dough prior to or during oven baking. The agents consist of a CO_2-generating source, as a rule sodium bicarbonate, and an acid carrier, usually disodium dihydrogendiphosphate, sometimes monocalcium phosphate $[Ca(H_2PO_4)_2]$. Glucono-δ-lactone or tartar (acidic potassium tartrate) are used as acid carrier for phosphate-free baking powder. The phosphate-free leavening mixtures produced on an industrial scale contain citric acid or its acidic sodium salt. In baking powder, the two reactive constituents are blended with a filler which consists of corn, rice, wheat or tapioca starch or sometimes dried wheat flour. The filler content in the powder is up to 30%. The role of the filler is to prevent premature release of CO_2. The market also offers baking powders flavored with vanillin or ethyl vanillin.

For every 500 g of flour, baking powder should develop 2.35–2.85 g CO_2, equivalent to about 1.25 liters.

In individual cases, $NaHCO_3$ alone is used for some flat shelf-stable cookies and ammonium hydrogencarbonate (NH_4HCO_3) for many others. Ginger and honey cookies are leavened by NH_4HCO_3, mostly together with potash (K_2CO_3). To a small extent, a 1:1 mixture of ammonium hydrogencarbonate and ammonium carbamate $(H_2NCOONH_4)$ is used in some countries. Both decompose above 60 °C to NH_3, CO_2 and water.

15.4.2 Dough Preparation

15.4.2.1 Addition of Yeast to Wheat Dough

15.4.2.1.1 Direct Addition

Flour, water, yeast, salt and other ingredients are directly mixed into the dough.

15.4.2.1.2 Indirect Addition

Yeast is propagated at 25–27 °C in a well-aerated liquid pre-ferment which contains flour, water and some sugar. After a given time, the liquid is blended with the bulk of flour and water and other ingredients and then made into a dough in a mixer.

For continuous indirect addition of yeast, special liquid starters (sponges) with a pH of 5.0–5.3 are also used with incubation at 38 °C to develop aroma. Such matured fermented sponge is then metered continuously into a kneader which handles the dough.

Table 15.48. Amount of yeast used in bread and other baked products

Baked product	Yeast added[a] (%)
Rye bread	0.5–1.5
Rye mix bread	1.0–2.0
Wheat mix bread	1.5–2.5
Wheat bread	2.0–4.0
Breakfast rolls	4.0–6.0
Rusk ("Zwieback")	6.0–10.0

[a] Based on flour content.

15.4.2.2 Sour Dough Making

In sour dough making (lowering the pH to 4.0–4.3) rye flour acquires the aroma and taste properties so typical of rye bread (cf. 15.1.5).

Yeast (*Saccharomyces cerevisae, Saccharomyces minor* and others), which are mainly responsible for dough leavening, and a complex bacterial flora in which lactic acid-forming organisms dominate (*Lactobacillus plantarum, Lactobac. San Francisco* and *Lactobacillus brevis*) are present in sour dough.

Sour dough is prepared by various procedures which differ considerably in the length of time required (Fig. 15.38). A three-stage procedure takes into account the optimum temperature and humidity needs of yeast and bacteria. Yeast prefer to grow at 26 °C, while the bacteria of interest grow best at 35 °C.

In setting up a three-stage process, initially an aqueous flour suspension is inoculated. This is the first "full sour" build-up stage (Fig. 15.38). After maturation, further amounts of flour and water are added and the process is continued with a "basic sour" stage at 35 °C and then, in a similar way, continued with an additional "full sour" third stage at 26 °C.

The incubation conditions given in Fig. 15.38 are only the essential outline. Temperature deviations influence the spectrum of fermentation products.

At warmer temperatures (30–35 °C) lactic acid is preferentially formed (Fig. 15.39), while at cooler temperatures (20–25 °C) more acetic acid is produced. The desirable lactic acid: acetic acid ratio, called the "fermentation ratio", is close to 80:20. A ratio with a higher acetic acid concentration gives too sharp an acid taste. The portion of rye flour in the end-product determines the amount of rye sour (full sour) to be added to the dough in the preparation stage. Thus, for rye bread the sour dough to be added is 35–45%, while for a rye mix bread it is 40–60% (on the basis of rye flour). In the short sour method the growth of yeast is negligible. Only a single sour stage, which lasts about 3 h, is involved, yeast is added and the dough is ready for use (Fig. 15.38). However, this short method requires a relatively high content of starter saved from a previous ripe sour. Additional time can be saved by using dough acidifiers (cf. 15.4.1.5.2 and Fig. 15.38). In short sour processes all the organic acids needed for the sour taste of the rye endproduct are present. However, there is a lack of aroma compounds and precursors from which odorants can be generated during baking. In a three-stage rye sour procedure, part of the flour proteins is hydrolyzed by proteinases of the microflora into free amino acids which then

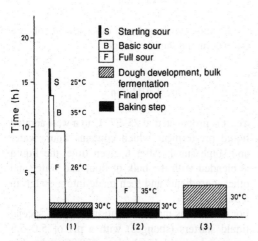

Fig. 15.38. Time requirement for various sour dough development methods (according to *Rothe*, 1974). *1* A three step process, *2* short sour, *3* dough souring agents used

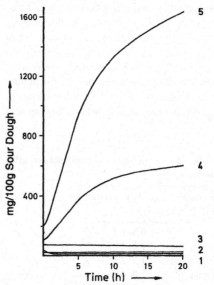

Fig. 15.39. Acid formation in sour dough versus time at 30 °C (according to *Rabe*, 1980). *1* Malate, *2* pyruvate, *3* citrate, *4* acetate, and *5* lactate

participate in *Maillard* reactions during baking, providing the more intense aroma.

15.4.2.3 Kneading

The kneading process is characterized by the following stages: mixing of the ingredients and seasonings; dough development and dough plastification.

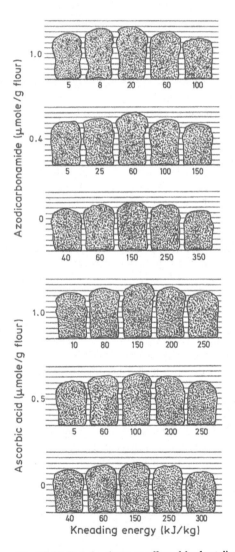

Fig. 15.40. Bread volume as affected by kneading energy input (according to *Frazier* et al., 1979)

The energy input into dough kneading, the dough properties and baking volumes are interrelated. For each dough the baking volume passes through an optimum which is dependent on kneading energy input (Fig. 15.40). This optimum shifts towards lower energy input with a flour of weak gluten content and towards higher energy input with flours of strong gluten content; and, as expected, the position of the optimum can be influenced by flour improvers. Increased additions, especially of azodicarbonamide, to the dough result in a successive drop in kneading energy input (Fig. 15.40).

As the kneading energy moves away from the optimum, the dough becomes wetter, it starts to stick to trough walls and its gasholding ability drops (cf. 15.4.2.5 and Fig. 15.44, *14* and *56*). Dough development of wheat flours requires close to double the kneading time of rye flours.

The machines used for kneading are grouped according to their performance based on kneading time: fast, intensive, and high power kneaders and mixers (Table 15.49). However, the groups are not sharply divided. As the kneading speed increases, the temperature of the dough rises (Table 15.49). Hence, cooling must be used during kneading to keep the temperature at 22–30 °C or, with high speed mixers, at 26–33 °C. The mixer, in a true sense, does not knead the dough, but rips or ruptures it. This could reduce the stability of the dough to such an extent that it could be baked only as panbread (in which case the pan walls support the dough) but not as bread made from selfsupporting dough.

Table 15.49. Examples for kneading conditions in white bread dough making

Dough mixer/ kneader	Speed (rpm)	Kneading time (min)	Dough heat[a] ΔT(°C)
Rapid kneader	60–75	20	2
Intensive kneader	120–180	10	5
High power kneader	450	3–5	
Mixer	1440	1	9
Mixer	2900	0.75	14

[a] Temperature rise during kneading time.

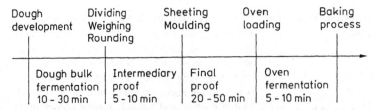

Fig. 15.41. Fermentation process for biologically leavened dough; temperature 26–32 °C (according to *Bueskens*, 1978)

15.4.2.4 Fermentation

Dough passes through several stages of fermentation in the presence of growing yeast, a biological leavening agent (Fig. 15.41). After initial fermentation, the dough is divided and scaled, then the dough pieces are rounded-off. A short fermentation is followed by sheeting and moulded dough fermentation. The dough acquires its enlarged final volume in the oven. The yeast produces CO_2 and ethanol which, as long as they do not dissolve in the aqueous phase of the dough, expand the air bubbles (10^2–10^5 /mm^3) that arise in the dough during kneading. The volume of a square white loaf increases 4 to 5 fold and more during initial, intermediate, and moulded dough fermentation and 5 to 7 fold during oven fermentation. The length of time of the fermentation varies. It depends on flour type (cf. Fig. 15.42), seasonings incorporated, the amount of yeast and oven temperature. The flour character determines the fermentation tolerance, i.e. the minimum or maximum time after which the fermentation has to be stopped and the dough loaded into the oven. Dough fermentation of a weak gluten flour is rapid, but its fermentation tolerance is low.

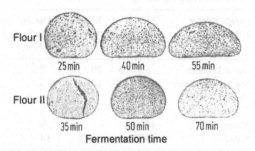

Fig. 15.42. The effect of fermentation time on baking results. (rye-mix bread with two flours which differ in baking quality; according to *Bueskens*, 1978)

The main dough fermentation step (cf. Fig. 15.41) can be substantially shortened by kneading the dough energetically and/or by incorporating fast-acting additives (for example, a mixture of bromates, ascorbic acid and cysteine) into the dough. This provides a favorable dough structure, able to accommodate large amounts of yeast. This is the basis for "no-time" dough making procedures, which provide a continuous flow of dough.

In continuously operated baking processes, the resting times required during the working of dough (intermediate and final fermentation) are realized in air conditioned fermentation rooms. The resting dough forms pass through these rooms with a defined speed.

15.4.2.5 Events Involved in Dough Making and Dough Strengthening

15.4.2.5.1 Dough Making

Bread dough is prepared by mixing water and flour (70:30 w/w). Water uptake, which depends on flour type, predetermines most of the subsequent reactions. A high water uptake favors the mobility of all the constituents involved in reactions, e.g., enzymatic degradation of starch into reducing sugars (Fig. 15.43).

Observation of wheat dough development by light or scanning electron microscopy reveals that a sequence of forceful changes occurs in the arrangement of the water-insoluble flour proteins.

When a light microscope is used to look at a wheat flour particle under water, practically no protein structure is discernible (Fig. 15.44, *1a*). If the particle is stretched in one direction by moving the slide cover glass against the microscope slide, numerous protein strands with inserted starch granules become visible. These strands are

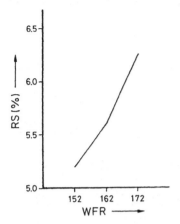

Fig. 15.43. The reducing sugar content in wheat bread crumb as affected by water content of dough (according to *Wassermann* and *Doerfner*, 1971).

$$\text{Water flour ratio (WFR)} = \frac{(\text{flour} + \text{water}) \times 100}{\text{flour}}$$

RS: Reducing sugar expressed as maltose

oriented in the direction of stretching (*1b*, *1c*) and partially adhere to the glass at one end (*2*). If circular movements are made with the cover glass, the protein strands are two-dimensionally stressed and most of the starch is released (*3* and *4*). As a result of the stickiness of the protein, the strands can be easily aggregated to a ball by further rotary movements. Another way of representing the protein structure is to spread flour particles on the water surface (*5*). The protein strands which radially grow out of the flour particle during hydration are linked by protein films and, thus, bent. After appropriate fixation of these structures, the protein films can be selectively removed with 60% ethanol and the strands lose their taut structure (*8*). The ethanol-soluble gliadins and the strand-shaped insoluble glutenins possibly exist separately even in the grain. Under a scanning electron microscope at higher magnification, a flour particle, after the removal of starch with amylase, looks like a protein sponge (*6*) in which starch granules were inserted. One-dimensional stretching gives strands (*7*). Similar gluten structures are detected in dough as in flour particles, but the proteins form differently arranged aggregates, which are more resistant to tear, because of the strong mechanical treatment. In ripe, dry grain, the gluten proteins are stored as particles in the endosperm cells. The diameter of these particles is 1–10 μm, depending on the

wheat cultivar. In addition, these particles can still fuse together in the cell to form aggregates with a diameter of up to 50 μm. At the start of the kneading process, the particles and the aggregates are hydrated and they form net-like structures (*10*) as a result of their cohesive properties. The exceptional cohesiveness of the gluten proteins is due to their high glutamine content, which allows the formation of innumerable hydrogen bridges. Due to the mechanical processing in the kneader, the proteins are increasingly brought into close contact so that they aggregate to larger networks (*12*). Strong shear forces are present in the dough because of the low amount of free water. Thus, like in a ball mill, the proteins are mixed with other flour constituents and can react with them. With increasing kneading time, the interactions between the gluten proteins become stronger and stronger, making the structures denser and denser (*14*) until the kneading resistance reaches a maximum, which can be measured in a farinograph. As a result of the high content of starch (70% of the dough), which is homogenously distributed in the dough, the net-like structures are still very thin (Fig. 15.46a). These structures are partially broken down (*56*) again by overkneading, weakening the support function of the gluten. If the gluten is extended two-dimensionally to a thin membrane, it starts to perforate (*15*) and with increasing relaxation forms strands, which are as round as possible. The energetic state of these strands is lower than that of the membranes because of the low surface area.

The connected gluten framework formed in this way is responsible for the gas retention capacity of wheat dough. In fact, stands which are as thick as possible but easily extensible under the pressure of the fermentation gases are of advantage for the stability of the dough.

With the help of transmission electron microscopy, it can be shown at still higher magnification that the surface of unstretched protein strands has an irregular globular structure (*18*). As a result of the washing out of gliadin with a large excess of water, these strands should essentially consist of glutenin. On twodimensional stretching, the globular surface is flattened (*19*) and platelet-like forms appear (*20*) which are arranged parallel to the plane of stretching and are less than 10 nm in thickness. The globular surface structures are probably highly tangled, strand-shaped proteins which are unfolded due to

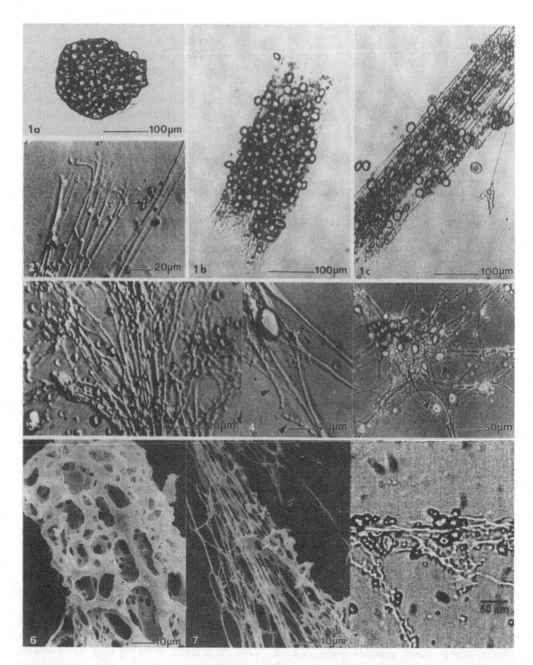

Fig. 15.44. *1–5 Light microscopy. 1a*: Individual flour particles in water. *1b*: Flour particles, slightly extended by moving the cover glass. *1c*: Flour particles, highly extended; *2*: Extended protein strands with one end adhering to the glass. *3*: Network of protein strands after two-dimensional extension of a flour particle. *4*: Protein film (*arrows*) between bent protein strands. *5*: Flour particles stretched on the water surface. Protein films between bent protein strands (*arrows*). *6–17, 56–57 Scanning electron microscopy. 6*: Flour particle unstretched, holes from enzymatically removed starch. *7*: Flour particle, extended (starch removed). *8*: Spread preparation (*5*) washed with 60% ethanol.

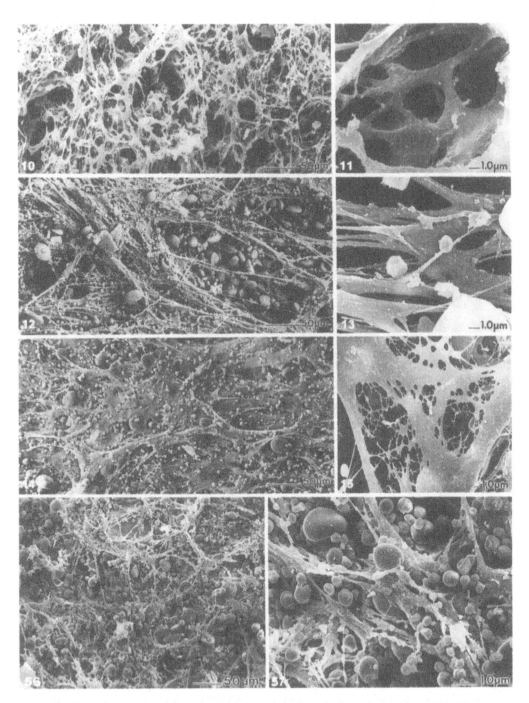

Fig. 15.44. (continued) *10*: Dough after addition of water (starch removed), not kneaded, practically unstretched, connected protein network. *11*: Detail from *10*. *12*: Kneaded dough with extended network. *13*: Detail from *12*, beginning film formation between protein strands. *14*: Optimally kneaded dough. *15*: Detail from *14* with partially perforated protein film. *56*: Highly overkneaded dough with short irregular protein strands. *57*: Detail from *56*.

Fig. 15.44. (continued) *18–22*: *Transmission electron microscopy.* *18*: Slightly extended gluten protein strand with rough surface, enlarged section. *19*: Protein strand, more extended and with smoother surface compared with *18*. *20*: Platelet-like structures on a highly extended gluten film. *21*: Protein threads on a gluten film in a) water, b) triethylamine solution, c) dithioerythritol solution. *22*: a) Highly extended protein filbrils with thickenings, b,c) enlarged

mechanical stress and are stabilized in the form of superimposed layers due to intermolecular interactions.

These protein threads have a diameter of 10–30 nm and all look similar, irrespective of the type of preparation, e. g., in water (*21a*), in triethylamine (*21b*) or in dithioerythritol solution (*21c*). One-dimensional stretching causes individual protein threads to be partly stretched into fibrils, which, including the metal layers vapor deposited for stabilization, have a diameter of only 3 nm (*22a, b, c*).

Based on the microscope pictures, dough formation can be summarized as follows. The individual flour particles consist of a sponge-like protein matrix in which starch is embedded. After addition of water, the matrix protein becomes sticky and causes the flour particles to form a continuous structure on kneading. At the same time, the protein matrix is extended and protein films are formed at the branch points

of the strands. In an optimally kneaded dough, the protein films are the predominant structural element and should contribute to the gas-holding capacity. Further kneading causes increased perforation of the films with formation of short, irregular protein strands, which are characteristic of overkneaded dough.

15.4.2.5.2 Dough Strengthening

A wheat dough is kneaded to the optimum and pressed, rolled or formed after a resting time of, e. g., up to 3 minutes or longer. This dough is subjected to a relatively weak shear compared with kneading. In this case, the resistance to extension is increased in tensile tests in the extensograph (Fig. 15.45).

Microscopic studies show that an unmixing of starch and gluten occurs. While starch and gluten are homogeneously distributed in freshly kneaded dough (Fig. 15.46a), the gluten relaxes in the sub-

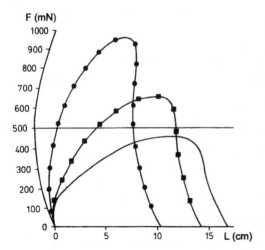

F (mN)

Fig. 15.45. Tensile tests with dough made of flour from the wheat cultivar Soisson (according to *Kieffer* and *Stein*, 1999). Dough after 45 min rest without shear (■–■), dough with shear after 135 min (●–●), dough without shear after 135 min (—)

sequent dough resting stage and contracts to form islands which are only slightly connected to each other (Fig. 15.46b). In shearing (Fig. 15.46c), these islands aggregate to form a network made of thicker gluten strands. The extent of the effect depends on the variety of wheat. In cultivars with weak gluten, the gluten is finely distributed even after shearing and the net is only weakly formed. The driving force behind the unmixing of starch and gluten is the tendency of gluten proteins to aggregate via intermolecular interactions, e. g., hydrogen bridges and hydrophobic interactions. Rye contains fewer gluten proteins than wheat. In dough development, its aggregation is additionally hindered by pentosans so that no gluten network can be formed.

Baking experiments have shown that the dough stability during fermentation is better, the baking form rounder and the baking volume larger if a clear unmixing of starch and gluten occurs due to the shearing of dough.

15.4.3 Baking Process

15.4.3.1 Conditions

The oven temperature and time of baking for some baked products are summarized in Table 15.50. Conditions for baking of rye and

rye mix bread sometimes deviate from these values. They are prebaked at higher temperatures, for instance at 400 °C for 1–3 min, and then post-baked at 150 °C (for the effect on quality see Table 15.51). In a continuous process, tunnel-type ovens with circulation heaters are used. Gratings frequently serve as the conveyor band.

In an oven with the temperatures given in Table 15.50, since heat transfer occurs slowly in dough, there is a steep temperature gradient, $200 \rightarrow 120\,°C$, inward from the crust of the dough piece. By the end of baking, a temperature of 96 °C is attained within the product. Higher temperatures up to 106 °C are found when the crust is able to resist the rise in inner steam pressure. The water evaporates only in the crust region during dough baking. Water diffusion towards the center of the bread can give the fresh crumb a higher moisture content than the dough. The steam concentration in the oven also affects the baking results. A steam header is provided in most oven designs to regulate oven moisture.

A baking weight loss is experienced as a result of water evaporation during crust formation. The extent of the loss is related to the form and size of the baked bread and is 8–14% of the fresh dough weight.

15.4.3.2 Chemical and Physical Changes – Formation of Crumb

The foamy texture of dough is changed into the spongy texture of crumb by baking. The following processes are involved in this conversion.

Up to ca. 50 °C, yeast produces CO_2 and ethanol at a rate that initially increases. At the same time, water and ethanol evaporate and, together with the liberated CO_2, expand the exisiting gas bubbles, further increasing the volume of the baked product. Parallel to this, the viscosity of the dough falls rapidly in the lower temperature range, reaches a minimum at ca. 60 °C, and then increases rapidly (Fig. 15.47). The increase is caused, on the one hand, by the swelling of starch and the accompanying release of amylose and, on the other hand, by protein denaturation. These processes result in a tremendous increase in the tensile stress of the dough and in the pressure in the gas bubbles at temperatures

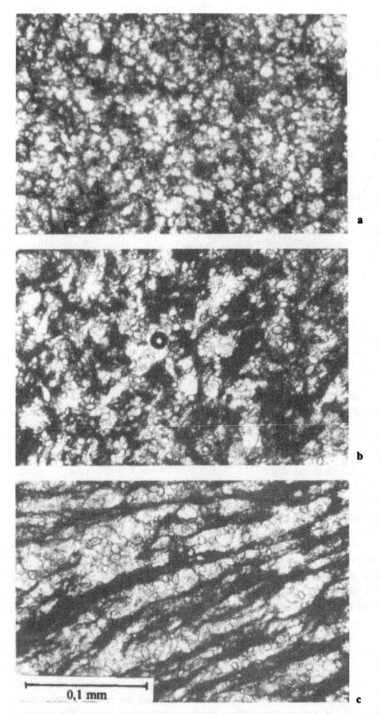

Fig. 15.46. Frozen section (thickness: 40 μm) of dough made of the flour Soisson (according to *Kieffer* and *Stein*, 1999). a) dough, freshly kneaded; b) dough, relaxed for 45 min; c) dough after 135 min and shearing

Table 15.50. Baking times and temperatures

Baked product	Weight (g)	Baking time (min)	Oven temperature (°C)
Buns, rolls and other small baked products	45	18–20	250–240
Wheat bread (self-supported dough)[a]	500	25–30	240–230
Wheat bread (pan-baked)[b]	500	35–40	240–230
Wheat bread (self-supported dough)	1000	40–50	240–220
Rye mix bread (self-supported dough)	1500	55–65	250–200
Rye bread (self-supported dough)	1500	60–70	260–200
Pumpernickel (pan-baked)	3000	16–14 hrs.	180–100

[a] Hearth bread.
[b] Pan bread.

Table 15.51. The effect of baking time and temperature on the quality of rye whole meal bread

Baking time (min)	90	180	270
Baking temperature (°C)	240–210	210–185	185–160
Bread yield (ml)	142	142	140
Crust strength (mm)	4	5	6
Taste	raw, slightly aromatic	aromatic	strongly aromatic

above ca. 60 °C (Fig. 15.47). The membranes give way and become permeable, allowing CO_2, ethanol, and water vapor to escape. The baking volume decreases slightly until the denatured proteins, with swollen and partially gelatinized starch, form a stable crumb framework, which contains pores down to 3 µm in diameter. Thin-walled membranes which can stand a greater increase in temperature on stretching, without becoming gas permeable; this is the prerequisite for a baked product with a large volume and uniform fine pores. A relatively large amount of high-molecular glutelins in gluten has a favorable effect. Dough made from wheat varieties with poor baking properties becomes gas permeable at a relatively low temperature and the baking volume remains correspondingly low. The extent of starch swelling depends on the available water. The water in dough is preferentially bound by prolamins, glutelins and pentosans. Part of this water becomes available to swell the starch during baking. Limited starch swelling results in a brittle crumb, whereas extensive swelling makes the crumb greasy or gluey.

In contrast to the crumb, the starch granules of the crust surface gelatinize almost completely. This is expecially the case when the oven humidity is high, e. g., when baking occurs below a steam header. Investigations involving gluten and starch mixtures to which the emulsifier stearyl-2-lactylate was added revealed that lipid transfer occurs from gluten to starch during heating of the mixture above 50 °C (Table 15.52). Apparently, the high swelling and gelatinization of the starch granules, which occurs above 50 °C (cf. Table 4.20), promotes lipid binding.

The specific volume of white bread is higher than that of rye bread (Table 15.53). The rye crumb is stronger and less elastic, suggesting that the pentosans can not fully compensate for the lack

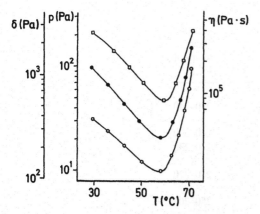

Fig. 15.47. Viscosity (η, □—□) and tensile stress (δ, o—o) of a wheat dough as well as pressure (p, ●—●) in the gas bubbles as a function of temperature during the baking process (according to *Bloksma*, 1990)

of texturizing quality of rye proteins (cf. 15.1.5). Heating of a dough accelerates enzymatic reactions, e. g., starch degradation (cf. 15.4.2.4). Above the "temperature optimum" (cf. 2.5.4.3) the reactions are inhibited by denaturation of the enzymes.

The vitamins of the B group are lost to different extents during baking. In white bread, the losses amount to 20% (flour type 550)−50% (flour type 1150) of thiamine, 6−14% of riboflavin and 0−15% of pyridoxine.

Starch degrades to dextrins, mono- and disaccharides at the relatively high temperatures to which the outer part of the dough is exposed. Caramelization and nonenzymatic browning reactions also occur, providing the sweetness and color of the crust. The thickness of the crust is dependent on temperature and baking time (Table 15.51) and type of baked product (Table 15.54). The composition of some types of bread is presented in Table 15.55.

15.4.3.3 Aroma

15.4.3.3.1 White Bread Crust

The substances which produce the aroma profile of a loaf of French bread (baguette) (Table 15.56) originate from the crust. They are listed in Table 15.57. 2-Acetyl-1-pyrroline

Table 15.52. The effect of temperature on stearyl-2-lactylate (SSL) binding in a blend of gluten and starch[a]

T (°C)	SSL free[b]	SSL bound[b] to	
		gluten	starch
30	22.0	64	14
40	20.0	66	14
50	22.0	62	16
60	20.0	6	74
70	16.0	6	78
80	12.0	8	80
90	12.0	2	86

[a] Blends of 17.9 g starch, 2.7 g gluten and 0.103 g SSL.
[b] Values in % of total SSL.

Table 15.53. Specific volumes[a] of bread

Bread variety	ml/g
Toast bread	3.5–4.0
White bread	3.3–3.7
White mix bread[b]	2.5–3.0
Rye mix bread[b]	2.1–2.6
Rye bread	1.9–2.4

[a] Specific volume = volume/weight.
[b] cf. Table 15.63.

Table 15.54. Crumb and crust ratios in different bread varieties

Bread variety	Crumb (%)	Crust (%)
Buns, rolls (50 g)	72.5	27.5
Stick (French) white bread	68.5	31.5
White bread, pan-baked (500 g)	75.0	25.0
White bread (self-supported dough, 500 g)	73.8	26.2
Rye mix bread (self-supported dough, 1000 g)	73.3	26.7
Rye mix bread (pan-baked, 1000 g)	84.5	15.5

produces the roasty note in the aroma profile and the *Strecker* aldehydes methylpropanal, 2- and 3-methylbutanal the malty note. The compounds (E)-2-nonenal and 1-octen-3-one are primarily responsible for the fatty impression.

The aroma of a baguette is not stable. Even four hours after the bread has left the oven, the intensities of the malty and sweet notes in the

Table 15.55. Chemical composition of some types of bread

Bread	Water	Protein (N × 5.8)	Available carbohydrates	Dietary fiber	Lipid	Minerals
	(%)	(%)	(%)	(%)	(%)	(%)
White bread	38.3	7.6	49.7	3.2	1.2	1.6
White mix bread	37.6	6.2	47.7	4.6	1.1	1.5
Rye mix bread	39.1	6.4	43.7	6.1	1.1	1.8
Rye bread	38.1	6.2	45.8	6.5	1.0	1.6
Rye whole grain bread	42.0	6.8	38.7	8.1	1.2	1.5
Crisp bread	7.0	9.4	66.1	14.6	1.4	2.3

Table 15.56. Aroma profile of baguette after 1 h and 4 h[a]

Aroma quality	Intensity[b] after	
	1 h	4 h
Roasty	1.8	1.9
Malty	1.9	0.4
Sweet	1.2	0
Fatty	1.0	2.3

[a] After leaving the oven the baguettes were cooled for 1 h at room temperature, then wrapped in aluminium foil and stored at room temperature for another 3 h.
[b] The intensity of each aroma quality was evaluated from 0–3 (strong). Average values of five testers.

Table 15.57. Odorants of the baguette crust[a]

Compound	Concentration (µg/kg)	Aroma-value[b]
Methylpropanal	1750	31
2-Methylbutanal	962	18
3-Methylbutanal	335	26
Methional	29	107
Dimethyl trisulfide	5.1	59
2,3-Butandione	1320	203
2-Acetyl-1-pyrroline	16	2191
2-Ethyl-3,5-dimethylpyrazine	3.2	19
Hexanal	214	7
1-Octen-3-one	6.6	150
(E)-2-Nonenal	87	164
(E,E)-2,4-Decadienal	56	21

[a] After leaving the oven the baguettes were cooled for 1 h at room temperature, then the crust was separated, frozen with liquid nitrogen and analyzed.
[b] On the basis of the odor threshold of the compound on starch.

aroma profile are greatly reduced and the fatty note predominates (Table 15.56). These changes are caused by the differences in the evaporation rates of the aroma substances. To determine these rates, the amounts of odorants which evaporate from a baguette in 15 min are collected at the times given in Table 15.58 and measured. The results in Table 15.58 show the evaporation rates of the malty smelling aldehydes methylpropanal, 2- and 3-methylbutanal decrease continuously in accordance with the changes in the aroma profile. This also results in the fact that odorants like the fatty smelling (E)-2-nonenal are more detectable with time. However, the intensity of the fatty note also increases because the evaporation rate of (E)-2-nonenal increases (Table 15.58), possibly due to the decomposition of the linoleic acid hydroperoxides formed during baking.

2-Acetyl-1-pyrroline evaporates uniformly (Table 15.58); the intensity of the roasty note is correspondingly stable during storage for 4 hours (Table 15.56). The losses of this odorant in the white bread crust on longer storage is shown in Table 15.59.

The aroma is influenced by the recipe but also by the fermentation, e. g., the *Strecker* aldehydes increase and those from lipid peroxidation decrease if the dough matures at lower temperatures (Table 15.60). An extension of the kneading process

Table 15.58. Concentrations of the odorants in the headspace of baguettes as a function of storage at room temperature[a]

Odorants	Concentration (ng/l air) after		
	1 h	2.5 h	4 h
Methylpropanal	830	536	400
2-Methylbutanal	320	230	170
3-Methylbutanal	150	85	68
2,3-Butandione	980	705	670
2-Acetyl-1-pyrroline	3.7	3.3	3.7
Hexanal	216	237	254
(E)-2-Nonenal	28	36	44
(E,E)-2,4-Decadienal	7.8	6.5	6.6

[a] To determine the concentration, the air above the baguette was collected for 15 min.

Table 15.59. Decrease (%) in 2-acetyl-1-pyrroline in the crust of white bread during storage

Time (h)	2-Acetyl-1-pyrroline
0	0
3	46
24	77
168	89

Table 15.60. Influence of the fermentation time and temperature on the concentrations of odorants in the baguette crust[a]

Odorant	Concentration (μg/kg)	
	Baguette I	Baguette II
2-Acetyl-1-pyrroline	16	14
Methylpropanal	1733	4331
2-Methylbutanal	1147	1487
3-Methylbutanal	426	680
Methional	31	49
1-Octen-3-one	3.8	2.1
(E)-2-Nonenal	61.8	40.4

[a] Dough I was fermented for 2 h 40 min at 26 °C, dough II for 2 h at 22 °C and then 18 h at 4 °C.

and the resulting increase in the dough temperature result in a decrease in the *Strecker* aldehydes in the crust (Fig. 15.48).

15.4.3.3.2 White Bread Crumb

In dilution analyses, 3-methylbutanol, 2-phenylethanol, methional, (E)-2-nonenal and (E,E)-2,4-decadienal were identified as the most important odorants in the baguette crumb. The odorants which produce the intense aroma note were detected in a comparison of two baguettes which were baked with two differently composed pre-fermented doughs. The crumb of baguette I had a pleasant intense odor, but in II this note was weak and a rancid aroma defect appeared. Table 15.61 shows that the concentrations of 2-phenylethanol and 3-methylbutanol, which have a flowery/intense and alcoholic odor, are higher in I than in II. The higher concentration of the sweaty 2-/3-methylbutyric acid in II produces the rancid aroma defect. A low yeast concentration in a liquid preliminary dough, which is only 1.5% based on the finished dough I compared with 4.6% in dough II, is the prerequisite for an optimal formation of both the alcohols shown in Fig. 15.49. The curves show that the concentrations of the two aroma substances reach a plateau after eight hours under the selected conditions. Their precursors phenylalanine and leucine, which originate in the flour and are degraded by yeast via the *Ehrlich* pathway to give the odorants (cf. Formula 15.8) are converted after this time or the bioconversion stops because important yeast nutrients are increasingly lacking.

Table 15.61. Odorants of white bread crumb – comparison of two kinds of bread subjected to different dough making[a]

Compound	Concentration (mg/kg)	
	Bread I	Bread II
2-Phenylethanol	11.8	2.87
3-Methylbutanol	18.1	9.7
2-/3-Methylbutyric acid	0.55	1.5

[a] Recipe (kg): flour (I: 4.15; II: 4.9), water (I: 2.27; II: 2.825), salt (I, II: 0.11), yeast (I: 0.125; II: 0.325), pre-ferment (I: 2.005 of A; II: 0.5 of B). Pre-ferment A: a suspension of flour (1 kg), water (1 kg) and yeast (5 g) was incubated for 15 h at 30 °C. Pre-ferment B: dough made of flour (250 g), water (175 g) and yeast (75 g) was incubated as in A.

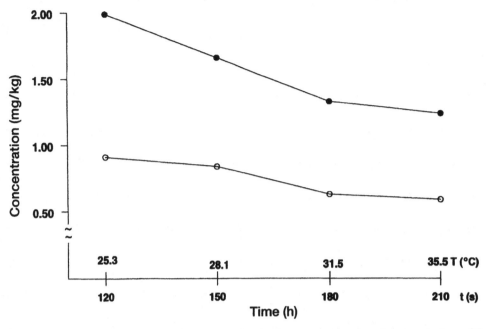

Fig. 15.48. Influence of the kneading intensity on the dough temperature and on the concentrations of 2- and 3-methylbutanal in the baguette crust. 2-Methylbutanal (•—•), 3-methylbutanal (o—o). *Abscissa*: kneading time (s) and dough temperature T (°C) (according to *Zehentbauer* and *Grosch*, 1998)

The crumb contains the precursors of the roasty odorants 2-acetyl-1-pyrroline and 2-acetyltetrahydropyridine (cf. 5.3.1.6), but the temperature in the baking process is sufficient to form these substances only in the crust. If white bread is toasted, the two odorants are formed with increasing browning of the toast (Fig. 15.50), whereby 2-acetyl-1-pyrroline increases much more. Of the odorants from the *Maillard* reaction, 4-hydroxy-2,5-dimethyl-3(2H)-furanone also appears with a high aroma value. In the case of fiber-enriched toast bread 2-/3-methylbutyric acid is a critical odorant. The addition of oat bran is preferable to wheat bran because it yields less 2-/3-methylbutyric acid and, consequently, a rancid/sweaty aroma defect is avoided.

Important precursors of 2-acetyl-1-pyrroline are ornithine and 2-oxopropanal (cf. 5.3.1.7), mainly originate from yeast metabolism. The concentrations of 2-acetyl-1-pyrroline and 2-acetyltetrahydropyridine in the crumb are less than those in the crust by a factor of about 30. The reason being that only in the crust area is the temperature high enough to release aroma substances or their precursors from yeast.

15.4.3.3.3 Rye Bread Crust

Dilution analyses show that the following compounds are involved in the aroma of rye bread

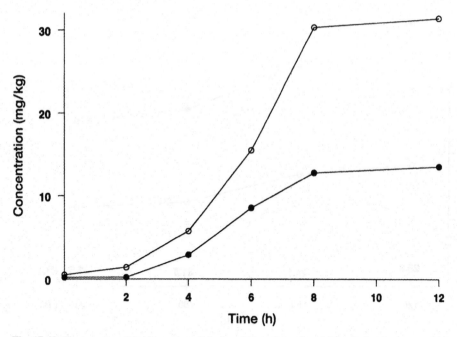

Fig. 15.49. Time curves of the formation of 2-phenylethanol (•–•) and 3-methylbutanol (○–○) in a prefermented dough (according to *Gassenmeier* and *Schieberle*, 1995): flour (59 g), water (50 g) and yeast (0.25 g) were incubated at 30 °C

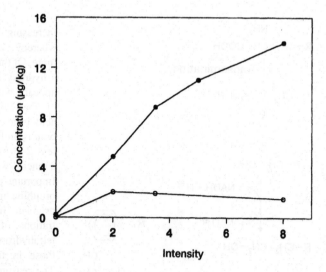

Fig. 15.50. Formation of 2-acetyl-1-pyrroline (•–•) and 2-acetyltetrahydropyridine (○–○) on toasting white bread (according to *Rychlik* and *Grosch*). Abscissa: intensity of browning (8: very strong).

crust with high FD values: methional, 2-/3-methylbutanal, 4-hydroxy-2,5-dimethyl-3(2H)-furanone (HD3F), 2-furfurylthiol, (Z)-4-heptenal, 1-octen-3-one, (Z)-1,5-octadien-3-one, phenylacetaldehyde, 2,3-diethyl-5-methylpyrazine and (E)-β-damascenone. The difference in the crust aroma of white and rye bread is due to the fact that the rye bread crust contains much more 3-methylbutanal, methional and HD3F but less 2-acetyl-1-pyrroline (Table 15.62). 2-Furfurylthiol also contributes only to the aroma of the rye bread crust.

Table 15.62. Concentrations of odorants in the crusts of white bread and rye bread

Compound	Concentration (µg/kg)	
	White bread	Rye bread
2-Acetyl-1-pyrroline	19	0.8
3-Methylbutanal	1406	3295
Methional	51	480
(E)-2-Nonenal	56	45
4-Hydroxy-2,5-dimethyl-3(2H)-furanone	1920	4310

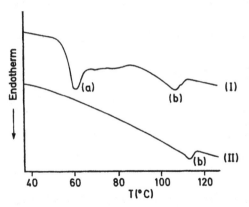

Fig. 15.51. DSC thermograms of wheat starch in water (45:55, g/g) I: native starch, II: gelatinized starch (according to *Slade, Levine*, 1991)

15.4.4 Changes During Storage

Bread quality changes rapidly during storage. Due to moisture adsorption, the crust loses its crispiness and glossyness. The aroma compounds of freshly baked bread evaporate or are entrapped preferentially by amylose helices which occur in the crumb. Repeated heating of aged bread releases these compounds. Very labile aroma compounds also contribute to the aroma of bread, e. g., 2-acetyl-1-pyrroline. They decrease rapidly on storage due to oxidation or other reactions (Table 15.59).

The crumb structure also changes, although at a lower rate. The crumb becomes firm, its elasticity and juiciness are lost, and it crumbles more easily. The so-called staling defect of the crumb is basically a starch retrogradation phenomenon (cf. 4.4.4.14.2) which proceeds at different rates with amylose and amylopectin. On cooling bread, the high-molecular amylose very rapidly forms a three-dimensional network and the crystalline states of order of amylose/lipid complexes increase. These processes stabilize the crumb.

On the other hand, the amylopectin is in an amorphous state because the crystalline regions present in flour melt on baking. This is in contrast to the behavior of crystalline amylose/lipid complexes. Thermograms of an aqueous starch suspension (Fig. 15.51) show the differences in the melting points. In comparison with native starch (I), the endotherm peak a at 60 °C caused by the melting of crystalline amylopectin is absent in the thermogram of gelatinized starch (II). However, the melting point of amylose/lipid complexes (ca. 110 °C, peak b in curve II) is not reached in the crumb on baking.

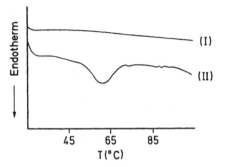

Fig. 15.52. DSC thermograms of white bread: I: fresh from the oven, II: after storage for 1 week at room temperature (according to *Slade, Levine*, 1991)

Staling of white-bread crumb begins with the formation of crystalline structures in amylopectin. The endotherm peak at 60 °C appears again in the thermogram of stored white bread (Fig. 15.52). A state of order arises which corresponds to that of B starch (cf. 4.4.4.14.2) and binds up to 27% of crystal water, which is withdrawn from amorphous starch and proteins. The crumb loses its elasticity and becomes stale. On storage of white bread, the amount of water that can freeze decreases corresponding to the conversion to non-freezing crystal water (Fig. 15.53).

The formation of crystal nuclei, which proceeds very rapidly at 0 °C and does not occur at temperatures below −5 °C (Fig. 15.54), determines the rate of amylopectin retrogradation. The nuclei grow most rapidly shortly before the melting point (60 °C) is reached (Fig. 15.54).

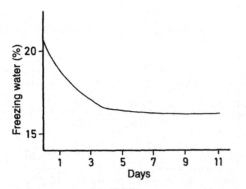

Fig. 15.53. Decrease in freezing water in the storage of white bread. The bread was stored at room temperature encapsulated to prevent drying (according to *Slade, Levine*, 1991)

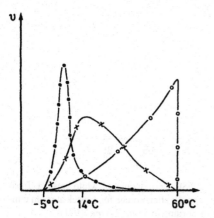

Fig. 15.54. Rate of crystallization of B starch as a function of temperature, $(-\bullet-\bullet)$ formation of crystal nuclei, $(-\circ-\circ)$ crystal growth, $(-\times-\times-)$ total crystal formation (according to *Slade, Levine*, 1991)

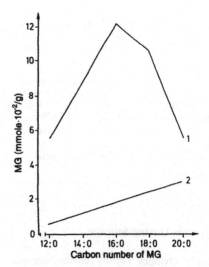

Fig. 15.55. Complex formation between monoacyl-glycerides (MG) and amylose (1) or amylopectin (2) (according to *Knightly*, 1977). *x-axis*: carbon number of the saturated acyl residue. *y-axis*: tendency of MG to form a complex with amylose or amylopectin (mmole $\cdot 10^{-2}$ MG/g polysaccharide)

The ageing process resulting from these events reaches a maximum at ca. 14 °C. As a result of this course, the ageing of white-bread crumb can be prevented by storage at <-5 °C. But the temperature must very quickly fall below the critical temperature for nucleation.

Temperatures above 14 °C also inhibit staling, e. g., increasing the storage temperature from 21 to 35 °C decreases the rate of amylopectin retrogradation by a factor of 4 and improves freshness of the crumb, but the aroma is dissipated. Increased protein or pentosan content slows retrogradation. A choice – actually a rule – to extend the shelf life or freshness of the baked product is the use of emulsifiers, such as mono-acylglycerides or stearyl-2-lactylate. During baking the emulsifier will be complexed with both starch constituents, though to a different extents (Fig. 15.55). Such complexes retard starch retrogradation. Fewer carbohydrates can be extracted from starch/monoacylglyceride adducts than from starch alone. This effect probably contributes to the increase in the cooking stability of pasta after addition of monoacylglycerides (cf. 15.5).

The staling of crumb is also delayed by bacterial α-amylase. From amylopectin, this enzyme cleaves branched oligosaccharides that consist of 19–24 glucose units. Consequently, the formation of crystalline structures in amylopectin is hindered.

15.4.5 Bread Types

Only those bread types of significant economic importance are listed in Table 15.63. Corresponding data on chemical composition are given in Table 15.55.

Crisp bread (Knaeckebrot) and Pumpernickel are special rye breads.

Table 15.63. Bread varieties

No.	Bread variety	Formulation
1.	Wheat bread (white bread)	At least 90% wheat; middlings less than 10%; occasionally with addition of dairy products, sugar, shortenings.
2.	Wheat mix bread	50–89% wheat, the rest rye milling products and other ingredients as under 1.
3.	Rye mix bread	50–89% rye, the rest wheat milling products and others as under 1.
4.	Rye bread	At least 90% rye flour, up to 10% wheat flour; other ingredients as under 1.
5.	Rye whole grain bread	From whole rye meal including also whole kernels, other rye and wheat products less than 10%.

The flat crisp bread is produced mostly from whole rye meal with low α-amylase activity. The dough is ice-cooled and mixed using compressors until foaming occurs, then sheeted and baked for 8–10 min in a tunnel-type oven. Additional drying reduces moisture by 10–20% to a level of 5%. In addition to this mechanically leavened bread, made by mixing air or nitrogen into the dough, there are crisp breads in which biological leavening (yeast or rye sour) is used. Flat bread is also produced in fully automatic cooker-extruders. The heart of these systems is represented by single-screw or double-screw extruders with co- or counter-rotating screws. This is mainly a high-temperature, short-time heating process. The material is degraded to some extent (partial starch gelatinization amongst others) by a combination of pressure, temperature, and shear forces and then deformed by the nozzle head plate. The sudden drop in pressure at the nozzle mouth results in expansion. Water then evaporates and causes the formation of the desired light and bubbly structure.

Pumpernickel bread originates from Westphalia. The sour rye dough, heated in sealed ovens, is more steam-cooked than baked (cf. Table 15.50). Prolonged heating considerably degrades the starch into dextrins and maltose, which are responsible for the sweet taste. The increased buildup of melanoidin pigments accounts for the dark color.

15.4.6 Fine Bakery Products

Until a few years ago, the production of fine bakery products was the domain of confectioners. Today, the importance of the industrial produc-

tion of these products has grown substantially. In general, the process techniques described for the production of bread can be adapted for fine bakery products. Thus, the relevant machine-building companies offer practically automatic production lines for various fine and stable bakery products.

15.5 Pasta Products

15.5.1 Raw Materials

Pasta products are made of wheat semolina and grist (cf. 15.3.1.3), in which the flour extraction grade is less than 70%, and may incorporate egg. The preferred ingredient is durum wheat semolina rather than the soft wheat counterpart (farina) since the former has better cooking and biting strengths and also has a higher content of carotenoids (cf. 15.2.5) which provide the yellow color of pasta products. In wheat mixtures, the soft wheat characteristics emerge when the soft wheat content is higher than 30%. In egg-pasta products (chemical composition in Table 15.64), 2–4 eggs/kg semolina provide a pasta with improved cooking strength and color.

Table 15.64. Composition of pasta products containing eggs (4 eggs per 1 kg flour)

Constituent	%	Constituent	%
Water	11.1	Available carbohydrates	70.0
Protein (N × 5.8)	12.3	Dietary fiber	3.4
Fat	2.9	Minerals	1.0

15.5.2 Additives

Cysteine hydrochloride (about 0.01%) lowers the mixing/kneading time by 15–20% (cf. 15.4.1.4.4). The cysteine also inhibits melanoidin build-up due to nonenzymatic browning, and suppresses the greyish-brown pigmentation. Addition of monoglycerides (about 0.4%) brings about amylose and amylopectin complexing, thereby increasing cooking strength (cf. 15.4.4). Through competitive inhibition, ascorbic acid prevents the action of lipoxygenase (Fig. 15.56). Although the enzyme is a lipoxygenase with high regio- and stereospecificity (cf. 3.7.2.2) and only slowly cooxidizes carotenoids, the low enzyme activity can still destroy the pigments because pasta production is a relatively long process. Addition of ascorbic acid inhibits this cooxidation (Fig. 15.57).

15.5.3 Production

Pasta products are manufactured continuously by a vacuum extruder, which consists of a mixing trough and press segments. The vacuum is used to retard oxidative degradation of carotenoids.

The semolina and added water (30%) and, when necessary, egg or egg powder are mixed in a mixing trough to form a crumb dough (diameter 1–3

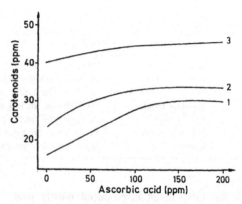

Fig. 15.57. Carotenoid stability in pasta products made of three Durum wheat cultivars as affected by added ascorbic acid (according to *Walsh* et al., 1970). 1–3 wheat cv. Durum

mm), pressed at 150–200 bar into a uniform paste and then pressed through an extruder pressure head die to provide the familiar pasta strings.

Drying is the most demanding stage of pasta manufacturing. The surface of a pasta product must not be allowed to harden prior to the interior core, otherwise cracks, fractures or bursts develop. The freshly extruded strings are initially dried from the outside until they are no longer sticky, then drying is continued at 45–60 °C, either very slowly or stepwise. The moisture content drops to 20–24% after such a predrying process. The moisture is then allowed to equilibrate between inner and outer parts, which brings the content of the final dried product to 11–13%.

15.6 References

Acker, L.: Phospholipases of cereals. In: Advances in cereal science and technology (Ed.: Pomeranz, Y.), Vol. VII, p. 85, American Association of Cereal Chemists: St. Paul. Minn. 1985

Ali, M. R., D'Appolonia, B.L.: Einfluß von Roggenpentosanen auf die Teig- und Broteigenschaften. Getreide Mehl Brot *33*, 334 (1979)

Aman, P., Graham, H.: Analysis of total and insoluble mixed-linked $(1 \rightarrow 3), (1 \rightarrow 4) - \beta$-D-Glucans in barley and oats. J. Agric. Food Chem. *35*, 704 (1987)

Amend, T., Belitz, H.-D.: Gluten formation studied by the transmission electron microscope. Z. Lebensm. Unters. Forsch. *191*, 184 (1990)

Amend, T., Belitz, H.-D.: Electron microscopic studies on protein films from wheat and other sources at the

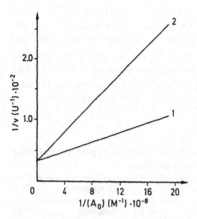

Fig.   15.56.   Competitive   inhibition   of   wheat lipoxy-genase   by   ascorbic   acid   (according   to *Walsh* et al., 1970). Activity assay with linoleic acid as a substrate (1) without, and (2) in the presence of ascorbic acid ($2 \cdot 10^{-6}$mol/l)

air/water interface. Z. Lebensm. Unters. Forsch. *190*, 217 (1990)

Amend, T., Belitz, H.-D.: The formation of dough and gluten – a study by scanning electron microscopy. Z. Lebensm. Unters. Forsch. *190*, 401 (1990)

Amend, T., Belitz, H.-D.: Mikroskopische Untersuchungen von Mehl/Wasser-Systemen. Getreide Mehl Brot *43*, 296 (1989)

Amend, T., Belitz, H.-D.: Microstructural studies of gluten and a hypothesis on dough formation. Food Structure *10*, 277 (1991)

Autorenkollektiv: Lebensmittel-Lexikon. VEB Fachbuchverlag: Leipzig. 1979

Barnes, P.J. (Ed.): Lipids in cereal technology. Academic Press: New York. 1983

Bernardin, J.E.: Gluten protein interaction with small molecules and ions – the control of flour properties. Bakers Dig. *52*, Aug. 1978, p. 20

Bhattacharya, K.R., Ali, S.Z.: Changes in rice during parboiling, and properties of parboiled rice. In: Advances in cereal science and technology (Ed.: Pomeranz, Y.), Vol. VII, p. 105, American Association of Cereal Chemists: St. Paul, Minn. 1985

Bietz, J.A., Wall, J.S.: Identity of high molecular weight gliadin and ethanol-soluble glutenin subunits of wheat: relation to gluten structure. Cereal Chem. *57*, 415 (1980)

Biliaderis, C.G., Tonogai, J.R.: Influence of lipids on the thermal and mechanical properties of concentrated starch gels. J. Agric. Food Chem. *39*, 833 (1991)

Bleukx, W., Roels, S.P., Delcour, J.A.: On the presence and activities of proteolytic enzymes in vital wheat gluten. J. Cereal Sci. *26*, 183 (1997)

Bloksma, A.H.: Dough structure, dough rheology, and baking quality. Cereal Foods World *35*, 237 (1990)

Bloksma, A.H.: Rheology of the breadmaking process. Cereal Foods World *35*, 228 (1990)

Brose, E., Becker, G., Bouchain, W.: Chemische Backtriebmittel. Chemische Fabrik Budenheim, Rudolf A. Oethker, Budenheim, 1996

Bushuk, W. (Ed.): Rye: Production, chemistry, and technology. American Association of Cereal Chemists: St. Paul, Minn. 1976

Büskens, H.: Die Backschule – Fachkunde für Bäcker. 7th edn., W. Girardet: Essen. 1978

Buttery, R.G., Ling, L.C., Mon, T.R.: Quantitative analysis of 2-acetyl-1-pyrroline in rice. J. Agric. Food Chem. *34*, 112 (1986)

Dubreil, L., Compoint, J.-P., Marion, D.: Interaction of puroindolines with wheat flour polar lipids determines their foaming properties. J. Agric. Food Chem. *45*, 108 (1997)

Finney, K.F., Tsen, C.C., Shogren, M.D.: Cysteine's effect on mixing time, water absorption, oxidation requirement, and loaf volume of Red River 68. Cereal Chem. *48*, 540 (1971)

Fox, P.F., Mulvihill, D.M.: Enzymes in wheat, flour, and bread. In: Advances in cereal science and technology (Ed.: Pomeranz, Y.), Vol. V, p. 107, American Association of Cereal Chemists: St. Paul, Minn. 1982

Gassenmeier, K., Schieberle, P.: Potent aromatic compounds in the crumb of wheat bread (French-type). Influence of preferments and studies on the formation of key odorants during dough processing. Z. Lebensm. Unters. Forsch. *201*, 241 (1995)

Greiner, R., Jany, K.-D.: Ist Phytat ein unerwünschter Inhaltsstoff in Getreideprodukten? Getreide Mehl Brot *50*, 368 (1996)

Groesaert, H., Brijs, K., Veraverbeke, W.S., Courtin, C.M., Gegruers, K., Delcour, J.A.: Wheat flour constituents: how they impact bread quality, and how to impact their functionality. Trends Food Sci. Technol. *16*, 12 (2005)

Grosch, W., Schieberle, P.: Flavor of cereal products – a review. Cereal Chem. *74*, 91 (1997)

Grosch, W., Wieser, H.: Redox reactions in wheat dough as affected by ascorbic acid. J. Cereal Sci. *29*, 1 (1999)

Hebeda, R.E., Zobel, H.F.: Baked good freshness. Marcel Dekker, New York, 1996

Hoseney, R.C.: Principles of cereal science and technology. American Association of Cereal Chemists St. Paul, Minn. 1986

Houston, D.F. (Ed.): Rice, chemistry and technology. American Association of Cereal Chemists: St. Paul, Minn. 1972

ICC-Standards: Standard-Methoden der Internationalen Gesellschaft für Getreidechemie (ICC). Verlag Moritz Schäfer: Detmold

Inglett, G.E. (Ed.): Maize, recent progress in chemistry and technology. Academic Press: New York. 1982

Inglett, G.E., Munck, L. (Eds.): Cereals for food and beverages, recent progress in cereal chemistry and technology. Academic Press: New York. 1980

Jackson, E.A., Holt, L.M., Payne, P.I.: Characterization of high-molecular weight gliadin and low-molecular weight glutenin subunits of wheat endosperm by two-dimensional electrophoresis and the chromosomal location of their controlling genes. Theor. Appl. Genet. *66*, 29 (1983)

Juliano, B.O.: Rice. Chemistry and technology. American Association of Cereal Chemists: St. Paul. Minn. 1985

Kasarda, D.D., Okita, T.W., Bernardin, J.E., Baecker, P.A., Nimmo, C.C., Lew, E.J.-L., Dietler, M.D., Greene, F.C.: Nucleic acid (cDNA) and amino acid sequences of α-type gliadins from wheat (*Triticum aestivum*). Proc. Natl. Acad. Sci. USA *81*, 4712 (1984)

Kieffer, R., Grosch, W.: Verbesserung der Back-
eigenschaften von Weizenmehlen durch die Typ
II-Lipoxygenase aus Sojabohnen. Z. Lebensm.
Unters. Forsch. *170*, 258 (1980)

Kieffer, R., Stein, N.: Demixing in wheat doughs –
Its influence on dough and gluten rheology. Cereal
Chem. *76*, 688 (1999)

Kirchhoff, E., Schieberle, P.: Einfluß der Sauerteig-
trocknung auf die Konzentrationen wichtiger Aromastoffe in Sauerteig und Brotkrume. Getreide Mehl
und Brot *52*, 273 (1998)

Köhler P., Belitz, H.-D., Wieser, H.: Disulphide bonds
in wheat gluten: further cystine peptides from high
molecular weight (HMW) and low molecular weight
(LMW) subunits of glutenin and from γ-gliadin. Z.
Lebensm. Unters. Forsch. *196*, 239 (1993)

Krause, J., Müller, U., Belitz, H.-D.: Charakterisierung
von Weizensorten durch SDS-Polyacryl-amid-
Elektrophorese (SDS-PAGE) und zwei-dimensionale
Elektrophorese (2D-PAGE) der Glutenine. Z.
Lebensm. Unters. Forsch. *186*, 398 (1988)

Kruger, J.E., Marchylo, B.A.: A comparison of the
catalysis of starch components by iso-enzymes
from the two major groups of germinated wheat-α-
amylase. Cereal Chem. *62*, 11 (1985)

Lorenz, K.: The history, development, and utilization of
Triticale. Crit. Rev. Food Technol. *5*, 175 (1974)

MacRitchie, F.: Physicochemical aspects of some problems in wheat research. In: Advances in cereal science and technology (Ed.: Pomeranz, Y.), Vol. III, p.
271, American Association of Cereal Chemists: St.
Paul, Minn. 1980

Maga, J.A.: Bread staling. Crit. Rev. Food Technol. *5*,
443 (1974)

Marchylo, B.A., Kruger, J.E., Hatcher, D.W.: Quantitative reversed-phase high-performance liquid chromatographic analysis of wheat storage proteins as
a potential quality prediction tool. J. Cereal Sci. *9*,
113 (1989)

Morris, C.F., Greenblatt, G.A., Bettge, A.D., Malkawi,
H.I.: Isolation and characterization of multiple forms
of friabilin. J. Cereal Sci. *21*, 167 (1994)

Morrison, W.R.: Lipide in Mehl, Teig und Brot. Getreide Mehl Brot *30*, 244 (1976)

Morrison, W.R.: Cereal lipids. In: Advances in cereal
science and technology (Ed.: Pomeranz, Y.), Vol. II,
p. 221, American Association of Cereal Chemists:
St. Paul, Minn. 1978

Morrison, W.R.: Recent progress in the chemistry and
functionality of flour lipids. In: Wheat enduse properties (Ed.: Salovaara, H.) Proc. Symp. ICC '89. University of Helsinki, Helsinki, pp. 131–149 (1989)

Müller, S., Wieser, H.: The location of disulphide bonds
in monomeric γ-type gliadins. J. Cereal Sci. *26*, 169
(1997)

Payne, P.I., Harris, P.A., Law, C.N., Holt, L.M., Blackman, J.A.: The high-molecular-weight sub-units

of glutenin: structure, genetics and relationship to
bread-making-quality. Ann. Technol. Agric. *29*, 309
(1980)

Payne, P.I., Corfield, K.G., Holt, L.M., Blackman,
J.A.: Correlation between the inheritance of certain
high-molecular-weight subunits of glutenin and
bread-making quality in progenies of six crosses of
bread wheat. J. Sci. Food Agric. *32*, 51 (1981a)

Payne, P.J., Holt, L.M., Lano, C.N.: Structural and ge-
netical studies on the high-molecular weight subunits
of wheat glutenin. Part 1: Allelic variation in subunits amongst varieties of wheat (*Triticum aestivum*).
Theor. Appl. Genet. *60*, 229 (1981b)

Payne, P.I., Holt, L.M., Jackson, E.A., Law, C.N.:
Wheat storage proteins: their genetics and their
potential for manipulation by plant breeding. Philos.
Trans. R. Soc. London Ser. B *304*, 359 (1984)

Payne, P.I., Holt, L.M., Jarvis, M.G., Jackson, E. A.:
Two-dimensional fractionation of the endosperm
proteins of bread wheat (*Triticum aestivum*): biochemical and genetic studies. Cereal Chem. *62*, 319
(1985)

Pomeranz, Y.: Cereals and cereal products. In: Ullmann's encyclopedia of industrial chemistry, 5th
Edition, Volume A6, p. 93, Verlag VCH, Weinheim,
1986

Pomeranz, Y. (Ed.): Wheat: Chemistry and technology.
Volumes I and II. American Association of Cereal
Chemists: St. Paul, Minn. 1988

Pomeranz, Y.: Wheat is unique. American Association
of Cereal Chemists: St. Paul. Minn. 1989

Rabe, E.: Organische Säuren in unterschiedlich
geführten Broten. Getreide Mehl Brot *34*, 90 (1980)

Raunum, P.: Potassium bromate in bread making. Cereal Foods World *37*, 253 (1992)

Rohrlich, M., Brückner, G.: Das Getreide. I. und II.
Teil, 2. Aufl., Verlag Paul Parey: Berlin. 1966/1967

Rohrlich, M., Thomas, B.: Getreide und Getreide-
mahlprodukte. In: Handbuch der Lebensmittelchemie (Ed.: Schormüller, J.), Vol. V/1st Part,
Springer-Verlag: Berlin. 1967

Rychlik, M., Grosch, W.: Identification and quantification of potent odorants formed by toasting of wheat
bread. Lebensm. Wiss. Technol. *29*, 515 (1996)

Sauer, D.B.: Storage of cereal grains and their products.
American Association of Cereal Chemists, Inc., St.
Paul, Minn., 1992

Schieberle, P., Grosch, W.: Quantitative analysis of
aroma compounds in wheat and rye bread crusts
using a stable isotope dilution assay. J. Agric. Food
Chem. *35*, 252 (1987)

Schieberle, P., Grosch, W.: Potent odorants of the wheat
bread crumb. Differences to the crust and effect of
a longer dough fermentation. Z. Lebensm. Unters.
Forsch. *192*, 130 (1991)

Schuh, C., Schieberle, P.: Characterization of (E,E,Z)-
2,4,6-nonatrienal as a character impact aroma com-

pound of oat flakes. J. Agric. Food Chem. *53*, 8699 (2005)

Seibel, W. (Ed.): Feine Backwaren. Paul Parey: Berlin, 1991

Seilmeier, W., Belitz, H.-D., Wieser, H.: Separation and quantitative determination of high-molecular subunits of glutenin from different wheat varieties and genetic variants of the variety Sicco. Z. Lebensm. Unters. Forsch. *192*, 124 (1991)

Shewry, P.R., Miflin, B.J.: Seed storage proteins of economically important cereals. In: Advances in cereal science and technology (Ed.: Pomeranz, Y.), Vol. VII, p. 1, American Association of Cereal Chemists: St. Paul, Minn. 1985

Shewry, P.R., Tatham, A.S.: Disulphide bonds in wheat gluten proteins. J. Cereal Sci. *25*, 207 (1997)

Slade, W., Levine, H.: Beyond water activity – Recent advances based on an alternative approach to the assessment of food quality and safety. Crit. Rev. Food Sci. Nutr. *30*, 115 (1991)

Spicher, G., Pomeranz, Y.: Bread and other baked products. In: Ullmann's encyclopedia of industrial chemistry, 5th Edition, Volume A4, p. 331, Verlag VCH, Weinheim, 1985

Sprößler, B.: Wirkung von Proteinasen beim Zusatz zum Mehl. Getreide Mehl Brot *35*, 60 (1981)

Stear, C.A.: Handbook of breadmaking technology. Elsevier Applied Science, London. 1990

Tan, S.L., Morrison, W.R.: The distribution of lipids in the germ, endosperm, pericarp and tip cap of amylomaize, LG-11 hybrid maize and waxy maize. J. Am. Oil Chem. Soc. *56*, 531 (1979)

Tatham, A.S., Shewry, P.R.: The conformation of wheat gluten proteins. The secondary structures and thermal stabilities of α-, β-, γ- and ω-gliadins. J. Cereal Sci. *3*, 103 (1985)

Walsh, D.E., Youngs, V.L., Gilles, K.A.: Inhibition of durum wheat lipoxygenase with L-ascorbic acid. Cereal Chem. *47*, 119 (1970)

Wassermann, L., Dörffner, H.H.: Der Einfluß des Wasser-Mehl-Verhältnisses in Brotteigen auf die Zusammensetzung und Eigenschaften der daraus hergestellten Brote. Brot Gebäck *25*, 148 (1971)

Watson, S.A., Ramstad, P.E.: Corn. Chemistry and technology. American Association of Cereal Chemists: St. Paul, Minn. 1987

Webster, F.H. (Ed.): Oats: Chemistry and technology. American Association of Cereal Chemists: St. Paul, Minn. 1986

Wieser, H., Belitz, H.-D.: Amino acid compositions of avenins separated by reversed-phase high-performance chromatography. J. Cereal Sci. *9*, 221 (1989)

Wieser, H., Seilmeier, W., Belitz, H.-D.: Characterization of ethanol-extractable reduced subunits of glutenin separated by reversed-phase high-performance liquid chromatography. J. Cereal Sci. *12*, 63 (1990)

Wieser, H., Seilmeier, W., Belitz, H.-D.: Use of RP-HPLC for a better understanding of the structure and the functionality of wheat gluten proteins. In: High-Performance Liquid Chromatography of Cereal and Legume Proteins (Eds.: Kruger, J.E., Bietz, J.A.), p. 235, American Association of Cereal Chemists: St. Paul. Minn., 1994

Wood, P.J., Siddiqui, I.R., Paton, D.: Extraction of high-viscosity gums from oats. Cereal Chem. *55*, 1038 (1978)

Youngs, V.L., Peterson, D.M., Brown, C.M.: Oats. In: Advances in cereal science and technology (Ed.: Pomeranz, Y.), Vol. V. p. 49, American Association of Cereal Chemists: St. Paul, Minn. 1982

Zehentbauer, G., Grosch, W.: Apparatus for quantitative headspace analysis of the characteristic odorants of baguettes. Z. Lebensm. Unters. Forsch. *205*, 262 (1997)

Zehentbauer, G., Grosch, W.: Crust aroma of baguettes I and II. J. Cereal Sci. *28*, 81 u. 93 (1998)

16 Legumes

16.1 Foreword

Ripe seeds of the plant family *Fabaceae*, known commonly as "legumes" or "pulses", are an important source of proteins for much of the world's population[*]. The extent of the production of major legumes is illustrated in Table 16.1. Legumes contain relatively high amounts of protein (Table 16.2). Hence, they are an indispensable supply of protein for the "third world". Soybeans and peanuts are oil seeds (cf. 14.3.2.2.5) and, even in industrialized countries, are used as an important source of raw proteins.

With regard to the biological value, legume proteins are somewhat deficient in the S-containing amino acids (Table 16.3 and 1.8).

Antinutritive substances, e. g., allergenic proteins, proteinase inhibitors, lectins and cyanogenic glycosides, are found in food raw materials. These substances will be described in this chapter since a large variety have been identified in legumes.

16.2 Individual Constituents

16.2.1 Proteins

About 80% of the proteins from soybean can be extracted at pH 6.8. A large number of these proteins can be precipitated by acidification at pH 4.5 (cf. Figure 1.54). This pH-dependent solubility is used in large-scale preparations of soy proteins. Fractionation of legume proteins using solubility procedures, as applied to cereals by *Osborne* (cf. 15.2.1.2), yields three fractions: albumins, globulins, and glutelins, with globulins being predominant (Table 16.4).

---

[*] Semi-ripe peas and beans are considered as vegetables (cf. Chapter 17).

16.2.1.1 Glubulines

The high content of globulins in seeds indicates that they function mostly as storage proteins, which are mobilized during the course of germination.

The globulin fraction can be separated by ultracentrifugation or chromatography into two major components present in all the legumes: *vicilin* ($\sim$7S) and *legumin* ($\sim$11S). Legumin from soybeans is called glycinin and from peanuts is called arachin. Molecular weights and sedimentation coefficients for the 7 S and 11 S globulins isolated from various legumes are presented in Table 16.5.

The 11 S globulins originate from a protein precursor ($M_r \sim 60,000$) which is split into an acidic α-polypeptide (pI $\sim$ 5) and a basic β-polypeptide (pI $\sim$ 8.2) by cleaving the peptide bond between Asn (417) and Gly (418) (cf. Table 16.6). These two polypeptides are connected by a disulfide bridge between Cys (92) and Cys (424) and are regarded as one subunit. Six such subunits join to give 11 S globulin, the hydrophobic β-polypeptides evidently forming the core of the subunits and of the entire structure. Little is known about the tertiary and quaternary structure. On the other hand, the amino acid sequences of the subunits of the 11 S globulins of a number of legumes are known. They were mainly derived from the nucleotide sequences of the coding nucleic acids. As an example, Table 16.6 shows the sequences of legumin J from the pea (*Pisum sativum*) and glycinin A_2B_{1a} from the soybean (*Glycine max*). Homology exists between the 11 S globulins of different legumes. Variable regions are primarily found in the acidic α-polypeptide, while the basic β-polypeptide is conservative with slight variability in the region of the C-terminal. Conserved residues are uniformly distributed throughout the sequence. The α/β cleavage site is conserved in all the proteins studied until now (cf. Table 16.7). Thus,

H.-D. Belitz · W. Grosch · P. Schieberle, *Food Chemistry*
© Springer 2009

Table 16.1. World production of seed legumes, 2006 (1,000 t)

Continent	Pulses Legumes total[a]	Beans[b]	Broad beans	Peas
World	60,194	19,559	4577	10,563
Africa	11,111	2856	1321	382
America, Central	2185	1853	37	8
America, North	6025	1430	18	3405
America, South and Caribbean	6865	6153	158	94
Asia	28,505	8701	2256	2392
Europe	6841	404	719	3898
Oceania	846	15	104	392

Continent	Chick peas	Lentils	Soybeans	Groundnuts Peanuts[c]
World	8241	3455	221,501	47,768
Africa	324	105	1417	8967
America, Central	163	7	124	175
America, North	231	928	91,203	1479
America, South and Caribbean	170	16,759	98,885	1026
Asia	7365	2316	26,334	36,258
Europe	43	48	3607	9
Oceania	108	41	55	29

Country	Pulses Legumes, grand total	Country	Beans	Country	Broad beans
India	14,264	Brazil	3437	China	2100
China	5557	India	3174	Ethiopia	599
Canada	4072	China	2007	Egypt	315
Brazil	3448	Myanmar	1700	France	290
Nigeria	3091	Mexico	1375	Morocco	181
Myanmar	2571	USA	1057	Sudan	138
USA	1953	Kenya	532	United Kingdom	130
Russian Fed.	1764	Uganda	424	Australia	104
Mexico	1663	Canada	373	Italy	83
Turkey	1550	Indonesia	327	Spain	76
France	1347	Argentina	323	Σ (%)[d]	90
Ethiopia	1265	Σ (%)[d]	75		
Iran	1007				
UK	830				
Australia	803				
Pakistan	803				
Σ (%)[d]	76				

[a] Without soybeans and peanuts.
[b] Without broad beans.
[c] With hull included.
[d] World production = 100%.

Table 16.1. (Continued)

Country	Peas	Country	Chick peas	Country	Lentils
Canada	2806	India	5600	India	950
Russian Fed.	1158	Turkey	552	Canada	693
China	1140	Pakistan	480	Turkey	623
France	1010	Iran	293	USA	235
India	800	Canada	182	Syria	165
USA	599	Myanmar	172	Nepal	158
Ukraine	485	Mexico	163	China	150
Australia	360	Ethiopia	125	Bangladesh	120
Germany	288	Australia	108	Iran	113
Iran	265	Morocco	66	Ethiopia	65
Σ (%)[d]	84	Σ (%)[d]	94	Σ (%)[d]	95

Country	Soybeans	Country	Groundnuts Peanuts
USA	87,670	China	14,722
Brazil	52,356	Indonesia	14,700
Argentina	40,467	India	4980
China	15,500	Nigeria	3825
India	8270	USA	1479
Paraguay	3800	Myanmar	910
Canada	3533	Sudan	540
Bolivia	1350	Ghana	520
Ukraine	889	Argentina	496
Russian Fed.	807	Viet Nam	465
Σ (%)[d]	97	Σ (%)[d]	89

Table 16.2. Chemical composition of legumes[a]

Name	Systematic name	Crude protein[b] (%)	Lipid (%)	Available carbohydrates (%)	Dietary fiber (%)	Minerals (%)
Soybeans	*Glycine hyspida max*	41.0	19.6	7.6	24.0	5.5
Peanuts	*Arachis hypogaea*	31.4	50.7	7.9	12.3	2.7
Peas	*Pisum sativum*	25.7	1.4	53.7	18.7	3.0
Garden beans	*Phaseolus vulgaris*	24.1	1.8	54.1	19.2	4.4
Runner beans	*Phaseolus coccineus*	23.1	2.1	n.a.	n.a.	3.9
Black gram	*Phaseolus mungo*	26.9	1.6	46.3	n.a.	3.6
Green gram (mungo beans)	*Phaseolus aureus*	26.7	1.3	51.7	21.7	3.8
Lima beans	*Phaseolus lunatus*	25.0	1.6	n.a.	n.a.	3.9
Chick peas	*Cicer arietinum*	22.7	5.0	54.6	10.7	3.0
Broad beans	*Vicia faba*	26.7	2.3	n.a.	n.a.	3.6
Lentils	*Lens culinaris*	28.6	1.6	57.6	11.9	3.6

[a] The result are average values given as weight-%/dry matter.
[b] N $\times$ 6.25.
n.a.: not analyzed.

cleavage of the protein precursor evidently occurs through the same, very specific, but not yet characterized proteinase. With a few exceptions, the 11 S globulins are not glycosylated.

Table 16.3. Essential amino acids in legumes (g/16 g N)

Amino acid	Soybean	Broad bean
Cystine	1.3	0.8
Methionine	1.3	0.7
Lysine	6.4	6.5
Isoleucine	4.5	4.0
Leucine	7.8	7.1
Phenylalanine	4.9	4.3
Tyrosine	3.1	3.2
Threonine	3.9	3.4
Tryptophan	1.3	n.a.
Valine	4.8	4.4

n.a.: not analyzed.

Table 16.4. Legumes: protein distribution (%) by *Osborne* fractions

Fraction	Soybeans	Peanuts	Peas	Mungo beans	Broad beans
Albumin	10	15	21	4	20
Globulin	90	70	66	67	60
Glutelin	0	10	12	29	15

Table 16.5. Molecular weight and sedimentation coefficient of the 7 S and 11 S globulins from legumes

Legume	7 S globulin		11 S globulin	
	Sedimentation coefficient	Mol. weight (kdal)	Sedimentation coefficient	Mol. weight (kdal)
Soybeans	7.9 ($S_{20,w}$)	193	12.3 ($S_{20,w}$)	360
Peanuts	8.7 (S_{20})	190	13.2 ($S_{20,w}$)	340
Peas	8.1 (S_{20})		13.1 (S_{20})	398
Garden beans	7.6 ($S_{20,w}$)	140	11.6 ($S_{20,w}$)	340
Broad beans	7.1 ($S_{20,w}$)	150	11.4 ($S_{20,w}$)	328

The 7 S globulins are made of three subunits ($M_r \sim 50,000$) which can be identical or different (homo- and heteropolymeric forms). There is little information available on the tertiary and quaternary structure. The subunits can consist of up to three polypeptides (α, β, γ) which are formed from the intact subunit (precursor protein) by proteolysis. Since the amino acid sequences of the α/β (239/240 in Table 16.8) and β/γ (376/377 in Table 16.8) cleavage sites are variable (Table 16.9), intact subunits and subunits

with only one cleaved bond are observed, unlike the behavior of the 11 S globulin subunits. Thus, the bond between N (376) and D (377) in vicilin 47k is cleaved, but corresponding ED bonds in other vicilins are evidently not split (cf. Table 16.9).

The amino acid sequences of the 7 S globulin subunits of a number of legumes are known and were mainly derived from the nucleotide sequences of the coding nucleic acids. Table 16.8 shows the sequences of phaseolin from the garden bean (*Phaseolus vulgaris*), vicilin from the pea (*Pisum sativum*), and β-conglycinin (β) from the soybean (*Glycine max*). Sequence homology, which is more pronounced than in the 11 S globulins, exists between the proteins of various legumes. Variable domains are found in the N- and C-terminal regions, but not inside the structure.

The 7 S globulins are glycosylated to different extents. The carbohydrate content is 0.5–1.4% in vicilin from peas, 1.2–5.5% in phaseolin from garden beans, and 2.7–5.4% in β-conglycinin from soybeans. The structures of the oligosaccharide residues are partly known. In β-conglycinin, for example, 6–8 mannose residues are bound to Asn in a branched structure via two N-acetylglucosamine residues.

Under non-denaturing conditions, the 11 S and 7 S globulins exhibit a tendency towards reversible dissociation/association, which greatly depends on the pH value and the ionic strength. According to their behavior, they can be attributed to different types. The 11 S globulins are relatively more stable than the 7 S globulins. They noticeably associate only in the region of the isoelectric point, isoelectric precipitation occurring at low ionic strength (cf. 16.3.1.2.1). If at pH 7.6, the ionic strength is reduced from $\mu = 0.5$ to $\mu < 0.1$, soybean 11 S globulin dissociates stepwise (α, β: acidic and basic proteins):

$$11S(6\alpha\beta) \rightarrow 7.5S(3\alpha\beta) \rightarrow 3S(\alpha\beta) \quad (16.1)$$

Complete dissociation occurs when the disulfide bonds are reduced in the presence of protein-unfolding agents, such as urea or SDS:

$$(\alpha\beta) \rightarrow \alpha + \beta \quad (16.2)$$

Soybean 7 S globulin has similar properties, as illustrated in Fig. 16.1. Hence, its molecular weight is also strongly dependent on pH and ionic strength.

Table 16.6. Amino acid sequences of the α/β subunits[a] of 11 S globulins, 1) legumin J (*Pisum sativum*) and 2) glycinin A_2B_{1a} (*Glycine max*)

```
                 10            20            30            40            50            60
1 L A T S S E F D R L - - N Q C Q L D S I N A L E P D H R V V S E A G L T E T W N P N H P E L K C A G V S L I R R T I D
2 - - - - L R E Q A Q Q N E C Q I Q K L N A L K P D N R I E S E G G F I E T W N P N N K P F Q C A G V A L S R C T L N
 61              70            80            90           100           110           120
1 P N G L H L P S F S P S P Q L I F I I Q G K G V L G L S F P G C P E T Y E E P R S S Q S R Q - - E S R Q - - - - P Q
2 R N A L R R P S Y T N G P Q E I Y I Q Q G N G I F G M I F P G C P S T Y Q E P Q E S Q - - Q R G - R S Q - - - R - P Q
121             130           140           150           160           170           180
1 Q G - - - - - - - - - - - - - - - - - - - - - - - - - - D S H Q K V R R F R K G D I I A I P S G I P Y W T Y N H G
2 - - - - - - - - - - - - - - - - - - - - - - - - - - - D R H Q K V H R F R E G D L I A V P T G V A W W M Y N N E
181             190           200           210           220           230           240
1 D E P L V A I S L L D T S N I A N Q L D S T P R V F Y L G G N P E T E F P E T Q E E Q Q G R H R - Q K H S Y P V G R R S
2 D T P V V A V S I I D T N S L E N Q L D Q M P R R F Y L A G N Q E Q E F - L K Y Q Q Q Q Q G - - - - G S Q S Q K
241             250           260           270           280           290           300
1 G H H - Q Q E E E S E Q Q N E G N S V L S G F S S E F L A Q T F N T E E D T A K R L R S P R D E - R S - Q I V R V E G G
2 G K - Q Q E E E - - - N E G S N I L S G F A P E F L K E A F G V N M Q I V R N L Q G E N E E D S G A I V T V K G G
301             310           320           330           340           350           360
1 L R I I K G - - R - - T - - - - - - E E - E K - E Q - - S H - - - - S H S H R E E K E E - - - - - -
2 L R V T A P A M R K P - - - Q Q - - - E - E D D D D E E E Q - P - - Q - C V E T - D K G C Q - - - - - -
361             370           380           390           400           410           420
1 - - - - - - - - - - E E E E E E D E E E - - - - - - - - - - E E - R - - - - - K N G L E
2 R - - - O S K R S R - - - - - - - - - - K Q - R S - - - - - - - - - N G I D
421             430           440           450           460           470           480
1 E T I C S A K I R E N I A D A A R A D L Y N P R A G R I S T A N S L T L P V L R Y L R L S A E Y V R L Y R N G I Y A P H
2 E T I C T M R L R Q N I G Q N S S P D I Y N P Q A G S I T T A T S L D F P A L W L L K L S A Q Y G S L R K N A M F V P H
481             490           500           510           520           530           540
1 W N I N A N S L L Y V I R G E G R V R I R S C E L P T N T M F D N K L R K G H L V V V P Q N F V V A E Q A G E E E G L E
2 Y T L N A N S I I Y A L N G R A L V Q V V N C N - G - E R V F D G E L Q E G G V L I V P Q N F A V A A K S Q S D N - F E
541             550           560           570           580           590           600
1 Y V V F K T N D R A A V S - H V Q Q - - V F R A T P S E V L A N A F G L R Q R Q V T E L K L S G N R G P M - V H - P -
2 Y V S F K T N D R P - S I G N L A G A N S L L N A L P E E V I Q H T F N L K S Q Q A R Q V K - - N N - P F S F L V P P
601             610           620
1 R - S Q S Q S H
2 Q E S Q - - R R A V A
```

[a] The α-polypeptide ends with Asn (417) and the β-polypeptide starts with Gly (418). The two polypeptides are connected by a disulfide bond between Cys (92) and Cys (424). The two remaining Cys residues of the α-polypeptide (16, 49) probably form an intramolecular disulfide bridge. –: space to maximize homology.

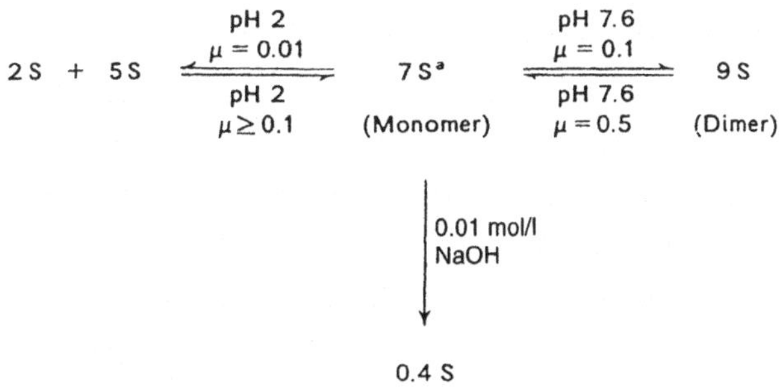

$$2S \; + \; 5S \; \underset{\substack{pH\;2 \\ \mu \geq 0.1}}{\overset{\substack{pH\;2 \\ \mu = 0.01}}{\rightleftharpoons}} \; \underset{\text{(Monomer)}}{7S^a} \; \underset{\substack{pH\;7.6 \\ \mu = 0.5}}{\overset{\substack{pH\;7.6 \\ \mu = 0.1}}{\rightleftharpoons}} \; \underset{\text{(Dimer)}}{9S}$$

$$\downarrow \begin{array}{l} 0.01 \text{ mol/l} \\ \text{NaOH} \end{array}$$

0.4 S

Fig. 16.1. Dissociation and aggregation of the soybean 7 S globulin

---

[a]Molecular weight: 193 kdal.

Table 16.7. Amino acid sequences in the vicinity of the α/β cleavage site (417/418 in Table 16.6) of sub-units of various 11 S globulins (–: space to maximize homology, $\cdots$ sequence not known)

Protein	420
Legumin J (*Pisum sativum*)	– –KNGLEETI CS
Legumin A (*Pisum sativum*)	D– – NGLEETVCT
Glycinin A_2B_{1a}(*Glycine max*)	– – – NGI DETI CT
Glycinin $A_5A_4B_3$ (*Glycine max*)	ETRNGVEENI CT
Glycinin A_3B_4(*Glycine max*)	QTRNGVEENI CT
Glycinin $A_{1a}B_{1b}$(*Glycine max*)	– – – NGI DETI CT
Glycinin $A_{1b}B_2$(*Glycine max*)	. . . NGI DETI CT
Cruciferin(*Brassica napus*)	– – – NGLEETI CS
Legumin β_1(*Vicia faba*)	D– – NGLEETVCT
Legumin B(*Vicia faba*)	– –RNGLEETI CS
Avenin(*Avena sativa*)	– – – NGLEENFCD
Glutelin(*Oryza sativa*)	NGLDETFCT

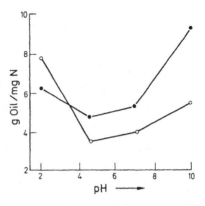

Fig. 16.2. Soybean globulin as an emulsifier. (According to *Aoki* et al., 1980). The capacity of an o/w-emulsion after addition of 11 S globulin (–o–) and 7 S globulin (–•–) is plotted versus pH

The thermal stability of the 11 S and 7 S globulins varies. While 7 S globulin coagulates in a 10% salt solution at 99 °C, 11 S globulin remains in solution. The opposite is true at $\mu = 0.001$. Since dissociated proteins are more easily coagulable thermally than associated proteins, it follows that 11 S globulin is destabilized by dissociation at low ionic strength, as shown above. Under these conditions, however, the 7 S globulin is stabilized by association.

The amino acid compositions of both major soybean proteins, with the exception of methionine, are very similar (Table 16.10). However, large differences exist in their carbohydrate contents. The 7 S globulin contains 5% carbohydrate and the 11 S globulin less than 1% carbohydrate.

Legume proteins exhibit a marked gelling capacity. The gel properties depend on the protein used and on the production conditions (pH value, ionic species, ionic strength, and temperature). They are suitable for the production of foams and emulsions.

In the pH range of 4–10, the 7 S globulin is a better emulsifier than the 11 S globulin, when the capacity (Fig. 16.2) and the stability of an o/w emulsion are compared. Partial acid hydrolysis improves the emulsifier properties.

16.2.1.2 Allergens

A food allergy is a diseased state caused by immunologic reactions which are induced by the intake of food. All other reactions which are

Table 16.8. Amino acid sequences of subunits[a] of the 7 S globulins, 1) β-conglycinin α' (*Glycine max*), 2) phaseolin (*Phaseolus vulgaris*) and 3) vicilin (*Pisum sativum*)

```
       1              10              20              30              40              50
1b–    – – – – D E D E E Q D K E S   Q E S E G   S E S   Q R E P R R   H K N K N P   F H F N S   K R –   F Q T L F K N Q
2  A   T S L R E E E E – – – – S   Q D – – –   – – – – – – – – – – – – –   N P F Y F N S   D N S   W N T L F K N Q
3  –   – – – – – – – – – – R – – – – – – –   – S D P   Q – – – – – – – –   N P F I   F K S N K – F   Q T L F E N E
      51             60              70              80              90             100
1  Y   G H V R V L Q R F N K R S   Q Q L Q N   L R D Y R I   L E F   N S K P N T L L L P   H H A D   A D Y L I   V I   L
2  Y   G H I R V L Q R F D Q Q S   K R L Q N   L E D Y R L V E F   R S K P E T L L L P   Q Q A D   A E L L L   V V R
3  N   G H I R L L Q K F D Q R S   K I   F E N – Q N Y R   L L E Y K S   K P H T I   F L P Q H T D A   D Y I   L V V L
     101            110             120             130             140             150
1  N   G T A I   L T L V N N D D R   D S Y – N   L Q S   G D A – L   – – – – – R V P A G T T F Y V   V N P D N D E N
2  S   G S A I   L V L V K P D D R   R E Y F F   L T Q G D N P I   F S D N Q K I   P – – – – – – –   – – – – – – – –
3  S   G K A I   L T V L K P D D R   N S F – N   – E R   G D T – I   – – – – – K L P – – – – – – –   – – – – – – – –
     151            160             170             180             190             200
1  L   R M I A G T T F Y V V N P   D N D E N   L R M I   T L A I   P   V N K P E R F E S F F L S S   T Q A Q Q S Y L
2  –   – – – A G T I   F Y L V N P   D P K E D   L R I   I   Q L A M P   V N N P Q – I   H E F F L S S   T E A Q Q S Y L
3  –   – – – A G T I   A Y L V N R   D D N E E   L R V L   D L A I   P   V N R P G Q L Q S F L L S G   N Q N Q Q N Y L
     201            210             220             230             240             250
1  Q   G F S K N I   L E A S Y D T   K F E E I   N K V L   F G R E   E G Q Q – Q – – – G – – – E E   R – – – – L Q E
2  Q   E F S K H I   L E A S F N S   K F E E I   N R V L   F – E E   E G Q Q – – – – – – – – – – –   – – E E G Q Q E
3  S   G F S K N I   L E A S F N T   D Y E E I   E K V L   L – E E   H E K E T Q H R R S L K D – K   R – Q Q S Q E E
     251            260             270             280             290             300
1  S   V I   V E I   S K K Q I   R E L S K H A K   S S S R K T I   S S E D K P F N L G S R D P I   Y   S N K L G K L F
2  G   V I   V N I   D S E Q I   E E L S K H A K   S S S R K S H S – – – K Q – D – – – – N T I –   G N E F G N L T
3  N   V I   V K L S R G Q I   E E L S K N A K   S T S K K S V S S E S E P F N L R S R G P I   Y   S N E F G K F F
     301            310             320             330             340             350
1  E   I T – Q R N – P Q L R D L   D V F L S   V V D M N E G A L   F L P H F N S K A I   V V L V   I   N E G E A N I
2  E   R T – D – N – – – – S – L   N V L I S   S I   E M K E G A L   F V P H Y Y S K A I   V I L V   V N E G E A H V
3  E   I T P E K N – P Q L Q D L   D I   F V N   S V E I   K E G S   L L L P H Y N S R A I   V I V T   V N E G K G D F
     351            360             370             380             390             400
1  E   L V G I   K – – – – – – – – – – – E Q   Q Q R Q Q Q – – – E E Q – – P – – – – – L E V   R K Y R A E L S
2  E   L V G P K – – – – – – – – – – – G N   – – – – K E – – – – – – – – – – – – T – L E F   E S Y R A E L S
3  E   L V G Q R – – – – – – – – – N E N   Q Q E Q R K E D D E E E E Q G E E E I   N K Q V   Q N Y K A K L S
     401            410             420             430             440             450
1  E   Q D I   F V I   P A G Y P V M V N A T S   D L N F – F – A F   G – – – – – I   N A E N N Q R   N F L A G S K D
2  K   D D V F V I   P A A Y P V A I   K A T S   N V N F – – T G F   G – – – – – I   N A N N N N R   N L L A G K T D
3  S   G D V F V I   P A G H P V A L K A S S   N L D L – L – G F   G – – – – – I   N A E N N Q R   N F L A G D E D
     451            460             470             480             490             500
1  N   V I   S Q I   P S Q V – – Q E – – – L A   F P R S   A K D I   E N L I   K S Q – S E S Y F V D   A – – – – Q P Q
2  X   V I   S S I   G R A L D G K D V L G L T   F S G S   G E E V M K L I   N K Q – S G S Y F V D   G H H H Q Q E Q
3  N   V I   S Q V Q R P V – – K E – – – L A   F P G S   A Q E V D R I   L E N Q – K Q S H F A D   A – – – – Q P Q
     501            510             520             530
1  Q   K E E – G N – – – – – – K G R K G P   L S S I   L R A F – Y
2  Q   K – – – G S H Q Q E Q Q K G R K G –   – – – – – – A F V Y
3  Q   R E – R G S – – R E – T R D R – – –   L S S V
```

[a] α/β cleavage site: 239/240, β/γ cleavage site: 376/377; – – space to maximize homology.

[b] In order to show the homology clearly, the N-terminal of β-conglycin was omitted in the table. It consists of the following sequence:

V E E E E C E E G Q I P R P R P Q H P E R E R Q Q H G E K E E D E G E Q P R P F P F P R P R Q P H Q
E E E H E Q K E E H E W H R K E E K H G G K G S E E E Q D E R E H P R P H Q P H Q K E E E K H E W
H K Q E Q K H Q G K E S E E E E E D Q

Table 16.9. Amino acid sequences in the vicinity of the α/β cleavage site (239/240 in Table 16.8) and the β/γ cleavage site (376/377 in Table 16.8) of subunits of various 7 S globulins (–: space to maximize the homology)

Protein	α/β 240		β/γ 377	
Phaseolin (*Phaseolus vulgaris*)	–	–	–	–
Vicilin (*Pisum sativum*)	K	D	E	D
Convicilin (*Pisum sativum*)	R	D	E	D
Vicilin 50k (*Pisum sativum*)	R	D	E	D
Vicilin 47k (*Pisum sativum*)	K	D	N	D
β-Conglycinin α′ (*Glycine max*)	–	–	–	–
β-Conglycinin β (*Glycine max*)	–	–	–	–
α-Gossypulin B (*Gossypium sp.*)	–	–	–	–

Table 16.10. Amino acid composition of 7 S and 11 S globulins from soybeans

Amino acid	g amino acid/100 g protein	
	7 S globulin	11 S globulin
Asx	11.18	13.10
Thr	3.14	3.37
Ser	4.79	4.16
Glx	17.54	18.03
Pro	5.21	5.40
Gly	3.37	3.97
Ala	3.66	3.55
Cys	1.52	1.44
Val	4.68	5.05
Met	0.43	1.84
Ile	4.99	4.69
Leu	8.15	7.17
Tyr	3.51	4.05
Phe	5.55	5.73
His	2.32	2.22
Lys	6.26	4.88
Arg	7.37	7.75

not based on specific immunologic mechanisms are classified as food intolerance. It has been estimated that 1–2% of adults and up to 8% of children suffers from a food allergy. Persons with a genetic predisposition develop a food allergy in two stages. In the first stage, sensitization, the allergen (= antigen, cf. 2.6.3) initiates reactions which lead to the formation of allergen-specific antibodies of the immunoglobulin class E (IgE). This process proceeds without discernible changes in the condition of the affected person. In the second stage, the allergen binds to the IgE molecules, leading to the release of phar-

macologically active mediators, e. g., histamine, leucotriens, prostaglandin D_2. These mediators can start inflammations. The specificity of IgE applies to certain structural features of the allergen. Accordingly, the IgE can react in a cross reaction not only with the antigen which generated the IgE, but also with other antigens which have partial structures that correspond with those of the sensitizer. For example, birch pollen, which enters the body via the respiratory tract, produces IgE which can cross react with proteins from apples, hazelnuts, celery or carrots, triggering an allergy.

A region of 5–7 amino acids of the antigen is responsible for the binding to IgE. It is called the epitope. It is either a section of the sequence (linear epitope) or, which is more often the case, amino acid residues which have come together as a result of the folding of the protein (conformational epitope).

Sensitization can occur not only by the inhalation of allergens, but also in the digestive tract. This is the cause of the allergenic effect of milk, egg and fish proteins and of some proteins of plant origin on sensitive persons. Allergens are proteins or glycoproteins occurring in food. Among plant foods, primarily peanuts and other legumes, hazel nuts and other nuts as well as celery and some spices can be allergenic. Examples of some well characterized plant food allergens are listed in Table 16.11. Some allergens, e. g., Mal dl (Table 16.11), show a high degree of correspondence in their sequence with the main allergens of birch pollen, many kinds of stone fruit and pomaceous fruit, celery (Api g1) and carrots. The thermal stability of the allergens varies (Table 16.11). While the allergens of soybeans survive cooking by microwaves (25 min), the allergen Api g1 is so thermolabile that it is no longer detectable in celery after 30 min (100 °C).

16.2.2 Enzymes

Various forms of lipoxygenase (cf. 3.7.2.2) are of interest in food chemistry since they strongly affect the legume aroma.

Urease, which hydrolyses urea,

$$NH_2 - CO - NH_2 + H_2O \rightarrow CO_2 + 2NH_3$$

$$(16.3)$$

Table 16.11. Examples of allergens in plant foods

Food	Allergen	Molecular weight	Characteristics	Stability[a]
Peanuts	Ara h1	$6.3–6.6 \times 10^4$	7 S Globulin (vicilin), glycoprotein	High
	Ara h2	17,000	Glycoprotein	Medium
Soybeans	Glycinin	$3.5–3.6 \times 10^5$		Medium
	β-Conglycinin	156,000	Glycoprotein	Unknown
	2 S Globulin	18,000		Medium
	Kunitz trypsin inhibitor	21,500		Medium
Mustard	Sin a1	14,000	2 S Albumin	High
Rice	16 kDa allergen	16,000	Albumin	Medium
Celery root	Api g1	ca. 16,000		Low
Celery root	Profilin	$1.5–1.6 \times 10^4$	2 Isoforms	Medium
Apple	Mal d1	$1.7–1.8 \times 10^4$		Low
Apple	Mal d3	11,410		High
Peach	Pru p1	9178		High
Apricot	Pru ar3	9178		High
Cherry	Pru av3	9200		High

[a] Thermal stability.

occurs in soybeans in relatively high concentration. Heat treatments of soy preparations can be detected by measuring the activity of this enzyme.

16.2.3 Proteinase and Amylase Inhibitors

16.2.3.1 Occurrence and Properties

Inhibitors of hydrolases, themselves proteins, form stoichiometric inactive complexes with the enzymes and are distributed in microorganisms, plants, and animals. Apart from the thoroughly examined group of proteinase inhibitors, some proteins that inhibit amylases are known.

Of the large number of known proteinase inhibitors, only those compounds found in foods are of interest in food chemistry. These include in particular the inhibitors in egg white, plant seeds, and plant nodules. Table 16.12 shows the most important sources of proteinase inhibitors, which have molecular weights between 6000 and 50,000. The specificity for proteinases varies considerably. Some inhibitors inhibit only trypsin, many act on both trypsin and chymotrypsin, and others inhibit microbial or plant proteinases as well, e. g., subtilisin or papain. Proteinase inhibitors are often located in,

but not limited to, the seeds of plants. The seeds of legumes (soybeans ca. 20 g/kg, white beans ca. 3.6 g/kg, chick peas ca. 1.5 g/kg, mungo beans ca. 0.25 g/kg), the tubers of Solanaceae (potatoes ca. 1–2 g/kg), and cereal grains (ca. 2–3 g/kg) contain especially high concentrations. The inhibitor content greatly depends on the variety, degree of ripeness, and storage time.

Often, several different inhibitors are found in plant materials. They differ in their isoelectric points and also in their specificity for proteinases, specific activities and thermal stabilities. For example, in the more than 30 legumes analyzed so far, nine inhibitors have been identified and five partially purified.

Food which contains inhibitors might cause nutritional problems. For example, feeding rats and chickens with raw soymeal leads to reversible pancreatic blistering. A consequence of excessive secretion of pancreatic juice is increased secretion of nitrogen in the feces. Furthermore, growth inhibition occurs which can be eliminated by incorporating methionine, threonine and valine into the diet.

These findings indicate that the poor growth rate might be due to some amino acid deficiencies, which are a result of increased N-excretion. All the possible effects of proteinase inhibitors are not fully understood.

Table 16.12. Proteinase inhibitors of aninal and plant origin

Source/Inhibitor	Molecular weight	Inhibition of[a]						
		T	CT	P	Bs	AP	SG	PP
Animal tissues								
Bovine pancreas								
Kazal inhibitor	6153	+	−	−				
Kunitz inhibitor	6512	+	+	−	−	−		+
Chicken egg								
Ovomucoid	27–31,000	+	−	−				
Ovoinhibitor	44–52,000	+	+	−	+	+		
Ficin-papain-inhibitor	12,700	−	−	+	−			
Plant tissues								
Cruciferae								
Raphanus sativus[b]	8–11,200	+	±	−	+	+		
Brassica juncea[b]	10–20,000	+	±					
Leguminosae								
Arachis hypogaea[b]	7500–17,000	+	+					
Cicer arietinum[b]	12,000	+	+					
Glycine max								
Kunitz inhibitor	21,500	+	+	−	−			
Bowman–Birk inhibitor	8000	+	+	−	+			
P. coccineus[c]	8800–10,700	+	+					
P. lunatus[c]	8300–16,200	+	+	−	−	±		
P. vulgaris[c]	8–10,000	+	+	−	−			
Pisum sativum[b]	8–12,800	+	+					
Vicia faba[b]	23,000	+	+					
Convolvulaceae								
Ipomoea batatas[b]	23–24,000	+	−	−	−	−		
Solanaceae								
Solanum tuberosum[b]	22–42,000[c]	±	±	−	±	±	±	±
Bromeliaceae								
Ananas comosus[b]	5500	+	+					
Gramineae								
Hordeum vulgare[b]	14–25,000	±	−	−	±	±	±	
Oryza sativa[b]		±	−	+				
Secale cereale[b]	10–18,700	+	+	−				
Triticum aestivum[b]	12–18,500	±	−	−				
Zea mays[c]	7–18,500	+	+	−				

[a] T: trypsin, CT: α-chymotrypsin, P: papain, Bs: Bacillus subtilis proteases, AP: Aspergillus spp. proteases, SG: Streptomyces griseus proteases, PP: Penicillium spp. proteases, +: inhibited, −: not inhibited, ±: inhibited by some inhibitors of the particular source.
[b] The properties of different inhibitors are combined.
[c] Subunits 6–10,000.

16.2.3.2 Structure

Many proteinase inhibitors have been isolated and their structures elucidated. The active center often contains a peptide bond specific for the inhibited enzyme, e. g., Lys-X or Arg-X in trypsin inhibitors and Leu-X, Phe-X or Tyr-X in chymotrypsin inhibitors (Table 16.13, 16.14). In addition, inhibitors are known that inhibit trypsin and chymotrypsin and contain only one trypsin-specific peptide bond at the active center, e. g., *Kunitz* inhibitors from bovine pancreas and soybeans (cf. Table 16.14). Some double-headed inhibitors contain two different active

Table 16.13. Active centers in *Bowman–Birk* type inhibitors

Inhibited enzyme	Active center	Occurrence
Trypsin	Lys-X	Adzuki bean (API II)
		Chickpea
		Garden bean (GBI I)
		Lima bean (LBI IV)
		Soybean (BBI)
		Wisteria (inhibitor II)
	Arg-X	Garden bean (GBI II)
		Soybean (inhibitor C-II)
		Soybean (inhibitor D-II)
α-Chymotrypsin	Leu-X	Lima bean (LBI IV)
		Soybean (BBI)
	Tyr-X	Adzuki bean (API II)
		Chickpea
	Phe-X	Garden bean (PVI 3)
		Lima bean (LBI IV')
Elastase	Ala-X	Garden bean (GBI II)
		Soybean (inhibitor C-II)

Table 16.14. Amino acid sequences in the region of the active centers of proteinase inhibitors

Inhibitor	Sequences at the active center[a]	Inhibited enzyme[b]
Bovine pancreas *Kazal* inh.	18 NGCP**RI**YNPVCG	T
Kunitz inh.	15 TGPCK**AR**IIRYF	T,CT
Soybean *Kunitz* inh.	63 SPSY**RI**RFIAFG	T,CT
Bowman–Birk inh.	16 CACT**KS**NPPQCR	T
	43 CICA**LS**YPAQCF	CT
Lima bean inhibior IV	26 CLACT**KS**IPPQCR	T
	53 CICT**LS**IPAQCV	CT
Potato, subunit A	60 PVVG**MD**FRCDRV	CT
Corn	25 GIPG**RL**PPLZKT	T

[a] Active center underlined.
[b] T: trypsin, CT: α-chymotrypsin.

centers, which, e. g., are both directed towards trypsin or towards trypsin and chymotrypsin. An example of the latter type is represented by the *Bowman–Birk* inhibitors found in legumes (cf. Table 16.14). Their reactive centers are localized in two homologous domains of the peptide chain, each of which form a 29 membered ring via a disulfide bridge (cf. 1.4.2.3.2). In this way, the centers are exposed for contact with the enzyme. An active center can also be exposed by another suitable conformation, as is the case with the *Kunitz* inhibitor from soybeans.

X-ray analyses of the trypsin inhibitor complex show that 12 amino acid residues of the inhibitor are involved in enzyme contact, including the sequence Ser(61)-Phe(66) with the active center Arg(63)-Ile(64).

The double-headed *Bowman–Birk* inhibitor from soybeans was cleaved into two fragments by cyanogen bromide (Met(27)-Arg(28)) and pepsin (Asp(56)-Phe(57)) (cf. Fig. 1.25). Each of these fragments contained an active center and, therefore, inhibited only one enzyme with remaining activities of 84% (trypsin) and 16% (chymotrypsin) compared with the native inhibitor.

Modifications of the active center of an inhibitor result in changes in the properties. For example, Arg(63) of the *Kunitz* inhibitor from soybeans can be replaced by Lys without changing the inhibitory behavior, while substitution by Trp abolishes the inhibition of trypsin and increases the inhibition of chymotrypsin. Indeed, Ile(64) can be replaced by Ala, Leu, or Gly without change in activity, while the insertion of an amino acid residue, e. g., Arg(63)-Glu(63a)-Ile(64), abolishes all inhibition and makes the inhibitor a normal substrate of trypsin.

16.2.3.3 Physiological Function

The biological functions of most proteinase inhibitors of plant origin are unknown. During germination of seeds or bulbs, an increase as well as a decrease in the inhibitor concentration has been observed, but only in a few cases were endogenous enzymes inhibited. It is probable that the inhibitors act against damage to plants by higher animals, insects, and microorganisms. This is indicated by the inhibition of proteinases of the genera *Tribolium* and *Tenebrio* and by the increase in inhibitor concentration in tomato and potato leaves after infection with the potato bug or its larvae. Proteinase inhibitors from the potato also inhibit the proteinases of microorganisms found in rotting potatoes, e. g., *Fusarium solani*.

16.2.3.4 Action on Human Enzymes

Inhibitor activity is normally determined with commercial animal enzymes, e. g., bovine trypsin or bovine chymotrypsin. The evaluation of a potential effect of the inhibitors on human health assumes that the inhibition of human enzymes is known. Present data show that inhibitors from legumes generally inhibit human trypsin to the same extent or a little less than bovine trypsin. On the other hand, human chymotrypsin is inhibited to a much greater extent by most legumes. Ovomucoid and ovoinhibitor from egg white as well as the *Kazal* inhibitor from bovine pancreas do not inhibit the human enzymes. The *Kunitz* inhibitor from bovine pancreas inhibits human trypsin but not chymotrypsin. The data obtained greatly depend not only on the substrate used, but also on the enzyme preparation and the reaction conditions, e. g., on the ratio enzyme/inhibitor. The stability of an inhibitor as it passes through the stomach must also be taken into account in the evaluation of a potential effect (cf. Table 16.15). The *Kunitz* inhibitor of soybeans, for

example, is completely inactivated by human gastric juice, but the *Bowman–Birk* inhibitor from the same source is not. The available data show that the average amount of trypsin and chymotrypsin produced by humans per day can be completely inhibited by extracts from 100 g of raw soybeans or 200 g of lentils or other legumes.

16.2.3.5 Inactivation

The inactivation of proteinase inhibitors in the course of food processing has been the subject of many studies. In general, the inhibitors are thermolabile and can be more or less extensively inactivated by suitable heating processes. In these processes, both the starting material as well as the process parameters are of great importance (time, temperature, pressure, and water content of the sample) (Table 16.16). Steaming of soybeans for 9 minutes at 100 °C causes an 87% destruction of inhibitors (Table 16.17).

A decrease in the inhibitor activity can also be achieved by soaking. A thermal step can then follow under gentler conditions. Although the processing of soybeans into protein isolates, textured protein, or meat surrogates causes a decrease in the inhibitor activity against trypsin, noticeable activity can still be present (Table 16.18).

Soybeans promote the growth of rats to the same extent as casein when about 90% of the inhibitor activity is eliminated (Table 16.19).

Table 16.15. Resistance of inhibitors[a] to pepsin at pH2

Source/Inhibitor	Remaining activity[b](%)
Soybean, *Kunitz* inhibitor	0
Bowman–Birk inhibitor (BBI)	100
extract	30–40
Lima bean, BBI-type inhibitor	70–93
Kidney bean, BBI-type inhibitor	100
Kintoki bean, BBI-type inhibitor	100
Lentil, BBI-type inhibitor	83–100
Chick pea, inhibitors	100
Broad bean, trypsin-chymotrypsin inhibitor	100
Moth bean, trypsin inhibitor	91
Broad bean, trypsin inhibitor	100

[a] Different incubation times.
[b] Against bovine and human trypsin and chymotrypsin.

16.2.3.6 Amylase Inhibitors

Relatively thermostable proteins, which have an inhibitory effect on pancreatic amylase, are found in aqueous extracts of navy beans, wheat, and rye. As a result of the high thermostabilty, inhibitor activity is also detectable in breakfast cereals and bread.

The amylase inhibitor of navy beans is instable in the stomach and becomes active only after preincubation with the enzyme in the absence of starch. As a result, it has no measurable influence on the digestion of starch by human beings. Moreover, the average amounts of inhibitor ingested with the food are small compared to the amylase activity present.

Table 16.16. Destruction of trypsin inhibitors by heating

Sample	Process	Destruction (%)
Soy flour	Live steam, 100 °C, 9 min	87
Soy bean	10% Ca(OH)$_2$, 80 °C, 1 h	100
Navy bean	Autoclaving, 121 °C, 5 min	80
	Autoclaving, 121 °C, 30 min	100
	Dry roasting, 196–204 °C, 20–25 s	75
Navy bean	Pressure cooking, 15 min	89
Winged bean	Autoclaving	92
	Soaking + autoclaving	95
Chickpea	Autoclaving	54
Broad bean	Autoclaving, 120 °C, 20 min	90
Horse bean	Autoclaving	100
Black gram	Cooking, 100 °C, 10 min	15
	Autoclaving, 108 °C, 10 min	27
	Autoclaving, 116 °C, 10 min	38
Cow pea	Cooking, 90–95 °C, 45 min	52
	Autoclaving, 121 °C, 15 min	11
	Toasting, 210 °C, 30 min	44
	Toasting, 240 °C, 30 min	22
	Extrusion cooking	19
Peanut	Moist heat, 100 °C, 15 min	100

Table 16.17. Inactivation of soybean trypsin inhibitors by steaming (100 °C)

Steaming (min)	Trypsin inhibitor	
	Concentration (mg/g soy meal)	Inactivation (%)
0	40	0
3	30	25
6	16.5	59
9	5.2	87
12	1.7	96
15	0.9	98

Table 16.19. Effect of soybean trypsin inhibitors on the growth of rats

Trypsin inhibitor Amount (mg/100 g diet)	Inactivation (%)	Body weight (g)	Protein efficiency ratio (PER)
887	0	79	1.59
532	40	111	2.37
282	68	121	2.78
119	87	148	3.08
Control (casein)		145	3.35

Table 16.18. Inhibition of bovine trypsin activity by some soya products[a]

Product	Extracted with	
	0.125 mol/l H$_2$SO$_4$	0.01 mol/l NaOH
Untreated soybean (cv. Caloria)	51.5	33.7
Supro G 10	6.8	15.6
Soyflour	1.1	8.7
TVPU 110 chunks	0	4.1
Flocosoya	0	1.9

[a] A 50% inhibition of mg trypsin/g product; substrate: N$^\alpha$-benzoyl-L-arginine-p-nitroanilide.

16.2.3.7 Conclusions

In summary, it can be concluded that many foods in the raw state contain inhibitors of hydrolases. The heating processes normally used in the home and in industry generally inactivate the inhibitors more or less completely, so that damage to human health is not to be expected. As a result of the greatly varying thermal stability of the inhibitors, constant and careful control of raw materials and products is required, especially when new materials and processes are applied.

16.2.4 Lectins

Lectins are sugar-binding proteins which differ from antibodies and enzymes. They are widely distributed in plants, e. g., in more than 600 species of legumes. One method of detection is based on the fact that lectins attach themselves to erythrocytes and cause their precipitation (agglutination). It should be taken into account that some lectins (e. g., those from pinto beans) only agglutinate erythrocytes which have been treated with pronase or trypsin. Alternatively, lectins can be detected by the precipitation of polysaccharides and glycoproteins. The examples in Table 16.20 show that biopolymers which contain N-acetyl galactosamine are preferentially bound, but other sugar specificities also exist.

Most lectins are glycoproteins. In general, they consist of several subunits (Table 16.20), which readily dissociate by a change in pH or ionic strength. A characteristic feature of their amino acid composition is the high content of acidic and hydroxy amino acids and the absence or low content of methionine.

Animal tests have demonstrated that their toxicity often does not parallel hemagglutination activity. Thus, lectins from soybeans and garden beans are toxic, but not those from peas and lentils. These and other observations suggest that it is not the hemagglutination activity but other activities of lectins which are responsible for their toxicity. One toxic effect originates in the, at least partial,

resistance of lectins to proteolysis *in vivo*. After reaching the intestinal tract, some lectins attach themselves to the epithelial cells of the intestinal villi and enter the intercellular space, resulting in severe metabolic damage.

After cooking or dry heating, the activities of legume lectins and the associated toxic effects are destroyed. After heating to 100 °C for 10 minutes, e. g., soybeans were free of lectin activity. However, the lectins in some legumes are much more stable.

16.2.5 Carbohydrates

The carbohydrates which are present in legumes are listed in Table 16.21. The major carbohydrate is starch, amounting to 75–80%. The soybean is an exception (Table 16.21), but it contains arabinoxylans and galactans (3.6 and 2.3% respectively). In peanuts, about onethird of the total carbohydrate is starch.

Oligosaccharides in legumes are present in higher concentration than in cereals. Predominant in this fraction are sucrose, stachyose and verbascose (Table 16.21).

After legume consumption, oligosaccharides might cause flatulency, a symptom of gas accumulation in the stomach or intestines. It is a result of the growth of anaerobic microorganisms in the intestines, which hydrolyze the

Table 16.20. Occurrence of lectins in food

Source	Molecular weight (kdal)	Subunits	Glycan-component		Specificity[a]
			% Carbo-hydrate	Building blocks	
Soybean	122	4	6.0	D-Man, D-GlcNAc	D-GalNAc, D-Gal
Garden beans	98–138	4	4.1	GlcN, Man	D-GalNAc
Jack beans[b]	112	4	0		α-D-Man
Lentils	52	2	2.0	GlcN, Glc	α-D-Man, α-D-Glc
Peas	53	4	0.3		α-D-Man, α-D-Glc
Peanuts	11	4	0		α-D-Gal
Potato	20		5.2	Ara	D-GlcNAc
Wheat	26		4.5	Glc, Xyl, Hexosamine	D-GlcNAc

[a] Precipitates biopolymers that contain the given building blocks (polysaccharides, glycoproteins, lipopolysaccharides).

[b] *Canavalia ensiformis*.

Table 16.21. Carbohydrates in legume flours[a]

Flour	Glucose	Saccharose	Raffinose	Stachyose	Verbascose	Starch
Garden beans	0.04	2.23	0.41	2.59	0.13	51.6
Broad beans	0.34	1.55	0.24	0.80	1.94	52.7
Lentils	0.07	1.81	0.39	1.85	1.20	52.3
Green gram (mungo beans)	0.05	1.28	0.32	1.65	2.77	52.0
Soybean[b]	0.01	4.5	1.1	3.7		0.62

[a] Weight-% of the dry matter.
[b] Defatted flour.

oligo- into monosaccharides and cause their further degradation to CO_2, CH_4 and H_2. Model feeding tests have demonstrated that phenolic ingredients, such as ferulic and syringic acids, inhibit microorganism metabolism and the related flatulency.

16.2.6 Cyanogenic Glycosides

Cyanogenic glycosides (Table 16.22) are present in lima beans and in some other plant foods. Precursors of cyanogenic glycosides are the amino acids listed in Table 16.22. As in the biosynthesis of glucosinolates (cf. 17.1.2.6.5), an aldoxime is initially formed, which is then transformed into a cyanogenic glycoside by means of the postulated reaction pathway shown in Fig. 16.3.

Seeds are ground and moistened in order to detoxify them. This initiates glycoside degradation with formation of HCN (cf. Table 16.23) which, after incubation, is expelled by heating.

The cyanogenic glycoside degradation is initiated by β-glucosidase (Fig. 16.4) which in the cells is separated from its substrate. Once the cell structure is ruptured by seed grinding, the enzyme and the substrate are brought together and the reaction starts.

The substrate specificity of β-glucosidase is governed by an aglycon moiety. Thus, the enzymes present in *"emulsin"*, a glycosidase mixture from bitter almonds, hydrolyze not only amygdalin but also other cyanogenic glycosides which are de-

Table 16.22. Cyanogenic glycosides in fruit and some field crops

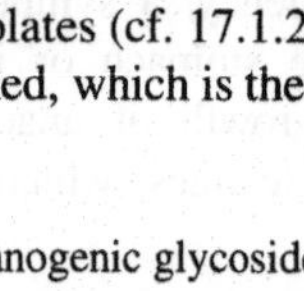

Glycoside Name	Structure R₁	Structure R₂	Sugar	Amino acid precursor	Occurrence (seeds)
Linamarin	CH_3	CH_3	Glucose	Val	Lima bean Linseed (flax) Cassava
(R)-Lotaustralin	C_2H_5	CH_3	Glucose	Ile	like Linamarin
(R)-Prunasin	Phenyl	H	Glucose	Phe	Prunes
(R)-Amygdalin	Phenyl	H	Gentiobiose	Phe	Bitter almond Apricots Peaches Apples
(S)-Dhurrin	HO–Phenyl	H	Glucose	Tyr	Sorghum sp.

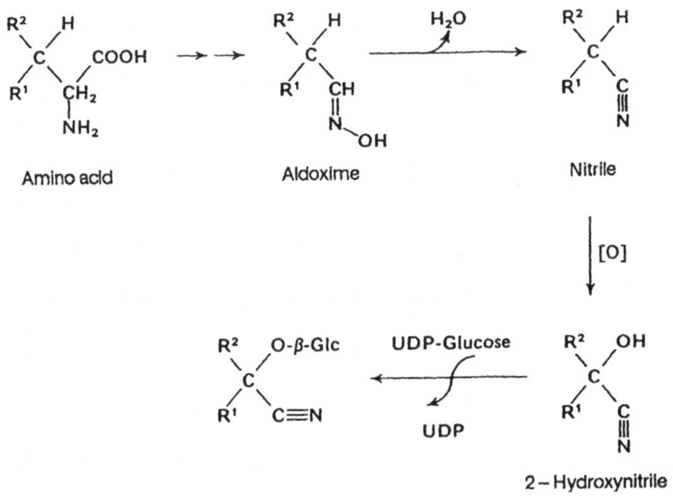

Fig. 16.3. Biosynthesis of cyanogenic glucosides

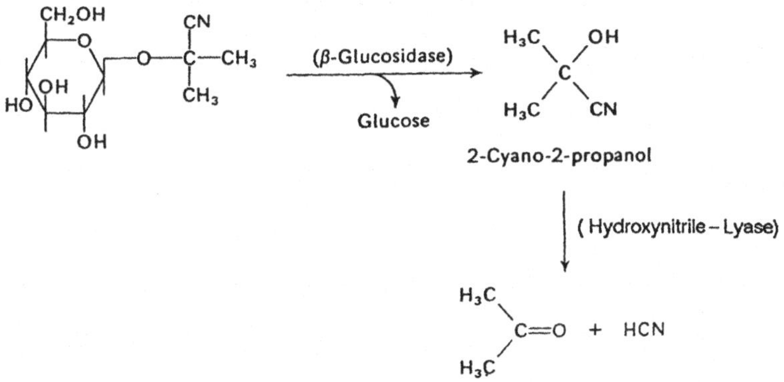

Fig. 16.4. Lima beans: linamarin degradation, resulting in a release of hydrocyanic acid

Table 16.23. Amount of glycoside-bound hydrocyanic acid in food

Food	HCN (mg/100 g)
Lima bean[a]	210–310
Bitter almond	280–310
Sorghum sp.	250
Cassava	110
Pea	2.3
Bean	2.0
Chick pea	0.8

[a] In the United States new cultivars have been developed that contain only 10 mg HCN/100 g seed.

As shown in Fig. 16.4, β-glucosidase hydrolysis produces an unstable hydroxynitrile which slowly degrades into the corresponding carbonyl compound and HCN. However, most legume seeds contain a hydroxynitrile lyase which accelerates this reaction.

16.2.7 Lipids*

With the exception of soybeans and peanuts, the lipid content of most legumes is so low (cf. Table 16.2) that they can not be considered

rived from phenylalanine or tyrosine, but not linamarin.

---

* The composition of soy and peanut lipids is covered in Chapters 3 and 14.

Table 16.24. Fatty acid composition of legume lipids (weight-%)[a]

Fatty acid	Garden beans	Chick peas	Broad beans	Lentils
14:0	0.22	1.3	0.6	0.85
16:0	21.8	8.9	9.3	23.2
18:0	4.7	1.6	4.9	4.6
20:0	0.53	0.03	0.7	2.3
22:0	2.9	0	0.42	2.7
24:0	1.1	0	0	0.85
16:1 (9)	0.21	0.05	0	0.15
18:1 (9)	11.6	35.4	33.8	36.0
18:2 (9,12)	29.8	51.1	42.1	20.6
18:3 (9,12,15)	27.4	1.7	6.4	1.6
20:1	0.02	0	0.7	1.9

[a] In Table 14.11 the fatty acid compositions are provided for soya oil and peanut butter.

as a source of fats or oils. Examples of their fatty acid composition are listed in Table 16.24.

16.2.8 Vitamins and Minerals

Vitamin and mineral content of some legumes is presented in Table 16.25. In addition to B-vitamins, the two oilseeds are rich in tocopherols.

16.2.9 Phytoestrogens

The isoflavones daidzein (Ia in Formula 16.4), genistein (Ib) and glycitein (Ic) as well as coumestrol (II) together with the lignans (cf. 18.1.2.5.7) are called phytoestrogens because they can dock onto estrogen receptors. Accordingly, they are competitors of endogenous estrogen, but have lower activity. Sources of isoflavones are soybeans and soybean products (Table 16.26). In addition, they occur in traces in many plant foods.

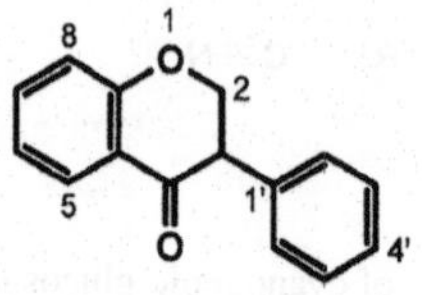

(Ia)	Daidzein:	4' - OH, 7 - OH
(Ib)	Genistein:	4' - OH, 5 - OH, 7 - OH
(Ic)	Glycitein:	4'- OH, 6 - OCH₃, 7 - OH

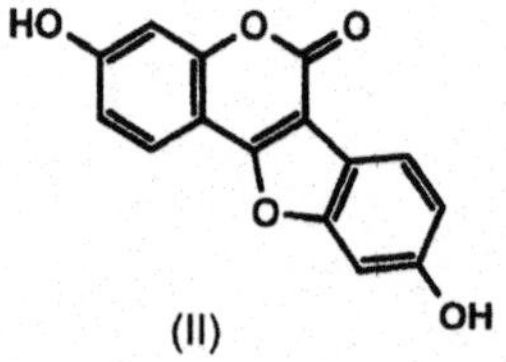

(II)   (16.4)

Table 16.25. Vitamin and mineral composition of legumes[a]

	Soybeans	Peas	Garden beans	Peanuts
Vitamins				
Tocopherols	127	12		202
B_1	8.2	1.2	4.5	9.0
B_2	4.3	0.64	1.6	1.5
Nicotinamide	20.8	9.5	20.8	153
Pantothenic acid	15.9	2.9	9.7	26
B_6	9.9	0.64	2.8	3.0
Minerals				
Na	33	8.0	20	52
K	1.4×10^4	1.2×10^3	1.3×10^4	7.1×10^3
Mg	2.1×10^3	132	1.3×10^3	1.6×10^3
Ca	2.1×10^3	96	1.1×10^3	590
Fe	71	7.4	60.4	21.1
Zn	42	10.6	26	30.7
P	4.9×10^3	432	4.3×10^3	3.7×10^3
Cl	58	160	248	70

[a] Results are given in mg/kg.

Table 16.26. Daidzein (Dai), genistein (Gen), glycitein (Gly) and coumestrol (Cou) in soybeans[a]

Foods	Dai	Gen	Gly	Cou
Soybeans	566	442	28.1	0.015
Soy milk	9.2	18	1.7	0.006
Soybean sprouts	2.7	5.1	0.045	n.n.
Tempeh	69.7	107	5.7	0.006
Tofu	93.4	170	7.3	0.007
Miso	44.2	59	8	0.024
Soybean protein	25.3	59.7	3.1	0.005

[a] values in mg/kg
n.n.: not detected

Table 16.27. Saponin content in foods

Food	Saponin (g/kg solids)
Chick peas	56
Soybeans	43
Garden beans	4.5–21
Peanuts	6.3
Lentils	3.7–4.6
Broad beans	3.5
Peas	11
Spinach	47
Asparagus	15
Oat bran	1.0

16.2.10 Saponins

Saponins are surface-active plant constituents, which are broken down into a carbohydrate portion and an aglycone on acid hydrolysis (Formula 16.5). The carbohydrate chain consists of 1 to 8 monosaccharides or uronic acids. It is usually branched and often terminated by a pentose, e. g., arabinose. There are saponins with one carbohydrate chain (mono-desmosides) and with two chains which are independent of each other (bisdesmosides).

$$\text{Saponin} \xrightarrow{\text{H}_2\text{O(H}^{\oplus})} \text{Sapogenin} + \text{Monosaccharide}$$
(16.5)

According to the structure of the aglycone, there are two groups: pentacyclic triterpenes and steroid sapogenins. The first mentioned group is found in legumes, the main source of saponins in food (Table 16.27). A steroid sapogenin is found, e. g., in oats, which also contain representatives of the first group. The saponins of soybeans have been most intensively studied; more than one dozen have been identified. As examples, the structures of two soybean saponins of the A (bisdesmosides) and B (monodesmosides) series are shown in Formula 16.6 and 16.7. Saponins of the B series contain a 2,3-dihydro-2,5-dihydroxy-6-methyl-4H-pyran-4-one residue at C_{22}. (Formula 16.7)

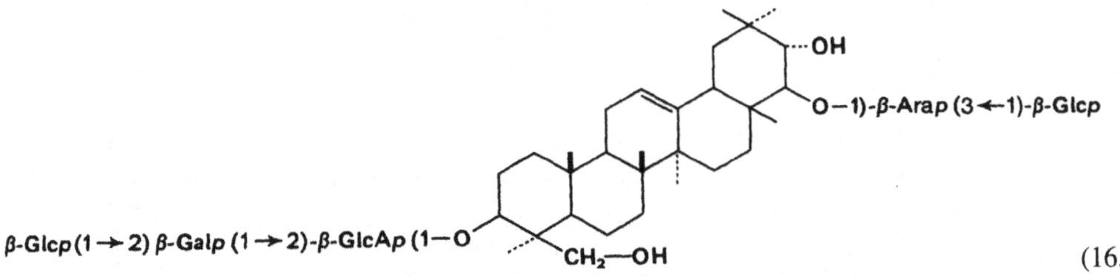

O—1)-β-Ara$p$ (3◄—1)-β-Glc$p$

β-Glc$p$(1→2) β-Gal$p$ (1→2)-β-GlcAp (1—O

CH$_2$—OH

(16.6)

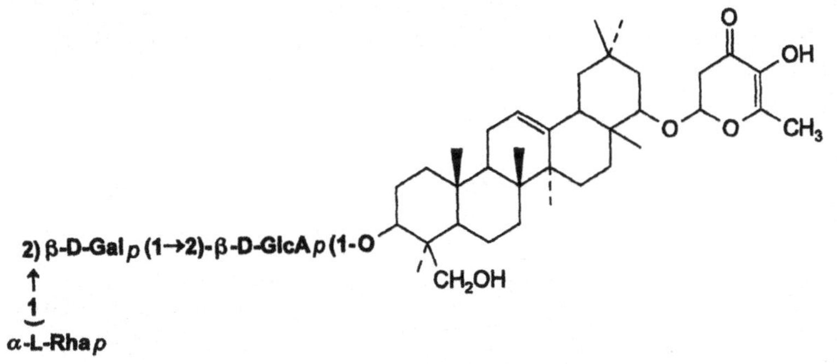

2) β-D-Gal$p$ (1→2)-β-D-GlcAp (1-O

CH$_2$OH

α-L-Rha$p$

(16.7)

Saponins contribute to the characteristic taste of soybeans and other legumes. They are heat stable in the neutral pH range. Since a considerable portion of the saponins occurs in the seed coat and in the hypocotyl, the taste of soybean products, e. g., tofu, improves when these parts are removed.

A series of saponins are hemolytically active, the aglycone as well as the sugar residue playing a role. Monodesmoside triterpene saponins are more active than the bisdesmosides, a longer sugar residue and a branch weaken the effect.

Steroid saponins and, to a smaller extent, triterpene saponins complex cholesterol, ergosterol and 7-dehydrocholesterol but not vitamin D. Since saponins are very poorly absorbed, their toxic effect is negligible. Even in vegetarians who ingest higher amounts of saponins with their food, no negative symptoms have been observed.

16.2.11 Other Constituents

The meadow pea, *Lathyrus sativus*, which is cultivated in India in periods of drought, contains β-N-oxalyl-α, β-diaminopropionic acid (cf. XXXVII in Table 17.5). Possibly due to its structural similarity to glutamic acid, this compound causes the disease known as neurolathyrism, which is characterized by paralysis of the lower limbs. More than 100,000 cases of this disease were described in 1975 alone. The diaminopropionic acid derivative can be largely eliminated by cooking the seeds in excess water, which is then discarded, or by soaking the seeds overnight, followed by steaming, roasting, or drying in the sun. The flour obtained from dried seeds has 24–28% of protein and a high lysine content. It can be used to make unleavened Indian bread ("chapatis").

The horse bean, *Vicia faba*, contains the glucosides vicin (Formula 16.8, I) and convicin (II).

The aglycones of these compounds divicin (III) and isouramil (IV) can be released by the β-glycosidases of the digestive tract. In the oxidized form, they cause quick oxidation of glutathione in erythrocytes (cf. Formula 16.8) which have a hereditary deficiency of glucose-6-phosphate dehydrogenase. Consequently, these erythrocytes are incapable of re-producing reduced glutathione with the help of glutathione reductase for lack of NADPH. The lack of reduced glutathione causes a hemolytic anemia called favism. This genetic defect is found especially in people from the Middle East. Since *Vicia faba* plays a big role in the protein supply of people in this region, attempts are being made to cultivate variants which do not contain these toxic glucosides or to develop suitable methods for its removal (soaking, heating).

16.3 Processing

16.3.1 Soybeans and Peanuts

16.3.1.1 Aroma Defects

Preparation and storage of products from both oilseeds is often inhibited by rancidity and bitter aroma defects caused mostly by volatile aroma active carbonyl compounds, e. g., (Z)-3-hexenal, (Z)-1,5-octadien-3-one and 3-methyl-2,4-nonandione. The rancidity-causing compounds are formed through peroxidation of linolenic acid, accelerated by the enzyme lipoxygenase and/or by hem(in) proteins (cf. 3.7.2.2). Furan fatty acids are the precursors in the case of the dione (cf. 14.3.2.2.5). Lipid peroxidation is also involved in the formation of another very potent odorant, 2-pentylpyridine, which produces grassy aroma defects in soybean products. Defatted soybean protein isolates contained 60–510 µg/kg of this compound, which with an odor threshold

I : R = NH_2
II : R = OH

III (ox.) : R = NH_2
IV (ox.) : R = OH

III (red.) : R = NH_2
IV (red.) : R = OH   (16.8)

of $0.012\,\mu g/kg$ (water), corresponds to aroma values of 5×10^3–4.25×10^4. One way to increase quality is to thermally inactivate enzymes or hem(in) catalysts. Table 16.28 illustrates steam heating of peanuts for a prolonged time in order to inactivate peroxidase activity. Lipoxygenase denaturation, under the conditions given in Table 16.28, occurs after 2 min, but this alone does not yield a satisfactory storage stability. Peroxidase and probably other catalysts should be excluded as well (Fig. 16.5).

The complete removal of lipids is used as an additional precautionary measure in order to obtain an off-flavor-free product, particularly in the case of production of protein isolates. For example, the lipid residue which remains in soy flakes after hexane solvent extraction (cf. 14.3.2.2.1) is removed by extraction with hexane-ethanol 82:18 v/v.

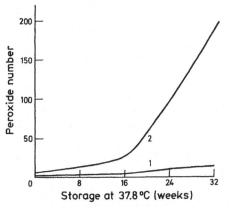

Fig. 16.5. Storage stability of peanut flakes. (according to *Mitchel* and *Malphrus*, 1977) Peanut flakes treated with steam at $100\,°C$ for 30 min (1) and 5 min (2)

Table 16.28. Thermal inactivation of lipoxygenase and peroxidase in peanuts

Heat treatment			Enzyme activity (%)	
Type	°C	Time (min)	Per-oxidase	Lipoxy-genase
Control			100	100
Dry heat	110	60	48	7
Steam	100	2	35	0
Steam	100	6	8	0
Steam	100	30	1	0

16.3.1.2 Individual Products

Protein preparations and milk-like products are processed from soybeans and peanuts. Alone or together with cereals, soybeans are processed into a large number of fermented products in Asia. The following products are made from soybeans.

16.3.1.2.1 Soy Proteins

Figure 16.6 gives an overview of the most important process steps in soybean processing. Soy protein concentrate is usually obtained from the flaked and defatted soy meal that is left after oil extraction (cf. 14.3.2.2.1). The process involves soaking of flakes in water, acidification of the aqueous extract to pH 4–5 (cf. 16.2.1) and separation of the precipitate from solubilized ingredients by centrifugation followed by washing and drying of the sediment collected.

Soy meal isolates enriched in protein are obtained by a preliminary extraction of soluble soy constituents with water or diluted alkali, pH 8–9, followed by protein precipitation from the aqueous extract by adjusting the pH to 4–5. Such protein isolates, texturized and flavored (cf. 1.4.7) are used as meat substitutes. The compositions of protein concentrates and isolates are compared in Table 16.29. For both products, the essential amino acid content corresponds to that of soybeans (cf. Table 16.3). Soy protein is added as an ingredient to baked and meat products and to baby food preparations to raise their protein level and to improve their processing qualities, such as increased water binding capacity or stabilization of o/w emulsions. These properties are required for processing at higher temperatures. The addition of soy protein to beverages at a pH of 3 results in better solubility of beverage constituents. Soy protein market value may be increased by its partial hydrolysis with papain (cf. 2.7.2.2.1).

Table 16.29. Composition of soya protein concentrate and isolate (%)

Product	Protein	Crude fiber	Ash
Concentrate	72	3.5	5.5
Isolate	95.6	0.2	4.0

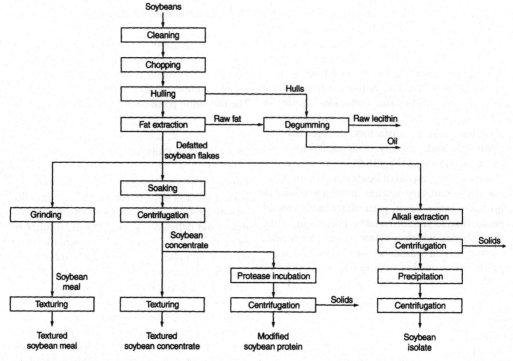

Fig. 16.6. Soybean processing

16.3.1.2.2 Soy Milk

Soybeans are swollen and ground in the presence of a 10-fold excess of water. Heating the suspension close to its boiling point for 15–20 min pasteurizes the suspension and inactivates lipoxygenase enzyme and proteinase inhibitors. A soy milk preparation enriched with calcium and vitamins is of importance in infant nutrition as a replacement for cow's milk, which close to 7% of infants in the USA are unable to tolerate.

16.3.1.2.3 Tofu

When calcium sulfate (3 g salt/kg milk) is added to soy milk at 65 °C, a gel (called soy "curd") slowly precipitates. The curd is separated from excess fluid by gentle squeezing in a special wooden filter box. A washing procedure then follows. The water content of the product is about 88%. Tofu contains 55% protein and 28% fat dry weight. In China and some other Asian countries, tofu is the largest source of food protein. It is consumed fresh or dried, or fried in fat and seasoned with soy sauce.

16.3.1.2.4 Soy Sauce (Shoyu)

Defatted soy meal is used as a starting material in the production of this seasoning sauce (Fig. 16.7). The meal is moistened, then mixed with roasted and crushed wheat and heated in an autoclave for 45 min. The mix ratio in Japan is fixed at 1:1, while in China it varies up to 4:1. Increasing the amount of soy decreases the quality of the endproduct. The mix, with a water content of 26%, is then inoculated with *Aspergillus oryzae* and *Aspergillus soyae*. Initial incubation is at 30 °C for 24 h and then at 40 °C for an additional 48 h. This fermentation starter, called "koji", is then salted to 18% by addition of 22.6% NaCl solution. Inoculation with *Lactobacillus delbrueckii* and with *Hansenula* yeast species results in lactic acid fermentation, which proceeds under gentle aeration in order to prevent the growth

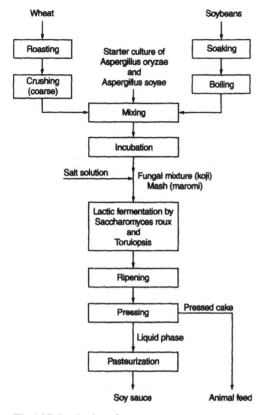

Fig. 16.7. Production of soy sauce

of undesired anaerobic microorganisms. It is a long and tedious fermentation carried out in stepwise fashion: for example, starting at 15 °C for one month, followed by 28 °C for four months, and finishing at 15 °C for an additional month. Highly-valued products ripen for several years. After the fermentation is completed, the soy sauce of pH 4.6 is filtered, pasteurized at 65–80 °C and preserved with benzoic acid for the export market.

During fermentation the microorganisms produce extracellular hydrolases which decompose the main components of the raw material: proteins, carbohydrates and nucleic acids. Soy sauce contains 1.5% N (of which 60% corresponds to amino-N) and 4.4% reducing sugar. The N-containing fraction consists of 40–50% amino acids (glutamic acid predominates at 1.2% of the product), 40–50% peptides, 10–15% ammonia and less than 1% protein. In addition, soy sauce contains by-products of microorganism

metabolism, such as ethanol (1.2%) and lactic, succinic and acetic acids.

Soy sauce products of lower quality are blended with spices and are prepared by acid hydrolysis of the above mix of raw materials (cf. 12.7.3.5).

The compound 2(5)-ethyl-4-hydroxy-5(2)-methyl-3(2H)-furanone (EHM3F) is responsible for the sweetish caramel-like aroma note. It is formed by the yeast *Zygosaccharomyces rouxii* from D-sedoheptulose-7-phosphate, which originates from the pentose phosphate cycle. Apart from EHM3F, 4-hydroxy-2,5-dimethyl-3(2H)-furanone (HD3F) and 3-hydroxy-4,5-dimethyl-2(5H)-furanone (HD2F) contribute to the aroma. 5-Ethyl-3-hydroxy-4-methyl-2(5H)-furanone (EHM2F) is also present, but is of secondary importance because of its lower concentration compared to that of HD2F.

16.3.1.2.5 Miso

Miso is a fermented soybean paste. To produce this substance, rice is soaked, heated, and incubated with *Aspergillus oryzae* at 28–35 °C for 40–50 hours. At the same time, whole soybeans are soaked, heated, and mixed with the incubated rice (60:30) with the addition of salt (4–13%). The mixture is allowed to ferment for several months at 25–30 °C in the presence of lactic acid bacteria and yeasts. The product is then pasteurized and packed. The aroma of miso can be enhanced by the addition of EHMF (cf. 16.3.1.2.4).

16.3.1.2.6 Natto

Various types of natto, a fermented soybean product, are known. For production (Itohiki type), soybeans are soaked in water, boiled and after cooling, incubated with *Bacillus nato*, a variant of *Bacillus subtilis*, for 16–20 hours at 40–45 °C. The surface of natto has a characteristic viscous texture caused by a polyglutamic acid produced by *B. natto*.

16.3.1.2.7 Sufu

Sufu is soy cheese made from tofu. Tofu is cut into cubes (3 cm edge length), treated with an acidified salt solution (6% NaCl, 2.5% citric acid), heated (100 °C, 15 min) and inoculated with *Actinomucor elegans*. After incubation at 12–25 °C for 2–7 days, sufu is placed in a 5–10% salt solution which contains fermented soybean paste and ethanol, if necessary, and allowed to ripen for 1–12 months.

16.3.2 Peas and Beans

Peas and beans are consumed only when cooked. In order to shorten the cooking time which, even after preliminary soaking in water overnight (preliminary swelling), is several hours, the legumes are precooked or parboiled by the process described in 15.3.2.2.1.

Additionally, seed hull removal provides about a 40% reduction in cooking time which, for peas, involves seed steaming at 90 °C, followed by drying and subsequent dehulling.

The softening of legumes during cooking is due to the disintegration of the cotyledonous tissue in individual cells. This is caused by the conversion of native protopectin to pectin, which quickly depolymerizes on heating. The middle lamella of the cell walls, which consists of pectins and strengthens the tissue, disintegrates in this process.

Conversely, the hardening of legumes during cooking is due to cross linkage of the cell walls. The following reactions which can start even during storage at higher temperatures are under discussion as the cause of cross linkage. Calcium and magnesium phytates included in the middle lamellae are hydrolyzed by the phytase present (cf. 15.2.2.4). Apart from meso-inositol and phosphoric acid, Ca^{2+} and Mg^{2+} ions also released cross link the pectic acids and thus strengthen the middle lamellae. Pectin esterases, which demethylate pectin to the acid, promote the hardening of the tissue. In the case of legumes that are relatively rich in phenolic compounds and polyphenol oxidases, the formation of complexes between proteins and polyphenols should contribute to the strengthening of the tissue.

Similar to soybeans, a number of beans are processed into fermented products in Asia.

16.4 References

Angelo, A.J.S., Ory, R.L.: Effects of lipoperoxides on proteins in raw and processed peanuts. J. Agric. Food Chem. *23*, 141 (1975)

Aoki, H., Taneyana, O., Inami, M.: Emulsifying properties of soy protein: characteristics of 7S and 11S proteins. J. Food Sci. *45*, 534 (1980)

Badley, R.A., Atkinson, D., Hauser, H., Oldani, D., Green, J.P., Stubbs, J.M.: The structure, physical and chemical properties of the soybean protein glycinin. Biochim. Biophys. Acta *412*, 214 (1975)

Belitz, H.-D.: Vegetable proteins as human food. FEBS 11th Meeting Copenhagen 1977, Vol. 44, Symposium A3, Pergamon Press: Oxford–New York. 1978

Belitz, H.-D., Kaiser, K.-P., Santarius, K.: Trypsin and chymotrypsin inhibitors from potatoes: isolation and some properties. Biochem. Biophys. Res. Commun. *42*, 420 (1971)

Belitz, H.-D., Weder, J.K.P.: Protein inhibitors of hydrolases in plant foodstuffs. Food Rev. Int. *6*, 151 (1990)

Beuchat, L.R.: Indigenous fermented foods. In: Biotechnology (Eds.: Rehm, H.-J., Reed, G.), Vol. 6, p. 477, Verlag Chemie: Weinheim. 1983

Boatright, W.L., Crum, A.D., Lei, Q.: Effect of prooxidants on the occurrence of 2-pentyl pyridine in soy protein isolate. J. Am. Oil Chem. Soc. *75*, 1379 (1998)

Boulter, D., Derbyshire, E.: The general properties, classification and distribution of plant proteins. In: Plant proteins (Ed.: Norton G.), p. 3, Butterworths: London. 1978

Derbyshire, E., Wright, D.J., Boulter, D.: Legumin and vicilin, storage proteins of legume seeds. Phytochemistry *15*, 3 (1976)

Friedman, M. (Ed.): Nutritional and toxicological significance of enzyme inhibitors in foods. Adv. Exp. Med. Biol. 199, Plenum Press: *New York*. 1986

Gallaher, D., Schneeman, B.O.: Nutritional and metabolic response to plant inhibitors of digestive enzymes. Adv. Exp. Med. Biol. *177*, 299 (1984)

Grant, G., van Driessche, E.: Legume lectins: physico-chemical and nutritional properties. In: Recent advances of research in antinutritional factors in legume seeds. Proc. 2nd Int. Workshop Antinutritional Factors (ANFs) in Legume Seeds (Eds.: A.F.P. van der Poel, J. Huisman, H.S. Saini) Wageningen Pers, Wageningen, 1993, pp. 219

Guegnuen, J., van Oort, M.G., Quillien, L., Hessing, M.: The composition, biochemical characteristics and analysis of proteinaceous antinutritional factors in legume seeds. In: Recent advances of research in antinutritional factors in legume seeds.

Proc. 2nd Int. Workshop Antinutritional Factors (ANFs) in Legume Seeds (Eds.: A.F.P. van der Poel, J. Huisman, H.S. Saini) Wageningen Pers, Wageningen, 1993, pp. 9

IFST: Current Hot Topics: Phytoestrogens (2001) www.ifst.org/hottop 34.htm

Lasztity, R., Hidvegi, M., Bata, A.: Saponins in food. Food Rev. Int. 14, 371 (1998)

Le Guen, M.P., Birk, Y.: Protein protease inhibitors from legume seeds: nutritional effects, mode of action and structure-relationship. In: Recent advances of research in antinutritional factors in legume seeds. Proc. 2nd Int. Workshop Antinutritional Factors (ANFs) in Legume Seeds (Eds.: A.F.P. van der Poel, J. Huisman, H.S. Saini) Wageningen Pers, Wageningen, 1993, pp. 157

Liener, I.E. (Ed.): Toxic constituents of plant foodstuffs. 2nd. ed. Academic Press: New York. 1980

Melcion, J.-P., van der Poel, A.F.B.: Process technology and antinutritional factors: principles, adequacy and process optimization. In: Recent advances of research in antinutritional factors in legume seeds. Proc. 2nd Int. Workshop Antinutritional Factors (ANFs) in Legume Seeds (Eds.: A.F.P. van der Poel, J. Huisman, H.S. Saini) Wageningen Pers, Wageningen, 1993, pp. 419

Mills, E.N.C., Jenkins, J.A., Alcocer, M.J.C., Shewry, P.: Structural, biological, and evolutionary relationships of plant food allergens sensitizing via the gastrointestinal tract. Crit. Rev. Food Sci. Nutr. 44, 379 (2004)

Mills, E.N.C., Madsen, C., Shewry, P.R., Wichers, H.J.: Food allergens of plant origin – their molecular and evolutionary relationships. Trend Food Sci Technol 14, 145 (2003)

Mitchell, J.H., Malphrus, R.K.: Lipid oxidation in spanish peanuts: the effect of moist heat treatments. J. Food Sci. 42, 1457 (1977)

Mossor, G., Skupin, J., Romanowska, B.: Plant inhibitors of proteolytic enzymes. Nahrung 28, 93 (1984)

Naivikul, O., D'Appolonia, B.L.: Comparison of legume and wheat flour carbohydrates. I. Sugar analysis. Cereal Chem. 55, 913 (1978)

Pernollet, J.-C., Mossé, J.: Structure and location of legume and cereal seed storage proteins. In: Seed proteins (Eds.: Daussant, J., Mossé, J., Vaughan, J.), p. 155, Academic Press: London. 1983

Preinerstorfer, B., Sonntag, G.: Determination of isoflavones in commercial soy products by HPLC and coulometric electrode array detection. Eur. Food Res. Technol. 219, 305 (2004)

Salunkhe, D.K., Kadam, S.S. (Eds.): CRC Handbook of World Food Legumes: Nutritional Chemistry, Processing Technology and Utilization, Vol. I–III. CRC Press: Boca Raton, FL, 1989

Sasaki, M., Nunomura, N., Matsudo, T.: Biosynthesis of 4-hydroxy-2(or5)-ethyl-5(or2)-methyl-3-(2H)-furanone by yeast. J. Agric. Food Chem. 39, 934 (1991)

Sathe, S.K., Salunkhe, D.K.: Technology of removal of unwanted components of dry beans. CRC Crit. Rev. Food Sci. Nutr. 21, 263 (1984)

Smith, A.K., Circle, S.J. (Eds.): Soybeans: Chemistry and technology. Vol. 1, AVI Publ. Co.: Westport, Conn. 1972

Stanley, D.W., Aguilera, J.M.: A review of textural defects in cooked reconstituted legumes – The influence of structure and composition. J. Food Biochem. 9, 277 (1985)

Thompson, L.U., Boucher, B.A., Liu, Z., Cotterchio, M., Krieger, N.: Phytoestrogen content in foods consumed in Canada, including isoflavones, lignans and coumestan. Nutrition and Cancer 54, 184 (2006)

Vieths, S. Haustein, D., Hoffmann, A., Jankiewics, A., Schöning, B.: Labile und stabile Allergene in Lebensmitteln pflanzlicher Herkunft. GIT Fachz. Lab. 4, 360 (1996)

Warchalewski, J.R.: Present-day studies on cereals protein nature α-amylase inhibitors. Nahrung 27, 103 (1983)

Weder, J.K.P.: Proteinaseinhibitoren in Lebensmitteln. Analytische Aspekte, Spezifität und Bedeutung. GIT Fachz. Lab. 4, 350 (1996)

Wright, D.J.: The seed globulins. In: Development of Food Proteins-5; (Ed.: Hudson, B.J.F.), p. 81, Elsevier Applied Science: London. 1987

Wright, D.J.: The seed globulins – Part. II. In: Development in Food Proteins-6; (Ed.: Hudson, B. J. F.), p. 119, Elsevier Applied Science: London. 1987

Wüthrich, B.: Lebensmittelallergien und -intoleranzen. Lebensmittelchemie 50, 155 (1996)

17 Vegetables and Vegetable Products

17.1 Vegetables

17.1.1 Foreword

Vegetables are defined as the fresh parts of plants which, either raw, cooked, canned or processed in some other way, provide suitable human nutrition. Fruits of perennial trees are not considered to be vegetables. Ripe seeds are also excluded (peas, beans, cereal grains, etc.). From a botanical point of view, vegetables can be divided into algae (seaweed), mushrooms, root vegetables (carrots), tubers (potatoes, yams), bulbs and stem or stalk (kohlrabi, parsley), leafy (spinach), inflorescence (broccoli), seed (green peas) and fruit (tomato) vegetables. The most important vegetables, with data relating to their botanical classification and use, are presented in Table 17.1. Information about vegetable production follows in Tables 17.2 and 17.3.

17.1.2 Composition

The composition of vegetables can vary significantly depending upon the cultivar and origin. Table 17.4 shows that the amount of dry matter in most vegetables is between 10 and 20%. The nitrogen content is in the range of 1–5%, carbohydrates 3–20%, lipids 0.1–0.3%, crude fiber about 1%, and minerals close to 1%. Some tuber and seed vegetables have a high starch content and therefore a high dry matter content. Vitamins, minerals, flavor substances and dietary fibers are important secondary constituents.

17.1.2.1 Nitrogen Compounds

Vegetables contain an average of 1–3% nitrogen compounds. Of this, 35–80% is protein, the rest is amino acids, peptides and other compounds.

17.1.2.1.1 Proteins

The protein fraction consists to a great extent of enzymes which may have either a beneficial or a detrimental effect on processing. They may contribute to the typical flavor or to formation of undesirable flavors, tissue softening and discoloration. Enzymes of all the main groups are present in vegetables:

- *Oxidoreductases* such as lipoxygenases, phenoloxidases, peroxidases;
- *Hydrolases* such as glycosidases, esterases, proteinases;
- *Transferases* such as transaminases;
- *Lyases* such as glutamic acid decarboxylase, alliinase, hydroperoxide lyase.
- *Ligases* such as glutamine synthetase.

Enzyme inhibitors are also present, e. g., potatoes contain proteins which have an inhibitory effect on serine proteinases, while proteins from beans and cucumbers inhibit pectolytic enzymes. Protein and enzyme patterns, as obtained by electrophoretic separation, are often characteristic of species or cultivars and can be used for analytical differentiation. Figure 17.1 shows typical protein and proteinase inhibitor patterns for several potato cultivars.

17.1.2.1.2 Free Amino Acids

In addition to protein-building amino acids, nonprotein amino acids occur in vegetables as well as in other plants. Tables 17.5 and 17.6 present data on the occurrence and structure of these amino acids. Information about their biosynthetic pathways is given below.
The higher homologues of amino acids, such as homoserine, homomethionine and aminoadipic acid, are generally derived from a reaction sequence which corresponds to that of oxalacetate

Table 17.1. List of some important vegetables

Number	Common name	Latin name	Class, order, family	Consumed as
Mushrooms (cultivated or wildly grown edible species)				
1	Ringed boletus	*Suillus luteus*	Basidiomycetes/Boletales	
2	Saffron milk cap	*Lactarius deliciosus*	Basidiomycetes/Agaricales	
3	Field champignon	*Agaricus campester*	Basidiomycetes/Agaricales	
4	Garden champignon	*Agaricus hortensis*	Basidiomycetes/Agaricales	
5	Cep	*Xerocomus badius*	Basidiomycetes/Boletales	
6	Truffle	*Tuber melanosporum*	Ascomycetes/Tuberales	Steamed, fried, dried,
7	Chanterelle	*Cantharellus cibarius*	Basidiomycetes/Aphyllophorales	pickled or salted
8		*Xerocomus chrysenteron*	Basidiomycetes/Boletales	
9	Morel	*Morchella esculenta*	Ascomycetes/Pezizales	
10	Edible boletus	*Boletus edulis*	Basidiomycetes/Boletales	
11	Goat's lip	*Xerocomus subtomentosus*	Basidiomycetes/Boletales	
Algae (seaweed)				
12	Sea lettuce	*Ulva lactuca*		Eaten raw as a salad, cooked in soups (Chile, Scotland, West Indies)
13	Sweet tangle	*Laminaria saccharina*		Eaten raw or cooked (Scotland)
14		*Laminaria sp.*		Eaten dried ("combu") or as a vegetable (Japan)
15		*Porphyra laciniata*		Eaten raw in salads, cooked as a vegetable (England, America)
16		*Porphyra sp.*		Dried or cooked ("nari" products, Japan and Korea)
17		*Undaria pinnatifida*		Eaten dried ("wakami"') and as a vegetable (Japan)
Rooty vegetables				
18	Carrot	*Daucus carota*	Apiaceae	Eaten raw or cooked
19	Radish (white elongated fleshy root)	*Raphanus sativus var. niger*	Brassicaceae	The pungent fleshy root eaten raw, salted
20	Viper's grass, scorzonera	*Scorzonera hispanica*	Asteraceae	Cooked as a vegetable
21	Parsley	*Petroselinum crispum ssp. tuberosum*	Apiaceae	Long tapered roots cooked as a vegetable, or used for seasoning
Tuberous vegetables (sprouting tubers)				
22	Arrowroot	*Tacca leontopetaloides*	Taccaceae	Cooked or milled into flour for breadmaking

Table 17.1. (Continued)

Number	Common name	Latin name	Class, order, family	Consumed as
23	White (Irish) potato	*Solanum tuberosum*	Solanaceae	Cooked, fried or deep fried in many forms, or unpeeled baked, also for starch and alcohol production
24	Celery tuber	*Apium graveolens, var. rapaceum*	Apiaceae	Cooked as salad, and cooked and fried as a vegetable
25	Kohlrabi, turnip cabbage	*Brassica oleracea convar. acephala var. gongylodes*	Brassicaceae	Eaten raw or cooked as a vegetable
26	Rutabaga	*Brassica napus var. naprobrassica*	Brassicaceae	Cooked as a vegetable
27	Radish (reddish round root)	*Raphanus sativus var. sativus/var. niger*	Brassicaceae	The pungent fleshy root is eaten raw, usually salted
28	Red beet, beetroot	*Beta vulgaris spp. vulgaris var. conditiva*	Chenopodiaceae	Cooked as a salad
Tuberous (rhizomatic) vegetables				
29	Sweet potatoes	*Ipomoea batatas*	Convolvulaceae	Cooked, fried or baked
30	Cassava (manioc)	*Manihot esculenta*	Euphorbiaceae	Cooked or roasted
31	Yam	*Dioscorea*	Dioscoreaceae	Cooked or roasted
Bulbous rooty vegetables				
32	Vegetable fennel	*Foeniculum vulgare var. azoricum*	Apiaceae	Eaten raw as salad, cooked as a vegetable
33	Garlic	*Allium sativum*	Liliaceae	Raw, cooked as seasoning
34	Onion	*Allium cepa*	Liliaceae	Eaten raw, fried as seasoning, cooked as a vegetable
34a	Leek	*Allium porrum*	Liliaceae	The pungent succulent leaves and thick cylindrical stalk are cooked as a vegetable
Stem (shoot) vegetables				
35	Bamboo roots	*Bambusa vulgaris*	Poaceae	Cooked for salads
36	Asparagus	*Asparagus officinalis*	Liliaceae	Young shoots cooked as a vegetable or eaten as salad
Leafy (stalk) vegetables				
37	Celery	*Apium graveolens var. dulce*	Apiaceae	Leafy crispy stalks eaten raw as salad, or are cooked as vegetable
38	Rhubarb	*Rheum rhabarbarum, Rheum rhaponticum*	Polygonaceae	Large thick and succulent petioles are cooked as preserves or baked; used as a pie filling

Table 17.1. (Continued)

Number	Common name	Latin name	Class, order, family	Consumed as
Leafy vegetables				
39	Watercress	*Nasturtium officinale*	Brassicaceae	Moderately pungent leaves are eaten raw in salads or used as garnish
40	Endive (escarole, *chicory*)	*Cichorium intybus L. var. foliosum*	Cichoriaceae	Eaten raw as a salad, or is cooked as a vegetable
41	Chinese cabbage	*Brassica chinensis*	Brassicaceae	Eaten raw in salads, or is cooked as a vegetable
42	Lamb's salad (lettuce or corn salad)	*Valerianella locusta*	Valerianaceae	Eaten raw in salads
43	Garden cress	*Lepidium sativum*	Brassicaceae	Eaten raw in salads
44	Kale (borecole)	*Brassica oleracea convar. acephala var. sabellica*	Brassicaceae	Cooked as a vegetable
45	Head lettuce	*Lactuca capitata var. capitata*	Cichoriaceae	Juicy succulent leaves are eaten raw in salads
46	Mangold (mangel-wurzel, beet root)	*Beta vulgaris spp. vulgaris var. vulgaris*	Chenopidiaceae	Cooked as a vegetable
47	Chinese (Peking) cabbage	*Brassica pekinensis*	Brassicaceae	Cooked as a vegetable
48	Brussels sprouts	*Brassica oleracea convar. oleracea var. gemmifera*	Brassicaceae	Cooked as a vegetable
49	Red cabbage	*Brassica oleracea convar. capitata var. capitata f. rubra*	Brassicaceae	Eaten raw in salads or is cooked as a vegetable
50	Romaine lettuce	*Lactuca capitata var. crispa*	Cichoriaceae	Eaten raw as a salad
51	Spinach	*Spinacia oleracea*	Chenopodiaceae	Cooked as a vegetable or is eaten raw as a salad
52	White (common) cabbage	*Brassica oleracea convar. capitata var. capitata f.alba*	Brassicaceae	Juicy succulent leaves are eaten raw in salads, or are fermented (sauerkraut), steamed or cooked as a vegetable
53	Winter endive	*Cichoricum endivia*	Cichoriaceae	Eaten raw as a salad
54	Savoy cabbage	*Brassica oleracea convar. capitata, var. sabauda*	Brassicaceae	Cooked as a vegetable
Flowerhead (calix) vegetables				
55	Artichoke	*Cynara scolymus*	Asteraceae	Flowerhead is cooked as a vegetable
56	Cauliflower	*Brassica oleracea convar. botrytis var botrytis.*	Brassicaceae	Cooked as a vegetable or used in salads (raw or pickled)
57	Broccoli	*Brassica oleracea convar. botrytis var. italica*	Brassicaceae	The tight green florets are cooked as a vegetable

Table 17.1. (Continued)

Number	Common name	Latin name	Class, order, family	Consumed as
	Seed vegetables			
58	Chestnut	*Castanea sativa*	Fagaceae	Cooked as a vegetable, roasted, or milled into a flour and used in soups and bread doughs
59	Green beans	*Phaseolus vulgaris*	Fabaceae	The immature pod is cooked as a vegetable or is steamed or pickled for salads
60	Green peas	*Pisum sativum ssp. sativum*	Fabaceae	The rounded smooth or (wrinkled) Green seeds are cooked as a vegetable or are steamed/cooked for salads
	Fruity vegetables			
61	Eggplant	*Solanum melongena*	Solanaceae	Steamed as a vegetable
62	Garden squash	*Cucurbita pepo*	Cucurbitaceae	Cooked as a compote or as a vegetable
63	Green bell pepper	*Capsicum annuum*	Solanaceae	Eaten raw in salads, or is cooked, steamed or baked
64	Cucumber	*Cucumis sativus*	Cucurbitaceae	Eaten raw in salads, cooked as a vegetable or pickled
65	Okra	*Abelmoschus esculentus*	Malvaceae	Its mucilaginous green pods are cooked as a vegetable in soups or stewed, or eaten as a salad
66	Tomato	*Lycopersicon lycopersicum*	Solanaceae	The reddish pulpy berry is eaten raw, in salads, cooked as a vegetable, used as a paste or seasoned puree; immature green tomatoes are pickled and then eaten as salad
67	Zucchini	*Cucurbita pepo convar. giromontiina*	Cucurbitaceae	The cylindrical dark green fruits are peeled and cooked as a vegetable

Table 17.2. Production of vegetables in 2006 (1000 t)

Continent	Vegetables + melons, grand total	Cabbages	Artichokes	Tomatoes
World	903,405	68,991	1270	125,543
Africa	56,498	2038	167	14,336
America, Central	14,192	441	1	3331
America, North	39,296	1262	38	11,829
America, South and Caribbean	39,220	1023	190	10,559
Asia	667,827	52,200	122	66,990
Europe	97,200	12,426	752	21,326
Oceania	3365	42	–	503

Continent	Cauliflower	Pumpkin, squash and gourds	Cucumbers and gherkins	Eggplants (aubergines)
World	18,141	21,003	43,887	31,930
Africa	299	1669	1163	1497
America, Central	365	89	582	50
America, North	1324	924	1173	75
America, South and Caribbean	452	1335	859	88
Asia	13,544	13,168	35,405	29,364
Europe	2325	3672	5271	900
Oceania	196	235	17	4

Continent	Chilies[a] and peppers, green	Onions, air dried	Garlic	Green beans
World	25,924	61,637	15,184	6424
Africa	2468	5441	367	553
America, Central	1732	1322	49	55
America, North	940	3575	211	140
America, South and Caribbean	2252	5140	386	141
Asia	17,056	38,842	13,396	4574
Europe	3154	8383	823	976
Oceania	54	256	1	39

Continent	Green peas	Carrots and turnips	Watermelons	Cantaloupes and other melons (muskmelons)
World	7666	26,830	100,602	27,977
Africa	607	1230	4412	1432
America, Central	65	450	1410	1345
America, North	905	1892	1728	1221
America, South and Caribbean	271	1536	3704	2070
Asia	4599	12,799	85,735	20,827
Europe	1193	8992	4905	2340
Oceania	90	381	119	86

Table 17.2. (Continued)

Country	Vegetables + melons grand total	Country	Cabbages	Country	Artichokes
China	448,446	China	34,826	Italy	469
India	81,947	India	6148	Spain	200
USA	37,052	Russian Fed.	4073	Argentina	89
Turkey	25,723	Korea Rep.	3068	Egypt	70
Egypt	16,165	Japan	2287	Peru	68
Russian Fed.	15,930	Ukraine	1465	China	60
Iran	15,760	Indonesia	1293	Morocco	55
Italy	15,133	Poland	1249	France	54
Spain	12,513	Romania	1113	USA	38
Japan	11,624	USA	1100	Turkey	35
Σ (%)[b]	75	Σ (%)[b]	82	Σ (%)[b]	90

Country	Tomatoes	Country	Cauliflower	Country	Pumpkin, squash and gourds
China	32,540	China	8083	China	6060
USA	11,250	India	4508	India	3678
Turkey	9855	USA	1288	Russian Fed.	1185
India	8638	Spain	460	Ukraine	1064
Egypt	7600	Italy	438	USA	862
Italy	6351	France	362	Egypt	690
Iran	4781	Mexico	305	Iran	591
Spain	3679	Poland	250	Italy	512
Brazil	3278	UK	219	Cuba	447
Mexico	2878	Pakistan	209	Philippines	371
Russian Fed.	2415	Σ (%)[b]	89	Turkey	365
Greece	1712			Σ (%)[b]	75
Σ (%)[b]	76				

Country	Cucumbers and gherkins	Country	Eggplants (aubergines)	Country	Chilies[a] and peppers, green
China	27,357	China	17,530	China	13,031
Turkey	1800	India	8704	Turkey	1842
Iran	1721	Egypt	1000	Mexico	1681
Russian Fed.	1423	Turkey	924	Spain	1074
USA	982	Japan	372	USA	894
Ukraine	685	Italy	338	Indonesia	871
Japan	628	Sudan	272	Nigeria	722
Egypt	600	Indonesia	252	Egypt	460
Indonesia	553	Philippines	192	Korea, Rep.	395
Spain	500	Spain	175	Italy	345
Σ (%)[b]	83	Σ (%)[b]	93	Σ (%)[b]	82

Table 17.2. (Continued)

Country	Onions, air dried	Country	Garlic	Country	Green beans
China	19,600	China	11,587	China	2431
India	6435	India	647	Indonesia	830
USA	3346	Korea, Rep.	331	Turkey	564
Pakistan	2056	Russian Fed.	256	India	420
Russian Fed.	1789	USA	211	Egypt	215
Turkey	1765	Egypt	162	Spain	215
Iran	1685	Spain	148	Italy	191
Egypt	1302	Ukraine	145	Morocco	142
Brazil	1175	Argentina	116	Belgium	110
Japan	1158	Myanmar	104	USA	97
Mexico	1151	Σ (%)[b]	90	Σ (%)[b]	81
Spain	1151				
Netherlands	983				
Korea, Rep.	890				
Morocco	882				
Indonesia	809				
Σ (%)[b]	76				

Country	Green peas	Country	Carrots and turnips	Country	Watermelons
China	2408	China	8700	China	71,220
India	1918	Russian Fed.	1918	Turkey	3805
USA	859	USA	1588	Iran	3259
France	354	Poland	833	USA	1719
Egypt	290	UK	833	Brazil	1505
Morocco	147	Japan	762	Egypt	1500
UK	133	Uzbekistan	745	Russian Fed.	986
Turkey	90	France	693	Mexico	969
Italy	88	Ukraine	640	Algeria	785
Hungary	85	Italy	615	Korea, Rep.	778
Σ (%)[b]	83	Spain	600	Σ (%)[b]	86
		Germany	504		
		Netherlands	487		
		Indonesia	440		
		Turkey	402		
		Mexico	383		
		Σ (%)[b]	75		

Country	Cantaloupes and other melons
China	15,525
Turkey	1766
USA	1208
Iran	1126
Spain	1042
India	653
Morocco	648
Italy	625
Mexico	570
Egypt	565
Σ (%)[b]	85

[a] Data including other Capsicum species.
[b] World production = 100 %.

Table 17.3. Production of starch containing roots, rhizomes and tubers in 2006 (1000 t)

Continent	Roots and tubers grand total	Potato	Sweet potato	Cassava (manioc)
World	736,748	315,100	123,510	226,337
Africa	216,059	16,446	12,904	122,088
America, Central	2759	1951	63	508
America, North	25,447	24,709	737	–
America, South and Caribbean	57,276	16,015	1846	37,042
Asia	307,396	129,624	107,320	67,011
Europe	126,869	126,515	77	–
Oceania	3700	1792	626	196

Country	Roots and tubers grand total	Country	Potato	Country	Sweet potato
China	176,433	China	70,338	China	100,222
Nigeria	92,214	Russian Fed.	38,573	Nigeria	3462
Russian Fed.	38,573	India	23,910	Uganda	2628
India	32,485	USA	19,713	Indonesia	1852
Brazil	30,602	Ukraine	19,467	Vietnam	1455
Indonesia	23,139	Germany	10,031	Tanzania	1056
Thailand	22,842	Poland	8982	Japan	989
USA	20,451	Belarus	8329	India	955
Ukraine	19,467	Netherlands	6500	Burundi	835
Congo	15,523	France	6354	Kenya	809
Ghana	14,988	UK	5684	Σ (%)[a]	93
Mozambique	11,615	Canada	4995		
Angola	10,088	Iran	4830		
Germany	10,031	Turkey	4397		
Vietnam	9539	Bangladesh	4161		
Poland	8982	Σ (%)[a]	75		
Belarus	8329				
Uganda	8182				
Σ (%)[a]	75				

Country	Cassava (manioc)
Nigeria	45,721
Brazil	26,713
Thailand	22,584
Indonesia	19,928
Congo	14,974
Mozambique	11,458
Ghana	9638
Angola	8810
Vietnam	7714
India	7620
Σ (%)[a]	77

[a] World production = 100%.

Table 17.4. Average composition of vegetables (as % of fresh edible portion)

Vegetable	Dry matter	N-Compounds (N ×6.25)	Available carbo-hydrates	Lipids	Dietary fiber	Ash
Mushrooms						
Champignon (cultivated Agaricus arvensis, campestris)	9.0	4.1	0.6	0.3	2.0	1.0
Chanterelle	8.5	2.6	0.2	0.5	3.3	1.6
Edible boletus (*Boletus edulis*)	11.4	5.4	0.5	0.4	6.0	0.9
Rooty vegetables						
Carrots	11.8	1.1	4.8	0.2	3.6	0.8
Radish (*Raphanus sativus*, elongated white fleshy root)	7.0	1.0	2.4	0.2	2.5	0.8
Viper's grass, *scorzonera*	23.2	1.4	2.2	0.4	18.3	1.0
Parsley	16.1	2.9	6.1	0.6		1.6
Tuberous vegetables (sprouting tubers)						
White (Irish) potato	22.2	2.0	14.8[a]	0.1	2.1	1.1
Celery (root)	11.6	1.6	2.3	0.3	4.2	1.0
Kohlrabi	8.4	2.0	3.7	0.2	1.4	1.0
Rutabaga	10.7	1.1	5.7	0.2	2.9	0.8
Radish (*Raphanus sativus*, reddish fleshy root)	5.6	1.1	2.1	0.1	1.6	0.9
Red beet, beetroot	13.8	1.6	8.4	0.1	2.5	1.1
Tuberous root vegetables						
Sweet potato	30.8	1.6	24.1[b]	0.6	3.1	1.1
Cassava (manioc)	36.9	0.9	32.0	0.2	2.9	0.7
Yam	31.1	2.0	22.4	0.1	5.6	1.0
Bulbous root vegetables						
Onion	11.4	1.2	4.9	0.3	1.8	0.6
Leek	12.1	2.2	3.3	0.3	2.3	0.9
Vegetable fennel	7.6	1.4	3.0	0.2	2.0	1.0
Stem (shoot) vegetables						
Asparagus	6.5	1.9	2.0	0.2	1.3	0.6
Leafy (stalk) vegetables						
Rhubarb	7.3	0.6	1.4	0.1	3.2	0.6
Leafy vegetables						
Endive (escarole)	5.6	1.3	2.3	0.2	1.3	0.8
Kale (curly cabbage)	14.1	4.3	2.5	0.9	4.2	1.5
Head lettuce	5.1	1.2	1.1	0.2	1.4	0.9
Brussels sprouts	15.0	4.5	3.3	0.3	4.4	1.2
Red cabbage	9.0	1.5	3.5	0.2	2.5	0.7
Spinach	8.5	2.6	0.6	0.3	2.6	1.5
Common (white) cabbage	9.6	1.3	4.2	0.2	3.0	0.7
Flowerhead (calix) vegetables						
Artichoke	17.5	2.4	2.6	0.1	10.8	1.3
Cauliflower	9.0	2.5	2.3	0.3	2.9	0.9
Broccoli	10.9	3.6	2.7	0.2	3.0	1.1

[a] Starch content 14.1%. [b] Starch and saccharose contents 19.6 and 2.8%, respectively.

Table 17.4. (Continued)

Vegetable	Dry matter	N-Com-pounds (N ×6.25)	Available carbo-hydrates	Lipids	Dietary fiber	Ash
Seed vegetables						
Chestnut	55.1	2.4	41.2	1.9	8.4	1.2
Green beans	10.5	2.4	5.1	0.2	1.9	0.7
Green peas	24.8	6.6	12.4	0.5	4.3	0.9
Fruity vegetables						
Eggplant	7.4	1.2	2.5	0.2	2.8	0.6
Squash	9.0	1.1	4.6	0.1	2.2	0.8
Green bell pepper	7.7	1.1	2.9	0.2	3.6	0.4
Cucumber	4.0	0.6	1.8	0.2	0.5	0.5
Tomato	5.8	1.0	2.6	0.2	1.0	0.5

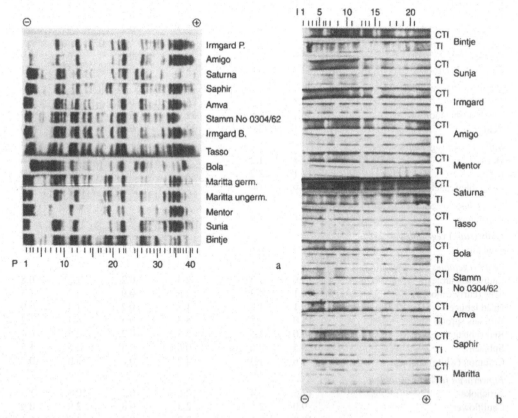

Fig. 17.1. Protein patterns of different potato cultivars obtained by isoelectric focussing on polyacrylamide gel pH 3–10. **a** Protein bands stained with Coomassie Blue; **b** Staining of trypsin and chymotrypsin inhibitors (TI, CTI): Incubation with trypsin or chymotrypsin, N-acetylphenylalanine-β-naphthyl ester and diazo blue B: inhibitor zones appear white on a red-violet background. (according to *Kaiser, Bruhn* and *Belitz*, 1974)

to ketoglutarate in the *Krebs* cycle:

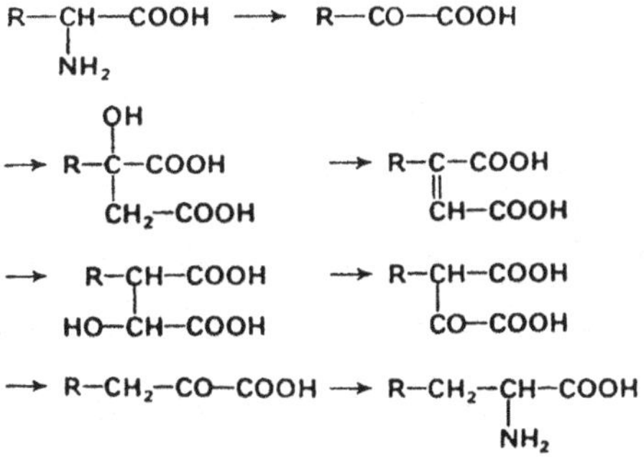

$$R\text{—}CH\text{—}COOH \longrightarrow R\text{—}CO\text{—}COOH$$
$$\quad\ \ | $$
$$\quad NH_2$$

$$OH$$
$$\longrightarrow R\text{—}\overset{|}{C}\text{—}COOH \quad \longrightarrow R\text{—}\overset{\|}{C}\text{—}COOH$$
$$\quad\ \ CH_2\text{—}COOH \qquad\ \ CH\text{—}COOH$$

$$\longrightarrow R\text{—}CH\text{—}COOH \quad \longrightarrow R\text{—}CH\text{—}COOH$$
$$\quad\ \ HO\text{—}CH\text{—}COOH \qquad\ \ CO\text{—}COOH$$

$$\longrightarrow R\text{—}CH_2\text{—}CO\text{—}COOH \longrightarrow R\text{—}CH_2\text{—}CH\text{—}COOH$$
$$\qquad\qquad\qquad\qquad\qquad\qquad\quad\ NH_2$$

(17.1)

4-Methyleneglutamic acid (Table 17.5: XXXI) is formed from pyruvic acid:

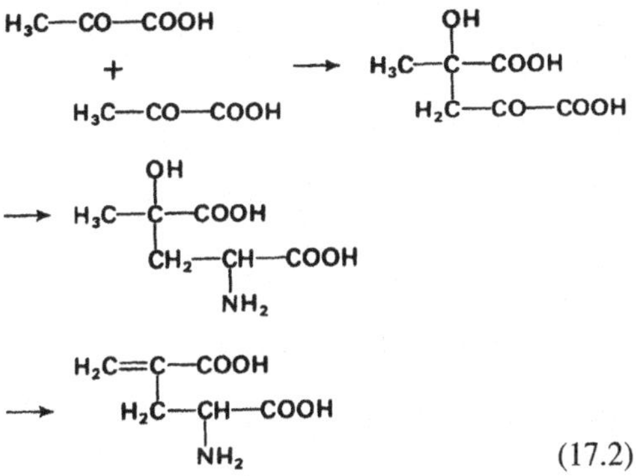

$$H_3C\text{—}CO\text{—}COOH$$
$$\qquad +$$
$$H_3C\text{—}CO\text{—}COOH$$

$$\longrightarrow \ H_3C\text{—}\overset{OH}{\underset{\ }{C}}\text{—}COOH$$
$$\qquad\qquad\ H_2C\text{—}CO\text{—}COOH$$

$$\longrightarrow \ H_3C\text{—}\overset{OH}{\underset{\ }{C}}\text{—}COOH$$
$$\qquad\qquad CH_2\text{—}CH\text{—}COOH$$
$$\qquad\qquad\qquad\ \ NH_2$$

$$\longrightarrow \ H_2C\text{=}C\text{—}COOH$$
$$\qquad\ H_2C\text{—}CH\text{—}COOH$$
$$\qquad\qquad\ \ NH_2$$

(17.2)

$$\text{Cysteine} \xrightarrow[H_2S]{HCN} NC\text{—}CH_2\text{—}CH\text{—}COOH$$
$$\qquad\qquad\qquad\qquad\qquad\qquad NH_2$$

$$-CO_2 \qquad\qquad +H_2O \qquad\qquad Red.$$

$$NC\text{—}CH_2\text{—}CH_2\text{—}NH_2 \quad H_2NOC\text{—}CH_2\text{—}CH\text{—}COOH \quad H_2N\text{—}CH_2\text{—}CH_2\text{—}CH\text{—}COOH$$
$$\qquad\qquad\qquad\qquad\qquad\qquad\qquad\qquad NH_2 \qquad\qquad\qquad\qquad\qquad\qquad NH_2$$

γ-Glutamyl-β-aminopropionitrile

$$OHC\text{—}CH_2\text{—}CH\text{—}COOH$$
$$\qquad\qquad\qquad NH_2$$

XXI

(17.4)

The important precursors of onion flavor, the S-alkylcysteine sulfoxides, are formed as follows:

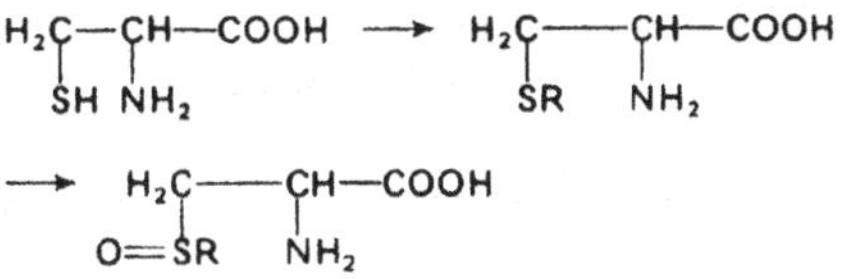

$$H_2C\text{—}CH\text{—}COOH \longrightarrow H_2C\text{———}CH\text{—}COOH$$
$$\ \ SH \ NH_2 \qquad\qquad\qquad\quad SR \quad NH_2$$

$$\longrightarrow \ H_2C\text{———}CH\text{—}COOH$$
$$\qquad\ O\text{=}SR \quad NH_2$$

(17.3)

2,4-Diaminobutyric acid and some other compounds are derived from cysteine (cf. Reaction 17.4).

The aspartic acid semi-nitrile formed initially can be decarboxylated to β-amino propionitrile which, just as its γ-glutamyl derivative, is responsible for osteolathyrism in animals. Hydrolysis of the semi-nitrile yields aspartic acid, hydrolysis and reduction yield 2,4- diaminobutyric acid, the oxalyl derivative of which, like oxalyldiaminopropionic acid, is a human neurotoxin. The main symptoms of neurolathyrism are paralysis of the limbs and muscular rigidity. 2,4-Diaminobutyric acid can be converted via the aspartic acid semialdehyde into 2-azetidine carboxylic acid (XXI), which occurs, for example, in sugar beets (Table 17.5).

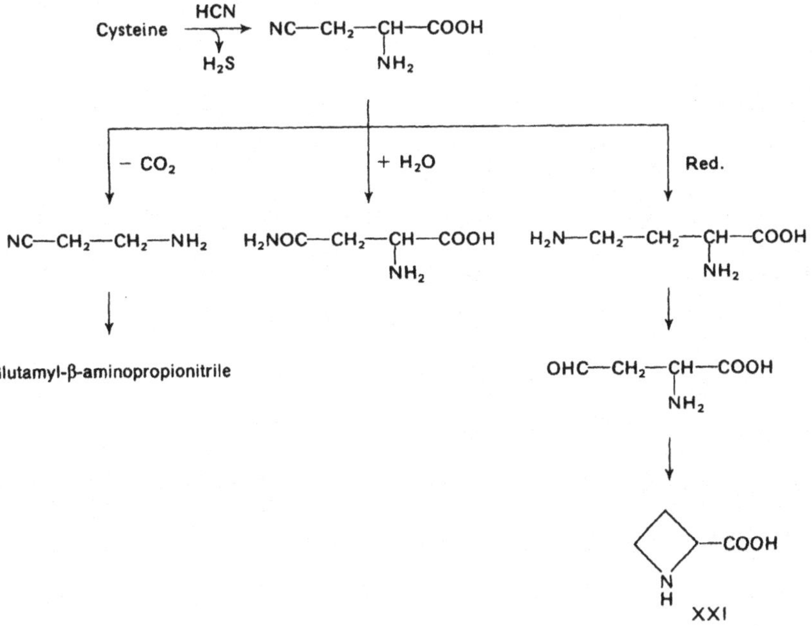

Table 17.5. Occurrence of nonprotein amino acids in plants (the Roman numerals refer to Table 17.6)

Amino acid		Plant		Family
Neutral aliphatic amino acids				
I	2-(Methylenecyclopropyl)-glycine	litchi	*Litchi chinensis*	Sapidaceae
II	3-(Methylenecyclopropyl)-L-alanine (Hypoglycine A)	akee	*Bligia sapida*	Sapidaceae
III	3-Cyano-L-alanine	common vetch	*Vicia sativa*	Fabaceae
IV	L-2-Aminobutyric acid	garden sage	*Salvia officinalis*	Lamiaceae
V	L-Homoserine	garden pea	*Pisum sativum*	Fabaceae
VI	O-Acetyl-L-homoserine	garden pea		
VII	O-Oxalyl-L-homoserine	vetchling	*Lathyrus sativum*	Fabaceae
VIII	5-Hydroxy-L-norvaline	jackbean	*Canavalia ensiformis*	Fabaceae
IX	4-Hydroxy-L-isoleucine	fenugreek	*Trigonella foenum-graecum*	Fabaceae
X	1-Amino-cyclopropane-1-carboxylic acid	apple pear	*Malus sylvestris* *Pyrus communis*	Rosaceae Rosaceae
Sulfurcontaining amino acids				
XI	S-Methyl-L-cysteine	garden bean	*Phaseolus vulgaris*	Fabaceae
XII	S-Methyl-L-cysteinesulfoxide	radish, cabbage cauliflower, broccoli	*Brassica oleracea*	Brassicaceae
XIII	S-(Prop-1-enyl)cysteine	garlic	*Allium sativum*	Liliaceae
XIV	S-(Prop-1-enyl)cysteinesulfoxide	onion	*Allium cepa*	Liliaceae
XV	γ-Glutamyl-S-(prop-1-enyl)cysteine	chive	*Allium schoenoprasum*	Liliaceae
XVI	S-(Carboxymethyl)cysteine	radish	*Raphanus sativus*	Brassicaceae
XVII	3,3′-(Methylenedithio)dialanine (Djenkolic acid)	djenkol bean	*Pithecolobium lobatum*	Fabaceae
XVIII	3,3′(-2-Methylethenyl-1,2-dithio)-dialanine (as γ-Glutamyl derivative)	chive	*Allium schoenoprasum*	Liliaceae
XIX	S-Methylmethionine	jackbean white cabbage asparagus	*Canavalia ensiformis* *Brassica oleracea* *Asparagus officinalis*	Fabaceae Brassicaceae Liliaceae
XX	Homomethionine	white cabbage	*Brassica oleracea*	Brassicaceae
Imino acids				
XXI	Azetidine-2-carboxylic acid	sugar beet	*Beta vulgaris ssp.*	Chenopodiaceae
XXII	tr-4-Methyl-L-proline	apple	*Malus sylvestris*	Rosaceae
XXIII	cis-4-Hydroxymethyl-L-proline	apple peel	*Malus sylvestris*	Rosaceae
XXIV	trans-4-Hydroxymethyl-L-proline	loquat	*Eriobotrya japonica*	Rosaceae
XXV	trans-4-Hydroxymethyl-D-proline	loquat	*Eriobotrya japonica*	Rosaceae
XXVI	4-Methylene-D,L-proline	loquat	*Eriobotrya japonica*	Rosaceae
XXVII	cis-3-Amino-L-proline	morel	*Morchella esculenta*	Ascomycetes
XXVIII	Pipecolic acid	many plants		
XXIX	3-Carboxy-6,7-dihydroxy-1,2,3,4-tetrahydroisoquinoline	cowage	*Mucuna sp.*	Fabaceae
XXX	1-Methyl-3-carboxy-6,7-dihydroxy-1,2,3,4-tetrahydroisoquinoline	cowage	*Mucuna sp.*	Fabaceae
Acidic amino acids and related compounds				
XXXI	4-Methyleneglutamic acid	peanut	*Arachis hypogaea*	Fabaceae
XXXII	4-Methyleneglutamine	peanut	*Arachis hypogaea*	Fabaceae
XXXIII	N^5-Ethyl-L-glutamine (L-Theanine)	tea	*Thea sinensis*	Theaceae
XXXIV	L-threo-4-Hydroxyglutamic acid			

Table 17.5. (continued)

Amino acid		Plant		Family
XXXV	3,4-Dihydroxyglutamic acid	garden cress	*Lepidium sativum*	Brassicaceae
		rhubarb	*Rheum rhabarbarum*	Polygonaceae
		carrot	*Daucus carota*	Apiaceae
		currant	*Ribis rubrum*	Saxifragaceae
		spinach	*Spinacia oleracea*	Chenopodiaceae
		longwort	*Angelica archangelica*	Apiaceae
XXXVI	L-2-Aminoadipic acid	many plants		
Basic amino acids and related compounds				
XXXVII	N²-Oxalyl-diaminopropionic acid	vetchling	*Lathyrus sativus*	Fabaceae
XXXVIII	N³-Oxalyl-diaminopropionic acid	vetchling	*Lathyrus sativus*	Fabaceae
XXXIX	2,4-Diaminobutyric acid (as N⁴-Lactyl compound)	sugar beet	*Beta vulgaris ssp.*	Chenopodiaceae
XL	2-Amino-4-(guanidinooxy)butyric acid (Canavanine)	jackbean	*Canavalia ensiformis*	Fabaceae
		soybean	*Glycine max*	Fabaceae
XLI	4-Hydroxyornithine	common vetch	*Vicia sativa*	Fabaceae
XLII	L-Citrulline	watermelon	*Citrullus lanatus*	Cucurbitaceae
XLIII	Homocitrulline	horse bean	*Vicia faba*	Fabaceae
XLIV	4-Hydroxyhomocitrulline	horse bean	*Vicia faba*	Fabaceae
XLV	4-Hydroxyarginine	common vetch	*Vicia sativa*	Fabaceae
XLVI	4-Hydroxylysine	garden sage	*Salvia officinalis*	Lamiaceae
XLVII	5-Hydroxylysine	lucern	*Medicago sativa*	Fabaceae
XLVIII	N⁶-Acetyl-L-lysine	sugar beet	*Beta vulgaris*	Chenopodiaceae
XLIX	N⁶-Acetyl-allo-5-hydroxy-L-lysine	sugar beet	*Beta vulgaris*	Chenopodiaceae
Heterocyclic amino acids				
L	3-(2-Furoyl)-L-alanine	buck wheat	*Fagopyrum esculentum*	Polygonaceae
LI	3-Pyrazol-l-ylalanine	watermelon	*Citrullus lanatus*	Cucurbitaceae
LII	1-Alanyluracil (Willardin)	cucumber	*Cicumis sativus*	Cucurbitaceae
		garden pea	*Pisum sativum*	Fabaceae
LIII	3-Alanyluracil (Isowillardin)	garden pea	*Pisum sativum*	Fabaceae
LIV	3-Amino-3-carboxypyrrolidine	musk melon	*Cucurbita monlata*	Cucurbitaceae
LV	3-(2,6-Dihydroxypyrimidine-5-yl)-alanine	garden pea	*Pisum sativum*	Fabaceae
LVI	3-(Isoxazoline-5-one-2-yl)alanine	garden pea	*Pisum sativum*	Fabaceae
LVII	3-(2-β-D-Glucopyranosyl-isoxazoline-5-one-4-yl)alanine	garden pea	*Pisum sativum*	Fabaceae
Aromatic amino acids				
LVIII	N-Carbamoyl-4-hydroxy-phenylglycine	horse bean	*Vicia faba*	Fabaceae
LIX	L-3,4-Dihydroxyphenylalanine	horse bean	*Vicia fabea*	Fabaceae
		cowage	*Mucuna sp.*	Fabaceae
Other amino acids				
LX	γ-Glutamyl-L-β-phenyl-β-alaninie	adzuki bean	*Phaseolus angularis*	Fabaceae
LXI	Saccharopine	yeast	*Saccharomyces cerevisiae*	Saccharomy-cetaceae

Freshly harvested mushrooms contain aprox. 0.1% agaritin, β-N-(γ-L(+)-glutamyl)-4 hydroxymethylphenylhydrazine. Enzymes present can hydrolyze agaritin and oxidize the released 4-hydroxymethyl-phenylhydrazine to the diazonium salt.

Table 17.6. Structures of nonprotein amino acids in plants (structures and Roman numerals refer to Table 17.5)

I

II

III

IV

V: R = H
VI: R = H_3C—CO
VII: R = HOOC—CO

VIII

IX

X

XI: R = CH_3
XIII: R = H_3C—CH═CH
XVI: R = HOOC—CH_2

XII: R = CH_3
XIV: R = H_3C—CH═CH

XV

XVII

XVIII

XIX

XX

XXI

XXII: R = CH_3
XXIII, XXIV, XXV: R = $HOCH_2$

XXVII

XXVI

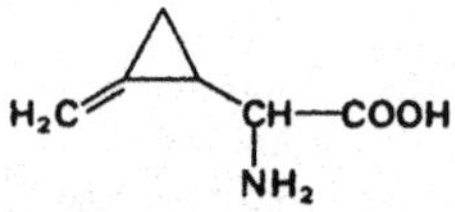

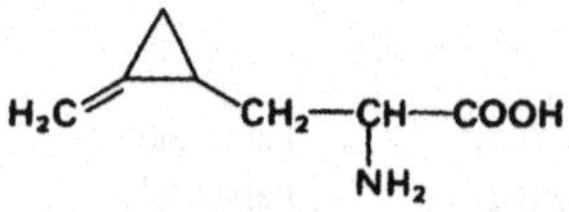

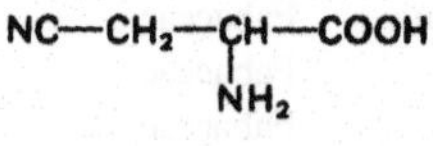

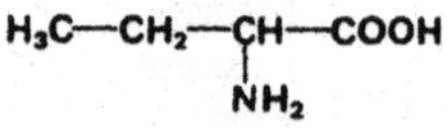

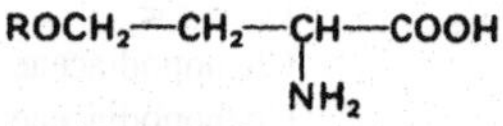

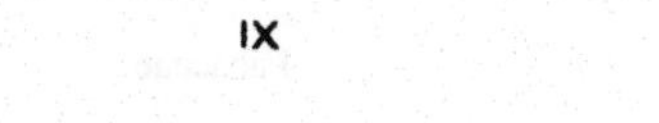

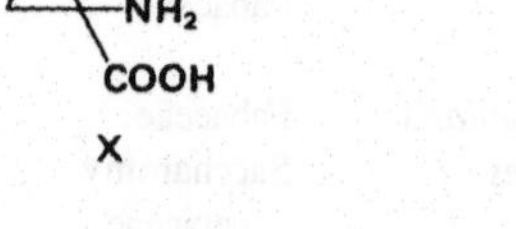

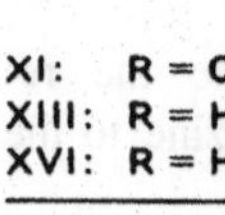

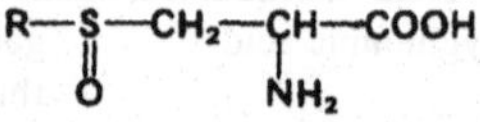

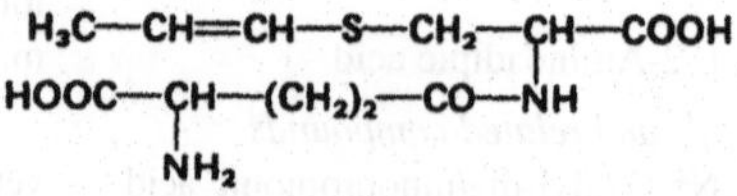

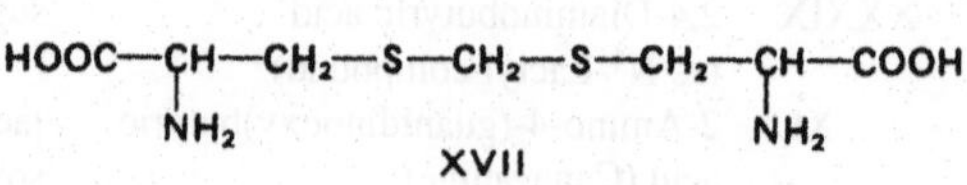

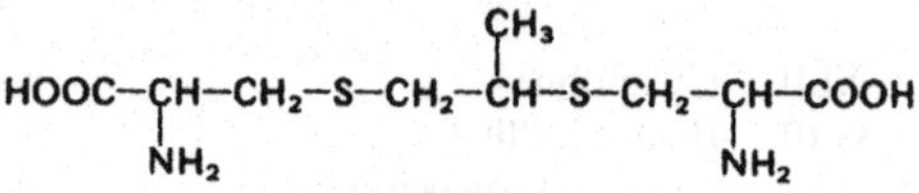

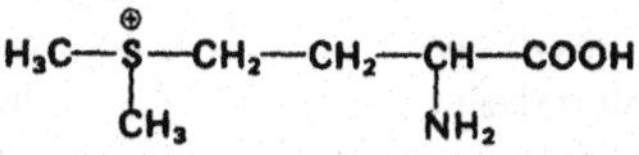

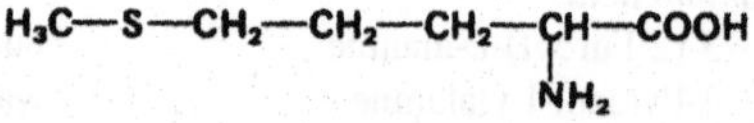

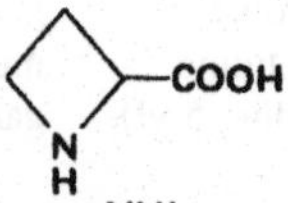

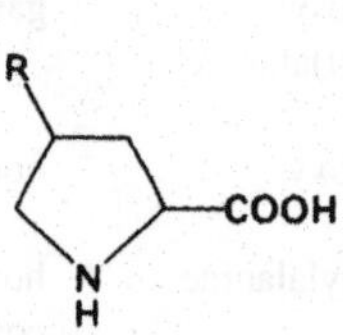

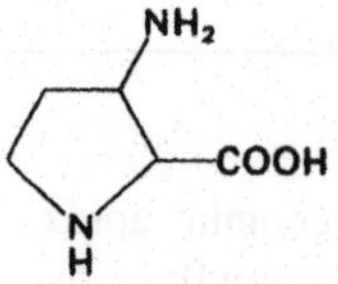

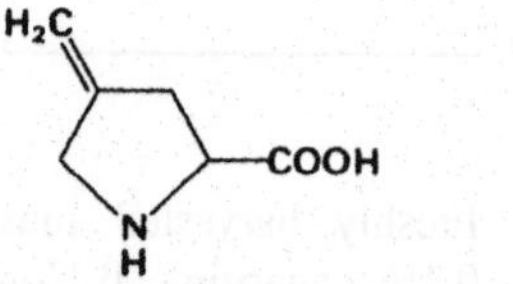

Table 17.6. (Continued)

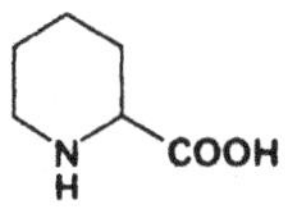

XXVIII

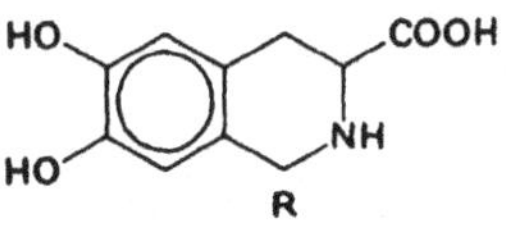

HO—, COOH structure
XXIX: R = H
XXX: R = CH₃

H_2N—C—NH—(CH₂)ₙ—CH—CH₂—CH—COOH

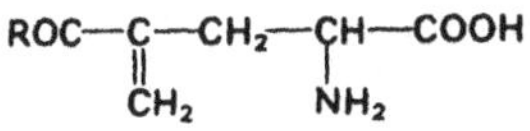

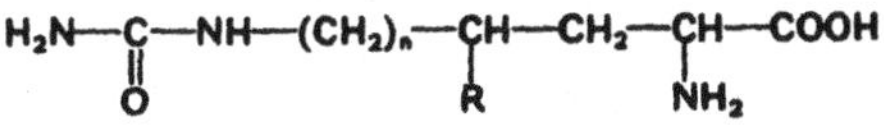

XLII: n = 1, R = H
XLIII: n = 2, R = H
XLIV: n = 2, R = OH

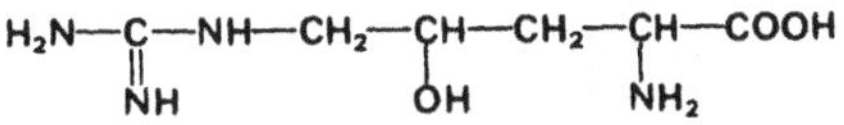

ROC—C—CH₂—CH—COOH
with CH₂ and NH₂

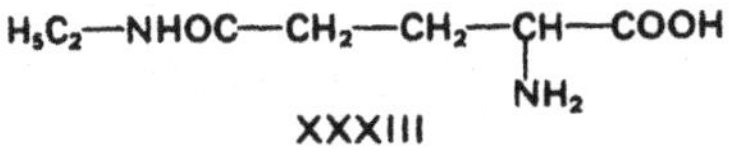

XXXI: R = OH   XXXII: R = NH₂

H_2N—C—NH—CH₂—CH—CH₂—CH—COOH
XLV

H₅C₂—NHOC—CH₂—CH₂—CH—COOH
XXXIII

R²—HN—CH₂—CH—CH—CH₂—CH—COOH

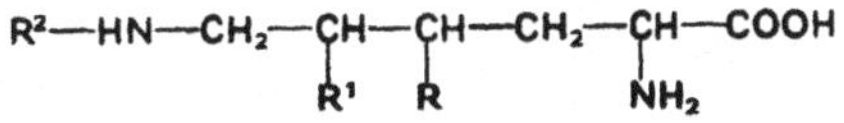

XLVI: R = OH, R¹, R² = H
XLVII: R¹ = OH, R, R² = H
XLVIII: R, R¹ = H, R² = CH₃CO
XLIX: R = H, R¹ = OH, R² = CH₃CO

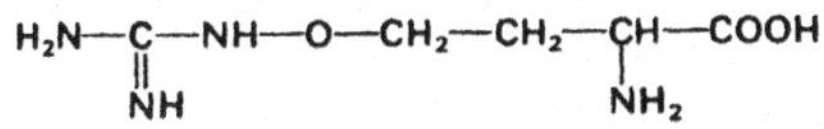

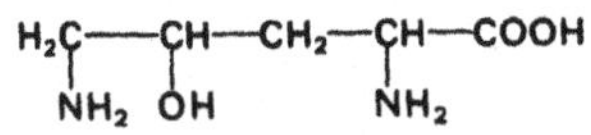

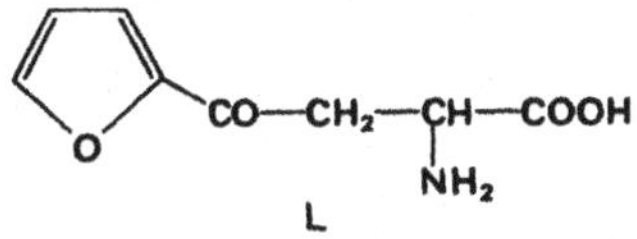

HOOC—CH—CH—CH—COOH

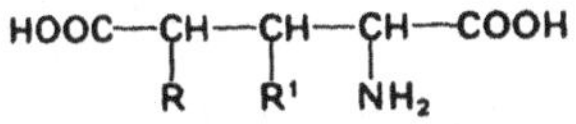

XXXIV: R = HO, R¹ = H
XXXV: R, R¹ = HO

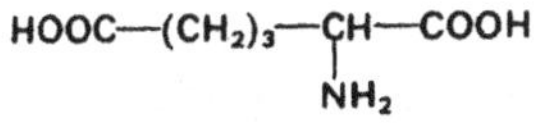

L

HOOC—(CH₂)₃—CH—COOH
XXXVI

LI

H₂C—CH—COOH
NHR¹ NHR

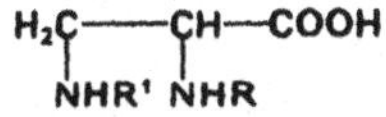

XXXVII: R = HOOC—CO, R¹ = H
XXXVIII: R¹ = HOOC—CO, R = H

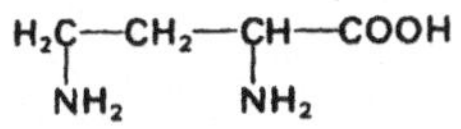

LII

H₂C—CH₂—CH—COOH
NH₂   NH₂
XXXIX

LIII

H_2N—C—NH—O—CH₂—CH₂—CH—COOH
XL

LIV

H₂C—CH—CH₂—CH—COOH
NH₂ OH   NH₂
XLI

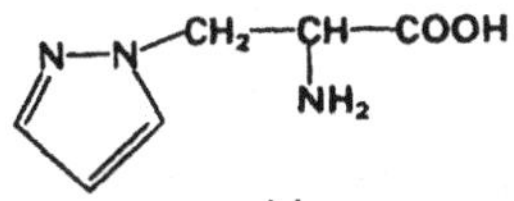

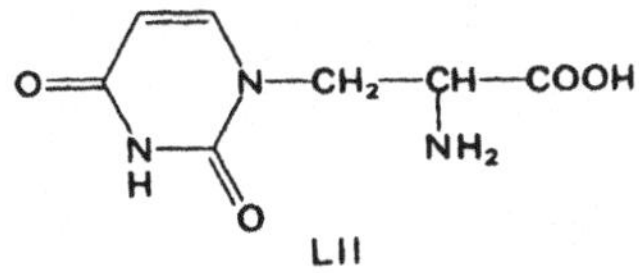

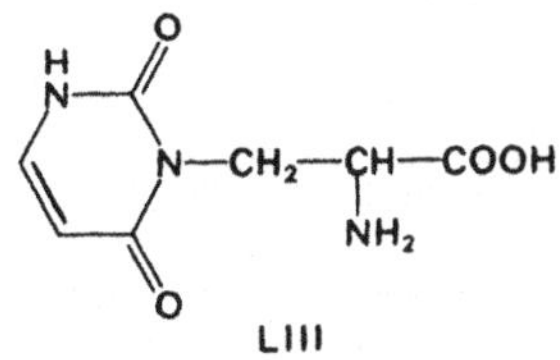

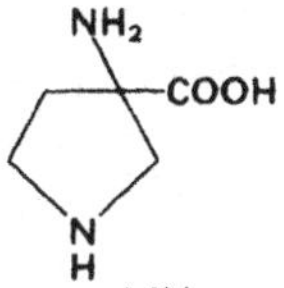

Table 17.6. (Continued)

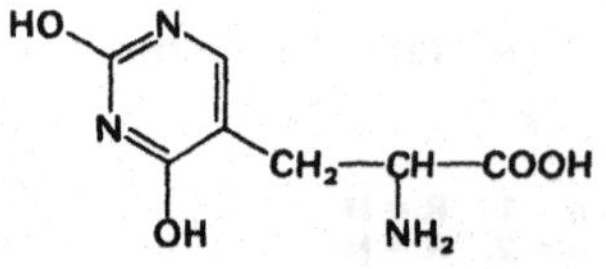

17.1.2.1.3 Amines

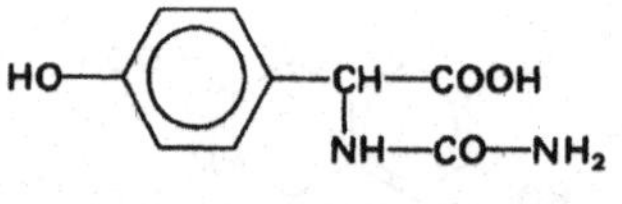

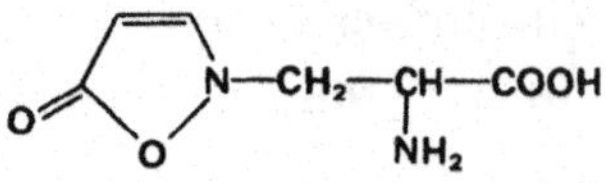

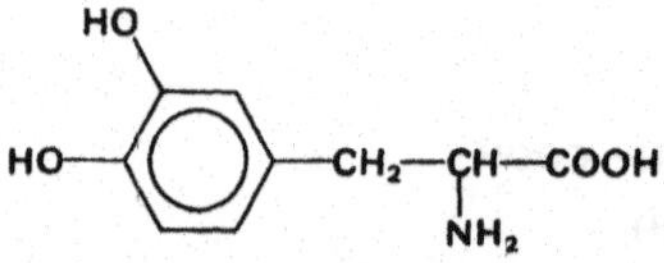

The presence of amines has been confirmed in various vegetables; e. g., histamine, N-acetylhistamine and N,N-dimethylhistamine in spinach; and tryptamine, serotonin, melatonin and tyramine in tomatoes and eggplant (cf. 18.1.2.1.3).

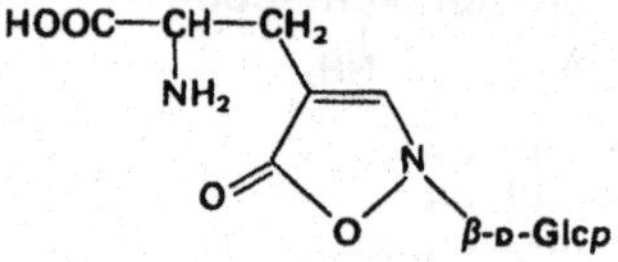

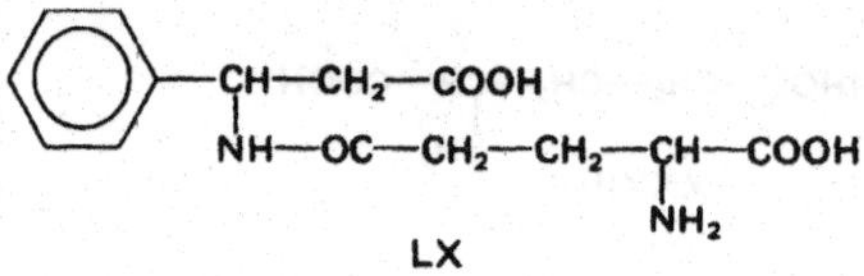

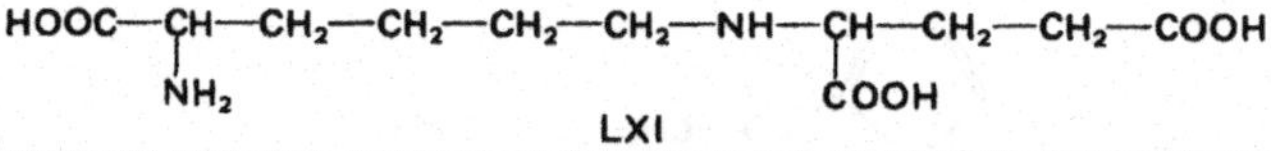

17.1.2.2 Carbohydrates

17.1.2.2.1 Mono- and Oligosaccharides, Sugar Alcohols

The predominant sugars in vegetables are glucose and fructose (0.3–4%) as well as sucrose (0.1–12%). Other sugars occur in small amounts; e. g. glycosidically bound apiose in *Umbelliferae* (celery and parsley); 1^F-β- and 6^G-β-fructosylsaccharose in the allium group (onions, leeks); raffinose, stachyose and verbascose in *Fabaceae*; and mannitol in *Brassicaceae* and *Cucurbitaceae*.

17.1.2.2.2 Polysaccharides

Starch occurs widely as a storage carbohydrate and is present in large amounts in some root and tuber vegetables. In *Compositae* (e. g., artichoke, viper's grass, bot. *Scorzonera*), inulin, rather than starch, is the storge carbohydrate.

Other polysaccharides are cellulose, hemicelluloses and pectins. The pectin fraction has a distinct role in the tissue firmness of vegetables. Tomatoes become firmer as the total pectin content and the content of some minerals (Ca, Mg) increases, and as the degree of esterification of the pectin decreases. In processing cauliflower (cf. 17.2.3), 70 °C is favorable for preserving tissue firmness. The reason for this effect is the presence of pectinmethylesterase which, in vegetables, is fully inactivated only at temperatures above 88 °C, while at 70 °C it is active and provides a build-up of insoluble pectates. For the conversion of protopectin to pectin during plant tissue maturation or ripening see 18.1.3.3.1.

Table 17.7. Carotenoids[a] in vegetables[b]

	Green bell pepper	Red pepper (paprika)	Tomato	Watermelon
Total carotenoids[b]	0.9–3.0	12.7–28.4	5.1–8.5	5.5
Phytoene (I)	–	0.03	1.3	–
Phytofluene (II)	0.01	0.56	0.7	
α-Carotene (VI)	0.01	0.1	–	–
β-Carotene (VII)	0.54	2.7	0.59	0.23
γ-Carotene (V)	–	–	–	0.09
ζ-Carotene (III)	0.01	0.45	0.84	–
Lycopene (IV)			4.7	4.5
α-Cryptoxanthin } β-Cryptoxanthin }	0.7	1.3	0.5	0.46
Lutein (IX)	0.6	–	0.12	0.01
Zeaxanthin (VIII)	0.02	3.9	–	–
Violaxanthin (XIII)	0.6	2.4	–	–
Capsanthin (X)		9.4	–	–
Neoxanthin (XX)	0.23	0.16	–	–

[a] Roman numerals refer to structural formula presented in Chapter 3.8.4.1.
[b] Values in mg carotene/100 g fresh weight.

17.1.2.3 Lipids

The lipid content of vegetables is generally low (0.1–0.9%). In addition to triacylglycerides, glyco- and phospholipids are present. Carotenoids are occasionally found in large amounts (cf. 18.1.2.3.2). Table 17.7 provides data on carotenoid compounds in green bell and paprika peppers, tomato and watermelon. For the occurrence of bitter cucurbitacins in *Cucurbitaceae*, see 18.1.2.3.3.

17.1.2.4 Organic Acids

The organic acids present in the highest concentration in vegetables are malic and citric acids (Table 17.8). The content of free titratable acids is 0.2–0.4 g/100 g fresh tissue, an amount which is low in comparison to fruits. Accordingly, the pH, with several exceptions such as tomato or rhubarb, is relatively high (5.5–6.5). Other acids of the citric acid cycle are present in negligible amounts. Oxalic acid occurs in larger amounts in some vegetables (Table 17.8).

Table 17.8. Organic acids in vegetables (mg/100 g fresh weight)

Vegetable	Malic acid	Citric acid	Oxalic acid
Artichoke	170	100	8.8
Eggplant	170	10	9.5
Cauliflower	201	20	–
Green beans	177	23	20–45
Broccoli	120	210	–
Green peas	139	142	–
Kale	215	220	7.5
Carrot	240	12	0–60
Leek	–	59	0–89
Rhubarb	910	137	230–500
Brussels sprouts	200	350	6.1
Red beet	37	195	181
Sorrel	–	–	360
White common cabbage	159	73	–
Onion	170	20	5.5
Potato	92	520	–
Tomato	51	328	–
Spinach	42	24	442

17.1.2.5 Phenolic Compounds

The phenolic compounds in plant material are dealt with in detail in 18.1.2.5. Hydroxybenzoic and hydroxycinnamic acids, flavones and flavonols also occur in vegetables. Table 17.9 provides data on the occurrence of anthocyanins in some vegetables.

17.1.2.6 Aroma Substances

Characteristic aroma compounds of several vegetables will be dealt with in more detail. The number following each vegetable corresponds to that given in Table 17.1. For aroma biosynthesis see 5.3.2.

17.1.2.6.1 Mushrooms (4)

The aroma in champignons originates from (R)-l-octen-3-ol derived from enzymatic oxidative degradation of linoleic acid (cf. 3.7.2.3). A small part of the alcohol is oxidized to 1-octen-3-one in fresh champignons. This compound has a mushroom-like odor when highly diluted and a metallic odor in higher concentrations. It contributes to the mushroom odor because its threshold value is lower by two powers of ten. Heating of champignons results in the complete oxidation of the alcohol to the ketone. Dried morels are a seasoning agent. The following compounds were identified as typical taste-compounds: (S)-morelid, (mixture of (S)-malic acid 1-O-α- and (S)-malic acid 1-O-β-D-glucopyranoside), L-glutamic acid, L-aspartic acid, γ-aminobutyric acid, malic acid, citric acid, acetic acid. (S)-Morelid intensifies the taste of L-glutamic acid and of NaCl. The mushroom *Lentium ediodes*, which is widely consumed in China and Japan, has a very intense aroma. The presence of 1,2,3,5,6-pentathiepane (lenthionine) has been confirmed, and it is a typical impact compound:

$$(17.5)$$

Its threshold values are 0.27–0.53 ppm (in water) or 12.5–25 ppm (in edible oil) It is derived biosynthetically from an S-alkyl cysteine sulfoxide, lentinic acid. Truffles, edible potato-shaped fungi, contain approx. 50 ng/g 5α-androst-16-ene-3 α-ol, which has a musky odor that contributes to the typical aroma (cf. 3.8.2.2.1).

17.1.2.6.2 Potatoes (23)

3-Isobutyl-2-methoxypyrazine and 2,3-diethyl-5-methylpyrazine belong to the key aroma substances in raw potatoes. These two pyrazines are also essential for the aroma of boiled potatoes. The substances responsible for the aroma of boiled potatoes are shown in Table 17.10.

The potato aroma note can be reproduced with an aqueous solution (pH 6) of methanethiol, dimethylsulfide, 2,3-diethyl-5-methylpyrazine, 3-isobutyl-2-methoxypyrazine and methional in the concentrations given in Table 17.10. Although it smells of boiled potatoes, methional only rounds off this aroma quality. In the drying of blanched potatoes to give a granulate, the concentrations of the two pyrazines decrease and, therefore, the intensity of the potato note also decreases.

17.1.2.6.3 Celery Tubers (24)

Celery aroma is due to the occurrence of phthalides in leaves, root, tuber and seeds. The

Table 17.9. Anthocyanins in vegetables

Vegetable	Anthocyanin
Eggplant	Delphinidin-3-(p-coumaroyl-L-rhamnosyl-D-glucosyl)-5-D-glucoside
Radish	Pelargonidin-3-[glucosyl(1 → 2)-6-(p-coumaroyl)-β-D-glucosido]-5-glucoside
	Pelargonidin-3-[glucosyl(1 → 2)-6-(feruloyl)-β-D-glucosido]-5-glucoside
Red cabbage	Cyanidin-3-sophorosido-5-glucoside (sugar moiety esterified with sinapic acid, 1–3 moles)
Onion (red shell)	Cyanidin glycoside Peonidin-3-arabinoside

Table 17.10. Odorants in boiled potatoes[a]

Odorants	Concentration[b] (µg/kg)
Methylpropanal	4.4
2-Methylbutanal	5.7
3-Methylbutanal	2.6
Hexanal	102.0
(E,E)-2,4-Decadienal	7.3
trans-4,5-Epoxy-(E)-2-decenal	58.0
Methional	65.0
Dimethyltrisulfide	1.0
Methanethiol	15.4
Dimethylsulfide	8.8
2,3-Diethyl-5-methylpyrazine	0.17
3-Isobutyl-2-methoxypyrazine	0.07
4-Hydroxy-2,5-dimethyl-3(2H)-furanone (HD3F)	67.0
3-Hydroxy-4,5-dimethyl-2(5H)-furanone (HD2F)	2.2
Vanillin	1000

[a]Potatoes, boiled in water for 40 min, then peeled.
[b]Reference: fresh weight; water content: 78%.

main compound 3-butyl-4,5-dihydrophthalide (sedanolide: I, Formula 17.6) occurs in amounts of 3–20 mg/kg. In addition, 3-butylphthalide-(II, 0.6–1.6 mg/kg), 3-butyl-3a,4,5,6-tetrahydrophthalide (III, 1.0–4.4 mg/kg), 3-butylhexahydrophthalide (IV) and (Z)-3-butyliden-4,5-dihydrophthalide (Z-ligustilide: V, 0.6–2 mg/kg) have been identified. The (S)-enantiomer of II plays a big part in the aroma and it not only predominates, but also has a much lower odor threshold when compared with the (R)-enantiomer (S: 0.01 µg/kg; R: 10 µg/kg, water). Of the

eight possible stereoisomers of the phthalide IV, the enantiomers 3R,3aR,7aS and 3S,3aR,7aS dominate in celery. But their contribution to the aroma must be low because of the high odor threshold (>125 µg/kg). Apart from the phthalides, the participation of (E,Z)-1,3,5-undecatriene in the aroma is under discussion.

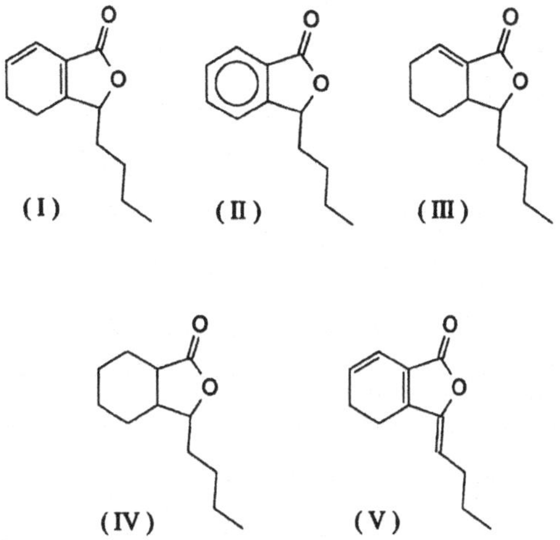

(17.6)

17.1.2.6.4 Radishes (27)

The sharp taste of the radish is due to 4-methylthio-trans-3-butenyl-isothiocyanate, which is released from the corresponding glucosinolate after the radish is sliced. Glucosinolates are widely distributed among *Brassicaceae* and some other plant families. Their occurrence in some types of cabbage is presented in Table 17.11.

Table 17.11. Glucosinolates in different types of cabbage (mg/kg fresh weight)

Compound[a]	Broccoli	Red cabbage	Brussels sprouts	Cauliflower	Savoy cabbage	White cabbage
Glucobrassicin (Ia)	20	16	31	21	46	22
4-Hydroxy-glucobrassicin (Ib)	5					
4-Methoxy-glucobrassicin (Ic)	4					
Glucoiberin (II)	4	11	24	16	52	23
Gluconapin (III)	n.d.	8	5	0.1	0.3	2
Glucoraphanin (IV)	21	21	4	0.7	1	4
Progoitrin (R-V)	n.d.	18	11	3	2	8
Sinigrin (VI)	n.d.	14	44	17	46	30

[a] The chemical structures are shown in Formula 17.7 und 17.10.
n.d.: not detected.

Glucosinolates are hydrolyzed by myrosinase, a thioglucosidase enzyme, to the corresponding isothiocyanates (mustard oils) on disintegration of the tissue (Formula 17.7). The residue R for the glucosinolates presented in Table 17.11 is shown in Formula 17.8.

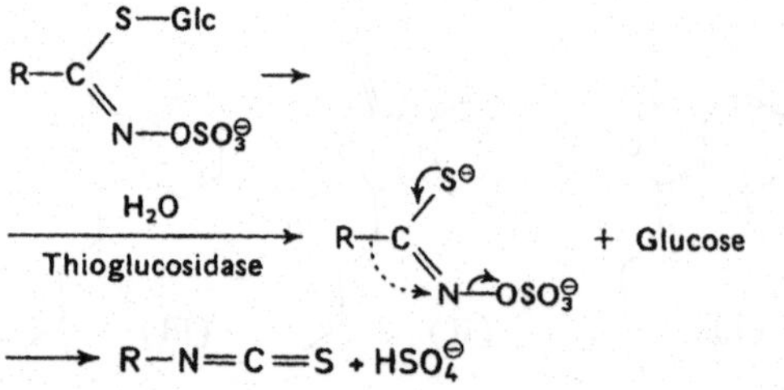

$$(17.7)$$

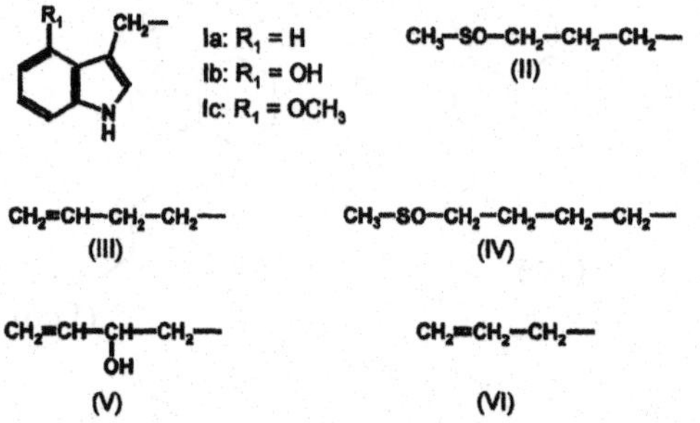

$$(17.8)$$

The decomposition corresponds to a *Lossen's* rearrangement of a hydroxamic acid. In addition to isothiocyanates, rhodanides and nitriles have been observed among the reaction products.

The isothiocyanates can react further, e. g., with hydroxy compounds or thiols, to form thiourethanes or dithiourethanes. In the presence of amines, thioureas result; while hydrolysis yields the corresponding amines and releases CO_2 and H_2S:

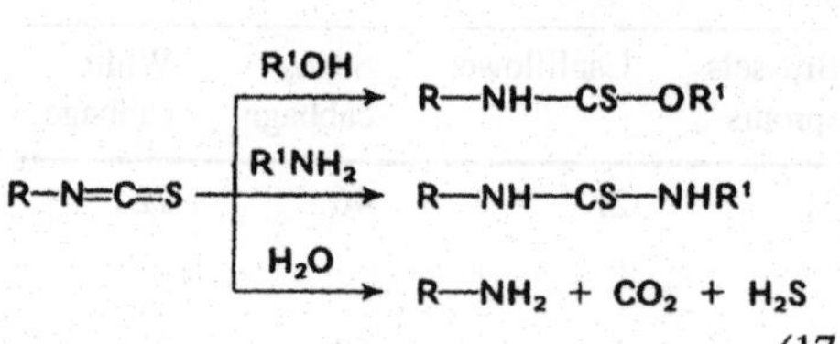

$$(17.9)$$

Biosynthesis of glucosinolates (reaction 17.10) starts from the corresponding amino acids, and proceeds via an oxime (I) and thiohydroximic acid (III). The intermediate reactions between

steps I and III are not yet clarified. Tests with [14]C- and ^{35}S-labelled compounds suggest that the aci-form of the corresponding nitro-compound (II) functions as a thiol acceptor. Cysteine may be involved as a thiol donor. The sulfation is achieved by 3'-phosphoadenosine-5'-phosphosulfate (PAPS). The biosynthetic pathway for cyanogenic glycosides branches at the aldoxime (I) intermediate (cf. 16.2.6).

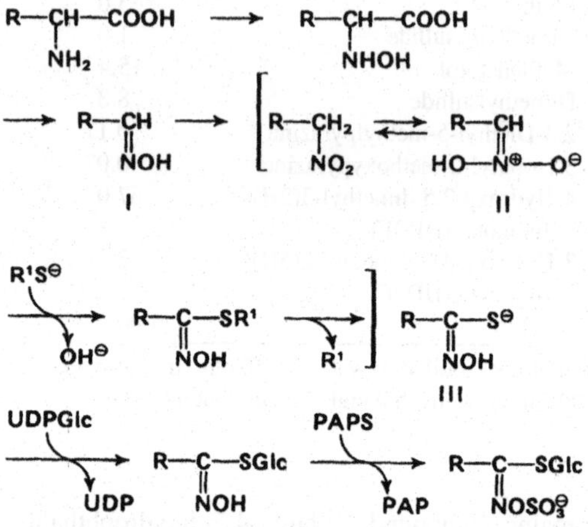

$$(17.10)$$

17.1.2.6.5 Red Beets (28)

Geosmin (structure cf. 5.1.5) is the character impact compound of the red beet.

17.1.2.6.6 Garlic (33) and Onions (34)

The compound which causes tears (the lachrymatory factor) is (Z)-propanethial-S-oxide (II) which, once the onion bulb is sliced, is derived from trans-(+)-S-(1-propenyl)-L-cysteine sulfoxide (I) by the action of the enzyme alliinase. Alliinase has pyridoxalphosphate as its coenzyme (cf. reaction sequence 17.11). Chopping of onions releases 3-mercapto-2-methylpentan-1-ol, which, with its very low threshold of $0.0016\,\mu g/l$ (water), smells of meat broth and onions. Raw onions contain $8–32\,\mu g/kg$, and onions which have been cut, stored for 30 minutes and then cooked contain

34–246 µg/kg. The formation involves the attachment of H_2S to the aldol condensation product of propanal and enzymatic reduction of the carbonyl group.

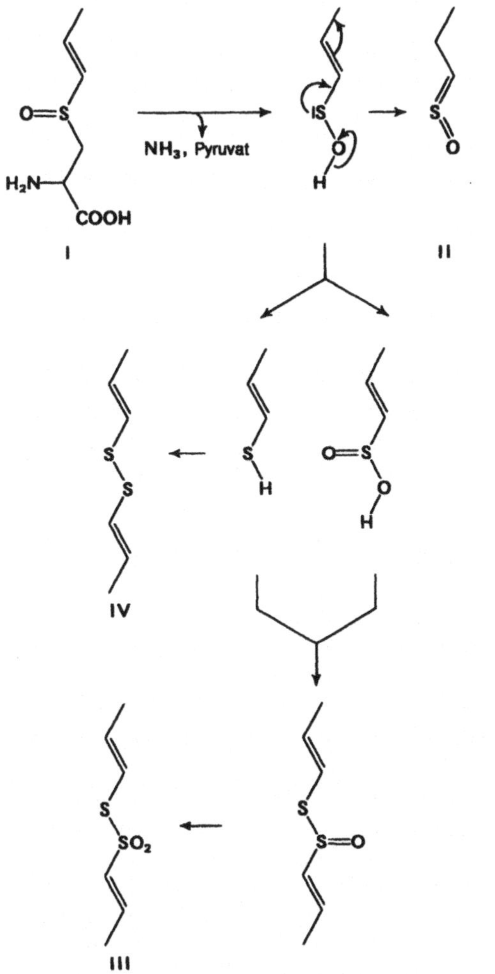

I

II

IV

III

$$(17.11)$$

Alkylthiosulfonates (III) are also responsible for the aroma of raw onions, while propyl- and propenyl disulfides (IV) and trisulfides are also supposed to play a role in the aroma of cooked onions. The aroma of fried onions is derived from dimethyl-thiophenes.

Precursors of importance for the aroma of onions, other than compound I, are S-methyl and S-propyl-L-cysteine sulfoxide. Precursor I is biosynthesized from valine and cysteine (cf. reaction sequence 17.12).

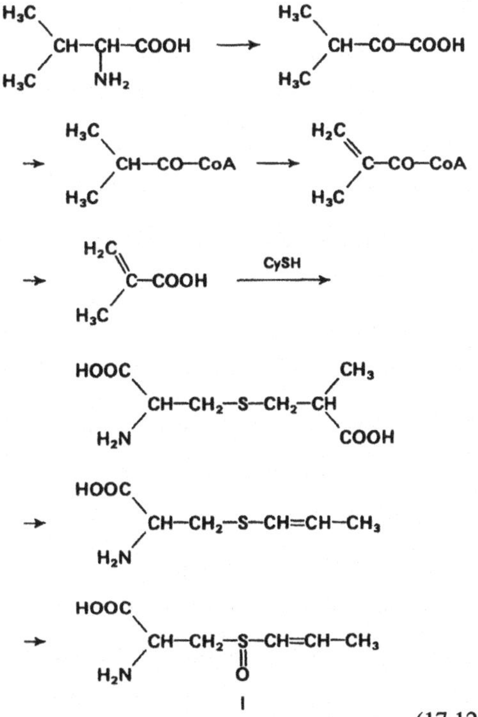

$$(17.12)$$

The key precursor for garlic aroma is S-allyl-L-cysteine sulfoxide (alliin) which, as in onions, occurs in garlic bulbs together with S-methyl- and S-propyl-compounds. The allyl and propyl-compounds are assumed to be synthesized from serine and corresponding thiols:

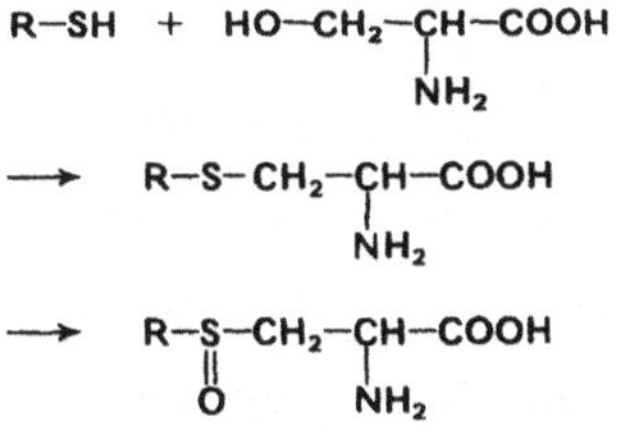

$$R–SH \; + \; HO–CH_2–CH–COOH$$
$$\qquad\qquad\qquad\qquad |$$
$$\qquad\qquad\qquad\qquad NH_2$$

$$\longrightarrow \; R–S–CH_2–CH–COOH$$
$$\qquad\qquad\qquad\qquad |$$
$$\qquad\qquad\qquad\qquad NH_2$$

$$\longrightarrow \; R–S–CH_2–CH–COOH$$
$$\qquad\qquad\;\; || \qquad\quad |$$
$$\qquad\qquad\;\; O \qquad\quad NH_2 \qquad (17.13)$$

Diallylthiosulfinate (allicin) and diallyldisulfide are formed from the main component by means of the enzyme alliinase. Both are character impact compounds of garlic.

17.1.2.6.7 Watercress (39)

Phenylethylisothiocyanate is responsible for the aroma of this plant of the mustard family (*Brassicaceae*). Decomposition of the corresponding glucosinolate gives phenylpropionitrile, the main component, and some other nitriles, e.g., 8-methylthiooctanonitrile and 9-methylthiononanonitrile.

17.1.2.6.8 White Cabbage, Red Cabage and Brussels Sprouts (52, 49, 48)

Mustard oil is more than 6% of the total volatile fraction of cooked white and red cabbages. There is such a high proportion of allylisothiocyanate (I, Formula 17.14) present that it participates in the aroma of boiled white cabbage in spite of its high odor threshold of 375 µg/kg (water). In addition, 2-phenylethylthiocyanate (II, odor threshold 6 µg/kg, water), 3-methylthiopropylisothiocyanate (III, 5 µg/kg) and 2-phenylethylcyanide (IV, 15 µg/kg) could be involved in the aroma.
Dimethylsulfide is another important odorant formed during the cooking of cabbage and other vegetables. It also appears that 3-alkyl-2-methoxypyrazine plays a role in cabbage aroma.

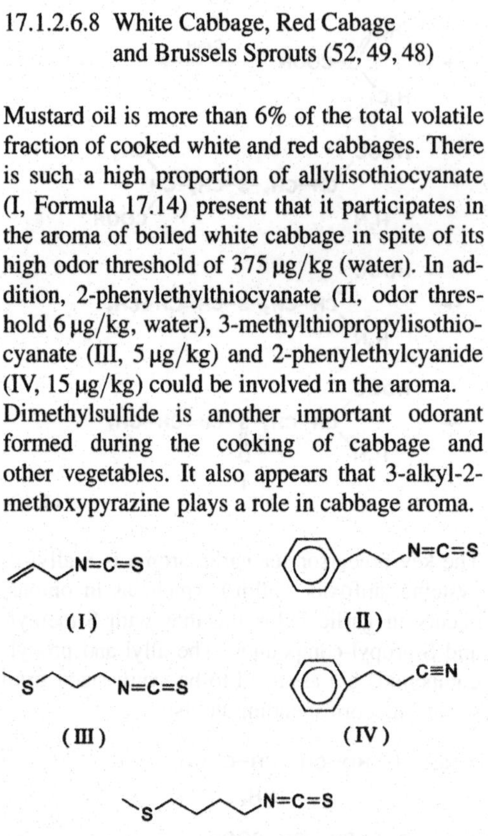

(17.14)

The total impact of the aroma in cooked frozen Brussels sprouts is less satisfactory than in cooked fresh material. In the former case, analysis has revealed comparatively little allyl mustard oil and more allylnitrile. Isothiocyanates in low concentrations are pleasant and appetite-stimulating, while nitriles are reminiscent of garlic odor. The shift in the concentration ratio of the two compounds is attributed to myrosinase enzyme inactivation during blanching prior to freezing. As a consequence of this, allylglucosinolate in frozen Brussels sprouts is thermally degraded only on subsequent cooking, preferentially forming nitriles. Goitrin is responsible for

the bitter taste that can occur in Brussels sprouts (cf. 17.1.2.9.3).

17.1.2.6.9 Spinach (51)

The compounds (Z)-3-hexenal, methanethiol, (Z)-1,5-octadien-3-one, dimethyltrisulfide, 3-isopropyl-2-methoxypyrazine and 3-*sec*-butyl-2-methoxypyrazine contribute to the aroma of the fresh vegetable. In cooked spinach, (Z)-3-hexenal decreases and dimethylsulfide, methanethiol, methional and 2-acetyl-1-pyrroline are dominant.

17.1.2.6.10 Artichoke (55)

1-Octen-3-one, the herbaceous smelling 1-hexen-3-one (odor threshold 0.02 µg/kg, water) and phenylacetaldehyde contribute to the aroma of boiled artichokes with high aroma values.

17.1.2.6.11 Cauliflower (56), Broccoli (57)

In cooked cauliflower and broccoli, the aroma compounds of importance are the sulfur compounds mentioned for white cabbage. 3-Methylthiopropylisothiocyanate, 3-methylthiopropylcyanide (odor threshold 82 µg/kg, water) and nonanal contribute to the typical aroma of cauliflower and 4-methylthiobutylisothiocyanate (V, cf. Formula 17.14), 4-methylthiobutylcyanide as well as II and IV to the aroma of broccoli.
During blanching of these vegetables, cystathionine-β-lyase (EC 4.4.1.8, cystine lyase) must be inactivated because this enzyme, which catalyzes the reaction shown in formula 17.15, produces an aroma defect. The undesirable aroma substances are formed by the degradation of the homocysteine released.

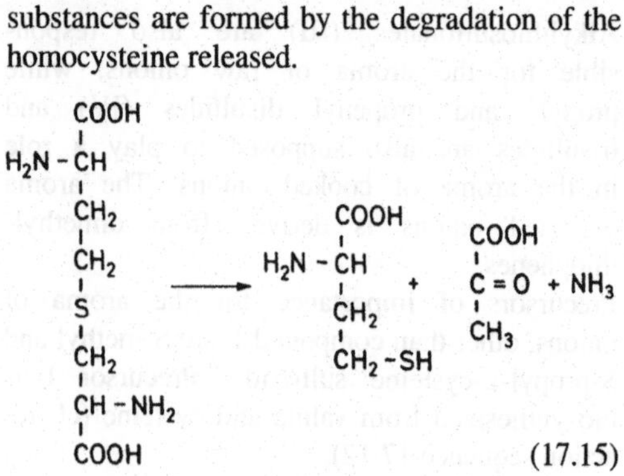

(17.15)

17.1.2.6.12 Green Peas (60)

The aroma of green peas is derived from aldehydes and pyrazines (3-isopropyl-, 3-*sec*-butyl- and 3-isobutyl-2-methoxypyrazine).

17.1.2.6.13 Cucumbers (64)

The following aldehydes play an important role in cucumber aroma: (E,Z)-2,6-nonadienal and (E)-2-nonenal. Linoleic and linolenic acids, as shown in Fig. 3.31, are the precursors for these and other aldehydes (Z)-3-hexenal, (E)-2-hexenal, (E)-2-nonenal.

17.1.2.6.14 Tomatoes (66)

Among a large number of volatile compounds, (Z)-3-hexenal, β-ionone, hexanal, β-damascenone, 1-penten-3-one, and 3-methylbutanal are of special importance for the aroma of tomatoes (cf. Table 17.12).

Table 17.12. Odorants in tomatoes and tomato paste

Compound	Aroma value[a]	
	Tomato	Tomato-paste
(Z)-3-Hexenal	5×10^4	<30
β-Ionone	6.3×10^2	_[b]
Hexanal	6.2×10^2	–
(E)-β-Damascenone	5×10^2	5.7×10^3
1-Penten-3-one	5×10^2	–
3-Methylbutanal	130	152
(E)-2-Hexenal	16	–
2-Isobutylthiazole	10	–
Dimethylsulfide	–	1.4×10^3
Methional	–	650
3-Hydroxy-4,5-dimethyl-5(2H)-furanone (HD2F)	–	213
4-Hydroxy-2,5-dimethyl-3(2H)-furanone (HD3F)	–	138
Eugenol	–	95
Methylpropanal	–	40

[a] The aroma values were calculated on the basis of the odor threshold in water.
[b] The compound does not contribute to the aroma here.

In tomato paste, for example (cf. Table 17.12), it was found that the changes in aroma caused by heating are primarily due to the formation of dimethylsulfide, methional, the furanones HD2F and HD3F and the increase in β-damascenone, and a substantial decrease in (Z)-3-hexenal and hexanal.

17.1.2.7 Vitamins

Table 17.13 provides data on the vitamin content of some vegetables. The values given may vary significantly with vegetable cultivar and climate. In spinach, for example, the ascorbic acid content varies from 40–155 mg/100g fresh weight. Freshly harvested potatoes contain 15–20 mg/100 g of vitamin C. The content drops by 50% on storage (4 °C) for 6–8 months and by 40–60% on peeling and cooking.

17.1.2.8 Minerals

Table 17.14 reviews the mineral content of some vegetables. Potassium is by far the most abundant constituent, followed by calcium, sodium and magnesium. The major anions are phosphate, chloride and carbonate. All other elements are present in much lower amounts. For nitrate content see 9.8.

17.1.2.9 Other Constituents

Plant pigments other than carotenoids and anthocyanins, e. g., chlorophyll and betalains, are also of great importance in vegetables and are covered in this section together with goitrogenic compounds occurring in *Brassicaceae*.

17.1.2.9.1 Chlorophyll

The green color of leaves and unripe fruits is due to the pigments chlorophyll a (blue-green) and chlorophyll b (yellow-green), occurring together in a ratio shown in Table 17.15 (see Formula 17.16). Figure 17.2 shows the absorption spectra of chlorophylls a and b. Removal of magnesium

Table 17.13. Vitamin content in vegetables (mg/100 g fresh weight)

Vegetable	Ascorbic acid	Thiamine	Riboflavin	Nicotinicacid	Folacid	α-Tocopherol	β-Carotene
Artichoke	8	0.14	0.01	1.0	–	0.19	0.10
Eggplant	5	0.05	0.05	0.6	0.03	0.03	0.04
Cauliflower	78	0.09	0.10	0.7	0.09	0.07	0.01
Broccoli	100	0.10	0.18	0.9	0.11	0.61	0.9
Kale	105	0.10	0.26	2.1	0.19	1.7	5.2
Cucumber	8	0.02	0.03	0.2	0.02	0.06	0.4
Head lettuce	10	0.06	0.09	0.3	0.06	0.6	1.1
Carrot	8	0.06	0.05	0.6	0.03	0.4	7.6
Green bell pepper	138	0.05	0.04	0.3	0.06	2.5	0.5
Leek	26	0.09	0.06	0.5	0.10	0.5	0.7
Radish	26	0.03	0.03	0.4	0.02	–	0.01
Brussels sprouts	102	0.10	0.16	0.7	0.10	0.6	0.5
Red beet	10	0.03	0.05	0.2	0.08	0.04	0.01
Red cabbage	61	0.06	0.04	0.4	0.04	1.7	0.02
Celery	8	0.05	0.06	0.7	0.01	–	2.9
Asparagus	20	0.11	0.10	1.0	0.11	2.0	0.5
Spinach	51	0.10	0.20	0.6	0.15	1.3	4.8
Tomato	23	0.06	0.04	0.5	0.02	0.8	0.6

Table 17.14. Minerals in vegetables (mg/100 g fresh weight)

Vegetable	K	Na	Ca	Mg	Fe	Mn	Co	Cu	Zn	P	Cl	F	I
Potato	418	2.7	6.4	21	0.4	0.15	0.001	0.09	0.3	50	50	0.01	0.003
Spinach	554	69	60	117	3.8	0.6	0.002	0.1	0.6	46	54	0.08	0.012
Carrot	321	61	37	13	0.4	0.2	0.001	0.05	0.3	35	59	0.02	0.002
Cauliflower	328	16	20	17	0.6	0.2	–	0.05	0.2	54	19	0.01	0.006
Green beans	256	1.7	51	26	0.8	0.2	–	0.1	0.3	37	13	0.01	0.003
Green peas	296	2	26	33	1.9	0.4	0.003	0.2	0.9	119	40	0.02	0.004
Cucumber	141	8.5	15	8	0.5	0.1	–	0.04	0.2	17	37	0.01	0.003
Red beet	336	86	29	1.4	0.9	0.2	0.01	0.08	0.4	45	0.2	0.01	0.005
Tomato	297	6.3	14	20	0.5	0.1	0.01	0.06	0.2	26	30	0.02	0.002
White common cabbage	227	13	46	23	0.5	0.2	0.01	0.03	0.2	36	37	0.01	0.005

Table 17.15. Chlorophylls a and b in vegetables and fruit

Food	Chlorophyll a	Chlorophyll b
	(mg/kg)[a]	
Green beans	118	35
Kale	1898	406
White cabbage	8	2
Cucumber	64	24
Parsley	890	288
Green bell pepper	98	33
Green peas	106	22
Spinach	946	202
Kiwi	17	8
Gooseberry	5	1

[a] Refers to fresh weight.

from the chlorophylls gives pheophytins a and b, both of which are olive-brown. Replacing magnesium by metal ions such as Sn^{2+} or Fe^{3+} likewise yields greyish-brown compounds. If, however, $Mg^{2\oplus}$ is replaced by $Zn^{2\oplus}$ and $Cu^{2\oplus}$ (weight ratio 10:1), a green colored complex is formed, which is very stable at pH 5.5. Upon removal of the phytol group, for example by the action of the chlorophyllase enzyme, the chlorophylls are converted into chlorophyllides a and b, while the hydrolysis of pheophytins yields pheophorbides a and b.

Chlorophylls and pheophytins are lipophilic due to the presence of the phytol group, while chlorophyllides and pheophorbides, without phytol,

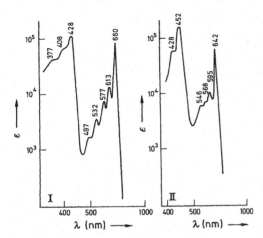

Chlorophyll a: $R^1 = CH_3$
Chlorophyll b: $R^1 = CHO$ (17.16)

are hydrophilic. Conversion of chlorophylls to pheophytins, which is accompanied by a color change, occurs readily upon heating plant material in weakly acidic solutions and, less readily, at pH 7. Color changes are encountered most visibly in processing of green peas, green beans, kale, Brussels sprouts and spinach. Table 17.16 shows that higher temperatures and shorter heating times provide better color retention than prolonged heating at lower temperatures.

Chlorophyllase is mostly inactivated when vegetables are blanched, hence chlorophyllides and pheophorbides are rarely detected. However, in the fermentation of cucumbers, chlorophyllase is active. The result is a color change from dark-green to olive-green, caused by large amounts of pheophorbides.

On stronger heating (sterilization, drying), a part of the pheophytins undergoes hydrolysis, releasing carbonic acid monomethylester which decomposes into CO_2 and methanol:

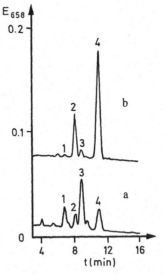

(17.17)

The corresponding pyropheophytins are formed which can be determined next to the pheophytins by using HPLC (Fig. 17.3). For example, Table 17.17 shows the changes in the chloropigments of spinach as a function of the duration of heat sterilization.

A change in color occurs during storage of dried vegetables, its extent increases with increasing water content. The conversion of chlorophylls to pheophytins continues in blanched vegeta-

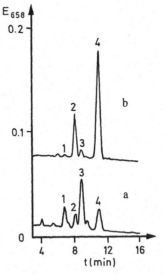

Fig. 17.2. Absorption spectra of chlorophylls a (I) and b (II). Solvent: diethyl ether (I) or diethyl ether +1% CCl_4 (II)

Fig. 17.3. HPLC of chloro-pigments from sterilized cans. Green beans *(a)*, spinach *(b)* (according to *Schwartz* and *von Elbe*, 1983). *1* Pheophytin b, *2* pyropheophytin b, *3* pheophytin a, *4* pyropheophytin a

Table 17.16. Changes in the chlorophyll fraction during processing (values in % of the total pigment content of unprocessed vegetables)

Vegetable	Process	Chlorophylls		Chlorophyllides		Pheophytins		Pheophorbides	
		a	b	a	b	a	b	a	b
Green beans	Untreated	49	25	0	0	18	8	0	0
	Blanched, 4 min/100 °C	37	24	0	0	19	10	0	0
Cucumbers	Untreated	51	30	0	0	15	5	0	0
	Blanched, 4 min/100 °C	34	24	6	3	22	1	5	7
Cucumbers	Untreated	67	33	0	0	0	0	0	0
	Fermented (pickled), 6 days	4	7	3	5	10	3	47	15
	Fermented (pickled), 24 days	0	0	0	0	16	7	57	28

Table 17.17. Effects of the heat sterilization of spinach on the composition of chloropigments (mg/g solids)

Heating to 121 °C (min)	Chlorophyll		Pheophytin		Pyropheophytin	
	a	b	a	b	a	b
Control	6.98	2.49	0	0	0	0
2	5.72	2.46	1.36	0.13	0	0
4	4.59	2.21	2.20	0.29	0.12	0
7	2.81	1.75	3.12	0.57	0.35	0
15	0.59	0.89	3.32	0.78	1.09	0.27
30	0	0.24	2.45	0.66	1.74	0.57
60			1.01	0.32	3.62	1.24

bles even during frozen storage. In beans and Brussels sprouts, immediately after blanching (2 min at 100 °C), the pheophytin content amounts to 8–9%, while after storage for 12 months at −18 °C it increases to 68–83%. Pheophytin content rises from 0% to only 4–6% in paprika peppers and peas under the same conditions.

17.1.2.9.2 Betalains

Pigments known as betalains occur in centrospermae, e. g., in red beet and also in some mushrooms (the red cap of fly amanita). They consist of red-violet betacyanins ($\lambda_{max} \sim 540$ nm) and yellow betaxanthins ($\lambda_{max} \sim 480$ nm). They have the general structure:

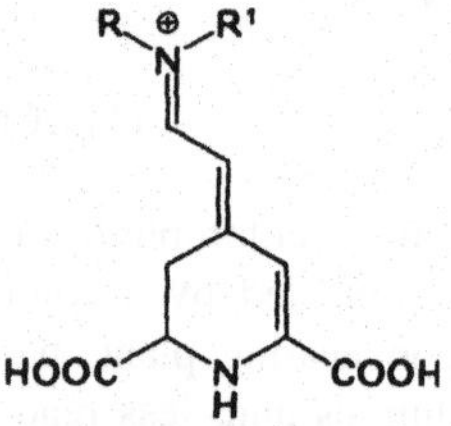

$$(17.18)$$

About 50 betalains have been identified. The majority have an acylated sugar moiety. The acids involved are sulfuric, malonic, caffeic, sinapic, citric and p-coumaric acids. All betacyanins are derived from two aglycones: betanidin (I) and isobetanidin (II), the latter being the C-15 epimer of betanidin:

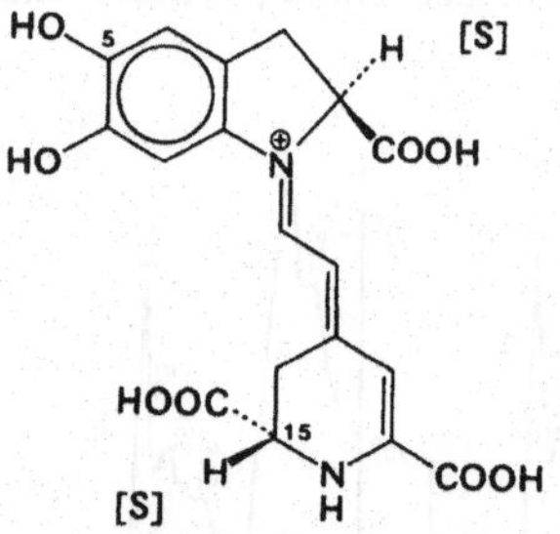

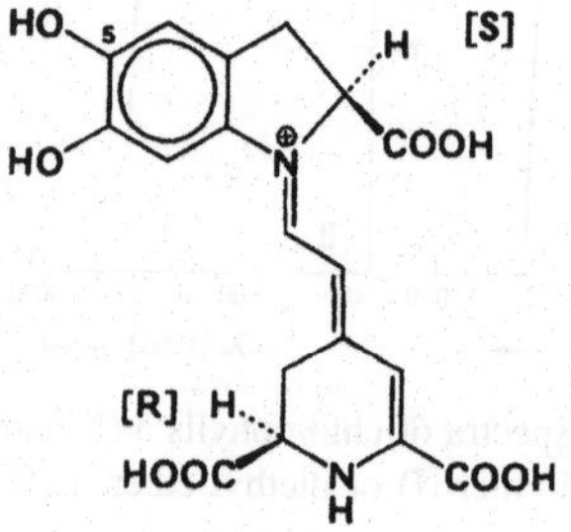

$$(17.19)$$

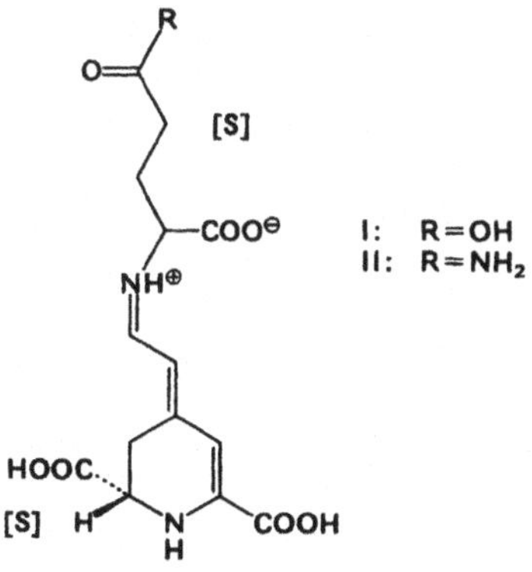

(17.20)

Betanin is the main pigment of red beet. It is a betanidin 5-0-β-glucoside. The betaxanthins have only the dihydropyridine ring in common. The other structural features are more variable than in betacyanins. Examples of betaxanthins are natural vulgaxanthins I and II, also from red beet (*Beta vulgaris*):

Betalain biosynthesis starts with dopa by opening of its benzene ring, followed by cyclization to a dihydropyridine. The (S)-betalamic acid which is formed undergoes condensation with (S)-cyclodopa to betacyanins or with some other amino acids to betaxanthins (cf. reaction sequence 17.21).

Red betanin is water soluble and is used to color food. Its application is, however, limited because it hydrolytically decomposes into the colorless cyclodopa-5-0-β-glucoside and the yellow (S)-betalamic acid. This reaction is reversible. Since the activation energy of the forward reaction $(72 \text{ kJ} \times \text{mol}^{-1})$, greatly exceeds that of the back reaction $(2.7 \text{ kJ} \times \text{mol}^{-1})$, a part of the betanin is regenerated at higher temperatures. Betanin is also sensitive to oxygen.

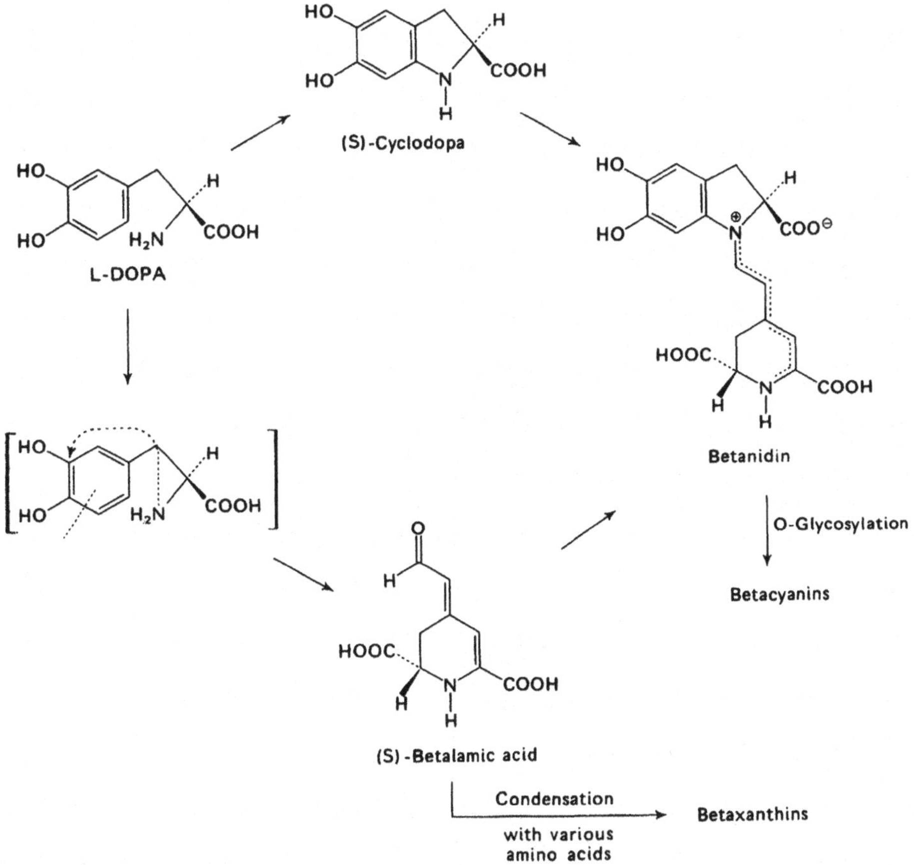

(17.21)

17.1.2.9.3 Goitrogenic Substances

Brassicaceae contain glucosinolates which decompose enzymatically, e. g., into rhodanides. For example, in savoy cabbage the rhodanide content is 30 mg/100 g fresh weight, while in cauliflower it is 10 mg and in kohlrabi 2 mg. Since rhodanide interferes with iodine uptake by the thyroid gland, large amounts of cabbage together with low amounts of iodine in the diet may cause goiter.

Oxazolidine-2-thiones are also goitrogenic. They occur as secondary products in the enzymatic hydrolysate of glucosinolates when the initially formed mustard oils contain a hydroxy group in position 2:

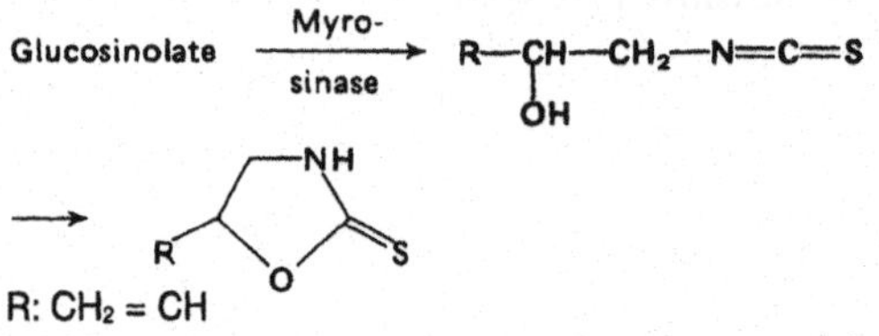

$$\text{Glucosinolate} \xrightarrow[\text{sinase}]{\text{Myro-}} \underset{\overset{|}{\text{OH}}}{\text{R—CH—CH}_2\text{—N=C=S}}$$

$$\longrightarrow$$

R: $CH_2 = CH$

$$(17.22)$$

The levels of the corresponding glucosinolates are up to 0.02% in yellow and white beets and up to 0.8% in seeds of *Brassicaceae* (all members of the cabbage family; kohlrabi, turnip; rapeseed). The leaves contain only negligible amounts of these compounds.

There are 3–15 mg/kg of 5-vinyloxazolidine-2-thione in sliced turnips. Direct intake of thiooxazolidones by humans is unlikely since the vegetable is generally consumed in cooked form. Consequently, the myrosinase enzyme is inactivated and there is no release of goitrogenic compounds. However, brussels sprouts are exceptions, as higher amounts (70–110 mg/kg) of bitter tasting goitrin is formed from progoitrin

during cooking. An indirect intake is possible through milk when such plants are used as animal feed, resulting in a goitrogenic compound content of 50–100 µg/l of milk. The oxazolidine-2-thiones inhibit the iodination of tyrosine, an effect unlike that of rhodanides, which may be offset not by intake of iodine but only by intake of thyroxine.

17.1.2.9.4 Steroid Alkaloids

Steroid alkaloids are plant constituents having a C_{27} steroid skeleton and nitrogen content. *Solanaceae* contain these compounds, their occurrence in potatoes being the most interesting from a food chemistry point of view.

The main compounds in the potato tuber are α-solanine (Formula 17.23) and α-chaconine, which differs from the former compound only in the structure of the trisaccharide (substitution of galactose and glucose with glucose and rhamnose). α-Solanine and α-chaconine and their aglycone solanidine have a bitter/burning taste (Table 17.18) and these sensations last long. The taste thresholds have to be determined in the presence of lactic acid due to a lack of water

Table 17.18. Taste of the steroid alkaloids occurring in potatoes

Compound[a]	Taste threshold (mg/kg)	
	Bitter	Burning
α-Solanine	3.1	6.25
α-Chaconine	0.78	3.13
Solanidine	3.1	–
Caffeine	12.5	–

[a] Dissolved in 0.02% lactic acid.

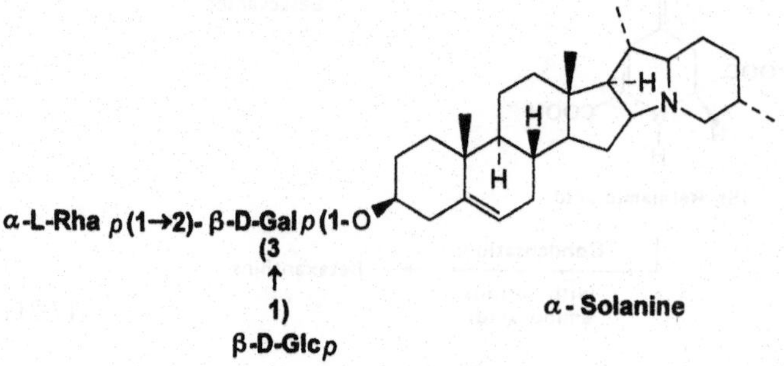

α-L-Rha $p(1\rightarrow2)$- β-D-Gal $p(1$-O

(3
↑
1)

β-D-Glc$p$

α - Solanine

$$(17.23)$$

solubility. Caffeine was used as a comparison. In potatoes, the bitter taste appears if the concentration of the steroid alkaloids exceeds 73 mg/kg. Stress during growth and the exposure of the potatoes to light after harvesting stimulate the formation of these bitter substances.

17.1.3 Storage

The storability of vegetables varies greatly and depends mostly on type, but also on vegetable quality. While some leafy vegetables, such as lettuce and spinach as well as beans, peas, cauliflower, cucumbers, asparagus and tomatoes have limited storage time, root and tuber vegetables, such as carrots, potatoes, kohlrabi, turnips, red table beets, celery, onions and late cabbage cultivars, can be stored for months. Cold storage at high air humidity is the most appropriate. Table 17.19 lists some common storage conditions. The relative air humidity has to be 80–95%. The weight loss experienced in these storage times is 2–10%. Ascorbic acid and carotene contents generally decrease with storage. Starch and protein degradation also occurs and there can be a rise in the free acid content of vegetables such as cauliflower, lettuce and spinach.

Table 17.19. Effect of cold storage temperature on vegetable shelf life

Vegetable	Temperature range (°C)	Shelf life (weeks)
Cauliflower	−1/0	4–6
Green beans	+3/+4	1–2
Green peas[a]	−1/0	4–6
Kale	−2/−1	12
Cucumber	+1/+2	2–3
Head lettuce	+0.5/+1	2–4
Carrot	−0.5/+0.5	8–10
Green bell pepper	−1/0	4
Leek	−1/0	8–12
Brussels sprouts	−3/−2	6–10
Red beet	−0.5/+0.5	16–26
Celery	−0.5/+1	26
Asparagus	+0.5/+1	2–4
Spinach	−1/0	2–4
Tomato	+1/+2	2–4
Onion	−2.5/−2	40

[a] Kept in pods.

17.2 Vegetable Products

A number of processing techniques provide vegetable products which have a substantially higher storage stability compared to fresh vegetables, and are readily converted into a consumable form. As is the case with dairy products, unique vegetable products can be produced by fermentation.

17.2.1 Dehydrated Vegetables

Vegetable dehydration reduces the natural water content of the plant below the level critical for the growth of microorganisms (12–15%) without being detrimental to important nutrients. Also, it is aimed at preserving flavor, aroma and appearance, and the ability to regain the original shape or appearance by swelling when water is added. The dehydration process is accompanied by significant changes. First, there is a concentration of major ingredients such as proteins, carbohydrates and minerals. This occurs along with some chemical changes. Fats are oxidatively degraded and, although present in low amounts in vegetables, this oxidation often diminishes odor and flavor. Amino compounds and carbohydrates interact in a *Maillard* reaction, resulting in a darker color and development of new aroma substances (cf. 4.2.4.4). Vitamin levels may also drop sharply. The original volatile aroma and flavor compounds are lost to a great extent.

In the production of the dehydrated product, the vegetable is first washed, peeled or cleaned, and may be sliced or diced. Blanching for 2–7 min to inactivate the enzymes is then done in hot water or steam. Vegetables may also be treated with SO_2.

Dehydration is performed in a conveyor or tube dryer at 55–60 °C to a residual moisture content of 4–8%. Liquid or paste forms, such as tomato or potato mash, are dried in a spray or drum dryer or, in the case of some special products, in a fluidized bed dryer. Dehydration by freeze-drying provides high quality products (good shape retention) with a spongy and porous structure that is readily rehydrated. Some vegetables used in soup powders, e. g., peas and cauliflower, are prepared in this way. For production of dehydrated potato products (Fig. 17.4), tubers are peeled, cleaned, sliced into strings or chips or diced and, after

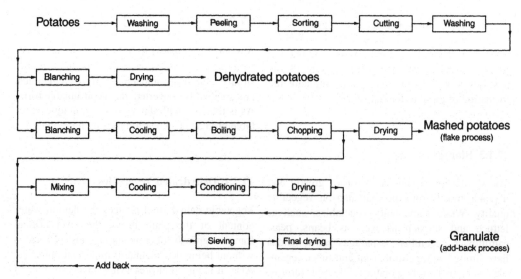

Fig. 17.4. Production of dehydrated potatoes, mashed potato flakes and potato granulate

steam-cooking, dried. For production of dehydrated mashed potato flakes or potato granulate, the steamed slices are squeezed between rollers into a mash with the least possible damage to cell walls. Cell wall damage allows the gelatinized starch to escape from the ruptured cells and to later impart a gluey-sticky texture to the final product. The mashed potato is dried on rollers for the production of flakes and in a pneumatic dryer for the production of granulate. Since the latter drying process requires a flowable product, the mash is mixed with dried powder containing 12–15% of water in a ratio of 1:2 (*add-back* process). The mixture obtained is then brought to a final water content of 6–8% in a fluidized bed dryer.

Dehydrated vegetables are light, air and moisture sensitive and therefore require careful packaging. Wax-impregnated paper or cardboard, multi-layer foils, metal cans or glass containers are commonly used and, occasionally, the packaging is done under nitrogen or vacuum. Also, the dehydrated product may be pressed prior to packaging.

17.2.2 Canned Vegetables

Canning, which involves heat sterilization, is one of the most important processes in vegetable preservation. The selected and sorted freshly harvested products are trimmed and blanched as outlined for dehydrated vegetables. Blanching here serves not only to inactivate the enzymes, but to remove both undesirable flavoring compounds (cabbages), and the air present in plant tissue, and to induce shrinkage or softening of the product, thereby increasing packaging density.

Brine (1–2% NaCl solution) often serves as a filling liquid. Sugar (peas, red table beets, tomato, sweet corn), citric acid (up to 0.05%, used for example for celery, cauliflower and horse beans), calcium salts for firming the plant tissue (tomato, cauliflower) or monosodium glutamate (100–150 mg per kg filling) are also added to round-off the flavor.

Sterilization is performed in autoclaves. The autoclaves can be classified according to the heat transfer into water and steam autoclaves and according to the mode of operation into vertical and rotation autoclaves. Rotation autoclaves can be used in a continuous operation only when the cans enter and exit via locks without loss of pressure and steam. The advantage of rotation heating lies in the quicker and more uniform heating of the product. After the required sterilization effect is achieved, the product is quickly cooled to avoid excessive after-heating. As with other foods, vegetable sterilization processes tend toward higher temperatures and shorter times (HTST sterilization) since, in this way, the products retain a better quality (texture, aroma, color).

The nutritional/physiological value of the main constituents of vegetables (proteins and carbohydrates) is not diminished by this common heat sterilization process. Damage due to interaction of amino acids with reducing sugars, which occurs to a small extent, is also negligible. However, there is often a negative effect on vitamins (cf. 6.1). Carotene, a fatsoluble provitamin A, is not affected by the washing and blanching steps, but it is moderately destroyed (5–30%) during actual canning. Vitamin B_1 in carrots and tomatoes does not decrease significantly, while losses are 10–50% for other vegetables (green beans, peas and asparagus). Vitamin B_1 losses are high in spinach (66%) due to the large surface area. Vitamin B_2 is lost (5–25%) by leaching during blanching, but not significantly during further processing. Nicotinic acid losses are similar. Vitamin C losses are due to its water solubility and its enzymatic and chemical degradation, particularly in the presence of traces of heavy metal ions. Vitamin C retention is 55–90% during the canning of asparagus, peas and green beans. Storage of canned vegetables for several years generally results in an additional 20% vitamin loss.

17.2.3 Frozen Vegetables

Beans, peas, paprika peppers, Brussels sprouts, edible mushrooms (*Boletus edulis*), tomato pulp and carrots are particularly suitable for freezing. Radishes, lettuce or whole tomatoes are unsuitable. High quality fresh vegetables are treated with boiling water for 1.5–4 min or steam for 2–5 min for enzyme inactivation. The blanching time is generally shorter than that used in canning, and varies according to type, ripeness and size of vegetable. It is kept as short as possible to prevent leaching. Steam blanching is generally more advantageous than blanching in hot water. The blanching time required for enzyme inactivation is determined by measuring the rate of inactivation of an indicator enzyme (cf. 2.5.4.4). Immediately after blanching, the vegetable is cooled, frozen at −40 °C or lower, then stored at −18 to −20 °C. Freezing is mainly conducted using conventional freezing techniques by indirect cold-transfer in plate or air freezers. At present, cryogenic freezing techniques play no appreciable part in vegetable processing.

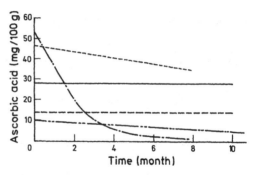

Fig. 17.5. Changes in vitamin C content in frozen vegetables kept at −21 °C. —— Peas precooked, − − − beans precooked, − · − · · − · · − beans raw, − · − · − spinach raw, − − − − spinach precooked. (according to *Heimann*, 1958)

Freezing preserves vegetable nutrients to a great extent. Vitamin A and its provitamin, carotene, are well preserved in spinach, peas and beans, or are moderately lost (asparagus) after proper blanching, freezing and deepfreeze storage and even after thawing to room temperature. Losses in the Vitamin B group depend mostly on the conditions of the primary processing steps (washing, blanching). The other steps have no effect on B vitamins. Vitamin C leaching by water or steam is detrimental. It is generaly preserved during freezing and thawing. Careful blanching and low temperature storage are critical for vitamin C preservation (Figs. 17.5 and 17.6).

Irreversible textural changes can occur in deep-frozen vegetables. Typical symptoms are soften-

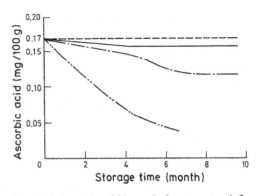

Fig. 17.6. Ascorbic acid losses in frozen peas as influenced by storage temperature. − − − −40 °C, —— −18 °C, − · − · − −12 °C − · · − · · − −9 °C. (according to *Schormuller*, 1966)

ing, ductile stickiness, or looseness or flaccidity (beans, cucumbers, carrots); build-up of a sticky, ductile, gum-like structure (asparagus), or pasty, soggy structure (celery, kohlrabi); or hull hardening (peas).

17.2.4 Pickled Vegetables

Pickled vegetables are produced by spontaneous lactic acid fermentation (white cabbage, green beans, cucumbers, etc.). The fermentation lowers the pH, inhibits the growth of undesirable acid-sensitive microorganisms and, simultaneously, affects the enzymatic softening of cells and their tissues, thus improving digestibility and wholesomeness. The use of salt also has a preservative effect. The acidic pH of the medium stabilizes vitamin C.

While the preservation techniques outlined in earlier sections were aimed at retention of the original odor and flavoring substances of the raw material, including regeneration of lost aroma constituents, this is not important in pickled vegetables since a new typical aroma is developed.

17.2.4.1 Pickled Cucumbers (Salt and Dill Pickles)

Unripe cucumbers, after addition of dill herb and, if necessary, other flavoring spices (vine leaf, garlic or bay leaf), are placed into 4–6% NaCl solution or are sometimes salted dry. Usually, the salt solution is poured on the cucumbers in a barrel and then allowed to ferment and, if necessary, glucose is added. Fermentation takes place at 18–20 °C and yields lactic acid, CO_2, some volatile acids, ethanol and small amounts of various aroma substances. Homo- and heterofermentative lactic acid bacteria like *Lactobacillus plantarum*, *L. brevis* and *Pediococ-*

cus cerevisiae are involved in the fermentation of pickled cucumbers. In contrast to sauerkraut, *Leuconostoc mesenteroides* does not play a role. The lactic acid (0.5–1%) initially formed is later metabolized partly by film yeast or oxidative yeasts that grow on the surface of the brine. Thus, the original pH value of the fermenting medium (3.4–3.8) is slightly increased.

Apart from spontaneous fermentation, controlled fermentatioin on inoculation with *Lactobacillus plantarum* and *Pediococcus cerevisiae* is also used.

17.2.4.2 Other Vegetables

Green beans, carrots, kohlrabi, celery, asparagus, turnips and others are processed similarly to cucumbers. Sliced green beans, for example, are treated with salt (2.5–3%), subjected to lactic acid fermentation at about 20 °C, and marketed in barrels, cans or glass jars. Some pickled vegetables, mostly those that were not blanched or precooked, will not soften during later cooking.

17.2.4.3 Sauerkraut

Lactic acid fermentation has been used for millenia for the production of sauerkraut (Fig. 17.7). It was also customary earlier to place the cabbage into acidified wine or vinegar. White cabbage heads are cut into 0.75–1.5 mm thick shreds, then mixed with salt at 1.8–2.5% by weight. The shreds are then packed into tanks of wood or reinforced concrete, coated with synthetics. After the shreds have been packed in layers, they are tamped and weighted down so that a layer of expressed brine juice covers the surface. The lactic acid fermentation initiated by starter cultures occurs spontaneously at 18–24 °C for

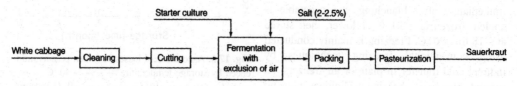

Fig. 17.7. Production of sauerkraut

3–6 weeks. During the first 48 h of fermentation the pH falls from 6.2 to the range of 3.7 to 4.2. The acid formed inhibits the growth of competing interfering microorganisms. *Leuconostoc mesenteroides* and in addition *Lactobacillus brevis* are the predominating microorganisms during the initial phase of fermentation. Homofermentative bacteria like *Lactobacillus plantarum* and *Pediococcus cerevisiae* appear later. The amount of acid formed depends on the initial sugar content of the cabbage. Hence, sugar is sometimes added (to 1%) to cabbage which does not ferment readily. In addition to *Lactobacillus* ssp., yeasts are also involved in fermentation. The products are lactic and acetic acids (in ratios of 4:1 to 6:1), ethanol (0.2–0.8%), CO_2, mannitol (from fructose) and, most importantly, aroma substances which appear in the prefermentation phase. After fermentation is complete, the sauerkraut pH is about 3.6. Lactic acid values of less than 6 g/l indicate unsatisfactorily fermented cabbage. The end-product is kept in barrels under brine. The sauerkraut is also packaged or canned in retail containers. The cans are filled at 70 °C, then exhausted, sealed and sterilized at 95–100 °C. In addition, sauerkraut is packed and distributed in plastic foils and containers. Mildly acidic sauerkraut, preferred in South Germany, is produced by stopping the fermentation before all the sugar is degraded. After pasteurization, the product can be stored for a longer time and still retains a clearly sour taste. Sauerkraut is flavored and spiced to some extent by addition of sugar, juniper berries, caraway or dill seeds. For wine sauerkraut at least 1 liter of wine per 50 kg sauerkraut is added after fermentation.

Drained sauerkraut contains on the average 90.7% water, 1.5% nitrogen compounds, 0.3% crude fat, 3.9% carbohydrates, 1.1% crude fiber, 0.6% minerals (excluding NaCl), 0.8–3.3% NaCl, 1.4–1.9% titratable acid (calculated as lactic acid; 0.28–0.42% is acetic acid) and 0.29–0.61% ethanol. There are small amounts of formic, *n*-heptanoic and *n*-octanoic acids, methanol, and compounds important for palatability, i.e., dextran and mannitol. Vitamin C content (10–38 mg/100 g) is not changed when sauerkraut is heated in a pressure cooker. However, after several reheatings about 30% is destroyed.

17.2.4.4 Eating Olives

Eating olives include not only the green, lactic-fermented olives, but also the black, lactic-fermented ones and the black, unfermented ones. Table 17.20 shows the composition of the flesh of fresh and green lactic-fermented olives.

For the production of green lactic-fermented olives, the fruit is harvested in a yellow-green to yellow state and placed in 1.3–2.6% NaOH for 6–10 h. During this time, most of the bitter substance oleuropein (Formula 17.24) is hydrolyzed.

$$O-\beta-D-Glcp \qquad (17.24)$$

The olives are then washed with water and allowed to undergo spontaneous lactic fermentation in a 10–12% NaCl solution. Fermentation is carried out in concrete containers coated with epoxide resin or in polyesters tanks reinforced with glass fibers. In addition to yeasts, *Pediococcus* and *Leuconostoc spp.* are involved in the first fermentation stages and *Lactobacillus spp.* (*L. plantarum*) in the later stages. After fermentation, the olives are left in the brine or filled into small packs with fresh salt solution and pasteurized. Before packing, the olives are usually stoned and filled (paprika, anchovies, almonds, capers, and onions). The final product has a pH value of 3.8–4.2 and contains 0.8–1.2% of lactic acid. The salt

Table 17.20. Composition[a] of the flesh of fresh (1) and green lactic-fermented olives 2)

Component	1	2
Water	50–75	61–81
Lipids	6–30	9–28
Reducing sugar	2–6	
Non-reducing sugar	0.1–0.3	
Raw protein	1–3	1–1.5
Raw fiber	1–4	1.4–2.1
Ash	0.6–1	4.2–5.5
Other components	6–10	

[a] Percentage by weight.

concentration should be at least 7% and at least 8% in products with a longer shelf life.

For the production of black lactic-fermented olives, the ripe, violet to black fruit is washed and directly allowed to undergo spontaneous lactic fermentation in a 8–10% salt solution. Lactobacilli and yeasts are involved, but the yeasts dominate normally. Fermentation proceeds slowly because the olive skin is not as permeable as after alkali treatment. After fermentation, the olives are packed into glass or plastic containers and pasteurized. The final product has a pH value of 4.5–4.8 and contains 0.1–0.6% of lactic acid. The salt concentration is 6–9%.

For the production of black unfermented olives, the ripe fruit is placed 3–5 times in 1–2% NaOH. In between the fruit is washed and well aired to ensure that the flesh is uniformly dyed black by intensive phenol oxidation. Iron gluconate is added to the last wash water to stabilize the color. The olives are then packed in a 3% NaCl solution and sterilized. The product has a pH value of 5.8–7.9 and contains 1–3% of common salt.

17.2.4.5 Faulty Processing of Pickles

Pickled cucumbers are often softened due to the effects of their own or microbial pectolytic enzymes. Brown-to-black discoloration is caused by iron sulfide build-up or by black pigments formed by microorganisms (*Bacillus nigrificans*). Hollowness is caused by gasforming microorganisms, i. e. gaseous fermentation, and can be prevented readily by pickling in the presence of sorbic acid.

Sauerkraut is darkened by chemical or enzymatic oxidations when the brine does not cover the surface. Reddish color is caused by yeasts. Sauerkraut softening occurs when fermentation takes place at too high a temperature, when the cabbage is exposed to air, too little salt is added; or by faulty fermentation when the lactic acid content remains too low. In addition to faulty fermentation, the kraut can be ruined by infections caused by molds and other flora of the surface film and by rotting (insufficient brine for full protection).

Small chain fatty acids like propionic acid and butyric acid cause an aroma defect.

17.2.5 Vinegar-Pickled Vegetables

These products are prepared by pouring preboiled and still hot vinegar onto the vegetables. Vegetables used are cucumbers, red table beets, pearl and silver onions, paprika peppers, mixed vegetables, which also include cauliflower, carrots, onions, peas, mushrooms (in particular the table mushroom, *Boletus edulis*), asparagus, tender corncobs, celery, parsley root, parsnip, kohlrabi, pumpkin and pepperoni peppers.

The raw vegetable is covered with a solution of 2.5% vinegar. Salt, spices and herbs, herb extracts, sugar and chemical preservatives are usually added. Depending on the vegetable and its preparation method, there are "single pickles" in vinegar (vinegar cucumbers, chili pepper-flavored cucumbers or gherkins, mustard cucumbers, sterilized deli and spiced garlic, dill-flavored cucumbers) and "mixed pickles" in vinegar, which are made partly from fresh and partly from precanned vegetables (unsliced cucumbers, cauliflower, onions, delicate and tender corncobs, paprika peppers).

17.2.6 Stock Brining of Vegetables

Salting is a practical method for preserving some vegetables in bulk until further processing. Usually the vegetable is salted with table salt after being blanched. Brined vegetables are kept for the production of other products. Salted asparagus, for example, is obtained by addition of ∼20% by weight of salt and used for the preparation of "Leipzig medley" and mixed fresh vegetables. Stock brining of beans is also important. Blanched or nonblanched beans are soaked in salt brine or are treated with dry salt to 10–20% by weight (added by hand or by machine spreading or dusting) and kept in brine prior to the manufacture of other products. As with other vegetables, the beans are thoroughly drained of brine and rinsed in a stream of hot water before further processing. In the same way, vegetables such as cauliflower, cabbage, carrots, pearly onions and gherkins are stock brined. Mushrooms and morels are also salted; a practice primarily found in Poland and Russia.

17.2.7 Vegetable Juices

The vegetable is cleaned, washed, then blanched and disintegrated in a mill. In some instances, e. g., the tomato, it is first disintegrated and the slurry heated to >70°C for some time. The juice is then separated in presses or by centrifuging and salt is usually added to 0.25–1%. Nonsour juices are mixed with lactic or citric acid. For storage stability, such products are subjected to pasteurization in plate heat exchangers. Mostly tomatoes and occasionally other vegetables such as cucumbers, carrots, red beets, radishes, sauerkraut, celery or spinach are used for processing into juice.

17.2.8 Vegetable Paste

A vegetable purée or paste is a finely dispersed slurry from which skins and seeds have been removed by passing the slurry through a pulper or finisher. The most important product is tomato purée which, depending on the brand, has a dry matter content of 14–36%, and contains 0.8–2% NaCl. Tomato ketchup is made by the intensive premixing of tomato paste (28% or 38%) with vinegar, water, sugar, spices, and stabilizers, followed by fine homogenization via colloid mills, if necessary. Each charge, which is usually made batchwise, is fed via a plate-type heat exchanger (90 °C) and via a degassing device to a hot-filling apparatus with subsequent cooling. If the heat treatment is too long, defects such as caramelization, color change, and bitter taste can be caused. Since the product tends to separate, especially at air bubbles when degassing is inadequate, it is important that the viscosity is sufficient. If the natural pectin content is well preserved (e. g., by hot break tomato puree), the use of thickening agents is unnecessary. The filled bottles are often stored upside down to prevent a relatively frequent defect called "black neck", a browning at the neck of the bottle due to a high proportion of air in the headspace.

Some other vegetable purées are important primarily as baby foods.

17.2.9 Vegetable Powders

Vegetable powders are obtained by drying the corresponding juice with or without addition of a drying enhancer, such as starch or a starch degradation product, to a residual moisture content of about 3%. Drying processes used are spray-drying, vacuum drum drying, and freeze-drying. The most important product is tomato powder. Other powders, such as those of spinach or red beets, are in part used in food colorings.

17.3 References

Adler, G.: Kartoffeln und Kartoffelerzeugnisse. Verlag Paul Barey: Berlin. 1971

Bötticher, W.: Technologie der Pilzverwertung. Verlag Eugen Ulmer: Stuttgart. 1974

Buttery, R.G., Teranishi, R., Flath, R. A., Ling, L. C: Fresh tomato volatiles. Composition and sensory studies. ACS Symposium Series 388, American Chemical Society, Washington, DC 1989, p. 213

Elbe, J.H. von: Influence of water activity on pigment stability in food products. In: Water Activity: Theory and Applications to Food (Eds.: Rockland, L.B., Beuchat, L.R.) Marcel Dekker, Inc.: New York. 1987

Fenwick, G.R., Griffiths, N.M.: The identification of the goitrogen, (-)5-vinyloxazolidine-2-thione (goitrin), as a bitter principle of cooked Brussels sprouts (*Brassica oleracea L. var. gemmifer*). Z. Lebensm. Unters. Forsch. *172*, 90 (1981)

Fernández Diez, M.J.: Olives. In: Biotechnology (Eds.: Rehm, H.-J., Reed, G.), Vol. 5, p. 379, Verlag Chemie: Weinheim. 1983

Fischer, K.-H., Grosch, W.: Volatile compounds of importance in the aroma of mushrooms (*Psalliota bispora*). Lebensm: Wiss. Technol. *20*, 233 (1987)

Granvogl, M., Christlbauer, M., Schieberle, P.: Quantitation of the intense aroma compound 3-mercapto-2-methylpentan-1-ol in raw and processed onions (*Allium cepa*) of different origins and in other Allium varieties using a stable isotope dilution assay. J. Agric. Food Chem. *52*, 2797 (2004)

Grosch, W.: Aromen von gekochten Kartoffeln, Trockenkartoffeln und Pommes frites. Kartoffelbau *50* (9/10), 362 (1999)

Hadar, Y., Dosoretz, C.G.: Mushroom mycelium as a potential source of food flavour. Trends in Food Science & Technology 2, 214 (1991)

Maga, J. A.: Potato flavor. Food Reviews International *10*, 1 (1994)

Mutti, B., Grosch, W.: Potent odorants of boiled potatoes. Nahrung/Food *43*, 302 (1999)

Ross, A.E., Nagel, D.L., Toth, B.: Evidence for the occurrence and formation of diazonium ions in the *Agaricus bisporus* mushroom and its extracts. J. Agric. Food Chem. *30*, 521 (1982)

Rotzoll, N., Dunkel, A., Hofmann, T.: Quantitative studies, taste reconstitution, and omission experiments on the key taste compounds in morel mushrooms (*morchella deliciosa* Fr.). J. Agric. Food Chem. *54*, 2705 (2006)

Salunkhe, D.K., Do, J.Y.: Biogenesis of aroma constituents of fruits and vegetables. Crit. Rev. Food Sci. Nutr. *8*, 161 (1977)

Schobinger, U.: Frucht- und Gemüsesäfte, Verlag Eugen Ulmer: Stuttgart. 1978

Shah, B.M., Salunkhe, D.K., Olson, L.E.: Effects of ripening processes on chemistry of tomato volatiles. J. Am. Soc. Hort. Sci. *94*, 171 (1969)

Stamer, J.R.: Lactic acid fermentation of cabbage and cucumbers. In: Biotechnology (Eds.: Rhem, H.-J., Reed, G.), Vol. 5, p. 365, Verlag Chemie: Weinheim. 1983

Talburt, W.F., Smith, O.R.A.: Potato Processing. AVI Publ., Westport, CT 1975

Tiu, C.S., Purcell, A.E., Collins, W. W.: Contribution of some volatile compounds to sweet potato aroma. J. Agric. Food Chem. *33*, 223 (1985)

Whitfield, F.B., Last, J.H.: Vegetables. In: Volatile compounds in foods and beverages (Ed.: Maarse, H.) Marcel Dekker, Inc.: New York, 1991

18 Fruits and Fruit Products

18.1 Fruits

18.1.1 Foreword

Fruits include both true fruits and spurious fruits, as well as seeds of cultivated and wild perennial plants. Fruits are commonly classified as pomaceous fruits, stone fruits, berries, tropical and subtropical fruits, hard-shelled dry fruits and wild fruits. The most important fruits are presented in Table 18.1 with pertinent data on botanical classification and use. Table 18.2 provides data about fruit production.

18.1.2 Composition

Fruit composition can be strongly influenced by the variety and ripeness, thus data given should be used only as a guide. Table 18.3 shows that the dry matter content of fruits (berries and pomme, stone, citrus and tropical fruits) varies between 10–20%. The major constituents are sugars, polysaccharides and organic acids, while N-compounds and lipids are present in lesser amounts. Minor constituents include pigments and aroma substances of importance to organoleptic quality, and vitamins and minerals of nutritional importance. Nuts are highly variable in composition (Table 18.4). Their moisture content is below 10%, N-compounds are about 20% and lipids are as high as 50%.

18.1.2.1 N-Containing Compounds

Fruits contain 0.1–1.5% N-compounds, of which 35–75% is protein. Free amino acids are also widely distributed. Other nitrogen compounds are only minor constituents. The special value of nuts, with their high protein content, has already been outlined.

18.1.2.1.1 Proteins, Enzymes

The protein fraction varies greatly with fruit variety and ripeness. This fraction is primarily enzymes. Besides those involved in carbohydrate metabolism (e. g., pectinolytic enzymes, cellulases, amylases, phosphorylases, saccharases, enzymes of the pentose phosphate cycle, aldolases), there are enzymes involved in lipid

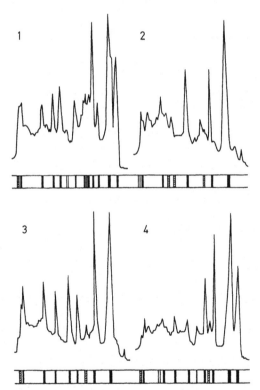

Fig. 18.1. Protein patterns of various wine cultivars obtained by isoelectric focussing (pH 3–10) using Sephadex G-75 as a gel support medium. Staining was done by Coomassie Blue. The figures show the electropherograms and the corresponding densitograms. Cultivation region South Palatinate; *1* Morio Muscat, *2* Mueller-Thurgau, *3* Rulaender, *4* Sylvaner (according to *Drawert* and *Mueller*, 1973)

Table 18.1. Eddible fruits: a classification

Number	Common name	Latin name	Family/ subfamily	Form of consumption
Pomme fruits				
1	Apple	*Malus sylvestris*	Rosaceae	Fresh, dried, purée, jelly, juice, apple cider, brandy
2	Pear	*Pyrus communis*	Rosaceae	Fresh, dried, compote, brandy
3	Quince apple shaped pear shaped	*Cydonia oblonga* var. *maliformis* var. *pyriformis*	Rosaceae	Jelly, ingredient of apple purée
Stone fruits				
4	Apricot	*Prunus armeniaca*	Rosaceae	Fresh, dried, compote, jam, juice, seed for persipan, brandy
5	Peach	*Prunus persica*	Rosaceae	Fresh, compote, juice, brandy
6	Prune/plum	*Prunus domestica*	Rosaceae	Fresh, dried, compote, jam, brandy
7	Sour cherry	*Prunus cerasus*	Rosaceae	Fresh, compote, jam, juice, brandy
8	Sweet cherry	*Prunus avium*	Rosaceae	Fresh, candied, compote
Berry fruits				
9	Blackberry	*Rubus fruticosus*	Rosaceae	Fresh, jam, jelly, juice, wine, liqueur
10	Strawberry	*Fragaria vesca*	Rosaceae	Fresh, compote, jam, brandy
11	Bilberry	*Vaccinium myrtillus*	Ericaceae	Fresh, compote, jam, brandy
12	Raspberry	*Rubus idaeus*	Rosaceae	Fresh, jam, jelly syrup, brandy
13	Red currant	*Ribes rubrum*	Saxifragaceae	Fresh, jelly, juice, brandy
14	Black currant	*Ribes nigrum*	Saxifragaceae	Fresh, juice, liqueur
15	Cranberry	*Vaccinium vitis-idaea*	Ericaeae	Compote
16	Gooseberry	*Ribes uva-crispa*	Saxifragaceae	Unripe: compote; ripe: fresh, jam, juice
17	Grapes	*Vites vinifera* ssp. *vinifera*	Vitaceae	Fresh, dried (raisins) juice, wine, brandy
Citrus fruits				
18	Orange	*Citrus sinensis*	Rutaceae	Fresh, juice, marmelade
19	Grapefruit	*Citrus paradisi*	Rutaceae	Fresh, juice
20	Kumquat	*Fortunella margarita*	Rutaceae	Fresh, compote, jam
21	Mandarine	*Citrus reticulata*	Rutaceae	Fresh, compote
22	Pomelo	*Citrus maxima*	Rutaceae	Fresh, juice
23	Seville orange	*Citrus aurantium* ssp. *aurantium*	Rutaceae	Candied, marmalade
24	Lemon	*Citrus limon*	Rutaceae	Juice
25	Citron	*Citrus medica*	Rutaceae	Peel candied (citronat)
Other tropical/ subtropical fruits				
26	Acerola	*Malpighia emarginata*	Malpighiaceae	Fresh, compote, juice
27	Pineapple	*Ananas comosus*	Bromeliaceae	Fresh, compote, jam, juice
28	Avocado	*Persea americana*	Lauraceae	Fresh
29	Banana	*Musa*	Musaceae	Fresh, dried, cooked, baked
30	Cherimoya	*Annona cherimola*	Annonaceae	Fresh
31	Date	*Phoenix dactylifera*	Arecaceae	Fresh, dried
32	Fig	*Ficus carica*	Moraceae	Fresh, dried, jam, dessert wine
33	Indian fig	*Opuntia ficus-indica*	Cactaceae	Fresh
34	Guava	*Psidium guajava*	Myrtaceae	Compote, juice
35	Persimmon	*Diospyros kaki*	Ebenaceae	Fresh, candied, compote

Table 18.1. (Continued)

Number	Common name	Latin name	Family/ subfamily	Form of consumption
36	Kiwi	*Actinidia chinensis*	Actinidiaceae	Fresh, compote
37	Litchi	*Litchi chinensis*	Sapindaceae	Fresh, dried, compote
38	Mango	*Mangifera indica*	Anacardiaceae	Fresh, compote, juice
39	Melons			
	cantaloups	*Cucumis melo*	Cucurbitaceae	Fresh
	watermelon	*Citrullus lanatus*	Cucurbitaceae	Fresh
40	Papaya	*Carica papaya*	Caricaceae	Fresh, compote, juice
41	Passion fruit	*Passiflora edulis*	Passifloraceae	Fresh, juice
42	Golden shower	*Cassia fistula*	Caesalpiniaceae	Fresh
Shell(nut) fruits				
43	Cashew nut	*Anacardium occidentale*	Anacardiaceae	Roasted
44	Peanut	*Arachis hypogaea*	Fabaceae	Roasted, salted
45	Hazel-nut (Filbert)	*Corylus avellana*	Betulaceae	Fresh, baked and confectionary products (nougat, crocant)
46	Almond sweet bitter	*Prunus dulcis* var. *dulcis* var. *amara*	Rosaceae	Baked and confectionary products (marzipan); flavoring of baked and confectionary products
47	Brazil nut	*Bertholletia excelsa*	Lecythidaceae	Fresh
48	Pistachio	*Pistacia vera*	Anacardiaceae	Fresh, salted, sausage flavoring, decoration of baked products
49	Walnut	*Juglans regia*	Juglandaceae	Fresh, baked and confectionary products, unripe fruits in vinegar and sugar-containing preserves
Wild fruits				
50	Rose hips	*Rosa sp.*	Rosaceae	Jam, wine
51	Elderberry	*Sambucus nigra*	Caprifoliaceae	Juice, jam
52	Seabuckthorn	*Hippophae rhamnoides*	Elaeagnaceae	Jam, juice

metabolism (e. g. lipases, lipoxygenases, enzymes involved in lipid biosynthesis), and in the citric acid and glyoxylate cycles, and many other enzymes such as acid phosphatases, ribonucleases, esterases, catalases, peroxidases, phenoloxidases and O-methyl transferases.

Protein and enzyme patterns, which can be obtained, for example, by electrophoretic separation, are generally highly specific for fruits and can be utilized for analytical differentiation of the species and variety. Figure 18.1 shows protein patterns of various grape species and Fig. 18.2 presents enzyme patterns of various species and cultivars of strawberries.

18.1.2.1.2 Free Amino Acids

Free amino acids are on average 50% of the soluble N-compounds. The amino acid pattern is typical of a fruit and hence can be utilized for the analytical characterization of a fruit product. Table 18.5 provides some relevant data.

In addition to common protein-building amino acids, there are nonprotein amino acids present in fruits, as in other plant tissues. Examples are the toxic 2-(methylene cyclopropyl)-glycine (I) in litchi fruits (*Litchi sinensis*), the toxic hypoglycine A (II) in akee (*Blighia sapida*), 1-aminocyclopropane-1-carboxylic acid (X) in apples and pears, trans-4-methylproline (XXII), 4-hydroxymethylprolines (XXIII–XXV) and 4-methyleneproline (XXVI) in apples and in loquat fruits (*Eviobotrya japonica*), 3,4-dihydroxyglutamic acid (XXXV) in red currants, 4-methyleneglutamic acid (XXXI) and 4-methyleneglutamine (XXXII) in peanuts and 3-amino-3-carboxypyrrolidine (LIV) in cashew. The nonprotein amino acids are discussed in more detail in Section 17.1.2.1.2. The Roman

Table 18.2. Production of fruits in 2006 (1000 t)

Continent	Fruits[a], grand total	Nuts, grand total	Grapes	Raisins	Dates
World	526,496	11,106	68,953	1189	6704
Africa	67,848	1676	3822	41	2413
America, Central	106,024	208	260	4	3
America, North	25,316	1305	6172	320	15
America, South and Caribbean	106,024	599	6999	85	3
Asia	244,643	6403	20,740	635	4261
Europe	73,058	1070	29,097	79	13
Oceania	6965	52	2123	30	–

Continent	Apples	Pears	Peaches + nectarines	Plums + sloes	Oranges
World	63,805	19,540	17,189	9431	64,795
Africa	2006	651	810	244	5620
America, Central	632	30	223	74	5260
America, North	4909	770	941	415	9000
America, South and Caribbean	4368	827	1166	505	26,118
Asia	36,768	13,906	9639	5505	16,721
Europe	14,952	3212	4515	2727	6757
Oceania	800	174	119	34	580

Continent	Mandarins[b]	Lemons and limes	Grapefruit	Apricots	Avocado
World	25,660	12,990	4563	3251	3317
Africa	1483	832	665	502	396
America, Central	342	2087	465	2	1201
America, North	417	942	1118	41	247
America, South and Caribbean	2800	5295	1093	54	165
Asia	18,008	4324	1616	1735	429
Europe	2843	1552	60	895	94
Oceania	109	46	13	23	57

Continent	Guavas, mangoes	Pineapples	Bananas	Papaya
World	30,541	18,261	70,756	6591
Africa	3003	2598	7755	1422
America, Central	2342	2191	7137	964
America, North	3	192	9	13
America, South and Caribbean	5098	6060	24,897	134
Asia	22,383	9273	36,457	1926
Europe	–	2	393	–
Oceania	55	135	1244	12

Table 18.2. (Continued)

Continent	Strawberries	Raspberries and other berries	Currants	Almonds	Pistachios
World	4082	1155	760	1766	576
Africa	227	3	–	214	1
America, Central	202	8	–	–	–
America, North	1279	109	–	716	122
America, South and Caribbean	297	14	–	10,651	–
Asia	711	510	2	418	440
Europe	1539	411	748	396	12
Oceania	29	108	9	12	–

Continent	Hazelnuts	Chestnuts	Cashew nuts	Walnuts
World	961	1180	3103	1664
Africa	–	–	1059	42
America, Central	–	–	8	80
America, North	41	–	–	308
America, South and Caribbean	–	42	247	107
Asia	737	1013	1797	891
Europe	183	125	–	316
Oceania	–	–	–	–

Country	Fruits, grand total	Country	Nuts, grand total	Country	Grapes
China	93,410	China	1625	Italy	8326
India	43,525	USA	1305	France	6693
Brazil	37,736	India	1092	Spain	6402
USA	27,328	Turkey	1000	China	6375
Italy	17,812	Vietnam	950	USA	6094
Spain	16,514	Nigeria	727	Turkey	4000
Indonesia	15,406	Iran	507	Iran	2964
Mexico	15,385	Italy	328	Argentina	2881
Iran	13,848	Brazil	271	Chile	2250
Philippines	13,582	Spain	268	Japan	2098
Turkey	12,563	Indonesia	264	Australia	1981
Nigeria	9874	Σ (%)^c	75	South Africa	1550
Uganda	9731			Σ (%)^c	75
France	9682				
Thailand	8648				
Argentina	8351				
Egypt	8196				
Colombia	7910				
Ecuador	7536				
Pakistan	6379				
Vietnam	5691				
South Africa	5690				
Σ (%)^c	75				

Table 18.2. (Continued)

Country	Raisins	Country	Dates	Country	Apples
Turkey	376	Egypt	1170	China	26,066
USA	320	Iran	997	USA	4569
Iran	146	Saudi Arabia	970	Iran	2662
Greece	77	Pakistan	507	Poland	2305
Chile	64	Algeria	491	Italy	2113
South Africa	40	Sudan	328	Turkey	2002
Uzbekistan	39	Oman	259	India	1739
Australia	30	Libya	181	France	1705
Argentina	17	China	125	Russian Fed.	1617
Syria	12	Tunisia	125	Chile	1350
$\Sigma\ (\%)^c$	94	$\Sigma\ (\%)^c$	77	Argentina	1272
				Germany	948
				$\Sigma\ (\%)^c$	76

Country	Pears	Country	Peaches + nectarines	Country	Plums + sloes
China	11,988	China	7510	China	4535
Italy	907	Italy	1665	Romania	599
USA	758	Spain	1256	Serbia, Rep.	556
Spain	590	USA	916	USA	412
Argentina	510	Greece	864	Chile	255
Korea, Rep.	431	Turkey	553	France	230
Japan	319	Iran	456	Turkey	214
Turkey	318	France	401	Italy	180
South Africa	316	Egypt	360	Spain	160
Netherlands	222	Chile	315	Iran	146
$\Sigma\ (\%)^c$	84	$\Sigma\ (\%)^c$	83	$\Sigma\ (\%)^c$	83

Country	Oranges	Country	Mandarins	Country	Lemons and limes
Brazil	18,059	China	13,240	Mexico	1866
USA	9000	Spain	2000	India	1618
Mexico	3980	Brazil	1233	Argentina	1393
India	3469	Japan	842	Brazil	1031
Spain	3211	Turkey	791	USA	942
China	2765	Iran	702	Spain	868
Italy	2356	Thailand	670	China	783
Iran	2253	Egypt	665	Turkey	710
Indonesia	2214	Argentina	660	Iran	615
Egypt	1789	Pakistan	639	Italy	583
$\Sigma\ (\%)^c$	76	$\Sigma\ (\%)^c$	84	$\Sigma\ (\%)^c$	80

numerals given in brackets above correspond to Tables 17.5 and 17.6.

18.1.2.1.3 Amines

A number of aliphatic and aromatic amines are found in various fruits and vegetables (Tables 18.6 and 18.7 and 18.1.4.2.1). They are formed in part by amino acid decarboxylation such as in apples, or by amination (cf. Reaction 18.1) or transamination of aldehydes (cf. Reaction 18.2).

$$R\text{---}CHO \xrightarrow[2\,[H]]{NH_3} R\text{---}CH_2\text{---}NH_2 \tag{18.1}$$

Table 18.2. (Continued)

Country	Grapefruit	Country	Apricots	Country	Avocado
USA	1118	Turkey	460	Mexico	1137
China	505	Iran	276	USA	247
South Africa	415	Uzbekistan	236	Indonesia	228
Mexico	380	Italy	222	Colombia	186
Syria	282	Pakistan	190	Brazil	169
Israel	266	France	180	Chile	163
Argentina	191	Algeria	167	Dominican Rep.	114
Turkey	180	Spain	141	Peru	103
Cuba	170	Morocco	129	China	90
India	154	Japan	120	Ethiopia	83
$\Sigma\,(\%)^c$	80	Syria	101	$\Sigma\,(\%)^c$	76
		China	85		
		South Africa	84		
		Greece	73		
		Egypt	73		
		$\Sigma\,(\%)^c$	83		

Country	Guavas, mangoes	Country	Pineapples	Country	Bananas
India	11,140	Thailand	2705	India	11,710
China	3550	Brazil	2487	Brazil	7088
Pakistan	2243	Philippines	1834	China	7053
Mexico	2050	China	1400	Philippines	6795
Thailand	1800	India	1229	Ecuador	6118
Indonesia	1413	Costa Rica	1200	Indonesia	5178
Brazil	1348	Indonesia	925	Costa Rica	2353
Philippines	937	Nigeria	895	Mexico	2197
Nigeria	732	Mexico	628	Thailand	1865
Egypt	380	Kenya	600	Colombia	1765
$\Sigma\,(\%)^c$	84	$\Sigma\,(\%)^c$	76	Burundi	1539
				$\Sigma\,(\%)^c$	76

Country	Papaya	Country	Strawberries
Brazil	1574	USA	1259
Mexico	806	Spain	334
India	783	Russian Fed.	235
Nigeria	759	Turkey	211
Indonesia	549	Korea, Rep.	205
Ethiopia	259	Poland	194
Congo	218	Mexico	192
Peru	171	Japan	191
Philippines	157	Germany	173
Venezuela	151	Italy	131
$\Sigma\,(\%)^c$	82	$\Sigma\,(\%)^c$	77

$$R-CHO + R'-\underset{\underset{NH_2}{|}}{C}H-COOH$$

$$\longrightarrow R-CH_2-NH_2 + R'-\underset{\underset{O}{\|}}{C}-COOH \qquad (18.2)$$

Some amines are derived from tyramine (e. g., hordenine, synephrine, octopamine, dopamine and noradrenaline; cf. Formula 18.3) and others from tryptophan (serotonin, tryptamine, melatonin; Formula 18.4). The occurrence of

Table 18.2. (Continued)

Country	Raspberries and other berries	Country	Currants	Country	Almonds
Iran	212	Russian Fed.	435	USA	716
Russian Fed.	184	Poland	195	Spain	220
Vietnam	106	UK	21	Syria	120
USA	99	Austria	19	Italy	113
Serbia, Rep.	80	Ukraine	16	Iran	109
Turkey	55	Czech Rep.	12	Morocco	83
Poland	53	Germany	11	Algeria	54
China	40	Denmark	10	Tunisia	50
Ukraine	19	New Zealand	8	Greece	47
Armenia	12	Lithuania	6	Turkey	43
UK	10	Σ (%)^c	96	Σ (%)^c	88
Σ (%)^c	75				

Country	Pistachios	Country	Hazelnuts	Country	Chestnuts
Iran	230	Turkey	661	China	850
USA	122	Italy	142	Korea, Rep.	76
Turkey	110	USA	41	Turkey	54
Syria	60	Spain	25	Italy	52
China	36	Azerbaijan, Rep.	25	Bolivia	41
Greece	9	Iran	18	Portugal	29
Italy	3	Georgia	17	Japan	23
Tunisia	1	China	14	Greece	21
Pakistan	1	France	6	France	10
Σ (%)^c	100	Poland	3	Spain	10
		Σ (%)^c	99	Σ (%)^c	99

Country	Cashewnuts	Country	Walnuts
Vietnam	942	China	499
Nigeria	636	USA	308
India	573	Iran	150
Brazil	236	Turkey	130
Indonesia	122	Mexico	80
Philippines	113	Ukraine	60
Côte d'Ivoire	94	Romania	38
Tanzania	90	France	36
Guinea-Bissau	85	India	36
Mozambique	68	Egypt	32
Σ (%)^c	95	Σ (%)^c	80

[a] Without melons and nuts.
[b] Inclusive tangerines, clementines and satsumas.
[c] World production = 100%.

Table 18.3. Average chemical composition of fruits (as % of fresh edible portion)

Fruit	Dry matter	Total sugar	Titratable acidity[a]	Dietary fiber	Pectin[b]	Ash	pH
Apple	16.0	11.1	0.6	2.1	0.6	0.3	3.3
Pear	17.5	12.4	0.2	3.1	0.5	0.4	3.9
Apricot	14.7	8.5	1.4	1.6	1.0	0.6	3.7
Sour cherry	14.7	9.9	1.8	1.0	0.3	0.5	3.4
Sweet cherry	17.2	13.3	1.0	1.3	0.3	0.6	4.0
Peach	12.9	8.5	0.6	1.9	0.5	0.5	3.7
Plum/prune	16.3	10.2	1.5 (Ä)[c]	1.6	0.9	0.5	3.3
Blackberry	15.3	6.2	1.7	3.2	0.5	0.5	3.4
Strawberry	10.2	5.7	1.1	1.6	0.5	0.5	
Currant, red	15.3	4.8	2.3 (C)[c]	3.5	0.9	0.6	3.0
Currant, black	18.7	6.3	2.6 (C)[c]	6.8	1.7	0.8	3.3
Raspberry	15.5	4.5	2.1 (C)[c]	4.7	0.4	0.5	3.4
Grapes	18.9	15.2	0.9 (W)[c]	1.5	0.3	0.5	3.3
Orange	14.3	8.3	1.1	1.6		0.5	3.3
Grapefruit	11.4	7.4	1.5	1.6		0.4	3.3
Lemon	9.8	3.2	4.9			0.5	2.5
Pineapple	15.4	12.3	0.7	1.0		0.4	3.4
Banana	26.4	20.0	0.6 (Ä)[c]	1.8	0.9	0.8	4.7
Cherimoya	25.9	13	0.2			0.9	
Date	80	65.1	1.3	8.7		1.8	
Fig	20	13	0.4	2.0	0.6	0.7	
Guava	17	5.8	0.9	5.2	0.7	0.7	
Mango	19	12.5	0.3	1.7	0.5	0.5	
Papaya	11	7.1	0.1	1.7	0.6	0.6	

[a] Sum: citric acid + malic acid + tartaric acid.
[b] Results are expressed as calcium pectate.
[c] Calculated as malic (M), citric (C), or tartaric acid (T).

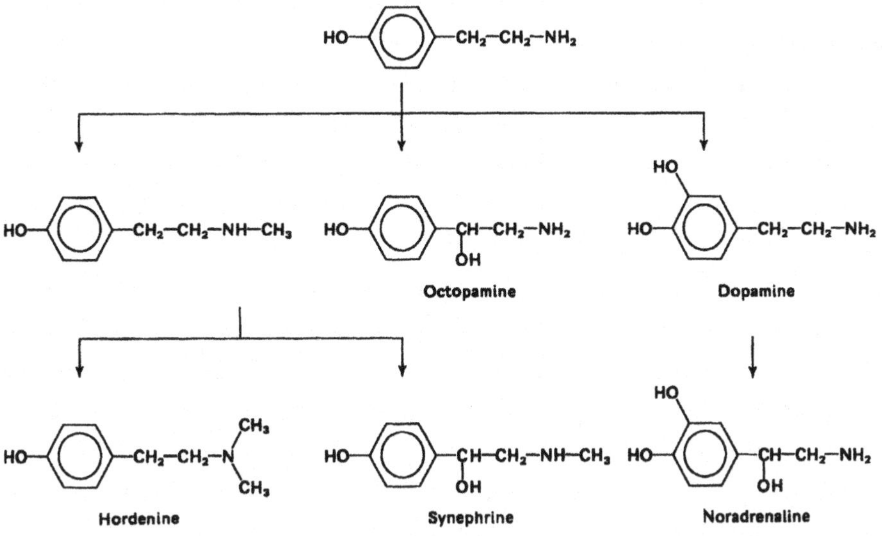

(18.3)

Table 18.4. Proximate composition of shell-nut fruit (as % of fresh edible portion)

Fruit	Mois-ture	N-Com-pounds (N × 5.3)	Lipids	Available carbo-hydrates	Ash	Dietary fiber
Cashewnut	4.0	16	42.2	30.5	2.9	2.9
Peanut	5.0	25.3	47.5	7.5	2.2	11.7
Hazelnut	5.2	12.0	66.0	10.5	2.5	8.2
Pistachio	5.9	17.6	53.5	11.6	2.7	10.6
Almond	5.7	20.5	56.0	5.4	2.7	13.5
Walnut	4.4	15.0	64.4	10.6	2.0	6.1

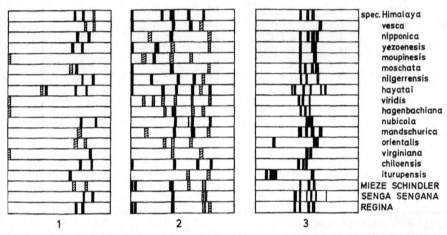

spec. Himalaya
vesca
nipponica
yezoenesis
moupinesis
moschata
nilgerrensis
hayatai
viridis
hagenbachiana
nubicola
mandschurica
orientalis
virginiana
chiloensis
iturupensis
MIEZE SCHINDLER
SENGA SENGANA
REGINA

1   2   3

Fig. 18.2. Enzyme patterns of some strawberry species (*Fragaria sp.*) and (*Fragaria ananas*) obtained by PAGE disc gel electrophoresis. Large pore concentrating gel pH 6.7, small pore separating gel, pH 8.9. *1* Peroxidase: incubation with o-toluidine/H_2O_2 at pH 7. *2* Esterase: incubation with α-naphthylacetate at pH 7, the released α-naphthol is diazotized and then coupled with p-chloroaniline. *3* Malate-dehydrogenase: incubation with malate, nitro-blue-tetrazolium chloride and NAD at pH 7.5. (according to *Drawert* et al., 1974)

Table 18.5. Free amino acids in fruits (as % of total free amino acids)

Fruit	Asp	Asn	Glu	Gln	Ser	Thr	Pro	Ala	Abu[b]	His	Arg	Pip[c]
Apple (juice)	21	17	15		10	3	2	7	5			
Pear (juice)	10	9	10		11	2	14	9	3			
Grapes	3		13			6	31	9	6		27	
Currant black		7	17	24	5		8	17	12			
Orange[a]	7–115	20–188	6–93	3–63	4–37		8–79	3–26	4–73		0–23	
Grapefruit[a]	34–99		8–90		310	10		4–27		76	20–45	
Lemon[a]	19–60		6–35		12–28			1–31	4–20		25–106	
Banana	5–10	15		10–15				5–10	10–15			5–10

[a] Values in mg/100 ml juice.
[b] γ-Aminobutyric acid.
[c] Pipecolinic acid.

Table 18.6. Amines in fruit

Fruit	Amines
Apple	Methylamine, ethylamine, propylamine, butylamine, hexylamine, octylamine, dimethylamine, spermine, spermidine
Plum/prune	Dopamine
Orange	Feruloylputrescine, methyltyramine, synephrine
Grapefruit	Feruloylputrescine
Lemon	Tyramine, synephrine, octopamine
Pineapple	Tyramine, serotonin
Avocado	Tyramine, dopamine
Banana	Methylamine, ethylamine, isobutylamine, isoamylamine, dimethylamine, putrescine, spermidine, ethanolamine, propanolamine, histamine, 2-phenyl-ethylamine, tyramine, dopamine, noradrenaline, serotonin

these biologically active amines in fruits and vegetables (Table 18.7) could influence their concentrations in human serum.

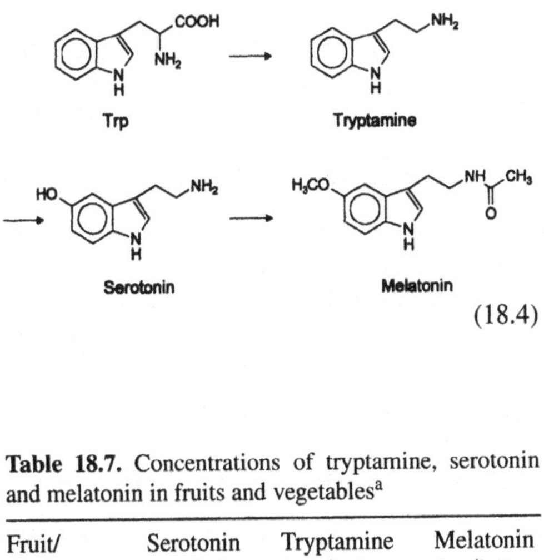

(18.4)

Table 18.7. Concentrations of tryptamine, serotonin and melatonin in fruits and vegetables[a]

Fruit/ Vegetable	Serotonin (mg/kg)	Tryptamine (mg/kg)	Melatonin (ng/kg)
Banana	11.7	0.03	466
Kiwi	6.2	4.2	
Pineapple	29.0	1.4	
Cherry			2–15
Walnut	278.9		
Cucumber			86
Tomato		2.9	112–506

[a] Reference: fresh weight.

18.1.2.2 Carbohydrates

18.1.2.2.1 Monosaccharides

In addition to glucose and fructose, the ratios of which vary greatly in various fruits (Table 18.8), other monosaccharides occur only in trace amounts. For example, arabinose and xylose have been found in several fruits. An exceptional case is avocado in which a number of higher sugars are present at 0.2 to 5.0% of the fresh weight (D-manno-heptulose, D-talo-heptulose, D-glycero-D-galacto-heptose, D-glycero-D-manno-octulose, D-glycero-L-galacto-octulose, D-erythro-L-gluco-nonulose and D-erythro-L-galacto-nonulose). Small amounts of heptuloses have been found in the fruit flesh of apples, peaches and strawberries, and in the peels of grapefruit, peaches and grapes.

18.1.2.2.2 Oligosaccharides

Saccharose (sucrose) is the dominant oligosaccharide. Other disaccharides do not have quantitative importance. Maltose occurs in small amounts in grapes, bananas and guava. Melibiose, raffinose and stachyose have also been detected in grapes. 6-Kestose has been identified in ripe bananas.

Other oligosaccharides occur only in trace amounts. The proportion of reducing sugars to saccharose can vary greatly (Table 18.8). Some fruits have no saccharose (e. g., cherries, grapes and figs), while in some the saccharose content is significantly higher than the reducing sugar content (e. g., apricots, peaches and pineapples).

18.1.2.2.3 Sugar Alcohols

D-Sorbitol is abundant in *Rosaceae* fruits (pomme fruits, stone fruits). For example, its concentration is 300–800 mg/100 ml in apple juice. Since fruits such as berries, citrus fruits, pineapples or bananas do not contain sorbitol, its detection is of analytical importance in the evaluation of wine and other fruit products. Meso-inositol also occurs in fruits; in orange juice it ranges from 130–170 mg/100 ml.

Table 18.8. Sugar content in various fruits (as % of the edible portion)

Fruit	Glucose	Fructose	Saccharose
Apple	1.8	5.7	2.4
Pear	1.8	6.7	1.8
Apricot	1.9	0.9	5.1
Cherry	6.9	6.1	0.2
Peach	1.0	1.2	5.7
Plum/prune	3.5	2.0	3.4
Blackberry	3.2	2.9	0.2
Strawberry	2.2	2.3	1.3
Currant, red	2.0	2.5	0.3
Currant, black	2.4	3.1	0.7
Raspberry	1.8	2.1	1.0
Grapes	7.2	7.4	0.4
Orange	2.4	2.4	3.4
Grapefruit	2.0	2.1	2.9
Lemon	1.4	1.4	0.4
Pineapple	2.3	2.4	7.9
Banana	3.5	3.4	10.3
Date	25.0	24.9	13.8
Fig	5.5	4.0	0.0

18.1.2.2.4 Polysaccharides

All fruits contain cellulose, hemicellulose (pentosans) and pectins. The building blocks of these polysaccharides are glucose, galactose, mannose, arabinose, xylose, rhamnose, fucose and galacturonic and glucuronic acids. The pectin fractions of fruits are particularly affected by ripening. A decrease in insoluble pectin is accompanied by an increase in the soluble pectin fraction. The total pectin content can also decrease. Starch is present primarily in unripe fruits and its content decreases to a negligible level as ripening proceeds. Exceptions are bananas, in which the starch content can be 3% or more even in ripe bananas, and various nuts such as cashew and Brazil nuts.

18.1.2.3 Lipids

The lipid content of fruits is generally low, 0.1–0.5% of the fresh weight. Only fruit seeds and nuts contain significantly higher levels of lipids (cf. Table 18.4). The fruit flesh of avocado is also rich in fat. The lipid fraction of fruits consists of triacylglycerols, glyco- and phospholipids, carotenoids, triterpenoids and waxes.

Table 18.9. Lipids of apple flesh (as % of the total lipids)

Triacylglycerols	5	Sterols	15
Glycolipids	17	Sterol esters	2
Phospholipids	47	Sulfolipids	1
		Others	15

18.1.2.3.1 Fruit Flesh Lipids (Other than Carotenoids and Triterpenoids)

Table 18.9 presents the lipid fractions of apple flesh. Phospholipids, about 50% of the lipid fraction, are predominant. The most abundant fatty acids are palmitic, oleic and linoleic acids (Table 18.10).

18.1.2.3.2 Carotenoids

Carotenoids are widespread in fruits and, in a number of fruits, such as citrus fruits, peaches and sweet melons, their presence is the main factor determining color. The most important carotenoids found in fruits are compiled in Table 18.11, while Table 18.12 gives the carotenoid composition of some fruits.

Fruits can be divided into various classes according to the content and distribution pattern of carotenoids:

- Fruits with low content of carotenoids (occurring mostly in chloroplasts) such as β-carotene, lutein, violaxanthin, neoxanthin (e. g., pineapples, bananas, figs and grapes).

Table 18.10. Fatty acid composition of some fruit flesh lipids (as % of the total fatty acids)

Fatty acid	Avocado	Apple	Banana
12:0	+[a]	0.6	+
14:0	+	0.6	0.6
16:0	15	30	58
16:1	4	0.5	8.3
18:0	+	6.4	2.5
18:1	69	18.5	15
18:2	11	42.5	10.6
18:3	+	1	3.6

[a] Traces.

Table 18.11. Carotenoids occurring in fruit (Roman numerals refer to their structures shown in 3.8.4)

Number	Carotenoid
1	Phytoene (I)
2	Phytofluene (II)
3	ζ-Carotene (III)
4	Lycopene (IV)
5	α-Carotene (VI)
6	β-Carotene (VII)
7	β-Zeacarotene (Va)
8	Lycoxanthin (16-hydroxylycopene)
9	α-Cryptoxanthin (3-hydroxy-α-carotene)
10	β-Cryptoxanthin (3-hydroxy-β-carotene)
11	β-Carotene-5,6-epoxide
12	Mutatochrome (β-carotene-5,8-epoxide)
13	Lutein (IX)
14	Zeaxanthin (VIII)
15	Cryptoflavin (α-cryptoxanthin-5,8-epoxide)
16	β-Carotene-5,6,5',6'-diepoxide
17	Antheraxanthin (zeaxanthin-5,6-epoxide)
18	Lutein-5,6-epoxide
19	Mutatoxanthin (XVI)
20	Lutein-5,8-epoxide
21	Cryptoxanthin-5,8,5',8'-diepoxide
22	Violaxanthin (XIII)
23	Luteoxanthin (XIV)
24	Auroxanthin(zeaxanthin-5,8,5',8'-diepoxide)
25	Neoxanthin (XX)
26	Capsanthin (X)

Table 18.12. Carotenoid patterns of various fruits

Fruit	Carotenoid Content[a]	Compounds[b]
Pineapple		6, 13
Orange	24	1, 2, 3, 4, 6, 10, 11, 12, 15, 17, 20, 21, 22, 23, 24
Banana		6, 13
Pear	0.3–1.3	2, 3, 6, 7, 8, 11, 12, 13, 14, 16, 18, 20, 24, 25
Fig	8.5	1, 2, 5, 6, 13, 14, 22, 23, 25
Guava		5, 6
Peach	27	1, 2, 3, 5, 6, 9, 10, 13, 14, 17, 18, 19, 22, 23, 24
Plum/ prune		1, 2, 3, 5, 6, 9, 10, 12, 14, 15, 17, 18, 19, 20, 22, 23, 25
Grapes	1.8	1, 2, 4, 5, 13, 14, 22, 23
Cantaloupe	20–30	1, 2, 3, 5, 6, 13, 14, 22, 23

[a] mg/kg fresh weight.
[b] The numerals refer to Arabic numerals in Table 18.11.

- Fruits with relatively high contents of lycopene, phytoene, phytofluene, ζ-carotene and neurosporene, e. g., peaches.
- Fruits with relatively high contents of β-carotene, cryptoxanthin and zeaxanthin. This class includes oranges, pears, peaches and sweet melons.
- Fruits with high amounts of epoxides, e. g., oranges and pears.
- Fruits which contain unusual carotenoids, e. g., oranges.

The compositional pattern of carotenoids which can be readily analyzed by HPLC is important for analytical characterization of fruit products.

18.1.2.3.3 Triterpenoids

This fraction contains bitter compounds of special interest, limonoids and cucurbitacins. Limonoids are found in the flesh and seeds of *Rutaceae* fruits. For example, limonin (II) is present in seeds, juice, and fruit flesh of oranges and grapefruit. The limonin content decreases with fruit ripening in oranges but remains constant in grapefruit. Development of a bitter taste in heated orange juice is a processing problem. Limonin monolactone (I), a nonbitter compound which is stable in the neutral pH range, is present in orange albedo and endocarp. During production of orange juice it is transferred to the juice in which, due to the lower pH, it is transformed

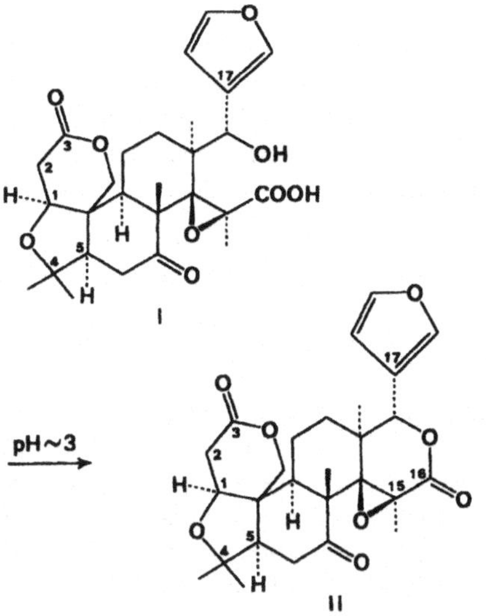

$$pH \sim 3$$

(18.5)

into the bitter tasting dilactone, limonin (II; cf. Formula 18.5).

Bitter and nonbitter forms of the many *Cucurbitaceae* are known. The bitter forms contain cucurbitacins (III) in fruits and seeds. For example, *Citrullus lanatus* (watermelon) contains IIIE in glycosidic form; while *Cucumis sativus* (cucumber) contains IIIC and *Cucurbita* spp. (pumpkin) contains IIIB, D, E and I (cf. Formula 18.6).

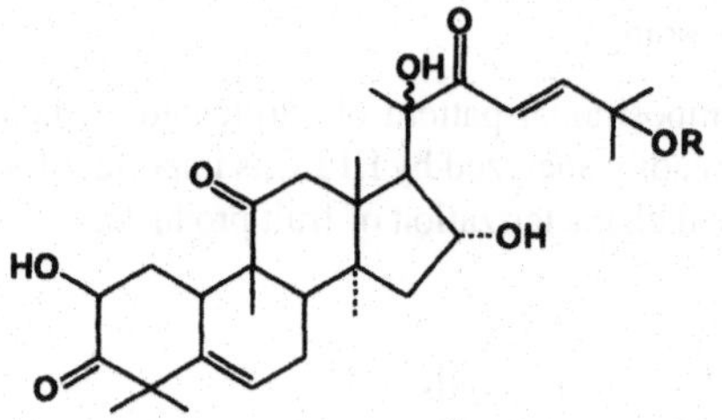

III B: R=Ac
III D: R=H

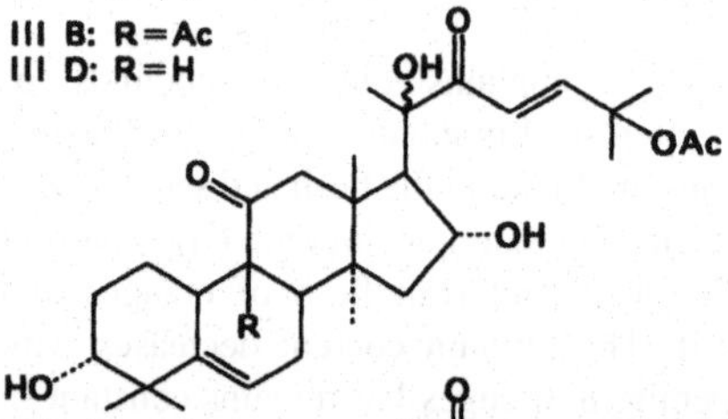

III C: R=CH₂OH

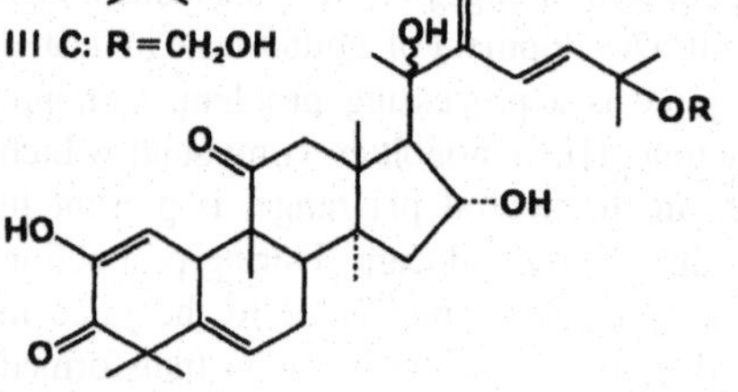

III E: R=Ac
III I: R=H   (18.6)

The common precursor in the biosynthesis of limonoids and cucurbitacins is squalene-2,3-oxide (IV). Based on some identified intermediary compounds, the biosynthetic pathway is probably as postulated in Reaction 18.7.

18.1.2.3.4 Fruit Waxes

The fruit peel is often coated with a waxy layer. In addition to the esters of higher fatty acids with higher alcohols, these waxes contain hydrocarbons, free fatty acids, free alcohols, ketones and aldehydes. The ester fraction in apples and grapes predominantly consists of alcohols of 24, 26 and 28 carbons, but their fatty acid patterns differ. Ap-

ples contain mostly 18:1, 18:2, 16:0 and 18:0 fatty acids, while grapes contain 20:0, 18:0, 22:0 and 24:0 fatty acids.

18.1.2.4 Organic Acids

L-Malic and citric acids are the major organic acids of fruits (Table 18.13). Malic acid is predominant in pomme and stone fruits, while citric acid is most abundant in berries, citrus and tropical fruits. (2R:3R)-Tartaric acid occurs only in grapes. Many other acids, including the acids in the citric acid cycle, occur only in low amounts. Examples are cis-aconitic, succinic, pyruvic, citramalic, fumaric, glyceric, glycolic, glyoxylic, isocitric, lactic, oxalacetic, oxalic and 2-oxoglutaric acids. In fruit juices, the ratio of citric acid to isocitric acid (examples in Table 18.14) serves as an indicator of dilution with an aqueous solution of citric acid.

Table 18.13. Organic acids in various fruits (milliequivalents/100 g fresh weight)

Fruit	Major acid	Other acids
Apple	Malic 3–19	Quinic (in unripe fruits)
Pear	Malic 1–2	Citric
Apricot	Malic 12	Citric 12, quinic 2–3
Cherry	Malic 5–9	Citric, quinic, and shikimic
Peach	Malic 4	Citric 4
Plum/prune	Malic 4–6	Quinic (especially in unripe fruits)
Strawberry	Citric 10–18	Malic 1–3, quinic 0.1, succinic 0.1
Raspberry	Citric 24	Malic 1
Currant, red	Citric 21–28	Malic 2–4, succinic, oxalic
Currant, black	Citric 43	Malic 6
Gooseberry	Citric 11–14	Malic 10–13, shikimic 1–2
Grapes	Tartaric 1.5–2	Malic 1.5–2
Orange	Citric 15	Malic 3, quinic
Lemon	Citric 73	Malic 4, quinic
Pineapple	Citric 6–20	Malic 1.5–7
Banana	Malic 4	
Fig	Citric 6	Malic, acetic
Guava	Citric 10–20	Malic

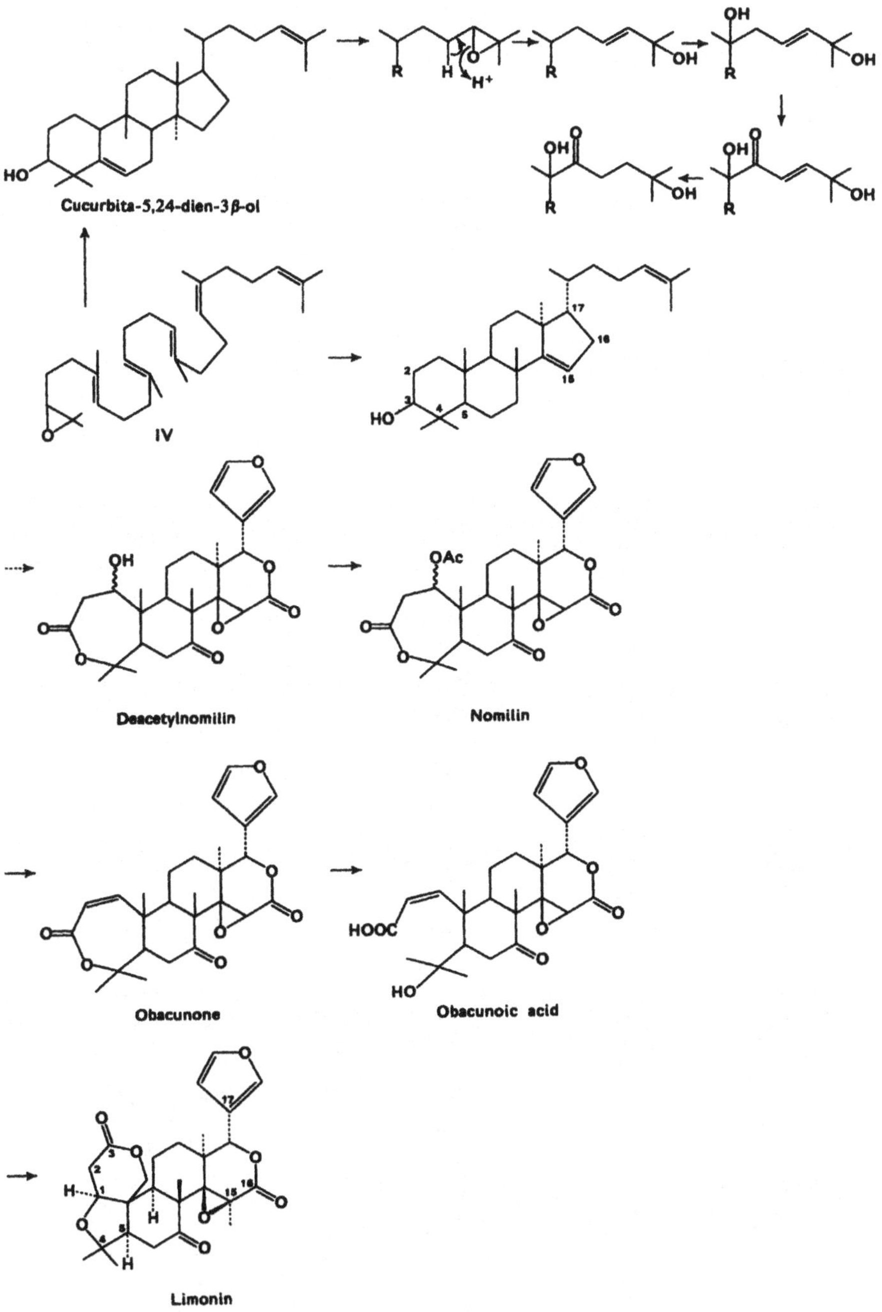

Cucurbita-5,24-dien-3β-ol

IV

Deacetylnomilin

Nomilin

Obacunone

Obacunoic acid

Limonin

(18.7)

Important phenolic acids, dealt with in Section 18.1.2.5.1, are quinic, caffeic, chlorogenic and shikimic acids. Galacturonic and glucuronic acids are also found.

Tartaric acid biosynthesis in *Vitis* spp. starts from glucose or fructose and probably proceeds through 5–oxogluconic acid or ascorbic acid respectively:

$$
\begin{array}{c}
\text{CHO} \\
\text{H}-\text{C}-\text{OH} \\
\text{HO}-\text{C}-\text{H} \\
\text{H}-\text{C}-\text{OH} \\
\text{H}-\text{C}-\text{OH} \\
\text{CH}_2\text{OH}
\end{array}
\;\longrightarrow\longrightarrow\;
\begin{array}{c}
\text{COOH} \\
\text{H}-\text{C}-\text{OH} \\
\text{HO}-\text{C}-\text{H} \\
\text{H}-\text{C}-\text{OH} \\
\text{C}=\text{O} \\
\text{CH}_2\text{OH}
\end{array}
\;\longrightarrow
$$

$$
\begin{array}{c}
\text{COOH} \\
\text{H}-\text{C}-\text{OH} \\
\text{HO}-\text{C}-\text{H} \\
\text{CHO}
\end{array}
\;\longrightarrow\;
\begin{array}{c}
\text{COOH} \\
\text{H}-\text{C}-\text{OH} \\
\text{HO}-\text{C}-\text{H} \\
\text{COOH}
\end{array}
$$

$$+$$

$$(2\,\text{R},\,3\,\text{R})\text{-Tartaric acid}$$

$$
\begin{array}{c}
\text{CHO} \\
\text{CH}_2\text{OH}
\end{array}
\tag{18.8}
$$

Human and animal metabolism oxidatively degrade (2R:3R)-tartaric and meso-tartaric acids into glyoxylic and hydroxypyruvic acids respectively:

$$
\begin{array}{c}
\text{COOH} \\
\text{CHOH} \\
\text{CHOH} \\
\text{COOH}
\end{array}
\;\longrightarrow\;
\begin{array}{c}
\text{COOH} \\
\text{C}-\text{OH} \\
\text{C}-\text{OH} \\
\text{COOH}
\end{array}
\;\rightleftharpoons\;
\begin{array}{c}
\text{COOH} \\
\text{C}=\text{O} \\
\text{CHOH} \\
\text{COOH}
\end{array}
$$

$$
\begin{array}{c}
\text{COOH} \\
\text{CO} \\
\text{CO} \\
\text{COOH}
\end{array}
$$

$$
\begin{array}{c}
\text{COOH} \\
\text{CO} \\
\text{CH}_2\text{OH}
\end{array}
\;+\;\text{CO}_2
$$

$$
2\;\begin{array}{c}
\text{COOH} \\
\text{CHO}
\end{array}
\tag{18.9}
$$

$$
\begin{array}{l}
\text{I:} \quad \text{R}^1,\ \text{R}^2=\text{H} \\
\text{II:} \quad \text{R}^1=\text{H},\ \text{R}^2=\text{OCH}_3 \\
\text{III:} \quad \text{R}^1=\text{H},\ \text{R}^2=\text{OH} \\
\text{IV:} \quad \text{R}^1=\text{R}^2=\text{OCH}_3
\end{array}
\tag{18.10}
$$

$$
\begin{array}{l}
\text{V:} \quad \text{R}^3=\text{Caf, R}^1,\ \text{R}^4,\ \text{R}^5=\text{H} \\
\text{VI:} \quad \text{R}^4=\text{Caf, R}^1,\ \text{R}^3,\ \text{R}^5=\text{H} \\
\text{VII:} \quad \text{R}^5=\text{Caf, R}^1,\ \text{R}^3,\ \text{R}^4=\text{H}
\end{array}
\tag{18.11}
$$

Table 18.14. Ratio of citric acid (C) to isocitric acid (I) in fruit juices

Fruit juice	C/I
Orange	80–130
Currant	60–140
Grapefruit	50–90
Raspberry	80–200
Lemon	≤ 250
Blackberry	ca. 0.25

18.1.2.5 Phenolic Compounds

In plant foods, several hundred polyphenols have been identified, which are classified according to the number of phenol rings and their linkage, as presented in Fig. 18.3. The flavonoids exhibit a great multiplicity. According to Fig. 18.4, they are divided into six subclasses. The polyphenols are mostly present as glycosides, and partly also as esters. They are antioxidatively active, their activity depending on the number and position of the OH-groups and on the pH (cf. 3.7.3.2.1). As antioxidants, they are of interest from

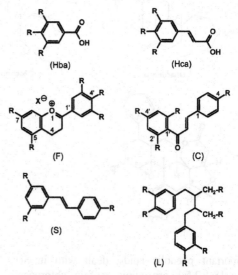

Fig. 18.3. Chemical structures of polyphenols Hydroxybenzoic acids (Hba), Hydroxycinammic acids (Hca), Flavonoids (F), Chalcones (C), Stilbenes (S, cf. 20.2.6.6), Lignans (L). R: H, OH or OCH$_3$

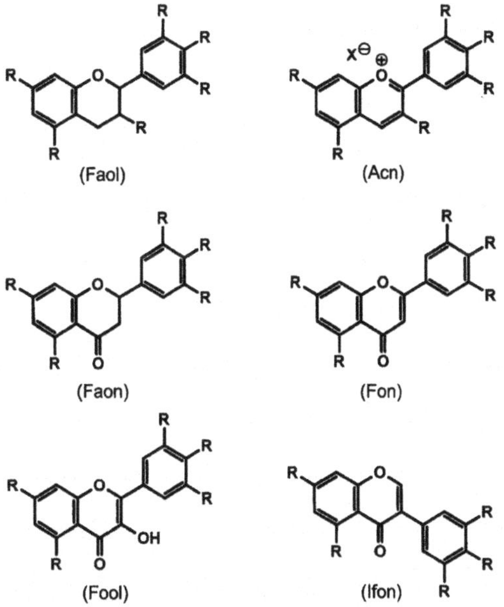

Fig. 18.4. Chemical structures of flavonoids
Flavanols (Faol), Anthocyanidines (Acn), Flavanones
(Faon), Flavones (Fon), Flavonols (Fool), Isoflavones
(Ifon, cf. 16.2.9). R: H, OH or OCH$_3$

a nutritional and physiological viewpoint. Up
to now, however, unambiguous evidence of
a health-building effect is lacking (literature
till 2006). The phenolic compounds contribute
to the color and taste of many types of fruit. In
processing, they can cause discoloration by the
formation of metal complexes and turbidity by
complexation of proteins. Table 18.15 shows the
polyphenol content of foods, which is strongly
dependent on the variety, climate and degree of
ripeness, among other factors. For instance, their
content in apples varies from 1.3 g/kg (*Golden
Delicious*) to 6 g/kg (*Jeanne Renard*) and can
even rise to 10 g/kg in this variety.

18.1.2.5.1 Hydroxycinnammic Acids, Hydroxycoumarins and Hydroxybenzoic Acids

p-Coumaric (I), ferulic (II), caffeic (III) and
sinapic (IV) acids are widespread in fruits and
vegetables.
These hydroxycinnamic acids are present
mainly as derivatives. The most common are

Table 18.15. Polyphenol content of foods

Compound[a]	Food	Content[b]
Hydroxybenzoic	Blackberries	80–270
acids (Hba)	Raspberries	60–100
	Black currants	40–130
	Strawberries	20–90
Hydroxycinnammic	Bilberries	2000–2200
acids (Hca)	Kiwi	600–1000
	Cherry	180–1150
	Plum	140–1140
	Apple	50–600
	Pear	15–600
	Aubergine	600–660
	Artichoke	450
	Chicory	200–500
	Potato	100–190
Flavanols[c]	Apricot	100–250
(monomer) (Faol)	Cherry	50–250
	Grape	30–175
	Peach	50–140
	Blackberry	130
	Apple	20–120
Anthocyans	Blackberry	1000–4000
(Acn-glycoside)	Black currant	1300–4000
	Bilberry	250–5000
	Red grape	300–7500
	Cherry	350–4500
	Rhubarb	2000
	Strawberry	150–750
	Plum	20–250
	Red cabbage	250
Flavanones[c]	Orange (juice)	215–685
(Faon)	Grapefruit (juice)	100–650
	Lemon (juice)	50–300
Flavones[c]	Parsley	240–1850
(Fon)	Celery	20–140
Flavonols[c]	Bilberry	30–160
(Fool)	Black currant	30–70
	Apricot	25–50
	Apple	20–40
	Red grape	15–40
	Kale	300–600
	Leek	30–225
	Broccoli	40–100
	Bean, green	10–50
	Tomato	2–15

[a] The numbers and abbreviations refer to the formulas
in Fig. 18.5 and 18.6.
[b] Values in mg/kg fresh weight or mg/l juice.
[c] Concentration of the aglycone.

Table 18.16. Derivatives of hydroxycinnamic acid in pomme and stone fruits[a]

Compound	Apple	Pear	Sweet cherry	Sour cherry	Plum	Peach	Apricot
5-Caffeoylquinic acid	62–385[b]	64–280	11–40	50–140	15–142	43–282	37–123
4-Caffeoylquinic acid	2	–	+	+	9	–	–
3-Caffeoylquinic acid	–	–	73–628	82–536	88–771	33–142	26–132
3-p-Coumaroylquinic acid	–	–	81–450	40–226	4–40	2	2–9
3-Feruloylquinic acid	–	–	4	1	13	1	7
p-Coumaroylglucose	4	+	–	–	15	–	–
Feruloylglucose	3	–	–	–	5	–	–

[a] mg/kg of fresh weight,
[b] variety "Boskop": 400–500 mg/kg.

Table 18.17. Derivatives of hydroxycinnamic acid in berries[a]

Compound	Straw-berry	Rasp-berry	Black-berry	Red currant	Black currant	Goose-berry	Cultured blueberry
Caffeoylquinic acid	–	1	45–53	1	45–52	3	1860–2080
p-Coumaroylquinic acid	–	1	2–5	+	14–23	1	2–5
Feruloylquinic acid	–	+	2–4	2	4	1	8
Caffeoylglucose	1	3–7	3–6	2–5	19–30	5–13	+
p-Coumaroylglucose	14–17	6–14	4–11	1	10–14	7	+
Feruloylglucose	1	4–7	2–6	+	11–15	1–6	+
Caffeic acid-4-O-glucoside	–	–	–	2	2	2	3
p-Coumaric acid-O-glucoside	+	5–10	2–5	5–16	4–10	6–8	3–15
Ferulic acid-O-glucoside	–	+	–	–	3	2–7	8–10

[a] mg/kg of fresh weight. The hydroxycinnamoylquinic acids are present mostly as the 3-isomers, but in blueberries as the 5-isomer.

esters of caffeic, coumaric, and ferulic acids with D-quinic and, in addition, with D-glucose (Tables 18.16 and 18.17). Since quinic acid has four OH groups, four bonding possibilities exist (R^1, R^3–R^5 in Formula 18.11), the 3- and 5-isomers being preferred. According to IUPAC nomenclature for cyclitols, the 3-, 4-, and 5-caffeoylquinic acids are identical to neochlorogenic acid (V), cryptochlorogenic acid (VI), and chlorogenic acid (VII). Isochlorogenic acid is a mixture of di-O-caffeoylquinic acids. Apart from quinic acid and glucose, other alcoholic components are shikimic, malic and tartaric acids and meso-inositol. Sinapine, which is found in mustard seed, is the counter ion of the glucosinolate sinalbin and is the choline ester of sinapic acid, which is as bitter as caffeine. Hydroxycinnamic acid amides are also found in plants.

Scopoletin (VIII, Formula 18.12) in esterified form is the only hydroxycoumarin (Table 18.18)

Table 18.18. Hydroxycoumarins in fruit (VIII)[a]

Compound	Substitution pattern		
	R^1	R^2	R^3
Coumarin	H	H	H
Umbelliferone	H	OH	H
Herniarin	H	OCH_3	H
Aesculetin	OH	OH	H
Scopoletin	OCH_3	OH	H
Fraxetin	OCH_3	OH	OH

[a] See Formula 18.12.

which has been found, in small amounts, in plums and apricots.

VIII

(18.12)

Table 18.19. Occurrence of hydroxybenzoic acids in fruit[a]

Type of fruit	4-Hydroxy-benzoic acid	Protoca-techuic acid	Gallic acid
Blackberry	10–16	68–189	8–67
Black currant	0–6	10–52	30–62
Raspberry	15–27	25–37	19–38
Red currant	10–23	3–08	
Strawberry	10–36		11–44
White currant	5–19		3–38

[a] After hydrolysis; values in mg/kg fresh weight.

The hydroxybenzoic acids that are found in various fruits and occur mostly as esters include: salicylic acid (2-hydroxybenzoic acid), 4-hydroxybenzoic acid, gentisic acid (2,4-dihydroxybenzoic acid), protocatechuic acid (3,4-dihydroxybenzoic acid), gallic acid (3,4,5-trihydroxybenzoic acid), vanillic acid (3-methoxy-4-hydroxybenzoic acid) and ellagic acid (IX, Formula 18.13), the dilactone of hexahydroxydiphenic acid (Table 18.19). Table 18.19 shows the most important sources of 4-hydroxybenzoic acid, protocatechuic acid and gallic acid. Strawberries (0.2–0.5), raspberries (1.2) and blackberries (1.9–2.0) contain higher concentrations of free and bound ellagic acid (g/kg).

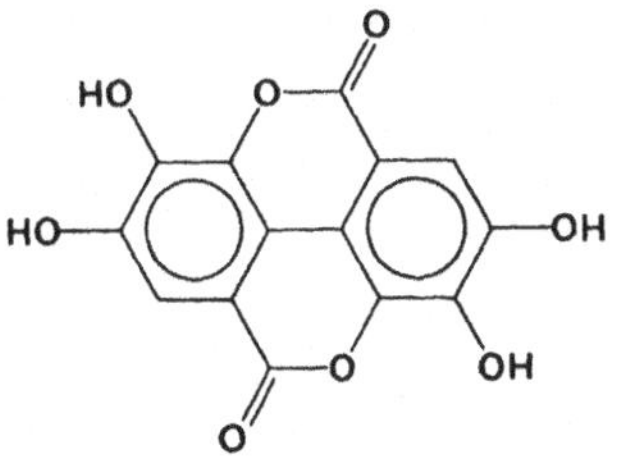

IX   (18.13)

Apart from the proanthocyanidins (cf. 18.1.2.5.2), esters of gallic acid and hexahydroxydiphenic acid form one of the two main classes of plant tanning agents, the "hydrolyzable tanning agents" or tannins. In addition to simple esters with different hydroxy components, such as β-D-glucogallin (X in Formula 18.14), theogallin (XI) and the flavan-3-olgallates XII and XIII, found, e.g., in tea leaves, complex polyesters with D-glucose are known.
They have molecular weights of M_r 500–3000, are generally readily soluble, and contribute their astringent properties to the taste of foods of plant origin.
Apart from gallic acid, most of the tanning agents of this type contain as acyl residues intermolecular gallic acid esters (depsides XIV), their ethers (depsidones, XV), and hexahydroxydiphenic acid (XVI) formed by oxidative coupling of two gallic acids. Some of the polyphenols derived from β-pentagalloyl-D-glucose are shown

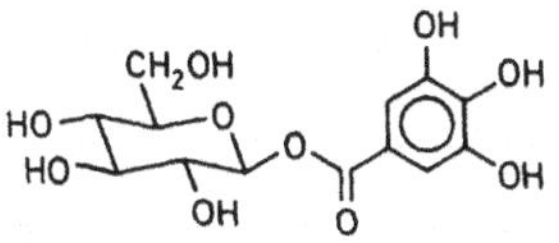

X 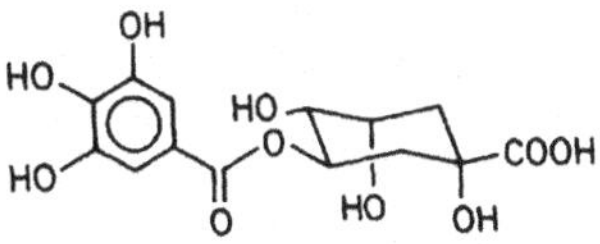   XI 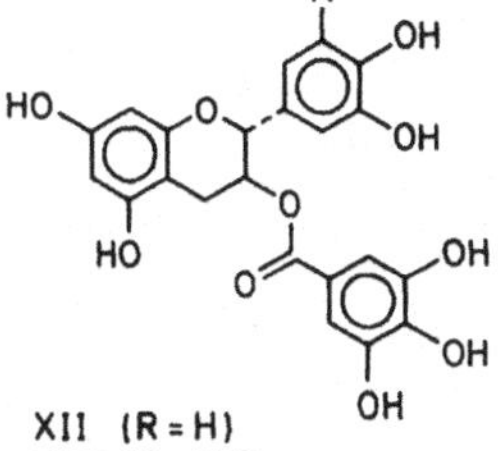   XII (R = H)
                                          XIII (R = OH)

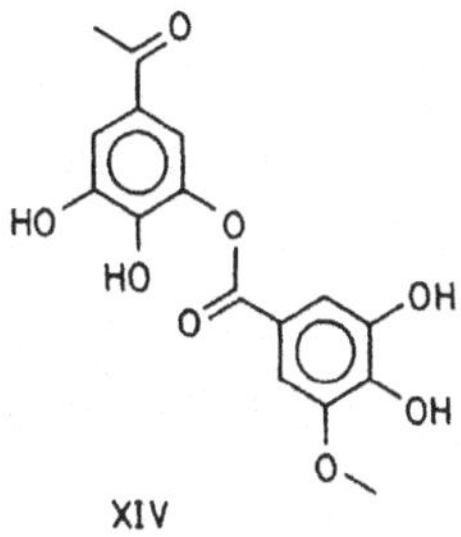

XIV 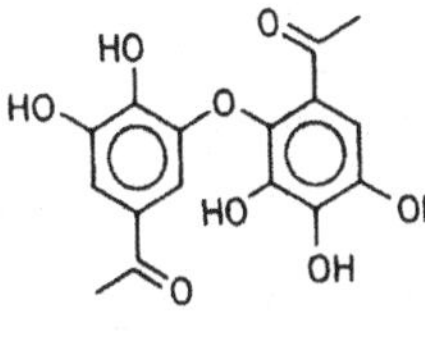   XV 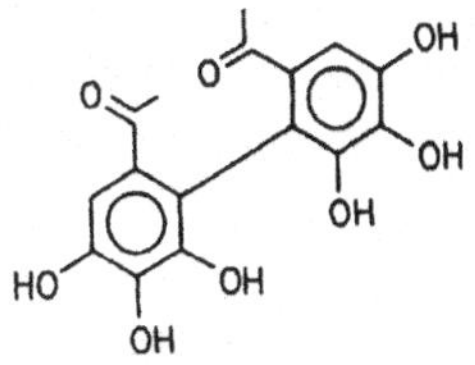   XVI   (18.14)

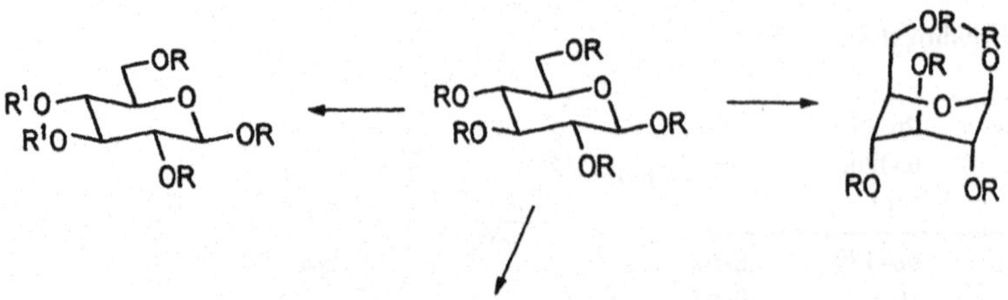

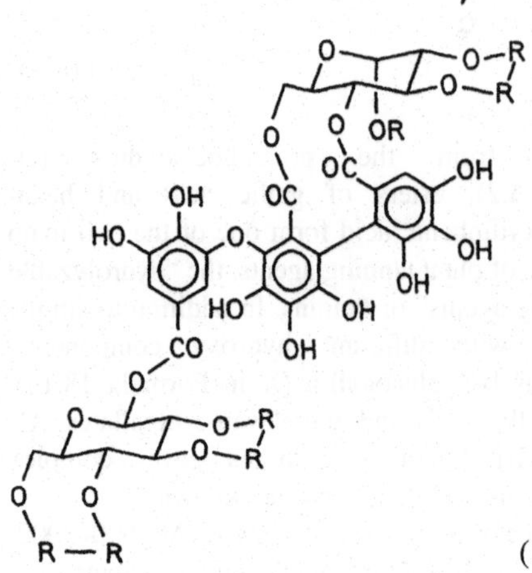

(18.15)

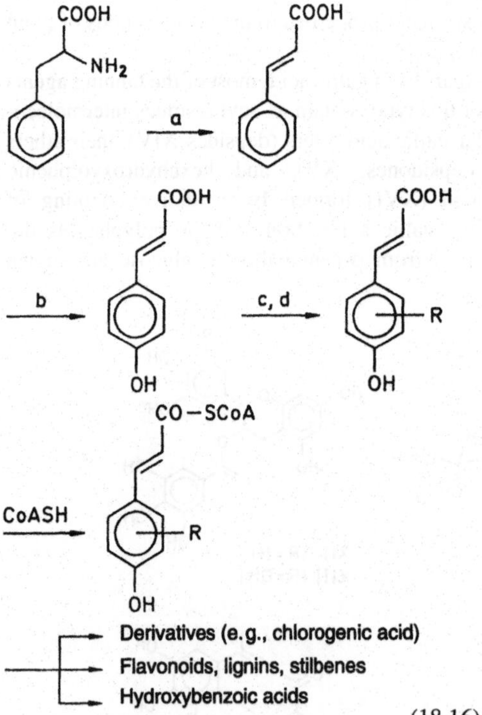

(18.16)

in Formula 18.15. They have been found in various *Rosaceae*, e. g., raspberries and blackberries, and contain the structural elements mentioned above.

The phenolic acids develop differently during the ripening of fruits. In apples, they reach a maximum in June and then decrease, e. g., from mid-June to the beginning of October (mg/kg): p-coumaric acid (460 → 15), caffeic acid (1270 → 85), ferulic acid (95 → 4). Subsequent storage results in further losses. In contrast, the concentrations in strawberries of, e. g., β-coumaric acid (69 → 175), caffeic acid (15→39), gallic acid (80 →121) and vanillic acid (3 → 34) increase from the beginning of June till ripening in the beginning of July (mg/kg). Hydroxycinnamic acid biosynthesis starts with phenylalanine [cf. Reaction route 18.16: a) phenylalanine-ammonia lyase; b) cinnamic acid 4-hydroxylase; c) phenolases; d) methyl transferases, R: OH and OCH_3 in various positions].

The hydroxybenzoic acids are derived from hydroxycinnamic acids by a pathway analogous to β-oxidation of fatty acids:

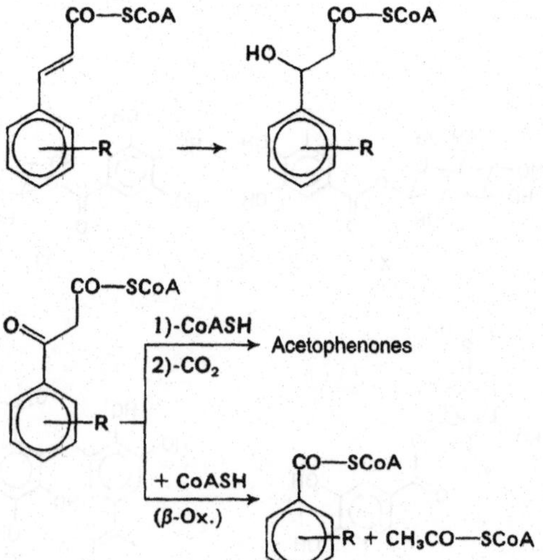

(18.17)

Reduction of the benzoic acid carboxyl groups yields the corresponding aldehydes and alcohols, as for instance vanillin and vanillyl alcohol (XVII and XVIII respectively in Formula 18.18) from 3-methoxy-4-hydroxybenzoic acid and coniferyl alcohol from ferulic acid.

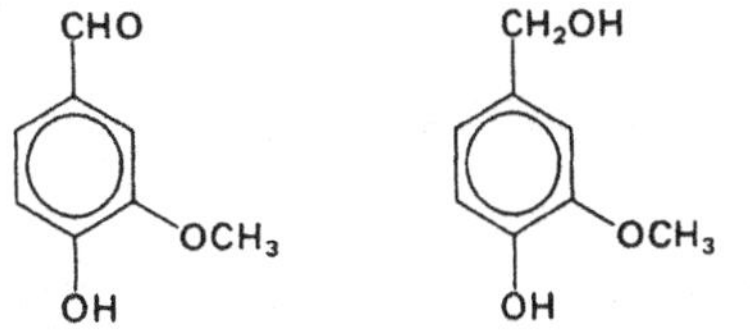

CHO   CH₂OH

$$\text{XVII} \qquad \text{XVIII} \qquad (18.18)$$

The glucosides of cis-o-coumaric acid are the precursors of coumarins. Disintegration of plant tissue releases the free acids from the glucosides. The acids then close spontaneously to the ring forms (R: OH and OCH₃ in various positions):

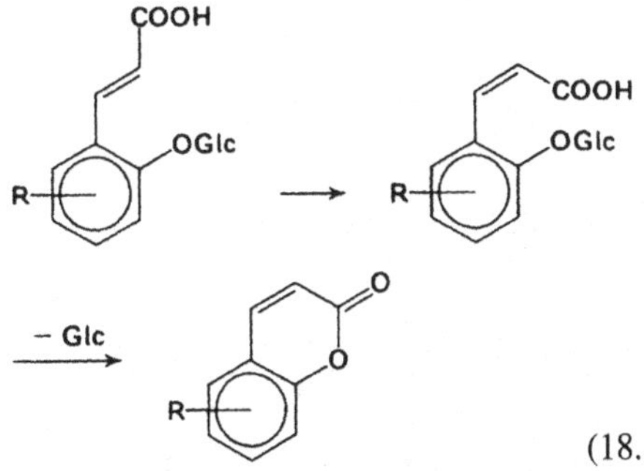

$$(18.19)$$

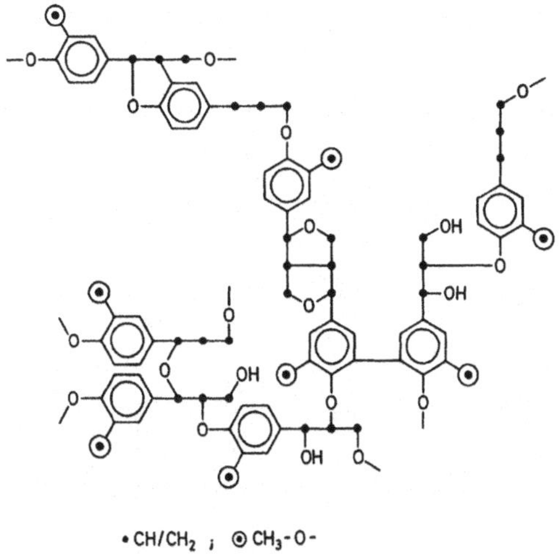

$$\bullet \, CH/CH_2 \; ; \; \odot \, CH_3-O-$$

Fig. 18.5. Section of the structure of a lignin (according to *Kindl* and *Wöber*, 1975)

Lignin is formed by the dehydrogenative polymerization of coniferyl, sinapyl, and p-coumaryl alcohol, which is catalyzed by a peroxidase and requires H_2O_2. A section of the structure of lignin formed by the polymerization of coniferyl alcohol is shown in Fig. 18.5. Lignins strengthen the walls of plant cells. They play a role as fiber in foods (cf. 15.2.4.2).

18.1.2.5.2 Flavan-3-ols (Catechins), Flavan-3,4-diols, and Proanthocyanidins (Condensed Tanning Agents)

These colorless compounds are the following: [R, R¹ = H: a) catechin, b) epicatechin; R = H, R¹ = OH: gallocatechin, epigallocatechin; R = OH: flavan-3,4-diols]:

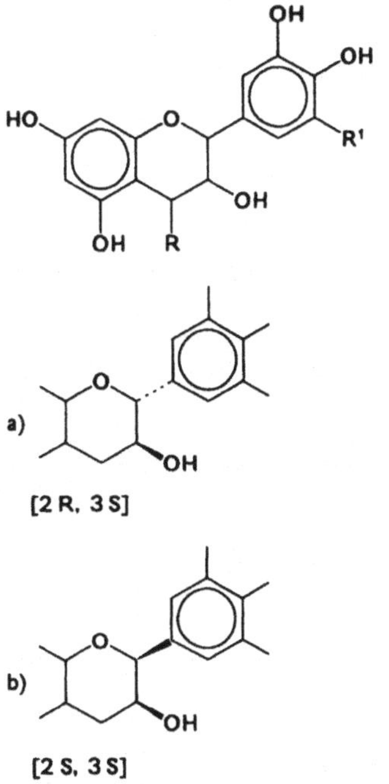

a) [2 R, 3 S]

b) [2 S, 3 S] (18.20)

Table 18.20 shows the occurrence of catechin and epicatechin in fruit.

In flavonoid biosynthesis, flavan-3,4-diols can be converted to flavan-3-ols (cf. 18.1.2.5.7). The intermediate is assumed to be a carbocation which is reduced to flavan-3-ol (Formula 18.21). When the reducing agent, e. g., NADPH, is limited, the cation can react with flavan-3-ol to

(18.21)

Table 18.20. Occurrence of catechin and epicatechin in foods[a]

Food	(+)Catechin	(-)Epicatechin
Apple	4–15	67–103
Apricot	50	61
Blackberry	7	181
Bilberry	n.d.	11
Sweet cherry	22	95
Black currant	7	5
Red currant	12	n.d.
Kiwi	n.d.	5
Mango	17	n.d.
Peach	23	n.d.
Pear	1–2	29–37
Plum	33	28
Strawberry	44	n.d.

[a] Values in mg/kg fresh weight.
n.d.: not detected.

give dimers and higher oligomers, which are called proanthocyanidins. As "condensed tanning agents", they contribute to the astringent taste of fruits. The spectrum of oligomers depends on the ratio of the rate of formation to the rate of reduction of the cation. The proanthocyanidins are soluble up to $M_r \sim 7000$, corresponding to ca. 20 flavanol units. Plant tissues also contain insoluble polymeric forms which frequently even predominate and can be covalently bound to the polysaccharide matrix of the cell. Procyanidins (Formula 18.21, R = H) are the most common group of proanthocyanidins; prodelphinidins (R = OH) also occur.

The name proanthocyanidins, previously called leucoanthocyanidins, implies that these are colorless precursors of anthocyanidins. On heating in acidic solution, the C—C bond made during formation is cleaved and terminal flavan units are released from the oligomers as carbocations, which are then oxidized to colored anthocyanidins (cf. 18.1.2.5.3) by atmospheric oxygen (Formula 18.21). Base-catalyzed cleavage via the quinone method is also possible.

18.1.2.5.3 Anthocyanidins

These red, blue or violet colored benzopyrylium and flavylium salts (Formula 18.22) occur in the form of glycosides, the anthocyanins, in most

commonly grown fruit varieties (Table 18.21) and also in tropical fruits; approximately 70 have been identified up to now.

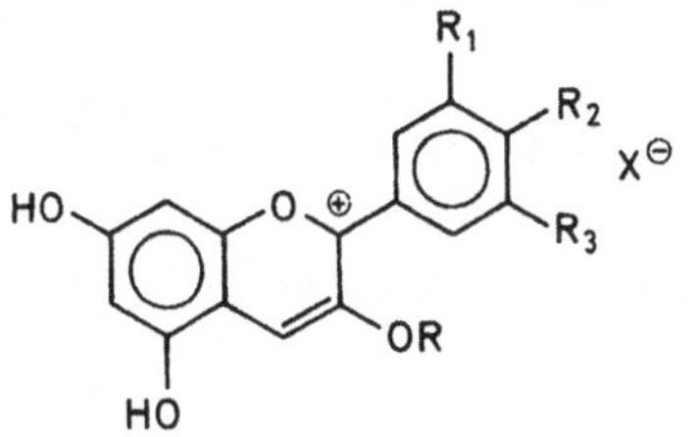

$$(18.22)$$

The cation is to be regarded as a resonance hybrid of the following oxonium and carbenium forms:

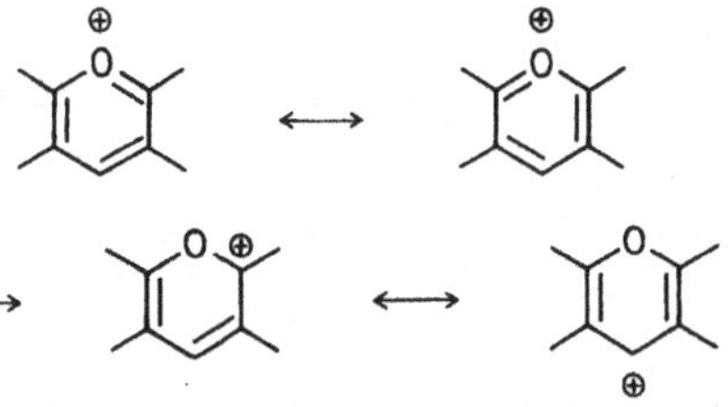

$$(18.23)$$

The sugar residues at the 3 or 5 position are easily cleaved in an acid catalyzed reaction with the formation of the corresponding aglycones (anthocyanidins).

Table 18.22 provides data on the structure and absorption maxima of the most important antho-

Table 18.21. Anthocyan content of fruit[a,b]

Type of fruit	Dp	Cy	Pt	Pg	Pn	Mv
Apple						
Gala	–	2.3	–	–	–	–
Red Delicious	–	12	–	–	0.2	–
Blackberry	–	244	–	0.7	–	–
Bilberry	120	29	72	–	34	131
Cherry, sweet	–	113	–	1.4	7.5	–
Black currant	333	133	7.3	1.9	1	–
Red currant	0.1	13	–	–	–	–
Red grape	1.1	3.9	1.1	–	10	10
Peach	–	4.8	–	–	–	–
Plum	–	19	–	–	–	–
Raspberry	–	90	–	1.9	–	–
Strawberry	–	1.2	–	20	–	–

[a] Values in mg/kg fresh weight.
[b] Dp, delphinidin; Cy, cyanidin; Pt, petunidin; Pg, pelargonidin; Pn, peonidin; Mv, malvidin.
– not detected.

cyanidins. Increasing hydroxylation results in a shift towards blue color (pelargonidin → cyanidin → delphinidin), whereas glycoside formation and methylation results in a shift towards red color (pelargonidin → pelargonidin-3-glucoside; cyanidin → peonidin). The visual detection threshold (VDT) is also influenced by the glycoside residue, as shown by the following examples (VDT in mg/l water): cyanidin-3-sophorosyl-5-glucoside (3.6), cyanidin-3-glucoside (1.3), cyanidin-3-xylosylgalactoside (0.9).

The color of an anthocyanin changes with the pH of the medium (R = sugar moiety; cf. Formula 18.24).

Table 18.22. Anthocyanidins: absorption maxima in the visible spectrum

Compound (Formula 18.22)	R^1	R^2	R^3	λ_{max} (nm)[a] R = H	R = Glc[b]
Pelargonidin	H	OH	H	520	506
Cyanidin	OH	OH	H	535[c]	525[c]
Peonidin	OCH₃	OH	H	532	523
Delphinidin	OH	OH	OH	544[c]	535[c]
Petunidin	OCH₃	OH	OH	543[c]	535[c]
Malvidin	OCH₃	OH	OCH₃	542	535

[a] In methanol with 0.01% HCl.
[b] 3-Glucoside.
[c] AlCl₃ shifts the absorption towards the blue region of the spectrum by 14 to 23 nm.

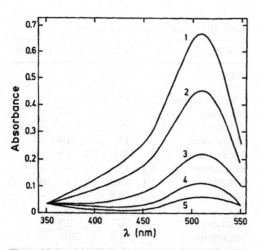

I: pH≤1, red

⇅

II: pH=4−5, colorless

⇅

III: pH=6−7, purple

⇅

IV: pH=7−8, deep blue

⇅

V: pH=7−8, yellow

(18.24)

Fig. 18.6. Absorption spectra of cyanidin-3-rhamnoglucoside (16 mg/l) in aqueous buffered solution at pH 0.71 (1), pH 2.53 (2), pH 3.31 (3), pH 3.70 (4), and pH 4.02 (5). (According to *Jurd*, 1964)

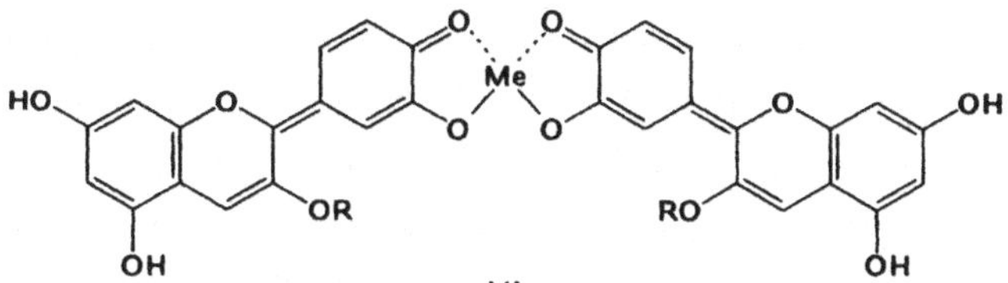

VI   (18.25)

The flavylium cation (I) is stable only at very low pH. As the pH increases it is transformed into colorless chromenol (II). Figure 18.6 shows the decrease in absorption in the visible spectrum at various pH's, reflecting these transformations. Formation of a quinoidal (III) and ionic anhydro base (IV) at pH 6–8 intensifies the color. At pH 7–8 structure IV is transformed through ring opening to yellow chalcone (V). At higher pH's the color can be stabilized by the presence of multivalent metal ions (Me: Al^{3+}, Fe^{3+}). The complexes formed are deep blue (cf. Formula 18.25). Figure 18.7 illustrates the shift in absorption maximum from 510 to 558 nm for cyanidin-3-glucoside over the pH range of 1.9–5.4. Readings were taken in the presence of aluminium chloride. At higher pH's free anthocyanidins (VII, Formula 18.26) are degraded via chromenols (VIII) and α-diketones (IX) to aldehydes (X) and carboxylic acids (XI):

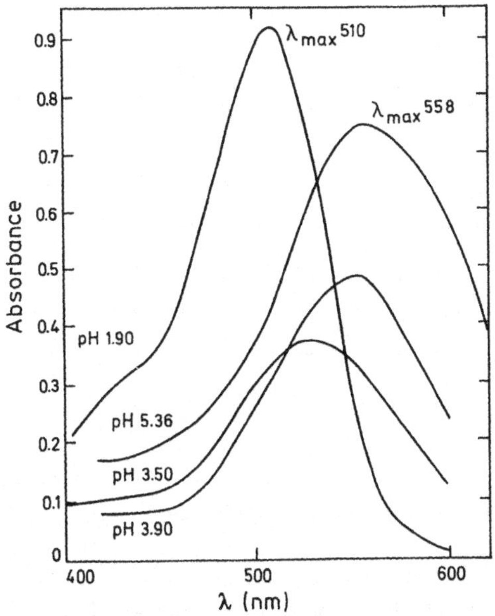

Fig. 18.7. Absorption spectra of cyanidin-3-glucoside (35 µmole/l + 830 µmole/l $AlCl_3$) in aqueous buffered solutions at pH 1.90, pH 3.50, pH 3.90, and pH 5.36. (According to *Jurd* and *Asen*, 1966)

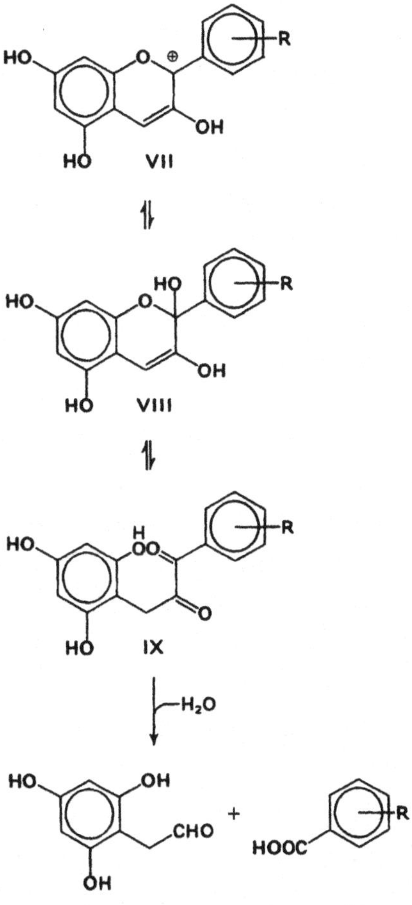

Addition of SO_2 bleaches anthocyanins. The flavylium cation reacts to form a carbinol base corresponding to compounds XII or XIII (Formula 18.27). The color is restored by acidification to pH 1 or by addition of a carbonyl compound (e. g. ethanal). Since compounds of type XIV ($R^1 = CH_3$, C_2H_5) are not affected by SO_2, it appears that compound XIII is involved in such bleaching reactions.

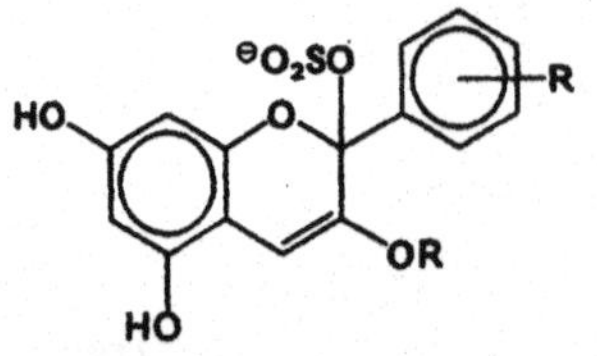

XII

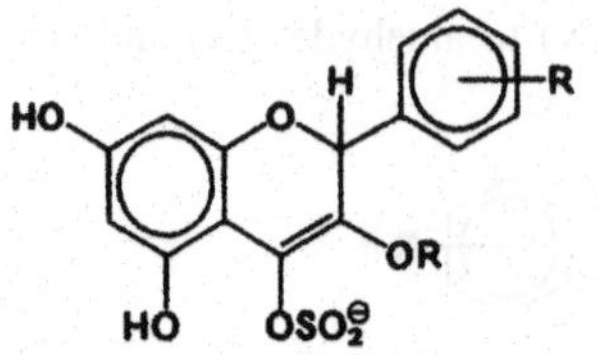

XIII

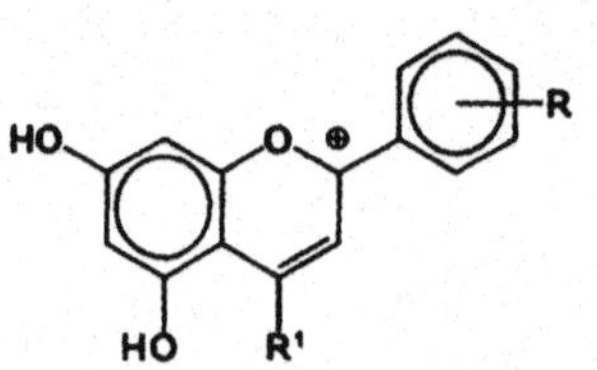

XIV   (18.27)

18.1.2.5.4 Flavanones

Flavanones (Formula 18.28: R^1 = H, R^2 = OCH_3: isosacuranetin; R^1 = H, R^2 = OH: naringenin; R^1 = OH, R^2 = OCH_3: hesperitin; R^1, R^2 = OH: eriodictyol) occur mostly as glycosides in citrus fruits (Table 18.23):

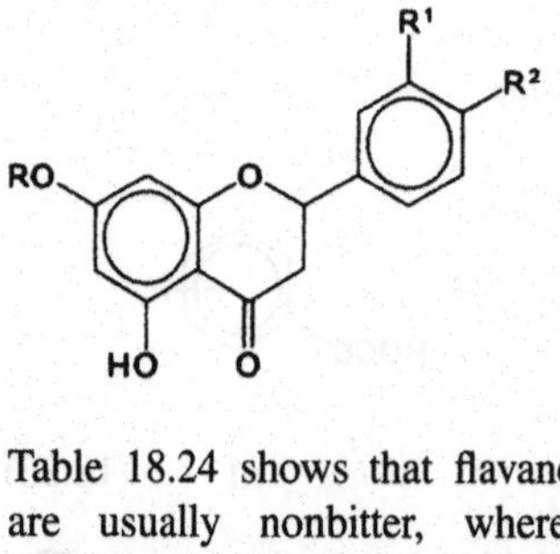

(18.28)

Table 18.24 shows that flavanone-7-rutinosides are usually nonbitter, whereas flavanone-7-neohesperidosides are generally bitter. The intensity of the bitter taste is influenced by the substitution pattern. Compounds with R^1 = H, R^2 = OH, OCH_3 (e. g., naringin, poncirin) are an order of magnitude more bitter than those with R^1 = OH, R^2 = OH, OCH_3 (e. g., neohesperidin, neoeriocitrin). Naringenin-7-neohesperidoside (naringin) is the bitter constituent of grapefruit. Hesperetin-7-neohesperidoside (neohesperidin)

is the bitter compound of bitter oranges (*Citrus auranticus*). The nonbitter isomer, hesperitin-7-rutinoside (hesperidin) occurs in oranges (*Citrus sinensis*) (cf. Table 18.23).

Removal of the bitter taste of citrus juices and citrus fruit pulps is possible by enzymatic cleavage of the sugar moiety using a mixture of α-rhamnosidase and β-glucosidase. These enzymes are isolated from microorganisms such as *Phomopsis citri*, *Cochliobolus miyabeanus* or *Rhizoctonia solanii*:

Naringin → Naringenin + Rhamnose
+ Glucose

A number of neutral or bitter flavanone glycosides can be converted through ring opening to sweet chalcones (II) which, by additional hydrogenation, can be stabilized as sweet dihydrochalcones (III):

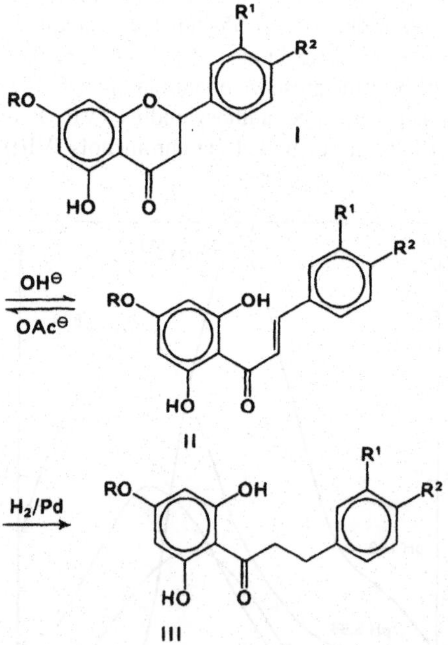

(18.29)

The presence of a free OH-group in position R^1 or R^2 is necessary for a sweet taste. Table 18.25 shows that the dihydrochalcone of naringin corresponds to saccharin in sweetness intensity, whereas the dihydrochalcone of neohesperidin is sweeter than saccharin by a factor of 20.

Conversion of naringin to highly sweet neohesperidin dihydrochalcone (VII) is possible by al-

Table 18.23. Flavanones and flavones in citrus fruits[a]

Polyphenol glucoside[b]	Orange, sweet[c]		Bitter orange		Grapefruit		Lemon	
	Fruit	Peel	Fruit	Peel	Fruit	Peel	Fruit	Peel
Flavanones								
Eriocitrin (Eri-7-rut)	159	59	49	38	183	92	1020	1320
Neoeriocitrin (Eri-7-neo)	27	n.d.	2100	2200	n.d.	n.d.	n.d.	n.d.
Narirutin (Nar-7-rut)	1660	665	170	220	1700	1900	114	225
Naringin (Nar-7-neo)	n.d.[d]	n.d.	9790	14,700	13,600	21,000	n.d.	n.d.
Hesperidin (Hes-7-rut)	9620	14,100	n.d.	n.d.	n.d.	n.d.	3560	711
Neohesperidin (Hes-7-neo)	n.d.	n.d.	6840	10,900	210	203	n.d.	n.d.
Neoponcirin (Isa-7-rut)	571	421	16	27	53	84	n.d.	n.d.
Poncirin (Isa-7-neo)	n.d.	n.d.	2820	5670	3040	462	n.d.	n.d.
Flavones								
Rutin (Que-3-rut)	108	n.d.	290	473	51	51	n.d.	n.d.
Isorhoifolin (Api-7-rut)	3	11	20	37	n.d.	n.d.	158	355
Rhoifolin (Api-7-neo)	15	58	566	1080	95	184	13	29
Diosmin (Dio-7-rut)	14	55	16	38	n.d.	n.d.	208	432
Neodiosmin (Dio-7-neo)	77	30	173	438	185	110	n.d.	n.d.

[a] Values in mg/kg fresh weight; n.d., not detected.
[b] Api: apigenin, Dio: diosmetin, Eri: eriodictyol, Hes: hesperitin, Isa: isosakuranetin, Nar: naringenin, Que: quercetin, rut: rutinose (O-α-L-Rha$_p$-(1 → 6)-D-Glc$_p$), neo: neophesperidose (O-α-L-Rha$_p$(1 → 2)-D-Glc$_p$)
[c] *C. sinensis* cv. Valencia
[d] *C. sinensis* var Brasiliensis cv. Morita contained 14 mg/kg naringin

Table 18.24. Taste of flavanone glycosides[a]

Compound	R	R[1]	R[2]	Taste	
				quality	intensity[b]
Naringenin-rutinoside	rut[c]	H	OH	neutral	–
Naringin	neo[d]	H	OH	bitter	20
Isosacuranetin-rutinoside	rut	H	OCH$_3$	neutral	–
Poncirin	neo	H	OCH$_3$	bitter	20
Hesperidin	rut	OH	OCH$_3$	neutral	–
Neohesperidin	neo	OH	OCH$_3$	bitter	2
Eriocitrin	rut	OH	OH	neutral	–
Neoeriocitrin	neo	OH	OH	bitter	2

[a] Data for R, R[1] and R[2] refer to Formula 18.28.
[b] Relative bitterness refers to quinine hydrochloride = 100.
[c] Rutinosyl.
[d] Neohesperidosyl.

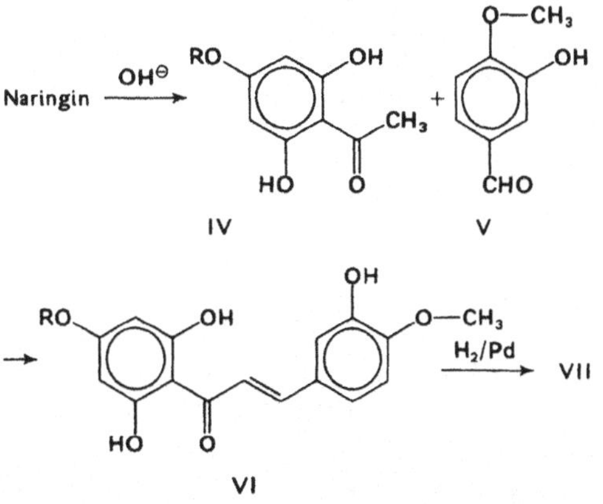

$$(18.30)$$

A sweet compound can be obtained from the neutral-tasting hesperidin of oranges by first converting hesperidin to another neutraltasting compound, hesperidin dihydrochalcone. The latter can then be hydrolyzed, by acidic or enzymatic catalysis, to remove the rhamnose residue, yielding hesperidin dihydrochalcone glucoside, which is sweet. The use of dihydrochalcones as sweeteners is discussed in Section 8.8.11.

kali fragmentation to a methylketone (IV), condensation with isovanillin (V) to the corresponding chalcone (VI), then hydrogenation:

Table 18.25. Taste of dihydrochalcones

Dihydrochalcone from	Tastes		
	quality	intensity[a] (μmole/l)	Relative intensity[b]
Naringin	sweet	200	1
Neohesperidin	sweet	10	20
Neoeriocitrin	slightly sweet	–	–
Poncirin	slightly bitter	–	–
Saccharin (Sodium salt)	sweet	200	1

[a] Concentration of iso-sweet solutions.
[b] Related to saccharin.

The dihydrochalcone glycoside phloridzin (Formula 18.31) occurs in apples.

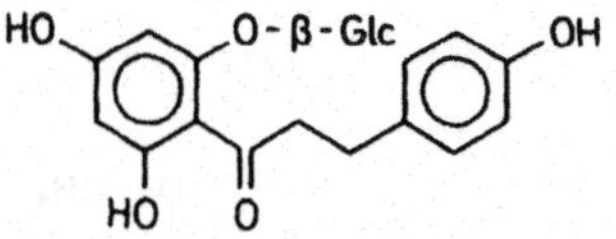

$$(18.31)$$

Table 18.26. Occurrence of flavonols in fruit

Fruit	Flavonols
Apple	Que-3-gal, Que-3-glc, Que-3-rha, Que-3-rha-glc, Que-3-ara, Que-3-xyl
Pear	Que-3-glc
Peach	Que-3-glc
Apricot	Que-3-glc, Kaem-gly
Plum/prune	Que-3-glc, Que-3-rha, Que-3-ara
Sour cherry	
Sweet cherry	Que-3-glc
Blackberry	Que-3-glc
Strawberry	
Currant, black	Que-3-glc, Kaem-3-glc, Myr-3-glc, further Que-glc, Kaem-glc
Raspberry	
Grapes	Que-3-rha, Que-3-glc, Que-3-rha-glc

Kaem: kaempferol, Myr: myricetin, Que: quercetin. ara: arabinoside, gal: galactoside, glc: glucoside, gly: glycoside, rha: rhamnoside, and xyl: xyloside.

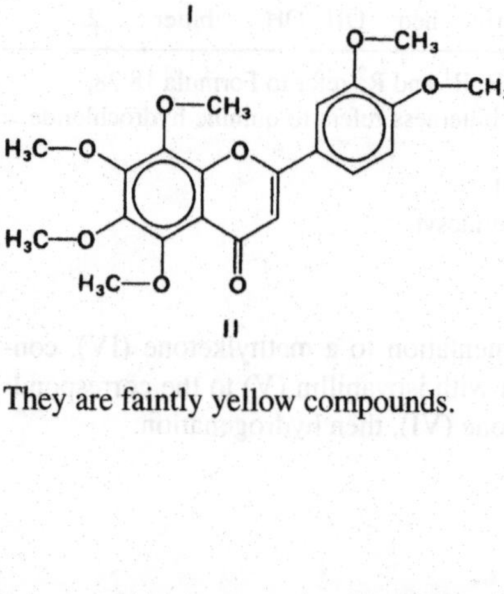

18.1.2.5.5 Flavones, Flavonols

Flavones (Formula 18.32: I, R = H; R^1, R^2 = H, R^3 = OH: apigenin; R^1, R^2 = OH, R^3 = H: luteolin; R^1 = OH, R^2 = H, R^3 = OCH_3: diosmetin; R^1 = OCH_3, R^2 = H, R^3 = OH: chrysoeriol; II: nobiletin) and flavonols (Formula 18.32: I, R, R^3 = OH; R^1, R^2 = H: kaempferol; R^1 = OH, R^2 = H: quercetin; R^1, R^2 = OH: myricetin; R^1 = OCH_3, R^2 = H: isorhamnetin) occur in all common fruits and citrus and tropical fruits as the 3-glycosides and, less frequently, as the 7-glycosides (Tables 18.23 und 18.26). Quercetin is a very effective antioxidant (cf. 3.7.3.2.1). Higher concentrations (mg/kg) have been found in quince (180), elderberry (170), cranberry (130), raspberry (70), apple (49), cherry (14) and red/blackcurrant (13).

$$(18.32)$$

They are faintly yellow compounds.

Table 18.27. Lignans in foods[a,b]

Food	Pino (I)	Lari (II)	Seco (III)	Mat (IV)	Sum
Linseed	33	30	2942	5	3011
Sesame	293	94	0.7	5	393
Bread, wholemeal wheat	0.3	0.7	0.2	n.d.	1.2
Bread, Rye, dark	1.7	1.2	0.1	0.1	3.2
Broccoli	3.2	9.7	0.4	n.d.	13.3
Garlic	2.0	2.9	0.5	n.d.	5.4
Apricot	3.1	1.1	0.3	n.d.	4.5
Strawberry	2.1	1.2	0.05	n.d.	3.3
Peach	1.9	0.8	0.3	n.d.	2.9

[a] Values in mg/kg fresh weight.
[b] Structures see Formula 18.33, Name: pinoresinol (I), lariciresinol (II), secoisolariciresinol (III), matairesinol (IV)
n.d.: not detected

18.1.2.5.6 Lignans

Lignans are polyphenols which belong to the group of phytoestrogens (cf. 16.2.9). Four compounds (I–IV in Formula 18.33) will be presented here. Pinoresinol (I) is a dimer of coniferyl alcohol. Lignans widely occur in low concentrations in food (examples in Table 18.29), linseed and sesame being especially rich.

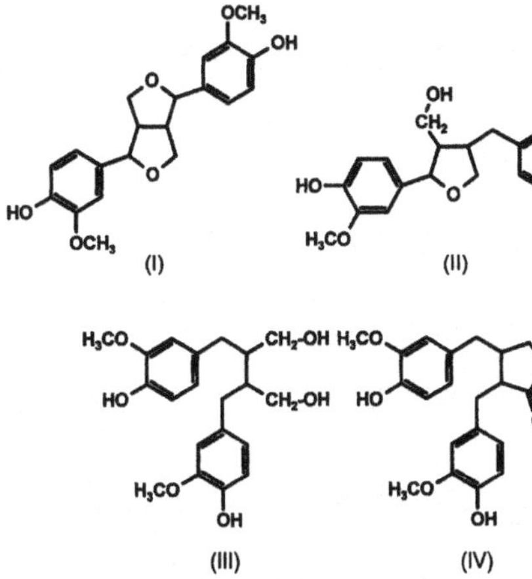

(18.33)

18.1.2.5.7 Flavonoid Biosynthesis

Flavonoid biosynthesis (cf. Formula 18.34) occurs through the stepwise condensation of activated hydroxycinnamic acid (I) with three activated malonic acid molecules (II). The primary condensation product, a chalcone (III), is in equilibrium with a flavanone (IV) with the equilibrium shifted toward product IV. The condensation directly yields a flavanone, hence chalcone is not an obligatory intermediary product.
A 2,7-cyclization yields stilbenes (IIIa).
One pathway converts flavanones (IV) to flavones (V) and, through another pathway, flavanones are converted to flavanonols (VI). The latter compounds are converted to flavandiols (VII), flavanols (VIII) and flavonols (IX), as well as anthocyanidins (XII) via endiols (X) and enols (XI).

18.1.2.5.8 Technological Importance of Phenolic Compounds

The taste of fruits is influenced by phenolic compounds. The presence of tannins yields an astringent, harsh taste, similar to an unripe apple (or an apple variety suitable only for processing). Table quality apples are low in phenolic compounds. Flavanones (naringin, neohesperidin) are the bitter compounds of citrus fruits.
Phenolic compounds are substrates for polyphenol oxidases. These enzymes hydroxylate monophenols to o-diphenols and also oxidize o-diphenols to o-quinones (cf. 2.3.3.2).
o-Quinones can enter into a number of other reactions, thus giving the undesired brown discoloration of fruits and fruit products. Protective measures against discoloration include inactivation of enzymes by heat treatment, use of reductive agents such as SO_2 or ascorbic acid, or removal of available oxygen.
Polyvalent phenols form colored complexes with metal ions. For example, at pH > 4, Fe^{3+} forms

$$(18.34)$$

complexes which are bluish-gray or bluish-black in color. Al^{3+} and Sn^{2+} also form intensely colored complexes. Leucoanthocyanins, when heated in the presence of an acid, are converted into anthocyanins. The red color of apples and pears, which is formed during cooking, is derived from leucoanthocyanins.

Phenolic compounds can also form complexes with proteins. These complexes increase the turbidity of fruit juices, beer and wine. The tendency to form complexes of this type increases with increasing degree of polymerization of the phenols; even dimeric procyanidins are active, e. g., procyanidin B2 (epicatechin-epicatechin) in apple juice. Based on model experiments, it is thought that especially the amino acid proline should be involved in complex formation, its ring system forming a π-complex with that of the phenols. Hydrogen bridges are also supposed to contribute to stabilization of the complexes. In the pH range 4.0–4.2, the amount of precipitate is maximum, being 7 times higher than at pH 3.0. In a similar manner as proteins and peptides, polyvinylpolypyrrolidone (PVPP) binds polyphenols. Therefore, it is especially suitable for the separation of haze active polyphenols.

18.1.2.6 Aroma Compounds

Aroma compounds contribute significantly to the importance of fruits in human nutrition. The aroma substances of selected fruits will be outlined below in more detail. The structures and synthesis pathways of common aroma substances are explained in Chapter 5.

The aroma of fruits can change on heating due to the liberation of aroma substances from glycosidic precursors (cf. 5.3.2.4), oxidation, water addition, and cyclization of individual compounds (cf. 5.5.4).

18.1.2.6.1 Bananas

The characteristic aroma compound of bananas is isopentyl acetate. Some esters of pentanol, such as those of acetic, propionic and butyric acids, also contribute to the typical aroma of bananas, while esters of butanol and hexanol with acetic and butyric acids are generally fruity

in character. An important contribution to the complete, mild banana aroma is supposed to be provided by eugenol (1), O-methyleugenol (II) and elemicin (III):

I: R^1, $R^2 = H$
II: $R^1 = H$, $R^2 = CH_3$
III: $R^1 = OCH_3$, $R^2 = CH_3$ (18.35)

18.1.2.6.2 Grapes

The compounds responsible for the typical aromas of different grape varieties has not been clarified in each case. Esters contribute to the fruity notes. The flowery-fruity aroma note of American grapes (*Vitis labrusca*) is based on 2-aminobenzoic acid methylester (methyl anthranilate), which is not found in European varieties, 2-Isobutyl-3-methoxypyrazine is responsible for the green paprika-like aroma note of Cabernet Sauvignon grapes.

18.1.2.6.3 Citrus Fruits

The aroma of the most important citrus fruit, the orange, has been analyzed in detail. The potent aroma substances identified in the freshly pressed juice of the variety *Valencia late* by dilution analyses are shown in Table 18.28.

On the basis of high orthonasal aroma values, it is expected that (S)-2-methylbutyric acid ethylester, ethyl butyrate, (Z)-3-hexenal, isobutyric acid ethylester, acetaldehyde and (R)-limonene are especially important for the aroma of orange juice (Table 18.28). Based on the retronasal odor threshold, the group of important compounds is enlarged to include 1-octen-3-one, trans-4,5-epoxy-(E)-2-decenal and ethyl caproate. A mixture of the odorants listed in Table 18.28, in which only wine lactone was missing, reproduced and aroma of orange juice. Omission experiments showed that the key aroma substances of the orange are acetalde-

Table 18.28. Odorants in fresh orange juice[a]

Compound	Concentration (µg/kg)	Aroma value[b]	
		Ortho-nasal	Retro-nasal
Acetaldehyde	8305	332	831
Isobutyric acid ethylester	8.8	440	293
(R)-α-Pinene	308	62	9
Ethyl butyrate	1192	1192	11,920
(S)-2-Methylbutyric acid ethylester	48	8000	12,000
Hexanal	197	19	19
(Z)-3-Hexenal	187	747	6227
Myrcene	594	42	36
(R)-Limonene	85,598	228	1339
3-Methylbutanol	639	<1	2.6
2-Methylbutanol	270	<1	n.b.
Ethyl caproate	63	13	125
Octanal	25	3.2	<1
1-Octen-3-one	4.1	4.1	410
Nonanal	13	2.7	3.8
Methional	0.4	<1	10
Decanal	45	9	6
(E)-2-Nonenal	0.6	<1	8
(S)-Linalool	81	13	54
3-Hydroxyhexanoic acid ethylester	1136	4	18
(E,E)-2,4-Decadienal	1.2	6	24
trans-4,5-Epoxy-(E)-2-decenal	4.3	36	287
Wine lactone	0.8	n.b.	94
Vanillin	67	3	2

[a] Variety: *Valencia late.*
[b] Aroma value: the ratio of the concentration to ortho- or retronasal odor threshold value of the substance in water.
n.d.: not determined.

hyde, (Z)-3-hexenal, decanal, (R)-limonene and trans-4,5-epoxy-(E)-2-decenal. It is remarkable that these substances include decanal although its aroma value is fairly low (Table 18.28). Esters are also indispensable, but the aroma of the recombined mixture is not impaired when a member of this group is missing. The contribution of (R)-α-pinene and myrcene is negligible.

The concentrations of the odorants in juice differ depending on the variety. Thus, the weaker citrus note of *Navel* oranges compared with the variety *Valencia late* is due to a 70% lower content of (R)-limonene.

The aroma of oranges changes on storage. In the juice of oranges stored for three weeks at 4°C, the concentrations of esters and especially aldehydes were much lower than in the fresh juice. For instance, the content of (Z)-3-hexenal was only 15%.

Orange juice from rediluted concentrate differs in its aroma. This can be the result of big losses of acetaldehyde and (Z)-3-hexenal, the formation of carvone by peroxidation of limonene and a large increase in the vanillin concentration, probably due to the degradation of ferulic acid.

Dilution analyses of grapefruit juice gave high FD factors (definition in 5.2.2.1) for ethyl butyrate, (Z)-3-hexenal, 1-hepten-3-one, 4-mercapto-4-methylpentan-2-one and 1-p-menthene-8-thiol (IV, probably the R enantiomer). The concentrations of the two sulfur compounds in juices were 0.4–0.8 µg/l and 0.007–0.1 µg/l respectively. Omission experiments (cf. 5.2.7) indicate that the grapefruit aroma note is produced by 4-mercapto-4-methylpentan-2-one. 1-p-Menthene-8-thiol, which occurs in even lower concentrations in oranges, contributes to the aroma but is not typical. It is possibly formed by the addition of H_2S to limonene. Traces of hydrogen sulfide occur in all citrus juices.

Grapefruit juice differs from orange juice also in the considerably lower limonene content. (+)-Nootkatone (V) only contributes to the aroma of grapefruit peel-oil, but not to that of the juice.

IV   V   (18.36)

Citral, which is actually a mixture of two stereoisomers, geranial (VIa) and neral (VIb), is the character impact compound of lemon oil (cf. 5.5.1.5):

VIa   VIIb   (18.37)

Linalool, myrcene, and limonene also have high aroma values.

18.1.2.6.4 Apples, Pears

The potent odorants identified in two apple varieties with a fruity/green (*Elstar*) and fruity/sweet/aromatic (*Cox Orange*) odor are shown in Table 18.29.

The fruity note in the aroma profile of both varieties is produced by acetic acid esters. On the other hand, there is a decrease in the ethyl esters, which are more odor active than the acetates (cf. 5.3.2.2) and dominate in some other fruits, e.g., oranges and olives. Hexanal, (Z)-3-hexenal and (Z)-3-nonenal are responsible for the green/apple-like note. (E)-β-Damascenone, which smells of cooked apples, has the highest aroma value in both varieties due to its much lower odor threshold. Eugenol and (E)-anethol contribute to the aniseed-like note which is a characteristic especially of the aroma of the peel of the *Cox Orange*.

The aroma of the pear Williams Christ is characterized by esters produced by the degradation of unsaturated fatty acids (example in 5.3.2.2): ethyl esters of (E,Z)-2,4-decadienoic acid, (E)-2-octenoic acid, and (Z)-4-decenoic acid, as well as hexyl acetate. In fact, butyl acetate and ethyl butyrate are also involved in the fruity odor note.

18.1.2.6.5 Raspberries

The character impact compound is the "raspberry ketone", i.e. 1-(p-hydroxyphenyl)-3-butanone (VII). Its concentration is 2 mg/kg and its odor threshold is 5 µg/kg (water). The starting point for the biosynthesis of VII is the condensation reaction of p-cumaroyl-CoA with malonyl-CoA (cf. Formula 18.38). Additional aroma notes are provided by (Z)-3-hexenol, α- and β-ionone.

In addition, the ethyl esters of 5-hydroxyoctanoic acid and 5-hydroxydecanoic acid should contribute to the aroma. A part of the esters hydrolyzes during cooking and the hydroxy acids released cyclize to the corresponding lactones.

18.1.2.6.6 Apricots

The following compounds are discussed as contributors to the aroma: myrcene, limonene, p-cymene, terpinolene, α-terpineol, geranial, geraniol, linalool, acids (acetic and 2-methylbutyric acids) alcohols, e.g., trans-2-hexenol; and a number of γ- and δ-lactones, e.g., γ-caprolactone,

Table 18.29. Odorants of the apple varieties *Elstar* and *Cox Orange*

Compound	Elstar		Cox Orange	
	Concentration (µg/kg)	Aroma value[a]	Concentration (µg/kg)	Aroma value[a]
(E)-β-Damascenone	1.4	1813	0.99	1320
2-Methylbutyric acid methylester	0.2	<1	1.8	7
Ethyl butyrate	0.7	7	0.3	3
Hexyl acetate	5595	112	1500	30
Butyl acetate	4640	93	1595	32
Acetic acid 2-methylbutylester	240	48	217	43
Hexanal	85	19	48	11
(E)-2-Hexenal	77	2	114	2
(Z)-3-Hexenal	6.4	25	30	120
(Z)-2-Nonenal	8.8	440	1.1	55
Butanol	4860	10	975	2
Hexanol	1390	3	350	<1
Linalool	9.3	19	4.5	9

[a] Aroma value: ratio of the concentration to the orthonasal odor threshold value of the substance in water.

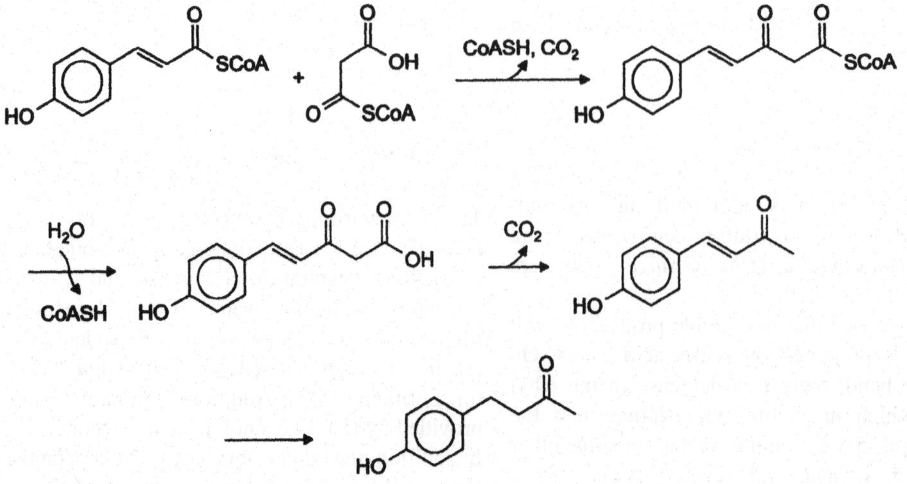

(18.38)

γ-octalactone, γ-decalactone, γ-dodecalactone, and hexyl hexanoate (69.6).
δ-octalactone and δ-decalactone.

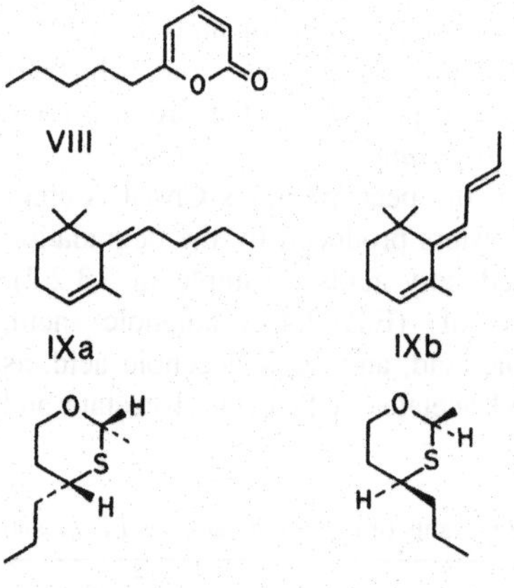

18.1.2.6.7 Peaches

The aroma of peaches is characterized by
γ-lactones (C_6–C_{12}) and δ-lactones (C_{10} and
C_{12}). The main compound in the lactone fraction
is (R)-1,4-decanolide, which has a creamy,
fruity, peach-like odor. Other important compounds should be benzaldehyde, benzyl alcohol,
ethyl cinnamate, isopentyl acetate, linalool,
α-terpineol, α- and β-ionone, 6-pentyl-α-pyrone
(Formula 18.39, VIII), hexanal, (Z)-3-hexenal,
and (E)-2-hexenal. Aroma differences in different
varieties of peaches are correlated with the different proportions of the esters and monoterpenes.
In the case of nectarines (*Prunus persica* L.,
Batsch var. nucipersica Schneid), the lactones
γ-C_8–C_{12} and δ-C_{10} belong to the compounds
with the highest aroma values.

18.1.2.6.8 Passion Fruit

The aroma of the yellow fruit (*Passiflora edulis*
var. *flavica*) is supposed to be superior to that of
the crimson fruit (*Passiflora edulis* var. *edulis*).
Compounds that contribute to the aroma of both
varieties are β-ionone and the following esters
(% of the volatile fraction): ethyl butyrate (1.4),
ethyl hexanoate (9.7), hexyl butyrate (13.9),

(18.39)

Four stereoisomeric megastigmatrienes have
been found in crimson passion fruit. A mixture of
the isomers IXa and IXb (Formula 18.39) gives
a rose-like aroma, that has a hint of strawberries
in it (threshold = 100 µg/kg; water).

The following S-containing aroma substances
have been isolated from the yellow fruit: 3-
methylthiohexane-1-ol which probably gives rise
to 2-methyl-4-propyl-1,3-oxathianes (cis/trans
isomers in the ratio of 10:1) (Xa, b: Formula 18.39). Of the two cis isomers, only the
2R,4S-isomer (Xb), which has a sulfurous
herb-like odor (threshold = 4 µg/kg; water), has
been found in the fruit. However, the aroma note
more typical of passion fruit is exhibited by the
2S,4R-isomer (Xa).

18.1.2.6.9 Strawberries

The concentrations of the odorants in strawberry juice are given in Table 18.30. With a solution of the first 11 compounds listed in the table, the original aroma was largely approximated. If HD3F was missing, the mixture smelt green and fruity and if (Z)-3-hexenal was missing, the caramel-like/sweetish note of HD3F dominated.

The HD3F concentration depends on the variety of strawberry. Values between 1.1 and 33.8 mg/kg were found. The red parts of the fruit are richer in HD3F than the white parts. The strawberry aroma changes on heating due to an increase in HD3F, formation of (E)-β-damascenone, (E,E)-2,4-decadienal and guaiacol as well as great losses of (Z)-3-hexenal and the esters (Table 18.30). A freeze/thaw process also changes the aroma as a result of a great increase in HD3F and degradation of (Z)-3-hexenal.

18.1.2.6.10 Pineapples

A model on the basis of the compounds listed in Table 18.31 reproduces the aroma of pineapples.

Table 18.30. Concentrations of odorants in fresh and heated strawberry juice

Compound	Concentration (mg/kg)	
	Fresh	Heated[a]
4-Hydroxy-2,5-dimethyl-3(2H)-furanone (HD3F)	16.2	29.4
(Z)-3-Hexenal	0.333	0.025
Methyl butyrate	5.0	1.0
Ethyl butyrate	0.41	0.048
Isobutyric acid ethylester	0.043	0.012
2-/3-Methylbutyric acid methylester	0.048	0.007
2-/3-Methylbutyric acid ethylester	0.007	0.0012
Acetic acid	74.5	74.9
2,3-Butandione	1.29	0.85
Butyric acid	1.83	1.79
2-/3-Methylbutyric acid	2.24	2.20
(E)-β-Damascenone	<0.1	5.4
(E,E)-2,4-Decadienal	<0.1	4.1
Guaiacol	0.8	2.8

[a] 100°C, 30 min (reflux).

Table 18.31. Odorants of pineapples

Compound	Concentration (μg/kg)	Aroma value[a]
4-Hydroxy-2,5-dimethyl-3(2H)-furanone	26,800	2680
2-Methylpropionic acid ethyl ester	48	1400
2-Methylbutyric acid ethyl ester	157	1050
2-Methylbutyric acid methyl ester	1190	595
(E,Z)-1,3,5-Undecatriene	8.9	445
β-Damascenone	0.083	111
Butyric acid ethyl ester	75	75
2-Methylpropionic acid methyl ester	154	24
Octanal	19	2
δ-Ocalactone	78	<1
δ-Decalactone	32.7	<1
Vanillin	6	<1

[a] Aroma value: quotient of the concentration and orthonasal odor threshold value in water.

Omission experiments have shown the five odorants with the highest aroma values shown in Table 18.31 are the key aroma substances.

18.1.2.6.11 Cherries, Plums

The compounds essentially involved in the aroma of cherries are benzaldehyde, linalool, hexanal, (E)-2-hexenal, phenylacetaldehyde, (E,Z)-2,6-nonadienal, and eugenol (Table 18.32). On heating cherry juice or in the making of jams, the concentration of benzaldehyde increases due to the hydrolysis of amygdalin and prunasin

Table 18.32. Odorants of cherry juice and jams made from it[a]

Aroma substance	Juice (μg/kg)	Jam (μg/kg)
Benzaldehyde	202	1510
Linalool	1.1	13.1
Hexanal	5.6	0.2
(E)-2-Hexenal	8.5	3.8
Eugenol	10.0	4.9

[a] Fruit content: 50 w/w per cent.

(cf. 16.2.6) and the concentration of linalool increases due to the hydrolysis of the corresponding glycoside (Table 18.32). Since the C_6-aldehydes and nonadienal decrease simultaneously, the fruity-flowery aroma notes are enhanced and the "green" notes diminished.

The important compounds in plums are linalool, benzaldehyde, methyl cinnamate, and γ-decalactone together with the C_6-aldehydes. Benzaldehyde, nonanal, and benzyl acetate contribute to the aroma of canned plums.

18.1.2.6.12 Litchi

The compounds acetic acid isobutylester, guaiacol, cis-rose oxide, 2-acetylthiazoline, (E)-β-damascenone, 4-hydroxy-2,5-dimethyl-3(2H)-furanone, linalool, geraniol and 2-phenylethanol exhibit the highest aroma activity.

18.1.2.7 Vitamins

Many fruits are important sources of vitamin C (Table 18.33). Its biosynthesis in plants starts from hexoses, e. g., glucose. It is postulated that following C-1 oxidation and cyclization to 1,4-lactone (II), the 5-keto compound (III) appears as an intermediary product which is oxidized to the 2,3-endiol (IV) then reduced stereospecifically to L-ascorbic acid (V) (cf. Formula 18.40).

Table 18.33. Ascorbic acid in various fruits (mg/100 g edible portion)

Fruit	Ascorbic acid	Fruit	Ascorbic acid
Apple	3–35	Currant, black	177
Pear	1–4		
Apricot	5–15	Orange	50
Cherry	8–37	Grapefruit	40
Peach	5–29	Lemon	50
Plum/prune	2–14	Acerola	1000–2000
Blackberry	17	Pineapple	25
Strawberry	60	Banana	7–21
Raspberry	25	Guava	300
Currant, red	40	Melons	6–32

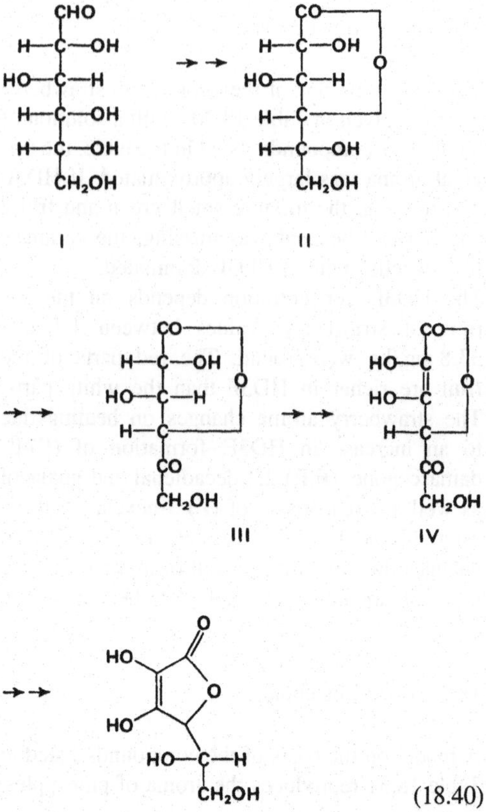

(18.40)

Industrial-scale production of ascorbic acid also starts with glucose. The sugar is first reduced to sorbitol (VI) and then oxidized with *Acetobacter suboxidans* to L-sorbose (VII) which after cyclization and conversion to the diisopropylidene derivative (VIII) is oxidized to the corresponding derivative of L-2-oxogulonic acid (IX). After removal of the protecting isopropylidene groups, L-ascorbic acid (vitamin C) is obtained via L-2-oxogulonic acid (X; cf. Reaction 18.41).

The synthesis can be shortened with a genetically modified strain of the bacterium *Erwinia herbicola* which directly converts D-glucose to L-2-oxogulonic acid (X).

β-Carotene (provitamin A) occurs in large amounts in apricots, cherries, cantaloups and peaches. B-vitamins present in some fruits (apricots, citrus fruits, figs, black currants and gooseberries) are pantothenic acid and biotin. Other B-group vitamins occur at levels of no nutritional significance. Vitamins B_{12} and D and tocopherols are found in no more than trace amounts.

Glucose →

VI

VII

VIII

IX

→ L-Ascorbic acid

X   (18.41)

18.1.2.8 Minerals

Table 18.34 gives the composition of the ash of orange juice and apples. The most important cation is potassium and the most important inorganic anion is phosphate.

18.1.3 Chemical Changes During Ripening of Fruit

Ripening of fruit involves highly complex changes in physical and chemical properties.

Table 18.34. Minerals in fruit

Element	Orange juice (% in ash)	Apple (mg/100g dry matter)
Potassium	40	840
Sodium	0.3	7.9
Calcium	2.8	38
Magnesium	3.0	40
Iron	0.06	1.6
Aluminium	0.12	0.43
Phosphorus	3.8	73
Sulfur	0.8	
Chlorine	1.0	

Zinc, titanium, barium,			Zinc	0.65
copper, manganese, tin	≤ 0.03		Manganese	0.3
Boron	≤ 0.01		Copper	0.35

Softening, increasing sweetness, aroma and color changes are among the most striking phenomena related to ripening. Some changes will be outlined below in more detail.

18.1.3.1 Changes in Respiration Rate

The respiration rate is affected by the development stage of the fruit. A rise in respiration rate occurs with growth. This is followed by a slow decrease in respiration rate until the fruit is fully ripe. In a number of fruits ripening is associated with a renewed rise in respiration rate, which is often denoted as a climacteric rise. Maximal CO_2 production occurs in the climacteric stage. De-

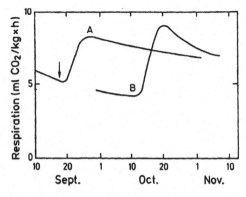

Fig. 18.8. Respiration rise in apples, Bramsley's seedlings (according to *Hulme*, 1963.) A, apple picked →, B, left on tree to ripen

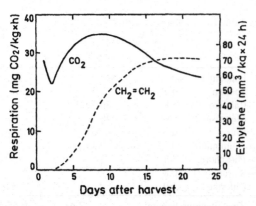

Fig. 18.9. Respiration rise in tomatoes ——— : CO_2, – – – : Ethylene

pending on the fruit, this can occur before or after harvesting. Figures 18.8 and 18.9 show that such a rise occurs a short time after harvest for apples and tomatoes and is accompanied by increased ethylene production.

The climacteric respiration rise is so specific that fruits can be divided into:

- *Climacteric types*, such as apples, apricots, avocados, bananas, pears, mangoes, papaya, passion fruit, peaches, plums/prunes and tomatoes; and
- *Nonclimacteric types*, which include pineapples, oranges, strawberries, figs, grapefruit, cucumbers, cherries, cantaloupes, melons, grapes and lemons.

It should be emphasized that nonclimacteric fruits generally ripen on the plants and contain no starch. The differing effects of ethylene on the two types of fruits are covered in Section 18.1.4.2.

Fruits can also be classified according to respiration behavior after harvesting. Three fruit types are distinguished:

Type 1: A slow drop in CO_2 production during ripening (as illustrated by citrus fruits).

Type 2: A temporary rise in CO_2 production. The fruits are fully ripe after this increase reaches a maximum (e. g., avocados, bananas, mangoes or tomatoes).

Type 3: Maximum CO_2 production in the fully ripe stage, until the fruit is overripe (e. g. strawberries and peaches).

The reason for the increase in CO_2 production is not yet fully elucidated. Physical and chemical factors are involved. For example, a change in permeability for gases occurs in fruit peels. With increasing age the peel cuticle becomes thicker and is more strongly impregnated with fluid waxes and oils. Thus, the total permeability drops, while the CO_2 concentration within the fruit increases. Three possibilities are usually considered for the rise in CO_2 production. The first is related to increased protein biosynthesis coupled with increased ATP consumption thus stimulating enhanced respiration. Secondly, since the respiratory quotient (RQ) increases from 1 to 1.4–1.6, it is assumed that the additional CO_2 source is not due to respiration but to decarboxylation of malate and pyruvate, i.e. there is a switch from the citric acid cycle to malate degradation. Another possibility is the partial uncoupling of respiration from phosphorylation by an unknown decoupler.

New concepts involving structural factors suggest that fruit flesh possesses marked photosynthetic activity which is then associated with CO_2 uptake. With the onset of ripening, an increased disorganization occurs in chloroplasts and other cell organelles. Photosynthetic activity decreases and finally stops completely. The same is the case for other synthetic activities. Catabolic processes, catalyzed by cytoplasmic enzymes, become dominant. Based on such a perception (*Phan et al.*, 1975) the "climacteric is seen as an indication of the natural end of a period of active synthesis and maintenance, and the beginning of the actual senescence of the fruit".

18.1.3.2 Changes in Metabolic Pathways

Metabolic shifts may occur in several fruits during ripening. For example, during ripening of bananas, there is a marked rise in aldolase and carboxylase activities and thus it appears that at this stage the *Embden–Meyerhoff* pathway becomes dominant and the pentose-phosphate pathway is suppressed.

An increase in malate and pyruvate decarboxylase activities is observed in apples during the climacteric stage. The activities drop as

CO_2 production decreases. This provides an explanation for the change in RQ during the climacteric stage. CO_2 production increases more rapidly than O_2 uptake, thus the RQ is greater than 1. The shift from the citric acid cycle to malate degradation in apples is also reflected by the effect of citrate and malate on succinate production. As ripening proceeds, production of succinate from citrate drops to zero. An increase in succinate content after addition of malate in the initial stage of ripening is probably a feedback reaction. In this case, a decrease is also observed later on, suggesting a greater change in metabolic patterns.

18.1.3.3 Changes in Individual Constituents

18.1.3.3.1 Carbohydrates

During ripening of fruits, significant changes occur in the carbohydrate fraction. For example, between picking and onset of decay in apples about 20% of the available carbohydrates have been utilized.

During the growth of apples on trees, the starch content rises and then drops to a negligible level by the time of harvest. This drop appears to be related to the increase in climacteric respiration. Contrary to starch, the sugar content rises. Other sources in addition to starch should be available for conversion to sugars. A decrease in hemicelluloses suggests that they are a possible source. Organic acids may also be an additional source of sugars.

A marked decrease in starch in bananas parallels an increase in the contents of glucose, fructose and saccharose. Biosynthesis of the latter occurs by two pathways:

1) $UDPG + Fru-6-p \rightarrow UDP + Sac-6^F-P$

     $\rightarrow Sac + P_{in}$

2) $UDPG + Fru \rightarrow UDP + Sac$ $\qquad$ (18.42)

The content of hemicelluloses drops from 9% to 1–2% (relative to fresh weight), hence they act as a storage pool in carbohydrate metabolism. There is also a drop in the sugar content in bananas during the post-climacteric stage.

Differences in various fruits can be remarkable. In oranges and grapefruits the acid content drops during ripening while the sugar level rises. In lemons, however, there is an increase in acids.

Decreases in arabinans, cellulose and other polysaccharides are found in pears during ripening. Cellulase enzyme activity has been confirmed in tomatoes.

Remarkable changes occur in the pectin fractions during ripening of many fruits (e. g., bananas, citrus fruits, strawberries, mangoes, cantaloupes and melons). The molecular weight of pectins decreases and there is a decrease in the degree of methylation. Insoluble protopectin is increasingly transformed into soluble forms. Protopectin is tightly associated with cellulose in the cell wall matrix. Its galacturonic acid residues are acetylated at OH-groups in positions 2 and 3 or are bound to polysaccharides as lignin ($R^1 = H$, CH_3, polysaccharide: arabinan, galactan and possibly cellulose; $R^2 = H$, CH_3CO, polysaccharide, lignin):

$$\downarrow \text{Protopectinase}$$

Polysaccharides + soluble Pectins

$\qquad\qquad\qquad\qquad\qquad\qquad\qquad$ (18.43)

Soluble pectins bind polyphenols, quench their astringent effect and, thus, contribute to the mild taste of ripe fruits.

After prolonged storage there is a decrease in soluble pectins in apples. This drop is associated with a mealy, soft texture. Similar events occur in pears, but much more rapidly and with more extensive demethylation of pectin. Generally, the degree of pectin esterification drops from 85% to about 40% during ripening of pears, peaches and avocados. This drop is due to a remarkable increase in activities of polygalacturonases and pectin esterases. The rise in free galacturonic acid is negligible; therefore it appears that the release of uronic acid is associated with its simultaneous conversion through other reactions.

18.1.3.3.2 Proteins, Enzymes

During ripening of some fruits, although the total content of nitrogen is constant, there is an increase in protein content, an increase assigned primarily to increased biosynthesis of enzymes. For example, during ripening of fruit there is increased activity of hydrolases (amylases, cellulases, pectinolytic enzymes, glycolytic enzymes, enzymes involved in the citric acid cycle, transaminases, peroxidases and catalases). Proteinaceous enzyme inhibitors which inhibit the activities of amylases, peroxidases and catalases are found in unripe bananas and mangoes. The activities of these inhibitors appear to decrease with increasing ripeness.

The ratios of NADH/NAD$^+$ or NADPH/NADP$^+$ pass through a maximum during ripening of fruit. For example, the values for mangoes are 0.32–0.67 in the unripe stage, 1.44–6.50 in the semiripe stage and 0.57–0.93 in the ripe stage. During ripening of fruit, shifts also occur in the amino acid and amine fractions. The shifts are not uniform and are affected by type and ripening stage of fruits.

18.1.3.3.3 Lipids

Little is known about changes in the lipid fraction. Shifts in composition and quantity have been found, especially in the phospholipid fraction.

18.1.3.3.4 Acids

There is a drop in acid content during ripening of fruits. Lemons, as already mentioned, are an exception. The proportion of various acids can change. In ripe apples malic acid is the major acid, while in young, unripe apples, quinic acid is dominant. In the various tissues of any single fruit, various acids can be dominant. For example, apple peels contain citramalic acid (I, cf. Formula 18.44) which is formed from pyruvic acid, and can produce acetone through acetoacetic acid. Acetone is formed abundantly during ripening:

$$O=\overset{\overset{\displaystyle COOH}{|}}{\underset{\underset{\displaystyle CH_3}{|}}{C}} + H_3C-CO-COOH$$

$$\longrightarrow HO-\overset{\overset{\displaystyle COOH}{|}}{\underset{\underset{\displaystyle CH_3}{|}}{C}}-CH_2-CO-COOH$$

$$\longrightarrow HO-\overset{\overset{\displaystyle COOH}{|}}{\underset{\underset{\displaystyle CH_3}{|}}{C}}-CH_2-COOH \quad I$$

$$\longrightarrow \overset{\overset{\displaystyle OC}{}}{\underset{\underset{\displaystyle CH_3}{|}}{}}-CH_2-COOH \longrightarrow H_3C-CO-CH_3$$

$$(18.44)$$

The synthesis of ascorbic acid is also of importance. It takes place in many fruits during ripening (cf. 18.1.2.7)

18.1.3.3.5 Pigments

The ripening of fruit is usually accompanied by a change in color. The transition of green to another color is due to the degradation of chlorophyll and the appearance of concealed pigments. Furthermore, the synthesis of other pigments plays a big role. For example, the lycopene content of the tomato increases greatly during ripening. The same applies to the carotenoid content of citrus fruits and mangoes. The formation of anthocyanin is frequently enhanced by light.

18.1.3.3.6 Aroma Compounds

The formation of typical aromas takes place during the ripening of fruit. In bananas, for example, noticeable amounts of volatile compounds are formed only 24 h after the climacteric stage has passed. The aroma build-up is affected by external factors such as temperature and day/night variations. Bananas, with a day/night rhythm of 30 °C/20 °C, produce about 60% more volatiles than those kept at a constant temperature of 30 °C. The synthesis of aroma substances is discussed in section 5.3.2.

18.1.4 Ripening as Influenced by Chemical Agents

18.1.4.1 Ethylene

Fruit ripening is coupled with ethylene biosynthesis:

$$CH_3-S-CH_2-CH_2-CH-COO^\ominus$$
$$\overset{|}{\underset{\oplus}{NH_3}}$$

$$\text{Ad} \quad \overset{\curvearrowleft ATP}{\underset{\curvearrowright PP_i + P_i}{}}$$

$$CH_3-\overset{|}{\underset{\oplus}{S}}-CH_2-CH_2-CH-COO^\ominus$$
$$\overset{|}{NH_2}$$

$$\overset{\curvearrowleft R-CHO}{}$$

$$\text{Ad} \qquad H$$
$$CH_3-\overset{|}{\underset{\oplus}{S}}-CH_2-CH_2-\overset{|}{C} \overset{}{\smile} N \overset{}{=} CH \smile R$$
$$\overset{|}{COO^\ominus}$$

$$\text{Ad} \qquad\qquad COO^\ominus$$
$$CH_3-\overset{|}{\underset{\oplus}{S}} \overset{}{\to} CH_2-CH_2-C$$
$$\qquad\qquad N \overset{}{\smile} CH \overset{}{=} R$$

$$\overset{\curvearrowleft CH_3-S-Ad}{}$$

$$\begin{array}{ccccc}
H_2C-CH_2 & & H_2O & R-CHO & H_2C-CH_2 \\
\diagdown C\diagup & & & & \diagdown C\diagup \\
\ominus_{OOC}\quad N=CH-R & & & & \ominus_{OOC}\quad \overset{\oplus}{N}H_3
\end{array}$$

$$\text{Ox.} \xrightarrow{} \begin{array}{cc} H_2C-CH_2 & H_2O \\ \diagdown C\diagup & \\ \ominus_{OOC}\quad N-OH & \\ \qquad\qquad H \end{array} \xrightarrow{} \left\{\begin{array}{l} CH_2\!=\!CH_2 \\ CO \\ NH_3 \\ HCOOH \end{array}\right.$$

$$(18.45)$$

Table 18.35. Ethylene production in ripening fruits

	Fruit	Ethylene (µg/l)
Climacteric maximum	Avocado	500
	Banana	40
	Mango	3
	Pear	40
	Tomato	27
Nonclimacteric stationary state	Lemon	0.2–0.2
	Orange	0.1–0.3
	Pineapple	0.2–0.4

Climacteric and nonclimacteric fruits respond differently to external ethylene (Fig. 18.10). Depending on the ethylene level, the respiratory increase sets in earlier in unripe climacteric fruits, but its height is not influenced. In contrast, in nonclimacteric fruits there is an increase in respiration rate at each ripening stage which is clearly dependent on ethylene concentration. The reaction pathway 18.45 is suggested for the biosynthesis of ethylene (R–CHO: pyridoxal phosphate; Ad: adenosine).
Ethylene and compounds capable of releasing ethylene under suitable conditions are utilized commercially for enhancing the ripening process. A number of such compounds are known, e. g., 2-chloroethylphosphonic acid (ethephon; R=H or CH$_2$–CH$_2$Cl) (Formula 18.46).

$$Cl-CH_2-CH_2-\overset{\overset{O}{\|}}{\underset{OR}{P}}-OH \xrightarrow{pH>4}$$

$$H_2C\!=\!CH_2 + H_2PO_4^\ominus + Cl^\ominus \qquad (18.46)$$

Ethylene increases rapidly but differently in the case of climacteric fruits. The maximum values for some fruits are given in Table 18.35. However, nonclimacteric fruits produce only a little ethylene (Table 18.35). This gaseous compound increases membrane permeability and thereby probably accelerates metabolism and fruit ripening. With mango fruits, for example, it has been demonstrated that before the climacteric stage, ethylene stimulates oxidative and hydrolytic enzymes (catalase, peroxidase and amylase) and inactivates inhibitors of these enzymes.

The use of ethylene before picking fruit (as with pineapples, figs, mangoes, melons, cantaloups and tomatoes) results in faster and more uniform ripening. Its utilization after harvesting accelerates ripening (e. g., with bananas, citrus fruits and mangoes). Ethylene can induce blossoming in the pineapple plant and facilitate detachment of stone fruits and olives. Vine defoliation can also be achieved.
The activity of propylene is only 1% of that of ethylene. Acetylene also accelerates ripening but only at substantially higher concentrations.

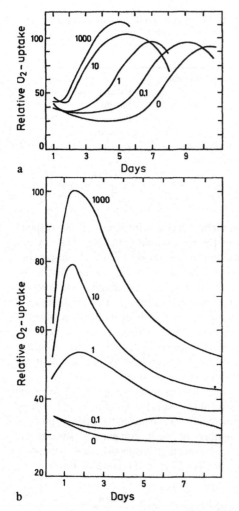

Fig. 18.10. The effect of ethylene on fruit respiration. (a) climacteric, (b) nonclimacteric. Numerals on the curves: ethylene in air, ppm (according to *Biale*, 1994)

18.1.4.2 Anti-Senescence Agents

In order to keep fruit fresh, the ripening is slowed down by cold storage (cf. 18.1.5) and/or with additives which inhibit the formation or action of ethylene.

18.1.4.2.1 Polyamines

The polyamines putrescine (butane-1,4-diamine), spermidine   [N-(3-aminopropyl)butane-1,4-diamine] and spermine [N,N'-bis-(3-aminopropyl)-

butane-1,4-diamine] belong to the group of naturally occurring anti-senescence agents. They increase during the early phase of growth of the fruit, when intensive cell division takes place, and, apart from exceptions, decrease during ripening. Studies have shown that the treatment of fruit with polyamines results in a slowing down of the ethylene production and respiration, which in turn has a positive effect on the firmness of the fruit flesh and on the color.

18.1.4.2.2 1-Methylcyclopropene (MCP)

As a result of structural similarities, MCP docks on to receptor proteins for ethylene, causing it to lose activity. The texture and color of apples and pears (sensitive varieties) do not change through months of cool storage. The sugar content also remains constant while the acid content increases. Gaseous MCP adsorbed on dextran is used for the treatment of fruit. The MCP is liberated by the addition of water. The concentrations applied are in the range 300–1000 ppb.

18.1.5 Storage of Fruits

18.1.5.1 Cold Storage

The suitability, duration and required conditions of fruit storage are dependent on variety and quality. Commonly used conditions are −1 °C to +2 °C at 80–90% relative humidity. The storage time varies from 4–8 months for apples, 2–6 months for pears, 2–3 months for grapes, 1–2 weeks for strawberries and raspberries, and 4–5 days for cherries. Efficient aeration is required during storage. Air circulation is often combined with purging to remove ethylene, the volatile promoter of fruit ripening. Weight losses occur during fruit storage due to moisture losses of 3–10%.

18.1.5.2 Storage in a Controlled (Modified) Atmosphere

This term is applied to an atmosphere which, in comparison to air, has a lowered oxygen con-

Table 18.36. Minimal O_2 and maximal CO_2 concentrations in the atmosphere during storage of fruits (temperature 0–5 °C)

Fruit	Minimal O_2 concentration (%)	Minimal CO_2 concentration (%)
Pear	1–2	2
Apple, kiwi	1–2	5
Peach, plum		
Pineapples	2	10
Sour cherry	2	15
Citrus fruits	5	10

centration and an increased CO_2 concentration. Common conditions for storage of many fruits are shown in Table 18.36. For each fruit variety it is important that optimal conditions for controlled atmosphere storage be maintained. For example, a high O_2 concentration accelerates ripening, while an overly low O_2 concentration results in high production of CO_2. An overly high concentration of CO_2 promotes glycolysis, which can cause off-flavors due to the formation of acetaldehyde and ethanol. Discoloration can also occur.

18.2 Fruit Products

The short shelf life of most fruits and the frequent need to store and spread out the surplus of a harvest for a prolonged period of time has brought about a number of processes which provide more durable and stable fruit products.

18.2.1 Dried Fruits

Like many other food products, moisture removal from fruits by a suitable drying process results in a product in which microbial growth is retarded and, with a suitable pretreatment, the enzymes present are largely inactivated. Fruit drying is probably the oldest procedure for preservation. It was originally performed in a rather primitive way (spreading the fruit in the hot air of a fireplace or hearth, kitchen stove or oven), thus providing dark "baked products". Solar drying is still a common process in southern and tropical

countries for obtaining dried apple slices, apricots, peaches or pears or tropical fruits such as dates, figs or raisins. Predrying is often achieved in sunshine and additional drying by artifical heat in drying installations. The temperature in drying chambers, flat or tunnel dryers is between 75 °C (incoming air) and 65 °C (temperature of the exit air) at a relative humidity of 15–20%. Vacuum drying at about 60 °C is particularly gentle.

Carefully washed and trimmed fruits of suitable varieties are pretreated in various ways: *Pomme fruits* (apples, pears) are initially peeled mechanically and freed from the core and calix (seed compartment). Apples are then cut preferentially into 5–7 mm thick slices, and dried in rings (a yield of 10–20% of the unpeeled fresh weight). Sulfite treatment is used to prevent browning during processing and storage. The sulfurous acid prevents both enzymatic and nonenzymatic browning reactions, stabilizes vitamin C and prevents microbial contamination during storage of the end product. The utilization of dilute solutions of citric acid is also suitable for preventing browning. Whole or sliced pears are heated with steam to achieve a translucent appearance and then are dried at 60–65 °C. The yield is 13–14% of the fresh weight.

The *stone fruits* usually dried are plums/prunes, apricots and peaches. Plums are first dipped for 5–15 s into a hot, diluted solution of sodium hydroxide, or into 0.7% aqueous K-carbonate and then rinsed and dried at 70–75 °C or dried in the sun. Plum peels are often fissured to facilitate drying. In order to clean and to provide a black, glossy surface, dried plums are steamed additionally at 80–85 °C for a short time. The plum yield is 25–30% at a moisture content of not more than 19%. Apricots and peaches are treated alternately with cold and hot water, then are halved, the stone seed is removed and the fruit is dried in the sun or in drying installations at 65–70 °C. The yield, depending on fruit size, is 10–15%. SO_2 (sulfurous acid) treatment is common for apricots and peaches. Cherries play a less important role as dried fruit. To avoid substantial aroma losses, cherries are dried slowly and with a number of precautions.

Grapes are the most commonly dried *berry fruits*. Raisins are dark-colored, dried grapes which contain seeds, whereas sultana raisins are

Table 18.37. Composition of some dried fruits (g/100 g edible portion)

Fruit	Moisture	N-containing compounds (N × 6.25)	Lipid	Available carbohy- drates	Dietary fiber	Minerals	Vitamin C
Apricots	17.6	5.0	0.4	48	17.7	3.5	0.011
Dates	20.2	1.9	0.5	65	8.7	1.8	0.003
Figs	23.7	3.5	1.3	55	12.9	2.4	0–0.005
Peaches	24.0	3.0	0.6	53	12.8	3.0	0.017
Plums/ prunes	24.0	2.3	0.6	47	17.8	2.1	0.004
Raisins	15.7	2.5	0.5	68	5.2	2.0	0.001

seedless, light-colored, dried grapes. Currants, with or without seeds, are dark and are much smaller in size than the other two raisin products. The surface treatment of raisins, with the exception of currants, involves the use of acetylated monoglycerides to prevent caking or sticking.

The compositions of some dried fruits are presented in Table 18.37. Dried fruits are exceptionally rich in calories and they supply significant amounts of minerals. Of the vitamins found in fruit, β-carotene and the B-vitamins remain intact. Vitamin C is lost to a great extent. Sulfite treatment destroys vitamin B_1. However, fruit color and vitamin C content are retained and stabilized.

18.2.2 Canned Fruits

Since the middle of the 19th century, heat sterilization in cans and glass jars has been the most important process for fruit preservation.

Undamaged, aroma-rich and not overripe fruits are suitable for heat sterilization. Aseptic canning is applicable only for fruit purées. Canned fruits used are primarily stone fruits, pears, pineapples and apples (usually apple purée). Strawberries and gooseberries are canned to a lesser extent.

Canned fruits are produced in a large volume by the food industry and also in individual households. Cherries are freed from stone seeds and stems, plums/prunes, apricots and peaches are halved and the stone seeds are removed, strawberry calix is removed, gooseberry and red currant stems are removed, apples and pears are peeled and sliced. Specialized equipment has been developed for these procedures.

With a few exceptions (raspberries and blackberries) all fruits are washed or rinsed. Apricots are readily peeled after alkali treatment at 65 °C. Fruits sterilized unpeeled, e. g., prunes or yellow plums, are first fissured to prevent later bursting. To avoid aroma loss and to prevent floating in the can, fruits which shrink considerably (such as cherries, yellow plums, strawberries and gooseberries) are dipped prior to canning into a hot 30% sugar solution and then covered with a sugar solution, with a sugar concentration approximately twice the desired final concentration in the can. Finally, the can is vacuum sealed at 77–95 °C for 4–6 min and, according to the fruit species, heat sterilized under the required conditions. For example, a 1 liter can of strawberries is sterilized in a boiling water bath at 100 °C for 18 min, while pears, peaches and apricots are heated at 100 °C for 22 min. Additions of ascorbic and citric acids for stabilization of color and calcium salts for the preservation of firm texture have been accepted as standard procedures for canned fruits consumed as desserts.

Canned fruits used for bakery products, confections or candies are produced like canned dessert fruits, however, the fruits are covered with water instead of sugar solutions.

18.2.3 Deep-Frozen Fruits

Fruits are frozen and stored either as an end product or for further processing. The choice of suitable varieties of fruit at an optimal ripening stage is very important. Pineapples, apples, apricots, grapefruit, strawberries and dark-colored cherries are highly suitable. Light-colored cherries, plums, grapes and many subtropical or tropical fruits are of low suitability.

Rapid chilling is important (air temperature $\leq -30\,^\circ$C, freezing time about 3 h) to avoid microbial growth, large concentration shifts in fruit tissues, and formation of large ice crystals which damage tissue structure. A blanching step prior to freezing is commonly used only for few fruits, such as pears, and occasionally for apples, apricots and peaches. Some fruits are covered, prior to freezing, with a 30–50% sugar solution or with solid granulated sugar (1 part per 4–10 parts by weight) and are left to stand until the sap separates. In both instances oxygen is eliminated, enzymatic browning is prevented, and the texture and aroma of the fruit are better preserved. Addition of ascorbic acid or citric acid is also common.

Frozen fruit which is stored at -18 to $24\,^\circ$C is stable for two to four years.

18.2.4 Rum Fruits, Fruits in Sugar Syrup, etc.

Rum fruits are produced by steeping the fruit in dilute spirits in the presence of sufficient sugar.

Fruits preserved in vinegar, mostly pears and plums, are prepared by poaching in wine vinegar sweetened with sugar and spiced with cinnamon and cloves.

Fruits in sugar syrup are prepared by treating raw or precooked fruits or fruit portions (may be precooked under a vacuum) with highly concentrated sucrose solutions which also contain starch syrup. The latter is added to enhance translucency, smoothness and tractability of the product. Candied lemon or orange peels are products of this kind.

Other varieties provide intermediary products processed further into fruit confections: glazed fruits (these are washed fruits treated with a sugar solution containing gum arabic and then subsequently dried at 30–35 $^\circ$C) or candied fruits in which the dried, glazed fruit is also immersed in a concentrated sugar solution and then dried to form a candied hull. Another product is crystallized fruit in which the dried, glazed fruits are rolled over icing or granulated sugar (sucrose), then dried aditionally and, to achieve a shiny, glossy appearance, are exposed to steam for a short time.

18.2.5 Fruit Pulps and Slurries

Fruit pulp is not suitable for direct consumption. The pulp is in the form of slurried fresh fruit or pieces of fruit either split or whole, and, when necessary, stabilized by chemical preservatives. The minimum dry matter content of various pulps is 7–11%. For pulp production the fruit, which has been washed in special machines, is lightly steamed in steam conduits or precooking retorts. The fruit slurry is an intermediary product, also not suitable for direct consumption. The production steps are similar to those for pulp. However, there is an additional step: slurrying and straining, i.e. passing the slurry through sieves. Both the pulp and the slurry can be stored frozen.

18.2.6 Marmalades, Jams and Jellies

18.2.6.1 Marmalades

Marmalade is a spreadable preparation made from pulp, slurry, juice, aqueous extracts or peels of citrus fruits and sugars. The product (1 kg) has to contain at least 200 g of citrus fruit (of which 75 g endocarp) and 60% by weight of soluble solids. The addition of fruit pectin and starch syrup are customary.

For the production of marmalade, the fresh fruits or intermediary products, such as fruit pulps or slurries, are boiled in an open kettle at atmospheric pressure (T up to 105 $^\circ$C) or in a closed vacuum boiler at reduced pressure (T: 65–80 $^\circ$C) with the addition of sugar (usually added in two batches). In general, the latter process is used industrially. The aroma substances are recovered from the vapor and returned in concentrated form before filling. The solids content and pH value are usually controlled automatically during boiling. Other ingredients (gelling agents, starch syrup and acids) are added before the thickening is completed by boiling. The end of boiling is determined by refractometer readings (the total boiling time is usually 15–30 min). The hot (70–75 $^\circ$C) marmalade is then poured into appropriate containers.

18.2.6.2 Jams

Jams are produced similarly to marmalades but usually from one kind of fruit. They are thickened by boiling and constant stirring of the whole or sliced fresh or fresh stored raw material, or of fruit pulp. Ordinary jams are also made from fruit slurry. Boiling under a vacuum at 65–80 °C offers the advantage of preserving the aroma and color. The disadvantages are the absence of sucrose inversion and the low caramelization. These reactions produce the characteristic taste of jams boiled in an open kettle (T: 105 °C). Table 18.38 provides compositional data for some commercial jams. The optimal pH of 3.0 required for gelling is adjusted by the addition of lactic, citric or tartaric acid, if necessary.

18.2.6.3 Jellies

Jellies are gelatinous, spreadable preparations made from the juice or aqueous extract of fresh fruits by boiling down with sugar. The addition of fruit pectin (0.5% as calcium pectate) and tartaric acid or lactic acid (0.5%) is normal. In general, the water content is 42%, and the sugar content between 50% and 70%. The juice is boiled down in open kettles or in vacuum kettles with sugar (about half the weight of the fruit), pectin, if necessary, and the substances mentioned above.

Table 18.38. Composition of various jams (average values in %)

Jam from	Moisture	Total sugar	Total acid[a]	Ash	Dietary fiber
Strawberries	35.0	58.7	0.89	0.23	0.80
Apricots	36.9	51.3	1.14	0.28	0.60
Cherries	36.6	57.3	1.26	0.28	0.50
Blackberries	34.2	58.0	0.37	0.24	1.20
Raspberries	35.9	54.6	1.03	0.23	1.20
Bilberries	35.1	55.8	0.60	0.22	0.37[b]
Plums/prunes	31.1	59.1	0.42	0.24	0.43[b]

[a] Sum of malic and citric acid.
[b] Pectin as calcium pectate.

The scum is carefully skimmed off and the mixture is boiled further until a moisture content of about 42% is reached.

18.2.7 Plum Sauce (Damson Cheese)

Plum/prune sauce is produced by thickening through boiling of fresh fruit pulps or fruit slurries. The use of dried plums is also common. Normally, the product has no added sugar, but sweetened products or products with other ingredients added are also produced. The soluble solids have to be at least 60% by weight.

18.2.8 Fruit Juices

Fruit juices are usually obtained directly from fruit by mechanical means, and also from juice concentrates (cf. 18.2.10) by dilution with water. The solid matter content is generally 5–20%. The juices are consumed as such or are used as intermediary products, e. g., for the production of syrups, jellies, lemonades, fruit juice liqueurs or fruit candies. Fruit juice production is regulated in most countries.

Juices from acidic fruits are usually sweetened by adding sucrose, glucose or fructose. Juices used for further processing usually contain chemical preservatives to inhibit fermentation. Some juices from berries and stone fruits, because of their high acid content, are not suitable for direct consumption. Addition of sugar and subsequent dilution with water provides fruit nectars or sweet musts (cf. 18.2.9). Since 1990, the per capita consumption of fruit juice and fruit nectar in Germany has been fairly constant at 40 l. In the case of fruit juices, the products presented in Table 18.39 are predominant.

Table 18.39. Per capita consumption of fruit juices in Germany (2004)

Product	Amount (l)
Apple juice	12.8
Orange juice	8.9
Multivitamin juice	3.8
Grape juice	1.3

Table 18.40. Composition of fruit juices and nectars (g/l)

	Total sugar	Volatile acid	Ash	Total acid[a]	Vitamin C
Apple juice	72–102	0.15–0.25	2.2–3.1	1.4	0–0.03
Grape juice	120–180	–	2.1–3.2	3.6–11.7	0.017–0.02
Blackcurrant nectar	95–145	–	2.25–3.2	9.15–12.75	0.2–0.56
Raspberry juice	2.7–69.6	–	4.1–5.2	–	0.12–0.49
Orange juice	60–110	0.13	2.2.–4.0	5–18	0.28–0.86
Lemon juice	7.7–40.8	–	3.0–4.3	42–83.3	0.37–0.63
Grapefruit juice	50–83	0.16	2.5–5.6	5–27	0.25–0.5

[a] Calculated as the sum of malic and citric acid (and tartaric acid in the case of grape juice.

Table 18.40 lists data on the composition of some juices and nectars. Multivitamin juices are produced from orange and apple juice with the addition of banana slurry, passion fruit, mango, pineapple and papaya as well as a mixture of vitamins C, E, B_1, folic acid, niacin and panthothenic acid.

Production of fruit juice involves the processing steps: fruit preparationn and the extraction, treatment and preservation of the juice.

18.2.8.1 Preparation of the Fruit

Preparation of the Fruit involves washing, rinsing and trimming, i. e. the faulty and unripe fruits are removed. The stone seeds and stalks, stems or calyx are then removed. Disintegration is accomplished mechanically in mills, thermally by heating (thermobreak at about 80 °C) or by freezing (less than −5 °C). The yield can be increased to 90% by enzymatic pectin degradation ("mash fermentation", particularly of stone fruits and of berries) or by applying procedures such as ultrasound or electropermeabilization. In the last mentioned process, the raw material is subjected to preliminary disintegration, the cells are then opened up by means of electrical impulses of high field strength, e. g., 2–5 kV/cm.

18.2.8.2 Juice Extraction

Separation of the juice is achieved using continuous or discontinuous presses or processes such as vacuum filtration or extraction.

Before pressing, the fruit tissue is digested with pectinolytic and cellulolytic enzymes at 50 °C to increase the yield. In this way, especially fruit with a soft texture can be directly converted into drinkable juices without the addition of water according to the scheme: preparation – washing – mashing – enzyme treatment – filtration – pasteurization – filling.

18.2.8.3 Juice Treatment

The juice treatment step involves fining and clarification, i. e. removal of turbidity, and stabilization to prevent additional turbidity. The former step commonly involves treatment with enzymes, mostly pectinolytic, and, if necessary, removal of starch and polyphenols using gelatin, alone or together with colloidal silicic acid or tannin, or polyvinylpyrrolidone. Finally, proteins are removed by adsorption on bentonite.

Clarification of juice is achieved by filtration through porous pads or layers of cellulose, asbestos or kieselguhr, or by centrifugation.

Since juice production provides juices which are well-saturated with air oxygen-sensitive products are deaerated. This is achieved by an evacuation step or by purging the juice with an inert gas such as N_2 or CO_2.

Fruit juices (with the exception of citrus juices) are produced as transparent, clarified products, although some turbid juices are available. In the latter case, measures are required to obtain a stable, turbid suspension. This is achieved with stone fruit juices by a short treatment with polygalacturonase preparations which have a low pectin esterase activity and which then partially degrade and, thus, stabilize the ingredients required for turbidity. Citrus juices (lemons, oranges, grapefruits) are heat-treated to inactivate the endogenous pectin esterase, which would otherwise provide pectic acid which can aggregate and flocculate in the presence of Ca^{2+} ions. However, since heat treatment damages fruit aroma, the use of polygalacturonase is preferred. This enzyme degrades the pectic acid to such an extent that flocculation does not occur in the presence of divalent cations.

18.2.8.4 Preservation

Finally, the fruit juice preservation step involves pasteurization, preservation by freezing, storage under an inert atmosphere, or concentration (cf. 18.2.10) and drying (cf. 18.2.12).

Pasteurization kills the microflora and inactivates the enzymes, particularly the phenol oxidases. Since a longer heating time is detrimental to the quality, a short, high-temperature heat treatment is the preferred process, using plate heat exchangers (clear juices 85–92 °C, 10–15 s; fruit slurries up to 105 °C and up to 30 s) with subsequent rapid cooling. The juice is stored in germ-free tanks. Filling operations for the retail market can lead to reinfection, hence a second pasteurization is required. It is achieved by filling preheated containers with the heated juice, or by heating the filled and sealed containers in chambers or tunnel pasteurizers.

Preservation by freezing generally involves transforming the juice or juice concentrate into an ice slurry (at −2.5 °C to −6.5 °C), then packing and cooling to the retail market storage temperature. The product is stable for 5–10 months in a temperature range of −18 °C to −23 °C.

Storage in an inert atmosphere makes use of the fact that filtered, sterilized juices are microbiologically stable at temperatures below 10 °C and in an atmosphere of more than 14.6 g CO_2/l. To attain such a concentration of CO_2, the filled storage tank has to be at a pressure of 0.59 MPa at 10 °C, or 0.47 MPa at 5 °C.

Fruit juices are poured into retail containers, i.e. glass bottles, synthetic polyethylene pouches, aluminum cans, or aluminum-lined cardboard containers.

18.2.8.5 Side Products

Pomace is the residue from the production of fruit juices. Citrus fruits and apple pomace are used for the recovery of pectins. Other fruit residues are used as animal feed, as organic fertilizer, or are incinerated.

18.2.9 Fruit Nectars

Fruit nectars are produced from fruit slurries or whole fruits by homogenization in the presence of sugar, water and, when necessary, citric and ascorbic acids. The fruit content (as fresh weight) is 25–50% and is regulated in most countries, as is the minimum total acid content. Apricots, pears, strawberries, peaches and sour cherries are suitable for nectar production. The fruits are washed, rinsed, disintegrated and heated to inactivate the enzymes present. The fruit mash is then treated with a suitable mixture of pectinolytic and cellulolytic enzymes. The treatment degrades protopectin and, thus, separates the tissue into its individual intact cells ("*maceration*").

High molecular weight and highly esterified pectin formed from protopectin provides the high viscosity and the required turbidity for the nectar. Finally, the disintegrated product is filtered hot, then saturated with the usual additives, homogenized and pasteurized.

Fruit products from citrus fruits (comminuted bases) are obtained by autoclaving (2–3 min at 0.3 MPa) and then straining the fruits through sieves, followed by homogenization. Fruit nectars also include juices or juice concentrates from berries or stone fruits, adjusted by addition of water and sugar.

18.2.10 Fruit Juice Concentrates

Fruit juice concentrates are chemically and microbiologically more stable than fruit juices and their storage and transport costs are lower. The solid content (dry matter) of the concentrates is 60–75%. Intermediary products, less stable concentrates with a dry matter content of 36–48%, are also produced. These semiconcentrates are pasteurized at 87 °C. Fruit juice concentration is achieved by evaporation, freezing, or by a process involving high pressure filtration. Initially, the pectin is degraded to avoid high viscosity and gel setting (undesired properties).

18.2.10.1 Evaporation

Concentration by evaporation is the preferred industrial process. Since the process leads to losses in volatile aroma constituents, it is combined with an aroma recovery step. The aroma of the juice is enriched 100 to 200 times by a counter-current distillation. This aroma is stored and recombined with the juice only at the dilution stage. In order to maintain quality, the residence time in evaporators is as short as possible. In a high-temperature, short-time heating installation, e.g., in a 3- and 4-fold stepwise gradient-type evaporator, the residence time is 3–8 min at an evaporating temperature of 100 °C in the first step and about 40 °C in the fourth step. The concentrate is then cooled to 10 °C. Recovery of the aroma is achieved by rectifying the condensate of the first evaporation stage. A short-time treatment of juices is also possible in thin-layer falling film evaporators. These are particularly suitable for concentrating highly viscous products such as fruit slurries.

18.2.10.2 Freeze Concentration

Concentration of juice by freezing is less economical than evaporation. Hence, it is utilized mostly for products containing sensitive aroma constituents, e.g., orange juice. The juice is cooled continuously below its freezing point in a scraper-type cooler. The ice crystals are separated from the resultant ice slurry by pressing or by centrifugation. The obtainable solid content of the end product is 40–50%. This content is a function of freezing temperature, as illustrated with apple juice in Fig. 18.11.

18.2.10.3 Membrane Filtration

Concentration of juice by filtration using semipermeable membranes and high pressure (0.1–1 MPa) is known as ultrafiltration. When the membrane is permeable for water and only to a limited extent for other small molecules ($M_r < 500$, e.g., salts, sugars, aroma compounds), the process is called reverse osmosis. Concentration of juice is possible only to about 25% dry matter content.

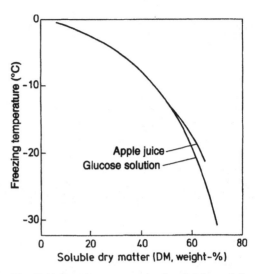

Fig. 18.11. Freezing temperature of apple juice and glucose solution as affected by soluble dry matter (DM) (according to *Schobinger*, 1978)

18.2.11 Fruit Syrups

Fruit syrups are thick, fluid preparations made by boiling one kind of fruit with an excess of sugar. They are sometimes prepared without heating by directly treating fresh fruit or fruit juice with sugar, occasionally also using small amounts of tartaric or lactic acids. Fruit syrups from citrus fruits often contain small amounts of peel aromas.

Fruit syrups are rapidly cooled to avoid aroma losses and caramelization of sugar. The boiling process partially inverts sucrose, preventing subsequent sucrose crystallization. Low-acid fruits are treated with tartaric or lactic acid. Boiling in closed kettles permits recovery of vaporized aroma compounds which can be added back to the end product. As in marmalade production, the boiling is occasionally done under vacuum (50 °C starting temperature, 65–70 °C final temperature) in order to retain the aroma. Syrup production by a cold process is particularly gentle. The raw juice flows over the granulated sucrose in the cold until the required sugar concentration has been achieved. Aroma-sensitive syrups which contain turbidity-causing substances, e.g., citrus fruit syrups, are made by adding sugar to the mother liquor with vigorous stirring. Fruit syrup can contain at most 68% of sugar (calculated as invert sugar) and has to contain at least 65% of soluble solids.

18.2.12 Fruit Powders

Fruit powders are produced by drying juices, juice concentrates or slurries. The hygroscopic powders contain less than 3–4% moisture. Addition of drying aids (such as glucose, maltose or starch syrup) in amounts greater than 50% of the dry matter can efficiently control clumping or caking due to the presence of fructose in the drying process. Freeze-drying, vacuum foam-drying (0.1–1 kPa, 40–60 °C) and foam-mat drying are suitable drying processes. In the last mentioned process, the material to be dried is foamed with foam stabilizers and inert gas and then dried. Spray drying is also applied. It has the disadvantage that often strong color and aroma changes occur.

18.3 Alcohol-Free Beverages

18.3.1 Fruit Juice Beverages

Fruits mostly provide only the taste in fruit juice beverages. These drinks are prepared from fruit juices or their mixtures or from fruit concentrates, with or without addition of sucrose or glucose, and are diluted with water or soda or mineral water. A minimum amount of fruit juice is stipulated: 30% of seed fruit juices or grape juice, 6% of citrus juice or citrus juice mixtures, and 10% of other juices or juice mixtures.

18.3.2 Lemonades, Cold and Hot Beverages

These drinks are prepared with or without the use of fruit juice or fruit extracts by the addition of natural fruit essences and sugar (sucrose or glucose), fruit acids and soda or mineral water. They are also consumed without added carbon dioxide, either cold or warmed. The drinks are usually colored. Lemonades made with the addition of fruit juices contain at least half the amount of fruit juice normally contained in fruit juice drinks.

The sugar added to lemonades has to be at least 7% based on the finished beverage.

Tonic water is also considered a lemonade. It contains about 80 mg quinine/l to provide the characteristic bitter taste.

18.3.3 Caffeine-Containing Beverages

These are also considered as "lemonades" (particularly in Europe). The most popular are the cola drinks, which contain extracts from the cola nut (*Cola nitida*) or aromatic extracts from ginger, orange blossoms, carob and tonka beans or lime peels. Caffeine is often added (6.5–25 mg/100 ml). Phosphoric acid is sometimes used as an acidulant (70 mg/100 ml). The sugar content of cola drinks averages 10–11%. The deep-brown color of the drink is adjusted with caramel.

18.3.4 Other Pop Beverages

Some effervescent pop drinks are imitations of fruit juices and lemonade-type drinks, however, their sugar content is fully or partially replaced by artificial sweeteners and the natural essence of flavoring ingredients are replaced by artificial or artificially-enhanced essences. Coloring substances are usually added.

18.4 Analysis

As a result of the numerous raw materials and processes involved, the analysis of fruit products is difficult and tedious. Information on the following is important for an evaluation:

- Type, amount, and origin, if necessary, of the fruit and additives used (e. g., acids, sugar).
- Constituents that determine quality (e. g., aroma substances, vitamins).
- Method of processing.

Information of this type is provided by the quantitative analysis of various constituents, determination of species-specific compounds, and by the determination of abundance ratios of isotopes.

18.4.1 Various Constituents

Since the composition of the raw materials varies greatly, deviations from the standard can be recognized only by collective changes in the concentration of as many components as possible.

For orange juice, Table 18.41 shows that when the guide values for certain components are exceeded or fallen short of, information is provided on the proportion of fruit, the use of expressed residues, acidification, sweetening, and microbiological spoilage.

Table 18.41. Guide values for orange juice

Quantity being measured/Component	Mean	Range of variation		Guide value[a]	Indicator[b]
Specific gravity at 20/20 °C	1.046	1.045–1.055	x	1.0450	1
Extract (°Brix)[c]	11.41	11.18–13.54	x	11.18	1
Soluble solids (g/L)	119.4	116.8–142.9	x	116.8	1
Titratable acids (pH 7.0)					
calc. as tartaric acid (g/L)	9.5	8.0–12	x	8.0	1
Ethanol (g/L)	– –	– –	o	3.0	3
Volatile acids					
calc. as acetic acid (g/L)	–	–	o	0.4	3
Lactic acid (g/L)	–	–	o	0.5	3
L(+)-Ascorbic acid (mg/L)	350	–	x	200	2
Peel oil (g/L)	–	–	o	0.3	7
Glucose (g/L)	28	20	x	22	1
Fructose (g/L)	30	22	x	24	1
Ratio of glucose to fructose	0.92	0.85–1.0	o	1.0	5
Sucrose (g/L)	33	47	o	45	5
Sucrose (% of total sugar)	–	–	o	50	5
Ash (g/L)	4.0	2.9–4.8	x	3.5	1
Sodium (mg/L)	14	–	o	30	1
Potassium (mg/L)	1900	1400–2300	x	1700	1
Calcium (mg/L)	80	60–120	o	110	1
Magnesium (mg/L)	100	70–150	x	90	1
Chloride (mg/L)	–	–	o	60	1
Nitrate (mg/L)	–	–	o	10	1
Phosphate (mg/L)	460	350–600	x	400	1
Sulfate (mg/L)	–	–	o	150	1
Citric acid (g/L)	9.4	7.6–11.5	x	8.0	4
Isocitric acid (mg/L)	90	65–130	x	70	4
Ratio of citric acid to isocitric acid[d]	105	80–130	o	130	4
L-Malic acid (g/L)	1.7	1.1–2.9	o	2.5	1
Prolin (mg/L)	800	450–1300	x	575	1
Formol value (0.1 mol/L NaOH per 100 mL)[e]	20	15–26	x	18	1
Flavonoid glycosides					
calc. as hesperidin (mg/L)	800	500–1000	o	1000	6
Water soluble pectins calc. as					
galacturonic acid anhydride (mg/L)	300	–	o	500	6

[a] Minimal (x) maximal (o) guide value.
[b] Indicator for: 1 fruit content, 2 heat or oxidation damage, 3 microbiological spoilage, 4 acidification, 5 sweetening, 6 extract of expressed residue, 7 as 6 but aromatized with peel oil.
[c] 1°Brix = 1 g of extract in 100 g of solution.
[d] cf. 18.1.2.4.
[e] Formol titration: after the addition of formaldehyde to a solution of the sample at pH 8–9, the free amino acids are determined by titration with sodium hydroxide solution.

Table 18.42. Phenolic compounds as indicator substances for the detection of adulteration of fruit products

Compound	Occurrence	Detection
Quercetin-3-rutinoside	Common, but not in strawberries	Elderberry juice in strawberry juice
Quercetin-3-O-(2″-O-α-L-rhamnosyl-6″-O-α-L-rhamnosyl)-β-D-glucoside	Red currants	Red currants in products from black currants
Naringin or naringenin	Grapefruits	Grapefruit juice in orange juice
Apigenin-6-C-β-D-glucopyranosyl-8-C-α-L-arabinopyranoside (schaftoside)	Figs	Fig juice in grape juice

18.4.2 Species-Specific Constituents

The occurrence of species-specific constituents is also analytically useful. The composition of the plant phenols of individual fruits can be analyzed quickly and very accurately by using HPLC. These data have shown that certain compounds are suitable indicators of adulteration (Table 18.42). These indicators must be fixed with great care. In fact, phloretin-2-glucoside (phloridzine) and isorhamnetinglucoside have been proposed as markers for apples and pears. Improvements in the analyses, however, showed that phloridzine and isorhamnetinglucoside widely occur in low concentrations in fruit, and the last mentioned glucoside also occurs in apples, among other fruit.

It must be guaranteed that the selected indicator substance is stable under the production conditions for the particular fruit product. Therefore, anthocyanins are generally not suitable. For fermented products, O-glycosides are not suitable because they are degraded by yeast enzymes. Suitable compounds are C-glycosidically bound flavonoids which are resistant to enzymatic hydrolysis and common chemical hydrolysis, e. g., schaftoside (cf. Table 18.42) can be detected even in wine and champagne when the must is adulterated with fig juice.

The analytical importance of amino acid (cf. 18.1.2.1.2), protein, enzyme (cf. 18.1.2.1.1), and carotinoid patterns (cf. 18.1.2.3.2) have already been mentioned.

Adulteration of orange juice by the addition of an aqueous extract of the pulp, which remains after pressing of the juice (pulp wash), is detected by the marker N,N-dimethylproline. The levels of this amino acid are higher in pulp wash than in juice.

18.4.3 Abundance Ratios of Isotopes

The content of the isotopes ^{2}H and ^{13}C is a criterion of the origin of the food or of individual constituents, e. g., sugar used to sweeten fruit juice. The method is based on the fact that isotopomeric molecules, e. g., $^{12}CO_2$ and $^{13}CO_2$, react at different rates in biochemical and chemical reactions (kinetic isotope effect). In general, the molecules with the heavier isotope react slower, so that this isotope is enriched in the products.

The resulting change in the abundance ratio is expressed as the δ-value, based on an international standard (Table 18.43).

$$\delta = \frac{R_{sample} - R_{standard}}{R_{standard}} \times 1000 [\text{‰}] \qquad (18.47)$$

$$R = \frac{C_1}{C_2} \qquad (18.48)$$

c_1/c_2: concentrations of heavy/light isotopes. The δ(^{13}C) value, which is -8 ± 1‰ for atmospheric CO_2, increases during CO_2 fixation as a function of the type of photosynthesis of the

Table 18.43. Abundance of important isotopes and international standards for their determination

Isotope	Rel. mean natural abundance [atom %]	International standard Name	R^a
^{1}H	99.9855	V-SMOW^b	0.00015576
^{2}H	0.0145		
^{12}C	98.8920	PDB^c	0.0112372
^{13}C	1.108		

^{a} Abundance ratio (Formula 18.48).
^{b} Vienna Standard Mean Ocean Water.
^{c} Pee Dee Belemnite (CaCO$_3$ from the Pee Dee formation in South Carolina).

Table 18.44. Isotope discrimination in primary photosynthetic CO_2 binding

Plant group	CO_2 Acceptor	$\delta(^{13}C)$ value ‰	Foods
C_3-Plant	D-Ribulose-1,5-bis-phosphate carboxylase (RuPB-C)	−32 to −24	Wheat, rice, oats, rye, potatoes, barley, batata, soybean, orange, sugar beet, grapes
C_4-Plant	Phosphoenolpyruvate carboxylase (PEP-C)	−16 to −10	Corn, millet, sugar cane
CAM-Plant[a]	RuBP-C/REP-C	−30 to −12	Pineapples, vanilla, cactaceae, agave

[a] CAM: Crassulacean acid metabolism.

plant (Table 18.44). The discrimination in C_3-plants is the greatest and is caused by the kinetic isotope effect in the reaction catalyzed by ribulose-1,5-biphosphate carboxylase. It is considerably less in C_4-plants. CAM plants occupy an intermediate position (Table 18.44) because the C_3- or the C_4-path is taken depending on the growth conditions.

The large differences in the masses of ^1H_2O, $^2H^1HO$, and ^2H_2O result in considerable thermodynamic isotope effects on phase transitions. On evaporation, deuterium (^2H) correspondingly decreases in the volatile phase, so that surface-, ground-, and rain-water contains less ^2H than the oceans. The ^2H enrichment in the oceans is greatest at the equator and decreases with increasing latitude because the amount of water evaporating depends on the temperature.

The hydrogen of plant foods comes from precipitation and from the ground-water in that particular location. Therefore, plants of the same type of photosynthesis, which are cultivated at different places, differ in their $\delta(^2H)$ values. Kinetic isotope effects in plant metabolism, which due to the mass difference $^2H/^1H$ are much higher than in the case of $^{13}C/^{12}C$, also have an effect on the $\delta(^2H)$ values.

For isotope analysis, the sample is subjected to catalytic combustion to give CO_2 and H_2O. After drying, the $^{13}C/^{12}C$ ratio in CO_2 is determined by mass spectrometry. The $^2H/^1H$ ratio is determined in hydrogen, which is formed by reducing the water obtained from catalytic combustion. The $^2H/^1H$ ratio can change by $^2H/^1H$ exchange, e. g., as undergone by OH groups. Therefore, such groups are eliminated before combustion. For example, only the $\delta(^2H)$ values of the CH-skeleton in carbohydrates are determined after conversion to the nitrate ester.

Table 18.45. $\delta(^{13}C)$ and $\delta(^2H)$ values for orange juice and sugar of different origins

Food	$\delta(^{13}C)(‰)$	$\delta(^2H)(‰)$
Orange juice, freeze-dried	−25.6±0.8	n.a.
Sucrose isolated from orange juice	−25.5±2.5	−22±10
Beet sugar	−25.6±1.0	−135±25
Cane sugar	−11.5±0.5	−50±20
Glucose-fructose syrup (corn)	−10.8±0.9	−31

n.a.: not analyzed.

Sweetening orange juice with cane sugar or glucose-fructose syrup from corn starch lowers the $\delta(^{13}C)$ value of sugar, which is −25.5‰ in the native juice (Table 18.45). On the other hand, the addition of beet sugar (C_3-plant) can be recognized only via the $\delta(^2H)$ value. The addition of synthetic products from petrochemicals ($\delta(^{13}C)$: −27±5‰) to foods from C_3-plants cannot be detected via the $\delta(^{13}C)$ value, but via the $\delta(^2H)$ value in many cases.

Apart from the global ^{13}C and ^2H contents of food constituents, the intramolecular distributions of these isotopes are typical of the origin and, therefore, of great analytical importance. They can be measured after chemical decomposition of the substance or with ^{13}C or ^2H NMR spectroscopy (example in 5.5.1.5).

18.5 References

Bell, E.A., Charlwood, B.Y. (Eds.): Secondary plant products. Springer-Verlag: Berlin. 1980

Berger, R.G., Shaw, P.E., Latrasse, A., Winterhalter, P: Fruits I–IV. In: Volatile compounds in foods and beverages (Ed.: Maarse, H.) Marcel Dekker, New York, 1991

Bricout, J., Koziet, J.: Control of authenticity of orange juice by isotopic analysis. J. Agric. Food Chem. *35*, 758 (1987)

Buettner, A., Schieberle, P.: Characterization of the most odor-active volatiles in fresh, hand-squeezed juice of grapefruit (*Citrus paradisi Macfayden*). J. Agric. Food Chem. *47*, 5189 (1999)

Demole, E., Enggist, P., Ohloff, G.: 1-p-Menthene-8-thiol: A powerful flavor impact constituent of grapefruit juice (*Citrus paradisi* MacFayden). Helv. Chim. Acta *65*, 1785 (1982)

Drawert, F., Görg, A., Staudt, G.: Über die elektrophoretische Differenzierung und Klassifizierung von Proteinen. IV. Disk-Elektrophorese und isoelektrische Fokussierung in Poly-Acrylamid-Gelen von Proteinen und Enzymen aus verschiedenen Erdbeerarten, -sorten und Artkreuzungsversuchen. Z. Lebensm. Unters. Forsch. *156*, 129 (1974)

Flügel, R., Carle, R., Schieber, A.: Quality and authenticity control of fruit purees, fruit preparations and jams – a review. Trends Food Sci. Technol. *16*, 433 (2005)

Fowden, L., Lea, P.J., Bell, E.A.: The nonprotein amino acids of plants. Adv. Enzymol. *50*, 117 (1979)

Frankel, E.N., Bruce, G.J.: Antioxidants in foods and health: problems and fallacies in the field. J. Sci. Food Agric. *86*, 1999 (2006)

Fuchs, G., Knechtel, W.: Fruchtsäfte und Gemüsesäfte. In: Taschenbuch für Lebensmittelchemiker und -technologen. Editor D. Osteroth, Band 2, S. 287, Springer-Verlag, Berlin, 1991

Fuhrmann, E., Grosch, W.: Character impact odorants of the apple cultivars Elstar and Cox Orange. Nahrung/Food *3*, 187 (2002)

Handwerk, R.L., Coleman, R.L.: Approaches to the citrus browning problem. A review. J. Agric. Food Chem. *36*, 231 (1988)

Haslam, E., Lilley, T.H.: Natural astringency in foodstuffs – a molecular interpretation. Crit. Rev. Food Sci. Nutr. *27*, 1 (1988)

Herderich, M., Gutsche, B.: Tryptophan-derived bioactive compounds in food. Food Rev. Int. *13*, 103 (1997)

Hermann, K.: Occurrence and content of hydroxycinnamic and hydroxybenzoic acid compounds in foods. Crit. Rev. Food Sci. Nutr. *28*, 315 (1989)

Herrmann, K.: Obst, Gemüse und deren Dauerwaren und Erzeugnisse. In: Taschenbuch für Lebensmittelchemiker und -technologen. Editor W. Frede, Band 1, S. 337, Springer-Verlag, Berlin, 1991

Hinterholzer, A., Schieberle, P.: Identification of the most odour-active volatiles in fresh, hand-extracted juice of Valencia late oranges by odour dilution techniques. Flavour Fragrance J. *13*, 49 (1998)

Jurd, L.: Reactions involved in sulfite bleaching of anthocyanins. J. Food Sci. *29*, 16 (1964)

Jurd, L., Asen, S.: Formation of metal and copigment complexes of cyanidin 3-d-glucoside. Phytochemistry *5*, 1263 (1966)

Kader, A.A., Zagory, D., Kerbel, E.L.: Modified atmosphere packaging of fruits and vegetables. Crit. Rev. Food Sci. Nutr. *28*, 1 (1989)

Larsen, M., Poll, L.: Odour thresholds of some important aroma compounds in raspberries. Z. Lebensm. Unters. Forsch. *191*, 129 (1990)

Linskens, H.F., Jackson, J.F. (Eds.): Analysis of non-alcoholic beverages. In: Modern methods of plant analysis. Volume 8. Springer-Verlag: Berlin. 1988

Manach, C.: Polyphenols: food sources and bioavability. Am. J. Clin. Nutr. *79*, 727 (2004)

Nagy, S., Attaway, J.A. (Eds.): Citrus nutrition and quality. ACS Symposium Series 143, American Chemical Society: Washington, D.C. 1980

Nogata, Y., Sakamoto, K. Shiratsuchi, H., Ishi, T., Yano, M., Ohta, H.: Flavonoid composition of fruit tissues of citrus species. Biosci. Biotechnol. Biochem. *70*, 178 (2006)

Ong, P.K.C., Acree, T.E.: Gas chromatography/olfactometry analysis of Lichee (*Litchi chinensis*). J. Agric. Food Chem. *46*, 2282 (1998)

Schieberle, P., Buettner, A.: Comparison of key odorants in fresh, hand-squeezd juices of orange (Valencia late) and grapefruit (White marsh). In: Frontiers of Flavour Science (P. Schieberle, K.-H. Engel, eds.) p. 10, Deutsche Forschungsanstalt für Lebensmittelchemie, Garching, 2000

Schieberle, P., Hofmann, T.: Evaluation of the character impact odorants in fresh strawberry juice by quantitative measurements and sensory studies on model mixtures. J. Agric. Food Chem. *45*, 227 (1997)

Schmid, W., Grosch, W.: Quantitative Analyse flüchtiger Aromastoffe mit hohen Aromawerten in Sauerkirschen, Süßkirschen und Kirschkonfitüren. Z. Lebensm. Unters. Forsch. *183*, 39 (1986)

Seymour, G.B., Taylor, J.E., Tucker, G.A.: Biochemistry of fruit ripening. Chapman & Hall, London, 1993

Shahidi, F., Naczk, M.: Phenolics in food and neutraceuticals, CRC Press, Boca Raton, 2004

Siebert, K.J.: Protein-polyphenol haze in beverages. Food Technol. *53* (no. 1), 54 (1999)

Siewek, F., Galensa, R., Herrmann, K.: Nachweis eines Zusatzes von Feigensaft zu Traubensaft und daraus hergestellten alkoholischen Erzeugnissen über die HPLC-Bestimmung von Flavon-C-glykosiden. Z. Lebensm. Unters. Forsch. *181*, 391 (1985)

Takeoka, G., Buttery, R.G., Flath, R.A., Teranishi, R., Wheeler, E.L., Wieczorek, R.L., Guentert, M.: Volatile constituents of pineapple (*Ananas comosus*). ACS Symp. Ser. *388*, 223 (1989)

Tokitomo, Y., Steinhaus, M., Büttner, A., Schieberle, P.: Odor-active constituents in fresh pineapple (*Ananas cosmosus* [L.] Merr.) by quantitative and sensory evaluation. Biosci. Biotechnol. Biochem. *69*, 1323 (2005)

Valero, D., Martinez-Romero, D., Serrano, M.: The role of polyamines in the improvement of the shelf life of fruit. Trends Food Sci. Technol. *13*, 228 (2002)

Weiss, H.O.: Konfitüren, Gelees, Marmeladen. In: Taschenbuch für Lebensmittelchemiker und -technologen. Editor D. Osteroth. Band 2, S. 427, Springer-Verlag, Berlin, 1991

Williams, A.A.: The flavour of non-citrus fruits and their products. In: International symposium on food flavors (Eds.: Adda, J., Richard, H.), p. 46, Technique et Documentation (Lavoisier): Paris. 1983

Winkler, F.J., Schmidt, H.-L.: Einsatzmöglichkeiten der ^{13}C-Isotopen-Massenspektrometrie in der Lebensmitteluntersuchung. Z. Lebensm. Unters. Forsch. *171*, 85 (1980)

Wu, X., Beecher, G.R., Holden, J.M., Haytowitz, D. B., Gebhardt, S.E., Prior, R.L.: Concentrations of anthoxyanins in common foods in the United States and estimation of normal consumption. J. Agric. Food Chem. *54*, 4069 (2006)

19 Sugars, Sugar Alcohols and Honey

19.1 Sugars, Sugar Alcohols and Sugar Products

19.1.1 Foreword

Only a few of the sugars occurring in nature are used extensively as sweeteners. Besides sucrose (saccharose), other important sugars are: glucose (starch sugar or starch syrup); invert sugar (equimolar mixture of glucose and fructose); maltose; lactose; and fructose. In addition, some other sugars and sugar alcohols (polyhydric alcohols) are used in diets or for some technical purposes. These include sorbitol, xylitol, mannitol, maltulose, isomaltulose, maltitol, isomaltitol, lactulose and lactitol. Some are used commonly in food and pharmaceutical industries, while applications for others are being developed. Food-grade oligosaccharides, which can be economically produced, are physiologically and technologically interesting. This group includes galacto-, fructo-, malto- and isomalto-oligosaccharides. Table 19.1 reviews relative sweetness, source and means of production, and Table 19.2 gives nutritional and physiological properties. Whether compounds will be successful as a sweetener depends on nutritional, physiological and processing properties, cariogenicity as compared to sucrose, economic impact, and the quality and intensity of the sweet taste.

19.1.2 Processing Properties

The potential of a compound for use as a sweetener depends upon its physical, processing and sensory properties. Important physical properties are solubility, viscosity of the solutions, and hygroscopicity. Figure 19.1 shows that the solubility of sugars and their alcohols in water is variable and affected to a great extent by temperature.

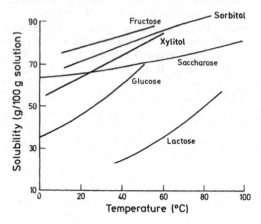

Fig. 19.1. Solubility of sugars and sugar alcohols in water (according to *Koivistoinen*, 1980)

There are similar temperature and concentration influences on the viscosity of aqueous solutions of many sugars and sugar alcohols. As an example, Fig. 19.2 shows viscosity curves for sucrose as a function of both temperature and concentration.

The viscosity of glucose syrup depends on its composition. It increases as the proportion of the high molecular weight carbohydrates increases (Fig. 19.3).

Figure 19.4 shows the water absorption characteristics of several sweeteners. Sorbitol and fructose are very hygroscopic, while other sugars absorb water only at higher relative humidities. Chemical reactions of sugars were covered in detail in Chapter 4. Only those reactions important from a technological viewpoint will be emphasized here.

All sugars with free reducing groups are very reactive. In mildly acidic solutions monosaccharides are stable, while disaccharides hydrolyze to yield monosaccharides. Fructose is maximally stable at pH 3.3; glucose at pH 4.0. At lower pH's dehydration reactions prevail, while the *Lobry de Bruyn–van Ekenstein* rearrangement oc-

H.-D. Belitz · W. Grosch · P. Schieberle, *Food Chemistry*
© Springer 2009

Table 19.1. Sweeteners of carbohydrate origin

Name	Relative sweetness[a]	Starting material, applied process
Saccharose	1.00	Isolation from sugar beet and sugar cane
Glucose	0.5−0.8	Hydrolysis of starch with acids and/or enzymes (α-amylase + glucoamylase)
Fructose	1.1−1.7	a) Hydrolysis of saccharose, followed by separation of the hydrolysate by chromatography. b) Hydrolysis of starch to glucose, followed by isomerization and separation by chromatography
Lactose	0.2−0.6	Isolation from whey
Mannitol	0.4−0.5	Hydrogenation of fructose
Sorbitol	0.4−0.5	Hydrogenation of glucose
Xylitol	1.0	Hydrogenation of xylose
Galactose	0.3−0.5	Hydrolysis of lactose, followed by separation of hydrolysate
Glucose syrup (starch syrup)	0.3−0.5[b]	Hydrolysis of starch with acids and/or enzymes; hydrolysate composition is strongly affected by process parameters (percentage of glucose, maltose, maltotriose and higher oligosaccharides)
Maltose	0.3−0.6	Hydrolysis of starch
Maltose syrup		As glucose syrup; process parameters adjusted for higher proportion of maltose in hydrolysate (amylase from *Aspergillus oryzae*)
Glucose/fructose syrup (isoglucose, high fructose syrup)	0.8−0.9	Isomerization of glucose to glucose/fructose mixture with glucose isomerase
Invert sugar		Hydrolysis of saccharose
Hydrogenated glucose syrup	0.3−0.8	Hydrogenation of starch hydrolysate (glucose syrup); composition is highly dependent on starting material (content of sorbitol, maltitol and hydrogenated oligosaccharides)
Maltitol syrup		Hydrogenation of maltose syrup
Galacto-oligosaccharides	0.3−0.6	Transgalactosylation of lactose by β-galactosidase (lactase)
Lactitol	0.3	Hydrogenation of lactose
Lactulose	ca. 0.6	Alkaline isomerization of lactose
Lactosucrose	0.3−0.6	Fructose from sucrose is transferred to lactose by β-fructofuranosidase
Maltitol	ca. 0.9	Hydrogenation of maltose
Isomaltitol	0.5	Hydrogenation of isomaltose
Fructo-oligosaccharides	0.3−0.6	Controlled enzymatic hydrolysis of inulin by inulase
Palatinose	0.3−0.6	Enzymatic isomerization of sucrose
Palatinose oligosaccharides	0.3−0.6	Intermolecular condensation of palatinose
Palatinit	0.45	Hydrogenation of palatinose
Glucosyl sucrose	0.5	Glucose from maltose is transferred to sucrose by cyclomaltodextrin-glucotransferase
Malto-oligosaccharides	0.3−0.6	a) Hydrolysis of the 1,6-α-glycosidic bonds in starch (debranching) by pullulanase; b) controlled hydrolysis by α-amylase
Isomalto-oligosaccharides	0.3−0.6	a) Hydrolysis of starch by α- and β-amylase; b) transglucosylation by α-glucosidase
Gentio-oligosaccharides	0.3−0.6	From glucose syrup by enzymatic transglucosylation
L-Sorbose	0.6−0.8	From glucose
Xylitol	1.0	Hydrogenation of xylose
Xylo-oligosaccharides	0.3−0.6	Controlled hydrolysis of xylan by endo-1,4-β-xylanase

[a] Sweetness is related to saccharose sweetness (= 1); the values are affected by sweetener concentration.
[b] Sweetness value is strongly influenced by syrup composition.

Table 19.2. Nutritional/physiological properties of carbohydrate-derived sweeteners

Sweetener	Resorption in metabolism	Utilization	Effect on blood sugar level and insulin secretion	Other properties
Sucrose	Effective after being hydrolyzed	Hydrolysis to fructose and glucose	Moderately high	Cariogenic
Glucose	Effective	Insulin-dependent in all tissues	High	Less cariogenic than sucrose
Fructose	Faster than by diffusion process	In liver to an extent of 80%	Low	Accelerates alcohol conversion in liver
Lactose	Effective after being hydrolyzed	Hydrolysis to glucose and galactose	High	Intolerance by humans lacking lactase enzyme; laxative effect
Sorbitol	Diffusion	Oxidation to fructose	Low	Slightly cariogenic and laxative
Mannitol	Diffusion	Partially utilized by liver	Low	Slightly cariogenic and laxative
Xylitol	Diffusion	Utilized preferentially by liver and red blood cells	Low	Not cariogenic, available data indicate an anticariogenic effect; mildly laxative
Hydrogenated glucose syrup	After hydrolysis glucose effective; sorbitol by diffusion	Variable depending on composition	Variable, composition dependent	Slightly cariogenic; mildly laxative
Arabinitol	Diffusion	Not metabolized by humans	None	Side effects unknown; probably laxative
Galactose	Effective	Isomerization to glucose	High	Forms cataracts in the eyes in feeding trials with rats; probably laxative
Isomaltitol	None	Probably not metabolized	None	Side effects unknown; strongly laxative
Lactitol	None	Partial hydrolysis to galactose and sorbitol	None	Side effects unknown; strongly laxative
Lactulose	None	No hydrolysis	None	Effects the N-balance; strongly laxative, bifidogenic
Maltitol	Effective as glucose after hydrolysis; sorbitol by diffusion	Hydrolysis to glucose and sorbitol	Probably slight	Side effects unknown; laxative
Maltose	Effective after hydrolysis	Hydrolysis to glucose	High	Cariogenic; intravenously given it appears to be utilized directly and, like glucose, it is insulin-dependent
L-Sorbose	Diffusion	Utilized preferentially by liver	Probably slight	Feeding trials with dogs revealed hemolytic anemia at a higher dosage; probably laxative
D-Xylose	Diffusion	Not metabolized by humans	None	Forms cataract in the eyes in feeding trials with rats; probably laxative

Table 19.2. continued

Sweetener	Resorption in metabolism	Utilization	Effect on blood sugar level and insulin secretion	Other properties
Palatinit		Partial hydrolysis to glucose, sorbitol, and mannitol	Probably slight	Side effects unknown
Galacto-oligosaccharides	Active after hydrolysis	Partial hydrolysis	Moderate	Bifidogenic, slightly cariogenic
Lactosucrose	Active after hydrolysis	Partial hydrolysis	Moderate	Bifidogenic, slightly cariogenic
Fructo-oligosaccharides	Active after hydrolysis	Partial hydrolysis	Slight	
Glucosyl-sucrose	Active after hydrolysis	Partial hydrolysis	Slight	Slightly cariogenic
Malto-oligosaccharides	Active after hydrolysis	Hydrolysis in the small intestine	High	Reduction of undesirable bacteria in intestine
Isomalto-oligosaccharides	Active after hydrolysis	Slight hydrolysis	Low	Bifidogenic
Gentio-oligosaccharides	None		None	Bifidogenic
Xylo-oligosaccharides	None	No change	None	Bifidogenic

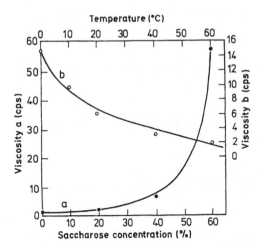

Fig. 19.2. Viscosity of aqueous saccharose solutions as a function of (a) saccharose concentration (20 °C) and temperature (40% saccharose) (according to *Shallenberger* and *Birch*, 1975)

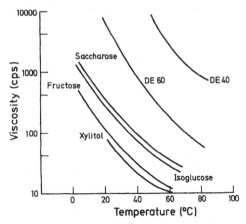

Fig. 19.3. Viscosity of some sugar solutions. Glucose syrup DE40: 78 weight-%; glucose syrup DE60: 77 weight-%; all other sugar solutions: 70 weight-% (according to *Koivistoinen*, 1980)

curs at higher pH's. Reducing sugars are unstable in mildly alkaline solutions, while nonreducing disaccharides, e. g., sucrose, have their stability maxima in this pH region.

The thermal stability of sugars is also quite variable. Sucrose and glucose can be heated in neutral solutions up to 100 °C, but fructose decomposes at temperatures as low at 60 °C.

Sugar alcohols are very stable in acidic or alkaline solutions. Relative taste intensity values for various sweeteners are found in Table 19.1. Taste intensity within a food can depend on a series of parameters, e. g., aroma, pH or food texture.

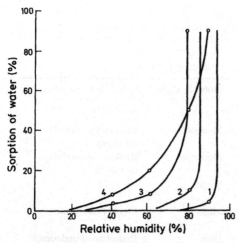

Fig. 19.4. Sorption of water by sugars at room temperature. *1* Saccharose, *2* xylitol, *3* fructose, *4* sorbitol (according to *Koivistoinen*, 1980)

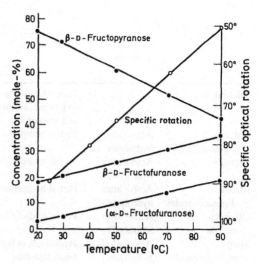

Fig. 19.6. Fructose mutarotation equilibrium as affected by temperature (according to *Shallenberger*, 1975)

Creams and gels with the same amounts of sweetener are often less sweet than the corresponding aqueous solutions. The sweet taste intensity may also depend on temperature (Fig. 19.5),

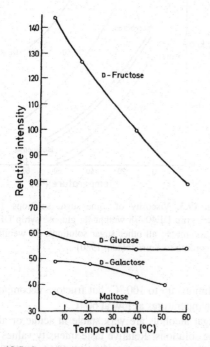

Fig. 19.5. Sugar sweetness intensity versus temperature. At all temperatures the saccharose taste intensity is 100 (according to *Shallenberger*, 1975)

an effect which is particularly pronounced with fructose – hot fructose solutions are less sweet than cold ones. The cause of such effects is the mass equilibrium of sugar isomers in solution. At higher temperatures the concentration of the very sweet β-D-fructopyranose drops in favor of both the less sweet α-D-fructofuranose and the β-D-fructofuranose (Fig. 19.6). Such strong shifts in isomer concentrations do not occur with glucose, hence its sweet taste intensity is relatively unchanged in the range of 5–50 °C.

19.1.3 Nutritional/Physiological Properties

19.1.3.1 Metabolism

The role of carbohydrates in metabolism is primarily determined by the ability of disaccharides to be hydrolyzed in the gastrointestinal tract and by the mechanisms of monosaccharide absorption.

The human organism hydrolyzes sucrose, lactose and oligosaccharides of the maltose and isomaltose type. The enzyme lactase, which is responsible for lactose hydrolysis, is lacking in some adults. Glucose and galactose are actively transported, while all other monosaccharides are transported only by diffusion. Sugar phosphorylation occurs preferentially in the liver. All monosaccharides which are metabolized can be interconverted. Sugar alcohols are oxidized: sorbitol → fructose, xylitol → xylulose. However, only glucose can enter the insulin-regulated and -dependent energy metabolism and be utilized by all tissues. Galactose is rapidly transformed into glucose and is therefore nutritionally equal to glucose. Oral intake of glucose and galactose causes a rapid increase in blood sugar levels and, as a consequence, insulin secretion. All other monosaccharides are primarily metabolized by the liver and do not directly affect glucose status or insulin release.

19.1.3.2 Glycemic Index

The glycemic index (GI) was introduced for the quantification of the blood sugar raising effect of carbohydrates. To determine the GI, the duration and the extent of the increase in blood sugar after consumption of 50 g of carbohydrate from food are measured. The reference value is the increase in blood sugar after the intake of 50 g of glucose (GI = 100%). The GI of maltose (105) is higher, but the GI of sucrose (65), lactose (46) and fructose (23) is lower.

The glycemic load (GL) was introduced to take into account the quantity of food consumed. This value refers to the glycemic total load of a portion of food consumed. Results are to be found on the Internet. The consumer, especially diabetics, should favour carbohydrate-containing foods with a low GL value.

19.1.3.3 Functional Food

Some oligosaccharides are bifidogenic (Table 19.2) because they enter the large intestine and promote the growth of Bifidobacteria there. This is desirable because potential pathogenic microorganisms (*Enterobacteriaceae, Clostridia*), which cannot metabolize these oligosaccharides, are simultaneously repressed. Apart from vitamins, natural substances with an antioxidative effect, minerals and trace elements, *n*-3 and *n*-6 polyunsaturated fatty acids and phytosterols (cf. 3.8.2.3), bifidogenic oligosaccharides belong to the components of functional foods. These are

products which not only have a pure nutritional value, but also offer a physiological advantage which is supposed to promote health. This definition is naturally unclear because it includes many traditional foods, such as water, which prevents the formation of kidney and bladder stones. Functional foods contain, e. g., substances which inhibit cancer or reduce cholesterol, protect against infections of the gastro-intestinal tract, reduce blood pressure etc. Whether a product actually meets these requirements must be properly checked so that the consumer is not disappointed (Katan and DeRoos, 2004). Bifidogenic oligosaccharides and inulin (cf. 4.4.4.22) belong to the group of *prebiotics*. These are indigestible substances which promote the growth in the intestine of bifidobacteria or possibly also other microorganisms. In this way, they should have positive effects on health (cf. Probiotocs, 10.2.1.2).

19.1.4 Individual Sugars and Sugar Alcohols

19.1.4.1 Sucrose (Beet Sugar, Cane Sugar)

19.1.4.1.1 General Outline

Sucrose is widely distributed in nature, particularly in green plants, leaves and stalks (sugar cane 12−26%; sweet corn 12−17%; sugar millet 7−15%; palm sap 3−6%); in fruits and seeds (stone fruits, such as peaches; core fruits, such as sweet apples; pumpkins; carobs or St. John's bread; pineapples, coconuts; walnuts; chestnuts); and in roots and rhizomes (sweet potatoes 2−3%; peanuts 4−12%; onions 10−11%; beet roots and selected breeding forms 3−20%). The two most important sources for sucrose production are sugar cane (*Saccharum officinarum*) and sugar beet (*Beta vulgaris ssp. vulgaris* var. *altissima*). Cane sugar and beet sugar are distingushed by the spectrum of accompanying substances and by the $^{13}C/^{12}C$ ratio, which can be used for identification (cf. 18.4.3).

Sucrose is the most economically significant sugar and is produced industrially in the largest quantity. Table 19.3 provides an overview of the annual world production of beet and cane sugar. Table 19.4 lists the main producers and Table 19.5 gives the sugar consumption in some countries.

Table 19.3. World production of sugar (beet/cane)

Year	Total production 10^6 t	Cane sugar %
1900/01	11.3	47.0
1920/21	16.4	70.5
1940/41	30.9	62.3
1960/61	61.1	60.3
1965/66	71.1	61.8
1970/71	82.3	64.2
1975/76	92.2	64.6
1980/81	98.4	66.6
1981/82	108.5	66.2
1982/83	98.6	62.9
1996	124	
1999	133	
2003/04	146	76.0

Honey is the oldest known sweetener and has relatively recently been displaced by cane sugar. Cane sugar was brought to Europe from Persia by the Arabs. After the Crusades, it was imported by Cyprus and Venice and, later, primarily by Holland, from Cuba, Mexico, Peru and Brazil.

In 1747 *Marggraf* discovered sucrose in beets and in 1802 *Achard* was the first to produce sucrose commercially from sugar beets. The

Table 19.5. Sugar consumption in selected countries in 2003

Country	Consumption[a]
Brazil	53.4
Mexico	52
Australia	50.5
Germany	36.2
EU-15	34.2
Former USSR	33.2
USA	27.9
Turkey	22.4
India	16.7
China	7.8

[a] kg/year and head.

Table 19.4. Production of sugar beet, sugar cane and saccharose in 2006 (1000 t)

Continent	Sugar cane	Sugar beet	Saccharose[a]
World	1,392,365	256,407	2264
Africa	92,540	5682	16
America, Central	91,417	–	5
America, North	26,835	29,751	11,000
America, South and Caribbean	661,456	2225	1732
Asia	569,852	36,224	365
Europe	60	182,525	128
Oceania	41,622	–	12

Country	Sugar cane	Country	Sugar beet	Country	Saccharose[a]
Brazil	455,291	Russian Fed.	30,861	Colombia	1722
India	281,170	France	29,879	Indonesia	205
China	100,684	USA	28,880	China	46
Mexico	50,597	Ukraine	22,421	Bangladesh	36
Thailand	47,658	Germany	20,647	Myanmar	31
Pakistan	44,666	Turkey	14,452	UK	27
Colombia	39,849	Poland	11,475	Lithuania	26
Australia	38,169	Italy	10,641	Belgium	22
Indonesia	30,150	China	10,536	Korea, Rep.	18
USA	26,835	UK	7150	Thailand	18
Σ (%)[b]	78	Spain	6045	Σ (%)[b]	75
		Σ (%)[b]	75		

[a] As raw (centrifuged) sugar.
[b] World production = 100%.

new sugar source had great economic impact; the more so when sucrose accumulation in the beets was increased by selection and breeding.

19.1.4.1.2 Production of Beet Sugar

The isolation of beet sugar will be described first because the processes used in material preparation and sugar separation have been developed to perfection. These processes were later transferred to the production of cane sugar from the clear juice concentration stage onwards. In fact, cane sugar was processed fairly primitively for a long time.

Prolonged selection efforts have led to sugar beets which reach their maximum sucrose content of 15–20% in the middle of October. The average for 1980–85 in FR Germany was 16.3%. The early yield achieved by *Achard* of 4.5 kg/100 kg beets has been increased to about 14 kg. Currently, beet varieties have a high sugar content and small amounts of nonsugar substances. Anatomically they have a favorable shape, i.e. are small and slim with a smooth surface, and have a firm texture. Since the sugar accumulation in beets peaks in October and since sugar decomposition due to respiration occurs during subsequent storage of beets, they are rapidly processed from the end of September to the middle of December.

The beet sugar extract contains about 17% sucrose, 0.5% inorganic and 1.4% organic nonsucrose matter. Invert sugar and raffinose content is 0.1% (in molasses this may be as high as 2%). The trisaccharide kestose (cf. 4.3.2), which is present in the extract, is an artifact generated in the course of beet processing. In addition to pectic substances, beet extract contains saponins which are responsible for foaming of the extract and binding with sugars. N-containing, nonsugar constituents of particular importance are proteins, free amino acids, and their amides, (e.g., glutamine) and glycine betaine ("betaine"). These constituents are 0.3% of beets and about 5% of molasses. Beet ash averages 28% potassium, 4% sodium, 5% calcium and 13% phosphoric acid, and contains numerous trace elements. The nonsugar constituents of the sugar extract also include steam-distillable odorous compounds, phenolic acids, e.g., ferulic acid, and numerous

beet enzymes which are extensively inactivated during extract processing. These enzymes, e.g., polyphenol oxidase, can induce darkening through melanin build-up, with the color being transferred during beet extraction into the raw sugar extract.

Beet processing involves the following steps:

- *Flushing and cleaning* in flushing chutes and whirlwashers.
- *Slicing* with machines into thin shreds (cossettes) with the shape of "shoestrings" 2–3 mm thick and 4–7 mm wide.
- *Extraction* by leaching of beet slices. The extraction water is adjusted to pH 5.6–5.8 and to 30–60 °dH with $CaCl_2$ or $CaSO_4$ to stabilize the skeletal substances of the slices in the following pressing step. To denature the cells, the slices are first heated to 70–78 °C for ca. 5 minutes (preliminary scalding) and then extracted at 69–73 °C for 70 to 85 minutes. To eliminate thermophilic microorganisms in the extraction system, 30–40% formaldehyde solution is intermittently added to the raw material at intervals of 8–24 hours in amounts of 0.5–1% of the raw juice accumulating hourly. This was once performed in a so-called diffusion battery of 12–14 bottom sievo-equipped cylindrical containers (diffusers) connected in series and operating discontinuously on a countercurrent principle. Today this battery operation has been replaced to a great extent by a continuous and automatically operated extraction tower into which the shreds are introduced at the bottom while the extraction fluid flows from the top. The extracted shreds (pulp) are discharged at the top. The pulp contains residual sugar of approx. 0.2% of the beet dry weight.

The pulp is pressed, dried on band dryers, and pelleted. It serves as cattle feed. Before drying, 2–3% of molasses and, for nitrogen enrichment, urea is sometimes added.

- *Raw Sugar Extract Purification* (liming and carbonatation). Juice purification results in the removal of 30–40% of the nonsugar substances and has the following objectives:

   – Elimination of fibers and cell residue
   – Precipitation of proteins and polysaccharides (pectins, arabans, galactans)
   – Precipitation of inorganic (phosphate, sulfate) and organic anions (citrate, malate,

oxalate) as calcium salts and precipitation of magnesium ions as $Mg(OH)_2$

- Degradation of reducing sugar (invert sugar, galactose) and, therefore, suppression of the *Maillard* reaction during evaporation
- Conversion of glutamine to pyrrolidone carboxylic acid and asparagine to aspartic acid. However, these reactions proceed only partially under the usual conditions of juice purification.
- Adsorption of pigments on the $CaCO_3$ formed.

Moreover, the sludge formed must be easily settleable and filtrable.

The raw juice from the extraction tower is turbid and greyish black in color due to the enzymatic oxidation of phenols, especially of tyrosine, and to the presence of phenoliron complexes. The raw juice has a pH of 6.2 and contains on an average 15% of solids, of which sucrose accounts for 13.5%. It is first mechanically filtered and then treated with lime milk in two steps (preliming and main liming). Preliming is generally conducted at $60-70\,°C$ up to a pH of $10.8-11.9$ with a residence time of at least 20 minutes. Main liming is conducted at $80-85\,°C$ with a residence time of ca. 30 minutes up to a total CaO content of $2-2.5\%$ in the juice. A number of organic acids and phosphate are precipitated as calcium salts and colloids flocculate.

In order to remove excess calcium, decompose the calcium saccharate $(C_{12}H_{22}O_{11} \times 3CaO)$ formed, and transform the precipitated turbidity-causing solids into a more filtrable form, the solution is quickly gassed with an amount of carbon dioxide required for the formation of calcium carbonate. Carbonatation is also performed in two steps. In the 1st. carbonatation step at $85\,°C$, the pH is adjusted to $10.8-11.9$. The sludge formed ($50-60\,g$ solids/l) is separated at $90-95\,°C$ via decanters and filters and washed on the filters up to a residual sugar content of $0.1-1\%$. In the 2nd. carbonatation step, a pH of $8.9-9.2$ is reached at $94-98\,°C$. The small amount of sludge ($1-3\,g$ solids/l) is filtered off. To lighten and stabilize the color during subsequent evaporation, $50\,g/m^3$ of SO_2 (sulfitation) are frequently added to the thin syrup (juice). Subsequently, the solution is again clarified by filtration, finally producing a clear light-colored thin juice with a solids content of 15 to 18%.

Apart from the classical juice purification processes, different variants are known, which have both advantages and disadvantages. They yield carbonatation juices that can be decanted and filtered more easily. However, these juices are frequently more thermolabile because of incomplete destruction of invert sugar and consequently discolor on evaporation.

Ion exchangers have become important in juice purification. They soften the thin syrup and prevent the formation of hardness scale on the evaporator coils. The substitution of alkaline earth ions (Mg) for the alkali ions is beneficial because it decreases the sugar lost in the molasses by ca. 30% due to the stronger hydration of the alkaline earth ions. Bleaching of thin syrup is possible with activated carbon or with large-pore ion exchangers, which bind the pigments mainly by adsorption.

Extensive elimination (ca. 85%) of nonsugar substances with a corresponding increase in the sugar yield can be achieved by a combination of cation exchangers ($H^\ominus$ form) and anion exchangers ($OH^\ominus$ form) (complete desalting). To suppress inversion during the temporary pH drop, the operation must be conducted at low temperatures ($14\,°C$). Higher temperatures ($60\,°C$) can be used if the cations are first replaced by ammonium ions which are then eliminated as ammonia with the help of an anion exchanger or fixed on a mixed-bed exchanger. In comparison with the lime-carbon dioxide treatment, however, complete desalting has not yet gained acceptance.

- *Evaporation of the thin syrup* ($15-18\%$ solids) is achieved in multiple-stage evaporators (falling film evaporators, natural or forced circulation evaporators). Mildly alkaline conditions (pH 9) are maintained to prevent sucrose inversion. The boiling temperatures decrease in the range of $130-90\,°C$. The resultant thick syrup (yield of $25-30\,kg/100\,kg$ beet) is once more filtered. The syrup contains $68-72\%$ solids and sucrose content is $61-67\%$. The raw, thin and thick syrups have purity quotients of approx. 89, 92 and 93, respectively, i.e. the percentage sucrose on a dry matter basis.

During evaporation, calcium salts precipitate, glutamine still present is converted to pyrrolidone carboxylic acid with lowering of the pH, alkaline degradation of sugar occurs to a small extent, and darkening of the syrup occurs due to *Maillard* reaction and caramelization, depending on the process management (temperature, residence time in the evaporation stages).

• *Crystallization.* Multistage crystallization can be used to isolate 85–90% of the sucrose contained in the thick syrup. The remaining sucrose and practically all the non-sugar substances are found in the last mother liquor called molasses. The crystallization process is predominantly a discontinuous operation. However, efforts are being made to introduce a continuous process (evaporative crystallization and centrifugation).

The thick syrup is evaporated in a boiling apparatus at 0.2–0.3 bar and 65–80 °C until slight supersaturation is achieved (evaporative crystallization). Crystallization is then initiated by seeding, e.g., by adding a dispersion of sucrose crystals (0.5–30 µm) in isopropanol. The mixture is further boiled until the crystals acquire the required size. In this process, the formation of both new crystals and crystal conglomerates is to be carefully prevented by intensive circulation (steam generation, stirring). The crystal paste (magma) with a crystal content of 50–60% is discharged into mashers for homogenization with constant stirring at a constant temperature.

A further crystallization occurs in part on very slow cooling to 35–40 °C (cooling crystallization). In this process, the viscosity of the mash must be maintained constant by the addition of water or mother syrup. Today, cooling crystallization is generally used only for after-product magma, but it will be of importance for raw sugar and white sugar.

Subsequently, the crystalline sugar from the mashers or massecuite is centrifuged in centrifugal baskets, eliminating the mother liquor called green syrup, which is returned to the process. The sugar (with the exception of raw sugar) is then freed from adhering syrup by washing with hot water and steam in the centrifuge. The resulting sugar solution (wash syrup) is fed back to the crystallization process. The presence of higher concentrations of raffinose in the magma (>1%, based on dry matter) reduces the rate of crystallization of sucrose and produces needle-shaped crystals. For this reason, raffinose is cleaved by α-galactosidase.

In this manner, thick syrup can be processed into raw sugar or consumer sugar (white sugar and refined sugar), depending on the process operation. The different crystallization schemes are simplified in Fig. 19.7. Raw sugar contains 1–1.2% of organic and 0.8–1% of inorganic nonsugar substances

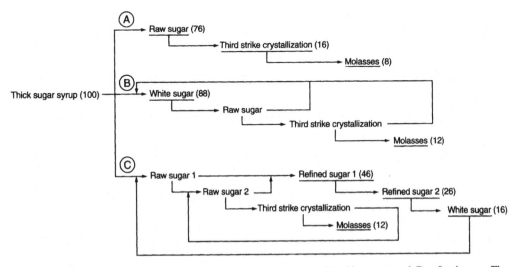

Fig. 19.7. Crystallization scheme for the production of A) raw sugar, B) white sugar, and C) refined sugar. The yields of sucrose (%), based on the amount of sucrose added with the thick syrup, are given in brackets behind the final products (*underlined*)

and 1−2% of water. It is light yellow to dark brown in color due to the adhering syrup. Like the after-product sugar (3−4% of organic and 1.5−2.5% of inorganic nonsugar substances and 2−3% of water) obtained in the last crystallization stage, raw sugar is generally not suitable for direct use. It is processed to consumer sugar in refineries.

In the refinery, the sugar is mashed into a magma with a suitable syrup, centrifuged, and washed with water and steam (affination). Thus, it directly yields a consumer sugar called affinated sugar. Another possibility is to dissolve the sugar and feed the resulting syrup (liquor) to a crystallization process which then yields refined sugar, a consumer sugar of the highest quality.

A simplified crystallization scheme for the production of white sugar is presented in Fig. 19.8. After affination and dissolving, the raw sugar and after-product sugar accumulating in the course of the process are boiled down together with the thick syrup, and the

Table 19.6. Production losses[a] during saccharose recovery from sugar beet

Processing step	1950	1974
Beet slice extraction	0.4−0.5	0.15−0.25
Sugar extract purification	0.1−0.2	0.02−0.05
Other steps	0.6−0.8	0.25−0.90
Total process	1.1−1.5	0.42−0.60

[a] Sugar amount in % based on the processed beet weight.

main part of the sugar finally crystallizes out of the supersaturated solution as white sugar. Centrifugation at 40–45 °C yields not only crystals of 2−4 mm (first-product sugar), but also centrifugal syrup (green syrup) which is subjected to two further crystallization steps. The last discharge, a highly viscous brown syrup, is molasses. In the processing of thick syrup to refined sugar, first raw sugar is isolated exclusively. It is then dissolved and fed back into the crystallization process. In this way, the process is independent of variations in the quality of the thick syrup.

Processing losses in sucrose recovery from beets in 1974 were 0.4−0.9% (sugar determined polarimetrically; and based on processed beet weight) and, when compared to 1950, represent a significant improvement of the sucrose yield (Table 19.6). This technological progress is also reflected in a rise of work productivity (work min/t beets), which was 130−150 in 1950 but only 12−30 in 1974.

19.1.4.1.3 Production of Cane Sugar

Sugar cane processing starts with squeezing out the sweet sap from thoroughly washed cane. For this purpose, the cane moves to a shredding machine where knives shred the stalks and then moves to crushing machines where a series of revolving heavy steel rollers squeeze the cane under high pressure. After the first roller, more than 60% of the cane weight is removed in the form of sap which contains 70% or more of the cane sucrose content. Repeated squeezing provides a sucrose yield of 93–97.5%. The squeezing may be combined with extraction by mixing the "bagasse" (the pressed cane) with hot

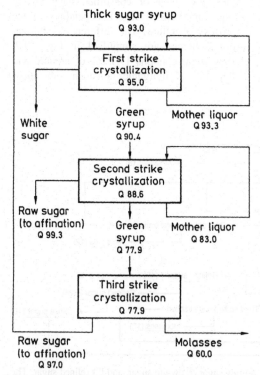

Fig. 19.8. White sugar evaporation and crystallization. Quotient Q: % saccharose in dry matter

water or dilute hot cane juice, followed by a final pressing. The experience gained in continuous thin beet syrup production is applied to sugar cane production, with a resultant energy saving and a rise in sugar yields.

Clarification and neutralization of the mildly acidic, raw extract (pH 4.8−5.0) is done by treatment with lime or lime and carbon dioxide. Further processing of the clarified pure syrup parallels that of sugar beet processing. The yield of raw cane sugar is 6−11% of the cane weight. The "bagasse" is used as fuel, made into wall-board or used as insulation.

19.1.4.1.4 Other Sources for Sucrose Production

Some plants other than sugar beet or sugar cane can serve as sources of sucrose:

Date Sugar is obtained from the sweet, fleshy fruit of the date palm (Algeria, Iraq), which contains up to 81% sucrose in its solids.

Palm Sugar originates from various palm species, e. g., palmyra, saga or Toddy palm, coconut and Nipa palm grown in India, Sri Lanka, Malaysia and the Philippines, respectively.

Maple Sugar is obtained from the maple tree (*Acer saccharum*), found solely in North America (USA and Canada) and Japan. The sap, which drips from holes drilled in the maple tree trunk, flows down metal spiles into metal pails. This sap contains about 5% sucrose, minute amounts of raffinose and several other oligosaccharides of unknown structures. It is marketed in concentrated form either as maple syrup or as maple sugar. Aroma substances are important constituents of these products. The syrup also contains various acids, e. g., citric, malic, fumaric, glycolic and succinic acids. The main component of maple sugar is sucrose (88–99% of the total solids). Aroma constituents include vanillin, syringic aldehyde, dihydroconiferyl alcohol, vanilloyl methyl ketone and furfural.

Sorghum Sugar. Sugar sorghum (*Sorghum dochna*) stalks contain 12% sucrose. This source was important earlier in the USA. Sugar sorghum is processed into sorghum syrup on a small scale on individual farms in the Midwestern United States.

19.1.4.1.5 Packaging and Storage

Sucrose is packaged in paper, jute or linen sacks, in cardboard boxes, paper bags or cones, in glass containers and in polyethylene foils; the latter serving as lining in paper, jute or wooden containers.

Sugar is stored at a relative humidity of 65−70% in loose form in bins or by stacking the paper or jute sacks. The unbagged, loose or bulk sugar is distributed to industry and wholesalers in bins on trucks or rail freight cars.

19.1.4.1.6 Types of Sugar

Sucrose is known under many trade and popular names. These may be related to its purity grade (raffinade, white, consumer's berry, raw or yellow sugar), to its extent of granulation or crystal size (icing, crystal, berry and candy sugar, and cube and cone sugar) and to its use (canning, confectionery or soft drink sugar). Liquid sugar is a sucrose solution in water with at least 62% solids (of which a maximum of 3% is invert sugar). The invert sugar content is high in liquid invert sugars and invert sugar syrups. Such solutions are easily stored, handled and transported. They are dosed by pumps and are widely used by the beverage industry (soft drinks and spirits), the canning industry and ice cream makers, confectionery and baking industries, and in production of jams, jellies and marmalades. Use of liquid sugar avoids the additional crystallization steps of sugar processing and problems associated with packaging of sugar.

Criteria for the analytical determination of sugars are: (a) color; (b) color extinction coefficient (absorbance) of a 50% sugar solution, expressed in ICUMSA-units; (c) ash content determined from conductivity measurements of a 28% aqueous sugar solutions; (d) moisture content; (e) optical rotation; and (f) criteria based on the content of invert sugar.

19.1.4.1.7 Composition of some Sugar Types

The chemical composition of a given type of sugar depends on the extent of sugar raffination. A raffinade, as mentioned above, consists of practically 100% sucrose. Washed raw beet sugar has about 96% sucrose, <1.4% moisture, 0.9% ash and 1.5% nonsugar organic substances. Berry sugar consists of 98.8% sucrose, 0.70% moisture, 0.20% ash and 0.29% nonsugar organic substances. The presence of raffinose, a trisaccharide, is detected by high optical rotation readings or by the presence of needle- or spearlike crystals.

19.1.4.1.8 Molasses

The molasses obtained after sugar beet processing contains about 60% sucrose and 40% other components (both on dry basis). The nonsucrose substances, expressed as percent weight of molasses, include: 10% inorganic salts, especially those of potassium; raffinose (about 1.2%); the trisaccharide kestose, an artifact of processing; organic acids (formic, acetic, propionic, butyric and valeric); and N-containing compounds (amino acids, betaine, etc.). The main amino acids are glutamic acid and its derivative, pyrrolidone carboxylic acid. Molasses is used in the production of baker's yeast; in fermentation technology for production of ethanol and citric, lactic and gluconic acids, as well as glycerol, butanol and acetone; as an ingredient of mixed feeds; or in the production of amino acids.

The residual molasses after cane sugar processing contains about 4% invert sugar, 30–40%

sucrose, 10–25% reducing substances, a very low amount of raffinose and no betaine, but unlike beet molasses, contains about 5% aconitic acid. Cane sugar molasses is fermented to provide arrack and rum.

19.1.4.2 Sugars Produced from Sucrose

Hydrolysis of sucrose with acids, or enzymes (invertase or saccharase) results in invert sugar which, after chromatographic separation, can provide *glucose* and *fructose*. Invert sugar syrup is a commercially available liquid sugar. Invert sugar also serves as a raw material for production of sorbitol and mannitol. Isomerization of sucrose with isomaltulose synthase (EC 5.4.99.11) gives isomaltulose. Apart from 6-O-α-glucopyranosidofructose (palatinose, Ia, Formula 19.1), 1-O-α-glucopyranosidofructose (Ib) is also formed, the ratio depending on the reaction conditions. The process is operated continuously with the immobilized enzyme. The fructose component of palatinose is present as furanose, the anomer ratio being $\alpha/\beta = 0.25$ (34 °C). The sweetening strength is 0.4, based on a 10% sucrose solution. Palatinose is not cleaved by human mouth flora; it undergoes delayed cleavage by the glucosidases in the wall of the small intestines.

Catalytic hydrogenation yields isomaltol (palatinit), a mixture of the disaccharide alcohols

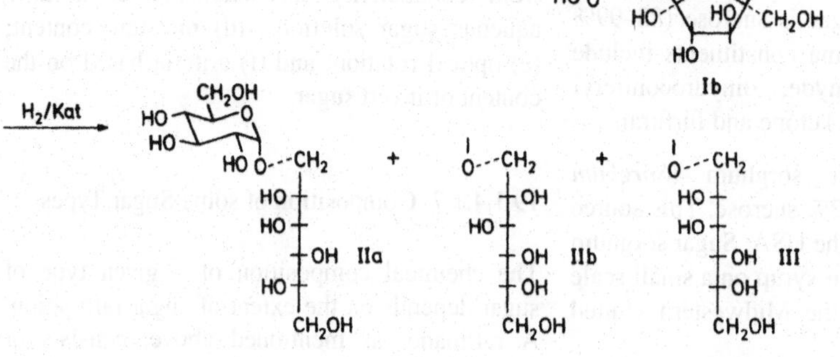

$$(19.1)$$

6-O-α-D-glucopyranosidosorbitol (IIa, Formula 19.1), 1-O-α-D-glucopyranosidosorbitol (IIb) (isomaltitol), and 1-O-α-D-glucopyranosidomannitol (III).

This mixture of sugar alcohols can be separated by fractional crystallization. Palatinit is a sugar substitute.

Isomalto-oligosaccharides [α-D-Glu-(1 → 6)-]$_n$, $n = 2 - 5$, produced by the intermolecular condensation of palatinose, can pass through the small intestine.

Enzymatic isomerization of sucrose with the help of *Leuconostoc mesenteroides* gives an α-D-glucopyranosido (1 → 5)-D-fructopyranose called *leucrose*. This sugar is fully metabolized but is non-cariogenic.

The transfer of glucose residues from maltose or soluble starch to sucrose with the help of a cyclodextrin glucosyltransferase gives mixtures of oligosaccharides [α-D-Glu-(1 → 4)-α-D-Glu-(1 → 2)-β-D-Fru], which are called glucosyl sucrose, and are only slightly cariogenic. The transfer of fructose residues to sucrose catalyzed by a fructosyltransferase gives fructo-oligosaccharides of the general formula α-D-Glu-(1 → 2)-[β-D-Fru-(1 → 2)-]$_n$ with $n = 2 - 4$, β-D-Fru-(1 → 2)-[β-D-Fru-(1 → 2)-]$_n$ with $n = 1 - 9$ and α-D-Glu-(1 → 2)-[β-D-Fru-(1 → 2)-]$_n$ with $n = 1-9$. Alternatively, the controlled hydrolysis of inulin is used in the production of fructo-oligosaccharides.

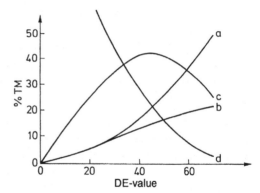

Fig. 19.9. Composition of starch syrups (acid hydrolysis). *a* Glucose, *b* maltose (disaccharide), *c* oligosaccharide (degree of polymerization DP = 3–7), *d* higher saccharides

19.1.4.3 Starch Degradation Products

19.1.4.3.1 General Outline

In principle, either starch or cellulose could be used as a source for saccharification, but only starch hydrolysis is currently of economic importance. Most of the enzymes used for this purpose are derived from genetically modified microorganisms.

19.1.4.3.2 Starch Syrup
(Glucose or Maltose Syrup)

Starch saccharification is achieved by either acidic or enzymatic hydrolysis. Controlled processing conditions yield products of widely different compositions to suit the diversified fields of application. Acid hydrolysis is conducted with hydrochloric acid or sulfuric acid, mainly in a continuous process, and yields glucose syrups with dextrose equivalents (DE value) between 20 and 68. The composition is constant for each DE value (Fig. 19.9).

The raw juice is neutralized and passes through various purification steps. Proteins and lipids from starch flocculate at a suitable pH value and are separated as sludge. Pigments are eliminated with activated carbon and minerals with ion exchangers. The purified juice is evaporated under vacuum (falling-film evaporator) up to a solids content of 70–85%.

During acid hydrolysis, a number of side reactions occur (cf. 4.2.4.3.1). Reversion products are formed in amounts of 5–6% of the glucose used. These are predominantly isomaltose (68–70%) and gentiobiose (17–18%), and, in addition, other di- and trisaccharides. Furthermore, degradation products of glucose are formed, e.g., 5-hydroxymethylfurfural and other compounds typical of caramelization and the *Maillard* reaction (cf. 4.2.4.4).

In enzymatic processes, α-amylases, β-amylases, glucoamylases, and pullulanases are used. First, starch liquefaction is conducted with acid, with α-amylase, or with a combination acid/enzyme process.

The enzyme most commonly used is α-amylase isolated from, for example, *Bacillus subtilis* or *B. licheniformis*. Optimal pH and temperature are 6.5 and 70–90 °C, respectively. The enzyme from *B. licheniformis* is active even at 110 °C. Hydrolysis can be carried out to obtain a product consisting mostly of maltose and, in addition, maltotriose and small amounts of glucose. When, for instance, starch is subjected to combined degradation with bacterial α-amylase and β-amylase or fungal α-amylase, the product obtained has 5% of glucose, 55% of maltose, 15% of maltotriose, 5% of maltotetraose, and 20% of dextrins in the dry matter. Maltose contents of up to 95% (dry matter) can be attained by using pullulanases (cf. 2.7.2.2.4).

A suitable combination of enzymes gives rise to products that cannot be obtained by acid hydrolysis alone.

The extent of starch conversion into sugars is generally expressed as dextrose equivalents (DE value), i.e. the amount of reducing sugars produced, calculated as glucose (DE value: glucose = 100, starch = 0).

The sweet taste intensity of the starch hydrolysates depends on the degree of saccharification and ranges from 25–50% of that of sucrose. Table 19.7 provides data on some hydrolysis products. The wide range of starch syrups starts with those with a DE value of 10–20 (maltodextrins) and ends with those with a DE value of 96. Starch syrups are used in sweet commodity products. They retard sucrose crystallization (hard caramel candies) and act as softening agents, as in soft caramel candies, fondants and chewing gum. They are also used in ice cream manufacturing, production of alcoholic beverages and soft drinks, canning and processing of fruits and in the baking industry.

19.1.4.3.3 Dried Starch Syrup (Dried Glucose Syrup)

Dried starch syrups with a moisture content of 3–4% are produced by spray drying of starch hydrolysates. The products are readily soluble in water and dilute alcohol and are used, for example, in sausage production as a red color enhancer. The average composition of dried starch syrups is 50% dextrin, 30% maltose and 20% glucose.

19.1.4.3.4 Glucose (Dextrose)

The raw source for glucose production is primarily starch isolated from corn, potatoes or wheat. The starch is first liquefied with thermostable α-amylases of microbial origin at 90 °C and pH 6.0 or by partial acid hydrolysis. The dextrins are then hydrolyzed by amyloglucosidase. The enzyme from *Aspergillus niger*, at pH 4.5 and 60 °C, provides a hydrolysate with 94–96% glucose. After a purification step, the hydrolysate is evaporated and crystallized. Glucose crystallizes as α-D-glucose monohydrate. Water-free α-D-glucose is obtainable from the monohydrate by drying in a stream of warm air or

Table 19.7. Average composition of starch hydrolysates[a]

DE-Value[b]	Glucose	Maltose	Maltotriose	Higher oligo-saccharides
Acid hydrolysis				
30	10	9	9	72
40	17	13	11	59
60	36	20	13	31
Enzymatic hydrolysis[c]				
20	1	5	6	88
45	5	50	20	25
65	39	35	11	15
97	96	2	–	2

[a] All values expressed as % of starch hydrolysate (dry weight basis).
[b] cf. 19.1.4.3.2
[c] Occasionally it involves a combined acid/enzymatic hydrolysis.

by crystallization from ethanol, methanol or glacial acetic acid. Dextrose, due to its great and rapid resorption, is used as an invigorating and strengthening agent in many nourishing formulations and medicines. Like dried glucose syrup, crystalline dextrose is used as a red color enhancer of meat and frying sausages.

19.1.4.3.5 Glucose-Fructose Syrup (High Fructose Corn Syrup, HFCS)

Glucose-fructose syrup is made by the enzymatic isomerization of glucose, which is derived from the process given in 19.1.4.3.4. The conversion occurs at pH 7.5 and 60 °C in a reactor with an isomerase of microbial origin fixed to a carrier. The pH is finally adjusted to 4–5 to avoid the *Maillard* reaction (browning). Since an isomerization of only 42% is achieved, the production of higher concentrations (e. g., 55%) requires the addition of fructose. The fructose is obtained from the syrup by chromatographic enrichment.As a result of a comparable sweetening strength, HFCS replaces sugar in many sweet foods. For example, HFCS accounts for 55% of the total sugar consumption in the USA and sugar for only 45%.

19.1.4.3.6 Starch Syrup Derivatives

Hydrogenation of glucose syrups results in products which, since they are nonfermentable and are less cariogenic, are used in manufacturing of sweet commodity products. Alkaline isomerization of maltose gives *maltulose*, which is sweeter than maltose, while hydrogenation yields *maltitol* in a mixture with maltotriit. This mixture of sugar alcohols is not crystallizable but, after addition of suitable polysaccharides (alginate, methylcellulose), can be spray-dried into a powder.
Enzymatic transglucosylation of glucose syrup gives bifidogenic gentio-oligosaccharides. They consist of some glucose residues with a $\beta(1 \rightarrow 6)$ linkage.

19.1.4.3.7 Polydextrose

When D-glucose is melted in the presence of small amounts of sorbitol and citric acid,

a cross-linked polymer called polydextrose is formed, which contains primarily 1,6-glucosidic bonds but also other bonds. The caloric value is $\geq 4.2 \, kJ/g$. For this reason, the use of polydextrose as a sweetener for diabetics and for the production of low-calorie baked products and candies is under discussion.

19.1.4.4 Milk Sugar (Lactose) and Derived Products

19.1.4.4.1 Milk Sugar

Lactose is produced from whey and its concentrates. The whey is adjusted to pH 4.7 and then heated directly with steam at 95–98 °C to remove milk albumins. The deproteinated filtered fluid is further concentrated in a multistage evaporator and then the separated salts are removed. The desalted concentrate yields a yellow raw sugar with a moisture content of 12–14%. The remaining mother liquor still contains an appreciable amount of lactose, so it is recirculated through the process or is used for the production of ethanol or lactic or propionic acids. The raw lactose is raffinated by solubilization, filtration and several crystallizations. The snow-white α-lactose monohydrate is pulverized in a pin mill and separated according to particle size in a centrifugal classifier. Spray drying of lactose is gaining in importance. To increase lactose digestibility, sweetness and solubility, a 60% lactose solution can be heated to 93.5 °C and the crystallizate discharged to a vacuum drum dryer. β-Lactose (cf. 10.1.2.2) is formed. Its moisture content is not more than 1% and it is more soluble than α-lactose. Uses of β-lactose include: a nutrient for children; a filler or diluter in medicinal preparations (tablets); and an ingredient of nutrient solutions used in microbial production of antibiotics.

19.1.4.4.2 Products from Lactose

Enzymatic or acidic hydrolysis of lactose provides a glucose-galactose mixture which is twice as sweet as lactose. A further increase in taste intensity is achieved by enzymatic isomerization of glucose. Such enzyme-treated products contain about 50% galactose, 29% glucose and 21% fructose.

Lactulose is obtained by the alkaline isomerization of lactose. It is sweeter than lactose. Hydrogenation of lactose yields *lactitol*, while hydrogenation of lactulose yields a mixture of lactilol and β-D-galactopyranosido-1,4-mannitol.

Bifidogenic galacto-oligosaccharides (α-D-Glu-(1 → 4)-[β-D-Gal-(1 → 6)-]$_n$, n = 2−5) and lactosucrose (β-D-Gal-(1 → 4)-α-D-Glu-(1 → 2)-β-D-Fru) are produced from lactose by transgalactosylation and transfructosylation (Table 19.1).

19.1.4.5 Fruit Sugar (Fructose)

Fructose is obtainable from its natural polymer, inulin, which occurs in: topinambur tubers (India) or its North American counterpart, Jerusalem artichoke (*Helianthus tuberosus*); chicory; tuberous roots of dahlia plants; and in flowerheads of globe or true artichoke (*Cyanara scolymus*), grown extensively in France. Fructose is obtained by acidic hydrolysis of inulin or by chromatographic separation of a glucose-fructose mixture (invert sugar, isomerized glucose syrup). Only the latter process has commercial significance. Fructose is present in the crystallized state as β-pyranose. Sweeter than sucrose, fructose is used as a sugar substitute for diabetics. It can be partly converted to glucose on boiling for longer periods due to the acid in fruit products.

19.1.4.6 L-Sorbose and Other L-Sugars

L-Sorbose can be formed from glucose via sorbitol. Sorbitol is oxidized by *Acetobacter*

xylium into L-sorbose, an intermediary product for commercial synthesis of ascorbic acid (cf. 18.1.2.7). Sorbose is under discussion as a sucrose substitute for diabetics and as an ingredient with neglible cariogenicity in low calorie foods. It is resorbed only slowly on oral administration.

Until now, other L-sugars have been available only in small amounts. It is assumed that they are metabolized not at all or only to a small extent by human beings and even in low concentrations, they are capable of inhibiting the glycosidases of the small intestine. Therefore, economic methods of synthesis are of interest. A suitable educt is L-arabinose, which yields a L-glucose/L-mannose mixture by chain extension. This mixture can be oxidized directly to L-fructose or after reduction via L-sorbitol/L-mannitol. The isomerization of L-sorbose to L-idose and L-gulose is also under discussion.

19.1.4.7 Sugar Alcohols (Polyalcohols)

Sugar alcohols serve as sweetening agents for diabetics and are used as sugar substitutes in sugar-free candies and confectionery. Candies, bread and cakes contain these alcohols as moisturizers and softeners. Table 19.8 shows data on their use. Sugar alcohols have a low physiological calorific value.

19.1.4.7.1 Isomaltol (Palatinit)

Palatinit is produced as described in 19.1.4.2. The sweetening strength of a 10% solution is 0.45 with reference to a 10% sucrose solution (sweet-

Table 19.8. Polyalcohols in candies, bread and cakes[a]

Product	Sorbitol	Xylitol	Mannitol	Lactitol	Maltitol	Isomaltitol
Cake	2.2–38.7	–	1.2–3.0	–	2.9–4.9	–
Sugar-free confectionery	342–864	–	23–41	–	–	487
Confectionery	1.5–101	–	1.4–1.7	0.9–2.7	5–360	165
Chocolate	2.9–19.5	–	–	53–122	46.5–109	–
Chewing gum	328–593	63.4–290	2.5–47.5	–	7.1–16.3	–

[a] Values in g/kg.

ening strength $= 1.0$). Palatinit is practically not hygroscopic.

19.1.4.7.2 Sorbitol

Sorbitol, a hygroscopic alcohol, is approximately half as sweet as sucrose. It is used as a sweetener for diabetics and in food canning. Sorbitol can be produced on a commercial scale by catalytic hydrogenation of glucose. Acid-catalyzed elimination of water yields a mixture of 1,4-sorbitan (85%, I) and 3,6-sorbitan (15%, II). Under more drastic conditions (action of concentrated acids), 1,4:3,6-dianhydrosorbitol (isosorbid III) is formed (Formula 19.2).

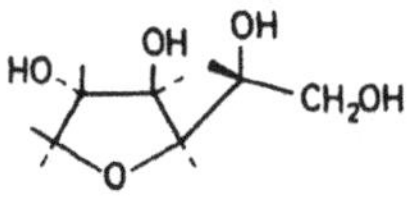

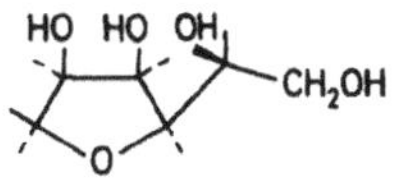

$$(19.2)$$

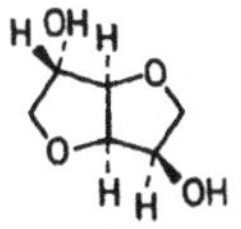

19.1.4.7.3 Xylitol

Xylose is obtained by hydrolysis of hemicelluloses. Catalytic hydrogenation of xylose yields xylitol. Xylitol is as sweet as sucrose. Due to its high heat of solution of $-23.27\,\text{kJ/mol}$ (sucrose: $6.21\,\text{kJ/mol}$), it produces a cooling effect in the mouth when it dissolves. This effect is utilized in some candies.

19.1.4.7.4 Mannitol

Mannitol can be made by the hydrogenation of invert sugar. As a result of its lower solubility, it is separated from sorbitol, which is also produced in the process, by chromatography.

19.1.5 Candies

19.1.5.1 General Outline

Candies represent a subgroup of sweet commodities generally called confectionery. Products such as long-storage cookies, cocoa and chocolate products, ice cream and invert sugar cream are also confections.

Candies are manufactured from all forms of sugar and may also incorporate other foods of diverse origin (dairy products, honey, fat, cocoa, chocolate, marmalade, jellies, fruit juices, herbs, spices, malt extract, seed kernels, rigid or elastic gels, liqueurs or spirits, essences, etc.). The essential and characteristic component of all types of candy is sugar, not only sucrose, but also other forms of sugar such as starch sugar, starch syrup, invert sugar, maltose, lactose, etc.

The important product groups include hard and soft caramels (bonbons, toffees), fondant, coconut flakes, foamy candies, gum candies, licorice products, dragees, pastilles, fruit pastes, chewing gum, croquant, effervescent powders, and products made of sugar and almonds, nuts and other protein-rich oil-containing seeds (marzipan, persipan, nougat).

19.1.5.2 Hard Caramel (Bonbons)

For the production of these candies, a sucrose solution is mixed with starch syrup and boiled down to the desired water content either batchwise or continuously (Fig. 19.10). Generally used are vacuum pans ($120-160\,^\circ\text{C}$) and film boiling machines in which evaporation takes place in a rotating cylinder ($110\,^\circ\text{C} \rightarrow 142\,^\circ\text{C}$, 5 s). Volatile labile components (aroma substances) are added after cooling. This applies to acids as well in order to prevent inversion. Air is incorporated into the mass, if necessary. Subsequently, the mass is formed into a cord and processed into bonbons with the help of stamping or casting machines that require a slightly thinner mass. Modern plants have a capacity of $0.6-1.5\,\text{t/h}$.

The composition of hard caramels is presented in Table 19.9.

Table 19.9. Composition[a] of some candies

Component	Hard caramel	Soft caramel	Fondant	Marzipan filler	Marzipan
Sucrose	40–70	30–60	65–80[d]	≤35	≤67.5
Starch syrup[b]	30–60	20–50	10–20	0	3.5
Invert sugar	1–8	1–10		0–10	0–20
Lactose		0–6			
Sorbitol				0	0–5
Fat		2–15		28–33	14–16
Acids[c]	0.5–2				
Milk protein		0–5			
Gelatin		0–0.5			
Aroma	0.1–0.3				
Water	1–34	4–8	10–15	15–17	7–8.5
Minerals	0.1–0.2	0.5–1.5		1.4–1.6	0.7–0.8

[a] Orientation values in %.
[b] Dry matter.
[c] Citric acid or tartaric acid.
[d] Glucose is also used, if necessary.

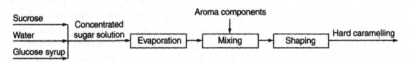

Fig. 19.10. Production of hard caramels

19.1.5.3 Soft Caramel (Toffees)

Milk, starch syrup, and fat are homogenized, mixed with sucrose solution, and boiled down as described above (cf. 19.1.5.2). Labile components are added after cooling. The fat content and, compared to hard caramel, slightly higher water content produce a plastic, partially elastic consistency, which is further improved by the incorporation of air in drawing machines. The mixing of powdered sugar or fondant filler during drawing produces a crumbly consistency due to partial sucrose crystallization. The cooled mass is formed into cords and cut.
The average composition is presented in Table 19.9.

19.1.5.4 Fondant

A sucrose or glucose solution is mixed with starch syrup and boiled down to a water content of 10–15%. The mass is rapidly cooled while it is subjected to intensive mechanical treatment. With partial crystallization (crystal diameter 3–30 mm), a dispersion of sucrose in a saturated sugar solution is formed. The mass solidifies on further cooling and is meltable and pourable on heating. It is aromatized and processed into various products, e. g., chocolate fillings.
The composition is presented in Table 19.9.

19.1.5.5 Foamy Candies

For the production of these candies, a hot sugar solution (sucrose/starch syrup) is carefully mixed into a stable protein foam (egg white, digested milk protein, gelatin). Apart from conventional beaters, pressure beaters are also used in which all the components are first mixed at 2–9 bar and then foamed by subsequent expansion. The light mass is shaped in a pressing step and possibly coated with chocolate.

19.1.5.6 Jellies, Gum and Gelatine Candies

For the production of these products, an aromatized sugar solution is heated with polysaccharides (agar, pectin, gum arabic, thin-boiling starch, amylopectin) and gelatine, poured into starch moulds, and removed with powder after hardening. Typical products are jelly fruit and gum bears.

19.1.5.7 Tablets

Powdered sugar and dextrose are aromatized, granulated with the addition of binding agents (fat, gelatine, gum arabic, tragacanth gum, starch) and lubricants (magnesium stearate), and tabletted under pressure.

19.1.5.8 Dragées

The core (almond, nut, sugar crystal etc.) is moistened with a sugar solution in a rotating boiler and then covered with a layer of sugar by the subsequent addition of powdered sugar. This process is repeated until the desired layer thickness is attained. Chocolate, if necessary, is applied in a corresponding manner.

Sugared or burnt almonds are a well-known product. They consist of raw or roasted almonds which are covered with a hot-saturated and caramelized sugar syrup. The rough crispy surface is formed by the blowing of hot air. Burnt almonds also contain spice and flavoring matter, like vanillin etc. In a larger average sample, the ratio of sugar to almonds should not exceed 4:1. The sugar covering of burnt almonds produced in the dragée process can also be colored.

19.1.5.9 Marzipan

In the traditional production of marzipan raw filler, sweet almonds are scalded, peeled on rubber-covered rolls, coarsely chopped, and then ground with the addition of not more than 35% of sucrose. The mixture is then roasted, e. g., in open roasting pans (95 °C; 45 min) or continually. After cooling, an equal amount of powdered

sucrose is worked into this semi-finished product, possibly with the addition of starch syrup and/or sorbitol, to give the actual marzipan.

Almonds which contain the cyanogenic glycoside amygdalin (cf. 16.2.6) are scalded, peeled and then debittered by leaching with flowing water. The HCN content decreases by 80% in 24 h and the water content in the almonds increases to 38%. An extension of the process reduces the HCN content only slightly.

For reasons of rationalization and bacteriological stability, efforts are being made in modern processes to conduct all the process steps in one hermetically sealed reaction chamber, e. g., in a combined vacuum/boiling/cutting/mixing machine (high speed cooker), to prevent infections. After heating, partial drying occurs here by the application of vacuum. The cooker is aerated with germ-free filtered air. Koenigsberg marzipan is briefly baked on top after shaping. Marzipan potatoes are rolled in cocoa.

19.1.5.10 Persipan

As with marzipan, a raw paste is initially prepared, but instead of almonds, the seed kernels of apricots, peaches or bitter almonds (with bitterness removed) are used. Commercial persipan is a mix of raw persipan filler and sucrose, the latter not more than half of the mix weight. Sucrose can be partially replaced by starch syrup and/or sorbitol.

19.1.5.11 Other Raw Candy Fillers

These are produced from dehulled nuts such as cashews or peanuts. They correspond in composition to raw persipan paste. They are designated according to the oilseed component.

19.1.5.12 Nougat Fillers

Nougat paste serves as a soft or firm candy filling. It contains up to 2% water and roasted dehulled filberts (hazelnuts) or roasted dehulled almonds, finely ground in the presence of sugar and cocoa products. Cocoa products used are cocoa beans; cocoa liquor and butter; pulverized defatted cocoa; chocolate; baking, cream and milk chocolate; chocolate icing; cream and milk chocolate icings; and chocolate powders. The filler may contain a small amount of flavoring and/or lecithin. Also, part of the sugar may be replaced by cream or milk powder. Sweet nougat fillings can also be produced without cocoa ingredients and cream or milk powders. The kneaded nougat paste is often designated just as nougat or noisette.

Recently the trans- and cis-isomers of 5-methyl-4-hepten-2-one have been detected as character impact compounds for the flavor of filberts. The aroma threshold of the trans-isomer (Filbertone) is extremely low: 5 ng/kg (water as solvent).

Of special importance for the aroma of nougat is Filbertone, 5-methyl-(E)-2-hepten-3-one (odor threshold 5 ng/kg oil), which is produced during the roasting of hazelnuts. In a model experiment, the concentration of this compound increased at 180 °C from 1.4 to 660 µg/kg in 9 min and to 1150 µg/kg in 15 min. In fact, 315 µg/kg were found in a commercially produced oil from roasted nuts. Oil from unroasted nuts contains less than 10 µg/kg Filbertone.

19.1.5.13 Croquant

Croquant serves generally as a filling for candy. It is made of molten sucrose, which has been at least partly caramelized, and ground and roasted almonds or nuts. It is occasionally mixed with marzipan, nougat, stable dairy products, fruit constituents and/or starch syrup. Croquant can be formulated to a brittle or soft consistency.

19.1.5.14 Licorice and its Products

To manufacture licorice products, flour dough is mixed with sugar, starch syrup, concentrated flavoring of the licorice herb root and gelatin, and the mix is evaporated to a thick consistency. It is then molded into sticks, bands, figurines, etc. and dried further. The characteristic and flavor-determining ingredient derived from the perennial licorice herb is the diglucuronide of β-glycyrrhetinic acid (cf. 8.8.10).

Simple licorice products contain starch (30–45%), sucrose (30–40%) and at least 5% licorice extract. Better quality products have an extract content of at least 30%. The aroma is enhanced, usually, with anise seed oil in conjunction with low amounts of ammonium chloride.

19.1.5.15 Chewing Gum

Chewing gum is made of a natural or a synthetic gum base impregnated with nutrients and flavoring constituents, mostly sugars and aroma substances, which are gradually released by chewing. The gum base is a blend of latex products from rubber trees that grow in tropical forests or plantations. The most important sources are chicle latex from the *Sapodilla* tree of Mexico, Indonesia and Malaysia; jelutongs; and rubber latex. Natural (mastic tree) and synthetic resins and waxes are also used. Synthetic thermoplastic resins are polyvinyl esters and ethers, polyethylene, polyisobutylene, butadiene-styrene copolymerisates, paraffin, microcrystalline waxes etc. The gum base may also contain cellulose as a filler and a break-up agent. The wax portion predominates in normal chewing gum and the gum-like substances predominate in bubble gum. To be able to process this base into a homogeneous plastic mass, it must be heated to ca. 60 °C before it is kneaded with the sugar components. The mass is made more malleable by the addition of small amounts of glycerol or glycerol triacetate. The mass is cooled to ca. 30 °C before it is rolled out. In fact, very strong kneaders must be used because of the high viscosity of the mass. Recently extruders have been used increasingly in continuous production lines. The production of chewing gum is summarized in Fig. 19.11.

19.1.5.16 Effervescent Lemonade Powders

The powder or compressed tablets (effervescent bonbons) are used for preparation of artificial sparkling lemonades. They contain sodium bicarbonate and an acid component (lactic, tartaric or citric acid). When dissolved in water, they generate carbon dioxide. Other constituents of the product are sucrose or another sweetener, and nat-

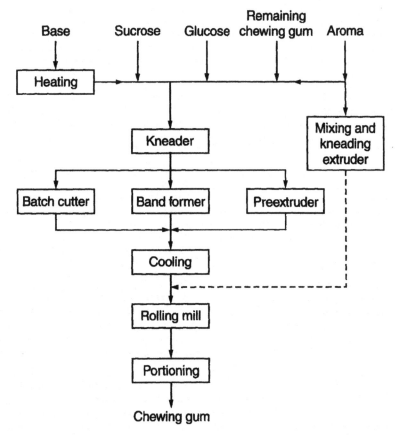

Fig. 19.11. Production of chewing gum

ural or artificial flavoring substances. Sodium bicarbonate and acids are often packaged and marketed separately in individual capsules or in two separate containers.

19.2 Honey and Artificial Honey

19.2.1 Honey

19.2.1.1 Foreword

Honey is produced by honeybees. They suck up nectar from flowers or other sweet saps found in living plants, store the nectar in their honey sac, and enrich it with some of their own substances to induce changes. When the bees return to the hive, they deposit the nectar in honeycombs for storage and ripening.

Honey production starts immediately after the flower pollen, nectar and honeydew are collected and deposited in the bee's pouch (honey sac). The mixture of raw materials is then given to worker bees in the hive to deposit it in the six-sided individual cells of the honeycomb. The changing of nectar into honey proceeds in the cell in the following stages: water evaporates from the nectar, which then thickens; the content of invert sugar increases through sucrose hydrolysis by acids and enzymes derived from bees, while an additional isomerization of glucose to fructose occurs in the honey sac; absorption of proteins from plant and bees, and acids from the bee's body; assimilation of forage minerals, vitamins and aroma substances; and absorption of enzymes from the bees' salivary glands and honey sacs. When the water content of the honey drops to 16–19%, the cells are closed with a wax lid and ripening continues, as reflected by a continued hydrolysis of sucrose by the enzyme invertase and by the synthesis of new sugars.

19.2.1.2 Production and Types

In the production and processing of honey, it is important to preserve the original composition, particularly the content of aroma substances, and to avoid contamination. The following kinds of honey are differentiated according to recovery techniques:

Comb Honey (honey with waxy cells), i. e. honey present in freshly-built, closed combs devoid of brood combs (young virgin combs). Such honey is produced in high amounts, but is not readily found in Germany. In other countries, primarily the USA, Canada and Mexico, it is widely available. Darker colored honey is obtained from covered virgin combs not more than one year old and from combs which include those used as brood combs.

Extracted Honey is obtained with a honey extractor, i. e. by centrifugation at somewhat elevated temperatures of brood-free comb cells. This recovery technique provides the bulk of the honey found on the market. Gentle warming up to 40 °C facilitates the release of honey from the combs.

Pressed Honey is collected by compressing the brood-free honey combs in a hydraulic press at room temperature.

Strained Honey is collected from brood-free, pulped or unpulped honey combs by gentle heating followed by pressing.

Beetle Honey is recovered by pulping honey combs which include brood combs. This type of honey is used only for feeding bees.

Based on its use, honey is distinguished as:

Honey for Domestic use. This is the highest quality product, and is consumed and enjoyed in pure form.

Baking Honey. This type of honey is not of high quality and is used in place of sugar in the baking industry. Such honey has spontaneously fermented, to a certain degree has absorbed or acquired other foreign odors and flavors, or was overheated. This category includes caramelized honey.

According to the recovery (harvest) time, honey is characterized as: early (collected until the end of May); main (June and July); and late (August and September).

Honey can be classified according to geographical origin, e. g., German (Black Forest or Allgäu honey), Hungarian, Californian, Canadian, Chilean, Havanan, etc.

The flavor and color of honey are influenced by the kinds of flowers from which the nectar originates. The following kinds of honey are classified on the basis of the type of plant from which they are obtained.

Flower Honey, e. g., from: heather; linden; acacia; alsike, sweet and white clovers; alfalfa; rape; buckwheat and fruit tree blossoms. When freshly manufactured, these are thick, transparent liquids which gradually granulate by developing sugar crystals. Flower honey is white, light-to-dark, greenish-yellow or brownish. Maple tree honey is light amber; alfalfa honey, dark-red; clover honey, light amber-to-reddish; and meadow flower honey, amber-to-brown. Flower honey has a typical sweet and highly aromatic flavor that is dependent on the flavor substances which together with the nectar are collected by the bees; it sometimes has a flavor reminiscent of molasses. This is especially true of honey derived from heather (alfalfa and buckwheat honeys).

Honeydew Honey (pine, spruce or leaf honeydew). This type of honey solidifies with difficulty. It is less sweet, dark colored, and may often have a resinous terpene-like odor and flavor.

19.2.1.3 Processing

Honey is marketed as a liquid or semisolid product.

It is usually oversaturated with glucose, which granulates, i. e. crystallizes, within the thick syrup in the form of glucose hydrate. To stabilize liquid honey, it has to be filtered under pressure to remove the sugar crystals and other crystallization seeds. Heating of honey decreases its viscosity during processing and filling, and provides complete glucose solubilization and pasteurization. Heating has to be gentle since the low pH of honey and its high fructose content make it sensitive to heat treatment. As with other foods, continuous, high temperature-short time processing (e. g., 65 °C for 30 s followed by rapid cooling) is advantageous.

Processing of honey into a semisolid product involves seeding of liquid honey with fine crystalline honey to 10% and storing for one week at 14 °C to allow full crystallization. This product is marketed as creamed honey.

19.2.1.4 Physical Properties

Honey density (at 20 °C) depends on the water content and may range from 1.4404 (14% water) to 1.3550 (21% water). Honey is hygroscopic and hence is kept in airtight containers. Viscosity data at various temperatures are given in Table 19.10. Most honeys behave like *Newtonian* fluids. Some, however, such as alfalfa honey, show thixotropic properties which are traceable to the presence of proteins, or dilating properties (as with opuntia cactus honey) due to the presence of trace amounts of dextran.

19.2.1.5 Composition

Honey is essentially a concentrated aqueous solution of invert sugar, but it also contains a very

Table 19.10. Viscosity of honey at various temperatures

	Temperature (°C)	Viscosity (Poise)
Honey 1[a]	13.7	600.0
	20.6	189.6
	29.0	68.4
	39.4	21.4
	48.1	10.7
	71.1	2.6
Honey 2[b]	11.7	729.6
	20.2	184.8
	30.7	55.2
	40.9	19.2
	50.7	9.5

[a] Melilot honey (*Melilotus officinalis*; 16.1% moisture).
[b] Sage honey (*Salvia officinalis*; moisture content 18.6%).

Table 19.11. Composition of honey (%)

Constituent	Average value	Variation range
Moisture	17.2	13.4–22.9
Fructose	38.2	27.3–44.3
Glucose	31.3	22.0–40.8
Saccharose	2.4	1.7–3.0
Maltose	7.3	2.7–16.0
Higher sugars	1.5	0.1–8.5
Others	3.1	0–13.2
Nitrogen	0.06	0.05–0.08
Minerals (ash)	0.22	0.20–0.24
Free acids[a]	22	6.8–47.2
Lactones[a]	7.1	0–18.8
Total acids[a]	29.1	8.7–59.5
pH value	3.9	3.4–6.1
Diastase value	20.8	2.1–61.2

[a] mequivalent/kg.

complex mixture of other carbohydrates, several enzymes, amino and organic acids, minerals, aroma substances, pigments, waxes, pollen grains, etc. Table 19.11 provides compositional data. The analytical data correspond to honey from the USA, nevertheless, they basically represent the composition of honey from other countries.

19.2.1.5.1 Water

The water content of honey should be less than 20%. Honey with higher water content is readily susceptible to fermentation by osmophilic yeasts. Yeast fermentation is negligible when the water contents is less than 17.1%, while between 17.1 and 20% fermentation depends on the count of osmophilic yeast buds.

19.2.1.5.2 Carbohydrates

Fructose (averaging 38%) and glucose (averaging 31%) are the predominant sugars in honey. Other monosaccharides have not been found. However, more than 20 di- and oligosaccharides have been identified (Table 19.12), with maltose predominating, followed by kojibiose (Table 19.13). The composition of disaccharides depends largely on

Table 19.12. Sugars identified in honey

Common name	Systematic name
Glucose	
Fructose	
Saccharose	α-D-glucopyranosyl-β-D-fructo-furanoside
Maltose	O-α-D-glucopyranosyl-(1 → 4)-D-glucopyranose
Isomaltose	O-α-D-glucopyranosyl-(1 → 6)-D-glucopyranose
Maltulose	O-α-D-glucopyranosyl-(1 → 4)-D-fructose
Nigerose	O-α-D-glucopyranosyl-(1 → 3)- D-glucopyranose
Turanose	O-α-D-glucopyranosyl-(1 → 3)-D-fructose
Kojibiose	O-α-D-glucopyranosyl-(1 → 2)-D-glucopyranose
Laminaribiose	O-β-D-glucopyranosyl-(1 → 3)-D-glucopyranose
α, β-Trehalose	α-D-glucopyranosyl-β-D-glucopyranoside
Gentiobiose	O-β-D-glucopyranosyl-(1 → 6)-D-glucopyranose
Melezitose	O-β-D-glucopyranosyl-(1 → 3)-O-β-D-fructofuranosyl-(2 → 1)-α-D-glucopyranoside
3-α-Isomaltosylglucose	O-α-D-glucopyranosyl-(1 → 6)-O-α-D-glucopyranosyl-(1 → 3)-D-glucopyranose
Maltotriose	O-α-D-glucopyranosyl-(1 → 4)-O-α-D-glucopyranosyl-(1 → 4)-D-gluco-pyranose
1-Kestose	O-α-D-glucopyranosyl-(1 → 2)-β-D-α-fructofuranosyl-(1 → 2)-β-D-fructofuranoside
Panose	O-α-D-glucopyranosyl-(1 → 6)-O-α-D-glucopyranosyl-(1 → 4)-D-glucopyranose
Isomaltotriose	O-α-D-glucopyranosyl-(1 → 6)-O-α-D-glucopyranosyl-(1 → 6)-D-glucopyranose
Erlose	O-α-D-glucopyranosyl-(1 → 4)-α-D-glucopyranosyl-β-D-fructofuranoside
Theanderose	O-α-D-glucopyranosyl-(1 → 6)-α-D-glucopyranosyl-β-D-fructofuranoside
Centose	O-α-D-glucopyranosyl-(1 → 4)-O-α-D-glucopyranosyl-(1 → 2)-D-glucopyranose
Isopanose	O-α-D-glucopyranosyl-(1 → 4)-O-α-D-glucopyranosyl-(1 → 6)-D-gluco-pyranose
Isomaltotetraose	O-α-D-glucopyranosyl-(1 → 6)-[O-α-D-glucopyranosyl-(1 → 6)]₂-D-gluco-pyranose
Isomaltopentaose	O-α-D-glucopyranosyl-(1 → 6)-[O-α-D-glucopyranosyl-(1 → 6)]₃-D-gluco-pyranose

the plants from which the honey was derived, while geographical and seasonal effects are negligible. The content of sucrose varies appreciably with the honey ripening stage.

19.2.1.5.3 Enzymes

The most prominent enzymes in honey are α-glucosidase(invertase or saccharase), α- and β-amylases (diastase), glucose oxidase, catalase and acid phosphatase. Average enzyme activities are presented in Table 19.14. Invertase and diastase activities, together with the hydroxymethyl furfural content, are of significance for assessing whether or not the honey was heated.

For α-glucosidase, 7−18 isoenzymes are known. In a wide pH optimum between 5.8−6.5 the enzyme hydrolyzes maltose and other α-glucosides. The K_M with sucrose as substrate is 0.030 mol/l. It also possesses transglucosylase activity. During the first stage of sucrose hydrolysis the trisaccharide erlose (α-maltosyl-β-D-fructofuranoside)

plus other oligosaccharides are formed (E = enzyme, S = sucrose, G = glucose, F = fructose):

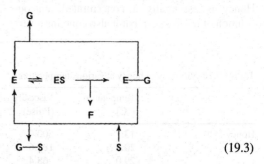

$$(19.3)$$

As the hydrolysis proceeds, most of these oligosaccharides are cleaved into monosaccharides.

Thermal inactivation of invertase in honey and its half-life values at various temperatures have been thoroughly investigated. These data are presented in Figs. 19.12 and 19.13. Practically all invertase activity is derived from bees.

Honey α- and β-*amylases (diastase)* also originate from bees. Their pH optimum range

Table 19.13. Oligosaccharide composition of honey

Sugar	Content[a] (%)
Disaccharides	
Maltose	29.4
Kojibiose	8.2
Turanose	4.7
Isomaltose	4.4
Saccharose	3.9
Maltulose (and two unidentified ketoses)	3.1
Nigerose	1.7
α-, β-Trehalose	1.1
Gentiobiose	0.4
Laminaribiose	0.09
Trisaccharides	
Erlose	4.5
Theanderose	2.7
Panose	2.5
Maltotriose	1.9
1-Kestose	0.9
Isomaltotriose	0.6
Melezitose	0.3
Isopanose	0.24
Gentose	0.05
3-α-Isomaltosylglucose	+[b]
Higher Oligosaccharides	
Isomaltotetraose	0.33
Isomaltopentaose	0.16
Acidic fraction	6.51

[a] Values are based on oligosaccharide total content (= 100%) which in honey averages 3.65%. Only the most important sugars are presented.
[b] Traces.

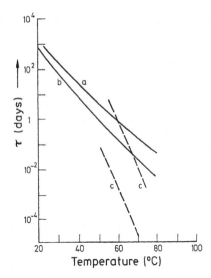

Fig. 19.13. Half-life activity ("τ") of diastase (a), invertase (b), and glucose oxidase (c) in honey at various temperatures (according to *White*, 1978)

Table 19.14. Average enzyme activity in honey

Number	Enzyme	Activity[a]
1	α-Glucosidase (saccharase)	7.5–10
2	Diastase (α- and β-amylase)	16–24
3	Glucose oxidase	80.8–210
4	Catalase	0–86.8
5	Acid phosphatase	5.07–13.4

[a] **1**: g saccharose hydrolyzed by 100 g honey per hour at 40 °C; **2**: g starch degraded by 100 g honey per hour at 40 °C; **3**: μg H_2O_2 formed per g honey/h; **4**: catalytic activity/g honey, and **5**: mg P/100 g honey released in 24 h.

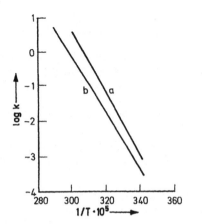

Fig. 19.12. Inactivation rate of (a) invertase and (b) diastase in honey (according to *White*, 1978)

is 5.0–5.3. Diastase activity is somewhat more thermally stable than invertase activity (Figs. 19.12 and 19.13).

Glucose oxidase presence in honey is also derived from bees. Its optimum pH is 6.1. The enzyme oxidizes glucose (100%) and mannose (9%). The enzymatic oxidation by-product, hydrogen peroxide, is partly responsible for a bacteriostatic effect of nonheated honey, an effect earlier ascribed to a so-called "inhibine". The enzymatic oxidation yields gluconic acid, the main acid in honey. Glucose oxidase activity and thermal stability in honey vary widely (limit values were given in Table 19.13), hence this enzyme is not a suitable indicator of the thermally treatment of honey.

Catalase in honey most probably originates from pollen, which, unlike flower nectar, has a high activity of this enzyme. Similarly, honey *acid phosphatase* originates mainly from pollen, although some activity comes from flower nectars.

19.2.1.5.4 Proteins

Honey proteins are derived partly from plants and partly from honeybees. Figure 19.14 shows that bees fed on sucrose provide proteins with less complex patterns than, for example, cottonflower honey.

19.2.1.5.5 Amino Acids

Honey contains free amino acids at a level of 100 mg/100 g solids. Proline, which might originate from bees, is the prevalent amino acid and is 50–85% of the amino acid fraction (Table 19.15). Based on several amino acid ratios, it is possible

Table 19.15. Free amino acids in honey

Amino acid	mg/100 g honey (dry weight basis)	Amino acid	mg/100 g honey (dry weight basis)
Asp	3.44	Tyr	2.58
Asn + Gln	11.64	Phe	14.75
Glu	2.94	β-Ala	1.06
Pro	59.65	γ-Abu	2.15
Gly	0.68	Lys	0.99
Ala	2.07	Orn	0.26
Cys	0.47	His	3.84
Val	2.00	Trp	3.84
Met	0.33	Unidentified	
Met-O	1.74	AA's (6)	24.53
Ile	1.12		
Leu	1.03		
Arg	1.72		
		Total	118.77

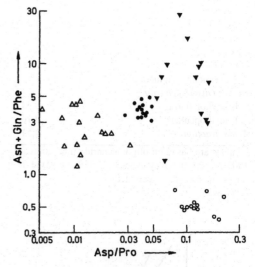

Fig. 19.15. Regional origin of honey as related to its amino acid composition. (according to *White*, 1978) Honey origin: △ Australia, ● Canada, ▼ United States (clover), ○ Yucatan

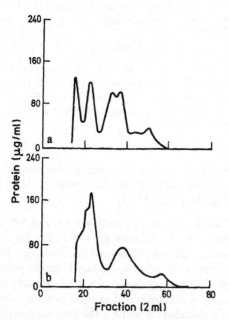

Fig. 19.14. Protein profiles of two honey varieties as revealed by gel filtration on Sephadex G-200. (a) Cottonflower honey (b), honey from sugar-fed bees (according to *White*, 1978)

to identify the geographical or regional origin of honeys (Fig. 19.15).

19.2.1.5.6 Acids

The principal organic acid in honey is gluconic acid, which results from glucose oxidase activity.

In honey gluconic acid is in equilibrium with its gluconolactone. The acid level is mostly dependent on the time elapsed between nectar collection by bees and achievement of the final honey density in honeycomb cells. Glucose oxidase activity drops to a negligible level in thickened honey. Other acids present in honey only in small amounts are: acetic, butyric, lactic, citric, succinic, formic, maleic, malic and oxalic acids.

19.2.1.5.7 Aroma Substances

About 300 volatile compounds are present in honey and more than 200 have been identified. There are esters of aliphatic and aromatic acids, aldehydes, ketones and alcohols. Of importance are especially β-damascenone and phenylacetaldehyde, which have a honey-like odor and taste. Methyl anthranilate is typical of the honey from citrus varieties and lavender and 2,4,5,7a-tetrahydro-3, 6-dimethylbenzofuran (Formula 19.4, linden ether) is typical of linden honey.

$$(19.4)$$

19.2.1.5.8 Pigments

Relatively little is known about honey color pigments. The amber color appears to originate from phenolic compounds and from products of the nonenzymic browning reactions between amino acids and fructose.

19.2.1.5.9 Toxic Constituents

Poisonous honey (pontius or insane honey) has been known since the time of the Greek historian and general, *Xenophon*, and the Roman writer, *Plinius*. It comes mostly from bees collecting their nectar from: rhododendron species (Asia Minor, Caucasus Mountains); some plants of the

family *Ericacea*; insane ("mad") berries; *Kalmia* evergreen shrubs; *Eurphorbiaceae*; and honey collected from other sweet substances, e.g., honeydew exudates of grasshoppers. Rhododendrons contain the poisonous compounds, andromedotoxin (an acetylandromedol) and grayanotoxins I, II and III (a tetra-cyclic diterpene) used in medicine as a muscle relaxant (I: R^1 = OH, R^2 = CH_3, R^3 = $COCH_3$; II: R^1, R^2 = CH_2, R^3 = H; III: R^1 = OH, R^2 = CH_3, R^3 = H) (see Formula 19.5).

$$(19.5)$$

The poisonous nature of New Zealand honey is a result of tutin and hyenanchin (mellitoxin) toxins from the tutu shrub (tanner shrub plant, *Coriaria arbora*). Poisonous flowers of tobacco, oleander, jasmine, henbane (*Datura metel*) and of hemlock (*Conium maculatum*) provide nonpoisonous honeys. The production of these honeys is negligible in Europe.

19.2.1.6 Storage

Honey color generally darkens on storage, the aroma intensity decreases and the content of hydroxymethyl furfural increases, depending on pH,

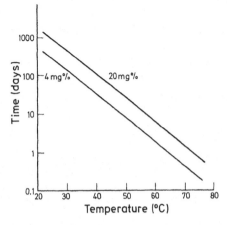

Fig. 19.16. Hydroxymethyl furfural formation in honey versus temperature and time (according to *White*, 1978)

storage time and temperature (Fig. 19.16). The enzymatic inversion of sucrose also continues at a low level even when honey has reached its final density.

Honey should be protected from air moisture and kept at temperatures lower than 10 °C when stored. The desired temperature range for use is 18 – 24 °C.

19.2.1.7 Utilization

Honey use goes back to prehistoric times. Beeswax and honey played an important role in ancient civilizations. They were placed into tombs as food for deceased spirits, while the Old Testament describes the promised land as "a land flowing with milk and honey". In the Middle Ages honey was used as an excellent energy food and, up to the introduction of cane sugar, served as the only food sweetener. Besides being enjoyed as honey, it is used in baking (honey cookies, etc.) or in the manufacturing of alcoholic beverages by mixing with alcohol (honey liqueur, "beartrag") or by fermentation into honey flavored wine (Met). Preparations containing honey, in combination with milk and cereals, are processed for children. Tobacco products are occasionally flavored with honey. In medicine, honey is used in pure form or prescribed in preparations such as honey milk, fennel honey and ointments for wounds. It is incorporated into cosmetics in glycerol-honey gels and tanning cream products. The importance of honey as a food and as a nutrient is based primarily on its aroma constituents and the high content and fast absorption of its carbohydrates.

19.2.2 Artificial Honey

19.2.2.1 Foreword

Artificial honey is mostly inverted sucrose from beet or cane sugar and is produced with or without starch sugar or starch syrup. It is adjusted in appearance, odor and flavor to imitate true honey. Depending on the production method, such creams contain nonsugar constituents, minerals, sucrose and hydroxymethyl furfural.

19.2.2.2 Production

Sucrose (75% solution) is cleaved into glucose and fructose by acidic hydrolysis using hydrochloric, sulfuric, phosphoric, carbonic, formic, lactic, tartaric or citric acid or, less frequently, enzymatically using invertase. The acid used for inversion is then neutralized with sodium carbonate or bicarbonate, calcium carbonate, etc. The inverted sugar is then aromatized, occasionally with strongly flavored natural honey. To facilitate crystallization, it is seeded with an invert sugar mixture that has already solidified, then packaged with automated machines. During inversion, an oligosaccharide (a "reversion dextrin") is also formed, mostly from fructose. Overinversion by prolonged heating results in dark coloring of the product and in some bitter flavor. Moreover, glucose and fructose degradation forms a noticeable level of hydroxymethyl furfural – this could be used for identification of artificial honey.

Liquid artificial honey is made from inverted and neutralized sucrose syrup. To prevent crystallization, up to 20% of a mildly degraded, dextrin-enriched starch syrup is added (the amount added is proportional to the end-product weight).

19.2.2.3 Composition

Artificial honey contains invert sugar ($\geq$50%), sucrose ($\leq$38.5%) water ($\leq$22%), ash ($\leq$0.5%) and, when necessary, saccharified starch products ($\leq$38.5%). The pH of the mixture should be $\geq$2.5. The aroma carrier is primarily phenylacetic acid ethyl ester and, occasionally, diacetyl, etc. Hydroxymethyl furfural content is 0.08–0.14%. The product is often colored with certified food colors.

19.2.2.4 Utilization

Artificial honey is used as a sweet spread for bread and for making Printen (honey cookies covered with almonds), gingerbread and other baked products.

19.3 References[a]

Birch, G.G., Green, L.F. (Eds.): Molecular structure and function of food carbohydrates. Applied Science Publ.: London. 1973

Blank, I., Fischer, K.-H., Grosch, W.: Intensive neutral odorants of linden honey. – Differences from honeys of other botanical origin. Z. Lebensm. Unters. Forsch. *189*, 426 (1989)

Crane, E. (Ed.): Honey. Heinemann. London. 1979

Crittenden, R.G., Playne, M.J.: Production, properties and applications of food-grade oligosaccharides. Trends Food Sci. Technol. *7*, 353 (1996)

Fincke, A.: Zuckerwaren. In: Ullmanns Encyklopädie der technischen Chemie, 4. edn., Vol. 24, p. 795, Verlag Chemie: Weinheim. 1983

Hoffmann, H., Mauch, W., Untze, W.: Zucker und Zuckerwaren. Verlag Paul Parey: Berlin. 1985

Hough, C.A.M., Parker, K.J.; Vlitos, A.J. (Eds.): Developments in sweeteners-1 ff. Applied Science Publ.: London. 1979 ff

Jeanes, A., Hodge, J. (Eds.): Physiological effects of food carbohydrates. ACS Symposium Series 15, American Chemical Society: Washington, D.C. 1975

Katan, M.B., DeRoos, N.: Promises and problems of functional foods. Crit. Rev. Food Sci. Nutr. *44*, 369 (2004)

Koivistoinen, P., Hyvönen, L. (Eds.): Carbohydrate sweeteners in foods and nutrition. Academic Press: New York. 1980

Lehmann, J., Tegge, G., Huchette, M., Pritzwald-Stegmann, B.F., Reiff, F., Raunhardt, O., Van Velthuysen, J.A., Schiweck, H., Schulz, G.: Zucker, Zuckeralkohole und Gluconsäure. In: Ullmanns Encyklopädie der technischen Chemie, 4. edn., Vol. 24, p. 749, Verlag Chemie: Weinheim. 1983

Pancoast, H.M., Junk, W.R.: Handbook of sugars, 2nd edn., AVI Publ. Co.: Westport, Conn. 1980

Pfnuer, P., Matsui, T., Grosch, W., Guth, H., Hofmann, T., Schieberle, P.: Development of a stable isotope dilution assay for the quantification of 5-methyl-(E)-2-hepten-4-one: Application to hazelnut oils and hazelnuts. J. Agric. Food Chem. *47*, 2044 (1999)

Quax, W.J.: Thermostable glucose isomerases. Trends Food Sci. Technol. *4*, 31 (1993)

Rymon Lipinski, G.-W. von, Schiweck, H.: Handbuch Süßungsmittel. Behr's Verlag: Hamburg. 1991

Schiweck, H.: Disaccharidalkohole. Süßwaren *22* (14), 13 (1978)

Schiweck, H., Clarke, M.: Sugar. In: Ullmann's Encyclopedia of Industrial Chemistry, 5th Edition, Volume A25, p. 345, Verlag Chemie, Weinheim, 1994

Schiweck, H., Bär, A., Vogel, R., Schwarz, E., Kunz, M.: Sugar alcohols. In: Ullmann's Encyclopedia of Industrial Chemistry, 5th Edition, Volume A25, p. 413, Verlag Chemie, Weinheim, 1994

Shallenberger, R.S., Birch, G.G.: Sugar chemistry. AVI Publ. Co.: Westport, Conn. 1975

Sturm, W., Hanssen, E.: Über Cyanwasserstoff in Prunoideensamen und einigen anderen Lebensmitteln. Z. Lebensm. Unters. Forsch. *135*, 249 (1967)

Tegge, G., Riehm, T., Sinner, M., Puls, J., Sahm, H.: Verzuckerung von Stärke und cellulosehaltigen Materialien. In: Ullmanns Encyklopädie der technischen Chemie, 4. edn., Vol. 23, p. 555, Verlag Chemie: Weinheim. 1983

White jr., J.W.: Honey. Adv. Food Res. *24*, 287 (1978)

---

[a] cf. 4.5

20 Alcoholic Beverages

Alcoholic beverages are produced from sugar-containing liquids by alcoholic fermentation. Sugars, fermentable by yeasts, are either present as such or are generated from the raw material by processing, i. e. by hydrolytic cleavage of starches and dextrins, yielding simple sugars. The most important alcoholic beverages are beer, wine and brandy. Beer and wine were known to early civilizations and were produced by a well-developed industry. The distillation process for liquor production was introduced much later. The nutritional energy value of ethanol is high (29 kJ/g or 7 kcal/g).

Figure 20.1 illustrates the *Embden–Meyerhoff–Parnas* scheme of alcoholic fermentation and glycolysis. For related details about the reactions and enzymes involved, the reader is referred to a textbook of biochemistry.

20.1 Beer

20.1.1 Foreword

Beer making or brewing involves the use of germinated barley (malt), hops, yeast and water. In addition to malt from barley, other starch- and/or sugar-containing raw materials have a role, e. g., other kinds of malt such as wheat, unmalted cereals called adjuncts (barley, wheat, corn, rice), starch flour, starch degradation products and fermentable sugars. The use of additional raw materials may necessitate in part the use of microbial enzyme preparations.

Beer owes its invigorating and intoxicating properties to ethanol; its aroma, flavor and bitter taste to hops, kiln-dried products and aroma constituents formed during fermentation; its nutritional value to the content of unfermented solubilized extracts (carbohydrates, protein); and, lastly, its refreshing effect to carbon dioxide, a major constituent. Data on beer production and consumption are given in Table 20.1 and a schematic representation of the production of beer is given in Fig. 20.2.

20.1.2 Raw Materials

20.1.2.1 Barley

Barley is the most important of the raw materials used for beer production. Different cultivars of the spring barley (*Hordeum vulgare convar. distichon*) with exceptionally suitable properties are used as brewing and malting barley in Germany. In addition, six-row winter barley has an increasing role. Barley of high brewing value provides ample quantities of extract from the resultant malt, and has a high starch but moderate protein (9–10%) content, a high degree of germination (at least 95% of kernels), high germination vigor and good swelling ability. Sensory assay (hand appraisal) should also be included in the evaluation of a barley.

20.1.2.2 Other Starch- and Sugar-Containing Raw Materials

20.1.2.2.1 Wheat Malt

Wheat malt is mixed with barley malt in a ratio of 40:60 in the production of top fermented beer.

20.1.2.2.2 Adjuncts

In addition to barley malt, supplementary sources of starch are used in the form of unmalted cereals (adjuncts) in order to dilute the mash by 15–50%. The adjuncts are barley, wheat, corn and rice (cracked rice) in the form of whole meal, grits, flakes or flour.

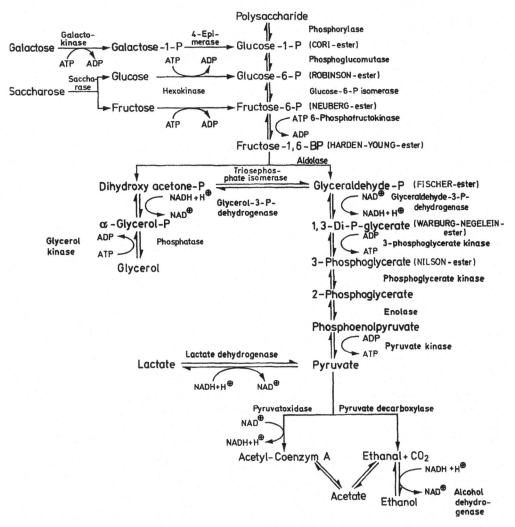

Fig. 20.1. *Embden–Meyerhoff–Parnas*-scheme of glycolysis and alcoholic fermentation

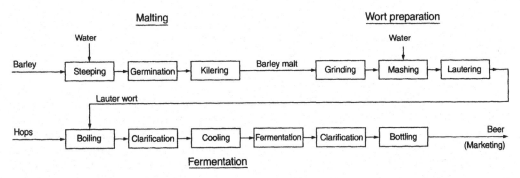

Fig. 20.2. Production of beer

Table 20.1. Production and consumption of beer in 1980, 1997 and 2004

Country	Production (10^6 hl)			Consumption (l/capita)		
	1980	1997	2004	1980	1997	2004
Belgium	14.3	14.0	15.4	131	101	93
Denmark	8.1	9.2	8.4	131	117	90
Germany	92.3[a]	114.8	106.0	146[a]	131	116
Finland	2.8	4.8	4.6	54	67	84
France	21.7	19.5	18.1	57		33
Greece	4.1[b]	3.9	4.4	41[b]	39	
Ireland	6.0	8.1	8.0	122	124	108
Italy	8.6	11.5	13.7	17	25	30
Luxembourg	0.7	0.5	0.4	116	115	
Holland	15.7	24.7	25.1	86	86	78
Austria	7.6	9.4	8.9	102	113	109
Portugal	3.6	6.6	7.3	35	64	62
Sweden	3.7	4.9	4.0	47	62	52
Spain	20.0	24.9	30.6	54	67	
United Kingdom	64.8	59.1	58.0	118	104	101
Czech Rep.			18.5			161

[a]Without GDR.
[b]1990.

Adjuncts are low in enzyme activity, hence their use may necessitate the addition of microbial enzyme preparations with α-amylase and proteinase activities.

Unmalted barley contains about three times more β-glucans than malted barley. In order to decrease the viscosity of unmalted barley extract to values similar to those of malted barley, β-glucans must be degraded with the enzyme β-glucanase, which is present in microbial enzyme preparations.

20.1.2.2.3 Syrups, Extract Powders

Since adjunct processing may result in undesirable changes, extracts from enzyme- or acid-treated barley, wheat or corn have recently been introduced in the form of syrup or powder. The use of syrup from barley to as much as 45% of the total mash is possible.

20.1.2.2.4 Malt Extracts, Wort Concentrates

For production of hop-free malt extracts or hopped wort concentrates, the usual worts are evaporated in vacuum or concentrated by freeze drying. Such concentrates are diluted prior to use. The content of bitter substances and the tendency to produce cloudiness or turbidity are decreased in such concentrates, since tannins and proteins are removed during the evaporation step.

20.1.2.2.5 Brewing Sugars

Sucrose, invert sugar and starch-sugar are introduced at the stage of hopping or before the beer is bottled.

20.1.2.3 Hops

20.1.2.3.1 General Outline

Hops are a very important and indispensable ingredient in beer production. They act as a clarifier, since they precipitate the proteins in wort, change the wort character to give a specific aroma and bitter taste and, together with ethanol and carbon dioxide, their active antibiotic properties contribute to the stability of beer. Lastly, the pectin content of hops enhances the foam-building ability of beer. The hop (*Humulus lupulus*) is a tall,

Table 20.2. Production of hops in 2006 (1000 t)

Continent	Hops	Country	Hops
World	129	Germany	34
		USA	26
Africa	23	Ethiopia	23
America, Central	–	China	22
America, North	26	Czech Rep.	5
America, South		Poland	3
and Caribbean	–	UK	2
Asia	25	Korea	2
Europe	53	Slovenia	2
Oceania	2	Spain	1
		Australia	1
		France	1
		Albania	1
		Σ (%)[a]	98

[a] World production = 100 %.

hardy, perennial climbing vine. The flowers of the female plants, though lacking pollination, grow well and cluster into a conical blossom which has large thin scales or bracts. This cone, when ripe, is harvested and used commercially. The plant is propagated vegetatively by planting cuttings from fleshy roots. The hop cones are picked in August or September and are dried and pressed into bales. The lupulin gland in the upper and lower portion of bracts contains, in addition to essential oils, bitter constituents. Data on hop production are given in Table 20.2.

20.1.2.3.2 Composition

Table 20.3 presents data on the composition of hops. The constituents of utmost importance are

Table 20.3. Composition of hops

Constituent	Content (%)[a]	Constituent	Content (%)[a]
Bitter compounds	18.3	Crude fiber	15.0
Essential oil	0.5	Ash	8.5
Polyphenols	3.5	N-free extract-	
Crude protein	20.0	able mater	34.0

[a] As % dry matter; moisture content approx. 11%.

the bitter substances. In fresh hops they occur mostly in the form of α-acids (cf. Formula 20.1): humulon (I), cohumulon (II), adhumulon (III); and in the form of β-acids: lupulon (IV), colupulon (V) and adlupulon (VI). These compounds are susceptible to changes during drying, storage and processing of hops. The changes usually involve isomerization, oxidation and/or polymerization. As a consequence, a great number of secondary products are found.

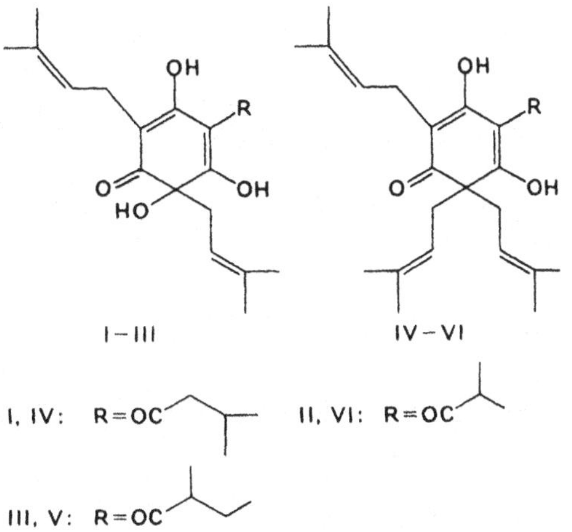

$$(20.1)$$

The quality and intensity of the bitter taste derived from these secondary products are different. Evaluation of hops is therefore based on a determination of composition of individual α- and β-acids, rather than of the total content of bitter substances. As seen in Table 20.4, the composition varies greatly with hop origin. During the boiling of hops, humulons isomerize into isohumulons (cis-compounds, VII; trans-compounds, VIII; cf. Formula 20.2), which are more soluble and bitter than the initial compounds. The isohumulons can be further transformed into humulinic acids (IX,X), which have only about 30% of the bitterness of isohumulons.

Hulupons (XI) and luputrions (XII) are the secondary products of the lupulons. They possess an exceptionally pleasant and mild bitter taste which is much less bitter than the compounds from which they are derived. Hence the bitter taste of beer is primarily due to compounds of the humulon fraction.

Table 20.4. Content of humulons and lupulons in hops from various sources (values in %)

Hops	α-Acids			β-Acids		
	humulon	cohumulon	adhumulon	lupulon	colupulon	adlupulon
Japan	46	41	13	21	68	11
America	54	34	12	32	57	11
Hallertau	59	27	14	45	43	12
Northern						
Brewer	64	24	12	46	43	11
Saaz	67	21	12	51	37	12

Table 20.5 shows the odorants of dried hops. The occurrence of undecatriene and -tetraene, which have a balsam-like, aromatic and pine-like odor, is remarkable. These hydrocarbons, and particularly myrcene and linalool belong to the compounds which produce the characteristic odor of hops.

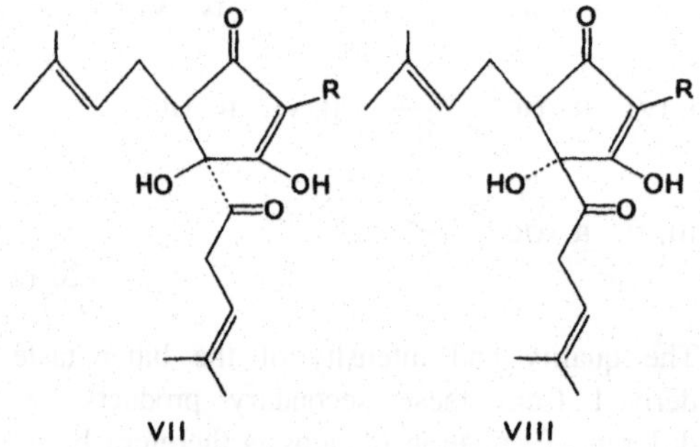

VII   VIII

IX   X

XI   XII

(20.2)

Table 20.5. Potent odorants in hops

Compound	Concentration (mg/kg solids)
Myrcene	3200
R-Linalool	68
Butyric acid	28
Hexanoic acid	21
Pentanoic acid	20
Nonanal	3.6
2-Methylpropionic acid	2.4
3-Methylbutyric acid	2.3
2-Methylbutyric acid	1.5
Hexanal	1.5
4-Vinylguaiacol	1.5
2-Methylbutyric acid methylester	0.15
(E,Z)-1,3,5-Undecatriene	0.076
(E,Z,E)-1,3,5,9-Undecatetraene	0.045
(Z)-3-Hexenal	0.029
1,3,5,8-Undecatetraene[a]	0.024
Isobutyric acid ethylester	0.023
1-Octen-3-one	0.022
2-Methylbutyric acid propylester	0.0018

[a] Steriochemistry unknown.

20.1.2.3.3 Processing

Freshly harvested hops are dried in a hop kiln in a stream of warm air (30–65 °C) to 8–10% moisture, followed by a readjustment of moisture content to 11–12%.

In addition to hop cones, which are prone to quality loss even under proper storage conditions, processed products from hops are acceptable and utilized.

Hop powder (water content 3–8%) is obtained by drying and grinding the cones, which makes the active aroma ingredients more extractable. Prior to grinding, part of the inert material is separated and thus lupulin-enriched concentrates are obtained.

Hops are extracted with a mixture of water and an organic solvent (e. g., alcohol, diethylether), giving extracts of varying compositions. Recently, a hops extraction process using supercritical carbon dioxide has become important.

Isomerized extracts, in which humulon has been converted into isohumulon by heat treatment, are suitable for a cold hopping procedure. In traditional beer hopping this conversion is achieved by boiling the wort for a long time. Isomerized extracts are used in the main fermentation or at a later step in brewing.

Boiling of hops results in the loss of a large portion of oil constituents with the steam. The addition of hops shortly before the end of the boiling process or the use of hop resins or concentrates may greatly enhance the hop aroma of the product. Phenolic constituents in hops contribute to protein coagulation during wort boiling. A part of protein-tannin complexes formed may precipitate at low temperatures after long storage, resulting in turbidity in the beer.

20.1.2.4 Brewing Water

The water used for wort preparation in a brewery has a great influence on beer quality and character. The salt constituents of water can change the pH of the mash and wort. Bicarbonate ions cause a pH increase, while Ca^{2+} and Mg^{2+} ions cause a pH decrease. Heating of water which contains bicarbonates increases the alkalinity according to the equation:

$$HCO_3^\ominus + H^\oplus \rightleftharpoons CO_2 + H_2O \qquad (20.3)$$

in which the equilibrium is shifted to the left since, during heating, the CO_2 component escapes as a gas. Ca and Mg ions react with secondary phosphates in wort to form insoluble tertiary phosphates, releasing protons which add to the acidity of the water:

$$3Ca^{2\oplus} + 2HPO_4^{2\ominus} \rightleftharpoons Ca_3(PO_4)_2 + 2H^\oplus \qquad (20.4)$$

Magnesium sulfate in high concentrations imparts an unpleasant bitter taste to beer. Manganese and iron salts induce turbidity, discoloration and taste deterioration. Concentrations of NaCl or nitrate (>300 mg/l) which are too high interfere with fermentation. During fermentation, nitrate is reduced to nitrite, which is toxic for yeast.

The unique character of different kinds of beer (Pilsen, Dortmund, Munich, Burton-on-Trent), without doubt, can historically be ascribed to the brewing water used in those places, with residual alkalinity playing the major role. Water, low in soluble bicarbonates of calcium, magnesium, sodium or potassium, and soluble carbonates and hydroxides, is suitable for strongly-hopped light beers, such as Pilsener, while alkaline water is suitable for dark beers, such as those from Munich.

Preparation of brewing water is mainly directed to the removal of carbonates. Precipitation by heating with lime is customary. Furthermore, when lime water is used without heating, water softening occurs. Removal of excess salt by ion-exchange resins is also advantageous. Today any water can be treated to match the requirement of a desired type of beer.

20.1.2.5 Brewing Yeasts

Brewing yeasts are exclusively strains of *Saccharomyces*. Two types are recognized: top fermenting yeasts for temperatures $>10\,°C$, and bottom fermenting yeasts used down to $0\,°C$. The top fermenting yeasts, e. g., *Saccharomyces cerevisiae Hansen*, rise to the surface during fermentation in the form of large budding ("sprouting") associations. They ferment raffinose only partially since they lack the enzyme melibiase. The bottom fermenting yeasts, e. g., *Saccharomyces carlsbergensis Hansen*, settle to the bottom during fermentation and completely ferment all sugars including raffinose. There are yeasts with high fermentation ability which remain suspended for a long time, giving a high fermentation rate. Yeasts with low fermentation ability flocculate early and settle to the bottom (super-flocculent yeasts) and hence are unable to continue active fermentation. Pure cultures of many yeast strains currently in use are derived from a single yeast cell and are used as "starter yeast" in plant operations. After the main fermentation, a part of the yeast is harvested for use in freshly-prepared worts, until the yeast becomes useless due to contamination or degeneration. In this way, it is possible to continuously select suitable yeast strains for a defined goal.

20.1.3 Malt Preparation

The cereals are soaked (steeped) in water and then allowed to germinate. The product, green malt, is dried and mildly roasted into a more or less dark and aroma-rich kiln-dried malt. During processing, the rootlets are removed from the malt. The loss due to malting is 11–13% of the dry weight. Prior to use, the malt is stored for 4–6 weeks.

20.1.3.1 Steeping

Cereal kernels are steeped in water to raise their moisture content to induce germination. The water content is 42–44% for light and 44–46% for dark malt. Usually the steeping water is alternately added and removed. Barley is first steeped for only 4–6 h at 12–15 °C water temperature so that the water content is adjusted to ca. 30%. In the following dry period, which lasts 18–20 h, the grains swell and enzymatic processes start rapidly. In the second wet steeping at ca. 18 °C, a water content of 38% is obtained in 2 h. Good aeration is needed in all phases to remove the CO_2 produced by respiration. The normal steeping temperature is 12–24 °C. Alkali treatment (CaO, NaOH) of steeping water serves to reduce microbial contamination and to remove undesirable polyphenols from the hulls.

20.1.3.2 Germination

When the cereals reach the desired moisture content (after ca. 26 h) and germination is started, they are allowed to germinate in germinator chests or less often in drums. The removal of CO_2 and heat is achieved by blowing in moist air (500 m^3/t). The sprouts appear at 16–18 °C in 16–20 h. The water content of barley is first

increased to ca. 41% by spraying and then in steps to 47% to further support germination. The growth of the rootlet continues up to 1.5 times the grain length. At the end of the process which on the whole lasts ca. 40–50 h, the temperature is reduced to 11–13 °C. The law in some countries allows the addition of growth substances to accelerate germination, e. g., gibberellic acid.

20.1.3.3 Kilning

The germinated cereals, termed green malt, contain 43–47% moisture. They are dried in a kiln to give a storable malt with a water content from 2.5% (dark) to 4.5% (lager).

Light malt requires fast drying so that the *Maillard* reaction does not get a look-in. The process is carried out in high-performance kilns at a temperature which is raised from 50 to 65 °C. The barley heats up and germination stops above 40 °C at a water content which is reduced to 20%. However, the activities of hydrolases (endopeptidases, α-amylases) still increase, as desired. The final drying is carried out at 82–85 °C, leading to unavoidable enzyme losses.

In the production of dark malt, the moisture is withdrawn so slowly that the material temperature is higher than with light malt. Although this results in an inhibition of germination, there is an extension of the period in which the activities of the hydrolases increase. The degradation of proteins and carbohydrates to precursors of the *Maillard* reaction is correspondingly extensive. Finally, the malt is rapidly dried at 100 to 105 °C, the *Maillard* reaction providing intensive color and aroma substances.

20.1.3.4 Continuous Processes

Several kinds of installations have been developed which provide continuous steeping, germination and, occasionally, also kilning, offering substantial savings in time. Steeping in this case is performed as a single washing followed by water spraying and continuous transferring to the germination stage. The process conditions are regulated by means of forced air. In some installations the malt is moved for different stages, while in others it remains in the same container from steeping to kilning.

20.1.3.5 Special Malts

Special malts are prepared for many purposes. Dark caramelized malt is held briefly at 60–80 °C to saccharify its starch and is then roasted at 150–180 °C for the desired degree of color. Such color-rich malt is free of diastase enzyme activity, is a good foam builder, and is mostly used for aromatizing malt beers and strong bock beers. Light caramelized malt is made in a similar way, but is treated at lower temperatures after the saccharification step. This preserves the enzyme activities. It is lightly colored and when used gives beer an increased foaming capacity and full-bodied properties. Colored malt is obtained by roasting the kiln-dried malt at 190–220 °C, omitting the prior saccharification step. It can be used to intensify the color of dark beers.

20.1.4 Wort Preparation

The coarsely ground malt is dispersed in water. During this time, the malt enzymes hydrolyze starch and other ingredients. A clear fermentable solution, the so-called wort, is obtained by filtration. When boiled with added hops, the wort takes on the typical beer flavors.

20.1.4.1 Ground Malt

Malt is disintegrated by passing it through several grinding rolls and sifters. The ground products, hull, middlings and flour, are then combined in the desired proportions. By using finely ground meal, the extraction yield increases, but problems arise in wort filtration. Wet milling is commonly preferred for better filtration as it yields a higher proportion of intact hulls. In addition, it provides the desired high extraction yield. For wet milling the water content of the malt is adjusted to 25–30%. It is then ground by a set of rollers and processed immediately into wort.

Continuous wet meal steepers have been developed to guarantee a defined steeping time and,

thus, to prevent the malt grains from becoming slippery and gelatinized by overly long steepage.

20.1.4.2 Mashing

In the mashing step, the malt meal is made into a paste with brewing water (heatable mixing vessel) and partially degraded and solubilized with malt enzymes.

For 100 kg of malt, 4–5 hl of water for light beers and 3–3.5 hl for dark beers are needed. This amount of water is divided into a major portion for production of the mash, and into one or several post-mashing rinses used to wash out extract from the hulls. The course of pH and temperature during mashing are of utmost importance for determining wort composition and, hence, the type and quality of beer. The optimum activity of malt α-amylases is from 70–75 °C at pH 5.6–5.8, and of malt β-amylases from 60–65 °C at pH 5.4–5.6, while that of malt endopeptidases is from 50–60 °C at pH 5.0–5.2. Hence, wort with a pH near 6 will not, without prior pH adjustment, provide optimal conditions for the action of enzymes. The methods used for temperature control in mashing are of two types: decoction and infusion. In the decoction method, the initial temperature of the total mash is raised by removing an aliquot of mash, heating this to boiling and then returning it to the main mash in the mash tun. In general one-, two- or three-mash return procedures are used commercially. The latter is used exclusively for dark beer brewing; the two-mash return for light beer; and the one-mash return procedure for brewing all types of beer. The three-mash return procedure will be briefly described as an example: The crushed malt is mixed in the mash tun with water at 37 °C; the first aliquot is drawn, heated to boiling and returned to the mash tun. In this way the total mash temperature is raised to 52 °C. Two repetitions raise the total mash temperature stepwise to 64 and then to 75 °C. The mashing process is completed at a terminal mash temperature of 74–78 °C.

In the case of poorly "solubilized" malt in which the starch-containing membranes have not ruptured, enzymatic degradation and the extract yield can be improved by stopping briefly at 47–50 °C before further temperature increase. This delays enzyme inactivation. On the other hand, when a low alcohol beer is desired, the malt mashed at 37 °C is drained into boiling water, increasing the temperature to 70 °C and resulting in extensive enzyme inactivation.

In infusion mashing, used mostly in England for brewing top fermented beer, the terminal mashing temperature is achieved not by stepwise increases, but by live steam injection or addition of hot water. As in the decoction method, the temperature program used can vary greatly.

20.1.4.3 Lautering

The separation of wort from hulls and insoluble residues of the grain is done by a classical procedure in a lauter tun, a vessel with a slotted false bottom. The hull and other residues form a ca. 35 cm deep layer in the bottom which acts as a filter through which the extract, or wort, is strained. The initial turbid liquid (turbid wort) with 16–20% extract is pumped back to the tun. Finally, to obtain more wort, the spent grains are rinsed or sparged 3 to 4 times with water.

Modern installations for lautering use strain masters or discontinuous or continuous mash filters. The draff, the lautering residue, is used for animal feed.

20.1.4.4 Wort Boiling and Hopping

Wort boiling with hops or hop products is done in a brew kettle (hop kettle) in which the initial and subsequent worts from the lautering step are collected. Addition of hops is adjusted according to the type and quality of beer desired. The quantity (in hop cones/hectoliter) for light lager beer is 130–150 g; for Dortmund-type beer, 180–220 g; for Pilsener beer, 250–400 g; for dark Munich beer, 130–170 g; and for malt beer and dark bock beer, 50–90 g. The critical factor is the content of bitter substances in the hops selected. The utilization of the bitter substances (α-acids) is only 30–35%.

Boiling for 70 to 120 min concentrates the wort, coagulates protein ("break forming"), solubilizes hop ingredients and converts the bitter components to their isoforms and, lastly, inactivates en-

zymes. The hot wort is then chilled, filtered, aerated and, finally, "pitched" with yeast.

In modern processes, the classical brew kettle is replaced by a whirlpool kettle with external cooker. Shorter boiling times and a better quality of beer are achieved with this system. Moreover, separation from the spent hops can be conducted in the same vessel.

Processes that use pressure boiling (high-temperature wort boiling up to 150 °C) can produce beer with an unpleasant cooked taste.

20.1.4.5 Continuous Processes

Efforts are being made to introduce continuous processes via heat exchangers and to save energy and make the process environmentally friendly with heat recovery from the exhaust steam.

Wort treatment, i. e., removal of the trub formed during boiling (protein-polyphenol complexes, cf. 18.1.2.5.8), is generally conducted in whirlpool vats (possibly combined with wort drying) or via continuous centrifuges. After cooling to the pitching temperature (6–8 °C), the cooling trub obtained is separated by filtration or centrifugation.

20.1.5 Fermentation

20.1.5.1 Bottom Fermentation

Bottom fermentation involves a primary and a secondary step. In the primary fermentation step, the cooled wort with about 6.5–18% dry mass extracted from malt ("stemwort") is pumped into fermenting tanks, located in fermentation cellars cooled to 5–6 °C. The tanks are made of plastic-lined concrete, enamelcoated steel, aluminum or V_2A steel. The wort is inoculated ("pitched") with yeast in the form of a thick yeast slurry of *Saccharomyces Carlsbergensis* (0.5–1 l/hl) and fermented at 8–14 °C until more than 90% of the fermentable extract has been converted. The primary fermentation is completed in 7–8 days, at which point the yeast "breaks", i. e., flocculates and settles to the bottom. The beer is transferred to large clean tanks. With a wort extract content of 12%, 4% of ethanol is produced during fermentation.

The young "green" beer is stored for 1–2 months in tanks at 0–1 °C for secondary fermentation. The beer is clarified by settling of the yeast and separation of protein-polyphenol complexes (cf. 18.1.2.5.8). The yeast multiplies attaining 4 to 5 times the original quantity and is harvested after fermentation. It is used several times until it is no longer biologically pure or it loses fermenting power.

20.1.5.2 Top Fermentation

Primary fermentation proceeds in fermentation tanks, but at higher temperatures (18–24 °C) than bottom fermentation, and requires a total time of ca. 3 days. The yeast builds a solid cap at the top of the tank. It is skimmed off into individual fractions (hops flock, yeast flock, post-flock). The secondary fermentation is a very slow process and may continue in tanks or bottles. Top fermentation is used mostly in England and Belgium, while in Germany it is used in the production of "Kölsch", "Altbier" and "Weiss" beer, a light tart ale made from wheat.

20.1.5.3 Continuous Processes, Rapid Methods

Several continuous processing methods provide accelerated fermentation. They make use of thermophilic yeasts, higher fermentation temperatures and more intensive wort aeration.

20.1.6 Bottling

After ageing, beer is filtered through cotton filter pads and some silicates, often having been preclarified through a kieselguhr pad or by centrifugation. Then, with the aid of a special cask/keg filling apparatus, it is foamlessly filled into transportable casks or metal cisterns. In addition to impregnated oakwood casks, specially-lined iron, aluminum or V_2A steel containers are also acceptable. Bottle filling proceeds from a "bottle tank" in a fully automated process. Tin-plated or aluminum cans are also used.

Pasteurization gives the beer biological stability for overseas export. To avoid cloudiness due to protein precipitation and changes in flavor, the beer is heated to 60–70 °C in a water bath or by steam. The beer is often pasteurized at 62 °C for 20 min. For sterile filling the beer is heated to 70 °C for 30 s or is passed through microfilters (with pore size less than the size of bacteria) and then poured into sterilized bottles or cans.

Temperature fluctuations during storage and transport must be avoided if beer quality is to be preserved.

20.1.7 Composition

20.1.7.1 Ethanol

The ethanol content, which has a very important influence on the aroma, is 1.0–1.5% by weight for a low fermented extract-rich beer, 1.5–2.0% for a weak or thin beer, 3.5–4.5% for a full beer, and 4.8–5.5% for a strong beer. Higher alcohols, such as 2-methylbutanol, 3-methylbutanol, methylpropanol and 2-phenylethanol, are also present in very small quantities.

20.1.7.2 Extract

The nonalcoholic constituents of beer vary within a wide range from 2–3% for plain beers to 8–10% for strong beers. These constituents are the beer solids and consist of to 80% carbohydrate, mostly dextrins. It is possible to calculate the solids content of the original wort before fermentation from the solids content (E, weight %) and alcohol content (A, weight %) of the beer product. The calculation is based on the fermentation equation: 2 parts by weight of sugar equal 1 part by weight of alcohol. The initial solids content of wort, which actually represents a measure of malt utilization, is designated as "stemwort" (St) and can be calculated by the formula:

$$St = \frac{100\,(E + 2.0665\,A)}{100 + 1.0665\,A} \tag{20.5}$$

Thus, for example, if the solids content (E) of a beer is 3% (w/v) and the alcohol content (A) is 5.0% (v/v), then the solids content of the wort before fermentation was 12.6% (w/v). The stem-wort content in Germany is 2–5.5% for plain beers, 7–8% for draft beers, 11–14% for full beers and above 16% for strong beers.

20.1.7.3 Acids

Carbon dioxide is responsible to a substantial extent for the refreshing value and stability of beer. CO_2 is 0.36–0.44% in bottom fermented beers, while in Weiss beer the CO_2 content is up to 0.6–0.7%. A CO_2 content below 0.2% gives flat and dull beers. Apart from small amounts of lactic, acetic, formic, and succinic acids, beer contains 9,10,13- and 9,12,13-tri-hydroxyoctadecenoic acid. In fact, 9.9 ± 2.1 mg/l were found in five types of beer and 9(S), 12(S), 13(S)-trihydroxy-10(E)-octadecenoic acid was the main compound and accounted for 50–55% of the 16 stereoisomers. The pH of beer is between 4.7 (dark, strong beer) and 4.1 (Weiss beer).

20.1.7.4 Nitrogen Compounds

The N-compounds in beer (0.15–0.75%) originate primarily from proteins in the raw materials and from yeast. They consist mainly of proteins plus high molecular weight protein degradation products; both being responsible for cloudiness in beer during cold storage. The free amino acids found in malt are also present in beer. It appears that glutamic acid contributes to beer taste. The presence of volatile amines has also been confirmed.

20.1.7.5 Carbohydrates

The carbohydrate content is approximately 3–5%, while in some strong beers or malt beers it may be considerably higher. Pentosans are also present in addition to dextrins, mono- and oligosaccharides (maltotriose, maltose, etc.). Glycerol normally is 0.2–0.3% of beer.

20.1.7.6 Minerals

Minerals make up 0.3–0.4% of beer and consist mostly of potassium and phosphate. Calcium, magnesium, iron, chloride, sulfate and silicates are also present.

20.1.7.7 Vitamins

Vitamins of the B-group (vitamins B_1 and B_2, nicotinic acid, pyridoxine and pantothenic acid) are present in various beers, often in significant amounts.

20.1.7.8 Aroma Substances

The odorants, e. g., for Pilsener beer are shown in Table 20.6. The aroma is reproduced by a suitable mixture of these substances dissolved in water, the pH of which is adjusted to 4.3 with carbonic acid. This emphasizes the fact that the key odorants of this type of beer can be analytically identified. (R)-Linalool and ethyl-4-methylpentanoate are derived from hops and pass into the beer on boiling the wort.

The odor- and taste-active substances essentially determine the type of beer. The bitter taste of Pilsener beers is produced by relatively high concentrations of isohumulons, and humulenes (including oxidation products), while larger amounts of furaneol are responsible for the caramel note of dark beers.

In the production of alcohol-free beer, the concentrations of important aroma substances drop (Table 20.7).

20.1.7.9 Foam Builders

The foam building properties of beer are due to proteins, polysaccharides and bitter constituents. The β-glucans stabilize the foam through their ability to increase viscosity. Addition of semisynthetic polysaccharides, e. g., propyleneglycol alginate (40 g/hectoliter), to beer provides a very stable foam although the addition is judged as unfavorable.

Table 20.6. Odorants in Pilsener Bier

Compound	Concentration (mg/l)	Aroma value[a]
Ethanol	4080	1639
(E)-β-Damascenone	0.0023	575
(R)-Linalool	0.045	321
Acetaldehyde	5.1	204
Ethyl butanoate	0.198	198
Ethyl-2-methylpropanoate	0.0032	160
Ethyl-4-methylpentanoate	0.00028	93
Dimethylsulfide	0.059	59
3-Methylbutanol	49.6	50
2-Methylbutanol	14.4	45
Ethylhexanoate	0.205	41
4-Hydroxy-2(or 5)-ethyl-5(or 2)-methyl-3(2H)-furanone	0.019	17
2-Phenylethanol	15.1	15
4-Hydroxy-2,5-dimethyl-3(2H)-furanone	0.312	13
Diethoxyethane	0.050	10
3-Methylbutanal	0.004	10
3-Methyl-2-buten-1-thiol	0.00001	8
3-(Methylthio)propanol	0.991	4
3-Hydroxy-4,5-dimethyl-2(5H)-furanone	0.001	3
Butyric acid	1.8	2
Ethyloctanoate	0.160	2
3-Methylbutyric acid	0.855	1
4-Vinyl-2-methoxyphenol	0.137	1

[a] Aroma value: quotient of the concentration to the orthonasal odor threshold value of the substance in water.

Table 20.7. Odorants in lager beer and alcohol-free beer

Compound	Lager beer (mg/l)	Alcohol-free beer (mg/l)
3-Methylbutanol	49.6	6.7
2-Phenylethanol	17.5	2.3
Ethyl hexanoate	0.15	0.01
Ethyl butanoate	0.06	0.01
4-Hydroxy-2,5-dimethyl-3(2H)-furanone (HD3F)	0.35	0.19
4-Vinylguaiacol	0.52	0.13

Lysophosphatidyl cholines (LPC), which occur in cereal as amylose inclusion compounds (starch lipids: cf. 15.2.5), reduce the foam stability. The temperature management during the mashing process regulates the LPC concentration because it determines the activity ratio of α-amylase, which contributes to the release of LPC from amylose, to phospholipase B, which catalyzes the degradation of LPC. Temperatures above 65 °C favor the more stable α-amylase, increasing the LPC concentration.

20.1.8 Kinds of Beer

There is a distinction between top and bottom fermented beers.

20.1.8.1 Top Fermented Beers

Selected examples of top fermented beers from Germany are: Berlin weiss beer, brewed from a wort having 7–8% solids from barley and wheat malts and inoculated at fermentation with yeast and lactic acid bacteria; Bavarian weiss beer brewed from weakly-smoked barley malt with a little wheat malt and fermented only with yeast; Graetzer beer made from wheat malt with a smoky flavor and with a stemwort content of

7–8%; malt beer (caramel beer), a dark, sweet and slightly hopflavored full beer; the bitter beers such as those from Cologne or Duesseldorf (Altbier) which are strongly hop-flavored full beers; top fermented plain beers (Jungbier or Frischbier) with a low stemwort content and often artificially sweetened; Braunschweig's mumme, an unfermented, non-hop flavored malt extract, hence not a true beer or a beer-like beverage. English beers have a stemwort content up to 11–13%. Stout is a very darkly colored and alcohol-rich beer made from concentrated boiled wort (up to 25% stemwort; alcohol content >6.5%). Milder varieties of stout are known as Porter beer.

Pale ale is strongly hopped light beer, whereas mild ale is mildly hopped dark beer. Incorporation of ginger root essence into these beers yields ginger-flavored ale.

Top fermented beers from Belgium, which are stored for a longer time, are called Lambic and Faro beers.

20.1.8.2 Bottom Fermented Beers

These beers show a significantly increased storage stability and are brewed as light, mildly colored or dark beers.

Pilsener beer, an example of a light colored beer, is typically hop flavored, containing 11.8–12.7% stemwort. In contrast, Dortmunder-type beer is made from a more concentrated wort which is fermented longer and thereby has a higher alcohol content. Lager beer (North German Lager) is similar to Dortmunder in hop flavoring, while the stemwort content is close to a Pilsener beer. Munich beers are dark, lightly hop flavored and contain 0.5–2% colored malt and often a little caramel malt. They taste sweet, have a typical malt aromatic flavor, and are fermented with a stemwort content of 11–14%. Beers with a high content of extract are designated as export beers. Traditional dark beers and currently produced special light beers, are the bock beers (Salvator, Animator, etc.). They are also strong beers with more than 16% stemwort. The dark Nuernberg and Kulmbacher beers are even higher in colored malt extracts and thereby are darker than Munich beers. An example of mildly colored beer is the Maerzen beer (averaging 13.8% stemwort). It is

produced from malt of Munich in which the use of colored malt is omitted.

20.1.8.3 Diet Beers

Diet beers exhibit a high degree of fermentation and contain almost no carbohydrates, which are a burden for diabetics. They are produced by special fermentation processes and contain a relatively high alcohol content. Subsequently, the alcohol level is frequently reduced to values typical of normal beer.

20.1.8.4 Alcohol-Free Beers

In the production of alcohol-free beers, the alcohol content of normal beer (top or bottom fermented, light or dark) is largely removed (≤0.5% by volume) by reverse osmosis (cf. 18.2.10.3) or distillation under vacuum at ca. 40 °C. The influence on the aroma is presented in 20.1.7.8.

20.1.8.5 Export Beers

These originate from widely different kinds of beer. They are mostly pasteurized and additionally treated with flocculating or adsorption agents (tannin, bentonite) or with proteolytic enzyme preparations to remove most of the proteins. The proteolytic enzymes split the large protein molecules into soluble products. Such beers are free of cloudiness or turbidity (chill-proofed beers) even after prolonged transport and cold storage.

20.1.9 Beer Flavor and Beer Defects

The taste and odor profile of a beer, including possible aroma defects, can be described in detail with the help of 44 terms grouped into 14 general terms, as shown in Fig. 20.3. Apart from a great variety of terms for odor notes, the terms bitter, salty, metallic, and alkaline are used only for taste and the terms sour, sweet, "body" etc. are applied to both taste as well as odor.

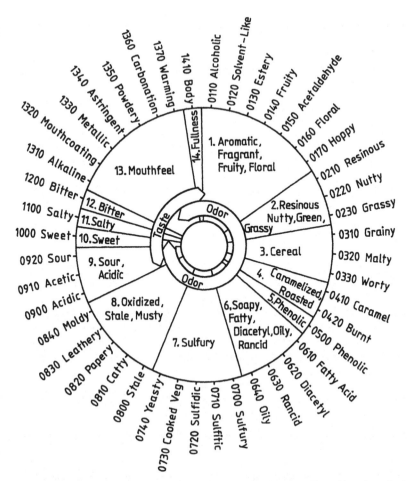

Fig. 20.3. Terminology for the description of odor and taste notes of beer (*American Society of Brewing Chemists*, according to *Meilgaard*, 1982)

Nine of the terms given in Fig. 20.3 describe the most important odor and taste characteristics of a good beer (Table 20.8). They are also suitable for the differentiation of different types of beer (Table 20.8).

Foaming is an important criterion of the taste of beer. A distinction is made between foam volume (produced by the content of carbon dioxide), foam density, and especially foam stability (caused by protein degradation products, bitter hop compounds, and pentosans). Lower fatty acids that are present in beer bouquet act as defoamers.

Beer defects detract from the odor and taste and are caused by improper production and storage. An example of a taste defect is the harsh, hard, bitter taste produced by the oxidation of polyphenols and some hop constituents. A flat taste, as already mentioned, comes from a low content of carbon dioxide. Diacetyl and ethanal in concentrations greater than 0.13 mg/l and 25 mg/l respectively, produce a taste defect. Acceleration of fermentation caused, e. g., by intensive stirring of the wort, raises the content of diacetyl and higher alcohols in the beer and lowers the content of esters and acids. On the whole, the aroma is negatively influenced. Higher concentrations of ethanal can arise, e. g., at higher fermentation temperatures and higher yeast concentrations.

Beer is very sensitive to light and oxidation. The "light" taste is due to the formation of 3-methyl-

Table 20.8. Main characteristics of the odor and taste of various types of beer

Flavor group	Intensity[a]					
	Munich	Pilsner	Pale ale	US lager	Stout[b]	Lambic
Bitterness	3–6	6–10	5–8	2–4	6–10	3–6
Alcoholic flavor	2–4	3–4	3–4	3–5	3–5	3–6
Carbonation	3–4	3–4	1–3	4	3–4	3–5
Hop character	2–6	6–10	5–8	0.5–4	6–10	3–6
Caramel flavor	4–8	0.5–2	3–5	0.5–1	6–100	1–3
Fruity/estery flavor	1–2	1–1.5	1–2	2–3	2–3	3–5
Sweetness	2–3	1–2	1–2	2–3	1–2	1–2
Acidity	1–2	1–2	1–2	1–2	2–3	3–20
Cabbage-like	1–2	1–3	0.2–0.8	1–3	0.2–0.8	1–10

[a] Semiquantitative values on the basis of aroma values.
[b] Top fermented English strong beer with a stemwort content of up to 25%.

2-buten-l-thiol (cf. Table 5.5). This substance becomes unpleasantly noticeable at concentrations higher than 0.3 µg/l. It is one of the characteristic aroma substances below this concentration. Enzymatic peroxidation of lipids contained in the wort and nonenzymatic secondary reactions during wort boiling give rise to the aroma defects listed as No. 8 in Fig. 20.3.

A sweetish off-flavor formed during storage of beer is caused by an increase in 3-methylbutanal, methional, phenylacetaldehyde, ethyl methylpropanoate and ethyl 2-methylbutanoate.

The addition of ascorbic acid or glucose oxidase/catalase (cf. 2.7.2.1.1) is recommended to overcome color and flavor defects caused by oxidation. Therefore, low-oxygen bottling is of great importance. Bottled beer should not contain more than 1 mg O_2/l.

Unwanted carbonyl compounds, which can produce an off-flavor in stored beer, are bound by sulfite derived from yeast metabolism. Yeast reduces the sulfate present in the wort to sulfite and sulfide, which is then consumed in the biosynthesis of sulfur-containing amino acids. If the growth of yeast comes to a standstill, excess sulfite is eliminated, increasing the stability of the beer to oxidative processes.

The very potent aroma substance 3-methyl-3-mercaptobutyl formate (cf. 5.3.2.5) can produce an off-aroma called "catty" (0810 in Fig. 20.3). The concentration of phenylacetaldehyde can also increase to such an extent on the storage of beer that it becomes noticeable in the aroma.

On storage, beer can become cloudy and form a sediment. Proteins and polypeptides make up 40–75% of the turbidity-causing solids. They become insoluble due to the formation of intermolecular disulfide bonds, complex formation with polyphenols, or reactions with heavy metals ions (Cu, Fe, Sn). Other components of the sediment are carbohydrates (2–25%), mainly α- and β-glucans. For measures used to prevent cloudiness, see 20.1.8.5. Undesirable microorganisms, e. g., thermophilic lactic acid bacteria, acetic acid bacteria (*Acetobacter, Gluconobacter*) and yeasts, can cause disturbances and defects in various process steps (mashing, fermentation, finished product).

20.2 Wine

20.2.1 Foreword

Wine is a beverage obtained by full or partial alcoholic fermentation of fresh, crushed grapes or grape juice (must). The woody vine grape has thrived in the Mediterranean region since ancient times and Italy, France and Spain are still among the leading wine-producing countries in the world. Other major producers are USA, Argentina, Chile, Germany and South Africa. Table 20.9 provides data on wine production and consumption in some countries. An overview of the individual process steps in wine production is presented in Fig. 20.4.

Table 20.9. Wine production 1999 and 2006 (1000 t), vineyard area (1993) and wine consumption (l/capita)

Continent	Production 1999	2006	Vineyard area (1993)			
World	27,944	27,772	82.81			
Africa	1005	1174	3.47			
America, North, Central	2217	2282	7.93			
America, South	2077	3046				
Asia	1250	1908	13.85			
Europe	20,593	17,850	56.86			
Oceania	802	1512	0.7			

Country	Production 1999	2006	Vineyard area (1993)	Consumption (1971)	(1993)	(1997)
France	5806	5349	9.42	107	64	54
Italy	6265	4712	9.81	111	61	60
Spain	3298	3644	13.7	60	39	38
USA	2045	2232	3.25	5	6	–
Argentina	1255	1540	2.05	85	48	–
Germany	1229	891	1.06	18	23	23
South Africa	878	1013	–	–	–	–
Australia	679	1410	0.63	–	–	–
Portugal	742		3.7	91	55	53
Romania	650		2.51	23	55	–
China		1400				
Chile		977				
$\Sigma(\%)^a$	82					

^{a} World production = 100%.

20.2.2 Grape Cultivars

Among the cultivated species of *Vitis*, the most important is the grapevine *Vitis vinifera*, L. ssp. *vinifera* in its many forms; more than 8000 cultivars are known. The size, shape and color of the grapes vary: there are round, elongated, large or small grape clusters. Grapes are either wine-type grapes, for white or red wine making, or table grapes, which are even grown in greenhouses in some northern countries. The cultivars are different in sugar content and aroma. Table 20.10 provides information about the major grape cultivars of Germany, with some of their characteristics. Table 20.11 shows the share of the major cultivars in vine growing areas. Table 20.12 gives data on the grape cultivars of some other countries. The European *V. vinifera* and the American vines (*V. labrusca*) have been crossed in order to produce pest-resistant

forms (hybrids, "direct producers"), giving plants with pest resistance and good quality must production, although the hybrids still leave much to be desired. The wines are considered rather ordinary, with less character and a more obtrusive flavor than the parent plants. Grape cultivars providing top quality white wines are:

- *Riesling* – native to Germany; a hardy cultivar grown in the Pfalz (Rhine Palatinate) and along the Mosel (Moselle), Rhine and Nahe rivers.
- *Traminer* – cultivated extensively in Alsace, Baden and Pfalz, and in Austria.
- *Rulaender* (grey burgundy, Pinot gris) – from Alsace and Burgundy regions in the Kaiserstuhl district, and from Hungary.
- *Kerner* – an early ripening cultivar, which comes close to the balance of Riesling.

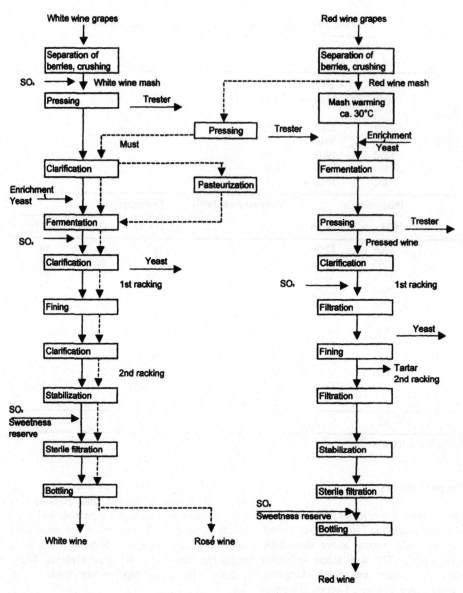

Fig. 20.4. Production of wine

- *Semillon Blanc* – together with Sauvignon and sometimes with Muscatel, provides Sauternes from the Bordeaux region.
- *Sauvignon* – used for Sauternes, and processed into its own types of wine, such as in the Loire region.
- *White Burgundy* (Pinot blanc) – yields the white wines from Burgundy (Chablis, Meursault, Puligny-Montrachet).

- *Chardonnay* – related to white burgundy, cultivated for example in Champagne.
- *Auxerrois* – also related to white burgundy.

Grape cultivars providing good white wines are:

- *Muscatel* and *Muscat-Ottonel* – cultivars with an exceptionally rich bouquet.
- *Furmint* – the grape cultivar of Hungarian Tokay wines.

Table 20.10. Important German grape cultivars

Cultivar	Wine type[a]	Acid[b]	Must weight[c]	Maturation characteristics[d]	Yield[e]	Comments about wine[f]
White wine cultivars						
Auxerrois		2	2	4	3	A vivacious wine with distinquished bouquet
Bacchus	M	2	2	3	3	Flowery with a muscat note fragrance
Burgundy, white	S	3	2	4		A full-bodied wine, pleasantly aromatic and more neutral as Rulaender
Ehrenfelser	R	2	2	5	2	Fruity, mildly acidic, a Riesling-like wine
Elbling, white	S	3	1	6	3	A light wine, devoid of rounded body and bouquet
Faber	M		2	2	3	A refined, refreshing and fruity flavored wine
Gutedel, white	M	1	1	3	3	Light wine, pleasing and captivating, mildly aromatic
Huxelrebe	T	2	2	2	3	A mellow wine with muscat-like bouquet
Kerner	R	2	2	4	2	A refreshing wine with a fine Riesling-like bouquet
Morio-Muscat	B	2	1	3	3	A wine with strong captivating muscat aroma
Mueller-Thurgau	M	2	1	2	3	Mild and refreshing wine with fine muscat flavor
Muscatel-yellow	B	2	2	5	1–2	A superior wine with a strong muscat-like aroma
Muscat-Ottonel	B	2	2		2	A pleasing wine with a strong refined muscat bouquet
Nobling	S	2	2	2	2	A full-bodied wine with a fruity flavor and fine bouquet
Optima	A	2	3	2	2	A refined, captivating wine with a fragrant aroma
Ortega	B	2	3	1	1	A wine with refined peach-like aroma
Perle	T	1	2	3	3	A mellow wine with flowery bouquet
Riesling, white	R	3	1	5	2	A superior refreshing and pleasing wine, with a fruity and flowery flavor
Rulaender (grey burgundy)	T	2	2	3	3	A body-rich wine with burning and passionate perception, and a pleasing bouquet
Scheurebe	T	3	2	4	3	A strong fruity flavored body-rich wine with a bouquet reminiscent of black currants
Siegerrebe	B	1	3	1	1	A wine with highly intensive refined bouquet
Sylvaner, green	S	2	1	4	3	A mellow pleasing wine with a delicately fruity flavor
Traminer, reddish (Clevner)	T	2	2	4	1	A wine with an exceptionally strong persisting bouquet

Table 20.10. (continued)

Cultivar	Wine type[a]	Acid[b]	Must weight[c]	Matura-tion charac-teristics[d]	Yield[e]	Comments about wine[f]
Red wine cultivars						
Burgundy, blue, late	2–3	2	4	2–3	Full-bodied, strongly flavored with a rounded bouquet, dark red mellow wine
Heroldrebe					A superior neutral wine with a tannin-like astringency
Limberger, blue	2	2	5	2	Characteristically fruity, a somewhat herbaceous, tarty and finely astringent bluish-red wine
Muellerrebe (black riesling)	2	2	4	2	Reminiscent of late Burgundy, but of lower quality
Portuguese, blue	2	1–2	1	3	A neutral mellow bluish-red wine with a bouquet deficiency
Trollinger, blue	2	2		3	A mellow refreshing light wine with a pungent flavor and light-red in color

[a] Quality German wines are classified as table wines (Tafelwein, Oechsle degrees less than 60), quality wines (with all the required characteristics of the growing region and an Oechsle degree of at least 60) and the special high quality wines (Oechsle degrees at least 73). The latter are denoted according to increasing quality as Kabinett, Spaetlese, Auslese, Beerenauslese and for the top quality as Trockenbeerenauslese. In addition to the rating, the label might carry a designation as Eiswein (ice-wine, see text).
R: Riesling group of wine (superior, fruity wine with distinct acidity)
S: Sylvaner group (neutral wine devoid of a distinct bouquet)
M: Mueller-Thurgau group (light, flowery with discrete bouquet)
T: Traminer group (wine with a fine bouquet)
B: Bouquet group of wine (wine with strong and aromatic bouquet)
A: Auslese group of wines (fullbodied great wines).
[b] 1: Low (approx. 50 g/l), 2: medium (approx. 5–10 g/l), and 3: high acidity (10–15 g/l).
[c] 1: 60–70 Oechsle degrees, 2: 70–85 °C, and 3: >85 Oechsle degrees.
[d] 1: Very early maturing (beginning–middle of September), 2: early (middle–end of September, 3: early-medium (end of September, beginning of October), 4: medium late (beginning–middle of October), 5: late (middle–end of October), and 6: very late maturing cultivar (end of October beginning of November).
[e] 1: Low (60 hl/ha), 2: average (60–80 hl/ha), and 3: high yielding cultivar ($\geq$90 hl/ha).
[f] The wine organoleptic quality description has its own wine dictionary. Terms classify and refer to wine (1) aroma or bouquet, (2) body, (3) sweetness and acids, (4) variety or cultivar, (5) age and (6) wine taste harmony (i.e. to which extent are the constituents of wine agreeably blended or related).

- *Sylvaner* – grown in Pfalz, Rheinhessen and Franken regions of Germany.
- *Mueller-Thurgau* – grown widely in east Switzerland and in Germany; it is a cross between Riesling and Sylvaner.
- *Gutedel* (Chasselas, Fendant, Dorin) – often found in Baden, Alsace, West Switzerland, and France.
- *Scheurebe* – a favored cultivar in Germany, obtained by crossing Sylvaner and Riesling.
- *Morio-Muscat*, a cultivar of exceptional bouquet.
- *Veltliner* – of significance in Austria, as is
- *Zierfandler*.

Grape cultivars providing top quality red wines are:

- *Pinot Noir* – the famous red vine cultivated in the Cote d'Or region of Burgundy, and also in Germany along the river Ahr and in Baden.

Table 20.11. Cultivation of important grape cultivars in Germany

Grape cultivar	Vineyard area in ha, 2005	
White grape cultivars	64,500	(63.2%)
Riesling	20,794	(20.4%)
Müller-Thurgau	14,346	(14.1%)
Silvaner	5383	(6.3%)
Kerner	4253	(4.2%)
Grey burgundy		
(Rulaender)	4211	(4.1%)
White burgundy	3335	(3.3%)
Bacchus	2205	(2.2%)
Scheurebe	1864	(1.8%)
Gutedel	1129	(1.1%)
Chardonnay	1018	(1.0%)
Other	5962	(9.2%)
Red grape cultivars	37,537	(36.8%)
Blue late burgundy	11,660	(11.4%)
Dornfelder	8259	(8.1%)
Blue Portuguese	4818	(4.7%)
Blue trollinger	2543	(2.5%)
Black riesling	2459	(2.4%)
Regent	2158	(2.1%)
Lemberger	1612	(1.6%)
Other	4028	(10.7%)

- *Cabernet-Sauvignon*,
- *Cabernet-Franc*, and
- *Merlot* – are cultivated together and provide the famous red wines of the Bordeaux region.

Other red grape cultivars are:

- *Gamay* – from the southern part of Burgundy and from Beaujolais and Maconnais.
- *Pinot Meunier* – black Riesling; of importance in Champagne, Wuerttemberg and Baden.
- *Portuguese* – found in Pfalz, Rheinhessen, and Wuerttemberg.
- *Trollinger* (Vernatsch) – cultivated in south Tyrol and in Wuerttemberg.
- *Limberger* – found in Wuerttemberg and Austria.
- *Blue Aramon* – the cultivar which provides the wines from Midi, France.
- *Rossary* – widely cultivated in south Tyrol.

Grape vine cultivation requires an average annual temperature of 10–12 °C. The average monthly temperature from April to October should not fall below 15 °C. The northern limit for cultivating the grape vine is close to 50 °C latitude. The per-

missible altitude for cultivation is dependent on the climate (plains in Italy, Spain and Portugal; sunny slopes of Germany; up to 13,000 m on Mt. Aetna in Sicily; up to 2700 m in the Himalayas). Soil cultivability and quality and weather are of decisive importance.

20.2.3 Grape Must

20.2.3.1 Growth and Harvest

After blooming and fruit formation, the grape berry continues to grow until the middle or the end of August, but remains green and hard. The acid content is high, while the sugar content is low. As ripening proceeds, the berry color changes to yellow-green or blue-red. The sugar content rises abruptly, while both the acid and water contents drop (Fig. 20.5).

The harvest (picking the berry clusters from the vines) is performed as nearly as possible when the grape is fully ripe, about the middle of September until the end of November, or it may be delayed until the grapes are overripe. In the USA and Europe, machines are being increasingly used for this very laborious harvesting, e. g., grape harvesters. However, they cannot sort the grapes according to the degree of ripeness. Terms which relate to the time of harvest include "vorlese", early harvest, "normallese", normal harvest, and "spaetlese", late harvest. The latter term, when applied to German wines, identifies excellent, top quality wines. Particularly well-developed grapes of the best cultivars from selected locations are picked separately and processed into a wine called "Auslese". When the grapes are left on the vine stock, they become overripe and dry – this provides the raisins or dried berries for "Beerenauslese", "Trockenbeerenauslese", or "Ausbruch" wine (fortified wine). In some districts, such as Tyrol and Trentino, the grapes are spread on straw or on reed mats to obtain shrivelled berries – this provides the so-called straw wines, Grapes that are botrytised (a state of "dry rot" caused by the mold *Botrytis cynerea*, the noble rot) have a high sugar content and a must of superior quality, consequently also producing a superior, fortified wine. Frozen grapes left on the vine stock (harvested at −6 °C to −8 °C and pressed in the frozen state) provide

Table 20.12. Major grape cultivars of selected countries

Country	Grape cultivar	Comments about cultivation area and quality
France	*White wine cultivars*	
	Aligote	Bourgogne, a "vin ordinaire", modest quality wine
	Chardonnay	Cultivated in Champagne and Bourgogne area (Chablis, Montrachet, Pouilly), a very good quality wine
	Chemin blanc	Cultivated in regions of Tourraine, Anjou and Loire
	Folle blanche	Wine used for brandy production in Cognac and Armagnac
	Grenach blanc	Midi
	Melon blanc	
	(Muscadet)	Mellow refreshing wine with a slight muscat bouquet
	Muscadelle	Cultivated in Bordeaux and Charente regions, 5–10% blended into Sauternes and Graves wines
	Pinot blanc	Cultivated in Alsace, Champagne, Loire and Cote d'Or
	Pinot gris	Alsace wine
	Roussane (Rouselle)	Cultivated in Rhone region, a full-bodied, pleasing fragrancy wine
	Sauvignon	Wine of Bordeaux, Loire and Cher regions, a full-bodied fragrant wine, with Semillon used for production of Sauternes wine
	Semillon blanc	As Sauvignon, used for production of Sauternes wine
	Red wine cultivars	
	Cabernet Franc	Spread in Bordeaux and Loire regions, a superior, strong pleasing wine, with Cabernet Sauvignon and Merlot is an ingredient of Bordeaux wines
	Cabernet Sauvignon	As Cabernet Franc, aroma rich, a superior quality wine
	Carignan	Grown in Rhone, Midi and Provence regions
	Cot (Malbec)	Bordeaux, one of the best grape cultivars
	Cinsaut	Grown in Southern France
	Grenach noir	Grown in Southern France
	Gamay noir	Beaujolais, Maconnais; fruity pleasant, refreshing wine
	Merlot	A Bordeaux wine, full-bodied, rich and mellow; as Cabernet Franc and Cabernet Sauvignon is an ingredient of Bordeaux wines
	Petit Verdot	Grown in Bordeaux region, component of Bordeaux wines
	Pinot noir	Bourgogne, and wine of Cote d'Or
	Syrah	Grown in Southern France
Italy	*White wine cultivars*	
	Malvasia, bianca	An important cultivar across Italy
	Mascato, bianco	Grown mostly in Northern Italy, a wine of Asti region
	Trebbiano	Widely grown across Italy
	Vermentino	Grown along Italian riviera, a very good white wine
	Weissterlaner	Wine of South Tyrol
	Red wine cultivars	
	Aleatico	Widely grown in Italy
	Barbera	One of the most important cultivars
	Freisa	Grown in Piemont and Vercelli regions, one of the best Italian cultivars
	Gross-Vernatsch	
	(Trollinger)	The wine of Bolzano, Trento and Como
	Lagrein	Grown in South Tyrol
	Merlot	
	Nebbiolo	A prefered cultivar of Piemont and Lombardy regions
	Pinot Nero	Grown in Northern Italy with Rome as Southern limit
	San Giovese	Spread from Toscana till Latium; major constituent of Chianti wine

Table 20.12. (continued)

Country	Grape cultivar	Comments about cultivation area and quality
Austria	*White wine cultivars*	
	Mueller-Thurgau	
	Muscat-Ottonel	
	Neuburger	A pronounced cultivar bouquet, pleasantly acidic
	Rheinriesling	
	Rotgipfler	Fruity, aroma rich, full-bodied; together with Zierfandler an ingredient of Gumpoldskirchner wines
	Sylvaner	
	Traminer	
	Veltiliner, green	A pleasant pleasing bouquet refreshing wine
	Veltiner, red	Fruity wine with a fine bouquet
	Veltliner, (early red, Malvasier)	
	Welschriesling	A mellow wine with fine bouquet
	Zierfandler, red	A wine with burning and passionate perception, fragrant aromatic, with a cultivar specific bouquet
	Red wine cultivars	
	Burgundy, blue, late	
	Blaufraenkisch (Limberger)	
	Portuguese, blue	
	Sankt Laurent	A strong wine, dark red colored, with a fine Bordeaux-like bouquet
Switzerland	*White wine cultivars*	
	Gutedel (Chasselas, Fendant, Dorin)	A major Swiss grape cultivar
	Marsanne Blanche (Hermitage)	A mellow wine with a refined bouquet
	Riesling	
	Mueller-Thurgau	Major cultivar of Eastern Switzerland
	Red wine cultivars	
	Burgundy, blue	
	Gamay	Grown in Western Switzerland
	Merlot	The wine of Tessin (Ticino)
Hungary	*White wine cultivars*	
	Furmint, yellow	Used for Tokay wine production
	Red wine cultivars	
	Kadarka	The most important Hungarian red wine cultivar

ice-must which, because of freezing, is enriched in sugar and acid and, as such, is a source of high quality wines (ice-wines).

20.2.3.2 Must Production and Treatment

The grape clusters cut from vine stocks using grape shears are cleaned of rotten and dried berries and then, as fast as possible, separated from the stems. This is done in a roller crusher which consists of two fluted horizontal rolls by which the berries are crushed without breaking the seeds or grinding the stems. The latter are separated out by a stemmer. The crushed grapes are then subjected to pressing to release their juice, the must. The mechanical and partly continuously operated presses are basket-type screw-presses (extruder-like tapered screw), hydraulic or pneumatic presses. The free-running must is collected prior to pressing (first run) and, after mild pressing, the major portion (pressed-must)

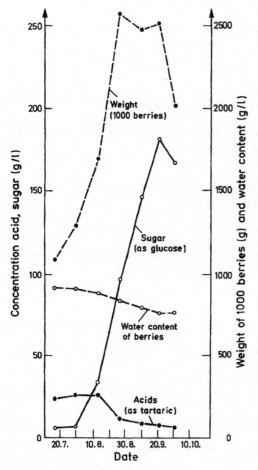

Fig. 20.5. Sylvaner wine grape ripening with measurement of the content of acid (as tartaric), sugar (as glucose), weight of 1000 berries and water content of the berries

is produced. The remaining grape skins and seeds (pomace) are loosened or shaken-up and pressed again. This provides the second or post-extract. In red wine making the crushed berries (the mash) are fermented without prior removal of the pomace, i.e. the must is fermented together with the skin. This is done in order to extract the red pigments localized in the skin, which are released only during fermentation. When blue grapes are processed in the same manner as white ones, or blends of blue and white grapes are combined and then processed, pink wines are obtained. They are designated as rosé wines. In red wine making the extraction of red pigments is sometimes facilitated by raising the temperature

to 50 °C prior to fermentation of the mash, or to 30 °C after the main fermentation, followed by a short additional fermentation.

The left over stems, skins and seeds provide the pomace. It is used as feed or fertilizer, or is fermented to provide pomace wine. This is consumed as a homemade drink and is not marketed. Pomace brandy is obtained by distillation of fermented pomace. The average must yield is 75 l/100 kg grapes. Of this, 60% is free juice (must), 30% press-must and 10% must from the second pressing.

The fresh, sweet must can be treated with sulfur dioxide (50 mg SO_2/l) to suppress oxidative discoloration and the growth of undesirable microorganisms. In order to remove undesirable odors or off-tastes, the must is treated with activated charcoal and, when necessary, is clarified by separators or filters. In general, sulfurization before fermentation is dispensed with if the material is faultless and pure culture yeast is used. If required, the must is pasteurized by a short heat treatment (87 °C/2 min).

The addition of sugar to and deacidification of must will be discussed in 20.2.5.4.

20.2.3.3 Must Composition

Table 20.13 provides data on the average composition of grape musts. For the quality assessment of grape must, its relative density at 20 °C is decisive. This is measured with a special aerometer (must balance). The must weight M, expressed in Oechsle degrees, is directly read off.

$$M\,[°Oe] = (D-1) \times 10^3 \qquad (20.6)$$

Accordingly, a must with $D = 1.080$, has an M of 80 °Oe. In Germany, the quality levels for

Table 20.13. Average composition of grape must

Constituent	Content (g/l)
Water	780–850
Sugar (as glucose)	120–250
Acids (as tartaric acid)	6–14
N-Compounds	0.5–1
Minerals	2.5–3.5

Table 20.14. Quality levels and natural minimum alcohol content of German wines

Quality level	Minimum alcohol content[a]	
	Zone A[b]	Zone B[b]
Table wine	5.0	6.0
Country wine	5.5	6.5
Quality wine	7.0[c]	8.0
Quality wine with vintage		
- Cabinet	9.5[d]	10.0
- "Spaetlese"	10.0	11.4
- "Auslese"	11.1	13.4
- "Beerenauslese"	15.3	17.5
- Ice wine	15.3	17.5
- "Trockenbeeren"	21.5	21.5

[a] in % vol.
[b] Vine growing areas: Germany without Baden (Zone A), Baden (Zone B).
[c] Partly 6.0.
[d] Partly 9.0.

wine are traditionally defined through the must weight, e. g., (°Oe): Cabinet (70–73), "Spaetlese" (85–90), "Auslese" (92–100), "Beerenauslese" (120). Internationally, the natural alcohol content is a characteristic feature of quality. The corresponding values for German wine are presented in Table 20.14.

Since the density of the must is primarily dependent on the sugar content c, it can be estimated using the following equation:

$$c[\%] = (0.25 \times ° \text{Oe}) - 3 \qquad (20.7)$$

Hence, a must of 100 °Oe contains about 22% sugar.

20.2.3.3.1 Carbohydrates

Ripe grapes contain equal amounts of glucose and fructose, while fructose predominates in overripe or botrytised berries.

In addition L-arabinose (ca. 1 g/l), rhamnose (up to ca. 400 mg/l), galactose (up to ca. 200 mg/l), D-ribose (ca. 100 mg/l), D-xylose (ca. 100 mg/l) and mannose (up to ca. 50 mg/l) are present. Saccharose (ca. 10 g/l) is detectable only if the saccharase is inhibited during pressing. Other oligosaccharides present are: raffinose (up to ca. 200 mg/l), maltose (ca. 20 mg/l), melezitose (ca.

100 mg/l) and stachyose (ca. 150 mg/l). Pectins (0.12–0.15%) and small amounts of pentosans are present.

20.2.3.3.2 Acids

The major acids of must are L-tartaric and L-malic acids. Succinic, citric and some other acids are minor constituents. In a good vintage, tartaric acid is 65–70% of the titratable acidity, but in years when unripe grapes are fermented, its content is only 35–40% and malic acid predominates. The good vintage year of 1911, for example, yielded grapes with 3.1 g/l malic acid and 6.4 g/l tartaric acid; in the inferior vintage year of 1912, on the other hand, malic acid was 10.7 g/l and tartaric acid 6.0 g/l.

20.2.3.3.3 Nitrogen Compounds

Proteins, which include various enzymes, peptides and amino acids, are present in low amounts (cf. 18.1.2.1)

20.2.3.3.4 Lipids

The lipid content of must is about 0.01 g/l.

20.2.3.3.5 Phenolic Compounds

Tannins occur primarily in stems, skin and seeds. In a carefully prepared white must, the tannin content is no more than 0.2 g/l. In contrast, red wines contain high levels of tannin, 1–2.5 g/l or even higher. In white grapes, quercetin, its 3-rhamnoside quercitrin and carotinoids contribute to the color. The main part of the color pigments of European red-wine vines are free (unesterified) anthocyanidin-3-glucosides with malvidin-3-glucoside (40–90% of the anthocyans) as the dominating compound. Apart from the 3-monoglucosides, anthocyanidin-3,5-diglucosides also occur on crossing with American cultivars (hybrids).

β-Glucosidases, which come from yeast, hydrolyze the free anthocyanidin glucosides to the instable aglycones during fermentation. Of special analytical interest are the anthocyan glucosides which are not attacked by the hydrolases and can easily be separated by RP-HPLC. These glucosides occur as side products and are acylated with acetic acid, p-cumaric acid or caffeic acid. The spectrum of the pigments depends on the grape cultivar, e. g., Cabernet Sauvignon contains about three times as much malvidin-3-acetylglucoside as malvidin-3-cumarylglucoside. However, the acylated anthocyans also decrease with time due to oxidation and condensation reactions. Consequently, their detection in wines that are more than 2–3 years old becomes increasingly difficult.

Cyanidin-3-glucoside is a suitable indicator of cherry wines which have been added to a wine to intensify the red color.

20.2.3.3.6 Minerals

Must contains predominantly potassium, followed by calcium, magnesium, sodium and iron. Important anions are phosphate, sulfate, silicate and chloride.

20.2.3.3.7 Aroma Substances

The must aroma substances will be discussed together with wine aroma substances (cf. 20.2.6.9).

20.2.4 Fermentation

Wine fermentation may occur spontaneously due to the presence of various desirable wine yeasts and wild yeasts found on the surface of grapes. Fermentation can also be conducted after must pasteurization by inoculation of the must with a pure culture of a selected strain of wine yeast. Wild yeasts include Saccharomyces apiculatus and exiguus, while the pure selected yeasts are derived from Saccharomyces cerevisiae var. ellipsoides or pastorianus. The pure wine yeast possesses various desirable fermentation properties. High fermenting strains are used to give high alcohol wines (up to 145 g/l) and those which are resistant to tannin and high alcohol levels are used

in red wine fermentation. Other types of yeast are "sulfite yeast" with little sensitivity to sulfurous acid (sulfur dioxide solutions), "cold fermentation yeasts", which are active at low temperatures and, finally, special yeasts for sparkling wines, which are able to form a dense, coarse-grained cloudiness that is readily removed from the wine. The desired yeasts (5–10 g of dried yeast per hectoliter of must) are added to must held in fermenters (vats made of oak, or chromiumnickel steel tanks lined with glass, enamel or plastic). The must is then fermented slowly for up to 21 days below 20 °C for white wines or 20–24 °C for red wines. The course of fermentation is influenced by sulfurous acid: 100 mg/l SO_2 delay the start of fermentation by 3 days, 200 mg/l SO_2 by 3 weeks.

As a safeguard against air (discoloration), bacterial spoilage (acetic acid bacteria) and also to retain carbon dioxide, the liquid loss in the fermenter is compensated for by topping up with the same wine. After the end of main (primary) fermentation, which lasts 5–7 days, the sugar has been largely converted to alcohol while the protein, pectin and tannins, along with tartrate and cell debris, settle with the yeast cells at the bottom of the fermenter. This sediment is called bottom mud, dregs or lees.

Partial precipitation of tartaric acid as cream of tartar (mixture of K hydrogentartrate and Ca tartrate) is affected by temperature, alcohol content and pH (Fig. 20.6). The crystallization of tartar can be retarded by the addition of metatartaric acid (up to 100 m/l), obtained by heating tartaric acid above the melting point. The addition is carried out directly before bottling. A tartar stability of 6–9 months is achieved. After this period, the metatartaric acid is slowly converted to tartaric acid. The unfermented residual sugar (residual sweetness) may be retained when necessary, if the secondary fermentation is suppressed by addition of sulfurous acid. Fermentation stops at an ethanol concentration of 12–15% (v/v), depending on the type of yeast.

The young wine, which is drunk with the yeast in some regions of Germany and Austria ("Federweisser" or "Sauser"), is usually withdrawn from the fermentation tank via clarifying separators after the primary fermentation. Red wine mash is fermented at somewhat higher temperatures by using various procedures, often in closed

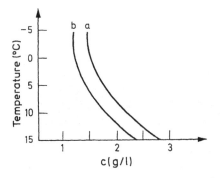

Fig. 20.6. The effect of temperature and ethanol concentration on cream of tartar solubility in wine. (a) 8 vol-%, (b) 12 vol-%. (according to *Vogt*, 1974)

double-walled enamellined tanks. The wine initially drawn off is the better quality free-run wine, followed by the pressed wine, an astringent and dry fraction ("*press-wine*"). The young wine should not stay on the pomace longer than necessary to extract the pigments, otherwise it will become tannin-enriched and hence harsh and astringent. In industrial production the extraction of the red pigments is not done by fermentation of the mash but by heat treatment of the mash (cf. 20.2.3.2).

The fermentation residue or pomace is processed into yeast-pressed wine or yeastbrandy, into wine oil (for brandy essence) and into tartaric acid. The left-over pomace is used as a feed or fertilizer. Pomace wine, obtained by fermenting a sugar solution containing the dispersed pressed-out pomace, is made only into a household drink and is not marketed.

20.2.5 Cellar Operations After Fermentation; Storage

The following cellar operations develop a particular character in the wine and give it stability and durability.

20.2.5.1 Racking, Storing and Aging

Racking of young wine is required to get rid of the sediment. The wine is drawn-off or decanted into large sulfur-treated vats, with or without aer-

ation. The time for racking is determined by the cellar master's experience. The wine racking is repeated as required. Racking should be carried but as early as possible. When necessary, 5–10% of unfermented sterile grape must is blended with the young wine to round-off and sweeten its flavor.

The objective of wine aging/storage is to further build up the aroma and flavor constituents. Aging requires various lengths of time. In general the wine is removed from vats after 3–9 months and poured into bottles in which aging continues. Duration of aging and storability differ and depend on wine quality. The great Burgundy and Bordeaux wines require at least 4–8 years in order to develop while for an average German wine, maximum development is achieved well within 5–7 years. Only great quality wines endure aging lasting 10–12 years or more without quality loss. Changes induced during wine maturation are not yet well understood. Reactions between wine ingredients, such as ethanol, acids and carbonyl compounds, which form the typical aroma components of wine, are covered in 20.2.6.9.

20.2.5.2 Sulfur Treatment

Crushed grapes (mash) or must are treated with sulfur immediately after grape crushing to preserve the constituents that are sensitive to oxidation, prevent enzymatic browning via phenol oxidation and suppress the growth of undesirable microorganisms (acetic acid bacteria, wild yeast, molds). Sulfur treatment of wines prior to the first racking serves the same purpose: wine stabilization (cf. 8.12.6). Futhermore, a very important effect is the suppression of undesirable aroma notes ("air", "oxidation", "ageing", "sherry" notes) by the binding of carbonyl compounds, especially of ethanal, as hydroxysulfonic acids. Sulfur treatment is achieved by the addition of sulfites, an aqueous solution of sulfurous acid or by adding liquid SO_2. The maximum quantities are stipulated by law. Only a part of the added sulfurous acid remains as free acid. A portion is oxidized to sulfate, while another binds to sugars and carbonyl compounds. The rapid oxidation of sulfurous acid can be partially reversed by the addition of L-ascorbic acid. Use of the right amount of SO_2 is important for fermentation, aging and stability and hence for the quality of the wine. Efforts are made to achieve 30–50 mg of free SO_2/l of finished wine.

20.2.5.3 Clarification and Stabilization

Suitable measures should not only eliminate any turbidity present, but also prevent its formation during storage (fining).

Turbidity-causing solids are mostly proteins as well as oxidized and condensed polyphenols. Furthermore, multivalent metal ions can cause discoloration and sediments. Wine clarification is usually achieved by precipitation reactions, filtration or centrifugation. In blue-fining the excess metal ions which are responsible for metal-induced cloudiness (iron, copper and zinc) are precipitated by precisely calculated amounts of potassium ferrocyanide. In this process, soluble Berlin blue is formed first,

$$KFe(CN)_6 + FePO_4 \rightarrow KFe_4 \cdot$$
$$Fe(CN)_6 + K_3PO_4 \tag{20.8}$$

which is then converted to insoluble Berlin blue.

$$3KFe_4(CN)_6 + 3FePO_4 \rightarrow Fe_4 \cdot$$
$$Fe(CN)_6 + K_3PO_4 \tag{20.9}$$

The blue turbidity formed helps to eliminate the persistent protein turbidity (grayish and black casse). The treated wine is tested for excess cyanoferrate and for free cyanide to be on the safe side. In other fining procedures, edible gelatin, isinglass (beluga dried bladder gelatin) combined with casein, egg albumen, tannin, iron-free bentonite, kaoline, agar-agar and purified or activated charcoals are added to the wine. This results in adsorption or precipitation of the substances causing cloudiness and unpleasant taste, the interaction products all being quick-settling coagulums. Phenolic compounds are removed from wine by polyvinylpyrrolidone (detanninizing) and undesirable sulfur compounds by cupric sulfate.

The clarification by filtering involves pads of asbestos, cellulose, infusorial earth, and filter aids such as Hyflo Super Cel and Filter Cel. The filters are built either as sheet filters or as washable filter presses. Sterile filtration has achieved great importance for the stability of wine and sweet must. Sterilizing filters made of asbestos or membrane sheets retain not only yeast cells, but also the much smaller spores of fungi and even bacteria. Sterilizing filters are also suitable for stopping fermentation and thus retaining a desired level of unfermented sugar (residual sweetness) at a selected stage of fermentation.

Suitable measures to prevent crystalline sediments in the bottle are, e. g., cooling the wine for a few days to 0–4 °C, addition of metatartaric acid (cf. 20.2.4), and reducing the concentrations of potassium, calcium, and tartaric acid by electrodialysis. Excessive concentrations of calcium produced by deacidification measures (cf. 20.2.5.4) can also result in additional crystal sedimentation (calcium tartrate, calcium mucate, and calcium oxalate). The elimination of excess calcium with D-tartaric acid is recommended as a counter-measure.

20.2.5.4 Amelioration

Must and wine amelioration is required when unfavorable weather in some years results in grapes with an excess of acids and a low sugar content. Such grapes would provide a must which could not be processed directly into a drinkable, palatable wine. The ameliorated wine should contain neither more alcohol nor less acid than the wine of the same type and origin from a good vintage year. The usual procedures involved are the addition of sugar, deacidification and wine blending.

The addition of sugar (*enrichment*), for which regulations exist in most countries, can be carried out before or during fermentation. Sucrose (dry sweetening) or grape must concentrates are added. To improve the quality, the *sweetness reserves* of the wine can be raised by the addition of grape must. The fermentation of this must is prevented by cold sterile storage, short-time heating (87 °C) or impregnation with CO_2 (15 g/l, pressure tank). The bouquet (aroma) is not improved. Poor or inferior wine is not improved by amelioration. Deacidification is achieved primarily by adding calcium carbonate, which may give either a precipitate of calcium tartrate or a mixture of calcium tartrate and calcium malate. Unsulfured wines which still have an excessively high acid content can be subjected to biological acid degradation (malolactic fermentation). In this process, lactic acid bacteria (e. g., *Leuconostoc oenos*) convert L-malic acid (10 g) to lactic acid (6.7 g):

$$HOOC-CHOH-CH_2-COOH$$
$$\rightarrow HOOC-CHOH-CH_3 + CO_2 \qquad (20.10)$$

In addition, the residual sugar, aldehydes and pyruvate are degraded so that less SO_2 is required in a subsequent sulfur treatment step. The multiplication of the lactic acid bacteria is promoted by increasing the temperature to 20 °C and stirring up the yeast settlings.

Wine blending is a suitable way of rectifying defects, refreshing old wines, deepening the color of red wines (table wines) and enhancing the bouquet or readjusting the low acid content, thus producing a uniform quality wine for the market.

Tartaric or citric acid can be added to low-acid wines from southern European countries. The addition of gypsum or phosphate treatment to enhance the color of red wines, which is used in the case of certain southern wines (e. g., Malaga, Marsala) is based on the increase in the color yield caused by lowering the pH with $CaSO_4$ or $CaHPO_4$.

20.2.6 Composition

The chemical composition of wine varies over a wide range. It is influenced by environmental factors, such as climate, weather and soil, as well as by cultivar and by storage and handling of the grapes, must and wine.

Within the scope of wine analysis, wine extract, alcohol, sugar, acids, ash, tannins, color pigments, nitrogen compounds and bouquet-forming substances are important. Hence, the value and quality of a wine is assessed through the content of ethanol, extract, sugar, glycerol, acids and bouquet substances. With the large number of quality-determining constituents, the evaluation and classification of wine are possible only by a combination of chemical analysis and sensory testing.

20.2.6.1 Extract

The extract includes all the components of wine mentioned above, except the volatile, distillable ones. Many of the extract components are present in must and are described in that section; others are typical fermentation and degradation products. The extract content of 85% of all German white wines is about 20–30 g/l (average about 22 g/l), while the extract content of red wines is somewhat higher – German "Auslese" wines contain about 60 g/l; other sweet wines, 30–40 g/l.

Since the sugar content can be manipulated, the "sugar-free extract" (extract in g/l minus reducing sugar in g/l plus 1 g/l for arabinose, which is also detected in the reductometric determination, but is not fermentable) is of greater importance for an evaluation of quality.

20.2.6.2 Carbohydrates

Carbohydrates (0.03–0.5%) present in fully fermented wines are small amounts of the hexoses glucose and fructose and of nonfermentable pentoses. Incompletely fermented wines contain higher concentrations of both hexoses, but substantially more of the slower fermenting fructose. The average ratio of glucose to fructose in the residual sugar of wine is 0.58:1, but it varies to a great extent. The pentose sugars which are present in fermented wines consist of 0.05–0.13% arabinose, 0.02–0.04% rhamnose, and xylose in trace amounts.

20.2.6.3 Ethanol

The ethanol content of wine varies over a wide range. It serves as a quality feature (cf. 20.2.3.3). An alcohol level above 144 g/l indicates addition of ethanol.

The extent to which ethanol is derived from added sugar in fermentation can be determined by the NMR spectroscopic measurement of the ratio of the hydrogen isotopes ^1H to ^2H. The method is based on the fact that the plant-specific $^2H/^1H$ ratio (R value, cf. 18.4.3) of the sugar also appears in ethanol: about 2.24 (corn sugar), about 2.70 (beet sugar), about 2.45 (wine). The detection limit is 6–8 °C Oe for a wine of unknown origin and it falls to 2–3°Oe (corresponding to ca. 0.5% v/v of ethanol) when the age, origin and grape cultivar are known.

20.2.6.4 Other Alcohols

Methanol occurs in wines at a very low level (38–200 mg/l), but much more is present in the fermentation of pomace as a product of pectin hydrolysis. Brandy distilled from pomace often contains 1–2% methanol. Higher alcohols in wine are propyl, butyl and amyl alcohols which, together, constitute 99% of the wine fusel oil. Hexyl, heptyl and nonyl alcohol and other alcohols including 2-phenylethanol (up to 150 mg/l) are present in small amounts. The average butylene glycol (2,3-butanediol) content is 0.4–0.7 g/l and is derived from diacetyl by yeast fermentation. Glycerol, 6–10 g/l, originates from sugars and gives wine its body and round taste. Added glycerol can be revealed by the determination of the glycerol factor (GF):

$$GF = \frac{Glycerol\,(g/l) \times 100}{Alcohol\,(g/l)} \qquad (20.11)$$

The natural variation range of GF in wines, as long as they have not been produced from noble rot material, lies between 8 and 10. Values above 12 indicate an additive. Low amounts of additive cannot be safely detected using the GF. More suitable is then the use of GC-MS to test the wine for the by-products of the technical synthesis of glycerol, e. g. 3-methoxypropanediol or cyclic diglycerols. Sorbitol is found in very low amounts. D-Mannitol is not present in healthy wines, but is present in spoiled, bacteria-infected wines at levels up to 35 g/l.

20.2.6.5 Acids

The pH of grape wine is between 2.8 and 3.8. Titratable acidity in German wines is between 4 and 9 g/l (expressed as tartaric acid). Acid degradation and cream of tartar precipitation decrease the acid content of ripe wines. Red wines generally contain less acids than white wines. The wines from Mediterranean countries and often high-grade wines ("Beerenauslese", "Trockenbeerenauslese") are low in acid content. Wine acids from grapes are tartaric, malic and citric acids and acids from fermentation and acid degradation are succinic, carbonic (carbon dioxide) and lactic acids and low amounts of some volatile acids. The presence of acetic and propionic acids as well as an anomalous amount of lactic acid is an indication of diseased wine.

Botrytis cinerea can form gluconic acid in concentrations of up to 2 g/l of must. Therefore, this acid is found in the corresponding wines.

20.2.6.6 Phenolic Compounds

Red wines contain phenols in considerably higher concentrations than white wines (Table 20.15). Exceptions are gentisic and ferulic acid, but relatively high concentrations of the last mentioned compound are characteristic of Riesling.

In the maturation of red wine, tanning agents polymerize (proanthocyanidins, cf. 18.1.2.5.2) in two ways and become insoluble, reducing the astringent taste. Acidcatalyzed polymerization proceeds via the carbocation shown in Formula 18.21. In addition, proanthocyanidins are cross-linked by acetaldehyde, which is formed

Table 20.15. Phenols in white and red wine[a]

Compound	White wine	Red wine
Gentisic acid	0.15–1.07	0.44–0.046
Vanillic acid	0.09–0.38	2.3–3.7
Ferulic acid	0.05–4.40	0.05–2.9
p-Coumaric acid	1.57–3.20	2.6–4.5
Caffeic acid	1.50–5.20	3.15–13
Gallic acid	0.50–2.80	13–30
cis-Reservatol	<0.10	0.27–0.88
trans-Reservatrol[b]	<0.25	0.71–2.5
cis-Polydatin[c]		0.02–0.68
trans-Polydatin[c]		0.02–0.98
(+)-Catechin	3.8–4.20	60–213
(−)-Epicatechin	1.7–3.8	25–82
Quercetin		0.5–2.6

[a] Concentration in mg/l.
[b] trans-3,4′,5-trihydroxystilbene (cf. Formula 20.12).
[c] cis-or trans-3,4′,5-trihydroxystilbene-3-β-D-glucoside.

by the slight oxidation of ethanol that occurs on storage of red wine.

$$(20.12)$$

20.2.6.7 Nitrogen Compounds

The nitrogen compounds in must precipitate to a smaller extent by binding to tannins during grape crushing and mashing, while most (70–80%) of them are metabolized by the growing yeast during fermentation. Free amino acids, especially proline (about 200–800 mg/l) are the major nitrogen compounds which remain in wine. Tryptophan, which is present in must in concentrations of 1–30 mg/l, and acetaldehyde, which is provided by yeast, are precursors of 1-methyl-1,2,3,4-tetrahydro-β-carbolin-3-carboxylic acid (MTCA). In fact, 0–18 mg/l of MTCA have been detected in wine. Its formation (cf. Formula 20.13) is inhibited by SO_2, which traps the precursors. On distillation, MTCA apparently remains in the residue because only traces, if at all, are present in brandy and whiskey. However, MTCA is not restricted to fermented products like wine, beer and soy sauce as the precursors are widely found, e. g., in milk, cheese and smoked foods.

20.2.6.8 Minerals

The mineral content of wine is lower than that of the must since a part of the minerals is primarily removed by precipitation as salts of tartaric acid. The ash content of wines is about 1.8–2.5 g/l, while that of must is 3–5 g/l. The average composition of ash in %, is: K_2O, 40; MgO, 6; CaO, 4; Na_2O, 2; Al_2O_3, 1; CO_2; 18; P_2O, 16; SO_3, 10; Cl, 2; SiO_2, 1.

The iron (as Fe_2O_3) content of wine is 5.7–13.4 mg/l, but it can rise to much higher levels (20–30 mg/l) through improper processing of grapes.

20.2.6.9 Aroma Substances

Most of the volatiles in wine, more than 800 compounds, with a total concentration of 0.8–1.2 g/l, has been identified. For the wines Gewürztraminer and Scheurebe, it has been found that the compounds listed in Tables 20.16 are so odor active that they can produce the aroma in each case. This could be confirmed for Gewürztraminer in a model experiment. A synthetic mixture of odor and taste compounds in the concentrations given in Table 20.16 and 20.17 reproduced the aroma and the taste of Gewürztraminer.

The two cis-rose oxides and 4-mercapto-4-methylpentan-2-one, which has an exceptionally low odor threshold (cf. 5.3.2.5), have been identified as the cultivar-specific odorants in Gewürztraminer and Scheurebe. In addition, ethyloctanoate, ethylhexanoate, 3-methylbutylacetate, ethylisobutanoate, linalool, (E)-β-damascenone and wine lactone (cf. structure in 5.2.5) exhibit high but different aroma values in the two types of wine. Other typical odorants are listed in Table 20.18. Some red wines, e. g. from Shiraz grapes contain the bicyclic terpene (−)-rotundone, the key aroma compound of pepper (cf. 22.1.1.2.1). Its concentration is high (up to 145 ng/l) in samples showing an intense "peppery" aroma note. Ethanol is essential for the aroma of wine. Since the odor thresholds of many volatiles increase in the presence of ethanol, e. g., those of ethyl-2- and -3-methylbutanoate increase by a factor of 100 (Table 20.19), it influences the bouquet of wine. Correspondingly, the intensity of the fruity note increased in an aroma model for Gewürztraminer when the alcohol content was lowered. The odorants partly originate from the grape (primary aroma) and are partly formed on

$$(20.13)$$

Table 20.16. Odorants in Gewürztraminer and Scheurebe

Aroma substance	Concentration (mg/l)	
	Gewürz-traminer[a]	Scheurebe[b]
Acetaldehyde	1.86	1.97
3-Methylbutylacetate	2.9	1.45
Ethylhexanoate	0.49	0.28
(2S,4R)-Rosenoxide	0.015	0.003
(2R,4S)-Rosenoxide	0.006	
Ethyloctanoate	0.63	0.27
(E)-β-Damascenone	0.00083	0.00098
Geraniol	0.221	0.038
3-Hydroxy-4,5-dimethyl-2(5H)-furanone (HD2F)	0.0054	0.0033
(3S,3aS,7aR)-Tetrahydro-3,6-dimethyl-2(3)-benzofuranone (wine lactone)	0.0001	0.0001
Ethanol	90,000	90,000
Ethylisobutanoate	0.150	0.480
Ethylbutanoate	0.210	0.184
Linalool	0.175	0.307
Ethylacetate	63.5	22.5
1,1-Diethoxyethane	0.375	n.a.
Diacetyl	0.150	0.180
Ethyl-2-methylbutanoate	0.0044	0.0045
Ethyl-3-methylbutanoate	0.0036	0.0027
2-Methylpropanol	52	108
3-Methylbutanol	128	109
Dimethyltrisulfide	0.00025	0.00009
4-Mercapto-4-methyl-pentan-2-one	<0.00001	0.0004
(3-Methylthio)-1-propanol (Methionol)	1.415	1.040
Hexanoic acid	3.23	2.47
2-Phenylethanol	18	21.6
trans-Ethylcinnamate	0.002	0.023
Eugenol	0.0054	0.0005
(Z)-6-Dodecenoic acid-γ-lactone	0.00027	0.00014
Vanillin	0.045	n.a.
Sulfur dioxide	7.3	30

[a] Gewürztraminer, dry, year 1992.
[b] Scheurebe, Cabinet semi-dry, year 1993.
n.a., not analyzed.

Table 20.17. Taste substances in Gewürztraminer and Scheurebe

Compound/Ion	Concentration (mg/l)	
	Gewürz-traminer[a]	Scheurebe[b]
Group I: sour, astringent		
Acetic acid	280	255
Tartaric acid	1575	1260
Citric acid	875	594
Malic acid	377	4790
Lactic acid	1680	980
Succinic acid	590	480
Oxalic acid	100	<50
γ-Aminobutyric acid	21	23
Group II: sweet		
D-Glucose	870	13,040
D-Fructose	575	13,500
Proline	760	320
Group III: salty		
Chloride	20	135
Phosphate	270	245
Sulfite	35	120
Potassium	1240	1100
Calcium	32	231
Magnesium	55	81
Glutamic acid	54	18
Group IV: bitter		
Lysine	27	16

[a,b] cf. Table 20.16

Table 20.18. Cultivar-specific aroma substances of wine

Compound	Cultivar
Ethyl cinnamate	Muscatel wines
β-Ionone	
Linalool	
Geraniol	
Nerol	
β-Damascenone	Riesling, Chardonnay
cis-Rose oxide	Gewürztraminer
2-sec-Butyl-3-methoxypyrazine	Sauvignon
2-Isobutyl-3-methoxypyrazine	
4-Mercapto-4-methylpentan-2-one	Scheurebe

fermentation (secondary aroma). A large part of the grapes used is neutral in aroma (e. g., white Burgundy, Silvaner, Chardonnay). However, there are also aroma-rich grapes, e. g., Muscatel, Gewürztraminer, Morio-Muskat and Sauvignon.

Table 20.19. Odor thresholds of the odorants of wine in water (I) and in 10% (w/w) ethanol (II)

Compound	Threshold value (µg/l)	
	I	II
Acetaldehyde	10	500
Ethylacetate	7500	7500
Ethyl-2-methylbutanoate	0.06	1
Ethyl-3-methylbutanoate	0.03	3
3-Methylbutylacetate	3	30
Ethylhexanoate	0.5	5
Ethyloctanoate	0.1	2
Acetic acid	22,000	200,000
cis-Rose oxide	0.1	0.2
4-Mercapto-4-methylpentan-2-one	0.0001	0.0006
Wine lactone	0.008	0.01
(E)-β-Damascenone	0.001	0.05
Linalool	1.5	15
Geraniol	7.5	30

Table 20.20. Esters in wine with sensory relevance

Compound	White wine (mg/l)	Red wine (mg/l)
Ethyl acetate	0.15–150	9–257
Ethyl propanoate	0–0.9	0–20
Ethyl pentanoate	1.3	5–10
Ethyl hexanoate	0.03–1.3	0–3.4
Ethyl octanoate	0.05–2.3	0.2–3.8
Ethyl decanoate	0–2.1	0–0.3
Hexylacetate	0–3.6	0–4.8
2-Phenylethyl acetate	0–18.5	0.02–8
3-Methylbutyl acetate	0.03–0.5	0–23
Ethyl lactate	0.17–378	12–382

Table 20.21. Effect of fermentation conditions on formation of higher alcohols and esters

Temperature (°C)	pH	Higher alcohols total (mg/l)	Fatty acid esters total (mg/l)
20	3.4	201	10.8
20	2.9	180	9.9
30	3.4	188	7.8
30	2.9	148	5.4

as an example, Table 20.22 shows that terpenes as well as esters and alcohols increase rapidly on fermentation. In addition, the monoterpenes also increase on aging of the wine in stainless steel tanks. On the one hand, terpene glycosides are hydrolyzed by must glycosidases, and on the other, the nonenzymatic hydrolysis of these precursors are promoted by heat treatment of the must and the low pH.

In addition a broad pattern of aroma-active monoterpenes (e. g., nerol oxide, hotrienol) is formed by cyclization and dehydration reactions of di- and polyhydroxylated monoterpenes (examples, cf. 5.3.2.4), e. g., the cis-rose oxides can be formed by the cyclization of 3,7-dimethylocta-6-en-1,5-diol.

Wine extracts the quercus lactone (structure cf. 5.3.2.3) on storage in oak barrels. The difference in aroma compared with maturation in steel tanks also results from oxidative processes, which cause an increase in aldehydes in oak bar-

The concentration ranges of esters found in a larger number of white and red wines are presented in Table 20.20. The high variability of the ester fraction has an effect on the intensity of fruity notes in the aroma profile.

The content and composition of the ester fraction is greatly influenced by fermentation conditions. The higher the temperature and the lower the pH during fermentation, the lower the ester concentration (Table 20.21).

Terpenes mainly contribute to the aroma of Muscatel wines and, to a smaller extent, other wines. In the must, however, these terpenes are still largely present as odorless glycosides, di- and polyols (cf. 5.3.2.4). Using Gewürztraminer

Table 20.22. Concentration changes in odorants in the production of Gewürztraminer[a]

Odorants	Concentration (mg/l)		
	I	II	III
Ethyl-2-methyl-butanoate	<0.0001	0.0023	0.0026
Ethyl hexanoate	0.0035	0.465	0.345
cis-Rose oxide	0.0011	0.0053	0.011
Linalool	0.0026	0.029	0.043
Geraniol	0.0087	0.035	0.045
3-Methylbutanol	0.440	64.0	61.0
(E)-β-Damascenone	0.00003	0.0063	0.0017

[a] I, pressed juice; II, after malolactic fermentation (cf. 20.2.5.4); III, after aging in a steel tank.

Table 20.23. Aging of Gewürztraminer in a steel stank (I) and in an oak barrel (II) – changes in the concentrations of important aroma substances[a]

Compounds	Concentration (mg/l)	
	I	II
Acetaldehyde	1.86	4.32
3-Methylbutanal	<0.001	0.051
3-Methylbutylacetate	2.9	0.450
Methional	<0.0005	0.0099
β-Damascenone	0.00084	0.0028
Guaiacol	0.0036	0.056
Vanillin	0.045	0.335
Quercus lactone	n.a.	0.134

[a] Storage 14 months; n.a. not analyzed.

rels, as shown in Table 20.23 for Gewürztraminer. β-Damascenone and vanillin also increase.

The aroma substances formed on storage of bottles include 1,1,6-trimethyl-1,2-dihydro-naphthalene (TDN). After longer storage, it exceeds the aroma threshold (ca. 20 μg/l water) and contributes a kerosine-like aroma note to the aroma profile in particular of old Riesling wines. In Riesling wines from southern European countries, this aroma substance can increase to such an extent on aging that they acquire a very unpleasant taste after even a short storage time (kerosine/petrol note, Table 20.24). As a result of the intensive sunshine and high temperatures, the precursor carotinoids are formed in relative high concentrations and are then degraded to TDN in this cultivar.

The monoterpene pattern can be used to differentiate cultivars. For example, a clear distinction can be made between wines from the grape cultivar "White Riesling" and wines from other grape cultivars which are also sold as "Riesling". As shown in Fig. 20.7, the monoterpene concentrations (especially of linalool, hotrienol, α-terpineol, and 3,7-dimethylocta-1,5-*trans*-dien-3,7-diol) in "White Riesling" are considerably higher than in the other "Riesling" wines.

Methoxypyrazines (Table 20.18) in concentrations of 10–20 ng/l are characteristic of Sauvignon wines. They are exceptionally odor active (cf. 5.3.1.7) and produce a paprika note in the odor profile.

20.2.7 Spoilage

As with beer, defects in wine are reflected in appearance, odor and taste and, if not controlled, result in complete spoilage. A full explanation of all defects is beyond the scope of this book; hence only a general outline will be provided.

Of importance is browning due to oxidative reactions of phenolic compounds which, in red wine, may result in complete flocculation of the color pigments. This oxidative darkening process is as much chemical as enzymatic (polyphenoloxidases). Sulfurous acid is the preferred agent to prevent browning. Once the wine is affected by browning, it may be lightened by treatment with activated charcoal. The charcoal treatment can also remove other defects, such as the taste of mash or rotten grapes. Iron-induced turbidity (white or greyish casse) appears as a white, greyish-white or greyish haze or cloudiness and consists mostly of ferric phosphate ($FePO_4$). It is formed by the oxidation of ferrous compounds in wine. Proteins, tannins or pectins can participate in the build-up of such cloudiness (black casse). The so-called copper casse or turbidity is based on the formation of Cu_2S and other compounds with monovalent copper. It originates from the Cu^{2+} ions present in wine and their reduction in the presence of excess SO_2. Other taste defects compiled in Table 20.24 can be divided into:

- those produced by the cultivar (e. g., strawberry note, fox note)
- those produced in fermentation by other microbial processes (e. g., "boeckser", mousy note, medicine note)
- those formed during wine storage and aging in wood barrels or by contamination (e. g., cork note, musty note, kerosine note, untypical aging note).

A medicine note is detected when the phenols listed in Table 20.24 are formed in excessively high concentrations on the degradation of ferulic and p-coumaric acid. This aroma defect has been observed especially in the cultivar "Kerner" when the grapes were exposed to intensive sunlight.

The "untypical aging note" (Table 20.24) is produced by stress during ripening of the berries. Dryness, low nitrogen uptake with a high yield can result in the formation of the unwanted odorant 2-aminoacetophenone during fermentation.

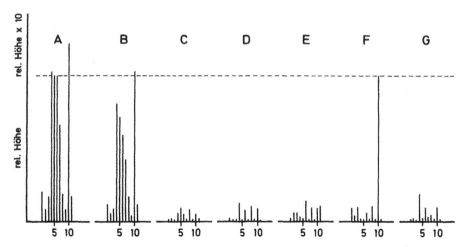

Fig. 20.7. Monoterpenes in wines from the grape cultivar "White Riesling" and in wines from other grape cultivars which are also sold as "Riesling" (according to *Rapp* et al., 1985). A: White Riesling (Rheinpfalz), B: White Riesling (France), C: Welschriesling (Austria), D: Welschriesling (Italy), E: Laski Rizling (Yugoslavia), F: Hunter Valley Riesling (Australia), G: Emerald Riesling (USA). Monoterpenes: **1** *trans*-furan linalool oxide, **2** *cis*-furan linalool oxide, **3** neroloxide, **4** linalool, **5** hotrienol, **6** α-terpineol, **7** not identified, **8** *trans*-pyran linalool oxide, **9** *cis*-pyran linalool oxide, **10** 3,7-dimethylocta-1,5-*trans*-dien-3,7-diol, **11** 3,7-dimethyl-1-octen-3,7-diol

Table 20.24. Aroma defects in wine

Aroma defect	Key aroma substances	Cause
Mousy note	2-Ethyl-3,4,5,6-tetrahydropyridine, 2-acetyl-3,4,5,6-tetrahydropyridine, 2-acetyl-1,2,5,6-tetrahydropyridine	Lactic acid bacteria in combination with yeasts of the genus *Brettanomyces*
Strawberry note	4-Hydroxy-2,5-dimethyl-3(2H)-furanone (HD3F): >1500 µg/l	Characteristic of the cultivar
Medicine note	4-Vinylphenol + 4-vinylguaiacol: >800 µg/l	Climate, microbiological processes
Medicinal, woody, smoky, horse sweat	4-Ethylphenol + 4-ethylguaiacol: >400 µg/l	Climate, microbiological processes
Kerosine/petrol note	1,1,6-Trimethyl-1,2-dihydro-naphthaline (TDN): >300 µg/l	Southern climate, excessively high carotinoid concentration in Riesling
"Boeckser"	Hydrogen sulfide	Fermentation
Cork taste/musty note	2,4,6-Trichloroanisole, geosmin, 2-methylisoborneol, 1-octen-3-one, 4,5-dichloroguaiacol, chlorovanillin	Contamination during wine storage
Untypical aging note (naphthaline note, fox note, hybrid note)	2-Aminoacetophenone: >0.5 µg/l	Stress reaction of the vine

The so-called "boeckser" is caused by the smell of hydrogen sulfide. The very unpleasant, rotten and yeasty "boeckser" (= mercaptan) odor is most objectionable and lingers for a long time. It is due to ethylthiol, which can be removed by activated charcoal. The volatile sulfur compounds originate from sulfite which is reduced to H_2S by yeast, later reacting with ethanol to form ethylthiol. Additional wine taste defects are the odd and disagreeable cork tastes which is due to the formation of the odorants listed in Table 20.24 and 2,4,6-trichloroanisole. Above concentrations of 15–20 ng/l, it contributes to the cork flavor of wine. Geosmin, 1-octen-3-one, 4,5-di-chloroguaiacol and chlorovanillin may cause musty off-flavors.

Additional yeast spoilage is induced by species of the genera *Candida* (*Mycoderma*), *Pischia* and *Hansenula* (*Willia*). Other microorganisms are involved in the formation of viscous, moldy and ropy wine flavor defects. Bacterial spoilage may involve acetic acid and lactic acid bacteria. In this case vinegar or lactic acid souring is detectable. It has usually been associated with mannitol fermentation which may result in considerable amounts of mannitol.

Sorbic acid can be converted to 2-ethoxy-3,5-hexadiene by heterofermentative lactic acid bacteria. In concentrations of 0.1 m/l, this compound produces a "geranium" note.

A "mousy" taint is occasionally detected in fruit and berry wines and, less often, in grape wines. It is thought that the tetrahydropyridines given in Table 20.24, which have been also identified as important flavor compounds of toasted bread (cf. 15.4.3.3.2), contribute to the "mousy" taint. These compounds might be formed by microorganisms in wine.

Likewise, red wines, particularly color-deficient wines, show a microbiologically-induced change reflected in a substantial increase in volatile acids and the degradation of tartaric acid and glycerol. The bitter taste of red wines is caused by bacteria, mold and yeast. The bitter taste is usually a result of glycerol conversion to divinyl glycol. Cloudiness of red wines appears to be due either to bacterial or yeast spoilage or to physical reasons alone, such as the precipitation of cream of tartar. The latter occurs frequently and mostly in bottled wines. Cream of tartar precipitates as a result of oversaturation of the salt solution, as appears to be the case with protein-tannin interaction products. With oversaturation, they sediment as a fine greyish-yellow haze. Cloudiness caused by mucic acid salts also occurs.

20.2.8 Liqueur Wines

In contrast to wine, liqueur wines (older term "dessert wines") are not exclusively made from fresh or mashed grapes or grape must. The alcohol content is at least 15% by volume and at most 22% by volume. The production proceeds according to two different processes, which are partly also combined:

- *Concentrated liqueur wines* are produced by the fermentation of concentrated grape juices which are very rich in sugar (e. g., from dry grapes) or by the addition of concentrated grape juice to wine.
- *Mixed liqueur wines* (e. g., Sherry/Malaga, Port wine/Madeira, Samos, Marsala) are produced from partly fermented must with the addition of alcohol or mixed thickened must. The addition of alcohol stops the fermentation.

The extract, alcohol and sugar contents of dessert wines are given in Table 20.25.

At least 2–5 years are needed to make dessert wines. In the production of sherry the wine is stored in partially filled butts, i. e. in the presence of excess air. Flor yeasts develop on the wine surface in the form of a continuous film or wine cover (sherry yeast). The typical sherry flavor is derived from the aerobic conditions of maturation. During this time the concentrations of the following compounds increase at the expense of alcohol and volatile acids: ethanal, acetals, esters, sotolon (cf. 5.3.1.3) and 2,3-butylene glycol. In port wine production the wine is drawn off to casks before the end of fermentation and is fortified with wine distillates. The fortifying procedure is repeated several times ("multiple addition") until the desired alcohol content is reached. Sotolon is the key aroma substance of Port wine. Its odor threshold in this wine is 19 µg/l. Its concentration increases linearly during storage. Port stored for one year and for 60 years contained 5 and 958 µg/l sotolon respectively.

20.2.9 Sparkling Wine

Experience has shown that carbon dioxide imparts a refreshing, prickling and lively character to wine (as already mentioned for young wines). Hence, the production of a refined form of wine, enriched with carbon dioxide (sparkling wine) was developed and used in the early 18th Century, originally in the Champagne region of France ("Champagne" wine).

Table 20.25. Composition of liqueur wines

	Extract (g/l)	Alcohol (g/l)	Sugar (g/l)	Glycerol (g/l)	Tritratable acid (g/l)[a]
Malaga	159.2	143.4	135.8	5.0	5.3
Portwine	67.6	166.5	47.0	2.8	4.5
Madeira	129.0	149.5	107.5		
Marsala	81.0	150.4	52.2	6.2	5.9
Samos	119.0	152.0	82.0	7.5	6.8
Tokay essence	257.5	84.4	225.3	4.1	6.5
Rheingauer top quality	140.6	107.7	99.4	14.3	10.2
Pfaelzer (Palatinate) top quality	171.6	86.7	121.3	10.5	11.6
Sauternes top quality	127.8	101.2	82.7		0.3

[a] Expressed as tartaric acid.

20.2.9.1 Bottle Fermentation ("*Méthode Champenoise*")

In the production of sparkling wine, young wines from suitable regions are used since fermentation of their grape juice in casks provides the special, fresh, fruity bouquet ("cuvé") desired. Blending of wines ("coupage") from different localities, often with older wines, is aimed at obtaining a uniform end-product. In this way clarified wine is then converted into an effervescent beverage by subjecting it to a second fermentation. Sugar is added (about 20–25 g/l) to wine, together with a pure yeast culture, for the purpose of attaining the desired final alcohol content (85–108 g/l) and carbon dioxide pressure (4.5 bar at 20 °C). Special yeasts are selected which, in addition to being good fermenters and insensitive to carbonic acid, sediment as a firm, grainy precipitate ("depot") after fermentation is complete.

The wine is bottled ("tirage") in such a way as to leave a small headspace of air and is then corked with a natural or plastic cork or, very often, with "crown" caps. The cork is finally and firmly secured with an iron clamp ("agrafe"). The bottles are stacked in cellars at normal temperatures (9–12 °C). The fermentation lasts several months while the build-up of carbonation may go on longer, perhaps for up to 3–5 years. During this time, the carbon dioxide pressure within the bottle rises considerably. The sparkling wines are classified in France depending on the pressure: "grand mousseux", high pressure (4.5–5 bar); "mousseux", intermediate pressure (4–4.5 bar) and "cremant", low pressure (below 4 bar).

At this stage the sparkling wine is ready for yeast removal (disgorging). The bottles are restacked upside-down. Then the contents are repeatedly shaken until the yeast is loosened and settles on the cork. After 6–8 weeks the bottles are placed upright and the cork is removed using disgorging pliers. Simultaneously, the yeast is pushed out by the pressure from within the bottle. In order to simplify this production step, which is considered the most difficult step, the neck of the bottle is frozen to about −20 °C and the yeast is forced out as an "ice" plug. Because of this time-consuming, costly procedure, the loss of wine, and other problems inherent in clearing the bottle of its yeast deposit, a "transfer system" has been introduced. The raw wine which has fermented in the bottle is emptied into a tank. The measured wine is filtered from the tank under pressure into a shipping/export bottle. The sparkling, yeast-free wine ("vin brut", dry wine) is then, depending on the market demand, supplemented with "liqueur" (dosage), quite commonly a plain solution of candy in wine. The bottle, with a headspace volume of 15 ml, is then corked and the cork is wired down. For further build-up of CO_2, the finished sparkling wine needs to be stored for an additional 3–6 months.

20.2.9.2 Tank Fermentation Process ("*Produit en Cuve Close*")

With the aim of simplifying the costly and time-consuming classical process, much of the sparkling wine production is now based on fermentation of wine in pressurized steel tanks instead of in bottles. The carbon dioxide-saturated wine is clarified and filtered and then chilled thoroughly and bottled. Fermentation is carried out at a pressure of about 7 bar over a 3 to 4 week period.

20.2.9.3 Carbonation Process

The carbonation of wine ("vin mousseux gacéi-fié") involves artificial saturation of wine with carbon dioxide, instead of the natural CO_2 developed during fermentation. Thus, the process is identical to the production of carbonated mineral water. The second fermentation, sugar addition and disgorging are omitted. However, sweetening with liqueur, corking and cork wiring are all retained. Perl wine is also a wine with artificially added carbon dioxide, which has a pressure limited to 2.5 bar, in comparison with "sekt".

20.2.9.4 Various Types of Sparkling Wines

Champagne is obtained by the classical bottle fermentation of wine from the French grape which grows in the region of Champagne. Sparkling wines produced in this region are the only ones that may be sold under the name of "Champagne". German sparkling wines are called "Schaumwein" and are commonly sold as "Sekt"; such Italian wines are "Spumante"; while in Spain and Portugal they are "Espumante".
According to the residual sugar content (g/l), sekt is classified as extra brut (0–6), brut (0–15), extra dry (12–20), dry (17–35), semi-dry (35–50) or mild (>50). Sparkling wine for diabetics is sweetened with sorbitol. Sparkling wines are also made from fruit and berry wines (apple, pear, white and red currant, bilberry). The process is that described above for carbonation.

20.2.10 Wine-Like Beverages

Compositions of some typical wine-like products are given in Table 20.26.

Table 20.26. Composition of some wine-like beverages[a]

Beverage	Alcohol	Extract	Acids[b]	Sugar	Minerals
Apple cider	58.4	23.4	3.8[+]	1.7	2.8
Cidre	51.0	29.7	2.8[+]	10.4	2.6
Pear wine	49.3	53.7	6.5[+]	9.0	4.1
Red currant cider	62.1	39.8	18.6[*]	1.8	4.0
Gooseberry cider	96.3	78.6	7.5[*]	55.8	1.8
Sour cherry cider	101.4	62.7	11.7[*]	3.8	3.61
Malt wine	70.6	24.5	4.6[+]	4.9	1.36
Malton sherry	123.0	115.2	8.1[+]	55.9	2.3
Mead	51.4	242.4	3.9[+]	208.0	1.34
Sake	121.2	28.6	5.7[+]	5.5	1.0

[a] Results are given in g/l.
[b] Acids are calculated as malic ([+]) or citric acid ([*]).

20.2.10.1 Fruit Wines

For the production of fruit wine, pressed juice (fruit must) is made from apples, pears, cherries, plums, peaches, red currants, gooseberries, bilberries, cranberries, raspberries, hip berries and rhubarb. In general, the process used is the same as that for making wine from grapes. Apple and pear mash are first pressed and the pressed juice (must) is fermented, while berry mash is fermented directly in order to extract the color pigments. Natural fermentation is suppressed by inoculation with pure, cultured yeast (cold-fermenting yeast). The vigor of the fermentation of berry musts, which are nitrogen deficient, is increased by addition of small amounts of ammonium salts (fermentation salts). Lactic acid (3 g/l) is added to acid-deficient musts, such as that from pears, in order to achieve a clear ferment and, often, sucrose solutions are added to berry and fruit musts to alleviate acidity. The yield and quality of pome fruit must is improved by mixing 9 parts of fruit residue with 1 part of water and adding sucrose to raise the density of the must to 55 Oechsle degrees.
Fruit wines are produced industrially in many countries, e. g., apple wine, which is called cider in France, the UK and the USA, and pear

wine, known as "poiré" in France. In Germany fruit wine is made along the Mosel river, around Frankfurt and in the state of Baden-Wuerttemberg. It is a popular beverage and is commonly called "plain must".

20.2.10.2 Malt Wine; Mead

Malt wine is made from fermented malt extract (the hot water extract of whole meal malt). Malton wine is made in the same way, except that sucrose is added at 1.8-times the amount of malt in order to increase the sugar and alcohol content of the wine. The wort is then soured by the action of lactic acid bacteria (0.6–0.8% lactic acid, final concentration). The acid fermentation is stopped by heating the wort to 78 °C and, after inoculation with a pure yeast culture, the wort is fermented to an alcohol content of 10–13%. The beverage thus formed has the character of a dessert wine, but is different because of its high content of lactic acid and its malt extract flavor. Mead is an alcoholic liquor made of fermented honey, malt and spices, or just of honey and water (not more than 21 water per kg of honey). Since early times, mead has been widely consumed in Europe and, even today, it is enjoyed the most of all the wine beverages in eastern and northern Europe.

20.2.10.3 Other Products

Other wine-like products include palm and agave wines ("Pulque"), maple and tamarind (Indian date) wines, and sake, the Japanese alcoholic drink made from fermented rice, which resembles sherry and is enjoyed as a warm drink.

20.2.11 Wine-Containing Beverages

Wine-containing beverages are made with wine, liquor wines or sparkling wines and, hence, they are alcoholic beverages.

20.2.11.1 Vermouth

Vermouth was first produced in the late 18th century in Italy (Vermouth di Torino, Vino Ver-

mouth) and later in Hungary, France, Slovenia and Germany. For the production of vermouth, wormwood (*Artemisia absynthium*) is extracted with the fermenting must or with wine, or it is made from a concentrate of plant extracts added to wine. Other herbs or spices are additionally used, such as seeds, bark, leaves or roots, as is the case with thyme, gentian or calamus, the sweet flag plant.

20.2.11.2 Aromatic Wines

These wines are similar to vermouth aperitif wines. They are flavored by different herbs and spices. Ginger-flavored wine is an example of this type of wine.

20.3 Spirits

20.3.1 Foreword

Spirits or liquors are alcoholic beverages in which the high alcohol concentration is achieved by distillation of a fermented sugar-containing liquid. Examples are distilled wines (brandies), liqueurs, punch extracts and alcohol-containing mixed drinks. Table 20.27 compares the alcohol consumption with respect to spirits, wine and beer in selected countries.

20.3.2 Liquor

The term liquor includes all liquids, even pure alcohol, which are obtained by fermentation followed by distillation. Some types of liquors contain flavorings.

20.3.2.1 Production

Liquors are produced by removing alcohol from an alcohol-containing liquid by distillation. Such liquids may already contain the alcohol, or alcohol is produced by the fermentation of a sugar-containing mash. The mash may include fermentable forms of sugars (D-glucose, D-fructose, D-mannose and D-galactose), or those forms are prepared by prior hydrolysis of di-

Table 20.27. Alcohol consumption with respect to the type of drink in l per inhabitant in 2003

Country	Spirits	Wine	Beer	Total
Luxembourg	1.6	6.7	4.3	12.6
Hungary	3.5	3.9	4.0	11.4
Czech Rep.	3.8	1.0	6.2	11.0
Ireland	2.0	2.7	6.1	10.8
Germany[a]	1.0	2.6	5.6	10.1
Spain	2.4	3.2	4.4	10.0
UK	1.8	2.2	5.6	9.6
Denmark	1.1	3.5	4.9	9.5
France	2.4	4.9	2.0	9.3
Austria	1.4	3.2	4.7	9.3
Switzerland	1.6	4.1	3.3	9.0
Slowakia	3.5	1.2	3.8	8.5
Lettland	6.1	0.5	1.5	8.1
Greece	1.6	3.4	2.7	7.7
Sweden	0.9	1.7	2.3	4.9

[a] In 2004.

and oligosaccharides (sucrose, lactose, raffinose, gentianose, melecitose, etc.) or polysaccharides. The main raw materials are:

- alcohol-containing liquids (wine, beer, fruit wines, fermented milk);
- sugar-containing sources, such as sugar cane and beet, molasses, fruit and fruit products, fruit pomace, whey, palm extract and sugar-rich parts of tropical plants;
- starch- and inulin-containing raw materials (fruit, cereal, potato, topinambur, sweet potato, cassava, tapioca or chicory).

Saccharification of the starch-containing material is achieved with malt (green malt or kiln-dried malt), or by microbial amylases e. g., from the molds Aspergillus niger and A. oryzae. Fermentation is achieved with *Saccharomyces cerevisiae*, which converts sucrose and hexoses (glucose, galactose, mannose, fructose). Other substrates can be fermented, e. g., with *Saccharomyces uvarum* (raffinose), *Kluyveromyces fragilis* (lactose), and *Kluyveromyces marxianus* (inulin). Distillation is performed in various ways, depending on the source and desired end-product. For the distillation of rum, arrack, fruit brandies and cereals, and brandy from wine, the apparatus is often a relatively simple still, used in such a way as to obtain a distillate which contains several

other products of fermentation besides ethanol, or which contains the aroma substances of the starting raw material. These aroma substances are alcohols, esters, aldehydes, acids, essential oils and hydrogen cyanide. Repeated distillation is needed to obtain an alcohol-enriched distillate. In the production of pure or absolute alcohol the aim is the opposite: the final product being free from materials other than ethanol.

20.3.2.2 Alcohol Production

Alcohol used for drinks is made primarily from potatoes, cereals and molasses. Distiller's yeast, especially the top fermenting culture (cf. 20.3.2.1), is used for fermentation. Since the fermentation proceeds in an unsterilized mash and at elevated temperatures and since the growth of yeast occurs in mash acidified with lactic or sulfuric acid (pH 2.5–5.5), the yeast must be highly fermentative, tolerant of elevated temperatures ($\leq 43\,°C$) and resistant to acids and alcohol. In addition to saccharification by malt which contains mainly β-amylase, high-activity microbial α-amylases are also used. Molasses does not require saccharification. The saccharified mash is cooled to $30\,°C$ and then inoculated with a yeast starter which has been cultured on a sulfuric or lactic acid medium of the mash or directly with distiller's yeast. After 48 h of fermentation, the ethanol present at 6–10% by volume in the mash is distilled off along with the other volatile constituents. This step and the following rectification of the crude alcohol are achieved by continuous processes.

To facilitate the removal of the fusel oils, the crude alcohol is diluted to 15% by volume prior to rectification. The head product obtained from the rectification column consists nearly of pure ethanol (96.6% by volume) which is used for production of alcohol-fortified beverages. Large amounts of acetaldehyde, methanol and low boiling esters are present in the first runnings of the distillate, while the last runnings contain primarily fusel oil, other high alcohols, furfural and esters. These runnings combined with other intermediate fractions provide technical alcohol. The fusel oil, obtained in amounts of 0.1–0.5 l per 100 l alcohol, is used for technical purposes, while the distillation residue (the wash or stillage) is frequently used as animal feed. The yield of alcohol from 100 kg of mash starch is 62–64 l, i.e. about 89% of the theoretical value.

Technical alcohol is denatured or embittered to prevent its use for other than technical purposes, e.g., for drinking. Burning alcohol is denatured by addition of a mixture of methylethylketone and pyridine and alcohol for industrial use with other solvents, such as petroleum ether, camphor, diethyl ether or dyes.

20.3.2.3 Liquor from Wine, Fruit, Cereals and Sugar Cane

These beverages have a distinct taste and odor and contain at least 38% ethanol by volume. They are called natural, genuine or true liquors. The distillate resulting from a single distillation has a low alcohol content and often contains the specific odor and taste components of the starting material (harsh raw grain or harsh raw juniper liquor-gin). In the production of liquor, the ultimate aim is to collect most of the desirable, specific fragrance and aroma substances (esters, essential oils) or to develop them (hydrogen cyanide, fermentation products, yeast oil) by using suitable mashing, fermentation and distillation processes. The freshly distilled liquor has a hard, burning taste and unpleasant odor. It is improved by aging, which gives it a new, desirable aroma and flavor. Therefore, aging of liquor is of the utmost importance.

20.3.2.3.1 Wine Liquor (Brandy)

Brandy is distilled wine which contains at least 38% by volume of alcohol. Brandy to which alcohol is added is designated as a brandy blend or adulterated brandy.

The term "cognac" is restricted to brandy made in France in the region of Charente. The brandy produced in southern France, called Armagnac, is close in quality to cognac. Brandy production originated in France. Fermented grape juices (must) are distilled in very simple copper-pot stills on an open fire, often without prior removal of the yeast. The primary distillate (sectionnement) with a harsh, unpleasant odor is refined by repeated distillations ("repasse"). Brandy pro-

duction soon spread to other countries (Germany, Russia, Spain, Hungary, the USA, Australia) and today brandy is frequently distilled by a continuous process and its production has become a large-scale industry. In Germany imported wines serve as starting material and are increasingly obtained from raw distillates. Distilled wine is a wine without residual sugar, to which a non-rectfied wine distillate with maximum 86 per cent by volume of alcohol has been added. It contains 18–24% (v/v) of alcohol and max. 1.5 g/l of volatile acids (calculated as acetic acid).

The primary wine distillate contains 52–86% by volume ethanol and is considered as an intermediate product. It is used as the raw ingredient in the production of adulterated brandy by aging from 6 months to several years in wooden casks. Hard oak wood is used predominantly (barrels are made from "limousin" wood, holding about 300 l). Wild chestnut and other woods are also used. During aging, the wine distillate extracts phenolic compounds and colors of the wood, thus acquiring the typical golden-yellow and, occasionally, greenish-yellow color of brandy. Simultaneously, oxidation and esterification reactions mellow and polish the flavor and aroma. In order to improve quality, it is common to add an essence prepared by extraction of oakwood, plums, green walnuts or deshelled almond with a wine distillate and also sugar, burnt sugar ("couleur") and 1% dessert wine to sweeten the brandy. In addition, treatment of brandy with clarifying agents and filtering agents is also common. The desired alcohol content is obtained by dilution of brandy with water.

20.3.2.3.2 Fruit Liquor (Fruit Brandy)

Fruit liquors are also called cherry or plum waters or bilberry or raspberry spirits. Production of fruit liquor will be illustrated by cherry and plum liquors. Kirschwasser is made mostly in southern Germany (Black Forest's cherry water), France and Switzerland (Chriesiwasser). Whole fruits of the various sweet cherry cultivars are partly crushed together with the seeds and are pounded into a pulp. The fruit is left to ferment for several weeks, using a pure yeast culture. The fermented mash is then distilled in a copper still on an open fire or is heated with steam.

During distillation the first and last fractions are separated. The main distillate contains 60% by volume or more alcohol. It is usually diluted with water to about 40–50% by volume alcohol and is marketed as clear, colorless brandy. The low levels of benzaldehyde and hydrogen cyanide which both contribute to the flavor are derived from the enzymatic cleavage of seed amygdalin. Kirschwasser, as is the case with Marasca from Dalmatia or Italy, is often used as an admixture in liqueur or cordial production (curacao, cherry brandy, maraschino, etc.).

Plum brandy is produced from fully-ripe plums in a similar way to Kirschwasser, though mostly no seed crushing is involved. Besides Germany and Switzerland (Pfluemli water), major producers are the Balkan states, Czech Republic and France. In addition to the common plum, the highly aromatic yellow plum, mirabelle, is also fermented. Mirabelle liquor is a desirable admixture to liqueurs containing fruit extract.

Fruit spirits are obtained from fresh or frozen fruit pulp or juice to which alcohol has been added prior to distillation. Fruits and berries used for this purpose are apricot, peach, bilberry, raspberry, strawberry, red currant, etc. "Williams" is a pear brandy made exclusively from the pear variety "Williams Christ". (E,Z)-2,4-Decadienoic acid ethylester (formation, cf. 5.3.2.2) has been identified as the characteristic aroma substance.

Pome fruit liquor is obtained from freshly fermented apple or other pome fruits, either whole or crushed, or their juices, without prior addition of sugar-containing materials, sucrose or alcohol of some other origin. The alcohol content of liquor from pome fruits is at least 38% by volume. Hydrogen cyanide plays an important role in the chemical composition of fruit liquors of either stone or pome fruit. The cherry liquor sold on the market contains about 0.3–60 mg of hydrogen cyanide per liter of alcohol. In the same range are the concentrations of benzaldehyde (at least 20 mg/l) and the bouquet substances (about 7–15 mg/100 ml). Plum brandy contains less hydrogen cyanide (0.6–21.3 mg/l).

20.3.2.3.3 Gentian Liquor ("*Enzian*")

Gentian brandy is a product obtained by distilling the fermented mash of gentian roots, or in which gentian distillate is used. The raw materials are the roots of many plants of the gentian family which, in the fresh state, contain substantial amounts of sugars (6–13%) in addition to the bitter glycoside-type compounds, such as gentiopicrin, amarogentin and others. The major production regions are the Alps (Tyrol, Bavaria, Switzerland) as well as the French and Swiss Jura mountains.

20.3.2.3.4 Juniper Liquor (*Brandy*) and Gin

Juniper brandy is obtained from pure alcohol and/or grain distillate by the addition of juniper distillate or its harsh, raw brandy. The use of juniper oil is uncommon. Juniper spirit is made exclusively from the distillate of whole juniper berries or from a fermented aqueous extract of juniper. The berries of *Juniper communis* are processed into brandy in Germany, Hungary, Austria, France and Switzerland. Pure juniper brandy is also used as an intermediate product for the production of alcoholic beverages with a juniper flavor as, for example, in Geneva gin. The alcohol of this gin is obtained by distillation of a cereal mash prepared from kiln-dried smoked malt. Juniper brandy also flavors the Bommerlunder from the state of Schleswig-Holstein and the Doornkaat of East Friesland in Germany. Common gin is made from juniper distillates and spices, and contains at least 38% by volume alcohol. Dry gin has an alcohol content of at least 40% by volume.

20.3.2.3.5 Rum

Major rum-producing countries are in the West Indies (Jamaica, Cuba, Barbados, Puerto Rico, Guyana and Martinique) and also Brazil and Mauritius.

Rum production in sugar cane-cultivating regions uses the sugar syrup or the freshly pressed extract, often with the addition of such by-products as foam skimmings, molasses, press-skimmings and their extracts, and distiller's wash ("dunder"), the residue leftover from a previous distillation. The sugar-containing solutions are diluted and allowed to ferment spontaneously at a maximum temperature of 36 °C and then are usually distilled in simple pot stills. Parts of aromatic plants are occasionally added to increase the aroma of the fermenting mash. This results in rum brands with different aromas. The quality of individual products fluctuates greatly. Especially highly regarded is Jamaican rum, which is marketed in various quality grades. A general classification divides them into drinking and blending types. Export rums have an alcohol content of about 76–80% by volume ("original rum"). Rum has the most intense aroma of all the distilled spirits enjoyed as drinks. This is acquired only after long aerobic aging in casks, by absorption of extracted substances from oakwood, and by formation of esters and other aroma constituents during aging. Original rum contains about 80–150 mg acids per 100 ml, calculated as acetic acid. A large part occurs in free form as acetic and formic acids, the rest, along with other low molecular weight fatty acids, is esterified. The ester content and composition are of utmost importance for the assessment of aroma quality.

20.3.2.3.6 Arrack

Arrack is made from rice, sugar cane molasses, or sugar-containing plant juices (primarily from sweet coconut palm extract or its bloom spadix) by fermentation and subsequent distillation. Dates are used for the same purpose in the Middle East.

Countries which produce arrack are Indonesia (Java), Sri Lanka, India (Malabar coast) and Thailand. Well-known brands are Batavia and Goa arracks. In comparison to rum, arrack is not available in very many varieties. It is imported as the "original arrack" with an alcohol content of 56–60% by volume, from which "true arrack" is obtained by dilution with water to 38–50% by volume alcohol. At least a tenth of the alcohol in arrack blends must be from genuine arrack. Arrack is used for hot drink preparations, for Swedish punch, as an admixture for liqueurs, and in baking and as a flavoring ingredient in candy manufacture. Batavia brand arrack, with an alcohol content of about 57% by volume, contains on the average 92 mg acids, 189 mg esters, 21 mg aldehydes and 174 mg higher alcohols per 100 ml of ethanol.

20.3.2.3.7 Liquors from Cereals

Typical products are grain alcohol and whiskey (American and Irish brands are usually spelled with an "e", while Scottish and Canadian brands tend to use "whisky"). Different cereals (rye, wheat, buckwheat, oats, barley, corn, millet) are used. The cereals are first ground, mixed with acidified water, and made into an uniform mash by starch gelatinization. Saccharification is then accomplished by incorporating 15% kiln-dried malt in a premashing vat and stirring constantly at 56 °C. Saccharification proceeds rapidly through the action of malt diastase enzymes. The enzymes are inactivated by heating the mash to 62 °C. This step is followed by rapid cooling of the mash to 19–23 °C. The sweet mash is fermented by a special yeast and is then distilled. Grain liquors are obtained by distilling the mash, while malt liquors commonly are produced by distillation of the wort. Simple stills are used for distillation in small plants, while both distillation and rectification are achieved on highly efficient, continuously run column stills in industrial-scale production. According to the process used, the yield is 30–35 l of alcohol per 100 kg of cereal (e. g., rye), while the quality and character of the spirits vary greatly. Simple stills, with an unsophisticated separation of head and tail fractions, provide characteristic products rich in grain fusel oils. A modern distillery is able to remove the fusel oils to a great extent, yielding a high percentage grain alcohol, from which it is then possible to make a mellow, tasty, pure grain brandy with a subtle aroma. The final flavor of all these products is dependent on well-conducted aging in wooden casks.

Whiskey, depending on the kind, is made by different processes. The raw material for Scotch single malt whiskey is barley malt which has been exposed to peat moss or coal smoke during kiln drying. Such smoked malt is mashed at 60 °C and filtered. The resulting wort is then fermented at 20–32 °C after the addition of yeast (*Saccharomyces cerevisiae*). Irish whiskey is never made from smoked malt. The distillation is conducted in two steps, sometimes in simple pot stills. The harsh, raw liquor is collected in the first distillation step. The undesirable harsh components are removed in the head and tail fractions in the second distillation.

In the production of Scotch grain whiskey the saccharified starch is distilled in continuous column stills. The character of the distillate is neutral, with less aroma than malt whiskey. In both Scotch whiskey processes, the distillates, with about 63% by volume ethanol, have to be stored/aged in order to develop their full aroma. This is best achieved by aging in old sherry casks or in charred casks. At the end of processing, the alcohol content is reduced to a drinkable level, about 43% by volume. Depending on the desired flavor or current preferences, the malt whiskey might be blended with grain whiskey ("blended whiskey"). American whiskey is made from corn, rye or wheat by saccharification with malt enzymes, fermentation of the wort, followed by doubledistillation in column stills and aging, usually in charred oakwood casks. The corn distillate content of bourbon whiskey is at least 51% by volume and that of corn whiskey is at least 80% by

Table 20.28. Odorants of whisky

Aroma substance	Concentration (mg/l)	
	Malt whisky[a]	Bourbon whisky[b]
Ethanol	316,000	316,000
2-Methoxyphenol	0.025	0.056
5-Methyl-2-methoxyphenol	0.0019	0.0011
4-Methyl-2-methoxyphenol	0.01	0.016
4-Ethyl-2-methoxyphenol	0.017	0.059
2-Methylphenol	0.034	0.003
4-Methylphenol	0.017	0.008
3-Methylphenol	0.008	0.004
4-Ethylphenol	0.016	0.166
3-Ethylphenol	0.0033	n.d.
Eugenol	0.027	0.24
3-Methylbutanol	568.1	1062.1
2-Methylbutanol	194.2	423.8
2-Phenylethanol	11.2	13.87
Ethylbutanoate	0.76	0.55
Ethylhexanoate	2.07	1.99
Ethyloctanoate	12.3	8.35
Ethyl-2-methylpropanoate	0.52	0.13
Ethyl-3-methylbutanoate	0.21	0.052
(S)-Ethyl-2-methylbutanoate	0.092	0.030
(S)-Ethyl-2-hydroxy-3-methylbutanoate	0.005	0.003
(2S,3S)-Ethyl-2-hydroxy-3-methylpentanoate	0.004	0.003
3-Methylbutylacetate	4.02	2.59
2-Phenylethylacetate	4.10	1.94
Methylpropanal	1.74	0.23
3-Methylbutanal	0.65	0.34
Vanillin	0.68	2.13
(3S,4S)-cis-Whisky lactone	0.39	2.49
(3S,4R)-trans-Whisky lactone	0.07	0.34
γ-Nonalactone	0.11	0.12
γ-Decalactone	0.010	0.002
2,3-Butandione	0.39	0.033
(E)-β-Damascenone	0.024	0.011
(E)-2-Nonenal	0.024	0.009
(E,E)-2,4-Decadienal	0.006	0.039
1,1-Diethoxyethane	21.86	15.33

[a] Single malt whisky from Scotland, stored for 8 years in an oak cask.
[b] American Kentucky straight bourbon whisky, stored for at least 3 years in a charred oak cask.

volume. Rye whiskey contains at least 51% by volume distillate from rye, while wheat whiskey must contain mostly distillate from wheat.

Table 20.28 presents data on the composition of the aromas of malt whisky (MW) and bourbon whisky (BW). A comparison shows that the same odorants are present in both drinks, but in different amounts. The concentrations of the esters and of 2,3-butanedione are higher in MW and the concentrations of 2- and 3-methylbutanol, eugenol, vanillin and whisky lactone are higher in BW. The extraction yield of the three last mentioned aroma substances clearly increases on storage of BW in charred casks. On longer storage of MW, this effect is also achieved in addition to an increase in the esters, e.g., an 18 year old MW contains (mg/l): eugenol (0.09), vanillin (2.2), cis- and trans-whisky lactone (1.1), ethylhexanoate (3.7) and ethyloctanoate (22.2). In MW (stored for 10 years) made from strongly smoked malt, the phenol fraction (mg/l) is higher than in the MW shown in Table 20.28: 2-methoxyphenol (1.5), 4-methyl-2-methoxyphenol (0.6) and 2-methylphenol (1.3) 4-methylphenol (1.2).

20.3.2.4 Miscellaneous Alcoholic Beverages

Many liquors are made "cold" by simply mixing the purified alcohols of various brands with water and are named according to the place of origin: Klarer, Weisser, East-German, etc. Such mixes often contain flavorings (seasonings, spices), e.g., freshly distilled or aged grain liquor, extracts of caraway, anise, fennel, etc., as well as sugar, essence, essential oils or other flavoring substances. These products are designated as aromatized liquors. Some examples are:

Vodka (in Russian = diminutive of water) is made of alcohol and/or grain distillate by a special process. In all cases the characteristic smoothness and flavor must be achieved. The flavor should be neutral. The extract content is 0.3 g/100 ml and the alcohol content is at least 37.5% by volume.

Aquavit is a liquor flavored primarily with caraway or dill seed. It is made from a distillate of herbs, spices or drugs and contains at least 37.5% by volume alcohol (potato alcohol or grain distillate). It is a favorite type of liquor in the Scandinavian countries.

Bitters are made from alcohol and bitter and aromatic plant or fruit extracts and/or their distillates, fruit saps and natural essential oils, with or without sugar, i. e. starch syrup. This group of products includes Boonekamp, bitter drops, English and Spanish bitters, and Angostura. The so-called "Aufgesetzter" is made of black currants and spirit or grain alcohol.

Absinthe is a liqueur flavored with aromatic constituents of wormwood and other aromatic plants. It becomes turbid after dilution with water.

Other Products. Some special liquors of regional importance should be mentioned: tequila and mescal from Mexico and South America, made from fermented sap of the agave cactus; and liquors from the Middle East, made of sultana raisins, figs or dates.

20.3.3 Liqueurs (Cordials)

Liqueurs are alcoholic beverages with at least 15% (advocaat 14%) by volume alcohol and at least 150 g/l of sugar (expressed as invert sugar) and flavored with fruit, spices, extracts or essences.

20.3.3.1 Fruit Sap Liqueurs

Fruit liqueurs contain the sap of fruits which give the liqueur its name. The lowest concentration of sap is 20 l per 1000 l of end-product (25% by volume alcohol). Cherry brandy, a special type of cherry liqueur, consists of cherry sap, cherry-water, sucrose or starch syrup, wine essence and water.

20.3.3.2 Fruit Aroma Liqueurs

These liqueurs are alcoholic beverages made of natural fruit essences, distillates or fruit extracts.

20.3.3.3 Other Liqueurs

Other liqueurs include:
Crystal liqueur, which contains sugar crystals (e. g., "crystal caraway").

Allasch, a special aromatic alcohol- and sugarrich caraway liqueur with at least 40% by volume alcohol.

Ice liqueur, which is mixed and drunk with ice (e. g., lemon ice liqueur), and has an extract content of at least 30 g/100 ml and a minimum alcohol content of 35% by volume.

Gold water, a spice liqueur containing gold leaf as a characteristic ingredient.

Fragrant vanilla liqueur, the aroma of which is derived exclusively from pod-like vanilla capsules (vanilla beans).

Honey liqueur ("Baerenfang", "Petzfang", the "bear traps") has at least 25 kg of honey in 100 l of end-product.

Swedish punch is made of arrack and spices and has an alcohol content of at least 25%. Cocoa, coffee and tea liqueurs are made from the corresponding extracts of raw materials. Emulsion liqueurs are chocolate, cream and milk liqueurs, mocca with cream liqueur, egg liqueur (the egg cream, "Advokat"), egg wine brandy, and other liqueurs with eggs added. The widespread and common egg liqueur is made from alcohol, sucrose and egg yolk. Herb, spice and bitter liqueurs are made from fruit saps and/or plant parts, natural essential oils or essences, and sugar. Examples are anise, caraway, curacao, peppermint, ginger, quince and many other liqueurs.

20.3.4 Punch Extracts

Punch extracts or punch syrups, known simply as punch, are concentrates which are diluted before they are drunk. Rum or arrack punches contain 5% rum or 10% arrack, calculated relative to the total alcohol content. Aromatization with artificial rum or arrack essences, or with fruit ethers or esters, is not commonly done.

20.3.5 Mixed Drinks

Mixed drinks or cocktails are mixtures of liquors, liqueurs, wines, essences, fruit and plant extracts, etc.

Mixed drinks include alcopops, i. e., sweet drinks made by mixing lemonade with distilled alcohol, beer or alcoholic drinks. The alcohol content is between 2.5 and 7% by volume. This category also includes instant drink powders, which mainly consist of sugar, acidifiers, aromas and coloring substances. The alcohol is adsorbed on a sugar/dextrose matrix. If the powder is dissolved in water as instructed, a drink with an alcohol content of 4.8% by volume is obtained.

20.4 References

Bluhm, L.: Distilled beverages. In: Biotechnology (Eds.: Rehm, H.J., Reed, G.), Vol. 5, p. 447, Verlag Chemie: Weinheim. 1983

Fritsch, H.T., Schieberle, P.: Identification based on quantitative measurements and aroma recombination of the character impact odorants in a Bavarian Pilsner-type beer. J. Agric. Food Chem. 53, 7544 (2005)

Guth, H.: Identification of character impact odorants of different white wine varieties. J. Agric. Food Chem. 45, 3022 (1997)

Guth, H.: Quantitation and sensory studies of character impact odorants of different white wine varieties. J. Agric. Food Chem. 45, 3027 (1997)

Guth, H.: Comparison of different white wine varieties in odor profiles by instrumental analysis and sensory studies. In: Chemistry of Wine Flavor. ACS Symp. Ser. 714, 39 (1998)

Hamberg, M.: Trihydrooctadecenoic acids in beer: qualitative and quantitative analysis. J. Agric. Food Chem. 39, 1568 (1991)

Hardwick, W.A.: Beer. In: Biotechnology (Eds.: Rehm, H.-J., Reed, G.), Vol. 5, p. 165, Verlag Chemie: Weinheim. 1983

Herraiz, T, Ough, C.S.: Chemical and techological factors determining tetrahydro-β-carboline-3-carboxylic acid content in fermented alcoholic beverages. J. Agric. Food Chem. 41, 959 (1993)

Hillebrand, W.: Taschenbuch der Rebsorten. 5. edn., Zeitschriftenverlag Dr. Bilz and Dr. Fraund KG.: Wiesbaden. 1978

Hoffmann, K.M.: Weinkunde in Stichworten. 3. edn, Verlag Ferdinand Hirt: Unterägeri. 1987

Jounela-Eriksson, P.: The aroma composition of distilled beverages and the perceived aroma of whisky. In: Flavor of foods and beverages (Eds.: Charalambous, G., Inglett, G.E.), p. 339, Academic Press: New York. 1978

Kreipe, H.: Getreide- und Kartoffelbrennerei. 3. edn., Verlag Eugen Ulmer: Stuttgart. 1981

Lafon-Lafourcade, S.: Wine and brandy. In: Biotechnology (Eds.: Rehm, H.-J., Reed, G.), Vol. 5, p. 81, Verlag Chemie: Weinheim. 1983

Meilgaard, M.C.: Prediction of flavor differences between beers from their chemical composition. J. Agric. Food Chem. *30*, 1009 (1982)

Molyneux, R.J., Wong, Yen-i.: High-pressure liquid chromatography in the separation and detection of bitter compounds. J. Agric. Food. Chem. *21*, 531 (1973)

Narziß, L.: Abriß der Bierbrauerei. 5. edn., Ferdinand Enke Verlag: Stuttgart. 1986

Narziß, L.: Malz. In: Lebensmitteltechnologie (Ed.: R. Heiss) Springer, Berlin, 1988, pp. 286

Narziß, L.: Bier. In: Lebensmitteltechnologie (Ed.: R. Heiss) Springer, Berlin 1988, pp. 294

Nykänen, L., Suomalainen, H.: Aroma of beer, wine and distilled alcoholic beverages. D. Reidel Publ. Co.: Dordrecht. 1983

Palamand, S.R., Aldenhoff, J.M.: Bitter tasting compounds of beer. Chemistry and taste properties of some hop resin compounds. J. Agric. Food Chem. *21*, 535 (1973)

Pieper, H.J., Bruchmann, E.-E., Kolb, E.: Technologie der Obstbrennerei. Verlag Eugen Ulmer: Stuttgart. 1977

Poisson, L., Schieberle, P.: Characterization of the most odor-active compounds in Bourbon Whisky. J. Agric. Food Chem. (submitted)

Pollock, J.R.A. (Ed.): Brewing Science, Vol. 1, 2. Academic Press: London. 1979/81

Rapp, A., Güntert, M., Heimann, W.: Beitrag zur Sortencharakterisierung der Rebsorte Weißer Riesling. Z. Lebensm. Unters. Forsch. *181*, 357 (1985)

Rapp, A., Markowetz, A.: NMR-Spektroskopie in der Weinanalytik. Chem. unserer Zeit *27*, 149 (1993)

Rapp, A.: Technologie des Weines. In: Taschenbuch für Lebensmittelchemiker und -technologen, Vol. 2 (Ed.: D. Osteroth) Springer, Berlin, 1991, pp. 315

Rapp, A.: Volatile flavour of wine: Correlation between instrumental analysis and sensory perception. Nahrung *42*, 351 (1998)

Schieberle, P.: Primary odorants of pale lager beer. Differences to other beers and changes during storage. Z. Lebensm. Unters. Forsch. *193*, 558 (1991)

Schieberle, P., Komarek, D.: Changes in key aroma compounds during natural beer ageing. In: Freshness and Shelf Life of Foods. ACS Symp. Ser. *836*, 70 (2002)

Soleas, G.J., Dam, J., Carey, M., Goldberg, D.M.: Toward the fingerprinting of wines: Cultivarrelated patterns of polyphenolic constituents in Ontario wines. J. Agric. Food Chem. *45*, 3871 (1997)

Tressel, R., Friese, L., Fendesack, F., Köppler, H.: Gas chromatography – mass spectrometric investigation of hop aroma constituents in beer. J. Agric. Food Chem. *26*, 1422 (1978)

Williams, P.J., Strauss, C.R., Wilson, B., Dimitriadis. E.: Recent studies into grape terpene glycosides. In: Progress in flavour research 1984 (Ed.: Adda, J.) p. 349. Elsevier Science Publ.: Amsterdam. 1985

Wilson, B., Strauss, C.R., Williams, P.J.: Changes in free and glycosidically bound monoterpenes in developing muscat grapes. J. Agric. Food. Chem. *32*, 919 (1984)

Zimmerli, B., Baumann, U., Nägeli, P., Battaglia, R.: Occurrence and formation of ethylcarbamate (urethane) in fermented foods. Some preliminary results. Proc. Euro Food Tox II, Zürich. October 1986, p. 243.

21 Coffee, Tea, Cocoa

21.1 Coffee and Coffee Substitutes

21.1.1 Foreword

Coffee (coffee beans) includes the seeds of crimson fruits from which the outer pericarp is completely removed and the silverskin (spermoderm) is occasionally removed. The seeds may be raw or roasted, whole or ground, and should be from the botanical genus *Coffea*. The drink prepared from such seeds is also called coffee.

Coffee is native to Africa (Ethiopia). From there it reached Arabia, then Constantinople and Venice. Regardless of the prohibition of use and medical warnings, coffee had spread all over Europe by the middle of the 17th century. The coffee tree or shrub belongs to the family *Rubiaceae*. Depending on the species, it can grow from 3–12 m in height. The shrubs are pruned to keep them at 2–2.5 m height and thus facilitate harvesting. The evergreen shrubs have leathery short-stemmed leaves and white, jasmin-like fragrant flowers from which the stone fruit, cherry-like berries, develop with a diameter of about 1.5 cm. The fruit or berry (Fig. 21.1) has a green outer skin which, when ripe, turns red-violet or deep red and encloses the sweet mesocarp or the pulp and the stone-fruit bean. The latter consists of two elliptical hemispheres with flattened adjacent sides. A yellowish transparent spermoderm, or silverskin, covers each hemisphere. Covering both hemispheres and separating them from each other is the strong fibrous endocarp, called the "parchment". Occasionally, 10–15% of the fruit berries consist of only one spherical bean ("peaberry" or "caracol"), which often brings a premium price.

The coffee shrub thrives in high tropical altitudes (600–1200 m) with an annual average temperature of 15–25 °C and moderate moisture and cloudiness. The shrubs start to bloom 3–4 years after planting and after six years of growth they provide a full harvest. The shrubs can bear fruit for 40 years, but the maximum yield is attained after 10–15 years. Fruit ripening occurs within 8–12 months after flowering. Only 3 of the 70 species of coffee are cultivated: *Coffea arabica*, which provides 75% of the world's production; *C. canephora*, about 25%; and *C. liberica* and others, less than 1%. The quantity (in

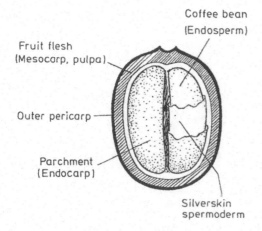

Fig. 21.1. Longitudinal section of a coffee fruit (according to *Vitzthum*, 1976)

Coffee bean (Endosperm)

Fruit flesh (Mesocarp, pulpa)

Outer pericarp

Parchment (Endocarp)

Silverskin spermoderm

Table 21.1. Production of coffee beans in 2006 (1000 t)

Continent	Raw coffee	Country	Raw coffee
World	7843	Brazil	2593
		Viet Nam	854
Africa	922	Colombia	696
America, Central	1020	Indonesia	653
America, North	3	Mexico	288
America, South		India	274
and Caribbean	4782	Ethiopia	260
Asia	2069	Guatemala	257
Europe	–	Honduras	191
Oceania	68	Peru	175
		Σ (%)[a]	75

[a] World production = 100%.

H.-D. Belitz · W. Grosch · P. Schieberle, *Food Chemistry*
© Springer 2009

kg) of fresh coffee cherries which yields 1 kg of marketable coffee beans is for *C. arabica* 6.38, *C. canephora* 4.35, and *C. liberica* 11.5. The most important countries providing the world's coffee harvest in 1996 are listed in Table 21.1.

21.1.2 Green Coffee

21.1.2.1 Harvesting and Processing

The coffee harvest occurs from about December until February from the Equator north to the Tropic of Cancer, while south of the Equator to the Tropic of Capricorn harvest occurs from May until August. Harvesting is done by hand-picking of each ripe berry or by strip-picking all of the berries from three branches after most of the berries (often present as clusters) have matured. Harvesting may also be done by sweeping under the tree, i. e. collecting the ripe berries from the ground. Processing commences with removal of the fleshy pulp by using one of the two following processes:

The dry or natural process used in Brazil involves rapid transport of the harvested berries to a central processing plant, where the whole fruit is spread out on sun-drying terraces and dried until the beans separate by shrinking from the surrounding parchment layer.

Dehulling machines – conical screws with a helical pitch increasing toward the discharge end – remove the dried husks and parchment from the dried berries and, as much as possible, the silverskin. The dehulled and cleaned coffee beans are then classified according to size and packed in 60 kg bags. Often, the fresh cherries, instead of being spread on the drying terrace, are piled up, left for 3–4 days under their own heat to ferment the fruity pulp, and are then processed as outlined below. In both cases unwashed beans are obtained.

The wet (washing) process is more sophisticated than the dry process, and by general consent leads to better quality coffee. The method is generally used for Arabica coffee (except in Brazil) in Central America, Colombia and Africa. The freshly harvested berries are brought to a pulper in which the soft fruit is squeezed between a rotating cylinder or disc and a slotted plate, the gap of which

is adjustable. The passage of the fruit produces a rubbing action which detaches the skin and the pulp from the beans without damaging the seed. The removed pulp is used as fertilizer.

The pulped beans still have the silver-skin, the parchment and a very adhesive mucilaginous layer (mucilage). Hence, such coffee is carried into water stream fermentation tanks made of concrete, the water is drained off and the beans are left to ferment for 12–48 h. During this time, the mucilaginous layer, which consists of 84.2% water, 8.9% protein, 4.1% sugar, 0.91% pectic subtances and 0.7% ash, is hydrolyzed by enzymes of the coffee and by similar enzymes produced by microorganisms found on the fruit skins. The mucilage is degraded to an extent which can be readily dispersed by washing with water. The beans are then collected, sun-dried on concrete floors or dried in mechanical dryers in a stream of hot air (65–85 °C). Beans dried in this way are still covered with the parchment shell ("pergament" coffee or "cafe pergmino") and are further processed by dehulling machines as in the dry process. This yields the green coffee beans. Premium-priced coffee beans are often polished to a smooth, glossy surface and the silverskin, except that retained in the centrecut of the beans, is removed.

21.1.2.2 Green Coffee Varieties

About 80 varieties of the three coffee bean species mentioned above are known. The most important of the species *Coffea arabica* are *typica, bourbon, maragogips* and *mocca*; and of *Coffea canephora* are *robusta* (the most common), *typica, uganda* and *quillon*. All varieties of *Coffea canephora* are marketed under the common name *"robusta"*.

The names of green coffees may be characteristic of the place of origin; i. e. the country and the port of export. Important washed Arabica coffees are, for example, Kenyan, Tanzanian, Colombian, Salvadorian, Guatemalon or Mexican.

Unwashed Arabica beans are the mild Santos and the hard Rio and Bahia beans. All three are from Brazil. Robusta coffees, mostly unwashed, are, for example, those from Angola, Uganda, the Ivory Coast and Madagascar.

Arabica coffees, particularly those from Kenya, Colombia and Central America, have a soft, rich, clean flavor or "fine acid" and "good body". The Arabica Santos from Brazil is an important ingredient of roasted coffee blends because of its strong but mellow flavor. Robusta coffee, on the other hand, is stronger but harsh and rough in aroma.

The quality assessment of green coffee is based on odor and taste assays, as well as on the size, shape, color, hardness and cross-section of the bean. Major defects or imperfections are primarily due to objectionable off-flavored blemished beans, which are removed by careful hand sorting. Blemished beans consist of: unripe seeds (grassy beans) which stay light colored during roasting; overfermented beans with an off-flavor due to the presence of acetic acid, diacetyl, butanol and isobutanol; frost-bitten and cracked beans; insect and rainfall-damaged beans; and excessively withered beans. Even a single blemished bean can spoil the whole coffee infusion. Additional imperfections are the moldy, musty flavor of insufficiently dried and prematurely sacked coffee and earthy or haylike off-flavors. Coffee varieties grown at high altitudes are generally more valuable than those from the plains or lowlands.

21.1.2.3 Composition of Green Coffee

The composition of green coffee is dependent on variety, origin, processing and climate. A review of the differences between Arabica and Robusta coffee is provided in Table 21.2. The constituents will be covered in more detail in the section dealing with roasted coffee.

21.1.3 Roasted Coffee

21.1.3.1 Roasting

Green beans smell green-earthy, so they must be heat treated in a process called roasting to bring about their truly delightful aroma. Roasting in the temperature range between 100 and the final temperature of ca. 200 °C causes profound changes. The beans increase in volume (50–80%) and change their structure and color. The green is replaced by a brown color, a 11–20%

loss in weight occurs, and there is a build-up of the typical roasted flavor of the beans. Simultaneously, the specific gravity falls from 1.126–1.272 to 0.570–0.694, hence the roasted coffee floats on water and the green beans sink. The horny, tough and difficult-to-crack beans become brittle and mellow after roasting.

Four major phases are distinguished during the roasting process: drying, development, decomposition and full roasting. The initial changes occur at or above 50 °C when the protein in the tissue cells denatures and water evaporates. Browning occurs above 100 °C due to pyrolysis of organic compounds, accompanied by swelling and an initial dry distillation; at about 150 °C there is a release of volatile products (water, CO_2, CO) which results in an increase in bean volume. The decomposition phase, which begins at 180–200 °C, is recognizable by the beans being forced to pop and burst (bursting by cracking along the groove or furrow); formation of bluish smoke; and the release of coffee aroma. Lastly, under optimum caramelization, the full roasting phase is achieved, during which the moisture content of the beans drops to its final level of 1.5–3.5%.

The roasting process is characterized by a decrease in old and formation of new compounds. This is covered in section 21.1.3.3, which deals with the composition of roasted coffee. The running of a roasting process requires skill and experience to achieve uniform color and optimum aroma development and to minimize the damage through over-roasting, scorching or burning.

During roasting, heat is transferred by contact of the beans with the walls of the roasting apparatus or by hot air or combusted gases (convection). Actual *contact roasting* is no longer of importance because heat transfer is uneven and the roasting times required are long (20–40 min). In the *contact-convection roasting* process (roasting time 6–15 min), efforts are made to increase the convection component as much as possible by suitable process management. Centrifugal roasters (rotating flat pans), revolving tube roasters, fluid-bed roasters (ca. 90% convection) etc. are used either batchwise or continuously. In the new *short-time roasting* process (roasting time 2 to 5 min), the heating-up phase is significantly shortened by improved heat transfer. Water evaporation proceeds by puffing, producing

Table 21.2. Composition of green Arabica and Robusta coffee[a,b]

Constituent	Arabica	Robusta	Components
Soluble carbohydrates	9–12.5	6–11.5	
Monosaccharides	0.2–0.5		Fructose, glucose, galactose, arabinose (traces)
Oligosaccharides	6–9	3–7	Sucrose (>90%), raffinose (0–0.9%), stachyose (0–0.13%)
Polysaccharides	3–4		Polymers of galactose (55–65%), mannose (10–20%), arabinose (20–35%), glucose (0–2%)
Insoluble polysaccharides	46–53	34–44	
Hemicelluloses	5–10	3–4	Polymers of galactose (65–75%), arabinose (25–30%), mannose (0–10%)
Cellulose, β(1–4)mannan	41–43	32–40	
Acids and phenols			
Volatile acids	0.1		
Nonvolatile aliphatic acids	2–2.9	1.3–2.2	Citric acid, malic acid, quinic acid
Chlorogenic acid[c]	6.7–9.2	7.1–12.1	Mono-, dicaffeoyl- and feruloylquinic acid
Lignin	1–3		
Lipids	15–18	8–12	
Wax	0.2–0.3		
Oil	7.7–17.7		Main fatty acids: 16:0 and 18:2 (9,12)
N Compounds	11–15		
Free amino acids	0.2–0.8		Main amino acids: Glu, Asp, Asp-NH$_2$
Proteins	8.5–12		
Caffeine	0.8–1.4	1.7–4.0	Traces of theobromine and theophylline
Trigonelline	0.6–1.2	0.3–0.9	
Minerals	3–5.4		

[a] Values in % of solids.

[b] Water content of raw coffee: 7–13%.

[c] Main components: 5-caffeoylquinic acid (chlorogenic acid: Arabica 3.0–5.6%; Robusta 4.4–6.6%).

a greater bean volume increase than conventional roasting processes. Therefore, the density in the ground state of coffee roasted by this process is 15–25% lower.

The roasting process is controlled electronically or by sampling roasted beans. The end-product is discharged rapidly to cooling sifters or is sprinkled with water in order to avoid over-roasting or burning and aroma loss. During roasting, vapors formed and cell fragments (silverskin particles) are removed by suction of an exhauster and, in larger plants, incinerated.

There are different roasting grades desired. In the USA and Central Europe, beans are roasted to a light color (200–220 °C, 3–10 min, weight loss 14–17%), and in France, Italy and the Balkan states, to a dark color (espresso, 230 °C, weight loss 20%).

21.1.3.2 Storing and Packaging

Roasted coffee is freed of faulty beans either by hand picking on a sorting board or, at large plants, automatically by using photo cells. Commercially available roasted coffee is a blend of 4–8 varieties which, because of their different characteristics, are normally roasted separately. Especially strong blends are usually designated as mocca blends.

While green coffee can be stored for 1–3 years, roasted coffee, commercially packaged (can, plastic bags, pouches, bottles), remains fresh for only 8–10 weeks. The roasting aroma decreases, while a stale, rancid taste or aroma appears. Ground coffee packaged in the absence of oxygen (vacuum packaging) keeps for 6–8 months but, as soon as the package is opened, this drops to 1–2 weeks. Little is known of the nature of the changes involved in aroma and flavor damage. The changes are retarded by storing coffee at low temperatures, excluding oxygen and water vapor.

21.1.3.3 Composition of Roasted Coffee

Table 21.3 provides information about the composition of roasted coffee. This varies greatly, depending on variety and extent of roasting.

21.1.3.3.1 Proteins

Protein is subjected to extensive changes when heated in the presence of carbohydrates. There is a shift of the amino acid composition of coffee protein acid hydrolysates before and after bean roasting (Table 21.4). The total amino acid content of the hydrolysate drops by about 30% because of considerable degradation.

Arginine, aspartic acid, cystine, histidine, lysine, serine, threonine and methionine, being especially reactive amino acids, are somewhat decreased in roasted coffee, while the stable amino acids, particularly alanine, glutamic acid and leucine, are relatively increased. Free amino acids occur only in traces in roasted coffee.

Table 21.3. Composition of roasted coffee (medium degree of roasting)

Component	Content (%)[a]	
	Arabica	Robusta
Caffeine	1.3	2.4
Lipids	17.0	11.0
Protein[b]	10.0	10.0
Carbohydrates	38.0	41.5
Trigonelline, niacin	1.0	0.7
Aliphatic acids	2.4	2.5
Chlorogenic acids	2.7	3.1
Volatile compounds	0.1	0.1
Minerals	4.5	4.7
Melanoidins[c]	23.0	23.0

[a] Based on solids. Water content varies between 1 and 5%.
[b] Calculated as the sum of the amino acids after acid hydrolysis.
[c] Calculated as the difference.

Table 21.4. Amino acid composition of the acid hydrolysate of Colombia coffee beans prior to and after roasting

Amino acid	Green coffee (%)	Roasted coffee[a] (%)
Alanine	4.75	5.52
Arginine	3.61	0
Aspartic acid	10.63	7.13
Cystine	2.89	0.69
Glutamic acid	19.80	23.22
Glycine	6.40	6.78
Histidine	2.79	1.61
Isoleucine	4.64	4.60
Leucine	8.77	10.34
Lysine	6.81	2.76
Methionine	1.44	1.26
Phenylalanine	5.78	6.32
Proline	6.60	7.01
Serine	5.88	0.80
Threonine	3.82	1.38
Tyrosine	3.61	4.35
Valine	8.05	8.05

[a] A loss due to roasting amounts to 17.6%.

21.1.3.3.2 Carbohydrates

Most of the carbohydrates present, such as cellulose and polysaccharides consisting of mannose, galactose and arabinose, are insoluble. During roasting a proportion of the polysaccharides are degraded into fragments which are soluble. Sucrose (cf. Table 21.2) present in raw coffee is decomposed in roasted coffee up to concentrations of 0.4–2.8%. Monosaccharides also hardly occur.

21.1.3.3.3 Lipids

The lipid fraction appears to be very stable and survives the roasting process with only minor changes. Its composition is given in Table 21.5. Linoleic acid is the predominant fatty acid, followed by palmitic acid. The raw coffee waxes, together with hydroxytryptamide esters of various fatty acids (arachidic, behenic and lignoceric) originate from the fruit epicarp. These compounds are 0.06–0.1% of normally roasted coffee. The diterpenes present are cafestol (I, R = H), 16-O-methylcafestol (I, R = CH$_3$), and kah-

Table 21.5. Lipid composition of roasted coffee beans (coffee oil)

Constituent	Content (%)	Constituent	Content (%)
Triacylglycerols	78.8	Triterpenes	
Diterpene esters	15.0	(sterols)	0.34
Diterpenes	0.12	Unidentified	
Triterpene esters	1.8	compounds	4.0

Table 21.6. Chlorogenic acid content as a function of the degree of roasting

Raw/degree of roasting	Arabica	Robusta
Raw	6.9%	8.8%
Light	2.7%	3.5%
Medium	2.2%	2.1%
Dark	0.2%	0.2%

weol (II). Cafestol and kahweol are degraded by the roasting process.

Since 16-O-methylcafestol is found only in Robusta coffee (0.6–1.8 g/kg of dry weight, green coffee), it is a suitable indicator for the detection of the blending of Arabica with Robusta coffee, even in instant coffee.

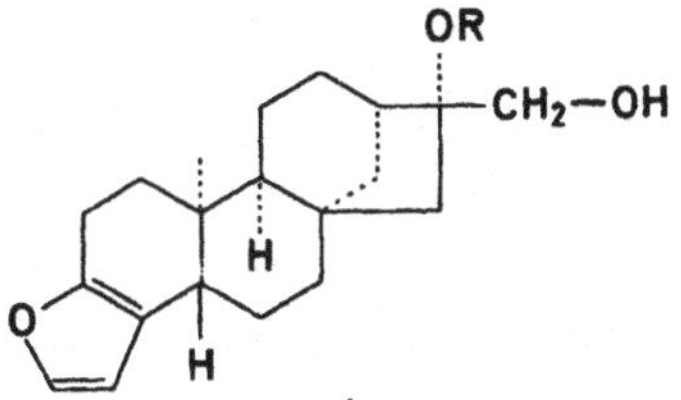

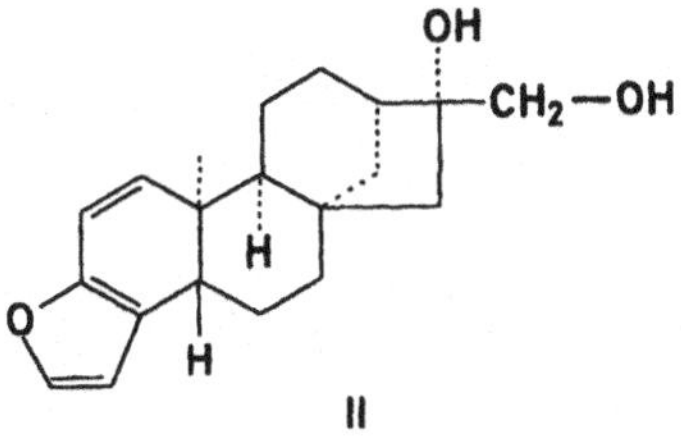

(21.1)

A diterpene glycoside is atractyloside and its aglycon, atractyligenin:

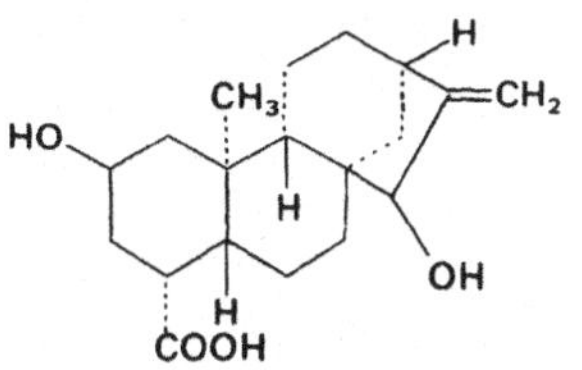

(21.2)

Sitosterol and stigmasterol are major compounds of the sterol fraction.

21.1.3.3.4 Acids

Formic and acetic acids predominate among the volatile acids, while nonvolatile acids are lac-

tic, tartaric, pyruvic and citric. Higher fatty acids and malonic, succinic, glutaric and malic acids are only minor constituents. Itaconic (I), citraconic (II) and mesaconic acids (III) are degradation products of citric acid, while fumaric and maleic acids are degradation products of malic acid:

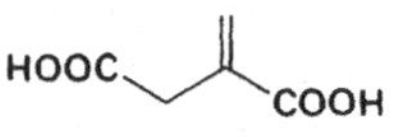

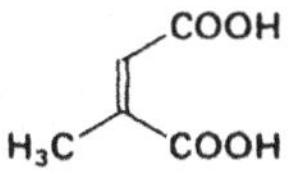

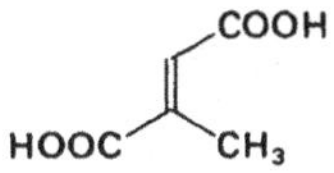

(21.3)

Chlorogenic acids are the most abundant acids of coffee (Tables 21.2 and 21.3). The content of these acids drops on roasting as shown in Table 21.6.

21.1.3.3.5 Caffeine

The best known N-compound is caffeine (1,3,7-trimethylxanthine) because of its physiological effects (stimulation of the central nervous system, increased blood circulation and respiration). It is mildly bitter in taste (threshold value in water is 0.8–1.2 mmole/l), crystallizes with one molecule of water into silky, white needles, which melt at 236.5 °C and sublime without decomposition at 178 °C. The caffeine content of raw Arabica coffee is 0.9–1.4%, while in the Robusta variety, it is 1.5–2.6%. In contrast there are caffeine-free Coffea varieties. Santos, an Arabica coffee, is on the low side, while Robusta from Angola is at the top of the range given for caffeine content. Other purine alkaloids are theobromine (Arabica: 36–40 mg/kg, Robusta: 26–82 mg/kg) and theophylline (Arabica: 7–23 µg/kg, Robusta: 86–344 µg/kg).

Caffeine forms, in part, a hydrophobic π-complex with chlorogenic acid in a molar ratio of 1:1. In a coffee drink, 10% of the caffeine and about 6% of the chlorogenic acid present occur in this form. The caffeine level in beans is only slightly decreased during roasting. Caffeine obtained by the decaffeination process and synthetic caffeine are used by the pharmaceutical and soft drink industries. Synthetic caffeine is obtained by methylation of xanthine which is synthesized from uric acid and formamide.

21.1.3.3.6 Trigonelline, Nicotinic Acid

Trigonelline (N-methylnicotinic acid) is present in green coffee up to 0.6% and is 50% decomposed during roasting. The degradation products include nicotinic acid, pyridine, 3-methyl pyridine, nicotinic acid methyl ester, and a number of other compounds.

21.1.3.3.7 Aroma Substances

The volatile fraction of roasted coffee has a very complex composition. Dilution analyses (cf. 5.2.2) have shown that of the 850 volatile compounds identified until now, only the 40 listed in Table 21.7 contribute to the aroma. Indeed, 28 aroma substances in the concentrations present in a medium roasted Arabica coffee drink (Table 21.8) can largely approximate its aroma. The correspondence becomes even better by the addition of 4-methoxy-2-methylbutan-2-thiol (cf.5.3.2.5), which has a concentration of 0.022 µg/kg in the drink.

The aroma profile of coffee is composed of the following notes: sweet/caramel-like, earthy, sulfurous/roasty and smoky/phenolic. Table 21.8 shows that most of the odorants can be assigned to these notes. The remaining odorants have a fruity or spicy odor. In the aroma profile, they are discretely detectable if their concentrations are considerably higher than shown in Table 21.8. Omission experiments (cf. 5.2.7) show that 2-furfurylthiol makes the most important contribution to the aroma of coffee.

Table 21.7. Odorants of roasted coffee – results of dilution analyses

Aroma substance

Acetaldehyde, methanethiol, propanal, methylpropanal, 2-/3-methylbutanal, 2,3-butandione, 2,3-pentandione, 3-methyl-2-buten-1-thiol, 2-methyl-3-furanthiol, 2-furfurylthiol, 2-/3-methylbutyric acid, methional, 2,3,5-trimethylthiazole, trimethylpyrazine, 3-mercapto-3-methyl-1-butanol, 3-mercapto-3-methylbutylformiate, 2-(1-mercaptoethyl)-furan, 2-methoxy-3-isopropylpyrazine, 5-ethyl-2, 4-dimethylthiazole, 2-ethyl-3, 5-dimethylpyrazine, phenylacetaldehyde, 2-ethenyl-3, 5-dimethylpyrazine, linalool, 2,3-diethyl-5-methylpyrazine, 3,4-dimethyl-2-cyclopentenol-1-one, guaiacol, 4-hydroxy-2, 5-dimethyl-3(2H)-furanone, 3-isobutyl-2-methoxypyrazine, 2-ethenyl-3-ethyl-5-methylpyrazine, 6,7-dihydro-5-methyl-5H-cyclopentapyrazine, (E)-2-nonenal, 5-ethyl-4-hydroxy-2-methyl-3(2H)-furanone, 3-hydroxy-4,5-dimethyl-2(5H)-furanone, 4-ethylguaiacol, p-anisaldehyde, 5-ethyl-3-hydroxy-4-methyl-2(5H)-furanone, 4-vinylguaiacol, (E)-β-damascenone, bis(2-methyl-3-furyl)disulfide, vanillin

Its precursors are polysaccharides containing arabinose, e. g., arabinogalactans, as well as cysteine in the free and bound form. A considerable part of furfurylthiol and the other thiols listed in Table 21.8 is present in roasted coffee as disulfide bound to cysteine, SH-peptides and proteins. On roasting, the formation of furfurylthiol is promoted by the water content and the slightly acidic pH value of the beans because under these conditions, the precursor arabinose in the polysaccharides is released by partial hydrolysis.

Robusta coffees contain alkylpyrazines and phenols in significantly higher concentrations than Arabica (Table 21.9). Correspondingly, the earthy and smoky/phenolic notes in the aroma profile are more intensive. Arabica coffees are usually richer in the odorants of the sweet/caramel-like group.

The pea-like, potato-like aroma note of raw coffee is produced by 3-alkyl-2-methoxypyrazines, 3-isobutyl-2-methoxypyrazine having the highest aroma value. Being very stable compounds, they easily survive the roasting process. However, this process yields very intensively smelling odor-

Table 21.8. Concentrations of potent odorants in Arabica coffee from Colombia[a] – Yields of odorants in the production of the beverage[b]

No.	Group/odorant	Concentration (mg/kg)	Yield (%)
	Sweet/caramel-like group		
1	Methylpropanal	28.2	59
2	2-Methylbutanal	23.4	62
3	3-Methylbutanal	17.8	62
4	2,3-Butandione	49.4	79
5	2,3-Pentandione	36.2	85
6	4-Hydroxy-2,5-dimethyl-3(2H)-furanone (HD3F)	120	95
7	5-Ethyl-4-hydroxy-2-methyl-3(2H)-furanone (EHM3F)	16.7	93
8	Vanillin	4.1	95
	Earthy group		
9	2-Ethyl-3,5-dimethylpyrazine	0.326	79
10	2-Ethenyl-3,5-dimethylpyrazine	0.053	35
11	2,3-Diethyl-5-methylpyrazine	0.090	67
12	2-Ethenyl-3-ethyl-5-methylpyrazine	0.017	25
13	3-Isobutyl-2-methoxy-pyrazine	0.087	23
	Sulfurous/roasty group		
14	2-Furfurylthiol	1.70	19
15	2-Methyl-3-furanthiol	0.064	34
16	Methional	0.239	74
17	3-Mercapto-3-methylbutyl-formiate	0.112	81
18	3-Methyl-2-butene-1-thiol	0.0099	85
19	Methanethiol	4.55	72
20	Dimethyltrisulfide	0.028	n.a.
	Smoky/phenolic group		
21	Guaiacol	3.2	65
22	4-Ethylguaiacol	1.6	49
23	4-Vinylguaiacol	55	30
	Fruity group		
24	Acetaldehyde	130	73
25	Propanal	17.4	n.a.
26	(E)-β-Damascenone	0.226	11
	Spicy group		
27	3-Hydroxy-4,5-dimethyl-3(5H)-furanone (HD2F)	1.58	78
28	5-Ethyl-3-hydroxy-4-methyl-2(5H)-furanone (EHM2F)	0.132	n.a.

[a] Degree of roasting: medium.
[b] Yield of the aroma substances in the production of the beverage (11) by percolation of coffee powder (54 g) with water (ca. 90 °C).
n.a.: not analyzed.

Table 21.9. Key odorants for the difference between Arabica and Robusta coffee

Aroma substance	Concentration (mg/kg)	
	Arabica	Robusta
2-Ethyl-3,5-dimethylpyrazine	0.326	0.940
2,3-Diethyl-5-methylpyrazine	0.090	0.310
Guaiacol	3.2	28.2
4-Ethylguaiacol	1.61	18.1
4-Vinylguaiacol	55	178

ants so that the odor of the methoxypyrazines is largely suppressed. An aroma defect, the potato taste, (Table 21.10) is produced in roasted coffee only if the concentrations of the alkyl-methoxypyrazines increase excessively. These compounds are synthesized by bacteria which penetrate into the coffee fruit after insects have done the groundwork. In particular, 2-furfurylthiol and guaiacol increase with increasing degree of roasting (Fig. 21.2).

The aroma of coffee is not stable, the fresh note is rapidly lost. Of the highly volatile odorants,

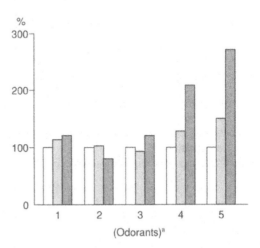

Fig. 21.2. Changes in the concentration of potent odorants in the roasting process (according to *Mayer* et al. 1999). Arabica coffee from Colombia was slightly (☐), moderately (▨) and strongly (■) roasted. *1*, 2,3-Butandione; *2*, 4-Hydroxy-2,5-dimethyl-3(2H)-furanone; *3*, 2-ethyl-3,5-dimethylpyrazine; *4*, 2-furfurylthiol; *5*, guaiacol

Table 21.10. Aroma defects in coffee

Aroma defect	Key aroma substance	Cause
Phenolic, musty, medicinal	2,4,6-Trichloroanisole	Degradation of fungicides
Mouldy	2-Methylisoborneol	Microorganisms
Potato taste	Alkylmethoxypyrazines	Combination of insects and bacteria
Fruity, silage-like	Cyclohexanecarboxylic acid ethylester	Uncontrolled fermentation

Table 21.11. Losses of odorants in ground and open stored coffee

Odorants	Loss (%)[a]
Methanethiol	66
Acetaldehyde	45
2-Methylbutanal	32
3-Methylbutanal	27
2-Furfurylthiol	23
3-Isobutyl-2-methoxypyrazine	21
Guaiacol	18
2-Ethyl-3,5-dimethylpyrazine	12
4-Vinylguaiacol	5
4-Hydroxy-2,5-dimethyl-3(2H)-furanone (HD3F)	1.4
3-Hydroxy-4,5-dimethyl-2(5H)-furanone (HD2F)	1.1

[a] Loss in 30 minutes at room temperature.

methanethiol evaporates the fastest, followed by acetaldehyde (Table 21.11). The aroma profile changes because especially the slow-evaporating furanones remain (Table 21.11). As a result, the aroma balance can be destroyed by the spicy odor of HD2F (cf. 12.7.3.5) because it is individually detectable. In the case of open storage of intact beans, losses of the highly volatile aroma substances are significantly lower, e. g., evaporation of methanethiol is only 11% in 15 minutes at room temperature instead of 43%.

21.1.3.3.8 Minerals

As with all plant materials, potassium is predominant in coffee ash (1.1%), followed by calcium (0.2%) and magnesium (0.2%). The predominant anions are phosphate (0.2%) and sulfate (0.1%). Many other elements are present in trace amounts.

21.1.3.3.9 Other Constituents

Brown compounds (melanoidins) are present in the soluble fraction of roasted coffee. They have a molecular weight range of 5–10 kdal and are derived from *Maillard* reactions or from carbohydrate caramelization. The structures of these compounds have not yet been elucidated. Apparently, chlorogenic acid is also involved in such browning reactions since caffeic acid has been identified in alkali hydrolysates of melanoidins.

21.1.3.4 Coffee Beverages

In order to obtain an aromatic brewed coffee with a high content of flavoring and stimulant constituents, a number of prerequisites must be fulfilled. The brewing, leaching and filtration procedures used give rise to a variety of combinations. While in our society brewed coffee is enjoyed as a transparent, clear drink, in the Orient brewed coffee is prepared from pulverized beans (roasted beans ground to a fine powder) and water brought to a boil, and is drunk as a turbid beverage with the sediment (Turkish mocca). Coffee extract is made by boiling the coffee for 10 min in water and then filtering. In the boiling-up procedure the coffee is added to hot water, brought to a boil within a short time and then filtered. The steeping method involves pouring hot water on a bag filled with ground coffee and occasionally swirling the bag in a pot for 10 min. In the filtration-percolation method, ground coffee is placed on a support grid (filter paper, muslin, perforated plastic filter, sintered glass, etc.) and extracted by dripping or spraying with hot water, i. e. by slow gravity percolation. This procedure, in principle, is the method used in most coffee machines. In an expresso machine, which was developed in Italy, coffee is extracted briefly by superheated water (100–110 °C), while filtration is accelerated by steam at a pressure of 4–5 bar. The exceptionally strong drink is usually turbid and is made of freshly ground, darkly roasted coffee. The water temperature should not exceed 85–95 °C in order to obtain an aromatic drink with most of the volatile substances retained. Water quality obviously plays a role, especially water with an unusual composition (some mineral spring waters, excessively hard water, and chlorinated water) might reduce the quality of the coffee brew. Brewed coffee allowed to stand for a longer time undergoes a change in flavor.

For regular brewed coffee, 50 g of roasted coffee/l (7.5 g/150 ml cup) is used; for mocca, 100 g/l; and for Italian espresso, 150 g/l. Depending on the particle size and brewing procedure, 18–35% of the roasted coffee is solubilized. The dry matter content of coffee beverages is 1–3%. The composition is presented in Table 21.12.

The taste of coffee depends greatly on the pH of the brew. The pH using 42.5 g/l of mild roasted coffee should be 4.9–5.2. At pH < 4.9 the coffee tastes sour; at pH > 5.2 it is flat and bitter. Coffees of different origins provide extracts with different pH's. Generally, the pH's of Robusta var-

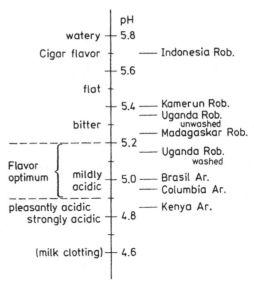

Fig. 21.3. The flavor of roasted coffee brew as related to pH value (according to *Vitzthum*, 1976)

ieties are higher than those of Arabica varieties. Figure 21.3 shows the relationship between pH and extract taste for some coffees of known origin.

The difference between the aroma of the beverage and that of ground coffee is the more intensive phenolic, buttery, caramel-like note and a weaker roasty note. These changes are caused by shifts in the concentrations of the aroma substances during brewing (Table 21.8). Compounds like 2,3-butandione, the furanones 6, 7 and 27, 2-ethyl-3,5-pyrazine, the thiols 17 and 18 are extracted with yields of >75%, while only 25% or less of 2-ethenyl-3-ethyl-5-methylpyrazine, 3-iso-butyl-2-methoxypyrazine, 2-furfurylthiol and β-damascenone pass into the beverage. The low yield of 2-furfurylthiol is partly due to reactions which occur during percolation of the coffee powder.

Caffeine and the quinic acid lactones listed in Table 21.13 are the bitter substances in the coffee drink. Accordingly, these lactones are almost exclusively responsible for the bitter note of a decaffeinated coffee drink (Table 21.13). Although the concentrations of the lactones III–VII, IX and X in the drink are lower than their threshold concentrations (cf. Table 21.13), they still additively contribute to the bitter taste (cf. 5.1.2: additive effect).

Table 21.12. Composition of coffee beverages[a]

Constituent	Content (% dry weight basis)
Protein[b]	6
Polysaccharides	24
Saccharose	0.8
Monosaccharides	0.4
Lipids	0.8
Volatile acids	1.4
Nonvolatile acids	1.6
Chlorogenic acids	14.8
Caffeine	4.8
Trigonelline	1.6
Nicotinic acid	0.08
Volatile aroma compounds	0.4
Minerals	14
Unidentified constituents (pigments, bitter compounds etc.)	29.4

[a] Arabica-coffee, medium roast, 50 g/l.
[b] Calculated as sum of the amino acids after acid hydrolysis.

$$(21.4)$$

Table 21.13. Bitter quinic acid lactones in a decaffeinated coffee drink[a]

No.	Quinic acid lactone (quinides)[b]	Threshold[c]	Concentration
		(mg/l)	
I	3-O-Caffeoyl-γ-	13.4	33.15
II	4-O-Caffeoyl-γ-	12.1	19.68
III	4-O-Caffeoyl-*muco*-γ-	11.2	8.27
IV	5-O-Caffeoyl-*muco*-γ-	9.7	6.12
V	5-O-Caffeoyl-*epi*-δ-	60.5	3.28
VI	3-O-Feruloyl-γ-	13.7	6.75
VII	4-O-Feruloyl-γ-	13.7	3.03
VIII	3,4-Dicaffeoyl-γ-	4.9	5.40
IX	4,5-Dicaffeoyl-*muco*-γ-	4.9	1.65
X	3,5-Dicaffeoyl-*epi*-δ-	24.9	0.80

[a] Made by the percolation of coffee powder (54 g) with water (80 °C, 1.1 l).
[b] The structures of the lactones I, III and V are presented in Formula 21.4.
[c] Threshold for the bitter taste.

21.1.4 Coffee Products

The coffee products which will be discussed are instant coffee, decaffeinated coffee and those containing additives.

21.1.4.1 Instant Coffee

Instant (soluble) coffee is obtained by the extraction of roasted coffee. The first technically sound process was developed by *Morgenthaler*

in Switzerland in 1938. Ground coffee is batchwise extracted under pressure in percolator batteries or continuously in extractors. The water temperature may be as high as 200 °C while the temperature of the extract leaving the last extraction cell is 40–80 °C. The extracts exhibit a concentration of ca. 15% and are evaporated in vacuum film evaporators to a solids content of 35–70%. To minimize aroma losses, the extraction can be conducted in two stages. In a gentle stage, the first extract is obtained with a solids content of 25–27% and carries the main portion of the aroma. Without concentration, it is mixed with a second extract which was obtained under stronger conditions and concentrated. In addition, aroma concentrates can be isolated by stripping; they can be added back before or after drying. The technical extraction yields are 36–46%. Further processing involves spray or freeze drying. In the latter method, the liquid extract is foamed and frozen in a stream of cold air or an inert gas (−40 °C), then granulated (grain size of 2–3 mm), sifted and dried in vacuum in the frozen state. Spray-dried coffee extract can be agglomerated in vibration fluid beds by steam or spray.

The resultant extract powder is hygroscopic and unstable. It is packaged in glass jars, vacuum packed in cans, aluminum foil-lined bags, flexible polyethylene, laminated pouches or bags, or packaged in air-tight plastic beakers or mugs, often under vacuum or under an inert gas.

Like roasted coffee, instant coffee is marketed in different varieties, e. g., regular roasted or as a dark, strongly-roasted espresso, or caffeine free. Instant coffee contains 1.0–6.0% moisture. The dry matter consists of 7.6–14.6% minerals, 3.2–

13.1% reducing sugars (calculated as glucose), 2.4–10.5% galactomannan, 12% low molecular organic acids, 15–28% brown pigments, 2.5–5.4% caffeine and 1.56–2.65% trigonelline. The products are used not only for the preparation of coffee beverages but also as flavorings for desserts, cakes, sweet cookies and ice cream.

21.1.4.2 Decaffeinated Coffee

The physiological effects of caffeine are not beneficial nor are they tolerated by everyone. Hence, many processes have been developed to remove caffeine (<0.1%) from coffee. The following process steps are normally used:

- Swelling of the raw coffee with water or steam at 22–100 °C up to a water content of 30–40%,
- Extraction of the caffeine-potassium-chorogenate complex with a water-saturated solvent (methylene chloride, ethyl acetate) at 60–150 °C,
- Treatment with steam at 100–110 °C to remove the solvent (deodorization),
- Drying with warm air or under vacuum at 40–80 °C.

In another indirect process, used in the USA, initially all the water-soluble compounds including caffeine are extracted from the green beans. The aqueous extract is decaffeinated with an organic solvent (e. g., dichloroethane), then added back to the green beans and evaporated to dryness with the beans.

Swollen raw coffee can also be decaffeinated with supercritical CO_2 (crit. point: 31.06 °C; 73.8 bar) at 40–80 °C and a pressure of 200–300 bar. The high vapor pressure of carbon dioxide under normal conditions guarantees a product that is free from solvent residues. Apart from the extraction of caffeine, this process can also be applied in the extraction of odor- and taste-active substances from hops and other plant materials.

21.1.4.3 Treated Coffee

The "roast" compounds, the phenolic acids and the coffee waxes, are irritating substances in roasted coffee. Various processes have been developed to separate these constituents to make roasted coffee tolerable for sensitive people.

Lendrich (1927) investigated the effect of steaming green beans, without caffeine extraction, on the removal of some substances (e. g., waxes) and hydrolysis of chlorogenic acid. In a process developed by *Bach* (1957), roasted coffee beans are washed with liquid carbon dioxide. In another process, the surface waxes of the raw beans are first removed by a lowboiling organic solvent, followed by steaming, as used by *Lendrich*. The extent of wax removal can be monitored by the analysis of fatty acid tryptamides, which have already been mentioned (cf. 21.1.3.3.3).

21.1.5 Coffee Substitutes and Adjuncts

21.1.5.1 Introduction

Coffee substitutes, or surrogates, are the parts of roasted plants and other sources which are made into a product which, with hot water, provides a coffee-like brew and serves as a coffee substitute or as a coffee blend.

Coffee adjuncts (coffee spices) are roasted parts of plants or material derived from plants, mixed with sugar, or a blend of all three sources and, when other ingredients are added, are used as an additive to coffee or as coffee substitutes. The starting materials for manufacturing such products vary: barley, rye, milo (a sorghum-type grain) and similar starch-rich seeds, barley and rye malts and other malted cereals, chicory, sugar beets, carrots and other roots, figs, dates, locust fruit (St. John's bread) and similar sugar-rich fruits, peanuts, soybeans and other oilseeds, fully or partially defatted acorns and other tannin-free plant parts, and, lastly, various sugars.

Coffee substitutes have been known for a long time, as exemplified by the coffee brew made of chicory roots (*Cichoricum intybus* var. *sativum*) or by clear drinks prepared from roasted cereals.

21.1.5.2 Processing of Raw Materials

The raw materials are stored as such (all cereals, figs), or are stored until processing as dried slices (e. g., root crops such as chicory or sugar beet). After careful cleaning, steeping, malting and steaming in steaming vats, pots or pressure vats take place. Roasting follows, with a final temperature of 180–200 °C, and then the grains may be polished or coated with sugar.

For the manufacture of substitutes and adjunct essences, liquid sugar juice (cane or beet molasses, syrup or starch-sugar plant extracts) is caramelized in a cooker by heating above 160 °C under atmospheric pressure. The dark, brown-black product solidifies to a glassy, strongly hygroscopic mass which is then ground.

Pulverized coffee substitutes are obtained from the corresponding starting materials, as with true coffee, by a spray, drum, conveyor or other drying process.

The starch present in the raw materials is diastatically degraded to readily-caramelized, water-soluble sugars in the manufacture of coffee substitutes during the steeping, steaming and, particularly, the malting steps. This is especially the case with malt coffee. Caramel substances ("bitter roast") formed in the roasting step, which provide the color and aroma of the brew, are derived from carbohydrate-rich raw materials (starch, inulin or sucrose). Since oilseeds readily develop rancidity, processing of carbohydrate-rich materials is preferred to oil- or protein-rich raw materials.

As aroma carriers, the oils from roasted products have been analyzed in detail, specially for malt and chicory coffees. From the volatiles identified in the coffee aroma, numerous constituents are also found in these oils. However, a basic difference appears to be that the numerous sulfur-containing substances, e. g., 2-furfurylthiol, that are present in roasted beans appear in considerably lower amounts.

21.1.5.3 Individual Products

21.1.5.3.1 Barley Coffee

Barley (or rye, corn or wheat) coffee is obtained by roasting the cleaned cereal grains after steeping or steaming. The products contain up to 12% moisture and have about 4% ash.

21.1.5.3.2 Malt Coffee

Malt coffee is made from barley malt by roasting, with or without an additional steaming step. It contains 4.5% moisture, 2.6% minerals, 74.7% carbohydrates (calculated), 1.8% fat, 10.8% crude protein, 5.6% crude fiber and provides an extract which is 42.4% soluble in water. Polycyclic aromatic hydrocarbons are also detected. Rye and wheat malt coffees are manufactured from their respective malts in the same way.

21.1.5.3.3 Chicory Coffee

Chicory coffee is manufactured by roasting the cleaned roots of the chicory plant possibly with addition of sugar beet, low amounts of edible fats or oils, salt and alkali carbonates. This is followed by grinding of the roasted product, with or without an additional steaming step or treatment with hot water. Chicory contains on the average 13.3% moisture, 4.4% minerals, 68.4% carbohydrates, 1.6% fat, 6.8% crude protein, 5.5% crude fiber andprovides an extract which is 64.6% soluble in water.

21.1.5.3.4 Fig Coffee

Fig coffee is made from figs by roasting and grinding, with or without an additional steaming step or treatment with hot water. It contains 11.4% moisture, 70.2% carbohydrates and 3.0% fat and provides an extract which is 67.9% soluble in water.

21.1.5.3.5 Acorn Coffee

This product is made from acorns, freed from fruit hull and the bulk of the seed coat, by the same process as used for coffee. It contains an average of 10.5% moisture, 73.0% carbohydrates and provides an extract which is 28.9% soluble in water.

21.1.5.3.6 Other Products

Coffee substitute blends and similarly designated products are blends of the above-outlined coffee substitutes, coffee adjuncts and coffee beans. Caffeine-containing coffee substitutes or adjuncts are made by incorporating plant caffeine extracts into substitutes before, during or after the roasting step. The content of caffeine never exceeds 0.2% in such products.

21.2 Tea and Tea-Like Products

21.2.1 Foreword

Tea or tea blends are considered to be the young, tender shoots of tea shrubs, consisting of young leaves and the bud, processed in a way traditional to the country of origin. The tea shrub was cultivated in China and Japan well before the time of Christ. Plantations are now also found in India, Pakistan, Sri Lanka, Indonesia, Taiwan, East Africa, South America, etc. Table 21.14 shows some data on the production of tea.

The evergreen tea shrub (*Camellia sinensis*, synonym *Thea sinensis*) has three principal varieties, of which the Chinese (var, *sinensis* small leaves) and the Assam varieties (var. *assamica*, large leaves) are the more important and widely cultivated. Grown in the wild, the shrub reaches a height of 9 m but, in order to facilitate harvest on plantations and in tea gardens, it is kept pruned as a low spreading shrub of 1–1.5 m in height. The plant is propagated from seeds or by vegetative propagation using leaf cuttings. It thrives in tropical and subtropical climates with high humidity. The first harvest is obtained after 4–5 years. The shrub can be used for 60 to 70 years. The harvesting season depends upon the region and climate and lasts for 8–9 months per

year, or leaves can be plucked at intervals of 6–9 days all year round. In China there are 3–4 harvests per year.

The younger the plucked leaves, the better the tea quality. The white-haired bud and the two adjacent youngest leaves (the famous "two leaves and the bud" formula) are plucked, but plucking of longer shoots containing three or even four to six leaves is not uncommon. Further processing of the leaves provides black or green tea.

21.2.2 Black Tea

The bulk of harvested tea leaves is processed into black tea. First, the leaves are withered in trays or drying racks in drying rooms, or are drum dried. This involves dehydration, reducing the moisture content of the fresh leaves from about 75% to about 55–65% so that the leaves become flaccid, a prerequisite for the next stage of processing: rolling without cracking of the leaves. Withering at 20–35 °C lasts about 4–18 h. During this time the thinly spread leaves lose about 50% of their weight in air or in a stream of warm air as in drum drying. In the next stage of processing, the leaves are fed into rollers and are lightly, without pressure, conditioned in order to attain a uniform distribution of polyphenol oxidase enzymes. These enzymes are present in epidermis tissue cells, spatially separated from their substrates. This is followed by a true rolling step in which the tea leaf tissue is completely macerated by conventional crank rollers under pressure. The cell sap is released and subjected to oxidation by oxygen from the air. The rolling process is regarded as fermentation and proceeds at 25 °C for tea leaves spread thinly in layers 3.5–7 cm thick. The traditional fermentation takes about 2–3 h. The fermented tea is dried in belt dryers counter-currently with hot air at ca. 90 °C to a water content of 3 to 4%. In this process the leaf material is heated to 80 °C, which is sufficient to inactivate the polyphenol oxidases. The sap released during rolling and fermentation solidifies during drying on the fine little hairs on the surface of the leaf. This tea extract has a gold or silver color. These are the "tips", which are a sign of good quality. They dissolve on brewing. During drying, aroma substances are formed and the coppery-red color is changed to black (hence "black tea").

Table 21.14. Production of tea in 2006 (1000 t)

Continent	Tea	Country	Tea
World	3649	China	1050
		India	893
Africa	486	Sri Lanka	311
America, Central	1	Kenya	311
America, North	–	Turkey	205
America, South		Indonesia	171
and Caribbean	95	Vietnam	142
Asia	3058	Japan	92
Europe	1	Argentina	68
Oceania	9	Iran	59
		Σ (%)[a]	90

[a] World production = 100 %.

India and Sri Lanka tea factories use both rollers and machines of continuous operation – the socalled CTC machines (crushing, tearing and curling). They provide a simultaneous crushing, grinding, and rolling of the tea leaf, thus reducing the rolling and fermentation time to 1 to 2 hours. Earl Grey tea is black tea perfumed with bergamot oil.

21.2.3 Green Tea

In the green tea manufacture, the development of oxidative processes is regarded as an adverse factor. The fresher the tea leaf used in manufacture, the better the tea produced. Since oxidative processes catalyzed by the leaf enzymes are undesirable, the enzymes are inactivated at an early stage and their reactions are replaced by thermochemical processes. In contrast to black tea manufacture, withering and fermentation stages are omitted in green tea processing.

There are two methods of manufacturing green tea: Japanese and Chinese. The Japanese method involves steaming of the freshly plucked leaf at 95 °C, followed by cooling and drying. Then the leaf undergoes high-temperature rolling at 75 to 80 °C. In the Chinese method the fresh leaves are placed into a roaster which is heated by smokeless charcoal, and roasted. After rolling and sifting, firing is the final step in the production of green tea. During the processing of green tea the content of tannin, chlorophyll, vitamin C and organic acids decreases only slightly as a consequence of enzyme inactivation.

Green tea provides a very light, clear, bitter tasting beverage. In China and Japan it is often aromatized by flowers of orange, rose or jasmin. Yellow tea and red tea (*Oolong*) occupy an intermediate position between the black and green teas, yellow tea being closer to green teas, and red tea to black teas.

Yellow tea production does not include fermentation. Nevertheless, in withering, roasting, and firing, a portion of tannins undergoes oxidation, and, therefore, dry yellow tea is darker than green tea.

Red tea is a partially fermented tea. Its special flavor which is free from the grassy note of green tea is formed during roasting and higher-temperature rolling.

21.2.4 Grades of Tea

The numerous grades of tea found in the trade are defined by origin, climate, age, processing method, and leaf grade. They can be classified somewhat arbitrarily:

- According to leaf grade (tea with full, intact leaves), such as Flowery Orange Pekoe and Orange Pekoe (made from leaf buds and the two youngest, hairy, silver leaves with yellowish tips); Pekoe (the third leaf); Pekoe Souchong (with the coarsest leaves, fourth to sixth, on the young twig).
- Broken-tea, with broken or cut leaves similar to the above grades, in which the fine broken or cut teas with the outermost golden leaf tips are distinguished from coarse, broken leaves. Broken/cut tea (loose tea) is the preferred product in world trade since it provides a finer aroma which, because of increased surface area, produces larger amounts of the beverage.
- Fannings and the fluff from broken/cut leaves, freed from stalks or stems, are used preferentially for manufacturing of tea bags.
- Tea dust, which is not used in Europe.
- Brick tea is also not available on the European market. It is made of tea dust by sifting, steaming and pressing the dust in the presence of a binder into a stiff, compact teabrick.

With regard to the origin, teas of especially high quality are those from the Himalayan region Darjeeling and from the highlands of Sri Lanka.

All over the world there is blending of teas (e. g., Chinese, Russian, East-Friesen blends, household blends) to adjust the quality and flavor of the brewed tea to suit consumer taste, acceptance or trends and to accommodate regional cultural practices for tea-water ratios. Like coffee, tea extracts are dried and marketed in the form of a soluble powder, often called instant tea.

21.2.5 Composition

The chemical composition of tea leaves varies greatly depending on their origin, age and the type of processing. Table 21.15 provides data on the constituents of fresh and fermented tea leaves. In fermented teas 38–41% of the dry matter is sol-

Table 21.15. Composition (%, dry weight basis) of fresh and fermented tea leaves and of tea brew

Constituent	Fresh flush	Black tea	Black tea brew[a]
Phenolic compounds[b]	30	5	4.5
Oxidized phenolic compounds[c]	0	25	15
Protein	15	15	+[d]
Amino acids	4	4	3.5
Caffeine	4	4	3.2
Crude fiber	26	26	0
Other carbohydrates	7	7	4
Lipids	7	7	+
Pigments[e]	2	2	+
Volatile compounds	0.1	0.1	0.1
Minerals	5	5	4.5

[a] Brewing time 3 min. [b] Mostly flavanols. [c] Mostly thearubigins. [d] Traces. [e] Chlorophyll and carotenoids.

Table 21.16. Phenolic compounds in fresh tea leaves (% dry matter)

Compound	Content
(−)-Epicatechin	1–3
(−)-Epicatechin gallate	3–6
(−)-Epicatechin digallate	+[a]
(−)-Epigallocatechin	3–6
(−)-Epigallocatechin gallate	9–13
(−)-Epigallocatechin digallate	+
(+)-Catechin	1–2
(+)-Gallocatechin	3–4
Flavonols and flavonolglycosides (quercetin, kaempherol, etc.)	+
Flavones (vitexin, etc.)	+
Leucoanthocyanins	2–3
Phenolic acids and esters (gallic acid, chlorogenic acids) p-Coumaroylquinic acid, theogallin	~5
Phenols, grand total	25–35

[a] Quantitative data are not available.

uble in hot water; this is significantly more than for roasted coffee.

21.2.5.1 Phenolic Compounds (cf. 18.1.2.5)

Phenolic compounds make up 25–35% of the dry matter content of young, fresh tea leaves. Flavanol compounds (Table 21.16) are 80% of the phenols, while the remainder is proanthocyanidins, phenolic acids, flavonols and flavones. During fermentation the flavanols are oxidized enzymatically to compounds which are responsible for the color and flavor of black tea. The reddish-yellow color of black tea extract is largely due to theaflavins and thearubigins (cf. 21.2.6). The astringent taste is caused primarily by flavonol-3-glycosides. Quercetin-3-O-[α-L-rhamnopyranosyl-(1→6)-β-D-glucopyranside] is especially active with a threshold value of 0.001 μmol/l. Also of importance are (threshold values): kaempferol-3-O-[α-L-rhamnopyranosyl-(1→6)-O-β-D-glucopyranoside] (0.25 μmol/l), quercetin-3-O-β-15-D-galactopyranoside (0.43 μmol/l), quercetin-3-O-β-D-glucoside (0.65 μmol/l) and kaempferol-3-O-β-D-glucopyranoside (0.67 μmol/l). The enzymes are inactivated in green tea, hence flavanol oxidation is prevented. The greenish or yellowish color of green tea is due to the presence of flavonols and flavones. Thus, tea which is processed into green or black tea is chemically readily distinguishable mainly by the composition of phenolic compounds. Green tea contains 17.5% and black tea 14.4% of polyphenols (expressed in gallic acid equivalents). The main components in green tea are the catechins (90% of the polyphenol fraction), which account for only 25% in black tea.

Changes in the content of the phenols occur during tea leaf growth on the shrub: the concentration decreases and the composition of this fraction is altered. Therefore, good quality tea is obtained only from young leaves. Among the remaining phenolic compounds theogallin (XI in Formula 18.14) plays a special role, since it is found only in tea and is correlated with tea quality.

21.2.5.2 Enzymes

A substantial part of the protein fraction in tea consists of enzymes.

The *polyphenol oxidases*, which are located mainly within the cells of leaf epidermis, are of great importance for tea fermentation. Their activity rises during the leaf withering and rolling process and then drops during the fermentation stage, probably as a consequence of reactions of some products (e. g., o-quinones) with the enzyme proteins.

5-Dehydroshikimate reductase which reversibly interconverts dehydroshikimate and shikimate is a key enzyme in the biosynthesis of phenolic compounds via the phenylalanine pathway.

Phenylalanine ammonia-lyase which catalyzes the cleavage of phenylalanine into transcinnamate and NH_3, is equally important for the biosynthesis of phenols. Its activity in tea leaves parallels the content of catechins and epicatechins.

Proteinases cause protein hydrolysis during withering, resulting in a rise in peptides and free amino acids.

The observed oxidation of linolenic acid to (Z)-3-hexenal, which then partly isomerizes to (E)-2-hexenal, is catalyzed by a *lipoxygenase* and a *hydroperoxide lyase* (cf. 3.7.2.3) and also occurs by autoxidation. (Z)-3-Hexenal contributes to the aroma of green tea.

Chlorophyllases participate in the degradation of chlorophyll and *transaminases* in the production of precursors for aroma constituents.

Demethylation of pectins by *pectin methyl esterase* (cf. 4.4.5.2) results in the formation of a pectic acid gel, which affects cell membrane permeability, thus resulting in a drop in the rate of oxygen diffusion into leaves during fermentation.

21.2.5.3 Amino Acids

Free amino acids constitute about 1–3% of the dry matter of the tea leaf. Of this, 50% is theanine (5-N-ethylglutamine) and the rest consists of protein-forming amino acids; β-alanine is also present.

Green tea contains more theanine than black tea. Generally, there is a characteristic difference in amino acid content as well as difference in phenolic compounds between the two types of tea (Table 21.17).

The contribution of theanine to the taste of green tea is discussed. Theanine biosynthesis occurs in the plant roots from glutamic acid and ethy-

Table 21.17. Amino acids and phenolic compounds in green and black tea (% dry matter)

Tea	Phenolic compounds	Amino acids
Green tea		
Prime quality (Japan)	13.2	4.8
Consumer quality (Japan)	22.9	2.1
Consumer quality (China)	25.8	1.8
Black tea		
Highlands (Sri Lanka)	28.0	1.6
Plains (Sri Lanka)	30.2	1.7

lamine, the latter being derived from alanine. The compound is then transported into the leaves. The analogous compounds, 4-N-ethylasparagine and 5-N-methylglutamine, are present at very low levels in tea leaves.

21.2.5.4 Caffeine

Caffeine constitutes 2.5–5.5% of the dry matter of tea leaves. It is of importance for the taste of tea. Theobromine (0.07–0.17%) and theophylline (0.002–0.013%) are also preset but in very low amounts. The biosynthesis of these two compounds involves methylation of hypoxanthine or xanthine:

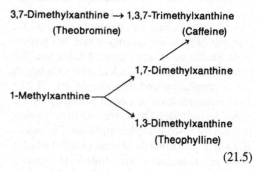

$$ \tag{21.5} $$

21.2.5.5 Carbohydrates

Glucose (0.72%), fructose, sucrose, arabinose and ribose are among sugars present in tea leaves. Rhamnose and galactose are bound to glycosides. Polysaccharides found include cellulose, hemicelluloses and pectic substances. Inositol occurs also in tea leaves.

21.2.5.6 Lipids

Lipids are present only at low levels. The polar fraction (glycerophospholipids) in young tea leaves is predominant, while glycolipids predominate in older leaves.

Triterpene alcohols, such as β-amyrin, butyrospermol and lupeol are predominant in the unsaponifiable fraction. The sterol fraction contains only Δ^7-sterols, primarily α-spinasterol and Δ^7-stigmasterol.

21.2.5.7 Pigments (Chlorophyll and Carotenoids)

Chlorophyll is degraded during tea processing. Chlorophyllides and pheophorbides (brownish in color) are present in fermented leaves, both being converted to pheophytines (black) during the firing step.

Fourteen carotenoids have been identified in tea leaves. The main carotenoids are xanthophylls, neoxanthin, violaxanthin and β-carotene (cf. 3.8.4.1). The content decreases during the processing of black tea. Degradation of neoxanthin (cf. 3.8.4.4), as an example, yields β-damascenone, a significant contributor to tea aroma (Table 21.18).

21.2.5.8 Aroma Substances

The aroma substances of black tea are shown in Table 21.18. A number of aroma substances greatly increase when the drink is brewed. It has been proposed that a modified Strecker reaction (cf. 4.2.4.4.7) contributes to an increase in 2-methylpropanal, 2- and 3-methylbutanal. The o-diquinones, which are produced by the oxidation of the numerous phenolic compounds present in tea, then take on the role of the dicarbonyl compound. The increase in geraniol is probably due to the hydrolysis of the corresponding glycosides. Some aroma substances which are produced by

Table 21.18. Concentrations of potent odorants in black tea (Darjeeling Gold Selection) – yields in the making of the drink[a]

Aroma substance	Concentration (mg/kg)	Yield (%)
2-Methyllpropanal	0.25	2300
3-Methylbutanal	0.32	1105
2-Methylbutanal	0.54	1262
Hexanal	1.60	289
(E)-2-Hexenal	0.27	2406
(Z)-4-Heptenal	0.051	108
(Z)-3-Hexen-1-ol	1.60	500
(E)-2-Nonenal	0.032	103
R/S-Linalool	6.60	180
(E,Z)-2,6-Nonadienal	0.038	122
Phenylacetaldehyde	0.65	731
(E,E)-2,4-Nonadienal	0.087	45
3-Methylnonan-2,4-dione	0.062	65
(E,E)-2,4-Decadienal	0.073	330
(E)-β-Damascenone	0.0098	125
Geraniol	0.37	3227
(E,E,Z)-2,4,6-Nonatrienal	0.16	58
β-Ionone	0.17	75
4-Hydroxy-2,5-dimethyl-3(2H)-furanone	0.10	2167

[a] Yield of the aroma substances obtained in the making of the drink from 12 g tea and 1 l water (95°).

Table 21.19. Concentrations of important aroma substances in the powder and brew of green tea

Compound	Amount[a]	
	Powder	Brew[b]
(Z)-1,5-Octadien-3-one	1.8	0.012
3-Hydroxy-4,5-dimethyl-2(5H)-furanone (HD2F)	49	0.6
3-Methyl-2,4-nonandione (MND)	83	0.56
(Z)-4-Heptenal	112	0.63
(Z)-3-Hexenal	101	0.28
(E,Z)-2,6-Nonadienal	61	0.48
1-Octen-3-one	6	0.03
(E,E)-2,4-Decadienal	127	0.9
(E)-β-Damascenone	9	0.01
4-Hydroxy-2,5-dimethyl-3(2H)-furanone (HD3F)	276	n.a.
2-/3-Methylbutyric acid	5280	63
2-Phenylethanol	1140	10.5
Linalool	206	1.0

[a] Values in μg/kg.
[b] Brew (1 kg) prepared from 10 g of the powder.
n.a., not analyzed.

the peroxidation of unsaturated fatty acids, play a role in black tea and are even more important in green tea (Table 21.19). Thus, (Z)-l,5-octadien-3-one, (Z)-3-hexenal and 3-methyl-2,4-nonandione (MND) are responsible for the green and hay-like notes in the aroma profile of this tea. Linolenic acid is the precursor of the first two carbonyl compounds. MND is a degradation product of furan fatty acids (cf. 3.7.2.1.5) and is present in tea in the concentrations shown in Table 21.19. A comparison of the values for tea and for the beverage made from it (Table 21.19) shows that the extraction yield for most of the aroma substances is >50%. β-Damascenone is an exception with a yield of 11%.

21.2.5.9 Minerals

Tea contains about 5% minerals. The major element is potassium, which is half the total mineral content. Some tea varieties contain fluorine in higher amounts (0.015–0.03%).

21.2.6 Reactions Involved in the Processing of Tea

Changes in tea constituents begin during the *withering* step of processing. Enzymatic protein hydrolysis yields amino acids of which a part is transaminated to the corresponding keto acids. Both types of acids provide a precursor pool for aroma substances. The induced chlorophyll degradation has significance for the appearance of the end-product. A more extensive conversion of chlorophyll into chlorophyllide, a reaction catalyzed by the enzyme chlorophyllase (cf. 17.1.2.9.1) is undesirable since it gives rise to pheophorbides (brown) and not to the desired oliveblack pheophytins. Increased cell permeability during withering favors the fermentation procedure. As already mentioned, a uniform distribution of polyphenol oxidases in tea leaves is achieved during the *conditioning* step of processing.

During *rolling*, the tea leaf is macerated and the substrate and enzymes are brought together; a prerequisite for fermentation. The subsequent enzymatic oxidative reactions are designated as

"fermentation". This term is a misnomer and originates from the time when the participation of microorganisms was assumed. In this processing step, the pigments are formed primarily as a result of phenolic oxidation by the polyphenol oxidases. In addition, oxidation of amino acids, carotenoids and unsaturated fatty acids, preferentially by oxidized phenols, is of importance for the formation of odorants.

Harler (1963) described tea aroma development during processing: "The aroma of the leaf changes as fermentation proceeds. Withered leaf has the smell of apples. When rolling (or leaf maceration) begins, this changes to one of pears, which then fades and the acrid smell of the green leaf returns. Later, a nutty aroma develops and, finally, a sweet smell, together with a flowery smell if flavor is present."

The enzymatic oxidation of flavanols via the corresponding o-quinones gives theaflavins (Formula 21.6, IX–XII: bright red color, good solubility), bisflavanols (XIII–XV: colorless), and epitheaflavic acids (XVI, XVII: bright red color, excellent solubility). The theaflavins and epitheaflavic acids are important ben-zotropolone derivatives that impart color to black tea.

A second, obviously heterogenous group of compounds, found in tea after the enzymatic oxidation of flavanols, are the thearubigins (XVIII, XIX), a group of compounds responsible for the characteristic reddish-yellow color of black tea extracts (cf. 18.1.2.5.2, Formula 18.21). On the whole, the phenol fraction of black tea consists of the following main components (g/kg): thearubigens (59.5), epigallocatechingallates (16.5), epigallocatechin (10.5), epicatechingallate (8.0) and theaflavin gallate (6.6).

Aroma development during fermentation is accompanied by an increase in the volatile compounds typical of black tea. They are produced by *Strecker* degradation reactions of amino acids with oxidized flavanols (Formula 21.7) and by oxidation of unsaturated fatty acids and the carotinoid neoxanthin.

During the *firing* step of tea processing, there is an initial rise in enzyme activity (10–15% of the theaflavins are formed during the first 10 min), then all the enzymes are inactivated. Conversion of chlorophyll into pheophytin is involved in reactions leading to the black color of tea. A prerequisite for these reactions is high temperature

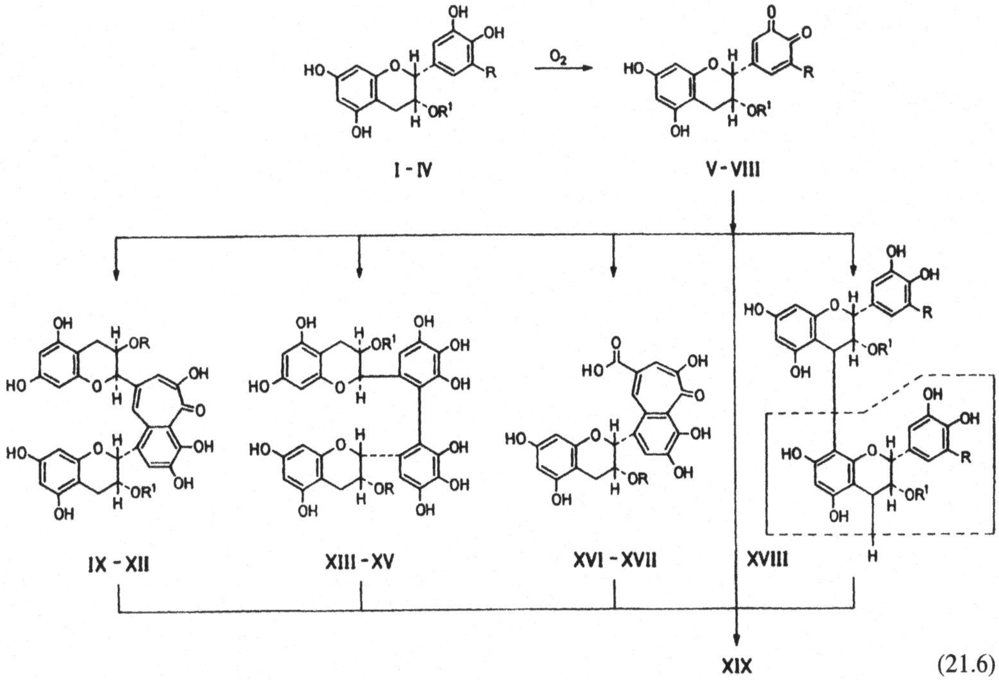

(21.6)

I: (−)-epicatechin, R^1, R^2 = H
II: (−)-epicatechin-3-gallate, R = H, R^1 = 3,4,5-trihydroxybenzoyl
III: (−)-epigallocatechin, R = OH, R^1 = H
IV: (−)-epigallocatechin-3-gallate, R = OH, R^1 = 3,4,5-trihydroxybenzoyl
V–VIII: o-quinones of compounds I–IV
IX: theaflavin, R, R^1 = H
X: theaflavin gallate A, R = H, R^1 = 3,4,5-trihydroxybenzoyl
XI: theaflavin gallate B, R = 3,4,5-trihydroxybenzoyl, R^1 = H
XII: theaflavin digallate, R, R^1 = 3,4,5-trihydroxybenzoyl
XIII: bisflavanol A, R = R^1 = 3,4,5-trihydroxybenzoyl
XIV: bisflavanol B, R = 3,4,5-trihydroxybenzoyl, R^1 = H
XV: bisflavanol C, R = R^1 = H
XVI: epitheaflavic acid, R = H
XVII: 3-galloyl epitheaflavic acid, R = 3,4,5-trihydroxybenzoyl
XVIII: thearubigins (proanthocyanidin-type), R = H, OH; R^1 = H, 3,4,5-trihydroxybenzoyl
XIX: thearubigins (polymeric catechins of unknown structure)

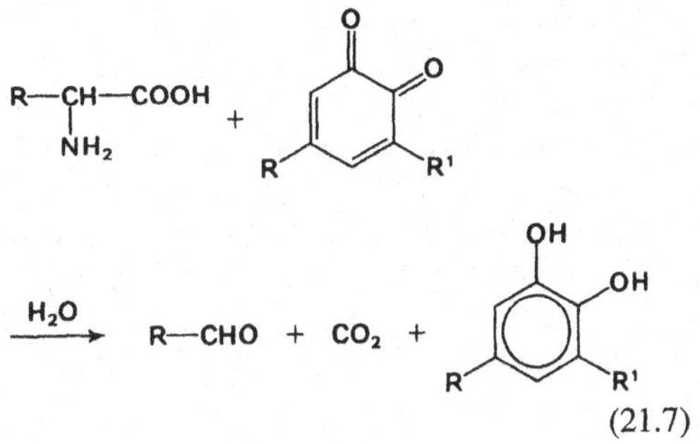

(21.7)

and an acidic environment. The undesired brown color is obtained at higher pH's. The astringent character of teas is decreased by the formation of complexes between phenolic compounds and proteins. The firing step also affects the balance of aroma substances. On the one hand there is a loss of volatile compounds, on the other hand, at high temperatures, an enhancement of the build-up of typical aroma constituents occurs, e. g., as a result of sugar-amino acid interactions.

21.2.7 Packaging, Storage, Brewing

In the country in which it is grown, the tea is cleaned of coarse impurities, graded according to leaf size, and then packed in standard plywood chests of 20–50 kg lined with aluminum, zinc or plastic foils. To preserve tea quality, the foils are sealed, soldered or welded. China, glass or metal containers are suitable for storing tea. Bags made of pergament or filter papers and filled with metered quantities of tea are also very common.

During storage, the tea is protected from light, heat (T < 30 °C) and moisture, otherwise its aroma becomes flat and light. Other sources of odor should be avoided during storage.

To prepare brewed tea, hot water is usually poured on the leaves and, with occasional swirling, left for 3–5 min. An initial tea concentrate or extract is often made, which is subsequently diluted with water. Usually 4–6 g of tea leaves per liter are required, but stronger extracts need about 8 g. The stimulating effect of tea is due primarily to the presence of caffeine.

21.2.8 Maté (Paraguayan Tea)

Maté, or Paraguayan tea, is made from leaves of a South American palm, *Ilex paraguariensis*. The palm grows in Argentina, Brazil, Paraguay and Uruguay, either wild or cultivated, and reaches a height of 8–12 m. To obtain maté, the palm leaves, petioles, flower stems and young shoot tips are collected and charred slightly on an open fire or in a woven wire drum. During such firing, oxidase enzymes are inactivated, the green color is fixed and a specific aroma is formed. The dried product is then pounded into burlap sacks or is ground to a fine powder (maté pulver, maté en pod). Maté may also be prepared by an alternative process: brief blanching of the leaf in boiling water, followed by drying on warm floors and disintegration of the leaves to rather coarse particles. In the countries in which it is grown, maté is drunk as a hot brew (yerva) from a gourd (maté = bulbshaped pumpkin fruit) using a special metal straw called a bombilla, or it is enjoyed simply in a powdered form. Maté stimulates the appetite and, because of its caffeine content (0.5–1.5%), it has long been the most important alkaloid-containing brewed plant product of South America. It contains on the average 12% crude protein, 4.5% ether-soluble material, 7.4% polyphenols and 6% minerals. About one third of the total dry matter of the leaves is solubilized in a maté brew, except for caffeine, which solubilizes to the extent of only 0.019–0.028%, and is 50% bound in leaves.

21.2.9 Products from Cola Nut

Cola (kola) nuts, called guru, goora and bissey nuts by Africans, are not nuts but actually seeds of an evergreen tree of the *Sterculiacea* family, genus *Cola*, species *verticillata, nitida* or *acuminata*, which grows wild in West Africa up to a height of 20 m. The tree is indigenous to Africa, but plantations of Cola are found on Madagascar, in Sri Lanka, Central and South America. Each fruit borne by the tree contains several red or yellow-white cola nuts, shaped like horse chestnuts. The nuts change color to brownish-red when dried, with the typical cola-red color resulting from the action of polyphenol oxidase enzymes. The nuts are on the average 5 cm long and 3 cm wide and have a bitter, astringent taste. The fresh nuts, wrapped in cola leaves and moistened with water, are the most enjoyed plant product of Western and Central Africa. They are consumed mostly in fresh form but are also chewed as dried nuts or ground to a powder and eaten with milk or honey. Cola nuts are used in the making of tinctures, extracts or medical stimulants in tablet or pastille form. They are also used in the liqueur, cocoa and chocolate industries and, especially, in the making of alcohol-free soft drinks, colawines, etc. The stimulating effect of cola nuts is due to the presence of caffeine (average content 2.16%), the main portion of which is in bound form. In addition, cola nuts contain on the average 12.2% moisture, 9.2% nitrogen compounds, 0.05% theobromine, 1.35% crude fat (ether extract), 3.4% polyphenols, 1.25% red pigments, 2.8% sugar, 43.8% starch, 15% other N-free extractable substances, 7.9% crude fiber and 3% ash.

21.3 Cocoa and Chocolate

21.3.1 Introduction

Cocoa, as a drink, is different from coffee or tea since it is consumed not in the form of an aqueous extract, i. e. a clear brew, but as a suspension. In addition to stimulating alkaloids, particularly theobromine, cacao products contain substantial amounts of nutrients: fats, carbohydrates and proteins. Unlike coffee and tea, cocoa has to be consumed in large amounts in order to experience a stimulating effect.

Cacao beans were known in Mexico and Central America for more than a thousand years before America was discovered by *Columbus*. They were enjoyed originally in the form of a slurry of roasted cocoa beans and corn which was seasoned with paprika, vanilla or cinnamon. In the first half of the 17th century, cacao beans were introduced into Germany. Cocoa became popular in the Old World only after sugar was added to the chocolate preparation. Initially, cocoa was treated as a luxury item, until the 19th century, when production of pulverized chocolate and defatted cocoa was established and they were distributed extensively as a food commodity.

The world production of cacao was 31,000 t in 1870/80, 103,000 t in 1900 and 1,585,000 t in 1979. The production in 2006 and the main cacaoproducing countries are listed in Table 21.20. The processing of cacao beans into cocoa powder and chocolate is presented schematically in Fig. 21.4.

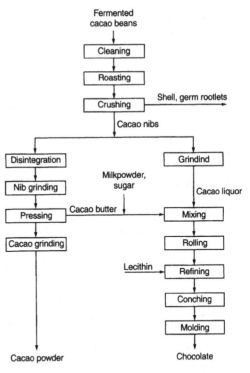

Fig. 21.4. Production of cocoa powder and chocolate

21.3.2 Cacao

21.3.2.1 General Information

Cacao beans are the seeds of the tropical cacao tree, *Theobroma cacao*, family *Sterculiaceae*. Originating in the northern part of South America and currently grown within 20 °C latitude of the Equator, the tree flourishes in warm, moist climates with an average annual temperature of 24–28 °C and at elevations up to 600 m. The tree, because of its sensitivity to sunshine and wind, is often planted and cultivated under shade trees ("cacao mothers"), such as forest trees, coconut palms and banana trees. The perennial tree grows in the wild to a height of 10–15 m, but on plantations it is kept at 2–4 m by pruning. The tree blooms all year round and the small red or white flowers bear 20–50 ripe fruits per tree. The ripe fruit or pod resembles a cantaloupe, 15–25 cm long and 7–10 cm wide. The pod is surrounded by a strong 10–15 mm thick shell. Embedded within the pod are pulpa, i. e. a sweet, mucilaginous pulp containing 10% glucose and fructose. The

Table 21.20. Production cacao bean in 2006 (1000 t)

Continent	Cacao beans	Country	Cacao beans
World	4059	Côte d'Ivoire	1400
		Ghana	734
Africa	2922	Indonesia	580
America, Central	43	Nigeria	485
America, North	–	Brazil	199
America, South		Cameroon	165
and Caribbean	462	Ecuador	94
Asia	628	Togo	73
Europe	–	Mexico	38
Oceania	48	Colombia	37
		Σ (%)[a]	94

[a] World production = 100%.

pulp surrounds 20–50 almond-shaped seeds (cacao beans). The seed is oval and flattened, about 2 cm long and 1 cm wide, and weighs close to 1 g after drying. The embryo, with two thick cotyledons (nibs) and a germ rootlet, 5 mm long and 1 mm thick, is under a thin, brittle seed coat. The colors in the cross-section of a nib range from white to light brown, to greyish-brown or brown-violet, to deep violet.

The fruit is harvested year round but, preferentially, twice a year. The main harvest time in Mexico is from March through April; in Brazil, February and, in particular, July. The summer harvest is larger and of higher quality. After the tree is planted (progagation by seed or by vegetative methods), it begins to bear pods after five or six years, giving a maximum yield after 20–30 years, while it is nearly exhausted after 40 years of growth. After reaching full beaning capacity, a cacao tree provides only 0.5–2 kg of fermented and dried beans per year. Harvesting at the right time is of great importance for the aroma of cacao and its products. The fruit is harvested fully ripe but not overripe, avoiding damage to the seed during its removal from the fruit.

The tree species *Theobroma cacao* (the only one of commercial importance) is divided into two major groups. The "Criollo" tree (criollo = native) is sensitive to climatic changes and to attack by diseases and pests. It bears highly aromatic beans, hence their commercial name "flavor beans", but they are relatively low yielding. The second group of trees, "Forastero" (forastero = strange, inferior), is characterized by great vigor and the trees are more resistant to climatic changes and to diseases and are higher yielding. The purple-red Forastero beans are less flavorful than Criollo varieties. Nevertheless, the Forastero bean is by far the most important commercial type of cacao and accounts for the bulk of world cacao production (Bahia and Accra cacaos).

Other varieties worth mentioning are the resistant and productive Calabacillo and the Ecuadorian Amelonado varieties.

Cacao beans are differentiated by their geographical origin, grade of cleanliness and the number of preparation steps to which they are subjected prior to shipment. "Flavor beans" come from Ecuador, Venezuela, Trinidad, Sri Lanka and Indonesia, while "commercial beans" are exported by the leading cacao-growing countries of West Africa (Ghana, Nigeria, Ivory Coast and Cameroon), and by Brazil (the port of Bahia) and the Dominica Republic.

21.3.2.2 Harvesting and Processing

At harvest the fully ripe pods are carefully cut from trees, gathered into heaps, cut open and the seeds scooped out with the surrounding pulp. Only rarely are the seeds dried in the sun without a prior fermentation step (*Arriba* and *Machala* varieties from South America). The bulk of the harvest is fermented before being dried. In this fermentation step the seeds with the adhering pulp are transferred to heaps, ditches or fermentation floors, baskets, boxes or perforated barrels and, depending on the variety, are left to ferment for 2–8 days. From time to time the seeds are mixed to make the oxygen in the air accessible to the fermentation process. During this time the temperature of the material rises rapidly to 45–50 °C and the germination ability of the seeds is lost. First, alcoholic fermentation occurs, which later turns into the production of acetic acid. Flavor and color formation and partial conversion of astringent phenolic compounds also occur. The adhering pulp is decomposed enzymatically and becomes liquid. It drains away as a fermentation juice. In addition, there are reactions between constituents of the seeds and pulp. After fermentation is completed, the seeds may be washed (Java, Sri Lanka), and are dried to a moisture content of 6–8%.

Well-fermented seeds, called cocoa beans from this step, provide uniformly colored, dark-brown beans which are readily separated into their cotyledons. Inadequate or unripe fermented beans are smooth in appearance (violetas) and are of low quality.

The cocoa imported by consuming countries is processed further. The cocoa beans are cleaned by a series of operations and separated according to size in order to facilitate uniform roasting in the next processing step. Roasting is being performed more and more as a two-step process. Roasting reduces the moisture content of the beans to 3%, contributes to further oxidation of phenolic compounds and the removal of acetic acid, volatile esters and other undesirable aroma components. In addition the eggs and larvae of pests are destroyed. The aroma of the beans is enhanced, the color deepens, the seed hardens and becomes more brittle and the shell is loosened and made more readily removable because of enzymatic and thermal reactions. The ripeness, moisture content, variety and size of the beans and preliminary processing steps done in the country of origin determine the extent and other parameters of the bean roasting process. This process should be carried out in two stages. First, a drying phase and then a phase in which important aroma substances are formed. For instance, African cocoa is heated to between 120 and 130 °C and high-quality cocoa to less than 130 °C for 30 minutes. Losses induced by roasting are 5–8%. As with coffee, roasted beans are immediately cooled to avoid overroasting. The roasters are batch or continuous. Heat transfer occurs either directly through heated surfaces or by a stream of hot air, without burning the shell of the beans. Roasting lasts 10–35 min, depending on the extent desired.

Roasted beans are transferred, after cooling, to winnowing machines to remove the shells and germ rootlets (these have a particularly unpleasant flavor and impart other undesirable properties to cocoa drinks). During winnowing the beans are lightly crushed in order to preserve the nibs and the shells in larger pieces and to avoid dust formation.

The winnowing process provides on the average 78–80% nibs, 10–12% shells, with a small amount of germ and about 4% of fine cocoa particles as waste. All yields are calculated on the basis of the weight of the raw beans.

The whole nibs, dried or roasted, dehulled and degermed or cracked, are still contaminated with 1.5–2% shell, seed coats and germ. The debris fraction, collected by purifying the cocoa waste, consists of fine nib particles and contains up to 10% shell, seed coating and germ. Although the cocoa shell is considered as waste material of little value, it can be used for recovery of theobromine, production of activated charcoal, or as a feed, cork substitute or tea substitute (cocoa shell tea) and, after extraction of fat, as a fertilizer or a fuel. In the adulteration of cocoa, the detection of cocoa shells is promising if based on the indicators lignoceric acid tryptamide (LAT, Formula 21.8) and behenic acid tryptamide (BAT), which are present in the ratio

of 2:1 in cocoa shells. These two tryptamides can be separated by HPLC with fluorescence detection and very exactly quantified. Cocoa shells contain 330–395 µg/g of LAT plus BAT, but the cotyledons only 7–10 µg/g.

$$R: \quad CH_3\text{-}(CH_2)_{19} \quad (BAT)$$
$$CH_3\text{-}(CH_2)_{21} \quad (LAT) \tag{21.8}$$

21.3.2.3 Composition

The compositions of fermented and air-dried cacao nib, cacao shell and germ are presented in Table 21.21.

21.3.2.3.1 Proteins and Amino Acids

About 60% of the total nitrogen content of fermented beans is protein. The nonprotein nitrogen is found as amino acids, about 0.3% in amide form, and 0.02% as ammonia, which is formed during fermentation of the beans.

Among the various enzymes, α-amylase, β-fructosidase, β-glucosidase, β-galactosidase, pec-

Table 21.21. Composition (%) of fermented and air dried cacao beans (1), cacao shells (2) and cacao germs (3)

Constituent	1	2	3
Moisture	5.0	4.5	8.5
Fat	54.0	1.5	3.5
Caffeine	0.2		
Theobromine	1.2	1.4	
Polyhydroxyphenols	6.0		
Crude protein	11.5	10.9	25.1
Mono- and oligosaccharides	1.0	0.1	2.3
Starch	6.0		
Pentosans	1.5	7.0	
Cellulose	9.0	26.5	4.3
Carboxylic acids	1.5		
Other compounds	0.5		
Ash	2.6	8.0	6.3

tinesterase, polygalacturonase, proteinase, alkaline and acid phosphatases, lipase, catalase,
peroxidase and polyphenol oxidase activities
have been detected in fresh cacao beans. These
enzymes are inactivated to a great extent during
processing.

21.3.2.3.2 Theobromine and Caffeine

Theobromine (3,7-dimethylxanthine), which is
1.2% in cocoa, provides a stimulating effect,
which is less than that of caffeine in coffee.
Therefore, it is of physiological importance.
Caffeine is also present, but in much lower
amounts (average 0.2%). A cup of cocoa contains 0.1 g of theobromine and 0.01 g of caffeine.
Theobromine crystallizes in the form of small
rhombic prisms which sublime at 290 °C without
decomposition. In cocoa beans theobromine is
often weakly bound to tannins and is released by
the acetic acid formed during fermentation of the
beans. Part of this theobromine then diffuses into
the shell.

21.3.2.3.3 Lipids

Cocoa fat (cocoa butter), because of its abundance
and value, is the most significant ingredient of cacao beans, and is dealt with in detail elsewhere
(cf. 14.3.2.2.3).

21.3.2.3.4 Carbohydrates

Starch is the predominant carbohydrate. It is
present in nibs but not in shells, a fact useful in
the microscopic examination of cocoa powders
in methods based on the occurrence of starch
as a characteristic constituent. Components of
the dietary fiber are amongst others pentosans,
galactans, mucins containing galac–turonic acid,
and cellulose. Soluble carbohydrates present
include stachyose, raffinose and sucrose (0.08–
1.5%), glucose and fructose. Sucrose hydrolysis,
which occurs during fermentation of the beans,
provides the reducing sugar pool important for
aroma formation during the roasting process.
Mesoinositol, phytin, verbascotetrose, and some
other sugars are found in cocoa nib.

21.3.2.3.5 Phenolic Compounds

The nib cotyledons consist of two types of
parenchyma cells (Fig. 21.5). More than 90% of
the cells are small and contain protoplasm, starch
granules, aleurone grains and fat globules. The
larger cells are scattered among them and contain
all the phenolic compounds and purines. These
polyphenol storage cells (pigment cells) make up
11–13% of the tissue and contain anthocyanins
and, depending on their composition, are white
to dark purple. Data on the composition of these
cells and that of the total tissue are given in
Table 21.22.

The content of phenolic compounds is also of interest from the standpoint of chemoprevention.
With 84 mg/g gallic acid equivalents (GAE) and

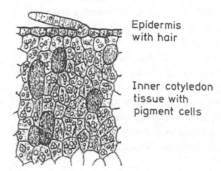

Epidermis
with hair

Inner cotyledon
tissue with
pigment cells

Fig. 21.5. A cross-section of cocoa cotyledon tissue

Table 21.22. Composition of polyphenol storage cells
of cacao tissue

Constituent	Polyphenol storage cell (%)	Cotyle-dons (%)[a]
Catechins	25.0	3.0
Leucocyanidins	21.0	2.5
Polymeric leucocyanidins	17.5	2.1
Anthocyanins	3.0	0.4
Total phenols	66.5	8.0
Theobromine	14.0	1.7
Caffeine	0.5	0.1
Free sugars	1.6	
Polysaccharides	3.0	
Other compounds	14.4	

[a] As % of dry matter.

77 mg/g epicatechin equivalents (ECE), cocoa powder contains very high concentrations compared to, e. g., green (83 mg/g GAE, 24 mg/g ECE) and black tea (62 mg/g GAE, 17 mg/G ECE).

Three groups of phenols are present in cocoa: catechins (about 37%), anthocyanins (about 4%) and leucoanthocyanins (about 58%).

The main catechin is (−)-epicatechin, besides (+)-catechin, (+)-gallocatechin and (−)-epigal-locatechin. The anthocyanin fraction consists mostly of cyanidin-3-arabinoside and cyanidin-3-galactoside.

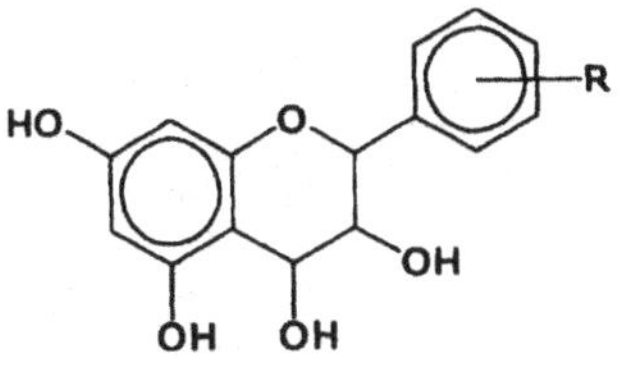

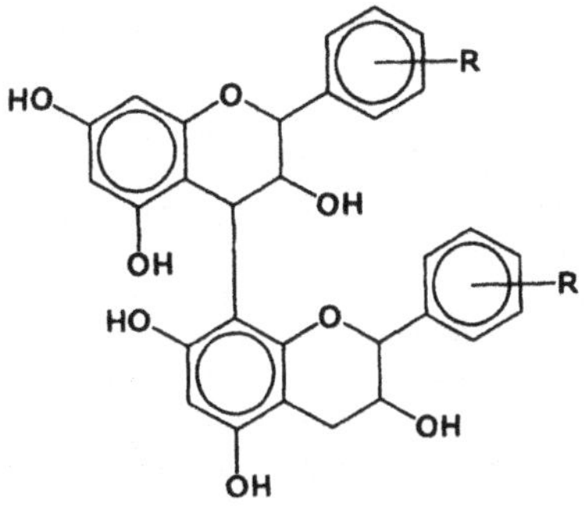

Pro- or leucoanthocyanins are compounds which, when heated in acidic media, yield anthocyanins and catechins or epicatechins, respectively. The form present in the greatest amount is flavan-3-4-diol (I in Formula 21.9) which, through 4 → 8 (II) or 4 → 6 (III) linkages, condenses to form dimers, trimers or higher oligomers (cf. 18.1.2.5.2, Formula 18.20).

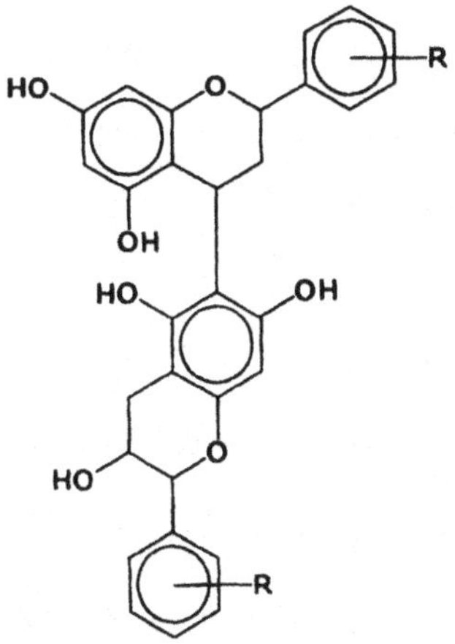

Leucoanthocyanins occur in fruits of various plants in addition to cacao; e. g., apples, pears and cola (kola) nuts.

21.3.2.3.6 Organic Acids

Of the organic acids (1.2–1.6%), citric, acetic, succinic and malic acid contribute to the taste of cocoa (Table 21.23). They are formed during fermentation. The amount of acetic acid released by the pulp and partly retained by the bean cotyledons depends on the duration of fermentation and on the drying method used. Eight brands of cocoa were found to contain 1.22–1.64% total acids, 0.79–1.25% volatile acids and 0.19–0.71% acetic acid.

(21.9)

Table 21.23. Taste substances of roasted cocoa nibs

Compound	Concentration (mmol/kg)
Bitter Group	
cis-cyclo (L-Pro-L-Val)	8.9
Theobromine	63.6
cis-cyclo (L-Val-L-Leu)	0.82
cis-cyclo (L-Ala-L-Ile)	0.64
cis-cyclo (L-Ile-L-Pro)	0.54
Astringent Group	
N-[3′,4′-Dihydroxy-(E)-cinnamoyl]-3-hydroxy-L-tyrosine	0.9
(−)-Epicatechin	8.6
Quercetin-3-0-β-D-glucopyranoside	0.10
Quercetin-3-0-β-D-galactopyranoside	0.034
γ-Aminobutyric acid	5.0
Sour Group	
Citric acid	31
Acetic acid	17
Succinic acid	1.7
Malic acid	3.6

21.3.2.3.7 Volatile Compounds and Flavor Substances

Cocoa aroma is crucially dependent on harvesting, fermentation, drying and roasting. The fresh beans have the odor and taste of vinegar. The characteristic bitter and astringent taste and the residual sweet taste of fermented beans might be impaired by various faults, such as processing of unripe or overripe fruit, insufficient aeration, lack of mixing of the fruit, infection with foreign organisms and/or smoke damage as a result of improper drying.

The odorants of cocoa powder are listed in Table 21.24. A model (cf. 5.2.7) made by using the 24 aroma substances on the basis of deodorized cocoa powder reproduced the aroma of cocoa very closely.

The taste of cocoa is described by the attributes bitter, astringent and sour. It can be reproduced by mixing 41 constituents dissolved in water (pH 5.5). The key compounds for the individual notes are the substances listed in Table 21.23. Apart from theobromine, a series of diketopiperazines are involved in the bitter taste (cf. Table 21.23), which are formed during the thermal degradation of proteins during roasting (Formula 21.10):

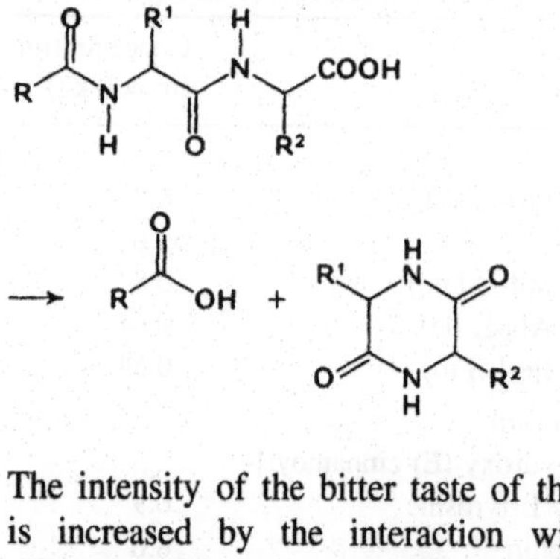

$$(21.10)$$

The intensity of the bitter taste of theobromine is increased by the interaction with certain diketopiperazines, a molar ratio of 2:1 giving the highest intensity. However, only those complexes are synergistically active in which the hydrogen bridges shown in Formula 21.11 can be formed. Thus, synergism does not occur when, e. g., in caffeine, the N(1)-atom in the purine ring is methylated.

Apart from the compounds mentioned in Table 21.23, epicatechin contributes to the astringent taste and also influences the bitter note.

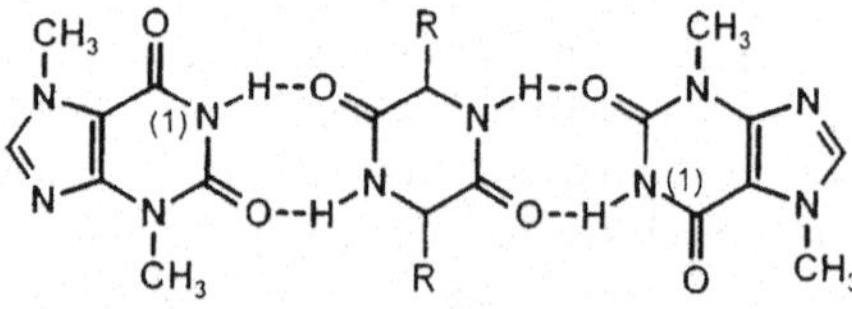

$$(21.11)$$

Table 21.24. Concentrations of potent odorants in a cocoa powder[a]

Compound	Concentration (mg/kg)
Acetic acid	332
3-Methylbutanal	25.8
2-Methylbutanal	14.3
Phenylacetaldehyde	6.60
3-Methylbutyric acid	8.55
2-Phenylacetic acid	7.70
Methylpropionic acid	2.80
2-Methylbutyric acid	1.75
4-Hydroxy-2,5-dimethyl-3(2H)-furanone	0.62
2-Phenylethanol	0.59
Butyric acid	0.32
2-Phenylethylacetate	0.32
2-Methoxyphenol	0.23
4-Methylphenol	0.12
Linalool	0.072
2-Ethyl-3,6-dimethylpyrazine	0.070
3-Methylindole	0.055
2-Ethyl-3,5-dimethylpyrazine	0.031
3-Hydroxy-4,5-dimethyl-2(5H)-furanone	0.015
2,3-Diethyl-5-methylpyrazine	0.008
Dimethyltrisulfide	0.007
2-Acetyl-1-pyrroline	0.006
2-Methyl-3-(methylthio)-furan	0.0005
2-Isobutyl-3-methoxypyrazine	0.0009

[a] Partially defatted cocoa powder (fat content: 20%), treated with alkali.

21.3.2.4 Reactions During Fermentation and Drying

The reactions occurring within the pulp during fermentation of whole cacao fruit can be distinguished from those occurring in the nibs or cotyledons. The pulp sugar is fermented by yeast to alcohol and CO_2 on the first day. Lactic acid fermentation may also occur to a small extent. Pectolytic enzymes and other glycosidases affect the degradation of polysaccharides. This is reflected in the fruit pulp becoming liquid and draining away. This improves aeration, resulting in oxidation of alcohol to acetic acid by acetic acid bacteria during the second to fourth days. The pH drops from about 6.5 to about 4.5 and the temperature increases to 45–50 °C. The seed cell walls become permeable, the living cacao seed is killed and an oxidative process takes over the entire mass. From the fifth to the seventh day, the oxidation and condensation reactions of phenolic compounds predominate. Amino acids and peptides react with the oxidation products of the phenolic compounds, giving rise to water-insoluble brown or brown-violet phlobaphenes (cacao-brown and red), which confer the characteristic color to fermented cacao beans. A decrease in the content of soluble phenols mellows the original harsh and astringent cacao flavor. Finally, the oxidation reactions are terminated by drying the seeds to a moisture content of less than 8%.

The hydrolysis of the proteins and peptides during fermentation yields with the free amino acids the precursors of aroma substances. Table 21.25 shows the increase in free amino

Table 21.25. Formation of free amino acids, accompanying *Strecker* aldehydes and amines in cocoa

Compound	Process		
	Without fermentation[a]	After fermentation[b]	After roasting[c]
Amino acid (mg/kg)			
L-Phenylalanine	190	1120	700
L-Leucine	170	1240	760
L-Isoleucine	140	390	280
Aldehydes (µg/kg)			
Phenylacetaldehyde	16	34	202
3-Methylbutanal	116	1636	8470
2-Methylbutanal	143	2075	3791
Amines (µg/kg)			
2-Phenylethylamine	227	1168	10,216
2-/-3-Methylbutylamine	129	1219	17,070

[a] After washing the pulp and drying in the sun.
[b] Fermentation (7 days) and drying in the sun.
[c] As in "b", then roasting of the nibs (15 min at 95°, increase in temperature in 20 min to 115 °C, cooling).

acids during fermentation and the extent of their degradation to aldehydes and amines. The decisive step for the degradation is the roasting, not the fermentation. Hence, the *Strecker* reaction (cf. 4.2.4.4.7) has a considerably higher share in the formation of these aroma substances (exception 2-methylbutanal) than the corresponding enzymatic degradation reactions.

The proper running of the fermentation process prevents the growth of detrimental microorganisms, such as molds, butyric acid bacteria and putrefaction-inducing bacteria.

21.3.2.5 Production of Cocoa Liquor

After roasting and drying, the cocoa nib is disintegrated and milled in order to rupture the cell walls of aggregates and expose the cocoa butter. Knife-hammer mills or crushing rolls usually serve for disintegration, while rollerball, horizontal "stone", steel disc or disc attrition mills are used for fine disintegration of cocoa particles. The resultant product is a homogeneous mobile paste, a flowing cocoa mass or cocoa liquor.

21.3.2.6 Production of Cocoa Liquor with Improved Dispersability

The cocoa nib or the cocoa mass is subjected to an alkalization process in order to mellow the flavor by partial neutralization of free acids, improve the color and enhance the wettability of cocoa powder, improve dispersability and lengthen suspension-holding ability, thus preventing formation of a sediment in the cocoa drink. The process involves the use of solutions or suspensions of magnesium oxide or hydroxide, potassium or sodium carbonate or their hydroxides. It is occasionally performed at elevated temperature and pressure, usually using steam. In this process, introduced by *C.I. van Houten* in 1828 (hence the term "Dutch cocoa process"), the roasted nibs are treated with a dilute 2–2.5% alkali solution at 75–100 °C, then neutralized, if necessary, by tartaric acid, and dried to a moisture content of about 2% in a vacuum dryer or by further kneading of the mass at a temperature above 100 °C. This treatment, in addition to acid neutralization, causes swelling of starch and an overall spongy and porous cell structure of the cocoa mass. Cocoa so treated is often incorrectly designated as "soluble cocoa" – the process does not increase solubility. Finally, the cocoa is disintegrated with fine roller mills. The "alkalized" cocoa generally contains 52–58% cocoa butter, up to 5% ash and up to 7% alkalized mass or liquor.

21.3.2.7 Production of Cocoa Powder by Cocoa Mass Pressing

To convert the cocoa mass/liquor into cocoa powder, the cocoa fat (54% of nib weight on the average) has to be reduced by pressing, usually by means of a hydraulic, mechanical or, preferentially, horizontally-run expeller press at a pressure of 400–500 bar and a temperature of 90–100 °C. To remove the contaminating cell debris, the hot cocoa butter is passed through a filter press, then molded and cooled. The bulk of the cocoa butter produced is used in chocolate manufacturing. The "stone hard" cocoa press cake, with a residual fat content of 10–24%, is disintegrated by a cook breaker, i.e. rollers with intermashing teeth. It is then ground in a peg mill and separated into a fine and a coarse fraction by an air sifter, the coarse fraction being recycled and milled repeatedly. Cocoa powders are divided according to the extent of defatting into lightly defatted powder, with 20–22% residual cocoa butter, and extensively-defatted powder, which contains less than 20% but more than 10% butter. Lightly defatted powder is darker in color and milder in flavor. Cocoa powder is widely used in the manufacture of other products, e.g., cake fillings, icings, pudding powders, ice creams and cocoa (chocolate) beverages.

21.3.3 Chocolate

21.3.3.1 Introduction

Switzerland has the highest per capita consumption of chocolate at 10.2 kg (2004), followed by Norway (9.2), Belgium (9.1), Germany (9.0), Ireland (8.8), Great Britain (8.8). The consumption of chocolate is low in Italy (3.5), Greece (2.5), Japan (1.8), Spain (1.6) and Brazil (1.0).

Chocolates were originally made directly from cocoa nibs by grinding them in the presence of sugar. Chocolate is now made from nonalkalized cocoa liquor by incorporating sucrose, cocoa butter, aroma or flavoring substances and, occasionally, other constituents (milk ingredients, nuts, coffee paste, etc.). The ingredients are mixed, refined, thoroughly conched and, finally, the chocolate mass is molded. To obtain a highly aromatic, structurally homogeneous and stable form and a product which "melts in the mouth", a set of chocolate processing steps is required, as described below.

21.3.3.2 Chocolate Production

21.3.3.2.1 Mixing

Mixing is a processing step by which ingredients such as cocoa liquor, high grade crystalline sucrose, cocoa butter and, for milk chocolate, milk powder are brought together in a mixer ("melangeur") or paster. A homogeneous, coarse chocolate paste is formed after intense mixing.

21.3.3.2.2 Refining

The refining step is performed by single or multiple refining rollers which disintegrate the chocolate paste into a smooth-textured mass made up of much finer particles. The rollers are hollow and can be adjusted to the desired temperature by water cooling. The refined end-product has a particle size of less than 30 to 40 µm. Its fat content should be 23–28%.

21.3.3.2.3 Conching

The refined chocolate mass is dry and powdery at room temperature and has a harsh, sour flavor. It is ripened before further processing by keeping it in warm chambers at 45–50 °C for about 24 h. Ripening imparts a doughy consistency to the chocolate and it may be used for the production of baking or other commercial chocolates. An additional conching step is required to obtain fine chocolates of extra smoothness. It is performed in oblong or round conche pots with roller or rotary conches. The chocolate mass is mixed, ground and kneaded.

This process is usually run in three stages. The temperature is maximum 65 °C with milk-containing chocolate and 75 °C with milk-free chocolate. In the first, the mass is treated, depending on the recipe and process, for more than 6–12 h. Loss of moisture occurs (dry conching) during heating, a protion of the volatiles is removed (ethanal, acetone, diacetyl, methanol, ethanol, isopropanol, isobutanol, isopentanol and acetic acid ethyl ester) and the fat becomes uniformly distributed, so that each cocoa particle is covered with a film of fat. The temperature at this stage is not allowed to rise since important aroma substances, e. g., pyrazines (cf. 21.3.2.3.7), may be lost. In the second stage, the mass is liquefied by the addition of residual cocoa butter and at a higher stirring speed homogenized further. Here, too, the time required greatly depends on the desired product quality: about 6 to 40 h. In the third phase, which starts 2 to 3 h before the end of the conching process, lecithin and other ingredients are added. Up to a limit of about 1.5%, lecithin lowers the flow rate and the viscosity of the mass; 1 part of lecithin can replace about 8 to 10 parts of cocoa butter. Chemical processes involved in conching are only partially understood.

Efforts have been made to shorten this time-, energy- and space-consuming final refinement in conche pots. Processes have been developed that are based on the separate pre-refinement of cocoa nibs or cocoa mass. The spray-film technique uses a cocoa mass with its natural water content or, in the case of highly acidic cocoa varieties, with the continuous addition of 0.5–2% of water. In a turbulent film with direct heat transfer, the cocoa mass is continuously dehumidified, deacidified, degassed, and roasted in counterflow with hot air (up to 130 °C). For the final refinement, apart from the time-tested conche pots, newly developed intensive refiners can be used. They reduce the conching time to 8 hours. The development of continuously operated conche pots is also being expedited.

21.3.3.2.4 Tempering and Molding

Before molding, the mass must be tempered to initiate crystallization. For both the structure (hard nibs, filling the mold) and appearance (glossy surface that is not dull), this is an important operation in which crystal nuclei are produced under controlled conditions (pre-crystallization). Molten chocolate is initially cooled from 50 °C to 18 °C within 10 min with constant stirring. It is kept at this lower temperature for 10 min to form the stable β-modification of cocoa butter (cf. 3.3.1.2). The temperature of the chocolate is then raised within 5 min to 29–31 °C. The process conditions vary according to composition. Regardless of processing variables, tempering serves to provide a great abundance of small fat crystals with high melting points. During the cooling step, the bulk of the molten chocolate develops a solid, homogeneous, finely crystalline, heat-stable fat structure characterized by good melting properties and a nice glossy surface.

Before molding, the chocolate is kept at 30–32 °C and delivered to warmed plastic or metal molds with a metering pump. The filled molds pass over a vibrating shaker to let the trapped air escape. They then pass through a cooling channel where, by slow cooling, the mass hardens and, finally, at 10 °C, the final chocolate product falls out of the mold. Tempering, metering, filling, cooling, wrapping and packaging machines now provide nearly fully mechanized and automated production of chocolate.

21.3.3.3 Kinds of Chocolate

In a strict sense, chocolate represents a food commodity which may be molded and which consists of cocoa nibs, nib particles, or cocoa liquor and sucrose, with or without added cocoa butter, natural herbs or spices, vanillin or ethyl vanillin. Chocolate contains at least 40% cocoa liquor or a blend of liquor and cocoa butter, and up to 60% sugar. The content of cocoa butter is at least 21% and, when cocoa liquor is blended with cocoa butter, at least 33%.

The composition of the more important kinds of chocolates and confectionery coatings are shown in Table 21.26.

Table 21.26. Composition of some chocolate products

Product	Cocoa mass %	Skim milk powder %	Cocoa butter %	Total fat %	Butter fat (milk) %	Sugar %
Baking chocolate	33–50	–	5–7	22–30	–	50–60
Chocolate for coating	35–60	–	to 15	28–35	–	38–50
Milk cream chocolate	10–20	8–16	10–22	33–36	5.5–10	35–60
Whole milk chocolate	10–30	9.3–23	12–20	28–32	3.2–7.5	32–60
Skim milk chocolate	10–35	12.5–25	15–25	22–30	0–2	30–60
Icings	33–65		5–25	35–46		25–50

Baking chocolate is made by a special process. Other kinds of chocolates include: cream; full or skim milk; filled; fruit, nut, almond; and those containing coffee or candied orange peels. Cola-chocolate is a caffeine-containing product (maximum of 0.25% caffeine) prepared by mixing with extracts obtained from coffee, cola or other caffeine-containing plants. Diabetic- or diet-chocolates are made by replacing sucrose with fructose, mannitol, sorbitol or xylitol. Information about chocolate coatings is presented in Table 21.26. Chocolates can also contain nuts and almonds whose oil contents are occasionally reduced by pressing to reach $\frac{2}{3}$ of the original amount. This is because the oil has a melting point lower than that of cocoa butter. In filled chocolates, the filler is first placed into a chocolate cup and then closed with a chocolate lid or cover. Fine crumbs of chocolate are made by pressing low-fat chocolate through a plate with orifices. Hollow figures are made in two-part molds, by a hollow press or by gluing together the individually molded parts.

The term "praline" originates from the name of the French Marshal *Duplessis-Praslin*, whose cook covered sweets with chocolate. Only a few of the many processing options will be mentioned. For pralines with a hard core, the hot, supersaturated sugar syrup (fondant) is poured into molds dusted with wheat powder and left to cool. The congealed core (korpus) is dipped into molgen kuverture and, in this way, covered with a chocolate coat (creme-praline). The fondant can be fully or partly replaced by fruit pastes like marzipan, jams, nuts, almonds, etc. (dessert-pralines). Such pralines are prepared with or without a sugar crust. Products with a sugar crust are made from a mixture of thick sugar solution and liqueur by pouring the mixture into mold cavities. The solid crust crystallizes on the outer walls, while the inner portion of the mixture remains liquid. The core so obtained is then dipped into melted chocolate, as described above. For pralines without a sugar crust (brandy or liqueur), the processing involves hollow-body machines in which the chocolate shell is formed, then filled with, e. g., brandy, and covered with a lid in a second machine. The fondant may also contain invertase and, thereby, the praline filling liquefies after several days. Plastic pastes are made by preliminary pulverization of the ingredients in a mill and then refiner by rollers. The oil content of the ingredients (nuts, almonds, peanuts) provides the consistency for a workable paste after grinding. Chocolate for beverages or drinks (chocolate powder or flour) is made from cocoa liquor or cocoa powder and sucrose. It is customary to incorporate seasonings, especially vanillin. The sugar content in chocolate drink powders is at most 65%.

Chocolate syrups are made in the USA by adding bacterial amylase. The enzyme prevents the syrup from thickening or setting by solubilizing and dextrinizing cocoa starch. A fat coating is a glazing like chocolate coatings made from a fat other than cocoa butter (fat from peanuts, coconuts, etc.). It is often used on baked or confectionery products. Tropical chocolates contain high melting fats or are specially prepared to make the chocolate resistant to heat. The melting point of cocoa butter can be raised by a controlled pre-crystallization procedure. Another option is based on the formation of a coherent sugar skeleton in which the fat is deposited in hollow or void spaces. In this case, in contrast to regular chocolate, there is no continuous fat phase to collapse during heating.

21.3.4 Storage of Cocoa Products

All products, from the raw cacao to chocolate, demand careful storage – dry, cool, well aerated space, protected from light and sources of other odors. A temperature of 10–12 °C and a relative humidity of 55–65% are suitable. Chocolate products are readily attacked by pests, particularly cacao moths (*Ephestia elutella* and *Cadra cauteila*), the flour moth (*Ephestia kuhniella*) and also beetles (*Coleoptera*), cockroaches (*Dictyoptera*) and ants (order *Hymenoptera*). Chocolates not properly stored are recognized by a greyish matte surface. Sugar bloom is caused by storage of chocolate in moist conditions (relative humidity above 75–80%) or by deposition of dew, causing the tiny sugar particles on the surface of the chocolate to solubilize and then, after evaporation, to form larger crystals. A fat bloom arises from chocolate fat at temperatures above 30 °C. At these temperatures the liquid fat is separated and, after repeated congealing, forms a white and larger spot. This may also occur as a result of improper precrystallization or tempering during chocolate production. The defect may be prevented or rectified by posttempering at 30 °C for 6 h.

21.4 References

Bokuchava, M.A., Skobeleva, N.I.: The biochemistry and technology of tea manufacture. Crit. Rev. Food Sci. Nutr. *12*, 303 (1979/80)

Castelein, J., Verachtert, H.: Coffee fermentation. In: Biotechnology (Eds.: Rehm, H.-J., Reed, G.), Vol. 5, p. 587, Verlag Chemie: Weinheim. 1983

Clarke, R.J., Vitzthum, O.G.: Coffee. Recent Developments. Blackwell Science Ltd., Oxford, 2001, p. 1, 50 and 68

Clifford, M.N., Willson, K.C. (Eds.): Coffee, botany, biochemistry and production of beans and beverage. The AVI Publishing Comp. Inc., Westport, Conn. 1985

Engelhardt, U.H., Lakenbrink, C., Lapczynski, S.: Antioxidative phenolic compounds in green-black tea and other methylxanthine-containing beverages. In: Caffeinated beverages. ACS Symposium Series *754*, 111 (2000)

Frank, O., Zehentbauer, G., Hofmann, T.: Bioresponse-guided decomposition of roast coffee beverage and identification of key bitter taste compounds. Eur. Food Res. Technol. *222*, 492 (2006)

Frauendorfer, F., Schieberle, P.: Identification of the key aroma compounds in cocoa powder based on molecular sensory correlations. J. Agric. Food Chem. *54*, 5521 (2006)

Garloff, H., Lange, H.: Kaffee. In: Lebensmitteltechnologie (Ed.: R. Heiss) Springer, Berlin, 1988, p. 355

Granvogl, M., Bugan, S., Schieberle, P.: Formation of amines and aldehydes from parent amino acids during thermal processing of cocoa and model systems: new insights into pathways of the Strecker reaction. J. Agric. Food Chem. *54*, 1730 (2006)

Guth, H., Grosch, W.: Furanoid fatty acids as precursors of a key aroma compound of green tea. In: Progress in Flavour Precursor Studies (Eds.: P. Schreier, P. Winterhalter) Allured Publishing Corporation, 1993, p. 189

Hatanaka, A.: The fresh green odor emitted by plants. Food Rev. Int. *12*, 303 (1996)

Ho, C.-T., Zhu, N.: The chemistry of tea. In: Caffeinated beverages. ACS Symposium Series *754*, 316 (2000).

Lange, H., Fincke, A.: Kakao und Schokolade. In: Handbuch der Lebensmittelchemie, Bd. VI (Ed.: Schormüller, J.), p. 210, Springer-Verlag: Berlin, 1970

Lee, K.W., Kim, Y.J., Lee, H.J., Lee, C.Y.: Cocoa has more phenolic phytochemicals and a higher antioxidant capacity than teas and red wine. J. Agric. Food Chem. *51*, 7292 (2003)

Maier, H.G.: Kaffee. Verlag Paul Parey: Berlin. 1981

Münch, M., Schieberle, P.: A sensitive and selective method for the quantitative determination of fatty acid tryptamides as shell indicators in cocoa products. Z. Lebensm. Unters. Forsch. A *208*, 39 (1999)

Poisson, L., Kerler, J., Liardon, R.: Assessment of the contribution of new aroma compounds found in coffee to the aroma of coffee brews. In: State of the Art in Flavour Chemistry and Biology. T. Hofmann, M. Rothe, P. Schieberle (eds.) Deutsche Forschungsanstalt für Lebensmittelchemie, Garching, 2004, p. 495

Rizzi, G.P.: Formation of sulfur-containing volatiles under coffee roasting conditions. In: Caffeinated beverages. ACS Symposium Series *754*, 210 (2000)

Sanderson, G.W.: Black tea aroma and its formation. In: Geruch- and Geschmackstoffe (Ed.: Drawert, F.), p. 65, Verlag Hans Carl: Nürnberg. 1975

Scharbert, S., Holzmann, N., Hofmann, T.: Identification of astringent taste compounds by combining instrumental analysis and human bioresponse. J. Agric. Food Chem. *52*, 3498 (2004)

Schieberle, P.: The chemistry and technology of cocoa. In: Caffeinated beverages. ACS Symposium Series *754*, 262 (2000)

Schuh, C., Schieberle, P.: Characterization of the key aroma compounds in the beverage prepared from Darjeeling black tea – quantitative difference between tea leaves and infusion. J. Agric. Food Chem. *54*, 916 (2006)

Speer, K., Tewis, R., Montag, A.: 16-O-Methyl-cafestol a quality indicator for coffee. Fourteenth International Conference on Coffee Science, San Francisco, July 14–19, 1991. ASIC 91, p. 237

Stark, T., Bareuther, S., Hofmann, T.: Molecular definition of the taste of roasted cocoa nibs (Theobroma cocoa) by means of quantitative studies and sensory experiments. J. Agric. Food Chem. *54*, 5530 (2006)

Täupmann, R.: Tee. In: Lebensmitteltechnologie (Ed.: R. Heiss) Springer, Berlin, 1988, p. 364

Viani, R.: Coffee. In: Ullmann's encyclopedia of industrial chemistry. 5th Edition, Volume A7, p. 315, Verlag VCH, Weinheim, 1986

Zürcher, K.: Kakao. In: Lebensmitteltechnologie, (Ed.: R. Heiss) Springer, Berlin, 1988, p. 341

22 Spices, Salt and Vinegar

22.1 Spices

Some plants with intensive and distinctive flavors and aromas are used dried or in fresh form as seasonings or spices. Table 22.1 lists the most important spice plants together with the part of the plant used for seasoning.

22.1.1 Composition

22.1.1.1 Components of Essential Oils

Most spices contain an essential or volatile oil (Table 22.2), which can be isolated by steam distillation. The main oil constituents are either

Table 22.1. Spices used in food preparation/processing

Number	Common name	Latin name	Class/order family (bot)	Cultivation region
Fruits				
1	Pepper, black	*Piper nigrum*	Piperaceae	Tropical and subtropical regions
2	Vanilla	*Vanilla planifolia*	Orchidaceae	Madagascar, Comore Island,
		Vanilla fragans		Mexico, Uganda
		Vanilla tahitensis		
		Vanilla pompona		
3	Allspice	*Pimenta dioica*	Myrtaceae	Caribbean Islands, Central America
4	Paprika (bell pepper)	*Capsicum annuum. var. annuum*	Solanaceae	Mediterranean and Balkan region
	Chili (Tabasco)	*Capsicum frutescens*		
	Brown pepper	*Capsicum baccatum, var. pendulum*		
5	Bay tree[a]	*Laurus nobilis*	Lauraceae	Mediterranean region
6	Juniper berries	*Juniperus communis*	Cupressaceae	Temperate climate region
7	Aniseed	*Pimpinella anisum*	Apiaceae	
8	Caraway	*Carum carvi*	Apiaceae	Temperate climate region
9	Coriander	*Coriandrum sativum*	Apiaceae	
10	Dill[a]	*Anethum graveolens*	Apiaceae	
Seeds				
11	Fenugreek	*Trigonella foenum greacum*	Leguminosae	Mediterranean region, temperate climate region
12	Mustard	*Sinapsis alba*[b]	Brassicaceae	
		Brassica nigra[c]	Brassicaceae	Temperate climate region
13	Nutmeg	*Myristica fragrans*	Myristicaceae	Indonesia, Sri Lanka, India
14	Cardamom	*Elettaria cardamomum*	Zingiberaceae	India, Sri Lanka
Flowers				
15	Cloves	*Syzygium aromaticum*	Myrtaceae	Indonesia, Sri Lanka, Madagascar
16	Saffron	*Crocus sativus*	Iridaceae	Mediterranean region, India, Australia
17	Caper	*Capparis spinosa*	Capparidaceae	Mediterranean region
Rhizomes				
18	Ginger	*Zingiber officinale*	Zingiberaceae	South China, India, Japan, Caribbean Islands, Africa
19	Turmeric	*Curcuma longa*	Zingiberaceae	India, China, Indonesia

Table 22.1. (continued)

Number	Common name	Latin name	Class/order family (bot)	Cultivation region
Barks				
20	Cinnamon	*Cinnamomum zeylanicum, C. aromaticum, C. burmanii*	Lauraceae	China, Sri Lanka, Indonesia, Caribbean Islands
Roots				
21	Horseradish	*Armoracia rusticana*	Brassicaceae	Temperate climate region
Leaves				
22	Basil	*Ocimum basilicum*	Labiate	Mediterranean region, India
23	Parsley	*Petroselinum crispum*	Apiaceae	Temperate climate region
24	Savory	*Satureia hortensis*	Labiate	Temperate climate region
25	Tarragon	*Artemisia dracunculus*	Compositae	Temperate climate region, Mediterranean region
26	Marjoram	*Origanum majorana*	Lamiaceae	Temperate climate region
27	Origano	*Origanum heracleoticum, O. onites*	Lamiaceae	Temperate climate region
28	Rosemary	*Rosmarinus officinalis*	Lamiaceae	Mediterranean region
29	Sage	*Salvia officinalis*	Lamiaceae	Mediterranean region
30	Chives	*Allium schoenoprasum*	Liliaceae	Temperate climate region
31	Thyme	*Thymus vulgaris*	Lamiaceae	Temperate climate region

[a] Fruits and leaves, [b] white mustard, [c] black mustard.

mono- and sesquiterpenes or phenols and phenol-ethers. Examples of the latter two classes of compounds are eugenol (I), carvacrol (II), thymol (III), estragole (IV), anethole (V), safrole (VI) and myristicin (VII):

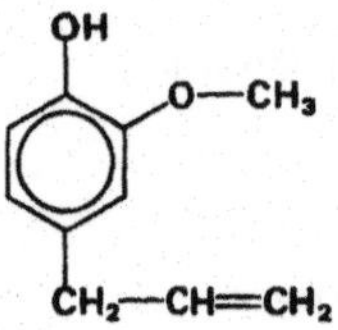

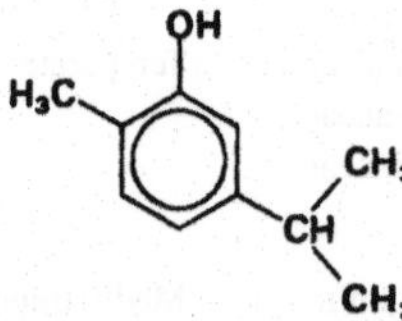

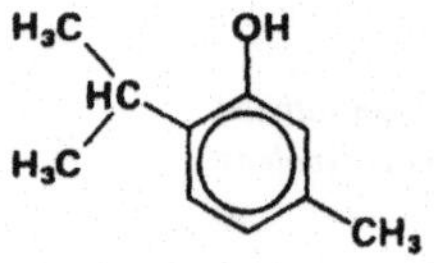

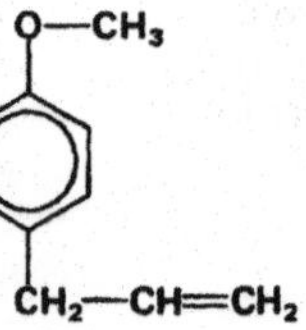

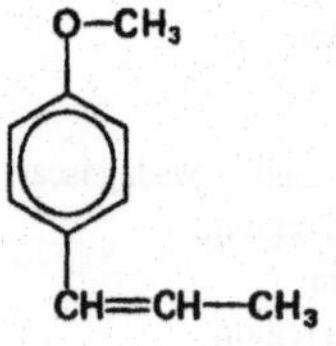

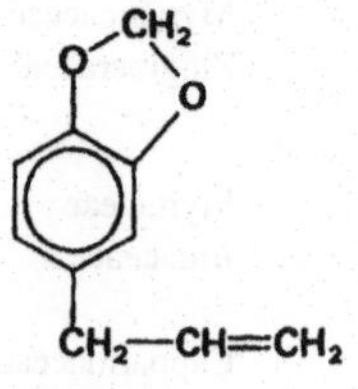

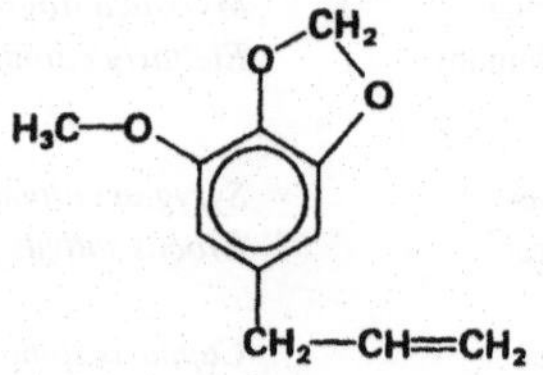

$$(22.1)$$

Biosynthesis of cinnamaldehyde (VIII) and also of eugenol (I) an safrole (VI) originates from phenylalanine (compare biosynthesis of other plant phenols in 18.1.2.5.1). The following reaction sequence is assumed:

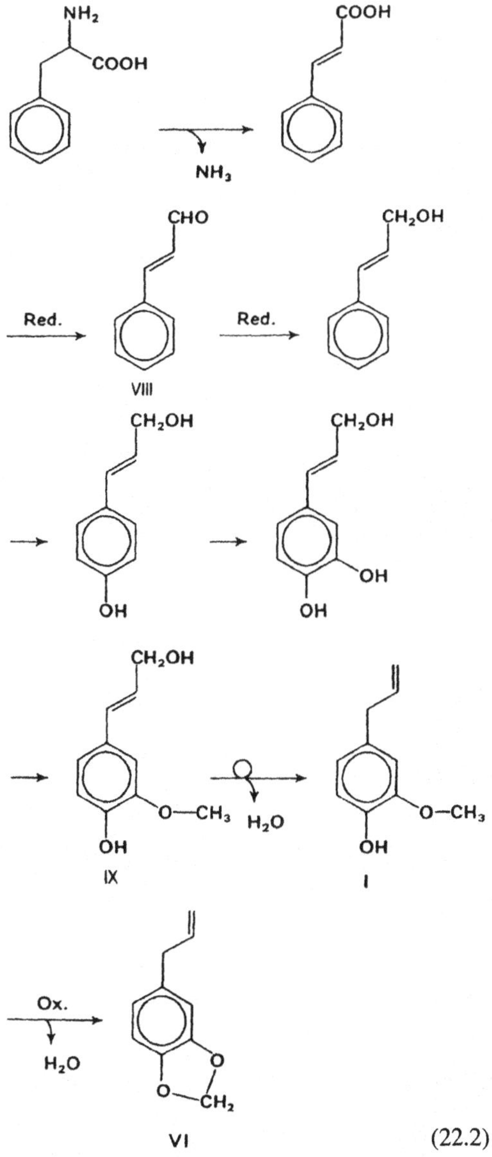

(22.2)

Table 22.2. Content of essential oils in some spices[a]

Spice	% Vol./Weight
Black pepper	2.0–4.5
White pepper	1.5–2.5
Aniseed	1.5–3.5
Caraway	2.7–7.5
Coriander	0.4–1.0
Dill	2.0–4.0
Nutmeg	6.5–15
Cardamom	4–10
Ginger	1–3
Turmeric	4–5
Marjoram	0.3–0.4
Origano	1.1
Rosemary	0.72
Sage	0.7–2.0

[a] For leaf spices, the values refer to the weight of the fresh material.

Some aromatic hydrocarbons are probably generated in spices by terpene oxidation. Examples are: 1-methyl-4-isopropenylbenzene (XI, Formula 22.3) derived from p-men-tha-1,3,8-triene (X) and (+)-ar-curcumene (XIV) from zingiberene (XII) or β-sesquiphel-landrene (XIII) [cf. Formula 22.4].

The formation of (+)-ar-curcumene from the above-mentioned precursor was detected during storage of ginger oil.

Another aromatic hydrocarbon present in significant amounts in essential oils of some spices (Table 22.3) is p-cymene (XV, Formula 22.3).

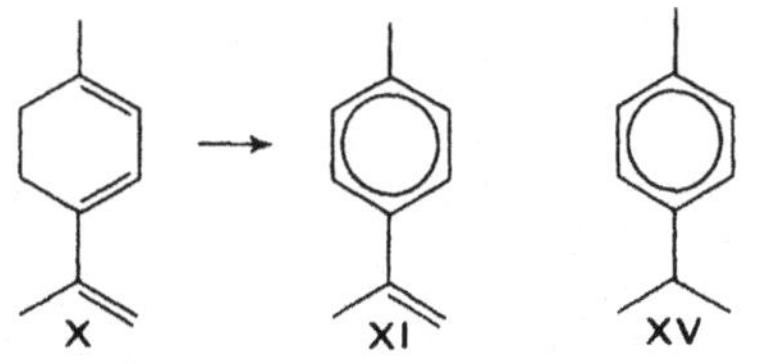

(22.3)

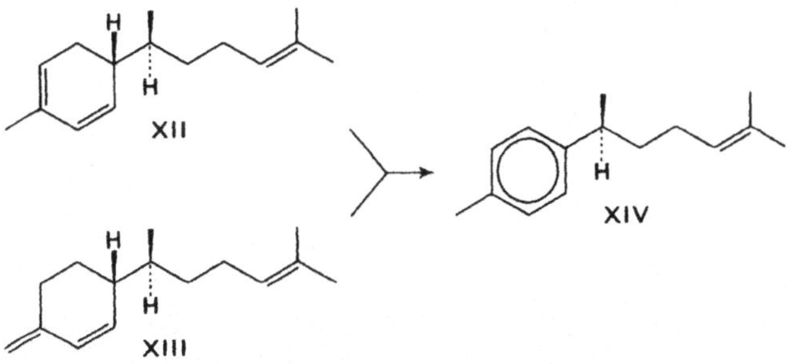

(22.4)

Table 22.3. Volatile compounds of spices[a]

Spice[b]	Components[c]
Pepper (1)	1–16% α-Pinene (XXIX*), 0.2–19% sabinene (XXV*), 9–30% β-caryophyllene (XLIX*), 0–20% Δ³-carene (XXXII*), 16–24% limonene (IX*), 5–14% β-pinene (XXX*)
Vanilla (2)	Vanillin (1.3–3.8%, dry matter), (R)(+)-trans-α-ionone, p-hydroxybenzylmethylether (XVII)
Allspice (3)	50–80% Eugenol (I), 4–7% β-caryophyllene (XLIX*), 3–28% methyleugenol, 1,8-cineole (XXIII*), α-phellandrene (XI*)
Bay leaf (5)	50–70%, 1,8-Cineole (XXIII*), α-pinene (XXIX*), β-pinene (XXX*), α-phellandrene (XI*), linalool (IV*)
Juniper berries (6)	36% α-Pinene (XXIX*), 13% myrcene (I*), β-pinene (XXX*), Δ³-carene (XXXII*)
Aniseed (7)	80–95% (E)-anethole (V)
Caraway (8)	55% (S)(+)-Carvone (XXI*), 44% limonene (IX*)
Coriander (9)	(S)(+)- and (R)(-)-Linalool (IV*), linalyl acetate, citral[d], 2-alkenales C₁₀–C₁₄
Dill (fruit, 10)	20–40% (S)(+)-Carvone (XXI*), 30–50% (R)(+)-limonene (IX*)
Dill (herb, 10)	70% (S)(+)-Phellandrene (XI*), 17% (3R,4S,8S)(+)-epoxy-p-menth-1-ene (XVIII), myristicin (VII), (R)-limonene (IX)
Fenugreek (11)	Linalool, 3-isobutyl-2-methoxypyrazine, 2-methoxy-3-isopropylpyrazine, 3-hydroxy-4,5-dimethyl-2(5H)-furanone (HD2F)
Nutmeg (13)	27% α-Pinene (XXIX*), 21% β-pinene (XXX*), 15% sabinene (XXV*), 9% limonene (IX*), 0.1–3.3% safrole (VI), 0.5–14% myristicin (VII), 1.5–4.2% 1,8-cineole (XXII*)2
Cardamom (14)	20–40%, 1,8-Cineole (XXIII*), 28–34% α-terpinyl acetate, 2–14% limonene (IX*), 3–5% sabinene (XXV*)
Clove (15)	73–85% Eugenol (I), 7–12% β-caryophyllene (XLIX*), 1.5–11% eugenol acetate
Saffron (16)	47% Safranal (XIV), 14% 2,6,6-trimethyl-4-hydroxy-1-cyclohexen-1-formaldehyde (XXIII)
Ginger (18)	30% (−)-Zingiberene (XLII*), 10–15% β-bisabolene (XLI*), 15–20% (−)-sesquiphellandrene (XLIII*), (+)-ar-curcumene (XIV), citral[c], citronellyl acetate
Tumeric (19)	30% Turmerone (XVIa), 25% ar-turmerone (XVIb), 25% zingiberene (XLII*)
Cinnamon (20)	50–80% Cinnamaldehyde (VIII), 10% eugenol (I), 0–11% safrole (VI), 10–15% linalool (IV*), camphor (XXXIII*)
Parsley (23)	p-Mentha-1,3,8-triene (X), myristicin (VII), 2-sec-butyl-3-methoxypyrazine, 2-isopropyl-3-methoxypyrazine, (Z)-6-decenal, (E,E)-2,4-decadienal, myrcene (I*)
Marjoram (26)	3–18% cis-Sabinenehydrate (XXVII*), 1–7% trans-sabinenehydrate, 16–36% 1-terpinen-4-ol
Origano (27)	60% Carvacrol (II), thymol (III)
Rosemary (28)	1,8-Cineole (XXIII*), camphor (XXXIII*), β-pinene (XXX*), camphene (XXXI*)
Sage (29)	1,8-Cineole (XXIII*), camphor (XXXIII*), thujone (XXVI*)
Thyme (31)	Thymol (III), p-cymene (XV), carvacrol (II), linalool (IV*)

[a] With the exception of vanillin and dill (herb), the quantitative values refer to the composition of the essential oil.

[b] The number in brackets refers to Table 22.1.

[c] Roman numerals with an asterisk refer to the chemical structures of the terpenes presented in Table 5.33. Roman numerals without an asterisk refer to chemical structures shown in Chapter 22.

[d] A mixture of neral and geranial (cf. footnote "b" in Table 5.33).

The concentrations given in Table 22.3 are guide values which can vary greatly depending on the variety and cultivation conditions.

22.1.1.2 Aroma Substances

In some spice plants, the odor corresponds with that of the main components of the volatile fraction. These include aniseed with (E)-anethole, caraway with (S)-carvone, clove with eugenol and cinnamon with cinnamalde-hyde (cf. Table 22.3). In the case of the following spice plants, further details about the important aroma substances are known.

22.1.1.2.1 Pepper

Black and white pepper are available commercially. Black pepper is harvested before it is fully ripe and then dried. After removal of the flesh, the seed of the ripe fruit gives white pepper, which has a milder aroma.

In the concentration range of 1 to 2 mg/kg (−)-rotundone (cf. 5.3.2.4) is the key odorant of black and white pepper. Further important odorants of black pepper are given in Table 22.4. White pepper contains the same typical aroma substances, but usually in lower concentrations.

The aroma of ground pepper is not stable due to losses of important aroma substances, the extent of which is shown in Fig. 22.1.

Table 22.4. Odorants in black pepper[a]

No.	Compound	Odor threshold (mg/kg)[b]	Concentration (mg/kg)
1	Methylpropanal	0.056	1.03
2	2-Methylbutanal	0.053	1.99
3	3-Methylbutanal	0.032	4.18
4a	(−)-α-Pinene	3.4	2070
4b	(+)-α-Pinene	2.1	486
5a	(−)-Sabinene	50	4470
5b	(+)-Sabinene	6.3	285
6a	(−)-β-Pinene	2.9	3950
6b	(+)-β-Pinene	2.1	298
7	Myrcene	1.9	870
8a	(R)-α-Phellandrene	1.4	227
8b	(S)-α-Phellandrene	1.1	1390
9a	(S)-Limonene	2.8	4000
9b	(R)-Limonene	1.8	3280
10	1,8-Cineol	0.084	22.4
11	(±)-Linalool	0.069	231
12	Butyric acid	0.10	1.28
13	2-/3-Methylbutyric acid		4.27

[a] Origin: India.
[b] Odor threshold on starch.

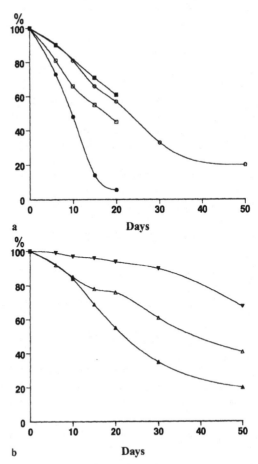

Fig. 22.1. Storage of ground black pepper at room temperature – changes in the concentrations of odorants. (a) (•—•) 3-methylbutanal, (o—o) α-pinene, (■—■) myrcene, (□—□) α-phellandrene (b) (▲—▲) limonene, (△—△) 1,8-cineol, (▼—▼) linalool

Musty/mouldy aroma defects in black pepper are caused by a mixture of 2,3-diethyl-5-methyl-pyrazine and 3-isopropyl-2-methoxy-pyrazine. Some samples of white pepper contain up to 2.5 mg/kg of skatole (odor threshold on starch: 0.23 µg/kg), which together with 3- and 4-methylphenol can cause a fecal aroma defect. This aroma defect arises during fermentation (degradation of amino acids, e. g., tryptophan → 3-methylindole), which is carried out to remove the flesh. On longer storage, this defect becomes more noticeable because intensive aroma substances, which disguise it in fresh white pepper, volatilize.

22.1.1.2.2 Vanilla

In the capsular fruit of vanilla, incorrectly called vanilla bean, 170 volatile compounds have been identified. However, the only fact that is certain is that apart from the main aroma substance vanillin, which is released from the glucoside on fermentation of the fruits, and (R)(+)-trans-α-ionone, the p-hydroxybenzyl-methylether (XVII) contributes to the aroma since its concentration (115–187 mg/kg) greatly exceeds the odor threshold (0.1 mg/kg, water). A mixture of 99% of sugar and 1% of ground vanilla is sold as vanilla sugar and a mixture of 98% of sugar and 2% of vanillin is sold as vanillin sugar.

22.1.1.2.3 Dill

AEDA and sensory investigations show that (S)-α-phellandrene in combination with (3S,3aS, 7aR)-3,6-dimethyl-2,3,3a,4,5,7a-hexahy-droben-zo[b]furan (XVIII, dill ether, cf. Formula 22.5) produce the aroma of dill. Both compounds are not stable and are largely lost on drying (Table 22.5).

The most important aroma substance in dill fruit is (S)-carvone, which smells of caraway. In fact, dill seeds were used as a substitute for caraway in the past.

22.1.1.2.4 Fenugreek

The most important odorants of seasoning (cf. 12.7.3.5) is 3-hydroxy-4,5-dimethyl-2(5H)-furanone (HD2F, XXI in Formula 22.6), 95% of which is present in the S form. This compound is also the outstanding odorant of fenugreek. Correspondingly, the seeds or the seed extract serves as the starting material for the production of seasoning. Other odorants are 3-amino-4,5-dimethyl-3,4-dihydro-2(5H)-furanone (XX in Formula 22.6), 1-octen-3-one, linalool and eugenol.

The concentration of HD2F varies in fenugreek between 3 and 12 mg/kg. However, this aroma substance is absent in *Trigonella* varieties.

If an extract of fenugreek is heated (100 °C, 60 min) at pH 2.4, there is a ca. 10 fold increase in the HD2F concentration. Under these conditions, the precursor (2S,3R,4S)-4-hydroxy-

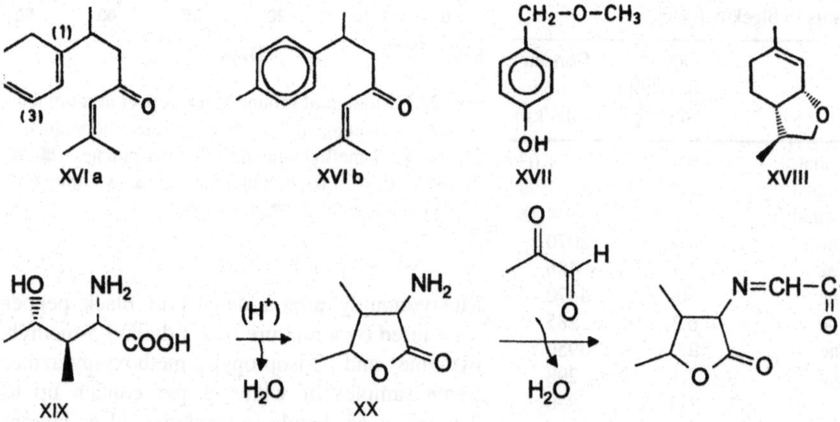

XVI a   XVI b   XVII   XVIII   (22.5)

XIX   XX

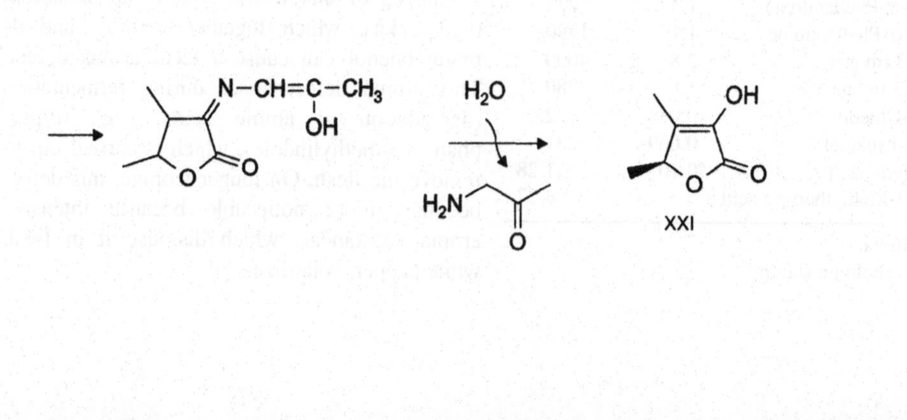

(22.6)

Table 22.5. Changes in aroma substances in the drying of dill (leaves)

| | Fresh | Dried (air) | | Freeze dried | |
		25 °C/4 h	50 °C/4 h	−25 °C/59 h	−25 °C/65 h
Water (w/w %)	90	11	12	16	2
Volatile compounds[a]	326	49	37	188	83
			Volatile compounds[a]		
α-Pinene	5.8	1.2	1.4	3.1	0.6
α-Phellandrene[b]	198.1	13.3	8.1	41.6	14.9
Limonene	10.0	0.7	0.4	2.0	0.7
β-Phellandrene	27.5	2.2	1.1	6.5	1.8
p-Cymene	5.5	1.1	0.4	4.0	0.1
3,9-Epoxy-p-ment-1-ene[b]	39.8	0.5	Traces	8.9	1.4
Myristicin	4.4	0.6	0.3	4.3	1.5
Neophytadiene	1.0	6.3	2.6	38.2	26.0

[a] Values in mg per 100 g of dry weight.
[b] Aroma substances that determine quality.

L-isoleucine (XIX in Formula 22.6) is cyclized to the amine XX, which is then converted to HD2F via the *Strecker* reaction, e. g., with methylglyoxal.

22.1.1.2.5 Saffron

In aroma extract dilution analyses (cf. 5.2.2), a compound with a saffron and hay-like odor, which could be 2-hydroxy-4,4,6-tri-methyl-2,5-cyclohexadien-1-one, gave the highest FD factor. This was followed by the terpene aldehyde safranal and an unknown compound, both of which have a saffron odor. Safranal (XXIV) is probably obtained from the bitter substance picrocrocin (XXII) by hydrolysis and elimination of water (Formula 22.6).

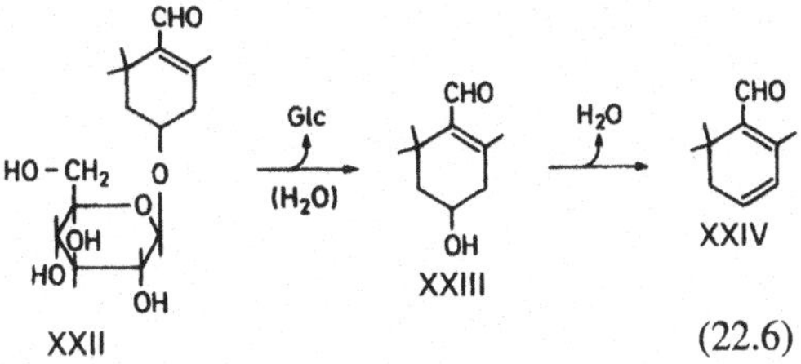

$$(22.6)$$

22.1.1.2.6 Mustard, Horseradish

Mustard and horseradish contain glucosinolates (Table 22.6) which, after cell rupture, are ex-

posed to the action of a *thioglucosidase* enzyme (cf. 17.1.2.6.5), yielding isothiocyanates (mustard oil). Allyl isothiocyanate is obtained from the glucoside sinigrin, a compound responsible for the pungent burning odor and taste of both spices. p-Hydroxybenzyl isothiocyanate obtained from sinalbin is only slightly volatile and contributes significantly to the sharp pungent taste of mustard.

The aroma of horseradish is also influenced by methyl, ethyl, isopropyl and 4-pentenyl iso-thiocyanates which, however, are present only in very small amounts in comparison to allyl isothiocyanate.

22.1.1.2.7 Ginger

The fresh ginger root has a citrus and camphor-like, flowery, musty, fatty and green odor. In a column chromatographic preliminary separation of an extract, the characteristic aroma substances appeared in the fraction of the oxidized hydrocarbons. The highest FD factors in dilution analyses were obtained for geraniol, linalool, geranial, citronellyl acetate, borneol, 1,8-cineol and neral.

22.1.1.2.8 Basil

The aroma profile of basil is characterized by green/fresh, flowery, clover- and pepper-

Table 22.6. The most important glucosinolates of mustard and horseradish

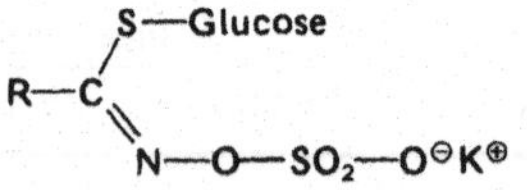

R	Name	Occurrence
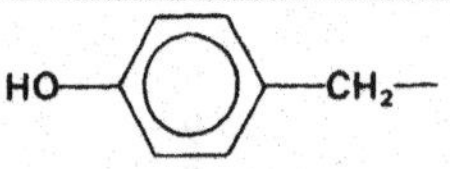 HO—⟨○⟩—CH₂—	Sinalbin	Mustard
H₂C=CH—CH₂—	Sinigrin	Mustard, horseradish
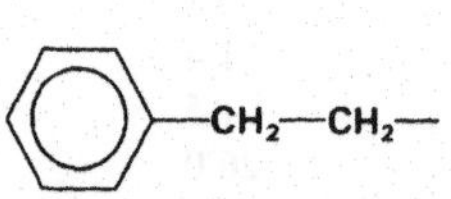 ⟨○⟩—CH₂—CH₂—	Gluconasturtiin	Horseradish

Table 22.7. Concentrations of odorants in fresh and dried basil

Compound	Concentration[a]		
	Fresh	Freeze-dried	Dried at 60 °C
(Z)-3-Hexenal	124	0.5	<0.01
1,8-Cineol	640	112	610
4-Mercapto-4-methylpentan-2-one	0.10	0.006	<0.01
Linalool	602	33	1210
4-Allyl-1,2-dimethoxybenzene	4950	1600	9540
Eugenol	890	214	391
3a,4,5,7a-Tetrahydro-3,6-dimethyl-benzofuran-2(3)-one (Winelactone)[b]	0.034	0.015	n.a.
Methylcinnamate	25	n.a.	n.a.
Estragol	12	n.a.	n.a.
α-Pineol	18	14.2	11.9
Decanal	0.39	n.a.	n.a.

[a] In mg/kg solids.
[b] cf. 5.2.5.
n.a: not analyzed.

like/spicy notes. The compounds given in Table 22.7 produce the aroma, which has been reproduced by a successful simulation. Omission experiments (cf. 5.2.7) show that eugenol, (Z)-3-hexenal, α-pinene, 4-mercapto-4-methylpentan-2-one, linalool and 1,8-cineol make the largest contributions to the aroma.

Drying damages the aroma considerably. (Z)-Hexenal and 4-mercapto-4-methylpentan-2-one are still detectable in freeze-dried basil (Table 22.7) and the green/fresh note is still perceptible. This note is absent in an air-dried sample, and the increase in linalool (Table 22.7), possibly through the enzymatic hydrolysis of the corresponding glycosides, causes the flowery note to become undesirably evident. The intensity of the pepper-like/spicy note also greatly decreases on drying.

22.1.1.2.9 Parsley

The most important odorants of parsley leaves are listed in Table 22.8. Sensory evaluations have shown that p-mentha-1,3,8-triene (X in Formula 22.3) and myrcene contribute to the characteristic aroma. (Z)-6-Decenal and (Z)-3-hexenal are responsible for the green notes. Myristicin, 2-sec-butyl-3-methoxypyrazine, (E,E)-2,4-decadienal, methanethiol and β-phellandrene also

Table 22.8. Concentrations of potent odorants in fresh and dried parsley[a]

Compound	Concentration[b]		
	Fresh		Dried[c]
	Cultivar I[d]	Cultivar II[d]	Cultivar II[e]
Methanthiol	1.2	0.972	0.067
Myrcene	83.6	133.8	135
2-Methoxy-3-isopropylpyrazine	0.007	0.01	n.a.
2-sec-Butyl-3-methoxypyrazine	0.036	0.056	0.085
Myristicin	269	991	2770
1-Octen-3-one	0.014	0.047	0.010
p-Mentha-1,3,8-triene	1829	313	1026
(Z)-3-Hexenal	0.93	1.378	0.139
(Z)-3-Hexenylacetate	0.763	0.328	n.a.
(Z)-1,5-Octadien-3-one	0.005	0.005	<0.001
(E,E)-2,4-Decadienal	4.9	4.7	0.27
(Z)-6-Decenal	27.4	16.5	2.75
Linalool	3.2	1.6	0.42
β-Phellandrene	949	1026	200
2-Methylbutanal	n.a.	n.a.	2.0
3-Methylbutanal	n.a.	n.a.	1.3
Methylpropanal	n.a.	n.a.	2.5
Dimethylsulfide	n.a.	n.a.	22.5
Methional	n.a.	n.a.	0.065
Acetaldehyde	n.a.	n.a.	8.0
Propanal	n.a.	n.a.	19.0
3-Methyl-2,4-nonandione	n.a.	n.a.	0.029

[a] Only those aroma substances which showed high FD factors in dilution analyses were quantified.
[b] In mg/kg based on solids.
[c] Dried at 70 °C (80 to 120 min), then stored for 3 months at −20 °C in nitrogen.
[d] Cultivar I: Hamburger cut, cultivar II: "Mooskrause".
[e] The fresh and dried cultivar II are of different origin.

exhibit high aroma values. The two cultivars of parsley compared in Table 22.8 differ considerably in the concentrations of some aroma substances, e. g., cultivar I contains 6 times more p-mentha-1,3,8-triene.

Drying of parsley on exposure to air leads to a large decrease in (Z)-3-hexenal and (Z)-6-decenal (Table 22.8), resulting in a reduction of the green note. In addition, sulfurous/cabbage-like and hay-like aroma defects appear due to the formation of dimethylsulfide and 3-methyl-2,4-nonandione. If drying proceeds at a higher temperature, methylpropanal, 2- and 3-methylbutanal, which do not play a role in the aroma of fresh parsley, also increase to such an extent that their malty aroma quality can assert itself in the aroma profile.

22.1.1.3 Substances with Pungent Taste

The hot, burning pungent taste of paprika (red pepper), pepper (black pepper) and ginger is caused by the nonvolatile compounds listed in Table 22.9.

Black pepper contains 3–8% of piperine (XXV) as the most important pungent substance. Pepper is sensitive to light since the trans,trans-diene system of piperine isomerizes to the cis,trans-diene system of the almost tasteless isochavicin on exposure to light.

In the processing and storage of ginger, gingerol easily dehydrates to shogaol, increasing the pungency (Table 22.9). A retroaldol cleavage of shogaol can also occur with the formation of sweet-spicy zingerone and hexanal (Formula 22.7). Above a certain concentra-

Table 22.9. Compounds present in spices causing a hot burning organoleptic perception

Name	Structure	Occurrence[a]	Relative pungency[b]
Piperine[c]	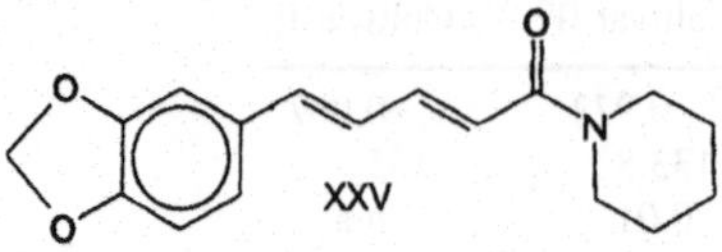 XXV	Pepper (1)	1.0
Piperanine	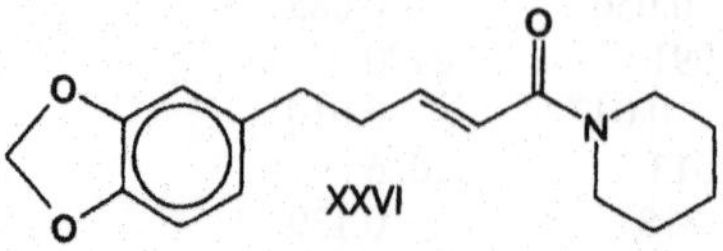 XXVI	Pepper (1)	0.5
Piperylin	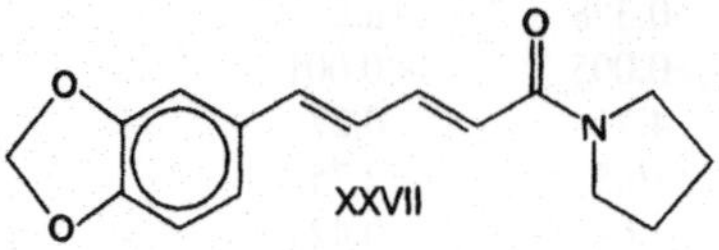 XXVII	Pepper (1)	0–1[d]
Gingerol	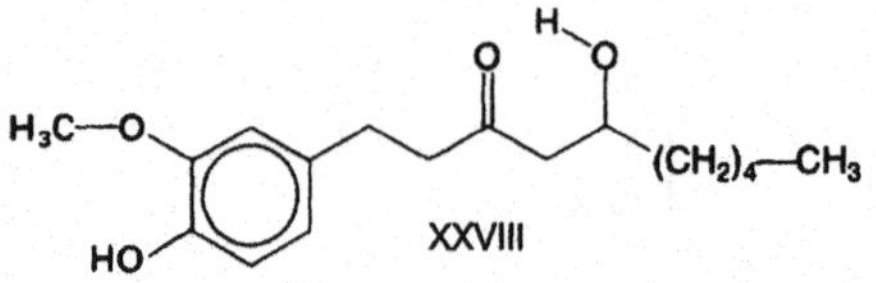 XXVIII	Ginger (17)	0.8
Shogaol	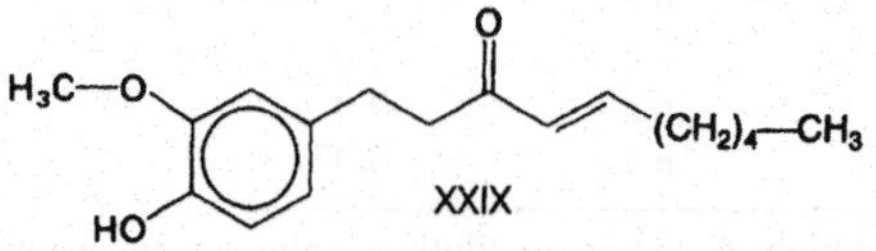 XXIX	Ginger (17)	1.6
Capsaicin	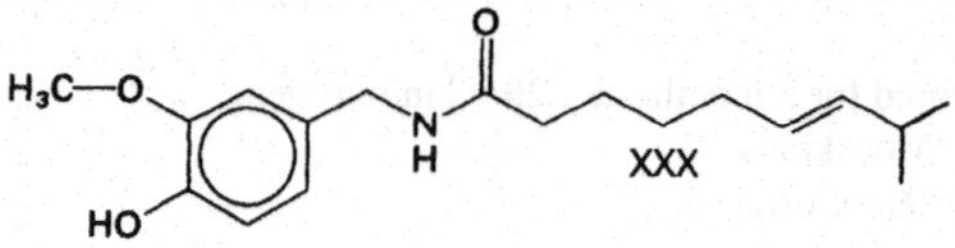 XXX	Capsicum (4; 7)	150–300[d]
Dihydro-capsain	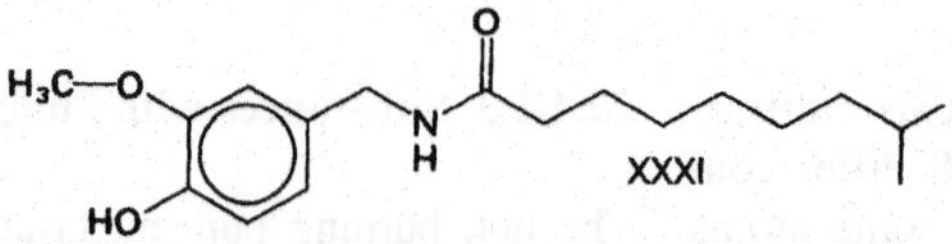 XXXI	Capsicum (4; 7)	like Capsaicin
Nordihydro-caicin	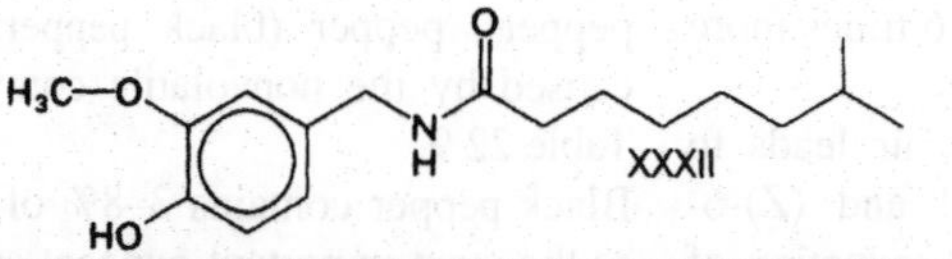 XXXII	Capsicum (4; 7)	75% Capsaicin

[a] The numerals in brackets refer to Table 22.1.
[b] Reference: pungency of piperine = 1.
[c] The corresponding *cis,trans*-compound is devoid of pungent taste.
[d] Literature data are within the range of values presented.

tion, hexanal causes an aroma defect in ginger oleoresins.

The concentration of the capsaicinoids XXX, XXXI, and XXXII (Table 22.9) in the fruits of capsicum or in various other pepper plants depends on the variety, cultivation, drying and storage conditions, and varies between 0.01 and 1.2%. These compounds are the most pungent spice constituents. Their concentrations are at the upper limit in chillies and tobasco varieties and at the lower limit in sweet varieties.

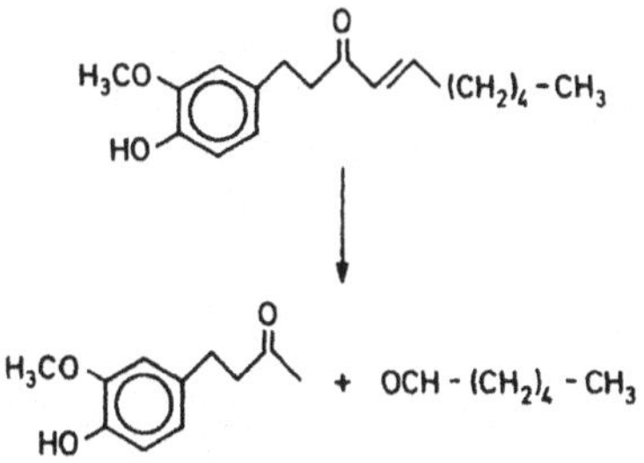

(22.7)

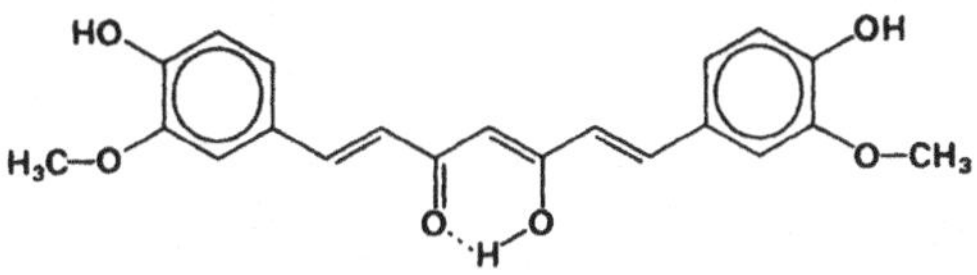

(22.8)

Investigations of the structure/effect relationship show that the intensity of the pungency does not change when 8-methyl-*trans*-6-nonenoic acid in capsaicin is replaced by nonanoic acid (9:0). However, it decreases when shorter, e.g., 8:0 (75%), 7:0 (25%), 6:0 (5%), or longer fatty acids, e.g., 10:0 (50%), 11:0 (25%), are introduced.

22.1.1.4 Pigments

Paprika (red pepper) and curcuma pigments are used as food colorants. Paprika pigments are carotenoids, with capsanthin as the main compound (cf. 3.8.4.1.2 and Fig. 3.47). Curcumin (cf. Formula 22.8) is the main pigment of curcuma, a tropical plant of the ginger family.

22.1.1.5 Antioxidants

Extracts of several spices, particularly of sage and rosemary, have the ability to prevent the autoxidation of unsaturated triacylglycerols. Among the most effective antioxidant constituents of both spices, the cyclic diterpene diphenols, carnosolic acid (XXXIII in formula 22.9) and carnosol (XXXIV) have been identified.

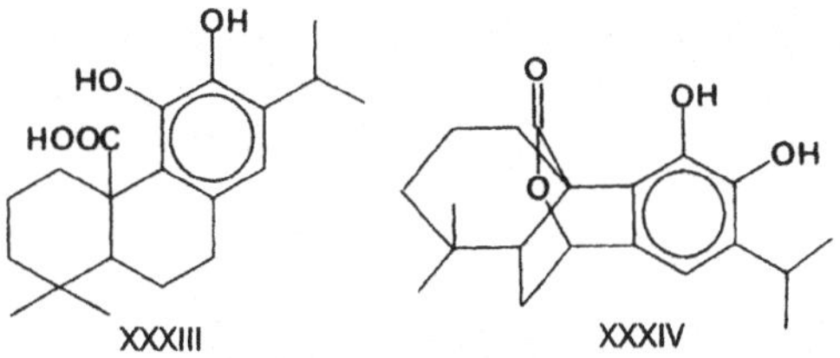

(22.9)

The high antioxidative activity of the two compounds is probably based on the fact that they are o-diphenols (cf. 3.7.3.2.1).

22.1.2 Products

22.1.2.1 Spice Powders

Spices are marketed unground or as coarsely or finely ground powders. The flavor is improved when the spices are ground using a cryogenic mill. After grinding the shelf life of the spices is limited. Favorable storage conditions are the absence of air, a relative humidity less than 60% and a temperature less than 20 °C. Crushed spices rapidly lose their aroma and absorb aromas from other sources. Leaf and herb spices are dried before they are crushed. The loss of aroma substances depends on the spice and on the drying conditions (for examples, see 22.1.2.3, 22.1.1.2.8 and 22.1.1.2.9). In comparison with air-drying at a raised temperature, there are no changes in aroma caused by the *Maillard* reaction in freeze-drying. However, gentle drying leads to increased hydrolysis of chlorophylls and to dehydration of the phytol released to phytadienes, e.g., to neophytadiene (7,11,15-trimethyl-3-methylene-1-hexadecene):

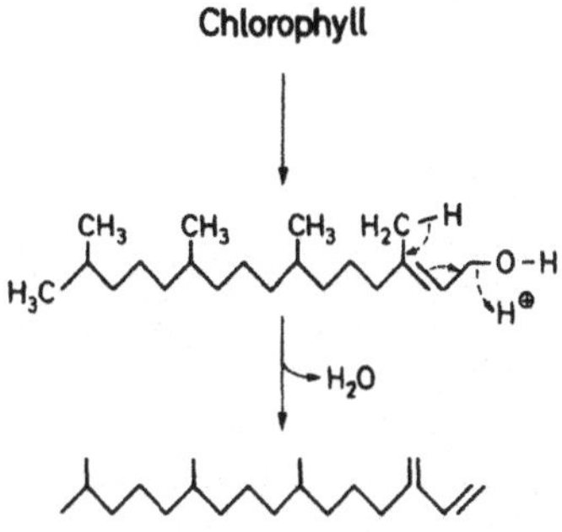

(22.10)

Contamination of spice powders with microorganisms is often very high, hence the addition of ground spices to food preparations may accelerate microbial food spoilage.

22.1.2.2 Spice Extracts or Concentrates (Oleoresins)

Spice extracts are being used in increasing amounts in industrial-scale food preparation since they are easier to handle than spice powders and are free of microorganisms. The production of these extracts is outlined in 5.5.1.2. The flavor quality depends on the solvent used and also on the raw material.

22.1.2.3 Blended Spices

Specially blended spices are offered commercially for some food preparations, such as liver sausage which uses a spice blend consisting of sweet marjoram, mace, nutmeg, cardamom, ginger, pepper and a little cinnamon.
Smoked, saveloy sausage spice blend consists of coriander, ginger, mustard kernels, paprika and pepper. Common spices for bread are aniseed, fennel and caraway. Gingerbread spice blend consists of aniseed, clove, coriander, cardamom, allspice and cinnamon.

22.1.2.4 Spice Preparations

Spice preparations are obtained by the addition of spices and blended spices to other substances, such as salt, sugar, glutamate, yeast extract and starch flour.

22.1.2.4.1 Curry Powder

A spice preparation containing a spice blend of turmeric as the main ingredient and paprika, chili, ginger, coriander, cardamom, clove, allspice and cinnamon, mixed together with up to 10% legume meal, starch and glucose, and with less than 5% salt.

22.1.2.4.2 Mustard Paste

A dark yellow paste used as a pungent seasoning for food. It is made from finely ground, often defatted mustard seeds, mixed into a slurry with water, vinegar, salt, oil and some other spices (pepper, clove, coriander, curcuma, ginger, paprika, etc.) and ground repeatedly or refined. During processing, lasting 1–4 h at a temperature not exceeding 60 °C, the mustard oil is released from its glucoside, as outlined in 22.1.1.2.6. "Extra strong" mustard is primarily made from dehulled black mustard seed, while the "medium hot" or "hot" types are made from seeds with hull, using varying proportions of black and white mustards.

22.1.2.4.3 Sambal

A spice preparation from Asia used for seasoning rice dishes. Its base is Sambal oelek, which consists mainly of crushed or pulverized saltpreserved chili.

22.2 Salt (Cooking Salt)

Common salt occupies a special position among the spices. Salt is used in greater amounts than all other spices to enhance the flavor and taste of food. Also, some foods are preserved when salted with large amounts of NaCl (cf. 0.3.1).
Humans require a certain constant level of intake of sodium and chloride ions to maintain their vital concentrations in plasma and extracellular fluids. The daily requirement is about 5 g of NaCl; an excessive intake is detrimental to health.
The salty taste is stimulated by ions. In comparison with the sour taste (cf. 8.10), the cation and the anion are significantly involved. The pure salt taste is only produced by NaCl. In fact, the very next chemical relative, KCl, has a sour/bitter aftertaste.

22.2.1 Composition

Common (cooking or kitchen) salt is nearly entirely NaCl. Impurities are moisture (up to 3%)

and other salts, not exceeding 2.5% (magnesium and calcium chloride; magnesium, calcium and sodium sulfates). Salt also contains trace elements.

22.2.2 Occurrence

Salt is abundant in sea water (2.7–3.7%) and in various landlocked seas (7.9% in the Dead Sea; 15.1% in the Great Salt Lake in Utah) and also in salt springs (Lueneburg, Reichenhall) and, above all, in salt beds formed in various geological periods, e.g., the European Zechstein salt deposits.

22.2.3 Production

In Germany salt is mainly mined as rock salt. It is selected, crushed and finely ground. Salt springs are also an important source. Saturated brine is recovered by tapping underground brine springs or by dissolving the salt out of beds with freshwater. For purification, magnesium is first eliminated as the hydroxide with lime milk and then calcium is removed as calcium carbonate with soda. Gypsiferous brine is treated with sodium sulfate containing mother liquor from the evaporation process. Evaporative crystallization occurs in multistage systems at 50–150 °C. The salt is centrifuged and dried. Salt obtained in such a manner is called "boiling" salt.

In warm countries sea water is concentrated in shallow flat basins by the sun, heat and wind until it crystallizes ("solar salt").

The addition of 0.25–2.0% calcium or magnesium carbonate, calcium silicate, or silicic acid improves the flowability. Indeed, 20 ppm of potassium ferrocyanide prevents the formation of lumps in the salt. The latter compound modifies the crystallization process of NaCl during the evaporation of salt spring water. In the presence of potassium ferrocyanide, the salt builds dendrites, which have strongly reduced volume, density and inclination to agglomerate.

In 1975 the worldwide production of NaCl was 162.2×10^6 t and 240×10^6 t in 2006. In 1974 only 5% of the NaCl produced in FR Germany was used for consumption; the remainder, 95%, was used in industry or trade (raw materials, salt for regeneration of ion-exchange resins, etc.).

22.2.4 Special Salt

Iodized salt is produced as a preventive measure against goiter, a disease of the thyroid gland (cf. 17.1.2.9.3). It contains 5 mg/kg of sodium-, potassium- or calcium iodide.

Nitrite salts are used for pickling and dry curing of meat (cf. 12.6.2.4). They consist of common salt and sodium nitrite (0.4–0.5%), with or without additional potassium nitrate.

22.2.5 Salt Substitutes

Some human diseases make it necessary to avoid excessive intake of sodium ions, so attempts have been made to eliminate the use of added salt as a spice or flavoring, without attempting to achieve completely salt-free nutrition. This "low salt" nutrition is actually only related to reduced sodium levels, hence a "low sodium" diet is a more relevant designation.

The compounds listed in Table 22.10 are used as salt substitutes. Their blends are marketed as "diet salts". Peptide hydrochlorides with a salty taste are discussed in Section 1.3.3.

22.3 Vinegar

Vinegar was known in old Oriental civilizations and was used as a poor man's drink and later as a remedy in ancient Greece and Rome. Vinegar is the most important single flavoring used to

Table 22.10. Substitutes for common salt

Potassium, calcium and magnesium salts of adipic, succinic, glutamic, carbonic, lactic, hydrochloric, tartaric and citric acids;
Monopotassium phosphate, adipic and glutamic acids and potassium sulfate;
Choline salt of acetic, carbonic, lactic, hydrochloric, tartaric and citric acids;
Potassium salt of guanylic and inosinic acids

provide or enhance the sour, acidic taste of food (cf. 8.12.5).

22.3.1 Production

Vinegar is produced microbiologically from ethanol or by dilution of acetic acid.

$$CH_3CH_2OH + O_2$$
$$\longrightarrow \quad CH_3COOH + H_2O + 494\,kJ \quad (22.11)$$

22.3.1.1 Microbiological Production

Acetobacter species are cultivated in aqueous ethanol solution or, to a lesser extent, in wine, fermented apple juice, malt mash or fermented whey. Ethanol, as shown in Fig. 22.2, is dehydrogenated stepwise to acetic acid; the resulting reduced form of the cosubstrate methoxatin ($PQQH_2$) is oxidized via the respiratory chain. Part of the energy formed by oxidation is released as heat which has to be removed by cooling during the processing of vinegar. If there is an insufficient supply of oxygen, the microorganisms disproportionate a proportion of the acetaldehyde, the intermediate compound (cf. Fig. 22.2) in this aerobic reaction pathway:

$$2CH_3CHO \longrightarrow CH_3COOH + CH_3CH_2OH \quad (22.12)$$

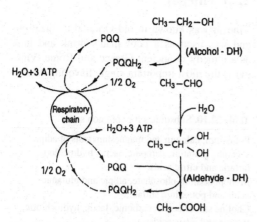

Fig. 22.2. Oxidation of ethanol to acetic acid by *Acetobacter* species (according to *Rehm*, 1980)

Fermentation of ethanol is conducted as a top fermentation and increasingly as a submerged oxidative process. In top fermentation the bacteria are cultivated on spongy, porous laminated carriers (usually beechwood shavings) with the alcoholic solution trickling down over carrier surfaces while a plentiful supply of air is provided from below. The fermentation is stopped at a 0.3% by volume residual ethanol level to avoid overoxidation, i.e., oxidation of acetic acid to CO_2 and water.

22.3.1.2 Chemical Synthesis

Acetic acid is usually synthesized by catalytic oxidation of acetaldehyde:

$$CH_3CHO + \tfrac{1}{2}O_2 \xrightarrow{\text{cat.}} CH_3COOH \quad (22.13)$$

Acetaldehyde is obtained by the catalytic hydration of acetylene or by the catalytic dehydrogenation of ethanol. Formic acid and formaldehyde are by-products of acetic acid synthesis. They are removed by distillation. Chemically pure acetic acid is diluted with water to 60–80% by volume to obtain the vinegar essence. The essence is a strongly corrosive liquid and is sold with special precautions. It is diluted further with water for production of food grade vinegar.

22.3.2 Composition

There are 5–15.5 g acetic acid in 100 g of vinegar. The blending (or adulteration) of fermented vinegar with synthetic acid can be detected by mass spectrometric determination of the $^{13}C/^{12}C$-isotope ratio (cf. 18.4.3); fermented vinegar has 5‰ more ^{13}C isotope than acetic acid synthesized petrochemically. In addition fermented vinegar can be distinguished from synthetic vinegar by analyzing the accompanying compounds. With this method fermented vinegars of different origin can also be distinguished from each other; e.g. spirit vinegar (fermented from aqueous ethanol) from wine, apple, malt and/or whey vinegar. The fermented vinegars contain metabolic by-products of *Acetobacter* strains, such as amino acids, 2,3-butylene glycol and acetyl methyl carbinol, in addition to substances derived from the raw materials used in vinegar production.

22.4 References

Blank, I. Lin, J., Devaud, S., Fumeaux, R., Fay, L. B.: The principal flavor components of fenugreek (*Trigonella foenum-graecum L.*) ACS Symposium Ser. *660*, 12 (1996)

Boelens, M.H., Richard, H.M.J.: Spieces and condiments I and II. In: Volatile compounds in foods and beverages (Ed.: Maarse, H.), Marcel Dekker, Inc.: New York. 1991

Chadwallader, K.R., Baek, H.H., Cai, M.: Characterization of saffron flavor by aroma extract dilution analysis. ACS Symposium Ser. *660*, 66 (1996)

Ebner, H., Follmann, H.: Acetic acid. In: Biotechnology (Eds.: Rehm, H.-J., Reed, G.), Vol. 3, p. 387, Verlag Chemie: Weinheim. 1983

Gerhardt, U.: Gewürze in der Lebensmittelindustrie. Eigenschaften, Technologien, Verwendung. B. Behr's Verlag: Hamburg. 1990

Gottschalk, G.: Bacterial Metabolism. 2nd edn., Springer-Verlag: Heidelberg. 1985

Guth, H., Murgoci, A.-M.: Identification of the key odorants of basil (*Ocimum basilicum L.*) – Effect of different drying procedures on the overall flavour. In: Flavour Perception, Aroma Evaluation (Eds. H.-P. Kruse, M. Rothe) Universität Potsdam, 1997, p. 232

Hall, G., Siewek, F., Gerhart, U.: Handbuch Aromen und Gewürze, Behr's Verlag, Hamburg, 1999

Huopalahti, R., Kesälahti, E., Linko, R.: Effect of hot air and freeze drying on the volatile compounds of dill (*Anethum graveolens L.*) herb. J. Agric. Sci. Finland *57*, 133 (1985)

Jagella, T., Grosch, W.: Flavour and off-flavour compounds of black and white pepper. Eur. Food Res. Technol. *209*, 27 (1999)

Maga, J.A.: Capsicum. Crit. Rev. Food Sci. Nutr. *6*, 177 (1975)

Masanetz, C., Grosch, W.: Hay-like off-flavour of dry parsley. Z. Lebensm. Unters. Forsch. *206*, 114 (1998)

Masanetz, C., Grosch, W.: Key odorants of parsley leaves (*Petroselinum crispum* [Mill.] *Nym. ssp. crispum*) by odour activity values. Flavour Fragrance J. *13*, 115 (1998)

Nishimura, O.: Identification of the characteristic odorants in fresh rhizomes of ginger (*Zingiber officinale* Roscoe) using aroma extract dilution analysis and modified multidimensional gas chromatography-mass spectrometry. J. Agric. Food Chem. *43*, 2941 (1995)

Rehm, H.-J.: Industrielle Mikrobiologie. 2nd. edn., Springer-Verlag: Berlin. 1980

Risch, S.J., Ho, C.-T.: Spices. Flavor chemistry and antioxidant properties. ACS Symposium Ser. 660, American Chemical Society, Washington DC, 1996

Salzer, U.-J.: The analysis of essential oils and extracts (oleoresins) from seasonings – a critical review. Crit. Rev. Food. Sci. Nutr. *9*, 345 (1977)

Sampathu, S.R., Shivashankar, S., Lewis, Y.S.: Saffron (*Crocus sativus* Linn.) – Cultivation, processing, chemistry and standardization. Crit. Rev. Food Sci. Nutr. *20*, 123 (1984)

Schmid, E.R., Fogy, I., Schwarz, P.: Beitrag zur Unterscheidung von Gärungsessig und synthetischem Säureessig durch die massenspektrometrische Bestimmung des $^{13}C/^{12}C$-Isotopenverhältnisses. Z. Lebensm. Unters. Forsch. *166*, 89 (1978)

Siewek, F.: Exotische Gewürze. Herkunft, Verwendung, Inhaltsstoffe. Birkhäuser Verlag: Basel. 1990

Steinhaus, M., Schieberle, P.: Role of the fermentation process in off-odorant formation in white pepper: on-site trial in Thailand. J. Agric. Food Chem. *53*, 6056 (2005)

Wijesekera, R.O.B.: The chemistry and technology of cinnamon. Crit. Rev. Food Sci. Nutr. *10*, 1 (1978)

23 Drinking Water, Mineral and Table Water

23.1 Drinking Water

Drinking water should be clear, cool, colorless and odorless, free from pathogens (low in microorganisms), perfect with regard to taste, cause no materials corrosion, and contain soluble substances only in narrow limits and minerals normally in concentrations of less than 1 g/l. In individual countries, criteria have been defined by law for the quality of drinking water, especially limiting values for microorganisms and contamination. As an example, limiting values stipulated in the German decree on drinking water are presented in Table 23.1.

Drinking water is recovered from springs, groundwater, and surface water. In sparsely populated areas, springs and brooks provide water that can be used without further pretreatment. Frequently, however, the water available does not fulfil the requirements and must be laboriously purified.

In dry areas, drinking water is obtained by desalting brackish or sea water. The usual processes applied are reverse osmosis with the use of semipermeable membranes for slightly saline brackish water and multistage evaporation, mainly as flash evaporation, for sea water.

23.1.1 Treatment

To remove suspended particles, the water is first filtered through gravel and sand layers of different grain size. Humic acids, which may color water yellow to brown, are flocculated with aluminium sulfate. After clarification, the quality of the water is improved still further, if required, by the application of the following processes.

Water should not contain more than 0.2 mg/l of iron, which is present as the bicarbonate, and 0.05 mg/l of manganese (Table 23.1). The iron can be eliminated as iron (III) hydroxide by aeration. In this process, manganese also precipitates as MnO_2 if the pH is higher than 8.5. Biological processes have also been developed for deferrization and demanganization. Free carbonic acid must be removed because it attacks pipes. The deacidification process applied depends on the hardness of the water and on the concentration of free carbonic acid. The usual process involves aeration and filtration through carbonate rock (e. g., marble or magnesite).

The disinfection of water is mostly achieved by chlorination or ozonation. At a pH of 6–8, the chlorine gas passed into the water forms practically only HClO and ClO$^-$ which, together with the dissolved Cl_2, are expressed as free chlorine. In the case of superchlorination for the killing of very resistant microorganisms, the excess chlorine (>0.1 mg/l of free chlorine) must be withdrawn with the help of SO_2, Na_2SO_3, $Na_2S_2O_3$ and filtration through calcium sulfite or coal. Disinfection with ozone has the advantage that due to its decomposition into oxygen, no chemicals remain in the water. Interfering odor- and taste-active substances are eliminated by filtration through activated carbon.

Overly high conventrations of nitrate (limiting value in Table 23.1) can be reduced by bacterial denitrification, ion exchange, or reverse osmosis. The fluoridation of drinking water is discussed in 7.3.2.10.

23.1.2 Hardness

The total water hardness refers to the total concentration of alkaline earths calcium and magnesium in mmol/l. The concentrations of strontium and barium, which are usually very low, are not considered. The following is valid for conversion to German degress of hardness (°d): 1 mmol/l hardness = 5.61 °d. Factors for conversion to the degree of hardness of other countries are given in Table 23.2.

H.-D. Belitz · W. Grosch · P. Schieberle, *Food Chemistry*
© Springer 2009

Table 23.1. Chemical and physical analysis of drinking water

Parameter	Limiting value[a]
General values to be measured	
Temperature	25 °C
pH Value	6.5–9.5
Electrical conductivity at 25 °C	2000 μS · cm^{-1}
Oxidizability[b]	5 mg O$_2$/l
Hardness	–[c]
Individual Constituents	mg/l
Sodium	150
Potassium	12
Calcium	–[c]
Magnesium	50
Iron	0.2
Manganese	0.05
Aluminium	0.2
Ammonium	0.5
Silver	0.01
Sulfate	240
Arsenic	0.04
Lead	0.04
Cadmium	0.005
Chromium	0.05
Nickel	0.05
Mercury	0.001
Cyanide	0.05
Fluoride	1.5
Nitrate	50
Nitrite	0.1
Polycyclic aromatic hydrocarbons, calculated as carbon	0.0002
Chlorine-containing solvents, sum of 1,1,1-trichloroethane, trichloroethylene, tetrachloro-ethylene, dichloromethane	0.025
Carbon tetrachloride	0.003
Pesticides, biphenyls, terphenyls	0.0001[d]
Surfactants	0.2

[a] The limiting values have been taken from the decree on drinking water, Dec. 5, 1990 (BGBL. I. p. 2612)/Jan. 23, 1991 (BGBL. I. p. 277).
[b] Organic substances are detected on the whole by oxidation, e. g., with permanganate.
[c] No limiting value required.
[d] Per individual substance.

The assessment of water involves an evaluation in accordance with the steps of hardness presented in Table 23.3.

Table 23.2. Conversion factors for degrees of hardness

Value	Alkaline earth metal ions (mmol/l)
Hardness[a]	1.00
1 German degree of hardness (°d)	0.18
1 English degree of hardness (°e)	0.14
1 French degree of hardness (°f)	0.10
1 USA degree of hardness (°US)[b]	0.01

[a] Hardness is now expressed as the concentration of the amount of substance (mmol/l). The following correspond: 1 mg/l Ca$^{2\oplus}$ = 0.025 mmol/l; 1 mg/l Mg$^{2\oplus}$ = 0.041 mmol/l.
[b] 1°US = 1 ppm CaCO$_3$.

Table 23.3. Classification in steps of hardness

Step	Range of hardness (mmol/l)	Degree of hardness (°d)	Characteristics
1	<1.3	<7	Soft
2	1.3–2.5	7–14	Medium-hard
3	2.5–3.8	14–21	Hard
4	>3.8	>21	Very hard

On heating, the hydrogen carbonates dissolved in water are converted to carbonates. On boiling, a part of the calcium salts precipitates out as slightly soluble CaCO$_3$. This part of the hardness is called carbonate hardness.

23.1.3 Analysis

The extent and frequency of the analysis of drinking water are regulated by law in many countries. Apart from monitoring the hygienic state of the water resources and of the treated drinking water, maintenance of limiting values is controlled. The data given in Table 23.1 show that extensive analysis of drinking water is a very laborious process. The question of whether the drinking water supply is possibly endangered by drug residues has risen recently. In spot checks, the concentrations of persistent drugs, e. g., chlofibrinic acid, detected in drinking water have been far below the human therapeutic activity threshold. From a hygienic viewpoint, however, this situation is not tolerable in the long run.

Table 23.4. Classification of mineral water

Description	Requirement
With low mineral content	Solid residue = mineral matter content ≤ 500 mg/l
With very low mineral content	Solid residue ≤ 50 mg/l
With high mineral content	Solid residue > 1500 mg/l
Bicarbonate containing	Hydrogen carbonate > 600 mg/l
Sulfate containing	Sulfate > 200 mg/l
Chloride containing	Chloride > 200 mg/l
Calcium containing	Calcium > 150 mg/l
Magnesium containing	Magnesium > 50 mg/l
Fluoride containing	Fluoride > 1 mg/l
Iron containing	Divalent iron > 1 mg/l
Sodium containing	Sodium > 200 mg/l
Suitable for preparation of infant food	Sodium ≤ 20 mg/l, nitrate ≤ 10 mg/l, nitrite ≤ 0.02 mg/l fluoride ≤ 1.5 mg/l
Suitable for low-sodium nutrition	Sodium < 20 mg/l
"Säuerling"	Carbon dioxide of natural origin > 250 mg/l

23.2 Mineral Water

Mineral water comes from a hygienically fault-less spring that is protected from contamination. It has a nutritional and physiological effect due to its mineral content.

In many countries, the recovery and composition of mineral water are controlled by the state and only a few processes for quality improvement are permitted. These are: separation of iron and sulfur compounds, complete or partial removal of free carbonic acid, and addition of carbon dioxide. Mineral water is bottled directly at the place of the spring. With regard to the heavy metal content and possible contamination, limiting values have been stipulated by law. The classification of mineral water is presented in Table 23.4.

In Germany, water used for therapeutic purposes (medicinal waters), because of its chemical composition, is subject to the law governing the manufacture and prescription of drugs.

23.3 Table Water

Table water is made from mineral water, drinking water, and/or sea water by using $NaCl$, $CaCl_2$, Na_2CO_3, $NaHCO_3$, $CaCO_3$, $MgCO_3$, and CO_2. If it contains at least 570 mg/l of $NaHCO_3$ and carbon dioxide, it can be called soda water. Selters is a soda water that comes from Selters on the Lahn.

23.4 References

Heberer, T., Stan, H.-J.: Arzneimittelrückstände im aquatischen System. Wasser & Boden 50(4), 20 (1998)

Höll, K.: Wasser, Walter de Gruyter, Berlin, 1979

Quentin, K.-E.: Trinkwasser. Springer-Verlag: Berlin. 1988

Weingärtner, H. et al.: Water. In: Ullmann's encyclopedia of industrial chemistry. 5th Edition, Volume A28, p. 1, VCH Verlag, Weinheim, 1996

Index

Page number in italics: formula
Page number (F): figure
Page number (T): table

Abhexon (see furanone),
  5-ethyl-3-hydroxy-4-methyl-2(5H)-
Absinthe 935
Absolue 394
Abundance ratio, isotope 858, 859(T)
Acaricide, chemical structure 477(F)
–, trade name and activity 480(T)
Acceptable Daily Intake (ADI) 467, 468
Acesulfame 440, 440(T)
–, taste, synergistic effect 433, 435(F)
Acetaldehyde, sensory property 204(T)
–, wine 922(T)
Acetem 461(T)
Acetic acid 443
–, breakdown of monosaccharide 267, 268
–, cacao bean 963
–, formation in sour dough 724(F)
–, production, microbiological 984, 984(F)
–, synthesis 984, 984
Acetic acid isopentyl ester, aroma substance 837
Acetoin, biosynthesis 378(F)
Acetone, fruit 846, 846
Acetyl-1,4,5,6-tetrahydropyridine, 2-
–, wine defect 925(T), 926
Acetyl-1-pyrroline, 2-
–, aroma quality 340(T)
–, aromatic rice 710
–, bread aroma 734, 735, 735, 736(T)
–, bread toasting 737, 738(F)
Acetyl-2-thiazoline, 2-
–, formation 368, 368(F)
–, –, time course 369(F)
–, sensory property 367(T)
Acetylamino acid, N- 017
Acetylcholine esterase, active serine, mechanism 115
–, rate constant 120(T)
Acetylene, fruit ripening 847
Acetylformoin, formation 277, 278, 282, 282
Acetylfuran, 2- 264
Acetylglycosyl halogenide 289, 290
Acetylimidazole, reaction with tyrosine 070
Acetylmethionine, N- 017
Acetyltetrahydropyridine, 2-
–, formation, bread toasting 737, 738(F)
Acetylthreonine, N- 017

Acid, additives 443
–, cacao bean 963
–, fruit 815(T)
–, fruit juice 853(T)
–, honey 888
–, jam 852(T)
–, organic, fruit 820, 820(T)
–, –, vegetable 787, 787(T)
–, rye dough 722
–, volatile, formation by heating fat 221(T)
–, wine 920
Acid generator, glucono-δ-lactone 448
Acidity, wheat flour 711
Acorn coffee 950
Acrylamide, formation 025, 025, 026
–, occurrence 490, 491(F), 491(T)
Actin 570
–, amino acid composition 578(T)
–, fibrillar 570, 572(F)
–, globular 570, 572(F)
Actinine 019(T)
Actinins, α-,β-,γ- 571(T)
Activation energy, chemical reaction 132
–, enzyme 131, 131
–, enzyme catalysis 132, 134(F)
–, killing of microorganism 132
–, protein denaturation 057(T), 058, 132
Activation entropy, protein denaturation 058, 058(T)
Activator, enzyme 124
Active site, enzyme 106
Actomyosin 573
Acyl lipid, building block 158, 158(T)
–, unsaturated, peroxidation 191
Acylglycerol 169
Acylhydrolase, potato, specificity 191(T)
Add-back process 800
Added protein, meat, detection 613
Added water, meat 613
Additive, food 429
–, utilization 429(T)
Additive effect, aroma substance 341
Adenine 047, 143
Adenosine triphosphate (ATP) 102, 102
Adhumulon 886(T), 895, 895
ADI utilization 486, 486(T)

ADI-value 467, 468
Adipic acid 447, *447*
Adjunct, beer production 892
Adlupulon 886(T), 895, *895*
ADP-deaminase, meat 573
Adulteration, fruit product, detection 857,
    857–859(T)
AEDA 350, 351(F)
–, coffee aroma 944, 944(T)
Aerosol 456(T)
Aesculetin 824(T)
Affination, production of sugar 872
Aflatoxin 472, *473*, 473, 474(T)
Agar 302, *302*
Agaritin 783
Agaropectin 303
Agarose 303
Aggregated dispersion 062, 063
Aging, meat 589, 590(T)
Alanine 009, *010*
Alanine, β- 039
Albumen (see also egg white) 548
Albumin, blood plasma 594
–, cereal 677, 678(T)
–, legume 746, 749(T)
Alcohol, beer aroma 901
–, consumption 929, 930(T)
–, denaturation 931
–, energy value 892
–, enzymatic formation 379
–, higher, wine 920
–, production 930
–, rectification 930
–, substrate for alcohol dehydrogenase 132(T)
–, technical 930
–, wine aroma 921
–, wine quality 915, 915(T)
Alcohol (see also ethanol)
Alcohol dehydrogenase, activation energy 132(T)
–, formation of volatile alcohol 379, *379*
–, rate constant 120(T)
–, reaction 139(T)
–, stereospecificity 099, 108, 109(F)
–, substrate 132(T)
–, substrate binding 122
–, zinc 099, *099*, 105
Alcoholic beverage 892
Alcoholic fermentation, scheme 893(F)
Alcopop 936
Aldehyde, *Strecker* reaction, sensory property 360(T)
–, autoxidation 205, 206(F)
–, formation by heating fat 222, *222*, 222(T)
–, formation by lipoxygenase 209
–, formation during cheese ripening 534, *534*
–, odor threshold in air 399(T)
–, reaction with protein 213(F), 214

–, sensory property 204(T)
–, stability 397, 397(T)
–, structure and odor 399, 399(T)
–, substrate for alcohol dehydrogenase 132(T)
Aldehyde dehydrogenase 149, 150(T)
–, *Michaelis* constant 149, 150(T)
Aldol condensation 268
Aldolase, active lysine 108
–, mechanism 116, 116(F)
Aldonitrile acetate 290, *290*, 291
Aldose, cyclization 249, *249*
–, equilibrium 255(T)
–, family tree 251(F)
–, occurrence 250(T)
–, rearrangement to ketose 266, *266*
Aldosylamine 271
Aleurone protein, wheat 671
Alfa process, churning 526
Algin 301(T)
Alginate 303, *303*
–, conformation *297*
–, viscosity 304(F), 308(T)
Alginate gel, egg box, 304(F)
–, thermo-reversibility, calcium 304
Alitame 442, *442*
Alkali-baked product 448
Alkane, formation by heating fat 221(T)
–, radiolysis 225(F)
Alkohol, formation by heating fat 221(T)
Alkohol dehydrogenase, mechanism 099
Alkoxy radical, formation 192, 192(F), *199*, 211,
    213(F)
Alkoxylipid 187
Alkyl cellulose 329, *329*
–, gelling behavior 329(F)
Alkylating agent, cysteine 024, *024*
Alkylcysteine sulfoxide, S- *781*
–, biosynthesis 781, 790
Alkylthiosulfonate 791, *791*
Allene oxide, formation 210, *210*
Allene oxide cyclase 210, *210*
Allene oxide synthetase (AOS) 210, *210*
Allene, oxide, reaction 210, *210*
Allergen, food 751, 754(T)
–, reaction 753
–, thermal stability 753, 754(T)
Allicin 791
Alliin 791, *791*
Allis shad 621
Allitol 261
Alloisoleucine 072, *072*, 073(F)
Allose *251*
Allosteric effector 125
Allosteric inhibitor 125
Allosterically regulated enzyme 123
Allspice, essential oil, chemical composition 974(T)

Allyl phenoxyacetate, structure, sensory property 396(T)
Allylcysteine sulfoxide 791, *791*
–, biosynthesis *791*
Allylisothiocyanate, odor threshold value 792
Almond, bitter, hydrocyanic acid 761(T)
–, chemical composition 816(T)
–, debittering 881
Almond oil, tocopherol content 234(T)
Altbier 904
Altritol 261
Altrose *251*
Aluminium 427
Alveogram 714(F)
Alveograph 714, 714(F)
Amadori compound, occurrence 271
–, reaction 272, *272, 273*
–, stability 272
Amadori rearrangement *270*, 271
Amidation, protein 067
Amine, fish 626, 627(F)
–, formation 373
–, formation during cheese ripening 534, *534*
–, formation in fruit 812, *812, 813*
–, fruit 812, 817(T)
–, occurrence 495(T)
–, precursor 374(T)
–, product of *Strecker* reaction 373, 374(T)
–, volatile 373
Amino acetophenone, o-
–, aroma quality 539
–, occurrence 539
–, odor threshold value 539
Amino acid, acylation 016
–, alkylation 018
–, analysis 019
–, arylation 018
–, bitter taste 034, 035(T)
–, cacao bean 961
–, carbamoylation 020
–, classification 009
–, cleavage of racemate 014, 015(F)
–, configuration 013
–, decarboxylation *534*
–, –, mechanism 104, 104(F)
–, degradation by lipid peroxidation 214, 215(T)
–, degradation during cheese ripening 534, *534*
–, discovery 009, 009(F)
–, dissociation 012, *012*
–, dissociation constant 013(T)
–, enzymatic analysis 139(T)
–, enzymatic degradation, carbonyl compound 377, *377, 378*
–, essential 009
–, –, legume 749(T)
–, –, requirement 030(T)

–, esterification 016
–, formation of aroma substance 377
–, free, fruit 626, 809, 816(T)
–, –, honey 888, 888(T)
–, –, vegetable 770, 782(T), *784*(T)
–, helix breaker 054, 054(T)
–, helix former 054, 054(T)
–, HPLC 019, 022
–, isoelectric point 012, *012*, 013(T)
–, N-terminal, analysis 018, 019
–, nonessential 009
–, nonprotein 770, 782(T), *784*(T)
–, –, biosynthesis 770, *781*
–, –, fruit 809
–, occurrence 009
–, photometric determination 021
–, physical property 012
–, production figure 031(T)
–, pyrolysis product 025, 027, 028(T)
–, reaction 016
–, reaction with carbonyl compound 021
–, resolution of racemate 022
–, side chain, charged 009
–, –, uncharged, nonpolar 009
–, –, uncharged, polar 009
–, solubility 015, 015(T)
–, specific rotation 014(T)
–, structure and taste 034, 035(T)
–, sweet taste 034, 035(T)
–, symbol 010(T)
–, synthesis 017
–, taste quality 035(T)
–, taste threshold value 035(T)
–, tea 954, 954(T)
–, thermal decomposition 025, *025*, 027(T)
–, –, mutagenic and carcinogenic product 026, *026*, 027, 027(T)
–, thiocarbamoylation 020
–, titration curve 013(F)
–, transamination 021
–, –, mechanism 103, 104, 104(F)
–, use as additive 430
–, utilization 031(T)
–, UV absorption 015, *015*, 016(F)
Amino acid analyser 041
Amino acid analysis 041, 042(F)
–, precolumn derivatization, reagent 041
Amino acid composition, *Osborne* fraction, cereal 678(T)
–, actin 578(T)
–, barley 677(T)
–, broad bean 749(T)
–, casein 502(T), 624(T)
–, cereal 677(T)
–, collagen (calf skin) 578(T)
–, corn 677(T)

–, egg 548(T)
–, egg white 548(T)
–, egg yolk 548(T)
–, elastin 578(T), 584
–, fish (cod) 624, 624(T)
–, fish protein concentrate 087(T)
–, gluten protein 683(T)
–, lipovitellin, α- 554(T)
–, lipovitellin, β- 554(T)
–, meat (beef) 578(T), 624(T)
–, meat (poultry) 578(T)
–, milk 502(T)
–, millet 677(T)
–, myosin 578(T)
–, oat 677(T)
–, plastein 085(T), 087(T)
–, rice 677(T)
–, rye 677(T)
–, soy bean 749(T)
–, soya protein isolate 087(T)
–, wheat 677(T)
–, whey protein 502(T)
–, zein 085(T)
Amino acid decarboxylase, enzymatic analysis
    139(T)
Amino acid ester, cyclization, polymerization 016
Amino acid pyrolyzate, mutagenic compound 027,
    028(T)
Amino acid sequence, *Bowman-Birk* inhibitor,
    soybean 055(F)
–, avidin 552(T)
–, casein 503(T)
–, –, genetic variant 505(T)
–, collagen, chain, α¹- 580(T)
–, globulin, 11S- 750, 751(T)
–, globulin, 7S- 752(T)
–, HMW subunit 687(T)
–, lactalbumin, α- 504(T)
–, –, genetic variant 505(T)
–, lactoglobulin, β- 504(T)
–, –, genetic variant 505(T)
–, lysozyme 552(T)
–, monellin 436(T)
–, N-terminal, gliadin 685(T)
–, –, glutenin 685(T)
–, thaumatin I 437(T)
–, wheat prolamin 689(T)
Amino acid, D-
–, analysis 014
–, formation 072
–, indicator 022
Amino aicd, D-
–, bacterial contamination 022
Amino-1-deoxyketose, 1- 271, *271*
Amino-2-deoxyaldose, 2- 271, *271*
Amino-3-hydroxy-6-methyl-heptanoic acid, 4-

–, aspartic proteinase, inhibitor 079
Aminoacetophenone, o-
–, formation 389, *389*
–, sensory property 389
Aminoacetophenone, p-
–, sensory property 389
Aminoacrylic acid, 2- 071
Aminoadipic acid 770, *781*, 782(T)
Aminoalanine, β- 071, *071*, 072(T)
Aminoaldose 271, *272*
Aminobutyric acid, mouth feeling 023
Aminobutyric acid, α- 023
Aminocrotonic acid, 2- 071
Aminocycloalkane carboxylic acid, taste 034, 035(T)
Aminocyclopropane-1-carboxylic acid, 1- 809
Aminoketone, α- 282, *283*
Aminoketose 271, *271*
Aminomalonic acid derivative, sweet taste 037,
    038(T)
Aminopeptidase 042, 077(T)
Aminopropanol 603
Aminopropionamide, 3-
–, formation 025, *025, 026*
Aminopropionitrile, β- 781, *781*
Aminoreductone, formation 280, *280*
Aminos acid, D- 072(T)
Ammonium carbamate, leavening agent 723
Amphiphilic lipid, definition 158
Amygdalin 760(T)
–, aroma precursor 841
Amyl cinnamicaldehyde, α-
–, structure, sensory property 396(T)
Amylase 333
–, cereal 695
–, chocolate syrup 968
Amylase inhibitor 757
Amylase, α- 333
–, application 150
–, calcium 105
–, dough making 721, 722(T)
–, honey 886
–, inactivation 133, 134(F), 134(T)
–, inhibition, pressure 136, 137(F)
–, temperature optimum 133, 134(F)
–, thermal stability 152(F)
–, wheat 695, 696(T)
–, –, pH optimum 696(T)
Amylase, β- 333
–, honey 886
–, mechanism 114(T), 115, 116(F)
–, pH optimum 129(T)
–, wheat, pH optimum 696(T)
Amylo corn starch, gelatinization behavior 320(F)
Amyloglucosidase, reaction 139(T)
Amylogram 715(F)
Amylograph 715

Amylopectin 323
–, adhesive 325
–, application 325(T)
–, conformation 324
–, monoacylglyceride, complexation 740(F)
–, retrogradation 739, 740(F)
–, structural model 324(F)
–, structure *323*
Amylose 317(T), 321
–, A-conformation 321, 322(F)
–, B-conformation 321, 322, 323(F)
–, conformation 297, *297*
–, film for packaging 324
–, gel formation 320, 320(F)
–, inclusion compound 322
–, monoacylglyceride, complexation 740(F)
–, retrogradation 320
–, structure *321*
–, V-conformation 322, 323(F)
Amylose-lipid complex, crumb 739
Amyrine, α-
–, detection in cocoa butter 233, 233(F)
Amyrine, β- *232*
Anabolic steroid, meat, detection 612, 612(F)
Analogne cheese (see imitation cheese)
Analysis, added protein, meat 613
–, added water, meat 613
–, amino acid 019
–, amino acid composition of protein 041, 042(F)
–, amino acid sequence 021, 044
–, amino acid sequence of protein 041
–, amino acid, D- 014
–, anabolic compound, meat 612, 612(F)
–, animal origin 608, 609, 610(F)
–, antibiotic, meat 612
–, carbonyl compound 667
–, carotinoid 244
–, cholesterol 232
–, cocoa butter 173
–, cocoa butter substitute 173
–, collagen 613
–, connective tissue 613
–, deep fried fat 220(T), 221
–, dietary fiber 336
–, egg yolk content 233
–, enzymatic 120, 137
–, fat 662
–, fish 627(F)
–, fish freshness 629, 630
–, frozen meat *611*, 611(F), 612
–, fruit 856, 857–859(T)
–, fruit product 857, 857, 858(T)
–, frying oil 668
–, heating of milk 519, 520
–, lipid 182
–, lipoprotein 184

–, margarine, indicator 661
–, marzipan 233
–, meat 608
–, meat extract 039
–, milk, heat treatment 518(F)
–, N-terminal amino acid 018, 019
–, NIR 706, 706(F), 706(T)
–, nitrosamine, meat 614
–, odorant 345
–, peptide mixture, HPLC 044
–, persipan 233
–, photometric, amino acid 022
–, –, linoleic acid 168
–, –, linolenic acid, α- 168
–, –, protein 016, 020
–, polysaccharide 335
–, protein, cross-linking 071, *071*
–, radiolysis of food 224
–, rye flour 715
–, soybean oil 233
–, stereospecific, triacylglycerol 175, 175(F)
–, sterol 666(F)
–, sterol ester 666(F)
–, sunflower oil 233
–, tenderized meat 611
–, thickening agent 335
–, tocopherol 233
–, triacylglycerol 173
–, wax ester 666(F)
–, wheat flour 711
–, wheat germ oil 233
Anatto, isolation 244
Anatto extract, pigment 240, *240*
Anchosen 635
Anchovy 618
Androst-16-ene-3α-ol, 5α-
–, truffle 788
Anethol, aroma, apple 839
Anethole 972, *972*
Anhydro-D-galactose, 3,6-
–, carrageenan 304, 305(T)
Anhydro-D-galactose-2-sulfate, 3,6- 304, 305(T)
Anhydro-L-galactose, 3,6- 302
Anhydroglucopyranose, 1,6- *263*
Anhydroidopyranose, 1,6- 263, *263*
Anhydrosugar 263
Aniseed, aroma substance 974(T)
–, essential oil, chemical composition 974(T)
Anomer 251
Anomeric effect 254(T), 255
Anserine *039*, 040(T)
–, fish 626
Anthelmintica 487, 487(T), 488(F)
Antheraxanthine, occurrence in orange 240(T)
Anthocyan, bleaching with SO_2 831, *832*
–, color, pH dependency 830, *830*, 830, 831(F)

–, fruit, content 829(T)
–, vegetable 788, 788(T)
Anthocyanidin 829, 829, 830(T)
–, absorption maxima 830(T)
–, biosynthesis 835, 836
–, chemical structure 823
–, formation 828, 829
–, metal complex 831, 831
–, wine 915
Anthocyanin, cacao bean 963
Anti-senescene agent, fruit storage 848
Antibiotic, additive 454
–, meat, detection 612
–, veterinary medicine 487, 487(T), 488(F)
Antibody, catalytic activity 112
Antifreeze glycoprotein, fish, blood serum 626, 626
Antimicrobial agent 448
Antioxidant, action 215, 215(F)
–, additives 455
–, combined effect 219, 219, 219(T)
–, natural 215
–, nitrosylmyoglobin 577, 577
–, prooxidative effect 220
–, spice 981
–, stoichiometric factor 215
–, synthetic 218
Antioxidative factor, definition 219, 219
Antiserum, enzyme immunoassay 141
Apigenin 834, 834
Apiose, occurrence 250(T)
Apple 815(T)
–, aroma formation 211(T)
–, aroma substance 839
–, fatty acid composition 818(T)
–, hydroxycinammic acid derivative 824(T)
–, lipid 818(T)
–, lipid composition 178(T)
–, ripening, respiration rise 843(F)
–, temperature of phase transition 006(T)
–, wine 928(T)
Apricot 815(T)
–, aroma substance 839
–, carotinoid content 235(T)
–, hydroxycinammic acid derivative 824(T)
–, kernel oil, tocopherol content 234(T)
Aquavit 935
Arabinase 335
Arabinitol, nutritional/physiological property 864(T)
Arabinogalactan from larch 313, 313
Arabinoic acid, formation 267, 267
Arabinose 251
–, occurrence 250(T)
–, specific rotation 258(T)
–, wine 919
Arabinose, L- 878
Arabinoxylan 330

–, endoxylanase 703
–, hydrolase, wheat 698
–, structure 702(F)
Arabitol 261
Arachidic acid, structure, melting point 161(T)
Arachidonic acid, biosynthesis 169, 169(F)
–, configuration 163(T)
–, molecule geometry 165
–, occurrence 162
–, photometric determination 168
–, structure, melting point 163(T)
Arachidonic aicd, taste 164(T)
Arachin 746
Argentation chromatography, fatty acid 168
–, triacylglycerol 173
Arginine 010
–, alkali cleavage 072, 073(F)
–, discovery 009
–, formation of NO 492
–, Maillard reaction 270
–, modification, Maillard reaction 286, 288
–, precursor of nitrate 492
–, reaction of the guanidyl group 023
–, reaction with dicarbonyl compound 066
Armagnac 931
Aroma, AEDA 350, 351(F)
–, analysis 345
–, –, enrichment 351, 353(T)
–, –, GC olfactometry 352, 352(F)
–, beer, reproduction 902, 903(T)
–, boiled beef 605, 606(T)
–, boiled pork 605, 606(T)
–, Camembert 542, 542(T)
–, CHARM analysis 350
–, cod 629, 629(T)
–, cream 539, 540(T)
–, Effect of binding 389, 390(F), 390(T)
–, Emmentaler 542, 542(T)
–, enantiomer, odor threshold value 341(T), 355(T)
–, encapsulation 398
–, pepper, key odorant 975
–, roasted chicken 605, 606(T)
–, salmon 628, 628(T)
–, tea 955, 955(T)
–, white bread 367
Aroma analysis, SPME 349
Aroma defect, beer 904
–, benzoic acid esterification 450
–, causes 345(F)
–, detergent in milk 541
–, fish 388
–, linolenic acid, α-
–, –, autoxidation 203
–, meat 595, 596(T)
–, milk product 388, 543
–, orange juice 838

–, partial fat hydrogenation 205
–, perfume rancidity 225
–, pyrazine 388
–, rancid, lipolysis 187
–, sexual odor 228, 229(T)
–, sorbic acid degradation 452, 452
–, soybean oil 198
Aroma defect (see also off-flavor)
Aroma distillate 394
Aroma extract 394
Aroma extract dilution analysis 350, 351(F)
Aroma extract dilution analysis (see also AEDA)
Aroma fixative, gum arabic 308
–, polyvinylpyrrolidone 333
Aroma model, omission experiment 358, 359(F)
–, preparation 357, 359(T)
Aroma note, beer 905(F), 906(T)
Aroma profile, coffee 944, 945(T)
Aroma recovery, fruit juice, evaporation 855
Aroma substance, Brussels sprout 792
–, aniseed 974(T)
–, apple 839, 839(T)
–, apricot 839
–, artichoke 792
–, banana 837
–, beef broth 605, 605(T)
–, beer 902, 903(T)
–, –, alcohol-free 902, 903(T)
–, binding, model experiment 389, 390(F), 390(T)
–, binding by lipid 389, 390
–, binding by polysaccharide 391, 392(T)
–, binding by protein 391, 392(F), 393(T)
–, black tea 955, 955(T)
–, broccoli 792
–, butter 541, 541(T)
–, cacao 964, 964(T)
–, cacao bean 964
–, caraway 974(T)
–, cauliflower 792
–, celery 788
–, change during isolation 345, 346(T)
–, cheese 541, 542(T)
–, cherry 841, 841(T)
–, cherry jam 841, 841(T)
–, cinnamon 972, 974, 974(T)
–, citrus fruit 837
–, clove 974(T)
–, coffee beverage 945(T), 947
–, condensed milk 539
–, cooked mutton 607, 607(T)
–, cucumber 793
–, distillation 340, 346, 347(F), 347(T)
–, dried milk product 539
–, egg 557
–, enantiomeric excess 353, 355(T)
–, enantioselective analysis 353, 355(T), 356(F)

–, enzymatic formation 374
–, enzymatic lipid peroxidation 209, 210(F), 211(T)
–, extract 394
–, fenugreek 976, 976
–, fish 628
–, formation 360
–, –, hydroperoxide lyase 209, 210(F), 211(T)
–, –, nonenzymatic 360
–, fruit 837
–, furanone 361, 362(T)
–, garlic 790
–, gas extraction 348, 348(T)
–, grape 837
–, grape must 916
–, grapefruit 838
–, honey 889
–, hop 896, 896(T)
–, hydrolysis of glycoside 382, 382
–, identification 353, 354(F)
–, impact compound 340
–, inclusion complex 398
–, isolation 345, 346(T), 347(F)
–, –, hydrolysis of glycoside 346, 346(T)
–, key compound 340(T)
–, lager beer 902, 903(T)
–, lipid peroxidation 191, 203, 204(T)
–, litchi 842
–, maple syrup 873
–, meat 605
–, milk 539, 540(T)
–, mushroom 788
–, mustard 978(T)
–, natural, definition 395
–, nectarine 840
–, odor, influence of structure 361
–, odor intensity of mixture 341, 341(T), 343(F)
–, odor quality of mixture 342, 343, 343(F)
–, odor threshold value 341, 341(T)
–, onion 790
–, orange 836(T), 837
–, parsley 978, 979(T)
–, passion fruit 840
–, pea 793
–, peach 840
–, pear 839
–, pepper 975, 975(F), 975(T)
–, pineapple 841
–, plum 841
–, potato 788, 789(T)
–, quantitative analysis 356, 357, 358(F), 359(T)
–, radish 789
–, raspberry 839
–, raw mutton 607, 607(T)
–, red beet 790
–, red cabbage 792
–, rice 710

–, saffron 974(T), 977, *977*
–, sensory identification 350
–, sensory relevance 349
–, sorption 391, 391(F)
–, –, binding constant 391
–, sour milk product 540
–, spinach 792
–, stability 397
–, strawberry 841, 841(T)
–, structure and odor 398
–, synthetic, definition 395
–, –, example 396(T)
–, synthetic natural, definition 395
–, tomato 793, 793(T)
–, turmeric 974(T)
–, type of bread 738, 739(T)
–, type of cabbage 792
–, use as additive 430
–, vegetable 788
–, watercress 791
–, whisky 934(T)
–, white cabbage 792
–, white-bread crumb 736, 736(T)
–, white-bread crust 735, 735, 736(T)
–, wine 921, 922–924(T)
–, yoghurt 540
Aroma substance (see also odorant)
Aroma type, economic importance 394(T)
Aroma value, calculation 356
–, definition 342
–, identification of aroma substance 356
–, tomato aroma substance 793(T)
Aromatization 394(T)
–, detection 356
–, food 393
–, invert sugar cream 890
–, lactone 380
–, margarine 661
Arrack 933
Arrhenius equation 131, *131*
–, crystallization of water 007, 007(F)
–, growth of microorganism 134, 134(F)
Arrhenius factor 131, *131*
Arsenic 428
–, toxicity 468
Artichoke, aroma substance 792
Artificial honey 890
Artificial honey (see also invert sugar cream)
Ascorbic acid, anaerobic degradation 418, *418*
–, antioxidant 217
–, beer 906
–, biosynthesis 842, *842*
–, browning reaction 419, *419*
–, enzymatic browning 122
–, flour improvement 716, 716(F), 717, 717(F),
  717(T), 718, 718(T), 725(F)

–, inhibitor for lipoxygenase 742, 742(F)
–, loss 419, 419(F)
–, Maillard reaction 419, *419*
–, metal catalysis 418, *418*
–, occurrence in fruit 842(T)
–, oxidation 418, *418*
–, –, dough 717(F)
–, prooxidant 217
–, stereoisomer 717(F)
–, storage of vegetable 799, 801, 801(F)
–, synergist for α-tocopherol 217
–, synthesis 418, 842, *843*
–, UV absorption 418, 418(T)
Ascorbic acid (see also vitamin C) 417, *418*
Ascorbic acid oxidase, copper 105
–, reaction 097
–, systematic name 097
–, wheat 698, 699(T)
Ascorbyl palmitate *218*
–, synergist for α-tocopherol 217
ASF, safety factor 468
Asparagine *010*
–, discovery 009
–, precursor of acrylamide 025, *025, 026*
–, sweet taste, AH/B/X-model 261
Asparagus, aroma formation 387, *387*
–, biosynthesis of asparagus acid 387, *387*
–, saponin content 763(T)
Asparagus acid, biosynthesis 387, *387*
Aspartame 037, 038(T), 441
–, analog compound 037, 038(T)
–, degradation reaction *442*
–, structure and taste 037, 038(T)
–, sweet taste 441(T)
–, sweetening strength, relative 441(T)
–, synthesis *442*
–, taste, synergistic effect 431, 435(F)
Aspartase 032, *032*
Aspartic acid *010*
–, discovery 011
–, reaction 023
–, synthesis 032, *032*
Aspartic endopeptidase 077(T)
Aspartic peptidase, specificity 078(T)
Aspartic proteinase 076
–, inhibition 079
–, mechanism 079
Aspartyl-L-phenylalanine-methylester, α-L- 037
Aspartylphenylalanine methyl ester 441, *442*
Aspergillopeptidase, specificity 078(T)
Assay, enzyme activity 140
–, lipase 189
Astaxanthine 238, *238*
Astringent taste 825, 829
–, quenching 845
Atmosphere, fruit storage 848, 849(T)

ATPase, myosin 569
Atractyligenin, coffee 943, *943*
Atractyloside, coffee 943, *943*
Auroxanthine *239*
–, occurrence in orange 240(T)
Autoxidation, fatty acid, saturated 221
–, hexenal, 2(E)- 397, *397*
–, lipid 191
–, –, start 196
Auxiliary reaction, enzymatic analysis 138
Avenasterol, Δ-5
–, occurrence 229(T)
Avenasterol, Δ-5- *231*
Avenasterol, Δ-7- *231*
Avidin 229(T), 549(T), 551
–, amino acid sequence 552(T)
–, biotin complex 551
Avocado, fatty acid composition 818(T)
–, ketose 250(T)
–, sugar 817
Azetidine carboxylic acid 781, *781*
Azlactone 017, *017*
Azodicarbonamide, flour improvement 719, 725(F)
Azomethine, amino acid 021

Bacillus cereus
–, food poisoning 470, 471(T)
Bacon 598
Bacterial count, decrease, rate equation 131, *131*
–, milk 133(F)
Bacterial food poisoning 470, 471(T)
Bacterial protein, plastein reaction 085(F)
Bacterial toxin 471(T)
Bactotherm process 518
Bagasse 872
BAI value 584
Baked product, agar 301(T), 302
–, alginate 301(T), 304
–, baking time 733(T)
–, carboxymethyl cellulose 330, 330(T)
–, carrageenan 301(T), 306
–, dextran 332
–, fine 741
–, –, definition 711(T)
–, gum arabic 308
–, gum tragacanth 310
–, raw materials 711
–, ropiness, propionic acid 452
–, staling, lipase 154
–, starch 324
–, type 711(T)
Baking chocolate 968, 968(T)
Baking honey 884
Baking margarine 660(T)
Baking powder, additive 447, 448
–, chemical composition 723

Baking process, bread 731
–, –, temperature and time 731, 733(T)
–, crumb formation 731, 734(F)
Baking property, addition of shortening 721
–, HMW subunit 712
–, lipid 705, 705(F)
–, wheat, lipid 704
–, wheat flour 712(T)
–, –, additive 716
–, –, ascorbic acid 716, 716(F), 717, 717(F), 717(T), 718, 718(T)
–, –, bromate 718(T), 719
Baking quality, addition of shortening 721(T)
Baking test 714
–, lipoxygenase 719(F)
Baking volume, protein content of wheat flour 712, 713(F)
Balenine *039*, 040(T)
Banana 815(T)
–, amine 817(T)
–, aroma substance 837
–, fatty acid composition 818(T)
–, temperature of phase transition 006(T)
Barley, *Osborne* fraction 676, 677(T)
–, beer production 892
–, lipoxygenase, reaction specificity 207(T)
–, origin 670, 670(F)
–, production data 672(T)
Barley coffee 950
Basil, aroma substance 977, 978(T)
–, dried, aroma 977, 978(T)
Batyl alcohol *187*
Baudouin reaction 664(T)
Bay leaf, essential oil, chemical composition 974(T)
Bean, chlorophyll 794(T)
–, processing, discoloration 796(T)
–, seed, cooking 768
Bean (see green or French bean)
Bearded spelt 670
Beaumaris shark 618
Beef, age, determination 607
–, boiled, aroma 605, 606(T)
–, –, furaneol 363(T)
–, –, odorant 605, 606(T)
–, consumption data 565(T)
–, glutathione 038
–, progesterone 228(T)
–, roasted 162(T)
–, sexual origin, detection 610
–, tenderness 581
Beef extract 040(T)
Beef fat, fatty acid composition 643(T)
Beef meat, category 592
–, composition 569(T)
Beef tallow, melting property 173
–, triacylglycerol 173(T)

Beer 892
–, acid 902
–, alcohol-free 904
–, –, aroma substance 902, 903(T)
–, –, production 904
–, alginate 301(T), 304
–, application of α-amylase 150
–, aroma, reproduction 902, 903(T)
–, aroma defect 904
–, aroma note 905(F)
–, aroma substance 902, 902, 903(T)
–, Berlin weiss 903
–, bottling 901
–, bottom fermented 903
–, carbohydrate 902
–, chemical composition 901
–, cold turbidity 150
–, consumption data 894(T)
–, continuous process 900
–, defect, bacteria 906
–, diacetyl 905
–, –, reduction 149
–, ethanal 905
–, ethanol content 901
–, extract content 901
–, fermentation 900
–, filtration 900
–, flavor 904
–, foam builder 902
–, foam stability, phospholipase 696
–, furaneol 363(T)
–, furcellaran 301(T), 307
–, glucose oxidase 149
–, hydrolysis of β-glucan 152
–, light taste 905
–, low-alcohol 904
–, lysophosphatidyl choline 903
–, maltol 361(T)
–, mash 899
–, mineral 902
–, nitrogen compound 902
–, oxidation 906
–, –, inhibition 906
–, pH-value 902
–, polyvinylpyrrolidone 333
–, production 893(F)
–, –, β-glucan 703
–, production data 894(T)
–, protein, cloudiness 902
–, stemwort 901, 901
–, sunlight off-flavor 344(T)
–, taste 905(F), 906(T)
–, top fermented 903
–, trihydroxyoctadecenoic acid, 9,12,13- 902
–, turbidity 906
–, type 903, 906(T)

–, vitamin 902
–, volatile sulfur compound, synthesis 387, 388
–, wort 899, 900
Beet sugar, production 867
–, production data 868(T)
–, ratio $^{13}C/^{12}C$ 859
Beetle honey 884
Behenic acid, structure, melting point 161(T)
Behenic acid tryptamide 961, 961
Benedict's reaction 267
Bentonite 464
Benzaldehyde, aroma quality 340(T)
–, binding by protein 393(T)
–, cherry aroma 841, 841(T)
–, fruit brandy 932
–, odor threshold value 341(T)
–, plum aroma 841
Benzil, reaction with arginine 066, 066
Benzo(a)pyrene 490, 490
Benzoic acid, antimicrobial action 449, 450(F)
–, antimicrobial agent 449, 450(F)
Benzoyl-D-glucose, 6- 291, 291
Benzylidenelysine, ε-N- 023
Benzyloxycarbonylamino acid, N- 017
Beri-beri 411
Betacyan 797, 797
Betaine 019, 019(T)
–, fish 626
–, sugar beet 869
Betalain, biosynthesis 797, 797
Betalamic acid 797, 797
Betanidin 797
–, hydrolysis 797
Betanin 796, 796, 797
–, stability 797
Betaxanthin 797
Betonicine 019(T)
Beverage, alcohol-free 856
–, caffeine containing 856
–, carrageenan 301(T), 306
–, clarification, polyvinylpyrrolidone 333
–, dextran 332
–, gum arabic 308
–, nonalcoholic, consumption 852(T)
–, pectin 315
–, wine-like 928, 928(T)
–, –, chemical composition 928(T)
BHA 218, 218, 219, 219
BHT 218, 218, 219, 219
Bicarbonate, addition to condensed milk 527
–, additive 449
Bifidogenic sweetener 864(T), 867
Bifidus pathway, glucose 521(F)
Bifunctional reagent 070
Biogenic amine, cheese 534, 536(T)
–, fish 626

–, meat 584
–, precursor
Biological value, protein 029, 030(T)
Biosynthesis, flavnonoid 835, *836*
Biotin 415, *415*
–, fruit 842
–, occurrence 409(T)
–, stability 425
Bis(2-methyl-3-furyl)disulfide, formation 364, *365*
–, sensory property 364(T)
Bisabolene, β- *385*
Biscuit, maltol 361(T)
Bitter acid, α- 886(T), 895, *895*
Bitter acid, β- 886(T), 895, *895*
Bitter compound, cocoa 963(T), 964
–, cucumber 820
–, cucurbitaceae 820
–, grapefruit 819
–, orange juice 819
Bitter liquor 935
Bitter orange, taste 832
Bitter substance, *Maillard* reaction 280, *280*
–, hop 895, 896(T)
Bitter taste, *Brussels* sprout 798
–, amino acid 034
–, bitter orange 832
–, cacao bean, formation 964, *964*
–, cheese, glutamic acid 543
–, divinyl glycol, wine 926
–, elimination 085
–, enzymatic protein hydrolysis 150
–, formation in oat 696, 697, 697(F)
–, grapefruit 832
–, oat 211
–, olive 803
–, oxidized fatty acid 212(T)
–, peptide 036
–, proteolysis 037
–, roasted coffee 947, *948*, 948(T)
–, steroid alkaloid 798, 798(T)
–, synergism 964, *964*
Bixin *240*
Black gram, chemical composition 748(T)
Blackberry, ellagic acid 825
–, hydroxycinammic acid derivative 824(T)
Blanching, loss of ascorbic acid 420(F)
–, vegetable 795, 799, 801
Blanching process, peroxidase 134
Bleaching, refining of oil 655, *655*
–, wheat dough 719
–, wheat flour 716, 719
Bleaching agent 464
Bleaching of flour, mechanism 209, 209(F)
Blended spice 982
Blood 594, 594(T)
Blood plasma 594

–, dried 595
Blood serum, lipoprotein 184, 185(T)
Blue cheese, aroma 542
–, ripening 533(F), 534
Blue No. 1 445(T), 447(T)
Blue No. 2 445(T), 447(T)
BOC-amino acid, reaction 081, *081*
Bohr effect, muscle, oxygen binding 574
Boiling sausage 601
–, chemical composition 599(T)
–, production 601(F)
Booser process, churning 526
Borneo tallow, triacylglycerol, composition 173(T)
Boron 427
Bottom fermentation, beer 900
Botulism 454
Bouillon, aroma substance 605, 605(T)
–, taste substance 605(T)
Bourbon whisky 933
Bovine serum albumin, binding of aroma substance 393(T)
Bowman-Birk inhibitor 055(F), 756, 756(T)
Brain, veal, chemical composition 594(T)
Bran, cereal 675(T)
–, composition 673, 706, 707(F)
Brandy 931
–, pear aroma 932
Brandy blend 931
Brassicasterol, occurrence 229(T)
Bread, agar 301(T), 302
–, aroma profile 735, 735(T)
–, aroma substance 734, 735, 736(T), 737, 738(F), 739(T)
–, baking process 731
–, baking time 733(T)
–, chemical composition 735(T)
–, freezing water 740(F)
–, kneading condition 725, 725(F), 725(T)
–, pyrraline 286(T)
–, ratio crumb/crust 734(T)
–, reheating 739
–, specific volume 733, 734(T)
–, staling, thermogram 739(F)
–, staling defect 739
–, storage 739
–, toasting, aroma 737, 738(F)
–, vitamin loss 734
–, white, aroma 367
Bread crumb, reducing sugar content 727(F)
Bread crust, aroma, stability 735, 736(T)
–, aroma analysis 346, 347(F)
–, aroma formation, fermentation 734, 736(T), 737(F)
–, furaneol 363(T)
–, odorant, headspace 735, 736(T)
Brewing water 897
Brewing yeast 897

Brisling 618
Broad bean, chemical composition 748(T)
–, essential amino acid 749(T)
–, production data 747(T)
–, saponin content 763(T)
Broccoli, aroma substance 792
–, glutathione 038
–, temperature of phase transition 006(T)
Bromate, flour improvement 718(T), 719
Bromelain 076, 077(T)
–, active cysteine 115
–, meat, tenderizing 598
Bromosuccinimide, N- 070
Brown algae 302
Brown rice 710
Browning, enzymatic 105
–, –, fruit 835
–, nonenzymatic 270
–, –, inhibition 007(T)
–, –, water activity 004, 004(F)
Browning reaction, inhibition 040
Brussels sprout, aroma substance 792
–, bitter taste 798
BSE 055
Bubble gum 882
Bucherer reaction 032, 033
Buffalo milk, chemical composition 501(T)
–, production data 499(T)
Burnt almond 881
Butanal, biosynthesis 377(F)
Butanal, 2-methyl-
–, biosynthesis 378(F)
–, sensory property 360(T)
Butanal, 3-methyl-
–, biosynthesis 378(F)
–, sensory property 360(T)
Butanediol, formation 541(F)
Butanediol dehydrogenase 149
Butenal, (E)-2-
–, linolenic acid, α-
–, –, autoxidation 203(T)
Butter 162(T)
–, aroma defect 541
–, aroma profile 540, 541(T)
–, aroma substance 540, 541(T)
–, chemical composition 524
–, consumption 501(T)
–, crystalline shell of a fat grain 525(F)
–, detection 665(T)
–, freeze-fracture 524(F)
–, furan fatty acid 164(T)
–, heated, maltol 361(T)
–, heptenal, (Z)-4- 205
–, lipolysis 190(T)
–, melted 526
–, production 524, 525(F)

–, production data 500(T)
–, progesterone 228(T)
–, solid/fluid ratio 525
–, starter culture 525
Butter granule 526
Butterfat, fractionated 526
–, melting characteristic 514(T)
Buttermilk, metallic taste 543
Buttermilk powder, detection 515
Butyl acetate, aroma, apple 839(T)
Butyloxycarbonylamino acid, N-tert.- 017
Butyric acid, butter aroma 541, 541(T)
–, sensory property 160(T)
–, structure, melting point 161(T)
–, variation range, milk fat 665

Cabbage, aroma substance 792
–, cooking, aroma 792
–, glucosinolate 789(T)
Cacao, aroma substance 964, 964(T)
–, bitter taste 964
Cacao bean, acid 963
–, anthocyanin 962(T), 963
–, carbohydrate 962
–, catechin 962(T), 963
–, chemical composition 961(T)
–, cotyledon tissue, morphology 962(F)
–, crushing 961
–, dioxopiperazine 964, 964
–, enzyme 961
–, fermentation 960
–, –, reaction 964
–, leucoanthocyanin 962(T), 963, 963
–, morphology 962(F)
–, phenolic compound 962
–, pigment 965
–, pigment cell 962, 962(F)
–, polyphenol storage cell 962, 962(F)
–, proanthocyanin 963
–, processing 959(F), 960
–, production data 959(T)
–, protein 961
–, roasting process 961
–, saccharose 962
–, shell, detection 961
–, variety 960
Cacao germ, chemical composition 961(T)
Cacao shell, chemical composition 961(T)
–, detection 961
Cadaverine 584
Cadinene, β- 385
Cadmium 469, 469(T)
Cafestol, coffee 942, 943
Caffeic acid 822
–, antioxidative activity 217, 217(T)
Caffeine, cacao bean 962

–, coffee 943
–, cola nut 958
–, cola-chocolate 968
–, green coffee 941(T)
–, maté 958
–, taste threshold value 034, 035(T)
–, tea 954
Caffeine-containing beverage 856
Caffeoylquinic acid, fruit 824(T)
Cake, polyalcohol 878(T)
Calcidiol 406
Calciferol (see also vitamin D) 406, *406*
Calcitriol 406
Calcium, content, human body 421(T)
–, enzyme cofactor 104
–, occurrence in food 422(T), 429
Calcium phosphate, formation of casein micelle
    510(F)
Calmar 639
Calpain, meat 590, 590(T)
CAM plant 859, 859(T)
Camel milk, chemical composition 501(T)
Camembert, aroma 541, 542(T)
–, plasmin 517
–, taste compound 541, 542(T)
Campesterol *230*
–, identification of cocoa butter 231(T)
–, occurrence 229(T)
Camphene *384*
Camphor *384*
Can, fruit 850
Candied fruit 851
Candied lemon peel 851
Candied orange peel 851
Candy 878
–, application of invertase 153
–, carboxymethyl cellulose 330
–, chemical composition 880(T)
–, dextran 332
Cane sugar, production 872
–, production data 868(T)
–, ratio ^{13}C/^{12}C 859
Canned fish 635
Canned meat 600, 600(T)
–, carrageenan 306
Canned vegetable 800
–, vitamin loss 801
Cannizzaro reaction 268
Canthaxanthine *238*
–, electronic spectrum 241(F)
Capric acid, sensory property 160(T)
–, structure, melting point 161(T)
Caproic acid, crystal structure 166(F)
–, sensory property 160(T)
–, structure, melting point 161(T)
Caprolactam, lysine precursor 032, *032*

Caprylic acid, sensory property 160(T)
–, structure, melting point 161(T)
–, titration curve 165(F)
Capsaicin *980*
Capsaicinoid, pungent taste 980
Capsanthin *238*
–, degradation in paprika 243(F)
Capsanthone *243*
Caramel aroma 361, 362(T)
Caramelization 270
Caraway, aroma substance 974(T)
–, essential oil, chemical composition 974(T)
Carbamate 476, 480(T)
Carbamoylamino acid 020
Carbazole 027(T)
Carbohydrate 248
–, available, dried fruit 850(T)
–, cacao bean 962
–, egg white 552
–, egg yolk 556
–, energy value 008
–, fat substitute 463
–, fruit 817, 818(T)
–, fruit ripening 845
–, green coffee 941(T)
–, honey 885, 885–887(T)
–, legume 748(T), 759, 760(T)
–, milk 512
–, tea 954
–, vegetable 779(T), 786
–, wheat 701, 702(T)
–, wine 919
Carbohydrate polyester 464
Carboline 026, *026*
Carbon-13
–, abundance 858, 858(T)
Carbon-14 470
Carbonatation, sugar extract 870
Carbonyl compound, analysis 667
–, enzymatic formation 378
–, formation 361
–, reaction with amino acid 021
Carboxcyclic acid, formation from monosaccharide
    267, *267*
Carboxyanhydride 064
Carboxycathepsin 077(T)
Carboxyglutamic acid, γ-
–, biosynthesis 408, *411*
Carboxylester hydrolase 188
Carboxymethyl cellulose 329
Carboxymethyl starch 326, *326*
Carboxymethyllysine, formation 286, *287*
Carboxypeptidase, tertiary structure 056(F)
Carboxypeptidase A 077(T)
–, active site 079(F)
–, specificity 043

Carboxypeptidase B 077(T)
–, specificity 043
Carboxypeptidase C 077(T)
–, specificity 043
Carcinogenicity, contaminant 467
Cardamom, essential oil, chemical composition
    974(T)
Cardiolipin 179
Carene 384
Carnitine 019(T), 585, 585
Carnosine 039, 040(T)
–, fish 626
Carnosol 981
Carnosolic acid 981
Carob bean 312
Carotenal, β-apo-8'- 240
Carotene, definition 236
Carotene, α- 237
–, aroma precursor 242(T)
–, electronic spectrum 241(F)
–, occurrence in palm oil 244
Carotene, β-
–, antioxidant activity 217, 217
–, aroma precursor 242(T)
–, electronic spectrum 240(T), 241(F)
–, fruit 842
–, occurrence in tomato cultivar 236(T)
Carotene, β-, 237
Carotene, γ- 237
–, electronic spectrum 240(T), 241(F)
Carotene, zeta-
–, occurrence in orange 240(T)
Carotinal, 3-keto-β-apo-8'- 243
Carotinoid 234
–, analysis 244
–, aroma precursor 241, 242(T)
–, chemical property 241
–, chromatography 244, 244(T)
–, co-oxidation 209, 209(F)
–, electronic spectra 240, 240(T), 241(F)
–, epoxide rearrangement 245
–, esterified with fatty acid 240
–, fat hardening 658
–, food pigment 244
–, fruit 818, 819(T)
–, main group 235
–, nomenclature 235
–, occurrence 235(T)
–, occurrence in palm oil 244
–, photooxygenation, inhibition 198
–, physical property 240
–, solubility 240
–, stability in pasta 742, 742(F)
–, vegetable 787, 787(T)
–, wheat 705

Carrageenan 298, 298, 299, 301(T), 304, 305(T),
    306(F)
–, biosynthesis 299
–, conformation 298, 298, 305(F)
–, gel formation 299
–, protein, coagulation 306
–, thermally reversible gel 306
–, viscosity 306(F), 308(T)
Carrot, Maillard reaction, inhibition 289, 289(F)
–, carotinoid content 235(T)
–, temperature of phase transition 006(T)
Carubin 306
Carvacrol 972, 972
Carvone, aroma, dill fruit 976
–, formation in orange juice 344(T)
–, sensory property 386, 386(T)
–, stereochemistry 386
Caryophyllene 385
Caryophyllene, β-
–, sensory property 386(T)
Casein, amino acid composition 502(T)
–, amino acid sequence 503(T)
–, –, genetic variant 505(T)
–, calcium binding 506(F)
–, enrichment with methionine 081, 082(T)
–, gel formation 509
–, genetic variant 502(T)
–, –, amino acid sequence 505(T)
–, heat coagulation 518(F), 519
–, micelle formation 508, 508(F)
–, molecular weight 502(T)
–, PER-value 085(T)
–, phosphorus content 502(T)
–, precipitation, proteinase 150
–, production 537(F)
–, reductive methylation 081(F)
–, succinylation 081, 081(F)
Casein fraction, composition 502(T)
–, electrophoretic separation 501
Casein micelle 510(F)
–, aggregation 510, 510(F)
–, chemical composition 509(T)
–, coagulation 510, 511(F)
–, distribution of casein 509(T)
–, surface hydrophobicity 510
Casein, α_{s1}- 502, 502(T)
–, amino acid sequence 503(T)
Casein, α_{s2}- 502, 504(T)
–, amino acid sequence 503(T)
Casein, β- 502(T), 504
–, acetylated, association 081, 082(T)
–, acylation 082, 082(T)
–, amino acid sequence 503(T)
–, emulsifying effect 063
–, enzymatic dephosphorylation 083, 083(F)
Casein, γ- 502(T), 507

Casein, κ- 502(T), 504
–, amino acid sequence 503(T)
–, –, genetic variant 505(T)
–, lactoglobulin, interaction 510
–, solubility 506
–, solubility of casein 506(F)
Caseinate, production 537(F)
–, recovery 537
–, utilization 537
Cashew nut, chemical composition 816(T)
Cassava, hydrocyanic acid 761(T)
Castor bean, production data 641(T)
Castor oil, detection 665(T)
–, ricinoleic acid 164
Cat milk, chemical composition 501(T)
Catalase, activation energy 094(T)
–, honey 888
–, milk, inactivation 135(F)
–, pH optimum 698
–, preservation 149
–, prosthetic group 103
–, rate constant 120(T)
–, reaction 149
–, wheat 698
Catalysis, energy profile 093(F)
Catalyst, activity 093
–, fat hardening 657
Catechin 827, 827
–, antioxidative activity 217, 217(T)
–, cacao bean 963
–, fruit, content 829(T)
–, tea 953
–, wine 920(T)
Catecholase 105
Catecholase activity 105
Cathepsin 077(T), 079
–, meat 590, 590(T)
Catty odorant 387, 388(T)
Cauliflower, aroma substance 792
–, glutathione 038
–, off-flavor 792
Caviar 635
Caviar substitute 636
Celery, aroma substance 788
Celery seed, aldose 250(T)
Celiac disease 674
–, prolamin 674
Cellobiase 335, 337(T)
Cellobiose 295(T)
–, conformation 293
–, hydrolysis 294
–, specific rotation 258(T)
Cellulase 335, 337(T)
–, application 153
Cellulose 327
–, conformation 296

–, crystal 328, 328(F)
–, microcrystalline 328
Cellulose derivative 328
–, utilization 330(T)
Cellulose ether, gel formation 329, 329(F)
Centose, honey 886(T)
Cephalin 179
Ceramide 181
Cereal, Osborne fraction 675, 677(T)
–, chemical composition 675(T)
–, crude fiber 675(T)
–, fatty acid composition 704(T)
–, kernel fraction 675(T)
–, lipid content 675(T)
–, milling 707(F)
–, origin 670, 670(F)
–, pest control 707
–, phenoloxidase 698
–, phylogeny 670(F)
–, production data 672(T)
–, prolamin, RP-HPLC 684(F)
–, protein, amino acid composition 674, 677(T)
–, protein content 675(T)
–, starch 317(T)
–, starch content 675(T)
–, storage 707
–, thousand kernel weight 674(T)
–, vitamin 675(T)
–, world production 673(T)
–, yield per hectare 674(T)
Cerotic acid, structure, melting point 161(T)
Ceryl cerotate 187
Cesium-134 470
Cesium-137 470
Cetyl alcohol 186
Chaconine 798, 798(T)
Chair model, triacylglycerol 171, 171(F)
Chalcone 822(F), 830, 831, 832, 832
–, chemical structure 822
–, sweet taste 832
Champagne 926
Champignon, aroma 376
–, aroma formation 211(T)
–, aroma substance 788
Chapatis 764
Cheese 162(T)
–, agar 302
–, antibiotic 454
–, aroma, lipase 154
–, aroma substance 541, 542(T)
–, biogenic amine 534, 536(T)
–, classification 529
–, fruity off-flavor 543
–, locust bean gum 312
–, melting salt 536
–, production 529, 532(F)

–, –, application of lysozyme 153
–, production data 500(T)
–, progesterone 228(T)
–, ripening 532
–, unriped 530, 537
–, –, aroma 541
–, variety 529, 531(T)
–, –, microflora 530(T)
Cheese aroma, concentrate 394
Cheese ripening, fat degradation 533
–, methyl ketone 225, 226(F)
–, peptide pattern 534, 534(T), 535(F)
–, propionic acid 532, 533(F)
–, protein degradation 534
Cheese surrogate 536, 536(T)
Chelating agent, additive 455, 455(T)
–, association constant 455(T)
Chemical composition, coffee whitener 528(T)
–, vegetable 770, 779(T)
Chemometry 666
Chemoprevention, phenolic compound 962
Cherry, aroma change 841, 841(T)
–, aroma substance 841
–, sour 815(T)
–, sweet 815(T)
Cherry jam, aroma substance 841, 841(T)
Cherry water 931
Chewing gum 882
–, polyalcohol 878(T)
–, production 883(F)
Chick pea, chemical composition 748(T)
–, hydrocyanic acid 761(T)
–, production data 747(T)
–, saponin content 763(T)
Chicken, egg, fatty acid distribution 176(T)
–, –, progesterone 228(T)
–, extract 040(T)
–, glutathione 038
–, meat, composition 569(T)
–, –, radiation 225(F)
–, progesterone 228(T)
–, roasted, aroma 606, 606(T)
Chicory coffee 950
Chili, formation of pyrazine 388, 388
–, pungent substance 980(T)
Chimyl alcohol 187
Chinese restaurant syndrome 431
Chloride 424
–, content, human body 421(T)
Chlorinated Maillard compound 603
Chlorination, drinking water 986
Chlorine, bleaching agent 464
Chlorine dioxide, bleaching agent 464
Chloro-3-tosylamido-4-phenylbutan-2-one, 1-,(TPCK)
   069

Chloro-3-tosylamido-7-aminoheptan-2-one, 1-,
   (TLCK) 069
Chloro-4-nitrobenzo-2-oxa-1,3-diazol,7- 019
Chloro-pigment, HPLC 795(F)
–, vegetable 793, 794(T), 795, 796(T)
Chlorogenic acid 822, 824, 824(T)
–, coffee 942(T), 943, 943(T)
–, green coffee 941(T)
Chlorohydrin, formation 453, 453
Chlorophyll 793, 795
–, absorption spectra 795(F)
–, bleaching of fat 655, 655(F), 655(T)
–, degradation 981, 981
–, degradation in vegetable processing 795, 796(T)
–, dried herb 981
–, Zn/Cu-complex 794
Chlorophyll a, occurrence 794(T)
Chlorophyll b, occurrence 794(T)
Chlorophyllase 795
Chlorophyllide 794, 796(T)
Chloropropanol 603
Chlortetracycline 454
Chocolate, chemical composition 968(T)
–, conching 966
–, consumption 966
–, crystallization 967
–, diabetic 968
–, emulsifier PGPR 462
–, fat bloom 969
–, kind 967, 968(T)
–, lecithin 967
–, maltol 361(T)
–, molding 967
–, polyalcohol 878(T)
–, production 959(F), 966
–, –, bicarbonate 449
–, refining 966
–, sugar bloom 969
Chocolate coating 968(T)
Chocolate milk, carrageenan 306
Chocolate powder 968
Chocolate syrup 968
Cholecalciferol 406, 406
Cholesterol 228, 228
–, analysis 232
–, autoxidation 227
–, determination of egg yolk content 233
–, egg yolk 556
–, milk 514, 514(T)
–, occurrence in food 228, 229(T)
–, oil, refining 656
Cholesterol reduction, milk 539, 539(T)
Choline 585, 585
Chromanoxyl radical, stability 217, 217
Chromenol 830
Chromium 425(T), 426

Chrysoeriol 834, *834*
Chylomicron 184, 185(T)
Chymopapain, specificity 078(T)
Chymosin 077(T)
–, retardation of coagulation 519
–, specificity 506, 507(T)
Chymotrypsin 075, 077(T)
–, active serine 107
–, acylation 115(F)
–, deacylation 115(F)
–, entropy effect 112
–, inhibition 055, 069
–, mechanism 115, 115(F)
–, –, steric effect 113(F)
–, pH optimum 129(T)
–, specificity 043, 045(F), 078(T)
–, structure 115(F)
–, substrate analog inhibitor 107, *107*
–, substrate binding 110(F)
Cidre 928(T)
Cineol, 1,8-
–, basil 978, 978(T)
–, sensory property 386(T)
Cineole, 1,8- *384*
Cinnamaldehyde, biosynthesis *973*
Cinnamon, aroma substance *973*, 974(T)
–, essential oil, chemical composition 974(T)
Citaurin, β-, *243*
Citraconic acid, coffee 943, *943*
–, synergistic effect with antioxidant 219
Citral *838*
–, degradation 397, *397*
–, lemon aroma 838, *838*
–, synthesis 395, *395*
Citramalic acid, biosynthesis 846, *846*
Citranaxanthin *557*
Citrate, addition to condensed milk 527
Citrem 461(T), *462*
Citric acid, additive 448
–, fruit 820, 820(T)
–, synergistic effect with antioxidant 220, 220(T)
Citronellol *383*
–, sensory property 386(T)
Citrus fruit, aroma substance 837
Citrus fruit juice, glucose oxidase 149
–, pectic acid, flocculation 153
Citrus juice, debitterizing 832
–, polygalacturonase 854
Citrus oil, autoxidation 397, *398*
CLA 162
Clarifying agent 464
Cleansing agent 495
Climacteric stage, fruit 843
–, –, ethylene 848(F)
Clipped fish 633
Clorine, disinfectant 495

Clostridium botulinum, food poisoning 470, 471(T)
Clostridium perfringens
–, food poisoning 470, 471(T)
Cloudiness, beer 901
Clove, aroma substance 974(T)
–, essential oil, chemical composition 974(T)
Clupeine, protein foam 062
Co-oxidation 209, 209(F)
–, degradation of carotinoid 241
Coating chocolate 968(T)
Cobalt 425(T), 426
Coberine, detection in cocoa butter 233, 233(F)
–, identification of cocoa butter 231(T)
Coccidiostatica 487, 487(T), 489(F)
–, chemical structure 487(T), 488(F)
Cocktail 936
Cocoa, acetic acid 964
–, adstringent taste 963(T), 964
–, aroma, odorant 964, 964(T)
–, GAE 962
–, shell, tryptamide 961, *961*
–, taste 963, 963(T)
Cocoa bean, fermentation, aldehyde 965, 965(T)
–, –, amine 965, 965(T)
–, –, free amino acid 965, 965(T)
Cocoa beverage, caffeine 962
–, theobromine 962
Cocoa butter, adulteration, detection 233, 233(F)
–, detection 649
–, detection of cocoa butter substitute 231, 233, 233(F)
–, fatty acid composition 648(T)
–, fatty acid distribution 176(T)
–, melting property 172
–, polymorphism 172
–, production 966
–, sterol 229(T)
–, tocopherol content 234(T)
–, triacylglycerol 173(T)
Cocoa butter substitute 648
–, analysis 173
–, detection 231(T), 233, 233(F)
–, sterol 231(T)
–, tocopherol content 234(T)
Cocoa liquor 965
Cocoa mass, disintegrated 965
–, production 965
Cocoa nib 960, 962
Cocoa powder, defatted 965
–, production 959(F), 966
Cocoa product, storage 969
Coconut fat, differential thermal analysis 667(F)
–, polymorphism 172(T)
Coconut oil 647, 648(T)
–, detection 665(T)
–, perfume rancidity 225

1006   Index

-, sterol 229(T)
-, unsaponifiable component 226(T)
Cod, aroma 629, 629(T)
Cod fish 621
Cofactor 098, 098(F)
Cofee, roasting, aroma change 945, 946(F)
Coffee 938
-, Arabica 939
-, Arabica/Robusta, aroma difference 944, 945(T)
-, aroma, AEDA 944, 944(T)
-, -, stability 945, 946(T)
-, aroma profile 944, 945(T)
-, caffeine 943
-, contact roasting 940
-, contact-convection roasting 940
-, decaffeinated 949
-, extract, pH value and taste 947, 947(F)
-, green, amino acid composition 942(T)
-, -, caffeine 941(T)
-, -, carbohydrate 941(T)
-, -, chemical composition 941, 941(T)
-, -, chlorogenic acid 941(T)
-, -, fatty acid 941(T)
-, -, polysaccharide 941(T)
-, -, protein 941(T)
-, instant 948
-, maltol 361(T)
-, off-flavor 945, 946(T)
-, packaging 941
-, potato note 945, 946(T)
-, potent odorant 944, 944, 945(T)
-, powder, natural radionuclide 470
-, production data 938(T)
-, roasted, acid 943
-, -, amino acid composition 942(T)
-, -, analysis 353, 353(T)
-, -, aroma substance 944, 944–946(T)
-, -, atractyligenin 943, 943
-, -, bitter taste 947, 948, 948(T)
-, -, cafestol 942, 943
-, -, carbohydrate 942
-, -, chemical composition 942, 942(T)
-, -, chlorogenic acid 942(T), 943, 943(T)
-, -, differentiation, Arabica/Robusta 943
-, -, diterpene 942, 943
-, -, kahweol 942, 943
-, -, lipid 942, 943(T)
-, -, methylcafestol,16-O- 942, 943
-, -, mineral 946
-, -, protein 942
-, -, storage 941
-, roasting 940
-, -, arginine loss 286(T)
-, -, weight loss 940
-, roasting grade 941, 943(T)
-, Robusta 939

-, short-time roasting process 940
-, species 938
-, storage, odorant loss 945, 946(T)
-, theobromine 943
-, theophylline 944
-, treated 949
-, trigonelline 944
Coffee bean 938, 938(F)
Coffee beverage 946, 947(T)
-, aroma substance 945(T), 947
-, chemical composition 947, 947(F)
-, pH value 947, 947(F)
Coffee cream 524
Coffee fruit, morphology 938, 938(F)
Coffee product 948
Coffee shrub 938
Coffee substitute 949
-, production 949
-, raw material 949
Coffee whitener 528, 528(T)
Coffee, green, harvest, processing 939
Cognac 931
Cohumulon 895, 895, 896(T)
Cola beverage 856
Cola nut, chemical composition 958
-, product 958
Cola-chocolate, caffeine 968
Colamine 585
Cold hopping 897
Cold instant pudding, alginate 304
Collagen 053, 571(T), 577
-, amino acid composition 578(T)
-, amino acid sequence 578, 580(T)
-, -, chain, α^1- 580(T)
-, biosynthesis 582, 582(F)
-, conformation 578, 578(F)
-, cross link 579
-, detection 613
-, enzymatic hydrolysis 583
-, fiber structure 579(F)
-, fish, shrinkage temperature 626
-, gelatin formation 583, 583(F)
-, mammal, shrinkage temperature 626
-, shrinkage 583
-, type 579(T)
Collagen fiber 579(F)
Collagenase 077(T), 583
Colling effect, threshold 431
Color change, meat 576
Colorant 443, 444(T), 446(T)
-, egg yolk 557, 557
-, margarine 661
Colored product, Maillard reaction 284, 284
Colupulon 895, 895, 896(T)
Comb honey 884
Comminuted base 854

Competitive immunoassay 141, 141(F)
Competitive inhibition 126
–, lipoxygenase 742(F)
–, pectin esterase 153, 154(F)
Compounded flour 708(T)
Conalbumin 549(T), 550
–, metal complex 550(T)
Conching, chocolate production 966
Condensed milk 526, 527(F)
–, aroma substance 539
–, carrageenan 306
–, polysaccharide 301(T)
–, production 526, 527(F)
–, production data 500(T)
Conditioning, oil raw material 647
Confectionary, polyalcohol 878(T)
Configuration, amino acid 013
Conformation, alginate 297
–, amino acid in protein 049(F)
–, amylopectin 324
–, amylose 297, 321, 322, 323(F), 333(F)
–, carrageenan 298
–, cellobiose 293
–, furanose 255
–, glucan, β-D- 296
–, lactose 293
–, lichenin 297
–, lysozyme 551(F)
–, maltose 293
–, monosaccharide 254
–, oligosaccharide 293
–, pectin 297
–, polysaccharide 296
–, protein 048, 054, 056(F)
–, pyranose 254
–, saccharose 293
Conglycinin, β-
–, amino acid sequence 752(T)
Coniferyl alcohol, biosynthesis 826, 826
Conjugate, enzyme immunoassay 141
Conjugated fatty acid 168
–, formation by heating fat 223, 223
–, occurrence 162
–, UV absorption 167(F)
Conjugated linoleic acid, milk 513
–, occurrence 162, 162(T)
–, structure 162
Connectin 571, 571(T)
Connective tissue 564, 577
–, detection 613
–, protein, determination 613
Consumption, alcohol 929, 930(T)
–, chocolate 966
–, fish 617(T), 618
–, food, national study 475
Consumption study 492

Contamination, bacterial, Indicator 022
–, food 467
–, prevention 467
Convicin 764
Cooked fish product 634
Cooked ham 598
Cooked sausage, chemical composition 599(T)
–, production 600(F)
Cooking cheese 531(T)
Cooking salt 982
Cooking salt (see also salt)
Cooling, fish 631
–, meat 595
–, vegetable, shelf life 799(T)
Cooling crystallization, saccharose 871
Cooling effect 431
–, Maillard reaction, product 431
–, xylitol 879
Coordination number, water 002, 002(T)
Copper 425, 425(T)
–, complexing by synergist 220(T)
–, enzyme cofactor 105
–, lipid peroxidation 199
–, taste threshold 425
Copra 648
–, production data 641(T)
Coprecipitate, production 537, 537(T)
Coriander, essential oil, chemical composition 974(T)
Corilagin 291, 291
Cork taste, wine 925, 925(T)
Corn, Osborne fraction 676, 677(T)
–, carotinoid 705
–, chemical composition 675(T)
–, flake 710
–, origin 670, 670(F)
–, product 710
–, production data 672(T)
–, zeaxanthine 237
Corn oil 649(T), 650
–, detection 665(T)
–, furan fatty acid 164(T)
–, polymorphism 172(T)
–, progesterone 228(T)
–, sterol 229(T)
–, tocopherol content 234(T)
Corned beef, fat content determination 663(T)
Cosubstrate 099
Cottage cheese, production 532
Cotton effect 060, 060(F)
Cottonseed, production data 641(T)
Cottonseed oil 649, 649(T)
–, detection 665(T)
–, sterol 229(T)
–, tocopherol content 234(T)
Coumestrol 762, 762
Coupled assay 138

Crabs, color retention 149
Crayfish 637, 637(T)
Cream 524
–, aroma substance 539, 540(T)
–, pasteurization, aroma 539, 540(T)
–, preparation, aroma 539, 540(T)
–, progesterone 228(T)
Cream cheese 529
–, propellant 465
Cream chocolate 968(T)
Cream of tartar 916
–, solubility 917(F)
Cream powder 528
Creamy flavor, milk product 543
Creamy margarine 660(T)
Creatine 585, 585
–, enzymatic analysis 139(T)
–, fish 626
Creatine kinase, inhibition 127(T)
–, reaction 139(T)
–, substrate binding 121
Creatine phosphate 585, 585
Creatine phosphokinase, meat 573
Creatininase, reaction 139(T)
Creatinine 585, 585
–, enzymatic analysis 139(T)
–, milk indicator 515(T)
Cresol, p-
–, formation 389
–, sensory property 389
Cresol, para- 375(T)
Cresolase 105
Crisp bread 740
Crocetin 239
Crocin 239
Croquant 882
Cross-link, protein, Maillard reaction 286
–, –, example 287, 288
–, –, pyrraline 287, 288
Cross-linkage, protein 086
Cross-linking, protein 070
Crude lecithin 654
Crumb, aroma substance 736(T)
–, bread, formation 731
Crust, proportion, bread 734(T)
Crustacean 636
–, chemical composition 637(T)
–, pigment 238
Crustacyanin 238
Cryptochlorogenic acid 822, 824
Cryptoxanthine, ester 240
–, occurrence in orange 240(T)
Crystal, cellulose 328(F)
Crystal lattice, triacylglycerol 171, 171(F)
Crystal modification, margarine 660
Crystal structure, fatty acid 165

Crystallization, inhibition 261
–, saccharose 871
Crystallization scheme, saccharose 872(F)
C-terminus, protein, determination 042
Cucumber, aroma formation 211(T)
–, aroma substance 793
–, bitter taste 820
–, color change during fermentation 795
–, pickled 802
–, –, faulty 804
–, processing, discoloration 796(T)
Cucurbitaceae, bitter taste 820
Cucurbitacin 820, 820
–, biosynthesis 820, 821
Cultivar differentiation, fruit 807(F), 816(F)
Cumaric acid, p- 822
Cumarin 824(T)
–, biosynthesis 826
Cumarin, 6-methyl-
–, sensory property 396(T)
Cumaroylquinic acid, fruit 824(T)
Curculin 438, 438(T)
–, amino acid sequence 438(T)
Curcumene, -ar- 973, 973
Curcumin 981
Curing, color, stability 576
–, meat 576, 577(F), 597
Currant, black 815(T)
–, red 815(T)
Curry powder 982
Cutin 187, 187(F)
Cyanidin 829, 830(T)
Cyanidinglycoside, visual detection threshold 830
Cyanocobalamin (see also vitamin B_{12}) 416, 417
Cyanogenic glucoside, biosynthese 761(F)
Cyanogenic glycoside, occurrence 760, 760(T)
Cyanosis 492
Cyclamate 433, 435
–, structure and taste 436(T)
Cycloartenol 232
Cyclobutanone, detection of radiolysis 224
Cyclocitral, β-, 242
Cyclodextrin 294, 294
Cycloheptaglucan 295(T)
Cyclohexaglucan 295(T)
Cyclohexanedione, 1,2- 066
Cyclomaltodextringlucanotransferase 294
Cyclooctaglucan 295(T)
Cyclopentanone, formation 283, 283
Cyclopentenolone, formation 268, 269
Cymene, p- 973
–, aroma, apricot 839
Cystathionine-β-lyase 792, 792
Cystatin C, egg white 551
Cysteic acid 024
Cysteine 010

–, *Strecker* degradation 365(F)
–, alkylation 068
–, discovery 011
–, disulfide interchange 700
–, dough, ascorbic acid effect 717(T), 718, 718(T)
–, flour improvement 717(T), 719, 720(F)
–, pasta 742
–, reaction 024, 068
–, reaction with monohydroperoxide 212, 213(F), 215(T)
–, S-alkylation 024
–, S-aminoethylation 043
–, S-methylation 068
–, S-sulfo derivative 067
Cysteine endopeptidase 076, 077(T)
–, specificity 078(T)
Cysteine peptidase, mechanism 076
Cysteine, free, wheat flour 699, 699(T)
Cystine, discovery 011
–, electrophilic cleavage 068
–, nucleophilic cleavage 067
–, reaction 024, 067
–, reduction 024
Cystine lyase 792, *792*
Cytochrome oxidase, inhibition 127(T)
Cytoskeletal protein 568

DABITC 044
DABMA 069, *069*
DABS-Cl 018
Daidzein *762*
Damascenone, honey aroma 889
Damascenone, β- *242*
–, aroma, tea 955, 955(T)
–, formation 243, *243*
Damascenone, (E)-β-
–, aroma, apple 839, 839(T)
Damascone α- *242*
–, sensory property 242
Damascone β- *242*
DANS-Cl 018
Dansylamino acid, N- 018, *018*
Dansylchloride 018, *018*
Date sugar 873
Datem 461
Datem 461, *461*, 461(T)
–, structure and activity 461, *461*
–, synthesis 461, *461*
DDT 476
–, half life 476
DE value, definition 876
Deacidification 654
Debittering, almond 881
–, citrus juice 832
–, cyclodextrin 296
–, protein hydrolysate 085, 086(T)

Decadienal, (E,E)-2,4-
–, sensory property 204(T)
Decadienal, (E,Z)-2,4-
–, linoleic acid, autoxidation 203(T)
–, sensory property 204(T)
Decadienal, 2,4-
–, formation 205, *205*
Decaffeination, coffee 949
Decalactone, butter aroma 541, 541(T)
–, formation 381, 386(F)
Decalactone, δ-
–, aroma, apricot 840
–, enantiomeric excess 355(T)
Decalactone, γ-
–, enantiomeric excess 355(T)
Decanal, aroma, orange 837, 838(T)
–, oleic acid, autoxidation 203(T)
–, sensory property 204(T)
Decarboxylase, substrate specificity 377(T)
Decarboxylation of amino acid, mechanism 104, 104(F)
Decatrienal, (E,E)-2,4-
–, deep-fried flavor 222
Decatrienal, (E,Z,Z)-2,4,7-
–, sensory property 204(T)
Decatrienal, 2,4,7-
–, linolenic acid, α-
–, –, autoxidation 203(T)
Decenal, (E)-2-
–, odor profile 343(F)
–, oleic acid, autoxidation 203(T)
–, sensory property 204(T)
Decenal, (Z)-2-
–, linoleic acid, autoxidation 203(T)
Decenal, (Z)-6-
–, parsley 978, 979(T)
Decoction process, beer 899
Decontamination, radiolysis, dose 225
Deep fried fat, analysis 220(T), 221
–, gel permeation chromatoraphy 221, 221(F)
–, hydroxyl number 220(T), 221
–, iodine number 220(T), 221
–, peroxide value 220(T), 221
–, unsaturated aldehyde, degradation 222, *222*
Deep fried taste, degradation of linoleic acid 222
Deep frying, reaction of fat 221, 221(T)
–, tocopherol loss 408(T)
Deep-frying, fat, composition 661
Deferrization, drinking water 986
Degree of hardness, English 987(T)
–, French 987(T)
–, German 986, 987(T)
–, USA 987(T)
Degumming 654
Dehydrated milk, aroma defect 344(T)
Dehydrated potato product, *Add-back* process 800

–, production 799, 800(F)

Dehydrated vegetable, off-flavor 799

Dehydro-β-methylalanine, nisin 039

Dehydroalanine 039, 071

–, formation 285

Dehydroaminobutyric acid 071

Dehydroascorbate reductase (see glutathione dehydrogenase)

Dehydroascorbic acid 418, *418*

–, biological activity 418, 428

Dehydrocholesterol, 7-

–, photochemical reaction *229*, 230(F)

Delphinidin 829, 830(T)

Demanganization, drinking water 986

Denaturation, protein 056

Deodorization, refining of oil 655

Deoxydicarbonyl compound 272

Deoxyfructosyl-1-lysine, ε-N- 071

Deoxylactulosyl-1-lysine, ε-N- 071

Deoxynivalenol, reference value, utilization 475(T)

Deoxyosone 272

–, detection 271, *274*

Deoxyosone, 1-

–, formation 272, *273*

–, secondary product 276, *277*

Deoxyosone, 3-

–, formation 272, *273*

–, secondary product 274, *275*

Deoxyosone, 4-

–, formation 272, *274*

–, secondary product 280

Deoxyribonucleic acid, nucleotide sequence 046

Depside 825, *825*

Depsidone 825, *825*

Desmethylsterol 229

Desmin 571, 571(T)

Desmosine 581, *581*

Desoxypentosone, 4-

–, formation 419, *419*

Dessert, alginate 301(T), 304

–, carrageenan 301(T), 306

Dessert wine (see liqueur wine)

Detection, heat treatment 285

Detection odor threshold, definition 341

Detergent, aroma defect in milk 541

Deterioration, fat, analysis 667

Deuterium, abundance 858, 858(T)

Dextran 332, *332*

Dextrin 326

Dextrin value according to *Lemmerzahl* 713

Dextrose 876

Dextrose equivalent, definition 876

DHA, structure, melting point 163(T)

Dhurrin 760(T)

Di-D-ketosylamino acid, formation 272, *272*

Di-tert-butyl-4-hydroxyanisole, antioxidative activity 216, 216(T)

Diabetes, *Maillard* reaction 271

Diacetyl, beer 905

–, –, aroma defect 149

–, biosynthesis 378(F)

–, butter aroma 540, 541(T)

–, formation 541(F)

–, off-flavor, soybean oil 198

–, soya oil, revesion flavor 651

Diacylglyceride, emulsifier 460

–, isomerization 645

Diagram, protein, helix parameter 052(F)

Diagram, φ, Ψ

–, protein 049, 050(F)

Diallyldisulfide, garlic 791

Diallylthiosulfinate, garlic 791

Diaminobutyric acid, 2,4- 781, *781*, 782(T)

Dianhydrosorbitol, 1,4:3,6- *879*

Diastase, honey 886, 887(F)

–, –, inactivation 887(F)

Diastatic activity, flour 713

Diazoacetamide, reaction with protein 067

Dibenzodioxin, polychlorinated 496, *496*

Dibromphenol, 2,6-

–, aroma, fish 629

–, odor threshold value 629

Dicarboxylic acid, oxidation of aldose 262

Dideoxyosone, 1,5- *279*

Dideoxyosone, 3,4-

–, formation 274, *275*

Dideoxyosone, 4,5- *280*

Diels-Alder adduct 223, *223*

Diet beer 904

Diet food, alkyl cellulose 330

Diet salt 983, 983(T)

Dietary fiber, analysis 335, 336

–, cereal 702, 709(T)

–, dried fruit 850(T)

–, fruit 815(T)

–, jam 852(T)

–, legume 748(T)

–, lignin 827, 827(F)

–, type of bread 735(T)

–, vegetable 779(T)

Dietary fiber (see fiber)

Diethyl carbonate 453, *453*

Diethyl dicarbonate 453, *453*

Diethyl-5-ethylpyrazine, 2,3-

–, potato aroma 788, 789(T)

Differential thermal analysis, fat 666, 667(F)

Differentiation, fruit 809

Difructosylamino acid 272

Digalactosyl diacylglycerol *180*

Digallate, enzymatic hydrolysis *154*

Dihydro-1,1,6-trimethylnaphthalene, 1,2- 242, *243*

Dihydrocapsaicin *980*
Dihydrochalcone, sweet taste 439, *439*, 834(T)
Dihydroisocumarin, taste 438, 439(T)
Dihydropyrazine 373, *373*
Dihydroxy-2-methyl-5,6-dihydropyran-4-one, formation 265, *266*
Dihydroxyacetone, from fructose 267, *267*
Dihydroxycholecalciferol, 1,25-
–, formation 229
Dihydroxyflavone, antioxidative activity 217, 217(T)
Dihydroxyglutamic acid, 3,4- 809
Dihydroxylysinonorleucine 582
Diisopropylfluorophosphate 075
–, enzyme inhibitor 107, *107*
Diketene 066
Diketogulonic acid, 2,3- 418, *418*
Dill, aroma substance 976, 977(T)
–, dried, aroma 976, 977(T)
–, herb, aroma substance 974(T)
–, –, dill ether 976, *976*, 977(T)
–, seed, essential oil, chemical composition 974(T)
Dill ether, dill, aroma 976, *976*, 977(T)
Dimethylamino acid, N- 019
Dimethylaminoazobenzene isothiocyanate 044
Dimethylaminoazobenzenemaleic imide,N- 069, *069*
Dimethylaminoazobenzenesulfonyl chloride 018, *018*
Dimethylaminonaphthaline-1-sulfonyl chloride, 5- 018
Dimethyldisulfide, formation 363, 367(F)
–, sensory property 364(T)
Dimethylproline, N,N-
–, orange juice, adulteration 858
Dimethylsterol 231
Dimethylsulfide, cabbage aroma 792
–, formation 363, *365*
–, sensory property 364(T)
Dimethyltrisulfide, formation 363, 367(F)
–, sensory property 364(T)
Dinitrofluorbenzene, enzyme inhibitor 107
Dinitrogen oxide, propellant 465
Dinitrogen tetraoxide, bleaching agent 464
Dinitrophenylamino acid, N-2,4- 019
Dinkel 670
Diol lipid 186
Diosmetin 834, *834*
Dioxin 496, *496*
–, formation 490
–, intake, average daily 496(T)
–, occurrence 496(T)
–, risk assessment 496(T)
Dioxoimidazolidine, 2,4- 020
Dioxopiperazine, cocoa, bitter taste 963(T), 964
–, formation, cocoa aroma 964, *964*
Dioxopiperazine, 2,5-
–, electron density 048(F)
Dipeptidase 077(T)

Dipeptide ester, sweet taste 441
Dipeptidylpeptidase 077(T)
Dipetide amide, sweet taste 441(T), 448
Diphenyl, antimicrobial action 454
Diphosphatidyl glycerol *179*
Disaccharide 295(T)
–, conformation 293
–, nonreducing 292
–, occurrence 295(T)
–, reducing 292
–, stability 866
Disinfectant 495
Dispersion 456(T)
–, aggregated 062, 063
Distillative deacidification 656
Disulfide bond, cleavage 041
–, gliadin 691, 692, 693(F)
–, gluten protein 692–694(F)
–, HMW subunit 691, 694(F)
Disulfide exchange, protein 068
Disulfide interchange, glutathione 699, *699, 700*
–, wheat dough 717(F)
Dithiobis-(2-nitrobenzoic acid), 5,5'- 068, *069*
Dithiothreitol 024
Dityrosine 087, 088(T)
Divicin *764*
Divinyl glycol, bitter taste, wine 926
DNA, dideoxy process 047
–, polymerase reaction 047
–, sequencing 046, *047*, 048(F)
DNP-Amino acid 019
Docosahexaenoic acid 162, 163(T)
Dodecen-γ-lactone, (Z)-6-
–, formation 380(F), 381
Dodecyl gallate *218*
Dog milk, chemical composition 501(T)
Dogfish 618
Domain, protein 055, 055(F)
Donkey milk, chemical composition 501(T)
Dopamine, fruit 813, *815*
Double cream cheese 529
Dough, fermentation 726, 726(F)
–, glutathione, reaction 699, *699, 700*, 700(F)
–, lipid binding 704
–, methylbutanol, formation 736, 736(T), 738(F)
–, microscopic study 726, 731, 732(F)
–, phenylethanol, formation 736, 736(T), 738(F)
–, production, addition of yeast 723
–, –, kneading process 725, 725(F), 725(T)
–, –, no-time process 726
–, rye flour, sour dough making 724, 724(F)
–, structure 726, 727(F)
–, viscosity 734(F)
–, water binding 733
Dough leavening, additive 722
Draff, beer 900

Dragée 881
Dried fruit, *Amadori* compound 271
–, vitamin C 849, 850(T)
–, water activity 004(T)
Dried vegetable, *Amadori* compound 271
Drinking chocolate 968
Drinking water 986
–, analysis 987, 987(T)
–, chlorination 986
–, deferrization 986
–, demanganization 986
–, disinfection 986
–, drug 987
–, fluoridation 426
–, hardness 986, 987(T)
–, hardness step 987(T)
–, limiting value 987(T)
–, ozonation 986
–, treatment 986
Drug, drinking water 987
Dry matter, fruit 815(T)
Dry product, alkyl cellulose 330
Dry sauce, production 604(F)
Dry soup, production 604(F)
Drying, fish 632
–, fruit 849
–, meat 597
–, milk 527, 528(T)
–, vegetable 799
Dulcitol, relative sweetness 259(T)
Durum wheat, pasta 741
–, semolina 710
D-value 135(F)
–, definition 131
–, lipase 189(T)

E. coli
–, enterotoxic strain 471(T), 472
Eating olive, bitter taste 803
–, black 804
–, chemical composition 803(T)
–, green 803
–, production 803
Edestin, aminoacylation 082(F)
Edible beef fat 642
Edible fruit, overview 808(T)
Edman degradation 021, *021*, 045
EDTA, additive 455(T)
ee-Value, definition 353
Eels 623
E-Factor, definition 664
–, linoleic acid 665(T)
–, palmitic acid 664
Effervescent lemonade powder 882
Egg 546
–, amino acid composition 548(T)

–, aroma defect 388
–, aroma substance 557
–, chemical composition 547(T)
–, dried product 558(F), 559, 560(T)
–, emulsifying property 559
–, fishy off-flavor 557
–, foam formation 558
–, frozen product 560, 560(F), 561(T)
–, liquid product 561(F), 561(T)
–, off-flavor 557
–, phospholipid, occurrence 179(T)
–, production data 547(T)
–, protein, foam formation 062
–, shell 547
–, storage 557
–, structure 546, 551(F)
–, thermal coagulation 558
–, trimethylamine 557
Egg box, alginate gel 304(F)
Egg powder 558(F), 559, 560(T)
Egg product 557
–, production 558(F)
Egg protein, biological valence 030
Egg white 548
–, carbohydrate 552
–, glycoprotein 549(T)
–, mineral 552(T)
–, ovomacroglobulin 549(T)
–, protein 548, 549(T)
–, viscosity 549(F)
–, vitamin 556(T)
Egg yolk 553
–, apoprotein 553, 553(T)
–, carbohydrate 556
–, cholesterol 556
–, colorant 557, *557*
–, content, determination 233
–, electrophoresis 553
–, fatty acid composition 555, 556(T)
–, –, feed 555, 556(T)
–, fractionation 553, 553(F)
–, frozen product 558(F)
–, –, gel formation 560, 560(F)
–, –, viscosity 560(F)
–, HDL 553
–, LDL 553, 553(F)
–, lipid 555, 555(T)
–, lipoprotein 185(T), 553, 553(T)
–, livetin 555
–, lutein 238
–, mineral 552(T), 556
–, phosvitin 554, 554(T), *555*
–, plasma 553, 553(T)
–, protein 553, 553(T)
–, –, analysis 553, 553(F), 553(T)
–, vitamin 556, 556(T)

EHMF (see Ethylfuraneol)
Ehrlich pathway *737*
Eicosapentaenoic acid 162, 163(T)
Elaeostearic acid, α-
–, UV absorption 167(F)
Elaidic acid 163(T), *165*
–, melting point 166(T)
Elastase 075, 584
–, active serine, mechanism 115
–, substrate binding 110(F)
Elastin 571(T), 584
–, amino acid composition 578(T), 584
–, enzymatic hydrolysis 584
Electrophoresis, egg yolk protein 553, 553(T)
–, hemoglobin 605(F), 610(F)
–, meat protein 609–611(F)
–, myoglobin 605(F), 610(F)
–, protein, vegetable species 770, 780(F)
Elemicin, banana 837, *837*
Eleostearic acid, α-
–, structure, melting point 163(T)
Eleostearic acid, β-
–, structure, melting point 163(T)
ELISA 142, 142(F)
–, noncompetitive 142, 142(F)
Ellagic acid 825, *825*
–, occurrence 825(T)
Emmental cheese, furaneol 363(T)
Emmentaler, aroma 542, 542(T)
–, taste compound 542, 542(T)
Emulsifier 456, 456, 457(T)
–, capacity 457
–, casein, β- 063
–, critical micelle concentration 458, 458(F)
–, critical micelle temperature 458, 458(F), 458(T)
–, crystalline mesophase 458, 458(F)
–, dough making 721
–, HLB-value 460(T)
–, phase diagram 459, 459(F)
–, protein 063
–, structure and activity 457, 457(F)
–, synthetic 460
Emulsin, glucosidase, β- 760
Emulsion 456, 456(T), 457(F)
–, phase diagram 459, 459(F)
–, sausage 599, 599(F)
–, stability 457(F)
Emulsion, O/W, protein 063
Enantiomeric excess 353, 355(T)
Encapsulation, aroma 398
End-point method, enzymatic analysis 138
Endomysium 564, 566(F)
Endopeptidase 074, 076, 077(T)
–, aspartic acid 079
–, industrial application 150, 151(T)
–, metal-containing, mechanism 076, 079(F)

–, –, specificity 078(T)
–, rennin-type 079
–, specificity 075, 078(T)
Endoperoxide 196, 206, *207*
Endoproteinase Glu-C 043
Endotoxin, food poisoning 470
Endoxohexahydrophthalic acid anhydride, Exo-cis-
   3,6- 066, *066*
Endoxylanase, arabinoxylan 703
Eneaminol, 2,3-
–, *retro-Michael* reaction 272, *273*
Energy value, alcohol 892
–, carbohydrate 008
–, lipid 158
–, protein 008
–, triacylglycerol 171
Enterotoxin, food poisoning 470
Enzymatic analysis 137
–, competitive inhibitor 140
–, coupled assay 138
–, determination of enzyme activity 140
–, determination of substrate concentration 120
–, end-point method 138
–, enzyme immunoassay 141, 141(F), 142, 142(F)
–, inhibitor 125, 140
–, kinetic method 140
–, substrate determination 138
–, two-substrate reaction 140
Enzymatic browning 105
–, potato 122
Enzymatic liquefaction, fruit 855
Enzyme, activation energy, denaturation 132
–, active conformation 109, 110
–, active site 106
–, activity curve 118, 118(F)
–, activity determination 098, 117, 140
–, –, value to be measured 098
–, allosteric effector 125
–, aroma defect, prevention 130(T)
–, cacao bean 961
–, catalytic activity 110
–, chemical modification 107
–, cofactor 098, 098(F)
–, collagenolytic 583
–, cosubstrate 098(F), 099
–, covalent catalysis 114, 114(T)
–, definition 093
–, dissociation 129
–, filter acid 853
–, first order reaction 120, 140
–, food technology 144
–, fruit 807
–, fruit ripening 845, 846
–, general acid-base catalysis 113
–, glutathione dehydrogenase, wheat 698, *698*, 698(T)
–, honey 886, 887(T)

–, immobilization 147, 147, 148(F)
–, immobilized 141, 145
–, inactivation 112
–, –, rate equation 131, *131*
–, inactivation rate, pH value 135
–, industrial application 146(T)
–, industrial preparation, isolation 145
–, inhibition, *Lineweaver-Burk* plot 128(F)
–, irreversible inhibition 126
–, isolation 096, 096(T)
–, kinetics 117
–, –, *Hofstee* plot 121(F)
–, –, *Lineweaver-Burk* plot 120, 121(F)
–, –, *Michaelis* constant 119
–, –, maximum velocity 119
–, –, pH dependency 128
–, –, single-substrate reaction 117
–, –, two-substrate reaction 121
–, metal ion 104
–, milk 516, 516(T)
–, molar catalytic activity, definition 098
–, multiple form 097
–, muscle 571(T), 573
–, nomenclature 097
–, pH optimum 129(T)
–, ping-pong mechanism 122
–, prochiral substrate, binding 108, 108(F)
–, prosthetic group 098, 098(F), 102
–, purification 096, 096(T)
–, pyruvate at active center 117(F)
–, pyruvate at active site 114(T)
–, rate constant 120(T)
–, reaction, retardation 007(T)
–, reaction rate 110
–, –, influencing fractor 117
–, –, temperature 141
–, reaction specificity 095, 095(F)
–, redox catalysis 102
–, regulatory specificity 094, 124
–, reversible inhibition 126, 127(T)
–, –, competitive 126, 127(T)
–, –, non-competitive 127, 127(F)
–, –, uncompetitive 128
–, sarcoplasma 573
–, specific catalytic activity 098
–, specificity 094
–, stereospecificity 099, 110
–, structure 095
–, substrate analog inhibitor 107
–, substrate binding 108, 108, 109(F)
–, –, order 121
–, –, reaction rate 109
–, substrate specificity 094
–, systematic classification 100(T)
–, systematic numbering 097
–, tea 953

–, thermal inactivation 130, 131(T)
–, thermal stability 134
–, transition state 110
–, transition state analog 112, 112(F)
–, vegetable 770
–, wheat 695
–, zero order reaction 119
Enzyme activity, D-value 131
–, detection 117
–, detection of heat treatment 093
–, frozen food 135
–, pressure 136
–, water content 137, 137(T)
–, z-value 133
Enzyme catalysis, activation energy 093, 094(T)
–, activator 124
–, electrophilic reaction 104, 115, 117, *117*
–, hydrogen ion concentration 128
–, inhibitor 125
–, initial reaction rate 118
–, nucleophilic reaction 114
–, pre-steady state 118
–, reaction mechanism 110
–, reaction rate 094, 094(T)
–, steady state 119
–, temperature dependency 130
–, temperature optimum 133
–, theory 106
Enzyme denaturation, activation energy 132, 133
Enzyme immunoassay 141
–, example 142(T)
–, principle 141(F)
Enzyme inhibitor, occurrence 126
Enzyme preparation, industrial 144, 146(T)
Enzyme unit, definition 098
Enzyme-substrate complex, binary 123, 124(F)
–, bond deformation 109
–, covalent binding 114(T)
–, dissociation constant 119
–, entropy effect 112
–, induced-fit model 109, 110(F)
–, lock and key hypothesis 109
–, ordered mechanism 122
–, orientation effect 111, 111(T)
–, random mechanism 121
–, steric effect 111
Enzyme-substrate-complex, ternary 121
EPA, structure, melting point 163(T)
Epicatechin 827, *827*
–, cacao bean 963
–, cocoa taste 963, 963(T), 964
–, fruit, content 829(T)
–, wine 920(T)
Epicatechin equivalent (ECE) 963
Epimysium 564, 566(F)
Epitheaflavic acid 956, *957*

Epitope, antigen 753
Epoxy fatty acid, formation in lipid peroxidation 212(T), 213(F)
Epoxy-(E)-2-decenal, trans-4,5-
–, autoxidation, linoleic acid 203(T)
Epoxy-2-decenal, 4,5-
–, aroma, orange 837, 838(T)
Epoxy-p-menth-1-ene, 3,9-
–, aroma, dill herb 976, 977(T)
Ergocalciferol 406, 406
Ergosterol 229, 407
Ergot 472
–, ergotism 472
Ergot alkaloid 472, 472, 474(T)
Eriodictyol 832, 832
Erlose, honey 886(T)
Erucic acid, occurrence 651(T), 652
–, structure, melting point 163(T)
Erythrodiol 646, 646(T)
Erythrose 249, 249, 251
Escherichia coli, food poisoning 470
Essence 396
Essential amino acid, requirement 030(T)
Essential fatty acid 162
–, biosynthesis 169
Essential oil, concentration in spice 973(T)
–, constituent 971, 974(T)
–, preparation 394
Ester, biosynthesis 379, 379
–, odor threshold value 381(T)
–, wine 921, 922, 923(T)
Ester oil 176
–, detection 176, 664
Esterase, active serine, detection 107
–, differentiation from lipase 188
–, wheat 696
Estradiol, 3,17- 228, 228
Estragole, spice 972, 972
Estrone, 17- 228, 228
Ethanal, oxidation to acetic acid 984
–, sensory property 360(T)
–, yoghurt 540
Ethane, linolenic acid, α-
–, –, autoxidation 207
Ethanol, content, wine 919
–, denaturation 931
–, enzymatic analysis 139(T)
–, odor threshold value 341(T)
–, production 930
–, sugar fermentation, detection 919
Ethephon 847, 847
Ethoxyquin 218
Ethyl 2-methylbutyrate, aroma, orange 837, 838(T)
–, olive oil 646(T)
Ethyl butyrate, aroma, grapefruit 838
–, odor threshold value 341(T)

Ethyl carbamate 453, 453
Ethyl cyclohexanoate, olive oil 646(T)
Ethyl decadienoate, (E,Z)-2,4-
–, formation 379(F)
Ethyl isobutyrate, olive oil 646(T)
Ethyl urethane 453, 453
Ethyl vanillin, structure, sensory property 396(T)
Ethyl-3-hydroxy-5-methyl-3(2H)-furanone, 2-
–, soy sauce 767
Ethylasparagine, 4-N- 954
Ethylene, biosynthesis 847, 847
–, fruit ripening 844(F), 847, 848(F)
–, production, fruit 847, 847(T)
Ethylene oxide 453
–, antimicrobial action 453
Ethylenediamine tetraacetic acid, additive 455(T)
Ethylfuraneol, formation 268, 269
–, swiss cheese 542(T)
Ethylglutamine, 5-N- 954
Ethylmaleic imide, N-
–, SH-reagent 068
Ethylmaleic imide, N-, 068
Ethylmaltol 361
Ethyloctanoic acid, 4-
–, mutton odor 607, 607(T)
Ethylthiol, wine defect 925
Eugenol 375(T)
–, banana 837
–, basil 978, 978(T)
Exopeptidase 074, 076, 077(T)
Exotoxin, food poisoning 470
Extension test, HMW subunit 692
Extensograph 714(F)
Extract content, beer 901
Extraction, oilseed 647
Extrusion, pasta 742
–, protein 089
–, starch 325

Falling number, rye flour 715
Falling number according to Hagberg and Perten 713
Farinograph 713, 713(F)
Farnesene, β- 385
Farnesene, cis-α- 385
Farnesene, tr-α- 385
Farnesol 385
Fat, analysis 662
–, –, minor constituent 665
–, –, oxidation state 667
–, animal, detection 665(T)
–, –, recovery 640
–, bleaching, detection 669(T)
–, characteristic value 663
–, consumption 640, 643(T)
–, deep-frying, composition 661
–, detection 664, 664(T)

–, –, lipolysis 667
–, determination, NIR 706(T)
–, differential thermal analysis 666, 667(F)
–, E-factor 665
–, fatty acid, oxidation 221, 222(F)
–, fractionation 659
–, hardened, detection 665(T)
–, heating (deep frying) 220
–, hydrocarbon 224
–, hydrogenation 656
–, –, principle 169, 169
–, identification 665(T)
–, indigestion 177
–, interesterification, process 658
–, peroxide value 667
–, plant, fatty acid composition 159(T)
–, –, fatty acid distribution, rule 176
–, –, resistance to autoxidation 215
–, polymerization 223
–, production data 640, 641(T), 643(T)
–, raffination, loss of tocopherol 233
–, raw material, production 641(T)
–, refining 653, 654(T)
–, –, detection 669, 669(T)
–, smoke point 668
–, stability in deep frying 224, 224(T)
–, sterol 229(T)
–, storage stability 668
–, tailored 659
–, unsaponifiable component 225, 226(T)
Fat cheese 529
Fat content, determinaton 662
Fat degradation, cheese ripening 533
Fat globule, membrane, composition 514
–, milk 514, 514(T)
Fat hardening, trans-fatty acid 658
Fat mimetics 463
Fat powder 661
–, production 662(F)
Fat replacer (see fat substitute)
Fat subsitute, retrofat 464
Fat substitute 463
–, carbohydrate 463
–, carbohydrate polyester 464
–, microparticulated protein 463
–, synthetic 463
Fatty acid, bitter taste 212(T)
–, branched 160, 161(T)
–, –, analysis 664
–, carboxyl group, dissociation 165
–, –, formation of hydrogen bond 165
–, cereal 704, 704(T)
–, chemical property 167
–, conformation 1645
–, crystal structure 165
–, fractional crystallization 167

–, free, analysis 667
–, –, fat 667
–, –, recovery 654
–, green coffee 941(T)
–, isolation 172
–, low-molecular, occurrence 159
–, –, sensory property 159, 160(T)
–, melting point 165, 166(T)
–, methylation 167, 167
–, nomenclature 159
–, odd numbered 160, 161(T)
–, oxidized, formation 211
–, physical property 165
–, saturated, autoxidation 221
–, –, melting point 161(T)
–, –, sensory property 160(T)
–, –, structure 161(T)
–, sensory property, pH value 160, 160(T)
–, shorthand designation 159
–, solubility 167, 167(F)
–, tocopherol requirement 407(T)
–, unsaturated, argentation chromatography 168
–, –, autoxidation 191
–, –, autoxidation, heavy metal 199
–, –, autoxidation, secondary product 203
–, –, autoxidation, start 196
–, –, biosynthesis 169, 169(F)
–, –, common structural feature 162
–, –, double bond, configuration 159, 162, 163(T)
–, –, halogen addition 168, 168
–, –, hydrogenation 169, 169
–, –, melting point 163(T)
–, –, photometric determination 168
–, –, reaction 168
–, –, rearrangement in conjugated acid 168
–, –, taste 164(T)
–, urea adduct 166
–, UV absorption 167, 167(F)
Fatty acid composition, analysis 664
–, legume 762(T)
–, sunflower oil, variation range 665(T)
Fatty acid methyl ester, production 172
–, synthesis 167
Fatty alcohol 186
Favism 764
FD-chromatogram, definition 350
–, example 347(F), 351, 352(F)
FD-factor, definition 350
FDNB 019
Feder's method, meat 613
Feed additive, chemical structure 488(F)
Fehling's reaction 267
Fenchone 384
Fenton reaction 202
Fenugreek, aroma substance 974(T), 976, 976
Fermentation, alcoholic 893(F)

–, beer 900
–, dough 726, 726(F)
–, grape must 916
–, red wine 917
–, white wine 917
Ferri-protoporphyrin IX *103*
Ferulic acid *822*
–, antioxidative activity 217, 217(T)
–, thermal degradation 374(F)
Ferulic acid sterol ester, antioxidant 662
FFI 281, *281*
Fibrillar protein 053
Fibrin, formation 594
Fibrinogen, blood plasma 594
Fibronectin, meat 590
Fibrous protein 053
Ficin 076, 077(T)
–, active cysteine 115
–, meat, tenderizing 598
–, specificity 078(T)
Ficin inhibitor 549(T)
–, egg white 551
Fig 815(T)
Fig coffee 950
Filament, G-actin 570, 570(F)
–, thick 567, 568, 569(F)
–, thin 566, 568, 569(F), 572, 572(F)
Filbertone, aroma quality 340(T)
–, halzelnut, roasting 882
–, identification 353, 354(F)
–, nougat aroma 882
–, odor threshold value 882
Filled milk 520
Filtration enzyme 853
Fining, wine 918
First order reaction, enzyme 131
Fischer's indole synthesis 033
Fish 617
–, amine 626, 627(F)
–, amine oxide 626, 627(F)
–, amino acid composition 624(T)
–, ammonia odor 618
–, aroma 628, 628(T)
–, aroma defect 388
–, aroma substance 628
–, betaine 626
–, category 617
–, chemical composition 622(T)
–, consumption 617(T), 618
–, contractile protein 625
–, cooling 631
–, creatine 626
–, dark muscle 623, 624(F)
–, drying 632
–, electric resistance 627(F)
–, formaldehyde, formation 626

–, free amino acid 626
–, freezing 631, 631(F), 632(T)
–, freshness 629, 630
–, fried 634
–, frozen, shelf life 632(T)
–, glycogen 627
–, guanophore 623
–, heating 634
–, hemoglobin 625
–, histidine 626
–, K-value 630
–, kamboko 635
–, light muscle 623, 624(F)
–, lipid 627
–, mineral 628, 628(T)
–, muscle fiber 623, 624(F)
–, myofibril, enzymatic hydrolysis 625(F)
–, myofibrillar protein 625, 625(T)
–, myoglobin 625
–, off-flavor 629
–, pickling 634
–, polyunsaturated fatty acid, ω-3- 627, 627(T)
–, post-mortem change 627(F), 629
–, processing 630, 631(F)
–, product, chemical composition 632(T)
–, protein 624
–, –, heat stability 625
–, –, solubility 634(T)
–, protein concentrate 636
–, quality criteria 627(F), 630
–, salting 633
–, sarcoplasma protein 625
–, shelf life 627(F), 632, 633(F)
–, –, bacteria 623
–, skin, component 623
–, smoking 633
–, species 619(T)
–, storage 627(F), 630
–, –, aroma 628(T), 629
–, –, off-flavor 629
–, surimi 635
–, toxic constituent 629
–, trimethylamine 626
–, urea 627
–, urea content 618
–, vitamin 628
–, world catch data 617(T)
Fish egg 635
Fish liver oil, furan fatty acid 164
Fish muscle, structure 623, 624(F)
Fish oil, wax 187
Fish product 630
Fish protein, enzymatic liquefaction 150
–, solubility 633(F)
Fish protein concentrate, plastein reaction 086, 087(T)

Fish sperm 636
Fishy off-flavor 629, 629(T)
Fitelson reaction 664(T)
Flat fish 622
Flatulency, legume 759
Flavan-3,4-diol, cocoa 963, *963*
Flavan-3-ol 827, *828*
Flavanol, biosynthesis 835, *836*
–, enzymatic oxidation, tea 956, *957*
–, occurrence 823(T)
Flavanone 832, *832, 833*, 833(T)
–, biosynthesis 835, *836*
–, chemical structure *823*
–, citrus fruit 833(T)
–, conversion to dihydrochalcone *832*
Flavin adenine dinucleotide (FAD) 102, *102*
–, reaction *102*
Flavin mononucleotide (FMN) 102, *102*
Flavone 834, *834*
–, biosynthesis 835, *836*
–, citrus fruit 833(T)
–, fruit 834
–, occurrence 823(T)
Flavonoid, biosynthesis 835, *836*
–, chemical structure *822*
Flavonol 834, *834*
–, fruit 834(T)
–, occurrence 823(T)
Flavononol, biosynthesis 835, *836*
Flavoprotein, egg 551
–, egg white 549(T)
Flavor, definition 340
Flavor dilution factor 350
Flavor enhancer 430
Flavylium cation *830*
Flounder 622
Flour, triglyceride, spherosome 705
Flower honey 884
Fluorenylmethylchloroformate 018, *018*
Fluorescamine 022
Fluorine 425(T), 426
Fluoro-2,4-dinitrobenzene, 1-, 019, 065
Fluoro-3-nitrobenzene sulfonic acid, 4-, 065
Fluoro-4-nitrobenzo-2-oxa-1,3-diazol,7- 019
FMOC 018
Foam 456(T)
–, egg protein 062
–, puroindolin 695
Foam builder, beer 902
Foam formation, protein 062
Foamy candy 880
Folic acid 415, *415*
–, bioavailability 416
–, deficiency 416
–, occurrence 409(T)
–, supplementation 416

Folic acid conjugase 416
Folin reagent 020
Fondant 880
Fondant mass, chemical composition 880(T)
Food, allergen 751, 754(T)
–, –, example 753
–, analysis, chemometry 666
–, aroma, analysis 345
–, isotope analysis 858, 858, 859(T)
–, lignan, content 835(T)
–, mutagenic compound 027, 028(T)
–, origin, determination 857, 860(T)
–, phase transition 005
–, radiation detection 073
–, saponin content 763(T)
–, taint 343, 345(F)
–, viscosity, temperature dependency 006
Food color 443, 444(T), *446(T)*
Food consumption, national study 475
Food contamination 467
Food monitoring 492
–, mycotoxin 472
Food Orange 8 444(T), 446(T)
Food poisoning, bacterial 470, 471(T)
Food processing, reaction of protein 070
Food seasoning, protein hydrolysate 603
Formaldehyde, oxidation of glucose 267, *267*
–, sensory property 360(T)
Formic acid, oxidation of glucose 267, *267*
Fortification, food 029
–, lysine 023
Fraxetin 824(T)
Freeze concentration, fruit juice 855
Freeze-thaw stability, modified starch 326
–, xanthan gum 331
Freezing, enzyme activity 135
–, fish 631, 631(F), 632(T)
–, fruit 850
–, meat 595
–, vegetable 801
French fries, aroma analysis 350, 351, 352(F)
–, aroma model 357, 359(T)
–, aroma value 356, 359(T)
–, odorant, concentration 356, 359(T)
Fresh cheese 531(T), 532
Freshwater fish 623
Friabilin, wheat 695
Frogdrum 639
Frozen meat, detection *611*, 611(F), 612
–, shelf life 595(T), 596, 596(F)
Fructo-oligosaccharide, production 875
Fructofuranose, α-D- 255(T)
Fructofuranose, β-D- 255(T)
Fructofuranosidase, β-D-
–, application 152
Fructopyranose, α-D- 255(T)

Fructopyranose, β-D- 255(T)
Fructose 251
–, alkaline degradation 268, 268(T)
–, aqueous solution, viscosity 865(F)
–, cyclization 249
–, enzymatic analysis 139(T)
–, equilibrium mixture 255(T)
–, fruit 818(T)
–, glycemic index 867
–, isomerization 266, 266
–, Maillard reaction 271, 271
–, mutarotation, temperature dependence 866(F)
–, nutritional/physiological property 864(T), 866
–, occurrence 250(T)
–, production 863(T), 878
–, relative sweetness 259(T), 513(T), 863(T)
–, solubility 862(T)
–, specific rotation 258(T)
–, stability 862
–, sweet taste, AH/B/X-model 261
–, –, temperature dependence 866(F)
–, taste threshold value 259(T)
–, temperature dependency, mutarotation equilibrium
    259, 260(F)
–, –, sweet taste 259, 259(F)
–, water absorption 866(F)
Fructose, D-
–, production 332
Fructose, L- 878
Fructoselysine 285, 285, 286
Fructosidase, β-
–, enzymatic analysis 138
Fruit 807
–, amine 812, 817(T)
–, analysis 856, 857–859(T)
–, anthocyan, content 829(T)
–, aroma formation 846
–, aroma substance 837
–, ascorbic acid content 842(T)
–, astringent taste 825, 829
–, biotin 842
–, candied 851
–, –, protective coating 324
–, canned 850
–, –, bakery product 851
–, carbohydrate 817, 818(T)
–, carotene, β- 842
–, carotinoid 818, 819(T)
–, catechin, content 829(T)
–, chemical composition 807, 815, 816(T)
–, climacteric stage 843
–, consumption form, overview 808(T)
–, cultivar differentiation 807(F), 809, 816(F)
–, dietary fiber 815(T)
–, dried, chemical composition 850(T)
–, drying 849

–, electropermeabilization 853
–, enzymatic browning 835
–, enzymatic liquefaction 854
–, enzyme 807
–, epicatechin, content 829(T)
–, ester formation 379, 379
–, ethylene, time course 844(F)
–, ethylene production 847, 847(T)
–, extraction 853
–, flavanone 833(T)
–, flavone 833(T), 834
–, flavonol 834(T)
–, free amino acid 809, 816(T)
–, freezing 850
–, heating 849–851
–, lignin 827, 827(F)
–, lipid 818, 818(T)
–, melatonin 813, 817, 817(T)
–, mineral 843, 843(T)
–, organic acid 820, 820(T)
–, pantothenic acid 842
–, pH value 815(T)
–, phenolic compound 822
–, pigment formation 846
–, production data 810(T)
–, promotion of ripening 847
–, protein 807
–, protopectin 845, 845
–, quercetin, content 834
–, retardation of aging 848
–, ripening 843
–, –, phenolcarboxylic acid 826
–, steeping in alcohol 851
–, storage, controlled atmosphere 848, 849(T)
–, –, retardation of aging 848
–, sugar 818(T)
–, sugar alcohol 817
–, sulfurous acid, treatment 849, 850
–, sweetening with sugar 851
–, temperature of phase transition 006(T)
–, triterpenoids 819
–, tryptamine 813, 817, 817(T)
–, vitamin 842
Fruit aroma liqueur 935
Fruit juice 852
–, adulteration 820
–, beverage 856
–, chemical composition 853(T)
–, citric acid/isocitric acid 822(T)
–, clarification 853
–, concentrate, natural radionuclide 470
–, concentration 855
–, consumption 852, 852(T)
–, enzymatic clarification 153
–, freeze concentration 855
–, freezing temperature 855(F)

–, haze, PVPP 837
–, membrane filtration 855
–, pasteurization 854
–, polyvinylpyrrolidone 333
–, protein complex 837
–, storage 854
–, treatment 853
–, turbidity 153
Fruit liqour 931
Fruit liquor, benzaldehyde 932
–, chriesiwasser 931
–, hydrogen cyanide 932
Fruit nectar 854
–, chemical composition 853(T)
–, production 854
Fruit paste, production, cellulase 153
Fruit powder 856
Fruit product 849
–, adulteration, detection 857, 857–859(T)
–, deep frozen fruit 850
–, fruit juice 852
–, fruit juice beverage 856
–, fruit juice concentrate 854
–, fruit powder 856
–, fruit pulp 851
–, fruit slurry 851
–, fruit syrup 855
–, jam 851
–, jelly 851
–, marmalade 851
–, nectar 854
–, plum sauce 852
–, sensory evaluation 260, 260(F)
–, sterile can 850
Fruit pulp 851
Fruit ripening, ethylene 847, 848(F)
Fruit sap liqueur 935
Fruit slurry 851
Fruit spirit 932
Fruit sugar (fructose)
–, production 878
Fruit syrup 855
Fruit wax 820
Fruit wine 928, 928(T)
Frying, change in fat 220, 220(T)
–, oil, smoke point 668
Fucose, occurrence 250(T)
Fucosidolactose 295(T)
Fumarase, rate constant 120(T)
Fumaric acid 447, 448
–, synergistic effect with antioxidant 219
Fumosin, mycotoxin 473, 474(T)
Functional food 523, 867
–, definition 867
Fungicide 473
–, chemical structure 477(F)

–, LD$_{50}$ 480(T)
–, trade name and activity 480(T)
Furan 490
–, formation 490, 491(F)
–, occurrence 490
Furan fatty acid 164, 164, 164(T)
–, enrichment 664
–, oxidation 652(T)
–, photooxidation 198, 199(F), 650, 652(T)
Furaneol, aroma, pineapple 841, 841(T)
–, –, strawberry 841, 841(T)
–, –, tea 955, 955(T)
–, caramel-like odor 277
–, formation 268, 269, 277, 282, 282
–, odor threshold value, effect of odorant 341, 341(T)
–, potato aroma 789(T)
–, swiss cheese 542(T)
–, warmed over flavor 608, 608(T)
Furaneol (see furanone, 4-hydroxy-2,5-dimethyl-3(2H)-)
Furanone 362(T)
–, hydroxy-2,5-dimethyl, 4-
–, –, quantification 362
Furanone,
–, hydroxy-2,5-dimethyl, 4-
–, –, glycoside 361, 361
Furanone, 2-ethyl-4-hydroxy-5-methyl-3(2H)-
–, occurrence 362(T)
Furanone, 3-hydroxy-4,5-dimethyl-2(5H)- 361, 362(T)
–, degradation 363
–, occurrence 362(T)
–, sensory property 362(T)
Furanone, 4-hydroxy-2,5-dimethyl-3(2H)- 362
–, aroma quality 340(T)
–, occurrence 363(T)
–, odor threshold value 363(T)
–, sensory property 363(T)
Furanone, 4-hydroxy-5-methyl-
–, occurrence 362, 363(T)
Furanone, 4-methoxy-2,5-dimethyl-3(2H)-
–, occurrence 363(T)
Furanone, 5-ethyl-3-hydroxy-4-methyl-2(5H)- 362(T), 363(F)
–, formation 362, 363(F)
–, sensory property 362(T)
Furanose, conformation 255
Furcellaran 306
Furfural, formation 274
–, odor threshold value 341(T)
Furfurylthiol, 2-
–, coffee aroma 944, 944(T), 945(F), 945, 946(T)
–, –, formation 944
–, formation 364, 365
–, quantitative analysis 356, 358(F), 359(T)
–, release from disulfide 364

–, sensory property 364(T)
Furosine 070
–, formation 279, *279*, 284, *284*
–, indicator, heat treatment 285
–, occurrence 285(T)
Fusariotoxin F$_2$ 474(T)
Fusariotoxin T$_2$ *473*, 474(T)
Fusel oil, alcohol production 930

Galactaric acid 262, *262*
Galactitol 261
Galacto-oligosaccharide, production 878
Galactose *251*
–, enzymatic analysis 139(T)
–, equilibrium mixture 255(T)
–, nutritional/physiological property 864(T), 866
–, occurrence 250(T)
–, production 863(T)
–, relative sweetness 259(T), 863(T)
–, specific rotation 258(T)
–, sweet taste, temperature dependence 866(F)
Galactose dehydrogenase, reaction 139(T)
Galactose sulfate 304, 305(T)
Galactose, α-D-
–, energy content 255(T)
Galactose, β-D-
–, energy content 255(T)
Galactosidase, α-
–, saccharose production 871
Galactosidase, β-D-
–, application 152
Galacturonic acid 262
–, competitive inhibitor 153, 154(F)
–, synthesis *262*
Gallic acid 825
–, fruit, content 825(T)
–, occurrence 825(T)
–, wine 920(T)
Galllic acid equivalent (GAE) 962
Gallocatechin, cacao bean 963
Game 593
Ganglioside 181
–, milk 181, *182*, 515
Garden bean, hydrocyanic acid 761(T)
–, production data 747(T)
–, proteinase inhibitor, stability 758(T)
–, saponin content 763(T)
Garlic, aroma substance 790
Gas extraction, aroma analysis 348, 348(T)
Gaseous sterilant 453
GC olfactometry 350, 352(F)
Gel, formation, amylose 320(F)
–, –, cellulose ether 329, 329(F)
–, –, polysaccharide 299, 299(F)
–, –, protein 062
–, –, xanthan gum 331

–, heat resistant 303
–, ionic strength 063
–, permeation chromatography, deep fried fat 221, 221(F)
–, thermally reversible 306
–, thermoplastic 063
–, thermoreversible 063
Gel builder 464
Gelatin, candy 881
–, gel formation 583, 583(F)
–, HLB-value 460(T)
–, production, lipase 154
Gelatinization temperature, starch 317(T)
Genetic code 046, 046(F)
Genistein *762*
Gentian liquor 932
Gentianose 295(T)
–, specific rotation 258(T)
Gentio-oligosaccharide, production 877
Gentiobiose *263*, 295(T)
–, component of crocin 239
–, honey 886(T)
–, specific rotation 258(T)
Gentisic acid 825
Geosmin
–, aroma quality 340(T)
–, odor threshold value *343*
Geranial, aroma quality 340(T)
–, sensory property 386(T)
Geraniol *383*
–, sensory property 386(T)
Germ, cereal 675(T), 676(F), 676(T)
German degree of hardness, drinking water 986, 987(T)
Gin 932
Ginger, aroma substance 977
–, essential oil, chemical composition 974(T)
–, pungent substance 980(T)
Ginger ale 904
Gingerol *980*
Glandular product 595
Glazed fruit 851
Gliadin, amino acid composition 683(T)
–, aminoacylation 082(F)
–, baking quality 692
–, rheological property 695(F)
–, RP-HPLC, wheat cultivar 682(F)
Gliadin, α- 681, 683(T), 685
–, amino acid sequence 689(T)
Gliadin, γ- 681, 683(T), 685
–, amino acid sequence 689(T)
Gliadin, ω1,2- 683(T), 684, 685(T)
Gliadin, ω5- 683(T), 684, 685(T)
Globular protein 053
Globulin, blood plasma 594
–, cereal 676(T), 678(T)

1022   Index

–, egg white 549(T)
–, legume 746, 749(T)
–, –, amino acid composition 753(T)
–, –, molecular weight 749(T)
–, –, sedimentation coefficient 749(T)
–, –, subunit 746, 749
Globulin, 11S-
–, legume 746
Globulin, 7S-
–, legume 749
–, soybean, dissociation 751(F)
Glucan, β- 330
–, cereal 703
–, determination, NIR 706(T)
Glucan, β-D-
–, conformation 297(F)
–, solubility 703
Glucitol 261, 261
Glucanase, β- 335
–, relative sweetness 259(T)
Glucoamylase 333
Glucobrassicin 789(T)
Glucofructan, wheat 703, 703
Glucofuranose, α-D- 252, 255(T)
Glucofuranose, β-D- 252, 255(T)
Glucogallin, β-D- 825
Glucoiberin 789(T)
Glucokinase, inhibition 127(T)
Gluconapin 789(T)
Gluconate, additive 455(T)
Gluconic acid 262
–, honey 889
Glucono-δ-lactone 262, 449
Glucono-γ-lactone 262
Glucopyranose, α-D- 252, 255(T)
–, borate complex 252, 253
Glucopyranose, β-D- 255(T)
Glucopyranosidomannitol 874
Glucopyranosidosorbitol, 1-0-α-D- 874
Glucopyranosidosorbitol, 6-0-α-D- 874
Glucose 251
–, Bifidus pathway 521(F)
–, cyclization 249
–, enolization 264, 265
–, enzymatic analysis 138, 138(F), 139(T)
–, equilibrium mixture 255(T)
–, fruit 818(T)
–, glycemic index 867
–, heterofermentation 521(F)
–, homofermentation 521(F)
–, isomerization 266, 266
–, legume 760(T)
–, Maillard reaction 270, 271
–, metabolism, lactic acid bacteria 521(F)
–, nutritional/physiological property 864(T), 866
–, occurrence 250(T)

–, oxidation with cupric ion 267, 267
–, production 863(T), 876
–, reaction with glucose oxidase 139(T)
–, relative sweetness 259(T), 513(T), 863(T)
–, solubility 862(T)
–, solution, freezing temperature 855(F)
–, specific rotation 258(T)
–, stability 862
–, sweet taste, AH/B/X-model 261
–, –, temperature dependence 866(F)
–, sweetness receptor 433, 434(F)
–, taste threshold value 259(T)
–, utilization 877
Glucose 6-phosphate dehydrogenase, enzymatic
    analysis 138
–, inhibition 127(T)
–, reaction 139(T)
Glucose dehydrogenase, inhibition 127(T)
Glucose isomerase, industrial application 154
Glucose oxidase, application 149
–, honey 887, 887(F)
–, mechanism 102, 102
–, phase transition temperature 137
–, reaction 139(T)
Glucose phosphate isomerase, reaction 139(T)
Glucose syrup, hydrogenated 877
–, –, production 863(T), 877
–, –, relative sweetness 863(T)
–, production 863(T), 875
–, relative sweetness 863(T)
–, viscosity 865(F)
Glucose, α-D-
–, conformation $^1C_4$ 254
–, conformation $^4C_1$ 254
–, energy content 255(T)
Glucose, β-D-
–, conformation $^1C_4$ 254
–, conformation $^4C_1$ 254
–, energy content 255(T), 863(T)
Glucose, D- 252
–, proton resonance spectrum 256, 257(F)
Glucose, L- 878
Glucose/fructose syrup, production 877
–, relative sweetness 863(T)
Glucosidase, α-
–, honey 886
–, inhibition 127(T)
–, isolation 096(T)
–, pH optimum 129(T)
–, reaction 139(T)
–, substrate specificity 095(T)
Glucosidase, β- 335, 337(T)
–, emulsin 760
–, substrate analog inhibitor 108, 108
Glucosidase, β-D-
–, immobilization, stability 148(F)

Glucosidase, Exo-1,4-α-D-
–, application 151
Glucoside, toxic, legume 764
–, toxicity 760
Glucosinolate 798, *798*
–, biosynthesis *790*
–, cabbage 789(T)
–, enzymatic degradation 153
–, horseradish 977, 978(T)
–, mustard 977, 978(T)
–, rapeseed 652
–, vegetable 789, 789(T), *790*
Glucosone *267*
Glucosyl sucrose, production 863(T), 875
–, relative sweetness 863(T)
Glucosylamine *270, 272*
Glucuronic acid *262, 263*
–, synthesis *263*
Glutamate, canned vegetable 800
–, cheese, bitter taste 542, 542(T)
Glutamate dehydrogenase, molecular weight 097
Glutamic acid *010*
–, decarboxylation 023
–, discovery 011
–, molasses 874
–, reaction 023
–, synthesis 032, *032*
–, taste 034, 035(T)
–, use as a flavor enhancer 430
Glutaminase 154
Glutamine *010*
–, discovery 011
–, sugar beet 869, 870
Glutathione, disulfide interchange 699, *699, 700*
–, dough, ascorbic acid effect 717(T), 718, 718(T)
–, nomenclature 036
–, occurrence 038
–, protein-bound, wheat 699, 700(T)
–, structure 038, *038*
–, wheat dough, rheology 699, 700(F)
–, wheat flour 699, 699, 700(T), 717(F)
Glutathione dehydrogenase, flour improvement with
    ascorbic acid 717, 717(F)
–, substrate specificity 699(T)
–, wheat 698, *698*, 698(T)
Glutathione peroxidase 426, *426*
Glutelin, cereal 676(T), 679(T)
–, legume 746, 749(T)
–, wheat 679(T)
–, –, electropherogram 690(F)
–, –, electrophoresis 680(F)
–, wheat (see also glutenin)
Gluten 673
–, aggregation 730, 732(F)
–, aminoacylation 082(F)
–, baking quality 691

–, bread volume 057(F)
–, disulfide bridge 083, 083(F)
–, enrichment with lysine 083(F)
–, formation, wheat 691
–, protein, aggregation 691
–, –, amino acid composition 683(T)
–, –, classification 683(T)
–, –, cystein residue 683(T)
–, –, genes 680, 682
–, –, intermolecular disulfide bond 691
–, –, molecular masses 683(T)
–, –, property 683(T)
–, –, wheat 680
–, reduction and reoxidation 082, 083(F)
–, sedimentation value 712
–, solubility 057(F)
–, structure 729, 729(F)
–, succinylation 081, 081(F)
–, tensile test 695(F)
–, thermal denaturation 057(F)
–, yield 711
Gluten-free baked product, alkyl cellulose 329
Glutenin, rheological property 692, 694, 695(F)
–, wheat, RP-HPLC 686(F)
Glycemic index, definition 867
–, example 867
Glycemic load 867
Glyceraldehyde *251*
–, from fructose 267, *267*
Glyceraldehyde-3-phosphate dehydrogenase, muscle
    573
Glyceric acid *292*
Glyceroglycolipid 180
Glycerokinase, reaction 139(T)
Glycerol, enzymatic analysis 139(T), 140, 140(T)
Glycerol ether 187
Glycerol factor, wine 920, *920*
Glycine *010*
–, discovery 011
Glycinin, amino acid sequence 750(T)
Glycitein *762*
Glycogen, fish 627
–, meat 587
Glycol cleavage 292, *292*
Glycolipid 180
–, wheat flour 704, 705(T)
Glycolipid hydrolase 190
Glycolysis, scheme 893(F)
Glycoprotein 041, *041*
–, asparagine 009
–, casein 505, *506*
–, collagen 577
–, egg white 553, 554(T)
–, fish, blood serum 626, *626*
–, lectin 759, 759(T)
–, ovalbumin 548, *550*

–, ovomucin 551
–, ovomucoid 550
–, phosvitin 554, *554*
–, serine 012
Glycosidase, β-
–, improvement of rye flour 152
Glycoside 250(T)
Glycoside, N- 270
–, isomerization 270, *271*
–, mutarotation 272
Glycoside, O- 289
–, formation 289
–, hydrolysis 290, 290(T)
–, occurrence 290
–, stereospecific synthesis 289, *289*
Glycosylamine *270, 272*
Glycyrrhizin 439, *439*
Glyoxylic acid *292*
GMO 143
–, detection 142(T), 144
GMP 431
–, taste, synergistic effect 431
Goat meat 593
Goat milk, chemical composition 501(T)
–, production data 499(T)
Goiter 427
–, rhodanide 797
–, thiooxazolidone 798
Goitrin 798
Goitrogenic substance, cabbage family 798
–, milk 798
Goose fat 644
–, fatty acid composition 643(T)
Gooseberry, hydroxycinammic acid derivative 824(T)
Gossypol *649*
GOT, frozen meat, detection 610, 611(F)
Grain alcohol 933
Gramisterol *231*
Granule, egg yolk 553, 553(F)
Grape 815(T)
–, acid content 911, 914(F)
–, aroma formation 211(T)
–, aroma substance 837
–, botrytised 911
–, cultivar differentiation 816(F)
–, methyl anthranilate 837
–, peppery note 921
–, ripening 911, 914(F)
–, rotundone 921
–, sugar content 911, 911(F)
Grape cultivar, German 909(T)
–, –, cultivation 911(T)
–, international 912(T)
–, red wine 910
–, white wine 907
Grape juice, adulteration 858(T)

Grape must 913
–, acid 915
–, aroma substance 916
–, carbohydrate 915
–, chemical composition 914, 914(T)
–, fermentation 916
–, mineral 916
–, phenolic compound 915
–, production 913
–, sulfur treatment 914, 917
–, tannin content 915
Grapefruit 815(T)
–, aroma substance 838
–, bitter taste 819, 832
–, flavanone content 833(T)
–, flavone content 833(T)
–, nootkatone 838
Grashopper ketone 243, *243*
Gravy, taste compound 605(T)
Grayanotoxin, honey *889*
Green malt 898
Groat 708(T)
–, barley 711
Guaiacol 375(T)
–, formation from ferulic acid 374(F)
Guaiacol, 4-vinyl- 375(T)
Guanidine derivative, sweet taste 440, *440*, 440(T)
Guanine *047, 143*
Guanosine monophosphate, 5'- 431
Guar flour 311
Guaran gum 308(T), 311, *311*
–, viscosity 312(F)
Gulose *251*
Gulose, L- 878
Guluronic acid, alginate 303, *303*
Gum arabic 307, *308*
–, viscosity 308(F), 308(T)
Gum bear 881
Gum candy 881
Gum ghatti 309, *309*
Gum tragacanth *310*
–, viscosity 310(F)
Gymnema silvestre
–, miraculin 438

Haddock 621
Hake 621
Half life, insecticide 476
Halibut 623
Halphen reaction 650, 664, 664(T)
Ham 598
–, chemical composition 599(T)
Hapten 141
Hard caramel 879, 880(F)
–, chemical composition 880(T)
–, production 880(F)

Hard cheese 529, 531(T)
–, ripening, yield 532
Hardened off-flavor, soybean oil 205
Harman 026, *029*
Harvest loss, pest control 475
Haze, fruit juice 837
Hazel nut, chemical composition 816(T)
Hazelnut, aroma analysis 353, 354(F)
–, roasting, filbertone 882
HD$_3$F (see Furaneol)
HDL 184
Headspace analysis 347(F), 348, 350, 352(F)
–, dynamic 350
–, static 347(F), 348, 349, 349(F), 352(F)
Health hazard, mycotoxin 475
–, PPA 486
Heart, chemical composition 594(T)
Heat treatment, detection 285
Heating, fish 634
–, meat 597
–, milk 133(F)
–, vegetable 799, 800
–, –, discoloration 795
Heavy metal, acceleration of lipid peroxidation 199
–, reaction with hydroperoxide 199
Helical structure, protein 051
Hemagglutination 759
Heme *574*, 574(F)
Heme(in)
–, formation of oxidized fatty acid, 212(T)
–, reaction with monohydroperoxide 200, *200*
Hemicellulase 335
Hemicellulose 250(T), 330
–, fruit ripening 845
Hemin *103*
Hemoglobin 573, 574(F)
–, electrophoresis 605(F), 610(F)
–, fish 625
–, oxygen binding 574(F)
–, tertiary structure 056(F)
Heptadienal, (E,E)-2,4-
–, linolenic acid, α-
–, –, autoxidation 203(T)
–, sensory property 204(T)
Heptadienal, (E,Z)-2,4-
–, linolenic acid, α-
–, –, autoxidation 203(T)
–, sensory property 204(T)
Heptadienal, 2,4-
–, autoxidation 206(F)
Heptanal, linoleic acid, autoxidation 203(T)
–, oleic acid, autoxidation 203(T)
–, sensory property 204(T)
Heptanone, 2-
–, sensory property 226(T)
Heptenal, (E)-2-

–, linoleic acid, autoxidation 203(T)
–, linolenic acid, α-
–, –, autoxidation 203(T)
–, sensory property 204(T)
Heptenal, (Z)-4-
–, formation 222
–, precursor 203
–, sensory property 204(T)
–, stability 397(T)
Heptulose, occurrence 250(T)
Heptulose, D-manno-2- 250(T)
–, specific rotation 258(T)
Herbicide 483
–, chemical structure 477(F)
–, trade name and activity 480(T)
Hercynine 019(T)
Hernandulcin 442, *443*
–, structure and taste 443
–, sweet taste 443
Herniarin 824(T)
Herring 618
Herring oil 644, 644(T)
–, unsaponifiable component 226(T)
Hesperitin 832, *832*
Heterofermentation, glucose 521(F)
Heterogeneous immunoassay 142
Hexadienoic acid, 2,4- 452
Hexahydroxydiphenic acid 825, *825*
Hexametaphosphate, additive 455(T)
Hexamethyldisilazane 291
Hexanal, autoxidation of linoleic acid 205
–, formation, enzymatic 210(F), 211(T)
–, indicator for aroma defect 203
–, indicator of fat deterioration 668
–, linoleic acid, autoxidation 203(T)
–, sensory property 204(T)
–, stability 397(T)
–, warmed over flavor 608, 608(T)
Hexanol, odor threshold value 341(T)
Hexanone, 2-
–, sensory property 226(T)
Hexen-3-one, 1-
–, artichoke aroma 792
–, odor threshold value 792
Hexenal, (E)-2-
–, autoxidation 397, *397*
–, formation, enzymatic 210
–, linolenic acid, α-
–, –, autoxidation 203(T)
–, odor profile 343(F)
–, sensory property 204(T)
–, stability 397(T)
Hexenal, (E)-3-
–, formation, enzymatic 211(T)
–, linolenic acid, α-
–, –, autoxidation 203(T)

Hexenal, (Z)-3-
–, aroma, apple 839, 839(T)
–, –, grapefruit 838
–, –, orange 838
–, –, tea, green 955, 955(T)
–, basil 978, 978(T)
–, formation, enzymatic 210(F)
–, linolenic acid, α-
–, –, autoxidation 203(T)
–, olive oil 646(T)
–, parsley 978, 979(T)
–, sensory property 204(T)
–, stability 397(T)
Hexokinase, enzymatic analysis 138
–, mechanism 104, 104
–, rate constant 120(T)
–, reaction 139(T)
–, substrate binding 122
Hexopyranose, energy content 255(T)
Hexopyranose, α-D- 253
Hexopyranose, α-L- 253
Hexopyranose, β-D- 253
Hexopyranose, β-L- 253
Hexose diphosphatase, inhibition 127(T)
Heyns compound 271
Heyns rearrangement 271, 271
High density lipoprotein (HDL) 184
High fructose syrup 877
High-temperature short-time process 134
Hill coefficient 125
Hill equation 124, 124, 125(F)
Histamine, meat 584
Histidine 010
–, acid-base catalysis 113, 114
–, discovery 011
–, fish 626
–, reaction 069
Histidine decarboxylase, mechanism 117, 117(F)
Histidine, 1-methyl-
–, fish 626
Histidino-hydroxylysino-norleucine 582, 582
HLB value, lysolecithin 180
HLB-value 459, 460(F), 460(T)
–, calculation 459, 460, 460(T)
–, lecithin 180
HMF, heating of milk 518, 520(F)
HMF (see Hydroxymethylfurfural)
HMW subunit, baking quality 692, 694(F)
–, disulfide bond 691, 694(F)
–, glutenin 681, 686(F)
–, –, amino acid sequence 687(T)
–, –, type, x- 681, 683(T)
–, –, type, y- 681, 683(T)
–, stretching resistance of wheat dough 692, 694(F)
Hofmeister series 061
Hog fat 643

–, detection 644
–, fatty acid composition 643(T)
Holoenzyme 098(F)
Homobetaine 019(T)
Homocysteine, folate deficiency, marker 416
Homofermentation, glucose 521(F)
Homogenization, milk 514, 518
Homomethionine, biosynthesis 781
Homoserine, biosynthesis 781
Honey 883
–, acid 888
–, aroma substance 889
–, carbohydrate 885, 885–887(T)
–, chemical composition 885, 885–887(T)
–, enzyme 886, 887(T)
–, free amino acid 888, 888(T)
–, gluconic acid 888
–, hydroxymethyl furfural 889, 889(F)
–, ketose 250(T)
–, oligosaccharide 886(T)
–, physical property 885
–, pigment 889
–, processing 884
–, production 884
–, protein 888
–, regional origin 888, 888(F)
–, storage 889
–, toxic constituent 889
–, transglucosylase activity 886
–, type 884
–, utilization 890
–, viscosity 885(T)
–, water activity 004(T)
Honeydew honey 884
Hop 894
–, aroma substance 896, 896(T)
–, bitter substance 895, 896(T)
–, chemical composition 895(T)
–, processing 896
–, production data 895(T)
Hop extract 897
Hop isoextract 897
Hordenine, fruit 813, 815
Horse meat 593
Horseradish, aroma substance 977
Hot beverage 856
Hotrienol 382, 383
–, formation 386, 582
HPLC, amino acid 022
–, lecithin 182(F)
–, lipid 182(F)
–, peptide 044
–, tocopherol 234
–, triacylglycerol 173, 174(F)
HTST 134
HTST sterilization 800

Hulupone 895, *896*
Human milk 250(T)
–, chemical composition 501(T)
–, palmitic acid 177
Humectant 004, 261, 464
Humulene *385*
Humulinic acid 895, *896*
Humulon 895, *895*
–, isomerization 895
HVP (Hydrolyzed Vegetable Protein) 602
Hydantoin 020
Hydration, enzyme activity 137, 137(T)
–, protein 726, 727(F)
Hydration shell, Structure 003
Hydrocarbon, aroma substance 379
–, fat 227
–, formation on radiation 224, 224(F)
Hydrochloric acid, additive 449
Hydrocyanic acid, formation, legume 761(F)
–, occurrence 761(T)
Hydrogen bond, water 001, 001(F), 002
Hydrogen cyanide, fruit brandy 932
Hydrogen peroxide, enzymatic degradation 149
–, milk, pasteurization 149
Hydrogen sulfide, citrus fruit juice 838
–, formation 365(F)
–, sensory property 364(T)
Hydrogenation, fat 656
–, –, off-flavor 205
–, –, principle 169, *169*
Hydrolase, substrate specificity 094
–, systematic, example 100(T)
Hydrolysis, enzymatic, potato lipid 188(T)
Hydrolytic rancidity 189, 190(T)
Hydroperoxide, bleaching of fat 655, *655*
–, degradation, bimolecular 192, 192(F)
–, –, monomolecular 192, 192(F)
Hydroperoxide isomerase (see allene oxide synthetase)
Hydroperoxide lyase, mechanism 209, 210(F)
–, product 211(T)
Hydroperoxyepidioxide 195
Hydrophilic-lipophilic balance 459
Hydroxamic acid, *Lossen* rearrangement *790*
Hydroxy fatty acid, bitter taste 212(T)
–, cutin 187
–, formation in lipid peroxidation 212(T), 213(F)
–, occurrence 164
–, precursor of γ- and δ-lactone 164
Hydroxy radical, lipid peroxidation 202
–, reaction 201(T)
Hydroxy,2,5-dimethyl-3(2H)-furanone, 4- (see Furaneol)
Hydroxy-2-methyl-3(2H)-furanone, 4- (see Norfuraneol)
Hydroxy-4,5-dimethyl-2(5H)-furanone, 3- (see Sotolon)

Hydroxy-L-isoleucine, sotolon precursor 976, *976*
Hydroxyacetylfuran, 2-
–, formation 264, *265*
Hydroxyacyl-CoA-dehydrogenase, β-
–, frozen meat 610, *611*
Hydroxyaldehyde, retroaldol condensation 222, *222*
Hydroxyalkyl cellulose 329
Hydroxyamino acid, hydrolysis 023
Hydroxybenzoic acid, biosynthesis *826*
–, fruit, content 823(T)
–, occurrence 823(T)
Hydroxybenzoic acid ester, p-
–, antimicrobial action 450
Hydroxybenzoic acid, p- 825(T)
–, occurrence 825(T)
Hydroxybenzylmethylether, p-
–, aroma, vanilla 976, *976*
Hydroxycinamic acid, occurrence 823(T)
Hydroxycinammic acid *822*, *823*
–, biosynthesis 826, *826*
Hydroxycitronellal 396(T)
Hydroxycumarin 824
Hydroxyethyl starch 326, *326*
Hydroxyglucobrassicin, 4- 789(T)
Hydroxyisoleucine, sotolon precursor 361
Hydroxyl ion, mobility 002
Hydroxyl number, deep fried fat 220(T), 221
Hydroxyl value, definition 663
Hydroxylysine, 5- *010*, 577, 580(F)
–, discovery 011
Hydroxylysino-norleucine 582
Hydroxymethyl furfural, honey 889, 889(F)
–, invert sugar cream, detection 890
Hydroxymethylfurfural, 5-
–, formation 264, *265*, 274, *275*
Hydroxymethylproline, 4- 809
Hydroxyphenyl-3-butanone, 1-p- 839, *840*
Hydroxyphenylalanine, o-
–, food radiation 073
Hydroxyproline, 3-
–, collagen 580(T)
Hydroxyproline, 4- *010*, 577, 580(F)
–, biosynthesis *583*
–, connective tissue, detection 577, 578(T)
–, discovery 011
–, indicator for connective tissue 613
Hydroxypropyl starch 326, *326*
Hydroxypyruvic acid *292*
Hypoglycine A 809
Hypothiocynite, milk 517, *517*
Hypoxanthine, oxidation 105, *105*
Hysteresis, sorption isotherm 003

Ice, structure 002
Ice cream 162(T), 529
–, agar 301(T), 303

–, alginate 301(T), 304
–, carboxymethyl cellulose 330
–, carrageenan 301(T), 306
–, dextran 332
–, guaran gum 312
–, gum arabic 309
–, gum tragacanth 310
–, locust bean gum 312
–, pectin 315
–, rate of crystallization 007, 007(F)
Icing 968(T)
IDA 356
Iditol 261
Idofuranose, α-D- 255(T)
Idofuranose, β-D- 255(T)
Idopyranose, α-D- 255(T)
Idopyranose, β-D- 255(T)
Idose 251
–, equilibrium mixture 255(T)
Idose, α-D-
–, conformation $^1C_4$ 254
–, conformation $^4C_1$ 254
–, energy content 255(T)
Idose, β-D-
–, energy content 255(T)
Idose, L- 878
IEDI 468
Illipé butter 648(T), 649
Imidazole, product of Maillard reaction 274, 276
Imidazoloquinoline (IQ) 026, 028(T), 029(F)
Imidazoloquinoxaline (IQx) 026, 028(T), 029(F)
Imidazoquinoline 026, 027(T)
Imidazoquinoxaline 026, 027(T)
Imidoester, bifunctional, protein 064
Imitation cheese 536, 536(T)
–, chemical composition 536(T)
Immobilized enzyme, kinetics 148
–, pH optimum 148
–, production 145, 148(F)
–, stability 148, 148(F)
Immunoassay, competitive 141, 141(F)
IMP 431
–, taste, synergistic effect 431, 431(F)
Impact compound, aroma 340
Inactivation, enzyme, rate equation 131, 131
–, thermal, enzyme 130, 133, 134
Inclusion compound, polysaccharide 297
Indicator, cherry wine 915
–, fat deterioration 667
–, radiolysis 224
–, refining of oil 669
Indicator reaction, enzymatic analysis 137
Induced fit 109
Induction period, fatty acid, unsaturated 192(T)
Infant food, milk preparation 528
–, palmitic acid 177

Infant nutrition, lactulose 266
Infusion process, beer 900
Inhibition, enzyme activity 125, 133(F)
–, –, thermal treatment 130, 131(T)
–, lipid peroxidation 214
–, photooxygenation 198
Inhibitor, enzyme 125
–, substrate analog 107
Inhibitor constant, definition 126, 126
–, determination 127
Inosine monophosphate, 5'- 431
Insecticide 476
–, chemical structure 477(F)
–, LD$_{50}$ 480(T)
–, natural 476
–, persistence 476
–, resistance of pest population 476
–, trade name and activity 480(T)
Instant coffee 948
Instant pudding, amylose 324
–, xanthan gum 331
Instant sauce, amylose 324
–, production 604(F)
Interesterification, directed 173, 173
–, fat, process 658
–, random rearrangement 173, 173
–, triacylglycerol, reaction 172, 172
Intermediary substrate 099
Intermediate moisture food 004
Inulin 332
–, solubility 332
–, structure 332
–, utilization 332
Inversion, saccharose 294
Invert sugar, production 863(T)
–, relative sweetness 259(T)
Invert sugar cream 890
–, chemical composition 890
–, production 890
Invertase, candy 153
–, honey 887(F)
–, –, inactivation 886
–, pH optimum 129(T)
Iodine 425(T), 427
Iodine number, deep fried fat 220(T), 221
–, reaction 168
Iodine value, definition, example 663, 664(T)
–, fat 665(T)
Iodine-131 470
Iodine-deficiency induced goiter 427
Iodization, salt 427
Iodoacetic acid 069
–, enzyme inhibitor 107
–, –, irreversible inhibition 126
Ionone, α- 242
–, enantiomeric excess 355(T)

–, enantioselective analysis 356, 356(F)
Ionone, β- *242*
Iron 422(T), 424, 425(T)
–, complexing by synergist 220(T)
–, enzyme cofactor 105
–, lipid peroxidation 199, 200(T), 213(F)
–, meat 587(T)
Iron(II)/iron(III)-redox system 106(T)
Irreversible inhibition, enzyme catalysis 126
Iso-sweet concentration, sugar 259(T)
Isoamylase (see pullulanase)
Isobetanidin 797, *797*
Isobutyl acetamide, N-
–, Camembert 543
Isobutyl-3-methoxy-pyrazine, 2-
–, grape 837
Isochavicin 979
Isochlorogenic acid 824
Isoelectric point, amino acid 012, *012*, 013(T)
–, peptide 036, 036(T)
–, protein 059
–, –, estimation 059
Isoelectric precipitation, protein 059, 061
Isoenzyme, definition 097
Isoflavone, chemical structure *823*
Isoglucose, production 863(T)
Isohumulon 895, *896*
Isoindole, amino acid derivative 022
Isoionic point, protein 059
Isokestose 295(T)
Isoleucine *010*
–, biosynthesis 377(F)
–, discovery 011
Isomaltitol *874*
–, production 863(T), 874
–, relative sweetness 863(T)
Isomalto-oligosaccharide, production 875
Isomaltol *264*
–, formation 278, *279*
Isomaltol, β-galactosyl-
–, heated milk 279
Isomaltopentaose, honey 886(T)
Isomaltose *263*, 295(T)
–, honey 886(T)
Isomaltosylglucose, 3-α-
–, honey 886(T)
Isomaltotetraose, honey 886(T)
Isomaltotriose, honey 886(T)
Isomerase, definition 097
–, example 100(T)
Isomerization, monohydroperoxide 195, *195*
Isoooxazolium salt, amidation of protein 067
Isopanose, honey 886(T)
Isopeptide bond 023, 072
Isoprene fatty acid, milk fat 160
Isopropyl-2-methoxypyrazine, 3-

–, potato aroma 788, 789(T)
Isorhamnetin 834, *834*
Isosaccharinic acid *268*
Isosacuranetin 832, *832*
Isosorbid *879*
Isothiocyanate 790, *790*
–, odor threshold value 792
–, reaction 790, *790*
–, vegetable 789, 789(T), *790*
Isotope analysis 858, 858, 859(T)
Isotope dilution analysis, aroma substance 356, 357,
    358(F), 359(T)
Isotope dilution assay, patulin 475
Isotope effect, kinetic 858, 859
–, thermodynamc 859
Isotopic analysis, vanillin 395, 395(T)
Isouramil *764*
Itaconic acid, coffee 943, *943*

Jam 852, 852(T)
–, chemical composition 852(T)
–, furcellaran 307
–, pectin 315
Jelly 881
–, alginate 301(T), 304
–, fruit 852
–, pectin 315
Jelly fruit 881
Juniper berry, essential oil, chemical composition
    974(T)
Juniper liquor 932

Kaempferol 834, *834*
Kaempferol-3-glycoside, taste threshold 953
Kahweol, coffee 942, *943*
Kamboko, production 635
Karaya gum 310, *311*
Kefir 524
Kernel hardness, measurement 708
Kestose 295(T)
–, honey, 886(T)
–, molasses 874
–, specific rotation 258(T)
–, sugar extract 869
Kestose, 1- 295(T)
Kestose, 6- 295(T)
Ketchup, propellant 465
Ketol fatty acid 210, *210*
–, enzymatic formation 210, *210*
Keton enamine, α-
–, cooling effect 431, *431, 432*
Ketose, cyclization 249, *249*
–, equilibrium 255(T)
–, family tree 252(F)
–, occurrence 251(T)
–, rearrangement to aldose 266, *266*

Key odorant, example 340(T)
Kidney, chemical composition 594(T)
Kiln-dried malt 898
Kinetic method, enzymatic analysis 140
Kinetics, enzyme 117
Kneading process, dough making 725, 725(F), 725(T)
Koji 766
Kojibiose, honey 885, 886, 887(T)
Krafft point 458, 458(F), 458(T)
Kumiss 524
Kunitz inhibitor 755, 756(T)
K-value, fish 630

Lactalbumin 502(T)
–, genetic variant 505(T)
Lactalbumin, α- 502(T), 512
–, amino acid sequence 504(T)
–, genetic variant, amino acid sequence 505(T)
–, lactose synthetase 512
–, thermal denaturation 057, 058(F), 058(T)
Lactase, application in milk technology 152
Lactate, enzymatic analysis 139(T)
Lactate 2-monooxygenase, reaction 095(F)
Lactate dehydrogenase, isoenzyme 097
–, mechanism 099
–, reaction 095(F)
Lactate malate transhydrogenase, reaction 095(F)
Lactate racemase, reaction 095(F)
Lactem 461(T)
Lactic acid, additive 448, *448*
–, formation in sour dough 724(F)
–, lactic acid bacteria 522(T)
–, milk product 522(T)
–, synthesis *448*
–, wine 919, 919(T)
Lactic acid bacteria 522(T)
–, glucose metabolism 521(F)
Lactic acid fermentation, sauerkraut 802
–, scheme 521(F)
–, vegetable 802
Lactic acid, D-
–, formation 521, 522(T)
Lactic acid, L-
–, formation 521, 522(T)
Lactide, formation 448, *448*
Lactitol 878
–, nutritional/physiological property 864(T), 866
–, production 863(T)
–, relative sweetness 863(T)
Lactoglobulin 502(T)
–, genetic variant 505(T)
Lactoglobulin B, β-
–, denaturing 510, 511(F)
Lactoglobulin, β- 502(T), 511
–, amino acid sequence 504(T)
–, genetic variant, amino acid sequence 505(T)

–, solubility 061(F)
–, thermal denaturation 057, 057(F), 058(T)
Lactol 249, 251
Lactone, aroma substance 380, 381(T)
–, biosynthesis 380(F), 381, 386(F)
–, chirality 382
Lactone, δ-
–, precursor 164
Lactone, γ-
–, formation by heating fat 221(T)
–, precursor 164
Lactoperoxidase, milk 517
Lactose 295(T), *512*
–, conformation *293*
–, degradation during cheese ripening 532, 533(F)
–, enzymatic analysis 138, 138(F)
–, enzymatic hydrolysis 152
–, furanoid-1-deoxyosone, formation 279
–, glycemic index 867
–, graininess 528
–, hydrolysis *294*
–, hydrolysis in milk product 152
–, mutarotation, pH dependency 512(F)
–, nutritional/physiological property 864(T)
–, physical data 513(T)
–, production 539, 863(T), 877
–, relative sweetness 259(T), 513(T), 863(T)
–, solubility 513(T), 862(F)
–, specific rotation 258(T)
–, stereochemistry *266*
–, structure 512
–, taste threshold value 259(T)
Lactose fatty acid ester 463
Lactose intolerance 513
Lactose synthetase, lactalbumin, α- 512
Lactosucrose, nutritional property 865(T)
–, production 863(T)
–, relative sweetness 863(T)
Lactosylceramide 181
Lactulose 295(T), 878
–, formation 266, *266*
–, milk product 513
–, nutritional/physiological property 864(T)
–, production 863(T)
–, relative sweetness 863(T)
Lager beer, aroma substance 902, 903(T)
Lamb 593
Lamella 186(F)
Laminaribiose, honey 886(T)
Lampante 645, 646(T)
Lanthionine 039
Lard 643, 643(T)
–, detection 644
–, effect of antioxidant 219(T)
–, improvement, interesterification 659
–, unsaponifiable component 226(T)

Lard (see hog fat)
Lariciresinol *835*
Lathyrism, neuro- 764, 781
Lathyrism, osteo- 781
Lauric acid, sensory property 160(T)
–, structure, melting point 161(T)
LD$_{50}$ 467
–, PPA 480(T)
LDL 184
Lead 468, 469(T)
Leaf protein, plastein reaction 085(F)
Lecithin 179, *179*
–, analysis 182
–, chemical hydrolysis 180
–, chocolate, production 967
–, HLB value 180
–, hydrolysis, enzymatic 190, *190*
–, hydroxylated 180
–, occurrence 179(T)
–, production 647(F)
–, raw, composition 182(F)
–, –, HPLC 182(F)
–, removal, oil processing 654
–, synergistic effect with antioxidant 220
Lectin, occurrence 759
–, specificity 759(T)
–, stability 759
–, structure 759, 759(T)
–, toxicity 759
Legume, carbohydrate 748(T), 759, 760(T)
–, chemical composition 748(T)
–, cyanogenic glycoside 760
–, dietary fiber 748(T)
–, fatty acid composition 762(T)
–, flatulency 759
–, hardening 768
–, lectin 759, 759(T)
–, mineral 762(T)
–, pectin esterase 768
–, phenoloxidase 768
–, phytase 768
–, phytoestrogen 762
–, production data 747(T)
–, protein 746
–, –, *Osborne* fraction 746, 749(T)
–, stachyose, hydrolysis 152
–, vitamin 762(T)
Legumin 746
Legumin J, amino acid sequence 750(T)
Lemon, aroma substance 838
–, carotinoid content 235(T)
–, flavanone content 833(T)
–, flavone content 833(T)
Lemonade 856
Lenthionine *788*
Lentil, chemical composition 748(T)

–, production data 747(T)
–, saponin content 763(T)
Leucine *010*
–, biosynthesis 378(F)
–, degradation, *Ehrlich* pathway *737*
–, discovery 011
Leucoanthocyanidin (see also proanthocyanidin)
Leucoanthocyanin, cacao bean 962(T), 963, *963*
Leucrose 875
Lichenin, conformation *297*
–, oat, barley 703
Licorice 882
Licorice root, sweet taste 439, *439*
Liebermann-Burchard reaction 232, 232(F)
Ligase, definition 097
Light taste, beer 905
–, milk product 543
Lignan 835
–, chemical structure *822*
–, content, fruit 835(T)
Lignin 827, 827(F)
Lignoceric acid, structure, melting point 161(T)
Lignoceric acid tryptamide 961, *961*
Ligustilide 789, *789*
Likens-Nickerson apparatus 348, 348(F), 348(T)
Lima bean, chemical composition 748(T)
–, hydrocyanic acid 761(T)
–, hydrolysis of linamarin 761(F)
Liming, sugar extract 870
Limonene *383*
–, odor threshold value 341(T)
–, oxidation 344(T)
–, sensory property 386(T)
Limonene, (R)-
–, aroma, orange 837, 838(T)
Limonin 819, *819*
–, biosynthesis 820, *821*
Limonin monolactone 819, *820*
Linalool *382*
–, aroma, cherry 841
–, –, lemon oil 839
–, basil 978, 978(T)
–, glycoside *382*
–, hop 896, 896(T)
–, sensory property 386(T)
Linalool oxide *386*
–, formation 386, *386*
–, sensory property 386(T)
Linamarin 760(T), 761(F)
–, formation of hydrocyanic acid 761(F)
Lindane 476
Linden ether, honey aroma 889, *889*
Lineweaver-Burk plot, enzyme inhibition 128(F)
–, pH dependency 129(F)
–, single-substrate reaction 121(F)
–, two-substrate reaction 124(F)

Linoleic acid, autoxidation 193
–, –, carbonyl compound 203(T)
–, –, monohydroperoxide 194(F), 195(T)
–, –, pentane 205, *205*, 207
–, autoxidation by peroxidase 200
–, biohydrogenation, milk 513, 513(F)
–, biosynthesis 169, 169(F)
–, conjugated 162, 162(T)
–, decrease on deep frying 220(T)
–, degradation to ethyl decadienoate, (E,Z)-2,4-
   379(F)
–, degradation to lactone 381, 386(F)
–, E-factor 665(T)
–, enzymatic-oxidative cleavage 207
–, essential fatty acid 162
–, heating, product 223
–, induction period 192(T)
–, iodine value 663
–, isomer, fat hardening 658
–, milk fat 513
–, photometric determination 168
–, photooxygenation 195(T), 198, 198(F)
–, pK value 165, 165(F)
–, plant fat *159*, 159(T)
–, rate of autoxidation 192(T), 200
–, reactant for co-oxidation 209(F)
–, reaction with lipoxygenase 208(F)
–, structure, melting point 163(T)
–, taste 164(T)
–, titration curve 165(F)
Linolelaidic acid, melting point 166(T)
Linolenic acid methyl ester, rate of autoxidation
   206(F)
Linolenic acid, α-
–, autoxidation, carbonyl compound 203(T)
–, –, ethane 207
–, –, monohydroperoxide 193, 195(T)
–, biosynthesis 169, 169(F)
–, enzymatic-oxidative cleavage 207
–, hydroperoxyepidioxide 196
–, induction period 192(T)
–, iodine value 663
–, nutritional role 162
–, partial hydrogenation, aroma defect 205
–, photometric determination 168
–, photooxygenation 195(T)
–, plant fat *159*, 159(T)
–, rate of autoxidation, relative 192(T)
–, structure, melting point 163(T)
–, taste 164(T)
Linolenic acid, γ-
–, biosynthesis 169, 169(F)
–, structure, melting point 163(T)
–, taste 164(T)
Linseed, lignan 835, 835(T)
Linseed oil 651(T), 653

Lipase 188
–, *Penicillium roqueforti*
–, –, substrate specificity 534(T)
–, activation, Ca^{2+}-ions 189
–, activation energy 094(T)
–, active center 188, 188(F)
–, assay 189
–, differentiation from esterase 188
–, heat inactivation 189, 189(T)
–, industrial application 154
–, milk 515
–, –, inactivation 135(F)
–, mode of action 188
–, occurrence 188, 189(T)
–, pH optimum 129(T)
–, porcine pancreas, property 188
–, potato, inactivation 135(F)
–, specificity 094, 188, 189(T)
–, wheat 696
Lipid, analysis 182
–, apple 818(T)
–, autoxidation 191
–, binding of aroma substance 390, 390, 391(F)
–, building block 158, 158(T)
–, cereal starch 701(T)
–, classification 158(T)
–, density 184
–, egg yolk 555, 555(T)
–, –, fatty acid composition 555, 556(T)
–, energy value 158
–, extraction 182
–, fish 627
–, formation of aroma substance 376
–, fruit 818, 818(T)
–, HPLC 182(F)
–, legume 748(T)
–, milk 513, 513(T)
–, nutritional property 158
–, peroxidation 191
–, –, acceleration by heavy metal 199
–, –, acceleration by heme(in) 200
–, –, degradation of amino acid 214, 215(T)
–, –, detection 214
–, –, detection *in vivo* 207
–, –, formation of aldehyde 205, 205(F)
–, –, indicator 203
–, –, inhibition 214
–, –, malondialdehyde 206, *207*
–, –, start by superoxide radical anion 201
–, –, water activity 199
–, property 158
–, protein foam, stability 062
–, radiolysis 224, 224(F)
–, rye flour 709(T)
–, simple 158(T)
–, solubility 158

–, spherosome, wheat flour 705
–, tea 955
–, thin layer chromatography 182, 184(F)
–, type of bread 735(T)
–, vegetable 787
–, wheat, baking property 704
–, –, baking quality 705(F)
–, –, composition 701(T), 705(T)
–, –, solubility 704, 704(F)
–, wheat flour 709(T)
Lipid binding, wheat dough 704
Lipid double layer 185
Lipochrome 240
Lipolysis, butter 190(T)
–, Ca^{2+} ion 189
–, cheese ripening 533, 533(F), 534(T)
–, –, aroma 187
–, chocolate, aroma 188
–, detection 189, 667
–, milk 515
–, milk fat 189
–, prediction of storage stability 189
Lipoperoxidase 210
–, oat 697, 697(F)
Lipoprotein 184
–, analysis 184
–, blood serum 185(T)
–, centrifugation 184, 185(F)
–, classification 184
–, composition 183, 185(T)
–, egg yolk, composition 185(T)
–, milk 185(T)
Lipoprotein, α- 184, 185(T)
Lipoprotein, β- 184
Lipoprotein, pre-ψ 184, 185(T)
Lipovitellenin 554
Lipovitellin 554, 554(T)
Lipoxygenase 207
–, activation energy 094(T)
–, blanching process 135
–, bleaching agent 464
–, co-oxidation of carotinoid 209, 209(F)
–, competitive inhibition 742(F)
–, flour improvement 719, 719(F)
–, indicator enzyme 135
–, inhibition, pressure 137
–, mechanism of catalysis 207, 208(F)
–, oat 697, 697(F)
–, pea, inactivation 135(F)
–, –, thermal stability 135, 135(F)
–, peanut, inactivation 765(T)
–, potato 135(F)
–, reaction specificity 207(T)
–, specificity 207, 208(F)
–, wheat 697
Lipoxygenase, non-specific, 209

–, –, reaction with linoleic acid 207(T), 209(F)
Lipoxygenase, type II, 207(T), 209, 209(F)
Liqueur 935
Liqueur wine 926, 927(T)
–, chemical composition 927(T)
Liquid egg 561, 561(T)
Liquid sugar 873
Liquor 929
–, bitter 935
–, production 929
–, raw materials 929
Litchi, aroma substance 842
Liver, chemical composition 594(T)
Liverwurst, water activity 004(T)
Livetin 553(T), 555
LMW subunit, amino acid composition 683(T)
–, baking quality 692
–, glutenin 685, 685(F)
–, –, amino acid sequence 689(T)
Lobry de Bruyn-van Ekenstein rearrangement 266, *266*
Lobster 637
–, chemical composition 637(T)
Locust bean gum 312
–, polysaccharide structure 312
–, viscosity 308(T)
Lotaustralin 760(T)
Low density lipoprotein (LDL) 184
Low-calorie food, cellulose 328
Lugduname, sweetness receptor 433, 434(F)
Lumiflavin 413, *413*
Lumisterol 230(F)
Lung, chemical composition 594(T)
Lupeol *232*
Lupulon 895, *895*
Luputrions 895, *896*
Lutein *237*
Luteolin 834, *834*
Luteoxanthine *239*
–, occurrence in orange 240(T)
Lyase, definition 097
–, example 100(T)
Lycopene *236*
–, aroma precursor 242(T)
–, electronic spectrum 240(T), 241(F)
–, tomato cultivar 236(T)
Lysine *010*
–, acetoacetylation 066
–, acylation 064, 065
–, –, reversible 066
–, alkylation 064
–, aminoacylation 064
–, arylation 064
–, biologically-available 064
–, cereal 677(T)
–, deamination, peroxidase 087, *087*

–, discovery 011
–, guanidination 064
–, loss by lipid peroxidation 215(T)
–, maleylation 066
–, *Maillard* reaction 270
–, modification, *Maillard* reaction 285, *285*, 286
–, reaction of the ε-amino group 017, 023
–, reaction with reducing sugar 071
–, succinylation 065
–, synthesis 032
Lysine aldolase 116
Lysine lyase 108
Lysine peptide 040, 072(T)
–, browning reaction 040(F)
–, supplementation of food 040
Lysinoalanine 072(T), 073, 073(F)
–, occurrence in food 075(T)
Lysinonorleucine 581
Lysolecithin 179
–, HLB value 180
Lysophosphatide, substrate for phospolipase 190
Lysophosphatidyl choline, beer 903
Lysophospholipase 190
Lysozyme 549(T), 550
–, amino acid sequence 552(T)
–, conformation 551(F)
–, hydration 137, 137(T)
–, industrial application 153
–, turn, β- 053(T)
Lyxose 250(T), *251*

Maceration, fruit 854
Magnesium 424
–, content, human body 421(T)
–, enzyme cofactor 104
Maillard compound, chlorinated 603
Maillard reaction, *Amadori* compound *270*, 271
–, *Heyn's* compound 271, *271*
–, *Strecker* degradation 282, *288*
–, acetylformoin 277, *278*
–, aminoreductone 280, *280*
–, arginine 270
–, bis-pyrraline 288, *288*
–, bisarg 288, *288*
–, bitter substance 280, *280*
–, carboxymethyllysine 285, *286*
–, colored product 284, *284*
–, cooling effect 431
–, deoxyosone 272, 274, 276
–, formation of antioxidant 218
–, fructoselysine 285, *285*, *286*
–, furaneol 282, *282*
–, furosine 280, *280*, 285, *285*
–, GLARG, formation 286, *287*
–, heating of milk 519
–, inhibition 082, 149, 289, 289(F)

–, initial phase 271
–, isomaltol 278, *279*
–, lactic acid ester, formation 279, *279*
–, lysine 070, 270
–, maltol 278, *279*, *281*, 282
–, melanoidin 284
–, ornithino-imidazolinone 286(T), 287, *287*
–, pentosidine 287, *288*, 288(T)
–, protein modification 285, *285*
–, pyridosine 285, *286*
–, pyrraline *285*, 286, 286(T)
–, redox reaction 282, *282*
–, reductone 277, *278*
–, vegetable drying 289(F)
–, water activity 004, 004(F)
–, volatile compound 274
Main element 421
Malate, enzymatic analysis 139(T)
Malate dehydrogenase, reaction 139(T)
Maleic acid anhydride 066
–, reaction with protein 066
Malic acid, additive 448
–, degradation in wine 919, *919*
–, fruit 820, 820(T)
Malic acid-1-O-glucopyranoside 788
Malondialdehyde, formation 206, *207*
–, reaction with protein 214
Malossol 636
Malt, β-amylase 150
–, caramelized 899
–, colored 899
–, dark 899
–, dark beer 899
–, light 899
–, light beer 898
–, production 898
Malt coffee 950
Malt extract, beer production 894
Malt wine 929
Malted flour, dough making 721
Maltitol 261
–, nutritional/physiological property 864(T)
–, production 877
Maltitol syrup, production 863(T)
Maltol, formation 278, *279*, *281*, 282
–, occurrence 361(T)
–, odor threshold value 341(T)
–, use as an additive 431
Malton wine 929
Maltopentaose 295(T)
–, specific rotation 258(T)
Maltose 295(T)
–, conformation *293*
–, enzymatic analysis 139(T)
–, glycemic index 867
–, glycol cleavage *292*

–, honey 885, 886(T)
–, hydrolysis *294*
–, nutritional/physiological property 864(T)
–, production 863(T)
–, pyranoid-1-deoxyosone, formation 279
–, relative sweetness 259(T), 863(T)
–, specific rotation 258(T)
–, sweet taste, temperature dependence 866(F)
–, taste threshold value 259(T)
Maltose syrup, production 875
Maltotetraose 295(T)
–, specific rotation 258(T)
Maltotriit 877
Maltotriose 295(T)
–, honey 886(T)
–, specific rotation 258(T)
Maltoxazine, formation 274, *275*
Maltulose 295(T), 877
–, honey 886(T)
–, specific rotation 258(T)
Malvalia acid *649*
Malvidin 829, 830(T)
Manganese 425, 425(T)
Mango 815(T)
Manninotriose 295(T)
–, specific rotation 258(T)
Mannitol 261, *261*
–, nutritional/physiological property 864(T)
–, production 863(T), 879
–, relative sweetness 259(T), 863(T)
Mannitol, D-
–, wine 920
Mannofuranose, α-D- 255(T)
Mannofuranose, β-D- 255(T)
Mannopyranose, α-D- 255(T)
Mannopyranose, β-D- 255(T)
Mannose *251*
–, equilibrium mixture 255(T)
–, isomerization 266, *266*
–, occurrence 250(T)
–, relative sweetness 259(T)
–, specific rotation 258(T)
Mannose, α-D-
–, energy content 255(T)
Mannose, β-D-
–, energy content 255(T)
Mannose, L- 878
Mannuronic acid 303
Maple sugar 873
Maple syrup 873
–, aroma substance 873
Mare milk, chemical composition 501(T)
Margaric acid 160
–, structure, melting point 161(T)
Margarine, "sandy" texture 660
–, aroma substance 661

–, coloring with palm oil 244
–, composition 660
–, detection 661
–, interesterification of raw material 659
–, modification of fat crystal 660
–, phytosterol 229
–, production 661
–, sterol ester 229
–, trans fatty acid 162
–, vitaminization 661
Margarine for pastry production 660(T)
Marinaded fish 634
Marine oil 644, 644(T)
–, detection 665(T)
Marjoram, essential oil, chemical composition 974(T)
Marmalade 851
–, water activity 004(T)
Marzipan 881
–, chemical composition 880(T)
–, detection of persipan 233
Marzipan raw filler, chemical composition 880(T)
Masa corn, aroma 389
Mash, beer 899
Mashed potato powder, production 800, 800(F)
Mass chromatogram, odorant 358(F)
Matairesinol *835*
Maté
–, chemical composition 958
Maximum residue limit, definition 468
Mayonnaise 661
Mead 928(T), 929
Meat 563
–, added protein, detection 613
–, added water, determination 613
–, aging 589, 590(T)
–, –, proteinase 150
–, amine 584
–, amino acid composition 578(T)
–, anabolic compound, detection 612, 612(F)
–, analog 088
–, animal origin, determination 608, 609, 610(F)
–, antibiotic, detection 612
–, aroma 605
–, –, model system 607(T)
–, aroma defect 595, 596(T)
–, biogenic amine 584
–, canned 598, 598(T)
–, carnitine 585, *585*
–, chemical composition 569(T)
–, color 575
–, color change 576
–, connective tissue, detection 613
–, consumption data 565(T)
–, contractile apparatus, protein 568, 571(T)
–, cooling 595
–, curing 576, 577(F), 597

–, defect 588
–, DFD 588, 589(T)
–, drying 597
–, endopeptidase 590, 590(T)
–, extender 088
–, extract 040(T), 601, 602(T)
–, –, analysis 039
–, –, chemical composition 602(T)
–, F-actin 570
–, free amino acid 584
–, freeze dried, proteinase 150
–, freezing 595
–, freshness, evaluation 611
–, frozen, detection 611, 611(F), 612
–, –, fluid loss 596
–, –, water holding capacity 596
–, G-actin 570, 570(F)
–, glycogen 587
–, heating 597
–, –, color change 576
–, IMP 586
–, iron content 587(T)
–, kind 592
–, microbial quality, indicator 584
–, mineral 587, 587(T)
–, myofibril, enzymatic hydrolysis 625(F)
–, nitrosamine, detection 614
–, organ, chemical composition 593, 594(T)
–, organic acid 586
–, packaging, color 576
–, peptide 584
–, phospholipid, occurrence 179(T)
–, preservation, processing 595
–, processing 592
–, product 598
–, –, agar 303
–, –, furcellaran 307
–, –, locust bean gum 312
–, production data 563(T)
–, proteolysis 590, 590(T)
–, PSE 588, 589(T)
–, purine 586, 586(T)
–, pyrimidine 586, 586(T)
–, salting 597
–, sexual odor 228, 229(T)
–, smoking 597
–, soya protein, detection 614
–, storage 595
–, surrogate 088
–, swelling 591, 592(F)
–, tenderizing 598
–, –, detection 611
–, thaw rigor 596
–, titin 569, 570(F)
–, variety, chemical composition 593, 594(T)
–, vitamin 586, 587(T)

–, warmed over flavor 595, 596(T)
–, water holding capacity 590, 591, 592(F), 598(F)
Medicinal water 988
Megastigmatriene, passion fruit 840, *840*
Meisenheimer adduct, sterol 233
Meisenheimer complex 020
Melanoidin 271
–, formation 284
Melatonin, fruit 813, *817*, 817(T)
Melezitose 295(T)
–, honey 886(T)
–, specific rotation 258(T)
Melibiose 295(T)
Melting point, fatty acid, influence of structure 166, 166(T)
–, –, saturated 161(T), 166(T)
–, –, unsaturated 163(T), 166, 166(T)
–, triacylglycerol 171(T)
Melting salt, processed cheese 536
Membrane, biological 185, 186(F)
Menhaden 621
Menthadien-7-al, 1,3-p- *384*
Menthane-8-thiol, 1-p- 388(T)
Menthantriene, 1,3,8-p-
–, aroma, parsley *973*, 978, 979(T)
Menthen-8-thiol, 1-p- 838, *838*
Menthene-8-thiol, 1-p-
–, aroma, grapefruit 838, *838*
Menthol *383*
–, absolute configuration 386
–, cooling effect 431
–, odor threshold 431
–, sensory property 386
–, stereochemistry 386
–, synthesis 395, *395*
Menthone *383*
Mercapto-2-butanone, 3-
–, sensory property 364(T)
Mercapto-2-methylpentan-1-ol, 3-
–, formation 791
–, onion 790
–, sensory property 364(T)
Mercapto-2-pentanone, 3-
–, formation 364
–, quantitative analysis 356, 358(F), 359(T)
Mercapto-3-methylbutylformate, 3- 388(T)
Mercapto-3-pentanone, 2-
–, formation 364, *365*
–, sensory property 364(T)
Mercapto-4-methylpenta-3-one, 4-
–, basil 978, 978(T)
Mercapto-4-methyl-2-pentanone, 4- 388(T)
–, grapefruit 838
–, wine 921, 922(T)
Mercaptoethanal, 2-
–, formation 365(F)

Mercuribenzoate, p- 069, *069*
Mercury 468, 469(T)
Mercury compound, organic 468
Merodesmosine 581
Meromyosin, H- 571(T)
Meromyosin, L- 571(T)
Mesaconic acid, coffee 943, *943*
Mesifuran (see furanone, 4-methoxy-2,5-dimethyl-3(2H)-)
Met, honey wine 890
Metal ion, enzyme cofactor 104
Metallic taste, buttermilk 543
Metallo-aldolase 117, *117*
Metalo carboxypeptidase 076
Metalo peptidase 076, 077(T)
–, mechanism 076, 079(F)
–, specificity 078(T)
Metalocarboxypeptidase 077(T)
Metasaccharinic acid *269*
Methanethiol, oxidation 363
Methanol, wine 920
Methanolysis, triacylglycerol 172, *172*
Methional,
–, enzymatic formation 387, *387*
–, formation 365(F)
–, milk, sunlight off-flavor 344(T)
–, potato aroma 788, 789(T)
–, sensory property 364(T)
–, *Strecker* degradation, off-flavor 017
Methionine *010*
–, cereal 677(T)
–, cyanogen bromide cleavage 043, *044*
–, discovery 011
–, fruit ripening 847, *847*
–, loss by lipid peroxidation 215(T)
–, loss in food processing 024
–, reaction 069
–, *Strecker* degradation 365(F)
–, sulfonium derivative 069
–, synthesis 033
Methionine sulfone 024
Methionine sulfoxide 024
–, formation 069, 215(T)
–, protein 072, *072*
Methionine sulfoximide 012, *012*
Methionol, enzymatic formation 387, *387*
–, sensory property 364(T)
Methoxy-2-methyl-2-butanethiol, 4- 388(T)
–, coffee aroma 944
–, olive oil 646, 646(T)
Methoxyglucobrassicin, 4- 789(T)
Methyl anthranilate, grape 837
–, honey aroma 888
Methyl ketone, blue cheese 534(T)
–, formation by heating fat 222, 222(F), 222(T)
–, formation by microorganism 225, 226(F)

–, sensory property 226(T)
Methyl-2,4-nonandione, 3-
–, aroma, tea 955, 955(T)
–, formation 198, 198(F), 199(F)
–, sensory property 204(T)
–, soya oil, reversion flavor 650, 652(T)
–, tea aroma 198
Methyl-2-buten-1-thiol, 3-
–, aroma defect, beer 905
Methyl-2-butene-1-thiol, 3-
–, formation, coffee 366
–, sensory property 364(T)
Methyl-2-hepten-4-one, 5-
–, aroma quality 340(T)
–, enantiomeric excess 355(T)
Methyl-3-furanthiol, 2-
–, formation 364, *365*
–, quantitative analysis 356, 358(F), 359(T)
–, sensory property 364(T)
–, –, difference to the disulfide 365
–, stability 364
–, thiyl radical formation 364, *365*
Methyl-3-mercaptobutyl formate, 3-
–, aroma defect, beer 906
Methyl-5-hepten-2-one, 6- *242*
Methyl-p-nitrobenzene sulfonate 068
Methylamino acid, N- 018, *018*
Methylation, fatty acid 167, *167*
Methylbutanal, 2-/3-
–, formation, crust 735, 737(F)
Methylbutanol, 3-
–, formation, dough 736, 736(T), 738(F)
Methylcafestol, 16-O-
–, coffee 942, *943*
Methylcellulose, viscosity 308(T)
Methylcyclopropene, 1-
–, anti-senescence agent, fruit storage 848
Methylene cycloartenol, occurrence 229(T)
Methylene cyclopropylglycine, 2- 809
Methylene reductic acid, formation 277, *278*
Methyleneglutamic acid, 4-
–, biosynthesis *781*
Methyleneproline, 4- 809
Methyleugenol, banana *837*
Methylfurfural, 5- *264*
Methylhistidine, 1- 039
Methylhistidine, 3- 039
Methylisoborneol, odor threshold value *343*
Methyllanthionine 071, 072(T)
Methyllanthionine, β- 039, *039*
Methyloctanoic acid, 4-
–, mutton odor 607, 607(T)
Methylproline, 4- 809
Methylpropanal, biosynthesis 378(F)
–, sensory property 360(T)
Methylpropanal,

–, odor threshold value 341(T)
Methylsterol 231
Methylthiol, formation 365(F)
–, odor threshold value 341(T)
–, sensory property 364(T)
Methyltridecanal, 12-
–, formation 606, *607*
Metmyoglobin 574, 575, 575(F), 576
Micelle, lipid 186(F)
Michaelis constant 119
–, competitive inhibition 126
–, determination 120
–, enzymatic analysis 140, 140(T)
–, immobilized enzyme 148
–, pH dependency 129, 129(F)
Michaelis-Menten
–, enzyme kinetics 117
Michaelis-Menten kinetics, allosteric effect 124
–, allosterism 124
Microbial aroma 394
Microbial quality, meat, indicator 584
Microbial rennin 150
Microorganism, enzyme preparation, industrial
    application 146(T)
–, growth 131
–, –, temperature 133, 134, 134(F)
–, pressure 136
Middling 707(F)
Milk 498
–, alkaline phosphatase, indicator 134
–, amino acid composition 502(T)
–, aroma defect, 367
–, aroma substance 539, 540(T)
–, carbohydrate 512
–, casein fraction 502
–, casein micelle 508, 509(F)
–, –, composition 509(T)
–, certified 517
–, cholesterol reduction 539, 539(T)
–, citric acid 515
–, conjugated linleic acid 162(T)
–, consumption 501(T)
–, cream separation 498
–, creaming 518
–, curd formation 529
–, dehydrated product 527, 528(T)
–, dried product, aroma substance 539
–, dry matter, calculation 498, *500*
–, drying 527, 528(T)
–, electrical conductivity 500
–, enzyme 516, 516(T)
–, –, thermal stability 135(F)
–, fat globule 514, 514(T)
–, fatty acid composition 514(T)
–, freezing point 500
–, ganglioside 181, *182*, 515

–, goitrogenic substance 798
–, heated, galactosylisomaltol, β- 279
–, heating 518
–, –, aroma substance 539, 540(T)
–, –, bacterial reduction 133(F)
–, –, coagulation 519, 519(F)
–, –, killing of microorganism 518(F)
–, –, reaction 519, 520(F)
–, –, reaction rate 520, 520(F)
–, –, thiamine degradation 133(F)
–, homogenization 514, 518
–, indicator 515, 515(T)
–, ketone formation 519
–, lactone formation 519
–, lactoperoxidase 517
–, lactulose 513
–, lipase 515
–, lipid 514(T)
–, lipid composition 178(T)
–, lipolysis 515
–, lipoprotein 185, 185(T)
–, lipoprotein lipase 515
–, low fat 520
–, main structural element 498, 501(T)
–, mineral 515, 516(T)
–, natural radionuclide 470
–, off-flavor 515
–, orotic acid 515, *515*
–, pasteurization 518, 518(F)
–, pasteurization with H_2O_2 149
–, pH value 500
–, phospholipid, occurrence 179(T)
–, physico-chemical property 498
–, plasmalogen 179
–, plasmin 516
–, processing 517
–, –, overview 518(F), 521(F)
–, product, furosine 285(T)
–, –, malty off-flavor 543
–, –, phenolic taste 543
–, –, pyrraline 286(T)
–, –, rancidity 543
–, production data 500(T)
–, progesterone 228(T)
–, protein 501, 502(T)
–, purification 518
–, raw 520
–, reconstituted 520
–, redox potential 501
–, refractive index 500
–, skimming 520
–, sour 521, 523
–, specific mass 498
–, sterilization 518, 518(F)
–, –, vitamin loss 520
–, sunlight off-flavor 344(T)

–, thermization 518
–, thiocyanate 517, *517*
–, UHT, aroma substance 539, 540(T)
–, ultrahigh temperature treatment 518, 518(F)
–, unclean taste 543
–, various mammals 501(T)
–, vitamin 515, 516(T)
–, –, degradation 519
–, whey protein 502(T)
–, –, composition 511
–, –, genetic variant 505(T)
–, yield, somatropin 498
Milk chocolate 968(T)
–, lipolysis 154
Milk fat, adulteration 665
–, aroma defect 344(T)
–, branched fatty acid 160
–, detection 665(T)
–, foreign fat, detection 665
–, oxofatty acid 164
–, perfume rancidity 225
Milk powder, natural radionuclide 470
Milk product 521
–, aroma defect 388, 543
–, dehydrated 527, 528(T)
–, furcellaran 307
–, infant food 528
–, lactic acid content 522(T)
Milk sugar 512, *512*
Milk sugar (lactose)
–, production 877
Millet,
–, chemical composition 675(T)
–, origin 670, 670(F)
–, *Osborne* fraction 676, 677(T)
–, production data 672(T)
Milling extraction grade 708
Mineral 421, 422(T), 425(T)
–, bioavailability 421
–, determination, NIR 706(T)
–, fish 628, 628(T)
–, food processing 428
–, fruit 843, 843(T)
–, grape must 916
–, legume 762(T)
–, loss in food processing 423(T)
–, milk 515, 516(T)
–, occurrence in food 422(T)
–, occurrence in the human body 421(T)
–, use as additive 430
–, vegetable 793, 794(T)
–, wheat 676(T), 708(F), 709
–, whey 538, 538(F)
–, wine 921
Mineral water 988
–, classification 988(T)

Miraculin 438
Miso 767
–, phytoestrogen 763(T)
Mixed drink 936
–, alcopop 936
Mixograph 714
Modified polysaccharide 302
Modified protein 070
Modified starch, freeze-thaw stability 326
Molasse, psicose 251(T)
Molasses 874
–, chemical composition 874
–, saccharose production 871(F)
Mold, formation of methyl ketone 225
Mollusk 638
Mollusks, paramyosin 625, 625(T)
Molybdenum 425(T), 426
–, enzyme cofactor 105, *105*
Monellin 436
–, amino acid sequence 436(T)
–, conformation 436, 437(F)
–, stabilization by genetic engineering 437(F)
Monoacylglyceride 460
–, complex formation with amylopectin 740(F)
–, complex formation with amylose 740(F)
–, pasta 741
Monoacylglycero, physical property 178
Monoacylglycerol, hydrolysis, enzymatic 190, 191(T)
–, melting point 178
–, production 177, *177*
–, raisin 849
Monogalactosyl diacylglycerol *180*
Monohydroperoxide, analysis, HPLC 193, 195(F)
–, degradation, enzymatic 209
–, –, nonvolatile product 212(T), 213(F)
–, formation 192, 192(F)
–, fragmentation, β- 205(F)
–, isomerization 195
–, linoleic acid 194(F), 195(T)
–, linolenic acid, α- 195(T)
–, lipoxygenase, activation 207
–, oleic acid 194(F), 195(T)
–, proton catalyzed fragmentation 206(F)
–, reaction with heavy metal 199, 200(T)
–, reaction with heme(in) compound 200, *200*
–, reaction with iron ions 213(F)
–, reaction with protein 212, 213
Monohydroxyacetone, formation 267, *268*
Monophenol, enzymatic hydroxylation 106, 106(F)
Monosaccharide 248
–, acetylation *290*
–, acid-catalyzed reaction 263
–, anomeric effect 255
–, chain lengthening 249, *249*, *251*
–, chain shortening *251*
–, configuration 249

–, conformation 254
–, constitution 248
–, cyanohydrin synthesis 249, *249*
–, dicarbonyl cleavage 267, *268*
–, ester 290
–, esterification *290*
–, ether 291, *291*
–, etherification 291
–, family tree 251(F)
–, hemiacetal formation 249
–, –, diastereomer 251
–, hygroscopicity 256
–, melting point 005(T)
–, methylether 291
–, nitroalkane synthesis 248, *250*
–, nomenclature 248, 251, *252*
–, occurrence 250, 251(T)
–, oxidation with cupric ion 267, *267*
–, proton resonance spectrum 256, 257(F)
–, reaction with amino compound 270
–, reduction 261
–, *Reeves* formula 254
–, retro-aldol reaction 267, *268*
–, reversion, mechanism 263, *264*
–, reversion product 263, *263*
–, solubility 256
–, specific rotation 258(T)
–, stability 263, 862
–, strong alkaline, reaction 266, *266*
–, sweet taste 259, 259(T)
–, temperature of phase transition 005, 006(T)
–, trimethylsilylether 291, *292*
–, wheat 703(T)
Monosodium glutamate 430
Morel, dried, taste compound 788
Morelid 788
Mouth feeling, aminobutyric acid 023
Mowrah butter 648(T)
Mozzarella, aroma 541
MSG 430
–, taste, synergistic effect 431, 431(F)
–, *umami* 430
MTCA, wine 921
Mucic acid 262, *262*
Muconic acid 452, *452*
Multidimensional GC 353
Mumme, beer 904
Mungo bean, chemical composition 748(T)
Murein, hydrolysis 153
–, structure 153
Muscle, amine 584
–, amino acid composition 578(T)
–, ATP degradation 586
–, connective tissue 577
–, contraction 572, 572(F)
–, enzyme 571(T), 573

–, fish 623, 624(F)
–, free amino acid 584
–, heart 567
–, insoluble protein 571(T), 577
–, membrane material 577
–, myofibrillar protein 568
–, myosin, filament 568, 569(F), 570, 570(F), 572(F)
–, peptide 584
–, pigment 573
–, post-mortem change 580(F), 587
–, purine, post-mortem change 586, 586(T)
–, relaxation 572, 572(F)
–, rigor mortis 587
–, smooth 567
–, soluble protein 571(T), 573
–, striation 566, 568(F)
Muscle fiber, fiber 564, 566–568(F)
–, –, light absorption/light scattering 574
Mushroom, agaritin 783
–, aroma substance 788
Musk ambrette 396(T)
Mussel 638
–, chemical composition 637(T)
Must, yield 914
Must weight 914, *914*
Mustard, aroma substance 977, 978(T)
Mustard oil, cabbage aroma 792
–, vegetable 789(T), 790, *790*
Mustard paste 982
Mutagenic compound 026, 027, 028(T)
Mutagenicity 467
Mutarotation 253, 257
–, fructose, temperature dependence 866(F)
–, lactose 512(F)
–, mechanism 253
–, reaction rate 253, 253(T), 257
Mutatoxanthine *239*
–, occurrence in orange 240(T)
Mutton 593
–, cooked, aroma substance 607, 607(T)
Mutton odor 607, 607(T)
Mycotoxin 472, *472*, 474(T)
–, carcinogen 473
–, food monitoring 472
–, health hazard 475
Myocommata, fish muscle 626
Myofibril 566, 567(F)
–, beef, enzymatic hydrolysis 625(F)
–, fish, enzymatic hydrolysis 625(F)
–, swelling 590
Myoglobin, dissociation 576, *576*
–, electrophoresis 605(F), 610(F)
–, fish 625
–, light absorption 574, 575(F)
–, NO complex 576, *577*, 577(F)
–, oxygen binding 574(F), *576*

–, reaction with monohydroperoxide 200, 200(T)
–, reaction with nitrite 576, *576*, 577(F)
Myomere, fish muscle 623, 624(F)
Myomesin 571, 571(T)
Myosepta, fish muscle 623, 624(F)
Myosin 568, 570(F), 571(T)
–, amino acid composition 578(T)
Myrcene *382*, 896, 896(T)
–, sensory property 386(T)
Myricetin *834*, 834(T)
Myristic acid, plant fat *159*, 159(T)
–, sensory property 160(T)
–, structure, melting point 161(T)
Myristicin 972, *972*
Myristoleic acid, structure, melting point 163(T)
Myrosinase 790, *790*, 798
–, *Brussels* sprout, blanching 792

Naphthoquinone-4-sulfonic acid, 1,2- 020
Naringenin *832*, 832, 833(T)
–, indicator 858(T)
Naringin 832, 833(T)
–, indicator 858(T)
Natamycin 454, *454*
Natto 767
NBD-Cl 019
NBD-F 019
Nectarine, aroma substance 840
NEDI 468
Nef-reaction *250*
Nematicide, chemical structure 477(F)
–, trade name and activity 480(T)
Neochlorogenic acid 824, *824*
Neohesperidin 832, 833(T)
Neohesperidose 295(T)
Neokestose 295(T)
Neophytadiene, chlorophyll degradation 977(T), 981, *981*
Neotrehalose 295(T)
Neoxanthine *239*
Neral, aroma, lemon oil 838, *838*
–, aroma quality 340(T)
Nerol *383*
–, sensory property 386(T)
Nerol oxide *383*
–, formation *386*
Nervonic acid, structure, melting point 163(T)
Net protein utilization 030
–, determination 030
Network, polymeric 062
Neuraminic acid *182*
Neurolathyrism 764, 781
Neurosporene, absorption maxima 240(T)
Neutral fat 169
Neutral lipid, definition 158
Niacin 414, *414*

–, loss during food processing 403(T)
–, rice 710(T)
–, stability 425
Nickel 426
Nickel contact, production 657
Nickel, 425(T)
Nicotinamide, loss during food processing 403(T)
–, occurrence 409(T)
Nicotinamide adenine dinucleotide (NAD,NADH) 099
–, electronic spectrum 099(F)
–, fruit ripening 846
–, reaction 099, *108*
Nicotinamide adenine dinucleotide phosphate (NADP) 099(F)
Nigerose 295(T)
–, honey 886(T)
Ninhydrin reaction 021, *021*
NIR 706, 706(F), 706(T)
Nisin 039, *039*, 454
Nitgrate, biochemical production 492
Nitrate, antimicrobial action 454
–, arginine, precursor 492
–, cheese ripening 532
–, meat 576
–, nitrite pickling salt 983
–, occurrence 493, 493(T)
–, ultilization of ADI 492, 493(T)
Nitrite, antimicrobial action 454
–, formation 492
–, meat 576
–, occurrence 493(T)
–, reaction with myoglobin 576, *576*, 577(F)
Nitrite pickling salt 983
Nitrogen, protective gas 465
Nitrogen compound, wine 921
Nitrogen dioxide, bleaching agent 464
Nitrosamide 492
Nitrosamine 454
–, formation 492, *493, 494*
–, meat, detection 614
–, occurrence 494(T)
Nitrosodimethylamine, occurrence 494(T)
Nitrosopiperidine, occcurrence 494(T)
Nitrosopyrrolidine 494, 494(T)
–, formation *494*
Nitrosylchloride 464
Nitrosylmyoglobin 576
–, antioxidant 577, *577*
Nizo process, churning 526
N-line protein 571, 571(T)
No Observed Adverse Effect Level (NOAEL) 467
No-Time process, dough making 726
Nobiletin *834*, 834(T)
Noisette 882
Nomilin, biosynthesis *821*

Non-competitive inhibition 127
Nonadienal, (E)-2,6-
–, cucumber aroma 793
Nonadienal, (E,Z)-2,4-
–, linoleic acid, autoxidation 203(T)
–, sensory property 204(T)
Nonadienal, (E,Z)-2,6-
–, linolenic acid, α-
–, –, autoxidation 203(T)
–, sensory property 204(T)
Nonadienal, (Z,Z)-3,6-
–, formation, enzymatic 210(F), 211(T)
–, sensory property 204(T)
Nonanal, oleic acid, autoxidation 203(T)
–, rate of autoxidation 206(F)
–, sensory property 204(T)
Nonanone, 2-
–, sensory property 226(T)
Nonatrienal, 2,4,6-
–, oat flake, aroma 710
Nonen-3-one, precursor 523
Nonen-3-one, 1-
–, sensory property 204(T)
–, yoghurt 523
Nonenal, (E)-2-
–, linoleic acid, autoxidation 203(T)
–, sensory property 204(T)
Nonenal, (E)-3-
–, linoleic acid, autoxidation 203(T)
Nonenal, (E)-6-
–, hardened flavor 205
Nonenal, (Z)-2-
–, aroma, apple 839, 839(T)
–, olive oil 646(T)
–, sensory property 204(T)
Nonenal, (Z)-3-
–, linoleic acid, autoxidation 203(T)
–, sensory property 204(T)
Nonenal, 2-
–, rate of autoxidation 206(F)
Nonenal, 6-tr-
–, aroma defect 344(T)
Nonenzymatic browning 270, 284, 284
Nonulose, occurrence 251(T)
Nonulose, D-erythro-L-gluco-2- 251(T)
Nootkatone 385
–, grapefruit peel aroma 838, 838
–, odor threshold value 341(T)
Noradrenaline, fruit 813, 815
Norbixin 244
Nordihydrocapsaicin 980
Norfuraneol, formation 276, 277
Norfuraneol (see furanone, 4-hydroxy-5-methyl-
  3(2H)-)
Norharmane 026, 029
Norisoprenoid, C$_{13}$-

–, glycoside 243
Norisoprenoid, C$_{13}$-, 242
Nougat, aroma, filbertone 882
Nougat filler 881
NPU 030, 030
N-terminus, protein, determination 042
Nuclear magnetic resonance spectroscopy (^{1}H-NMR)
–, determination of fat 663
Nucleotide, 5'- 431
Nutra sweet 037
Nutraceutical 867
Nutritional property, sweetness 864(T), 866
Nylander's reaction 267

Oat,
–, bitter substance, formation 697, 697(F)
–, bitter taste 211
–, chemical composition 675(T)
–, flake, aroma 710
–, –, production 710
–, lipoperoxidase 210
–, origin 670(F), 671
–, Osborne fraction 676, 677(T)
–, production data 672(T)
Oat bran, saponin content 763(T)
Obacunone, biosynthesis 821
Obtusifoliol 231
Ochratoxin A 473, 474(T)
–, reference value, utilization 475(T)
Ocimene, cis- 382
Ocimene, trans- 382
Octadien-2-one, (E,E)-3,5-
–, sensory property 204(T)
Octadien-2-one, (E,Z)-3,5-
–, sensory property 204(T)
Octadien-2-one, 3,5-
–, linolenic acid, α-
–, –, autoxidation 203(T)
Octadien-3-hydroperoxide, (Z)-1,5-, 203(T)
Octadien-3-ol, (Z)-1,5-
–, formation, enzymatic 210(F), 211(T)
Octadien-3-one, (Z)-1,5-
–, aroma, tea, green 955, 955(T)
–, autoxidation, linolenic acid, α- 203(T)
–, fish aroma 628(T), 629
–, sensory property 204(T)
Octadienal, (Z)-2,5-
–, linolenic acid, α-
–, –, autoxidation 203(T)
Octanal, linoleic acid, autoxidation 203(T)
–, oleic acid, autoxidation 203(T)
–, sensory property 204(T)
Octanone, 2-
–, sensory property 226(T)
Octen-1-ol, Camembert 542, 542(T)

Octen-1-one, Camembert 542, 542(T)
Octen-3-hydroperoxide, 1-
–, autoxidation, linoleic acid 203(T)
–, sensory property 204(T)
Octen-3-ol, 1-
–, aroma quality 340(T)
–, mushroom aroma 788
Octen-3-ol,1-
–, enantiomeric excess 355(T)
–, formation, enzymatic 210(F), 211(T)
Octen-3-one, 1-
–, linoleic acid, autoxidation 203(T)
–, mushroom aroma 788
–, sensory property 204(T)
Octenal, (E)-2-
–, linoleic acid, autoxidation 203(T)
–, sensory property 204(T)
Octenal, (Z)-2-
–, linoleic acid, autoxidation 203(T)
–, sensory property 204(T)
Octopamine, fruit 813, 815
Octopus 639
Octulose, occurrence 251(T)
Octulose, D-glycero-D-manno-2- 251(T)
Octyl gallate 218
Odor, definition 340
–, fatty acid 160(T)
–, law of Stevens 342, 342(F)
Odor intensity, dependency on stimulant concentration
    342(F)
Odor quality, aroma substance 342, 343(F)
–, key odorant 340(T)
Odor threshold value,
–, air and water 351(T)
–, aldehyde 204(T)
–, amino acetophenone, o- 539
–, aroma substance 341(T)
–, catty odorant 388(T)
–, definition 341
–, diastereomer 354, 355(T)
–, ester 381(T)
–, fatty acid 160, 160(T)
–, fatty acid, 160, 160(T)
–, filbertone 882
–, furanone 362, 363(T)
–, hexen-3-one, 1- 792
–, hydroxybenzylmethylether, p- 976
–, influence of the matrix 342(T)
–, interaction of odorant 341, 341(T)
–, isothiocyanate 792
–, lactone 381(T)
–, pentylpyridine, 2- 370, 764
–, phenol 375(T)
–, pyrazine 372(T)
–, pyrroline,2-acetyl-1- 369(T)
–, saturated fatty acid 160(T)

–, steroid, C$_{19}$- 229(T)
–, Strecker aldehyde 360(T)
–, sulfur compound 364(T)
–, trimethylamine 557
–, wine odorant 921, 923(T)
Odor-bearing substance, definition 340
Odorant, boiled beef 605, 606(T)
–, composition, green tea 955, 955(T)
–, formation 211(T)
–, –, hydroperoxide lyase 209, 210(F)
–, lipid peroxidation 191, 203, 204(T)
–, olive oil 646, 646(T)
Odorant (see also aroma substance)
Oechsle degree 914, 914
Oenanthic acid, structure, melting point 161(T)
Off-flavor 343, 345(F)
–, cauliflower 792
–, dehydrated vegetable 799
–, egg 557
–, enzyme, inactivation 131(T)
–, example 344(T)
–, fecal, pepper 975
–, fish 629
–, fishy 628(T), 629
–, indicator 203
–, linoleic acid 203
–, linolenic acid, α-
–, –, autoxidation 203
–, lipolysis 189, 190(T)
–, milk 515
–, mouldy, pepper 975
–, partial fat hydrogenation 205
–, rancid, lipolysis 188
–, sexual odor 228, 229(T)
–, soybean oil 198
–, soybean product 764
–, swiss cheese 543
Off-flavor (see also aroma defect)
Oil, fruit pulp 645, 645(T)
–, plant 644
–, seed 647
Oilseed, oil recovery 647
Oily taste 158
Oleanolic acid 232
Oleic acid, autoxidation 193
–, –, carbonyl compound 203(T)
–, –, monohydroperoxide 194(F), 195(T)
–, configuration 165
–, degradation to lactone 380(F)
–, induction period 192(T)
–, iodine value 663
–, melting point 166(T)
–, molecule geometry 165
–, photooxygenation 195(T)
–, plant fat 159, 159(T)
–, rate of autoxidation 192(T)

–, structure, melting point 163(T)
–, taste 164(T)
Oleomargarine, fatty acid composition 160
–, production 643
Oleoresin, from paprika 244
Oleostearine, production 643
Oleuropein *803*
Oleyl alcohol *186*
Oligofructose 332
Oligosaccharide, abbreviated designation 293
–, cacao bean 962
–, conformation 293
–, glycol cleavage *292*
–, honey 886, 887(T)
–, hydrolysis 294
–, hygroscopicity 256
–, nomenclature 292
–, occurrence 295(T)
–, solubility 256
–, specific rotation 258(T)
–, structure 292
–, sweet taste 259, 259(T)
–, temperature of phase transition 005, 006(T)
–, wheat 703(T)
Oliseed, pressing 647
Olive, alkali treatment 449
–, production data 641(T)
Olive (see also table olive)
Olive oil 645, 645(T)
–, adulteration 666
–, analysis 645, 646(T), 666, 666(T)
–, –, HPLC 174(F)
–, detection 665(T)
–, ester oil, detection 664
–, fatty acid distribution 176(T)
–, heat treatment, detection 645
–, odorant 646, 646(T)
–, oleanolic acid 232
–, pectinolytic enzyme 153
–, polymorphism 172(T)
–, production data 641(T), 643(T)
–, quality difference, analysis 666, 666(F)
–, squalene 227
–, sterol 229(T)
–, storage, detection 645
–, tocopherol content 234(T)
–, type 645
–, unsaponifiable component 226(T)
Onion, aroma substance 790
–, key odorant, 790
OPA 022, *022*
Optimum, pH-
–, catalase 698
–, peroxidase 698
Optimum,pH-
–, amylase, wheat 696(T)

Orange 815(T)
–, aroma substance 837, 838(T)
–, bitter, flavanone content 833(T)
–, –, flavone content 833(T)
–, bitter compound 819
–, carotinoid 240(T)
–, flavanone content 833(T)
–, flavone content 833(T)
–, oxocarboxylic acid decarboxylase 377(T)
Orange juice, adulteration 857, 857(T)
–, –, detection 240, 858, 858(T)
–, aroma change 838
–, aroma defect 344(T), 838
–, aroma value 837, 838(T)
–, bitter taste 819
–, composition, guide value 857(T)
–, rearrangement of epoxycarotinoid 245
–, sweetening, detection 859, 859(T)
Organophosphate compound 476, 480(T)
Origano, essential oil, chemical composition 974(T)
Origin, determination 857
Ornithine 071, *072*, 073(F)
–, *Strecker* degradation *370*
Ornithinoalanine 071, 072(T)
Orotic acid, biosynthesis 515, *515*
–, milk 515, *515*
–, milk indicator 515(T)
Oryzanal, γ-
–, antioxidant 662
Osborne fraction, cereal 675
–, legume 746, 749(T)
Osteolathyrism 781
Osteomalacia 406
Ovalbumin 548, 549(T), *550*
–, plastein reaction 085(F)
Ovalbumin, S- 548
Ovoglobulin 551
Ovoinhibitor 549(T), 551
Ovomacroglobulin 549(T)
Ovomucin 551
Ovomucoid 550
Oxalyl-2,3-diaminopropionic acid 549(T), 550, 764, 781, 782(T)
Oxalyl-2,4-diaminobutyric acid 781, 782(T)
Oxathiane, passion fruit 840, *840*
Oxathiazinone dioxide, structure and taste 440, *440*
Oxazolinone, amino acid 017
Oxene 200, *201*
Oxidation, phenolic compound, enzymatic 835
Oxidation, β-
–, pathway 225, 226(F)
Oxidoreductase, example 100(T)
–, nomenclature 097
Oxime, taste 440
Oxo fatty acid, formation in lipid peroxidation 212(T), 213(F)

Oxocarboxylic acid decarboxylase, substrate
  specificity 377(T)
Oxofatty acid 164, *164*
Oxopropanal, 2-
–, formation 267, *267, 370*
Oxygen, activated, reaction 201, 202(T)
–, electron configuration 196, 197(F)
–, enzymatic removal 149
Oxymyoglobin 574, 575(F)
–, stability 576
Oxystearin, additive 455(T)
Oxytetracycline 454
Oyster, chemical composition 637(T)
Ozonation, drinking water 986
Ozone, disinfectant 495

Paddy rice 710
PAH 490
Pain 601
Pal oil, bleaching 647
Palatinit 874, *874*
–, nutritional/physiological property 865(T)
–, production 863(T)
Palatinol, biotechnology 262
Palatinose 295(T)
–, production 863(T), 874, *874*
–, relative sweetness 863(T)
–, specific rotation 258(T)
Palm kernel, production data 641(T)
Palm kernel oil 648, 648(T)
–, perfume rancidity 225
–, polymorphism 172(T)
Palm oil 645(T), 646
–, adulteration, detection 647
–, detection 665(T)
–, fatty acid distribution 176(T)
–, fractionation 659(T)
–, production data 641(T)
–, sterol 229(T)
–, thermal bleaching 656
–, tocopherol content 234(T)
–, unsaponifiable component 226(T)
Palm sugar 873
Palmitic acid, human milk 177
–, infant food 177
–, plant fat *159*, 159(T)
Palmitoleic acid, structure, melting point 161(T),
  163(T)
Panose 295(T)
–, honey 886(T)
–, specific rotation 258(T)
Panthothenic acid, stability 425
Pantothenic acid 415, *416*
–, fruit 842
–, occurrence 409(T)
Papain 076, 077(T)

–, active cysteine 115
–, meat, tenderizing 598
–, pH optimum 129(T)
–, specificity 078(T)
Paprika, degradation of capsanthin 243(F)
–, formation of pyrazine 388
–, oleoresin 244
–, pungent substance 979, 980(T)
Para-κ-casein, formation 506
Paraguayan tea 958
Paramyosin, molluslk 625, 625(T)
–, surimi 635
Parasorbic acid 451, *451*
Parboiling process, rice 710
Parinaric acid, structure, melting point 163(T)
–, UV absorption 167(F)
Parsley, aldose 250(T)
–, aroma substance 978, 979(T)
–, chlorophyll 794(T)
–, dried, aroma 979, 979(T)
–, essential oil, chemical composition 974(T)
–, glutathione 038
Passion fruit, aroma defect 344(T)
–, aroma substance 840, *840*
–, odorant 242(T)
Pasta, additive 742
–, chemical composition 741(T)
–, egg content 741
–, production 742
–, raw material 741
Paste 601
Pasteurization, fruit juice 854
–, gaseous sterilant 453
–, milk 518, 518(F)
Patulin *473*, 474(T)
–, isotope dilution assay 475
Pauly reagent, reaction of tyrosine 024, *024*
PCB 489
–, contamination, milk 490
PCDD 496, *496*
PCDF 496, *496*
PCR 142, 143, 143(F)
PDB 858(T)
Pea, aroma defect 344(T)
–, aroma substance 793
–, blanching process, lipoxygenase 135, 136(F)
–, carotinoid content 235(T)
–, chemical composition 748(T)
–, chlorophyll 794(T)
–, cooking process 768
–, hardening 768
–, hydrocyanic acid 761(T)
–, lipoxygenase, reaction specificity 207(T)
–, parboiling 768
–, production data 747(T)
–, proteinase inhibitor, stability 758(T)

–, saponin content 763(T)
–, softening 768
–, temperature of phase transition 006(T)
Peach 815(T)
–, aroma substance 840
–, carotinoid content 235(T)
–, hydroxycinammic acid derivative 824(T)
–, temperature of phase transition 006(T)
Peach kernel oil, tocopherol content 234(T)
Peanut, chemical composition 748(T), 816(T)
–, lipoxygenase, inactivation 765(T)
–, peroxidase, inactivation 765(T)
–, phytosphingolipid 182, 182
–, production data 749(T)
–, saponin content 763(T)
Peanut butter 652
–, fatty acid distribution 176(T)
–, polymorphism 172(T)
–, unsaponifiable component 226(T)
Peanut flake, stability 766(F)
Peanut oil 651(T), 652
–, detection 665(T)
–, production data 643(T)
–, sterol 229(T)
–, tocopherol content 234(T)
Peanut raw filler 881
Pear 815(T)
–, aroma formation 211(T)
–, aroma substance 839
–, formation of ethyl decadienoate, (E,Z)-2,4- 379(F)
–, hydroxycinammic acid derivative 824(T)
Pecctinolysis, fruit nectar 854
Pectic acid, formation in citrus fruit juice 153
Pectin 250(T), 314, 314
–, binding of polyphenol 845
–, conformation 297
–, elimination reaction 314
–, fruit 815(T)
–, fruit ripening 845
–, gel formation 314
–, gel strength 313(F)
–, gelling time 315(T)
–, legume, cooking 768
–, stability 314, 314
Pectin esterase 334, 334, 334(T)
–, citrus fruit juice 153, 154(F)
–, competitive inhibition 153, 154(F)
–, legume 768
Pectin lyase 334, 334, 334(T)
Pectin methylesterase, inhibition, pressure 136
Pectinolytic enzyme 334
–, industrial application 153
–, overview 334(T)
Pelargonic acid, structure, melting point 161(T)
Pelargonidin 829, 830(T)
Pellagra 414

Pentadiene, 1,3-
–, aroma defect 452
Pentagalloyl-D-glucose 825, 826
Pentanal, linoleic acid, autoxidation 203(T)
–, sensory property 204(T)
Pentane, linoleic acid, autoxidation 205, 205, 207
Pentanone, 2-
–, sensory property 226(T)
Penten-3-one, 1-
–, linolenic acid, α-
–, –, autoxidation 203(T)
–, sensory property 204(T)
Pentenal, (E)-2-
–, linolenic aicd, α-
–, –, autoxidation 203(T)
–, sensory property 204(T)
Pentenal, (Z)-2-
–, linolenic acid, α-
–, –, autoxidation 203(T)
Pentofuranose, α-D- 253
Pentofuranose, α-L- 253
Pentofuranose, β-D- 253
Pentofuranose, β-L- 253
Pentosan, cereal 702, 702(F)
–, determination, NIR 706(T)
–, enzymatic degradation 153
–, oxidative cross linking 703(F)
–, solubility 702, 703
–, structure 702(F)
Pentosidine, collagen 582, 582
–, formation 287, 288
–, occurrence 288(T)
Pentyl-α-pyrone, 6-
–, peach aroma 840, 840
Pentylfuran, sensory property 204(T)
Pentylpyridine, 2- 370, 370
–, occurrence 764
–, odor threshold 370
–, odor threshold value 764
Peonidin 829, 830(T)
Pepper, aroma, key odorant 975
–, –, storage 975, 975(F)
–, aroma defect 979
–, aroma substance 975, 975(F), 975(T)
–, essential oil, chemical composition 974(T)
–, fecal off-flavor 975
–, light sensitivity 979
–, mouldy off-flavor 975
–, pungent substance 980(T)
–, white, off-flavor 388
–, –, skatole 388
Pepsin 077(T), 079
–, pH optimum 129(T)
–, specificity 044, 045(F), 078(T), 079
Pepstatin 079
Peptidase 074, 077(T)

–, classification 077(T)
–, meat, aging 590, 590(T)
–, metal-containing 076
Peptide 034
–, antiparallel chain 051(F)
–, basicity 036
–, binding to carrier 044, *044*
–, bitter taste 036, 036(T)
–, dissociation constant 036(T)
–, HPLC 044
–, isoelectric point 036(T)
–, mass spectrometric analysis 046
–, nomenclature 034
–, parallel chain 051(F)
–, salty taste 038, 038(T)
–, sequence analysis 021, 044
–, structure and taste 036
–, sweet taste 037, 038(T), 441
–, taste threshold value 036(T)
–, vapor phase sequence analysis 046
Peptide bond, configuration 048
Peptide chain, conformation 048, 048(F)
–, extended 048, 048(F)
–, folding 053
–, torsion angle 049, 049(F)
Peptide pattern, cheese ripening 534, 534(T), 535(F)
Peptide synthesis 034
–, protective group 017, 018, 034
Peptidyl-prolyl-cis/trans-isomerase 049
Peptidyldipeptidase 077(T)
PER-value 030, *030*
–, modified casein 081, 082(T), 085(T)
–, plastein 085(T)
Perch-like fish 622
Perfume rancidity 225
–, occurrence 648
Perilla alcohol *383*
Perillaldehyde, oxime 440, *440*
Perimysium 564, 566(F)
Periodate, glycol cleavage 292, *292*
Perlwein 928
Peroxidase, autoxidation of linoleic acid 200
–, blanching process 134
–, electronic spectrum 103(F)
–, inhibition, pressure 136
–, mechanism 103, 103(F)
–, milk 135(F)
–, potato 135(F)
–, prosthetic group 103
–, protein cross-linking 087, *087*
–, rate constant 120(T)
–, reaction 139(T)
–, reactivation 135
–, thermal inactivation, peanut 765(T)
–, wheat 698
Peroxide value, deep fried fat 220(T), 221

–, fat 667
Peroxy nitrite 201
Peroxy radical, cyclization 195, *195*
–, formation 192, 192(F)
–, reaction 192, 192(F), 193, 195, *195*, 201(T),
–, rearrangement 195, *195*, 196, *196*
Persipan 881
Pest control, cereal 707
Pesticide, natural 486
–, –, example 486
–, removal, oil refining 656(T)
Pesticide (see also PPA)
Petunidin 830(T)
Pfluemli water 932
PGPR 462, *462*
pH optimum, enzyme 127, 129(T)
pH Value, fruit 815(T)
Phase transition, kinetics 005
Phase transition temperature, glucose oxidase 137
Phaseolin, amino acid sequence 752(T)
PHB-Ester, antimicrobial action 450
Phellandrene, α- *383*
–, pepper 975, 975(T)
–, sensory property 386(T)
Phellandrene, β- *383*
Phenol, black tea 956, *957*
–, chemoprevention 962
Phenol oxidase, potato 135(F)
Phenol oxidase (see also polyphenol oxidase)
Phenol, 2-methoxy-4-vinyl- 375(T)
Phenol, 4-ethyl- 375(T)
Phenol, 4-vinyl- 375(T)
Phenolase 105
Phenolcarboxylic acid, fruit 823
Phenolic compound, cacao bean 962(T), 963
–, fruit 822, 823(T)
–, –, taste 835
–, indicator, fruit 857(T), 858, 858(T)
–, metal complex, fruit 835
–, oxidation 835
–, polymerization, red wine 920
–, protein complex, fruit juice 837
–, tea 953, 953(T)
–, wine 920, 920(T)
Phenoloxidase, cresolase activity 698
–, legume 768
Phenyl-2-thiohydantoin, 3- 021
Phenylacetaldehyde, aroma defect, beer 906
–, honey aroma 888
Phenylacetic acid ethyl ester, invert sugar cream,
    aroma 890
Phenylalanine *010*
–, degradation, *Ehrlich* pathway 737
–, discovery 012
–, synthesis 033
–, UV absorption 015(F)

Phenylalanine-free diet 086, 087(F), 087(T)
Phenylene diisothiocyanate, p- 045, *046*
Phenylenediamine, reaction with deoxyosone 273, *274*
Phenylethanal, 2-
–, odor threshold value 341(T)
–, sensory property 360(T)
Phenylethanol, 2-
–, formation, dough 736, 736(T), 738(F)
Phenylisothiocyanate 021, 044
Phenylmethanesulfonyl fluoride 075
Phenylphenol, o-
–, antimicrobial action 454
Pheophorbide 794, 796(T)
Pheophytin 794, 796(T)
–, bleaching of fat 655, 655(F), 655(T)
–, HPLC 795(F)
–, photooxidans 655, 655(F)
Phlobaphene 965
Phloridzin *834*
Phosphatase, acid, honey 886
–, –, milk 518(F)
–, alkaline, milk 135(F), 518(F)
–, milk, reactivation 135
Phosphatase,
–, alkaline, indicator 134
Phosphate, addition to condensed milk 527
–, occurrence in food 422(T)
Phosphate starch 326, *326*
Phosphatide, definition 158(T)
Phosphatidyl choline *179*
Phosphatidyl ethanolamine *179*
Phosphatidyl glycerol *179*
Phosphatidyl inositol *179*
Phosphatidyl serine *179*
–, occurrence 179(T)
Phosphofructokinase, allosteric regulation 124
Phospholipase, cereal 696
–, specificity 190
Phospholipase A$_1$ 190
Phospholipase A$_2$ 190
Phospholipase B 190
Phospholipase C 190
Phospholipase D 190
Phospholipid, dissociation 180
–, hydrolysis 180, 190
–, milk 514
–, occurrence 179(T)
–, solubility 180
–, synergistic effect with antioxidant 220
–, wheat flour 704, 705(T)
Phosphoprotein 012, 040
–, casein 502
–, –, amino acid sequence 503(T)
–, phosvitin 554, *554*
Phosphoric acid, acid salt, additive 449

–, cola drink 449
–, synergistic effect with antioxidant 220, 220(T)
Phosphorous 424
Phosphorus, content, human body 421(T)
Phosphytidyl ethanolamine, occurrence 179(T)
Phosvitin 554, 554(T), *555*
Photooxidation, furan fatty acid 650, 652(T)
–, reaction rate 198
–, soya oil 650, 652(T)
Photooxidation (see photooxygenation)
Photooxygenation 196
–, furan fatty acid 198, 199(F)
–, inhibition 198
–, linoleic acid 198, 198(F)
–, –, monohydroperoxide 195(T)
–, linolenic acid, α-
–, –, monohydroperoxide 195(T)
–, oleic acid, monohydroperoxide 195(T)
–, type-1 reaction 197
–, type-2 reaction 197
Photosynthesis, plant group 858, 859(T)
Phthaldialdehyde, O-
–, amino acid derivative 022, *022*
Phthalide, celery 788, *789*
Phthalylamino acid, N- 017
Phyllodulcin 439, *439*
Physical refining, oil 656
Phytadiene, chlorophyll degradation 977(T), 981, *981*
Phytane 227
Phytanic acid 160
–, structure 161(T)
Phytase, legume 768
–, wheat 696, *697*, 697(T)
Phytate 455(T)
–, wheat *697*, 697(T)
Phytoene *236*
–, absorption maxima 240(T)
–, orange 240(T)
–, tomato cultivar 236(T)
Phytoestrogen 762
–, lignan 835
–, occurrence 762, 763(T)
Phytofluene *236*
–, absorption maxima 240(T)
Phytol, dehydration 981, *981*
Phytomenadione (see also vitamin K$_1$) 408, *411*
Phytosphingolipid 182
Phytosphingosine 181
–, structure determination 183, *183*
Phytosterol 229, *231*
–, nutritional value 229
–, oil refining 655
Pickled vegetable 802
Pickling, fish 634
Picrocrocin 977, *977*
Pigment, carotinoid 244

–, vegetable 793, 794(T)
–, wine 915
Pilsener beer, odorant 902, 903(T)
Pimaricin 454, *454*
Pineapple 815(T)
–, aroma substance 841, 841(T)
–, furaneol 363(T)
Pinene, α- *384*
–, basil 978, 978(T)
–, pepper 975, 975(T)
Pinene, β- *384*
–, pepper 975, 975(T)
Ping-pong mechanism, enzyme 122
Pinoresinol *835*
Piperanine *980*
Piperine *980*
–, isomerization 979
Piperonyl isobutyrate, sensory property 396(T)
Piperylin *980*
Pistachio, chemical composition 816(T)
Plant food, enzymatic digestion 153
Plant gums, aldose 250(T)
Plant phenol 822
–, biosynthesis 835
Plant, C$_3$- 859, 859(T)
Plant, C$_4$- 859, 859(T)
Plant-protective agent 475
Plant-protective agent (PPA) 467, 468
Plasmalogen 179, *180*
Plasmin 075
–, milk 516
Plastein, amino acid composition 085(T)
–, phenylalanine free 086, 087(T)
–, taste 085, 086(T)
Plastein reaction 083, 083(F), 083(T), 085(T), 087(F)
–, single-stage 087(F)
–, taste 085
Pleated-sheet structure, protein 050
Plum 815(T)
–, aroma substance 841
–, hydroxycinammic acid derivative 824(T)
Plum sauce 852
Plum water 932
Polar lipid, definition 158
Pollack 621
Polyalcohol 878
–, content 878, 878(T)
Polyamide, clarifying agent 464
Polyamine, anti-senescene agent, fruit storage 848
Polychlorinated biphenyls (PCB) 489
Polycyclic aromatic compound, removal, oil refining 656(T)
Polycyclic aromatic hydrocarbon (PAH) 490
Polydextrose 877
Polygalacturonase 334, 334(T)
–, citrus juice 854

Polyglycerol-polyricinolate 462, *462*
Polyglycine, conformation 050(T)
Polylysine, conformation 060(F)
Polymerase chain reaction 142
Polymeric network 062
Polymerization, fat 223
Polymorphism, fat 171, 172(T)
–, triacylglyceride 171
Polyoxyethylene sorbitan ester 460(T), 462, *462*
Polyphenol 825, *826*
–, chemical structure *822*
–, occurrence 823(T)
Polyphenol oxidase, cereal 698
–, copper 105
–, fruit 835
–, inhibition, pressure 137
–, mechanism 106(F)
–, ordered mechanism 122
–, substrate binding 122
Polyphenol oxidase (see phenol oxidase)
Polyproline, conformation 050(T)
Polysaccharide 296
–, analysis 335
–, aperiodic sequence 296, 299
–, binding of aroma substance 391, 392(T)
–, branched type 300
–, classification 296
–, conformation 296
–, containing carboxyl group 301
–, crumpled-type conformation *298*
–, double helix 298
–, effective volume 302(F)
–, egg box type *297*
–, egg-box-type conformation *297*
–, enzymatic degradation 333
–, gel formation 299, 299(F)
–, green coffee 941(T)
–, helical conformation 297, *297*, 298(F)
–, hollow helix type 297, *297*
–, inclusion compound 297
–, interchain aggregation 299(F)
–, interchain interaction 298
–, linearly branched type 301
–, loosely joint type *298*
–, modified 302
–, nomenclature 296
–, perfectly linear type 300
–, periodic sequence 296, 298
–, phosphoric acid ester 302
–, property 300
–, reductive polymerization 335, 336(F)
–, ribbon type 296, *296*
–, ribbon-type conformation 296
–, stretched conformation 296
–, structure 296
–, sulfuric acid ester 302

1050   Index

–, temperature of phase transition 006, 006(T), 007
–, triple helices 298, 299(F)
–, use in food 301(T)
Polysaccharide other than starch, cereal 702
Polyunsaturated fatty acid, ω-3-
–, fish 627, 627(T)
Polyvinyl pyrrolidone 333, *333*, 464
Pomace 914
Pomace wine 917
Poncirin 832, 833(T)
Pop beverage 856
Pop corn aroma 368, 369(T)
Poppy oil 651(T), 653
Pork 593
–, boiled, aroma 605, 606(T)
–, consumption data 565(T)
–, meat, composition 569(T)
–, progesterone 228(T)
Porter, beer 904
Portwine 926, 927(T)
–, sotolon 926
Post-mortem change, fish 627(F), 629
–, meat 587
Potassium 422(T)
–, content, human body 421(T)
–, occurrence in food 422(T)
Potassium chloride, taste 982
Potassium ferrocyanide, addition to salt 983
Potassium-40 470
Potato, aroma substance 788, 789(T)
–, boiled, aroma 788, 789(T)
–, cultivar differentiation 780(F)
–, enzymatic browning 122
–, enzyme, inactivation 134, 135(F)
–, –, thermal stability 135(F)
–, fatty acid with dienyl ether structure 210, *210*
–, glutathione 038
–, lectin 759(T)
–, lipolysis 191(T)
–, lipoxygenase, reaction specificity 207(T)
–, phospholipid, occurrence 179(T)
–, proteinase inhibitor 780(F)
–, steroid alkaloid 798, *798*, 798(T)
–, temperature of phase transition 006(T)
Potato protein, biological valence 030
Potato starch, gelatinization 320(F)
Poultry 593
–, consumption data 565(T)
–, meat extract 602
–, production data 563(T)
PPA 480(T)
–, ADI utilization 486, 486(T)
–, amount used 482, 482(T)
–, analysis 483
–, chemical structure 477(F)
–, LD$_{50}$ 480(T)

–, market share 475(T)
–, metabolite 484
–, monitoring program 482(T), 485
–, MPL 485
–, natural 486
–, –, consumption 486
–, plant-protective agent 475
–, residue 475, 480(T), 482(T), 484(T), 485
–, resistance of pest population 476
–, results ofd analysis, example 482, 482(T), 484(T)
–, risk assessment 485, 486(T)
Praline 968
Prebiotics 867
Precalciferol 229
Pregelatinized flour 722
Pregelatinized starch 326
Premier jus 642
Preservation, catalase 149
Preservative 449
Pressed honey 884
Pressing, oilseed 647
Pressure, enzyme activity 136
–, microorganism 136
Primary structure, protein 040
Prime beef fat 642
Prion, pathogenic 055
Pristane 227
Pristanic acid 160
–, structure 161(T)
Proanthocyanidin 828, *829*
Proanthocyanin, cacao bean 963
Probiotics 523
Process flavor 607, 607(T)
Processed cheese, alkali treatment 449
–, carboxymethyl cellulose 330
Processed meat, additive, determination 612
Procyanidin *828*
Prodelphidin *828*
Progesterone 228, *228*
–, occurrence 228(T)
Progoitrin 789(T), 798
Prolamin, barley 680, 684(F)
–, celiac disease 674
–, cereal 676(T), 679(T)
–, oat 684(F)
–, rye 680, 684(F)
–, wheat 679(T), 680, 684(F)
–, –, amino acid sequence 689(T)
–, wheat (see also gliadin)
Prolinase 076
Proline *010*
–, discovery 012
–, ninhydrin reaction 022
–, precursor of odorant 368, *370*
Promotion of ripening, fruit 847
Pronase, specificity 086

Prooxidants, rates of lipid autoxidation 191(F)
Prooxidative effect 220
Propanal, formation from threonine *376*
–, linolenic acid, α-
–, –, autoxidation 203(T)
–, sensory property 204(T)
Propanethial-S-oxide, (Z)-
–, lachrymatory factor 790, *791*
Propellant 465
Propenylcysteine sulfoxide 790, *791*
Propenylguaethol, sensory property 396(T)
Propiolactone, β- 069
Propionic acid 452
–, antimicrobial agent 443
–, Ca-salt, sweet taste 543
–, fermentation 532, 533(F)
–, Mg-salt, sweet taste 543
–, swiss cheese 542, 542(T)
–, titration curve 165(F)
Propionyl-2-thiazoline, 2- 367(T)
Propyl gallate *218*
Propylene, fruit ripening 847
Propylene glycol alginate 304
–, beer 902
Propylene oxide, antimicrobial action 453
Prosthetic group 098, 098(F), 102
Protective coating, amylose 324
–, scleroglucan 331
Protective gase 465
Protein 040
–, *Edman* degradation 044
–, 3₁₀-helix 050(T)
–, acetoacetylation 066
–, acidic polysaccharide, interaction 306
–, acylation 065, 066, 080(T)
–, –, reversible 066
–, alkali treatment 071
–, alkylation 064, 082
–, amidation 067
–, amino acid composition 030(T), 041, 085(T)
–, amino acid sequence 041
–, aminoacylation 081, 082(F)
–, binding of aroma substance 391, 392(F), 393(T)
–, binding of ion 059, 059(F)
–, binding to carrier 044, *044*
–, biological value 029
–, C-terminus, hydrazinolysis 042, *042*
–, –, titration 042, *042*
–, cacao bean 961
–, carbamoylation 064
–, circular dichroism 060
–, cleavage of disulfide bridge 067
–, complex, fruit juice 837
–, conformation 048
–, cross-linking 071, 086
–, cyanogen bromide cleavage 043, *044*

–, cytoskeletal 568
–, deamination 065
–, degradation during cheese ripening 534
–, denaturation 056
–, –, activation energy 058, 058(T)
–, –, activation entropy 058, 058(T)
–, density 184
–, determination, NIR 706(T)
–, dissociation 058, 059(T)
–, disulfide bridge 055, 055(F)
–, –, bond strength 054(T)
–, –, reduction und reoxidation 082, 083(F)
–, disulfide exchange 068
–, domain 055, 055, 056(F)
–, egg white 548, 549(T)
–, electrostatic interactions, bond strength 054(T)
–, emulsifier 063
–, emulsifying property 082(T)
–, energy value 008
–, enrichment 008
–, enrichment with essential amino acid 081, 082(F),
    085(F)
–, enzymatic dephosphorylation 083, 083(F)
–, enzymatic hydrolysis, bitter taste 150
–, –, industrial application 150
–, enzyme-catalyzed reaction 074, 076(T)
–, –, overview 076(T)
–, esterification 067, 080(T)
–, extrusion 089
–, fibrillar 053
–, fish 624
–, foam formation 062
–, fruit 807
–, fruit ripening 846
–, gel formation 062
–, globular 053, 056(T)
–, green coffee 941(T)
–, guanidination 064
–, helical structure 050(T), 051, 052(F)
–, helix, α-
–, –, frequenzy of amino acid 054(T)
–, helix, α-, 050(T), 052(F)
–, helix, π- 050(T)
–, honey 888
–, hydration 061
–, hydrazinolysis 042, *042*
–, hydrogen bond 054, 054(T)
–, hydrolysis 080(T)
–, hydrolysis at aspartic acid residue 044
–, hydrolysis at serine and threonine 043, *044*
–, hydrophobic bond 054, 054(T)
–, hydrophobicity, calculation 507, *508*
–, ionic bond, bond strength 054(T)
–, isoelectric point 059
–, –, estimation 059
–, isoelectric precipitation 059, 061

–, isoionic point 059, 059(F)
–, legume 746, 748(T)
–, maleylation 066
–, microbial 008
–, microparticulated 463
–, modification 064, 079, 080(T)
–, molecular weight 056(T)
–, –, determination 041
–, muscle 571(T)
–, N,O-acyl migration 043, *044*
–, N-terminus, determination 042
–, net charge 059
–, optical activity 060
–, optical rotatory dispersion (ORD) 060, 060(F)
–, overlapping cleavage 045(F)
–, oxidation with peroxidase/H_2O_2 087
–, oxidative change 072
–, partial hydrolysis, chemical 043
–, –, enzymatic 043
–, phenylalanine free 086, 087(T)
–, photometric determination 016, 020
–, pK value of the side chain 059(T)
–, pleated-sheet structure 050, 050(T), 051(F)
–, primary structure 040
–, quaternary structure 056, 056(T)
–, reaction 064
–, reaction in food processing 070
–, reaction with aldehyde 214, 214(F)
–, reaction with amino acid ester 067
–, reaction with bifunctional reagent 070
–, reaction with diazoacetamide 067
–, reaction with malondialdehyde 214
–, reaction with monohydroperoxide 212, 213
–, reaction with pyridoxal phosphate 065
–, reductive alkylation 080(T)
–, reductive methylation 083(F)
–, regular structure element 049, 050(T)
–, –, frequency of amino acid 054(T)
–, rye flour 709(T)
–, S-sulfo derivative 067
–, salting-in effect 061
–, salting-out effect 061
–, secondary structure 049
–, selective cleavage at cysteine residue 067
–, selective cleavage at methionine residue 043, *044*
–, sequence analysis 021, 041, 044
–, sequence analysis via nucleotide sequence 046
–, sheet, β- 050, 050(T), 051(F)
–, single cell 008
–, solubility 060, 061(F), 085, 086(F)
–, –, fish 634(T)
–, –, frozen fish 633(F)
–, source 008
–, spinning 088
–, structural domain 055, 055, 056(F)
–, structure, β-

–, –, frequenzy of amino acid 054(T)
–, subunit 041
–, succinylation 065, 081, 081(F)
–, super-secondary structure 052, 053(F)
–, surface denaturation 062, 063
–, sweet 436
–, swelling 061
–, technological property 080(T)
–, temperature of phase transition 006, 006(T)
–, terminal group 042
–, tertiary structure 053, 056(F)
–, texturization 088
–, texturized 087
–, titration curve 060(F)
–, total charge 059
–, turn, β-
–, –, frequenzy of amino acid 054(T)
–, turn, β-, 052, 052(T), 053(F)
–, tyrosine cross-linking with peroxidase/H_2O_2 086
–, vegetable 770, 780(F)
–, water binding capacity 061
–, wheat, hydration 727, 728(F)
–, wheat flour 709(T)
–, X-ray analysis 048
Protein concentrate 008
–, fish 636
Protein content, type of bread 735(T)
Protein cross-link, fish 626
Protein denaturation, activation energy 132
Protein efficiency ratio 030
Protein film, dough 727, 728(F)
–, stability 062
Protein foam 062
Protein gel 062
Protein hydrolysate 602, 602(F)
–, debitterizing 085, 086(T)
–, production 602(F)
Protein hydrolysis, loss 023
Protein isolate 008
Protein modification, *Maillard* reaction 285, *285*
Protein nature, enzyme 095
Protein radical, formation 214, *214*
Protein strand, dough 729, 729(F)
Proteinase 074, 077(T)
–, active serine, detection 108
–, application in meat ripening 150
–, flour improvement 720, 720(F)
–, industrial application 150
–, industrial preparation 150, 151(T)
–, pH optimum 151(T)
–, stability, pH range 151(T)
–, wheat 696
Proteinase (see also endopeptidase)
Proteinase inhibitor 055, 055(F), 075
–, active center 756(T)
–, amount, legume 754

–, inactivation 757, 758(T)
–, nutritional test 757, 757(T)
–, nutritional-physiological effect 754, 757, 757(T), 758
–, occurrence 755(T)
–, property 753
–, soybean, inactivation 758(T)
–, soybean product 758(T)
–, specificity 755, 755(T)
–, structure 755
Proteinase K 056
Proteose peptone 507
Proteus spp.
–, food poisoning 470
Protocatechuic acid 825
–, fruit, content 825(T)
–, occurrence 825(T)
Proton, mobility 002
Protopectin 845, *845*
–, fruit ripening 845
Protopectinase, fruit ripening 845, *845*
Protoporphyrin, Fe^{2+}- *574*, 574(F)
Protoporphyrin, Fe^{3+}- 574, 576(F)
Prunasin 760(T)
–, aroma precursor 841
Pseudo ionone *242*
Pseudolysine 043
Pseudomonas spp.
–, food poisoning 470
Psicose 251(T), *252*
PTC-peptide 021
Pulegol *383*
Pulegone *383*
Pullulanase 334
–, application 152
Pumpernickel 741
Pumpkin seed oil 649(T), 650
Punch extract 936
Pungent substance 980(T)
Pungent taste, capsaicin 979
–, paprika 980
–, relative 980(T)
–, structure/effect relationship 979
Purity quotient, sugar extract 873
Puroindolin, foam formation 695
–, wheat 695
Putrescine 584
Pyranose, conformation 254
Pyrazine, biosynthesis 388
–, cabbage aroma 792
–, ethenyl-3,5-dimethyl-, 2- 372(T)
–, ethenyl-3-ethyl-5-methyl-, 2- 372(T)
–, ethyl-3,6-dimethyl-, 2- 372(T)
–, formation 371, 373, *373*, 373(T), *388*
–, odor threshold in air 400(F)
–, potato 788, 789(T)

–, product of *Maillard* reaction 274, *275*
–, sensory property 372(T)
–, structure and odor 399, 400(F)
–, trimethyl 372(T)
Pyrazine, 2,3-diethyl-5-methyl- 372(T)
Pyrazine, 2-acetyl- 372(T)
Pyrazine, 2-ethyl-3,5-dimethyl- 372(T)
Pyrazine, 2-isobutyl-3-methoxy- 372(T)
–, odor threshold value 341(T)
Pyrazine, 2-isopropyl-3-methoxy- 372(T)
Pyrazine, 2-sec-butyl-3-methoxy- 372(T)
Pyrethrin 476, *476*
Pyrethrum 476
Pyridine, product of *Maillard* reaction 274, 275, 280
Pyridine, 2-acetyl-
–, odor threshold value 369(T)
Pyridinoline, beef, tenderness 581, *581*
Pyridocarbazole 027(T)
Pyridoimidazole 026, 027(T)
Pyridoindole 026, 027(T)
Pyridosine 071
Pyridoxal (see also vitamin B$_6$) 413, *414*
Pyridoxal phosphate 065, 103, *103*, 413
–, coenzyme of alliinase 790
Pyridoxamine (see also vitamin B$_6$) *103*
Pyridoxine (see also vitamin B$_6$) 413, *414*
Pyridoxol (see also vitamin B$_6$) 413, *414*
Pyrolysis product, amino acid 025, 027, 028(T)
Pyropheophytin 795, *795*
–, HPLC 795(F)
Pyrraline, occurrence 286(T)
Pyrrole, aroma substance 367
–, product of *Maillard* reaction 274, *275*
Pyrrolidone carboxylic acid 011, *011*
–, molasses 874
–, sugar extract 869, 874
Pyrroline, 2-acetyl-1-
–, formation 369, 371(F)
–, odor threshold value 369(T)
Pyrroline, 2-propionyl-1- 369(T)
Pyruvate kinase 139(T)

Quaternary structure, protein 056, 056(T)
Quercetin 834, *834*
–, antioxidative activity 217, 217(T)
–, content, fruit 834
–, red wine 920(T)
Quercetin-3-glycoside, taste threshold 953
Quercus lactone. See whisky lactone
Quinic acid 820(T), *822*
Quinic acid lactone, coffee 947, *948*, 948(T)
Quinide *948*, 948(T)
Quinit acid lactone, coffee, bitter taste 947, *948*, 948(T)
Q$_{10}$-value, definition 133, *133*

Racemate, cleavage 014

Radiation detection, food 073
Radical chain reaction, autoxidation of lipid 192, 192(F)
Radical scavenger 215(F), 216(T)
Radiolysis, alkyl cyclobutanone, formation 224, *225*
–, decontamination, dose 225
–, detection 224
–, lipid 224, 224(F)
Radionuclide 470
–, natural 470
Radish, aroma substance 789
Raffination, fat, loss of tocopherol 233
Raffinose 295(T)
–, enzymatic degradation 152
–, legume 760(T)
–, relative sweetness 259(T)
–, saccharose production 871
–, specific rotation 258(T)
–, sugar beet 869
Raisin, chemical composition 850(T)
Ramachandran diagram, protein 049(F)
Rancidity, hydrolytic 189, 190(T)
Random-random hypothesis, triacylglycerol 174
Rapeseed, glucosinolate 652
–, production data 641(T)
Rapeseed oil 651(T), 652
–, bleaching 655, 655(T)
–, detection 665(T)
–, furan fatty acid 164(T)
–, lecithin removal 654
–, sensitizer 655, 655(T)
–, unsaponifiable component 226(T)
Raspberry 815(T)
–, aroma, cooking process 839
–, ellagic acid 825
–, hydroxycinammic acid derivative 824(T)
–, odorant 242(T), 839
Raspberry ketone, aroma quality 340(T)
–, biosynthesis 839, *840*
–, odor threshold 839
–, odor threshold value 341(T)
Rate constant, activated oxygen 201, 202(T)
–, enzyme 120(T)
Raw milk 517
Raw sausage 600
–, production 600(F)
Raw sugar 872, 872(F)
Reaction aroma 607, 607(T)
Reaction rate, photooxidation 198
–, temperature dependency 131, *131*
Reaction rate constant, temperature dependency 131, *131*
Reaction specificity
–, enzyme 095, 095(F)
Recognition odor threshold, definition 341
Red beet, aroma substance 790

–, pigment 796
Red cabbage, aroma substance 792
Red color enhancer, meat 877
Red currant, hydroxycinammic acid derivative 824(T)
Red No. 2 444(T), 446(T)
Red No. 3 445, 446(T)
Red wine, fermentation 916
–, grape cultivar 910
–, phenolic compound 920, 920(T)
Redox lipid 158, 233, 234(F)
Redox reaction, reductone 277, *278*
Reductive depolymerisation, polysaccharide 335, 336(F)
Reductone, *Maillard* reaction 278, *278*
–, redox reaction 278, *278*
Reeves formula, monosaccharide 254
Reference Dose (RfD) 467
Refined sugar 872, 872(F)
Refining, fat 653, 654(F)
–, –, detection 669(T)
–, loss of chloro pigment 655(T)
–, oil, cholesterol 656
–, –, hydroperoxide degradation 655, *655*
–, physical, oil 656
Refining fatty acid 656
Regulatory specificity, enzyme 094, 124
Reindeer milk, chemical composition 501(T)
Rennin 077(T), 078
–, from microorganism 150
–, specificity 078(T)
Rennin (see chymosin)
Resistant starch 325
–, determination 325
–, formation 325
–, structure 325
Respiratory quotient, fruit ripening 844
Resveratol, wine 920(T), 921
Retinal, 11-cis *404*, 404(F)
Retinol (see also vitamin A) *404*
–, loss during food processing 403
Retroaldol condensation 222, *222*
–, enzyme catalyzed 117, *117*
Retrofat, fat substitute 464
Retrogradation, amylose 320
Reversible inhibition, enzyme catalysis 126
Reversion, glucose 263
Reversion dextrin 890
Reversion flavor, soya oil 650, 652(T)
Rhamnose 250(T)
–, relative sweetness 259(T)
–, specific rotation 258(T)
Rhamnosidase, α-L- 153
Rheological property, gliadin 692
–, HMW subunit 692, 694(F)
–, LMW 692
Rhodanide, goiter 798

RIA 142
Ribitol 261
Riboflavin 102, *102*
–, rice 710(T)
Riboflavin (see also vitamin B₂) 413, *413*
Ribofuranose, α-D- 255(T)
Ribofuranose, β-D- 255(T)
Ribonuclease, conformation 049, 060(F)
–, mechanism 114(F)
Ribopyranose, α-D- 255(T)
Ribopyranose, β-D- 255(T)
Ribose 250(T), *251*
–, equilibrium mixture 255(T)
–, specific rotation 258(T)
Ribulose *252*
Rice 710
–, aroma substance 710
–, chemical composition 675(T)
–, oil, squalene content 227
–, origin 670, 670(F)
–, *Osborne* fraction 676, 677(T)
–, parboiled, production 710
–, –, vitamin content 710(T)
–, production data 672(T)
–, yield per hectare 674(T)
Ricinoleic acid 164, *164*
Rigor mortis 587
Ripening, fruit 843
–, –, phenolcarboxylic acid 826
–, grape 911, 914(F)
Roasted aroma 367, 367(T)
Rock salmon 618
Rodenticide, chemical structure 477(F)
–, trade name and activity 480(T)
Rose oxide *383*
–, sensory property 386(T)
Rosemary, antioxidant 981
–, essential oil, chemical composition 974(T)
Rosenoxide, wine 921, 922(T)
–, –, formation 923
Rosé wine 914
Rotation, molecular 257
–, specific 257
Rotundone *385*
–, pepper aroma 975
–, sensory property 386(T)
–, shiraz grape 921
Rum 932
Rum fruit 851
Runner bean, chemical composition 748(T)
Rutinose 295(T)
Rye,
–, chemical composition 675(T)
–, milling 707
–, origin 670(F)
–, *Osborne* fraction 676, 677(T)

–, yield per hectare 674(T)
Rye bread, aroma substance 737, *737*, 739(T)
Rye flour, amylase, α- 715
–, amylogram 715(F)
–, baking property, acid 722
–, –, pregelatinized flour 722
–, chemical composition 709(T)
–, extraction rate 709(T)
–, milling extraction rate, mineral 708(F)
–, –, vitamin 708(F)
–, physical assay 715
–, production 707
–, sour dough 724, 724(F)
–, storage 716
–, treatment with β-glycosidase 153
–, type 709, 709(T)
Rye mix bread 741(T)

Sabinene *384*
Sabinene hydrate, cis- *384*
Sabinene hydrate, trans- *384*
Saccharase, honey 886
Saccharification degree, starch degradation 875
Saccharin 433, *435*
–, synthesis *435*
Saccharinic acid *269*
Saccharose 295(T)
–, aqueous solution, viscosity 865(F)
–, biosynthesis 845, *845*
–, cacao bean 962
–, conformation 293
–, enzymatic analysis 138, 138(F)
–, enzymatic isomerization 875
–, fruit 818(T)
–, glycemic index 867
–, glycol cleavage 292, *292*
–, hydrolysis *294*
–, legume 760(T)
–, nutritional/physiological property 864(T), 866
–, occurrence 868
–, production 863(T), 869
–, –, crystallization 871, 871, 872(F)
–, –, evaporation of thin syrup 870
–, –, extract purification 869
–, –, extraction 869
–, –, loss 872, 872(T)
–, production data 868(T)
–, secondary product 874
–, solubility 862(F)
–, specific rotation 258(T)
–, stability 865
–, taste threshold value 035(T), 259(T)
–, water absorption 866(F)
Saccharose fatty acid ester 291
Safflower oil 651(T), 653
–, fatty acid composition, climate 178(F)

Safflower seed, production data 641(T)
Saffron, aroma substance 974(T), 977, *977*
–, extract 244
–, pigment 239
Safranal 977, *977*
Safrole *972*
–, biosynthesis *973*
Sage, antioxidant 981
–, essential oil, chemical composition 974(T)
Saithe 621, 635
Sake 928(T)
Salad sauce, alginate 304
–, guaran gum 312
–, gum tragacanth 310
–, locust bean gum 312
–, xanthan gum 331
Salami, water activity 004(T)
Salicylic acid 825
Salicylidenelysine, ε-N- 023
Salmon 623
–, aroma 628, 628(T)
Salmonella spp.
–, food poisoning 470
Salt 982
–, additive 983
–, drying 983
–, effect in dough 721
–, iodization 427
–, iodized 983
–, potassium ferrocyanide 983
–, production 983
–, requirement 982
Salt substitute 983, 983(T)
Salted vegetable 802, 804
Salting, fish 633
–, meat 597
–, vegetable 804
Salting-in effect, protein 061
Salting-out effect, protein 061
Salty taste 982
–, cheese 543
–, peptide 038, 038(T)
Sambal 982
Sandwich ELISA 142, 142(F)
Sansa 645
Sapogenin 763
Saponification 172
Saponification value, definition, example 663, 664(T)
Sansa 645
Saponin, chemical structure 763, *763*
–, cholesterol complex 764
–, hemolysis 764
–, legume 763, *763*
–, toxicity 764
Sarcolemma 565
Sarcomere 566, 568, 569(F)

–, titin 569, 570(F)
Sarcoplasm 565
–, muscle 572
Sarcoplasmic reticulum 572
Sardine 621
Sauce 603
Sauerkraut 802, 802(F)
–, chemical composition 803
–, faulty 804
–, reducing agent 804
Sausage 598
–, chemical composition 600(T)
–, cooked 600, 600(F)
–, emulsion 599, 599(F)
–, production 600, 601(F)
Sauser 916
Schaal test 668
Schaftoside, indicator 858, 858(T)
Schardinger dextrin 295(T)
–, specific rotation 258(T)
Scleroglucan 331, *332*
Scleroprotein 053
Scopoletin 824, *824*, 824(T)
Scorbamic acid 419, *419*
Scorpaenidae 622
SCP 008
Scurvy 417
Sea fish 618
Seal oil 644, 644(T)
Seasoning, genotoxic compound 603
–, protein hydrolysate 603
–, –, aroma 602
Seaweed, aldose 250(T)
Secoisolariciresinol *835*
Secondary product of metabolism, formation of aroma
    substance 360
Secondary structure, protein 049
Secunda beef fat 643
Sedanolide 789, *789*
Sedimentation value, gluten 712
Sedoheptulose-7-phosphate, aroma precursor 767
Selachyl alcohol *187*
Selenium 426, 426(T)
Selinene β- *385*
Selters 988
Semi-fat cheese 529
Semi-fat margarine 660, 660(T)
Semolina, durum wheat 710
–, wheat 707(F), 710
Sensitizer, edible oil 655(F), 655(T)
–, photooxygenation 197
Sensory property, amino acid 034, 035(T)
–, peptide 036
Sepia 639
Sequence analysis, peptide 021
–, protein 021

Sequene analysis, DNA 046, *047*, 048(F)
Sequestrant 455, 455(T)
Serine *010*
–, discovery 012
–, reaction 024
Serine endopeptidase 075, 077(T)
–, inhibition 075, *075*
–, specificity 075, 078(T)
Serine peptidase 077(T)
Serotonin, fruit 813, *817*, 817(T)
Sesame oil 651(T), 653
–, detection 664(T)
–, detection of margarine 661
Sesame seed, production data 641(T)
Sesamol, formation 653(F)
Sesamolin 653(F)
Sesquiphellandrene *385*
Shark 618
Shark liver oil, squalene 227
Shea butter 648(T), 649
Shea fat, unsaponifiable component 226(T)
Sheep milk, chemical composition 501(T)
–, production data 499(T)
Sheep tallow 643
Sheet, β-
–, protein 050
Shelf life, fish 627(F), 629, 632(T)
Shelfish, consumption 617(T), 618
Shellfish, chemical composition 637(T)
Sherry 926, 927(T)
Shigella spp.
–, food poisoning 470
Shikimic acid 820(T), 821
Shogaol, retroaldol cleavage 981, *981*
Short sour, rye bread 724, 724(F)
Shortening, baking quality 721, 722(T)
Shoyu 766
Shrimp 636
–, chemical composition 637(T)
–, removal of shell 153
Shrinkage temperature, collagen 626
Sialic acid 181, *182*
Silicon 428
Silk fibroin 053
Sinapic acid 825
–, thermal degradation 376(T)
Sinensal, α- *385*
Sinensal, (all-E)-α-
–, sensory property 386(T)
Single cell protein 008
Single-phase interesterification 173, *173*
Single-substrate reaction, kinetics 117
Singlet oxygen, formation 197, *197*
–, reaction 201(T)
–, reaction with unsaturated fatty acid 197, *197*
Sinigrin 789(T)

Sitosterol *230*
–, olive oil 646(T)
Sitosterol, β-
–, occurrence 229(T)
Skatole, formation *389*
–, odor threshold value *343*
–, sensory property 388
Skeletal muscle 564, 566–568(F)
–, contractile apparatus 568
Skim cheese 529
Skim milk 520
–, chemical composition 538(T)
Skim milk powder 527, 528(T)
–, analysis 515
–, production data 500(T)
Slaughtering, process 592
Slicing cheese 531(T)
Smoke point, frying oil 668
Smoking, fish 633
–, formation of antioxidant 218
–, meat 597
SMOW 858(T)
Snail 638
–, helix garden 638
Soap, isolation 172
Soda water 988
Sodium 421
–, content, human body 421(T)
–, occurrence in food 422(T)
Sodium hydroxide, additive 449
Soft caramel 880
–, chemical composition 880(T)
Soft cheese 529, 531(T)
–, ripening, yield 532
Soft roe 636
Softener 261, 464
Soja, taste 764
Solanidine 798, 798(T)
Solanine *798*, 798(T)
Sole 623
Solubility, amino acid 015, 015(T)
–, fatty acid 167, 167(F)
–, lactose 513(T)
–, monosaccharide 256
–, oligosaccharide 256
–, protein 060
–, sugar 862(F)
–, sugar alcohol 862(F)
Solvent, extraction of aroma substance 394
–, extraction of oilseeds 647
Somatropin, milk yield 498
Sorbic acid, antimicrobial action 451(F), 452
–, degradation, aroma defect 452, *452*
–, synthesis *451*
–, wine, bacterial degradation 926
Sorbitan, production 879, *879*

Sorbitan fatty acid ester 291, 460(T), 462, *462*
Sorbitol 261, *261*
–, enzymatic analysis 139(T)
–, fruit 817
–, nutritional/physiological property 864(T)
–, production 863(T), 879
–, relative sweetness 863(T)
–, solubility 862(F)
–, water absorption 866(F)
Sorbitol dehydrogenase, reaction 139(T)
Sorbitol, D-
–, wine 920
Sorbitol, L- 879
Sorbose *252*
–, nutritional/physiological property 864(T)
–, production 863(T)
–, relative sweetness 863(T)
Sorbose, L- 878
–, production 878
Sorghum, hydrocyanic acid 761(T)
Sorghum sugar 873
Sorption isotherm 003, 004(F)
Sotolon, degradation 362, *363*
–, enantiomeric excess 355(T)
–, formation 268, *269*
–, –, fenugreek 976, *976*
–, odor threshold, wine 926
–, portwine 926
–, seasoning aroma 603
–, soy sauce 767
Sotolon (see furanone), 3-hydroxy-4,5-dimethyl-
    2(5H)-
Soup 603
Soup turtle 639
Sour cherry, hydroxycinammic acid derivative 824(T)
Sour cream butter 526
Sour dough, acid formation 724, 724(F)
–, fermentation ratio 724
–, production 724, 724(F)
–, short sour method 724, 724(F)
Sour milk 523
Sour milk cheese 531(T)
Sour milk product 521
–, aroma substance 540
Soya milk, phytoestrogen 763(T)
Soya sauce, aroma substance 767
–, chemical composition 767
–, production 766, 766(F)
Soya, *Bowman-Birk* inhibitor 055, 055(F), 756
–, chemical composition 748(T)
–, coumestrol 762, *762*
–, essential amino acid 749(T)
–, flour, phytoestrogen 762
–, globulin, amino acid composition 753(T)
–, globulin fraction 746
–, –, emulsifying effect 751, 751(F)

–, –, subunit 746
–, heat treatment, detection 754
–, *Kunitz* inhibitor 755
–, lipid composition 178(T)
–, lipoxygenase, reaction specificity 207(T), 208(F)
–, meat, detection 614
–, milk 766
–, miso 767
–, oil 651, 651(T)
–, –, change on deep frying 220, 220(T)
–, –, detection 665(T)
–, –, differentiation from sunflower oil 233
–, –, fatty acid distribution 176(T)
–, –, furan fatty acid 164(T)
–, –, hardened flavor 205
–, –, high oleic 650, 652(T)
–, –, high stearic 650(T)
–, –, HPLC-analysis 174(F)
–, –, hydrogenation 658(T)
–, –, lecithin removal 654, 654(F)
–, –, low linolenic 650, 652(T)
–, –, low palmitic 650, 652(T)
–, –, low saturate 650, 652(T)
–, –, production 647(F)
–, –, production data 643(T)
–, –, reversion flavor 650, 652(T)
–, –, sensitizer 655(F)
–, –, sterol 229(T)
–, –, tocopherol content 234(T)
–, –, unsaponifiable component 226(T)
–, phospholipid, occurrence 179(T)
–, phytoestrogen 763(T)
–, processing 765, 766(F)
–, product, aroma defect, removal 149
–, –, natto 767
–, –, off-flavor 370, 764
–, –, soy milk 766
–, –, soy sauce 766
–, –, sufu 767
–, –, tofu 766
–, production data 748(T)
–, protein, aminoacylation 083(F)
–, –, binding of aroma substance 393(T)
–, –, detection 142(T), 144
–, –, enrichment with glutamic acid 085(F)
–, –, enrichment with methionine 083(F)
–, –, enrichment with tryptophan 083(F)
–, –, phytoestrogen 763(T)
–, –, plastein reaction 086(F), 087(T)
–, –, plastein reaction, solubility 085, 086(F)
–, –, production 765, 766(F)
–, protein concentrate 765, 765(T)
–, protein isolate 765, 765(T)
–, rancidity 764
–, saponin content 763(T)
–, urease 753, *753*

Soybean genetically modified, detection 142(T), 144
Soybean genotype, fatty acid composition 650, 652(T)
Sparkling wine 926
–, bottle fermentation 927
–, carbonation process 928
–, classification 928
–, disgorging 927
–, dosage 927
–, tank fermentation 928
Special flour 708(T)
Special salt 983
Specific activity, enzyme 098
Spelt, origin 670
Spelt wheat 670
Sperm, fish 636
Sperm whale extract 040(T)
Spermidine 585
Spermine 585
Spherosome, lipid, wheat flour 705
Spherosomes 177
Sphingoglycolipid 181
–, acidic 181
–, neutral 181
Sphingolipid 181
Sphingomyelin *181*
Sphingophospholipid 181
Sphingosine 181
Spice 971
–, aroma change 981
–, chemical composition 971
–, class 971(T)
–, content of essential oil 973(T)
–, extract, propellant 465
–, plant 971(T)
–, powder 981
–, preparation 981
–, volatile 971, 974(T)
Spinach, aroma substance 792
–, carotinoid content 235(T)
–, chlorophyll 794(T)
–, furan fatty acid 164(T)
–, glutathione 038
–, heating, chloro-pigment 796(T)
–, processing, discoloration 796(T)
–, saponin content 763(T)
–, temperature of phase transition 006(T)
Spinning, protein 088
Spirit 929
Spirit vinegar 984
Spleen, chemical composition 594(T)
SPME 349
Sprat 618
Sprouting, cereal 695
–, –, detection 695, 696
Squalene 227

Stability in deep frying, fat 224, 224(T)
Stabilizer 464
Stachydrine 019(T)
Stachyose 295(T)
–, enzymatic degradation 153
–, legume 760(T)
–, specific rotation 258(T)
Staling, bread, carboxymethyl cellulose 330
Staling defect, bread 739
Standard flour 708(T)
Stanol 229
Staphylococcus aureus, food poisoning 470, 471(T)
Starch 315
–, amorphous region 315
–, amylose content 317(T)
–, binding of aroma substance 392(T)
–, cacao bean 962
–, cereal 701
–, cross linked 327, 327(F)
–, crystalline region 318, 318(F)
–, crystallinity 316, 317(T)
–, crystallization, kinetics 740(F)
–, damage, measurement 708
–, degradation, α–amylase 152(F)
–, –, enzymatic 151, 152(F)
–, –, product 875
–, –, saccharification degree 876
–, degree of cross linking and viscosity 327(F)
–, determination, NIR 706(T)
–, E-type 326
–, enzymatic analysis 139(T)
–, extruded 325
–, extrusion, anhydroglucopyranose, 1,6- 325
–, fruit ripening 845
–, gelatinization 316, 318, 319(F)
–, gelatinization behavior 320(F)
–, gelatinization temperature 317(T), 318, 320, 321(F)
–, legume 759, 760(T)
–, mechanical damage 708
–, mechanically damaged 325
–, modified 325
–, oxidized 327
–, production 315
–, raw material 315(T)
–, resistant 325
–, retrogradation 739, 740(F)
–, rye flour 709(T)
–, saccharification, enzymatic 151, 152(F)
–, swelling 316, 317(T)
–, temperature of phase transition 005(F), 006(T)
–, thermal modification 318, 319(F), 319(T)
–, thermogram 739(F)
–, thin boiling 326
–, unfrozen water 005
–, wheat flour 709(T)
–, X-ray diffraction diagram 316, 318, 319(F)

Starch ester 326
Starch ether 326
Starch granule, structure 316, 317(T)
Starch lipid, composition 701(T), 702
Starch phosphate 326
Starch syrup 875
–, chemical composition 875(F), 876(T)
–, hydrogenated 877
–, maltose enriched 152
–, production 863(T), 875
Starchy endosperm, cereal 673, 675(T)
Steam-rendered lard 643
Stearic acid, plant fat 159, 159(T)
–, structure, melting point 161(T)
Stearyl alcohol 186
Stearyl-2-lactylate 463, 463
–, binding in baking process 734(T)
Stemwort, beer 901, 901
Stereospecific analysis 175
–, triacylglycerol 175, 175(F)
Sterigmatocystin 473, 474(T)
Sterilization, canned vegetable 800
–, milk 518, 518(F)
Steroid, chlorinated 603, 603
–, odorant 228, 229(T)
Steroid alkaloid, bitter taste 798, 798(T)
–, potato 798, 798, 798(T)
Steroid hormone, beef, sexual origin 610
Steroid, C$_{19}$-
–, odor threshold 229(T)
Sterol 227
–, analysis 232, 666(F)
–, fat hardening 658
–, Meisenheimer adduct 233
–, vegetable fat 229, 229(T)
Sterol ester, analysis 666(F)
–, margarine 229
Steven'slaw 342
Stevioside 438, 438
Stiebene, chemical structure 822
Stigmasterol 230
–, identification of cocoa butter 231(T)
–, occurrence 229(T)
Stigmasterol, Δ-7
–, occurrence 229(T)
Stigmasterol, Δ-7- 231
Stock fish 633
Storage stability, temperature of phase transition 006, 006(T)
Stout 904
Strained honey 884
Strawberry 815(T)
–, aroma change 841, 841(T)
–, aroma substance 841, 841(T)
–, cultivar differentiation 816(F)
–, ellagic acid 825

–, furaneol 363(T)
–, heat treatment, flavor change 841
–, hydroxycinammic acid derivative 824(T)
–, temperature of phase transition 006(T)
Strawberry juice, adulteration 858(T)
Strecker acid, formation 283, 283
Strecker aldehyde 021, 282
–, sensory property 360(T)
Strecker degradation, tea 956, 957
Strecker degradation 021
–, cysteine 363(F)
–, methionine 363(F)
–, ornithine 370
–, proline 368, 370
Strecker reaction 017, 032, 282, 288
Strontium-90 470
Structure and odor, aroma substance 398
Structure and taste, amino acid 034, 261
–, aspartame 037, 038(T)
–, cyclamate 436(T)
–, dipeptide amide 441(T), 448
–, hernandulcin 443
–, oxathiazinone dioxide 440, 440(T)
–, peptide 036
–, sweet peptide 037, 038(T)
–, sweet substance 432
Structure und taste, sugar 260
Substrate determination, enzymatic analysis 138
Substrate specificity, decarboxylase 377(T)
–, enzyme 094
–, glutathione dehydrogenase 699(T)
Subtilin 039
Subtilisin, active serine, mechanism 115
–, enzymatic hydrolysis 045(F)
–, specificity 078(T)
Succinate dehydrogenase, inhibition 127(T)
Succinic acid 443
Succinic acid anhydride, additive 447
–, reaction with protein 065
Sucrose, sweetness receptor 433, 434(F)
Sucrose fatty acid ester 462
Sufu 767
Sugar 862
–, affinated 872
–, caramelization 270
–, composition 873
–, consumption data 868(T)
–, fruit 818(T)
–, fruit juice 853(T)
–, iso-sweet concentration 259(T)
–, jam 852(T)
–, metabolism 866
–, must 915
–, nutritional/physiological property 864(T), 866
–, production from sugar beet 869
–, production from sugar cane 872

-, reducing, bread crumb 727(F)
-, refined 871
-, relative sweetness 863(T)
-, sensory property 863(T), 866
-, solubility 862(F)
-, sweet taste, AH/B/X model 260
-, -, temperature dependency 259, 259(F)
-, type 873
Sugar alcohol 261
-, fruit 817
-, nutritional/physiological property 864(T), 867
-, production 878, 879
-, solubility 862(F)
Sugar anhydride 263, 263
Sugar beet, extract, chemical composition 869
-, -, nonsugar substance 869, 872
-, extraction 869
-, production data 868(T)
Sugar cane, production data 868(T)
Sugar couleur 270
Sugar ester 291, 462
Sugar ether 291
Sugar fatty acid ester 291
Sugar substitute 261, 432
Sugar, L-
-, synthesis 878
Sulfhydrylation, reductive 364, 365
Sulfitation, sugar extract 870
Sulfite, antimicrobial action 452
-, reaction with food constituent 452
-, sugar coleur, stabilization 270, 270
Sulfolipid 181
Sulfonamide 487(T), 488(F)
Sulfur compound, odor threshold value 364(T)
-, volatile 387, 388(T)
Sulfur dioxide, antimicrobial action 452
-, bleaching of anthocyan 831, 832
Sulfur treatment, grape must 917
-, wine 917
Sulfuric acid, additive 449
Sulfurous acid, treatment, fruit 849, 850
Sunflower oil 650, 651(T)
-, analysis, HPLC 174(F)
-, detection 665(T)
-, differentiation from soybean oil 233
-, fatty acid composition, climate 178(F)
-, -, variation 665(T)
-, fatty acid distribution 176(T)
-, production data 643(T)
-, tocopherol content 234(T)
-, unsaponifiable component 226(T)
Sunflower seed, production data 641(T)
Sunshine vitamin 406
Suosan 439, 440
Super-secondary structure, protein 052
Superaspartame 037, 442, 442

Superoxide dismutase 201
Superoxide radical anion 201, 201(F)
-, formation 201, 201
-, lipid peroxidation 201
-, reaction 201, 202(T)
-, reaction with superoxide dismutase 201
-, xanthine oxidase 105
Surface-active agent 456, 456(T)
-, occurrence in food 456(T)
-, use in food 456(T)
Surimi, production 635
Suspension 456(T)
Sweet cherry, hydroxycinammic acid derivative
    824(T)
Sweet compound, AH/B-system, AH/B/X-system
    260, 261
Sweet must 854
-, definition 855
Sweet taste 432
-, AH/B/X-system 432, 433(F)
-, alitame 442, 442
-, amino acid 034, 035(T)
-, aspartame 441
-, chalcone 834
-, dihydrochalcone 439, 439, 834(T)
-, dipeptide amide 441
-, dipeptide ester 441
-, guanidine derivative 440, 440, 440(T)
-, n/e-system 432
-, peptide 037, 038(T)
-, relative, sugar 259(T)
-, structural requirement 260
-, sugar, temperature dependence 866, 866(F)
-, superaspartame 442, 442
-, threshold detection value 432
-, urea derivative 439
Sweetener 432
-, nutritional property 864(T), 866
Sweetening strength, relative 433, 435(F)
Sweetening with sugar, fruit 851
Sweetish-off-flavor, beer 906
Sweetness, relative, sugar 513(T)
Sweetness receptor, glucose 433, 434(F)
-, lugduname 433, 434(F)
-, model 433, 434(F)
-, sucrose 433, 434(F)
Swelling, protein 061
Swift test 668
Swiss cheese, aroma 542, 542(T)
-, taste compound 542, 542(T)
Synemin 571
Synephrine, fruit 813, 815
Synergist, antioxidative effect 219
Synthetic, chewing gum 882

Table water 988

Table wine 919
Tablet 881
Tachysterol 230(F)
Taco shell, aroma 389
Taette, milk 524
Tagatose 252
Taint, food 343, 345(F)
Tall oil 229
Talose 251
Tamarind flour 312
–, gel strength 313(F)
–, polysaccharide structure 312
Tannase, reaction 154
Tannin 825
–, clarifying agent 464
Tanning agent 825, 828, 829
–, condensed 829
–, enzymatic hydrolysis 154
Tartaric acid, additive 448, 448
–, biosynthesis 822
–, fruit 820(T)
–, metabolism 822, 822
–, wine 916, 917(F)
Taste, amino acid 034, 035(T)
–, astringent 825, 829
–, definition 340
–, fatty acid 160(T)
–, –, unsaturated 164(T)
–, oily 158
–, peptide 036
–, salty 982
–, synergistic effect 431, 431(F), 433, 435(F)
Taste compound, bouillon 605(T)
–, coffee 947, 948, 948(T)
–, dried morel 788
–, tea 953
Taste intensity, definition 034
Taste modifier 438
Taste substance, potato 798(T)
–, steroid alkaloid 798(T)
–, wine 921, 922(T)
Taste threshold, fatty acid, 160, 160(T)
Taste threshold value, amino acid 034, 035(T)
–, sugar 259(T)
Taste-bearing substance, definition 340
Taste;
–, cocoa 963, 963(T)
Taurine 584, 584
–, biosynthesis 584
TBA test 668
TBHQ 218
Tea 951
–, amino acid 954, 954(T)
–, astringent taste 953
–, black 951
–, –, aroma 955, 955(T)

–, –, GAE 963
–, catechin, difference green/black 953
–, chemical composition 952, 953(T)
–, chlorophyllase reaction 955
–, enzyme 953
–, epitheaflavic acid 956, 957
–, flavonol-3-glycoside 953
–, formation of pigment 956, 957
–, furan fatty acid 164(T)
–, green 952
–, –, aroma 955, 955(T)
–, –, aroma of brew 955, 955(T)
–, –, GAE 963
–, mineral 956
–, odorant 242(T)
–, packaging 958
–, phenol, composition 956, 957
–, phenol oxidation, enzymatic 953
–, phenolic compound 953, 953(T)
–, pheophytin 955
–, polyphenol oxidase 956
–, processing, reaction 956
–, production, reaction 956
–, production data 951(T)
–, seed oil, detection 646
–, storage 958
–, Strecker degradation 956, 957
–, theaflavin 956, 957
–, thearubigen 956, 957
–, theogallin 954
–, turbidity, tannase 154
Tea grade 952
Teaseed oil, detection 664(T)
Tempeh, phytoestrogen 763(T)
Temperature, growth of microorganism 130, 132(F), 133, 134(F)
Temperature of phase transition 005, 005, 006(F), 006(T)
–, storage stability 006, 006(T)
Temperature optimum, enzyme 133
Tenderizing, meat 598
–, –, detection 611
Tenkawang fat, detection in cocoa butter 231(T)
Tenside 456
Teratogenicity 467
Terpene, chemical structure 382(T)
–, grape 924, 925(F)
–, –, cultivar differentiation 924, 925(F)
–, sensory property 386, 386(T)
Terpene glucoside, wine, hydrolysis 923, 923(T)
Terpinene, α- 383
Terpinene, γ- 383
Terpineol, α- 383
–, sensory property 386(T)
Tertiary structure, carboxypeptidase 056(F)
–, hemoglobin 056(F)

–, protein 053
–, triosephosphate isomerase 056(F)
Testosterone 228, *228*
Tetrahydropyridine, 2-acetyl, formation 369, 371(F)
–, odor threshold value 369(T)
Tetranitromethane 070
Tetrasaccharide 295(T)
Tetrulose *252*
Texturized protein 087
–, extrusion process 089
–, spin process 088
Thaumatin 437
Thaumatin I, amino acid sequence 438(T)
–, conformation 436, 437(F), 437(T)
Theaflavin 956, *957*
Theanderose, honey 886(T)
Theanine, tea 954
Thearubigen 956, *957*
Theobromine, bitter taste, synergism 963(T), 964, *964*
–, cacao bean 961, 961(T)
–, coffee 943
–, cola nut 958
Theogallin 825, *954*
Theophylline, coffee 944
Thermization, milk 518
Thermogram 739(F)
Thermolysine, specificity 078(T), 079
Thermoplastic gel 063
Thermoreversible gel 063
Thiabendazole, antimicrobial action 454, *455*
Thiamine, degradation to aroma substance 365, 366(F)
–, determination 413, *413*
–, loss 412(T)
–, rice 710(T)
Thiamine (see also vitamin B$_1$) 411, *411*
Thiazole 367
Thiazole 2-acetyl-
–, odor threshold value 367(T)
Thiazole, 2-acetyl- 367(T)
Thiazole, 2-isobutyl- 367(T), 387
–, odor threshold value 367(T)
Thiazole, benzo- 367(T)
Thiazolylbenzimidazole, antimicrobial action 454, *455*
Thickening agent 464
–, analysis 335
Thiobarbituric acid test 668
Thiocarbamylamino acid 020
Thiocyanate, milk 517, *517*
Thiocyanate (see also rhodanide)
Thioether, formation 024
Thioglucosidase 790
–, industrial application 153
Thiol-disulfide exchange, milk protein 519

Thiooxazolidone, goiter 798
Thiosulfate 456(T)
Thiourethane 790
Thousand kernel weight, cereal 674(T)
Threonine *010*
–, aroma substance precursor 362
–, degradation to propanal *376*
–, discovery 012
–, reaction 024
–, synthesis 033
Threose 249, *249, 251*
Threshold value, taste 034
Thrombin 075
Thujone *384*
Thyme, essential oil, chemical composition 974(T)
Thymine *047, 143*
Thymol *972*
Thymus, veal, chemical composition 594(T)
Thyroxine 427
Tin 427
Titin, meat 569, 570(F)
Titration curve, amino acid 013(F)
TLCK 069
Tocopherol, analysis 233
–, antioxidative activity, mechanism 217, *217*
–, biological activity 407(T)
–, fat hardening 658
–, HPLC 234, 235(F)
–, loss, deep frying 408(T)
–, loss in raffination 233
–, photometrc determination 234
–, reaction with peroxy radical 215, 216, *216*
–, requirement, fatty acid 407(T)
–, stereochemistry 407(T)
–, wheat, part of kernel 706, 706(T)
Tocopherol (see also vitamin E) 407, *408*
Tocopherol, α-
–, antioxidant activity 216, *216*, 217(T)
–, margarine 661
–, reaction product 216, *216*
Tocopherol, β-
–, indicator for wheat germ oil 233
–, occurrence 234(T)
–, structure 234(F)
Tocopherol, δ-
–, occurrence 234(T)
–, structure 234(F)
Tocopherol, γ-
–, antioxidant activity 216, *216*, 217(T)
–, occurrence 234(T)
–, reaction product 216, *216*
–, structure 234(F)
Tocotrienol, α-
–, occurrence 234(T)
–, structure 234(F)
Tocotrienol, β-

–, occurrence 234(T)
–, structure 234(F)
Tocotrienol, δ-
–, occurrence 234(T)
–, structure 234(F)
Tocotrienol, γ-
–, occurrence 234(T)
–, structure 234(F)
Toffee 880
Tofu 766
–, phytoestrogen 763(T)
Toluenesulfonyl-L-phenylalanine ethylester, N- 107
Tomato, aroma formation 211(T)
–, aroma substance 793, 793(T)
–, carotene, differences in cultivar 236(T)
–, lipoxygenase, reaction specificity 207(T), 208(F)
–, lycopene 236
–, odorant 242(T)
–, ripening, respiration rise 844(F)
–, temperature of phase transition 006(T)
Tomato genetically modified, detection 142(T), 144
Tomato ketchup 805
Tomato paste 805
–, aroma substance 793, 793(T)
Tomato product, aromatization 367
Toned milk 520
Tongue, chemical composition 594(T)
Tonic Water 856
Top fermentation, beer 901
Top quality, chemical composition 927(T)
Tortilla, aroma 389
Tosyl-L-lysine chlormethyl ketone, N-,(TLCK) 107
Tosyl-L-phenylalanine chlormethyl ketone,
  N-,(TPCK) 107
Tosylamino acid, N- 018, 018
Toxic compound, arsenic 468
–, tolerable concentration 467, 468
Toxic trace element 468
Toxicity, acute 467
–, chronic 467
–, dioxin 496(T)
–, Reference Dose (RfD) 467
–, subacute 467
Toxin, bacterial 471(T)
TPCK 069
Trace element 421, 424, 425(T)
–, toxic 468
Tragacanth, viscosity 308(T)
Trans fatty acid 162, 163(T)
–, fat hardening 658
Transamination, mechanism 104, 104(F)
Transferase 097, 100(T)
–, industrial application 154, 155(T)
Transglucosidase, purification of glucoamylase 151
Transglucosylase, honey 886
Transglutaminase, industrial application 154, 155(T)

–, microbial, property 154
–, protein crosslink 155
–, reaction 155
Transhydrogenase, mechanism 099
Transition state analog 112, 112(F)
Transport metabolite 099
Trehalose 295(T)
Trehalose, α-, β-
–, honey 886(T)
Triacylglycerol 169
–, autoxidation rate 191
–, biosynthesis 177, 177(F)
–, chemical property 172
–, chirality 170
–, composition, calculation 174
–, crystal lattice 171, 171(F)
–, energy value 171
–, HPLC 173, 174(F)
–, hydrolysis 172, 172
–, –, enzymatic 188
–, interesterification 172, 173
–, melting point 171(T)
–, methanolysis 172, 172
–, modification, α- 171
–, modification, β- 171
–, modification, ψ'- 171
–, molecule geometry 171
–, nomenclature 170
–, number of isomer, calculation 170
–, polymorphism 171
–, position isomer, analysis 173
–, saponification 172, 172
–, stereospecific analysis 175, 175(F)
–, structure determination 173
Trichloroanisole, 2,4,6-
–, formation 343, 343
–, odor threshold value 343
Trichloroanisole,2,4,6-
–, wine 925, 925(T)
Trideoxy-2,5-hexulose, 1,3,5-
–, formation 268
Trifluoroacetylamino acid, N-, 017
Trigalloyl-D-glucose, 1,3,6- 291
Trigonelline, coffee 944
Trigonelline amide 414, 414
Trihydroxyoctadecenoic acid, 9,12,13-
–, beer 902
Triiodothyronine 427
Trilaurin, melting point 171(T)
Trimethylamine, egg 557
–, fish 626, 629
Trimethylamine oxide, fish 626
Trimethylamino acid, N- 019, 019(T)
Trimyristin, melting point 171(T)
Trinitrobenzene sulfonic acid 019, 020
Trinitrophenylamino acid, N-, 019

olein, melting point 171(T)
iosephosphate isomerase, mechanism 112, 112(F), 114, *114*
tertiary structure 056(F)
ipalmitin, melting point 171(T)
iphosphate 455(T)
iple helix, collagen 578(F)
isaccharide 295(T)
istearin, heating product 222(T)
  melting point 171(T)
riterpene alcohol 645
riterpenoids, fruit 819
rithioacetaldehyde 366, 366(F)
riticale 671
ritium 470
ritylamino acid, N- 019
ritylchloride 019
, reaction with amino acid 019
ropomyosin 571, 571(T), 572(F)
roponin 571, 571(T), 572(F)
rub stabilization, xanthan gum 331
ruffle, aroma substance 788
rypsin 075, 077(T)
-, activation energy 094(T)
-, active serine, mechanism 115
-, inhibition 054, 069
-, specificity 043, 045(F), 078(T)
-, substrate analog inhibitor 107
-, substrate binding 110(F)
Tryptamide, cocoa shell 961, *961*
Tryptamine 584
-, fruit 813, *817*, 817(T)
Tryptophan *010*
-, discovery 012
-, loss by lipid peroxidation 215(T)
-, reaction 070
-, synthesis 033
-, UV absorption 015(F)
Tryptophan, D-
-, sweet taste 034, 035(T)
Tryptophan, L-
-, bitter taste 034
Tuna 622
-, meaty aroma note 629
Turanose, honey 886(T)
Turbidity, beer 906
Turkey, progesterone 228(T)
Turmeric, essential oil, chemical composition 974(T)
Turmerone 974(T), *976*
Turn, β-
-, lysozyme 053(T)
-, protein 052, 052(F), 052(T), 053(F)
Turn, β-, 052
Turnover number, definition 098
Turtle 639
Two-substrate reaction, rate equation 122

-, substrate binding, order 121
Type of bread 741(T)
-, amount of yeast 723(T)
Type of margarine 660(T)
Type of milk 520
Tyramine 584
-, fruit 813, *815*
Tyrosinase 105
Tyrosine *010*
-, acylation 070
-, discovery 012
-, loss by lipid peroxidation 215(T)
-, nitration 070
-, reaction 024, 070
-, UV absorption 015, 015(F)
-, -, pH dependency 016(F)
Tyrosine, o-
-, food radiation 073

UHT milk, vitamin loss 520
Ultra high temperature treatment (UHT)
-, milk 518, 518(F)
Ultra-trace element 421, 427
Ultrafiltration, fruit juice 855
Umami
-, fifth taste quality 430
Umbelliferone 295(T), 824(T)
Uncompetitive inhibition 127
Undecatetraene, (E,Z,Z),1,3,5,8-
-, sensory property 379
Undecatriene, hop 896, 896(T)
Undecatriene, (E,Z)-1,3,5-
-, formation *379*
-, sensory property 379
Undecenal, (E)-2-
-, oleic acid, autoxidation 203(T)
Unripened cheese, carrageenan 301(T), 306
Unsaponifiable component, fat 225, 226(T)
Urea, fish 627
Urea adduct, fatty acid 166
Urea derivative, sweet taste 439
Urease, rate constant 120(T)
-, reaction *753*
Uric acid, enzymatic analysis 140(T)
-, milk indicator 515(T)
Uronic acid, occurrence 263
-, synthesis 262
Uvaol *646*, 646(T)

Vaccinin 291, *291*
Valencene *385*
-, oxidation 344(T)
Valeric acid, structure, melting point 161(T)
Valine *010*
-, biosynthesis 378(F)
-, discovery 012

Vanilla, aroma substance 974, 974(T)
–, vanillin 976
Vanilla sugar 976
Vanillin 375(T)
–, antioxidative effect 218
–, aroma defect, orange 838
–, biosynthesis 826, *826*
–, formation from ferulic acid 374(F)
–, isotopic analysis 395, 395(T)
–, odor threshold value 341(T), 375(T)
–, synthesis 395, *395*
Vanillyl alcohol, biosynthesis 826, *826*
VDT 830
Veal 592
Vegetable 770
–, amine 786
–, aroma substance 788
–, blanching 795, 801
–, carbohydrate 779(T), 786
–, carotinoid 787, 787(T)
–, chemical composition 770, 779(T)
–, color change during drying 795
–, cooling 799(T)
–, cultivar differentiation 770, 780(F)
–, deep freezing 801
–, –, ascorbic acid 801(F)
–, dehydrated 799
–, dietary fiber 779(T)
–, discoloration 795
–, drying 799
–, enzyme 770
–, fermentation 802
–, fermentation product, faulty 804
–, free amino acid 770, 782(T), *784*(T)
–, heating 799, 800
–, lipid 787
–, mineral 793, 794(T)
–, organic acid 787, 788(T)
–, pectin 786
–, pickling 804
–, pigment 793, 794(T)
–, production data 775(T)
–, promotion of ripening 847
–, protein 770
–, salting 804
–, species 771(T)
–, species differentiation 770, 780(F)
–, storage 799
–, tissue firmness 786
–, vitamin 793, 794(T)
Vegetable drying, *Maillard* reaction 289(F)
Vegetable juice 805
–, enzymatic clarification 153
Vegetable paste 805
Vegetable powder 805
Vegetable product 799

–, deep frozen 801
–, eating olive 803
–, pickled cucumber 802
–, pickled vegetable 802
–, salted vegetable 804
–, sauerkraut 802
–, sterile can 800
–, vegetable juice 805
–, vegetable paste 805
–, vegetable powder 805
–, vinegar-pickled vegetable 804
Verbascose, legume 760(T)
Verdoperoxidase 103
Vermouth 929
Very low density lipoprotein (VLDL) 184
Veterinary medicine 486, 487(T), 488(F)
–, analysis 487
–, chemical structure 488(F)
Vicilin 746
–, amino acid sequence 752(T)
Vicin *764*
Vierge type, olive oil 645
Vimentin 571
Vinegar, chemical composition 984
–, differentiation fermented/synthetic vinegar 984
–, production, microbiological 983, 984(F)
Vinegar essence 984
Vinegar-pickled vegetable 804
Vinegar-preserved fruit 851
Vinyloxazolidine-2-thione, 5- 798
Violaxanthine *238*
–, orange 240(T)
Virgin oil 645
Viscosity, temperature dependency 006
Visual cycle 404, 404(F)
Visual detection threshold, cyanidinglycoside 830
Vitamin 403
–, beer 902
–, daily requirement 405(T)
–, egg 556(T)
–, enrichment, example 430, 430(T)
–, fish 628
–, fruit 842
–, legume 762(T)
–, loss during canning 800
–, loss during food processing 403(T)
–, loss in the baking process 734
–, margarine 661
–, milk 515, 516(T)
–, occurrence 409(T)
–, storage of vegetable 799, 801, 801(F)
–, use as additive 430
–, vegetable 793, 794(T)
–, wheat 675, 676(T)
Vitamin A 403(T), *404*
–, loss during food processing 403(T), 404

Vitamin A concentrate, production 655
Vitamin B$_{12}$ 416, *417*
–, coenzyme, mechanism 532, *533*
–, occurrence 409(T)
Vitamin B$_1$,
–, degradation reaction 412, *412*
–, determination 413, *413*
–, loss 412(T)
–, loss during food processing 403(T)
–, occurrence 409(T)
Vitamin B$_2$ 102, *102*
–, loss during food processing 403(T)
–, stability 413
Vitamin B$_6$ 413, *414*
–, occurrence 409(T)
–, stability 424
Vitamin C 417, *418*
–, dried fruit 849, 850, 850(T)
–, fruit juice 853(T)
–, loss 419, 419, 420(F)
–, loss during food processing 403(T)
–, occurrence 409(T)
–, pickled vegetable 802
–, sauerkraut 803
Vitamin D 406, *406*
–, determination 233
–, occurrence 409(T)
Vitamin D$_2$ 406, *406*
Vitamin D$_3$ 406, *406*
–, formation from 7-dehydrocholesterol 229, 230(F)
Vitamin E 407, *408*
–, occurrence 409(T)
Vitamin K, biological activity 408, *411*
Vitamin K$_1$ *411*
–, occurrence 409(T)
Vitispiran 243, *244*
VLDL 184
Vodka 935
Volatile, spice 971, 971(T)
Volatile compound, *Maillard* reaction 274
Vomitoxin *473*, 474(T)
Vulgaxanthin 797, *797*

Walnut, chemical composition 816(T)
–, oil 651(T), 653
Warmed over flavor, formation 608, 608(T)
–, meat 595, 596(T)
Water, coordination number 002, 002(T)
–, density 002
–, determination, NIR 706(T)
–, dissociation 002
–, H-bridge 002
–, molecular geometry 001(F)
–, orbital model 001(F)
–, structure 001
–, unfrozen 005, 006(T)

Water (see also drinking water)
Water activity, browning, nonenzymatic 004, 004(F)
–, food preservation 003
–, food quality 003, 004(T)
–, lipid peroxidation 004, 004(F)
–, moisture content 004(T)
–, range 004
–, storage stability 003(F)
Water binding, desorption isotherm 004(F)
–, sorption isotherm 004(F)
Water binding capacity, protein 061
Water content, enzyme activity 137, 137(T)
–, food 001(T)
Water holding capacity, meat 590, 591, 592(F), 598(F)
Water molecule, coordination 001(F)
–, hydrogen bond 001, 001(F), 002
Water retention, protein 061, *061*
Watercress, aroma substance 791
Wax 186
–, fruit 820
Wax ester, analysis 666(F)
Waxy corn 317(T)
Waxy corn starch, gelatinization behavior 317(T)
Weiss beer, Bavarian 903
–, Berlin 903
Whale 636
Whale meat extract 040(T), 602
Whale oil 644
–, fatty acid composition 644(T)
Wheat,
–, amylase 695, 696(T)
–, arabinoxylan hydrolase 698
–, ascorbic acid oxidase 698, 699(T)
–, baking quality, gluten protein 690(F)
–, carbohydrate 701
–, carotinoid 705
–, catalase 698
–, cellulose 703
–, chemical composition 675(T)
–, cultivar difference, HMW subunit 692, 694(F)
–, cysteinyl glycine 699, 699(T)
–, differentiation of wheat cultivar 676(F), 682(F)
–, dough making, microscopy 726, 728(F)
–, dough strengthening 723(F), 730, 731(F)
–, enzyme 695
–, flour, air classification 709
–, flour extraction rate 709(T)
–, friabilin 695
–, glucofructan 703
–, glutamyl cysteine 699, 699(T)
–, glutathione, protein-bound 699, 700(T)
–, glutathione dehydrogenase 698, *698*, 699(T)
–, gluten 673
–, gluten protein 680
–, glutenin 679

–, –, electropherogram 680(F)
–, lectin 759(T)
–, lipase 696
–, lipid 701(T), 704, 705(T)
–, lipid composition 178(T)
–, lipoxygenase 697
–, –, reaction specificity 207(T)
–, milling 707, 707(F)
–, milling product 708(T)
–, monosaccharide 703(T)
–, NIR absorption 706(F)
–, oligosaccharide 703(T)
–, origin 670, 670(F)
–, *Osborne* fraction 676, 677(T)
–, pentosan 702
–, peroxidase 698
–, phytase 696, 697(T)
–, production data 672(T)
–, progesterone 228(T)
–, prolamin 680
–, –, electropherogram 680(F)
–, protein, electropherogram 680(F), 690(F)
–, proteinase 696
–, puroindolin 695
–, soy sauce 766
–, yield per hectare 674(T)
Wheat dough, emulsifier 721
–, production 723
–, rheological property 039
–, rheology, glutathione 699, 700(F)
Wheat flour, acidity 711
–, amylase, α-
–, –, flour improvement 721
–, ascorbic acid, flour improvement 725(F)
–, azodicarbonamide, flour improvement 725(F)
–, baking test 714
–, –, lipoxygenase 719(F)
–, bromate, flour improvement 718(T), 719
–, characterization of baking property 711
–, chemical assay 711
–, chemical bleaching 716
–, chemical composition 709(T)
–, cultivar difference 712(T), 713(F)
–, dextrin value 713
–, disulfide group 712, 712(T)
–, event involved in dough making 726, 728(F)
–, extensogram 714(F), 715(T)
–, falling number 713
–, farinogram 714(F), 714(T)
–, free cysteine 699, 699(T)
–, glutathione 699, 699, 700(T)
–, gluten content 711
–, improvement, ascorbic acid 716, 716, 717(F), 717(T)
–, –, azodicarbonamide 719
–, lipoxygenase, flour improvement 719, 719(F)

–, maltose value 713
–, milling extraction grade 708
–, milling extraction rate 709(T)
–, particle size, protein content 709(T)
–, physical assay 713
–, protein content, baking volume 712, 713(F)
–, proteinase, flour improvement 720, 720(F)
–, starch damage 708
–, storage 716
–, thiol group 712(T)
–, type 709, 709(T)
Wheat germ oil 650(T), 653
–, analysis, HPLC 174(F)
–, detection 665(T)
–, furan fatty acid 164(T)
–, tocopherol as indicator, β- 233
–, tocopherol content 234(T)
Wheat gluten, succinylation 081, 081(F)
Wheat grain, anatomy 676(F)
–, chemical composition 671, 675, 676(T)
–, mineral 676(T)
–, vitamin 676(T)
Wheat malt 892
Wheat mix bread 741(T)
Whey, chemical composition 538(T)
–, demineralization 538
–, electrodialysis 538(F)
Whey powder 528(T), 537
–, production data 500(T)
Whey product 537
–, chemical composition 538(T)
Whey protein 502(T)
–, amino acid composition 502(T)
–, composition 511
–, denaturation 518, 519(F)
–, genetic variant 505(T)
–, thermal denaturation 057, 058(F), 058(T)
Whey syrup, hydrolysis 539
Whipped cream 524
–, propellant 465
Whisky 933
–, blended 934
–, Bourbon, odorant 934(T), 935
–, malt, odorant 934(T), 935
–, storage, aroma 934(T)
Whisky lactone 381(T)
–, formation 382
White bread, aroma substance 734, 736, 736, 737(T)
White cabbage, aroma substance 792
White sugar 872, 872(F)
–, production 872(F)
White wine, fermentation 916
–, grape cultivar 907
–, phenolic compound 920, 920(T)
Whole grain groat 708(T)
Whole milk 520

Whole milk chocolate 968(T)
Whole milk powder 527, 528(T)
Whole-fat cheese 529
Whole-meal flour, phenoloxidase 698
Wiliams, pear brandy 932
Williams, Landel, Ferry-equation 006
Wine 906
–, acid 920
–, aging 917
–, –, aroma 923, 924(T)
–, alcohol reduced, aroma 921, 923(T)
–, arabinose 919
–, aroma, ageing 243
–, aroma substance 921, 922–924(T)
–, –, fermentation condition 923, 923(T)
–, bitter taste, divinyl glycol 926
–, blending 919
–, blue fining 918
–, boekser 925, 925(T)
–, butanediol, 2,3- 920
–, carbohydrate 919
–, cellar operation 917
–, Chardonnay, aroma 922(T)
–, chemical composition 919
–, clarification 918
–, consumption data 907(T)
–, cork taste 925, 925(T)
–, cultivar-specific aroma 921, 922(T)
–, deacidification 918
–, degradation of malic acid 919, *919*
–, diacetyl 920
–, distilled, alcohol content 931
–, ethanol 919
–, –, sugar added 919
–, ethanol content 919
–, extract 919
–, fining 918
–, flavored 929
–, fortified 911
–, geranium note 926
–, glycerol 920
–, glycerol factor 920, *920*
–, grades 915(T)
–, grape 907
–, higher alcohol 920
–, iron content 921
–, kerosine note 924, 925(T)
–, lactic acid, formation 919, *919*
–, mannitol, D- 920
–, medicine note 924, 925(T)
–, methanol 920
–, mineral 921
–, minimum alcohol content 915, 915(T)
–, mousy note 925(T), 926
–, MTCA formation 921, *921*
–, Muscatel, aroma 922(T)

–, odorant, odor threshold value 921, 923(T)
–, off-flavor 924, 925(T)
–, pH-value 920
–, pigment 915
–, production data 907(T)
–, production scheme 908(F)
–, proline 921
–, resveratol 920(T), *921*
–, sauvignon type, aroma 922, 922(T)
–, sorbitol 920
–, strawberry note 925(T)
–, sugar added, detection 919
–, sugar addition 919
–, sugar-free extract 919
–, sulfur treatment 917
–, tartaric acid 916, 917(F)
–, taste substance 921, 922(T)
–, top quality 911, 915(T)
–, tryptophan 921
–, turbidity 924, 926
–, untypical aging note 924, 925(T)
–, volatile sulfur compound, biosynthesis 387
–, young 916
Wine distillate 931
Wine lactone, diastereomer, odor threshold value 354,
   355(T)
–, –, separation 354, 355(F)
–, wine 922(T)
Wine sauerkraut 803
Wine-like beverage 928
–, chemical composition 928(T)
Winterization, cottonseed oil, corn oil 650
–, fat, process 659
WLF-equation 006
Woodward reagent 067
Wool keratin 053
–, plastein reaction 085(F)
Wort, beer 899
Wort concentrate, beer production 894

Xanthan gum 331, *331*
–, viscosity 331(F)
Xanthine, oxidation 105
Xanthine oxidase, mechanism 105, *105*
–, milk 135(F)
–, pH optimum 129(T)
–, superoxide radical anion 105
Xanthophyll 237
Xylanase 335
Xylane 250(T)
Xylitol 261
–, aqueous solution, viscosity 865(F)
–, nutritional/physiological property 864(T)
–, production 863(T), 879
–, relative sweetness 259(T), 863(T)
–, sensory propertiy 879

−, solubility 862(F)
−, water absorption 866(F)
Xyloglucan 330, *330*
Xylopyranose, α-D- 255(T)
Xylopyranose, β-D- 255(T)
Xylose 250(T), *251*
−, equilibrium mixture 255(T)
−, nutritional/physiological property 864(T)
−, relative sweetness 259(T)
−, specific rotation 258(T)
Xylosone, formation 419, *419*
Xylulose *252*

Yak milk, chemical composition 501(T)
Yeast, amount, type of bread 723(T)
−, beer 897
−, dough leavening 723
−, extract 602
−, nucleic acid 250(T)
−, ornithine 368
−, protein, succinylation 081, 082(T)
Yellow No. 5 444(T), 446(T)
Yellow No. 6 444(T), 446(T)

Yoghurt 162(T), 523
−, agar 303
−, aroma substance 540
−, consistence 510, 511(F)
−, nonen-3-one, 1- 523, *523*
−, probiotics 523
−, type, production 523(F)
Yolk droplet 553

Zeacarotene, β- *237*
Zearalenone *473*, 474(T)
Zeaxanthine *237*
Zebu milk, chemical composition 501(T)
Zein, enrichment with lysine, threonine and tryptophan 084(F), 085(T)
Zinc 425, 426(T)
−, enzyme cofactor 104
Zingerone, formation 979, *981*
Zingiberene *385*
−, oxidation 973, *973*
z-value, definition 133, *133*
−, temperature dependency 133

Ba: 625965

008694118

Lightning Source UK Ltd.
Milton Keynes UK
UKOW04f1330220617

303722UK00021B/353/P

9 783540 699354